Advanced Engineering Mathematics

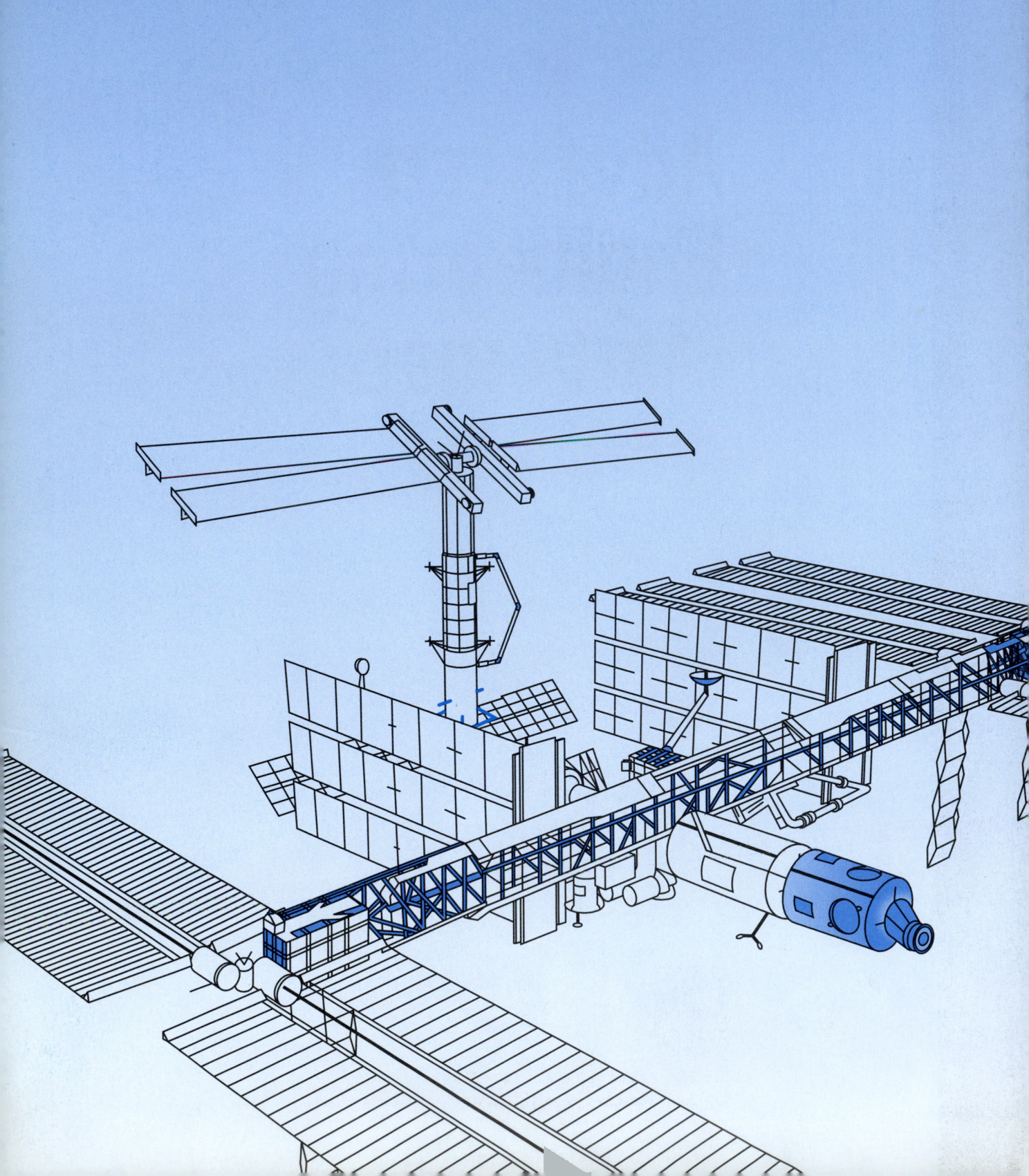

Advanced Engineering Mathematics

Alan Jeffrey

University of Newcastle-upon-Tyne

San Diego   San Francisco   New York   Boston
London   Toronto   Sydney   Tokyo

Sponsoring Editor	Barbara Holland
Production Editor	Julie Bolduc
Promotions Manager	Stephanie Stevens
Cover Design	Monty Lewis Design
Text Design	Thompson Steele Production Services
Front Matter Design	Perspectives
Copyeditor	Kristin Landon
Composition	TechBooks
Printer	RR Donnelley & Sons, Inc.

This book is printed on acid-free paper. ⊗

Harcourt/Academic Press
A Harcourt Science and Technology Company
200 Wheeler Road, Burlington, Massachusetts 01803, USA
http://www.harcourt-ap.com

Academic Press
A Harcourt Science and Technology Company
525 B Street, Suite 1900, San Diego, California 92101-4495, USA
http://www.academicpress.com

Academic Press
Harcourt Place, 32 Jamestown Road, London NW1 7BY, UK
http://www.academicpress.com

Library of Congress Catalog Card Number: 00-108262

International Standard Book Number: 0-12-382592-X

PRINTED IN THE UNITED STATES OF AMERICA
01 02 03 04 05 06 DOC 9 8 7 6 5 4 3 2 1

To Lisl and our family

CONTENTS

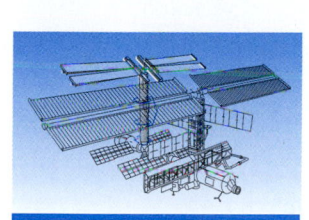

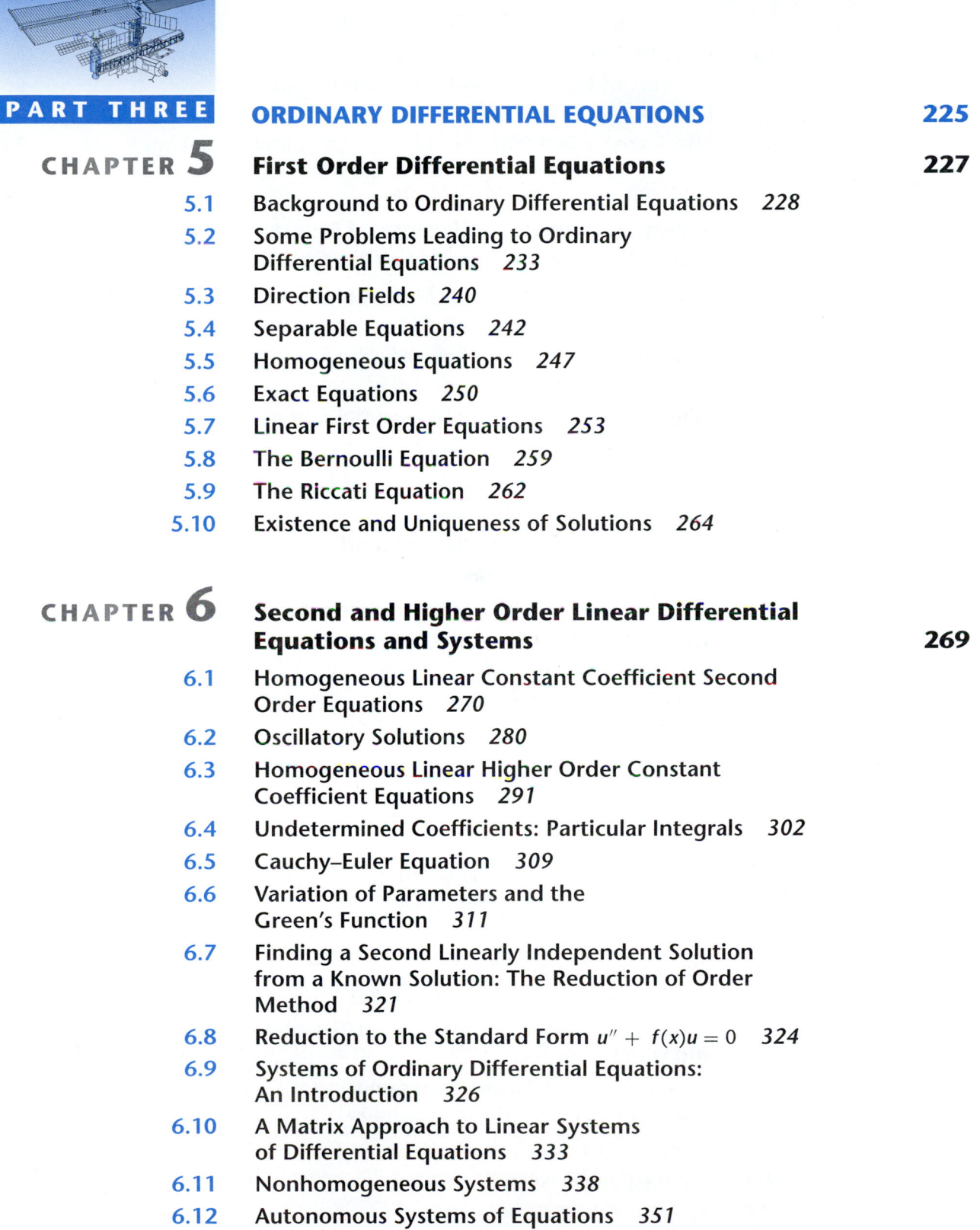

PART EIGHT

CHAPTER 19

NUMERICAL MATHEMATICS   1043

Numerical Mathematics   1045

PREFACE

This book has evolved from lectures on engineering mathematics given regularly over many years to students at all levels in the United States, England, and elsewhere. It covers the more advanced aspects of engineering mathematics that are common to all first engineering degrees, and it differs from texts with similar names by the emphasis it places on certain topics, the systematic development of the underlying theory before making applications, and the inclusion of new material. Its special features are as follows.

Prerequisites

The opening chapter, which reviews mathematical prerequisites, serves two purposes. The first is to refresh ideas from previous courses and to provide basic self-contained reference material. The second is to remove from the main body of the text certain elementary material that by tradition is usually reviewed when first used in the text, thereby allowing the development of more advanced ideas to proceed without interruption.

Worked Examples

The numerous worked examples that follow the introduction of each new idea serve in the earlier chapters to illustrate applications that require relatively little background knowledge. The ability to formulate physical problems in mathematical terms is an essential part of all mathematics applications. Although this is not a text on mathematical modeling, where more complicated physical applications are considered, the essential background is first developed to the point at which the physical nature of the problem becomes clear. Some examples, such as the ones involving the determination of the forces acting in the struts of a framed structure, the damping of vibrations caused by a generator and the vibrational modes of clamped membranes, illustrate important mathematical ideas in the context of practical applications. Other examples occur without specific applications and their purpose is to reinforce new mathematical ideas and techniques as they arise.

A different type of example is the one that seeks to determine the height of the tallest flagpole, where the height limitation is due to the phenomenon of buckling. Although the model used does not give an accurate answer, it provides a typical example of how a mathematical model is constructed. It also illustrates the reasoning used to select a physical solution from a scenario in which other purely mathematical solutions are possible. In addition, the example demonstrates how the choice of a unique physically meaningful solution from a set of mathematically possible ones can sometimes depend on physical considerations that did not enter into the formulation of the original problem.

Exercise Sets

The need for engineering students to have a sound understanding of mathematics is recognized by the systematic development of the underlying theory and the provision of many carefully selected fully worked examples, coupled with their reinforcement through the provision of large sets of exercises at the ends of sections. These sets, to which answers to odd-numbered exercises are listed at the end of the book, contain many routine exercises intended to provide practice when dealing with the various special cases that can arise, and also more challenging exercises, each of which is starred, that extend the subject matter of the text in different ways.

Although many of these exercises can be solved quickly by using standard computer algebra packages, the author believes the fundamental mathematical ideas involved are only properly understood once a significant number of exercises have first been solved by hand. Computer algebra can then be used with advantage to confirm the results, as is required in various exercise sets. Where computer algebra is either required or can be used to advantage, the exercise numbers are in blue. A comparison of computer-based solutions with those obtained by hand not only confirms the correctness of hand calculations, but also serves to illustrate how the method of solution often determines its form, and that transforming one form of solution to another is sometimes difficult. It is the author's belief that only when fundamental ideas are fully understood is it safe to make routine use of computer algebra, or to use a numerical package to solve more complicated problems where the manipulation involved is prohibitive, or where a numerical result may be the only form of solution that is possible.

New Material

Typical of some of the new material to be found in the book is the matrix exponential and its application to the solution of linear systems of ordinary differential equations, and the use of the Green's function. The introductory discussion of the development of discontinuous solutions of first order quasilinear equations, which are essential in the study of supersonic gas flow and in various other physical applications, is also new and is not to be found elsewhere. The account of the Laplace transform contains more detail than usual. While the Laplace transform is applied to standard engineering problems, including control theory, various nonstandard problems are also considered, such as the solution of a boundary value problem for the equation that describes the bending of a beam and the derivation of the Laplace transform of a function from its differential equation. The chapter on vector integral calculus first derives and then applies two fundamental vector transport theorems that are not found in similar texts, but which are of considerable importance in many branches of engineering.

Series Solutions of Differential Equations

Understanding the derivation of series solutions of ordinary differential equations is often difficult for students. This is recognized by the provision of detailed examples, followed by carefully chosen sets of exercises. The worked examples illustrate all of the special cases that can arise. The chapter then builds on this by deriving the most important properties of Legendre polynomials and Bessel functions, which are essential when solving partial differential equations involving cylindrical and spherical polar coordinates.

Complex Analysis

Because of its importance in so many different applications, the chapters on complex analysis contain more topics than are found in similar texts. In particular, the inclusion of an account of the inversion integral for the Laplace transform makes it possible to introduce transform methods for the solution of problems involving ordinary and partial differential equations for which tables of transform pairs are inadequate. To avoid unnecessary complication, and to restrict the material to a reasonable length, some topics are not developed with full mathematical rigor, though where this occurs the arguments used will suffice for all practical purposes. If required, the account of complex analysis is sufficiently detailed for it to serve as a basis for a single subject course.

Conformal Mapping and Boundary Value Problems

Sufficient information is provided about conformal transformations for them to be used to provide geometrical insight into the solution of some fundamental two-dimensional boundary value problems for the Laplace equation. Physical applications are made to steady-state temperature distributions, electrostatic problems, and fluid mechanics. The conformal mapping chapter also provides a quite different approach to the solution of certain two-dimensional boundary value problems that in the subsequent chapter on partial differential equations are solved by the very different method of separation of variables.

Partial Differential Equations

An understanding of partial differential equations is essential in all branches of engineering, but accounts in engineering mathematics texts often fall short of what is required. This is because of their tendency to focus on the three standard types of linear second order partial differential equations, and their solution by means of separation of variables, to the virtual exclusion of first order equations and the systems from which these fundamental linear second order equations are derived. Often very little is said about the types of boundary and initial conditions that are appropriate for the different types of partial differential equations. Mention is seldom if ever made of the important part played by nonlinearity in first order equations and the way it influences the properties of their solutions. The account given here approaches these matters by starting with first order linear and quasilinear equations, where the way initial and boundary conditions and nonlinearity influence solutions is easily understood. The discussion of the effects of nonlinearity is introduced at a comparatively early stage in the study of partial differential equations because of its importance in subjects like fluid mechanics and chemical engineering. The account of nonlinearity also includes a brief discussion of shock wave solutions that are of fundamental importance in both supersonic gas flow and elsewhere.

Linear and nonlinear wave propagation is examined in some detail because of its considerable practical importance; in addition, the way integral transform methods can be used to solve linear partial differential equations is described. From a rigorous mathematical point of view, the solution of a partial differential equation by the method of separation of variables only yields a formal solution, which only becomes a rigorous solution once the completeness of any set of eigenfunctions that arises has been established. To develop the subject in this manner would take the text far beyond the level for which it is intended and so the completeness of any set of eigenfunctions that occurs will always be assumed. This assumption can be fully justified when applying separation of variables to the applications considered here and also in virtually all other practical cases.

Technology Projects

To encourage the use of technology and computer algebra and to broaden the range of problems that can be considered, technology-based projects have been added wherever appropriate; in addition, standard sets of exercises of a theoretical nature have been included at the ends of sections. These projects are not linked to a particular computer algebra package: Some projects illustrating standard results are intended to make use of simple computer skills while others provide insight into more advanced and physically important theoretical questions. Typical of the projects designed to introduce new ideas are those at the end of the chapter on partial differential equations, which offer a brief introduction to the special nonlinear wave solutions called solitons.

Numerical Mathematics

Although an understanding of basic numerical mathematics is essential for all engineering students, in a book such as this it is impossible to provide a systematic account of this important discipline. The aim of this chapter is to provide a general idea of how to approach and deal with some of the most important and frequently encountered numerical operations, using only basic numerical techniques, and thereafter to encourage the use of standard numerical packages. The routines available in numerical packages are sophisticated, highly optimized and efficient, but the general ideas that are involved are easily understood once the material in the chapter has been assimilated. The accounts that are given here purposely avoid going into great detail as this can be found in the quoted references. However, the chapter does indicate when it is best to use certain types of routine and those circumstances where routines might be inappropriate.

The details of references to literature contained in square brackets at the ends of sections are listed at the back of the book with suggestions for additional reading. An instructor's *Solutions Manual* that gives outline solutions for the technology projects is also available.

Acknowledgments

I wish to express my sincere thanks to the reviewers and accuracy readers, those cited below and many who remain anonymous, whose critical comments and suggestions were so valuable, and also to my many students whose questions when studying the material in this book have contributed so fundamentally to its development. Particular thanks go to:

Chun Liu, Pennsylvania State University
William F. Moss, Clemson University
Donald Hartig, California Polytechnic State University at San Luis Obispo
Howard A. Stone, Harvard University
Donald Estep, Georgia Institute of Technology
Preetham B. Kumar, California State University at Sacramento
Anthony L. Peressini, University of Illinois at Urbana-Champaign
Eutiquio C. Young, Florida State University
Colin H. Marks, University of Maryland
Ronald Jodoin, Rochester Institute of Technology
Edgar Pechlaner, Simon Fraser University
Ronald B. Guenther, Oregon State University
Mattias Kawski, Arizona State University
L. F. Shampine, Southern Methodist University

In conclusion, I also wish to thank my editor, Barbara Holland, for her invaluable help and advice on presentation; Julie Bolduc, senior production editor, for her patience and guidance; Mike Sugarman, for his comments during the early stages of writing; and, finally, Chuck Glaser, for encouraging me to write the book in the first place.

PART ONE

REVIEW MATERIAL

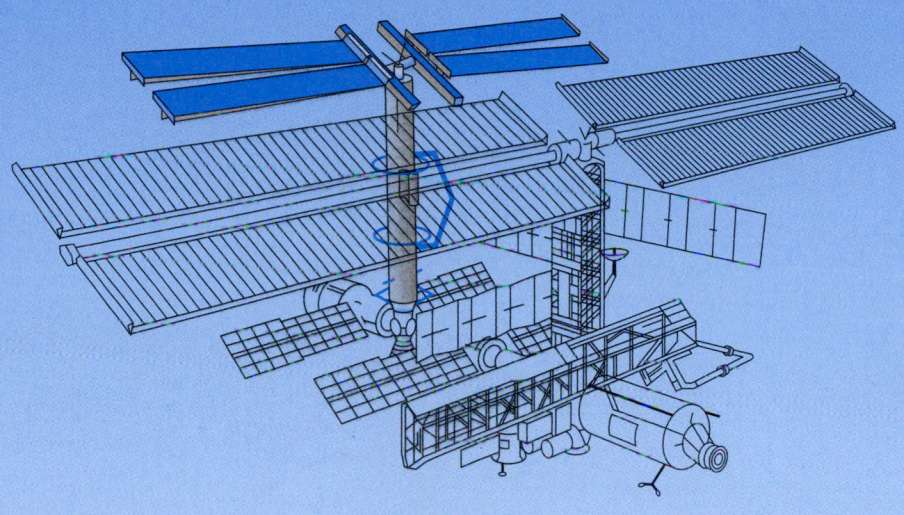

Chapter 1   Review of Prerequisites

CHAPTER 1

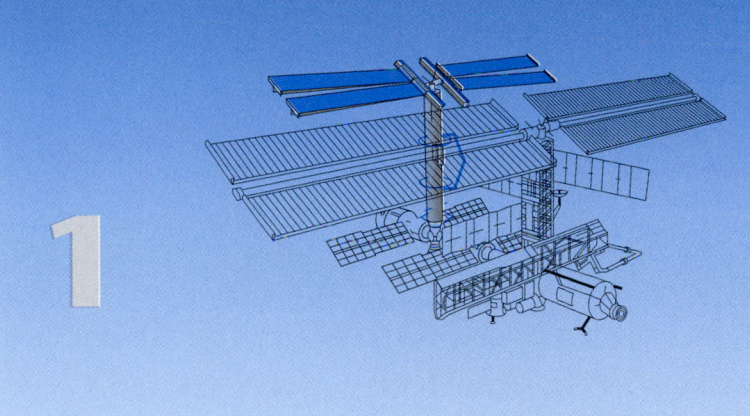

Review of Prerequisites

Every account of advanced engineering mathematics must rely on earlier mathematics courses to provide the necessary background. The essentials are a first course in calculus and some knowledge of elementary algebraic concepts and techniques. The purpose of the present chapter is to review the most important of these ideas that have already been encountered, and to provide for convenient reference results and techniques that can be consulted later, thereby avoiding the need to interrupt the development of subsequent chapters by the inclusion of review material prior to its use.

Some basic mathematical conventions are reviewed in Section 1.1, together with the method of proof by mathematical induction that will be required in later chapters. The essential algebraic operations involving complex numbers are summarized in Section 1.2, the complex plane is introduced in Section 1.3, the modulus and argument representation of complex numbers is reviewed in Section 1.4, and roots of complex numbers are considered in Section 1.5. Some of this material is required throughout the book, though its main use will be in Part 5 when developing the theory of analytic functions.

The use of partial fractions is reviewed in Section 1.6 because of the part they play in Chapter 7 in developing the Laplace transform. As the most basic properties of determinants are often required, the expansion of determinants is summarized in Section 1.7, though a somewhat fuller account of determinants is to be found later in Section 3.3 of Chapter 3.

The related concepts of limit, continuity, and differentiability of functions of one or more independent variables are fundamental to the calculus, and to the use that will be made of them throughout the book, so these ideas are reviewed in Sections 1.8 and 1.9. Tangent line and tangent plane approximations are illustrated in Section 1.10, and improper integrals that play an essential role in the Laplace and Fourier transforms, and also in complex analysis, are discussed in Section 1.11.

The importance of Taylor series expansions of functions involving one or more independent variables is recognized by their inclusion in Section 1.12. A brief mention is also made of the two most frequently used tests for the convergence of series, and of the differentiation and integration of power series that is used in Chapter 8 when considering series solutions of linear ordinary differential equations. These topics are considered again in Part 5 when the theory of analytic functions is developed.

The solution of many problems involving partial differential equations can be simplified by a convenient choice of coordinate system, so Section 1.13 reviews the theorem for the change of variable in partial differentiation, and describes the cylindrical polar and spherical polar coordinate systems that are the two that occur most frequently in practical problems.

Because of its fundamental importance, the implicit function theorem is stated without proof in Section 1.14, though it is not usually mentioned in first calculus courses.

1.1   Real Numbers, Mathematical Induction, and Mathematical Conventions

Numbers are fundamental to all mathematics, and real numbers are a subset of complex numbers. A real number can be classified as being an **integer**, a **rational** number, or an **irrational** number. From the set of positive and negative integers, and zero, the set of positive integers $1, 2, 3, \ldots$ is called the set of **natural numbers**. The rational numbers are those that can be expressed in the form m/n, where m and n are integers with $n \neq 0$. Irrational numbers such as π, $\sqrt{2}$, and $\sin 2$ are numbers that cannot be expressed in rational form, so, for example, for no integers m and n is it true that $\sqrt{2}$ is equal to m/n. Practical calculations can only be performed using rational numbers, so all irrational numbers that arise must be approximated arbitrarily closely by rational numbers.

Collectively, the sets of integers and rational and irrational numbers form what is called the set of all **real numbers**, and this set is denoted by **R**. When it is necessary to indicate that an arbitrary number a is a real number a shorthand notation is adopted involving the symbol $\in$, and we will write $a \in \mathbf{R}$. The symbol $\in$ is to be read "belongs to" or, more formally, as "is an element of the set." If a is not a member of set **R**, the symbol $\in$ is negated by writing $\notin$, and we will write $a \notin \mathbf{R}$ where, of course, the symbol $\notin$ is to be read as "does not belong to," or "is not an element of the set." As real numbers can be identified in a unique manner with points on a line, the set of all real numbers **R** is often called the **real line**. The set of all complex numbers **C** to which **R** belongs will be introduced later.

One of the most important properties of real numbers that distinguishes them from other complex numbers is that they can be arranged in numerical order. This fundamental property is expressed by saying that the real numbers possess the **order property**. This simply means that if $x, y \in \mathbf{R}$, with $x \neq y$, then

$$\text{either } x < y \quad \text{or} \quad x > y,$$

where the symbol $<$ is to be read "is less than" and the symbol $>$ is to be read "is greater than." When the foregoing results are expressed differently, though equivalently, if $x, y \in \mathbf{R}$, with $x \neq y$, then

$$\text{either } x - y < 0 \quad \text{or} \quad x - y > 0.$$

It is the order property that enables the graph of a real function f of a real variable x to be constructed. This follows because once length scales have been chosen for the axes together with a common origin, a real number can be made to correspond to a unique point on an axis. The graph of f follows by plotting all possible points $(x, f(x))$ in the plane, with x measured along one axis and $f(x)$ along the other axis.

absolute value

The **absolute value** $|x|$ of a real number x is defined by the formula

$$|x| = \begin{cases} x & \text{if } x \geq 0 \\ -x & \text{if } x < 0. \end{cases}$$

This form of definition is in reality a concise way of expressing two separate statements. One statement is obtained by reading $|x|$ with the top condition on the right and the other by reading it with the bottom condition on the right. The absolute value of a real number provides a measure of its magnitude without regard to its sign so, for example, $|3| = 3$, $|-7.41| = 7.41$, and $|0| = 0$.

Sometimes the form of a general mathematical result that only depends on an arbitrary natural number n can be found by experiment or by conjecture, and then the problem that remains is how to prove that the result is either true or false for all n. A typical example is the proposition that the product

$$(1 - 1/4)(1 - 1/9)(1 - 1/16) \ldots [1 - 1/(n+1)^2]$$
$$= (n+2)/(2n+2), \quad \text{for } n = 1, 2, \ldots.$$

This assertion is easily checked for any specific positive integer n, but this does not amount to a proof that the result is true for all natural numbers.

A powerful method by which such propositions can often be shown to be either true or false involves using a form of argument called **mathematical induction**. This type of proof depends for its success on the order property of numbers and the fact that if n is a natural number, then so also is $n + 1$. The steps involved in an inductive proof can be summarized as follows.

mathematical induction

Proof by Mathematical Induction

Let $P(n)$ be a proposition depending on a positive integer n.

STEP 1   Show, if possible, that $P(n)$ is true for some positive integer n_0.

STEP 2   Show, if possible, that if $P(n)$ is true for an arbitrary integer $n = k \geq n_0$, then the proposition $P(k + 1)$ follows from proposition $P(k)$.

STEP 3   If Step 2 is true, the fact that $P(n_0)$ is true implies that $P(n_0 + 1)$ is true, and then that $P(n_0 + 2)$ is true, and hence that $P(n)$ is true for all $n \geq n_0$.

STEP 4   If no number $n = n_0$ can be found for which Step 1 is true, or if in Step 2 it can be shown that $P(k)$ does not imply $P(k + 1)$, the proposition $P(n)$ is false.

The example that follows is typical of the situation where an inductive proof is used. It arises when determining the nth term in the Maclaurin series for $\sin ax$ that involves finding the nth derivative of $\sin ax$. A result such as this may be found intuitively by inspection of the first few derivatives, though this does not amount to a formal proof that the result is true for all natural numbers n.

EXAMPLE 1.1   Prove by mathematical induction that

$$d^n/dx^n[\sin ax] = a^n \sin(ax + n\pi/2), \quad \text{for } n = 1, 2, \ldots.$$

Solution   The proposition $P(n)$ is that

$$d^n/dx^n[\sin ax] = a^n \sin(ax + n\pi/2), \quad \text{for } n = 1, 2, \ldots.$$

STEP 1   Differentiation gives

$$d/dx[\sin ax] = a \cos ax,$$

but setting $n = 1$ in $P(n)$ leads to the result

$$d/dx[\sin ax] = a \sin(ax + \pi/2) = a \cos ax,$$

showing that proposition $P(n)$ is true for $n = 1$ (so in this case $n_0 = 1$).

STEP 2   Assuming $P(k)$ to be true for $k > 1$, differentiation gives

$$d/dx\{d^k/dx^k[\sin ax]\} = d/dx[a^k \sin(ax + k\pi/2)],$$

so

$$d^{k+1}/dx^{k+1}[\sin ax] = a^{k+1} \cos(ax + k\pi/2).$$

However, replacing k by $k + 1$ in $P(k)$ gives

$$\begin{aligned}
d^{k+1}/dx^{k+1}[\sin ax] &= a^{k+1} \sin[ax + (k+1)\pi/2] \\
&= a^{k+1} \sin[(ax + k\pi/2) + \pi/2] \\
&= a^{k+1} \cos(ax + k\pi/2),
\end{aligned}$$

showing, as required, that proposition $P(k)$ implies proposition $P(k+1)$, so Step 2 is true.

STEP 3   As $P(n)$ is true for $n = 1$, and $P(k)$ implies $P(k+1)$, it follows that the result is true for $n = 1, 2, \ldots$ and the proof is complete. ■

The **binomial theorem** finds applications throughout mathematics at all levels, so we quote it first when the exponent n is a positive integer, and then in its more general form when the exponent α involved is any real number.

Binomial theorem when n is a positive integer

If a, b are real numbers and n is a positive integer, then

$$(a + b)^n = a^n + na^{n-1}b + \frac{n(n-1)}{2!}a^{n-2}b^2$$
$$+ \frac{n(n-1)(n-2)}{3!}a^{n-3}b^3 + \cdots + b^n,$$

binomial coefficient

or more concisely in terms of the **binomial coefficient**

$$\binom{n}{r} = \frac{n!}{(n-r)!r!},$$

we have

$$(a + b)^n = \sum_{r=0}^{n} \binom{n}{r}a^{n-r}b^r,$$

where $m!$ is the factorial function defined as $m! = 1 \cdot 2 \cdot 3 \cdots m$ with $m > 0$ an integer, and $0!$ is defined as $0! = 1$. It follows at once that

$$\binom{n}{0} = \binom{n}{n} = 1.$$

The binomial theorem involving the expression $(a+b)^\alpha$, where a and b are real numbers with $|b/a| < 1$ and α is an arbitrary real number takes the following form.

General form of the binomial theorem when α is an arbitrary real number

If a and b are real numbers such that $|b/a| < 1$ and α is an arbitrary real number, then

$$
(a+b)^\alpha = a^\alpha \left(1 + \frac{b}{a}\right)^\alpha = a^\alpha \left(1 + \frac{\alpha}{1!}\left(\frac{b}{a}\right) + \frac{\alpha(\alpha-1)}{2!}\left(\frac{b}{a}\right)^2 \right.
$$
$$
\left. + \frac{\alpha(\alpha-1)(\alpha-2)}{3!}\left(\frac{b}{a}\right)^3 + \cdots \right).
$$

The series on the right only terminates after a finite number of terms if α is a positive integer, in which case the result reduces to the one just given. If α is a negative integer, or a nonintegral real number, the expression on the right becomes an infinite series that diverges if $|b/a| > 1$.

EXAMPLE 1.2

Expand $(3+x)^{-1/2}$ by the binomial theorem, stating for what values of x the series converges.

Solution   Setting $b/a = \frac{1}{3}x$ in the general form of the binomial theorem gives

$$
(3+x)^{-1/2} = 3^{-1/2}\left(1 + \frac{1}{3}x\right)^{-1/2} = \frac{1}{\sqrt{3}}\left(1 - \frac{1}{6}x + \frac{1}{24}x^2 - \frac{5}{432}x^3 + \cdots\right).
$$

The series only converges if $|\frac{1}{3}x| < 1$, and so it is convergent provided $|x| < 3$. ∎

Some standard mathematical conventions

Use of combinations of the $\pm$ and $\mp$ signs

The occurrence of two or more of the symbols $\pm$ and $\mp$ in an expression is to be taken to imply two separate results, the first obtained by taking the upper signs and the second by taking the lower signs. Thus, the expression $a \pm b\sin\theta \mp c\cos\theta$ is an abbreviation for the two separate expressions

$$
a + b\sin\theta - c\cos\theta \quad \text{and} \quad a - b\sin\theta + c\cos\theta.
$$

Multi-statements

multi-statement

When a function is defined sectionally on n different intervals of the real line, instead of formulating n separate definitions these are usually simplified by being combined into what can be considered to be a single **multi-statement**. The following example is typical of a multi-statement:

$$
f(x) = \begin{cases} \sin x, & x < \pi \\ 0, & \pi \leq x \leq 3\pi/2 \\ -1, & x > 3\pi/2. \end{cases}
$$

It is, in fact, three statements. The first is obtained by reading $f(x)$ in conjunction with the top line on the right, the second by reading it in conjunction with the second line on the right, and the third by reading it in conjunction with the third line on the right. An example of a multi-statement has already been encountered in the definition of the absolute value $|x|$ of a number x. Frequent use of multi-statements will be made in Chapter 9 on Fourier series, and elsewhere.

Polynomials

polynomials

A **polynomial** is an expression of the form $P(x) = a_0 x^n + a_1 x^{n-1} + \cdots + a_{n-1} x + a_n$. The integer n is called the **degree** of the polynomial, and the numbers a_i are called its **coefficients**. The **fundamental theorem of algebra** that is proved in Chapter 14 asserts that $P(x) = 0$ has n **roots** that may be either real or complex, though some of them may be repeated. ($a_0 \neq 0$ is assumed.)

Notation for ordinary and partial derivatives

If $f(x)$ is an n times differentiable function then $f^{(n)}(x)$ will, on occasion, be used to signify $d^n f / dx^n$, so that

$$f^{(n)}(x) = \frac{d^n f}{dx^n}.$$

suffix notation for partial derivatives

If $f(x, y)$ is a suitably differentiable function of x and y, a concise notation used to signify partial differentiation involves using suffixes, so that

$$f_x = \frac{\partial f}{\partial x}, \; f_{yx} = (f_y)_x = \frac{\partial}{\partial x}\left(\frac{\partial f}{\partial y}\right) = \frac{\partial^2 f}{\partial y \partial x}, \; f_{yy} = \frac{\partial^2 f}{\partial y^2}, \ldots,$$

with similar results when f is a function of more than two independent variables.

Inverse trigonometric functions

The periodicity of the real variable trigonometric sine, cosine, and tangent functions means that the corresponding general inverse trigonometric functions are *many valued*. So, for example, if $y = \sin x$ and we ask for what values of x is $y = 1/\sqrt{2}$, we find this is true for $x = \pi/4 \pm 2n\pi$ and $x = 3\pi/4 \pm 2n\pi$ for $n = 0, 1, 2, \ldots$. To overcome this ambiguity, we introduce the *single valued* inverses, denoted respectively by $x = \text{Arcsin } y$, $x = \text{Arccos } y$, and $x = \text{Arctan } y$ by restricting the domain and range of the sine, cosine, and tangent functions to one where they are either strictly increasing or strictly decreasing functions, because then one value of x corresponds to one value of y and, conversely, one value of y corresponds to one value of x.

In the case of the function $y = \sin x$, by restricting the argument x to the interval $-\pi/2 \leq x \leq \pi/2$ the function becomes a strictly increasing function of x. The corresponding single valued inverse function is denoted by $x = \text{Arcsin } y$, where y is a number in the domain of definition $[-1, 1]$ of the Arcsine function and x is a number in its range $[-\pi/2, \pi/2]$. Similarly, when considering the function $y = \cos x$, the argument is restricted to $0 \leq x \leq \pi$ to make $\cos x$ a strictly decreasing function of x. The corresponding single valued inverse function is denoted by $x = \text{Arccos } y$, where y is a number in the domain of definition $[-1, 1]$ of the Arccosine function and x is a number in its range $[0, \pi]$. Finally, in the case of the function $y = \tan x$, restricting the argument to the interval $-\pi/2 < x < \pi/2$ makes the tangent function a strictly increasing function of x. The corresponding single valued inverse function is denoted by $x = \text{Arctan } y$ where y is a number in the domain of definition $(-\infty, \infty)$ of the Arctangent function and x is a number in its range $(-\pi/2, \pi/2)$.

As the inverse trigonometric functions are important in their own right, the variables x and y in the preceding definitions are interchanged to allow consideration of the inverse functions $y = \text{Arcsin } x$, $y = \text{Arccos } x$, and $y = \text{Arctan } x$, so that now x is the independent variable and y is the dependent variable.

With this interchange of variables the expression $y = \arcsin x$ will be used to refer to any single valued inverse function with the *same* domain of definition as Arcsin x, but with a *different* range. Similar definitions apply to the functions $y = \arccos x$ and $y = \arctan x$.

Double summations

An expression involving a double summation like

$$\sum_{m=1}^{\infty} \sum_{n=1}^{\infty} a_{mn} \sin mx \sin ny,$$

double summation

means sum the terms $a_{mn} \sin mx \sin ny$ over all possible values of m and n, so that

$$\sum_{m=1}^{\infty} \sum_{n=1}^{\infty} a_{mn} \sin mx \sin ny = a_{11} \sin x \sin y + a_{12} \sin x \sin 2y$$

$$+ a_{21} \sin 2x \sin y + a_{22} \sin 2x \sin 2y + \cdots.$$

A more concise notation also in use involves writing the double summation as

$$\sum_{m=1,n=1}^{\infty} a_{mn} \sin mx \sin nx.$$

The signum function

signum function

The **signum function**, usually written $\text{sign}(x)$, and sometimes $\text{sgn}(x)$, is defined as

$$\text{sign}(x) = \begin{cases} 1 & \text{if } x > 0 \\ -1 & \text{if } x < 0. \end{cases}$$

We have, for example, $\text{sign}(\cos x) = 1$ for $0 < x < \pi/2$, and $\text{sign}(\cos x) = -1$ for $\pi/2 < x < \pi$ or, equivalently,

$$\text{sign}(\cos x) = \begin{cases} 1, & 0 < x < \frac{1}{2}\pi \\ -1, & \frac{1}{2}\pi < x < \pi. \end{cases}$$

Products

Let $\{u_k\}_{k=1}^{n}$ be a sequence of numbers or functions $u_1, u_2, \ldots$; then the product of the n members of this sequence is denoted by $\prod_{k=1}^{n} u_k$, so that

$$\prod_{k=1}^{n} u_k = u_1 u_2 \cdots u_n.$$

infinite product

When the sequence is infinite,

$$\lim_{n \to \infty} \prod_{k=1}^{n} u_k = \prod_{k=1}^{\infty} u_k$$

is called an **infinite product** involving the sequence $\{u_k\}$. Typical examples of infinite products are

$$\prod_{k=2}^{\infty}\left(1-\frac{1}{k^2}\right)=\frac{1}{2} \quad \text{and} \quad \prod_{k=1}^{\infty}\left(1-\frac{x^2}{k^2\pi^2}\right)=\frac{\sin x}{x}.$$

More background information and examples can be found in the appropriate sections in any of references [1.1], [1.2], and [1.5].

Logarithmic functions

the functions ln and Log

The notation $\ln x$ is used to denote the **natural logarithm** of a real number x, that is, the logarithm of x to the base e, and in some books this is written $\log_e x$. In this book logarithms to the base 10 are not used, and when working with functions of a complex variable the notation $\text{Log } z$, with $z = re^{i\theta}$ means $\text{Log } z = \ln r + i\theta$.

EXERCISES 1.1

1. Prove that if $a > 0, b > 0$, then $a/\sqrt{b}+b/\sqrt{a} \geq \sqrt{a}+\sqrt{b}$.

Prove Exercises 2 through 6 by mathematical induction.

2. $\sum_{k=0}^{n-1}(a+kd)=(n/2)[2a+(n-1)d]$
(sum of an arithmetic series).

3. $\sum_{k=0}^{n-1} r^k = (1-r^n)/(1-r) \quad (r \neq 1)$
(sum of a geometric series).

4. $\sum_{k=1}^{n} k^2 = (1/6)n(n+1)(2n+1)$ (sum of squares).

5. $d^n/dx^n[\cos ax] = a^n \cos(ax+n\pi/2)$, with n a natural number.

6. $d^n/dx^n[\ln(1+x)] = (-1)^{n+1}(n-1)!/(1+x)^n$, with n a natural number.

7. Use the binomial theorem to expand $(3+2x)^4$.

8. Use the binomial theorem and multiplication to expand $(1-x^2)(2+3x)^3$.

In Exercises 9 through 12 find the first four terms of the binomial expansion of the function and state conditions for the convergence of the series.

9. $(3+2x)^{-2}$.

10. $(2-x^2)^{1/3}$.

11. $(4+2x^2)^{-1/2}$.

12. $(1-3x^2)^{3/4}$.

1.2   Complex Numbers

Mathematical operations can lead to numbers that do not belong to the real number system **R** introduced in Section 1.1. In the simplest case this occurs when finding the roots of the quadratic equation

$$ax^2+bx+c=0 \quad \text{with } a,b,c \in \mathbf{R}, a \neq 0$$

by means of the quadratic formula

$$x=\frac{-b\pm\sqrt{b^2-4ac}}{2a}.$$

discriminant of a quadratic

The **discriminant** of the equation is $b^2 - 4ac$, and if $b^2 - 4ac < 0$ the formula involves the square root of a negative real number; so, if the formula is to have meaning, numbers must be allowed that lie outside the real number system.

The inadequacy of the real number system when considering different mathematical operations can be illustrated in other ways by asking, for example, how to find the three roots that are expected of a third degree algebraic equation as simple as $x^3 - 1 = 0$, where only the real root 1 can be found using $y = x^3 - 1$, or by seeking to give meaning to $\ln(-1)$, both of which questions will arise later.

Difficulties such as these can all be overcome if the real number system is extended by introducing the **imaginary unit** i defined as

$$i^2 = -1,$$

so expressions like $\sqrt{(-k^2)}$ where k a positive real number may be written $\sqrt{(-1)}\sqrt{(k^2)} = \pm ik$. Notice that as the real number k only *scales* the imaginary unit i, it is immaterial whether the result is written as ik or as ki.

The extension to the real number system that is required to resolve problems of the type just illustrated involves the introduction of **complex numbers**, denoted collectively by **C**, in which the general complex number, usually denoted by z, has the form

$$z = \alpha + i\beta, \quad \text{with } \alpha, \beta \text{ real numbers.}$$

real and imaginary part notation

The real number α is called the **real part** of the complex number z, and the real number β is called its **imaginary part**. When these need to be identified separately, we write

$$\text{Re}\{z\} = \alpha \quad \text{and} \quad \text{Im}\{z\} = \beta,$$

so if $z = 3 - 7i$, $\text{Re}\{z\} = 3$ and $\text{Im}\{z\} = -7$.

If $\text{Im}\{z\} = \beta = 0$ the complex number z reduces to a real number, and if $\text{Re}\{z\} = \alpha = 0$ it becomes a purely imaginary number, so, for example, $z = 5i$ is a purely imaginary number. When a complex number z is considered as a variable it is usual to write it as

$$z = x + iy,$$

where x and y are now real variables. If it is necessary to indicate that z is a general complex number we write $z \in \mathbf{C}$.

When solving the quadratic equation $az^2 + bz + c = 0$ with $a, b,$ and c real numbers and a discriminant $b^2 - 4ac < 0$, by setting $4ac - b^2 = k^2$ in the quadratic formula, with $k > 0$, the two roots z_1 and z_2 are given by the complex numbers

$$z_1 = -(b/2a) + i(k/2a) \quad \text{and} \quad z_2 = -(b/2a) - i(k/2a).$$

Algebraic rules for complex numbers

Let the complex numbers z_1 and z_2 be defined as

$$z_1 = a + ib \quad \text{and} \quad z_2 = c + id,$$

with $a, b, c,$ and d arbitrary real numbers. Then the following rules govern the arithmetic manipulation of complex numbers.

Equality of complex numbers

The complex numbers z_1 and z_2 are **equal**, written $z_1 = z_2$ if, and only if, $\text{Re}\{z_1\} = \text{Re}\{z_2\}$ and $\text{Im}\{z_1\} = \text{Im}\{z_2\}$. So $a + ib = c + id$ if, and only if,

$$a = c \quad \text{and} \quad b = d.$$

EXAMPLE 1.3

(a) $3 - 9i = 3 + bi$ if, and only if, $b = -9$.

(b) If $u = -2 + 5i$, $v = 3 + 5i$, $w = a + 5i$, then
$u = w$ if, and only if, $a = -2$ but $u \neq v$, and
$v = w$ if, and only if, $a = 3$.   ∎

Zero complex number

The **zero** complex number, also called the **null** complex number, is the number $0 + 0i$ that, for simplicity, is usually written as an ordinary zero 0.

EXAMPLE 1.4

If $a + ib = 0$, then $a = 0$ and $b = 0$.   ∎

Addition and subtraction of complex numbers

The **addition (sum)** and **subtraction (difference)** of the complex numbers z_1 and z_2 is defined as

$$z_1 + z_2 = \text{Re}\{z_1\} + \text{Re}\{z_2\} + i[\text{Im}\{z_1\} + \text{Im}\{z_2\}]$$

and

$$z_1 - z_2 = \text{Re}\{z_1\} - \text{Re}\{z_2\} + i[\text{Im}\{z_1\} - \text{Im}\{z_2\}].$$

So, if $z_1 = a + ib$ and $z_2 = c + id$, then

$$\begin{aligned} z_1 + z_2 &= (a + ib) + (c + id) \\ &= (a + c) + i(b + d), \end{aligned}$$

and

$$\begin{aligned} z_1 - z_2 &= (a + ib) - (c + id) \\ &= (a - c) + i(b - d). \end{aligned}$$

EXAMPLE 1.5

If $z_1 = 3 + 7i$ and $z_2 = 3 + 2i$, then the sum

$$z_1 + z_2 = (3 + 3) + (7 + 2)i = 6 + 9i,$$

and the difference

$$z_1 - z_2 = (3 - 3) + (7 - 2)i = 5i.$$ ∎

Multiplication of complex numbers

The **multiplication (product)** of the two complex numbers $z_1 = a + ib$ and $z_2 = c + id$ is defined by the rule

$$z_1 z_2 = (a + ib)(c + id) = (ac - bd) + i(ad + bc).$$

An immediate consequence of this definition is that if k is a real number, then $kz_1 = k(a + ib) = ka + ikb$. This operation involving multiplication of a complex number by a real number is called *scaling* a complex number. Thus, if $z_1 = 3 + 7i$ and $z_2 = 3 + 2i$, then $2z_1 - 3z_2 = (6 + 14i) - (9 + 6i) = -3 + 8i$.

In particular, if $z = a + ib$, then $-z = (-1)z = -a - ib$. This is as would be expected, because it leads to the result $z - z = 0$.

In practice, instead of using this formal definition of multiplication, it is more convenient to perform multiplication of complex numbers by multiplying the bracketed quantities in the usual algebraic manner, replacing every product i^2 by -1, and then combining separately the real and imaginary terms to arrive at the required product.

EXAMPLE 1.6

(a) $5i(-4 + 3i) = -15 - 20i$.

(b) $(3 - 2i)(-1 + 4i)(1 + i) = (-3 + 12i + 2i - 8i^2)(1 + i)$
$$= [(-3 + 8) + (12 + 2)i](1 + i) = (5 + 14i)(1 + i)$$
$$= 5 + 14i + 5i + 14i^2 = (5 - 14) + (5 + 14)i = -9 + 19i.$$ ∎

Complex conjugate

If $z = a + ib$, then the **complex conjugate** of z, usually denoted by $\bar{z}$ and read "z bar," is defined as $\bar{z} = a - ib$. It follows directly that

$$\overline{(\bar{z})} = z \quad \text{and} \quad z\bar{z} = a^2 + b^2.$$

In words, the complex conjugate operation has the property that taking the complex conjugate of a complex conjugate returns the original complex number, whereas the product of a complex number and its complex conjugate always yields a real number.

If $z = a + ib$, then adding and subtracting z and $\bar{z}$ gives the useful results

$$z + \bar{z} = 2\text{Re}\{z\} = 2a \quad \text{and} \quad z - \bar{z} = 2i\,\text{Im}\{z\} = 2ib.$$

These can be written in the equivalent form

$$\text{Re}\{z\} = a = \frac{1}{2}(z + \bar{z}) \quad \text{and} \quad \text{Im}\{z\} = b = \frac{1}{2i}(z - \bar{z}).$$

Quotient (division) of complex numbers

Let $z_1 = a + ib$ and $z_2 = c + id$. Then the **quotient** z_1/z_2 is defined as

$$\frac{z_1}{z_2} = \frac{(ac + bd) + i(bc - ad)}{c^2 + d^2}, \ z_2 \neq 0.$$

In practice, division of complex numbers is not carried out using this definition. Instead, the quotient is written in the form

$$\frac{z_1}{z_2} = \frac{z_1 \bar{z}_2}{z_2 \bar{z}_2},$$

where the denominator is now seen to be a real number. The quotient is then found by multiplying out and simplifying the numerator in the usual manner and dividing the real and imaginary parts of the numerator by the real number $z_2 \bar{z}_2$.

EXAMPLE 1.7   Find z_1/z_2 given that $z_1 = (3 + 2i)$ and $z_2 = 1 + 3i$.

Solution

$$\frac{3 + 2i}{1 + 3i} = \frac{(3 + 2i)(1 - 3i)}{(1 + 3i)(1 - 3i)} = \frac{3 - 9i + 2i - 6i^2}{10} = \frac{9}{10} - \frac{7i}{10}. \qquad \blacksquare$$

Modulus of a complex number

The **modulus** of the complex number $z = a + ib$ denoted by $|z|$, and also called its **magnitude**, is defined as

$$|z| = (a^2 + b^2)^{1/2} = (z\bar{z})^{1/2}.$$

It follows directly from the definitions of the modulus and division that

$$|z| = |\bar{z}| = (a^2 + b^2)^{1/2},$$

and

$$z_1/z_2 = z_1 \bar{z}_2 / |z_2|^2.$$

EXAMPLE 1.8   If $z = 3 + 7i$, then $|z| = |3 + 7i| = (3^2 + 7^2)^{1/2} = \sqrt{58}$. $\qquad \blacksquare$

It is seen that the foregoing rules for the arithmetic manipulation of complex numbers reduce to the ordinary arithmetic rules for the algebraic manipulation of real numbers when all the complex numbers involved are real numbers. Complex numbers are the most general numbers that need to be used in mathematics, and they contain the real numbers as a special case. There is, however, a fundamental difference between real and complex numbers to which attention will be drawn after their common properties have been listed.

Properties shared by real and complex numbers

Let z, u, and w be arbitrary real or complex numbers. Then the following properties are true:

1. $z + u = u + z$.   This means that the order in which complex numbers are added does not affect their sum.
2. $zu = uz$.   This means that the order in which complex numbers are multiplied does not affect their product.

3. $(z + u) + w = z + (u + w)$.   This means that the order in which brackets are inserted into a sum of finitely many complex numbers does not affect the sum.

4. $z(uw) = (zu)w$.   This means that the terms in a product of complex numbers may be grouped and multiplied in any order without affecting the resulting product.

5. $z(u + w) = zu + zw$.   This means that the product of z and a sum of complex numbers equals the sum of the products of z and the individual complex numbers involved in the sum.

6. $z + 0 = 0 + z = z$.   This result means that the addition of zero to any complex number leaves it unchanged.

7. $z \cdot 1 = 1 \cdot z = z$.   This result means that multiplication of any complex number by unity leaves the complex number unchanged.

Despite the properties common to real and complex numbers just listed, there remains a fundamental difference because, unlike real numbers, complex numbers have *no* natural order. So if z_1 and z_2 are any complex numbers, a statement such as $z_1 < z_2$ has no meaning.

EXERCISES 1.2

Find the roots of the equations in Exercises 1 through 6.

1. $z^2 + z + 1 = 0$.

2. $2z^2 + 5z + 4 = 0$.

3. $z^2 + z + 6 = 0$.

4. $3z^2 + 2z + 1 = 0$.

5. $3z^2 + 3z + 1 = 0$.

6. $2z^2 - 2z + 3 = 0$.

7. Given that $z = 1$ is a root, find the other two roots of $2z^3 - z^2 + 3z - 4 = 0$.

8. Given that $z = -2$ is a root, find the other two roots of $4z^3 + 11z^2 + 10z + 8 = 0$.

9. Given $u = 4 - 2i$, $v = 3 - 4i$, $w = -5i$ and $a + ib = (u + iv)w$, find a and b.

10. Given $u = -4 + 3i$, $v = 2 + 4i$, and $a + ib = uv^2$, find a and b.

11. Given $u = 2 + 3i$, $v = 1 - 2i$, $w = -3 - 6i$, find $|u + v|$, $u + 2v$, $u - 3v + 2w$, uv, uvw, $|u/v|$, v/w.

12. Given $u = 1 + 3i$, $v = 2 - i$, $w = -3 + 4i$, find uv/w, uw/v and $|v|w/u$.

1.3   The Complex Plane

Complex numbers can be represented geometrically either as *points*, or as *directed line segments* (*vectors*), in the **complex plane**. The complex plane is also called the **z-plane** because of the representation of complex numbers in the form $z = x + iy$. Both of these representations are accomplished by using rectangular cartesian coordinates and plotting the complex number $z = a + ib$ as the point (a, b) in the plane, so the x-coordinate of z is $a = \text{Re}\{z\}$ and its y-coordinate is $b = \text{Im}\{z\}$. Because of this geometrical representation, a complex number written in the form $z = a + ib$ is said to be expressed in **cartesian form**. To acknowledge the Swiss amateur mathematician Jean-Robert Argand, who introduced the concept of the complex plane in 1806, and who by profession was a bookkeeper, this representation is also called the **Argand diagram**.

cartesian representation of z

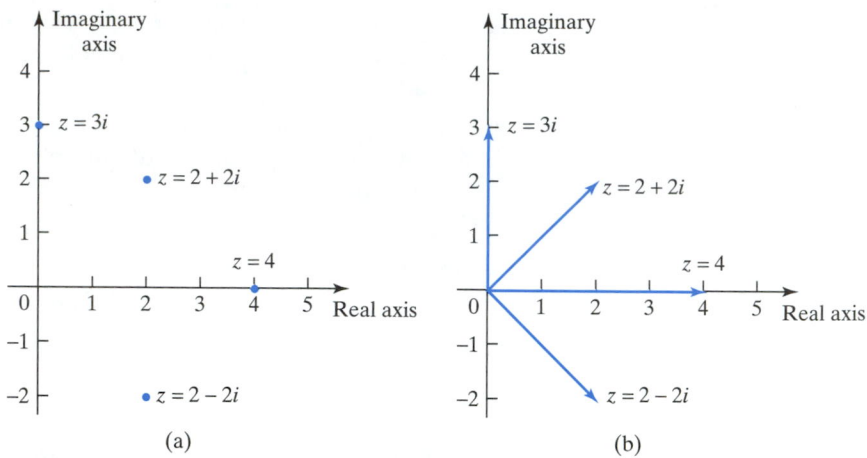

FIGURE 1.1   (a) Complex numbers as points. (b) Complex numbers as vectors.

For obvious reasons, the x-axis is called the **real axis** and the y-axis the **imaginary axis**. Purely real numbers are represented by points on the real axis and purely imaginary ones by points on the imaginary axis. Examples of the representation of typical points in the complex plane are given in Fig. 1.1a, where the numbers $4, 3i, 2 + 2i$, and $2 - 2i$ are plotted as points. These same complex numbers are shown again in Fig. 1.1b as directed line segments drawn from the origin (vectors). The arrow shows the *sense* along the line, that is, the direction from the origin to the tip of the vector representing the complex number. It can be seen from both figures that, when represented in the complex plane, a complex number and its complex conjugate (in this case $2 + 2i$ and $2 - 2i$) lie symmetrically above and below the real axis. Another way of expressing this result is by saying that a complex number and its complex conjugate appear as **reflections** of each other in the real axis, which acts like a mirror.

The addition and subtraction of two complex numbers have convenient geometrical interpretations that follow from the definitions given in Section 1.2. When complex numbers are added, their respective real and imaginary parts are added, whereas when they are subtracted, their respective real and imaginary parts are subtracted. This leads at once to the **triangle law** for addition illustrated in Fig. 1.2a, in which the directed line segment (vector) representing z_2 is translated without rotation or change of scale, to bring its base (the end opposite to the arrow) into coincidence with the tip of the directed line element representing z_1 (the end at which the arrow is located). The sum $z_1 + z_2$ of the two complex numbers is then represented by the directed line segment from the base of the line segment representing z_1 to the tip of the newly positioned line segment representing z_2.

triangle and parallelogram laws

The name *triangle law* comes from the triangle that is constructed in the complex plane during this geometrical process of addition. Notice that an immediate consequence of this law is that addition is *commutative*, because both $z_1 + z_2$ and $z_2 + z_1$ are seen to lead to the same directed line segment in the complex plane. For this reason the addition of complex numbers is also said to obey the **parallelogram law** for addition, because the commutative property generates the parallelogram shown in Fig. 1.2a.

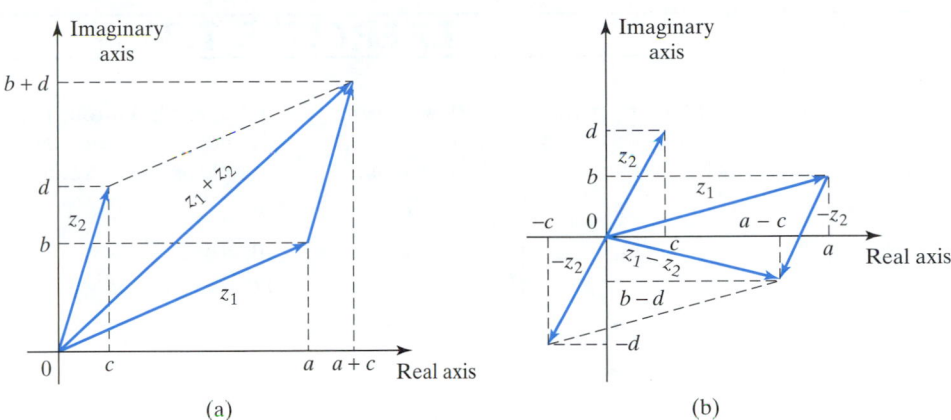

FIGURE 1.2   Addition and subtraction of complex numbers using the triangle/parallelogram law.

The geometrical interpretation of the subtraction of z_2 from z_1 follows similarly by adding to z_1 the directed line segment $-z_2$ that is obtained by reversing of the sense (arrow) along z_2, as shown in Fig. 1.2b.

It is an elementary fact from Euclidean geometry that the sum of the lengths of the two sides $|u|$ and $|v|$ of the triangle in Fig. 1.3 is greater than or equal to the length of the hypotenuse $|u + v|$, so from geometrical considerations we can write

$$|u + v| \leq |u| + |v|.$$

triangle inequality   This result involving the moduli of the complex numbers u and v is called the **triangle inequality** for complex numbers, and it has many applications.

An algebraic proof of the triangle inequality proceeds as follows:

$$\begin{aligned} |u + v|^2 = (u + v)\overline{(u + v)} &= u\bar{u} + v\bar{u} + u\bar{v} + v\bar{v} \\ &= |u|^2 + |v|^2 + (u\bar{v} + \overline{u\bar{v}}) \leq |u^2| + |v^2| + 2|u\bar{v}| \\ &= (|u| + |v|)^2. \end{aligned}$$

The required result now follows from taking the positive square root.

A similar argument, the proof of which is left as an exercise, can be used to show that $\|u| - |v\| \leq |u + v|$, so when combined with the triangle inequality we have

$$\|u| - |v\| \leq |u + v| \leq |u| + |v|.$$

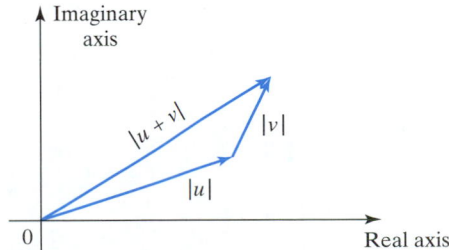

FIGURE 1.3   The triangle inequality.

EXERCISES 1.3

In Exercises 1 through 8 use the parallelogram law to form the sum and difference of the given complex numbers and then verify the results by direct addition and subtraction.

In Exercises 9 through 11 use the parallelogram law to verify the triangle inequality $|u + v| \leq |u| + |v|$ for the given complex numbers u and v.

1. $u = 2 + 3i, v = 1 - 2i$.

5. $u = 3 + 6i, v = -4 + 2i$.

9. $u = -4 + 2i, v = 3 + 5i$.

2. $u = 4 + 7i, v = -2 - 3i$.

6. $u = -3 + 2i, v = 6i$.

10. $u = 2 + 5i, v = 3 - 2i$.

3. $u = -3, v = -3 - 4i$.

7. $u = -4 + 2i, v = -4 - 10i$.

11. $u = -3 + 5i, v = 2 + 6i$.

4. $u = 4 + 3i, v = 3 + 4i$.

8. $u = 4 + 7i, v = -3 + 5i$.

1.4   Modulus and Argument Representation of Complex Numbers

polar representation of z

When representing $z = x + iy$ in the complex plane by a point P with coordinates (x, y), a natural alternative to the cartesian representation is to give the *polar coordinates* (r, θ) of P. This polar representation of z is shown in Fig. 1.4, where

$$OP = r = |z| = (x^2 + y^2)^{1/2} \quad \text{and} \quad \tan \theta = y/x. \tag{1}$$

The radial distance OP is the **modulus** of z, so $r = |z|$, and the angle θ measured counterclockwise from the positive real axis is called the **argument** of z. Because of this, a complex number expressed in terms of the polar coordinates (r, θ) is said to be in **modulus–argument** form. The argument θ is indeterminate up to a multiple of 2π, because the polar coordinates (r, θ), and $(r, \theta + 2k\pi)$, with $k = \pm 1, \pm 2, \ldots$, identify the *same* point P. By convention, the the angle θ is called the **principal value** of the argument of z when it lies in the interval $-\pi < \theta \leq \pi$. To distinguish the principal value of the argument from all of its other values, we write

$$\text{Arg } z = \theta, \quad \text{when } -\pi < \theta \leq \pi. \tag{2}$$

The values of the argument of z that differ from this value of θ by a multiple of 2π are denoted by arg z, so that

$$\arg z = \theta + 2k\pi, \quad \text{with } k = \pm 1, \pm 2, \ldots . \tag{3}$$

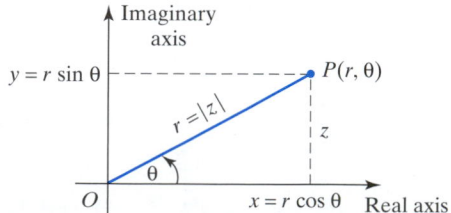

FIGURE 1.4   The complex plane and the (r, θ) representation of z.

The significance of the multivalued nature of arg z will become apparent later when the roots of complex numbers are determined.

The connection between the cartesian coordinates (x, y) and the polar coordinates (r, θ) of the point P corresponding to $z = x + iy$ is easily seen to be given by

$$x = r \cos \theta \quad \text{and} \quad y = r \sin \theta.$$

modulus–argument representation of z

This leads immediately to the representation of $z = x + iy$ in the alternative *modulus–argument form*

$$z = r(\cos \theta + i \sin \theta). \tag{4}$$

A routine calculation using elementary trigonometric identities shows that

$$(\cos \theta + i \sin \theta)^2 = (\cos 2\theta + i \sin 2\theta).$$

An inductive argument using the above result as its first step then establishes the following simple but important theorem.

THEOREM 1.1   **De Moivre's theorem**

$$(\cos \theta + i \sin \theta)^n = (\cos n\theta + i \sin n\theta), \quad \text{for } n \text{ a natural number.} \qquad \blacksquare$$

EXAMPLE 1.9   Use de Moivre's theorem to express $\cos 4\theta$ and $\sin 4\theta$ in terms of powers of $\cos \theta$ and $\sin \theta$.

Solution   The result is obtained by first setting $n = 4$ in de Moivre's theorem and expanding $(\cos \theta + i \sin \theta)^4$ to obtain

$$\cos^4 \theta + 4i \cos^3 \theta \sin \theta - 6 \cos^2 \theta \sin^2 \theta - 4i \cos \theta \sin^3 \theta + \sin^4 \theta = \cos 4\theta + i \sin 4\theta.$$

Equating the respective real and imaginary parts on either side of this identity gives the required results

$$\cos 4\theta = \cos^4 \theta - 6 \cos^2 \theta \sin^2 \theta + \sin^4 \theta$$

and

$$\sin 4\theta = 4 \cos^3 \theta \sin \theta - 4 \cos \theta \sin^3 \theta. \qquad \blacksquare$$

As the complex number $z = \cos \theta + i \sin \theta$ has unit modulus, it follows that all numbers of this form lie on the unit circle (a circle of radius 1) centered on the origin, as shown in Fig. 1.5.

Using (5), we see that if $z = r(\cos \theta + i \sin \theta)$, then

$$z^n = r^n(\cos n\theta + i \sin n\theta), \text{ for } n \text{ a natural number.} \tag{5}$$

The relationship between e^θ, $\sin \theta$, and $\cos \theta$ can be seen from the following well-known series expansions of the functions

$$e^\theta = \sum_{n=0}^{\infty} \frac{\theta^n}{n!} = 1 + \theta + \frac{\theta^2}{2!} + \frac{\theta^3}{3!} + \frac{\theta^4}{4!} + \frac{\theta^5}{5!} + \frac{\theta^6}{6!} + \cdots;$$

$$\sin \theta = \sum_{n=0}^{\infty} (-1)^n \frac{\theta^{2n+1}}{(2n+1)!} = \theta - \frac{\theta^3}{3!} + \frac{\theta^5}{5!} - \frac{\theta^7}{7!} + \cdots;$$

$$\cos \theta = \sum_{n=0}^{\infty} (-1)^n \frac{\theta^{2n}}{(2n)!} = 1 - \frac{\theta^2}{2!} + \frac{\theta^4}{4!} - \frac{\theta^6}{6!} + \cdots.$$

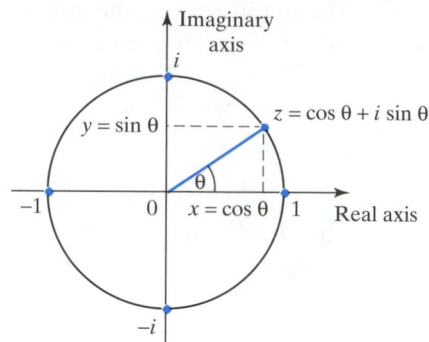

FIGURE 1.5   Point $z = \cos\theta + i\sin\theta$ on the unit circle centered on the origin.

By making a formal power series expansion of the function $e^{i\theta}$, simplifying powers of i, grouping together the real and imaginary terms, and using the series representations for $\cos\theta$ and $\sin\theta$, we arrive at what is called the real variable form of the

Euler formula   **Euler formula**

$$e^{i\theta} = \cos\theta + i\sin\theta, \quad \text{for any real } \theta. \tag{6}$$

This immediately implies that if $z = re^{i\theta}$, then

$$z^{\alpha} = r^{\alpha} e^{i\alpha\theta}, \quad \text{for any real } \alpha. \tag{7}$$

When θ is restricted to the interval $-\pi < \theta \leq \pi$, formula (6) leads to the useful results

$$1 = e^{i0}, \quad i = e^{i\pi/2}, \quad -1 = e^{i\pi}, \quad -i = e^{-i\pi/2}$$

and, in particular, to

$$1 = e^{2k\pi i} \quad \text{for } k = 0, \pm 1, \pm 2, \ldots.$$

The Euler form for complex numbers makes their multiplication and division very simple. To see this we set $z_1 = r_1 e^{i\alpha}$ and $z_2 = r_2 e^{i\beta}$ and then use the results

$$z_1 z_2 = r_1 r_2 e^{i(\alpha+\beta)} \quad \text{and} \quad z_1/z_2 = r_1/r_2 e^{i(\alpha-\beta)}. \tag{8}$$

These show that when complex numbers are multiplied, their moduli are *multiplied* and their arguments are *added*, whereas when complex numbers are divided, their moduli are *divided* and their arguments are *subtracted*.

EXAMPLE 1.10   Find $uv, u/v$, and u^{25} given that $u = 1 + i, v = \sqrt{3} - i$.

Solution $u = 1 + i = \sqrt{2}e^{i\pi/4}$, $v = \sqrt{3} - i = 2e^{-i\pi/6}$, so $uv = 2\sqrt{2}e^{i\pi/12}, u/v = (1/\sqrt{2})e^{i5\pi/12}$ while $u^{25} = (\sqrt{2}e^{i\pi/4})^{25} = (\sqrt{2})^{25}(e^{i\pi/4})^{25} = 4096\sqrt{2}(e^{i(6+1/4)\pi}) = 4096\sqrt{2}(e^{i6\pi})(e^{i\pi/4}) = 4096\sqrt{2}(e^{i\pi/4}) = 4096\sqrt{2}(1 + i)$. ∎

To find the principal value of the argument of a given complex number z, namely Arg z, use should be made of the signs of $x = \text{Re}\{z\}$, and $y = \text{Im}\{z\}$ together with the results listed below, all of which follow by inspection of Fig. 1.5.

Signs of x and y	_Arg z = θ_
$x < 0, \ y < 0$	$-\pi < \theta < -\pi/2$
$x > 0, \ y < 0$	$-\pi/2 < \theta < 0$
$x > 0, \ y > 0$	$0 < \theta < \pi/2$
$x < 0, \ y > 0$	$\pi/2 < \theta < \pi$

EXAMPLE 1.11   Find $r = |z|$, Arg z, arg z, and the modulus–argument form of the following values of z.

(a) $-2\sqrt{3} - 2i$ (b) $-1 + i\sqrt{3}$ (c) $1 + i$ (d) $2 - i2\sqrt{3}$.

Solution   **(a)** $r = \{(-2\sqrt{3})^2 + (-2)^2\}^{1/2} = 4$, $\text{Arg} z = \theta = -5\pi/6$ and $\arg z = -5\pi/6 + 2k\pi, k = \pm1, \pm2, \ldots, z = 4(\cos(-5\pi/6) + i\sin(-5\pi/6))$.

(b) $r = \{(-1)^2 + (\sqrt{3})^2\}^{1/2} = 2$, $\text{Arg} z = \theta = 2\pi/3$ and $\arg z = 2\pi/3 + 2k\pi$, $k = \pm1, \pm2, \ldots, z = 2(\cos(2\pi/3) + i\sin(2\pi/3))$.

(c) $r = \{(1)^2 + (1)^2\}^{1/2} = \sqrt{2}$, $\text{Arg} z = \theta = \pi/4$ and $\arg z = \pi/4 + 2k\pi$, $k = \pm1, \pm2, \ldots, z = \sqrt{2}(\cos(\pi/4) + i\sin(\pi/4))$.

(d) $r = \{(2)^2 + (-2\sqrt{3})^2\}^{1/2} = 4$, $\text{Arg} z = \theta = -\pi/3$ and $\arg z = -\pi/3 + 2k\pi$, $k = \pm1, \pm2, \ldots, z = 4(\cos(-\pi/3) + i\sin(-\pi/3))$. ∎

EXERCISES 1.4

1. Expand $(\cos\theta + i\sin\theta)^2$ and then use trigonometric identities to show that

$$(\cos\theta + i\sin\theta)^2 = (\cos 2\theta + i\sin 2\theta).$$

2. Give an inductive proof of de Moivre's theorem

$$(\cos\theta + i\sin\theta)^n = (\cos n\theta + i\sin n\theta),$$

for n a natural number.

3. Use de Moivre's theorem to express $\cos 5\theta$ and $\sin 5\theta$ in terms of powers of $\cos\theta$ and $\sin\theta$.

4. Use de Moivre's theorem to express $\cos 6\theta$ and $\sin 6\theta$ in terms of powers of $\cos\theta$ and $\sin\theta$.

5. Show by expanding $(\cos\alpha + i\sin\alpha)(\cos\beta + i\sin\beta)$ and using trigonometric identities that

$$(\cos\alpha + i\sin\alpha)(\cos\beta + i\sin\beta)$$
$$= \cos(\alpha + \beta) + i\sin(\alpha + \beta).$$

6. Show by expanding $(\cos\alpha + i\sin\alpha)/(\cos\beta + i\sin\beta)$ and using trigonometric identities that

$$(\cos\alpha + i\sin\alpha)/(\cos\beta + i\sin\beta)$$
$$= \cos(\alpha - \beta) + i\sin(\alpha - \beta).$$

7. If $z = \cos\theta + i\sin\theta = e^{i\theta}$, show that when n is a natural number,

$$\cos(n\theta) = \frac{1}{2}\left(z^n + \frac{1}{z^n}\right) \quad \text{and} \quad \sin(n\theta) = \frac{1}{2i}\left(z^n - \frac{1}{z^n}\right).$$

Use these results to express $\cos^3\theta\sin^3\theta$ in terms of multiple angles of θ. Hint: $\bar{z} = 1/z$.

8. Use the method of Exercise 7 to express $\sin^6\theta$ in terms of multiple angles of θ.

9. By expanding $(z + 1/z)^4$, grouping terms, and using the method of Exercise 7, show that

$$\cos^4\theta = (1/8)(3 + 4\cos 2\theta + \cos 4\theta).$$

10. By expanding $(z - 1/z)^5$, grouping terms, and using the method of Exercise 7, show that

$$\sin^5\theta = (1/16)(\sin 5\theta - 5\sin 3\theta + 10\sin\theta).$$

11. Use the method of Exercise 7 to show that

$$\cos^3\theta + \sin^3\theta = (1/4)(\cos 3\theta + 3\cos\theta$$
$$- \sin^3\theta + 3\sin\theta).$$

In Exercises 12 through 15 express the functions of u, v, and w in modulus-argument form.

12. uv, u/v, and v^5, given that $u = 2 - 2i$ and $v = 3 + i3\sqrt{3}$.

13. uv, u/v, and u^7, given that $u = -1 - i\sqrt{3}$, $v = -4 + 4i$.

14. uv, u/v, and v^6, given that $u = 2 - 2i$, $v = 2 - i2\sqrt{3}$.

15. uvw, uw/v, and w^3/u^4, given that $u = 2 - 2i$, $v = 3 - i3\sqrt{3}$ and $w = 1 + i$.

16. Express $[(-8 + i8\sqrt{3})/(-1 - i)]^2$ in modulus–argument form.

17. Find in modulus–argument form $[(1 + i\sqrt{3})^3/(-1 + i)^2]^3$.

18. Use the factorization

$$(1 - z^{n+1}) = (1 - z)(1 + z + z^2 + \cdots + z^n) \qquad (z \neq 1)$$

with $z = e^{i\theta} = \exp(i\theta)$ to show that

$$\sum_{k=1}^{n} \exp(ik\theta) = \frac{\exp(in\theta) - 1}{1 - \exp(-i\theta)}.$$

19. Use the final result of Exercise 18 to show that

$$\sum_{k=1}^{n} \exp(ik\theta) = \frac{\exp[i(n + 1/2)\theta] - \exp(i\theta/2)}{\exp(i\theta/2) - \exp(-i\theta/2)},$$

and then use the result to deduce the **Lagrange identity**

$$1 + \cos\theta + \cos 2\theta + \cdots + \cos n\theta$$
$$= 1/2 + \frac{\sin[(n + 1/2)\theta]}{2\sin(\theta/2)}, \quad \text{for } 0 < \theta < 2\pi.$$

1.5   Roots of Complex Numbers

It is often necessary to find the n values of $z^{1/n}$ when n is a positive integer and z is an arbitrary complex number. This process is called finding the **nth roots** of z. To determine these roots we start by setting

$$w = z^{1/n}, \quad \text{which is equivalent to } w^n = z.$$

Then, after defining w and z in modulus–argument form as

$$w = \rho e^{i\phi} \quad \text{and} \quad z = re^{i\theta}, \tag{9}$$

we substitute for w and z in $w^n = z$ to obtain

$$\rho^n e^{in\phi} = re^{i\theta}.$$

It is at this stage, in order to find all n roots, that use must be made of the many-valued nature of the argument of a complex number by recognizing that $1 = e^{2k\pi i}$ for $k = 0, \pm 1, \pm 2, \dots$. Using this result we now multiply the right-hand side of the foregoing result by by $e^{2k\pi i}$ (that is, by 1) to obtain

$$\rho^n e^{in\phi} = re^{i\theta} e^{2k\pi i} = re^{i(\theta + 2k\pi)}.$$

Equality of complex numbers in modulus–argument form means the equality of their moduli and, correspondingly, the equality of their arguments, so applying this to the last result we have

$$\rho^n = r \quad \text{and} \quad n\phi = \theta + 2k\pi,$$

showing that

$$\rho = r^{1/n} \quad \text{and} \quad \phi = (\theta + 2k\pi)/n.$$

Here $r^{1/n}$ is simply the nth positive root of r: $\rho = \sqrt[n]{r}$.

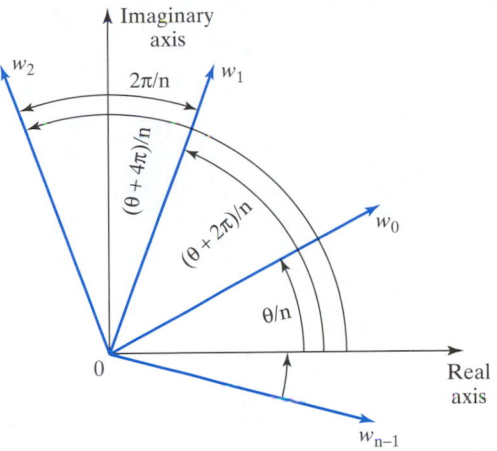

FIGURE 1.6   Location of the roots of $z^{1/n}$.

nth roots of a complex number z

Finally, when we substitute these results into the expression for w, we see that the n values of the roots denoted by $w_0, w_1, \ldots, w_{n-1}$ are given by

$$w_k = r^{1/n}\{\cos[(\theta + 2k\pi)/n] + i \sin[(\theta + 2k\pi)/n]\},$$
$$\text{for } k = 0, 1, \ldots, n-1. \tag{10}$$

Notice that it is only necessary to allow k to run through the successive integers $0, 1, \ldots, n-1$, because the period of the sine and cosine functions is 2π, so allowing k to increase beyond the value $n-1$ will simply repeat this *same* set of roots. An identical argument shows that allowing k to run through successive negative integers can again only generate the same n roots $w_0, w_1, \ldots, w_{n-1}$.

Examination of the arguments of the roots shows them to be spaced uniformly around a circle of radius $r^{1/n}$ centered on the origin. The angle between the radial lines drawn from the origin to each successive root is $2\pi/n$, with the radial line from the origin to the first root w_0 making an angle θ/n to the positive real axis, as shown in Fig. 1.6. This means that if the location on the circle of any one root is known, then the locations of the others follow immediately.

Writing unity in the form $1 = e^{i0}$ shows its modulus to be $r = 1$ and the principal value of its argument to be $\theta = 0$. Substitution in formula (10) then shows the n roots of $1^{1/n}$, called the **nth roots of unity**, to be

$$w_0 = 1, \quad w_1 = e^{i\pi/n}, \quad w_2 = e^{i2\pi/n}, \ldots, w_{n-1} = e^{i(n-1)\pi/n}. \tag{11}$$

By way of example, the fifth roots of unity are located around the unit circle as shown in Fig. 1.7.

If we set $\omega = w_1$, it follows that the nth roots of unity can be written in the form

$$1, \omega, \omega^2, \ldots, \omega^{n-1}.$$

As $\omega^n = 1$ and $\omega^n - 1 = (\omega - 1)(1 + \omega + \omega^2 + \cdots + \omega^{n-1}) = 0$, as $\omega_1 \neq 1$ we see that the the nth roots of unity satisfy

$$1 + \omega + \omega^2 + \cdots + \omega^{n-1} = 0. \tag{12}$$

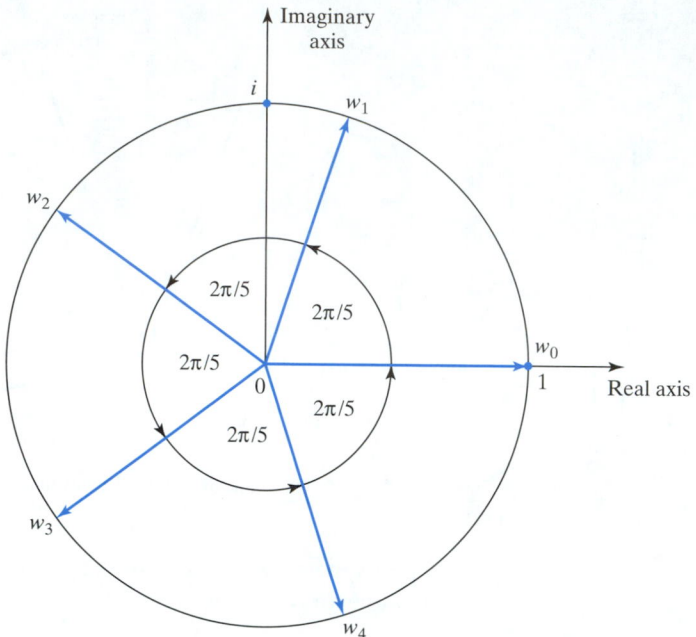

FIGURE 1.7   The fifth roots of unity.

This result remains true if ω is replaced by any one of the other nth roots of unity, with the exception of 1 itself.

EXAMPLE 1.12   Find $w = (1 + i)^{1/3}$.

Solution   Setting $z = 1 + i = \sqrt{2}e^{i\pi/4}$ shows that $r = |z| = \sqrt{2}$ and $\theta = \pi/4$. Substituting these results into formula (1) gives

$$w_k = 2^{1/6}\{\cos[(1/12)(1 + 8k)\pi] + i\sin[(1/12)(1 + 8k)\pi]\}, \quad \text{for } k = 0, 1, 2. \quad \blacksquare$$

The square root of a complex number $\zeta = \alpha + i\beta$ is often required, so we now derive a useful formula for its two roots in terms of $|\zeta|$, α and the sign of β. To obtain the result we consider the equation

$$z^2 = \zeta, \quad \text{where } \zeta = \alpha + i\beta,$$

and let Arg $\zeta = \theta$. Then we may write

$$z^2 = |\zeta|e^{i\theta},$$

and taking the square root of this result we find the two square roots z_- and z_+ are given by

$$z_\pm = \pm|\zeta|^{1/2}e^{i\theta/2}$$
$$= \pm|\zeta|^{1/2}\{\cos(\theta/2) + i\sin(\theta/2)\}.$$

Now $\cos\theta = \alpha/|\zeta|$, but

$$\cos^2(\theta/2) = (1/2)(1 + \cos\theta), \quad \text{and} \quad \sin^2(\theta/2) = (1/2)(1 - \cos\theta),$$

so

$$\cos^2(\theta/2) = (1/2)(1 + \alpha/|\zeta|), \quad \text{and} \quad \sin^2(\theta/2) = (1/2)(1 - \alpha/|\zeta|).$$

As $-\pi < \theta \leq \pi$, it follows that in this interval $\cos(\theta/2)$ is nonnegative, so taking the square root of $\cos^2(\theta/2)$ we obtain

$$\cos(\theta/2) = \left(\frac{|\zeta| + \alpha}{2|\zeta|}\right)^{1/2}.$$

However, the function $\sin(\theta/2)$ is negative in the interval $-\pi < \theta < 0$ and positive in the interval $0 < \theta < \pi$, and so has the same sign as β. Thus, the square root of $\sin^2(\theta/2)$ can be written in the form

$$\sin(\theta/2) = \text{sign}\,(\beta)\left(\frac{|\zeta| - \alpha}{2|\zeta|}\right)^{1/2}.$$

Using these expressions for $\cos(\theta/2)$ and $\sin(\theta/2)$ in the square roots $z_\pm$ brings us to the following useful rule.

Rule for finding the square root of a complex number

Let $z^2 = \zeta$, with $\zeta = \alpha + i\beta$. Then the square roots z_+ and z_- of ζ are given by

$$z_+ = \left(\frac{|\zeta| + \alpha}{2}\right)^{1/2} + i\,\text{sign}\,(\beta)\left(\frac{|\zeta| - \alpha}{2}\right)^{1/2}$$

$$z_- = -\left(\frac{|\zeta| + \alpha}{2}\right)^{1/2} - i\,\text{sign}\,(\beta)\left(\frac{|\zeta| - \alpha}{2}\right)^{1/2}.$$

EXAMPLE 1.13   Find the square roots of (a) $\zeta = 1 + i$ and (b) $\zeta = 1 - i$.

Solution   (a) $\zeta = 1 + i$ so $|\zeta| = \sqrt{2}$, $\alpha = 1$ and $\text{sign}(\beta) = 1$, so the square roots of $\zeta = 1 + i$ are

$$z_\pm = \pm\left\{\left(\frac{\sqrt{2} + 1}{2}\right)^{1/2} + i\left(\frac{\sqrt{2} - 1}{2}\right)^{1/2}\right\}.$$

(b) $\zeta = 1 - i$, so $|\zeta| = \sqrt{2}$, $\alpha = 1$ and $\text{sign}(\beta) = -1$, from which it follows that the square roots of $\zeta = 1 - i$ are

$$z_\pm = \pm\left\{\left(\frac{\sqrt{2} + 1}{2}\right)^{1/2} - i\left(\frac{\sqrt{2} - 1}{2}\right)^{1/2}\right\}.$$

The theorem that follows provides information about the roots of polynomials with *real* coefficients that proves to be useful in a variety of ways.

THEOREM 1.2   **Roots of a polynomial with real coefficients**   Let

$$P(z) = z^n + a_1 z^{n-1} + a_2 z^{n-2} + \cdots a_{n-1} z + a_n$$

be a polynomial of **degree** n in which all the coefficients $a_1, a_2, \ldots, a_n$ are real. Then either all the n **roots** of $P(z) = 0$ are real, that is, the n **zeros** of $P(z)$ are all real, or any that are complex must occur in complex conjugate pairs.

Proof   The proof uses the following simple properties of the complex conjugate operation.

1. If a is real, then $\bar{a} = a$. This result follows directly from the definition of the complex conjugate operation.
2. If b and c are any two complex numbers, then $\overline{b + c} = \bar{b} + \bar{c}$. This result also follows directly from the definition of the complex conjugate operation.
3. If b and c are any two complex numbers, then $\overline{bc} = \bar{b}\bar{c}$ and $\overline{b^r} = (\bar{b})^r$.

We now proceed to the proof. Taking the complex conjugate of $P(z) = 0$ gives

$$\bar{z}^n + \overline{a_1 z^{n-1}} + \overline{a_2 z^{n-2}} + \cdots + \overline{a_{n-1} z} + \overline{a_n} = 0,$$

but the a_r are all real so $\overline{a_r z^{n-r}} = \overline{a_r}\,\overline{z^{n-r}} = a_r \overline{z^{n-r}} = a_r (\bar{z})^{n-r}$, allowing the preceding equation to be rewritten as

$$(\bar{z})^n + a_1 (\bar{z})^{n-1} + a_2 (\bar{z})^{n-2} + \cdots + a_{n-1}\bar{z} + a_n = 0.$$

This result is simply $P(\bar{z}) = 0$, showing that if z is a complex root of $P(z)$, then so also is $\bar{z}$. Equivalently, z and $\bar{z}$ are both zeros of $P(z)$.

If, however, z is a real root, then $z = \bar{z}$ and the result remains true, so the first part of the theorem is proved. The second part follows from the fact that if $z = \alpha + i\beta$ is a root, then so also is $z = \alpha - i\beta$, and so $(z - \alpha - i\beta)$ and $(z - \alpha + i\beta)$ are factors of $P(z)$. The product of these factors must also be a factor of $P(z)$, but

$$(z - \alpha - i\beta)(z - \alpha + i\beta) = z^2 - 2\alpha z + \alpha^2 + \beta^2,$$

and the expression on the right is a quadratic in z with real coefficients, so the final result of the theorem is established.   ∎

EXAMPLE 1.14   Find the roots of $z^3 - z^2 - z - 2 = 0$, given that $z = 2$ is a root.

Solution   If $z = 2$ is a root of $P(z) = 0$, then $z - 2$ is a factor of $P(z)$, so dividing $P(z)$ by $z - 2$ we obtain $z^2 + z + 1$. The remaining two roots of $P(z) = 0$ are the roots of $z^2 + z + 1 = 0$. Solving this quadratic equation we find that $z = (-1 \pm i\sqrt{3})/2$, so the three roots of the equation are 2, $(-1 + i\sqrt{3})/2$, and $(-1 - i\sqrt{3})/2$.   ∎

For more background information and examples on complex numbers, the complex plane and roots of complex numbers, see Chapter 1 of reference [6.1], Sections 1.1 to 1.5 of reference [6.4], and Chapter 1 of reference [6.6].

EXERCISES 1.5

In Exercises 1 through 8 find the square roots of the given complex number by using result (10), and then confirm the result by using the formula for finding the square root of a complex mumber.

1. $-1+i$.

2. $3+2i$.

3. i.

4. $-1+4i$.

5. $2-3i$.

6. $-2-i$.

7. $4-3i$.

8. $-5+i$.

In Exercises 9 through 14 find the roots of the given complex number.

9. $(1+i\sqrt{3})^{1/3}$.

10. $i^{1/4}$.

11. $(-1)^{1/4}$.

12. $(-1-i)^{1/3}$.

13. $(-i)^{1/3}$.

14. $(4+4i)^{1/4}$.

15. Find the roots of $z^3 + z(i-1) = 0$.

16. Find the roots of $z^3 + iz/(1+i) = 0$.

17. Use result (12) to show that
$$1 + \cos(2\pi/n) + \cos(4\pi/n) + \cdots$$
$$+ \cos[(2(n-1)\pi/n)] = 0$$
and
$$\sin(2\pi/n) + \sin(4\pi/n) + \cdots + \sin[(2(n-1)\pi/n)] = 0.$$

18. Use Theorem 1.1 and the representation $z = re^{i\theta}$ to prove that if a and b are any two arbitrary complex numbers, then $\overline{ab} = \overline{a}\,\overline{b}$ and $\overline{(a^r)} = (\overline{a})^r$.

19. Given $z = 1$ is a zero of the polynomial $P(z) = z^3 - 5z^2 + 17z - 13$, find its other two zeros and verify that they are complex conjugates.

20. Given that $z = -2$ is a zero of the polynomial $P(z) = z^5 + 2z^4 - 4z - 8$, find its other four zeros and verify that they occur in complex conjugate pairs.

21. Find the two zeros of the quadratic $P(z) = z^2 - 1 + i$, and explain why they do not occur as a complex conjugate pair.

1.6   Partial Fractions

Let $N(x)$ and $D(x)$ be two polynomials. Then a **rational function** of x is any function of the form $N(x)/D(x)$. The method of **partial fractions** involves the decomposition of rational functions into an equivalent sum of simpler terms of the type

$$\frac{P_1}{ax+b}, \frac{P_2}{(ax+b)^2}, \ldots \quad \text{and} \quad \frac{Q_1x+R_1}{Ax^2+Bx+C}, \frac{Q_2x+R_2}{(Ax^2+Bx+C)^2}, \ldots,$$

where the coefficients are all real together with, possibly, a polynomial in x.

The steps in the reduction of a rational function to its partial fraction representation are as follows:

STEP 1   Factorize $D(x)$ into a product of linear factors and quadratic factors with real coefficients with complex roots, called **irreducible** factors. This is the hardest step, and real quadratic factors will only arise when $D(x) = 0$ has pairs of complex conjugate roots (see Theorem 1.2). Use the result to express $D(x)$ in the form

$$D(x) = (a_1x+b_1)^{r_1} \ldots (a_mx+b_m)^{r_m}(A_1x^2+B_1x+C_1)^{s_1}$$
$$\ldots (A_kx^2+B_kx+C_k)^{s_k},$$

where r_i is the number of times the linear factor (a_ix+b_i) occurs in the factorization of $D(x)$, called its **multiplicity**, and s_j is the corresponding multiplicity of the quadratic factor $(A_jx^2+B_jx+C_j)$.

STEP 2   Suppose first that the degree n of the numerator is *less* than the degree d of the denominator. Then, to every different linear factor $(ax + b)$ with multiplicity r, include in the partial fraction expansion the terms

$$\frac{P_1}{(ax + b)} + \frac{P_2}{(ax + b)^2} + \cdots + \frac{P_r}{(ax + b)^r},$$

where the constant coefficients P_i are unknown at this stage, and so are called **undetermined coefficients**.

partial fraction undetermined coefficients

STEP 3   To every quadratic factor $(Ax^2 + Bx + C)^s$ with multiplicity s include in the partial fraction expansion the terms

$$\frac{Q_1 x + R_1}{(Ax^2 + Bx + C)} + \frac{Q_2 x + R_2}{(Ax^2 + Bx + C)^2} + \cdots + \frac{Q_s x + R_s}{(Ax^2 + Bx + C)^s},$$

where the Q_j and R_j for $j = 1, 2, \ldots, s$ are undetermined coefficients.

STEP 4   Take as the partial fraction representation of $N(x)/D(x)$ the sum of all the terms in Steps 2 and 3.

STEP 5   Multiply the expression

$$N(x)/D(x) = \text{Partial fraction representation in Step 4}$$

by $D(x)$, and determine the unknown coefficients by equating the coefficients of corresponding powers of x on either side of this expression to make it an identity (that is, true for all x).

STEP 6   Substitute the values of the coefficients determined in Step 5 into the expression in Step 4 to obtain the required partial fraction representation.

STEP 7   If $n \geq d$, use long division to divide the denominator into the numerator to obtain the sum of a polynomial of degree $n - d$ of the form

$$T_0 + T_1 x + T_2 x^2 + \cdots + T_{n-d} x^{n-d},$$

together with a remainder term in the form of a rational function $R(x)$ of the type just considered. Find the partial fraction representation of the rational function $R(x)$ using Steps 1 to 6. The required partial fraction representation is then the sum of the polynomial found by long division and the partial fraction representation of $R(x)$.

EXAMPLE 1.15   Find the partial fraction representations of

(a) $F(x) = \dfrac{x^2}{(x + 1)(x - 2)(x + 3)}$ and   (b) $F(x) = \dfrac{2x^3 - 4x^2 + 3x + 1}{(x - 1)^2}.$

Solution   **(a)** All terms in the denominator are linear factors, so by Step 1 the appropriate form of partial fraction representation is

$$\frac{x^2}{(x + 1)(x - 2)(x + 3)} = \frac{A}{x + 1} + \frac{B}{x - 2} + \frac{C}{x + 3}.$$

Cross multiplying, we obtain

$$x^2 = A(x - 2)(x + 3) + B(x + 1)(x + 3) + C(x + 1)(x - 2).$$

Setting $x = -1$ makes the terms in B and C vanish and gives $A = -1/6$. Setting $x = 2$ makes the terms in A and C vanish and gives $B = 4/15$, whereas setting $x = -3$ makes the terms in A and B vanish and gives $C = 9/10$, so

$$\frac{x^2}{(x+1)(x-2)(x+3)} = \frac{-1}{6(x+1)} + \frac{4}{15(x-2)} + \frac{9}{10(x+3)}.$$

(b) The degree of the numerator exceeds that of the denominator, so from Step 7 it is necessary to start by dividing the denominator into the numerator longhand to obtain

$$\frac{2x^3 - 4x^2 + x + 3}{(x-1)^2} = 2x + \frac{3-x}{(x-1)^2}.$$

We now seek a partial fraction representation of $(3-x)/(x-1)^2$ by using Step 1 and writing

$$\frac{3-x}{(x-1)^2} = \frac{A}{x-1} + \frac{B}{(x-1)^2}.$$

When we multiply by $(x-1)^2$, this becomes

$$3 - x = A(x-1) + B.$$

Equating the constant terms gives $3 = -A + B$, whereas equating the coefficients of x gives $-1 = A$ so that $B = 2$. Thus, the required partial fraction representation is

$$\frac{2x^3 - 4x^2 + x + 3}{(x-1)^2} = 2x + \frac{1}{1-x} + \frac{2}{(x-1)^2}. \qquad ■$$

An examination of the way the undetermined coefficients were obtained in (a) earlier, where the degree of the numerator is less than that of the denominator and linear factors occur in the denominator, leads to a simple rule for finding the undetermined coefficients called the "cover-up rule."

The cover-up rule

Let a partial fraction decomposition be required for a rational function $N(x)/D(x)$ in which the degree of the numerator $N(x)$ is less than that of the denominator $D(x)$ and, when factored, let $D(x)$ contain some linear factors (factors of degree 1).

Let $(x - \alpha)$ be a linear factor of $D(x)$. Then the unknown coefficient K in the term $K/(x - \alpha)$ in the partial fraction decomposition of $N(x)/D(x)$ is obtained by "covering up" (ignoring) all of the other terms in the partial fraction expansion, multiplying the remaining expression $N(x)/D(x) = K/(x - \alpha)$ by $(x - \alpha)$, and then determining K by setting $x = \alpha$ in the result.

To illustrate the use of this rule we use it in case (a) given earlier to find A from the representation

$$\frac{x^2}{(x+1)(x-2)(x+3)} = \frac{A}{x+1} + \frac{B}{x-2} + \frac{C}{x+3}.$$

We "cover up" (ignore) the terms involving B and C, multiply through by $(x + 1)$, and find A from the result

$$\frac{x^2}{(x-2)(x+3)} = A$$

by setting $x = -1$, when we obtain $A = -1/6$. The undetermined coefficients B and C follow in similar fashion.

Once a partial fraction representation of a function has been obtained, it is often necessary to express any quadratic $x^2 + px + q$ that occurs in a denominator in the form $(x + A)^2 + B$, where A and B may be either positive or negative real numbers. This is called **completing the square**, and it is used, for example, when integrating rational functions and when finding inverse Laplace transforms.

completing the square

To find A and B we set

$$x^2 + px + q = (x + A)^2 + B$$
$$= x^2 + 2Ax + A^2 + B,$$

and to make this an identity we now equate the coefficients of corresponding powers of x on either side of this expression:

(coefficients of x^2) $1 = 1$ (this tells us nothing)
(coefficients of x) $p = 2A$
(constant terms) $q = A^2 + B.$

Consequently $A = (1/2)p$ and $B = q - (1/4)p^2$, and so the result obtained by completing the square is

$$x^2 + px + q = [x + (1/2)p]^2 + q - (1/4)p^2.$$

If the more general quadratic $ax^2 + bx + c$ occurs, all that is necessary to reduce it to this same form is to write it as

$$ax^2 + bx + c = a[x^2 + (b/a)x + c/a],$$

and then to complete the square using $p = b/a$ and $q = c/a$.

EXAMPLE 1.16

Complete the square in the following expressions:

(a) $x^2 + x + 1$.
(b) $x^2 + 4x$.
(c) $3x^2 + 2x + 1$.

Solution   (a) $p = 1, q = 1$, so $A = 1/2$, $B = 3/4$, and hence

$$x^2 + x + 1 = (x + 1/2)^2 + 3/4.$$

(b) $p = 4$, $q = 0$, so $A = 2$, $B = -4$, and hence

$$x^2 + 4x = (x + 2)^2 - 4.$$

(c) $3x^2 + 2x + 1 = 3[x^2 + (2/3)x + 1/3]$ and so $p = 2/3, q = 1/3$, from which it follows that $A = 1/3$ and $B = 2/9$, so

$$3x^2 + 2x + 1 = 3\{(x + 1/3)^2 + 2/9\}.$$

Further information and examples of partial fractions can be found in any one of references [1.1] to [1.7]. ∎

EXERCISES 1.6

Express the rational functions in Exercises 1 through 8 in terms of partial fractions using the method of Section 1.6, and verify the results by using computer algebra to determine the partial fractions.

1. $(3x + 4)/(2x^2 + 5x + 2)$.
2. $(x^2 + 3x + 5)/(2x^2 + 5x + 3)$.
3. $(3x - 7)/(2x^2 + 9x + 10)$.
4. $(x^2 + 3x + 2)/(x^2 + 2x - 3)$.
5. $(x^3 + x^2 + x + 1)/[(x + 2)^2(x^2 + 1)]$.

6. $(x^2 - 1)/(x^2 + x + 1)$.
7. $(x^3 + x^2 + x + 1)/\{(x + 2)^2(x + 1)\}$.
8. $(x^2 + 4)/(x^3 + 3x^2 + 3x + 1)$.

Complete the square in Exercises 9 through 14.

9. $x^2 + 4x + 5$.
10. $x^2 + 6x + 7$.
11. $2x^2 + 3x - 6$.
12. $4x^2 - 4x - 3$.
13. $2 - 2x + 9x^2$.
14. $2 + 2x - x^2$.

1.7   Fundamentals of Determinants

A **determinant** of **order n** is a single number associated with an array $\mathbf{A}$ of n^2 numbers arranged in n rows and n columns. If the number in the ith row and jth column of a determinant is a_{ij}, the determinant of $\mathbf{A}$, denoted by det $\mathbf{A}$ and sometimes by $|\mathbf{A}|$, is written

$$\det \mathbf{A} = |\mathbf{A}| = \begin{vmatrix} a_{11} & a_{12} & \dots & a_{1n} \\ a_{21} & a_{22} & \dots & a_{2n} \\ \dots & \dots & \dots & \dots \\ a_{n1} & a_{n2} & \dots & a_{nn} \end{vmatrix}. \tag{13}$$

It is customary to refer to the entries a_{ij} in a determinant as its *elements*. Notice the use of vertical bars enclosing the array $\mathbf{A}$ in the notation $|\mathbf{A}|$ for the *determinant* of $\mathbf{A}$, as opposed to the use of the square brackets in $[\mathbf{A}]$ that will be used later to denote the *matrix* associated with an array $\mathbf{A}$ of quantities in which the number of rows need not be equal to the number of columns.

The value of a first order determinant det $\mathbf{A}$ with the single element a_{11} is defined as a_{11} so that $\det[a_{11}] = a_{11}$ or, in terms of the alternative notation for a determinant, $|a_{11}| = a_{11}$. This use of the notation $|.|$ to signify a determinant should not be confused with the notation used to signify the absolute value of a number.

The second order determinant associated with an array of elements containing two rows and two columns is defined as

$$\det \mathbf{A} = \begin{vmatrix} a_{11} & a_{12} \\ a_{21} & a_{22} \end{vmatrix} = a_{11}a_{22} - a_{12}a_{21}, \tag{14}$$

so, for example, using the alternative notation for a determinant we have

$$\begin{vmatrix} 9 & 3 \\ -7 & -4 \end{vmatrix} = 9(-4) - (-7)3 = -15.$$

Notice that **interchanging** two rows or columns of a determinant changes its sign.

We now introduce the terms *minor* and *cofactor* that are used in connection with determinants of all orders, and to do so we consider the third order determinant

$$\det \mathbf{A} = \begin{vmatrix} a_{11} & a_{12} & a_{13} \\ a_{21} & a_{22} & a_{23} \\ a_{31} & a_{32} & a_{33} \end{vmatrix}. \tag{15}$$

minors and cofactors

The **minor** M_{ij} associated with a_{ij}, the element in the ith row and jth column of det $\mathbf{A}$, is defined as the second order determinant obtained from det $\mathbf{A}$ by deleting the elements (numbers) in its ith row and jth column. The **cofactor** C_{ij} of an element in the ith row and jth column of the det $\mathbf{A}$ in (15) is defined as the **signed minor** using the rule

$$C_{ij} = (-1)^{i+j} M_{ij}. \tag{16}$$

With these ideas in mind, the determinant det $\mathbf{A}$ in (15) is defined as

$$\det \mathbf{A} = \sum_{j=1}^{3} a_{1j}(-1)^{1+j} \det M_{1j}$$
$$= a_{11}M_{11} - a_{12}M_{12} + a_{13}M_{13}.$$

If we introduce the cofactors C_{ij}, this last result can be written

$$\det \mathbf{A} = a_{11}C_{11} + a_{12}C_{12} + a_{13}C_{13}, \tag{17}$$

and more concisely as

$$\det \mathbf{A} = \sum_{j=1}^{3} a_{1j}C_{1j}. \tag{18}$$

Result (18), or equivalently (17), will be taken as the definition of a third order determinant.

EXAMPLE 1.17   Evaluate the determinant

$$\begin{vmatrix} 1 & 3 & -3 \\ 2 & 1 & 0 \\ -2 & 1 & 1 \end{vmatrix}.$$

Solution

The minor $M_{11} = \begin{vmatrix} 1 & 0 \\ 1 & 1 \end{vmatrix} = (1)(1) - (0)(1) = 1$, so the cofactor $C_{11} = (-1)^{(1+1)} M_{11} = 1$.
The minor $M_{12} = \begin{vmatrix} 2 & 0 \\ -2 & 1 \end{vmatrix} = (2)(1) - (0)(-2) = 2$, so the cofactor $C_{12} = (-1)^{(1+2)} M_{12} = -2$.
The minor $M_{13} = \begin{vmatrix} 2 & 1 \\ -2 & 1 \end{vmatrix} = (2)(1) - (1)(-2) = 4$, so the cofactor $C_{13} = (-1)^{(1+3)} M_{13} = 4$.

Using (17) we have

$$\begin{vmatrix} 1 & 3 & -3 \\ 2 & 1 & 0 \\ -2 & 1 & 1 \end{vmatrix} = (1)C_{11} + (3)C_{12} + (-3)C_{13} = (1)(1) + (3)(-2) + (-3)(4) = -17.$$

When expanded, (17) becomes

$$\det \mathbf{A} = a_{11}a_{22}a_{33} - a_{11}a_{32}a_{23} - a_{12}a_{21}a_{33} + a_{12}a_{31}a_{23} + a_{13}a_{21}a_{32} - a_{13}a_{31}a_{22},$$

and after regrouping these terms in the form

$$\det \mathbf{A} = -a_{21}a_{12}a_{33} + a_{21}a_{32}a_{13} + a_{22}a_{11}a_{33} - a_{22}a_{31}a_{13} - a_{23}a_{11}a_{32} + a_{23}a_{31}a_{12},$$

we find that

$$\det \mathbf{A} = a_{21}C_{21} + a_{22}C_{22} + a_{23}C_{23}.$$

Proceeding in this manner, we can easily show that det **A** may be obtained by forming the sum of the products of the elements of **A** and their cofactors in *any* row or column of det **A**. These results can be expressed symbolically as follows.

Expanding in terms of the elements of the *i*th row:

$$\det \mathbf{A} = a_{i1}C_{i1} + a_{i2}C_{i2} + a_{i3}C_{i3} = \sum_{j=1}^{3} a_{ij}C_{ij}. \tag{19}$$

Laplace expansion theorem

Expanding in terms of the elements of the *j*th column:

$$\det \mathbf{A} = a_{1j}C_{1j} + a_{2j}C_{2j} + a_{3j}C_{3j} = \sum_{i=1}^{3} a_{ij}C_{ij}. \tag{20}$$

Results (19) and (20) are the form taken by the **Laplace expansion theorem** when applied to a third order determinant. The extension of the theorem to determinants of any order will be made later in Chapter 3, Section 3.3.

EXAMPLE 1.18

Expand the following determinant (a) in terms of elements of its first row, and (b) in terms of elements of its third column:

$$|\mathbf{A}| = \begin{vmatrix} 1 & 2 & 4 \\ 1 & 0 & 2 \\ 1 & 2 & 1 \end{vmatrix}.$$

Solution (a) Expanding in terms of the elements of the first row requires the three cofactors $C_{11} = M_{11}$, $C_{12} = -M_{12}$, and $C_{13} = M_{13}$, where

$$M_{11} = \begin{vmatrix} 0 & 2 \\ 2 & 1 \end{vmatrix} = -4, \quad M_{12} = \begin{vmatrix} 1 & 2 \\ 1 & 1 \end{vmatrix} = -1, \quad M_{13} = \begin{vmatrix} 1 & 0 \\ 1 & 2 \end{vmatrix} = 2,$$

so $C_{11} = (-1)^{(1+1)}(-4) = -4$, $C_{12} = (-1)^{(1+2)}(-1) = 1$, $C_{13} = (-1)^{(1+3)}(2) = 2$, and so

$$|\mathbf{A}| = (1)(-4) + (2)(1) + (4)(2) = 6.$$

(b) Expanding in terms of the elements of the third column requires the three cofactors $C_{13} = M_{13}$, $C_{23} = -M_{23}$, and $C_{33} = M_{33}$, where

$$M_{13} = \begin{vmatrix} 1 & 0 \\ 1 & 2 \end{vmatrix} = 2, \quad M_{23} = \begin{vmatrix} 1 & 2 \\ 1 & 2 \end{vmatrix} = 0, \quad M_{33} = \begin{vmatrix} 1 & 2 \\ 1 & 0 \end{vmatrix} = -2,$$

so $C_{13} = (-1)^{(1+3)}(2) = 2$, $C_{23} = 0$, $C_{33} = (-1)^{(3+3)}(-2) = -2$ and so

$$|\mathbf{A}| = (4)(2) + (2)(0) + (1)(-2) = 6.$$

Two especially simple *third order* determinants are of the form

$$\det \mathbf{A} = \begin{vmatrix} a_{11} & a_{12} & a_{13} \\ 0 & a_{22} & a_{23} \\ 0 & 0 & a_{33} \end{vmatrix} \quad \text{and} \quad \det \mathbf{A} = \begin{vmatrix} a_{11} & 0 & 0 \\ a_{21} & a_{22} & 0 \\ a_{31} & a_{32} & a_{33} \end{vmatrix}.$$

The first of these determinants has only zero elements below the diagonal line drawn from its top left element to its bottom right one, and the second determinant has only zero elements above this line. This diagonal line in every determinant is called the **leading diagonal**. The value of each of the preceding determinants is easily seen to be given by the product $a_{11}a_{22}a_{33}$ of the terms on its leading diagonal.

Simpler still in form is the third order determinant

$$\det \mathbf{A} = \begin{vmatrix} a_{11} & 0 & 0 \\ 0 & a_{22} & 0 \\ 0 & 0 & a_{33} \end{vmatrix} = a_{11}a_{22}a_{33},$$

whose value $a_{11}a_{22}a_{33}$ is again the product of the elements on the leading diagonal.

For another approach to the elementary properties of determinants, see Appendix A16 of reference [1.2], and Chapter 2 of reference [2.1].

EXERCISES 1.7

Evaluate the determinants in Exercises 1 through 6 (a) in terms of elements of the first row and (b) in terms of elements of the second column.

1. $\begin{vmatrix} 1 & 5 & 7 \\ 1 & -1 & 1 \\ 1 & 2 & 1 \end{vmatrix}.$

2. $\begin{vmatrix} 2 & 1 & -1 \\ 2 & 6 & -1 \\ 5 & 1 & -1 \end{vmatrix}.$

3. $\begin{vmatrix} 5 & 2 & 4 \\ 1 & 2 & 1 \\ 3 & 1 & 5 \end{vmatrix}.$

4. $\begin{vmatrix} -1 & 3 & 6 \\ 2 & 1 & 4 \\ -1 & 3 & 1 \end{vmatrix}.$

5. $\begin{vmatrix} 1 & 0 & -6 \\ 2 & 1 & 3 \\ 4 & 3 & 21 \end{vmatrix}.$

6. $\begin{vmatrix} 1 & 5 & -1 \\ 2 & 1 & -3 \\ -4 & 1 & 1 \end{vmatrix}.$

7. On occasion the elements of a matrix may be functions, in which case the determinant may be a function. Evaluate the *functional* determinant

$$\begin{vmatrix} 1 & 0 & 0 \\ 0 & \sin x & -\cos x \\ 0 & \cos x & \sin x \end{vmatrix}.$$

8. Determine the values of λ that make the following determinant vanish:

$$\begin{vmatrix} 3 - \lambda & 2 & 2 \\ 2 & 2 - \lambda & 0 \\ 2 & 0 & 4 - \lambda \end{vmatrix}.$$

Hint: This is a polynomial in λ of degree 3.

9. A matrix is said to be **transposed** if its first row is written as its first column, its second row is written as its second

column ..., and its last row is written as its last column. If the determinant is $|\mathbf{A}|$, the determinant of $\mathbf{A}^{\mathrm{T}}$, the transpose matrix $\mathbf{A}$, is denoted by $|\mathbf{A}^{\mathrm{T}}|$. Write out the expansion of $|\mathbf{A}|$ using (17) and reorder the terms to show that

$$|\mathbf{A}| = |\mathbf{A}^{\mathrm{T}}|.$$

10. Use elimination to solve the system of linear equations

$$a_{11}x_1 + a_{12}x_2 = b_1$$
$$a_{21}x_1 + a_{22}x_2 = b_2$$

for x_1 and x_2, in which not both b_1 and b_2 are zero, and show that the solution can be written in the form

$$x_1 = D_1/|\mathbf{A}| \quad \text{and} \quad x_2 = D_2/|\mathbf{A}|, \quad \text{provided } |\mathbf{A}| \neq 0,$$

where $|\mathbf{A}|$ is the determinant of the matrix of *coefficients* of the system

$$|\mathbf{A}| = \begin{vmatrix} a_{11} & a_{12} \\ a_{21} & a_{22} \end{vmatrix}, \quad D_1 = \begin{vmatrix} b_1 & a_{12} \\ b_2 & a_{22} \end{vmatrix}, \quad \text{and} \quad D_2 = \begin{vmatrix} a_{11} & b_1 \\ a_{21} & b_2 \end{vmatrix}.$$

Notice that D_1 is obtained from $|\mathbf{A}|$ by replacing its *first* column by b_1 and b_2, whereas D_2 is obtained from $|\mathbf{A}|$ by replacing its *second* column by b_1 and b_2. This is **Cramer's rule** for a system of two simultaneous equations. Use this method to find the solution of

$$x_1 + 5x_2 = 3$$
$$7x_1 - 3x_2 = -1.$$

11. Repeat the calculation in Exercise 10 using the system of equations

$$a_{11}x_1 + a_{12}x_2 + a_{13}x_3 = b_1$$
$$a_{21}x_1 + a_{22}x_2 + a_{23}x_3 = b_2$$
$$a_{31}x_1 + a_{32}x_2 + a_{33}x_3 = b_3,$$

in which not all of b_1, b_2, and b_3 are zero, and show that provided $|\mathbf{A}| \neq 0$,

$$x_1 = D_1/|\mathbf{A}|, \quad x_2 = D_2/|\mathbf{A}|, \quad \text{and} \quad x_3 = D_3/|\mathbf{A}|,$$

where $|\mathbf{A}|$ is the determinant of the matrix of coefficients and D_i is the determinant obtained from $|\mathbf{A}|$ by replacing its ith column by b_1, b_2, and b_3 for $i = 1, 2, 3$. This is **Cramer's rule** for a system of three simultaneous equations, and the method generalizes to a system of n linear equations in n unknowns. Use this method to find the solution of

$$x_1 + 2x_2 - x_3 = 2$$
$$x_1 - 3x_2 - 2x_3 = -1$$
$$2x_1 + x_2 + 2x_3 = 1.$$

1.8   Continuity in One or More Variables

If the function $y = f(x)$ is defined in the interval $a \leq x \leq b$, the interval is called the **domain of definition** of the function. The function f is said to have a **limit** at a point c in $a \leq x \leq b$, written $\lim_{x \to c} f(x) = L$, if for every arbitrarily small number $\varepsilon > 0$ there exists a number $\delta > 0$ such that

$$|f(x) - L| < \varepsilon \quad \text{when} \quad |x - c| < \delta. \tag{21}$$

This technical definition means that as x either increases toward c and becomes arbitrarily close to it, or decreases toward c and becomes arbitrarily close to it, so $f(x)$ approaches arbitrarily close to the value L. Notice that it is not necessary for $f(x)$ to be defined at $x = c$, or, if it is, that $f(c)$ assumes the value L. If $f(x)$ has a limit L as $x \to c$ and in addition $f(c) = L$, so that

$$\lim_{x \to c} f(x) = f(c) = L, \tag{22}$$

then the function f is said to be **continuous** at c. It must be emphasized that in this definition of continuity the limiting operation $x \to c$ must be true as x tends to c from *both* the left and right. It is convenient to say that x approaches c from the *left* when it increases toward c and, correspondingly, to say that x approaches c from the *right* when it decreases toward it.

continuity from the right

The function f is **continuous from the right** at $x = c$ if

$$\lim_{x \to c^+} f(x) = f(c), \tag{23}$$

where the notation $x \to c^+$ means that x decreases toward c, causing x to tend to c from the *right*. Similarly, f is **continuous from the left** at $x = c$ if

continuity from the left

$$\lim_{x \to c^-} f(x) = f(c), \tag{24}$$

where now $x \to c^-$ means that x increases toward c, causing x to tend to c from the *left*. The relationship among definitions (22), (23), and (24) is that f is continuous at the point c if

continuity at $x = c$

$$\lim_{x \to c^-} f(x) = \lim_{x \to c^+} f(x) = f(c). \tag{25}$$

When expressed in words, this says that f is continuous at $x = c$ if the limits of f as x tends to c from both the left and right exist and, furthermore, the limits equal the functional value $f(c)$.

continuous function

A function f that is continuous at all points of $a \leq x \leq b$ is said to be a **continuous function** on that interval. Graphically, a continuous function on $a \leq x \leq b$ is a function whose graph is unbroken but not necessarily smooth. A function f is said

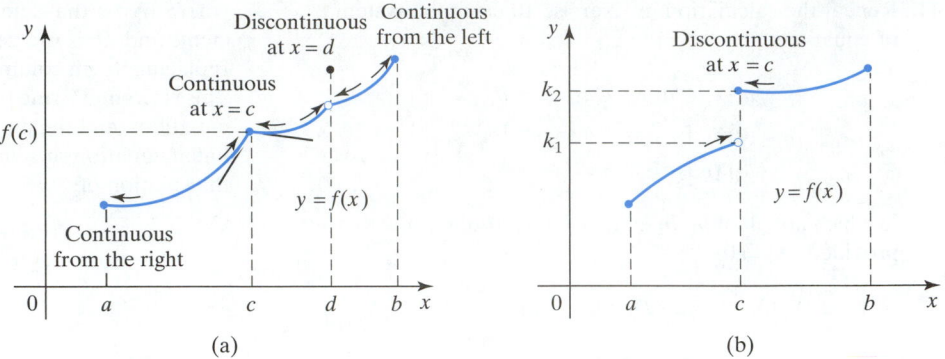

(a)   (b)

FIGURE 1.8   (a) A continuous function for $a < x < b$. (b) A discontinuous function.

smooth function

continuous and piecewise smooth function

discontinuous function

piecewise continuity

to be **smooth** over an interval if at each point of the graph the tangent lines to the left and right of the point are the same. Figure 1.8a shows the graph of a continuous function that is smooth over the intervals $a \leq x < c$ and $c < x < b$ but has different tangent lines to the immediate left and right of $x = c$ where the function is *not* smooth. A function such as this is said to be **continuous and piecewise smooth** over the interval $a \leq x \leq b$.

A function f is said to be **discontinuous** at a point c if it is not continuous there. For a jump discontinuity we have

$$\lim_{x \to c^-} f(x) = k_1 \quad \text{and} \quad \lim_{x \to c^+} f(x) = k_2, \quad \text{but } k_1 \neq k_2. \tag{26}$$

A function f is said to have a **removable discontinuity** at a point c if $k_1 = k_2$ in (26), but $f(c) \neq k_1$, as at the point c_2 in Fig. 1.9.

An example of a discontinuous function is shown in Fig. 1.8b where a jump discontinuity occurs at $x = c$.

A function f is said to be **piecewise continuous** on an interval $a \leq x \leq b$ if it is continuous on a finite number of adjacent subintervals, but discontinuous at the end points of the subintervals, as shown in Fig. 1.9.

The notion of continuity of a function of several variables is best illustrated by considering a function $f(x, y)$ of the two independent variables x and y. The function f defined in some region of the (x, y)-plane D, say, is said to be **continuous**

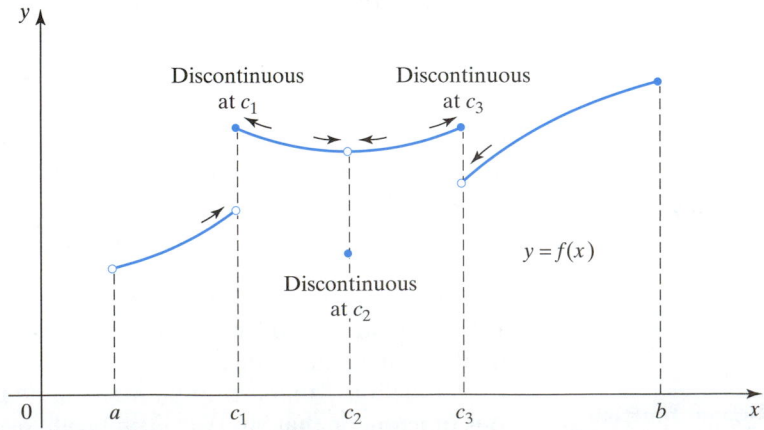

FIGURE 1.9   A piecewise continuous function.

at the point (a, b) in D if

continuity of $f(x, y)$

$$\lim_{x \to a, y \to b} f(x, y) = f(a, b), \tag{27}$$

and to be discontinuous otherwise.

In this definition of continuity, it is important to recognize that a general point P at (x, y) is allowed to tend to the point (a, b) in D along *any* path in the (x, y)-plane that lies in D. Expressed differently, f will only be continuous at (a, b) if the limit in (27) is independent of the way in which the point (x, y) approaches the point (a, b). When this is true for all points in D, the function f is said to be **continuous in D**.

discontinuity of $f(x, y)$ The function f is, for instance, **discontinuous** at (a, b) if

$$\lim_{x \to a, y \to b} f(x, y) = k, \quad \text{but } f(a, b) \neq k.$$

Sufficient for showing that a function f is discontinuous at a point (a, b) is by demonstrating that two *different* limiting values of f are obtained if the point P at (x, y) is allowed to tend to (a, b) along two *different* straight-line paths. This approach can be used to show that the function

$$f(x, y) = \frac{xy}{x^2 + a^2 y^2}$$

has no limit at the origin. If we allow the point P at (x, y) to tend to the origin along the straight line $y = kx$, with k an arbitrary constant, the function f becomes

$$f(x, kx) = \frac{k}{1 + a^2 k^2},$$

and it is seen from this that f is constant along each such line. However, the value of f on each line, and hence at the origin, depends on k, so f has no limit at the origin and so is discontinuous at that point, though f is defined and continuous at all other points of the (x, y)-plane.

An example of a function $f(x, y)$ that is continuous everywhere except at points along a curve Γ in the (x, y)-plane is shown in Fig. 1.10.

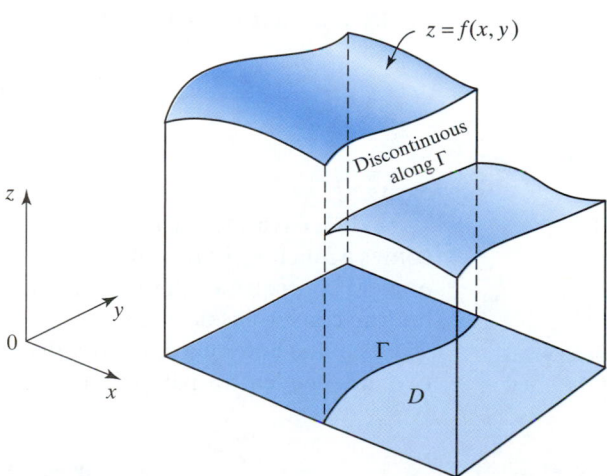

FIGURE 1.10 A function $f(x, y)$ continuous everywhere except at points on Γ.

The extension of these definitions to functions of n variables is immediate and so will not be discussed.

Discussions on continuity and its consequences can be found in any one of references [1.1] to [1.7].

1.9 Differentiability of Functions of One or More Variables

The function $f(x)$ defined in $a \leq x \leq b$ is said to be **differentiable** with the **derivative** $f'(c)$ at a point c inside the interval if the following limit exists:

$$\lim_{h \to 0} \frac{f(c+h) - f(c)}{h} = f'(c). \tag{28}$$

Here, as in the definition of continuity, for f to be differentiable at point c the limit must remain unchanged as h tends to zero through both positive and negative values. The function f is said to be **differentiable** in the interval $a \leq x \leq b$ if it is differentiable at every point in the interval. When f is differentiable at a point c with derivative $f'(c)$, the number $f'(c)$ is the gradient, or slope, of the tangent line to the graph at the point $(c, f(c))$. A function with a continuous derivative throughout an interval is said to be a **smooth** function over the interval. The function f will be said to be **nondifferentiable** at any point c where the limit in (28) does not exist.

differentiability of f(x)

Even when a function f is nondifferentiable at a point, it is possible that a special form of derivative can still be defined to the left and right of the point if the requirement that the limit in (28) exists as $h \to 0$ through both positive and negative values is relaxed. The function f has a **right-hand derivative** at a if the limit

left- and right-hand derivatives of f(x)

$$\lim_{h \to 0^+} \frac{f(a+h) - f(a)}{h} \tag{29}$$

exists, and a **left-hand derivative** at b if the limit

$$\lim_{h \to 0^-} \frac{f(b+h) - f(b)}{h}. \tag{30}$$

exists.

When c is a specific point, $f'(c)$ is a *number*, but when x is a variable, $f'(x)$ becomes a function. Left- and right-hand derivatives are illustrated in Fig. 1.11. An important consequence of differentiability is that **differentiability implies continuity**, but the converse is not true.

first order partial derivatives of f(x, y)

The **first order partial derivative with respect to x** of the function $f(x, y)$ of the two independent variables x and y at the point (a, b) is the number defined by

$$\lim_{h \to 0} \frac{f(a+h, b) - f(a, b)}{h}, \tag{31}$$

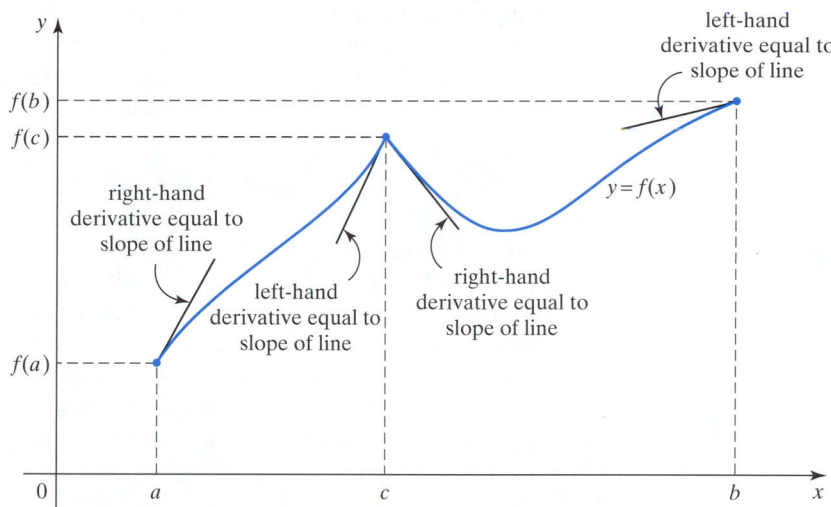

FIGURE 1.11   Left- and right-hand derivatives as tangent lines.

provided the limit exists. The value of this partial derivative is denoted either by $\partial f/\partial x$ at (a, b), or by $f_x(a, b)$. The corresponding partial derivative at a general point (x, y) is the function $f_x(x, y)$.

Similarly, the **first order partial derivative with respect to y** of the function $f(x, y)$ at the point (a, b) is the number defined by the limit

$$\lim_{k \to 0} \frac{f(a, b + k) - f(a, b)}{k}, \tag{32}$$

provided the limit exists. The value of this partial derivative is denoted either by $\partial f/\partial y$ at (a, b), or by $f_y(a, b)$. At a general point (x, y) this partial derivative becomes the function $f_y(x, y)$. Higher order partial derivatives are defined in a similar

second order partial derivatives of $f(x, y)$

fashion leading, for example, to the **second order partial derivatives**

$$\partial^2 f/\partial x^2 = \partial/\partial x(\partial f/\partial x), \; \partial^2 f/\partial y^2 = \partial/\partial y(\partial f/\partial y),$$

$$\partial^2 f/\partial x\partial y = \partial/\partial y(\partial f/\partial x), \quad \text{and} \quad \partial^2 f/\partial y\partial x = \partial/\partial x(\partial f/\partial y).$$

A more compact notation for these same derivatives is

$f_{xx}, \; f_{yy}, \; f_{xy},$ and f_{yx}, so that, for example $f_{yx} = \partial^2 f/\partial y\partial x$ and $f_{yy} = \partial^2 f/\partial y^2$.

mixed partial derivatives

The derivatives f_{xy} and f_{yx} are called **mixed partial derivatives**, and their relationship forms the statement of the next theorem, the proof of which can be found in any one of references [1.1] to [1.7].

THEOREM 1.3   **Equality of mixed partial derivatives**   Let $f, f_x, f_{xy},$ and f_{yx} all be defined and continuous at a point (a, b) in a region. Then

$$f_{xy}(a, b) = f_{yx}(a, b).$$
∎

This result, given conditions for the *equality* of mixed partial derivatives, is an important one, and use will be made of it on numerous occasions as, for example, in Chapter 18 when second order partial differential equations are considered.

total differential

If $z = f(x, y)$, the **total differential** dz of f is defined as

$$dz = (\partial f/\partial x)\, dx + (\partial f/\partial y)\, dy, \tag{33}$$

where dz, dx, and dy are *differentials*. Here, a **differential** means a small quantity, and the differential dz is determined by (33) when the differentials dx and dy are specified. When $\partial f/\partial x$ and $\partial f/\partial y$ are evaluated at a specific point (a, b), result (33) provides a linear approximation to $f(x, y)$ near to the point (a, b). Although finite, the limits of the quotients of the differentials $dz \div dx$ and $dy \div dx$ as the differential $dx \to 0$ are such that they become the values of the derivatives dz/dx and dy/dx, respectively, at a point (x, y) where $\partial f/\partial x$ and $\partial f/\partial y$ are evaluated.

1.10   Tangent Line and Tangent Plane Approximations to Functions

tangent line approximation

Let $y = f(x)$ be defined in the interval $a \le x \le b$ and be differentiable throughout it. Then a **tangent line (linear) approximation** to f near a point x_0 in the interval is given by

$$y_T = f(x_0) + (x - x_0)\, f'(x_0). \tag{34}$$

This linear expression approximates the function f close to x_0 by the tangent to the graph of $y = f(x)$ at the point $(x_0, f(x_0))$.

This simple approximation has many uses; one will be in the Euler and modified Euler methods for solving initial value problems for ordinary differential equations developed in Chapter 19.

EXAMPLE 1.19

Find a tangent line approximation to $y = 1 + x^2 + \sin x$ near the point $x = \alpha$.

Solution   Setting $x_0 = \alpha$ and substituting into (34) gives

$$y \approx 1 + \alpha^2 + \sin\alpha + (x - \alpha)(2\alpha + \cos\alpha) \text{ for } x \text{ close to } \alpha. \qquad \blacksquare$$

tangent plane approximation

Let the function $z = f(x, y)$ be defined in a region D of the (x, y)-plane where it possesses continuous first order partial derivatives $\partial f/\partial x$ and $\partial f/\partial y$. Then a **tangent plane (linear) approximation** to f near any point (x_0, y_0) in D is given by

$$z_T = f(x_0, y_0) + (x - x_0)\, f_x(x_0, y_0) + (y - y_0)\, f_y(x_0, y_0). \tag{35}$$

This linear expression approximates the function f close to the point (x_0, y_0) by a plane that is tangent to the surface $z = f(x, y)$ at the point $(x_0, y_0, f(x_0, y_0))$. The tangent plane approximation in (35) is an immediate extension to functions of two variables of the tangent line approximation in (34), to which it simplifies when only one independent variable is involved.

Both of these approximations are derived from the appropriate Taylor series expansions of functions discussed in Section 1.12 by retaining only the linear terms.

EXAMPLE 1.20

Find the tangent plane approximation to the function $z = x^2 - 3y^2$ near the point $(1, 2)$.

Solution   Setting $x_0 = 1$, $y_0 = 2$ and substituting into (35) gives

$$z \approx -11 + 2(x - 1) - 12(y - 2) \text{ for } (x, y) \text{ close to } (1, 2).$$ ∎

1.11   Integrals

A differentiable function $F(x)$ is called an **antiderivative** of the function $f(x)$ on some interval if at each point of the interval $dF/dx = f(x)$. If $F(x)$ is any antiderivative of $f(x)$, the **indefinite integral** of $f(x)$, written $\int f(x)\,dx$, is

indefinite and definite integrals

$$\int f(x)\,dx = F(x) + c,$$

where c is an arbitrary constant called the *constant of integration*. The function $f(x)$ is called the **integrand** of the integral. Thus, an indefinite integral is a function, and an antiderivative and an indefinite integral can only differ by an arbitrary additive constant.

The expression $\int_a^b f(x)\,dx$, called a **definite integral**, is a number and may be interpreted geometrically as the area between the graph of $f(x)$ and the lines $x = a$ and $x = b$, for $b > a$, with areas above the x-axis counted as positive and those below it as negative.

The relationship between definite integrals that are *numbers* and indefinite integrals that are *functions* is given in the next theorem, included in which is also the mean value theorem for integrals. See the references at the end of the chapter for proofs and further information.

THEOREM 1.4

Fundamental theorem of integral calculus and the mean value theorem for integrals
If $F'(x)$ is continuous in the interval $a \leq x \leq b$, throughout which $F'(x) = f(x)$, then

$$\int_a^b f(x)\,dx = F(b) - F(a).$$

Another result is

$$\int_a^b f(x)\,dx = (b - a)\,f'(\xi),$$

if f is differentiable, where the number ξ, although unknown, lies in the interval $a < \xi < b$. In this form the result is called the **mean value theorem for integrals**. ∎

An **improper integral** is a definite integral in which one or more of the following cases arises: (a) the integrand becomes infinite inside or at the end of the interval of integration, or (b) one (or both) of the limits of integration is infinite.

Types of Improper Integrals
Case (a)

If the integrand of an integral becomes infinite at a point c inside the interval of integration $a \leq x \leq b$ as shown in Fig. 1.12a, the improper integral is said to exist if the limits in (36) exist. When the improper integral exists it is said to **converge** to the (finite) value of the following limit:

convergence and divergence of improper integrals

$$\int_a^b f(x)\,dx = \lim_{h \to 0} \int_a^{c-h} f(x)\,dx + \lim_{k \to 0} \int_{c+k}^b f(x)\,dx. \tag{36}$$

In this definition $h > 0$ and $k > 0$ are allowed to tend to zero *independently* of each other. If, when the limit is taken, the integral is either infinite or indeterminate, the integral is said to **diverge**.

Some integrals of this type diverge when h and k are allowed to tend to zero independently of each other, but converge when the limit is taken with $h = k$, in which case the result of the limit is called the **Cauchy principal value** of the integral. Integrals of this type arise frequently when certain types of definite integral are evaluated in the complex plane by means of contour integration (see Chapter 15, Section 15.5).

Cauchy principal value

Case (b)

If a limit of integration in a definite integral is infinite, say the upper limit as shown in Fig. 1.12b, then, when it exists, the improper integral is said to **converge** to the value of the limit

$$\int_a^\infty f(x)\,dx = \lim_{R \to \infty} \int_a^R f(x)\,dx, \tag{37}$$

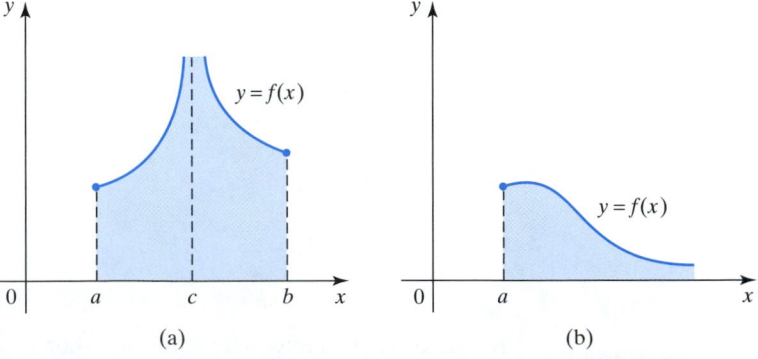

FIGURE 1.12   (a) $f(x)$ is infinite inside the interval of integration. (b) The interval of integration is infinite in length.

and the integral is **divergent** if the limit is either infinite or indeterminate. If both limits are infinite, the improper integral is said to **converge** to the value of the limit

$$\int_{-\infty}^{\infty} f(x)\,dx = \lim_{R\to\infty, S\to\infty} \int_{-S}^{R} f(x)\,dx \tag{38}$$

when it exists, and the integral is said to be **divergent** if the limit is either infinite or indeterminate.

In (38) R and S are allowed to tend to infinity *independently* of each other. Integrals of this type also have Cauchy principal values if the foregoing process leads to divergence, but the integrals are convergent when the limit is taken with $R = S$. Integrals of this type also occur when certain real integrals are evaluated by means of contour integration (see Chapter 15, Section 15.5).

Elementary examples of convergent improper integrals of the types shown in (36) to (38) are

$$\int_{0}^{1} \frac{x^p - x^{-p}}{x - 1}\,dx = \frac{1}{p} - \pi\cot p\pi, \quad (p^2 < 1),$$

$$\int_{0}^{\infty} \exp(-x)\sin x\,dx = 1/2 \quad \text{and} \quad \int_{-\infty}^{\infty} \frac{dx}{1 + x^2} = \pi.$$

THEOREM 1.5   **Differentiation under the integral sign — Leibniz' rule**   If $\xi(t)$, $\eta(t)$, $d\xi/dt$, $d\eta/dt$, $f(x,t)$, and $\partial f/\partial t$ are continuous for $t_0 \le t \le t_1$ and for x in the interval of integration, then

$$\frac{d}{dt}\int_{\xi(t)}^{\eta(t)} f(x,t)\,dx = \int_{\xi(t)}^{\eta(t)} \frac{\partial f(x,t)}{\partial t}\,dx + f(\eta(t),t)\frac{d\eta}{dt} - f(\xi(t),t)\frac{d\xi}{dt}. \qquad \blacksquare$$

This theorem is used, for example, in Chapter 18 when discussing discontinuous solutions of a class of partial differential equations called *conservation laws*. Extensions of the theorem to functions of more variables are developed in Chapter 12, Section 12.3, where certain vector integral theorems are developed, and applications of the results of that section to fluid mechanics are to be found in Chapter 12, Section 12.4.

An application of Theorem 1.5 that is easily checked by direct calculation is

$$\frac{d}{dt}\int_{2t}^{t^2} (x^2 + t)\,dx = \int_{2t}^{t^2} dx + (t^4 + t)\cdot 2t - (4t^2 + t)\cdot 2 = 2t^5 - 5t^2 - 4t.$$

A proof of Leibniz' rule can be found, for example, in Chapter 12 of reference [1.6].

1.12   Taylor and Maclaurin Theorems

THEOREM 1.6   **Taylor's theorem for a function of one variable**   Let a function $f(x)$ have derivatives of all orders in the interval $a < x < b$. Then for each positive integer n and each x_0 in the interval

$$f(x) = f(x_0) + (x - x_0)f^{(1)}(x_0) + \frac{(x - x_0)^2}{2!} f^{(2)}(x_0) + \cdots$$
$$+ \frac{(x - x_0)^n}{n!} f^{(n)}(x_0) + R_{n+1}(x),$$

where $f^{(r)}(x) = d^r f/dx^r$, and the **remainder term** $R_{n+1}(x)$ is given by

$$R_{n+1}(x) = \frac{(x - x_0)^{n+1}}{(n + 1)!} f^{(n+1)}(\xi),$$

for some ξ between x_0 and x.   ∎

Taylor's theorem becomes the **Taylor series** for $f(x)$ when n is allowed to become infinite, and if the remainder term is neglected in Taylor's theorem the result is called the **Taylor polynomial approximation** to $f(x)$ of **degree** n. The Taylor polynomial of degree 1 is simply the tangent line approximation to f at x_0 given in (34).

Taylor polynomial

Taylor's theorem reduces to **Maclaurin's theorem** if $x_0 = 0$, and if we allow n to become infinite in Maclaurin's theorem, it becomes the **Maclaurin series** for $f(x)$.

Maclaurin's theorem

A special case of Theorem 1.6 arises when Taylor's theorem is terminated with the term $R_1(x)$, corresponding to $n = 0$, because the result can be written

$$\frac{f(x) - f(x_0)}{x - x_0} = f'(\xi), \tag{39}$$

with ξ between x_0 and x, and in this form it is called the **mean value theorem for derivatives** (see the last result of Theorem 1.4).

mean value theorem

A Taylor series is an example of an infinite series called a **power series**, the general form of which is

$$\sum_{n=0}^{\infty} a_n(x - x_0)^n = a_0 + a_1(x - x_0) + a_2(x - x_0)^2 + \cdots. \tag{40}$$

In (40) the quantity x is a variable, the numbers a_i are the **coefficients** of the power series, the constant x_0 is called the **center** of the series, or the point about which the series is **expanded**, and unless otherwise stated, x, x_0, and the a_i are real numbers, so the power series is a function of x.

A power series is said to **converge** for a given value of x if the sum of the infinite series for this value of x is *finite*. If the sum is *infinite*, or is *not defined*, the power series will be said to **diverge** for that value of x. Power series converge in an interval $x_0 - R < x < x_0 + R$, where the number R is called the **radius of convergence** of the series. Expressions for R are derived in Section 15.1.

The interval $x_0 - R < x < x_0 + R$ is called the **interval of convergence** of the power series. A power series converges for all x inside the interval of convergence and diverges for all x outside it, and the series may, or may not, converge at the end points of the interval. The convergence properties of power series are shown diagramatically in Fig. 1.13, and results (40) and combining expressions for R with

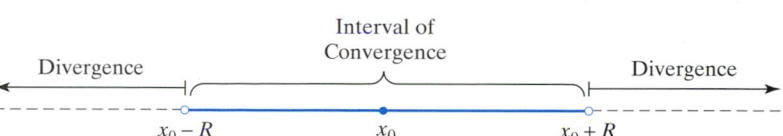

FIGURE 1.13   Interval of convergence of a power series with center x_0.

(40) gives the following theorem (see the references at the end of the chapter for real variable proofs of the following results and for more information).

THEOREM 1.7   **Ratio test and nth root test for the convergence of power series**   The power series

$$\sum_{n=0}^{\infty} a_n(x - x_0)^n = a_0 + a_1(x - x_0) + a_2(x - x_0)^2 + \cdots$$

radius and interval of convergence

converges in the *interval of convergence* $x_0 - R < x < x_0 + R$, where the *radius of convergence R* is determined by either of the formulas

(a) $R = 1/ \lim_{n \to \infty} |a_{n+1}/a_n|$ or   (b) $R = 1/ \lim_{n \to \infty} |a_n|^{1/n}$.

The power series will diverge outside the interval of convergence, and its behavior at the ends of the interval of convergence must be determined separately. ∎

A simple result on the convergence of a series that is often useful is the alternating series test. An **alternating series** is so named because the signs of successive terms of the series alternate in sign.

THEOREM 1.8   **The alternating series test for convergence**   The alternating series $\sum_{n=1}^{\infty}(-1)^{n+1}a_n$ converges if $a_n > 0$ and $a_{n+1} < a_n$ for all n and $\lim_{n \to \infty} a_n = 0$. ∎

The following theorem on the differentiation and integration of power series is often needed, and it is a real variable form of a result proved later in Chapter 15 when complex power series are studied.

THEOREM 1.9   **Differentiation and integration of power series**   Let a power series have an interval of convergence $x_0 - R < x < x_0 + R$. Then the series may be differentiated and integrated term by term, and in each case the resulting series will have the same interval of convergence as the original series. In addition, within an interval of convergence common to any two power series, the series may be scaled by a constant and added or subtracted term by term and the resulting power series will have the same common interval of convergence. ∎

The simplest form of Taylor's theorem for a function of two variables that finds many applications is given in the next theorem.

THEOREM 1.10   **Taylor's theorem for a function of two variables**   Let $f(x, y)$ be defined for $a < x < b$ and $c < y < d$ and have continuous partial derivatives up to and including those of order 2. Then for x_0 and y_0 any points such that $a < x_0 < b$ and $c < y_0 < d$,

$$
\begin{aligned}
f(x, y) = {} & f(x_0, y_0) + (x - x_0) f_x(x_0, y_0) + (y - y_0) f_y(x_0, y_0) \\
& + \frac{1}{2!}[(x - x_0)^2 f_{xx}(x_0 + \xi, y_0 + \eta) + 2(x - x_0)(y - y_0) \\
& \times f_{xy}(x_0 + \xi, y_0 + \eta)(y - y_0)^2 f_{yy}(x_0 + \xi, y_0 + \eta)],
\end{aligned}
$$

where the numbers ξ and η are unknown, but ξ lies between x_0 and x and η lies between y_0 and y. ∎

The group of second order partial derivatives in Theorem 1.10 forms the remainder term, and when these derivatives are ignored, the result reduces to the tangent plane approximation to $f(x, y)$ at the point (x_0, y_0) given in (35).

More information on Taylor's theorem and series can be found, for example, in reference [1.2].

1.13 Cylindrical and Spherical Polar Coordinates and Change of Variables in Partial Differentiation

Mathematical problems formulated using a particular coordinate system, such as cartesian coordinates, often need to be reexpressed in terms of a different coordinate system in order to simplify the task of finding a solution. When partial derivatives occur in the formulation of problems, it becomes necessary to know how they transform when a different coordinate system is used. The fundamental theorem governing the transformation of partial derivatives under a change of variables takes the following form (see the references at the end of the chapter for the proof of Theorem 1.11 and for more examples of its use).

THEOREM 1.11

Change of variables in partial differentiation Let $f(x_1, x_2, \ldots, x_n)$ be a differentiable function with respect to the n independent variables $x_1, x_2, \ldots, x_n$, and let the n new independent variables $u_1, u_2, \ldots, u_n$ be determined in terms of $x_1, x_2, \ldots, x_n$ by

$$
x_1 = X_1(u_1, u_2, \ldots, u_n), \quad x_2 = X_2(u_1, u_2, \ldots, u_n), \ldots, \quad x_n = X_n(u_1, u_2, \ldots, u_n),
$$

where $X_1, X_2, \ldots, X_n$ are differentiable functions of their arguments. Then, if as a result of the change of variables the function $f(x_1, x_2, \ldots, x_n)$ becomes the function $F(X_1, X_2, \ldots, X_n)$, and using chain rules we have

$$
\begin{aligned}
\frac{\partial F}{\partial u_1} &= \frac{\partial f}{\partial x_1}\frac{\partial X_1}{\partial u_1} + \frac{\partial f}{\partial x_2}\frac{\partial X_2}{\partial u_1} + \cdots + \frac{\partial f}{\partial x_n}\frac{\partial X_n}{\partial u_1} \\
\frac{\partial F}{\partial u_2} &= \frac{\partial f}{\partial x_1}\frac{\partial X_1}{\partial u_2} + \frac{\partial f}{\partial x_2}\frac{\partial X_2}{\partial u_2} + \cdots + \frac{\partial f}{\partial x_n}\frac{\partial X_n}{\partial u_2} \\
&\qquad\qquad\cdots\cdots\cdots\cdots\cdots\cdots\cdots \\
\frac{\partial F}{\partial u_n} &= \frac{\partial f}{\partial x_1}\frac{\partial X_1}{\partial u_n} + \frac{\partial f}{\partial x_2}\frac{\partial X_2}{\partial u_n} + \cdots + \frac{\partial f}{\partial x_n}\frac{\partial X_n}{\partial u_n}.
\end{aligned} \tag{41}
$$
∎

To find higher order partial derivatives it is necessary to express the relationships between the *operations* of differentiation in the two coordinate systems, rather than between the actual derivatives themseves. This can be accomplished by rewriting the results of Theorem 1.11 in the form of **partial differential operators** as follows:

$$
\begin{aligned}
\frac{\partial}{\partial u_1} &\equiv \frac{\partial X_1}{\partial u_1}\frac{\partial}{\partial x_1} + \frac{\partial X_2}{\partial u_1}\frac{\partial}{\partial x_2} + \cdots + \frac{\partial X_n}{\partial u_1}\frac{\partial}{\partial x_n} \\
\frac{\partial}{\partial u_2} &\equiv \frac{\partial X_1}{\partial u_2}\frac{\partial}{\partial x_1} + \frac{\partial X_2}{\partial u_2}\frac{\partial}{\partial x_2} + \cdots + \frac{\partial X_n}{\partial u_2}\frac{\partial}{\partial x_n} \\
&\cdots\cdots\cdots\cdots\cdots\cdots\cdots\cdots\cdots\cdots \\
\frac{\partial}{\partial u_n} &\equiv \frac{\partial X_1}{\partial u_n}\frac{\partial}{\partial x_1} + \frac{\partial X_2}{\partial u_n}\frac{\partial}{\partial x_2} + \cdots + \frac{\partial X_n}{\partial u_n}\frac{\partial}{\partial x_n}.
\end{aligned}
\tag{42}
$$

When expressed in this form the relationships between the partial differentiation operations $\partial/\partial x_1, \partial/\partial x_2, \ldots, \partial/\partial x_n$ and $\partial/\partial u_1, \partial/\partial u_2, \ldots, \partial/\partial u_n$ become clear. This interpretation is needed when finding higher order partial derivatives such as $\partial^2 F/\partial u_2 \partial u_1$, because

$$
\frac{\partial^2 F}{\partial u_2 \partial u_1} = \frac{\partial}{\partial u_1}\left(\frac{\partial F}{\partial u_2}\right) = \left(\frac{\partial X_1}{\partial u_1}\frac{\partial}{\partial x_1} + \frac{\partial X_2}{\partial u_1}\frac{\partial}{\partial x_2} + \cdots + \frac{\partial X_n}{\partial u_1}\frac{\partial}{\partial x_n}\right)\left(\frac{\partial F}{\partial u_2}\right).
$$

An important combination of partial derivatives that occurs throughout physics and engineering is called the **Laplacian** of a function. When a twice differentiable function $f(x, y, z)$ of the cartesian coordinates x, y, and z is involved, the Laplacian of f, denoted by Δf and sometimes by $\nabla^2 f$, read "del squared f," takes the form

$$
\Delta f = \nabla^2 f = \frac{\partial^2 f}{\partial x^2} + \frac{\partial^2 f}{\partial y^2} + \frac{\partial^2 f}{\partial z^2}.
\tag{43}
$$

Cylindrical Polar Coordinates (r, θ, z)

The cylindrical polar coordinate system (r, θ, z) is illustrated in Fig. 1.14, and its relationship to cartesian coordinates is given by

$$
x = r\cos\theta, \quad y = r\sin\theta, \quad z = z, \quad \text{with } 0 \le \theta < 2\pi \text{ and } r \ge 0.
\tag{44}
$$

Spherical Polar Coordinates (r, ϕ, θ)

The spherical polar coordinate system (r, ϕ, θ) shown in Fig. 1.15 is related to cartesian coordinates by

$$
\begin{aligned}
x = r\sin\theta\cos\phi, \quad y = r\sin\theta\sin\phi, \quad z = r\cos\theta, \\
\text{with } 0 \le \theta \le \pi,\ 0 \le \phi < 2\pi.
\end{aligned}
\tag{45}
$$

The derivation of the formulas for the change of variables in functions of several variables can be found in any one of references [1.1] to [1.7], where cylindrical and

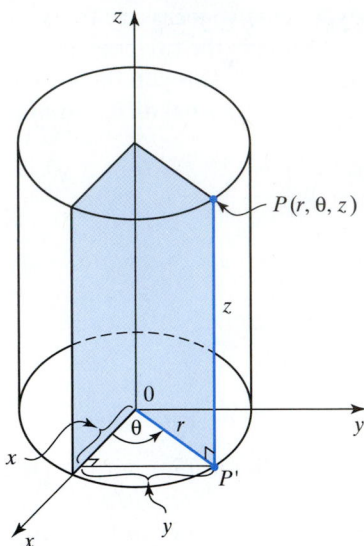

FIGURE 1.14   Cylindrical polar coordinates (r, θ, z).

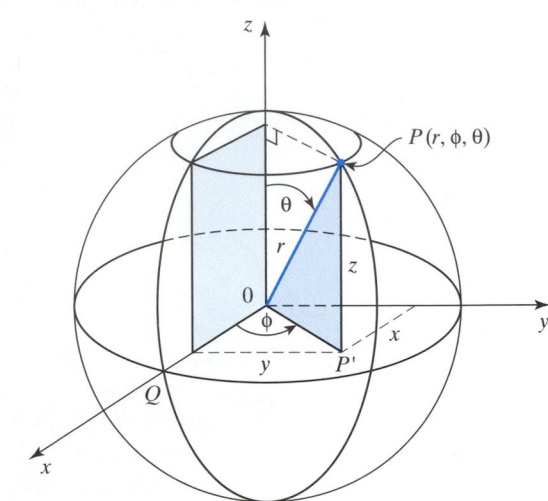

FIGURE 1.15   Spherical polar coordinates (r, ϕ, θ).

spherical polar coordinates are also discussed. Information on general orthogonal coordinate systems can be found in references [G.3] and [2.3].

EXERCISES 1.13

1. By making the change of variables $x = r \cos \theta$, $y = r \sin \theta$, $z = z$, in the function $f(x, y, z)$, when it becomes the function $F(r, \theta, z)$, show that in cylindrical polar coordinates

$$\frac{\partial F}{\partial r} = \cos \theta \frac{\partial f}{\partial x} + \sin \theta \frac{\partial f}{\partial y},$$

$$\frac{\partial F}{\partial \theta} = -r \sin \theta \frac{\partial f}{\partial x} + r \cos \theta \frac{\partial f}{\partial y}, \quad \frac{\partial F}{\partial z} = \frac{\partial f}{\partial z}.$$

2. Use the results of Exercise 1 to show that in cylindrical polar coordinates the Laplacian

$$\Delta f = \frac{\partial^2 f}{\partial x^2} + \frac{\partial^2 f}{\partial y^2} + \frac{\partial^2 f}{\partial z^2} \quad \text{becomes}$$

$$\Delta F = \frac{\partial^2 F}{\partial r^2} + \frac{1}{r} \frac{\partial F}{\partial r} + \frac{1}{r^2} \frac{\partial^2 F}{\partial \theta^2} + \frac{\partial^2 F}{\partial z^2},$$

and hence that an equivalent form of ΔF is

$$\Delta F = \frac{1}{r} \left[\frac{\partial}{\partial r} \left(r \frac{\partial F}{\partial r} \right) + \frac{1}{r} \frac{\partial}{\partial \theta} \left(\frac{\partial F}{\partial \theta} \right) + \frac{\partial}{\partial z} \left(r \frac{\partial F}{\partial z} \right) \right].$$

3. By making the change of variable $x = r \sin \theta \cos \phi$, $y = r \sin \theta \sin \phi$, $z = r \cos \theta$ in the function $f(x, y, z)$, when it

becomes $F(r, \phi, \theta)$, show that in spherical polar coordinates

$$\frac{\partial F}{\partial r} = \sin \theta \cos \phi \frac{\partial f}{\partial x} + \sin \phi \sin \theta \frac{\partial f}{\partial y} + \cos \phi \frac{\partial f}{\partial z}$$

$$\frac{\partial F}{\partial \phi} = r \cos \phi \cos \theta \frac{\partial f}{\partial x} + r \cos \phi \sin \theta \frac{\partial f}{\partial y} - r \sin \phi \frac{\partial f}{\partial z}$$

$$\frac{\partial F}{\partial z} = -r \sin \phi \sin \theta \frac{\partial f}{\partial x} + r \sin \phi \cos \theta \frac{\partial f}{\partial y}.$$

4. Use the results of Exercise 3 to show that in spherical polar coordinates the Laplacian

$$\Delta f = \frac{\partial^2 f}{\partial x^2} + \frac{\partial^2 f}{\partial y^2} + \frac{\partial^2 f}{\partial z^2}$$

becomes

$$\Delta F = \frac{1}{r^2} \frac{\partial}{\partial r} \left(r^2 \frac{\partial F}{\partial r} \right) + \frac{1}{r^2 \sin^2 \theta} \left(\frac{\partial^2 F}{\partial \phi^2} \right)$$

$$+ \frac{1}{r^2 \sin \theta} \frac{\partial}{\partial \theta} \left(\sin \theta \frac{\partial F}{\partial \theta} \right).$$

1.14   Inverse Functions and the Inverse Function Theorem

In mathematics and its applications it is often necessary to find the inverse of a function $y = f(x)$ so x can be expressed in the form $x = g(y)$, and when this can be done the function g is called the **inverse** of f and is such that $y = f(g(y))$. When f is an arbitrary function its inverse is often denoted by f^{-1}, and this superscript notation is also used to denote the inverse of trigonometric functions so if, for example, $y = \sin x$, the inverse sine function is written $\sin^{-1}$, so that $x = \sin^{-1} y$. However, the notation $y = \arcsin y$ is also used with the understanding that the notations $\arcsin$ and $\sin^{-1}$ are equivalent.

A trivial example of a function whose inverse can be found unambiguously is $y = ax + b$, because provided $a \neq 0$ we can write $x = (y - b)/a$ for all x and y. This is not the case, however, when trigonometric functions are involved, because the function $y = \sin x$ will give a unique value of y for any given x, but given y there are infinitely many values of x for which $y = \sin x$. This and similar inverse trigonometric functions are considered in elementary calculus courses. There the multivalued nature of the inverse sine function is resolved by restricting it to make y lie in a specific interval chosen so that one y corresponds to one x and, conversely, one x corresponds to one y. This situation is described by saying that the relationship between x and y is **one-to-one**. Specifically, in the case of the sine function, this is accomplished by requiring that if $x = \sin y$, the inverse function $y = \text{Arcsin } x$ is restricted so its **principal value** lies in the interval $-\pi/2 \leq \text{Arcsin} x \leq \pi/2$, where the domain of definition of the inverse function is $-1 \leq x \leq 1$.

A different possibility that arises frequently is when x and y are related by an equation of the form $f(x, y) = 0$ from which it is impossible to extract either x as a function of y, or y as a function of x in terms of known functions. A typical example of this type is $f(x, y) = x^2 - 2y^2 - \sin xy$. To make matters precise, if x and y are related by an equation $f(x, y) = 0$, then if a function $y = g(x)$ exists such that $f(x, g(x)) = 0$, the function $y = g(x)$ is said to be defined **implicitly** by $f(x, y) = 0$.

Although it is often not possible to find the function $g(x)$, it is still necessary to know when, in a neighborhood of a point (x_0, y_0), given a value of x, a unique value of y can be found, sometimes only numerically. The *implicit function theorem* that follows is seldom mentioned in first calculus courses because its proof involves certain technicalities, but it is quoted here in the simplest possible form because of its fundamental importance and the fact that is it frequently used by implication.

THEOREM 1.12

The implicit function theorem   Let $f(x, y)$ and $f_y(x, y)$ be continuous in a region D of the (x, y)-plane and let (x_0, y_0) be a point inside D, where $f(x_0, y_0) = 0$ and $f_y(x_0, y_0) \neq 0$. Then

(i) There is a rectangle R inside D containing (x_0, y_0) at all points of which there can be found a unique y such that $f(x, y) = 0$.

(ii) If the value of y is denoted by $g(x)$, then $y_0 = g(x_0)$, with $f(x, g(x)) = 0$, and $g(x)$ is continuous inside R.

(iii) If, in addition, $f_x(x, y)$ is continuous in D then $g(x)$ is differentiable in R and $g'(x) = -\frac{f_x(x,g(x))}{f_y(x,g(x))}$. ∎

In general terms, the implicit function theorem gives conditions that ensure the existence of an inverse function that is continuous and smooth enough to be differentiable. The theorem has a more general form involving functions $f(x_1, x_2, \ldots, x_n)$ of n variables, though this will not be given here. The interested reader can find accounts of the implicit function theorem and some of its generalizations in references [1.4], [1.6], and [5.1].

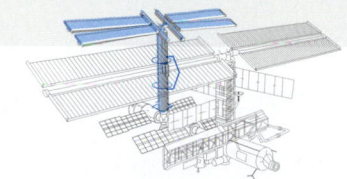

CHAPTER 1
TECHNOLOGY PROJECTS

Project 1

Linear Difference Equations and the Fibonacci Sequence

In Italy in 1202, Leonardo of Pisa, also known as Fibonacci, posed the following question. Let a newly born pair of rabbits produce two offspring each month, with breeding starting when they are 2 months old. Assuming that the pair of offspring start breeding in the same fashion when 2 months old, and that the process continues thereafter in a similar manner with no deaths, how many pairs of rabbits will there be after n months?

If u_n, is the number of pairs of rabbits after n months, the production of rabbits can be represented by the **linear difference equation**, or **recurrence relation**,

$$u_{n+2} = u_{n+1} + u_n,$$

where the sequence of numbers u_r with $r = 1, 2, \ldots$ is generated by setting $u_1 = 1$ and $u_2 = 1$, since this represents the initial pair of rabbits that began the breeding process. A simple calculation using this difference equation shows that the sequence of numbers generated in this manner that represents the number of pairs of rabbits present each month is

$$1, 1, 2, 3, 5, 8, \ldots,$$

and this is called the **Fibonacci** sequence. This sequence is found to occur in the study of regular solids, in numerical analysis, and elsewhere in mathematics.

A linear difference equation of the form

$$u_{n+2} = au_{n+1} + bu_n,$$

with a and b real numbers, can be solved by substituting $u_n = A\lambda^n$ into the difference equation and finding the two roots λ_1 and λ_2 of the resulting quadratic equation in λ. When $\lambda_1 \neq \lambda_2$, the general solution is $u_n = A_1\lambda_1^n + A\lambda_2^n$, and when $\lambda_1 = \lambda_2 = \lambda$, say, the general solution is $u_n = (A_1 + nA_2)\mu^n$. The arbitrary constants A_1 and A_2 are found by requiring u_n to satisfy some given conditions of the form $u_1 = \alpha$ and $u_2 = \beta$,

where the numbers α and β specify the way the sequence starts (the **initial conditions**).

Use this method to show that the solution u_n for the Fibonacci sequence is

$$u_n = \frac{1}{\sqrt{5}}\left[\left(\frac{1+\sqrt{5}}{2}\right)^n - \left(\frac{1-\sqrt{5}}{2}\right)^n\right],$$

$$\text{for } n = 1, 2, \ldots.$$

Make use of computer algebra to generate the first 30 terms of the Fibonacci sequence directly from the difference equation, and verify that the results are in agreement with the preceding formula.

Use computer algebra to show that $\lim_{n\to\infty}(u_n/u_{n-1}) = \frac{1}{2}(\sqrt{5}+1)$. This number is called the **golden mean**, and in art and architecture it represents the ratio of the sides of a rectangle that is considered to have the most pleasing appearance.

Project 2

Erratic Behavior of a Sequence Generated by a Difference Equation

1. Not all difference equations generate sequences of numbers that evolve steadily as happens with the Fibonacci sequence. Use computer algebra to generate the first 20 terms of the sequence produced by the difference equation

   $$u_{n+2} = 2u_{n+1} - 5u_n \quad \text{with } u_1 = 1, u_2 = -3,$$

   and observe its erratic behavior. Use the method of Project 1 to determine the analytical solution, and by means of computer algebra confirm that the two results are in agreement. Examine the analytical solution and explain why the behavior of the sequence of terms is so erratic.

2. Construct a difference equation of your own in which the roots λ_1 and λ_2 are equal. Find the analytical solution and use computer algebra to determine the first 20 terms of the sequence. Verify that these terms are in agreement with the ones generated directly from the difference equation.

PART TWO

VECTORS AND MATRICES

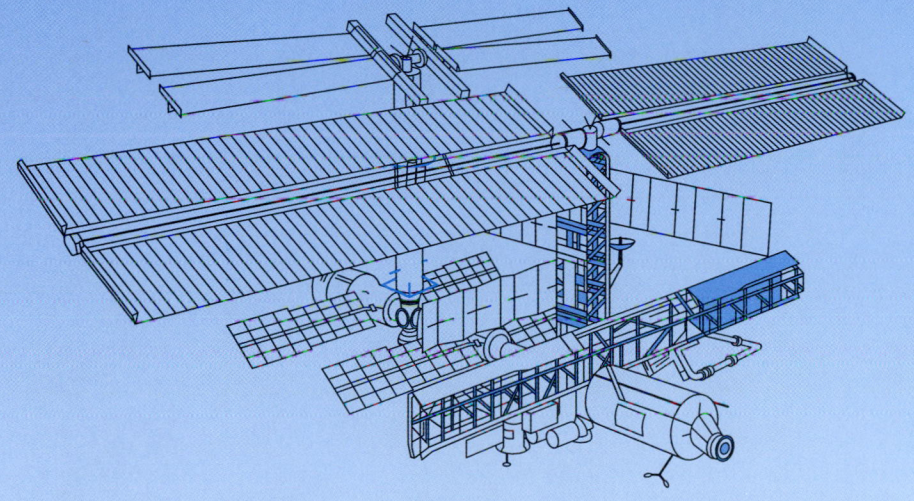

CHAPTER 2

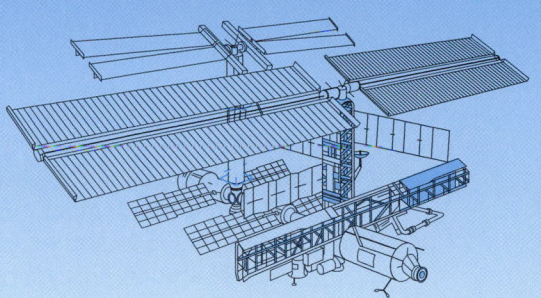

Vectors and Vector Spaces

Engineers, scientists, and physicists need to work with systems involving physical quantities that, unlike the density of a solid, cannot be characterized by a single number. This chapter is about the algebra of important and useful quantities called vectors that arise naturally when studying physical systems, and are defined by an ordered group of three numbers (a, b, c). Vectors are of fundamental importance and they play an essential role when the laws governing engineering and physics are expressed in mathematical terms.

A scalar quantity is one that is completely described when its magnitude is known, such as pressure, temperature, and area. A vector is a quantity that is completely specified when both its magnitude and direction are given, such as force, velocity, and momentum. A vector can be described geometrically as a directed straight line segment, with its length proportional to the magnitude of the vector, the line representing the vector parallel to the line of action of the vector, and an arrow on the line showing the direction along the line, or the sense, in which the vector acts.

This geometrical interpretation of a vector is valuable in many ways, as it can be used to add and subtract vectors and to multiply them by a scalar, since this merely involves changing their magnitude and sense, while leaving the line to which they are parallel unchanged. However, to perform more general algebraic operations on vectors some other form of representation is required. The one that is used most frequently involves describing a vector in terms of what are called its components along a set of three mutually orthogonal axes, which are usually taken to be the axes O{x, y, z} in the cartesian coordinate system. Here, by the component of a vector along a given line l, we mean the length of the perpendicular projection of the vector onto the line l.

We will see later that this cartesian representation of a vector identifies it completely in terms of three components and enables algebraic operations to be performed on it. In particular, it allows the introduction of the scalar product, or dot product, of two vectors that results in a scalar, and a vector product, or cross product, of two vectors that leads to a vector.

Finally, vectors and their algebra will be generalized to n space dimensions, leading to the concept of a vector space and to some related ideas.

2.1   Vectors, Geometry, and Algebra

Many quantities are completely described once their magnitude is known. A typical example of a physical quantity of this type is provided by the temperature at a given point in a room that is determined by the number specifying its value measured on a temperature scale, such as degrees F or degrees C. A quantity such as this is called a **scalar** quantity, and different examples of mathematical and physical scalar quantities are real numbers, length, area, volume, mass, speed, pressure, chemical concentration, electrical resistance, electric potential, and energy.

scalar

Other physical quantities are only fully specified when both their magnitude and direction are given. Quantities like this are called **vector** quantities, and a typical example of a vector quantity arises when specifying the instantaneous motion of a fluid particle in a river. In this case both the particle speed and its direction must be given if the description of its motion is to be complete. Speed in a given direction is called **velocity**, and velocity is a vector quantity. Some other examples of vector quantities are force, acceleration, momentum, the heat flow vector at a point in a block of metal, the earth's magnetic field at a given location, and a mathematical quantity called the gradient of a scalar function of position that will be defined later. By definition, the magnitude of a vector quantity is a nonnegative number (a scalar) that measures its size without regard to its direction, so, for example, the magnitude of a velocity is a speed.

vector

A convenient geometrical representation of a vector is provided by a straight line segment drawn in space parallel to the required direction, with an arrowhead indicating the **sense** in which the vector acts along the line segment, and the length of the line segment proportional to the magnitude of the vector. This is called a **directed straight line segment**, and by definition all directed straight line segments that are parallel to one another and have the same sense and length are regarded as equal. Expressed differently, moving a directed straight line segment parallel to itself so that its length remains the same and its arrow still points in the same direction leaves the vector it represents unchanged. A shift of a directed straight line segment of this type is called a **translation** of the vector it represents. For this reason the terms *directed straight line segment* and *vector* can be used interchangeably. Some examples of vectors that are equal through translation are shown in Fig. 2.1.

directed straight line segment

translation

It must be emphasized that geometrical representations of vectors as directed straight line segments in space are defined without reference to a specific coordinate system. This purely geometrical interpretation of vectors finds many applications, though a different form of representation is necessary if an effective vector algebra is to be developed for use with the calculus. An analytical representation of vectors that allows a vector algebra to be constructed with this purpose in mind can be based on a general coordinate system. However, throughout this chapter only rectangular cartesian coordinates will be used because they provide a simple and natural way of representing vectors.

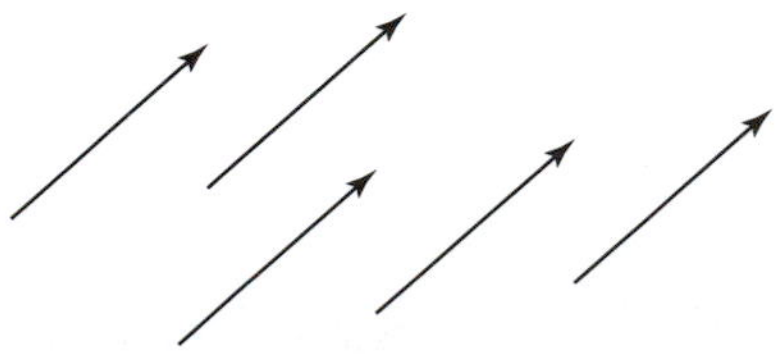

FIGURE 2.1   Equal geometrical vectors.

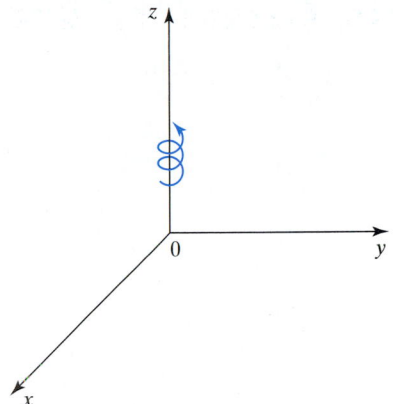

FIGURE 2.2   A right-handed
rectangular cartesian coordinate system.

In rectangular cartesian coordinates the x-, y-, and z-axes are all mutually orthogonal (perpendicular), and the positive sense along the axes is taken to be in the direction of increasing x, y, and z. The orientation of the axes will always be such that the positive direction along the z-axis is the one in which a right-handed screw (such as a corkscrew) aligned with the z-axis will advance when rotated from the positive x-axis to the positive y-axis, as shown in Fig. 2.2. A system of axes with this property is called a **right-handed system**.

right-handed system

The end of a vector toward which the arrow points will be called the **tip** of the vector, and the other end its **base**. Because a vector is invariant under a translation, there is no loss of generality in taking its base to be located at the origin O of the coordinate system, and its tip at a point P with the coordinates (a_1, a_2, a_3), say, as shown in Fig. 2.3. An application of the Pythagoras theorem to the triangle OPP'

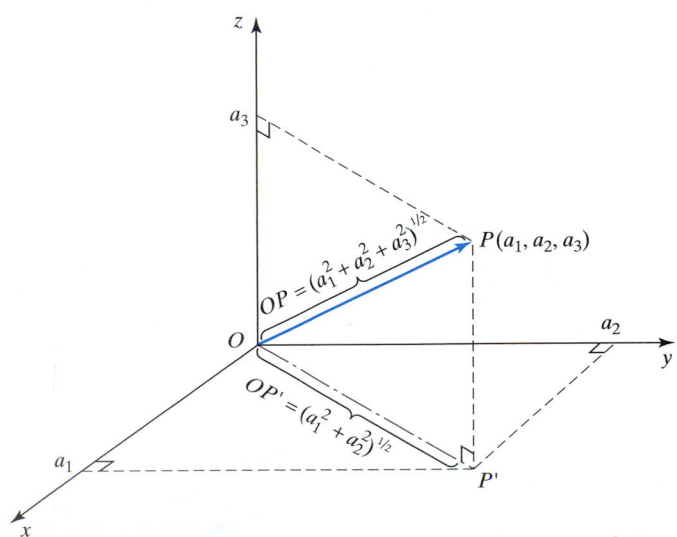

FIGURE 2.3   The vector from O to P and its components a_1, a_2, and a_3 in the x-, y-, z-coordinate system.

magnitude, unit vector, and components

shows the length of the line from O to P to be $(a_1^2 + a_2^2 + a_3^2)^{1/2}$. This length is *proportional* to the **magnitude** of the vector it represents, and as the base of the vector is at O, the sense of the vector is from O to P. For convenience, the constant of proportionality will be taken to be 1, so a directed straight line segment of unit length will represent a vector of magnitude 1 and so will be called a **unit vector**. Using this convention, the vector represented by the line from O to P in Fig. 2.3 has magnitude $(a_1^2 + a_2^2 + a_3^2)^{1/2}$. The three numbers a_1, a_2, and a_3, in this order, that define the vector from O to P are called its **components** in the x, y, and z directions, respectively.

ordered number triple

A set of three numbers a_1, a_2, and a_3 in a given order, written (a_1, a_2, a_3), is called an **ordered number triple**. As the coordinates (a_1, a_2, a_3) of point P in Fig. 2.3 completely define the vector from O to P, this ordered number triple may be taken as the definition of the vector itself. In general, changing the order of the numbers in an ordered number triple changes the vector it defines.

Sometimes it is necessary to consider a vector whose base does not coincide with the origin. Suppose that when this occurs the base C is at the point (c_1, c_2, c_3) and the tip D is at the point (d_1, d_2, d_3). Then Fig. 2.4 shows the components of this vector in the x, y, and z directions to be $d_1 - c_1$, $d_2 - c_2$, and $d_3 - c_3$. These components determine both the magnitude and direction of the vector. The vector is described by the ordered number triple $(d_1 - c_1, d_2 - c_2, d_3 - c_3)$, and the length of CD that is equal to the magnitude of the vector is $[(d_1 - c_1)^2 + (d_2 - c_2)^2 + (d_3 - c_3)^2]^{1/2}$.

norm and modulus

For convenience, it is usual to represent a vector by a single boldface character such as $\mathbf{a}$, and its **magnitude** (length) by $\|\mathbf{a}\|$, called the **norm** of $\mathbf{a}$. It is necessary to say here that in applications of vectors to mechanics, and in some purely geometrical applications of vectors, the norm of vector $\mathbf{r}$ is often called its **modulus** and written $|\mathbf{r}|$. When this convention is used, because $|\mathbf{r}|$ is a scalar it is usual to denote it by the corresponding ordinary italic letter r, so that $r = |\mathbf{r}|$.

If the base and tip of a vector need to be identified by letters, a vector such as the one from C to D in Fig. 2.4 is written $\underline{CD}$, with underlining used to indicate that a vector is involved, and the ordering of the letters is such that the first shows the

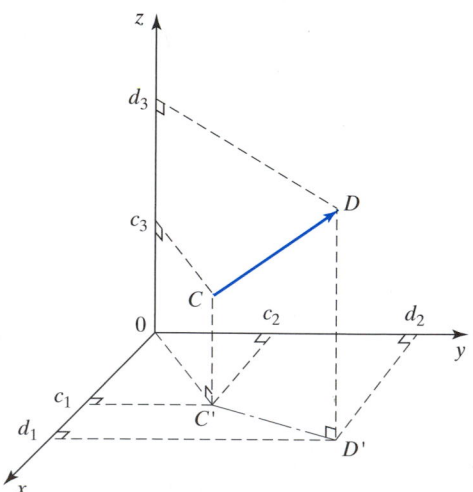

FIGURE 2.4   Vector directed from point C at (c_1, c_2, c_3) to point D at (d_1, d_2, d_3).

base and the second the tip of the vector. Thus, $\underline{CD}$ and $\underline{DC}$ are vectors of equal magnitude but opposite sense, and when these vectors are represented by arrows, the arrows are parallel and of equal length, but point in opposite directions.

EXAMPLE 2.1   If, in Fig. 2.4, C is the point $(-3, 4, 9)$ and D the point $(2, 5, 7)$, the vector $\underline{CD}$ has components $2 - (-3) = 5$, $5 - 4 = 1$, and $7 - 9 = -2$, and so is represented by the ordered number triple $(5, 1, -2)$, whereas vector $\underline{DC}$ has components -5, -1, and 2 and is represented by the ordered number triple $(-5, -1, 2)$. ∎

Having illustrated the concepts of scalars and vectors using some familiar examples, we now develop the algebra of vectors in rather more general terms.

Vectors

A **vector** quantity $\mathbf{a}$ is an ordered number triple (a_1, a_2, a_3) in which a_1, a_2, and a_3 are real numbers, and we shall write $\mathbf{a} = (a_1, a_2, a_3)$. The numbers a_1, a_2, and a_3, in this order, are called the first, second, and third **components** of vector $\mathbf{a}$ or, equivalently, its x-, y-, and z-components.

Null vector

The **null (zero)** vector, written $\mathbf{0}$, has neither magnitude nor direction and is the ordered number triple $\mathbf{0} = (0, 0, 0)$.

Equality of vectors

Two vectors $\mathbf{a} = (a_1, a_2, a_3)$ and $\mathbf{b} = (b_1, b_2, b_3)$ are **equal**, written $\mathbf{a} = \mathbf{b}$, if, and only if, $a_1 = b_1$, $a_2 = b_2$, and $a_3 = b_3$.

EXAMPLE 2.2   If $\mathbf{a} = (a_1, -5, 6)$, $\mathbf{b} = (3, b_2, b_3)$ and $\mathbf{c} = (3, -5, 1)$, then $\mathbf{a} = \mathbf{b}$ if $a_1 = 3$, $b_2 = -5$ and $b_3 = 6$, and $\mathbf{b} = \mathbf{c}$ if $b_2 = -5$ and $b_3 = 1$, but $\mathbf{a} \neq \mathbf{c}$ for any choice of a_1 because $6 \neq 1$. ∎

Norm of a vector

The **norm** of vector $\mathbf{a} = (a_1, a_2, a_3)$, denoted by $\|\mathbf{a}\|$, is the non-negative real number

$$\|\mathbf{a}\| = \left(a_1^2 + a_2^2 + a_3^2\right)^{1/2},$$

and in geometrical terms $\|\mathbf{a}\|$ is the *length* of vector $\mathbf{a}$. The norm of the null vector $\mathbf{0}$ is $\|\mathbf{0}\| = 0$. For example, if $\mathbf{a}$ is in m/sec, "length" of $\mathbf{a}$ is in m/sec.

EXAMPLE 2.3   If $\mathbf{a} = (1, -3, 2)$, then $\|\mathbf{a}\| = [1^2 + (-3)^2 + 2^2]^{1/2} = \sqrt{14}$, as illustrated in Fig. 2.5. ∎

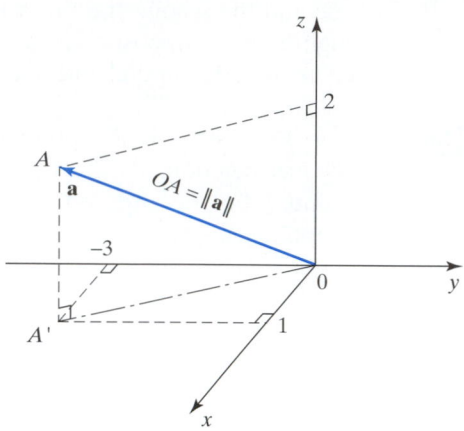

FIGURE 2.5   Vector **a** and its norm $\|\mathbf{a}\|$.

The sum of two vectors

If $\mathbf{a} = (a_1, a_2, a_3)$ and $\mathbf{b} = (b_1, b_2, b_3)$ have the same dimensions, say, both are m/sec, their **sum**, written $\mathbf{a} + \mathbf{b}$, is defined as the ordered number triple (vector) obtained by adding corresponding components of **a** and **b** to give

$$\mathbf{a} + \mathbf{b} = (a_1 + b_1, a_2 + b_2, a_3 + b_3).$$

EXAMPLE 2.4

If $\mathbf{a} = (1, 2, -5)$ and $\mathbf{b} = (-2, 2, 4)$, then

$$\mathbf{a} + \mathbf{b} = (1 + (-2), 2 + 2, -5 + 4) = (-1, 4, -1). \qquad \blacksquare$$

Multiplying a vector by a scalar

Let $\mathbf{a} = (a_1, a_2, a_3)$ and λ be an arbitrary real number. Then the product $\lambda\mathbf{a}$ is defined as the vector

$$\lambda\mathbf{a} = (\lambda a_1, \lambda a_2, \lambda a_3).$$

EXAMPLE 2.5

Let $\mathbf{a} = (2, -3, 5)$, $\mathbf{b} = (-1, 2, 4)$. Then $2\mathbf{a} = (4, -6, 10)$, $4\mathbf{b} = (-4, 8, 16)$, and $2\mathbf{a} + 4\mathbf{b} = (4 + (-4), -6 + 8, 10 + 16) = (0, 2, 26)$. $\qquad \blacksquare$

This definition of the product of a vector and a scalar, called **scaling** a vector, shows that when vector **a** is multiplied by a scalar λ, the norm of **a** is multiplied by $|\lambda|$, because

$$\|\lambda\mathbf{a}\| = \left(\lambda^2 a_1^2 + \lambda^2 a_2^2 + \lambda^2 a_3^2\right)^{1/2} = |\lambda| \cdot \|\mathbf{a}\|.$$

It also follows from the definition that the sense of vector **a** is reversed when it is multiplied by -1, though its norm is left unaltered. The definition of the **difference**

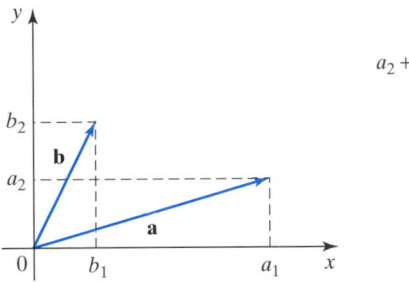
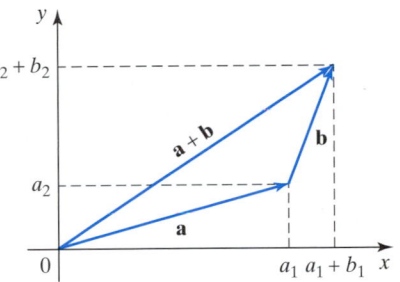

FIGURE 2.6   The vector sum $\mathbf{a} + \mathbf{b}$.

of two vectors is seen to be contained in the definition of their sum, because $\mathbf{a} - \mathbf{b} = \mathbf{a} + (-\mathbf{b})$. In particular, when $\mathbf{a} = \mathbf{b}$, we find that that $\mathbf{a} - \mathbf{a} = \mathbf{0}$, showing that $-\mathbf{a}$ is the *additive inverse* of $\mathbf{a}$.

The geometrical interpretations of the sum $\mathbf{a} + \mathbf{b}$, the difference $\mathbf{a} - \mathbf{b}$, and the **scaled** vector $\lambda\mathbf{a}$ in terms of their components are shown in Figs. 2.6 to 2.8, though to simplify the diagrams only the two-dimensional cases are illustrated. This involves no loss of generality, because it is always possible to choose the (x, y)-plane to coincide with the plane containing the vectors $\mathbf{a}$ and $\mathbf{b}$.

Vector Addition by the Triangle Rule

Consideration of Fig. 2.6 shows that the addition of vector $\mathbf{b}$ to vector $\mathbf{a}$ is obtained geometrically by translating vector $\mathbf{b}$ until its base is located at the tip of vector $\mathbf{a}$, and then the vector representing the sum $\mathbf{a} + \mathbf{b}$ has its base at the base of vector $\mathbf{a}$ and its tip at the tip of the repositioned vector $\mathbf{b}$. Because of the triangle involving vectors $\mathbf{a}$, $\mathbf{b}$, and $\mathbf{a} + \mathbf{b}$, this geometrical interpretation of a vector sum is called the **triangle rule** for vector addition. The triangle rule also applies to the difference of two vectors, as may be seen by considering Fig. 2.7, because after obtaining $-\mathbf{b}$ from $\mathbf{b}$ by reversing its sense, the difference $\mathbf{a} - \mathbf{b}$ can be written as the vector sum $\mathbf{a} + (-\mathbf{b})$, where $-\mathbf{b}$ is added to vector $\mathbf{a}$ by means of the triangle rule.

triangle rule for addition

The algebraic results discussed so far concerning the addition and scaling of vectors, together with some of their consequences, are combined to form the following theorem.

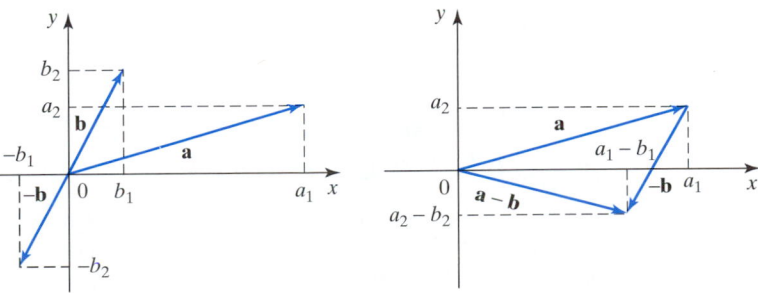

FIGURE 2.7   The vector difference $\mathbf{a} - \mathbf{b}$.

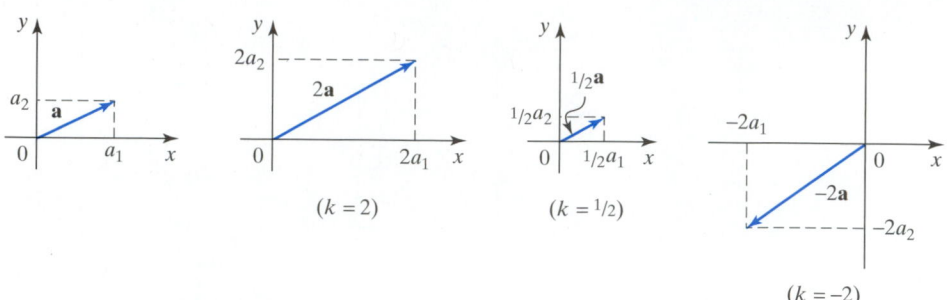

FIGURE 2.8 The vector $k\mathbf{a}$ for different values of k.

THEOREM 2.1

Addition and scaling of vectors   Let $\mathbf{P}, \mathbf{Q}$, and $\mathbf{R}$ be arbitrary vectors and let α and β be arbitrary real numbers. Then:

1. $\mathbf{P} + \mathbf{Q} = \mathbf{Q} + \mathbf{P}$ (vector addition is **commutative**);

2. $\mathbf{P} + \mathbf{0} = \mathbf{0} + \mathbf{P} = \mathbf{P}$ ($\mathbf{0}$ is the **identity element** in vector addition);

3. $(\mathbf{P} + \mathbf{Q}) + \mathbf{R} = \mathbf{P} + (\mathbf{Q} + \mathbf{R})$ (vector addition is **associative**);

4. $\alpha(\mathbf{P} + \mathbf{Q}) = \alpha\mathbf{P} + \alpha\mathbf{Q}$ (multiplication by a scalar is **distributive** over **vector addition**);

5. $(\alpha\beta)\mathbf{P} = \alpha(\beta\mathbf{P}) = \beta(\alpha\mathbf{P})$ (multiplication of a vector by a product of scalars is **associative**);

6. $(\alpha + \beta)\mathbf{P} = \alpha\mathbf{P} + \beta\mathbf{P}$ (multiplication of a vector by a sum of scalars is **distributive**);

7. $\|\alpha\mathbf{P}\| = |\alpha| \cdot \|\mathbf{P}\|$ (scaling $\mathbf{P}$ by α scales the norm of $\mathbf{P}$ by $|\alpha|$).

Proof   The results of this theorem are all immediate consequences of the above definitions so as the proofs of results 1 to 6 are all very similar, and result 7 has already been established, we only prove result 4.

$$\text{Let } \mathbf{P} = (p_1, p_2, p_3) \text{ and } \mathbf{Q} = (q_1, q_2, q_3); \text{ then}$$
$$\alpha(\mathbf{P} + \mathbf{Q}) = \alpha(p_1 + q_1, p_2 + q_2, p_3 + q_3)$$
$$= \alpha[(p_1, p_2, p_3) + (q_1, q_2, q_3)]$$
$$= \alpha(p_1, p_2, p_3) + \alpha(q_1, q_2, q_3)$$
$$= \alpha\mathbf{P} + \alpha\mathbf{Q},$$

as was to be shown. ∎

The Representation of Vectors in Terms of the Unit Vectors i, j, and k

The components of a vector, together with vector addition, can be used to describe vectors in a very convenient way. The idea is simple, and it involves using the standard convention that $\mathbf{i}, \mathbf{j}$, and $\mathbf{k}$ are vectors of unit length that point in the positive sense along the x-, y-, and z-axes, respectively. Vectors such as $\mathbf{i}, \mathbf{j}$, and $\mathbf{k}$ that have a unit norm (length) are called **unit vectors**, so $\|\mathbf{i}\| = \|\mathbf{j}\| = \|\mathbf{k}\| = 1$.

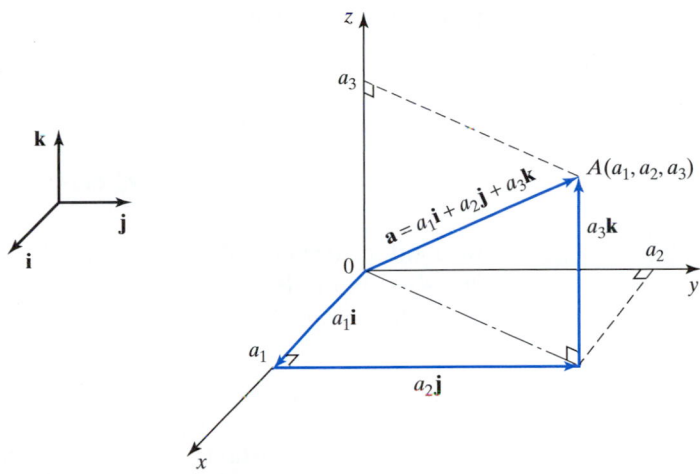

FIGURE 2.9   Vector **a** in terms of the unit vectors **i**, **j**, and **k**.

An arbitrary vector **a** can be represented by an "arrow," with its base at the origin and its tip at the point A with cartesian coordinates (a_1, a_2, a_3) where, of course, a_1, a_2, and a_3 are also the components of **a**. Consequently, scaling the unit vectors **i**, **j**, and **k** by the respective x, y, and z components a_1, a_2, and a_3 of **a**, followed by vector addition of these three vectors, shows that **a** can be written

$$\mathbf{a} = a_1\mathbf{i} + a_2\mathbf{j} + a_3\mathbf{k}, \tag{1}$$

as can be seen from Fig. 2.9. The representation of vector **a** in terms of the unit vectors **i**, **j**, and **k** in (1), and the ordered triple notation, are equivalent, so

$$\mathbf{a} = a_1\mathbf{i} + a_2\mathbf{j} + a_3\mathbf{k} = (a_1, a_2, a_3). \tag{2}$$

position vector

In some applications a vector defines a point in space, so vectors of this type are called **position vectors**. The symbol **r** is normally used for a position vector, so if point P with coordinates (x, y, z) is a general point in space, as in Fig. 2.10, its

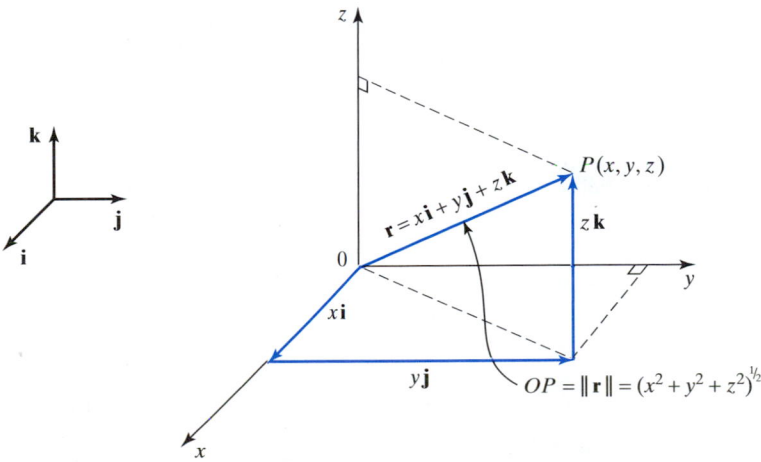

FIGURE 2.10   Position vector of a general point P in space.

position vector relative to the origin is

$$\mathbf{r} = x\mathbf{i} + y\mathbf{j} + z\mathbf{k}, \tag{3}$$

and its norm (length) is

$$\|\mathbf{r}\| = (x^2 + y^2 + z^2)^{1/2}. \tag{4}$$

EXAMPLE 2.6

(a) Find the distance of point P from the origin given that its position vector is $\mathbf{r} = 2\mathbf{i} + 4\mathbf{j} - 3\mathbf{k}$. (b) If a general point P in space has position vector $\mathbf{r} = x\mathbf{i} + y\mathbf{j} + z\mathbf{k}$, describe the surface defined by $\|\mathbf{r}\| = 3$ and find its cartesian equation.

Solution **(a)** As $\mathbf{r}$ is the position vector of P relative to the origin, the distance of point P from the origin is $\|\mathbf{r}\| = [2^2 + 4^2 + (-3)^2]^{1/2} = \sqrt{29}$.
(b) As $\|\mathbf{r}\| = 3$ (constant), it follows that the required surface is one for which every point lies at a distance 3 from the origin, so the surface must be a sphere of radius 3 centered on the origin. As $\mathbf{r} = x\mathbf{i} + y\mathbf{j} + z\mathbf{k}$ is the general position vector of a point on this sphere, the result $\|\mathbf{r}\| = 3$ is equivalent to $(x^2 + y^2 + z^2)^{1/2} = 3$, so the cartesian equation of the sphere is $x^2 + y^2 + z^2 = 9$. ∎

Because of the equivalence of the ordered number triple notation and the representation of vectors in terms of the unit vectors $\mathbf{i}$, $\mathbf{j}$, and $\mathbf{k}$ given in (2), both systems obey the same rules governing the addition and scaling of vectors in terms of their components. Thus, the following rules apply to the combination of any two vectors $\mathbf{a} = a_1\mathbf{i} + a_2\mathbf{j} + a_3\mathbf{k}, \mathbf{b} = b_1\mathbf{i} + b_2\mathbf{j} + b_3\mathbf{k}$ expressed in terms of $\mathbf{i}, \mathbf{j},$ and $\mathbf{k}$, and an arbitrary real number λ.

The sum $\mathbf{a} + \mathbf{b}$ is given by

$$\mathbf{a} + \mathbf{b} = (a_1 + b_1)\mathbf{i} + (a_2 + b_2)\mathbf{j} + (a_3 + b_3)\mathbf{k}. \tag{5}$$

The product $\lambda\,\mathbf{a}$ is given by

$$\lambda\,\mathbf{a} = \lambda a_1\,\mathbf{i} + \lambda a_2\,\mathbf{j} + \lambda a_3\,\mathbf{k}. \tag{6}$$

The norm of scaled vector $\lambda\,\mathbf{a}$ is given by

$$\|\lambda\,\mathbf{a}\| = |\lambda| \cdot \|\mathbf{a}\|$$
$$= |\lambda|\left(a_1^2 + a_2^2 + a_3^2\right)^{1/2}. \tag{7}$$

EXAMPLE 2.7

If $\mathbf{a} = 5\mathbf{i} + \mathbf{j} - 3\mathbf{k}$ and $\mathbf{b} = 2\mathbf{i} - 2\mathbf{j} - 7\mathbf{k}$, find (a) $\mathbf{a} + \mathbf{b}$, (b) $\mathbf{a} - \mathbf{b}$, (c) $2\mathbf{a} + \mathbf{b}$, and (d) $|-2\mathbf{a}|$.

Solution

(a)
$$\begin{aligned}
\mathbf{a} + \mathbf{b} &= (5\mathbf{i} + \mathbf{j} - 3\mathbf{k}) + (2\mathbf{i} - 2\mathbf{j} - 7\mathbf{k}) \\
&= (5 + 2)\mathbf{i} + (1 - 2)\mathbf{j} + (-3 - 7)\mathbf{k} \\
&= 7\mathbf{i} - \mathbf{j} - 10\mathbf{k}.
\end{aligned}$$

(b)
$$\begin{aligned}
\mathbf{a} - \mathbf{b} &= (5\mathbf{i} + \mathbf{j} - 3\mathbf{k}) - (2\mathbf{i} - 2\mathbf{j} - 7\mathbf{k}) \\
&= (5 - 2)\mathbf{i} + (1 - (-2))\mathbf{j} + (-3 - (-7))\mathbf{k} \\
&= 3\mathbf{i} + 3\mathbf{j} + 4\mathbf{k}.
\end{aligned}$$

(c)
$$2\mathbf{a} + \mathbf{b} = 2(5\mathbf{i} + \mathbf{j} - 3\mathbf{k}) + (2\mathbf{i} - 2\mathbf{j} - 7\mathbf{k})$$
$$= (10\mathbf{i} + 2\mathbf{j} - 6\mathbf{k}) + (2\mathbf{i} - 2\mathbf{j} - 7\mathbf{k})$$
$$= (10 + 2)\mathbf{i} + (2 + (-2))\mathbf{j} + (-6 + (-7))\mathbf{k}$$
$$= 12\mathbf{i} - 13\mathbf{k}.$$

(d)
$$|-2\mathbf{a}| = [(-10)^2 + (-2)^2 + 6^2]^{1/2} = 2\sqrt{35}$$

or, equivalently,

$$|-2\mathbf{a}| = |-2| \cdot \|\mathbf{a}\| = 2\|\mathbf{a}\| = 2[5^2 + 1^2 + (-3)^2]^{1/2} = 2\sqrt{35}. \qquad \blacksquare$$

Finding a Unit Vector in the Direction of an Arbitrary Vector

It is often necessary to find a unit vector in the direction of an arbitrary vector $\mathbf{a} = a_1\mathbf{i} + a_2\mathbf{j} + a_3\mathbf{k}$. This is accomplished by dividing $\mathbf{a}$ by its norm $\|\mathbf{a}\|$, because the vector $\mathbf{a}/\|\mathbf{a}\|$ has the same sense as $\mathbf{a}$ and its norm is 1. It is convenient to use a symbol related to an arbitrary vector $\mathbf{a}$ to indicate the unit vector in its direction, so from now on such a vector will be denoted by $\hat{\mathbf{a}}$, read "a hat." So if $\mathbf{a} = a_1\mathbf{i} + a_2\mathbf{j} + a_3\mathbf{k}$,

$$\hat{\mathbf{a}} = \mathbf{a}/\|\mathbf{a}\| = (a_1\mathbf{i} + a_2\mathbf{j} + a_3\mathbf{k})/\left(a_1^2 + a_2^2 + a_3^2\right)^{1/2}$$
$$= (a_1/a)\mathbf{i} + (a_2/a)\mathbf{j} + (a_3/a)\mathbf{k}, \quad \text{with } a = \left(a_1^2 + a_2^2 + a_3^2\right)^{1/2}. \qquad (8)$$

As the symbols $\mathbf{i}, \mathbf{j}$, and $\mathbf{k}$ are used exclusively for the unit vectors in the x-, y-, and z-directions, it is not necessary to write $\hat{\mathbf{i}}, \hat{\mathbf{j}}$, and $\hat{\mathbf{k}}$.

The relationship between $\mathbf{a}, \hat{\mathbf{a}}$, and $\|\mathbf{a}\|$ can be put in the useful form

$$\mathbf{a} = \|\mathbf{a}\|\hat{\mathbf{a}}, \qquad (9)$$

showing that a general vector $\mathbf{a}$ can always be written as the unit vector $\hat{\mathbf{a}}$ scaled by $\|\mathbf{a}\|$. Unless otherwise stated, $\mathbf{a} \neq \mathbf{0}$.

EXAMPLE 2.8   Find a unit vector in the direction of $\mathbf{a} = 3\mathbf{i} + 2\mathbf{j} + 5\mathbf{k}$.

Solution   As $\|\mathbf{a}\| = (3^2 + 2^2 + 5^2)^{1/2} = \sqrt{38}$, it follows that

$$\hat{\mathbf{a}} = \mathbf{a}/\|\mathbf{a}\| = (3/\sqrt{38})\mathbf{i} + (2/\sqrt{38})\mathbf{j} + (5/\sqrt{38})\mathbf{k}. \qquad \blacksquare$$

EXAMPLE 2.9   It is known from experiments in mechanics that forces are vector quantities and so combine according to the laws of vector algebra. Use this fact to find the sum and difference of a force of 9 units in the direction of $2\mathbf{i} + \mathbf{j} - 2\mathbf{k}$ and a force of 10 units in the direction of $4\mathbf{i} - 3\mathbf{j}$, and determine the magnitudes of these forces.

Solution   We will use the convention that a unit vector represents a force of 1 unit. Let $\mathbf{F}$ be the force of 9 units. Then as $\|2\mathbf{i} + \mathbf{j} - 2\mathbf{k}\| = [2^2 + 1^2 + (-2)^2]^{1/2} = 3$, the unit vector in the direction of $\mathbf{F}$ is

$$\hat{\mathbf{F}} = (1/3)(2\mathbf{i} + \mathbf{j} - 2\mathbf{k}) = (2/3)\mathbf{i} + (1/3)\mathbf{j} - (2/3)\mathbf{k},$$

so $\mathbf{F} = 9\hat{\mathbf{F}} = 6\mathbf{i} + 3\mathbf{j} - 6\mathbf{k}$ units.

Similarly, let $\mathbf{G}$ be the force of 10 units. Then as $\|4\mathbf{i} - 3\mathbf{j}\| = 5$, the unit vector in the direction of $\mathbf{G}$ is

$$\hat{\mathbf{G}} = (1/5)(4\mathbf{i} - 3\mathbf{j}) = (4/5)\mathbf{i} - (3/5)\mathbf{j},$$

so $\mathbf{G} = 10\hat{\mathbf{G}} = 8\mathbf{i} - 6\mathbf{j}$ units.

Combining these results shows that $\mathbf{F} + \mathbf{G} = 14\mathbf{i} - 3\mathbf{j} - 6\mathbf{k}$ units, and $\mathbf{F} - \mathbf{G} = -2\mathbf{i} + 9\mathbf{j} - 6\mathbf{k}$ units, from which it follows that the magnitudes of the forces are given by

$$\|\mathbf{F} + \mathbf{G}\| = \sqrt{241} \text{ units and } \|\mathbf{F} - \mathbf{G}\| = 11 \text{ units.} \qquad \blacksquare$$

Equality of vectors expressed in terms of unit vectors

As the difference of two equal and opposite vectors is the null vector $\mathbf{0}$, this shows that if $\mathbf{a} = \mathbf{b}$, where $\mathbf{a} = a_1\mathbf{i} + a_2\mathbf{j} + a_3\mathbf{k}$, and $\mathbf{b} = b_1\mathbf{i} + b_2\mathbf{j} + b_3\mathbf{k}$, then the respective components of vectors $\mathbf{a}$ and $\mathbf{b}$ must be equal, leading to the result that

$$\mathbf{a} = \mathbf{b} \text{ if, and only if, } a_1 = b_1, a_2 = b_2, \text{ and } a_3 = b_3. \qquad (10)$$

Simple Geometrical Applications of Vectors

Although our use of vectors will be mainly in connection with the calculus, the following simple geometrical applications are helpful because they illustrate basic vector arguments and properties.

Although we have seen how an arbitrary vector can be expressed in terms of unit vectors associated with a cartesian coordinate system, it must be remembered that the fundamental concept of a vector and its algebra is independent of a coordinate system. Because of this, it is often possible to use the rules governing elementary vector algebra given in Theorem 2.1 to establish equations in a purely vectorial manner, without the need to appeal to any coordinate system. Once a general vector equation has been established, the representation of the vectors involved in terms of their components and the unit vectors $\mathbf{i}$, $\mathbf{j}$, and $\mathbf{k}$ can be used to convert the vector equation into the equivalent cartesian equations.

The purely vectorial approach to geometrical problems is well illustrated by finding the vector $\underline{AB}$ in terms of the position vectors of points A and B, and then using the result to find the position vector of the mid-point of $\underline{AB}$. After this, the purely vectorial derivation of a geometrical result followed by its interpretation in cartesian form will be illustrated by finding the equation of a straight line in three space dimensions.

Vector $\underline{AB}$ in terms of the position vectors of A and B

Let $\mathbf{a}$ and $\mathbf{b}$ be the position vectors of points A and B relative to an origin O, as shown in Fig. 2.11.

An application of the triangle rule for the addition of vectors gives

$$\underline{OA} + \underline{AB} = \underline{OB},$$

but $\underline{OA} = \mathbf{a}$ and $\underline{OB} = \mathbf{b}$, so

$$\mathbf{a} + \underline{AB} = \mathbf{b},$$

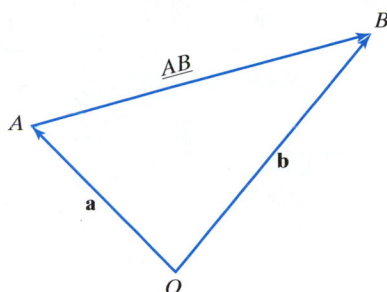

FIGURE 2.11 Vectors **a**, **b**, and $\underline{AB}$.

giving

$$\underline{AB} = \mathbf{b} - \mathbf{a}. \tag{11}$$

When expressed in words, this simple but useful result asserts that vector $\underline{AB}$ is obtained by subtracting the position vector **a** of point A from the position vector **b** of point B.

EXAMPLE 2.10

Find the position vector of the mid-point of $\underline{AB}$ if point A has position vector **a** and point B has position vector **b** relative to an origin O.

Solution   Let point C, with position vector **c** relative to origin O, be the mid-point of $\underline{AB}$, as shown in Fig. 2.12.
By the triangle rule,

$$\underline{OA} + \underline{AC} = \underline{OC},$$

but $\underline{OA} = \mathbf{a}$, and from (11) $\underline{AC} = (1/2)(\mathbf{b} - \mathbf{a})$, so

$$\underline{OC} = \mathbf{a} + (1/2)(\mathbf{b} - \mathbf{a}),$$

so the required result is

$$\mathbf{c} = \underline{OC} = (1/2)(\mathbf{b} + \mathbf{a}).$$ ■

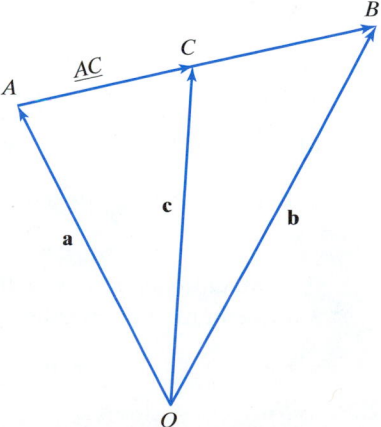

FIGURE 2.12 C is the mid-point of $\underline{AB}$.

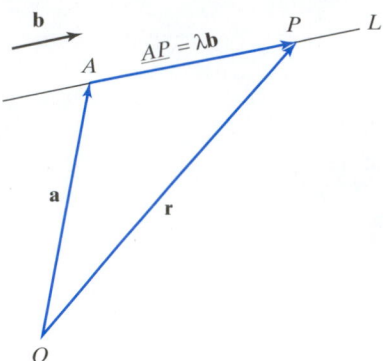

FIGURE 2.13 The straight line L.

The vector and cartesian equations of a straight line

Let line L be a straight line through point A with position vector $\mathbf{a}$ relative to an origin O, and let the line be parallel to a vector $\mathbf{b}$. If P is an arbitrary point on line L with position vector $\mathbf{r}$ relative to O, an application of the triangle rule for vector addition to the vectors shown in Fig. 2.13 gives

$$\mathbf{r} = \underline{OA} + \underline{AP}.$$

vector equation of straight line

But $\underline{OA} = \mathbf{a}$, and as $\underline{AP}$ is parallel to $\mathbf{b}$, a number λ can always be found such that $\underline{AP} = \lambda\mathbf{b}$, so the **vector equation** of line L becomes

$$\mathbf{r} = \mathbf{a} + \lambda\mathbf{b}. \tag{12}$$

Notice that result (12) determines all points P on L if λ is taken to be a number in the interval $-\infty < \lambda < \infty$.

The cartesian equations of line L follow by setting $\mathbf{a} = a_1\mathbf{i} + a_2\mathbf{j} + a_3\mathbf{k}$, $\mathbf{b} = b_1\mathbf{i} + b_2\mathbf{j} + b_3\mathbf{k}$, and $\mathbf{r} = x\mathbf{i} + y\mathbf{j} + z\mathbf{k}$ in result (12), and then using the definition of equality of vectors given in (10) to obtain the corresponding three scalar cartesian equations. Proceeding in this way we find that

$$x\mathbf{i} + y\mathbf{j} + z\mathbf{k} = a_1\mathbf{i} + a_2\mathbf{j} + a_3\mathbf{k} + \lambda(b_1\mathbf{i} + b_2\mathbf{j} + b_3\mathbf{k}),$$

cartesian and standard form of straight line

so equating corresponding components of $\mathbf{i}, \mathbf{j}$, and $\mathbf{k}$ on each side of this equation brings us to the required **cartesian equations** for L in the form

$$x_1 = a_1 + \lambda b_1, \quad x_2 = a_2 + \lambda b_2, \quad x_3 = a_3 + \lambda b_3. \tag{13}$$

An equivalent form of these equations is obtained by solving each equation for λ and equating the results to get

$$\frac{x - a_1}{b_1} = \frac{y - a_2}{b_2} = \frac{z - a_3}{b_3} = \lambda. \tag{14}$$

This is the **standard form** (also called the **canonical form**) of the cartesian equations of a straight line. It is important to notice that when written in standard form the coefficients of x, y, and z are all *unity*. Once the equation of a straight line is written in standard form, equating each numerator to zero determines the components (a_1, a_2, a_3) of a position vector of a point on the line, while the denominators in the order (b_1, b_2, b_3) determine the components of a vector parallel to the line.

EXAMPLE 2.11

A straight line L is given in the form

$$\frac{2x - 3}{4} = \frac{3 - y}{2} = \frac{z + 1}{3}.$$

Find the position vector of a point on L and a vector parallel to L.

Solution   When the equation is written in standard form it becomes

$$\frac{x - 3/2}{2} = \frac{y - 3}{-2} = \frac{z + 1}{3} = \lambda.$$

Comparing these equations with (14) shows that $(a_1, a_2, a_3) = (3/2, 3, -1)$ and $\mathbf{b} = (b_1, b_2, b_3) = (2, -2, 3)$. So the position vector of a point on the line is $\mathbf{a} = (3/2)\mathbf{i} + 3\mathbf{j} - \mathbf{k}$, and a vector parallel to the line is $\mathbf{b} = 2\mathbf{i} - 2\mathbf{j} + 3\mathbf{k}$.

Neither of these results is unique, because $\mu\mathbf{b}$ is also parallel to the line for any scalar $\mu \neq 0$, and any other point on L would suffice. For example, the vector $14\mathbf{i} - 14\mathbf{j} + 21\mathbf{k}$ is also parallel to the line, while setting $\lambda = 2$ leads to the result $(a_1, a_2, a_3) = (11/2, -1, 5)$, corresponding to a different point on the same line, this time with position vector $\mathbf{a} = (11/2)\mathbf{i} - \mathbf{j} + 5\mathbf{k}$.   ■

Summary

This section has introduced vectors both as geometrical quantities that can be represented by directed line segments and, using a right-handed system of cartesian axes, as ordered number triples. Definitions of the scaling, addition, and subtraction of vectors have been given, and a general vector has been defined in terms of the set of three unit vectors $\mathbf{i}$, $\mathbf{j}$, and $\mathbf{k}$ that lie along the orthogonal cartesian axes O$\{x, y, z\}$. Finally, the vector and cartesian equations of a straight line in space have been derived, and the standard form of the cartesian equations has been introduced from which a vector parallel to the line may be found by inspection.

EXERCISES 2.1

1. Prove Results 1, 3, and 6 of Theorem 2.1.

2. Given that $\mathbf{a} = 2\mathbf{i} + 3\mathbf{j} - \mathbf{k}$, $\mathbf{b} = \mathbf{i} - \mathbf{j} + 2\mathbf{k}$, and $\mathbf{c} = 3\mathbf{i} + 4\mathbf{j} + \mathbf{k}$, find (a) $\mathbf{a} + 2\mathbf{b} - \mathbf{c}$, (b) a vector $\mathbf{d}$ such that $\mathbf{a} + \mathbf{b} + \mathbf{c} + \mathbf{d} = 0$, and (c) a vector $\mathbf{d}$ such that $\mathbf{a} - \mathbf{b} + \mathbf{c} + 3\mathbf{d} = 0$.

3. Given $\mathbf{a} = \mathbf{i} + 2\mathbf{j} + 3\mathbf{k}$, $\mathbf{b} = 2\mathbf{i} - 2\mathbf{j} + \mathbf{k}$, find (a) a vector $\mathbf{c}$ such that $2\mathbf{a} + \mathbf{b} + 2\mathbf{c} = \mathbf{i} + \mathbf{k}$, (b) a vector $\mathbf{c}$ such that $3\mathbf{a} - 2\mathbf{b} + \mathbf{c} = \mathbf{i} + \mathbf{j} - 2\mathbf{k}$.

4. Given that $\mathbf{a} = 3\mathbf{i} + 2\mathbf{j} - 3\mathbf{k}$, $\mathbf{b} = 2\mathbf{i} - \mathbf{j} + 5\mathbf{k}$, and $\mathbf{c} = 2\mathbf{i} + 5\mathbf{j} + 2\mathbf{k}$, find (a) $2\mathbf{a} + 3\mathbf{b} - 3\mathbf{c}$, (b) a vector $\mathbf{d}$ such that $\mathbf{a} + 3\mathbf{b} - 2\mathbf{c} + 3\mathbf{d} = 0$, and (c) a vector $\mathbf{d}$ such that $2\mathbf{a} - 3\mathbf{d} = \mathbf{b} + 4\mathbf{c}$.

5. Given that A and B have the respective position vectors $2\mathbf{i} + 3\mathbf{j} - \mathbf{k}$ and $\mathbf{i} + 2\mathbf{j} + 4\mathbf{k}$, find the vector $\underline{AB}$ and a unit vector in the direction of $\underline{AB}$.

6. Given that A and B have the respective position vectors $3\mathbf{i} - \mathbf{j} + 4\mathbf{k}$ and $2\mathbf{i} + \mathbf{j} + \mathbf{k}$, find the vector $\underline{AB}$ and the position vector $\mathbf{c}$ of the mid-point of $\underline{AB}$.

7. Given that A and B have the respective position vectors $\mathbf{a}$ and $\mathbf{b}$, find the position vector of a point P on the line AB located between A and B such that

$$(\text{length } AP)/(\text{length } PB) = m/n, \quad \text{where } m, n > 0$$

are any two real numbers.

8. Find the position vector **r** of a point P on the straight line joining point A at $(1, 2, 1)$ and point B at $(3, -1, 2)$ and between A and B such that

$$\text{(length } AP)/\text{(length } PB) = 3/2.$$

9. It is known from Euclidean geometry that the medians of a triangle (lines drawn from a vertex to the mid-point of the opposite side) all meet at a single point P, and that P is two-thirds of the distance along each median from the vertex through which it passes. If the vertices A, B, and C of a triangle have the respective position vectors **a**, **b**, and **c**, show that the position vector of P is $(1/3)(\mathbf{a} + \mathbf{b} + \mathbf{c})$.

10. Forces of 1, 2, and 3 units act through the origin along, and in the positive directions of, the respective x-, y-, and z-axes. Find the vector sum **S** of these forces, the magnitude $\|\mathbf{S}\|$ of the sum of the vectors, and a unit vector in the direction of **S**.

11. Forces of 2, 1, and 4 units act through the origin along, and in the positive directions of, the respective x-, y-, and z-axes. Find the vector sum **S** of these forces, the magnitude $\|\mathbf{S}\|$ of the sum of the vectors, and a unit vector in the direction of **S**.

12. A straight line L is given in the form

$$\frac{3x - 1}{4} = \frac{2y + 3}{2} = \frac{2 - 3z}{1}.$$

Find the position vectors of two different points on L and a unit vector parallel to L.

13. A straight line L is given in the form

$$\frac{2x + 1}{3} = \frac{3y + 2}{4} = \frac{2 - 4z}{-1}.$$

Find position vectors of two different points on L and a unit vector parallel to L.

14. Given that a straight line L_1 passes through the points $(-2, 3, 1)$ and $(1, 4, 6)$, find (a) the position vector of a point on the line and a vector parallel to it, and (b) a straight line L_2 parallel to L_1 that passes through the point $(1, 2, 1)$.

15. Given that a straight line L_1 passes through the points $(3, 2, 4)$ and $(2, 1, 6)$, find (a) the position vector of a point on the line and a vector parallel to it, and (b) a straight line L_2 parallel to L_1 that passes through the point $(-2, 1, 2)$.

16. A straight line has the vector equation $\mathbf{r} = \mathbf{a} + \lambda \mathbf{b}$, where $\mathbf{a} = 3\mathbf{j} + 2\mathbf{k}$, and $\mathbf{b} = 2\mathbf{i} + \mathbf{j} + 2\mathbf{k}$. Find the cartesian equations of the line and the coordinates of three points that lie on it.

17. A straight line passes through the point $(3, 2, -3)$ parallel to the vector $2\mathbf{i} + 3\mathbf{j} - 3\mathbf{k}$. Find the cartesian equations of the line and the coordinates of three points that lie on it.

18. In mechanics, if a point A moves with velocity $\mathbf{v}_A$ and point B moves with velocity $\mathbf{v}_B$, the velocity $\mathbf{v}_R$ of A **relative** to B (the **relative velocity** of A with respect to B) is defined as $\mathbf{v}_R = \mathbf{v}_A - \mathbf{v}_B$. Power boat A moves northeast at 20 knots and power boat B moves southeast at 30 knots. Find the velocity of boat A relative to boat B, and a unit vector in the direction of the relative velocity.

2.2   The Dot Product (Scalar Product)

A product of two vectors **a** and **b** can be formed in such a way that the result is a scalar. The result is written $\mathbf{a} \cdot \mathbf{b}$ and called the **dot product** of **a** and **b**. The names **scalar product** and **inner product** are also used in place of the term *dot product*.

Dot Product

Let **a** and **b** be any two vectors that after a translation to bring their bases into coincidence are inclined to one another at an angle θ, as shown in Fig. 2.14, where $0 \le \theta \le \pi$. Then the **dot product** of **a** and **b** is defined as the number

dot or scalar product

$$\mathbf{a} \cdot \mathbf{b} = \|\mathbf{a}\| \cdot \|\mathbf{b}\| \cos \theta.$$

This geometrical definition of the dot product has many uses, but when working with vectors **a** and **b** that are expressed in terms of their components in the **i**, **j**, and

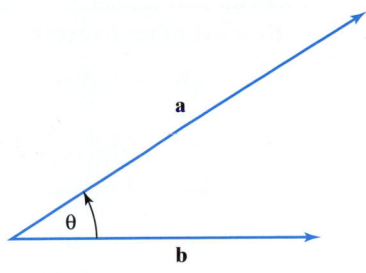

FIGURE 2.14 Vectors **a** and **b** inclined at an angle θ.

k directions, a more convenient form is needed. An equivalent definition that is easier to use is given later in (23).

properties of the dot product

Properties of the dot product

The following results, in which **a** and **b** are any two vectors and λ and μ are any two scalars, are all immediate consequences of the definition of the dot product.

The dot product is commutative

$$\mathbf{a} \cdot \mathbf{b} = \mathbf{b} \cdot \mathbf{a} \quad \text{and} \quad \lambda\mathbf{a} \cdot \mu\mathbf{b} = \mu\mathbf{a} \cdot \lambda\mathbf{b} = \lambda\mu\mathbf{a} \cdot \mathbf{b} \tag{15}$$

The dot product is distributive and linear

$$\mathbf{a} \cdot (\mathbf{b} + \mathbf{c}) = \mathbf{a} \cdot \mathbf{b} + \mathbf{a} \cdot \mathbf{c} \quad \text{and} \quad \mathbf{a} \cdot (\lambda\mathbf{b} + \mu\mathbf{c}) = \lambda\mathbf{a} \cdot \mathbf{b} + \mu\mathbf{a} \cdot \mathbf{c}. \tag{16}$$

The angle between two vectors

The angle θ between vectors **a** and **b** is given by

$$\cos\theta = \frac{\mathbf{a} \cdot \mathbf{b}}{\|\mathbf{a}\| \cdot \|\mathbf{b}\|}, \quad \text{with } 0 \le \theta \le \pi. \tag{17}$$

Parallel vectors ($\theta = 0$)

If vectors **a** and **b** are parallel, then

$$\mathbf{a} \cdot \mathbf{b} = \|\mathbf{a}\| \cdot \|\mathbf{b}\| \quad \text{and, in particular,} \quad \mathbf{a} \cdot \mathbf{a} = \|\mathbf{a}\|^2. \tag{18}$$

Orthogonal vectors ($\theta = \pi/2$)

If vectors **a** and **b** are orthogonal, then

$$\mathbf{a} \cdot \mathbf{b} = 0. \tag{19}$$

Product of unit vectors

If $\hat{\mathbf{a}}$ and $\hat{\mathbf{b}}$ are unit vectors, then

$$\hat{\mathbf{a}} \cdot \hat{\mathbf{b}} = \cos\theta, \quad \text{with } 0 \leq \theta \leq \pi. \tag{20}$$

An immediate consequence of properties (15), (19), and (20) is that

$$\mathbf{i} \cdot \mathbf{i} = \mathbf{j} \cdot \mathbf{j} = \mathbf{k} \cdot \mathbf{k} = 1, \tag{21}$$

and

$$\mathbf{i} \cdot \mathbf{j} = \mathbf{j} \cdot \mathbf{i} = \mathbf{i} \cdot \mathbf{k} = \mathbf{k} \cdot \mathbf{i} = \mathbf{j} \cdot \mathbf{k} = \mathbf{k} \cdot \mathbf{j} = 0. \tag{22}$$

We now use results (21) and (22) to arrive at a simple expression for the dot product in terms of the components of $\mathbf{a}$ and $\mathbf{b}$. To arrive at the result we set $\mathbf{a} = a_1\mathbf{i} + a_2\mathbf{j} + a_3\mathbf{k}$, $\mathbf{b} = b_1\mathbf{i} + b_2\mathbf{j} + b_3\mathbf{k}$ and form the dot product

$$\mathbf{a} \cdot \mathbf{b} = (a_1\mathbf{i} + a_2\mathbf{j} + a_3\mathbf{k}) \cdot (b_1\mathbf{i} + b_2\mathbf{j} + b_3\mathbf{k}).$$

dot product in terms of components

Expanding this product using (15) and (16) and making use of results (21) and (22) brings us to the following *alternative definition* of the **dot product** expressed in terms of the components of $\mathbf{a}$ and $\mathbf{b}$:

$$\mathbf{a} \cdot \mathbf{b} = a_1 b_1 + a_2 b_2 + a_3 b_3. \tag{23}$$

Using (23) in (17) produces the following useful expression that can be used to find the angle θ between $\mathbf{a}$ and $\mathbf{b}$:

$$\cos\theta = \frac{a_1 b_1 + a_2 b_2 + a_3 b_3}{\left(a_1^2 + a_2^2 + a_3^2\right)^{1/2} \left(b_1^2 + b_2^2 + b_3^2\right)^{1/2}} \quad \text{where } 0 \leq \theta \leq \pi. \tag{24}$$

EXAMPLE 2.12

Find $\mathbf{a} \cdot \mathbf{b}$ and the angle between the vectors $\mathbf{a}$ and $\mathbf{b}$, given that $\mathbf{a} = \mathbf{i} + 2\mathbf{j} + 3\mathbf{k}$ and $\mathbf{b} = 2\mathbf{i} - \mathbf{j} - 2\mathbf{k}$.

Solution $\|\mathbf{a}\| = \sqrt{14}$, $\|\mathbf{b}\| = 3$, and $\mathbf{a} \cdot \mathbf{b} = 1 \cdot 2 + 2 \cdot (-1) + 3 \cdot (-2) = -6$. Using these results in (24) gives

$$\cos\theta = -6/(3\sqrt{14}) = -2/\sqrt{14},$$

so as $0 \leq \theta \leq \pi$ we see that $\theta = 2.1347$ radians, or $\theta = 122.3°$. ∎

projecting a vector onto a line

The projection of a vector onto the line of another vector

The projection of vector $\mathbf{a}$ onto the line of vector $\mathbf{b}$ is a scalar, and it is the *signed* length of the geometrical projection of vector $\mathbf{a}$ onto a line parallel to $\mathbf{b}$, with the sign positive for $0 \leq \theta < \pi/2$ and negative for $\pi/2 < \theta \leq \pi$. This is illustrated in Fig. 2.15, from which it is seen that the signed length of the projection of $\mathbf{a}$ onto the line of vector $\mathbf{b}$ is ON, where $ON = \|\mathbf{a}\| \cos\theta$.

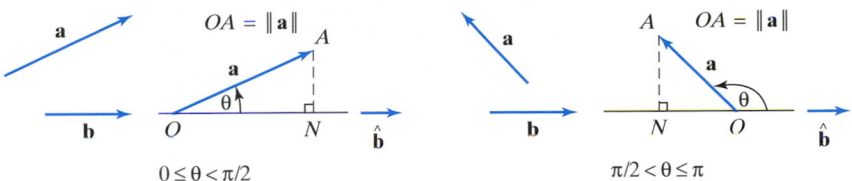

FIGURE 2.15   The projection of vector **a** onto the line of vector **b**.

If $\hat{\mathbf{b}}$ is the unit vector along **b**, then as $\mathbf{a} = \hat{\mathbf{a}}\|\mathbf{a}\|$, and $\hat{\mathbf{a}} \cdot \hat{\mathbf{b}} = \cos\theta$, the projection $ON = \|\mathbf{a}\|\cos\theta$ can be written as the dot product

$$ON = \|\mathbf{a}\|\hat{\mathbf{a}} \cdot \hat{\mathbf{b}} = \mathbf{a} \cdot \hat{\mathbf{b}} = \frac{\mathbf{a} \cdot \mathbf{b}}{\|\mathbf{b}\|} \tag{25}$$

EXAMPLE 2.13   Find the strength of the magnetic field vector $\mathbf{H} = 5\mathbf{i} + 3\mathbf{j} + 7\mathbf{k}$ in the direction of $2\mathbf{i} - \mathbf{j} + 2\mathbf{k}$, where a unit vector represents one unit of magnetic flux.

Solution   We are required to find the projection of vector **H** in the direction of the vector $2\mathbf{i} - \mathbf{j} + 2\mathbf{k}$. Setting $\mathbf{b} = 2\mathbf{i} - \mathbf{j} + 2\mathbf{k}$, $\|\mathbf{b}\| = 3$, so $\hat{\mathbf{b}} = (1/3)(2\mathbf{i} - \mathbf{j} + 2\mathbf{k})$, so the strength of the vector **H** in the direction of **b** is

$$\mathbf{H} \cdot \hat{\mathbf{b}} = (1/3)(5\mathbf{i} + 3\mathbf{j} + 7\mathbf{k}) \cdot (2\mathbf{i} - \mathbf{j} + 2\mathbf{k}) = 7. \quad\blacksquare$$

Direction cosines and direction ratios

If $\mathbf{a} = a_1\mathbf{i} + a_2\mathbf{j} + a_3\mathbf{k}$ is an arbitrary vector, the unit vector $\hat{\mathbf{a}}$ in the direction of **a** is

$$\hat{\mathbf{a}} = (a_1\mathbf{i} + a_2\mathbf{j} + a_3\mathbf{k})/\|\mathbf{a}\|$$
$$= (a_1\mathbf{i} + a_2\mathbf{j} + a_3\mathbf{k})/\left(a_1^2 + a_2^2 + a_3^2\right)^{1/2}. \tag{26}$$

Taking the dot product of **a** with **i**, **j**, and **k**, and setting $l = a_1/(a_1^2 + a_2^2 + a_3^2)^{1/2}$, $m = a_2/(a_1^2 + a_2^2 + a_3^2)^{1/2}$, and $n = a_3/(a_1^2 + a_2^2 + a_3^2)^{1/2}$ gives

$$l = \mathbf{i} \cdot \hat{\mathbf{a}}, \quad m = \mathbf{j} \cdot \hat{\mathbf{a}}, \quad \text{and} \quad n = \mathbf{k} \cdot \hat{\mathbf{a}},$$

so we may write

$$\hat{\mathbf{a}} = l\mathbf{i} + m\mathbf{j} + n\mathbf{k}. \tag{27}$$

The dot product $\hat{\mathbf{a}} \cdot \hat{\mathbf{a}} = l^2 + m^2 + n^2 = (a_1^2 + a_2^2 + a_3^2)/\|\mathbf{a}\|^2$, but $\|\mathbf{a}\|^2 = a_1^2 + a_2^2 + a_3^2$, so

$$l^2 + m^2 + n^2 = 1. \tag{28}$$

The number l is the cosine of the angle β_1 between **a** and the x-axis, the number m is the cosine of the angle β_2 between **a** and the y-axis, and the number n is the cosine of the angle β_3 between **a** and the z-axis, as shown in

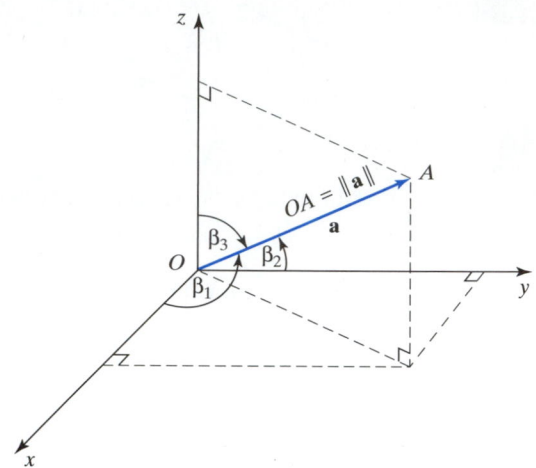

FIGURE 2.16   The angles β_1, β_2, and β_3.

direction cosines

Fig. 2.16. The numbers (l, m, n) are called the **direction cosines** of **a**, because they determine the direction of the unit vector $\hat{\mathbf{a}}$ that is parallel to **a**.

Notice that when any two of the three direction cosines l, m, and n of a vector **a** are given, the third is related to them by

$$l^2 + m^2 + n^2 = 1.$$

Because of result (27) it is always possible to write

$$\mathbf{a} = \|\mathbf{a}\|(l\mathbf{i} + m\mathbf{j} + n\mathbf{k}), \tag{29}$$

where l, m, and n are the direction cosines of **a**.

direction ratios

As the components a_1, a_2, and a_3 of **a** are *proportional* to the direction cosines, they are called the **direction ratios** of **a**.

EXAMPLE 2.14

Find the direction cosines and direction ratios of $\mathbf{a} = 3\mathbf{i} + \mathbf{j} - 2\mathbf{k}$.

Solution   As $\|\mathbf{a}\| = \sqrt{14}$, the direction cosines are $l = 3/\sqrt{14}$, $m = 1/\sqrt{14}$, and $n = -2/\sqrt{14}$. The direction ratios of **a** are 3, 1, and -2, or any nonnegative multiple of these three numbers such as $15/\sqrt{14}$, $5/\sqrt{14}$, and $-10/\sqrt{14}$.   ∎

The triangle inequality

The following result will be needed in the proof of the triangle inequality that is to follow. The absolute value of $\mathbf{a} \cdot \mathbf{b} = \|\mathbf{a}\| \cdot \|\mathbf{b}\| \cos\theta$ is

$$|\mathbf{a} \cdot \mathbf{b}| = \|\mathbf{a}\| \cdot \|\mathbf{b}\||\cos\theta|,$$

but $|\cos\theta| \leq 1$, so using this in the above result we obtain the **Cauchy–Schwarz inequality**,

$$|\mathbf{a} \cdot \mathbf{b}| \leq \|\mathbf{a}\| \cdot \|\mathbf{b}\|. \tag{30}$$

| THEOREM 2.2 |

The triangle inequality   If **a** and **b** are any two vectors, then

$$\|\mathbf{a} + \mathbf{b}\| \le \|\mathbf{a}\| + \|\mathbf{b}\|.$$

Proof   From (18) we have

$$\|\mathbf{a} + \mathbf{b}\|^2 = (\mathbf{a} + \mathbf{b}) \cdot (\mathbf{a} + \mathbf{b}) = \mathbf{a} \cdot \mathbf{a} + 2\mathbf{a} \cdot \mathbf{b} + \mathbf{b} \cdot \mathbf{b}$$
$$= \|\mathbf{a}\|^2 + 2\mathbf{a} \cdot \mathbf{b} + \|\mathbf{b}\|^2,$$

but $\mathbf{a} \cdot \mathbf{b} \le |\mathbf{a} \cdot \mathbf{b}|$, so from the Cauchy–Schwarz inequality (30)

$$\|\mathbf{a} + \mathbf{b}\|^2 \le \|\mathbf{a}\|^2 + 2\|\mathbf{a}\| \cdot \|\mathbf{b}\| + \|\mathbf{b}\|^2$$
$$= (\|\mathbf{a}\| + \|\mathbf{b}\|)^2.$$

Taking the positive square root of this last result, we obtain the triangle inequality

$$\|\mathbf{a} + \mathbf{b}\| \le \|\mathbf{a}\| + \|\mathbf{b}\|. \qquad \blacksquare$$

The triangle inequality will be generalized in Section 2.5, but in its present form it is the vector equivalent of the Euclidean theorem that "the sum of the lengths of any two sides of a triangle is greater than or equal to the length of the third side," and it is from this theorem that the inequality derives its name.

Equation of a Plane

When working with the vector calculus it is sometimes necessary to consider a plane that is locally tangent to a point on a surface in space so it will be useful to derive the general equation of a plane in both its vector and cartesian forms.

A plane Π can be defined by specifying a fixed point belonging to the plane and a vector **n** that is perpendicular to the plane. This follows because if **n** is perpendicular at a point on the plane, it must be perpendicular at every point on the plane. Any vector **n** that is perpendicular to a plane is called a **normal** to the plane. Clearly a normal to a plane is not unique, because a plane has two sides, so if a normal **n** is directed away from one side of the plane, the vector −**n** is a normal directed away from the other side. Both **n** and −**n** can be scaled by any nonzero number and still remain normals; consequently, if **n** is a normal to a plane, so also are all vectors of the form λ**n**, with λ ≠ 0 any real number.

Let a fixed point A on plane Π with normal **n** have position vector **a** relative to an origin O, and let P be a general point on plane Π with position vector **r** relative to O. Then, as may be seen from Fig. 2.17, the vector **r** − **a** lies in the plane, and so is perpendicular (normal) to **n**. Forming the dot product of **n** and **r** − **a**, and using (19), shows that the **vector equation** of plane Π is

vector equation of a plane

$$\mathbf{n} \cdot (\mathbf{r} - \mathbf{a}) = 0, \tag{31}$$

or, equivalently,

$$\mathbf{n} \cdot \mathbf{r} = \mathbf{n} \cdot \mathbf{a}. \tag{32}$$

cartesian equation of a plane

The **cartesian form** of this equation follows by considering a general point with coordinates (x, y, z) on plane Π, setting $\mathbf{r} = x\mathbf{i} + y\mathbf{j} + z\mathbf{k}$, $\mathbf{a} = a_1\mathbf{i} + a_2\mathbf{j} + a_3\mathbf{k}$,

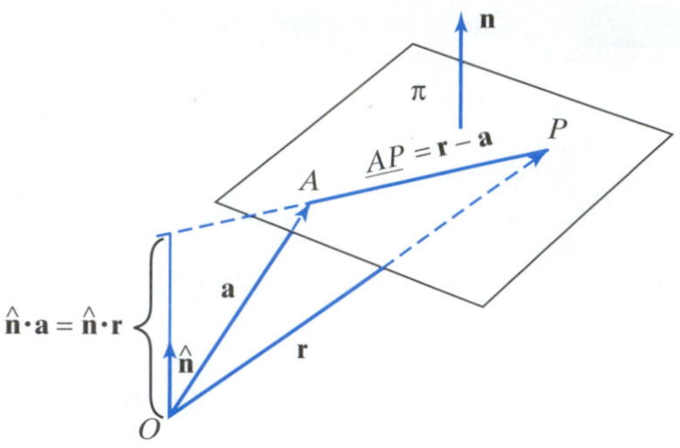

FIGURE 2.17   Plane Π with normal **n** passing through point A.

and $\mathbf{n} = n_1\mathbf{i} + n_2\mathbf{j} + n_3\mathbf{k}$, and then substituting into (32) to get

$$(n_1\mathbf{i} + n_2\mathbf{j} + n_3\mathbf{k}) \cdot (x\mathbf{i} + y\mathbf{j} + z\mathbf{k}) = (n_1\mathbf{i} + n_2\mathbf{j} + n_3\mathbf{k}) \cdot (a_1\mathbf{i} + a_2\mathbf{j} + a_3\mathbf{k}).$$

Taking the dot products and using results (21) and (22) show the cartesian equation of plane Π to be

$$n_1x + n_2y + n_3z = n_1a_1 + n_2a_2 + n_3a_3 = d, \text{ a constant.} \qquad (33)$$

EXAMPLE 2.15   Find the cartesian equation of the plane through the point $(2, 5, 3)$ with normal $3\mathbf{i} + 2\mathbf{j} - 7\mathbf{k}$.

Solution   Here $n_1 = 3$, $n_2 = 2$, $n_3 = -7$ and $a_1 = 2$, $a_2 = 5$, and $a_3 = 3$, so substituting into (33) shows the plane has the equation

$$3x + 2y - 7z = -5.$$ ■

Summary

This section has introduced the dot or scalar product of two vectors in geometrical terms and, more conveniently for calculations, in terms of the components of the two vectors involved. The applications given include the important operation of projecting a vector onto the line of another vector and the derivation of the vector equation and cartesian equation of a plane.

EXERCISES 2.2

1. Find the dot products of the following pairs of vectors:
   (a) $\mathbf{i} - \mathbf{j} + 3\mathbf{k}, 2\mathbf{i} + 3\mathbf{j} + \mathbf{k}$.
   (b) $2\mathbf{i} - \mathbf{j} + 4\mathbf{k}, -\mathbf{i} + 2\mathbf{j} + 2\mathbf{k}$.
   (c) $\mathbf{i} + \mathbf{j} - 3\mathbf{k}, 2\mathbf{i} + \mathbf{j} + \mathbf{k}$.

2. Find the dot products of the following pairs of vectors:
   (a) $\mathbf{i} - 2\mathbf{j} + 4\mathbf{k}, \mathbf{i} + 2\mathbf{j} + 3\mathbf{k}$.
   (b) $3\mathbf{i} + \mathbf{j} + 2\mathbf{k}, 4\mathbf{i} - 3\mathbf{j} + \mathbf{k}$.
   (c) $5\mathbf{i} - 3\mathbf{j} + 3\mathbf{k}, 2\mathbf{i} - 3\mathbf{j} + 5\mathbf{k}$.

3. Find which of the following pairs of vectors are orthogonal:
   (a) $3\mathbf{i} + 2\mathbf{j} - 6\mathbf{k}, -9\mathbf{i} - 6\mathbf{j} + 18\mathbf{k}$.
   (b) $3\mathbf{i} - \mathbf{j} + 7\mathbf{k}, 3\mathbf{i} + 2\mathbf{j} + \mathbf{k}$.

   (c) $2\mathbf{i} + \mathbf{j} + \mathbf{k}, \mathbf{i} + \mathbf{j} - \mathbf{k}$.
   (d) $\mathbf{i} + \mathbf{j} - 3\mathbf{k}, 2\mathbf{i} + \mathbf{j} + \mathbf{k}$.

4. Find which, if any, of the following pairs of vectors are orthogonal:
   (a) $2\mathbf{i} + \mathbf{j} + \mathbf{k}, 8\mathbf{i} + 2\mathbf{j} + 2\mathbf{k}$.
   (b) $\mathbf{i} + 2\mathbf{j} + 3\mathbf{k}, 2\mathbf{i} - 2\mathbf{j} - 3\mathbf{k}$.
   (c) $\mathbf{i} + 2\mathbf{j} + 4\mathbf{k}, 2\mathbf{i} + \mathbf{j} + 3\mathbf{k}$.
   (d) $\mathbf{i} + \mathbf{j}, 2\mathbf{j} + 3\mathbf{k}$.

5. Given that $\mathbf{a} = 2\mathbf{i} + 3\mathbf{j} - 2\mathbf{k}$, $\mathbf{b} = \mathbf{i} + 3\mathbf{j} + \mathbf{k}$ and $\mathbf{c} = 3\mathbf{i} + \mathbf{j} - \mathbf{k}$, find (a) $(\mathbf{a} + \mathbf{b}) \cdot \mathbf{c}$. (b) $(2\mathbf{b} - 3\mathbf{c}) \cdot \mathbf{a}$. (c) $\mathbf{a} \cdot \mathbf{a}$. (d) $\mathbf{c} \cdot (\mathbf{a} - 2\mathbf{b})$.

6. Given that $\mathbf{a} = 3\mathbf{i} + 2\mathbf{j} - 3\mathbf{k}, \mathbf{b} = 2\mathbf{i} + \mathbf{j} + 2\mathbf{k}$, and $\mathbf{c} = 5\mathbf{i} + 2\mathbf{j} - 2\mathbf{k}$, find (a) $\mathbf{b} \cdot (\mathbf{b} + (\mathbf{a} \cdot \mathbf{c})\mathbf{c})$. (b) $(\mathbf{a} + 2\mathbf{b}) \cdot (2\mathbf{b} - 3\mathbf{c})$. (c) $(\mathbf{c} \cdot \mathbf{c})\mathbf{b} - (\mathbf{a} \cdot \mathbf{a})\mathbf{c}$.

7. Find the angle between the following pairs of vectors:
   (a) $\mathbf{i} + \mathbf{j} + \mathbf{k}, 2\mathbf{i} + \mathbf{j} - \mathbf{k}$.
   (b) $2\mathbf{i} - \mathbf{j} + 3\mathbf{k}, 2\mathbf{i} + \mathbf{j} + 3\mathbf{k}$.
   (c) $3\mathbf{i} - \mathbf{j} + \mathbf{k}, \mathbf{i} - 2\mathbf{j} + 3\mathbf{k}$.
   (d) $\mathbf{i} - 2\mathbf{j} + \mathbf{k}, 4\mathbf{i} - 8\mathbf{j} + 16\mathbf{k}$.

8. Given $\mathbf{a} = 2\mathbf{i} - 3\mathbf{j} - 3\mathbf{k}$, $\mathbf{b} = \mathbf{i} + \mathbf{j} + 2\mathbf{k}$, and $\mathbf{c} = 3\mathbf{i} - 2\mathbf{j} - \mathbf{k}$, find the angles between the following pairs of vectors:
   (a) $\mathbf{a} + \mathbf{b}, \mathbf{b} - 2\mathbf{c}$. (b) $2\mathbf{a} - \mathbf{c}, \mathbf{a} + \mathbf{b} - \mathbf{c}$. (c) $\mathbf{b} + 3\mathbf{c}, \mathbf{a} - 2\mathbf{c}$.

9. Find the component of the force $\mathbf{F} = 4\mathbf{i} + 3\mathbf{j} + 2\mathbf{k}$ in the direction of the vector $\mathbf{i} + \mathbf{j} + \mathbf{k}$.

10. Find the component of the force $\mathbf{F} = 2\mathbf{i} + 5\mathbf{j} - 3\mathbf{k}$ in the direction of the vector $2\mathbf{i} + \mathbf{j} - 2\mathbf{k}$.

11. Given that $\mathbf{a} = \mathbf{i} + 2\mathbf{j} + 2\mathbf{k}$ and $\mathbf{b} = 2\mathbf{i} - 3\mathbf{j} + \mathbf{k}$, find (a) the projection of $\mathbf{a}$ onto the line of $\mathbf{b}$, and (b) the projection of $\mathbf{b}$ onto the line of $\mathbf{a}$.

12. Given that $\mathbf{a} = 3\mathbf{i} + 6\mathbf{j} + 9\mathbf{k}$ and $\mathbf{b} = \mathbf{i} + 2\mathbf{j} + 3\mathbf{k}$, (a) find the projection of $\mathbf{a}$ onto the line of $\mathbf{b}$ and (b) compare the magnitude of $\mathbf{a}$ with the result found in (a) and comment on the result.

13. Find the direction cosines and corresponding angles for the following vectors:
    (a) $\mathbf{i} + \mathbf{j} + \mathbf{k}$. (b) $\mathbf{i} - 2\mathbf{j} + 2\mathbf{k}$. (c) $4\mathbf{i} - 2\mathbf{j} + 3\mathbf{k}$.

14. Find the direction cosines and corresponding angles for the following vectors:
    (a) $\mathbf{i} - \mathbf{j} - \mathbf{k}$. (b) $2\mathbf{i} + 2\mathbf{j} - 5\mathbf{k}$. (c) $-4\mathbf{j} - \mathbf{k}$.

15. Verify the triangle inequality for vectors $\mathbf{a} = \mathbf{i} + 2\mathbf{j} + 3\mathbf{k}$ and $\mathbf{b} = 2\mathbf{i} + \mathbf{j} + 7\mathbf{k}$.

16. Verify the triangle inequality for vectors $\mathbf{a} = 2\mathbf{i} - \mathbf{j} - 2\mathbf{k}$ and $3\mathbf{i} + 2\mathbf{j} + 3\mathbf{k}$.

17. Find the equation of the plane with normal $2\mathbf{i} - 3\mathbf{j} + \mathbf{k}$ that contains the point $(1, 0, 1)$.

18. Find the equation of the plane with normal $\mathbf{i} - 2\mathbf{j} + 2\mathbf{k}$ that contains the point $(2, -3, 4)$.

19. Given that a plane passes through the point $(2, 3, -5)$, and the vector $2\mathbf{i} + \mathbf{k}$ is normal to the plane, find the cartesian form of its equation.

20. The equation of a plane is $3x + 2y - 5z = 4$. Find a vector that is normal to the plane, and the position vector of a point on the plane.

21. Explain why if the vector equation of plane Π in (32) is divided by $\|\mathbf{n}\|$ to bring it into the form $\mathbf{r} \cdot \mathbf{n} = \mathbf{a} \cdot \mathbf{n}$, the number $|\mathbf{a} \cdot \mathbf{n}|$ is the perpendicular distance of origin O from the plane. Explain also why if $\mathbf{a} \cdot \mathbf{n} > 0$ the plane lies to the side of O toward which $\mathbf{n}$ is directed, as in Fig. 2.15, but that if $\mathbf{a} \cdot \mathbf{n} < 0$ it lies on the opposite side of O toward which $-\mathbf{n}$ is directed.

22. Use the result of Exercise 21 to find the perpendicular distance of the plane $2x - 4y - 5z = 5$ from the origin.

23. The angle between two planes is defined as the angle between their normals. Find the angle between the two planes $x + 3y + 2z = 4$ and $2x - 5y + z = 2$.

24. Find the angle between the two planes $3x + 2y - 2z = 4$ and $2x + y + 2z = 1$.

25. Let $\mathbf{a}$ and $\mathbf{b}$ be two arbitrary skew (nonparallel) vectors, and set $\mathbf{a} = \mathbf{a}_b + \mathbf{a}_p$, where $\mathbf{a}_b$ is parallel to $\mathbf{b}$ and $\mathbf{a}_p$ is perpendicular to $\mathbf{b}$ and lies in the plane of $\mathbf{a}$ and $\mathbf{b}$. Find $\mathbf{a}_b$ and $\mathbf{a}_p$ in terms of $\mathbf{a}$ and $\mathbf{b}$.

26. The **law of cosines** for a triangle with sides of length a, b, and c, in which the angle opposite the side of length c is C, takes the form

$$c^2 = a^2 + b^2 - 2ab \cos C.$$

Prove this by taking vectors $\mathbf{a}$, $\mathbf{b}$, and $\mathbf{c}$ such that $\mathbf{c} = \mathbf{a} - \mathbf{b}$ and considering the dot product $\mathbf{c} \cdot \mathbf{c} = (\mathbf{a} - \mathbf{b}) \cdot (\mathbf{a} - \mathbf{b})$.

27. The work units W done by a constant force $\mathbf{F}$ when moving its point of application along a straight line L parallel to a vector $\mathbf{a}$ are defined as the product of the component of $\mathbf{F}$ in the direction of $\mathbf{a}$ and the distance d moved along line L. Express W in terms of $\mathbf{F}$, $\mathbf{a}$, and d.

28. If $\mathbf{a}$ and $\mathbf{b}$ are arbitrary vectors and λ and μ are any two scalars, prove that

$$\|\lambda\mathbf{a} + \mu\mathbf{b}\|^2 \leq \lambda^2\|\mathbf{a}\|^2 + 2\lambda\mu\mathbf{a} \cdot \mathbf{b} + \mu^2\|\mathbf{b}\|^2.$$

29. Verify the result of Exercise 28 by setting $\lambda = 2, \mu = -3$, $\mathbf{a} = 3\mathbf{i} + \mathbf{j} - 4\mathbf{k}$, and $\mathbf{b} = 2\mathbf{i} + 3\mathbf{j} + \mathbf{k}$.

2.3   The Cross Product

A product of two vectors $\mathbf{a}$ and $\mathbf{b}$ can be defined in such a way that the result is a vector. The result is written $\mathbf{a} \times \mathbf{b}$ and called the **cross product** of $\mathbf{a}$ and $\mathbf{b}$. The name **vector product** is also used in place of the term *cross product*.

Before defining the cross product we first formulate what is called the right-hand rule. Given any two skew vectors $\mathbf{a}$ and $\mathbf{b}$, the right-hand rule is used to determine the sense of a third vector **c** that is required to be normal to the plane containing vectors **a** and **b**.

right-hand rule

The Right-Hand Rule

Let **a** and **b** be two arbitrary skew vectors with the same base point, with **c** a vector normal to the plane containing them. If the fingers of the right hand are curled in such a way that they point from vector **a** to vector **b** through the angle θ between them, with $0 < \theta < \pi$, then when the thumb is extended away from the palm it will point in the direction of vector **c**.

When applying the right-hand rule, the order of the vectors is important. If vectors **a**, **b**, and **c** obey the right-hand rule, they will always be written in the order **a**, **b**, **c**, with the understanding that **c** is normal to the plane of **a** and **b**, with its sense determined by the right-hand rule. Figure 2.18 illustrates the right-hand rule.

An important special case of the right-hand rule has already been encountered in connection with the unit vectors **i**, **j**, and **k** that obey the rule, and because the vectors are mutually orthogonal the vectors **j**, **k**, **i** and **k**, **i**, **j** also obey the right-hand rule.

geometrical definition of a cross product

The cross product (a geometrical interpretation)

Let **a** and **b** be two arbitrary vectors, with **n̂** a unit vector normal to the plane of **a** and **b** chosen so that **a**, **b**, and **n̂**, in this order, obey the right-hand rule. Then the **cross product** of vectors **a** and **b**, written **a** × **b**, is defined as the vector

$$\mathbf{a} \times \mathbf{b} = \|\mathbf{a}\|.\|\mathbf{b}\| \sin\theta\,\hat{\mathbf{n}}. \tag{34}$$

This geometrical definition of the cross product is useful in many situations, but when the vectors **a** and **b** are specified in terms of their cartesian components a different form of the definition will be needed.

The cross product can be interpreted as a *vector area*, in the sense that it can be written $\mathbf{a} \times \mathbf{b} = S\hat{\mathbf{n}}$, where $S = OA \cdot BN = \|\mathbf{a}\| \cdot \|\mathbf{b}\| \sin\theta$ is the geometrical area

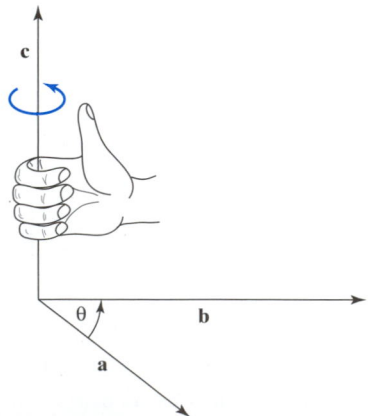

FIGURE 2.18   The right-hand rule.

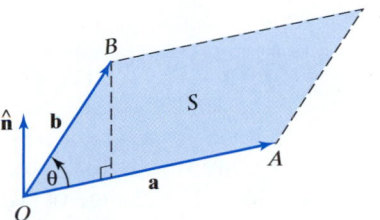

FIGURE 2.19   The cross product interpreted as the vector area of a parallelogram.

of the parallelogram in Fig. 2.19, and the unit vector $\hat{\mathbf{n}}$ is normal to the area. This shows that the geometrical area S of the vector parallelogram with sides $\mathbf{a}$ and $\mathbf{b}$ is simply the modulus of the cross product $\mathbf{a} \times \mathbf{b}$, so $S = \|\mathbf{a} \times \mathbf{b}\|$.

Properties of the cross product

properties of the cross product

The following results are consequences of the definition of the cross product.

The cross product is anticommutative

$$\mathbf{a} \times \mathbf{b} = -\mathbf{b} \times \mathbf{a} \tag{35}$$

The cross product is associative

$$\mathbf{a} \times (\mathbf{b} + \mathbf{c}) = \mathbf{a} \times \mathbf{b} + \mathbf{a} \times \mathbf{c}. \tag{36}$$

Parallel vectors ($\theta = 0$)

If vectors $\mathbf{a}$ and $\mathbf{b}$ are parallel, then

$$\mathbf{a} \times \mathbf{b} = \mathbf{0}. \tag{37}$$

Orthogonal vectors ($\theta = \pi/2$)

If vectors $\mathbf{a}$ and $\mathbf{b}$ are orthogonal, then

$$\mathbf{a} \times \mathbf{b} = \|\mathbf{a}\| . \|\mathbf{b}\| \hat{\mathbf{n}}. \tag{38}$$

Product of unit vectors

If $\mathbf{a}$ and $\mathbf{b}$ are unit vectors, then

$$\mathbf{a} \times \mathbf{b} = \sin\theta \hat{\mathbf{n}}. \tag{39}$$

An immediate consequence of properties (34), (35), and (37) is that

$$\mathbf{i} \times \mathbf{i} = \mathbf{j} \times \mathbf{j} = \mathbf{k} \times \mathbf{k} = \mathbf{0}, \tag{40}$$

and

$$\mathbf{i} \times \mathbf{j} = \mathbf{k}, \quad \mathbf{j} \times \mathbf{i} = -\mathbf{k}, \quad \mathbf{j} \times \mathbf{k} = \mathbf{i}, \quad \mathbf{k} \times \mathbf{j} = -\mathbf{i}, \quad \mathbf{k} \times \mathbf{i} = \mathbf{j}, \quad \mathbf{i} \times \mathbf{k} = -\mathbf{j}.$$

(41)

Only results (35) and (36) require some comment, as the other results are obvious. The change of sign in (35) that makes the cross product *anticommutative* occurs because when the vectors **a** and **b** are interchanged, the right-hand rule causes the direction of $\hat{\mathbf{n}}$ to be reversed. Result (36) can be proved in several ways, but we shall postpone its proof until a different expression for the cross product has been derived.

To obtain a more convenient expression for the cross product that can be used when **a** and **b** are known in terms of their components, we proceed as follows. Let $\mathbf{a} = a_1\mathbf{i} + a_2\mathbf{j} + a_3\mathbf{k}$ and $\mathbf{b} = b_1\mathbf{i} + b_2\mathbf{j} + b_3\mathbf{k}$, and consider the cross product $\mathbf{a} \times \mathbf{b} = (a_1\mathbf{i} + a_2\mathbf{j} + a_3\mathbf{k}) \times (b_1\mathbf{i} + b_2\mathbf{j} + b_3\mathbf{k})$. Expanding this expression term by term is justified because of the associative property given in (36), and it leads to the result

$$\mathbf{a} \times \mathbf{b} = a_1 b_1 \mathbf{i} \times \mathbf{i} + a_1 b_2 \mathbf{i} \times \mathbf{j} + a_1 b_3 \mathbf{i} \times \mathbf{k} + a_2 b_1 \mathbf{j} \times \mathbf{i} + a_2 b_2 \mathbf{j} \times \mathbf{j}$$
$$+ a_2 b_3 \mathbf{j} \times \mathbf{k} + a_3 b_1 \mathbf{k} \times \mathbf{i} + a_3 b_2 \mathbf{k} \times \mathbf{j} + a_3 b_3 \mathbf{k} \times \mathbf{k}.$$

cross product in terms of components

Results (40) cause three terms on the right-hand side to vanish, and results (41) allow the remaining six terms to be collected into three groups as follows to give

$$\mathbf{a} \times \mathbf{b} = (a_2 b_3 - a_3 b_2)\mathbf{i} - (a_1 b_3 - a_3 b_1)\mathbf{j} + (a_1 b_2 - a_2 b_1)\mathbf{k}.$$

(42)

This alternative expression for the cross product in terms of the cartesian components of vectors **a** and **b** can be further simplified by making formal use of the third-order determinant,

$$\mathbf{a} \times \mathbf{b} = \begin{vmatrix} \mathbf{i} & \mathbf{j} & \mathbf{k} \\ a_1 & a_2 & a_3 \\ b_1 & b_2 & b_3 \end{vmatrix},$$

because a formal expansion in terms of elements of the first row generates result (42). We take this result as an alternative but equivalent definition of the cross product.

practical definition of a cross product using a determinant

The cross product (cartesian component form)

Let $\mathbf{a} = a_1\mathbf{i} + a_2\mathbf{j} + a_3\mathbf{k}$ and $\mathbf{b} = b_1\mathbf{i} + b_2\mathbf{j} + b_3\mathbf{k}$. Then

$$\mathbf{a} \times \mathbf{b} = \begin{vmatrix} \mathbf{i} & \mathbf{j} & \mathbf{k} \\ a_1 & a_2 & a_3 \\ b_1 & b_2 & b_3 \end{vmatrix},$$

(43)

When expressing $\mathbf{a} \times \mathbf{b}$ as the determinant in (43), purely *formal* use was made of the method of expansion of a determinant in terms of the elements of its first row, because (43) is not a determinant in the ordinary sense as its elements are a mixture of vectors and numbers.

EXAMPLE 2.16

Given that $\mathbf{a} = 3\mathbf{i} - 2\mathbf{j} - \mathbf{k}$ and $\mathbf{b} = \mathbf{i} + 4\mathbf{j} + 2\mathbf{k}$, find $\mathbf{a} \times \mathbf{b}$ and a unit vector $\hat{\mathbf{n}}$ normal to the plane containing $\mathbf{a}$ and $\mathbf{b}$ such that $\mathbf{a}, \mathbf{b}$, and $\mathbf{n}$, in this order, obey the right-hand rule.

Solution   Substitution into expression (43) gives

$$\mathbf{a} \times \mathbf{b} = \begin{vmatrix} \mathbf{i} & \mathbf{j} & \mathbf{k} \\ 3 & -2 & -1 \\ 1 & 4 & 2 \end{vmatrix}$$

$$= [(-2) \cdot 2 - 4 \cdot (-1)]\mathbf{i} - [3 \cdot 2 - 1 \cdot (-1)]\mathbf{j} + [3 \cdot 4 - 1 \cdot (-2)]\mathbf{k}$$

$$= -7\mathbf{j} + 14\mathbf{k}.$$

The required unit vector $\hat{\mathbf{n}}$ is simply the unit vector in the direction of $\mathbf{a} \times \mathbf{b}$, so

$$\hat{\mathbf{n}} = (\mathbf{a} \times \mathbf{b})/\|\mathbf{a} \times \mathbf{b}\| = (-7\mathbf{j} + 14\mathbf{k})/(7\sqrt{5}).$$
$$= (-1/\sqrt{5})\mathbf{j} + (2/\sqrt{5})\mathbf{k}. \qquad \blacksquare$$

We now return to the proof of the associative property stated in (35) and establish it by means of result (43).

Setting $\mathbf{a} = a_1\mathbf{i} + a_2\mathbf{j} + a_3\mathbf{k}$, $\mathbf{b} = b_1\mathbf{i} + b_2\mathbf{j} + b_3\mathbf{k}$, and $\mathbf{c} = c_1\mathbf{i} + c_2\mathbf{j} + c_3\mathbf{k}$, we have

$$\mathbf{a} \times (\mathbf{b} + \mathbf{c}) = \begin{vmatrix} \mathbf{i} & \mathbf{j} & \mathbf{k} \\ a_1 & a_2 & a_3 \\ (b_1 + c_1) & (b_2 + c_2) & (b_3 + c_3) \end{vmatrix}.$$

Expanding the determinant in terms of elements of its first row and grouping terms gives

$$\mathbf{a} \times (\mathbf{b} + \mathbf{c}) = (a_2 b_3 - a_3 b_2)\mathbf{i} - (a_1 b_3 - a_3 b_1)\mathbf{j} + (a_1 b_2 - a_2 b_1)\mathbf{k}$$
$$+ (a_2 c_3 - a_3 c_2)\mathbf{i} - (a_1 c_3 - a_3 c_1)\mathbf{j} + (a_1 c_2 - a_2 c_1)\mathbf{k}$$
$$= \mathbf{a} \times \mathbf{b} + \mathbf{a} \times \mathbf{c},$$

and the result is proved.

Summary

This section first introduced the vector or cross product of two vectors in geometrical terms and then used the result to show that the vector product is anticommutative, in the sense that $\mathbf{a} \times \mathbf{b} = -\mathbf{b} \times \mathbf{a}$. Important results involving the vector product are given in terms of the components of the two vectors that are involved. Finally, the vector product was expressed in a form that is most convenient for calculations by writing it in determinantal form, the rows of which contain the unit vectors $\mathbf{i}, \mathbf{j}$, and $\mathbf{k}$ and the components of the respective vectors.

EXERCISES 2.3

In Exercises 1 through 6 use (43) to find $\mathbf{a} \times \mathbf{b}$.

1. For $\mathbf{a} = 2\mathbf{i} - \mathbf{j} - 4\mathbf{k}, \mathbf{b} = 3\mathbf{i} - \mathbf{j} - \mathbf{k}$.
2. For $\mathbf{a} = -3\mathbf{i} + 2\mathbf{j} + 4\mathbf{k}, \mathbf{b} = 2\mathbf{i} + \mathbf{j} - 2\mathbf{k}$.
3. For $\mathbf{a} = 7\mathbf{i} + 6\mathbf{k}, \mathbf{b} = 3\mathbf{j} + \mathbf{k}$.
4. For $\mathbf{a} = 3\mathbf{i} + 7\mathbf{j} + 2\mathbf{k}, \mathbf{b} = \mathbf{i} - \mathbf{j} + \mathbf{k}$.
5. For $\mathbf{a} = 2\mathbf{i} + \mathbf{j} + \mathbf{k}, \mathbf{b} = 2\mathbf{i} - \mathbf{j} + \mathbf{k}$.

6. For $\mathbf{a} = 3\mathbf{i} - 2\mathbf{j} + 6\mathbf{k}, \mathbf{b} = 2\mathbf{i} + \mathbf{j} + 3\mathbf{k}$.

In Exercises 7 through 10 verify the equivalence of the definitions of the cross product in (34) and (43) by first using (43) to calculate $\mathbf{a} \times \mathbf{b}$, and hence $\|\mathbf{a} \times \mathbf{b}\|$ and $\hat{\mathbf{n}}$, and then calculating $\|\mathbf{a}\|$ and $\|\mathbf{b}\|$ directly, using result (17) to find $\cos \theta$ and hence $\sin \theta$, and using the results to find $\mathbf{a} \times \mathbf{b}$ from (34).

7. For $\mathbf{a} = \mathbf{i} + \mathbf{j} + 3\mathbf{k}$ and $\mathbf{b} = 3\mathbf{i} + 2\mathbf{j} + \mathbf{k}$.
8. For $\mathbf{a} = \mathbf{i} + \mathbf{j} + \mathbf{k}$ and $\mathbf{b} = 4\mathbf{i} + 2\mathbf{j} + 2\mathbf{k}$.
9. For $\mathbf{a} = 2\mathbf{i} + \mathbf{j} - 3\mathbf{k}$ and $\mathbf{b} = 5\mathbf{i} - 2\mathbf{k}$.
10. For $\mathbf{a} = -2\mathbf{i} - 3\mathbf{j} + \mathbf{k}$ and $\mathbf{b} = 3\mathbf{i} + \mathbf{j} + 2\mathbf{k}$.

In Exercises 11 through 14, verify by direct calculation that $(\mathbf{b} + \mathbf{c}) \times \mathbf{a} = -\mathbf{a} \times (\mathbf{b} + \mathbf{c})$.

11. $\mathbf{a} = 3\mathbf{j} + 2\mathbf{k}$, $\mathbf{b} = \mathbf{i} - 4\mathbf{j} + \mathbf{k}$, and $\mathbf{c} = 5\mathbf{i} - 2\mathbf{j} + 3\mathbf{k}$.
12. $\mathbf{a} = -\mathbf{i} + 5\mathbf{j} + 2\mathbf{k}$, $\mathbf{b} = 4\mathbf{i} + \mathbf{k}$, and $\mathbf{c} = -2\mathbf{i} - 4\mathbf{j} + 3\mathbf{k}$.
13. $\mathbf{a} = \mathbf{i} + \mathbf{k}$, $\mathbf{b} = 3\mathbf{i} - \mathbf{j} - 2\mathbf{k}$, and $\mathbf{c} = 3\mathbf{i} + \mathbf{j} + \mathbf{k}$.
14. $\mathbf{a} = 5\mathbf{i} + \mathbf{j} + \mathbf{k}$, $\mathbf{b} = 2\mathbf{i} - \mathbf{j} - \mathbf{k}$, and $\mathbf{c} = 4\mathbf{i} + 2\mathbf{j} + 3\mathbf{k}$.

In Exercises 15 through 18 find a unit vector normal to a plane containing the given vectors.

15. $3\mathbf{i} + \mathbf{j} + \mathbf{k}$ and $\mathbf{i} + 2\mathbf{j} + \mathbf{k}$.
16. $2\mathbf{i} - \mathbf{j} + 2\mathbf{k}$ and $2\mathbf{i} + 3\mathbf{j} + \mathbf{k}$.
17. $\mathbf{i} + \mathbf{j} + \mathbf{k}$ and $2\mathbf{i} + 3\mathbf{j} - \mathbf{k}$.
18. $2\mathbf{i} + 2\mathbf{j} - \mathbf{k}$ and $3\mathbf{i} + \mathbf{j} + 4\mathbf{k}$.
19. Find a unit vector normal to a plane containing vectors $\mathbf{a} + \mathbf{b}$ and $\mathbf{a} + \mathbf{c}$, given that $\mathbf{a} = \mathbf{i} + 2\mathbf{j} + \mathbf{k}$, $\mathbf{b} = 2\mathbf{i} + \mathbf{j} - 2\mathbf{k}$, and $\mathbf{c} = 3\mathbf{i} + 2\mathbf{j} + 4\mathbf{k}$.
20. Given that $\mathbf{a} = 3\mathbf{i} + \mathbf{j} + \mathbf{k}$, $\mathbf{b} = 2\mathbf{i} - \mathbf{j} + 2\mathbf{k}$, and $\mathbf{c} = \mathbf{i} + \mathbf{j} + \mathbf{k}$, find (a) a vector normal to the plane containing the vectors $\mathbf{a} + (\mathbf{a} \cdot \mathbf{b})\mathbf{b}$ and $\mathbf{c}$ and, (b) explain why the normal to a plane containing the vectors $\mathbf{a}$ and $\mathbf{b}$ and the normal to a plane containing the vectors $(\mathbf{a} \cdot \mathbf{b})\mathbf{a}$ and $(\mathbf{b} \cdot \mathbf{c})\mathbf{b}$ are parallel.

In Exercises 21 through 24, find the cartesian equation of the plane that passes through the given points.

21. $(1, 3, 2)$, $(2, 0, -4)$, and $(1, 6, 11)$.
22. $(1, 4, 3)$, $(2, 0, 1)$, and $(3, 4, -6)$.
23. $(1, 2, 3)$, $(2, -4, 1)$, and $(3, 6, -1)$.
24. $(1, 0, 1)$, $(2, 5, 7)$, and $(2, 3, 9)$.

Three points with position vectors $\mathbf{a}$, $\mathbf{b}$, and $\mathbf{c}$ will be **collinear** (lie on a line) if the parallelogram with adjacent sides $\mathbf{a} - \mathbf{b}$ and $\mathbf{a} - \mathbf{c}$ has zero geometrical area. Use this result in Exercises 25 through 28 to determine which sets of points are collinear.

25. $(2, 2, 3)$, $(6, 1, 5)$, $(-2, 4, 3)$.
26. $(1, 2, 4)$, $(7, 0, 8)$, $(-8, 5, -2)$.
27. $(2, 3, 3)$, $(3, 7, 5)$, $(0, -5, -1)$.
28. $(1, 3, 2)$, $(4, 2, 1)$, $(1, 0, 2)$.
29. A vector $\mathbf{N}$ normal to the plane containing the skew vectors $\mathbf{a}$ and $\mathbf{b}$ can be found as follows. $\mathbf{N}$ is normal to $\mathbf{a}$ and $\mathbf{b}$, so $\mathbf{a} \cdot \mathbf{N} = 0$ and $\mathbf{b} \cdot \mathbf{N} = 0$. If a component of $\mathbf{N}$ is assigned an arbitrary nonzero value c, say, the other two components can be found from these two equations as multiples of c, and $\mathbf{N}$ will then be determined as a multiple of c. A suitable choice of c will make $\mathbf{N}$ a unit normal $\hat{\mathbf{N}}$. Apply this method to vectors $\mathbf{a}$ and $\mathbf{b}$ in Exercise 7 to find a vector $\hat{\mathbf{N}}$. Compare the result with the unit vector

$$\hat{\mathbf{n}} = (\mathbf{a} \times \mathbf{b})/\|\mathbf{a} \times \mathbf{b}\|$$

found from (43). Explain why although both $\hat{\mathbf{n}}$ and $\hat{\mathbf{N}}$ are normal to the plane containing $\mathbf{a}$ and $\mathbf{b}$ they may have opposite senses.

2.4   Linear Dependence and Independence of Vectors and Triple Products

The dot and cross products can be combined to provide a simple test that determines whether or not an arbitrary set of three vectors possesses a property of fundamental importance to the algebra of vectors. First, however, some introductory remarks are necessary.

Given a set of n vectors $\mathbf{a}_1, \mathbf{a}_2, \ldots, \mathbf{a}_n$, and a set of n constants $c_1, c_2, \ldots, c_n$, the sum

$$c_1\mathbf{a}_1 + c_2\mathbf{a}_2 + \cdots + c_n\mathbf{a}_n$$

linear combination of vectors

is called a **linear combination** of the vectors. Linear combinations of the vectors $\mathbf{i}$, $\mathbf{j}$, and $\mathbf{k}$ were used in Section 2.1 to express *every* vector in three-dimensional space as a linear combination of these three vectors. A triad of vectors such as $\mathbf{i}$, $\mathbf{j}$, and $\mathbf{k}$ with the property that *all* vectors in three-dimensional space can be represented as linear combinations of these three vectors is said to form a **basis** for the space.

basis

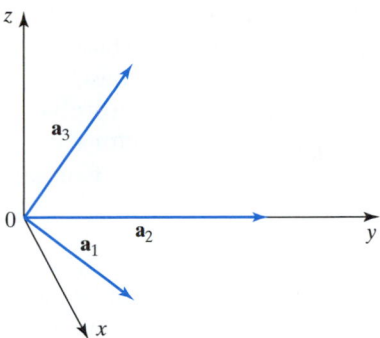

FIGURE 2.20   Nonorthogonal triad forming a basis in three-dimensional space.

It is a fundamental property of three-dimensional space that a basis for the space comprises a set of three vectors $\mathbf{a}_1, \mathbf{a}_2$, and $\mathbf{a}_3$, with the property that the linear combination

$$c_1\mathbf{a}_1 + c_2\mathbf{a}_2 + c_3\mathbf{a}_3 = \mathbf{0} \qquad (44)$$

is *only* true when $c_1 = c_2 = c_3 = 0$. Vectors $\mathbf{a}_1, \mathbf{a}_2$, and $\mathbf{a}_3$ satisfying this condition are said to be **linearly independent** vectors, and a vector $\mathbf{d}$ of the form

linear independence and linear dependence

$$\mathbf{d} = c_1\mathbf{a}_1 + c_2\mathbf{a}_2 + c_3\mathbf{a}_3,$$

where *not all* of c_1, c_2, and c_3 are zero, is said to be **linearly dependent** on the vectors $\mathbf{a}_1, \mathbf{a}_2$, and $\mathbf{a}_3$. The vectors $\mathbf{i}, \mathbf{j}$, and $\mathbf{k}$ that form a basis for three-dimensional space are linearly independent vectors, but the position vector $\mathbf{r} = 2\mathbf{i} - 3\mathbf{j} + 5\mathbf{k}$ is linearly dependent on vectors $\mathbf{i}, \mathbf{j}$, and $\mathbf{k}$.

Clearly, vectors $\mathbf{i}, \mathbf{j}$, and $\mathbf{k}$ do not form the only basis for three-dimensional space, because any triad of linearly independent vectors $\mathbf{a}_1, \mathbf{a}_2$, and $\mathbf{a}_3$ will serve equally well, as, for example, the nonorthogonal set of vectors shown in Fig. 2.20.

The dot and cross products will now be combined to develop a test for linear dependence and independence based on the elementary geometrical idea of the volume of the parallelepiped shown in Fig. 2.21, three edges $\mathbf{a}, \mathbf{b}$, and $\mathbf{c}$ of which meet at the origin.

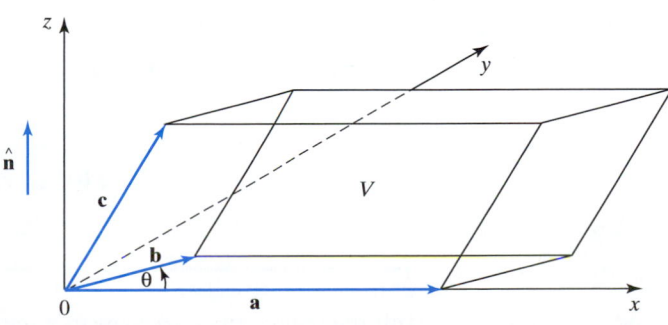

FIGURE 2.21   Volume V of a parallelepiped.

The volume V of a parallelepiped is a nonnegative number given by the product of the area of its base and its height. Suppose vectors $\mathbf{a}$ and $\mathbf{b}$ are chosen to form two sides of the base of the parallelepiped. Then the vector area of this base has already been interpreted as $\mathbf{a} \times \mathbf{b}$. The vertical height of the parallelepiped is the projection of vector $\mathbf{c}$ in the direction of the unit vector $\hat{\mathbf{n}}$ normal to the base, and so is given by $\hat{\mathbf{n}} \cdot \mathbf{c}$. Consequently, as $\mathbf{a} \times \mathbf{b} = \|\mathbf{a}\| \cdot \|\mathbf{b}\| \sin\theta\hat{\mathbf{n}}$, it follows that

$$V = |(\mathbf{a} \times \mathbf{b}) \cdot \mathbf{c}|. \tag{45}$$

The absolute value of the right-hand side of (45) has been taken because a volume must be a nonnegative quantity, whereas the dot product $(\mathbf{a} \times \mathbf{b}) \cdot \mathbf{c}$ may be of either sign.

If vectors $\mathbf{a}$, $\mathbf{b}$, and $\mathbf{c}$ form a basis for three-dimensional space, vector $\mathbf{c}$ cannot be linearly dependent on vectors $\mathbf{a}$ and $\mathbf{b}$, and so the parallelepiped in Fig. 2.21 with these vectors as its sides must have a nonzero volume. If, however, vectors $\mathbf{a}$, $\mathbf{b}$, and $\mathbf{c}$ are **coplanar** (all lie in the same plane), and so cannot form a basis for the space, the volume of the parallelepiped will be zero. These simple geometrical observations lead to the following test for the linear independence of three vectors in three-dimensional space.

THEOREM 2.3

a test for linear independence

Test for linear independence of vectors in three-dimensional space   Let $\mathbf{a}$, $\mathbf{b}$, and $\mathbf{c}$ be any three vectors.   Then the vectors are linearly independent if $(\mathbf{a} \times \mathbf{b}) \cdot \mathbf{c} \neq 0$, and they are linearly dependent if $(\mathbf{a} \times \mathbf{b}) \cdot \mathbf{c} = 0$. ∎

A product of the type $(\mathbf{a} \times \mathbf{b}) \cdot \mathbf{c}$ is called a **scalar triple product**. The name arises because the result is a scalar. It is also called a mixed triple product since both $\cdot$ and $\times$ appear. Three vectors are involved in this dot (scalar) product, one of which is the vector $\mathbf{a} \times \mathbf{b}$ and the other is the vector $\mathbf{c}$.

scalar triple product

Scalar triple products are easily evaluated, because taking the dot product of $\mathbf{a} \times \mathbf{b}$ in the form given in (42) with $\mathbf{c} = c_1\mathbf{i} + c_2\mathbf{j} + c_3\mathbf{k}$ gives

$$(\mathbf{a} \times \mathbf{b}) \cdot \mathbf{c} = (a_2b_3 - a_3b_2)c_1 - (a_1b_3 - a_3b_1)c_2 + (a_1b_2 - a_2b_1)c_3.$$

The right-hand side of this expression is simply the value of a determinant with successive rows given by the components of $\mathbf{a}$, $\mathbf{b}$, and $\mathbf{c}$, so we have arrived at the following convenient formula for the scalar triple product.

scalar triple product as a determinant

Scalar triple product

Let $\mathbf{a} = a_1\mathbf{i} + a_2\mathbf{j} + a_3\mathbf{k}$, $\mathbf{b} = b_1\mathbf{i} + b_2\mathbf{j} + b_3\mathbf{k}$, and $\mathbf{c} = c_1\mathbf{i} + c_2\mathbf{j} + c_3\mathbf{k}$. Then

$$(\mathbf{a} \times \mathbf{b}) \cdot \mathbf{c} = \begin{vmatrix} a_1 & a_2 & a_3 \\ b_1 & b_2 & b_3 \\ c_1 & c_2 & c_3 \end{vmatrix}. \tag{46}$$

Interchanging any two rows in a matrix changes the sign but not the value of its determinant. Two such switches in (46) leave the value unchanged, so the dot product is commutative and so we arrive at the useful result

$$(\mathbf{a} \times \mathbf{b}) \cdot \mathbf{c} = \mathbf{a} \cdot (\mathbf{b} \times \mathbf{c}). \tag{47}$$

So, in a scalar triple product the dot and cross may be *interchanged* without altering the result.

EXAMPLE 2.17   Given the two sets of vectors (a) $\mathbf{a} = \mathbf{i} + 2\mathbf{j} - 5\mathbf{k}, \mathbf{b} = \mathbf{i} + \mathbf{j} + 2\mathbf{k}, \mathbf{c} = \mathbf{i} + 4\mathbf{j} - 19\mathbf{k}$ and (b) $\mathbf{a} = 2\mathbf{i} + \mathbf{j} + \mathbf{k}, \mathbf{b} = 3\mathbf{i} + 4\mathbf{k}, \mathbf{c} = \mathbf{i} + \mathbf{j} + \mathbf{k}$, find if the vectors are linearly independent or linearly dependent.

Solution   We apply Theorem 2.3 to each set, using result (46) to evaluate the scalar triple products.

(a)
$$(\mathbf{a} \times \mathbf{b}) \cdot \mathbf{c} = \begin{vmatrix} 1 & 2 & -5 \\ 1 & 1 & 2 \\ 1 & 4 & -19 \end{vmatrix} = 0,$$

so the set of three vectors in (a) is linearly dependent. In fact this can be seen from the fact that $\mathbf{c} = 3\mathbf{a} - 2\mathbf{b}$.

(b)
$$(\mathbf{a} \times \mathbf{b}) \cdot \mathbf{c} = \begin{vmatrix} 2 & 1 & 1 \\ 3 & 0 & 4 \\ 1 & 1 & 1 \end{vmatrix} = -4 \neq 0,$$

so the set of three vectors in (b) is linearly independent. Although not required, the volume V of the parallelepiped formed by these three vectors is $V = |(\mathbf{a} \times \mathbf{b}) \cdot \mathbf{c}| = |-4| = 4$. ∎

Another notation for the scalar triple product of vectors $\mathbf{a}, \mathbf{b}$, and $\mathbf{c}$ is $[\mathbf{a}, \mathbf{b}, \mathbf{c}]$, so

$$[\mathbf{a}, \mathbf{b}, \mathbf{c}] = (\mathbf{a} \times \mathbf{b}) \cdot \mathbf{c}, \tag{48}$$

or, in terms of a determinant,

$$[\mathbf{a}, \mathbf{b}, \mathbf{c}] = \begin{vmatrix} a_1 & a_2 & a_3 \\ b_1 & b_2 & b_3 \\ c_1 & c_2 & c_3 \end{vmatrix}. \tag{49}$$

alternative forms of a scalar triple product   Using this definition of $[\mathbf{a}, \mathbf{b}, \mathbf{c}]$ with the row interchange property of determinants (see Section 1.7) shows that

$$[\mathbf{a}, \mathbf{b}, \mathbf{c}] = [\mathbf{b}, \mathbf{c}, \mathbf{a}] = [\mathbf{c}, \mathbf{a}, \mathbf{b}], \tag{50}$$

because two row interchanges are needed to arrive at $[\mathbf{b}, \mathbf{c}, \mathbf{a}]$ from $[\mathbf{a}, \mathbf{b}, \mathbf{c}]$, leaving the sign of the determinant unchanged, whereas two more are required to arrive at $[\mathbf{c}, \mathbf{a}, \mathbf{b}]$ from $[\mathbf{b}, \mathbf{c}, \mathbf{a}]$, again leaving the sign of the determinant unchanged.

The order of the vectors in results (46), or in the equivalent notation of (48), is easily remembered when the results are abbreviated to

$$\begin{matrix} \mathbf{a} & \mathbf{b} & \mathbf{c} \\ \mathbf{b} & \mathbf{c} & \mathbf{a} \\ \mathbf{c} & \mathbf{a} & \mathbf{b} \end{matrix}$$

In this pattern, row two follows from row one when the first letter is moved to the end position, and row three follows from row two by means of the same process. The effect of applying this process to the third row is simply to regenerate the first row. Rearrangements of this kind are called **cyclic permutations** of the three vectors.

Again making use of the row interchange property of determinants (see Section 1.7), it follows that

$$[\mathbf{a}, \mathbf{b}, \mathbf{c}] = -[\mathbf{a}, \mathbf{c}, \mathbf{b}],$$

because this time only one row interchange is needed to produce the result on the right from the one on the left, so that a sign change is involved.

A different product involving the three vectors $\mathbf{a}$, $\mathbf{b}$, and $\mathbf{c}$ that this time generates another vector is of the form

$$\mathbf{a} \times (\mathbf{b} \times \mathbf{c}),$$

and products of this type are called **vector triple products** since the results are vectors. In these products it is essential to include the brackets because, in general, $\mathbf{a} \times (\mathbf{b} \times \mathbf{c}) \neq (\mathbf{a} \times \mathbf{b}) \times \mathbf{c}$. The most important results concerning vector triple products are given in the following theorem.

THEOREM 2.4

Vector triple products   If $\mathbf{a}$, $\mathbf{b}$, and $\mathbf{c}$ are any three vectors, then

(a)
$$\mathbf{a} \times (\mathbf{b} \times \mathbf{c}) = (\mathbf{a} \cdot \mathbf{c})\mathbf{b} - (\mathbf{a} \cdot \mathbf{b})\mathbf{c}$$

vector triple product   and

(b)
$$(\mathbf{a} \times \mathbf{b}) \times \mathbf{c} = (\mathbf{a} \cdot \mathbf{c})\mathbf{b} - (\mathbf{b} \cdot \mathbf{c})\mathbf{a}.$$

Proof   The proof of the results in Theorem 2.4 both follow in similar fashion, so we only prove result (a) and leave the proof of result (b) as an exercise. We write the cross product $\mathbf{a} \times (\mathbf{b} \times \mathbf{c})$ in the form of the determinant in (43), with the components of $\mathbf{a}$ in the second row and those of $\mathbf{b} \times \mathbf{c}$ (obtained from (42)) in the third row when we find that

$$\mathbf{a} \times (\mathbf{b} \times \mathbf{c}) = \begin{vmatrix} \mathbf{i} & \mathbf{j} & \mathbf{k} \\ a_1 & a_2 & a_3 \\ (b_2 c_3 - b_3 c_2) & (b_3 c_1 - b_1 c_3) & (b_1 c_2 - b_2 c_1) \end{vmatrix}.$$

Expanding this determinant in terms of the elements of its first row and grouping terms gives

$$\mathbf{a} \times (\mathbf{b} \times \mathbf{c}) = [(a_2 c_2 + a_3 c_3)b_1 - (a_2 b_2 + a_3 b_3)c_1]\mathbf{i} + [(a_1 c_1 + a_3 c_3)b_2$$
$$- (a_1 b_1 + a_3 b_3)c_2]\mathbf{j} + [(a_1 c_1 + a_2 c_2)b_3 - (a_1 b_1 + a_2 b_2)c_3]\mathbf{k}.$$

As it stands, this result is not yet in the form that is required, but adding and subtracting $a_1 b_1 c_1$ to the coefficient of $\mathbf{i}$, $a_2 b_2 c_2$ to the coefficient of $\mathbf{j}$, and $a_3 b_3 c_3$ to the coefficient of $\mathbf{k}$ followed by grouping terms give

$$\mathbf{a} \times (\mathbf{b} \times \mathbf{c}) = (\mathbf{a} \cdot \mathbf{c})\mathbf{b} - (\mathbf{a} \cdot \mathbf{b})\mathbf{c},$$

and the result is established.   ∎

| EXAMPLE 2.18 | Find $\mathbf{a} \times (\mathbf{b} \times \mathbf{c})$ and $(\mathbf{a} \times \mathbf{b}) \times \mathbf{c}$, given that $\mathbf{a} = 3\mathbf{i} + \mathbf{j} - 4\mathbf{k}, \mathbf{b} = 2\mathbf{i} + \mathbf{j} + 3\mathbf{k}$, and $\mathbf{c} = \mathbf{i} + 5\mathbf{j} - \mathbf{k}$. |

Solution $\mathbf{a} \cdot \mathbf{b} = -5, \mathbf{a} \cdot \mathbf{c} = 12$, and $\mathbf{b} \cdot \mathbf{c} = 4$, so

$$\mathbf{a} \times (\mathbf{b} \times \mathbf{c}) = (\mathbf{a} \cdot \mathbf{c})\mathbf{b} - (\mathbf{a} \cdot \mathbf{b})\mathbf{c} = 12\mathbf{b} + 5\mathbf{c} = 29\mathbf{i} + 37\mathbf{j} + 31\mathbf{k},$$

and

$$(\mathbf{a} \times \mathbf{b}) \times \mathbf{c} = (\mathbf{a} \cdot \mathbf{c})\mathbf{b} - (\mathbf{b} \cdot \mathbf{c})\mathbf{a} = 12\mathbf{b} - 4\mathbf{a} = 12\mathbf{i} + 8\mathbf{j} + 52\mathbf{k}. \qquad ■$$

Accounts of geometrical vectors can be found, for example, in references [2.1], [2.3], [2.6], and [1.6].

Summary

This section introduced the two fundamental concepts of linear dependence and independence of vectors. It then showed how the scalar triple product involving three vectors, that gives rise to a scalar quantity, provides a simple test for the linear dependence or independence of the vectors involved. A simple and convenient way of calculating a scalar triple product was shown to be in terms of a determinant with the elements in its rows formed by the components of the three vectors involved in the product. Finally a vector triple product was defined that gives rise to a vector quantity, and it was shown that to avoid ambiguity it is necessary to bracket a pair of vectors in such a product. A rule for the expansion of a vector triple product was derived and shown to involve a linear combination of two of the vectors multiplied by scalar products so that, for example, $\mathbf{a} \times (\mathbf{b} \times \mathbf{c}) = (\mathbf{a} \cdot \mathbf{c})\mathbf{b} - (\mathbf{a} \cdot \mathbf{b})\mathbf{c}$.

EXERCISES 2.4

In Exercises 1 through 4 use the vectors $\mathbf{a}$, $\mathbf{b}$, and $\mathbf{c}$ to find (a) the scalar triple product $\mathbf{a} \cdot (\mathbf{b} \times \mathbf{c})$, and (b) the volume of the parallelepiped determined by these three vectors directed away from a corner.

1. $\mathbf{a} = 2\mathbf{i} - \mathbf{j} - 3\mathbf{k}, \ \mathbf{b} = 3\mathbf{i} - 2\mathbf{k}, \ \mathbf{c} = \mathbf{i} + \mathbf{j} - 4\mathbf{k}$.
2. $\mathbf{a} = \mathbf{i} - \mathbf{j} + 2\mathbf{k}, \ \mathbf{b} = \mathbf{i} + \mathbf{j} + 3\mathbf{k}, \ \mathbf{c} = 2\mathbf{i} - \mathbf{j} + 3\mathbf{k}$.
3. $\mathbf{a} = -\mathbf{i} - \mathbf{j} + \mathbf{k}, \ \mathbf{b} = 2\mathbf{i} + 2\mathbf{j} + 3\mathbf{k}, \ \mathbf{c} = -4\mathbf{i} + \mathbf{j} + 3\mathbf{k}$.
4. $\mathbf{a} = 5\mathbf{i} + 3\mathbf{k}, \ \mathbf{b} = 2\mathbf{i} - \mathbf{j}, \ \mathbf{c} = -2\mathbf{i} + 3\mathbf{j} - 2\mathbf{k}$.

In Exercises 5 through 10 find which sets of vectors are coplanar.

5. $\mathbf{i} + 3\mathbf{j} + 2\mathbf{k}, \ 2\mathbf{i} + \mathbf{j} + 4\mathbf{k}, \ 4\mathbf{i} + 7\mathbf{j} + 8\mathbf{k}$.
6. $2\mathbf{i} + \mathbf{j} + 4\mathbf{k}, \ \mathbf{i} + 2\mathbf{j} + \mathbf{k}, \ 4\mathbf{i} + 3\mathbf{j} + 6\mathbf{k}$.
7. $2\mathbf{i} + \mathbf{k}, \ \mathbf{i} + 4\mathbf{j} + 2\mathbf{k}, \ 3\mathbf{i} + 12\mathbf{j} + 7\mathbf{k}$.
8. $\mathbf{i} + \mathbf{j} + \mathbf{k}, \ 2\mathbf{i} + \mathbf{j} + 2\mathbf{k}, \ 4\mathbf{i} + 3\mathbf{j} + \mathbf{k}$.
9. $2\mathbf{i} + \mathbf{j} - \mathbf{k}, \ 3\mathbf{i} + \mathbf{j} + 2\mathbf{k}, \ 5\mathbf{i} + \mathbf{j} + 8\mathbf{k}$.
10. $2\mathbf{i} + \mathbf{j} - \mathbf{k}, \ \mathbf{i} + 2\mathbf{j} + 2\mathbf{k}, \ 5\mathbf{i} + 4\mathbf{j} + \mathbf{k}$.

In Exercises 11 through 15 use computer algebra to verify that $[\mathbf{a}, \mathbf{b}, \mathbf{c}] = [\mathbf{c}, \mathbf{a}, \mathbf{b}] = -[\mathbf{a}, \mathbf{c}, \mathbf{b}]$.

11. $\mathbf{a} = \mathbf{i} + \mathbf{j} + \mathbf{k}, \mathbf{b} = 2\mathbf{i} + \mathbf{j} - \mathbf{k}, \ \mathbf{c} = 3\mathbf{i} - \mathbf{j} + \mathbf{k}$.
12. $\mathbf{a} = \mathbf{i} - \mathbf{j} - \mathbf{k}, \ \mathbf{b} = -5\mathbf{i} + 2\mathbf{j} - 3\mathbf{k}, \ \mathbf{c} = 2\mathbf{i} + 3\mathbf{j} - 2\mathbf{k}$.

13. $\mathbf{a} = -3\mathbf{i} - 4\mathbf{j} + \mathbf{k}, \ \mathbf{b} = 9\mathbf{i} + 12\mathbf{j} - 3\mathbf{k}, \ \mathbf{c} = \mathbf{i} + 2\mathbf{j} + \mathbf{k}$.
14. $\mathbf{a} = 3\mathbf{i} + 4\mathbf{k}, \ \mathbf{b} = \mathbf{i} + 5\mathbf{k}, \ \mathbf{c} = 2\mathbf{j} + \mathbf{k}$.
15. Prove that if $\mathbf{a}$, $\mathbf{b}$, $\mathbf{c}$, and $\mathbf{d}$ are any four vectors, and λ, μ are arbitrary scalars $[\lambda\mathbf{a} + \mu\mathbf{b}, \mathbf{c}, \mathbf{d}] = \lambda[\mathbf{a}, \mathbf{c}, \mathbf{d}] + \mu[\mathbf{b}, \mathbf{c}, \mathbf{d}]$. Use computer algebra with vectors $\mathbf{a}$, $\mathbf{b}$, $\mathbf{c}$, $\mathbf{d}$ from Exercise 12 with $\mathbf{d} = 4\mathbf{c} - 2\mathbf{j} + 6\mathbf{k}$, and scalars λ, μ of your choice, to verify this result.

In Exercises 16 through 20 find (a) the cartesian equation of the plane containing the given points, and (b) a unit vector normal to the plane.

16. $(1, 2, 1), (3, 1, -2), (2, 1, 4)$.
17. $(2, 0, 3), (0, 1, 0), (2, 4, 5)$.
18. $(-1, 2, -3), (2, 4, 1), (3, 0, 1)$.
19. $(1, 2, 5), (-2, 1, 0), (0, 2, 0)$.
20. Prove result (b) of Theorem 2.4.
21. Show that

$$\mathbf{a} \times (\mathbf{b} \times \mathbf{c}) + \mathbf{b} \times (\mathbf{c} \times \mathbf{a}) + \mathbf{c} \times (\mathbf{a} \times \mathbf{b}) = \mathbf{0}.$$

22. The **law of sines** for a triangle with angles A, B, and C opposite sides with the respective lengths a, b, and c

takes the form

$$\frac{a}{\sin A} = \frac{b}{\sin B} = \frac{c}{\sin C}.$$

Prove this by considering a vector triangle with sides **a**, **b**, and **c**, where $\mathbf{c} = \mathbf{a} + \mathbf{b}$, and taking the cross product of $\mathbf{c} = \mathbf{a} + \mathbf{b}$ first with **a**, then with **b**, and finally with **c**.

In Exercises 23 through 26 use the fact that four points with position vectors **p**, **q**, **r**, and **s** will be coplanar if the vectors $\mathbf{p} - \mathbf{q}$, $\mathbf{p} - \mathbf{r}$, and $\mathbf{p} - \mathbf{s}$ are coplanar to find which sets of points all lie in a plane.

23. $(1, 1, -1), (-3, 1, 1), (-1, 2, -1), (1, 0, 0)$.
24. $(1, 2, -1), (2, 1, 1), (0, 1, 2), (1, 1, 1)$.
25. $(0, -4, 0), (2, 3, 1), (3, -4, -2), (4, -2, -2)$.
26. $(1, 2, 3), (1, 0, 1), (2, 1, 2), (4, 1, 0)$.
27. The volume of a tetrahedron is one-third of the product of the area of its base and its vertical height. Show the volume V of the tetrahedron in Fig. 2.22, in which three edges formed by the vectors **a**, **b**, and **c** are directed away from a vertex, is given by

$$V = (1/6)|\mathbf{a} \cdot (\mathbf{b} \times \mathbf{c})|$$

28. Let **a**, **b**, **c**, and **d** be vectors and λ, μ, ν be scalars satisfying the equation

$$\lambda(\mathbf{b} \times \mathbf{c}) + \mu(\mathbf{c} \times \mathbf{a}) + \nu(\mathbf{a} \times \mathbf{b}) + \mathbf{d} = \mathbf{0}.$$

Show that if **a**, **b**, and **c** are linearly independent, then

$$\lambda = -(\mathbf{a} \cdot \mathbf{d})/[\mathbf{a} \cdot (\mathbf{b} \times \mathbf{c})], \quad \mu = -(\mathbf{b} \cdot \mathbf{d})/[\mathbf{a} \cdot (\mathbf{b} \times \mathbf{c})],$$
$$\nu = -(\mathbf{c} \cdot \mathbf{d})/[\mathbf{a} \cdot (\mathbf{b} \times \mathbf{c})].$$

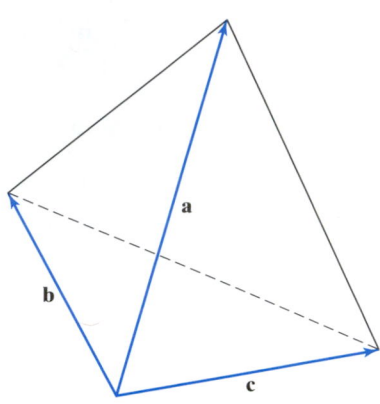

FIGURE 2.22   Tetrahedron.

29. Let **a**, **b**, **c**, and **d** be vectors and λ, μ, ν be scalars satisfying the equation

$$\lambda \mathbf{a} + \mu \mathbf{b} + \nu \mathbf{c} + \mathbf{d} = \mathbf{0}.$$

By taking the scalar products of this equation first with $\mathbf{b} \times \mathbf{c}$, then with $\mathbf{a} \times \mathbf{c}$, and finally with $\mathbf{a} \times \mathbf{b}$, show that if **a**, **b**, and **c** are linearly independent, then

$$\lambda = -\mathbf{d} \cdot (\mathbf{b} \times \mathbf{c})/[\mathbf{a} \cdot (\mathbf{b} \times \mathbf{c})],$$
$$\mu = -\mathbf{d} \cdot (\mathbf{c} \times \mathbf{a})/[\mathbf{a} \cdot (\mathbf{b} \times \mathbf{c})],$$
$$\nu = -\mathbf{d} \cdot (\mathbf{a} \times \mathbf{b})/[\mathbf{a} \cdot (\mathbf{b} \times \mathbf{c})].$$

30. Show that $\mathbf{a} = \mathbf{i} + 2\mathbf{j} + \mathbf{k}$, $\mathbf{b} = 2\mathbf{i} - \mathbf{j} - \mathbf{k}$, and $\mathbf{c} = 4\mathbf{i} + 3\mathbf{j} + i\mathbf{k}$ are linearly independent vectors, and use them with a vector **d** of your choice to verify the results of Exercises 28 and 29.

31. Prove the **Lagrange identity**

$$(\mathbf{a} \times \mathbf{b}) \cdot (\mathbf{c} \times \mathbf{d}) = (\mathbf{a} \cdot \mathbf{c})(\mathbf{b} \cdot \mathbf{d}) - (\mathbf{a} \cdot \mathbf{d})(\mathbf{b} \cdot \mathbf{c}).$$

2.5   *n*-Vectors and the Vector Space R^n

There are many occasions when it is convenient to generalize a vector and its associated algebra to spaces of more than three dimensions. A typical situation occurs in mechanics, where it is sometimes necessary to consider both the position and the momentum of a particle as functions of time. This leads to the study of a 6-vector, three components of which specify the particle position and three its momentum vector at a time t.

Sets of n numbers $(x_1, x_2, \ldots, x_n)$ in a given order, that can be thought of either as n-vectors or as the coordinates of a point in n-dimensional space are called **ordered *n*-tuples** of real numbers or, simply, ***n*-tuples**.

n-tuples

n-vector

An *n*-vector

If $n \geq 2$ is an integer, and $x_1, x_2, \ldots, x_n$ are real numbers, an **n-vector** is an ordered *n*-tuple

$$(x_1, x_2, \ldots, x_n).$$

components and dimension

The numbers $x_1, x_2, \ldots, x_n$ are called the **components** of the *n*-vector, x_i is the *i*th component of the vector, and *n* is called the **dimension** of the space to which the *n*-vector belongs. For any given *n*, the set of all vectors with *n* real components is called a **real *n*-space** or, simply, an **n-space**, and it is denoted by the symbol R^n. A corresponding space exists when the *n* numbers $x_1, x_2, \ldots, x_n$ are allowed to be complex numbers, leading to a **complex *n*-space** denoted by C^n. In this notation R^3 is the three-dimensional space used in previous sections.

In R^3 the length of a vector was taken as the definition of its norm, so if $\mathbf{r} = x_1\mathbf{i} + x_2\mathbf{j} + x_3\mathbf{k}$, then $\|\mathbf{r}\| = (x_1^2 + x_2^2 + x_3^2)^{1/2}$. A generalization of this norm to R^n leads to the following definition.

norm in R^n

The norm in R^n

The **norm** of the *n*-vector $(x_1, x_2, \ldots, x_n)$, denoted by $\|(x_1, x_2, \ldots, x_n)\|$ is

$$\begin{aligned}
\|(x_1, x_2, \ldots, x_n)\| &= \sqrt{\left(x_1^2 + x_2^2 + \cdots + x_n^2\right)} \\
&= \left[\sum_{i=1}^{n} x_i^2\right]^{1/2}.
\end{aligned} \tag{51}$$

The laws for the equality, addition, and scaling of vectors in R^3 in terms of the components of the vector generalize to R^n as follows.

Equality of *n*-vectors

Let $(x_1, x_2, \ldots, x_n)$ and $(y_1, y_2, \ldots, y_n)$ be two *n*-vectors. Then the vectors will be **equal**, written $(x_1, x_2, \ldots, x_n) = (y_1, y_2, \ldots, y_n)$, if, and only if, corresponding components are equal, so that

algebraic rules for equality, addition, and scaling using components

$$x_1 = y_1, \quad x_2 = y_2, \ldots, x_n = y_n. \tag{52}$$

Addition of *n*-vectors

Let $(x_1, x_2, \ldots, x_n)$ and $(y_1, y_2, \ldots, y_n)$ be any two *n*-vectors. Then the **sum** of these vectors, written $(x_1, x_2, \ldots, x_n) + (y_1, y_2, \ldots, y_n)$, is defined as the vector whose *i*th component is the sum of the corresponding *i*th components of the vectors for $i = 1, 2, \ldots, n$, so that

$$(x_1, x_2, \ldots, x_n) + (y_1, y_2, \ldots, y_n) = (x_1 + y_1, x_2 + y_2, \ldots, x_n + y_n). \tag{53}$$

Scaling an *n*-vector

Let $(x_1, x_2, \ldots, x_n)$ be an arbitrary *n*-vector and λ be any scalar. Then the result of **scaling** the vector by λ, written $\lambda(x_1, x_2, \ldots, x_n)$, is defined as the vector whose *i*th component is λ times the *i*th component of the original vector, for $i = 1, 2, \ldots, n$, so that

$$\lambda(x_1, x_2, \ldots, x_n) = (\lambda x_1, \lambda x_2, \ldots, \lambda x_n). \tag{54}$$

The **null (zero)** vector in R^n is the vector $\mathbf{0}$ in which every component is zero, so that

$$\mathbf{0} = (0, 0, \ldots, 0). \tag{55}$$

As with vectors in R^3, so also with *n*-vectors in R^n, it is convenient to use a single boldface symbol for a vector and the corresponding italic symbols with suffixes when it is necessary to specify the components. So we will write

$$\mathbf{x} = (x_1, x_2, \ldots, x_n) \quad \text{and} \quad \mathbf{y} = (y_1, y_2, \ldots, y_n).$$

The reasoning that led to the interpretation of Theorem 2.1 on the algebraic rules for the addition and scaling of vectors in R^3 leads also the following theorem for *n*-vectors.

THEOREM 2.5 **Algebraic rules for the addition and scaling of *n*-vectors in R^n**   Let $\mathbf{x}$, $\mathbf{y}$, and $\mathbf{z}$ be arbitrary *n*-vectors, and let λ and μ be arbitrary real numbers. Then:

(i) $\mathbf{x} + \mathbf{y} = \mathbf{y} + \mathbf{x}$;

(ii) $\mathbf{x} + \mathbf{0} = \mathbf{0} + \mathbf{x} = \mathbf{x}$;

(iii) $(\mathbf{x} + \mathbf{y}) + \mathbf{z} = \mathbf{x} + (\mathbf{y} + \mathbf{z})$;

(iv) $\lambda(\mathbf{x} + \mathbf{y}) = \lambda\mathbf{x} + \lambda\mathbf{y}$;

(v) $(\lambda\mu)\mathbf{x} = \lambda(\mu\mathbf{x}) = \mu(\lambda\mathbf{x})$;

(vi) $(\lambda + \mu)\mathbf{x} = \lambda\mathbf{x} + \mu\mathbf{x}$;

(vii) $\|\lambda\mathbf{x}\| = |\lambda|\|\mathbf{x}\|$. ∎

Because of this similarity between vectors in R^3 and in R^n, the space R^n is called a **real vector space**, though because the symbol R indicates *real* numbers this is usually abbreviated a **vector space**. Analogously, when the elements of the *n*-vectors are allowed to be complex, the resulting space is called the **complex vector space** C^n.

So far there would seem to be little difference between vectors in R^3 and R^n, but major differences do exist, and they are best appreciated when geometrical analogies are sought for vector operations in R^n.

dot product of *n*-vectors

The dot product of *n*-vectors

Let $\mathbf{x} = (x_1, x_2, \ldots, x_n)$ and $\mathbf{y} = (y_1, y_2, \ldots, y_n)$ be any two *n*-vectors. Then the **dot product** of these two vectors, written $\mathbf{x} \cdot \mathbf{y}$ and also called their **inner**

product, is defined as the sum of the products of corresponding components, so that

$$(x_1, x_2, \ldots, x_n) \cdot (y_1, y_2, \ldots, y_n) = x_1 y_1 + x_2 y_2 + \ldots + x_n y_n. \quad (56)$$

The following properties of this dot product are strictly analogous to those of the dot product in R^3 and can be deduced directly from (56).

THEOREM 2.6

Properties of the dot product in R^n   Let $\mathbf{x}$, $\mathbf{y}$, and $\mathbf{z}$ be any three *n*-vectors and λ be any scalar. Then:

(i) $\mathbf{x} \cdot \mathbf{y} = \mathbf{y} \cdot \mathbf{x}$;
(ii) $\mathbf{x} \cdot (\mathbf{y} + \mathbf{z}) = \mathbf{x} \cdot \mathbf{y} + \mathbf{x} \cdot \mathbf{z}$;
(iii) $(\lambda \mathbf{x}) \cdot \mathbf{y} = \mathbf{x} \cdot (\lambda \mathbf{y}) = \lambda (\mathbf{x} \cdot \mathbf{y})$;
(iv) $\mathbf{x} \cdot \mathbf{x} = \|\mathbf{x}\|^2$;
(v) $\mathbf{x} \cdot \mathbf{0} = 0$;
(vi) $\|\mathbf{x}\|^2 = 0$ if, and only if, $\mathbf{x} = \mathbf{0}$. ∎

The existence of a dot product in R^n allows the Cauchy–Schwarz and triangle inequalities to be generalized, both of which play a fundamental role in the study of vector spaces. Various forms of proof of these inequalities are possible, but the one given here has been chosen because it makes full use of the properties of the dot product listed in Theorem 2.6.

THEOREM 2.7

The Cauchy–Schwarz and triangle inequalities   Let $\mathbf{x} = (x_1, x_2, \ldots, x_n)$ and $\mathbf{y} = (y_1, y_2, \ldots, y_n)$ be any two *n*-vectors. Then

generalized inequalities for *n*-vectors

(a)
$$|\mathbf{x} \cdot \mathbf{y}| \leq \|\mathbf{x}\| \cdot \|\mathbf{y}\| \quad \text{(\textbf{Cauchy–Schwarz inequality}}),$$

and

(b)
$$\|\mathbf{x} + \mathbf{y}\| \leq \|\mathbf{x}\| + \|\mathbf{y}\| \quad \text{(\textbf{triangle inequality}}).$$

Proof   We start by proving the Cauchy–Schwarz inequality in (a). The inequality is certainly true if $\mathbf{x} \cdot \mathbf{y} = 0$, so we need only consider the case $\mathbf{x} \cdot \mathbf{y} \neq 0$. Let $\mathbf{x}$ and $\mathbf{y}$ be any two *n*-vectors, and λ be a scalar. Then, using properties (ii) to (iv) of Theorem 2.6,

$$\|\mathbf{x} + \lambda \mathbf{y}\|^2 = (\mathbf{x} + \lambda \mathbf{y}) \cdot (\mathbf{x} + \lambda \mathbf{y}),$$
$$= \|\mathbf{x}\|^2 + \lambda \mathbf{x} \cdot \mathbf{y} + \lambda \mathbf{y} \cdot \mathbf{x} + \lambda^2 \|\mathbf{y}\|^2.$$

However, by result (1) of Theorem 2.6, $\mathbf{y} \cdot \mathbf{x} = \mathbf{x} \cdot \mathbf{y}$, so

$$\|\mathbf{x} + \lambda \mathbf{y}\|^2 = \|\mathbf{x}\|^2 + 2\lambda \mathbf{x} \cdot \mathbf{y} + \lambda^2 \|\mathbf{y}\|^2.$$

We now set $\lambda = -\|x\|^2/(\mathbf{x} \cdot \mathbf{y})$ to obtain

$$\|\mathbf{x} + \lambda \mathbf{y}\|^2 = -\|\mathbf{x}\|^2 + (\|\mathbf{x}\|^4 \|\mathbf{y}\|^2)/|\mathbf{x} \cdot \mathbf{y}|^2,$$

where we have used the fact that $(\mathbf{x} \cdot \mathbf{y})^2 = |\mathbf{x} \cdot \mathbf{y}|^2$. As $\|\mathbf{x} + \lambda \mathbf{y}\|^2$ is nonnegative, this result is equivalent to

$$-\|\mathbf{x}\|^2 + (\|\mathbf{x}\|^4 \cdot \|\mathbf{y}\|^2)/|\mathbf{x} \cdot \mathbf{y}|^2 \geq 0.$$

Cancelling the nonnegative number $\|\mathbf{x}\|^2$, which leaves the inequality sign unchanged; rearranging the terms; and taking the square root of the remaining nonnegative result on each side of the inequality yields the Cauchy–Schwarz inequality

$$|\mathbf{x} \cdot \mathbf{y}| \leq \|\mathbf{x}\| \cdot \|\mathbf{y}\|.$$

To prove the triangle inequality (b) we set $\lambda = 1$ and start from the result

$$\|\mathbf{x} + \mathbf{y}\|^2 = \|\mathbf{x}\|^2 + 2\mathbf{x} \cdot \mathbf{y} + \|\mathbf{y}\|^2.$$

As $\mathbf{x} \cdot \mathbf{y}$ may be either positive or negative, $\mathbf{x} \cdot \mathbf{y} \leq |\mathbf{x} \cdot \mathbf{y}|$, so making use of the Cauchy–Schwarz inequality shows that

$$\|\mathbf{x} + \mathbf{y}\|^2 \leq \|\mathbf{x}\|^2 + 2\|\mathbf{x}\| \cdot \|\mathbf{y}\| + \|\mathbf{y}\|^2$$
$$= (\|\mathbf{x}\| + \|\mathbf{y}\|)^2.$$

The *triangle inequality* follows from taking the square root of each side of this inequality, which is permitted because both are nonnegative numbers.   ∎

The dot product in R^3 allowed the angle between vectors to be determined and, more importantly, it provided a test for the orthogonality of vectors. These same geometrical ideas can be introduced into the vector space R^n if the Cauchy–Schwarz inequality is written in the form

$$-\|\mathbf{x}\| \cdot \|\mathbf{y}\| \leq \mathbf{x} \cdot \mathbf{y} \leq \|\mathbf{x}\| \cdot \|\mathbf{y}\|.$$

After division by the nonnegative number $\|\mathbf{x}\| \cdot \|\mathbf{y}\|$, this becomes

$$-1 \leq \frac{\mathbf{x} \cdot \mathbf{y}}{\|\mathbf{x}\| \cdot \|\mathbf{y}\|} \leq 1.$$

This enables the angle θ between the two n-vectors $\mathbf{x}$ and $\mathbf{y}$ to be defined by the result

$$\cos \theta = \frac{\mathbf{x} \cdot \mathbf{y}}{\|\mathbf{x}\| \cdot \|\mathbf{y}\|}.$$

orthogonality of n-vectors

On account of this result, two n-vectors $\mathbf{x}$ and $\mathbf{y}$ in R^n will be said to be **orthogonal** when $\mathbf{x} \cdot \mathbf{y} = 0$.

unit n-vector

By analogy with R^3 we will call $\mathbf{x} = (x_1, x_2, \ldots, x_n)$ a **unit n-vector** if $\|\mathbf{x}\| = 1$. If we define the unit n-vectors $\mathbf{e}_1, \mathbf{e}_2, \ldots, \mathbf{e}_n$ as

$$\mathbf{e}_1 = (1, 0, 0, 0, \ldots, 0), \ \mathbf{e}_2 = (0, 1, 0, 0, \ldots, 0), \ldots, \mathbf{e}_n = (0, 0, 0, 0, \ldots, 1),$$

we see that

$$\mathbf{e}_i \cdot \mathbf{e}_j = \begin{cases} 1 & \text{for } i = j \\ 0 & \text{for } i \neq j, \end{cases}$$

showing that the $\mathbf{e}_i$ are mutually orthogonal unit n-vectors in R^n. As a result of this the vectors $\mathbf{e}_1, \mathbf{e}_2, \ldots, \mathbf{e}_n$ play the same role in R^n as the vectors $\mathbf{i}, \mathbf{j}$, and $\mathbf{k}$ in R^3. This allows the vector $\mathbf{x} = (x_1, x_2, \ldots, x_n)$ to be written as

$$\mathbf{x} = x_1\mathbf{e}_1 + x_2\mathbf{e}_2 + \cdots + x_n\mathbf{e}_n,$$

where x_i is the ith component of $\mathbf{x}$.

Now suppose that for $n > 3$, we set

$$\mathbf{u}_1 = (1, 0, 0, 0, \ldots, 0), \quad \mathbf{u}_2 = (0, 1, 0, 0, \ldots, 0), \quad \mathbf{u}_3 = (0, 0, 1, 0, \ldots, 0),$$

and all other $\mathbf{u}_i$ identically zero, so that $\mathbf{u}_i = (0, 0, 0, 0, \ldots, 0)$ for $i = 4, 5, \ldots, n$. Then it is not difficult to see that $\mathbf{u}_1, \mathbf{u}_2$, and $\mathbf{u}_3$ behave like the unit vectors $\mathbf{i}, \mathbf{j}$, and $\mathbf{k}$, so that, in some sense the vector space R^3 is embedded in the vector space R^n with vectors in both spaces obeying the same algebraic rules for addition and scaling. This is recognized by saying that R^3 is a **subspace** of R^n.

subspaces

Subspace of R^n

A subset S of vectors in the vector space R^n is called a **subspace** of R^n if S is itself a vector space that obeys the rules for the addition and the scaling of vectors in R^n.

EXAMPLE 2.19

Find the condition that the set S of vectors of the form $(x, mx + c, 0)$, for any m and all real x forms a subspace of the vector space R^3, and give a geometrical interpretation of the result.

Solution   The set S can only contain the null vector $(0, 0, 0)$ if $c = 0$, so if $c \neq 0$ the vectors in S cannot form a subspace of R^3. Now let $c = 0$, so that S contains the null vector. The vector addition law holds, because if $(x, mx, 0)$ and $(x', mx', 0)$ are vectors in S, the sum

$$(x, mx, 0) + (x', mx', 0) = (x + x', m(x + x'), 0)$$

is also a vector in S. The scaling $\lambda(x, mx, 0) = (\lambda x, m\lambda x, 0)$ also generates a vector in S, so the scaling law for vectors also holds, showing that S is a subspace of R^3 provided $c = 0$.

If the three components of vectors in S are regarded as the x-, y-, and z-components of a vector in R^3, the vectors can be interpreted as points on the straight line $y = mx$ passing through the origin and lying in the plane $z = 0$. This subspace is a one-dimensional vector space embedded in the three-dimensional vector space R^3. ∎

EXAMPLE 2.20

Test the following subsets of R^n to determine if they form a subspace.

(a) S is the set of vectors $(x_1, x_1 + 1, \ldots, x_n)$ with all the x_i real numbers.

(b) S is the set of vectors $(x_1, x_2, \ldots, x_n)$ with $x_1 + x_2 + \cdots + x_n = 0$ and all the x_i are real numbers.

Solution   **(a)** The set S does not contain the null vector and so cannot form a subspace of R^n. This result is sufficient to show that S is not a subspace, but to see what properties of a subspace the set S possesses we consider both the summation and scaling of vectors in S. If $(x_1, x_1 + 1, \ldots, x_n)$ and $(x_1', x_1' + 1, \ldots, x_n')$ are two vectors in S, their sum

$$(x_1, x_1 + 1, \ldots, x_n) + (x_1', x_1' + 1, \ldots, x_n') = (x_1 + x_1', x_1 + x_1' + 2, \ldots, x_n + x_n')$$

is *not* a vector in S, so the summation law is not satisfied.

The scaling condition for vectors is not satisfied, because if λ is an arbitrary scalar,

$$\lambda(x_1, x_1 + 1, \ldots, x_n) = (\lambda x_1, \lambda x_1 + \lambda, \ldots, \lambda x_n) \neq (a, a + 1, \ldots), \quad (\lambda n_1 = a)$$

showing that scaling generates another a vector in S. We have proved that the vectors in S do *not* form a subspace of R^n.

(b) The set S does contain the null vector, because $x_1 = x_2 = \cdots = x_n = 0$ satisfies the constraint condition $x_1 + x_2 + \cdots + x_n = 0$. Both the summation law and the scaling law for vectors are easily seen to be satisfied, so this set S does form a subspace of R^n. ∎

EXAMPLE 2.21

Let $C(a, b)$ be the space of all real functions of a single real variable x that are continuous for $a < x < b$, and let $S(a, b)$ be the set of all functions belonging to $C(a, b)$ that have a derivative at every point of the interval $a < x < b$. Show that $S(a, b)$ forms a subspace of $C(a, b)$.

Solution In this case a vector in the space is simply any real function of a single real variable x that is continuous in the interval $a < x < b$. The null vector corresponds to the continuous function that is identically zero in the stated interval, so as the derivative of this function is also zero, it follows that the set $S(a, b)$ must also contain the null vector. The sum of continuous functions in $a < x < b$ is a continuous function, and the sum of differentiable functions in this same interval is a differentiable function, so the summation law for vectors is satisfied. Similarly, scaling continuous functions and differentiable functions does not affect either their continuity or their differentiability, so the scaling law for vectors is also satisfied. Thus, $S(a, b)$ forms a subspace of $C(a, b)$. Think of the dimension of these spores as infinite; norm and inner product are easy to define. ∎

Summary

This section generalized the concept of a three-dimensional vector to a vector with n components in R^n. It was shown that the magnitude of a vector in three space dimensions generalizes to the norm of a vector in R^n and that in terms of components, the equality, addition, and scaling of vectors in R^n follow the same pattern as with three space dimensions. The dot product was generalized and two fundamental inequalities for vectors in R^n were derived. The concept of orthogonality of vectors was generalized and the notion of a subspace of R^n was introduced.

EXERCISES 2.5

In Exercises 1 through 8 find the sum of the given pairs of vectors, their norms, and their dot product.

1. $(2, 1, 0, 2, 2), (1, -1, 2, 2, 4)$.

2. $(3, -1, -1, 2, -4), (1, 2, 0, 0, 3)$.

3. $(2, 1, -1, 2, 1), (-2, -1, 1, -2, -1)$.

4. $(3, -2, 1, 1, 2, 0, 1), (1, -1, 1, -1, 1, 0, 1)$.

5. $(3, 0, 1, 0), (0, 2, 0, 4)$.

6. $(1, -1, 2, 2, 0, 1), (2, -2, 1, 1, 1, 0)$.

7. $(-1, 2, -4, 0, 1), (2, -1, 1, 0, 2)$.

8. $(3, 1, 2, 4, 1, 1, 1), (1, 2, 3, -1, -2, 1, 3)$.

In Exercises 9 through 12 find the angle between the given pairs of n-vectors and the unit n-vector associated with each vector.

9. $(3, 1, 2, 1), (1, -1, 2, 2)$.

10. $(4, 1, 0, 2), (2, -1, 2, 1)$.

11. $(2, -2, -2, 4), (1, -1, -1, 2)$.

12. $(2, 1, -1, 1), (1, -2, 2, 2)$.

In Exercises 13 through 18 determine if the set of vectors S forms a subspace of the given vector space. Give reasons why S either is or is not a subspace.

13. S is the set of vectors of the form $(x_1, x_2, \ldots, x_n)$ in R^n, with the x_i real numbers and $x_2 = x_1^4$.

14. S is the set of vectors of the form $(x_1, x_2, \ldots, x_n)$ in R^n, with the x_i real numbers and $x_1 + 2x_2 + 3x_3 + \cdots + nx_n = 0$.

15. S is the set of vectors of the form $(x_1, x_2, \ldots, x_n)$ in R^n, with the x_i real numbers and $x_1 + x_2 + x_3 + \cdots + x_n = 2$.

16. S is the set of vectors of the form $(x_1, x_2, \ldots, x_6)$ in R^6, with the x_i real numbers and $x_1 = 0$ or $x_6 = 0$.

17. S is the set of vectors of the form $(x_1, x_2, \ldots, x_6)$ in R^6, with the x_i real numbers and $x_1 - x_2 + x_3 \cdots + x_6 = 0$.

18. S is the set of vectors of the form $(x_1, x_2, \ldots, x_5)$ in R^5, with the x_i real numbers and $x_2 < x_3$.

In Exercises 19 to 23 determine if the given set S is a subspace of the space $C[0, 1]$ of all real valued functions that are continuous on the interval $0 \le x \le 1$. Give reasons why either S is a subspace, or it is not.

19. S is the set of all polynomials of degree two.

20. S is the set of all polynomial functions.

21. S is the set of all continuous functions such that $f(0) = f(1) = 0$.

22. S is the set of all continuous functions such that $f(0) = 0$ and $f(1) = 2$.

23. S is the set of all continuous once differentiable functions such that $f(0) = 0$ and $f'(x) > 0$.

24. Prove that the set S of all vectors lying in any plane in R^3 that passes through the origin forms a subspace of R^3.

25. Explain why the set S of all vectors lying in any plane in R^3 that does not pass through the origin does not form a subspace of R^3.

26. Consider the polynomial $P(\lambda)$ defined as

$$P(\lambda) = \|\mathbf{x} + \lambda\mathbf{y}\|^2,$$

where $\mathbf{x}$ and $\mathbf{y}$ are vectors in R^n. Show, provided not both $\mathbf{x}$ and $\mathbf{y}$ are null vectors, that the graph of $P(\lambda)$ as a function of λ is nonnegative, so $P(\lambda) = 0$ cannot have real roots. Use this result to prove the Cauchy–Schwarz inequality

$$|\mathbf{x} \cdot \mathbf{y}| \le \|\mathbf{x}\| \cdot \|\mathbf{y}\|.$$

27. Let $\mathbf{x}$ and $\mathbf{y}$ be vectors in R^n and λ be a scalar. Prove that

$$\|\mathbf{x} + \lambda\mathbf{y}\|^2 + \|\mathbf{x} - \lambda\mathbf{y}\|^2 = 2(\|\mathbf{x}\|^2 + \lambda^2\|\mathbf{y}\|^2).$$

28. If $\mathbf{x}$ and $\mathbf{y}$ are orthogonal vectors in R^n, prove that the Pythagoras theorem takes the form

$$\|\mathbf{x} + \mathbf{y}\|^2 = \|\mathbf{x}\|^2 + \|\mathbf{y}\|^2.$$

29. What conditions on the components of vectors $\mathbf{x}$ and $\mathbf{y}$ in the Cauchy–Schwarz inequality cause it to become an equality, so that

$$\sum_{i=1}^{n} x_i y_i = \left(\sum_{i=1}^{n} x_i^2 \right)^{1/2} \left(\sum_{i=1}^{n} y_i^2 \right)^{1/2} ?$$

30. Modify the method of proof used in Theorem 2.7 to prove the complex form of the Cauchy–Schwarz inequality

$$\left| \sum_{i=1}^{n} x_i y_i \right| \le \left(\sum_{i=1}^{n} |x_i|^2 \right)^{1/2} + \left(\sum_{i=1}^{n} |y_i|^2 \right)^{1/2},$$

where the x_i and y_i are complex numbers.

2.6   Linear Independence, Basis, and Dimension

The concept of the linear independence of a set of vectors in R^3 introduced in Section 2.4 generalizes to R^n and involves a linear combination of n-vectors.

Linear combination of n-vectors

Let $\mathbf{x}_1, \mathbf{x}_2, \ldots, \mathbf{x}_m$ be a set of n-vectors in R^n. Then a **linear combination** of the n-vectors is a sum of the form

$$c_1\mathbf{x}_1 + c_2\mathbf{x}_2 + \cdots + c_m\mathbf{x}_m,$$

where $c_1, c_2, \ldots, c_m$ are nonzero scalars.

An example of a linear combination of vectors in R^5 is provided by the vector sum ($m = 3, n = 5$)

$$2\mathbf{x}_1 + \mathbf{x}_2 + 3\mathbf{x}_3,$$

where $\mathbf{x}_1 = (1, 2, 3, 0, 4)$, $\mathbf{x}_2 = (2, 1, 4, 1, -3)$, and $\mathbf{x}_3 = (6, 0, 2, 2, -1)$. The vector in R^5 formed by this linear combination is

$$2\mathbf{x}_1 + \mathbf{x}_2 + 3\mathbf{x}_3 = 2(1, 2, 3, 0, 4) + (2, 1, 4, 1, -3) + 3(6, 0, 2, 2, -1),$$
$$= (22, 5, 16, 7, 2).$$

A linear combination of n-vectors is the most general way of combining n-vectors, and the definition of a linear combination of vectors contains within it the definition of the scaling of a single n-vector as a special case. This can be seen by setting $m = 1$, because this reduces the linear combination to the single scaled n-vector $c_1\mathbf{x}_1$.

linear dependence and independence of n-vectors

Linear dependence of n-vectors

Let $\mathbf{x}_1, \mathbf{x}_2, \ldots, \mathbf{x}_m$ be a set of n-vectors in R^n. Then the set is said to be **linearly dependent** if, and only if, one of the n-vectors can be expressed as a linear combination of the remaining n-vectors.

An example of linear dependence in R^4 is provided by the vectors $\mathbf{x}_1 = (1, 0, 2, 5)$, $\mathbf{x}_2 = (2, 1, 2, 1)$, $\mathbf{x}_3 = (3, 2, 1, 0)$, and $\mathbf{x}_4 = (-1, -1, -1, 7)$, because

$$\mathbf{x}_4 = 2\mathbf{x}_1 - 3\mathbf{x}_2 + \mathbf{x}_3.$$

Linear independence of n-vectors

Let $\mathbf{x}_1, \mathbf{x}_2, \ldots, \mathbf{x}_m$ be a set of n-vectors in R^n. Then the set is said to be **linearly independent** if, and only if, the n-vectors are not linearly dependent.

A simple example of a set of linearly independent vectors in R^4 is provided by the vectors $\mathbf{e}_1 = (1, 0, 0, 0)$, $\mathbf{e}_2 = (0, 1, 0, 0)$, and $\mathbf{e}_3 = (0, 0, 1, 0)$. The linear independence of these 4-vectors can be seen from the fact that for no choice of c_1 and c_2 can the vector $c_1\mathbf{e}_1 + c_2\mathbf{e}_2$ be made equal to $\mathbf{e}_3$.

To make effective use of the concept of linear independence, and to understand the notion of the *basis* and *dimension* of a vector space, it is necessary to have a test for linear independence. Such a test is provided by the following theorem.

THEOREM 2.8   **Linear dependence and independence**   Let S be a set of non-zero n-vectors $\mathbf{x}_1, \mathbf{x}_2, \ldots, \mathbf{x}_m$, with $m \geq 2$. Then:

(a) Set S is linearly dependent if the vector equation

$$c_1\mathbf{x}_1 + c_2\mathbf{x}_2 + \cdots + c_m\mathbf{x}_m = \mathbf{0}$$

is true for some set of scalars (constants) $c_1, c_2, \ldots, c_m$ that are not all zero;

(b) Set S is linearly independent if the vector equation

$$c_1\mathbf{x}_1 + c_2\mathbf{x}_2 + \cdots + c_m\mathbf{x}_m = \mathbf{0}$$

is only true when $c_1 = c_2 = \cdots = c_m = 0$.

Proof   To establish result (a) it is necessary to show that the conditions of the definition of linear dependence are satisfied. First, if the set S of n-vectors is linearly dependent, scalars $d_1, d_2, \ldots, d_m$ exist such that

$$d_1\mathbf{x}_1 + d_2\mathbf{x}_2 + \cdots + d_m\mathbf{x}_m = \mathbf{0}.$$

There is no loss of generality in assuming that $d_1 \neq 0$, because if this is not the case a renumbering of the vectors can always make this possible. Consequently,

$$\mathbf{x}_1 = (-d_2/d_1)\mathbf{x}_2 + (-d_3/d_1)\mathbf{x}_3 + \cdots + (-d_m/d_1)\mathbf{x}_m,$$

which shows, as claimed, that the set S is linearly dependent, because $\mathbf{x}_1$ is linearly dependent on $\mathbf{x}_2, \mathbf{x}_3, \ldots, \mathbf{x}_m$. A similar argument applies to show that $\mathbf{x}_r$ is linearly dependent on the remaining n-vectors in S provided $d_r \neq 0$, for $r = 2, 3, \ldots, m$.

Conversely, if one of the n-vectors in set S, say $\mathbf{x}_1$, is linearly dependent on the remaining n-vectors in the set, scalars $d_1, d_2, \ldots, d_m$ can be found such that

$$\mathbf{x}_1 = d_2\mathbf{x}_2 + \cdots + d_m\mathbf{x}_m,$$

so that

$$\mathbf{x}_1 - d_2\mathbf{x}_2 - \cdots - d_m\mathbf{x}_m = 0.$$

This result is of the form given in definition of linear dependence with $c_1 = 1$, $c_2 = -d_2, \ldots, c_m = -d_m$, not all of which constants are zero, so again the set of n-vectors in S is seen to be linearly dependent.

To establish result (b), suppose, if possible, that the set S of vectors is linearly independent, but that some scalars $d_1, d_2, \ldots, d_m$ that are not all zero can be found such that

$$d_1\mathbf{x}_1 + d_2\mathbf{x}_2 + \cdots + d_m\mathbf{x}_m = \mathbf{0}.$$

Then if $d_1 \neq 0$, say, is one of these scalars, it follows that

$$\mathbf{x}_1 = (-d_2/d_1)\mathbf{x}_2 + (-d_3/d_1)\mathbf{x}_3 + \cdots + (-d_m/d_1)\mathbf{x}_m,$$

which is impossible because this shows that, contrary to the hypothesis, $\mathbf{x}_1$ is linearly dependent on the remaining n-vectors in S. So we must have $c_1 = c_2 = \cdots = c_m = 0$. ∎

A systematic and efficient computational method for the application of Theorem 2.8 to vectors in R^n will be developed in the next chapter for the three separate cases that arise, (a) $m < n$, (b) $m = n$, and (c) $m > n$. However, when n and m are small, a straightforward approach is possible, as illustrated in the next example.

EXAMPLE 2.22

Test the following sets of vectors in R^4 for linear dependence or independence.

(a) $\mathbf{x}_1 = (2, 1, 1, 0)$, $\mathbf{x}_2 = (0, 2, 0, 1)$, $\mathbf{x}_3 = (1, 1, 0, 2)$, $\mathbf{x}_4 = (0, 2, 1, 1)$.

(b) $\mathbf{x}_1 = (4, 0, 2)$, $\mathbf{x}_2 = (2, 2, 0)$, $\mathbf{x}_3 = (1, 1, 0)$, $\mathbf{x}_4 = (5, 1, 2)$.

Solution   In both (a) and (b) it is necessary to consider the vector equation

$$c_1 \mathbf{x}_1 + \cdots + c_m \mathbf{x}_m = \mathbf{0}.$$

If the equation is only satisfied when $c_1 = c_2 = \cdots = c_m = 0$, the set of vectors will be linearly independent, whereas if a solution can be found in which not all of the constants c_1, c_2, c_3, c_4 vanish, the set of vectors will be linearly dependent.

(a) Substituting for $\mathbf{x}_1, \mathbf{x}_2, \mathbf{x}_3, \mathbf{x}_4$ in the preceding equation and equating corresponding components show the coefficients c_i must satisfy the following equations

$$2c_1 + c_3 = 0$$
$$c_1 + 2c_2 + c_3 + 2c_4 = 0$$
$$c_1 + c_4 = 0$$
$$c_2 + 2c_3 + c_4 = 0.$$

The third equation shows that $c_4 = -c_1$, so the equations can be rewritten as

$$2c_1 + c_3 = 0$$
$$-c_1 + 2c_2 + c_3 = 0$$
$$c_2 - c_1 + 2c_3 = 0.$$

Adding twice the third equation to the first equation shows that $c_3 = 0$, so $c_1 = 0$, and it then follows that $c_2 = c_3 = c_4 = 0$. This has established the linear independence of the set of vectors in (a).

(b) Proceeding in the same manner with the set of vectors in (b) leads to the following equations for the coefficients c_i:

$$4c_1 + 2c_2 + c_3 + 5c_4 = 0$$
$$2c_2 + c_3 + c_4 = 0$$
$$2c_1 + 2c_4 = 0.$$

The third equation shows that $c_4 = -c_1$, so using this result in the first two equations reduces the first one to

$$-c_1 + 2c_2 + c_3 = 0$$

and the second to

$$-c_1 + 2c_2 + c_3 = 0.$$

There is only one equation connecting $c_1, c_2,$ and c_3, and hence also c_4. This means that if c_2 and c_3 are given arbitrary values, not both of which are zero, the constants c_1 and c_4 will be determined in terms of them. Thus, a set of constants c_1, c_2, c_3, c_4 that are not all zero can be found that satisfy the vector equation, showing that the set of vectors in (b) is linearly dependent. This set of constants is not unique, but this does affect the conclusion that the set of vectors is linearly dependent, because to establish linear dependence it is sufficient that at least one such set of constants can be found.   ∎

Example 2.22 has shown one way in which Theorem 2.8 can be implemented for vectors in R^n, but it also illustrates the need for a systematic approach to the solution of the system of equations for the coefficients when n is large.

A trivial case of Theorem 2.8 arises when the set of vectors S contains the null vector $\mathbf{0}$, because then the set of vectors in S is always linearly dependent. This can be seen by assuming that $\mathbf{x}_1 = \mathbf{0}$, because then the vector equation in the theorem becomes

$$c_1\mathbf{0} + c_2\mathbf{x}_2 + \cdots + c_m\mathbf{x}_m = \mathbf{0}.$$

This vector equation is satisfied if $c_1 \neq 0$ (arbitrary) and $c_2 = c_3 = \cdots = c_m = 0$, so, as not all of the coefficients are zero, the set of vectors must be linearly dependent.

We conclude this introduction to the vector space R^n by defining the *span*, a *basis*, and the *dimension* of a vector space.

span of a vector space

Span of a vector space

Let the set of non-zero vectors $\mathbf{x}_1, \mathbf{x}_2, \ldots, \mathbf{x}_m$ belonging to a vector space V have the property that every vector in V can be expressed as a linear combination of these vectors. Then the vectors $\mathbf{x}_1, \mathbf{x}_2, \ldots, \mathbf{x}_m$ are said to **span** the vector space V.

EXAMPLE 2.23   All vectors $\mathbf{v}$ in the (x, y)-plane are spanned by the vectors $\mathbf{i}$ and $\mathbf{j}$, because any vector $\mathbf{v} = (v_1, v_2)$ can always be written $\mathbf{v} = v_1\mathbf{i} + v_2\mathbf{j}$. This is an example of vectors spanning the space R^2.   ∎

EXAMPLE 2.24   The vector space R^n is spanned by the unit n-vectors

$$\mathbf{e}_1 = (1, 0, 0, 0, \ldots, 0), \quad \mathbf{e}_2 = (0, 1, 0, 0, \ldots, 0), \ldots, \quad \mathbf{e}_n = (0, 0, 0, 0, \ldots, 1).$$ ∎

EXAMPLE 2.25   The subspace R^3 of the vector space R^5 is spanned by the unit vectors

$$\mathbf{e}_1 = (1, 0, 0, 0, 0), \quad \mathbf{e}_2 = (0, 1, 0, 0, 0), \quad \mathbf{e}_3 = (0, 0, 1, 0, 0),$$

because all vectors $\mathbf{v} = (v_1, v_2, v_3)$ in R^3 can be written in the form of the linear combination $\mathbf{v} = v_1\mathbf{e}_1 + v_1\mathbf{e}_2 + v_3\mathbf{e}_3$.   ∎

basis of a vector space in R^n

Basis of a vector space

Let $\mathbf{x}_1, \mathbf{x}_2, \ldots, \mathbf{x}_n$ be vectors in R^n. Then the vectors are said to form a **basis** for the vector space R^n if:

(i) The vectors $\mathbf{x}_1, \mathbf{x}_2, \ldots, \mathbf{x}_n$ are linearly independent.
(ii) Every vector in R^n can be expressed as a linear combination of the vectors $\mathbf{x}_1, \mathbf{x}_2, \ldots, \mathbf{x}_n$.

dimension of a vector space

Dimension of a vector space

The **dimension** of a vector space is the number of vectors in its basis.

EXAMPLE 2.26

A basis for the space of ordinary vectors in three dimensions is provided by the vectors $\mathbf{i}$, $\mathbf{j}$, and $\mathbf{k}$, so the dimension of the space is 3.   ∎

EXAMPLE 2.27

A basis for R^n is provided by the n vectors

$$\mathbf{e}_1 = (1,0,0,0,\ldots,0), \quad \mathbf{e}_2 = (0,1,0,0,\ldots,0),\ldots, \quad \mathbf{e}_n = (0,0,0,0,\ldots,1),$$

so its dimension is n.   ∎

EXAMPLE 2.28

It was shown in Example 2.20 (b) that the set S of vectors $(x_1, x_2, \ldots, x_n)$ with $x_1 + x_2 + \cdots + x_n = 0$ forms a subspace of R^n. The dimension of R^n is n, but the constraint condition $x_1 + x_2 + \cdots + x_n = 0$ implies that only $n-1$ of the components $x_1, x_2, \ldots, x_n$ can be specified independently, because the constraint itself determines the value of the remaining component. This in turn implies that the basis for the subspace S can only contain $n-1$ linearly independent vectors, so S must have dimension $n-1$.   ∎

More information on linear vector spaces can be found in references [2.1] and [2.5] to [2.12].

Summary

In this section the concepts of linear dependence and independence were generalized to vectors in R^n, and the span of a vector space was defined as a set of vectors in R^n with the property that every vector in R^n can be expressed as a linear combination of these vectors. Naturally in R^n, as in R^3, a set of vectors spanning the space is not unique. The smallest set of n vectors spanning a vector space is said to form a basis for the vector space, and the dimension of a vector space is the number of vectors in its basis. This corresponds to the fact that the unit vectors $\mathbf{i}$, $\mathbf{j}$, and $\mathbf{k}$ form a basis for the ordinary three-dimensional space R^3, because every vector in this space can be represented as a linear combination of $\mathbf{i}$, $\mathbf{j}$, and $\mathbf{k}$.

EXERCISES 2.6

In Exercises 1 through 12 determine if the set of m vectors in three-dimensional space is linearly independent by solving for the scalars $c_1, c_2 \ldots c_m$ in Theorem 2.8. Where appropriate, verify the result by using Theorem 2.3.

1. $\mathbf{a} = \mathbf{i} + 2\mathbf{j} + \mathbf{k}$, $\mathbf{b} = \mathbf{i} - \mathbf{j} + \mathbf{k}$, $\mathbf{c} = 2\mathbf{i} + \mathbf{k}$.
2. $\mathbf{a} = 3\mathbf{i} - \mathbf{j} + \mathbf{k}$, $\mathbf{b} = \mathbf{i} + 3\mathbf{k}$, $\mathbf{c} = 5\mathbf{i} - \mathbf{j} + 7\mathbf{k}$.
3. $\mathbf{a} = 2\mathbf{i} - \mathbf{j} + \mathbf{k}$, $\mathbf{b} = 3\mathbf{i} + \mathbf{j} - \mathbf{k}$, $\mathbf{c} = 8\mathbf{i} + \mathbf{j} + 7\mathbf{k}$.
4. $\mathbf{a} = 3\mathbf{i} + 2\mathbf{k}$, $\mathbf{b} = \mathbf{i} + \mathbf{j} + 2\mathbf{k}$, $\mathbf{c} = 11\mathbf{i} + 2\mathbf{j} - 2\mathbf{k}$.
5. $\mathbf{a} = 4\mathbf{i} - \mathbf{j} + 3\mathbf{k}$, $\mathbf{b} = \mathbf{i} + 4\mathbf{j} - 2\mathbf{k}$, $\mathbf{c} = 3\mathbf{i} - \mathbf{j} - \mathbf{k}$.
6. $\mathbf{a} = \mathbf{i} + \mathbf{j} - \mathbf{k}$, $\mathbf{b} = \mathbf{i} - \mathbf{j} + \mathbf{k}$, $\mathbf{c} = -\mathbf{i} + \mathbf{j} + \mathbf{k}$.
7. $\mathbf{a} = \mathbf{i} + 2\mathbf{j} + \mathbf{k}$, $\mathbf{b} = \mathbf{i} + 3\mathbf{j} - \mathbf{k}$, $\mathbf{c} = 3\mathbf{i} + 10\mathbf{j} - 5\mathbf{k}$.
8. $\mathbf{a} = 2\mathbf{i} + 3\mathbf{j} + \mathbf{k}$, $\mathbf{b} = \mathbf{i} - 3\mathbf{j} + 2\mathbf{k}$, $\mathbf{c} = \mathbf{i} + 15\mathbf{j} - 4\mathbf{k}$.
9. $\mathbf{a} = 3\mathbf{i} - \mathbf{j} + 2\mathbf{k}$, $\mathbf{b} = \mathbf{i} + \mathbf{j} + \mathbf{k}$ $(m = 2)$.
10. $\mathbf{a} = \mathbf{i} + \mathbf{j} + \mathbf{k}$, $\mathbf{b} = \mathbf{i} + 2\mathbf{j} + \mathbf{k}$, $\mathbf{c} = \mathbf{i} + 3\mathbf{j} + \mathbf{k}$, $\mathbf{d} = \mathbf{i} - 4\mathbf{j} + \mathbf{k}$ $(m = 4)$.
11. $\mathbf{a} = \mathbf{i} - \mathbf{j} + 3\mathbf{k}$, $\mathbf{b} = 2\mathbf{i} - \mathbf{j} + 2\mathbf{k}$, $\mathbf{c} = \mathbf{i} + \mathbf{k}$, $\mathbf{d} = 3\mathbf{i} + \mathbf{j} + \mathbf{k}$ $(m = 4)$.

12. $\mathbf{a} = \mathbf{i} + \mathbf{j}$, $\mathbf{b} = \mathbf{j} + \mathbf{k}$, $\mathbf{c} = \mathbf{i} - \mathbf{k}$.

In Exercises 13 through 16, determine if the set of vectors in R^4 is linearly independent by using the method of Example 2.22.

13. $(1, 3, -1, 0), (1, 2, 0, 1), (0, 1, 0, -1), (1, 1, 0, 1)$.
14. $(1, -2, 1, 2), (4, -1, 0, 2), (2, 1, -1, 1), (1, 0, 0, -1)$.
15. $(2, 1, 0, 1), (1, 0, 1, 1), (4, 1, 2, -1), (1, 0, 1, -1)$.
16. $(1, 2, 1, 1), (1, -2, 0, -1), (1, 1, 1, 2), (1, -1, 0, 0)$.

In Exercises 17 through 20, find a basis and the dimension of the given subspace S.

17. The subspace S of vectors in R^5 of the form $(x_1, x_2, x_3, x_4, x_5)$ with $x_1 = x_2$.
18. The subspace S of vectors in R^4 of the form (x_1, x_2, x_3, x_4) with $x_1 = 2x_2$.
19. The subspace S of vectors in R^5 of the form $(x_1, x_2, x_3, x_4, x_5)$ with $x_1 = x_2 = 2x_3$.

20. The subspace S of vectors in R^6 of the form $(x_1, x_2, x_3, x_4, x_5, x_6)$ with $x_1 = 2x_2$ and $x_3 = -x_4$.

21. Let $\mathbf{u} = \cos^2 x$ and $\mathbf{v} = \sin^2 x$ form a basis for a vector space V. Find which of the following can be represented in terms of $\mathbf{u}$ and $\mathbf{v}$, and so lie in V.

(a) 2. (b) $\sin 2x$. (c) 0. (d) $\cos 2x$. (e) $2 + 3x$. (f) $3 - 4\cos 2x$.

22. Given that $r \leq n$, prove that any subset S of r vectors selected from a set of n linearly independent vectors is linearly independent.

2.7   Gram–Schmidt Orthogonalization Process

A set of vectors forming a basis for a vector space is not unique, and having obtained a basis by some means, it is often useful to replace it by an equivalent set of orthogonal vectors. The **Gram–Schmidt orthogonalization process** accomplishes this by means of a sequence of simple steps that have a convenient geometrical interpretation. We now develop the Gram–Schmidt orthogonalization process for geometrical vectors in R^3, though in Section 4.2 the method will be extended to vectors in R^n to enable orthogonal matrices to be constructed from a set of eigenvectors associated with a symmetric matrix.

Let us now show how any basis for R^3, comprising three nonorthogonal linearly independent vectors $\mathbf{a}_1$, $\mathbf{a}_2$, and $\mathbf{a}_3$, can be used to construct an equivalent basis involving three linearly independent orthogonal vectors $\mathbf{u}_1$, $\mathbf{u}_2$, and $\mathbf{u}_3$. It is essential that the vectors $\mathbf{a}_1$, $\mathbf{a}_2$, and $\mathbf{a}_3$ be linearly independent, because if not, the vectors $\mathbf{u}_1$, $\mathbf{u}_2$, and $\mathbf{u}_3$ generated by the Gram–Schmidt orthogonalization process will be linearly dependent and so cannot form a basis for R^3. The derivation of the method starts by setting

$$\mathbf{u}_1 = \mathbf{a}_1,$$

where the choice of $\mathbf{a}_1$ instead of $\mathbf{a}_2$ or $\mathbf{a}_3$ is arbitrary.

The component of $\mathbf{a}_2$ in the direction of $\mathbf{u}_1$ is $\mathbf{u}_1 \cdot \mathbf{a}_2$, so the vector component of $\mathbf{a}_2$ in this direction is

$$(\mathbf{u}_1 \cdot \mathbf{a}_2)\mathbf{u}_1 = \frac{(\mathbf{u}_1 \cdot \mathbf{a}_2)\mathbf{u}_1}{\|\mathbf{u}_1\|^2},$$

and this always exists because $\|\mathbf{u}_1\|^2 > 0$. Subtracting this vector from $\mathbf{a}_2$ gives a vector $\mathbf{u}_2$ that is normal to $\mathbf{u}_1$, so

$$\mathbf{u}_2 = \mathbf{a}_2 - \frac{(\mathbf{u}_1 \cdot \mathbf{a}_2)\mathbf{u}_1}{\|\mathbf{u}_1\|^2}.$$

Similarly, to find a vector normal to both $\mathbf{u}_1$ and $\mathbf{u}_2$ involving $\mathbf{a}_3$, it is necessary to subtract from $\mathbf{a}_3$ the components of vector $\mathbf{a}_3$ in the direction of $\mathbf{u}_1$ and also in the direction of $\mathbf{u}_2$, so that

$$\mathbf{u}_3 = \mathbf{a}_3 - \frac{(\mathbf{u}_1 \cdot \mathbf{a}_3)\mathbf{u}_1}{\|\mathbf{u}_1\|^2} - \frac{(\mathbf{u}_2 \cdot \mathbf{a}_3)\mathbf{u}_2}{\|\mathbf{u}_2\|^2},$$

and this also always exists, because $\|\mathbf{u}_1\|^2 > 0$ and $\|\mathbf{u}_2\|^2 > 0$.

If an orthonormal basis is required, it is necessary to normalize the vectors $\mathbf{u}_1$, $\mathbf{u}_2$, and $\mathbf{u}_3$ by dividing each by its norm.

Rule for the Gram–Schmidt orthogonalization process in R^3

A set of nonorthogonal linearly independent vectors $\mathbf{a}_1$, $\mathbf{a}_2$, and $\mathbf{a}_3$ that form a basis in R^3 can be used to generate an equivalent orthogonal basis involving the vectors, $\mathbf{u}_1$, $\mathbf{u}_2$, and $\mathbf{u}_3$ by setting

$$\mathbf{u}_1 = \mathbf{a}_1, \quad \mathbf{u}_2 = \mathbf{a}_2 - \frac{(\mathbf{u}_1 \cdot \mathbf{a}_2)\mathbf{u}_1}{\|\mathbf{u}_1\|^2}, \quad \text{and}$$

$$\mathbf{u}_3 = \mathbf{a}_3 - \frac{(\mathbf{u}_1 \cdot \mathbf{a}_3)\mathbf{u}_1}{\|\mathbf{u}_1\|^2} - \frac{(\mathbf{u}_2 \cdot \mathbf{a}_3)\mathbf{u}_2}{\|\mathbf{u}_2\|^2}.$$

As already remarked, the choice of $\mathbf{a}_1$ as the vector with which to start the orthogonalization process was arbitrary, and the process could equally well have been started by setting $\mathbf{u}_1 = \mathbf{a}_2$ or $\mathbf{u}_1 = \mathbf{a}_3$. Using a different vector will produce a different set of orthogonal vectors $\mathbf{u}_1$, $\mathbf{u}_2$, and $\mathbf{u}_3$, but any basis for R^3 is equivalent to any other basis, so unless there is a practical reason for starting with a particular vector, the choice is immaterial.

EXAMPLE 2.29

Given the nonorthogonal basis $\mathbf{a}_1 = \mathbf{i} - \mathbf{j} - \mathbf{k}$, $\mathbf{a}_2 = \mathbf{i} + \mathbf{j} + \mathbf{k}$, and $\mathbf{a}_3 = -\mathbf{i} + 2\mathbf{k}$, use the Gram–Schmidt orthogonalization process to find an equivalent orthogonal basis, and then find the corresponding orthonormal basis.

Solution   Using the preceding rule we start with $\mathbf{u}_1 = \mathbf{i} - \mathbf{j} - \mathbf{k}$, and to find $\mathbf{u}_2$ we need to use the results $\mathbf{u}_1 \cdot \mathbf{a}_2 = -1$ and $\|\mathbf{u}_1\|^2 = 3$, so that

$$\mathbf{u}_2 = \mathbf{i} + \mathbf{j} + \mathbf{k} - (-1/3)(\mathbf{i} - \mathbf{j} - \mathbf{k}) = (4/3)\mathbf{i} + (2/3)\mathbf{j} + (2/3)\mathbf{k}.$$

To find $\mathbf{u}_3$ we need to use the results $\mathbf{u}_1 \cdot \mathbf{a}_3 = -3$, $\|\mathbf{u}_1\|^2 = 3$, $\mathbf{u}_2 \cdot \mathbf{a}_3 = 0$, and $\|\mathbf{u}_2\|^2 = 24/9$, so that

$$\mathbf{u}_3 = -\mathbf{i} + 2\mathbf{k} - (-3/3)(\mathbf{i} - \mathbf{j} - \mathbf{k}) = -\mathbf{j} + \mathbf{k}.$$

So the required equivalent orthogonal basis is

$$\mathbf{u}_1 = \mathbf{i} - \mathbf{j} - \mathbf{k}, \quad \mathbf{u}_2 = (4/3)\mathbf{i} + (2/3)\mathbf{j} + 2/3\mathbf{k}, \quad \text{and} \quad \mathbf{u}_3 = -\mathbf{j} + \mathbf{k}.$$

The corresponding orthonormal basis obtained by dividing each of these vectors by its norm (modulus) is

$$\hat{\mathbf{u}}_1 = (1/\sqrt{3})\mathbf{u}_1, \quad \hat{\mathbf{u}}_2 = (1/2)\sqrt{(3/2)}\mathbf{u}_2 \quad \text{and} \quad \hat{\mathbf{u}}_3 = (1/\sqrt{2})\mathbf{u}_3. \qquad \blacksquare$$

Other accounts of the Gram–Schmidt orthogonalization process are to be found in references [2.1] and [2.7] to [2.12].

Summary

In this section it is shown how in R^3 the Gram–Schmidt orthogonalization process converts any three nonorthogonal linearly independent vectors $\mathbf{a}_1$, $\mathbf{a}_2$, and $\mathbf{a}_3$ into three orthogonal vectors $\mathbf{u}_1$, $\mathbf{u}_2$, and $\mathbf{u}_3$. If necessary, the vectors $\mathbf{u}_1$, $\mathbf{u}_2$, and $\mathbf{u}_3$ can then be normalized in the usual manner to form an orthogonal set of unit vectors.

EXERCISES 2.7

In Exercises 1 through 6, use the given nonorthogonal basis for vectors in R^3 to find an equivalent orthogonal basis by means of the Gram–Schmidt orthogonalization process.

1. $\mathbf{a}_1 = \mathbf{i} + 2\mathbf{j} + \mathbf{k}$, $\mathbf{a}_2 = \mathbf{i} - \mathbf{j}$, $\mathbf{a}_3 = 2\mathbf{j} - \mathbf{k}$.
2. $\mathbf{a}_1 = \mathbf{j} + 3\mathbf{k}$, $\mathbf{a}_2 = \mathbf{i} + \mathbf{j} - \mathbf{k}$, $\mathbf{a}_3 = \mathbf{i} + 2\mathbf{k}$.
3. $\mathbf{a}_1 = 2\mathbf{i} + \mathbf{j}$, $\mathbf{a}_2 = 2\mathbf{j} + \mathbf{k}$, $\mathbf{a}_3 = \mathbf{k}$.
4. $\mathbf{a}_1 = \mathbf{i} + 3\mathbf{k}$, $\mathbf{a}_2 = \mathbf{i} - \mathbf{j} + \mathbf{k}$, $\mathbf{a}_3 = 2\mathbf{i} + \mathbf{j}$.
5. $\mathbf{a}_1 = -\mathbf{i} + \mathbf{k}$, $\mathbf{a}_2 = 2\mathbf{j} + \mathbf{k}$, $\mathbf{a}_3 = \mathbf{i} + \mathbf{j} + \mathbf{k}$.

6. $\mathbf{a}_1 = \mathbf{i} + \mathbf{k}$, $\mathbf{a}_2 = -\mathbf{j} + \mathbf{k}$, $\mathbf{a}_3 = \mathbf{i} + \mathbf{j} + 2\mathbf{k}$.

In Exercises 7 and 8, find two different but equivalent sets of orthogonal vectors by arranging the same three nonorthogonal vectors in the orders indicated.

7. (a) $\mathbf{a}_1 = 3\mathbf{j} - \mathbf{k}$, $\mathbf{a}_2 = \mathbf{i} + \mathbf{j}$, $\mathbf{a}_3 = \mathbf{i} + 2\mathbf{k}$.

    (b) $\mathbf{a}_1 = \mathbf{i} + \mathbf{j}$, $\mathbf{a}_2 = 3\mathbf{j} - \mathbf{k}$, $\mathbf{a}_3 = \mathbf{i} + 2\mathbf{k}$.

8. (a) $\mathbf{a}_1 = \mathbf{j} - \mathbf{k}$, $\mathbf{a}_2 = \mathbf{i} + \mathbf{k}$, $\mathbf{a}_3 = -\mathbf{i} - \mathbf{j} + \mathbf{k}$.

    (b) $\mathbf{a}_1 = -\mathbf{i} - \mathbf{j} + \mathbf{k}$, $\mathbf{a}_2 = \mathbf{i} + \mathbf{k}$, $\mathbf{a}_3 = \mathbf{j} - \mathbf{k}$.

CHAPTER 3

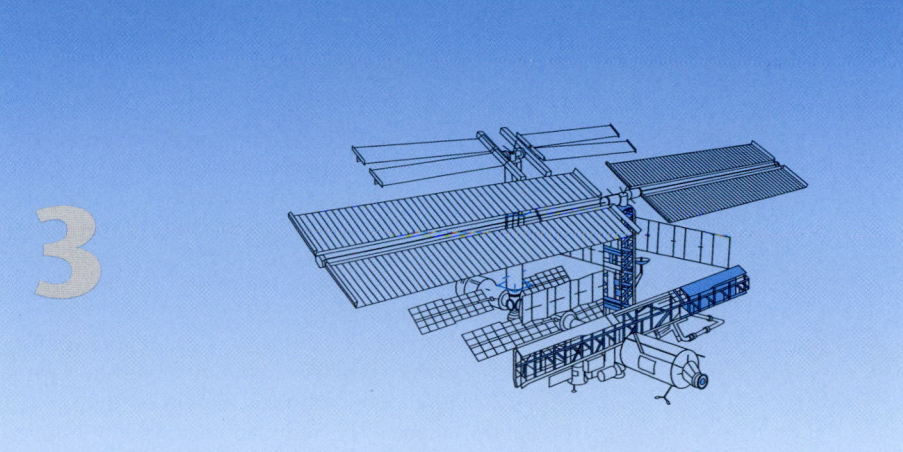

Matrices and Systems of Linear Equations

Many types of problems that arise in engineering and physics give rise to linear algebraic simultaneous equations. A typical engineering example involves the determination of the forces acting in the struts of a pin-jointed structure like a truss that forms the side of a bridge supporting a load. The determination of the forces in a strut is important in order to know when it is in compression or tension, and to ensure that no truss exceeds its safe load. The analysis of the forces in structures of this type gives rise to a set of linear simultaneous equations that relate the forces in the struts and the external load.

It is necessary to know when systems of linear equations are consistent so a solution exists, when they are inconsistent so there is no solution, and whether when a solution exists it is unique or nonunique in the sense that it involves a number of arbitrary parameters. In practical problems all of these mathematical possibilities have physical meaning, and in the case of a truss, the inability to determine the forces acting in a particular strut indicates that it is redundant and so can be removed without compromising the integrity of the structure.

A more complicated though very similar situation occurs when linearly vibrating systems are coupled together, as may happen when an active vibration damper is attached to a spring-mounted motor. However, in this case it is a system of simultaneous linear ordinary differential equations determining the amplitudes of the vibrations of the motor and vibration damper that are coupled together. The analysis of this problem, which will be considered later, also gives rise to a linear system of simultaneous algebraic equations.

Linear ordinary differential equations are also coupled together when working with linear control systems involving feedback. When such systems are solved by means of the Laplace transform to be described later, linear algebraic systems again arise and the nature of the zeros of the determinant of a certain quantity then determines the stability of the control system.

Linear systems of simultaneous algebraic equations also play an essential role in computer graphics, where at the simplest level they are used to transform images by translating, rotating, and stretching them by differing amounts in different directions.

Although each equation in a system of linear algebraic equations can be considered separately, such can be discovered about the properties of the physical problem that gave rise to the equations if the system of equations can be studied as a whole. This can be accomplished by using the algebra of matrices that provides a way of analyzing systems as a single entity, and it is the purpose of this chapter to introduce and develop this aspect of what is called linear algebra.

After defining the notion of a matrix, this chapter develops the fundamental matrix operations of equality, addition, scaling, transposition, and multiplication. Various applications of matrices are given, and the brief review of determinants given in Chapter 1 is developed in greater detail, prior to its use when considering the solution of systems of linear algebraic equations.

The concept of elementary row operations is introduced and used to reduce systems of linear algebraic equations to a form that shows whether or not a unique solution exists. When a solution does exist, which is either unique or determined in terms of some of the remaining variables, this reduction enables the solution to be found immediately.

The inverse of an $n \times n$ matrix is defined and shown only to exist when the determinant of the matrix is nonvanishing, and, finally, the derivative of a matrix whose elements are functions of a variable is introduced and some of its most important properties are derived.

3.1   Matrices

Matrices arise naturally in many different ways, one of the most common being in the study of systems of linear equations such as

$$
\begin{aligned}
a_{11}x_1 + a_{12}x_2 + \cdots + a_{1n}x_n &= b_1 \\
a_{21}x_1 + a_{22}x_2 + \cdots + a_{2n}x_n &= b_2 \\
&\cdots \cdots \cdots \\
a_{m1}x_1 + a_{m2}x_2 + \cdots + a_{mn}x_n &= b_m.
\end{aligned}
\tag{1}
$$

In system (1) the numbers a_{ij} are the **coefficients** of the equations, the numbers b_i are the **nonhomogeneous terms**, and the number of equations m may equal, exceed, or be less than n, the number of unknowns $x_1, x_2, \ldots, x_n$.

System (1) is said to be **homogeneous** when $b_1 = b_2 = \cdots = b_m = 0$, and to be **nonhomogeneous** when at least one of the b_i is nonvanishing. The algebraic properties of the system are determined by the array of coefficients a_{ij}, the nonhomogeneous terms b_i and the numbers m and n. From now on, the array of coefficients and the nonhomogeneous terms on the right will be denoted by the single symbols **A** and **b**, respectively, where

$$
\mathbf{A} = \begin{bmatrix} a_{11} & a_{12} & \cdots & a_{1n} \\ a_{21} & a_{22} & \cdots & a_{2n} \\ & \cdots \cdots \cdots & \\ a_{m1} & a_{m2} & \cdots & a_{mn} \end{bmatrix} \quad \text{and} \quad \mathbf{b} = \begin{bmatrix} b_1 \\ b_2 \\ \cdot \\ b_m \end{bmatrix}.
\tag{2}
$$

The array of mn coefficients a_{ij} in m rows and n columns that form **A** is an example of an $m \times n$ **matrix**, where $m \times n$ is read "m by n." The array **b** is an example of an $m \times 1$ **matrix**, and it is called an m element **column vector**. We will use the convention that an array such as **A**, with two or more rows and two or more columns, will be denoted by a boldface capital letter. An array with a single row, or a column such as **b**, will be denoted by a boldface lowercase letter.

Each entry in a matrix is called an **element** of the matrix, and entries may be numbers, functions, or even matrices themselves. The suffixes associated with an element show its position in the matrix, because the first suffix is the **row number** and the second is the **column number**. Because of this convention, the element a_{35} in a matrix belongs to the third row and the fifth column of the matrix. So, for example, if $\mathbf{A}$ is a 3×2 matrix and its general element $a_{ij} = i + 3j$, then as i may only take the values 1, 2, and 3, and j the values 1 and 2, it follows that

$$\mathbf{A} = \begin{bmatrix} 4 & 7 \\ 5 & 8 \\ 6 & 9 \end{bmatrix}.$$

In a *column* vector $\mathbf{c}$ with elements $c_{11}, c_{21}, c_{31}, \ldots, c_{m1}$, as only a single column is involved, it is usual to vary the suffix convention by omitting the second suffix and instead numbering the elements sequentially as $c_1, c_2, c_3, \ldots, c_m$, so that

$$\mathbf{c} = \begin{bmatrix} c_1 \\ c_2 \\ \vdots \\ \vdots \\ c_m \end{bmatrix}.$$

Later it will be necessary to introduce **row vectors**, and in an s element row vector $\mathbf{r}$ with elements $r_{11}, r_{12}, r_{13}, \ldots, r_{1s}$, the notation is again simplified, this time by omitting the first suffix and numbering the elements sequentially as $r_1, r_2, \ldots, r_s$, so

$$\mathbf{r} = [r_1, \ r_2, \ldots, r_s]. \tag{3}$$

In general, row and column vectors will be denoted by boldface lowercase letters such as $\mathbf{a}$, $\mathbf{b}$, $\mathbf{c}$, and $\mathbf{x}$, and matrices such as the coefficient matrix in (2) will be denoted by boldface capital letters such as $\mathbf{A}$, $\mathbf{B}$, $\mathbf{P}$, and $\mathbf{Q}$.

A different convention that is also used to denote a matrix involves enclosing the array between curved brackets instead of the square ones used here. Thus,

$$\begin{pmatrix} 1 & 5 & 9 \\ -3 & 2 & 4 \end{pmatrix} \quad \text{and} \quad \begin{bmatrix} 1 & 5 & 9 \\ -3 & 2 & 4 \end{bmatrix} \tag{4}$$

denote the same 2×3 matrix. A matrix should never be enclosed between two vertical rules in order to avoid confusion with the determinant notation because

$$\begin{bmatrix} 3 & -4 \\ 5 & 2 \end{bmatrix} \ \text{is a matrix, but} \ \begin{vmatrix} 3 & -4 \\ 5 & 2 \end{vmatrix} = 26 \ \text{is a determinant.}$$

Definition of a matrix

An $m \times n$ **matrix** is an array of mn entries, called **elements**, arranged in m rows and n columns. If a matrix is denoted by $\mathbf{A}$, then the element in its ith row and jth column is denoted by a_{ij} and

$$\mathbf{A} = [a_{ij}] = \begin{bmatrix} a_{11} & a_{12} & \cdots & a_{1n} \\ a_{21} & a_{22} & \cdots & a_{2n} \\ \cdots & \cdots & \cdots & \cdots \\ a_{m1} & a_{m2} & \cdots & a_{mn} \end{bmatrix}.$$

some typical matrices

The following are typical examples of matrices:

A 1×1 matrix: $[3]$; a single element may be regarded as a matrix.

A 3×4 matrix: $\begin{bmatrix} 1 & 3 & 5 & 0 \\ 2 & -1 & 4 & 3 \\ 7 & 2 & 1 & 6 \end{bmatrix}$; a matrix with real numbers as elements.

A 2×2 matrix: $\begin{bmatrix} 1+i & 1-i \\ 3+4i & 2-3i \end{bmatrix}$; a matrix with complex numbers as elements.

A 2×2 matrix: $\begin{bmatrix} \cos\theta & \sin\theta \\ -\sin\theta & \cos\theta \end{bmatrix}$; a matrix with functions as elements.

A 1×3 matrix: $[2, \ -5, \ 7]$; a three-element row vector.

A 2×1 matrix: $\begin{bmatrix} 11 \\ 9 \end{bmatrix}$; a two-element column vector.

A **square matrix** is a matrix in which the number of rows m equals the number of columns n. A typical square matrix is the 3×3 matrix

$$\begin{bmatrix} 2 & 0 & 5 \\ 1 & -3 & 4 \\ 3 & 1 & 7 \end{bmatrix}.$$

Definition of the equality of matrices

Let $\mathbf{A} = [a_{ij}]$ be an $m \times n$ matrix and $\mathbf{B} = [b_{ij}]$ be a $p \times q$ matrix. Then matrices $\mathbf{A}$ and $\mathbf{B}$ will be **equal**, written $\mathbf{A} = \mathbf{B}$, if, and only if:

(a) $\mathbf{A}$ and $\mathbf{B}$ have the same number of rows, and the same number of columns, so that $m = p$ and $n = q$, and

(b) $a_{ij} = b_{ij}$, for each i and j.

Equality of matrices means that if $\mathbf{A}$ and $\mathbf{B}$ are equal, then each is an identical copy of the other.

EXAMPLE 3.1

If $\mathbf{A} = \begin{bmatrix} 2 & 3 & a \\ b & 6 & 1 \end{bmatrix}$, $\mathbf{B} = \begin{bmatrix} 2 & 3 & 9 \\ -3 & 6 & 1 \end{bmatrix}$, and $\mathbf{C} = \begin{bmatrix} 2 & 3 & 9 \\ -3 & 6 & 1 \\ 0 & 0 & 0 \end{bmatrix}$,

then $\mathbf{A} = \mathbf{B}$ if and only if $a = 9$ and $b = -3$, but $\mathbf{A} \neq \mathbf{C}$ and $\mathbf{B} \neq \mathbf{C}$. ∎

Definition of matrix addition

The addition of matrices $\mathbf{A}$ and $\mathbf{B}$ is only defined if the matrices each have the same number of rows and the same number of columns. Let $\mathbf{A} = [a_{ij}]$ and $\mathbf{B} = [b_{ij}]$ be $m \times n$ matrices. Then the the $m \times n$ matrix formed by adding $\mathbf{A}$ and $\mathbf{B}$, called the **sum** of $\mathbf{A}$ and $\mathbf{B}$ and written $\mathbf{A} + \mathbf{B}$, is the matrix whose element in the ith row and jth column is $a_{ij} + b_{ij}$, for each i and j, so that

$$\mathbf{A} + \mathbf{B} = [a_{ij} + b_{ij}].$$

Matrices that can be added are said to be **conformable** for addition.

It is an immediate consequence of this definition that $\mathbf{A} + \mathbf{B} = \mathbf{B} + \mathbf{A}$, so matrix addition is **commutative**.

Definition of the transpose of a matrix

Let $\mathbf{A} = [a_{ij}]$ be an $m \times n$ matrix. Then the **transpose** of $\mathbf{A}$, denoted by $\mathbf{A}^T$ (and sometimes by $\mathbf{A}'$), is the matrix obtained from $\mathbf{A}$ by interchanging rows and columns to produce the $n \times m$ matrix

$$\mathbf{A}^T = [a_{ij}]^T = [a_{ji}].$$

The definition of the transpose of a matrix means that the first *row* of $\mathbf{A}$ becomes the first *column* of $\mathbf{A}^T$, the second *row* of $\mathbf{A}$ becomes the second *column* of $\mathbf{A}^T$, ..., and, finally, the mth *row* of $\mathbf{A}$ becomes the mth *column* of $\mathbf{A}^T$. In particular, if $\mathbf{A}$ is a row vector, then its transpose is a column vector, and conversely.

EXAMPLE 3.2

If $\mathbf{A} = \begin{bmatrix} 2 & 6 & 3 \\ 1 & 0 & 4 \end{bmatrix}$ then $\mathbf{A}^T = \begin{bmatrix} 2 & 1 \\ 6 & 0 \\ 3 & 4 \end{bmatrix}$, and if $\mathbf{A} = [7, 3, 2]$ then $\mathbf{A}^T = \begin{bmatrix} 7 \\ 3 \\ 2 \end{bmatrix}$. ∎

Definition of scaling a matrix by a number

Let $\mathbf{A} = [a_{ij}]$ be an $m \times n$ matrix and λ be a scalar (real or complex). Then if $\mathbf{A}$ is **scaled** by λ, written $\lambda\mathbf{A}$, every element of $\mathbf{A}$ is multiplied by λ to yield the $m \times n$ matrix

$$\lambda\mathbf{A} = [\lambda a_{ij}].$$

EXAMPLE 3.3

If $\lambda = 2$ and $\mathbf{A} = \begin{bmatrix} 2 & -6 & 7 \\ 1 & 4 & 15 \end{bmatrix}$, then $\lambda\mathbf{A} = 2\mathbf{A} = \begin{bmatrix} 4 & -12 & 14 \\ 2 & 8 & 30 \end{bmatrix}$,

and if $\lambda = -1$, then

$$\lambda\mathbf{A} = (-1)\mathbf{A} = -\mathbf{A} = \begin{bmatrix} -2 & 6 & -7 \\ -1 & -4 & -15 \end{bmatrix}.$$ ∎

Taken together, the definitions of the addition and scaling of matrices show that if the matrices $\mathbf{A}$ and $\mathbf{B}$ are conformable for addition, then the subtraction of matrix $\mathbf{B}$ from $\mathbf{A}$, called their **difference** and written $\mathbf{A} - \mathbf{B}$, is to be interpreted as

difference (subtraction) of matrices

$$\mathbf{A} - \mathbf{B} = \mathbf{A} + (-1)\mathbf{B}.$$

EXAMPLE 3.4

If $\mathbf{A} = \begin{bmatrix} 2 & 5 & 8 \\ 1 & -4 & 5 \end{bmatrix}$ and $\mathbf{B} = \begin{bmatrix} 2 & 4 & 5 \\ 2 & -4 & 1 \end{bmatrix}$, then $\mathbf{A} - \mathbf{B} = \begin{bmatrix} 0 & 1 & 3 \\ -1 & 0 & 4 \end{bmatrix}$. ∎

negative of a matrix

The **null** or **zero** matrix **0** is defined as any matrix in which every element is zero. The introduction of the null matrix makes it appropriate to call $-\mathbf{A}$ the **negative** of **A**, because

$$\mathbf{A} - \mathbf{A} = \mathbf{A} + (-1)\mathbf{A} = \mathbf{0}.$$

When working with the null matrix the number of its rows and columns is never stated, because these are always taken to be whatever is appropriate for the equation that is involved.

Definition of the product of a row and a column vector

Let $\mathbf{a} = [a_1, a_2, \ldots, a_r]$ be an r-element row vector, and $\mathbf{b} = [b_1, b_2, \ldots, b_r]^{\mathrm{T}}$ be an r-element column vector. Then the product $\mathbf{ab}$, in this order, is the number defined as

$$\mathbf{ab} = a_1 b_1 + a_2 b_2 \cdots + a_r b_r.$$

Notice that this product is *only* defined when the number of elements in the row vector **A** equals the number of elements in the column vector **B**.

EXAMPLE 3.5

Find the product $\mathbf{ab}$ given that $\mathbf{a} = [1,\ 4,\ -3,\ 10]$ and $\mathbf{b} = [2,\ 1,\ 4,\ -2]^{\mathrm{T}}$.

Solution

$$\mathbf{ab} = [1,\ 4,\ -3,\ 10] \begin{bmatrix} 2 \\ 1 \\ 4 \\ -2 \end{bmatrix}$$
$$= (1)\cdot(2) + (4)\cdot(1) + (-3)\cdot(4) + (10)\cdot(-2)$$
$$= -26.$$

Definition of the product of matrices

Let $\mathbf{A} = [a_{ij}]$ be an $m \times n$ matrix in which the rth row is the row vector $\mathbf{a}_r$, and let $\mathbf{B} = [b_{ij}]$ be a $p \times q$ matrix in which the sth column is the column vector $\mathbf{b}_s$. The matrix product $\mathbf{AB}$, in this order, is only defined if the number of columns in **A** equals the number of rows in **B**, so that $n = p$. The product is then an $m \times q$ matrix with the element in its rth row and sth column defined as $\mathbf{a}_r \mathbf{b}_s$. Thus, if $c_{rs} = \mathbf{a}_r \mathbf{b}_s$, as $c_{rs} = a_{r1}b_{1s} + a_{r2}b_{2s} + \cdots + a_{rn}b_{ns}$,

$$\mathbf{AB} = [c_{rs}] = [a_{r1}b_{1s} + a_{r2}b_{2s} + \cdots + a_{rn}b_{ns}],$$

for $1 \leq r \leq m$ and $1 \leq s \leq q$, or, equivalently,

$$\mathbf{AB} = \begin{bmatrix} \mathbf{a}_1\mathbf{b}_1 & \mathbf{a}_1\mathbf{b}_2 & \mathbf{a}_1\mathbf{b}_3 & \ldots & \mathbf{a}_1\mathbf{b}_q \\ \mathbf{a}_2\mathbf{b}_1 & \mathbf{a}_2\mathbf{b}_2 & \mathbf{a}_2\mathbf{b}_3 & \ldots & \mathbf{a}_2\mathbf{b}_q \\ \cdots\cdots\cdots\cdots\cdots\cdots\cdots \\ \mathbf{a}_m\mathbf{b}_1 & \mathbf{a}_m\mathbf{b}_2 & \mathbf{a}_m\mathbf{b}_3 & \ldots & \mathbf{a}_m\mathbf{b}_q \end{bmatrix}.$$

When a matrix product **AB** is defined, the matrices are said to be **conformable** for matrix multiplication in the given order.

in general, matrix multiplication is noncommutative

It is important to notice that when the product **AB** is defined, the product **BA** may or may not be defined, and even when **BA** is defined, in general $\mathbf{AB} \neq \mathbf{BA}$. This situation is recognized by saying that, in general, matrix multiplication is **noncommutative**.

Provided matrices **A** and **B** are conformable for multiplication, the above rule for finding their product **AB**, in this order, is best remembered by saying that the element in the ith row and jth column of **AB** is the product of the ith row of **A** and the jth column of **B**.

EXAMPLE 3.6

Form the matrix products **AB** and **BA** given that

$$\mathbf{A} = \begin{bmatrix} 1 & 4 & -3 \\ 2 & 5 & 4 \end{bmatrix} \quad \text{and} \quad \mathbf{B} = \begin{bmatrix} 4 & 1 \\ 2 & 6 \\ 0 & 3 \end{bmatrix}.$$

Solution   Let us calculate the matrix product **AB**. The first and second row vectors of **A** are $\mathbf{a}_1 = [1, 4, -3]$ and $\mathbf{a}_2 = [2, 5, 4]$, and the first and second column vectors of **B** are $\mathbf{b}_1 = [4, 2, 0]^T$ and $\mathbf{b}_2 = [1, 6, 3]^T$. As **A** is a 2×3 matrix and **B** is a 3×2 matrix, the product **AB** is conformable for multiplication and yields a 2×2 matrix

$$\mathbf{AB} = \begin{bmatrix} \mathbf{a}_1\mathbf{b}_1 & \mathbf{a}_1\mathbf{b}_2 \\ \mathbf{a}_2\mathbf{b}_1 & \mathbf{a}_2\mathbf{b}_2 \end{bmatrix} = \begin{bmatrix} (1\cdot 4 + 4\cdot 2 + (-3)\cdot 0) & (1\cdot 1 + 4\cdot 6 + (-3)\cdot 3) \\ (2\cdot 4 + 5\cdot 2 + 4\cdot 0) & (2\cdot 1 + 5\cdot 6 + 4\cdot 3) \end{bmatrix}$$

$$= \begin{bmatrix} 12 & 16 \\ 18 & 44 \end{bmatrix}.$$

The product **BA** is also conformable for multiplication and yields a 3×3 matrix, where now we must use the *row* vectors of **B** that with an obvious change of notation are $\mathbf{b}_1 = [4, 1], \mathbf{b}_2 = [2, 6], \mathbf{b}_3 = [0, 3]$, and the *column* vectors of **A** that are $\mathbf{a}_1 = [1, 2]^T, \mathbf{a}_2 = [4, 5]^T$, and $\mathbf{a}_3 = [-3, 4]^T$, so that

$$\mathbf{BA} = \begin{bmatrix} \mathbf{b}_1\mathbf{a}_1 & \mathbf{b}_1\mathbf{a}_2 & \mathbf{b}_1\mathbf{a}_3 \\ \mathbf{b}_2\mathbf{a}_1 & \mathbf{b}_2\mathbf{a}_2 & \mathbf{b}_2\mathbf{a}_3 \\ \mathbf{b}_3\mathbf{a}_1 & \mathbf{b}_3\mathbf{a}_2 & \mathbf{b}_3\mathbf{a}_3 \end{bmatrix} = \begin{bmatrix} (4\cdot 1 + 1\cdot 2) & (4\cdot 4 + 1\cdot 5) & (4\cdot (-3) + 1\cdot 4) \\ (2\cdot 1 + 6\cdot 2) & (2\cdot 4 + 6\cdot 5) & (2\cdot (-3) + 6\cdot 4) \\ (0\cdot 1 + 3\cdot 2) & (0\cdot 4 + 3\cdot 5) & (0\cdot (-3) + 3\cdot 4) \end{bmatrix}$$

$$= \begin{bmatrix} 6 & 21 & -8 \\ 14 & 38 & 18 \\ 6 & 15 & 12 \end{bmatrix}.$$

This is an example of two matrices **A** and **B** that can be combined to form the products **AB** and **BA**, but $\mathbf{AB} \neq \mathbf{BA}$. ∎

EXAMPLE 3.7

Write the system of simultaneous equations (1) in matrix form.

Solution   Using the matrices **A** and **b** in (2) and setting $\mathbf{x} = [x_1, x_2, \ldots, x_n]^T$ allows the system of equations (1) to be written

$$\mathbf{Ax} = \mathbf{b}.$$

Here, as is usual, to save space the transpose operation has been used to display the elements of column vector **x** in the more convenient form $\mathbf{x} = [x_1, x_2, \ldots, x_n]^T$. ∎

The definitions of matrix multiplication and addition lead almost immediately to the results of the following theorem, so the proof is left as an exercise.

THEOREM 3.1

some important
properties of matrices

Associative and distributive properties of matrices   Let $\mathbf{A}$, $\mathbf{B}$, and $\mathbf{C}$ be matrices that are conformable for the operations that follow, and let λ be a scalar. Then:

(i)   If $\mathbf{AB}$ and $\mathbf{BA}$ are both defined, in general $\mathbf{AB} \neq \mathbf{BA}$;

(ii) $\mathbf{A}(\mathbf{BC}) = (\mathbf{AB})\mathbf{C} = \mathbf{ABC}$;

(iii) $(\lambda\mathbf{A})\mathbf{B} = \mathbf{A}(\lambda\mathbf{B}) = \lambda\mathbf{AB}$;

(iv) $\mathbf{A}(\mathbf{B} + \mathbf{C}) = \mathbf{AB} + \mathbf{AC}$;

(v) $(\mathbf{A} + \mathbf{B})\mathbf{C} = \mathbf{AC} + \mathbf{BC}$.   ∎

THEOREM 3.2

Transposition of a product   If matrices $\mathbf{A}$ and $\mathbf{B}$ are conformable to form the product $\mathbf{AB}$, then

$$(\mathbf{AB})^{\mathrm{T}} = \mathbf{B}^{\mathrm{T}}\mathbf{A}^{\mathrm{T}}.$$

Proof   The products $(\mathbf{AB})^{\mathrm{T}}$ and $\mathbf{B}^{\mathrm{T}}\mathbf{A}^{\mathrm{T}}$ are both defined, and each is an $m \times q$ matrix. Introduce the notation $[\mathbf{M}]_{ij}$ to denote the element of $\mathbf{M}$ in row i and column j. Then from the transpose operation and the rule for matrix multiplication, for all permissible i, j,

$$[\mathbf{AB}]^{\mathrm{T}}_{i,j} = [\mathbf{AB}]_{j,i} = (\text{product of } j\text{th row of } \mathbf{A} \text{ with } i\text{th column of } \mathbf{B}) = \sum_{k=1}^{n} a_{jk}b_{ki}.$$

Similarly,

$$[\mathbf{B}^{\mathrm{T}}\mathbf{A}^{\mathrm{T}}]_{i,j} = (\text{product of } i\text{th row of } \mathbf{B}^{\mathrm{T}} \text{ with } j\text{th column of } \mathbf{A}^{\mathrm{T}})$$

$$= (\text{product of } i\text{th column of } \mathbf{B} \text{ with } j\text{th row of } \mathbf{A}) = \sum_{k=1}^{n} a_{jk}b_{ki}.$$

So $[\mathbf{AB}]^{\mathrm{T}}_{ij} = [\mathbf{B}^{\mathrm{T}}\mathbf{A}^{\mathrm{T}}]_{ij}$ for all permissible i, j, showing that $(\mathbf{AB})^{\mathrm{T}} = \mathbf{B}^{\mathrm{T}}\mathbf{A}^{\mathrm{T}}$.   ∎

raising a matrix to a
power

It is an immediate consequence of Theorem 3.1(ii) that if $\mathbf{A}$ is a square matrix and m and n are positive integers,

$$\underbrace{\mathbf{A} \cdot \mathbf{A} \cdot \mathbf{A} \cdot \ldots \cdot \mathbf{A}}_{n \text{ times}} = \mathbf{A}^{n} \quad \text{and} \quad \mathbf{A}^{m} \cdot \mathbf{A}^{n} = \mathbf{A}^{m+n}.$$

A useful result from the definition of addition is

$$(\mathbf{A} + \mathbf{B})^{\mathrm{T}} = \mathbf{A}^{\mathrm{T}} + \mathbf{B}^{\mathrm{T}},$$

while from Theorem 3.2

$$(\mathbf{ABC})^{\mathrm{T}} = \mathbf{C}^{\mathrm{T}}\mathbf{B}^{\mathrm{T}}\mathbf{A}^{\mathrm{T}}.$$

As the order in which a sequence of permissible matrix multiplications is performed influences the product, it is necessary to introduce a form of words that makes the order unambiguous. This is accomplished by saying that if matrix $\mathbf{A}$ multiplies matrix $\mathbf{B}$ from the *left*, as in $\mathbf{AB}$, then $\mathbf{B}$ is **premultiplied** by $\mathbf{A}$, while if $\mathbf{A}$ multiplies $\mathbf{B}$ from the *right*, as in $\mathbf{BA}$, then $\mathbf{B}$ is **postmultiplied** by $\mathbf{A}$. Equivalently, in the product $\mathbf{AB}$, we can say that $\mathbf{A}$ is *postmultiplied* by $\mathbf{B}$, or that $\mathbf{B}$ is *premultiplied* by $\mathbf{A}$.

pre- and post-multiplication of matrices

Important Differences Between Ordinary Algebraic Equations and Matrix Equations

(i) The algebraic equation $ab = 0$, in which a and b are numbers, not both of which are zero, implies that either $a = 0$ or $b = 0$. However, if the matrix product $\mathbf{AB}$ is defined and is such that $\mathbf{AB} = \mathbf{0}$, then it does *not* necessarily follow that either $\mathbf{A} = \mathbf{0}$ or $\mathbf{B} = \mathbf{0}$.

(ii) The algebraic equation $ab = ac$ in which a, b, and c are numbers, with $a \neq 0$, allows cancellation of the factor a leading to the conclusion that $b = c$. However, if the matrix products $\mathbf{AB}$ and $\mathbf{AC}$ are defined and are such that $\mathbf{AB} = \mathbf{AC}$, this does *not* necessarily imply that $\mathbf{B} = \mathbf{C}$, so that cancellation of matrix factors is *not* permissible.

The validity of these two statements can be seen by considering the following simple examples.

EXAMPLE 3.8

Consider matrices $\mathbf{A}$ and $\mathbf{B}$ given by

$$\mathbf{A} = \begin{bmatrix} 1 & 4 \\ 3 & 12 \end{bmatrix} \quad \text{and} \quad \mathbf{B} = \begin{bmatrix} 4 & -8 \\ -1 & 2 \end{bmatrix}.$$

Then $\mathbf{AB} = \mathbf{0}$, but neither $\mathbf{A}$ nor $\mathbf{B}$ is a null matrix. ∎

EXAMPLE 3.9

Consider the matrices $\mathbf{A}$, $\mathbf{B}$, and $\mathbf{C}$ given by

$$\mathbf{A} = \begin{bmatrix} 1 & -1 \\ 2 & -2 \end{bmatrix}, \quad \mathbf{B} = \begin{bmatrix} 2 & 4 & 6 \\ 2 & 3 & 4 \end{bmatrix}, \quad \text{and} \quad \mathbf{C} = \begin{bmatrix} 3 & 6 & 8 \\ 3 & 5 & 6 \end{bmatrix}.$$

Then

$$\mathbf{AB} = \mathbf{AC} = \begin{bmatrix} 0 & 1 & 2 \\ 0 & 2 & 4 \end{bmatrix},$$

but $\mathbf{B} \neq \mathbf{C}$. ∎

In a square $n \times n$ matrix $\mathbf{A} = [a_{ij}]$, the elements on a line extending from top left to bottom right is called the **leading diagonal** of $\mathbf{A}$, and it contains the n elements $a_{11}, a_{22}, \ldots, a_{nn}$.

So the leading diagonal of the 2×2 matrix $\mathbf{A}$ in Example 3.8 contains the elements 1 and 12, and the leading diagonal of the 2×2 matrix $\mathbf{B}$ contains the elements 4 and 2. Symbolically, the leading diagonal of the $n \times n$ matrix $\mathbf{A} = [a_{ij}]$ shown below comprises the n elements in the shaded diagonal strip, though these

leading diagonal and trace of a matrix

n elements do *not* form an n element vector.

$$\mathbf{A} = \begin{bmatrix} a_{11} & a_{12} & a_{13} & \cdot & \cdot & \cdot & a_{1n} \\ a_{21} & a_{22} & a_{23} & \cdot & \cdot & \cdot & a_{2n} \\ a_{31} & a_{32} & a_{33} & & \cdot & \cdot & a_{3n} \\ \cdot & \cdot & \cdot & & \cdot & & \cdot \\ \cdot & \cdot & \cdot & & & \cdot & \cdot \\ a_{n1} & a_{n2} & a_{n3} & \cdot & \cdot & \cdot & a_{nn} \end{bmatrix}.$$

The **trace** of a square matrix $\mathbf{A}$, written $\operatorname{tr}(\mathbf{A})$, is the sum of the terms on its leading diagonal, so for the foregoing matrix $\operatorname{tr}(\mathbf{A}) = a_{11} + a_{22} + \cdots + a_{nn}$.

Square matrices in which all elements away from the leading diagonal are zero, but not every element on the leading diagonal is zero, are called **diagonal matrices**. Of the class of diagonal matrices, the most important are the **unit** matrices, also called **identity** matrices, in which every element on the leading diagonal is the number 1. These $n \times n$ matrices are usually all denoted by the symbol $\mathbf{I}$, with the value of n being understood to be appropriate to the context in which they arise. If, however, the value of n needs to be indicated, the symbol $\mathbf{I}$ can be replaced by $\mathbf{I}_n$. It is easily seen from the definition of matrix multiplication that for any $m \times n$ matrix $\mathbf{A}$ it follows that

$$\mathbf{I}_m \mathbf{A} = \mathbf{A} \mathbf{I}_n \text{ or, more simply, } \mathbf{IA} = \mathbf{AI} = \mathbf{A},$$

and that when $\mathbf{A}$ is an $n \times n$ matrix,

$$\mathbf{IA} = \mathbf{AI} = \mathbf{A}.$$

identity or unit matrix

When working with matrices, the unit matrix $\mathbf{I}$ plays the part of the unit real number, and it is because of this that $\mathbf{I}$ is called either the *unit* or the *identity* matrix.

An example of a 4×4 diagonal matrix is

$$\mathbf{D} = \begin{bmatrix} 3 & 0 & 0 & 0 \\ 0 & 2 & 0 & 0 \\ 0 & 0 & 0 & 0 \\ 0 & 0 & 0 & 1 \end{bmatrix}, \quad \text{with the trace given by } \operatorname{tr}(\mathbf{D}) = 3 + 2 + 0 + 1 = 6.$$

The 3×3 unit matrix is the diagonal matrix

$$\mathbf{I} = \mathbf{I}_3 = \begin{bmatrix} 1 & 0 & 0 \\ 0 & 1 & 0 \\ 0 & 0 & 1 \end{bmatrix}, \quad \text{and its trace is } \operatorname{tr}(\mathbf{I}) = 1 + 1 + 1 = 3.$$

Various special square $n \times n$ matrices occur sufficiently frequently for them to be given names, and some of the most important of these are the following:

Upper triangular matrices are matrices in which all elements below the leading diagonal are zero. A typical example of a 4×4 upper triangular matrix is

some special matrices

$$\mathbf{U} = \begin{bmatrix} 1 & 3 & -1 & 0 \\ 0 & 2 & -6 & 1 \\ 0 & 0 & -3 & 2 \\ 0 & 0 & 0 & 4 \end{bmatrix}.$$

Lower triangular matrices are matrices in which all elements above the leading diagonal are zero. A typical example of a 4×4 lower triangular matrix is

$$\mathbf{L} = \begin{bmatrix} 2 & 0 & 0 & 0 \\ 1 & 0 & 0 & 0 \\ 3 & -2 & 5 & 0 \\ -2 & 4 & 7 & 3 \end{bmatrix}.$$

Symmetric matrices $\mathbf{A} = [a_{ij}]$ are matrices in which $a_{ij} = a_{ji}$ for all i and j. If $\mathbf{A}$ is symmetric, then $\mathbf{A} = \mathbf{A}^{\mathrm{T}}$. A typical example of a symmetric matrix is

$$\mathbf{M} = \begin{bmatrix} 1 & 5 & -3 \\ 5 & 4 & 2 \\ -3 & 2 & 7 \end{bmatrix}.$$

Skew-symmetric matrices $\mathbf{A} = [a_{ji}]$ are matrices in which $a_{ij} = -a_{ji}$ for all i and j. From the definition of an $n \times n$ skew-symmetric matrix we have $a_{ii} = -a_{ii}$ for $i = 1, 2, \ldots, n$, so the elements on the leading diagonal must all be zero. An equivalent definition of a skew-symmetric matrix $\mathbf{A}$ is that $\mathbf{A}^{\mathrm{T}} = -\mathbf{A}$. A typical example of a skew-symmetric matrix is

$$\mathbf{S} = \begin{bmatrix} 0 & 3 & -5 & 6 \\ -3 & 0 & 2 & -4 \\ 5 & -2 & 0 & -1 \\ -6 & 4 & 1 & 0 \end{bmatrix}.$$

An **orthogonal** matrix $\mathbf{Q}$ is a matrix such that $\mathbf{Q}\mathbf{Q}^{\mathrm{T}} = \mathbf{Q}^{\mathrm{T}}\mathbf{Q} = \mathbf{I}$. A typical orthogonal matrix is

$$\mathbf{Q} = \begin{bmatrix} \dfrac{1}{\sqrt{2}} & -\dfrac{1}{\sqrt{2}} \\ \dfrac{1}{\sqrt{2}} & \dfrac{1}{\sqrt{2}} \end{bmatrix}.$$

More special than the preceding real valued matrices are matrices $\mathbf{A} = [a_{ij}]$ in which the elements a_{ij} are complex numbers. We will write $\overline{\mathbf{A}}$ to denote the matrix obtained from $\mathbf{A}$ by replacing each of its elements a_{ij} by its complex conjugate $\overline{a}_{ij}$, so that

$$\overline{\mathbf{A}} = [\overline{a}_{ij}].$$

Then matrix $\mathbf{A}$ is said to be **Hermitian** if

$$\overline{\mathbf{A}}^{\mathrm{T}} = \mathbf{A}.$$

A typical Hermitian matrix is

$$\mathbf{A} = \begin{bmatrix} 7 & 1 - 4i \\ 1 + 4i & 3 \end{bmatrix}.$$

The matrix $\mathbf{A}$ is said to be **skew-Hermitian** if

$$\overline{\mathbf{A}}^{\mathrm{T}} = -\mathbf{A}.$$

A typical skew-Hermitian matrix is

$$\mathbf{A} = \begin{bmatrix} 3i & 5 + 2i \\ -5 + 2i & 0 \end{bmatrix}.$$

More will be said later about some of these special square matrices and the ways in which they arise.

Finally, we mention that every $m \times n$ matrix $\mathbf{A}$ can be represented differently as a **block matrix**, in which each element is itself a matrix. This is accomplished by **partitioning** the matrix $\mathbf{A}$ into **submatrices** by considering horizontal and vertical lines to be drawn through $\mathbf{A}$ between some of its rows and columns, and then identifying each group of elements so defined as a **submatrix** of $\mathbf{A}$. Clearly there is more than one way in which a matrix can be partitioned. As an example of matrix partitioning, let us consider the 3×3 matrix

block matrices

$$\mathbf{A} = \begin{bmatrix} 3 & -1 & 2 \\ 1 & 2 & 0 \\ 2 & 1 & 0 \end{bmatrix}.$$

One way in which this matrix can be partitioned is as follows:

$$\mathbf{A} = \left[\begin{array}{cc:c} 3 & -1 & 2 \\ \hdashline 1 & 2 & 0 \\ 2 & 1 & 0 \end{array} \right].$$

This can now be written in block matrix form as

$$\mathbf{A} = \begin{bmatrix} \mathbf{A}_{11} & \mathbf{A}_{12} \\ \mathbf{A}_{21} & \mathbf{A}_{22} \end{bmatrix},$$

where the submatrices are

$$\mathbf{A}_{11} = [3 \ -1], \quad \mathbf{A}_{12} = [2], \quad \mathbf{A}_{21} = \begin{bmatrix} 1 & 2 \\ 2 & 1 \end{bmatrix}, \quad \text{and} \quad \mathbf{A}_{22} = \begin{bmatrix} 0 \\ 0 \end{bmatrix}.$$

The addition and scaling of block matrices follow the same rules as those for ordinary matrices, but care must be exercised when multiplying block matrices. To see how multiplication of block matrices can be performed, let us consider the product of matrix $\mathbf{A}$ above and the 3×4 matrix

$$\mathbf{B} = \left[\begin{array}{c:ccc} 1 & 2 & 2 & 1 \\ 3 & 1 & 1 & 0 \\ \hdashline 2 & 3 & 0 & 2 \end{array} \right],$$

which are conformable for the product $\mathbf{AB}$ that is itself a 3×4 matrix. If $\mathbf{B}$ is partitioned as indicated by the dashed lines, it can be written as

$$\mathbf{B} = \begin{bmatrix} \mathbf{B}_{11} & \mathbf{B}_{12} \\ \mathbf{B}_{21} & \mathbf{B}_{22} \end{bmatrix},$$

where the submatrices are

$$\mathbf{B}_{11} = \begin{bmatrix} 1 \\ 3 \end{bmatrix}, \quad \mathbf{B}_{12} = \begin{bmatrix} 2 & 2 & 1 \\ 1 & 1 & 0 \end{bmatrix}, \quad \mathbf{B}_{21} = [2], \quad \text{and} \quad \mathbf{B}_{22} = [3, \ 0, \ 2].$$

Consideration of the definition of the product of matrices shows that we may now write the matrix product $\mathbf{AB}$ in the condensed form

$$\mathbf{AB} = \begin{bmatrix} \mathbf{A}_{11}\mathbf{B}_{11} + \mathbf{A}_{12}\mathbf{B}_{21} & \mathbf{A}_{11}\mathbf{B}_{12} + \mathbf{A}_{12}\mathbf{B}_{22} \\ \mathbf{A}_{21}\mathbf{B}_{11} + \mathbf{A}_{22}\mathbf{B}_{21} & \mathbf{A}_{21}\mathbf{B}_{12} + \mathbf{A}_{22}\mathbf{B}_{22} \end{bmatrix},$$

where the partitioned matrices have been multiplied as though their elements were ordinary elements. This result follows because of correct partitioning, as each product of submatrices is conformable for multiplication and all of the matrix sums are conformable for addition.

In this illustration, routine calculations show that

$$\mathbf{A}_{11}\mathbf{B}_{11} + \mathbf{A}_{12}\mathbf{B}_{21} = [4], \quad \mathbf{A}_{11}\mathbf{B}_{12} + \mathbf{A}_{12}\mathbf{B}_{22} = [11,\ 5,\ 7],$$

$$\mathbf{A}_{21}\mathbf{B}_{11} + \mathbf{A}_{22}\mathbf{B}_{21} = \begin{bmatrix} 7 \\ 5 \end{bmatrix}, \quad \text{and} \quad \mathbf{A}_{21}\mathbf{B}_{12} + \mathbf{A}_{22}\mathbf{B}_{22} = \begin{bmatrix} 4 & 4 & 1 \\ 5 & 5 & 2 \end{bmatrix},$$

so

$$\mathbf{AB} = \begin{bmatrix} [4] & [11,\ 5,\ 7] \\ \begin{bmatrix} 7 \\ 5 \end{bmatrix} & \begin{bmatrix} 4 & 4 & 1 \\ 5 & 5 & 2 \end{bmatrix} \end{bmatrix} = \begin{bmatrix} 4 & 11 & 5 & 7 \\ 7 & 4 & 4 & 1 \\ 5 & 5 & 5 & 2 \end{bmatrix}.$$

This result is easily confirmed by direct matrix multiplication.

The calculation of a matrix product **AB** using partitioned matrices applies in general, provided the partitioning of **A** and **B** is performed in such a way that the products of all the submatrices involved are defined.

Matrix partitioning has various uses, one of which arises in machine computation when a very large fixed matrix **A** needs to be multiplied by a sequence of very large matrices **P**, **Q**, **R**, ... If it happens that **A** can be partitioned in such a way that some of its submatrices are null matrices, the computational time involved can be drastically reduced, because the product of a submatrix and a null matrix is a null matrix, and so need not be computed. The economy follows from the fact that in machine computation multiplications occupy most of the time, so any reduction in their number produces a significant reduction in the time taken to evaluate a matrix product, and the result is even more significant when the same partitioned matrix with null blocks is involved in a sequence of calculations.

Block matrices are also of significance when describing complex oscillation problems governed by a large system of simultaneous ordinary differential equations. Their importance arises from the fact that the matrix of coefficients of the equations often contains many null submatrices, and when this happens the structure of the nonnull blocks provides useful information about the fundamental modes of oscillation that are possible, and also about their interconnections.

For other accounts of elementary matrices see the appropriate chapters in references [2.1], [2.5], and [2.7] to [2.12].

Summary

This section defined $m \times n$ matrices, and the special cases of column and row vectors, and it introduced the fundamental algebraic operations of equality, addition, scaling, transposition, and multiplication of matrices. It was shown that, in general, matrix multiplication is not commutative, so that even when both of the products **AB** and **BA** are defined, it is usually the case that $\mathbf{AB} \neq \mathbf{BA}$.

Pre- and postmultiplication of matrices was defined, and some important special types of matrices were introduced, such as the unit matrix **I**. It was also shown how a matrix **A** can be subdivided into blocks, and that a matrix operation performed on **A** can be interpreted in terms of matrix operations performed on block matrices obtained by subdivision of **A**.

EXERCISES 3.1

In Exercises 1 through 4 find the values of the constants a, b, and c in order that $\mathbf{A} = \mathbf{B}$.

1. $\mathbf{A} = \begin{bmatrix} a^2 & 1 & c \\ 2 & 3 & a \end{bmatrix}$, $\quad \mathbf{B} = \begin{bmatrix} -a & 1 & 4 \\ 2 & b & -1 \end{bmatrix}$.

2. $\mathbf{A} = \begin{bmatrix} 1 & 4 & 3 \\ a & 2 & 4 \\ 9 & 1 & c \end{bmatrix}$, $\quad \mathbf{B} = \begin{bmatrix} 1 & 4 & 3 \\ 2 & 2 & 4 \\ b & 1 & 0 \end{bmatrix}$.

3. $\mathbf{A} = \begin{bmatrix} a^2 & a & 1 \\ b & 1 & 2 \\ 1+a & 2+c & 6 \end{bmatrix}$, $\quad \mathbf{B} = \begin{bmatrix} a^2 & a & 1 \\ 3 & 1 & 2 \\ 2 & 4 & 6 \end{bmatrix}$.

4. $\mathbf{A} = \begin{bmatrix} 1 & 3+a & 2 \\ 1+b & a & 5 \\ b^2 & 1 & a^2 \end{bmatrix}$, $\quad \mathbf{B} = \begin{bmatrix} 1 & -1 & c \\ 4 & a & 5 \\ b^2 & 1 & a^2 \end{bmatrix}$.

In Exercises 5 through 8 find $\mathbf{A} + \mathbf{B}$ and $\mathbf{A} - \mathbf{B}$.

5. $\mathbf{A} = \begin{bmatrix} 1 & 4 & 3 & 6 \\ 2 & 1 & 0 & 2 \\ 1 & -1 & 0 & 1 \end{bmatrix}$, $\quad \mathbf{B} = \begin{bmatrix} 2 & 0 & 1 & -2 \\ 1 & 1 & -3 & 1 \\ 0 & 1 & 1 & 0 \end{bmatrix}$.

6. $\mathbf{A} = \begin{bmatrix} 1 & 7 & 6 \\ 0 & 2 & 4 \\ -1 & 0 & 1 \end{bmatrix}$, $\quad \mathbf{B} = \begin{bmatrix} 2 & -1 & 6 \\ 1 & -2 & 3 \\ 2 & 1 & 2 \end{bmatrix}$.

7. $\mathbf{A} = \begin{bmatrix} 1 & 2 & 4 \\ 3 & 1 & 0 \\ 1 & 1 & 0 \\ 2 & 2 & 4 \end{bmatrix}$, $\quad \mathbf{B} = \begin{bmatrix} 0 & 2 & 3 \\ 3 & -1 & 1 \\ 0 & 1 & 1 \\ 1 & 3 & 2 \end{bmatrix}$.

8. $\mathbf{A} = \begin{bmatrix} 1 & 4 & 3 & 6 \\ 0 & 2 & 1 & 4 \\ 0 & 0 & 3 & 1 \\ 0 & 0 & 0 & 2 \end{bmatrix}$, $\quad \mathbf{B} = \begin{bmatrix} 1 & 0 & 0 & 0 \\ 3 & 1 & 0 & 0 \\ 1 & 2 & 4 & 0 \\ 1 & 1 & 1 & 3 \end{bmatrix}$.

In Exercises 9 through 12 form the sum $\lambda\mathbf{A} + \mu\mathbf{B}$.

9. $\lambda = 1$, $\mu = 3$, $\mathbf{A} = \begin{bmatrix} 1 & 4 & 2 \\ 2 & 1 & 4 \\ 3 & 2 & 2 \end{bmatrix}$,

$\mathbf{B} = \begin{bmatrix} 2 & 3 & -1 \\ 1 & 2 & 4 \\ 1 & 0 & 3 \end{bmatrix}$.

10. $\lambda = -1$, $\mu = 2$, $\mathbf{A} = \begin{bmatrix} 1 & 4 & 1 \\ 2 & 4 & 0 \end{bmatrix}$,

$\mathbf{B} = \begin{bmatrix} 2 & 1 & 1 \\ 0 & 2 & 4 \end{bmatrix}$.

11. $\lambda = 4$, $\mu = -2$, $\mathbf{A} = \begin{bmatrix} 4 & 3 & 1 \\ 2 & 1 & 1 \\ 1 & 2 & 1 \end{bmatrix}$,

$\mathbf{B} = \begin{bmatrix} 6 & 1 & 0 \\ 2 & 4 & 2 \\ 1 & 1 & 2 \end{bmatrix}$.

12. $\lambda = 3$, $\mu = -3$, $\mathbf{A} = \begin{bmatrix} 3 & 1 & 4 \\ 2 & 2 & 1 \\ 3 & 6 & 2 \end{bmatrix}$,

$\mathbf{B} = \begin{bmatrix} 3 & 2 & 1 \\ 4 & 2 & 3 \\ 2 & 1 & 1 \end{bmatrix}$.

In Exercises 13 through 16 find the product $\mathbf{AB}$.

13. $\mathbf{A} = [1, \quad 4, \quad -2, \quad 3]$, $\quad \mathbf{B} = [2, \quad 1, \quad -1, \quad 2]^{\mathrm{T}}$.

14. $\mathbf{A} = [2, \quad 3, \quad 1, \quad 4]$, $\quad \mathbf{B} = [3, \quad 1, \quad 1, \quad 3]^{\mathrm{T}}$.

15. $\mathbf{A} = [1, \quad 4, \quad 3, \quad 7, \quad 5]$,
$\mathbf{B} = [2, \quad 2, \quad -1, \quad -1, \quad 3]^{\mathrm{T}}$.

16. $\mathbf{A} = [1, \quad 3, \quad -1, \quad 2, \quad 0]$,
$\mathbf{B} = [-1, \quad 2, \quad 13, \quad 4, \quad 1]^{\mathrm{T}}$.

In Exercises 17 through 22 find the product $\mathbf{AB}$ and, when it exists, the product $\mathbf{BA}$.

17. $\mathbf{A} = \begin{bmatrix} 1 & 4 \\ 2 & 0 \end{bmatrix}$, $\quad \mathbf{B} = \begin{bmatrix} 2 & 1 & 3 \\ 1 & 4 & 1 \end{bmatrix}$.

18. $\mathbf{A} = [1, 4, 6, -7]$, $\quad \mathbf{B} = [2, 3, -2, 3]^{\mathrm{T}}$.

19. $\mathbf{A} = \begin{bmatrix} 1 & 0 & 0 \\ 0 & 1 & 0 \\ 0 & 0 & 1 \end{bmatrix}$, $\quad \mathbf{B} = \begin{bmatrix} 3 & 1 & 4 \\ 2 & 1 & -5 \\ 7 & 2 & 0 \end{bmatrix}$.

20. $\mathbf{A} = \begin{bmatrix} 2 & 0 & 0 \\ 0 & -3 & 0 \\ 0 & 0 & 5 \end{bmatrix}$, $\quad \mathbf{B} = \begin{bmatrix} 9 & -1 & 4 \\ 1 & 6 & -2 \\ 2 & 2 & 3 \end{bmatrix}$.

21. $\mathbf{A} = \begin{bmatrix} 2 & 3 & 1 \\ 4 & 1 & 2 \\ 2 & 2 & 6 \\ 1 & 5 & 2 \end{bmatrix}$, $\quad \mathbf{B} = \begin{bmatrix} 5 & 2 & 3 \\ 2 & 0 & 4 \\ 1 & 4 & 7 \end{bmatrix}$.

22. $\mathbf{A} = \begin{bmatrix} 1 & 2 & 1 & 0 \\ 2 & 1 & 1 & 4 \\ 1 & 0 & 2 & 1 \\ 1 & 1 & 2 & 1 \end{bmatrix}$, $\quad \mathbf{B} = \begin{bmatrix} 3 & 1 \\ 4 & 2 \\ 6 & -2 \\ -1 & 4 \end{bmatrix}$.

23. Given

$$\mathbf{A} = \begin{bmatrix} 2 & 5 & -3 \\ 5 & 1 & 4 \\ -3 & 4 & 6 \end{bmatrix} \quad \text{and} \quad \mathbf{B} = \begin{bmatrix} 4 & 2 & 1 \\ 2 & 5 & 6 \\ 1 & 6 & 3 \end{bmatrix},$$

show that $(\mathbf{AB})^{\mathrm{T}} = \mathbf{BA}$.

In Exercises 24 through 28 write the given systems of equations in the matrix form $\mathbf{Ax} = \mathbf{b}$, where $\mathbf{A}$ is the coefficient matrix, $\mathbf{x}$ is the vector of unknowns, and $\mathbf{b}$ is the nonhomogeneous vector term.

24. $3x + 5y - 6z = 7$
$x - 7y + 4z = -3$
$2x + 4y - 5z = 4.$

25. $4u + 5v - w + 7z = 25$
$3u + 2v + 3z = 6$
$v + 6w - 7z = 0.$

26. $5x + 3y - 6z = 14$
$6x - 5y + 11z = 20$
$x - 4y + 3z = 2$
$9x - 3y + 2z = 35.$

27. $3x + 4y - 2z = \lambda x$
$2x - 7y + 6z = \lambda y$
$8x + 3y + 5z = \lambda z.$

28. $2x + 3y + 6z = \lambda(3x + 2y + 3z)$
$3x - 4y + 2z = \lambda(x - 5y + 2z)$
$4x + 9y + 2z = \lambda(x - 2y + 4z).$

29. If
$$\mathbf{A} = \begin{bmatrix} 1 & 3 & 6 \\ 1 & 2 & 0 \\ 0 & 1 & 3 \end{bmatrix}, \quad \mathbf{B} = \begin{bmatrix} 2 & 0 & 1 \\ 4 & 2 & 3 \\ 0 & -1 & 1 \end{bmatrix}, \quad \text{and}$$
$$\mathbf{X} = \begin{bmatrix} x_1 & x_2 & x_3 \\ y_1 & y_2 & y_3 \\ z_1 & z_2 & z_3 \end{bmatrix},$$
solve for $\mathbf{X}$ given that
$$3\mathbf{X} + \mathbf{A} = \mathbf{A}^T\mathbf{B} - \mathbf{X} + 3\mathbf{B}.$$

30. If
$$\mathbf{A} = \begin{bmatrix} 2 & 1 & 4 \\ 1 & 2 & 1 \\ 3 & 0 & 2 \end{bmatrix}, \quad \mathbf{B} = \begin{bmatrix} 1 & 4 & 1 \\ 2 & 1 & 2 \\ 1 & 1 & 2 \end{bmatrix}, \quad \text{and}$$
$$\mathbf{X} = \begin{bmatrix} x_1 & x_2 & x_3 \\ y_1 & y_2 & y_3 \\ z_1 & z_2 & z_3 \end{bmatrix},$$
solve for $\mathbf{X}$ given that
$$2\mathbf{AB}^T + \mathbf{X} - 2\mathbf{I} = 3\mathbf{X} + 4\mathbf{B} - 2\mathbf{A}.$$

31. Given that
$$\mathbf{A} = \begin{bmatrix} 3 & 2 & 2 \\ 2 & 2 & 0 \\ 2 & 0 & 4 \end{bmatrix},$$
show that
$$\mathbf{A}^3 - 9\mathbf{A}^2 + 18\mathbf{A} = \mathbf{0}.$$

32. Given that
$$\mathbf{A} = \begin{bmatrix} 0 & 1 & 0 \\ 0 & 0 & 1 \\ 2 & 1 & -2 \end{bmatrix},$$
show that
$$\mathbf{A}^3 + 2\mathbf{A}^2 - \mathbf{A} - 2\mathbf{I} = \mathbf{0}.$$

33. Prove the second result in Theorem 3.1 that $\mathbf{A}(\mathbf{BC}) = (\mathbf{AB})\mathbf{C} = \mathbf{ABC}$.

34. Prove the third result in Theorem 3.1 that $(\lambda\mathbf{A})\mathbf{B} = \mathbf{A}(\lambda\mathbf{B}) = \lambda\mathbf{AB}$.

35. Prove the fourth result in Theorem 3.1 that $\mathbf{A}(\mathbf{B} + \mathbf{C}) = \mathbf{AB} + \mathbf{AC}$.

In Exercises 36 through 39 verify that $(\mathbf{AB})^T = \mathbf{B}^T\mathbf{A}^T$.

36. $\mathbf{A} = \begin{bmatrix} 3 & 1 & 4 \\ 2 & 1 & 2 \\ 4 & 2 & 3 \end{bmatrix}, \quad \mathbf{B} = \begin{bmatrix} 2 & 1 & 3 \\ 1 & 2 & 5 \\ 0 & 2 & 1 \end{bmatrix}.$

37. $\mathbf{A} = \begin{bmatrix} 2 & 1 & 4 & 3 \\ 1 & 6 & 2 & 1 \\ 1 & 1 & -2 & 4 \end{bmatrix}, \quad \mathbf{B} = \begin{bmatrix} 1 & 4 & 3 \\ 2 & 1 & 5 \\ -1 & 3 & 2 \\ 1 & 7 & 3 \end{bmatrix}.$

38. $\mathbf{A} = \begin{bmatrix} 1 & 4 & 2 \\ 7 & 3 & -1 \\ 0 & 2 & 5 \end{bmatrix}, \quad \mathbf{B} = \begin{bmatrix} 3 & 1 & -5 \\ 1 & 3 & 4 \\ 2 & 0 & 8 \end{bmatrix}.$

39. $\mathbf{A} = \begin{bmatrix} 1 & 4 & 6 & 2 \\ 2 & 1 & 4 & 1 \\ 3 & 0 & 0 & 2 \end{bmatrix}, \quad \mathbf{B} = \begin{bmatrix} 1 & 2 & 1 \\ -2 & 1 & 4 \\ 2 & 2 & 5 \\ 1 & 1 & 1 \end{bmatrix}.$

40. Verify that $(\mathbf{ABC})^T = \mathbf{C}^T\mathbf{B}^T\mathbf{A}^T$ given that
$$\mathbf{A} = \begin{bmatrix} 1 & 5 \\ 3 & 1 \end{bmatrix}, \quad \mathbf{B} = \begin{bmatrix} 3 & -2 \\ 4 & 5 \end{bmatrix}, \quad \text{and} \quad \mathbf{C} = \begin{bmatrix} -2 & 3 \\ 5 & 7 \end{bmatrix}.$$

41. Prove that if $\mathbf{D}$ is the $n \times n$ diagonal matrix
$$\mathbf{D} = \begin{bmatrix} k_1 & 0 & 0 & \cdots & 0 \\ 0 & k_2 & 0 & \cdots & 0 \\ 0 & 0 & k_3 & \cdots & 0 \\ \cdots & \cdots & \cdots & \cdots & \cdots \\ 0 & 0 & 0 & \cdots & k_n \end{bmatrix}, \quad \text{then}$$
$$\mathbf{D}^m = \begin{bmatrix} k_1^m & 0 & 0 & \cdots & 0 \\ 0 & k_2^m & 0 & \cdots & 0 \\ 0 & 0 & k_3^m & \cdots & 0 \\ \cdots & \cdots & \cdots & \cdots & \cdots \\ 0 & 0 & 0 & \cdots & k_n^m \end{bmatrix},$$
where m is a positive integer.

42. Find $\mathbf{A}^2$, $\mathbf{A}^3$, and $\mathbf{A}^4$, given that
$$\mathbf{A} = \begin{bmatrix} 1 & 2 & 7 \\ 2 & 5 & 6 \\ 1 & 0 & -1 \end{bmatrix}.$$

43. Find $\mathbf{A}^2$, $\mathbf{A}^4$, and $\mathbf{A}^6$, given that
$$\mathbf{A} = \begin{bmatrix} 1/2 & -(\sqrt{3})/2 \\ (\sqrt{3})/2 & 1/2 \end{bmatrix}.$$

44. Use the matrix **A** in Exercise 42 to find $\mathbf{A}^3$, $\mathbf{A}^5$, and $\mathbf{A}^7$.

45. A square matrix **A** such that $\mathbf{A}^2 = \mathbf{A}$ is said to be **idempotent**. Find the three idempotent matrices of the form

$$\mathbf{A} = \begin{bmatrix} 1 & p \\ q & r \end{bmatrix}.$$

46. A square matrix **A** such that for some positive integer n has the property that $\mathbf{A}^{n-1} \neq \mathbf{0}$, but $\mathbf{A}^n = \mathbf{0}$ is said to be **nilpotent of index n** $(n \geq 2)$. Show that the matrix

$$\mathbf{A} = \begin{bmatrix} 0 & 0 & 0 \\ 4 & 0 & 0 \\ 1 & -1 & 0 \end{bmatrix}$$

is nilpotent and find its index.

47. A **quadratic form** in the variables $x_1, x_2, x_3, \ldots, x_n$ is an expression of the form $ax_1^2 + bx_1x_2 + cx_2^2 + dx_1x_3 + \cdots + fx_{n-1}x_n + gx_n^2$ in which some of the coefficients $a, b, c, d, \ldots, f, g$ may be zero. Explain why $\mathbf{x}^T\mathbf{A}\mathbf{x}$ is a quadratic form and find the quadratic form for which

$$\mathbf{A} = \begin{bmatrix} 3 & 4 & 0 & 3 \\ 4 & 2 & 2 & 6 \\ 0 & 2 & 5 & 1 \\ 3 & 6 & 1 & 7 \end{bmatrix} \quad \text{and} \quad \mathbf{x} = \begin{bmatrix} x_1 \\ x_2 \\ x_3 \\ x_4 \end{bmatrix}.$$

48. Find the quadratic form $\mathbf{x}^T\mathbf{A}\mathbf{x}$ when

$$\mathbf{A} = \begin{bmatrix} 4 & 1 & 3 & 6 \\ 2 & 3 & 5 & 4 \\ 1 & 4 & 1 & 2 \\ 2 & 0 & 4 & 1 \end{bmatrix} \quad \text{and} \quad \mathbf{x} = \begin{bmatrix} x_1 \\ x_2 \\ x_3 \\ x_4 \end{bmatrix}.$$

49. Explain why the matrix **A** in the general expression for a quadratic form $\mathbf{x}^T\mathbf{A}\mathbf{x}$ can always be written as a *symmetric* matrix.

In Exercises 50 through 52 find the symmetric matrix **A** for the given quadratic form when written $\mathbf{x}^T\mathbf{A}\mathbf{x}$, with $\mathbf{x} = [x, y, z]^T$.

50. $x^2 + 3xy - 4y^2 + 4xz + 6yz - z^2$.

51. $2x^2 + 4xy + 6y^2 + 7xz - 9z^2$.

52. $7x^2 + 7xy - 5y^2 + 4xz + 2yz - 9z^2$.

53. A square matrix **P** is called a **stochastic matrix** if all its elements are nonnegative and the sum of the elements in each row is 1. Thus, the matrix

$$\mathbf{P} = \begin{bmatrix} p_{11} & p_{12} & \cdots & p_{1n} \\ p_{21} & p_{22} & \cdots & p_{2n} \\ \cdots & \cdots & \cdots & \cdots \\ p_{n1} & p_{n2} & \cdots & p_{nn} \end{bmatrix}$$

will be a stochastic matrix if $p_{ij} \geq 0$ for $0 \leq i \leq n$, $0 \leq j \leq n$, and

$$\sum_{j=1}^{n} p_{ij} = 1 \quad \text{for } i = 1, 2, \ldots, n.$$

Let the n element column vector $\mathbf{E} = [1, 1, 1, \ldots, 1]^T$. By considering the matrix product $\mathbf{PE}$, and using mathematical induction, prove that $\mathbf{P}^m$ is a stochastic matrix for all positive integral values of m.

54. Construct a 3×3 stochastic matrix **P**. Find $\mathbf{P}^2$ and $\mathbf{P}^3$, and by showing that all elements of these matrices are nonnegative and that all their row-sums are 1, verify the result of Exercise 53 that each of these matrices is a stochastic matrix.

3.2   Some Problems That Give Rise to Matrices

(a) Electric Circuits with Resistors and Applied Voltages

A simple electric circuit involving five resistors and three applied voltages is shown in Fig. 3.1. The directions of the currents i_1, i_2, and i_3 flowing in each branch of the circuit are shown by arrows. The currents themselves can be determined by an application of *Ohm's* law and the *Kirchhoff* laws that can be stated as follows:

(a) Voltage = current × resistance (Ohm's law);

(b) The algebraic sum of the potential drops around each closed circuit is zero (Kirchhoff's second law);

(c) The current entering each junction must equal the algebraic sum of the currents leaving it (Kirchhoff's first law).

equations and matrices for electric circuits

An application of these laws to the circuit in Fig. 3.1, where the potentials are in volts, the resistances are in ohms, and the currents are in amps, leads to the following

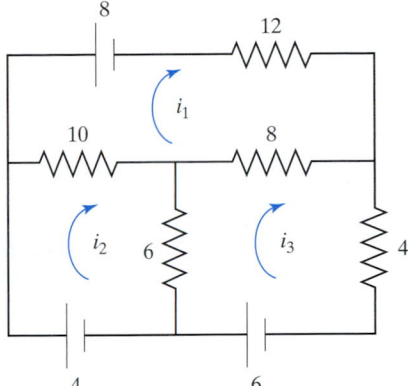

FIGURE 3.1   An electric circuit with resistors and applied voltages.

set of simultaneous equations:

$$8 = 12i_1 + 10(i_1 - i_2) + 8(i_1 - i_3)$$
$$4 = 10(i_2 - i_1) + 6(i_2 - i_3)$$
$$6 = 8(i_3 - i_1) + 6(i_3 - i_2) + 4i_3.$$

After collecting terms this system can be written as the matrix equation $\mathbf{Ax} = \mathbf{b}$, with

$$\mathbf{A} = \begin{bmatrix} 30 & -10 & -8 \\ -10 & 16 & -6 \\ -8 & -6 & 18 \end{bmatrix}, \quad \mathbf{x} = \begin{bmatrix} i_1 \\ i_2 \\ i_3 \end{bmatrix}, \quad \mathbf{b} = \begin{bmatrix} 8 \\ 4 \\ 6 \end{bmatrix}.$$

The directions assumed for the currents i_r for $r = 1, 2, 3$ are shown by the arrows in Fig. 3.1, but if after the system of equations is solved, the value of the current is found to be negative, the direction of its arrow must be reversed.

The circuit in Fig. 3.1 is simple, so in this example the currents can be found by routine elimination between the three equations. When many coupled circuits are involved a matrix approach is useful, and it then becomes necessary to develop a method for solving for $\mathbf{x}$ the matrix equation $\mathbf{Ax} = \mathbf{b}$, the elements of which are the required currents. If the number of equations is small, $\mathbf{x}$ can be found by making use of the matrix $\mathbf{A}^{-1}$, inverse to $\mathbf{A}$, that will be introduced later, though the most computationally efficient approach is to use one of the numerical methods for solving systems of linear simultaneous equations described in Chapter 19.

(b) Combinatorial Problems: Graph Theory

Matrices play an important role in combinatorial problems of many different types and, in particular, in graph theory. The purpose of the brief account offered here will be to illustrate a particular application of matrices, and no attempt will be made to discuss their subsequent use in the solution of the associated problems.

Combinatorial problems involve dealing with the possible arrangements of situations of various different kinds, and computing the number and properties of such arrangements. The arrangements may be of very diverse types, involving at one extreme the ordering of matches that are to take place in a tennis tournament,

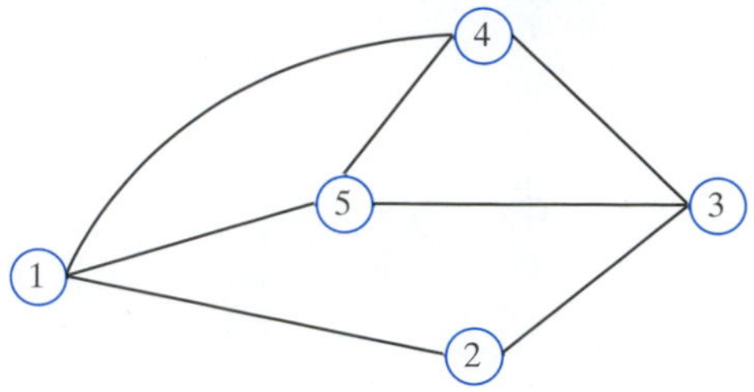

FIGURE 3.2 The graph representing routes.

and at the other extreme finding an optimum route for a delivery truck or for the most efficient routing of work through a machine shop.

The ideas involved are most easily illustrated by means of examples, the first of which involves the delivery from a storage depot of a consumable product to a group of supermarkets in a large city where it is important that daily deliveries be made as rapidly as possible. One possibility involves a delivery truck making a delivery to each supermarket in turn and returning to the storage depot between each delivery before setting out on the next delivery.

An alternative is to travel between supermarkets after each delivery without returning to the storage depot. The question that then arises is which approach to routing is the best, and how it is to be determined.

A typical situation is illustrated in Fig. 3.2, in which supermarkets numbered 1 to 5 are involved, with circles representing supermarkets and lines and arcs representing the routes.

The representation in Fig. 3.2 is called a **graph**, and it is to be regarded as a set of points represented by the circles called **vertices** of the graph, and **edges** of the graph represented by the lines and arcs. In Fig. 3.2 the vertices are the circles 1, $2, \ldots, 5$ and the seven edges are the lines and arcs connecting the vertices.

A special type of matrix associated with such a graph is an **adjacency matrix**, that is, a matrix whose only entries are 0 or 1. The rules for the entries in an adjacency matrix $\mathbf{A} = [a_{ij}]$ are that

$$a_{ij} = \begin{cases} 1, & \text{if vertices } i \text{ and } j \text{ are joined by an edge} \\ 0, & \text{otherwise.} \end{cases}$$

The adjacency matrix for the graph in Fig. 3.2 is seen to be the symmetric matrix

$$\mathbf{A} = \begin{bmatrix} 0 & 1 & 0 & 1 & 1 \\ 1 & 0 & 1 & 0 & 0 \\ 0 & 1 & 0 & 1 & 1 \\ 1 & 0 & 1 & 0 & 1 \\ 1 & 0 & 1 & 1 & 0 \end{bmatrix}.$$

It is to be expected that an adjacency matrix is symmetric, because if i is adjacent to j, then j is adjacent to i.

Although we shall not attempt to do so here, the interconnection properties of the problem represented by the graph in Fig. 3.2 can be analyzed in terms of its adjacency matrix $\mathbf{A}$. The optimum routing problem can then be resolved once the traveling times along roads (lines or arcs) are known.

Sometimes it happens that the edges in a graph represent connections that only operate in one direction, so then arrows must be added to the graph to indicate these

graphs, vertices, edges, and adjacency matrix

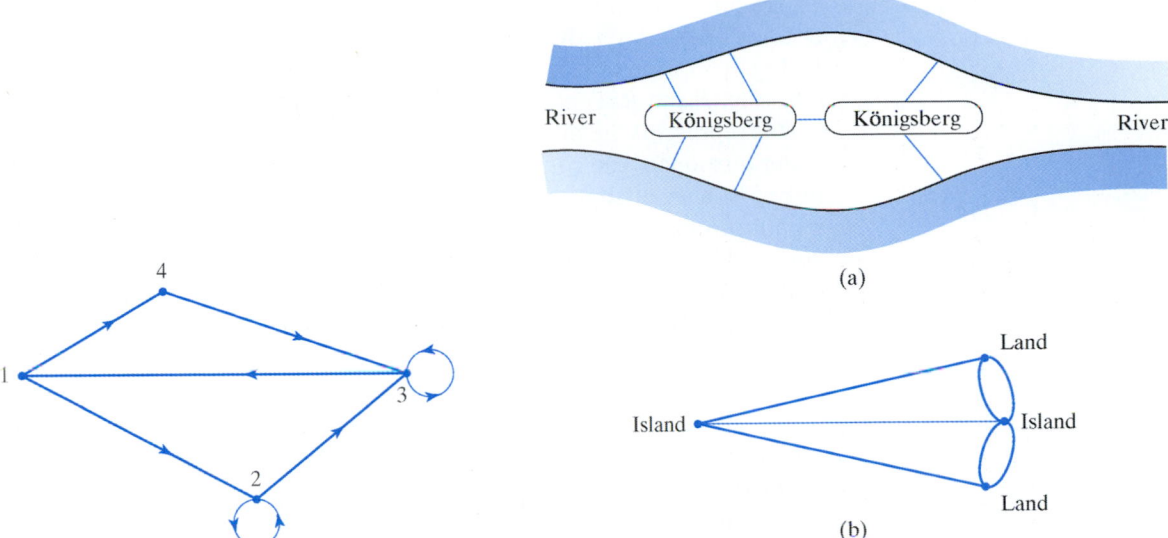

FIGURE 3.3   A typical digraph.

FIGURE 3.4   The Königsberg bridge problem.

digraph

directions. A graph of this type is called a **digraph** (directed graph). The rules for the entries in the adjacency matrix $\mathbf{A} = [a_{ij}]$ of a digraph are that

$$a_{ij} = \begin{cases} 1, & \text{if vertices } i \text{ and } j \text{ are joined by an edge with an arrow from } i \text{ to } j \\ 0, & \text{otherwise.} \end{cases}$$

A typical digraph is shown in Fig. 3.3, and it has the associated adjacency matrix

$$\mathbf{A} = \begin{bmatrix} 0 & 1 & 0 & 1 \\ 0 & 0 & 1 & 0 \\ 1 & 0 & 0 & 0 \\ 0 & 0 & 1 & 0 \end{bmatrix}.$$

The adjacency matrix $\mathbf{A}$ characterizes all the possible interconnections between the four vertices and, as with the previous example, an analysis of the properties of any situation capable of representation in terms of this digraph can be performed using the matrix $\mathbf{A}$. Problems of this type can arise in transportation problems in cities with one-way streets, and in chemical processes where a fluid is piped to different parts of a plant through an interconnecting network of pipes through which fluid may only flow in a given direction.

Before closing this brief introduction to graph theory, mention should be made of a problem of historical significance, since it represented the start of graph theory as it is known today. The problem is called the **Königsberg bridge problem**, and it was solved by Euler (1707–1783). During the early 18th century the Prussian town of Königsberg was established on two adjacent islands in the middle of the river Pregel. The islands were linked to the land on either side of the river, and to one another, by seven bridges, as shown in Fig. 3.4a. It was suggested to Euler that he should resolve the conjecture that it ought to be possible to walk through the town, starting and ending at the same place, while crossing each of the seven bridges only once.

Königsberg bridge problem

Euler replaced the picture in Fig. 3.4a by the graph in Fig. 3.4b, though it was not until much later that the term *graph* in the sense used here was introduced. In Fig. 3.4b the vertices S and Q represent the two islands and, using the same lettering, P and R represent the riverbanks. The number of edges incident on each vertex represents the number of bridges connected to the corresponding land mass. Euler introduced the concept of a *connected graph*, in which each pair of vertices is linked by a set of edges, and also what is now called an *eulerian circuit*, comprising a path through all vertices that starts and ends at the same vertex and uses every edge only once. He called the number of edges incident upon a vertex the *degree* of the vertex, and by using these ideas he was able to prove the impossibility of the conjecture. The arguments involved are not difficult, but their details would be out of place here.

Many more practical problems are capable of solution by graph theory, which itself belongs to the branch of mathematics called *combinatorics*. In elementary accounts, graph theory and related combinatorial issues are usually called *discrete mathematics*. More information about combinatorics and its connection with matrices can be found in References [2.2] and [2.13].

(c) Translations, Rotations, and Scaling of Graphs: Computer Graphics

matrices and computer graphics

The simplest operations in computer graphics involve copying a picture to a different location, rotating a picture about a fixed point, and scaling a picture, where the scaling can be different in the horizontal and vertical directions. These operations are called, respectively, a **translation**, a **rotation**, and a **scaling** of the picture. Operations of this nature can all be represented in terms of matrices, and they involve what are called **linear transformations** of the original picture.

Translation

A translation of a two-dimensional picture involves copying it to a different location without either rotating it or changing its horizontal and vertical scales. Figure 3.5 shows the original cartesian axes $O(x, y)$ and the shifted axes $O'(x', y')$, where the respective axes remain parallel to their original directions and the origin O' is located at the point (h, k) relative to the $O(x, y)$ axes.

The relationship between the two sets of coordinates is given by

$$x = x' + h \quad \text{and} \quad y = y' + k.$$

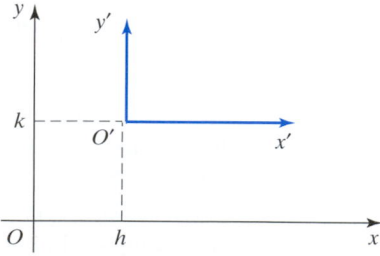

FIGURE 3.5   A translation.

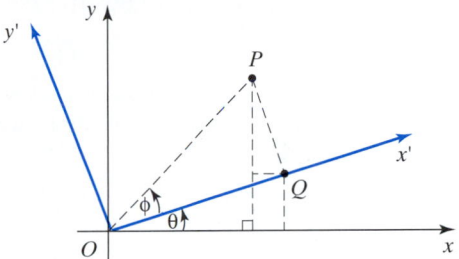

FIGURE 3.6   A rotation through an angle θ.

If $\mathbf{x} = [x, y]^{\mathrm{T}}$, $\mathbf{x}' = [x', y']^{\mathrm{T}}$, and $\mathbf{b} = [h, k]^{\mathrm{T}}$, the coordinate transformation can be written in matrix form as

$$\mathbf{x} = \mathbf{x}' + \mathbf{b},$$

where matrix $\mathbf{b}$ represents the translation.

Rotation

A rotation of the coordinate axes through an angle θ is shown in Fig. 3.6, where $P(x, y)$ is an arbitrary point. The coordinates of P in the (x, y) reference frame and the (x', y') reference frame are related as

$$x = OR = OP\cos(\phi + \theta) = OP\cos\phi\cos\theta - OP\sin\phi\sin\theta$$
$$= OQ\cos\theta - PQ\sin\theta = x'\cos\theta - y'\sin\theta,$$

and

$$y = PR = OP\sin(\phi + \theta) = OP\sin\phi\cos\theta + OP\cos\phi\sin\theta$$
$$= PQ\cos\theta + OQ\sin\theta = y'\cos\theta + x'\sin\theta,$$

so

$$x = x'\cos\theta - y'\sin\theta \quad \text{and} \quad y = y'\cos\theta + x'\sin\theta.$$

Defining the matrices $\mathbf{x}$, $\mathbf{x}'$, and $\mathbf{R}$ as

$$\mathbf{x} = \begin{bmatrix} x \\ y \end{bmatrix}, \quad \mathbf{x}' = \begin{bmatrix} x' \\ y' \end{bmatrix}, \quad \text{and} \quad \mathbf{R} = \begin{bmatrix} \cos\theta & -\sin\theta \\ \sin\theta & \cos\theta \end{bmatrix}$$

allows the coordinate transformation to be written as

$$\mathbf{x} = \mathbf{R}\mathbf{x}'.$$

Scaling

If $\mathbf{S}$ is a matrix of the form

$$\mathbf{S} = \begin{bmatrix} k_x & 0 \\ 0 & k_y \end{bmatrix},$$

where k_x and k_y are positive constants, and $\mathbf{x}' = \mathbf{Sx}$, it follows that

$$x = k_x x' \quad \text{and} \quad y = k_y y',$$

showing that x is obtained by scaling x' by k_x, while y is obtained by scaling y' by k_y. This form of scaling is represented by premultiplication of $\mathbf{x}$ by $\mathbf{S}$, and if, for example,

$$\mathbf{S} = \begin{bmatrix} 4 & 0 \\ 0 & 3 \end{bmatrix},$$

the effect of this transformation on a circle of radius a will be to map it into an ellipse with semimajor axis of length $4a$ parallel to the x-axis and a semiminor axis of length $3a$ parallel to the y-axis.

Composite transformations

By combining the preceding matrix operations to form a **composite transformation**, it is possible to carry out several transformations simultaneously. As an example, the effect of a rotation $\mathbf{R}$ followed by a translation $\mathbf{b}$ when performed on a vector $\mathbf{x}'$ are seen to be described by the matrix equation

$$\mathbf{x} = \mathbf{Rx}' + \mathbf{b},$$

the effect of which is shown in Fig. 3.7.

If a scaling $\mathbf{S}$ is performed before the rotation and translation, the effect on a vector $\mathbf{x}'$ is described by the matrix equation

$$\mathbf{x} = \mathbf{RSx}' + \mathbf{b}.$$

This is illustrated in Fig. 3.8b, which shows the effect when a transformation of this type is performed on the circle of radius a centered on the origin shown in Fig. 3.8a, with

$$\mathbf{b} = \begin{bmatrix} h \\ k \end{bmatrix}, \quad \mathbf{R} = \begin{bmatrix} \cos \pi/3 & -\sin \pi/3 \\ \sin \pi/3 & \cos \pi/3 \end{bmatrix}, \quad \text{and} \quad \mathbf{S} = \begin{bmatrix} 3 & 0 \\ 0 & 2 \end{bmatrix}.$$

It is seen that the circle has first been scaled to become an ellipse with semiaxes $3a$ and $2a$, after which the ellipse has been rotated through an angle $\pi/3$, and finally its center has been translated to the point (h, k).

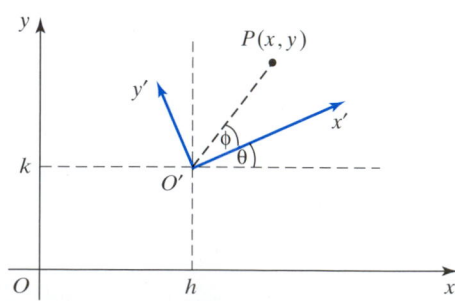

FIGURE 3.7 A rotation and a translation.

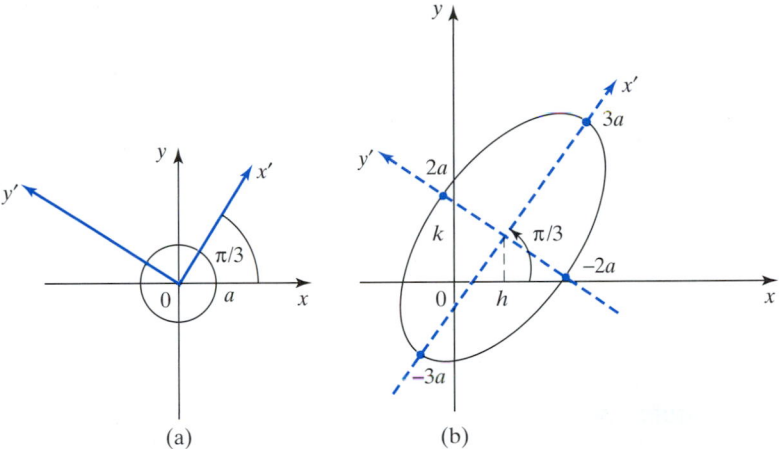

FIGURE 3.8   The composite transformation $\mathbf{x} = \mathbf{RSx'} + \mathbf{b}$.

It is essential to remember that the *order* in which transformations are per-formed will, in general, influence the result. This is easily seen by considering the two transformations $\mathbf{x} = \mathbf{RSx'} + \mathbf{b}$ and $\mathbf{x} = \mathbf{SRx'} + \mathbf{b}$. If the first of these is per-formed on the circle in Fig. 3.8a, it produces Fig. 3.8b, but when the second is performed on the same circle, it first converts it into an ellipse with its major axis horizontal, and then translates the center of the ellipse to the point (h, k). In this case the effect of the rotation cannot be seen, because the circle is symmetric with respect to rotations.

A relationship of the form $\mathbf{x} = \mathbf{F(x')}$ can be interpreted geometrically in two distinct ways which are equally valid:

1. As the change in the way we describe the location of a point P. Then the relationship is called a transformation of coordinates (Figs. 3.5, 3.6, 3.7).
2. As a mapping of a point P from one location to a new one.

(d) Matrix Analysis of Framed Structures

A **framed structure** is a network of straight struts joined at their ends to form a rigid three-dimensional structure. A typical framed structure is the steel work for a large building before the walls and floors have been added. A simple example of a framed structure, called a **truss**, is a plane construction in which the struts are joined together to form triangles, as in the side section of the small bridge shown in Fig. 3.9.

For safety, to ensure that no strut fails when the bridge carries the largest permitted load, it is necessary to determine the force experienced by each strut in the truss when the bridge supports its maximum load in several different positions. Typically, the largest load could be due to a heavy truck crossing the bridge. The analysis of trusses is usually simplified by making the following assumptions:

- The structure is in the vertical plane;
- The weight of each strut can be neglected;
- Struts are rigid and so remain straight;

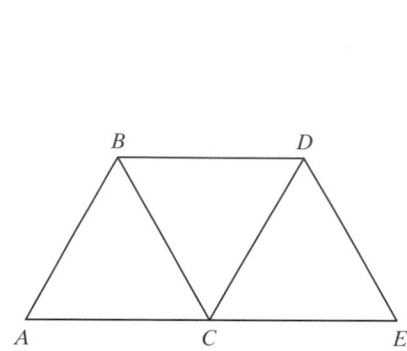

FIGURE 3.9 A typical truss found in a side section of a bridge.

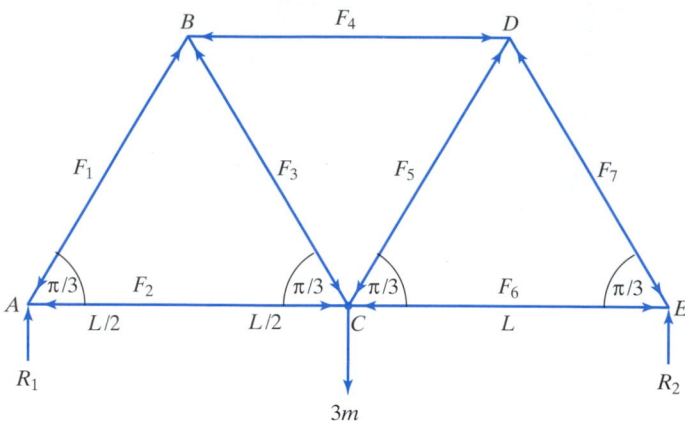

FIGURE 3.10 A truss supporting a concentrated load.

- Each joint is considered to be hinged, so the only forces acting at a joint are along the struts meeting at the joint if forces are applied at joints only.
- There are no redundant struts, so that removing a strut will cause the truss to collapse.

We now write down the simultaneous equations that must be solved to find the forces acting in the seven struts of length L that form the truss shown in Fig. 3.10, when a concentrated load $3m$ is located at point C midway between A and E. This load could be considered to be a heavily laden truck standing in the center of the bridge.

To determine the reactions at the support points A and E, we use the fact that for equilibrium the turning moments about these two points must be zero. The turning moment of the load $3m$ about the point A must be cancelled by the turning moment of the reaction R_2 at E, so $3m(L) = R_2(2L)$, showing that $R_2 = 3m/2$. Similarly, the turning moment of the load $3m$ about the point E must be cancelled by the turning moment of the reaction R_1 at A, so $3m(L) = R_1(2L)$, showing that $R_1 = 3m/2$.

The directions in which the forces F_1 to F_7 are assumed to act are shown by arrows, and if later a force is found to be negative, the direction of the associated arrows must be reversed. For equilibrium the sum of the vertical components of all forces acting at each joint must be zero, as must be the sum of the horizontal components of all forces acting at each joint. The equations representing the balance of forces at each joint are as follows, where when resolving the forces acting at joint C, the effect of the load $3m$ which acts vertically downwards must be taken into account:

equations and matrices for a framed structure

Joint A(vertical)	$F_1 \sin \pi/3 - 3m/2 = 0$
Joint A(horizontal)	$F_1 \cos \pi/3 + F_2 = 0$
Joint B(vertical)	$F_1 \sin \pi/3 + F_3 \sin \pi/3 = 0$
Joint B(horizontal)	$F_1 \cos \pi/3 - F_3 \cos \pi/3 - F_4 = 0$
Joint C(vertical)	$F_3 \sin \pi/3 + F_5 \sin \pi/3 + 3m = 0$
Joint C(horizontal)	$F_2 + F_3 \cos \pi/3 - F_5 \cos \pi/3 - F_6 = 0$
Joint D(vertical)	$F_5 \sin \pi/3 + F_7 \sin \pi/3 = 0$

Joint D(horizontal) $F_4 + F_5 \cos \pi/3 - F_7 \cos \pi/3 = 0$
Joint E(vertical) $F_7 \sin \pi/3 - 3m/2 = 0$
Joint E(horizontal) $F_6 + F_7 \cos \pi/3 = 0.$

After substituting for $\sin \pi/3$ and $\cos \pi/3$, these equations can be written in the matrix form $\mathbf{A}\mathbf{x} = \mathbf{b}$, where

$$
\mathbf{A} = \begin{bmatrix}
\frac{1}{2}\sqrt{3} & 0 & 0 & 0 & 0 & 0 & 0 \\
\frac{1}{2} & 1 & 0 & 0 & 0 & 0 & 0 \\
\frac{1}{2}\sqrt{3} & 0 & \frac{1}{2}\sqrt{3} & 0 & 0 & 0 & 0 \\
\frac{1}{2} & 0 & -\frac{1}{2} & -1 & 0 & 0 & 0 \\
0 & 0 & \frac{1}{2}\sqrt{3} & 0 & \frac{1}{2}\sqrt{3} & 0 & 0 \\
0 & 1 & \frac{1}{2} & 0 & -\frac{1}{2} & -1 & 0 \\
0 & 0 & 0 & 0 & \frac{1}{2}\sqrt{3} & 0 & \frac{1}{2}\sqrt{3} \\
0 & 0 & 0 & 1 & \frac{1}{2} & 0 & -\frac{1}{2} \\
0 & 0 & 0 & 0 & 0 & 0 & \frac{1}{2}\sqrt{3} \\
0 & 0 & 0 & 0 & 0 & 1 & \frac{1}{2}
\end{bmatrix},
\quad
\mathbf{x} = \begin{bmatrix} F_1 \\ F_2 \\ F_3 \\ F_4 \\ F_5 \\ F_6 \\ F_7 \end{bmatrix},
\quad
\mathbf{b} = \begin{bmatrix} 3m/2 \\ 0 \\ 0 \\ -3m \\ 0 \\ 0 \\ 0 \\ 3m/2 \\ 0 \end{bmatrix}.
$$

These are 10 equations for the 7 unknown forces F_1 to F_7, so unless 3 of the equations represented in $\mathbf{A}\mathbf{x} = \mathbf{b}$ are combinations of the remaining 7 equations, we cannot expect there to be a solution. When the **rank** of a matrix is introduced in Section 3.6, we will see how systems of this type can be checked for consistency and, when appropriate, simplified and solved.

In this case the equations are sufficiently simple that they can be solved sequentially, without the use of matrices. The solution is seen to be

$$F_1 = m\sqrt{3}, \quad F_2 = -m/(\sqrt{3}/2), \quad F_3 = -m\sqrt{3}, \quad F_4 = m\sqrt{3},$$
$$F_5 = -m\sqrt{3}, \quad F_6 = -m(\sqrt{3}/2), \quad F_7 = m\sqrt{3}.$$

The signs show that the arrows in Fig. 3.10 associated with forces F_2, F_3, F_5, and F_6 should be reversed, so these struts are in tension, while the others are in compression.

Notice that matrix $\mathbf{A}$ is determined by the geometry of the truss, and so does not change when forces are applied to more than one of the joints on the truss (bridge). This means that after the 10 equations have been reduced to seven, the same modified matrix $\mathbf{A}$ can be use to find the forces in the struts for *any* form of concentrated loading. Had a more complicated struss been involved, many more equations would have been involved, so that a matrix approach becomes necessary. This approach also identifies any redundant struts in a structure, because the force in a redundant strut is indeterminate.

(e) A Compound Mass–Spring System

Matrices can have variables as elements, and an analysis of the compound mass–spring system shown in Fig. 3.11 shows one way in which this can arise. Figure 3.11 represents a mass m_1 suspended from a rigid support by a spring of negligible mass with spring constant k_1, and a mass m_2 suspended from mass m_1 by a spring of negligible mass with spring constant k_2. The vertical displacement of m_1 from its

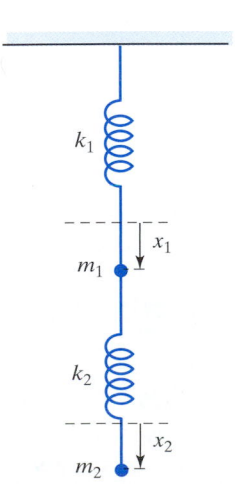

FIGURE 3.11 A compound mass–spring system.

equilibrium position is x_1, and the vertical displacement of m_2 from its equilibrium position is x_2. Each spring is assumed to be linearly elastic, so the restoring force exerted by a spring is equal to the product of the displacement from its equilibrium position and the spring constant.

The product of the mass m_1 and its acceleration is $m_1 d^2 x_1 / dt^2$, and the restoring force due to spring k_1 is $k_1 x_1$, while the restoring force due to spring k_2 is $k_2(x_1 - x_2)$, so the equation of motion of m_1 is

equations of motion of a coupled mass–spring system

$$m_1 \frac{d^2 x_1}{dt^2} = -k_1 x_1 - k_2(x_1 - x_2).$$

Similarly, the equation of motion of m_2 is

$$m_2 \frac{d^2 x_2}{dt^2} = -k_2(x_2 - x_1),$$

where the negative signs are necessary because the springs act to restore the masses to their original positions.

This system can be written as the matrix differential equation $\ddot{\mathbf{x}} + \mathbf{A}\mathbf{x} = \mathbf{0}$, by defining $\mathbf{A}$ and $\mathbf{x}$ as

$$\mathbf{A} = \begin{bmatrix} \dfrac{(k_1 + k_2)}{m_1} & -\dfrac{k_2}{m_1} \\ -\dfrac{k_2}{m_2} & \dfrac{k_2}{m_2} \end{bmatrix}, \quad \mathbf{x} = \begin{bmatrix} x_1 \\ x_2 \end{bmatrix}, \quad \text{and} \quad \ddot{\mathbf{x}} = \begin{bmatrix} \dfrac{d^2 x_1}{dt^2} \\ \dfrac{d^2 x_2}{dt^2} \end{bmatrix}.$$

The solution of this system will not be considered here as ordinary differential equations and systems of the type derived here are discussed in detail in Chapter 6, where matrix methods are also developed. Chapter 7 develops Laplace transform methods for the solution of differential equations and systems. It will suffice to mention here that the dynamical behavior of the compound mass–spring system in Fig. 3.11 is completely characterized by matrix $\mathbf{A}$.

(f) Stochastic Processes

Certain problems arise that are not of a deterministic nature, so that both the formulation of the problem and its outcome must be expressed in terms of probabilities. The probability p that a certain event occurs is a number in the interval $0 \le p \le 1$. An event with probability $p = 0$ is one that never occurs, and an event with probability $p = 1$ is one that is certain to occur. So, for example, when tossing a coin N times and recording each outcome as an H (head) or a T (tail), if the number of heads is N_H and the number of tails is N_T, so that $N = N_H + N_T$, the numbers N_H/N and N_T/N will be approximations to the respective probabilities that a head or a tail occurs when the coin is tossed. If the coin is *unbiased*, it is reasonable to expect that as N increases both N_H/N and N_T/N will approach the value 1/2. This will mean, of course, that the chances of either a head or a tail occurring on each occasion are equal.

The example we now outline is called a **stochastic process** and is illustrated by considering a process that evolves with time and is such that at any given moment it may be in precisely one of N different situations, usually called **states**, where N is finite. We shall denote the N states in which the process may find itself at any given time t_m by $S_1, S_2, \ldots, S_N$, with $m = 0, 1, 2, \ldots$, and $t_{m-1} < t_m$, it being assumed that the outcome at each time depends on probabilities, and so is *not* deterministic.

To formulate the problem we assume that what are called the **conditional probabilities** p_{ki} (also called **transition probabilities**) that determine the probability with which the process will be in state S_j at time t_m are all known, given that it was in state S_k at time t_{m-1}, and that these probabilities are the same from t_1 to t_2 as from t_{m-1} to t_m for $m = 0, 1, 2, \ldots$. This last assumption means that the probability with which the transition from state S_k to S_j occurs is *independent* of the time at which the process was in state S_k.

The conditional probabilities can be arranged as the $N \times N$ matrix $\mathbf{P} = [p_{jk}]$, so as probabilities are involved, all the p_{jk} are nonnegative, and as each stage must have an outcome, the sum of the elements in every row of matrix $\mathbf{P}$ must equal 1. A matrix $\mathbf{P}$ with these properties, namely that

$$0 \leq p_{jk} \leq 1, \quad 0 \leq j \leq N, \quad 0 \leq k \leq N, \quad \text{and} \quad \sum_{j=1}^{N} p_{jk} = 1,$$

is called a **stochastic matrix** (see Exercise 53, Section 3.1).

stochastic matrix and a Markov process

Processes like these, whose condition at any subsequent instant does not depend on how the process arrived at its present state, are called **Markov processes**. Simple but typical examples of such processes involving only two states are gambling wins and losses, the reliability of machines that may either be operational or under repair, shells fired from a gun that either hit or miss the target and errors that introduce an incorrect digit 1 or 0 when transferring binary coded information.

To develop the argument a little further, let us now consider a process that can be in one of two states, and that the matrix $\mathbf{P}$ describing its transitions is given by

$$\mathbf{P} = \begin{bmatrix} 2/3 & 1/3 \\ 1/4 & 3/4 \end{bmatrix}.$$

Now suppose that initially the probability distribution is given by the row matrix $\mathbf{E}(0) = [p, q]$, where, of course, $p + q = 1$. Then if $\mathbf{E}(m)$ denotes the probability distribution of the states at time t_m, it follows that $\mathbf{E}(1) = \mathbf{E}(0)\mathbf{P}$, but as $\mathbf{P}$ is independent of the time we conclude that after m transitions the general result must be

$$\mathbf{E}(m) = \mathbf{E}(0)\mathbf{P}^m,$$

so in this case

$$\mathbf{E}(m) = [p, q] \begin{bmatrix} 2/3 & 1/3 \\ 1/4 & 3/4 \end{bmatrix}^m.$$

Direct calculation shows that

$$\mathbf{E}(3) = [0.470p + 0.398q, 0.530p + 0.602q],$$
$$\mathbf{E}(6) = [0.432p + 0.426q, 0.568p + 0.574q],$$

and

$$\mathbf{E}(10) = [0.429p + 0.429q, 0.571p + 0.571q],$$

so it is reasonable to ask if $\mathbf{E}(m)$ tends to a limiting vector as $m \to \infty$ and, if so, what this is? As this problem is simple, an analytical answer is possible, though it involves using a *diagonalizing* matrix $\mathbf{P}$ which will be discussed later.

We will see later that $\mathbf{P}$ can be written as $\mathbf{ADB}$, where $\mathbf{D}$ is a diagonal matrix and $\mathbf{AB} = \mathbf{I}$. In this case

$$\mathbf{A} = \begin{bmatrix} 1 & 4 \\ 1 & -3 \end{bmatrix}, \quad \mathbf{D} = \begin{bmatrix} 1 & 0 \\ 0 & 5/12 \end{bmatrix}, \quad \text{and} \quad \mathbf{B} = \begin{bmatrix} 3/7 & 4/7 \\ 1/7 & -1/7 \end{bmatrix},$$

so

$$\mathbf{P} = \begin{bmatrix} 1 & 4 \\ 1 & -3 \end{bmatrix} \begin{bmatrix} 1 & 0 \\ 0 & 5/12 \end{bmatrix} \begin{bmatrix} 3/7 & 4/7 \\ 1/7 & -1/7 \end{bmatrix}.$$

In what follows we will need to make repeated use of the fact that

$$\mathbf{BA} = \begin{bmatrix} 3/7 & 4/7 \\ 1/7 & -1/7 \end{bmatrix} \begin{bmatrix} 1 & 4 \\ 1 & -3 \end{bmatrix} = \begin{bmatrix} 1 & 0 \\ 0 & 1 \end{bmatrix} = \mathbf{I}.$$

Using this last property we find that

$$\mathbf{P}^2 = \begin{bmatrix} 1 & 4 \\ 1 & -3 \end{bmatrix} \begin{bmatrix} 1 & 0 \\ 0 & 5/12 \end{bmatrix} \begin{bmatrix} 3/7 & 4/7 \\ 1/7 & -1/7 \end{bmatrix} \begin{bmatrix} 1 & 4 \\ 1 & -3 \end{bmatrix} \begin{bmatrix} 1 & 0 \\ 0 & 5/12 \end{bmatrix} \begin{bmatrix} 3/7 & 4/7 \\ 1/7 & -1/7 \end{bmatrix}$$

$$= \begin{bmatrix} 1 & 4 \\ 1 & -3 \end{bmatrix} \begin{bmatrix} 1 & 0 \\ 0 & 5/12 \end{bmatrix}^2 \begin{bmatrix} 3/7 & 4/7 \\ 1/7 & -1/7 \end{bmatrix}.$$

However, when a diagonal matrix is raised to a power, each of its elements is raised to that same power (see Problem 41, Section 3.1), so

$$\mathbf{P}^2 = \begin{bmatrix} 1 & 4 \\ 1 & -3 \end{bmatrix} \begin{bmatrix} 1 & 0 \\ 0 & (5/12)^2 \end{bmatrix} \begin{bmatrix} 3/7 & 4/7 \\ 1/7 & -1/7 \end{bmatrix}$$

and, in general,

$$\mathbf{P}^m = \begin{bmatrix} 1 & 4 \\ 1 & -3 \end{bmatrix} \begin{bmatrix} 1 & 0 \\ 0 & (5/12)^m \end{bmatrix} \begin{bmatrix} 3/7 & 4/7 \\ 1/7 & -1/7 \end{bmatrix}.$$

Thus,

$$\mathbf{P}^m = \begin{bmatrix} \dfrac{3 + 4(5/12)^m}{7} & \dfrac{4 - 4(5/12)^m}{7} \\[2ex] \dfrac{3 - 3(5/12)^m}{7} & \dfrac{4 + 3(5/12)^m}{7} \end{bmatrix},$$

showing that as $m \to \infty$, so

$$\lim_{m \to \infty} \mathbf{E}(m)\mathbf{P}^m = [3(p+q)/7, 4(p+q)/7] = [3/7, 4/7],$$

and we have found the limiting state of the system.

Stochastic processes also occur that involve more than two states. The problem of determining the probability with which such processes will be in a given state, and when a limiting state exists, the limiting values of the probabilities involved, is of considerable practical importance. An introduction to stochastic process can be found in reference [2.4].

Summary

This section has introduced some of the many areas in which matrices play an essential role. These range from electric circuits needing the application of Kirchhoff's laws, through routing problems involving the concepts of directed graphs and adjacency matrices, to the classical Königsberg bridge problem, computer graphic operations performed by linear transformations, the matrix analysis of forces in a framed structure, the oscillations of a coupled mass–spring system, and stochastic processes.

EXERCISES 3.2

1. State which of the following matrices is a stochastic matrix, giving a reason when this is not the case.

(a) $\begin{bmatrix} 0.5 & 0.3 & 0.2 \\ 0.25 & 0 & 0.75 \\ 0.5 & 0.5 & 0 \end{bmatrix}$.

(c) $\begin{bmatrix} 0.5 & 0.2 & 0.3 \\ 0.7 & 0.3 & 0.2 \\ 0.4 & 0.2 & 0.4 \end{bmatrix}$.

(b) $\begin{bmatrix} 1.2 & 0 & -0.2 \\ 0 & 0.8 & 0.2 \\ 0.6 & 0.3 & 0.1 \end{bmatrix}$.

(d) $\begin{bmatrix} 0.3 & 0.1 & 0.6 \\ 0.8 & 0 & 0.2 \\ 0 & 1 & 0 \end{bmatrix}$.

2. Given the stochastic matrix

$$\mathbf{P} = \begin{bmatrix} 3/4 & 1/4 \\ 1/2 & 1/2 \end{bmatrix}$$

and the initial probability distribution $\mathbf{E}(0) = [p, q]$, with $p, q \geq 0$ and $p + q = 1$, the probability distribution of the two states at time t_m is given by

$$\mathbf{E}(m) = \mathbf{E}(0)\mathbf{P}^m.$$

Find $\mathbf{E}(2)$, $\mathbf{E}(4)$, and $\mathbf{E}(6)$, together with their values when $p = 1/4$, $q = 3/4$.

In Exercises 3 through 6 find the adjacency matrices for the given graphs and digraphs.

3.

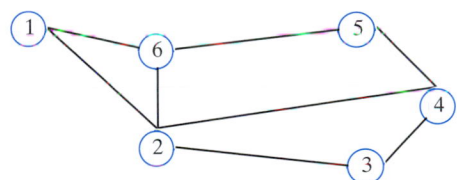

FIGURE 3.12

4.

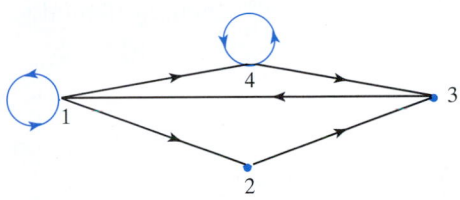

FIGURE 3.13

5.

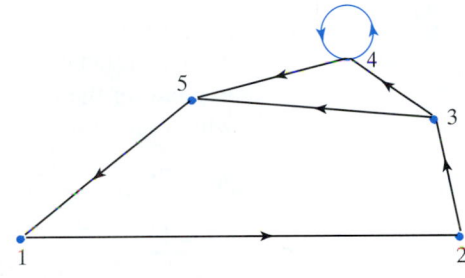

FIGURE 3.14

6.

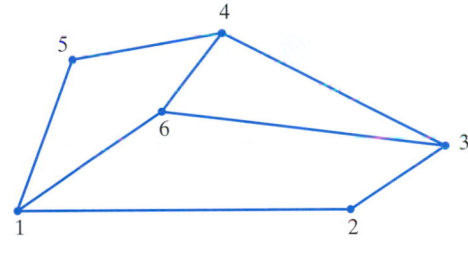

FIGURE 3.15

3.3   Determinants

Every square matrix $\mathbf{A}$ with numbers as elements has associated with it a single unique number called the *determinant* of $\mathbf{A}$, which is written $\det\mathbf{A}$. If $\mathbf{A}$ is an $n \times n$ matrix, the determinant of $\mathbf{A}$ is indicated by displaying the elements a_{ij} of $\mathbf{A}$ between two vertical bars as follows:

notation for a determinant

$$\det\mathbf{A} = \begin{vmatrix} a_{11} & a_{12} & \cdots & a_{1n} \\ a_{21} & a_{22} & \cdots & a_{2n} \\ \cdots & \cdots & \cdots & \cdots \\ a_{n1} & a_{n2} & \cdots & a_{nn} \end{vmatrix}. \tag{5}$$

The number n is called the **order** of determinant $\mathbf{A}$, and in (5) the vertical bars are used to distinguish $\det\mathbf{A}$, that is a number, from matrix $\mathbf{A}$ that is an $n \times n$ array of numbers.

A general definition of the value of $\det\mathbf{A}$ in terms of its elements a_{ij} will be given later, so for the moment we define only the value of first and second order determinants (see Section 1.7). If $\mathbf{A}$ only contains a single element a_{11} so $\mathbf{A} = [a_{11}]$ then, by definition, $\det\mathbf{A} = a_{11}$, and if $\mathbf{A}$ is the 2×2 matrix

$$\mathbf{A} = \begin{bmatrix} a_{11} & a_{12} \\ a_{21} & a_{22} \end{bmatrix},$$

then, by definition,

$$\det\mathbf{A} = \begin{bmatrix} a_{11} & a_{12} \\ a_{21} & a_{22} \end{bmatrix} = a_{11}a_{22} - a_{21}a_{12}. \tag{6}$$

Notice that in (6) the numerical value of $\det\mathbf{A}$ is obtained by forming the product of the two terms a_{11} and a_{22} on the leading diagonal, and subtracting from it the product of the two terms a_{21} and a_{12} on the cross diagonal. This process, called **expanding** the determinant, is easily remembered by representing the method by which the determinant is expanded as

$$\begin{matrix} a_{11} & a_{12} \\ & \times & \\ a_{21} & a_{22} \end{matrix} = a_{11}a_{22} - a_{21}a_{12},$$

where the product involving the downward arrow generates the first pair of terms on the right and the product involving the upward arrow indicates that the product of the associated pair of terms is to be subtracted.

EXAMPLE 3.10   Find $\det\mathbf{A}$ given

(a) $\det\mathbf{A} = \begin{vmatrix} 3 & -1 \\ 2 & 6 \end{vmatrix}$ and   (b) $\det\mathbf{A} = \begin{vmatrix} 1+i & i \\ -3i & 2 \end{vmatrix}$.

Solution   **(a)** Using (5) we have

$$\det\mathbf{A} = \begin{vmatrix} 3 & -1 \\ 2 & 6 \end{vmatrix} = 3 \cdot 6 - 2 \cdot (-1) = 20.$$

(b) Again using (5) we have

$$\det\mathbf{A} = \begin{vmatrix} 1+i & i \\ -3i & 2 \end{vmatrix} = (1+i) \cdot 2 - (-3i) \cdot i = -1 + 2i. \qquad \blacksquare$$

To provide some motivation for the introduction of determinants, we solve by elimination the two linear simultaneous algebraic equations

$$\begin{aligned} a_{11}x_1 + a_{12}x_2 &= b_1 \\ a_{21}x_1 + a_{22}x_2 &= b_2. \end{aligned} \tag{7}$$

To eliminate x_2 we multiply the first equation by a_{22} and the second equation by a_{12}, and then subtract the results to obtain

$$(a_{11}a_{22} - a_{21}a_{12})x_1 = a_{22}b_1 - a_{12}b_2.$$

This shows that when $a_{11}a_{22} - a_{21}a_{12} \neq 0$,

$$x_1 = \frac{a_{22}b_1 - a_{12}b_2}{a_{11}a_{22} - a_{21}a_{12}}.$$

This result can be expressed in terms of det$\mathbf{A}$ as

$$x_1 = (a_{22}b_1 - a_{12}b_2)/\det\mathbf{A}. \tag{8}$$

Similarly, when x_1 is eliminated from equations (7) we find that

$$x_2 = (a_{11}b_2 - a_{21}b_1)/\det\mathbf{A}. \tag{9}$$

Cramer's rule for a system of two equations

Examination of (8) and (9) shows that their numerators can be written in terms of determinants that are closely related to det$\mathbf{A}$, because

$$x_1 = \frac{D_1}{D} \quad \text{and} \quad x_2 = \frac{D_2}{D}, \tag{10}$$

where

$$D = \det\mathbf{A}, \quad D_1 = \begin{vmatrix} b_1 & a_{12} \\ b_2 & a_{22} \end{vmatrix}, \quad \text{and} \quad D_2 = \begin{vmatrix} a_{11} & b_1 \\ a_{21} & b_2 \end{vmatrix}. \tag{11}$$

The form of solution of equations (7) in terms of the determinants in (10) and (11) is called **Cramer's rule**. The rule itself says that $x_i = D_i/D$ for $i = 1, 2$, where determinant D_1 is obtained from $D = \det\mathbf{A}$ by replacing the *first* column of $\mathbf{A}$ by the nonhomogeneous terms b_1 and b_2 on the right of equations (7), and determinant D_2 is obtained from D by replacing the *second* column of $\mathbf{A}$ by these same two terms.

EXAMPLE 3.11

Use Cramer's rule to solve the equations

$$3x_1 + 5x_2 = 4$$
$$2x_1 - 4x_2 = 1.$$

Solution   The three determinants required by Cramer's are

$$D = \det\mathbf{A} = \begin{vmatrix} 3 & 5 \\ 2 & -4 \end{vmatrix} = -22, \quad D_1 = \begin{vmatrix} 4 & 5 \\ 1 & -4 \end{vmatrix} = -21, \quad D_2 = \begin{vmatrix} 3 & 4 \\ 2 & 1 \end{vmatrix} = -5,$$

so $x_1 = D_1/D = 21/22$ and $x_2 = D_2/D = 5/22$. ∎

This example shows how determinants enter naturally into the solution of a system of equations. As determinants of order $n > 2$ occur in the study of differential equations, analytical geometry, throughout linear algebra, and elsewhere, it is necessary to generalize the definition of a determinant of order 2 given in (6) to determinants of any order n.

With this objective in mind, we first define the *minors* and *cofactors* of a determinant of order n. The **minor** M_{ij} associated with the element a_{ij} in the ith row and jth column of the nth order determinant in (5) is the determinant of order $n-1$ formed from det$\mathbf{A}$ by deleting the elements in the ith row and jth column. As each element of det$\mathbf{A}$ has an associated minor, a determinant of order n has n^2 minors.

minors and cofactors

By way of example, the minor M_{3j} of the nth order determinant in (5) is the determinant of order $n-1$

$$M_{3j} = \begin{vmatrix} a_{11} & a_{12} & \cdots & a_{1j-1} & a_{1j+1} & \cdots & a_{1n} \\ a_{21} & a_{22} & \cdots & a_{2j-1} & a_{2j+1} & \cdots & a_{2n} \\ a_{41} & a_{42} & \cdots & a_{4j-1} & a_{4j+1} & \cdots & a_{4n} \\ \cdots & \cdots & \cdots & \cdots & \cdots & \cdots & \cdots \\ a_{n1} & a_{n2} & \cdots & a_{nj-1} & a_{nj+1} & \cdots & a_{nn} \end{vmatrix}. \tag{12}$$

The **cofactor** C_{ij} associated with the element a_{ij} in determinant (5) is defined in terms of the minor M_{ij} as

$$C_{ij} = (-1)^{i+j} M_{ij} \quad \text{for } i, j = 1, 2, \ldots, n, \tag{13}$$

so an nth order determinant has n^2 cofactors.

EXAMPLE 3.12   Find the minors and cofactors of

$$\det\mathbf{A} = \begin{vmatrix} 2 & -3 \\ 1 & 4 \end{vmatrix}.$$

Solution   Inspection shows that $M_{11} = 4$, $M_{12} = 1$, $M_{21} = -3$, and $M_{22} = 2$. Using definition (12), the cofactors are seen to be

$$C_{11} = (-1)^{1+1} M_{11} = 4, \quad C_{12} = (-1)^{1+2} M_{12} = -1, \quad C_{21} = (-1)^{2+1} M_{21} = 3,$$

and $C_{22} = (-1)^{2+2} M_{22} = 2$. ∎

Recognizing that the cofactors of the second order determinant

$$\det\mathbf{A} = \begin{vmatrix} a_{11} & a_{12} \\ a_{21} & a_{22} \end{vmatrix} \quad \text{are } C_{11} = a_{22}, \; C_{12} = -a_{21}, \; C_{21} = -a_{12}, \; \text{and } C_{22} = a_{11},$$

expanding a second order determinant in terms of rows or columns

we see from the definition $\det\mathbf{A} = a_{11}a_{22} - a_{21}a_{12}$ that $\det\mathbf{A}$ can be expressed in terms of these cofactors in four different ways:

$\det\mathbf{A} = a_{11}C_{11} + a_{12}C_{12}$, using elements and cofactors from the first row of $\mathbf{A}$;

$\det\mathbf{A} = a_{21}C_{21} + a_{22}C_{22}$, using elements and cofactors from the second row of $\mathbf{A}$;

$\det\mathbf{A} = a_{11}C_{11} + a_{21}C_{21}$, using elements and cofactors from the first column of $\mathbf{A}$;

$\det\mathbf{A} = a_{12}C_{12} + a_{22}C_{22}$, using elements and cofactors from the second

column of $\mathbf{A}$.

This has proved by direct calculation that the value of the general second order determinant $\det\mathbf{A}$ is given by the sum of the products of the elements and their associated cofactors in any row or column of the determinant. When the definition of a determinant is extended to the case $n > 2$ it will be seen that this same property remains true.

There are various ways of defining an nth order determinant, and from among these we have chosen to use one that involves a recursive process. More will be said about this recursive process, and how it can be used to evaluate the determinant, once the definition has been formulated.

Definition of a determinant of order n

The nth order determinant $\det\mathbf{A}$ in which the element a_{ij} has the associated cofactor C_{ij} for $i, j = 1, 2, \ldots, n$ is defined as

$$\det\mathbf{A} = \begin{vmatrix} a_{11} & a_{12} & \cdots & a_{1n} \\ a_{21} & a_{22} & \cdots & a_{2n} \\ \cdots & \cdots & \cdots & \cdots \\ a_{n1} & a_{n2} & \cdots & a_{nn} \end{vmatrix} = \sum_{j=1}^{n} a_{1j} C_{1j}. \tag{14}$$

Recalling the different ways in which a second order determinant can be evaluated, we see that the expansion of det$\mathbf{A}$ in (14) is in terms of the elements and cofactors of the first row, so for conciseness this expansion is said to be in terms of the *elements of the first row*.

The recursive process enters this definition through the fact that each cofactor C_{1j} is a determinant of order $n-1$, as can be seen from (12), so each cofactor in turn can be expanded in terms of determinants of order $n-2$, and the process continued until determinants of order 2 are obtained that can then be calculated using (6).

EXAMPLE 3.13   Expand

$$\det\mathbf{A} = \begin{vmatrix} 1 & 4 & -1 \\ 2 & 0 & 3 \\ 1 & 2 & 1 \end{vmatrix}.$$

Solution   To expand this third order determinant using (14), we must find the cofactors of the elements of the first row, so to do this we first find the minors and then use (13) to find the cofactors, as a result of which we find that

$$M_{11} = \begin{vmatrix} 0 & 3 \\ 2 & 1 \end{vmatrix} = -6, \quad \text{so } C_{11} = (-1)^{1+1}(-6) = -6$$

$$M_{12} = \begin{vmatrix} 2 & 3 \\ 1 & 1 \end{vmatrix} = -1, \quad \text{so } C_{12} = (-1)^{1+2}(-1) = 1$$

$$M_{13} = \begin{vmatrix} 2 & 0 \\ 1 & 2 \end{vmatrix} = 4, \quad \text{so } C_{13} = (-1)^{1+3}(4) = 4.$$

As the elements of the first row are $a_{11} = 1$, $a_{12} = 4$, and $a_{13} = -1$, we find from (12) that

$$\det\mathbf{A} = (1)C_{11} + (4)C_{12} + (-1)C_{13} = (1)(-6) + (4)(1) + (-1)(4) = -6. \quad \blacksquare$$

The determinant associated with either an upper or a lower triangular matrix $\mathbf{A}$ of any order is easily expanded, because repeated application of (12) shows that it reduces to the product of the terms on the leading diagonal, so the expansion of the nth order upper triangular determinant with elements $a_{11}, a_{22}, \ldots, a_{nn}$ on its leading diagonal

$$\det\mathbf{A} = \begin{vmatrix} a_{11} & a_{12} & \cdots & a_{1n} \\ 0 & a_{22} & \cdots & a_{2n} \\ 0 & 0 & \cdots & \cdots \\ 0 & 0 & 0 & a_{nn} \end{vmatrix} = a_{11}a_{22}\ldots a_{nn}, \tag{15}$$

and a corresponding result is true for a lower triangular matrix.

Definition (14) can be used to prove that nth order determinants, like second order determinants, have the property that their value is given by the sum of the products of the elements and their cofactors in *any* row or column. This result, together with a generalization concerning the vanishing of the sum of the products of the elements in any row (or column) and the corresponding cofactors in a different row (or column), forms the next theorem. The details of the proof can be found in linear algebra texts, for example, [2.1], [2.5], [2.7], [2.9], but the method used has no other application in what is to follow, so the proof will be omitted.

THEOREM 3.3

Laplace expansion theorem

Laplace expansion theorem and an extension   Let $\mathbf{A}$ be an $n \times n$ matrix with elements a_{ij}. Then,

(i) $\det\mathbf{A}$ can be expanded in terms of elements of its ith row and the cofactors C_{ij} of the ith row as

$$\det\mathbf{A} = a_{i1}C_{i1} + a_{i2}C_{i2} + \cdots + a_{in}C_{in} = \sum_{j=1}^{n} a_{ij}C_{ij}$$

for any fixed i with $1 \leq i \leq n$.

(ii) $\det\mathbf{A}$ can be expanded in terms of elements of its jth column and the cofactors C_{ij} of the jth column as

$$\det\mathbf{A} = a_{1j}C_{1j} + a_{2j}C_{2j} + \cdots + a_{nj}C_{nj} = \sum_{j=1}^{n} a_{ij}C_{ij}$$

for any fixed j with $1 \leq j \leq n$.

(iii) The sum of the products of the elements of the ith row with the corresponding cofactors of the kth row is zero when $i \neq k$.

(iv) The sum of the products of the elements in the jth column with the corresponding cofactors of the kth column is zero when $j \neq k$. ∎

Results (i) and (ii) are often used to advantage when a row or column contains many zeros, because if the determinant is expanded in terms of the elements of that row or column, the cofactors associated with each zero element need not be calculated.

Results (iii) and (iv) simply say that the sum of the products of the elements in any row (or column) with the corresponding cofactors in a *different* row (or column) is zero.

PIERRE SIMON LAPLACE (1749–1827)
A French mathematician of remarkable ability who made contributions to analysis, differential equations, probability, and celestial mechanics. He used mathematics as a tool with which to investigate physical phenomena, and made fundamental contributions to hydrodynamics, the propagation of sound, surface tension in liquids, and many other topics. His many contributions had a wide-ranging effect on the development of mathematics.

EXAMPLE 3.14

Verify Theorem 3.3(i) by expanding the determinant in Example 3.13 in terms of the elements of its second row. Use the determinant to check the result of Theorem 3.3(iii).

Solution   The second row contains a zero element in its mid position, so the cofactor C_{22} associated with the zero element need not be calculated. The necessary cofactors in the second row that follow from the minors are

$$M_{21} = \begin{vmatrix} 4 & -1 \\ 2 & 1 \end{vmatrix} = 6 \quad \text{so } C_{21} = (-1)^{2+1}(6) = -6$$

$$M_{23} = \begin{vmatrix} 1 & 4 \\ 1 & 2 \end{vmatrix} = -2 \quad \text{so } C_{23} = (-1)^{2+3}(-2) = 2.$$

As $a_{21} = 2$ and $a_{23} = 3$, it follows from Theorem 3.3(i) that when det$\mathbf{A}$ is expanded in terms of elements of its second row,

$$\det\mathbf{A} = (2)(-6) + (3)(2) = -6,$$

confirming the result obtained in Example 3.13.

As a particular case of Theorem 3.3(iii), let us show that the sum of the products of the cofactors in the first row of det$\mathbf{A}$ and the corresponding elements in the third row is zero.

In Example 3.13 it was found that $C_{11} = -6$, $C_{12} = 1$, and $C_{13} = 4$, so as the elements of the third row are $a_{31} = 1$, $a_{32} = 2$, and $a_{33} = 1$, we have

$$a_{31}C_{11} + a_{32}C_{12} + a_{33}C_{13} = (-6)(1) + (2)(1) + (1)(4) = 0,$$

confirming the result of Theorem 3.3(iii) when the elements of row 3 and the cofactors of row 1 are used.   ■

Determinants have a number of special properties that can be used to simplify their expansion, though their main uses are found elsewhere in mathematics, where determinants often characterize some important theoretical feature of a problem. The most important and useful of these properties are contained in the next theorem.

THEOREM 3.4

basic properties of determinants

Properties of determinants   A determinant det$\mathbf{A}$ has the following properties:

(i) If any row or column of a determinant det$\mathbf{A}$ only contains zero elements, then det$\mathbf{A} = 0$.

(ii) If $\mathbf{A}$ is a square matrix with the transpose $\mathbf{A}^{\mathrm{T}}$, then det$\mathbf{A} = \det\mathbf{A}^{\mathrm{T}}$.

(iii) If each element of a row or column of a square matrix $\mathbf{A}$ is multiplied by a constant k, then the value of the determinant is $k\det\mathbf{A}$.

(iv) If two rows (or columns) of a square matrix are interchanged, the sign of the determinant is changed.

(v) If any two rows or columns of a square matrix $\mathbf{A}$ are proportional, then det$\mathbf{A} = 0$.

(vi) Let the square matrix $\mathbf{A}$ be such that each element a_{ij} of the ith row (or the jth column) can be written as $a_{ij} = a_{ij}^{(1)} + a_{ij}^{(2)}$. Then if $\mathbf{A}_1$ is the matrix derived from $\mathbf{A}$ by replacing its ith row (or jth column) by the elements $a_{ij}^{(1)}$ and $\mathbf{A}_2$ is the matrix derived from $\mathbf{A}$ by replacing its ith row (or jth column) by the elements $a_{ij}^{(2)}$,

$$\det\mathbf{A} = \det\mathbf{A}_1 + \det\mathbf{A}_2.$$

(vii) The addition of a multiple of a row (or column) of a determinant to another row (or column) of the determinant leaves the value of the determinant unchanged.

(viii) Let $\mathbf{A}$ and $\mathbf{B}$ be two $n \times n$ matrix, then

$$\det(\mathbf{AB}) = \det\mathbf{A}\,\det\mathbf{B}.$$

Proof

(i) The result follows by expanding the determinant in terms of the row or column that only contains zero elements.

(ii) The result follows from the fact that expanding $\det\mathbf{A}$ in terms of the elements of its first row is the same as expanding $\det\mathbf{A}^{\mathrm{T}}$ in terms of the elements of its first column.

(iii) The result follows by expanding the determinant in terms of the row or column in which each element has been multiplied by the constant k, because k appears as a factor in each term, so the result becomes $k\det\mathbf{A}$.

(iv) The proof is by induction, starting with a second order determinant for which the result can be seen to be true from definition (6). To proceed with an inductive proof we assume the results to be true for a determinant of order $r - 1$, and show it must be true for a determinant of order r. Expand a row of a determinant of order r in terms of the elements of a row (or column) that has not been interchanged. Then, by hypothesis, as the cofactors are determinants of order $r - 1$, their signs will all be reversed. This establishes that if the hypothesis is true for a determinant of order $r - 1$ it must also be true for a determinant of order r. As the result is true for $r = 2$, it follows by induction that it is true for all integers $r > 2$, and the result is proved.

(v) If the value of the determinant is $\det\mathbf{A}$, and one row is k times another, then from (ii) by removing the factor k from the row the value of the determinant will be $k\det\mathbf{A}_1$, where $\mathbf{A}_1$ is now a determinant with two identical rows. From (ii), interchanging two rows changes the sign of the determinant, but the rows are identical, leaving the determinant invariant, so $\det\mathbf{A}_1 = 0$. A similar argument shows the result to be true when two columns are proportional, so the result is proved.

(vi) The result is proved directly by expanding the determinant in terms of the elements of the ith row (or the jth column).

(vii) Let the square matrix $\mathbf{B}$ be obtained from $\mathbf{A}$ by adding k times the ith row (or a column) to the jth row (or column). Then from (iii) and (vi),

$$\det\mathbf{B} = \det\mathbf{A} + k\det\mathbf{C},$$

where $\mathbf{C}$ is obtained from $\mathbf{A}$ by replacing the ith row (or column) by the jth row (or column). As $\det\mathbf{C}$ has two identical rows (or columns), it follows from (v) that $\det\mathbf{C} = 0$, so $\det\mathbf{B} = \det\mathbf{A}$ and the result is proved.

(viii) A proof of this result will be given later after the introduction of elementary row operation matrices. ∎

Cramer's rule, which was first encountered when seeking the solution of the two equations in (7), can be extended to a system of n equations in a very straightforward manner, and it takes the following form.

Cramer's rule

Cramer's rule for a system of n equations in n unknowns

The solution of the system of n equations in the n unknowns $x_1, x_2, \ldots, x_n$

$$a_{11}x_1 + a_{12}x_2 + \cdots + a_{1n}x_n = b_1$$
$$a_{21}x_1 + a_{22}x_2 + \cdots + a_{2n}x_n = b_2$$
$$\cdots \cdots$$
$$a_{n1}x_1 + a_{n2}x_2 + \cdots + a_{nn}x_n = b_n$$

is given by

$$x_i = \det\mathbf{A}_i / \det\mathbf{A} \quad \text{for } i = 1, 2, \ldots, n,$$

where $\det\mathbf{A}$ is the determinant of the coefficient matrix with elements a_{ij}, and $\det\mathbf{A}_i$ is the determinant obtained from the coefficient matrix by replacing its ith column by the column containing the number $b_1, b_2, \ldots, b_n$.

The justification for Cramer's rule in this more general form will be postponed until after the introduction of inverse matrices, when a simple proof can be given. Cramer's rule is mainly of theoretical importance and, in general, it should not be used to solve equations when $n > 3$. This is because the number of multiplications required to evaluate a determinant of order n is $(n-1)n!$, so to solve for n unknowns $(n+1)$ determinants must be evaluated leading to a total of $(n^2 - 1)n!$ multiplications, and this calculation becomes excessive when $n > 3$. An efficient way of solving large systems by means of elimination is given in Chapter 19.

EXAMPLE 3.15   Use Cramer's rule to solve

$$x_1 - 2x_2 + x_3 = 1$$
$$2x_1 + x_2 - 2x_3 = 3$$
$$-x_1 + 3x_2 + 4x_3 = -2.$$

Solution   The determinants involved are

$$\det\mathbf{A} = \begin{vmatrix} 1 & -2 & 1 \\ 2 & 1 & -2 \\ -1 & 3 & 4 \end{vmatrix} = 29, \quad \det\mathbf{A}_1 = \begin{vmatrix} 1 & -2 & 1 \\ 3 & 1 & -2 \\ -2 & 3 & 4 \end{vmatrix} = 37$$

$$\det\mathbf{A}_2 = \begin{vmatrix} 1 & 1 & 1 \\ 2 & 3 & -2 \\ -1 & -2 & 4 \end{vmatrix} = 1, \quad \det\mathbf{A}_3 = \begin{vmatrix} 1 & -2 & 1 \\ 2 & 1 & 3 \\ -1 & 3 & -2 \end{vmatrix} = -6,$$

so $x_1 = 37/29$, $x_2 = 1/29$, and $x_3 = -6/29$.   ∎

A purely algebraic approach to the study of determinants and their properties is to be found in reference [2.8], and many examples of their applications are given in references [2.11] and [2.12].

Summary

This section has extended to an nth order determinant the basic notion of a second order determinant that was reviewed in Chapter 1, and then established its most important properties. The Laplace expansion formulas that were established are of theoretical importance, but it will be seen later that the practical evaluation of a determinant is most easily performed by reducing the $n \times n$ matrix associated with a determinant to its echelon form.

EXERCISES 3.3

In Exercises 1 through 4 find $\det\mathbf{A}$.

1. $\det\mathbf{A} = \begin{vmatrix} 2 & 1 & -1 \\ 0 & 4 & 3 \\ 3 & 2 & -2 \end{vmatrix}$.

2. $\det\mathbf{A} = \begin{vmatrix} -1 & 2 & 1 \\ 1 & 3 & 2 \\ -4 & 1 & 2 \end{vmatrix}$.

3. $\det\mathbf{A} = \begin{vmatrix} 2 & 4 & -3 \\ -2 & 1 & 0 \\ 5 & -2 & 4 \end{vmatrix}$.

4. $\det\mathbf{A} = \begin{vmatrix} 4 & 0 & 0 \\ -2 & \cos x & -\sin x \\ 5 & \sin x & \cos x \end{vmatrix}$.

5. Given that
$$\det\mathbf{A} = \begin{vmatrix} -3 & 1 & 4 \\ 2 & -1 & 5 \\ 4 & 2 & 5 \end{vmatrix} = 87,$$
confirm by direct calculation that (a) interchanging the first and last rows changes the sign of $\det\mathbf{A}$ and (b) interchanging the second and third columns changes the sign of $\det\mathbf{A}$.

6. Given that
$$\det\mathbf{A} = \begin{vmatrix} 2 & 1 & 3 \\ 5 & -2 & 2 \\ -1 & 1 & 3 \end{vmatrix} = -24,$$
confirm by direct calculation that (a) adding twice row two to row three leaves $\det\mathbf{A}$ unchanged and (b) subtracting three times column three from column one leaves $\det\mathbf{A}$ unchanged.

Establish the results in Exercises 7 through 12 without a direct expansion of the determinant by using the properties listed in Theorem 3.4.

7. $\begin{vmatrix} 1+a & a & a \\ b & 1+b & b \\ c & c & 1+c \end{vmatrix} = (1 + a + b + c)$.

8. $\begin{vmatrix} 1 & a & b+c \\ 1 & b & c+a \\ 1 & c & a+b \end{vmatrix} = 0$.

9. $\begin{vmatrix} a^2 & b^2 & c^2 \\ a & b & c \\ 1 & 1 & 1 \end{vmatrix} = (a-b)(a-c)(b-c)$.

10. $\begin{vmatrix} x^2+a^2 & ab & ac \\ ab & x^2+b^2 & bc \\ ac & cb & x^2+c^2 \end{vmatrix} = x^4(x^2 + a^2 + b^2 + c^2)$.

11. $\begin{vmatrix} 1 & a & b \\ a & 1 & b \\ a & b & 1 \end{vmatrix} = (a + b + 1)(a - 1)(b - 1)$.

12. $\begin{vmatrix} k & 1 & 1 & 1 \\ 1 & k & 1 & 1 \\ 1 & 1 & k & 1 \\ 1 & 1 & 1 & k \end{vmatrix} = (k+3)(k-1)^3$.

In Exercises 13 and 14 use Cramer's rule to solve the system of equations.

13. $2x_1 - 3x_2 + x_3 = 4$
$x_1 + 2x_2 - 2x_3 = 1$
$3x_1 + x_2 - 2x_3 = -2$.

14. $3x_1 + x_2 + 2x_3 = 5$
$2x_1 - 4x_2 + 3x_3 = -3$
$x_1 + 2x_2 + 4x_3 = 2$.

15. Let $P(\lambda)$ be given by
$$P(\lambda) = \begin{vmatrix} 3-\lambda & 0 & 1 \\ 2 & 2-\lambda & 2 \\ 4 & 2 & 1-\lambda \end{vmatrix},$$
where λ is a parameter. Expand the determinant to find the form of the polynomial $P(\lambda)$ and use the result to find for what values of λ the determinant vanishes.

16. Let $P(\lambda)$ be given by
$$P(\lambda) = \begin{vmatrix} 4-\lambda & 0 & 1 \\ 1 & -\lambda & 1 \\ -1 & -2 & 2-\lambda \end{vmatrix},$$
where λ is a parameter. Expand the determinant to find the form of the polynomial $P(\lambda)$ and use the result to find for what values of λ the determinant vanishes.

17. Given that
$$\mathbf{A} = \begin{bmatrix} -3 & 0 & 4 \\ 1 & 2 & -1 \\ 1 & 0 & 1 \end{bmatrix} \quad \text{and} \quad \mathbf{B} = \begin{bmatrix} 1 & 2 & 3 \\ 2 & 3 & 1 \\ 3 & 1 & 2 \end{bmatrix},$$
calculate $\det(\mathbf{AB})$, $\det\mathbf{A}$, $\det\mathbf{B}$, and hence verify that $\det(\mathbf{AB}) = \det\mathbf{A}\det\mathbf{B}$.

3.4   Elementary Row Operations, Elementary Matrices, and Their Connection with Matrix Multiplication

To motivate what is to follow we will examine the processes involved when solving by elimination the system of linear equations

$$
\begin{aligned}
a_{11}x_1 + a_{12}x_2 + \cdots + a_{1n}x_n &= b_1 \\
a_{21}x_1 + a_{22}x_2 + \cdots + a_{2n}x_n &= b_2 \\
&\;\;\cdots\cdots\cdots \\
a_{m1}x_1 + a_{m2}x_2 + \cdots + a_{mn}x_n &= b_m,
\end{aligned}
\tag{16}
$$

though later more will need to be said about the details of this important problem, and how it is influenced by the number of equations m and the number of unknowns n.

Elementary Row Operations

The three types of elementary row operations used when solving equations (16) by elimination are:

the three basic types of elementary row operation

TYPE I   The interchange of two equations
TYPE II   The scaling of an equation by a nonzero constant
TYPE III   The addition of a scalar multiple of an equation to another equation

In matrix notation the system of equations (16) becomes

$$
\mathbf{Ax} = \mathbf{b},
\tag{17}
$$

where $\mathbf{A} = [a_{ij}]$ is an $m \times n$ matrix, $\mathbf{x} = [x_1, x_2, \ldots, x_n]^{\mathrm{T}}$, and $\mathbf{b} = [b_1, b_2, \ldots, b_m]^{\mathrm{T}}$. The three elementary row operations of types I to III that can be performed on the *equations* in (16) can be interpreted as the corresponding operations performed on the *rows* of the matrices $\mathbf{A}$ and $\mathbf{b}$. This is equivalent to performing these same operations on the rows of the new matrix denoted by $(\mathbf{A}, \mathbf{b})$, defined as

$$
(\mathbf{A}, \mathbf{b}) =
\left[
\begin{array}{cccc:c}
a_{11} & a_{12} & \cdots & a_{1n} & b_1 \\
a_{21} & a_{22} & \cdots & a_{2n} & b_2 \\
& & \cdots\cdots & & \\
a_{m1} & a_{m2} & \cdots & a_{mn} & b_m
\end{array}
\right],
\tag{18}
$$

that has m rows and $n+1$ columns and is obtained by inserting the column vector $\mathbf{b}$ containing the nonhomogeneous terms on the right of matrix $\mathbf{A}$.

When considering the system of linear equations in (16), matrix $(\mathbf{A}, \mathbf{b})$ is called the **augmented matrix** associated with the system. The separation of the last column in (18) by a vertical dashed line is to indicate *partitioning* of the matrix to show that the elements of the last column are not elements of the coefficient matrix $\mathbf{A}$.

the augmented matrix

We are now in a position to introduce a notation for the three *elementary row operations* that are necessary when using an elimination process to find the solution of a system of equations in matrix form (ordinary or augmented).

Elementary row operations

The three **elementary row operations** that may be performed on a matrix are:

(i) The interchange of the ith and jth rows, which will be denoted by $R\{i \to j, j \to i\}$.

(ii) The replacement of each element in the ith row by its product with a nonzero constant α, which will be denoted by $R\{(\alpha)i \to i\}$.

(iii) The replacement of each element of the jth row by the sum of β times the corresponding element in the ith row and the element in the jth row, which will be denoted by $R\{(\beta)i + j \to j\}$.

EXAMPLE 3.16

To illustrate the elementary row operations, we consider the matrix

$$\mathbf{A} = \begin{bmatrix} 1 & 6 & 4 & -3 & 2 \\ 2 & 0 & 1 & 7 & 4 \\ 5 & 2 & 8 & 2 & 3 \end{bmatrix}.$$

An example of an elementary row operation of type (i) performed on $\mathbf{A}$ is provided by $R\{1 \to 3, 3 \to 1\}$. This requires rows 1 and 3 to be interchanged to give the new matrix

$$R\{1 \to 3, 3 \to 1\}\mathbf{A} = \begin{bmatrix} 5 & 2 & 8 & 2 & 3 \\ 2 & 0 & 1 & 7 & 4 \\ 1 & 6 & 4 & -3 & 2 \end{bmatrix}.$$

An example of an elementary row operation of type (ii) performed on $\mathbf{A}$ is provided by $R\{(-3)1 \to 1\}$. This requires each element in row 1 to be multiplied by -3 to give the new matrix

$$R\{(-3)1 \to 1\}\mathbf{A} = \begin{bmatrix} -3 & -18 & -12 & 9 & -6 \\ 2 & 0 & 1 & 7 & 4 \\ 5 & 2 & 8 & 2 & 3 \end{bmatrix}.$$

An example of an elementary row operation of type (iii) performed on $\mathbf{A}$ is provided by $R\{(4)1 + 2 \to 2\}$, which requires the elements of row 1 to be multiplied by 4 and then added to the corresponding elements of row 2 to give the new matrix

$$R\{(4)1 + 2 \to 2\}\mathbf{A} = \begin{bmatrix} 1 & 6 & 4 & -3 & 2 \\ 6 & 24 & 17 & -5 & 12 \\ 5 & 2 & 8 & 2 & 3 \end{bmatrix}. \qquad \blacksquare$$

A sequence of elementary row operations performed on the augmented matrix $(\mathbf{A}, \mathbf{b})$ will lead to a different augmented matrix $(\mathbf{A}', \mathbf{b}')$. However, as this is equivalent to performing the corresponding sequence of operations on the actual equations in (16), although $(\mathbf{A}, \mathbf{b})$ and $(\mathbf{A}', \mathbf{b}')$ will look different, the interpretation of $(\mathbf{A}', \mathbf{b}')$ in terms of the solution of the system of equations in (16) will, of course, be the same as that of $(\mathbf{A}, \mathbf{b})$. It will be seen later that the purpose of carrying out these operations on a matrix is to simplify it while leaving its essential algebraic structure unaltered, e.g., without changing the solution $x_1, \ldots, x_n$ of the corresponding system of equations.

The definition that now follows is a consequence of the equivalence, in terms of equations (16), of matrix $(\mathbf{A}, \mathbf{b})$ and any matrix $(\mathbf{A}', \mathbf{b}')$ that can be derived from it by means of a sequence of elementary row operations, though the definition applies to matrices in general, and not only to augmented matrices.

Row equivalence of matrices

Two $m \times n$ matrices will be said to be **row equivalent** if one can be obtained from the other by means of a sequence of elementary row operations. Row equivalence between matrices $\mathbf{A}$ and $\mathbf{B}$ is denoted by writing $\mathbf{A} \sim \mathbf{B}$.

The row equivalence of matrices has the useful properties listed in the following theorem.

THEOREM 3.5   **Reflexive, symmetric, and transitive properties of row equivalence**

(i) Every $m \times n$ matrix $\mathbf{A}$ is row equivalent to itself (*reflexive* property).

(ii) Let $\mathbf{A}$ and $\mathbf{B}$ be $m \times n$ matrices. Then if $\mathbf{A}$ is row equivalent to $\mathbf{B}$, $\mathbf{B}$ is row equivalent to $\mathbf{A}$ (*symmetric* property).

(iii) Let $\mathbf{A}$, $\mathbf{B}$, and $\mathbf{C}$ be $m \times n$ matrices. Then if matrix $\mathbf{A}$ is row equivalent to $\mathbf{B}$ and $\mathbf{B}$ is row equivalent to $\mathbf{C}$, $\mathbf{A}$ is row equivalent to $\mathbf{C}$ (*transitive* property).

Proof

(i) The property is self-evident.

(ii) To establish this property we must show the three elementary row operations involved are reversible. In the case of elementary row operations of type (i) the result follows from the fact that if an application of the operation $R\{i \to j, j \to i\}$ to matrix $\mathbf{A}$ yields a new matrix $\mathbf{B}$, an application of the operation $R\{j \to i, i \to j\}$ to matrix $\mathbf{B}$ generates the original matrix $\mathbf{A}$.

Similarly, in the case of elementary row operations of type (ii), if an application of the operation $R\{(\alpha)i \to i\}$ to matrix $\mathbf{A}$ yields a new matrix $\mathbf{B}$, an application of the operation $R\{(1/\alpha)i \to i\}$ to matrix $\mathbf{B}$ reproduces the original matrix $\mathbf{A}$.

Finally we consider the case of elementary row operations of type (iii). If an application of the operation $R\{(\beta)i + j \to j\}$ to matrix $\mathbf{A}$ yields a new matrix $\mathbf{B}$, an application of the operation $R\{(-\beta)i + j \to j\}$ to $\mathbf{B}$ returns the original matrix $\mathbf{A}$. Taken together these results establish property (ii).

(iii) Using property (ii) in (iii) establishes the row equivalence first of $\mathbf{A}$ and $\mathbf{B}$, and then of $\mathbf{B}$ and $\mathbf{C}$, and hence of $\mathbf{A}$ and $\mathbf{C}$, so property (iii) is proved.   ∎

Let us now define what are called *elementary matrices* and examine the effect they have when used to premultiply a matrix.

Elementary matrices

An $n \times n$ **elementary matrix** is any matrix that is obtained from an $n \times n$ unit matrix $\mathbf{I}$ by performing a single elementary row operation.

The following concise notation will be used to identify the elementary matrices that correspond to each of the three elementary row operations.

the three basic types of elementary matrix

TYPE I $\mathbf{E}_{ij}$ will denote the elementary matrix obtained from the unit matrix $\mathbf{I}$ by interchanging its ith and jth rows.

TYPE II $\mathbf{E}_i(c)$ will denote the matrix obtained from the unit matrix $\mathbf{I}$ by multiplying its ith row by the nonzero scalar c.

TYPE III $\mathbf{E}_{ij}(c)$ will denote the matrix obtained from the unit matrix $\mathbf{I}$ by adding c times its ith row to its jth row.

EXAMPLE 3.17

Let $\mathbf{I}$ be the 3×3 unit matrix. Then

$$\mathbf{I} = \begin{bmatrix} 1 & 0 & 0 \\ 0 & 1 & 0 \\ 0 & 0 & 1 \end{bmatrix}, \quad \mathbf{E}_{23} = \begin{bmatrix} 1 & 0 & 0 \\ 0 & 0 & 1 \\ 0 & 1 & 0 \end{bmatrix}, \quad \mathbf{E}_3(4) = \begin{bmatrix} 1 & 0 & 0 \\ 0 & 1 & 0 \\ 0 & 0 & 4 \end{bmatrix}, \quad \text{and}$$

$$\mathbf{E}_{13}(5) = \begin{bmatrix} 1 & 0 & 0 \\ 0 & 1 & 0 \\ 5 & 0 & 1 \end{bmatrix}. \qquad \blacksquare$$

Determinants of Elementary Matrices

It follows directly from the definitions of elementary matrices that:

(a) The determinant of an elementary matrix of Type I is -1, because two rows of a unit matrix have been interchanged so, in terms of $\mathbf{E}_{ij}$, we have $\det(\mathbf{E}_{ij}) = -1$.

(b) The determinant of an elementary matrix of Type II in which a row is multiplied by a nonzero constant c is c, because a row of a unit matrix has been multiplied by c so, in terms of $\mathbf{E}_i(c)$, we have $\det(\mathbf{E}_i(c)) = c$.

(c) The determinant of an elementary matrix of Type III in which c times one row has been added to another row is 1, because the addition of a multiple of a row of a unit matrix to another row leaves its value unchanged so, in terms of $\mathbf{E}_{ij}(c)$, we have $\det(\mathbf{E}_{ij}(c)) = 1$.

The next theorem shows that premultiplication of a matrix $\mathbf{A}$ by an elementary matrix $\mathbf{E}$ that is conformable for multiplication performs on $\mathbf{A}$ the same elementary row operation that was used to generate $\mathbf{E}$ from $\mathbf{I}$.

THEOREM 3.6

Row operations performed by elementary matrices   Let $\mathbf{E}$ be an $m \times m$ elementary matrix produced by performing an elementary row operation on the unit matrix $\mathbf{I}$, and let $\mathbf{A}$ be an $m \times n$ matrix. Then the matrix product $\mathbf{EA}$ is the matrix that is obtained when the row operation that generated $\mathbf{E}$ from $\mathbf{I}$ is performed on $\mathbf{A}$.

Proof   The proof of the theorem follows directly from the definition of a matrix product and the fact that, with the exception of the ith element in the ith row of $\mathbf{I}$, which is 1, all the other elements in that row are zero. So if $\mathbf{E}$ is the elementary matrix obtained from $\mathbf{I}$ by replacing the element 1 in its ith row by α, the result of the matrix product $\mathbf{EA}$ will be that the elements in the ith row of $\mathbf{A}$ will be multiplied by α. As the form of argument used to establish the effect on $\mathbf{A}$ of premultiplication by $\mathbf{P}$ to form $\mathbf{PA}$ can also be employed when the other two elementary row operations are used to generate an elementary matrix $\mathbf{E}$, the details will be left as an exercise. $\blacksquare$

EXAMPLE 3.18   Let $\mathbf{A}$ be the matrix

$$\mathbf{A} = \begin{bmatrix} 2 & 4 & 5 \\ 1 & 3 & 7 \\ 6 & 1 & 2 \end{bmatrix}.$$

If we use the notation for elementary matrices, and introduce the elementary matrix $\mathbf{E}_{23}$ from Example 3.17 obtained by interchanging the last two rows of $\mathbf{I}_3$, a routine calculation shows that

$$\mathbf{E}_{23}\mathbf{A} = \begin{bmatrix} 1 & 0 & 0 \\ 0 & 0 & 1 \\ 0 & 1 & 0 \end{bmatrix} \begin{bmatrix} 2 & 4 & 5 \\ 1 & 3 & 7 \\ 6 & 1 & 2 \end{bmatrix} = \begin{bmatrix} 2 & 4 & 5 \\ 6 & 1 & 2 \\ 1 & 3 & 7 \end{bmatrix},$$

so the product $\mathbf{E}_{23}\mathbf{A}$ has indeed interchanged the last two rows of $\mathbf{A}$.

Similarly, again using the elementary matrices in Example 3.17, it is easily checked that $\mathbf{E}_3(4)\mathbf{A}$ multiplies the elements in the third row of $\mathbf{A}$ by 4, while $\mathbf{E}_{13}(5)\mathbf{A}$ adds five times the first row of $\mathbf{A}$ to the last row. ∎

The main use of Theorem 3.6 is to be found in the theory of matrix algebra, and in the justification it provides for various practical methods that are used when working with matrices. This is because when solving purely numerical problems the necessary row operations need only be performed on the rows of the augmented matrix instead of on the system of equations itself.

Typical uses of the theorem will occur later after a discussion of the linear independence of equations, the definition of what is called the *rank* of a matrix, and the introduction of the inverse of an $n \times n$ matrix $\mathbf{A}$. In this last case, the results of the theorem will be used to provide an elementary method by which what is called the inverse matrix of an $n \times n$ matrix can be obtained when n is small.

Summary   This section introduced the three types of elementary row operations that are used when manipulating matrices together with the corresponding three types of elementary matrix that can be used to perform elementary row operations.

3.5   The Echelon and Row-Reduced Echelon Forms of a Matrix

We now use the row equivalence of matrices to reduce a matrix $\mathbf{A}$ to one of two slightly different but related standard forms called, respectively, its *echelon form* and its *row-reduced echelon form*. It is helpful to introduce these two new concepts by considering the solution of the system of m equations in n unknowns introduced in (16) and written in an equivalent but more condensed form as $(\mathbf{A}, \mathbf{b})$, where

$$(\mathbf{A}, \mathbf{b}) = \begin{bmatrix} a_{11} & a_{12} & \cdots & a_{1n} & b_1 \\ a_{21} & a_{22} & \cdots & a_{2n} & b_2 \\ & & \cdots \cdots & & \\ a_{m1} & a_{m2} & \cdots & a_{mn} & b_m \end{bmatrix}, \tag{19}$$

because this is equivalent to the full matrix equation $\mathbf{Ax} = \mathbf{b}$.

Echelon and row-reduced echelon forms of a matrix

echelon and
row-reduced
echelon forms

A matrix $\mathbf{A}$ is said to be in **echelon form** if:

(i) The first nonzero element in each row, called its *leading entry*, is 1;

(ii) In any two successive rows i and $i+1$ that do not consist entirely of zeros the leading element in the $(i+1)$th row lies to the right of the leading element in ith row;

(iii) Any rows that consist entirely of zeros lie at the bottom of the matrix. Matrix $\mathbf{A}$ is said to be in **row-reduced echelon form** if, in addition to conditions (i) to (iii), it is also true that

(iv) In a column that contains the leading entry of a row, all the other elements are zero.

In summary, this definition means that a matrix $\mathbf{A}$ is in *echelon form* if the first nonzero entry in any row is a 1, the entry appears to the right of the first nonzero entry in the row above, and all rows of zeros lie at the bottom of the matrix. Furthermore, matrix $\mathbf{A}$ is in *row-reduced echelon form* if, in addition to these conditions, the first nonzero entry in any row is the only nonzero entry in the column containing that entry.

EXAMPLE 3.19

The following matrices are in *echelon form*:

$$\begin{bmatrix} 1 & 0 & 5 & 7 \\ 0 & 0 & 1 & 0 \\ 0 & 0 & 0 & 0 \end{bmatrix} \quad \text{and} \quad \begin{bmatrix} 1 & 1 & 1 & 1 & 1 & 1 \\ 0 & 0 & 1 & 2 & 0 & 1 \\ 0 & 0 & 0 & 1 & 5 & 2 \\ 0 & 0 & 0 & 0 & 1 & 3 \\ 0 & 0 & 0 & 0 & 0 & 0 \end{bmatrix}.$$

The matrices

$$\begin{bmatrix} 0 & 1 & 0 & 2 & 0 & 5 & 0 \\ 0 & 0 & 1 & 1 & 0 & 3 & 2 \\ 0 & 0 & 0 & 0 & 1 & 1 & 0 \\ 0 & 0 & 0 & 0 & 0 & 0 & 0 \end{bmatrix}, \quad \begin{bmatrix} 1 & 0 & 0 & 9 & 2 \\ 0 & 1 & 0 & 2 & 3 \\ 0 & 0 & 1 & 1 & 0 \end{bmatrix}, \quad \text{and} \quad \begin{bmatrix} 1 & 0 & 0 & 5 \\ 0 & 1 & 0 & 2 \\ 0 & 0 & 1 & 1 \end{bmatrix}$$

are in *row-reduced echelon form*. ∎

Rules for the reduction of a matrix to echelon form

rules for finding the
echelon form

The reduction of the $m \times n$ matrix to its echelon form is accomplished by means of the following steps:

1. Find the row whose first nonzero element is furthest to the left and, if necessary, move it into row 1; if there is more than one such row, choose the row whose first nonzero element has the largest absolute value.
2. Scale row 1 to make its leading entry 1.
3. Subtract multiples of row 1 from the $m-1$ rows below it to reduce to zero all entries that lie below the leading entry in the first column.
4. In the $m-1$ rows below row 1, find the row whose first nonzero entry is furthest to the left and, if necessary, move it into row 2; if there is more

than one such row, choose the row whose first nonzero entry has the largest absolute value.

5. Scale row 2 to make its leading entry 1.

6. Subtract multiples of row 2 from the $m - 2$ rows below it to reduce to zero all entries in the column below the leading entry in row 2.

7. Continue this process until either the first nonzero entry in the mth row is 1, or a stage is reached at which all subsequent rows consist entirely of zeros.

8. The matrix is then in its echelon form.

Remark

The selection in Step 1, and the steps corresponding to Step 4, of a row whose first nonzero entry has the largest magnitude is made to reduce computational errors, and is not necessary mathematically. This criterion is introduced to ensure that the elimination procedure does not use an unnecessary scaling of a nonzero entry of small absolute magnitude to reduce to zero an entry of large absolute magnitude.

rules for finding the row-reduced echelon form

Rules for the reduction of a matrix to row-reduced echelon form

1. Proceed as in the reduction of a matrix to echelon form, but when steps equivalent to Step 6 are reached, in addition to subtracting multiples of the row containing a leading entry 1 from the rows below to reduce to zero all elements in the column below the leading entry, this same process must be repeated to reduce to zero all elements in the column above the leading entry.

2. An equivalent approach is first to reduce the matrix to echelon form and then, starting with row 2 and working downwards, to subtract multiples of successive rows from the rows above to generate columns with leading entries to ones with the single nonzero entry 1.

Each of these methods reduces a matrix to its row-reduced echelon form.

The row equivalence of a matrix with either its echelon or its row-reduced echelon form means that the different-looking systems of equations represented by these three matrices all have identical solution sets. The simplified structure of the row echelon and row-reduced echelon forms of the original augmented matrix makes the solution of the associated system of equations particularly easy, as can be seen from the following examples.

EXAMPLE 3.20   Reduce the following matrix to its echelon and its row-reduced echelon form:

$$\begin{bmatrix} 0 & 1 & 2 & 0 & 3 \\ 2 & 4 & 8 & 2 & 4 \\ 1 & 2 & 4 & 2 & 2 \\ 1 & 3 & 6 & 1 & 5 \end{bmatrix}.$$

Solution
$$\begin{bmatrix} 0 & 1 & 2 & 0 & 3 \\ 2 & 4 & 8 & 2 & 4 \\ 1 & 2 & 4 & 2 & 2 \\ 1 & 3 & 6 & 1 & 5 \end{bmatrix}$$
switch rows
2 and 1
$$\sim \begin{bmatrix} 2 & 4 & 8 & 2 & 4 \\ 0 & 1 & 2 & 0 & 3 \\ 1 & 2 & 4 & 2 & 2 \\ 1 & 3 & 6 & 1 & 5 \end{bmatrix}$$

divide row 1 by 2
$$\sim \begin{bmatrix} 1 & 2 & 4 & 1 & 2 \\ 0 & 1 & 2 & 0 & 3 \\ 1 & 2 & 4 & 2 & 2 \\ 1 & 3 & 6 & 1 & 5 \end{bmatrix}$$
subtract row 1
from rows 3 and 4
$$\sim \begin{bmatrix} 1 & 2 & 4 & 1 & 2 \\ 0 & 1 & 2 & 0 & 3 \\ 0 & 0 & 0 & 1 & 0 \\ 0 & 1 & 2 & 0 & 3 \end{bmatrix}$$

subtract row 2
from row 4
$$\sim \begin{bmatrix} 1 & 2 & 4 & 1 & 2 \\ 0 & 1 & 2 & 0 & 3 \\ 0 & 0 & 0 & 1 & 0 \\ 0 & 0 & 0 & 0 & 0 \end{bmatrix}$$

and the matrix is now in echelon form.

Having already obtained the echelon form of the matrix, we now use it to obtain the row-reduced echelon form. We already have

$$\begin{bmatrix} 0 & 1 & 2 & 0 & 3 \\ 2 & 4 & 8 & 2 & 4 \\ 1 & 2 & 4 & 2 & 2 \\ 1 & 3 & 6 & 1 & 5 \end{bmatrix} \sim \begin{bmatrix} 1 & 2 & 4 & 1 & 2 \\ 0 & 1 & 2 & 0 & 3 \\ 0 & 0 & 0 & 1 & 0 \\ 0 & 0 & 0 & 0 & 0 \end{bmatrix}$$
subtract twice row 2
from row 1

$$\sim \begin{bmatrix} 1 & 0 & 0 & 1 & -4 \\ 0 & 1 & 2 & 0 & 3 \\ 0 & 0 & 0 & 1 & 0 \\ 0 & 0 & 0 & 0 & 0 \end{bmatrix}$$
subtract row 3
from row 1
$$\sim \begin{bmatrix} 1 & 0 & 0 & 0 & -4 \\ 0 & 1 & 2 & 0 & 3 \\ 0 & 0 & 0 & 1 & 0 \\ 0 & 0 & 0 & 0 & 0 \end{bmatrix},$$

and the matrix is now in its row-reduced echelon form. ∎

EXAMPLE 3.21   Solve the system of equations

$$x_2 + 2x_3 = 3$$
$$2x_1 + 4x_2 + 8x_3 + 2x_4 = 4$$
$$x_1 + 2x_2 + 4x_3 + 2x_4 = 2$$
$$x_1 + 3x_2 + 6x_3 + x_4 = 5.$$

Solution   The augmented matrix $(\mathbf{A}, \mathbf{b})$ for this system is the matrix in Example 3.20 that was shown to be equivalent to the row-reduced echelon form

$$\begin{bmatrix} 1 & 0 & 0 & 0 & -4 \\ 0 & 1 & 2 & 0 & 3 \\ 0 & 0 & 0 & 1 & 0 \\ 0 & 0 & 0 & 0 & 0 \end{bmatrix}.$$

If we recall that the first four columns of this matrix contain the coefficients of x_1, x_2, x_3, and x_4, while the last column contains the nonhomogeneous terms, the matrix implies the much simpler system of equations

$$x_4 = 0, \quad x_2 + 2x_3 = 3, \quad \text{and} \quad x_1 = -4.$$

As there are only three equations connecting four unknowns, it follows that in the second equation either x_2 or x_3 can be assigned arbitrarily, so if we choose to set $x_3 = k$ (an arbitrary number), the solution set of the system in terms of the parameter k becomes

$$x_1 = -4, \quad x_2 = 3 - 2k, \quad x_3 = k, \quad \text{and} \quad x_4 = 0.$$

The same solution could have been obtained from the echelon form of the matrix

$$\begin{bmatrix} 1 & 2 & 4 & 1 & | & 2 \\ 0 & 1 & 2 & 0 & | & 3 \\ 0 & 0 & 0 & 1 & | & 0 \\ 0 & 0 & 0 & 0 & | & 0 \end{bmatrix},$$

because this implies the system of equations

$$x_1 + 2x_2 + 4x_3 + x_4 = 2, \quad x_2 + 2x_3 = 3, \quad \text{and} \quad x_4 = 0.$$

Starting from the last equation we find $x_4 = 0$, and setting $x_3 = k$ in the middle equation gives, as before, $x_2 = 3 - 2k$. Finally, substituting x_2, x_3, and x_4 in the first equation gives $x_1 = -4$. This process of arriving at a solution of a system of equations whose coefficient matrix is in upper triangular form is called **back substitution**.

back substitution

It should be noticed that the system of equations would have had *no solution* if the row-reduced echelon form had been

$$\begin{bmatrix} 1 & 0 & 0 & 0 & | & -4 \\ 0 & 1 & 2 & 0 & | & 3 \\ 0 & 0 & 0 & 1 & | & 0 \\ 0 & 0 & 0 & 0 & | & 5 \end{bmatrix}.$$

This is because although the equations corresponding to the first three rows of this matrix would have been the same as before, the fourth row implies $0 = 5$, which is impossible. This corresponds to a system of equations where one equation contradicts the others, so that no solution is possible. ∎

Summary

This section defined two related types of fundamental matrix that can be obtained from a general matrix by means of elementary row operations. The first was a reduction to echelon form and the second, derived from the first form, was a reduction to row-reduced echelon form. Each of the reduced forms retains the essential properties of the original matrix, while simplifying the task of solving the associated system of linear algebraic equations.

EXERCISES 3.5

Let $\mathbf{P}$, $\mathbf{Q}$, and $\mathbf{R}$ be the matrices

$$\mathbf{P} = \begin{bmatrix} 3 & 0 & 0 \\ 0 & 1 & 0 \\ 0 & 0 & 1 \end{bmatrix}, \quad \mathbf{Q} = \begin{bmatrix} 0 & 0 & 1 \\ 0 & 1 & 0 \\ 1 & 0 & 0 \end{bmatrix}, \quad \mathbf{R} = \begin{bmatrix} 1 & 2 & 0 \\ 0 & 1 & 0 \\ 0 & 0 & 1 \end{bmatrix}.$$

In Exercises 1 through 4 verify by direct calculation that (a) premultiplication by $\mathbf{P}$ multiplies row 1 by 3; (b) premultiplication by $\mathbf{Q}$ interchanges rows 1 and 3; and (c) premultiplication by $\mathbf{R}$ adds twice row 2 to row 1.

1. $\begin{bmatrix} 2 & 1 & 1 \\ 1 & 3 & 0 \\ 1 & 2 & 4 \end{bmatrix}.$

2. $\begin{bmatrix} 1 & -1 & 2 \\ 2 & 1 & 3 \\ 3 & 0 & 7 \end{bmatrix}.$

3. $\begin{bmatrix} 4 & 0 & 1 \\ 2 & 0 & 3 \\ 1 & 2 & 5 \end{bmatrix}.$

4. $\begin{bmatrix} 9 & 1 & 3 \\ 2 & 4 & 7 \\ 1 & 2 & 2 \end{bmatrix}.$

In Exercises 5 and 6 write down the required elementary matrices.

5. When $\mathbf{I}$ is the 3×3 unit matrix, write down $\mathbf{E}_{12}$, $\mathbf{E}_2(3)$, and $\mathbf{E}_{12}(6)$.

6. When $\mathbf{I}$ is the 4×4 unit matrix, write down $\mathbf{E}_{41}$, $\mathbf{E}_4(3)$, and $\mathbf{E}_{23}(4)$.

In Exercises 7 through 12, reduce the given matrices to their row-reduced echelon form.

7. $\begin{bmatrix} 0 & 3 & 4 & 1 \\ 3 & 1 & 2 & 2 \\ 1 & 5 & 2 & 1 \end{bmatrix}$.

8. $\begin{bmatrix} 4 & 1 & 3 & 1 & 3 \\ 2 & 1 & 1 & 2 & 0 \\ 3 & 2 & 1 & 1 & 0 \end{bmatrix}$.

9. $\begin{bmatrix} 4 & -2 & 2 & 3 & 1 \\ 2 & 0 & 0 & 3 & 2 \\ 4 & 1 & 2 & 5 & 1 \end{bmatrix}$.

10. $\begin{bmatrix} 3 & 2 & 1 & 1 \\ 2 & 5 & 1 & 2 \\ 3 & 1 & 1 & 3 \\ 0 & 1 & 3 & 4 \\ 2 & 1 & 3 & 1 \end{bmatrix}$.

11. $\begin{bmatrix} 2 & 2 & 4 & 1 & 4 \\ 1 & 1 & 3 & 2 & 1 \\ 3 & 2 & 5 & 1 & 4 \\ 1 & 0 & 3 & 1 & 2 \end{bmatrix}$.

12. $\begin{bmatrix} 3 & 2 & 3 & 2 \\ 3 & 7 & 1 & -1 \\ 5 & 1 & 1 & 3 \end{bmatrix}$.

In Exercises 13 through 18, reduce the given augmented matrices to their row-reduced echelon form and, where appropriate, use the result to solve the related system of equations in terms of an appropriate number of the unknowns $x_1, x_2, \ldots$.

13. $\left[\begin{array}{ccc|c} 2 & 3 & 1 & 0 \\ 1 & 3 & 1 & 4 \\ 6 & 9 & 4 & 8 \end{array}\right]$.

14. $\left[\begin{array}{ccc|c} 2 & 1 & 1 & 0 \\ 2 & 3 & 1 & 4 \\ 4 & 9 & 4 & 8 \end{array}\right]$.

15. $\left[\begin{array}{cccc|c} 0 & 2 & 1 & 1 & 1 \\ 1 & 3 & 1 & 2 & 1 \\ 3 & 9 & 4 & 3 & 0 \end{array}\right]$.

16. $\left[\begin{array}{cccc|c} 2 & 1 & 0 & 2 & 1 \\ 1 & 3 & 1 & 4 & 2 \\ 2 & 1 & 2 & 3 & 1 \\ 4 & 7 & 4 & 11 & 7 \end{array}\right]$.

17. $\left[\begin{array}{ccccc|c} 1 & 0 & 1 & 0 & 2 & 0 \\ 2 & 2 & 6 & 0 & 6 & 0 \\ 1 & 0 & 1 & 1 & 6 & 0 \\ 3 & 2 & 7 & 0 & 8 & 2 \end{array}\right]$.

18. $\left[\begin{array}{cccc|c} 3 & 0 & 6 & 0 & 6 \\ 1 & 1 & 5 & 1 & 9 \\ 2 & 0 & 4 & 2 & 10 \end{array}\right]$.

3.6 Row and Column Spaces and Rank

The reduction of an $m \times n$ matrix $\mathbf{A}$ to either its echelon or its row-reduced echelon form will produce a row of zeros whenever the row is a linear combination of some (or all) of the rows above it. So if an echelon form contains $r \leq m$ nonzero rows, it follows that these r rows are linearly independent, and hence that the remaining $m - r$ rows are linearly dependent on the first r rows. The number r is called the *row rank* of matrix $\mathbf{A}$.

This means that if the r nonzero rows of an echelon form $\mathbf{u}_1, \mathbf{u}_2, \ldots, \mathbf{u}_r$ are regarded as n element row vectors belonging to a vector space $\mathbf{R}^n$, the r vectors will *span* a subspace of $\mathbf{R}^n$. Consequently, as these vectors form a *basis* for this subspace, every vector in it can be expressed as a linear combination of the form

$$a_1\mathbf{u}_1 + a_2\mathbf{u}_2 + \cdots + a_r\mathbf{u}_r,$$

where the $a_1, a_2, \ldots, a_r$ are scalar constants. This subspace of $\mathbf{R}^n$ is called the **row space** of matrix $\mathbf{A}$.

row and column ranks and spaces

It should be remembered that the vectors forming a basis for a space are not unique, and that any basis can be transformed to any other one by means of suitable linear combinations of the vectors involved. So although the r nonzero rows of the echelon form of $\mathbf{A}$ and those of its row-reduced echelon form look different, they are equivalent, and each forms a basis for the row space of $\mathbf{A}$.

Just as there may be linear dependence between the rows of $\mathbf{A}$, so also may there be linear dependence between its columns. If s of the n columns of an $m \times n$ matrix $\mathbf{A}$ are linearly independent, the number s is called the **column rank** of matrix $\mathbf{A}$. When the s nonzero columns $\mathbf{v}_1, \mathbf{v}_2, \ldots, \mathbf{v}_s$ are regarded as m element column vectors belonging to a vector space $\mathbf{R}^m$, these vectors will span a subspace of $\mathbf{R}^m$.

Consequently, as these vectors form a basis for this subspace, every vector in it can be expressed as a linear combination of the form

$$b_1 \mathbf{v}_1 + b_2 \mathbf{v}_2 + \cdots + b_s \mathbf{v}_s,$$

where the $b_1, b_2, \ldots, b_s$ are scalar constants. This subspace of R^m is called the **column space** of matrix $\mathbf{A}$.

The connection between the row and column ranks of a matrix is provided by the following theorem.

THEOREM 3.7

equality of the rank of a matrix and its transpose

The equality of the row and column ranks Let $\mathbf{A}$ be any matrix. Then the row rank and column rank of $\mathbf{A}$ are equal.

Proof Let an $m \times n$ matrix $\mathbf{A}$ have row rank r. Then in its row-reduced echelon form it must contain r columns $\mathbf{v}_1, \mathbf{v}_2, \ldots, \mathbf{v}_r$, in each of which only the single nonzero entry 1 appears. Call these columns the *leading columns* of the row-reduced echelon form, and let them be arranged so that in the ith column $\mathbf{v}$, the entry 1 appears in the ith row.

The row-reduced echelon form of $\mathbf{A}$ will comprise the leading columns arranged in numerical order with, possibly, columns between the ith and the $(i + 1)$th leading columns in which zero elements lie below the ith row but nonzero elements may occur above it. Furthermore, there may be columns to the right of column $\mathbf{v}_r$ in which zero elements lie below the rth row but nonzero elements may lie above it.

By subtracting suitable multiples of the leading columns from any columns that lie between them or to the right of $\mathbf{v}_r$, it is possible to reduce all entries in such columns to zero. Consequently, at the end of this process, the only remaining nonzero columns will be the r linearly independent leading columns $\mathbf{v}_1, \mathbf{v}_2, \ldots, \mathbf{v}_r$. This establishes the equality of the row and column ranks. ∎

Rank

The **rank** of matrix $\mathbf{A}$, denoted by rank $(\mathbf{A})$, is the value common to the row and column ranks of $\mathbf{A}$.

THEOREM 3.8

Rank of A and $\mathbf{A}^{\mathrm{T}}$ Let $\mathbf{A}$ be any matrix. Then

$$\text{rank } (\mathbf{A}) = \text{rank } (\mathbf{A}^{\mathrm{T}}).$$

Proof The columns of $\mathbf{A}$ are the rows of $\mathbf{A}^{\mathrm{T}}$, so the column rank of A is the row rank of $\mathbf{A}^{\mathrm{T}}$. However, by Theorem 3.7 these two ranks are equal, so the result is proved. ∎

EXAMPLE 3.22

Let

$$\mathbf{A} = \begin{bmatrix} 1 & 0 & 3 & 0 & 4 & 0 \\ 2 & 1 & 7 & 0 & 10 & 1 \\ 1 & 0 & 3 & 2 & 6 & 4 \\ 1 & 0 & 3 & 0 & 4 & 0 \end{bmatrix}.$$

Then the row-reduced echelon form of **A** is **B** (**B** ∼ **A**)

$$\mathbf{B} = \begin{bmatrix} 1 & 0 & 3 & 0 & 4 & 0 \\ 0 & 1 & 1 & 0 & 2 & 1 \\ 0 & 0 & 0 & 1 & 1 & 2 \\ 0 & 0 & 0 & 0 & 0 & 0 \end{bmatrix},$$

showing that the number of leading columns is 3, so the row rank of **A** is 3, and hence its *rank* is 3. Three row vectors spanning a subspace of R^6, and so forming a basis for this subspace, are the three nonzero row vectors in this 4×6 matrix,

$$\mathbf{u}_1 = [1, 0, 3, 0, 4, 0], \quad \mathbf{u}_2 = [0, 1, 1, 0, 2, 1], \quad \text{and} \quad \mathbf{u}_3 = [0, 0, 0, 1, 1, 2].$$

The row-reduced echelon form of $\mathbf{A}^T$ is

$$\begin{bmatrix} 1 & 0 & 0 & 1 \\ 0 & 1 & 0 & 0 \\ 0 & 0 & 1 & 0 \\ 0 & 0 & 0 & 0 \\ 0 & 0 & 0 & 0 \\ 0 & 0 & 0 & 0 \end{bmatrix},$$

showing that the number of leading columns is 3, confirming as would be expected that the column rank of **A** (the row rank of $\mathbf{A}^T$) is 3. The three *row* vectors of $\mathbf{A}^T$ spanning a subspace of R^4, and so forming a basis for this subspace, are the three nonzero rows in this 6×4 matrix, namely,

$$[1, 0, 0, 1], \quad [0, 1, 0, 0], \quad \text{and} \quad [0, 0, 1, 0].$$

The three linearly independent *column* vectors of **A** are obtained by transposing these vectors to obtain

$$\mathbf{v}_1 = \begin{bmatrix} 1 \\ 0 \\ 0 \\ 1 \end{bmatrix}, \quad \mathbf{v}_2 = \begin{bmatrix} 0 \\ 1 \\ 0 \\ 0 \end{bmatrix}, \quad \text{and} \quad \mathbf{v}_3 = \begin{bmatrix} 0 \\ 0 \\ 1 \\ 0 \end{bmatrix}. \quad \blacksquare$$

Summary

This section introduced the important algebraic concepts of the rank of a matrix, and of the row and column spaces of a matrix. The equality of the row and column ranks of a matrix was then proved. It will be seen later that the rank of a matrix plays a fundamental role when we seek a solution of a linear algebraic system of equations.

EXERCISES 3.6

In Exercises 1 through 14 find the row-reduced echelon form of the given matrix, its rank, a basis for its row space, and a basis for its column space.

1. $\begin{bmatrix} 1 & 3 & 1 & 0 & 1 & 1 \\ 2 & 2 & 1 & 0 & 0 & 1 \\ 0 & 2 & 1 & 4 & 1 & 3 \end{bmatrix}.$

2. $\begin{bmatrix} 1 & 3 & 2 & 1 \\ 2 & 0 & 2 & 1 \\ 1 & 0 & 4 & 5 \\ 0 & 1 & 2 & 4 \end{bmatrix}.$

3. $\begin{bmatrix} 3 & 0 & 2 & 6 & 0 \\ 4 & 1 & 0 & 11 & 3 \\ 2 & 0 & 2 & 4 & 0 \\ 3 & 0 & 0 & 6 & 3 \end{bmatrix}.$

4. $\begin{bmatrix} 2 & 3 & 1 & 0 & 0 & 2 & 4 \\ 1 & 2 & 1 & 0 & 4 & 1 & 2 \end{bmatrix}.$

5. $\begin{bmatrix} 1 & 2 & 3 \\ 2 & 3 & 1 \\ 3 & 2 & 1 \end{bmatrix}.$

6. $\begin{bmatrix} 3 & 2 & 4 \\ 1 & 2 & 2 \\ 8 & 8 & 12 \end{bmatrix}.$

7. $\begin{bmatrix} 1 & 3 & 4 \\ 3 & 0 & 4 \\ 2 & 3 & 1 \\ 0 & 3 & 5 \end{bmatrix}.$

8. $\begin{bmatrix} 2 & 1 & 3 & 1 \\ 1 & 1 & 0 & 3 \\ 1 & 2 & 1 & 0 \\ 3 & 3 & 4 & 1 \\ 2 & 3 & 1 & 3 \end{bmatrix}.$

9. $\begin{bmatrix} 1 & 2 & 1 & 4 & 5 & 7 \\ 2 & 1 & 0 & 1 & 2 & 1 \\ 3 & 3 & 1 & 5 & 7 & 8 \end{bmatrix}.$

10. $\begin{bmatrix} 2 & 4 & 0 & 10 & 8 \\ 0 & 2 & 1 & 3 & 1 \\ 2 & 6 & 1 & 13 & 9 \end{bmatrix}.$

11. $\begin{bmatrix} 0 & -1 & 4 & 3 \\ 0 & 0 & 1 & 2 \\ 0 & 0 & 0 & -1 \\ 1 & 0 & 0 & 0 \end{bmatrix}.$

12. $\begin{bmatrix} 1 & 0 & 0 & 0 \\ 1 & -1 & 0 & 0 \\ 2 & 5 & -1 & 0 \\ 1 & 3 & 2 & 1 \end{bmatrix}.$

13. $\begin{bmatrix} 1 & 7 & 2 & 4 \\ 0 & 0 & 5 & 7 \\ 0 & 0 & 0 & 3 \end{bmatrix}.$

14. $\begin{bmatrix} 1 & 5 & 0 & 3 \\ 2 & 1 & 1 & 1 \\ 1 & 2 & 3 & 2 \\ 3 & 3 & 4 & 3 \\ 4 & 5 & 7 & 5 \end{bmatrix}.$

3.7 The Solution of Homogeneous Systems of Linear Equations

Having now introduced the echelon and row-reduced echelon forms of an $m \times n$ matrix $\mathbf{A}$, we are in a position to discuss the nature of the solution set of the system of linear equations

homogeneous and nonhomogeneous systems of equations

$$\begin{aligned} a_{11}x_1 + a_{12}x_2 + \cdots + a_{1n}x_n &= b_1 \\ a_{21}x_1 + a_{22}x_2 + \cdots + a_{2n}x_n &= b_2 \\ &\cdots \\ a_{m1}x_1 + a_{m2}x_2 + \cdots + a_{mn}x_n &= b_m, \end{aligned} \tag{20}$$

which will be **nonhomogeneous** when at least one of the terms b_i on the right is nonzero, and **homogeneous** when $b_1 = b_2 = \cdots = b_m = 0$. In this section we will only consider homogeneous systems.

Rather than working with the full system of homogeneous equations corresponding to $b_i = 0$, $i = 1, 2, \ldots, m$ in (20), it is more convenient to work with its coefficient matrix

$$\mathbf{A} = \begin{bmatrix} a_{11} & a_{12} & \cdots & a_{1n} \\ a_{21} & a_{22} & \cdots & a_{2n} \\ & & \cdots & \\ a_{m1} & a_{m2} & \cdots & a_{mn} \end{bmatrix}, \tag{21}$$

which contains all the information about the system. The coefficients in the first column of $\mathbf{A}$ are multipliers of x_1, those in the second column are multipliers of $x_2, \ldots,$ and those in the nth column are multipliers of x_n.

Denote by $\mathbf{A}_E$ either the echelon or the row-reduced echelon form of the coefficient matrix $\mathbf{A}$. Then, as elementary row operations performed on a coefficient matrix are equivalent in all respects to performing the same operations on the corresponding full system of equations, the solution set of the matrix equation

$$\mathbf{Ax} = \mathbf{0} \tag{22}$$

will be the same as the solution set of an echelon form of the homogeneous equations

$$\mathbf{A}_E\mathbf{x} = \mathbf{0}. \tag{23}$$

trivial solution

It is obvious that $\mathbf{x} = \mathbf{0}$, corresponding to $\mathbf{x} = [0, 0, \ldots, 0]^T$, is always a solution of (22) and, of course of (23), and it is called the **trivial solution** of the homogeneous system of equations. To discover when nontrivial solutions exist it is necessary to work with the equivalent echelon form of the equations given in (23).

If $\text{rank}(\mathbf{A}) = r$, the first r rows of $\mathbf{A}_E$ will be nonzero rows, and the last $m - r$ rows will be zero rows. As there are m rows in $\mathbf{A}$, we must consider the three separate cases (a) $m < n$, (b) $m = n$, and (c) $m > n$.

Case (a): $m < n$. In this case there are more variables than equations. As $\text{rank}(\mathbf{A}) = r$, and there are m equations, it follows that $r = \text{rank}(\mathbf{A}) \leq m$. The system in (22) will thus contain only r linearly independent equations corresponding to the first r rows of $\mathbf{A}_E$. So working with system (23), we see that r of the variables $x_1, x_2, \ldots, x_n$ will be determined in terms of the remaining $m - r$ variables regarded as parameters (see Example 3.23).

Case (b): $m = n$. In this case the number of variables equals the number of equations. If $\text{rank}(\mathbf{A}) = r < n$ we have the same situation as in Case (a), and the variables $x_1, x_2, \ldots, x_n$ will be determined by the system of equations in (23) in terms of the remaining $m - r$ variables regarded as parameters. However, if $r = n$, only the trivial solution $\mathbf{x} = \mathbf{0}$ is possible, because in this case $\mathbf{A}_E$ becomes the unit matrix $\mathbf{I}_n$, from which it follows directly that $\mathbf{x} = \mathbf{0}$.

Case (c): $m > n$. In this case the number of equations exceeds the number of variables and $r = \text{rank}(\mathbf{A}) \leq n$. This is essentially the same situation as in Case (b), because if $r = \text{rank}(\mathbf{A}) < n$, the variables $x_1, x_2, \ldots, x_n$ will be determined by the system of equations in (22) in terms of the remaining $m - r$ variables regarded as parameters, while if $\text{rank}(\mathbf{A}) = n$ only the trivial solution $\mathbf{x} = \mathbf{0}$ is possible.

The practical determination of solution sets to homogeneous systems of linear equations is illustrated in the next example.

EXAMPLE 3.23

Find the solution sets of the homogeneous systems of linear equations with coefficient matrices given by:

(a) $\mathbf{A} = \begin{bmatrix} 1 & 2 & 1 & 7 & 0 \\ 3 & 6 & 4 & 24 & 3 \\ 1 & 4 & 4 & 12 & 3 \end{bmatrix}$, (b) $\mathbf{A} = \begin{bmatrix} 1 & 3 & 2 \\ 2 & 1 & 0 \\ 1 & 2 & 1 \end{bmatrix}$, (c) $\mathbf{A} = \begin{bmatrix} 2 & 3 & 6 & 1 \\ 1 & 4 & 2 & 2 \\ 4 & 11 & 10 & 5 \\ 1 & 0 & 1 & 1 \end{bmatrix}$,

(d) $\mathbf{A} = \begin{bmatrix} 1 & 4 & 1 & 2 \\ 1 & 3 & 0 & 1 \\ 2 & 1 & 1 & 1 \\ 4 & 9 & 3 & 5 \\ 5 & 5 & 2 & 3 \end{bmatrix}$, (e) $\mathbf{A} = \begin{bmatrix} 1 & 2 & 3 & 1 & 4 & 3 \\ 0 & 1 & 3 & 0 & 1 & 5 \\ 3 & 1 & 2 & 3 & 1 & 4 \end{bmatrix}$

Solution

(a) The row-reduced echelon form of the matrix is

$$\mathbf{A}_E = \begin{bmatrix} 1 & 0 & 0 & 8 & 3 \\ 0 & 1 & 0 & -2 & -3 \\ 0 & 0 & 1 & 3 & 3 \end{bmatrix},$$

showing that rank($\mathbf{A}$) = 3. This corresponds to the following three equations between the five variables x_1, x_2, x_3, x_4, and x_5:

$$x_1 + 8x_4 + 3x_5 = 0, \quad x_2 - 2x_4 - 3x_5 = 0, \quad \text{and} \quad x_3 + 3x_4 + 3x_5 = 0.$$

Letting $x_4 = \alpha$ and $x_5 = \beta$ be arbitrary numbers (parameters) allows the solution set to be written

$$x_1 = -8\alpha - 3\beta, \quad x_2 = 2\alpha + 3\beta, \quad x_3 = -3\alpha - 3\beta, \quad x_4 = \alpha, \quad x_5 = \beta.$$

(b) The row-reduced echelon form of the matrix is

$$\mathbf{A}_E = \begin{bmatrix} 1 & 0 & 0 \\ 0 & 1 & 0 \\ 0 & 0 & 1 \end{bmatrix},$$

showing that rank($\mathbf{A}$) = 3. This corresponds to the trivial solution $x_1 = x_2 = x_3 = 0$.

(c) The row-reduced echelon form of the matrix is

$$\mathbf{A}_E = \begin{bmatrix} 1 & 0 & 0 & 20/13 \\ 0 & 1 & 0 & 5/13 \\ 0 & 0 & 1 & -7/13 \\ 0 & 0 & 0 & 0 \end{bmatrix},$$

showing that rank($\mathbf{A}$) = 3. This corresponds to the solution set $x_1 + (20/13)x_4 = 0$, $x_2 + (5/13)x_4 = 0$, and $x_3 - (7/13)x_4 = 0$. Setting $x_4 = k$, an arbitrary number (a parameter), shows the solution set to be given by

$$x_1 = -(20/13)k, \quad x_2 = -(5/13)k, \quad x_3 = (7/13)k, \quad \text{and} \quad x_4 = k.$$

(d) The row-reduced echelon form of the matrix is

$$\mathbf{A}_E = \begin{bmatrix} 1 & 0 & 0 & 0 \\ 0 & 1 & 0 & 1/3 \\ 0 & 0 & 1 & 2/3 \\ 0 & 0 & 0 & 0 \\ 0 & 0 & 0 & 0 \end{bmatrix},$$

showing that rank($\mathbf{A}$) = 3. This corresponds to the following three equations for the four variables x_1, x_2, x_3, and x_4:

$$x_1 = 0, \quad x_2 + (1/3)x_4 = 0, \quad \text{and} \quad x_3 + (2/3)x_4 = 0.$$

Setting $x_4 = k$, an arbitrary number (a parameter), shows the solution set to be given by

$$x_1 = 0, \quad x_2 = -k/3 = 0, \quad x_3 = -2k/3, \quad \text{and} \quad x_4 = k.$$

(e) The row-reduced echelon form of the matrix is

$$\mathbf{A}_E = \begin{bmatrix} 1 & 0 & 0 & 1 & -1/4 & 1/2 \\ 0 & 1 & 0 & 0 & 13/4 & -5/2 \\ 0 & 0 & 1 & 0 & -3/4 & 5/2 \end{bmatrix},$$

showing that rank($\mathbf{A}$) = 3. This corresponds to the following three equations for the six variables x_1 to x_6:

$$x_1 + x_4 - (1/4)x_5 + (1/2)x_6 = 0, \quad x_2 + (13/4)x_5 - (5/2)x_6 = 0$$
$$x_3 - (3/4)x_5 + (5/2)x_6 = 0.$$

Setting $x_4 = \alpha$, $x_5 = \beta$, and $x_6 = \gamma$, where α, β, and γ are arbitrary numbers (parameters), shows the solution set to be given by

$$x_1 = -\alpha + (1/4)\beta - (1/2)\gamma, \quad x_2 = -(13/4)\beta + (5/2)\gamma, \quad x_3 = (3/4)\beta - (5/2)\gamma$$
$$x_4 = \alpha, \quad x_5 = \beta, \quad \text{and} \quad x_6 = \gamma. \qquad \blacksquare$$

Summary

This section made use of the rank of a matrix to determine when a nontrivial solution of a linear system of homogeneous linear algebraic equations exists and, when it does, its precise form.

EXERCISES 3.7

In Exercises 1 through 10, use the given form of the matrix **A** to find the solution set of the associated homogeneous linear system of equations $\mathbf{Ax} = \mathbf{0}$.

1. $\begin{bmatrix} 1 & 3 & 2 & 1 & 1 \\ 1 & 1 & 0 & 1 & 2 \\ 0 & 1 & 2 & 1 & 3 \end{bmatrix}$.

2. $\begin{bmatrix} 1 & 2 & 0 & 1 & 1 \\ 0 & 3 & 1 & 0 & 1 \\ 2 & 0 & 2 & 0 & 1 \\ 1 & 0 & 3 & 1 & 1 \end{bmatrix}$.

3. $\begin{bmatrix} 1 & 2 & 4 & 1 \\ 0 & 3 & 1 & 3 \\ 1 & 4 & 1 & 3 \\ 2 & 6 & 5 & 4 \end{bmatrix}$.

4. $\begin{bmatrix} 1 & 2 & 1 & 0 \\ 2 & 1 & 0 & 1 \\ 0 & 3 & 5 & 1 \\ 1 & 0 & 1 & 5 \end{bmatrix}$.

5. $\begin{bmatrix} 1 & 3 & 4 \\ 2 & 1 & 3 \\ 1 & 0 & 2 \\ 3 & 1 & 1 \\ 2 & 3 & 1 \end{bmatrix}$.

6. $\begin{bmatrix} 2 & 1 & 1 & 3 \\ 1 & 2 & 3 & 0 \\ 0 & 1 & 4 & 2 \\ 1 & 3 & 1 & 2 \\ 0 & 4 & 1 & 1 \end{bmatrix}$.

7. $\begin{bmatrix} 1 & 5 & 2 & 2 & 1 & 3 & 2 \\ 0 & 1 & 4 & 1 & 0 & 1 & 1 \\ 1 & 2 & 1 & 0 & 0 & 2 & 0 \\ 2 & 3 & 0 & 1 & 1 & 0 & 2 \end{bmatrix}$.

8. $\begin{bmatrix} 1 & 4 & 1 & 0 \\ 2 & 1 & 3 & 1 \\ 5 & 6 & 7 & 2 \\ 2 & 1 & 0 & 1 \end{bmatrix}$.

9. $\begin{bmatrix} 1 & 1 & 5 & 0 & 0 & 1 \\ 2 & 3 & 1 & 2 & 1 & 3 \\ 0 & 1 & 0 & 1 & 3 & 0 \end{bmatrix}$.

10. $\begin{bmatrix} 1 & 3 & 2 & 1 & 1 \\ 2 & 5 & 1 & 0 & 2 \\ 0 & 1 & 2 & 0 & 3 \\ 1 & 0 & 3 & 1 & 2 \end{bmatrix}$.

3.8 The Solution of Nonhomogeneous Systems of Linear Equations

We now turn our attention to the solution of the nonhomogeneous system of equations in (20) that may be written in the matrix form

$$\mathbf{Ax} = \mathbf{b}, \qquad (24)$$

where **A** is an $m \times n$ matrix and **b** is an $m \times 1$ nonzero column vector. In many respects the arguments we now use parallel the ones used when seeking the form of the solution set for a homogeneous system, but there are important differences. This time, rather than working with the matrix **A**, we must work with the augmented matrix $(\mathbf{A}, \mathbf{b})$ and use elementary row operations to transform it into either an echelon or a row-reduced echelon form that will be denoted by $(\mathbf{A}, \mathbf{b})_E$. When this is done, system (24) and the echelon form corresponding to $(\mathbf{A}, \mathbf{b})_E$ will, of course, each have the same solution set.

It is important to recognize that rank(**A**) is not necessarily equal to rank $(\mathbf{A}, \mathbf{b})_E$, so that in general rank(**A**) $\leq$ rank$((\mathbf{A}, \mathbf{b})_E)$. The significance of this observation will become clear when we seek solutions of systems like (24).

Case (a): $m < n$. In this case there are more variables than equations, and it must follow that $\operatorname{rank}((\mathbf{A}, \mathbf{b})_E) \le m$. If $\operatorname{rank}(\mathbf{A}) = \operatorname{rank}((\mathbf{A}, \mathbf{b})_E) = r$, it follows that r of the equations in (24) are linearly independent and $m - r$ are linear combinations of these r equations. This means that the first r rows of $(\mathbf{A}, \mathbf{b})_E$ are linearly independent while the last $m - r$ rows are rows of zeros. Thus, r of the variables x_1 to x_n will be determined by the equations corresponding to these r nonzero rows, in terms of the remaining $m - r$ variables as parameters. It can happen, however, that $\operatorname{rank}(\mathbf{A}) = r < \operatorname{rank}((\mathbf{A}, \mathbf{b})_E)$, and then the situation is different, because one or more of the rows following the rth row will have zeros in its first n entries and nonzero numbers for their last entries. When interpreted as equations, these will imply contradictions, because they will assert expressions such as $0 = c$ with $c \neq 0$ that are impossible. Thus, no solution will exist if $\operatorname{rank}(\mathbf{A}) \neq \operatorname{rank}((\mathbf{A}, \mathbf{b})_E)$.

Case (b): $m = n$. In this case the number of variables equals the number of equations, and it must follow that $\operatorname{rank}((\mathbf{A}, \mathbf{b})_E) \le n$. The situation now parallels that of Case (a), because if $\operatorname{rank}(\mathbf{A}) = \operatorname{rank}((\mathbf{A}, \mathbf{b})_E) = r < m$, then r of the equations in (24) will be linearly independent, while $m - r$ will be linear combinations of these r equations. So, as before, the first r rows of $(\mathbf{A}, \mathbf{b})_E$ will be linearly independent while the last $m - r$ rows will be rows of zeros. Thus, r of the variables x_1 to x_n will be determined by the equations corresponding to these r nonzero rows in terms of the remaining $m - r$ variables as parameters. In the case $r = n$, the solution will be unique, because then $\mathbf{A}_E = \mathbf{I}$. Finally, if $\operatorname{rank}(\mathbf{A}) \neq \operatorname{rank}((\mathbf{A}, \mathbf{b})_E)$, it follows, as in Case (a), that no solution will exist.

Case (c): $m > n$. In this case there are more equations than variables, and it must follow that $\operatorname{rank}((\mathbf{A}, \mathbf{b})_E) \le n$. If $\operatorname{rank}(\mathbf{A}) = \operatorname{rank}((\mathbf{A}, \mathbf{b})_E) = r$, it follows, as in Case (b), that r of the equations in (24) are linearly independent while $m - r$ are linear combinations of these r equations. Thus, again, the first r rows of $(\mathbf{A}, \mathbf{b})_E$ will be linearly independent while the last $m - r$ rows will be rows of zeros. Consequently, r of the variables x_1 to x_n will be determined by the equations corresponding to these r nonzero rows in terms of the remaining $m - r$ variables as parameters. If $\operatorname{rank}(\mathbf{A}) \neq \operatorname{rank}((\mathbf{A}, \mathbf{b})_E)$, then as before no solution will exist.

These considerations bring us to the definition of consistent and inconsistent systems of nonhomogeneous equations, with consistent systems having solutions, sometimes in terms of parameters, and inconsistent systems have no solution.

consistent and inconsistent systems

Consistent and inconsistent nonhomogeneous systems

The nonhomogeneous system $\mathbf{Ax} = \mathbf{b}$ is said to be **consistent** when it has a solution; otherwise, it is said to be **inconsistent**.

As with homogeneous systems, the practical determination of solution sets of nonhomogeneous systems of linear equations will be illustrated by means of examples.

EXAMPLE 3.24 Find the solution sets for each of the following augmented matrices $(\mathbf{A}, \mathbf{b})$, where the matrices $\mathbf{A}$ are those given in Example 3.23.

(a) $(\mathbf{A}, \mathbf{b}) = \begin{bmatrix} 1 & 2 & 1 & 7 & 0 & | & 1 \\ 3 & 6 & 4 & 24 & 3 & | & 0 \\ 1 & 4 & 4 & 12 & 3 & | & 3 \end{bmatrix}$ **(b)** $(\mathbf{A}, \mathbf{b}) = \begin{bmatrix} 1 & 3 & 2 & | & 2 \\ 2 & 1 & 0 & | & 1 \\ 1 & 2 & 1 & | & -3 \end{bmatrix}$

(c) $(\mathbf{A}, \mathbf{b}) = \begin{bmatrix} 2 & 3 & 6 & 1 & | & 2 \\ 1 & 4 & 2 & 2 & | & 3 \\ 4 & 11 & 10 & 5 & | & 1 \\ 1 & 0 & 1 & 1 & | & 2 \end{bmatrix}$ **(d)** $(\mathbf{A}, \mathbf{b}) = \begin{bmatrix} 1 & 4 & 1 & 2 & | & 2 \\ 1 & 3 & 0 & 1 & | & 0 \\ 2 & 1 & 1 & 1 & | & 3 \\ 4 & 9 & 3 & 5 & | & 7 \\ 5 & 5 & 2 & 3 & | & 0 \end{bmatrix}$

(e) $(\mathbf{A}, \mathbf{b}) = \begin{bmatrix} 1 & 2 & 3 & 1 & 4 & 3 & | & -2 \\ 0 & 1 & 3 & 0 & 1 & 5 & | & 0 \\ 3 & 1 & 2 & 3 & 1 & 4 & | & 1 \end{bmatrix}$.

Solution

(a) In this case,

$$(\mathbf{A}, \mathbf{b})_E = \begin{bmatrix} 1 & 0 & 0 & 8 & 3 & | & -7 \\ 0 & 1 & 0 & -2 & -3 & | & 11/2 \\ 0 & 0 & 1 & 3 & 3 & | & -3 \end{bmatrix}.$$

As rank$(\mathbf{A}, \mathbf{b})_E = 3$, and the rank of matrix $\mathbf{A}$ is the rank of the matrix formed by deleting the last column of $(\mathbf{A}, \mathbf{b})_E$, it follows that rank$(\mathbf{A}) = 3$. So rank$(\mathbf{A}, \mathbf{b})_E =$ rank$(\mathbf{A})$, showing the equations to be consistent, so they have a solution.

If we remember that the first column contains the coefficients of x_1, the second column the coefficients of $x_2, \ldots$, and the fifth column the coefficients of x_5, while the last column contains the nonhomogeneous terms, we can see that the matrix $(\mathbf{A}, \mathbf{b})_E$ is equivalent to the three equations

$$x_1 + 8x_4 + 3x_5 = -7, \quad x_2 - 2x_4 - 3x_5 = 11/2, \quad x_3 + 3x_4 + 3x_5 = -3.$$

So, if we set $x_4 = \alpha$ and $x_5 = \beta$, with α and β arbitrary numbers (parameters), the solution set becomes

$$x_1 = -8\alpha - 3\beta - 7, \quad x_2 = 2\alpha + 3\beta + 11/2, \quad x_3 = -3\alpha - 3\beta - 3,$$
$$x_4 = \alpha \quad \text{and} \quad x_5 = \beta.$$

(b) In this case,

$$(\mathbf{A}, \mathbf{b})_E = \begin{bmatrix} 1 & 0 & 0 & | & 9 \\ 0 & 1 & 0 & | & -17 \\ 0 & 0 & 1 & | & 22 \end{bmatrix}.$$

Here $\mathbf{A}$ is a 3×3 matrix and rank$(\mathbf{A}) =$ rank$((\mathbf{A}, \mathbf{b})_E) = 3$, so the equations are consistent and the solution is unique. The solution set is seen to be

$$x_1 = 9, \quad x_2 = -17, \quad \text{and} \quad x_3 = 22.$$

(c) In this case,

$$(\mathbf{A}, \mathbf{b})_E = \begin{bmatrix} 1 & 0 & 0 & 20/13 & \vdots & 0 \\ 0 & 1 & 0 & 5/13 & \vdots & 0 \\ 0 & 0 & 0 & -7/13 & \vdots & 0 \\ 0 & 0 & 0 & 0 & \vdots & 1 \\ 0 & 0 & 0 & 0 & \vdots & 0 \end{bmatrix}.$$

This system has no solution because the equations are inconsistent. This follows from the fact that $\text{rank}(\mathbf{A}) = 3$, as can be seen from the first four columns, while the five columns show that $\text{rank}((\mathbf{A}, \mathbf{b})_E) = 4$, so that $\text{rank}(\mathbf{A}) \neq \text{rank}((\mathbf{A}, \mathbf{b})_E)$. The inconsistency can be seen from the contradiction contained in the last row, which asserts that $0 = 1$.

(d) In this case

$$(\mathbf{A}, \mathbf{b})_E = \begin{bmatrix} 1 & 0 & 0 & 0 & \vdots & 0 \\ 0 & 1 & 0 & 1/3 & \vdots & 0 \\ 0 & 0 & 1 & 2/3 & \vdots & 0 \\ 0 & 0 & 0 & 0 & \vdots & 1 \\ 0 & 0 & 0 & 0 & \vdots & 0 \end{bmatrix}.$$

This system also has no solution because the equations are inconsistent. This follows from the fact that $\text{rank}(\mathbf{A}) = 3$ and $\text{rank}((\mathbf{A}, \mathbf{b})_E) = 4$, so that $\text{rank}(\mathbf{A}) \neq \text{rank}((\mathbf{A}, \mathbf{b})_E)$. The inconsistency can again be seen from the contradiction in the last row, which again asserts that $0 = 1$.

(e) In this case

$$(\mathbf{A}, \mathbf{b})_E = \begin{bmatrix} 1 & 0 & 0 & 1 & -1/4 & 1/2 & \vdots & 5/8 \\ 0 & 1 & 0 & 0 & 13/4 & -5/2 & \vdots & -21/8 \\ 0 & 0 & 1 & 0 & -3/4 & 5/2 & \vdots & 7/8 \end{bmatrix},$$

showing that $\text{rank}(\mathbf{A}) = \text{rank}((\mathbf{A}, \mathbf{b})_E) = 3$, so the equations are consistent.

Reasoning as in (a) and setting $x_4 = \alpha, x_5 = \beta$, and $x_6 = \gamma$, with α, β, and γ arbitrary numbers (parameters), shows the solution set to be given by

$$x_1 = -\alpha + (1/4)\beta - (1/2)\gamma + 5/8, \quad x_2 = -(13/4)\beta + (5/2)\gamma - 21/8,$$
$$x_3 = (3/4)\beta - (5/2)\gamma + 7/8, \quad x_4 = \alpha, \quad x_5 = \beta, \quad x_6 = \gamma. \quad \blacksquare$$

A comparison of the corresponding solution sets in Examples 3.23 and 3.24 shows that whenever the nonhomogeneous system has a solution, it comprises the sum of the solution set of the corresponding homogeneous system, containing arbitrary parameters, and numerical constants contributed by the nonhomogeneous terms. This is no coincidence, because it is a fundamental property of nonhomogeneous linear systems of equations. The combination of solutions comprising the sum of a solution of the homogeneous system $\mathbf{Ax} = \mathbf{0}$ containing arbitrary constants, and a particular fixed solution of the nonhomogeneous system $\mathbf{Ax} = \mathbf{b}$ that is free from arbitrary constants, is called the **general solution** of a nonhomogeneous system. The result is important, so it will be recorded as a theorem.

general solution of a nonhomogeneous system

THEOREM 3.9

General solution of a nonhomogeneous system   The nonhomogeneous system of equations

$$\mathbf{Ax} = \mathbf{b}$$

for which $\text{rank}(\mathbf{A}) = \text{rank}\,((\mathbf{A}, \mathbf{b})_E)$ has a general solution of the form

$$\mathbf{x} = \mathbf{x}_H + \mathbf{x}_P,$$

where $\mathbf{x}_H$ is the general solution of the associated homogeneous system $\mathbf{Ax}_H = \mathbf{0}$ and $\mathbf{x}_P$ is a particular (fixed) solution of the nonhomogeneous system $\mathbf{Ax}_P = \mathbf{b}$.

Proof   Let $\mathbf{x}$ be any solution of the nonhomogeneous system $\mathbf{Ax} = \mathbf{b}$, and let $\mathbf{x}_P$ be a solution of the nonhomogeneous system $\mathbf{Ax}_P = \mathbf{b}$ that contains no arbitrary constants (a *fixed* solution). Then, as the equations are linear,

$$\mathbf{A}(\mathbf{x} - \mathbf{x}_P) = \mathbf{Ax} - \mathbf{Ax}_P = \mathbf{b} - \mathbf{b} = \mathbf{0},$$

showing that the difference $\mathbf{x}_D = \mathbf{x} - \mathbf{x}_P$ is itself a solution of the homogeneous system. Consequently, all solutions of the nonhomogeneous system are contained in the solution set of the homogeneous system to which $\mathbf{x}_D$ belongs, and the theorem is proved.   ∎

Summary

This section used the rank of a matrix to determine when a solution of a linear system of nonhomogeneous equations exists and to determine its precise form. If the ranks of a matrix and an augmented matrix are equal, it was shown that a solution exists, furthermore, if there are n equations and the rank $r < n$, then r unknowns can be expressed in terms of arbitrary values assigned to the remaining $n - r$ unknowns. The system was shown to have a unique solution when $r = n$, and no solution if the ranks of the matrix and the augmented matrix are different.

EXERCISES 3.8

In Exercises 1 through 10 write down a system of equations with an appropriate number of unknowns $x_1, x_2, \ldots$ corresponding to the augmented matrix. Find the solution set when the equations are consistent, and state when the equations are inconsistent.

1. $\begin{bmatrix} 1 & -2 & 1 & 3 & | & 11 \\ 0 & 3 & -2 & 1 & | & 11 \\ 2 & 1 & 0 & 4 & | & 23 \\ 3 & 2 & -1 & 2 & | & 21 \\ 1 & -1 & 3 & 2 & | & 4 \end{bmatrix}$.

2. $\begin{bmatrix} 2 & 1 & 3 & 1 & | & 1 \\ 0 & 1 & 4 & 1 & | & 1 \\ 3 & 0 & 0 & 2 & | & 1 \end{bmatrix}$.

3. $\begin{bmatrix} 1 & 3 & 1 & 1 & | & 0 \\ 1 & 1 & 3 & 2 & | & 1 \\ 1 & 1 & 0 & 3 & | & 1 \\ 2 & 0 & 2 & 1 & | & 0 \end{bmatrix}$.

4. $\begin{bmatrix} 1 & 4 & 2 & 3 & | & 4 \\ 2 & 0 & 3 & 1 & | & 2 \\ 5 & 4 & 8 & 5 & | & 8 \end{bmatrix}$.

5. $\begin{bmatrix} 1 & -1 & 2 & -1 & | & -4 \\ 2 & 3 & 1 & 2 & | & 12 \\ 1 & 2 & -2 & 3 & | & 15 \\ 3 & 1 & -1 & 1 & | & 11 \\ 1 & 1 & -1 & 2 & | & 3 \end{bmatrix}$.

6. $\begin{bmatrix} 1 & 2 & 3 & | & 1 \\ 2 & 1 & 1 & | & 3 \\ 0 & 2 & 1 & | & 3 \\ 2 & 6 & 7 & | & 5 \\ 1 & -2 & 1 & | & 0 \end{bmatrix}$.

7. $\begin{bmatrix} 1 & 2 & 3 & 0 & | & 1 \\ 0 & 1 & 0 & 2 & | & 1 \\ 2 & 1 & 3 & 1 & | & 0 \\ 1 & 4 & 1 & 5 & | & 2 \end{bmatrix}$.

8. $\begin{bmatrix} 2 & 1 & 0 & 0 & 3 & | & 1 \\ 1 & 2 & 1 & 1 & 3 & | & 0 \\ 0 & 1 & 2 & 5 & 1 & | & 2 \end{bmatrix}$.

9. $\begin{bmatrix} 1 & 2 & 1 & | & 4 \\ 1 & 1 & 2 & | & 0 \\ 2 & 1 & 1 & | & 4 \\ 0 & 3 & 5 & | & 1 \end{bmatrix}$.

10. $\begin{bmatrix} 1 & 3 & 1 & 1 & 2 & | & 1 \\ 1 & -2 & 1 & 3 & 1 & | & 0 \\ 2 & 0 & 1 & 0 & 3 & | & 0 \end{bmatrix}$.

3.9   The Inverse Matrix

multiplicative inverse matrix

The operation of division is not defined for matrices. However, we will see that $n \times n$ matrices **A** for which $\det \mathbf{A} \neq 0$ have associated with them an $n \times n$ matrix **B**, called its **multiplicative inverse**, with the property that

$$\mathbf{AB} = \mathbf{BA} = \mathbf{I}.$$

The purpose of this section will be to develop ways of finding the multiplicative inverse of a matrix, which for simplicity is usually called the **inverse** matrix, but first we give a formal definition of the inverse of a matrix.

The inverse of a matrix

Let **A** and **B** be two $n \times n$ matrices. Then matrix **A** is said to be invertible and to have an associated **inverse** matrix **B** if

$$\mathbf{AB} = \mathbf{BA} = \mathbf{I}.$$

Interchanging the order of **A** and **B** in this definition shows that if **B** is the inverse of **A**, then **A** must be the inverse of **B**.

To see that not all $n \times n$ matrices have inverses, it will be sufficient to try to find a matrix **B** such that the product $\mathbf{AB} = \mathbf{I}$, where

$$\mathbf{A} = \begin{bmatrix} 1 & 2 \\ 1 & 2 \end{bmatrix} \quad \text{and} \quad \mathbf{B} = \begin{bmatrix} a & b \\ c & d \end{bmatrix}.$$

The product **AB** is

$$\mathbf{AB} = \begin{bmatrix} 1 & 2 \\ 1 & 2 \end{bmatrix} \begin{bmatrix} a & b \\ c & d \end{bmatrix} = \begin{bmatrix} a + 2c & b + 2d \\ a + 2c & b + 2d \end{bmatrix},$$

so if this product is to equal the 2×2 unit matrix **I**, it is necessary that

$$\begin{bmatrix} a + 2c & b + 2d \\ a + 2c & b + 2d \end{bmatrix} = \begin{bmatrix} 1 & 0 \\ 0 & 1 \end{bmatrix}.$$

Equating corresponding elements in the first columns shows that this can only hold if $a + 2c = 1$ and $a + 2c = 0$, while equating corresponding elements in the second columns shows that $b + 2d = 0$ and $b + 2d = 1$, which is impossible, so matrix **A** has no inverse. In this case $\det \mathbf{A} = 0$, and we will see later why the nonvanishing of $\det \mathbf{A}$ is necessary if **A** is to have an inverse.

Nonsingular and singular matrices

singular and nonsingular $n \times n$ matrices

An $n \times n$ matrix is said to be **nonsingular** when its inverse exists, and to be **singular** when it has no inverse.

EXAMPLE 3.25

We have already seen that the matrix

$$\mathbf{A} = \begin{bmatrix} 1 & 2 \\ 1 & 2 \end{bmatrix},$$

for which $\det\mathbf{A} = 0$, has no inverse and so is *singular*. However, in the case of matrix $\mathbf{A}$ that follows, a simple matrix multiplication confirms that it has associated with it an inverse $\mathbf{B}$, where

$$\mathbf{A} = \begin{bmatrix} 1 & 0 & 1 \\ -1 & 2 & 0 \\ 0 & 1 & 1 \end{bmatrix} \quad \text{and} \quad \mathbf{B} = \begin{bmatrix} 2 & 1 & -2 \\ 1 & 1 & -1 \\ -1 & -1 & 2 \end{bmatrix},$$

because $\mathbf{AB} = \mathbf{BA} = \mathbf{I}$. Furthermore, $\det\mathbf{A} \neq 0$, so $\mathbf{A}$ is *nonsingular*, as is $\mathbf{B}$, and each is the inverse of the other. ∎

Before proceeding further it is necessary to establish that, when it exists, the inverse matrix is unique.

THEOREM 3.10

Uniqueness of the inverse matrix   A nonsingular matrix $\mathbf{A}$ has a unique inverse.

Proof   Suppose, if possible, that the nonsingular $n \times n$ matrix $\mathbf{A}$ has the two different inverses $\mathbf{B}$ and $\mathbf{C}$. Then as $\mathbf{AC} = \mathbf{I}$, we have

$$\mathbf{B} = \mathbf{BI} = \mathbf{B}(\mathbf{AC}) = (\mathbf{BA})\mathbf{C} = \mathbf{IC} = \mathbf{C},$$

showing that $\mathbf{B} = \mathbf{C}$, so the inverse matrix is unique. ∎

It is convenient to denote the inverse of a nonsingular $n \times n$ matrix $\mathbf{A}$ by the symbol $\mathbf{A}^{-1}$. This is suggested by the exponentiation notation (raising to a power), because if for the moment we write $\mathbf{A} = \mathbf{A}^1$, then $\mathbf{AA}^{-1} = \mathbf{A}^1\mathbf{A}^{-1} = \mathbf{I}$, showing that exponents may be combined in the usual way, with the understanding that $\mathbf{A}^1\mathbf{A}^{-1} = \mathbf{A}^{(1-1)} = \mathbf{A}^0 = \mathbf{I}$.

THEOREM 3.11

basic properties of the inverse matrix

Basic properties of inverse matrices

(i) The unit matrix $\mathbf{I}$ is its own inverse, so $\mathbf{I} = \mathbf{I}^{-1}$.

(ii) If $\mathbf{A}$ is nonsingular, so also is $\mathbf{A}^{-1}$, and $(\mathbf{A}^{-1})^{-1} = \mathbf{A}$.

(iii) If $\mathbf{A}$ is nonsingular, so also is $\mathbf{A}^T$, and $(\mathbf{A}^{-1})^T = (\mathbf{A}^T)^{-1}$.

(iv) If $\mathbf{A}$ and $\mathbf{B}$ are nonsingular $n \times n$ matrices, so is $\mathbf{AB}$, and

$$(\mathbf{AB})^{-1} = \mathbf{B}^{-1}\mathbf{A}^{-1}.$$

(v) If $\mathbf{A}$ is nonsingular, then $(\mathbf{A}^{-1})^m = (\mathbf{A}^m)^{-1}$ for $m = 1, 2, \ldots$.

Proof   We prove only (i) and (iv), and leave the proofs of (ii), (iii), and (v) as exercises. The proof of (i) is almost immediate, because $\mathbf{I}^2 = \mathbf{I}$, showing that $\mathbf{I} = \mathbf{I}^{-1}$. To prove (iv) we premultiply $\mathbf{B}^{-1}\mathbf{A}^{-1}$ by $\mathbf{AB}$ to obtain

$$\mathbf{ABB}^{-1}\mathbf{A}^{-1} = \mathbf{AIA}^{-1} = \mathbf{AA}^{-1} = \mathbf{I},$$

which shows that $(\mathbf{AB})^{-1}$ is $\mathbf{B}^{-1}\mathbf{A}^{-1}$, so the proof is complete. ∎

A simple method of finding the inverse of an $n \times n$ matrix is by means of elementary row operations, but to justify the method we first need the following theorem.

THEOREM 3.12

Elementary row operation matrices are nonsingular   Every $n \times n$ matrix $\mathbf{E}$ that represents an elementary row operation is nonsingular.

Proof   Every $n \times n$ matrix $\mathbf{E}$ that represents an elementary row operation is derived from the unit matrix $\mathbf{I}$ by means of one of the three operations defined at the start of Section 3.4. So, as rank$(\mathbf{I}) = n$ and $\mathbf{E}$ and $\mathbf{I}$ are row similar, it follows that rank$(\mathbf{E}) = n$, and so $\mathbf{E}$ is also nonsingular.   ∎

finding an inverse matrix using elementary row operations

We can now describe an elementary way of finding an inverse matrix by means of elementary row transformations. Let $\mathbf{A}$ be a nonsingular $n \times n$ matrix, and let $\mathbf{E}_1, \mathbf{E}_2, \ldots, \mathbf{E}_m$ represent a sequence of elementary row operations of Types I, II, and III that reduces $\mathbf{A}$ to $\mathbf{I}$, so that

$$\mathbf{E}_m \mathbf{E}_{m-1} \ldots \mathbf{E}_2 \mathbf{E}_1 \mathbf{A} = \mathbf{I}.$$

Then postmultiplying this result by $\mathbf{A}^{-1}$ gives

$$\mathbf{E}_m \mathbf{E}_{m-1} \ldots \mathbf{E}_2 \mathbf{E}_1 \mathbf{I} = \mathbf{A}^{-1},$$

so $\mathbf{A}^{-1}$ is given by

$$\mathbf{A}^{-1} = \mathbf{E}_m \mathbf{E}_{m-1} \cdots \mathbf{E}_2 \mathbf{E}_1 \mathbf{I},$$

where the product of the first m matrices on the right is nonsingular because of Theorem 3.11. Expressed in words, this result states that when a sequence of elementary row operations is used to reduce a nonsingular matrix $\mathbf{A}$ to the unit matrix $\mathbf{I}$, performing the same sequence of elementary row operations on $\mathbf{I}$, in the same order, will generate the inverse matrix $\mathbf{A}^{-1}$. If matrix $\mathbf{A}$ is singular, this will be indicated by the generation of either a complete row or a complete column of zeros before $\mathbf{I}$ is reached.

If $\mathbf{A}$ is nonsingular, it is reducible to the unit matrix $\mathbf{I}$, and clearly det$\mathbf{A} \neq 0$. However, if $\mathbf{A}$ is singular, the attempt to reduce it to $\mathbf{I}$ will generate either a row or a column of zeros, so that then det$\mathbf{A} = 0$. The vanishing or nonvanishing of det$\mathbf{A}$ provides a simple and convenient test for the singularity or nonsingularity of $\mathbf{A}$ whenever n is *small*, say $n \leq 3$, because only then is it a simple matter to calculate det$\mathbf{A}$.

The practical way in which to implement this result is not to use the matrices $\mathbf{E}_i$ to reduce $\mathbf{A}$ to $\mathbf{I}$, but to perform the operations directly on the rows of the partitioned matrix $(\mathbf{A}, \mathbf{I})$, because when $\mathbf{A}$ in the left half of the partitioned matrix has been reduced to $\mathbf{I}$, the matrix $\mathbf{I}$ in the right half will have been transformed into $\mathbf{A}^{-1}$.

EXAMPLE 3.26

Use elementary row operations to find $\mathbf{A}^{-1}$ given that

$$\mathbf{A} = \begin{bmatrix} 1 & 0 & 1 \\ -1 & 2 & 0 \\ 0 & 1 & 1 \end{bmatrix}.$$

Solution   We form the augmented matrix $(\mathbf{A}, \mathbf{I})$ and proceed as described earlier.

$$(\mathbf{A}, \mathbf{I}) = \begin{bmatrix} 1 & 0 & 1 & \vdots & 1 & 0 & 0 \\ -1 & 2 & 0 & \vdots & 0 & 1 & 0 \\ 0 & 1 & 1 & \vdots & 0 & 0 & 1 \end{bmatrix} \begin{array}{c} \sim \\ \text{add row 1} \\ \text{to row 2} \end{array} \begin{bmatrix} 1 & 0 & 1 & \vdots & 1 & 0 & 0 \\ 0 & 2 & 1 & \vdots & 1 & 1 & 0 \\ 0 & 1 & 1 & \vdots & 0 & 0 & 1 \end{bmatrix}$$

$$\begin{array}{c} \sim \\ \text{subtract row 3} \\ \text{from row 2} \end{array} \begin{bmatrix} 1 & 0 & 1 & \vdots & 1 & 0 & 0 \\ 0 & 1 & 0 & \vdots & 1 & 1 & -1 \\ 0 & 1 & 1 & \vdots & 0 & 0 & 1 \end{bmatrix} \begin{array}{c} \sim \\ \text{subtract row 2} \\ \text{from row 3} \end{array} \begin{bmatrix} 1 & 0 & 1 & \vdots & 1 & 0 & 0 \\ 0 & 1 & 0 & \vdots & 1 & 1 & -1 \\ 0 & 0 & 1 & \vdots & -1 & -1 & 2 \end{bmatrix}$$

$$\begin{array}{c} \sim \\ \text{subtract row 3} \\ \text{from row 1} \end{array} \begin{bmatrix} 1 & 0 & 0 & \vdots & 2 & 1 & -2 \\ 0 & 1 & 0 & \vdots & 1 & 1 & -1 \\ 0 & 0 & 1 & \vdots & -1 & -1 & 2 \end{bmatrix}.$$

The 3×3 matrix on the left of this row-equivalent partitioned matrix is now the unit matrix $\mathbf{I}$, so the required inverse matrix is the one to the right of the partition, namely,

$$\mathbf{A}^{-1} = \begin{bmatrix} 2 & 1 & -2 \\ 1 & 1 & -1 \\ -1 & -1 & 2 \end{bmatrix}.$$

Once $\mathbf{A}^{-1}$ has been obtained, it is always advisable to check the result by verifying that $\mathbf{A}\mathbf{A}^{-1} = \mathbf{I}$. ∎

Before proceeding further we will use elementary matrices to provide the promised proof of Theorem 3.4(viii).

the proof that det(AB) = detA detB

Proof that det(AB) = detA detB   Let $\mathbf{E}_1$ be a row matrix of Type I. Then if $\mathbf{A}$ is a nonsingular matrix, $\det(\mathbf{E}_I\mathbf{A}) = -\det\mathbf{A}$, because only a row interchange is involved. However, $\det(\mathbf{E}_I) = -1$, so $\det(\mathbf{E}_I\mathbf{A}) = \det\mathbf{E}_I\det\mathbf{A}$. Similar arguments show this to be true for elementary row operation matrices of the other two types, so if $\mathbf{E}$ is an elementary row operation of any type, then

$$\det(\mathbf{E}\mathbf{A}) = \det\mathbf{E}\det\mathbf{A}.$$

If $\det\mathbf{A} \neq 0$, premultiplication by a sequence of elementary row operation matrices $\mathbf{E}_1, \mathbf{E}_2, \ldots, \mathbf{E}_r$ will reduce $\mathbf{A}$ to $\mathbf{I}$, so performing them on $\mathbf{I}$ in the reverse order allows us to write

$$\mathbf{A} = \mathbf{E}_1\mathbf{E}_2 \ldots \mathbf{E}_r\mathbf{I} = \mathbf{E}_1\mathbf{E}_2 \ldots \mathbf{E}_r.$$

A repetition of the result $\det(\mathbf{E}\mathbf{A}) = \det\mathbf{E}\det\mathbf{A}$ shows that

$$\det\mathbf{A} = \det\mathbf{E}_1\det\mathbf{E}_2 \ldots \det\mathbf{E}_r.$$

If $\mathbf{B}$ is conformable for multiplication with $\mathbf{A}$, using the preceding result we have

$$\det(\mathbf{A}\mathbf{B}) = \det(\mathbf{E}_1\mathbf{E}_2 \ldots \mathbf{E}_r\mathbf{B})$$
$$= \det\mathbf{E}_1\det\mathbf{E}_2 \ldots \det\mathbf{E}_r\det\mathbf{B},$$

but

$$\det\mathbf{E}_1\det\mathbf{E}_2 \ldots \det\mathbf{E}_n = \det\mathbf{A}, \quad \text{and so} \quad \det(\mathbf{A}\mathbf{B}) = \det\mathbf{A}\det\mathbf{B}.$$

To complete the proof we must show this result remains true if **A** is singular, in which case $\det\mathbf{A} = 0$. When $\det\mathbf{A} = 0$, the attempt to reduce it to the unit matrix **I** by elementary row operation matrices will fail because at one stage it will produce a determinant in which a row will contain only zero elements. Consequently, a determinant $\det\mathbf{E}_m$, say, will be zero, which is impossible, so $\det(\mathbf{AB}) = 0$. However, if $\det\mathbf{A} = 0$, then $\det\mathbf{A}\det\mathbf{B} = 0$, so that once again $\det(\mathbf{AB}) = \det\mathbf{A}\det\mathbf{B}$, and the result is proved.   ∎

EXAMPLE 3.27

Use (a) elementary row operations and (b) the determinant test to show matrix **A** is singular, given that

$$\mathbf{A} = \begin{bmatrix} 1 & 1 & 0 \\ 1 & 0 & 1 \\ 4 & 3 & 1 \end{bmatrix}.$$

Solution

(a) Using elementary row operations on the augmented matrix gives

$$(\mathbf{A}, \mathbf{I}) = \left[\begin{array}{ccc|ccc} 1 & 1 & 0 & 1 & 0 & 0 \\ 1 & 0 & 1 & 0 & 1 & 0 \\ 4 & 3 & 1 & 0 & 0 & 1 \end{array}\right] \begin{array}{c} \sim \\ \text{subtract row 1} \\ \text{from row 2} \end{array} \left[\begin{array}{ccc|ccc} 1 & 1 & 0 & 1 & 0 & 0 \\ 0 & -1 & 1 & -1 & 1 & 0 \\ 4 & 3 & 1 & 0 & 0 & 1 \end{array}\right]$$

$$\begin{array}{c} \sim \\ \text{subtract 4 times} \\ \text{row 1 from row 3} \end{array} \left[\begin{array}{ccc|ccc} 1 & 1 & 0 & 1 & 0 & 0 \\ 0 & -1 & 1 & -1 & 1 & 0 \\ 0 & -1 & 1 & -4 & 0 & 1 \end{array}\right]$$

$$\begin{array}{c} \sim \\ \text{subtract row 2} \\ \text{from row 3} \end{array} \left[\begin{array}{ccc|ccc} 1 & 1 & 0 & 1 & 0 & 0 \\ 0 & -1 & 1 & -1 & 1 & 0 \\ 0 & 0 & 0 & -3 & -1 & 1 \end{array}\right].$$

The reduction is terminated at this stage by the appearance of a row of zeros on the matrix to the left of the partition, showing that **A** cannot be reduced to **I**, and hence that **A** is singular.

(b) Applying the determinant test to **A**, we find that $\det\mathbf{A} = 0$, showing that **A** is singular. Although in this case this is by far the quickest way to establish the singularity of **A**, this would not have been so had the order of $\det\mathbf{A}$ been much greater than 3. This is because when $n > 3$, the effort involved in performing the elementary row operations in an attempt to reduce **A** to **I** is considerably less than the effort involved when calculating $\det\mathbf{A}$.   ∎

The following very different way of finding the inverse of an $n \times n$ matrix **A** is mainly of theoretical importance, though it is a practical method when n is small. The method is based on the properties of the sum of products of elements and cofactors of a determinant.

Let $\mathbf{A} = [a_{ij}]$ be an $n \times n$ matrix, $\mathbf{C} = [C_{ij}]$ be the associated $n \times n$ matrix of cofactors and form the matrix product

$$\mathbf{AC}^\mathrm{T} = \begin{bmatrix} a_{11} & a_{12} & \ldots & a_{1n} \\ a_{21} & a_{22} & \ldots & a_{2n} \\ \multicolumn{4}{c}{\cdots\cdots\cdots\cdots} \\ a_{n1} & a_{n2} & \ldots & a_{nn} \end{bmatrix} \begin{bmatrix} C_{11} & C_{21} & \ldots & C_{n1} \\ C_{12} & C_{22} & \ldots & C_{n2} \\ \multicolumn{4}{c}{\cdots\cdots\cdots\cdots} \\ C_{1n} & C_{2n} & \ldots & C_{nn} \end{bmatrix}.$$

If we write $\mathbf{B} = \mathbf{A}\mathbf{C}^{\mathrm{T}}$, with $\mathbf{B} = [b_{ij}]$, it follows from the rule for matrix multiplication that

$$b_{ij} = a_{i1}C_{1j} + a_{i2}C_{2j} + \cdots + a_{in}C_{nj}.$$

Thus, b_{ij} is seen to be the sum of the product of the elements of the ith row of $\mathbf{A}$ and the corresponding cofactors of the elements of the jth row of $\mathbf{A}$. It then follows from the Laplace expansion theorem for determinants that

$$b_{ij} = \det\mathbf{A}, \ \ \text{for } i = j = 1, 2, \ldots, n$$

and

$$b_{ij} = 0, \ \ \text{for } i \neq j.$$

Using these results in the matrix product, we find that

$$\mathbf{A}\mathbf{C}^{\mathrm{T}} = \begin{bmatrix} \det\mathbf{A} & 0 & 0 & \ldots & 0 \\ 0 & \det\mathbf{A} & 0 & & \\ 0 & 0 & \det\mathbf{A} & \ldots & 0 \\ & & \cdots & & \\ 0 & 0 & 0 & \ldots & \det\mathbf{A} \end{bmatrix}$$

$$= \det\mathbf{A} \ \mathbf{I}.$$

Consequently, provided $\det\mathbf{A} \neq 0$, it follows that

$$(1/\det\mathbf{A})\mathbf{A}\mathbf{C}^{\mathrm{T}} = \mathbf{I}.$$

Writing this as

$$\mathbf{A}\{(1/\det\mathbf{A})\mathbf{C}^{\mathrm{T}}\} = \mathbf{I}$$

shows that

$$\mathbf{A}^{-1} = (1/\det\mathbf{A})\mathbf{C}^{\mathrm{T}}.$$

adjoint matrix

The matrix $\mathbf{C}^{\mathrm{T}}$, called the **adjoint** of $\mathbf{A}$ and written adj$\mathbf{A}$, is the *transpose* of the matrix of cofactors of $\mathbf{A}$. So the formula for the inverse of $\mathbf{A}$ becomes

$$\mathbf{A}^{-1} = (1/\det\mathbf{A})\text{adj}\mathbf{A}. \tag{25}$$

We have arrived at the following definition and theorem.

Adjoint matrix

If $\mathbf{A}$ is an $n \times n$ matrix, and $\mathbf{C}$ is the associated matrix of cofactors, the transpose $\mathbf{C}^{\mathrm{T}}$ of the matrix of cofactors is called the adjoint of $\mathbf{A}$ and is written adj$\mathbf{A}$.

THEOREM 3.13

formal definition of an inverse matrix

The inverse matrix in terms of the adjoint of A   Let $\mathbf{A}$ be a nonsingular $n \times n$ matrix. Then the inverse of $\mathbf{A}$ is given by

$$\mathbf{A}^{-1} = (1/\det\mathbf{A})\text{adj}\mathbf{A}.$$

∎

| **EXAMPLE 3.28** |

Use Theorem 3.13 to find $\mathbf{A}^{-1}$, given that

$$\mathbf{A} = \begin{bmatrix} 1 & 3 & 0 \\ 2 & 1 & 1 \\ 1 & 0 & 1 \end{bmatrix}.$$

Solution   The matrix of cofactors

$$\mathbf{C} = \begin{bmatrix} 1 & -1 & -1 \\ -3 & 1 & 3 \\ 3 & -1 & -5 \end{bmatrix}, \quad \text{so } \mathbf{C}^{\mathrm{T}} = \begin{bmatrix} 1 & -3 & 3 \\ -1 & 1 & -1 \\ -1 & 3 & -5 \end{bmatrix}.$$

Expanding $\det\mathbf{A}$ in terms of the elements of its first row (we already have its associated cofactors in the first row of $\mathbf{C}$) gives $\det\mathbf{A} = 1 \cdot 1 + (-1) \cdot 3 + 1 \cdot 0 = -2$, so from Theorem 3.13,

$$\mathbf{A}^{-1} = (-1/2)\mathbf{C}^{\mathrm{T}} = \begin{bmatrix} -\frac{1}{2} & \frac{3}{2} & -\frac{3}{2} \\ \frac{1}{2} & -\frac{1}{2} & \frac{1}{2} \\ \frac{1}{2} & -\frac{3}{2} & \frac{5}{2} \end{bmatrix}. \qquad \blacksquare$$

Although the result of Theorem 3.13 is of considerable theoretical importance, unless n is small, the task of evaluating the determinants involved makes it impractical for the determination of inverse matrices. In general, for large n, an inverse matrix is found by means of a computer using elementary row operations to reduce $\mathbf{A}$ to $\mathbf{I}$.

General Proof of Cramer's Rule

proof of Cramer's rule for a system of n equations

In conclusion, we will use Theorem 3.13 to arrive at a simple proof of Cramer's rule for the system of equations

$$a_{11}x_1 + a_{12}x_2 + \cdots + a_{1n}x_n = b_1$$
$$a_{21}x_1 + a_{22}x_2 + \cdots + a_{2n}x_n = b_2$$
$$\cdots\cdots\cdots\cdots$$
$$a_{n1}x_1 + a_{n2}x_2 + \cdots + a_{nn}x_n = b_n.$$

If we write the system as $\mathbf{Ax} = \mathbf{b}$, then, provided $\det\mathbf{A} \neq 0$, the solution can be written

$$\mathbf{x} = \mathbf{A}^{-1}\mathbf{b} = (1/\det\mathbf{A})(\mathrm{adj}\mathbf{A})\mathbf{b} = (1/\det\mathbf{A})\mathbf{C}^{\mathrm{T}}\mathbf{b},$$

where $\mathbf{C}^{\mathrm{T}}$ is the transpose of the matrix of cofactors of $\mathbf{A}$. If $\mathbf{x} = (x_1, x_2, \ldots, x_n)^{\mathrm{T}}$ and $\mathbf{b} = (b_1, b_2, \ldots, b_n)^{\mathrm{T}}$, the ith element of $\mathbf{x}$ is given by

$$x_i = (1/\det\mathbf{A})(C_{1i}b_1 + C_{2i}b_2 + \cdots + C_{ni}b_n) \quad \text{for } i = 1, 2, \ldots, n.$$

This is simply the expansion of $\det\mathbf{A}_i$ in terms of the elements of its ith column, where $\mathbf{A}_i$ is the matrix obtained from $\mathbf{A}$ by replacing the elements of the ith column by the elements of $\mathbf{b}$. This has established that

$$x_i = \det\mathbf{A}_i/\det\mathbf{A}, \quad \text{for } i = 1, 2, \ldots, n,$$

and the proof is complete. $\qquad \blacksquare$

More information about the material in Sections 3.4 to 3.9 is to be found in the appropriate chapters of references [2.1], [2.5], and [2.7] to [2.12].

> **GABRIEL CRAMER (1704–1752):**
> A Swiss mathematician who made many contributions to algebra and geometry. The result called Cramer's rule was, in fact, first formulated by Maclaurin around 1729 and published posthumously in his *Treatise on Algebra* (1748). The form of the rule attributed to Cramer appeared in his book *Traité des courbes algebraiques* (1750), which became a standard reference work during the remainder of the century. The work was so well written and so often quoted that after his death Cramer was, on occasions, considered to be the originator of the rule.

Summary

Division by matrices is not defined, but the introduction of a multiplicative inverse $\mathbf{A}^{-1}$ of a nonsingular $n \times n$ matrix $\mathbf{A}$, called the inverse of $\mathbf{A}$, enables certain operations that in some sense are similar to matrix division to be performed. This section gave the formal definition of the inverse of a matrix and established its most important algebraic properties. The inverse matrix was used to prove Cramer's rule for a general system of n nonhomogeneous linear algebraic equations when the determinant of the coefficient matrix is nonsingular.

EXERCISES 3.9

In Exercises 1 through 8, construct a suitable augmented matrix and find the inverse of the given matrix using elementary row operations.

1. $\begin{bmatrix} 1 & 3 & 7 \\ 2 & 1 & -1 \\ 2 & 1 & 5 \end{bmatrix}$.

2. $\begin{bmatrix} -4 & 1 & 0 \\ 1 & -3 & 1 \\ 2 & 1 & 4 \end{bmatrix}$.

3. $\begin{bmatrix} 1 & 1 & 3 \\ 5 & 2 & 1 \\ 1 & 6 & 2 \end{bmatrix}$.

4. $\begin{bmatrix} 2 & -6 & 1 \\ 1 & 3 & 4 \\ 0 & -2 & 1 \end{bmatrix}$.

5. $\begin{bmatrix} 2 & 3 & 1 \\ 1 & 2 & 0 \\ 2 & 4 & 1 \end{bmatrix}$.

6. $\begin{bmatrix} 3 & 0 & 1 \\ 1 & -1 & 1 \\ 0 & 4 & 5 \end{bmatrix}$.

7. $\begin{bmatrix} 1 & 2 & 0 & 1 \\ 1 & 0 & -3 & 4 \\ 0 & 1 & 2 & 5 \\ 2 & -1 & 2 & 2 \end{bmatrix}$.

8. $\begin{bmatrix} 0 & 1 & 2 & 3 \\ 2 & 2 & -4 & 2 \\ 1 & 3 & 0 & 1 \\ 3 & 1 & 1 & 0 \end{bmatrix}$.

9. Given that
$$\mathbf{A} = \begin{bmatrix} 3 & -1 & 1 \\ 1 & 4 & 0 \\ 2 & 1 & -3 \end{bmatrix} \text{ and } \mathbf{B} = \begin{bmatrix} 1 & -3 & 1 \\ 2 & 0 & 5 \\ 3 & 1 & 2 \end{bmatrix},$$
verify that $(\mathbf{AB})^{-1} = \mathbf{B}^{-1}\mathbf{A}^{-1}$.

10. Given that
$$\mathbf{A} = \begin{bmatrix} 4 & 1 & 2 \\ 3 & 1 & 0 \\ 3 & 2 & 1 \end{bmatrix}, \text{ verify that } (\mathbf{A}^{-1})^{\mathrm{T}} = (\mathbf{A}^{\mathrm{T}})^{-1} \text{ and}$$
$(\mathbf{A}^{-1})^2 = (\mathbf{A}^2)^{-1}$.

In Exercises 11 through 16, use Theorem 3.13 to find the inverse of the given matrix, and check the result by showing that $\mathbf{AA}^{-1} = \mathbf{I}$.

11. $\begin{bmatrix} 2 & 4 & -5 \\ 2 & 7 & 1 \\ 1 & 3 & 4 \end{bmatrix}$.

12. $\begin{bmatrix} 3 & -7 & 8 \\ 1 & 4 & 3 \\ 0 & -5 & 1 \end{bmatrix}$.

13. $\begin{bmatrix} 9 & 2 & 1 \\ 1 & 4 & 10 \\ 3 & 1 & 2 \end{bmatrix}$.

14. $\begin{bmatrix} -3 & 2 & 6 \\ 2 & -1 & 7 \\ 5 & 4 & -2 \end{bmatrix}$.

15. $\begin{bmatrix} 2 & 0 & 1 & 2 \\ 3 & 1 & 3 & 4 \\ 1 & 0 & -2 & 3 \\ 1 & -2 & 2 & 7 \end{bmatrix}$.

16. $\begin{bmatrix} 0 & 1 & -4 & 1 \\ 3 & 7 & 5 & 2 \\ 1 & -2 & 6 & 0 \\ 0 & 1 & 3 & 1 \end{bmatrix}$.

In the following two exercises, use the determinant test to show the given matrix is singular, and then verify this by using elementary row operations applied to a suitable augmented matrix, as in Example 3.27. Compare the effort involved in each case.

17. $\begin{bmatrix} 0 & 2 & 1 & 0 \\ 1 & 1 & 3 & 0 \\ 2 & 1 & 4 & 2 \\ 4 & 3 & 10 & 2 \end{bmatrix}$.

18. $\begin{bmatrix} 1 & 3 & 0 & 1 \\ 1 & 1 & 2 & 1 \\ 1 & 1 & 2 & 5 \\ 0 & -1 & 1 & 2 \end{bmatrix}$.

3.10   Derivative of a Matrix

When the elements of matrix $\mathbf{A}$ are differentiable functions of a single variable, say t, so that $\mathbf{A} = \mathbf{A}[a_{ij}(t)]$, calculus can be performed on matrices, so it becomes necessary to define the derivative of a matrix. An illustration of the need for this was given in Section 3.2(e), where the matrix differential equation $\ddot{\mathbf{x}} + \mathbf{A}\mathbf{x} = \mathbf{0}$ was obtained as the system of second order differential equations determining the motion of a compound mass–spring system.

Derivative of a matrix

Let the $m \times n$ matrix $\mathbf{A}$ have elements $a_{ij}(t)$ that are differentiable functions of the variable t. Then the **first order derivative** of $\mathbf{A}$ with respect to t, written $d\mathbf{A}/dt$, is defined as

fundamental definition of dA/dt

$$d\mathbf{A}/dt = [d(a_{ij})/dt],$$

and its ***n*th order derivative** with respect to t is defined recursively as

$$d^n\mathbf{A}/dt^n = d/dt[d^{n-1}\mathbf{A}/dt^{n-1}], \quad \text{for } n = 1, 2, \ldots,$$

with the convention that $d^0(a_{ij})/dt^0 = a_{ij}$, so that $d^0\mathbf{A}/dt^0 = \mathbf{A}$. The derivative of a constant matrix is the null (zero) matrix $\mathbf{0}$.

EXAMPLE 3.29

Find $d\mathbf{A}/dt$ and $d^2\mathbf{A}/dt^2$ given that

(a) $\mathbf{A} = \begin{bmatrix} t^2 & 3t & \cosh t \\ 2t+1 & e^t & \sin 2t \end{bmatrix}$, (b) $\mathbf{A} = \begin{bmatrix} te^t \\ \cos 3t \end{bmatrix}$.

Solution

(a) By definition,

$$d\mathbf{A}/dt = \begin{bmatrix} 2t & 3 & \sinh t \\ 2 & e^t & 2\cos 2t \end{bmatrix} \quad \text{and} \quad d^2\mathbf{A}/dt^2 = \begin{bmatrix} 2 & 0 & \cosh t \\ 0 & e^t & -4\sin 2t \end{bmatrix}.$$

(b) $d\mathbf{A}/dt = \begin{bmatrix} e^t + te^t \\ -3\sin 3t \end{bmatrix} \quad \text{and} \quad d^2\mathbf{A}/dt^2 = \begin{bmatrix} 2e^t + te^t \\ -9\cos 3t \end{bmatrix}.$ ∎

THEOREM 3.14

derivative of a sum, a product, and an inverse matrix

Derivative of the sum of two matrices   Let $\mathbf{A}(t)$ and $\mathbf{B}(t)$ be an $m \times n$ matrices, each with differentiable elements. Then

$$d/dt\{\mathbf{A} + \mathbf{B}\} = d\mathbf{A}/dt + d\mathbf{B}/dt.$$

Proof   The result follows immediately from the definition of the sum of two matrices. ∎

THEOREM 3.15

Derivative of a matrix product Let $\mathbf{A}(t)$ be an $m \times n$ matrix and $\mathbf{B}(t)$ be an $n \times q$ matrix, each with differentiable elements. Then, if the $m \times q$ matrix $\mathbf{C}(t) = \mathbf{A}(t)\,\mathbf{B}(t)$,

$$d\mathbf{C}/dt = \{d\mathbf{A}/dt\}\mathbf{B} + \mathbf{A}\{d\mathbf{B}/dt\}.$$

Proof It follows from the definition of the matrix product of two matrices $\mathbf{A}$ and $\mathbf{B}$ that are conformable for multiplication that $c_{rs} = a_{r1}b_{1s} + a_{r2}b_{2s} + \cdots + a_{rn}b_{ns}$, so each term in c_{rs} is a product of two differentiable functions. Differentiating c_{rs} establishes the theorem in which the order of the matrix products must be as shown. ∎

THEOREM 3.16

Derivative of an inverse matrix Let $\mathbf{A}(t)$ be an $n \times n$ nonsingular matrix with differentiable elements. Then

$$d\mathbf{A}^{-1}/dt = -\mathbf{A}^{-1}\{d\mathbf{A}/dt\}\mathbf{A}^{-1}.$$

Proof As $\mathbf{A}$ is nonsingular, its inverse $\mathbf{A}^{-1}$ exists and $\mathbf{A}\mathbf{A}^{-1} = \mathbf{I}$. Differentiating the matrix product $\mathbf{A}\mathbf{A}^{-1} = \mathbf{I}$ gives

$$\{d\mathbf{A}/dt\}\mathbf{A}^{-1} + \mathbf{A}d\mathbf{A}^{-1}/dt = \mathbf{0}.$$

Premultiplication by $\mathbf{A}^{-1}$ followed by a rearrangement establishes the theorem. ∎

EXAMPLE 3.30

Find $d\mathbf{A}^{-1}/dt$ given that

$$\mathbf{A} = \begin{bmatrix} \cos t & -\sin t \\ \sin t & \cos t \end{bmatrix}.$$

Solution We have

$$d\mathbf{A}/dt = \begin{bmatrix} -\sin t & -\cos t \\ \cos t & -\sin t \end{bmatrix} \quad \text{and} \quad \mathbf{A}^{-1} = \begin{bmatrix} \cos t & \sin t \\ -\sin t & \cos t \end{bmatrix},$$

so from Theorem 3.16

$$d\mathbf{A}^{-1}/dt = -\mathbf{A}^{-1}\{d\mathbf{A}/dt\}\mathbf{A}^{-1} = \begin{bmatrix} -\sin t & \cos t \\ -\cos t & -\sin t \end{bmatrix}.$$

In this case the result is easily checked by direct differentiation of $\mathbf{A}^{-1}$. ∎

Applications of the derivative of a matrix are to be found in reference [2.11] and, for example, in connection with systems of ordinary differential equations in reference [3.15].

Summary

Matrices can occur with functions as their elements as, for example, when a matrix describes a rotation through an angle θ about the origin of a cartesian coordinate system $O\{x, y\}$, or when a column vector contains the unknown functions $u_1(t), u_2(t), \ldots, u_n(t)$ that form the solution set of a system of linear differential equations with independent variable t. Because of this, it is necessary to understand how to differentiate a matrix with respect to an independent variable that is present in functions forming its elements. This section addressed this matter by first defining the fundamental operation of differentiation of a matrix, and then establishing the way in which it is to be applied to the sum and product of two matrices and to the inverse matrix.

EXERCISES 3.10

In Exercises 1 through 4, find dC/dt and d^2C/dt^2.

1. $C = A + B$, where $A = \begin{bmatrix} t^3 & t & t\sin t \\ t^2 & \cos t & \sin 2t \end{bmatrix}$ and

$B = \begin{bmatrix} 1 & 2t^2 & \cosh t \\ t & 3 & \cos t \end{bmatrix}$.

2. $C = A - B$, where $A = \begin{bmatrix} e^{2t} & 1 & \tan t \\ t & \sin t & \cos 3t \end{bmatrix}$ and

$B = \begin{bmatrix} 2 & 2t & \sinh t \\ t & t & \sin t \end{bmatrix}$.

3. $C = A - 2B$, where $A = \begin{bmatrix} t+2 & 2t & t^3 \\ 3 & 3t & e^{2t} \end{bmatrix}$ and

$B = \begin{bmatrix} e^{2t} & t & t^3 \\ 1 & t^2 & \sinh t \end{bmatrix}$.

4. $C = A + 3B$, where $A = \begin{bmatrix} (t+1)^2 & t & t^2 \\ 2t & 1 & \ln t \end{bmatrix}$ and

$B = \begin{bmatrix} t\sin t & 4 & t \\ t & t & \cosh t \end{bmatrix}$.

In Exercises 5 and 6, use Theorem 3.15 to find dC/dt, where $C = AB$, and check the result by direct differentiation of C.

5. $A = \begin{bmatrix} \sin t & -\cos 3t \\ \cos t & \sin t \end{bmatrix}$ and $B = \begin{bmatrix} 1+2t & 2\sin t \\ 2 & \cos t \end{bmatrix}$.

6. $A = \begin{bmatrix} \cosh t & \cos t \\ \sinh t & \sin t \end{bmatrix}$ and $B = \begin{bmatrix} \ln(2t) & t \\ t & \cos t \end{bmatrix}$.

In Exercises 7 and 8 find dA^{-1}/dt by means of Theorem 3.16 and then verify the result by direct differentiation of A^{-1}.

7. $A = \begin{bmatrix} \cos t & \sin t & 0 \\ -\sin t & \cos t & 0 \\ t^2 & t & 1 \end{bmatrix}$.

8. $A = \begin{bmatrix} t^2 & 2t \\ -t & 3t \end{bmatrix}$.

9. Find an expression for

$$d^2\{A^{-1}\}/dt^2$$

in terms of A^{-1}, dA/dt, and d^2A/dt^2. Apply the result to

$$A = \begin{bmatrix} \cos t & -\sin t \\ \sin t & \cos t \end{bmatrix}$$

and verify it by direct differentiation of A^{-1}.

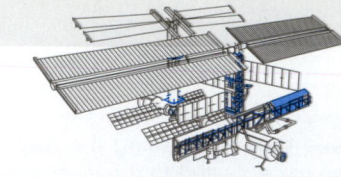

CHAPTER 3
TECHNOLOGY PROJECTS

Project 1

Simplification of det C When $C = [c_{ij} + d_{ij}]$

The purpose of this project is to provide practice with the computer algebra of determinants and to extend the result of Theorem 3.4(vi) to the case when each element of a determinant is the sum of two numbers.

1. Let $\mathbf{a}_1, \mathbf{a}_2, \mathbf{a}_3, \mathbf{b}_1, \mathbf{b}_2, \mathbf{b}_3$ be arbitrary 3×1 element column vectors. Then, by repeated application of Theorem 3.4(vi), extend its result to the case when $\mathbf{C} = [\mathbf{a}_1 + \mathbf{b}_1, \mathbf{a}_2 + \mathbf{b}_2, \mathbf{a}_3 + \mathbf{b}_3]$ by expressing det $\mathbf{C}$ as a sum of 3×3 determinants with columns formed from $\mathbf{a}_1, \mathbf{a}_2, \mathbf{a}_3, \mathbf{b}_1, \mathbf{b}_2,$ and $\mathbf{b}_3$.

2. Define an arbitrary matrix $\mathbf{C}$ of the form $\mathbf{C} = [\mathbf{a}_1 + \mathbf{b}_1, \mathbf{a}_2 + \mathbf{b}_2, \mathbf{a}_3 + \mathbf{b}_3]$, and with the aid of a computer algebra determinant package find det $\mathbf{C}$ by using the result of Step 1. Confirm the result by applying the computer algebra package directly to find det $\mathbf{C}$.

Project 2

The Row-Reduced Echelon Form of a Matrix and Its Rank

The purpose of this project is to provide practice with elementary row operations performed by means of computer algebra. It involves reducing a matrix step by step, using the rules given in Section 3.5, to its row-reduced echelon form, from which its rank can then be determined by inspection.

1. Let $\mathbf{A}$ be the matrix

$$\mathbf{A} = \begin{bmatrix} 0 & 1 & 3 & 2 & 4 & 2 \\ 1 & 2 & 1 & -3 & 1 & 1 \\ -4 & 0 & 1 & 2 & 0 & 1 \\ 0 & -3 & -4 & 5 & 0 & -3 \\ 2 & 1 & -2 & -1 & 2 & -1 \end{bmatrix}.$$

Using computer algebra, apply sequentially the steps in the rule in Section 3.5 to reduce $\mathbf{A}$ to its row-reduced echelon form, and hence find rank $(\mathbf{A})$.

2. Confirm the result obtained in Step 1 by using a computer algebra package to find directly the row-reduced echelon form of $\mathbf{A}$. Take note that in some computer algebra packages the row-reduced echelon form of a matrix $\mathbf{A}$ is called the *Gauss–Jordan form* of $\mathbf{A}$.

Project 3

A Theorem on the Rank of a Matrix Product ABC

The purpose of this project is to provide practice with matrix multiplication and the reduction of matrices to their row-reduced echelon forms using computer algebra.

1. If $\mathbf{A}$, $\mathbf{B}$, and $\mathbf{C}$ are arbitrary rectangular matrices, it can be shown that when the matrix product $\mathbf{ABC}$ exists, then

$$\text{Rank}(\mathbf{AB}) + \text{Rank}(\mathbf{BC}) \leq \text{Rank}(\mathbf{B}) + \text{Rank}(\mathbf{ABC}).$$

2. Define three arbitrary rectangular matrices $\mathbf{A}$, $\mathbf{B}$, and $\mathbf{C}$ for which the product $\mathbf{ABC}$ is defined. Using computer algebra matrix multiplication and computer algebra row-reduction to echelon form, find the ranks of $\mathbf{AB}$, $\mathbf{BC}$, $\mathbf{B}$, and $\mathbf{ABC}$, and hence confirm the inequality in Step 1 for this particular case.

Project 4

Consistency of Augmented Coefficient Matrices, Solution by Back Substitution and Cramer's Rule

The purpose of this project is to use computer algebra to determine the consistency of two 6×7 augmented coefficient matrices. The solution for the corresponding consistent set of linear equations is then found after the reduction of its augmented coefficient matrix to row-reduced echelon form followed by back substitution. Finally, the solution is checked using Cramer's rule, which, despite the large determinants involved, becomes feasible when computer algebra is used.

1. Use computer algebra to determine which of the augmented coefficient matrices **A** and **B** is consistent, given that

$$\mathbf{A} = \begin{bmatrix} 1 & 4 & 7 & 3 & 0 & 2 & 4 \\ 3 & 1 & 0 & 2 & -3 & 4 & 1 \\ 1 & 2 & 1 & 1 & 4 & 3 & 2 \\ 2 & 4 & 0 & 0 & 1 & 6 & 3 \\ 0 & 1 & 2 & 1 & -2 & 1 & 0 \\ 2 & 5 & 2 & 1 & -1 & 7 & 5 \end{bmatrix} \quad \text{and}$$

$$\mathbf{B} = \begin{bmatrix} 4 & -1 & 3 & 0 & 1 & 4 & 2 \\ 1 & 1 & -1 & -3 & 2 & 1 & -1 \\ 0 & 1 & -1 & -2 & 2 & 1 & 3 \\ 4 & 0 & 1 & -1 & 2 & 3 & -4 \\ 1 & -1 & 3 & 2 & -4 & 2 & -1 \\ 0 & 4 & 3 & -3 & 1 & 2 & 0 \end{bmatrix}.$$

2. In the case of the consistent set of equations, using the reduction of the coefficient matrix to its row-reduced echelon form, find the solution by back substitution.

3. Using computer algebra, apply Cramer's rule to the consistent set of equations to find the solution, and so confirm the result found in step 2.

Project 5

A One-Way Traffic Flow Problem

The diagram shows the pattern of one way traffic flow at six road intersections at the corners of two city blocks. The arrows show the directions of traffic flow, and the associated numbers are the traffic flow rates in vehicles per hour at peak traffic time.

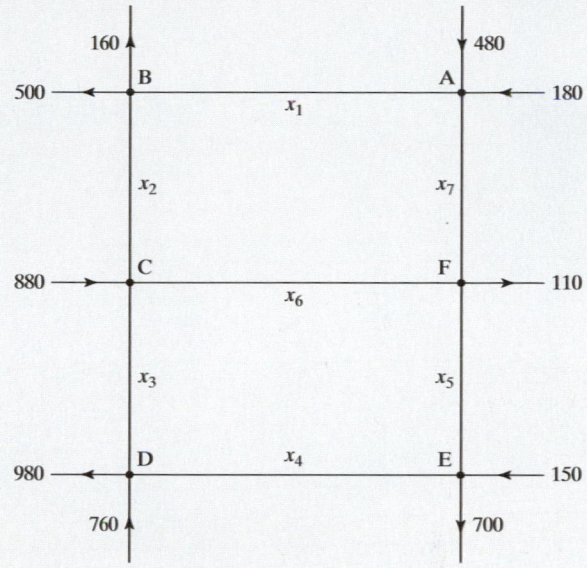

By equating the flow rate of traffic into an intersection to the flow rate out of it (no parking is allowed), find equations relating the traffic flow rates $x_1, x_2, \ldots, x_7$ along each of the roads. Explain why with the given peak flow rates it is impossible to close road DE, and comment on the effect on traffic flow if road CD is closed for repairs.

Project 6

Forces in Bridge Struts

Use matrix methods to find the forces in the pin-jointed framed bridge section shown in Fig. 3.10, given that a concentrated load m acts vertically downwards at joint B.

Give a simple example of a pin-jointed framed structure that contains a redundant strut, and prove its redundancy by attempting to determine the forces acting in the strut when the structure is loaded.

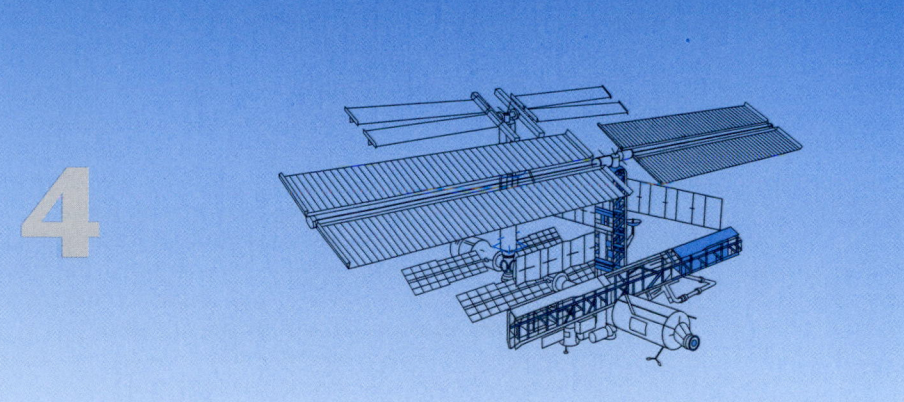

CHAPTER 4

Eigenvalues, Eigenvectors, and Diagonalization

In engineering and physics, problems involving n linear algebraic equations in n independent variables with a constant coefficient matrix $\mathbf{A}$ often arise where a solution vector $\mathbf{x}$ is required to be proportional to $\mathbf{Ax}$. Setting the constant of proportionality equal to λ, this means that $\mathbf{x}$ must be a solution of the equation $\mathbf{Ax} = \lambda\mathbf{x}$ or, equivalently, of the equation $(\mathbf{A} - \lambda\mathbf{I})\mathbf{x} = \mathbf{0}$. The numbers λ_i for which nonzero solutions $\mathbf{x}_i$ exist are called the eigenvalues of matrix $\mathbf{A}$, and the corresponding vectors $\mathbf{x}_i$ are called the eigenvectors of $\mathbf{A}$.

Eigenvalues and eigenvectors arise, for example, when studying vibrational problems, where the eigenvalues represent fundamental frequencies of vibration and the eigenvectors characterize the corresponding fundamental modes of vibration.

They also occur in many other ways; in mechanics, for example, the eigenvalues can represent the principal stresses in a solid body, in which case the eigenvectors then describe the corresponding principal axes of stress caused by the body being subjected to external forces. Also in mechanics, the moment of inertia of a solid body about lines through its center of gravity can be represented by an ellipsoid, with the length of a line drawn from its center to the surface of the ellipsoid proportional to the moment of inertia of the body about an axis through the center of gravity of the body drawn parallel to the line. In this case the eigenvalues represent the principal moments of inertia of the body about the principal axes of inertia, that are then determined by the eigenvectors.

More precisely, if $\mathbf{A}$ is an $n \times n$ matrix, the polynomial $P_n(\lambda)$ of degree n in the scalar λ defined as $P_n(\lambda) = \det(\mathbf{A} - \lambda\mathbf{I})$ is called the characteristic polynomial of $\mathbf{A}$. The roots of the equation $P_n(\lambda) = 0$ are called the eigenvalues of matrix $\mathbf{A}$, and the column vectors $\mathbf{x}_1, \mathbf{x}_2, \ldots, \mathbf{x}_n$ satisfying the matrix equation $(\mathbf{A} - \lambda_i\mathbf{I})\mathbf{x}_i = \mathbf{0}$ are called the eigenvectors of matrix $\mathbf{A}$.

This chapter explains how eigenvalues and eigenvectors are determined and establishes important properties of eigenvectors. The eigenvectors of an $n \times n$ matrix $\mathbf{A}$ with n linearly independent eigenvectors are then used to simplify the structure of $\mathbf{A}$ by means of a process called diagonalization. An important application of diagonalization will arise later when considering the solution of linear systems of ordinary differential equations that arise from the study of mechanical, electrical, and chemical reaction problems. Diagonalization is also an important tool when working with partial differential equations, different types of which describe the temperature distribution in a metal, electromagnetic wave propagation, and diffusion processes, to name a few examples.

After a brief discussion of some special $n \times n$ matrices with complex elements, real quadratic forms are defined and the properties of eigenvectors are used to reduce a general quadratic form to a sum of squares. This is a process that finds many different applications, one of which occurs later when classifying the partial differential equations of engineering and physics in order to know the type of auxiliary conditions that must be imposed in order for them to give rise to physically meaningful solutions.

The chapter ends with the introduction of the matrix exponential $e^{\mathbf{A}}$, where $\mathbf{A}$ is a real $n \times n$ matrix, and it is shown how this enters into the solution of a linear first order matrix differential equation of the form $d\mathbf{x}/dt = \mathbf{A}\mathbf{x}$.

4.1 Characteristic Polynomial, Eigenvalues, and Eigenvectors

Throughout this chapter we will be considering the solutions of the homogeneous system of algebraic equations

$$\mathbf{A}\mathbf{x} = \lambda \mathbf{x}, \tag{1}$$

where $\mathbf{A}[a_{ij}]$ is an $n \times n$ matrix, $\mathbf{x}$ is an n element column vector with elements $x_1, x_2, \ldots, x_n$, and λ is a scalar. For $\mathbf{A}$ given we wish to find x and λ. Introducing the $n \times n$ unit matrix by $\mathbf{I}$ allows (1) to be written

$$(\mathbf{A} - \lambda \mathbf{I})\mathbf{x} = \mathbf{0}, \tag{2}$$

showing that $\mathbf{x}$ is a solution of a homogeneous system of equations with the coefficient matrix $\mathbf{A} - \lambda \mathbf{I}$. It was seen in Chapter 3 that nontrivial solutions $\mathbf{x}$ of (2) are only possible if one or more rows of the coefficient matrix $\mathbf{A} - \lambda \mathbf{I}$ are linearly dependent on its remaining rows. This means that nontrivial solutions $\mathbf{x}$ will exist if $\operatorname{rank}(\mathbf{A} - \lambda \mathbf{I}) < n$, but this, in turn, is equivalent to the more convenient condition $\det(\mathbf{A} - \lambda \mathbf{I}) = 0$. This is a polynomial equation for λ.

Let $P_n(\lambda)$ be the polynomial of degree n in λ defined by the determinant

$$P_n(\lambda) = \begin{vmatrix} a_{11} - \lambda & a_{12} & a_{13} & a_{14} & \cdot & \cdot & \cdot & \cdot & a_{1n} \\ a_{21} & a_{22} - \lambda & a_{23} & a_{24} & \cdot & \cdot & \cdot & \cdot & a_{2n} \\ a_{31} & a_{32} & a_{33} - \lambda & a_{34} & \cdot & \cdot & \cdot & \cdot & a_{3n} \\ & & \cdot & \cdot & \cdot & \cdot & \cdot & & \\ a_{n1} & a_{n2} & a_{n3} & a_{n4} & & \cdot & \cdot & \cdot & a_{nn} - \lambda \end{vmatrix}. \tag{3}$$

Inspection of the determinant defining $P_n(\lambda)$ shows the coefficient of λ^n is $(-1)^n$, so the polynomial is of the form

$$P_n(\lambda) = (-1)^n [\lambda^n + c_1 \lambda^{n-1} + c_2 \lambda^{n-2} - \cdots + c_{n-1}\lambda + c_0]. \tag{4}$$

characteristic
polynomial, equation,
and eigenvalue

The polynomial $P_n(\lambda)$ is called the **characteristic polynomial** of $\mathbf{A}$ and the associated polynomial equation $P_n(\lambda) = 0$ is the **characteristic equation** of $\mathbf{A}$. As the characteristic equation of $\mathbf{A}$ is of degree n in λ, it will have n roots, some of which may be repeated. The roots of $P_n(\lambda) = 0$, or equivalently the zeros of $P_n(\lambda)$, are called the **eigenvalues** of $\mathbf{A}$ or, sometimes, the **characteristic values** of $\mathbf{A}$.

Eigenvalues (characteristic values) of A

The eigenvalues of an $n \times n$ matrix $\mathbf{A}$ are the n zeros of the polynomial $P(\lambda) = \det(\mathbf{A} - \lambda \mathbf{I})$, or, equivalently, the n roots of the nth degree polynomial equation $\det(\mathbf{A} - \lambda \mathbf{I}) = 0$.

In general, a matrix with complex coefficients will have complex eigenvalues, though even when the coefficients of $\mathbf{A}$ are all real it is still possible for complex eigenvalues to arise. This is because then the characteristic equation will have real coefficients, so if complex roots occur they must do so in complex conjugate pairs.

If an eigenvalue λ^* is repeated r times, corresponding to the presence of a factor $(\lambda - \lambda^*)^r$ in the characteristic polynomial $P_n(\lambda)$, the number r is called the **algebraic multiplicity** of the eigenvalue λ^*. The set of all eigenvalues $\lambda_1, \lambda_2, \ldots, \lambda_n$ of $\mathbf{A}$ is called the **spectrum** of $\mathbf{A}$, and the number $R = \max\{|\lambda_1|, |\lambda_2|, \ldots, |\lambda_n|\}$, equal to the largest of the moduli of the eigenvalues, is called the **spectral radius** of $\mathbf{A}$. The name comes from the fact that when the spectrum of $\mathbf{A}$ is plotted as points in the complex plane, they all lie inside or on a circle of radius R centered on the origin.

spectrum and spectral radius

An **eigenvector** of an $n \times n$ matrix $\mathbf{A}$, corresponding to an eigenvalue $\lambda = \lambda_i$, is a nonzero n-element column vector $\mathbf{x}_i$ that satisfies the matrix equation

eigenvectors and eigenvalues

$$\mathbf{A}\mathbf{x}_i = \lambda_i \mathbf{x}_i$$

or, equivalently, that is a solution of the homogeneous system of n algebraic equations

$$(\mathbf{A} - \lambda_i \mathbf{I})\mathbf{x}_i = \mathbf{0}. \tag{5}$$

Eigenvectors of A

The eigenvector $\mathbf{x}_i$ of the $n \times n$ matrix $\mathbf{A}$, corresponding to the eigenvalue $\lambda = \lambda_i$, is a solution of the homogeneous equation $(\mathbf{A} - \lambda_i \mathbf{I})\mathbf{x}_i = \mathbf{0}$.

It is important to recognize that because system (5) is homogeneous, the elements of an eigenvector can only be determined as multiples of one of its nonzero elements as a parameter. This means that if for some choice of the parameter $\mathbf{x}$ is an eigenvalue, then $k\mathbf{x}$ will also be an eigenvalue for any $k \neq 0$.

The next theorem is fundamental to the use of eigenvectors and shows that when an $n \times n$ matrix $\mathbf{A}$ has n distinct (different) eigenvalues, its n eigenvectors form a basis for the vector space associated with the matrix $\mathbf{A}$.

THEOREM 4.1

eigenvectors are linearly independent

Linear independence of eigenvectors   The eigenvectors $\mathbf{x}_1, \mathbf{x}_2, \ldots, \mathbf{x}_m$, corresponding to m distinct eigenvalues $\lambda_1, \lambda_2, \ldots, \lambda_m$, of an $n \times n$ matrix $\mathbf{A}$, are linearly independent. Furthermore, if $m = n$, the set of eigenvectors $\mathbf{x}_1, \mathbf{x}_2, \ldots, \mathbf{x}_n$ forms a basis for the n-dimensional vector space associated with $\mathbf{A}$.

Proof   The proof will be by induction, starting with two vectors, and it uses the fact that $\mathbf{A}\mathbf{x}_i = \lambda_i \mathbf{x}_i$ for $i = 1, 2, \ldots, m$.

Let $\mathbf{x}_1$ and $\mathbf{x}_2$ correspond to distinct eigenvalues λ_1 and λ_2, and let constants k_1 and k_2 be such that

$$k_1\mathbf{x}_1 + k_2\mathbf{x}_2 = \mathbf{0}.$$

Then

$$\mathbf{A}(k_1\mathbf{x}_1 + k_2\mathbf{x}_2) = \mathbf{0},$$

but $\mathbf{A}\mathbf{x}_i = \lambda_i\mathbf{x}_i$, so this is equivalent to

$$k_1\lambda_1\mathbf{x}_1 + k_2\lambda_2\mathbf{x}_2 = \mathbf{0}.$$

Subtracting λ_2 times the first equation from the last result gives

$$(\lambda_1 - \lambda_2)k_1\mathbf{x}_1 = \mathbf{0}.$$

By hypothesis, $\lambda_1 \neq \lambda_2$, so as $\mathbf{x}_1 \neq \mathbf{0}$ it follows that $k_1 = 0$. Using this result in $k_1\mathbf{x}_1 + k_2\mathbf{x}_2 = \mathbf{0}$ shows that $k_2 = 0$, so we have established the linear independence of $\mathbf{x}_1$ and $\mathbf{x}_2$.

To proceed with an inductive proof we now assume that linear independence has been proved for the first $r - 1$ vectors, and show that the rth vector must also be linearly independent. To accomplish this we consider the equation

$$k_1\mathbf{x}_1 + k_2\mathbf{x}_2 + \cdots + k_r\mathbf{x}_r = \mathbf{0}.$$

Premultiplying this equation by $\mathbf{A}$ and reasoning as before, we arrive at the result

$$k_1\lambda_1\mathbf{x}_1 + k_2\lambda_2\mathbf{x}_2 + \cdots + k_r\lambda_r\mathbf{x}_r = \mathbf{0}.$$

Subtracting λ_r times the first equation from the last one gives

$$(\lambda_1 - \lambda_r)k_1\mathbf{x}_1 + (\lambda_2 - \lambda_r)k_2\mathbf{x}_2 + \cdots + (\lambda_{r-1} - \lambda_r)k_{r-1}\mathbf{x}_{r-1} = \mathbf{0}.$$

By the inductive hypothesis $\mathbf{x}_1, \mathbf{x}_2, \ldots, \mathbf{x}_{r-1}$ are linearly independent, so as $\mathbf{x}_r \neq \mathbf{0}$,

$$(\lambda_1 - \lambda_r)k_1 = (\lambda_2 - \lambda_r)k_2 = \cdots = (\lambda_{r-1} - \lambda_r)k_{r-1} = 0.$$

The eigenvalues are distinct, so the last result can only be true if $k_1 = k_2 = \cdots = k_{r-1} = 0$. Thus $k_r = 0$, and so the vector $\mathbf{x}_r$ is linearly independent of the vectors $\mathbf{x}_1, \mathbf{x}_2, \ldots, \mathbf{x}_{r-1}$. It has been shown that $\mathbf{x}_1$ and $\mathbf{x}_2$ are linearly independent, so by induction we conclude that the set of vectors $\mathbf{x}_i$ is linearly independent for $i = 1, 2, \ldots, m$.

A matrix $\mathbf{A}$ can have no more than n linearly independent eigenvectors, so when $m = n$ the set of eigenvectors $\mathbf{x}_1, \mathbf{x}_2, \ldots, \mathbf{x}_n$ spans the n-dimensional vector space associated with matrix $\mathbf{A}$ and forms a basis for this space. The proof is complete. ∎

algebraic and geometric multiplicity

It can happen that an eigenvalue with **algebraic multiplicity** $r > 1$ only has s different eigenvectors associated with it, where $s < r$, and when this occurs the number s is called the **geometric multiplicity** of the eigenvalue. The set of all eigenvectors associated with an eigenvalue with geometric multiplicity s together with the null vector $\mathbf{0}$ forms what is called the **eigenspace** associated with the eigenvalue. When one or more eigenvalues has a geometric multiplicity that is less than its algebraic multiplicity, it follows directly that the vector space associated with $\mathbf{A}$ must have dimension less than n.

EXAMPLE 4.1

Find the characteristic polynomial, the eigenvalues, and the eigenvectors of the matrix

$$\mathbf{A} = \begin{bmatrix} 2 & 1 & -1 \\ 3 & 2 & -3 \\ 3 & 1 & -2 \end{bmatrix}.$$

Solution   The characteristic polynomial $P_3(\lambda)$ is given by

$$P_3(\lambda) = \begin{vmatrix} 2-\lambda & 1 & -1 \\ 3 & 2-\lambda & -3 \\ 3 & 1 & -2-\lambda \end{vmatrix},$$

and after expanding the determinant we find that

$$P_3(\lambda) = -\lambda^3 + 2\lambda^2 + \lambda - 2.$$

The characteristic equation $P_3(\lambda) = 0$ is

$$\lambda^3 - 2\lambda^2 - \lambda + 2 = 0,$$

and inspection shows it has the roots 2, 1, and -1. So the *eigenvalues* of **A** are $\lambda_1 = 2$, $\lambda_2 = 1$, and $\lambda_3 = -1$, and as these roots are all distinct (there are no repeated roots), each has an algebraic and geometric multiplicity of 1 (each is a single root). The set of numbers $-1, 1, 2$ forms the *spectrum* of matrix **A**. As the *spectral radius* R of a matrix is defined as the largest of the moduli of the eigenvalues, we see that $R = 2$.

To find the eigenvectors $\mathbf{x}_i$ of **A** corresponding to the eigenvalues $\lambda = \lambda_i$, for $i = 1, 2, 3$, it will be necessary to solve the homogeneous system of algebraic equations

$$(\mathbf{A} - \lambda_i \mathbf{I})\mathbf{x}_i = \mathbf{0} \quad \text{for } i = 1, 2, 3,$$

where $\mathbf{x}_i = [x_1, x_2, x_3]^{\mathsf{T}}$.

Case $\lambda_1 = 2$

The system of equations to be solved is

$$\begin{bmatrix} 2-2 & 1 & -1 \\ 3 & 2-2 & -3 \\ 3 & 1 & -2-2 \end{bmatrix} \begin{bmatrix} x_1 \\ x_2 \\ x_3 \end{bmatrix} = \begin{bmatrix} 0 \\ 0 \\ 0 \end{bmatrix},$$

and this matrix equation is equivalent to the set of three linear algebraic equations

$$x_2 - x_3 = 0, \quad 3x_1 - 3x_3 = 0, \quad \text{and} \quad 3x_1 + x_2 - 4x_3 = 0.$$

The first two equations are equivalent, so only one of the first two equations and the third equation are linearly independent. Solving the last two equations for x_1 and x_2 in terms of x_3, we find that $x_1 = x_2 = x_3$, so setting $x_3 = k_1$ where k_1 is an arbitrary real number (a parameter) shows that the eigenvector $\mathbf{x}_1$ corresponding to the eigenvalue $\lambda_1 = 2$ is given by

$$\mathbf{x}_1 = \begin{bmatrix} k_1 \\ k_1 \\ k_1 \end{bmatrix} = k_1 \begin{bmatrix} 1 \\ 1 \\ 1 \end{bmatrix}.$$

As k_1 is an arbitrary parameter, for convenience we set $k_1 = 1$ and as a result obtain the eigenvector

$$\mathbf{x}_1 = \begin{bmatrix} 1 \\ 1 \\ 1 \end{bmatrix}.$$

Case $\lambda_2 = 1$

This time the system of equations to be solved to find the eigenvector $\mathbf{x}_2$ is

$$\begin{bmatrix} 2-1 & 1 & -1 \\ 3 & 2-1 & -3 \\ 3 & 1 & -2-1 \end{bmatrix} \begin{bmatrix} x_1 \\ x_2 \\ x_3 \end{bmatrix} = \begin{bmatrix} 0 \\ 0 \\ 0 \end{bmatrix},$$

and this is equivalent to the three linear algebraic equations

$$x_1 + x_2 - x_3 = 0, \quad 3x_1 + x_2 - 3x_3 = 0, \quad \text{and} \quad 3x_1 + x_2 - 3x_3 = 0.$$

The last two equations are identical, so we must solve for x_1, x_2, and x_3 using the first two equations. It is easily seen from these two equations that $x_2 = 0$ and $x_1 = x_3$, so setting $x_1 = k_2$, where k_2 is an arbitrary real number (a parameter), gives

$$\mathbf{x}_2 = k_2 \begin{bmatrix} 1 \\ 0 \\ 1 \end{bmatrix}.$$

Making the arbitrary choice $k_2 = 1$ shows that the eigenvector $\mathbf{x}_2$ corresponding to $\lambda_2 = 1$ is

$$\mathbf{x}_2 = \begin{bmatrix} 1 \\ 0 \\ 1 \end{bmatrix}.$$

Case $\lambda_3 = -1$

Setting $\lambda = \lambda_3$, and proceeding as before, shows that the elements of the eigenvector $\mathbf{x}_3$ must satisfy the three equations

$$3x_1 + x_2 - x_3 = 0, \quad 3x_1 + 3x_2 - 3x_3 = 0, \quad \text{and} \quad 3x_1 + x_2 - x_3 = 0,$$

with the solution $x_1 = 0, x_2 = x_3 = k_3$, where k_3 is an arbitrary real number (a parameter). Making the arbitrary choice $k_3 = 1$ allows the eigenvector $\mathbf{x}_3$ to be written as

$$\mathbf{x}_3 = \begin{bmatrix} 0 \\ 1 \\ 1 \end{bmatrix}.$$

We have shown that matrix $\mathbf{A}$ has the three distinct eigenvalues $\lambda_1 = 2, \lambda_2 = 1$, and $\lambda_3 = -1$, corresponding to which there are the three eigenvectors

$$\mathbf{x}_1 = \begin{bmatrix} 1 \\ 1 \\ 1 \end{bmatrix}, \quad \mathbf{x}_2 = \begin{bmatrix} 1 \\ 0 \\ 1 \end{bmatrix}, \quad \text{and} \quad \mathbf{x}_3 = \begin{bmatrix} 0 \\ 1 \\ 1 \end{bmatrix}.$$

These three eigenvectors form a basis for the three-dimensional vector space associated with $\mathbf{A}$. ∎

As the eigenvectors $\mathbf{x}$ of matrix $\mathbf{A}$ satisfy the homogeneous equation (2), they can be multiplied by an arbitrary nonzero number K, which is either positive or negative, and still remain an eigenvector. This property is used to *scale* the eigenvectors of $\mathbf{A}$ to produce what are called **normalized** eigenvectors. This scaling is used in numerical calculations involving the iteration of eigenvectors, because without normalization the elements of $\mathbf{x}$ may either grow or diminish in absolute value after each stage of the calculation, leading to a progressive loss of accuracy.

Normalization of eigenvectors

a frequently used way of normalizing eigenvectors

Various normalizations are in use. The most common one for eigenvectors with real elements involves scaling the eigenvector so that the square root of the sum of the squares of its elements is 1. So, for example, if

$$\mathbf{x} = \begin{bmatrix} a \\ b \\ c \end{bmatrix}, \quad \text{the normalizing factor} \quad K = \frac{1}{(a^2 + b^2 + c^2)^{1/2}} \quad (6)$$

and the normalized eigenvector $\hat{\mathbf{x}}$ becomes

$$\hat{\mathbf{x}} = \begin{bmatrix} a/(a^2 + b^2 + c^2)^{1/2} \\ b/(a^2 + b^2 + c^2)^{1/2} \\ c/(a^2 + b^2 + c^2)^{1/2} \end{bmatrix}. \quad (7)$$

When the eigenvectors in Example 4.1 are normalized in this way, they become

$$\hat{\mathbf{x}}_1 = \begin{bmatrix} 1/\sqrt{3} \\ 1/\sqrt{3} \\ 1/\sqrt{3} \end{bmatrix}, \quad \hat{\mathbf{x}}_2 = \begin{bmatrix} 1/\sqrt{2} \\ 0 \\ 1/\sqrt{2} \end{bmatrix}, \quad \text{and} \quad \hat{\mathbf{x}}_3 = \begin{bmatrix} 0 \\ 1/\sqrt{2} \\ 1/\sqrt{2} \end{bmatrix}.$$

EXAMPLE 4.2   Find the characteristic polynomial, eigenvalues, and eigenvectors of the matrix

$$\mathbf{A} = \begin{bmatrix} 0 & 0 & 1 & 1 \\ -1 & 2 & 0 & 1 \\ -1 & 0 & 2 & 1 \\ 1 & 0 & -1 & 0 \end{bmatrix}.$$

Solution   The determinant defining the characteristic polynomial is

$$P_4(\lambda) = \begin{bmatrix} -\lambda & 0 & 1 & 1 \\ -1 & 2-\lambda & 0 & 1 \\ -1 & 0 & 2-\lambda & 1 \\ 1 & 0 & -1 & -\lambda \end{bmatrix},$$

and after the determinant is expanded the characteristic equation $P_4(\lambda) = 0$ is found to be

$$P_4(\lambda) = \lambda(\lambda^3 - 4\lambda^2 + 5\lambda - 2) = 0.$$

Clearly, $\lambda = 0$ is a root of $P_4(\lambda) = 0$, and inspection shows the other three roots to be 1, 1, and 2. So the eigenvalues of $\mathbf{A}$ are $\lambda_1 = 0$, $\lambda_2 = 1$, $\lambda_3 = 1$, and $\lambda_4 = 2$. In this case $\lambda_2 = \lambda_3 = 1$, so the eigenvalue 1 has algebraic multiplicity 2, and the remaining two eigenvalues each have an algebraic multiplicity of 1. To find the eigenvectors corresponding to these eigenvalues we proceed as in Example 4.1.

Case $\lambda_1 = 0$

Setting $\lambda = \lambda_1 = 0$ in $(\mathbf{A} - \lambda\mathbf{I})\mathbf{x} = 0$ leads to the four equations

$$x_3 + x_4 = 0, \quad -x_1 + 2x_2 + x_4 = 0, \quad -x_1 + 2x_3 + x_4 = 0, \quad \text{and} \quad x_1 - x_3 = 0.$$

Proceeding as before we find that $x_1 = x_2 = x_3 = -x_4$, so solving for x_1, x_2, and x_3 in terms of x_4, and setting $x_4 = 1$ (an arbitrary choice), shows the eigenvector $\mathbf{x}_1$ to be

$$\mathbf{x}_1 = \begin{bmatrix} -1 \\ -1 \\ -1 \\ 1 \end{bmatrix}.$$

Case $\lambda_2 = \lambda_3 = 1$

The eigenvalue 1 has algebraic multiplicity 2, so we must attempt to find two *different* eigenvectors that correspond to the single eigenvalue $\lambda = 1$. Setting $\lambda = 1$ in $(\mathbf{A} - \lambda\mathbf{I})\mathbf{x} = \mathbf{0}$ leads to the four equations

$$-x_1 + x_3 + x_4 = 0, \quad -x_1 + x_2 + x_4 = 0, \quad -x_1 + x_3 + x_4 = 0, \quad x_1 - x_3 - x_4 = 0.$$

The first, third, and fourth equations are identical, so x_1, x_2, x_3, and x_4 must be determined from the two equations

$$-x_1 + x_3 + x_4 = 0 \quad \text{and} \quad -x_1 + x_2 + x_4 = 0.$$

As there are four unknown quantities x_1, x_2, x_3, and x_4, and only two equations relating them, it will only be possible to solve for two of these quantities in terms of the remaining two. The equations show that $x_2 = x_3$ and $x_4 = x_1 - x_3$, so choosing to solve for x_3 and x_4 in terms of x_1 and x_2 by setting $x_1 = \alpha$ and $x_2 = \beta$, with α and β arbitrary constants, shows that the eigenvectors $\mathbf{x}_2$ and $\mathbf{x}_3$ are both of the form

$$\mathbf{x}_{2,3} = \begin{bmatrix} \alpha \\ \beta \\ \beta \\ \alpha - \beta \end{bmatrix}.$$

It is possible to obtain two *different* eigenvectors from this last result by choosing two different pairs of values for the arbitrary parameters α and β. We will define $\mathbf{x}_2$ by setting $\alpha = 1$ and $\beta = 1$, and $\mathbf{x}_3$ by setting $\alpha = 1$ and $\beta = 0$, and as a result we find that

$$\mathbf{x}_2 = \begin{bmatrix} 1 \\ 1 \\ 1 \\ 0 \end{bmatrix} \quad \text{and} \quad \mathbf{x}_3 = \begin{bmatrix} 1 \\ 0 \\ 0 \\ 1 \end{bmatrix}.$$

Had other choices of the parameters α and β been made, two different eigenvectors would have been produced.

Case $\lambda_4 = 2$

Setting $\lambda = \lambda_4 = 2$ in $(\mathbf{A} - \lambda\mathbf{I})\mathbf{x} = 0$ leads to the four equations

$$-2x_1 + x_3 + x_4 = 0, \quad -x_1 + x_4 = 0, \quad -x_1 + x_4 = 0, \quad x_1 - x_3 - 2x_4 = 0.$$

These equations have the solution $x_1 = x_3 = x_4 = 0$, with no condition being imposed on x_2. For simplicity we choose to set $x_2 = 1$ to obtain

$$\mathbf{x}_4 = \begin{bmatrix} 0 \\ 1 \\ 0 \\ 0 \end{bmatrix}.$$

In this example, the eigenvalue 1 has algebraic multiplicity 2, and two different eigenvectors can be associated with it, so the geometric multiplicity of the eigenvalue is also 2. The four eigenvectors $\mathbf{x}_1, \mathbf{x}_2, \mathbf{x}_3$, and $\mathbf{x}_4$ form a basis for the four-dimensional vector space associated with matrix $\mathbf{A}$.

Had different values been used for α and β, the basis vectors for this vector space would have been different, though the vector space itself would have remained the same because linear combinations of basis vectors will produce an equivalent set of basis vectors.

The *spectrum* of $\mathbf{A}$ is the set of numbers 0, 1, 2, and the *spectral radius* of $\mathbf{A}$ is seen to be $R = 2$. ∎

EXAMPLE 4.3   Show that the matrix

$$\mathbf{A} = \begin{bmatrix} 1 & 1 & 0 \\ 0 & 1 & 0 \\ 0 & 0 & 0 \end{bmatrix}$$

has three eigenvalues, but only two linearly independent eigenvectors.

Solution   The characteristic polynomial

$$P_3(\lambda) = \begin{vmatrix} 1-\lambda & 1 & 0 \\ 0 & 1-\lambda & 0 \\ 0 & 0 & -\lambda \end{vmatrix},$$

and after expanding the determinant the characteristic equation $P_3(\lambda) = 0$ becomes

$$P_3(\lambda) = -\lambda(1-\lambda)^2 = 0.$$

The eigenvalue $\lambda_1 = 0$ occurs with algebraic multiplicity 1 and the eigenvalue $\lambda_2 = \lambda_3 = 1$ occurs with algebraic multiplicity 2.

The equations determining the eigenvector $\mathbf{x}_1$, corresponding to the eigenvalue $\lambda = \lambda_1 = 0$, are

$$x_1 + x_2 = 0 \quad \text{and} \quad x_2 = 0,$$

so $x_1 = x_2 = 0$ and x_3 is arbitrary. Setting $x_3 = 1$ gives

$$\mathbf{x}_1 = \begin{bmatrix} 0 \\ 0 \\ 1 \end{bmatrix}.$$

The equations determining $\mathbf{x}_2$ and $\mathbf{x}_3$, corresponding to $\lambda = \lambda_2 = \lambda_3 = 1$, are

$$x_1 = k(\text{arbitrary}) \quad \text{and} \quad x_2 = x_3 = 0,$$

so setting $k = 1$, we find that the eigenvalue $\lambda_2 = \lambda_3 = 1$ with algebraic multiplicity 2 only has associated with it the *single* eigenvector

$$\mathbf{x}_{2,3} = \begin{bmatrix} 1 \\ 0 \\ 0 \end{bmatrix}.$$

So the algebraic multiplicity of the eigenvalue $\lambda = 1$ is 2, but its geometric multiplicity is 1. The *spectrum* of $\mathbf{A}$ is the set of numbers 0, 1, so the *spectral radius* of $\mathbf{A}$ is $R = 1$. ■

The eigenvalues of a diagonal matrix can be found immediately, and the corresponding eigenvectors take on a particularly simple form. Let $\mathbf{D}$ be the $n \times n$ diagonal matrix

$$\mathbf{D} = \begin{bmatrix} a_1 & 0 & 0 & \cdot & \cdot & \cdot & \cdot & 0 \\ 0 & a_2 & 0 & \cdot & \cdot & \cdot & \cdot & 0 \\ & & \cdot & \cdot & \cdot & \cdot & \cdot & \\ & & \cdot & \cdot & \cdot & \cdot & \cdot & \\ 0 & 0 & 0 & \cdot & \cdot & \cdot & \cdot & a_n \end{bmatrix},$$

with entries $a_1, a_2, \ldots, a_n$ on its leading diagonal, not all of which are zero, and zeros elsewhere. Then it is easily seen that the eigenvalues of $\mathbf{D}$ are $\lambda_1 = a_1, \lambda_2 = a_2, \ldots,$ $\lambda_n = a_n$. The eigenvector $\mathbf{x}_i$ corresponding to the eigenvalue $\lambda_i = a_i$ becomes an n-element column vector in which only the ith element is nonzero. It is not difficult to show that this result remains true whatever the algebraic multiplicity of an eigenvalue, so *every* diagonal $n \times n$ matrix has n eigenvectors of this form. For convenience, the ith element in $\mathbf{x}_i$ is usually taken to be 1 so, for example, the matrix

$$\mathbf{A} = \begin{bmatrix} 3 & 0 & 0 \\ 0 & -5 & 0 \\ 0 & 0 & 4 \end{bmatrix}$$

has eigenvalues $\lambda_1 = 3$, $\lambda_2 = -5$, and $\lambda_3 = 4$ and eigenvectors

$$\mathbf{x}_1 = \begin{bmatrix} 1 \\ 0 \\ 0 \end{bmatrix}, \quad \mathbf{x}_2 = \begin{bmatrix} 0 \\ 1 \\ 0 \end{bmatrix}, \quad \text{and} \quad \mathbf{x}_3 = \begin{bmatrix} 0 \\ 0 \\ 1 \end{bmatrix}.$$

Similarly, the diagonal matrix

$$\mathbf{A} = \begin{bmatrix} -2 & 0 & 0 \\ 0 & 4 & 0 \\ 0 & 0 & 4 \end{bmatrix}$$

has an eigenvalue $\lambda_1 = -2$ with multiplicity 1 and a double eigenvalue $\lambda_2 = \lambda_3 = 4$ with multiplicity 2, but the matrix still has the three distinct eigenvectors

$$\mathbf{x}_1 = \begin{bmatrix} 1 \\ 0 \\ 0 \end{bmatrix}, \quad \mathbf{x}_2 = \begin{bmatrix} 0 \\ 1 \\ 0 \end{bmatrix}, \quad \text{and} \quad \mathbf{x}_3 = \begin{bmatrix} 0 \\ 0 \\ 1 \end{bmatrix}.$$

When the degree of the characteristic equation of a matrix exceeds 2, its roots must usually be found by means of a numerical technique. In such circumstances the next theorem provides a simple and useful check for the values of the eigenvalues that have been computed.

THEOREM 4.2

a check on the sum
of the eigenvectors

The sum of eigenvalues   Let the $n \times n$ matrix $\mathbf{A}[a_{ij}]$ have the n eigenvalues λ_1, $\lambda_2, \ldots, \lambda_n$, which may be either real or complex. Then

$$\lambda_1 + \lambda_2 + \cdots + \lambda_n = (-1)^{n-1}(a_{11} + a_{22} + \cdots + a_{nn}) = (-1)^{n-1}\mathrm{tr}(\mathbf{A}).$$

Proof   As the multiplication of a column of a matrix by a number k is equivalent to multiplication of its determinant by k, we can write

$$P_n(\lambda) = \det(\mathbf{A} - \lambda\mathbf{I}) = (-1)^n \det(\lambda\mathbf{I} - \mathbf{A}).$$

Expanding the determinant on the right in terms of the elements of the first column and separating out the factors that can give rise to the terms in λ^n and λ^{n-1}, we arrive at the result

$$P_n(\lambda) = (-1)^n\{(\lambda - a_{11})(\lambda - a_{22})\cdots(\lambda - a_{nn}) + Q_{n-2}(\lambda)\},$$

where $Q_{n-2}(\lambda)$ is a polynomial in λ of degree $n-2$.

Identifying the coefficients of λ^n and λ^{n-1} in the expression for $P_n(\lambda)$ shows that

$$P_n(\lambda) = (-1)^n\{\lambda^n - (a_{11} + a_{22} + \cdots + a_{nn})\lambda^{n-1} + \cdots + \text{constant} + Q_{n-2}(\lambda)\}.$$

An equivalent expression for $P_n(\lambda)$ can be obtained by expanding it in terms of its factors $(\lambda - \lambda_1), (\lambda - \lambda_2), \ldots, (\lambda - \lambda_n)$ to obtain

$$P_n(\lambda) = (-1)^n(\lambda - \lambda_1)(\lambda - \lambda_2)\cdots(\lambda - \lambda_n)$$
$$= (-1)^n\{\lambda^n - (\lambda_1 + \lambda_2 + \cdots + \lambda_n)\lambda^{n-1} + \cdots + \text{constant}\}.$$

The statement of the theorem then follows by comparing the coefficients of λ^{n-1} in the two different expressions for $P_n(\lambda)$, where it will be recalled that the **trace** of an $n \times n$ matrix $\mathbf{A}[a_{ij}]$, written $\mathrm{tr}(\mathbf{A})$, is the sum of the elements on its leading diagonal, so that $\mathrm{tr}(\mathbf{A}) = a_{11} + a_{22} + \cdots + a_{nn}$.   ∎

EXAMPLE 4.4

Use Theorem 4.2 to check the eigenvalues of the matrices in Examples 4.1 and 4.2.

Solution   In Example 4.1, $\lambda_1 = 2$, $\lambda_2 = 1$, and $\lambda_3 = -1$, so $\lambda_1 + \lambda_2 + \lambda_3 = 2$, and $\mathrm{tr}(\mathbf{A}) = 2 + 2 - 2 = 2$, so the result of Theorem 4.2 is verified. Similarly, in Example 4.2, $\lambda_1 = 0$, $\lambda_2 = 1$, $\lambda_3 = 1$, and $\lambda_4 = 2$, so $\lambda_1 + \lambda_2 + \lambda_3 + \lambda_4 = 4$, and $\mathrm{tr}(\mathbf{A}) = 0 + 2 + 2 + 0 = 4$, showing that the result of Theorem 4.2 is again verified.   ∎

EXAMPLE 4.5

Find the characteristic polynomial, eigenvalues, and eigenvectors of

$$\mathbf{A} = \begin{bmatrix} -1-2i & -1-i & 2+2i \\ -4i & -i & 4i \\ -1-3i & -1-i & 2+3i \end{bmatrix},$$

and use Theorem 4.2 to check the eigenvalues.

Solution   This matrix has complex elements. Expanding $\det(\mathbf{A} - \lambda\mathbf{I}) = 0$ shows that the characteristic polynomial $P_3(\lambda)$ is

$$P_3(\lambda) = \lambda^3 - \lambda^2 + \lambda - 1.$$

Inspection shows the eigenvalues determined by $P_3(\lambda) = 0$ to be $\lambda_1 = 1$, $\lambda_2 = i$, and $\lambda_3 = -i$. Finding the eigenvectors, as in Example 4.1, gives

$$(\lambda_1 = 1)\ \mathbf{x}_1 = \begin{bmatrix} 1 \\ 0 \\ 1 \end{bmatrix}, \quad (\lambda_2 = i)\ \mathbf{x}_2 = \begin{bmatrix} 0 \\ 1 \\ 1/2 \end{bmatrix}, \quad \text{and} \quad (\lambda_3 = -i)\ \mathbf{x}_3 = \begin{bmatrix} 1 \\ 1 \\ 1 \end{bmatrix}.$$

In this example, although the matrix $\mathbf{A}$ has complex elements, the characteristic polynomial has real coefficients, and one of its zeros (an eigenvalue) is real and its other two zeros (eigenvalues) are complex conjugates. The test in Theorem 4.2 is satisfied because $\mathrm{tr}(\mathbf{A}) = \lambda_1 + \lambda_2 + \lambda_3 = \mathrm{tr}(\mathbf{A}) = 1$. ■

Complex eigenvalues arise in numerous applications of matrices, and when this happens it is often useful to have qualitative information about a region in the complex plane that contains all of the eigenvalues, without the necessity of computing their actual values. This form of approach is particularly useful when the coefficients of a polynomial are not specific, and all that is known is that they lie within given intervals or, if complex, that the modulus of each is bounded by a given number.

Another need for this type of information occurs when working with systems of linear differential equations, because it will be seen in Chapter 6 that the roots of a characteristic polynomial equation determine the form of the general solution of a homogeneous system. Roots of the form $\alpha + i\beta$ will be seen to lead to real solutions of the form $e^{\alpha t} \sin \beta t$ and $e^{\alpha t} \cos \beta t$, and these solutions will only remain bounded (stable) as $t \to +\infty$ if the real part of every root is negative. This means that the qualitative knowledge that all of the roots lie to the left of the imaginary axis will be sufficient to ensure that the solution remains finite (is stable) as $t \to +\infty$.

The theorem that follows is the simplest of many similar results that are available, all of which provide information about regions in the complex plane where all of the zeros of a characteristic polynomial are located. Two other results are to be found in the exercise set at the end of this section; the one called the **Routh–Hurwitz stability criterion** is particularly useful when working with systems of linear differential equations.

Although the theorem to be proved in this section identifies a region less precisely than many similar theorems, it has been included to illustrate how such regions can be found, and also because the derivation of the result is elementary. The proof only uses the basic properties of complex numbers extending as far as the triangle inequality.

THEOREM 4.3

finding a region that contains all the eigenvalues

The Gerschgorin circle theorem   Let $\mathbf{A}[a_{ij}]$ be an $n \times n$ matrix, and define the circles $C_1, C_2, \ldots, C_n$ in the complex plane such that circle C_r has its center at a_{rr} and the radius

$$\rho_r = \sum_{j=1, j \neq r}^{n} |a_{rj}| = |a_{r1}| + |a_{r2}| + \cdots + |a_{r,r-1}| + |a_{r,r+1}| + \cdots + |a_{rn}|.$$

Then each of the eigenvalues of $\mathbf{A}$ lies in at least one of these circles.

Proof   The rth equation of $\mathbf{Ax} = \lambda\mathbf{x}$ is

$$a_{r1}x_1 + \cdots + a_{r,r-1}x_{r-1} + (a_{rr} - \lambda)x_r + a_{r,r+1}x_{r+1} + \cdots + a_{rn}x_n = 0.$$

Solving for $(a_{rr} - \lambda)$, taking the modulus of the result, and making repeated use of the triangle inequality $|a + b| \le |a| + |b|$, where a and b are arbitrary complex numbers, leads to the inequality

$$|\lambda - a_{rr}| < \sum_{j=1,\, j\neq r}^{n} |a_{rj}||x_j|/|x_r|, \quad \text{for } r = 1, 2, \ldots, n.$$

We now choose x_r to be the element of $\mathbf{x}$ with the largest modulus, so that $|x_j|/|x_r| \le 1$ for $r = 1, 2, \ldots, n$. The statement of the theorem is obtained from the inequality involving $|\lambda - a_{rr}|$ by replacing each term $|x_j|/|x_r|$ on the right by 1, and then repeating the argument for $r = 1, 2, \ldots, n$.   ∎

EXAMPLE 4.6

Apply the Gerschgorin circle theorem to Example 4.1.

Solution   Circle C_1 has its center at the point $a_{11} = (2, 0)$ and its radius $\rho_1 = |a_{12}| + |a_{13}| = 1 + 1 = 2$. Circle C_2 has its center at the point $a_{22} = (2, 0)$ and its radius $\rho_2 = |a_{21}| + |a_{23}| = 3 + 3 = 6$, while circle C_3 has its center at the point $a_{33} = (-2, 0)$ and its radius $\rho_3 = |a_{31}| + |a_{32}| = 3 + 1 = 4$.

Consequently, the Gerschgorin circle theorem asserts that all the eigenvalues of $\mathbf{A}$ lie in the region of the complex plane enclosed by these three circles. The circles are shown in Fig. 4.1 together with the locations of the three eigenvalues 2, 1, and −1.   ∎

Physical problems that give rise to matrices with real coefficients often do so in the form of real valued symmetric matrices. These matrices have a number of useful properties that we will examine after first introducing the notions of the *inner product* and *norm* of a matrix vector, and then *orthogonal* and *orthonormal* sets of matrix vectors.

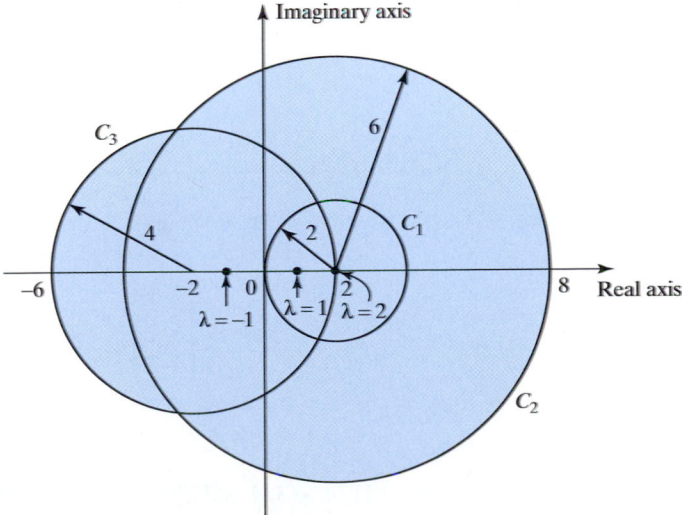

FIGURE 4.1   The Gerschgorin circles for Example 4.1.

Inner product of vectors

Let $\mathbf{u}$ and $\mathbf{v}$ be two n-element matrix vectors (row or column) with the respective elements $u_1, u_2, \ldots, u_n$ and $v_1, v_2, \ldots, v_n$. Then their **dot** or **inner product**, denoted here by $\mathbf{u} \cdot \mathbf{v}$ but elsewhere often by $\langle \mathbf{u}, \mathbf{v} \rangle$, is defined as

$$\mathbf{u} \cdot \mathbf{v} = u_1 v_1 + u_2 v_2 + \cdots + u_n v_n. \tag{8}$$

Norm of a vector

The **norm** of an n-element vector $\mathbf{w}$ (row or column) with elements $w_1, w_2, \ldots, w_n$, written $\|\mathbf{w}\|$, is defined as $(\mathbf{w} \cdot \mathbf{w})^{1/2}$, and so is given by

$$\|\mathbf{w}\| = \left(w_1^2 + w_2^2 + \cdots + w_n^2\right)^{1/2}. \tag{9}$$

We now use the matrix norm to introduce the idea of the *orthogonality* of sets of matrix vectors, and then to show how such sets can be replaced by an equivalent *orthonormal* set of vectors.

Orthogonal and orthonormal sets of vectors

Let $\mathbf{u}_1, \mathbf{u}_2, \ldots, \mathbf{u}_n$ be a set of n-element vectors (row or column). Then the set is said to be **orthogonal** if

$$\mathbf{u}_i \cdot \mathbf{u}_j = \begin{cases} 0 & \text{for } i \neq j, \\ \|\mathbf{u}_i\|^2 & \text{for } i = j, \end{cases} \tag{10}$$

and to be **orthonormal** if, in addition to being orthogonal, the norm of each vector is 1, so that $\|\mathbf{u}_i\| = 1$ for $i = 1, 2, \ldots, n$. This means that the set of vectors $\mathbf{u}_1, \mathbf{u}_2 \ldots, \mathbf{u}_n$ will form an *orthonormal* set if

$$\mathbf{u}_i \cdot \mathbf{u}_j = \begin{cases} 0 & \text{for } i \neq j, \\ \|\mathbf{u}_i\|^2 = 1 & \text{for } i = j. \end{cases} \tag{11}$$

EXAMPLE 4.7

Given the sets of vectors
(a)

$$\mathbf{u}_1 = \begin{bmatrix} 1 \\ 2 \\ -2 \end{bmatrix}, \quad \mathbf{u}_2 = \begin{bmatrix} 2 \\ 1 \\ 2 \end{bmatrix} \text{ and } \mathbf{u}_3 = \begin{bmatrix} -2 \\ 2 \\ 1 \end{bmatrix},$$

and
(b)

$$\mathbf{u}_1 = [1/4, \sqrt{3}/4, \sqrt{3}/2], \quad \mathbf{u}_2 = [\sqrt{3}/2, -1/2, 0], \quad \mathbf{u}_3 = [\sqrt{3}/4, 3/4, -1/2],$$

show the vectors in set (a) are orthogonal and convert them to an orthonormal set, and that those in set in (b) are orthonormal.

Solution

(a) $\mathbf{u}_1 \cdot \mathbf{u}_2 = 1.2 + 2.1 - 2.2 = 0$ and, similarly, $\mathbf{u}_1 \cdot \mathbf{u}_3 = \mathbf{u}_2 \cdot \mathbf{u}_3 = 0$, and $\|\mathbf{u}_1\| = \|\mathbf{u}_2\| = \|\mathbf{u}_3\| = \sqrt{9} = 3$. So the set is orthogonal but *not* orthonormal, because the vector norms are not all equal to 1. To convert the set into an orthonormal set, it is only necessary to divide each vector by its norm to arrive at the equivalent *orthonormal* set

$$\hat{\mathbf{u}}_1 = \begin{bmatrix} 1/3 \\ 2/3 \\ -2/3 \end{bmatrix}, \quad \hat{\mathbf{u}}_2 = \begin{bmatrix} 2/3 \\ 1/3 \\ 2/3 \end{bmatrix}, \quad \text{and} \quad \hat{\mathbf{u}}_3 = \begin{bmatrix} -2/3 \\ 2/3 \\ 1/3 \end{bmatrix}.$$

(b) Proceeding as in (a) we have $\mathbf{u}_1 \cdot \mathbf{u}_2 = \mathbf{u}_1 \cdot \mathbf{u}_3 = \mathbf{u}_2 \cdot \mathbf{u}_3 = 0$, showing that the set is orthogonal. However, $\|\mathbf{u}_1\| = \|\mathbf{u}_2\| = \|\mathbf{u}_3\| = 1$, so the set is also orthonormal. ∎

THEOREM 4.4

properties of eigenvalues and eigenvectors of symmetric matrices

Eigenvalues and eigenvectors of a symmetric matrix   Let $\mathbf{A}$ be an $n \times n$ real symmetric matrix. Then

(i) the eigenvalues of $\mathbf{A}$ are all real;

(ii) the eigenvectors of $\mathbf{A}$ corresponding to distinct eigenvalues are mutually orthogonal.

Proof   We start by observing that if $\mathbf{x}$ and $\mathbf{y}$ are two n-element column vectors the product $\mathbf{y}^T\mathbf{A}\mathbf{x}$ is a scalar, and so is equal to its transpose. Thus, $\mathbf{y}^T\mathbf{A}\mathbf{x} = (\mathbf{y}^T\mathbf{A}\mathbf{x})^T = \mathbf{x}^T\mathbf{A}^T\mathbf{y}$, but as $\mathbf{A}$ is symmetric $\mathbf{A}^T = \mathbf{A}$, so that $\mathbf{y}^T\mathbf{A}\mathbf{x} = \mathbf{x}^T\mathbf{A}^T\mathbf{y}$.

To prove (i), let λ be an eigenvalue of $\mathbf{A}$ with the corresponding eigenvector $\mathbf{x}$. Then

$$\mathbf{A}\mathbf{x} = \lambda\mathbf{x}.$$

Taking the complex conjugate of this result and using the fact that $\mathbf{A}$ is real valued, so that $\overline{\mathbf{A}} = \mathbf{A}$, gives

$$\mathbf{A}\overline{\mathbf{x}} = \overline{\lambda}\overline{\mathbf{x}}.$$

This shows that $\overline{\lambda}$ is an eigenvalue of $\mathbf{A}$ with the associated eigenvector $\overline{\mathbf{x}}$. If we now premultiply this result by $\mathbf{x}^T$, we obtain the scalar equation

$$\mathbf{x}^T\mathbf{A}\overline{\mathbf{x}} = \overline{\lambda}\mathbf{x}^T\overline{\mathbf{x}},$$

but premultiplying the original eigenvalue equation by $\overline{\mathbf{x}}^T$ gives

$$\overline{\mathbf{x}}^T\mathbf{A}\mathbf{x} = \lambda\overline{\mathbf{x}}^T\mathbf{x}.$$

Using the result $\mathbf{x}^T\mathbf{A}\overline{\mathbf{x}} = \overline{\mathbf{x}}^T\mathbf{A}\mathbf{x}$ then shows that $\lambda\overline{\mathbf{x}}^T\mathbf{x} = \overline{\lambda}\mathbf{x}^T\overline{\mathbf{x}}$, but $\overline{\mathbf{x}}^T\mathbf{x} = \mathbf{x}^T\overline{\mathbf{x}}$ so $\lambda = \overline{\lambda}$, which is only possible if λ is real. This has established the first part of the theorem.

To prove (ii) we must show that if $\mathbf{x}_r$ and $\mathbf{x}_s$ are eigenvectors of $\mathbf{A}$ corresponding to the distinct eigenvalues λ_r and λ_s, with $r \neq s$, then $\mathbf{x}_r \cdot \mathbf{x}_s = 0$, which is equivalent to the condition $\mathbf{x}_r^T\mathbf{x}_s = 0$. The eigenvalues λ_r and λ_s and the corresponding eigenvectors $\mathbf{x}_r$ and $\mathbf{x}_s$ satisfy the equations

$$\mathbf{A}\mathbf{x}_r = \lambda_r\mathbf{x}_r \quad \text{and} \quad \mathbf{A}\mathbf{x}_s = \lambda_s\mathbf{x}_s,$$

from which, after premultiplication by $\mathbf{x}_s^T$ and $\mathbf{x}_r^T$, respectively, we obtain the two scalar equations

$$\mathbf{x}_s^T \mathbf{A} \mathbf{x}_r = \lambda_r \mathbf{x}_s^T \mathbf{x}_r \quad \text{and} \quad \mathbf{x}_r^T \mathbf{A} \mathbf{x}_s = \lambda_s \mathbf{x}_r^T \mathbf{x}_s.$$

Again, using the fact that the transpose of a scalar leaves it unchanged, we see that the preceding results are identical, so subtracting them we arrive at the condition

$$(\lambda_r - \lambda_s)\mathbf{x}_r^T \mathbf{x}_s = 0.$$

As $\lambda_r \neq \lambda_s$ for $r \neq s$, this is only possible if $\mathbf{x}_r^T \mathbf{x}_s = 0$, so the eigenvectors are mutually orthogonal and the proof is complete. ∎

It can be shown that even when some of the eigenvalues of a real symmetric $n \times n$ matrix $\mathbf{A}$ are repeated, the matrix $\mathbf{A}$ will still have n linearly independent eigenvectors, though this result will not be proved here. See, for example, references [2.1], [2.5], [2.8], [2.9], and [2.10].

Orthogonal matrices

orthogonal matrices and rotations

An $n \times n$ real matrix $\mathbf{Q}$ will be said to be an **orthogonal** matrix if

$$\mathbf{Q}^T \mathbf{Q} = \mathbf{I} \tag{12}$$

so, if $\mathbf{Q}$ is an orthogonal matrix, it follows that

$$\mathbf{Q}^T = \mathbf{Q}^{-1}.$$

When interpreted geometrically in terms of the cartesian geometry of two or three space dimensions, premultiplication of a linear transformation by an orthogonal matrix corresponds to a pure rotation (or a reflection or both; rotation only if $\det Q = +1$) in space that preserves the lengths between any two points in space, and also the angles between any two straight lines.

A typical geometrical interpretation of a two-dimensional transformation performed by an orthogonal matrix has already been encountered in Section 3.2(c), where the transformation considered was $\mathbf{x}' = \mathbf{R}\mathbf{x}$, with

$$\mathbf{R} = \begin{bmatrix} \cos\theta & -\sin\theta \\ \sin\theta & \cos\theta \end{bmatrix}, \quad \mathbf{x} = \begin{bmatrix} x \\ y \end{bmatrix}, \quad \text{and} \quad \mathbf{x}' = \begin{bmatrix} x' \\ y' \end{bmatrix}.$$

When this transformation was considered in Section 3.2(c), the column vector $\mathbf{x}$ represented a point P in the (x, y)-plane with coordinates (x, y), and $\mathbf{x}'$ represented the same point with coordinates (x', y') in the (x', y')-plane, which was obtained by rotating the $O\{x, y\}$ axes counterclockwise through an angle θ about the origin, as shown in Fig. 4.2.

The transformation (interpreted as a mapping of points) shows that every point in the $O\{x', y'\}$ plane experiences the same rotation through an angle θ about the origin. To show that lengths are preserved, let points P_1 and P_2 have coordinates (x_1, y_1) and (x_2, y_2) in the $O\{x, y\}$ plane and their image points P_1' and P_2' have the coordinates (x_1', y_1') and (x_2', y_2') in the $O\{x', y'\}$ plane. Then the square of the distance d between P_1 and P_2 is given by $d^2 = (x_1 - x_2)^2 + (y_1 - y_2)^2$, and the square

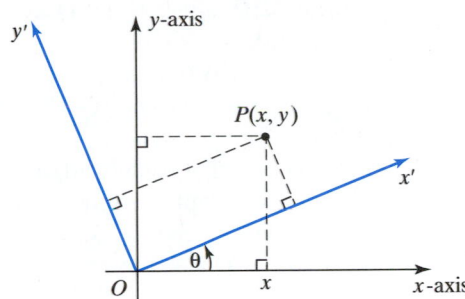

FIGURE 4.2   A rotation of axes about the origin through the angle θ.

of the distance $(d')^2$ between P'_1 and P'_2 is given by $(d')^2 = (x'_1 - x'_2)^2 + (y'_1 - y'_2)^2$. However, from the linear transformation $\mathbf{x}' = \mathbf{R}\mathbf{x}$ we find that

$$x_1 = x'_1 \cos\theta - y'_1 \sin\theta, \quad x_2 = x'_2 \cos\theta - y'_2 \sin\theta$$

and

$$y_1 = x'_1 \sin\theta + y'_1 \cos\theta, \quad y_2 = x'_2 \sin\theta + y'_2 \cos\theta,$$

from which, after substituting for x'_1, x'_2, y'_1, and y'_2, it follows that $(d')^2 = d^2$, showing that distances are preserved. The angles between straight lines in the plane will be preserved because the points on each line will be rotated about the origin through the same angle without changing their distance from the origin.

EXAMPLE 4.8   Show that the matrix

$$\mathbf{R} = \begin{bmatrix} \cos\theta & -\sin\theta \\ \sin\theta & \cos\theta \end{bmatrix}$$

is orthogonal.

Solution   We have

$$\mathbf{R}^{\mathrm{T}} = \begin{bmatrix} \cos\theta & \sin\theta \\ -\sin\theta & \cos\theta \end{bmatrix},$$

but $\mathbf{R}^{\mathrm{T}}\mathbf{R} = \mathbf{I}$, so $\mathbf{R}$ is orthogonal.   ■

THEOREM 4.5

main properties of orthogonal matrices

Properties of orthogonal matrices

   (i) If $\mathbf{Q}$ is orthogonal then $\det\mathbf{Q} = \pm 1$;
   (ii) The product of $n \times n$ orthogonal matrices is an orthogonal matrix;
   (iii) The eigenvalues of an orthogonal matrix are all of unit modulus;
   (iv) The rows (columns) of an orthogonal matrix form an orthonormal set of vectors.

Proof   To prove (i) we start from the fact that $\det\mathbf{Q} = \det\mathbf{Q}^{\mathrm{T}}$. This follows directly from the Laplace expansion of a determinant, because expanding $\det\mathbf{Q}$ in terms of the elements of its ith row is the same as expanding $\det\mathbf{Q}^{\mathrm{T}}$ in terms of the elements of its ith column. From (12), $\mathbf{Q}\mathbf{Q}^{\mathrm{T}} = \mathbf{1}$, so as $\det(\mathbf{AB}) = \det\mathbf{A}\det\mathbf{B}$ we can write $\det\mathbf{Q}\det\mathbf{Q}^{\mathrm{T}} = 1$, but $\det\mathbf{Q}^{\mathrm{T}} = \det\mathbf{Q}$ by Theorem 3.4 so $\det\mathbf{Q}\det\mathbf{Q}^{\mathrm{T}} = (\det\mathbf{Q})^2 = 1$, and so det$\mathbf{Q} = \pm 1$. If det $\mathbf{Q} = +1$, rotation. If det $\mathbf{Q} = -1$, rotation plus reflection in general.

Result (ii) follows from the fact that if $\mathbf{Q}_1$ and $\mathbf{Q}_2$ are two $n \times n$ orthogonal matrices, then $(\mathbf{Q}_1\mathbf{Q}_2)^\mathrm{T}\mathbf{Q}_1\mathbf{Q}_2 = \mathbf{Q}_2^\mathrm{T}\mathbf{Q}_1^\mathrm{T}\mathbf{Q}_1\mathbf{Q}_2 = \mathbf{Q}_2^\mathrm{T}\mathbf{Q}_2 = \mathbf{1}$, and the result is established.

The proof of Result (iii) is similar to the proof of (i) in Theorem 4.3. If $\mathbf{Q}$ is real, taking the complex conjugate of $\mathbf{Qx} = \lambda\mathbf{x}$ gives $\mathbf{Q}\overline{\mathbf{x}} = \overline{\lambda}\overline{\mathbf{x}}$, so taking the transpose of this we find that $\overline{\mathbf{x}}^\mathrm{T}\mathbf{Q}^\mathrm{T} = \overline{\lambda}\overline{\mathbf{x}}^\mathrm{T}$. Forming the product of these two results gives $\overline{\mathbf{x}}^\mathrm{T}\mathbf{Q}^\mathrm{T}\mathbf{Qx} = \lambda\overline{\lambda}\overline{\mathbf{x}}^\mathrm{T}\mathbf{x}$, but $\mathbf{Q}^\mathrm{T}\mathbf{Q} = \mathbf{I}$, so $\overline{\mathbf{x}}^\mathrm{T}\mathbf{x} = \lambda\overline{\lambda}\overline{\mathbf{x}}^\mathrm{T}\mathbf{x}$, showing that $\lambda\overline{\lambda} = 1$. Result (iii) follows from this last result because $\lambda\overline{\lambda} = |\lambda|^2 = 1$.

Finally, Result (iv) follows from the definition of an orthogonal matrix, because $\mathbf{QQ}^\mathrm{T} = \mathbf{1}$, and if $\mathbf{u}_i$ is the ith row of $\mathbf{Q}$ and $\mathbf{v}_j$ is the jth column of $\mathbf{Q}^\mathrm{T}$ (the jth column of $\mathbf{Q}$), then $\mathbf{u}_i\mathbf{v}_j = 0$ for $i \neq j$, and $\mathbf{u}_i\mathbf{v}_j = 1$ for $i = j$, confirming that the vectors form an orthonormal set. ∎

Summary

After definition of the eigenvalues of an $n \times n$ matrix $\mathbf{A}$ in terms of its characteristic polynomial, the associated eigenvectors were defined. An eigenvalue that is repeated r times was said to have the algebraic multiplicity r, and the set of all eigenvalues of $\mathbf{A}$ was called the spectrum of $\mathbf{A}$. The spectral radius of $\mathbf{A}$ was defined in terms of the eigenvalues $\lambda_1, \lambda_2, \ldots, \lambda_n$ as the number $R = \max\{|\lambda_1|, |\lambda_2|, \ldots |\lambda_n|\}$, and the linear independence of the set of all eigenvectors was established. The most frequently used method of normalizing eigenvectors was introduced, and examples were worked showing how to determine eigenvectors once the eigenvalues are known.

A simple test was given to check the sum of all eigenvalues, and the Gerschgorin circle theorem was proved that determines a region inside which all eigenvalues must lie, though the region determined in this manner is far from optimal. Inner products, the norm, and systems of orthogonal and orthonormal vectors were introduced, and the most important eigenvalue and eigenvector properties of symmetric matrices and orthogonal matrices were derived.

EXERCISES 4.1

In Exercises 1 through 8, find the characteristic polynomial of the given matrix.

1. $\begin{bmatrix} 2 & 1 & 3 \\ 1 & 0 & 1 \\ 0 & 1 & 1 \end{bmatrix}$.

2. $\begin{bmatrix} 2 & 1 & 3 \\ 1 & 1 & 1 \\ 1 & 0 & 1 \end{bmatrix}$.

3. $\begin{bmatrix} 1 & 0 & 2 \\ -1 & 1 & -1 \\ 0 & 2 & 1 \end{bmatrix}$.

4. $\begin{bmatrix} 3 & 1 & 1 \\ -2 & 2 & 1 \\ 1 & -1 & 2 \end{bmatrix}$.

5. $\begin{bmatrix} -1 & 0 & 1 \\ 3 & 2 & 1 \\ 1 & 2 & 3 \end{bmatrix}$.

6. $\begin{bmatrix} 4 & 1 & -1 \\ 1 & 0 & 2 \\ -1 & 1 & 2 \end{bmatrix}$.

7. $\begin{bmatrix} 1 & 1 & -1 & 0 \\ 1 & -1 & 1 & 0 \\ 1 & -3 & 3 & 0 \\ -1 & 2 & -1 & -1 \end{bmatrix}$.

8. $\begin{bmatrix} -1 & 1 & 0 & 1 \\ -1 & 2 & -1 & 1 \\ 5 & -3 & 4 & -5 \\ 3 & -2 & 3 & -3 \end{bmatrix}$.

In Exercises 9 through 24 find the eigenvalues and eigenvectors of the given matrix.

9. $\begin{bmatrix} 3 & -2 & 2 \\ 6 & -4 & 6 \\ 2 & -1 & 3 \end{bmatrix}$.

10. $\begin{bmatrix} 3 & -1 & 1 \\ 4 & -1 & 4 \\ 2 & -1 & 4 \end{bmatrix}$.

11. $\begin{bmatrix} -3 & 2 & -2 \\ 4 & -1 & 4 \\ 8 & -4 & 7 \end{bmatrix}$.

12. $\begin{bmatrix} 3 & -2 & 4 \\ -4 & 5 & -4 \\ -4 & 4 & -5 \end{bmatrix}$.

13. $\begin{bmatrix} -5 & 4 & -1 \\ -3 & 2 & -1 \\ 6 & -4 & 2 \end{bmatrix}$.

14. $\begin{bmatrix} 0 & 1 & -2 \\ 2 & -1 & 2 \\ 2 & -2 & 4 \end{bmatrix}$.

15. $\begin{bmatrix} -5 & 8 & 1 \\ -3 & 6 & 1 \\ 6 & -8 & 0 \end{bmatrix}$.

16. $\begin{bmatrix} -1 & 0 & -2 \\ -1 & 2 & -1 \\ 4 & 0 & 5 \end{bmatrix}$.

17. $\begin{bmatrix} -1 & 0 & 2 \\ -1 & 2 & 0 \\ -1 & 0 & 2 \end{bmatrix}$.

18. $\begin{bmatrix} 6 & 0 & 4 \\ 3 & 1 & 3 \\ -8 & 0 & -6 \end{bmatrix}$.

19. $\begin{bmatrix} 0 & 0 & 2 \\ -1 & 1 & 2 \\ -1 & 0 & 3 \end{bmatrix}.$

20. $\begin{bmatrix} 4 & 0 & 2 \\ 2 & 2 & 2 \\ -4 & 0 & 2 \end{bmatrix}.$

21. $\begin{bmatrix} 4 & 0 & -4 \\ 2 & 2 & -4 \\ 2 & 0 & -2 \end{bmatrix}.$

22. $\begin{bmatrix} 3 & 0 & 1 \\ 2 & 1 & 1 \\ -2 & 0 & 0 \end{bmatrix}.$

23. $\begin{bmatrix} -1 & -1 & 1 & 0 \\ 1 & 1 & 1 & -1 \\ 1 & 3 & -1 & -1 \\ -2 & 2 & -2 & 1 \end{bmatrix}.$

24. $\begin{bmatrix} 0 & 1 & 0 & -1 \\ 1 & 0 & 0 & -1 \\ 1 & -2 & 0 & -1 \\ -3 & 3 & 0 & 2 \end{bmatrix}.$

25. Prove that the eigenvalues of upper and lower triangular matrices are equal to the elements on the leading diagonal. Show by example that, unlike the case of diagonal matrices, an eigenvalue of an upper or lower triangular matrix with algebraic multiplicity r has fewer than r eigenvectors.

26. Apply the Gerschgorin circle theorem to one or more of the matrices in Exercises 9 through 24 to verify that the eigenvalues lie within or on the circles determined by the theorem.

27. It can be shown that all the zeros of the polynomial

$$P_n(\lambda) = a_0 + a_1\lambda + a_2\lambda^2 + \cdots + a_n\lambda^n, \quad a_n \neq 0,$$

lie in the circle

$$|\lambda| < 1 + \max\left|\frac{a_k}{a_n}\right|, \quad k = 0, 1, 2, \ldots, n-1.$$

Verify this result by applying it to one or more of the characteristic equations associated with the matrices in Exercises 9 through 24.

The Routh–Hurwitz stability criterion

Let the real polynomial $P_n(\lambda)$ be given by

$$P_n(\lambda) = \lambda^n + a_1\lambda^{n-1} + a_2\lambda^{n-2} + \cdots + a_n$$

and form the determinants

$$\Delta_1 = a_1, \quad \Delta_2 = \begin{vmatrix} a_1 & a_3 \\ 1 & a_2 \end{vmatrix}, \quad \Delta_3 = \begin{vmatrix} a_1 & a_3 & a_5 \\ 1 & a_2 & a_4 \\ 0 & a_1 & a_3 \end{vmatrix}, \ldots,$$

$$\Delta_n = \begin{vmatrix} a_1 & a_3 & a_5 & \cdots & a_{2n-1} \\ 1 & a_2 & a_4 & \cdots & a_{2n-2} \\ 0 & a_1 & a_3 & \cdots & a_{2n-3} \\ \cdots & \cdots & \cdots & \cdots & \cdots \\ 0 & 0 & 0 & 0 & a_n \end{vmatrix} \quad \text{with } a_k = 0 \text{ for } k > n.$$

Then, $\Delta_r > 0$ for $r = 1, 2, \ldots, n$, if and only if every zero of $P_n(\lambda)$ has a negative real part.

28.

(a) Numerical computation shows that the matrix

$$\mathbf{A} = \begin{bmatrix} -2 & 1 & 5 \\ 2 & 3 & 1 \\ 0 & 4 & 2 \end{bmatrix}$$

has the eigenvalues 5.7238, $-1.3619 + 1.9328i$, and $-1.3619 - 1.9328i$. Apply the Routh–Hurwitz stability criterion to confirm that not every zero of the characteristic polynomial has a negative real part.

(b) Numerical computation shows that the matrix

$$\mathbf{A} = \begin{bmatrix} -2 & -2 & -3 \\ 3 & -1 & 0 \\ -4 & 0 & -3 \end{bmatrix}$$

has the eigenvalues -5.4873, $-0.2563 - 1.4564i$, and $-0.2563 + 1.4564i$. Apply the Routh–Hurwitz stability criterion to confirm that every zero of the characteristic polynomial has a negative real part.

An $n \times n$ matrix $\mathbf{A}$ is said to be **similar** to an $n \times n$ matrix $\mathbf{B}$ if there exists a nonsingular $n \times n$ matrix $\mathbf{M}$ such that $\mathbf{B} = \mathbf{M}^{-1}\mathbf{AM}$. The relationship between $\mathbf{A}$ and $\mathbf{B}$ is said to constitute a **similarity transformation** between the two matrices.

29. If $\mathbf{A}$ and $\mathbf{B}$ are similar, show that $\det\mathbf{A} = \det\mathbf{B}$, and by substituting $\mathbf{B} = \mathbf{M}^{-1}\mathbf{AM}$ in $\det\mathbf{B}$ and expanding the result, show that similar matrices have the same eigenvalues.

30. Verify the result of Exercise 29 by direct calculation by using

$$\mathbf{A} = \begin{bmatrix} 3 & 1 & -1 \\ 4 & 0 & -1 \\ 4 & -2 & 1 \end{bmatrix} \quad \text{and} \quad \mathbf{M} = \begin{bmatrix} 1 & 4 & 1 \\ 1 & 0 & 1 \\ 2 & 1 & 0 \end{bmatrix}$$

to show that both $\mathbf{A}$ and $\mathbf{B}$ have the eigenvalues $-1, 2$, and 3.

31. Let the $n \times n$ elementary matrix $\mathbf{E}$ be obtained from the unit matrix $\mathbf{I}$ by interchanging its ith and jth rows (columns). By considering the product $\mathbf{EQ}$, where $\mathbf{Q}$ is an $n \times n$ orthogonal matrix, prove that an orthogonal matrix remains orthogonal when its rows (columns) are interchanged.

4.2   **Diagonalization of Matrices**

Our purpose in this section will be to examine the possibility of diagonalizing an $n \times n$ matrix $\mathbf{A}$. The reason for this is to try to simplify the structure of $\mathbf{A}$ so that, in some ways, it reflects the simple properties of a diagonal matrix. Diagonalization finds many applications, some of which will be discussed later.

diagonal matrix

Let $\mathbf{D}$ be the general $n \times n$ diagonal matrix

$$\mathbf{D} = \begin{bmatrix} \lambda_1 & 0 & 0 & . & . & . & . & 0 \\ 0 & \lambda_2 & 0 & . & . & . & . & 0 \\ & . & . & . & . & . & . & \\ & . & . & . & . & . & . & \\ 0 & 0 & 0 & . & . & . & . & \lambda_n \end{bmatrix}. \tag{13}$$

Then, as already seen in Section 4.1, the eigenvalues of $\mathbf{D}$ are the entries $\lambda_1, \lambda_2, \ldots, \lambda_n$ on its leading diagonal, and the corresponding n linearly independent eigenvectors can be taken to be

$$\mathbf{x_1} = \begin{bmatrix} 1 \\ 0 \\ 0 \\ . \\ . \\ 0 \end{bmatrix}, \quad \mathbf{x_2} = \begin{bmatrix} 0 \\ 1 \\ 0 \\ . \\ . \\ 0 \end{bmatrix}, \ldots, \mathbf{x_n} = \begin{bmatrix} 0 \\ 0 \\ 0 \\ . \\ . \\ 1 \end{bmatrix}. \tag{14}$$

The rule for matrix multiplication shows that

$$\mathbf{D}^m = \begin{bmatrix} \lambda_1^m & 0 & 0 & . & . & . & . & 0 \\ 0 & \lambda_2^m & 0 & . & . & . & . & 0 \\ & . & . & . & . & . & . & \\ & . & . & . & . & . & . & \\ 0 & 0 & 0 & . & . & . & . & \lambda_n^m \end{bmatrix}, \tag{15}$$

for any positive integer m, so $\mathbf{D}^m$ is easily computed and will have the same set of eigenvectors as $\mathbf{D}$, though its eigenvalues will be $\lambda_1^m, \lambda_2^m, \ldots, \lambda_n^m$.

In addition to these properties, it is obvious that $\det \mathbf{D} = \lambda_1 \cdot \lambda_2 \cdots \lambda_n$, so $\mathbf{D}$ will be nonsingular provided no entry on its leading diagonal is zero. As a result, when $\mathbf{D}$ is nonsingular, the rule for matrix multiplication shows that $\mathbf{D}\mathbf{D}^{-1} = \mathbf{I}$, where

$$\mathbf{D}^{-1} = \begin{bmatrix} 1/\lambda_1 & 0 & 0 & . & . & . & . & 0 \\ 0 & 1/\lambda_2 & 0 & . & . & . & . & 0 \\ & . & . & . & . & . & . & \\ & . & . & . & . & . & . & \\ 0 & 0 & 0 & 0 & & & & 1/\lambda_n \end{bmatrix}. \tag{16}$$

We now state and prove the fundamental theorem on the diagonalization of $n \times n$ matrices.

THEOREM 4.6

how to diagonalize a matrix

Diagonalization of an $n \times n$ matrix   Let the $n \times n$ matrix $\mathbf{A}$ have n eigenvalues $\lambda_1, \lambda_2, \ldots, \lambda_n$, not all of which need be distinct, and let there be n corresponding distinct eigenvectors $\mathbf{x}_1, \mathbf{x}_2, \ldots, \mathbf{x}_n$, so that

$$\mathbf{A}\mathbf{x}_i = \lambda_i \mathbf{x}_i, \quad i = 1, 2, \ldots, n.$$

Define the matrix $\mathbf{P}$ to be the $n \times n$ matrix in which the ith column is the eigenvector $\mathbf{x}_i$, with $i = 1, 2, \ldots, n$, so that in partitioned form $\mathbf{P} = [\mathbf{x}_1 \ \ \mathbf{x}_2 \ \ \cdots \ \ \mathbf{x}_n]$, and let $\mathbf{D}$ be the diagonal matrix

$$
\mathbf{D} = \begin{bmatrix} \lambda_1 & 0 & 0 & . & . & . & . & 0 \\ 0 & \lambda_2 & 0 & . & . & . & . & 0 \\ . & . & . & . & . & . & . & . \\ . & . & . & . & . & . & . & . \\ 0 & 0 & 0 & . & . & . & . & \lambda_n \end{bmatrix},
$$

where the eigenvalue λ_i is in the ith position in the ith row. Then

$$
\mathbf{P}^{-1}\mathbf{A}\mathbf{P} = \mathbf{D}.
$$

Proof   Consider the product $\mathbf{B} = \mathbf{A}\mathbf{P}$. Then, by expressing $\mathbf{P}$ in partitioned form, we can write $\mathbf{B}$ as

$$
\mathbf{B} = [\mathbf{A}\mathbf{x}_1 \ \ \mathbf{A}\mathbf{x}_2 \ \ \ldots \ \ \mathbf{A}\mathbf{x}_n].
$$

Using the fact that $\mathbf{A}\mathbf{x}_i = \lambda_i \mathbf{x}_i$ allows this to be rewritten as

$$
\mathbf{B} = [\lambda_1 \mathbf{x}_1 \ \ \lambda_2 \mathbf{x}_2 \ \ \ldots \ \ \lambda_n \mathbf{x}_n] = \mathbf{P}\mathbf{D},
$$

showing that

$$
\mathbf{P}\mathbf{D} = \mathbf{A}\mathbf{P}.
$$

As the columns of $\mathbf{P}$ are linearly independent, $\mathbf{P}$ is nonsingular, so $\mathbf{P}^{-1}$ exists and we can premultiply by $\mathbf{P}^{-1}$ to obtain

$$
\mathbf{D} = \mathbf{P}^{-1}\mathbf{A}\mathbf{P},
$$

and the theorem is proved.   ∎

General Remarks About Diagonalization

(i) An $n \times n$ matrix can be diagonalized provided it possesses n linearly independent eigenvectors.

(ii) A symmetric matrix can always be diagonalized.

(iii) The diagonalizing matrix for a real $n \times n$ matrix $\mathbf{A}$ may contain complex elements. This is because although the characteristic polynomial of $\mathbf{A}$ has real coefficients, its zeros either will be real or will occur in complex conjugate pairs.

(iv) A diagonalizing matrix is not unique, because its form depends on the order in which the eigenvectors of $\mathbf{A}$ are used to form its columns.

A useful consequence of the diagonalized form of a matrix is that it enables it to be raised to a positive integral power with the minimum of effort. This property will be used later when the matrix exponential is introduced.

To see the ease with which an $n \times n$ matrix can be raised to a power when it is diagonalizable, we start by writing $\mathbf{A}$ in the form $\mathbf{A} = \mathbf{P}\mathbf{D}\mathbf{P}^{-1}$. We then have

$$
\mathbf{A}^2 = (\mathbf{P}\mathbf{D}\mathbf{P}^{-1})(\mathbf{P}\mathbf{D}\mathbf{P}^{-1}) = \mathbf{P}\mathbf{D}\mathbf{P}^{-1}\mathbf{P}\mathbf{D}\mathbf{P}^{-1} = \mathbf{P}\mathbf{D}\mathbf{D}\mathbf{P}^{-1} = \mathbf{P}\mathbf{D}^2\mathbf{P}^{-1},
$$

so that, in general,

$$\mathbf{A}^m = \mathbf{PD}^m\mathbf{P}^{-1}, \quad \text{for } m = 1, 2, \dots.$$

As evaluating $\mathbf{D}^m$ simply involves raising each entry on its leading diagonal to the power m, the evaluation of $\mathbf{A}^m$ only involves three matrix multiplications.

This last result was used without justification in Section 3.2(f) when a stochastic matrix was raised to the power m (do not confuse the stochastic matrix $\mathbf{P}$ in that section with the orthogonalizing matrix $\mathbf{P}$ just defined).

EXAMPLE 4.9   Diagonalize the matrix

$$\mathbf{A} = \begin{bmatrix} 2 & 1 & -1 \\ 3 & 2 & -3 \\ 3 & 1 & -2 \end{bmatrix},$$

and use the result to find $\mathbf{A}^5$.

Solution   Matrix $\mathbf{A}$ was examined in Example 4.1 and shown to have the eigenvalues $\lambda_1 = 2$, $\lambda_2 = 1$, and $\lambda_3 = -1$, and the corresponding eigenvectors

$$\mathbf{x}_1 = \begin{bmatrix} 1 \\ 1 \\ 1 \end{bmatrix}, \quad \mathbf{x}_2 = \begin{bmatrix} 1 \\ 0 \\ 1 \end{bmatrix}, \quad \text{and} \quad \mathbf{x}_3 = \begin{bmatrix} 0 \\ 1 \\ 1 \end{bmatrix}.$$

Theorem 4.5 shows that a diagonalizing matrix $\mathbf{P}$ is given by

$$\mathbf{P} = \begin{bmatrix} 1 & 1 & 0 \\ 1 & 0 & 1 \\ 1 & 1 & 1 \end{bmatrix},$$

and a routine calculation shows that

$$\mathbf{P}^{-1} = \begin{bmatrix} 1 & 1 & -1 \\ 0 & -1 & 1 \\ -1 & 0 & 1 \end{bmatrix}.$$

Before finding $\mathbf{A}^5$, and although it is unnecessary for what is to follow, it is instructive to check that when the matrix $\mathbf{P}^{-1}\mathbf{AP}$ is formed, the eigenvalues appearing in the diagonal matrix $\mathbf{D}$ do so in the order in which the corresponding eigenvectors of $\mathbf{A}$ have been used to form the columns of $\mathbf{P}$. This is seen to be so in this case because

$$\mathbf{D} = \mathbf{P}^{-1}\mathbf{AP} = \begin{bmatrix} 2 & 0 & 0 \\ 0 & 1 & 0 \\ 0 & 0 & -1 \end{bmatrix}.$$

Returning to the calculation of $\mathbf{A}^5$ and using the expressions for $\mathbf{P}$, $\mathbf{P}^{-1}$, and $\mathbf{D}$ in $\mathbf{A}^5 = \mathbf{PD}^5\mathbf{P}^{-1}$ gives

$$\mathbf{A}^5 = \begin{bmatrix} 1 & 1 & 0 \\ 1 & 0 & 1 \\ 1 & 1 & 1 \end{bmatrix} \begin{bmatrix} 2^5 & 0 & 0 \\ 0 & 1^5 & 0 \\ 0 & 0 & (-1)^5 \end{bmatrix} \begin{bmatrix} 1 & 1 & 0 \\ 1 & 0 & 1 \\ 1 & 0 & 1 \end{bmatrix} = \begin{bmatrix} 32 & 31 & -31 \\ 33 & 32 & -33 \\ 33 & 31 & -32 \end{bmatrix}.$$

Had the eigenvectors been arranged in a different order when constructing $\mathbf{P}$, a different but equivalent diagonal matrix would have been obtained. For example, if **P** had been written

$$\mathbf{P} = \begin{bmatrix} 1 & 1 & 0 \\ 0 & 1 & 1 \\ 1 & 1 & 1 \end{bmatrix},$$

D would have become

$$\mathbf{D} = \begin{bmatrix} 1 & 0 & 0 \\ 0 & 2 & 0 \\ 0 & 0 & -1 \end{bmatrix},$$

though after $\mathbf{P}^{-1}$ was found and $\mathbf{A}^5 = \mathbf{P}\mathbf{D}^5\mathbf{P}^{-1}$ was computed, the matrix $\mathbf{A}^5$ would, of course, remain the same. ∎

EXAMPLE 4.10   Diagonalize the matrix

$$\mathbf{A} = \begin{bmatrix} 0 & 0 & 1 & 1 \\ -1 & 2 & 0 & 1 \\ -1 & 0 & 2 & 1 \\ 1 & 0 & -1 & 0 \end{bmatrix}.$$

Solution   Matrix **A** was considered in Example 4.2, which showed that it had the eigenvalues $\lambda_1 = 0, \lambda_2 = 1, \lambda_3 = 1,$ and $\lambda_4 = 2,$ and that although the eigenvalue 1 occurred with algebraic multiplicity 2, the matrix still had the four linearly independent eigenvectors

$$(\lambda_1 = 0) \quad \mathbf{x}_1 = \begin{bmatrix} -1 \\ -1 \\ -1 \\ 1 \end{bmatrix}, \quad (\lambda_2 = 1) \quad \mathbf{x}_2 = \begin{bmatrix} 1 \\ 1 \\ 1 \\ 0 \end{bmatrix}, \quad (\lambda_3 = 1) \quad \mathbf{x}_3 = \begin{bmatrix} 1 \\ 0 \\ 0 \\ 1 \end{bmatrix},$$

and

$$(\lambda_4 = 2) \quad \mathbf{x}_1 = \begin{bmatrix} 0 \\ 1 \\ 0 \\ 0 \end{bmatrix}.$$

Using these eigenvectors to form **P** gives

$$\mathbf{P} = \begin{bmatrix} -1 & 1 & 1 & 0 \\ -1 & 1 & 0 & 1 \\ -1 & 1 & 0 & 0 \\ 1 & 0 & 1 & 0 \end{bmatrix},$$

from which it follows that

$$\mathbf{P}^{-1} = \begin{bmatrix} -1 & 0 & 1 & 1 \\ -1 & 0 & 2 & 1 \\ 1 & 0 & -1 & 0 \\ 0 & 1 & -1 & 0 \end{bmatrix}.$$

Because of the ordering of the eigenvectors, the diagonal matrix $\mathbf{D}$ will be

$$\mathbf{D} = \begin{bmatrix} 0 & 0 & 0 & 0 \\ 0 & 1 & 0 & 0 \\ 0 & 0 & 1 & 0 \\ 0 & 0 & 0 & 2 \end{bmatrix},$$

where

$$\mathbf{P}^{-1}\mathbf{A}\mathbf{P} = \mathbf{D}. \qquad\qquad \blacksquare$$

We saw in Theorem 4.4 that a real symmetric $n \times n$ matrix $\mathbf{A}$ with distinct eigenvalues has a set of n mutually orthogonal linearly independent eigenvectors. It follows at once that if when constructing the diagonalizing matrix for $\mathbf{A}$ the normalized eigenvectors of $\mathbf{A}$ are used to form the columns of $\mathbf{P}$, the resulting diagonalizing matrix will be an *orthogonal* matrix. This is often advantageous, because the properties of orthogonal matrices can simplify subsequent calculations that may arise. However, if an eigenvalue is repeated, the corresponding eigenvectors will not, in general, be orthogonal to the other eigenvectors, so although there will still be a set of n linearly independent eigenvectors, the set will no longer form an orthogonal set.

Because of the frequency with which symmetric matrices arise in applications, and the fact that symmetric matrices with repeated eigenvalues are not unusual, it is reasonable to ask if it is possible for symmetric matrices always to be diagonalized by an orthogonal matrix and, if so, how this can be achieved. The answer to the question about the possibility of diagonalization by an orthogonal matrix is in the affirmative. The method of arriving at an orthonormal set of vectors to be used when constructing $\mathbf{P}$ involves using a generalization of the Gram–Schmidt orthogonalization process introduced in Section 2.7 in the context of geometrical vectors in R^3.

As an n element matrix vector is simply a vector in a vector space, an extension of the Gram–Schmidt orthogonalization process to include n-element matrix vectors can be used to construct an *orthonormal* set of n vectors from any set of n linearly independent eigenvectors that are always associated with an $n \times n$ symmetric matrix $\mathbf{A}$. The required generalization of the orthogonalization process that leads to an **orthonormal system** is an immediate extension of the one derived in Section 2.7, so the details of its derivation will be omitted.

Rule for the Gram–Schmidt orthogonalization process for matrix vectors

orthogonalization of a set of linearly independent vectors

Let $\mathbf{x}_1, \mathbf{x}_2, \ldots, \mathbf{x}_n$ be a set of n element linearly independent nonorthogonal matrix column vectors. Then an equivalent **orthonormal set** of vectors $\mathbf{p}_1, \mathbf{p}_2, \ldots, \mathbf{p}_n$ can be constructed from the vectors $\mathbf{x}_1, \mathbf{x}_2, \ldots, \mathbf{x}_n$, via an intermediate set of orthogonal nonnormalized vectors $\mathbf{v}_2, \mathbf{v}_2, \ldots, \mathbf{v}_n$. The steps involved in the determination of the vectors $\mathbf{p}_1, \mathbf{p}_2, \ldots, \mathbf{p}_n$ are as follows:

$$\begin{aligned} \mathbf{p}_1 &= \mathbf{x}_1/\|\mathbf{x}_1\|, \\ \mathbf{v}_2 &= \mathbf{x}_2 - (\mathbf{p}_1 \cdot \mathbf{x}_2)\mathbf{p}_1, \\ \mathbf{p}_2 &= \mathbf{v}_2/\|\mathbf{v}_2\|, \\ \mathbf{v}_r &= \mathbf{x}_r - \{(\mathbf{p}_1 \cdot \mathbf{x}_r)\mathbf{p}_1 + (\mathbf{p}_2 \cdot \mathbf{x}_r)\mathbf{p}_2 + \cdots + (\mathbf{p}_{r-1} \cdot \mathbf{x}_r)\mathbf{p}_{r-1}\} \\ \mathbf{p}_r &= \mathbf{v}_r/\|\mathbf{v}_r\|, \quad \text{for } r = 2, 3, \ldots, n. \end{aligned}$$

When the Gram–Schmidt orthogonalization process is applied to the eigenvectors of a real symmetric matrix $\mathbf{A}$ with repeated eigenvalues, the diagonalizing matrix $\mathbf{P}$ is constructed by using the vectors $\mathbf{p}_1, \mathbf{p}_2, \ldots, \mathbf{p}_n$, obtained from the preceding scheme after starting with any linearly independent set of eigenvectors $\mathbf{x}_1, \mathbf{x}_2, \ldots, \mathbf{x}_n$ of $\mathbf{A}$. Then, in partitioned form,

$$\mathbf{P} = [\mathbf{p}_1 \quad \mathbf{p}_2 \quad \cdots \quad \mathbf{p}_n]$$

and, as before,

$$\mathbf{D} = \mathbf{P}^{-1}\mathbf{A}\mathbf{P},$$

where $\mathbf{D}$ is again a diagonal matrix with its diagonal elements equal to the eigenvalues of $\mathbf{A}$ arranged in the same order as the corresponding columns of $\mathbf{P}$. This time, however, entries on the leading diagonal will be repeated as many times as the multiplicity of the eigenvalues concerned.

EXAMPLE 4.11

Use the Gram–Schmidt orthogonalization process to construct an orthonormal set of vectors from the vectors

$$\mathbf{x}_1 = \begin{bmatrix} 1 \\ 1 \\ 1 \end{bmatrix}, \quad \mathbf{x}_2 = \begin{bmatrix} 1 \\ 0 \\ -1 \end{bmatrix}, \quad \text{and} \quad \mathbf{x}_3 = \begin{bmatrix} 1 \\ 2 \\ 0 \end{bmatrix}.$$

Solution   In this case the Gram–Schmidt orthogonalization process involves the three vectors $\mathbf{x}_1$, $\mathbf{x}_2$, and $\mathbf{x}_3$, so a set of orthonormal vectors $\mathbf{p}_1$, $\mathbf{p}_2$, and $\mathbf{p}_3$ is given by the scheme

$$\mathbf{p}_1 = \mathbf{x}_1/\|\mathbf{x}_1\|$$
$$\mathbf{v}_2 = \mathbf{x}_2 - (\mathbf{p}_1 \cdot \mathbf{x}_2)\mathbf{p}_1$$
$$\mathbf{p}_2 = \mathbf{v}_2/\|\mathbf{v}_2\|$$
$$\mathbf{v}_3 = \mathbf{x}_3 - \{(\mathbf{p}_1 \cdot \mathbf{x}_3)\mathbf{p}_1 + (\mathbf{p}_2 \cdot \mathbf{x}_3)\mathbf{p}_2\}$$
$$\mathbf{p}_3 = \mathbf{v}_3/\|\mathbf{v}_3\|.$$

A series of straightforward calculations gives

$$\mathbf{p}_1 = \begin{bmatrix} 1/\sqrt{3} \\ 1/\sqrt{3} \\ 1/\sqrt{3} \end{bmatrix}, \quad \text{and} \quad \mathbf{v}_2 = \begin{bmatrix} 1 \\ 0 \\ -1 \end{bmatrix} - 0\,\mathbf{p}_1 = \begin{bmatrix} 1 \\ 0 \\ -1 \end{bmatrix}, \quad \text{so } \mathbf{p}_2 = \begin{bmatrix} 1/\sqrt{2} \\ 0 \\ -1/\sqrt{2} \end{bmatrix},$$

and, finally,

$$\mathbf{v}_3 = \begin{bmatrix} 1 \\ 2 \\ 0 \end{bmatrix} - \sqrt{3}\begin{bmatrix} 1/\sqrt{3} \\ 1/\sqrt{3} \\ 1/\sqrt{3} \end{bmatrix} - 1/\sqrt{2}\begin{bmatrix} 1/\sqrt{2} \\ 0 \\ -1/\sqrt{2} \end{bmatrix} = \begin{bmatrix} -1/2 \\ 1 \\ -1/2 \end{bmatrix},$$

so

$$\mathbf{p}_3 = \begin{bmatrix} -1/\sqrt{6} \\ \sqrt{(2/3)} \\ -1/\sqrt{6} \end{bmatrix}.$$

EXAMPLE 4.12

Construct an orthogonal diagonalizing matrix for the symmetric matrix

$$\mathbf{A} = \begin{bmatrix} 4 & 0 & 0 \\ 0 & 1 & 2 \\ 0 & 2 & 1 \end{bmatrix}.$$

Solution This has the *distinct* eigenvalues $\lambda_1 = -1$, $\lambda_2 = 3$, and $\lambda_1 = 4$, so the corresponding eigenvectors $\mathbf{x}_1, \mathbf{x}_2$, and $\mathbf{x}_3$ are orthogonal. Simple calculations show that

$$\mathbf{x}_1 = \begin{bmatrix} 0 \\ -1 \\ 1 \end{bmatrix}, \quad \mathbf{x}_2 = \begin{bmatrix} 0 \\ 1 \\ 1 \end{bmatrix}, \quad \text{and} \quad \mathbf{x}_3 = \begin{bmatrix} 1 \\ 0 \\ 0 \end{bmatrix}.$$

The normalized eigenvectors are

$$\hat{\mathbf{x}}_1 = \begin{bmatrix} 0 \\ -1/\sqrt{2} \\ 1/\sqrt{2} \end{bmatrix}, \quad \hat{\mathbf{x}}_2 = \begin{bmatrix} 0 \\ 1/\sqrt{2} \\ 1/\sqrt{2} \end{bmatrix}, \quad \text{and} \quad \hat{\mathbf{x}}_3 = \begin{bmatrix} 1 \\ 0 \\ 0 \end{bmatrix},$$

so the diagonalizing matrix $\mathbf{P}$ and the corresponding diagonal matrix $\mathbf{D}$ are

$$\mathbf{P} = \begin{bmatrix} 0 & 0 & 1 \\ -1/\sqrt{2} & 1/\sqrt{2} & 0 \\ 1/\sqrt{2} & 1/\sqrt{2} & 0 \end{bmatrix} \quad \text{and} \quad \mathbf{D} = \begin{bmatrix} -1 & 0 & 0 \\ 0 & 3 & 0 \\ 0 & 0 & 4 \end{bmatrix}. \qquad \blacksquare$$

EXAMPLE 4.13

Construct an orthogonal diagonalizing matrix for the real symmetric matrix

$$\mathbf{A} = \begin{bmatrix} -1 & 2 & 4 \\ 2 & 2 & -2 \\ 4 & -2 & -1 \end{bmatrix}.$$

Solution This has the eigenvalues $\lambda_1 = -6$, $\lambda_2 = 3$, and $\lambda_3 = 3$, so as the eigenvalue 3 has multiplicity 2, the corresponding set of eigenvectors $\mathbf{x}_1, \mathbf{x}_2$, and $\mathbf{x}_3$ will *not* be orthogonal. The eigenvectors $\mathbf{x}_1, \mathbf{x}_2$, and $\mathbf{x}_3$ are easily shown to be

$$\mathbf{x}_1 = \begin{bmatrix} -2 \\ 1 \\ 2 \end{bmatrix}, \quad \mathbf{x}_2 = \begin{bmatrix} 1 \\ 2 \\ 0 \end{bmatrix}, \quad \text{and} \quad \mathbf{x}_3 = \begin{bmatrix} 0 \\ -2 \\ 1 \end{bmatrix}.$$

Applying the Gram–Schmidt orthogonalization process to vectors $\mathbf{x}_1, \mathbf{x}_2$, and $\mathbf{x}_3$, as in Example 4.11, after some straightforward calculations we arrive at the orthonormal set

$$\mathbf{p}_1 = \begin{bmatrix} -2/3 \\ 1/3 \\ 2/3 \end{bmatrix}, \quad \mathbf{p}_2 = \begin{bmatrix} 1/\sqrt{5} \\ 2/\sqrt{5} \\ 0 \end{bmatrix}, \quad \text{and} \quad \mathbf{p}_3 = \begin{bmatrix} 4/(3\sqrt{5}) \\ -2/(3\sqrt{5}) \\ \sqrt{5}/3 \end{bmatrix}.$$

In this case an orthogonal diagonalizing matrix is

$$\mathbf{P} = \begin{bmatrix} -2/3 & 1/\sqrt{5} & 4/(3\sqrt{5}) \\ 1/3 & 2/\sqrt{5} & -2/(3\sqrt{5}) \\ 2/3 & 0 & \sqrt{5}/3 \end{bmatrix},$$

and the corresponding diagonal matrix is

$$\mathbf{D} = \begin{bmatrix} -6 & 0 & 0 \\ 0 & 3 & 0 \\ 0 & 0 & 3 \end{bmatrix}.$$

To close this section we state the important Cayley–Hamilton theorem, which is true for *all* square matrices, though before considering the theorem we first define a matrix polynomial.

A **matrix polynomial** involving an $n \times n$ matrix $\mathbf{A}$ is an expression of the form

$$\mathbf{A}^m + b_1\mathbf{A}^{m-1} + b_2\mathbf{A}^{m-2} + \cdots + b_{m-1}\mathbf{A} + b_m\mathbf{I},$$

in which m is an integer and $b_1, b_2, \ldots, b_m$ are real or complex numbers.

THEOREM 4.7

a matrix satisfies its own characteristic equation

The Cayley–Hamilton theorem   Let $P_n(\lambda)$ be the characteristic polynomial of an arbitrary $n \times n$ square matrix $\mathbf{A}$. Then $\mathbf{A}$ satisfies its own characteristic equation, and so is a solution of the matrix polynomial equation $P_n(\mathbf{A}) = \mathbf{0}$.

Proof   For simplicity, we only prove the theorem for real symmetric matrices, though it is true for every $n \times n$ matrix. If $\mathbf{A}$ is a real $n \times n$ symmetric matrix, then from Theorem 4.6 we may write $\mathbf{A} = \mathbf{PDP}^{-1}$. Let the characteristic polynomial of $\mathbf{A}$ be

$$P_n(\lambda) = (-1)^n \{\lambda^n + c_1\lambda^{n-1} + \cdots + c_{n-1}\lambda + c_n\}.$$

Then replacing λ by $\mathbf{A}$ converts $P_n(\lambda)$ to the matrix polynomial

$$P_n(\mathbf{A}) = (-1)^n \{\mathbf{A}^n + c_1\mathbf{A}^{n-1} + \cdots + c_{n-1}\mathbf{A} + c_n\mathbf{I}\},$$

but $\mathbf{A}^r = \mathbf{PD}^r\mathbf{P}^{-1}$, so

$$P_n(\mathbf{A}) = (-1)^n \{\mathbf{P}\{\mathbf{D}^n + c_1\mathbf{D}^{n-1} + \cdots + c_{n-1}\mathbf{D} + c_n\mathbf{I}_n\}\mathbf{P}^{-1}\}.$$

The ith row of the matrix polynomial $\mathbf{D}^n + c_1\mathbf{D}^{n-1} + \cdots + c_{n-1}\mathbf{D} + c_n\mathbf{I}$ is simply $\lambda_i^n + c_1\lambda_i^{n-1} + \cdots + c_{n-1}\lambda_i + c_n$, but this is $P_n(\lambda_i)$, and it must vanish for $i = 1, 2, \ldots, n$ because λ_i is an eigenvalue of $\mathbf{A}$. Thus, $\mathbf{D}^n + c_1\mathbf{D}^{n-1} + \cdots + c_{n-1}\mathbf{D} + c_n\mathbf{I} = \mathbf{0}$, showing that $P_n(\mathbf{A}) = \mathbf{P}\{\mathbf{0}\}\mathbf{P}^{-1} = \mathbf{0}$, and the result is proved. ∎

EXAMPLE 4.14

Verify the Cayley–Hamilton theorem for the matrix

$$\mathbf{A} = \begin{bmatrix} 2 & 1 \\ 5 & 2 \end{bmatrix}.$$

Solution   The characteristic polynomial is $P_2(\lambda) = \lambda^2 - 4\lambda - 1$, and

$$\mathbf{A}^2 = \begin{bmatrix} 9 & 4 \\ 20 & 9 \end{bmatrix}, \quad \text{so } P_2(\mathbf{A}) = \begin{bmatrix} 9 & 4 \\ 20 & 9 \end{bmatrix} - 4\begin{bmatrix} 2 & 1 \\ 5 & 2 \end{bmatrix} - \begin{bmatrix} 1 & 0 \\ 0 & 1 \end{bmatrix} = \begin{bmatrix} 0 & 0 \\ 0 & 0 \end{bmatrix}.$$

Finding $\mathbf{A}^{-1}$ from the Cayley–Hamilton theorem

If the $n \times n$ matrix $\mathbf{A}$ is nonsingular, the following interesting result can be obtained directly from the Cayley–Hamilton theorem. Let the characteristic polynomial of $\mathbf{A}$ be $P_n(\lambda) = (-1)^n\{\lambda^n + c_1\lambda^{n-1} + \cdots + c_{n-1}\lambda + c_n\}$, so from Theorem 4.7

$$\mathbf{A}^n + c_1\mathbf{A}^{n-1} + \cdots + c_{n-1}\mathbf{A} + c_n\mathbf{I} = 0.$$

The matrix $\mathbf{A}^{-1}$ exists because by hypothesis $\mathbf{A}$ is nonsingular, so premultiplication of the preceding equation by $\mathbf{A}^{-1}$, followed by a rearrangement of terms, allows $\mathbf{A}^{-1}$ to be expressed in terms of powers of $\mathbf{A}$ through the result

$$\mathbf{A}^{-1} = (-1/c_n)\{\mathbf{A}^{n-1} + c_1\mathbf{A}^{n-2} + \cdots + c_{n-1}\mathbf{I}\}. \tag{17}$$

EXAMPLE 4.15   Use the result of equation (17) to find $\mathbf{A}^{-1}$ for the nonsingular matrix

$$\mathbf{A} = \begin{bmatrix} 2 & 1 \\ 5 & 2 \end{bmatrix}.$$

Solution   Matrix $\mathbf{A}$ was considered in Example 4.14, where it was found that the characteristic polynomial $P_2(\lambda) = \lambda^2 - 4\lambda - 1$, so in terms of (17) we see that $c_1 = -4$ and $c_2 = -1$. Thus,

$$\mathbf{A}^{-1} = -1/(-1)\left\{ \begin{bmatrix} 2 & 1 \\ 5 & 2 \end{bmatrix} - 4\begin{bmatrix} 1 & 0 \\ 0 & 1 \end{bmatrix} \right\} = \begin{bmatrix} -2 & 1 \\ 5 & -2 \end{bmatrix}. \quad\blacksquare$$

Summary

This section has described how an $n \times n$ matrix can be diagonalized when it possesses n linearly independent eigenvectors. The diagonalization was shown not to be unique, since its form depends on the order in which the eigenvectors are used to construct the diagonalizing matrix $\mathbf{P}$.

Sometimes, when a linearly independent set of n vectors has been obtained, it is desirable to replace it by an equivalent set of n orthogonal or orthonormal vectors. The section closed by showing how this can be accomplished by means of the Gram–Schmidt orthogonalization procedure.

EXERCISES 4.2

In Exercises 1 through 12, find a diagonalizing matrix $\mathbf{P}$ for the given matrix, in each case using the fact that the zeros of the characteristic polynomial are small integers that can be found by trial and error.

1. $\begin{bmatrix} -2 & -3 & -1 \\ 1 & 2 & 1 \\ 3 & 3 & 2 \end{bmatrix}.$

2. $\begin{bmatrix} 3 & 1 & 4 \\ -4 & -2 & -4 \\ -1 & -1 & 2 \end{bmatrix}.$

3. $\begin{bmatrix} 3 & 1 & -2 \\ 6 & 2 & -6 \\ 4 & 1 & -3 \end{bmatrix}.$

4. $\begin{bmatrix} -6 & -10 & -4 \\ 2 & 3 & 2 \\ 7 & 10 & 5 \end{bmatrix}.$

5. $\begin{bmatrix} -1 & 2 & -2 \\ 2 & -1 & 2 \\ 2 & -2 & 3 \end{bmatrix}.$

6. $\begin{bmatrix} 14 & 2 & 8 \\ -8 & -3 & -4 \\ -26 & -4 & -15 \end{bmatrix}.$

7. $\begin{bmatrix} 5 & -2 & 2 \\ 2 & 1 & 2 \\ -2 & 2 & 1 \end{bmatrix}.$

8. $\begin{bmatrix} 12 & 4 & 6 \\ -6 & -2 & -3 \\ -22 & -8 & -11 \end{bmatrix}.$

9. $\begin{bmatrix} 2 & 0 & 0 \\ 1 & -1 & 2 \\ -2 & 0 & 1 \end{bmatrix}.$

10. $\begin{bmatrix} 12 & -4 & 8 \\ -6 & 2 & -4 \\ -20 & 8 & -14 \end{bmatrix}.$

11. $\begin{bmatrix} -6 & 2 & -4 \\ -4 & 0 & -4 \\ 4 & -2 & 2 \end{bmatrix}.$

12. $\begin{bmatrix} -7 & 0 & -6 \\ 3 & -1 & 3 \\ 9 & 0 & 8 \end{bmatrix}.$

In Exercises 13 through 16 use the Gram–Schmidt orthogonalization process with the given set of vectors to find (a) an equivalent set of orthogonal vectors and (b) an orthonormal set.

13. $\begin{bmatrix} 1 \\ 1 \\ 1 \end{bmatrix}, \begin{bmatrix} 0 \\ 1 \\ 1 \end{bmatrix}, \begin{bmatrix} 0 \\ 0 \\ 1 \end{bmatrix}.$ **15.** $\begin{bmatrix} -1 \\ 1 \\ 0 \end{bmatrix}, \begin{bmatrix} 2 \\ 1 \\ -1 \end{bmatrix}, \begin{bmatrix} 1 \\ -2 \\ 2 \end{bmatrix}.$

14. $\begin{bmatrix} 2 \\ 1 \\ 1 \end{bmatrix}, \begin{bmatrix} 1 \\ -1 \\ 1 \end{bmatrix}, \begin{bmatrix} 0 \\ 2 \\ 1 \end{bmatrix}.$ **16.** $\begin{bmatrix} -1 \\ 2 \\ 0 \end{bmatrix}, \begin{bmatrix} 1 \\ 1 \\ -1 \end{bmatrix}, \begin{bmatrix} 1 \\ -1 \\ 1 \end{bmatrix}.$

In Exercises 17 through 22 find an orthogonal diagonalizing matrix **P** for the given symmetric matrix.

17. $\begin{bmatrix} 3 & 0 & 0 \\ 0 & 3 & 1 \\ 0 & 1 & 3 \end{bmatrix}.$ **19.** $\begin{bmatrix} 4 & 1 & 0 \\ 1 & 4 & 0 \\ 0 & 0 & 3 \end{bmatrix}.$

18. $\begin{bmatrix} 5 & 1 & 0 \\ 1 & 5 & 0 \\ 0 & 0 & 2 \end{bmatrix}.$ **20.** $\begin{bmatrix} 2 & 1 & 1 \\ 1 & 2 & 1 \\ 1 & 1 & 2 \end{bmatrix}.$

21. $\begin{bmatrix} 4 & 2 & 0 \\ 2 & 4 & 0 \\ 0 & 0 & 2 \end{bmatrix}.$ **22.** $\begin{bmatrix} 4 & 1 & 1 \\ 1 & 4 & 1 \\ 1 & 1 & 4 \end{bmatrix}.$

23. Verify by direct calculation that the matrix in Exercise 1 satisfies the Cayley–Hamilton theorem.

24. Verify by direct calculation that the matrix in Exercise 7 satisfies the Cayley–Hamilton theorem.

In Exercises 25 through 28 use (17) to find $\mathbf{A}^{-1}$ and check the result by showing that $\mathbf{AA}^{-1} = \mathbf{I}$.

25. $\mathbf{A} = \begin{bmatrix} 2 & 3 \\ -1 & 4 \end{bmatrix}.$ **27.** $\mathbf{A} = \begin{bmatrix} 2 & 1 & 0 \\ -2 & 1 & 2 \\ 0 & -1 & -2 \end{bmatrix}.$

26. $\mathbf{A} = \begin{bmatrix} 5 & 1 \\ 3 & -2 \end{bmatrix}.$ **28.** $\mathbf{A} = \begin{bmatrix} 1 & 0 & 2 \\ 3 & 1 & 0 \\ 0 & 2 & 4 \end{bmatrix}.$

4.3   Special Matrices with Complex Elements

In the previous section it was seen that one way in which matrices with complex elements can occur is when the eigenvectors of an arbitrary $n \times n$ matrix are used to construct a diagonalizing matrix. This is not the only reason for considering $n \times n$ matrices with complex elements, because the following three special types of matrices arise naturally in applications of mathematics to physics and engineering, and elsewhere.

Hermitian, skew-Hermitian, and unitary matrices

Let $\mathbf{A} = [a_{ij}]$ be an $n \times n$ matrix with possibly complex elements. Then:

$\mathbf{A}$ is called an **Hermitian** matrix if $\overline{\mathbf{A}}^\mathrm{T} = \mathbf{A}$, so that $\overline{a}_{kj} = a_{jk}$;

$\mathbf{A}$ is called a **skew-Hermitian** matrix if $\overline{\mathbf{A}}^\mathrm{T} = -\mathbf{A}$, so that $\overline{a}_{kj} = -a_{jk}$;

$\mathbf{U}$ is called a **unitary** matrix if $\overline{\mathbf{U}}^\mathrm{T} = \mathbf{U}^{-1}$.

The basic properties of these three types of matrices follow almost directly from their definitions.

Basic Properties of Hermitian, Skew-Hermitian, and Unitary Matrices

1. The elements on the leading diagonal of an Hermitian matrix are real, because $\overline{a}_{ii} = a_{ii}$, and this is only possible if a_{ii} is real.
2. The elements on the leading diagonal of a skew-Hermitian matrix are either purely imaginary or 0. This follows from the fact that $\overline{a}_{ii} = -a_{ii}$, so the real part of a_{ii} must equal its negative, and this is only possible if a_{ii} is purely imaginary or 0.

3. If the elements of an Hermitian matrix are real, then the matrix is a real symmetric matrix, because then $\overline{\mathbf{A}}^{\mathrm{T}} = \mathbf{A}^{\mathrm{T}}$, and the definition of an Hermitian matrix reduces to the definition of a real symmetric matrix.

4. If the elements of a skew-Hermitian matrix are real, then the matrix is a skew-symmetric matrix, because then the definition of a skew-Hermitian matrix reduces to the definition of a skew-symmetric matrix.

5. Any $n \times n$ matrix $\mathbf{A}$ of the form $\mathbf{A} = \mathbf{B} + i\mathbf{C}$, where $\mathbf{B}$ is a real symmetric matrix and $\mathbf{C}$ is a real skew-symmetric matrix, is an Hermitian matrix. This follows directly from Properties 3 and 4.

6. Any $n \times n$ matrix $\mathbf{A}$ can be written in the form $\mathbf{A} = \mathbf{B} + \mathbf{C}$, where $\mathbf{B}$ is Hermitian and $\mathbf{C}$ is a skew-Hermitian. To see this we write $\mathbf{A} = (1/2)(\mathbf{A} + \overline{\mathbf{A}}^{\mathrm{T}}) + (1/2)(\mathbf{A} - \overline{\mathbf{A}}^{\mathrm{T}})$, and then set $\mathbf{B} = (1/2)(\mathbf{A} + \overline{\mathbf{A}}^{\mathrm{T}})$ and $\mathbf{C} = (1/2)(\mathbf{A} - \overline{\mathbf{A}}^{\mathrm{T}})$. Then $\overline{\mathbf{B}}^{\mathrm{T}} = (1/2)(\overline{\mathbf{A}^{\mathrm{T}} + \overline{\mathbf{A}}}) = (1/2)(\mathbf{A} + \overline{\mathbf{A}}^{\mathrm{T}}) = \mathbf{B}$ and $\mathbf{C}^{\mathrm{T}} = (1/2)(\overline{\mathbf{A}^{\mathrm{T}} - \overline{\mathbf{A}}}) = -(1/2)(\mathbf{A} - \overline{\mathbf{A}}^{\mathrm{T}}) = -\mathbf{C}$, showing that $\mathbf{B}$ is Hermitian and $\mathbf{C}$ is skew-Hermitian.

7. A real unitary matrix is an orthogonal matrix, because in that case $\overline{\mathbf{A}}^{\mathrm{T}} = \mathbf{A}^{\mathrm{T}}$, causing the definition of a unitary matrix to reduce to the definition of an orthogonal matrix.

8. The determinant of a unitary matrix is ± 1. This result is established in essentially the same way as the result of Theorem 4.4(i), so the argument will not be repeated.

EXAMPLE 4.16

The following are examples of Hermitian, skew-Hermitian, and unitary matrices.

Hermitian matrix:

$$\mathbf{A} = \begin{bmatrix} 3 & 2+5i & -7+3i \\ 2-5i & 0 & 1-i \\ -7-3i & 1+i & 4 \end{bmatrix}.$$

Skew-Hermitian matrix:

$$\mathbf{B} = \begin{bmatrix} 4i & -3-2i & -6-4i \\ 3-2i & -2i & 5 \\ 6-4i & -5 & 0 \end{bmatrix}.$$

Unitary matrix:

$$\mathbf{U} = \begin{bmatrix} \dfrac{1+i}{2} & \dfrac{-1+i}{2} & 0 \\ \dfrac{1+i}{2} & \dfrac{1-i}{2} & 0 \\ 0 & 0 & 1 \end{bmatrix}.$$

It can be seen from Properties 3, 4, and 7 that Hermitian, skew-Hermitian, and unitary matrices are, respectively, generalizations of symmetric, skew-symmetric, and orthogonal real-valued matrices. Accordingly, it is to be expected that some of the properties exhibited by these real-valued matrices are shared by their complex generalizations, and this is indeed the case as we now show.

| THEOREM 4.8 | **Eigenvalues of Hermitian, skew-Hermitian, and unitary matrices** |

(i) The eigenvalues of an Hermitian matrix are real.

(ii) The eigenvalues of a skew-Hermitian matrix are either purely imaginary or 0.

(iii) The eigenvalues λ of a unitary matrix are all such that $|\lambda| = 1$.

Proof

(i) Apart for the need to introduce the complex conjugate operation, the proof is essentially the same as that of Theorem 4.4 for symmetric matrices, and so it is omitted.

(ii) Let $\mathbf{x}$ be the eigenvector of $\mathbf{A}$ corresponding to the eigenvalue λ, so $\mathbf{A}\mathbf{x} = \lambda\mathbf{x}$. Then $\bar{\mathbf{x}}^{\mathrm{T}}\mathbf{A}\mathbf{x} = \lambda\bar{\mathbf{x}}^{\mathrm{T}}\mathbf{x}$, from which we have

$$\lambda = \bar{\mathbf{x}}^{\mathrm{T}}\mathbf{A}\mathbf{x}/\bar{\mathbf{x}}^{\mathrm{T}}\mathbf{x},$$

but $\bar{\mathbf{x}}^{\mathrm{T}}\mathbf{x} = x_1\bar{x}_1 + x_2\bar{x}_2 + \cdots + x_n\bar{x}_n$ is real. However, $\overline{\mathbf{A}} = -\mathbf{A}^{\mathrm{T}}$, so $\bar{\mathbf{x}}^{\mathrm{T}}\mathbf{A}\mathbf{x} = -\overline{\bar{\mathbf{x}}^{\mathrm{T}}\mathbf{A}\mathbf{x}}$, so we can write

$$\lambda = \bar{\mathbf{x}}^{\mathrm{T}}\mathbf{A}\mathbf{x}/\bar{\mathbf{x}}^{\mathrm{T}}\mathbf{x} = -\overline{\bar{\mathbf{x}}^{\mathrm{T}}\mathbf{A}\mathbf{x}/\bar{\mathbf{x}}^{\mathrm{T}}\mathbf{x}}.$$

The product $\bar{\mathbf{x}}^{\mathrm{T}}\mathbf{x}$ is real, so this last result shows that the complex number λ equals the negative of its complex conjugate, and this is only possible if λ is purely imaginary or 0, so the proof is complete.

(iii) Apart from the need to introduce the complex conjugate operation, the proof is essentially that of Theorem 4.5(iii), so it will be omitted. ∎

The location of the eigenvalues of these complex matrices and of their corresponding real forms are illustrated in Fig. 4.3.

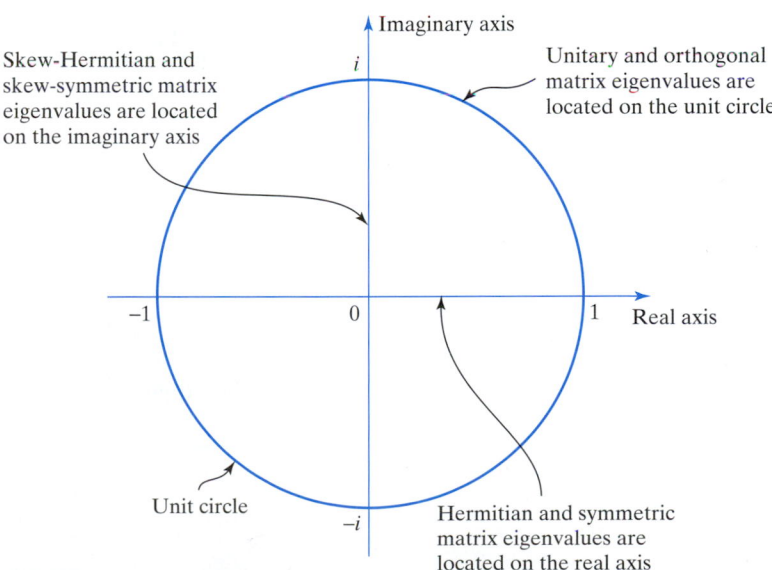

FIGURE 4.3 The location of the eigenvalues of Hermitian, skew-Hermitian, and unitary matrices in the complex plane.

If the definitions of an inner product and a norm are generalized, the concept of orthogonality can be extended to include vectors with complex elements. These generalizations have many applications, but they will only be used here to prove the orthogonality of the rows and columns of unitary matrices.

As the norm of a vector is essentially its *length* and so must be nonnegative, the previous definition of a norm in terms of an inner product must be modified in such a way that the inner product and norm of a complex vector coincide with those for a real vector when purely real vectors are considered. This is achieved by introducing the complex conjugate operation into the definition of an inner product.

Inner product of complex vectors

Let $\mathbf{w} = [w_1, w_2, \ldots, w_n]^{\mathrm{T}}$ and $\mathbf{z} = [z_1, z_2, \ldots, z_n]^{\mathrm{T}}$ be two column vectors with complex elements. Then the **inner product** of the column vectors $\mathbf{w}$ and $\mathbf{z}$, again denoted by $\mathbf{w} \cdot \mathbf{z}$, is defined as $\mathbf{w} \cdot \mathbf{z} = \overline{\mathbf{w}}^{\mathrm{T}} \mathbf{z}$, so that

$$\mathbf{w} \cdot \mathbf{z} = \overline{w}_1 z_1 + \overline{w}_2 z_2 + \cdots + \overline{w}_n z_n. \tag{18}$$

Norm of complex vectors

The **norm** of a vector $\mathbf{z}$, again denoted by $\|\mathbf{z}\|$, is defined as the nonnegative number

$$\begin{aligned}
\|\mathbf{z}\| &= (\mathbf{z} \cdot \mathbf{z})^{1/2} = (\overline{\mathbf{z}}^{\mathrm{T}} \mathbf{z})^{1/2} \\
&= (\overline{z}_1 z_1 + \overline{z}_2 z_2 + \cdots + \overline{z}_n z_n)^{1/2} \\
&= (|z_1|^2 + |z_2|^2 + \cdots + |z_n|^2)^{1/2}.
\end{aligned} \tag{19}$$

It can be seen from the preceding definition that the inner product of two arbitrary complex vectors is a complex number. However, the definition of the norm of a complex vector $\mathbf{z}$ is a real nonnegative number, as would be expected.

EXAMPLE 4.17   If $\mathbf{w} = [1 + 2i, \ 3 - i, \ i]^{\mathrm{T}}$ and $\mathbf{z} = [2 + i, \ 1 - i, \ 1 + 3i]^{\mathrm{T}}$, find $\mathbf{w} \cdot \mathbf{z}$ and $\|\mathbf{z}\|$.

Solution $\mathbf{w} \cdot \mathbf{z} = (\overline{1 + 2i})(2 + i) + (\overline{3 - i})(1 - i) + \overline{i}(1 + 3i) = 11 - 6i$, and $\|\mathbf{z}\| = [|2 + i|^2 + |1 - i|^2 + |1 + 3i|^2]^{1/2} = 17^{1/2}$. ∎

We are now in a position to generalize the concept of an orthonormal system of real vectors to a system of complex vectors that will be called a *unitary* system if the vectors satisfy the following conditions.

A unitary system

A set of complex vectors $\mathbf{z}_1, \mathbf{z}_2, \ldots, \mathbf{z}_n$ is said to form a **unitary system** if

$$\mathbf{z}_i \cdot \mathbf{z}_j = \overline{\mathbf{z}}_i^{\mathrm{T}} \mathbf{z}_j = \begin{cases} 0 & \text{if } i \neq j \\ 1 & \text{if } i = j. \end{cases} \tag{20}$$

THEOREM 4.9

The eigenvectors of a unitary matrix   The rows and columns of a unitary matrix each form a unitary system of vectors.

Proof   By definition the $n \times n$ matrix $\mathbf{U}$ is unitary if $\overline{\mathbf{U}}^{\mathrm{T}} = \mathbf{U}^{-1}$, so that $\overline{\mathbf{U}}^{\mathrm{T}}\mathbf{U} = \mathbf{I}$. The element in the ith row and jth column of $\mathbf{I}$ is the inner product $\mathbf{x}_i \cdot \mathbf{x}_j = \overline{\mathbf{x}}_i^{\mathrm{T}}\mathbf{x}_j$, where $\mathbf{x}_i$ and $\mathbf{x}_j$ are the ith and jth columns of $\mathbf{U}$. Consequently,

$$\overline{\mathbf{x}}_i^{\mathrm{T}}\mathbf{x}_j = \begin{cases} 0 & \text{if } i \neq j \\ 1 & \text{if } i = j, \end{cases}$$

showing that the columns of $\mathbf{U}$ form a unitary system. The rows also form a unitary system, because taking the transpose of $\overline{\mathbf{U}}^{\mathrm{T}}\mathbf{U}$ we find that $(\overline{\mathbf{U}}^{\mathrm{T}}\mathbf{U})^{\mathrm{T}} = \mathbf{U}^{\mathrm{T}}\overline{\mathbf{U}} = \mathbf{I}^{\mathrm{T}} = \mathbf{I}$. ∎

Summary

Matrices with complex elements arise in a variety of different applications, and from among these matrices, the most important are Hermitian, skew-Hermitian, and unitary matrices. Hermitian and skew-Hermitian matrices are the complex analogues of real symmetric and skew-symmetric matrices, respectively, and unitary matrices are the complex analogue of real orthogonal matrices. This section derived and illustrated by means of examples the most important properties of these matrices, and then introduced the inner product and norm of matrices with complex elements.

EXERCISES 4.3

In Exercises 1 through 4 write the given matrix as the sum of an Hermitian and a skew-Hermitian matrix.

1. $\begin{bmatrix} 1+i & 3+i & 3+2i \\ -1+3i & 2 & 4+i \\ -3-2i & 2+3i & 4+2i \end{bmatrix}$.

2. $\begin{bmatrix} 0 & 3+i & 1+2i \\ 1-5i & 1+i & 2 \\ 1+4i & -2i & 3 \end{bmatrix}$.

3. $\begin{bmatrix} 4-2i & 1+i & 2+2i \\ -1-3i & 1+2i & 4 \\ 0 & 2 & 0 \end{bmatrix}$.

4. $\begin{bmatrix} 3+i & 4-i & 5+2i \\ 2+i & 1+2i & 2 \\ -1 & 2i & 4-i \end{bmatrix}$.

In Exercises 5 through 8 find the eigenvalues of the Hermitian matrices and hence confirm the result of Theorem 4.8(a) that they are real.

5. $\begin{bmatrix} 1 & 2-i \\ 2+i & 2 \end{bmatrix}$.

6. $\begin{bmatrix} 2 & 2+2i \\ 1-2i & 3 \end{bmatrix}$.

7. $\begin{bmatrix} 3 & 2-3i \\ 2+3i & 1 \end{bmatrix}$.

8. $\begin{bmatrix} -4 & 2-2i \\ 2+2i & 3 \end{bmatrix}$.

In Exercises 9 through 12 find the eigenvalues of the skew-Hermitian matrices and hence confirm the result of Theorem 4.8(b) that they are purely imaginary.

9. $\begin{bmatrix} i & 3+i \\ -3+i & 2i \end{bmatrix}$.

10. $\begin{bmatrix} 3i & 2-i \\ -2-i & 0 \end{bmatrix}$.

11. $\begin{bmatrix} 0 & 3+2i \\ -3+2i & 0 \end{bmatrix}$.

12. $\begin{bmatrix} 4i & 2+3i \\ -2+3i & i \end{bmatrix}$.

13. Show the following matrix is unitary:

$$\begin{bmatrix} 1/\sqrt{2} & -i/\sqrt{2} \\ i/\sqrt{2} & 1/\sqrt{2} \end{bmatrix}.$$

In Exercises 14 and 15 show the matrices are unitary, find their eigenvalues and eigenvectors, and confirm that the eigenvalues all lie on the unit circle.

14. $\begin{bmatrix} (i-1)/\sqrt{2} & (1-i)/\sqrt{2} \\ (i-1)/\sqrt{2} & (i-1)/\sqrt{2} \end{bmatrix}$.

15. $\begin{bmatrix} (1+i)/\sqrt{2} & -(1+i)/\sqrt{2} \\ (1+i)/\sqrt{2} & (1+i)/\sqrt{2} \end{bmatrix}$.

4.4   Quadratic Forms

A homogeneous polynomial $P(\mathbf{x})$ of degree two of the form

$$
\begin{aligned}
P(\mathbf{x}) \equiv\ & a_{11}x_1^2 + a_{22}x_2^2 + \cdots + a_{nn}x_n^2 + 2a_{12}x_1x_2 \\
& + 2a_{13}x_1x_3 + \cdots + 2a_{n-1,n}x_{n-1}x_n,
\end{aligned}
\tag{21}
$$

real quadratic form

in which the coefficients a_{ij} and the variables in $\mathbf{x}(x_1, x_2, \ldots, x_n)$ are real numbers, is called a **real quadratic form** in the variables $x_1, x_2, \ldots, x_n$. The term *homogeneous* of degree two or, more precisely, *algebraically* homogeneous of degree two, means that each term in P is quadratic in the sense that it involves a product of precisely two of the variables $x_1, x_2, \ldots, x_n$. The terms involving the products $x_i x_j$ with $i \neq j$ are called the **mixed product** or **cross-product terms**.

Real quadratic forms

A real quadratic form $P(\mathbf{x})$ is a homogeneous polynomial in the real variables $x_1, x_2, \ldots, x_n$ of the form shown in (21). If $\mathbf{A}$ is a real symmetric $n \times n$ matrix and $\mathbf{x}$ is an n-element column vector defined as

$$
\mathbf{x} = \begin{bmatrix} x_1 \\ x_2 \\ \cdot \\ \cdot \\ x_n \end{bmatrix} \quad \text{and} \quad \mathbf{A} = \begin{bmatrix} a_{11} & a_{12} & \cdots & a_{1n} \\ a_{12} & a_{22} & \cdots & a_{2n} \\ \cdot & \cdot & \cdots & \cdot \\ \cdot & \cdot & \cdots & \cdot \\ a_{1n} & a_{2n} & \cdots & a_{nn} \end{bmatrix},
\tag{22}
$$

then $P(\mathbf{x})$ can be written in the matrix form

$$
P(\mathbf{x}) \equiv \mathbf{x}^{\mathrm{T}}\mathbf{A}\mathbf{x}.
\tag{23}
$$

There is no loss of generality in requiring $\mathbf{A}$ to be a symmetric matrix, because if the coefficient of a cross-product term $x_i x_j$ equals b_{ij}, this can always be rewritten as $b_{ij} = 2a_{ij}$ allowing the terms a_{ij} to be positioned symmetrically about the leading diagonal, as shown in the matrix $\mathbf{A}$ in (22). Exercise 30 at the end of this section shows how the definition of a real quadratic form can be extended to any real $n \times n$ matrix.

EXAMPLE 4.18

Express the quadratic form

$$
P(\mathbf{x}) \equiv 3x_1^2 - 2x_2^2 + 4x_3^2 + x_1x_2 + 3x_1x_3 - 2x_2x_3
$$

as the matrix product $P(\mathbf{x}) = \mathbf{x}^{\mathrm{T}}\mathbf{A}\mathbf{x}$.

Solution   By defining $\mathbf{x}$ and $\mathbf{A}$ as

$$
\mathbf{x} = \begin{bmatrix} x_1 \\ x_2 \\ x_3 \end{bmatrix}, \quad \mathbf{A} = \begin{bmatrix} 3 & 1/2 & 3/2 \\ 1/2 & -2 & -1 \\ 3/2 & -1 & 4 \end{bmatrix},
$$

we can write $P(\mathbf{x}) = \mathbf{x}^{\mathrm{T}}\mathbf{A}\mathbf{x}$.   ∎

Quadratic forms arise in various ways; for example, in mechanics a quadratic form can describe the ellipsoid of inertia of a solid body, the angular momentum of a solid body rotating about an axis, and the kinetic energy of a system of moving particles. Other areas in which quadratic forms occur include the geometry of conics in two space dimensions and of quadrics in three space dimensions, optimization problems, crystallography, and in the classification of partial differential equations (see Chapter 18).

We now give a general definition of a quadratic form that allows both the matrix $\mathbf{A}$ and the vector $\mathbf{x}$ to contain complex elements.

General quadratic forms

quadratic form and vectors with complex elements

Let the elements of an $n \times n$ matrix $\mathbf{A} = [a_{ij}]$ and an n-element column vector $\mathbf{z}$ be complex numbers. Then a **quadratic form** $P(\mathbf{z})$ involving the variables $z_1, z_2, \ldots, z_n$ of vector $\mathbf{z}$ is an expression of the form

$$P(\mathbf{z}) = \bar{\mathbf{z}}^{\mathrm{T}} \mathbf{A} \mathbf{z} = \sum_{i=1, j=1}^{n} a_{ij} \bar{z}_i z_j. \tag{24}$$

This definition is seen to include real quadratic forms, because when the elements of $\mathbf{A}$ and $\mathbf{z}$ are real, result (24) reduces to the real quadratic form defined in (23).

The structure of a quadratic form becomes clearer if a change of variables is made that removes the mixed product terms, leaving only the squared terms. This is called the **reduction** of the quadratic form to its **standard form**, also known as its **canonical form**. The next theorem shows how such a simplification can be achieved.

THEOREM 4.10

how to reduce a quadratic form to a sum of squares

Reduction of a quadratic form   Let the $n \times n$ real symmetric matrix $\mathbf{A}$ have the eigenvalues $\lambda_1, \lambda_2, \ldots, \lambda_n$, and let $\mathbf{Q}$ be an orthogonal matrix that diagonalizes $\mathbf{A}$, so that $\mathbf{Q}^{\mathrm{T}} \mathbf{A} \mathbf{Q} = \mathbf{D}$, where $\mathbf{D}$ is a diagonal matrix with the eigenvalues of $\mathbf{A}$ as the elements on its leading diagonal. Then the change of variable $\mathbf{x} = \mathbf{Q}\mathbf{y}$, involving the column vectors $\mathbf{x} = [x_1, x_2, \ldots, x_n]^{\mathrm{T}}$ and $\mathbf{y} = [y_1, y_2, \ldots, y_n]^{\mathrm{T}}$, transforms the real quadratic form $P(\mathbf{x}) \equiv \mathbf{x}^{\mathrm{T}} \mathbf{A} \mathbf{x}$ into the **standard form**

$$P(\mathbf{x}) \equiv \sum_{i=1, j=1}^{n} a_{ij} x_i x_j = \lambda_1 y_1^2 + \lambda_2 y_2^2 + \cdots + \lambda_n y_n^2.$$

Proof   The proof uses the fact that because $\mathbf{Q}$ is an orthogonal matrix, $\mathbf{Q}^{\mathrm{T}} \mathbf{A} \mathbf{Q} = \mathbf{D}$. Substituting $\mathbf{x} = \mathbf{Q}\mathbf{y}$ into the real quadratic form $\mathbf{x}^{\mathrm{T}} \mathbf{A} \mathbf{x}$ gives

$$P(\mathbf{x}) \equiv \mathbf{x}^{\mathrm{T}} \mathbf{A} \mathbf{x} = (\mathbf{Q}\mathbf{y})^{\mathrm{T}} \mathbf{A} \mathbf{Q} \mathbf{y}$$
$$= \mathbf{y}^{\mathrm{T}} \mathbf{Q}^{\mathrm{T}} \mathbf{A} \mathbf{Q} \mathbf{y}$$
$$= \mathbf{y}^{\mathrm{T}} \mathbf{D} \mathbf{y} = \lambda_1 y_1^2 + \lambda_2 y_2^2 + \cdots + \lambda_n y_n^2. \qquad \blacksquare$$

It follows immediately from Theorem 4.10 that the standard form of $P(\mathbf{x})$ is determined once the eigenvalues of $\mathbf{A}$ are known and, when needed, the transformation of coordinates between $\mathbf{x}$ and $\mathbf{y}$ is given by $\mathbf{x} = \mathbf{Q}\mathbf{y}$ or, equivalently, by $\mathbf{y} = \mathbf{Q}^{\mathrm{T}} \mathbf{x}$.

The next example provides a geometrical interpretation of Theorem 4.10 in the context of rigid body mechanics. In order to understand its implications it is necessary to know that if an origin O is taken at an arbitrary point inside a solid body, and an orthogonal set of axes O$\{x_1, x_2, x_3\}$ is located at O, nine moments and products of inertia of the body can be defined relative to these axes and displayed in the form of a 3×3 inertia matrix. The moment of inertia of the body about any line passing through the origin O is proportional to the length of the segment of the line that lies between O and the point where it intersects a three-dimensional surface defined by a quadratic form determined by the inertia matrix.

When the surface determined by the inertia matrix is scaled so the length of the line from O to its point of intersection with the surface equals the reciprocal of the moment of inertia about that line, the surface is called the **ellipsoid of inertia**. If the orientation of the O$\{x_1, x_2, x_3\}$ axes is chosen arbitrarily, the resulting quadratic form will be complicated by the presence of mixed product terms, but a suitable rotation of the axes can always remove these terms and lead to the most convenient orientation of the new system of axes O$\{y_1, y_2, y_3\}$. In the geometry of both conics and quadrics, and also in mechanics, new axes obtained in this way that lead to the elimination of mixed product terms are called the **principal axes**, and it is because of this that Theorem 4.10 is often known as the **principal axes theorem**.

quadratic forms and principal axes

EXAMPLE 4.19

The ellipsoid of inertia of a solid body is given by

$$P(\mathbf{x}) \equiv 4x_1^2 + 4x_2^2 + x_3^2 - 2x_1x_2.$$

Find its standard form in terms of a new orthogonal set of axes O$\{y_1, y_2, y_3\}$, and find the linear transformation that connects the two sets of coordinates.

Solution   The quadratic form $P(\mathbf{x})$ can be written as $\mathbf{x}^T \mathbf{A}\mathbf{x}$ by defining

$$\mathbf{x} = \begin{bmatrix} x_1 \\ x_2 \\ x_3 \end{bmatrix} \quad \text{and} \quad \mathbf{A} = \begin{bmatrix} 4 & -1 & 0 \\ -1 & 4 & 0 \\ 0 & 0 & 1 \end{bmatrix}.$$

The eigenvalues of $\mathbf{A}$ are $\lambda_1 = 1$, $\lambda_2 = 5$, and $\lambda_3 = 3$, so the standard form of $P(\mathbf{x})$ is

$$P(\mathbf{x}) \equiv y_1^2 + 5y_2^2 + 3y_3^2.$$

The eigenvalues and corresponding normalized eigenvectors of $\mathbf{A}$ are

$$\lambda_1 = 1, \ \hat{\mathbf{x}}_1 = \begin{bmatrix} 0 \\ 0 \\ 1 \end{bmatrix}, \quad \lambda_2 = 5, \quad \hat{\mathbf{x}}_2 = \begin{bmatrix} -1/\sqrt{2} \\ 1/\sqrt{2} \\ 0 \end{bmatrix}, \quad \lambda_3 = 3, \quad \hat{\mathbf{x}}_3 = \begin{bmatrix} 1/\sqrt{2} \\ 1/\sqrt{2} \\ 0 \end{bmatrix},$$

so the orthogonal diagonalizing matrix for $\mathbf{A}$ is

$$\mathbf{Q} = \begin{bmatrix} 0 & -1/\sqrt{2} & 1/\sqrt{2} \\ 0 & 1/\sqrt{2} & 1/\sqrt{2} \\ 1 & 0 & 0 \end{bmatrix},$$

and the change of variables between $\mathbf{x}$ and $\mathbf{y}$ determined by $\mathbf{x} = \mathbf{Q}\mathbf{y}$ becomes

$$x_1 = (-y_2 + y_3)/\sqrt{2}, \quad x_2 = (y_2 + y_3)/\sqrt{2}, \quad x_3 = y_1.$$

The equation $P(\mathbf{x}) = $ constant is seen to be an *ellipsoid* for which O$\{y_1, y_2, y_3\}$ are the *principal axes*.

EXAMPLE 4.20

Reduce the quadratic part of the following expression to its standard form involving the principal axes $O\{y_1, y_2\}$, and hence find the form taken by the complete expression in terms of y_1 and y_2:

$$x_1^2 + 4x_1x_2 + 4x_2^2 + x_1 - 2x_2.$$

Solution   The quadratic part of the expression is $x_1^2 + 4x_1x_2 + 4x_2^2$, and this can be expressed in the form $\mathbf{x}^T\mathbf{A}\mathbf{x}$ by setting

$$\mathbf{x} = \begin{bmatrix} x_1 \\ x_2 \end{bmatrix} \quad \text{and} \quad \mathbf{A} = \begin{bmatrix} 1 & 2 \\ 2 & 4 \end{bmatrix}.$$

The eigenvalues and eigenvectors of $\mathbf{A}$ are

$$\lambda_1 = 5, \quad \mathbf{x}_1 = \begin{bmatrix} 1 \\ 2 \end{bmatrix} \quad \text{and} \quad \lambda_2 = 0, \quad \mathbf{x}_2 = \begin{bmatrix} -2 \\ 1 \end{bmatrix},$$

so the orthogonal diagonalizing matrix is

$$\mathbf{Q} = \begin{bmatrix} 1/\sqrt{5} & -2/\sqrt{5} \\ 2/\sqrt{5} & 1/\sqrt{5} \end{bmatrix} \quad \text{and} \quad \mathbf{D} = \begin{bmatrix} 5 & 0 \\ 0 & 0 \end{bmatrix}.$$

Making the variable change $\mathbf{x} = \mathbf{Q}\mathbf{y}$ shows the standard form of the quadratic terms to be $5y_1^2$. The variables x_1 and x_2 are related to y_1 and y_2 by the expressions $x_1 = y_1/\sqrt{5} - 2y_2/\sqrt{5}$ and $x_2 = 2y_1/\sqrt{5} + y_2/\sqrt{5}$, so $x_1 - 2x_2 = -(3y_1 + 4y_2)/\sqrt{5}$. In terms of the principal axes involving the coordinates y_1 and y_2, the complete expression $x_1^2 + 4x_1x_2 + 4x_2^2 + x_1 - 2x_2$ reduces to

$$x_1^2 + 4x_1x_2 + 4x_2^2 + x_1 - 2x_2 = 5y_1^2 - (3y_1 + 4y_2)/\sqrt{5}. \qquad \blacksquare$$

Quadratic forms $P(\mathbf{x})$ are classified according to the behavior of the sign of $P(\mathbf{x})$ when $\mathbf{x}$ is allowed to take all possible values. In terms of vector spaces, this amounts to saying that if the vector $\mathbf{x}$ in $P(\mathbf{x})$ is an n vector, then $\mathbf{x} \in R^n$.

**how to classify
quadratic forms**

Classification of quadratic forms

Let $P(\mathbf{x})$ be a quadratic form. Then:

1. $P(\mathbf{x})$ is said to be **positive definite** if $P(\mathbf{x}) > 0$ for all $\mathbf{x} \neq \mathbf{0}$ in R^n, with $P(\mathbf{x}) = 0$ if, and only if, $\mathbf{x} = \mathbf{0}$. $P(\mathbf{x})$ is said to be **negative definite** if in this definition the inequality sign $>$ is replaced by $<$.
2. $P(\mathbf{x})$ is said to be **positive semidefinite** if $P(\mathbf{x}) \geq 0$ for all $\mathbf{x} \neq \mathbf{0}$ in R^n, and to be **negative semidefinite** if in this definition the inequality sign $\geq$ is replaced by $\leq$.
3. $P(\mathbf{x})$ is said to be **indefinite** if it satisfies none of the above conditions.

It is an immediate consequence of Theorem 4.10 that if $P(\mathbf{x})$ is associated with a real symmetric matrix $\mathbf{A}$, then:

(a) $P(\mathbf{x})$ is positive definite if all the eigenvalues of $\mathbf{A}$ are positive, and it is negative definite if all the eigenvalues of $\mathbf{A}$ are negative.

(b) $P(\mathbf{x})$ is positive semidefinite if all the eigenvalues of $\mathbf{A}$ are nonnegative, and it is negative semidefinite if all the eigenvalues of $\mathbf{A}$ are nonpositive. So, in each semidefinite case, one or more of the eigenvalues may be zero.

(c) $P(\mathbf{x})$ is indefinite if at least one eigenvalue is opposite in sign to the others. In this case, depending on the choice of $\mathbf{x}$, $P(\mathbf{x})$ may be either positive or negative.

EXAMPLE 4.21 The following are examples of different types of standard forms associated with a 3×3 matrix:

$x_1^2 + 2x_2^2 + 5x_3^2$ is positive definite;

$-(2x_1^2 + 7x_2^2 + 4x_3^2)$ is negative definite;

$4x_1^2 + 3x_3^2$ is positive semidefinite (it is positive, but irrespective of the value of $x_2 \neq 0$ it can vanish when $\mathbf{x} \neq \mathbf{0}$);

$-(2x_1^2 + x_3^2)$ is negative semidefinite (it is negative, but irrespective of the value of $x_2 \neq 0$ it can vanish when $\mathbf{x} \neq \mathbf{0}$);

$3x_1^2 - 2x_2^2 + x_3^2$ is indefinite (it can be positive or negative). ∎

Further, and more detailed, information relating to the material in Sections 4.1 to 4.4 is to be found in the appropriate chapters of references [2.1] and [2.5] to [2.12].

Summary

A real quadratic form involving the n real variables $x_1, x_2, \ldots, x_n$ is a homogeneous polynomial of degree two in these variables. Such forms arise in many different ways, one of which occurs in optimization problems where a reduction to a sum of squares simplifies the task of finding an optimum least squares solution. In this section it was shown that a real quadratic form arises when studying the mechanics of solid bodies, since there a set of principal axes $O\{x_1, x_2, x_3\}$ is used to simplify the description of the body in terms of its inertia about each of the three axes. The reduction of a quadratic form to a sum of squares both simplifies the analysis of its properties and also enables it to be classified as being positive or negative definite, semipositive or seminegative, or of indefinite type, all of which classifications have important implications in applications.

EXERCISES 4.4

In Exercises 1 through 6 find the symmetric matrix $\mathbf{A}$ that is associated with the given quadratic form.

1. $x_1^2 + 4x_1x_3 - 6x_2x_3 + 3x_2^2 - 2x_3^2$.

2. $5x_1^2 - 2x_2^2 - 5x_3^2 - 4x_2x_3$.

3. $-2x_1^2 + 3x_2^2 - 2x_1x_3 + 4x_2x_3$.

4. $x_1^2 + 3x_2^2 - 2x_1x_2 + 4x_2x_4 - 2x_3x_4 + x_3^2 + 6x_4^2$.

5. $3x_1^2 - 4x_1x_2 - 6x_2x_3 - 2x_2x_4 + 2x_3^2 + 8x_4^2$.

6. $x_1^2 + x_2^2 + 4x_3^2 - 3x_4^2 - x_1x_2 + 2x_2x_4 + 2x_3x_4$.

In Exercises 7 through 10 write down the quadratic form associated with the given matrix.

7. $\begin{bmatrix} 2 & 4 & 4 & 0 \\ 4 & 1 & 2 & 1 \\ 4 & 2 & -1 & 2 \\ 0 & 1 & 2 & 3 \end{bmatrix}$.

8. $\begin{bmatrix} 1 & -3 & 2 & 1 \\ -3 & 2 & 0 & 2 \\ 2 & 0 & -3 & 0 \\ 1 & 2 & 0 & 4 \end{bmatrix}$.

9. $\begin{bmatrix} 0 & 2 & -4 & 2 \\ 2 & 3 & 1 & 0 \\ -4 & 1 & 2 & 1 \\ 2 & 0 & 1 & 7 \end{bmatrix}$.

10. $\begin{bmatrix} 1 & -2 & 4 & 3 \\ -2 & 3 & 1 & 2 \\ 4 & 1 & 5 & 0 \\ 3 & 2 & 0 & 3 \end{bmatrix}$.

In Exercises 11 through 18 use hand computation to reduce the quadratic form to its standard form, and use the reduction to classify it. Confirm the reduction by using computer algebra.

11. $(5/2)x_1^2 + x_1x_3 + x_2^2 + (5/2)x_3^2$.

12. $4x_1^2 + x_2^2 + 2x_2x_3 + x_3^2$.

13. $4x_1^2 + 4x_2^2 + 2x_2x_3 + 4x_3^2$.

14. $(3/2)x_1^2 - x_1x_3 + x_2^2 + (3/2)x_3^2$.

15. $(3/2)x_1^2 + x_1x_3 - x_2^2 + (3/2)x_3^2$.

16. $(1/2)x_1^2 + x_1x_3 + 2x_2^2 + (1/2)x_3^2$.

17. $2x_1^2 + x_2^2 - 4x_2x_3 + x_3^2$.

18. $2x_1^2 + 2x_2^2 + 2x_2x_3 + 2x_3^2$.

In Exercises 19 through 24 use computer algebra to reduce the quadratic form on the left to its standard form. Use the result to identify the conic section described by the equation as a circle, an ellipse, or a hyperbola.

19. $3x_1^2 - 6x_1x_2 + 9x_2^2 = 3$.

20. $8x_2^2 - x_1^2 + 20x_1x_2 = 12$.

21. $5x_1^2 + 4x_1x_2 - 10x_2^2 = 1$.
22. $10x_1^2 + 2x_1x_2 + 5x_2^2 = 4$.
23. $13x_1^2 + 18x_1x_2 + 10x_2^2 = 9$.
24. $2x_1^2 + 16x_1x_2 + 5x_2^2 = 4$.

In Exercises 25 through 29 use hand computation to reduce the quadratic part of the expression to its standard form involving the principal axes $O\{y_1, y_2\}$, and find the form taken by the complete expression in terms of y_1 and y_2. Confirm the reduction by using computer algebra.

25. $x_1^2 + 8x_1x_2 + x_2^2 + 3x_1 - 2x_2$.
26. $x_1^2 - 8x_1x_2 + x_2^2 + 2x_1 + 3x_2$.
27. $-2x_1^2 + 4x_1x_2 + x_2^2 + 4x_1 - x_2$.
28. $(8/5)x_1^2 - (8/5)x_1x_2 + (2/5)x_2^2 + 2x_1 + 4x_2$.
29. $(35/17)x_1^2 + (8/17)x_1x_2 + (50/17)x_2^2 + 4x_2$.

30. By using the definitions of a symmetric and a skew-symmetric matrix, generalize the definition of a quadratic form by proving that the quadratic form associated with any real $n \times n$ matrix $\mathbf{A}$ can be written $\mathbf{x}^T\mathbf{B}\mathbf{x}$, where $\mathbf{B}$ is the symmetric part of $\mathbf{A}$.

4.5   The Matrix Exponential

It is shown in Chapter 6 that the matrix exponential can be used when solving systems of linear first order differential equations. As this approach uses matrix diagonalization when determining what is called the *matrix exponential* involving an arbitrary $n \times n$ diagonalizable matrix, it is convenient to introduce the matrix exponential in this chapter.

To motivate what is to follow, we notice that the first order homogeneous linear differential equation

$$dx/dt = ax \quad (a = \text{constant}) \tag{25}$$

has the general solution

$$x = ce^{at} \tag{26}$$

where c is an arbitrary constant.

Let us now consider the system of n linear first order homogeneous differential equations

$$
\begin{aligned}
dx_1/dt &= a_{11}x_1 + a_{12}x_2 + \cdots a_{1n}x_n \\
dx_2/dt &= a_{21}x_1 + a_{22}x_2 + \cdots a_{2n}x_n \\
& \cdot \quad \cdot \quad \cdot \quad \cdot \quad \cdot \quad \cdot \\
dx_n/dt &= a_{n1}x_1 + a_{n2}x_2 + \cdots a_{nn}x_n
\end{aligned}
\tag{27}
$$

Setting

$$
\mathbf{x} = \begin{bmatrix} x_1 \\ x_2 \\ \cdot \\ \cdot \\ x_n \end{bmatrix} \quad \text{and} \quad
\mathbf{A} = \begin{bmatrix} a_{11} & a_{12} & \cdot & \cdot & \cdot & \cdot & a_{1n} \\ a_{21} & a_{22} & \cdot & \cdot & \cdot & \cdot & a_{2n} \\ \cdot & \cdot & \cdot & \cdot & \cdot & \cdot & \cdot \\ \cdot & \cdot & \cdot & \cdot & \cdot & \cdot & \cdot \\ a_{n1} & a_{n2} & \cdot & \cdot & \cdot & \cdot & a_{nn} \end{bmatrix}
$$

allows the system of differential equations in (27) to be written in the matrix form

$$dx/dt = \mathbf{A}\mathbf{x}, \tag{28}$$

where $d\mathbf{x}/dt = [dx_1/dt, dx_2/dt, \ldots, dx_n/dt]^T$ (see Section 3.2(d)).

As the single differential equation (25) has the solution (26), it is reasonable to ask whether it is possible to express the solution of the system of differential equations in (28) in the form

$$\mathbf{x} = e^{\mathbf{A}t}\mathbf{C}. \tag{29}$$

the matrix exponential

For this to be possible it is necessary to give meaning to the expression $e^{\mathbf{A}t}$, which is called the **matrix exponential**, with t as a parameter. Our objective in the remainder of this section will be to give a brief introduction to the matrix exponential and to use the definition to determine its most important properties in preparation for their use in Chapter 6.

The starting point for this generalization of the exponential function is the familiar result

$$e^{at} = \sum_{m=0}^{\infty} \frac{a^m t^m}{m!}$$

$$= 1 + at + \frac{a^2 t^2}{2!} + \frac{a^3 t^3}{3!} + \cdots. \tag{30}$$

If $\mathbf{A}$ is an $n \times n$ constant matrix with real coefficients we take as an intuitive definition of the matrix exponential e^{At} the infinite series of matrices

$$e^{\mathbf{A}t} = \mathbf{I} + \mathbf{A}t + \mathbf{A}^2 \frac{t^2}{2!} + \mathbf{A}^3 \frac{t^3}{3!} + \cdots. \tag{31}$$

In adopting (31) as a possible definition of the matrix exponential, we have set $\mathbf{A}^0 = \mathbf{I}$ and chosen to vary the convention that a scalar multiplier of a matrix is placed in front of the matrix by writing $\mathbf{A}t$, $\mathbf{A}^2 t^2$, ..., instead of $t\mathbf{A}$, $t^2\mathbf{A}^2$, ... This notation has been adopted to make the appearance of the arguments that follow parallel as closely as possible those for the familiar single real variable case. Some books adopt this convention but make no mention of it, while others adhere strictly to the convention that a scalar multiplier is placed before a matrix and write

$$e^{t\mathbf{A}} = \mathbf{I} + t\mathbf{A} + \frac{t^2}{2!}\mathbf{A}^2 + \frac{t^3}{3!}\mathbf{A}^3 + \cdots.$$

The matrix exponential in (31) is an $n \times n$ matrix, each element of which is an ordinary infinite series. So to show that $e^{\mathbf{A}t}$ is convergent, it will be sufficient to show that an infinite sum of the required form containing the term of greatest absolute value in $\mathbf{A}$ is convergent. Let us consider the matrix product $\mathbf{A}^2$. Then the term $c_{rs}^{(2)}$ in the rth row and sth column of $\mathbf{A}^2$ is $c_{rs}^{(2)} = a_{r1}a_{1s} + a_{r2}a_{2s} + \cdots + a_{rn}a_{ns}$, so if the magnitude of the largest term in $\mathbf{A}$ is M, it follows that $|a_{rs}| \leq M$, and $|c_{rs}^{(2)}| \leq nM^2$. A similar argument shows that if $|c_{rs}^{(3)}|$ is the corresponding term in the matrix $\mathbf{A}^3$, then $c_{rs}^{(3)} = c_{r1}^{(2)}a_{1s} + c_{r2}^{(2)}a_{2s} + \cdots + c_{rn}^{(2)}a_{ns}$ and so $|c_{rs}^{(3)}| \leq n^2 M^3$. Either by induction or by inspection, we see that the magnitude of the term $c_{rs}^{(m)}$ in the rth row and sth column of $\mathbf{A}^m$ obeys the inequality $|c_{rs}^{(m)}| \leq n^{m-1} M^m$.

An overestimate of the magnitude of the term in the rth row and sth column of $e^{\mathbf{A}t}$ is provided by the series

$$1 + tM + t^2 n M^2/2! + t^3 n^2 M^3/3! + \cdots + t^m n^{m-1} M^m/m! + \cdots.$$

Setting $u_m = t^m n^{m-1} M^m / m!$ and applying the ratio test shows that for all fixed t

$$L = \lim_{m \to \infty} |u_{m+1}/u_m| = \lim_{m \to \infty} tnM/(m+1) = 0,$$

so the series is absolutely convergent for all fixed t. Thus, (26) serves as a satisfactory definition of the matrix exponential, and because it is absolutely convergent for all fixed t the series can be differentiated and integrated term by term with respect to t.

The matrix exponential

If $\mathbf{A}$ is an $n \times n$ constant matrix with real coefficients, the **matrix exponential** $e^{\mathbf{A}t}$ is defined by the infinite series

the formal
definition of $e^{\mathbf{A}t}$
and its properties

$$e^{\mathbf{A}t} = \mathbf{I}_n + \mathbf{A}t + \mathbf{A}^2 \frac{t^2}{2!} + \mathbf{A}^3 \frac{t^3}{3!} + \cdots, \qquad (32)$$

which is absolutely convergent for all fixed t.

The absolute convergence of the infinite series defining the matrix exponential allows it to be differentiated term by term, so

$$d[e^{\mathbf{A}t}]/dt = \mathbf{A} + \mathbf{A}^2 t + \mathbf{A}^3 \frac{t^2}{2!} + \cdots = \mathbf{A}\left\{ \mathbf{I} + \mathbf{A}t + \mathbf{A}^2 \frac{t^2}{2!} + \mathbf{A}^3 \frac{t^3}{3!} + \cdots \right\}$$

$$= \mathbf{A}e^{\mathbf{A}t}.$$

We have established the fundamental result that

$$d[e^{\mathbf{A}t}]/dt = \mathbf{A}e^{\mathbf{A}t}, \qquad (33)$$

and hence by repeated differentiation that

$$d^n[e^{\mathbf{A}t}]/dt^n = \mathbf{A}^n e^{\mathbf{A}t}. \qquad (34)$$

Setting $t = 1$ in (33) shows that

$$e^{\mathbf{A}} = \mathbf{I} + \mathbf{A} + \mathbf{A}^2 \frac{1}{2!} + \mathbf{A}^3 \frac{1}{3!} + \cdots, \qquad (35)$$

whereas setting $t = 0$ shows that $e^{\mathbf{0}} = \mathbf{I}$.

EXAMPLE 4.22   Find $e^{\mathbf{A}t}$ given that

$$\mathbf{A} = \begin{bmatrix} 3 & 0 & 0 \\ 0 & -2 & 0 \\ 0 & 0 & 4 \end{bmatrix}.$$

Solution   As $\mathbf{A}$ is a diagonal matrix

$$\mathbf{A}^m = \begin{bmatrix} 3^m & 0 & 0 \\ 0 & (-2)^m & 0 \\ 0 & 0 & 4^m \end{bmatrix},$$

so substituting into (32) gives

$$e^{\mathbf{A}t} = \begin{bmatrix} 1 & 0 & 0 \\ 0 & 1 & 0 \\ 0 & 0 & 1 \end{bmatrix} + \begin{bmatrix} 3 & 0 & 0 \\ 0 & -2 & 0 \\ 0 & 0 & 4 \end{bmatrix} t + \begin{bmatrix} 3^2 & 0 & 0 \\ 0 & (-2)^2 & 0 \\ 0 & 0 & 4^2 \end{bmatrix} \frac{t^2}{2!} + \cdots ,$$

showing that

$$e^{\mathbf{A}t} = \begin{bmatrix} \displaystyle\sum_{m=0}^{\infty} \frac{3^m t^m}{m!} & 0 & 0 \\ 0 & \displaystyle\sum_{m=0}^{\infty} \frac{(-2)^m t^m}{m!} & 0 \\ 0 & 0 & \displaystyle\sum_{m=0}^{\infty} \frac{4^m t^m}{m!} \end{bmatrix} = \begin{bmatrix} e^{3t} & 0 & 0 \\ 0 & e^{-2t} & 0 \\ 0 & 0 & e^{4t} \end{bmatrix}. \qquad \blacksquare$$

EXAMPLE 4.23   Find $e^{\mathbf{A}}$ and $e^{\mathbf{A}t}$, and show by direct differentiation that $d[e^{\mathbf{A}t}]/dt = \mathbf{A}e^{\mathbf{A}t}$, given that

$$\mathbf{A} = \begin{bmatrix} 0 & 2 & 1 & 1 \\ 0 & 0 & 3 & -2 \\ 0 & 0 & 0 & 1 \\ 0 & 0 & 0 & 0 \end{bmatrix}.$$

Solution

$$\mathbf{A}^2 = \begin{bmatrix} 0 & 0 & 6 & -3 \\ 0 & 0 & 0 & 3 \\ 0 & 0 & 0 & 0 \\ 0 & 0 & 0 & 0 \end{bmatrix}, \quad \mathbf{A}^3 = \begin{bmatrix} 0 & 0 & 0 & 6 \\ 0 & 0 & 0 & 0 \\ 0 & 0 & 0 & 0 \\ 0 & 0 & 0 & 0 \end{bmatrix}, \quad \text{and} \quad \mathbf{A}^n = \mathbf{0} \text{ for } n > 3.$$

Substituting into (32) and adding the scaled matrices gives

$$e^{\mathbf{A}t} = \begin{bmatrix} 1 & 2t & t+3t^2 & t-(3/2)t^2+t^3 \\ 0 & 1 & 3t & -2t+(3/2)t^2 \\ 0 & 0 & 1 & t \\ 0 & 0 & 0 & 1 \end{bmatrix}.$$

Setting $t = 1$ in this result, we find that

$$e^{\mathbf{A}} = \begin{bmatrix} 1 & 2 & 4 & 1/2 \\ 0 & 1 & 3 & -1/2 \\ 0 & 0 & 1 & 1 \\ 0 & 0 & 0 & 1 \end{bmatrix}.$$

Differentiation of the terms in the matrix $e^{\mathbf{A}t}$ gives

$$d[e^{\mathbf{A}t}]/dt = \begin{bmatrix} 0 & 2 & 1+6t & 1-3t+3t^2 \\ 0 & 0 & 3 & -2+3t \\ 0 & 0 & 0 & 1 \\ 0 & 0 & 0 & 0 \end{bmatrix},$$

and as this is equal to $\mathbf{A}e^{\mathbf{A}t}$, it confirms the result $d[e^{\mathbf{A}t}]/dt = \mathbf{A}e^{\mathbf{A}t}$. $\qquad \blacksquare$

It was possible to sum the infinite series of matrices in Example 4.22 because only a diagonal matrix was involved, so its powers could be determined immediately. The situation was different in Example 4.23 because $\mathbf{A}^n = \mathbf{0}$ for $n > 3$ so that only a finite sum of matrices was involved. Matrices such as those in Example 4.23, which vanish when raised to a finite power, are called **nilpotent** matrices.

If $\mathbf{A}$ is neither diagonal nor nilpotent, but is diagonalizable, in order to determine $\mathbf{A}^m$ it is first necessary to find the diagonalizing matrix $\mathbf{P}$ for $\mathbf{A}$. Then, if $\mathbf{D}$ is the diagonalized form of $\mathbf{A}$, so that $\mathbf{D} = \mathbf{P}^{-1}\mathbf{A}\mathbf{P}$, it follows that $\mathbf{A} = \mathbf{P}\mathbf{D}\mathbf{P}^{-1}$ and

$$\mathbf{A}^2 = (\mathbf{P}\mathbf{D}\mathbf{P}^{-1})(\mathbf{P}\mathbf{D}\mathbf{P}^{-1}) = \mathbf{P}\mathbf{D}^2\mathbf{P}^{-1}, \quad \mathbf{A}^3 = \mathbf{A}\mathbf{A}^2 = (\mathbf{P}\mathbf{D}\mathbf{P}^{-1})(\mathbf{P}\mathbf{D}^2\mathbf{P}^{-1})$$
$$= \mathbf{P}\mathbf{D}^3\mathbf{P}^{-1},$$

so that in general,

$$\mathbf{A}^m = \mathbf{P}\mathbf{D}^m\mathbf{P}^{-1}.$$

Using this result in the matrix exponential gives

$$e^{\mathbf{A}t} = \mathbf{I} + (\mathbf{P}\mathbf{D}\mathbf{P}^{-1})t + \mathbf{P}\mathbf{D}^2\mathbf{P}^{-1}\frac{t^2}{2!} + \cdots,$$

and writing $\mathbf{I} = \mathbf{P}\mathbf{P}^{-1}$ reduces this to

$$e^{\mathbf{A}t} = \mathbf{P}\left\{\mathbf{I}_n + \mathbf{D}t + \mathbf{D}^2\frac{t^2}{2!} + \mathbf{D}^3\frac{t^3}{3!} + \cdots\right\}\mathbf{P}^{-1}. \tag{36}$$

The form of $e^{\mathbf{A}}$ follows directly from this by setting $t = 1$.

EXAMPLE 4.24   Determine $e^{\mathbf{A}t}$ given that

$$\mathbf{A} = \begin{bmatrix} -2 & -3 \\ 6 & 7 \end{bmatrix},$$

and use the result to find $e^{\mathbf{A}}$.

Solution   The eigenvalues and eigenvectors of $\mathbf{A}$ are

$$\lambda_1 = 1, \quad \mathbf{x}_1 = \begin{bmatrix} -1 \\ 1 \end{bmatrix} \quad \text{and} \quad \lambda_2 = 4, \quad \mathbf{x}_2 = \begin{bmatrix} 1 \\ -2 \end{bmatrix},$$

so the diagonalizing matrix

$$\mathbf{P} = \begin{bmatrix} -1 & 1 \\ 1 & -2 \end{bmatrix} \quad \text{and} \quad \mathbf{P}^{-1} = \begin{bmatrix} -2 & -1 \\ -1 & -1 \end{bmatrix}, \quad \text{while } \mathbf{D} = \begin{bmatrix} 1 & 0 \\ 0 & 4 \end{bmatrix}.$$

Substituting these matrices into (36) gives

$$e^{\mathbf{A}t} = \mathbf{P}\left[\begin{bmatrix} 1 & 0 \\ 0 & 1 \end{bmatrix} + \begin{bmatrix} 1 & 0 \\ 0 & 4 \end{bmatrix}t + \begin{bmatrix} 1 & 0 \\ 0 & 4^2 \end{bmatrix}\frac{t^2}{2!} + \begin{bmatrix} 1 & 0 \\ 0 & 4^3 \end{bmatrix}\frac{t^3}{3!} + \cdots\right]\mathbf{P}^{-1}$$

$$= \mathbf{P}\begin{bmatrix} e^t & 0 \\ 0 & e^{4t} \end{bmatrix}\mathbf{P}^{-1} = \begin{bmatrix} (2e^t - e^{4t}) & (e^t - e^{4t}) \\ (2e^{4t} - 2e^t) & (2e^{4t} - e^t) \end{bmatrix}.$$

Finally, setting $t = 1$ we find that

$$e^{\mathbf{A}} = \begin{bmatrix} (2e - e^4) & (e - e^4) \\ (2e^4 - 2e) & (2e^4 - e) \end{bmatrix}.$$ ∎

So far, the properties of the matrix exponential have closely paralleled those of the ordinary exponential, but there are significant differences, one of the most important being that in general, even when $\mathbf{A} + \mathbf{B}$ is defined, $e^{\mathbf{A}}e^{\mathbf{B}} \neq e^{(\mathbf{A}+\mathbf{B})}$. To determine under what conditions the equality is true, we consider the matrix exponentials $e^{\mathbf{A}t}e^{\mathbf{B}t}$ and $e^{(\mathbf{A}+\mathbf{B})t}$ and require their derivatives to be equal when $t = 0$.

Differentiating each expression once with respect to t gives

$$d[e^{\mathbf{A}t}e^{\mathbf{B}t}]/dt = \mathbf{A}e^{\mathbf{A}t}e^{\mathbf{B}t} + e^{\mathbf{A}t}\mathbf{B}e^{\mathbf{B}t} \quad \text{and} \quad d[e^{(\mathbf{A}+\mathbf{B})t}]/dt = (\mathbf{A}+\mathbf{B})e^{(\mathbf{A}+\mathbf{B})t},$$

and these are seen to be equal when $t = 0$. Next, computing $d^2[e^{\mathbf{A}t}e^{\mathbf{B}t}]/dt^2$ and $d^2[e^{(\mathbf{A}+\mathbf{B})t}]/dt^2$, we obtain

$$d^2[e^{\mathbf{A}t}e^{\mathbf{B}t}]/dt^2 = \mathbf{A}^2 e^{\mathbf{A}t}e^{\mathbf{B}t} + 2\mathbf{A}e^{\mathbf{A}t}\mathbf{B}e^{\mathbf{B}t} + e^{\mathbf{A}t}\mathbf{B}^2 e^{\mathbf{B}t}$$

and

$$d^2[e^{(\mathbf{A}+\mathbf{B})t}]/dt^2 = (\mathbf{A}+\mathbf{B})^2 e^{(\mathbf{A}+\mathbf{B})t} = (\mathbf{A}^2 + \mathbf{AB} + \mathbf{BA} + \mathbf{B}^2)e^{(\mathbf{A}+\mathbf{B})t}.$$

Setting $t = 0$ shows that these two expressions are only equal if $\mathbf{AB} = \mathbf{BA}$; that is, the matrices $\mathbf{A}$ and $\mathbf{B}$ must *commute*, and the same condition applies when all higher order derivatives are considered. This has established the fundamental result that

when does
$e^{\mathbf{A}}e^{\mathbf{B}} = e^{(\mathbf{A}+\mathbf{B})}$

$$e^{\mathbf{A}}e^{\mathbf{B}} = e^{(\mathbf{A}+\mathbf{B})} \quad \text{if, and only if, } \mathbf{AB} = \mathbf{BA}. \tag{37}$$

Replacing $\mathbf{B}$ by $-\mathbf{A}$ in (37) gives

$$e^{\mathbf{A}}e^{-\mathbf{A}} = e^{\mathbf{0}} = \mathbf{I}, \tag{38}$$

from which we see, as would be expected, that $e^{-\mathbf{A}}$ is the inverse of $e^{\mathbf{A}}$, and also that as $e^{-\mathbf{A}}$ is nonsingular it always exists. This parallels the real variable situation, because e^{-x} exists for all finite x.

Having arrived at a satisfactory definition of $e^{\mathbf{A}t}$ and determined its derivatives, we are now in a position to define the **antiderivative** $\int e^{\mathbf{A}t}\, dt$ as the matrix obtained by integrating each element of $e^{\mathbf{A}t}$ with respect to t, it being understood that when this is done an arbitrary constant $n \times n$ matrix must always be added to the result representing the arbitrary additive constant of integration that arises when each term of $e^{\mathbf{A}t}$ is integrated.

EXAMPLE 4.25   Find $\int e^{\mathbf{A}t}\, dt$ given that $\mathbf{A}$ is the matrix in Example 4.21.

Solution   It was shown in Example 4.21 that if

$$\mathbf{A} = \begin{bmatrix} 3 & 0 & 0 \\ 0 & -2 & 0 \\ 0 & 0 & 4 \end{bmatrix} \text{ then } e^{\mathbf{A}t} = \begin{bmatrix} e^{3t} & 0 & 0 \\ 0 & e^{-2t} & 0 \\ 0 & 0 & e^{4t} \end{bmatrix},$$

so that

$$\int e^{\mathbf{A}t}\, dt = \begin{bmatrix} e^{3t}/3 + c_1 & 0 & 0 \\ 0 & -e^{-2t}/2 + c_2 & 0 \\ 0 & 0 & e^{4t}/4 + c_3 \end{bmatrix}$$

$$= \begin{bmatrix} e^{3t}/3 & 0 & 0 \\ 0 & -e^{-2t}/2 & 0 \\ 0 & 0 & e^{4t}/4 \end{bmatrix} + \begin{bmatrix} c_1 & 0 & 0 \\ 0 & c_2 & 0 \\ 0 & 0 & c_3 \end{bmatrix},$$

where c_1, c_2, and c_3 are arbitrary constants. ∎

Applications of the matrix exponential to ordinary differential equations are to be found in reference [3.15].

Summary

The matrix exponential $e^{\mathbf{A}t}$ arises as the natural extension of the exponential function when solving a system of linear first order constant coefficient differential equations in the matrix form $d\mathbf{x}/dt = \mathbf{Ax}$. This section has described how $e^{\mathbf{A}t}$ can be calculated in simple cases and shown that $e^{\mathbf{A}}e^{\mathbf{B}} = e^{\mathbf{A}+\mathbf{B}}$ if, and only if, $\mathbf{AB} = \mathbf{BA}$. A different way of finding $e^{\mathbf{A}t}$ using the Laplace transform is given later in Section 7.3(b).

EXERCISES 4.5

1. Given that

$$\mathbf{A} = \begin{bmatrix} 0 & 3 & 1 & 0 \\ 0 & 0 & 2 & 1 \\ 0 & 0 & 0 & 3 \\ 0 & 0 & 0 & 0 \end{bmatrix},$$

show that it is nilpotent and find the smallest power for which $\mathbf{A}^n = \mathbf{0}$.

2. Given that

$$\mathbf{A} = \begin{bmatrix} 0 & 1 & 2 & 2 \\ 0 & 0 & 3 & 1 \\ 0 & 0 & 0 & 1 \\ 0 & 0 & 0 & 0 \end{bmatrix},$$

find $e^{\mathbf{A}t}$.

3. Given that

$$\mathbf{A} = \begin{bmatrix} 0 & 2 \\ 0 & 0 \end{bmatrix} \quad \text{and} \quad \mathbf{B} = \begin{bmatrix} 0 & 0 \\ 3 & 0 \end{bmatrix},$$

show that $\mathbf{A}$ and $\mathbf{B}$ do not commute, and by finding $e^{\mathbf{A}t}$, $e^{\mathbf{B}t}$, and $e^{(\mathbf{A}+\mathbf{B})t}$, verify that $e^{\mathbf{A}t}e^{\mathbf{B}t} \neq e^{(\mathbf{A}+\mathbf{B})t}$.

In Exercises 4 through 9, find $e^{\mathbf{A}t}$.

4. $\mathbf{A} = \begin{bmatrix} 0 & 1 \\ 1 & 0 \end{bmatrix}$.

5. $\mathbf{A} = \begin{bmatrix} m & 0 \\ 0 & n \end{bmatrix}$.

6. $\mathbf{A} = \begin{bmatrix} 0 & -c \\ c & 0 \end{bmatrix}$.

7. $\mathbf{A} = \begin{bmatrix} -2 & 2 \\ 2 & 1 \end{bmatrix}$.

8. $\mathbf{A} = \begin{bmatrix} 3 & -2 & 2 \\ 6 & -4 & 6 \\ 2 & -1 & 3 \end{bmatrix}$.

9. $\mathbf{A} = \begin{bmatrix} 0 & 1 & -2 \\ 2 & -1 & 2 \\ 2 & -2 & 4 \end{bmatrix}$.

10. By considering the definition of $e^{\mathbf{A}t}$ show, provided the square matrices $\mathbf{A}$ and $\mathbf{B}$ commute, that

$$\mathbf{A}e^{\mathbf{B}t} = e^{\mathbf{B}t}\mathbf{A}.$$

11. By considering the definition of $e^{\mathbf{A}t}$ show that $\int e^{-\mathbf{A}t}\, dt = -\mathbf{A}^{-1}e^{-\mathbf{A}t} + \mathbf{C} = e^{-\mathbf{A}t}\mathbf{A}^{-1} + \mathbf{C}$, where $\mathbf{C}$ is an arbitrary constant matrix that is conformable for addition with $\mathbf{A}$.

12. Show that if the square matrices $\mathbf{A}$ and $\mathbf{B}$ commute, then the binomial theorem takes the form

$$(\mathbf{A} + \mathbf{B})^n = \sum_{k=0}^{n} \binom{n}{k} \mathbf{A}^k \mathbf{B}^{n-k}.$$

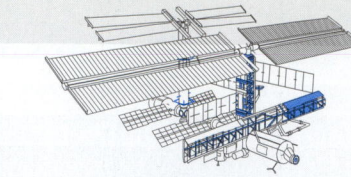

CHAPTER 4
TECHNOLOGY PROJECTS

Project 1

Verifying and Using the Cayley–Hamilton Theorem

The purpose of this project is to verify the Cayley–Hamilton theorem in a particular case by constructing an arbitrary 6×6 non-singular matrix $\mathbf{A}$ and, after finding its characteristic polynomial, showing by direct calculation that $\mathbf{A}$ satisfies its own characteristic matrix polynomial equation. The matrix polynomial equation is then to be used to compute the inverse matrix $\mathbf{A}^{-1}$, after which the inverse is to be checked by showing that the product $\mathbf{A}\mathbf{A}^{-1} = \mathbf{I}$. The project then explores the way in which this approach fails when $\mathbf{A}$ is singular.

1. Construct an arbitrary 6×6 matrix $\mathbf{A}$ and check that $\det \mathbf{A} \neq 0$ to ensure that it has an inverse $\mathbf{A}^{-1}$.
2. Find the characteristic polynomial for matrix $\mathbf{A}$.
3. Show by direct calculation that $\mathbf{A}$ satisfies its own characteristic matrix polynomial equation.
4. Use the characteristic matrix polynomial equation to find $\mathbf{A}^{-1}$, and check its correctness by showing that the product $\mathbf{A}\mathbf{A}^{-1} = \mathbf{I}$.
5. Replace the last row of $\mathbf{A}$ by the entries in the row above to form a matrix $\mathbf{B}$ that is singular, and find the characteristic polynomial for $\mathbf{B}$.
6. Try to use the characteristic matrix polynomial equation for $\mathbf{B}$ to find $\mathbf{B}^{-1}$, and comment on the way in which this approach fails.

Project 2

Diagonalization of a Matrix

This project involves the diagonalization of a 5×5 matrix $\mathbf{A}$ when two of its five eigenvalues are equal, but there are five linearly independent eigenvectors.

1. Find a diagonalizing matrix for

$$\mathbf{A} = \begin{bmatrix} 13 & 31 & 30 & 51 & -40 \\ 32 & 62 & 64 & 104 & -88 \\ -28 & -56 & -58 & -88 & 80 \\ -17 & -33 & -34 & -55 & 48 \\ -13 & -25 & -26 & -37 & 38 \end{bmatrix}.$$

2. Diagonalize the matrix $\mathbf{B} = \frac{1}{2}\mathbf{A}$, and comment on the relationship between the diagonalizing matrices for $\mathbf{A}$ and $\mathbf{B}$.

Project 3

Orthogonal Vectors Computed by the Gram–Schmidt Method

The purpose of this project is to develop a computer algebra procedure that generalizes the **Gram–Schmidt** process to n-dimensional vectors. The extension is almost immediate and follows from the fact that in the case of three-dimensional vectors one of them, say $\mathbf{a}_1$, was taken as the first vector $\mathbf{u}_1$ of an orthogonal basis, the second vector $\mathbf{u}_2$ was derived from $\mathbf{a}_2$ by subtracting from it the projection of $\mathbf{u}_1$ onto $\mathbf{a}_2$, and, finally, the third vector $\mathbf{u}_3$ was obtained from $\mathbf{a}_3$ by subtracting from it both the projection of $\mathbf{u}_1$ onto $\mathbf{a}_3$ and the projection of $\mathbf{u}_2$ onto $\mathbf{a}_3$.

Starting with a set of n linearly independent vectors $\{\mathbf{a}_1, \mathbf{a}_2, \ldots, \mathbf{a}_n\}$, an orthogonal basis $\{\mathbf{u}_1, \mathbf{u}_2, \ldots, \mathbf{u}_n\}$ for this space is obtained by extending the preceding method by setting

$$\mathbf{u}_1 = \mathbf{a}_1$$

$$\mathbf{u}_2 = \mathbf{a}_2 - \frac{\mathbf{a}_2 . \mathbf{u}_1}{\mathbf{u}_1 . \mathbf{u}_1}\mathbf{u}_1$$

$$\mathbf{u}_3 = \mathbf{a}_3 - \frac{\mathbf{a}_3 . \mathbf{u}_1}{\mathbf{u}_1 . \mathbf{u}_1}\mathbf{u}_1 - \frac{\mathbf{a}_3 . \mathbf{u}_2}{\mathbf{u}_2 . \mathbf{u}_2}\mathbf{u}_2$$

$$\vdots$$

$$\mathbf{u}_n = \mathbf{a}_n - \frac{\mathbf{a}_n . \mathbf{u}_1}{\mathbf{u}_1 . \mathbf{u}_1}\mathbf{u}_1 - \frac{\mathbf{a}_n . \mathbf{u}_2}{\mathbf{u}_2 . \mathbf{u}_2}\mathbf{u}_2 - \cdots - \frac{\mathbf{a}_n . \mathbf{u}_{n-1}}{\mathbf{u}_{n-1} . \mathbf{u}_{n-1}}\mathbf{u}_{n-1}.$$

Write a computer algebra procedure that reproduces these results step by step for four-dimensional vectors.

Check the procedure by applying it to the set of linearly independent vectors $\mathbf{a}_1 = [-1, -1, 1, 2]$, $\mathbf{a}_2 = [1, 0, 1, -2]$, $\mathbf{a}_3 = [0, 1, -1, -1]$, and $\mathbf{a}_4 = [2, -1, 1, 1]$, and showing that the corresponding set of orthogonal basis vectors is $\mathbf{u}_1 = [-1, -1, 1, 2]$, $\mathbf{u}_2 = [\frac{3}{7}, -\frac{4}{7}, \frac{11}{7}, -\frac{6}{7}]$, $\mathbf{u}_3 = [-\frac{11}{26}, \frac{3}{13}, \frac{3}{26}, -\frac{2}{13}]$, and $\mathbf{u}_4 = [\frac{2}{7}, \frac{4}{7}, \frac{2}{7}, \frac{2}{7}]$.

Define two other sets of linearly independent vectors and, after applying your procedure, verify that the resulting sets of vectors $\{\mathbf{u}_1, \mathbf{u}_2, \mathbf{u}_3, \mathbf{u}_4\}$ are orthogonal.

Project 4

Reduction of a Quadratic Form to Standard Form

The purpose of this project is to find a transformation that reduces a given quadratic form in four variables to a sum of squares.

1. Given the quadratic form $x_2^2 - x_1^2 - 2x_1x_2 - 2x_1x_3 + 2x_1x_4 - 2x_3x_4$, find a transformation that reduces it to a sum of squares.
2. Find the simplified quadratic form produced by the transformation in Step 1.

Project 5

The Hubble Space Telescope and Quadratic Forms

When the Hubble space telescope in orbit around the earth is required to photograph a particular nebula it has to be rotated until it is pointing in the correct direction. As it is a rigid body, the kinetic energy W required to rotate it at an angular velocity ω about a suitable axis is given by $W = \frac{1}{2}I\omega^2$, where I is the moment of inertia of the telescope about the axis of rotation. Because the telescope has an irregular shape, the moment of inertia I will depend on the axis of rotation, and a convenient way of representing the value of I about all possible axes through a given point in the telescope is by means of what is called the *ellipsoid of inertia*.

The ellipsoid of inertia for a given rigid body of mass m relative to a fixed point in the body is a three-dimensional plot of the moment of inertia relative to all possible axes of rotation passing through the point. It is shown in texts on mechanics that this plot is an ellipsoidal surface, with the property that the length of the straight line drawn from the center of the ellipsoid to its surface is inversely proportional to the radius of gyration k of the body about that line, where $I = mk^2$.

Given that an ellipsoid of inertia has the form

$$16x^2 - 4xy + 37y^2 - 12xz + 18yz + 11z^2 = 12,$$

use matrix methods to find a linear transformation from the variables x, y, and z to new variables X, Y, and Z that reduces the expression to one of the form

$$\frac{X^2}{a^2} + \frac{Y^2}{b^2} + \frac{Z^2}{c^2} = 1.$$

Hence find the radii of gyration $1/a$, $1/b$, and $1/c$ about the principal axes of the ellipsoid that form its three mutually orthogonal axes about which there is symmetry.

Project 6

Dynamical Systems and Logging Operations

Discrete dynamical systems are used to model situations in engineering, control theory, physics, ecology, and elsewhere that can be considered to evolve stage by stage, with each stage dependent on the previous one. For example, a logging operation to supply a saw mill in a specific area of forest, with tree replanting and the availability of a limited supply of logs from outside the area, can be described by a simple dynamical system that models the way the output of cut timber is influenced by the competition between the felling of trees, the importing of a limited amount of logs, and the regeneration of the forest.

In the simplest case the long-term behavior of a dynamical system can be represented mathematically by the matrix equation

$$\mathbf{x}_{k+1} = \mathbf{A}\mathbf{x}_k, \quad \text{for } k = 0, 1, 2, \ldots,$$

where $\mathbf{A}$ is an $n \times n$ matrix, and $\mathbf{x}_k$ is an n element column vector whose elements describe the physical characteristics of the system at the kth stage. In a logging operation $n = 2$, and $\mathbf{x}_k = [T_k, R_k]^{\mathrm{T}}$, where T_k is the amount of timber remaining after k years and R_k is the amount of replanted timber that has matured after k years.

In general, let $\mathbf{A}$ be diagonalizable with the real eigenvalues $\lambda_1, \lambda_2, \ldots, \lambda_n$, and let the corresponding linearly independent eigenvectors be $\mathbf{u}_1, \mathbf{u}_2, \ldots, \mathbf{u}_n$. Then, if $\mathbf{x}_0$ describes the initial state of the system, since the eigenvectors form a basis for the system we may set $\mathbf{x}_0 = c_1\mathbf{u}_1 + c_2\mathbf{u}_2 + \ldots + c_n\mathbf{u}_n$. Use the representation of $\mathbf{x}_0$ to find a general expression for $\mathbf{x}_k$ in terms of the eigenvalues, and comment on the approximate form taken by $\mathbf{x}_k$ as k becomes large.

Given that

$$\mathbf{A} = \begin{bmatrix} 0.4 & 0.8 \\ -0.1 & 1.12 \end{bmatrix} \quad \text{and} \quad \mathbf{x}_k = \begin{bmatrix} T_k \\ R_k \end{bmatrix},$$

interpret the meaning of the coefficients of $\mathbf{A}$ in the context of a logging operation. Starting with $\boldsymbol{x}_0 = [1, 0.9]^{\mathrm{T}}$, generate the first 15 vectors $\boldsymbol{x}_k$, com- pare the results with the approximation found ear- lier, and comment on the result in terms of a logging operation.

Suggest a physical dynamical system where $\mathbf{A}$ is a 3×3 matrix. Define a suitable numerical matrix $\mathbf{A}$ and initial vector $\boldsymbol{x}_0$, generate the first 15 vectors $\boldsymbol{x}_k$, and interpret the results in terms of the model.

PART THREE

ORDINARY DIFFERENTIAL EQUATIONS

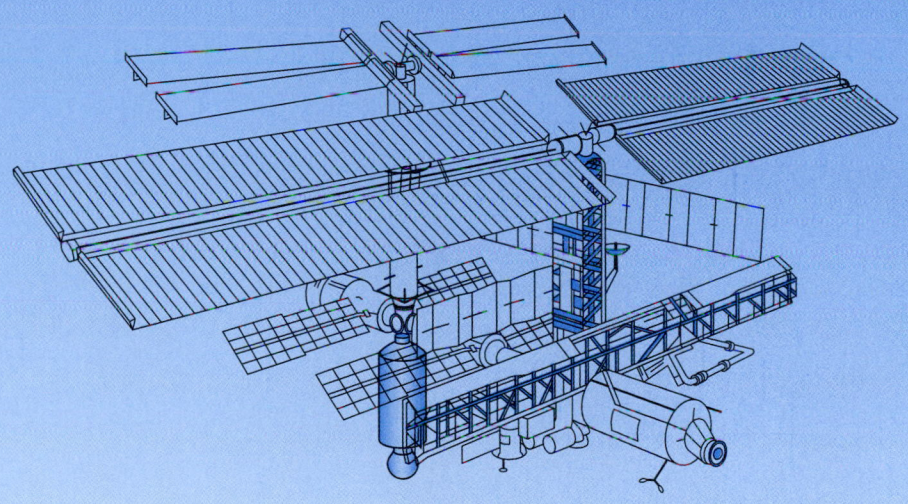

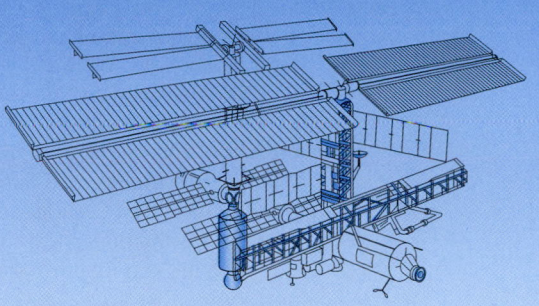

CHAPTER 5

First Order Differential Equations

Differential equations are fundamental to the study of engineering and physics, and this chapter marks the start of our discussion of this important topic. Typically, in an electrical problem, the dependent variable $i(t)$ in an ordinary differential equation might be the current flowing in a circuit at time t, in which case the independent variable would be the time. In all such examples, the nature of $i(t)$ depends on the current flow at the start, and the specification of information of this type is called an **initial condition** for the differential equation. Similarly, in chemical engineering, a dependent variable $m(t)$ might be the amount of a chemical produced by a reaction at time t. Here also the independent variable would be the time t, and to determine $m(t)$ in any particular case it would be necessary to specify the amount of $m(t)$ present at the start, that for convenience is usually taken to be when $t = 0$.

Many physical problems are capable of description in terms of a single first order ordinary differential equation, while other more complicated problems involve coupled first order differential equations, that after the elimination of all but one of the independent variables, can be replaced by a single higher order equation for the remaining dependent variable. This happens, for example, when determining the current in an R-L-C electrical circuit.

Thus first order ordinary differential equations can be considered as the building blocks in the study of higher order equations, and their properties are particularly important and easy to obtain when the equations are linear. The study and properties of the specially simple class of equations called **constant coefficient equations** is very important, as it forms the foundation of the study of higher order constant coefficient equations that will be developed later and have many and varied applications.

Motivation for the study of ordinary differential equations in general is provided by considering a number of typical problems that give rise to different types of differential equation. The first application involves the determination of orthogonal trajectories. A typical example of orthogonal trajectories arises in steady state two-dimensional temperature distributions, where one family of trajectories corresponds to the lines along which the temperature is constant, while the other family corresponds to lines along which heat flows. Other examples considered are the radioactive decay of a substance, the logistic equation and its connection with population growth, damped oscillations, the shape of a suspended power line, and the bending of beams.

The chapter starts by defining an ***m*th order** ordinary differential equation, of which a first order equation is a special case. Various important terms are defined, and the physical significance of **initial** and **boundary conditions** for differential equations are introduced and explained.

The geometrical interpretation of the derivative dy/dx as the slope of a curve is used in Section 5.3 to develop the concept of the **direction field** associated with the first order equation $dy/dx = f(x, y)$. This concept is particularly useful as it leads to a geometrical picture showing the qualitative behavior of all solutions of the differential equation. It will be seen later that the idea underlying a direction field forms the basis of the simple Euler method for the numerical solution of an initial value problem.

First order equations are considered, **separable equations** are defined and solved, and some other special types of equation are introduced that arise in applications, of which the most important is the general **linear first order differential equation**. Its solution is found by using what is called an **integrating factor**. The first order linear differential equation is important, because the structure of its solution is typical of linear differential equations of all orders.

Another special first order equation that is considered is the Bernoulli equation. The Bernoulli equation is an important type of nonlinear equation with many applications, and in a sense it stands on the border between linear and nonlinear first order differential equations. An application of the Bernoulli equation is outlined in the text, and another more detailed one is to be found in the Exercise set at the end of Section 5.8.

The chapter ends by considering the important and practical questions concerning the existence and uniqueness of solutions of $dy/dx = f(x, y)$.

5.1 Background to Ordinary Differential Equations

An **ordinary differential equation (ODE)** is an equation that relates a function $y(x)$ to some of its derivatives $y^{(r)}(x) = d^r y/dx^r$. It is usual to call x the **independent** variable and y the **dependent variable**, and to write the most general ordinary differential equation as

$$F\left(x, y, y^{(1)}, y^{(2)}, \ldots, y^{(n)}\right) = 0. \tag{1}$$

The number n in (1) is called the **order** of the ordinary differential equation, and it is the order of the highest derivative of y that occurs in the equation. A class of ODEs of particular importance in engineering and science, because of their frequency of occurrence and the extensive analytical methods that are available for their solution, are the linear ordinary differential equations.

The most general **nth order linear differential equation** can be written

nth order linear variable coefficient equation

$$a_0(x)\frac{d^n y}{dx^n} + a_1(x)\frac{d^{n-1}y}{dx^{n-1}} + \cdots + a_{n-1}(x)\frac{dy}{dx} + a_n(x)y = f(x), \tag{2}$$

with $a_0(x) \neq 0$ and we will consider it to be defined over some interval $a \leq x \leq b$. The functions $a_0(x), a_1(x), \ldots, a_n(x)$, called the **coefficients** of the equation, are known functions, and the known function $f(x)$ is called the **nonhomogeneous term**. The name **forcing function** is also sometimes given to $f(x)$, because in applications it represents the influence of an external input that drives a physical system represented by the differential equation. Equation (2) is called **homogeneous** if $f(x) \equiv 0$.

It will be seen later that the solution of the nonhomogeneous equation (2) is related in a fundamental manner to the solution of its associated homogeneous equation.

When one or more of the coefficients of (2) depend on x, it is called a **variable coefficient** equation. Simpler than **variable coefficient** linear equations, but still of considerable importance, are the linear equations in which the coefficients are the constants $a_0, a_1, \ldots, a_n$, so that (2) becomes

nth order linear constant coefficient equation

$$a_0\frac{d^n y}{dx^n} + a_1\frac{d^{n-1}y}{dx^{n-1}} + \cdots + a_{n-1}\frac{dy}{dx} + a_n y = f(x) \quad \text{for} \quad a \le x \le b. \qquad (3)$$

Equations of this type are called **constant coefficient** linear equations.

If the interval $a \le x \le b$ on which equations (2) and (3) are defined is not specified, it is to be understood to be the largest one for which the equations have meaning. Sometimes, in the case of (2), this interval is determined by the variable coefficients $a_r(x)$, whereas in applications it is often determined by the nature of the problem that restricts x to a specific interval.

nonlinear equation and degree

An ordinary differential equation that is not linear is said to be **nonlinear**. Nonlinearity arises in ordinary differential equations because of the occurrence of a nonlinear function of the dependent variable y that sometimes occurs in the form of a power or a radical. The terms homogeneous and nonhomogeneous have no meaning for nonlinear equations.

A term that is also in use, mainly as an indication of the complexity to be expected of a solution, is the *degree* of an equation. The **degree** is the greatest power to which the highest order derivative in the differential equation is raised after the radicals have been cleared from expressions involving the dependent variable y.

EXAMPLE 5.1

(a) The ODE

$$\frac{dy}{dx} + 2xy = \sin x$$

is a linear variable coefficient nonhomogeneous first order equation.

(b) The ODE

$$(1 - x^2)\frac{d^2 y}{dx^2} - 2x\frac{dy}{dx} + 6y = 0, \quad \text{with } -1 < x < 1,$$

is a linear variable coefficient homogeneous second order equation.

(c) The ODE

$$\frac{d^2 y}{dx^2} + a\frac{dy}{dx} + by = \sin \omega x, \quad \text{with } \omega = \text{constant},$$

is a linear constant coefficient nonhomogeneous second order equation.

(d) The ODE

$$\frac{d^2\theta}{dt^2} + k \sin\theta = 0, \quad \text{with } k = \text{constant}$$

is a nonlinear second order equation because θ occurs nonlinearly in the function $\sin\theta$.

(e) The ODE

$$k\frac{d^2y}{dx^2} = f(x)[1 + (dy/dx)^2]^{3/2}, \quad \text{with } k > 0 \text{ a constant}$$

is a nonlinear second order equation of degree 2 involving a power and a radical.

∎

A **solution** of an ordinary differential equation is a function $y = \Phi(x)$ that, when substituted into the equation, makes it identically zero over the interval on which the equation is defined. A solution of an nth order equation that contains n **general and particular solutions, and integral curves** arbitrary constants is called the **general solution** of the equation. If the arbitrary constants in the general solution are assigned specific values, the result is called a **particular solution** of the equation.

For obvious reasons the solution of an ordinary differential equation is also called an **integral curve**. A solution that cannot be obtained from the general solu- **singular solution** tion for any choice of its arbitrary constants is called a **singular solution**. In the case of linear equations *all* possible solutions of the equation can be obtained from the general solution, so linear equations have no singular solutions. Nonlinear equations possess a more complicated structure that often allows the existence of one or more singular solutions.

EXAMPLE 5.2   **(a)** The general solution of the linear constant coefficient nonhomogeneous equation

$$\frac{d^2y}{dx^2} - 4y = x$$

is $y = Ae^{2x} + Be^{-2x} - x/4$, where A and B are arbitrary constants. This is easily checked, because substituting for y in the equation leads to the identity $x \equiv x$.

(b) The nonlinear equation

$$\left(\frac{dy}{dx}\right)^2 + y^2 = 1$$

has the general solution $y = \sin(x + A)$. However, $y = \pm 1$ are also seen to be solutions, though as these cannot be obtained from the general solution for any choice of A, they are *singular solutions*.

∎

The linear equation (2) is often written in the more compact form

$$L[y] = f(x), \tag{4}$$

linear operator   where L is the **linear operator**

$$L[\cdot] \equiv a_0(x)\frac{d^n}{dx^n} + a_1(x)\frac{d^{n-1}}{dx^{n-1}} + \cdots + a_{n-1}(x)\frac{d}{dx} + a_n(x), \tag{5}$$

with coefficients that may or may not be functions of x. Only when $L[\cdot]$ acts on an n times differentiable function does it produce a function.

Equation (2) is called **linear** because if y_1 and y_2 are any two solutions of the homogeneous form of the equation $L[y] = 0$, the linear combination $y = C_1 y_1 + C_2 y_2$ where C_1 and C_2 are constants is also a solution. In terms of the differential operator $L[\cdot]$ this property becomes $L[C_1 y_1 + C_2 y_2] = C_1 L[y_1] + C_2 L[y_2]$, and it follows directly from the linearity of the differentiation operation, because

$$\frac{d^m}{dx^m}(y_1 + y_2) = \frac{d^m y_1}{dx^m} + \frac{d^m y_2}{dx^m},$$

for $m = 0, 1, \ldots, n$, with $d^0 y/dx^0 \equiv y$.

If $y_1(x), y_2(x), \ldots, y_m(x)$ are solutions of the nth order homogeneous equation $L[y] = 0$, with $m \leq n$ and $C_1, C_2, \ldots, C_m$ arbitrary constants, the linear combination

$$y(x) = C_1 y_1(x) + C_2 y(x) + \cdots + C_m y_m(x)$$

linear superposition

is called a **linear superposition** of the m solutions, and it is also a solution of the homogeneous equation.

Later we will define the linear independence of a set of functions over an interval and show that the homogeneous form of (2) has precisely n linearly independent solutions $y_1(x), y_2(x), \ldots, y_n(x)$, and that its general solution is

$$y_c(x) = C_1 y_1(x) + C_2 y(x) + \cdots + C_n y_n(x), \tag{6}$$

complementary solution, particular integral, and complete solution

where $C_1, C_2, \ldots, C_n$ are arbitrary constants. This general solution of the homogeneous form of equation (2) is called the **complementary function** or the **complementary solution** of (2). A function $y_p(x)$ that is a solution of the nonhomogeneous equation (2) but contains *no* arbitrary constants is called a **particular integral** of (2). The **complete solution** $y(x)$ of equation (2) is

$$y(x) = y_c(x) + y_p(x). \tag{7}$$

In applications of ordinary differential equations the values of the arbitrary constants in specific problems are obtained by choosing them so the solution satisfies auxiliary conditions that identify a particular problem.

Auxiliary conditions specified at a single point $x = a$, say, are called **initial conditions**, because x often represents the time so that conditions of this type describe how the solution starts. An **initial value problem (i.v.p.)** involves finding a solution of a differential equation that satisfies prescribed initial conditions.

A different type of problem arises when the auxiliary conditions are specified at two different points $x = a$ and $x = b$, say. Conditions of this type are called **boundary conditions**, because in such problems x usually represents a space variable, and the solution is required to be determined between two boundaries located at $x = a$ and $x = b$ where boundary conditions are prescribed. A **boundary value problem (b.v.p.)** involves finding a solution of a differential equation that satisfies prescribed boundary conditions.

boundary and initial conditions

EXAMPLE 5.3

(a) The linear nonhomogeneous ordinary differential equation

$$\frac{d^2y}{dx^2} + y = x$$

has the general solution $y = A\cos x + B\sin x + x$. This equation together with the initial conditions $y(0) = 0$, $y'(0) = 0$ specified at the point $x = 0$ constitutes an *initial value problem* for y. Choosing A and B to satisfy these initial conditions shows the unique solution of this i.v.p. to be $y = x - \sin x$ for $x \geq 0$.

(b) The linear homogeneous ordinary differential equation

$$\frac{d^2y}{dx^2} + y = 0$$

has the general solution $y = A\cos x + B\sin x$. This equation together with the conditions $y(0) = 0$, $y'(\pi/3) = 3$ specified at the two different points $x = 0$ and $x = \pi/3$ constitutes a *boundary value problem* for y. Choosing A and B to satisfy these conditions shows that this b.v.p. has the unique solution $y = 6\sin x$ for $0 < x < \pi/3$.

(c) Consider the linear homogeneous ordinary differential equation

$$\frac{d^2y}{dx^2} - y = 0 \quad \text{defined for } x \geq 0,$$

which is easily seen to have the general solution $y = Ae^x + Be^{-x}$. Imposing the boundary conditions $y(0) = 1$ and $y(+\infty) = 0$ constitutes a boundary value problem for y in which one condition is at $x = 0$ and the other is at plus infinity. The condition at infinity can only be satisfied if $A = 0$, so matching the solution $y = Be^{-x}$ to the condition $y(0) = 1$ shows that this b.v.p. has the **unique** (only) solution $y = e^{-x}$.

unique and nonunique solutions

(d) It is possible for a boundary value problem to have a unique solution as in (b), more than one solution, or no solution at all. More will be said about this later, but for the moment we give a simple example that shows why a boundary value problem may have many solutions or no solution.

The general solution of (b) is $y = A\cos x + B\sin x$, so if the boundary conditions $y(0) = 0$ and $y(\pi) = 0$ are imposed we find that $A = 0$ and B is indeterminate, so it may be assigned any value. In this case a solution certainly exists, as it is given by $y = B\sin x$, but B is arbitrary, so there is more than one solution. When more than one solution can be found that satisfies the auxiliary conditions, the solution is said to be **nonunique**.

If, in this example, the boundary conditions are replaced by $y(0) = 0$ and $y(\pi) = 1$, no choice of constants A and B can make the general solution satisfy the boundary conditions, so in this case there is no solution. ∎

Summary

This section introduced the concept of an *n*th order ordinary differential equation, and the initial and boundary conditions that such equations are often required to satisfy. Emphasis was placed on linear equations and, in particular, on the structure of the solution of a linear first order equation, because the structure of the solution of this fundamental type of equation is shared by the solutions of all higher order linear equations.

EXERCISES 5.1

In Exercises 1 through 10, determine the order and degree of the equation and classify it as homogeneous linear, non-homogeneous linear, or nonlinear.

1. $y''' + 3y'' + 4y' - y = 0$.
2. $y'' + 4y' + y = x \sin x$.
3. $y'' + x(y')^2 = \cosh x$.
4. $(y'')^{3/2} + xy' = [(1+x)y']$.

5. $y'' + 3y' + 2y = x^2 \sin y$.
6. $y^{(4)} + x^2 \sqrt{y} = 3 + x^3$.
7. $y' + 3xy = 1 + x^2$.
8. $y'' + y = \tan(y')$.
9. $(2 + x^2)y' + x(1 - y^2) = 0$.
10. $y'/y + \sin x = 3$.

5.2 Some Problems Leading to Ordinary Differential Equations

Before we develop methods for the solution of ordinary differential equations, it will be helpful to examine some simple geometrical and physical problems that lead to ODEs. There are many such problems, so we only consider some representative examples.

(a) A Geometrical Problem: Orthogonal Trajectories

The equation

$$F(x, y, c) = 0,$$

where the real variable c is a *parameter*, defines a **one-parameter** family of curves in the (x, y)-plane. This means that assigning a specific value to c determines a particular curve in the (x, y)-plane, and a different value of c will determine a different curve. It often happens that the equation $F(x, y, c) = 0$ defines y implicitly in terms of x, so that the equation cannot be solved explicitly as $y = f(x, c)$.

orthogonal trajectory

A curve that intersects every member of a one-parameter family of curves orthogonally (at right angles) is called an **orthogonal trajectory** of the family. A geometrical problem that often occurs is how to find a *family* of curves that form orthogonal trajectories to a given family.

When some applications of conformal mapping to two-dimensional physical problems are considered in Chapter 17, it will be seen that orthogonal trajectories arise in the study of steady state heat conduction, fluid dynamics, and electromagnetic theory. In heat conduction (see Chapter 18), one family of curves represents lines of constant temperature called **isotherms**, and their orthogonal trajectories then represent **heat flow** lines. In two-dimensional fluid dynamics, orthogonal trajectories express the relationship between the curves followed by fluid particles called **streamlines**, and the associated **equipotential lines** along which a function called the **fluid potential** is constant. In two-dimensional electromagnetic theory an analogous situation arises where one family of curves describes lines of constant electric potential, again called **equipotential lines**, and the family of orthogonal trajectories that describes what are then called **flux lines**.

isotherms, heat flow, streamlines, equipotentials, and flux lines

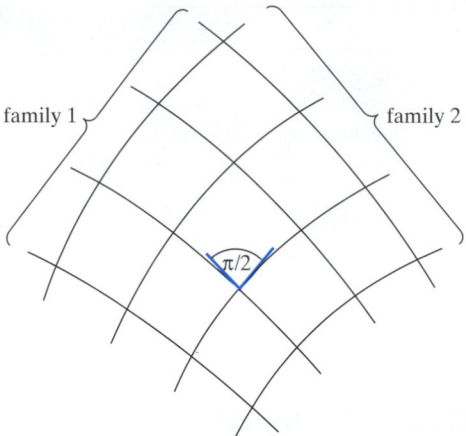

FIGURE 5.1   Two typical families of orthogonal trajectories.

Two typical families of orthogonal trajectories are illustrated in Fig. 5.1, and if these curves are related to steady state heat flow, family 1 could represent the isotherms and family 2 the heat flow lines.

Two specific examples of families of orthogonal trajectories are shown in Fig. 5.2, where in case (a) the curves are given by

$$x^2 + y^2 = c^2 \quad \text{and} \quad y = kx \qquad \text{(with } c \text{ and } k \text{ real).}$$

The first equation describes a family of concentric circles centered on the origin, and the second family that forms their orthogonal trajectories comprises all the straight lines that pass through the origin.

In case (b) the curves are given by

$$x^2 - y^2 = c \quad \text{and} \quad xy = k \qquad \text{(with } c \text{ and } k \text{ real),}$$

where the two families of curves are families of mutually orthogonal rectangular hyperbolas.

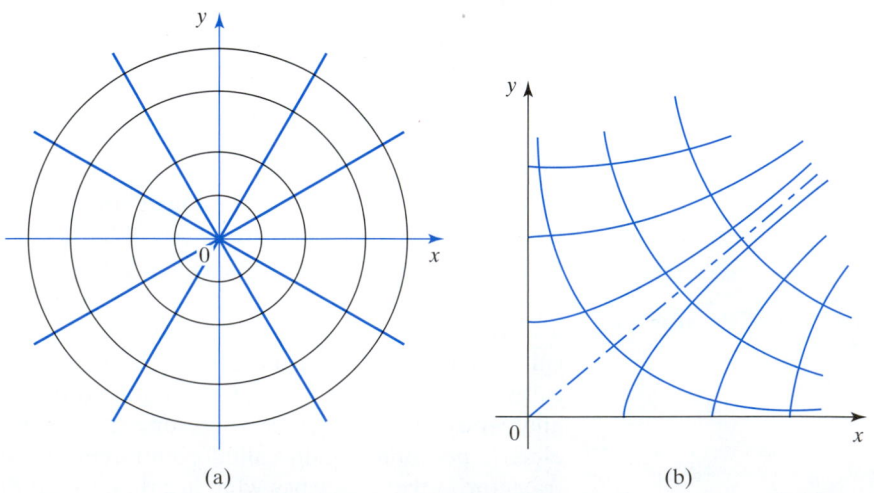

(a)   (b)

FIGURE 5.2   Specific examples of orthogonal trajectories.

In general the equation

$$F(x, y, c) = 0, \tag{8}$$

with c a parameter, describes a family of curves. To find their orthogonal trajectories we first need to obtain the differential equation for the family of curves determined by (8). This can be done by differentiating (8) with respect to x and then eliminating c between (8) and the equation with dy/dx to arrive at a differential equation of the form

$$\frac{dy}{dx} = f(x, y). \tag{9}$$

If the family of curves described by this differential equation is to be orthogonal to another family, the products of the gradients of every pair of intersecting curves must equal -1. So the gradient dy/dx of the family of curves that are mutually orthogonal to those of (9) must be such that

$$\frac{dy}{dx} = -\frac{1}{f(x, y)}. \tag{10}$$

This is the differential equation of the required family of orthogonal trajectories. In general (10) can often be solved by the method of separation of variables that will be discussed later.

(b) Chemical Reaction Rates and Radioactive Decay

In many circumstances, for a limited period of time, the rate of reaction of a chemical process can be considered to be proportional only to the amount Q of the chemical that is present at a given time t. The differential equation governing such a process then has the form

$$\frac{dQ}{dt} = kQ, \tag{11}$$

where $k \geq 0$ is a constant of proportionality. This is a homogeneous linear first order differential equation.

An analogous situation applies to the radioactive decay of an isotope for which the decay takes place at a rate proportional to the amount of radioactive isotope that is present at any given instant of time. The equation governing the amount Q of the isotope as a function of time t is also of the form shown in (11), but instead of the amount growing as in the previous case, it is decreasing, so as in this case the constant of proportionality is usually denoted by a positive number λ, the equation for radioactive decay takes the form

$$\frac{dQ}{dt} = -\lambda Q. \tag{12}$$

It is not difficult to see by inspection that the general solution of (12) is

$$Q = Q_0 e^{-\lambda t},$$

where Q_0 is the amount of the isotope present at the start when $t = 0$. The so-called **half-life** T_h of an isotope is the time taken for half of it to decay away, so setting $Q = (1/2)Q_0$ in the above result shows the half-life to be given by $T_h = (1/\lambda) \ln 2$.

(c) The Logistic Equation: Population Growth

In the study of phenomena involving the rate of increase of a quantity of interest, it often happens that the rate is influenced both by the amount of the quantity that is present at any given instant of time and by the limitation of a resource that is necessary to enable an increase to occur. Such a situation arises in a population of animals that compete for limited food resources, leading to the so-called *predator–prey* situations where an animal (the predator) feeds on another species (the prey) with the effect that overfeeding leads to starvation. This in turn leads to a reduction in the number of predators that in turn can lead to a recovery of the food stock. Similar situations arise in manufacturing when there is competition for scarce resources, and in a variety of similar situations.

To model the situation we let P represent the amount of the quantity of interest present at a given time t, and M represent the amount of resources available at the start. Then a simple model for this process is provided by the differential equation

$$\frac{dP}{dt} = kP(M - P), \tag{13}$$

in which k is a constant of proportionality. When constructing this equation the assumption has been made that the rate of increase dP/dt is proportional to both the amount P that is present at time t and to the amount $M - P$ that remains. Equation (13) is called the **logistic equation**, and it is nonlinear because of the presence of the term $-kP^2$ on the right, though it is easily integrated by the method of separation of variables to be described later.

(d) A Differential Equation that Models Damped Oscillations

Mechanical and electrical systems, and control systems in general, can exhibit oscillatory behavior that after an initial disturbance slowly decays to zero. The process producing the decay is a *dissipative* one that removes energy from the system, and it is called **damping**. To see the prototype equation that exhibits this phenomenon we need only consider the following very simple mechanical model. A mass M rests on a rough horizontal surface and is attached by a spring of negligible mass to a fixed point. The mass–spring system is caused to oscillate along the line of the spring by being displaced from its equilibrium position by a small amount and then released. Figure 5.3a shows the system in its equilibrium configuration, and Fig. 5.3b shows it when the mass has been displaced through a distance x from its rest position.

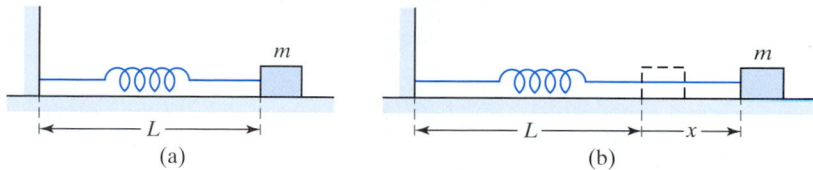

FIGURE 5.3 Mass–spring system.

If t is the time, the acceleration of the mass is d^2x/dt^2, so the force acting due to the motion is Md^2x/dt^2. The forces opposing the motion are the spring force, assumed to be proportional to the displacement x from the equilibrium position, and the frictional force, assumed to be proportional to the velocity dx/dt of the mass M. If the spring constant of proportionality is p and the frictional constant of proportionality is k, the two opposing forces are kdx/dt due to friction and px due to the spring. Equating the forces acting along the line of the spring and taking account of the fact that the spring and frictional forces oppose the force due to the acceleration shows the equation of motion to be the homogeneous second order linear equation

$$M\frac{d^2x}{dt^2} = -k\frac{dx}{dt} - px,$$

or

$$\frac{d^2x}{dt^2} + a\frac{dx}{dt} + bx = 0, \tag{14}$$

where $a = k/M$ and $b = p/M$.

If an external force $Mf(t)$ is applied to the spring, the equation governing the damped oscillations becomes the linear nonhomogeneous second order equation

$$\frac{d^2x}{dt^2} + a\frac{dx}{dt} + bx = f(t).$$

An equation of the same form as (14) governs the oscillation of the charge q in the R–L–C electric circuit shown in Fig. 5.4. The open circuit is shown in Fig. 5.4a with the plates of the capacitor C carrying initial charges Q and $-Q$, while Fig. 5.4b shows the circuit when the switch S has been closed, causing a current i to flow due to a charge is q at time t.

The respective potential drops in the direction of the arrow across the resistor R, the inductance L, and the capacitor C are $V = iR$, where $i = dq/dt$, Ldi/dt,

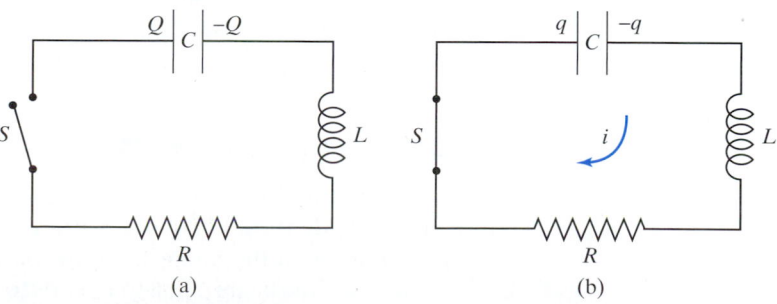

FIGURE 5.4 An R–L–C circuit.

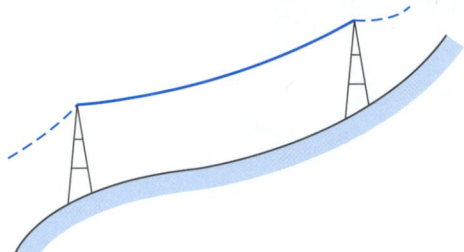

FIGURE 5.5   Suspended cable.

and q/C. Applying Kirchhoff's law, which requires the sum of the potential drops around the circuit to be zero, gives

$$L\frac{di}{dt} + Ri + \frac{q}{C} = 0.$$

Eliminating i by using the result $i = dq/dt$ leads to the following homogeneous linear second order equation for q:

$$LC\frac{d^2q}{dt^2} + RC\frac{dq}{dt} + q = 0.$$

This ODE is of the same form as (14) with $a = R/L$ and $b = 1/LC$.

(e) The Shape of a Suspended Power Line: The Catenary

An analysis of the forces acting on a power line attached to pylons as shown in Fig. 5.5, or on the suspension cable of a cable car, shows the shape of the cable to be determined by the solution $y(x)$ of the nonlinear differential equation

$$\frac{d^2y}{dx^2} = a\sqrt{1 + (dy/dx)^2}.$$

The shape taken by the cable is called a **catenary**, after the Latin word *catena*, meaning chain. Although this equation will not be solved here, it is not difficult to show that its solution is a hyperbolic cosine curve.

(f) Bending of Beams

An analysis of the forces and moments acting on a horizontal beam of uniform construction made from a material with Young's modulus E and supported at its two end points, with the moment of inertia of its cross-section about the central horizontal axis of the beam equal to I, leads to the following equation for the vertical

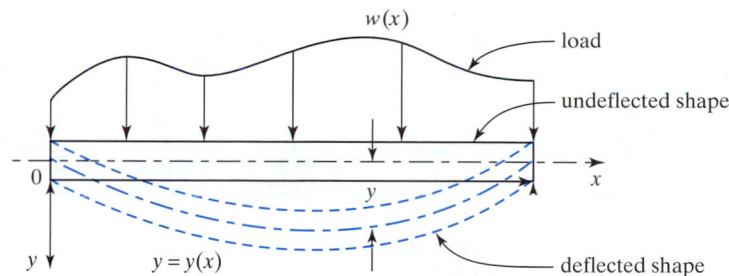

FIGURE 5.6   Deflection of a loaded beam.

deflection y caused by the weight of the beam and any loads it is supporting:

$$\frac{EI\, d^2y/dx^2}{[1+(dy/dx)^2]^{3/2}} = M(x). \tag{15}$$

Here $M(x)$ is the bending moment that acts to one side of a point x in the beam. If a distributed load of line density $w(x)$ acts along the beam creating a load $\int_a^b w(x)dx$ on the segment from $x = a$ to $x = b$, as represented in Fig. 5.6, it can be shown that $M(x)$ and $w(x)$ are related by the result

$$\frac{d^2 M}{dx^2} = -w(x). \tag{16}$$

Using this result in (15) shows that the deflection $y(x)$ is determined by the solution of the nonlinear fourth order equation

$$\frac{d^2}{dx^2}\left\{ \frac{EI d^2 y/dx^2}{[1+(dy/dx)^2]^{3/2}} \right\} = w(x), \tag{17}$$

flexural rigidity   in which the product EI is called the **flexural rigidity** of the beam. If the bending is small and the term $(dy/dx)^2$ can be neglected, (17) simplifies to the linear fourth order constant coefficient equation

$$\frac{d^4 y}{dx^4} = \frac{w(x)}{EI},$$

which can be solved by direct integration.

Many applications of ordinary differential equations to physical problems are to be found in reference [3.6].

Summary

This section has provided mathematical and physical examples of problems that give rise to ordinary differential equations, some with initial conditions and others with boundary conditions. The logistic equation was seen to be nonlinear and first order, whereas others such as the equation governing radioactive decay and the equation describing damped oscillations were seen to be linear and of first and second order, respectively. The beam equation is nonlinear, though when the bending is small it was seen to reduce to a simple linear fourth order equation that could be solved by direct integration.

EXERCISES 5.2

1. Derive the differential equation that describes the families of circles that are tangent to both the x- and y-axes.

2. Derive the differential equation satisfied by all curves such that the magnitude of the area under the curve between any two ordinates at $x = a$ and $x = b$ is proportional to the magnitude of the arc length of the curve from $x = a$ to $x = b$. Verify that the *catenary* $y(x) = k \cosh(x/k - K)$ is such a curve, with k and K parameters.

3.* A launch travels along the y-axis a constant speed U, starting from the origin, and a police launch starting from a point $a > 0$ on the x-axis pursues it at a constant speed $V > U$. If t is the time measured from the start of the pursuit, write down the differential equation that describes the pursuit path. At all times the police launch steers toward the first launch.

5.3   Direction Fields

In certain applications of mathematics it is necessary to know the qualitative behavior of solutions of a general first order equation

$$\frac{dy}{dx} = f(x, y) \tag{18}$$

global properties

over the entire (x, y)-plane, when either no analytical solution is available or, if one exists, it is too complicated to be useful. General properties of solutions of (18) that are known throughout the (x, y)-plane are called **global properties**. A typical global property might be that the solutions are known to be bounded for all x.

A numerical solution of (18) can always be obtained for any given initial condition (see Chapter 19), but it is impracticable to obtain such solutions for a large enough set of initial conditions simply to enable general the behavior of solutions all over the (x, y)-plane to be understood.

A convenient answer to this problem involves constructing a graphical representation of what is called the *direction field* of (18) at a conveniently chosen mesh of points covering a region R of interest in the (x, y)-plane.

The idea involved is simple and starts by dividing the interval $a \leq x \leq b$ into m subintervals of equal length $\Delta x = (b - a)/m$, and the interval $c \leq y \leq d$ into n subintervals of equal length $\Delta y = (d - c)/n$. The mesh of points to be used to cover R are then located at the points (x_r, y_s), where $x_r = a + r\Delta x$ and $y_s = c + s\Delta y$ with $r = 0, 1, \ldots, m$ and $s = 0, 1, \ldots, n$.

Once the mesh has been chosen, the function $f(x, y)$ is evaluated at each of the points (x_r, y_s). It follows directly that the number $f(x_r, y_s)$ associated with the point (x_r, y_s) is the *gradient* (slope) of the integral curve (solution curve) that passes through that point. Accordingly, the next step is to construct through each point (x_r, y_s), a small straight line segment making an angle $\theta_{rs} = \text{Arctan } f(x_r, y_s)$ with the x-axis, as in Fig. 5.7a.

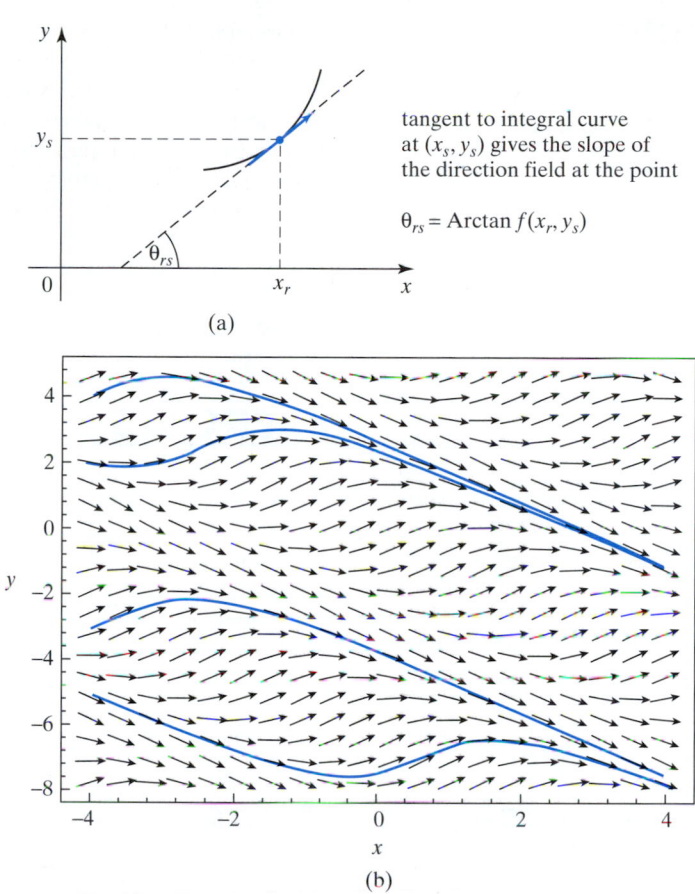

FIGURE 5.7   (a) The construction of a direction field vector at the point (x_r, y_s). (b) The direction field and integral curves for $dy/dx = \cos(x + y)$.

By the nature of their construction, each line segment that is drawn in this manner is tangent to the integral curve that passes through the point through which the segment is drawn. An examination of the pattern of the line segments indicates the overall pattern of behavior of all of the integral curves passing through region R. The assignment of a gradient $f(x, y)$ to each point of R is said to define the **direction field** of the ODE in (18) over R, and the method just described is its geometrical interpretation at a finite number of points of R.

direction field

The graphical interpretation of a direction field can be used to obtain an approximation to the integral curve that passes through an initial point (x_0, y_0) in R. This is accomplished by starting with the line segment through the point (x_0, y_0) and then joining up successive line segments as they intersect one another. As the construction of a direction field over a large region involves many calculations, it is usual to construct them with the aid of a computer.

The direction field for the nonlinear first order equation

$$\frac{dy}{dx} = \cos(x + y)$$

over the region $-4 \le x \le 4$ and $-8 \le y \le 5$ is shown in Fig. 5.7b, to which have been added some integral curves to show their relationship to the direction field.

Summary

The concept of a direction field of a first order differential equation $dy/dx = f(x, y)$ was introduced in this section. It is a graphical representation of the slope (gradient) of solution curves of the differential equation where they pass through a rectangular mesh of points inside a region of the (x, y)-plane where the solution of the differential equation is of interest. It involves plotting at each mesh point (x_i, y_i) a short segment of the tangent to the solution curve with slope $f(x_i, y_i)$ that passes through that point, to which is added an arrow showing the direction in which the solution is changing as x increases. A direction field provides a geometrical representation of the global nature of the solution inside the region of interest, and tracing successive line segments from one to another, starting from any mesh point, provides a rough picture of the solution curve that originates from the initial condition represented by that mesh point.

EXERCISES 5.3

In each of the following exercises, with the aid of a computer algebra package: (a) Construct the direction field for the given equation at a suitable number of mesh points, (b) use the results of (a) to sketch some representative integral curves, and (c) compare an approximate integral curve through a chosen initial point (x_0, y_0) with the exact solution found by requiring the given general solution to pass through that point.

1. $dy/dx = y + 2x$; $\quad y = Ce^x - 2 - 2x$.
2. $dy/dx = y + 2\cos x$; $\quad y = Ce^x - \cos x + \sin x$.
3. $dy/dx = 2x - y$; $\quad y = Ce^{-x} - 2 + 2x$.
4. $dy/dx = x(1 + y/2)$; $\quad y = C\exp(x^2/4) - 2$.
5. $dy/dx = y + x^2$; $\quad y = Ce^x - 2 - 2x - x^2$.

5.4 Separable Equations

Sometimes the function $f(x, y)$ in the first order differential equation

$$\frac{dy}{dx} = f(x, y) \tag{19}$$

can be written as the product of a function $F(x)$ depending only on x and a function $G(y)$ depending only on y, so that $f(x, y) = F(x)G(y)$, allowing (19) to be written

$$\frac{dy}{dx} = F(x)G(y). \tag{20}$$

two forms of a separable equation

When (19) can be expressed in this simple form, its variables x and y are said to be **separable**, and the equation itself to be of **variables separable** type. If we use differential notation, (20) becomes

$$\frac{1}{G(y)}dy = F(x)dx, \tag{21}$$

so provided $G(y) \neq 0$, equation (21) can be solved by routine integration of the left side with respect to y and of the right side with respect to x. Thus, in principle, the solution of a first order differential equation in which the variables are separable can always be found, though in practice the integrals involved may be difficult or sometimes impossible to evaluate analytically.

Separable first order equations

The differential equation

$$\frac{dy}{dx} = f(x, y)$$

is said to be **separable** if it can be written in the form

$$\frac{dy}{dx} = F(x)G(y),$$

or, in differential form,

$$\frac{1}{G(y)}dy = F(x)dx.$$

EXAMPLE 5.4

examples of
separable equations

Solve the logistic equation

$$\frac{dP}{dt} = kP(M - P)$$

given in equation (13) of Section 5.2(c), assuming $k > 0$ and $0 \leq P \leq M$. Find the solution of the initial value problem in which $P = P_0$ when $t = 0$, and draw some typical integral curves.

Solution   The equation is separable and can be written in the differential form

$$\frac{dP}{P(M - P)} = kdt.$$

If we write the left-hand side in partial fraction form, the equation becomes

$$\frac{dP}{P} + \frac{dP}{(M - P)} = Mkdt,$$

and after integration we find that

$$\ln\left|\frac{P}{M - P}\right| = Mkt + C,$$

where C is an arbitrary constant of integration. As the solution for P must lie in the interval $0 \leq P \leq M$, this result simplifies to

$$P = \frac{MA}{A + \exp(-Mkt)},$$

where A is an arbitrary constant.

The arbitrary constant A is related to C by $A = e^C$, but as C is arbitrary, the constant A is also arbitrary, so for simplicity we denote the arbitrary constant in this last result by A without mentioning how it is related to C. In general, arithmetic is not usually performed on arbitrary constants, so after algebraic manipulations, either constants are renamed or the same symbol is used for a related constant.

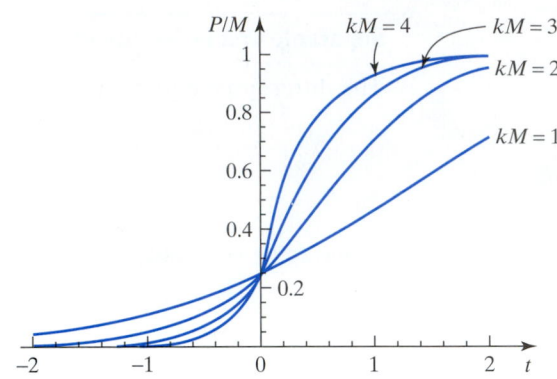

FIGURE 5.8 Integral curves for the logistic equation.

To solve the initial value problem we must find A such that $P = P_0$ when $t = 0$, from which it is easily seen that $A = P_0/(M - P_0)$. The required particular solution is thus

$$P = \frac{MP_0}{P_0 + (M - P_0)\exp(-Mkt)}.$$

Representative integral curves of $P(t)/M$ obtained from this expression using $P_0/M = \tfrac{1}{4}$ and $kM = 1, 2, 3$, and 4 are shown in Fig. 5.8 for $-2 \leq t \leq 2$. ■

EXAMPLE 5.5

Solve the initial value problem for the equation expressed in differential form

$$x^2 y^2 dx - (1 + x^2)dy = 0, \quad \text{given that } y(0) = 1.$$

Solution   The equation is separable because it can be written

$$\frac{dy}{y^2} = \frac{x^2}{(1 + x^2)}dx.$$

Integration gives

$$\int \frac{dy}{y^2} = \int \frac{x^2}{(1 + x^2)}dx,$$

and after the integrations have been performed this becomes

$$-1/y = x - \operatorname{Arctan} x + C,$$

where C is an arbitrary constant of integration. This general solution will satisfy the initial condition $y(0) = 1$ if $C = -1$, so the required solution is seen to be

$$y = 1/(\operatorname{Arctan} x - x + 1).$$ ■

EXAMPLE 5.6

Derive the differential equation that determines the orthogonal trajectories of the one parameter family of curves $y = Cxe^x$, and solve it to find the equation of these trajectories.

Solution   The differential equation describing the family of curves $y = Cxe^x$ is found by first calculating $y'(x)$, and then using the original equation to eliminate C

from the result. We have

$$y'(x) = Ce^x(1 + x),$$

but from the original equation $C = y/xe^x$, so eliminating C between these two results shows that the required differential is

$$y'(x) = y(1 + x)/x.$$

The product of the gradient $y'(x)$ of curves belonging to this family and the gradient of the family of orthogonal trajectories must equal -1 (see Section 5.2(a)), so the differential equation of the orthogonal trajectories is the separable equation

$$\frac{dy}{dx} = -\frac{x}{y(1 + x)}.$$

After separation of the variables and integration, this becomes

$$\int y\,dy = -\int \frac{x}{1 + x}dx,$$

so that

$$y^2 = \ln(1 + x)^2 - 2x + C. \qquad\blacksquare$$

EXAMPLE 5.7   A circular metal radiator pipe has inner radius R_1 and outer radius R_2 $(R_2 > R_1)$. When operating under steady conditions the radial temperature distribution $T(r)$ in the metal wall of the pipe is known to be a solution of the ordinary differential equation (see the heat equation in cylindrical polar coordinates in Section 18.5)

$$r\frac{d^2 T}{dr^2} + \frac{dT}{dr} = 0.$$

(i) Find the radial temperature distribution in the pipe wall when the inner surface is maintained at a constant temperature T_1 and the outer surface is maintained at a constant temperature T_2.

(ii) Find the radial temperature distribution in the pipe wall when the inner surface is maintained at a constant temperature T_1 and heat is lost by radiation from the outer surface according to Newton's law of cooling that requires the heat flux across the outer surface to be proportional to the difference in temperature between the surface and the surrounding air at a temperature T_2.

Solution

(i) Setting $u = dT/dr$ the equation becomes the separable equation

$$r\frac{du}{dr} + u = 0 \quad \text{and so} \quad \frac{du}{u} = -\frac{dr}{r},$$

from which it follows that

$$\ln u = -\ln r + \ln A,$$

where for convenience the arbitrary integration constant has been written $\ln A$. Thus $ur = A$, so after substituting for u and again separating variables we have

$$\frac{dT}{dr} = \frac{A}{r}.$$

A final integration gives the general solution

$$T(r) = A \ln r + B,$$

where B is another arbitrary integration constant.

Matching the arbitrary constants A and B to the required conditions $T(R_1) = T_1$ and $T(R_2) = T_2$ then gives the required solution

$$T(r) = \frac{T_1 \ln(R_2/r) + T_2 \ln(r/R_1)}{\ln(R_2/R_1)}.$$

(ii) The heat flux across the surface $r = R_2$ is proportional to dT/dr at $r = R_2$, and this in turn is proportional to the temperature difference $T(R_2) - T_2$, so the required boundary condition on the outer surface of the pipe is of the form

$$\left(\frac{dT}{dr}\right)_{r=R_2} = -h[T(R_2) - T_2],$$

where the negative sign is necessary because heat is being lost across the surface $r = R_2$, and h is a constant depending on the metal in the pipe and the heat transfer condition at its surface.

The general solution is still $T(r) = A \ln r + b$, but now the arbitrary constants A and B must be matched to the condition $T(R_1) = T_1$ on the inside wall of the pipe, and to the above condition derived from Newton's law of cooling. When this is done the temperature distribution in the pipe is found to be

$$T(r) = T_1 + \frac{h R_2(T_2 - T_1)}{1 + h R_2 \ln(R_2/R_1)} \ln\left(\frac{r}{R_1}\right). \qquad \blacksquare$$

Summary

This section introduced the important class of separable differential equations $dy/dx = F(x)G(y)$, so called because when written in the form $dy/G(y) = F(x)dx$ the variables are separated by the $=$ sign; they can be integrated immediately provided antiderivatives (indefinite integrals) of $1/G(y)$ and $F(x)$ can be found. This method was used to integrate the nonlinear logistic equation and to obtain the equation of some orthogonal trajectories.

EXERCISES 5.4

In Exercises 1 through 4 solve the given differential equation by hand and confirm the result by using computer algebra.

1. $2yy' = x(1 - 2y)$ with $y(1) = 1$.

2. $2x^2y^2y' + y^4 = 4$ with $y(1) = 3$.

3. $(x^2 - 4)y' = x(1 - 2y)$ with $y(\sqrt{5}) = 1$.

4. $2\sqrt{(1 + x^2)}y' = \sqrt{(1 - y^2)}$ with $y(1) = 1$.

In Exercises 5 through 14 find the general solution of the given differential equation.

5. $\sqrt{(1 + x^2)}y' - 3x\sqrt{(y^2 - 1)} = 0$.

6. $e^{-3x}y' + x \sin 2y = 0$.

7. $2(1 + x)(1 + y)y' + (y + 2)^2 = 0$.

8. $2(x - 1)y' + (x^2 - 2x + 3)\cos^2 y = 0$.

9. $(1 + 3y^2)y' + 2y \ln|1 + x| = 0$.

10. $2(1 - \cos x)y' + 3 \sin y = 0$.

11. $(1 + x^2)yy' - x(y^2 + y + 1) = 0$.

12. $(x^2 + 9)y^2y' - \sqrt{(4 - y^2)} = 0$.

13. $y'\text{ctg }x + 2y = 4$.

14. $(x + 1)y^2y' = x(y^2 + 4)$.

In Exercises 15 through 17 derive and then solve the differential equation that determines the orthogonal trajectories to the given one parameter family of curves.

15. $y = b + k(x - a)$ with a and b constants and k a parameter.

16. $x^2 - 4y^2 + y = c$ with c a parameter.

17. $y = Cx^2e^{2x}$ with C a parameter.

18. A snowball of radius 2 inches is brought into a warm room at a constant temperature above freezing point, and it is found that after 6 hours it has melted to a radius of 1.5 inches. Assuming the melting occurs at a rate proportional to the surface area, write down the differential equation determining the radius as a function of time t in hours, and find the general expression for the radius as a function of time. Comment on any deficiency exhibited by this mathematical model.

19. A simple model called *Malthus' law* for the change in a bacterial population $N(t)$ as a function of time t involves assuming the rate of change is proportional to the population present at time t. Write down the differential equation governing $N(t)$ if the constant of proportionality is $\lambda > 0$, and find an expression for $N(t)$ given that initially $N(0) = N_0$. Find λ if $N(t_1) = N_1$ when $t = t_1$ and $N(t_2) = N_2$ when $t = t_2$, with $N_1 > N_2$ and $t_2 > t_1$. Give a reason why this model is unrealistic when t is large.

20. When a beam of light enters a parallel slab of transparent material at right angles to its plane surface, its intensity I decreases at a rate proportional to the intensity $I(x)$ at a perpendicular distance x into the material. Given a slab of material where the intensity at a distance h into the slab is 40% of the initial intensity, write down the differential equation for $I(x)$. Solve the equation for $I(x)$ and find the distance at which the intensity is 10% of its initial value.

21. The dating of a fossilized bone is based on the amount of radioactive isotope carbon-14 present in the bone.

The method uses the fact that the isotope is produced in the atmosphere at a steady rate by bombardment of nitrogen by cosmic radiation when it is absorbed into the living bone. The process stops when the bone is dead, after which the C-14 present in the bone decays exponentially. Assuming the half-life of C-14 is 5600 years, and a bone is found to contain 1/500th of the original amount of C-14 that was present originally, determine its age. This approach is called **radioactive carbon dating**.

22. A cylindrical tank of cross-sectional area A standing in a vertical position is filled with water to a depth h. At time $t = 0$ a circular hole of radius a in the bottom of the tank is opened and water is allowed to drain away under gravity. It is known from *Torricelli's law* that the speed of flow of the water through the hole when the water in the tank has depth x is equal to $\sqrt{2gx}$, this being the speed attained by a particle falling freely from rest under gravity through a distance x, where g is the acceleration due to gravity. Write down the differential equation determining the water height $x(t)$ in the tank when $t > 0$, and solve the equation for $x(t)$. If water is added to the tank at a rate $V(t)$, write down the modified equation governing the water height. If $V(t) = V_0$ is constant, and the flow into and out of the tank reaches equilibrium, find the equilibrium height of the water in the tank. Remark: In applications the expression $\sqrt{2gx}$ is replaced by $k\sqrt{2gx}$, with $0 < k < 1$ a constant. The factor k allows for the contraction of the jet after leaving the hole. In the case of water $k \approx 0.6$.

5.5   Homogeneous Equations

homogeneous equation of degree n

A function $f(x, y)$ is said to be **algebraically homogeneous of degree n**, or simply **homogeneous of degree n**, if $f(tx, ty) = t^n f(x, y)$ for some real number n and all $t > 0$, for $(x, y) \neq (0, 0)$.

EXAMPLE 5.8

(a) If $f(x, y) = x^2 + 3xy + 4y^2$, then $f(tx, ty) = t^2(x^2 + 3xy + 4y^2) = t^2 f(x, y)$, so $f(x, y)$ is homogeneous of degree 2.

(b) If $f(x, y) = \ln|y| - \ln|x|$ for $(x, y) \neq (0, 0)$, then $f(x, y) = \ln|y/x|$, so $f(tx, ty) = f(x, y)$, showing that $f(x, y)$ is homogeneous of degree 0.

(c) If

$$f(x, y) = \frac{x^{3/2} + x^{1/2}y + 3y^{3/2}}{2x^{3/2} - xy^{1/2}}, \text{ then } f(tx, ty) = t^0 f(x, y),$$

showing that $f(x, y)$ is homogeneous of degree 0.

(d) If

$$f(x, y) = x^2 + 4y^2 + \sin(x/y), \text{ then } f(tx, ty) = t^2(x^2 + 4y^2) + \sin(x/y),$$

so $f(x, y)$ is *not* homogeneous, because although both the first group of terms and the last term are homogeneous functions of x and y, they are not both homogeneous of the same degree.

(e) If $f(x, y) = \tan(xy + 1)$, then $f(tx, ty) = \tan(t^2 xy + 1)$, so $f(x, y)$ is *not* homogeneous.   ■

Homogeneous differential equations

The first order ODE in differential form

$$P(x, y)dx + Q(x, y)dy = 0$$

is called **homogeneous** if P and Q are homogeneous functions of the same degree or, equivalently, if when written in the form

$$\frac{dy}{dx} = f(x, y), \quad \text{the function } f(x, y) \text{ can be written as} \quad f(x, y) = g(y/x).$$

The substitution $y = ux$ will reduce either form of the homogeneous equation to an equation involving the independent variable x and the new dependent variable u in which the variables are separable. As with most separable equations the solution can be complicated, and it is often the case that y is determined implicitly in terms of x.

EXAMPLE 5.9   Solve

$$(y^2 + 2xy)dx - x^2\, dy = 0.$$

Solution   Both terms in the differential equation are homogeneous of degree 2, so the equation itself is homogeneous. Differentiating the substitution $y = ux$ gives

$$\frac{dy}{dx} = u + x\frac{du}{dx}, \quad \text{or} \quad dy = udx + xdu.$$

After substituting for y and dy in the differential equation and cancelling x^2, we obtain the variables separable equation

$$u(u + 1)dx = xdu, \quad \text{or} \quad \frac{du}{u(u+1)} = \frac{dx}{x}.$$

This has the general solution

$$u = \frac{Cx}{1 - Cx}, \quad \text{but} \quad y = ux \quad \text{and so } y = \frac{Cx^2}{1 - Cx},$$

where C is an arbitrary constant. In this case the general solution is simple and y is determined explicitly in terms of x.   ■

EXAMPLE 5.10   Solve

$$\frac{dy}{dx} = \frac{y^2}{xy - x^2}.$$

Solution   The equation is homogeneous because it can be written

$$\frac{dy}{dx} = \frac{(y/x)^2}{(y/x) - 1}.$$

Making the substitution $y = ux$, and again using the result $dy/dx = u + x\,du/dx$, reduces this to the separable equation

$$u + x\frac{du}{dx} = \frac{u^2}{u-1}, \text{ or } \left(1 - \frac{1}{u}\right)du = \frac{dx}{x}.$$

Integration gives

$$u - \ln|u| = \ln|x| + \ln|C|,$$

where C is an arbitrary integration constant. Finally, substituting $u = y/x$ and simplifying the result we arrive at the following implicit solution for y:

$$y = Ce^{y/x}. \qquad \blacksquare$$

An equation of the form

$$\frac{dy}{dx} = \frac{ax + by + c}{px + qy + r}$$

near-homogeneous

is called **near-homogeneous**, because it can be transformed into a homogeneous equation by means of a variable change that shifts the origin to the point of intersection of the two lines

$$ax + by + c = 0 \quad \text{and} \quad px + qy + r = 0.$$

EXAMPLE 5.11   Solve the initial value problem

$$\frac{dy}{dx} = \frac{y+1}{x+2y} \quad \text{with } y(2) = 0.$$

Solution   The equation is near-homogeneous and the lines $y + 1 = 0$ and $x + 2y = 0$ intersect at the point $x = 2$ and $y = -1$, so we make the variable change $x = X + 2$ and $y = Y - 1$, as a result of which the equation becomes the homogeneous equation

$$\frac{dY}{dX} = \frac{Y}{X + 2Y}.$$

Solving this as in Example 5.9 by setting $Y = uX$ leads to the equation

$$-\left(\frac{1 + 2u}{2u^2}\right)du = \frac{dX}{X},$$

with the solution

$$1/u = 2\ln|CuX|,$$

where C is an arbitrary integration constant. If we set $u = Y/X$, this becomes

$$X = 2Y \ln |CY|,$$

where C is an arbitrary constant. Returning to the original variables by substituting $X = x - 2$, $Y = y + 1$, we arrive at the required general solution

$$x = 2 + 2(y + 1) \ln |C(y + 1)|.$$

Although this is an implicit solution for y, if we regard y as the independent variable and x as the dependent variable, solution curves (integral curves) are easily graphed. Substituting the initial condition $y = 0$ when $x = 2$ in the general solution shows that $C = 1$, so the solution of the initial value problem is

$$x = 2 + 2(y + 1) \ln |y + 1|. \qquad \blacksquare$$

Summary

This section introduced the special type of first order ordinary differential equation known as an algebraically homogeneous equation. This name is frequently shortened to the term homogeneous equation, though this must not be confused with the sense in which the term homogeneous is used in Section 5.1. After showing how such equations can be solved, it was shown how a simple linear change of variables changes a near-homogeneous equation to a homogeneous equation that can then be solved.

EXERCISES 5.5

In Exercises 1 through 14 find by hand calculation the general solution of the given homogeneous or near-homogeneous equations and confirm the result by using computer algebra.

1. $y' = y/(2x + y)$.
2. $y' = (2xy + y^2)/(3x^2)$.
3. $y' = (2x^2 + y^2)/xy$.
4. $y' = (2xy + y^2)/x^2$.
5. $y' = (x - y)/(x + 2y)$.

6. $y' = (x + 4y)/x$.
7. $y' = (2x + y \cos^2(y/x))/(x \cos^2(y/x))$.
8. $y' = 3y^2/(1 + x^2)$.
9. $y' = (x + y \sin^2(y/x))/(x \sin^2(y/x))$.
10. $y' = 3x \exp(x + 2y)/y$.
11. $y' = (y + 2)/(x + y + 2)$.
12. $y' = (y + 1)/(x + 2y + 2)$.
13. $y' = (x + y + 1)/(x - y + 1)$.
14. $y' = (x - y + 1)/(x + y)$.

5.6   Exact Equations

The so-called *exact* equations have a simple structure, and they arise in many important applications as, for example, in the study of thermodynamics. After definition of an exact equation, a test for exactness will be derived and the general solution of such an equation will be found.

Exact equations

definition of an
exact equation

The first order ODE

$$M(x, y)dx + N(x, y)dy = 0$$

is said to be **exact** if a function $F(x, y)$ exists such that the total differential

$$d[F(x, y)] = M(x, y)dx + N(x, y)dy.$$

It follows directly that if

$$M(x, y)dx + N(x, y)dy = 0 \tag{22}$$

is exact, then the total differential

$$d[F(x, y)] = 0,$$

so the general solution of (22) must be

$$F(x, y) = \text{constant}. \tag{23}$$

EXAMPLE 5.12   The total differential of $F(x, y) = 3x^3 + 2xy^2 + 4y^3 + 2x$ is

$$
\begin{aligned}
d[F(x, y)] &= (\partial F/\partial x)dx + (\partial F/\partial y)dy \\
&= (9x^2 + 2y^2 + 2)dx + (4xy + 12y^2)dy,
\end{aligned}
$$

so the exact differential equation

$$(9x^2 + 2y^2 + 2)dx + (4xy + 12y^2)dy = 0$$

has the general solution

$$3x^3 + 2xy^2 + 4y^3 + 2x = \text{constant}. \qquad \blacksquare$$

Three questions now arise:

 (i) Is there a test for exactness?
 (ii) If an equation is exact, is it possible to find its general solution?
 (iii) If an equation is not exact, is it possible to modify it to make it exact?

There are satisfactory answers to the first two questions, and a less satisfactory answer to the third question. We deal with the last question first.

It can be shown that an equation of the form (21) that is *not* exact can always be made exact if it is multiplied by a suitable factor $\mu(x, y)$, called an **integrating factor**, though there is no general method by which such an integrating factor can be found. Fortunately, however, an integrating factor can always be found for a variable coefficient linear first order ODE, and in the next section the integrating factor will be derived for such an ODE and then used to find its general solution.

We now turn our attention to the first question. If $F(x, y) = \text{constant}$ is a solution of the exact differential equation

$$M(x, y)dx + N(x, y)dy = 0, \tag{24}$$

then $M(x, y) = \partial F/\partial x$ and $N(x, y) = \partial F/\partial y$. So, provided the derivatives $\partial F/\partial x$, $\partial F/\partial y$, $\partial^2 F/\partial x \partial y$, and $\partial^2 F/\partial y \partial x$ are defined and continuous in the region within which the differential equation is defined, the mixed derivatives will be equal so that $\partial^2 F/\partial x \partial y = \partial^2 F/\partial y \partial x$. This last result is equivalent to requiring that $\partial M/\partial y = \partial N/\partial x$ in order that (24) is exact, so this provides the required test for exactness.

THEOREM 5.1

a simple test for exactness

Test for exactness   The differential equation

$$M(x, y)dx + N(x, y)dy = 0$$

is exact if and only if $\partial M/\partial y = \partial N/\partial x$. ∎

EXAMPLE 5.13

Test for exactness the differential equations

(a) $\{\sin(xy + 1) + xy\cos(xy + 1)\}dx + x^2\cos(xy + 1)dy = 0$.

(b) $(2x + \sin y)dx + (2x\cos y + y)dy = 0$.

Solution   In case (a) $M(x, y) = \sin(xy + 1) + xy\cos(xy + 1)$ and $N(x, y) = x^2\cos(xy + 1)$, and $\partial M/\partial y = \partial N/\partial x$, so the equation is exact.

In case (b) $M(x, y) = 2x + \sin y$ and $N(x, y) = 2x\cos y + y$ but $\partial M/\partial y \neq \partial N/\partial x$, so the equation is not exact. ∎

Having established a test for exactness, it remains for us to determine how the general solution of an exact equation can be found. The starting point is the fact that if $F(x, y) = $ constant is a solution of the exact equation

$$M(x, y)dx + N(x, y)dy = 0,$$

then $\partial F/\partial x = M(x, y)$ and $\partial F/\partial y = N(x, y)$.

Two expressions for $F(x, y)$ can be obtained from these results by integrating M with respect to x while regarding y as a constant, and integrating N with respect to y while regarding x as a constant, because this reverses the process of partial differentiation by which M and N were obtained. However, after integrating M it will be necessary to add not only an arbitrary constant, but also an arbitrary function $f(y)$ of y, because this will behave like a constant when F is differentiated partially with respect to x to obtain M. Similarly, after integrating N it will be necessary to add not only an arbitrary constant, but also an arbitrary function $g(x)$ of x, because this will behave like a constant when F is differentiated partially with respect to y to obtain N.

These two expressions for F will look different but must, of course, be identical. The arbitrary function $f(y)$ can be found by identifying it with any function only of y that occurs in the expression for F obtained by integrating N, while the arbitrary function $g(x)$ can be found by identifying it with any function only of x that occurs in the expression for F found by integrating M, where, of course, the true constants introduced after each integration must be identical.

EXAMPLE 5.14

Show the following equation is exact and find its general solution:

$$\{3x^2 + 2y + 2\cosh(2x + 3y)\}dx + \{2x + 2y + 3\cosh(2x + 3y)\}dy = 0.$$

Solution   In this equation $M(x, y) = 3x^2 + 2y + 2\cosh(2x + 3y)$, and $N(x, y) = 2x + 2y + 3\cosh(2x + 3y)$, so as $M_y = N_x = 2 + 6\sinh(2x + 3y)$ the equation is exact:

$$F(x, y) = \int M(x, y)dx = \int \{3x^2 + 2y + 2\cosh(2x + 3y)\}dx$$

$$= x^3 + 2xy + \sinh(2x + 3y) + f(y) + C,$$

and

$$F(x, y) = \int N(x, y)dy = \int \{2x + 2y + 3\cosh(2x + 3y)\}dy$$

$$= 2xy + y^2 + \sinh(2x + 3y) + g(x) + D.$$

For these two expressions to be identical, we must set $f(y) \equiv y^2$, $g(x) \equiv x^3$, and $D = C$, so $F(x, y)$ is seen to be

$$F(x, y) = x^3 + 2xy + y^2 + \sinh(2x + 3y) + C,$$

and so the general solution is

$$x^3 + 2xy + y^2 + \sinh(2x + 3y) = C,$$

where as C is an arbitrary constant we have chosen to write C rather than $-C$ on the right of the solution.   ∎

Summary

This section introduced the class of first order ordinary differential equations known as exact equations that arise in many different applications. It was then shown how the equality of mixed derivatives yields a simple test for exactness.

EXERCISES 5.6

In Exercises 1 through 8 test the equation for exactness, and when an equation is exact, find its general solution.

1. (a) $\{\sin(3y) + 4x^2 y\}dx + \{3x\cos(3y) + y + 2x^3\}dy = 0$;
   (b) $\{4x^3 + 3y^2 + \cos x\}dx + \{6xy + 2\}dy = 0$.
2. (a) $\{(2x + 3y^2)^{-1/2} + 4y^3 + 2x\}dx + \{3y/(2x + 3y^2) + 12xy^2\}dy = 0$;
   (b) $\{\cos(x + 3y^2) + 4xy^3\}dx + \{6y\cos(x + 3y^2) + 3x^2 y^2 + 2y\}dy = 0$.
3. (a) $\{\sin x + x\cos x + \cosh(x + 2y)\}dx + \{3y^2 + 2\cosh(x + 2y)\}dy = 0$;
   (b) $\{6x(2x^2 + y^2)^{1/2} + x^2\}dx + 2y(2x^2 + y^2)^{1/2}dy = 0$.
4. (a) $\{6x/(3x^2 + y) + 4xy^3\}dx + \{1/(3x^2 + y) + 6x^2 y^2 + 3y^2\}dy = 0$;
   (b) $\{\sin(xy) + xy\cos(xy) + y^2\sin(xy)\}dx + \{x^2\cos(xy) + \cos(xy) - xy\sin(xy)\}dy = 0$.

5. (a) $\dfrac{3x^2}{2\sqrt{x^3 + y^2}}dx + \left\{\dfrac{y}{\sqrt{x^3 + y^2}} + 6y\right\}dy = 0$;
   (b) $\{y/x + 2x\sinh(y^2)\}dx + \{\ln x + 2x^2 y\cosh(y^2)\}dy = 0$.
6. (a) $\{4xy + 1/x\}dx + \{2x^2 - 1/y\}dy = 0$;
   (b) $\{6xy - 2/(x^2 y)\}dx + \{3x^2 - 2/(xy^2)\}dy = 0$.
7. (a) $\{2xy + 6/x\}dx + \{x^2 + 4/y\}dy = 0$;
   (b) $\{2x/(2x + 3y^2) - 2x^2/(2x + 3y^2)^2 + 2\}dx - 6x^2 y/(2x + 3y^2)^2 dy = 0$.
8. (a) $\{(5/2)x^{3/2} + 14y^3\}dx + \{(3/2)\sqrt{y} + 42xy^2\}dy = 0$;
   (b) $(y/x^2)\cos(y/x)dx + \{(1/x)\cos(y/x) + 6y\exp(y^2)\}dy = 0$.

5.7   Linear First Order Equations

The **standard form** of the **linear first order differential equation** is

standard form of linear first order equation

$$\frac{dy}{dx} + P(x)y = Q(x), \tag{25}$$

where $P(x)$ and $Q(x)$ are known functions. An **initial value problem** (**i.v.p**) for a linear first order ODE involves the specification of an initial condition

$$y(x_0) = y_0, \tag{26}$$

where this last condition means that $y = y_0$ when $x = x_0$. Thus, the solution of the initial value problem will evolve away from the point (x_0, y_0) in the (x, y)-plane as x increases from x_0.

To find the general solution of (25) we multiply the equation by a function $\mu(x)$, still to be determined, to obtain

$$\mu \frac{dy}{dx} + \mu P(x)y = \mu Q(x), \tag{27}$$

and seek a choice for μ that allows the left-hand side of (26) to be written as $d(\mu y)/dx$.

With this choice of μ, equation (27) becomes

$$\frac{d(\mu y)}{dx} = \mu Q(x), \tag{28}$$

so integrating with respect to x and dividing by μ shows the general solution of (25) to be

$$y(x) = \frac{C}{\mu(x)} + \frac{1}{\mu(x)} \int \mu(x)Q(x)dx, \tag{29}$$

where C is an arbitrary integration constant. Notice that it is essential to include the arbitrary integration constant *immediately* after the integration $\int \mu(x)Q(x)dx$ has been performed, and *before* dividing by $\mu(x)$; otherwise, the form of the general solution will be incorrect.

To make use of (29) it is necessary to determine the function $\mu(x)$ called the **integrating factor** for the linear first order ODE in (24). By definition

integrating factor

$$\frac{d(\mu y)}{dx} = \mu \frac{dy}{dx} + \mu P(x)y,$$

so after expanding the left-hand side this becomes

$$\mu \frac{dy}{dx} + y \frac{d\mu}{dx} = \mu \frac{dy}{dx} + \mu P(x)y.$$

Cancelling the terms $\mu dy/dx$ and dividing by y gives the following variables separable equation for the integrating factor $\mu(x)$:

$$\frac{d\mu}{dx} = \mu P(x).$$

This has the solution

$$\mu(x) = A \exp\left\{ \int P(x)dx \right\},$$

where A is an arbitrary integration constant. As μ multiplies the entire equation (27), the choice of A is immaterial, so for simplicity we will always set $A = 1$ and take the **integrating factor** to be

finding the integrating factor

$$\mu(x) = \exp\left\{\int P(x)dx\right\}. \tag{30}$$

Inserting (30) into (29) shows the **general solution** of (25) to be

$$y(x) = C \exp\left\{-\int P(x)dx\right\} + \exp\left\{-\int P(x)dx\right\} \int \exp\left\{\int P(x)dx\right\} Q(x)dx. \tag{31}$$

If an initial value problem is involved in which the solution of (25) is required subject to the initial condition $y(x_0) = y_0$, the value of the arbitrary constant C in (31) must be chosen accordingly.

The form of the general solution in (31) is mainly of importance for theoretical reasons, because it shows that the general solution is the sum of a **complementary function**

complementary function, particular integral, and general solution

$$y_c(x) = C \exp\left\{-\int P(x)dx\right\} \tag{32}$$

that contains the arbitrary constant belonging to the general solution of (25), and a **particular integral**

$$y_p(x) = \exp\left\{-\int P(x)dx\right\} \int \exp\left\{\int P(x)dx\right\} Q(x)dx \tag{33}$$

that contains no arbitrary constant and is determined by the nonhomogeneous term $Q(x)$.

Substitution of $y_c(x)$ into the homogeneous form of (25) given by

$$\frac{dy}{dx} + P(x)y = 0$$

shows that $y_c(x)$ is its general solution. The general solution of the nonhomogeneous equation (25) is now seen to be the sum of the general solution of the homogeneous form of the equation, and a particular integral determined by the nonhomogeneous term. It will be shown later that this is the pattern of the general solution for all linear nonhomogeneous differential equations, no matter what their order.

Rather than trying to remember the form of general solution given in (31), it is better to obtain the solution by starting from the integrating factor $\mu(x)$ in (30) and integrating result (28), while not forgetting to include the arbitrary constant immediately after the integration before dividing by $\mu(x)$. For convenience, the steps in the determination of the general solution of (25) can be listed as follows.

**steps used when
solving a linear first
order equation**

Rule for solving linear first order equations

STEP 1   If the equation is not in standard form and is written

$$a(x)\frac{dy}{dx} + b(x)y = c(x),$$

divide by $a(x)$ to bring it to the standard form

$$\frac{dy}{dx} + P(x)y = Q(x),$$

with $P(x) = b(x)/a(x)$ and $Q(x) = c(x)/a(x)$

STEP 2   Find the integrating factor

$$\mu(x) = \exp\left\{\int P(x)dx\right\}.$$

STEP 3   Rewrite the original differential equation in the form

$$\frac{d(\mu y)}{dx} = \mu Q(x).$$

STEP 4   Integrate the equation in Step 3 to obtain

$$\mu(x)y(x) = \int \mu(x)Q(x)dx + C.$$

STEP 5   Divide the result of Step 4 by $\mu(x)$ to obtain the required general solution of the linear first order differential equation in Step 1.

STEP 6   If an initial condition $y(x_0) = y_0$ is given, the required solution of the i.v.p. is obtained by choosing the arbitrary constant C in the general solution found in Step 5 so that $y = y_0$ when $x = x_0$.

EXAMPLE 5.15   Solve the initial value problem

$$\cos x \frac{dy}{dx} + y = \sin x, \text{ subject to the initial condition } y(0) = 2.$$

Solution   We follow the steps in the above rule.

STEP 1   When written in standard form the equation becomes

$$\frac{dy}{dx} + \frac{1}{\cos x}y = \tan x,$$

so $P(x) = 1/\cos x$ and $Q(x) = \tan x$.

STEP 2   The integrating factor

$$\mu(x) = \exp\left\{\int \frac{dx}{\cos x}\right\} = \exp\{\ln |\sec x + \tan x|\}$$

$$= \sec x + \tan x = \frac{1 + \sin x}{\cos x}.$$

STEP 3   The original differential equation can now be written

$$\frac{d}{dx}\left[\left(\frac{1 + \sin x}{\cos x}\right)y(x)\right] = \left(\frac{1 + \sin x}{\cos x}\right)\tan x.$$

STEP 4   Integrating the result of Step 3 gives

$$\left(\frac{1 + \sin x}{\cos x}\right)y(x) = \int \left(\frac{1 + \sin x}{\cos x}\right)\tan x\, dx + C$$

$$= \int \sec x \tan x\, dx + \int \tan^2 x\, dx + C$$

$$= \sec x + \tan x - x + C = \frac{1 + \sin x}{\cos x} - x + C.$$

STEP 5   Dividing the result of Step 4 by the integrating factor $\mu(x) = (1 + \sin x)/\cos x$ shows that the required general solution is

$$y(x) = \frac{C \cos x}{1 + \sin x} + 1 - \frac{x \cos x}{1 + \sin x},$$

for x such that $1 + \sin x \neq 0$.

The *complementary function* is seen to be

$$y_c(x) = \frac{C \cos x}{1 + \sin x},$$

and the *particular integral* is

$$y_p(x) = 1 - \frac{x \cos x}{1 + \sin x}.$$

STEP 6   The initial condition requires that $y = 2$ when $x = 0$, and the general solution is seen to satisfy this condition if $C = 1$, so the solution of the i.v.p. is

$$y(x) = 1 + \frac{(1 - x) \cos x}{1 + \sin x}.$$ ■

EXAMPLE 5.16

An R–L circuit contains an inductor and resistor in series, and a current is made to flow through them by applying a voltage across the ends of the circuit. If the inductance varies linearly with time in such a way that $L(t) = L_0(1 + kt)$, find the current $i(t)$ flowing in the circuit when $t > 0$, given that a constant voltage V_0 is applied at time $t = 0$ when $i(t) = 0$.

Solution   The voltage change due to a current $i(t)$ flowing through the inductance is $d(L(t)i)/dt$, and from Ohm's law the corresponding voltage change across the resistance R is Ri, so as the sum of the these voltage changes must equal the imposed constant voltage V_0, the differential equation determining the current becomes

$$\frac{d}{dt}(L(t)i) + Ri = V_0 \quad \text{for} \quad t > 0.$$

Substituting for $L(t)$ and rearranging terms we arrive at the following linear first order variable coefficient nonhomogeneous equation for $i(t)$

$$\frac{di}{dt} + \left(\frac{kL_0 + R}{L_0(1 + kt)}\right)i = \frac{V_0}{L_0(1 + kt)},$$

subject to the initial condition $i(0) = 0$.

In the notation of this section $P(t) = \left(\dfrac{kL_0 + R}{L_0(1 + kt)}\right)$ and $Q(t) = \dfrac{V_0}{L_0(1 + kt)}$, so the integrating factor in Step 2 becomes

$$\mu(t) = \exp\left\{\int P(t)dt\right\} = (1 + kt)^{[kL_0+R]/kL_0}.$$

Using $\mu(t)$ and $Q(t)$ in Step 4 and applying the initial condition $i(0) = 0$ then shows that the current $i(t)$ at a time $t > 0$ is determined by

$$i(t) = \left(\frac{V_0}{kL_0 + R}\right)\left(1 - (1 + kt)^{\left(\frac{kL_0+R}{kL_0}\right)}\right). \qquad \blacksquare$$

Summary

The study of the linear first order differential equation considered in this section is important in its own right, and it also provides the key to understanding the nature of the solution of linear higher order differential equations. It was shown how, after an equation is written in standard form, it can be solved by means of an integrating factor that can be found directly from the coefficient of y in the equation.

EXERCISES 5.7

In Exercises 1 through 10 find the general solution for the linear first order differential equation, and check your result by using computer algebra.

1. $dy/dx + 2y = 1.$
2. $dy/dx + (1/x)y = x.$
3. $(x + 1)dy/dx + y = 2x(x + 1).$
4. $x^2 dy/dx + xy = x^2 \sin x.$
5. $x^2 dy/dx - 2xy = 1 + x.$
6. $\sin x\, dy/dx - y\cos x = 2\sin^2 x.$
7. $x\, dy/dx + 2y = x^2.$
8. $(x + 3)dy/dx - 2y = x + 3.$
9. $\sin x\, dy/dx - y = 2\sin x.$
10. $\sin x\, dy/dx + y = \sin x.$

In Exercises 11 through 16 solve the initial value problem for the linear first order differential equation, and check your result by using a computer algebra package.

11. $x\, dy/dx - y = x^2\cos x,$ with $y(\pi/2) = \pi.$
12. $x^2 dy/dx + 2xy = 2 + x,$ with $y(1) = 1.$
13. $x\, dy/dx - 2y = 2 + x,$ with $y(1) = 0.$
14. $x\, dy/dx + 2y = 2x^4,$ with $y(1) = 1.$
15. $\sin x\, dy/dx + y\cos x = 2\sin^2 x,$ with $y(\pi/2) = 0.$
16. $2\, dy/dx + y = x^2,$ with $y(0) = 1.$
17. A 25-liter gas cylinder contains 80% oxygen and 20% helium. If helium is added at a rate of 0.2 liters a second, and the mixture is drawn off at the same rate, how long will it be before the cylinder contains 80% helium?
18. If in Exercise 17 the volume of the gas cylinder is 20 liters and initially it contains 90% oxygen and 10% helium, and the rate of supply of helium is q liters a second, what must be the value of q if the cylinder is to become 80% full of helium in 1 minute?

19. A particle of unit mass moves horizontally in a resisting medium with velocity $v(t)$ at time t with a resistance opposing the motion given by $kv(t)$, with $k > 0$. If the particle is also subject to an additional resisting force kt, write down the differential equation for $v(t)$, and hence find the value of k if the motion starts with $v(0) = v_0$, and at time $t = 1/k$ its velocity is $v(1/k) = \frac{1}{4}v_0$.

5.8   The Bernoulli Equation

The **Bernoulli equation** is a nonlinear first order differential equation with the standard form

standard form of the Bernoulli equation

$$\frac{dy}{dx} + P(x)y = Q(x)y^n, \quad (n \neq 1). \tag{34}$$

The substitution

$$u = y^{1-n} \tag{35}$$

reduces (34) to the linear first order ODE

$$\frac{1}{(1-n)}\frac{du}{dx} + P(x)u = Q(x), \tag{36}$$

and this can be solved by the method described in Section 5.7. Once the general solution $u(x)$ of (36) has been found, the general solution $y(x)$ of (34) follows by returning to the original dependent variable by making the substitution $u = y^{1-n}$.

When using the general solution in (36) it is important to write the Bernoulli equation in standard form before identifying $P(x)$, $Q(x)$, and n. However, if the form of the equation corresponding to (36) is derived directly, starting from the substitution $u = y^{1-n}$, there is no need for the equation to be in standard form.

The Bernoulli equation occurs in various applications of mathematics that involve some form of nonlinearity. It occurs, for example, in solid and fluid mechanics, where it is found to describe an important characteristic of special types of wave that propagate through space as time increases. To appreciate how this ODE enters into these problems, we consider a simple application to solid mechanics involving a long bar made of a composite material or a polymer whose properties are such that the extension caused by a force does not obey Hooke's law, and so is *not* proportional to the force. Materials of this type are said to be **nonlinearly elastic**. If such a bar receives a blow at one end a disturbance will propagate along it at a finite speed, so that at any instant of time there will be a region in the bar through which the disturbance has passed, and a region ahead of the disturbance through which it has still to pass. When the blow is not large, the propagating boundary between these two regions is called a **wavefront** and t the function representing the displacement at position x at any given time t will be continuous along the bar, though its derivative with respect to x will be discontinuous across the wavefront.

wavefront and acceleration wave

The propagating jump in the derivative of the displacement with respect to x at the wavefront as a function of time is called an **acceleration wave**, and we will denote it by $a(t)$. For many nonlinear materials the magnitude $a(t)$ of the acceleration wave obeys a Bernoulli equation of the form

$$\frac{da}{dt} + \mu(t)a = \beta(t)a^2. \tag{37}$$

It was shown by P. J. Chen (*Selected Topics in Wave Propagation*, Noordhoff, Leyden, 1976, p. 29) that $\mu(t)$ depends on the material properties of the medium through which the disturbance propagates and also the geometry involved, which in a one-dimensional case may be plane, cylindrically, or spherically symmetric, but that the function $\beta(t)$ depends only on the material properties of the medium. This same equation governs the behavior of acceleration waves in three space dimensions and time.

Because of the effects of nonlinearity, in many materials it is possible for the acceleration wave to strengthen as it propagates to the point at which the continuity of the displacement function breaks down and what is called a **shock wave** forms. When this occurs, the speed of propagation of disturbances and other physical quantities become discontinuous across the shock wave, and this in turn can lead to the fracture of the material. Once the material properties of such a medium are specified together with the nature of the initial disturbance, the Bernoulli equation in (37) can be used to determine whether or not a shock wave will form and, if it does, the point along the bar where this occurs.

EXAMPLE 5.17

examples of the Bernoulli equation

Solve the Bernoulli equation

$$\frac{da}{dt} + a = ta^2,$$

and find a condition that determines when the solution becomes unbounded.

Solution   The equation is in standard form with $P(t) = 1$, $Q(t) = t$, and $n = 2$. Making the substitution $u = 1/a$ corresponding to (35) and substituting into (36) leads to the linear first order equation

$$\frac{du}{dt} - u = -t.$$

Solving this by the method described in Section 5.7 gives

$$u(t) = Ce^t + 1 + t,$$

so transforming back to the variable $a(t)$, we find that

$$a(t) = 1/(Ce^t + 1 + t).$$

The solution $a(t)$ of the Bernoulli equation will become unbounded at $t = t_c$ if t_c is a solution of the equation $C\exp(t_c) + 1 + t_c = 0$. This result shows that an acceleration wave starting at time $t = 0$ will decay instead of evolving into a shock wave if $C > 0$, because then the equation for t_c has no positive solution, whereas a shock wave will always form if $C < 0$.

Had $a(t)$ represented the magnitude of an acceleration wave, the development of an infinite gradient in the displacement corresponding to $a(t_c) = \infty$ would indicate shock formation. ∎

EXAMPLE 5.18   Find the general solution of

$$\frac{dy}{dx} - 2y = xy^{1/2}.$$

Solution   In terms of the standard form of the Bernoulli equation given in (34), $P(x) = -2$, $Q(x) = x$, and $n = 1/2$. However, rather than substituting into equation (36) to obtain a linear differential equation for $u(x)$, we will derive it directly starting from the substitution $u = y^{1/2}$, and differentiating it to find du/dx in terms of dy/dx. We have

$$\frac{du}{dx} = \frac{1}{2}y^{-1/2}\frac{dy}{dx} = \frac{1}{2u}\frac{dy}{dx}, \quad \text{so} \quad \frac{dy}{dx} = 2u\frac{du}{dx}.$$

Substituting for y and dy/dx in the Bernoulli equation and cancelling a factor $2u$ gives the following linear equation (compare it with (36) after substituting for $P(x)$, $Q(x)$ and n):

$$\frac{du}{dx} - u = \frac{1}{2}x.$$

The method of Section 5.6 shows this equation to have the general solution

$$u(x) = Ce^x - (1/2)(1 + x),$$

so as $u = y^{1/2}$, the required general solution of the Bernoulli equation is

$$y(x) = [Ce^x - (1/2)(1 + x)]^2. \qquad \blacksquare$$

> **JACOB BERNOULLI (1654–1705)**
> A Swiss mathematician born in Basel where he was professor of mathematics until his death. He was a member of one of the most distinguished families of mathematicians in all of the history of mathematics. His most important contributions were to the theory of probability and the calculus and theory of elasticity. Other members of the family contributed to many different parts of mathematics including hydrodynamics and the calculus of variations.

Summary   In a sense, the Bernoulli equation, which is a nonlinear first order differential equation, stands on the boundary between linear and nonlinear first order differential equations, so for this and other reasons it is important in applications. It arises in different applications, many of which themselves arise from problems bordering on linear and nonlinear regimes. This section showed how a straightforward change of variable transforms a Bernoulli equation into a linear first order differential equation that can then be solved by the method of Section 5.6.

EXERCISES 5.8

In Exercises 1 through 8 find the general solution of the Bernoulli equation.

1. $dy/dx + 2y = 2xy^{1/2}$.
2. $dy/dx + y = 3y^2$.
3. $dy/dx - y = 2xy^{3/2}$.
4. $x\,dy/dx + y = xy^2$.
5. $dy/dx + 2y\sin x = 2y^2\sin x$.
6. $x\,dy/dx + y = 2xy^{1/2}$.
7. $x\,dy/dx - 2y = xy^{3/2}$.
8. $dy/dx + 4xy = xy^3$.

9. A model for the variation of a finite amount of stock $n(t)$ in a warehouse as a function of the time t caused by the supply of fresh stock and its removal by demand is

$$\frac{dn}{dt} = (a - bn)n \quad \text{with the constants } a, b > 0,$$

where $n(0) = n_0$. Find $n(t)$ and discuss the nature of the change in the stock level as a function of time according as n_0 is less than a/b, equal to a/b, or greater than a/b.

10.* This exercise concerns water in a canal of variable depth with the x-axis taken along the canal in the equilibrium surface of the water, and the y-axis vertically downwards. Let the equilibrium depth of water in a channel be $h(x)$, and the cross-sectional area of water in the canal be a slowly varying function $W(x)$. When a water wave advances along the channel into water at rest there will be a change of acceleration across the advancing line (wavefront) that separates the disturbed water from the undisturbed water. Such an advancing disturbance is called an **acceleration wave**. If the change in acceleration across the wavefront at point x along the channel is $a(x)$, it can be shown that the strength $a(x)$ of the acceleration wave obeys the Bernoulli equation

$$\frac{da}{dx} + \left(\frac{3h'}{4h} + \frac{W'}{2W} \right) a + \frac{3a^2}{2h} = 0.$$

If the initial condition for $a(x)$ is $a(0) = a_0$, then a wave of **elevation** wave is one for which $a_0 < 0$, and a wave of **depression** is one for which $a_0 > 0$. In this approximation the wave will **break**, due to the water surface becoming vertical at the wavefront if, after propagating a critical distance x_c along the channel, the strength of the acceleration $a(x_c) = \infty$.

(i) Find $a(x)$ in terms of $a_0 = a(0), h_0 = h(0)$ and $W_0 = W(0)$.

(ii) Discuss the breaking and non-breaking of waves of elevation and depression.

(iii) If the water shelves to zero at $x = l$, so that $h(l) = 0$, find a condition that ensures the wave breaks before $x = l$.

5.9   The Riccati Equation

The **Riccati equation** is an important nonlinear equation with the standard form

standard form of the Riccati equation

$$\frac{dy}{dx} + P(x)y + R(x)y^2 = Q(x). \tag{38}$$

Its significance derives from the fact that it stands at the boundary between linear and nonlinear equations, and it occurs in various applications of mathematics that involve nonlinear problems. The Riccati equation reduces to a linear first order equation when $R(x) \equiv 0$, and to a Bernoulli equation when $Q(x) \equiv 0$.

Obtaining the general solution of a Riccati equation is difficult, but the task is simplified if a particular solution is known, or can be found by inspection. If a particular solution is $y_1(x)$ is known, then

(i) The substitution $y = y_1 + 1/u$ reduces the equation to a linear first order equation.

substitutions that simplify the Riccati equation

(ii) The substitution $y = y_1 + u$ reduces the equation to a Bernoulli equation.

(iii) The general substitution

$$y = \frac{1}{R(x)z}\frac{dz}{dx}$$

reduces the Riccati equation to the linear homogeneous second order ODE

$$\frac{d^2z}{dx^2} + \left\{ P(x) - \frac{R'(x)}{R(x)} \right\} \frac{dz}{dx} - R(x)Q(x)z = 0$$

discussed in Chapters 6 and 8.

Substitution (i) is often the most convenient one to use, as will be seen from the next example.

EXAMPLE 5.19   Find the general solution of the Riccati equation

$$\frac{dy}{dx} + x^2y - xy^2 = 1.$$

Solution   Inspection shows that $y_1(x) = x$ is a particular solution, so we make the substitution $y = x + 1/u$, from which it follows that

$$\frac{dy}{dx} = 1 - \frac{1}{u^2}\frac{du}{dx},$$

and after substitution for y and dy/dx in the Riccati equation it reduces to the linear ODE

$$\frac{du}{dx} + x^2u = -x.$$

Solving this by the method of Section 5.6 gives

$$u(x) = C\exp(-x^3/3) - \exp(-x^3/3)\int x\exp(x^3/3)dx,$$

where the integral in the last term cannot be expressed in terms of elementary functions. Transforming back to the variable $y(x)$ shows the general solution of the Riccati equation to be

$$y(x) = x + \frac{\exp(x^3/3)}{C - \int x\exp(x^3/3)dx}.$$

It is not unusual for solutions of ODEs to give rise to functions such as $\int x\exp(x^3/3)dx$ that have no representation in terms of known functions, because not all functions have antiderivatives that are expressible in terms of elementary functions. ∎

JACOPO FRANCESCO (COUNT) RICCATI (1676–1754)
An Italian mathematician whose main contributions to mathematics were in the field of differential equations, though he also contributed to geometry and the study of acoustics.

Additional information relevant to the material in Sections 5.4 to 5.9 is to be found in the appropriate chapters of any one of references [3.3] to [3.5], [3.15], [3.16], and [3.19]. A sophisticated and extremely enlightening discussion of ordinary differential equations is to be found in reference [3.1] that considers not only first order equations, but also higher order equations and systems.

Summary   This section introduced the Riccati equation, of which the Bernoulli equation is a special case. Solving the Riccati equation is difficult, but some substitutions were given that simplify this task when one solution of the Riccati equation is already known, possibly by inspection.

EXERCISES 5.9

1. Show that the substitution $y = y_1 + 1/u$ reduces the Riccati equation in (38) to a linear first order equation.

2. Show that the substitution $y = y_1 + u$ reduces the Riccati equation in (38) to a Bernoulli equation.

In Exercises 3 through 6 verify that $y_1(x)$ is a solution of the Riccati equation and use it to find the general solution of the equation.

3. $dy/dx + 2x^2 y - 2xy^2 = 1$, with $y_1(x) = x$.

4. $dy/dx + 2y^2 - y = 1$, with $y_1(x) \equiv 1$.

5. $dy/dx - 2y^2 + 3y = 1$, with $y_1(x) \equiv 1$.

6. $dy/dx - 3x^2 y + 3xy^2 = 1$, with $y_1(x) = x$.

7. Verify that the substitution

$$y = \frac{1}{R(x)z}\frac{dz}{dx}$$

reduces the Riccati equation (38) to the linear homogeneous second order ODE

$$\frac{d^2 z}{dx^2} + \left\{ P(x) - \frac{R'(x)}{R(x)} \right\} \frac{dz}{dx} - R(x)Q(x)z = 0.$$

5.10   Existence and Uniqueness of Solutions

The questions of whether a solution to an initial value problem for a first order differential equation can be found and, when a solution does exist, whether it is the only solution are of fundamental importance in the theory of differential equations, and also in their applications. Establishing that a solution to an initial value problem

existence and uniqueness

can be found is called the **existence** problem, while ensuring that when a solution exists it is the only one is called the **uniqueness** problem. To show that the questions of existence and uniqueness arise even with very simple initial value problems we examine the following two examples.

Let us consider the initial value problem

$$\frac{dy}{dx} = \frac{4}{3}y^{1/4}, \quad \text{with } y(0) = -1,$$

involving a variables separable equation. Integration shows the general solution to be

$$y^3 = (x + C)^4,$$

from which it can be seen that y is essentially nonnegative. Clearly there can be no solution to this equation such that $y = -1$ when $x = 0$, so this is an example of an initial value problem that has *no* solution. Had the initial condition been $y(0) = 1$ the unique solution would have been

$$y^3 = (x + 1)^4.$$

In fact this equation has a solution for any initial condition in which $y(x)$ is *positive*, but no solution when it is *negative*. This is hardly surprising, because had we examined the function $y^{1/4}$ carefully before proceeding with the integration we would have seen that it is a complex number whenever y is negative. Sometimes, as here, an inspection of the initial condition and the equation can show in advance whether or not the condition is appropriate, but more frequently constraints on an initial condition that allow a solution to the differential equation to exist only emerge when the form of the solution is known.

To illustrate nonuniqueness, we need only consider the differential equation

$$\frac{dy}{dx} = 3y^{2/3}, \text{ subject to the initial condition } y(0) = 0.$$

The equation is variables separable, and integration shows it has the solution $y = x^3$, but this is not the only solution because it also has the singular, though somewhat uninteresting, solution $y = 0$.

However, these are not the only two solutions, because for any $a > 0$ the function

$$y(x) = \begin{cases} 0, & x < a \\ (x - a)^3, & x \geq a \end{cases}$$

is continuous, has a continuous first derivative, and satisfies both the differential equation and the initial condition, showing that it also is a solution. As $a > 0$ is arbitrary, we see that $y(x)$ is a one-parameter family of solutions, so clearly this initial value problem does not have a unique solution.

The following theorem on existence and uniqueness is stated without proof (see, for example, references [3.1], [3.3], [3.4], [3.10] and [3.12]). It is important to appreciate that though the conditions in the theorem are *sufficient* to ensure existence and uniqueness, they are not *necessary* conditions, as examples can be constructed that fail to satisfy the conditions of the theorem, but nevertheless have a unique solution.

THEOREM 5.2

conditions that definitely ensure existence and uniqueness

Existence and uniqueness of solutions   Let $f(x, y)$ be a continuous and bounded function of x and y in a rectangular region R of the (x, y)-plane that contains a given point (x_0, y_0). Then for some suitably small positive number h the initial value problem

$$\frac{dy}{dx} = f(x, y), \qquad \text{with} \qquad y(x_0) = y_0$$

has at least one solution within the open interval $x_0 - h < x < x_0 + h$. If, in addition, $\partial f / \partial y$ is continuous and bounded in R, the solution is unique in an open interval centered on x_0 that may lie within the interval $x_0 - h < x < x_0 + h$. ∎

Let us apply this theorem to the initial value problem

$$\frac{dy}{dx} = 3y^{2/3}, \quad \text{with } y(0) = 0,$$

that we have just shown does not have a unique solution. The function $f(x, y) = 3y^{2/3}$ is continuous in any neighborhood of the origin where the initial condition is given, but $\partial f / \partial y = 2y^{-1/3}$ is unbounded at the origin. So the first condition of Theorem 5.2 is satisfied but the second is not, showing that although this initial value problem has a solution, it is not unique.

Summary
This section described what is meant by the existence of a solution of a differential equation, and the uniqueness of a solution that is usually expected in applications to physical problems. A theorem, stated without proof, was given that guarantees both the existence and uniqueness of a solution. However, the conditions of the theorem are more restrictive than necessary, so equations can be found that while not satisfying the conditions of the theorem nevertheless have a solution, and it is unique.

EXERCISES 5.10

In Exercises 1 through 6, find any points at which the imposition of initial conditions will not lead to a unique solution.

1. $dy/dx = (1 - x)^{1/2}$.

2. $dy/dx = xy + 1$.

3. $dy/dx = x^2 + y^2$.

4. $dy/dx = (x^2 + y^2 - 1)^{-1/2}$.

5. $dy/dx = -y/x$.

6. $dy/dx = x \ln|1 - y^2|$.

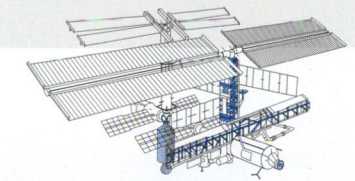

CHAPTER 5
TECHNOLOGY PROJECTS

Project 1

Solution of First Order Linear Differential Equation

The purpose of this project is to use computer algebra to solve a first order equation step by step from first principles, and then to obtain the same result by means of a computer software ODE solver.

1. Given the linear first order differential equation

$$y' + (3x^2 \sin x)y = 2x^2 \sin x,$$

   use computer integration to find the general solution by reproducing the steps in the rule for the solution by means of an integrating factor given in Section 5.6, and check the result by substitution into the differential equation.

2. Use a computer ODE solver to find the general solution and confirm that it is the same as the result obtained in step 1.

Project 2

Direction Fields and Integral Curves

The purpose of the following project is to gain insight into the relationship between direction fields and integral curves by using a computer package to plot the direction fields for two nonlinear first order differential equations, and then to add to the direction field plots some typical integral curves obtained by using a standard numerical ODE solver package.

1. Construct the direction field for the nonlinear ODE

$$y' = \sin\left(\frac{1}{2}x\right)\cos\left(\frac{1}{2}x + y\right) \text{ for } -6 \le x \le 6,$$
$$-6 \le y \le 6.$$

2. Use a standard ODE numerical solver package to find the solutions (the integral curves) through the points $(-6, -4)$, $(-6, -2)$, $(-6, 2)$,

$(-6, 4)$. Superimpose the integral curves on the direction field and compare them with the arrows in the direction field.

3. Repeat Steps 1 and 2, but this time using the nonlinear ODE

$$y' = x \sin(y - 1)/(3 + \cos x) \text{ for } -6 \le x \le 6,$$
$$-6 \le y \le 6.$$

Project 3

Direction Fields and Isoclines

An *isocline* is a curve in the direction field of the differential equation $y' = f(x, y)$ at each point of which the slope of the direction field has the same constant value. This means that wherever a solution curve of the equation intersects an isocline, its tangent will have the same slope. The isoclines of the differential equation $y' = f(x, y)$ are the curves $k = f(x, y)$, where k is the slope (gradient) of all solution curves at the points where they intersect the isocline. In general an isocline is not a solution curve and, depending on the function $f(x, y)$, there may be no isoclines for some values of the constant k. The purpose of this project is to construct the direction field for an ODE, and to superimpose on it some representative isoclines and solution curves to illustrate their interrelationship.

1. Use computer algebra to construct the direction field for the ordinary differential equation

$$y' = x^2 - y - 1 \text{ for } -2 \le x \le 2, -2 \le y \le 2,$$

   and superimpose on the direction field the isoclines corresponding to $k = -1, 0, 1, 2$. Verify that all arrows intersecting an isocline are parallel.

2. Use a standard ODE numerical solver package to find the solutions through the points $(-2, -1.5)$, $(-2, -0.5)$, $(-2, 0.5)$, $(-2, 1.5)$. Superimpose the solution curves on the isoclines found in Step 1 and confirm that the tangents to solution curves where they intersect an isocline are all parallel.

CHAPTER 6

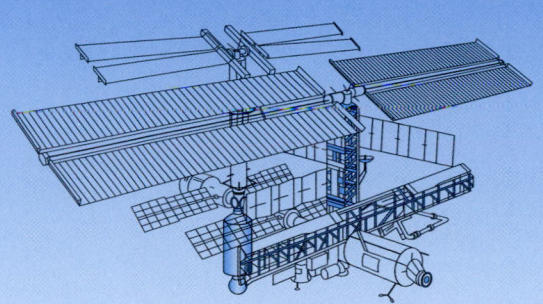

Second and Higher Order Linear Differential Equations and Systems

Linear second order differential equations with constant coefficients are the simplest of the higher order differential equations, and they have many applications. They are of the general form $y'' + Ay' + By = F(x)$ with A and B constants and $F(x)$, called the nonhomogeneous term, a known function of x. The equation is called nonhomogeneous when $F(x)$ is not identically zero; otherwise, it is called homogeneous. All general solutions are shown to be the sum of two quite different parts, one being a solution of the homogeneous equation called the complementary function that contains the expected two arbitrary constants of integration, and the other a special solution called a particular integral that depends only on $F(x)$ and contains no arbitrary constants.

Methods are developed for the solution of homogeneous and nonhomogeneous second order equations and for the solution of associated initial value problems. Particular attention is paid to the second order equations that describe oscillatory phenomena, because equations of this type arise in practical problems involving oscillations in electrical circuits, in the description of many types of mechanical vibration, and elsewhere. It is shown that in stable oscillatory motions the particular integral describes the start-up of an initial value problem, after which it decays, leaving only the complementary function that describes the long-term behavior known as the steady state solution.

The methods of solution for second order equations developed in this chapter include the simplest one, called the method of undetermined coefficients; the powerful method of variation of parameters; and a related method involving a function called the Green's function that is independent of the nonhomogeneous term $F(x)$.

Various useful special cases of second order equations are considered, after which higher order linear differential equations and first order systems are introduced and solved, the solutions of which have the same general structure as the second order equations. Matrix methods are introduced for the description and solution of first order systems of equations. The chapter concludes with a discussion of linear autonomous systems of equations, followed by a brief introduction to nonlinear autonomous systems that arise in many practical problems and can lead to oscillatory solutions of a nonlinear nature. The general behavior of solutions of both types of autonomous system is described in an interesting and useful geometrical manner involving what are called trajectories in the phase plane.

6.1   Homogeneous Linear Constant Coefficient Second Order Equations

The simplest general higher order homogeneous differential equation that occurs in applications is the **linear constant coefficient second order equation**

$$\frac{d^2y}{dx^2} + A\frac{dy}{dx} + By = 0. \tag{1}$$

Equations like this were derived in Section 5.2(d), where they were shown to describe the motion of a mass–spring system subject to frictional resistance, and also the variation of charge in an R–L–C electric circuit. The equation also describes the pendulum-like motion of a load suspended from a crane that is set in motion when the crane rotates to a new position and soon stops. The motion can be modeled as shown in Fig. 6.1, where ℓ is the length of the crane cable, m is the load, F is the resisting frictional force exerted by the air due to motion, and θ is the angular deflection of the cable from the vertical.

The angular momentum of the load about a line through the support point of the cable at O normal to the plane of motion is $m\ell^2(d\theta/dt)$, so the rate of change of angular momentum about O is $m\ell^2(d^2\theta/dt^2)$. The moments acting to restore the load to its equilibrium position at Q are due to the air resistance F opposing the motion and the turning moment of the gravitational force mg about O. If the air resistance acting on the load is proportional to the speed of the load, and the constant of proportionality is μ, the resisting frictional force is $F = \mu\ell(d\theta/dt)$, so the restoring moment exerted by F about O is $\ell F = \mu\ell^2(d\theta/dt)$. The turning moment exerted by the gravitational force mg about O is $mg\ell\sin\theta$, so equating the rate of change of angular momentum to the sum of the two restoring moments gives

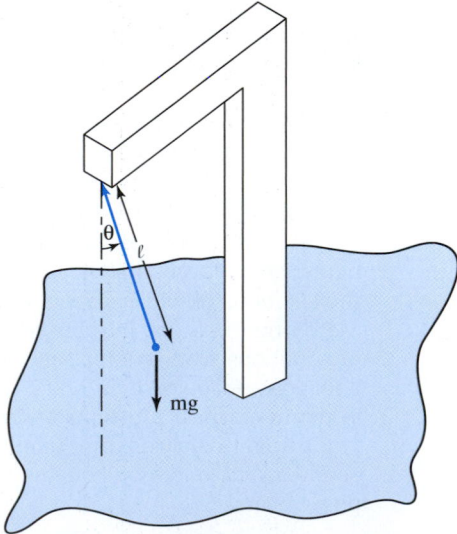

FIGURE 6.1   A deflected load supported by a crane cable.

the equation of motion

$$m\ell^2 \frac{d^2\theta}{dt^2} = -\mu\ell^2 \frac{d\theta}{dt} - mg\ell\sin\theta.$$

The negative signs on the right are necessary because the restoring moments act in the opposite sense to that of the rate of change of angular momentum.

When the angle of swing is small $\sin\theta$ can be approximated by θ, and the equation of motion simplifies to

$$\frac{d^2\theta}{dt^2} + \frac{\mu}{m}\frac{d\theta}{dt} + \frac{g}{\ell}\theta = 0.$$

Because of its many applications we start our discussion of higher order equations by examining the properties and general solution of equation (1).

Let $y_1(x)$ and $y_2(x)$ be any two solutions of (1). Then because each function satisfies the differential equation, it follows that

$$\frac{d^2 y_1}{dx^2} + A\frac{dy_1}{dx} + By_1 = 0 \quad \text{and} \quad \frac{d^2 y_2}{dx^2} + A\frac{dy_2}{dx} + By_2 = 0. \tag{2}$$

Now consider the linear combination of the two solutions

$$y(x) = c_1 y_1(x) + c_2 y_2(x), \tag{3}$$

where c_1 and c_2 are arbitrary constants. Substituting (3) into (1) and grouping terms gives

$$\frac{d^2[c_1 y_1 + c_2 y_2]}{dx^2} + A\frac{d[c_1 y_1 + c_2 y_2]}{dx} + B[c_1 y_1 + c_2 y_2]$$

$$= c_1\left[\frac{d^2 y_1}{dx^2} + A\frac{dy_1}{dx} + By_1\right] + c_2\left[\frac{d^2 y_2}{dx^2} + A\frac{dy_2}{dx} + By_2\right] = 0,$$

because each of the bracketed groups of terms vanishes on account of (2). This has shown that $y(x) = c_1 y_1(x) + c_2 y_2(x)$ is also a solution of (1).

This last result is described by saying equation (1) allows the **linear superposition** of solutions and it means that the sum of solutions is again a solution. Later we will see that linear superposition of solutions is a fundamental property of all homogeneous linear equations, including those with variable coefficients.

Two functions $y_1(x)$ and $y_2(x)$ are said to be **linearly independent** over an interval $a \leq x \leq b$ if the equation

linear superposition, dependence, and independence

$$c_1 y_1(x) + c_2 y_2(x) = 0 \tag{4}$$

is only true for all x in the interval if $c_1 = c_2 = 0$. The functions are said to be **linearly dependent** if (4) is true for some nonvanishing constants c_1 and c_2.

When the functions are linearly dependent, provided $c_1 \neq 0$, equation (4) can be written

$$y_1(x) = -\frac{c_2}{c_1} y_2(x),$$

with a corresponding result

$$y_2(x) = -\frac{c_1}{c_2} y_1(x),$$

if $c_2 \neq 0$, showing that in each case the linear dependence of the functions means they are proportional. We have established the following simple test.

---

simple test for linear independence

Test for linear independence of $y_1(x)$ and $y_2(x)$ over $a \leq x \leq b$

The two functions $y_1(x)$ and $y_2(x)$ will be linearly independent over $a \leq x \leq b$ if they are not proportional over the interval; otherwise, they will be linearly dependent.

EXAMPLE 6.1

Apply the test for linear independence to the following pairs of functions.

(a) e^x and e^{2x} are linearly independent for all x because $e^{2x}/e^x = e^x$ is defined for all x and e^x is *not* a constant.

(b) $\ln x^2$ and $\ln x^3$ are linearly dependent for $x > 0$, because $\ln x^2 = 2 \ln x$ and $\ln x^3 = 3 \ln x$, so $\ln x^2/\ln x^3 = 2/3$ is a constant, and the logarithmic function is defined for $x > 0$.

(c) $\sinh 2x$ and $\sinh x \cosh x$ are linearly dependent for all x because $\sinh 2x = 2 \sinh x \cosh x$. ∎

The notion of the linear independence of functions is of special significance when the functions are solutions of homogeneous differential equations. This is because it will be seen later that all particular solutions of such differential equations can be represented in the form of suitable linear combinations of as many linearly independent solutions as the equation allows. In fact, the number of linearly independent solutions is equal to the order of the differential equation, so the second order differential equation (1) has *two* linearly independent solutions. So, if $y_1(x)$ and $y_2(x)$ are linearly independent solutions of (1), and c_1 and c_2 are arbitrary constants, the **general solution** of (1) from which all particular solutions can be obtained can be written

general solution

$$y(x) = c_1 y_1(x) + c_2 y_2(x). \tag{5}$$

The justification of this assertion will be postponed until the nature of the linearly independent solutions of (1) has been established.

EXAMPLE 6.2

Direct substitution of the functions $y_1(x) = \sin 2x$ and $y_2(x) = \cos 2x$ into the second order differential equation

$$y'' + 4y = 0$$

confirms that they are solutions. The functions are linearly independent for all x because they are not proportional, so

$$y(x) = c_1 \cos 2x + c_2 \sin 2x$$

is the general solution of the differential equation. ∎

We will now find the general solution of (1), and when doing so use will be made of the fact that if $y(x) = ce^{\lambda x}$, with c and λ constants, then

$$\frac{dy}{dx} = \frac{d[ce^{\lambda x}]}{dx} = c\lambda e^{\lambda x} \quad \text{and} \quad \frac{d^2 y}{dx^2} = \frac{d^2[ce^{\lambda x}]}{dx^2} = c\lambda^2 e^{\lambda x}.$$

Substituting these results into (1) leads to the equation

$$(\lambda^2 + A\lambda + B)e^{\lambda x} = 0.$$

However, the factor $e^{\lambda x}$ is nonvanishing for all x, so after its cancellation this equation is seen to be equivalent to the quadratic equation for λ

$$\lambda^2 + A\lambda + B = 0. \tag{6}$$

When the quadratic equation (6) has two distinct (different) roots λ_1 and λ_2, the functions $y_1(x) = \exp(\lambda_1 x)$ and $y_2(x) = \exp(\lambda_2 x)$ will be linearly independent for all x, because $y_1(x)/y_2(x) = \exp[(\lambda_1 - \lambda_2)x]$ is not constant. Thus, then $\exp(\lambda_1 x)$ and $\exp(\lambda_2 x)$ are linearly independent solutions of (1), so the general solution is

$$y(x) = c_1 \exp(\lambda_1 x) + c_2 \exp(\lambda_2 x), \tag{7}$$

where c_1 and c_2 are arbitrary constants.

It is now necessary to introduce the type of initial conditions that are appropriate for (1). As (1) is a second order differential equation, it relates $y(x)$, $y'(x)$, and $y''(x)$, so it follows that suitable initial conditions will be the specification of $y(x)$ and $y'(x)$ at some point $x = a$. Then the value of $y''(a)$ cannot be assigned arbitrarily, because the differential equation itself will determine its value in terms of $y(a)$ and $y'(a)$. The solution of (1) satisfying these initial conditions can be found from the general solution (7) by determining c_1 and c_2 from the two equations:

Initial condition on y(x)

initial conditions

$$y(a) = c_1 \exp(\lambda_1 a) + c_2 \exp(\lambda_2 a),$$

Initial condition on y'(x)

$$\tag{8}$$

$$y'(a) = \lambda_1 c_1 \exp(\lambda_1 a) + \lambda_2 c_2 \exp(\lambda_2 a).$$

When we considered systems of linear algebraic equations in Chapter 3, it was shown that equations (8) will determine c_1 and c_2 uniquely if the determinant of the coefficients of c_1 and c_2 is nonvanishing. Thus, the specification of $y(a)$ and $y'(a)$ will be appropriate as initial conditions if

$$\Delta = \begin{vmatrix} \exp(\lambda_1 a) & \exp(\lambda_2 a) \\ \lambda_1 \exp(\lambda_1 a) & \lambda_2 \exp(\lambda_2 a) \end{vmatrix} \neq 0. \tag{9}$$

Expanding the determinant gives $\Delta = (\lambda_2 - \lambda_1)\exp[(\lambda_1 + \lambda_2)a]$. However, by hypothesis $\lambda_1 \neq \lambda_2$, while $\exp[(\lambda_1 + \lambda_2)a]$ never vanishes, so $\Delta \neq 0$. The particular solution satisfying the initial conditions follows by using the values of c_1 and c_2 found from (8) in the general solution (7).

EXAMPLE 6.3

Find the solution of the initial value problem

$$y'' + 4y = 0, \quad \text{if} \quad y(\pi/4) = 1 \quad \text{and} \quad y'(\pi/4) = 1.$$

Solution   In Example 6.2 direct substitution has already been used to show that $\cos 2x$ and $\sin 2x$ are linearly independent solutions of the differential equation, so its general solution is

$$y(x) = c_1 \cos 2x + c_2 \sin 2x,$$

from which it follows by differentiation that

$$y'(x) = -2c_1 \sin 2x + 2c_2 \cos 2x.$$

Imposing the initial condition on $y(x)$ at $x = \pi/4$ leads to the following equation that must be satisfied by c_1 and c_2:

$$1 = c_1 \cos \pi/2 + c_2 \sin \pi/2.$$

Similarly, imposing the initial condition on $y'(x)$ at $x = \pi/4$ leads to the second condition that must be satisfied by c_1 and c_2:

$$1 = -2c_1 \sin \pi/2 + 2c_2 \cos \pi/2.$$

These equations have the solution $c_1 = -1/2$ and $c_2 = 1$, so the particular solution satisfying the initial conditions $y(\pi/4) = 1$ and $y'(\pi/4) = 1$ is

$$y(x) = \sin 2x - \frac{1}{2} \cos 2x. \qquad \blacksquare$$

The quadratic equation determining the permissible values of λ in the exponential solutions $y_1(x) = \exp(\lambda_1 x)$ and $y_2(x) = \exp(\lambda_2 x)$ of differential equation (1), namely,

$$\lambda^2 + A\lambda + B = 0, \tag{10}$$

characteristic equation

is called the **characteristic equation** of the differential equation. Its two roots,

$$\lambda_1 = \frac{-A + \sqrt{A^2 - 4B}}{2} \quad \text{and} \quad \lambda_2 = \frac{-A - \sqrt{A^2 - 4B}}{2}, \tag{11}$$

are the values of λ to be used in the general solution (7). When the roots λ_1 and λ_2 are real and distinct, the functions

$$y_1(x) = \exp(\lambda_1 x) \quad \text{and} \quad y_2(x) = \exp(\lambda_2 x) \tag{12}$$

are said to form a **basis** for the solution space of (1). This means that the solution of every initial value problem for (1) can be obtained from the linear combination $y(x) = c_1 \exp(\lambda_1 x) + c_2 \exp(\lambda_2 x)$ by assigning suitable values to c_1 and c_2.

A comparison of differential equation (1) and its characteristic equation (10) shows the characteristic equation can be written down immediately from the differential equation by simply replacing y by 1, dy/dx by λ and d^2y/dx^2 by λ^2. It is usual to use this method when obtaining the characteristic equation, as it avoids the unnecessary intermediate steps involved when substituting $y(x) = \exp(\lambda x)$.

Three different cases must now be considered, according to whether (i) λ_1 and λ_2 are real and distinct ($\lambda_1 \neq \lambda_2$), (ii) λ_1 and λ_2 are complex conjugates, or (iii) the possibility, excluded so far, that λ_1 and λ_2 are real and equal, so $\lambda_1 = \lambda_2 = \mu$, say.

Case (I) (Real and Distinct Roots)

how a solution depends on the roots

This case corresponds to the condition $A^2 - 4B > 0$, with

$$\lambda_1 = \frac{-A + \sqrt{A^2 - 4B}}{2} \quad \text{and} \quad \lambda_2 = \frac{-A - \sqrt{A^2 - 4B}}{2}. \tag{13}$$

No more need be said about this case because it has already been established that the functions $\exp(\lambda_1 x)$ and $\exp(\lambda_2 x)$ form a basis for the solution space of (1), which thus has the general solution

$$y(x) = c_1 \exp(\lambda_1 x) + c_2 \exp(\lambda_2 x).$$

Case (II) (Complex Conjugate Roots)

This case corresponds to the condition $A^2 - 4B < 0$. A real solution $y(x)$ corresponding to complex conjugate roots λ_1 and λ_2 is only possible if the arbitrary constants c_1 and c_2 are themselves complex conjugates. A routine calculation shows that if $\lambda_1 = \alpha + i\beta$ and $\lambda_2 = \alpha - i\beta$, with

$$\alpha = -(1/2)A, \qquad \beta = (1/2)(4B - A^2)^{1/2}, \tag{14}$$

the two corresponding linearly independent solutions are

$$y_1(x) = e^{\alpha x} \cos \beta x \quad \text{and} \quad y_2(x) = e^{\alpha x} \sin \beta x. \tag{15}$$

A basis for the solution space of (1) is formed by the functions $e^{\alpha x} \cos \beta x$ and $e^{\alpha x} \sin \beta x$, corresponding to a general solution of the form

$$y_1(x) = e^{\alpha x}[c_1 \cos \beta x + c_2 \sin \beta x]. \tag{16}$$

The calculation required to establish the form of this result is left as an exercise.

Case (III) (Equal Real Roots)

This case corresponds to the condition $A^2 - 4B = 0$, with

$$\mu = \lambda_1 = \lambda_2 = -(1/2)A. \tag{17}$$

In this case only the one exponential solution

$$y_1(x) = e^{\mu x} \tag{18}$$

can be found.

However, substitution of the function

$$y_2(x) = xe^{\mu x} \tag{19}$$

into the differential equation shows that it is also a solution. The functions $y_1(x)$ and $y_2(x)$ are linearly independent because $y_2(x)/y_1(x) = x$ is not a constant, so in this case a basis for the solution space of (1) is formed by the functions $e^{\mu x}$ and $xe^{\mu x}$, with the corresponding general solution

$$y(x) = (c_1 + c_2 x)e^{\mu x}. \tag{20}$$

Summary of the forms of solution of $y'' + Ay' + By = 0$

summary of types of solution

Characteristic equation: $\lambda^2 + A\lambda + B = 0$

Case (I) $A^2 - 4B > 0$. The general solution is

$$y(x) = c_1 \exp(\lambda_1 x) + c_2 \exp(\lambda_2 x), \qquad \text{with}$$

$$\lambda_1 = \frac{-A + \sqrt{A^2 - 4B}}{2} \quad \text{and} \quad \lambda_2 = \frac{-A - \sqrt{A^2 - 4B}}{2}.$$

Case (II) $A^2 - 4B < 0$. The general solution is

$$y_1(x) = e^{\alpha x}[c_1 \cos \beta x + c_2 \sin \beta x], \qquad \text{with}$$

$$\alpha = -(1/2)A \quad \text{and} \quad \beta = (1/2)(4B - A^2)^{1/2}.$$

Case (III) $A^2 = 4B$. The general solution is

$$y(x) = (c_1 + c_2 x)e^{\mu x}, \qquad \text{with} \quad \mu = -(1/2)A.$$

EXAMPLE 6.4

Find the general solution and hence solve the stated initial value problem for

(i) $y'' + y' - 2y = 0$, with $y(0) = 1$ and $y'(0) = 2$;
(ii) $y'' + 2y' + 4y = 0$, with $y(0) = 2$ and $y'(0) = 1$;
(iii) $y'' + 4y' + 4y = 0$, with $y(0) = 3$ and $y'(0) = 1$.

Solution

(i) The characteristic equation is

$$\lambda^2 + \lambda - 2 = 0,$$

with the roots $\lambda_1 = 1, \lambda_2 = -2$, so this is Case (I). The general solution is

$$y(x) = c_1 e^x + c_2 e^{-2x}.$$

The initial condition $y(0) = 1$ is satisfied if

$$1 = c_1 + c_2,$$

while the initial condition $y'(0) = 2$ is satisfied if

$$2 = c_1 - 2c_2.$$

These equations have the solution $c_1 = 4/3$ and $c_2 = -1/3$, so the solution of the initial value problem is

$$y(x) = (4/3)e^x - (1/3)e^{-2x}.$$

(ii) The characteristic equation is

$$\lambda^2 + 2\lambda + 4 = 0,$$

with $A^2 - 4B = -12$, so this is Case (II) with $\alpha = -1$ and $\beta = \sqrt{3}$. The general solution is

$$y(x) = e^{-x}[c_1\cos(x\sqrt{3}) + c_2\sin(x\sqrt{3})].$$

The initial condition $y(0) = 2$ is satisfied if $2 = c_1$, while the initial condition $y'(0) = 1$ is satisfied if

$$1 = -2 + c_2\sqrt{3}.$$

Solving these equations gives $c_1 = 2$ and $c_2 = \sqrt{3}$, so the solution of the initial value problem is

$$y(x) = e^{-x}[\sqrt{3}\sin(x\sqrt{3}) + 2\cos(x\sqrt{3})].$$

(iii) The characteristic equation is

$$\lambda^2 + 4\lambda + 4 = 0,$$

with $A^2 - 4B = 0$, so this is Case (III) with $\mu = -2$. The general solution is

$$y(x) = (c_1 + c_2 x)e^{-2x}.$$

Using the initial condition $y(0) = 3$ shows that $3 = c_1$, whereas the initial condition $y'(0) = 1$ will be satisfied if

$$1 = -6 + c_2.$$

Solving these equations gives $c_1 = 3$ and $c_2 = 7$, so the solution of the initial value problem is

$$y(x) = (3 + 7x)e^{-2x}. \qquad \blacksquare$$

We now formulate the fundamental existence and uniqueness theorem for the homogeneous linear second order constant coefficient differential equation (1). This is a special case of a more general theorem that will be quoted later.

THEOREM 6.1

existence and
uniqueness of
solutions

Existence and uniqueness of solutions of homogeneous second order constant coefficient equations   Let differential equation (1) have two linearly independent solutions $y_1(x)$ and $y_2(x)$. Then, for any $x = x_0$ and numbers μ_1 and μ_2, a unique solution of (1) exists satisfying the initial conditions

$$y(x_0) = \mu_0, \quad y^{(1)}(x_0) = \mu_1.$$

Proof   The existence of the solutions $y_1(x)$ and $y_2(x)$ was established when the cases (I), (II), and (III) were examined. The nonvanishing of the determinant Δ in (9) showed c_1 and c_2 to be uniquely determined by the given initial conditions when the roots are real and distinct, so the solution of the initial value problem is also unique. An examination of the form of the determinant Δ in cases (II) and (III) establishes the uniqueness of the solution in the remaining two cases, though the details are left as an exercise.   ■

two-point boundary conditions

A different type of problem that can arise with second order equations occurs when the solution is required to satisfy a condition at two distinct points $x = a$ and $x = b$, instead of satisfying two initial conditions. Problems of this type are called **two-point boundary value problems**, because the points a and b can be regarded as boundaries between which the solution is required, and at which it must satisfy given **boundary conditions**. Problems of this type occur in the study of the bending of beams that are supported in different ways at each end, and elsewhere (see Section 8.10).

Typical two-point boundary value problems involve either the specification of $y(x)$ at $x = a$ and at $x = b$, or the specification of $y(x)$ at one boundary and $y'(x)$ at the other one. The most general two point boundary value problem involves finding a solution in the interval $a < x < b$ such that

$$y'' + Ay' + By = 0,$$

subject to the boundary condition at $x = a$

$$\alpha y(a) + \beta y'(a) = \mu,$$

and the boundary condition at $x = b$

$$\gamma y(b) + \delta y'(b) = K,$$

where $\alpha, \beta, \gamma, \delta, \mu$, and K are known constants.

EXAMPLE 6.5

Solve the two-point boundary value problem

$$y'' + 2y' + 17y = 0, \quad \text{with } y(0) = 1 \text{ and } y'(\pi/4) = 0.$$

Solution   The characteristic equation is

$$\lambda^2 + 2\lambda + 17 = 0$$

with the complex roots $\lambda_1 = -1 + 4i$ and $\lambda_2 = -1 - 4i$, so the general solution is

$$y(x) = e^{-x}[c_1 \cos 4x + c_2 \sin 4x].$$

At the boundary $x = 0$ the general solution reduces to $1 = c_1$, whereas at the boundary $x = \pi/4$ it reduces to $0 = -e^{-\pi/4} + 4c_2 e^{-\pi/4}$, showing that $c_2 = 1/4$. So the solution of the two-point boundary value problem is

$$y(x) = e^{-x}\left[\cos 4x + \frac{1}{4}\sin 4x\right], \quad \text{for } 0 < x < \pi/4. ■$$

Summary

This section introduced the homogeneous linear second order constant coefficient equation and explained the importance of the linear independence of solutions. It showed how for this second order equation the general solution can be expressed as a linear combination of the two linearly independent solutions that can always be found. The form of the two linearly independent solutions was shown to depend on the relationship between the roots of the characteristic equation. A fundamental existence and uniqueness theorem was given and the nature of a simple two-point boundary value problem was explained.

EXERCISES 6.1

In Exercises 1 through 4 test the given pairs of functions for linear independence or dependence over the stated intervals.

1. (a) $\sinh^2 x$, $\cosh^2 x$, for all x.
   (b) $x + \ln |x|$, $x + 2\ln |x|$, for $|x| > 0$.
   (c) $1 + x$, $x + x^2$, for all x.
2. (a) $\sin x$, $\cos x$, for all x.
   (b) $\sin x \cos x$, $\sin 2x$, for all x.
   (c) e^{2x}, xe^{2x}, for all x.
3. (a) $|x|x^2$, x^3, for $-1 < x < 1$.
   (b) $\sin x$, $\tan x$, for $-\pi/4 \le x \le \pi/4$.
   (c) $x|x|$, x^2, for $x \ge 0$.
4. (a) $\sin x$, $|\sin x|$, for $\pi \le x \le 2\pi$.
   (b) $x^3 - 2x + 4$, $-4x^3 + 8x - 16$, for all x.
   (c) $x + 2|x|$, $x - 2|x|$ for all x.

Find the general solution of the differential equations in Exercises 5 through 20.

5. $y'' + 3y' - 4y = 0$.
6. $y'' + 2y' + y = 0$.
7. $y'' - 2y' + 2y = 0$.
8. $y'' + 2y' + 2y = 0$.
9. $y'' + 2y' - 3y = 0$.
10. $y'' + 5y' + 4y = 0$.
11. $y'' + 6y' + 9y = 0$.
12. $y'' - 2y' + 4y = 0$.
13. $y'' - 4y' + 5y = 0$.
14. $y'' + 3y' + 3y = 0$.
15. $y'' + 6y' + 25y = 0$.
16. $y'' - 4y' + 20y = 0$.
17. $y'' + 5y' + 4y = 0$.
18. $y'' + 4y' + 5y = 0$.
19. $y'' - 3y' + 3y = 0$.
20. $y'' + y' + y = 0$.

Solve initial value problems in Exercises 21 through 28 using the method of this section, and confirm the solutions for even numbered problems by using computer algebra.

21. $y'' + 5y' + 6y = 0$, with $y(0) = 1$, $y'(0) = 2$.
22. $y'' + 4y' + 5y = 0$, with $y(0) = 1$, $y'(0) = 3$.
23. $y'' + 2y' + 2y = 0$, with $y(0) = 3$, $y'(0) = 1$.
24. $y'' + 6y' + 8y = 0$, with $y(0) = 1$, $y'(0) = 0$.
25. $y'' - 5y' + 6y = 0$, with $y(0) = 2$, $y'(0) = 1$.
26. $y'' - 3y' + 3y = 0$, with $y(0) = 0$, $y'(0) = 2$.
27. $y'' - 3y' - 4y = 0$, with $y(0) = -1$, $y'(0) = 2$.
28. $y'' - 2y' + 3y = 0$, with $y(0) = 1$, $y'(0) = 0$.

Solve the boundary value problems in Exercises 29 through 36 using the method of this section, and confirm the solutions for even-numbered problems by using computer algebra.

29. $y'' + 4y' + 3y = 0$, with $y(0) = 1$, $y'(1) = 0$.
30. $y'' + 4y' + 4y = 0$, with $y(0) = 2$, $y'(1) = 0$.
31. $y'' + 6y' + 9y = 0$, with $y(-1) = 1$, $y'(1) = 0$.
32. $y'' + 4y' + 5y = 0$, with $y(-\pi/2) = 1$, $y'(\pi/2) = 0$.
33. $y'' + 2y' + 26y = 0$, with $y(0) = 1$, $y'(\pi/4) = 0$.
34. $y'' + 2y' + 26y = 0$, with $y(0) = 0$, $y'(\pi/4) = 2$.
35. $y'' + 5y' + 6y = 0$, with $y(0) = 0$, $y'(1) = 1$.
36. $y'' + 2y' - 3y = 0$, with $y(0) = 1$, $y'(1) = 1$.

Theorem 6.1 ensures the existence and uniqueness of solutions of initial value problems for the differential equation in (1), but does not apply to two-point boundary value problems that may have no solution, a unique solution or infinitely many solutions. In Exercises 37 and 38 use the general solution of

$$y'' + y = 0$$

to find if a solution exists and is unique, exists but is nonunique, or does not exist for each set of boundary conditions.

37. (a) $y(0) = 0$, $y(\pi) = 0$.   (c) $y'(0) = 1$, $y(\pi/4) = \sqrt{2}$.
    (b) $y(0) = 1$, $y(2\pi) = 2$.
38. (a) $y(0) = 1$, $y(\pi/2) = 1$.   (c) $y'(0) = 0$, $y'(\pi) = 0$.
    (b) $y(0) = 0$, $y'(\pi) = 0$.
39. For what values of λ will the following two-point boundary value problem have infinitely many solutions, and what is the form of these solutions:

$$y'' + \lambda^2 y = 0, \quad \text{with } y(0) = 0, y(\pi) = 0.$$

40. A particle moves in a straight line in such a way that its distance x from the origin at time t obeys the differential equation $x'' + x' + x = 0$. Assuming it starts from the origin with speed 30 ft/sec, what will be its distance from the origin, its speed, and its acceleration after $\pi/\sqrt{3}$ seconds?

41. The angular displacement θ of a damped simple pendulum obeys the equation $\theta'' + 2\mu\theta' + (\mu^2 + p^2)\theta = 0$,

with $4p^2 > \mu^2$. Find the angular displacement $\theta(t)$, and the time t and angular displacement when it first comes to rest, given that it starts with $\theta = 0$ and $d\theta/dt = \alpha$.

42. The top of a vibration damper oscillates in a straight line in such a way that its position x from the origin at time t obeys the differential equation $x'' + 2x' + 4x = 0$. Given that it starts from the origin with speed U, find its position as a function of U and t and the distance from the origin when it first comes to rest.

43. The free oscillations of all physical systems giving rise to oscillatory solutions obey an equation of the form $x'' + 2\mu x' + (\mu^2 + p^2)x = 0$ with $p^2 > 0$. Given that $x(0) = 0$, and $(dx/dt)_{t=0} = Ap$, solve the equation and show that $x(t) = A\exp(-\mu t)\sin pt$. Use this result to prove that the ratio of the magnitude of successive extrema of $x(t)$ forms a geometric series with common ratio $r = \exp[-\mu\pi/(p)]$. The number $\mu\pi/(p)$ is called the **logarithmic decrement** of the oscillations.

6.2 Oscillatory Solutions

The nonhomogeneous constant coefficient second order equation

$$a_0\frac{d^2y}{dt^2} + a_1\frac{dy}{dt} + a_2y = f(t), \tag{21}$$

in which t can be regarded as the time and $f(t)$ as an external input to the system, is the simplest mathematical model capable of representing the oscillatory behavior of a physical system.

It was shown in Section 5.2(d) that one way this equation can arise is when describing the motion of a mass–spring system in which a mass moves on a rough horizontal surface, with the motion resisted by a frictional force proportional to the speed. Friction dissipates energy, so the motion will decay to zero as time increases unless it is sustained by some external input of energy in the form of a **forcing function** represented in (21) by the nonhomogeneous term $f(t)$. The dissipation of energy due to friction, or to a friction-like effect in other applications, is called **damping**, and in the R–L–C circuit considered in Section 5.2(d), where the charge q on the capacitor was shown to satisfy a homogeneous form of equation (21) with $a_0 = LC$, $a_1 = RC$, and $a_3 = 1$, the damping was due to the dissipative (friction-like) term $a_1 = RC$.

forcing function and damping

Another way in which equation (21) can arise is when a cylindrical mass with moment of inertia I about its axis of symmetry is mounted on a flexible shaft that can be twisted about its axis, with the resistance to **torsion** (twisting) proportional to the angle of twist θ, and damping proportional to the angular velocity $d\theta/dt$ about the shaft. This occurs, for example, in a torsional pendulum and also in heavy rotating machinery when a heavy flywheel is attached to a shaft. The equation governing the **torsional oscillations** $\theta(t)$ as a function of the time t becomes

$$I\frac{d^2\theta}{dt^2} + k\frac{d\theta}{dt} + \mu\theta = f(t),$$

where k and μ are constants and, as before, $f(t)$ is a forcing function.

A comparison of the second order constant coefficient differential equations that govern mechanical, electrical, and torsional oscillations leads to Table 6.1, which relates analogous physical quantities in each of the different systems.

Many other physical situations can be represented by this same constant coefficient second order differential equation with varying degrees of approximation. It does, for example, provide a simple model that describes the effect of a fluctuating vertical lift at the center of a flexible suspension bridge caused by gusts of wind. If

TABLE 6.1 A Comparison of Second Order Constant Coefficient Differential Equations		
Mechanical System	Electrical System with Elements in Series	Torsional System
Mass	Inductance	Moment of inertia
Damping constant	Resistance	Torsional damping constant
Spring constant	Reciprocal of capacitance	Shaft torsional constant
Applied force	Applied voltage	Applied torque

this effect is sustained, and the gusts come at the natural frequency of the bridge, the amplitude of the oscillations can become dangerously large. On November 7, 1940, in the state of Washington, this effect caused the failure of the Tacoma Narrows Bridge over Puget Sound. Powerful gusting winds at around the natural frequency of this excessively flexible bridge induced and then sustained vertical oscillations of the bridge that reached an amplitude of 28 feet before the bridge snapped and fell.

When analyzing the oscillatory nature of solutions of (21), and looking at the effect of *resonance* that occurs if the natural frequency of oscillation of the system coincides with the frequency of a periodic forcing function, it is helpful to have in mind a mathematical model of a simple but typical mechanical system. The mechanical system we will consider here is shown in Fig. 6.2, and it involves a piece of heavy machinery that vibrates vertically and is mounted on a spring and damper system to reduce the transmission of the vibrations to the foundations of the building. The damper is usually a piston that moves in a viscous fluid, with the resisting force considered to be proportional to the speed of the piston.

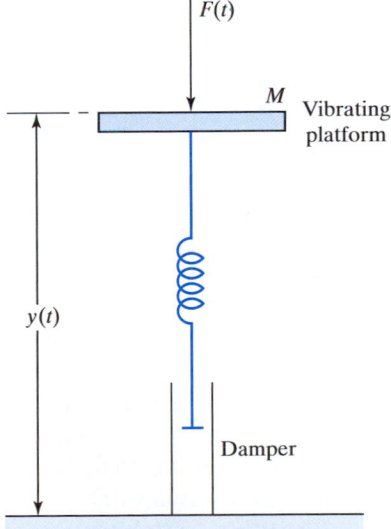

FIGURE 6.2   A vibrating machine mounted on a spring and damper system.

If the mass of the machine is M, the displacement of the machine from the floor at time t is $y(t)$, the spring constant is k, the constant of proportionality for the damper is μ, and the force exerted by the vibrating machine is $\tilde{F}(t)$, the rate of change of momentum $d/dt\{M(dy/dt)\}$ must be equated to the frictional resistance $-\mu\,dy/dt$, the restoring force of the spring $-ky$, and the external force $\tilde{F}(t)$. So this system, with the displacement $y(t)$ as its **one degree of freedom**, is seen to satisfy the differential equation

$$M\frac{d^2 y}{dt^2} = -\mu\frac{dy}{dt} - ky + \tilde{F}(t).$$

For convenience this will be written in the standard form

$$\frac{d^2 y}{dt^2} + a\frac{dy}{dt} + by = F(t), \tag{22}$$

with $a = \mu/M$, $b = k/M$, and $F(t) = \tilde{F}(t)/M$.

Differential equation (22) is nonhomogeneous, so its solution will be more complicated than the solution of the homogeneous equation considered in the previous section. The equation is linear, so as in Section 5.6 we will represent its general solution as the sum

$$y(t) = y_c(t) + y_p(t), \tag{23}$$

with $y_c(t)$ the general solution of the homogeneous form of equation (22)

$$\frac{d^2 y_c}{dt^2} + a\frac{dy_c}{dt} + by_c = 0, \tag{24}$$

and $y_p(t)$ a particular solution of

$$\frac{d^2 y_p}{dt^2} + a\frac{dy_p}{dt} + by_p = F(t) \tag{25}$$

that contains no arbitrary constants.

The justification for writing the general solution of (22) as $y(t) = y_c(t) + y_p(t)$ follows if we notice that (22) can be written

$$\frac{d^2[y_c + y_p]}{dt^2} + a\frac{d[y_c + y_p]}{dt} + b[y_c + y_p] = F(t)$$

or, equivalently, as

$$\left[\frac{d^2 y_c}{dt^2} + a\frac{dy_c}{dt} + by_c\right] + \frac{d^2 y_p}{dt^2} + a\frac{dy_p}{dt} + by_p = F(t),$$

where the group of terms in the bracket vanishes because of (24), while $y_p(t)$ satisfies the remainder of the equation because of (25).

complementary function and particular integral

As in Section 5.6, the solution $y_c(t)$ will be called the **complementary function**, and the solution $y_p(t)$ will be called a **particular integral**. It is important to recognize that the two arbitrary constants associated with the general solution of (22) occur in the complementary function $y_c(t)$, whereas the particular integral $y_p(t)$ contains no arbitrary constants.

We now introduce the notation $a = 2\zeta$ and $b = \Omega^2$, when the characteristic equation of (22) becomes

$$\lambda^2 + 2\zeta\lambda + \Omega^2 = 0, \tag{26}$$

with the roots

$$\lambda_1 = -\zeta + (\zeta^2 - \Omega^2)^{1/2} \quad \text{and} \quad \lambda_2 = -\zeta - (\zeta^2 - \Omega^2)^{1/2}. \tag{27}$$

The solution $y_c(t)$ of (22) will correspond to one of the Cases (I), (II), or (III) of Section 6.1, but before determining its form in each of these cases we further simplify the notation by setting $k^2 = \zeta^2 - \Omega^2$, so that

$$\lambda_1 = -\zeta + k \quad \text{and} \quad \lambda_2 = -\zeta - k. \tag{28}$$

Case (I): $k^2 > 0\,(\zeta^2 > \Omega^2)$

The complementary function $y_c(t)$ is nonoscillatory and given by

$$y_c(t) = \exp(-\zeta t)\{C_1 \exp(kt) + C_2 \exp(-kt)\}. \tag{29}$$

Case (II): $k^2 < 0\,(\zeta^2 < \Omega^2)$

If we set $k^2 = -\omega_0^2$ the complementary function is seen to be oscillatory and given by

$$y_c(t) = \exp(-\zeta t)\{C_1\cos\omega_0 t + C_2\sin\omega_0 t\}. \tag{30}$$

Case (III): $k^2 = 0\,(\Omega^2 = \zeta^2)$

The complementary function is nonoscillatory and given by

$$y_c(t) = \{C_1 + C_2 t\}\exp(-\zeta t). \tag{31}$$

critical damping and overdamping

Cases (I) and (III) exhibit no oscillatory behavior. Case (I) is said to be **overdamped** and Case (III) to be **critically damped**, because it marks the boundary between the overdamped behavior of Case (I) and the oscillatory behavior of Case (II). The parameter ζ entering into the exponential factor $\exp(-\zeta t)$ that is present in all three cases is called the **damping exponent** and, provided $\zeta > 0$, the factor $\exp(-\zeta t)$ will cause all three complementary functions to decay to zero as time increases. This property of a complementary function with $\zeta > 0$ has led to its being called the **transient solution** of the differential equation. Accordingly, after a suitable lapse of time, only the particular integral $y_p(t)$ will remain, and this property is recognized by calling $y_p(t)$ the **steady state solution**, with the understanding that it is the *time-dependent* solution that remains after the transient solution has become vanishingly small.

transient and steady state solutions

Typical transient solution behavior in the critically damped case is shown in Fig. 6.3 for different initial conditions, some of which can cause an initial increase in the amplitude of $y_c(t)$ before it decays to zero. The behavior in the overdamped case is similar to that in the critically damped case.

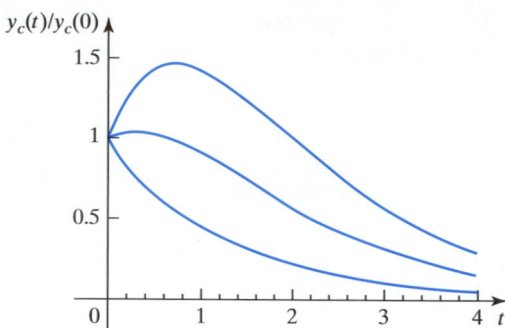

FIGURE 6.3 $y_c(t)$ in the critically damped case for different initial conditions.

It is now necessary to determine the form of the particular integral $y_p(t)$, and to do so the function $F(t)$ must be specified. A vibration is periodic in nature, so we shall model it by a nonhomogeneous term of the form

$$F(t) = A\cos \omega t, \tag{32}$$

in which the **amplitude** A will be considered to be fixed and the **angular frequency** ω will be regarded as a parameter. The angular frequency ω is expressed in terms of radians/unit time and corresponds to a time period of oscillation of $T = 2\pi/\omega$ time units, while the **frequency** of the vibration $1/T = \omega/2\pi$ measures the number of cycles (vibrations) occurring in one time unit. If the unit of time T is 1 sec, the frequency is measured in cycles/sec, called **hertz** (Hz), so 20 Hz is 20 cycles/sec.

> amplitude and angular frequency of a vibration

Setting $F(t) = A\cos \omega t$ in (22) shows that the differential equation to be considered is

$$\frac{d^2y}{dt^2} + 2\zeta \frac{dy}{dt} + \Omega^2 y = A\cos \omega t. \tag{33}$$

A systematic approach to the determination of particular integrals will be developed in the next section, but here we will proceed from first principles. As equation (33) has constant coefficients, and its nonhomogeneous term is $A\cos \omega t$, the way this nonhomogeneous term can be obtained by differentiating a particular integral $y_p(t)$ is if the particular integral is of the form

$$y_p(t) = C\sin \omega t + D\cos \omega t, \tag{34}$$

unless $\zeta = 0$ and $\Omega = \omega$ (then try $y_p = t(C\sin \omega t + D\cos \omega t)$).
Substituting (34) into (33) and collecting terms gives

$$[(\Omega^2 - \omega^2)C - 2\zeta \omega D]\sin \omega t + [(\Omega^2 - \omega^2)D + 2\zeta \omega C]\cos \omega t = A\cos \omega t.$$

This must be true for all t, but this will only be possible if the respective coefficients of $\sin \omega t$ and $\cos \omega t$ on either side of the equation are identical, so

$$(\Omega^2 - \omega^2)C - 2\zeta \omega D = 0 \quad \text{and} \quad (\Omega^2 - \omega^2)D + 2\zeta \omega C = A.$$

Solving for C and D gives

$$C = \frac{2A\omega\zeta}{(\Omega^2 - \omega^2)^2 + 4\zeta^2\omega^2} \quad \text{and} \quad D = \frac{A(\Omega^2 - \omega^2)}{(\Omega^2 - \omega^2)^2 + 4\zeta^2\omega^2}, \tag{35}$$

so the required particular integral is

$$y_p(t) = \frac{2A\omega\zeta}{(\Omega^2 - \omega^2)^2 + 4\zeta^2\omega^2}\sin\omega t + \frac{A(\Omega^2 - \omega^2)}{(\Omega^2 - \omega^2)^2 + 4\zeta^2\omega^2}\cos\omega t. \tag{36}$$

A better understanding of the nature of this particular integral can be obtained if it is rewritten. To accomplish this we return to (34) and write it as

$$y_p(t) = (C^2 + D^2)^{1/2}\left[\frac{C}{(C^2 + D^2)^{1/2}}\sin\omega t + \frac{D}{(C^2 + D^2)^{1/2}}\cos\omega t\right], \tag{37}$$

and then define an angle ϕ by the requirement that

$$\sin\phi = \frac{C}{(C^2 + D^2)^{1/2}}, \quad \text{and} \quad \cos\phi = \frac{D}{(C^2 + D^2)^{1/2}}, \tag{38}$$

or by the equivalent expression

$$\tan\phi = C/D = \frac{2\zeta\omega}{\Omega^2 - \omega^2}. \tag{39}$$

The trigonometric identity $\cos(\omega t - \phi) = \cos\omega t\cos\phi + \sin\omega t\sin\phi$ then allows $y_p(t)$ to be expressed in the more convenient form

$$y_p(t) = \frac{A}{[(\Omega^2 - \omega^2)^2 + 4\zeta^2\omega^2]^{1/2}}\cos(\omega t - \phi). \tag{40}$$

Using this form for $y_p(t)$ gives the simpler expression for the general solution

$$y(t) = y_c(t) + \frac{A}{[(\Omega^2 - \omega^2) + 4\zeta^2\omega^2]^{1/2}}\sin(\omega t - \phi), \tag{41}$$

where $y_c(t)$ is one of the Cases (I), (II), or (III), depending on the sign of $\zeta^2 - \Omega^2$.

The angle ϕ, which by convention is required to lie in the interval $0 < \phi < \pi$, is called the **phase angle** of the solution, and often the **phase lag**, because it represents the *delay* with which the steady-state solution (the *output* from the system) lags behind the *input* to the system determined by $F(t)$.

phase angle and phase lag

We have seen that provided $\zeta > 0$, the transient solution $y_c(t)$ decays to zero as t increases, leaving only the **steady state** solution $y_p(t)$. The steady state solution (41), illustrated in Fig. 6.4, is a sinusoid with the same angular frequency as the function $F(t) = A\sin(\omega t)$ that forces the oscillations, but with an amplitude that depends on both A and the angular forcing frequency ω. The effect of the phase lag ϕ is seen to shift the origin.

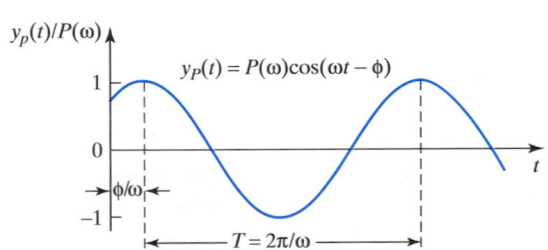

FIGURE 6.4   The steady state solution $y_p(t)/P(\omega)$.

If $\zeta < 0$, the general solution $y(t)$ will increase without bound as time increases, and in physical problems this behavior is called **instability**. In effect, when $\zeta < 0$, energy is fed into the system as time increases, instead of it being removed by friction.

The amplitude of the steady state solution is

$$P(\omega) = \frac{A}{[(\Omega^2 - \omega^2)^2 + 4\zeta^2\omega^2]^{1/2}}, \tag{42}$$

and $P(\omega)/A = [(\Omega^2 - \omega^2) + 4\zeta^2\omega^2]^{-1/2}$ is called the **amplification factor**, because it is the ratio of the amplitude of the solution (*response*) to the amplitude of the forcing function (*input*). The amplification factor attains its maximum value P_{max} when $\omega = \omega_c$, with $\omega_c^2 = \Omega^2 - 2\zeta^2$, in which case

$$P_{max} = \frac{A}{2\zeta(\Omega^2 - \zeta^2)^{1/2}}. \tag{43}$$

The angular frequency ω_c is called the **resonant angular frequency** of the system that is said to experience **resonance** at the the frequency ω_c. It is to avoid exciting resonance that troops marching across a flexible suspension bridge are told to break step. Conversely, it is for this same reason that when one pushes a swing, successive pushes need to be synchronized with each oscillation if the amplitude of the motion is to be built up. If $\zeta = 0$, result (42) shows that resonance occurs when $\omega = \Omega$, leading to an infinite amplification factor. The critical role played by damping in limiting the amplitude of oscillations can be seen from (43).

Figures 6.5a and 6.5b show the variation of the scaled amplification factor $\Omega^2 P(\omega)/A$ and the phase angle ϕ as functions of ω/Ω for a range of values of ζ/Ω.

Care must always be exercised when finding the phase angle ϕ, because the phase is required to lie in the interval $0 < \phi < \pi$, though the usual domain of definition of the inverse tangent function is $(-\pi/2, \pi/2)$.

The most extreme effect of resonance occurs when there is no damping ($\zeta = 0$), though this can never happen in physical problems because there are always some dissipative effects. In the absence of damping, the **natural angular frequency** of oscillations is Ω, and equation (42) shows that when the vibrations are forced by a

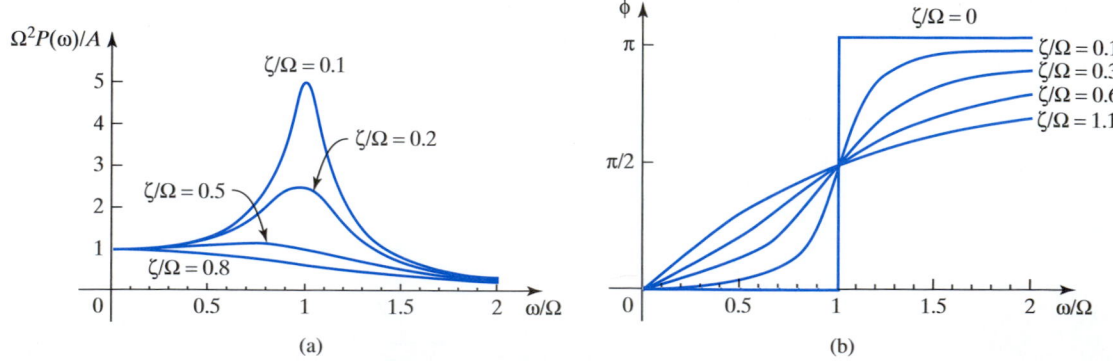

FIGURE 6.5   (a) Amplitude as a function of ω/Ω. (b) Phase angle as a function of ω/Ω.

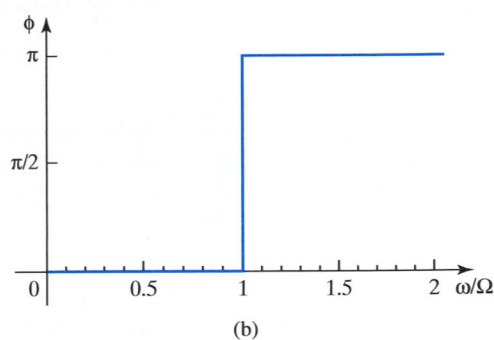

FIGURE 6.6 (a) Variation of amplitude. (b) Variation of phase.

sinusoidal input of angular frequency ω, the amplitude of the steady state solution is

$$P(\omega) = \frac{A}{|\Omega^2 - \omega^2|^{1/2}}.$$

This shows that $P(\omega)$ becomes infinite when the exciting angular frequency ω equals the natural angular frequency Ω. The variation of $\Omega P(\omega)/A$ as ω/Ω varies is shown in Fig. 6.6a, while the corresponding variation of the phase is shown in Fig. 6.6b for the limiting case $\omega \to \Omega$.

To understand how the solution becomes unstable when $\omega = \Omega$, it is necessary to consider the solution of

$$\frac{d^2 y}{dt^2} + \Omega^2 y = A \sin \Omega t, \quad \text{with } y(0) = 0, (dy/dt) = 0, t = 0.$$

We find that

$$y(t) = \frac{A}{2\Omega^2}(\sin \Omega t - \Omega t \cos \Omega t),$$

and the variation of $y(t)$ is shown in Fig. 6.7, from which it can be seen that when the damping is zero, forcing at the resonant angular frequency causes the amplitude of the oscillations to grow linearly with time.

An interesting and important property of oscillatory solutions under conditions **beats** that allow dissipation to be ignored is to be found in the occurrence of **beats** in the

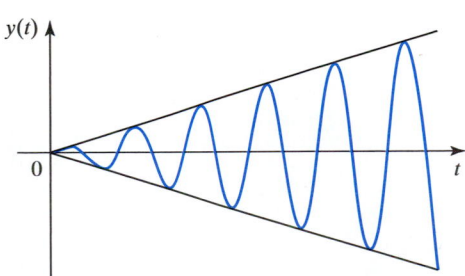

FIGURE 6.7 Linear growth of amplitude with time when $\zeta = 0$.

steady state solution. Consider a solution of the form

$$y(t) = \frac{A}{|\Omega^2 - \omega^2|^{1/2}}[\cos \omega t - \cos \Omega t].$$

Subtracting the trigonometric identities $\cos(C - D) = \cos C \cos D + \sin C \sin D$ and $\cos(C + D) = \cos C \cos D - \sin C \sin D$, and then setting $C = (\Omega + \omega)/2$ and $D = (\Omega - \omega)/2$, gives

$$\cos \omega t - \cos \Omega t = 2\sin\left(\frac{(\Omega + \omega)t}{2}\right)\sin\left(\frac{(\Omega - \omega)t}{2}\right),$$

so the solution becomes

$$y(t) = \frac{2A}{|\Omega^2 - \omega^2|^{1/2}}\sin\left(\frac{(\Omega + \omega)t}{2}\right)\sin\left(\frac{(\Omega - \omega)t}{2}\right).$$

This result can be written

$$y(t) = E(t)\sin\left(\frac{(\Omega + \omega)t}{2}\right), \quad \text{with } E(t) = \frac{2A}{|\Omega^2 - \omega^2|^{1/2}}\sin\left(\frac{(\Omega - \omega)t}{2}\right),$$

showing that when ω is close to Ω, the solution is in the form of a component with the "high angular frequency" $(\Omega + \omega)/2$, modulated by an amplitude

$$E(t) = \frac{2A}{|\Omega^2 - \omega^2|^{1/2}}\sin\left(\frac{(\Omega - \omega)t}{2}\right);$$

with the "low angular frequency" $(\Omega - \omega)/2$.

This solution is seen to be in the form of "pulses" at the higher angular frequency $(\Omega + \omega)/2$ modulated by the lower angular frequency $(\Omega - \omega)/2$. A typical physical example of *beats* can be experienced when listening to two sound waves with similar frequencies Ω_1 and Ω_2 that interact. Then, provided the amplitudes are similar, the sound at the higher frequency is heard as pulses that arrive at the lower frequency. Figure 6.8 shows a typical situation where beats occur, and when listening to such interacting sound waves the high frequency would be heard as a slow throbbing sound.

EXAMPLE 6.6   Solve the initial value problem

$$4\frac{d^2y}{dt^2} + 4\frac{dy}{dt} + 37y = 12\cos t, \quad \text{with } y(0) = 1, y'(0) = -2.$$

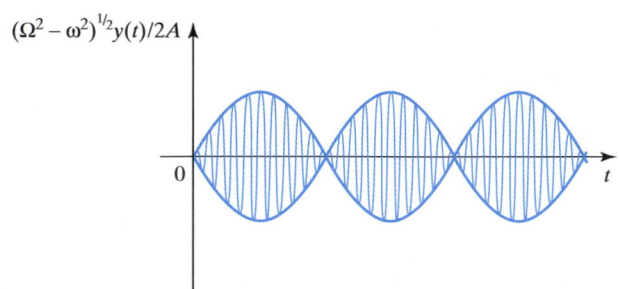

FIGURE 6.8 The phenomenon of beats produced when frequencies ω and Ω are close.

Solution   The characteristic equation is

$$4\lambda^2 + 4\lambda + 37 = 0,$$

an example showing the makeup of a typical solution

with the roots $\lambda_1 = -(1/2) + 3i$ and $\lambda_2 = -(1/2) - 3i$, so the complementary function is

$$y_c(t) = \exp[-t/2](C_1\sin 3t + C_2\cos 3t).$$

When written in the standard form the differential equation becomes

$$\frac{d^2y}{dt^2} + \frac{dy}{dt} + \frac{37}{4}y = 3\cos t.$$

Comparison with (33) shows that $\zeta = 1/2$, $\Omega^2 = 37/4$, $A = 3$, and $\omega = 1$, so $\omega_0 = (\Omega^2 - \zeta^2)^{1/2} = 3$. Substituting these results into equations (35) gives $C = 48/1105$ and $D = 396/1105$, so the general solution is

$$y(t) = \exp[-t/2](C_1\sin 3t + C_2\cos 3t) + \frac{48}{1105}\sin t + \frac{396}{1105}\cos t.$$

Imposing the initial condition $y(0) = 1$ on $y(t)$ gives

$$1 = C_2 + (396/1105), \quad \text{so} \quad C_2 = 709/1105.$$

Similarly, imposing the initial condition $y'(0) = -2$ on $y(t)$ gives

$$-2 = 48/1105 - (1/2)C_2 + 3C_1, \quad \text{so} \quad C_1 = -1269/2210.$$

Finally, substituting the values of C_1 and C_2 into the general solution shows that the solution of the initial value problem is

$$y(t) = \frac{1}{2210}\exp(-t/2)(1418\cos 3t - 1269\sin 3t) + \frac{1}{1105}(48\sin t + 396\cos t).$$

The steady state solution is

$$y_p(t) = \frac{1}{1105}(48\sin t + 396\cos t) = \frac{12}{\sqrt{1105}}\cos(t - \phi),$$

where the phase lag $\phi = \arctan C/D = \arctan(48/396) = 0.1206$ radians, and the transient solution is

$$y_c(t) = \frac{1}{2210}\exp(-t/2)(1418\cos 3t - 1269\sin 3t).$$

On the following page, the transient solution $y_p(t)$ is shown in Fig. 6.9a, the steady state solution $y_c(t)$ in Fig. 6.9b, and the complete solution $y(t)$ of the initial value problem in Fig. 6.9c. ∎

Summary

This section showed that the solution of a nonhomogeneous constant coefficient equation where the independent variable is the time comprises two parts: one called the transient solution, which describes the startup of the solution, and another called the steady state solution, which describes the nature of the time-dependent solution that remains when the transient solution has decayed sufficiently to become negligible.

The important case involving a sinusoidal forcing function was examined in detail, and the terms amplitude, frequency, and phase angle of the solution were explained, together with the important effect of resonance.

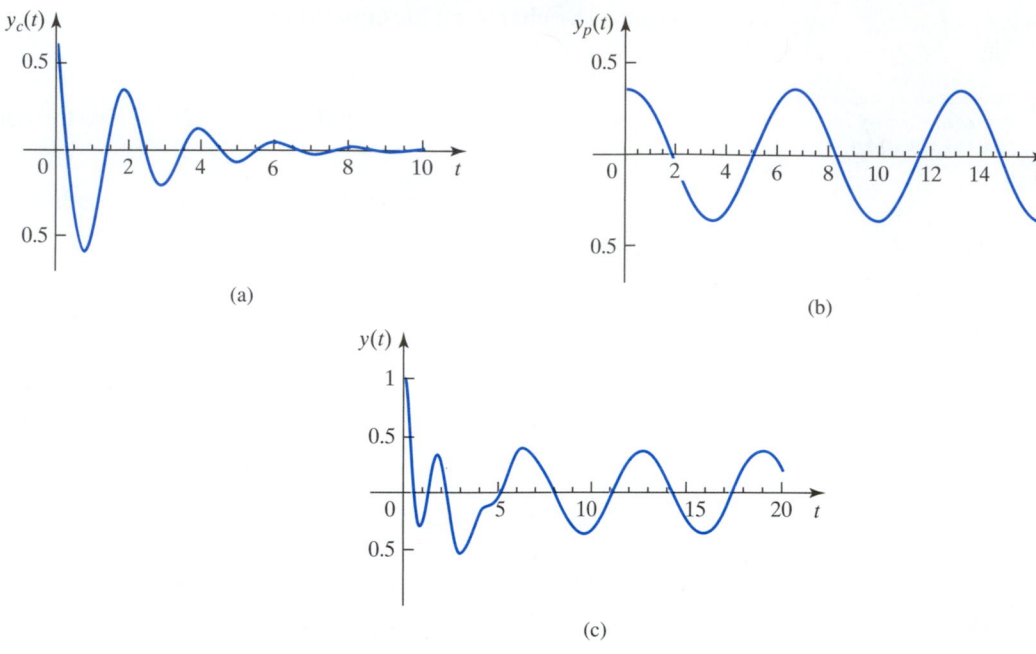

FIGURE 6.9   (a) The transient solution. (b) The steady state solution. (c) The complete solution.

EXERCISES 6.2

In Exercises 1 through 7 solve the initial value problem using the methods of this section, and identify the steady state and transient solutions. Confirm the results for the even numbered problems by computer algebra and plot their solutions for some interval $0 < t < T$.

1. $y'' + 2y' + 5y = 2\sin x$, with $y(0) = 1$, $y'(0) = 0$.

2. $y'' + 2y' + 5y = 3\sin x$, with $y(0) = 0$, $y'(0) = 0$.

3. $y'' + 2y' + y = \sin x$, with $y(0) = 1$, $y'(0) = 0$.

4. $y'' + 2y' + y = \sin 2x$, with $y(0) = 2$, $y'(0) = 0$.

5. $y'' + 3y' + 2y = \sin 3x$, with $y(0) = 0$, $y'(0) = 1$.

6. $y'' + 2y' + 5y = \sin x$, with $y(0) = 0$, $y'(0) = 1$.

7. $y'' + 5y' + 6y = A\sin x$, with $y(0) = 3$, $y'(0) = 1$.

8. Use the argument in Section 6.2 when establishing the results in (35) to show that if the forcing function on the right of (33) is replaced by $A\sin \omega t$, and the particular integral is written

$$y_p(t) = C\sin \omega t + D\cos \omega t,$$

the constants C and D are given by

$$C = \frac{A(\Omega^2 - \omega^2)}{(\Omega^2 - \omega^2)^2 + 4\zeta^2\omega^2} \quad \text{and} \quad D = \frac{2\zeta\omega A}{(\Omega^2 - \omega^2)^2 + 4\zeta^2\omega^2},$$

and that the phase angle ϕ is such that

$$\tan \phi = -\frac{2\zeta\omega}{\Omega^2 - \omega^2}.$$

In Exercises 9 through 14 use the results of Exercise 8 when solving the initial value problem. Find the phase angle and identify the steady state and transient solutions.

9. $y'' + 5y' + 6y = 2\cos x$, with $y(0) = 1$, $y'(0) = 1$.

10. $y'' + 7y' + 6y = 2\cos 3x$, with $y(0) = 2$, $y'(0) = 1$.

11. $y'' + 6y' + 9y = 2\cos 3x$, with $y(0) = 2$, $y'(0) = 2$.

12. $y'' + 2y' + 2y = \cos 4x$, with $y(0) = 0$, $y'(0) = 2$.

13. $y'' + 6y' + 8y = 3\cos 2x$, with $y(0) = 4$, $y'(0) = 1$.

14. $y'' + 2y' + 5y = 3\cos 3x$, with $y(0) = 2$, $y'(0) = 3$.

15. The fall of a loaded parachute is determined by the differential equation

$$m\frac{d^2y}{dt^2} + kg\frac{dy}{dt} + mg = 0,$$

where m is the weight of the payload in pounds, k is the drag coefficient of the parachute, $y(t)$ is its height above the ground at time t in seconds, and g is the acceleration due to gravity. Taking $g = 32$ ft/sec², $k = 10$ lb/ft/sec, and the initial speed of fall at time $t = 0$ when the parachute opens 2000 ft above the ground to be $dy/dt = -32$ ft/sec (remember $y(t)$ is measured upward but the speed is downward), find $y(t)$ and the speed of fall at time t as functions of m. Use the result to find the largest payload M in pounds if the speed of fall on landing is not to exceed 24 ft/sec. Plot $y(t)$ for $m = M$ and estimate the time of descent in this case.

16. Stokes' law $F = 6\pi a \eta u$ determines the drag F on a sphere of radius a moving slowly through a fluid with viscosity η at a speed u. Let the density of the sphere be ρ_1 and the fluid density be ρ_2 ($\rho_1 > \rho_2$). Find the equation of motion of the sphere in terms of the distance $x(t)$ from its point of release, if it falls from rest in the fluid at time $t = 0$. Solve the equation of motion to find $x(t)$, and hence the speed of fall, as functions of time. Suggest how this result could be used to determine the viscosity of oil in an experiment involving the release from rest of a ball bearing that is allowed to fall vertically through oil contained in a long glass cylinder.

17. A spherical container of radius a and density ρ_1 is released from rest on the sea bed at a depth h below the surface and allowed to float slowly upward in still water of density ρ_2, where $\rho_2 - \rho_1$ is small and $\rho_2 > \rho_1$. Assuming that Stokes' law in Exercise 16 applies, and the

viscosity of the water is η, find the distance $x(t)$ of the container from the sea bed as a function of time, and use it to write down the equation determining the time T when the container reaches the surface. Estimate this time, and suggest how a more accurate value of T could be obtained.

18. As $\omega \to 1$, from either above or below, so the solution $x(t)$ of $x'' + x = \sin \omega t$ subject to the initial conditions $x(0) = x'(0) = 0$ tends to the divergent resonance solution illustrated in Fig. 6.7. Use computer algebra to plot the solution for $\omega = 0.85$, 0.95, 0.99, 1.0, 1.05, and 1.1 to illustrate how the amplitude of the oscillations tends to a linear growth as $\omega \to 1$. Show that for $\omega = 1$, $x = \frac{1}{2}(\sin t - t \cos t)$.

19. Typically, beats occur when two slowly varying oscillations with equal amplitudes and almost equal frequencies are superimposed. Use computer algebra to plot $x(t) = \cos \omega_1 t + \cos \omega_2 t$, with suitable values of ω_1 and ω_2 and a sufficiently long time interval $0 \le t \le T$, to show a clear pattern of the beats. Find the equation determining the high-frequency oscillation and the equations forming the envelope of the high-frequency component.

6.3   Homogeneous Linear Higher Order Constant Coefficient Equations

A Typical Example Leading to a Fourth Order System

Linear nth order constant coefficient differential equations often arise as a result of the elimination of all but one of the unknowns in a system of simultaneous lower order differential equations. To see how this can happen, consider the longitudinal motion of three equal particles of mass m, coupled together by four identical springs each of unstrained length l with spring constant k, with the left and right ends of the system clamped, as illustrated in Fig. 6.10.

Now suppose that the system oscillates in the direction of the springs, with y_1, y_2, and y_3 the displacements of the masses from their equilibrium positions, as shown in Fig. 6.10. Equating the rate of change of momentum $d/dt\{m(dy_1/dt)\}$ of the mass with coordinate y_1 to the sum of the restoring force $k(y_2 - y_1)$ due to the second spring and the force $k(y_3 - y_1)$ due to the third spring shows that the equation of motion of the first mass is

$$m\frac{d^2 y_1}{dt^2} = k(l - y_1) + k(y_2 - y_1 - l) = k(y_2 - 2y_1).$$

Similar arguments applied to the second and third masses in this system with *three degrees of freedom* (the coordinates y_1, y_2, and y_3) gives the other two coupled equations of motion,

$$m\frac{d^2 y_2}{dt^2} = k(l + y_1 - y_2) + k(y_3 - y_2 - l) = k(y_1 + y_3 - 2y_2)$$

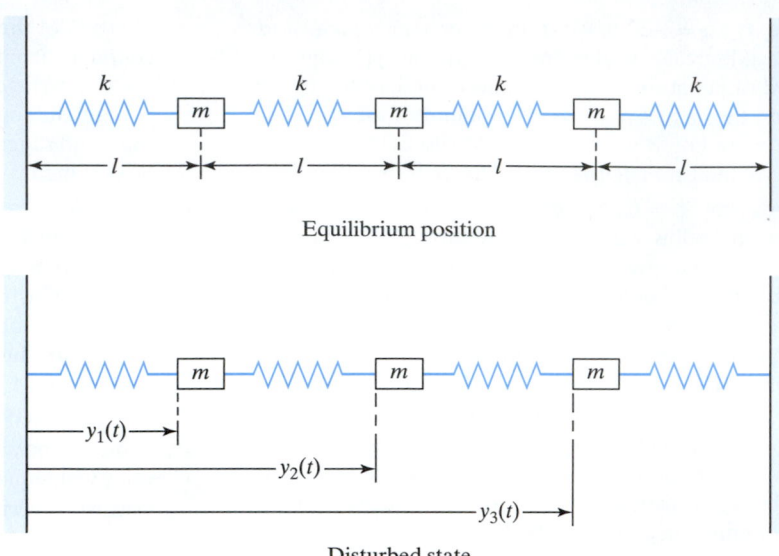

Equilibrium position

Disturbed state

FIGURE 6.10   A three-mass–spring system with its ends clamped.

and

$$m\frac{d^2 y_3}{dt^2} = k(l + y_2 - y_3) + k(4l - y_3 - l) = k(4l + y_2 - 2y_3).$$

Eliminating any two of the three unknowns y_1, y_2, and y_3 from these three equations of motion leads to a homogeneous sixth order constant coefficient differential equation for the remaining unknown. Initial conditions for the system are the values $y_i(0)$ and $y_i'(0)$ for $i = 1, 2$, and 3.

More complicated systems of this type are used to study one-dimensional waves in various types of periodic structure ranging from chains of low-pass electrical filters to the vibration of molecules in crystal lattices.

A different example that gives rise to a fourth order differential equation is the modeling of a two degree of freedom vibration damper for a motor generator of mass M. Unless damped, the vertical vibrations due to the periodic motion of the pistons are passed to the foundations of the building and can cause unacceptable vibrations throughout the building. One way of dealing with this problem is not only to mount the motor generator on a spring and damper system, but also to spring mount a smaller mass m on top of the motor generator, as in Fig. 6.11, and to adjust the two spring constants and the mass m so that the vertical oscillations of M are minimized and passed instead to the smaller mass m mounted on the motor generator.

Let the mass M be connected to the foundation by a spring with spring constant K, and let the spring constant of the spring supporting mass m be k. To make the model more realistic, suppose that in addition there is a viscous damper fitted between the mass M and the foundation that exerts a resistance proportional to the speed of its displacement with constant of proportionality μ, and let the displacements of the masses M and m from their equilibrium positions be x and y, respectively. Suppose also that the vibrational force acting on M due to the operation of the motor generator is $F(t)$.

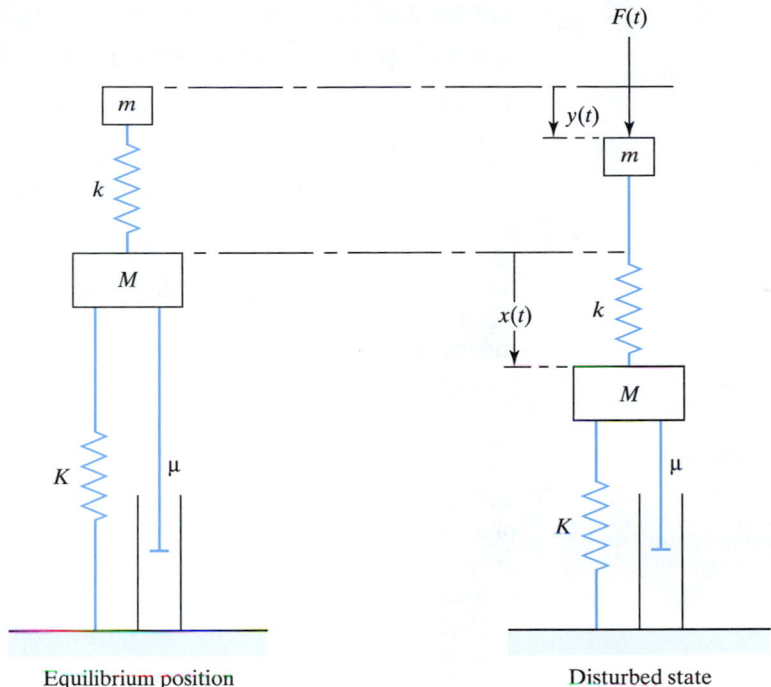

Equilibrium position   Disturbed state

FIGURE 6.11   A two degree of freedom vibration system with a viscous damper.

The equation of motion of the mass M obtained by equating its rate of change of momentum to the combined restoring forces of the two springs, the viscous damper, and the vibrational force $F(t)$ is

$$M\frac{d^2x}{dt^2} = -k(x - y) - Kx - \mu\frac{dx}{dt} + F(t),$$

and the equation of motion of the mass m obtained by equating its rate of change of momentum to the restoring force exerted by the top spring is

$$m\frac{d^2y}{dt^2} = -k(y - x).$$

Eliminating y between these two equations gives the fourth order constant coefficient equation for x

$$\frac{d^4x}{dt^4} + \alpha\frac{d^3x}{dt^3} + (\beta + \gamma + \gamma\delta/\beta)\frac{d^2x}{dt^2} + \alpha\beta\frac{dx}{dt} + \gamma\delta x = \frac{1}{M}\left(\gamma F(t) + \frac{d^2F}{dt^2}\right),$$

where $\alpha = \mu/M$, $\beta = k/m$, $\gamma = k/M$, and $\delta = K/m$.
Similarly, eliminating x between the two equations gives

$$\frac{d^4y}{dt^4} + \alpha\frac{d^3y}{dt^3} + (\beta + \gamma + \gamma\delta/\beta)\frac{d^2y}{dt^2} + \alpha\beta\frac{dy}{dt} + \gamma\delta y = \frac{\gamma}{M}F(t).$$

When $F(t)$ is a periodic force with frequency ω, and the constants k, K, and m are adjusted to take account of resonance in the spring and damper mounting, the system can be tuned so that the displacement $x(t)$ is reduced almost to zero, and the vibration is transferred instead to the mass m mounted on top of the motor generator.

General Homogeneous Higher Order Constant Coefficient Equations

The homogeneous **linear constant coefficient nth order equation**

$$\frac{d^n y}{dx^n} + a_1 \frac{d^{n-1} y}{dx^{n-1}} + \cdots + a_{n-1} \frac{dy}{dx} + a_n y = 0 \qquad (44)$$

has properties that are similar to those of second order equations.

If $y_1(x), y_2(x), \ldots, y_r(x)$ are any r solutions of (44), the linearity of the equation means that the linear combination of functions

$$y(x) = c_1 y_1(x) + c_2 y_2(x) + \cdots + c_r y_r(x),$$

linear superposition in higher order systems

with $c_1, c_2, \ldots, c_r$ arbitrary constants, is also a solution. This **linear superposition** property of solutions of the homogeneous equation is an extension of the same property encountered in Section 6.1 when considering homogeneous constant co-efficient second order equations. The proof of this property follows by substituting $y(x)$ into the left-hand side of (44), using the linearity of the differentiation operation

$$\frac{d^s}{dx^s}(c_1 y_1 + c_2 y_2 + \cdots + c_r y_r) = c_1 \frac{d^s y_1}{dx^s} + c_2 \frac{d^s y_2}{dx^s} + \cdots + c_r \frac{d^s y_r}{dx^s},$$

for $s = 0, 1, \ldots, n$, where $d^0 y/dx^0 \equiv y$ and grouping terms, to obtain r expressions of the form

$$c_i \left(\frac{d^n y_i}{dx^n} + a_1 \frac{d^{n-1} y_i}{dx^{n-1}} + \cdots + a_{n-1} \frac{dy_i}{dx} + a_n y_i \right).$$

Each of these expressions vanishes, because the $y_i(x)$ are solutions of the ho-mogeneous equation, so the result of substituting $y(x)$ into the left side of (44) is to reduce it to zero, showing that

$$y(x) = c_1 y_1(x) + c_2 y_2(x) + \cdots + c_r y_r(x)$$

is a solution.

basis, solution space, and general solutions

It will be shown later that the homogeneous equation (44) has n linearly in-dependent solutions $y_1(x), \ldots, y_n(x)$, and that these form a **basis** for its **solution space**. This means that every particular solution of (44) can be written as

$$y(x) = c_1 y_1(x) + c_2 y_2(x) + \cdots + c_n y_n(x), \qquad (45)$$

for some choice of constants $c_1, c_2, \ldots, c_n$. It is because of this property that (45) is called the **general solution** of (44).

A more general test for linear independence than the one in Section 6.1 is needed to ensure that the n solutions $y_1(x), y_2(x), \ldots, y_n(x)$ of (44) form a basis for the solution space. To obtain this test we must first extend the earlier definition of linear independence in a natural way to a set of functions $g_1(x), g_2(x), \ldots, g_n(x)$ defined over an interval $a \leq x \leq b$. The set of functions will be said to be **linearly**

independent over the interval if for all x in the interval,

$$k_1 g_1(x) + k_2 g_2(x) + \cdots + k_n g_n(x) = 0 \qquad (46)$$

linear independence and dependence

is only true if $k_1 = k_2 = \cdots = k_n = 0$; otherwise, the set of functions will be said to be **linearly dependent**.

As the test will be needed later for solutions of linear differential equations more general than (44), it will be derived for the variable coefficient differential equation

$$a_0(x)\frac{d^n y}{dx^n} + a_1(x)\frac{d^{n-1}y}{dx^{n-1}} + \cdots + a_{n-1}(x)\frac{dy}{dx} + a_n(x)y = 0, \qquad (47)$$

where the coefficients $a_i(x)$ are continuous functions of x for $a \le x \le b$. The test will also apply to solutions of (44), because a constant is a special case of a continuous function.

The derivation starts from the fact that if the functions $y_1(x), y_2(x), \ldots, y_n(x)$ are solutions of the nth order equation (47) with continuous coefficients over an interval $a \le x \le b$, then they will be everywhere continuous and differentiable at least $n - 1$ times over this same interval. By definition, the functions will be linearly independent over the interval $a \le x \le b$ if the equation

$$c_1 y_1(x) + c_2 y_2(x) + \cdots + c_n(x)y_n(x) = 0 \qquad (48)$$

is only true if $c_1 = c_2 = \cdots = c_n = 0$ for all x in the interval. Differentiating the equation $n - 1$ times gives

$$\begin{aligned}
c_1 y_1(x) + c_2 y_2(x) + \ldots + c_n(x)y_n(x) &= 0 \\
c_1 y_1^{(1)}(x) + c_2 y_2^{(1)}(x) + \cdots + c_n y_n^{(1)}(x) &= 0 \\
\cdots \cdots \cdots \cdots \cdots \cdots \cdots \cdots \cdots \\
c_1 y_1^{(n-1)}(x) + c_2 y_2^{(n-1)}(x) + \cdots + c_n y_n^{(n-1)}(x) &= 0.
\end{aligned} \qquad (49)$$

This homogeneous system of equations can only have the null solution $c_1 = c_2 = \cdots = c_n = 0$ that is necessary to ensure the linear independence of the functions $y_1(x), y_2(x), \ldots, y_n(x)$ if the determinant W of the coefficients is nonvanishing, for $a \le x \le b$. This shows that the required condition for linear independence is $W \ne 0$, for $a \le x \le b$, where

Wronskian determinant

$$W = \begin{vmatrix} y_1(x) & y_2(x) & \ldots & y_n(x) \\ y_1^{(1)}(x) & y_2^{(1)}(x) & \ldots & y_n^{(1)}(x) \\ \cdots & \cdots & \cdots & \cdots \\ y_1^{(n-1)}(x) & y_2^{(n-1)}(x) & \ldots & y_n^{(n-1)}(x) \end{vmatrix}. \qquad (50)$$

The determinant W is called the **Wronskian** of the set of functions $y_1(x)$, $y_2(x), \ldots, y_n(x)$, and it is named after the Polish mathematician who introduced the condition. We have proved the following theorem concerning the linear independence of solutions of homogeneous linear differential equations with continuous coefficients.

THEOREM 6.2

the Wronskian test for linear independence

Wronskian test for linear independence   Let $y_1(x), y_2(x), \ldots, y_n(x)$ be $n-1$ times differentiable solutions of a homogeneous linear nth order differential equation with continuous coefficients that is defined over an interval $a \le x \le b$. Then a necessary and sufficient condition for the functions to be linearly independent solutions of the differential equation is that their Wronskian W is nonvanishing over this interval. The solutions will be linearly dependent over the interval if W vanishes identically. ∎

EXAMPLE 6.7

(a) The set of continuous functions $\cosh x$, $\sinh x$, 1 is linearly independent, because the Wronskian

$$W = \begin{vmatrix} \cosh x & \sinh x & 1 \\ \sinh x & \cosh x & 0 \\ \cosh x & \sinh x & 0 \end{vmatrix} = \sinh^2 x - \cosh^2 x = -1, \quad \text{for all } x.$$

(b) The set of continuous functions $1, x, x^2, (1+x)^2$ is linearly dependent because the Wronskian

$$W = \begin{vmatrix} 1 & x & x^2 & (1+x)^2 \\ 0 & x & 2x & 2+2x \\ 0 & 1 & 2 & 2 \\ 0 & 0 & 0 & 0 \end{vmatrix} = 0 \quad \text{for all } x.$$

This result is obvious without appeal to Theorem 6.2, because setting $y_1 = 1$, $y_2 = x$, $y_3 = x^2$, and $y_4 = (1+x)^2$, we have $y_4 = y_1 + 2y_2 + y_3$, showing that y_4 is a linear combination of y_1, y_2, and y_3. ∎

It should be understood that when Theorem 6.2 is used as a general test for the linear independence of an arbitrary set of functions $u_1, u_2, \ldots u_n$ defined over an interval I, the vanishing of their Wronskian is a *necessary* condition for their linear independence over the interval, but it is *not* a sufficient condition if any of the functions involved are discontinuous within the interval.

It is the requirement in Theorem 6.2 that the functions be solutions of a homogeneous linear differential equation with *continuous* coefficients that ensures that the vanishing of the Wronskian is both a necessary and sufficient condition for their linear independence, though the details of the proof of this are omitted.

initial value problem and initial conditions

An **initial value problem** for the nth order linear differential equations (44) and (47) at a point $x = x_0$ involves specifying the **initial conditions** $y(x_0) = k_0, y^{(1)}(x_0) = k_1, \ldots, y^{(n-1)}(x_0) = k_n$ for $y(x)$, and its first $n-1$ derivatives at the point x_0, where the constants $k_1, k_2, \ldots, k_n$ can be specified arbitrarily. The derivative $y^{(n)}(x_0)$ cannot be specified as an initial condition, because it is determined by the differential equation itself once the stated initial conditions have been given.

The following is the fundamental existence and uniqueness theorem for linear higher order differential equations.

THEOREM 6.3

Existence and uniqueness of solutions   Let the coefficients of the homogeneous differential equation (47) be continuous functions over an interval $a < x < b$ that contains the point x_0 and $a_0(x) \neq 0$ in (a, b). Then a unique solution exists on this interval that satisfies the initial conditions

$$y(x_0) = k_0, \quad y^{(1)}(x_0) = k_1, \ldots, y^{(n-1)}(x_0) = k_n.$$

Proof A proof of the existence of solutions of initial value problems for linear higher order variable coefficient differential equations is beyond the level of this first account, and so will be omitted. However, the existence and uniqueness of solutions of initial value problems for constant coefficient equations will follow from the subsequent work in which the form of the general solution will be found and its constants matched so that it satisfies the initial conditions.

It remains for us to establish the uniqueness of the initial value problem for linear higher order variable coefficient equations with continuous coefficients. Let us consider equation (47), and write its general solution

$$y(x) = c_1 y_1(x) + c_2 y_2(x) + \cdots + c_n y_n(x).$$

Differentiating this result $n - 1$ times, and after each differentiation substituting the initial conditions, leads to the following system of simultaneous equations:

$$c_1 y_1(x_0) + c_2 y_2(x_0) + \cdots + c_n(x) y_n(x_0) = k_0$$

$$c_1 y_1^{(1)}(x_0) + c_2 y_2^{(1)}(x_0) + \cdots + c_n y_n^{(1)}(x_0) = k_1$$

$$\cdots \cdots \cdots \cdots \cdots \cdots \cdots \cdots \cdots \cdots$$

$$c_1 y_1^{(n-1)}(x_0) + c_2 y_2^{(n-1)}(x_0) + \cdots + c_n y_n^{(n-1)}(x_0) = k_{n-1}.$$

This nonhomogeneous system of linear equations will have a unique solution for the constant coefficients $c_1, c_2, \ldots, c_n$ provided the determinant of the coefficients does not vanish. The determinant is simply the Wronskian $W(x_0)$, and by hypothesis the n solutions are linearly independent, so $W(x_0) \neq 0$ for $a \leq x \leq b$. Consequently, the coefficients $c_1, c_2, \ldots, c_n$ are uniquely determined and, when substituted into the general solution, lead to a unique solution of the initial value problem. ∎

To solve the homogeneous constant coefficient equation

$$\frac{d^n y}{dx^n} + a_1 \frac{d^{n-1} y}{dx^{n-1}} + \cdots + a_{n-1} \frac{dy}{dx} + a_n y = 0, \tag{51}$$

we proceed as with a second order equation and seek solutions of the form $y(x) = ce^{\lambda x}$, with c and λ constants. Substituting $y(x)$ into (51) leads to the result

$$(\lambda^n + a_1 \lambda^{n-1} + a_2 \lambda^{n-2} + \cdots + a_n) e^{\lambda x} = 0,$$

after which cancellation of the nonvanishing factor $e^{\lambda x}$ shows λ may be any of the roots of the **characteristic equation**

$$\lambda^n + a_1 \lambda^{n-1} + a_2 \lambda^{n-2} + \cdots + a_n = 0. \tag{52}$$

This polynomial of degree n has n roots $\lambda_1, \lambda_2, \ldots, \lambda_n$ that either will all be real or, if some are complex, will occur in complex conjugate pairs. To each root λ_i there will correspond a solution $y_i(x)$, and the linearly independent solutions $y_1(x), y_2(x), \ldots, y_n(x)$ form a **basis** for the **solution space**. An arbitrary linear combination of the n basis functions forms the **complementary function** for (51).

Rules for constructing the complementary function of an nth order constant coefficient differential equation

how to construct the complementary function

The differential equation

$$\frac{d^n y}{dx^n} + a_1 \frac{d^{n-1}y}{dx^{n-1}} + \cdots + a_{n-1}\frac{dy}{dx} + a_n y = 0$$

with real coefficients $a_1, a_2, \ldots, a_n$ has the characteristic equation

$$\lambda^n + a_1\lambda^{n-1} + a_2\lambda^{n-2} + \cdots + a_n = 0,$$

with the n roots $\lambda_1, \lambda_2, \ldots, \lambda_n$.

1. To a single real root $\lambda = \alpha$ there corresponds the single solution $e^{\alpha x}$, with A an arbitrary constant.

2. Substitution shows that to a real root $\lambda = \alpha$ with multiplicity r (repeated r times) there correspond the r linearly independent solutions

$$e^{\alpha x},\ xe^{\alpha x}, \ldots, x^{r-1}e^{\alpha x}.$$

3. To a pair of complex conjugate roots $\lambda = \alpha \pm i\beta$ there correspond the two solutions

$$e^{\alpha x}\cos\beta x \quad \text{and} \quad e^{\alpha x}\sin\beta x.$$

4. To a pair of complex conjugate roots $\lambda = \alpha \pm i\beta$ repeated s times, there correspond the $2s$ solutions

$$e^{\alpha x}\cos\beta x,\ e^{\alpha x}\sin\beta x,\ e^{\alpha x}x\cos\beta x,\ e^{\alpha x}x\sin\beta x, \ldots$$
$$\ldots, e^{\alpha x}x^{s-2}\cos\beta x,\ e^{\alpha x}x^{s-2}\sin\beta x,\ e^{\alpha x}x^{s-1}\cos\beta x,$$
$$e^{\alpha x}x^{s-1}\sin\beta x.$$

5. The general solution of the differential equation is an arbitrary linear combination of all solutions produced by the preceding rules.

To see why the functions in Rules 2 and 4 are solutions of the differential equation, we consider a typical case in which the differential equation has a real root $\lambda = \mu$ with multiplicity 2. Removing the factor $(\lambda - \mu)^2$ from the characteristic polynomial allows it to be written

$$\lambda^n + a_1\lambda^{n-1} + a_2\lambda^{n-2} + \cdots + a_n = (\lambda - \mu)^2 \mathbf{Q}(\lambda),$$

where $\mathbf{Q}(\lambda)$ is a polynomial of degree $n - 2$ in λ that does not vanish when $\lambda = \mu$.

Differentiating this result with respect to λ gives

$$n\lambda^{n-1} + (n - 1)a_1\lambda^{n-2} + \cdots + a_{n-1} = 2(\lambda - \mu)\mathbf{Q}(\lambda) + (\lambda - \mu)^2\mathbf{Q}'(\lambda),$$

and setting $\lambda = \mu$ reduces this to

$$n\mu^{n-1} + (n - 1)a_1\mu^{n-2} + \cdots + a_{n-1} = 0.$$

As the multiplicity of the root is 2, and $e^{\mu x}$ is known to be a solution, it is necessary to show that $xe^{\mu x}$ is also a solution. This will follow if when $xe^{\mu x}$ is substituted into the differential equation the result becomes an identity.

Setting $y(x) = xe^{\mu x}$ and differentiating m times gives $y^{(m)} = m\mu^{m-1}e^{\mu x} + \mu^m xe^{\mu x}$. Substituting this into the left-hand side of the differential equation leads to the result

$$(n\mu^{n-1} + (n-1)a_1\mu^{n-2} + \cdots + a_{n-1})e^{\mu x} + (\mu^n + a_1\mu^{n-1} + \cdots + a_{n-1}\mu + a_n)xe^{\mu x},$$

but this is zero because we have shown that the coefficient of $e^{\mu x}$ is zero, and the coefficient of $xe^{\mu x}$ vanishes because μ is a root of the characteristic equation. Thus, $xe^{\mu x}$ satisfies the differential equation identically and so is a solution. The functions $e^{\mu x}$ and $xe^{\mu x}$ are linearly independent because they are not proportional.

The same form of argument can be extended to the case when $\lambda = \mu$ is a real root of arbitrary multiplicity, whereas the linear independence of the solutions follows from Theorem 6.2. A similar argument can be used when a pair of complex conjugate roots occurs with arbitrary multiplicity, though the details of these extensions are left as exercises.

EXAMPLE 6.8

some typical examples

Find the general solution of

(i) $y''' - 2y'' - 5y' + 6y = 0$;

(ii) $y''' + 2y'' + 4y' = 0$;

(iii) $y^{(iv)} + y'' - 2y = 0$.

Solution

(i) The characteristic equation is

$$\lambda^3 - 2\lambda^2 - 5\lambda + 6 = 0.$$

Inspection shows that $\lambda = 1$ is a root, so dividing the characteristic equation by the factor $(\lambda - 1)$ shows that the other two roots are the solutions of $\lambda^2 - \lambda - 6 = 0$, which are $\lambda = -2$ and $\lambda = 3$. Thus, from Rule 1 the general solution is

$$y(x) = C_1 e^x + C_2 e^{-2x} + C_3 e^{3x}.$$

(ii) The characteristic equation is

$$\lambda^3 + 2\lambda^2 + 4\lambda = 0 \quad \text{or} \quad \lambda(\lambda^2 + 2\lambda + 4) = 0,$$

from which we see that $\lambda = 0$, or $\lambda = -1 \pm i\sqrt{3}$.
Combining Rules 1 and 3 shows the general solution to be

$$y(x) = C_1 + e^{-x}(C_2\cos(x\sqrt{3}) + C_3\sin(x\sqrt{3})).$$

(iii) The characteristic equation is

$$\lambda^4 + \lambda^2 - 2 = 0.$$

This is a biquadratic equation, so if we set $m = \lambda^2$, this becomes $m^2 + m - 2 = 0$, with the solutions $m = -2$ and $m = 1$. Thus, λ can take the values $1, -1, i\sqrt{2}$, and $-i\sqrt{2}$. Combining Rules 1 and 3 shows the general solution to be

$$y(x) = C_1 e^x + C_2 e^{-x} + C_3\cos(x\sqrt{2}) + C_4\sin(x\sqrt{2}).$$

EXAMPLE 6.9   Find the general solution of a homogeneous equation with the characteristic equation

$$\lambda^3(\lambda + 4)^2(\lambda^2 + 2\lambda + 5)^2 = 0.$$

Solution   In this equation the real root $\lambda = 0$ occurs with multiplicity 3, the real root $\lambda = -4$ occurs with multiplicity 2, and the pair of complex conjugate roots ($\lambda = -1 + 2i$) and ($\lambda = -1 - 2i$) occur with multiplicity 2.

The terms to be included in the general solution corresponding to the repeated root $\lambda = 0$ follow by setting $\lambda = 0$ and $r = 3$ in Rule 2 to obtain

$$D_1 + D_2x + D_3x^2.$$

Similarly, the terms to be included corresponding to the repeated root $\lambda = -4$ follow by setting $\alpha = -4$ and $r = 2$ in Rule 2 to obtain

$$K_1e^{-4x} + K_2xe^{-4x},$$

where K_1 and K_2 are arbitrary constants.

Finally, the terms to be included because of the repeated complex conjugate roots follow by setting $\alpha = -1$, $\beta = 2$, and $s = 2$ in Rule 4 to obtain

$$e^{-x}\{E_1\cos 2x + F_1\sin 2x + E_2x\cos 2x + F_2x\sin 2x\}.$$

Collecting terms shows that the general solution is

$$y(x) = D_1 + D_2x + D_3x^2 + K_1e^{-4x} + K_2xe^{-4x}$$
$$+ e^{-x}\{E_1\cos(2x) + F_1\sin(2x) + E_2x\cos(2x) + F_2x\sin(2x)\}.$$

This general solution contains nine arbitrary constants, as would be expected because the characteristic polynomial is of degree 9.   ∎

EXAMPLE 6.10   Solve the initial value problem

$$y''' - 2y'' - 5y' + 6y = 0, \quad \text{with } y(0) = 1, \, y'(0) = y''(0) = 0.$$

Solution   The general solution was shown in Example 6.8 (i) to be

$$y(x) = C_1e^x + C_2e^{-2x} + C_3e^{3x}.$$

The initial conditions require that

$$\begin{array}{ll}(y(0) = 1) & 1 = C_1 + C_2 + C_3 \\ (y'(0) = 0) & 0 = C_1 - 2C_2 + 3C_3 \\ (y''(0) = 0) & 0 = C_1 + 4C_2 + 9C_3.\end{array}$$

The solution of this system of equations is $C_1 = 1, C_2 = 1/5, C_3 = -1/5$, so the solution of the initial value problem is

$$y(x) = e^x + \frac{1}{5}e^{-2x} - \frac{1}{5}e^{3x}.$$ ∎

EXAMPLE 6.11   Solve the initial value problem

$$y''' + 2y'' + 4y' = 0, \quad \text{with } y(0) = 0, \, y'(0) = 1, \, y''(0) = 0.$$

Solution   The general solution was found in Example 6.8 (ii) to be

$$y(x) = C_1 + e^{-x}(C_2\cos(x\sqrt{3}) + C_3\sin(x\sqrt{3})).$$

The initial conditions require that

$$(y(0) = 0) \quad C_1 + C_2 = 0$$
$$(y'(0) = 1) \quad 1 = -C_2 + C_3\sqrt{3}$$
$$(y''(0) = 0) \quad 0 = C_2 + C_3\sqrt{3}.$$

These equations have the solution $C_1 = \frac{1}{2}$, $C_2 = -\frac{1}{2}$, and $C_3 = \sqrt{3}/6$, so the solution is

$$y(x) = \frac{1}{2} + \frac{\sqrt{3}}{6}e^{-x}\sin x\sqrt{3} - \frac{1}{2}e^{-x}\cos x\sqrt{3}. \qquad \blacksquare$$

Summary

This section extended the discussion of linear second order constant coefficient equations to higher order equations, and showed how the characteristic equation again determines the nature of the solutions that enter into the complementary function. The concept of linearly independent functions was extended, and it was shown that the set of linearly independent functions associated with a higher order equation forms a basis for its solution space. The Wronskian was defined and shown to provide a test for the linear independence of a set of solutions of a higher order equation. Rules were given for construction of the complementary function of an nth order constant coefficient equation, and then applied to some typical examples.

EXERCISES 6.3

1. Use the Wronskian test to prove the linear independence of the functions
   e^x, xe^x, x^2e^x for $|x| < \infty$.

2. Use the Wronskian test to prove the linear independence of the functions
   $\sin x, e^x\sin x, e^x\cos x$.

3. Test the following functions for linear independence:
   $3, -x, x^2, (1+2x)^2$.

4. Test the following functions for linear independence:
   $1, \ln x, \ln x^{1/2}, e^x$ for $x = 0$.

In Exercises 5 through 12 show that the given functions form a basis for the associated differential equation. Write down the general solution, state the interval in which it is defined, and, where required, solve the given initial value problem.

5. $xy'' - y' - 4x^3y = 0$; $\cosh x^2$ and $\sinh x^2$.

6. $xy'' - y' + 4x^3y = 0$; $\sin x^2$ and $\cos x^2$.

7. $y''' + 3y'' + 9y' - 13y = 0$; e^x, $e^{-2x}\cos 3x$, $e^{-2x}\sin 3x$.

Solve the initial value problem for which $y(0) = 1$, $y'(0) = 0$, and $y''(0) = 0$.

8. $x^3y''' - x^2y'' + 2xy' - 2y = 0$; $x, x^2, x\ln|x|$.

Solve the initial value problem for which $y(1) = 1$, $y'(1) = 1$, and $y''(1) = 0$.

9. $(8x^2 + 1)y'' - 16xy' + 16y = 0$; $2x, 8x^2 - 1$.

10. $y'' - 16xy' + (64x^2 - 8)y = 0$; $\exp(4x^2), 2x\exp(4x^2)$.

11. $[4 - 2x\cot(x/2)]y'' - xy' + y = 0$; $x/2, \sin(x/2)$.

12. $3x^3y'' + xy' - y = 0$; $3x, 3x\exp[1/(3x)]$.

In Exercises 13 through 18 solve the initial value problems using the five stated rules for the construction of the complementary function and, when available, use computer algebra to check the results.

13. $y''' + y'' - 4y = 0$, with $y(0) = 1$, $y'(0) = 1$, $y''(0) = 0$.

14. $y''' + 3y'' - 4y = 0$, with $y(1) = -1$, $y'(1) = 0$, $y''(1) = 1$.

15. $y''' + 3y'' + 7y' + 5y = 0$, with $y(0) = 1$, $y'(0) = 0$, $y''(0) = 0$.

16. $y''' - 2y'' + 5y' + 26y = 0$, with $y(0) = 0$, $y'(0) = 1$, $y''(0) = 1$.

17. $y^{(iv)} - y'' - 2y = 0$, with $y(0) = 1$, $y'(0) = 0$, $y''(0) = 0$, $y'''(0) = 0$.

18. $y^{(iv)} - y'' - 6y = 0$, with $y(0) = 0$, $y'(0) = 1$, $y''(0) = 0$, $y'''(0) = 0$.

19.* A gyrostatic pendulum is a pendulum bob (mass) suspended by a light inextensible string from a fixed point, with the bob allowed to swing around its equilibrium position. If the displacement of the bob from its equilibrium position is small, the x and y coordinates of the bob as a function of time t can be shown to satisfy the coupled differential equations

$$\frac{d^2x}{dt^2} + a\frac{dy}{dt} + c^2x = 0 \quad \text{and} \quad \frac{d^2y}{dt^2} - a\frac{dx}{dt} + c^2y = 0,$$

with $a > 0$. Find the general solution for $x(t)$ and $y(t)$. By examination of the constants in the general solution identify two situation in which the motion of the bob will be in a circle (a circular pendulum), in each case commenting on the angular velocity of the bob.

20.* The discharge of capacitor in the primary circuit of an induction coil with a closed secondary circuit is oscillatory and governed by the equations

$$L\frac{dx}{dt} + M\frac{dy}{dt} + \frac{1}{C}\int x\,dt = f(t) \quad \text{and}$$

$$M\frac{dx}{dt} + N\frac{dy}{dt} = 0,$$

where L, M, N, and C are all positive constants and $f(t)$ is a forcing function. Find the differential equation satisfied by the discharge $x(t)$, and show that when $LN - M^2$ is small and positive the complementary function for the discharge $x(t)$ exhibits rapid oscillations.

Background material

21.* Let $y_1(x)$ and $y_2(x)$ be two linearly independent solutions of the differential equation

$$a_0(x)y'' + a_1(x)y' + a_2(x)y = 0,$$

defined on some interval I. Then the **Abel formula** for the Wronskian is

$$W(y_1(x), y_2(x)) = W(y_1(x_0), y_2(x_0))$$

$$\times \exp\left(-\int_{x_0}^{x} \frac{a_1(t)}{a_0(t)}dt\right),$$

where x_0 is any point in the interval I. Verify this result for the differential equation

$$x^2y'' - 2xy' - 4y = 0,$$

given that two linearly independent solutions over any interval that does not contain the origin are $1/x$ and x^4. Conclude that the choice of the point x_0 entering into the constant factor $W(y_1(x_0), y_2(x_0))$ is immaterial.

22.* Complete the details of the following outline proof of the Abel formula. Show that the derivative of the Wronskian of the functions in Exercise 21 can be written

$$W(y_1(x), y_2(x))' = y_1(x)y_2'' - y_2(x)y_1''(x).$$

Use the fact that $y_1(x)$ and $y_2(x)$ are solutions of the differential equation to show that

$$W' = -\frac{a_1(x)}{a_0(x)}W,$$

and by integrating over the interval $x_0 \le t \le x$ derive the result

$$W(y_1(x), y_2(x)) = W(y_1(x_0), y_2(x_0))$$

$$\times \exp\left(-\int_{x_0}^{x} \frac{a_1(t)}{a_0(t)}dt\right).$$

6.4 Undetermined Coefficients: Particular Integrals

Like the nonhomogeneous second order constant coefficient differential equation considered in Section 6.2, a **particular integral** $y_p(x)$ of the nonhomogeneous linear higher order constant coefficient differential equation

$$\frac{d^n y}{dx^n} + a_1\frac{d^{n-1}y}{dx^{n-1}} + \cdots + a_{n-1}\frac{dy}{dx} + a_n y = f(x) \tag{53}$$

is a solution of the equation that does not contain arbitrary constants, so

$$\frac{d^n y_p}{dx^n} + a_1\frac{d^{n-1}y_p}{dx^{n-1}} + \cdots + a_{n-1}\frac{dy_p}{dx} + a_n y_p = f(x).$$

particular integral, complementary function, and undetermined coefficients

The **complementary function** $y_c(x)$ associated with (53) is the general solution of the homogeneous form of the equation

$$\frac{d^n y_c}{dx^n} + a_1 \frac{d^{n-1} y_c}{dx^{n-1}} + \cdots + a_{n-1} \frac{dy_c}{dx} + a_n y_c = 0,$$

considered in Section 6.3. It follows from the definitions of $y_c(x)$ and $y_p(x)$ and the linearity of the equation that the general solution $y(x)$ of (53) can be written

$$y(x) = y_c(x) + y_p(x). \tag{54}$$

A particular integral of (53) can be found by the **method of undetermined coefficients** whenever the nonhomogeneous term $f(x)$ is a linear combination of elementary functions such as polynomials, exponentials, and sine or cosine functions.

The method depends for its success on recognizing the general form of a function that when substituted into the left-hand side of (53) yields the general form of the nonhomogeneous term $f(x)$ on the right-hand side. Undetermined coefficients are involved because although the general form of a particular integral $y_p(x)$ can be guessed from the function $f(x)$, any multiplicative constants (the undetermined coefficients) involved will not be known. Their values are found by substituting the possible form for $y_p(x)$ into the left-hand side of (53) and equating the undetermined coefficients of terms on the left of the equation to the known coefficients of corresponding terms in $f(x)$ on the right. The approach is illustrated in the following example.

EXAMPLE 6.12

Find the general solution of

$$y'' + 5y' + 6y = 4e^{-x} + 5\sin x.$$

Solution   The general solution is

$$y(x) = y_c(x) + y_p(x),$$

where $y_c(x)$ is the complementary function satisfying the homogeneous form of the equation

$$y_c'' + 5y_c' + 6y_c = 0,$$

and $y_p(x)$ is a particular integral that corresponds to the nonhomogeneous term $4e^{-x} + 5\sin x$.

The characteristic equation is

$$\lambda^2 + 5\lambda + 6 = 0,$$

with the roots $\lambda_1 = -2$ and $\lambda_2 = -3$ corresponding to the linearly independent solutions e^{-2x} and e^{-3x}, so the complementary function is

$$y_c(x) = C_1 e^{-2x} + C_2 e^{-3x},$$

where C_1 and C_2 are arbitrary constants.

To find a particular integral, we notice first that neither the term e^{-x} nor the term $\sin x$ is contained in the complementary function. This means that the only form of particular integral $y_p(x)$ that can produce the nonhomogeneous term $4e^{-x} + 5\sin x$ is

$$y_p(x) = Ae^{-x} + B\sin x + C\cos x,$$

undetermined coefficients

where A, B, and C are the *undetermined coefficients* that must be found.

Substituting this expression for $y_p(x)$ into the differential equation leads to the result

$$(Ae^{-x} - B\sin x - C\cos x) + 5(-Ae^{-x} + B\cos x - C\sin x)$$
$$+ 6(Ae^{-x} + B\sin x + C\cos x) = 4e^{-x} + 5\sin x.$$

When we collect terms involving e^{-x}, $\sin x$, and $\cos x$ this becomes

$$2Ae^{-x} + 5(B - C)\sin x + 5(B + C)\cos x = 4e^{-x} + 5\sin x.$$

If $y_p(x)$ is a particular integral, this expression must be an identity (true for all x), but this is only possible if the coefficients of corresponding functions of x on either side of the equation are identical. Equating corresponding coefficients gives

$$\text{(coefficients of } e^{-x}) \quad 2A = 4, \quad \text{so } A = 2$$
$$\text{(coefficient of } \sin x) \quad 5(B - C) = 5$$
$$\text{(coefficient of } \cos x) \quad 5(B + C) = 0.$$

Solving the last two equations for B and C gives $B = 1/2$, $C = -1/2$, so the particular integral is

$$y_p(x) = 2e^{-x} + (1/2)\sin x - (1/2)\cos x.$$

Substituting $y_c(x)$ and $y_p(x)$ into $y(x) = y_c(x) + y_p(x)$ shows that the general solution is

$$y(x) = C_1 e^{-2x} + C_2 e^{-3x} + 2e^{-x} + (1/2)\sin x - (1/2)\cos x. \quad \blacksquare$$

A complication arises if a term in the nonhomogeneous term $f(x)$ is contained in the complementary function, as illustrated in the next example.

EXAMPLE 6.13

Find a particular integral of the equation

$$y'' + y' - 12y = e^{3x}.$$

Solution This equation has the complementary function

$$y_c(x) = C_1 e^{3x} + C_2 e^{-4x},$$

so e^{3x} is contained in both the nonhomogeneous term and the complementary function.

An attempt to find a particular integral of the form $y_p(x) = Ae^{3x}$ will fail, because e^{3x} is a solution of the homogeneous form of the equation, so its substitution into the left-hand side of the differential equation will lead to the contradiction $0 = e^{3x}$. To overcome this difficulty we need to seek a more general particular integral that, when substituted into the differential equation, produces a multiple of e^{3x} whose scale factor can be equated to the coefficient of the nonhomogeneous term and other terms that cancel. As exponentials are involved, a natural choice is $y_p(x) = Axe^{3x}$.

Differentiation of $y_p(x)$ gives

$$y_p'(x) = Ae^{3x} + 3Axe^{3x} \quad \text{and} \quad y_p''(x) = 6Ae^{3x} + 9Axe^{3x}.$$

Substituting these results into the differential equation gives

$$6Ae^{3x} + 9Axe^{3x} + Ae^{3x} + 3Axe^{3x} - 12Axe^{3x} = e^{3x},$$

so after cancellation of the terms in Axe^{3x} this reduces to

$$7Ae^{3x} = e^{3x},$$

showing that $A = 1/7$. So the required particular integral is

$$y_p(x) = \frac{1}{7}xe^{3x}. \qquad \blacksquare$$

Table 6.2 lists the form of particular integral that correspond to the most common nonhomogeneous terms. Each of its entries can be constructed by using arguments similar to the one just given. When the nonhomogeneous term is a linear combination of terms in the table, the form of $y_p(x)$ is found by adding the forms of the corresponding particular integrals.

EXAMPLE 6.14

Find the general solution of

$$y''' - 5y'' + 6y' = x^2 + \sin x.$$

some typical examples

Solution   The characteristic equation is

$$\lambda^3 - 5\lambda^2 + 6\lambda = 0, \quad \text{or} \quad \lambda(\lambda^2 - 5\lambda + 6) = 0,$$

with the roots $\lambda_1 = 0$, $\lambda_2 = 2$, and $\lambda_3 = 3$, so the complementary function is

$$y_c(x) = C_1 + C_2 e^{2x} + C_3 e^{3x}.$$

The function x^2 on the right-hand side is not contained in the complementary function, but there is no undifferentiated term involving $y(x)$ in the equation, so from Step 2(b) in Table 6.2 the appropriate form of particular integral corresponding to this term is

$$Ax + Bx^2 + Cx^3.$$

The function $\sin x$ is not contained in the complementary function, so the form of particular integral appropriate to this term is seen from Step 4(a) to be

$$D\sin x + E\cos x.$$

Combining these two forms shows that the general form of $y_p(x)$ is

$$y_p(x) = Ax + Bx^2 + Cx^3 + D\sin x + E\cos x.$$

Substituting $y_p(x)$ into the differential equation gives

$$(6C - D\cos x + E\sin x) - 5(2B + 6Cx - D\sin x - E\cos x)$$
$$+ 6(A + 2Bx + 3Cx^2 + D\cos x - E\sin x) = x^2 + \sin x.$$

TABLE 6.2 Particular Integrals by the Method of Undetermined Coefficients

The method applies to the linear constant coefficient differential equation

$$\frac{d^n y}{dx^n} + a_1 \frac{d^{n-1} y}{dx^{n-1}} + \cdots + a_{n-1} \frac{dy}{dx} + a_n y = f(x),$$

which has the characteristic equation

$$\lambda^n + a_1 \lambda^{n-1} + \cdots + a_{n-1}\lambda + a_n = 0,$$

with the roots $\lambda_1, \lambda_2, \ldots, \lambda_n$, and the complementary function

$$y_c(x) = C_1 y_1(x) + C_2 y_2(x) + \cdots + C_n y_n(x),$$

where $y_1(x), y_2(x), \ldots, y_n(x)$ are the linearly independent solutions of the homogeneous equation appropriate to the nature of the roots.

1. $f(x) = $ constant. $\quad (\lambda \neq 0)$

   Include in $y_p(x)$ the constant term K.

2. $f(x) = a_0 + a_1 x + a_2 x^2 + \cdots + a_m x^m.$

   (a) If the left-hand side of the differential equation contains an undifferentiated term $y(x)$, include in $y_p(x)$ the polynomial

   $$A_0 x^m + A_1 x^{m-1} + \cdots + A_m.$$

   (b) If the left-hand side of the differential equation contains no undifferentiated function of $y(x)$, and the lowest order derivative is $d^s y/dx^s$, include in $y_p(x)$ the polynomial

   $$A_0 x^{m+s} + A_1 x^{m+s-1} + \cdots + A_m x^s.$$

3. $f(x) = Pe^{ax}.$

   (a) If e^{ax} is not contained in the complementary function, include in $y_p(x)$ the term

   $$Be^{ax}.$$

   (b) If the complementary function contains the terms $e^{ax}, xe^{ax}, \ldots, x^m e^{ax}$, include in $y_p(x)$ the term

   $$Bx^{m+1} e^{ax}.$$

4. $f(x)$ contains terms in $\cos px$ and/or $\sin px$.

   (a) If $\cos px$ and/or $\sin px$ are not contained in the complementary function, include in $y_p(x)$ the terms

   $$P\cos px + Q\sin px.$$

   (b) If the complementary function contains the terms $x \cos px$ and/or $x \sin px$, include in $y_p(x)$ terms of the form

   $$x^2(P\cos px + Q\sin px).$$

(continued)

TABLE 6.2   (*continued*)

   (c) If the complementary function contains the terms $x^2 \cos px$ and/or $x^2 \sin px$, include in $y_p(x)$ terms of the form

$$x^3(P\cos px + Q\sin px).$$

5. $f(x)$ contains terms in $e^{px}\cos qx$ and/or $e^{px}\sin qx$.

   (a) If $e^{px}\cos qx$ and/or $e^{px}\sin qx$ are not contained in the complementary function, include in $y_p(x)$ terms of the form

$$e^{px}(R\cos qx + S\sin qx).$$

   (b) If the complementary function contains $xe^{px}\cos qx$ and/or $xe^{px}\sin qx$, include in $y_p(x)$ terms of the form

$$x^2 e^{px}(R\cos qx + S\sin qx).$$

6. The required particular integral $y_p(x)$ is the sum of all the terms produced by identifying each term belonging to $f(x)$ with one of the types of term listed above.

7. The values of the undetermined coefficients K, A_0, A_1, ..., A_m, B, P, Q, R, and S are found by substituting $y_p(x)$ into the differential equation, equating the coefficients of corresponding functions on either side of the equation to make the result an identity, and then solving the resulting simultaneous equations for the undetermined coefficients.

   Equating coefficients of corresponding functions on each side of this expression to make it an identity, we have

(constant terms) $6C - 10B + 6A = 0,$

(terms in x) $-30C + 12B = 0,$

(terms in x^2) $18C = 1,$

(terms in $\sin x$) $5D - 5E = 1,$

(terms in $\cos x$) $5D + 5E = 0.$

Solving these simultaneous equations gives $A = 19/108$, $B = 5/36$, $C = 1/18$, $D = 1/10$, and $E = -1/10$, so the particular integral is

$$y_p(x) = \frac{19}{108}x + \frac{5}{36}x^2 + \frac{1}{18}x^3 + \frac{1}{10}\sin x - \frac{1}{10}\cos x.$$

Combining this with the complementary function shows the general solution to be

$$y(x) = C_1 + C_2 e^{2x} + C_3 e^{3x} + \frac{19}{108}x + \frac{5}{36}x^2 + \frac{1}{18}x^3 + \frac{1}{10}\sin x - \frac{1}{10}\cos x. \quad \blacksquare$$

   The existence and uniqueness of solutions of initial value problems for nonhomogeneous linear differential equations are guaranteed by the following theorem, which is a direct extension of Theorem 6.3.

THEOREM 6.4

more on existence and
uniqueness: this time
for nonhomogeneous
equations

Existence and uniqueness of solutions of nonhomogeneous linear equations Let the coefficients and nonhomogeneous term of differential equation (53) be continuous functions over an interval $a < x < b$ that contains the point x_0. Then a unique solution exists on this interval that satisfies the initial conditions

$$y(x_0) = k_0, \quad y^{(1)}(x_0) = k_1, \ldots, y^{(n-1)}(x_0) = k_{n-1}.$$

Proof As before, the proof of the existence of solutions of variable coefficient equations will be omitted, while the existence of solutions of constant coefficient equations has already been established. This only leaves the proof of uniqueness that follows along the same lines as those of Theorem 6.3, with $y(x)$ replaced by

$$y(x) = c_1 y_1(x) + c_2 y_2(x) + \cdots + c_n(x) + y_p(x),$$

and the system of equations determining $c_1, c_2, \ldots, c_n$ replaced by

$$c_1 y_1(x_0) + c_2 y_2(x_0) + \cdots + c_n(x) y_n(x_0) = k_0 - y_p(x_0)$$
$$c_1 y_1^{(1)}(x_0) + c_2 y_2^{(1)}(x_0) + \cdots + c_n y_n^{(1)}(x_0) = k_1 - y_p'(x_0)$$
$$\cdots \cdots \cdots \cdots \cdots$$
$$c_1 y_1^{(n-1)}(x_0) + c_2 y_2^{(n-1)}(x_0) + \cdots + c_n y_n^{(n-1)}(x_0) = k_{n-1} - y_p^{(n-1)}(x_0).$$

The constants $c_1, c_2, \ldots, c_n$ are uniquely determined by this system because, as with Theorem 6.3, the determinant of the coefficients is the Wronskian and so is nonvanishing for $x = x_0$. ∎

EXAMPLE 6.15

Solve the initial value problem

$$y'' + 4y' + 3y = e^{-x}, \quad \text{with } y(0) = 2, \quad y'(0) = 1.$$

Solution The characteristic equation is

$$\lambda^2 + 4\lambda + 3 = 0,$$

with the roots $\lambda_1 = -1$ and $\lambda_2 = -3$, so the complementary function is

$$y_c(x) = C_1 e^{-x} + C_2 e^{-3x}.$$

The nonhomogeneous term e^{-x} is contained in the complementary function, so by Step 3(b) in Table 6.2 we must seek a particular integral of the form

$$y_p(x) = Axe^{-x}.$$

Substituting the expression for $y_p(x)$ into the differential equation gives

$$(-2Ae^{-x} + Axe^{-x}) + 4(Ae^{-x} - Axe^{-x}) + 3Axe^{-x} = e^{-x}, \quad \text{or} \quad 2Ae^{-x} = e^{-x},$$

showing that $A = 1/2$. So, in this case, the particular integral is $y_p(x) = (1/2)xe^{-x}$ and the general solution is

$$y(x) = C_1 e^{-x} + C_2 e^{-3x} + (1/2)xe^{-x}.$$

The initial condition $y(0) = 2$ will be satisfied if

$$2 = C_1 + C_2,$$

and the initial condition $y'(0) = 1$ will be satisfied if

$$1/2 = -C_1 - 3C_2,$$

so $C_1 = 13/4$ and $C_2 = -5/4$. Substituting these values for C_1 and C_2 in the general solution gives the solution of the initial value problem

$$y(x) = \left(\frac{13}{4} + \frac{1}{2}x\right)e^{-x} - \frac{5}{4}e^{-3x}.$$

■

Summary

The determination of particular integrals for nonhomogeneous equations is important, and the method of undetermined coefficients that was described in this section is the simplest method by which they can be found. The method is only applicable to nonhomogeneous terms formed by a sum of polynomials, exponentials, trigonometric functions, and certain of their combinations. It depends for its success on recognizing the general form of function that, when substituted into the left of the differential equation, produces terms of the type found in the nonhomogeneous term on the right. The method involves substituting a linear combination of such terms with arbitrary constant multipliers (the undetermined coefficients) into the left of the equation and matching the constants so the terms that result are identical to the terms on the right.

EXERCISES 6.4

Find the general solutions of the following differential equations.

1. $y'' + 2y' - 3y = 4 + x + 4e^{2x}$.
2. $y'' + 4y' + 4y = 2 - \sin 3x$.
3. $y'' + 2y' + y = 5 + x^2 e^x$.
4. $y'' - 4y' + 4y = 3x^2 + 2e^{3x}$.
5. $y'' + 4y' + 4y = \sin x - 2\cos x$.
6. $y'' + 4y' + 5y = \sin x$.
7. $y'' + 2y' + 2y = 1 + x + e^{-x}$.
8. $y''' + 5y'' + 6y' = 3\sin x + 5x + x^2$.
9. $y''' + 2y'' + 2y' = 2 - 4x^2$.
10. $y'' + 2y' + 2y = \sin x$.
11. $y'' - 7y' + 12y = x + e^{2x} + e^{3x}$.
12. $y'' + 4y' + 5y = 3 + 2e^{-2x}$.
13. $y'' + 2y' - 8y = 3x\cos 4x$.
14. $y'' + 2y' - 15y = 3 + 2x\sin x$.
15. $y'' + 9y = 2\cos 3x + \sin 3x$.
16. $y'' - 4y = 3e^{2x} + 4e^{-2x}$.
17. $y'' + 3y' + 2y = x^2 + 3e^{-2x}$.
18. $y''' + y'' + 3y' - 5y = 4e^{-x}$.
19. $y'' + 4y' + 5y = e^{-2x}\sin x$.
20. $y'' + 4y' + 5y = x^2 - e^{-2x}\cos x$.

In Exercises 21 through 28 solve the initial value problems. Where the characteristic equation is of degree 3, at least one root is an integer and can be found by inspection.

21. $y'' + 6y' + 13y = e^{-3x}\cos x$, with $y(0) = 2$, $y'(0) = 1$.
22. $y'' - 4y' + 5y = e^{2x}\cos x$, with $y(0) = 0$, $y'(0) = 2$.
23. $y'' + 9y = 7 + 2\sin 3x - 4\cos 3x$, with $y(0) = -1$, $y'(0) = 1$.
24. $y'' + 4y' + 5y = x + \sin x$, with $y(0) = -1$, $y'(0) = 0$.
25. $y'' - 2y' + 5y = 1 + e^{-x}$, with $y(0) = 2$, $y'(0) = 1$.
26. $y'' + 4y' + 5y = 2 + e^{-2x}\sin x$, with $y(0) = 0$, $y'(0) = 0$.
27. $y''' + y'' - 2y = 3 + 2\cos x$, with $y(0) = 0$, $y'(0) = 1$, $y''(0) = -1$.
28. $y''' + y'' - y' - y = 2 + e^{-x}$, with $y(0) = 1$, $y'(0) = 1$, $y''(0) = 0$.

6.5   Cauchy–Euler Equation

Cauchy–Euler equation

One of the simplest linear variable coefficient differential equations is the homogeneous second order **Cauchy–Euler** equation, whose standard form is

$$x^2\frac{d^2y}{dx^2} + a_1 x\frac{dy}{dx} + a_2 y = 0. \tag{55}$$

The solution of this homogeneous equation can be reduced to a simple algebraic problem by seeking a solution of the form

$$y(x) = Ax^m, \tag{56}$$

where A is an arbitrary constant, and the permissible values of m are to be determined.

Differentiating $y(x)$ to obtain

$$\frac{dy}{dx} = mAx^{m-1} \quad \text{and} \quad \frac{d^2y}{dx^2} = m(m-1)Ax^{m-2} \tag{57}$$

and substituting these expressions into the Cauchy–Euler equation gives the following quadratic equation for m:

$$m(m-1) + a_1 m + a_2 = 0. \tag{58}$$

When this equation has two distinct real roots $m = \alpha$ and $m = \beta$, the general solution of (55) is

$$y(x) = C_1 x^\alpha + C_2 x^\beta, \tag{59}$$

but if the two roots are real and equal with $m = \mu$, the general solution of (55) is

$$y(x) = C_1 x^\mu + C_2 x^\mu \ln|x|, \tag{60}$$

where C_1 and C_2 are arbitrary real constants.

If the equation for m has the complex conjugate roots $m = \alpha \pm i\beta$, substitution confirms that the general solution of (55) is

$$y(x) = C_1 x^\alpha \cos(\beta \ln|x|) + C_2 x^\alpha \sin(\beta \ln|x|). \tag{61}$$

The second solution $x^\mu \ln|x|$ in (60) can be obtained from the method of Section 6.7 by using the known solution $y_1(x) = x^\mu$ to find a second linearly independent solution $y_2(x)$. The form of solution (61) follows from writing the general solution as $y(x) = A\exp(\alpha + i\beta) + B\exp(\alpha - i\beta)$, with A an arbitrary complex constant and B its complex conjugate so that $y(x)$ is real.

EXAMPLE 6.16   Find the general solution of

$$x^2 \frac{d^2y}{dx^2} + 3x \frac{dy}{dx} + 2y = 0 \quad \text{for } x \neq 0.$$

Solution   The equation for m is

$$m(m-1) + 3m + 2 = 0,$$

with the roots $m = -1 \pm i$. The general solution is thus

$$y(x) = C_1 x^{-1} \cos(\ln|x|) + C_2 x^{-1} \sin(\ln|x|).$$

Summary

The Cauchy–Euler equation is the simplest linear variable coefficient equation for which a closed form analytical solution can be found. The solution is obtained by recognizing that it must be of the form $y(x) = Ax^m$ and finding the permissible values of m.

EXERCISES 6.5

Find the general solutions of the following Cauchy–Euler equations.

1. $x^2 y'' + 3xy' - 3y = 0$.
2. $x^2 y'' + 3xy' + 5y = 0$.
3. $x^2 y'' + 5xy' + 9y = 0$.
4. $x^2 y'' - 3xy' - 5y = 0$.
5. $x^2 y'' + 3xy' - 8y = 0$.

6. $x^2 y'' + 2xy' + 4y = 0$.
7. $x^2 y'' + 6xy' + 4y = 0$.
8. $x^2 y'' + xy' + 4y = 0$.
9. $x^2 y'' + 4xy' + 4y = 0$.
10. $x^2 y'' + 3xy' + 6y = 0$.

11. With the change of variable $x = e^t$, we find using the chain rule that

$$\frac{dy}{dx} = \frac{1}{x}\frac{dy}{dt} \quad \text{and} \quad \frac{d^2 y}{dx^2} = \frac{1}{x^2}\left(\frac{d^2 y}{dt^2} - \frac{dy}{dt}\right).$$

Use these results to show that this change of variable transforms a Cauchy–Euler equation into a constant coefficient equation, and solve Exercise 3 by this method.

12. Use the substitution $y(x) = Ax^m$ to solve the third order Cauchy–Euler equation

$$x^3 y''' - 3x^2 y'' + 6xy' - 6y = 0.$$

13. Use the substitution of Exercise 11 to solve the Cauchy–Euler equation in Exercise 12.

14. Express dy/dx, $d^2 y/dx^2$, and $d^3 y/dx^3$ in terms of dy/dt, $d^2 y/dt^2$, and $d^3 y/dt^3$ if $ax + b = e^t$. Use the substitution to show that the general solution of

$$(2x + 3)^3 y''' + 3(2x + 3)y' - 6y = 0$$

is

$$y(x) = C_1(2x + 3) + C_2(2x + 3)^{1/2} + C_3(2x + 3)^{3/2}$$

for $x > 0$.

6.6   Variation of Parameters and the Green's Function

Variation of Parameters

The method of **variation of parameters**, perhaps more properly called **variation of constants**, is a powerful method used to find a particular integral of a linear differential equation once its complementary function is known. In what follows the method will be developed for a general linear second order variable coefficient differential equation, though it is easily extended to include linear variable coefficient differential equations of any order.

As linear constant coefficient equations are a special case of variable coefficient equations, the method enables particular integrals to be found for all linear equations. The method also has the advantage that no special cases arise due to the nonhomogeneous term being included in the complementary function.

Consider the general linear second order differential equation

$$\frac{d^2 y}{dx^2} + a(x)\frac{dy}{dx} + b(x)y = f(x), \tag{62}$$

idea underlying the method of variation of parameters

defined on some interval $\alpha \leq x \leq \beta$ over which $a(x)$, $b(x)$, and $f(x)$ are defined and continuous. Let $y_1(x)$ and $y_2(x)$ be two known linearly independent solutions of the homogeneous form of (62), so the complementary function is

$$y_c(x) = C_1 y_1(x) + C_2 y(x). \tag{63}$$

The idea underlying the method of variation of parameters, and from which it derives its name, is to replace the constants C_1 and C_2 by the unknown functions $u_1(x)$ and $u_2(x)$, and then to seek a particular integral of the form

$$y_p(x) = u_1(x)y_1(x) + u_2(x)y_2(x). \tag{64}$$

Two equations are needed in order to determine $u_1(x)$ and $u_2(x)$, and the first of these is obtained as follows. Differentiation of (64) gives

$$y_p'(x) = u_1(x)y_1'(x) + u_2(x)y_2'(x) + u_1'(x)y_1(x) + u_2'(x)y_2(x),$$

so by requiring $u_1(x)$ and $u_2(x)$ to be such that the last two terms vanish, we have

$$y_p'(x) = u_1(x)y_1'(x) + u_2(x)y_2'(x), \tag{65}$$

subject to the condition

$$u_1'(x)y_1(x) + u_2'(x)y_2(x) = 0. \tag{66}$$

Equation (66) is the *first* condition to be imposed on $u_1(x)$ and $u_2(x)$, and a second condition is obtained as follows. Differentiating (65) gives

$$y_p''(x) = u_1(x)y_1''(x) + u_2(x)y_2''(x) + u_1'(x)y_1'(x) + u_2'(x)y_2'(x), \tag{67}$$

so substituting (64), (65), and (67) into (62), followed by grouping terms, gives

$$u_1[y_1'' + a(x)y_1' + b(x)y_1] + u_2[y_2'' + a(x)y_2' + b(x)y_2] +$$
$$+ u_1'y_1' + u_2'y_2' = f(x). \tag{68}$$

As $y_1(x)$ and $y_2(x)$ are both solutions of differential equation (62) with $f(x) = 0$, the expressions multiplying $u_1(x)$ and $u_2(x)$ both vanish identically, reducing (68) to the *second* condition on $u_1(x)$ and $u_2(x)$,

$$u_1'y_1' + u_2'y_2' = f(x). \tag{69}$$

The functions $u_1(x)$ and $u_2(x)$ can now be found by solving equations (66) and (69). Solving these for $u_1'(x)$ and $u_2'(x)$ gives

$$u_1'(x) = \frac{-y_2(x)f(x)}{W(x)} \quad \text{and} \quad u_2'(x) = \frac{y_1(x)f(x)}{W(x)}, \tag{70}$$

where

$$W(x) = \begin{vmatrix} y_1 & y_2 \\ y_1' & y_2' \end{vmatrix} = y_1 y_2' - y_1' y_2 \tag{71}$$

is the Wronskian of $y_1(x)$ and $y_2(x)$ and so is never zero.

After integration, results (70) become

$$u_1(x) = -\int \frac{y_2(x) f(x)}{W(x)} dx \qquad \text{and} \qquad u_2(x) = \int \frac{y_1(x) f(x)}{W(x)} dx. \qquad (72)$$

the general solution

Finally, combining (64) and (72), we find that

$$y(x) = -y_1(x) \int \frac{y_2(x) f(x)}{W(x)} dx + y_2(x) \int \frac{y_1(x) f(x)}{W(x)} dx. \qquad (73)$$

This result represents the general solution of (62), because each indefinite integral has associated with it an additive arbitrary constant, and if these are $-C_1$ and C_2, say, they include in $y(x)$ the complementary function $y_c(x) = C_1 y_1(x) + C_2 y_2(x)$. When these constants are set equal to zero result (73) reduces to the particular integral $y_p(x)$.

Rule for the method of variation of parameters

1. Write the differential equation in the standard form

$$\frac{d^2 y}{dx^2} + a(x)\frac{dy}{dx} + b(x)y = f(x).$$

how to apply the method of variation of parameters

2. Find two linearly independent solutions $y_1(x)$ and $y_2(x)$ of the homogeneous form of the differential equation and construct the equations

$$u_1'(x)y_1(x) + u_2'(x)y_2(x) = 0 \qquad \text{and} \qquad u_1'y_1' + u_2'y_2' = f(x).$$

3. Solve the equations in Step 2 for $u_1'(x)$ and $u_2'(x)$ and integrate to find $u_1(x)$ and $u_2(x)$, each with an arbitrary additive constant of integration.
4. The general solution of the differential equation is then given by

$$y(x) = u_1(x)y_1(x) + u_2(x)y_2(x).$$

Or, alternatively, after finding $y_1(x)$ and $y_2(x)$:

5. Substitute into

$$y(x) = -y_1(x) \int \frac{y_2(x) f(x)}{W(x)} dx + y_2(x) \int \frac{y_1(x) f(x)}{W(x)} dx,$$

where

$$W(x) = \begin{vmatrix} y_1 & y_2 \\ y_1' & y_2' \end{vmatrix} = y_1 y_2' - y_1' y_2.$$

6. The result of Step 5 becomes the particular integral $y_p(x)$ if the arbitrary integration constants are set equal to zero.

The example that follows shows how the method of variation of parameters deals automatically with the presence of a nonhomogeneous term in the complementary function of a constant coefficient equation.

EXAMPLE 6.17

a simple example that could also be solved by undetermined coefficients

Find the general solution of the second order differential equation

$$y'' + 2y' + y = xe^{-x}.$$

Solution The characteristic equation is

$$\lambda^2 + 2\lambda + 1 = 0,$$

with the repeated root $\lambda = -1$. Thus, the complementary function is

$$y_c(x) = C_1 e^{-x} + C_2 x e^{-x}.$$

Two linearly independent solutions are thus

$$y_1(x) = e^{-x} \quad \text{and} \quad y_2(x) = xe^{-x},$$

while the nonhomogeneous term is $f(x) = xe^{-x}$. The Wronskian

$$W(x) = \begin{vmatrix} y_1 & y_2 \\ y_1' & y_2' \end{vmatrix} = e^{-x}(e^{-x} - xe^{-x}) + e^{-x}xe^{-x} = e^{-2x},$$

so substituting in (73) shows that the particular integral is

$$y_p(x) = -e^{-x} \int x^2 dx + xe^{-x} \int x dx = \frac{1}{6}x^3 e^{-x}.$$

The general solution is

$$y(x) = C_1 e^{-x} + C_2 x e^{-x} + \frac{1}{6}x^3 e^{-x}.$$

This result could, of course, have been found by the method of undetermined coefficients. ∎

The next example shows how the method of variation of parameters determines a particular integral for a constant coefficient equation whose particular integral could not have been found by using undetermined coefficients.

EXAMPLE 6.18

an example that could not be solved by undetermined coefficients

Find the general solution of the differential equation

$$y'' + y = \csc x$$

in any interval in which $x \neq n\pi$, for $n = 1, 2, \ldots$.

Solution It follows at once that the complementary function is

$$y_c(x) = C_1 \cos x + C_2 \sin x,$$

so two linearly independent solutions are

$$y_1(x) = \cos x \quad \text{and} \quad y_2(x) = \sin x.$$

The Wronskian $W(x) = y_1 y_2' - y_1' y_2 = \cos^2 x + \sin^2 x = 1$, and $f(x) = 1/\sin x$, so substituting into (73) shows that the particular integral is

$$y_p(x) = -\cos x \int dx + \sin x \int \cot x dx.$$

As $\int \cot x\, dx = \ln|\sin x|$,

$$y_p(x) = -x\cos x + \sin x \ln|\sin x|,$$

and the general solution is

$$y(x) = C_1\cos x + C_2\sin x - x\cos x + \sin x \ln|\sin x|,$$

in any interval in which $x \neq n\pi$, for $n = 1, 2, \ldots$, because $\ln|\sin n\pi| = \infty$. Although this is a constant coefficient equation, it is unlikely that its particular integral could have been found by the method of undetermined coefficients.   ■

The last example shows how the method of variation of parameters determines a particular integral for a linear second order variable coefficient equation.

EXAMPLE 6.19

Find the general solution of the second order variable coefficient equation

$$x^2 y'' - 3xy' + 4y = \ln x \quad (x > 0).$$

application to a variable coefficient equation

Solution   This is a Cauchy–Euler equation, and the method of Section 6.5 shows that its complementary function is

$$y_c(x) = C_1 x^2 + C_2 x^2 \ln x, \quad \text{for} \quad x > 0,$$

so two linearly independent solutions are

$$y_1(x) = x^2 \quad \text{and} \quad y_2(x) = x^2 \ln x \quad \text{for} \quad x > 0.$$

A routine calculation shows the Wronskian $W(x) = x^3$. Before identifying $f(x)$ the equation must be written in the standard form with the coefficient of y'' equal to 1. Dividing the differential equation by x^2 to bring it into the standard form shows that $f(x) = (\ln x)/x^2$.

Substitution into (73) then gives

$$y_p(x) = -x^2 \int \frac{(\ln x)^2}{x^3}dx + x^2 \ln x \int \frac{\ln x}{x^3}dx.$$

Integration by parts shows that

$$\int \frac{(\ln x)^2}{x^3} = -\frac{1}{2}\frac{(\ln x)^2}{x^2} - \frac{1}{2}\frac{\ln x}{x^2} - \frac{1}{4x^2} \quad \text{and} \quad \int \frac{\ln x}{x^3}dx = -\frac{1}{2}\frac{\ln x}{x^2} - \frac{1}{4x^2},$$

so using these results in the expression for $y_p(x)$ gives

$$y_p(x) = \frac{1}{4} + \frac{1}{4}\ln x \quad (x > 0).$$

The general solution is thus

$$y(x) = C_1 x^2 + C_2 x^2 \ln x + \frac{1}{4} + \frac{1}{4}\ln x \quad (x > 0).$$

Although the complementary function of a Cauchy–Euler equation is easily determined, a particular integral is usually sufficiently complicated that its general form cannot be guessed and so must be found by the method of variation of parameters.   ■

Finally, we remark that an application of the method of variation of parameters to the equation

what happens if an integral has no known antiderivative

$$y'' + y = (1 + x^2)^{1/2}$$

gives a particular integral in the form

$$y_p(x) = -\cos x \int (\sin x)(1 + x^2)^{1/2}dx + \sin x \int (\cos x)(1 + x^2)^{1/2}dx.$$

Neither of the two integrals involved can be evaluated in terms of known functions, so if an analytical solution is needed it must be obtained in series form. The Maclaurin series for the functions $(\sin x)(1 + x^2)^{1/2}$ and $(\cos x)(1 + x^2)^{1/2}$ are

$$(\sin x)(1 + x^2)^{1/2} = x + \frac{1}{3}x^3 - \frac{1}{5}x^5 + \cdots \quad \text{and}$$

$$(\cos x)(1 + x^2)^{1/2} = x - \frac{1}{3}x^4 - \frac{13}{90}x^6 + \cdots.$$

Integrating these results and substituting in the expression for $y_p(x)$ gives

$$y_p(x) = -(\cos x)\left(\frac{1}{2}x^2 + \frac{1}{12}x^4 - \frac{1}{30}x^6 + \cdots\right) + \sin x \left(x - \frac{1}{15}x^5 + \frac{13}{630}x^7 + \cdots\right).$$

Let $y(x)$ satisfy the differential equation

$$\frac{d^2y}{dx^2} + a(x)\frac{dy}{dx} + b(x)y = f(x), \tag{74}$$

defined on an interval $\alpha \leq x \leq \beta$, and let a be any point inside this interval. Then the general solution of (74) given in (73) can be put into a convenient form for solving the initial value problem for (74) when the initial conditions are $y(a) = 0$ and $y'(a) = 0$.

We start from the general solution in (73)

$$y(x) = -y_1(x) \int \frac{y_2(x)f(x)}{W(x)}dx + y_2(x) \int \frac{y_1(x)f(x)}{W(x)}dx. \tag{75}$$

Next, we rewrite the indefinite integral $\int \frac{y_2(x)f(x)}{W(x)}dx$ as the definite integral with a variable upper limit $\int_a^x \frac{y_2(t)f(t)}{W(t)}dt$ and an arbitrary fixed lower limit $x = a$. In this result, the additive arbitrary integration constant associated with the indefinite integral has been replaced by the arbitrary constant a in the lower integration limit. The implications of the lower limit will become apparent when an initial value problem is considered. A corresponding result holds for the second indefinite integral in (75). Using these results, taking the functions $y_1(x)$ and $y_2(x)$ under the respective integral signs as they are not involved in the integrations, and combining the integrals allows the general solution $y(x)$ to be written in the form

$$y(x) = \int_a^x \frac{y_1(t)y_2(x) - y_1(x)y_2(t)}{W(t)} f(t)dt. \tag{76}$$

Setting $x = a$ in this result shows that $y(a) = 0$. Differentiation of (76) with respect to x using Leibniz's rule

$$\frac{d}{dx}\int_{p(x)}^{q(x)} g(x,t)dt = \frac{dq}{dx}g(x,q) - \frac{dp}{dx}g(x,p) + \int_{p(x)}^{q(x)} \frac{\partial}{\partial x}g(x,t)dt$$

gives

$$y'(x) = \frac{y_1(x)y_2(x) - y_1(x)y_2(x)}{W(x)}f(x) + \int_a^x \frac{y_1(t)y_2'(x) - y_1'(x)y_2(t)}{W(t)}dt.$$

The first term on the right vanishes, and setting $x = a$ causes the integral to vanish, so we have shown that $y'(a) = 0$. Consequently, the integral

variation of parameters and initial value problems

$$y(x) = \int_a^x \frac{y_1(t)y_2(x) - y_1(x)y_2(t)}{W(t)}f(t)dt$$

solves the initial value problem

$$\frac{d^2y}{dx^2} + a(x)\frac{dy}{dx} + b(x)y = f(x), \quad \text{with} \quad y(a) = y'(a) = 0.$$

EXAMPLE 6.20   Use result (76) to solve the initial value problem

$$y'' + 4y = 1 + \cos 2x, \quad \text{with } y(0) = y'(0) = 0.$$

Solution   Two linearly independent solutions of the homogeneous equation are $y_1(x) = \sin 2x$ and $y_2(x) = \cos 2x$, so $W(t) = -2(\sin^2 2t + \cos^2 2t) = -2$. Substituting into (76) with $f(t) = 1 + \cos 2t$ gives

$$y(x) = \int_0^x \tfrac{1}{2}(\sin 2x \cos 2t - \sin 2t \cos 2x)(1 + \cos 2t)dt,$$

and so

$$y(x) = \tfrac{1}{4}(1 - \cos 2x + x\sin 2x). \qquad\blacksquare$$

The Green's Function

An important result that can be derived from the general solution of (74) when expressed in the form given in (75) is obtained by considering a boundary value problem for the equation written in the standard form

$$\frac{d^2y}{dx^2} + a(x)\frac{dy}{dx} + b(x)y = f(x), \tag{77}$$

and defined over the interval $a \le x \le b$.

Evaluating the first integral in (75) over the interval $b \le t \le x$, changing the sign by reversing the limits of integration, and then evaluating the second integral over the interval $a \le t \le x$ gives

$$y(x) = y_2(x) \int_a^x \frac{y_1(t)}{W(t)} f(t)dt + y_1(x) \int_x^b \frac{y_2(t)}{W(t)} f(t)dt. \tag{78}$$

As $y_2(x)$ is not involved in the first integral, and $y_1(x)$ is not involved in the second integral, they may be taken under the respective integral signs so that (78) becomes

$$y(x) = \int_a^x \frac{y_1(t)y_2(x)}{W(t)} f(t)dt + \int_x^b \frac{y_1(x)y_2(t)}{W(t)} f(t)dt. \tag{79}$$

This can be written

$$y(x) = \int_a^b G(x, t) f(t)dt, \tag{80}$$

the Green's function

where the function $G(x, t)$ is called the **Green's function** for differential equation (77) defined over the interval $a \le x \le b$ and is defined as

$$G(x, t) = \begin{cases} \dfrac{y_1(t)y_2(x)}{W(t)}, & a \le t \le x \\[2mm] \dfrac{y_1(x)y_2(t)}{W(t)}, & x \le t \le b. \end{cases} \tag{81}$$

Inspection of (81) shows $G(x, t)$ to be a continuous function of x for $a \le x \le b$. Differentiation of $G(x, t)$ with respect to x gives

$$G_x(x, t) = \begin{cases} \dfrac{y_1(t)y_2'(x)}{W(t)}, & a \le t \le x \\[2mm] \dfrac{y_1'(x)y_2(t)}{W(t)}, & x \le t \le b. \end{cases} \tag{82}$$

Examination of (82) shows that as t increases across $t = x$, the function $G_x(x, t)$ is discontinuous and experiences the jump

$$G_x(x, x_+) - G_x(x, x_-) = \frac{y_1'(x)y_2(x) - y_1(x)y_2'(x)}{W(x)} = -\frac{W(x)}{W(x)} = -1,$$

where x_+ is the limit at t decreases to x and x_- is the limit as t increases to x.

Now let $y_1(x)$ and $y_2(x)$ be two linearly independent solutions of the homogeneous differential equation, with $y_1(x)$ such that at $x = a$ it satisfies the homogeneous boundary condition

$$k_1 y_1(a) + K_1 y_1'(a) = 0,$$

and $y_2(x)$ such that at $x = b$ it satisfies the homogeneous boundary condition

$$k_2 y_2(b) + K_2 y_2'(b) = 0.$$

Then $G(x, t)$ is seen to satisfy these same homogeneous boundary conditions, and differentiation of (80) with respect to x, again using Leibniz's rule, shows that the solution $y(x)$ also satisfies these homogeneous boundary conditions. Combining results shows that

$$y(x) = \int_a^b G(x,t) f(t) dt \quad \text{with} \quad G(x,t) = \begin{cases} \dfrac{y_1(t) y_2(x)}{W(t)}, & a \le t \le x \\[2mm] \dfrac{y_1(x) y_2(t)}{W(t)}, & x \le t \le b \end{cases}$$

(83)

is the solution of the boundary value problem for the nonhomogeneous linear second order equation

$$\frac{d^2 y}{dx^2} + a(x)\frac{dy}{dx} + b(x) y = f(x),$$

subject to the homogeneous boundary conditions

$$k_1 y(a) + K_1 y'(a) = 0 \quad \text{with} \quad k_2 y(b) + K_2 y'(b) = 0.$$

When using this approach, unless the Green's function itself is required, it is usually more convenient to obtain the solution directly from result (78). The advantage of the Green's function is that it characterizes all the essential features of the differential equation without reference to the nonhomogeneous term $f(x)$, so that once it is known (80) solves the homogeneous boundary value problem for any function $f(x)$.

Properties of the Green's function defined over the interval $a \le x \le b$

Consider the boundary value problem

$$\frac{d^2 y}{dx^2} + a(x)\frac{dy}{dx} + b(x) y = 0,$$

fundamental properties of the Green's function

subject to the boundary conditions

$$k_1 y(a) + K_1 y'(a) = 0 \quad \text{and} \quad k_2 y(b) + K_2 y'(b) = 0$$

The Green's function in (81) has the following properties:

1. The piecewise defined Green's function $G(x,t)$ satisfies the differential equation in the respective intervals $a \le x < t$ and $t < x \le b$.
2. $G(x,t)$ is a continuous function of x for $a \le x \le b$.
3. $G(x,t)$ satisfies the homogeneous boundary conditions.
4. The function $G_x(x,t)$ is continuous for $a \le x < t$ and $t < x \le b$, but it is discontinuous across $x = t$ where it experiences the jump

$$G_x(x, x_+) - G_x(x, x_-) = -1.$$

EXAMPLE 6.21   Find the Green's function for the differential equation

$$x^2 y'' - 2xy' + 2y = 3x^2$$

and use it to solve the boundary value problem when $y(1) = 0$ and $y'(2) = 0$.

Solution   The homogeneous form of the equation is a Cauchy–Euler equation, and the method of Section 6.5 shows that it has the two linearly independent solutions $y_1(x) = x$ and $y_2(x) = x^2$, so the general solution is $y(x) = ax + bx^2$.

For the solution $y_1(x)$ we must use the form of this solution that satisfies the left boundary condition $y(1) = 0$, and this is easily seen to be $y_1(x) = x - x^2$. For the linearly independent solution $y_2(x)$ we must use the form of solution $y(x) = ax + bx^2$ that satisfies the right boundary condition $y'(2) = 0$. As $y'(x) = a + 2bx$, the condition $y'(2) = 0$ shows that $y_2(x) = 4x - x^2$. Using these results the Wronskian becomes $W(t) = 3t^2$.

The Green's function for this differential equation defined by (81) is

$$G(x, t) = \begin{cases} \dfrac{(t - t^2)(4x - x^2)}{3t^2}, & 1 \le t < x \\ \dfrac{(4t - t^2)(x - x^2)}{3t^2}, & x < t \le 2. \end{cases}$$

To find the function $f(x)$ we must write the equation in the standard form where the coefficient of y'' is 1, so

$$y'' - \frac{2}{x} y' + \frac{2}{x^2} y = 3,$$

showing that $f(x) = 3$. It now follows from (78), or from (80), that

$$y(x) = (4x - x^2) \int_1^x \frac{(t - t^2)}{3t^2} 3\,dt + (x - x^2) \int_x^2 \frac{(4t - t^2)3}{3t^2}\,dt,$$

and so

$$y(x) = x^2(3 \ln x - 2 - 4 \ln 2) + 2x(1 + 2 \ln 2).$$

It is easily checked that this is the required solution, because $y(1) = 0$, $y'(2) = 0$, and $y(x)$ satisfies the differential equation. ∎

More information and examples relating to the material in Sections 6.1 to 6.6 can be found in any one of the references [3.3], [3.4], [3.15], and [3.16].

Summary

This section described the powerful method of variation of parameters that enables the general solution of a linear nonhomogeneous equation to be found from the linearly independent solutions (the basis functions) that enter into its complementary function. It takes automatic account of nonhomogeneous terms that contain one or more basis functions, and it enables particular integrals, and hence general solutions, to be found where the method of undetermined coefficients fails. It was shown how the general solution obtained by the method of variation of parameters can be rewritten in terms of a Green's function that characterizes all of the essential features of the differential equation without reference to the nonhomogeneous term. Knowledge of the Green's function enables a homogeneous boundary value problem to be solved for any given nonhomogeneous term on the right of the equation.

EXERCISES 6.6

In Exercises 1 through 13 find the general solution.

1. $y'' + y' - 2y = xe^x$.

2. $y'' - 5y' + 6y = x^2 e^{3x}$.

3. $y'' + 5y' + 6y = x^2 e^{-2x}$.

4. $y'' + 4y' + 4y = x \sin x$.

5. $y'' - 2y + y = 2e^x/x$.

6. $y'' + 4y' + 5y = e^{-2x} \sin x$.

7. $y'' + 4y' + 5y = xe^{-2x} \cos x$.

8. $y'' - 4y' + 4y = e^{2x}/x$.

9. $y'' + 16y = x^2 e^x$.

10. $y'' + 16y = \sec x$.

11. $y'' + 3y' + 2y = 3/(1 + e^x)$.

12. $y'' + y = \tan x$.

13. $y'' + y = \sec^2 x$.

In Exercises 14 through 18 verify that the functions $y_1(x)$ and $y_2(x)$ are linearly independent solutions of homogeneous form of the stated differential equation, and use them to find a particular integral and a general solution of the given equation.

14. $x^2 y'' - 4xy' + 6y = 2x + \ln x$, where $y_1(x) = x^2$ and $y_2(x) = x^3$.

15. $x^2 y'' + 3xy' - 3y = \sqrt{x}$, where $y_1(x) = x$ and $y_2(x) = x^{-3}$.

16. $x^2 y'' + 3xy' - 8y = 2 \ln x$, where $y_1(x) = x^2$ and $y_2(x) = x^{-4}$.

17. $(1 - x^2)y'' - xy' + 4y = x$, where $y_1(x) = 2x^2 - 1$ and $y_2(x) = x(x^2 - 1)^{1/2}$.

18. $(1 - x^2)y'' - 2y' = 1$, where $y_1(x) = 1$ and $y_2(x) = x + 2 \ln(x - 1)$.

In Exercises 19 through 22 use result (76) to solve the stated initial value problem.

19. $x^2 y'' - 3xy' + 3y = 2x^2 \ln x$, with $y(1) = 0$ and $y'(1) = 0$.

20. $y'' + 5y' + 6y = xe^{-2x}$, with $y(1) = 0$ and $y'(1) = 0$.

21. $y'' + y = 2 \sec^2 x$, with $y(0) = 0$ and $y'(0) = 0$.

22. $y'' + 4y' + 5y = x$, with $y(0) = 0$ and $y'(0) = 0$.

In Exercises 23 through 26 find the Green's function for the given differential equation, subject to the associated homogeneous boundary conditions.

23. $y'' = f(x)$, with $y(0) = 0$ and $y(1) = 0$.

24. $y'' = f(x)$, with $y(0) = 0$ and $y'(1) = 0$.

25. $y'' + \lambda^2 y = f(x)$, with $y(0) = 0$ and $y(1) = 0$.

26. $y'' + \lambda^2 y = f(x)$, with $y(0) = 0$ and $y'(1) = 0$.

In Exercises 27 through 30 solve the given boundary value problem by means of a suitable Green's function.

27. $x^2 y'' + xy' - y = x^2 e^{-x}$, with $y(1) = 0$ and $y(2) = 0$.

28. $x^2 y'' + 2xy' - 2y = x^3$, with $y(1) = 0$ and $y(2) = 0$.

29. $x^2 y'' - 3xy' + 3y = x^2 \ln x$, with $y'(1) = 0$ and $y(2) = 0$.

30. $x^2 y'' - 3xy' = x^2$, with $y(1) = 0$ and $y(2) = 0$.

6.7 Finding a Second Linearly Independent Solution from a Known Solution: The Reduction of Order Method

reduction of order method

In working with homogeneous linear second order variable coefficient equations, it can happen that one solution $y_1(x)$ is known and it is necessary to find a second linearly independent solution $y_2(x)$. The method we now describe, called the **reduction of order** method, involves seeking a second solution of the form

$$y_2(x) = u(x)y_1(x), \tag{84}$$

where the function $u(x)$ is to be determined. Provided $u(x)$ is not constant, the solutions $y_1(x)$ and $y_2(x)$ will be linearly independent, because $y_1(x)$ and $y_2(x)$ will not be proportional.

The method will be developed using the homogeneous second order variable coefficient equation in the standard form

$$\frac{d^2 y}{dx^2} + a(x)\frac{dy}{dx} + b(x)y = 0. \tag{85}$$

Differentiating (84) gives

$$\frac{dy_2}{dx} = \frac{du}{dx}y_1 + u\frac{dy_1}{dx}, \quad \text{and} \quad \frac{d^2y_2}{dx^2} = \frac{d^2u}{dx^2}y_1 + 2\frac{du}{dx}\frac{dy_1}{dx} + u\frac{d^2y_1}{dx^2}. \tag{86}$$

Substituting (84) and (86) into (85) and grouping terms gives

$$y_1 u'' + (2y_1' + ay_1)u' + (y_1'' + ay_1' + by_1)u = 0. \tag{87}$$

As $y_1(x)$ is a solution of (85), the factor $y_1'' + ay_1' + by_1$ multiplying u is zero, causing the equation to be reduced to

$$\frac{d^2u}{dx^2} = -\left(\frac{2y_1'}{y_1} + a(x)\right)\frac{du}{dx}. \tag{88}$$

The substitution $v = du/dx$ reduces (88) to the first order variables separable equation

$$\frac{dv}{dx} = -\left(\frac{2y_1'}{y_1} + a(x)\right)v, \tag{89}$$

and it is from this reduction of order of the differential equation that the method derives its name.

Separating variables and integrating (89) we find that

$$\int \frac{dv}{v} = -\int \left(\frac{2y_1'}{y_1} + a(x)\right)dx + \ln C,$$

or

$$\ln(v/C) = -\int \left(\frac{2y_1'}{y_1} + a(x)\right)dx,$$

so

$$v(x) = C\frac{\exp\{-\int a(x)dx\}}{y_1^2}. \tag{90}$$

As $v = du/dx$, integration of (90) gives

$$u(x) = C\int \left[\frac{\exp\{-\int a(x)dx\}}{y_1^2}\right]dx + D,$$

where D is another arbitrary constant.

The arbitrary constant D can be set equal to zero, because when $u(x)$ is substituted in (84) the constant D will simply scale the solution $y_1(x)$. Furthermore, as any constant C that scales $u(x)$ will scale each term in the differential equation, its value is immaterial, so for convenience we set $C = 1$. Thus, the expression for $u(x)$ is given by

$$u(x) = \int \left[\frac{\exp\{-\int a(x)dx\}}{y_1^2}\right]dx. \tag{91}$$

Using this expression for $u(x)$ in (84) shows that the second linearly independent solution is

$$y_2(x) = y_1(x) \int \left[\frac{\exp\{-\int a(x)dx\}}{y_1^2} \right] dx. \tag{92}$$

Thus, in terms of $y_1(x)$, the general solution of (85) can be written

$$y(x) = C_1 y_1(x) + C_2 y_1(x) \int \left[\frac{\exp\{-\int a(x)dx\}}{y_1^2} \right] dx, \tag{93}$$

where C_1 and C_2 are arbitrary constants.

EXAMPLE 6.22   Given that $y_1(x) = e^{-3x}$ is a solution of $y'' + 6y' + 9y = 0$, find a second linearly independent solution, and hence find the general solution.

Solution   The equation is in standard form with $a(x) = 6$ and $y_1(x) = e^{-3x}$, so

$$u(x) = \int \left(\frac{\exp\{-\int 6\,dx\}}{\exp(-6x)} \right) dx = \int dx = x,$$

showing that

$$y_2(x) = xe^{-3x}.$$

This result is to be expected, because the linear constant coefficient equation corresponds to case (III) with $\mu = -3$. The general solution is thus

$$y(x) = (C_1 + C_2 x)e^{-3x}. \qquad \blacksquare$$

EXAMPLE 6.23   Given that $y_1(x) = x^2$ is a solution of $x^2 y'' - 3xy' + 4y = 0$ for $x > 0$, find a second linearly independent solution, and hence find the general solution.

Solution   Writing the equation in standard form (85) shows that $a(x) = -3/x$, so

$$u(x) = \int \frac{\exp\{-\int\{-3/x\}dx\}}{x^4} dx = \int \frac{\exp\{\ln x^3\}}{x^4} dx$$

$$= \int \frac{dx}{x} = \ln x,$$

from which it follows that the second linearly independent solution is

$$y_2(x) = x^2 \ln x \quad \text{for } x > 0.$$

The general solution is

$$y(x) = x^2(C_1 + C_2 \ln x). \qquad \blacksquare$$

The reduction of order method can lead to an expression for $u(x)$ that cannot be integrated analytically. In such cases, in order to find an analytical approximation to $y_2(x)$, the integrand in (92) must be expanded in powers of x and integrated term by term. This approach will be used in Chapter 8 in connection with series solutions of second order variable coefficient linear differential equations. See references [3.3] and [3.4].

Summary

It is often the case that one solution of a linear second order variable-coefficient homogeneous variable-coefficient equation can be found, often by inspection, though a second linearly independent solution cannot be found in similar fashion. This section showed how a known solution can be used to find a second linearly independent solution. It was shown that the second linearly independent solution of the original second order equation is determined in terms of a first order equation, and it is this feature that has caused this approach to be called the reduction of order method.

EXERCISES 6.7

In the following exercises, verify that $y_1(x)$ is a solution of the given differential equation and use it to find a second linearly independent solution.

1. $y'' - 5y' - 14y = 0$ with $y_1(x) = e^{7x}$.

2. $y'' + 4y = 0$, with $y_1(x) = \sin 2x$.

3. $y'' + 4y' + 5y = 0$, with $y_1(x) = e^{-x}\cos x$.

4. $x^2y'' + 3xy' + y = 0$, with $y_1(x) = 1/x$.

5. $x^2y'' - xy' + y = 0$, with $y_1(x) = x$.

6. $x^2y'' + xy' + y = 0$, with $y_1(x) = \cos(\ln x)$.

7. $xy'' + 2y' + xy = 0$, with $y_1(x) = \sin x/x$.

8. $x^2y'' + xy' + (x^2 - 1/4)y = 0$, with $y_1(x) = \sin x/\sqrt{x}$.

9. $x^2(\ln x - 1)y'' - xy' + y = 0$, with $y_1(x) = x$.

10. $(1 - x\cot x)y'' - xy' + y = 0$, with $y_1(x) = x$.

(Hint: When finding $\int -a(x)dx$, make the substitution $u = \sin x - x\cos$, and in the final integral make the substitution $v = \sin x/x$.)

6.8   Reduction to the Standard Form $u'' + f(x)u = 0$

When studying the properties of second order variable coefficient equations it is sometimes advantageous to reduce the equation

$$y'' + a(x)y' + b(x)y = 0 \qquad (94)$$

the standard form of a linear variable coefficient equation

to the *standard form* for a second order equation

$$u'' + f(x)u = 0, \qquad (95)$$

from which the first derivative term u' is missing. This reduction has many uses, one of which occurs in Section 8.6 when we derive the analytical form of Bessel functions of fractional order.

To accomplish the reduction we seek a solution of (94) of the form

$$y(x) = u(x)v(x), \qquad (96)$$

and then try to choose $v(x)$ so the first derivative term in u vanishes. Differentiation of $y = uv$ gives $y' = uv' + u'v$ and $y'' = u''v + 2u'v' + uv''$, so substitution into equation (94) gives

$$u''v + (2v' + av)u' + (v'' + av' + bv)u = 0. \qquad (97)$$

This result shows that the first derivative term u' will vanish if $v(x)$ is such that

$$2v' + av = 0, \qquad (98)$$

which has the solution

$$v(x) = \exp\left[-\frac{1}{2}\int a(x)dx\right]. \tag{99}$$

From (98) we have $v' = -(1/2)av$ and $v'' = -(1/2)(a'v + av')$, so eliminating v' and v'' from (97) gives

$$u'' + \left[-\frac{1}{2}a'(x) - \frac{1}{4}a^2(x) + b(x)\right]u = 0. \tag{100}$$

Because of its importance, we record this result in the form of a theorem.

THEOREM 6.5

how to perform the reduction

Reduction to the standard form $u'' + f(x)u = 0$ The substitution

$$y(x) = u(x)v(x) \quad \text{, with}$$

$$v(x) = \exp\left[-\frac{1}{2}\int a(x)dx\right],$$

reduces the differential equation

$$y'' + a(x)y' + b(x)y = 0$$

to

$$u'' + f(x)u = 0,$$

where

$$f(x) = -\frac{1}{2}a'(x) - \frac{1}{4}a^2(x) + b(x).$$ ∎

EXAMPLE 6.24   Reduce the equation

$$4x^2y'' + 4xy' + (16x^2 - 1)y = 0$$

to standard form and hence find the general solution.

Solution   Dividing the differential equation by $4x^2$ to reduce it to the form given in (94) shows that $a(x) = 1/x$ and $b(x) = 4 - 1/(4x^2)$. Applying the result of Theorem 6.5 then shows that

$$v(x) = \exp\left[-\frac{1}{2}\int (1/x)dx\right] = x^{-1/2} \quad \text{and} \quad f(x) = 4.$$

The equation for $u(x)$ is thus $u'' + 4u = 0$ with the general solution

$$u(x) = C_1\cos 2x + C_2\sin 2x,$$

but $y(x) = u(x)v(x) = x^{-1/2}u(x)$, so the general solution is

$$y(x) = C_1\sqrt{\frac{1}{x}}\cos 2x + C_2\sqrt{\frac{1}{x}}\sin 2x.$$

See references [3.3] and [3.4].

Summary

The study of the properties of some homogeneous linear variable coefficient equations of the form $y'' + a(x)y' + b(x)y = 0$ is simplified if a change of variable can be found that reduces them to an equivalent form $u'' + f(x)u = 0$. This section showed how such a change of variable can be found, and used it to solve a variable coefficient equation for which the two linearly independent functions entering into its general solution are by no means obvious.

EXERCISES 6.8

Reduce the equations in Exercises 1 and 2 to standard form, but do not attempt to find their general solutions.

1. $x^2y'' - xy' + 9xy = 0$.
2. $x^2y'' + xy' + (x^2 - 9)y = 0$.

In Exercises 3 through 7 reduce the equation to standard form and hence find its general solution.

3. $y'' - 2y' + y = 0$.
4. $y'' + 4y' + 3y = 0$.
5. $y'' - 4y' + 5y = 0$.
6. $x^2y'' + xy' + (36x^2 - 1)y = 0$.
7. $xy'' + 2y' + xy = 0$.

6.9 Systems of Ordinary Differential Equations: An Introduction

Physical problems that give rise to ordinary differential equations often do so in the form of coupled systems of first order linear differential equations, or systems of second order equations that are more easily treated if reduced to a first order system. A very simple example of this type was encountered in Section 5.2(d), where two first order equations were derived that linked the current i and the charge q flowing in an R–L–C circuit at time t. In that case it was convenient to eliminate the current i to obtain a simple second order equation for the current q that could be solved by the methods of Section 6.1 and 6.2.

Another example is the three-loop electric circuit shown in Fig. 6.12. In the circuit H is an inductance; C_1 and C_2 are capacitances; R_1, R_2, and R_3 are resistors; V_0 is an applied voltage; i_1, i_2, and i_3 are circulating currents; and q_2 and q_3 are the charges on the respective capacitances C_1 and C_2.

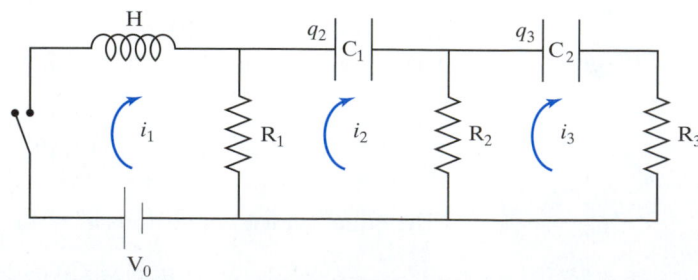

FIGURE 6.12 A three-loop electric circuit with an applied voltage.

an electrical problem
leading to a first
order system

Applying Kirchhoff's laws (see Section 5.2(d)) to each loop when the switch is closed leads to the three coupled equations

$$H\frac{di_1}{dt} + R_1(i_1 - i_2) = V_0$$
$$R_2 i_2 + R_1(i_2 - i_1) + q_2 C_1 = 0$$
$$R_3 i_3 + R_2(i_3 - i_2) + q_3 C_2 = 0.$$

Using the results $i_2 = dq_2/dt$ and $i_3 = dq_3/dt$ reduces these equations to the coupled system of first order equations

$$H\frac{di_1}{dt} + R_1 i_1 - R_1\frac{dq_2}{dt} = V_0$$
$$(R_1 + R_2)\frac{dq_2}{dt} - R_1 i_1 + q_2 C_1 = 0$$
$$(R_2 + R_3)\frac{dq_3}{dt} - R_2\frac{dq_2}{dt} + q_3 C_2 = 0$$

for $i_1, q_2,$ and q_3. When these are solved the currents i_2 and i_3 follow from $i_2 = dq_2/dt$ and $i_3 = dq_3/dt$.

An example of a different kind is provided by the two degree of freedom vibration system with a damper in Fig. 6.11 that was shown to lead to the two coupled second order equations

$$M\frac{d^2 x}{dt^2} = -k(x - y) - Kx - \mu\frac{dx}{dt} + F(t)$$

and

$$m\frac{d^2 y}{dt^2} = -k(y - x).$$

Instead of eliminating first y and then x to obtain two fourth order differential equations for x and y, respectively, a different approach is to reduce these two equations to a system of four first order equations by introducing first order derivatives of x and y as new variables.

To do this we set $w = dx/dt$ and $z = dy/dt$, and as a result obtain the simultaneous system of four first order equations

$$\frac{dx}{dt} = w$$
$$\frac{dy}{dt} = z$$
$$M\frac{dw}{dt} + (k + K)x - ky + \mu w = F(t)$$
$$m\frac{dz}{dt} + ky - kx = 0.$$

This reduction of a higher order differential equation, or a coupled system of differential equations, to a first order system is often useful. In Chapter 19 this approach is used when seeking the numerical solution of higher order differential equations by means of the Runge–Kutta method. This method provides accurate numerical solutions of first order differential equations that may be either linear or nonlinear, and it can be adapted to solve higher order differential equations by reducing them to a coupled system of first order equations.

a general
homogeneous
first order system

A general system of n first order linear variable coefficient differential equations involving the n dependent variables $x_1(t), x_2(t), \ldots, x_n(t)$ that are functions of the independent variable t (in applications t is often the time), the variable coefficients $a_{ij}(t)$, and the nonhomogeneous terms $f_1(t), f_2(t), \ldots, f_n(t)$ has the form

$$
\begin{aligned}
x_1'(t) &= a_{11}(t)x_1(t) + a_{12}(t)x_2(t) + \cdots + a_{1n}(t)x_n(t) + f_1(t) \\
x_2'(t) &= a_{21}(t)x_1(t) + a_{22}(t)x_2(t) + \cdots + a_{2n}(t)x_n(t) + f_2(t) \\
&\quad \cdots \cdots \cdots \cdots \\
x_n'(t) &= a_{n1}(t)x_1(t) + a_{n2}x_2(t) + \cdots + a_{nn}(t)x_n(t) + f_n(t).
\end{aligned} \tag{101}
$$

System (101) is said to be **homogeneous** when all the functions $f_i(t)$ are zero, and to be **nonhomogeneous** when at least one of them is nonzero. It is a linear system because it is linear in the functions $x_1(t), x_2(t), \ldots, x_n(t)$ and their derivatives, and it is a **variable coefficient** system whenever at least one of the coefficients $a_{ij}(t)$ is a function of t; otherwise, it becomes a **constant coefficient** system.

An **initial value problem** for system (101) involves seeking a solution of (101) such that at $t = t_0$ the variables $x_1(t), x_2(t), \ldots, x_n(t)$ satisfy the initial conditions

$$
x_1(t_0) = k_1, \; x_2(t_0) = k_2, \ldots, x_n(t_0) = k_n, \tag{102}
$$

where $k_1, k_2, \ldots, k_n$ are given constants.

Matrix notation allows system (101) to be written in the concise form

$$
\mathbf{x}'(t) = \mathbf{A}(t)\mathbf{x}(t) + \mathbf{b}(t), \tag{103}
$$

matrix notation for systems

or more simply as

$$
\mathbf{x}' = \mathbf{A}\mathbf{x} + \mathbf{b},
$$

where a prime again indicates differentiation with respect to t, and the matrices in (103) are defined as

$$
\mathbf{x}(t) = \begin{bmatrix} x_1(t) \\ x_2(t) \\ \vdots \\ \vdots \\ x_n(t) \end{bmatrix}, \quad \mathbf{x}'(t) = \begin{bmatrix} x_1'(t) \\ x_2'(t) \\ \vdots \\ \vdots \\ x_n'(t) \end{bmatrix},
$$

$$
\mathbf{A}(t) = \begin{bmatrix} a_{11}(t) & a_{12}(t) & \cdots & \cdots & a_{1n}(t) \\ a_{21}(t) & a_{22}(t) & \cdots & \cdots & a_{2n}(t) \\ \cdots & \cdots & \cdots & \cdots & \cdots \\ \cdots & \cdots & \cdots & \cdots & \cdots \\ a_{n1}(t) & a_{n2}(t) & \cdots & \cdots & a_{nn}(t) \end{bmatrix}, \quad \mathbf{b}(t) = \begin{bmatrix} f_1(t) \\ f_2(t) \\ \vdots \\ \vdots \\ f_n(t) \end{bmatrix}.
$$

$$\tag{104}$$

The $n \times 1$ vector $\mathbf{x}(t)$ is called the **solution vector**, the $n \times n$ matrix $\mathbf{A}(t)$ is called the **coefficient matrix**, and the $n \times 1$ vector $\mathbf{b}(t)$ is called the **nonhomogeneous term** of the system.

System (103) becomes an **initial value problem** for the solution $\mathbf{x}(t)$ when at $t = t_0$ the vector $\mathbf{x}(t)$ is required to satisfy the initial condition

$$\mathbf{x}(t_0) = \begin{bmatrix} k_1 \\ k_2 \\ \vdots \\ \vdots \\ k_n \end{bmatrix}, \tag{105}$$

where $\mathbf{x}(t_0)$ is the **initial vector** and $k_1, k_2, \ldots, k_n$ are given constants.

EXAMPLE 6.25

Express in matrix form the initial value problem

$$x_1' = 2x_1 - x_2 + 4 - t^2$$
$$x_2' = -x_1 + 2x_2 + 1, \quad \text{with } x_1(0) = 1 \quad \text{and} \quad x_2(0) = 0.$$

Solution   The system of equations can be written

$$\mathbf{x}'(t) = \mathbf{A}\,\mathbf{x}(t) + \mathbf{b}(t)$$

where

$$\mathbf{x}(t) = \begin{bmatrix} x_1 \\ x_2 \end{bmatrix}, \quad \mathbf{A} = \begin{bmatrix} 2 & -1 \\ -1 & 2 \end{bmatrix}, \quad \text{and} \quad \mathbf{b}(t) = \begin{bmatrix} 4 - t^2 \\ 1 \end{bmatrix},$$

and the initial vector is

$$\mathbf{x}(0) = \begin{bmatrix} 1 \\ 0 \end{bmatrix}.$$

As $\mathbf{A}$ is a constant matrix and $\mathbf{b}(t) \neq \mathbf{0}$, this is a constant coefficient nonhomogeneous system. ∎

solution by elimination: a first approach

In what follows, our main objective will be to develop matrix methods for the solution of initial value problems for systems of first order linear constant coefficient differential equations. Before developing a matrix approach, we first describe a simple way of solving system (102) when no more than three equations are involved. The method is straightforward and does not use matrix algebra, but it is often useful, and the examples that are solved show that systems can have oscillatory solutions even when no oscillatory term is present in the nonhomogeneous term.

The approach used is called **solution by elimination**, because it involves eliminating all but one of the dependent variables in order to arrive at a single higher order equation for the remaining variable, say $x_1(t)$. Once $x_1(t)$ has been found, it is used in the system of equations to determine sequentially the remaining variables $x_2(t), x_3(t), \ldots, x_n(t)$. The method will be illustrated by means of examples.

EXAMPLE 6.26

Solve by elimination the initial value problem of Example 6.25.

Solution   The equations involved are

$$x_1' = 2x_1 - x_2 + 4 - t^2$$
$$x_2' = -x_1 + 2x_2 + 1.$$

The method will be to eliminate the dependent variable x_2 between the two equations to obtain a single second order equation for x_1. After solving for x_1, the dependent variable x_2 will be found by substituting for x_1 in the first equation. Thus, the solution of this system of two first order equations will involve the solution of a single second order equation, and it will be through this equation that the two arbitrary constants expected to occur in the general solution of the system will enter.

Differentiation of the first equation belonging to the system gives

$$x_1'' = 2x_1' - x_2' - 2t,$$

and after substituting for x_2' from the second equation in the system, this becomes

$$x_1'' = 2x_1' + x_1 - 2x_2 - 1 - 2t.$$

Solving the first equation belonging to the system for x_2 gives

$$x_2 = 2x_1 + 4 - t^2 - x_1',$$

so using this result to eliminate x_2 from the second order equation for x_1 shows that x_1 satisfies the equation

$$x_1'' - 4x_1' + 3x_1 = 2t^2 - 2t - 9.$$

Solving this equation by any method, say by the method of undetermined coefficients, gives

$$x_1(t) = C_1 e^{3t} + C_2 e^t - \frac{53}{27} + \frac{10}{9}t + \frac{2}{3}t^2,$$

where C_1 and C_2 are arbitrary constants of integration.

It now remains for us to find x_2, and this is accomplished by substituting for x_1 in the first equation, which can be written in the form $x_2 = 2x_1 + 4 - t^2 - x_1'$. As a result we find that

$$x_2(t) = -C_1 e^{3t} + C_2 e^t - \frac{28}{27} + \frac{8}{9}t + \frac{1}{3}t^2,$$

so the general solution of the nonhomogeneous system is

$$x_1(t) = C_1 e^{3t} + C_2 e^t - \frac{53}{27} + \frac{10}{9}t + \frac{2}{3}t^2,$$

and

$$x_2(t) = -C_1 e^{3t} + C_2 e^t - \frac{28}{27} + \frac{8}{9}t + \frac{1}{3}t^2.$$

To solve the initial value problem, C_1 and C_2 must be chosen such that $x_1(0) = 1$ and $x_2(0) = 0$. Setting $t = 0$ in the general solution and using these initial conditions, we find that C_1 and C_2 must satisfy the equations

$$1 = C_1 + C_2 - \frac{53}{27} \quad \text{and} \quad 0 = -C_1 + C_2 - \frac{28}{27},$$

with the solution $C_1 = 26/27$ and $C_2 = 2$. Thus, the required solution of the initial value problem is

$$x_1(t) = \frac{26}{27}e^{3t} + 2e^t - \frac{53}{27} + \frac{10}{9}t + \frac{2}{3}t^2,$$

and

$$x_2(t) = -\frac{26}{27}e^{3t} + 2e^t - \frac{28}{27} + \frac{8}{9}t + \frac{1}{3}t^2. \qquad \blacksquare$$

Unlike first order linear differential equations whose complementary function can only contain an exponential function, systems of such equations can give rise to periodic solutions even when these do not occur in the nonhomogeneous term. This is illustrated by the next example.

EXAMPLE 6.27

Solve by elimination the system of differential equations

$$x_1' + 2x_1 - x_2 = 1 + e^{-t}, \quad x_2' + x_1 + 2x_2 = 3,$$

subject to the initial conditions $x_1(0) = 5/2$ and $x_2(0) = -1/2$.

Solution   Proceeding as in the previous example by differentiating the first equation with respect to t and substituting for x_2' from the second equation gives

$$x_1'' + 2x_1' + x_1 + 2x_2 - 3 = -e^{-t}.$$

Substituting for x_2 from the first equation belonging to the system then shows that x_1 must satisfy the second order differential equation

$$x_1'' + 4x_1' + 5x_1 = 5 + e^{-t},$$

with the general solution

$$x_1(t) = C_1 e^{-2t} \cos t + C_2 e^{-2t} \sin t + 1 + (1/2)e^{-t}.$$

Finally, solving the first equation belonging to the system for x_2 and substituting for x_1, we have

$$x_2(t) = -C_1 e^{-2t} \sin t + C_2 e^{-2t} \cos t + 1 - (1/2)e^{-t}.$$

Thus, the general solution of the system is

$$x_1(t) = C_1 e^{-2t} \cos t + C_2 e^{-2t} \sin t + 1 + (1/2)e^{-t}$$

and

$$x_2(t) = -C_1 e^{-2t} \sin t + C_2 e^{-2t} \cos t + 1 - (1/2)e^{-t}.$$

To satisfy the initial conditions, the arbitrary constants C_1 and C_2 must be chosen such that $x_1(0) = 5/2$ and $x_2(0) = -1/2$. Inserting these conditions into the preceding general solution leads to the equations

$$5/2 = C_1 + 3/2 \quad \text{and} \quad -1/2 = C_2 + 1/2, \quad \text{so that } C_1 = 1 \text{ and } C_2 = -1.$$

The solution of the initial value problem is then given by

$$x_1(t) = e^{-2t}(\cos t - \sin t) + 1 + \frac{1}{2}e^{-t}$$

and

$$x_2(t) = -e^{-2t}(\sin t + \cos t) + 1 - \frac{1}{2}e^{-t}.$$

This example illustrates the way in which oscillatory terms can enter into the solution through a higher order equation satisfied by one of the dependent variables, although they may not be present in the nonhomogeneous terms. $\blacksquare$

As a final example of the elimination method, we consider a homogeneous system of three equations to show how this simple method becomes more difficult when the number of equations is greater than two, and also to demonstrate how care must then be taken with the determination of the arbitrary constants of integration.

EXAMPLE 6.28

Find the general solution of the system of equations

$$x_1' = x_2 + x_3, \quad x_2' = x_1 + x_3, \quad \text{and} \quad x_3' = x_1 + x_2.$$

Solution   Differentiating the first equation with respect to t and substituting for x_2' and x_3' from the second and third equations and using the first equation shows that x_1 satisfies the second order equation

$$x_1'' - x_1' - 2x_1 = 0,$$

with the solution

$$x_1(t) = C_1 e^{-t} + C_2 e^{2t},$$

where C_1 and C_2 are arbitrary constants of integration.

Substituting for $x_1(t)$ in the second equation belonging to the system, differentiating the result with respect to t, and then substituting for x_3' from the third equation belonging to the system shows that x_2 satisfies the nonhomogeneous second order equation

$$x_2'' - x_2 = 3C_2 e^{2t},$$

with the solution

$$x_2(t) = C_2 e^{2t} + C_3 e^{-t} + C_4 e^{t}.$$

how to resolve the problem of the arbitrary constants

It now appears that an anomalous situation has arisen, because when seeking a solution of a system of *three* equations, *four* arbitrary integration constants have appeared. This apparent inconsistency will be resolved shortly, so for the moment we continue working with this form of solution for $x_2(t)$.

Subtracting the first two equations belonging to the system gives

$$x_1' - x_2' = x_2 - x_1.$$

After substituting for $x_1(t)$ and $x_2(t)$ in this equation and cancelling terms, this is seen to reduce to $-C_4 e^{t} = C_4 e^{t}$. As $e^{t} \neq 0$ for any t, it follows that $C_4 = 0$, and the apparent inconsistency has been resolved because now only the three arbitrary constants C_1, C_2, and C_3 appear in the general solutions for $x_1(t)$ and $x_2(t)$.

In fact, no further integration is required to determine $x_3(t)$, because substituting $x_1(t)$ and $x_2(t)$ into the first equation belonging to the system and solving for $x_3(t)$ gives

$$x_3(t) = -(C_1 + C_3)e^{-t} + C_2 e^{2t}.$$

Thus, the general solution of the system is given by

$$x_1(t) = C_1 e^{-t} + C_2 e^{2t}$$
$$x_2(t) = C_2 e^{2t} + C_3 e^{-t}$$
$$x_3(t) = -(C_1 + C_3)e^{-t} + C_2 e^{2t}.$$

Summary

This section has shown how a system of first order equations can arise from a typical electrical problem. A matrix notation for systems was introduced, and an elementary method for solving small systems of equations using elimination was described that avoided the use of matrices. This method was seen to lead to more arbitrary constants in the general solution than the number of equations involved, but a simple argument resolved this difficulty.

EXERCISES 6.9

Solve Exercises 1 through 6 by elimination.

1. $2x_1' = x_1 - x_2$, $2x_2' = 3x_1 + 5x_2$.

2. $x_1' = -10x_1 - 18x_2$, $x_2' = 6x_1 + 11x_2$.

3. $x_1' = 2x_1 - 12x_2$, $2x_2' = 3x_1 - 8x_2$, with $x_1(0) = 0$ and $x_2(0) = 1$.

4. $x_1' = 3x_2 + t$, $x_2' = 2x_1 + x_2 - 3$, with $x_1(0) = 1$ and $x_2(0) = 1$.

5. $x_1' = 2x_2 + 4x_3 + 3e^{-t}$, $x_2' = x_1 + x_2 - 2x_3 + 1$, $x_3' = -2x_1 + 5x_3$, with $x_1(0) = 1$, $x_2(0) = 0$, and $x_3(0) = 0$.

6. $x_1' = -2x_1 + 2x_2 + 2x_3 + 3e^t$, $x_2' = -x_1 - x_2 - 2x_3 + 1$, $x_3' = x_1 + 2x_2 + 3x_3 - 3$, with $x_1(0) = 1$, $x_2(0) = 1$, and $x_3(0) = 0$.

6.10   A Matrix Approach to Linear Systems of Differential Equations

We will now consider some general properties of the variable coefficient system

$$\mathbf{x}'(t) = \mathbf{A}(t)\mathbf{x}(t) + \mathbf{b}(t), \tag{106}$$

where the matrices $\mathbf{x}(t)$, $\mathbf{A}(t)$, and $\mathbf{b}(t)$ are as defined in (103).

A **solution** of system (106) is a vector $\mathbf{x}(t)$ with elements $x_1(t), x_2(t), \ldots, x_n(t)$ that when substituted in system (106) satisfies it identically. Thus, a solution of the initial value problem in Example 6.26 is the vector

a solution in matrix form

$$\mathbf{x}(t) = \begin{bmatrix} x_1(t) \\ x_2(t) \end{bmatrix} = \begin{bmatrix} \dfrac{26}{27}e^{3t} + 2e^t - \dfrac{53}{27} + \dfrac{10}{9}t + \dfrac{2}{3}t^2 \\[2mm] -\dfrac{26}{27}e^{3t} + 2e^t - \dfrac{28}{27} + \dfrac{8}{9}t + \dfrac{1}{3}t^2 \end{bmatrix}.$$

Structure of Solutions of Homogeneous Systems

(a) Linear superposition of solutions

The properties of linear homogeneous systems of differential equations are similar to those of a single linear higher order homogeneous differential equation. A most important property that is common to both is that a linear superposition of solutions of a linear homogeneous system of variable-coefficient first order differential equations is itself a solution of the homogeneous system.

This result is easily proved. Let $\mathbf{\Psi}_1(t), \mathbf{\Psi}_2(t), \ldots, \mathbf{\Psi}_m(t)$ be any m solutions of the linear homogeneous system $\mathbf{x}'(t) = \mathbf{A}(t)\mathbf{x}(t)$, and taking $C_1, C_2, \ldots, C_m$ to be any set of m arbitrary constants form the vector $\mathbf{\Psi}(t) = C_1\mathbf{\Psi}_1(t) + C_2\mathbf{\Psi}_2(t) + \cdots + C_m\mathbf{\Psi}_m(t)$. Then

$$\mathbf{\Psi}(t)' = (C_1\mathbf{\Psi}_1 + C_2\mathbf{\Psi}_2 + \cdots + C_m\mathbf{\Psi}_m)' = C_1\mathbf{\Psi}_1' + C_2\mathbf{\Psi}_2' + \cdots + C_m\mathbf{\Psi}_m',$$

so the system $\mathbf{\Psi}(t)' = \mathbf{A}(t)\mathbf{\Psi}(t)$ becomes

$$
\begin{aligned}
C_1\mathbf{\Psi}_1' + C_2\mathbf{\Psi}_2' + \cdots + C_m\mathbf{\Psi}_m' &= \mathbf{A}(C_1\mathbf{\Psi}_1 + C_2\mathbf{\Psi}_2 + \cdots + C_m\mathbf{\Psi}_m) \\
&= C_1\mathbf{A}\mathbf{\Psi}_1 + C_2\mathbf{A}\mathbf{\Psi}_2 + \cdots + C_m\mathbf{A}\mathbf{\Psi}_m.
\end{aligned}
$$

Consequently, as $\mathbf{\Psi}_i'(t) = \mathbf{A}(t)\mathbf{\Psi}_i(t)$, we have shown that $\mathbf{\Psi}(t)$ is also a solution of the homogeneous system, and the result is proved.

(b) Existence and uniqueness

We now state without proof the fundamental theorem on the existence and uniqueness of the solution to the initial value problem for a system of linear variable coefficient first order differential equations. (See, for example, references [3.4] and [3.5].)

THEOREM 6.5   **Existence and uniqueness of solutions of linear systems**   Let the vector $\mathbf{x}(t)$ with the n elements $x_i(t)$ $(i = 1, 2, \ldots, n)$ be the solution of the nonhomogeneous variable coefficient system of first order linear differential equations

$$\mathbf{x}'(t) = \mathbf{A}(t)\mathbf{x}(t) + \mathbf{b}(t),$$

where the functions $a_{ij}(t)$ $(i, j = 1, 2, \ldots, n)$ forming the elements of $\mathbf{A}(t)$ and the elements $f_i(t)$ $(i = 1, 2, \ldots, n)$ forming the elements of the vector $\mathbf{b}(t)$ are continuous functions in some interval $a < t < b$. Furthermore, let the elements of $\mathbf{x}(t)$ satisfy the initial conditions $x_i(t_0) = k_i$ $(i = 1, 2, \ldots, n)$, where the k_i are given constants and t_0 is any point such that $a < t_0 < b$. Then the solution of the initial value problem exists and is unique for all t such that $a < t < b$.   ∎

(c) Fundamental matrix and a test for linear independence of solutions

As with single higher order linear differential equations, the general solution of a homogeneous system will be constructed by the forming a linear combination of all possible linearly independent solutions of the system. For this reason it is necessary to know how many linearly independent solutions belong to a given homogeneous system, and how to test the linear independence of a set of solutions. The answers to these two fundamental questions are provided by the next two theorems, the results of which should be remembered. As the proofs of these theorems may be omitted at a first reading, they are given at the end of this section.

THEOREM 6.6   **Linearly independent solutions of a homogeneous system**   Let the elements $a_{ij}(t)$ $(i, j = 1, 2, \ldots, n)$ of the $n \times n$ matrix $\mathbf{A}(t)$ be continuous in the interval $a < t < b$. Then the linear homogeneous system

$$\mathbf{x}' = \mathbf{A}(t)\mathbf{x}$$

possesses n linearly independent solutions $\mathbf{\Psi}_1(t), \mathbf{\Psi}_2(t), \ldots, \mathbf{\Psi}_n(t)$, and every solution of the system is expressible as a linear combination of the form

$$\mathbf{\Psi}(t) = C_1 \mathbf{\Psi}_1(t) + C_2 \mathbf{\Psi}_2(t) + \cdots + C_n \mathbf{\Psi}_n(t)$$

for some choice of the constants $C_1, C_2, \ldots, C_n$. ∎

a fundamental matrix

An $n \times n$ matrix $\mathbf{\Phi}(t)$ whose columns are any n linearly independent solution vectors of the homogeneous system $\mathbf{x}' = \mathbf{A}(t)\mathbf{x}$ is called a **fundamental matrix** for the system, and Theorem 6.6 shows that the general solution of the system can always be written in the form

$$\mathbf{x}(t) = \mathbf{\Phi}(t)\mathbf{C},$$

where $\mathbf{C}$ is an n-element column vector with arbitrary constant elements C_1, $C_2, \ldots, C_n$.

Clearly, a fundamental matrix is not unique, because any of its columns may be replaced by a linear combination of its columns and the result will remain a fundamental matrix. This follows because if the columns of a determinant are replaced by linear combinations of its columns, the value of the determinant is unaltered, so if initially the determinant was nonsingular, it will remain nonsingular.

THEOREM 6.7

a determinant test for linear independence of solution vectors

Determinant test for the linear independence of solution vectors   Let the column vectors $\mathbf{\Psi}_m(t)\,(m = 1, 2, \ldots, n)$, whose elements $\Psi_1^{(m)}(t), \Psi_2^{(m)}(t), \ldots, \Psi_n^{(m)}(t)$, be n solutions of the homogeneous system

$$\mathbf{x}' = \mathbf{A}(t)\mathbf{x},$$

in which the elements $a_{ij}(t)\,(i, j = 1, 2, \ldots, n)$ of the $n \times n$ matrix $\mathbf{A}(t)$ are continuous functions for $a < t < b$. Then the n vectors $\mathbf{\Psi}_m(t)\,(m = 1, 2, \ldots, n)$ are linearly independent solutions for $a < t < b$ if, for some t_0 in the interval, the determinant

$$\Delta(t_0) = \begin{vmatrix} \Psi_1^{(1)}(t_0) & \Psi_1^{(2)}(t_0) & \cdots & \Psi_1^{(n)}(t_0) \\ \Psi_2^{(1)}(t_0) & \Psi_2^{(2)}(t_0) & \cdots & \Psi_2^{(n)}(t_0) \\ \vdots & \vdots & \vdots & \vdots \\ \Psi_n^{(1)}(t_0) & \Psi_n^{(2)}(t_0) & \cdots & \Psi_n^{(n)}(t_0) \end{vmatrix} \neq 0,$$

and the vectors $\mathbf{\Psi}_m(t)\,(m = 1, 2, \ldots, n)$ form a basis for solutions of the system. Furthermore, if $\Delta(t_0) \neq 0$, then $\Delta(t) \neq 0$, for all t in $a < t < b$. ∎

EXAMPLE 6.29

Find a set of linearly independent solution vectors for the system

$$x_1' = x_2 + x_3, \quad x_2' = x_1 + x_3 \quad \text{and} \quad x_3' = x_1 + x_2,$$

and construct a fundamental matrix.

Solution   In Example 6.28 the solution of this system was shown to be

$$x_1(t) = C_1 e^{-t} + C_2 e^{2t}$$
$$x_2(t) = C_2 e^{2t} + C_3 e^{-t}$$
$$x_3(t) = -(C_1 + C_3)e^{-t} + C_2 e^{2t}.$$

Writing this solution in the form $\mathbf{x}(t) = \Phi(t)\mathbf{C}$ determined by Theorem 6.7, we obtain

$$\begin{bmatrix} x_1(t) \\ x_2(t) \\ x_3(t) \end{bmatrix} = \begin{bmatrix} e^{-t} & e^{2t} & 0 \\ 0 & e^{2t} & e^{-t} \\ -e^{-t} & e^{2t} & -e^{-t} \end{bmatrix} \begin{bmatrix} C_1 \\ C_2 \\ C_3 \end{bmatrix}.$$

Thus, a fundamental matrix for the system, that is, a matrix whose columns are linearly independent solution vectors of the system, can be taken to be

$$\Phi(t) = \begin{bmatrix} e^{-t} & e^{2t} & 0 \\ 0 & e^{2t} & e^{-t} \\ -e^{-t} & e^{2t} & -e^{-t} \end{bmatrix},$$

provided the solution vectors corresponding to the columns of this matrix are linearly independent. The test for this is provided by Theorem 6.7, and as it is easily shown that det $\Phi(t) = -3$. So it follows from Theorem 6.7 that the three column vectors

$$\Psi_1(t) = \begin{bmatrix} e^{-t} \\ 0 \\ -e^{-t} \end{bmatrix}, \quad \Psi_2(t) = \begin{bmatrix} e^{2t} \\ e^{2t} \\ e^{2t} \end{bmatrix}, \quad \text{and} \quad \Psi_3(t) = \begin{bmatrix} 0 \\ e^{-t} \\ -e^{-t} \end{bmatrix}$$

are, indeed, linearly independent solution vectors.   ■

Proofs of Theorems 6.6 and 6.7

Proof of Theorem 6.6   Consider any set of n linearly independent column vectors $\mathbf{v}_1, \mathbf{v}_2, \ldots, \mathbf{v}_n$, each with constant elements, and for some t_0 in $a < t_0 < b$ use them as initial conditions in the set of initial value problems

$$\mathbf{x}' = \mathbf{A}(t)\mathbf{x} \quad \text{with} \quad \mathbf{x}(t_0) = \mathbf{v}_m, \quad \text{for } m = 1, 2, \ldots, n.$$

By the existence and uniqueness theorem, each of these initial value problems has a unique solution $\Psi_m(t)$ defined on $a < t < b$.

To establish the linear independence of these solutions on $a < t < b$, we suppose, if possible, that constants $C_1, C_2, \ldots, C_n$ can be found such that

$$C_1 \Psi_1(t) + C_2 \Psi_2(t) + \cdots + C_n \Psi_n(t) = \mathbf{0}$$

for every t in the interval. Setting $t = t_0$, this result becomes

$$C_1 \mathbf{v}_1 + C_2 \mathbf{v}_2 + \cdots + C_n \mathbf{v}_n = \mathbf{0},$$

but as the $\mathbf{v}_m$ are linearly independent, this can only be true if $C_1 = C_2 = \cdots = C_n = 0$, so we have proved that the solutions $\Psi_m(t)$ $(m = 1, 2, \ldots, n)$ are linearly independent over the interval.

We must now show that for some constants $C_1, C_2, \ldots, C_n$, not all of which are zero, every solution of the system $\mathbf{x}' = \mathbf{A}(t)\mathbf{x}$ can be written

$$\Psi(t) = C_1 \Psi_1(t) + C_2 \Psi_2(t) + \cdots + C_n \Psi_n(t),$$

and in particular this result must be true when $t = t_0$.

Define a matrix $\boldsymbol{\Phi}(t)$ whose columns are the n linearly independent vectors $\boldsymbol{\Psi}_1(t), \boldsymbol{\Psi}_2(t), \ldots, \boldsymbol{\Psi}_n(t)$, where the elements of $\boldsymbol{\Psi}_m(t)$ are $\Psi_1^{(m)}(t), \Psi_2^{(m)}(t), \ldots, \Psi_n^{(m)}(t)$, for $m = 1, 2, \ldots, n$, so

$$\boldsymbol{\Phi}(t) = \begin{bmatrix} \Psi_1^{(1)}(t) & \Psi_1^{(2)}(t) & \cdots & \Psi_1^{(n)}(t) \\ \Psi_2^{(1)}(t) & \Psi_2^{(2)}(t) & \cdots & \Psi_2^{(n)}(t) \\ \vdots & \vdots & \vdots & \vdots \\ \Psi_n^{(1)}(t) & \Psi_n^{(2)}(t) & \cdots & \Psi_n^{(n)}(t) \end{bmatrix}.$$

Now set $t = t_0$ and consider the matrix equation

$$\boldsymbol{\Phi}(t_0)\mathbf{C} = \boldsymbol{\Psi}(t_0),$$

where $\mathbf{C}$ is a column vector with the n elements $C_1, C_2, \ldots, C_n$. Expanding the expression on the left and grouping terms shows that

$$\boldsymbol{\Phi}(t_0)\mathbf{C} = C_1 \boldsymbol{\Psi}_1(t_0) + C_2 \boldsymbol{\Psi}_2(t_0) + \cdots + C_n \boldsymbol{\Psi}_n(t_0),$$

and so

$$C_1 \boldsymbol{\Psi}_1(t_0) + C_2 \boldsymbol{\Psi}_2(t_0) + \cdots + C_n \boldsymbol{\Psi}_n(t_0) = \boldsymbol{\Psi}(t_0).$$

The existence of a unique set of constants $C_1, C_2, \ldots, C_n$, not all of which are zero, follows from the fact that $\det \boldsymbol{\Phi}(t_0) \neq 0$, because of the linear independence of its columns.

As $\boldsymbol{\Psi}(t)$ and $C_1 \boldsymbol{\Psi}_1(t) + C_2 \boldsymbol{\Psi}_2(t) + \cdots + C_n \boldsymbol{\Psi}_n(t)$ are both solutions of the same initial value problem

$$\mathbf{x}' = \mathbf{A}(t)\mathbf{x} \quad \text{with} \quad \mathbf{x}(t_0) = \boldsymbol{\Psi}(t_0),$$

the existence and uniqueness theorem shows that $\boldsymbol{\Psi}(t) = C_1 \boldsymbol{\Psi}_1(t) + C_2 \boldsymbol{\Psi}_2(t)_1 + \cdots + C_n \boldsymbol{\Psi}_n(t)$ for all t such that $a < t < b$, and the theorem is proved.   ∎

Proof of Theorem 6.7   The proof is in two parts. First we show that if the vectors are linearly independent, then $\det \boldsymbol{\Phi}(t) \neq 0$ for all t in the interval. Then we assume the converse, namely that $\boldsymbol{\Phi}(t)$ is a fundamental matrix, and show this implies $\det \boldsymbol{\Phi}(t) \neq 0$ for all t in the interval. The fact that every solution of the system can be expressed as a linear combination of the n linearly independent solutions will then follow from Theorem 6.6.

If $\boldsymbol{\Phi}(t)$ is a matrix whose columns are solution vectors and $\det \boldsymbol{\Phi}(t) \neq 0$, then the vectors are linearly independent. To show this, suppose constants $C_1, C_2, \ldots, C_n$ can be found such that

$$C_1 \boldsymbol{\Psi}_1(t) + C_2 \boldsymbol{\Psi}_2(t) + \cdots + C_n \boldsymbol{\Psi}_n(t) = \mathbf{0}$$

for all t in the interval $a < t < b$. Then for any t_0 in the interval, setting $t = t_0$ the equation can be written

$$\boldsymbol{\Phi}(t_0)\mathbf{C} = \mathbf{0},$$

where $\mathbf{C}$ is a column matrix with elements $C_1, C_2, \ldots, C_n$. As $\det \boldsymbol{\Phi}(t_0) \neq 0$, the only solution of this homogeneous system of algebraic equations is $C_1 = C_2 = \cdots = C_n = 0$, so the column vectors must be linearly independent for all t in the interval.

We must now consider the converse situation and suppose that $\boldsymbol{\Phi}(t)$ is a fundamental matrix. Then, if $\boldsymbol{\Psi}(t)$ is a solution of the system, from the definition of a fundamental solution a unique constant vector $\mathbf{C}$ can always be found such that $\boldsymbol{\Psi}(t) = \boldsymbol{\Phi}(t)\mathbf{C}$ for all t in the interval. To find $\mathbf{C}$ we need only set $t = t_0$ in this last result, because as $\det \boldsymbol{\Phi}(t_0) \neq 0$ the homogeneous system of algebraic equations must have a unique solution. The result is true for each t_0 in the interval, and so it follows that $\det \boldsymbol{\Phi}(t) \neq 0$ over the interval $a < t < b$.

As the set of n vectors $\boldsymbol{\Psi}_m(t)$ $(m = 1, 2, \ldots, n)$ is linearly independent, it follows from Theorem 6.6 that every solution of the system is expressible as a linear combination of these vectors, so they form a basis for solutions of the system. ∎

For more information about the material in Sections 6.9 and 6.10 see, for example, references [3.3], [3.4], and [3.16].

Summary

The linear superposition of matrix vector solutions was shown to be permissible, and the concept of a fundamental matrix was introduced, the columns of which contained n linearly independent solution vectors of a linear system of n first order equations. The fundamental matrix had the property that the general solution of the system could be expressed in terms of its product with a column vector containing n arbitrary constants. A determinant test was then developed that established when a set of n solution vectors was suitable to form the columns of a fundamental matrix—that is, to form a basis for the solution set of the system.

EXERCISES 6.10

In Exercises 1 through 6, verify by substitution that the functions $x_1(t)$ and $x_2(t)$ are solutions of the given system of equations. By writing the solution in matrix form, find a fundamental matrix for the system and verify that its columns are linearly independent.

1. $x_1' = x_1 + x_2$, $x_2' = -x_1 + x_2$;
$x_1(t) = e^t(C_1 \cos t + C_2 \sin t)$, $x_2(t) = e^t(C_2 \cos t - C_1 \sin t)$.

2. $x_1' = 2x_1 + x_2$, $x_2' = -2x_1$;
$x_1(t) = e^t(C_1 \cos t + C_2 \sin t)$, $x_2(t) = (C_2 - C_1)e^t \cos t - (C_1 + C_2)e^t \sin t$.

3. $x_1' = x_1 - 2x_2$, $x_2' = x_1 - x_2$;
$x_1(t) = C_1 \cos t + C_2 \sin t$, $x_2(t) = (1/2)(C_1 - C_2) \cos t + (1/2)(C_1 + C_2) \sin t$.

4. $x_1' = -3x_1 - x_2$, $x_2' = 3x_1 + x_2$;
$x_1(t) = C_1 + C_2 e^{-2t}$, $x_2(t) = -3C_1 - C_2 e^{-2t}$.

5. $2x_1' = 2x_1 - x_2$, $x_2' = x_1 + 2x_2$;
$x_1(t) = C_1 e^{3t/2} \cos t/2 + C_2 e^{3t/2} \sin t/2$,
$x_2(t) = -(C_1 + C_2)e^{3t/2} \cos t/2 + (C_1 - C_2)e^{3t/2} \sin t/2$.

6. $2x_1' = -x_1 + x_2$, $x_2' = x_1 - x_2$;
$x_1(t) = C_1 + C_2 e^{-3t/2}$, $x_2(t) = C_1 - 2C_2 e^{-3t/2}$.

6.11 Nonhomogeneous Systems

A nonhomogeneous variable coefficient system of first order linear differential equations can be written

$$\mathbf{x}' = \mathbf{A}(t)\mathbf{x} + \mathbf{b}(t). \tag{107}$$

Its general solution can be expressed as the sum of the general solution of the associated homogeneous system $\mathbf{x}' = \mathbf{A}\mathbf{x}$ that will contain the arbitrary constants, and a *particular solution* free from arbitrary constants that can be taken to be any solution of the nonhomogeneous equation $\mathbf{x}' = \mathbf{A}\mathbf{x} + \mathbf{b}$. This result is recorded and proved in the next theorem.

The Structure of the Solution

THEOREM 6.8

nonhomogeneous system and the structure of the solution

Structure of the solution of $x' = A(t)x + b(t)$ Let $\Phi(t)$ be a fundamental matrix for the homogeneous linear first order system $x' = A(t)x$, and let $P(t)$ be any solution of the nonhomogeneous system $x' = A(t)x + b(t)$. Then the **general solution** of the nonhomogeneous system is $x(t) = \Phi(t)C + P(t)$, with C an n-element column matrix with arbitrary constants $C_1, C_2, \ldots, C_n$ as elements.

Proof   The result is almost immediate and follows by substitution. Setting $x = \Phi(t)C + P(t)$, we have $x' = \Phi'(t)C + P'(t)$, so after substitution into the system of differential equations we find that

$$\Phi'(t)C + P'(t) = A\Phi(t)C + AP(t) + b(t).$$

However, $\Phi'(t)C = A(t)\Phi(t)C$, and by definition $P(t)$ is any solution of $x' = A(t)x + b(t)$, so $P'(t) = AP(t) + b(t)$, showing that substitution of the general solution into the equation leads to an identity, so the theorem is proved. ∎

It is important to recognize that solutions of nonhomogeneous linear systems do not have the linear superposition property of solutions of homogeneous systems, and so they do *not* form a vector space.

EXAMPLE 6.30

Find the solution of the initial value problem for the nonhomogeneous system of equations

$$x_1' + 2x_1 + 4x_2 = 1 + 2t, \quad x_2' + x_1 - x_2 = 3t$$

subject to the initial conditions $x_1(0) = 56/9$ and $x_2(0) = -13/9$, and verify the results of Theorem 6.8.

Solution   Using the elimination method, the solution of the system can be shown to be

$$x_1(t) = 2/9 + (7/3)t + 2e^{2t} + 4e^{-3t} \quad \text{and} \quad x_2(t) = -4/9 - (2/3)t - 2e^{2t} + e^{-3t},$$

and in matrix form this becomes

$$\underbrace{\begin{bmatrix} x_1(t) \\ x_2(t) \end{bmatrix}}_{x(t)} = \underbrace{\begin{bmatrix} e^{2t} & 4e^{-3t} \\ -e^{2t} & e^{-3t} \end{bmatrix}}_{\Phi(t)} \underbrace{\begin{bmatrix} 2 \\ 1 \end{bmatrix}}_{C} + \underbrace{\begin{bmatrix} 2/9 + (7/3)t \\ -4/9 - (2/3)t \end{bmatrix}}_{P(t)}.$$

Inspection of this form of solution identifies the fundamental matrix $\Phi(t)$ containing exponentials, a column vector C with elements $C_1 = 2$ and $C_2 = 1$, and a particular solution $P(t)$ of the nonhomogeneous system represented by the last matrix vector. It is easily checked that the vector $P(t)$, which contains no constants, is a particular solution of the system. ∎

Matrix Methods of Solution

We now describe a number of matrix methods for the solution of both homogeneous and nonhomogeneous constant coefficient systems of linear first order differential equations.

(a) Solution by diagonalization when A has real eigenvalues

solution by
diagonalization
when eigenvalues
are real

Having already illustrated the elementary elimination method for solving a small system of equations, we now describe the powerful and systematic matrix diagonalization method that can be used with systems involving any number of differential equations.

Consider a general nonhomogeneous constant coefficient system

$$\mathbf{x}' = \mathbf{A}\mathbf{x} + \mathbf{b}(t) \tag{108}$$

where $\mathbf{A}$ is a constant coefficient $n \times n$ matrix with real eigenvalues and n linearly independent eigenvectors. The approach we will use is to try to find a transformation of the dependent variables $x_1, x_2, \ldots, x_n$ forming the elements of vector $\mathbf{x}$, which creates a new set of variables $u_1, u_2, \ldots, u_n$ that form the elements of a vector $\mathbf{u}$ with the property that system (108) can be written as

$$\mathbf{u}' = \mathbf{D}\mathbf{u} + \mathbf{h}, \tag{109}$$

where $\mathbf{D}$ is a diagonal matrix and $\mathbf{h}$ is an n-element column vector with elements that depend on the elements in the nonhomogeneous term $\mathbf{b}(t)$.

If such a transformation can be found, the equations in the system will have been *uncoupled*, because each equation for $u_1, u_2, \ldots, u_n$ can then be solved individually. When $u_1, u_2, \ldots, u_n$ are known, reversing the transformation will give the solution $x_1(t), x_2(t), \ldots, x_n(t)$ of system (108).

Such a transformation has already been provided by Theorem 4.6. It was shown there that if a matrix $\mathbf{P}$ is constructed with the n eigenvectors of $\mathbf{A}$ as its columns, then $\mathbf{P}^{-1}\mathbf{A}\mathbf{P} = \mathbf{D}$, where $\mathbf{D}$ is a diagonal matrix with the eigenvalues of $\mathbf{A}$ arranged along its leading diagonal in the same order as the corresponding eigenvectors appear in $\mathbf{P}$.

Adopting this approach, setting

$$\mathbf{x} = \mathbf{P}\mathbf{u}, \tag{110}$$

and substituting in (108) gives

$$\mathbf{P}\mathbf{u}' = \mathbf{A}\mathbf{P}\mathbf{u} + \mathbf{b}(t), \tag{111}$$

where when differentiating $\mathbf{x}(t)$ use has been made of the fact that $\mathbf{P}$ is a constant matrix. The linear independence of the n eigenvectors forming the columns of $\mathbf{P}$ ensures the existence of the inverse matrix $\mathbf{P}^{-1}$, so premultiplying (111) by $\mathbf{P}^{-1}$ gives

$$\mathbf{u}' = \mathbf{P}^{-1}\mathbf{A}\mathbf{P}\mathbf{u} + \mathbf{P}^{-1}\mathbf{b}(t),$$

but $\mathbf{P}^{-1}\mathbf{A}\mathbf{P} = \mathbf{D}$, so system (108) has been transformed into the uncoupled system

$$\mathbf{u}' = \mathbf{D}\mathbf{u} + \mathbf{P}^{-1}\mathbf{b}(t). \tag{112}$$

The required solution vector $\mathbf{x}(t)$ follows from the result $\mathbf{x}(t) = \mathbf{P}\mathbf{u}$.

Before giving an example, it is necessary to consider whether systems exist for which this method will fail. The answer to this question is not difficult to find, because the method depends for its success on the diagonalization of $\mathbf{A}$, and this in turn requires that $\mathbf{A}$ have n linearly independent eigenvectors. Consequently, we see that the method will fail if the $n \times n$ matrix $\mathbf{A}$ has fewer than n linearly independent eigenvectors, because then the diagonalizing matrix $\mathbf{P}$ cannot be constructed. This situation occurs when $\mathbf{A}$ has multiple eigenvalues but an eigenvalue with multiplicity r has associated with it fewer than r linearly independent eigenvectors. A typical matrix with this property is

$$\mathbf{A} = \begin{bmatrix} 1 & 5 & 7 \\ 0 & 1 & 1 \\ 0 & -1 & -1 \end{bmatrix}.$$

In this case the eigenvalue $\lambda = 1$ occurs with multiplicity 1 and the eigenvalue $\lambda = 0$ with multiplicity 2, but the matrix has only the two linearly independent eigenvectors

$$(\lambda = 1), \mathbf{x}_1 = \begin{bmatrix} 1 \\ 0 \\ 0 \end{bmatrix} \text{ and } (\lambda = 0, \text{twice}), \text{ the single eigenvector } \mathbf{x}_2 = \begin{bmatrix} -2 \\ -1 \\ 1 \end{bmatrix}.$$

EXAMPLE 6.31

Use diagonalization to solve the nonhomogeneous system

$$x_1'(t) + 2x_1 + 4x_2 = 2t - 1, \quad x_2'(t) + x_1 - x_2 = \sin t.$$

Solution The system can be written in the form $\mathbf{x}' = \mathbf{A}\mathbf{x} + \mathbf{b}(t)$ with

$$\mathbf{x} = \begin{bmatrix} x_1 \\ x_2 \end{bmatrix}, \quad \mathbf{A} = \begin{bmatrix} -2 & -4 \\ -1 & 1 \end{bmatrix}, \quad \text{and} \quad \mathbf{b}(t) = \begin{bmatrix} 2t - 1 \\ \sin t \end{bmatrix}.$$

Matrix $\mathbf{A}$ has the two eigenvalues and eigenvectors

$$\lambda_1 = 2, \ \mathbf{x}_1 = \begin{bmatrix} -1 \\ 1 \end{bmatrix}, \quad \lambda_2 = -3, \quad \mathbf{x}_2 = \begin{bmatrix} 4 \\ 1 \end{bmatrix}.$$

The diagonalizing matrix is thus

$$\mathbf{P} = \begin{bmatrix} -1 & 4 \\ 1 & 1 \end{bmatrix}, \quad \text{so} \quad \mathbf{P}^{-1} = \begin{bmatrix} -1/5 & 4/5 \\ 1/5 & 1/5 \end{bmatrix},$$

and from the order in which the eigenvectors have been entered as the columns of $\mathbf{P}$, it follows without further computation that

$$\mathbf{D} = \mathbf{P}^{-1}\mathbf{A}\mathbf{P} = \begin{bmatrix} 2 & 0 \\ 0 & -3 \end{bmatrix}.$$

We have

$$\mathbf{P}^{-1}\mathbf{b}(t) = \begin{bmatrix} 1/5 - (2/5)t + (4/5)\sin t \\ -1/5 + (2/5)t + (1/5)\sin t \end{bmatrix},$$

so, corresponding to (112) the transformed system becomes

$$\begin{bmatrix} u_1' \\ u_2' \end{bmatrix} = \begin{bmatrix} 2 & 0 \\ 0 & -3 \end{bmatrix} \begin{bmatrix} u_1 \\ u_2 \end{bmatrix} + \begin{bmatrix} 1/5 - (2/5)t + (4/5)\sin t \\ -1/5 + (2/5)t + (1/5)\sin t \end{bmatrix}.$$

In component form these are seen to be the uncoupled equations

$$u_1' = 2u_1 + 1/5 - (2/5)t + (4/5)\sin t$$

and

$$u_2' = -3u_2 - 1/5 + (2/5)t + (1/5)\sin t.$$

The solution of the uncoupled equations is easily shown to be

$$u_1(t) = C_1 e^{2t} - (4/25)\cos t - (8/25)\sin t + (1/5)t$$
$$u_2(t) = C_2 e^{-3t} - (1/50)\cos t + (3/50)\sin t + (2/15)t - 1/9,$$

where C_1 and C_2 are arbitrary constants. If we use these as the elements of the column vector $\mathbf{U}$, the required solution is given by $\mathbf{x}(t) = \mathbf{Pu}$, and so

$$\begin{bmatrix} x_1(t) \\ x_2(t) \end{bmatrix} = \begin{bmatrix} -1 & 4 \\ 1 & 1 \end{bmatrix} \begin{bmatrix} C_1 e^{2t} - (4/25)\cos t - (8/25)\sin t + (1/5)t \\ C_2 e^{-3t} - (1/50)\cos t + (3/50)\sin t + (2/15)t - 1/9 \end{bmatrix}.$$

In component form the solution becomes

$$x_1(t) = -4/9 + (1/3)t + (2/25)\cos t + (14/25)\sin t - C_1 e^{2t} + 4C_2 e^{-3t}$$
$$x_2(t) = -1/9 + (1/3)t - (9/50)\cos t - (13/50)\sin t + C_1 e^{2t} + C_2 e^{-3t}.$$ ■

(b) Solution by diagonalization when A has complex eigenvalues

solution by diagonalization when eigenvalues are complex

When the diagonalization method is used to solve a system in which $\mathbf{A}$ has pairs of complex conjugate eigenvalues, the approach only differs from the case involving real eigenvalues in that the arbitrary constants introduced at the integration stage are complex. When $\mathbf{A}$ has real coefficients and complex eigenvalues exist, they must do so in complex conjugate pairs, so after integrating an equation corresponding to the complex eigenvalue $\lambda = \alpha + i\beta$, we must introduce a complex integration constant $C_1 + iC_2$. Then, to make the solution real, when integrating the equation corresponding to the complex conjugate eigenvalue $\bar{\lambda} = \alpha - i\beta$ the complex conjugate integration constant $C_1 - iC_2$ must be introduced.

EXAMPLE 6.32   Use diagonalization to solve the system of nonhomogeneous equations

$$x_1'(t) = x_1 + 2x_2 + x_3 + 1$$
$$x_2'(t) = x_2 + x_3 + t$$
$$x_3'(t) = 2x_1 + x_3 + 2t.$$

Solution   The matrix $\mathbf{A}$ is

$$\mathbf{A} = \begin{bmatrix} 1 & 2 & 1 \\ 0 & 1 & 1 \\ 2 & 0 & 1 \end{bmatrix},$$

and its eigenvalues and eigenvectors are

$$\lambda_1 = 3, \quad \mathbf{x}_1 = \begin{bmatrix} 2 \\ 1 \\ 2 \end{bmatrix}, \quad \lambda_2 = i, \quad \mathbf{x}_2 = \begin{bmatrix} -i \\ 1 \\ -1+i \end{bmatrix}, \quad \lambda_3 = -i, \quad \mathbf{x}_3 = \begin{bmatrix} i \\ 1 \\ -1-i \end{bmatrix}.$$

The diagonalizing matrix

$$\mathbf{P} = \begin{bmatrix} 2 & -i & i \\ 1 & 1 & 1 \\ 2 & -1+i & -1-i \end{bmatrix} \quad \text{and}$$

$$\mathbf{P}^{-1} = \begin{bmatrix} 1/5 & 1/5 & 1/5 \\ -1/10+3i/10 & 2/5-i/5 & -1/10-i/5 \\ -1/10-3i/10 & 2/5+i/5 & -1/10+i/5 \end{bmatrix}.$$

The order in which the columns of $\mathbf{P}$ are arranged shows without further computation that when diagonalized, $\mathbf{A}$ will become the matrix

$$\mathbf{D} = \begin{bmatrix} 3 & 0 & 0 \\ 0 & i & 0 \\ 0 & 0 & -i \end{bmatrix}.$$

This is because $\mathbf{D}$ can be written down immediately without the need to calculate $\mathbf{D} = \mathbf{P}^{-1}\mathbf{A}\mathbf{P}$, because the order in which the eigenvalues are arranged along the leading diagonal of $\mathbf{D}$ is the order in which their corresponding eigenvectors form the columns of $\mathbf{A}$.

If we write the system as

$$\mathbf{x}' = \mathbf{A}\mathbf{x} + \mathbf{b}(t),$$

with

$$\mathbf{x} = \begin{bmatrix} x_1(t) \\ x_2(t) \\ x_3(t) \end{bmatrix} \quad \text{and} \quad \mathbf{b}(t) = \begin{bmatrix} 1 \\ t \\ 2t \end{bmatrix},$$

and set $\mathbf{x}(t) = \mathbf{P}\mathbf{u}$, the system becomes

$$\mathbf{P}\mathbf{u}' = \mathbf{A}\mathbf{P}\mathbf{u} + \mathbf{b}(t),$$

so

$$\mathbf{u}' = \mathbf{P}^{-1}\mathbf{A}\mathbf{P}\mathbf{u} + \mathbf{P}^{-1}\mathbf{b}(t) \quad \text{or} \quad \mathbf{u}' = \mathbf{D}\mathbf{u} + \mathbf{P}^{-1}\mathbf{b}(t).$$

A simple calculation then gives

$$\mathbf{P}^{-1}\mathbf{b}(t) = \begin{bmatrix} 1/5+3t/5 \\ -1/10+3i/10+t/5-3it/5 \\ -1/10-3i/10+t/5+3it/5 \end{bmatrix},$$

so writing $\mathbf{u}' = \mathbf{D}\mathbf{u} + \mathbf{P}^{-1}\mathbf{b}(t)$ in component form shows that the uncoupled equations become

$$u_1'(t) = 3u_1 + 1/5 + 3t/5$$
$$u_2'(t) = iu_2 - 1/10 + 3i/10 + t/5 - 3it/5$$
$$u_3'(t) = -iu_3 - 1/10 - 3i/10 + t/5 + 3it/5.$$

Solving the first equation involves no complex numbers and so gives rise to the solution

$$u_1(t) = -2/15 - t/5 + C_1 e^{3t}.$$

However, the other two equations are complex, so remembering that the complex integration constant in the third equation must be the complex conjugate of the one in the second equation leads to the results

$$u_2(t) = 3t/5 - 1/10 - 7i/10 + it/5 + (C_2 + iC_3)(\cos t + i\sin t)$$
$$u_3(t) = 3t/5 - 1/10 + 7i/10 - it/5 + (C_2 - iC_3)(\cos t - i\sin t).$$

Combining these results gives

$$\mathbf{u} = \begin{bmatrix} -2/15 - t/5 + C_1 e^{3t} \\ 3t/5 - 1/10 - 7i/10 + it/5 + (C_2 + iC_3)(\cos t + i\sin t) \\ 3t/5 - 1/10 + 7i/10 - it/5 + (C_2 - iC_3)(\cos t - i\sin t) \end{bmatrix},$$

so finally, using $\mathbf{x}(t) = \mathbf{Pu}$, we arrive at the required solution

$$x_1(t) = -5/3 + 2C_1 e^{3t} + 2C_2 \sin t + 2C_3 \cos t$$
$$x_2(t) = -1/3 + t + C_1 e^{3t} + 2C_2 \cos t - 2C_3 \sin t$$
$$x_3(t) = 4/3 - 2t + 2C_1 e^{3t} + (2C_3 - 2C_2)\sin t - (2C_2 + 2C_3)\cos t. \qquad \blacksquare$$

(c) Solution of a homogeneous system by the matrix exponential

solution using the matrix exponential

For the sake of completeness, we now show how, when $\mathbf{A}$ is diagonalizable, the solution of the homogeneous constant coefficient system $\mathbf{x}' = \mathbf{Ax}$ can be solved by means of the matrix exponential, and we indicate how the method can be extended to enable the solution to be found when $\mathbf{A}$ is *not* diagonalizable. As the Laplace transform method to be described later deals with the solution of initial value problems for linear equations automatically and is simpler to use, the ideas involved will only be outlined. Nevertheless, the matrix exponential is both useful and important when working with systems of equations, so it is necessary to make some mention of it here.

We consider the initial value problem

$$\mathbf{x}' = \mathbf{Ax} \quad \text{subject to the initial condition} \quad \mathbf{x}(t_0) = \mathbf{v}, \qquad (113)$$

where $\mathbf{A}$ is an $n \times n$ constant matrix and $\mathbf{v}$ is an arbitrary n-element constant column vector. Then the existence and uniqueness theorem guarantees that a solution certainly exists in some open interval containing t_0. If we define a vector $\mathbf{x}(t) = e^{t\mathbf{A}}\mathbf{v}$, and set

$$e^{t\mathbf{A}} = \mathbf{I}_n + t\mathbf{A} + \frac{t^2}{2!}\mathbf{A}^2 + \frac{t^3}{3!}\mathbf{A}^3 + \dots,$$

then

$$\begin{aligned} d\mathbf{x}/dt &= d(e^{t\mathbf{A}})/dt \, \mathbf{v} \\ &= \mathbf{A}e^{t\mathbf{A}}\mathbf{v} \\ &= \mathbf{Ax}, \end{aligned}$$

so the solution of the initial value problem in (113) can be represented in the form

$$\mathbf{x}(t) = e^{t\mathbf{A}}\mathbf{v}. \tag{114}$$

We saw in Section 4.5 that $e^{t\mathbf{A}}$ is easily computed when $\mathbf{A}$ is diagonalizable, but before using this result we first review the ideas that are involved. If $\mathbf{A}$ is diagonalizable to a matrix $\mathbf{D}$, a matrix $\mathbf{P}$ exists such that $\mathbf{A} = \mathbf{PDP}^{-1}$, where the columns of $\mathbf{P}$ are the eigenvectors of $\mathbf{A}$, and the elements of $\mathbf{D}$ are the corresponding eigenvalues of $\mathbf{A}$. Thus, $\mathbf{A}^2 = \mathbf{PDP}^{-1}\,\mathbf{PDP}^{-1} = \mathbf{PD}^2\mathbf{P}^{-1}$, and by extending this argument we have the general result $\mathbf{A}^m = \mathbf{PD}^m\mathbf{P}^{-1}$, for $m = 1, 2\ldots$. Using this property in the definition of the matrix exponential $e^{t\mathbf{A}}$ given above allows it to be written

$$e^{t\mathbf{A}} = \mathbf{P}[\mathbf{I}_n + t\mathbf{D} \frac{t^2}{2!}\mathbf{D}^2 + \frac{t^3}{3!}\mathbf{D}^3 + \ldots]\mathbf{P}^{-1}.$$

Consequently, if

$$\mathbf{D} = \begin{bmatrix} \lambda_1 & 0 & 0 & \ldots & 0 \\ 0 & \lambda_2 & 0 & \ldots & 0 \\ . & . & . & . & . \\ 0 & 0 & 0 & \ldots & \lambda_n \end{bmatrix},$$

then

$$\mathbf{D}^j = \begin{bmatrix} \lambda_1^j & 0 & 0 & \ldots & 0 \\ 0 & \lambda_2^j & 0 & \ldots & 0 \\ . & . & . & . & . \\ 0 & 0 & 0 & \ldots & \lambda_n^j \end{bmatrix},$$

and so

$$e^{t\mathbf{A}} = \mathbf{P} \begin{bmatrix} \sum_{j=0}^{\infty} \frac{\lambda_1^j t^j}{j!} & 0 & 0 & \ldots & 0 \\ 0 & \sum_{j=0}^{\infty} \frac{\lambda_2^j t^j}{j!} & 0 & \ldots & 0 \\ . & . & . & . & . \\ 0 & 0 & 0 & \sum_{j=0}^{\infty} \frac{\lambda_n^j t^j}{j!} \end{bmatrix} \mathbf{P}^{-1},$$

and this shows that

$$e^{t\mathbf{A}} = \mathbf{P} \begin{bmatrix} \exp(\lambda_1 t) & 0 & 0 & \ldots & 0 \\ 0 & \exp(\lambda_2 t) & 0 & \ldots & 0 \\ . & . & . & . & . \\ 0 & 0 & 0 & \ldots & \exp(\lambda_n t) \end{bmatrix} \mathbf{P}^{-1}. \tag{115}$$

We have shown that the matrix exponential $e^{t\mathbf{A}}$ is simply another way of representing a fundamental matrix for system (113). So, provided $\mathbf{A}$ can be diagonalized and has real eigenvalues, $e^{t\mathbf{A}}$ can be written down immediately by using result (115).

EXAMPLE 6.33   Use the matrix exponential to solve the system

$$x_1'(t) = -2x_1 + 6x_2$$
$$x_2'(t) = -2x_1 + 5x_2.$$

Solution   The matrix $\mathbf{A}$ is

$$\mathbf{A} = \begin{bmatrix} -2 & 6 \\ -2 & 5 \end{bmatrix},$$

and its eigenvalues and eigenvectors are

$$(\lambda_1 = 1)\, \mathbf{x}_1 = \begin{bmatrix} 2 \\ 1 \end{bmatrix}, \quad (\lambda_2 = 2)\, \mathbf{x}_2 = \begin{bmatrix} 3/2 \\ 1 \end{bmatrix}.$$

The diagonalizing matrix

$$\mathbf{P} = \begin{bmatrix} 2 & 3/2 \\ 1 & 1 \end{bmatrix}, \quad \mathbf{P}^{-1} = \begin{bmatrix} 2 & -3 \\ -2 & 4 \end{bmatrix},$$

and

$$\mathbf{D} = \begin{bmatrix} 1 & 0 \\ 0 & 2 \end{bmatrix}.$$

So from (115) we have

$$e^{t\mathbf{A}} = \begin{bmatrix} 2 & 3/2 \\ 1 & 1 \end{bmatrix} \begin{bmatrix} e^t & 0 \\ 0 & e^{2t} \end{bmatrix} \begin{bmatrix} 2 & -3 \\ -2 & 4 \end{bmatrix},$$

and after evaluating the matrix products we obtain

$$e^{t\mathbf{A}} = \begin{bmatrix} 4e^t - 3e^{2t}, & -6e^t + 6e^{2t} \\ 2e^t - 2e^{2t}, & -3e^t + 4e^{2t} \end{bmatrix}.$$

Defining a two-element column matrix $\mathbf{C}$ with the arbitrary constants C_1 and C_2 as elements allows the general solution to be written as

$$\mathbf{x}(t) = e^{t\mathbf{A}}\mathbf{C} = \begin{bmatrix} 4e^t - 3e^{2t}, & -6e^t + 6e^{2t} \\ 2e^t - 2e^{2t}, & -3e^t + 4e^{2t} \end{bmatrix} \begin{bmatrix} C_1 \\ C_2 \end{bmatrix},$$

so

$$\mathbf{x}(t) = \begin{bmatrix} (4C_1 - 6C_2)e^t + (6C_2 - 3C_1)e^{2t} \\ (2C_1 - 3C_2)e^t + (4C_2 - 2C_1)e^{2t} \end{bmatrix}.$$

In component form the solution is

$$x_1(t) = (4C_1 - 6C_2)e^t + (6C_2 - 3C_1)e^{2t}$$
$$x_2(t) = (2C_1 - 3C_2)e^t + (4C_2 - 2C_1)e^{2t}.$$ ∎

The method applies equally well to the situation in which matrix $\mathbf{A}$ is real but the eigenvalues occur in complex conjugate pairs, as shown by the next example.

EXAMPLE 6.34

Use the matrix exponential to solve the system

$$x_1'(t) = -3x_1 - 4x_2 \quad \text{and} \quad x_2'(t) = 2x_1 + x_2.$$

Solution   The matrix $\mathbf{A}$ is

$$\mathbf{A} = \begin{bmatrix} -3 & -4 \\ 2 & 1 \end{bmatrix}$$

and its eigenvalues λ_1, λ_2 and eigenvectors $\mathbf{x}_1$ and $\mathbf{x}_2$ are

$$\lambda_1 = -1 + 2i, \quad \mathbf{x}_1 = \begin{bmatrix} -1 + i \\ 1 \end{bmatrix}, \quad \lambda_2 = -1 - 2i, \quad \mathbf{x}_2 = \begin{bmatrix} -1 - i \\ 1 \end{bmatrix}.$$

So

$$\mathbf{P} = \begin{bmatrix} -1 + i & -1 - i \\ 1 & 1 \end{bmatrix}, \quad \mathbf{P}^{-1} = \begin{bmatrix} -i/2 & 1/2 - i/2 \\ i/2 & 1/2 + i/2 \end{bmatrix},$$

$$\mathbf{D} = \begin{bmatrix} -1 + 2i & 0 \\ 0 & -1 - 2i \end{bmatrix},$$

and consequently

$$e^{t\mathbf{A}} = \mathbf{P} \begin{bmatrix} e^{-t}(\cos 2t + i \sin 2t) & 0 \\ 0 & e^{-t}(\cos 2t - i \sin 2t) \end{bmatrix} \mathbf{P}^{-1}$$

$$= \begin{bmatrix} e^{-t}(\cos 2t - \sin 2t) & -2e^{-t} \sin 2t \\ e^{-t} \sin 2t & e^{-t}(\cos 2t + \sin 2t) \end{bmatrix}.$$

If we use this expression for $e^{t\mathbf{A}}$ in $\mathbf{x}(t) = e^{t\mathbf{A}}\mathbf{C}$, the general solution becomes

$$\mathbf{x}(t) = \begin{bmatrix} e^{-t}(\cos 2t - \sin 2t) & -2e^{-t} \sin 2t \\ e^{-t} \sin 2t & e^{-t}(\cos 2t + \sin 2t) \end{bmatrix} \begin{bmatrix} C_1 \\ C_2 \end{bmatrix}.$$

In component form this reduces to

$$x_1(t) = C_1 e^{-t} \cos 2t - (C_1 + 2C_2)e^{-t} \sin 2t$$
$$x_2(t) = (C_1 + C_2)e^{-t} \sin 2t + C_2 e^{-t} \cos 2t.$$ ∎

When $\mathbf{A}$ is not diagonalizable, it is still possible to compute $e^{\mathbf{A}}$ by writing $e^{\mathbf{A}} = e^{\mathbf{K}} e^{\mathbf{L}}$, where $\mathbf{A}$ is the sum of a *diagonal* matrix $\mathbf{K}$ and a *nilpotent* matrix $\mathbf{L}$ (a square matrix that when raised to a finite power becomes the null matrix), because under these circumstances the matrices $e^{\mathbf{K}}$ and $e^{\mathbf{L}}$ commute and $e^{\mathbf{K}+\mathbf{L}} = e^{\mathbf{K}} e^{\mathbf{L}}$. The next example illustrates this approach.

EXAMPLE 6.35

Find $e^{t\mathbf{A}}$ given that

$$\mathbf{A} = \begin{bmatrix} 4 & 1 \\ 0 & 4 \end{bmatrix}$$

and use it to solve the homogeneous system

$$x_1'(t) = 4x_1 + x_2 \quad \text{and} \quad x_2'(t) = 4x_2.$$

Solution   Matrix $\mathbf{A}$ is not diagonalizable, because the repeated eigenvalue $\lambda = 4$ only gives rise to a single eigenvector. However, $t\mathbf{A}$ can be written as the sum of the following diagonal matrix $t\mathbf{K}$ and nilpotent matrix $t\mathbf{L}$:

$$t\mathbf{A} = t\mathbf{K} + t\mathbf{L}, \quad \text{where } t\mathbf{K} = \begin{bmatrix} 4t & 0 \\ 0 & 4t \end{bmatrix} \quad \text{and} \quad t\mathbf{L} = \begin{bmatrix} 0 & t \\ 0 & 0 \end{bmatrix}.$$

It is easily checked that $(t\mathbf{L})^2 = 0$ and the matrices $t\mathbf{K}$ and $t\mathbf{L}$ commute, so $e^{t\mathbf{A}} = e^{t\mathbf{K}}e^{t\mathbf{L}}$. It follows from this that

$$e^{t\mathbf{K}} = \begin{bmatrix} e^{4t} & 0 \\ 0 & e^{4t} \end{bmatrix}, \quad \text{and} \quad e^{t\mathbf{L}} = \begin{bmatrix} 1 & 0 \\ 0 & 1 \end{bmatrix} + \begin{bmatrix} 0 & t \\ 0 & 0 \end{bmatrix} = \begin{bmatrix} 1 & t \\ 0 & 1 \end{bmatrix},$$

so we arrive at the result

$$e^{t\mathbf{A}} = \begin{bmatrix} e^{4t} & 0 \\ 0 & e^{4t} \end{bmatrix} \begin{bmatrix} 1 & t \\ 0 & 1 \end{bmatrix} = \begin{bmatrix} e^{4t} & te^{4t} \\ 0 & e^{4t} \end{bmatrix}.$$

The exponential matrix $e^{t\mathbf{A}}$ is a fundamental matrix for the system, so as the general solution is given by $\mathbf{x}(t) = e^{t\mathbf{A}}\mathbf{C}$,

$$\mathbf{x}(t) = \begin{bmatrix} e^{4t} & te^{4t} \\ 0 & e^{4t} \end{bmatrix} \begin{bmatrix} C_1 \\ C_2 \end{bmatrix} = \begin{bmatrix} C_1 e^{4t} + C_2 te^{4t} \\ C_2 e^{4t} \end{bmatrix}.$$

In component form the solution becomes

$$x_1(t) = C_1 e^{4t} + C_2 te^{4t} \quad \text{and} \quad x_2(t) = C_2 e^{4t}. \qquad \blacksquare$$

The nilpotent matrix $\mathbf{L}$ in the last example was seen to give rise to the second linearly independent solution te^{4t} corresponding to the eigenvalue $\lambda = 4$ that occurred with multiplicity 2. If in a larger system with a repeated eigenvalue λ and a nondiagonalizable matrix $\mathbf{A}$ it had been necessary to raise a nilpotent matrix to the power r before it became the null matrix, then in addition to a term of the form $e^{\lambda t}$ appearing in $e^{t\mathbf{A}}$, the repeated eigenvalue would also give rise to the linearly independent terms $te^{\lambda t}, t^2 e^{\lambda t}, \ldots, t^{(r-1)}e^{\lambda t}$.

(d) Variation of parameters

A particular integral can be found from the general solution of the homogeneous form of a constant coefficient system by a direct generalization of the method of **variation of parameters** described in Section 6.6. If the system is

the matrix exponential and variation of parameters

$$\mathbf{x}' = \mathbf{A}\mathbf{x} + \mathbf{b}(t), \tag{116}$$

to find a particular integral $\mathbf{x}_p(t)$ we set

$$\mathbf{x}_p(t) = e^{t\mathbf{A}}\mathbf{u}(t), \tag{117}$$

where the vector $\mathbf{u}(t)$ is to be determined.

Then, as $\mathbf{x}'_p(t) = \mathbf{A}e^{t\mathbf{A}}\mathbf{u}(t) + e^{t\mathbf{A}}\mathbf{u}'(t)$, substituting for $\mathbf{x}_p(t)$ in system (106) gives

$$\mathbf{A}e^{t\mathbf{A}}\mathbf{u}(t) + e^{t\mathbf{A}}\mathbf{u}'(t) = \mathbf{A}e^{t\mathbf{A}}\mathbf{u}(t) + \mathbf{b}(t),$$

so after cancelling the terms $\mathbf{A}e^{t\mathbf{A}}\mathbf{u}(t)$ and premultiplying the result by $e^{-t\mathbf{A}}$, the inverse of $e^{t\mathbf{A}}$ because $e^{t\mathbf{A}}$ and $e^{-t\mathbf{A}}$ commute, we find that

$$\mathbf{u}'(t) = e^{-t\mathbf{A}}\mathbf{b}(t), \tag{118}$$

from which $\mathbf{u}(t)$ now follows.

In the equation for $\mathbf{u}(t)$ the matrix exponential $e^{-t\mathbf{A}}$ is determined from $e^{t\mathbf{A}}$ by changing the sign of t. The expression on the right of (118) is simply a column vector with elements that are known functions of t, so the components of (118) can be integrated separately to find the elements $u_1(t), u_2(t), \ldots, u_n(t)$ of $\mathbf{U}(t)$. Then, when $\mathbf{U}(t)$ is known, the particular integral follows from (117).

The general solution of (116) is the sum of the solution of the homogeneous form of the system and the particular integral $\mathbf{x}_p(t)$.

EXAMPLE 6.36

Use the method of variation of parameters to solve the nonhomogeneous system.

$$x_1'(t) = -2x_1 + 6x_2 + t$$
$$x_2'(t) = -2x_1 + 5x_2 - 1.$$

Solution   The homogeneous form of this system was obtained in Example 6.33, where it was shown that

$$e^{t\mathbf{A}} = \begin{bmatrix} 4e^t - 3e^{2t} & -6e^t + 6e^{2t} \\ 2e^t - 2e^{2t} & -3e^t + 4e^{2t} \end{bmatrix},$$

so

$$e^{-t\mathbf{A}} = \begin{bmatrix} 4e^{-t} - 3e^{-2t} & -6e^{-t} + 6e^{-2t} \\ 2e^{-t} - 2e^{-2t} & -3e^{-t} + 4e^{-2t} \end{bmatrix}.$$

As

$$\mathbf{b}(t) = \begin{bmatrix} t \\ -1 \end{bmatrix}, \text{ we have } e^{-t\mathbf{A}}\mathbf{b}(t) = \begin{bmatrix} 2(3 + 2t)e^{-t} - 3(2 + t)e^{-2t} \\ (3 + 2t)e^{-t} - 2(2 + t)e^{-2t} \end{bmatrix},$$

but $\mathbf{u}'(t) = e^{-t\mathbf{A}}\mathbf{b}(t)$, so

$$u_1'(t) = 2(3 + 2t)e^{-t} - 3(2 + t)e^{-2t}$$
$$u_2'(t) = (3 + 2t)e^{-t} - 2(2 + t)e^{-2t}.$$

When these equations are integrated, the arbitrary constants of integration can be set equal to zero, because if they are nonzero the terms they introduce are of the same type as the solution of the homogeneous system, so they can be absorbed into it. As a result, integration gives

$$u_1(t) = -2(5 + 2t) + \frac{3}{2}\left(\frac{5}{2} + t\right)e^{-2t}$$

and

$$u_2(t) = -(5 + 2t)e^{-t} + \left(\frac{5}{2} + t\right)e^{-2t}.$$

Finally, if we set $\mathbf{x}_p(t) = e^{t\mathbf{A}}\mathbf{u}(t)$ with $u_1(t)$ and $u_2(t)$ taken as the elements of $\mathbf{u}(t)$, the particular integral becomes

$$\mathbf{x}_p(t) = \begin{bmatrix} -\dfrac{25}{4} - \dfrac{5}{2}t \\ -\dfrac{5}{2} - t \end{bmatrix}.$$

The solution $\mathbf{x}_c(t)$ of the homogeneous system (the complementary function) found in Example 6.33 was

$$\mathbf{x}_c(t) = \begin{bmatrix} (4C_1 - 6C_2)e^t + (6C_2 - 3C_1)e^{2t} \\ (2C_1 - 3C_2)e^t + (4C_2 - 2C_1)e^{2t} \end{bmatrix},$$

so the solution of the nonhomogeneous system $\mathbf{x}(t) = \mathbf{x}_c(t) + \mathbf{x}_p(t)$ is given by

$$x_1(t) = (4C_1 - 6C_2)e^t + (6C_2 - 3C_1)e^{2t} - \frac{25}{4} - \frac{5}{2}t$$

$$x_2(t) = (2C_1 - 3C_2)e^t + (4C_2 - 2C_1)e^{2t} - \frac{5}{2} - t. \qquad \blacksquare$$

General Remark

The way in which combinations of arbitrary constants appear when we multiply functions in the general solution of a homogeneous system of differential equations is determined by the method of solution. So, for example, when we solve a system by elimination, the choice of variable to be eliminated first will influence the form of the result, as will the ordering of the eigenvectors when diagonalizing the matrix **A**. A combination of arbitrary constants is simply an arbitrary constant, though the ratio of all similar combinations of constants multiplying corresponding functions in different forms of the solution must be the same.

This can be illustrated by considering the solution of the homogeneous form of the equation in Example 6.36 that was found to be

$$x_1(t) = (4C_1 - 6C_2)e^t + (6C_2 - 3C_1)e^{2t}$$
$$x_2(t) = (2C_1 - 3C_2)e^t + (4C_2 - 2C_1)e^{2t}.$$

This solution can be written in an equivalent but different-looking form by setting $K_1 = 2C_1 - 3C_2$ and $K_2 = 6C_2 - 3C_1$, where K_1 and K_2 are themselves arbitrary constants. After changing the constants in this manner the solution becomes

$$x_1(t) = 2K_1e^t + K_2e^{2t}$$

and

$$x_2(t) = K_1e^t + \frac{2}{3}K_2e^{2t},$$

and other equivalent forms are also possible.

The above remarks should be remembered when comparing solutions to problem sets with the solutions given at the end of the book. As a particular integral contains no arbitrary constants, its form remains the same irrespective of the manner in which it has been determined.

An account of the material in this section is to be found in references [3.5] and [3.15].

Summary

The structure of the solution of a linear nonhomogeneous system of equations was explained, and a matrix method of solution was developed for constant coefficient systems that depended on the diagonalization of the coefficient matrix. The cases of real and complex eigenvalues of the coefficient matrix were examined separately, and it was shown how systems of equations with real coefficient matrices can lead to solutions involving trigonometric functions. A different method of solution was then developed using the concept of the matrix exponential.

EXERCISES 6.11

In Exercises 1 through 6 find a fundamental matrix and the general solution of the system.

1. $x_1' = -x_2, x_2' = 2x_1$.
2. $x_1' = -x_1 - 5x_2, x_2' = x_1 - 5x_2$.
3. $x_1' = -3x_1 - 4x_2, x_2' = 2x_1 + x_2$.
4. $x_1' = -x_1 - 4x_2, x_2' = x_1 + 4x_2$.
5. $x_1' = 2x_2, x_2' = -2x_3, x_3' = 2x_2$.
6. $x_1' = -3x_2, x_2' = -3x_3, x_3' = 3x_2$.

In Exercises 7 through 18 find the general solution of the system by diagonalization.

7. $x_1' = -10x_1 - 18x_2 + t, x_2' = 6x_1 + 11x_2 + 3$.
8. $x_1' = -2x_2 + \sin t, x_2' = -2x_1 - t$.
9. $x_1' = x_1 - x_2 + \cos t, x_2' = -x_1 + x_2 + e^{3t}$.
10. $x_1' = x_2 + e^{-t}, x_2' = -x_1 + 2x_2 - 4$.
11. $x_1' = 2x_1 + 3x_2 - \sin t, x_2' = x_1 - 2x_2$.
12. $x_1' = -x_1 - 2x_2 + \cos t, x_2' = x_1 + x_2 + 4$.
13. $x_1' = -2x_1 + 2x_2 + 2x_3 + \sin t, x_2' = -x_2 + 3, x_3' = -2x_1 + 4x_2 + 3x_3$.
14. $x_1' = x_1 + 2x_2 + 3 + 2t, x_2' = x_2 + t, x_3' = 2x_1 + x_3 + 1$.
15. $x_1' = x_1 + 2x_2 + x_3 + t, \quad x_2' = x_2 - x_3 + 2, \quad x_3' = 2x_1 + x_3 + 2t$.
16. $x_1' = x_2 + t, x_2' = x_3, x_3' = x_2$.
17. $x_1' = x_1 + 2x_2 + x_3 + 2e^{-t}, x_2' = x_2 + x_3 + t, x_3' = 2x_1 + x_3 + 2t$.
18. $x_1' = x_2 + 5, x_2' = x_3 + t, x_3' = x_2 + 2t$.

Solve Exercises 19 through 26 by means of the matrix exponential.

19. $2x_1' = x_1 - x_2, 2x_2' = 3x_1 + 5x_2$.

20. $x_1' = -10x_1 - 18x_2, x_2' = 6x_1 + 11x_2$.
21. $x_1' = -x_2, x_2' = 2x_1$.
22. $x_1' = 2x_1 - 12x_2, 2x_2' = 3x_1 - 8x_2$.
23. $x_1' = 7x_1 - 34x_2, x_2' = 2x_1 - 9x_2$.
24. $x_1' = -x_1 - 5x_2, x_2' = x_1 - 5x_2$.
25. $x_1' = -3x_1 - 4x_2, x_2' = 2x_1 + x_2$.
26. $x_1' = -x_1 + 2x_2, x_2' = x_1 + x_2$.

Solve Exercises 27 through 30 by the method of variation of parameters.

27. $x_1' = 10x_1 + 18x_2 + \sin t, x_2' = -6x_1 - 11x_2 + t$.
28. $x_1' = -x_2 + 3e^{4t}, x_2' = -2x_1 + x_2 - 2$.
29. $x_1' = 3x_1 + 4x_2, x_2' = -2x_1 - x_2 - t^2$.
30. $x_1' = -x_2 + 5, x_2' = x_1 + 2x_2 - 1$.

Solve the initial value problems 31 through 36 by any of the methods in this chapter.

31. $x_1' = x_2 + 1, \ x_2' = 2x_1 - x_2 + t$, with $x_1(0) = 1$, $x_2(0) = 0$.
32. $x_1' = 3x_2 + t, \ x_2' = 2x_1 + x_2 - 3$, with $x_1(0) = 1$, $x_2(0) = 1$.
33. $x_1' = 2x_1 + x_2 - e^t, x_2' = -2x_1 - x_2 - 3$, with $x_1(0) = 0$, $x_2(0) = 1$.
34. $x_1' = -3x_1 - x_2 + 3t, \ x_2' = x_1 - x_2 - 3$, with $x_1(0) = 1$, $x_2(0) = 3$.
35. $x_1' = -3x_1 - 5x_2 - 12x_3 + \sin t, \quad x_2' = -2x_1 + 1, \quad x_3' = x_1 + x_2 + 2x_3 - t$, with $x_1(0) = 1, x_2(0) = 0, x_3(0) = -1$.
36. $x_1' = -2x_1 + 2x_2 + 2x_3 + 3e^t, \ x_2' = -x_1 - x_2 - 2x_3 + 1, x_3' = x_1 + 2x_2 + 3x_3 - 3$, with $x_1(0) = 1, \ x_2(0) = 1, x_3(0) = 0$.

6.12   Autonomous Systems of Equations

Autonomous Systems, the Phase Plane, Stability, and Linear Systems

The general form of a nonlinear system of two simultaneous first order differential equations for the functions $x(t)$, $y(t)$ that depend on the time t is

$$
\frac{dx}{dt} = f_1(x, y, t)
$$

$$
\frac{dy}{dt} = g_1(x, y, t). \tag{119}
$$

This system is linear and nonhomogeneous if

$$f_1(x, y, t) = a(t)x(t) + b(t)y(t) + h(t)$$

and

$$g_1(x, y, t) = c(t)x(t) + d(t)y(t) + k(t),$$

and homogeneous if, in addition, $h(t) = k(t) \equiv 0$.

If the dependence of the functions f_1 and g_1 on the time t is only through the functions $x(t)$ and $y(t)$, the time dependence is implicit and $f_1 = f(x, y)$ and $g_1 = g(x, y)$, causing the system of equations in (119) to become

$$\frac{dx}{dt} = f(x, y)$$

$$\frac{dy}{dt} = g(x, y). \tag{120}$$

autonomous and nonautonomous systems

Systems of this type are called **autonomous**, and they describe physical phenomena such as chemical reactions that, provided all conditions remain the same, will yield identical results whenever the reactions are repeated. It is because of this that autonomous systems are sometimes said to be **time invariant** systems. This situation should be contrasted with the **nonautonomous** behavior of an electrical circuit containing temperature-dependent elements that will cause its behavior to vary as the ambient temperature changes with time.

A point (x_0, y_0) where both of the derivatives dx/dt and dy/dt in (120) vanish, so that

equilibrium or critical point

$$\left(\frac{dx}{dt}\right)^2 + \left(\frac{dy}{dt}\right)^2 = 0,$$

is called an **equilibrium point** or a **critical point** of the system.

If the differential equations in (120) are solved subject to the initial conditions $x_0 = x(t_0)$, $y_0 = y(t_0)$ imposed at time $t = t_0$, it is convenient to regard $(x(t), y(t))$ as a point in the (x, y)-plane that traces out a curve as t increases. Such curves, along which the time t can be regarded as a parameter, are called **trajectories** or **paths**, and sometimes **orbits** in the (x, y)-plane. The (x, y)-plane itself is then called the **phase plane**. Associated with each trajectory is the direction in which the point $(x(t), y(t))$ moves as t increases, and in the phase plane these directions are usually indicated by adding arrows to trajectories. The pattern of trajectories associated with a given autonomous system of equations is called the **phase portrait** of the system.

trajectories or paths

phase portrait

The reason why in autonomous systems the time t can be regarded as a parameter can be seen by dividing the second equation in (120) by the first to obtain the differential equation

$$\frac{dy}{dx} = \frac{g(x, y)}{f(x, y)}, \tag{121}$$

in which t is absent. Had the **nonautonomous** system of equations in (119) been treated in similar fashion, dy/dx would have exhibited an explicit dependence on the time.

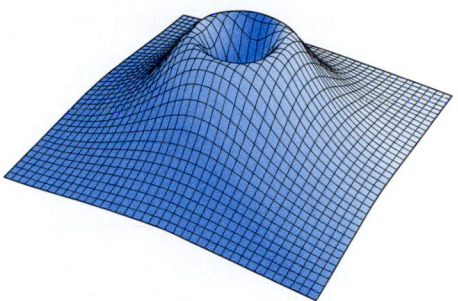

FIGURE 6.13   A depression in a surface surrounded by an elevated rim.

At an equilibrium point (x_0, y_0) of the system in (120), the vanishing of both f and g causes dy/dx in (121) to become *indeterminate* at that point, so initial conditions imposed at an equilibrium point cannot determine a unique solution. This has the effect that on passing through an equilibrium point, a point moving along one trajectory can move onto a different trajectory.

At an equilibrium point of an autonomous system, a physical system represented by the equations is in an equilibrium state. This state is said to be **stable** if, when the system is subjected to arbitrarily small disturbances, it always remains in the neighborhood of the same equilibrium state. If, however, the result of arbitrarily small disturbances is to make the system change to a different equilibrium state, to make the displacement grow unrestrictedly, or, depending on the displacement, to make the system sometimes return to the original equilibrium state and sometimes to cause the displacement increase unrestrictedly, the state is said to be **unstable**.

stability, instability, and asymptotic stability

A dynamical analogy illustrating stable and unstable situations is provided by considering Fig. 6.13, which represents a depression in a surface surrounded by an elevated rim, beyond which the level of the surface falls away steadily. A ball placed at the bottom of the depression is in a stable equilibrium state, because after any small displacement gravity will cause it to try to return to the equilibrium state. If, however, the displacement is large the motion will be unstable, because the ball will leave the depression and roll away indefinitely as time increases. Every point on the top of the rim represents an unstable equilibrium state because, depending on the direction of the displacement, the ball may move to another point on the rim, return to the depression, or roll away indefinitely. So this system has one stable equilibrium state at the bottom of the depression, and an infinite number of unstable states around the top of the rim.

Stability and asymptotic stability

The notion of stability can be made more precise by introducing the function $\Delta(t)$ that measures the distance in the phase plane of a point $(x(t), y(t))$ on a trajectory at time t from an equilibrium point at (x_0, y_0), where

$$\Delta(t) = \sqrt{(x(t) - x_0)^2 + (y(t) - y_0)^2}.$$

(i) The equilibrium point (x_0, y_0) is said to be **stable** if for every arbitrarily small number $\varepsilon > 0$, a number $\delta > 0$ can be found such that if $\sqrt{x_0^2 + y_0^2} < \delta$, then $\Delta(t) < \varepsilon$ for all time t.

(ii) The equilibrium point (x_0, y_0) is said to be **asymptotically stable** if it is stable in the sense of (i) and a number α can be found such that if $\sqrt{x_0^2 + y_0^2} < \alpha$, then $\Delta(t) \to 0$ as $t \to \infty$.

The implication of these definitions is that when an equilibrium point (x_0, y_0) is stable, a trajectory starting close to (x_0, y_0) will remain close to it, but if the point is asymptotically stable any trajectory starting close to (x_0, y_0) will eventually converge to the equilibrium point as $t \to \infty$. Asymptotically stable equilibrium points can be said to *attract* trajectories, so such points are called **attractors**, whereas equilibrium points from which the distance function $\Delta(t)$ increases without bound as t increases are said to **repel** trajectories.

In the dynamical example just given, in the absence of friction, the point at the bottom of the depression will be a *stable* state, because after a small displacement the ball will forever move around the lowest point. If, however, friction is present, the lowest point of the depression will be an *asymptotically stable* state, because after any small displacement the ball will eventually come to rest at the lowest point.

Interest in autonomous systems centers around the fact that trajectories in phase space provide qualitative information about the entire class of solutions of the system and, in particular, about properties of solutions when f and g are nonlinear and no analytical solution can be found.

predator–prey problem

A classical example of a nonlinear autonomous system is the **predator–prey** system of equations introduced and studied by Volterra and Lotka around 1930. They considered the ecological situation in which an isolated colony of foxes and rabbits coexist, with the foxes eating the rabbits and the rabbits feeding on a plentiful supply of vegetation. When the rabbits are numerous, the foxes are well fed and their numbers will grow, but when the number of foxes increases to the point where the rabbit population declines, the number of foxes will begin to fall, giving the rabbit population an opportunity to regenerate. This process, it was postulated, could explain the nonlinear cyclic variation in fox and rabbit populations that is observed in nature. This predator–prey model involving foxes and rabbits will become *nonautonomous* if some external factors are introduced that reduce the fox and rabbit populations by some other means.

To derive the predator–prey equations, let $x(t)$ be the number of rabbits present at time t. Then, as vegetation is plentiful, without foxes the rabbit population will grow at a rate proportional to the number of rabbits, so we can write

$$\frac{dx}{dt} = ax,$$

where $a > 0$ is a constant. Assuming that the rate at which foxes eat rabbits is proportional to the product of the number of rabbits $x(t)$ and the number of foxes $y(t)$ present at time t, the rabbit population described by the preceding equation must be modified to allow for this reduction, and so it becomes

$$\frac{dx}{dt} = ax - bxy,$$

where $b > 0$ is a constant.

The differential equation governing the fox population $y(t)$ is derived in a similar manner, but now the number of foxes *decreases* as the rabbit population decreases, leading to a differential equation of the form

$$\frac{dy}{dt} = -cy + dxy,$$

where $c > 0$ and $d > 0$ are constants. The classical **predator–prey** equations are the two nonlinear autonomous equations

$$\frac{dx}{dt} = x(a - by)$$

$$\frac{dy}{dt} = y(xd - c). \qquad (122)$$

This nonlinear autonomous system has no analytical solution, so either individual solutions must be found by numerical computation (see Section 19.7), or phase-plane methods must be used to determine the qualitative behavior of solutions of this system. An obvious feature of the predator–prey system of equations is that an equilibrium state exists when $dx/dt = dy/dt = 0$, and this occurs at the origin $(0, 0)$ and when $x = c/d$ and $y = a/b$. The first equilibrium state is of no interest because then neither rabbits nor foxes are present, but in the other equilibrium state the rabbit and fox populations will remain static, though deviations from this situation can be expected to initiate nonlinear oscillations in the population numbers.

The predator–prey model, although simple and developed initially for ecological reasons, can be modified and applied to other situations such as the spread of an infectious disease, the competition between industries for a raw material that is in limited supply, or when industries compete for the same market.

When the functions $f(x, y)$ and $g(x, y)$ in (120) are nonlinear, or are complicated in other ways, to help understand the behavior of the system the functions f and g are often **linearized** about an equilibrium point at (x_0, y_0) that is of interest. This involves expanding f and g about (x_0, y_0) as two-variable Taylor series expansions, and then replacing f and g in (120) by the linear terms in these expansions.

linearization

If, for example, (x_0, y_0) is an equilibrium point of the system of equations in (120), then $f(x_0, g_0) = 0$ and $g(x_0, y_0) = 0$, and expanding f and g about the point (x_0, y_0) gives

$$f(x, y) = f_x(x_0, y_0)(x - x_0) + f_y(x_0, y_0)(y - y_0) + \text{higher order terms}$$

and

$$g(x, y) = g_x(x_0, y_0)(x - x_0) + g_y(x_0, y_0)(y - y_0) + \text{higher order terms}.$$

Substituting only the first order terms from these expansions into system (120) simplifies it to the constant coefficient linear autonomous system

$$\frac{d(x - x_0)}{dt} = f_x(x_0, y_0)(x - x_0) + f_y(x_0, y_0)(y - y_0)$$

and

$$\frac{d(y - y_0)}{dt} = g_x(x_0, y_0)(x - x_0) + g_y(x_0, y_0)(y - y_0). \qquad (123)$$

Setting $X = x - x_0$ and $Y = y - x_0$, we can write these equations in the matrix form

$$\frac{d\mathbf{z}}{dt} = \mathbf{J}(x_0, y_0)\mathbf{z}, \tag{124}$$

where

$$\mathbf{z} = \begin{bmatrix} X \\ Y \end{bmatrix} \quad \text{and} \quad \mathbf{J}(x_0, y_0) = \begin{bmatrix} f_x(x_0, y_0) & f_y(x_0, y_0) \\ g_x(x_0, y_0) & g_y(x_0, y_0) \end{bmatrix}.$$

Jacobi matrix of the system

The matrix $\mathbf{J}(x_0, y_0)$ is called the **Jacobi matrix** of the system at the point (x_0, y_0), and we will see later how the eigenvalues of $\mathbf{J}(x_0, y_0)$ determine the nature of the equilibrium point at (x_0, y_0).

It is reasonable to suppose that when the neglected remainder terms in the Taylor series expansions of f and g are suitably small, the behavior of this linearized system of equations in some neighborhood of the equilibrium point at (x_0, y_0) will be qualitatively similar to that of the original nonlinear system.

As an illustration of the linearization process, let us now linearize the predator–prey equations in (122) about the equilibrium point at $x = c/d$ and $y = a/b$. Identifying $f(x, y)$ with $x(a - by)$ and $g(x, y)$ with $y(dx - c)$, substituting into the Jacobian $\mathbf{J}(x_0, y_0)$ with $x_0 = c/d$ and $y_0 = a/b$, and setting $X = x - c/d$ and $Y = y - a/b$ leads to the linearized predator–prey equations

$$\begin{aligned} \frac{dX}{dt} &= -\frac{bc}{d}Y \\ \frac{dY}{dt} &= \frac{ad}{b}X. \end{aligned} \tag{125}$$

These equations are easily integrated to give the following equation for the trajectories in the (X, Y) phase plane:

$$X^2 + (cb^2/ad^2)Y^2 = k^2, \quad \text{where } k \text{ is an integration constant.}$$

Reverting to the original variables shows that after linearization, each trajectory in the (x, y) phase plane that is close to the equilibrium point is a member of the family of ellipses

$$(x - c/d)^2 + (cb^2/ad^2)(y - a/b)^2 = k^2, \tag{126}$$

which have their common center at the point $(c/d, a/b)$ in the (x, y) phase plane. This shows that in a neighborhood of the equilibrium point the *phase portrait* of the predator–prey system can be expected to be approximated by this family of ellipses.

This result indicates that close to the equilibrium condition, the rabbit and fox populations can be expected to exhibit a cyclic variation with respect to time. This conclusion follows from the fact that as the time t increases, starting at an initial point on a trajectory where $x_0 = x(t_0)$, $y_0 = y(t_0)$ at a time $t = t_0$, the point $(x(t), y(t))$ will move around the ellipse that passes through this point until after a suitable interval of time it returns to its starting point. In this case linearization has produced elliptical trajectories centered on the equilibrium point, so in the nonlinear case the trajectories can be expected to be distorted ellipses.

Before considering nonlinear autonomous systems we will determine the nature of the equilibrium points associated with the general linear two variable autonomous system, which in standard notation can be written

$$\frac{dx}{dt} = ax + by$$
$$\frac{dy}{dt} = cx + yd,$$

(127)

where a, b, c, and d are constants, and the second term dy on the right of the second equation is *not* to be confused with the differential dy.

Setting $dx/dt = dy/dt = 0$ in (127) and solving for x and y shows the origin to be the *only* equilibrium point if

$$\begin{vmatrix} a & b \\ c & d \end{vmatrix} = ad - bc \neq 0.$$

(128)

When the (127) is integrated once, it yields what is called a **first integral** of the system. A first integral is *not* a solution of the system, because although it is an equation that connects $x(t)$ and $y(t)$, it does not express either function explicitly in terms of t. First integrals are useful because they are easier to obtain than solutions of general autonomous systems, and they provide qualitative information about the general behavior of the set of all solutions. This can be seen from the first integral of the linearized predator–prey system in (125) because, although this did not yield a solution in terms of t, it did confirm that the linearized system exhibits a periodic behavior of the two populations in a neighborhood of the equilibrium point.

A simple example of a linear autonomous system can be derived from any physical system, be it electrical, mechanical, or otherwise, that can be represented by the homogeneous constant-coefficient second order equation

$$\frac{d^2 y}{dt^2} + a\frac{dy}{dt} + by = 0.$$

(129)

Setting $dy/dt = x$, we can write the second order equation as the linear autonomous system

$$\frac{dx}{dt} = -ax - by$$
$$\frac{dy}{dt} = x,$$

(130)

with t as a parameter, or to the equivalent variables separable equation

$$\frac{dy}{dx} = -\frac{x}{ax + by},$$

(131)

where now only x and y are present.

As a special case, when $a = 0$ and $b = n^2$, result (131) becomes

$$\frac{dy}{dx} = -\frac{x}{n^2 y},$$

for which a first integral is seen to be

$$x^2 + n^2 y^2 = k^2.$$

This represents a family of elliptical trajectories all centered on the equilibrium point of the system that is located at the origin. The argument used earlier in connection with the linearized predator–prey equations shows that solutions of the system (131) when $a = 0$ and $b = n^2$ must be periodic. This is to be expected, because with these values of a and b equation (129) describes undamped simple harmonic oscillations. In this simple case, as $x = dy/dt$,

$$\frac{dy}{\sqrt{k^2 - n^2 y^2}} = dt,$$

and after integration this gives

$$y(t) = (k/n) \sin[n(t + t_0)],$$

which is the general solution of (129) for $a = 0$ and $b = x^2$.

When we considered the linearized predator–prey equations, the family of ellipses around the equilibrium point that were found represented an *approximation* to the phase portrait of the system in a *neighborhood* of the equilibrium point. In this case, however, system (130) is linear, so no linearization is involved and the family of elliptical trajectories forms the true phase portrait of system (130).

The linear autonomous system (127) can be written in the matrix form

$$d\mathbf{x}/dt = \mathbf{Jx}, \tag{132}$$

where

$$\mathbf{x} = \begin{bmatrix} x \\ y \end{bmatrix}, \quad \mathbf{J} = \begin{bmatrix} a & b \\ c & d \end{bmatrix}. \tag{133}$$

This system was studied in detail in Section 6.10, where it was seen that its solution depends on the eigenvalues of $\mathbf{J}$ determined by the characteristic equation

$$\begin{vmatrix} a - \lambda & b \\ c & d - \lambda \end{vmatrix} = \lambda^2 - (a + d)\lambda + (ad - bc) = 0. \tag{134}$$

Setting $\alpha = a + d$ and $\beta = ad - bc$, the characteristic equation in (134) becomes

$$\lambda^2 - \alpha\lambda + \beta = 0, \tag{135}$$

with the discriminant $\Delta = (a - d)^2 + 4bc$.

The pattern of the trajectories of the autonomous system in (132), equivalently (127), is determined completely by the eigenvalues λ_1 and λ_2 of $\mathbf{J}$ and their associated eigenvectors: that is to say, by the fundamental solutions of the system. If the eigenvalues are real and $\lambda_1 \neq \lambda_2$, a matrix $\mathbf{P}$ can always be found that simplifies the system by reducing $\mathbf{J}$ to a diagonal matrix $\mathbf{D}$ through the result $\mathbf{P}^{-1}\mathbf{JP} = \mathbf{D}$, with λ_1 and λ_2 the elements on the leading diagonal of $\mathbf{D}$ (see Section 4.2). The transformation $\mathbf{x} = \mathbf{Pu}$ with $\mathbf{u} = [u, v]^{\mathrm{T}}$ then reduces (132) to the simpler form $d\mathbf{u}/dt = \mathbf{Du}$, showing that $du/dt = \lambda_1 u$ and $dv/dt = \lambda_2 v$.

These equations have the general solution

$$u = Ae^{\lambda_1 t} \quad \text{and} \quad v = Be^{\lambda_2 t}, \tag{136}$$

so the form of the trajectories about the equilibrium point at the origin in the (u, v) phase plane is seen to depend on both the signs of the eigenvalues λ_1 and λ_2 and their magnitudes.

When the discriminant $\Delta > 0$, the eigenvalues λ_1 and λ_2 will be real, and then there are three cases to consider.

(i) Unstable nodes: λ_1 and λ_2 are positive

Examination of the solution in (136) shows that the trajectories must take one of the two forms illustrated in Figs. 6.14a and 6.14b. In this case the equilibrium point at

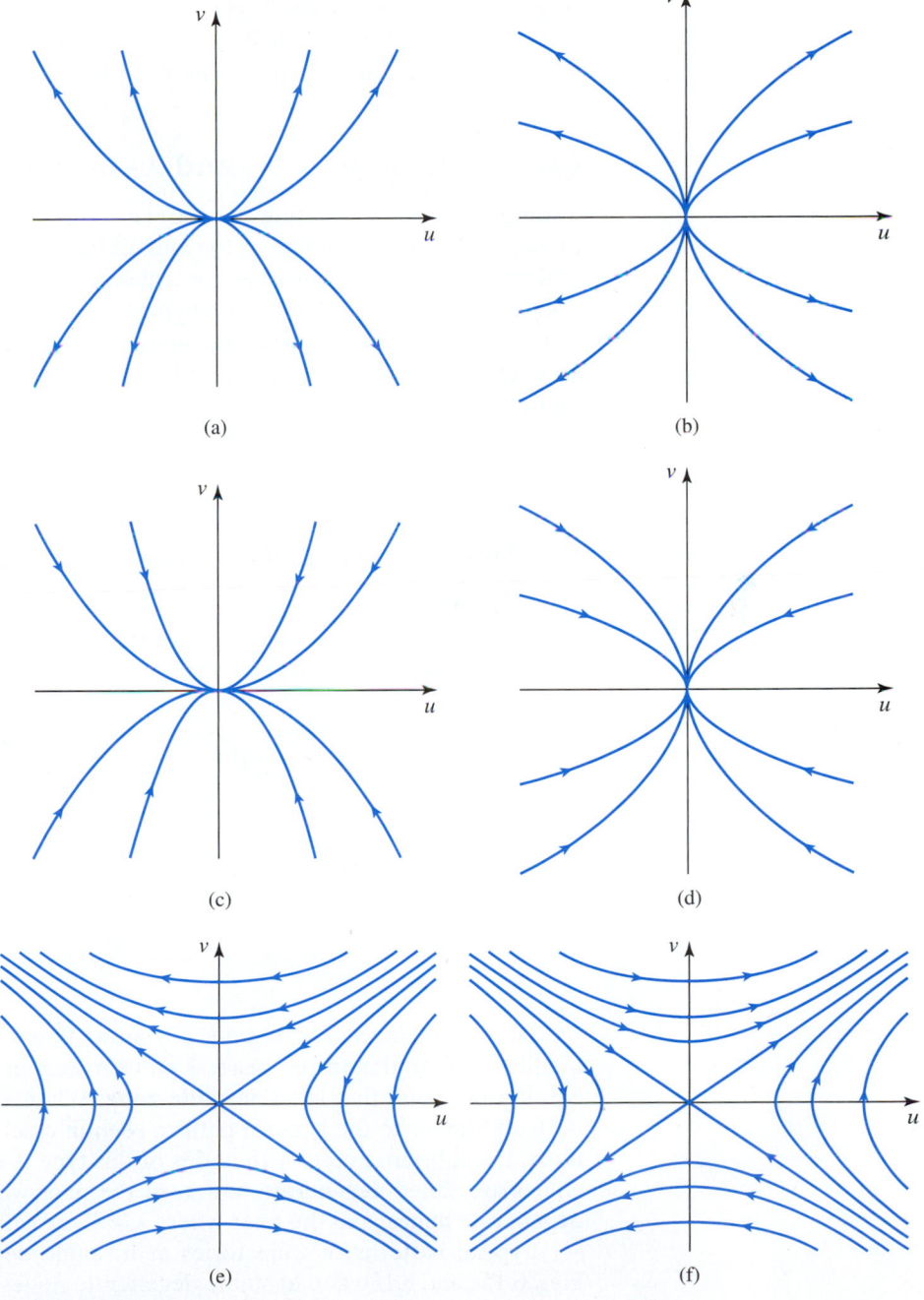

(a)

(b)

(c)

(d)

(e)

(f)

FIGURE 6.14   (a,b) Unstable nodes. (c,d) Stable nodes. (e,f) Saddle points.

the origin is called a **node**. As the eigenvalues are both positive, a point $(u(t), v(t))$ on a trajectory moves *away* from the origin as t increases, so this type of equilibrium point is called an **unstable node**.

(ii) Stable nodes: λ_1 and λ_2 are negative

Examination of the solution in (136) shows that the trajectories must take one of the two forms illustrated in Figs. 6.14c and 6.14d, where the equilibrium point at the origin is again called **node**. This time, as the eigenvalues are both negative, a point $(u(t), v(t))$ on a trajectory will move *toward* the origin as t increases, so in this case the equilibrium point is called a **stable node**.

(iii) Saddle points: λ_1 and λ_2 have opposite signs

Examination of the solution in (136) shows that the trajectories take one of the two forms illustrated in Figs. 6.14e and 6.14f, where the equilibrium point is called a **saddle point**. The eigenvalues are real and have opposite signs, so as t increases a point $(u(t), v(t))$ on a branch of a hyperbola will move toward the origin and then away again, showing that a **saddle point** represents an instability. The two diagonal straight lines that form degenerate hyperbolas are each called a **separatrix** in the phase portrait, because they separate the phase plane into four distinct regions, and a solution in any one of these regions cannot be related to a solution in a different region.

(iv) Degenerate node: Equal eigenvalues $\lambda = \lambda_1 = \lambda_2$

When the discriminant $\Delta = 0$ the eigenvalues coincide, so $\lambda = \lambda_1 = \lambda_2$. In this case the Jacobi matrix **J** cannot be diagonalized, but system (132) can always be reduced to the form $d\mathbf{u}/dt = \mathbf{Su}$, where

$$\mathbf{S} = \begin{bmatrix} \lambda & 0 \\ 1 & \lambda \end{bmatrix} \quad \text{and} \quad \mathbf{u} = \begin{bmatrix} u \\ v \end{bmatrix},$$

and this has the general solution

$$u = Ae^{\lambda t} \quad \text{and} \quad v = (At + B)e^{\lambda t}. \tag{137}$$

An examination of solution (137) shows that when $\lambda > 0$, the trajectories are qualitatively similar to the general pattern seen in case (i), corresponding to an equilibrium point that is an **unstable node**. When $\lambda < 0$ the trajectories are qualitatively similar to the general pattern seen in case (ii), corresponding to a **stable node**. Equilibrium points with nodes of this type that arise from coincident eigenvalues are called **degenerate nodes**, so the ones where $\lambda > 0$ are called **unstable degenerate nodes**, and the ones where $\lambda < 0$ are called **stable degenerate nodes**.

Typical patterns of trajectories at unstable degenerate nodes are shown in Figs. 6.15a and 6.15b and at stable degenerate nodes in Figs. 6.15c and 6.15d.

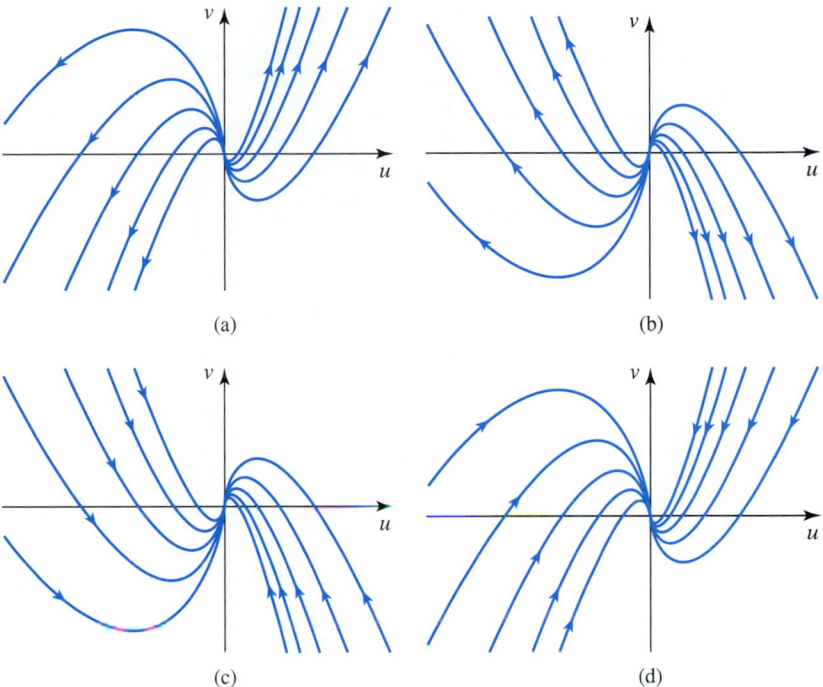

(a)   (b)

(c)   (d)

FIGURE 6.15   (a,b) Unstable degenerate nodes. (c,d) Stable degenerate nodes.

(v) Focus or spiral point: Complex conjugate eigenvalues

If the discriminant $\Delta < 0$, the eigenvalues will be the complex conjugates with $\lambda_1 = \xi + i\eta$ and $\lambda_2 = \xi - i\eta$. Diagonalization of $\mathbf{J}$ then produces a system of equations of the form $d\mathbf{u}/dt = \mathbf{C}\mathbf{u}$, where

$$\mathbf{C} = \begin{bmatrix} \xi & \eta \\ -\eta & \xi \end{bmatrix} \quad \text{and} \quad \mathbf{u} = \begin{bmatrix} u \\ v \end{bmatrix}.$$

This system is easily shown to have the general solution

$$u = e^{\xi t}(A \sin \eta t - B \cos \eta t) \quad \text{and} \quad v = e^{\xi t}(B \sin \eta t + A \cos \eta t),$$

(138)

which defines spiral trajectories about the equilibrium point. In this case the equilibrium point is called a **focus** or a **spiral point**. The direction in which a point $(u(t), v(t))$ along a spiral as t increases is determined by the sign of ξ. When $\xi > 0$ the point moves *away* from the origin as t increases, so the equilibrium point is then called either an **unstable focus** or an **unstable spiral point**. Conversely, when $\xi < 0$, the point moves toward the origin as t increases, so in this case the equilibrium point is called a **stable focus** or a **stable spiral point**. Figure. 6.16a shows an unstable focus and Figure. 6.16b a stable focus. Spirals may evolve in either a clockwise or a counterclockwise direction, and this can be determined by the direction of the vector with components $(dx/dt, dy/dt)$ at any point on the spiral (see Example 6.39).

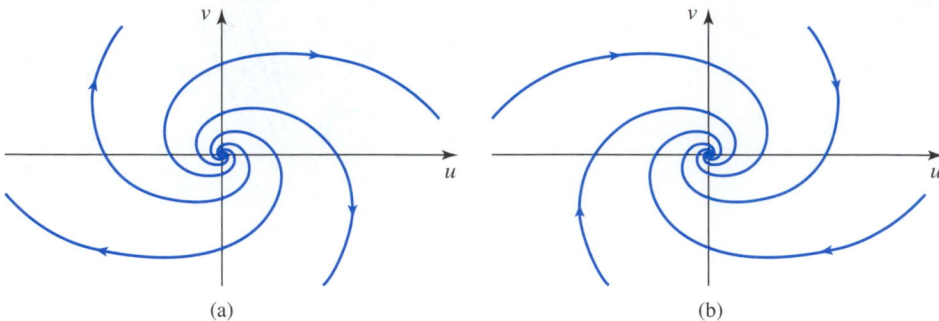

(a)   (b)

FIGURE 6.16   (a) An unstable focus. (b) A stable focus.

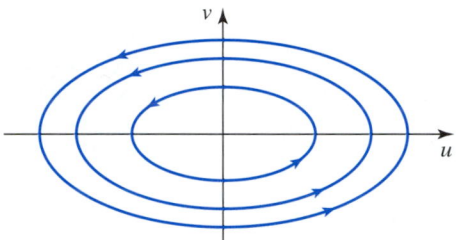

FIGURE 6.17   A center located at the origin.

(vi) Center: Purely imaginary complex conjugate eigenvalues

If in the characteristic equation (135) $\alpha = a - d = 0$ and the discriminant $\Delta < 0$, the eigenvalues will be purely imaginary complex conjugates. Setting $\xi = 0$ in (138) shows that the trajectories become a family of ellipses centered on the origin, as shown in Fig. 6.17. In this case the equilibrium point at the origin is called a **center**, and the corresponding solutions are considered to be *stable* because they remain bounded for all time. It follows from this that the equilibrium point in the linearized predator–prey system is a center.

EXAMPLE 6.37

Locate and identify the nature of the equilibrium point of the system

$$\frac{dx}{dt} = -x, \quad \frac{dy}{dt} = -x - 2y,$$

and draw some typical trajectories.

Solution   The equilibrium point is located at the origin, and its nature can be identified by examining the eigenvalues of the Jacobi matrix **J** that follows by setting $f(x, y) = -x$ and $g(x, y) = -x - 2y$. We have

$$\mathbf{J} = \begin{bmatrix} -1 & 0 \\ -1 & -2 \end{bmatrix},$$

and this has the eigenvalues $\lambda_1 = -1$ and $\lambda_2 = -2$. As the eigenvalues are real, and both are negative, it follows from Case (ii) that the equilibrium point at the origin is a *stable node*. To draw trajectories it is necessary to solve this system, and a routine

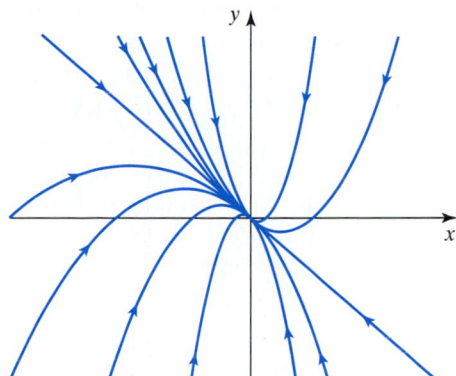

FIGURE 6.18 Trajectories in the neighborhood of the stable node at the origin.

calculation shows that

$$x = -C_1 e^{-t} \text{ and } y = C_1 e^{-t} + C_2 e^{-2t}.$$

Eliminating t, we find that the equation of the trajectories is

$$y = -x + (C_2/C_1^2)x^2.$$

This equation describes a family of parabolas that at the origin are all tangent to the degenerate parabola $y = -x$ that forms a separatrix marking a boundary between phase curves with different properties. Some typical trajectories are shown in Fig. 6.18, where the arrows indicate that the node is stable.

It is important to recognize that as the node is a singularity of the system where dy/dx is indeterminate, a point moving along a trajectory that passes through the node cannot leave it on a different trajectory. ■

EXAMPLE 6.38

Locate and identify the nature of the equilibrium point of the system

$$\frac{dx}{dt} = -x - y - 2, \quad \frac{dy}{dt} = -x + y - 4,$$

and draw some typical trajectories.

Solution   The equilibrium point occurs when $-x - y - 2 = 0$ and $-x + y - 4 = 0$, corresponding to $x = -3$, $y = 1$. For convenience we shift the equilibrium point to the origin in the (X, Y) phase plane by making the change of variables $X = x + 3$ and $Y = y - 1$, when the system becomes

$$\frac{dX}{dt} = -X - Y, \quad \frac{dY}{dt} = -X + Y.$$

The nature of the equilibrium point that is now located at the origin in the (X, Y) phase plane can be identified by examining the eigenvalues of the Jacobi matrix

$$\mathbf{J} = \begin{bmatrix} -1 & -1 \\ -1 & 1 \end{bmatrix},$$

which are easily seen to be $\lambda_1 = -\sqrt{2}$ and $\lambda_2 = \sqrt{2}$. As the eigenvalues are real, and opposite in sign, it follows from Case (iii) that the equilibrium point at the origin is a *saddle point*. To draw trajectories it is necessary to solve this system of equations.

After some calculations, the equation of the family of trajectories determined by $dY/dX = (X - Y)/(X + Y)$ is found to be given by

$$Y^2 + 2XY - X^2 = c,$$

where the constant c is determined by the point in the phase plane through which a trajectory is required to pass.

The general equation of a conic is

$$AX^2 + 2BXY + CY^2 + DX + EY + F = 0,$$

and this represents an ellipse if $B^2 - AC < 0$, a parabola if $B^2 - AC = 0$, and a hyperbola if $B^2 - AC > 0$. So comparing the equation of the trajectories with the general form of a conic, we see that $B^2 - AC > 0$, so it describes a family of hyperbolas.

This family of hyperbolas with parameter c is centered on the origin, and solving for Y gives

$$Y = -X + \sqrt{2X^2 + c} \quad \text{and} \quad Y = -X - \sqrt{2X^2 + c},$$

where for any given value of c, each equation represents one pair of hyperbolas.

Some typical hyperbolas are shown in Fig. 6.19, where the upper and lower branches correspond to different values of c in the first equation, and the left and right branches correspond to other values of c in the second equation. The asymptotes, which represent degenerate hyperbolas, are seen by inspection of these equations to be given by $Y = (\sqrt{2} - 1)X$ and $Y = -(\sqrt{2} + 1)X$. Each of these is a *separatrix* in the phase portrait of the system, and a solution in any one of the four regions into which these lines divides the phase plane cannot connect with a solution in any other region.

The simplest way to determine the direction along the upper and lower hyperbolic trajectories as t increases is to find the direction of the vector $(dX/dt, dY/dt)$ on a trajectory. For example, when $X = 0$, we see from the differential equations that the direction of the vector along a trajectory that crosses the Y-axis has the components $(-Y, Y)$. This shows that when $Y > 0$ the vector is directed upward and toward the left, whereas when $Y < 0$ it is directed downward and toward the

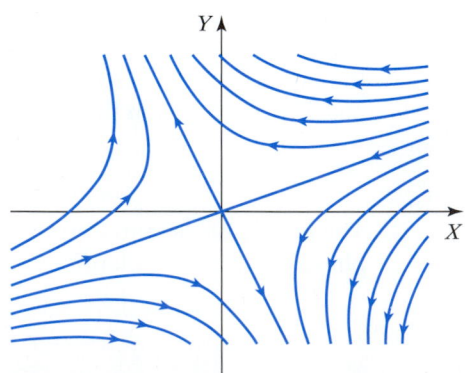

FIGURE 6.19 Trajectories around the saddle point at the origin in the (X, Y) phase plane.

right. The direction of the arrows on the left and right hyperbolic trajectories are determined in similar fashion by finding the direction of the vector $(dX/dt, dY/dt)$ that crosses the X-axis where $Y = 0$.

The pattern of the trajectories around the saddle point in the original coordinate system is obtained by translating the picture in Fig. 6.19 to the point $(-3, 1)$. ∎

EXAMPLE 6.39   Locate and identify the equilibrium point of the system

$$\frac{dx}{dt} = -x + 2y + 1, \quad \frac{dy}{dt} = -2x - y + 2,$$

and sketch some trajectories.

Solution   The equilibrium point occurs when $-x + 2y + 1 = 0$ and $-2x - y + 2 = 0$, corresponding to $x = 1$ and $y = 0$. For convenience we shift the equilibrium point to the origin in the (X, Y) phase-plane by making the change of variables $X = x - 1$ and $Y = y$, when the system becomes

$$\frac{dX}{dt} = -X + 2Y, \quad \frac{dY}{dt} = -2X - Y.$$

The nature of the equilibrium point that is now located at the origin in the (X, Y) phase plane can be identified by examining the eigenvalues of the Jacobi matrix

$$\mathbf{J} = \begin{bmatrix} -1 & 2 \\ -2 & -1 \end{bmatrix},$$

which follows from setting $f(X, Y) = -X + 2Y$ and $g(X, Y) = -2X - Y$.

The eigenvalues are $\lambda_1 = -1 + 2i$ and $\lambda_2 = -1 - 2i$, so as these are complex conjugates with negative real parts, it follows from Case (v) that the equilibrium point at the origin in the (X, Y) phase plane is a *stable focus*. This means that the trajectories spiral into the origin as t increases, so the only question that remains is whether the spiral is clockwise or counterclockwise.

Figure 6.20 shows two possible spirals, where in Fig. 6.20a the direction around the spiral is conterclockwise, while in Fig. 6.20b it is clockwise. Arguing as in Example 6.38, and considering the vector with components $(dX/dt, dY/dt)$ where the spiral crosses the X-axis, by setting $Y = 0$ we find that the vector has components $(-X, -2X)$. As this vector is directed downward and for $x > 0$ to the left, it

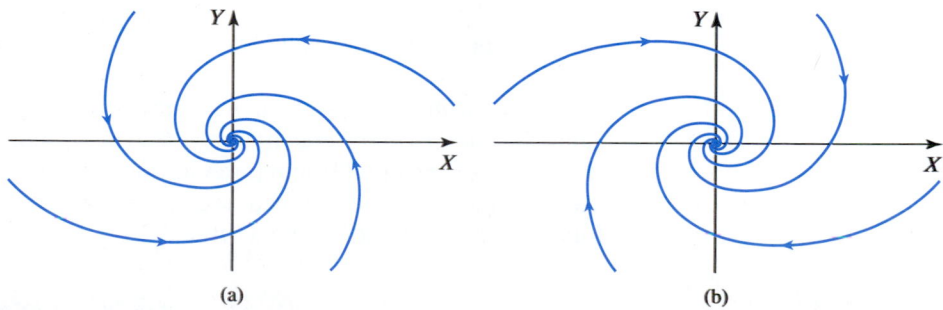

(a)   (b)

FIGURE 6.20   Two stable foci in the (X, Y) phase plane. (a) Counterclockwise spiral. (b) Clockwise spiral.

follows that the trajectories must spiral clockwise into the origin, so Fig. 6.20b is the only possible phase portrait for this system.

This information is sufficient to enable trajectories to be sketched, but as the general solution of the system is easily found to be

$$X(t) = e^{-t}(c_1 \sin 2t - c_2 \cos 2t), \quad Y(t) = e^{-t}(c_1 \cos 2t + c_2 \sin 2t),$$

it is not difficult to construct accurate spiral trajectories.

The pattern of trajectories for the original autonomous system is obtained by translating the pattern in Fig. 6.20b to the point $(1, 0)$ in the (x, y) phase plane. ∎

If it is only necessary to identify the nature of the equilibrium point at the origin belonging to the linear autonomous system,

$$\frac{dx}{dt} = ax + by$$

$$\frac{dy}{dt} = cx + dy,$$

identification of critical points

results (i) to (vi) can be summarized as follows:

(a) A **node** if $(a + d)^2 \geq 4(ad - bc) > 0$; stable if $a + d < 0$ and unstable if $a + d > 0$.

(b) A **saddle point** if $ad - bc < 0$.

(c) A **focus** if $(a + d)^2 < 4(ad - bc)$; stable if $a + d < 0$ and unstable if $a + d > 0$.

(d) A **center** if $a + d = 0$ and $ad - bc > 0$.

(vii) Nonlinear autonomous systems

nonlinear autonomous systems

If the nonlinear autonomous system

$$\begin{cases} \dfrac{dx}{dt} = f(x, y) \\[2mm] \dfrac{dy}{dt} = g(x, y) \end{cases} \tag{139}$$

has an equilibrium point at (x_0, y_0), the transformation $X = x - x_0$, $Y = y - y_0$ will shift it to the origin in the (X, Y) phase plane. Accordingly, when considering an equilibrium point of system (139), we will always assume that such a translation has been made.

It is plausible to expect that when the nonlinear system in (139) has an equilibrium point at the origin, and in some sense the system is *close* to a linear system, then the nature of the equilibrium point at the origin will be the same in both systems. To make more precise the meaning of the term *close*, we restrict consideration to functions f and g that can be written

$$\begin{aligned} f(x, y) &= ax + by + F(x, y) \\ g(x, y) &= cx + dy + G(x, y), \end{aligned} \tag{140}$$

where $ad - bc \neq 0$ and the nonlinear terms F and G are such that

$$\lim_{x \to 0, y \to 0} \frac{F(x, y)}{\sqrt{x^2 + y^2}} = 0 \quad \text{and} \quad \lim_{x \to 0, y \to 0} \frac{G(x, y)}{\sqrt{x^2 + y^2}} = 0. \tag{141}$$

This conjecture concerning the relationship between the equilibrium points of a nonlinear and a related linear autonomous system can be shown to be correct, subject only to a single qualification. Specifically, if the linearized system

$$\frac{dx}{dt} = ax + by$$

$$\frac{dy}{dt} = cx + dy \tag{142}$$

has a node, a saddle point, or a focus at the origin, then so also has the nonlinear system in (140). The qualification that must be added is that if the equilibrium point at the origin of the linearized system in (142) is a center, then the corresponding nonlinear system in (140) has an equilibrium point at the origin that is *either* a center *or* a focus.

The reason why a center of the linear system (142) may be either a center of a focus of the nonlinear system (140) is not difficult to understand. Conditions (c) and (d) at the end of section (vi) show that the criteria identifying a focus and a center in the linear case are closely related, and it is due to the insensitivity of the linearization process that it fails to distinguish between them when a nonlinear autonomous system is considered. No proof of these statements will be offered here, as this involves methods that do not belong to this first account of autonomous systems. However, a detailed proof of the nature of the relationship between the types of equilibrium points in nonlinear and linearized systems, together with other important results due to Liapunov, Poincaré, and others, can be found in the references at the end of the book.

Nonlinear autonomous systems possess an important property that is not shared by linear systems. This is that in the phase plane a curve Γ may exist, *not* enclosing an equilibrium point, with the property that a trajectory starting from a point either inside or outside Γ is attracted to Γ and spirals into it as t increases. A curve Γ of this type, to which trajectories are attracted, is called a **limit cycle** for the system. Clearly, although a limit cycle represents a stable oscillatory solution, it is *not* one that is asymptotically stable. This statement is essentially the substance of the **Poincaré–Bendixson** theorem, the details of which can be found in the references at the end of the book.

HENRI POINCARÉ (1854–1912)
An outstanding French mathematician who studied in the Ecole Polytechnique in France before proceeding to study in the Ecole Nationale Superieure des Mines in Paris and receiving his doctorate from the University of Paris in 1879. He was appointed to the chair of physical and experimental mechanics at the Sorbonne and later to the chairs of mathematical physics and then the chair of mathematical astronomy. He made fundamental contributions to almost all of mathematics and was probably the last of the mathematical geniuses about whom it could truly be said that he knew all that was then known about mathematics.

It was proved separately by Bendixson that if in system (139) the functions f and g have continuous partial derivatives for all x and y, and $f_x + f_y$ is either positive or negative in some region Ω of the phase plane, then the system has no limit cycle in Ω. Although the proof of this result is not difficult, it will not be given here. The result is useful for establishing the nonexistence of limit cycles in given regions of the phase plane.

A theorem that gives sufficient, though not necessary, conditions for the existence of a limit cycle for a special type of autonomous system is Liénard's theorem. The theorem is now stated without proof.

THEOREM 6.9

conditions identifying a limit cycle: Liénard's theorem

Liénard's theorem   Write the linear equation

$$\frac{d^2x}{dt^2} + f(x)\frac{dx}{dt} + g(x) = 0$$

as the first order Liénard system

$$\frac{dx}{dt} = y$$
$$\frac{dy}{dt} = -g(x) - f(x)y.$$

Let $f(x)$ and $g(x)$ satisfy the following conditions:

(i) $f(x)$ and $g(x)$ are continuous functions with continuous first derivatives for all x.

(ii) $g(x)$ is an odd function that is positive for $x > 0$ and $f(x)$ is an even function.

(iii) the function $F(x) = \int_0^x f(\xi)d\xi$, which is an odd function, has precisely one positive root at $x = \alpha$, with $F(x) < 0$ for $0 < x < \alpha$, $F(x) > 0$ and nondecreasing for $x > \alpha$, and $F(x) \to \infty$ as $x \to \infty$.

Then the Liénard system possesses a unique closed curve Γ enclosing the origin in the phase plane, with the property that every trajectory spirals toward Γ as $t \to \infty$. ∎

van der Pol equation and phase portraits

An application of this theorem will be made later to the **van der Pol equation**

$$\frac{d^2x}{dt^2} + \varepsilon(x^2 - 1)\frac{dx}{dt} + x = 0, \tag{143}$$

which provides a classical example of a limit cycle. The equation itself was derived in the 1920s by Balthazar van der Pol when studying self-sustained oscillations in vacuum tubes, and it was his work that prompted Liénard to study corresponding problems in nonlinear mechanics.

The task of finding the complete phase portrait of a nonlinear autonomous system, usually called the **global phase portrait**, can be difficult. This is because nonlinear systems may have more than one equilibrium point, and while linearization techniques provide information in a neighborhood of each of these points (with the exception of centers), they provide very little information about the general phase portrait or any separatrix that may occur, and no information at all about the existence of a limit cycle, though Liénard's theorem helps in the linear case.

The Predator–Prey Problem

The predator–prey equations have been shown to have a single physically meaning-ful equilibrium point at $(c/d, a/b)$ in the phase plane, where the linearized form of the equations has a center with elliptical trajectories surrounding it. In view of the fact that when the linearized form of a nonlinear system identifies an equilibrium point as a center, the associated nonlinear system may have either a center or a focus, a more careful examination is necessary in the predator–prey case before it is possible to state with certainty that $(c/d, a/b)$ is a center and that cyclic variations in the populations take place. In more advanced accounts of nonlinear autonomous systems, theorems exist that can resolve this ambiguity, but here we will make use of a simple device that in this and other straightforward cases will suffice to distinguish between the two possibilities.

more on the predator–prey problem

The idea is simple, and it involves asking how many times a trajectory will intersect a straight line drawn through the equilibrium point at $(c/d, a/b)$. If the equilibrium point is a center, a trajectory can only intersect this line twice, but if it is a focus (a spiral point) it will intersect it infinitely many times.

Dividing the second of the predator–prey equations in (122) by the first equation, rearranging terms, and integrating gives

$$\int \frac{(a - by)}{y} dy = \int \frac{(xd - c)}{x} dx,$$

and so

$$a \ln y + c \ln x - by - xd = k,$$

where k is an integration constant. To proceed further, we consider a typical case where $a = 1$, $b = 1$, $c = 2$, and $d = 1$, when the predator–prey system will have an equilibrium point at $(2, 1)$ in the phase plane, and the equation determining the trajectories becomes

$$\ln y + 2 \ln x - y - x = k.$$

Let us now select a convenient trajectory through any point in the first quadrant that does not coincide with the equilibrium point. It is convenient to choose the point $(1, 1)$, when it follows from the above equation that $k = -2$, so the equation of the trajectory through this point becomes

$$\ln y + 2 \ln x - x = -3.$$

We may choose any test line through the equilibrium point, but it is simplest to choose the line $y = 1$ that passes through the equilibrium point of the system at $(2, 1)$ in the phase plane. Setting $y = 1$ in the preceding equation reduces it to

$$2 \ln x - x = -3,$$

so if this equation has only two real roots the equilibrium point will be a center, but if it has infinitely many it will be a focus. Graphing $y = 2 \ln x - x$ and $y = -3$ to determine where they intersect, we find that only two intersections occur, with one at $x \approx 0.25$ and the other at $x \approx 6.85$. This shows that in this model of the predator–prey system the equilibrium point at $(2, 1)$ must be a center.

A similar argument applies to any other choice of nonnegative coefficients a, b, c, and d. This demonstrates that the equilibrium point of the predator–prey system located at $(2, 1)$ in the first quadrant of the phase plane is, indeed, a center.

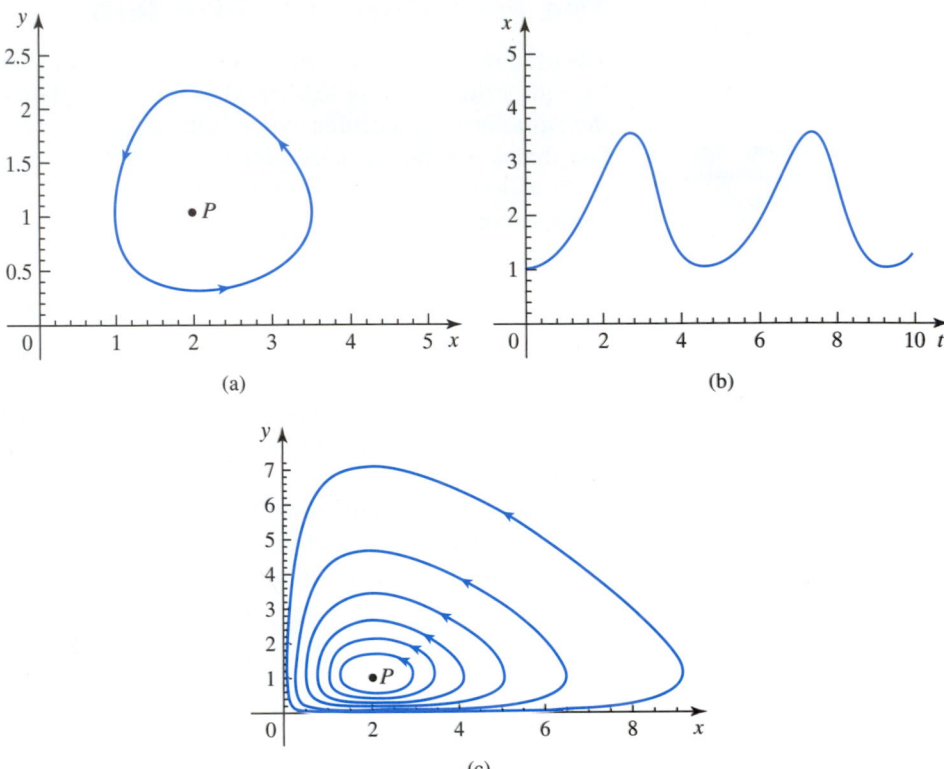

FIGURE 6.21   (a) The phase plane for the system through the point $(1, 1)$ with an equilibrium point at $(2, 1)$. (b) The variation of $x(t)$ showing the cycle time to be approximately 4.7 time units. (c) A general family of trajectories, each with the same equilibrium point.

Negative rabbit and fox populations have no physical significance, so no attention need be paid to the saddle point located at the origin of the phase plane, but notice that each axis is a separatrix belonging to the saddle point. Accordingly, the computer-generated phase portrait in the first quadrant is shown in Fig. 6.21a, with $a = 1, b = 1, c = 2$, and $d = 1$, the rabbit population along the horizontal axis, and the fox population along the vertical axis. The equilibrium point is shown as P. To find the period of this cycle of events, it is sufficient to find the period of either $x(t)$ or $y(t)$. The variation of $x(t)$ is shown in Fig. 6.21b with t along the horizontal axis and x along the vertical axis, from which the period is seen to be approximately $T \approx 4.7$ time units. Figure 6.21c shows a general family of trajectories for this system, each with a different period.

The Undamped and Damped Simple Pendulum

study of the undamped and damped pendulum

The geometry of the simple pendulum is illustrated in Fig. 6.22, where a mass m is attached to the end of a light rigid rod of length l that is pivoted at the end opposite to the mass and allowed to oscillate under gravity. The equation of motion, when damping proportional to $d\theta/dt$ is present, can be written

$$ml^2 \frac{d^2\theta}{dt^2} + 2mlk\frac{d\theta}{dt} + mgl\sin\theta = 0,$$

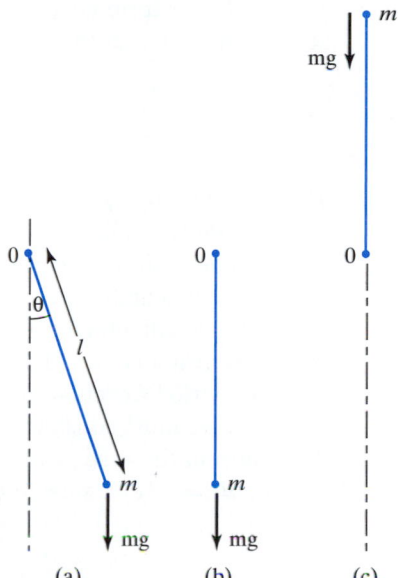

FIGURE 6.22   (a) Small oscillations.
(b) Stable equilibrium. (c) Inverted
pendulum—unstable equilibrium.

where $k > 0$ is a constant. Here, to simplify the associated characteristic equation, the constant of proportionality for wind resistance has been set equal to $2ml\,k$. This is equivalent to setting $\mu = 2mk/l$ in the equation of motion for a damped pendulum derived at the start of Section 6.1.

The undamped pendulum

Let us start by considering the undamped case $k = 0$. Introducing the new variable $x = d\theta/dt$, we see the nonlinear autonomous system determining the motion to be

$$\frac{dx}{dt} = -\left(\frac{g}{l}\right)\sin\theta \quad \text{and} \quad \frac{d\theta}{dt} = x,$$

with equilibrium points on the θ-axis where $\sin\theta = 0$. This shows there are infinitely many equilibrium points along the θ-axis at $\theta = \pm n\pi$, for $n = 0, 1, \ldots$. Accordingly, because of the periodicity of $\sin\theta$, only the interval $-\pi \leq \theta \leq \pi$ need be considered.

If we write $\sin\theta = \theta + (\sin\theta - \theta)$, the system becomes

$$\frac{dx}{dt} = -\left(\frac{g}{l}\right)\theta - \left(\frac{g}{l}\right)(\sin\theta - \theta)$$

$$\frac{d\theta}{dt} = x.$$

The nonlinear term $(g/l)\,(\sin\theta - \theta)$ satisfies the condition in (141), so when the equilibrium point at the origin is considered, the Jacobi matrix becomes

$$\mathbf{J} = \begin{bmatrix} 0 & -g/l \\ 1 & 0 \end{bmatrix}.$$

This has the purely imaginary eigenvalues $\lambda_1 = -i\sqrt{(g/l)}$ and $\lambda_2 = i\sqrt{(g/l)}$, so the equilibrium point of the linearized system located at the origin is a center. An argument similar to the one used with the predator–prey equations can be used to show that any trajectory starting at a point on the line $\theta = 0$ in the interval $-\pi < \theta < \pi$ will intersect the x-axis twice, so the equilibrium point of the nonlinear system is also a center. This confirms the expected result that the pendulum will perform periodic oscillations.

Next we must consider the equilibrium point at $(\pi, 0)$, and to do this we shift the origin of the system to this point by setting $u = \theta - \pi$. This causes the equation $dx/dt = -(g/l)\sin\theta$ to become $dx/dt = (g/l)\sin u$, so the system can now be written

$$\frac{du}{dt} = x$$

$$\frac{dx}{dt} = \left(\frac{g}{l}\right)u + \left(\frac{g}{l}\right)(\sin u - u).$$

The nonlinear term again satisfies the conditions in (141), so the nature of this equilibrium point is determined by the eigenvalues of the Jacobi matrix $\mathbf{J}$, which now becomes

$$\mathbf{J} = \begin{bmatrix} 0 & g/l \\ 1 & 0 \end{bmatrix}.$$

This has the real eigenvalues $\lambda_1 = -\sqrt{(g/l)}$ and $\lambda_2 = \sqrt{(g/l)}$, so as these are of opposite sign the equilibrium point at $(\pi, 0)$ is seen to be a saddle point. An analogous argument shows that the equilibrium point at $(-\pi, 0)$ is also a saddle point, so the nonlinear system also has saddle points at $(\pm\pi, 0)$.

A repetition of these arguments shows the equilibrium points at $(\pm 2n\pi, 0)$ all to be centers, and the equilibrium points at $((2n + 1)\pi, 0)$ all to be saddle points. A computer plot of some typical trajectories is shown in Fig. 6.23a.

An examination of Fig. 6.23a explains the significance of these centers and saddle points. As the angular displacement of the pendulum is indeterminate up to a multiple of 2π, each center represents the stable nonlinear oscillations that occur in Fig. 6.22a when the pendulum never becomes inverted. Similarly, each saddle point represents the unstable position of the inverted pendulum shown in Fig. 6.22c. As the oscillations are nonlinear, each different closed curve about a center represents a nonlinear oscillation with a different period. Each dashed curve is a separatrix forming a boundary between phase curves with different properties.

An important and useful result is obtained by writing

$$\frac{d^2\theta}{dt^2} = \frac{dx}{dt} = \frac{d\theta}{dt}\frac{dx}{d\theta} = x\frac{dx}{d\theta}.$$

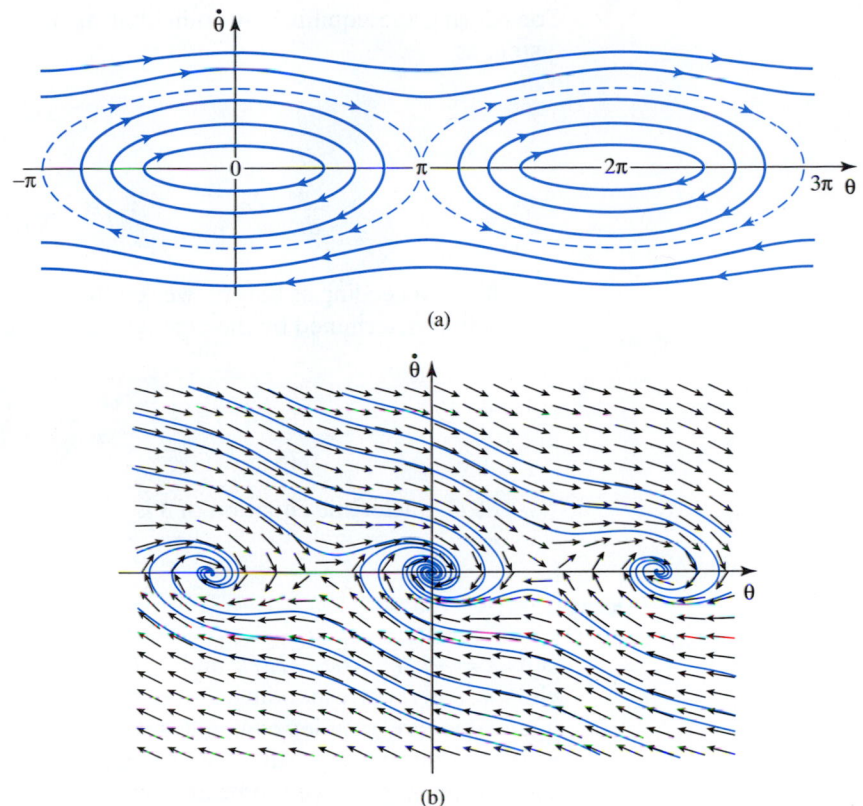

(a)

(b)

FIGURE 6.23   (a) The phase portrait for the undamped pendulum.
(b) The phase portrait for the damped pendulum.

Using this result the equation of motion becomes

$$ml^2 x \frac{dx}{d\theta} + mgl\sin\theta = 0,$$

so after integration we have

$$\frac{1}{2}ml^2 \left(\frac{d\theta}{dt}\right)^2 - mgl\cos\theta = C,$$

where C is an integration constant.

This first integral of the equation of motion expresses the *conservation of energy* in the system, which is possible because when $k = 0$ there is no dissipation of energy due to friction.

The damped pendulum

When damping occurs ($k > 0$), the nonlinear autonomous system governing the oscillations of the pendulum becomes

$$\frac{d\theta}{dt} = x \quad \text{and} \quad \frac{dx}{dt} = \frac{-2kx}{l} - \left(\frac{g}{l}\right)\sin\theta.$$

Considering the equilibrium point that again occurs at the origin, we write the system as

$$\frac{d\theta}{dt} = x$$

$$\frac{dx}{dt} = \frac{-2kx}{l} - \left(\frac{g}{l}\right)\theta - \left(\frac{g}{l}\right)(\sin\theta - \theta).$$

Then, proceeding as before, we see that the nature of the equilibrium point at the origin is determined by the eigenvalues of the Jacobi matrix

$$\mathbf{J} = \begin{bmatrix} -2k/l & -g/l \\ 1 & 0 \end{bmatrix}.$$

The characteristic equation of $\mathbf{J}$ is

$$\lambda^2 + (2k/l)\lambda + g/l = 0,$$

so as $\lambda = -k/l \pm \sqrt{k^2 - lg}/l$, and as $k > 0$, the eigenvalues are real and negative when $k > g/l$, corresponding to overdamped oscillations. When $(k/l)^2 < g/l$ the eigenvalues are complex conjugates with negative real parts, corresponding to the asymptotically stable oscillatory case. So, when friction is present, the equilibrium point at the origin is seen to be an asymptotically stable focus. In time, friction will cause the oscillations to decay to zero, causing the pendulum to come to rest in the positions shown in Fig. 6.23b.

EXAMPLE 6.40

Locate and classify the equilibrium points of the nonlinear autonomous system

$$\frac{dx}{dt} = 4 - x^2 - 4y^2 \quad \text{and} \quad \frac{dy}{dt} = xy.$$

Solution The equilibrium points occur when $4 - x^2 - 4y^2 = 0$ and $xy = 0$, so the points are located at $(0, -1)$, $(0, 1)$, $(2, 0)$, and $(-2, 0)$. Let us consider the equilibrium point at $(0, 1)$ and shift the origin to this point by setting $Y = y - 1$ and $X = x$. The system now becomes

$$\frac{dX}{dt} = -8Y - X^2 - 4Y^2 \quad \text{and} \quad \frac{dY}{dt} = X + XY.$$

Setting $X = r\cos\theta$, $Y = r\sin\theta$, we easily see that conditions (141) are satisfied, so the nature of the equilibrium point at $(0, 1)$ will be determined by the eigenvalues of the Jacobi matrix

$$\mathbf{J} = \begin{bmatrix} 0 & -8 \\ 1 & 0 \end{bmatrix}.$$

These satisfy the characteristic equation $\lambda^2 + 8 = 0$, so as they are purely imaginary, the equilibrium point of the linearized system that is located at $(0, 1)$ must be a center, and arguments similar to those used with the pendulum problem confirm that the nonlinear system also has a center at $(0, 1)$.

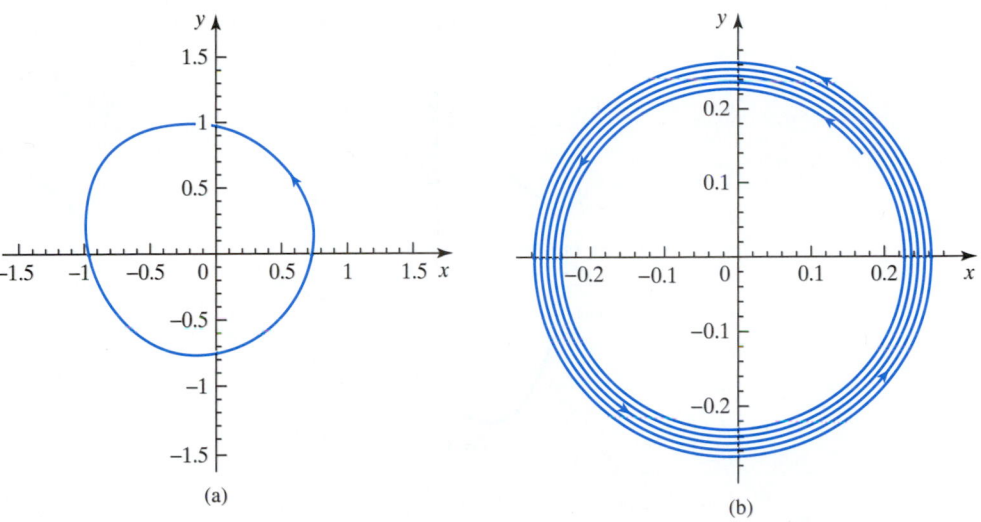

FIGURE 6.24 (a) The origin is a center. (b) The origin is an unstable focus.

It is left as an exercise to use similar arguments to show that the equilibrium point at $(0, -1)$ is also a center and the equilibrium points at $(-2, 0)$ and $(2, 0)$ are saddle points. ∎

The inability of a linearized system to reflect the difference between a center and a focus in the nonlinear system from which it is derived is best illustrated by means of computer-generated phase portraits. The following two systems only differ in the power of x associated with dx/dt, and each has the same linearized form that indicates the existence of a center at the origin of the phase plane:

$$(i) \qquad \frac{dx}{dt} = -4y + x^2 \quad \text{and} \quad \frac{dy}{dt} = 4x + y^2$$

and

$$(ii) \qquad \frac{dx}{dt} = -4y + x^3 \quad \text{and} \quad \frac{dy}{dt} = 4x + y^2.$$

However, the nonlinear phase portrait of system (i) in Fig. 6.24a shows that the system does, indeed, have a center located at the origin in the phase plane, but the nonlinear phase portrait of system (ii) in Fig. 6.24b shows that the system has an unstable focus at the origin.

A typical example of a limit cycle is provided by the van der Pol equation

$$\frac{d^2x}{dt^2} + \varepsilon(x^2 - 1)\frac{dx}{dt} + x = 0.$$

If we set $f(x) = \varepsilon(x^2 - 1)$ and $g(x) = x$ in Liénard's theorem, it is easily seen that the conditions of the theorem are satisfied provided $F(x) = \int_0^x \varepsilon(\xi^2 - 1)d\xi$ has precisely one positive root $x = \alpha$ with $F(x) < 0$ for $0 < x < \alpha$, and $F(x)$ is such that it is positive and nondecreasing for $x > \alpha$ with $F(x) \to \infty$ as $x \to \infty$. This

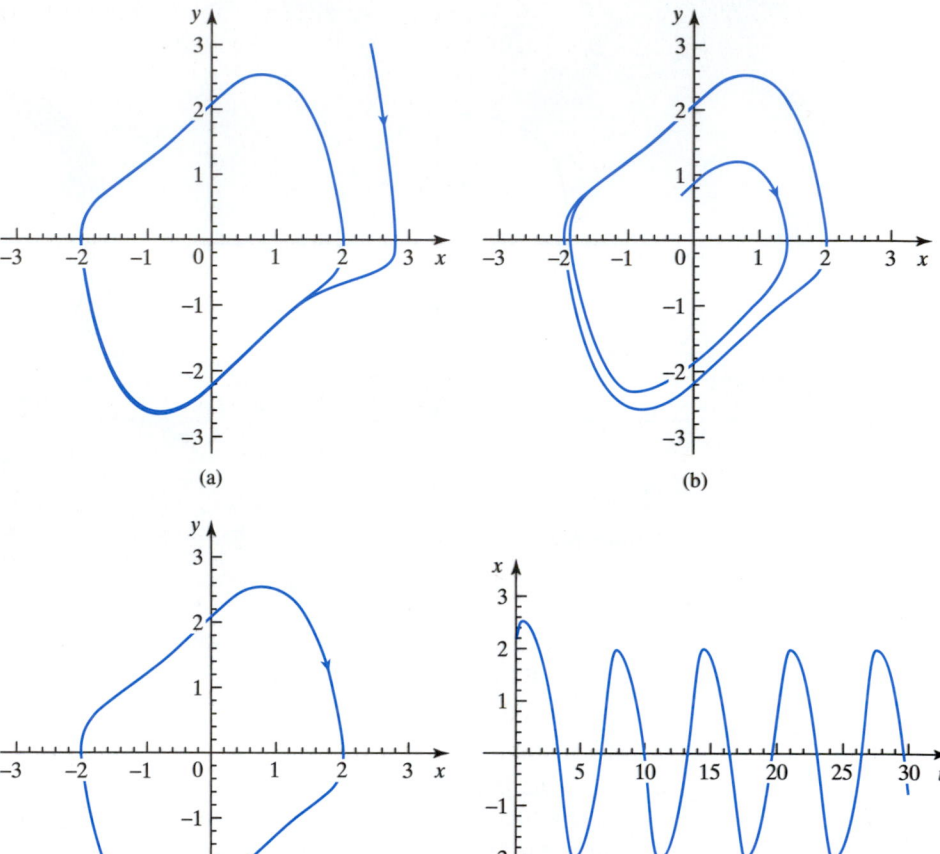

FIGURE 6.25 Phase portraits for the van der Pol equation with $\varepsilon = 0.9$ and the variation of $x(t)$ with t. (a) A trajectory starting outside the limit circle. (b) A trajectory starting inside the limit cycle. (c) The limit cycle. (d) The periodicity of $x(t)$ as a function of t.

is seen to be the case, because $F(x) = \frac{1}{3}\varepsilon(x^3 - 3x)$, so the theorem ensures the existence of a limit cycle for the van der Pol equation provided $\varepsilon > 0$.

Figure 6.25a shows a computer-generated phase portrait for the van der Pol equation with $\epsilon = 0.9$, where the trajectory starting from an initial point at $t = 0$ outside the limit cycle (the parallelogram-shaped closed curve) is attracted *inward* toward the limit cycle. Figure 6.25b shows the corresponding situation when the initial point lies inside the limit cycle, where here the trajectory is attracted *outward* toward the limit cycle. Figure 6.25c shows the limit cycle itself. A plot of $x(t)$ against t is shown in Fig. 6.25d, from which the solution is seen to become periodic, with a period of approximately 6.5 time units, after the time $t = 5$.

More examples of the phase plane are to be found in references [3.3] to [3.5], whereas a more extensive and advanced account is to be found in references [3.1], [3.2], and [3.13].

Summary

An autonomous system involving the variables $x(t)$ and $y(t)$, where the parameter t is usually the time, are systems of the form $dx/dt = f(x, y)$ and $dy/dt = g(x, y)$, where the dependence of the f and g on t is implicit. Critical points of such systems were defined and the concept of a trajectory, or path, was introduced leading to the notion of a phase portrait. Stability, instability, and asymptotic stability were defined, and the classical predator–prey problem was used to illustrate ideas. Linearization of the functions f and g led to the identification of different types of critical points for linear autonomous systems. These ideas were extended to nonlinear autonomous systems where it was possible for trajectories to spiral in or out until they entered a closed loop called a limit cycle, where the solution became periodic, though nonlinear. These ideas were illustrated by application to the full nonlinear predator–prey problem, the pendulum problem, and the van der Pol equation.

EXERCISES 6.12

In Exercises 1 through 6, locate and identify the nature of the equilibrium point and sketch the pattern of the trajectories.

1. $dx/dt = y$, $dy/dt = x$.
2. $dx/dt = x + 2$, $dy/dt = -x + 2y - 8$.
3. $dx/dt = x - 2y$, $dy/dt = 4x - 3y$.
4. $dx/dt = x - y$, $dy/dt = 2x - y$.
5. $dx/dt = x + 3y - 4$, $dy/dt = -6x - 5y + 22$.
6. $dx/dt = 2y - x$, $dy/dt = 3x + 6$.

In Exercises 7 through 9 locate the equilibrium points of the given nonlinear autonomous system and, where possible, use linearization to identify their nature.

7. $dx/dt = x^2 - y^2 - 4$, $dy/dt = y$.
8. $dx/dt = 2 + y - x^2$, $dy/dt = x^2 - xy$.

9. $dx/dt = x + y + y^2$, $dy/dt = 2x + y$.
10. Locate and identify the equilibrium points of

$$dx/dt = -x + xy, \quad dy/dt = 3y - 2xy + x.$$

11. Show that the only equilibrium point of the van der Pol equation

$$\frac{d^2x}{dt^2} + \varepsilon(x^2 - 1)\frac{dx}{dt} + x = 0$$

is located at the origin. By linearizing the equation about the origin, find conditions that must be imposed on ε in order that (a) the equilibrium point be an unstable spiral, (b) that it be an unstable node, and (c) that it be a center. Relate your results to the phase portraits in Fig. 6.25.

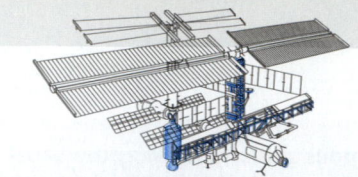

CHAPTER 6
TECHNOLOGY PROJECTS

The purpose of the first two projects is to use a computer algebra phase portrait package to construct the phase portraits for linear and nonlinear systems, and to examine the nature of the limit cycles in the van der Pol equation for different choices of the parameter ε and the initial conditions.

Project 1

Phase Portraits

Use a computer phase portrait package to construct the phase portraits for the following systems about the origin:

(a) $\dfrac{dx}{dt} = 2x^2 - 3y$, $\quad \dfrac{dy}{dt} = x + 2y$.

(b) $\dfrac{dx}{dt} = x + 2y$, $\quad \dfrac{dy}{dt} = x - 3y$.

(c) $\dfrac{dx}{dt} = 2x - 3y$, $\quad \dfrac{dy}{dt} = -x + 2y$.

(d) $\dfrac{dx}{dt} = 2x - 4y$, $\quad \dfrac{dy}{dt} = 4x - 2y$.

(e) $\dfrac{dx}{dt} = x + 3y^2$, $\quad \dfrac{dy}{dt} = x + 2y$.

Project 2

The Limit Cycle of the van der Pol Equation

Use a computer algebra phase portrait package to construct integral curves for the van der Pol equation

$$x'' + \varepsilon(x^2 - 1)x' + x = 0$$

for $\varepsilon = 0.5$, 1.0, and 1.5, starting trajectories from points inside and outside the limit cycle shown in Fig. 6.25.

Project 3

Period of Oscillation of a Nonlinear Pendulum

The nonlinear equation of motion of a simple pendulum when the mass of the pendulum rod is neglected is

$$m\phi'' + (mg/l)\sin\phi = 0,$$

where a prime denotes differentiation with respect to the time t, m is the mass of the pendulum bob, g is the acceleration due to gravity, l is the length of the pendulum, and ϕ is the angle of deflection of the pendulum from the vertical. When the maximum angle of deflection of the pendulum from the vertical is θ, the period of oscillation T is given by the complete elliptic integral

$$T = 4\sqrt{\frac{l}{g}} \int_0^{\pi/2} \frac{du}{(1 - \sin^2(u)\sin^2(\frac{1}{2}\theta))^{1/2}}. \qquad \text{(I)}$$

1. Use the numerical integration facility of MAPLE to find $(T/4)\sqrt{(g/l)}$ for some specific θ.

2. Expand the integrand of (I) as a Maclaurin series in u and integrate term by term to find a series representation for $(T/4)\sqrt{(g/l)}$ in terms of powers of $\sin\theta$.

3. Set $\theta = 2\pi/5$ and approximate the result in Part 2 by taking the first N terms, with $N = 2^m$ and $m = 1, 2, \ldots$. By repeatedly doubling N and comparing the estimate of $(T/4)\sqrt{(g/l)}$ with the result obtained in Part 1, find how many terms must be used in the approximation if the result is to agree to four decimal places.

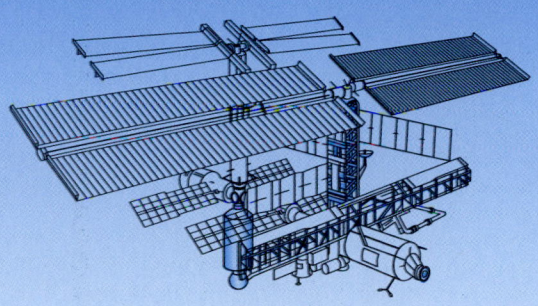

CHAPTER 7

The Laplace Transform

Many problems in engineering and physics can be described in terms of the evolution of solutions of linear differential equations subject to initial conditions. An important group of these problems involves constant coefficient differential equations, and equations like these can be solved very easily by using the Laplace transform.

The Laplace transform is an integral transform that changes a real variable function $f(t)$ into a function $F(s)$ of a variable s through

$$F(s) = \int_0^\infty e^{-st} f(t)\, dt,$$

where in general s is a complex variable.

The importance of the Laplace transform in the study of initial value problems for linear constant coefficient differential equations is that it replaces the operation of integrating a differential equation in $f(t)$ by much simpler algebraic operations involving $F(s)$. Unlike previous methods, where first a general solution is found, and then the constants in the complementary function are chosen to match the initial conditions, when the Laplace transform method is used the initial conditions are incorporated from the start. The task of finding the function $f(t)$ from its Laplace transform $F(s)$ is called inverting the transform, and when working with constant coefficient equations we can accomplish this by appeal to tables of Laplace transform pairs—that is, to a table listing a function $f(t)$ and its corresponding Laplace transform $F(s)$.

The fundamental ideas underlying the Laplace transform are derived, along with its operational properties, which are illustrated by examples. Initial value problems for ordinary differential equations are solved by the Laplace transform, which is then applied to systems of equations and to certain variable coefficient equations. The chapter concludes with applications of the Laplace transform to a variety of problems, the last of which is the heat equation.

7.1   Laplace Transform: Fundamental Ideas

Let the real function $f(t)$ be defined for $a \leq t \leq b$, and let the function $K(t, s)$ of the variables t and s be defined for $a \leq t \leq b$ and some s. When it exists, the integral $\int_a^b f(t)K(t,s)\,dt$ is a function of the single variable s, so denoting the integral by $F(s)$ we can write

$$F(s) = \int_a^b K(t,s)\,f(t)\,dt. \tag{1}$$

The function $F(s)$ in (1) is called an **integral transform** of $f(t)$, the function $K(t,s)$ is the **kernel** of the transform, and s is the **transform variable**. The limits a and b may be finite or infinite, and when at least one limit is infinite the integral in (1) becomes an improper integral.

When it exists, the **Laplace transform** $F(s)$ of a real function $f(t)$ with domain of definition $0 \le t < \infty$ is defined as the integral transform (1) with the kernel $K(t,s) = e^{-st}$, the interval of integration $0 \le t < \infty$, and s a complex variable such that $\mathrm{Re}\, s < c$ for some nonnegative constant c, so that

$$F(s) = \int_0^\infty e^{-st} f(t)\,dt. \tag{2}$$

Throughout the present chapter the transform variable s will be considered to be a real variable, and c will be chosen such that the integral in (2) converges. However, when the general problem of recovering a function $f(t)$ from its Laplace transform $F(s)$ is considered in Chapter 16, it will be seen that s must be allowed to be a complex variable. The advantage of restricting s to the real variable case in this chapter is that the recovery of many useful and frequently occurring functions $f(t)$ from their Laplace transforms $F(s)$ can be accomplished in a very simple manner without the use of complex variable methods.

The reason for interest in integral transforms in general, and the Laplace transform in particular, will become clear when the solution of initial value problems for differential equations is considered. It will then be seen that the Laplace transform replaces integrations with respect to t by simple algebraic operations involving $F(s)$, so provided $f(t)$ can be recovered from $F(s)$ in a simple manner, the solution of an initial value problem can be found by means of straightforward algebraic operations.

Clearly the kernel e^{-st} will only decrease as t increases if $s > 0$, and the Laplace transform of $f(t)$ will only be defined for functions $f(t)$ that decrease sufficiently rapidly as $t \to \infty$ for the integral in (2) to exist. In general, if the function to be transformed is denoted by a lowercase letter such as f, then its Laplace transform will be denoted by the corresponding uppercase letter F, as in (2). It is convenient to denote the Laplace transform operation by the symbol $\mathcal{L}$, so that symbolically $F(s) = \mathcal{L}\{f(t)\}$.

The Laplace transform

formal definition of the Laplace transform

Let $f(t)$ be defined for $0 \le t < \infty$. Then, when the improper integral exists, the Laplace transform $F(s)$ of $f(t)$, written symbolically $F(s) = \mathcal{L}\{f(t)\}$, is defined as

$$F(s) = \int_0^\infty e^{-st} f(t)\,dt.$$

EXAMPLE 7.1

Find $\mathcal{L}\{e^{at}\}$ where a is real.

Solution From (2) we have

$$\mathcal{L}\{e^{at}\} = \int_0^\infty e^{-st} e^{at}\, dt$$

$$= \left[\frac{-e^{-(s-a)t}}{s-a}\right]_0^{t\to\infty}$$

$$= \lim_{t\to\infty}\left[\frac{-e^{-(s-a)t}}{s-a}\right] + \frac{1}{s-a}$$

$$= \frac{1}{s-a},$$

provided $s > a$, for only then will the limit in the first term vanish. This has shown that $\mathcal{L}\{e^{at}\} = F(s) = 1/(s-a)$ for $s > a$, where it is necessary to include the inequality $s > a$ to ensure the convergence of the integral. ∎

PIERRE SIMON LAPLACE (1749–1827)
A French mathematician of remarkable ability who made contributions to analysis, differential equations, probability, and celestial mechanics. He used mathematics as a tool with which to investigate physical phenomena, and made fundamental contributions to hydrodynamics, the propagation of sound, surface tension in liquids, and many other topics. His many contributions had a wide-ranging effect on the development of mathematics.

Laplace transform pair and inverse transform

The two functions $f(t)$ and $F(s)$ are called a **Laplace transform pair**, and for all ordinary functions, given $F(s)$ the corresponding function $f(t)$ is determined uniquely, just as $f(t)$ determines $F(s)$ uniquely. This relationship is expressed symbolically by using the symbol $\mathcal{L}^{-1}$ to denote the operation of finding a function $f(t)$ with a given Laplace transform $F(s)$. This process is called finding the **inverse Laplace transform** of $F(s)$. In terms of the foregoing example, we have $\mathcal{L}\{e^{at}\} = 1/(s-a)$ and $\mathcal{L}^{-1}\{1/(s-a)\} = e^{at}$. This is a particular case of the general result that, by definition, the inverse Laplace transform acting on the Laplace transform of the function returns the original function, so we can write

$$\mathcal{L}^{-1}\{\mathcal{L}\{f(t)\}\} = f(t).$$

how to be sure a Laplace transform exists

A sufficient condition for the existence of the Laplace transform of a function $f(t)$ is that the absolute value of $f(t)$ can be bounded for all $t \geq 0$ by

$$|f(t)| \leq Me^{kt}, \tag{3}$$

for some constants M and k. This means that if numbers M and k can be found such that

$$|e^{-st} f(t)| \leq Me^{(k-s)t},$$

then

$$\mathcal{L}\{f(t)\} = \int_0^\infty e^{-st} f(t)\, dt \leq M\int_0^\infty e^{(k-s)t}\, dt = M/(s-k).$$

TABLE 7.1 Laplace Transform Pairs

$f(t)$	$F(s) = \mathcal{L}\{f(t)\}$	Condition on s
1. 1	$1/s$	$s > 0$
2. t	$1/s^2$	$s > 0$
3. t^n $(n = 1, 2, \ldots)$	$n!/s^{n+1}$	$s > 0$
4. t^a $(a > -1)$	$\Gamma(a+1)/s^{a+1}$	$s > a$
5. e^{at}	$1/(s-a)$	$s > a$
6. $t^n e^{at}$ $(n = 1, 2, \ldots)$	$n!/(s-a)^{n+1}$	$s > a$
7. $H(t-a)$	e^{-as}/s	$s \geq a$
8. $\delta(t-a)$	e^{-as}	$s > 0, a > 0$
9. $\sin at$	$a/(s^2 + a^2)$	$s > 0$
10. $\cos at$	$s/(s^2 + a^2)$	$s > 0$
11. $t \sin at$	$2as/(s^2 + a^2)^2$	$s > 0$
12. $t \cos at$	$(s^2 - a^2)/(s^2 + a^2)^2$	$s > 0$
13. $e^{at} \sin bt$	$b/[(s-a)^2 + b^2]$	$s > a$
14. $e^{at} \cos bt$	$(s-a)/[(s-a)^2 + b^2]$	$s > a$
15. $\dfrac{1}{2a^3} \sin at - \dfrac{1}{2a^2} t \cos at$	$1/(s^2 + a^2)^2$	$s > 0$
16. $\dfrac{1}{2a} \sin at + \dfrac{1}{2} t \cos at$	$s^2/(s^2 + a^2)^2$	$s > 0$
17. $1 - \cos at$	$a^2/[s(s^2 + a^2)]$	$s > 0$
18. $at - \sin at$	$a^3/[s^2(s^2 + a^2)]$	$s > 0$
19. $\sinh at$	$a/(s^2 - a^2)$	$s > |a|$
20. $\cosh at$	$s/(s^2 - a^2)$	$s > |a|$
21. $\dfrac{1}{2a^3} \sinh at + \dfrac{1}{2a^2} t \cosh at$	$1/(s^2 - a^2)^2$	$s > |a|$
22. $\dfrac{1}{2a} t \sinh at$	$s/(s^2 - a^2)^2$	$s > |a|$
23. $\dfrac{1}{2a} \sinh at + \dfrac{1}{2} t \cosh at$	$s^2/(s^2 - a^2)^2$	$s > |a|$
24. $\sinh at - \sin at$	$2a^3/(s^4 - a^4)$	$s > |a|$
25. $\cosh at - \cos at$	$2a^2 s/(s^4 - a^4)$	$s > |a|$

The integral on the right will be convergent provided $s > k > 0$, so when this is true the Laplace transform $F(s) = \mathcal{L}\{f(t)\}$ will exist. It should be clearly understood that (3) is only a *sufficient* condition for the existence of a Laplace transform, and *not* a necessary one, because Laplace transforms can be found for functions that do not satisfy condition (3). For example, the function $f(t) = t^{-1/4}$ does not satisfy condition (3), but its Laplace transform exists and is a special case of entry 4 in Table 7.1.

The preceding inequality implies that when $\mathcal{L}\{f(t)\}$ exists, $F(s)$ must be such that $\lim_{s \to \infty} F(s) = 0$. In addition, the condition $\mathcal{L}\{f(t)\} \leq M/(s - k)$ implies that $F(s)$ cannot be the Laplace transform of on ordinary function $f(t)$ unless $F(s) \to 0$ as $s \to \infty$. For example, $F(s) = (s^2 - 1)/(s^2 + 1)$ is not a Laplace transform of an ordinary function. Exceptions to this condition are functions like the *delta function*, which is defined in Section 7.2, though there the delta function will be seen to involve integration, and so it is not a *function* in the usual sense.

The Laplace transform is a linear operation, and the consequence of this important and useful property is expressed in the following theorem.

THEOREM 7.1

fundamental linearity property

Linearity of the Laplace transformation   Let the functions $f_1(t)$, $f_2(t)$, ..., $f_n(t)$ have Laplace transforms, and let $c_1, c_2, \ldots, c_n$ be any set of arbitrary constants. Then

$$\mathcal{L}\{c_1 f_1(t) + c_2 f_2(t) + \cdots + c_n f_n(t)\} = c_1\mathcal{L}\{f_1(t)\} + c_2\mathcal{L}\{f_2(t)\} + \cdots + c_n\mathcal{L}\{f_n(t)\}.$$

Proof   The proof is simple and follows directly from the fact that integration is a linear operation, so the integral of a sum of functions is the sum of their integrals. Thus,

$$\int_0^\infty e^{-st}\{c_1 f_1(t) + c_2 f_2(t) + \cdots + c_n f_n(t)\}dt$$

$$= c_1\int_0^\infty f_1(t)e^{-st}dt + c_2\int_0^\infty f_2(t)e^{-st}dt + \cdots + c_n\int_0^\infty f_n(t)e^{-st}dt$$

$$= c_1\mathcal{L}\{f_1(t)\} + c_2\mathcal{L}\{f_2(t)\} + \cdots + c_n\mathcal{L}\{f_n(t)\}. \qquad \blacksquare$$

This theorem has many applications and its use is essential when working with the Laplace transform.

EXAMPLE 7.2

Find the Laplace transform of $f(t) = c_1e^{at} + c_2e^{-at}$, and use the result to find $\mathcal{L}\{\sinh at\}$ and $\mathcal{L}\{\cosh at\}$.

some examples

Solution   Applying Theorem 7.1 and the result $\mathcal{L}\{e^{at}\} = 1/(s-a)$ from Example 7.1, we find that

$$\mathcal{L}\{c_1e^{at} + c_2e^{-at}\} = c_1\mathcal{L}\{e^{at}\} + c_2\mathcal{L}\{e^{-at}\} = c_1/(s-a) + c_2/(s+a).$$

As $\sinh at = (e^{at} - e^{-at})/2$ and $\cosh at = (e^{at} + e^{-at})/2$, $\mathcal{L}\{\sinh at\}$ is obtained from the preceding result by setting $c_1 = 1/2$ and $c_2 = -1/2$, and $\mathcal{L}\{\cosh at\}$ is obtained by setting $c_1 = c_2 = 1/2$, when we obtain

$$\mathcal{L}\{\sinh at\} = a/(s^2 - a^2) \quad \text{and} \quad \mathcal{L}\{\cosh at\} = s/(s^2 - a^2),$$

for $s > |a| \geq 0$. Notice that because s must be be positive, but in $\sinh at$ and $\cosh at$ the number a may be either positive or negative, the relationship between s and a necessary to ensure that the convergence of the integrals must be $s > |a| \geq 0$, and not $s > a > 0$.   ∎

The process of finding an inverse Laplace transformation involves reversing the foregoing argument and seeking a function $f(t)$ that has the required Laplace transform $F(s)$. Where possible, this is accomplished by simplifying the algebraic structure of $F(s)$ to the point at which it can be recognized as the sum of the Laplace transforms of known functions of t.

EXAMPLE 7.3

Find the inverse Laplace transform of

$$F(s) = \frac{4s + 10}{s^2 + 6s + 8}.$$

Solution   Expanding the Laplace transform in terms of partial fractions gives

$$\frac{4s+10}{s^2+6s+8} = \frac{1}{s+2} + \frac{3}{s+4},$$

so

$$\mathcal{L}^{-1}\{F(s)\} = \mathcal{L}^{-1}\left\{\frac{4s+10}{s^2+6s+8}\right\} = \mathcal{L}^{-1}\left\{\frac{1}{s+2}\right\} + 3\mathcal{L}^{-1}\left\{\frac{1}{s+4}\right\}.$$

Using the result of Example 7.1 we find that

$$f(t) = \mathcal{L}^{-1}\left\{\frac{4s+10}{s^2+6s+8}\right\} = e^{-2t} + 3e^{-4t}. \qquad \blacksquare$$

EXAMPLE 7.4   Find (a) $\mathcal{L}\{1\}$ and (b) $\mathcal{L}\{t\}$.

Solution

(a) By definition,

$$\mathcal{L}\{1\} = \int_0^\infty e^{-st}dt = \frac{1}{s}, \quad \text{for } s > 0.$$

(b) By definition,

$$\mathcal{L}\{t\} = \int_0^\infty e^{-st}t\,dt = \left(-\frac{t}{s}e^{-st} - \frac{e^{-st}}{s^2}\right)_{t=0}^\infty = \frac{1}{s^2}, \quad \text{for } s > 0. \qquad \blacksquare$$

EXAMPLE 7.5   Find $\mathcal{L}\{\sin at\}$.

Solution   By definition,

$$\mathcal{L}\{\sin at\} = \int_0^\infty e^{-st}\sin at\,dt = \lim_{k\to\infty}\int_0^k e^{-st}\sin at\,dt$$

$$= \lim_{k\to\infty}\left(\frac{-e^{-sk}(a\cos ak + s\sin ak)}{s^2+a^2}\right) + \frac{a}{s^2+a^2}$$

$$= \frac{a}{s^2+a^2} \quad \text{for } s > 0,$$

where the condition $s > 0$ is required to ensure that the limit is finite as $k \to 0$. This has shown that

$$\mathcal{L}\{\sin at\} = \frac{a}{s^2+a^2} \quad \text{for } s > 0. \qquad \blacksquare$$

In the next example we find $\mathcal{L}\{t^n\}$, and in the process introduce an integral that will be useful later in Chapter 8 when finding series solutions of linear second order variable coefficient differential equations.

EXAMPLE 7.6   Find $\mathcal{L}\{t^n\}$ for $n = 1, 2, \ldots$.

Solution   By definition

$$\mathcal{L}\{t^n\} = \int_0^\infty e^{-st}t^n dt.$$

To evaluate this integral we will make use of integration by parts to establish a recursion (recurrence) relation from which the result for arbitrary positive integral n can be found.

Accordingly, we define $I(n, s)$ as

$$I(n, s) = \int_0^\infty e^{-st} t^n dt = \lim_{k \to \infty} \int_0^k \frac{-t^n}{s} \frac{d}{dt}(e^{-st})dt$$

and use integration by parts to express this as

$$= \lim_{k \to \infty} \left[\frac{-t^n e^{-st}}{s} \right]_{t=0}^k + \frac{n}{s} \int_0^\infty t^{n-1} e^{-st} dt$$

$$= \left(\frac{n}{s} \right) I(n - 1, s), \quad \text{for } s > 0.$$

This has established the *recursion relation*

$$I(n, s) = (n/s)I(n - 1,\ s),$$

satisfied by the integral $I(n, s)$.

As $I(0, s) = \int_0^\infty e^{-st} dt = 1/s$, by setting $n = 1$ in the recursion relation we find that

$$I(1, s) = (1/s)I(0, s) = 1/s^2, \quad \text{for } s > 0.$$

Similarly, setting $n = 2$ in the recursion relation shows that

$$I(2, s) = (2/s)I(1, s) = 2 \cdot 1/s^3 = 2!/s^3, \quad \text{for } s > 0,$$

and an inductive argument shows that

$$I(n, s) = n!/s^{n+1}.$$

In terms of the Laplace transform notation, we have shown that

$$\mathcal{L}\{t^n\} = n!/s^{n+1} \quad \text{for } n = 0,\ 1,\ 2, \ldots, \quad \text{for } s > 0. \qquad \blacksquare$$

Notice that setting $s = 1$ in the general result of Example 7.3 enables $n!$ to be expressed as the integral

$$n! = \int_0^\infty e^{-t} t^n dt, \quad \text{for } n = 0,\ 1,\ 2, \ldots.$$

This provides a way of representing factorial n in terms of an integral, and it is our first encounter with a special case of the **Gamma function** that will be required later. The gamma function, denoted by $\Gamma(x)$ for $x > 0$, is defined by the integral

first encounter with the Gamma function

$$\Gamma(x) = \int_0^\infty e^{-t} t^{x-1} dt. \qquad (4)$$

In terms of the earlier notation, when the restriction that n is an integer is removed, and n is replaced by a positive real variable x, we can write

$$\Gamma(x + 1) = \int_0^\infty e^{-t} t^x dt = I(x, 1),$$

but

$$I(x, 1) = xI(x - 1, 1) = x\Gamma(x) \quad \text{for } x > 0,$$

so combining results shows that the gamma function satisfies the fundamental relation

$$\Gamma(x+1) = x\Gamma(x) \quad \text{for } x > 0. \tag{5}$$

It is easily seen from this that

$$\Gamma(n+1) = n! \quad \text{for } n = 0, 1, 2, \ldots,$$

so as $\Gamma(x)$ is defined for all positive x the gamma function provides a generalization of the factorial function $n!$ for positive non-integer values of n. It will be seen later that the gamma function, which belongs to the general class of functions called **higher transcendental functions**, occurs frequently throughout mathematics.

Discontinuous Functions

Because the Laplace transform is defined in terms of an integral, it is possible to find Laplace transforms of discontinuous functions. Suppose, for example, that a function $g(t)$ is discontinuous at $t = a$, as in Fig. 7.1. Then, provided it converges, the integral defining the Laplace transform of $g(t)$ is given by

$$\mathcal{L}\{g(t)\} = \lim_{\varepsilon \to 0} \int_0^{a-\varepsilon} e^{-st} g(t)dt + \lim_{\delta \to 0} \int_{a+\delta}^{\infty} e^{-st} g(t)dt, \tag{6}$$

where ε and δ are both positive. For simplicity, the upper limit in the first integral is usually denoted by a_- and the lower limit in the second integral by a_+. These are, respectively, the limits of integration to the left and right of $t = a$.

An important discontinuous function that finds numerous applications in connection with the Laplace transform, and elsewhere, is the **unit step function** $f(t) = H(t-a)$ with $a \geq 0$, known also as the **Heaviside step function**. The unit step function is defined as

Heaviside step function

$$H(t-a) = \begin{cases} 0 & \text{if } t < a \\ & \qquad\qquad (a \geq 0). \\ 1 & \text{if } t > a \end{cases} \tag{7}$$

A related function that is also of considerable importance is the **unit pulse function**,

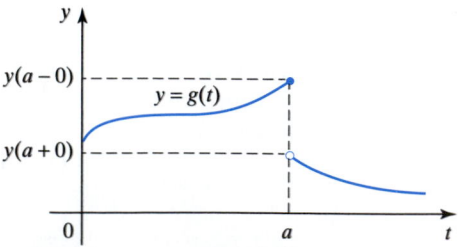

FIGURE 7.1   A discontinuous function $g(t)$.

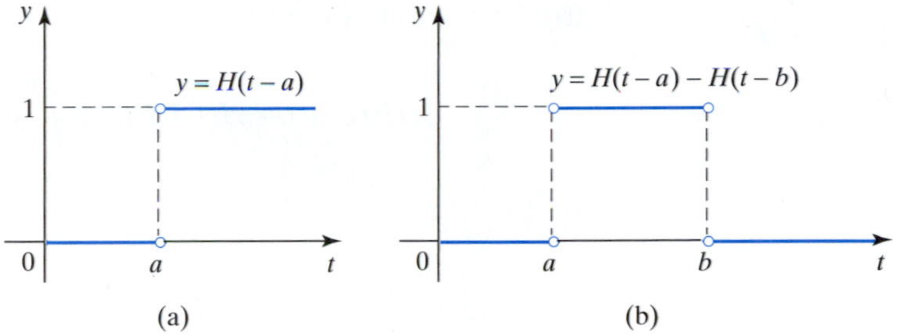

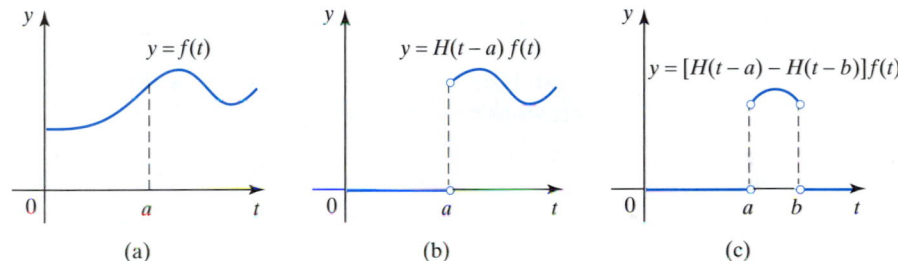

FIGURE 7.2   (a) The unit step function $y = H(t - a)$. (b) The unit pulse function $y = p(t) = H(t - a) - H(t - b)$.

FIGURE 7.3   The effect on $f(t)$ of multiplication by $H(t - a)$ and $H(t - a) - H(t - b)$.

defined as

$$p(t) = H(t - a) - H(t - b), \quad \text{with } b > a \geq 0. \tag{8}$$

The function $p(t)$ operates like a "switch," because it switches on at $t = a$ and off at $t = b$. Graphs of these two functions are shown in Fig. 7.2.

If a function $f(t)$ is multiplied by a unit step function, the function $f(t)$ can be considered to be "switched on" at time $t = a$, in the sense that the product $H(t - a) f(t)$ is zero for $t < a$ and $f(t)$ for $t > a$. Similarly, multiplication of $f(t)$ by a unit pulse function "switches on" the function $f(t)$ at time $t = a$ and "switches it off" at time $t = b$. This property is illustrated in Fig. 7.3, where Fig. 7.3(a) shows the original function $f(t)$, Fig. 7.3(b) shows the product $H(t - a) f(t)$, and Fig. 7.3(c) the product $\{H(t - a) - H(t - b)\} f(t)$.

In the next example we make use of result (6) to find the Laplace transforms of the unit step function and the unit pulse function.

> **switching functions on and off with the Heaviside step function**

EXAMPLE 7.7

Find (a) $\mathcal{L}\{H(t - a)\}$ and (b) $\mathcal{L}\{H(t - a) - H(t - b)\}$.

Solution

(a) By definition

$$\mathcal{L}\{H(t - a)\} = \int_a^\infty e^{-st} dt$$

$$= \left(-\frac{e^{-st}}{s}\right)_{t=a}^\infty = \frac{e^{-as}}{s} \quad \text{for } s > a \geq 0.$$

(b) Using result (a) we have

$$\mathcal{L}\{H(t-a) - H(t-b)\} = \int_a^b e^{-st}\, dt$$

$$= \int_a^\infty e^{-st}\, dt - \int_b^\infty e^{-st}\, dt$$

$$= \frac{e^{-as} - e^{-bs}}{s} \quad \text{for } s > b > a \geq 0. \qquad \blacksquare$$

EXAMPLE 7.8   Find (a) $\mathcal{L}\{t^3 - 4t + 5 + 3\sin 2t\}$ and (b) $\mathcal{L}^{-1}\{(s^4 + 5s^2 + 2)/[s^3(s^2+1)]\}$.

Solution

(a) Using Theorem 7.1 together with the Laplace transform pairs found in the previous examples, we have

$$\mathcal{L}\{t^3 - 4t + 5 + 2\sin 3t\} = \mathcal{L}\{t^3\} - 4\mathcal{L}\{t\} + \mathcal{L}\{5\} + 3\mathcal{L}\{\sin 2t\}$$

$$= 6/s^4 - 4/s^2 + 5/s + 6/(s^2 + 4)$$

$$= (5s^5 + 2s^4 + 20s^3 - 10s^2 + 24)/[s^4(s^2 + 4)].$$

(b) Simplifying the transform by means of partial fractions gives

$$\frac{s^4 + 5s^2 + 2}{s^3(s^2 + 1)} = \frac{2}{s^3} + \frac{3}{s} - 2\frac{s}{s^2 + 1}.$$

Taking the inverse Laplace transform of each term on the right and using the linearity property of the Laplace transform, we find that

$$\mathcal{L}^{-1}\left(\frac{s^4 + 5s^2 + 2}{s^3(s^2+1)}\right) = \mathcal{L}^{-1}\left\{\frac{2}{s^3}\right\} + \mathcal{L}^{-1}\left\{\frac{3}{s}\right\} - 2\mathcal{L}^{-1}\left\{\frac{s}{s^2+1}\right\}.$$

Finally, using the transform pairs established in the previous examples, we have

$$\mathcal{L}^{-1}\left\{\frac{s^4 + 5s^2 + 2}{s^3(s^2+1)}\right\} = t^2 + 3 - 2\cos t. \qquad \blacksquare$$

To make further progress with the Laplace transform it is necessary to have available a table of Laplace transform pairs for the most commonly occurring functions. Theorems to be developed later will enable such a table to be extended in a straightforward manner, so that transforms and inverse Laplace transforms of more complicated functions can be found.

Table 7.1 provides a list of the most useful Laplace transform pairs involving elementary functions. All of these entries can be established either by means of routine integration, or by the combination of simpler results, with the sole exception of the *delta function* $\delta(t-a)$ in entry 8. The derivation of this result is to be found in Section 7.2 after the delta function has been defined.

The example that now follows illustrates how entry 15 can be found from entries 9 through 12.

EXAMPLE 7.9   Find $\mathcal{L}^{-1}\{1/(s^2+a^2)^2\}$ by combining related entries in Table 7.1.

Solution   Our objective will be to use the linearity property of the Laplace transform to express $1/(s^2+a^2)^2$ as a linear combination of terms that we hope will be found listed in the column $F(s)$ of Table 7.1. If this is possible, the inverse Laplace transform can then be found by adding the inverse transform of each expression in partial fraction representation of $F(s)$. A routine calculation shows that $F(s)$ can be written as

$$\frac{1}{(s^2+a^2)^2} = \frac{1}{2a^3}\left(\frac{a}{s^2+a^2}\right) - \frac{1}{2a^2}\left(\frac{s^2-a^2}{(s^2+a^2)^2}\right),$$

so from using entries 9 and 12 in Table 7.1 we have

$$\mathcal{L}^{-1}\left\{\frac{1}{(s^2+a^2)^2}\right\} = \frac{1}{2a^3}\sin at - \frac{1}{2a^2}t\cos at,$$

and this is entry 15 in the table.   ∎

Summary   The Laplace transform of a function $f(t)$ has been defined. A condition has been given that ensures the existence of the transform, and the concept of a Laplace transform pair has been introduced. The transform has been shown to have the fundamental property of linearity, and some simple transform pairs have been found directly from the definition. The Heaviside unit step function $H(t-a)$, which jumps from zero for $0 \le t < a$ to unity for $t > a$, has been introduced and used. The section closed with a table of useful Laplace transform pairs.

EXERCISES 7.1

In Exercises 1 through 4 use the definition of the Laplace transform to obtain the stated result.

1. Show that $\mathcal{L}\{t^2\} = 2/s^3$ for $s > 0$.
2. Show that $\mathcal{L}\{te^{at}\} = 1/(s-a)^2$ for $s > a$.
3. Find $\mathcal{L}\{e^{iat}\}$, and by equating the real and imaginary parts show that $\mathcal{L}\{\sin at\} = a/(s^2+a^2)$ and $\mathcal{L}\{\cos at\} = s/(s^2+a^2)$ for $s > 0$.
4. Show that $\mathcal{L}\{\sinh at\} = a/(s^2-a^2)$ for $s > |a|$.

In Exercises 5 through 20 use Table 7.1 of Laplace transform pairs to find $\mathcal{L}\{f(t)\}$.

5. $f(t) = te^{2t}$.
6. $f(t) = 2\sin 3t - \cos 3t$.
7. $f(t) = t - t^2 + t^3$.
8. $f(t) = e^{3t}(\sin t - \cos t)$.
9. $f(t) = e^{-2t}(\cos 2t - \sin 2t)$.
10. $f(t) = t(\sin 2t - \cos 2t)$.
11. $f(t) = t\cosh 3t - \sinh 3t$.
12. $f(t) = \sinh t - t\cos t$.
13. $f(t) = e^{-t}\cos 2t - t$.
14. $f(t) = 2t^2 - 3t + 4\cos 3t$.

15. $f(t) = H(t-\pi/2)e^t\sin t$.
16. $f(t) = H(t-3\pi/2)(\sin t - 3\cos t)$.
17. $f(t) = [H(t-\pi/2) - H(t-\pi)]t$.
18. $f(t) = [1 - H(t-\pi/2)]t$.
19. $f(t) = H(t-\pi/2)e^{-t}\cos t$.
20. $f(t) = [1 - H(t-\pi/2)]e^{3t}$.

In Exercises 21 through 30 use Table 7.1 of Laplace transform pairs to find $\mathcal{L}^{-1}\{F(s)\}$.

21. $F(s) = (s^2-1)/[s(s^2+4)]$.
22. $F(s) = (s^2+3s+1)/[s(s^2-4)]$.
23. $F(s) = (3s+5)/[s(s^2+9)]$.
24. $F(s) = (s^2-4)/[(s^2+1)(s^2-1)]$.
25. $F(s) = (s^3-1)/[(s+2)^2(s^2-9)]$.
26. $F(s) = (s^2+s+1)/[(s^2+4)(s^2-9)]$.
27. $F(s) = s^2/[(s-1)^2(s+1)]$.
28. $F(s) = s/(s-1)^3$.
29. $F(s) = (s^2+4)/[(s^2-9)(s-1)]$.
30. $F(s) = (s^2+1)/[(s+1)(s+2)(s+3)]$.

In Exercises 31 through 36 find the Laplace transform of the function $f(t)$ shown in graphical form.

31.

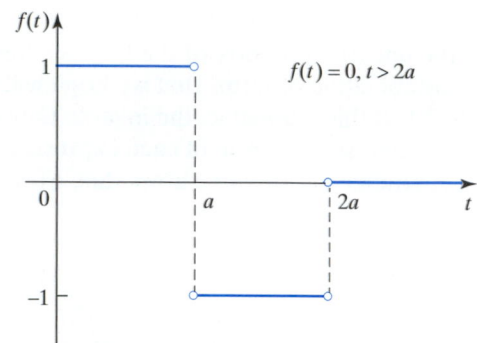

$f(t) = 0, t > 2a$

FIGURE 7.4

32.

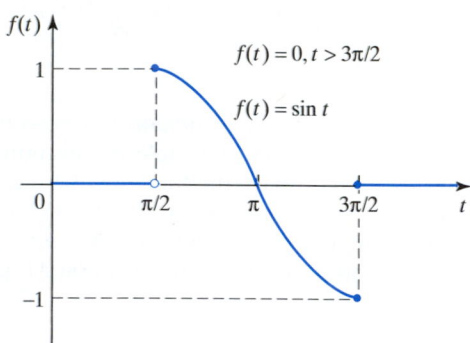

$f(t) = 0, t > 3\pi/2$

$f(t) = \sin t$

FIGURE 7.5

33.

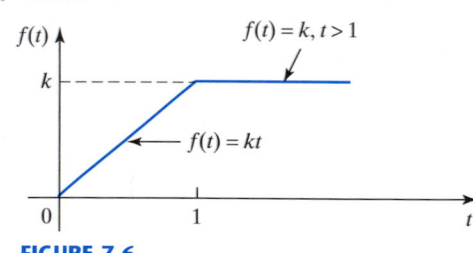

$f(t) = k, t > 1$

$f(t) = kt$

FIGURE 7.6

34.

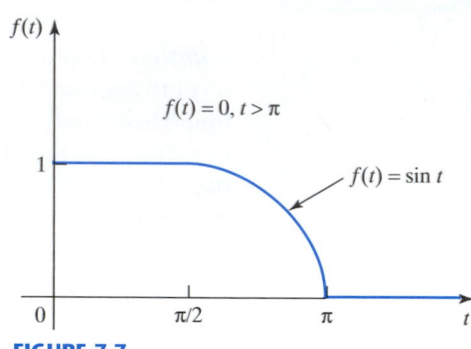

$f(t) = 0, t > \pi$

$f(t) = \sin t$

FIGURE 7.7

35.

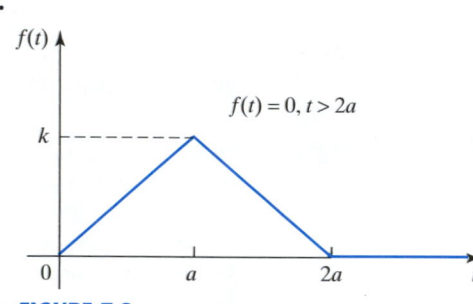

$f(t) = 0, t > 2a$

FIGURE 7.8

36.

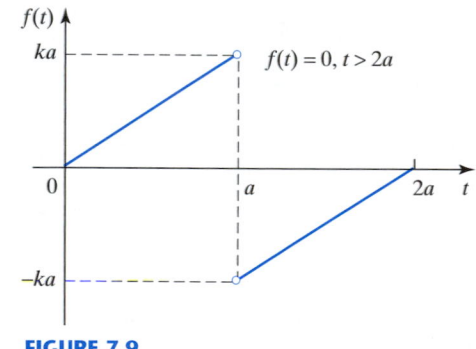

$f(t) = 0, t > 2a$

FIGURE 7.9

7.2   Operational Properties of the Laplace Transform

In the previous section the Laplace transform of a basic list of commonly occurring functions $f(t)$ was recorded as the list of Laplace transform pairs in Table 7.1. To use the Laplace transform to solve initial value problems for linear differential equations and systems it is necessary to establish a number of fundamental properties of the transform known as its **operational properties**. This name is given to properties of the transform itself that relate to the way it *operates* on any function $f(t)$ that is transformed, rather than to the effect these properties of the transform have on specific functions $f(t)$.

This means that operational properties are general properties of the Laplace transform that are not specific to any particular function $f(t)$ or to its transform

$F(s)$. An important example of an operational property has already been encountered in Theorem 7.1, where the linearity property of the transformation was established.

Some operational properties, such as the scaling and shift theorems that will be proved later, save effort when finding the Laplace transform of a function or inverting a transform, whereas others such as the transform of a derivative are essential when applying the Laplace transform to solve initial value problems for differential equations.

The way derivatives transform is used to find how the homogeneous part of a linear differential equation is transformed, and we will see later that it also shows how the initial conditions for the differential equation enter into the transformed equation. Table 7.1 of Laplace transform pairs is needed when transforming the nonhomogeneous term in the differential equation.

THEOREM 7.2

transforming
derivatives

Transform of a derivative   Let $f(t)$ be continuous on $0 \le t < \infty$, and let $f'(t)$ be piecewise continuous on every finite interval contained in $t \ge 0$. Then if $\mathcal{L}\{f(t)\} = F(s)$,

$$\mathcal{L}\{f'(t)\} = s F(s) - f(0).$$

Proof   Using integration by parts, and assuming that f satisfies the sufficiency condition for the existence of a Laplace transform, we have

$$\mathcal{L}\{f'(t)\} = \int_0^\infty e^{-st} f'(t)dt = \lim_{k \to \infty} \int_0^k e^{-st} f'(t)dt$$

$$= \lim_{k \to \infty} [e^{-st} f(t)]_0^k - \lim_{k \to \infty} \int_0^k -se^{-st} f(t)dt$$

$$= \lim_{k \to \infty} [e^{-sk} f(k) - f(0)] + s F(s)$$

$$= s F(s) - f(0),$$

where $\lim_{k \to \infty} e^{-sk} f(k) = 0$ because of condition (3).   ∎

THEOREM 7.3

Transform of a higher derivative   Let $f(t)$ be continuous on $0 \le t < \infty$, and let $f'(t), f''(t), \ldots, f^{(n-1)}(t)$ be piecewise continuous on every finite interval contained in $t \ge 0$. Then if $\mathcal{L}\{f(t)\} = F(s)$,

$$\mathcal{L}\{f^{(n)}(t)\} = s^n F(s) - s^{n-1} f(0) - s^{n-2} f'(0) - \cdots - s f^{(n-2)}(0) - f^{(n-1)}(0).$$

Proof   The proof uses repeated integration by parts, but otherwise is analogous to the one used in Theorem 7.2, so the details are left as an exercise.   ∎

The two most frequently used results are those of Theorem 7.2 and the result from Theorem 7.3 corresponding to $n = 2$, so for convenience we record these here.

The Laplace transform of first and second derivatives

$$\mathcal{L}\{f'(t)\} = s F(s) - f(0). \tag{9a}$$

$$\mathcal{L}\{f''(t)\} = s^2 F(s) - s f(0) - f'(0). \tag{9b}$$

THEOREM 7.4

Transform of f' when f is discontinuous at $t = a$ Let $f(t)$ be continuous on $0 \le t < a$ and on $a < t < \infty$, and let it have a simple jump discontinuity at $t = a$ with the value $f_-(a)$ to the immediate left of a at $t = a-$ and the value $f_+(a)$ to the immediate right of $t = a$ at $a+$. Then if $\mathcal{L}\{f(t)\} = F(s)$,

$$\mathcal{L}\{f'(t)\} = sF(s) - f(0) + [f_-(a) - f_+(a)]e^{-as}.$$

Proof   Using integration by parts, as in Theorem 7.2, we have

$$\mathcal{L}\{f'(t)\} = \int_0^{a-} e^{-st} f'(t)dt + \lim_{k \to \infty} \int_{a+}^{\infty} e^{-st} f'(t)dt$$

$$= [e^{-st} f(t)]_0^{a-} + \lim_{k \to \infty} [e^{-sk} f(k) - e^{-as} f_+(a)] + sF(s)$$

$$= sF(s) - f(0) + [f_-(a) - f_+(a)]e^{-as}. \qquad \blacksquare$$

The next example illustrates the application of results (8) and (9) to a simple initial value problem.

EXAMPLE 7.10

Solve the initial value problem

$$y'' + 3y' + 2y = \sin 2t, \quad \text{where } y(0) = 2 \quad \text{and} \quad y'(0) = -1.$$

Solution   Because of the linearity of the equation and of the Laplace transform operation, taking the Laplace transform of the differential equation we have

$$\mathcal{L}\{y''\} + 3\mathcal{L}\{y'\} + 2\mathcal{L}\{y\} = \mathcal{L}\{\sin 2t\}.$$

Setting $\mathcal{L}\{y(t)\} = Y(s)$, and using the initial conditions $y(0) = 2$ and $y'(0) = -1$, we find from (9a,b) that

$$\mathcal{L}\{y''\} = s^2 Y(s) - 2s + 1,$$

and

$$\mathcal{L}\{y'\} = sY(s) - 2.$$

Entry 9 in Table 7.1 shows that $\mathcal{L}\{\sin 2t\} = 2/(s^2 + 4)$, so combining these results enables the transformed differential equation to be written

$$s^2 Y(s) - 2s + 1 + 3[sY(s) - 2] + 2Y(s) = \frac{2}{s^2 + 4},$$

or as

$$(s^2 + 3s + 2)Y(s) = \frac{2s^3 + 5s^2 + 8s + 22}{s^2 + 4}.$$

Solving for the Laplace transform of the solution gives

$$Y(s) = \frac{2s^3 + 5s^2 + 8s + 22}{(s^2 + 4)(s^2 + 3s + 2)}.$$

When expressed in partial fraction form, $Y(s)$ becomes

$$Y(s) = \frac{-5}{4}\frac{1}{s+2} + \frac{17}{5}\frac{1}{s+1} - \frac{1}{20}\frac{2}{s^2+4} - \frac{3}{20}\frac{s}{s^2+4}.$$

Using the linearity property when taking the inverse Laplace transform, we have

$$\mathcal{L}^{-1}\{Y(s)\} = -\frac{5}{4}\mathcal{L}^{-1}\left\{\frac{1}{s+2}\right\} + \frac{17}{5}\mathcal{L}^{-1}\left\{\frac{1}{s+1}\right\}$$

$$-\frac{1}{20}\mathcal{L}^{-1}\left\{\frac{2}{s^2+4}\right\} - \frac{3}{20}\mathcal{L}^{-1}\left\{\frac{s}{s^2+4}\right\},$$

so using Table 7.1 to identify the four transforms involved shows that the solution of the initial value problem is

$$y(t) = -\frac{5}{4}e^{-2t} + \frac{17}{5}e^{-t} - \frac{1}{20}\sin 2t - \frac{3}{20}\cos 2t, \quad \text{for } t > 0. \qquad \blacksquare$$

This example illustrates a fundamental difference between the solution of an initial value problem obtained by using the Laplace transform and that obtained by the previous methods that have been developed. In the other methods, when solving an initial value problem, first a general solution was found, and then the arbitrary constants were matched to the initial conditions. However, in the Laplace transform approach the initial conditions are incorporated when the equation is transformed, so the inversion of $Y(s)$ gives the required solution of the initial value problem immediately.

As the *structure* of the solution in Example 7.10 is typical of the structure obtained when solving all initial value problems for ordinary differential equations by means of the Laplace transform, a closer examination of it will help understand how the solution is generated.

Returning to the point where the equation was transformed, the result can be rewritten as

$$\underbrace{(s^2 + 3s + 2)\,Y(s)}_{\substack{\text{Transformed homogeneous equation} \\ \text{with } y'', y', \text{ and } y \text{ replaced, respectively,} \\ \text{by } s^2, s, \text{ and } 1}} \quad = \quad \underbrace{2s + 5}_{\substack{\text{Transformed initial} \\ \text{conditions}}} \quad + \quad \underbrace{\frac{2}{s^2 + 2}}_{\substack{\text{Transformed nonhomogeneous} \\ \text{term}}}$$

Setting $G(s) = 1/(s^2 + 3s + 2)$, and denoting the transformed initial conditions by $I(s)$ and the transformed nonhomogeneous term by $R(s)$, the above result can be solved for $Y(s)$ and written in the form

$$Y(s) = G(s)I(s) + G(s)R(s). \qquad (10)$$

transfer function

This shows how the transform $G(s)$, called in engineering applications the **transfer function** associated with the differential equation, modifies the transform of the initial conditions and the transform of the nonhomogeneous term to arrive at the transform $Y(s)$ of the solution. The name *transfer function* comes from the fact that when all the initial conditions are zero, so $I(s) = 0$, the only term generating a solution is the forcing function (the nonhomogeneous term), so (10) describes how the effect of the *input* is *transferred* to the *output* (the solution). In terms of Example 7.10 we can write

$$G(s) = \frac{Y(s)}{R(s)} = \frac{\mathcal{L}\{y(t)\}}{\mathcal{L}\{\sin 2t\}}$$

$$= \frac{\mathcal{L}\{output\}}{\mathcal{L}\{input\}}. \qquad (11)$$

In control theory the transfer function of a system characterizes the behavior of the entire system.

We now develop the most important operational properties of the Laplace transform, starting with the first shift theorem, also called the *s*-shift theorem.

THEOREM 7.5

the s-shift theorem

The first shift theorem or the *s*-shift theorem   Let $\mathcal{L}\{f(t)\} = F(s)$ for $s > \gamma$. Then the Laplace transform of $e^{at} f(t)$ is obtained from $F(s)$ by replacing s by $s - a$, where $s - a > \gamma$. Thus,

$$\mathcal{L}\{e^{at} f(t)\} = F(s-a) \quad \text{for } s - a > \gamma.$$

Conversely, the inverse transform

$$\mathcal{L}^{-1}\{F(s-a)\} = e^{at} f(t).$$

Proof   From the conditions of the theorem, $\mathcal{L}\{f(t)\} = \int_0^\infty e^{-st} f(t)dt$ for $s > \gamma$, so

$$\mathcal{L}\{e^{-at} f(t)\} = \int_0^\infty e^{-st} e^{at} f(t)dt = \int_0^\infty e^{-(s-a)t} f(t)dt = F(s-a) \quad \text{for } s - a > \gamma.$$

The converse result follows by reversing this argument to arrive at the result

$$\mathcal{L}^{-1}\{F(s-a)\} = e^{at} f(t). \qquad \blacksquare$$

EXAMPLE 7.11

Use Theorem 7.5 to find $\mathcal{L}\{e^{at} t^n\}$, $\mathcal{L}\{e^{at} \cos bt\}$, and $\mathcal{L}\{e^{at} t \sin bt\}$.

Solution   Using the Laplace transforms of t^n, $\cos bt$, and $t \sin bt$ listed as entries 3, 10, and 11 in Table 7.1, with a replaced by b in entries 10 and 11, and then replacing s by $s - a$ we find that

$$\mathcal{L}\{e^{at} t^n\} = \frac{n!}{(s-a)^{n+1}} \quad \text{for } s > 0, \quad \mathcal{L}\{e^{at} \cos bt\} = \frac{(s-a)}{[(s-a)^2 + b^2]} \quad \text{for } s > a,$$

and

$$\mathcal{L}\{e^{at} t \sin bt\} = \frac{2b(s-a)}{[(s-a)^2 + b^2]^2} \quad \text{for } s > a. \qquad \blacksquare$$

EXAMPLE 7.12

Use Theorem 7.5 to find $\mathcal{L}^{-1}\{1/(s^2 + 4s + 13)\}$.

Solution   Completing the square in the denominator we have

$$\mathcal{L}^{-1}\left\{\frac{1}{s^2 + 4s + 13}\right\} = \mathcal{L}^{-1}\left\{\frac{1}{(s+2)^2 + 3^2}\right\}.$$

A comparison with entry 13 in Table 7.1 shows that

$$\mathcal{L}^{-1}\{1/(s^2 + 4s + 13)\} = \frac{1}{3} e^{-2t} \sin 3t. \qquad \blacksquare$$

We now derive the second shift theorem, also called the *t*-shift theorem, in which use will be made of the unit step function $H(t - a)$.

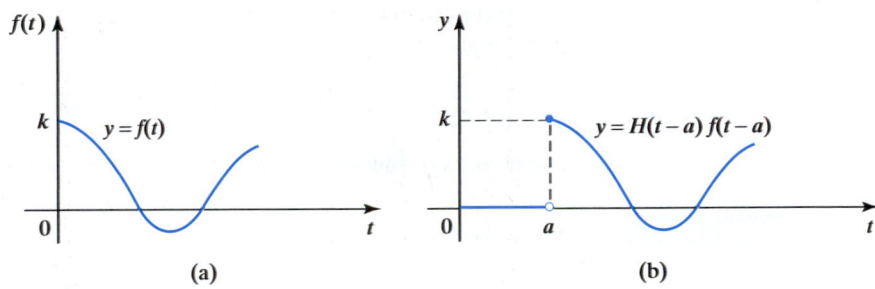

FIGURE 7.10 The relationship between $f(t)$ and $H(t-a)f(t-a)$.

THEOREM 7.6

the t-shift theorem

The second shift theorem or the t-shift theorem   Let $\mathcal{L}\{f(t)\} = F(s)$. Then

$$\mathcal{L}\{H(t-a)f(t-a)\} = e^{-as}F(s)$$

and, conversely,

$$\mathcal{L}^{-1}\{e^{-as}F(s)\} = H(t-a)f(t-a).$$

Proof   Before proving the theorem it is necessary to understand the precise meaning of $H(t-a)f(t-a)$. This can be seen by examining Fig. 7.10. The unit step function $H(t-a)$ is zero until $t = a$, when it jumps to the value 1 and thereafter remains constant for $t > a$. The function $f(t-a)$ is simply the function $f(t)$ with its origin shifted to $t = a$, so it can be considered to be the function $f(t)$ translated to the right by an amount a. Thus, $H(t-a)f(t-a)$ is a function that is zero until $t = a$, after which it reproduces the function $f(t)$ translated to the right by an amount a.

The result of the theorem is obtained as follows:

$$\mathcal{L}\{H(t-a)f(t-a)\} = \int_0^\infty e^{-st}H(t-a)f(t-a)dt = \int_a^\infty e^{-st}f(t-a)dt.$$

If we make the change of variable $\tau = t - a$, this becomes

$$\mathcal{L}\{H(t-a)f(t-a)\} = e^{-as}\int_0^\infty e^{-s\tau}f(\tau)d\tau$$

and so

$$\mathcal{L}\{H(t-a)f(t-a)\} = e^{-as}F(s).$$

The converse result follows by reversing this argument.   ∎

EXAMPLE 7.13

Use Theorem 7.6 to find (a) $\mathcal{L}\{H(t-4)\sin(t-4)\}$, (b) to show that $\mathcal{L}\{H(t-a)\} = e^{-as}/s$ in agreement with entry 7 in Table 7.1, and (c) to find $\mathcal{L}^{-1}\{se^{-as}/(s^2+b^2)\}$.

Solution   **(a)** From entry 9 in Table 7.1 we have $\mathcal{L}\{\sin t\} = 1/(s^2+1)$, so applying Theorem 7.6 with $a = 4$ gives

$$\mathcal{L}\{H(t-4)\sin(t-4)\} = e^{-4s}/(s^2+1).$$

(b) Setting $f(t) = 1$ in Theorem 7.6 and using the fact that $\mathcal{L}\{1\} = 1/s$ gives

$$\mathcal{L}\{H(t-a)\} = e^{-as}/s.$$

(c) Entry 10 in Table 7.1 shows that $\mathcal{L}\{\cos bt\} = s/(s^2 + b^2)$, so using this in Theorem 7.6 gives

$$\mathcal{L}^{-1}\{se^{-as}/(s^2 + b^2)\} = H(t - a)\cos[b(t - a)].$$

The next example makes use of Theorem 7.6 when solving an initial value problem.

EXAMPLE 7.14

Solve the initial value problem

$$y'' + 3y' + 2y = H(t - \pi)\sin 2t \quad \text{with} \quad y(0) = 1 \quad \text{and} \quad y'(0) = 0.$$

Solution   Setting $\mathcal{L}\{y(t)\} = Y(s)$, transforming the differential equation, and incorporating the initial conditions as in Example 7.10 gives

$$s^2 Y(s) - s + 3(sY(s) - 1) + 2Y(s) = \frac{2e^{-\pi s}}{s^2 + 4},$$

or

$$(s^2 + 3s + 2)Y(s) = s + 3 + \frac{2e^{-\pi s}}{s^2 + 4}.$$

As $s^2 + 3s + 2 = (s + 1)(s + 2)$, this last result can be written in the form

$$Y(s) = \frac{s + 3}{(s + 1)(s + 2)} + \frac{2e^{-\pi s}}{(s^2 + 4)(s + 1)(s + 2)}.$$

It is now necessary to invert $Y(s)$, and to accomplish this some algebraic manipulation will be necessary if we are to identify terms on the right with entries in Table 7.1. When expressed in terms of partial fractions, after a little manipulation $Y(s)$ becomes

$$Y(s) = \frac{2}{s + 1} - \frac{1}{s + 2} + e^{-\pi s}\left(\frac{2}{5}\frac{1}{s + 1} - \frac{1}{4}\frac{1}{s + 2} - \frac{1}{20}\frac{2}{s^2 + 4} - \frac{3}{20}\frac{s}{s^2 + 4}\right).$$

Each term can now be identified as the transform of an entry in Table 7.1, though as the last four terms are multiplied by $e^{-\pi s}$ their inverse Laplace transforms will need to be obtained by using Theorem 7.6. As a result, $y(t) = \mathcal{L}^{-1}\{Y(s)\}$ becomes

$$y(t) = 2e^{-t} - e^{-2t} + H(t - \pi)$$
$$\times \left(\frac{2}{5}e^{-(t-\pi)} - \frac{1}{4}e^{-2(t-\pi)} - \frac{1}{20}\sin 2(t - \pi) - \frac{3}{20}\cos 2(t - \pi)\right),$$

for $t > 0$. A graph of this solution is shown in Fig. 7.11, from which it can be seen that in the interval $0 < t < \pi$ the solution $y(t)$ only involves the first two terms, and so decays exponentially. At $t = \pi$ the forcing function $\sin 2t$ is switched on, after which all the exponential terms decay to zero as $t \to \infty$, leaving only the periodic steady state solution.

THEOREM 7.7

differentiating a transform

Differentiation of a transform: Multiplication of $f(t)$ by t^n   Let $\mathcal{L}\{f(t)\} = F(s)$. Then

$$\mathcal{L}\{t^n f(t)\} = (-1)^n \frac{d^n F(s)}{ds^n}.$$

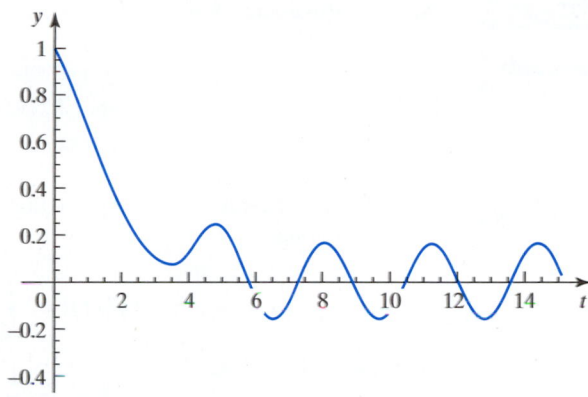

FIGURE 7.11   The solution $y(t)$ showing the influence of the forcing function after $t = \pi$.

Proof   By definition

$$\int_0^\infty e^{-st} f(t)dt = F(s),$$

so differentiating under the integral sign with respect to s gives

$$\frac{dF(s)}{ds} = \int_0^\infty \frac{\partial(e^{-st})}{\partial s} f(t)dt,$$

and so

$$\frac{dF(s)}{ds} = \int_0^\infty (-t)e^{-st} f(t)dt = -\int_0^\infty e^{-st} t f(t)dt,$$

which is the result of the theorem when $n = 1$. Each subsequent differentiation will introduce a further factor $(-t)$ into the integrand, leading the general result of the theorem.   ■

EXAMPLE 7.15

Use Theorem 7.7 to find (a) $\mathcal{L}\{t \sin at\}$ and (b) $\mathcal{L}\{t\, e^{at} \cos bt\}$.

Solution   **(a)** Entry 9 in Table 7.1 shows that $\mathcal{L}\{\sin at\} = a/(s^2 + a^2)$ for $s > 0$, so from Theorem 7.7

$$\mathcal{L}\{t \sin at\} = (-1)\frac{d}{ds}\frac{a}{(s^2 + a^2)} = \frac{2as}{(s^2 + a^2)^2} \quad \text{for } s > 0,$$

in agreement with entry 11 in Table 7.1.
(b) Entry 14 in Table 7.1 shows that $\mathcal{L}\{e^{at} \cos bt\} = (s - a)/[(s - a)^2 + b^2]$ for $s > a$, so from Theorem 7.7

$$\mathcal{L}\{t\, e^{at} \cos bt\} = (-1)\frac{d}{ds}\frac{(s - a)}{[(s - a)^2 + b^2]}$$

$$= \frac{(s - a)^2 - b^2}{[(s - a)^2 + b^2]^2} \quad \text{for } s > a.$$

These examples show that, in many cases, less effort is involved finding transforms by means of Theorem 7.7 than by direct use of the definition of the Laplace transform.

■

THEOREM 7.8

scaling a transform

Scaling theorem   Let $\mathcal{L}\{f(t)\} = F(s)$. Then if $k > 0$,

$$\mathcal{L}\{f(kt)\} = \frac{1}{k}F\left(\frac{s}{k}\right).$$

Proof   The result follows by setting $u = kt$ in the definition of the Laplace transform, because

$$\{f(kt)\} = \int_0^\infty e^{-st} f(kt)dt$$

$$= \frac{1}{k}\int_0^\infty e^{-s(u/k)} f(u)du$$

$$= \frac{1}{k}\int_0^\infty e^{-(s/k)u} du$$

$$= \frac{1}{k}F\left(\frac{s}{k}\right).$$ ∎

EXAMPLE 7.16

If $\mathcal{L}\{f(t)\} = e^{-3s}(1 - 2s)/(2s^2 - s + 1)$, find $\{f(3t)\}$.

Solution   In this case $k = 3 > 0$, so from Theorem 7.8, replacing s by $s/3$ in $\mathcal{L}\{f(t)\}$ and multiplying the result by $1/3$ gives

$$\mathcal{L}\{f(3t)\} = \frac{1}{3}\frac{e^{-s}(1 - 2s/3)}{(2(s/3)^2 - s/3 + 1)}$$

$$= \frac{e^{-s}(3 - 2s)}{2s^2 - 3s + 9}.$$ ∎

Many functions whose Laplace transform is required are periodic functions with period T, though they are not necessarily continuous functions for all $t > 0$. In the Laplace transform, where only the behavior of a function $f(t)$ for $t > 0$ is involved, a **periodic function** with **period** T is defined as a function $f(t)$ with the property that T is the smallest value for which

$$f(t + T) = f(t) \quad \text{for all } t > 0. \tag{12}$$

An example of a piecewise continuous function $f(t)$ with period T that is defined for $t > 0$ is shown in Fig. 7.12.

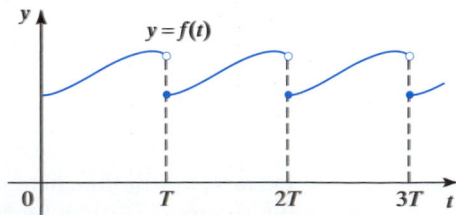

FIGURE 7.12   A function $f(t)$ with period T.

THEOREM 7.9

transforming a
periodic function

Transform of a periodic function with period T   Let $f(t)$ be a periodic function with period T such that $\int_0^T e^{-st} f(t)dt$ is finite. Then

$$\mathcal{L}\{f(t)\} = \frac{1}{1 - e^{-Ts}} \int_0^T e^{-st} f(t)dt \quad \text{for } s > 0.$$

Proof   In the definition of the Laplace transform we divide the interval of integration into subintervals of length T and write

$$\mathcal{L}\{f(t)\} = \int_0^T e^{-st} f(t)dt + \int_T^{2T} e^{-st} f(t)dt + \cdots.$$

Then, because of the periodicity of $f(t)$, the function $f(t)$ will be the same in each integral. Consequently, changing the variable in the $(r+1)$th integral to $t = \tau + rT$ with $r = 0, 1, 2, \ldots$ gives

$$\int_0^T e^{-s(\tau + rT)} f(\tau)d\tau = e^{-rsT} \int_0^T e^{-s\tau} f(\tau)d\tau \quad \text{for } r = 0, \ 1, 2, \ldots$$

$$= e^{-rsT} \int_0^T e^{-st} f(t)dt,$$

where the dummy variable τ has been replaced by t. Substituting this result into the original integral gives

$$\mathcal{L}\{f(t)\} = [1 + e^{-Ts} + e^{-2Ts} + \cdots] \int_0^T e^{-st} f(t)dt,$$

which is finite because we have assumed that $\int_0^T e^{-st} f(t)dt$ is finite. The bracketed terms form a geometrical series with the common ratio $e^{-Ts} < 1$, so its sum is $1/(1 - e^{-Ts})$, and thus

$$\mathcal{L}\{f(t)\} = \frac{1}{1 - e^{-Ts}} \int_0^T e^{-st} f(t)dt, \quad \text{for } s > 0,$$

and the proof is complete.   ∎

The necessity of the condition in Theorem 7.9 that $\int_0^T e^{-st} f(t)dt$ is finite arises because periodic functions exist for which this integral is divergent.

EXAMPLE 7.17

Find the Laplace transform of the square wave shown in Fig. 7.13.

Solution   As the function is discontinuous with period $2a$ we compute the integral in Theorem 7.9 in two parts as

$$\int_0^{2a} e^{-st} f(t)dt = \int_0^a ke^{-st}dt + \int_a^{2a} (-k)e^{-st}dt$$

$$= \frac{k}{s}(1 - e^{-as}) + \frac{k}{5}(e^{-2as} - e^{-as})$$

$$= \frac{k}{s}(1 + e^{-2as} - 2e^{-as}).$$

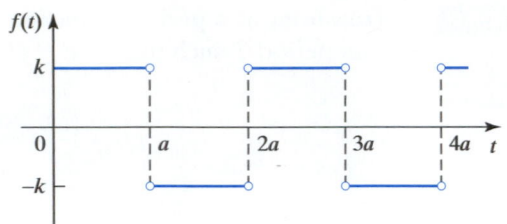

FIGURE 7.13 A square wave with period $2a$.

Then from Theorem 7.9 we have

$$\mathcal{L}\{f(t)\} = \frac{k(1 + e^{-2as} - 2e^{-as})}{s(1 - e^{-2as})}$$

$$= \frac{k(1 - e^{-as})}{s(1 + e^{-as})}$$

$$= \frac{k(e^{as/2} - e^{-as/2})}{s(e^{as/2} + e^{-as/2})}$$

$$= \frac{k\sinh(as/2)}{s\cosh(as/2)} = \frac{k}{s}\tanh(as/2) \quad \text{for } s > 0. \qquad \blacksquare$$

EXAMPLE 7.18

Use Theorem 7.9 to show that $\mathcal{L}\{\sin t\} = 1/(s^2 + 1)$ and Theorem 7.8 to show that $\mathcal{L}\{\sin at\} = a/(s^2 + a^2)$.

Solution The function $f(t) = \sin t$ is periodic with period 2π and $\int_0^{2\pi} e^{-st}\sin t\,dt$ is finite, so from Theorem 7.9 we have

$$\mathcal{L}\{\sin t\} = \frac{1}{(1 - e^{-2\pi s})}\int_0^{2\pi} e^{-st}\sin t\,dt$$

$$= \frac{1}{(1 - e^{-2\pi s})}\left(\frac{1}{s^2 + 1} - \frac{e^{-2\pi s}}{s^2 + 1}\right)$$

$$= \frac{1}{s^2 + 1} \quad \text{for } s > 0.$$

Setting $k = a$ in Theorem 7.8 and using the preceding result gives

$$\mathcal{L}\{\sin at\} = \frac{1}{a}\frac{1}{[(s/a)^2 + 1]}$$

$$= \frac{a}{s^2 + a^2} \quad \text{for } s > 0. \qquad \blacksquare$$

EXAMPLE 7.19

Find the Laplace transform of the solution of the initial value problem

$$y'' + 3y' + 2y = f(t), \quad \text{where } y(0) = y'(0) = 0$$

and $f(t)$ is the square wave in Example 7.17.

Solution Transforming the equation as in Examples 7.10 and 7.14 and using the result of Example 7.17 gives

$$s^2 Y(s) + 3s\,Y(s) + 2Y(s) = \frac{k}{s}\tanh(as/2),$$

so

$$Y(s) = \frac{k \tanh(as/2)}{s(s^2 + 3s + 2)}.$$ ∎

The convolution operation

Let the functions $f(t)$ and $g(t)$ be defined for $t \geq 0$. Then the **convolution** of the functions f and g denoted by $(f * g)(t)$, and in abbreviated form by $(f * g)$, is defined as the integral

$$(f * g)(t) = \int_0^t f(\tau)g(t - \tau)d\tau.$$

convolution and the convolution theorem

The change of variable $v = t - \tau$ followed by the replacement of the dummy variable v by t shows that the convolution operation is *commutative*, so

$$(f * g)(t) = (g * f)(t). \tag{13}$$

EXAMPLE 7.20

Find $(t^2 * \cos t)$ and $(\cos t * t^2)$ and hence confirm the equality of these two convolution operations. Compare the effort required in each case.

Solution   We have

$$(t^2 * \cos t) = \int_0^t \tau^2 \cos(t - \tau)d\tau$$

$$= \int_0^t \tau^2[\cos t \cos \tau + \sin t \sin \tau]d\tau$$

$$= \cos t \int_0^t \tau^2 \cos \tau \, d\tau + \sin t \int_0^t \tau^2 \sin \tau \, d\tau$$

$$= 2(t - \sin t).$$

Similarly,

$$(\cos t * t^2) = \int_0^t \cos \tau (t - \tau)^2 d\tau$$

$$= t^2 \int_0^t \cos \tau \, d\tau - 2t \int_0^t \tau \cos \tau \, d\tau + \int_0^t \tau^2 \cos \tau \, d\tau$$

$$= 2(t - \sin t).$$

While confirming that the convolution operation is commutative, this example also shows that sometimes calculating $(f * g)(t)$ is simpler than calculating $(g * f)(t)$. ∎

The convolution operation has various uses, one of the most important of which occurs in the following important theorem that expresses the relationship between the product of two Laplace transforms $F(s)$ and $G(s)$ and the convolution of their transform pairs $f(t)$ and $g(t)$.

THEOREM 7.10   **The convolution theorem**   Let $\mathcal{L}\{f(t)\} = F(s)$ and $\mathcal{L}\{g(t)\} = G(s)$. Then

$$\mathcal{L}\{(f * g)(t)\} = F(s)G(s)$$

or, equivalently,

$$\mathcal{L}\left\{\int_0^t f(\tau)g(t - \tau)d\tau\right\} = F(s)G(s).$$

Conversely,

$$\mathcal{L}^{-1}\{F(s)G(s)\} = \int_0^t f(\tau)g(t - \tau)d\tau.$$

Proof   From the definition of the Laplace transform and the convolution operation, we have

$$\mathcal{L}\{(f * g)(t)\} = \int_0^\infty e^{-st}\left[\int_0^t f(\tau)g(t - \tau)d\tau\right]dt.$$

Inspection of Fig. 7.14 shows that interchanging the order of integration allows the integral to be written as

$$\mathcal{L}\{(f * g)(t)\} = \int_0^\infty f(\tau)\left[\int_\tau^\infty e^{-st}g(t - \tau)dt\right]d\tau.$$

Using the second shift theorem reduces the inner integral to $e^{-st}G(s)$, so that

$$\mathcal{L}\{(f * g)(t)\} = \int_0^\infty G(s)e^{-s\tau}f(\tau)d\tau$$

$$= G(s)\int_0^\infty e^{-s\tau}f(\tau)d\tau$$

$$= G(s)F(s).$$

The converse result follows if we reverse the argument to find the inverse Laplace transform of $F(s)G(s)$.   ∎

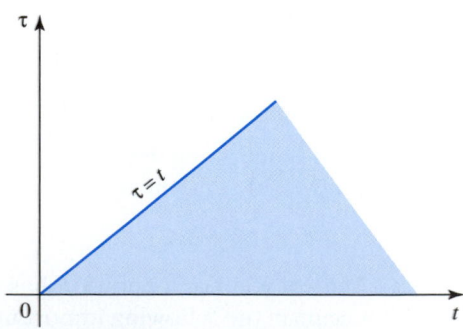

FIGURE 7.14   Region of integration for Theorem 7.10.

EXAMPLE 7.21

Use Theorem 7.10 to find (a) $\mathcal{L}\{t^2 * \cos t\}$ and (b) $\mathcal{L}^{-1}\{s/(s^2 + a^2)^2\}$.

Solution

(a) $\mathcal{L}\{t^2\} = 2/s^3$ and $\mathcal{L}\{\cos t\} = s/(s^2 + a^2)$, so from Theorem 7.10

$$\mathcal{L}\{t^2 * \cos t\} = \mathcal{L}\{t^2\}\,\mathcal{L}\{\cos t\} = \frac{2s}{(s^2 + a^2)}.$$

(b) Writing

$$\frac{s}{(s^2 + a^2)^2} = \frac{1}{(s^2 + a^2)}\frac{s}{(s^2 + a^2)}$$

shows that in Theorem 7.10 we may take

$$F(s) = \frac{1}{(s^2 + a^2)} \quad \text{and} \quad G(s) = \frac{s}{(s^2 + a^2)}.$$

So as $\mathcal{L}^{-1}\{F(s)\} = (1/a)\sin at$ and $\mathcal{L}^{-1}\{G(s)\} = \cos at$, it follows from Theorem 7.10 that

$$\mathcal{L}^{-1}\{s/(s^2 + a^2)^2\} = (1/a)(\sin at * \cos at)$$

$$= \frac{1}{a}\int_0^t \sin a\tau \cos a(t - \tau)d\tau$$

$$= \frac{1}{2a}t \sin at,$$

in agreement with entry 11 in Table 7.1.   ■

When evaluating convolution integrals of this type, instead of expanding a term such as $\cos a(t - \tau)$ and $\sin a(t - \tau)$ using integration by parts, it is often quicker to replace $\sin at$ and $\cos at$ by

$$\sin at = (e^{iat} - e^{-iat})/(2i) \quad \text{and} \quad \cos a(t - \tau) = \left(e^{i(t-\tau)} + e^{-i(t-\tau)}\right)/2$$

before performing the integrations, and again using these identities to interpret the result in terms of trigonometric functions.

EXAMPLE 7.22

Solve the initial value problem

$$y'' + 4y' + 13y = 2e^{-2t}\sin 3t \quad \text{with } y(0) = 1 \quad \text{and} \quad y'(0) = 0.$$

Solution   Before we solve this initial value problem, it should be noted that the complementary function is

$$y_c(t) = e^{-2t}(C_1 \cos 3t + C_2 \sin 3t),$$

so the nonhomogeneous term $2e^{-2t}\sin 3t$ is contained in $y_c(t)$. It will be seen that, unlike the special cases that arise when determining a particular integral by the method of undetermined coefficients, this situation does not give rise to a special case when the solution is obtained by means of the Laplace transform.

Transforming the equation in the usual way gives

$$s^2 Y(s) - s + 4(s Y(s) - 1) + 13Y(s) = \frac{6}{s^2 + 4s + 13},$$

and so

$$Y(s) = \frac{s+4}{s^2 + 4s + 13} + \frac{6}{(s^2 + 4s + 13)^2}.$$

Writing $s + 4 = s + 2 + (2/3)3$ allows $Y(s)$ to be rewritten as

$$Y(s) = \frac{s+2}{(s+2)^2 + 3^2} + \frac{2}{3}\frac{3}{(s+2)^2 + 3^2} + \frac{6}{[(s+2)^2 + 3^2]^2}.$$

Taking the inverse Laplace transform of $Y(s)$ and using entries 13 and 14 of Table 7.1 leads to the result

$$y(t) = e^{-2t}\left[\cos 3t + \frac{2}{3}\sin 3t\right] + \mathcal{L}^{-1}\{6/[(s+2)^2 + 3^2]^2\}.$$

To find $\mathcal{L}^{-1}\{6/[(s+2)^2 + 3^2]^2\}$, we first write this as

$$\frac{6}{[(s+2)^2 + 3^2]^2} = \frac{2}{3}\left(\frac{3}{(s+2)^2 + 3^2}\right)\left(\frac{3}{(s+2)^2 + 3^2}\right),$$

and then, from entry 13 in Table 7.1, we find that $\mathcal{L}^{-1}\{3/[(s+2)^2 + 3^2]\} = e^{-2t}\sin 3t$. An application of Theorem 7.10 shows that

$$\mathcal{L}^{-1}\{6/[(s+2)^2 + 3^2]^2\} = \frac{2}{3}(e^{-2t}\sin 3t * e^{-2t}\sin 3t)$$

$$= \frac{2}{3}\int_0^t e^{-2\tau}\sin 3\tau e^{-2(t-\tau)}\sin 3(t-\tau)d\tau$$

$$= \frac{2}{3}e^{-2t}\int_0^t \sin 3\tau \sin 3(t-\tau)d\tau$$

$$= \frac{2}{3}e^{-2t}\left(\frac{1}{6}\sin 3t - \frac{1}{2}t\cos 3t\right).$$

Substituting this result in the expression for $y(t)$ shows that the solution of the initial value problem is

$$y(t) = e^{-2t}\left(\cos 3t + \frac{7}{9}\sin 3t - \frac{1}{3}t\cos 3t\right), \quad \text{for } t > 0. \qquad \blacksquare$$

Although the previous example could have been solved by the method of undetermined coefficients, the next two examples cannot be solved in this manner. The first involves a special type of equation called an **integral equation**, and the second an **integro-differential** equation.

An equation of the form

$$y(t) = f(t) + \lambda\int_0^t K(t, \tau)y(\tau)d\tau \qquad (14)$$

is called a **Volterra integral equation**, where λ is a parameter and $K(t, \tau)$ is called the **kernel** of the integral equation. Equations of this type are often associated with the solution of initial value problems. The Laplace transform is well suited to the solution of such integral equations when the kernel $K(t, \tau)$ has a special form that depends on t and τ only through the difference $t - \tau$, because then $K(t, \tau) = K(t - \tau)$ and the integral in (14) becomes a convolution integral.

An examination of the Volterra integral equation in (14) shows it to be essentially the integral form of an initial value problem, and it relates the solution $y(t)$ at the current time t to an integral of the past history of the solution over the interval $[0, t]$.

The following is a simple example of a problem that leads to a Volterra integral equation. Determine the amount of a manufactured material contained in a store from time $t = 0$ until time t, if the only supply of material comes immediately from the manufacturer and it begins degrading exponentially with time from the moment it enters the store. Let the amount of material present at time $t = 0$ be Q and the amount present in the store at time t be $y(t)$, and suppose it degrades exponentially as e^{-kt} with $k > 0$. Then, by time t, the amount of material that entered the store at time τ but has not degraded is $e^{-k(t-\tau)}y(\tau)$. Thus the amount of material present at time t is determined by the solution of the Volterra integral equation

$$y(t) = Qe^{-kt} + \int_0^t e^{-k(t-\tau)}y(\tau)d\tau.$$

By using the method of solution explained in the next example, the solution of this problem is easily shown to be

$$y(t) = Qe^{-(k-1)t}.$$

EXAMPLE 7.23 Solve the Volterra integral equation

$$y(t) = 2e^{-t} + \int_0^t \sin(t - \tau)y(\tau)d\tau.$$

Solution   The Laplace transform of the integral equation is

$$Y(s) = \frac{2}{s+1} + \mathcal{L}\int_0^t \sin(t - \tau)y(\tau)d\tau,$$

and after applying Theorem 7.10 to the last term the equation for $Y(s)$ becomes

$$Y(s) = \frac{2}{s+1} + \frac{Y(s)}{s^2+1}.$$

Solving for $Y(s)$ and expanding the result in partial fractions shows that

$$Y(s) = \frac{2(s^2+1)}{s^2(s+1)} = \frac{2}{s^2} - \frac{2}{s} + \frac{4}{s+1}.$$

Taking the inverse Laplace transform shows the solution to be

$$y(t) = 2t - 2 + 4e^{-t}, \quad \text{for } t > 0. \qquad \blacksquare$$

The next example is a differential equation of an unusual type, because the function $y(t)$ occurs not only as the dependent variable in the differential equation, but also inside a convolution integral that forms the nonhomogeneous term. Equations of this type that involve both the integral of an unknown function and its derivative are called **integro-differential equations**. These equations occur in many applications of mathematics, one of which arises in the continuum mechanics of polymers, where the dynamical response $y(t)$ of certain types of material at time t depends on a derivative of $y(t)$ and the time-weighted cumulative effect of what has happened to the material prior to time t. For obvious reasons materials of this type are called *materials with memory*.

integro-differential equation

An example of an integro-differential equation was obtained in Section 5.3(d) when considering the R–L–C circuit in Fig. 5.4, though at the time this was not recognized. When the circuit was closed, and the charge q on the capacitor was allowed to flow causing a current $i(t)$ in the circuit, the equation determining $i(t)$ was shown to be

$$L\frac{di}{dt} + Ri + \frac{q}{C} = 0.$$

To recognize that this is an integro-differential equation, we use the result that at time t we have $q = \int_0^t i(\tau)d\tau$, so the equation determining $i(t)$ becomes the integro-differential equation

$$L\frac{di}{dt} + Ri + \frac{1}{C}\int_0^t i(\tau)d\tau.$$

In this case it was possible to reduce this to a second order constant coefficient differential equation for $i(t)$, but in other more complicated cases a reduction of this type may not be possible.

EXAMPLE 7.24

Solve the equation

$$y'' + y = \int_0^t \sin\tau\, y(t - \tau)d\tau,$$

subject to the initial conditions $y(0) = 1$ and $y'(0) = 0$.

Solution   Taking the Laplace transform in the usual way gives

$$s^2 Y(s) - s + Y(s) = \mathcal{L}\int_0^t \sin\tau\, y(t - \tau)d\tau.$$

The last term is the Laplace transform of a convolution integral, so from Theorem 7.10 it follows that

$$\mathcal{L}\left\{\int_0^t \sin\tau\, y(t - \tau)d\tau\right\} = \mathcal{L}\{\sin t\}\mathcal{L}\{y(t)\}$$

$$= \frac{Y(s)}{s^2 + 1}.$$

Using this result in the transformed equation, solving for $Y(s)$, and expanding the result using partial fractions gives

$$Y(s) = \frac{s^2 + 1}{s(s^2 + 2)} = \frac{1}{2}\frac{1}{s} + \frac{1}{2}\frac{s}{(s^2 + 2)}.$$

After the inverse Laplace transform is taken, the solution becomes

$$y(t) = \frac{1}{2}(1 + \cos\sqrt{2}t), \quad \text{for } t > 0. \qquad \blacksquare$$

THEOREM 7.11

transforming an integral

The transform of an integral   Let $f(t)$ be a piecewise continuous function such that $|f(t)| \leq Me^{kt}$ for $k > 0$ and all $t \geq 0$. Then, if $\mathcal{L}\{f(t)\} = F(s)$,

$$\mathcal{L}\left\{\int_0^t f(\tau)d\tau\right\} = \frac{F(s)}{s} \quad \text{for} \quad s > k,$$

and, conversely,

$$\mathcal{L}^{-1}\{F(s)/s\} = \int_0^t f(\tau)d\tau.$$

Proof   The condition $|f(t)| \leq Me^{kt}$ is sufficient to ensure the existence of the Laplace transform $F(s)$, so writing $h(t) = \int_0^t f(\tau)d\tau$ we have

$$|h(t)| \leq \int_0^t |f(\tau)|d\tau \leq M \int_0^t e^{k\tau}d\tau \leq M\frac{e^{kt}}{k} \quad \text{for} \quad t \geq 0.$$

This result shows that $|h(t)|$ grows no faster than $|f(t)|$ as $t \to \infty$, so the existence of the Laplace transform $Y(s)$ ensures the existence of the Laplace transform of $h(t)$. Using the fundamental result from the calculus that $h'(t) = f(t)$ together with Theorem 7.2 means that, apart from points where $f(t)$ is discontinuous,

$$F(s) = \mathcal{L}\{f(t)\} = \mathcal{L}\{h'(t)\} = s\mathcal{L}\{h(t)\} = s\mathcal{L}\left\{\int_0^t f(\tau)d\tau\right\},$$

and so

$$\mathcal{L}\left\{\int_0^t f(\tau)d\tau\right\} = \frac{F(s)}{s}.$$

The converse result follows by taking the inverse Laplace transform and the proof is complete.   ■

EXAMPLE 7.25   Find (a) $\mathcal{L}\{\int_0^t \tau \cos a\tau d\tau\}$ and (b) $\mathcal{L}^{-1}\{1/[s(s^2 + a^2)]\}$.

Solution   **(a)** As $\mathcal{L}\{t \cos at\} = (s^2 - a^2)/(s^2 + a^2)^2$ for $s > 0$, an application of Theorem 7.11 shows that

$$\mathcal{L}\left\{\int_0^t \tau \cos a\tau d\tau\right\} = \frac{s^2 - a^2}{s(s^2 + a^2)^2} \quad \text{for } s > 0.$$

(b) We can write

$$\frac{1}{s(s^2 + a^2)} = \frac{1}{s^2 + a^2}\frac{1}{s}.$$

So if we set $F(s) = 1/(s^2 + a^2)$, for which $f(t) = \mathcal{L}^{-1}F(s) = (1/a)\sin at$, it follows from Theorem 7.11 that

$$\mathcal{L}^{-1}\left\{\frac{F(s)}{s}\right\} = \mathcal{L}^{-1}\left\{\frac{1}{s(s^2 + a^2)}\right\} = \int_0^t \frac{1}{a}\sin a\tau d\tau$$

$$= \frac{1}{a^2}(1 - \cos at),$$

in agreement with entry 17 of Table 7.1.   ■

THEOREM 7.12

integrating a transform

The integral of a transform   Let $f(t)/t$ be piecewise continuous, defined for $t \geq 0$ and such that $|f(t)/t| \leq Me^{-kt}$ for $t \geq 0$. Then if $\mathcal{L}\{f(t)/t\} = G(s)$ for $s > k$, and $\mathcal{L}\{f(t)\} = F(s)$,

$$\mathcal{L}\left\{\frac{f(t)}{t}\right\} = \int_{s}^{\infty} F(u)du$$

and, conversely,

$$\mathcal{L}^{-1}\{G(s)\} = \frac{-1}{t}\mathcal{L}^{-1}\{G'(s)\}.$$

Proof   We have

$$G(s) = \int_{0}^{\infty} e^{-st}\frac{f(t)}{t} \quad \text{for} \quad s > k.$$

However, from Theorem 7.7,

$$G'(s) = \int_{0}^{\infty} e^{-st}(-t)\frac{f(t)}{t}dt = -\int_{0}^{\infty} e^{-st}f(t)dt = -F(s),$$

so after integration we have

$$\int_{s}^{\infty} F(u)du = -\int_{s}^{\infty} G'(u)du = G(s) - G(\infty)$$

To proceed further we now make use of the fact that the condition $|f(t)/t| \leq Me^{-kt}$ implies that $G(s)_{\lim s \to \infty} = 0$, showing that

$$G(s) = \mathcal{L}\{f(t)/t\} = \int_{s}^{\infty} F(u)du \quad \text{for } s > k.$$

The converse result follows by taking the inverse Laplace transform and using the fact that $\mathcal{L}^{-1}\{G(s)\} = f(t)/t$ together with the result $\mathcal{L}\{f(t)\} = F(s) = -G'(s)$. ∎

EXAMPLE 7.26

Find

$$\text{(a) } \mathcal{L}\left\{\frac{\sin at}{t}\right\} \quad \text{and} \quad \text{(b) } \mathcal{L}^{-1}\left\{\ln\left(\frac{s+a}{s+b}\right)\right\}.$$

Solution   **(a)** The function $(\sin at)/t$ is defined and finite for all $t > 0$, so Theorem 7.12 can be applied. If we use the fact that $\mathcal{L}\{\sin at\} = a/(s^2 + a^2)$, it follows from the first part of Theorem 7.12 that

$$\mathcal{L}\left\{\frac{\sin at}{t}\right\} = \int_{s}^{\infty} \frac{a}{u^2 + a^2}du$$

$$= \pi/2 - \text{Arctan}(s/a)$$

$$= \text{Arctan}(a/s).$$

(b)   If we set

$$G(s) = \ln\left(\frac{s+a}{s+b}\right),$$

differentiation gives

$$G'(s) = \frac{b-a}{(s+a)(s+b)} = \frac{1}{s+a} + \frac{1}{s+b},$$

from which we see that

$$\mathcal{L}^{-1}\{G'(s)\} = e^{-at} - e^{-bt}.$$

From the second part of Theorem 7.11 we have

$$\mathcal{L}^{-1}\{G(s)\} = \mathcal{L}^{-1}\left\{\ln\left(\frac{s+a}{s+b}\right)\right\} = \frac{-1}{t}\mathcal{L}^{-1}\{G'(s)\}$$
$$= (e^{-bt} - e^{-at})/t.$$

The conditions of Theorem 7.11 assert that method used to derive this result is permissible if $\mathcal{L}^{-1}\{G(s)\}$ is defined and finite for $t \geq 0$. We see from the preceding result that $\mathcal{L}^{-1}\{G(s)\}$ is defined and finite for $t > 0$ and $\lim_{t\to 0}[(e^{-bt} - e^{-at})/t] = a - b$, so the conditions of the theorem are satisfied and we have shown that

$$\mathcal{L}^{-1}\left\{\ln\left(\frac{s+a}{s+b}\right)\right\} = (e^{-bt} - e^{-at})/t. \qquad \blacksquare$$

The theorem that follows shows how the initial values $f(0)$, $f'(0)$, ..., of a suitably differentiable function $f(t)$ can be found directly from its Laplace transform $F(s)$. An example of the use of the theorem is to be found in Section 7.3(d) when determining the Laplace transform of a function known only as the solution of a differential equation.

THEOREM 7.13

relating initial values and the transform

The initial value theorem   Let $\mathcal{L}\{f(t)\} = F(s)$ be the Laplace transform of an n times differentiable function $f(t)$. Then

$$f^{(r)}(0) = \lim_{s\to\infty}\left\{s^{r+1}F(s) - s^r f(0) - s^{r-1}f'(0) - \cdots - sf^{(r-1)}(0)\right\},$$
$$r = 0, 1, \ldots, n.$$

In particular,

$$f(0) = \lim_{s\to\infty}\{sF(s)\}, \quad f'(0) = \lim_{s\to\infty}\{s^2 F(s) - sf(0)\}$$
$$f''(0) = \lim_{s\to\infty}\{s^3 F(s) - s^2 f(0) - sf'(0)\}.$$

Proof   The theorem follows directly from Theorem 7.3 by first replacing n by $r+1$ and rewriting the result as

$$f^{(r)}(0) = s^{r+1}F(s) - s^r f(0) - \cdots - sf^{(r-1)}(0) - \mathcal{L}\{f^{(r+1)}(t)\}.$$

Then, provided $f^{(r+1)}(t)$ satisfies the sufficiency condition for the existence of a Laplace transform given in (3), it follows that for some $M > 0$ and $k > 0$

$$\mathcal{L}\{f^{(r+1)}(t)\} < M/(s-k) \quad \text{for} \quad s > k \quad \text{and} \quad r = 0, 1, \ldots, n.$$

As a result,

$$\lim_{s\to\infty}\{f^{(r+1)}(t)\} = 0,$$

and the theorem is proved. $\blacksquare$

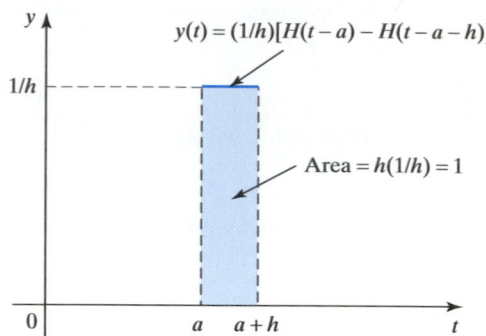

FIGURE 7.15 $\delta(t-a) = \lim_{h\to 0} y(t)$.

EXAMPLE 7.27

Given that $F(s) = 2as/(s^2 + a^2)^2$, use Theorem 7.13 to find $f(0)$, $f'(0)$, and $f''(0)$. Use $f(t) = \mathcal{L}^{-1}\{F(s)\} = t\sin at$ to confirm the results by direct differentiation.

Solution   From Theorem 7.13

$$f(0) = \lim_{s\to\infty}\{sF(s)\} = \lim_{s\to\infty}\frac{2as^2}{(s^2 + a^2)^2} = 0,$$

$$f'(0) = \lim_{s\to\infty}\{s^2 F(s) - sf(0)\} = \lim_{s\to\infty}\frac{2as^3}{(s^2 + a^2)^2} = 0,$$

$$f''(0) = \lim_{s\to\infty}\{s^3 F(s) - s^2 f(0) - sf'(0)\} = \lim_{s\to\infty}\frac{2as^4}{(s^2 + a^2)^2} = 2a.$$

These results are easily confirmed by differentiation of $f(t) = t\sin at$. ∎

The last operational property to be considered concerns the **Dirac delta function**, usually abbreviated to the **delta function** and sometimes called the **unit impulse function**. The Dirac delta function, named after the Oxford University Nobel laureate mathematical physicist P. A. M. Dirac and denoted by $\delta(t - a)$, is actually a limiting mathematical *operation*, and not a function as its name implies. For our purposes the delta function can be considered to be the limit of a rectangular "pulse" of height h and width $1/h$ in the limit as $h \to \infty$. Thus the area of the graph representing the pulse remains constant at 1 as $h \to \infty$, while its height increases to infinity and its width decreases to zero. The graphical representation of such a pulse $f(t) = \delta(t - a)$ located at $t = a$, before proceeding to the limit, is shown in Fig. 7.15.

We adopt the following definition of the delta function in terms of the unit step function.

The delta function

the delta or impulse function

The **delta function** located at $t = a$ and denoted by $\delta(t - a)$ is defined as the limit

$$\delta(t - a) = \lim_{h\to 0}\frac{1}{h}[H(t - a) - H(t - a - h)].$$

The operational property of the delta function, usually called its **filtering property** and sometimes its **sifting property**, is represented by the following theorem.

THEOREM 7.14

a useful property of the delta function

Filtering property of the delta function   Let $f(t)$ be defined and integrable over all intervals contained within $0 \leq t < \infty$, and let it be continuous in a neighborhood of a. Then for $a \geq 0$

$$\int_0^\infty f(t)\delta(t-a)dt = f(a).$$

Proof   From the definition of the delta function,

$$\int_0^\infty f(t)\delta(t-a)dt = \lim_{h \to 0} \int_a^{a+h} \frac{f(t)}{h}dt,$$

so applying the mean value theorem for integrals we have

$$\int_0^\infty f(t)\delta(t-a)dt = \lim_{h \to 0} \left[h\left(\frac{1}{h}\right)f(t_h) \right],$$

where $a < t_h < a + h$. In the limit as $h \to 0$ the variable $t_h \to a$, showing that

$$\int_0^\infty f(t)\delta(t-a)dt = f(a),$$

and the theorem is proved.   ∎

Consideration of the definition of the delta function suggests that, in a sense, $\delta(t-a)$ is the derivative of the unit step function $H(t-a)$, though the justification of this conjecture requires arguments involving **generalized functions** that are beyond the scope of this account.

In mechanical problems the delta function is used to represent an *impulse*, defined as the integral of a large force applied locally for a very short time. The delta function has many other applications, such as the distribution of point masses along a supporting beam, whereas in electrical systems it can be used to represent the brief application of a very large voltage, or the sudden discharge of energy contained in a capacitor.

A purely formal derivation of the Laplace transform of the delta function proceeds as follows. By definition,

$$\mathcal{L}\{\delta(t-a)\} = \int_0^\infty e^{-st}\delta(t-a)dt.$$

An application of the filtering property of Theorem 7.14 reduces this to

$$\mathcal{L}\{\delta(t-a)\} = e^{-as}. \tag{15}$$

As a special case we have

$$\mathcal{L}\{\delta(t)\} = 1. \tag{16}$$

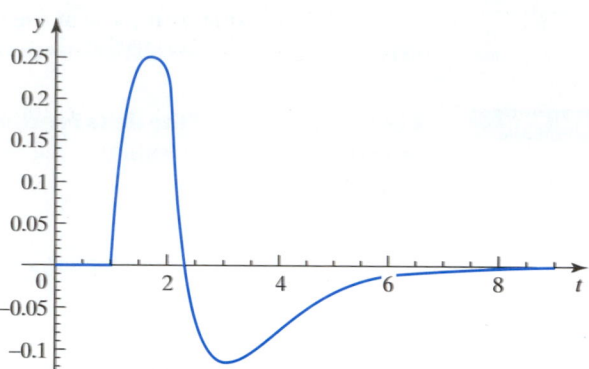

FIGURE 7.16   The solution $y(t)$ as a function of the time t.

Solve the initial value problem

$$y'' + 3y' + 2y = \delta(t-1) - \delta(t-2) \quad \text{with } y(0) = y'(0) = 0.$$

Solution   Taking the Laplace transform in the usual way and using result (15) gives

$$(s^2 + 3s + 2)Y(s) = e^{-s} - e^{-2s},$$

and so

$$Y(s) = \frac{e^{-s} - e^{-2s}}{s^2 + 3s + 2} = \frac{e^{-s} - e^{-2s}}{s+1} - \frac{e^{-s} - e^{-2s}}{s+2}.$$

Inverting the transform using Theorem 7.6 (the t-shift theorem) shows that

$$y(t) = H(t-1)[e^{1-t} - e^{2-2t}] - H(t-2)[e^{2-t} - e^{4-2t}].$$

A graph of this solution is given in Fig. 7.16. The graph shows that a physical system represented by the given differential equation subject to the equilibrium initial conditions $y(0) = y'(0) = 0$ is at rest until it is excited by the delta function at time $t = 1$ and then, after peaking just before $t = 2$, it is excited in the opposite sense by the delta function at time $t = 2$, after which the solution decays to zero as t increases, corresponding to the system returning to rest.

The Laplace transform is also discussed in references [3.4], [3.8], [3.9], [3.17], and [3.20]; tables of Laplace transform pairs are to be found in references [G.1], [G.3], [3.11], and [3.14]. An advanced account of the Laplace transform is to be found in reference [3.19].   ∎

PAUL ADRIEN DIRAC (1902–1984)
An English mathematical physicist who introduced the delta function in a fundamental paper on quantum mechanics presented to the Royal Society of London in 1927. Together with the German physicist **Erwin Schrodinger** he shared the Nobel Prize for physics because of contributions made to quantum mechanics.

Summary

This section has been concerned with what are known as the operational properties of the Laplace transform. These are general properties of the transform itself that can be applied to any function $f(t)$ that possesses a Laplace transform, or to any function $F(s)$ that is the Laplace transform of a function $f(t)$. It will be seen later that these properties can be used to extend the table of Laplace transforms given at the end of Section 7.1, and when using the Laplace transform to solve differential equations.

EXERCISES 7.2

Exercises involving the transformation of derivatives

1. Prove that $\mathcal{L}\{f''(t)\} = s^2 F(s) - sf(0) - f'(0)$.
2. Prove that $\mathcal{L}\{f'''(t)\} = s^3 F(s) - s^2 f(0) - sf'(0) - f''(0)$.
3. Given that $f(0) = 1$, $f'(0) = 0$, $f''(0) = 1$, find $\mathcal{L}\{f'''(t)\}$.
4. Given that $f(0) = 0$, $f'(0) = 2$, $f''(0) = 2$, $f'''(0) = -4$, find $\mathcal{L}\{f^{(4)}(t)\}$.
5. Given that $f(t) = \begin{cases} \sin t, & 0 \le t < \pi/2 \\ t = 0, & t \ge \pi/2 \end{cases}$, find $\mathcal{L}\{f(t)\}$.
6. Given that $f(t) = \begin{cases} \sin t, & 0 \le t < \pi/2 \\ 1, & t \ge \pi/2 \end{cases}$, find $\mathcal{L}\{f(t)\}$.
7. Solve $y'' - 3y' + 2y = \cos t$, with $y(0) = 1$, $y'(0) = -1$.
8. Solve $y'' + 5y' + 4y = \exp(-t)$, with $y(0) = 1$, $y'(0) = 0$.
9. Solve $y'' + 8y' - 9y = t$, with $y(0) = 2$, $y'(0) = 1$.
10. Solve $y'' + 5y' + 6y = 1 + t^2$, with $y(0) = 0$, $y'(0) = 0$.

Exercises involving the first shift theorem (s-shift)

11. Find $\mathcal{L}\{(2 + t^3)e^{-2t}\}$.
12. Find $\mathcal{L}\{e^{-3t} \cos 2t\}$.
13. Find $\mathcal{L}\{e^{-t} t \sin 2t\}$.
14. Find $\mathcal{L}\{(1 + t^2)e^{-4t}\}$.
15. Find $\mathcal{L}\{e^{2t} \sin 3t\}$.
16. Find $\mathcal{L}\{e^{-4t} \sinh 3t\}$.
17. Find $\mathcal{L}^{-1}\{1/(s^2 - 4s + 13)\}$.
18. Find $\mathcal{L}^{-1}\{s/(s^2 + 4s + 13)\}$.
19. Find $\mathcal{L}^{-1}\{(1 - 3s)/(s^2 + 2s + 5)\}$.
20. Find $\mathcal{L}^{-1}\{1/[s(s^2 - 2s + 5)]\}$.
21. Find $\mathcal{L}^{-1}\{s/[(s + 1)(s^2 - 4s + 13)]\}$.
22. Find $\mathcal{L}^{-1}\{3/(s^2 + 6s + 25)\}$.
23. Find $\mathcal{L}^{-1}\{3(s^2 + 4)/[s(s^2 + 4s + 8)]\}$.
24. Find $\mathcal{L}^{-1}\{2/[(s + 3)^2(s^2 + 8s + 20)]\}$.

Exercises involving graphing functions with a t-shift

25. Sketch $f(t) = H(t - 2)(1 + t)$.
26. Sketch $f(t) = H(t - \pi) \sin t + H(t - 2\pi)$.
27. Sketch $f(t) = [H(t - \pi) - H(t - 2\pi)] \cos t$.
28. Sketch $f(t) = \sum_{r=0}^{4} H(t - r)$.
29. Sketch $f(t) = H(t - \pi) \cos(t - \pi)$.
30. Sketch $f(t) = H(t - 1)(t - 1)^2$.
31. Sketch $f(t) = [H(t - 1) - H(t - 2)](t - 1)^2$.
32. Sketch $f(t) = H(t - \pi/2) \cos(t - \pi/2)$.

Exercises involving the second shift theorem (t-shift)

33. Find $\mathcal{L}\{H(t - 3)(t - 3)^3\}$.
34. Find $\mathcal{L}\{H(t - 1) \sin(t - 1)\}$.
35. Find $\mathcal{L}\{H(t - 3\pi/2) \sin 2(t - 3\pi/2)\}$.
36. Find $\mathcal{L}\{H(t - \pi/2)(t - \pi/2)^3 - H(t - 3\pi/2) \times (t - 3\pi/2)^3\}$.
37. Find $\mathcal{L}\{H(t - 4) \sinh 3(t - 4)\}$.
38. Find $\mathcal{L}\{H(t - 1)(t - 1) \sin(t - 1)\}$.
39. Find $\mathcal{L}^{-1}\{s\, e^{-2s}/(s^2 + 4)\}$.
40. Find $\mathcal{L}^{-1}\{e^{-\pi s/3}/(s^2 + 9)\}$.
41. Find $\mathcal{L}^{-1}\{e^{-\pi s/2}(s + 1)/(s^2 + 4s + 5)\}$.
42. Find $\mathcal{L}^{-1}\{e^{-2s}(s^2 + s + 1)/[s(s + 2)^2]\}$.
43. Find $\mathcal{L}^{-1}\{e^{-4s}(s + 3)/(s^2 + 4s + 13)\}$.
44. Find $\mathcal{L}^{-1}\{e^{-3s}s^2/[s(s^2 + 4s + 8)]\}$.
45. Solve $y'' + 5y' + 6y = H(t - \pi) \cos(t - \pi)$, with $y(0) = 1$, $y'(0) = 0$.
46. Solve $y'' - 5y' + 6y = t H(t - 1)$, with $y(0) = 0$, $y'(0) = 0$.
47. Solve $y'' - 5y' + 6y = 1 + t H(t - 2)$, with $y(0) = 0$, $y'(0) = 1$.
48. Solve $y'' - 6y' + 10y = t H(t - 3)$, with $y(0) = 1$, $y'(0) = 1$.
49. Solve $y'' + 2y' + 10y = e^{-t} H(t - 1)$, with $y(0) = -1$, $y'(0) = 0$.
50. Solve $y'' - y' - 2y = e^{-t} H(t - 1)$, with $y(0) = 1$, $y'(0) = 0$.

Exercises involving differentiation of transforms

51. Find $\mathcal{L}\{t^2 e^{3t} \sin t\}$.
52. Find $\mathcal{L}\{te^{-t} \sin 4t\}$.
53. Find $\mathcal{L}\{t^3 e^{2t} \sin 2t\}$.
54. Find $\mathcal{L}\{t^2 e^{3t} \cos 2t\}$.

Exercises involving scaling

55. If $\mathcal{L}\{f(t)\} = e^{-3s}(s^2 - 1)/(s^4 - a^4)$, find $\mathcal{L}\{f(2t)\}$.
56. If $\mathcal{L}\{f(t)\} = (s + 1)(s^2 + 2)/(s^2 + 4)^2$, find $\mathcal{L}\{f(3t)\}$.
57. If $\mathcal{L}\{f(t)\} = 1/[s^2(s^2 + 4)]$, find $\mathcal{L}\{f(t/3)\}$.
58. If $\mathcal{L}\{f(t)\} = (s^2 - 4)/[(s^2 + 4)^2]$, find $\mathcal{L}\{f(t/2)\}$.

Exercises involving the Laplace transform of periodic functions

In Exercises 59 through 66 find the Laplace transform of the periodic function $f(t)$.

59.

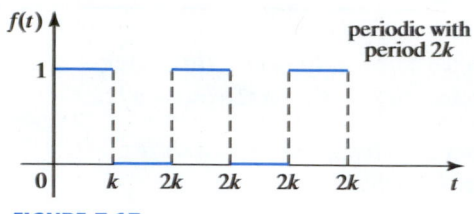

FIGURE 7.17

60.

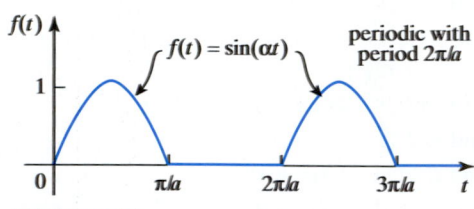

FIGURE 7.18

61.

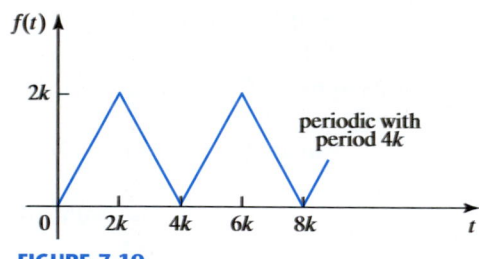

FIGURE 7.19

62.

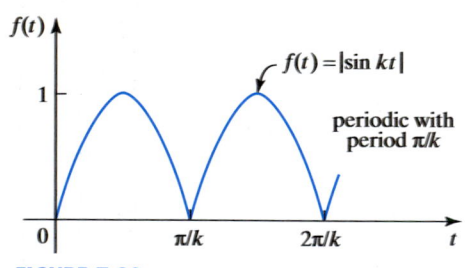

FIGURE 7.20

63.

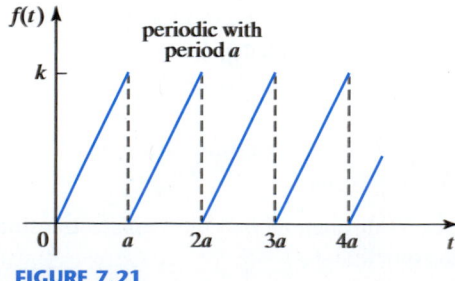

FIGURE 7.21

64.

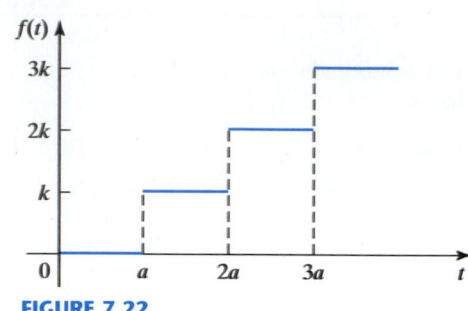

FIGURE 7.22

65.

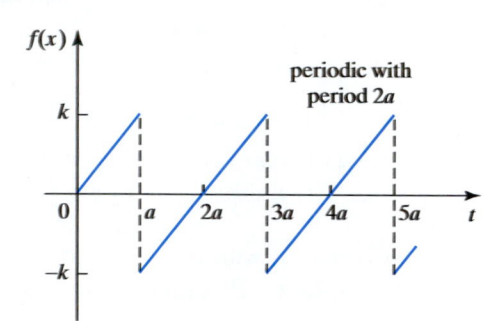

FIGURE 7.23

66.

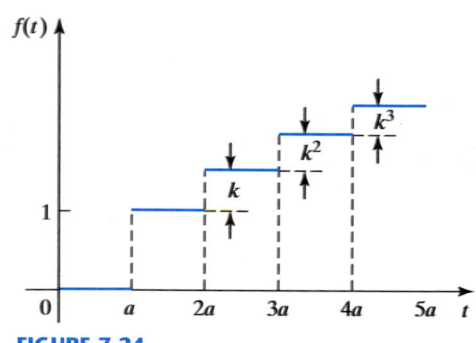

FIGURE 7.24

Exercises involving the convolution operation

67. Find $(e^{-t} * e^{-2t})$.

68. Find $(t * \sin t)$.

69. Find $(t^2 * \sin t)$.

70. Find $(t * e^{-t})$.

71. Find $(\cos t * \cos t)$.

72. Find $(\sin 2t * \sin 2t)$.

Exercises involving the convolution theorem

73. Find $\mathcal{L}\{t * e^{-2t}\}$.

74. Find $\mathcal{L}\{2t * \cos 2t\}$.

75. Find $\mathcal{L}\{e^{-t} \sin t * t\}$.

76. Find $\mathcal{L}\{e^{-2t} \cos t * e^t\}$.

77. Find $\mathcal{L}^{-1}\{1/[s^2(s^2 + 4)]\}$.

78. Find $\mathcal{L}^{-1}\{1/(s^2 - 9)^2\}$.

79. Find $\mathcal{L}^{-1}\{s^2/(s^2 - 1)^2\}$.

80. Find $\mathcal{L}^{-1}\{s/(s^2 - 4)^2\}$.

Exercises involving integral equations

81. Solve $y(t) = \sin t + \int_0^t \sin(t - \tau)y(\tau)d\tau$.

82. Solve $y(t) = \cos t + \int_0^t \sin[2(t - \tau)]y(\tau)d\tau$.

83. Solve $y(t) = t^2 + \int_0^t \cos(t - \tau)y(\tau)d\tau$.

84. Solve $y(t) = e^{-2t} + \int_0^t \cos(t - \tau)y(\tau)d\tau$.

Exercises involving integro-differential equations

85. Solve $y' + 4y = 4\int_0^t \sin \tau \, y(t - \tau)d\tau$, with $y(0) = 1$.

86. Solve $y' + y = \int_0^t e^{-2\tau} y(t - \tau)d\tau$, with $y(0) = 3$.

87. Solve $y'' - y = \int_0^t \sinh \tau \, y(t - \tau)d\tau$, with $y(0) = 1$, $y'(0) = 0$.

88. Solve $y'' - 4y = 2\int_0^t \sinh 2\tau \, y(t - \tau)d\tau$, with $y(0) = 1$, $y'(0) = 0$.

Exercises involving the transform of an integral

89. Find $\mathcal{L}\left\{\int_0^t \tau^2 \sin 2\tau \, d\tau\right\}$.

90. Find $\mathcal{L}\left\{\int_0^t e^{2\tau} \cos \tau \, d\tau\right\}$.

91. Find $\mathcal{L}^{-1}\{1/(s^2 + a^2)^2\}$.

92. Find $\mathcal{L}^{-1}\{s/(s^2 + a^2)\}$.

Exercises involving an integral of a transform

93. Find $\mathcal{L}\left\{\dfrac{\sinh 2t}{t}\right\}$.

94. Find $\mathcal{L}\left\{\dfrac{1 - \cos 3t}{t}\right\}$.

95. Find $\mathcal{L}^{-1}\left\{\ln\left(\dfrac{s^2 - a^2}{s^2}\right)\right\}$.

96. Find $\mathcal{L}^{-1}\left\{\ln\left(\dfrac{s^2 + a^2}{s^2}\right)\right\}$.

Exercises involving the initial value theorem

In Exercises 97 through 100 use the initial value theorem to find $f(0)$, $f'(0)$, and $f''(0)$ from $F(s)$, and verify the result by differentiation of $f(t) = \mathcal{L}^{-1}\{F(s)\}$.

97. $F(s) = (s^2 + 6)/\{s(s^2 + 9)\}$.

98. $F(s) = s/(s^2 + 6s + 9)$.

99. $F(s) = (s - 1)/(s^2 - 4s + 4)$.

100. $F(s) = (2s^2 + s - 12)/\{s(s + 2)(s + 3)\}$.

Exercises involving the delta function

101. Evaluate $\displaystyle\int_0^\infty \left(\frac{1 - 3\sin^2 t}{t}\right)\delta(t - \pi/2)dt$.

102. Evaluate $\displaystyle\int_0^4 \sin^2 t \delta(t - 2\pi)dt$.

103. Evaluate $\displaystyle\int_0^\infty \sum_{n=1}^3 \left\{\left(\frac{\sin nt}{t}\right)\delta\left[t - (2n + 1)\frac{\pi}{2}\right]\right\} dt$.

104. Evaluate $\displaystyle\int_0^\infty \{[H(t - 1) - H(t - 2)]t + \cos(t - 3\pi)\delta(t - 3\pi)\}dt$.

105. Solve $y'' + 9y = 1 + \delta(t - 1)$, with $y(0) = 0$, $y'(0) = 0$.

106. Solve $y'' + 4y' + 4y = \delta(t - 1)$, with $y(0) = 1$, $y'(0) = 1$.

107. Solve $y'' + 2y' + y = \sin t + \delta(t - \pi)$, with $y(0) = y'(0) = 0$.

108. Solve $y'' - 4y' + 3y = e^{-t} + 3\delta(t - 2)$, with $y(0) = y'(0) = 0$.

109. Solve $y'' + 4y = 1 - H(t - 1) + \delta(t - 2)$, with $y(0) = 1$, $y'(0) = 0$.

110. Solve $y'' + 3y' + 2y = \delta(t - 1)$, with $y(0) = 0$, $y'(0) = 1$.

7.3   Systems of Equations and Applications of the Laplace Transform

(a) Solution of Systems of Linear First Order Equations by the Laplace Transform

The Laplace transform can be used to solve initial value problems for systems of linear first order differential equations by introducing the Laplace transform of each dependent variable that is involved, solving the resulting algebraic equations for each transformed dependent variable, and then inverting the results.

As a system of linear higher order differential equations can always be reduced to a system of first order equations by introducing higher order derivatives as new dependent variables, the solution of a system of linear first order equations can be considered to be the most general case.

solving systems of equations

The example that follows, involving two simultaneous first order equations, illustrates the approach to be used in all cases, but by restricting the number of equations and using simple nonhomogeneous terms (forcing functions) the algebra is kept to a minimum.

EXAMPLE 7.29   Solve the initial value problem

$$x' - 2x + y = \sin t$$
$$y' + 2x - y = 1,$$

with $x(0) = 1$, $y(0) = -1$.

Solution   We define the transforms of the dependent variables $x(t)$ and $y(t)$ to be

$$\mathcal{L}\{x(t)\} = X(s), \quad \mathcal{L}\{y(t)\} = Y(s).$$

Transforming the system of equations in the usual way leads to the following system of linear algebraic equations for $X(s)$ and $Y(s)$:

$$s X(s) - 1 - 2X(s) + Y(s) = 1/(s^2 + 1)$$
$$s Y(s) + 1 + 2X(s) - Y(s) = 1/s.$$

Solving these for $X(s)$ and $Y(s)$ gives

$$X(s) = \frac{(s-1)(s^3 + s^2 + 2s + 1)}{s^2(s-3)(s^2+1)} \quad \text{and} \quad Y(s) = \frac{-(s^4 - s^3 + 3s^2 + s + 2)}{s^2(s-3)(s^2+1)}.$$

Expressing these results in terms of partial fractions, we find that

$$X(s) = \frac{4}{9}\frac{1}{s} + \frac{1}{3}\frac{1}{s^2} - \frac{1}{5}\frac{1}{s^2+1} - \frac{2}{5}\frac{s}{s^2+1} + \frac{43}{45}\frac{1}{s-3}$$

and

$$Y(s) = \frac{5}{9}\frac{1}{s} + \frac{2}{3}\frac{1}{s^2} + \frac{1}{5}\frac{1}{s^2+1} - \frac{3}{5}\frac{s}{s^2+1} - \frac{43}{45}\frac{1}{s-3}.$$

Finally, taking the inverse transform gives the solution

$$x(t) = \frac{4}{9} + \frac{1}{3}t - \frac{1}{5}\sin t - \frac{2}{5}\cos t + \frac{43}{45}e^{3t}$$

and

$$y(t) = \frac{5}{9} + \frac{2}{3}t + \frac{1}{5}\sin t - \frac{3}{5}\cos t - \frac{43}{45}e^{3t} \quad \text{for } t > 0. \qquad \blacksquare$$

This method can be used for any number of simultaneous linear differential equations, though the complexity of both the algebraic manipulation and the associated inversion problem increases rapidly when more than two equations are involved.

A typical example of the way systems of first order equations arise in practice is provided by considering a chemical reaction that converts a raw chemical into an end product, via several intermediate reactions. The simplest situation involves chemical reactions that are irreversible, so that once a product has been produced the chemical process cannot be reversed, causing the new product to revert to a previous one.

Let us derive the system of equations governing such a process when three intermediate reactions are involved, each of which is irreversible, with each reaction proceeding at a rate that is proportional to the amount of material to be converted from one stage to the next. Denote the raw chemical by A and the end product by E, with the intermediate products denoted by B, C, and D, and let the reaction rates (the constants of proportionality) from $A \rightarrow B$, $B \rightarrow C$, $C \rightarrow D$, and $D \rightarrow E$ be k_1, k_2, k_3, and k_4, respectively. Then if the amounts of chemicals A, B, C, D, and E present at time t are x, y, u, v, and w, the production and removal of the chemical products involved is described as follows.

Reaction	Reaction Rate of Removal	Reaction Rate of Production
$A \rightarrow B$	$\left(\dfrac{dx}{dt}\right)_{A \rightarrow B} = -k_1 x$	$\left(\dfrac{dy}{dt}\right)_{A \rightarrow B} = k_1 x$
$B \rightarrow C$	$\left(\dfrac{dy}{dt}\right)_{B \rightarrow C} = -k_2 y$	$\left(\dfrac{du}{dt}\right)_{B \rightarrow C} = k_2 y$
$C \rightarrow D$	$\left(\dfrac{du}{dt}\right)_{C \rightarrow D} = -k_3 u$	$\left(\dfrac{dv}{dt}\right)_{C \rightarrow D} = k_3 u$
$D \rightarrow E$	$\left(\dfrac{dv}{dt}\right)_{D \rightarrow E} = -k_4 v$	$\left(\dfrac{dw}{dt}\right)_{D \rightarrow E} = k_4 v$

Combining these results gives

$$\frac{dx}{dt} = \left(\frac{dx}{dt}\right)_{A \rightarrow B} = -k_1 x$$

$$\frac{dy}{dt} = \left(\frac{dy}{dt}\right)_{A \rightarrow B} + \left(\frac{dy}{dt}\right)_{B \rightarrow C} = k_1 x - k_2 y$$

$$\frac{du}{dt} = \left(\frac{du}{dt}\right)_{B \rightarrow C} + \left(\frac{du}{dt}\right)_{C \rightarrow D} = k_2 y - k_3 u$$

$$\frac{dv}{dt} = \left(\frac{dv}{dt}\right)_{C \rightarrow D} + \left(\frac{dv}{dt}\right)_{D \rightarrow E} = k_3 u - k_4 v.$$

If the amount of raw material A present at the start is Q, the initial conditions for the system are seen to be

$$x(0) = Q, \quad y(0) = 0, \quad u(0) = 0, \quad v(0) = 0, \quad \text{and} \quad w(0) = 0.$$

Provided no additional by-products are produced during the reactions, it follows from the conservation of mass that $x + y + u + v + w = Q$, and so

$$w = Q - x - y - u - v.$$

Taking the Laplace transform of this system of first order linear equations and using the stated initial conditions leads to the transformed system

$$s\,X(s) + k_1\,X(s) = Q$$
$$s\,Y(s) - k_1\,X(s) + k_2\,Y(s) = 0$$
$$s\,U(s) - k_2\,Y(s) + k_3\,U(s) = 0$$
$$s\,V(s) - k_3\,U(s) + k_4\,V(s) = 0,$$

where $\mathcal{L}\{x(t)\} = X(s)$, $\mathcal{L}\{y(t)\} = Y(s)$, $\mathcal{L}\{u(t)\} = U(s)$, and $\mathcal{L}\{v(t)\} = V(s)$.

Solving for the Laplace transforms, we have

$$X(s) = \frac{Q}{s + k_1}, \quad Y(s) = \frac{k_1 Q}{(s + k_1)(s + k_2)}, \quad U(s) = \frac{k_1 k_2 Q}{(s + k_1)(s + k_2)(s + k_3)},$$

and

$$V(s) = \frac{k_1 k_2 k_3 Q}{(s + k_1)(s + k_2)(s + k_3)(s + k_4)}.$$

After expressing these Laplace transforms in terms of partial fractions the required solutions are seen to be

$$x(t) = Q e^{-k_1 t}, \quad y(t) = \frac{k_1 Q}{k_1 - k_2}(e^{-k_1 t} - e^{-k_2 t})$$

and

$$u(t) = k_1 k_2 Q\left(\frac{1}{(k_2 - k_1)(k_3 - k_1)}e^{-k_1 t} + \frac{1}{(k_1 - k_2)(k_3 - k_2)}e^{-k_2 t}\right.$$
$$\left. + \frac{1}{(k_1 - k_3)(k_2 - k_3)}e^{-k_3 t}\right)$$

with $v(t)$ similarly defined. The amount of the end product $w(t)$ produced at time t follows from

$$w(t) = Q - x(t) - y(t) - u(t) - v(t).$$

solving systems of equations in matrix form

We now outline a matrix method of solution of initial value problems for systems of linear first order differential equations, of which Example 7.29 is a typical case. Let us consider the system

$$\frac{d}{dt}\mathbf{x}(t) = \mathbf{A}\mathbf{x}(t) + \mathbf{b}(t), \tag{17}$$

where

$$\mathbf{x}(t) = \begin{bmatrix} x_1(t) \\ x_2(t) \\ \cdot \\ \cdot \\ \cdot \\ x_n(t) \end{bmatrix}, \quad \mathbf{A} = \begin{bmatrix} a_{11} & a_{12} & \cdot & \cdot & \cdot & a_{1n} \\ a_{21} & a_{22} & & & \cdot & a_{2n} \\ & & \cdots\cdots & & \\ & & \cdots\cdots & & \\ a_{n1} & a_{n2} & \cdot & \cdot & \cdot & a_{nn} \end{bmatrix}, \quad \mathbf{b}(t) = \begin{bmatrix} b_1(t) \\ b_2(t) \\ \cdot \\ \cdot \\ \cdot \\ b_n(t) \end{bmatrix},$$

subject to the initial conditions $x_1(0) = x_1$, $x_2(0) = x_2, \ldots, x_n(0) = x_n$.

Define $\mathcal{L}\{x_1(t)\} = X_1(s)$, $\mathcal{L}\{x_2(t)\} = X_2(s)\ldots$, $\mathcal{L}\{x_n(t)\} = X_n(s)$, $\mathcal{L}\{b_1(t)\} = B_1(s)$, $\mathcal{L}\{b_2(t)\} = B_2(s), \ldots, \mathcal{L}\{b_n(t)\} = B_n(s)$, and set

$$\mathbf{Z}(s) = \begin{bmatrix} X_1(s) \\ X_2(s) \\ \cdot \\ \cdot \\ X_n(s) \end{bmatrix}, \quad \mathbf{c}(s) = \begin{bmatrix} B_1(s) \\ B_2(s) \\ \cdot \\ \cdot \\ B_n(s) \end{bmatrix} \quad \text{and} \quad \mathbf{v} = \begin{bmatrix} x_1 \\ x_2 \\ \cdot \\ \cdot \\ x_n \end{bmatrix}.$$

Then taking the Laplace transform of (17) and using the result $\mathcal{L}\{x_r'(t)\} = s\,X(s) - x_r$, for $r = 1, 2, \ldots, n$, we arrive at the system

$$s\mathbf{Z}(s) - \mathbf{v} = \mathbf{A}\mathbf{Z}(s) + \mathbf{c}(s)$$

or, equivalently,

$$(s\mathbf{I} - \mathbf{A})\mathbf{Z}(s) = \mathbf{v} + \mathbf{c}(s),$$

where $\mathbf{I}$ is the $n \times n$ unit matrix. Premultiplying this last result by $(s\mathbf{I} - \mathbf{A})^{-1}$ gives

$$\mathbf{Z}(s) = [s\mathbf{I} - \mathbf{A}]^{-1}[\mathbf{v} + \mathbf{c}(s)]. \tag{18}$$

Finally, taking the inverse Laplace transform of (18) we obtain the solution $\mathbf{x}(t)$ of the initial value problem in the form

$$\mathbf{x}(t) = \mathcal{L}^{-1}\{[s\mathbf{I} - \mathbf{A}]^{-1}[\mathbf{v} + \mathbf{c}(s)]\}. \tag{19}$$

EXAMPLE 7.30   Solve the initial value problem of Example 7.29 by using result (19).

Solution   Making the necessary identifications we have

$$\mathbf{I} = \begin{bmatrix} 1 & 0 \\ 0 & 1 \end{bmatrix}, \quad \mathbf{A} = \begin{bmatrix} 2 & -1 \\ -2 & 1 \end{bmatrix}, \quad \mathbf{v} = \begin{bmatrix} 1 \\ -1 \end{bmatrix}, \quad \mathbf{c}(s) = \begin{bmatrix} 1/(s^2+1) \\ 1/s \end{bmatrix},$$

so (18) becomes

$$\mathbf{Z}(s) = \left[s\begin{bmatrix} 1 & 0 \\ 0 & 1 \end{bmatrix} - \begin{bmatrix} 2 & -1 \\ -2 & 1 \end{bmatrix} \right]^{-1} \left[\begin{bmatrix} 1 \\ -1 \end{bmatrix} + \begin{bmatrix} 1/(s^2+1) \\ 1/s \end{bmatrix} \right],$$

or

$$\mathbf{Z}(s) = \begin{bmatrix} s-2 & 1 \\ 2 & s-1 \end{bmatrix}^{-1} \begin{bmatrix} (s^2+2)/(s^2+1) \\ (1-s)/s \end{bmatrix}.$$

The inverse of the first matrix in this product is

$$\begin{bmatrix} s-2 & 1 \\ 2 & s-1 \end{bmatrix}^{-1} = \begin{bmatrix} \dfrac{s-1}{s(s-3)} & \dfrac{-1}{s(s-3)} \\ \dfrac{-2}{s(s-3)} & \dfrac{s-2}{s(s-3)} \end{bmatrix},$$

so

$$\mathbf{Z}(s) = \begin{bmatrix} \dfrac{s-1}{s(s-3)} & \dfrac{-1}{s(s-3)} \\ \dfrac{-2}{s(s-3)} & \dfrac{s-2}{s(s-3)} \end{bmatrix} \begin{bmatrix} \dfrac{s^2+2}{s^2+1} \\ \dfrac{1-s}{s} \end{bmatrix}.$$

After forming the matrix product this becomes

$$\mathbf{Z}(s) = \begin{bmatrix} \dfrac{(s-1)(s^3 + s^2 + 2s + 1)}{s^2(s-3)(s^2+1)} \\[2ex] \dfrac{-(s^4 - s^3 + 3s^2 + s + 2)}{s^2(s-3)(s^2+1)} \end{bmatrix}.$$

The inverse transforms involved are, of course, the same as the ones in Example 7.29, so, as would be expected, the solution is the same as before, apart from a change of notation involving the replacement of $x(t)$ and $y(t)$ by $x_1(t)$ and $x_2(t)$ giving

$$x_1(t) = \frac{4}{9} + \frac{1}{3}t - \frac{1}{5}\sin t - \frac{2}{5}\cos t + \frac{43}{45}e^{3t}$$

and

$$x_2(t) = \frac{5}{9} + \frac{2}{3}t + \frac{1}{5}\sin t - \frac{3}{5}\cos t - \frac{43}{45}e^{3t} \quad \text{for } t > 0. \qquad \blacksquare$$

(b) Determination of $e^{t\mathbf{A}}$ by Means of the Laplace Transform

The matrix solution of system (17) given in (19) has an interesting and useful consequence, because it provides a different and efficient way of finding the matrix exponential $e^{t\mathbf{A}}$. To see how this comes about, notice that from equation (114) in Section 6.10(c) the solution of the homogeneous system of equations

$$\mathbf{x}' = \mathbf{A}\mathbf{x}, \tag{20}$$

subject to the initial condition $\mathbf{x}(0) = \mathbf{v}$, can be written

$$\mathbf{x}(t) = e^{t\mathbf{A}}\mathbf{v}. \tag{21}$$

Setting $\mathbf{c}(s) = \mathbf{0}$ (corresponding to $\mathbf{b}(t) = \mathbf{0}$) reduces solution (19) to

$$\mathbf{x}(t) = \mathcal{L}^{-1}\{[s\mathbf{I} - \mathbf{A}]^{-1}\}\mathbf{v}, \tag{22}$$

so comparison of (21) and (22) shows that

$$e^{t\mathbf{A}} = \mathcal{L}^{-1}\{[s\mathbf{I} - \mathbf{A}]^{-1}\}. \tag{23}$$

We have established the following theorem.

THEOREM 7.15

finding the matrix exponential by the Laplace transform

Determination of $e^{t\mathbf{A}}$ by means of the Laplace transform Let $\mathbf{A}$ be a real $n \times n$ matrix with constant elements. Then the exponential matrix

$$e^{t\mathbf{A}} = \mathcal{L}^{-1}\{[s\mathbf{I} - \mathbf{A}]^{-1}\}. \qquad \blacksquare$$

The following examples show how Theorem 7.15 determines $e^{t\mathbf{A}}$ in the cases when $\mathbf{A}$ is diagonalizable with real eigenvalues, when it is diagonalizable with complex conjugate eigenvalues, and also when it is not diagonalizable.

EXAMPLE 7.31

Use Theorem 7.15 to find $e^{t\mathbf{A}}$ when

$$\mathbf{A} = \begin{bmatrix} -2 & 6 \\ -2 & 5 \end{bmatrix}.$$

Solution   Matrix **A** has the distinct eigenvalues 1 and 2, and so is diagonalizable.

$$[s\mathbf{I} - \mathbf{A}] = \begin{bmatrix} s+2 & -6 \\ 2 & s-5 \end{bmatrix}$$

so

$$[s\mathbf{I} - \mathbf{A}]^{-1} = \begin{bmatrix} \dfrac{s-5}{s^2-3s+2} & \dfrac{6}{s^2-3s+2} \\ \dfrac{-2}{s^2-3s+2} & \dfrac{s+2}{s^2-3s+2} \end{bmatrix}.$$

Expressing each element of this matrix in terms of partial fractions and taking the inverse Laplace transform gives

$$e^{t\mathbf{A}} = \begin{bmatrix} 4e^t - 3e^{2t} & -6e^t + 6e^{2t} \\ 2e^t - 2e^{2t} & -3e^t + 4e^{2t} \end{bmatrix},$$

in agreement with the result in Example 6.33.   ∎

EXAMPLE 7.32   Use Theorem 7.14 to find $e^{t\mathbf{A}}$ when

$$\mathbf{A} = \begin{bmatrix} -3 & -4 \\ 2 & 1 \end{bmatrix}.$$

Solution   Matrix **A** has the complex conjugate eigenvalues $-1 \pm 2i$.

$$[s\mathbf{I} - \mathbf{A}] = \begin{bmatrix} s+3 & 4 \\ -2 & s-1 \end{bmatrix},$$

so

$$[s\mathbf{I} - \mathbf{A}]^{-1} = \begin{bmatrix} \dfrac{s-1}{s^2+2s+5} & \dfrac{-4}{s^2+2s+5} \\ \dfrac{2}{s^2+2s+5} & \dfrac{s+3}{s^2+2s+5} \end{bmatrix}.$$

Expressing each element of this matrix in terms of partial fractions and taking the inverse Laplace transform gives

$$e^{t\mathbf{A}} = \begin{bmatrix} e^{-t}(\cos 2t - \sin 2t) & -2e^{-t}\sin 2t \\ e^{-t}\sin 2t & e^{-t}(\cos 2t + \sin 2t) \end{bmatrix},$$

in agreement with the result of Example 6.34.   ∎

EXAMPLE 7.33   Use Theorem 7.14 to find $e^{t\mathbf{A}}$ when

$$\mathbf{A} = \begin{bmatrix} 4 & 1 \\ 0 & 4 \end{bmatrix}.$$

Solution   Matrix **A** has the repeated eigenvalue 4 and is not diagonalizable.

$$[s\mathbf{I} - \mathbf{A}] = \begin{bmatrix} s-4 & -1 \\ 0 & s-4 \end{bmatrix},$$

so

$$[s\mathbf{I} - \mathbf{A}]^{-1} = \begin{bmatrix} \dfrac{1}{s-4} & \dfrac{1}{(s-4)^2} \\ 0 & \dfrac{1}{s-4} \end{bmatrix}.$$

Taking the inverse of the elements of this matrix, we find that

$$e^{t\mathbf{A}} = \begin{bmatrix} e^{4t} & te^{4t} \\ 0 & e^{4t} \end{bmatrix},$$

in agreement with the result of Example 6.35.   ■

(c) The Weighting Function

To introduce the concept of a *weighting function*, which has important engineering applications, we consider the differential equation

$$a_0 \frac{d^n y}{dt^n} + a_1 \frac{d^{n-1} y}{dt^{n-1}} + \cdots + a_n y = f(t), \tag{24}$$

subject to the initial conditions $y(0) = y'(0) = \cdots = y^{(n-1)}(0) = 0$. We shall denote by $w(t)$ the solution of equation (24) when $f(t) = \delta(t)$, and call it the **weighting function** associated with the equation. Thus the solution $w(t)$ can be regarded as the *output* from a system described by equation (24) that is produced by the impulsive *input* (nonhomogeneous term) $\delta(t)$ applied at time $t = 0$ when the system is at rest. The weighting function $w(t)$ is the solution of the equation

weighting function and its uses

$$a_0 \frac{d^n w}{dt^n} + a_1 \frac{d^{n-1} w}{dt^{n-1}} + \cdots + a_n w = \delta(t), \tag{25}$$

with $w(t) = 0$ for $t < 0$.

Let us now consider the *output* $y(t)$ from a system described by (24) produced by an arbitrary *input* $f(t)$, subject to the homogeneous initial conditions $y(0) = y'(0) = \cdots = y^{(n-1)}(0) = 0$. Taking the Laplace transform of (24) we find that

$$G(s)Y(s) = F(s), \tag{26}$$

where

$$G(s) = a_0 s^n + a_1 s^{n-1} + \cdots + a_{n-1} s + a_n, \quad Y(s) = \mathcal{L}\{y(t)\} \quad \text{and} \quad F(s) = \mathcal{L}\{f(t)\}.$$

Setting $W(s) = \mathcal{L}\{w(t)\}$, taking the Laplace transform of (25), and using the fact that $w(t)$ and all its derivatives vanish for $t < 0$ leads to the result

$$G(s)W(s) = 1. \tag{27}$$

Eliminating $G(s)$ between (26) and (27) relates the Laplace transform of the output $Y(s)$ to the Laplace transform $F(s)$ of the input by the equation

$$Y(s) = W(s)F(s). \tag{28}$$

Taking the inverse Laplace transform of (28) and using the convolution theorem gives

$$y(t) = \int_0^t w(\tau) f(t - \tau) d\tau. \tag{29}$$

This form of the solution of (24) explains why $w(t)$ is called the *weighting function*, because (29) shows how the input $y(t - \tau)$ at time $t - \tau$ is *weighted* by the function $w(\tau)$ over the interval $0 \le \tau \le t$ in the integral determining $y(t)$.

The determination of the weighting function has the advantage that once it has been found, the solution of (24), subject to the conditions that $y(0) = y'(0) = \cdots = y^{(n-1)}(0) = 0$, is always expressible as result (29) for every nonhomogeneous term $f(t)$. It is instructive to compare this result, which applies to a linear differential equation of *any* order, to the one in (76) of Section 6.6, which was obtained by applying the method of variation of parameters to a second order equation with homogeneous initial conditions when $t = a$. The weighting function is also sometimes called the *Green's function* for an initial value problem for a homogeneous differential equation.

The modification that must be made to result (29) to take account of initial conditions for $y(t)$ that are not all zero at $t = 0$ is to be found in Exercise 25 at the end of this section.

EXAMPLE 7.34

Find the weighting function for the equation

$$y'' + 2y' + 5y = \sin t$$

and use it to solve the equation subject to the initial condition $y(0) = y'(0) = 0$.

Solution   The weighting function $w(t)$ is the solution of

$$w'' + 2w' + 5w = \delta(t)$$

with $w(0) = w'(0) = 0$. Taking the Laplace transform and setting $\mathcal{L}\{w(t)\} = W(s)$ gives

$$s^2 W(s) + 2s W(s) + 5W(s) = 1,$$

so

$$W(s) = \frac{1}{s^2 + 2s + 5}.$$

Taking the inverse Laplace transform, we find that

$$w(t) = \mathcal{L}^{-1}\{W(s)\} = \frac{1}{2} e^{-t} \sin 2t \quad \text{for } t \ge 0.$$

The solution of the differential equation with $y(0) = y'(0) = 0$ now follows from (29) as

$$y(t) = \int_0^t w(\tau) \sin(t - \tau) d\tau$$

$$= \frac{1}{2} \int_0^t e^{-\tau} \sin 2\tau \sin(t - \tau) d\tau$$

$$= \frac{1}{5} \sin t - \frac{1}{10} \cos t + \frac{e^{-t}}{20} (2\cos 2t - \sin 2t). \qquad \blacksquare$$

The concept of a weighting function can be generalized to include systems of equations, though then more than one weighting function must be introduced, and the solution of each dependent variable becomes the sum of convolution integrals of the type given in (29). The ideas involved are illustrated by considering the following system of equations involving $x(t)$ and $y(t)$:

$$x' + ax + by = f_1(t)$$
$$y' + cx + dy = f_2(t), \tag{30}$$

subject to the initial conditions $x(0) = y(0) = 0$.

It is necessary to introduce a weighting function for each of the variables $x(t)$ and $y(t)$ corresponding first to $f_1(t) = \delta(t)$ and $f_2(t) = 0$, and then to $f_1(t) = 0$ and $f_2(t) = \delta(t)$. Let $w_{x1}(t)$ and $w_{y1}(t)$ be the weighting functions corresponding to

$$w'_{x1} + aw_{x1} + bw_{y1} = \delta(t)$$
$$w'_{y1} + cw_{x1} + dw_{y1} = 0, \tag{31}$$

and $w_{x2}(t)$ and $w_{y2}(t)$ be the Green's functions corresponding to

$$w'_{x2} + aw_{x2} + bw_{y2} = 0$$
$$w'_{y2} + cw_{x2} + dw_{y2} = \delta(t), \tag{32}$$

where $w_{x1}(0) = w_{x2}(0) = w_{y1}(0) = w_{y2}(0) = 0$.

The notation used here indicates that $w_{x1}(t)$ is the x response and $w_{y1}(t)$ the y response to the input $f_1(t) = \delta(t)$ and $f_2(t) = 0$, and $w_{x2}(t)$ is the x response and $w_{y2}(t)$ the y response to the input $f_1(t) = 0$ and $f_2(t) = \delta(t)$. Then, because the equations are linear, to obtain the solution $x(t)$ subject to the initial conditions $x(0) = y(0) = 0$, it is necessary to add the contribution due to $w_{x1}(t)$ to the one due to $w_{x2}(t)$, and similarly for the solution $y(t)$.

This leads to the solution in the form

$$x(t) = \int_0^t w_{x1}(\tau) f_1(t - \tau)d\tau + \int_0^t w_{x2}(\tau) f_2(t - \tau)d\tau \tag{33a}$$

and

$$y(t) = \int_0^t w_{y1}(\tau) f_1(t - \tau)d\tau + \int_0^t w_{y2}(\tau) f_2(t - \tau)d\tau. \tag{33b}$$

Once the weighting functions have been found, equations (33) give the solution of system (30) for any choice of functions $f_1(t)$ and $f_2(t)$, subject to the initial conditions $x(0) = y(0) = 0$.

EXAMPLE 7.35   Find weighting functions for the equations

$$x' + 2x - y = f_1(t)$$
$$y' - 2x + y = f_2(t)$$

and use them to solve the system subject to the initial conditions $x(0) = y(0) = 0$ when (a) $f_1(t) = \sin t$ and $f_2(t) = 2$ and (b) $f_1(t) = \cos t$ and $f_2(t) = 0$.

Solution   **(a)** From (31) the functions $w_{x1}(t)$ and $w_{y1}(t)$ satisfy

$$w'_{x1} + 2w_{x1} - w_{y1} = \delta(t)$$
$$w'_{y1} - 2w_{x1} + w_{y1} = 0,$$

so taking the Laplace transform of these equations we have

$$(s+2)\mathcal{L}\{w_{x1}(t)\} - \mathcal{L}\{w_{y1}(t)\} = 1$$

$$(s+1)\mathcal{L}\{w_{y1}(t)\} - 2\mathcal{L}\{w_{x1}(t)\} = 0.$$

Solving for $\mathcal{L}\{w_{x1}(t)\}$ and $\mathcal{L}\{w_{y1}(t)\}$ gives

$$\mathcal{L}\{w_{x1}(t)\} = \frac{s+1}{s(s+3)} \quad \text{and} \quad \mathcal{L}\{w_{y1}(t)\} = \frac{2}{s(s+3)}.$$

Taking the inverse Laplace transforms, we find that

$$w_{x1}(t) = \frac{1}{3} + \frac{2}{3}e^{-3t} \quad \text{and} \quad w_{y1}(t) = \frac{2}{3} - \frac{2}{3}e^{-3t} \quad \text{for } t \geq 0.$$

Similarly, solving the equations for $w_{x2}(t)$ and $w_{y2}(t)$ corresponding to (32), we obtain

$$w_{x2}(t) = \frac{1}{3} - \frac{1}{3}e^{-3t} \quad \text{and} \quad w_{y2}(t) = \frac{2}{3} + \frac{1}{3}e^{-3t} \quad \text{for } t \geq 0.$$

The solution of the system subject to the initial conditions $x(0) = y(0) = 0$, $f_1(t) = \sin t$, and $f_2(t) = 2$ now follows from (33) as

$$x(t) = \int_0^t w_{x1}(\tau) \sin(t - \tau) d\tau + 2 \int_0^t w_{x2}(\tau) d\tau$$

and

$$y(t) = \int_0^t w_{y1}(\tau) \sin(t - \tau) d\tau + 2 \int_0^t w_{y2}(\tau) d\tau.$$

After the integrations are performed, the solution is found to be

$$x(t) = \frac{1}{9} + \frac{2}{3}t + \frac{13}{45}e^{-3t} + \frac{1}{5}\sin t - \frac{2}{5}\cos t$$

and

$$y(t) = \frac{8}{9} + \frac{4}{3}t - \frac{13}{45}e^{-3t} - \frac{1}{5}\sin t - \frac{3}{5}\cos t \quad \text{for } t > 0.$$

(b) Similarly, the solution when $f_1(t) = \cos t$ and $f_2(t) = 0$ is given by

$$x(t) = \int_0^t w_{x1}(\tau) \cos(t - \tau) d\tau$$

and

$$y(t) = \int_0^t w_{y1}(\tau) \cos(t - \tau) d\tau,$$

so after performing the integrations,

$$x(t) = -\frac{1}{5}e^{-3t} + \frac{2}{5}\sin t + \frac{1}{5}\cos t$$

and

$$y(t) = \frac{1}{5}e^{-3t} + \frac{3}{5}\sin t - \frac{1}{5}\cos t \quad \text{for } t > 0.$$ ■

(d) Differential Equations with Polynomial Coefficients

special variable coefficient differential equations

The Laplace transform can be applied to linear differential equations with polynomial coefficients to find the solution of an initial value problem in the usual way, and also to deduce the Laplace transform of a function from its defining differential equation. This last situation is useful when the integral defining the Laplace transform of a function $f(t)$ cannot be evaluated directly. First, however, we use Theorems 7.3 and 7.7 to find the transform of a product of a power of t and a derivative of $f(t)$.

THEOREM 7.16

$\mathcal{L}\{t^m f^{(n)}(t)\}$ Let $f(t)$ be n times differentiable with $\mathcal{L}\{f(t)\} = F(s)$. Then

$$\mathcal{L}\{t^m f^{(n)}(t)\} = (-1)^m \frac{d^m}{ds^m}[s^n F(s) - s^{n-1} f(0) - s^{n-2} f'(0)$$
$$- s^{n-3} f''(0) - \cdots - f^{(n-1)}(0)].$$

Useful special cases are:

(i) $\mathcal{L}\{t f(t)\} = -F'(s)$

(ii) $\mathcal{L}\{t f'(t)\} = -s F'(s) - F(s)$

(iii) $\mathcal{L}\{t f''(t)\} = -s^2 F'(s) - 2s F(s) + f(0)$

(iv) $\mathcal{L}\{t^2 f'(t)\} = s F'(s) + 2F(s)$

(v) $\mathcal{L}\{t^2 f''(t)\} = s^2 F''(s) + 4s F'(s) + 2F(s)$

Proof   The results of the theorem are direct consequences of Theorems 7.3 and 7.7. We prove the general result, from which the special cases all follow. From Theorem 7.3 we have

$$\mathcal{L}\{f^{(n)}(t)\} = s^n F(s) - s^{n-1} f(0) - s^{n-2} f'(0) - s^{n-3} f''(0) - \cdots - f^{(n-1)}(0),$$

whereas from Theorem 7.7 $\mathcal{L}\{t^m g(t)\} = (-1)^m \frac{d^m}{ds^m} G(s)$, where $\mathcal{L}\{g(t)\} = G(s)$. The main result of the theorem now follows by setting $g(t) = f^{(n)}(t)$ in this last result.   ∎

(i) $\mathcal{L}\{\exp(-t^2)\}$ and its connection with the error function

Laplace transform of the error function

We will use the differential equation satisfied by $y(t) = \exp(-t^2)$ to show that

$$\mathcal{L}\{\exp(-t^2)\} = \frac{1}{2}\sqrt{\pi} \exp(s^2/4)[1 - \operatorname{erf}(s/2)],$$

where

$$\operatorname{erf} s = \frac{2}{\sqrt{\pi}} \int_0^s \exp(-u^2) du$$

is a special function called the **error function**. The error function arises in the theory of heat conduction (see Section 7.3(f) and Chapter 18), in chemical diffusion processes, statistics, and elsewhere.

An attempt to find $\mathcal{L}\{\exp(-t^2)\}$ directly from the definition fails because the integral cannot be evaluated in terms of elementary functions, so some other method must be used. If we set $y(t) = \exp(-t^2)$, it is easily shown that $y(t)$ satisfies the first order variable coefficient equation

$$\frac{dy}{dt} + 2ty = 0,$$

subject to the initial condition $y(0) = \exp(0) = 1$.

Setting $\mathcal{L}\{y(t)\} = Y(s)$ and taking the Laplace transform of the differential equation gives

$$sY(s) - y(0) + 2\mathcal{L}\{ty(t)\} = 0.$$

However, $y(0) = 1$, and from result (i) of Theorem 7.15 (or directly from Theorem 7.7) $\mathcal{L}\{ty(t)\} = -Y'(s)$, so using these results in the preceding equation shows that the Laplace transform satisfies the differential equation

$$\frac{dY}{ds} - \frac{1}{2}sY = -\frac{1}{2}.$$

The integrating factor for this linear first order equation is $\mu(s) = \exp(-s^2/4)$, so after multiplication of the equation by $\mu(s)$ the result becomes

$$\frac{d}{ds}[\exp(-s^2/4)Y(s)] = -\frac{1}{2}\exp(-s^2/4).$$

Integrating over the interval $0 \le u \le s$ gives (after the introduction of the dummy variable u)

$$\int_0^s \frac{d}{du}[\exp(-u^2/4)Y(u)]du = -\frac{1}{2}\int_0^s \exp(-u^2/4)du,$$

or

$$\exp(-s^2/4)Y(s) - Y(0) = -\frac{1}{2}\int_0^s \exp(-u^2/4)du.$$

From the definition $Y(s) = \int_0^\infty e^{-st}\exp(-t^2)dt$, we find that $Y(0) = \int_0^\infty \exp(-t^2)dt$. The integral determining $Y(0)$ is a standard result, $\int_0^\infty \exp(-t^2)dt = \sqrt{\pi}/2$, so making use of this we find that

$$Y(s) = \frac{\sqrt{\pi}}{2}\exp(s^2/4)\left[1 - \frac{1}{\sqrt{\pi}}\int_0^s \exp(-u^2/4)du\right].$$

The change of variable $u = 2v$ brings this last result into the form

$$Y(s) = \frac{\sqrt{\pi}}{2}\exp(s^2/4)\left[1 - \frac{2}{\sqrt{\pi}}\int_0^{s/2} \exp(-v^2)dv\right].$$

If we now define the **error function as**

$$\text{erf}(x) = \frac{2}{\sqrt{\pi}}\int_0^x \exp(-v^2)dv,$$

the Laplace transform $Y(s)$ becomes

$$Y(s) = \mathcal{L}\{\exp(-t^2)\} = \frac{\sqrt{\pi}}{2} \exp(s^2/4)[1 - \text{erf}(s/2)].$$

The function erfc (x), defined as

$$\text{erfc}(x) = 1 - \text{erf}(x),$$

is called the **complementary error function**, so in terms of this function the transform $Y(s)$ becomes

$$Y(s) = \frac{\sqrt{\pi}}{2} \exp(s^2/4)\text{erfc}(s/2).$$

This method of determining the Laplace transform was successful because the differential equation satisfied by $Y(s)$ happened to be simpler than the differential equation satisfied by $y(t)$.

(ii) Laplace transform of the Bessel function $J_0(t)$ and the series expansion of $J_0(t)$

Laplace transform of a Bessel function

The following linear second order differential equation, called **Bessel's equation**,

$$t^2 \frac{d^2 y}{dt^2} + t \frac{dy}{dt} + (t^2 - v^2)y = 0,$$

contains a parameter v that is a constant. It has many applications, one of which is to be found in Chapter 18, where it enters into the solution of a vibrating circular membrane. The properties of its solutions are developed in some detail in Sections 8.6 and 8.7 of Chapter 8.

For each constant value v, Bessel's equation has two linearly independent solutions denoted by $J_v(t)$ and $Y_v(t)$, called, respectively, Bessel functions of order v of the first and second kind. We now use the Laplace transform to find $\mathcal{L}\{J_0(t)\}$, and then to find a power series expansion for $J_0(t)$ that will be obtained in a completely different way in Section 8.6. When $v = 0$, Bessel's equation reduces to

$$t \frac{d^2 J_0}{dt^2} + \frac{dJ_0}{dt} + t J_0 = 0,$$

and we will now find $\mathcal{L}\{J_0(t)\}$ subject to the initial condition $J_0(0) = 1$.

A second initial condition follows by setting $t = 0$ in the differential equation that gives $J_0'(0) = 0$, though this result will not be needed in what is to follow as the condition is implied later when the initial value Theorem 7.13 is used.

Taking the Laplace transform of Bessel's equation of order zero, setting $\mathcal{L}\{J_0(t)\} = Y(s)$, and using the results of Theorem 7.16, we obtain

$$-s^2 Y'(s) - 2s Y(s) + 1 + s Y(s) - 1 - Y'(s) = 0,$$

and after simplification this shows that $Y(s)$ satisfies the first order differential equation

$$\frac{dY}{ds} + \frac{s}{s^2 + 1} Y(s) = 0.$$

Separating the variables and integrating gives

$$\int \frac{dY}{Y} = -\int \frac{s}{s^2+1} ds,$$

and so

$$Y(s) = \frac{C}{(s^2+1)^{1/2}}.$$

We now know the form of $Y(s)$, apart from the magnitude of the constant C. To find the constant we use the initial value theorem (Theorem 7.13), which shows that we must have

$$J_0(0) = \lim_{s\to\infty} [s\,Y(s)],$$

but from the initial condition $J_0(0) = 1$, so

$$1 = \lim_{s\to\infty} \frac{sC}{(s^2+1)^{1/2}} = C,$$

and thus

$$\mathcal{L}\{J_0(t)\} = \frac{1}{(s^2+1)^{1/2}} \quad \text{for } s > 0.$$

This result can be used to obtain a series expansion for $J_0(t)$ by first writing it as

$$\mathcal{L}\{J_0(t)\} = \frac{1}{s}\left(1 + \frac{1}{s^2}\right)^{-1/2},$$

and then expanding the result by the binomial theorem to obtain

$$\mathcal{L}\{J_0(t)\} = \frac{1}{s} - \frac{1}{2}\frac{1}{s^3} + \frac{3}{8}\frac{1}{s^5} - \frac{5}{16}\frac{1}{s^7} + \cdots.$$

Finally, taking the inverse Laplace transform of each term and adding the results, we arrive at the series expansion of $J_0(t)$:

$$J_0(t) = 1 - \frac{t^2}{4} + \frac{t^4}{64} - \frac{t^6}{2304} + \cdots.$$

If the general term in the expansion of $\frac{1}{s}(1 + \frac{1}{s^2})^{-1/2}$ is found, and the result is combined with entry 3 of Table 7.1, it is not difficult to show that $J_0(t)$ can be written as

$$J_0(t) = \sum_{n=0}^{\infty} \frac{(-1)^n t^{2n}}{2^{2n}(n!)^2}.$$

(iii) $\mathcal{L}\{\sin \sqrt{t}\}$

We now show how $\mathcal{L}\{\sin \sqrt{t}\} = Y(s)$ can be found from the differential equation satisfied by the function $\sin \sqrt{t}$, and how in this case a different form of argument from the one used in (ii) must be employed to determine the constant of integration in the expression for $Y(s)$. It is easily seen that $y(t) = \sin\sqrt{t}$ is a solution of

$$4t\frac{d^2y}{dt^2} + 2\frac{dy}{dt} + y = 0,$$

and clearly $y(0) = 0$. Writing $\mathcal{L}\{y(t)\} = Y(s)$, transforming the equation using result (iii) of Theorem 7.16, and incorporating the initial condition $y(0) = 0$ leads to the following first order differential equation for $Y(s)$:

$$\frac{dY}{ds} = \left(\frac{1-6s}{4s^2}\right)Y.$$

Integration of this variables separable equation gives

$$Y(s) = Cs^{-3/2}\exp[-1/(4s)],$$

so it only remains to determine the value of the constant C.

In this case the initial value theorem is of no help in determining C, so to accomplish this we return to the definition of the Laplace transform:

$$\mathcal{L}\{\sin\sqrt{t}\} = Y(s) = \int_0^\infty e^{-st}\sin\sqrt{t}\,dt.$$

The intuitive argument we now use can be made rigorous, but as the details of its justification are not appropriate here, they will be omitted. Inspection of the integrand shows that as $|\sin\sqrt{t}| \leq 1$ for all t, when s is large and positive the exponential function will only be significant close to the origin where the function $\sin\sqrt{t}$ can be approximated by $\sqrt{t}$. So for large s the integral can be approximated by

$$\mathcal{L}\{\sin\sqrt{t}\} \approx \int_0^\infty e^{-st}t^{1/2}dt,$$

$$= \frac{\Gamma(3/2)}{s^{3/2}} = \frac{\sqrt{\pi}}{2s^{3/2}},$$

where entry 4 of Table 7.1 has been used together with the result $\Gamma(3/2) = \frac{1}{2}\Gamma(1/2) = \frac{1}{2}\sqrt{\pi}$ that will be proved later in Section 8.5 of Chapter 8.

Comparing the original expression for $Y(s)$ when s is large with this last result gives $C = \frac{1}{2}\sqrt{\pi}$, so

$$\mathcal{L}\{\sin\sqrt{t}\} = \frac{\sqrt{\pi}}{2s^{3/2}}\exp[-1/(4s)], \quad \text{for } s > 0.$$

This form of argument used to determine the behavior of the integral as $s \to \infty$, where the approximation approaches arbitrarily close to the exact value as s increases, is called an *asymptotic* argument (see, for example, reference [3.3]).

(e) Two-Point Boundary Value Problems: Bending of Beams

boundary value problems and the bending of beams

The Laplace transform is ideally suited to the solution of initial value problems because of the way the initial values of a function enter into the Laplace transform of its derivatives. It can, however, also be used to solve certain types of two-point boundary value problems, as we now show. It will be helpful to use a simple physical example to illustrate the method of approach, so we will consider the case of a

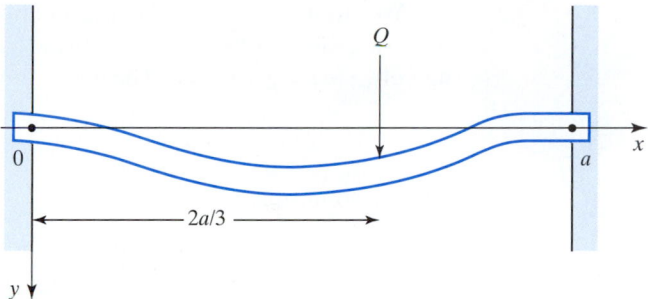

FIGURE 7.25   Clamped beam supporting a point load.

uniform horizontal beam of mass M and length a that is clamped at each end and supports a point load Q at a distance $2a/3$ from one end, as illustrated in Fig. 7.25. The beam equation was introduced in Section 5.2(f) and is

$$EI\frac{d^4y}{dx^4} = w(x).$$

Here x is measured along the axis of the undeflected beam, $y(x)$ is the vertical deflection, E is the Young's modulus of the material of the beam, I is the second moment of the area of the beam about an axis normal to the x- and y-axes, and $w(x)$ is the transverse load per unit length of the beam, which in this case is an isolated point mass Q located at $x = 2a/3$. The boundary conditions for a clamped beam are

$$y(0) = y'(0) = 0 \quad \text{and} \quad y(a) = y'(a) = 0,$$

because neither deflection nor bending can occur at the ends, so both $y(x)$ and $y'(x)$ vanish at $x = 0$ and $x = a$.

The function $w(x)$ can be expressed as

$$w(x) = \frac{M}{a} + Q\delta(x - 2a/3), \quad \text{for } 0 \leq x \leq a,$$

where the point load Q is represented by the delta function that only makes a contribution at $x = 2a/3$.

Transforming the equation, setting $\mathcal{L}\{y(x)\} = Y(s)$, and this time writing x in place of t, because it is conventional to denote a length by x, we find

$$EI[s^4Y(s) - s^3y(0) - s^2y'(0) - sy''(0) - y'''(0)] = \mathcal{L}\{w(x)\}.$$

However,

$$\mathcal{L}\{w(x)\} = \frac{M}{as} + Qe^{-2as/3},$$

so using this in the preceding equation, incorporating the two known initial conditions $y(0) = y'(0) = 0$, and rearranging terms, we find that

$$Y(s) = \frac{M}{a\,EI}\frac{1}{s^5} + \frac{Q}{EI}\frac{e^{-2as/3}}{s^4} + \frac{1}{s^3}y''(0) + \frac{1}{s^4}y'''(0).$$

Taking the inverse Laplace transform of this expression gives

$$y(x) = \frac{M}{24a\,EI}x^4 + \frac{Q}{6EI}(x - 2a/3)^3 H(x - 2a/3) + \frac{1}{2}x^2y''(0) + \frac{1}{6}x^3y'''(0).$$

We must now solve for the unknown initial conditions $y''(0)$ and $y'''(0)$ by requiring this expression to satisfy the two remaining boundary conditions at $x = a$, namely, $y(a) = y'(a) = 0$. The condition $y(a) = 0$ gives

$$0 = \frac{Ma}{4EI} + \frac{Qa}{27EI} + 3y''(0) + ay'''(0),$$

and the condition $y'(a) = 0$ gives

$$0 = \frac{Ma^2}{6EI} + \frac{Q}{18EI} + y''(0) + \frac{1}{2}ay'''(0),$$

so solving for $y''(0)$ and $y'''(0)$, we obtain

$$y''(0) = \frac{a}{108EI}(9M + 8Q) \quad \text{and} \quad y'''(0) = -\frac{1}{54EI}(27M + 14Q).$$

The required solution is then given by

$$y(x) = \frac{M}{24aEI}x^4 + \frac{Q}{6EI}(x - 2a/3)^3 H(x - 2a/3) + \frac{a}{216EI}(9M + 8Q)x^2$$
$$- \frac{1}{324EI}(27M + 14Q),$$

for $0 \leq x \leq a$.

This same form of approach can be used for other two-point boundary value problems, but its success depends on the ability to solve for the unknown initial values in terms of the given boundary conditions.

(f) An Application of the Laplace Transform to the Heat Equation

The Laplace transform can also be used to solve certain types of partial differential equation, involving two or more independent variables. Although the solution of partial differential equations (PDEs) forms the topic of Chapter 18, it will be instructive at this early stage to introduce a simple example that illustrates how the transform can be used for this purpose, and the way the result of Section 7.3d(i) enters into the solution.

The **one-dimensional heat equation** is the partial differential equation

$$\frac{1}{\kappa}\frac{\partial T}{\partial t} = \frac{\partial^2 T}{\partial x^2},$$

a first encounter with a partial differential equation: the heat equation

where $T(x, t)$ is the temperature in a one-dimensional heat-conducting solid at position x at time t, and κ is a constant that describes the thermal conductivity property of the solid. This is a *partial* differential equation because it is a differential equation that involves the partial derivatives of the dependent variable $T(x, t)$. The physical situation modeled by this equation can be considered to be a semi-infinite slab of metal with a plane face on which the origin of the x-axis is located, with the positive half of the axis directed into the slab. This situation is illustrated in Fig. 7.26.

We will consider the situation where for $t < 0$ all of the metal in the slab is at the temperature $T = 0$ and then, at time $t = 0$, the plane face of the slab is suddenly brought up to and maintained at the constant temperature $T = T_0$. The problem is to find the temperature inside the slab on any plane $x = $ constant at any time $t > 0$, knowing that physically the temperature must remain finite for all $x > 0$ and $t > 0$.

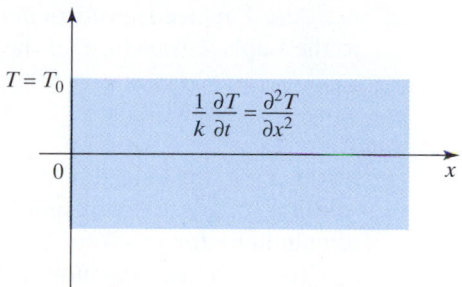

FIGURE 7.26   A semi-infinite metal slab.

The approach will be to take the Laplace transform of the dependent variable $T(x, t)$ in the heat equation with respect to the time t, as a result of which an ordinary differential equation with x as its independent variable will be obtained for the transformed variable that will then depend on both the Laplace transform variable s and x. After this ordinary differential equation has been solved for the transformed variable, the inverse Laplace transform will be used to recover the time variation, and so to arrive at the required solution as a function of x and t.

Before proceeding with this approach we notice first that if the Laplace transform is applied to the independent variable t in the function of two variables $T(x, t)$, the variable x will behave like a constant. Consequently, the rules for transforming derivatives of functions of a single independent variable also apply to a function of two independent variables. So, using the notation $\overline{T}(x, s) = {}_t\mathcal{L}\{T(x, t)\}$ to denote the Laplace transform of $T(x, t)$ with respect to the time t, it follows directly from the formula for the transform of a derivative in (9a) that

$${}_t\mathcal{L}\{\partial T(x, t)\} = s\overline{T}(x, s) - T(x, 0).$$

To proceed further we must now use the condition that at time $t = 0$ the material of the slab is at zero temperature, so $T(x, 0) = 0$, as a result of which

$${}_t\mathcal{L}\{\partial T(x, t)/\partial t\} = s\overline{T}(x, s).$$

Next, as x is regarded as a constant, we have

$${}_t\mathcal{L}\{\partial^2 T(x, t)/\partial x^2\} = \frac{\partial^2 \overline{T}(x, s)}{\partial x^2}.$$

Using these results when taking the Laplace transform of the heat equation with respect to t, and making use of the linearity property of the transform, gives

$$s\overline{T}(x, s) = \kappa \frac{d^2 \overline{T}(x, s)}{dx^2},$$

where we now use an ordinary derivative with respect to x because in this differential equation s appears as a parameter so x can be considered to be the only independent variable. When the differential equation is written

$$\overline{T}'' - \frac{s}{\kappa}\overline{T} = 0,$$

using a prime to denote a derivative with respect to x, it is seen to have the general solution

$$\overline{T}(x, s) = A \exp\left[\sqrt{\frac{s}{\kappa}}x\right] + B \exp\left[-\sqrt{\frac{s}{\kappa}}x\right].$$

As a Laplace transform must vanish in the limit $s \to +\infty$, we must set $A = 0$, so the Laplace transform of the temperature is seen to be given by

$$\overline{T}(x, s) = B \exp\left[-\sqrt{\frac{s}{\kappa}}x\right].$$

In this case, the rejection of the term with the positive exponent in the general solution for $\overline{T}(x, s)$ corresponds to the physical requirement that the temperature remain finite for $x > 0$ and $t > 0$.

To determine B we now make use of the boundary condition on the plane face of the slab that requires $T(0, t) = T_0$, from which it follows that $_t\mathcal{L}\{T(0, t)\} = T_0/s$. Thus, the Laplace transform of the solution with respect to the time t is seen to be

$$\overline{T}(x, s) = \frac{T_0}{s} \exp\left[-\sqrt{\frac{s}{\kappa}}x\right].$$

To recover the time variation from this Laplace transform it is necessary to find $_t\mathcal{L}^{-1}\{\overline{T}(x, s)\}$. As $\overline{T}(x, s)$ is not the Laplace transform of an elementary function listed in our table of transform pairs, the solution $T(x, t)$ must be found by means of the Laplace inversion integral. In Chapter 16 on the Laplace inversion integral, it is shown in Example 16.6 that

$$_t\mathcal{L}^{-1}\{e^{-k\sqrt{s}}\} = \frac{k}{2\sqrt{\pi t^3}} \exp\left\{-\frac{k^2}{4t}\right\}.$$

So, setting $k = x/\kappa^2$ in this result and using it with Theorem 7.11 to invert the Laplace transform $\overline{T}(x, s)$ shows that the solution is

$$T(x, t) = T_0 \operatorname{erfc}\left\{\frac{x}{2\sqrt{\kappa t}}\right\}, \qquad \text{for } x > 0, \ t > 0.$$

The use of integral transforms is discussed in reference [4.4].

Summary

The Laplace transform has been applied to systems of differential equations, and the results extended to systems in matrix form. Various applications have been made to some useful variable coefficient ordinary differential equations, and to the important partial differential equation that describes one-dimensional unsteady heat flow.

EXERCISES 7.3

(a) Exercises involving systems of equations

1. Solve

$$x' + 5x - 2y = 1 \quad \text{and} \quad y' - 5x + 2y = 3$$
$$\text{given } x(0) = 0, \ y(0) = 2.$$

2. Solve

$$x' - x - y = \cos t \quad \text{and} \quad y' + x + y = \cos t$$
$$\text{given } x(0) = 1, \ y(0) = 1.$$

3. Solve

$$x' + x + y = 2 \quad \text{and} \quad y' + x - y = 1$$
$$\text{given } x(0) = -1, \ y(0) = 1.$$

4. Solve

$$x' + x + 2y = e^{-t} \quad \text{and} \quad y' + 2x + y = 1 \quad \text{given}$$
$$x(0) = 0, \ y(0) = 0.$$

5. Solve

$$x' - x + 3y = 1 + t \quad \text{and} \quad y' + x - y = 2 \quad \text{given}$$
$$x(0) = 2, \ y(0) = -2.$$

6. Solve

$$x' + x + y = \sin 2t \quad \text{and} \quad y' + x - y = 1 \quad \text{given}$$
$$x(0) = 0, \ y(0) = 0.$$

7. Solve

$$x' + x - z = 1, \quad y' - x + y = 1, \quad z' + y - x = 0,$$
$$\text{given that } x(0) = 1, \ y(0) = 0, \ z(0) = 1.$$

8. Solve

$$x' + x - y = 1, \quad y' - y + 2z = 0, \quad z' + x - y = \sin t,$$
$$\text{given } x(0) = 1, \ y(0) = 0, \ z(0) = 2.$$

9. Solve

$$x' - z = e^t, \quad y' - z = 2, \quad z' - x = 1, \quad \text{given } x(0) = 0,$$
$$y(0) = 1, \ z(0) = 0.$$

10. Solve

$$x' + z = 3, \quad y' + x = 1, \quad z' - x = \sin t, \quad \text{given}$$
$$x(0) = 1, \ y(0) = 0, \ z(0) = 1.$$

(b) Exercises involving e^{tA}

In Exercises 11 through 24 find e^{tA} for the given matrix $\mathbf{A}$.

11. $\mathbf{A} = \begin{bmatrix} 1 & 3 \\ 1 & -1 \end{bmatrix}.$

12. $\mathbf{A} = \begin{bmatrix} -2 & 4 \\ 3 & 2 \end{bmatrix}.$

13. $\mathbf{A} = \begin{bmatrix} 3 & 6 \\ 2 & -1 \end{bmatrix}.$

14. $\mathbf{A} = \begin{bmatrix} 3 & 7 \\ 3 & -1 \end{bmatrix}.$

15. $\mathbf{A} = \begin{bmatrix} 4 & -5 \\ 4 & 0 \end{bmatrix}.$

16. $\mathbf{A} = \begin{bmatrix} 3 & 4 \\ 3 & -1 \end{bmatrix}.$

17. $\mathbf{A} = \begin{bmatrix} 2 & -4 \\ 1 & 2 \end{bmatrix}.$

18. $\mathbf{A} = \begin{bmatrix} -2 & 3 \\ 5 & 0 \end{bmatrix}.$

19. $\mathbf{A} = \begin{bmatrix} 6 & -1 \\ 0 & 6 \end{bmatrix}.$

20. $\mathbf{A} = \begin{bmatrix} 2 & 3 \\ 0 & 4 \end{bmatrix}.$

21. $\mathbf{A} = \begin{bmatrix} -2 & 4 \\ 0 & -2 \end{bmatrix}.$

22. $\mathbf{A} = \begin{bmatrix} 1 & 4 \\ 3 & 0 \end{bmatrix}.$

23. $\mathbf{A} = \begin{bmatrix} 5 & 10 & 7 \\ 0 & -1 & -1 \\ 0 & 2 & 2 \end{bmatrix}.$

24. $\mathbf{A} = \begin{bmatrix} 1 & 0 & 0 \\ 1 & 3 & 2 \\ 1 & 2 & 3 \end{bmatrix}.$

(c) Exercises involving the weighting function

In Exercises 26 through 32 find the weighting function when a single equation is involved, and the four weighting functions when a pair of equations is involved. Use the weighting function(s) to solve the given differential equation(s).

25. Show that if the initial conditions for equation (24) are $y(0) = y_0, \ y'(0) = y_1, \ \dots, \ y^{(n-1)} = y_{n-1}$, the solution

can be written in the form

$$y(t) = \int_0^t w(\tau)[y_0(t - \tau) - h(t - \tau)]d\tau.$$

Here $y_0(t)$ is the solution of the equation with the initial conditions $y(0) = y'(0) = \cdots = y^{(n-1)}(0) = 0$, and $h(t) = \{H(s)/G(s)\}$, with $H(s)$ the polynomial produced by the nonvanishing initial values of the derivatives, so that the transformed equation corresponding to (26) becomes

$$G(s)Y(s) + H(s) = F(s).$$

26. $y'' - 4y' + 3y = \cos t$, given $y(0) = 0$ and $y'(0) = 0$.

27. $y'' + 2y' + 2y = e^{2t}$, given $y(0) = 0$ and $y'(0) = 0$.

28. $y'' + 4y' + 13y = \cos 2t$, given $y(0) = 0$ and $y'(0) = 0$.

29. $y'' + 6y' + 5y = e^{-t}$, given $y(0) = 0$ and $y'(0) = 0$.

30. Use the result of Exercise 25 to solve

$$y'' - 2y' - 3y = 1 + \sin t, \quad \text{given } y(0) = 1$$
$$\text{and} \quad y'(0) = -1.$$

31. $x' - 3x + 2y = e^{-t}, \quad y' + 3x - 4y = 3, \quad \text{with } x(0) = y(0) = 0.$

32. $x' + 2x - y = \sin t, \quad y' - 2x + y = 2, \quad \text{with } x(0) = y(0) = 0.$

(d) Differential equations with polynomial coefficients

33. Use the fact that $y(x) = \sin ax$ satisfies the differential equation

$$y'' + a^2 y = 0 \quad \text{with } y(0) = 0, \ y'(0) = a$$

to derive $\mathcal{L}\{\sin ax\}$ from the differential equation.

34. Use the fact that $y(x) = 1 - \cos ax$ satisfies the differential equation

$$y'' + a^2 y = a^2 \quad \text{with } y(0) = 0, \ y'(0) = 0$$

to derive $\mathcal{L}\{1 - \cos ax\}$ from the differential equation.

35.* The **Laguerre equation**

$$xy'' + (1 - x)y' + ny = 0,$$

with $n = 0, \ 1, \ 2, \dots$ a parameter, has polynomial solutions $y(x) = L_n(x)$ called **Laguerre polynomials**. These polynomials are used in many branches of mathematics and physics, and also in connection with numerical integration. By taking the Laplace transform of the differential equation find $\mathcal{L}\{L_n(x)\}$ and hence show that

$$L_4(x) = 24 - 96x + 72x^2 - 16x^3 + x^4.$$

36.* The **Hermite equation**

$$y'' - 2xy' + 2ny = 0,$$

with $n = 0, 1, 2, \ldots$ a parameter, has polynomial solutions $y(x) = H_n(x)$ called **Hermite polynomials**. Like the Laguerre polynomials, these polynomials are also used in mathematics and physics, and in connection with numerical integration. By transforming the equation and using the initial conditions $y(0) = H_4(0) = 12$ and $y'(0) = 0$, find $\mathcal{L}\{H_4(x)\}$, and hence show that

$$H_4(x) = 16x^4 - 48x^2 + 12.$$

37.* The Bessel function $y(x) = J_0(ax)$ satisfies the differential equation

$$xy'' + y' + axy = 0$$

subject to the initial conditions $y(0) = J_0(0) = 0$. Derive $\mathcal{L}\{J_0(ax)\}$ from the differential equation and confirm the result by using $\mathcal{L}\{J_0(x)\} = 1/(s^2 + 1)^{1/2}$ in conjunction with the scaling theorem.

38.* The Bessel function $y(x) = J_1(x)$ satisfies the differential equation

$$x^2 y'' + xy' + (x^2 - 1)y = 0 \quad \text{with } J_1(0) = 0 \text{ and}$$
$$J_1'(0) = 1/2.$$

By taking the Laplace transform of the differential equation show that $\mathcal{L}\{J_1(x)\} = C\{1 - s/(s^2 + 1)^{1/2}\}$, and deduce that $C = 1$.

(e) Exercises involving two-point boundary value problems

39. Solve $x'' + x = \sin 2t$ with $x(0) = 0$ and $x(\pi/2) = 1$.
40. Using the notation of Section 7.3(e), solve the beam equation

$$EI \frac{d^4 y}{dx^4} = w(x)$$

for the uniform cantilevered beam of mass M and length a shown in Fig. 7.27, where a point mass Q is located at a distance $a/3$ from the clamped end. The boundary conditions to be used are

$$y(0) = y'(0) = 0 \quad \text{and} \quad y''(a) = y'''(a) = 0.$$

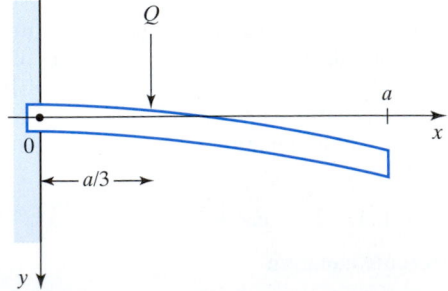

FIGURE 7.27 Cantilevered beam with a point load.

41. Using the notation of Section 7.3(e), solve the beam equation

$$EI \frac{d^4 y}{dx^4} = w(x)$$

for the uniform beam of mass M and length a with clamped ends shown in Fig. 7.28, where a point mass Q is located at a distance $3a/4$ from the left-hand end. The boundary conditions to be used are

$$y(0) = y'(0) = 0 \quad \text{and} \quad y(a) = y'(a) = 0.$$

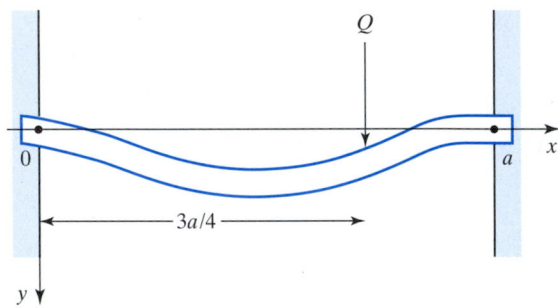

FIGURE 7.28 Supported beam with clamped ends and a point load.

42. Using the notation of Section 7.3(e), solve the beam equation

$$EI \frac{d^4 y}{dx^4} = w(x)$$

for the uniform beam of mass M and length a shown in Fig. 7.29 that is clamped at the end $x = 0$ and supported at the end $x = a$, where a point mass Q is located at a distance $a/4$ from the right-hand end. The boundary conditions to be used are

$$y(0) = y'(0) = 0 \quad \text{and} \quad y(a) = y''(a) = 0.$$

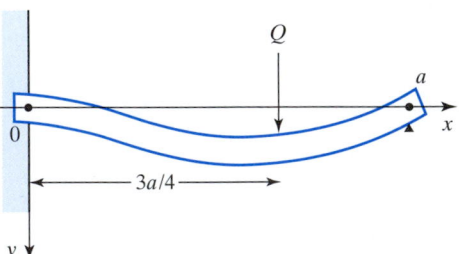

FIGURE 7.29 Beam clamped at one end and supported at the other with a point load.

(f) Physical problems to be solved by computer algebra

43. In an R–L–C circuit the current $i(t)$ and charge $q(t)$ resulting from a constant voltage E_0 applied at time

$t = 0$, when $i(0) = 0$ and $q(0) = 0$, are determined by the equations

$$L\frac{di}{dt} + Ri + \frac{q}{C} = E_0 \quad \text{and} \quad i = \frac{dq}{dt}.$$

Find $i(t)$, and comment on its form depending on the sign of $R^2C - 4L$. Choose representative values of R, L, C corresponding to each of the foregoing cases and plot $i(t)$ in a suitable interval $0 \le t \le T$.

44. Figure 6.10 in Section 6.3 illustrates three particles of equal mass joined by identical springs that oscillate in a straight line, with each end of the system clamped. In a representative case, the nondimensional equations determining the magnitudes of the displacements $y_1(t)$, $y_2(t)$, and $y_3(t)$ are

$$3\frac{d^2 y_1}{dt^2} = y_2 - 2y_1 + y_3, \quad 3\frac{d^2 y_2}{dt^2} = y_3 - 2y_2 + y_1,$$

$$3\frac{d^2 y_3}{dt^2} = y_1 - 2y_3 + y_2.$$

Find $y_1(t)$, $y_2(t)$, and $y_3(t)$ given that $y_1(0) = 1$, $y_1'(0) = 0$, $y_2(0) = 2$, $y_2'(0) = 1$, $y_3(0) = 3$, $y_3'(0) = 0$.

45. If, similar to the example in Section 7.3(a), an irreversible reaction converts a molecule of chemical A into a molecule of chemical D, via molecules of chemicals B and C, the governing equations in terms of the respec-

tive reaction rates k_1, k_2, and k_3 are

$$\frac{dx}{dt} = -k_1 x, \quad \frac{dy}{dt} = k_1 x - k_2 y, \quad \text{and} \quad \frac{dz}{dt} = k_2 y - k_3 z,$$

where x, y, and z are the number of molecules of A, B, and C present at time t. If Q molecules of A are present at time $t = 0$, the number of molecules of D present at time t is $w(t) = Q - x(t) - y(t) - z(t)$. Find $w(t)/Q$ as a function of t given that $k_1 = 2$, $k_2 = 3$, and $k_3 = 3$, and plot the result for $0 \le t \le 5$. Find the percentage of chemical A that has been transformed into chemical D at the instants of time $t = 1, 2$, and 3.

46. In the following nondimensional equations, $x(t)$ and $y(t)$ represent the magnitudes of the currents flowing in the primary and secondary windings of a transformer, when initially $x(0) = 0$, $y(0) = 0$ and at time $t = 0$ the primary winding is subjected to an exponentially decaying voltage of magnitude e^{-t}:

$$\frac{dx}{dt} + \frac{1}{3}\frac{dy}{dt} + 3x = e^{-t}, \quad \frac{dx}{dt} + 3\frac{dy}{dt} + 9y = 0.$$

Find $x(t)$ and $y(t)$, and by plotting the magnitudes of the currents show that $x(t)$ is always positive and after peaking decays to zero, while $y(t)$ is initially negative, but after becoming positive it decays to zero faster than $x(t)$.

7.4   The Transfer Function, Control Systems, and Time Lags

The study of engineering systems of all types whose behavior is determined by *linear* ordinary differential equations is often carried out by examining what is called the system **transfer function**. Typically, a system is governed by a linear nth order constant coefficient ordinary differential equation whose solution or **output**, also called the **response** of the system, we will denote by $u_0(t)$ and whose forcing function, or **input**, is a known function we will denote by $u_i(t)$, where t is the time.

A typical example of a simple system has already been encountered in Fig. 6.2, where the spring-mounted and damped vibrating machine has an input $F(t)$ and an output $y(t)$ that are related by

$$\frac{d^2 y}{dt^2} + a\frac{dy}{dt} + by = F(t).$$

An nth order system may be governed by the equation

$$a_n\frac{d^n u_0}{dt^n} + a_{n-1}\frac{d^{n-1} u_0}{dt^{n-1}} + \cdots + a_0 u_0 = u_i,$$

which can be represented graphically as in Fig. 7.30, where F[.] is the differential operator

$$F[.] \equiv a_n\frac{d^n}{dt^n} + a_{n-1}\frac{d^{n-1}}{dt^{n-1}} + \cdots + a_0. \tag{34}$$

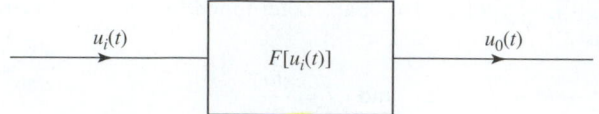

FIGURE 7.30 Block-diagram representation of equation (34).

More generally, in linear systems the input itself may be the solution of another linear differential equation, in which case the system relating the response $u_0(t)$ to the input $u_i(t)$ becomes

$$a_n \frac{d^n u_0}{dt^n} + a_{n-1} \frac{d^{n-1} u_0}{dt^{n-1}} + \cdots + a_0 u_0 = b_m \frac{d^m u_i}{dt^m} + b_{m-1} \frac{d^{m-1} u_i}{dt^{m-1}} + \cdots b_0 u_i, \qquad (35)$$

where $n \geq m$ and the coefficients a_r and b_s are constants.

The **transfer function** of a system is defined as the quotient of the Laplace transforms of the system output and the system input, when all of the initial conditions are taken to be *zero*. This last condition means that when the Laplace transform is used to transform a differential equation we may set $\mathcal{L}\{d^r u/dt^r\} = s^r U(s)$. So, after transforming (35), we obtain

$$(a_n s^n + a_{n-1} s^{n-1} + \cdots + a_0) U_0(s) = (b_m s^m + b_{m-1} s^{m-1} + \cdots b_0) U_i(s), \qquad (36)$$

where $U_0(s) = \mathcal{L}\{u_0(t)\}$ and $U_i(s) = \mathcal{L}\{u_i(t)\}$. The transfer function $G(s) = U_0(s)/U_i(s)$ becomes the rational function of the transform variable s

$$G(s) = \frac{b_m s^m + b_{m-1} s^{m-1} + \cdots b_0}{a_n s^n + a_{n-1} s^{n-1} + \cdots a_0}. \qquad (37)$$

Let us now set $G(s) = N(s)/D(s)$, where $N(s)$ is the polynomial in s of degree m in the numerator of $G(s)$, and $D(s)$ is the polynomial in s of degree n in the denominator. The polynomial $D(s)$ is called the **characteristic polynomial** of the system, and $D(s) = 0$ is called the **characteristic equation** of the system. The **order** of the system in (37) is the degree n of the polynomial $D(s)$.

As the coefficients of $D(s)$ are real, it follows that the roots of the characteristic equation, called the **poles** of the transfer function $G(s)$, either are all real or, if complex, they must occur in complex conjugate pairs. When $G(s)$ is expressed in partial fraction form, this last observation implies that the system will be **stable** provided all the roots of the characteristic equation have negative real parts. Here, by **stability**, we mean that any bounded input to a system that is stable will result in an output that is also bounded for all time, and this will be the case when every root of $D(s) = 0$ has a negative real part. The requirement that $n \geq m$ imposed on (35) is necessary in order to prevent unbounded behavior of the output caused by the occurrence of delta functions.

It is important to recognize that systems describing quite different physical phenomena can have the *same* transfer function, so transfer functions provide a means of examining a class of similar systems independently of their physical origin. It follows that for any given input with Laplace transform $U_i(s)$, the Laplace transform of the output $U_0(s)$ is given by

$$U_0(s) = G(s) U_i(s). \qquad (38)$$

The time variation of the output of the system then follows by taking the inverse Laplace transform of (38).

EXAMPLE 7.36

Find the transfer function of the system with input $u_i(t)$ and output $u_0(t)$ described by

$$4\frac{d^2 u_0(t)}{dt^2} + 16\frac{du_0(t)}{dt} + 25u_0(t) = 3\frac{du_i(t)}{dt} + 2u_i(t),$$

and show it is stable.

Solution Taking the Laplace transform of the governing equation and assuming all initial conditions to be zero gives

$$(4s^2 + 16s + 25)U_0(s) = (3s + 2)U_i(s),$$

so the system transfer function is

$$G(s) = \frac{U_0(s)}{U_i(s)} = \frac{3s + 2}{4s^2 + 16s + 25}.$$

The system is of order 2, and its characteristic equation is

$$4s^2 + 16s + 25 = 0.$$

The characteristic equation has the roots $s_1 = -2 - \frac{3}{2}i$ and $s_2 = -2 + \frac{3}{2}i$, so as their real parts are negative, the system is stable. ∎

Systems that compare the difference between an input and an output, and attempt to reduce the difference to zero to make the output *follow* the input, are called **control systems**. A typical example is a temperature control system for a chemical reactor in which the temperature is required to remain constant, but where as the reaction progresses heat is released at variable rates, causing cooling to become necessary.

A simple control system is illustrated in Fig. 7.31, where F is the system differential equation. The idea here is that an input u_i is compared with the output u_0, called the **feedback**, and the difference $\varepsilon = u_i - u_0$, called the **error signal**, is then used as an input to system F. The result is that $u_0 = u_i$ when $\varepsilon = 0$. It is often necessary to modify the feedback by passing u_0 through another system G with output $v = G[u_0]$, and then to use the the difference $v - u_i$ to drive F. The reason for this is to improve the overall performance of a system, whose physical characteristics may be difficult to alter, by using an easily modified feedback to make the system more responsive and to reduce any tendency it may have for excessive oscillation.

EXAMPLE 7.37

A steering mechanism for a small boat comprises an input heading θ_i from the helm, an amplifier for the error signal, and a servomotor to drive the rudder with moment of inertia I that produces a resisting torque proportional to the rate of change of the output angle θ_0. Derive the differential equation governing the system and find its transfer function given that the feedback is the unmodified output θ_0.

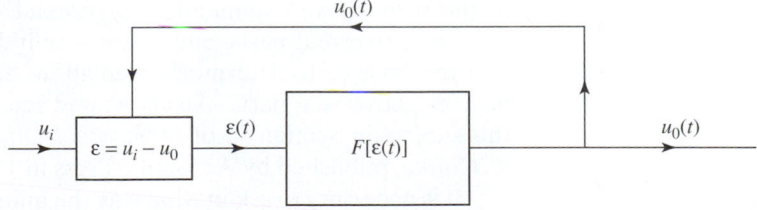

FIGURE 7.31   A typical feedback control system.

Solution If the resisting torque is $kd\theta_0/dt$ and the amplifier increases the magnitude of the error signal by a factor K, the system can be represented as in Fig. 7.31 with the governing differential equation

$$I\frac{d^2\theta_0}{dt^2} + k\frac{d\theta_0}{dt} = K(\theta_i - \theta_0).$$

Taking the Laplace transform of this equation gives

$$(Is^2 + ks + K)\mathcal{L}\{\theta_0\} = \mathcal{L}\{\theta_i\},$$

and so

$$\mathcal{L}\{\theta_0\} = \frac{1}{Is^2 + ks + K}\mathcal{L}\{\theta_i\}.$$

This result shows that the transfer function $G(s) = 1/(Is^2 + ks + K)$, so the system will be stable provided the roots of the characteristic equation $Is^2 + ks + K = 0$ have negative real parts. This will be the case since $I > 0$ and $K > 0$, but the steering will oscillate about the required heading if $4IK > k^2$.

As the design of the boat determines I and k, any improvement of the steering response can only be obtained by using a modified feedback signal instead of the direct feedback θ_0.   ∎

We close this section by mentioning an important consequence of the introduction of a delay into an equation governing the response of a system. Consider a vibrating system characterized by $y(t)$ in which instantaneous damping proportional to the velocity dy/dt occurs with coefficient of proportionality a_1, and where there is also present an additional time retarded damping of a similar type but with a time lag τ and a coefficient of proportionality a_2. Then, when a springlike restoring effect is present with constant of proportionality a_3, the governing equation takes the form

$$\frac{d^2y(t)}{dt^2} + a_1\frac{dy(t)}{dt} + a_2\frac{dy(t-\tau)}{dt} + a_3y(t) = 0. \tag{39}$$

Because of the presence of the time-delayed derivative $dy(t - \tau)/dt$, an equation of this type is called a **differential-difference equation**.

If we now seek a solution of this equation by using the Laplace transform (or by seeking solutions of the form $y(t) = A\exp(\lambda t)$, where A and λ are constants) we arrive at a characteristic equation of the form

$$s^2 + a_1s + a_2s\exp(-\tau s) + a_3 = 0. \tag{40}$$

This is called an **exponential polynomial** in s, and its root will determine both the stability and response of the system.

Without going into detail, by using Rouche's theorem from complex analysis it is not difficult to prove that exponential polynomials have an infinite number of zeros. Consequently, the response of a system with a characteristic polynomial in the form of an exponential polynomial will only be stable if all of its zeros have negative real parts, and this can only be shown analytically. Methods exist that can be used to determine when all the zeros of such exponential polynomials have negative real parts. An interested reader will find a valuable discussion of this subject in Section 13 of *Differential-Difference Equations* by R. Bellman and K. Cooke, published by Academic Press in 1963.

It is necessary to ask in what way the infinite number of zeros of an exponential polynomial of degree n approximate the n zeros of the ordinary polynomial of degree n when time lags are absent. This is a simpler question, and it can be answered by appeal to Hurwitz's theorem from complex analysis, though again the arguments used go beyond this first account of the subject.

A result on exponential polynomials

Let $P_\tau(s)$ be an exponential polynomial of degree n in s with a time lag τ, and let $P_0(s)$ be the corresponding constant coefficient polynomial when $\tau = 0$. Then, as $\tau \to 0$, so each of the n zeros s_i of $P_0(s)$ is approached arbitrarily closely by a number of zeros of $P_\tau(s)$ equal in number to its multiplicity, and the remaining infinite number of zeros of $P_\tau(s)$ can be made to lie outside a circle of arbitrarily large radius centered on the origin.

As this result says nothing about how the zeros move as $\tau \to 0$, it is possible for the system to be stable when τ lies in certain intervals and unstable otherwise.

EXERCISES 7.4

1. Find the transfer function for each of the following systems. Determine the order of each system and find which is stable.

(a)
$$\frac{d^3 u_0}{dt^3} + 3\frac{d^2 u_0}{dt^2} + 16\frac{du_0}{dt} - 20u_0$$
$$= 2\frac{d^2 u_i}{dt^2} + \frac{du_i}{dt} - 6u_i.$$

(b)
$$\frac{d^3 u_0}{dt^3} + 4\frac{d^2 u_0}{dt^2} + 14\frac{du_0}{dt} + 20u_0$$
$$= 6\frac{d^2 u_i}{dt^2} - 13\frac{du_i}{dt} + 6u_i.$$

(c)
$$9\frac{d^2 u_0}{dt^2} + 6\frac{du_0}{dt} + 10u_0 = 6\frac{d^2 u_i}{dt^2} + 5\frac{du_i}{dt} - 6u_i.$$

2.* For safety reasons, a control system is often duplicated, with the sensors for each system located in different positions, and in such cases the possibility of interaction between the control systems must be considered. A typical case is illustrated in Fig. 7.32, where two identical control systems are shown between which there is assumed to be linear **cross-coupling** of the error signals. This means that the respective actuating error signals are $\varepsilon_1' = a_{11}\varepsilon_1 + a_{12}\varepsilon_2$ and $\varepsilon_2' = a_{21}\varepsilon_1 + a_{22}\varepsilon_2$, with the coefficients a_{ij} constants. Derive and discuss the equations governing the response of the system when

$$F(u_0) = \frac{d^2 u_0}{dt^2} + 2\zeta\Omega\frac{du_0}{dt} + \Omega^2 u_0,$$

with $\zeta > 0$ and $\Omega > 0$.

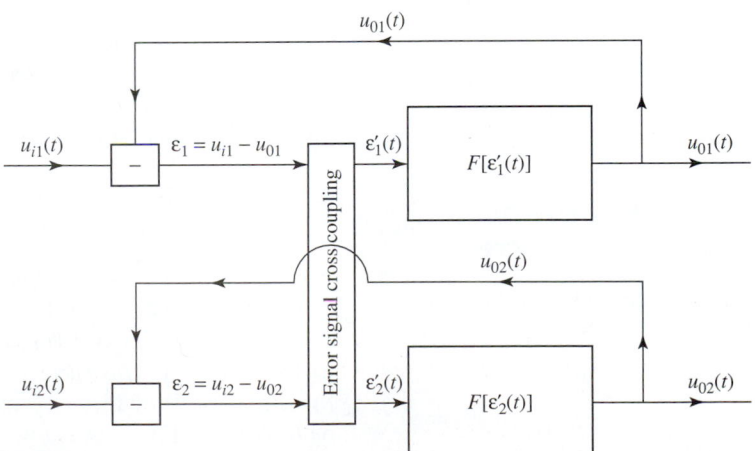

FIGURE 7.32   Two interacting control systems.

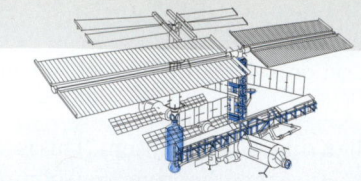

CHAPTER 7
TECHNOLOGY PROJECTS

The purpose of these projects is to use a computer algebra differential equation solver to find the analytical solutions of initial value problems involving linear constant coefficient differential equations, some of which contain either the Dirac delta function or the Heaviside step function. As all the initial conditions are given at $t = 0$, the Laplace transform can also be used to solve these problems.

Project 1

Solving a Third Order Initial Value Problem

Use a computer algebra Laplace solver to solve the initial value problem

$$x''' + 2x'' - x' - 2x = e^{-t} \sin t, \quad \text{with } x(0) = 1,$$
$$x'(0) = -1, \text{ and } x''(0) = 0.$$

Verify the result by using computer algebra (a) to take the Laplace transform of the equation, (b) to find the Laplace transform $X(s)$ of the solution, and (c) to invert the transform to find $x(t)$.

Project 2

Solving an Equation with the Heaviside Step Function in the Nonhomogeneous Term

Use a computer algebra Laplace solver to solve the initial value problem

$$x'' + 3x' + 2x = \{H(t - 1) - H(t - 2)\}t,$$
$$\text{with } x(0) = 1, \ x'(0) = -1.$$

Verify the result by using computer algebra (a) to take the Laplace transform of the equation, (b) to find the Laplace transform $X(s)$ of the solution, and (c) to invert the transform to find $x(t)$. Plot the solution for $0 \le t \le 6$.

Project 3

Solving an Equation with the Dirac Delta Function in the Nonhomogeneous Term

Use a computer algebra Laplace solver to solve the initial value problem

$$x'' + 3x' + 2x = 3e^{-t} + \delta(t - 2), \quad \text{with } x(0) = 1,$$
$$x'(0) = 2.$$

Verify the result by using computer algebra (a) to

take the Laplace transform of the equation, (b) to find the Laplace transform $X(s)$ of the solution, and (c) to invert the transform to find $x(t)$.

Project 4

Solving a System

Solve the initial value problem for the system

$$\frac{dx}{dt} = x(t) + 2y(t) + 3, \quad \frac{dy}{dt} = 1 - x(t) + y(t),$$
$$\text{with } x(0) = 1, \ y(0) = 0.$$

Verify the result by using computer algebra (a) to take the Laplace transform of the system, (b) to solve for the Laplace transforms $X(s)$ and $Y(s)$ of $x(t)$ and $y(t)$, and then (c) to invert the transforms to find $x(t)$ and $y(t)$.

Project 5

Examining the Properties of a Spring Damper System

In an experiment, a wheel of mass M is mounted vertically below a rigid plate to which it is attached by a spring with spring constant k and a damper whose resisting force is μ times the speed of its displacement. If at time t the vertical displacement of the wheel from its equilibrium position is $x(t)$, and a force $F(t)$ is applied to the wheel, its equation of motion is

$$M\frac{d^2x}{dt^2} + \mu\frac{dx}{dt} + kx = F(t).$$

Set $\Omega = (k/M)^{1/2}$, $\zeta = \frac{\mu}{2\sqrt{kM}}$, and assume the wheel is initially at rest, so that $x(0) = 0$ and $(dx/dt)_{t=0} = 0$. If a constant load $F(t) = F_0$ is suddenly applied to the wheel at the time $t = 0$, find an expression for $x(t)k/F_0$. Plot this expression for several values of ζ in the interval $0 < \zeta < 2$ and comment on the results.

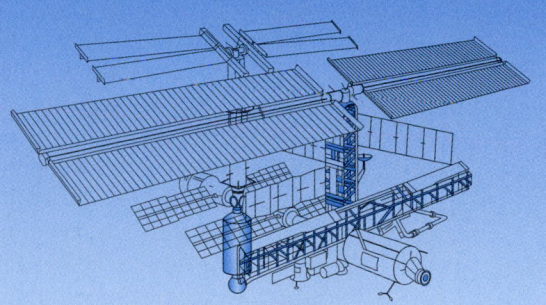

CHAPTER 8

Series Solutions of Differential Equations, Special Functions, and Sturm–Liouville Equations

Linear second order variable coefficient equations arise in many applications, but only in a few special cases is it possible to express their general solution of a finite linear combination of elementary functions. As analytical, rather than purely numerical, information about solutions is often essential, some other way must be found to represent the solutions of such equations. The approach developed in this chapter involves seeking solutions of certain types of equation in the form of power series, and in other cases using an approach due to Frobenius that involves seeking solutions in the form of power series multiplied by a factor x^c, where c is not an integer. Applications are made to a number of typical linear variable coefficient equations, and then to the important Legendre, Chebyshev, and Bessel equations that lead in turn to Legendre and Chebyshev polynomials and to Bessel functions.

Two-point boundary value problems, called Sturm–Liouville systems, that are defined over an interval $a \leq x \leq b$ and contain a parameter λ are introduced. It is shown that their solutions only exist for an infinite number of special values of the parameter $\lambda_1, \lambda_2, \ldots$, called the eigenvalues of the problem. Each solution $\varphi_n(x)$ corresponding to an eigenvalue λ_n is called an eigenfunction, and the eigenfunctions are shown to have the special property of orthogonality with respect to a function $w(x)$ called the weight function. This means that if the set of eigenfunctions is $\{\varphi_n(x)\}_{n=1}^{\infty}$, the integral $\int_a^b \varphi_m(x)\varphi_n(x)w(x)dx$ is positive when $n = m$ and zero when $n \neq m$. This property will be used extensively in Chapter 18 when solving partial differential equations.

Fundamental properties of eigenfunctions and eigenvalues are established for general Sturm–Liouville systems, after which a number of frequently occurring and important special cases are examined.

8.1 A First Approach to Power Series Solutions of Differential Equations

The solutions of many differential equations can be expressed in terms of elementary functions such as sine, cosine, exponential, and logarithm, all of whose mathematical properties are well known. When required, the analytical behavior of solutions that involve elementary functions can be explored by making use of

443

their familiar properties. Numerical solutions are obtained easily, either by using a pocket calculator to find the values of the elementary functions involved, or through the use of standard subroutines that form a part of all basic mathematical software packages. With either a pocket calculator or a software package, the method of calculating functional values is usually based on a series expansion of the function concerned.

Most differential equations cannot be solved in terms of elementary functions, yet some form of analytical solution is often needed rather than a purely numerical one, so the fundamental question that then arises is how to obtain a solution in the form of a series, when only the differential equation is known. It is the purpose of this chapter to answer this question, and in the process to show how the form of series solution obtained depends on what are called the *singular points* of the differential equation.

We begin our approach to this problem by showing how series solutions can be found for first and second order linear differential equations with initial conditions specified at $x = x_0$. The series we obtain will be in powers of $x - x_0$, and they will be said to be *expanded* about the point x_0. The first order linear differential equation will be assumed to be of the form

$$y' + p(x)y = r(x) \quad \text{with } y(x_0) = y_0, \tag{1}$$

and the second order linear differential equation will be assumed to be of the form

$$y'' + P(x)y' + Q(x)y = R(x) \quad \text{with } y(x_0) = y_0, \ y'(x_0) = y_1, \tag{2}$$

where the functions $p(x), r(x), P(x), Q(x)$, and $R(x)$ can all be expanded as Taylor series about the point x_0.

analytic in a neighborhood

Functions with this property are said to be **analytic** in a **neighborhood** of the point x_0 or, more simply, to be **analytic** at x_0. The method to be developed will be seen to be capable of extension to a higher order linear differential equation in an obvious manner, provided only that the coefficients of y and its derivatives that are involved and the nonhomogeneous term are analytic at x_0.

how to find a power series solution

The approach is best illustrated by considering equation (1), and seeking a solution about x_0 of the form

$$y(x) = y(x_0) + (x - x_0)y'(x_0) + \frac{(x - x_0)^2}{2!}y''(x_0) + \frac{(x - x_0)^3}{3!}y'''(x_0) + \cdots$$

$$= \sum_{n=0}^{\infty} \frac{(x - x_0)^n}{n!} y^{(n)}(x_0), \quad \text{with } y^{(n)}(x) = d^n y/dx^n.$$

$$\tag{3}$$

Setting $x = x_0$ in (1) gives

$$y^{(1)}(x_0) + p(x_0)y(x_0) = r(x_0),$$

but $y(x_0) = y_0$, so

$$y^{(1)}(x_0) = r(x_0) - p(x_0)y(x_0)$$
$$= r(x_0) - p(x_0)y_0.$$

To determine $y^{(2)}(x)$ we differentiate equation (1) once with respect to x to obtain

$$y^{(2)}(x) + p^{(1)}(x)y(x) + p(x)y^{(1)}(x) = r^{(1)}(x),$$

where $p^{(1)}(x) = p'(x)$ and $r^{(1)}(x) = r'(x)$. Then, after setting $x = x_0$ and using the fact that $y^{(1)}(x_0) = r(x_0) - p(x_0)y_0$, we find that

$$y^{(2)}(x_0) = r^{(1)}(x_0) - p^{(1)}(x_0)y_0 - p(x_0)[r(x_0) - p(x_0)y_0].$$

Higher order derivatives $y^{(n)}(x_0)$ can be computed in similar fashion by repeated differentiation of the original differential equation coupled with the use of lower order derivatives that have already been determined. Once the values of $y^{(k)}(x_0)$ have been found for $k = 1, 2, \ldots, N$, for some given integer N, substitution into series (3) provides the required approximation to the power series solution of the initial value problem for the differential equation up to terms of order $(x - x_0)^N$. The existence and uniqueness of the solution are guaranteed by Theorem 5.2.

This method generates the Taylor series expansion of $y(x)$ about the point x_0 when $x_0 \neq 0$, and its Maclaurin series expansion when $x_0 = 0$, though these series are often simply called power series about $x_0 \neq 0$ and $x_0 = 0$, respectively.

EXAMPLE 8.1

Find the first five terms in the series solution of

$$y' + (1 + x^2)y = \sin x, \quad \text{with } y(0) = a.$$

Solution   As the initial condition is specified at $x = 0$, the power series solution is an expansion about the origin and so is, in fact, a Maclaurin series. The functions $1 + x^2$ and $\cos x$ are analytic for all x, so the series expansion can certainly be found about the origin.

Setting $x = 0$ in the equation and substituting for the initial conditions shows that $y'(0) = y^{(1)}(0) = -a$. Differentiation of the differential equation gives

$$y^{(2)} + 2xy + (1 + x^2)y^{(1)} = \cos x,$$

where $y^{(2)} = y''$, so setting $x = 0$ this becomes

$$y^{(2)}(0) + y^{(1)}(0) = 1,$$

but $y^{(1)}(0) = -a$ and so $y^{(2)}(0) = 1 + a$. Repeating this process to find higher order derivatives leads to the results $y^{(3)}(0) = -(1 + 3a)$, $y^{(4)}(0) = 9a, \ldots$. Substituting these results into series (3) shows that, to terms of order x^4, the required solution takes the form

$$y(x) = a - ax + (1 + a)\frac{x^2}{2!} - (1 + 3a)\frac{x^3}{3!} + 9a\frac{x^4}{4!} + \cdots.$$ ∎

EXAMPLE 8.2

Find the first five terms in the series solution of

$$y' + 4xy = 3e^{x-1}, \quad \text{with } y(1) = 1.$$

Solution   In this case the functions x and e^{x-1} are analytic for all x, but as the expansion is about $x = 1$, the power series solution that is obtained will be a Taylor series expansion about the point $x = 1$. Setting $x = 1$ in the differential equation and using the initial condition $y(1) = 1$ shows that $y^{(1)}(1) = -1$.

Differentiation of the differential equation gives

$$y^{(2)} + 4y + 4xy^{(1)} = 3e^{x-1},$$

so setting $x = 1$ and using the result $y^{(1)}(1) = -1$ shows that $y^{(2)}(1) = 3$.

Repeating this process leads to the results that $y^{(3)}(1) = -1$ and $y^{(4)}(1) = -29$, so substituting into (3) shows that the Taylor series expansion of the solution up to terms of order $(x - 1)^4$ is

$$y(x) = 1 - (x - 1) + \frac{3}{2}(x - 1)^2 - \frac{1}{6}(x - 1)^3 - \frac{29}{24}(x - 1)^4 + \cdots. \qquad \blacksquare$$

This same method can be applied to a second order equation of the type shown in (2), though a more general approach will be developed later to deal with the case in which the first term is of the form $a(x)y''(x)$, and the expansion is about a point x_0 where $a(x_0) = 0$.

EXAMPLE 8.3

Find the terms up to x^5 in the series solution of

$$y'' + xy' + (1 - x^2)y = x \quad \text{with } y(0) = a, \ y'(0) = b.$$

Solution   The coefficients x and $(1 - x^2)$ and the nonhomogeneous term x are analytic for all x, so as the initial data is given at $x = 0$, a Maclaurin series solution can be found.

Setting $x = 0$ in the equation and using the initial conditions $y(0) = a$ and $y'(0) = b$ gives $y^{(2)}(0) = -a$. Differentiating the differential equation we have

$$y^{(3)} + y^{(1)} + xy^{(2)} - 2xy + (1 - x^2)y^{(1)} = 1,$$

so setting $x = 0$ and using the results $y^{(2)}(0) = -a$ and $y^{(1)}(0) = b$ shows that $y^{(3)}(0) = 1 - 2b$. A repetition of this process leads to the results $y^{(4)}(0) = 5a$, $y^{(5)}(0) = 14b - 4, \ldots$, so substituting into (3) shows that to terms of order x^5 the Maclaurin series expansion of the solution is

$$y(x) = a + bx - \frac{1}{2}ax^2 + \left(\frac{1 - 2b}{6}\right)x^3 + \frac{5a}{24}x^4 + \left(\frac{7b - 2}{60}\right)x^5 + \cdots. \qquad \blacksquare$$

Summary

Often a variable coefficient equation cannot be solved in terms of known functions, though some form of analytical solution is still required. This section has shown how to overcome this difficulty in some cases by finding a solution in terms of a power series expanded about a point of interest $x = a$. The method was seen to work provided the functions in the equation have Taylor series expansions about $x = a$. It will be shown later how to find series solutions in a systematic manner, and also how to generalize this approach to other types of equation.

EXERCISES 8.1

Find the first five terms in the power series solution of the following initial value problems.

1. $y' + (1 + x^2)y = x^2$, with $y(0) = 1$.
2. $2y' + xy = 1 - x$, with $y(0) = 2$.
3. $y' + (1 - 2x)y = x$, with $y(0) = -1$.
4. $4y' + (1 + x + x^2)y = x$, with $y(0) = 3$.

5. $y' + (x - 2x^2)y = 1$, with $y(0) = 1$.
6. $y' - 2xy = 1 - x$, with $y(0) = 2$.
7. $3y' + (1 - x^2)y = 1$, with $y(0) = 2$.
8. $y' + (1 + x)y = 1 + x^2$, with $y(0) = 1$.
9. $y'' - 2xy' + x^2y = 0$, with $y(0) = a$, $y'(0) = b$.
10. $2y'' + 2(1 + x)y' - y = 0$, with $y(0) = a$, $y'(0) = b$.

11. $(1 + x^2)y'' + 3xy' + (1 - x^2)y = 1 + x,$ with $y(0) = a,$ $y'(0) = b.$

12. $(1 + 3x^2)y'' + 2xy' + 2xy = 1,$ with $y(0) = a,$ $y'(0) = b.$

13. $y'' + 7y' + x^2y = 0,$ with $y(0) = a,$ $y'(0) = b.$

14. $xy'' + (1 + x)y' + xy = b,$ with $y(0) = a,$ $y'(0) = 0.$

15. $2y'' + 3x^2y' + (1 - x^2)y = 2x,$ with $y(0) = a,$ $y'(0) = b.$

16. $3y'' + 2xy' + (1 - 2x^2)y = 1 + 2x,$ with $y(0) = a,$ $y'(0) = b.$

8.2   A General Approach to Power Series Solutions of Homogeneous Equations

The method developed in Section 8.1 works satisfactorily if only the first few terms in a power series solution are required, but it has the disadvantage that a separate calculation is required each time a coefficient is determined. The present section shows how in many cases this difficulty can be overcome by introducing a systematic and simple way of generating arbitrarily many terms in a power series solution of the homogeneous linear differential equation

$$a(x)y'' + b(x)y' + c(x)y = 0 \qquad (4)$$

about a point x_0, when $a(x)$, $b(x)$, and $c(x)$ are polynomials with $a(x_0) \neq 0$.

The approach enables the coefficients of the power series solution to be determined by means of a *recurrence relation* that relates a few consecutive coefficients in the series. This has the advantage that once the first few coefficients in the series expansion have been found, the rest can be generated by means of the recurrence relation.

There will be no loss of generality if the approach is based on an expansion about the origin, because if one is required about an arbitrary point $x = x_0$, the change of variable $X = x - x_0$ will shift the point $x = x_0$ to $X = 0$. For example, suppose a solution of

$$y'' + (2 + 3x)y' + x^2y = 0$$

is required about the point $x = 1$, corresponding to the specification of the initial conditions for $y(1)$ and $y'(1)$ at $x = 1$. Setting $X = x - 1$ and $y(x) = Y(x - 1) = Y(X)$, it follows that $y(1) = Y(0)$, $dy/dx = dY/dX$, $d^2y/dx^2 = d^2Y/dX^2$, and $x = X + 1$, so in terms of the new variables X and Y the equation and initial conditions become

$$Y'' + (5 + 3X)Y' + (1 + X)^2Y = 0, \quad \text{with } Y(0) = y(1), \ Y'(0) = y'(1).$$

Setting $X = x - 1$ in the power series solution of this equation expanded about $X = 0$ reduces it to the solution of the original equation expanded about $x = 1$.

The approach we now describe involves seeking a solution in the form of a general power series

$$y(x) = \sum_{n=0}^{\infty} a_n x^n \qquad (5)$$

and finding a relationship between the coefficients a_n by substituting (5) into the homogeneous differential equation

$$a(x)y'' + b(x)y' + c(x)y = 0. \tag{6}$$

We will assume that the coefficients $a(x)$, $b(x)$, and $c(x)$ in the differential equation are polynomials in x, and so are analytic at $x = 0$, and also that $a(0) \neq 0$. If (5) is to be a solution of (6), it must satisfy the differential equation for all x, but this will only be possible if, after combining terms, the coefficient of each power of x in the new power series is zero. It will be seen later that it is this last requirement that leads to the determination of the coefficients a_n in terms of a recurrence relation.

Before illustrating the approach by means of an example, we first find expressions for the derivatives $y'(x)$ and $y''(x)$ that will be needed in the calculation. Writing out the first few terms of $y(x)$ in (5) gives

$$y(x) = a_0 + a_1 x + a_2 x^2 + a_3 x^3 + \cdots = \sum_{n=0}^{\infty} a_n x^n. \tag{7}$$

Differentiating this expression term by term with respect to x, which is permitted for x inside the interval of convergence of the series, we arrive at the result

$$y'(x) = a_1 + 2a_2 x + 3a_3 x^2 + \cdots = \sum_{n=1}^{\infty} n a_n x^{n-1}, \tag{8}$$

and after a further differentiation we have

$$y''(x) = 2a_2 + 2 \cdot 3a_3 x + 3 \cdot 4a_4 x^2 + \cdots = \sum_{n=2}^{\infty} n(n-1) a_n x^{n-2}. \tag{9}$$

In what is to follow it will be important to remember that the summation in (8) starts at $n = 1$, whereas the summation in (9) starts at $n = 2$.

EXAMPLE 8.4   Find the recurrence relation that must be satisfied by coefficients in the series solution of the differential equation

$$y'' + 2xy' + (1 + x^2)y = 0$$

when the expansion is about the origin. Solve the initial value problem for this differential equation given that $y(0) = 3$ and $y'(0) = -1$.

Solution   Substituting $y(x) = \sum_{n=0}^{\infty} a_n x^n$ into the differential equation and using (8) and (9) gives

$$\sum_{n=2}^{\infty} n(n-1) a_n x^{n-2} + 2x \sum_{n=1}^{\infty} n a_n x^{n-1} + (1 + x^2) \sum_{n=0}^{\infty} a_n x^n = 0.$$

Taking the factor $2x$ in the second term and the factor x^2 in the third term under their respective summation signs allows the equation to be written in the form

$$\sum_{n=2}^{\infty} n(n-1)a_n x^{n-2} + \sum_{n=1}^{\infty} 2na_n x^n + \sum_{n=0}^{\infty} a_n x^n + \sum_{n=0}^{\infty} a_n x^{n+2} = 0.$$

The powers of x in the first and last summations are different from those in the middle two summations, so before combining the summations in order to find the coefficient of each power of x, it will first be necessary to change the power of x in the first and last terms from $n-2$ and $n+2$ to n.

In the first summation we set $m = n - 2$, causing the summation to become

$$\sum_{m=0}^{\infty} (m+2)(m+1)a_{m+2} x^m.$$

However, m is simply a summation index that can be replaced by any other symbol, so we will replace it by n to obtain the equivalent expression

$$\sum_{n=0}^{\infty} (n+2)(n+1)a_{n+2} x^n.$$

Similarly, by setting $m = n + 2$ in the last summation, and then replacing m by n, we find that

$$\sum_{n=0}^{\infty} a_n x^{n+2} \quad \text{becomes} \quad \sum_{n=2}^{\infty} a_{n-2} x^n.$$

We now substitute these last two results into the series solution of the differential equation to obtain

$$\sum_{n=0}^{\infty} (n+2)(n+1)a_{n+2} x^n + \sum_{n=1}^{\infty} 2na_n x^n + \sum_{n=0}^{\infty} a_n x^n + \sum_{n=2}^{\infty} a_{n-2} x^n = 0,$$

where now each summation involves x^n, though not all summations start from $n = 0$.

Separating out the terms corresponding to $n = 0$ and $n = 1$, and collecting all the remaining terms under a single summation sign in which the summation starts from $n = 2$, this becomes

$$2a_2 + a_0 + (6a_3 + 3a_1)x + \sum_{n=2}^{\infty} [(n+2)(n+1)a_{n+2} + 3a_n + a_{n-2}]x^n = 0.$$

As already remarked, if this power series is to be a solution of the differential equation it must satisfy the equation identically for all x, but this will only be possible if in the foregoing expression the coefficient of each power of x vanishes. Applying this condition to the preceding series we find that for it to vanish identically for all x,

deriving and using a recurrence relation

$$\text{(coefficient of } x^0) \quad 2a_2 + a_0 = 0$$
$$\text{(coefficient of } x) \quad 6a_3 + 3a_1 = 0$$

and

$$\text{(coefficient of } x^n) \quad (n+2)(n+1)a_{n+2} + 3a_n + a_{n-2} = 0, \quad \text{for } n \geq 2.$$

The first condition shows that

$$a_2 = -\frac{1}{2}a_0,$$

while the second condition shows that

$$a_3 = -\frac{1}{2}a_1,$$

where a_0 and a_1 are arbitrary constants.

The third condition is a **recurrence relation** (also called a **recursion relation** or an **algorithm**) that in this case relates three coefficients whose indices differ by 2, so given a_{n-2} and a_n we can find a_{n+2} for $n = 2, 3, 4, \dots.$

We now show how to determine the first few coefficients a_n by writing the recursion relation in the form

$$a_{n+2} = -\frac{[(2n+1)a_n + a_{n-2}]}{(n+1)(n+2)}$$

and setting $n = 2, 3, 4, \dots.$

For $n = 2$, after using $a_2 = -\frac{1}{2}a_0$, we find that

$$a_4 = -\frac{(5a_2 + a_0)}{12} = \frac{a_0}{8},$$

whereas for $n = 3$, after using $a_3 = -\frac{1}{2}a_1$, we find that

$$a_5 = -\frac{(7a_3 + a_1)}{20} = \frac{a_1}{8}.$$

Continuing this process generates the coefficients

$$a_6 = -\frac{a_0}{48}, \quad a_7 = -\frac{a_1}{48}, \quad a_8 = \frac{a_0}{384}, \quad a_9 = \frac{a_1}{384}, \dots.$$

Thus, all the coefficients with even suffixes are determined in terms of the arbitrary constant a_0, whereas all the coefficients with odd suffixes are determined in terms of the arbitrary constant a_1.

Substituting these coefficients into the power series $y(x) = \sum_{n=0}^{\infty} a_n x^n$ and grouping terms gives

$$y(x) = a_0 \left(1 - \frac{1}{2}x^2 + \frac{1}{8}x^4 - \frac{1}{48}x^6 + \frac{1}{384}x^8 - \cdots\right)$$

$$+ a_1 \left(x - \frac{1}{2}x^3 + \frac{1}{8}x^5 - \frac{1}{48}x^7 + \frac{1}{384}x^9 - \cdots\right).$$

As the coefficients a_0 and a_1 are arbitrary, the functions represented by the series

$$y_1(x) = 1 - \frac{1}{2}x^2 + \frac{1}{8}x^4 - \frac{1}{48}x^6 + \frac{1}{384}x^8 - \cdots$$

and

$$y_2(x) = x - \frac{1}{2}x^3 + \frac{1}{8}x^5 - \frac{1}{48}x^7 + \frac{1}{384}x^9 - \cdots$$

are seen to be the two linearly independent solutions known to be associated with a homogeneous linear second order equation. So all possible solutions of the differential equation can be written in the form

$$y(x) = C_1 y_1(x) + C_2 y_2(x),$$

with C_1 and C_2 arbitrary constants, where to reconcile this result with our previous notation we notice that C_1 and C_2 have been written in place of a_0 and a_1.

To solve the initial value problem the constants C_1 and C_2 must be chosen such that $y(0) = 3$ and $y'(0) = -1$, so

$$3 = C_1 y_1(0) + C_2 y_2(0) \quad \text{and} \quad -1 = C_1 y_1'(0) + C_2 y_2'(0),$$

but $y_1(0) = 1$, $y_2(0) = 0$, and differentiation of the expressions for $y_1(x)$ and $y_2(x)$ shows that $y_1'(0) = 0$ and $y_2'(0) = 1$, so solving for C_1 and C_2 gives $C_1 = 3$ and $C_2 = -1$, showing that the required solution to the initial value problem is

$$y(x) = 3y_1(x) - y_2(x). \qquad \blacksquare$$

The coefficients of the power series expansions for $y_1(x)$ and $y_2(x)$ in the last example were sufficiently complicated that no attempt was made to deduce their general forms and they were merely generated from the recurrence relation. The next example is simpler, and we use it to illustrate the type of argument that is necessary when attempting to arrive at the form of the general term in a power series solution of a homogeneous linear differential equation. There are no specific rules to follow when seeking the form of a general term in a series, and success depends on experience and the ability to recognize the pattern of signs and numbers forming the coefficients.

EXAMPLE 8.5   Find two linearly independent solutions of

$$y'' + xy' + y = 0,$$

when the series expansion is about the origin, and hence solve the initial value problem for which $y(0) = 1$ and $y'(0) = 0$.

Solution   Substituting results (7) to (9) into the differential equation gives

$$\sum_{n=2}^{\infty} n(n-1)a_n x^{n-2} + x\sum_{n=1}^{\infty} na_n x^{n-1} + \sum_{n=0}^{\infty} a_n x^n = 0.$$

Shifting the summation index in the first term, taking the factor x under the second summation and separating out the constant term, as in Example 8.4, gives

$$2a_2 + a_0 + \sum_{n=1}^{\infty}[(n+2)(n+1)a_{n+2} + (n+1)a_n]x^n = 0.$$

Equating the coefficient of each power of x to zero, as in Example 8.4, shows that

$$2a_2 + a_0 = 0, \quad \text{so } a_2 = -\frac{a_0}{2},$$

and

$$(n+2)(n+1)a_{n+2} + (n+1)a_n = 0 \quad \text{for } n \geq 1,$$

but as $n+1 \neq 0$ this last condition reduces to the simpler *recurrence relation*

$$a_{n+2} = -\frac{a_n}{n+2}, \quad \text{for } n = 1, 2, \ldots.$$

It follows directly from the recurrence relation that all even coefficients are multiples of a_0 and all odd coefficients are multiples of a_1 with

$$a_3 = -\frac{a_1}{3}, \quad a_4 = -\frac{a_2}{4} = \frac{a_0}{2 \cdot 4}, \quad a_5 = -\frac{a_3}{5} = \frac{a_1}{3 \cdot 5}, \quad a_6 = -\frac{a_4}{6} = -\frac{a_0}{2 \cdot 4 \cdot 6},$$

$$a_7 = -\frac{a_5}{7} = -\frac{a_1}{3 \cdot 5 \cdot 7}, \quad a_8 = -\frac{a_6}{8} = \frac{a_0}{2 \cdot 4 \cdot 6 \cdot 8}, \quad a_9 = -\frac{a_7}{9} = \frac{a_1}{3 \cdot 5 \cdot 7 \cdot 9}, \dots,$$

where a_0 and a_1 are arbitrary constants.

It is apparent that the pattern of coefficients with even suffixes differs from the one for coefficients with odd suffixes, so each must be considered separately. Starting with the coefficients with even suffixes, we use the fact that if $m = 1, 2, \dots$, then $2m$ is an even number. A little experimentation shows that the signs of the terms with even suffixes are given by the factor $(-1)^m$.

Noticing that a_2, a_4, a_6, and a_8 can be written in the form

$$a_2 = \frac{(-1)a_0}{2}, \quad a_4 = \frac{1}{2 \cdot 4}\frac{(-1)^2 a_0}{2^2 2!}, \quad a_6 = \frac{-a_0}{2 \cdot 4 \cdot 6} = \frac{(-1)^3 a_0}{2^3 3!},$$

$$a_8 = \frac{(-1)^4 a_0}{2 \cdot 4 \cdot 6 \cdot 8} = \frac{(-1)^4 a_0}{2^4 4!}$$

suggests that if we set $n = 2m$, for $m = 0, 1, 2, \dots$, the even numbered terms can be written

$$a_{2m} = \frac{(-1)^m}{2^m m!} a_0.$$

A formal proof that this is the general coefficient in the series involving even powers of x can be obtained by mathematical induction, but we leave this as an exercise.

It is now necessary to consider the coefficients with odd suffixes, and to do this we use the fact that if $m = 1, 2, 3, \dots$, then $2m + 1$ is an odd number. Noticing that the coefficients a_3, a_5, a_7, and a_9 can be written

$$a_3 = \frac{-a_1}{3} = \frac{(-1)2a_1}{3!}, \quad a_5 = \frac{a_1}{3 \cdot 5} = \frac{(-1)^2 2 \cdot 4 a_1}{1 \cdot 2 \cdot 3 \cdot 4 \cdot 5} = \frac{(-1)^2 2^2 2!}{5!},$$

$$a_7 = \frac{-a_1}{3 \cdot 5 \cdot 7} = \frac{(-1)^3 2 \cdot 4 \cdot 6 a_1}{1 \cdot 2 \cdot 3 \cdot 4 \cdot 5 \cdot 6 \cdot 7} = \frac{(-1)^3 2^3 3! a_1}{7!},$$

$$a_9 = \frac{a_1}{3 \cdot 5 \cdot 7 \cdot 9} = \frac{(-1)^4 2 \cdot 4 \cdot 6 \cdot 8 a_1}{9!} = \frac{(-1)^4 2^4 4! a_1}{9!}$$

suggests that the coefficients in the series of odd powers of x can be written

$$a_{2m+1} = \frac{(-1)^m 2^m m!}{(2m+1)!} a_1.$$

Here again we leave as an exercise the task of giving an inductive proof that this is, indeed, the coefficient of the general term in the series involving odd powers of x.

The solution of the differential equation has now separated into two series, one multiplied by a_0 containing only even powers of x and the other multiplied by a_1 containing only odd powers of x, so the solution becomes

$$y(x) = a_0 \sum_{m=0}^{\infty} \frac{(-1)^m x^{2m}}{2^m m!} + a_1 \sum_{m=0}^{\infty} \frac{(-1)^m 2^m m! x^{2m+1}}{(2m+1)!}.$$

As a_0 and a_1 are arbitrary constants, and the two series are not proportional, it follows that two linearly independent solutions of the differential equation are

$$y_1(x) = \sum_{m=0}^{\infty} \frac{(-1)^m x^{2m}}{2^m m!} \quad \text{and} \quad y_2(x) = \sum_{m=0}^{\infty} \frac{(-1)^m 2^m m! x^{2m+1}}{(2m+1)!},$$

so the general solution is

$$y(x) = C_1 y_1(x) + C_2 y_2(x),$$

where C_1 and C_2 are arbitrary constants.

Using the series for $y_1(x)$ and $y_2(x)$, simple calculation gives $y_1(0) = 1$, $y_1'(0) = 0$, $y_2(0) = 0$, and $y_2'(0) = 1$, so the initial conditions $y(0) = 1$, $y'(0) = 0$ will be satisfied if the constants C_1 and C_2 are such that

$$1 = C_1 y_1(0) + C_2 y_2(0) \quad \text{and} \quad 0 = C_1 y_1'(0) + C_2 y_2'(0).$$

This pair of equations has the solution $C_1 = 1$ and $C_2 = 0$, so the solution of the initial value problem becomes

$$y(x) = \sum_{m=0}^{\infty} \frac{(-1)^m x^{2m}}{2^m m!}.$$

Rewriting this as

$$y(x) = \sum_{m=0}^{\infty} \frac{(-x^2/2)^m}{m!},$$

we recognize that the solution is simply $y(x) = \exp(-x^2/2)$, so this series is known to converge for all x.

Finally, to complete our examination of the two linearly independent solutions, let us find the radius of convergence of the second solution $y_2(x)$. The formula for the radius of convergence R based on the ratio test requires all powers of x to be present, whereas the series $y_2(x)$ only contains odd powers of x, so we must modify the series before using the test. All that is necessary is to set $z = x^2$ and to write the series in the form

$$y_2(x) = x \sum_{m=1}^{\infty} \frac{(-1)^m 2^m m!}{(2m+1)!} z^m,$$

for now the radius of convergence of the series in z can be found. The coefficient a_m of z^m is

$$a_m = (-1)^m \frac{2^m m!}{(2m+1)!},$$

so the radius of convergence R is given by

$$R = \lim_{m \to \infty} 1/|a_{m+1}/a_m| = \lim_{m \to \infty} \left| \frac{a_m}{a_{m+1}} \right| = \lim_{m \to \infty} \left(\frac{2^m m!}{(2m+1)!} \frac{(2m+3)!}{2^{m+1}(m+1)!} \right)$$

$$= \lim_{m \to \infty} (2m+3) = \infty.$$

As the series in z has an infinite radius of convergence, so also does the original series involving odd powers of x. This means that the general solution

$$y(x) = C_1 y_1(x) + C_2 y_2(x)$$

is valid for all real x. ∎

Legendre's equation

An important application of the power series method of solution is to the **Legendre differential equation**

$$(1 - x^2)y'' - 2xy' + \alpha(\alpha + 1)y = 0, \tag{10}$$

in which $\alpha \geq 0$ is a real parameter. The equation arises in a variety of applications, but mainly in connection with physical problems in which spherical symmetry is present. It will be seen later that the equation finds its origin in the study of Laplace's equation when expressed in spherical coordinates. Solutions of (10) are called **Legendre functions**, and they are examples of **special functions**, or so-called **higher transcendental functions**, as distinct from elementary functions such as sine, cosine, exponential, and logarithm. We first develop the series solutions for arbitrary $\alpha \geq 0$, and then consider the cases $\alpha = n = 0, 1, 2, \ldots$, which lead to a special class of polynomial solutions $P_n(x)$ called **Legendre polynomials** in which n is the degree of the polynomial. The important properties of Legendre polynomials will be examined later when the topic of orthogonal functions is introduced.

The coefficients of Legendre's equation are all analytic at the origin and the leading coefficient $(1 - x^2)$ only vanishes at $x = \pm 1$, so a power series solution can be expected to exist in the interval $-1 < x < 1$. Substituting (7) to (9) in (10) leads to the equation

$$(1 - x^2) \sum_{n=2}^{\infty} n(n-1)a_n x^{n-2} - 2x \sum_{n=1}^{\infty} na_n x^{n-1} + \alpha(\alpha + 1) \sum_{n=0}^{\infty} a_n x^n = 0.$$

Proceeding as in Example 8.4, this can be rewritten as

$$\sum_{n=0}^{\infty} (n+2)(n+1)a_{n+2}x^n - \sum_{n=2}^{\infty} n(n-1)a_n x^n - \sum_{n=1}^{\infty} 2na_n x^n + \alpha(\alpha + 1) \sum_{n=0}^{\infty} a_n x^n = 0,$$

so equating each coefficient to zero in the usual manner gives the following:
Coefficient of x^0:

$$2a_2 + \alpha(\alpha + 1)a_0 = 0,$$

Coefficient of x:

$$6a_3 - 2a_1 + \alpha(\alpha + 1)a_1 = 0,$$

Coefficient of x^n for $n \geq 2$:

$$(n+2)(n+1)a_{n+2} - n(n-1)a_n - 2na_n + \alpha(\alpha + 1)a_n = 0.$$

Solving the first two equations gives

$$a_2 = -\frac{\alpha(\alpha + 1)}{2}a_0 \quad \text{and} \quad a_3 = \frac{[2 - \alpha(\alpha + 1)]}{6}a_1,$$

whereas the third result gives the recurrence relation

$$a_{n+2} = -\frac{(\alpha - n)(\alpha + n + 1)}{(n+2)(n+1)}a_n \quad \text{for } n \geq 2. \tag{11}$$

Straightforward calculations show that the first few coefficients are given by

$$a_2 = -\frac{\alpha(\alpha + 1)}{2!}a_0, \quad a_3 = -\frac{(\alpha - 1)(\alpha + 2)}{3!}a_1,$$

$$a_4 = \frac{(\alpha - 2)\alpha(\alpha + 1)(\alpha + 3)}{4!}a_0, \quad a_5 = \frac{(\alpha - 3)(\alpha - 1)(\alpha + 2)(\alpha + 4)}{5!}a_1,$$

$$a_6 = -\frac{(\alpha - 4)(\alpha - 2)\alpha(\alpha + 1)(\alpha + 3)(\alpha + 5)}{6!}a_0.$$

Thus, the coefficients of the even powers of x are all multiples of a_0, whereas the coefficients of the odd powers of x are all multiples of a_1, where a_0 and a_1 are arbitrary real numbers. Substituting these coefficients into the series

$$y(x) = a_0 + a_1 x + a_2 x^2 + a_3 x^3 + \cdots = \sum_{n=0}^{\infty} a_n x^n$$

shows that the general solution of the Legendre differential equation can be written

$$y(x) = a_0 y_1(x) + a_1 y_2(x), \tag{12}$$

where

$$y_1(x) = 1 - \frac{\alpha(\alpha + 1)}{2!}x^2 + \frac{(\alpha - 2)\alpha(\alpha + 1)(\alpha + 3)}{4!}x^4 - \cdots, \tag{13}$$

and

$$y_2(x) = x - \frac{(\alpha - 1)(\alpha + 2)}{3!}x^3 + \frac{(\alpha - 3)(\alpha - 1)(\alpha + 2)(\alpha + 4)}{5!}x^5 - \cdots. \tag{14}$$

As the solutions $y_1(x)$ and $y_2(x)$ are not proportional, they must be linearly independent solutions of the Legendre equation (10). We leave as an exercise the task of showing that each series is convergent in the interval $-1 < x < 1$, so the general solution (12) has this same interval of convergence.

Examination of the recurrence relation (11) shows that if $\alpha = n$ is a nonnegative integer, the terms $a_{n+2} = a_{n+4} = a_{n+6} = \cdots$ all vanish. Thus, if $\alpha = n$ is even, the series $y_1(x)$ will reduce to a polynomial of degree n in even powers of x, whereas if $\alpha = n$ is odd the series $y_2(x)$ will reduce to a polynomial of degree n in odd powers of x.

The solution $y(x)$ reduces to the following polynomials when $n = 0, 1, 2, 3, 4$:

Case $n = 0$:

$$y(x) = a_0,$$

Case $n = 1$:

$$y(x) = a_1 x,$$

Case $n = 2$:

$$y(x) = a_0(1 - 3x^2),$$

Case $n = 3$:

$$y(x) = a_1 \left(x - \frac{5}{3} x^3 \right),$$

Case $n = 4$:

$$y(x) = a_0 \left(1 - 10x^2 + \frac{35}{3} x^4 \right).$$

When α is a nonnegative integer, after suitable scaling the foregoing polynomials are denoted by $P_n(x)$ and called **Legendre polynomials of degree n**. The standard scaling adopted involves choosing the arbitrary multiplier of each polynomial such that $P_n(1) = 1$ for $n = 0, 1, 2, \ldots$. When this is done the first few Legendre polynomials become

Legendre polynomials

Even polynomials	_Odd polynomials_
$P_0(x) = 1$	$P_1(x) = x$
$P_2(x) = \frac{1}{2}(3x^2 - 1)$	$P_3(x) = \frac{1}{2}(5x^3 - 3x)$
$P_4(x) = \frac{1}{8}(35x^4 - 30x^2 + 3)$	$P_5(x) = \frac{1}{8}(63x^5 - 70x^3 + 15x)$

A general expression for $P_n(x)$ can be obtained by writing the recurrence relation (11) in the form

$$a_r = \frac{(r+2)(r+1)}{(r-n)(n+r+1)} a_{r+2} \quad \text{for } r \leq n - 2$$

and finding that

$$a_n = \frac{1 \cdot 3 \cdot 4 \cdots (2n-1)}{n!} = \frac{(2n)!}{2^n (n!)^2} \quad \text{for } n = 1, 2, 3, \ldots,$$

in order to make $P_n(1) = 1$. As a result, the following expressions for $P_n(x)$ are obtained.

For *even* polynomials:

$$P_{2n}(x) = \sum_{r=0}^{n} (-1)^r \frac{(4n - 2r)!}{2^{2n} r! (2n-r)! (2n-2r)!} x^{2n-2r}, \quad n = 0, 1, 2, \ldots. \tag{15a}$$

For *odd* polynomials:

$$P_{2n+1}(x) = \sum_{r=0}^{n} (-1)^r \frac{(4n - 2r + 2)!}{2^{2n+1} r! (2n-r+1)! (2n-2r+1)!} x^{2n-2r+1},$$
$$n = 0, 1, 2, \ldots. \tag{15b}$$

Two alternative definitions of Legendre polynomials are to be found in Exercises 16 and 18 at the end of this section.

Results (15a, b) provide a general definition for a Legendre polynomial of any order, though when only a few low order polynomials are required it is often more convenient to generate them by means of the following recurrence relation that

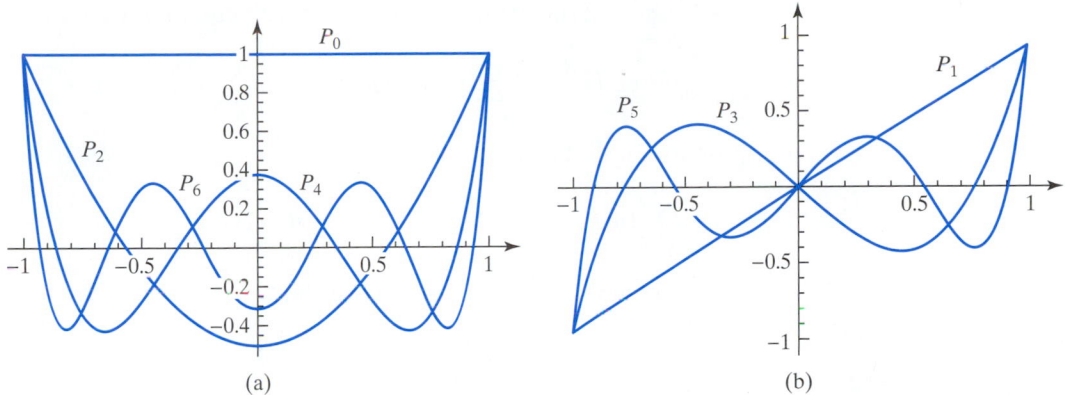

FIGURE 8.1   (a) Even Legendre polynomials. (b) Odd Legendre polynomials.

determines $P_{n+1}(x)$ in terms of $P_n(x)$ and $P_{n-1}(x)$;

$$(n+1)P_{n+1}(x) - (2n+1)xP_n(x) + nP_{n-1}(x) = 0, \tag{16}$$

recurrence relation for Legendre polynomials

for $n = 1, 2, 3, \ldots$. A derivation of this recurrence relation is to be found in Exercise 17 at the end of this section.

As an example of the use of (16) we set $n = 2$ to obtain

$$P_3(x) = \frac{1}{3}[5xP_2(x) - 2P_1(x)],$$

but $P_1(x) = x$ and $P_2(x) = \frac{1}{2}(3x^2 - 1)$, so substituting these expressions, we find $P_3(x) = \frac{1}{2}(5x^3 - 3x)$.

Graphs of the first few Legendre polynomials $P_n(x)$ are given in Fig. 8.1.

ADRIEN-MARIE LEGENDRE (1752–1833)
A French mathematician educated at a college in Paris whose remarkable mathematical ability enabled him to be appointed to the position of professor of mathematics at a military school in Paris. His work on the motion of projectiles in a resisting medium won him a prize offered by the Royal Academy in Berlin. He was subsequently appointed professor at the Normal School in Paris and his contributions as an analyst were second only to those of Laplace and Lagrange, who were his contemporaries. In addition to his contributions to the development of the calculus, he made major contributions to the study of elliptic functions.

For more information about Legendre polynomials, and for applications to boundary value problems, see Chapters 5 and 8 of reference [3.7]. Recurrence relations satisfied by Legendre polynomials and other orthogonal polynomials are to be found in Chapter 22 of reference [G.1], and also in Chapter 18 of reference [G.3].

Another important and useful differential equation with a power series solution

Chebyshev equation

is the **Chebyshev equation**,

$$(1 - x^2)y'' - xy' + \alpha y = 0. \tag{17}$$

The coefficients are all analytic functions and the leading coefficient $(1 - x^2)$ only vanishes at $x = \pm 1$, so a power series solution can be found in the interval

$-1 \le x \le 1$. Proceeding as with Legendre's equation we find

$$(1 - x^2) \sum_{n=2}^{\infty} n(n-1)a_n x^{n-2} - x \sum_{n=1}^{\infty} n a_n x^{n-1} + \alpha \sum_{n=0}^{\infty} a_n x^n = 0,$$

or after a shift of summation index,

$$\sum_{n=0}^{\infty}(n+2)(n+1)a_{n+2}x^n - \sum_{n=2}^{\infty} n(n-1)a_n x^n - \sum_{n=1}^{\infty} n a_n x^n + \alpha \sum_{n=0}^{\infty} a_n x^n = 0.$$

If we combine summations, this becomes

$$(1 \cdot 2a_2 + \alpha a_0) + [2 \cdot 3a_3 + (\alpha - 1)a_1]x$$
$$+ \sum_{n=2}^{\infty} \left[(n+1)(n+2)a_{n+2} + (\alpha - n^2)a_n\right]x^n = 0.$$

Equating the coefficients of each power of x to zero gives

$$a_2 = -\frac{\alpha}{2!}a_0, \quad a_3 = \frac{(1-\alpha)}{3!}a_1,$$

and the recurrence relation

$$a_{n+2} = \frac{(n^2 - \alpha)}{(n+1)(n+2)}a_n, \quad n = 2, 3 \ldots.$$

Thus,

$$a_4 = \frac{(2^2 - \alpha)}{3 \cdot 4}a_2 = -\frac{\alpha(2^2 - \alpha)}{4!}a_0$$

$$a_5 = \frac{(3^2 - \alpha)}{4 \cdot 5}a_3 = \frac{(1-\alpha)(3^2 - \alpha)}{5!}a_1$$

. . . .

Using these coefficients in the original power series $y(x) = \sum_{n=0}^{\infty} a_n x^n$ gives the solution of the Chebyshev equation in the form

$$y(x) = a_0 y_0(x) + a_1 y_1(x),$$

where

$$y_0(x) = a_0 \left[1 - \frac{\alpha}{2!}x^2 - \frac{\alpha(2^2 - \alpha)}{4!}x^4 - \frac{\alpha(2^2 - \alpha)(4^2 - \alpha)}{6!}x^6 - \cdots\right]$$

and

$$y_1(x) = a_1 \left[x + \frac{(1-\alpha)}{3!}x^3 + \frac{(1-\alpha)(3^2 - \alpha)}{5!}x^5 + \cdots\right].$$

In applications of this equation to approximation theory, numerical analysis, and elsewhere, it is usual that $\alpha = m^2$, where $m = 0, 1, 2, \ldots$. Inspection of $y_0(x)$ shows that when m is even, the solution reduces to a polynomial of degree m in even powers of x, whereas when m is odd $y_1(x)$ reduces to a polynomial of degree m in odd powers of x.

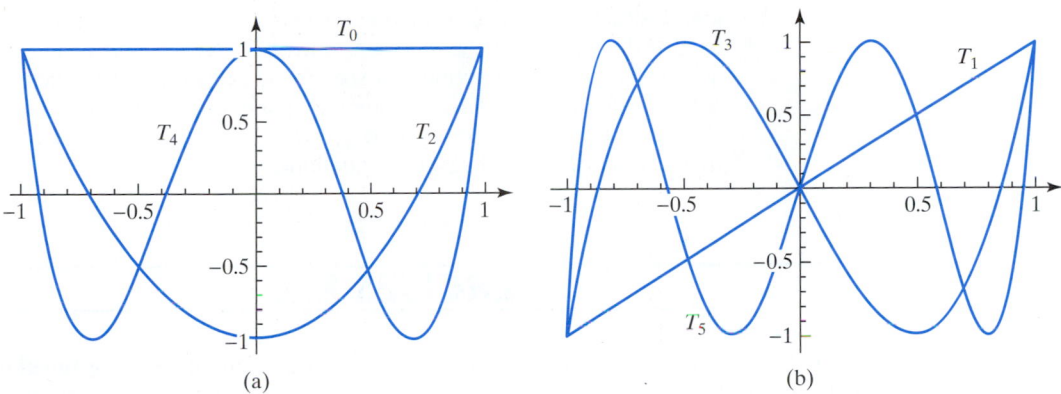

FIGURE 8.2   (a) Even Chebyshev polynomials. (b) Odd Chebyshev polynomials.

As the polynomials are solutions of a homogeneous differential equation, the scale factors for each polynomial can be chosen arbitrarily, so by convention they are chosen such that the term with the largest power of x is positive and the polynomial is free from fractional coefficients. These polynomials are called **Chebyshev polynomials**, and they are denoted by $T_n(x)$. The first six Chebyshev polynomials are:

Chebyshev polynomials

Even polynomials	*Odd polynomials*
$T_0(x) = 1$	$T_1(x) = x$
$T_2(x) = 2x^2 - 1$	$T_3(x) = 4x^3 - 3x$
$T_4(x) = 8x^4 - 8x^2 + 1$	$T_5(x) = 16x^5 - 20x^3 + 5x$

Using the forms for $T_{n+1}(x)$, $T_n(x)$ and $T_{n-1}(x)$ obtained from $y_0(x)$ and $y_1(x)$, it can be shown that Chebyshev polynomials obey the following recurrence relation:

recurrence relation for Chebyshev polynomials

$$T_{n+1}(x) - 2x\,T_n(x) + T_{n-1}(x) = 0. \qquad (18)$$

When used with the polynomials just listed, this recurrence relation is the simplest way of generating higher order polynomials. Graphs of the first six Chebyshev polynomials are shown in Fig. 8.2.

For applications of Chebyshev polynomials to numerical analysis see, for example, references [8.3] to [8.5].

PAFNUTI LIWOWICH CHEBYSHEV (1821–1894)
A distinguished Russian mathematician who was professor of mathematics at the University of Petrograd (now St. Petersburg). He made many contributions to analysis and number theory. There are many variations of the transliteration of his name, the most common probably being Tchebycheff.

Summary

This section showed how to find a series solution, expanded about the origin, of a homogeneous linear second order variable coefficient differential equation with polynomial coefficients, when the solution can be obtained in the form of a general power series with unknown coefficients. By substituting this series into the differential equation, grouping corresponding powers of x, and requiring the coefficient of each power of x to vanish identically, a recurrence relation connecting the unknown coefficients was obtained and used to find the coefficients of the power series in terms of two arbitrary constants a_0 and a_1. The general solution was seen to be the sum of two linearly independent power series with known coefficients, one multiplied by a_0 and the other by a_1. Two important special cases were considered that gave rise to polynomial solutions of the important and useful Legendre and Chebyshev equations.

EXERCISES 8.2

Find the first six terms in the power series expansion of each of the following initial value problems.

1. $y'' + (x - x^2)y' + y = 0$, with $y(0) = 2$, $y'(0) = -3$.

2. $2y'' + xy' + 2(1 + x)y = 0$, with $y(0) = -2$, $y'(0) = 1$.

3. $y'' + (1 + x^2)y' + xy = 0$, with $y(0) = 1$, $y'(0) = -3$.

4. $y'' - 3xy' + 2y = 0$, with $y(0) = 1$, $y'(0) = 1$.

5. $(1 - x^2)y'' + xy' - y = 0$, with $y(0) = 2$, $y'(0) = -1$.

6. $y'' + x^2y' + 2xy = 0$, with $y(0) = 3$, $y'(0) = -2$.

7. $y'' + 2(1 - x)y' - 3xy = 0$, with $y(0) = 1$, $y'(0) = -1$.

8. $(1 - x)y'' + 2xy' + (1 + x)y = 0$, with $y(0) = 4$, $y'(0) = -2$.

9. $(1 - 2x^2)y'' + 2y' + 3y = 0$, with $y(0) = 1$, $y'(0) = -1$.

10. $(1 + 2x^2)y'' + 3xy' + y = 0$, with $y(0) = 2$, $y'(0) = -2$.

11. $(2x^2 - 1)y'' + (1 + x)y' + 2y = 0$, with $y(0) = 1$, $y'(0) = 4$.

12. $y'' + (1 + 2x)y' + xy = 0$, with $y(2) = 1$, $y'(2) = 0$.

13. $(2 + x)y'' + 3(1 + x)y' + 2y = 0$, with $y(1) = 2$, $y'(1) = -3$.

14. $(x^2 - 2x + 2)y'' + (x - 1)y' - 3y = 0$, with $y(-1) = 1$, $y'(-1) = 2$.

15. $(1 - x)y'' + 2xy' - 2xy = 0$, with $y(2) = 1$, $y'(2) = 5$.

16. An alternative definition of the Legendre polynomial $P_n(x)$ is provided by the formula

$$P_n(x) = \frac{1}{2^n n!} \frac{d^n}{dx^n}(x^2 - 1)^n,$$

called the **Rodrigues formula**. Use the formula to compute $P_4(x)$ and $P_5(x)$.

17.* Set $u = (x^2 - 1)^n$ and use repeated differentiation of the Rodrigues formula to verify that $P_n(x)$ is a Legendre polynomial by showing it satisfies the Legendre differential equation

$$(1 - x^2)P_n''(x) - 2xP_n'(x) + n(n + 1)P_n(x) = 0.$$

18.* The function

$$G(x, t) = (1 - 2xt + t^2)^{-1/2}$$

is called the **generating function** for Legendre polynomials. It has the property that when expanded as a power series in t the coefficient of t^n is $P_n(x)$, so that

$$G(x, t) = P_0(x) + P_1(x)t + P_2(x)t^2 + \cdots.$$

Set $u = -2xt + t^2$ and expand $(1 + u)^{-1/2}$ by the binomial theorem. Collect all the terms in x multiplying t^5 and hence verify that the coefficient of t^5 is $P_5(x)$.

19.* Show that the generating function defined in Problem 18 satisfies the differential equation

$$(1 - 2xt + t^2)\frac{\partial G}{\partial t} - (x - t)G = 0$$

for arbitrary t. As the result must be an identity in t, the consequence of substituting

$$G(x, t) = P_0(x) + P_1(x)t + P_2(x)t^2 + \cdots$$

into the differential equation must be such that terms in x multiplying each power of t vanish. Collect the terms multiplying t^n, and hence establish the Legendre polynomial recurrence relation

$$(n + 1)P_{n+1}(x) - (2n + 1)xP_n(x) + nP_{n-1}(x) = 0$$

for $n = 1, 2, \ldots$. This result is called the **Bonnet recurrence relation**.

20.* The **electrostatic potential** ϕ at a point in a vacuum distant d from a charge Q is given by $\phi = Q/d$. Use the Legendre polynomial generating function

$$G(r, t) = \frac{1}{(1 - 2rt + t^2)^{1/2}},$$

together with the result from elementary trigonometry

$$r = \left(r_1^2 + r_2^2 - 2r_1r_2\cos\theta\right)^{1/2},$$

to show that the electrostatic potential at point A due to a charge Q at B in Fig. 8.3 is given by

$$\frac{Q}{r} = \frac{1}{r_2}\left[P_0(\cos\theta) + \left(\frac{r_1}{r_2}\right)P_1(\cos\theta)\right.$$

$$\left. + \left(\frac{r_1}{r_2}\right)^2 P_2(\cos\theta) + \cdots\right], \text{ for } r_1/r_2 < 1.$$

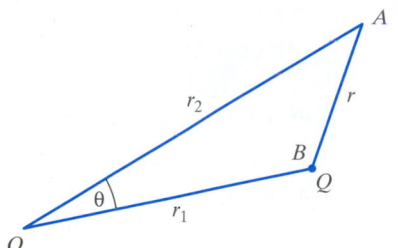

FIGURE 8.3   A point charge Q at B distant r from A.

8.3 Singular Points of Linear Differential Equations

In Section 8.2 the power series method was used to find a solution of a homogeneous variable coefficient differential equation of the form

$$a(x)y'' + b(x)y' + c(x)y = 0. \tag{19}$$

It was seen that the method could be applied about any point x_0 at which the coefficients of the differential equation are analytic and $a(x_0) \neq 0$. Expressed differently, when (19) is written in the standard form

$$y'' + P(x)y' + Q(x)y = 0, \tag{20}$$

with

$$P(x) = \frac{b(x)}{a(x)} \quad \text{and} \quad Q(x) = \frac{c(x)}{a(x)}, \tag{21}$$

the power series method can be applied to develop a solution about any point x_0 at which the functions $P(x)$ and $Q(x)$ are analytic.

regular and singular points

Points where $P(x)$ and $Q(x)$ are analytic are called **regular points** of the differential equation, and points where at least one is not analytic are called **singular points**.

Equation (20) will be said to have a **regular singular point** at x_0 if the functions

$$(x - x_0)P(x) \quad \text{and} \quad (x - x_0)^2 Q(x)$$

are analytic at x_0, and so have Taylor series expansions about x_0. If at least one of these functions is not analytic at x_0, the point will be said to be an **irregular singular point**.

EXAMPLE 8.6   Identify the nature of the singular points of the following equations:

(a) $x^2 y'' + xy' + (x^2 - n^2)y = 0$

(b) $(1 - x^2)y'' - 2xy' + n(n+1)y = 0, \quad (n = 0, 1, 2, \dots)$

(c) $(1 - x)y'' + 2(x - 1)y' + xy = 0$

(d) $(x - 1)^3 y'' + 3(x - 1)^2 y' + y = 0$

Solution

(a) This is *Bessel's equation* of order n in which the functions $P(x) = 1/x$ and $Q(x) = (x^2 - n^2)/x^2$. Neither of these functions is analytic at the origin, so the origin is a singular point of Bessel's equation. However, as the functions $x P(x) = 1$ and $x^2 Q(x) = x^2 - n^2$ are both analytic at the origin, it follows that $x = 0$ is a regular singular point of Bessel's equation.

(b) This is *Legendre's equation* of order n in which $P(x) = -2x/(1 - x^2)$ and $Q(x) = n(n + 1)/(1 - x^2)$. Neither of these functions is analytic at $x = \pm 1$, so these points are the singular points of the Legendre equation. Let us consider the singular point at $x = 1$. As the functions

$$(x - 1)P(x) = 2x/(1 + x) \quad \text{and} \quad (x - 1)^2 Q(x) = n(n + 1)(x - 1)/(1 + x)$$

are both analytic at $x = 1$, it follows that this is a regular singular point of Legendre's equation. A similar argument shows that $x = -1$ is also a regular singular point of the equation.

(c) In this case $P(x) = -2$ and $Q(x) = x/(1 - x)$, and while $P(x)$ is analytic for all x the function $Q(x)$ is not analytic at $x = 1$, so this is a singular point of the equation. The functions $(x - 1)P(x) = 2(1 - x)$ and $(x - 1)^2 Q(x) = x(1 - x)$ are both analytic at $x = 1$, so $x = 1$ is a regular singular point of the equation.

(d) In this equation $P(x) = 3/(x - 1)$ and $Q(x) = 1/(x - 1)^3$ and neither function is analytic at $x = 1$, so this is a singular point of the equation. We have

$$(x - 1)P(x) = 3 \quad \text{and} \quad (x - 1)^2 Q(x) = \frac{1}{x - 1},$$

and although the first of these functions is analytic for all x, the second is not analytic at $x = 1$, so $x = 1$ is an irregular singular point of the equation. ∎

In the next section the power series method will be generalized to arrive at what is called the **Frobenius method**, which always generates two linearly independent solutions about a *regular singular point* of equation (20). As the behavior of solutions in a neighborhood of an irregular singular point can be shown to be very erratic, no further consideration will be given to solutions near such points.

shifting a singular point

Sometimes it is more convenient to consider an equation with a regular singular point located at the origin rather at some other point $x_0 \neq 0$. In such cases a singular point located at x_0 can always be shifted to the origin by making the change of variable $X = x - x_0$, as in Section 8.2.

EXAMPLE 8.7

Shift the singular point of the following equation to the origin:

$$(x - 1)^2 y'' + 3(x + 2)y' + 2y = 0.$$

Solution   The equation has a regular singular point at $x = 1$, so we make the variable change $X = x - 1$ and set $y(x) = Y(x - 1) = Y(X)$. The equation then becomes

$$X^2 Y'' + 3(X + 3)Y' + 2Y = 0,$$

with a regular singular point now located at $X = 0$. ∎

To appreciate why an ordinary power series solution cannot be developed around a regular singular point, it will be sufficient to consider the Cauchy–Euler equation

$$x^2 y'' + 3xy' + 2y = 0,$$

which has a regular singular point at the origin. This Cauchy–Euler equation was solved analytically in Example 6.10, where its solution was found to be

$$y(x) = C_1 x^{-1} \cos(\ln|x|) + C_2 x^{-1} \sin(\ln|x|).$$

The reason that no power series solution exists in this case is seen to be the presence of the factor x^{-1} and the function $\ln|x|$ in the analytical solution, neither of which can be expanded in a power series about the origin.

Summary

The regular and singular points of a general homogeneous second order linear variable coefficient differential equation were defined and illustrated by example. It was shown how, if necessary, a singular point occurring at $x = a$ could be shifted to the origin, and an example was used to demonstrate why an ordinary power series solution cannot be developed around a regular singular point.

EXERCISES 8.3

Identify the nature of the singular points in each of the following equations.

1. $(1-x)^2 y'' + 2(x-1)y' + y = 0.$

2. $x^2 y'' + 3x^2 y' + (1+x^2)y = 0.$

3. $(1+x)^2 y'' + 2y' + y = 0.$

4. $xy'' + (1-x)y' + ny = 0 \quad (n > 0).$

5. $(x+4)^3 y'' + 2(x+4)y' + xy = 0.$

6. $(x^2 - 4)y'' + (x+3)y' - 5(x+1)y = 0.$

7. $(3-x)^2 y'' + 4y' + \cos x(3-x^2)y = 0.$

8. $x^2 y'' + 8y' + 3xy = 0.$

8.4   The Frobenius Method

A generalization of the power series method that was introduced by Frobenius (1849–1917) enables a solution of a homogeneous linear differential equation to be developed about a regular singular point. He considered the differential equation

$$a(x)y'' + b(x)y' + c(x)y = 0, \tag{22}$$

and established the following result that is stated without proof.

THEOREM 8.1   **Frobenius theorem**   Let x_0 be a regular singular point of (22). Then, in some interval $0 < x - x_0 < d$, the equation will always possess at least one solution of the form

$$y(x) = (x - x_0)^c \left(a_0 + a_1(x - x_0) + a_2(x - x_0)^2 + \cdots\right)$$

$$= (x - x_0)^c \sum_{n=0}^{\infty} a_n (x - x_0)^n,$$

where $a_0 \neq 0$ and c is a real or complex number. A second linearly independent solution of similar form will exist that may contain a logarithmic term, though with a different value of c and some other coefficients $b_0, b_1, b_2, \ldots$ in place of the coefficients $a_0, a_1, a_2, \ldots$. Taken together, these two solutions form a basis of solutions for the differential equation. ∎

GEORG FERDINAND FROBENIUS (1849–1917)
A German mathematician whose main research was in group theory and analysis. He worked in Zurich and Berlin and published his method for the series solution of linear ordinary differential equations in 1873.

For simplicity, and because of their frequent occurrence, in what follows we will develop the Frobenius method in terms of a slightly less general class of equations by setting $a(x) = x^2$ in (22). So we will consider the equation

$$x^2 y'' + b(x)y' + c(x)y = 0, \tag{23}$$

and write it in the standard form

$$y'' + P(x)y' + Q(x)y = 0, \tag{24}$$

where

$$P(x) = \frac{p(x)}{x} \quad \text{and} \quad Q(x) = \frac{q(x)}{x^2}, \tag{25}$$

and assume that $p(x)$ and $q(x)$ are analytic functions at $x = 0$. So we will only consider equations of the form (24) with regular singular points at the *origin*.

To determine the exponent c in Theorem 8.1 we substitute a solution of the form

$$y(x) = x^c \sum_{n=0}^{\infty} a_n x^n \tag{26}$$

into equation (24), where c is to be determined along with the coefficients a_n. When making this substitution we will need to use the following results obtained by differentiation of (26):

$$y'(x) = ca_0 x^{c-1} + (c+1)a_1 x^c + (c+2)a_2 x^{c+1} + \cdots = \sum_{n=0}^{\infty} (n+c)a_n x^{n+c-1} \tag{27}$$

and

$$y''(x) = c(c-1)a_0 x^{c-2} + (c+1)ca_1 x^{c-1} + (c+2)(c+1)a_2 x^c + \cdots$$

$$= \sum_{n=0}^{\infty} (n+c)(n+c-1)a_n x^{n+c-2}. \tag{28}$$

As the functions $p(x)$ and $q(x)$ are assumed to be analytic at the origin, they can be expanded as the Maclaurin series

$$p(x) = p_0 + p_1 x + p_2 x^2 + \cdots \quad \text{and} \quad q(x) = q_0 + q_1 x + q_2 x^2 + \cdots. \tag{29}$$

Substituting (27) to (29) into (24) leads to the result

$$x^{c-2}[c(c-1)a_0 + (c+1)ca_1 x + \cdots]$$

$$+ \left(p_0 + p_1 x + p_2 x^2 + \cdots\right)x^{c-2}(ca_0 + (c+1)a_1 x + \cdots)$$

$$+ x^{c-2}\left(q_0 + q_1 x + q_2 x^2 + \cdots\right)\left(a_0 + a_1 x + a_2 x^2 + \cdots\right) = 0.$$

If (26) is to be a solution of (24), the coefficient of each power of x in this last result must vanish to make it an identity. Collecting terms involving the same power of x and equating their coefficients to zero will lead to a sequence of equations connecting the coefficients a_n in (26), and equating the coefficient of the lowest power of x to zero will give an equation from which c can be determined.

The lowest power of x in the preceding result is x^{c-2}, so collecting terms involving x^{c-2} and equating the coefficient of x^{c-2} to zero gives

$$[c(c-1) + p_0 c + q_0]a_0 = 0.$$

As Theorem 8.1 requires $a_0 \neq 0$, it follows that c is determined by the equation

$$c(c-1) + p_0 c + q_0 = 0. \tag{30}$$

indicial equation

This equation is called the **indicial equation** associated with differential equation (24), because it determines the permissible values of the index c to be used in the solution given in Theorem 8.1.

The indicial equation of differential equation (24) can be constructed without the need to make the substitution (26), because it is easily seen that

$$p_0 = \lim_{x \to 0}[x P(x)] \quad \text{and} \quad q_0 = \lim_{x \to 0}[x^2 Q(x)]. \tag{31}$$

For the class of equations of type (24) that all have a regular singular point at the origin, the appropriate form of the Frobenius theorem follows from Theorem 8.1 if we set $x_0 = 0$.

It is important to notice that for a general equation (22) in which $a(x) \neq x^2$ the indicial equation does *not* take the form given in (30). When this situation arises the indicial equation must be obtained by substituting (26) into (22) and equating to zero the coefficient of the lowest power of x that occurs in the expansion.

As the indicial equation is a quadratic equation in c, the following relationships between its roots c_1 and c_2 are possible:

(a) Roots c_1 and c_2 are real and distinct and do not differ by an integer
(b) Roots c_1 and c_2 are real and differ by an integer
(c) Roots c_1 and c_2 are real and equal
(d) Roots c_1 and c_2 are complex conjugates

The reason for identifying these different cases is to be found in the following theorem, which is stated without proof in terms of a differential equation with a regular singular point located at the origin (see references [3.3] and [3.5]).

THEOREM 8.2

Forms of Frobenius solution depending on the nature of c_1 and c_2   Let a differential equation of the form

$$x^2 y'' + x[x P(x)]y' + [x^2 Q(x)]y = 0$$

have a regular singular point at $x = 0$. Let $x P(x)$ and $x^2 Q(x)$ each be capable of expansion as convergent power series in an interval $|x| < d$, where $d > 0$ is the smaller of the two radii of convergence, and suppose that

$$p_0 = \lim_{x \to 0}[x\,P(x)] \quad \text{and} \quad q_0 = \lim_{x \to 0}[x^2 Q(x)].$$

Then in terms of the exponent c in (26), and the coefficients p_0 and q_0, the indicial equation for the differential equation is

$$c(c-1) + p_0 c + q_0 = 0,$$

with two roots c_1 and c_2 that may be real or complex conjugates.

The two linearly independent solutions of the differential equation that exist depend on the relationship between the roots of the indicial equation, and they take the following forms.

Case (a) Real roots with $c_1 > c_2$ and $c_1 - c_2$ neither zero nor a positive integer

In the intervals $-d < x < 0$ and $0 < x < d$ the differential equation has two linearly independent solutions of the form

different forms of Frobenius solution and examples

$$y_1(x) = |x|^{c_1}\left[1 + \sum_{n=1}^{\infty} a_n x^n\right] \quad \text{and} \quad y_2(x) = |x|^{c_2}\left[1 + \sum_{n=1}^{\infty} b_n x^n\right],$$

where the coefficients a_n are obtained by substituting $c = c_1$ in the recurrence relation connecting coefficients and then setting $a_0 = 1$, and the coefficients b_n are obtained in similar fashion by substituting $c = c_2$ in the recurrence relation, replacing a_n by b_n and setting $b_0 = 1$.

Case (b) Real roots with $c_1 - c_2$ equal to a positive integer

In the intervals $-d < x < 0$ and $0 < x < d$ the differential equation has two linearly independent solutions of the form

$$y_1(x) = |x|^{c_1}\left[1 + \sum_{n=1}^{\infty} a_n x^n\right] \quad \text{and} \quad y_2(x) = A y_1(x) \ln |x| + |x|^{c_2} \sum_{n=1}^{\infty} \beta_n x^n,$$

where the coefficients a_n are determined as in Case (a), and the coefficients A and β_n are found by substituting $y(x) = y_2(x)$ in the differential equation. Some differential equations for which $c_1 - c_2$ is a positive integer have no logarithmic term in their solution $y_2(x)$, in which case $A = 0$.

Case (c) Real roots with $c_1 = c_2$

In the intervals $-d < x < 0$ and $0 < x < d$ the differential equation has two linearly independent solutions of the form

$$y_1(x) = |x|^{c_1}\left[1 + \sum_{n=1}^{\infty} a_n x^n\right] \quad \text{and} \quad y_2(x) = y_1(x) \ln |x| + |x|^{c_1} \sum_{n=1}^{\infty} \alpha_n x^n,$$

where the coefficients a_n are determined as in Case (a), and the coefficients α_n are found by substituting $y(x) = y_2(x)$ into the differential equation.

Case (d) Complex conjugate roots

If $c_1 = \lambda + i\mu$ and $c_2 = \lambda - i\mu$ with $\mu \neq 0$, then in the intervals $-d < x < 0$ and $0 < x < d$ the two linearly independent solutions of the differential equation are the real and imaginary parts of

$$y(x) = |x|^{\lambda + i\mu} \left[1 + \sum_{n=1}^{\infty} a_n x^n \right],$$

where the coefficients a_n are determined as in Case (a). ∎

It is important to recognize that the solutions in cases (a) to (d) of Theorem 8.2 all lie in intervals of the form $0 < x < d$ that do *not* contain the origin. A solution in the interval $-d < x < 0$ can be obtained from the above results by replacing x by $-x$ and, depending on the relationship between the roots c_1 and c_2, seeking a solution in the manner indicated in the illustrative examples that follow.

Case (a) Roots c_1 and c_2 Are Distinct and Do Not Differ by an Integer

EXAMPLE 8.8

Find the solution of

$$2xy'' + (x + 1)y' + y = 0$$

in some interval $0 < x < d$.

Solution   As the coefficient of y'' vanishes at $x = 0$ the origin must be a singular point of this equation. When the differential equation is written in standard form we find that $P(x) = (x + 1)/(2x)$ and $Q(x) = 1/(2x)$, so $p_0 = \lim_{x \to 0} x P(x) = 1/2$ and $q_0 = \lim_{x \to 0} x^2 Q(x) = 0$, showing that the origin is a regular singular point of the differential equation.

From (30) the indicial equation is seen to be

$$c(c - 1) + \frac{1}{2}c = 0, \quad \text{or} \quad c\left(c - \frac{1}{2}\right) = 0,$$

showing that the permissible values of c are $c = 0$ and $c = 1/2$. As these values of c are distinct and do not differ by an integer, the solution will be of the type given in Theorem 8.2(a).

Setting

$$y(x) = \sum_{n=0}^{\infty} a_n x^{n+c}$$

and substituting into the differential equation in the usual way leads to the result

$$2\sum_{n=0}^{\infty}(n + c)(n + c - 1)a_n x^{n+c-1} + \sum_{n=0}^{\infty}(n + c)a_n x^{n+c} + \sum_{n=0}^{\infty}(n + c)a_n x^{n+c-1}$$

$$+ \sum_{n=0}^{\infty} a_n x^{n+c} = 0.$$

Shifting the summation index in the first and third summations gives

$$2 \sum_{n=-1}^{\infty} (n+c+1)(n+c)a_{n+1}x^{n+c} + \sum_{n=0}^{\infty}(n+c)a_n x^{n+c} + \sum_{n=-1}^{\infty} (n+c+1)a_{n+1}x^{n+c}$$

$$+ \sum_{n=0}^{\infty} a_n x^{n+c} = 0,$$

and, finally, combining terms we arrive at the result

$$\sum_{n=-1}^{\infty} [2(n+c+1)(n+c) + (n+c+1)]a_{n+1}x^{n+c} + \sum_{n=0}^{\infty}(n+c+1)a_n x^{n+c} = 0.$$

Separating out the term corresponding to $n = -1$ allows this to be written

$$[2c(c-1) + c]a_0 x^{c-1} + \sum_{n=0}^{\infty}\{[2(n+c+1)(n+c) + (n+c+1)]a_{n+1}$$

$$+ (n+c+1)a_n\}x^{n+c} = 0.$$

To proceed further we must now equate to zero the coefficient of each power of x. Equating to zero the coefficient of x^{c-1} simply gives the indicial equation, but equating to zero the coefficient of x^{n+c} for $n = 0, 1, 2, \ldots$ gives

$$(n+c+1)(2n+2c+1)a_{n+1} + (n+c+1)a_n = 0.$$

As $n+c+1 \neq 0$ this recurrence relation can be written

$$a_{n+1} = -\frac{a_n}{2n+2c+1}.$$

Starting with the value $c = 0$, we find that

$$a_{n+1} = -\frac{a_n}{2n+1},$$

so

$$a_1 = -a_0, \quad a_2 = -\frac{a_1}{3} = \frac{a_0}{3}, \quad a_3 = -\frac{a_2}{5} = -\frac{a_0}{3 \cdot 5}, \quad a_4 = -\frac{a_3}{7} = \frac{a_0}{3 \cdot 5 \cdot 7},$$

$$a_5 = -\frac{a_4}{9} = -\frac{a_0}{3 \cdot 5 \cdot 7 \cdot 9}, \quad a_6 = -\frac{a_5}{11} = \frac{a_0}{3 \cdot 5 \cdot 7 \cdot 9 \cdot 11}, \ldots.$$

Examination of a_5 and a_6 shows they can be written

$$a_5 = -\frac{2 \cdot 4 \cdot 6 \cdot 8}{9!}a_0 = -\frac{2^4 \cdot 4!}{(2 \cdot 4 + 1)!}a_0$$

and

$$a_6 = \frac{2^5 \cdot 5!}{(2 \cdot 5 + 1)!}a_0.$$

These expressions suggest that the coefficient of the general term in the series is

$$a_{n+1} = \frac{(-1)^{n+1}2^n n!}{(2n+1)!}a_0 \quad \text{for } n = 0, 1, 2, \ldots,$$

and this is easily verified by mathematical induction. As we are considering the case in which $c = 0$, it follows from Theorem 8.2(a) that for some $d_1 > 0$ one solution is

$$y(x) = a_0 \left[1 + \sum_{n=0}^{\infty} \frac{(-1)^{n+1} 2^n n!}{(2n+1)!} x^{n+1} \right].$$

As the constant $a_0 \neq 0$ is arbitrary, we set $a_0 = 1$ and take for a fundamental solution of the differential equation

$$y_1(x) = 1 + \sum_{n=0}^{\infty} \frac{(-1)^{n+1} 2^n n!}{(2n+1)!} x^{n+1} \quad \text{for } 0 < x < d_1.$$

A second fundamental (linearly independent) solution follows by using the other value $c = 1/2$, for which the recurrence relation becomes

$$a_{n+1} = -\frac{a_n}{2n+2}.$$

Using this result and recognizing that the coefficients a_n are not the same as the ones in $y_1(x)$, we find that

$$a_1 = -\frac{a_0}{2}, \quad a_2 = -\frac{a_1}{2 \cdot 2} = \frac{a_0}{2^2 \cdot 2!}, \quad a_3 = -\frac{a_2}{2 \cdot 3} = -\frac{a_0}{2^3 \cdot 3!},$$

$$a_4 = -\frac{a_3}{2 \cdot 4} = \frac{a_0}{2^4 \cdot 4!}, \ldots$$

This pattern of coefficients suggests that the coefficient of the general term in the series is

$$a_n = \frac{(-1)^n}{2^n n!} a_0,$$

and this also is easily verified by using an inductive argument. Setting the arbitrary constant $a_0 = 1$, it follows from Theorem 8.2(a) that for some $d_2 > 0$ a second fundamental solution is given by

$$y_2(x) = x^{1/2} \sum_{n=0}^{\infty} \frac{(-1)^n}{2^n n!} x^n = x^{1/2} e^{-x/2}, \quad \text{for } 0 < x < d_2.$$

The solutions $y_1(x)$ and $y_2(x)$ form a basis for solutions of the differential equation in an interval of the form $0 < x < d$, where $d = \min\{d_1, d_2\}$. Thus, the general solution is

$$y(x) = C_1 y_1(x) + C_2 y_2(x), \quad \text{for } 0 < x < d,$$

where C_1 and C_2 are arbitrary constants. The value of d is

$$d = \min\{R_1, R_2\},$$

where R_1 and R_2 are the radii of convergence of the series solutions for $y_1(x)$ and $y_2(x)$, respectively. In this case $R_1 = R_2 = \infty$, so the general solution is valid for $x > 0$. ∎

Case (b) Roots c_1 and c_2 Are Real and Differ by an Integer

EXAMPLE 8.9

Find the solution of

$$x^2 y'' + x(2 + x) y' - 2y = 0$$

in some interval $0 < x < d$.

Solution   The equation has a singular point at the origin, and writing it in standard form shows that $P(x) = (2 + x)/x$ and $Q(x) = -2/x^2$. Thus,

$$p_0 = \lim_{x \to 0} x P(x) = 2 \quad \text{and} \quad q_0 = \lim_{x \to 0} x^2 Q(x) = -2,$$

so the equation has a regular singular point at the origin. It follows from (30) that the indicial equation is

$$c(c - 1) + 2c - 2 = 0, \quad \text{or} \quad c^2 + c - 2 = 0.$$

The permissible values of c are thus $c = -2$ and $c = 1$, and these differ by an integer. Substituting the series

$$y(x) = \sum_{n=0}^{\infty} a_n x^{n+c}$$

into the differential equation gives

$$\sum_{n=0}^{\infty}(n + c)(n + c - 1)a_n x^{n+c} + 2\sum_{n=0}^{\infty}(n + c)a_n x^{n+c} + \sum_{n=0}^{\infty}(n + c)a_n x^{n+c+1}$$

$$- 2\sum_{n=0}^{\infty} a_n x^{n+c} = 0.$$

Shifting the index in the third summation so it starts from $n = 1$ and separating out the terms multiplied by x^c enables the equation to be written

$$a_0(c^2 + c - 2)x^c + \sum_{n=1}^{\infty}\{[(n + c)(n + c + 1) - 2]a_n + (n + c - 1)a_{n-1}\}x^{n+c} = 0.$$

Proceeding as usual and equating the coefficient of x^c to zero simply gives the indicial equation, whereas equating the coefficient of x^{n+c} to zero gives the recurrence relation

$$a_n = \frac{(n + c - 1)}{[2 - (n + c)(n + c + 1)]}a_{n-1} \quad \text{for } n = 1, 2, \ldots.$$

Considering the larger root $c = 1$, as required by Theorem 8.2(b), we find that

$$a_n = \frac{n}{[2 - (1 + n)(2 + n)]}a_{n-1} \quad \text{for } n = 1, 2, \ldots.$$

So the first few coefficients are

$$a_1 = -\frac{a_0}{4}, \quad a_2 = \frac{2}{[2 - 3 \cdot 4]}a_1 = \frac{a_0}{4 \cdot 5}, \quad a_3 = -\frac{3a_2}{[2 - 4 \cdot 5]} = -\frac{a_0}{4 \cdot 5 \cdot 6},$$

$$a_4 = \frac{4a_3}{[2 - 5 \cdot 6]} = \frac{a_0}{4 \cdot 5 \cdot 6 \cdot 7}, \ldots.$$

As $c = 1$, setting the arbitrary constant $a_0 = 1$, it follows from Theorem 8.2(b) that for some $d_1 > 0$ a fundamental solution of the differential equation is

$$y_1(x) = x\left(1 - \frac{x}{4} + \frac{x^2}{4 \cdot 5} - \frac{x^3}{4 \cdot 5 \cdot 6} + \frac{x^4}{4 \cdot 5 \cdot 6 \cdot 7} - \cdots\right),$$

or

$$y_1(x) = x - \frac{x^2}{4} + \frac{x^3}{4 \cdot 5} - \frac{x^4}{4 \cdot 5 \cdot 6} + \frac{x^5}{4 \cdot 5 \cdot 6 \cdot 7} - \cdots,$$

with $0 < x < d$.

Theorem 8.2(b) asserts that, corresponding to the smaller root $c = -2$, a second fundamental solution is of the form

$$y_2(x) = Cy_1(x) \ln x + x^{-2} \sum_{n=0}^{\infty} b_n x^n$$

$$= Cy_1(x) \ln x + \sum_{n=0}^{\infty} b_n x^{n-2}.$$

To determine C and the coefficients b_n, we substitute this solution into the original differential equation, and because the result must be an identity in x, the coefficient of each power of x must vanish.

Differentiation of the foregoing result gives

$$y_2' = Cy_1'(x) \ln x + \frac{Cy_1(x)}{x} + \sum_{n=0}^{\infty} (n - 2)b_n x^{n-3}$$

and

$$y_2''(x) = Cy_1''(x) \ln x + \frac{2Cy_1'(x)}{x} - \frac{Cy_1(x)}{x^2} + \sum_{n=0}^{\infty} (n - 2)(n - 3)b_n x^{n-4}.$$

Substituting these results into the differential equation and collecting terms leads to the result

$$[x^2 y_1''(x) + x(2 + x)y_1'(x) - 2y_1(x)]C \ln x + C[y_1(x) + xy_1(x) + 2xy_1'(x)]$$

$$+ \sum_{n=0}^{\infty} (n - 3)(n - 2)b_n x^{n-2} + \sum_{n=0}^{\infty} 2(n - 2)b_n x^{n-2} + \sum_{n=0}^{\infty} (n - 2)b_n x^{n-1}$$

$$- \sum_{n=0}^{\infty} 2b_n x^{n-2} = 0.$$

The coefficient of the logarithmic term vanishes, because $y_1(x)$ is a solution of the differential equation, so the equation simplifies to

$$C[y_1(x) + xy_1(x) + 2xy_1'(x)]$$

$$+ \sum_{n=0}^{\infty} (n - 3)(n - 2)b_n x^{n-2} + \sum_{n=0}^{\infty} 2(n - 2)b_n x^{n-2} + \sum_{n=0}^{\infty} (n - 2)b_n x^{n-1}$$

$$- \sum_{n=0}^{\infty} 2b_n x^{n-2} = 0.$$

The terms corresponding to $n = 0$ cancel, and after shifting the summation index in the third summation, we have

$$C[y_1(x) + xy_1(x) + 2xy_1'(x)] + \sum_{n=1}^{\infty} (n - 3)(nb_n + b_{n-1})x^{n-2} = 0.$$

To find the form of the first group of terms $C[y_1(x) + xy_1(x) + 2xy_1'(x)]$, we must use the series solution for $y_1(x)$. As

$$y_1(x) = x - \frac{x^2}{4} + \frac{x^3}{4 \cdot 5} - \frac{x^4}{4 \cdot 5 \cdot 6} + \cdots,$$

differentiation gives

$$y_1'(x) = 1 - \frac{x}{2} + \frac{3x^2}{20} - \frac{x^3}{30} + \cdots,$$

and so

$$C[y_1(x) + xy_1(x) + 2xy_1'(x)] = 3Cx - \frac{Cx^2}{4} + \frac{Cx^3}{10} - \frac{Cx^4}{40} + \cdots.$$

Using this result in the equation and expanding the first few terms in the summation involving the unknown coefficients b_n shows that

$$\left(3Cx - \frac{Cx^2}{4} + \frac{Cx^3}{10} - \frac{Cx^4}{40} + \cdots\right) - (2b_1 + 2b_0)\frac{1}{x} - (2b_2 + b_1) + (4b_4 + b_3)x^2$$

$$+ (10b_5 + 2b_4)x^3 + (18b_6 + 3b_5)x^4 + (28b_7 + 4b_6)x^5 + (40b_8 + 5b_7)x^6 + \cdots = 0.$$

If we now equate to zero the coefficient of each power of x, we find that

$$b_1 = -b_0, \quad b_2 = -\frac{1}{2}b_1 = \frac{1}{2}b_0, \quad C = 0, \quad b_4 = -\frac{1}{4}b_3,$$

$$b_5 = -\frac{1}{5}b_4 = \frac{1}{4 \cdot 5}b_3, \quad b_6 = -\frac{1}{6}b_5 = -\frac{1}{4 \cdot 5 \cdot 6}b_3, \ldots.$$

The condition $C = 0$ shows that in this case the second linearly independent solution $y_2(x)$ does *not* contain a logarithmic term. The terms b_1 and b_2 are determined as multiples of b_0, and from Theorem 8.2(b) $b_0 \neq 0$, whereas for $n > 3$ all of the terms b_n are seen to be multiples of b_3, which is arbitrary because no equation connects it with b_0. Thus, the solution that has been generated appears to contain *two* arbitrary constants instead of the *one* that would have been expected. Substituting the b_n into the general form of the solution, which with $C = 0$ has reduced to

$$y_2(x) = \sum_{n=0}^{\infty} b_n x^{n-2},$$

gives

$$y_2(x) = b_0\left(\frac{1}{x^2} - \frac{1}{x} + \frac{1}{2}\right) + b_3 x\left(1 - \frac{x}{4} + \frac{x^2}{4 \cdot 5} - \frac{x^3}{4 \cdot 5 \cdot 6} + \frac{x^4}{4 \cdot 5 \cdot 6 \cdot 7} - \cdots\right).$$

The apparent incompatibility caused by the introduction of the two arbitrary constants b_0 and b_3 is now resolved, because the series multiplied by b_3 is simply the first linearly independent solution $y_1(x)$. So, in this case, when seeking the second linearly independent solution we have, in fact, generated a linear combination of the first linearly independent solution $y_1(x)$ and another linearly independent solution given by the expression

$$\frac{1}{x^2} - \frac{1}{x} + \frac{1}{2}.$$

Accordingly, we set $b_3 = 0$ and $b_0 = 1$, and take for the second linearly independent solution

$$y_2(x) = \frac{1}{x^2} - \frac{1}{x} + \frac{1}{2},$$

and since only three terms are involved we see that $y_2(x)$ is defined for $x > 0$.

When closed form solutions such as $y_2(x)$ are obtained, they should always be checked by substitution into the differential equation, and in this case it is easy to check that $y_2(x)$ is, indeed, a solution.

It is a simple matter to show the radius of convergence of the series solution $y_1(x)$ is infinite, so solutions $y_1(x)$ and $y_2(x)$ form a basis for the solution of the differential equation whose general solution is

$$y(x) = C_1 y_1(x) + C_2 y_2(x), \quad \text{for } x > 0,$$

where C_1 and C_2 are arbitrary constants.   ■

Case (c) Equal Real Roots $c_1 = c_2$

EXAMPLE 8.10

Find the solution of

$$x^2 y'' + (x^2 - x)y' + y = 0,$$

in some interval $0 < x < d$.

Solution   This equation has a singular point at the origin, and when expressed in standard form we see that $P(x) = (x - 1)/x$ and $Q(x) = 1/x^2$, so

$$p_0 = \lim_{x \to 0} x P(x) = -1 \quad \text{and} \quad q_0 = \lim_{x \to 0} x^2 Q(x) = 1.$$

Thus, the origin is a regular singular point, and from (30) the indicial equation is seen to be

$$c(c - 1) - c + 1 = 0, \quad \text{or} \quad (c - 1)^2 = 0,$$

so the roots are $c = 1$ (twice).

Substituting the series

$$y(x) = \sum_{n=0}^{\infty} a_n x^{n+c}$$

into the differential equation gives

$$\sum_{n=0}^{\infty}(n + c)(n + c - 1)a_n x^{n+c} + \sum_{n=0}^{\infty}(n + c)a_n x^{n+c+1}$$

$$- \sum_{n=0}^{\infty}(n + c)a_n x^{n+c} + \sum_{n=0}^{\infty} a_n x^{n+c} = 0.$$

Shifting the summation index in the second summation allows it to be written

$$\sum_{n=1}^{\infty}(n + c - 1)a_{n-1} x^{n+c},$$

so using this in the preceding equation and separating out the terms corresponding to $n = 0$ we find that

$$a_0[c(c-1) - c + 1]x^c + \sum_{n=1}^{\infty}\{[(n+c)(n+c-2) + 1]a_n + (n+c-1)a_{n-1}\}x^{n+c} = 0.$$

As usual, equating the coefficient of x^c to zero gives the indicial equation, and equating the coefficient of x^{n+c} to zero gives the recurrence relation

$$[(n+c)(n+c-2) + 1]a_n = -(n+c-1)a_{n-1} \quad \text{for } n = 1, 2, \ldots.$$

Setting $c = 1$ this becomes

$$a_n = -a_{n-1}/n,$$

so

$$a_1 = -a_0, \quad a_2 = -\frac{1}{2}a_0 = \frac{1}{2!}a_0, \quad a_3 = -\frac{1}{3}a_2 = \frac{1}{3!}a_0$$

and, in general,

$$a_n = \frac{(-1)^n}{n!} \quad \text{for } n = 0, 1, 2, \ldots.$$

Setting the arbitrary constant $a_0 = 1$ gives as a fundamental solution of the equation

$$y_1(x) = \sum_{n=0}^{\infty} \frac{(-1)^n}{n!} x^{n+1} = xe^{-x}.$$

The series for e^{-x} converges for $x > 0$, so this result is valid for all $x > 0$.

Continuing, we now illustrate two different methods by which a second linearly independent solution may be found.

Method 1. As the form of solution $y_1(x)$ is particularly simple, we will make use of result (35) of Section 6.3 that asserts that if $y_1(x)$ is a solution of the equation

$$y'' + P(x)y' + Q(x)y = 0,$$

an example using the reduction of order method

then a second linearly independent solution is given by the **reduction of order** formula

$$y_2(x) = y_1(x)\int \frac{\exp[-\int P(x)dx]}{[y_1(x)]^2}dx.$$

Substituting for $y_1(x)$ and $P(x)$ gives

$$\int P(x)dx = \int \frac{(x-1)}{x}dx = x - \ln x, \quad \text{so } \exp\left[-\int P(x)dx\right] = xe^{-x}.$$

Thus,

$$y_2(x) = y_1(x)\int \frac{xe^{-x}}{x^2e^{-2x}}dx = y_1(x)\int \frac{e^x}{x}dx.$$

To integrate this result we replace e^x by its series expansion and integrate term by term to obtain

$$y_2(x) = xe^{-x} \int \left(\frac{1 + x + \frac{x^2}{2!} + \frac{x^3}{3!} + \frac{x^4}{4!} + \cdots}{x} \right) dx$$

$$= xe^{-x} \left(\ln x + x + \frac{x^2}{4} + \frac{x^3}{18} + \frac{x^4}{96} + \frac{x^5}{600} + \cdots \right).$$

In order to compare this method with the one that is to follow, we rewrite this result by replacing e^{-x} by the first few terms of its series expansion to give

$$y_2(x) = xe^{-x} \ln x + x \left(1 - x + \frac{x^2}{2!} - \frac{x^3}{3!} + \cdots \right) \left(x + \frac{x^2}{4} + \frac{x^3}{18} + \frac{x^4}{96} + \frac{x^5}{600} + \cdots \right).$$

Multiplying the two series together then shows that for some d_2

$$y_2(x) = xe^{-x} \ln x + \left(x^2 - \frac{3x^3}{4} + \frac{11x^4}{36} - \frac{25x^5}{288} + \cdots \right), \quad \text{for } 0 < x < d_2,$$

where d_2 is the radius of convergence of the bracketed series.

Method 2.   Theorem 8.2(c) asserts that the second linearly independent solution has the form

$$y_2(x) = y_1(x) \ln x + x^2 \sum_{n=0}^{\infty} b_n x^n = y_1(x) \ln x + \sum_{n=0}^{\infty} b_n x^{n+2}.$$

Substituting this result into the differential equation and collecting terms gives

$$[x^2 y_1''(x) + (x^2 - x)y_1'(x) + y_1(x)] \ln x + 2xy_1'(x) + xy_1(x) - 2y_1(x)$$

$$+ \sum_{n=0}^{\infty} (n+2)(n+1)b_n x^{n+2} + \sum_{n=0}^{\infty} (n+2)b_n x^{n+3} - \sum_{n=0}^{\infty} (n+2)b_n x^{n+2}$$

$$+ \sum_{n=0}^{\infty} b_n x^{n+2} = 0.$$

Notice that the logarithmic term has vanished because $y_1(x)$ is a solution of the differential equation.

Shifting the summation index in the second summation, we obtain

$$2xy_1'(x) + xy_1(x) - 2y_1(x) + \sum_{n=0}^{\infty} (n+2)(n+1)b_n x^{n+2}$$

$$+ \sum_{n=1}^{\infty} (n+1)b_{n-1} x^{n+2} - \sum_{n=1}^{\infty} (n+2)b_n x^{n+2} + \sum_{n=0}^{\infty} b_n x^{n+2} = 0.$$

Separating out the terms corresponding to $n = 0$ allows this to be written as

$$2xy_1'(x) + xy_1(x) - 2y_1(x) + b_0 x^2 + \sum_{n=1}^{\infty} (n+1)[(n+1)b_n + b_{n-1}]x^{n+2} = 0.$$

The terms involving $y_1(x)$ are now obtained by differentiation of the series

$$y_1(x) = xe^{-x} = x - x^2 + x^3/3 - x^4/6 + x^5/24 - \cdots,$$

leading to

$$2xy_1'(x) + xy_1(x) - 2y_1(x) = -x^2 + x^3 - x^4/2 + x^5/6 - x^6/24 + \cdots.$$

Using this result in the above equation and expanding the terms involving b^n gives

$$(-x^2 + x^3 - x^4/2 + x^5/6 - x^6/24 + \cdots) + b_0x^2 + 2(2b_1 + b_0)x^3$$
$$+ 3(3b_2 + b_1)x^4 + 4(4b_3 + b_2)x^5 + 5(5b_4 + b_3)x^6 + \cdots = 0.$$

Finally, equating the coefficients of powers of x to zero gives

$$b_0 - 1 = 0, \quad 4b_1 + b_0 + 1 = 0, \quad 9b_2 + 3b_1 - 1/2 = 0, \ldots.$$

so that

$$b_0 = 1, \quad b_1 = -3/4, \quad b_2 = 11/36, \quad b_3 = -25/288, \ldots.$$

Substituting these coefficients into the general form of the solution again produces the second solution found by Method 1, though in this case Method 1 was simpler. ∎

When the indicial equation has either equal roots or roots differing by an integer, and only the leading terms (the most significant ones) are required in the second linearly independent solution $y_2(x)$, the reduction of order method is often the simplest one to use. This approach is illustrated in the following example, and it is typical of how best to proceed when the integrand in result (35) of Section 6.3 involves a quotient of polynomials.

EXAMPLE 8.11

Find the solution of

$$x^2y'' + (x^3 - x)y' + y = 0$$

in some interval $0 < x < d$.

Solution The equation has a singular point at the origin, and when it is written in standard form, we find that $P(x) = x - 1/x$ and $Q(x) = 1/x^2$. Thus,

$$p_0 = \lim_{x \to 0} xP(x) = -1 \quad \text{and} \quad q_0 = \lim_{x \to 0} x^2Q(x) = 1,$$

so the origin is a regular singular point and the indicial equation is

$$c(1 - c) - c + 1 = 0 \quad \text{or} \quad (c - 1)^2 = 0,$$

with the double root $c = 1$.

Making the substitution $y(x) = \sum_{n=0}^{\infty} a_n x^{n+c}$ in the differential equation gives

$$\sum_{n=0}^{\infty}(n+c)(n+c-1)a_n x^{n+c} + \sum_{n=0}^{\infty}(n+c)a_n x^{n+c+2} - \sum_{n=0}^{\infty}(n+c)a_n x^{n+c}$$
$$+ \sum_{n=0}^{\infty} a_n x^{n+c} = 0.$$

A shift of the summation index brings this to the form

$$(c^2 - 2c + 1)x^c + c^2 x^{c+1} + \sum_{n=2}^{\infty} (n+c)(n+c-1)a_n x^{n+c} + \sum_{n=2}^{\infty} (n+c-2)a_{n-2} x^{n+c}$$

$$- \sum_{n=2}^{\infty} (n+c)a_n x^{n+c} + \sum_{n=2}^{\infty} a_n x^{n+c} = 0,$$

and after combination of the summations this becomes

$$(c^2 - 2c + 1)a_0 x^c + c^2 a_1 x^{c+1} + \sum_{n=2}^{\infty} \{[(n+c)(n+c-2)+1]a_n$$

$$+ (n+c-2)a_{n-2}\} x^{n+c} = 0.$$

Equating the coefficient of x^c to zero gives the indicial equation with the double root $c = 1$, and equating the coefficient of x^{c+1} to zero shows that $a_1 = 0$, because $c = 1$. Equating the coefficient of x^{n+c} to zero leads to the recurrence relation

$$[(n+c)(n+c-2)+1]a_n + (n+c-2)a_{n-2} = 0 \quad \text{for } n \geq 2.$$

Setting $c = 1$ in the recurrence relation, we have

$$a_n = -\frac{(n-1)}{n^2} a_{n-2},$$

but as $a_1 = 0$, it follows immediately that $a_n = 0$ for all odd n. As a result we have

$$a_2 = -\frac{1}{2^2} a_0, \quad a_4 = -\frac{3}{4^2} a_2 = \frac{3}{2^2 \cdot 4^2} a_0, \quad a_6 = -\frac{5}{6^2} a_4 = -\frac{1 \cdot 3 \cdot 5}{2^2 \cdot 4^2 \cdot 6^2} a_0, \ldots,$$

so a fundamental solution is given by

$$y_1(x) = x \left(1 - \frac{1}{2^2} x^2 + \frac{1 \cdot 3}{2^2 \cdot 4^2} x^4 - \frac{1 \cdot 3 \cdot 5}{2^2 \cdot 4^2 \cdot 6^2} x^6 - \cdots \right),$$

or for $0 < x < d_1$, where d_1 is the radius of convergence of $y_1(x)$, by

$$y_1(x) = x - \frac{1}{2^2} x^3 + \frac{1 \cdot 3}{2^2 \cdot 4^2} x^5 - \frac{1 \cdot 3 \cdot 5}{2^2 \cdot 4^2 \cdot 6^2} x^7 + \cdots.$$

The reduction of order method in (35) of Section 6.3 shows that

$$y_2(x) = y_1(x) \int \frac{\exp\left[-\int P(x)dx\right]}{[y_1(x)]^2} dx,$$

but $\exp[-\int P(x)dx] = \exp(-x^2/2)$, so

$$y_2(x) = y_1(x) \int \frac{\exp(-x^2/2)}{[y_1(x)]^2} dx.$$

To find the leading terms in the expansion for $y_2(x)$ it is now necessary to replace $\exp(-x^2/2)$ and $[y_1(x)]^2$ by the first few terms of their series expansions and then to convert the integrand to a polynomial that can be integrated term by term. We have

$$y_2(x) = y_1(x) \int \frac{x \left(1 - \frac{1}{2}x^2 + \frac{1}{8}x^4 - \frac{1}{48}x^6 + \frac{1}{384}x^8 - \cdots \right)}{x^2 \left(1 - \frac{1}{4}x^2 + \frac{3}{64}x^4 - \frac{5}{768}x^6 + \frac{35}{49152}x^8 - \cdots \right)^2} dx.$$

If the bracketed term in the denominator is now squared, the integral becomes

$$y_2(x) = y_1(x) \int \frac{\left(1 - \frac{1}{2}x^2 + \frac{1}{8}x^4 - \frac{1}{48}x^6 + \frac{1}{384}x^8 - \cdots\right)}{x\left(1 - \frac{1}{2}x^2 + \frac{5}{32}x^4 - \frac{7}{192}x^6 + \frac{169}{24576}x^8 - \cdots\right)}\, dx.$$

Division of the two polynomials using long division, or writing the numerator as

$$\frac{1}{x}\left(1 - \frac{1}{2}x^2 + \frac{1}{8}x^4 + \cdots\right)\left(1 - \frac{1}{2}x^2 + \frac{5}{32}x^4 - \cdots\right)^{-1}$$

and multiplying the bracketed terms after using the binomial theorem to expand the second bracket, converts the expression for $y_2(x)$ to

$$y_2(x) = y_1(x) \int \frac{1}{x}\left(1 - \frac{1}{32}x^4 + \frac{5}{8192}x^8 - \cdots\right) dx.$$

Integrating term by term, we find that for some $d_2 > 0$, the first few terms of the series solution $y_2(x)$ are

$$y_2(x) = y_1(x)\left[\ln\, x - \frac{1}{128}x^4 + \frac{5}{65536}x^8 + \cdots\right],$$

or

$$y_2(x) = y_1(x)\ln x + x\left(1 - \frac{1}{4}x^2 + \frac{3}{64}x^4 - \frac{5}{768}x^6\right.$$
$$\left. + \frac{35}{4915}x^8 - \cdots\right)\left(-\frac{1}{128}x^4 + \frac{5}{65536}x^8 + \cdots\right).$$

After multiplication of the two series we obtain

$$y_2(x) = y_1(x)\ln\, x - \left(\frac{1}{128}x^5 - \frac{59}{65536}x^9 + \cdots\right)$$

in some interval of the form $0 < x < d_2$, where d_2 is the radius of convergence of the bracketed series. The general solution is thus

$$y(x) = C_1 y_1(x) + C_2 y_2(x), \quad \text{for } 0 < x < d,$$

where C_1 and C_2 are arbitrary constants and $d = \min\{d_1, d_2\}$.

When using this approach it is important to ensure that sufficient terms are retained in the intermediate calculations involving the polynomials for the final result to be accurate to the required power of x. ∎

Case (d) Complex Conjugate Roots

EXAMPLE 8.12

Find the solution of the Cauchy–Euler equation

$$x^2 y''(x) - xy'(x) + 10y(x) = 0$$

in some interval $0 < x < d$.

Solution   This equation has a singular point at the origin, and when expressed in standard form $P(x) = -1/x$ and $Q(x) = 10/x^2$. We have

$$\lim_{x \to 0} x P(x) = -1 \quad \text{and} \quad \lim_{x \to 0} x^2 Q(x) = 10,$$

so the origin is a regular singular point. From (30) the indicial equation is seen to be

$$c^2 - 2c + 10 = 0$$

with the complex conjugate roots $c = 1 \pm 3i$. Substituting

$$y(x) = \sum_{n=0}^{\infty} a_n x^{n+c}$$

into the differential equation leads to the result

$$\sum_{n=0}^{\infty}(n+c)(n+c-1)a_n x^{n+c} - \sum_{n=0}^{\infty}(n+c)a_n x^{n+c} + \sum_{n=0}^{\infty} 10 a_n x^{n+c} = 0.$$

After terms are collected under a single summation sign, this becomes

$$\sum_{n=0}^{\infty}[(n+c)(n+c-2)+10]a_n x^{n+c} = 0.$$

Equating to zero the coefficient of x^c, corresponding to $n = 0$, gives

$$(c^2 - 2c + 10)a_0 = 0,$$

but by hypothesis $a_0 \neq 0$, so this simply yields the indicial equation. Equating to zero the coefficient of x^{n+c} for $n = 1, 2, \ldots$ gives

$$(n+c)(n+c+10)a_n = 0,$$

but as $c = 1 \pm 3i$, the factor $(n+c)(n+c+10) \neq 0$ for any value of n, so it follows that $a_n = 0$ for $n = 1, 2, \ldots$. Thus, from Theorem 8.2(d), it follows that two linearly independent solutions of the differential equation are obtained by taking the real and imaginary parts of

$$y(x) = a_0 x^{1+3i} = a_0 x \exp\{\ln x^{3i}\} = a_0 x \exp\{3i \ln x\}$$

$$= a_0 x \{\cos(3 \ln x) + i \sin(3 \ln x)\}.$$

Setting the arbitrary constant $a_0 = 1$ and taking the real and imaginary parts of this last result shows that two linearly independent solutions are

$$y_1(x) = x \cos\{3 \ln x\} \quad \text{and} \quad y_2(x) = x \sin\{3 \ln x\},$$

each of which is defined for $x > 0$. These solutions form a basis for the solution of the differential equation whose general solution is

$$y(x) = C_1 x \cos\{3 \ln x\} + C_2 x \sin\{3 \ln x\}, \quad \text{for } x > 0,$$

where C_1 and C_2 are arbitrary constants. ■

More information about singular points and the Frobenius method can be found in references [3.3] to [3.6].

Summary

This section showed how the power series solutions considered previously must be modified if solutions are to be obtained in the form of expansions about regular singular points. The method due to Frobenius for obtaining such solutions was then developed systematically and illustrated by examples, with particular attention being given to the various special cases that arise depending on the relationship that exists between the roots of the indicial equation.

EXERCISES 8.4

In Exercises 1 and 2, shift the summation indices to combine the given expressions into the sum of a finite number of terms and a single summation.

1. (a) $2\sum_{n=0}^{\infty} a_n x^{n+c} + (1+x)\sum_{n=0}^{\infty} a_n x^{n+c-2}$.

(b) $3\sum_{n=0}^{\infty} a_n x^{n+c} + 2x^2 \sum_{n=0}^{\infty} a_n x^{n+c-1}$.

2. (a) $(x - x^3)\sum_{n=0}^{\infty} a_n x^{n+c} + 3\sum_{n=0}^{\infty} a_n x^{n+c-1}$.

(b) $(x^2 - x)\sum_{n=0}^{\infty} a_n x^{n+c} + 2\sum_{n=0}^{\infty} a_n x^{n+c-2}$.

In Exercises 3 through 6, use long division and multiplication of series to find the first four terms of the given expressions.

3. (a) $\dfrac{1}{\sum_{n=0}^{\infty} (-1)^n x^n/(n+1)}$.

(b) $(1 - x/2 + x^2/4 - x^3/8 + x^4/16 - x^5/32 + \cdots)\exp(x)$.

(c) $(1 - x/2 + x^2/3 - x^3/4 + x^4/5 - \cdots)(1 - x + x^2/2 - x^3/3 + x^4/4 - \cdots)$.

4. (a) $(1 + 2x + x^2)/(3 - x + 2x^4)$.

(b) $\left(\sum_{n=1}^{\infty} \dfrac{x^n}{n^2}\right)\left(\sum_{n=1}^{\infty} \dfrac{(-1)^{n+1}x^n}{(n+1)}\right)$.

5. (a) $\int \dfrac{1}{x}\left(\dfrac{1 - 3x + x^2}{2 - \exp(x)}\right)dx$. **(b)** $\int \dfrac{\exp x}{(x + x^2 + x^3)}dx$.

6. (a) $\int \dfrac{1}{x^2}\dfrac{(1 + 2x - x^2)}{(1 + x + 2x^3)}dx$. **(b)** $\int \dfrac{1}{x}\dfrac{\exp(-x)}{(1 - 2x + 2x^2)}dx$.

In Exercises 7 through 26, find two linearly independent solutions for $x > 0$, and determine at least the first four leading terms in the second solution $y_2(x)$.

7. $4x^2 y'' + 2xy' + (x - 2)y = 0$.
8. $3x^2 y'' - xy' + (x + 1)y = 0$.
9. $2x^2 y'' + xy' - (2x + 1)y = 0$.
10. $2x^2 y'' + xy' - (3x + 1)y = 0$.
11. $(x^2 - 1)y'' + 2xy' + y = 0$.
12. $2x^2 y'' + 2xy' + (x^2 - 2)y = 0$.
13. $x(1 - x)y'' + (1 - x)y' - y = 0$.
14. $2x^2 y'' - 2xy' + (x^2 + 2)y = 0$.
15. $x^2 y'' + (2x^2 - x)y' + y = 0$.
16. $x^2 y'' + 2(x^2 - x)y' + 2y = 0$.
17. $x^2 y'' + (x^2 - 2x)y' + 2y = 0$.
18. $x^2 y'' - xy' + (x^2 + 1)y = 0$.
19. $16x^2 y'' + 8xy' + (16x + 1)y = 0$.
20. $2x^2 y'' + 2xy' + (x - 2)y = 0$.
21. $x^2 y'' + (x^2 - x)y' - 3y = 0$.
22. $4x^2 y'' - 2x^2 y' + (2x + 1)y = 0$.
23. $x^2 y'' + (x^2 + x)y' - 4y = 0$.
24. $9x^2 y'' - 6xy' + 2y = 0$.
25. $x^2 y'' - 4xy' + 20y = 0$.
26. $4x^2 y'' + 8xy' + 5y = 0$.

27. By shifting the critical point to the origin, find two linearly independent solutions of the following equation in an interval of the form $0 < x + 1 < d$:

$$2(x+1)y'' + y' - (x+1)y = 0.$$

28. By shifting the critical point to the origin, find two linearly independent solutions of the following equation in an interval of the form $0 < x - 2 < d$:

$$(x-2)^2 y'' - (x-2)y' + (x^2 - 4x + 5)y = 0.$$

8.5   The Gamma Function Revisited

more about the Gamma function

The function $\Gamma(x)$, called the **gamma function**, was introduced in (4) of Section 7.1 in connection with the Laplace transform of t^a when a is not an integer, and it was defined in terms of the improper integral

$$\Gamma(x) = \int_0^\infty e^{-t} t^{x-1}\, dt \quad \text{for } x > 0. \tag{32}$$

a fundamental result

It was shown that $\Gamma(x)$ satisfies the recurrence relation

$$\Gamma(x+1) = x\Gamma(x) \quad \text{for } x > 0, \tag{33}$$

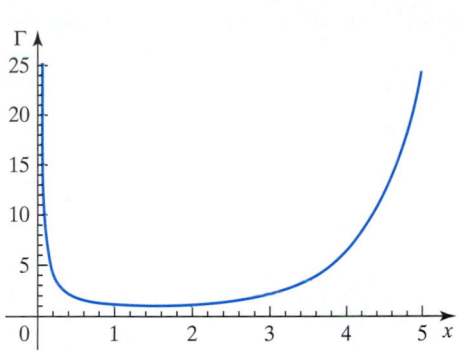

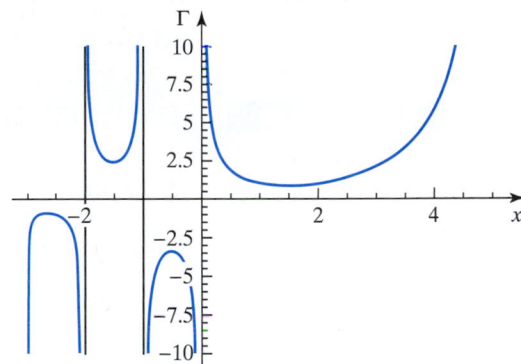

FIGURE 8.4   The function $\Gamma(x)$ in the interval $0 < x < 5$.

FIGURE 8.5   The function $\Gamma(x)$ in the interval $-3 < x < 4$.

and that when x is a positive integer n the gamma function reduces to

$$\Gamma(n+1) = n!. \tag{34}$$

Thus, for any real $x > 0$, the function $\Gamma(x)$ interpolates continuously between successive values of $n!$, and so generalizes the factorial function to nonintegral values of n. For obvious reasons the gamma function is sometimes called the **factorial function**. Figure 8.4 shows a graph of $\Gamma(x)$ in the interval $0 < x < 5$.

The gamma function can be extended to $x < 0$ for $x \neq -1, -2, \ldots$, at which point it becomes infinite. A graph of $\Gamma(x)$ in the interval $-3 < x < 4$ is shown in Fig. 8.5.

The value of $\Gamma(1/2)$ is often needed, and it can be found by means of the following method in which the integral defining $\Gamma(1/2)$ is squared and converted to a double integral that is easily evaluated. If the method used is unfamiliar the details can be omitted, though the result given in (35) is useful and should be remembered.

From (32) we have

$$[\Gamma(1/2)]^2 = \left(\int_0^\infty u^{-1/2} e^{-u} du \right) \left(\int_0^\infty v^{-1/2} e^{-v} dv \right),$$

where the two dummy variables u and v have been introduced to avoid confusion when the product of integrals is combined.

Writing $u = x^2$ and $v = y^2$ allows this product of integrals to be written as

$$[\Gamma(1/2)]^2 = \left(\int_0^\infty 2e^{-x^2} dx \right) \left(\int_0^\infty 2e^{-y^2} dy \right) = 4 \int_0^\infty \int_0^\infty e^{-(x^2+y^2)} dx dy.$$

As the integral in terms of cartesian coordinates is only evaluated over the first quadrant, changing to the polar coordinates (r, θ) by setting $x = r \cos \theta$, $y = r \sin \theta$, and using the result $r^2 = x^2 + y^2$ reduces this last integral to

$$[\Gamma(1/2)]^2 = \lim_{\rho \to \infty} 4 \int_0^{\pi/2} d\theta \int_0^\rho e^{-r^2} r \, dr = 4 \cdot (\pi/2) \lim_{\rho \to \infty} \left[-\frac{1}{2} e^{-r^2} \right]_0^\rho = \pi.$$

a useful special case

Taking the square root shows that

$$\Gamma(1/2) = \sqrt{\pi}. \tag{35}$$

When x is a multiple of $1/2$, repeated use of recurrence relation (33) combined with result (35) allows $\Gamma(x)$ to be simplified, as illustrated in the following example.

EXAMPLE 8.13

Find (a) $\Gamma(7/2)$ and (b) $\Gamma(-3/2)$.

Solution

(a) From (33) it follows that

$$\Gamma\left(\frac{7}{2}\right) = \frac{5}{2}\Gamma\left(\frac{5}{2}\right) = \frac{5}{2}\cdot\frac{3}{2}\Gamma\left(\frac{3}{2}\right) = \frac{5}{2}\cdot\frac{3}{2}\cdot\frac{1}{2}\Gamma\left(\frac{1}{2}\right) = \frac{15}{8}\sqrt{\pi}.$$

(b) Setting $x = -3/2$ in (33) gives

$$\left(-\frac{3}{2}\right)\Gamma\left(-\frac{3}{2}\right) = \Gamma\left(-\frac{1}{2}\right),$$

whereas setting $x = -1/2$ in (33) gives

$$\left(-\frac{1}{2}\right)\Gamma\left(-\frac{1}{2}\right) = \Gamma(1/2) = \sqrt{\pi}.$$

So, combining these two results, we find that

$$\Gamma\left(-\frac{3}{2}\right) = \left(-\frac{2}{3}\right)\left(-\frac{2}{1}\right)\Gamma(1/2) = \frac{4}{3}\sqrt{\pi}. \qquad \blacksquare$$

The reason for this re-examination of the gamma function is because it enables the coefficients of a series expansion to be expressed in a concise form. For example, it follows directly from (34) that the binomial coefficient

$$\binom{n}{m} = \frac{n!}{m!(n-m)!} = \frac{\Gamma(n+1)}{\Gamma(m+1)\Gamma(n-m+1)}. \qquad (36)$$

Expressing a binomial coefficient with integer entries in terms of the gamma function offers no particular advantage over the use of factorials, but the preceding result generalizes to the more useful result

$$\binom{\alpha}{m} = \frac{\Gamma(\alpha+1)}{\Gamma(m+1)\Gamma(\alpha-m+1)} \qquad (37)$$

when α is any nonnegative real number (not necessarily an integer). This expression is often useful when performing numerical calculations.

As another example of the use of (33) we notice that we can write

$$a(a+1)(a+2)\ldots(a+n) = \frac{\Gamma(a+n+1)}{\Gamma(a)}, \qquad (38)$$

where n is a positive integer and the real number $a > 0$. Thus, for example, in terms of the gamma function the following product becomes

$$\left(\frac{1}{2}\right)\left(\frac{3}{2}\right)\left(\frac{5}{2}\right)\left(\frac{7}{2}\right) = \frac{\Gamma(\frac{1}{2}+3+1)}{\Gamma(\frac{1}{2})} = \frac{\Gamma(\frac{9}{2})}{\Gamma(\frac{1}{2})}.$$

Result (38) generalizes further to provide a concise representation of the product of $n + 1$ factors $c(c + d)(c + 2d)\ldots(c + nd)$. By writing the product as

$$c(c+d)(c+2d)\ldots(c+nd) = d^{n+1}\left(\frac{c}{d}\right)\left(\frac{c}{d}+1\right)\left(\frac{c}{d}+2\right)\cdots\left(\frac{c}{d}+n\right),$$

and then setting $a = c/d$ in (38), we arrive at the useful result

$$c(c+d)(c+2d)\ldots(c+nd) = d^{n+1}\frac{\Gamma\left(\frac{c}{d}+n+1\right)}{\Gamma\left(\frac{c}{d}\right)}. \tag{39}$$

EXAMPLE 8.14

The nth coefficient of a series is given by

$$a_n = \frac{1\cdot5\cdot9\cdot13\ldots(4n+1)}{2^n}.$$

Express a_n in terms of the gamma function.

Solution   Comparing the numerator of a_n with result (39) shows that it contains $n + 1$ factors, and in the notation of (39) we have $c = 1$ and $d = 4$. Thus,

$$1\cdot5\cdot9\cdot13\ldots(4n+1) = 4^{n+1}\frac{\Gamma\left(n+\frac{5}{4}\right)}{\Gamma\left(\frac{1}{4}\right)},$$

so dividing by 2^n we find that

$$a_n = 4^{n+1}\frac{\Gamma\left(n+\frac{5}{4}\right)}{2^n\Gamma\left(\frac{1}{4}\right)} = 2^{n+2}\frac{\Gamma\left(n+\frac{5}{4}\right)}{\Gamma\left(\frac{1}{4}\right)}. \qquad\blacksquare$$

Two special products of this type arise when working with series as, for example, occurs in the case of Legendre polynomials. These products involve either the product of consecutive pairs of odd numbers or the product of consecutive pairs of even numbers. Although these products can be expressed in terms of the gamma function, a convenient and concise **double factorial** notation is used. We define the double factorial !! as follows:

the double factorial

$$1\cdot3\cdot5\cdots(2n+1) = (2n+1)!! \quad\text{and}\quad 2\cdot4\cdot6\cdots(2n) = (2n)!!. \tag{40}$$

Alternative expressions for these double factorials in terms of the usual factorial function are

$$(2n+1)!! = \frac{(2n+1)!}{2^n n!} \quad\text{and}\quad (2n)!! = 2^n n!. \tag{41}$$

The following relationship connecting gamma functions is sometimes useful:

$$\Gamma(x)\Gamma(1-x) = \frac{\pi}{\sin \pi x}. \tag{42}$$

However, this result will not be proved here as it requires the techniques of complex integration.

In passing, we mention a function $B(x, y)$ called the *beta function* that is related to the gamma function. The **beta function**, which has applications in statistics and elsewhere, is defined as the integral

the beta function

$$B(x, y) = \int_0^1 t^{x-1}(1 - t)^{y-1}dt \quad \text{with} \quad x > 0, y > 0. \tag{43}$$

The following are the most important properties of the beta function:

Symmetry:

$$B(x, y) = B(y, x) \tag{44}$$

relating gamma and beta functions

Connection with the gamma function:

$$B(x, y) = \frac{\Gamma(x)\Gamma(y)}{\Gamma(x + y)} \tag{45}$$

Relationship between beta functions:

$$B(x, y) = \left(\frac{y - 1}{x + y - 1}\right) B(x, y - 1) = \left(\frac{x + y}{y}\right) B(x, y + 1), \tag{46}$$

Special values:

$$B(1, 1) = 1 \quad \text{and} \quad B(1/2, 1/2) = \pi. \tag{47}$$

Outline proofs of results (42) to (44) will be found in the harder exercises at the end of this section.

The gamma function in the complex plane is discussed in reference [6.7], and general information about the gamma function and related functions is contained in Chapter 6 of reference [G.1] and Chapter 11 of reference [G.3].

Summary

The gamma function that was introduced earlier was seen to provide a natural extension to arbitrary values of x of the factorial function $n!$, where n is an integer. In this section the gamma function was examined in greater detail and some useful values were derived in terms of π. The beta function was then defined and related to the gamma function.

EXERCISES 8.5

1. Express $\Gamma(5/2), \Gamma(-5/2)$, and $\Gamma(9/2)$ in terms of $\sqrt{\pi}$.

2. Express $\Gamma(-9/2), \Gamma(11/2)$, and $\Gamma(-11/2)$ in terms of $\sqrt{\pi}$.

3. Express $\Gamma(5/4), \Gamma(-5/4)$, and $\Gamma(7/4)$ in terms of either $\Gamma(1/4)$ or $\Gamma(-1/4)$.

4. Express $\Gamma(-7/4), \Gamma(9/4)$, and $\Gamma(3/4)$ in terms of either $\Gamma(1/4)$ or $\Gamma(-1/4)$.

5. Express the product $6 \cdot 11 \cdot 16 \cdot 21 \ldots (5n + 6)$ in terms of the gamma function.

6. Express the product $1 \cdot 3 \cdot 5 \cdot 7 \cdot 11 \ldots . (2n + 1)$ in terms of the gamma function.

7. Express the product $5 \cdot 8 \cdot 11 \cdot 14 \ldots (3n+5)$ in terms of the gamma function.

8. Express the product $4 \cdot 8 \cdot 12 \cdot 16 \ldots (4n+4)$ in terms of the gamma function.

9. Show that
$$\Gamma\left(\frac{1}{2} - n\right) = \frac{(-1)^n \sqrt{\pi}}{\left(n - \frac{1}{2}\right)\left(n - \frac{3}{2}\right)\left(n - \frac{5}{2}\right)\cdots\left(\frac{1}{2}\right)}.$$

10. Show that
$$\Gamma\left(n + \frac{1}{2}\right) = \left(n - \frac{1}{2}\right)\left(n - \frac{3}{2}\right)\cdots\left(\frac{1}{2}\right)\sqrt{\pi}.$$

The following slightly harder exercises provide more information about the gamma function.

11.* Use the result $\Gamma(n + \frac{1}{2}) = (n - \frac{1}{2})\Gamma(n - \frac{1}{2})$ with the result of Exercise 9 to show that
$$\Gamma(2n) = \frac{2^{2n-1}\Gamma(n)\Gamma\left(n + \frac{1}{2}\right)}{\sqrt{\pi}}.$$

12.* Show that $\Gamma(x) = \int_0^1 (\ln\frac{1}{u})^{x-1} du$ for $x > 0$.

13.* Show that $\Gamma(x) = 2\int_0^\infty e^{-u^2} u^{2x-1} du$ for $x > 0$.

14.* The function $\psi(x)$, called the **psi function** or the **digamma function**, is defined as
$$\psi(x) = \frac{d}{dx}[\ln \Gamma(x)].$$

Show that
$$\psi(x+1) = \psi(x) + \frac{1}{x} \quad \text{for } x > 0.$$

15.* Use the result of Exercise 14 to show that
$$\psi(x+n) = \psi(x) + \sum_{k=0}^{n-1} \frac{1}{x+k} \quad \text{where } n > 1 \text{ is an integer.}$$

16.* By making the variable change $u = 1 - t$ in the integral defining $B(x, y)$, show that $B(x, y) = B(y, x)$.

17.* Integrate $B(x, y)$ by parts to obtain the result of (46) that
$$B(x, y) = \left(\frac{y - 1}{x + y - 1}\right) B(x, y - 1),$$
and use this result to obtain the second result of (46).

18.* Use the result of Exercise 17 to show that if m and n are integers,
$$B(m, n) = \frac{(m - n)!(n - 1)!}{(m + m - 1)!} = \frac{\Gamma(m)\Gamma(n)}{\Gamma(m + n)},$$
and so
$$B(m, n) = \frac{\Gamma(m)\Gamma(n)}{\Gamma(m + n)}.$$

8.6   Bessel Function of the First Kind $J_n(x)$

Bessel's equation

In standard form, **Bessel's equation** is written

$$x^2\frac{d^2y}{dx^2} + x\frac{dy}{dx} + (x^2 - v^2)y = 0, \tag{48}$$

where $v \geq 0$ is a real number. Another useful form of Bessel's equation that often arises in applications is

$$x^2\frac{d^2y}{dx^2} + x\frac{dy}{dx} + (\lambda^2 x^2 - v^2)y = 0. \tag{49}$$

This form of the equation is obtained from (48) by first making the change of variable $x = \lambda u$, and then replacing u by x.

When developing the properties of Bessel functions in this section the standard form of the equation given in (48) will be used. Applications of Bessel functions to partial differential equations are made in Chapter 18.

Bessel's equation has a singularity at the origin, and using the notation of Section 8.4 with $P(x) = 1/x$ and $Q(x) = (x^2 - v^2)/x^2$, we find that

$$p_0 = \lim_{x \to 0} x P(x) = 1 \quad \text{and} \quad q_0 = \lim_{x \to 0} x^2 Q(x) = -v^2,$$

showing that the origin is a regular singular point.

The indicial equation is seen to be

$$c^2 - v^2 = 0, \tag{50}$$

so the roots $c_1 = v$ and $c_2 = -v$ are distinct when $v \neq 0$, and there is a repeated zero root when $v = 0$. Thus, when $v = 0$, the second Frobenius solution will contain a logarithmic term, whereas when $c_1 - c_2$ is an integer the second Frobenius solution may or may not contain a logarithmic term. When $c_1 - c_2 \neq 0$ is not an integer, neither of the two linearly independent Frobenius solutions contains a logarithmic term.

Substituting $y(x) = \sum_{r=0}^{\infty} a_r x^{r+c}$ into (48) gives

$$\sum_{r=0}^{\infty}(r+c)(r+c-1)a_r x^{r+c} + \sum_{r=0}^{\infty}(r+c)a_r x^{r+c} + \sum_{r=0}^{\infty} a_r x^{r+c+2} - v^2 \sum_{r=0}^{\infty} a_r x^{r+c} = 0.$$

Shifting the summation index in the third summation and collecting terms under a single summation leads to the result

$$(c^2 - v^2)a_0 x^c + [(c+1)^2 - v^2]a_1 x^{c+1} + \sum_{r=2}^{\infty}[(r+c+v)(r+c-v)a_r + a_{r-2}]x^{r+c} = 0.$$

Equating the coefficients of powers of x to zero shows the following:

Coefficient of x^c:

$$(c^2 - v^2)a_0 = 0 \quad \text{(the indicial equation, because } a_0 \neq 0)$$

Coefficient of x^{c+1}:

$$[(c+1)^2 - v^2]a_1 = 0 \quad \text{(a condition on } a_1)$$

Coefficient of x^{r+c}:

$$[(r+c)^2 - v^2]a_r + a_{r-2} = 0 \quad \text{(a recurrence relation)} \tag{51}$$

As $(c+1)^2 - v^2 \neq 0$, it follows from the second result that $a_1 = 0$, and then from the recurrence relation (51) that $a_r = 0$ for all odd r. As only even indices r are involved in the recurrence relation, we set $r = 2m$ with $m = 0, 1, \ldots$, after which substituting $c = v$ in the recurrence relation reduces it to

$$a_{2m} = -\frac{1}{4m(m+v)}a_{2m-2}, \quad \text{for } m = 1, 2, \ldots. \tag{52}$$

As a_0 is arbitrary, we normalize the solution in the standard manner by setting

$$a_0 = \frac{1}{2^v \Gamma(1+v)},$$

after which the coefficients a_{2m} become

$$a_2 = -\frac{a_0}{2^2(1+v)} = -\frac{1}{2^{2+v}1!\Gamma(2+v)}, \quad a_4 = -\frac{a_2}{2^2 2(2+v)} = \frac{1}{2^{4+v}2!\Gamma(3+v)}, \ldots,$$

and, in general,

$$a_{2m} = -\frac{(-1)^m}{2^{2m+v}m!\Gamma(m+1+v)}, \quad \text{for } m = 1, 2, \ldots. \tag{53}$$

the Bessel function
$J_\nu(x)$

Using this result in the first Frobenius solution, which hereafter will be denoted by $J_\nu(x)$ and called a **Bessel function of the first kind of order** ν, gives

$$J_\nu(x) = x^\nu \sum_{m=0}^{\infty} \frac{(-1)^m x^{2m}}{2^{2m+\nu} m! \, \Gamma(m+1+\nu)} \quad \text{for } x \geq 0. \tag{54}$$

When $x < 0$ the corresponding expression for $J_\nu(x)$ follows from the preceding result by reversing the sign of x in the series and replacing x^ν by $|x|^\nu$. The ratio test shows the series for $J_\nu(x)$ to be absolutely convergent for all x.

So far ν has been an arbitrary nonnegative number, but the standard convention is that when ν is an integer it is denoted by n. Using the result that when $\nu = n$ the gamma function $\Gamma(m+1+n) = (m+n)!$ allows $J_n(x)$ to be written in the simpler form

$$J_n(x) = \sum_{m=0}^{\infty} \frac{(-1)^m x^{2m+n}}{2^{2m+n} m! (m+n)!}, \quad \text{for } n = 0, 1, 2, \ldots. \tag{55}$$

It was because of this use of n that, to avoid confusion, the summation index in the series was chosen to be m. The two most important special cases of (55) are:

Bessel functions
$J_0(x)$ and $J_1(x)$

Bessel function of the first kind of order zero:

$$J_0(x) = \sum_{m=0}^{\infty} \frac{(-1)^m x^{2m}}{2^{2m} (m!)^2} = 1 - \frac{x^2}{2^2 (1!)^2} + \frac{x^4}{2^4 (2!)^2} - \frac{x^6}{2^6 (3!)^2} + \cdots \tag{56}$$

Bessel function of the first kind of order 1:

$$J_1(x) = \sum_{m=0}^{\infty} \frac{(-1)^m x^{2m+1}}{2^{2m+1} m! (m+1)!} = \frac{x}{2} - \frac{x^3}{2^3 1! 2!} + \frac{x^5}{2^5 2! 3!} - \frac{x^7}{2^7 3! 4!} + \cdots.$$

$$\tag{57}$$

Graphs of $J_0(x)$, $J_1(x)$, and $J_2(x)$ are shown in Fig. 8.6.

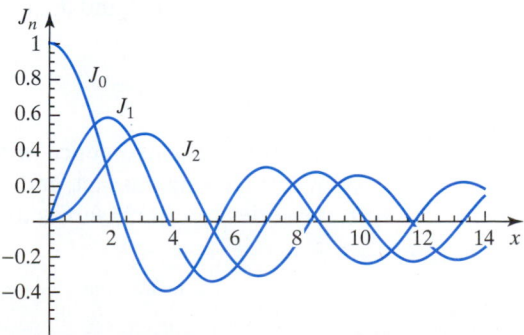

FIGURE 8.6   Graphs of the Bessel functions of the first kind $J_0(x)$, $J_1(x)$, and $J_2(x)$.

Having found $J_\nu(x)$, which is one solution of Bessel's equation (48), we must now find a second linearly independent solution in order to arrive at a basis for solutions of the equation, and hence to arive at the general solution. The nature of a second linearly independent solution will depend on the value of ν, and the simplest situation arises when ν is not an integer. In this case, because $c^2 = \nu^2$, a second linearly independent solution will follow from (54) by replacing ν by $-\nu$. Denoting this second solution by $J_{-\nu}(x)$ we find that

$$J_{-\nu}(x) = |x|^{-\nu} \sum_{m=0}^{\infty} \frac{(-1)^m x^{2m}}{2^{2m-\nu} m! \Gamma(m+1-\nu)} \quad \text{for } x \neq 0. \tag{58}$$

general solution of Bessel's equation

When ν is *not* an integer, the **general solution of Bessel's equation** (48) can be written

$$y(x) = C_1 J_\nu(x) + C_2 J_{-\nu}(x), \quad \text{for } x \neq 0, \tag{59}$$

with C_1 and C_2 arbitrary constants. The corresponding general solution of (49) is then

$$y(x) = C_1 J_\nu(\lambda x) + C_2 J_{-\nu}(\lambda x), \quad \text{for } x \neq 0. \tag{60}$$

The nature of the second linearly independent solution when $\nu = n$ will be considered later. In the meantime we will show that when $\nu = n$, the Bessel functions $J_n(x)$ and $J_{-n}(x)$ are linearly dependent. This is most easily seen by taking the limit of (58) as $\nu \to n$. Gamma functions with negative integer arguments are infinite, so the coefficients a_{2m} in which they occur will all vanish, causing the summation to start at the value $m = n$. Using the result $\Gamma(m+1-n) = (m-n)!$ then shows that the series for $J_{-n}(x)$ is

$$J_{-n}(x) = \sum_{m=n}^{\infty} \frac{(-1)^m x^{2m-n}}{2^{2m-n} m! (m-n)!},$$

and after a shift of the summation index this becomes

$$J_{-n}(x) = (-1)^n \sum_{m=n}^{\infty} \frac{(-1)^m x^{2m+n}}{2^{2m+n} m! (m+n)!}, \quad \text{for } n = 1, 2, \ldots. \tag{61}$$

A comparison of (55) and (61) shows that $J_{-n}(x)$ is a constant multiple of $J_n(x)$, so the two functions $J_n(x)$ and $J_{-n}(x)$ are linearly dependent. To be precise,

$$J_{-n}(x) = (-1)^n J_n(x) \quad \text{for } n = 1, 2, \ldots. \tag{62}$$

The absolute convergence of the series for $J_\nu(x)$ allows it to be differentiated term by term. Using this fact, and comparing of the derivative of the series for $J_0(x)$ with the series for $J_1(x)$, shows that

$$J_0'(x) = -J_1(x). \tag{63}$$

This result is the simplest example of the many relationships that exist between Bessel functions. The four most important results are the following:

Relationships between derivatives of $J_\nu(x)$:

$$\frac{d}{dx}\left[x^\nu J_\nu(x)\right] = x^\nu J_{\nu-1}(x) \tag{64}$$

$$\frac{d}{dx}\left[x^{-\nu} J_\nu(x)\right] = -x^{-\nu} J_{\nu-1}(x) \tag{65}$$

Recurrence relations involving $J_\nu(x)$:

$$J_{\nu-1}(x) + J_{\nu+1}(x) = \frac{2\nu}{x} J_\nu(x) \tag{66}$$

$$J_{\nu-1}(x) - J_{\nu+1}(x) = 2J_\nu'(x) \tag{67}$$

We show next that these results are easily verified by substituting the series solution for $J_\nu(x)$ given in (54) into each relationship, though the direct derivation of these relationships is a more complicated matter. An indication of one way in which to arrive at these results without appealing to the series solution (54) is to be found in the set of exercises at the end of this section.

To establish (64) we start by multiplying the series (54) for $J_\nu(x)$ by x^ν to obtain

$$x^\nu J_\nu(x) = \sum_{m=0}^\infty \frac{(-1)^m x^{2m+2\nu}}{2^{2m+\nu} m! \Gamma(m+1+\nu)}.$$

Differentiating this result and removing a factor x^ν from the summation gives

$$\frac{d}{dx}\left[x^\nu J_\nu(x)\right] = x^\nu \sum_{m=0}^\infty \frac{(-1)^m x^{2m+\nu-1}}{2^{2m+\nu-1}\Gamma(m+\nu)},$$

but the series on the right-hand side is simply $J_{\nu-1}(x)$, so we have shown that

$$\frac{d}{dx}\left[x^\nu J_\nu(x)\right] = x^\nu J_{\nu-1}(x).$$

Result (65) is established in similar fashion by differentiating $x^{-\nu} J_\nu(x)$.

The recurrence relations can be obtained as follows. Carrying out the indicated differentiations and cancelling a factor x^ν in (64) and (65) gives

$$J_\nu'(x) = J_{\nu-1}(x) - \frac{\nu}{x} J_\nu(x) \tag{64$'$}$$

and

$$J_\nu'(x) = \frac{\nu}{x} J_\nu(x) - J_{\nu+1}(x). \tag{65$'$}$$

Results (66) and (67) now follow first by subtraction and then by addition of these two results.

Result (66) is useful because it relates $J_\nu(x)$ to $J_{\nu-1}(x)$ and $J_{\nu+1}(x)$, whereas (64) and (65) can be used to evaluate certain integrals involving $J_\nu(x)$, because by integrating (64) and (65) we obtain

$$\int x^\nu J_{\nu-1}(x)dx = x^\nu J_\nu(x) + C \tag{68}$$

and

$$\int x^{-\nu} J_{\nu+1}(x)dx = -x^{-\nu} J_\nu(x) + C. \tag{69}$$

EXAMPLE 8.15

Express $J_4(x)$ in terms of $J_0(x)$ and $J_1(x)$, and use the result to compute $J_4(6.2)$ given that $J_0(6.2) = 0.20175$ and $J_1(6.2) = -0.23292$.

Solution   Rearranging (66) gives

$$J_{\nu+1}(x) = \frac{2\nu}{x} J_\nu(x) - J_{\nu-1}(x),$$

so setting $\nu = 3, 2,$ and 1 we have

$$J_4(x) = \frac{6}{x} J_3(x) - J_2(x), \quad J_3(x) = \frac{4}{x} J_2(x) - J_1(x), \quad \text{and} \quad J_2(x) = \frac{2}{x} J_1(x) - J_0(x).$$

Eliminating $J_2(x)$ and $J_3(x)$ between these results gives the required expression

$$J_4(x) = \left(\frac{48}{x^3} - \frac{8}{x}\right) J_1(x) + \left(1 - \frac{24}{x^2}\right) J_0(x).$$

Setting $x = 6.2$ and substituting the given values of $J_0(6.2)$ and $J_1(6.2)$ shows that $J_4(6.2) = 0.32941$. ∎

Numerical values of Bessel functions are extensively tabulated, and subroutines that enable their calculation for arbitrary values of their argument are found in most computer algebra packages. See the references at the end of the chapter for some of the most extensive tabulations of Bessel functions.

EXAMPLE 8.16

Evaluate

$$\int \left(x^2 + \frac{1}{x}\right) J_1(x)dx.$$

Solution   We write the integral as the sum of integrals

$$\int \left(x^2 + \frac{1}{x}\right) J_1(x)dx = \int x^2 J_1(x)dx + \int x^{-1} J_1(x)dx$$

and consider each separately. Setting $\nu = 2$ in (64) shows that

$$\frac{d}{dx}[x^2 J_2(x)] = x^2 J_1(x),$$

so it follows at once that

$$\int x^2 J_1(x)dx = x^2 J_2(x) + C.$$

The second integral is a little harder and requires the use of integration by parts. Writing it as

$$\int x^{-1} J_1(x)dx = \int x^{-2}[x J_1(x)]dx,$$

and noticing from (63) with $\nu = 1$ that $[xJ_1(x)]' = xJ_0(x)$, we find that

$$\int x^{-1}J_1(x)dx = \int x^{-2}[xJ_1(x)]dx = -J_1(x) + \int x^{-1}xJ_0(x)dx,$$

and so

$$\int x^{-1}J_1(x)dx = -J_1(x) + \int J_0(x)dx.$$

No further simplification is possible because $\int J_0(x)dx$ cannot be expressed in terms of simpler functions, though $\int_0^x J_0(u)du$ is available in tabular form and it is easily evaluated numerically on a computer. However, we will see later that $\int_0^\infty J_n(x)dx = 1$ for $n = 0, 1, 2, \ldots$. ∎

EXAMPLE 8.17

Evaluate $\int x^3 J_0(x)dx$.

Solution   Writing the integrand as the product $x^3 J_0(x) = x^2[xJ_0(x)]$ and using (64) with $\nu = 1$ gives

$$\int x^3 J_0(x)dx = \int x^2[xJ_0(x)]dx = \int x^2 \frac{d}{dx}[xJ_1(x)]dx.$$

Integration by parts then gives

$$\int x^3 J_0(x)dx = x^3 J_1(x) - 2x^2 J_2(x) + C.$$ ∎

It can be seen from Fig. 8.6 that the Bessel functions $J_0(x)$, $J_1(x)$, and $J_2(x)$ are oscillatory in nature and resemble damped sinusoids. The recurrence relation (66) implies that this same oscillatory property is true for all $J_n(x)$. Although these Bessel functions are not strictly periodic, in the sense that for any given n the zeros of $J_n(x)$ are not equally spaced along the x-axis, it can be shown that for fixed ν and large x the function $J_n(x)$ can be approximated by

$$J_\nu(x) \sim \sqrt{\frac{2}{\pi x}} \cos\left(x - \frac{\nu\pi}{2} - \frac{\pi}{4}\right), \tag{70}$$

where the symbol $\sim$ is to be read "is asymptotically equal to," with the understanding that the term *asymptotic* is used here in the technical sense and means that the ratio of the two sides of the expression tends to 1 as $x \to \infty$. This last result is an example of what is called an **asymptotic expansion** of the function $J_\nu(x)$, and asymptotic expansions have the property that the larger x becomes, the more accurate the asymptotic expansion becomes.

asymptotic expansion of $J_\nu(x)$

When the Bessel functions $J_\nu(x)$ are required in a computer program, the series solution (54) is used for small x, and different approximations are used for large x and in the intermediate region between small and large x. Corresponding approximations are used when the order ν of a Bessel function is large. The simplest approximation to $J_\nu(x)$ for small x, which follows from (54) by setting $m = 0$, is

$$J_\nu(x) \approx \frac{1}{\Gamma(1+\nu)}\left(\frac{x}{2}\right)^\nu. \tag{71}$$

The fact that the series for $J_\nu(x)$ is an alternating series means that the maximum magnitude of the error made when the series is truncated after n terms is the absolute

TABLE 8.1 Zeros $j_{n,r}$ of $J_n(x)$ for $n = 0, 1, 2, 3$				
r	$j_{0,r}$	$j_{1,r}$	$j_{2,r}$	$j_{3,r}$
1	2.40482	3.83171	5.13162	6.38016
2	5.52007	7.01559	8.41724	9.76102
3	8.65372	10.17347	11.61984	13.01520
4	11.79153	13.32369	14.79595	16.22347
5	14.93091	16.47063	17.95982	19.40942
6	18.07106	19.61586	21.11700	22.58273

value of the $(n + 1)$th term. So, if the series

$$J_0(x) = 1 - \frac{x^2}{2^2(1!)^2} + \frac{x^4}{2^4(2!)^2} - \frac{x^6}{2^6(3!)^2} + \cdots$$

is truncated after the term in x^4, the maximum error made is $|-x^6/[2^6(3!)^2]| = x^6/[2^6(3!)^2]$. Consequently, if $J_0(x)$ is approximated by

$$J_0(x) \approx 1 - \frac{x^2}{2^2(1!)^2} + \frac{x^4}{2^4(2!)^2},$$

then in the interval $0 \leq x \leq a$, the absolute value of the maximum error will not exceed $a^6/[2^6(3!)^2]$. When $J_\nu(x)$ is required to be accurate to a given number of decimal places in an interval $0 \leq x \leq a$, this simple estimate determines how many terms must be retained in the series approximation for $J_\nu(x)$.

zeros of Bessel functions $J_n(x)$

When using Bessel functions in applications, it is often necessary to know the location of the zeros of $J_n(x)$, so for future reference Table 8.1 lists the first six zeros of $J_n(x)$ for $n = 0, 1, 2, 3$. In the table the rth zero of $J_n(x)$ is denoted by $j_{n,r}$, where the first suffix indicates the order of the Bessel function and the second suffix the number of the zero. As $J_n(0) = 0$ for $n \geq 1$, the zeros $j_{1,r}$, $j_{2,r}$, and $j_{3,r}$ have been numbered so the first entry to appear in each column is the first nonvanishing zero of the function involved. Thus, although $J_1(0) = 0$, the first entry to appear in the column for $j_{1,r}$ is 3.83171, which it will be seen from Fig. 8.6 is the first nonvanishing zero of $J_1(x)$.

Bessel Functions $J_{\pm n/2}(x)$

The Bessel functions $J_{\pm n/2}(x)$ are particularly simple, despite the fact that the difference between the indices $c_1 = n/2$ and $c_2 = -n/2$ is an integer. The easiest way to find the form of $J_{\pm n/2}(x)$ is to use the reduction to standard form given in Lemma 6.1 of Section 6.3 to remove the first derivative term from Bessel's equation.

It follows from the lemma that the substitution $u = x^{1/2}y$ reduces Bessel's equation

$$x^2 y'' + xy' + (x^2 - \nu^2)y = 0$$

to the standard form for a second order equation

$$u'' + \left(1 - \frac{4\nu^2 - 1}{4x^2}\right)u = 0.$$

If we now consider the cases of $J_{1/2}(x)$ and $J_{-1/2}(x)$, corresponding to $\nu^2 = 1/4$, the differential equation simplifies to

$$u'' + u = 0,$$

with the general solution

$$u(x) = C_1 \sin x + C_2 \cos x.$$

As $y = x^{-1/2}u$, the general solution of Bessel's equation of order $\pm 1/2$ becomes

$$y(x) = C_1 \sqrt{\frac{1}{x}} \sin x + C_2 \sqrt{\frac{1}{x}} \cos x.$$

The two functions in the general solution for $y(x)$ are linearly independent, so we take for the solutions forming a basis for the differential equation with $\nu = \pm 1/2$ the functions $J_{1/2}(x)$ and $J_{-1/2}(x)$ given by

Bessel functions of fractional order

$$J_{1/2}(x) = C_1 \sqrt{\frac{1}{x}} \sin x \quad \text{and} \quad J_{-1/2}(x) = C_2 \sqrt{\frac{1}{x}} \cos x.$$

The constants C_1 and C_2 are arbitrary, but to make these results compatible with the normalization used for a_0 when developing the series solution for $J_\nu(x)$ we compare these expressions with the asymptotic formula (70), from which we see it is necessary to set $C_1 = C_2 = \sqrt{(2/\pi)}$, to obtain

$$J_{1/2}(x) = \sqrt{\frac{2}{\pi x}} \sin x \quad \text{and} \quad J_{-1/2}(x) = \sqrt{\frac{2}{\pi x}} \cos x. \tag{72}$$

Expressions for $J_{\pm n/2}(x)$ now follow by use of recurrence relation (66). Thus, for example, setting $\nu = 1/2$ in (66) gives

$$J_{3/2}(x) = \frac{1}{x} J_{1/2}(x) - J_{-1/2}(x) = \sqrt{\frac{2}{\pi x}} \left(\frac{\sin x}{x} - \cos x \right), \tag{73}$$

and, similarly, setting $\nu = -1/2$ gives

$$J_{-3/2}(x) = -\sqrt{\frac{2}{\pi x}} \left(\sin x + \frac{\cos x}{x} \right). \tag{74}$$

We have shown that all Bessel functions $J_{\pm n/2}(x)$ with n an odd integer are expressible in terms of elementary functions. The derivation of $J_{\pm 1/2}(x)$ directly from series (54) forms an exercise in the set at the end of this section.

FRIEDRICH WILHELM BESSEL (1784–1846)
A German mathematician who started his career as a clerk apprenticed to a mercantile office in Bremen where he remained for a number of years. Using published observations he calculated the orbit of Haley's comet and submitted his calculations to the astronomer H.W.M. Olbers who recognized his ability and, after recommending the work for publication, arranged for Bessel to become an assistant in the observatory in Lilienthal. His major mathematical contribution was the introduction, in a paper of 1824 devoted to planetary motions, of the class of transcendental functions now known as Bessel functions.

Summary

Bessel's equation was introduced and series solutions were obtained by the Frobenius method for the Bessel function $J_\nu(x)$ of the first kind of order ν. It was shown that Bessel functions of the first kind of fractional order $\pm n/2$, with n odd, could be expressed in terms of products of sines and cosines and $1/\sqrt{x}$.

EXERCISES 8.6

---

1. Write down the first six terms of the series expansion for $J_2(x)$.

2. Write down the first six terms of the series expansion for $J_3(x)$.

3. Derive result (65) by differentiating the product of $x^{-1/2}$ and the series for $J_\nu(x)$ given in (54).

4. Determine how many terms must be retained in the series for $J_0(x)$ for it to be accurate to four decimal places over the interval $0 \le x \le 4$.

5. Determine how many terms must be retained in the series for $J_0(x)$ for it to be accurate to four decimal places over the interval $0 \le x \le 2$.

6. Determine how many terms must be retained in the series for $J_0(x)$ for it to be accurate to six decimal places over the interval $0 \le x \le 1$.

7. Determine how many terms must be retained in the series for $J_0(x)$ for it to be accurate to six decimal places over the interval $0 \le x \le 2$.

8. Determine how many terms must be retained in the series for $J_1(x)$ for it to be accurate to four decimal places over the interval $0 \le x \le 2$.

9. Determine how many terms must be retained in the series for $J_1(x)$ for it to be accurate to four decimal places over the interval $0 \le x \le 3$.

10. Integrate the first four terms in the series for $J_0(x)$ term by term to obtain an approximation to

$$\int_0^x J_0(t)dt.$$

Estimate the maximum magnitude of the error when using the result in the interval $0 \le x \le a$.

11. Integrate the first four terms in the series for $J_1(x)$ term by term to obtain an approximation to

$$\int_0^x J_1(t)dt.$$

Estimate the maximum magnitude of the error when using the approximation in the interval $0 \le x \le a$. Integrate the integral analytically, and confirm that the analytical result and the approximation are in agreement.

The Bessel function $J_\nu(\lambda x)$ is a solution of $x^2 y'' + xy' + (\lambda^2 x^2 - \nu^2)y = 0$. Establish the following results by making

the change of variable $x = \lambda X$ in results (64) to (67), and then replacing X by x.

12. $\dfrac{d}{dx}[x^\nu J_\nu(\lambda x)] = \lambda x^\nu J_{\nu-1}(\lambda x).$

13. $\dfrac{d}{dx}[x^{-\nu} J_\nu(x)] = -\lambda x^{-\nu} J_{\nu+1}(\lambda x).$

14. $\dfrac{d}{dx}[J_\nu(\lambda x)] = \lambda\, J_{\nu-1}(\lambda x) - \dfrac{\nu}{x} J_\nu(\lambda x).$

15. $\dfrac{d}{dx}[J_\nu(\lambda x)] = -\lambda\, J_{\nu+1}(\lambda x) + \dfrac{\nu}{x} J_\nu(\lambda x).$

16. $\dfrac{d}{dx}[J_\nu(\lambda x)] = \dfrac{\lambda}{2}[J_{\nu-1}(\lambda x) - J_{\nu+1}(\lambda x)].$

17. $J_\nu(\lambda x) = \dfrac{\lambda x}{2\nu}[J_{\nu-1}(\lambda x) + J_{\nu+1}(\lambda x)].$

18. Use $(64)'$ and $(65)'$ to show that

$$\frac{d}{dx}[xJ_\nu(x)J_{\nu+1}(x)] = x\left[J_\nu^2(x) - J_{\nu+1}^2(x)\right].$$

19. Show that $\lim_{x\to 0} J_0(x) = 1$, $\lim_{x\to 0} J_n(x) = 0$ for $n = 1, 2, \ldots$ and, $\lim_{x\to\infty} J_n(x) = 0$ for $n = 0, 1, \ldots$, and prove that

$$\int_0^\infty J_1(x)dx = 1.$$

20. Use the results in Exercise 19 with (67) to show that

$$1 = \int_0^\infty J_1(x)ds = \int_0^\infty J_3(x)dx = \cdots$$

$$= \int_0^\infty J_{2n+1}(x)dx = \cdots \quad \text{for } n = 0, 1, \ldots.$$

21. In Section 7.3(d)(ii) it was shown that the Laplace transform of $J_0(x)$ was

$$\mathcal{L}\{J_0(x)\} = \frac{1}{(s^2 + 1)^{1/2}}.$$

Use this result to deduce the value of $\int_0^\infty J_0(x)dx$, and then use (67) together with the results of Exercise 20 to show that

$$1 = \int_0^\infty J_0(x)dx = \int_0^\infty J_1(x)dx = \int_0^\infty J_3(x)dx = \cdots$$

$$= \int_0^\infty J_n(x)dx = \cdots \quad \text{for } n = 0, 1, 2, \ldots.$$

22. Find (a) $\int x^3 J_2(x)dx$ and (b) $\int x^{-3} J_4(x)dx$.

23. Express $\int J_4(x)dx$ in terms of $\int J_0(x)dx$.

24. Express $\int J_5(x)dx$ in terms of $J_0(x)$, $J_2(x)$, and $J_4(x)$.

25. Express $\int x J_1(x)dx$ in terms of $\int J_0(x)dx$.

26. Express $\int x^2 J_0(x)dx$ in terms of $\int J_0(x)dx$.

The exercises that follow, some of which are slightly harder, provide background information about Bessel functions.

27.* By differentiating under the integral sign with respect to x, integrating by parts, and combining results using an elementary trigonometric identity, prove that

$$J_0(x) = \frac{1}{\pi} \int_0^\pi \cos(x \sin \theta) d\theta$$

is an integral representation of $J_0(x)$ by showing that it satisfies Bessel's equation of order zero

$$x J_0'' + J_0' + x J_0 = 0.$$

28.* The function $\exp[\frac{x}{2}(t - \frac{1}{t})]$ is the **generating function** for the Bessel functions $J_n(x)$, and it has the property that when it is expanded in powers of t (both positive and negative),

$$\exp\left[\frac{x}{2}\left(t - \frac{1}{t}\right)\right] = \sum_{n=-\infty}^{\infty} J_n(x)t^n.$$

Thus, $J_n(x)$ is the coefficient of t^n in the expansion of the generating function in powers of t. Expand the exponential as the product of the series for $\exp[xt/2]$ and $\exp[-x/(2t)]$, and hence derive the first three terms of the series expansion of $J_0(x)$.

29.* Differentiate the generating function partially with respect to x and equate the coefficients of t^n on each side

of the identity to prove that

$$2J_n'(x) = J_{n-1}(x) - J_{n+1}(x).$$

30.* Differentiate the generating function partially with respect to t and equate the coefficients of t^{n-1} on each side of the identity to prove that

$$\frac{2n}{x} J_n(x) = J_{n-1}(x) + J_{n+1}(x).$$

31.* Substitute $\nu = 1/2$ in (54) and (58), and hence show that $J_{1/2} = \sqrt{\frac{2}{\pi x}} \sin x$ and $J_{-1/2}(x) = \sqrt{\frac{2}{\pi x}} \cos x$.

32.* Use (66) together with results (73) and (74) to show that

$$J_{5/2}(x) = \sqrt{\frac{2}{\pi x}} \left[\left(\frac{3}{x^2} - 1 \right) \sin x - \frac{3}{x} \cos x \right]$$

$$J_{-5/2}(x) = \sqrt{\frac{2}{\pi x}} \left[\frac{3}{x} \sin x + \left(\frac{3}{x^2} - 1 \right) \cos x \right]$$

$$J_{9/2}(x) = \sqrt{\frac{2}{\pi x}} \left[\left(\frac{105}{x^4} - \frac{45}{x} + 1 \right) \sin x \right.$$
$$\left. - \left(\frac{105}{x^3} - \frac{10}{x} \right) \cos x \right]$$

and

$$J_{-9/2}(x) = \sqrt{\frac{2}{\pi x}} \left[\left(\frac{105}{x^3} - \frac{10}{x} \right) \sin x \right.$$
$$\left. + \left(\frac{105}{x^4} - \frac{45}{x^2} + 1 \right) \cos x \right].$$

8.7   Bessel Functions of the Second Kind $Y_\nu(x)$

It was shown in the previous section that, with the exception of $\nu = 1/2$, the two Bessel functions $J_\nu(x)$ and $J_{-\nu}(x)$ of the first kind are only linearly independent solutions of Bessel's equation when the roots of the indicial equation differ by an integer. So it remains for us to find a second linearly independent solution when $\nu = n$ and $n = 0, 1, 2, \ldots$. We begin by considering the case $n = 0$, corresponding to the repeated root $\nu = 0$, when it follows from Theorem 8.2(b) that the form of solution to be expected in the case of Bessel's equation of order zero

Bessel functions of the second kind

$$xy'' + y' + xy = 0 \qquad (75)$$

is

$$y_2(x) = J_0(x) \ln x + \sum_{r=0}^{\infty} b_r x^{r+1}. \qquad (76)$$

Differentiation of (76) gives

$$y_2'(x) = J_0'(x) \ln x + \frac{J_0(x)}{x} + \sum_{r=0}^{\infty} (r+1) b_r x^r$$

and

$$y_2''(x) = J_0''(x) \ln x + \frac{2 J_0'(x)}{x} - \frac{J_0(x)}{x^2} + \sum_{r=0}^{\infty} (r+1) r b_r x^{r-1}.$$

When these expressions are substituted into (75) the terms in $J_0(x)$ cancel, causing the equation to reduce to

$$[x J_0''(x) + J_0'(x) + x J_0(x)] \ln x + 2 J_0'(x) + \sum_{r=0}^{\infty} (r+1) r b_r x^r$$

$$+ \sum_{r=0}^{\infty} (r+1) b_r x^r + \sum_{r=0}^{\infty} b_r x^{r+2} = 0.$$

The logarithmic term vanishes because $J_0(x)$ is a solution of (75), so the coefficients b_r are determined by the equation

$$2 J_0'(x) + \sum_{r=0}^{\infty} (r+1) r b_r x^r + \sum_{r=0}^{\infty} (r+1) b_r x^r + \sum_{n=0}^{\infty} b_r x^{r+2} = 0.$$

To proceed further it is necessary to determine $J_0'(x)$, but this can be found by differentiating (56) in Section 8.6. After cancellation of a factor $2m$ from the numerator and denominator of the resulting expression, and noticing that the summation now starts from $m = 1$, it is found that

$$J_0'(x) = \sum_{m=1}^{\infty} \frac{(-1)^m x^{2m-1}}{2^{2m-1} (m-1)! m!}.$$

Combining this with the previous result gives

$$\sum_{m=1}^{\infty} \frac{(-1)^{m+1} x^{2m+1}}{2^{2m} (m+1)! m!} + \sum_{r=0}^{\infty} (r+1) r b_r x^r + \sum_{r=0}^{\infty} (r+1) b_r x^r + \sum_{r=0}^{\infty} b_r x^{r+2} = 0.$$

Shifting the summation index in the last term and combining the summations reduces this to

$$\sum_{m=1}^{\infty} \frac{(-1)^{m+1} x^{2m+1}}{2^{2m} (m+1)! m!} + b_0 + 4 b_1 x + \sum_{r=2}^{\infty} \{ (r+1)^2 b_r + b_{r-2} \} x^r = 0.$$

We now make use of the fact that terms may be rearranged in an absolutely convergent series in order to rewrite the last summation as a sum of even powers of x and a sum of odd powers of x before combining the results. The preceding equation then becomes

$$\sum_{m=1}^{\infty} \frac{(-1)^{m+1} x^{2m+1}}{2^{2m} (m+1)! m!} + b_0 + 4 b_1 x + \sum_{m=1}^{\infty} \{ (2m+1)^2 b_{2m} + b_{2m-2} \} x^{2m}$$

$$+ \sum_{m=2}^{\infty} [4 m^2 b_{2m-1} + b_{2m-3}] x^{2m-1} = 0.$$

Next we equate the coefficient of each power of x to zero in the usual manner. As there is no constant term in the first summation, it follows that $b_0 = 0$. The recurrence relation in the second summation is $(2m+1)^2 b_{2m} + b_{2m-2} = 0$, so together with the result $b_0 = 0$ this implies that $b_{2m} = 0$ for $m = 0, 1, 2, \ldots$. Setting the summation involving even powers of x to zero brings the equation into the form

$$\sum_{m=1}^{\infty} \frac{(-1)^{m+1} x^{2m+1}}{2^{2m}(m+1)!m!} + 4b_1 x + \sum_{m=2}^{\infty} \left[4m^2 b_{2m-1} + b_{2m-3}\right] x^{2m-1} = 0.$$

We now equate to zero the coefficients of each remaining power of x, and proceeding in this manner it is not difficult to show that the general coefficient b_{2m-1} can be written

$$b_{2m-1} = \frac{(-1)^{m-1}}{2^{2m}(m!)^2} \left(1 + \frac{1}{2} + \frac{1}{3} + \cdots + \frac{1}{m}\right), \quad \text{for } m = 1, 2, \ldots,$$

so the second linearly independent solution is

$$y_2(x) = J_0(x) \ln x + \sum_{m=1}^{\infty} \frac{(-1)^{m-1} x^{2m}}{2^{2m}(m!)^2} \left(1 + \frac{1}{2} + \frac{1}{3} + \cdots + \frac{1}{m}\right). \tag{77}$$

Defining h_m as

$$h_m = 1 + \frac{1}{2} + \frac{1}{3} + \cdots + \frac{1}{m} \tag{78}$$

allows $y_2(x)$ to be written in the more convenient form

$$y_2(x) = J_0(x) \ln x + \sum_{m=1}^{\infty} \frac{(-1)^{m-1} h_m x^{2m}}{2^{2m}(m!)^2}. \tag{79}$$

The series in (79) can be shown to converge, though as the logarithmic term becomes infinite at the origin, result (79) is only finite for $x > 0$.

As any linear combination of two linearly independent solutions of a differential equation is itself a solution, it proves to be convenient to take as the second solution of Bessel's equation of order zero the function $Y_0(x)$ defined as the linear combination

$$Y_0(x) = \frac{2}{\pi}[y_2(x) + (\gamma - \ln 2)J_0(x)], \tag{80}$$

where the constant γ, called the **Euler constant**, is defined as

$$\gamma = \lim_{m \to \infty} \left(1 + \frac{1}{2} + \frac{1}{3} + \cdots + \frac{1}{m} - \ln m\right), \tag{81}$$

where $\gamma = 0.577\ 215\ 664\ 901.\ldots$. This constant is also called the **Euler–Mascheroni constant**, and on occasion it is denoted by C and sometimes by $\ln \gamma$.

the Bessel functions
$Y_0(x)$ and $Y_\nu(x)$

The function $Y_0(x)$, called the **Bessel function of the second kind of order zero,** is defined as

$$Y_0(x) = \frac{2}{\pi} \left[J_0(x) \left(\ln \frac{x}{2} + \gamma \right) + \sum_{m=1}^{\infty} \frac{(-1)^{m-1} h_m}{2^{2m}(m!)^2} x^{2m} \right]. \tag{82}$$

The reason for choosing this particular combination of functions in the definition of $Y_0(x)$ is because of its convenient properties as $x \to \infty$. The function $Y_0(x)$ is also called the **Neumann** or **Weber function** of order zero and denoted by $N_0(x)$.

Some authors make a distinction in what they call a Bessel function of the second kind, so there may be a difference between the Weber function $Y_n(x)$ and the Neumann function $N_n(x)$. Because of this, care must be exercised when using these functions in software packages.

Bessel functions of the second kind of integral order can be defined in similar fashion, but to make them compatible with the functions $J_{-\nu}(x)$ introduced in Section 8.6 the following definition is adopted:

$$Y_\nu(x) = \frac{1}{\sin \nu\pi} [J_\nu(x) \cos \nu\pi - J_{-\nu}(x)] \tag{83}$$

with

$$Y_n(x) = \lim_{\nu \to n} Y_\nu(x). \tag{84}$$

Using this last result it is possible to show that for integral values of ν the function $Y_n(x)$ is given by

$$Y_n(x) = \frac{2}{\pi} J_n(x) \left(\ln \frac{x}{2} + \gamma \right) + \frac{x^n}{\pi} \sum_{m=0}^{\infty} \frac{(-1)^{m-1}(h_m + h_{m+n})}{2^{2m+n} \, m!(m+n)!} x^{2m}$$
$$- \frac{1}{\pi x^n} \sum_{m=0}^{n-1} \frac{(n-m-1)!}{2^{2m-n}m!} x^{2m} \tag{85}$$

where, by definition, $h_0 = 1$. It follows from this that the Bessel functions $Y_n(x)$ and $Y_{-n}(x)$ are linearly dependent, with

$$Y_{-n}(x) = (-1)^n Y_n(x).$$

Graphs of the first three Bessel functions of the second kind are shown in Fig. 8.7.

When x is small the following approximations are useful:

$$Y_0(x) \approx \frac{2}{\pi} \ln x \quad \text{and for } \nu > 0, \quad Y_\nu(x) \approx -\frac{\Gamma(\nu)}{\pi} \left(\frac{2}{x} \right)^\nu. \tag{86}$$

asymptotic form
for $Y_\nu(x)$

For large x, however, the asymptotic approximation to $Y_\nu(x)$ is

$$Y_\nu(x) \sim \sqrt{\frac{2}{\pi x}} \sin \left[x - \left(\frac{2\nu + 1}{4} \right) \pi \right]. \tag{87}$$

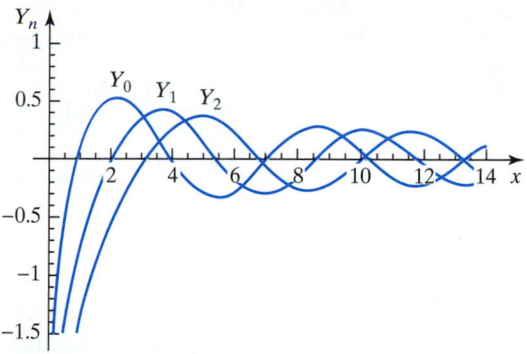

FIGURE 8.7 Bessel functions $Y_0(x)$, $Y_1(x)$, and $Y_2(x)$ of the second kind.

It follows from (86) and (87) that

$$\lim_{x\to 0} Y_v = -\infty \quad \text{and} \quad \lim_{x\to\infty} Y_v(x) = 0. \tag{88}$$

zeros of Bessel functions $Y_n(x)$

The zeros of $Y_n(x)$ are needed when working with Bessel functions, so the locations of the first six zeros of $Y_n(x)$ for $n = 0, 1, 2, 3$ are listed in Table 8.2. The rth zero of the Bessel function $Y_n(x)$ is denoted by $y_{n,r}$, so, for example, the second zero of $Y_1(x)$ is $y_{1,2} = 5.42968$.

It is a consequence of the definition of $Y_v(x)$ that for all v the general solution of Bessel's equation in the standard form

$$x^2 y'' + x y' + (x^2 - v^2) y = 0 \tag{89}$$

is

$$y(x) = C_1 J_v(x) + C_2 Y_v(x). \tag{90}$$

Similarly, the general solution of Bessel's equation in the form

$$x^2 y'' + x y' + (\lambda^2 x^2 - v^2) y = 0 \tag{91}$$

is

$$y(x) = C_1 J_v(\lambda x) + C_2 Y_v(\lambda x). \tag{92}$$

TABLE 8.2 Zeros $y_{n,r}$ of $Y_n(x)$ for $n = 0, 1, 2, 3$

r	$y_{0,r}$	$y_{1,r}$	$y_{2,r}$	$y_{3,r}$
1	0.89358	2.19714	3.38424	4.52702
2	3.95786	5.42968	6.79381	8.09755
3	7.08605	8.59601	10.02348	11.39647
4	10.22235	11.74915	13.20999	14.62308
5	13.36110	14.89744	16.37897	17.81846
6	16.50092	18.04340	19.53904	20.99728

Many differential equations can be solved in terms of Bessel functions after a suitable transformation of the dependent variable. In particular, the equation

$$y'' + \left(\frac{1-2a}{x}\right)y' + \left[b^2c^2x^{2c-2} + \left(\frac{a^2 - v^2c^2}{x^2}\right)\right]y = 0 \tag{93}$$

can be shown to have the solution

$$y(x) = x^a Z_v(bx^c), \tag{94}$$

where a, b, and c are numbers and Z_v is any linear combination of J_v and Y_v (see Exercise 16 at the end of this section).

The following is an application of Bessel functions to a simple physical problem. It illustrates how, in this case, the conditions of the problem only allow a Bessel function of the first kind to be retained in the solution. The problem, which is a classical one, can be stated as follows.

Find the radial temperature distribution $T(r)$ in a wire of circular cross-section with $0 \leq r \leq R$, when the electrical conductivity is σ, the thermal conductivity is K, and the wire carries a uniform current of density I amps per unit area of cross-section. Assume that the temperature at the center of the wire is T_0 and that the resistance of the wire varies linearly with the temperature as $\alpha T(r)$, with α a constant.

In order to formulate the problem in mathematical terms, we begin with the fact that the rate of heat generation in a unit volume of the wire is given by JI^2/σ heat units, where J is a physical constant (typically the number of calories in a joule). It follows from arguments given later in Chapter 18 that the equation determining the radial steady state temperature distribution is

$$K\frac{d^2T}{dr^2} + \frac{K}{r}\frac{dT}{dr} + \frac{\alpha JI^2}{\sigma}T = -\frac{JI^2}{\sigma},$$

where the last term on the left takes account of the linear variation of resistance with temperature, and the term on the right represents the heat generation due to the current.

When divided by K, this is seen to be Bessel's equation of order zero with a nonhomogeneous term $-JI^2/K\sigma$, and it is easily shown to have the general solution

$$T(r) = AJ_0\left(Ir\sqrt{\frac{\alpha J}{K\sigma}}\right) + BY_0\left(Ir\sqrt{\frac{\alpha J}{K\sigma}}\right) - \frac{1}{\alpha},$$

with A and B arbitrary constants. As the temperature must remain finite at the center of the wire, we must set $B=0$ to remove the infinite value of Y_0 when $r=0$. However, $T(0) = T_0$, so $A = T_0 + 1/\alpha$ and the required radial temperature distribution becomes

$$T(r) = \left(T_0 + \frac{1}{\alpha}\right)J_0\left(Ir\sqrt{\frac{\alpha J}{K\sigma}}\right) - \frac{1}{\alpha} \quad \text{for} \quad 0 \leq r \leq R. \quad \blacksquare$$

Summary

It was seen in the previous section that when n is an integer $J_n(x)$ and $J_{-n}(x)$ are linearly dependent. This section has shown how a second linearly independent solution $Y_v(x)$ can be constructed that for all v is linearly independent of $J_v(x)$, so the general solution of Bessel's equation can always be written $y(x) = AJ_v(x) + BY_v(x)$, where A and B are arbitrary constants. The function $Y_v(x)$ is called a Bessel function of the second kind of order v.

EXERCISES 8.7

In Exercises 1 through 10, find the general solution of the differential equation.

1. $x^2 y'' + xy' + (x^2 - 4)y = 0$.

2. $4x^2 y'' + 4xy' + (4x^2 - 1)y = 0$.

3. $xy'' + y' + xy = 0$.

4. $xy'' + y' + \lambda^2 xy = 0$.

5. $xy'' + y' + 4x^3 y = 0$; substitute $u = x^2$.

6. $x^2 y'' + 3xy' + (x^2 + 1)y = 0$; substitute $y = u/x$.

7. $x^2 y'' + xy' + 4(x^2 - 1)y = 0$.

8. $xy'' + y' + 9x^5 y = 0$; substitute $u = x^3$.

9. $4x^2 y'' + (16x^2 + 1)y = 0$; substitute $y = x^{1/2}u$.

10. $x^2 y'' + 5xy' + (x^2 + 4)y = 0$; substitute $y = u/x^2$.

Use (93) and (94) to find the solution of the differential equations in Exercises 11 through 15.

11. $x^2 y'' - xy' + (4x^4 - 3)y = 0$.

12. $xy'' - 3y' + xy = 0$.

13. $x^2 y'' - xy' + (9x^2 + 1)y = 0$.

14. $x^2 y'' - 5xy' + (16x^4 + 1)y = 0$.

15. $x^2 y'' - 3xy' + (64x^8 - 8)y = 0$.

16. Verify that $y(x) = x^a Z_\nu(bx^c)$ is a solution of (93) by substituting for $y(x)$ in the differential equation and showing that this leads to the equation

$$X^2 Z''_\nu(X) + XZ'_\nu(X) + (X^2 - \nu^2)Z_\nu(X) = 0,$$

with $X = bx^c$. Hence, conclude that $Z_\nu(X)$ is either $J_\nu(X)$ or $Y_\nu(X)$, and so, because of the linearity of the equation, $Z_\nu(X) = C_1 J_\nu(X) + C_2 Y_\nu(X)$ must be a solution.

17. Use the substitution $y(x) = x^{-\nu}u(x)$ to convert the equation

$$x^2 \frac{d^2 y}{dt^2} + ax\frac{dy}{dx} + (1 + k^2 x^2)y = 0,$$

in which a is a parameter, into an equation for $u(x)$. Find the values of a and ν that make the equation in $u(x)$ Bessel's equation of order zero. Use the result to find the general solution $y(x)$ that corresponds to this value of a.

8.8   Modified Bessel Functions $I_\nu(x)$ and $K_\nu(x)$

Replacing the independent variable x in Bessel's equation by ix changes the differential equation to

$$x^2 y'' + xy' - (x^2 + \nu^2)y = 0, \tag{95}$$

Bessel's modified equation

called **Bessel's modified equation of order** ν.

It follows directly from Section 8.7 that Bessel's modified equation has two linearly independent complex solutions $J_\nu(ix)$ and $Y_\nu(ix)$. These solutions are not convenient to use, so the process of scaling and combining linearly independent solutions of a linear differential equation to form other solutions is used to produce two real linearly independent solutions denoted by $I_\nu(x)$ and $K_\nu(x)$. These are called, respectively, **modified Bessel functions of the first and second kinds of order** ν.

The modification of $J_\nu(ix)$ is straightforward, because from (54)

$$J_\nu(ix) = \sum_{m=0}^{\infty} \frac{(-1)^m (ix)^{2m+\nu}}{2^{2m+\nu} m! \Gamma(m+1+\nu)} = i^\nu \sum_{m=0}^{\infty} \frac{x^{2m+\nu}}{2^{2m+\nu} m! \Gamma(\text{the first kind of } \nu)},$$

so the factor i^ν is removed and the **modified Bessel function** of the first kind of **order** ν is defined as the real function

$$I_\nu(x) = \sum_{m=0}^{\infty} \frac{x^{2m+\nu}}{2^{2m+\nu} m! \Gamma(}\tag{96}$$

Unlike the series for $J_\nu(x)$, the series for $I_\nu(x)$ in (96) is no longer an alternating series, though it converges rapidly. As with ordinary Bessel functions, provided ν is not an integer, the general solution of Bessel's modified equation (95) can be written

$$y(x) = C_1 I_\nu(x) + C_2 I_{-\nu}(x). \tag{97}$$

However, rather than use $I_{-\nu}(x)$, in its place it is usual to introduce the real function $K_\nu(x)$ defined as the linear combination of real functions

$$K_\nu(x) = \left(\frac{\pi}{2}\right)\left(\frac{I_{-\nu}(x) - I_\nu(x)}{\sin \nu\pi}\right), \tag{98}$$

and to call $K_\nu(x)$ the **modified Bessel function of the second kind of order** ν. It can be seen from (98) that the functions $I_\nu(x)$ and $K_\nu(x)$ are linearly independent.

The definition of $K_\nu(x)$ can be extended to the case in which ν is an integer n by defining the function $K_n(x)$ as

$$K_n(x) = \lim_{\nu \to n} \left(\frac{\pi}{2}\right)\left(\frac{I_{-\nu}(x) - I_\nu(x)}{\sin \nu\pi}\right). \tag{99}$$

Because of this extension of the definition of $K_\nu(x)$, the general solution of Bessel's modified equation (95) can always be written in the form

$$y(x) = C_1 I_\nu(x) + C_2 K_\nu(x), \tag{100}$$

with no restriction placed on ν. The function $K_\nu(x)$ is also sometimes called the Kelvin function.

Similarly, when Bessel's modified equation is written in the form

$$x^2 y'' + xy' - (\lambda^2 x^2 + \nu^2)y = 0, \tag{101}$$

its general solution is given by

$$y(x) = C_1 I_\nu(\lambda x) + C_2 K_\nu(\lambda x), \tag{102}$$

with no restriction placed on ν.

This definition of $K_0(x)$ leads to the expansion

$$K_0(x) = -\left[\ln\frac{x}{2} + \gamma\right] I_0(x) + \frac{x^2/4}{(1!)^2} + \left(1 + \frac{1}{2}\right)\frac{(x^2/4)^2}{(2!)^2}$$
$$+ \left(1 + \frac{1}{2} + \frac{1}{3}\right)\frac{(x^2/4)^3}{(3!)^2} + \cdots, \tag{103}$$

with similar though more complicated expansions for $K_n(x)$.

Graphs of $I_0(x)$ and $I_1(x)$ and of $K_0(x)$ and $K_1(x)$ are shown in Figs. 8.8 and 8.9, respectively.

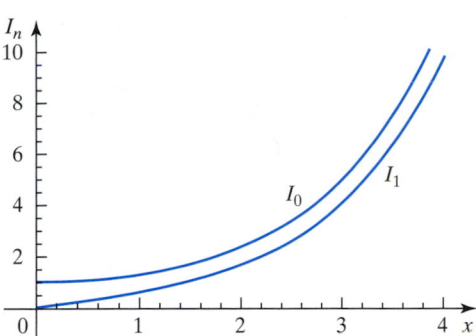

FIGURE 8.8   Graphs of $I_0(x)$ and $I_1(x)$.

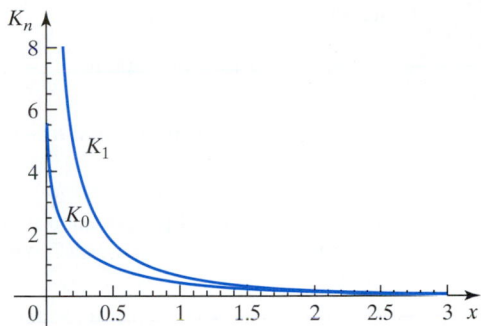

FIGURE 8.9   Graphs of $K_0(x)$ and $K_1(x)$.

The following are useful properties of $I_\nu(x)$ and $K_\nu(x)$:

$$I_0(0) = 1, \quad I_n(0) = 0 \quad \text{for } n = 1, 2, \ldots, \lim_{x \to 0} I_\nu(x) = 0,$$
$$K_n(0) = \infty, \quad \lim_{x \to \infty} K_n(x) = 0 \quad \text{for } n = 0, 1, 2, \ldots. \tag{104}$$

asymptotic expressions for modified Bessel functions

For small x

$$I_\nu(x) \sim \frac{1}{\Gamma(1 + \nu)} \left(\frac{x}{2}\right)^\nu, \quad K_0(x) = -\ln x \quad \text{and}$$
$$K_\nu(x) \sim \frac{\Gamma(\nu)}{2} \left(\frac{2}{x}\right)^\nu \quad \text{for } \nu > 0, \tag{105}$$

whereas for large x

$$I_\nu(x) \approx \frac{1}{\sqrt{2\pi x}} e^x \quad \text{and} \quad K_\nu(x) \approx \sqrt{\frac{\pi}{2x}} e^{-x}. \tag{106}$$

Results involving Bessel functions of the first and second kinds, together with applications, are to be found in Chapter 5 of reference [3.7]. Chapters 9 to 11 of Reference [G.1] and Chapter 17 of reference [G.3] give general information about all types of Bessel functions. The standard encyclopedic work covering all aspects of Bessel functions is reference [3.17].

Summary

Modified Bessel functions were introduced, their series solutions were obtained, the general solution was expressed in terms of $I_\nu(x)$ and $K_\nu(x)$, and asymptotic representations were given.

EXERCISES 8.8

1. By differentiating the series for $I_0(x)$, show that $I_0'(x) = I_1(x)$.

2. Use the definition of $I_\nu(x)$ to show that

$$I_{\nu-1}(x) - I_{\nu+1}(x) = \frac{2\nu}{x} I_\nu(x) \quad \text{for } \nu \geq 1.$$

3. Use the definition of $I_\nu(x)$ to show that

$$I_{\nu-1}(x) + I_{\nu+1}(x) = 2I_\nu'(x) \quad \text{for } \nu \geq 1.$$

4. Use Lemma 6.1 of Section 6.3 to reduce Bessel's modified equation of order $\nu = 1/2$ to standard form, and hence show that

$$I_{1/2}(x) \text{ is proportional to } \frac{\sinh x}{\sqrt{x}}, \quad \text{and}$$

$$I_{-1/2}(x) \text{ is proportional to } \frac{\cosh x}{\sqrt{x}}.$$

5. Use asymptotic result (106) for $I_\nu(x)$ when x is large to find the constants of proportionality in Exercise 4, and then use the result of Exercise 2 to find $I_{3/2}(x)$ and $I_{-3/2}(x)$.

6. Use Lemma 6.1 of Section 6.3 to reduce Bessel's modified equation of order $\nu = 1/2$ to standard form, and hence show that when x is large two linearly independent solutions of the equation are proportional to $e^x/\sqrt{x}$ and $e^{-x}/\sqrt{x}$.

7. Deduce the expressions for $I_{\pm 1/2}(x)$ and $I_{\pm 3/2}(x)$ from the corresponding results for $J_{\pm 1/2}(x)$ and $J_{\pm 3/2}(x)$ in (72) to (74) of Section 8.6.

8. Use Abel's formula in Exercise 6 of set 6.1 to show that if y_1 and y_2 are any two linearly independent solutions of Bessel's modified equation, then

$$y_1 y_2' - y_2 y_1' = C/x,$$

where C is a constant introduced through the Abel formula.

9. Set $y_1(x) = I_\nu(x)$ and $y_2(x) = I_{-\nu}(x)$ in the result of Exercise 8, where ν is not an integer. Substitute the series for $I_\nu(x)$ and $I_{-\nu}(x)$, and by finding the coefficient of $1/x$ on the left-hand side identify the coefficient C. Use the result

$$\Gamma(z)\Gamma(1-z) = \frac{\pi}{\sin \pi z}$$

to show that

$$I_\nu(x) I_{-\nu}'(x) - I_\nu'(x) I_{-\nu}(x) = -\frac{2}{\pi x} \sin \nu x.$$

10. Use the definition of $K_\nu(x)$ with the result of Exercise 9 to show that

$$I_\nu(x) K_\nu'(x) - I_\nu'(x) K_\nu(x) = -\frac{1}{x}.$$

11.* The amplitude $R(r)$ of the small symmetric vibrations of a flexible annular disc $a \le r \le b$ normal to its surface with its outer edge free and its inner edge fixed to a rod that oscillates along its length is governed by the equation

$$\frac{d^4 R}{dr^4} + \frac{2}{r}\frac{d^3 R}{dr^3} - \frac{1}{r^2}\frac{d^2 R}{dr^2} + \frac{1}{r^3}\frac{dR}{dr} - R = 0.$$

Show by expressing the equation as

$$\left(\frac{d^2}{dr^2} + \frac{1}{r}\frac{d}{dr} - 1 \right)\left(\frac{d^2}{dr^2} + \frac{1}{r}\frac{d}{dr} + 1 \right) R = 0$$

that its general solution is

$$R(r) = AJ_0(r) + BY_0(r) + CI_0(r) + DK_0(r),$$

where A, B, C, and D are arbitrary constants.

12.* In partial differential equations that govern physical phenomena with cylindrical and spherical polar coordinates, the following equation describes the radial variation $R(r)$ of the solution as a function of the radius r (see Chapter 18):

$$\frac{d^2 R}{dr^2} + \frac{1}{r}\frac{dR}{dr} + \left(\lambda^2 - \frac{n^2}{r^2} \right) R = 0.$$

Here, λ is a parameter and $n = 0, 1, 2, \ldots$. Show that the general solution of the equation is

$$R(r) = AJ_n(\lambda r) + BY_n(\lambda r).$$

Find the form of the solution of the following boundary value problems, given that $R(r)$ remains bounded, and determine the permissible values of the parameter λ.

(i) $0 \le r \le a$, for all n with the boundary conditions $R(a) = 0$.

(ii) $b \le r \le c$, for all n with the boundary conditions $R(b) = R(c) = 0$.

(iii) $0 \le r \le a$, for all n with the boundary conditions $R(a) + kR'(a) = 0 (k = \text{const})$.

(iv) $b \le r \le c$, for $n = 0$ with the boundary conditions $R(b) = R'(c) = 0$.

8.9 A Critical Bending Problem: Is There a Tallest Flagpole?

The implication of the question posed in the section heading will have been experienced by anyone who has tried holding a long, thin, flexible rod in a vertical position. If the rod is short, and its tip is given a small sideways displacement and released, the rod will perform transverse oscillations until it reaches an equilibrium position in a bent shape because of supporting its own weight. The longer the rod, the larger the amplitude of these oscillations, and the greater the bending under its own weight when in equilibrium, until at some critical length the rod will bend until its tip just touches the ground, after which it will remain in that position.

An idealization of this phenomenon can be modeled by a long, thin, flexible flagpole of uniform cross-section, the base of which is clamped in the ground so the pole is vertical. We then ask at what length will the pole become unstable, so that any displacement of the top of the pole will cause it to bend under its own weight until the top of the pole touches and remains in contact with the ground? This question can be posed in mathematical terms, and it is the one that will be answered here.

The solution to this question will involve the use of Bessel functions, but the linear differential equation involved will have to satisfy a two-point boundary condition instead of the initial conditions we have considered so far. This means that the existence and uniqueness of solutions to initial value problems guaranteed by Theorem 6.2 no longer applies, so even when a solution can be found it may not be unique — more will be said about this later.

Let us model the problem by considering a thin uniform flexible rod of length L with a constant cross-section that is constructed from material with a Young's modulus of elasticity E, with the moment of inertia of a cross-section about a diameter normal to the plane of bending equal to I. The line density along the rod will be assumed to be constant and equal to w. The x-axis will be taken to be vertical and to coincide with the undistorted axis of the rod, with its origin located at the base of the rod. The horizontal displacement of the rod at a position x will be taken to be y, as shown in Fig. 8.10.

It is known from Section 5.2(f) that if the moment acting on the rod at a position x is $M(x)$, the equation governing its transverse deflection y when in equilibrium is

$$EI\frac{d^2y}{dx^2} = M(x). \tag{107}$$

The **shear** on the rod at point x is the force exerted perpendicular to the axis of the rod at x due to the weight of the rod extending from x to the top at P. As the length of this part of the rod is $L - x$, and its line density is w, the weight of this section is

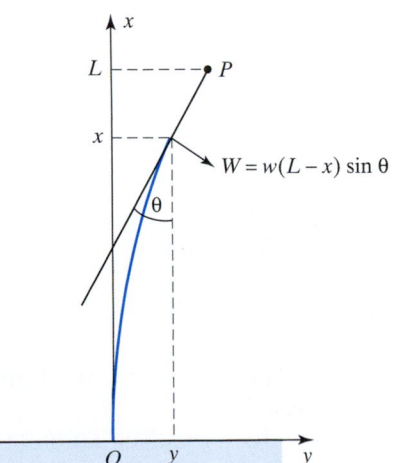

FIGURE 8.10 Equilibrium position of the rod when bent under its own weight.

given by $w(L - x)$, so the component W of this force normal to the axis of the rod at x is simply

$$W = w(L - x)\sin\theta, \tag{108}$$

where θ is the angle of deflection of the rod from the vertical at point x, as shown in Fig. 8.10.

It is known from mechanics that the shear on a rod is given in terms of the moment $M(x)$ by

$$\frac{dM}{dx} = -W(x). \tag{109}$$

We now make the approximation that the deflection at point x on the rod is small, so $\sin\theta \approx \tan\theta = dy/dx$, and by combining (107) to (109) we arrive that the governing equation for the deflection, which is the third order linear variable coefficient differential equation

$$EI\frac{d^3y}{dx^3} + w(L - x)\frac{dy}{dx} = 0. \tag{110}$$

Making the change of variable $z = L - x$ brings (110) to the more convenient form

$$\frac{d^3y}{dz^3} + \left(\frac{w}{EI}\right)z\frac{dy}{dz} = 0. \tag{111}$$

To apply this to our problem it is necessary to determine appropriate boundary conditions to be applied at the base and top of the rod. An obvious condition to be applied at the base is that due to clamping the pole in a vertical position at the origin, $(dy/dx)_{x=0} = (dy/dz)_{z=L} = 0$. To arrive at a second condition we notice that when the rod is bent and in equilibrium, there can be no bending moment at the top of the rod, so it can have no curvature at that point. Recalling that the radius of curvature ρ of a plane curve $y = y(x)$ is

$$\rho = \frac{(1 + (y')^2)^{2/3}}{y''}, \tag{112}$$

we see that the rod will have no curvature at $x = L$ (equivalently at $z = 0$) when $\rho = \infty$, corresponding to $(d^2y/dx^2)_{x=L} = (d^2y/dz^2)_{z=0} = 0$.

Setting $u(z) = dy/dz$, these two boundary conditions become

$$u(L) = 0 \quad \text{and} \quad (du/dz)_{z=0} = 0. \tag{113}$$

Equation (111) is third order, but in terms of $u(z)$ it is only second order, and we have found two conditions on $u(z)$ from which to determine u. Fortunately, we only need to work with $u(z)$ to solve our problem. This is because we will soon see that the two-point boundary conditions (113) applied to the differential equation for u

$$\frac{d^2u}{dz^2} + \left(\frac{w}{EI}\right)zu = 0 \tag{114}$$

will provide sufficient information for us to find the critical length at which bending occurs.

Identifying equation (114) with (93) from Section 8.7, with x replaced by z, shows that

$$1 - 2a = 0, \quad 2c - 2 = 1, \quad a^2 - v^2c^2 = 0, \quad \text{and} \quad b^2c^2 = w/EI, \tag{115}$$

so

$$a = 1/2, \quad c = 3/2, \quad \nu = 1/3, \quad \text{and} \quad b = \frac{2}{3}\sqrt{\frac{w}{EI}}. \tag{116}$$

Using this information in the solution (94) to equation (93) in Section 8.7 gives

$$u(z) = C_1\sqrt{z}\,J_{1/3}\left(\frac{2}{3}\sqrt{\frac{w}{EI}}z^{3/2}\right) + C_2\sqrt{z}\,J_{-1/3}\left(\frac{2}{3}\sqrt{\frac{w}{EI}}z^{3/2}\right). \tag{117}$$

Noticing from (71) of Section 8.6 that for small z

$$J_\nu(z) \approx \frac{1}{\Gamma(1+\nu)}\left(\frac{z}{2}\right)^\nu \quad \text{and} \quad J_{-\nu}(z) \approx \frac{1}{\Gamma(1-\nu)}\left(\frac{z}{2}\right)^{-\nu},$$

we see that close to the top of the rod, that is, for small z, $u(z)$ can be approximated by

$$u(z) \approx C_1\frac{z}{\Gamma(4/3)}\left(\frac{1}{3}\sqrt{\frac{w}{EI}}\right)^{1/3} + C_2\frac{1}{\Gamma(2/3)}\left(\frac{1}{3}\sqrt{\frac{w}{EI}}\right)^{-1/3}.$$

Differentiation of this result gives

$$u'(z) \approx C_1\frac{1}{\Gamma(4/3)}\left(\frac{1}{3}\sqrt{\frac{w}{EI}}\right)^{1/3},$$

but to satisfy the second boundary condition $(du/dz)_{z=0} = 0$, we must set $C_1 = 0$, causing solution (117) to reduce to

$$u(z) = C_2\sqrt{z}\,J_{-1/3}\left(\frac{2}{3}\sqrt{\frac{w}{EI}}z^{3/2}\right). \tag{118}$$

Applying the remaining boundary condition $u(L) = 0$ to (118) gives

$$0 = C_2\sqrt{L}\,J_{-1/3}\left(\frac{2}{3}\sqrt{\frac{w}{EI}}L^{3/2}\right), \tag{119}$$

and this will be satisfied if either $C_2 = 0$ or $J_{-1/3}(\frac{2}{3}\sqrt{\frac{w}{EI}}L^{3/2}) = 0$. The first condition $C_2 = 0$ corresponds to the unstable equilibrium configuration in which the rod is vertical, and so must be rejected, whereas the second condition corresponds to the required critical bending condition, and it will be satisfied when L is such that it causes $J_{-1/3}$ to vanish.

It is at this stage that we discover the boundary value problem does *not* have a unique solution, because the asymptotic behavior of $J_{-1/3}$ given in (70) of Section 8.6 shows that it has infinitely many zeros. To resolve this difficulty, and to find the length at which critical bending occurs, we must now seek a selection criterion for the length from *outside* the description of the physical situation provided by the differential equation.

Such a criterion is not hard to find, because critical bending must occur at the *smallest* value of L, say at L_c, that satisfies the condition

$$J_{-1/3}\left(\frac{2}{3}\sqrt{\frac{w}{EI}}L_c^{3/2}\right) = 0, \tag{120}$$

because if critical bending occurs when $L = L_c$, it will certainly occur at any larger value of L.

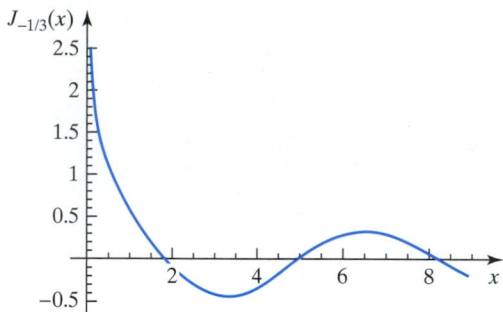

FIGURE 8.11   Graph of $J_{-1/3}(x)$ showing its first few zeros.

A graph of $J_{-1/3}(x)$ is shown in Fig. 8.11, from which it can be seen that the first zero α of $J_{-1/3}(x)$ occurs at around the value $\alpha \approx 1.87$, though numerical calculation provides the more accurate value $\alpha = 1.86635\ldots$. However, this accuracy is unnecessary, because the approximations made when modeling the physical situation introduce errors of sufficient magnitude that the value $\alpha \approx 1.87$ is adequate.

Using the value $\alpha = 1.87$ shows that the length L_c for critical bending must satisfy the formula

$$\frac{2}{3}\sqrt{\frac{w}{EI}}\,L_c^{3/2} \approx 1.87,$$

which is equivalent to

$$L_c \approx 1.99 \left(\frac{EI}{w}\right)^{1/3}.$$

This approximation shows, as would be expected, that if the rod is not cylindrically symmetric about its axis, the critical length L_c will depend on the plane in which bending occurs, because the moment of inertia will depend on the direction in which the rod bends. Thus, for example, the critical length of a rod with a rectangular cross-section that bends in a plane parallel to one pair of its faces will differ from the critical length when bending occurs in a plane parallel to its other pair of faces. In such cases the model used is too simple, because twisting (torsion) will be likely to occur, causing the rod always to buckle in such a way that L_c assumes its smallest possible value.

The simplest case arises when the rod has a circular cross-section of radius a, for then the moment of inertia of the cross-section about any diameter is $I = \pi a^4/4$. When this expression is substituted into the approximation for L_c, we obtain the approximation

$$L_c \approx 1.25 \left(\frac{Ea^4}{w}\right)^{1/3}.$$

Summary

In addition to involving Bessel functions, this idealization of a physical problem has illustrated the way in which a mathematical approach can sometimes lead to more than one solution, only one of which can be regarded as an approximation to the situation in the real world. The choice of the appropriate solution was seen to be based on an additional physical consideration that was outside the original formulation of the mathematical problem. This situation is not unusual in applied mathematics, where the choice of solution is often based on stability considerations, a physically possible solution being stable, whereas a nonphysical solution is unstable and so will not be observed. A different example occurs in the study of shock waves in air where two solutions are mathematically possible, though only one is physically realizable. In that case the selection principle is based on the thermodynamics of the problem, though it can also be based on stability considerations.

8.10   Sturm–Liouville Problems, Eigenfunctions, and Orthogonality

Mathematical models of physical situations arising in engineering and physics lead to two-point boundary value problems for a function $y(x)$ that is defined over an interval $a < x < b$ and satisfies a differential equation of the form

$$y''(x) + P(x)y'(x) + (Q(x) + \lambda R(x))y(x) = 0, \tag{121}$$

in which λ is a parameter. This equation always has the solution $y(x) \equiv 0$, called the **trivial solution**, but if it is to have nontrivial solutions (solutions that are not identically zero) satisfying boundary conditions at $x = a$ and $x = b$, the parameter λ cannot be arbitrary. In what follows our purpose will be to find constant values of λ for which nontrivial solutions exist satisfying given boundary conditions. It will be seen later how these nontrivial solutions can be used to generalize series expansions of arbitrary functions over the interval $a < x < b$ that, along with other uses, are needed in Chapter 18 when solving partial differential equations by the method of separation of variables.

To proceed further we will write (121) in a more convenient form, and to this end we simplify its first two terms using the method developed in Section 5.6 when finding an integrating factor for a linear first order equation. Defining the function $p(x)$ as

$$p(x) = \exp\left[\int P(x)\mathrm{d}x\right],$$

and multiplying (121) by $p(x)$ gives

$$p(x)[y''(x) + P(x)y'(x)] + p(x)(Q(x) + \lambda R(x))y(x) = 0.$$

However,

$$p(x)[y''(x) + P(x)y'(x)] = \frac{d}{dx}\left[p(x)\frac{dy(x)}{dx}\right],$$

so the equation becomes

$$\frac{d}{dx}\left[p(x)\frac{dy(x)}{dx}\right] + p(x)(Q(x) + \lambda R(x))y(x) = 0.$$

Finally, setting $q(x) = p(x)Q(x)$ and $r(x) = p(x)R(x)$ allows equation (121) to be written in the form

$$\frac{d}{dx}\left[p(x)\frac{dy(x)}{dx}\right] + [q(x) + \lambda r(x)]y(x) = 0. \tag{122}$$

In what follows $p(x), q(x), r(x)$, and $p'(x)$ will be assumed to be continuous functions defined on a closed interval $a \leq x \leq b$ on which $p(x) > 0, r(x) > 0$.

Differential equations with these properties and written in this form are called **Sturm–Liouville equations**, and the type of boundary conditions that are to be imposed will be introduced after the following typical examples of these equations.

> **JACQUES CHARLES FRANÇOIS STURM (1803–1855) AND JOSEPH LIOUVILLE (1809–1882)**
> Sturm, who was born in Geneva, Switzerland, was Poisson's successor in the Chair of Mechanics in the Sorbonne. Much of his work was in algebra, where he worked on the determination of intervals on the real line inside each of which was located one real root of a polynomial, though he also worked on the study of heat flow introduced by his contemporary Joseph Fourier. Liouville, a professor at the Collège de France, also studied algebraic problems and, in particular, quadratic forms, though he also made contributions to elliptic functions and to complex analysis. Sturm and Liouville, who were friends, collaborated on the eigenvalue and eigenfunction problems raised by the study of heat flow, and together their work led to what is now called the study of Sturm–Liouville systems.

examples of Sturm–Liouville equations

Simple harmonic motion equation

The differential equation describing undamped simple harmonic oscillations

$$y'' + n^2 y = 0 \tag{123}$$

follows from (122) by setting $p(x) = 1$, $q(x) = 0$, $r(x) = 1$, and $\lambda = n^2$.

The Legendre equation

The Legendre equation encountered in (10) of Section 8.2, usually written

$$(1 - x^2)y'' - 2xy' + \alpha(\alpha + 1)y = 0, \tag{124}$$

follows from (122) by setting $p(x) = 1 - x^2$, $q(x) = 0$, $r(x) = 1$, and $\lambda = \alpha(\alpha + 1)$.

Bessel's equation

When Bessel's equation of order ν is written in its more general form

$$x^2 y'' + xy' + (k^2 x^2 - \nu^2)y = 0, \tag{125}$$

the equation follows from (122) by setting $p(x) = x$, $q(x) = -\nu^2/x$, $r(x) = x$, and $\lambda = k^2$.

The Chebyshev equation

The Chebyshev equation of order ν is

$$(1 - x^2)y'' - xy' + n^2 y = 0, \tag{126}$$

and the equation follows from (122) by setting $p(x) = (1 - x^2)^{1/2}$, $q(x) = 0$, $r(x) = (1 - x^2)^{-1/2}$, and $\lambda = n^2$.

For future reference, Table 8.3 lists $p(x)$, $q(x)$, $r(x)$, and λ for the preceding equations, together with three other named equations that find applications in numerical analysis and elsewhere.

TABLE 8.3 $p(x), q(x), r(x)$ and λ for Some Named Equations

Name	$p(x)$	$q(x)$	$r(x)$	λ
Simple harmonic equation	1	0	1	n^2
Legendre's equation	$1 - x^2$	0	1	$\alpha(\alpha + 1)$
Bessel's equation	x	$-v^2/x$	x	k^2
Bessel's modified equation	x	$-v^2/x$	$-x$	k^2
Laguerre equation	xe^{-x}	0	e^{-x}	n
Chebyshev equation	$(1 - x^2)^{1/2}$	0	$(1 - x^2)^{-1/2}$	n^2
Hermite equation	e^{-x^2}	0	e^{-x^2}	$2n$

When the Sturm–Liouville equation (122) is associated with boundary condi-tions at $x = a$ and $x = b$, the equation itself together with the boundary conditions form what is called a **Sturm–Liouville problem**. The boundary conditions that will concern us here are the **homogeneous** boundary conditions,

$$A_1 y(a) + A_2 y'(a) = 0 \quad \text{and} \quad B_1 y(b) + B_2 y'(b) = 0, \tag{127}$$

where the term *homogeneous* is used in the sense that the linear combinations of $y(x)$ and $y'(x)$ at $x = a$ and $x = b$ are both equal to zero. There are three cate-gories of Sturm–Liouville problems that arise, called **regular**, **periodic**, and **singular** problems according to the nature of the boundary conditions and the behavior of $p(x)$ at the boundaries.

Regular Sturm–Liouville problems

Regular problems are those for which constant values of λ are sought corresponding to each of which a nontrivial solution can be found for the Sturm–Liouville equation

$$(py')' + (q + \lambda r)y = 0,$$

with $p(x) > 0$ continuous on $a \leq x \leq b$ and subject to the boundary conditions

$$A_1 y(a) + A_2 y'(a) = 0 \quad \text{and} \quad B_1 y(b) + B_2 y'(b) = 0,$$

where in neither of the boundary conditions do both constant coefficients vanish.

Periodic Sturm–Liouville problems

This class of problems arises when $p(x)$ and the boundary conditions involving $y(x)$ and $y'(x)$ are periodic over the interval $a \leq x \leq b$. In this case constant values of λ are sought corresponding to each of which a nontrivial solution can be found for the Sturm–Liouville problem

$$(py')' + (q + \lambda r)y = 0,$$

subject to the periodic boundary conditions

$$p(a) = p(b), \quad y(a) = y(b), \quad \text{and} \quad y'(a) = y'(b).$$

Singular Sturm–Liouville problems

In this class of problems constant values of λ are sought, corresponding to each of which a nontrivial solution can be found for the Sturm–Liouville equation

$$(py')' + (q + \lambda r)y = 0,$$

on a finite interval at one or or both ends of which $p(x)$ or $r(x)$ vanish, or on a semi-infinite or infinite interval. The most frequently occurring problem of this type, and the only one to be considered here, is the Sturm–Liouville problem defined on a finite interval $a \leq x \leq b$, where the singular point is located at either $x = a$ or $x = b$, so that either $p(a) = 0$ or $p(b) = 0$. In such cases the boundary condition that is often imposed at the singular point takes the form of the requirement that the solution remains bounded there. Typically, this happens when a bounded solution of Bessel's equation of the form $y(x) = AJ_0(x) + BY_0(x)$ is required over an interval $0 \leq x \leq a$, because then the requirement that the solution remains bounded at the singular point located at $x = 0$ means we must set $B = 0$ to exclude the infinite value of $Y_0(x)$ at $x = 0$.

When dealing with Sturm–Liouville problems, each value of λ for which a nontrivial solution can be found is called an **eigenvalue** of the problem, and the corresponding solution $y(x)$ is called an **eigenfunction** of the problem. Because the Sturm–Liouville equation (122) is homogeneous, it follows that an eigenfunction can be multiplied by any constant factor and still remain an eigenfunction. This simple but fundamental property will be used repeatedly, first when normalizing eigenfunctions and later when representing arbitrary functions defined over an interval $[a, b]$ in terms of series of eigenfunctions, as is done in Chapter 9 when working with Fourier series. Such representations of functions are called *eigenfunction expansions*.

In most practical situations an eigenvalue is associated with an important physical characteristic of the problem, such as the frequency of vibration of a string or of a metal plate. In such cases the eigenfunction can be considered to describe a "snapshot" of a particular mode of vibration of the string or plate when it vibrates at the frequency determined by the associated eigenvalue. This application, and others that lead to Sturm–Liouville problems, will be developed in detail when partial differential equations are discussed in the context of *separation of variables*.

A Regular Problem

EXAMPLE 8.18

Find the eigenvalues and eigenfunctions of the two-point boundary value problem

$$y'' + \lambda y = 0,$$

such that

$$y(0) = 0 \quad \text{and} \quad y'(\pi) = 0.$$

Solution The interval over which the eigenfunctions are defined is $0 \leq x \leq \pi$. We need to consider the three cases $\lambda = 0$, $\lambda < 0$, and $\lambda > 0$. The homogenous boundary conditions in this problem are of the type given in (127) with $A_2 = 0$ and $B_1 = 0$, where the values of the constants A_1 and B_2 are immaterial provided neither is zero.

Case $\lambda = 0$

When $\lambda = 0$ the equation has the general solution

$$y(x) = C_1 x + C_2,$$

so to satisfy the boundary condition $y(0) = 0$ we must have $C_2 = 0$, and to satisfy the boundary condition $y'(\pi) = 0$ we must have $C_1 = 0$, giving rise to the trivial solution $y(x) \equiv 0$. Thus, $\lambda = 0$ is not an eigenvalue of the problem.

Case $\lambda < 0$

If we set $\lambda = -\mu^2$, the general solution becomes

$$y(x) = C_1 e^{\mu x} + C_2 e^{-\mu x},$$

so the imposition of the boundary conditions requires that

$$0 = C_1 + C_2 \quad \text{and} \quad 0 = \mu C_1 e^{\mu \pi} - \mu C_2 e^{-\mu \pi}.$$

After the elimination of C_2, this last result can be written

$$0 = 2\mu C_1 \cosh \mu \pi,$$

but $\mu > 0$, so as $\cosh \mu \pi \neq 0$, this is only possible if $C_1 = 0$, so again we obtain the trivial solution showing that the problem has no negative eigenvalues.

Case $\lambda > 0$

As $\lambda > 0$, it is convenient to set $\lambda = \mu^2$, when the general solution of the equation becomes

$$y(x) = C_1 \cos \mu x + \mu C_2 \sin \mu x.$$

Applying the boundary condition $y(0) = 0$ to the general solution gives $C_1 = 0$, and applying the boundary condition $y'(\pi) = 0$ gives

$$\mu C_2 \cos \mu \pi = 0,$$

so either $C_2 = 0$ or $\cos \mu \pi = 0$. If we take $C_2 = 0$, then as $C_1 = 0$ we obtain the trivial solution, so we must take $C_2 \neq 0$. The condition $\cos \mu \pi = 0$ is satisfied if $\mu \pi$ is one of the zeros of the cosine function given by $\pm \frac{1}{2}(2n + 1)\pi$, for $n = 0, 1, 2, \dots$.

Denoting the permitted values of μ by μ_n we arrive at the condition

$$\mu_n = \pm \frac{1}{2}(2n + 1), \quad \text{with } n = 0, 1, 2, \dots.$$

The eigenvalues of this problem corresponding to the parameter $\lambda = \mu^2$ are thus

$$\lambda_n = \frac{(2n + 1)^2}{4}, \quad \text{with } n = 0, 1, 2, \dots,$$

and the corresponding eigenfunctions are

$$y_n(x) = \sin \frac{(2n + 1)x}{2} \quad \text{with } n = 0, 1, 2, \dots.$$

When writing down the form of the eigenfunction $y_n(x)$, we have set $C_2 = 1$ because, as has already been remarked, an eigenfunction can be multiplied by any constant nonzero factor and still remain an eigenfunction.

This example has shown the existence of an infinite increasing sequence of positive eigenvalues μ_n^2, corresponding to each of which there is a nontrivial solution of the Sturm–Liouville problem, namely the eigenfunction $y_n(x) = \sin \mu_n x$. If $\mu \neq \mu_n$, then the Sturm–Liouville problem only has the *trivial* solution $y(x) \equiv 0$.   ■

A Periodic Problem

Find the eigenvalues and eigenfunctions of the Sturm–Liouville equation

$$y'' + \lambda y = 0$$

subject to the conditions

$$y(0) = y(L), \quad y'(0) = y'(L).$$

Solution   The interval over which the eigenfunctions are defined is $0 \leq x \leq L$, and as in Example 8.18 we must again consider the three cases $\lambda = 0, \lambda < 0$, and $\lambda > 0$.

Case $\lambda = 0$

As in the previous problem, the general solution is

$$y(x) = C_1 x + C_2,$$

so applying the boundary condition $y(0) = y(L)$ leads to the result $C_2 = C_1 L + C_2$, from which it follows that $C_1 = 0$. As $y'(x) = C_1$ the boundary condition $y'(0) = y'(L)$ is automatically satisfied, showing that $y(x) = C_2$, with C_2 any nonzero constant. This shows that in this case $\lambda = 0$ is an eigenvalue, and that $y(x) = C_2$ (C_2 is an arbitrary nonzero constant) is the corresponding eigenfunction.

Case $\lambda < 0$

If we set $\lambda = -\mu^2$, the general solution becomes

$$y(x) = C_1 e^{\mu x} + C_2 e^{-\mu x}.$$

The boundary condition $y(0) = y(L)$ leads to the condition

$$C_1(1 - e^{\mu L}) = C_2(e^{-\mu L} - 1),$$

and the boundary condition $y'(0) = y'(L)$ leads to the condition

$$C_1(1 - e^{\mu L}) = -C_2(e^{-\mu L} - 1).$$

This last condition is only possible if $C_1 = 0$, but then $C_2 = 0$, so we again obtain the trivial solution. Consequently, we conclude that this problem has no negative eigenvalues.

Case $\lambda > 0$

Setting $\lambda = \mu^2$ the general solution of the equation becomes

$$y(x) = C_1 \cos \mu x + C_2 \sin \mu x.$$

The boundary condition $y(0) = y(L)$ leads to the condition

$$C_1(1 - \cos \mu L) = C_2 \sin \mu L,$$

and the boundary condition $y'(0) = y'(L)$ leads to the condition

$$C_2(1 - \cos \mu L) = -C_1 \sin \mu L.$$

Eliminating C_2 between these two equations and simplifying the result gives

$$2C_1(1 - \cos \mu L) = 0.$$

This condition is satisfied if either $C_1 = 0$, or if $\cos \mu L = 1$. If $C_1 = 0$, then $C_2 = 0$, and we obtain the trivial solution, so the only other possibility is that $\cos \mu L = 1$. This last condition will be satisfied if μL is zero or an integer multiple of 2π, so

$$\mu L = \pm 2n\pi \quad \text{for } n = 0, 1, 2, \ldots,$$

or

$$\mu_n = \pm 2n\pi / L \quad \text{for } n = 0, 1, 2, \ldots.$$

As $\lambda = \mu^2$ the eigenvalues are seen to be

$$\lambda_n = 4n^2\pi^2 / L^2, \quad \text{for } n = 0, 1, 2, \ldots.$$

The corresponding eigenfunctions are

$$y_n(x) = C_1 \cos \mu_n x + C_2 \sin \mu_n x,$$

or

$$y_n(x) = C_1 \cos(2n\pi x / L) + C_2 \sin(2n\pi x / L), \quad \text{for } n = 0, 1, 2, \ldots,$$

where not both constants C_1 and C_2 are zero. Because C_1 and C_2 are arbitrary, and both the cosine function and the sine function satisfy the Sturm–Liouville equation and the boundary conditions, by first setting $C_1 = 1$ and $C_2 = 0$ and then $C_1 = 0$ and $C_2 = 1$ it is seen that in this case the *single* eigenvalue $\lambda_n = 4n^2\pi^2 / L^2$ has associated with it the *two* distinct eigenfunctions

$$y_n^{(1)}(x) = \cos(2n\pi x / L) \quad \text{and} \quad y_n^{(2)}(x) = \sin(2n\pi x / L). \qquad \blacksquare$$

The eigenvalues in Sturm–Liouville problems are not always determined as easily as in the previous examples, and this is illustrated by the next example.

EXAMPLE 8.20   Find the eigenvalues and eigenfunctions of the Sturm–Liouville equation

$$y'' + \lambda y = 0,$$

subject to the conditions

$$y(0) - y'(0) = 0, \quad y(1) + y'(1) = 0.$$

Solution   The interval over which the eigenfunctions are defined is $0 \le x \le 1$, and as before we must again consider the cases $\lambda = 0$, $\lambda < 0$, and $\lambda > 0$.

Case $\lambda = 0$

The general solution is

$$y(x) = C_1 x + C_2,$$

so applying the boundary condition $y(0) - y'(0) = 0$ shows that $C_2 - C_1 = 0$, while applying the boundary condition $y(1) + y'(1) = 0$ gives the condition $2C_1 + C_2 = 0$.

The only solution for these equations is $C_1 = C_2 = 0$ corresponding to the trivial solution, so $\lambda = 0$ is not an eigenvalue of the problem.

Case $\lambda < 0$

Setting $\lambda = -\mu^2$ leads to the general solution

$$y(x) = C_1 e^{\mu x} + C_2 e^{-\mu x}.$$

Applying the boundary condition $y(0) - y'(0) = 0$ leads to the condition

$$C_1(1 - \mu) + C_2(1 + \mu) = 0,$$

and applying the boundary condition $y(1) + y'(1) = 0$ leads to the condition

$$C_1[(1 + \mu)e^{\mu} + C_2(1 - \mu)e^{-\mu}] = 0.$$

As a factor $\mu - 1$ appears, we must consider the cases $\mu = 1$ and $\mu \neq 1$ separately. If $\mu = 1$, the first equation gives $C_2 = 0$, and the second one gives $C_1 = 0$, corresponding to the trivial solution. So $\mu = 1$ is not an eigenvalue. If $\mu \neq 1$, eliminating C_2 between these two equations leads to the condition

$$C_1[(1 + \mu)^2 e^{\mu} - (1 - \mu)^2 e^{-\mu}] = 0.$$

As $\mu > 0$, $(\mu + 1)^2 e^{\mu} > (\mu - 1)^2 e^{-\mu}$, showing that the bracketed term is non-vanishing, from which we conclude that $C_1 = 0$, and so $C_2 = 0$, corresponding to the trivial solution. Thus, this Sturm–Liouville problem has no negative eigenvalues.

Case $\lambda > 0$

Setting $\lambda = \mu^2$ leads to the general solution

$$y(x) = C_1 \cos \mu x + C_2 \sin \mu x.$$

Applying the boundary condition $y(0) - y'(L) = 0$ shows that

$$C_1 - \mu C_2 = 0,$$

and applying the boundary condition $y(1) + y'(1) = 0$ gives

$$C_1 \cos \mu + C_2 \sin \mu - \mu C_1 \sin \mu + \mu C_2 \cos \mu = 0.$$

Eliminating C_1 between these two equations, we obtain

$$C_2[2\mu \cos \mu + (1 - \mu^2) \sin \mu] = 0.$$

The constant C_2 cannot be zero, because then $C_1 = 0$, corresponding to the trivial solution, so μ must be a solution of the equation

$$2\mu \cos \mu + (1 - \mu^2) \sin \mu = 0$$

or, equivalently, μ_n is a solution of the transcendental equation

$$\tan \mu_n = \frac{2\mu_n}{\mu_n^2 - 1}.$$

This equation can only be solved numerically, but approximate solutions can be found graphically. Figure 8.12(a) shows graphs of $y = \tan \mu$ and $y = 2\mu/(\mu^2 - 1)$, and the required solutions μ_n are the values of μ at which the graphs intersect. It has been shown that $\mu = 1$ is not an eigenvalue, so the permissible values of μ_n are all greater than 1. The vertical lines to the right of $x = 1$ are the asymptotes to the

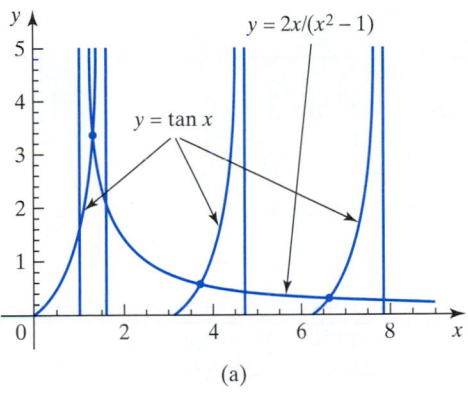
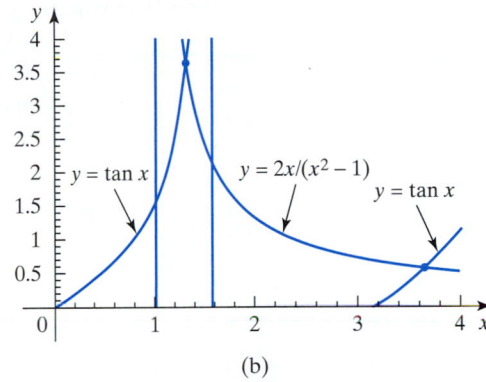

(a)   (b)

FIGURE 8.12   The roots of $\tan \mu = 2\mu/(\mu^2 - 1)$.

tangent function, and the vertical line at $x = 1$ is the asymptote to $2x/(x^2 - 1)$, to the right of which must lie all the solutions μ_n. The graph in Fig. 8.12b, drawn on a larger scale, shows that the first two values of μ are approximately $\mu_1 = 1.3$ and $\mu_2 = 3.7$. A numerical calculation using Newton's method described in Chapter 19 gives the better approximations $\mu_1 = 1.30654$ and $\mu = 3.67319$. It can be seen from Fig. 8.12a that when n is large $\mu_n \approx n\pi$. ■

A Singular Problem

EXAMPLE 8.21   Find the eigenvalues and eigenfunctions of Bessel's equation

$$x^2 y'' + xy' + (k^2 x^2 - n^2) y = 0$$

on the interval $0 \leq x \leq a$ on which the solution is bounded with $y(a) = 0$.

Solution   This is a singular Sturm–Liouville problem, because when Bessel's equation is written in the Sturm–Liouville form

$$\frac{d}{dx}\left[x \frac{dy}{dx} \right] + \left(k^2 x^2 - \frac{n^2}{x} \right) y = 0,$$

with $p(x) = x$, $q(x) = -n^2/x$, $r(x) = x$, and $\lambda = k^2$ (see Table 8.3), it is seen that $p(0) = 0$.

The general solution is

$$y(x) = C_1 J_n(kx) + C_2 Y_n(kx),$$

but $Y_n(kx)$ is infinite when $x = 0$, so for the solution to remain finite over the interval $0 \leq x \leq a$ we must set $C_2 = 0$.

The solution now reduces to

$$y(x) = C_1 J_n(kx),$$

so if the boundary condition $y(a) = 0$ is to be satisfied we must set

$$J_n(ka) = 0.$$

This condition will be satisfied if ka is one of the zeros of $J_n(x)$. If we denote the zeros of $J_n(x)$ by $j_{n,r}$, with $r = 1, 2, \ldots$, it follows that k must be such that it assumes one of the values

$$k_n = j_{n,r}/a, \quad \text{with } r = 1, 2, \ldots.$$

Thus, the eigenvalues $\lambda_n = k_n^2$ are given by

$$\lambda_n = j_{n,r}^2/a^2,$$

and the corresponding eigenfunctions are

$$y_r(x) = J_n(j_{n,r}x/a), \quad \text{with } r = 1, 2, \ldots,$$

where for convenience we have set $C_1 = 1$. Table 8.1 lists the first six zeros of $J_n(x)$ for $n = 0, 1, 2, 3$. Thus if, for example, we consider the case $n = 0$, the corresponding zeros are seen to be $j_{0,1} = 2.4048$, $j_{0,2} = 5.5201\ldots$, so the eigenvalues are $\lambda_1 = 5.7832/a^2$, $\lambda_2 = 30.4711/a^2, \ldots$, and the corresponding eigenfunctions are

$$y_1(x) = J_0(2.4048x/a), \quad y_2(x) = J_0(5.5201x/a), \ldots. \quad \blacksquare$$

Orthogonal and Orthonormal Systems of Functions

orthogonal and orthonormal systems

When working with eigenfunctions it is useful to introduce the notions of **orthogonal** and **orthonormal** systems of eigenfunctions that are defined as follows.

Let $\varphi_1(x)$, $\varphi_2(x)$, ... be an infinite sequence of functions defined over the interval $a \le x \le b$ on which a function $r(x) \ge 0$ is defined. Then the functions are said to be **orthogonal** with respect to the **weight function** $r(x)$ if

$$\int_a^b r(x)\varphi_m(x)\varphi_n(x)dx = 0 \quad \text{for } m \ne n.$$

Clearly, the integral $\int_a^b r(x)\varphi_m(x)\varphi_n(x)dx > 0$ when $m = n$, so we can define a number $\|\varphi_n(x)\|$, called the **norm** of $\varphi_n(x)$, where the square of the norm is defined as

$$\|\varphi_n(x)\|^2 = \int_a^b r(x)\varphi_n^2(x)dx.$$

Using this definition of the norm it is easy to see that the sequence of normalized functions $\hat{\varphi}_1(x) = \varphi_1(x)/\|\varphi_1(x)\|$, $\hat{\varphi}_2(x) = \varphi_2(x)/\|\varphi_2(x)\|, \ldots$ has the property that

$$\int_a^b \hat{\varphi}_m(x)\hat{\varphi}_n(x)r(x)dx = 0, \quad \text{for } m \ne n$$

and

$$\int_a^b \hat{\varphi}_m(x)\hat{\varphi}_n(x)r(x)dx = 1, \quad \text{for } m = n.$$

The sequence of functions $\hat{\varphi}_1(x)$, $\hat{\varphi}_2(x), \ldots$ derived from the sequence of orthogonal functions $\varphi_1(x)$, $\varphi_2(x), \ldots$ by normalization is said to form an **orthonormal** sequence of functions.

In what follows the orthogonality of eigenfunctions will be used extensively, but for the moment it will be sufficient to give a single elementary example of an orthogonal sequence of functions.

EXAMPLE 8.22

Show that the sequence of functions

$$1, \ \cos x, \ \sin x, \ \cos 2x, \ \sin 2x, \ \cos 3x, \ \sin 3x, \ldots$$

is orthogonal over the interval $-\pi \le x \le \pi$ with respect to the weight function $r(x) = 1$, and use it to construct an orthonormal sequence.

Solution   The functions in this sequence occur in the Fourier series representation of an arbitrary function $f(x)$ defined over the interval $-\pi \le x \le \pi$ that is discussed in Chapter 9. Routine calculation shows that for $m \ne n$,

$$\int_{-\pi}^{\pi} \sin mx \sin nx \, dx = 0, \quad \int_{-\pi}^{\pi} \cos mx \cos nx \, dx = 0, \quad \int_{-\pi}^{\pi} \sin mx \cos nx \, dx = 0,$$

and

$$\int_{-\pi}^{\pi} 1 dx = 2\pi, \quad \int_{-\pi}^{\pi} \sin^2 nx \, dx = \int_{-\pi}^{\pi} \cos^2 nx \, dx = \pi, \quad n = 1, 2, \ldots,$$

while $\int_{-\pi}^{\pi} 1 \cdot \cos mx \, dx = \int_{-\pi}^{\pi} 1 \cdot \sin mx \, dx = 0$.
So the functions

$$1, \ \cos x, \ \sin x, \ \cos 2x, \ \sin 2x, \ \cos 3x, \ \sin 3x, \ldots$$

are orthogonal over the interval $-\pi \le x \le \pi$ with respect to the weight function $r(x) = 1$. The respective norms are $\|1\| = \sqrt{2\pi}$ and $\|\sin nx\| = \|\cos nx\| = \sqrt{\pi}$, so the sequence of functions

$$1/\sqrt{2\pi}, \quad (\sin nx)/\sqrt{\pi}, \quad (\cos nx)/\sqrt{\pi}, \quad \text{with } n = 1, 2, \ldots,$$

forms an orthonormal sequence.   ∎

Fundamental Properties of Eigenvalues

The theorem that follows lists the most important properties of the eigenvalues and eigenfunctions of Sturm–Liouville problems. Apart from the important Rayleigh quotient that occurs in Theorem 8.3 (5), the other properties are all qualitative and their main use is to provide general information about eigenvalues that is often of considerable value when working with physical problems.

For convenience, the proofs of all results in Theorem 8.3 that can be established in a straightforward manner have been included in an appendix at the end of this chapter. The proofs of the other results can be found in the references listed at the end of the chapter. A reader who does not require the proofs that are given here may omit them, though the properties themselves should be understood.

THEOREM 8.3

A Sturm–Liouville theorem

1. Regular and periodic Sturm–Liouville problems have an infinite number of distinct real eigenvalues $\lambda_1, \lambda_2, \ldots$, that can be arranged in order so that

important properties of eigenvalues

$$\lambda_1 < \lambda_2 < \lambda_3 < \ldots,$$

where the smallest eigenvalue λ_1 is finite, and

$$\lim_{n\to\infty} \lambda_n = \infty.$$

2. To each eigenvalue of a regular Sturm–Liouville problem there corresponds only one eigenfunction that is unique apart from an arbitrary multiplicative constant.

3. Let the eigenfunctions of a Sturm–Liouville problem on an interval $a \le x \le b$ with weight function $r(x)$ be denoted by $\varphi_1, \varphi_2, \ldots$, with the corresponding eigenvalues $\lambda_1, \lambda_2, \ldots$. Then, if φ_m and φ_n are eigenfunctions corresponding to two distinct eigenvalues λ_m and λ_n ($\lambda_m \ne \lambda_n$ for $m \ne n$), the functions are orthogonal with respect to the weight function $r(x)$, so

$$\int_a^b r(x)\varphi_m(x)\varphi_n(x)dx = 0.$$

4. All the eigenvalues of a Sturm–Liouville problem are real.

5. Let λ_n be an eigenvalue of a regular Sturm–Liouville problem, with φ_n its associated eigenfunction defined on an interval $a \le x \le b$. Then λ_n is given in terms of the Sturm–Liouville functions p, q, r, and the boundary conditions by the **Rayleigh quotient**

$$\lambda_n = \frac{-[p\varphi_n\varphi_n']_a^b + \int_a^b p(\varphi_n')^2 dx - \int_a^b q\varphi_n^2\, dx}{\int_a^b r\varphi_n^2\, dx}.$$

6. Let λ_n be an eigenvalue and φ_n be the corresponding eigenfunction of a regular Sturm–Liouville problem defined on $a \le x \le b$. Then if $q(x) < 0$ and $[p(x)\varphi_n\varphi_n']_a^b \le 0$, all the eigenvalues are nonnegative.

7. The nth eigenfunction of a regular Sturm–Liouville problem defined on the interval $a \le x \le b$ has exactly $n-1$ zeros lying strictly inside the interval.

8. Let two regular Sturm–Liouville problems defined on an interval $a \le x \le b$ be such that $[p(x)\varphi_n\varphi_n']_a^b = 0$ and differ only in their coefficients $p(x)$. Furthermore, let the problem with the coefficient $p_1(x)$ have the eigenvalues $\lambda_1^{(1)}, \lambda_2^{(1)}, \ldots$, and the problem with the coefficient $p_2(x)$ have the eigenvalues $\lambda_1^{(2)}, \lambda_2^{(2)}, \ldots$. Then, if $p_1(x) > p_2(x)$,

$$\lambda_n^{(1)} > \lambda_n^{(2)} \quad \text{for } n = 1, 2, \ldots.$$

9. Let a regular Sturm–Liouville equation with $q(x) < 0$ be defined on an interval $a \le x \le b$ and have boundary conditions such that the first term in the numerator of the Rayleigh quotient in Property 5 is zero. Then reducing the length of the interval $a \le x \le b$ will not reduce the value of any eigenvalue. ■

Remarks about Theorem 8.3

Property 1 ensures that the eigenvalues are distinct ($\lambda_m \ne \lambda_n$ if $m \ne n$), that they are infinite in number, and, because $\lim_{n\to\infty} \lambda_n = \infty$, that there can be no clustering of eigenvalues about a finite limit point. If, for example, the eigenvalues represent the frequencies of vibration of a stretched string of finite length L, this means there is a lowest frequency of vibration, but no upper limit to the frequency of vibration of the string.

Property 2 says that to each distinct eigenvalue of a regular Sturm–Liouville problem there corresponds only one eigenfunction, and it is unique apart from a constant multiplicative factor. Notice that this only applies to *regular* Sturm–Liouville problems, because in periodic Sturm–Liouville problems an eigenvalue has associated with it two linearly independent eigenfunctions. This latter situation occurred in Example 8.19, where the *two* eigenfunctions

$$y_n^{(1)}(x) = \cos(2n\pi x/L) \quad \text{and} \quad y_n^{(2)}(x) = \sin(2n\pi x/L)$$

were seen to correspond to the *single* eigenvalue $\lambda_n = 4n^2\pi^2/L^2$. In such cases there can only be two eigenfunctions to each eigenvalue, because the equation is second order. The scaling of eigenfunctions by a constant is used repeatedly when representing arbitrary functions in terms of series of eigenfunctions.

Property 3 is of fundamental importance because of the part played by orthogonality when developing arbitrary functions in terms of series of eigenfunctions defined over some interval. It is the orthogonality of sine and cosine functions illustrated in Example 8.22 that is used when working with Fourier series.

It will be seen later that the representation (*expansion*) of arbitrary functions in terms of series of eigenfunctions is more general than in terms of power series. This is because, unlike Taylor series whose coefficients are determined by repeated differentiation of the function being expanded, the coefficients in series of eigenfunctions are determined in terms of integrals involving the function. This means that the function can have finite discontinuities at points within its interval of representation and still have an eigenfunction expansion.

Property 4 removes the necessity to check Sturm–Liouville problems for the possibility that negative eigenvalues occur. Had this property been known in advance of Examples 8.18 to 8.21, it would have been unnecessary to have examined the forms of solution corresponding to $\lambda < 0$.

Property 5 is useful when seeking qualitative properties of eigenvalues. The result is not directly useful when trying to determine an eigenvalue because the associated eigenfunction needs to be known. The main use of the Rayleigh quotient arises when it is used in the following rather different form.

Let a function $\Phi(x)$ containing some arbitrary constants $\alpha, \beta, \ldots$ satisfy the *boundary conditions* of a Sturm–Liouville problem. Then with any choice of the arbitrary constants, the Rayleigh quotient

$$\frac{-[p\Phi_n\Phi_n']_a^b + \int_a^b p(\Phi_n')^2 \, dx - \int_a^b q\Phi_n^2 \, dx}{\int_a^b r\Phi_n^2 \, dx} \tag{128}$$

provides an *upper bound* for the value of the smallest eigenvalue of the associated Sturm–Liouville problem. If the arbitrary constants $\alpha, \beta, \ldots$ are chosen to *minimize* this expression, its value becomes the best estimate of the smallest eigenvalue that can be obtained using that approximation. Furthermore, substituting the values of the constants that minimize the Rayleigh quotient into the function $\Phi(x)$ provides a corresponding approximation to the first eigenfunction. The actual value λ_1 is only attained when $\Phi(x) = \varphi_1(x)$.

Property 6, together with Property 4, ensures that under the given conditions the eigenvalues are both real and positive. In corresponding physical problems this result is usually to be expected on an intuitive basis, so the result provides the mathematical justification for making such an assumption on purely physical grounds.

Property 7 provides precise information about the number of zeros of a given eigenfunction within the interval over which it is defined. It is well illustrated by considering Figs. 8.1 showing graphs of Legendre polynomials. These show, for example, that $P_3(x)$ has precisely three zeros in the interval $-1 \leq x \leq 1$, whereas $P_4(x)$ has four zeros. It is important to recognize that these zeros lie strictly *inside* the interval, so that zeros that occur at either end of an interval are *not* counted.

Property 8 means that if in a Sturm–Liouville problem $p(x)$ is associated with a characteristic feature of a physical system, then increasing $p(x)$ increases each eigenvalue of the system. For example, if $p(x)$ is related to the density of a vibrating string, then *increasing* the density while keeping all other parameters constant will *decrease* the frequency of vibration, and increasing the tension will increase the frequency.

Property 9 means that reducing the length of the interval $a \leq x \leq b$ on which a Sturm–Liouville problem is set cannot reduce the values of the eigenvalues. In fact, it usually increases them. This is most easily understood in terms of a vibrating string for which the eigenvalues of the associated Sturm–Liouville problem represent its possible frequencies of vibration (see Chapter 18). In such a case *shortening* the string, while leaving other parameters unchanged, will *increase* the frequency, as any guitarist or violinist knows from experience.

EXAMPLE 8.23

orthogonality and weight functions

An orthogonal system of sine functions   The Sturm–Liouville problem considered in Example 8.18, namely

$$y'' + \lambda y = 0 \quad \text{with } y(0) = 0 \quad \text{and} \quad y'(\pi) = 0,$$

is such that $p(x) = 1$, $q(x) = 0$, and $r(x) = 1$. Its eigenvalues were shown to be $\lambda_n = (2n+1)^2/4$, and its corresponding eigenfunctions were

$$\varphi_n(x) = \sin \frac{(2n+1)x}{2}, \quad n = 0, 1, \ldots.$$

Thus, from Theorem 8.3 (3), the functions $\varphi_n(x)$ are orthogonal over the interval $0 \leq x \leq \pi$ with weight function $r(x) = 1$, and so

$$\int_0^\pi \varphi_m(x)\varphi_n(x)dx = 0 \quad \text{for } m \neq n.$$

The square of the norm is given by

$$\|\varphi_n(x)\|^2 = \left\| \sin \frac{(2n+1)x}{2} \right\|^2 = \int_0^\pi \left(\sin \frac{(2n+1)x}{2} \right)^2 dx = \frac{\pi}{2},$$

so $\|\varphi_n(x)\| = \sqrt{\pi/2}$.   ■

EXAMPLE 8.24

Orthogonality of Legendre polynomials   When written in Sturm–Liouville form, Legendre's equation becomes

$$[(1 - x^2)y']' + \lambda y = 0,$$

and it is defined over the interval $-1 \leq x \leq 1$, with $p(x) = 1 - x^2$, $q(x) = 0$, and $r(x) = 1$. The Legendre polynomial $P_n(x)$ corresponds to $\lambda = n(n+1)$, so from

Theorem 8.3 (3) we see that the Legendre polynomials are orthogonal with respect to the weight function $r(x) = 1$, so that

$$\int_{-1}^{1} P_m(x) P_n(x) dx = 0 \quad \text{for } m \neq n.$$

To determine the norm $\|P_n(x)\|$ we make use of recurrence relation (16),

$$(n+1) P_{n+1}(x) - (2n+1)x P_n(x) + n P_{n-1}(x) = 0.$$

Replacing n by $n-1$ and substituting for one of the factors $P_n(x)$ in the integral gives

$$\|P_n(x)\|^2 = \int_{-1}^{1} P_n(x) \left\{ \left(\frac{2n-1}{n} \right) x P_{n-1}(x) - \frac{n-1}{n} P_{n-2}(x) \right\} dx$$

$$= \left(\frac{2n-1}{n} \right) \int_{-1}^{1} x P_{n-1}(x) P_n(x) dx - \left(\frac{n-1}{n} \right) \int_{-1}^{1} P_n(x) P_{n-2}(x) dx$$

$$= \left(\frac{2n-1}{n} \right) \int_{-1}^{1} x P_{n-1}(x) P_n(x) dx,$$

where the second integral has been set equal to zero because of the orthogonality of $P_n(x)$ and $P_{n-2}(x)$. Using the recurrence relation to remove the term $x P_n(x)$ gives

$$\|P_n(x)\|^2 = \left(\frac{2n-1}{n} \right) \int_{-1}^{1} P_{n-1}(x) \left\{ \left(\frac{n+1}{2n+1} \right) P_{n+1}(x) + \left(\frac{n}{2n+1} \right) P_{n-1}(x) \right\} dx$$

$$= \left(\frac{2n-1}{2n+1} \right) \int_{-1}^{1} [P_{n-1}(x)]^2 dx,$$

where the first integral vanishes because of the orthogonality of $P_n(x)$ and $P_{n-1}(x)$.

This has established the recurrence relation for norms

$$\|P_n(x)\|^2 = \left(\frac{2n-1}{2n+1} \right) \|P_{n-1}(x)\|^2.$$

Using this result to relate $\|P_n(x)\|^2$ to $\|P_0(x)\|^2$ and cancelling factors shows that

$$\|P_n(x)\|^2 = \left(\frac{2n-1}{2n+1} \right) \left(\frac{2n-3}{2n-1} \right) \left(\frac{2n-5}{2n-3} \right) \cdots \left(\frac{3}{5} \right) \left(\frac{1}{3} \right) \|P_0(x)\|^2$$

$$= \left(\frac{1}{2n+1} \right) \|P_0(x)\|^2,$$

but $\|P_0(x)\|^2 = \int_{-1}^{1} 1 dx = 2$, so that

$$\|P_n(x)\|^2 = \frac{2}{2n+1}, \quad \text{and} \quad \|P_n(x)\| = \sqrt{\frac{2}{2n+1}} \quad \text{for } n = 0, 1, \ldots. \quad \blacksquare$$

EXAMPLE 8.25

Orthogonality of Bessel functions $J_n(x)$ When written in Sturm–Liouville form, Bessel's equation of order n becomes

$$[x J_n'(kx)]' + \left(k^2 x - \frac{n^2}{x} \right) J_n(kx) = 0,$$

where $p(x) = x$, $q(x) = -n^2/x$, $r(x) = x$, and $\lambda = k^2$.

The orthogonality of Bessel functions over an interval $0 \leq x \leq a$ takes a somewhat different form from that in the previous examples, because the orthogonality is between Bessel functions of the *same* order, but with *different* arguments, rather than between Bessel functions of different orders. If for fixed n the solution $J_n(kx)$ is required to satisfy the boundary condition

$$J_n(ka) = 0,$$

it follows, as in Example 8.21, that the permissible values of k are

$$k_r = j_{n,r}/a, \quad \text{with } r = 1, 2, \ldots,$$

where $j_{n,r}$ is the rth zero of $J_n(x)$, the first few of which are listed in Table 8.1.

Theorem 8.3 (3) then asserts that as the weight function $r(x) = x$, the orthogonality condition is

$$\int_0^a x J_n\left(\frac{j_{n,r}x}{a}\right) J_n\left(\frac{j_{n,s}x}{a}\right) dx = 0 \quad \text{for } r \neq s.$$

The square of the norm of $J_n(\frac{j_{n,r}x}{a})$ is

$$\left\| J_n\left(\frac{j_{n,r}x}{a}\right) \right\|^2 = \int_0^a x \left[J_n\left(\frac{j_{n,r}x}{a}\right) \right]^2 dx = \frac{a^2}{2}[J_{n+1}(j_{n,r})]^2.$$

A proof of this last result is given in Appendix 2 at the end of the chapter.   ∎

EXAMPLE 8.26

Orthogonality of Chebyshev polynomials   When written in Sturm–Liouville form, the Chebyshev equation for the polynomial $T_n(x)$ of degree n becomes

$$[(1 - x^2)^{1/2} y']' + n^2(1 - x^2)^{-1/2} y = 0.$$

As the weight function is $(1 - x^2)^{-1/2}$, the orthogonality relation becomes

$$\int_{-1}^1 \frac{T_m(x) T_n(x)}{\sqrt{1 - x^2}} dx = 0 \quad \text{for } m \neq n.$$

The square of the norm of $T_n(x)$ is given by

$$\| T_n(x) \|^2 = \int_{-1}^1 \frac{[T_n(x)]^2}{\sqrt{1 - x^2}} dx$$

where $\| T_0(x) \|^2 = \pi$ and $\| T_n(x) \|^2 = \pi/2$ for $n = 1, 2, \ldots$. As it is inappropriate to include the proof of this result here, an outline proof is given in Exercise 31 at the end of the section.

Accounts of Sturm–Liouville systems are to be found in references [3.3] and [3.4] and in Chapter 5 of reference [3.7].   ∎

Summary

The important idea of Sturm–Liouville systems was introduced, their relationship to eigenvalues and eigenfunctions was explained, and it was shown that the solutions of such systems comprise a system of functions that are orthogonal with respect to a suitable weight function. The examples of Sturm–Liouville systems that were given included trigonometric, Legendre, Chebyshev, and Bessel functions. Infinite sets of functions like these represent generalizations to an infinite dimensional space of the elementary notion of the orthogonality of vectors in the three-dimensional Euclidean space. The significance of the orthogonality of eigenfunctions will become clear later when arbitrary functions are expanded in terms of eigenfunctions.

EXERCISES 8.10

In Exercises 1 through 4, reduce the differential equation to Sturm–Liouville form by the method used when reducing equation (121) to the form in (122).

1. $xy'' + (1-x)y' + \lambda y = 0$.

2. $y'' - 2xy' + \lambda y = 0$.

3. $(1-x^2)y'' - xy' + \lambda y = 0$.

4. $(1-x^2)^2 y'' - 2x(1-x^2)y' + [\lambda(1-x^2) - m^2]y = 0$.

In Equations 5 through 14 find the eigenvalues and eigenfunctions of the differential equation.

5. $y'' + \lambda y = 0$, $y(0) = 0$, $y(L) = 0$.

6. $y'' + \lambda y = 0$, $y'(0) = 0$, $y'(L) = 0$.

7. $y'' + \lambda y = 0$, $y'(0) = 0$, $y(1) = 0$.

8. $y'' + \lambda y = 0$, $y(0) = 0$, $y'(2\pi) = 0$.

9. $y'' + \lambda y = 0$, $y(0) = 0$, $y'(1) - 2y(1) = 0$. Find numerical estimates for the first two eigenvalues.

10. $y'' + \lambda y = 0$, $y(0) = 0$, $y'(1) + y(1) = 0$. Find numerical estimates for the first two eigenvalues.

11. $y'' + \lambda y = 0$, $y(-1) = y(1)$, $y'(-1) = y'(1)$.

12. $y'' + \lambda y = 0$, $y(0) = y(1)$, $y'(0) = y'(1)$.

13. $x^2 y'' + xy' + k^2 y = 0$, $y(1) = 0$, $y(4) = 0$.
(Hint: This is a Cauchy–Euler equation)

14. $x^2 y'' + xy' + 9k^2 y = 0$, $y(1) = 0$, $y'(2) = 0$.
(Hint: This is a Cauchy–Euler equation)

In Exercises 15 through 18, verify that the sets of functions are orthogonal over their stated intervals with the weight function $r(x) = 1$, and find their norms.

15. $\varphi_n(x) = \sin\left(\dfrac{n\pi x}{L}\right)$, $n = 1, 2, \ldots$ $(0 \le x \le L)$.

16. $\varphi_n(x) = \cos\left(\dfrac{(2n-1)\pi x}{2}\right)$, $n = 1, 2, \ldots$ $(0 \le x \le 1)$.

17. $\varphi_n(x) = \cos\left(\dfrac{n\pi x}{L}\right)$, $n = 1, 2, \ldots$ $(0 \le x \le L)$.

18. $\varphi_n(x) = \sin\left(\dfrac{(2n-1)\pi x}{4}\right)$, $n = 1, 2, \ldots$ $(0 \le x \le 2\pi)$.

19.* It is known from Example 8.18 that the Sturm–Liouville problem

$$y'' + \lambda y = 0 \quad \text{with } y(0) = 0, y'(\pi) = 0$$

has for its first eigenvalue $\lambda_1 = 1/4$, and that the corresponding eigenfunction is $\varphi_1(x) = \sin x/2$. Verify that the function $\Phi(x) = x(2\pi - x)$ satisfies the boundary conditions for y. By using this expression in the form of the Rayleigh quotient given in (128), find the corresponding upper bound for λ_1 and compare it with the exact value. Why is it that replacing $\Phi(x)$ by

$\Phi(x) = Cx(2\pi - x)$, where C is any nonzero constant, leaves the estimate of the upper bound unchanged?

20.* Perform the calculation required in Exercise 19 using the function $\Phi(x) = x^2(1 - \frac{2x}{3\pi})$, after first showing that $\Phi(x)$ satisfies the boundary conditions. Compare the value of the upper bound so obtained with the exact value $\lambda_1 = 1/4$. Suggest a reason why this approximation is not likely to yield a better lower bound than the one obtained using the function $\Phi(x)$ in Exercise 19.

21.* The Sturm–Liouville form of Bessel's equation of order 1 is

$$[xy']' + \left(k^2 x - \frac{1}{x}\right)y = 0,$$

where $p(x) = x, q(x) = -1/x, r(x) = x$, and $\lambda = k^2$. The bounded solution of this equation on the interval $0 \le x \le 1$ subject to the condition $y(1) = 0$ is $y(x) = J_1(j_{1,1}x)$, where from Table 8.1 $j_{1,1} = 3.8317$ is the first zero of $J_1(x)$. The inverted parabola $\Phi(x) = x(1-x)$ provides a reasonable approximation to the shape of the required Bessel function for $0 \le x \le 1$. Use this expression in (128) to obtain an upper bound for the first eigenvalue λ_1 of the equation, and using the fact that $\lambda_1 = j_{1,1}^2$ find an upper bound for $j_{1,1}$. Compare this estimate with the correct result.

22.* The Sturm–Liouville form of Bessel's equation of order 2 is

$$[xy']' + \left(k^2 x - \frac{4}{x}\right)y = 0,$$

where $p(x) = x$, $q(x) = -4/x$, $r(x) = x$, and $\lambda = k^2$. The solution of this equation that is bounded on the interval $0 \le x \le 1$ and subject to the condition $y(1) = 0$ is $y(x) = J_2(j_{2,1}x)$, where from Table 8.1 $j_{2,1} = 5.1316$ is the first zero of $J_2(x)$. Use the approximation $\Phi(x) = x(1-x)$ to obtain an upper bound for the first eigenvalue of the equation, and using the fact that $\lambda_1 = j_{2,1}^2$, find an upper bound for $j_{2,1}$. Compare this estimate with the correct value.

23. The differential equation

$$L[y] = P(x)y'' + Q(x)y' + R(x)y = 0$$

has associated with it the **adjoint differential equation** defined by

$$M[w] = [P(x)w]'' - [Q(x)w]' + R(x)w = 0.$$

A differential equation is said to be **self-adjoint** if the differential equation and its adjoint are of the same form. When this occurs, the differential operator common to both equations is also said to be self-adjoint.

(a) Show that Bessel's equation of order ν

$$x^2 y'' + xy' + (x^2 - \nu^2)y = 0$$

is not self-adjoint.

(b) Find the value of α that makes the following equation self-adjoint

$$(\alpha \sin x)y'' + (\cos x)y' + 2y = 0.$$

24. Show that Legendre's equation

$$(1 - x^2)y'' - 2xy' - \lambda y = 0$$

is self-adjoint.

25. Show that Bessel's equation of order n in the form

$$x^2 y'' + xy' - (x^2 - n^2)y = 0$$

is not self-adjoint, but that it becomes so when multiplied by $1/x$.

26. Show that the Hermite equation in the form

$$y'' - 2xy' + \lambda y = 0$$

is not self-adjoint, but that it becomes so when multiplied by $\exp[-x^2]$.

27. Show that the Chebyshev equation in the form

$$(1 - x^2)y'' - xy' + \lambda y = 0$$

is not self-adjoint, but that it becomes self-adjoint when multiplied by $(1 - x^2)^{-1/2}$.

28.* Let $u(x)$ and $v(x)$ be any two solutions of

$$\frac{d}{dx}\left[p(x)\frac{dy}{dx}\right] + q(x)y = 0$$

defined over the interval $a \le x \le b$. Prove **Abel's identity**

$$p(x)[u(x)v'(x) - u'(x)v(x)] = \text{constant}$$

for all x in the interval. As $p(x) \ne 0$ in regular Sturm–Liouville problems, what conclusion can be drawn from Abel's identity if (a) the constant is not zero and (b) the constant is zero?

(Hint: Multiply the equations for u and v by suitable factors, subtract them, and integrate the resulting equation over the interval $a \le t \le x$.)

29.* The Chebyshev polynomial $T_n(x)$ can be defined as

$$T_n(x) = \cos(n \,\text{arc}\cos x), \quad n = 0, 1, \ldots.$$

Verify this by showing that this definition of $T_n(x)$ satisfies the Chebyshev differential equation

$$(1 - x^2)y'' - xy' + n^2 y = 0.$$

30.* Let $y = T_n(x) = \cos(n \,\text{arc}\cos x)$ and set $x = \cos\theta$. Use the fact that $y(\theta)$ satisfies the differential equation

$$\frac{d^2 y}{d\theta^2} + n^2 y = 0$$

together with a change of variable back from θ to x to show that this definition of $T_n(x)$ satisfies the Chebyshev equation

$$(1 - x^2)y'' - xy' + n^2 y = 0.$$

31.* Show that if $y_n(\theta) = \cos n\theta$ then

$$\int_0^\pi [y_n(\theta)]^2 d\theta = \begin{cases} \pi, & n = 0 \\ \frac{1}{2}\pi, & n \ge 1 \end{cases}.$$

By changing back from the variable θ to x, where $x = \cos\theta$ and using the definition of $T_n(x)$ in Problem 30, show that the square of the norm of $T_n(x)$ is given by

$$\| T_n(x) \|^2 = \int_{-1}^1 \frac{[T_n(x)]^2}{\sqrt{1 - x^2}}\,dx = \begin{cases} \pi, & n = 0 \\ \frac{1}{2}\pi, & n \ge 1 \end{cases}.$$

8.11   Eigenfunction Expansions and Completeness

The orthogonality of a set of functions $\varphi_0(x), \varphi_1(x), \ldots$ over the interval $a \le x \le b$ with respect to a weight function $r(x)$ allows them to be used to expand (represent) a function $f(x)$ over that same interval in terms of the functions $\varphi_i(x)$ by expressing it as the series

$$f(x) = \sum_{m=0}^\infty a_m \varphi_m(x) = a_0 \varphi_0(x) + a_1 \varphi_1(x) + \ldots, \tag{129}$$

where $a_0, a_1, \ldots$ are constants called the **coefficients** of the expansion.

The representation of functions in this manner is used in approximation theory, in numerical analysis, and in the solution of partial differential equations by the method of *separation of variables* to be described later (see Chapter 18). A series

eigenfunction expansions

such as (129) is called a **generalized Fourier series** representation of $f(x)$ or, when the functions $\varphi_n(x)$ are eigenfunctions, an **eigenfunction expansion** of $f(x)$.

To see how the coefficients a_m in (129) are derived for a specific function $f(x)$, it is necessary to recall that

$$\int_a^b r(x)\varphi_m(x)\varphi_n(x)dx = 0, \quad m \neq n, \tag{130}$$

and

$$\|\varphi_n(x)\|^2 = \int_a^b r(x)[\varphi_n(x)]^2\, dx. \tag{131}$$

If the expansion (129) is multiplied by $r(x)\varphi_n(x)$ and the result is integrated over the interval $a \leq x \leq b$, the orthogonality condition (130) causes every term on the right for which $m \neq n$ to vanish, leaving only the term involving a_n, so using (131) enables the result to be written

$$\int_a^b r(x)\varphi_n(x)f(x)dx = a_n \int_a^b r(x)[\varphi_n(x)]^2 dx = a_n\|\varphi_n(x)\|^2.$$

This has established that the coefficients a_n are given by the formula

$$a_n = \frac{\int_a^b r(x)\varphi_n(x)f(x)dx}{\|\varphi_n(x)\|^2}, \quad n = 0, 1, \ldots. \tag{132}$$

The term-by-term integration of series (129) leading to (132) requires justification, and this follows when the series is uniformly convergent.

Summary of Main Sets of Orthogonal Functions

1. Fourier series (see Chapter 9)

Interval of definition $\quad -\pi \leq x \leq \pi$

Set of functions $\quad \{1, \cos nx, \sin nx\}, n = 1, 2, \ldots$

Weight $\quad r(x) = 1$

Orthogonality $\quad \displaystyle\int_{-\pi}^{\pi} \sin mx \sin nx\, dx = 0, m \neq n$

$\displaystyle\int_{-\pi}^{\pi} \sin mx \cos nx\, dx = 0,$

$\displaystyle\int_{-\pi}^{\pi} \cos mx \cos nx\, dx = 0, m \neq n$

$\displaystyle\int_{-\pi}^{\pi} 1 \cdot \sin mx\, dx = 0$

$\displaystyle\int_{-\pi}^{\pi} 1 \cdot \cos mx\, dx = 0$

Norms $\quad \|1\|^2 = 2\pi, \|\cos nx\|^2 = \pi, \|\sin nx\|^2 = \pi$

2. Legendre polynomials

Interval of definition $-1 \leq x \leq 1$

Set of functions $P_0(x) = 1,\, P_1(x) = x,\, P_2(x) = \frac{1}{2}(3x^2 - 1), \ldots$

Recurrence relation $(n+1)P_{n+1}(x) - (2n+1)x P_n(x) + n P_{n-1}(x) = 0$

Weight $r(x) = 1$

Orthogonality $\displaystyle\int_{-1}^{1} P_m(x) P_n(x)\, dx = 0,\, m \neq n$

Norm $\displaystyle \|P_n(x)\|^2 = \frac{2}{2n+1},\, n = 0, 1, \ldots.$

3. Bessel functions

Interval of definition $0 \leq x \leq a$

Set of functions   There is a set of orthogonal functions for each fixed n:
$J_n(j_{n,r}x/a),\, r = 1, 2, \ldots$, with $j_{n,r}$ the nth zero of $J_n(x)$
(see Table 8.1)

Weight $r(x) = x$

Orthogonality $\displaystyle\int_{0}^{a} x J_n(j_{n,r}x/a) J_n(j_{n,s}x/a)\, dx = 0,\, r \neq s$

Norm $\displaystyle \|J_n(j_{n,r}x/a)\|^2 = \frac{1}{2}a^2 [J_{n+1}(j_{n,r})]^2,\, r = 1, 2, \ldots$

4. Chebyshev polynomials

Interval of definition $-1 \leq x \leq 1$

Set of functions $T_0(x) = 1,\, T_1(x) = x,\, T_2(x) = 2x^2 - 1, \ldots$

Recurrence relation $T_{n+1}(x) - 2x T_n(x) + T_{n-1}(x) = 0$

Weight $(1 - x^2)^{-1/2}$

Orthogonality $\displaystyle\int_{-1}^{1} \frac{T_m(x) T_n(x)}{\sqrt{1 - x^2}}\, dx = 0,\, m \neq n$

Norms $\displaystyle \|T_0(x)\|^2 = \pi,\ \|T_n(x)\|^2 = \frac{1}{2}\pi,\quad n = 1, 2, \ldots.$

(See Exercises 30 and 31 in Exercise Set 18.10 for the derivation of the norms.)

EXAMPLE 8.27

a first example of a
Fourier series

A Fourier series   Example 8.22 established the orthogonality of the set of functions

$$1,\ \cos x,\ \sin x,\ \cos 2x,\ \sin 2x,\ \ldots$$

over the interval $-\pi \leq x \leq \pi$ with weight $r(x) = 1$. It is left as a simple exercise to verify that these functions are the eigenfunctions of the Sturm–Liouville problem

$$y'' + \lambda y = 0, \quad y(-\pi) = y(\pi) = 0.$$

The **Fourier series** for a function $f(x)$ is

$$f(x) = a_0 + \sum_{n=1}^{\infty}(a_n \cos nx + b_n \sin nx),$$

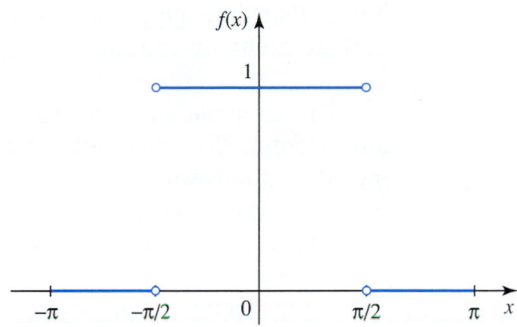

FIGURE 8.13 The rectangular pulse.

where from (132), the **Fourier coefficients** are

$$a_0 = \frac{1}{2\pi} \int_{-\pi}^{\pi} f(x)dx, \quad a_n = \frac{1}{\pi} \int_{-\pi}^{\pi} f(x) \cos nx \, dx$$

$$b_n = \frac{1}{\pi} \int_{-\pi}^{\pi} f(x) \sin nx dx, \quad n = 1, 2, \dots.$$

The formulas for the a_n and b_n are called the **Euler formulas** for the Fourier coefficients.

In anticipation of Chapter 9, let us use these results to find the Fourier series of the (discontinuous) rectangular pulse function

$$f(x) = \begin{cases} 0, & -\pi < x < -\pi/2 \\ 1, & -\pi/2 < x < \pi/2 \\ 0, & \pi/2 < x < \pi \end{cases}$$

shown in Fig. 8.13.

The discontinuities in $f(x)$ cause no problem when deriving the coefficients a_n and b_n because integrals of finite discontinuous functions are well defined:

$$a_0 = \frac{1}{2\pi} \int_{-\pi}^{\pi} f(x) \, dx = \frac{1}{2\pi} \int_{-\pi/2}^{\pi/2} 1 \, dx = \frac{1}{2}$$

$$a_n = \frac{1}{\pi} \int_{-\pi}^{\pi} f(x) \cos nx \, dx = \frac{1}{\pi} \int_{-\pi/2}^{\pi/2} \cos nx \, dx = \frac{2}{n\pi} \sin\left(\frac{1}{2}n\pi\right) = \begin{cases} 0 & \text{if } n \text{ is even} \\ \pm\frac{2}{n\pi} & \text{if } n \text{ is odd.} \end{cases}$$

A similar calculation shows that

$$b_n = \frac{1}{\pi} \int_{-\pi/2}^{\pi/2} \sin nx \, dx = \left[\frac{-1}{n\pi} \cos nx\right]_{-\pi/2}^{\pi/2} = 0, \quad n = 1, 2, \dots.$$

Substituting for the coefficients in the Fourier series gives

$$f(x) = \frac{1}{2} + \frac{2}{\pi} \sum_{n=1}^{\infty} (-1)^{n+1} \frac{\cos(2n-1)x}{2n-1},$$

and so

$$f(x) = \frac{1}{2} + \frac{2}{\pi}\left(\cos x - \frac{\cos 3x}{3} + \frac{\cos 5x}{5} - \cdots\right), \quad -\pi \leq x \leq \pi.$$

Notice that although $f(x)$ is discontinuous at $x = \pm\pi/2$, the Fourier series is defined at these points and has the value 1/2. ∎

This example illustrates the fact that a Fourier series expansion (and indeed any eigenfunction expansion) of $f(x)$ is defined for all x in its interval of definition, including points where $f(x)$ is discontinuous, or not even defined. Because of this it is necessary to question the use of the equality sign in (129) and to reinterpret its meaning at points of discontinuity of $f(x)$. More will be said about this in Chapter 9 in connection with Fourier series.

Some comments will be offered later about the convergence of eigenfunction expansions in general, and their behavior at points of discontinuity of $f(x)$ when the completeness of sets of orthogonal functions is discussed.

EXAMPLE 8.28

a Fourier–Legendre expansion

A Fourier–Legendre expansion   The expansion of a function $f(x)$ in terms of Legendre polynomials $P_n(x)$ over the interval $-1 \le x \le 1$ is called a **Fourier–Legendre expansion**, and it takes the form

$$f(x) = \sum_{n=0}^{\infty} a_n P_n(x) = a_0 + a_1 P_1(x) + \cdots. \tag{133}$$

From (135) the coefficients a_n are determined by

$$a_n = \frac{\int_a^b r(x)\varphi_n(x) f(x)dx}{\|\varphi_n(x)\|^2} = \left[\frac{2n+1}{2}\right]\int_{-1}^1 f(x)P_n(x)dx, \quad n = 0, 1, \ldots.$$

As any polynomial of degree m can be expressed as a linear combination of $P_0(x), P_1(x), \ldots, P_m(x)$, it follows from the orthogonality condition that

$$\int_{-1}^1 x^m P_n(x)dx = 0 \text{ for } n > m.$$

The Fourier–Legendre expansion of the discontinuous function

$$f(x) = \begin{cases} 0, & -1 < x < 0 \\ 1, & 0 < x < 1 \end{cases}$$

is determined as follows. From (133),

$$a_n = \left(\frac{2n+1}{2}\right)\int_{-1}^1 f(x)P_n(x)dx = \left(\frac{2n+1}{2}\right)\int_0^1 P_n(x)dx. \tag{134}$$

If we substitute for $P_n(x)$, it then follows that the first few coefficients in the expansion are

$$a_0 = \frac{1}{2}, a_1 = \frac{3}{4}, a_2 = 0, a_3 = -\frac{7}{16}, \ldots,$$

so the required expansion is

$$f(x) = \frac{1}{2}P_0(x) + \frac{3}{4}P_1(x) - \frac{7}{16}P_3(x) + \cdots.$$

Here also this Fourier-Legendre expansion attributes a value to $f(x)$ at its point of discontinuity at $x = 0$, and a closer examination shows that the value determined by the expansion is 1/2. ∎

Fourier–Bessel expansions   A function $f(x)$ can be expanded over the interval $0 \leq x \leq a$ in terms of the Bessel function J_n, with n fixed, to obtain a **Fourier–Bessel** expansion of the form

$$f(x) = \sum_{r=1}^{\infty} a_r J_n(j_{n,r}x/a) = a_1 J_n(j_{n,1}x/a) + a_2 J_n(j_{n,2}x/a) + \cdots, \qquad (135)$$

where

$$a_r = \left(\frac{2}{a^2}\right) \frac{\int_0^a J_n(j_{n,r}x/a)\, f(x)dx}{[J_{n+1}(j_{n,r})]^2} \qquad (136)$$

An expansion of this type will be used in Chapter 18 when solving the oscillations of a circular membrane, such as the membrane covering a circular drum head.   ∎

Fourier–Chebyshev expansions   The **Fourier–Chebyshev expansion** of a function $f(x)$ over the interval $-1 \leq x \leq 1$ takes the form

$$f(x) = \sum_{n=0}^{\infty} a_n T_n(x) = a_0 T_0(x) + a_1 T_1(x) + \cdots, \qquad (137)$$

where

$$a_n = \frac{\int_{-1}^{1} \dfrac{f(x)T_n(x)}{\sqrt{1-x^2}}dx}{\|T_n(x)\|^2}, \qquad (138)$$

with

$$\|T_0(x)\|^2 = \pi \quad \text{and} \quad \|T_n(x)\|^2 = \frac{1}{2}\pi, \quad n = 1, 2, \ldots.$$

Any polynomial of degree m can be expressed as a linear combination of $T_0(x)$, $T_1(x), \ldots, T_m(x)$, so it follows from the orthogonality conditions that

$$\int_{-1}^{1} \frac{x^m T_n(x)}{\sqrt{1-x^2}}dx = 0 \quad \text{for } n > m. \qquad ∎$$

It is now necessary to comment on the interpretation of the equality sign in (129) at points where $f(x)$ is discontinuous. For expansions in terms of orthogonal functions to be useful, they must be able to represent the class of functions that occur in practical applications. This means that an orthogonal set of functions defined over an interval $a \leq x \leq b$ must always be able to be used to expand functions that are piecewise continuous and differentiable at all but a finite number of points in the interval. For conciseness we will denote this set of functions by PC. In addition, the set of orthogonal functions must be sufficiently rich in functions that there is no function of practical importance that cannot be expanded in this manner.

Orthogonal (and orthonormal) sets of functions that have this property are said to be **complete**, and the ones introduced so far can all be shown to have this property of completeness. As sets of orthogonal functions are required to expand both continuous and piecewise continuous functions that belong to class PC, the convergence of these expansions must of necessity be more general in nature than ordinary convergence. It is this more general form of convergence, which will be introduced shortly, that will permit the equality sign in (129) to be interpreted in a special sense at points where $f(x)$ is discontinuous.

The special type of convergence we now introduce is called **convergence in the norm**, **mean-square convergence**, or **L^2 convergence**. This form of convergence is defined by requiring that if a sequence of functions $f_1(x), f_2(x), \ldots$ *converges in the mean* to a function $f(x)$, then

$$\lim_{n \to \infty} \| f_n(x) - f(x) \| = 0, \tag{139}$$

or, more explicitly,

$$\lim_{n \to \infty} \int_a^b r(x) [f_n(x) - f(x)]^2 \, dx = 0. \tag{140}$$

When interpreting (139) as (140) it is convenient to omit the square root in the definition of the norm, as this simplifies analysis and does not influence the limit.

The sequence of functions $f_n(x)$ in this definition can be taken to be the nth partial sum of the eigenfunction expansion (129),

$$f_n(x) = \sum_{m=0}^{n} a_m \varphi_m(x) = a_0 \varphi_0(x) + a_1 \varphi_1(x) + \cdots, \tag{141}$$

where from now on we will assume $\varphi_0(x), \varphi_1(x), \ldots$ to be an orthonormal set of functions so that $\| \varphi_n(x) \|^2 = 1, n = 0, 1, \ldots$. Such an orthonormal set of functions will be complete with respect to the functions $f(x)$ in C if every function in PC can be approximated by (141). Convergence in the norm and ordinary convergence are the same everywhere a function is continuous and differentiable.

We now state without proof the fundamental eigenfunction expansion theorem.

THEOREM 8.4

a fundamental eigenfunction expansion theorem

Eigenfunction expansion theorem   Let $f(x)$ and $f'(x)$ have at most a finite number of jump discontinuities in the interval $a \leq x \leq b$. Then the eigenfunction expansion (129) converges in the mean to $f(x)$ at every point of continuity of $f(x)$ inside this interval, and to the value $\frac{1}{2}[f(c-) + f(c+)]$ at any point c where $f(x)$ is discontinuous. ∎

This convergence property has already been demonstrated in Example 8.27, where the Fourier series converged to the value $1/2$ at the points where the function was discontinuous. Figure 8.14 shows the result in the general case.

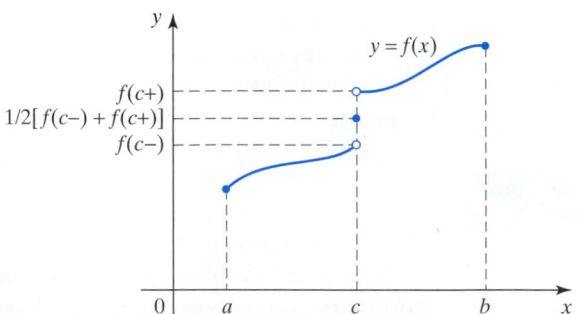

FIGURE 8.14   Convergence of an eigenfunction expansion at a point of discontinuity.

To develop the concept of completeness a little further, we substitute (129) into (140) to obtain

$$\int_a^b r(x)[f_n(x) - f(x)]^2 dx = \int_a^b r(x)[f_n(x)]^2 dx - 2 \int_a^b r(x)f(x)f_n(x)dx$$

$$+ \int_a^b r(x)[f(x)]^2 dx = \int_a^b r(x)\left[\sum_{s=0}^n a_s \varphi_s(x)\right]^2 dx$$

$$- 2 \sum_{s=0}^n a_s \int_a^b r(x)f(x)\varphi_s(x)dx + \int_a^b r(x)[f(x)]^2 dx.$$

The orthogonality property of the set of eigenfunctions $\varphi_s(x)$ reduces the first integral on the right to $\sum_{s=0}^n a_s^2$, while the definition of a_s shows that the second term on the right can be written $-2\sum_{s=0}^n a_s^2$, so the result becomes

$$\int_a^b r(x)[f_n(x) - f(x)]^2 dx = -\sum_{s=0}^n a_s^2 + \int_a^b r(x)[f(x)]^2 dx.$$

The integrands of both integrals are nonnegative, and the integral on the right is $\|f(x)\|^2$, so we have established the inequality

$$\sum_{s=0}^n a_s^2 \le \int_a^b r(x)[f(x)]^2 dx = \|f(x)\|^2 \quad \text{for all } n \ge 0. \tag{142}$$

Bessel's inequality

This result is called **Bessel's inequality**, and it shows that the sum $\sum_{s=0}^n a_s^2$ has the upper bound $\|f(x)\|^2$ as $n \to \infty$. As the terms of the series are nonnegative, the series increases as n increases, so it follows that $\sum_{s=0}^n a_s^2$ converges as $n \to \infty$.

If the system of orthonormal functions $\varphi_s(x)$ is complete, result (139) must be true for every function $f(x)$ in the class PC, so that then $\lim_{n\to\infty} \sum_{s=0}^n a_s^2 = \|f(x)\|^2$. Consequently, for complete orthonormal systems of functions

$$\sum_{s=0}^\infty a_s^2 = \|f(x)\|^2 = \int_a^b r(x)[f(x)]^2 dx. \tag{143}$$

Parseval relation

This result is called the **Parseval relation**.

THEOREM 8.5

Completeness of orthonormal systems   Let $\varphi_0(x), \varphi_1(x), \ldots$ be a complete orthonormal set of functions with respect to the set C to which the functions $f(x)$ belong. Then the only continuous function in C that is orthogonal to every function $\varphi_n(x)$ is the zero function $f(x) \equiv 0$. Furthermore, if the restriction of continuity is removed, the only functions that can be orthogonal to every function in the orthonormal set are those with zero norm.

Proof   In the first case the vanishing of the norm of $f(x)$ implies that $f(x) \equiv 0$. In the second case, the orthogonality of a function with respect to every eigenfunction implies that the function must be degenerate, and although not identically zero, must have a zero norm. ■

See Chapters 2 and 5 of reference [3.7] for information about eigenfunction expansions and orthonormal sets of functions.

Summary

Eigenfunction expansions have been introduced, and the most important sets of orthogonal functions summarized together with their intervals of definition, weight functions, and orthogonality relationships. Mean-square convergence has been defined and the fundamental eigenfunction theorem stated, and the notion of completeness of systems of orthogonal functions has been explained and related to the Parseval relation.

Appendix 1 (Proofs of Theorem 8.3)

The study of Sturm–Liouville problems is made more concise by the introduction of the notion of a **differential operator L** defined as

$$L \equiv \frac{d}{dx}\left[p(x)\frac{d}{dx}\right] + q(x), \tag{144}$$

with the understanding that if y is a suitably differentiable function,

$$L[y] \equiv \frac{d}{dx}\left[p(x)\frac{d}{dx}\right]y(x) + q(x)y(x). \tag{145}$$

Differential operators, of which L is a special case, have the property that when they operate on a function y they produce another function $L[y]$. For example, if

$$L \equiv \frac{d}{dx}\left[x\frac{d}{dx}\right] + 2,$$

and $y(x) = e^{-x}$, then

$$L[e^{-x}] = \frac{d}{dx}\left[x\frac{d[e^{-x}]}{dx}\right] + 2e^{-x} = \frac{d}{dx}[-xe^{-x}] + 2e^{-x} = (1+x)e^{-x}.$$

In terms of the differential operator L in (144), the Sturm–Liouville equation (122) with eigenvalue λ and corresponding eigenfunction φ becomes

$$L[\varphi] + \lambda r\varphi = 0, \tag{146}$$

where φ satisfies suitable boundary conditions.

The proof of the results of Theorem 8.3 that can be given here is simplified by appeal to the following theorem, which is important in its own right.

THEOREM 8.6   **One-dimensional form of Green's theorem**   Let L be the linear operator

$$L \equiv \frac{d}{dx}\left[p(x)\frac{d}{dx}\right] + q(x),$$

and, let u, v be any two twice differentiable functions defined on the interval $a \leq x \leq b$. Then,

(i)

$$\int_a^b uL[v]dx = [p(x)u(x)v'(x)]_a^b - \int_a^b pu'v'\,dx + \int_a^b quv\,dx$$

and

(ii)

$$\int_a^b \{u L[v] - v L[u]\} dx = [p(x)\{u(x)v'(x) - v(x)u'(x)\}]_a^b,$$

called the **Lagrange identity**. Furthermore, if u and v satisfy the boundary conditions

$$A_1\phi(a) + A_2\phi'(a) = 0 \quad \text{and} \quad B_1\phi(b) + B_2\phi'(b) = 0,$$

where ϕ may be either u or v, then

(iii)

$$\int_a^b \{u L[v] - v L[u]\} dx = 0.$$

Proof   Result (i) is the one-dimensional form of **Green's first theorem**, and result (ii) is the one-dimensional form of **Green's second theorem**. The three-dimensional forms of these theorems are derived in Chapter 12, Section 12.2. Result (iii) is the consequence of Green's second theorem when u and v satisfy the stated boundary conditions at the ends of the interval $a \leq x \leq b$.

The proof proceeds as follows. Differentiation of the product $u(pv')$ gives

$$[u(pv')]' = u(pv')' + u'(pv'),$$

so

$$u(pv')' = [puv']' - pu'v'.$$

Recalling the definition of L, we can write

$$u L[v] = [puv']' - pu'v' + quv,$$

so integrating over the interval $a \leq x \leq b$ gives

$$\int_a^b u L[v] dx = [p(x)u(x)v'(x)]_a^b - \int_a^b pu'v' dx + \int_a^b quv dx,$$

which is result (i).

Result (ii) follows if we interchange u and v in (i) and subtract the result from (i) to obtain

$$\int_a^b \{u L[v] - v L[u]\} dx = [p(x)\{u(x)v'(x) - v(x)u'(x)\}]_a^b.$$

Result (iii) follows from (ii) if we notice that, provided $A_2 \neq 0$, it follows from the boundary conditions at $x = a$ that

$$u'(a) = -(A_1/A_2)u(a) \quad \text{and} \quad v'(a) = -(A_1/A_2)v(a),$$

so

$$[p(uv' - vu')]_{x=a} = -(A_1/A_2)p(a)u(a)v(a) + (A_1/A_2)p(a)u(a)v(a) = 0,$$

and a similar argument shows that, provided $B_2 \neq 0$,

$$[p(uv' - vu')]_{x=b} = 0.$$

Thus, $[p(uv' - vu')]_a^b = 0$, reducing result (ii) to

$$\int_a^b \{uL[v] - vL[u]\}dx = 0,$$

which is result (iii).

Result (iii) is obviously true if the boundary conditions simplify to

$$\phi(a) = 0 \text{ and } \phi(b) = 0 \quad \text{or to } \phi'(a) = 0 \quad \text{and} \quad \phi'(b) = 0,$$

and the modification to the proof needed to show that the result remains true if A_2 and/or B_2 is zero is left as an exercise. ∎

JOSEPH LOUIS LAGRANGE (1736–1813)
Lagrange was born in Turin of French extraction and after working in Berlin for twenty years moved to Paris. His many fundamental contributions to mathematics have led to his being regarded as one of the most outstanding mathematicians of his time. He made contributions to algebra, calculus, differential equations, the calculus of variations, and also to mechanics.

We now prove the results in Theorem 8.3 that are straightforward, and refer to the references at the end of the chapter for details of the way in which the more complicated results can be established.

Property 1. The proof of this property is difficult and so will be omitted, but Examples 8.18 to 8.21 illustrate the existence of an ordered sequence of eigenvalues in specific cases.

Property 2. In a regular Sturm–Liouville problem suppose, if possible, that φ and ψ are eigenfunctions corresponding to the single eigenvalue λ. Then each of these functions satisfies the Sturm–Liouville equation, while φ and ψ both satisfy the boundary conditions at $x = a$ so that

$$A_1\varphi(a) + A_2\varphi'(a) = 0 \text{ and } A_1\psi(a) + A_2\psi'(a) = 0.$$

This pair of equations can be considered to determine A_1 and A_2 in terms of φ and ψ at $x = a$. The equations are homogeneous, so there can only be a nontrivial solution for A_1 and A_2 if the determinant of coefficients $W = \varphi(a)\psi'(a) - \varphi'(a)\psi(a)$ vanishes, but this determinant is the Wronskian of the solutions and can only vanish if φ is proportional to ψ, so the result is established.

Property 3. Let φ_m and φ_n be eigenfunctions corresponding to the two distinct eigenvalues λ_m and λ_n of the Sturm–Liouville problem

$$L[y] + \lambda r y = 0$$

defined on $a \le x \le b$ and satisfying homogeneous boundary conditions of the type given in (127). Then it follows that

$$L[\varphi_m] + \lambda_m r \varphi_m = 0 \text{ and } L[\varphi_n] + \lambda_n r \varphi_n = 0.$$

Multiplying the first equation by φ_n and the second by φ_m, subtracting the results, and integrating over the interval $a \le x \le b$ gives

$$\int_a^b \{\varphi_m L[\varphi_n] - \varphi_n L[\varphi_m]\}dx + (\lambda_n - \lambda_m)\int_a^b r\varphi_m\varphi_n\,dx = 0.$$

The first integral vanishes because of the result of Theorem 8.4 (iii), so

$$(\lambda_n - \lambda_m) \int_a^b r \varphi_m \varphi_n dx = 0.$$

The result now follows because $\lambda_n \neq \lambda_m$.

Property 4.   The proof is by contradiction. Suppose, if possible, that $\lambda = \alpha + i\beta$ is a complex eigenvalue associated with the complex eigenfunction $\Phi = \varphi + i\psi$. Then as Φ and λ satisfy the Sturm–Liouville equation, we have

$$[p(\varphi + i\psi)']' + [q + (\alpha + i\beta)r](\varphi + i\psi) = 0.$$

This can be written

$$[p\varphi']' + q\varphi + \alpha\varphi r - \beta\psi r + i\{[p\psi']' + q\psi + \beta\varphi r + \alpha\psi r\} = 0.$$

For this to be true, both real and imaginary parts of the equation must vanish, so

$$[p\varphi']' + q\varphi + \alpha\varphi r - \beta\psi r = 0 \quad \text{and} \quad [p\psi']' + q\psi + \beta\varphi r + \alpha\psi r = 0.$$

Multiplying the second equation by i, subtracting it from the first equation, and collecting terms gives

$$[p(\varphi - i\psi)']' + [q + (\alpha - i\beta)r](\varphi - i\psi) = 0,$$

showing that $\overline{\Phi} = \varphi - i\psi$ is an eigenfunction and $\overline{\lambda} = \alpha - i\beta$ is an eigenvalue. As Φ and $\overline{\Phi}$ are linearly independent eigenfunctions, it follows from Theorem 8.3 (3) that

$$\int_a^b r\Phi\overline{\Phi}dx = \int_a^b r(\varphi^2 + \psi^2)dx = 0,$$

but this is impossible because by hypothesis $r(x) \geq 0$ and $\varphi^2 + \psi^2 > 0$. Consequently the assumption that an eigenvalue can be complex is false.

Property 5.   Let λ_n be an eigenvalue and φ_n be the corresponding eigenfunction of the Sturm–Liouville equation

$$L[\varphi_n] + \lambda_n r \varphi_n = 0.$$

Multiplication of this equation by φ_n, followed by integration over the interval $a \leq x \leq b$, gives

$$\int_a^b \varphi_n L[\varphi_n]dx + \lambda_n \int_a^b r\varphi_n^2 dx = 0.$$

An application of Theorem 8.4 (i) with $u = v = \varphi_n$ then gives the result

$$\lambda_n = \frac{-[p\varphi_n\varphi_n']_a^b + \int_a^b p(\varphi_n')^2 dx - \int_a^b r\varphi_n^2 dx}{\int_a^b r\varphi_n^2 dx}.$$

Property 6.   This follows directly from Property 5 when $q(x) < 0$ and the condition $[p\varphi_n\varphi_n']_a^b \leq 0$ is satisfied.

Property 7.   We offer no proof of this result, though as already remarked it is well illustrated by graphs of the Legendre polynomials shown in Fig. 8.1.

Property 8.   This follows directly from Property 5 when the stated conditions are imposed, because increasing $p(x)$ will increase the numerator while leaving all other terms unchanged.

Property 9. No proof of this result is offered because it follows from the form of argument used to establish the upper bound property of the Rayleigh quotient given in (128).

Appendix 2 (Norm of $J_n(x)$)

The square of the norm of the Bessel function $J_n(j_{n,r}x/a)$ is the definite integral

$$\|J_n(j_{n,r}x/a)\|^2 = \int_0^a x[J_n(j_{n,r}x/a)]^2 dx = \frac{1}{2}a^2[J_{n+1}(j_{n,r})]^2,$$

and so the norm is

$$\|J_n(j_{n,r}x/a)\| = \frac{1}{\sqrt{2}}a[J_{n+1}(j_{n,r})]. \tag{147}$$

This result is most easily derived by considering the case $a = 1$, and then changing variables to obtain the foregoing more general result. Accordingly, we consider the two Bessel equations in Sturm–Liouville form,

$$[xu']' + (j_{n,r}^2 x - n^2/x)u = 0 \quad \text{and} \quad [xv']' + (k^2 x - n^2/x)v = 0,$$

defined on the interval $0 \le x \le 1$ with bounded solutions that satisfy the boundary conditions $u(1) = v(1) = 0$. These equations have the respective solutions $u(x) = J_n(j_{n,r}x)$ and $v(x) = J_n(kx)$.

Multiplying the first equation by u, the second by v, subtracting the second equation from the first, and integrating over the interval $0 \le x \le 1$ gives, after using Theorem 8.6 (ii) and the result $u'(x) = j_{n,r}J'(j_{n,r}x)$,

$$\int_0^1 xJ_n(j_{n,r}x)J_n(kx)dx = \frac{j_{n,r}J_n(k)J_n'(j_{n,r})}{k^2 - j_{n,r}^2}.$$

We now write this result as

$$\int_0^1 xJ_n(j_{n,r}x)J_n(kx)dx = \left(\frac{j_{n,r}}{k+j_{n,r}}\right)\left(\frac{J_n(k) - J_n(j_{n,r})}{k - j_{n,r}}\right)J_n'(j_{n,r}),$$

where the subtraction of $J_n(j_{n,r})$ in the bracketed term in the numerator leaves the result unchanged because $J_n(j_{n,r}) = 0$.

Taking the limit as $k \to j_{n,r}$, reduces this result to

$$\int_0^1 x[J_n(j_{n,r}x)]^2 dx = \frac{1}{2}[J_n'(j_{n,r})]^2, \quad r = 1, 2, \ldots.$$

It is inconvenient to work with $J_n'(j_{n,r})$, so we relate J_n to J_{n+1} by using recurrence relation (65)':

$$xJ_n'(x) = nJ_n(x) - xJ_{n+1}(x).$$

Setting $x = j_{n,r}$ causes this to simplify to $J_n'(j_{n,r}) = -J_{n+1}(j_{n,r})$, and so

$$\int_0^1 x[J_n(j_{n,r}x)]^2 dx = \frac{1}{2}[J_{n+1}(j_{n,r})]^2.$$

The more general result follows by making the change of variable $x = z/a$ and then replacing z by x.

EXERCISES 8.11

In Exercises 1 through 3 expand the given polynomials in terms of Legendre polynomials.

1. $4x^3 - 2x^2 + 1$.

2. $3x^3 + x^2 - 4x$.

3. $x^4 + 3x^2 + 2x$.

4. Represent x^2, x^3, and x^4 in terms of Legendre polynomials.

In Exercises 5 through 8 find the first four terms of the Fourier–Legendre expansions of the given functions. In each case graph the four term approximation to $f(x)$ and compare it with the graph of $f(x)$.

5. $f(x) = \begin{cases} 1, & -1 \le x \le 0 \\ x, & 0 < x \le 1. \end{cases}$

6. $f(x) = \begin{cases} 1 + x, & -1 \le x \le 0 \\ 1 - x, & 0 < x \le 1. \end{cases}$

7. $f(x) = \begin{cases} 0, & -1 \le x < -1/2 \\ 1, & -1/2 < x < 1/2 \\ 1/2, & 1/2 < x < 1. \end{cases}$

8. $f(x) = \begin{cases} -2x, & -1 \le x < 0 \\ x, & 0 \le x \le 1. \end{cases}$

9. Find the first four terms in the Fourier–Legendre expansion of e^x.

10. Find the first four terms in the Fourier–Legendre expansion of e^{-x}.

In Exercises 11 through 13 expand the given polynomials in terms of Chebyshev polynomials.

11. $3x^4 - 4x^2 - x$.

12. $4x^3 + x^2 - 3x + 1$.

13. $2x^4 - x^3 + x + 3$.

14. Represent x^2, x^3, and x^4 in terms of Chebyshev polynomials.

In Exercises 15 and 16 find the first four terms in the Fourier–Chebyshev expansion of the given function. In each case graph the four term approximation to $f(x)$ and compare it with the graph of $f(x)$.

15. $f(x) = \begin{cases} 2 + x, & -1 < x < 0 \\ 3, & 0 < x < 1. \end{cases}$

16. $f(x) = \begin{cases} -1, & -1 < x < 0 \\ 2x - 1, & 0 < x < 1. \end{cases}$

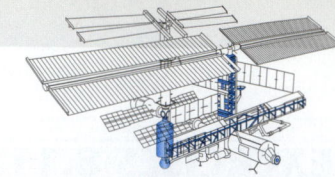

Project 1

The Asymptotic Formulas for $J_n(x)$ and $Y_n(x)$

The purpose of this project is to compare plots of the Bessel functions $J_n(x)$ and $Y_n(x)$ with the results obtained from the asymptotic formulas

$$J_n(x) \sim \sqrt{\frac{2}{\pi x}} \sin\left(x - \frac{1}{2}n\pi + \frac{1}{4}\pi\right) \quad \text{and}$$

$$Y_n(x) \sim \sqrt{\frac{2}{\pi x}} \sin\left(x - \frac{1}{4}\pi(2n+1)\right).$$

Make combined plots of $J_n(x)$ and its asymptotic form, and $Y_n(x)$ and its asymptotic form for $0 \le x \le 30$ for different values of n to illustrate the speed with which the asymptotic approximation tends to the function itself.

Project 2

Chebyshev Approximation

The purpose of this project is to make Chebyshev polynomial approximations of different orders to an asymmetric function $f(x)$ to illustrate the rapidity with which they converge to $f(x)$.

1. Let $f(x) = \sin(5x)(1 + x^2)^{1/4}$ for $-1 \le x \le 1$. Approximate $f(x)$ in terms of the Chebyshev polynomials $T_n(x)$ by the function $f_N(x)$:

$$f_N(x) = \sum_{n=0}^{N} a_n T_n(x).$$

Find the coefficients a_n numerically and make simultaneous plots of $f(x)$ and $f_N(x)$ for $N = 3, 5$, and 7 to show the convergence of $f_N(x)$ to $f(x)$ as N increases.

2. Repeat the calculations with a discontinuous function of your own choice and comment on the behavior of the approximation at the point of discontinuity in the cases when $N = 5, 10, 15, 20, 25$, and 30. Compare your observations with the remarks about the occurrence of the *Gibbs phenomenon* in Fourier series in Chapter 9.

Project 3

Legendre Approximation

The purpose of this project is to make Legendre polynomial approximations of different orders to the function $f(x)$ in Project 2 to illustrate the rapidity with which they converge to $f(x)$.

1. Let $f(x) = \sin(5x)(1 + x^2)^{1/4}$ for $-1 \le x \le 1$. Approximate $f(x)$ in terms of the Legendre polynomials $P_n(x)$ by the function $f_N(x)$:

$$f_N(x) = \sum_{n=0}^{N} a_n P_n(x).$$

Find the coefficients a_n numerically and make simultaneous plots of $f(x)$ and $f_N(x)$ for $N = 3, 5$, and 7 to show the convergence of $f_N(x)$ to $f(x)$ as N increases.

2. Repeat the calculations with a discontinuous function of your own choice and comment on the behavior of the approximation at the point of discontinuity for the cases $N = 5, 10, 15, 20, 25$, and 30. Compare your observations with the remarks about the occurrence of the oscillatory behavior of approximations near a finite jump discontinuity described in Chapter 9 on Fourier series, where the effect is called the *Gibbs phenomenon*.

Project 4

Bessel Function Approximation

The purpose of this project is to make Bessel function approximations of different orders to a function $f(x)$ over a given interval to illustrate the rapidity with which they converge to $f(x)$.

1. Approximate $f(x) = (1 + x^3)\sin x$ over the interval $0 \le x \le \pi$ in terms of the Bessel function $J_1(x)$ by the function $f_N(x)$

$$f_N(x) = \sum_{r=1}^{N} a_r J_1(j_{1,r} x/\pi),$$

where $j_{1,r}$ is the rth zero of $J_1(x)$ listed in Table 8.1. Find the coefficients a_n numerically and make simultaneous plots of $f(x)$ and $f_N(x)$ for $N = 3$, 5, and 7 to show the convergence of $f_N(x)$ to $f(x)$ as N increases.

2. Repeat the calculation with a continuous function $f(x)$ of your own choice. When making the series expansion in terms of the Bessel function $J_n(x)$, use the value $n = 0$ if $f(0) \neq 0$ and $n = 1$ if $f(0) = 0$.

PART FOUR

FOURIER SERIES, INTEGRALS, AND THE FOURIER TRANSFORM

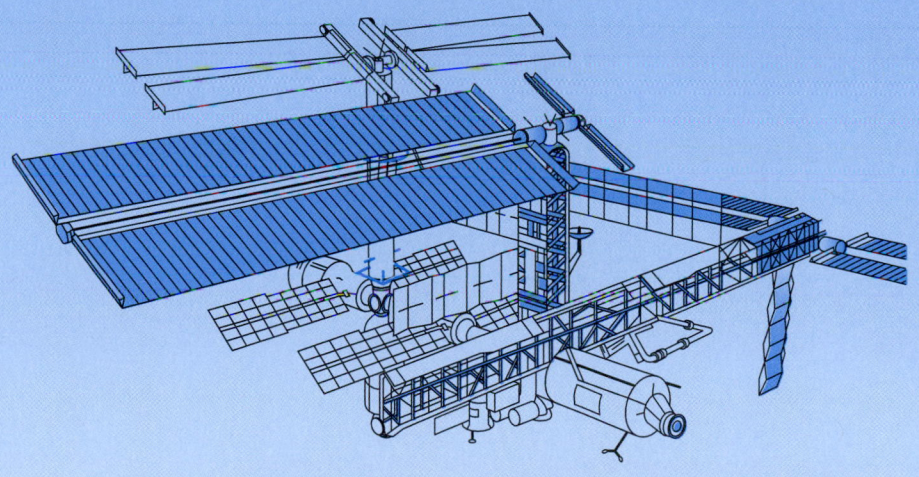

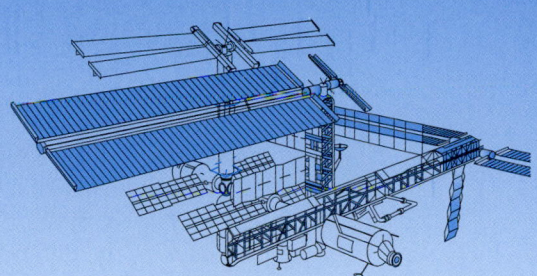

CHAPTER 9

Fourier Series

W hen analyzing situations as diverse as electrical oscillations, vibrating mechanical systems, longitudinal oscillations in crystals, and many other physical phenomena, Fourier series are found to arise naturally. Furthermore, the individual terms in a Fourier series often have an important physical interpretation. In a vibrating mechanical system, for example, each component of a Fourier series representation of the overall vibration represents a fundamental mode of vibration. The full Fourier series shows how each mode contributes to the solution, and which are the most significant modes. This information can often be used to advantage, either by showing how the modes can be utilized to achieve a desired effect, or by using the information to enable systems to be constructed that minimize undesirable vibrations. It is for these and other reasons that it is necessary for engineers and physicists to study the properties of Fourier series.

9.1 Introduction to Fourier Series

A **Fourier series** representation of a function $f(x)$ over the interval $-\pi \leq x \leq \pi$ is an expression of the form

$$f(x) = a_0 + \sum_{n=1}^{\infty}(a_n \cos nx + b_n \sin nx)$$

$$= a_0 + a_1 \cos x + b_1 \sin x + a_2 \cos 2x + b_2 \sin 2x + \cdots, \qquad (1)$$

where the coefficients $a_0, a_1, \ldots, b_1, b_2, \ldots$ are determined by the function $f(x)$.

It is important to notice that the Fourier series representation of $f(x)$ contains two infinite sums, one of even functions (the cosines) and the other of odd functions (the sines). It will be recalled that a function $f(x)$ defined in the interval $-L \leq x \leq L$ is said to be an **even function** in the interval if

$$f(-x) = f(x), \qquad (2)$$

545

and to be an **odd function** in the interval if

$$f(-x) = -f(x). \tag{3}$$

The cosine function is an even function because $\cos(-x) = \cos x$ in agreement with the definition in (2). As this is true for all x, the function $\cos x$ is an even function for $-\infty < x < \infty$. Similarly, $\sin x$ is an odd function because $\sin(-x) = -\sin x$ in agreement with the definition in (3). This also is true for all x, so the function $\sin x$ is an odd function for $-\infty < x < \infty$.

Most functions are neither even nor odd, but any function in an interval $-L \leq x \leq L$ can be expressed as the sum of an even function and an odd function defined over the interval. To see why this is, let $f(x)$ be an arbitrary function defined over the interval $-L \leq x \leq L$, and write it in the form

$$f(x) = \frac{1}{2}(f(x) + f(-x)) + \frac{1}{2}(f(x) - f(-x)) \quad \text{for } -L \leq x \leq L. \tag{4}$$

Then the function

$$h(x) = \frac{1}{2}(f(x) + f(-x)) \tag{5}$$

is seen to be an *even* function, because $h(-x) = h(x)$, whereas the function

$$g(x) = \frac{1}{2}(f(x) - f(-x)) \tag{6}$$

is seen to be an odd function, because $g(-x) = -g(x)$, so the assertion is proved.

EXAMPLE 9.1   Classify the following functions as even, odd, or neither.

(a) $\cosh x$. (b) $\sinh x$. (c) $x^2 + \sin x$. (d) $1 + x^2 + 3x^4$.

Solution   (a) As $\cosh(-x) = \cosh x$ for all x, the function $\cosh x$ is an even function for all x. (b) As $\sinh(-x) = -\sinh x$ for all x, the function $\sinh x$ is an odd function for all x. (c) $(-x)^2 = x^2$, so x^2 is an even function for all x, while $\sin x$ is an odd function for all x, so the function $x^2 + \sin x$ is neither even nor odd. In this case the function $x^2 + \sin x$ is already expressed as the sum of an even function and an odd function. (d) Set $f(x) = 1 + x^2 + 3x^4$. Then $f(-x) = 1 + (-x)^2 + (-x)^4 = f(x)$, so $f(x)$ is an even function. This result can be obtained by a different form of argument as follows. A constant does not change when the sign of x is changed, so all constants are even functions and, in particular, 1 is an even function. The function x^2 has already been shown to be an even function, and the function $3x^4$ is an even function because $3(-x)^4 = 3x^4$. Thus, as the function $1 + x^2 + 3x^4$ is a sum of three even functions, it must be an even function.   ∎

To arrive at a formula for the a_n in (1) corresponding to a given function $f(x)$, result (1) is first multiplied term by term by $\cos nx$ to obtain

deriving formulas for a_n and b_n

$$f(x) \cos nx = a_0 \cos nx + a_1 \cos x \cos nx + a_2 \cos 2x \cos nx + a_3 \cos 3x \cos nx$$
$$+ \cdots + a_{n-1} \cos(n-1)x \cos nx + a_n \cos^2 nx$$
$$+ a_{n+1} \cos(n+1)x \cos nx + \cdots + b_1 \sin x \cos nx$$
$$+ b_2 \sin 2x \cos nx + \cdots.$$

Integrating this result over the interval $-\pi \leq x \leq \pi$ gives

$$\int_{-\pi}^{\pi} f(x) \cos nx dx = a_0 \int_{-\pi}^{\pi} \cos nx dx + a_1 \int_{-\pi}^{\pi} \cos x \cos nx dx$$

$$+ a_2 \int_{-\pi}^{\pi} \cos 2x \cos nx dx + a_3 \int_{-\pi}^{\pi} \cos 3x \cos nx dx + \cdots$$

$$+ a_{n-1} \int_{-\pi}^{\pi} \cos(n-1)x \cos nx dx + a_n \int_{-\pi}^{\pi} \cos^2 nx dx$$

$$+ a_{n+1} \int_{-\pi}^{\pi} \cos(n+1)x \cos nx dx + \cdots + b_1 \int_{-\pi}^{\pi} \sin x \cos nx dx$$

$$+ b_2 \int_{-\pi}^{\pi} \sin 2x \cos nx dx + \cdots .$$

The orthogonality properties of the sine and cosine functions listed in entry 1 of the summary of main sets of orthogonal functions in Section 8.11 shows that all integrals on the right with the exception of the one with the integrand $\cos^2 nx$ vanish, giving rise to the result

$$\int_{-\pi}^{\pi} f(x) \cos nx dx = a_n \int_{-\pi}^{\pi} \cos^2 nx dx.$$

However, $\int_{-\pi}^{\pi} \cos^2 nx dx = \pi$, for $n \neq 0$ and $\int_{-\pi}^{\pi} 1.dx = 2\pi$, so

$$a_0 = \frac{1}{2\pi} \int_{-\pi}^{\pi} f(x) dx \quad \text{and} \quad a_n = \frac{1}{\pi} \int_{-\pi}^{\pi} f(x) \cos nx dx, \quad \text{for } n = 1, 2, \ldots .$$

A similar argument involving the multiplication of the Fourier series (1) by $\sin nx$ followed by integration over the interval $-\pi \leq x \leq \pi$ and use of the orthogonality properties of $\sin nx$ shows the coefficients b_n are given by

$$b_n = \frac{1}{\pi} \int_{-\pi}^{\pi} f(x) \sin nx dx, \quad \text{for } n = 1, 2, \ldots .$$

the Euler formulas

These results are the **Euler formulas** for the **Fourier coefficients** a_n and b_n, and for future reference they are now listed, together with the associated Fourier series representation of $f(x)$.

the Fourier series representation

Fourier series representation of $f(x)$ over the interval $-\pi \leq x \leq \pi$

Let the function $f(x)$ be defined on the interval $-\pi \leq x \leq \pi$. Then the Fourier coefficients a_n and b_n in the Fourier series representation of $f(x)$

$$f(x) = a_0 + \sum_{n=1}^{\infty} (a_n \cos nx + b_n \sin nx) \tag{7}$$

are given by the Euler formulas

$$a_0 = \frac{1}{2\pi} \int_{-\pi}^{\pi} f(x) dx, \quad a_n = \frac{1}{\pi} \int_{-\pi}^{\pi} f(x) \cos nx dx,$$

$$b_n = \frac{1}{\pi} \int_{-\pi}^{\pi} f(x) \sin nx dx, \ldots \quad \text{for } n = 1, 2, \ldots . \tag{8}$$

The arguments used to derive the Euler formulas in (8) are not rigorous, because the term by term integration needs to be justified and the convergence of the Fourier series representation of $f(x)$ to the function $f(x)$ itself has not been examined, so the use of an equality sign in (1) and (7) must be questioned.

> **JEAN BAPTISTE JOSEPH (BARON) FOURIER (1768–1830)**
> A remarkable French physicist who was also an outstanding mathematician. He was orphaned at eight, and educated in a military school run by the Benedictines who then gave him a lectureship in mathematics. He later moved to a chair at the Ecole Polytechnique in Paris, and later to Grenoble where he was appointed Prefect by Napoleon. His experiments on heat conduction while in Grenoble, suggested by Newton's Law of Cooling, led him to propose his law of heat conduction (Fourier's Law) and to the publication of his most important *Theorie Analytique de la Chaleur* in which he introduced the representation of arbitrary function over an interval in terms of trigonometric functions, now called Fourier series. He was created a Baron by Napoleon in 1808.

In fact, the preceding approach can be fully justified for all functions $f(x)$ that arise in practical situations, and we will see later that the equality sign can be used wherever $f(x)$ is continuous, whereas at points where $f(x)$ experiences a finite jump discontinuity the value assumed by the Fourier series representation is the average of the values to the immediate left and right of the jump. It is for these reasons that in more advanced accounts the equality sign in (7) is replaced by a tilde $\sim$, because this indicates that a relationship exists between a function $f(x)$ and its Fourier series representation without indicating that it is necessarily a strict equality. When this notation is used, the connection between $f(x)$ and its Fourier series is shown by writing

$$f(x) \sim a_0 + \sum_{n=1}^{\infty}(a_n \cos nx + b_n \sin nx). \tag{9}$$

fundamental interval, periodicity, and periodic extension

The interval of integration $-\pi \leq x \leq \pi$ used when deriving the Euler formulas is called the **fundamental interval** of the Fourier series, and the Fourier coefficients will always be defined provided the integral $\int_{-\pi}^{\pi} f(x)dx$ exists. Although Fourier series comprise only even and odd functions, results (4) to (6) allow a Fourier series to represent arbitrary functions that are neither even nor odd.

A Taylor series expansion of a function $f(x)$ about a point x_0 requires the function to be repeatedly differentiable at x_0. However, the coefficients of a Fourier series are defined in terms of definite integrals that are still defined when $f(x)$ has finite jump discontinuities in the fundamental interval, so the Euler formulas still remain valid when $f(x)$ is discontinuous. It is this property of a definite integral that makes a Fourier series representation of a function more general than a Taylor series expansion.

The properties of Fourier series reflect the *periodicity* of the sine and cosine functions used in the expansion, where the *period* of a periodic function is defined as follows. A function $g(x)$ is said to be **periodic** with **period** T if

$$g(x + T) = g(x) \tag{10}$$

for all x, and T is the *smallest* value for which (10) is true. A periodic function $g(x)$ may either be continuous or discontinuous, and an example of a continuous periodic function with period T is shown in Fig. 9.1.

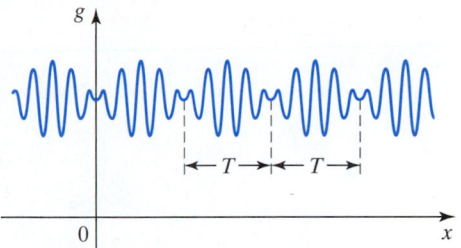

FIGURE 9.1   A continuous periodic function $g(x)$ with period T.

The functions 1, $\cos nx$, and $\sin nx$ in the Fourier series representation (7) of $f(x)$ are all periodic with period 2π, so the *Fourier series representation* of $f(x)$ defined on the interval $-\pi < x < \pi$ is also periodic with period 2π. It does not necessarily follow that outside the fundamental interval the function $f(x)$ coincides with its Fourier series representation, because the behavior of $f(x)$ outside the fundamental interval does not enter into the Euler formulas. Each representation of $f(x)$ in an interval of the form $(2n-1)\pi < x < (2n+1)\pi$, with $n = 0, \pm 1, \pm 2, \ldots$, is called a **periodic extension** of the fundamental interval for $f(x)$.

In Chapter 8, Example 8.22, the discontinuous rectangular pulse function

$$f(x) = \begin{cases} 0, & -\pi < x < -\pi/2 \\ 1, & -\pi/2 < x < \pi/2 \\ 0, & \pi/2 < x < \pi \end{cases}$$

was shown to be represented by the Fourier series

$$f(x) = \frac{1}{2} + \frac{2}{\pi}\left[\cos x - \frac{\cos 3x}{3} + \frac{\cos 5x}{5} - \frac{\cos 7x}{7} + \cdots\right] \quad \text{for all } x. \tag{11}$$

If this function $f(x)$ is defined for all x by the periodicity condition $f(x + 2\pi) = f(x)$, its graph takes the form shown in Fig. 9.2. Figure 9.3 shows the graph of the first five terms of the Fourier series representation (11) in the fundamental interval.

This simple example emphasizes two important issues that always arise when working with Fourier series representations of functions:

1. The need to interpret the equality sign in (7) at any point $x = x_0$ in the fundamental interval where $f(x)$ is discontinuous.

2. The fact that the Fourier series of a function and the periodic extensions of the function will only coincide when the function $f(x)$ is itself periodic with a period equal to the fundamental interval.

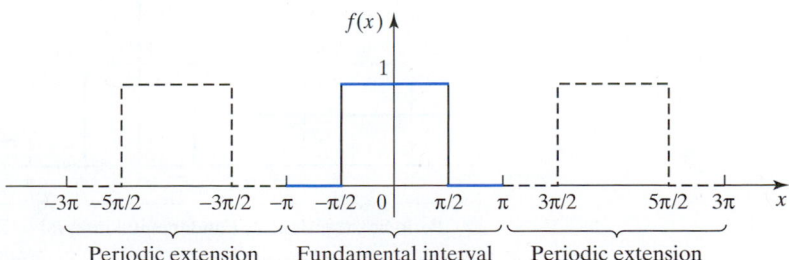

FIGURE 9.2   The periodic rectangular pulse function $f(x)$.

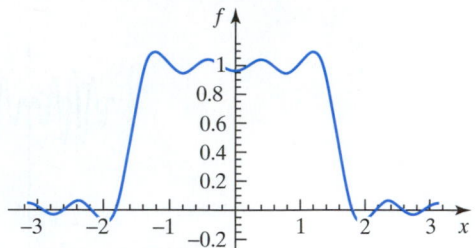

FIGURE 9.3   Graph of the first five terms of the Fourier series of $f(x)$.

An example of the difference that can arise between the behavior of a nonperiodic function $f(x)$ and its periodic extensions is illustrated in Fig. 9.4 in the case of the function

$$f(x) = \begin{cases} 1/2, & x < -\pi \\ 0, & -\pi < x < -\pi/2 \\ 1, & -\pi/2 < x < \pi/2 \\ 0, & \pi/2 < x < \pi \\ 1/4, & x > \pi. \end{cases}$$

The periodic extensions of $f(x)$ in its fundamental interval $-\pi \leq x \leq \pi$ shown as dashed lines are, of course, the same as those in Fig. 9.2, though in this case the behavior of $f(x)$ outside the fundamental interval is entirely different.

EXAMPLE 9.2

some illustrative examples

Find the Fourier series representation of

$$f(x) = \begin{cases} \sin 2x, & -\pi < x < -\pi/2 \\ 0, & -\pi/2 \leq x \leq 0 \\ \sin 2x, & 0 < x \leq \pi. \end{cases}$$

Solution   The function $f(x)$ is continuous over the fundamental interval $-\pi \leq x \leq \pi$, but it is defined in piecewise manner, so the Fourier coefficients must be determined by integrating the Euler equations (8) in a corresponding manner. We have

$$a_0 = \frac{1}{2\pi} \int_{-\pi}^{\pi} f(x)dx = \frac{1}{2\pi} \int_{-\pi}^{-\pi/2} \sin 2x\,dx + \frac{1}{2\pi} \int_0^{\pi} \sin 2x\,dx$$

$$= \frac{1}{2\pi}[-(1/2)\cos 2x]_{-\pi}^{-\pi/2} + \frac{1}{2\pi}[-(1/2)\cos 2x]_0^{\pi} = \frac{1}{2\pi} + 0 = \frac{1}{2\pi}.$$

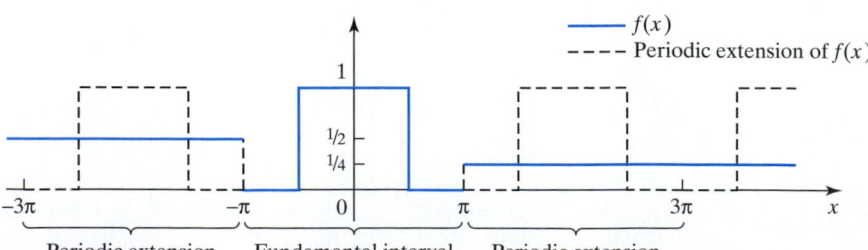

FIGURE 9.4   A nonperiodic function defined for all x, and the periodic extensions of the function in its fundamental interval.

Similarly,

$$a_n = \frac{1}{\pi} \int_{-\pi}^{\pi} f(x)\cos nx\, dx = \frac{1}{\pi} \int_{-\pi}^{-\pi/2} \sin 2x \cos nx\, dx + \frac{1}{\pi} \int_{0}^{\pi} \sin 2x \cos nx\, dx$$

$$= \frac{-2}{\pi}\left[\frac{\cos n\pi + \cos(n\pi/2)}{n^2 - 4}\right]_{-\pi}^{-\pi/2} + \frac{2}{\pi}\left[\frac{\cos n\pi - 1}{n^2 - 4}\right]_{0}^{\pi}, \quad \text{for } n \neq 2$$

$$= \frac{-2[1 + \cos(n\pi/2)]}{\pi(n^2 - 4)}, \quad \text{for } n \neq 2.$$

As the denominator in the expression for a_n is zero when $n = 2$, in order to find a_2 it is necessary to return to the Euler formula for a_n and set $n = 2$ *before* integrating, when we obtain

$$a_2 = \frac{1}{\pi} \int_{-\pi}^{-\pi/2} \sin 2x \cos 2x\, dx + \frac{1}{\pi} \int_{0}^{\pi} \sin 2x \cos 2x\, dx = 0 + 0 = 0.$$

The Euler formula for b_n becomes

$$b_n = \frac{1}{\pi} \int_{-\pi}^{\pi} f(x)\sin nx\, dx = \frac{1}{\pi} \int_{-\pi}^{-\pi/2} \sin 2x \sin nx\, dx + \frac{1}{\pi} \int_{0}^{\pi} \sin 2x \sin nx\, dx$$

$$= \frac{1}{2\pi}\left[\frac{\sin(n-2)x}{n-2} - \frac{\sin(n+2)x}{n+2}\right]_{-\pi}^{-\pi/2} + \frac{1}{2\pi}\left[\frac{\sin(n-2)x}{n-2} - \frac{\sin(n+2)x}{n+2}\right]_{0}^{\pi}$$

$$= \frac{2}{\pi}\frac{\sin(n\pi/2)}{(n^2 - 4)}, \quad \text{for } n \neq 2.$$

As the denominator in the expression for b_n is zero for $n = 2$, to find b_2 we must set $n = 2$ in the Euler formula for b_2 before integrating, as a result of which we find that

$$b_2 = \frac{1}{\pi} \int_{-\pi}^{-\pi/2} \sin^2 2x\, dx + \frac{1}{\pi} \int_{0}^{\pi} \sin^2 2x\, dx$$

$$= \frac{1}{4\pi}[2x - \sin 2x \cos 2x]_{-\pi}^{-\pi/2} + \frac{1}{4\pi}[2x - \sin 2x \cos 2x]_{0}^{\pi}$$

$$= \frac{1}{4} + \frac{1}{2} = \frac{3}{4}.$$

Combining the preceding results shows the first few Fourier coefficients to be

$$a_0 = \frac{1}{2\pi}, \quad a_1 = \frac{2}{3\pi}, \quad a_2 = 0, \quad a_3 = -\frac{2}{5\pi}, \quad a_4 = -\frac{1}{3\pi}, \quad a_5 = -\frac{2}{21\pi},$$

$$b_1 = -\frac{2}{3\pi}, \quad b_2 = \frac{3}{4}, \quad b_3 = -\frac{2}{5\pi}, \quad b_4 = 0, \quad b_5 = \frac{2}{21\pi}, \cdots.$$

When these coefficients are used, the first few terms of the Fourier series for $f(x)$ are seen to be

$$f(x) = \frac{1}{2\pi} + \frac{1}{\pi}\left(\frac{2}{3}\cos x - \frac{2}{5}\cos 3x - \frac{1}{3}\cos 4x - \frac{2}{21}\cos 5x + \cdots\right)$$

$$+ \frac{1}{\pi}\left(-\frac{2}{3}\sin x + \frac{3\pi}{4}\sin 2x - \frac{2}{5}\sin 3x + \frac{2}{21}\sin 5x + \cdots\right).$$

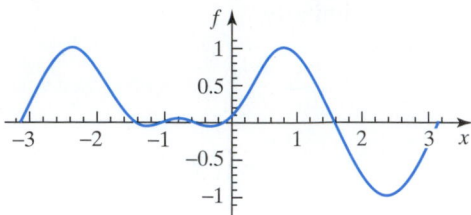

FIGURE 9.5 Fourier series approximation for $f(x)$.

This example illustrates how when a sine function (or a cosine function) with an argument mx with m an integer occurs in a piecewise defined function, its Fourier coefficients a_m and b_m must be found from the Euler formulas with n set equal to m before integration. Figure 9.5 shows a graph of this Fourier series approximation to $f(x)$ up to and including the terms in $\cos 5x$ and $\sin 5x$. ■

It is useful to have a special name for finite approximations to Fourier series such as the one used to construct the graph in Fig. 9.5. Because of this it is usual to call the approximation

$$S_N(x) = a_0 + \sum_{n=1}^{N}(a_n \cos nx + b_n \sin nx) \tag{12}$$

Nth partial sum

to the full Fourier series in (7) the **Nth partial sum** of the Fourier series. Thus, the graph in Fig. 9.5 shows the fifth partial sum $S_5(x)$ of the function $f(x)$ defined in Example 9.2. The Fourier series in (7) is related to its Nth partial sum $S_n(x)$ by the limit

$$f(x) = a_0 + \sum_{n=0}^{\infty}(a_n \cos nx + b_n \sin nx) = \lim_{N\to\infty} S_N(x). \tag{13}$$

Not every function has a Fourier series involving an infinite number of terms, as can be seen by considering the function $f(x) = 1 + 2\sin x \cos x$. When this is rewritten as $f(x) = 1 + \sin 2x$, it is recognized that it is, in fact, its own Fourier series.

There is nothing special about the choice of $-\pi \le x \le \pi$ as a fundamental interval, and it is often necessary to take the fundamental interval to be $-L \le x \le L$. Results (7) and (8) generalize immediately once it is recognized that the set of functions

$$1, \cos\frac{\pi x}{L}, \cos\frac{2\pi x}{L}, \cos\frac{3\pi x}{L}, \dots, \sin\frac{\pi x}{L}, \sin\frac{2\pi x}{L}, \sin\frac{3\pi x}{L}, \dots$$

form an orthogonal set over the interval $-L \le x \le L$. This can be seen by using routine integration to show that

$$\int_{-L}^{L} \sin\frac{m\pi x}{L} \cos\frac{n\pi x}{L} dx = 0 \quad \text{for all integers } m \text{ and } n, \tag{14}$$

$$\int_{-L}^{L} \sin\frac{m\pi x}{L} \sin\frac{n\pi x}{L} dx = \begin{cases} 0 & \text{for } m \ne n \\ L & \text{for } m = n \end{cases} \quad \text{for all integers } m \text{ and } n, \tag{15}$$

and

$$\int_{-L}^{L} \cos\frac{m\pi x}{L}\cos\frac{n\pi x}{L}dx = \begin{cases} 0 & \text{for } m \neq n \\ L & \text{for } m = n \neq 0 \text{ for all integers } m \text{ and } n \\ 2L & \text{for } m = n = 0. \end{cases}$$

(16)

The Fourier series of a function $f(x)$ defined on the interval $-L \leq x \leq L$ becomes

$$f(x) = a_0 + \sum_{n=1}^{\infty} \left(a_n \cos\frac{n\pi x}{L} + b_n \sin\frac{n\pi x}{L} \right),$$

(17)

and the corresponding Euler formulas for the a_n and b_n follow as before. The coefficients a_n are obtained by multiplying (17) by $\cos\frac{n\pi x}{L}$ and integrating over the interval $-L \leq x \leq L$, while the b_n follow by multiplying (17) by $\sin\frac{n\pi x}{L}$ and integrating over the same interval. The result is as follows, though the details are left as an exercise.

Fourier series representation of $f(x)$ over the interval $-L \leq x \leq L$

Fourier series over $-L \leq x \leq L$

Let the function $f(x)$ be defined on the interval $-L \leq x \leq L$. Then the Fourier coefficients a_n and b_n in the Fourier series representation of $f(x)$

$$f(x) = a_0 + \sum_{n=1}^{\infty} \left(a_n \cos\frac{n\pi x}{L} + b_n \sin\frac{n\pi x}{L} \right)$$

(18)

are given by the Euler formulas

$$a_0 = \frac{1}{2L}\int_{-L}^{L} f(x)dx, \quad a_n = \frac{1}{L}\int_{-L}^{L} f(x)\cos\frac{n\pi x}{L}dx,$$

$$b_n = \frac{1}{L}\int_{-L}^{L} f(x)\sin\frac{n\pi x}{L}dx, \quad \text{for } n = 1, 2, \ldots.$$

(19)

EXAMPLE 9.3

Find the Fourier series representation of $f(x) = x + 1$ for $-1 \leq x \leq 1$.

Solution   In this case $L = 1$, so using integration by parts we find that

$$a_0 = \frac{1}{2}\int_{-1}^{1}(x+1)dx = 1, \quad a_n = \int_{-1}^{1}(x+1)\cos n\pi x\, dx = \frac{\cos n\pi x}{n^2\pi^2} + \frac{x\sin n\pi x}{n\pi}$$

$$+ \frac{\sin n\pi x}{n\pi}\Big]_{-1}^{1} = 0$$

and

$$b_n = \int_{-1}^{1}(x+1)\sin n\pi x\, dx = \frac{\sin n\pi x}{n^2\pi^2} - \frac{x\cos n\pi x}{n\pi} - \frac{\cos n\pi x}{n\pi}\Big]_{-1}^{1} = \frac{2(-1)^{n+1}}{n\pi},$$

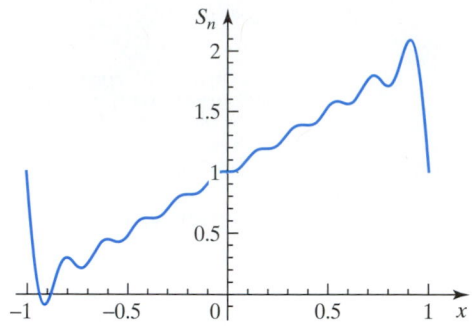

FIGURE 9.6   The partial sum approximation $S_{10}(x)$.

for $n = 1, 2, \ldots$, where we have used the fact that $\sin n\pi = 0$ and $\cos n\pi = (-1)^n$ for n a positive integer. Substituting these coefficients into (18) shows the required Fourier series representation to be

$$f(x) = 1 + \frac{2}{\pi} \sum_{n=1}^{\infty} \frac{(-1)^{n+1}}{n} \sin n\pi x, \quad \text{for } -1 \le x \le 1.$$

A graph of the partial sum approximation $S_{10}(x)$ to $f(x)$ is shown in Fig. 9.6.   ■

As cosines are even functions and sines are odd functions, it is to be expected that a Fourier series representation of an even function will only contain cosine terms, whereas a Fourier series representation of an odd function will only contain sine functions. These properties form the basis of the following result that simplifies the task of finding Fourier series representations of even and odd functions.

expanding even and odd functions

Fourier series of even and odd functions

If $f(x)$ is an even function defined on the interval $-L \le x \le L$, then

$$f(x) = a_0 + \sum_{n=1}^{\infty} a_n \cos \frac{n\pi x}{L}, \quad \text{with } a_0 = \frac{1}{L} \int_0^L f(x)\,dx,$$

$$a_n = \frac{2}{L} \int_0^L f(x) \cos \frac{n\pi x}{L}\,dx$$

for $n = 1, 2, \ldots$; if $f(x)$ is an odd function, then

$$f(x) = \sum_{n=1}^{\infty} b_n \sin \frac{n\pi x}{L}, \quad \text{with } b_n = \frac{2}{L} \int_0^L f(x) \sin \frac{n\pi x}{L}\,dx,$$
$$\text{for } n = 1, 2, \ldots,$$

The justification of these results is as follows. To find the form taken by the Fourier coefficients a_n of an even function, and why its Fourier coefficients b_n vanish, we will consider an even function $f(x)$ defined over the interval $-L \le x \le L$.

By definition,

$$a_0 = \frac{1}{2L}\int_{-L}^{L} f(x)dx = \frac{1}{2L}\int_{-L}^{0} f(x)dx + \frac{1}{2L}\int_{0}^{L} f(x)dx.$$

Setting $x = -u$ in the first integral on the right gives

$$\frac{1}{2L}\int_{-L}^{0} f(x)dx = -\frac{1}{2L}\int_{L}^{0} f(-u)du.$$

As f is an even function, $f(-u) = f(u)$, so using this result, changing the sign of the integral by interchanging its limits, and then replacing the dummy variable u by x gives

$$\frac{1}{2L}\int_{-L}^{0} f(x)dx = \frac{1}{2L}\int_{0}^{L} f(x)dx.$$

When this is combined with the original expression for a_0 we find that

$$a_0 = \frac{1}{L}\int_{0}^{L} f(x)dx,$$

and a strictly analogous argument shows that

$$a_n = \frac{2}{L}\int_{0}^{L} f(x)\cos n\pi x\,dx \quad \text{for } n = 1, 2, \ldots.$$

The Fourier coefficients b_n are given by

$$b_n = \frac{1}{L}\int_{-L}^{L} f(x)\sin\frac{n\pi x}{L}dx = \frac{1}{L}\int_{-L}^{0} f(x)\sin\frac{n\pi x}{L}dx + \frac{1}{L}\int_{0}^{L} f(x)\sin\frac{n\pi x}{L}dx.$$

Setting $x = -u$ in the integral taken over the interval $-L \le x \le 0$ gives

$$\frac{1}{L}\int_{-L}^{0} f(x)\sin\frac{n\pi x}{L}dx = -\frac{1}{L}\int_{L}^{0} f(-u)\sin\left(-\frac{n\pi u}{L}\right)du.$$

We now use the fact that f is an even function, so $f(-u) = f(u)$, together with the fact that the sine function is an odd function. Reversal of the limits coupled with changing the sign and replacing u by x gives

$$\frac{1}{L}\int_{-L}^{0} f(x)\sin\frac{n\pi x}{L}dx = -\frac{1}{L}\int_{0}^{L} f(x)\sin\frac{n\pi x}{L}dx.$$

Finally, using this result in the original expression for b_n gives

$$b_n = \frac{1}{L}\int_{0}^{L} f(x)\sin\frac{n\pi x}{L}dx - \frac{1}{L}\int_{0}^{L} f(x)\sin\frac{n\pi x}{L}dx = 0 \quad \text{for } n = 1, 2, \ldots,$$

and the result is proved.

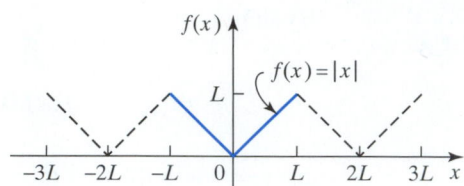

FIGURE 9.7   The function $f(x) = |x|$ in
$-L \le x \le L$ and two periodic extensions.

A similar argument shows that if $f(x)$ is an odd function over $-L \le x \le L$, then

$$a_n = 0 \quad \text{for } n = 0, 1, 2, \ldots,$$

and

$$b_n = \frac{2}{L} \int_0^L f(x) \sin \frac{n\pi x}{L} dx \quad \text{for } n = 1, 2, \ldots,$$

and the results have been established.

EXAMPLE 9.4   Find the Fourier series representation of $f(x) = |x|$ in the interval $-L \le x \le L$.

Solution   The graph of this even function, together with two of its periodic extensions outside the fundamental interval $-L \le x \le L$, is shown in Fig. 9.7.

The Euler formula for the coefficients a_n of the *even* function $|x|$ defined as

$$|x| = \begin{cases} -x & \text{for } < 0 \\ x & \text{for } x \ge 0 \end{cases}$$

gives

$$a_0 = \frac{1}{L} \int_0^L x\, dx = \frac{L}{2}$$

and

$$a_n = \frac{2}{L} \int_0^L x \cos \frac{n\pi x}{L} dx = \frac{2}{L} \left[\frac{L^2 \cos \frac{n\pi x}{L}}{n^2 \pi^2} + \frac{L n\pi x \sin \frac{n\pi x}{L}}{n^2 \pi^2} \right]_0^L, \quad \text{for } n = 1, 2, \ldots.$$

If we use the fact that $\sin n\pi = 0$ and $\cos n\pi = (-1)^n$ when n is a positive integer, it then follows that

a convenient representation of $\cos n\pi$

$$a_n = \frac{2L}{n^2 \pi^2}[(-1)^n - 1] \quad \text{for } n = 1, 2, \ldots,$$

and so

$$a_n = -\frac{4L}{n^2 \pi^2} \quad \text{when } n \text{ is odd}$$

and

$$a_n = 0 \quad \text{when } n \ne 0, \text{ is even.}$$

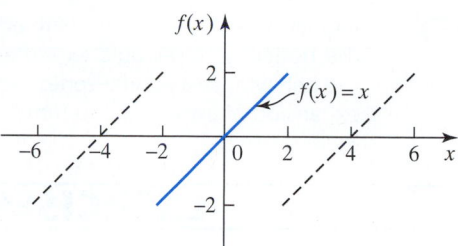

FIGURE 9.8   The function $f(x) = x$ in $-2 \le x \le 2$ and two periodic extensions.

Thus, the Fourier series representation of $f(x) = |x|$ for $-L \le x \le L$ is

$$f(x) = \frac{L}{2} - \frac{4L}{\pi^2} \left(\frac{\cos \dfrac{\pi x}{L}}{1^2} + \frac{\cos \dfrac{3\pi x}{L}}{3^2} + \frac{\cos \dfrac{5\pi x}{L}}{5^2} + \cdots \right).$$

The sequence of positive odd numbers can be written in the form $2n - 1$ with $n = 1, 2, \ldots$, so this last result can be expressed more concisely as

$$f(x) = \frac{L}{2} - \frac{4L}{\pi^2} \sum_{n=1}^{\infty} \frac{\cos \left(\dfrac{(2n-1)\pi x}{L} \right)}{(2n-1)^2} \quad \text{for } -L \le x \le L. \qquad \blacksquare$$

EXAMPLE 9.5   Find the Fourier series representation of $f(x) = x$ on the interval $-2 \le x \le 2$.

Solution   A graph of $f(x)$ and two of its periodic extensions outside the fundamental interval $-2 \le x \le 2$ is shown in Fig. 9.8.

Using the fact that $L = 2$, a straightforward calculation gives

$$b_n = \frac{1}{2} \int_{-2}^{2} x \sin \frac{n\pi x}{2} dx = \frac{2}{n^2 \pi^2} \left[\sin \frac{n\pi x}{2} - \frac{1}{2} n\pi x \cos \frac{n\pi x}{2} \right]_{-2}^{2}$$

$$= -\frac{4 \cos n\pi}{n\pi} = \frac{4(-1)^{n+1}}{n\pi},$$

and as the function is odd all the coefficients $a_n = 0$.

The required Fourier series representation is thus

$$f(x) = \frac{4}{\pi} \left(\frac{\sin \dfrac{\pi x}{2}}{1} - \frac{\sin \pi x}{2} + \frac{\sin \dfrac{3\pi x}{2}}{3} - \cdots \right),$$

which can be written in the more concise form

$$f(x) = \frac{4}{\pi} \sum_{n=1}^{\infty} \frac{(-1)^{n+1}}{n} \sin \frac{n\pi x}{2} \quad \text{for } -2 \le x \le 2. \qquad \blacksquare$$

Summary

Fourier series have been defined over more general intervals than $-\pi \leq x \leq \pi$ and the notion of a periodic extension has been introduced. Attention has been drawn to the behavior of a Fourier series representation at a point of discontinuity of $f(x)$, and the expansion of even and odd functions has been considered.

EXERCISES 9.1

Find the period of each of the functions in Exercises 1 through 6.

1. $\cos x + \sin 2x$.

2. $2 \sin 2x - 3 \cos \dfrac{x}{3}$.

3. $\sin x \cos x$.

4. $\cos 2x \sin x$.

5. $3 \sin \dfrac{x}{3} + \cos \dfrac{x}{2}$.

6. $\cos \dfrac{x}{3} + 5 \sin \dfrac{x}{4}$.

In Exercises 7 through 10 (a) sketch the given function in the interval $-3a < x < 3a$, and (b) in the intervals $-3a < x < -a$ and $a < x < 3a$, and state whether the function is periodic.

7. $f(x) = \begin{cases} 0, & x < a/2 \\ 1, & x > a/2. \end{cases}$

8. $f(x) = \begin{cases} -1, & -a < x < 0 \\ 2, & 0 < x < a, \end{cases} \quad f(x + 2a) = f(x).$

9. $f(x) = a - |x|$.

10. $f(x) = |\sin \pi x/a|$.

In Exercises 11 and 12 make use of the trigonometric identities $\sin(A \pm B) = \sin A \cos B \pm \cos A \sin B$ and $\cos(A \pm B) = \cos A \cos B + \sin A \sin B$ to transform the given functions into their (finite) Fourier series.

11. (a) $\sin x \cos x$. (b) $1 - 2 \sin^2 x$. (c) $\sin 3x \cos x$.

12. (a) $4 \cos 2x \cos 5x$. (b) $\sin x \sin 2x$. (c) $\cos^2 2x - 1/2$.

Verify the following definite integrals that were used when developing a Fourier series representation over the interval $-L < x < L$.

13. $\displaystyle \int_{-L}^{L} \sin \frac{m\pi x}{L} \cos \frac{n\pi x}{L} dx = 0$ for all integers m and n.

14. $\displaystyle \int_{-L}^{L} \sin \frac{m\pi x}{L} \sin \frac{n\pi x}{L} dx = \begin{cases} 0 & \text{for } m \neq n \\ L & \text{for } m = n, \end{cases}$
    with m, n integers.

15. $\displaystyle \int_{-L}^{L} \cos \frac{m\pi x}{L} \cos \frac{n\pi x}{L} dx$

$= \begin{cases} 0 & \text{for } m \neq n \\ L & \text{for } m = n \neq 0 \\ 2L & \text{for } m = n = 0 \end{cases}$ for all integers m and n.

16. Prove that the product of two even functions and of two odd functions is an even function, and that the product of an even and an odd function is an odd function.

17. Prove that the sum of two even functions is an even function and the sum of two odd functions is an odd function.

18. Prove that if $f(x)$ is an odd function all the Fourier coefficients $a_n = 0$.

19. Evaluate the following integrals that arise when finding the Fourier series expansion of x over the interval $-L < x < L$.

(a) $\displaystyle \int_{-L}^{L} x \sin \frac{\pi x}{L} dx$. (b) $\displaystyle \int_{-L}^{L} x \sin \frac{2\pi x}{L} dx$.

(c) $\displaystyle \int_{-L}^{L} x \sin \frac{3\pi x}{L} dx$.

20. Evaluate the following integrals that arise when finding the Fourier series expansion of x^2 over the interval $-L < x < L$.

(a) $\displaystyle \int_{-L}^{L} x^2 \sin \frac{\pi x}{L} dx$. (b) $\displaystyle \int_{-L}^{L} x^2 \sin \frac{2\pi x}{L} dx$.

(c) $\displaystyle \int_{-L}^{L} x^2 \sin \frac{3\pi x}{L} dx$.

The integrals in Exercises 21 and 22 arise when finding the Fourier series expansion of e^{ax} over the interval $-L < x < L$. Use the result $\cos n\pi = (-1)^n$ for integral values of n to establish the stated result.

21. $\displaystyle \int_{-\pi}^{\pi} e^{ax} \sin nx \, dx = (-1)^{n+1} \frac{n(e^{a\pi} - e^{-a\pi})}{(a^2 + n^2)}$ for integral values of n.

22. $\displaystyle \int_{-\pi}^{\pi} e^{ax} \cos nx \, dx = (-1)^n \frac{a(e^{a\pi} - e^{-a\pi})}{(a^2 + n^2)}$ for integral values of n.

In Exercises 23 through 35 find the Fourier series representation of the given function over the indicated fundamental interval and use a computer to plot the indicated partial sum $S_n(x)$ over the fundamental interval.

23. $f(x) = \begin{cases} a, & -\pi < x < 0 \\ b, & 0 < x < \pi. \end{cases}$ Plot $S_{10}(x)$ for $a = 3, b = 1$.

24. $f(x) = \begin{cases} x + 1, & -1 < x < 0 \\ x - 1, & 0 < x < 1. \end{cases}$ Plot $S_{10}(x)$.

25. $f(x) = 1 - |x|$, $-1 < x < 1$. Plot $S_{10}(x)$.

26. $f(x) = \begin{cases} 0, & -2 < x < 0 \\ x, & 0 \le x < 2. \end{cases}$ Plot $S_8(x)$.

27. $f(x) = |\sin x|, -\pi \le x \le \pi$ (a fully rectified sine wave). Plot $S_{10}(x)$.

28. $f(x) = \begin{cases} ax, & -\pi < x \le 0 \\ bx, & 0 \le x < \pi. \end{cases}$ Plot $S_8(x)$ for $a = 1, b = 3$.

29. $f(x) = \begin{cases} 0, & -\pi \le x \le 0 \\ \sin x, & 0 \le x \le \pi. \end{cases}$ Plot $S_8(x)$.

30. $f(x) = x^2, \quad -\pi \le x \le \pi$. Plot $S_8(x)$.

31. $f(x) = x^2, \quad -2\pi \le x \le 2\pi$. Plot $S_{10}(x)$.

32. $f(x) = \sin ax, \quad -\pi \le x \le \pi$ with a not an integer. Plot $S_{10}(x)$ for $a = 0.7$.

33. $f(x) = \cos ax, \quad -\pi \le x \le \pi$ with a not an integer. Plot $S_{10}(x)$ for $a = 0.7$.

34. $f(x) = e^{ax}, \quad -\pi \le x \le \pi$. Plot $S_7(x)$ for $a = 0.7$.

35. $f(x) = \begin{cases} 0, & -2\pi \le x < -\pi \\ \sin x, & -\pi \le x \le \pi \\ 0, & \pi \le x \le 2\pi. \end{cases}$ Plot $S_8(x)$.

9.2 Convergence of Fourier Series and Their Integration and Differentiation

The general theory of the convergence of Fourier series is complicated and still incomplete in some respects. Consequently, we will only derive some useful results that can be obtained in a straightforward manner, and then state without proof a convergence theorem due to the German mathematician P. G. L. Dirichlet (1805–1859) that is sufficient for all practical applications of Fourier series.

Let us consider the nth partial sum

$$S_n(x) = a_0 + \sum_{r=1}^{n}(a_r \cos rx + b_r \sin rx), \qquad (20)$$

of the Fourier series for $f(x)$ in (7) defined over the interval $-\pi \le x \le \pi$. Then, provided the integral $\int_{-\pi}^{\pi}[f(x)]^2 dx$ exists and is finite, we have the obvious result

$$\int_{-\pi}^{\pi}[f(x) - S_n(x)]^2 dx = \int_{-\pi}^{\pi}[f(x)]^2 dx - 2\int_{-\pi}^{\pi}f(x)S_n(x)dx + \int_{-\pi}^{\pi}[S_n(x)]^2 dx. \qquad (21)$$

From the definition of $S_n(x)$ in (20), it follows that

$$\int_{-\pi}^{\pi}[S_n(x)]^2 dx = \int_{-\pi}^{\pi}\left[a_0 + \sum_{r=1}^{n}(a_r \cos rx + b_r \sin rx)\right]^2 dx,$$

but the orthogonality of the sine and cosine functions reduces this to

$$\int_{-\pi}^{\pi}[S_n(x)]^2 dx = \int_{-\pi}^{\pi}a_0^2 dx + \sum_{r=1}^{n}\left[a_r^2\int_{-\pi}^{\pi}\cos^2 rx dx\right] + \sum_{r=1}^{n}\left[b_r^2\int_{-\pi}^{\pi}\sin^2 rx dx\right]$$

$$= \pi\left[2a_0^2 + \sum_{r=1}^{n}(a_r^2 + b_r^2)\right]. \qquad (22)$$

If $f(x)$ is replaced by its Fourier series, a similar argument shows that

$$\int_{-\pi}^{\pi}f(x)S_n(x)dx = \pi\left[2a_0^2 + \sum_{r=1}^{n}(a_r^2 + b_r^2)\right], \qquad (23)$$

so combining (21) to (23) gives

$$\int_{-\pi}^{\pi}[f(x) - S_n(x)]^2 dx = \int_{-\pi}^{\pi}[f(x)]^2 dx - \pi\left[2a_0^2 + \sum_{r=1}^{n}(a_r^2 + b_r^2)\right]. \qquad (24)$$

The integral on the left of (24) is nonnegative, because its integrand is a squared quantity, so it follows at once that for all n

$$2a_0^2 + \sum_{r=1}^{n}\left(a_r^2 + b_r^2\right) \le \frac{1}{\pi}\int_{-\pi}^{\pi}[f(x)]^2 dx,$$

so letting $n \to \infty$ we arrive at the inequality

$$2a_0^2 + \sum_{r=1}^{\infty}\left(a_r^2 + b_r^2\right) \le \frac{1}{\pi}\int_{-\pi}^{\pi}[f(x)]^2 dx. \tag{25}$$

Bessel's inequality

This is **Bessel's inequality** for Fourier series, and the restriction to functions $f(x)$ such that $\int_{-\pi}^{\pi}[f(x)]^2 dx$ exists and is finite implies that the series

$$2a_0^2 + \sum_{r=1}^{\infty}\left(a_r^2 + b_r^2\right)$$

is convergent, so the coefficients in the associated Fourier series (7) must be such that

$$\lim_{n\to\infty} a_n = 0 \quad \text{and} \quad \lim_{n\to\infty} b_n = 0. \tag{26}$$

the fundamental Riemann–Lebesgue lemma

This important result on the behavior of Fourier coefficients as $n \to \infty$ is called the **Riemann–Lebesgue** lemma, though its rigorous proof proceeds differently.

It is also a consequence of (24) that if the nth partial sum $S_n(x)$ converges to $f(x)$ in the sense that

$$\lim_{n\to\infty}\int_{-\pi}^{\pi}[f(x) - S_n(x)]^2 dx = 0,$$

which is true for all functions $f(x)$ encountered in applications, then

$$2a_0^2 + \sum_{r=1}^{\infty}\left(a_r^2 + b_r^2\right) = \frac{1}{\pi}\int_{-\pi}^{\pi}[f(x)]^2 dx. \tag{27}$$

Parseval relation

This is the **Parseval relation** for Fourier series.

EXAMPLE 9.6

Apply the Parseval relation to the Fourier series of $f(x) = |x|$ defined over the interval $-\pi \le x \le \pi$.

Solution   It follows from Example 9.4 with $L = \pi$ that the Fourier series representation of $f(x) = |x|$ over the interval $-\pi \le x \le \pi$ is

$$f(x) = \frac{\pi}{2} - \frac{4}{\pi}\sum_{n=1}^{\infty}\frac{\cos(2n-1)x}{(2n-1)^2},$$

so that

$$a_0 = \frac{\pi}{2}, \quad a_{2n-1} = -\frac{4}{\pi(2n-1)^2}, \quad \text{and} \quad a_{2n} = 0 \quad \text{for } n = 1, 2, \ldots.$$

We have

$$\int_{-\pi}^{\pi}[f(x)]^2 dx = \int_{-\pi}^{\pi} x^2 dx = \frac{2\pi^3}{3},$$

so as the integral is finite, provided $S_n(x)$ converges in the norm to $f(x)$, it follows from the Parseval relation in (27) that

$$\frac{1}{\pi}\left(\frac{2\pi^3}{3}\right) = 2\frac{\pi^2}{4} + \frac{16}{\pi^2}\sum_{n=1}^{\infty}\frac{1}{(2n-1)^4}.$$

After simplification this reduces to the well-known result

$$\frac{\pi^4}{96} = \sum_{n=1}^{\infty}\frac{1}{(2n-1)^4}$$

$$= \frac{1}{1^4} + \frac{1}{3^4} + \frac{1}{5^4} + \frac{1}{7^4} + \cdots.$$

The justification for applying the Parseval relation in this case is provided by the following theorem. It can be confirmed by summing a large number of terms and comparing the result with the known value of $\pi^4/96$. For example, using $n = 100$ leads to the result $\pi^4/96 \approx 1.01467801$, while a direct calculation shows that $\pi^4/96 = 1.01467803$, so the two results agree to seven decimal places. ■

THEOREM 9.1

Convergence of Fourier series   Let $f(x)$ be continuous over the interval $-L < x < L$ except possibly at a finite number of internal points $x_1, x_2, \ldots$, at each point x_n of which the function has a finite jump discontinuity $f(x_n+) - f(x_n-)$. Furthermore, let the left- and right-hand derivatives $f'(x_n-)$ and $f'(x_n+)$ exist for $n = 1, 2, \ldots$. Then at points of continuity of $f(x)$ its Fourier series converges uniformly to $f(x)$, and at each point of discontinuity it converges pointwise to

fundamental convergence theorem

$$\frac{1}{2}(f(x_n-) + f(x_n+)) \quad \text{for } n = 1, 2, \ldots.$$

If, in addition, $f(x)$ has a right-hand derivative $f'(-L+)$ at the left end point of the interval and a left-hand derivative $f'(L-)$ at the right end point of the interval, then at $x = \pm L$ the Fourier series converges pointwise to

$$\frac{1}{2}(f(-L+) + f(L-)).$$
■

In effect, this theorem says that if $f(x)$ is piecewise continuous and bounded over the interval $-L < x < L$ with derivatives defined to the left and right of each discontinuity, its Fourier series converges uniformly to $f(x)$ wherever it is continuous and to the mid-point of the jump where there is a discontinuity. If, in addition, one-sided derivatives exist at the ends of the interval, then at both $x = -L$ and $x = L$ the Fourier series converges to the average of the values of $f(x)$ at the two ends of the interval.

A consequence of this theorem that is sometimes useful is that it allows many numerical series to be summed in closed form. Results of this type follow by choosing a value of x for which the terms of the Fourier series take on a simple numerical form, and equating the result to the appropriate value of $f(x)$. At a point $x = x^*$ where $f(x)$ is continuous the series will converge to $f(x^*)$, and at a point $x = x^*$ where $f(x)$ is discontinuous the series will converge to the mid-point of the jump.

EXAMPLE 9.7

(a) Given that the step function

$$f(x) = \begin{cases} -1, & \text{for } -\pi < x < 0 \\ 1, & \text{for } 0 < x < \pi \end{cases}$$

has the Fourier series

$$f(x) = \frac{4}{\pi} \sum_{n=1}^{\infty} \frac{\sin(2n-1)x}{2n-1},$$

find a series for $\pi/4$.

(b) Given that

$$f(x) = \begin{cases} 0, & \text{for } -\pi < x < 0 \\ x^2, & \text{for } 0 \leq x < \pi \end{cases}$$

has the Fourier series

$$f(x) = \frac{\pi^2}{6} + \sum_{n=1}^{\infty} \left\{ \left[\frac{2(-1)^n}{n^2} \right] \cos nx + \frac{1}{\pi} \left[(-1)^n \left(\frac{2}{n^3} - \frac{\pi^2}{n} \right) - \frac{2}{n^3} \right] \sin nx \right\},$$

find a series for $\pi^2/6$.

Solution

(a) The function $f(x)$ graphed in Fig. 9.9 is seen to be discontinuous at $x = 0$ and to have different values at $x = \pm\pi$. The average of the values of $f(x)$ to the immediate left and right of the discontinuity at $x = 0$ is zero, so the Fourier series will converge to the value zero when $x = 0$. Setting $x = 0$ in the Fourier series causes every term to vanish, so equating this to the value to which the Fourier series converges at the origin yields the uninteresting result $0 = 0$.

how Fourier series can be used to sum series

To obtain a more interesting result, let us try setting $x = \pi/2$, which makes $\sin(2n-1)\frac{\pi}{2} = (-1)^{n+1}$. The function $f(x)$ is continuous at this point and equal to 1, so its Fourier series will converge to the value 1 when $x = \pi/2$. Inserting this value of x into the Fourier series and equating the result to 1 gives

$$1 = \frac{4}{\pi} \left(\frac{1}{1} - \frac{1}{3} + \frac{1}{5} - \cdots \right),$$

so

$$\frac{\pi}{4} = \frac{1}{1} - \frac{1}{3} + \frac{1}{5} - \cdots = \sum_{n=1}^{\infty} \frac{(-1)^{n+1}}{(2n-1)}.$$

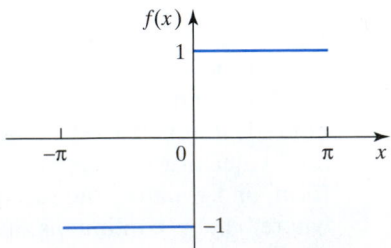

FIGURE 9.9 The step function $f(x)$.

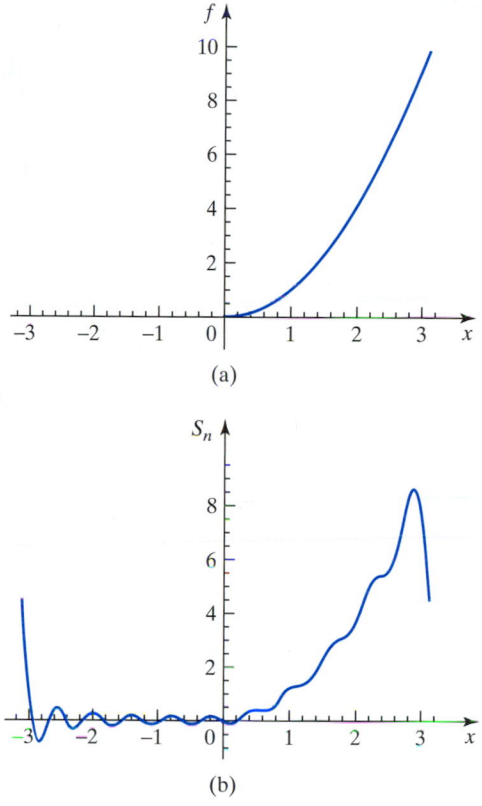

FIGURE 9.10   (a) The function $f(x)$ and (b) $S_{10}(x)$.

This series, known as Leibniz' formula, converges very slowly, so it is not useful for computing π.

(b) The function $f(x)$ is graphed in Fig. 9.10(a), and $S_{10}(x)$ in Fig. 9.10(b). The average of the values of $f(x)$ at the end points of the interval $-\pi < x < \pi$ is $\pi^2/2$, so setting $x = \pi$ in the Fourier series and equating the result to $\pi^2/2$ as required by the last part of Theorem 9.2 gives

$$\frac{\pi^2}{2} = \frac{\pi^2}{6} + 2\sum_{n=1}^{\infty}\frac{1}{n^2},$$

where we have used the fact that $\cos n\pi = (-1)^n$ and $\sin n\pi = 0$ for positive integers n.

This result simplifies to the series

$$\frac{\pi^2}{6} = 1 + \frac{1}{2^2} + \frac{1}{3^2} + \cdots = \sum_{n=1}^{\infty}\frac{1}{n^2},$$

which converges somewhat faster than the series in part (a).   ∎

Examination of Fig. 9.3 and also Fig. 9.6 in Section 9.1 shows that when $f(x)$ is discontinuous, the graph of the partial sum $S_n(x)$ of the Fourier series representation of the function exhibits over- and undershoots close to the discontinuities. This is called the **Gibbs phenomenon**, and it persists for all values of n. This behavior

Gibbs phenomenon

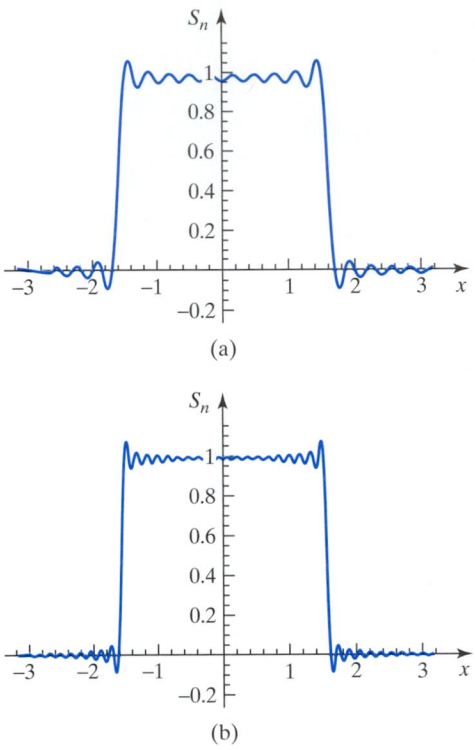

FIGURE 9.11   An example of the Gibbs phenomenon with (a) $n = 10$, and (b) $n = 20$.

reflects the way the continuous function $S_n(x)$ obtained from the Fourier series approximates the behavior of $f(x)$ at a point of discontinuity. Increasing n simply moves the under- and overshoots closer to the discontinuity while leaving their size approximately the same.

Figure 9.11 shows the Gibbs phenomena for the function

$$f(x) = \begin{cases} 0, & -\pi < x < -\pi/2 \\ 1, & -\pi/2 < x < \pi/2 \\ 0, & \pi/2 < x < \pi \end{cases}$$

for different partial sums $S_n(x)$. The results should be compared with Fig. 9.3, which shows the graph of $S_5(x)$.

We now state without proof two important theorems concerning the term-by-term integration and differentiation of Fourier series that are often useful, but before doing so we first define what are called Dirichlet conditions, which are satisfied by most functions of practical importance.

A function $f(x)$ is said to satisfy **Dirichlet conditions** on an interval $-L < x < L$ if it is bounded on the interval, has at most a finite number of maxima and minima, and is continuous apart from a finite number of discontinuities in the interval.

THEOREM 9.2

when a Fourier series can be integrated

Termwise integration of Fourier series   The integral of any function $f(x)$ satisfying Dirichlet conditions on the interval $-L \leq x \leq L$ can be obtained by term-by-term integration of the Fourier series representation of $f(x)$. So, if $f(x)$ has the Fourier series representation

$$f(x) = a_0 + \sum_{n=1}^{\infty} \left(a_n \cos\left(\frac{n\pi x}{L}\right) + b_n \sin\left(\frac{n\pi x}{L}\right) \right) \quad \text{for } -L \le x \le L,$$

then

$$\int_{-L}^{x} f(u)du = a_0(x + L)$$
$$+ \frac{L}{\pi} \sum_{n=1}^{\infty} \left[\frac{a_n}{n} \sin\left(\frac{n\pi x}{L}\right) - \frac{b_n}{n} \left(\cos\left(\frac{n\pi x}{L}\right) + (-1)^{n+1} \right) \right]$$
$$\text{for } -L \le x \le L.$$

■

THEOREM 9.3

when a Fourier series can be differentiated

Termwise differentiation of Fourier series   Let $f(x)$ be a continuous function on the interval $-L \le x \le L$ such that $f(-L) = f(L)$, and suppose also that $f'(x)$ is piecewise continuous. Then for any x strictly inside the interval at which $f''(x)$ exists, the derivative of $f(x)$ can be obtained by term-by-term differentiation of the Fourier series representation of $f(x)$. So, if $f(x)$ has the Fourier series representation

$$f(x) = a_0 + \frac{\pi}{L} \sum_{n=1}^{\infty} \left(a_n \cos\left(\frac{n\pi x}{L}\right) + b_n \sin\left(\frac{n\pi x}{L}\right) \right) \quad \text{for } -L \le x \le L,$$

then

$$f'(x) = \frac{\pi}{L} \sum_{n=1}^{\infty} \left(-na_n \sin\left(\frac{n\pi x}{L}\right) + nb_n \cos\left(\frac{n\pi x}{L}\right) \right) \quad \text{for } -L < x < L,$$

except for points at where $f'(x)$ and $f''(x)$ are not defined.   ■

EXAMPLE 9.8

Use the Fourier series representation of the function

$$f(x) = \begin{cases} -1, & -\pi < x < 0 \\ 1, & 0 < x < \pi \end{cases}$$

given in Example 9.7 to find a Fourier series representation of $F(x) = \int_{-\pi}^{x} f(t)dt$ in the interval $-\pi < x < \pi$, and relate the result to Example 9.4.

Solution   As $f(x)$ satisfies the conditions of Theorem 9.2, its Fourier series representation may be integrated term by term to obtain the Fourier series representation of

$$F(x) = \int_{-\pi}^{x} f(t)dt = \begin{cases} \int_{-\pi}^{x} -1 dt = -(x + \pi), & \text{for } -\pi < x < 0 \\ \int_{-\pi}^{0} -1 dt + \int_{0}^{x} 1 dt = x - \pi & \text{for } 0 < x < \pi. \end{cases}$$

From Example 9.7, the Fourier series representation of $f(x)$ is

$$f(x) = \frac{4}{\pi} \sum_{n=1}^{\infty} \frac{\sin(2n-1)x}{2n-1},$$

so replacing x by the dummy variable t and integrating over the interval $-\pi \le t \le x$ gives

$$F(x) = \frac{4}{\pi} \sum_{n=1}^{\infty} \int_{-\pi}^{x} \frac{\sin(2n-1)t}{2n-1} dt = -\frac{4}{\pi} \left[\sum_{n=1}^{\infty} \frac{\cos(2n-1)x}{(2n-1)^2} - \sum_{n=1}^{\infty} \frac{\cos(2n-1)\pi}{(2n-1)^2} \right].$$

As $\cos(2n-1)\pi = -1$ for $n = 1, 2, \ldots$, this reduces to

$$F(x) = -\frac{4}{\pi} \sum_{n=1}^{\infty} \frac{\cos(2n-1)x}{(2n-1)^2} - \frac{4}{\pi} \sum_{n=1}^{\infty} \frac{1}{(2n-1)^2}.$$

The numerical series on the right can be summed by applying the Parseval relation to the Fourier series representation of $f(x)$ to obtain

$$2 = \sum_{n=1}^{\infty} \left(\frac{4}{\pi(2n-1)} \right)^2, \quad \text{or} \quad \frac{\pi^2}{8} = \sum_{n=1}^{\infty} \frac{1}{(2n-1)^2}.$$

Replacing the numerical series in $F(x)$ by $\pi^2/8$ reduces it to

$$\int_{-\pi}^{x} f(t)dt = -\frac{4}{\pi} \sum_{n=1}^{\infty} \frac{\cos(2n-1)x}{(2n-1)^2} - \frac{4}{\pi} \frac{\pi^2}{8} = -\frac{\pi}{2} - \frac{4}{\pi} \sum_{n=1}^{\infty} \frac{\cos(2n-1)x}{(2n-1)^2},$$

and so the required Fourier series representation is

$$F(x) = \begin{cases} \displaystyle\int_{-\pi}^{x} -1 dt = -(x+\pi), & \text{for } -\pi < x < 0 \\[3mm] \displaystyle\int_{-\pi}^{0} -1 dt + \int_{0}^{x} 1 dt = x - \pi, & \text{for } 0 < x < \pi \end{cases}$$

$$= -\frac{\pi}{2} - \frac{4}{\pi} \sum_{n=1}^{\infty} \frac{\cos(2n-1)x}{(2n-1)^2}.$$

Examination of $F(x)$ shows that $F(x) = |x| - \pi$, so as a check we see that the Fourier series representation of the function $|x|$ in the interval $-\pi \le x \le \pi$ can be obtained by adding π to the Fourier series representation of $F(x)$ to obtain

$$|x| = \frac{\pi}{2} - \frac{4}{\pi} \sum_{n=1}^{\infty} \frac{\cos(2n-1)x}{(2n-1)^2}, \quad \text{for } -\pi \le x \le \pi,$$

in agreement with the result of Example 9.4 with $L = \pi$. ∎

EXAMPLE 9.9   Given

$$f(x) = \begin{cases} \sin 2x, & -\pi \le x < -\pi/2 \\ 0, & -\pi/2 \le x \le \pi/2 \\ \sin 2x, & \pi/2 < x \le \pi, \end{cases}$$

find $f'(x)$ by differentiation of the Fourier series representation of $f(x)$.

Solution   The function satisfies the conditions of Theorem 9.3, so its Fourier series representation may be differentiated term by term to find the Fourier series representation of $f'(x)$. It was shown in Example 9.2 that the Fourier series representation of $f(x)$ is

$$f(x) = \frac{1}{2\pi} + \frac{1}{\pi}\left(\frac{2}{3}\cos x - \frac{2}{5}\cos 3x - \frac{1}{3}\cos 4x - \cdots\right)$$

$$+ \frac{1}{\pi}\left(-\frac{2}{3}\sin x + \frac{3\pi}{4}\sin 2x - \frac{2}{5}\sin 3x + \cdots\right),$$

so differentiation shows the first few terms of the Fourier series for $f'(x)$ to be

$$f'(x) = \frac{1}{\pi}\left(-\frac{2}{3}\sin x + \frac{6}{5}\sin 3x + \cdots\right) + \frac{1}{\pi}\left(-\frac{2}{3}\cos x + \frac{3\pi}{2}\cos 2x - \cdots\right),$$

where from the definition of $f(x)$

$$f'(x) = \begin{cases} 2\cos 2x, & -\pi \le x < -\pi/2 \\ 0, & -\pi/2 \le x \le \pi/2 \\ 2\cos 2x, & \pi/2 < x \le \pi. \end{cases}$$ ■

Summary

The convergence of Fourier series has been examined, and it has been shown that where $f(x)$ is continuous its Fourier series representation converges to $f(x)$, but where it has a finite jump discontinuity it converges to the mid-point of the jump. The Bessel inequality and the Parseval relation have been established, and conditions given for the termwise integration and differentiation of a Fourier series.

EXERCISES 9.2

In Exercises 1 through 4, apply the Parseval relation to the given function and its Fourier series to obtain a series representation involving a power of π.

1. $f(x) = \begin{cases} -1, & -\pi < x < 0 \\ 1, & 0 < x < \pi \end{cases}$

   with $f(x) = \dfrac{4}{\pi}\displaystyle\sum_{n=1}^{\infty}\dfrac{\sin(2n-1)x}{2n-1}$.

2. $f(x) = x, -\pi < x < \pi$

   with $f(x) = 2\displaystyle\sum_{n=1}^{\infty}(-1)^{n+1}\dfrac{\sin nx}{n}$.

3. $f(x) = x^2, -\pi \le x \le \pi$,

   with $f(x) = \dfrac{\pi^2}{3} - 4\displaystyle\sum_{n=1}^{\infty}(-1)^{n+1}\dfrac{\cos nx}{n^2}$.

4. $f(x) = |\cos x|, -\pi \le x \le \pi$

   with $f(x) = \dfrac{2}{\pi} + \dfrac{4}{\pi}\displaystyle\sum_{n=1}^{\infty}(-1)^{n+1}\dfrac{\cos 2nx}{(4n^2-1)}$.

5. Show that the **Parseval relation** for a function $f(x)$ defined on the interval $-L < x < L$ takes the form

   $$\frac{1}{L}\int_{-L}^{L}[f(x)]^2 dx = 2a_0^2 + \sum_{n=1}^{\infty}\left(a_n^2 + b_n^2\right).$$

6. Find the Fourier series for the function

   $$f(x) = \begin{cases} 0, & -4 \le x < 0 \\ 4, & 0 \le x < 4 \end{cases}$$

   and apply the Parseval relation in Exercise 5 to the result.

7. Use the Fourier series in Example 10.6(b) for the function

   $$f(x) = \begin{cases} 0, & \text{for } -\pi \le x \le 0 \\ x^2, & \text{for } 0 < x < \pi \end{cases}$$

   to find a series for $\pi^2/12$.

8. Use the Fourier series for $f(x) = |\sin x|$, for $-\pi \le x \le \pi$, to find a series for $\pi/4$.

9. Use the Fourier series for

   $$f(x) = \begin{cases} 0, & \text{for } -1 < x < 0 \\ x, & \text{for } 0 \le x < 1 \end{cases}$$

   to find a series for $\pi^2/8$.

10. Integrate the Fourier series of $f(x)$ in Exercise 2 to find the Fourier series of x^2. What happens if the Fourier series of $f(x)$ is differentiated to find $f'(x)$?

11. Find the Fourier series of $f(x) = \pi^2 - x^2$ for $-\pi \leq x \leq \pi$ and use it with Theorems 10.2 and 10.3 to find the Fourier series of x and $x(\pi^2 - x^2)$.

Exercises 12 through 18 are optional. Exercises 12 through 14 show how the partial sum

$$S_n(x) = a_0 + \sum_{r=1}^{n} (a_r \cos rx + b_r \sin rx),$$

of the Fourier series of a function $f(x)$ defined over the fundamental interval $-\pi \leq x \leq \pi$, and by periodic extension outside it, can be expressed as an integral. Exercises 15 through 17 provide an intuitive justification of Theorem 9.1.

12. Starting from the trigonometric identity

$$\frac{1}{2} + \sum_{r=1}^{n} \cos rx = \frac{\sin\left[\left(n + \frac{1}{2}\right)x\right]}{2\sin\left(\frac{x}{2}\right)}$$

that formed Exercise 19 in Section 1.4, integrate the identity first over the interval $[-\pi, 0]$ and then over the interval $[0, \pi]$ to show that

$$\int_{-\pi}^{0} \frac{\sin\left[\left(n + \frac{1}{2}\right)x\right]}{\sin\left(\frac{x}{2}\right)} dx = \pi \quad \text{and}$$

$$\int_{0}^{\pi} \frac{\sin\left[\left(n + \frac{1}{2}\right)x\right]}{\sin\left(\frac{x}{2}\right)} dx = \pi.$$

13. Substitute the Euler formulas for a_r and b_r into $S_n(x)$, after first replacing the dummy variable x in each integral by the dummy variable u to avoid confusion with the variable x in $S_n(x)$. Combine all terms under a single integral sign and, after simplifying the result using the formula $\cos a \cos b + \sin a \sin b = \cos(a - b)$, use the results of Exercise 12 to show that

$$S_n(x) = \frac{1}{\pi} \int_{x-\pi}^{x+\pi} f(x-t) \frac{\sin\left[\left(n + \frac{1}{2}\right)t\right]}{2\sin\left(\frac{t}{2}\right)} dt.$$

14. Use the periodicity of the integrand of $S_n(x)$ in Exercise 13 to show that

$$S_n(x) = \frac{1}{\pi} \int_{0}^{\pi} [f(x-t) + f(x+t)] \frac{\sin\left[\left(n + \frac{1}{2}\right)t\right]}{2\sin\left(\frac{t}{2}\right)} dt.$$

The function $D_n(t) = \sin[(n + \frac{1}{2})t]/[2\sin(\frac{t}{2})]$ occurring in the integrand of $S_n(x)$ is called the **Dirichlet kernel**.

15. Use a computer to graph $D_n(t)$ in Exercise 14 in the interval $-\pi \leq t \leq \pi$, for $n = 10, 15, 30$. Confirm from the graphs that when n is large $D_n(t)$ only differs significantly from zero in the interval $-2\pi/(2n+1) \leq t \leq 2\pi/(2n+1)$.

16. Use the conclusion of Exercise 15 together with the result

$$\int_{-\pi}^{\pi} D_n(t) dt = \pi$$

established in Exercise 12 to give reasons why for large n the Dirichlet kernel $D_n(t)$ can be approximated by the rectangular pulse function

$$\Delta(t) = \begin{cases} 0, & -\pi \leq t < -2\pi(2n+1) \\ (2n+1)/4, & -2\pi/(2n+1) \leq t \leq 2\pi/(2n+1) \\ 0, & 2\pi/(2n+1) < t \leq \pi. \end{cases}$$

17. Use the result of Exercise 16, with

$$S_n(x) = \frac{1}{\pi} \int_{0}^{\pi} [f(x-t) + f(x+t)] D_n(t) dt$$

from Exercise 14, to suggest why in the limit as $n \to \infty$ this confirms the convergence properties of Fourier series stated in Theorem 9.1.

18. By first setting $f(x) = \sin mx$ and then $f(x) = \cos mx$ in the result of Exercise 17, with m a positive integer, and using the fact that the functions $\sin mx$ and $\cos mx$ are their own Fourier series on $-\pi \leq x \leq \pi$, deduce that

$$\int_{0}^{\pi} \sin mt\, D_n(t) dt = \int_{0}^{\pi} \cos mt\, D_n(t) dt$$

$$= \begin{cases} 0, & n = 1, 2, \ldots, m-1 \\ \pi/2, & n = m, m+1, \ldots. \end{cases}$$

9.3   Fourier Sine and Cosine Series on $0 \leq x \leq L$

A function $f(x)$ that is specified on the interval $0 \leq x \leq L$ can be represented in terms of a series either of sines or of cosines on the interval. These series are obtained by first extending the definition of the function to the interval $-L \leq x \leq L$ in a suitable manner, and then restricting the Fourier series representation of the extended function to the original interval $0 \leq x \leq L$.

Sine Series on $0 \leq x \leq L$

Let a function $f(x)$ specified on the interval $0 \leq x \leq L$ be extended to the interval $-L \leq x \leq L$ as an odd function by the requirement that $f(-x) = -f(x)$ for $-L \leq x \leq L$. Then the odd function $g(x)$ given by

$$g(x) = \begin{cases} -f(-x), & -L \leq x \leq 0 \\ f(x), & 0 \leq x \leq L, \end{cases}$$

and defined on the interval $-L \leq x \leq L$, coincides with the function $f(x)$ on the original interval $0 \leq x \leq L$.

It follows from Theorem 9.1 and the Fourier series representation of functions on the interval $-L \leq x \leq L$ that

$$f(x) = \sum_{n=1}^{\infty} b_n \sin \frac{n\pi x}{L}, \quad \text{for } -L \leq x \leq L, \tag{28}$$

where

$$b_n = \frac{2}{L} \int_0^L f(x) \sin \frac{n\pi x}{L} dx, \quad \text{for } n = 1, 2, \ldots. \tag{29}$$

As the functions $f(x)$ and $g(x)$ coincide for $0 \leq x \leq L$, we see that by restricting x to the interval $0 \leq x \leq L$, series (28) is the required sine series. Result (28) with the coefficients b_n defined by (29) is called the **sine series** representation of $f(x)$ on the interval $0 \leq x \leq L$, or sometimes the **half-range sine series expansion** of $f(x)$.

Cosine Series on $0 \leq x \leq L$

If $f(x)$ is extended to the interval $-L \leq x \leq L$ as an even function, by requiring that $f(-x) = f(x)$ for $-L \leq x \leq 0$, we can define an even function $g(x)$ by

$$g(x) = \begin{cases} f(-x), & -L \leq x \leq 0 \\ f(x), & 0 \leq x \leq L. \end{cases}$$

If we again use Theorem 9.1 with the Fourier series representation of functions on the interval $-L \leq x \leq L$, it follows that

$$f(x) = a_0 + \sum_{n=1}^{\infty} a_n \cos \frac{n\pi x}{L}, \quad \text{for } -L \leq x \leq L \tag{30}$$

where

$$a_0 = \frac{1}{L} \int_0^L f(x) dx \quad \text{and} \quad a_n = \frac{2}{L} \int_0^L f(x) \cos \frac{n\pi x}{L} dx, \quad \text{for } n = 1, 2, \ldots. \tag{31}$$

Here also the functions $f(x)$ and $g(x)$ coincide for $0 \leq x \leq L$, so by restricting x to this interval (30) is seen to provide required cosine series representation of $f(x)$ on the interval $0 \leq x \leq L$. Result (31) with the coefficients a_n defined by (32) is called the **cosine series** representation of $f(x)$ on the interval $0 \leq x \leq L$, or sometimes the **half-range cosine series expansion** of $f(x)$.

Fourier expansions only in terms of sines or cosines

Sine and cosine representations of $f(x)$ on $0 \le x \le L$

Let $f(x)$ be defined on the interval $0 \le x \le L$. Then the **sine series** representation of $f(x)$ is given by

$$f(x) = \sum_{n=1}^{\infty} b_n \sin \frac{n\pi x}{L}, \quad \text{for } 0 \le x \le L,$$

where

$$b_n = \frac{2}{L} \int_0^L f(x) \sin \frac{n\pi x}{L} dx, \quad \text{for } n = 1, 2, \ldots,$$

and the **cosine series** representation of $f(x)$ is given by

$$f(x) = a_0 + \sum_{n=1}^{\infty} a_n \cos \frac{n\pi x}{L}, \quad \text{for } 0 \le x \le L,$$

where

$$a_0 = \frac{1}{L} \int_0^L f(x) dx \quad \text{and} \quad a_n = \frac{2}{L} \int_0^L f(x) \cos \frac{n\pi x}{L} dx,$$

$$\text{for } n = 1, 2, \ldots.$$

EXAMPLE 9.10   Find the sine and cosine representations of $f(x) = x$ for $0 \le x \le \pi$.

Solution   The sine series representation is given by

$$f(x) = \sum_{n=1}^{\infty} b_n \sin nx,$$

where

$$b_n = \frac{2}{\pi} \int_0^\pi x \sin nx\, dx, \quad \text{for } n = 1, 2, \ldots.$$

Integrating this last result, we find that

$$b_n = (-1)^{n+1} \frac{2}{n},$$

so the required sine series representation is

$$f(x) = 2 \sum_{n=1}^{\infty} (-1)^{n+1} \frac{\sin nx}{n} \quad \text{for } 0 \le x \le \pi.$$

The cosine series representation is given by

$$f(x) = a_0 + \sum_{n=1}^{\infty} a_n \cos nx,$$

where

$$a_0 = \frac{1}{\pi}\int_0^\pi x\,dx \quad \text{and} \quad a_n = \frac{2}{\pi}\int_0^\pi x\cos nx\,dx \quad \text{for } n = 1, 2, \ldots.$$

Integration gives

$$a_0 = \frac{\pi}{2}, \quad \text{while } a_{2n-1} = -\frac{4}{\pi(2n-1)^2}, \quad \text{and} \quad a_{2n} = 0 \quad \text{for } n = 1, 2, \ldots,$$

so the cosine series representation is

$$f(x) = \frac{\pi}{2} - \frac{4}{\pi}\sum_{n=1}^\infty \frac{\cos(2n-1)x}{(2n-1)^2} \quad \text{for } 0 \le x \le \pi. \qquad \blacksquare$$

Summary

It has been shown how a function $f(x)$ defined on the interval $0 \le x \le L$ can be represented either in terms of a series involving only sine functions or as a series involving only cosine functions. These special Fourier series, called either half-range sine or cosine Fourier series, were obtained from the usual expansion over the interval $-L \le x \le L$ by extending the definition of $f(x)$ to the interval $-L \le x \le L$ in a suitable manner. As half-range Fourier series are derived from ordinary Fourier series, their convergence properties are the same as those of ordinary Fourier series.

EXERCISES 9.3

In Exercises 1 through 4 find the sine series for the given function defined on the interval $0 \le x \le \pi$.

1. $f(x) = x^2$.

2. $f(x) = |\cos x|$.

3. $f(x) = \begin{cases} \cos x, & 0 < x \le \pi/2 \\ 0, & \pi/2 < x \le \pi. \end{cases}$

4. $f(x) = (x - \pi)^2/\pi^2$.

In Exercises 5 through 8 find the cosine series for the given function defined on the interval $0 \le x \le \pi$.

5. $f(x) = \begin{cases} \cos x, & 0 < x \le \pi/2 \\ 0, & \pi/2 < x \le \pi. \end{cases}$

6. $f(x) = \sin x$.

7. $f(x) = \begin{cases} \sin x, & 0 < x \le \pi/2 \\ 0, & \pi/2 < x \le \pi. \end{cases}$

8. $f(x) = (x - \pi)^2/\pi^2$.

9. Use the sine series together with the orthogonality of the functions $\sin\frac{n\pi x}{L}$, for $n = 1, 2, \ldots$, on the interval $0 \le x \le L$ to show that the **Parseval relation** for the **sine series** takes the form

$$\frac{2}{L}\int_0^L [f(x)]^2\,dx = \sum_{n=1}^\infty b_n^2.$$

10. Use the cosine series together with the orthogonality of the functions $\cos\frac{n\pi x}{L}$, for $n = 1, 2, \ldots$, on the interval $0 \le x \le L$ to show that the **Parseval relation** for

the **cosine series** takes the form

$$\frac{2}{L}\int_0^L [f(x)]^2\,dx = 2a_0^2 + 0^2 + \sum_{n=1}^\infty a_n^2.$$

11. Find the sine series representation of

$$f(x) = e^{-x}, \quad 0 < x < \pi.$$

12. Find the sine and cosine series representations of $f(x) = \pi - x$ on the interval $0 \le x \le \pi$. Use them with the results of Exercises 9 and 10 to show that

$$\frac{\pi^2}{6} = \sum_{n=1}^\infty \frac{1}{n^2} \quad \text{and} \quad \frac{\pi^4}{96} = \sum_{n=1}^\infty \frac{1}{(2n-1)^4}.$$

Comment on which series representation converges most rapidly to $f(x)$.

13.* Explain why if $f(x)$ and $g(x)$ have Fourier series representations for $-\pi \le x \le \pi$, the Fourier series representations of $f(x) \pm g(x)$ can be obtained from those for $f(x)$ and $g(x)$ by term-by-term addition or subtraction. By adding and subtracting the Fourier series representations of

$$\int_{-\pi}^\pi [f(x) + g(x)]\,dx \quad \text{and} \quad \int_{-\pi}^\pi [f(x) - g(x)]\,dx,$$

obtain the **generalized Parseval relation**

$$\frac{1}{\pi}\int_{-\pi}^\pi f(x)g(x)\,dx = 2a_0 A_0 + \sum_{n=1}^\infty (a_n A_n + b_n B_n),$$

where the a_n, b_n are the Fourier coefficients of $f(x)$ and the A_n, B_n are the Fourier coefficients of $g(x)$.

14.* Let $f(x)$ defined for $-\pi \le x \le \pi$ be approximated by the nth partial sum of its Fourier series representation

$$S_n(x) = a_0 + \sum_{m=1}^{n}(a_m \cos mx + b_m \sin mx),$$

and let

$$\Phi(x) = A_0 + \sum_{m=1}^{n}(A_m \cos mx + B_m \sin mx)$$

be any other approximation to $f(x)$ with coefficients A_m and B_m. Show by expanding the square error

$$E_n = \int_{-\pi}^{\pi} [f(x) - \Phi_n(x)]^2 dx$$

in terms of the Fourier series representation of $f(x)$ that E_n is minimized when $A_m = a_m$ and $B_m = b_m$ for $m = 0, 1, 2, \ldots, n$. This establishes the fact that the Fourier series partial sum $S_n(x)$ provides the best trigonometric approximation to $f(x)$ in the least squares sense.

9.4   Other Forms of Fourier Series

In this section we introduce two other forms of Fourier series that prove useful. The first is the Fourier series of a function $f(x)$ defined over an interval $a - L \le x \le a + L$ with a an arbitrary real number, and by periodicity outside it. Frequently $a = L$, corresponding to the Fourier series over the interval $0 \le x \le 2L$. The second form of Fourier series considered uses the Euler identity $e^{ix} = \cos x + i \sin x$ to derive the **complex** form of the Fourier series, also often called the **exponential form** of the Fourier series.

Fourier Series over a Shifted Interval

Routine integration shows the set of functions

$$1, \quad \sin \frac{n\pi x}{L} \quad \text{and} \quad \cos \frac{n\pi x}{L} \quad \text{for } n = 1, 2, \ldots$$

form an orthogonal system over any interval of the form $a - L \le x \le a + L$, for any real number a, and that

$$\int_{a-L}^{a+L} \sin \frac{m\pi x}{L} \cos \frac{n\pi x}{L} dx = 0 \quad \text{for all integers } m \text{ and } n,$$

$$\int_{a-L}^{a+L} \sin \frac{m\pi x}{L} \sin \frac{n\pi x}{L} dx = \begin{cases} 0 & \text{for } m \ne n \\ L & \text{for } m = n, \end{cases} \text{ for all integers } m \text{ and } n,$$

$$\int_{a-L}^{a+L} \cos \frac{m\pi x}{L} \cos \frac{n\pi x}{L} dx = \begin{cases} 0 & \text{for } m \ne n \\ L & \text{for } m = n \ne 0 \\ 2L & \text{for } m = n = 0, \end{cases} \text{ for all integers } m \text{ and } n.$$

The following result is a direct consequence of these integrals, and it provides an extension of the definition of a Fourier series to the interval $-L \le x \le L$.

Fourier series over a shifted interval

Fourier series over the interval $a - L \le x \le a + L$

A function $f(x)$ defined on the interval $a - L \le x \le a + L$ has the Fourier series representation

$$f(x) = a_0 + \sum_{n=1}^{\infty}\left(a_n \cos \frac{n\pi x}{L} + b_n \sin \frac{n\pi x}{L}\right), \tag{32}$$

where

$$a_0 = \frac{1}{2L} \int_{a-L}^{a+L} f(x)dx, \quad a_n = \frac{1}{L} \int_{a-L}^{a+L} f(x)\cos\frac{n\pi x}{L}dx,$$

$$b_n = \frac{1}{L} \int_{a-L}^{a+L} f(x)\sin\frac{n\pi x}{L}dx, \quad \text{for } n = 1, 2, \ldots.$$

(33)

EXAMPLE 9.11

Find the Fourier series representation of

$$f(x) = \begin{cases} x, & 0 \le x \le \pi \\ \pi, & \pi \le x < 2\pi. \end{cases}$$

Solution A graph of the function $f(x)$ is shown in Fig. 9.12. Using (33) with $a = L = \pi$ gives

$$a_0 = \frac{1}{2\pi} \int_0^{2\pi} f(x)dx = \frac{3\pi}{4} \quad \text{and} \quad a_n = \frac{1}{\pi} \int_0^{2\pi} f(x)\cos nx\,dx,$$

from which it follows that

$$a_{2n-1} = -\frac{2}{\pi(2n-1)^2} \quad \text{and} \quad a_{2n} = 0 \quad \text{for } n = 1, 2, \ldots.$$

The Euler formula for b_n gives

$$b_n = \frac{1}{\pi} \int_0^{2\pi} f(x)\sin nx\,dx = -\frac{1}{n} \quad \text{for } n = 1, 2, \ldots,$$

so the required Fourier series is

$$f(x) = \frac{3\pi}{4} - \frac{2}{\pi}\sum_{n=1}^{\infty}\frac{\cos(2n-1)x}{(2n-1)^2} - \sum_{n=1}^{\infty}\frac{\sin nx}{n} \quad \text{for } 0 \le x < 2\pi. \qquad \blacksquare$$

Complex Fourier Series

The Euler identities $e^{ix} = \cos x + i\sin x$ and $e^{-ix} = \cos x - i\sin x$ allow us to write

$$\cos x = \frac{e^{ix} + e^{-ix}}{2} \quad \text{and} \quad \sin x = \frac{e^{ix} - e^{-ix}}{2i}.$$

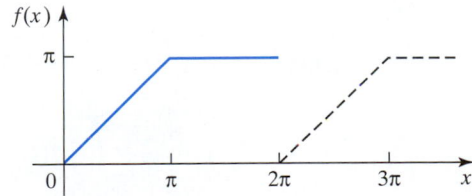

FIGURE 9.12 The function $f(x)$ defined for $0 \le x < 2\pi$.

When these results are used in the real variable Fourier series representation of $f(x)$ over the interval $-L \leq x \leq L$, it becomes

$$f(x) = a_0 + \sum_{n=1}^{\infty} \left[a_n \left(\frac{e^{in\pi x/L} + e^{-in\pi x/L}}{2} \right) + b_n \left(\frac{e^{in\pi x/L} - e^{-in\pi x/L}}{2i} \right) \right],$$

and after grouping terms we have

$$f(x) = a_0 + \sum_{n=1}^{n} \left(\frac{a_n - ib_n}{2} \right) e^{in\pi x/L} + \sum_{n=1}^{n} \left(\frac{a_n + ib_n}{2} \right) e^{-in\pi x/L}. \tag{34}$$

If we now define

$$c_0 = a_0, \quad c_n = \frac{a_n - ib_n}{2}, \quad \text{and} \quad c_{-n} = \frac{a_n + ib_n}{2} \quad \text{for } n = 1, 2, \ldots, \tag{35}$$

the Fourier series representation of $f(x)$ in (34) becomes

$$f(x) = \lim_{k \to \infty} \sum_{n=-k}^{k} c_n e^{in\pi x/L} \quad \text{for } -L \leq x \leq L. \tag{36}$$

This is the **complex** or **exponential** form of the Fourier series representation of $f(x)$.

If real functions $f(x)$ are considered, the Fourier coefficients a_n and b_n are real, and (35) then shows that c_n and c_{-n} are complex conjugates, because $c_{-n} = \bar{c}_n$. To proceed further we now make use of the fact that the functions $\exp(im\pi x/L)$ and $\exp(-in\pi x/L)$ are orthogonal over the interval $-L \leq x \leq L$, because integration shows that

$$\int_{-L}^{L} e^{im\pi x/L} e^{-in\pi x/L} dx = \begin{cases} 0, & \text{for } m \neq -n \\ 2\pi & \text{for } m = -n \end{cases} \text{ for } m, n \text{ positive integers.}$$

Multiplication of (36) by $\exp(-im\pi x/L)$, followed by integration over $-L \leq x \leq L$ and use of the above orthogonality condition gives

$$c_n = \frac{1}{2L} \int_{-L}^{L} f(x) e^{-in\pi x/L} dx, \quad \text{for } n = 0, \pm 1, \pm 2, \ldots. \tag{37}$$

Collecting these results we arrive at the following definition.

The complex form of a Fourier series

the complex or exponential form of a Fourier series

Let the real function $f(x)$ be defined on the interval $-L \leq x \leq L$. Then the complex Fourier series representation of $f(x)$ is

$$f(x) = \lim_{k \to \infty} \sum_{n=-k}^{k} c_n e^{in\pi x/L} \quad \text{for } -L \leq x \leq L,$$

where

$$c_n = \frac{1}{2L} \int_{-L}^{L} f(x) e^{-in\pi x/L} dx, \quad \text{for } n = 0, \pm 1, \pm 2, \ldots.$$

As the complex form of a Fourier series was derived directly from the real variable Fourier series, it follows directly that if $f(x)$ is defined for $a - L \leq x \leq a + L$, then

$$f(x) = \lim_{k \to \infty} \sum_{n=-k}^{k} c_n e^{in\pi x/L} \quad \text{for } a - L \leq x \leq a + L, \tag{38}$$

with

$$c_n = \frac{1}{2L} \int_{a-L}^{a+L} f(x) e^{-in\pi x/L} dx, \quad \text{for } n = 0, \pm 1, \pm 2, \ldots. \tag{39}$$

It is sometimes useful to separate out the coefficient c_0 from the summation in (36) (or in (38)) by writing

$$f(x) = c_0 + \lim_{k \to \infty} \sum_{n=-k}^{k}{}' c_n e^{in\pi x/L}, \tag{40}$$

with the understanding that Σ' indicates that the term corresponding to $n = 0$ has been omitted from the summation.

When $f(x)$ is real, so that $c_{-n} = \bar{c}_n$, result (40) becomes

$$f(x) = c_0 + \sum_{n=1}^{\infty} [c_n e^{in\pi x/L} + \bar{c}_n e^{-in\pi x/L}]. \tag{41}$$

Because the complex form of the Fourier series representation of a function is derived from its real variable definition, the convergence properties of complex Fourier series are the same as those already discussed for the real variable case. So at points of continuity of $f(x)$ the complex Fourier series converges uniformly to $f(x)$, while at points of discontinuity it converges to the mid-point of the jump discontinuity.

EXAMPLE 9.12   Find the complex Fourier series representation of

$$f(x) = \begin{cases} 0, & -\pi < x < -\pi/2 \\ 1, & -\pi/2 < x < \pi/2 \\ 0, & \pi/2 < x < \pi. \end{cases}$$

Solution   As the function $f(x)$ is defined on the interval $-\pi \leq x \leq \pi$, we have $L = \pi$, so the coefficients c_n are given by

$$c_0 = \frac{1}{2\pi} \int_{-\pi}^{\pi} f(x) dx = \frac{1}{2\pi} \int_{-\pi/2}^{\pi/2} 1 dx = \frac{1}{2}$$

and

$$c_n = \frac{1}{2\pi} \int_{-\pi}^{\pi} f(x) e^{-inx} dx = \frac{1}{2\pi} \int_{-\pi/2}^{\pi/2} e^{-inx} dx = \frac{1}{n\pi} \left(\frac{e^{in\pi/2} - e^{-in\pi/2}}{2i} \right)$$

$$\text{for } n = \pm 1, \pm 2, \ldots.$$

The coefficients c_n reduce to the real values

$$c_n = \frac{1}{n\pi} \sin \frac{n\pi}{2} \quad \text{for } n = \pm 1, \pm 2, \ldots,$$

so $c_n = c_{-n}$ because c_n is an even function of n. Consideration of the function

$\sin(n\pi/2)$ for integer values of n shows that

$$c_{2n-1} = \frac{(-1)^{n-1}}{\pi(2n-1)} \quad \text{and} \quad c_{2n} = 0 \quad \text{for } n = 1, 2, \ldots.$$

Thus, the complex Fourier series representation of $f(x)$ is

$$f(x) = \frac{1}{2} + \lim_{k \to \infty} \sum_{n=-k}^{k}{}' c_n(e^{inx} + e^{-inx}).$$

The real variable Fourier series representation of this function $f(x)$ was derived in Chapter 8, Example 8.22, and considered again at the start of Section 9.1. If c_n is used in the preceding result with $e^{inx} + e^{-inx} = 2\cos nx$, the complex Fourier series representation reduces to the real variable Fourier series representation

$$f(x) = \frac{1}{2} + \frac{2}{\pi} \sum_{n=1}^{\infty} (-1)^{n+1} \frac{\cos(2n-1)x}{(2n-1)}$$

that was obtained previously. This series, and the equivalent complex series, converges uniformly to $f(x)$ at points of continuity of $f(x)$ and to the value $1/2$ at the discontinuities located at $x = \pm\pi/2$. ∎

EXAMPLE 9.13   Find the complex Fourier series representation of

$$f(x) = \begin{cases} 0, & 0 < x < 1 \\ 1, & 1 < x < 4. \end{cases}$$

Solution   The function $f(x)$ is defined on the interval $0 \le x \le 2L$, with $2L = 4$, so $L = 2$. Thus, the complex Fourier coefficients c_n are given by

$$c_n = \frac{1}{4}\int_0^4 f(x)e^{-in\pi x/2}dx = \frac{1}{4}\int_1^4 e^{-in\pi x/2}dx, \quad \text{for } n = 0, \pm1, \pm2, \ldots.$$

Setting $n = 0$ gives

$$c_0 = \frac{3}{4},$$

whereas

$$c_n = \frac{i}{2\pi n}\left[1 - e^{-in\pi/2}\right], \quad \text{for } n = \pm1, \pm2, \ldots.$$

So the complex Fourier series representation of $f(x)$ is

$$f(x) = c_0 + \lim_{k \to \infty} \sum_{n=-k}^{k} c_n e^{in\pi x/2},$$

with c_0 and c_n defined as shown. ∎

Accounts of Fourier series and their general properties are to be found in references [3.3] to [3.5] and also in [3.7], [3.16], and [4.2]. An advanced and encyclopedic account of trigonometric series is given in reference [4.5].

Summary

Other forms of Fourier series have been derived, first by stretching and shifting the interval over which the expansion was required, and then by expressing the series in complex form. As both results were derived from the ordinary Fourier series, their convergence properties are the same as those of ordinary Fourier series.

EXERCISES 9.4

In Exercises 1 through 4 find the Fourier series representation of the function $f(x)$ over the given shifted interval.

1. $f(x) = \begin{cases} 0, & 0 < x < \pi \\ 1, & \pi < x < 2\pi. \end{cases}$

2. $f(x) = 1 - x, \quad 0 < x < 1.$

3. $f(x) = x, \quad 0 < x < \pi.$

4. $f(x) = x^2, \quad \pi < x < 3\pi.$

In Exercises 5 through 10 find the complex Fourier series

representations of the given function $f(x)$ over the stated interval.

5. $f(x) = e^x, \quad -1 < x < 1.$

6. $f(x) = x^2, \quad 0 < x < 2\pi.$

7. $f(x) = e^x, \quad 0 < x < 1.$

8. $f(x) = \sinh x, \quad -\pi < x < \pi.$

9. $f(x) = e^x, \quad -\pi < x < \pi.$

10. $f(x) = \cosh x, \quad -1 < x < 1.$

9.5 Frequency and Amplitude Spectra of a Function

When Fourier series are applied to periodic physical phenomena with period T, it is convenient to work in terms of the angular frequency ω_0 defined as

$$\omega_0 = \frac{2\pi}{T}, \tag{42}$$

where $1/T = \omega_0/2\pi$ measures the number of cycles (oscillations) occurring in one time unit. For example, the period of the function $\sin 2x$ is $T = \pi$, so in this case $\omega_0 = 2$.

interpreting Fourier series representations in a different way

The Fourier series representation of a function $f(x)$ defined on the interval $-L \leq x \leq L$ with the corresponding period $T = 2L$ has been shown to be

$$f(x) = a_0 + \sum_{n=1}^{\infty} \left(a_n \cos \frac{n\pi x}{L} + b_n \sin \frac{n\pi x}{L} \right),$$

so as $\omega_0 = \pi/L$ this can be written

$$f(x) = a_0 + \sum_{n=1}^{\infty} (a_n \cos n\omega_0 x + b_n \sin n\omega_0 x), \tag{43}$$

where

$$a_0 = \frac{1}{2L} \int_{-L}^{L} f(x)dx$$

$$a_n = \frac{1}{L} \int_{-L}^{L} f(x) \cos \frac{n\pi x}{L} dx = \frac{1}{L} \int_{-L}^{L} f(x) \cos n\omega_0 x dx \quad \text{for } n = 1, 2, \ldots, \tag{44}$$

and

$$b_n = \frac{1}{L} \int_{-L}^{L} f(x) \sin \frac{n\pi x}{L} dx = \frac{1}{L} \int_{-L}^{L} f(x) \sin n\omega_0 x dx \quad \text{for } n = 1, 2, \ldots. \tag{45}$$

In terms of these results (43) becomes

$$f(x) = a_0 + \sum_{n=1}^{\infty} \left(a_n^2 + b_n^2\right)^{1/2} \left[\frac{a_n}{\left(a_n^2 + b_n^2\right)^{1/2}} \cos n\omega_0 x + \frac{b_n}{\left(a_n^2 + b_n^2\right)^{1/2}} \sin n\omega_0 x \right].$$
(46)

Using the trigonometric identity $\cos(P + Q) = \cos P \cos Q - \sin P \sin Q$, and defining

$$A_n = \left(a_n^2 + b_n^2\right)^{1/2} \quad \text{and} \quad \delta_n = \text{Arctan}\,(-b_n/a_n),$$
(47)

with A_n the **amplitude** and δ_n the **phase**, allows (46) to be written more concisely in the **amplitude and phase angle** representation

$$f(x) = a_0 + \sum_{n=1}^{\infty} A_n \cos(n\omega_0 x + \delta_n).$$
(48)

When the Fourier series representation of $f(x)$ is expressed in this form, the set of numbers

$$\omega_0, \quad 2\omega_0, \quad 3\omega_0, \dots$$

is called the **frequency spectrum** of the function $f(x)$. The number $n\omega_0$ is called the **nth harmonic frequency** of $f(x)$, and the number δ_n the **nth phase angle** of $f(x)$. The set of numbers

$$A_0, \quad A_1, \quad A_2, \dots,$$

where $A_0 = |a_0|$, is called the **amplitude spectrum** of $f(x)$, and the function

$$\cos(n\omega_0 x + \delta_n)$$

is called the **nth harmonic** of the function $f(x)$. The amplitude spectrum can be displayed graphically by drawing lines of height $A_0, A_1, A_2, \dots$, against the respective harmonic frequencies $\omega_0, 2\omega_0, 3\omega_0, \dots$, as shown in the next example. This is called a **discrete spectrum**, because the amplitude is only defined at the discrete frequencies in the frequency spectrum.

Result (48) shows how $f(x)$ is representable in terms of a linear combination of harmonics, each weighted by an appropriate amplitude factor A_n.

EXAMPLE 9.14   Find the harmonics and amplitude spectrum of

$$f(x) = \begin{cases} \pi, & -\pi < x < 0 \\ \pi - x, & 0 \le x \le \pi. \end{cases}$$

Solution   In this case the function is defined on the interval $-\pi \le x \le \pi$, so $L = \pi$, $T = 2L = 2\pi$, and $\omega_0 = 2\pi/T = 1$. The frequency spectrum becomes $1, 2, 3, \dots$, and the Fourier series representation in terms of frequency is

$$f(x) = a_0 + \sum_{n=1}^{\infty} (a_n \cos nx + b_n \sin nx),$$

where

$$a_0 = \frac{1}{2\pi} \int_{-\pi}^{0} \pi \, dx + \frac{1}{2\pi} \int_{0}^{\pi} (\pi - x) dx = \frac{3\pi}{4},$$

and

$$a_n = \frac{1}{\pi} \int_{-\pi}^{0} \pi \cos nx dx + \frac{1}{\pi} \int_{0}^{\pi} (\pi - x) \cos nx dx = \frac{1}{\pi n^2} [1 - (-1)^n],$$

$$\text{for } n = 1, 2, \ldots.$$

This last result simplifies to

$$a_{2n-1} = \frac{2}{\pi (2n-1)^2}, \quad a_{2n} = 0, \quad \text{for } n = 1, 2, \ldots.$$

Similarly,

$$b_n = \frac{1}{\pi} \int_{-\pi}^{0} \pi \sin nx dx + \frac{1}{\pi} \int_{0}^{\pi} (\pi - x) \sin nx dx = \frac{(-1)^n}{n}, \quad \text{for } n = 1, 2, \ldots.$$

Substituting the coefficients a_n and b_n into the Fourier series gives

$$f(x) = \frac{3\pi}{4} + \frac{2}{\pi} \sum_{n=1}^{\infty} \frac{\cos(2n-1)x}{(2n-1)^2} + \sum_{n=1}^{\infty} \frac{(-1)^n \sin nx}{n} \quad \text{for } -\pi \le x \le \pi.$$

To find the harmonics and the amplitude spectrum, it is necessary to group together terms with corresponding frequencies. When this is done $f(x)$ becomes

$$f(x) = \frac{3\pi}{4} + \left(\frac{2}{\pi} \cos x - \sin x \right) + \frac{1}{2} \sin 2x + \left(\frac{2}{9\pi} \cos 3x - \frac{1}{3} \sin 3x \right)$$

$$+ \frac{1}{4} \sin 4x + \left(\frac{2}{25\pi} \cos 5x - \frac{1}{5} \sin 5x \right) + \cdots.$$

This shows, for example, that the fifth harmonic is proportional to

$$\frac{2}{25\pi} \cos 5x - \frac{1}{5} \sin 5x.$$

The amplitudes are

$$A_0 = |a_0| = \frac{3\pi}{4}, \quad A_1 = \left[\left(\frac{2}{\pi} \right)^2 + (-1)^2 \right]^{1/2},$$

$$A_2 = \frac{1}{2}, \quad A_3 = \left[\left(\frac{2}{9\pi} \right)^2 + \left(-\frac{1}{3} \right)^2 \right]^{1/2},$$

$$A_4 = \frac{1}{4}, \quad A_5 = \left[\left(\frac{2}{25\pi} \right)^2 + \left(-\frac{1}{5} \right)^2 \right]^{1/2}, \ldots.$$

In general

$$A_{2n-1} = \frac{1}{(2n-1)} \left[\frac{4}{(2n-1)^2 \pi^2} + 1 \right]^{1/2} \quad \text{and} \quad A_{2n} = \frac{1}{2n}, \quad \text{for } n = 1, 2, \ldots.$$

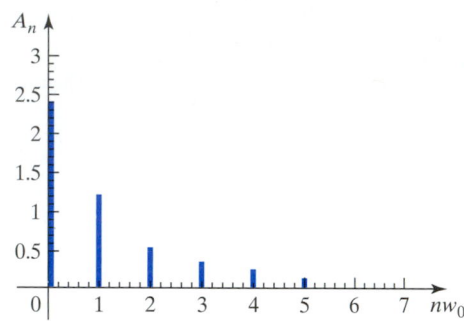

FIGURE 9.13   The amplitude spectrum of $f(x)$ as a function of frequency.

The first few numerical values of the amplitudes are

$$A_0 = 2.356, \quad A_1 = 1.185, \quad A_2 = 0.5, \quad A_3 = 0.341, \quad A_4 = 0.25, \quad A_5 = 0.202,$$
$$A_6 = 0.167, \ldots,$$

and the amplitude spectrum of $f(x)$ is shown in Fig. 9.13. In Fig. 9.13 the amplitudes $A_0, A_1, \ldots$, are represented by vertical lines of length $A_0, A_1, \ldots$, corresponding to the frequencies $0, 1, 2, \ldots$.

The phases $\delta_n = \text{Arctan}\,(-b_n/a_n)$ are seen to be given by

$$\delta_1 = \text{Arctan}\,(\pi/2), \quad \delta_2 = \text{Arctan}\,(-\infty), \quad \delta_3 = \text{Arctan}\,(3\pi/2),$$
$$\delta_4 = \text{Arctan}\,(-\infty), \quad \delta_5 = \text{Arctan}\,(5\pi/2), \ldots.$$

The negative sign is required in the arctangent functions associated with phases with even suffixes so that when the terms $A_{2n}\cos(2nx + \delta_{2n})$ are expanded, the functions $\sin 2nx$ have a positive sign.   ∎

Summary

It was shown how a Fourier series can be interpreted in a different way by introducing an angular frequency ω_0, combining sine and cosine terms with similar arguments into a single cosine term with a phase angle, and calling the magnitude of the multiplier of the cosine term the amplitude associated with the cosine term. A discrete plot of amplitude as a function of frequency was then called the amplitude spectrum of the representation. This form of representation is useful in many applications involving vibrations, because when the response of a system is represented in this way, the square of the amplitude is proportional to the energy in the system at that frequency, so the plot shows the distribution of energy as a function of frequency.

EXERCISES 9.5

In the following exercises find the frequency and amplitude spectrum of the given functions.

1. $f(x) = \begin{cases} 0, & -2\pi < x < 0 \\ x, & 0 < x < 2\pi. \end{cases}$

2. $f(x) = x, \quad -\pi/2 < x < \pi/2.$

3. $f(x) = \begin{cases} 1, & -\pi < x < 0 \\ -3, & 0 < x < \pi. \end{cases}$

4. $f(x) = \begin{cases} -1, & -\pi < x < 0 \\ 1, & 0 < x < \pi. \end{cases}$

5. $f(x) = x^2, \quad -\pi/4 < x < \pi/4.$

9.6   Double Fourier Series

Fourier series representations extend in a natural way to functions $f(x, y)$ of two real variables x and y over the intervals $-L_1 \leq x \leq L_1$ and $-L_2 \leq y \leq L_2$, provided f can be represented as a Fourier series in x when y is held constant, and as a Fourier series in y when x is held constant.

To arrive at a double Fourier series representation for $f(x, y)$, we first consider y to be a constant and write $f(x, y)$ as

$$f(x, y) = \sum_{m=0}^{\infty} \left(A_m(y) \cos \frac{m\pi x}{L_1} + B_m(y) \sin \frac{m\pi x}{L_1} \right), \tag{49}$$

extending Fourier series to function $f(x, y)$ of two variables

and then allow y to vary by replacing the Fourier coefficients $A_m(y)$ and $B_m(y)$ by their Fourier series representations

$$A_m(y) = \sum_{n=0}^{\infty} \left(a_{mn} \cos \frac{n\pi y}{L_2} + b_{mn} \sin \frac{n\pi y}{L_2} \right) \tag{50}$$

and

$$B_m(y) = \sum_{n=0}^{\infty} \left(c_{mn} \cos \frac{n\pi y}{L_2} + d_{mn} \sin \frac{n\pi y}{L_2} \right).$$

Substituting (50) into (49) shows $f(x, y)$ can be written as

$$f(x, y) = \sum_{m=0}^{\infty} \sum_{n=0}^{\infty} \left(a_{mn} \cos \frac{m\pi x}{L_1} \cos \frac{n\pi y}{L_2} + b_{mn} \cos \frac{m\pi x}{L_1} \sin \frac{n\pi y}{L_2} \right)$$
$$+ \sum_{m=0}^{\infty} \sum_{n=0}^{\infty} \left(c_{mn} \sin \frac{m\pi x}{L_1} \cos \frac{n\pi y}{L_2} + d_{mn} \sin \frac{m\pi x}{L_1} \sin \frac{n\pi y}{L_2} \right). \tag{51}$$

The Fourier coefficients a_{mn} for $m, n = 1, 2, \ldots$ are found by multiplying (51) by $\cos \frac{s\pi x}{L_1}$ and integrating over the interval $-L_1 \leq x \leq L_1$ to get

$$\int_{-L_1}^{L_1} f(x, y) \cos \frac{s\pi x}{L_1} dx = \sum_{m=0}^{\infty} \sum_{n=0}^{\infty} \left[a_{mn} \cos \frac{n\pi y}{L_2} \int_{-L_1}^{L_1} \cos \frac{m\pi x}{L_1} \cos \frac{s\pi x}{L_1} dx \right]$$

$$+ \sum_{m=0}^{\infty} \sum_{n=0}^{\infty} \left[b_{mn} \sin \frac{n\pi y}{L_2} \int_{-L_1}^{L_1} \cos \frac{m\pi x}{L_1} \cos \frac{s\pi x}{L_1} dx \right]$$

$$+ \sum_{m=0}^{\infty} \sum_{n=0}^{\infty} \left[c_{mn} \cos \frac{n\pi y}{L_2} \int_{-L_1}^{L_1} \sin \frac{m\pi x}{L_1} \cos \frac{s\pi x}{L_1} dx \right]$$

$$+ \sum_{m=0}^{\infty} \sum_{n=0}^{\infty} \left[d_{mn} \sin \frac{n\pi y}{L_2} \int_{-L_1}^{L_1} \sin \frac{m\pi x}{L_1} \cos \frac{s\pi x}{L_1} dx \right]. \tag{52}$$

The orthogonality of the functions $\cos \frac{m\pi x}{L_1}$ and $\sin \frac{s\pi x}{L_1}$ over the interval $-L_1 \leq x \leq L_1$ reduces (52) to

$$\int_{-L_1}^{L_1} f(x, y) \cos \frac{s\pi x}{L_1} dx = \sum_{n=0}^{\infty} \left(a_{sn} L_1 \cos \frac{n\pi y}{L_2} + b_{sn} L_1 \sin \frac{n\pi y}{L_2} \right). \tag{53}$$

Multiplication of (53) by $\cos \frac{t\pi y}{L_2}$ followed by integration over the interval $-L_2 \leq$

$y \le L_2$ reduces it further to

$$\int_{-L_2}^{L_2} \left[\int_{-L_1}^{L_1} f(x, y) \cos \frac{s\pi x}{L_1} \right] \cos \frac{t\pi y}{L_2} dy = a_{st} L_1 L_2,$$

so replacing s by m and t by n gives

$$a_{mn} = \frac{1}{L_1 L_2} \int_{-L_2}^{L_2} \int_{-L_1}^{L_1} f(x, y) \cos \frac{m\pi x}{L_1} \cos \frac{n\pi y}{L_2} dxdy \quad \text{for } m, n = 1, 2, \ldots.$$

(54)

The coefficient a_{00} follows by setting $m = n = 0$ in (51) and integrating over the intervals $-L_1 \le x \le L_1$ and $-L_2 \le y \le L_2$ to give

$$a_{00} = \frac{1}{4L_1 L_2} \int_{-L_2}^{L_2} \int_{-L_1}^{L_1} f(x, y)dxdy.$$

(55)

It remains to find the coefficients a_{m0} and a_{0n} for $m, n = 1, 2, \ldots$. Setting $n = 0$ in (53), integrating over $-L_2 \le y \le L_2$, and then replacing s by m gives

$$a_{m0} = \frac{1}{2L_1 L_2} \int_{-L_2}^{L_2} \int_{-L_1}^{L_1} f(x, y) \cos \frac{m\pi x}{L_1} dxdy.$$

(56)

The coefficients a_{0n} for $n = 1, 2, \ldots$ follow by multiplying (51) by $\cos \frac{t\pi y}{L_1}$, integrating over the interval $-L_2 \le y \le L_2$, and then replacing t by n to obtain

$$a_{0n} = \frac{1}{2L_1 L_2} \int_{-L_2}^{L_2} \int_{-L_1}^{L_1} f(x, y) \cos \frac{n\pi y}{L_2} dxdy.$$

(57)

Corresponding arguments show that for $m, n = 1, 2, \ldots$,

$$b_{mn} = \frac{1}{L_1 L_2} \int_{-L_2}^{L_2} \int_{-L_1}^{L_1} f(x, y) \cos \frac{m\pi x}{L_1} \sin \frac{n\pi y}{L_2} dxdy,$$

(58)

$$c_{mn} = \frac{1}{L_1 L_2} \int_{-L_2}^{L_2} \int_{-L_1}^{L_1} f(x, y) \sin \frac{m\pi x}{L_1} \cos \frac{n\pi y}{L_2} dxdy,$$

(59)

$$d_{mn} = \frac{1}{L_1 L_2} \int_{-L_2}^{L_2} \int_{-L_1}^{L_1} f(x, y) \sin \frac{m\pi x}{L_1} \sin \frac{n\pi y}{L_2} dxdy,$$

(60)

where

$$b_{m0} = 0, \quad c_{0n} = 0, \quad d_{0n} = 0 \quad \text{and} \quad d_{m0} = 0,$$

(61)

because the index zero causes the sine function to vanish in the integrands of the integrals defining these constants.

general and special double Fourier series representations

Thus, the general **double Fourier series representation** of $f(x, y)$ over the interval $-L_1 \le x \le L_1$ and $-L_2 \le y \le L_2$ is given by

$$f(x, y) = \sum_{m=0}^{\infty} \sum_{n=0}^{\infty} \left(a_{mn} \cos \frac{m\pi x}{L_1} \cos \frac{n\pi y}{L_2} + b_{mn} \cos \frac{m\pi x}{L_1} \sin \frac{n\pi y}{L_2} \right)$$

$$+ \sum_{m=0}^{\infty} \sum_{n=0}^{\infty} \left(c_{mn} \sin \frac{m\pi x}{L_1} \cos \frac{n\pi y}{L_2} + d_{mn} \sin \frac{m\pi x}{L_1} \sin \frac{n\pi y}{L_2} \right),$$

(62)

where the coefficients a_{mn}, b_{mn}, c_{mn}, and d_{mn} are given by expressions (54) to (61).

The following useful special cases arise according as the function $f(x, y)$ is even or odd in its variables.

Case (a) $f(x, y)$ Is Even in x and y

In this case $f(-x, y) = f(x, y)$ and $f(x, -y) = f(x, y)$, so only the coefficients a_{mn} are nonzero, leading to the **double Fourier cosine series representation**

$$f(x, y) = a_{00} + \sum_{m=1}^{\infty} a_{m0} \cos \frac{m\pi x}{L_1} + \sum_{n=1}^{\infty} a_{0n} \cos \frac{n\pi y}{L_2}$$

$$+ \sum_{m=1}^{\infty} \sum_{n=1}^{\infty} a_{mn} \cos \frac{m\pi x}{L_1} \cos \frac{n\pi y}{L_2}. \tag{63}$$

As $f(x, y)$ is even in both x and y, both limits of integration in the integrals defining the a_{mn} in (54) to (57) can be changed to give

$$a_{00} = \frac{1}{L_1 L_2} \int_0^{L_2} \int_0^{L_1} f(x, y)\,dxdy$$

$$a_{m0} = \frac{2}{L_1 L_2} \int_0^{L_2} \int_0^{L_1} f(x, y) \cos \frac{m\pi x}{L_1}\,dxdy, \quad m = 1, 2, \ldots$$

$$a_{0n} = \frac{2}{L_1 L_2} \int_0^{L_2} \int_0^{L_1} f(x, y) \cos \frac{n\pi y}{L_2}\,dxdy, \quad n = 1, 2, \ldots$$

$$a_{mn} = \frac{4}{L_1 L_2} \int_0^{L_2} \int_0^{L_1} f(x, y) \cos \frac{m\pi x}{L_1} \cos \frac{n\pi y}{L_2}\,dxdy, \quad m, n = 1, 2, \ldots. \tag{64}$$

Case (b) $f(x, y)$ Is Even in x and Odd in y

In this case $f(-x, y) = f(x, y)$ and $f(x, -y) = -f(x, y)$ so only the coefficients b_{mn} are nonzero, leading to the representation

$$f(x, y) = \sum_{n=1}^{\infty} b_{0n} \sin \frac{n\pi y}{L_2} + \sum_{m=1}^{\infty} \sum_{n=1}^{\infty} b_{mn} \cos \frac{m\pi x}{L_1} \sin \frac{n\pi y}{L_2}. \tag{65}$$

As $f(x, y)$ is even only in x, the limits of integration for x in integral (58) defining the coefficients b_{mn} can be changed to give

$$b_{mn} = \frac{2}{L_1 L_2} \int_{-L_2}^{L_2} \int_0^{L_1} f(x, y) \cos \frac{m\pi x}{L_1} \sin \frac{n\pi y}{L_2}\,dxdy$$

$$= \frac{4}{L_1 L_2} \int_0^{L_2} \int_0^{L_1} f(x, y) \cos \frac{m\pi x}{L_1} \sin \frac{n\pi y}{L_2}\,dxdy. \tag{66}$$

Case (c) $f(x, y)$ Is Odd in x and Even in y

In this case $f(-x, y) = -f(x, y)$ and $f(x, -y) = f(x, y)$, so only the coefficients c_{mn} are nonzero, leading to the representation

$$f(x, y) = \sum_{m=1}^{\infty} c_{m0} \sin \frac{m\pi y}{L_1} + \sum_{m=1}^{\infty} \sum_{n=1}^{\infty} c_{mn} \sin \frac{m\pi x}{L_1} \cos \frac{n\pi y}{L_2}. \tag{67}$$

As $f(x, y)$ is even only in y, the limits of integration for y in integral (59) defining the coefficients c_{mn} can be changed to give

$$\begin{aligned} c_{mn} &= \frac{2}{L_1 L_2} \int_{-L_2}^{L_2} \int_{-L_1}^{L_1} f(x, y) \sin \frac{m\pi x}{L_1} \cos \frac{n\pi y}{L_2} dx dy \\ &= \frac{4}{L_1 L_2} \int_{0}^{L_2} \int_{-L_1}^{L_2} f(x, y) \sin \frac{m\pi x}{L_1} \cos \frac{n\pi y}{L_2} dx dy. \end{aligned} \tag{68}$$

Case (d) $f(x, y)$ Is Odd in x and y

In this case $f(-x, y) = -f(x, y)$ and $f(x, -y) = -f(x, y)$ so only the coefficients d_{mn} are nonzero, leading to the **double Fourier sine series representation**

$$f(x, y) = \sum_{m=1}^{\infty} \sum_{n=1}^{\infty} d_{mn} \sin \frac{m\pi x}{L_1} \sin \frac{n\pi y}{L_2}. \tag{69}$$

As $f(x, y)$ is odd in both x and y, both limits of integration for x and y in integral (60) defining the coefficients d_{mn} can be changed to give

$$d_{mn} = \frac{4}{L_1 L_2} \int_{0}^{L_2} \int_{0}^{L_1} f(x, y) \sin \frac{m\pi x}{L_1} \sin \frac{n\pi y}{L_2} dx dy. \tag{70}$$

EXAMPLE 9.15   Find the double Fourier series representation of $f(x, y) = xy$ over $-2 \leq x \leq 2$ and $-4 \leq y \leq 4$.

Solution   The function $f(x, y)$ is odd in both x and y, so this corresponds to the double Fourier sine series representation of case (d) with $L_1 = 2$ and $L_2 = 4$. From (70) we have

$$\begin{aligned} d_{mn} &= \frac{4}{8} \int_{0}^{4} \int_{0}^{2} xy \sin \frac{m\pi x}{2} \sin \frac{n\pi y}{4} dx dy \\ &= \frac{1}{2} \left[\int_{0}^{2} x \sin \frac{m\pi x}{2} dx \right] \left[\int_{0}^{4} y \sin \frac{n\pi y}{4} dy \right] \\ &= \frac{1}{2} \left[\frac{-4(-1)^m}{m\pi} \right] \left[\frac{-16(-1)^n}{n\pi} \right] = (-1)^{m+n} \frac{32}{mn\pi^2}. \end{aligned}$$

Thus, the required double Fourier sine series representation is

$$f(x, y) = \frac{32}{\pi^2} \sum_{m=1}^{\infty} \sum_{n=1}^{\infty} (-1)^{m+n} \frac{1}{mn} \sin \frac{m\pi x}{2} \sin \frac{n\pi y}{4},$$

for $-2 \leq x \leq 2$ and $-4 \leq y \leq 4$. Notice that this same expression describes the representation of $f(x, y)$ for $0 \leq x \leq 2$ and $0 \leq y \leq 4$. ∎

By analogy with the half-range sine and cosine series of Section 9.3, a function $f(x, y)$ defined in a region $0 \leq x \leq a, 0 \leq y \leq b$ can be extended to the region $-a \leq x \leq a, -b \leq y \leq b$ either as a function that is odd in both x and y, or as one that is even in both x and y. If it is extended as an odd function, case (d) applies and the representation in the first quadrant follows by restricting the result to $0 \leq x \leq a$, $0 \leq y \leq b$, whereas if it is extended as an even function, case (a) applies, when the representation is again obtained by restricting the result to $0 \leq x \leq a, 0 \leq y \leq b$.

Suppose, for example, a double Fourier sine series representation of $f(x, y) = xy$ is required for $0 \leq x \leq 2$ and $0 \leq y \leq 4$. Then extending $f(x, y)$ to the region $-2 \leq x \leq 2, -4 \leq y \leq 4$ as a function that is odd in both x and y leads to Example 9.15, so the required representation is given by restricting the double Fourier sine series of Example 9.15 to $0 \leq x \leq 2$ and $0 \leq y \leq 4$. Similarly, $f(x, y) = xy$ can be represented by a double Fourier cosine series in $0 \leq x \leq 2$ and $0 \leq y \leq 4$ by extending it as $f(x, y) = |x||y|$ for $-2 \leq x \leq 2$ and $-4 \leq y \leq 4$. As $f(x, y)$ is even in both x and y, case (a) can be applied and the result again restricted so that $0 \leq x \leq 2$ and $0 \leq y \leq 4$.

A typical plot of a double Fourier series approximation to $f(x, y) = xy$ for $0 \leq x \leq 2$ and $0 \leq y \leq 4$ provided by a partial sum of the double Fourier sine series in Example 9.15 is shown in Fig. 9.14 for the case with $m = n = 10$. If, instead, the cosine approximation had been used (see Exercise 6), the plot of the corresponding approximation provided by the partial sum with $m = n = 10$ is shown in Fig. 9.15. The convergence of the double cosine series is seen to be the faster of the two.

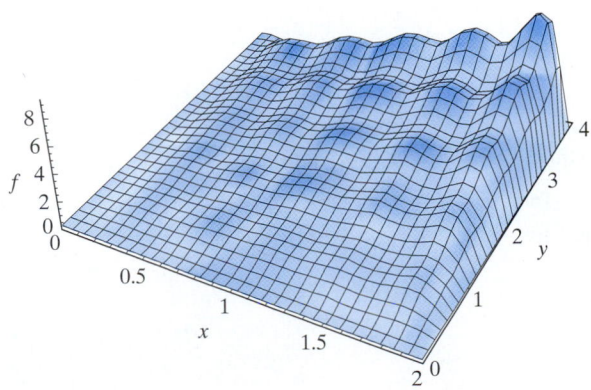

FIGURE 9.14   A double Fourier sine series approximation to $f(x, y) = xy$.

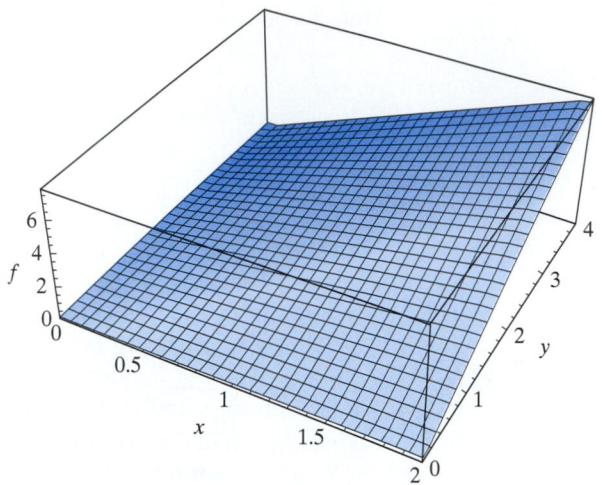

FIGURE 9.15 A double Fourier cosine series approximation to $f(x, y) = xy$.

Summary

It was shown how an ordinary Fourier series representation can be extended in a natural way to the expansion of functions $f(x, y)$ of two variables. After the derivation of the general expansion result, four useful special cases were examined and illustrated by example. Unless $f(x, y)$ is simple, the Fourier series approximation of functions of two variables can require numerical integration when finding the Fourier coefficients, and many terms are usually required to achieve good convergence, so in general it is necessary to perform such calculations and to plot the result by computer.

EXERCISES 9.6

1. By setting $y = 1$ in $f(x, y) = x^2 y$, with $-\pi \leq x \leq \pi$ and $-\pi \leq y \leq \pi$, show that the double Fourier series representation of $f(x, y)$ reduces to the ordinary Fourier series representation of $f(x) = x^2$ for $-\pi \leq x \leq \pi$ given by

$$f(x) = \frac{\pi^2}{3} + 4 \sum_{m=1}^{\infty} (-1)^m \frac{\cos mx}{m^2}$$

In Exercises 2 through 9 find and plot double Fourier series partial sum approximations to the given function.

2. $f(x, y) = xy^2$, for $-\pi \leq x \leq \pi$ and $-\pi \leq y \leq \pi$.

3. $f(x, y) = x^3 y$, for $-\pi \leq x \leq \pi$ and $-\pi \leq y \leq \pi$.

4. $f(x, y) = x^2 y^2$, for $-\pi \leq x \leq \pi$ and $-\pi \leq y \leq \pi$.

5.* $f(x, y) = \text{sign}(xy)$, for $-\pi \leq x \leq \pi$ and $-\pi \leq y \leq \pi$, where sign $u = 1$ if $u > 0$ and sign $u = -1$ if $u < 0$.

6.* $f(x, y) = |xy|$, for $-2 \leq x \leq 2$ and $-4 \leq y \leq 4$.

7.* $f(x, y) = \text{sign}(xy) + xy$, for $-\pi \leq x \leq \pi$ and $-\pi \leq y \leq \pi$.

8.* $f(x, y) = y|\sin x|$, for $-\pi \leq x \leq \pi$ and $-\pi \leq y \leq \pi$.

9.* Extend $f(x, y) = xy^2$, for $0 \leq x \leq \pi$ and $0 \leq y \leq \pi$, to $-\pi \leq x \leq \pi$ and $-\pi \leq y \leq \pi$ as an odd function, and hence find a double Fourier sine series representation of $f(x, y)$ for $0 \leq x \leq \pi$ and $0 \leq y \leq \pi$.

CHAPTER 9
TECHNOLOGY PROJECTS

The purpose of these projects is to use computer algebra to generate Fourier series for continuous and discontinuous functions, to use computer graphics to examine their convergence to the functions they represent, and to explore the nature of the Gibbs phenomenon.

Project 1

Finding Fourier Series and Plotting Partial Sums

Use computer algebra to find the first 11 terms $a_0, a_1, \ldots, a_5, b_1, b_2, \ldots, b_5$ of the Fourier series of

$$f(x) = (\pi^2 - x^2)e^{-x} \sin x \quad \text{for } -\pi \leq x \leq \pi.$$

Plot the approximation to $f(x)$ obtained by using (a) the terms involving $a_0, a_1, a_2, b_1,$ and b_2 and (b) the 11 terms involving $a_0, \ldots, a_5, b_1, \ldots, b_5$ in the partial sum approximation, and compare the results with the graph of $f(x)$.

Project 2

Examining the Gibbs Phenomenon

Use computer algebra to find the Fourier series representation of the function

$$f(x) = \begin{cases} \sin x - 1, & -\pi < x < 0 \\ \sin x + 1, & 0 < x < \pi. \end{cases}$$

By plotting the partial sum representations of $f(x)$ using different numbers of terms, demonstrate the persistence of the overshoot and undershoot caused by the Gibbs phenomenon as the number of terms in the approximation increases.

Project 3

The Complex Fourier Series

Use computer algebra with the complex Fourier series representation of a function to verify the coefficients c_n and c_{2n-1} found in Example 9.12. Plot different partial sum approximations to $f(x)$ and, as in Project 2, demonstrate the persistence of the Gibbs phenomena as the number of terms in the partial sum approximation increases.

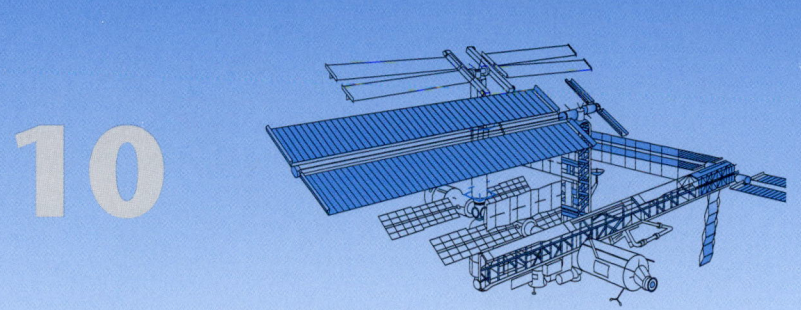

CHAPTER 10

Fourier Integrals and the Fourier Transform

ourier series enable functions and solutions of linear systems defined over a finite interval to be represented as an infinite series of sines and cosines. This suffices for many physical problems, but often the interval involved is either semi-infinite or infinite, in which case a somewhat different representation becomes necessary. This happens, for example, when working with the partial differential equations that describe heat conduction and diffusion in a half-space for which Fourier series cannot be used.

The Fourier integral can be regarded as the limiting case of a Fourier series representation of a function $f(x)$ defined over an interval $-L < x < L$ as $L \to \infty$. The meaning of the integral representation when the function to be represented is discontinuous is considered, and the special cases of the sine and cosine integral representations are introduced.

Fourier sine and cosine transforms are considered, tables of their transform pairs are given, and the transform of derivatives is discussed. In anticipation of Chapter 18, an application of the Fourier transform is made to the problem of the one-dimensional time dependent heat equation.

10.1 The Fourier Integral

A Fourier series has been shown to represent an arbitrary function $f(x)$ over an interval $-L \leq x \leq L$, and because the series is periodic with period $2L$ the representation of $f(x)$ in this fundamental interval is repeated by periodicity for all x outside the interval. However, even if $f(x)$ is defined outside the fundamental interval, it does not necessarily follow that the function and its periodic extensions coincide outside the interval. This means that if a nonperiodic function is to be represented over an arbitrarily large interval, some generalization of a Fourier series is required.

Letting $L \to \infty$ in a Fourier series leads to the introduction of a different type of representation called a **Fourier integral representation**, where the function $f(x)$ is defined for all x and need not be periodic. This representation forms the basis of an integral transform called the **Fourier transform** that is similar to the Laplace transform. As with the Laplace transform, one of the the main uses of the Fourier transform is in the solution of differential equations.

The derivation of the Fourier integral representation given here is heuristic, because a rigorous one requires techniques that are not needed elsewhere in the book. We start from the definition of a Fourier series of $f(x)$ over an interval $-L \leq x \leq L$ given in (18) and (19) of Section 9.1 by writing

$$f(x) = a_0 + \sum_{n=1}^{\infty} \left(a_n \cos \frac{n\pi x}{L} + b_n \sin \frac{n\pi x}{L} \right) \tag{1}$$

where

$$a_0 = \frac{1}{2L} \int_{-L}^{L} f(x) dx, \quad a_n = \frac{1}{L} \int_{-L}^{L} f(x) \cos \frac{n\pi x}{L} dx,$$

$$b_n = \frac{1}{L} \int_{-L}^{L} f(x) \sin \frac{n\pi x}{L} dx \quad \text{for } n = 1, 2, \ldots. \tag{2}$$

Substituting the Fourier coefficients (2) into Fourier series (1) allows it to be written in the integral form

$$f(x) = \frac{1}{2L} \int_{-L}^{L} f(u) du + \frac{1}{L} \sum_{n=1}^{\infty} \int_{-L}^{L} f(u) \cos \frac{n\pi (u - x)}{L} du. \tag{3}$$

To proceed further, if the representation is to remain valid as $L \to \infty$ the first term must not become either infinite or indeterminate. This will certainly be true if $\lim_{L \to \infty} \int_{-L}^{L} |f(x)| dx$ is finite, because then the integral involved in the first term will be *absolutely* convergent and the first term in (3) will vanish in the limit as $L \to \infty$. From now on we will assume this condition to be satisfied. We can now write (3) as

$$f(x) = \frac{1}{L} \sum_{n=1}^{\infty} \int_{-L}^{L} f(u) \cos \frac{n\pi (u - x)}{L} du. \tag{4}$$

It is from this point onward that our derivation of the Fourier integral representation becomes heuristic, because the arguments used to convert (4) to an integral over the interval $(-\infty, \infty)$ are merely intuitive. A careful examination of the convergence of the double integral involved would be necessary to provide a rigorous justification.

Setting $\Delta_n \omega = \pi / L$, and defining the frequency $\omega_n = n\pi / L$, allows (4) to be rewritten as

$$\frac{1}{\pi} \sum_{n=1}^{\infty} \Delta_n \omega \int_{-L}^{L} f(u) \cos[\omega_n (u - x)] du. \tag{5}$$

Examination of (5) suggests it is equivalent to the pre-limit sum approximation used in the definition of the definite (Riemann) integral of the function

$$F(u) = \frac{1}{\pi} \int_{-L}^{L} f(u) \cos \omega (u - x) du.$$

Using this last result in (5), and proceeding to the limit as $L \to \infty$, we obtain

$$f(x) = \frac{1}{\pi} \int_0^{\infty} d\omega \int_{-\infty}^{\infty} f(u) \cos \omega (u - x) du, \tag{6}$$

the Fourier integral representation

which is called the **Fourier integral representation** of $f(x)$.

By defining the functions $A(\omega)$ and $B(\omega)$ as

$$A(\omega) = \frac{1}{\pi} \int_{-\infty}^{\infty} f(u) \cos \omega u\, du \quad \text{and} \quad B(\omega) = \frac{1}{\pi} \int_{-\infty}^{\infty} f(u) \sin \omega u\, du, \qquad (7)$$

the Fourier integral representation in (6) can be written in the simpler form

$$f(x) = \int_{0}^{\infty} [A(\omega) \cos \omega x + B(\omega) \sin \omega x]\, d\omega. \qquad (8)$$

The convergence properties of Fourier series recorded in Theorem 9.1 can be shown to be transferred to the Fourier integral representation of $f(x)$ if, in addition to the integral of $f(x)$ being absolutely convergent over $(-\infty, \infty)$, it also satisfies certain other conditions. These conditions, called **Dirichlet conditions**, are as follows:

Dirichlet conditions

  (i) In any finite interval $f(x)$ has only a finite number of maxima and minima
  (ii) In any finite interval $f(x)$ has only a finite number of bounded jump discontinuities and no infinite jump discontinuities.

We now state the following theorem for the Fourier integral without proof.

PETER GUSTAV LEJEUNE DIRICHLET (1805–1859)
A German mathematician who studied under Gauss, was the son-in-law of Jacobi and succeeded Gauss as Professor of Mathematics at Göttingen. He did much to make some of the more abstruse contributions by Gauss better understood. His most important contributions to mathematics were his major contribution to the understanding of the convergence of Fourier series, and his work on number theory and the theory of potential.

THEOREM 10.1

the fundamental Fourier integral theorem

Fourier integral theorem Let $f(x)$ satisfy Dirichlet conditions, and suppose the (sufficiency) conditions that $f(x)$ be both integrable and absolutely integrable over the interval $-\infty < x < \infty$ are both satisfied, so each of the integrals $\int_{-\infty}^{\infty} f(x)dx$ and $\int_{-\infty}^{\infty} |f(x)|dx$ exists. Then

$$\frac{1}{2}[f(x+0) + f(x-0)] = \frac{1}{\pi} \int_{0}^{\infty} d\omega \int_{-\infty}^{\infty} f(u) \cos \omega(u-x) du$$

or, equivalently,

$$\frac{1}{2}[f(x+0) + f(x-0)] = \int_{0}^{\infty} [A(\omega) \cos \omega x + B(\omega) \sin \omega x]d\omega,$$

where

$$A(\omega) = \frac{1}{\pi} \int_{-\infty}^{\infty} f(u) \cos \omega u\, du \quad \text{and} \quad B(\omega) = \frac{1}{\pi} \int_{-\infty}^{\infty} f(u) \sin \omega u\, du. \qquad \blacksquare$$

EXAMPLE 10.1

Find the Fourier integral representation of $f(x) = e^{-|x|}$.

Solution The function $e^{-|x|}$ satisfies the Dirichlet conditions, and $\int_{-\infty}^{\infty} |e^{-|x|}|dx = 2$, so the integral of $f(x) = e^{-|x|}$ over $(-\infty, \infty)$ is absolutely convergent. This confirms that $f(x) = e^{-|x|}$ has a Fourier integral representation.

The function $e^{-|x|}$ is even in x, so $e^{-|u|} \cos \omega u$ is also even, and

$$A(\omega) = \frac{1}{\pi} \int_{-\infty}^{\infty} e^{-|u|} \cos \omega u \, du = \frac{2}{\pi} \int_{0}^{\infty} e^{-u} \cos \omega u \, du = \frac{2}{\pi(1 + \omega^2)}.$$

As the function $e^{-|u|} \sin \omega u$ is odd in u,

$$B(\omega) = \frac{1}{\pi} \int_{-\infty}^{\infty} e^{-|u|} \sin \omega u \, du = \frac{2}{\pi} \int_{0}^{\infty} e^{-u} \sin \omega u \, du = 0,$$

so from (8) the Fourier integral representation of $e^{-|x|}$ is seen to be

$$e^{-|x|} = \frac{2}{\pi} \int_{0}^{\infty} \frac{\cos \omega x}{1 + \omega^2} d\omega. \qquad \blacksquare$$

EXAMPLE 10.2   Find the Fourier integral representation of

$$f(x) = \begin{cases} e^{-x}, & x > 0 \\ 0, & x < 0 \end{cases}$$

and use Theorem 10.1 to find the value of the resulting integral when (a) $x < 0$, (b) $x = 0$, and (c) $x > 0$.

Solution   The function $f(x)$ satisfies the Dirichlet conditions and the integral $\int_{-\infty}^{\infty} |f(x)| dx = \int_{0}^{\infty} e^{-x} dx = 1$, so as the conditions of Theorem 10.1 are satisfied the function has a Fourier integral representation.

We have

$$A(\omega) = \frac{1}{\pi} \int_{-\infty}^{\infty} f(u) \cos \omega u \, du = \frac{1}{\pi} \int_{0}^{\infty} e^{-u} \cos \omega u \, du = \frac{1}{\pi(1 + \omega^2)}$$

and

$$B(\omega) = \frac{1}{\pi} \int_{-\infty}^{\infty} f(u) \sin \omega u \, du = \frac{1}{\pi} \int_{0}^{\infty} e^{-u} \sin \omega u \, du = \frac{\omega}{\pi(1 + \omega^2)}.$$

Substituting into (8) shows the Fourier integral representation to be

$$f(x) = \frac{1}{\pi} \int_{0}^{\infty} \frac{\cos \omega x + \omega \sin \omega x}{1 + \omega^2} d\omega \quad \text{for } -\infty < x < \infty.$$

Applying the results of Theorem 10.1 to this integral, we find that

$$\pi f(x) = \int_{0}^{\infty} \frac{\cos \omega x + \omega \sin \omega x}{1 + \omega^2} d\omega = \begin{cases} 0, & x < 0 \\ \pi/2, & x = 0 \\ \pi e^{-x}, & x > 0. \end{cases}$$

When $x = 0$, this last result is seen to reduce to the familiar definite integral

$$\int_{0}^{\infty} \frac{d\omega}{1 + \omega^2} = \frac{\pi}{2}. \qquad \blacksquare$$

Special forms of the Fourier integral representation arise according to whether $f(x)$ is even or odd. When $f(x)$ is an even function, $f(u) \sin \omega u$ is an odd function of u, so $B(\omega) \equiv 0$ and

$$A(\omega) = \frac{2}{\pi} \int_0^\infty f(u) \cos \omega u \, du, \tag{9}$$

so that (8) simplifies to the **Fourier cosine integral representation** of $f(x)$

$$f(x) = \int_0^\infty A(\omega) \cos \omega x \, d\omega. \tag{10}$$

Similarly, when $f(x)$ is an odd function, $f(u) \cos \omega u$ is an odd function of u, so $A(\omega) \equiv 0$ and

$$B(\omega) = \frac{2}{\pi} \int_0^\infty f(u) \sin \omega u \, du, \tag{11}$$

causing (8) to simplify to the **Fourier sine integral representation** of $f(x)$ given by

$$f(x) = \int_0^\infty B(\omega) \sin \omega x \, d\omega. \tag{12}$$

Summary of Fourier integral representations

    (a) An *arbitrary* function $f(x)$ satisfying the conditions of Theorem 10.1 has the **general Fourier integral representation**

different Fourier integral representations

$$\frac{1}{2}[f(x+0) + f(x-0)] = \int_0^\infty [A(\omega) \cos \omega x + B(\omega) \sin \omega x] d\omega.$$

$$\tag{13}$$

    (b) An *even* function $f(x)$ satisfying the conditions of Theorem 10.1 has the **Fourier cosine integral representation**

$$\frac{1}{2}[f(x+0) + f(x-0)] = \int_0^\infty A(\omega) \cos \omega x \, d\omega. \tag{14}$$

    (c) An *odd* function $f(x)$ satisfying the conditions of Theorem 10.1 has the **Fourier sine integral representation**

$$\frac{1}{2}[f(x+0) + f(x-0)] = \int_0^\infty B(\omega) \sin \omega x \, d\omega, \tag{15}$$

where

$$A(\omega) = \frac{1}{\pi} \int_{-\infty}^\infty f(u) \cos \omega u \, du \quad \text{and} \quad B(\omega) = \frac{1}{\pi} \int_{-\infty}^\infty f(u) \sin \omega u \, du.$$

$$\tag{16}$$

Summary   The Fourier integral representation of a function $f(x)$ was introduced as the natural extension of a Fourier series representation as the interval of the representation extends to become the interval $-\infty < x < \infty$. A fundamental representation theorem was given and illustrated by example, and some useful special cases of the theorem were considered.

EXERCISES 10.1

Find the Fourier integral representation of the given functions.

1. The rectangular pulse function $f(x) = \begin{cases} 1, & |x| < 1 \\ 0, & |x| > 1 \end{cases}$ (Fig. 10.1).

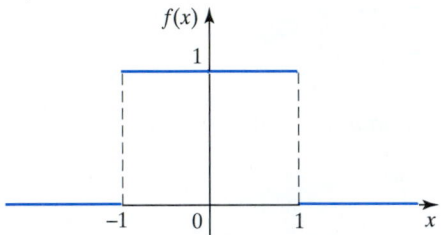

FIGURE 10.1   The rectangular pulse function.

2. The triangular function

$$f(x) = \begin{cases} 0, & |x| > a \\ b\left(1 + \dfrac{x}{a}\right), & -a \le x \le 0 \\ b\left(1 - \dfrac{x}{a}\right), & 0 \le x \le a \end{cases}$$ (Fig. 10.2).

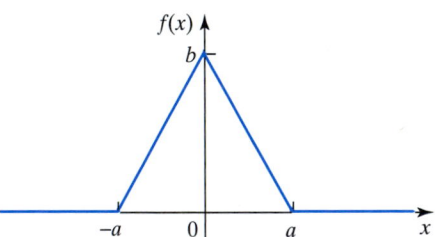

FIGURE 10.2   The triangular function.

3. $f(x) = \begin{cases} 0, & |x| > a \\ bx/a, & -a \le x \le a \end{cases}$ (Fig. 10.3).

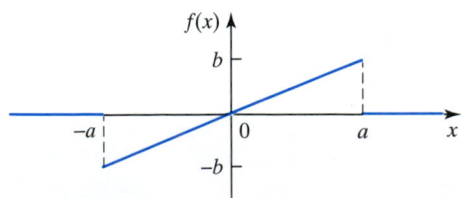

FIGURE 10.3   The truncated straight line function.

4. $f(x) = \begin{cases} 0, & x \le 0 \\ \sin x, & 0 \le x \le \pi \\ 0, & x \ge \pi \end{cases}$ (Fig. 10.4).

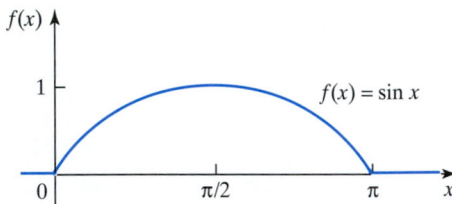

FIGURE 10.4   The asymmetric truncated sine function.

5. $f(x) = \begin{cases} (\pi/2)\cos x, & |x| < \pi/2 \\ 0, & |x| > \pi/2 \end{cases}$ (Fig. 10.5).

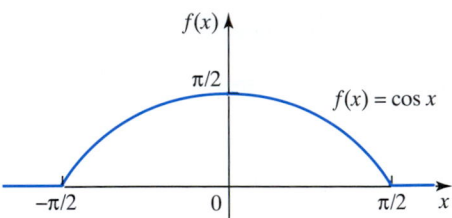

FIGURE 10.5   The truncated cosine function.

6. $f(x) = \begin{cases} (\pi/2)\sin x, & |x| < \pi/2 \\ 0, & |x| > \pi/2 \end{cases}$ (Fig. 10.6).

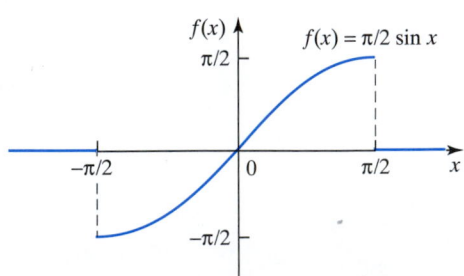

FIGURE 10.6   The truncated sine function.

7. $f(x) = \begin{cases} 0, & x < 0 \\ \cos x, & 0 < x < \pi \\ 0, & x > \pi \end{cases}$ (Fig. 10.7).

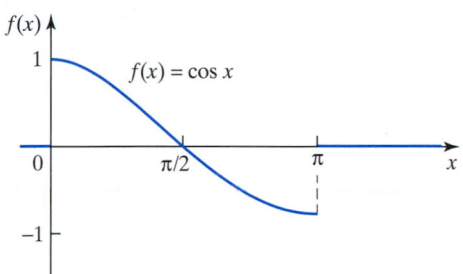

FIGURE 10.7   The asymmetric truncated cosine function.

8. The hump function $f(x) = 1/(1 + x^2)$ (Fig. 10.8). (Hint: Use the result of Example 10.16 with a change of notation.)

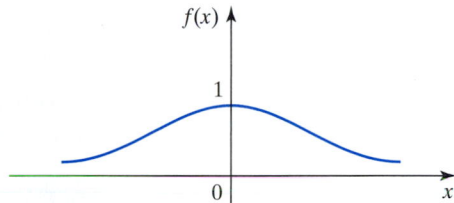

FIGURE 10.8   The hump function.

10.2   The Fourier Transform

The starting point for the development of the *Fourier transform* is the complex form of the Fourier integral representation of a function $f(x)$. To derive this representation in which $f(x)$ is defined over the interval $(-\infty, \infty)$, we substitute into (8) of Section 10.1 the expressions for $A(\omega)$ and $B(\omega)$ given in (7) to obtain

$$\frac{1}{2}[f(x+0) + f(x-0)] = \frac{1}{\pi} \int_0^\infty \left[\int_{-\infty}^\infty f(u)[\cos \omega u \cos \omega x + \sin \omega u \sin \omega x] du \right] d\omega$$

$$= \frac{1}{\pi} \int_0^\infty \left[\int_{-\infty}^\infty f(u) \cos\{\omega(u - x)\} du \right] d\omega$$

$$= \frac{1}{\pi} \int_0^\infty \left[\int_{-\infty}^\infty f(u) \cos\{\omega(x - u)\} du \right] d\omega,$$

where we have used the result $\cos \omega(u - x) = \cos \omega(x - u)$.

As the integrand in the last integral is an even function of ω, the interval of integration with respect to ω can be doubled and the result compensated by the introduction of a multiplicative factor $1/2$ to give

$$\frac{1}{2}[f(x+0) + f(x-0)] = \frac{1}{2\pi} \int_{-\infty}^\infty \left[\int_{-\infty}^\infty f(u) \cos \omega(x - u) du \right] d\omega. \tag{17}$$

The function $\sin \omega(x - u)$ is an odd function of ω, so it follows directly that

$$0 = \frac{1}{2\pi} \int_{-\infty}^\infty \left[\int_{-\infty}^\infty f(u) \sin\{\omega(x - u)\} du \right] d\omega. \tag{18}$$

the complex Fourier integral representation

Multiplying equation (18) by i, adding the result to equation (17), and using the Euler formula $e^{i\theta} = \cos \theta + i \sin \theta$, we arrive at the **complex Fourier integral**

representation

$$\frac{1}{2}[f(x+0)+f(x-0)] = \frac{1}{2\pi}\int_{-\infty}^{\infty}\left[\int_{-\infty}^{\infty}f(u)\exp\{i\omega(x-u)\}du\right]d\omega.$$

(19)

The brackets in (17) to (19) were retained to clarify the order in which the integrations are performed, but they are usually omitted in (19), which then becomes

$$\frac{1}{2}[f(x+0)+f(x-0)] = \frac{1}{2\pi}\int_{-\infty}^{\infty}\int_{-\infty}^{\infty}f(u)\exp\{i\omega(x-u)\}dud\omega.$$

(20)

Clearly, the left-hand side of (20) reduces to $f(x)$ wherever the function is continuous.

To arrive at the definitions of a Fourier transform and its inverse we write the factor $\exp\{i\omega(x-u)\}$ in (19) (equivalently (20)) as the product $\exp\{i\omega x\} \cdot \exp\{-i\omega u\}$. Then, as the inner integral only involves integration with respect to u, we rewrite (19) as

$$f(x) = \frac{1}{\sqrt{2\pi}}\int_{-\infty}^{\infty}\exp\{i\omega x\}\left[\frac{1}{\sqrt{2\pi}}\int_{-\infty}^{\infty}f(u)\exp\{-i\omega u\}du\right]d\omega, \qquad (21)$$

where the left-hand side is to be replaced by $(1/2)[f(x+0)+f(x-0)]$ whenever $f(x)$ is discontinuous.

If we now define the function $F(\omega)$ as

$$F(\omega) = \frac{1}{\sqrt{2\pi}}\int_{-\infty}^{\infty}f(u)\exp\{-i\omega u\}du,$$

then because u is a dummy variable it can be replaced by x and the result rewritten as

$$F(\omega) = \frac{1}{\sqrt{2\pi}}\int_{-\infty}^{\infty}f(x)\exp\{-i\omega x\}dx, \qquad (22)$$

so that (19) becomes

$$f(x) = \frac{1}{\sqrt{2\pi}}\int_{-\infty}^{\infty}F(\omega)\exp\{i\omega x\}d\omega. \qquad (23)$$

Fourier transforms and transform pairs

The function $F(\omega)$ in (22) is called the **Fourier transform** of $f(x)$, or sometimes the **exponential Fourier transform**, and because integral (23) recovers $f(x)$ from $F(\omega)$ it is called the **inversion integral** for the Fourier transform. As with the Laplace transform, when working with the Fourier transform the function $f(x)$ and the associated Fourier transform $F(\omega)$ are called a **Fourier transform pair**. A short table of Fourier transform pairs is to be found at the end of this section.

Various other notations are used to indicate the Fourier transform of $f(x)$, the most common of which involves representing it by $\bar{f}(\omega)$, so in terms of the notation used here, $\bar{f}(\omega) = F(\omega)$.

Another notation that is often useful involves representing the Fourier transform of $f(x)$ by $\mathcal{F}\{f(x)\}$, so that $\mathcal{F}\{f(x)\} = F(\omega)$, and when this notation is used the inverse Fourier transform is written $\mathcal{F}^{-1}\{F(\omega)\} = f(x)$. In what follows a function to be transformed is denoted by a lowercase letter, and the corresponding uppercase letter is then used to denote its Fourier transform. So, for example, $\mathcal{F}\{g(x)\} = G(\omega)$ and $\mathcal{F}\{h(x)\} = H(\omega)$.

The choice of the normalizing factors $1/\sqrt{2\pi}$ in integrals (22) and (23) is optional, and it is chosen here to introduce as much symmetry as possible into the definitions of a Fourier transform and its inverse. All that is required of the normalizing factors is that their product be $1/(2\pi)$, so in many reference works the factor $1/\sqrt{2\pi}$ in (22) is replaced by 1, while the factor $1/\sqrt{2\pi}$ in (23) is replaced by $1/(2\pi)$. It is impossible to achieve complete symmetry in the definitions of a Fourier integral and its inverse because the exponential factor occurs with opposite signs in (22) and (23).

When Fourier transforms listed in reference works are used, another source of confusion can arise because sometimes the signs in the exponential factors occurring in integrals (22) and (23) are interchanged. When this happens a Fourier transform obtained using this sign convention can be converted to the one used here by reversing the sign of ω. However, each definition of the Fourier transform and the corresponding inversion integral conform to the general pattern

$$\mathcal{F}\{f(x)\} = \frac{k}{2\pi}\int_{-\infty}^{\infty} f(x)\exp\{\pm i\omega x\}dx \quad \text{and}$$

$$\mathcal{F}^{-1}\{F(\omega)\} = \frac{1}{k}\int_{-\infty}^{\infty} F(\omega)\exp\{\mp i\omega x\}d\omega,$$

(24)

where k is an arbitrary scale factor.

In view of the different conventions that are in use, when working with Fourier transforms and referring to reference works, it is essential that the normalizing factor k and the sign convention employed in the exponential factors be established before any use is made of the results.

When we considered the convergence of Fourier series, the Riemann–Lebesgue lemma was established the results of which were that

$$\lim_{n\to\infty}\int_{-\pi}^{\pi} f(x)\cos nx\, dx = \lim_{n\to\infty}\int_{-\pi}^{\pi} f(x)\sin nx\, dx = 0. \tag{25}$$

A limiting argument similar to the one used in Section 10.1 when deriving the Fourier integral representation of $f(x)$ shows that, provided $f(x)$ has a Fourier transform,

$$\lim_{|\omega|\to\infty}\int_{-\infty}^{\infty} f(x)\cos\omega x\, dx = \lim_{|\omega|\to\infty}\int_{-\infty}^{\infty} f(x)\sin\omega x\, dx = 0. \tag{26}$$

As the Fourier transform $F(\omega)$ of $f(x)$ can be written

$$F(\omega) = \frac{1}{\sqrt{2\pi}}\left[\int_{-\infty}^{\infty} f(x)\cos\omega x\, dx - i\int_{-\infty}^{\infty} f(x)\sin\omega x\, dx\right], \tag{27}$$

an application of limits (26) in (27) establishes the important property of a Fourier transform that

$$\lim_{|\omega|\to\infty} F(\omega) = 0. \tag{28}$$

Find the Fourier transforms of

(a) $f(x) = \begin{cases} 1, & |x| < a \\ 0, & |x| > a, \end{cases}$ (b) $g(x) = \begin{cases} 1, & 0 < x < a \\ 0, & \text{otherwise}, \end{cases}$ (c) $p(x) = \dfrac{1}{x^2+a^2}$

by making use of the standard integral $\int_{-\infty}^{\infty} \frac{\cos\omega x}{x^2+a^2} dx = \frac{\pi}{a} e^{-|\omega|a}$ $(a > 0)$ and (d) $q(x) = \begin{cases} e^{iax}, & 0 < x < 1 \\ 0, & \text{otherwise} \end{cases}$. In each case confirm that the Fourier transform vanishes as $\omega \to \pm\infty$.

Solution

(a) $$F(\omega) = \frac{1}{\sqrt{2\pi}} \int_{-a}^{a} e^{-i\omega x} dx = \frac{1}{\omega\sqrt{2\pi}}\left[\frac{e^{i\omega a} - e^{-i\omega a}}{i}\right]$$

$$= \frac{1}{\omega}\sqrt{\frac{2}{\pi}}\left[\frac{e^{i\omega a} - e^{-i\omega a}}{2i}\right] = \sqrt{\frac{2}{\pi}}\frac{\sin\omega a}{\omega}.$$

As $\sin\omega a$ is bounded, it follows directly that $\lim_{|\omega|\to\infty} F(\omega) = 0$.

(b) $$G(\omega) = \frac{1}{\sqrt{2\pi}} \int_{0}^{a} e^{-i\omega x} dx = \frac{1}{\sqrt{2\pi}}\left(\frac{1 - e^{-i\omega a}}{i\omega}\right).$$

As the numerator of $G(\omega)$ is bounded, it follows that $\lim_{|\omega|\to\infty} G(\omega) = 0$. This example shows that although $f(x)$ may be real, its Fourier transform can be complex.

(c) $$P(\omega) = \frac{1}{\sqrt{2\pi}} \int_{-\infty}^{\infty} \frac{e^{-i\omega x}}{x^2+a^2} dx = \frac{1}{\sqrt{2\pi}} \int_{-\infty}^{\infty} \frac{\cos\omega x}{x^2+a^2} dx - \frac{i}{\sqrt{2\pi}} \int_{-\infty}^{\infty} \frac{\sin\omega x}{x^2+a^2} dx.$$

The integrand of the second integral is odd, so the value of the integral is zero. Using the standard result

$$\int_{-\infty}^{\infty} \frac{\cos\omega x}{x^2+a^2} dx = \frac{\pi}{a} e^{-|\omega|a}$$

in the remaining integral on the right, we find that

$$P(\omega) = \sqrt{\frac{\pi}{2}}\frac{e^{-|\omega|a}}{a} \quad (a > 0).$$

In this case the factor $e^{-|\omega|a}$ ensures that $\lim_{|\omega|\to\infty} P(\omega) = 0$.

(d) $$Q(\omega) = \frac{1}{\sqrt{2\pi}} \int_{-\infty}^{\infty} q(x) e^{-i\omega x} dx = \frac{1}{\sqrt{2\pi}} \int_{0}^{1} e^{-i(\omega-a)x} dx$$

$$= \frac{i}{\sqrt{2\pi}}\left(\frac{1 - e^{-i(\omega-a)}}{a - \omega}\right).$$

As the numerator of the Fourier transform is bounded, the denominator causes the transform to vanish as $|\omega| \to \infty$. This example shows that a complex function can also have a Fourier transform and, in general, that the transform will be complex.

the main operational properties of Fourier transforms

The fundamental properties contained in Theorems 10.2 to 10.8 that follow are called **operational properties** of the Fourier transform. Familiarity with these properties is essential, because they simplify calculations involving Fourier transforms and can lead to results that are difficult to obtain without their use.

THEOREM 10.2

Linearity of the Fourier transform   Let the functions $f(x)$ and $g(x)$ have the respective Fourier transforms $F(\omega)$ and $G(\omega)$, and let a and b be arbitrary constants. Then

$$\mathcal{F}\{af(x) + bg(x)\} = a\mathcal{F}\{f(x)\} + b\mathcal{F}\{g(x)\}.$$

Proof   As the Fourier integral involves the operation of integration, the linearity property of the transform follows directly from the linearity property of the definite integral. ∎

Theorem 10.2 is important when the Fourier transform of a sum of functions is required, because it is this result that allows each term involved in the sum to be transformed separately before the results are added.

EXAMPLE 10.4

Find the Fourier transform of $3\,f(x) - 2g(x)$, where $f(x)$ and $g(x)$ are the functions in (a) and (b) of Example 10.3.

Solution   Using the results of Example 10.3 and applying Theorem 10.2, we have

$$\mathcal{F}\{3\,f(x) - 2g(x)\} = 3\mathcal{F}\{f(x)\} - 2\mathcal{F}\{g(x)\}$$

$$= \sqrt{\frac{2}{\pi}}\left\{\frac{3\sin\omega a}{\omega} - \left(\frac{1 - e^{-i\omega a}}{i\omega}\right)\right\}. \quad ∎$$

THEOREM 10.3

Fourier transform of a derivative of $f(x)$ Let $f(x)$ be a continuous function of x with the property that $\lim_{|x|\to\infty} f(x) = 0$, and such that $f'(x)$ is absolutely integrable over $(-\infty, \infty)$. Then:

(a) $\mathcal{F}\{f'(x)\} = i\omega F(\omega).$

(b)   For all n such that the derivatives $f^{(r)}(x)$ with $r = 1, 2, \ldots, n$ satisfy Dirichlet conditions, are absolutely integrable over $(-\infty, \infty)$, and $\lim_{|x|\to\infty} f^{(n-1)}(x) = 0$,

$$\mathcal{F}\{f^{(n)}(x)\} = (i\omega)^n F(\omega),$$

where $f^{(n)}(x) = d^n f/dx^n$.

Proof

(a) Integration by parts coupled with the condition that $\lim_{|x|\to\infty} f(x) = 0$ gives

$$\mathcal{F}\{f'(x)\} = \frac{1}{\sqrt{2\pi}} \int_{-\infty}^{\infty} f'(x)e^{-i\omega x}dx$$

$$= \frac{1}{\sqrt{2\pi}}\left[f(x)e^{-i\omega x}\Big|_{-\infty}^{\infty} - (-i\omega) \int_{-\infty}^{\infty} f(x)e^{-i\omega x}dx \right]$$

$$= i\omega\,\mathcal{F}\{f(x)\} = i\omega F(\omega),$$

where the term $f(x)e^{-i\omega x}|_{-\infty}^{\infty}$ vanishes because of the condition $\lim_{|x|\to\infty} f(x) = 0$.

(b) The second part of the theorem follows by repeated application of result (a), and the conditions imposed on $f^{(n)}(x)$ are necessary to ensure that its Fourier transform exists. ∎

EXAMPLE 10.5

Find the Fourier transform of $p'(x)$ from the Fourier transform of $p(x)$, where $p(x)$ is the function in Example 10.3(c).

Solution   It was shown in Example 10.3(c) that $P(\omega) = \sqrt{\frac{\pi}{2}} \frac{e^{-|\omega|a}}{a}$, so it follows from Theorem 10.3 (a) that $\mathcal{F}\{p'(x)\} = i\omega P(\omega) = i\omega\sqrt{\frac{\pi}{2}}\frac{e^{-|\omega|a}}{a}$. ∎

THEOREM 10.4

Fourier transform of $x^n f(x)$ Let $f(x)$ be a continuous and differentiable function with an n times differentiable Fourier transform $F(\omega)$. Then

(a) $$\mathcal{F}\{xf(x)\} = i\frac{d}{d\omega}[F(\omega)]$$

and

(b) $$\mathcal{F}\{x^n f(x)\} = i^n \frac{d^n}{d\omega^n}[F(\omega)],$$

for all n such that $\lim_{|\omega|\to\infty} F^{(n)}(\omega) = 0$.

Proof   The proof of the theorem follows directly by the application of *Leibniz's rule* that governs differentiation under the integral sign. The rule may be stated as follows:

Leibniz' rule: Let $f(x, \omega)$ and $\partial f/\partial\omega$ be continuous functions of their variables with $-\infty < x < \infty$ and $-\infty < \omega < \infty$. Furthermore, let $\int_{-\infty}^{\infty} |f(x, \omega)| dx$ be finite and $|\partial f/\partial\omega| \le h(x)$ where $h(x)$ is piecewise continuous and such that $\int_{-\infty}^{\infty} h(x) dx$ is finite. Then

$$\frac{d}{d\omega}\int_{-\infty}^{\infty} f(x, \omega) dx = \int_{-\infty}^{\infty} \frac{\partial}{\partial\omega}[f(x, \omega)] dx.$$

(a) Using Leibniz' rule to differentiate the Fourier transform of $f(x)$, we obtain

$$\frac{d}{d\omega}[F(\omega)] = \frac{1}{\sqrt{2\pi}}\frac{d}{d\omega}\int_{-\infty}^{\infty} f(x)e^{-i\omega x} dx = \frac{-i}{\sqrt{2\pi}}\int_{-\infty}^{\infty} xf(x)e^{-i\omega x} dx.$$

The required result follows from this after multiplication by i, because the expression on the right is then $\mathcal{F}\{xf(x)\}$.

(b) The proof for the case when $n > 1$ follows by repeated application of result (a). The conditions imposed on $x^n f(x)$ and $F(\omega)$ are necessary to ensure the existence of the Fourier transform. ∎

THEOREM 10.5

Fourier transform of $x^m f^{(n)}(x)$ Let $f(x)$ be a continuous n times differentiable function. Furthermore, let $x^m f^{(r)}(x)$ for $r = 1, 2, \ldots, n$ satisfy Dirichlet conditions and be absolutely integrable over $(-\infty, \infty)$, and let $\omega^n F(\omega)$ possess an m times differentiable inverse Fourier transform. Then, provided $\lim_{|x|\to\infty} f^{(n-1)}(x) = 0$,

$$\mathcal{F}\{x^m f^{(n)}(x)\} = (i)^{m+n}\frac{d^m}{d\omega^m}[\omega^n F(\omega)].$$

Proof   The result follows directly by combining Theorems 10.3 and 10.4, because

$$\mathcal{F}\{x^m f^{(n)}(x)\} = (i)^m \frac{d^m}{d\omega^m}\mathcal{F}\{f^{(n)}(x)\} = (i)^{m+n}\frac{d^m}{d\omega^m}[\omega^n F(\omega)].$$

The conditions imposed on $x^m f^{(n)}(x)$ and $\omega^n F(\omega)$ are necessary to ensure the existence of the Fourier transform.   ∎

The examples that follow illustrate how Theorems 10.3 to 10.5 may be used to find the Fourier transforms of more complicated functions.

EXAMPLE 10.6

Find the Fourier transform of $f(x) = \exp(-a^2 x^2)(a > 0)$.

Solution   The function $f(x)$ is continuous and differentiable for all x and

$$\int_{-\infty}^{\infty} |\exp(-a^2 x^2)|dx = \int_{-\infty}^{\infty} \exp(-a^2 x^2)dx = \frac{1}{a}\int_{-\infty}^{\infty} \exp(-u^2)du = \frac{\sqrt{\pi}}{a},$$

where we have made use of the standard integral $\int_{-\infty}^{\infty} \exp(-u^2)du = \sqrt{\pi}$. This shows that $f(x)$ is absolutely integrable over the interval $(-\infty, \infty)$, and so $f(x)$ has a Fourier transform. A straightforward calculation establishes that $f(x)$ satisfies the differential equation

$$f' + 2a^2 xf = 0.$$

Taking the Fourier transform of this equation using Theorem 10.2 gives

$$\mathcal{F}\{f'(x)\} + 2a^2 \mathcal{F}\{xf(x)\} = 0.$$

Applying Theorem 10.3 to the first term and Theorem 10.4 to the second term and cancelling a factor i reduces this to the variables separable equation for $F(\omega)$,

$$2a^2 F' + \omega F = 0, \quad \text{where } F(\omega) = \frac{1}{\sqrt{2\pi}}\int_{-\infty}^{\infty}\exp(-a^2 x^2)e^{-i\omega x}dx.$$

When variables are separated, the equation becomes

$$\int \frac{F'}{F}d\omega = -\frac{1}{2a^2}\int \omega d\omega,$$

so

$$\ln F(\omega) = -\frac{\omega^2}{4a^2} + \ln A, \quad \text{or} \quad F(\omega) = A \exp\left[-\frac{\omega^2}{4a^2}\right],$$

where, for convenience, the arbitrary integration constant has been written in the form $\ln A$. To determine A we use the fact that $A = F(0)$, but

$$F(0) = \frac{1}{\sqrt{2\pi}}\int_{-\infty}^{\infty}\exp(-a^2 x^2)dx = \frac{1}{\sqrt{2\pi}}\frac{\sqrt{\pi}}{a} = \frac{1}{a\sqrt{2}},$$

and so

$$\mathcal{F}\{\exp(-a^2 x^2)\} = F(\omega) = \frac{1}{a\sqrt{2}}\exp\left\{-\frac{\omega^2}{4a^2}\right\} \quad (a > 0).$$ ∎

EXAMPLE 10.7

finding the Fourier
transform of a
function defined by a
differential equation

Find the Fourier transform of the Bessel function $J_0(x)$.

Solution   The Bessel function $J_0(x)$ does not satisfy the absolute integrability condition found in Theorem 10.1. However, this is merely a sufficient condition that ensures the existence of the Fourier transform of a function $f(x)$, though not a necessary one. Functions exist that possess a Fourier transform even though this condition is violated, and $J_0(x)$ is such a function. The function $f(x) = J_0(x)$ is an even function that is defined for all x and satisfies Bessel's differential equation of order zero

$$xf'' + f' + xf = 0.$$

Taking the Fourier transform of the differential equation by using Theorem 10.2 and then applying Theorem 10.5 to the first term, Theorem 10.3 to the second term, and Theorem 10.4 to the last term, we find, after the cancellation of a factor i and the combination of terms, that

$$(1 - \omega^2)F' - \omega F = 0, \quad \text{where } F(\omega) = \frac{1}{\sqrt{2\pi}} \int_{-\infty}^{\infty} J_0(x)e^{-i\omega x}dx.$$

This is a linear first order variables separable differential equation that can be written

$$\int \frac{F'}{F}d\omega = \int \frac{\omega}{1 - \omega^2}d\omega,$$

so integration gives

$$\ln F(\omega) = -\frac{1}{2}\ln(1 - \omega^2) + \ln A, \quad \text{or} \quad F(\omega) = \frac{A}{(1 - \omega^2)^{1/2}}, \quad \text{with } 0 < \omega^2 < 1.$$

In this equation, the arbitrary integration constant has again been written in the form $\ln A$, and the restriction on ω^2 is necessary because the real logarithmic function is not defined for negative arguments.

To determine A we use the fact that $A = F(0)$, together with the standard result $\int_0^{\infty} J_0(x)dx = 1$ and the fact that $J_0(x)$ is an even function, to obtain

$$A = F(0) = \frac{1}{\sqrt{2\pi}} \int_{-\infty}^{\infty} J_0(x)dx = \frac{2}{\sqrt{2\pi}} \int_0^{\infty} J_0(x)dx = \sqrt{\frac{2}{\pi}}.$$

Substituting A into $F(\omega)$ gives

$$\mathcal{F}\{J_0(x)\} = F(\omega) = \sqrt{\frac{2}{\pi}} \frac{1}{(1 - \omega^2)^{1/2}} H(1 - |\omega|),$$

where the Heaviside unit step function $H(1 - |\omega|)$ is necessary because of the restriction imposed by the real logarithmic function that requires ω to be such that $0 < \omega^2 < 1$. ∎

When working with Fourier integrals, as with the Laplace transform, it is useful to introduce the convolution operation to establish the relationship between the functions $f(x)$ and $g(x)$ and their respective Fourier transforms $F(\omega)$ and $G(\omega)$.

The **convolution** of functions $f(x)$ and $g(x)$ denoted by $f * g$ is a function of x, and if the dependence on a variable x in the convolution is to be emphasized, it is then denoted by $(f * g)(x)$. The convolution of $f(x)$ and $g(x)$ is defined as

$$(f * g)(x) = \int_{-\infty}^{\infty} f(t)g(x - t)dt = \int_{-\infty}^{\infty} f(x - t)g(t)dt. \tag{29}$$

A slightly different definition of the convolution operation for the Fourier transform is also to be found in the literature, where it is defined as

$$(f * g)(x) = \frac{1}{\sqrt{2\pi}} \int_{-\infty}^{\infty} f(t)g(x - t)dt.$$

When this definition is employed, the form taken by the next theorem (the convolution theorem for Fourier transforms) will require modification. This is because its form will depend on the factor $1/\sqrt{2\pi}$ and the way the constant 2π enters in the definition of the Fourier transform that is used.

THEOREM 10.6

relating the convolution of $f(x)$ and $g(x)$ and the product of their transforms

The convolution theorem for Fourier transforms   Let the functions $f(x)$ and $g(x)$ be piecewise continuous, bounded, and absolutely integrable over $(-\infty, \infty)$ with the respective Fourier transforms $F(\omega)$ and $G(\omega)$. Then

(a) $\mathcal{F}\{(f * g)(x)\} = 2\pi \, \mathcal{F}\{f(x)\}\mathcal{F}\{g(x)\}, \;\; \text{or} \;\; \mathcal{F}\{f * g\} = 2\pi \, F(\omega)G(\omega)$

and, conversely,

(b) $(f * g)(x) = \sqrt{2\pi} \int_{-\infty}^{\infty} F(\omega)G(\omega)e^{i\omega x} d\omega.$

Proof   **(a)** By definition,

$$\mathcal{F}\{(f * g)(x)\} = \frac{1}{2\pi} \int_{-\infty}^{\infty} \left[\frac{1}{\sqrt{2\pi}} \int_{-\infty}^{\infty} f(t)g(x - t)e^{-i\omega x} dt \right] dx$$

$$= \left[\frac{1}{2\pi} \int_{-\infty}^{\infty} \left[\int_{-\infty}^{\infty} f(t)g(x - t)e^{-i\omega x} dx \right] dt,$$

where the second result follows from the first by a change in the order of integration. If we set $v = x - t$, this becomes

$$\mathcal{F}\{(f * g)(x)\} = \frac{1}{2\pi} \int_{-\infty}^{\infty} \left[f(t)g(v)e^{-i\omega(t+v)} dt \right] dv$$

$$= \frac{1}{2\pi} \int_{-\infty}^{\infty} f(t)e^{-i\omega t} dt \int_{-\infty}^{\infty} g(v)e^{-i\omega v} dv.$$

However, t and v are dummy variables and so may be replaced by x, causing the preceding result to become

$$\mathcal{F}\{(f * g)(x)\} = \mathcal{F}\{f(x)\}2\pi \mathcal{F}\{g(x)\},$$

showing that

$$\mathcal{F}\{(f * g)(x)\} = 2\pi \, \mathcal{F}\{f(x)\} \, \mathcal{F}\{g(x)\}, \;\; \text{or} \;\; \mathcal{F}\{(f * g)(x)\} = 2\pi \, F(\omega)G(\omega).$$

Result (b) follows directly from the last result by taking the inverse Fourier transform that causes a factor $\sqrt{2\pi}$ to cancel.   ■

EXAMPLE 10.8

It was shown in Example 10.3(a) that the function $f(x) = \begin{cases} 1, & |x| < a \\ 0, & |x| > a \end{cases}$ has the Fourier transform $F(\omega) = \sqrt{\frac{2}{\pi}}\left(\frac{\sin \omega a}{\omega}\right)$, so by the convolution theorem it follows that

$$\mathcal{F}\{(f * f)(x)\} = \sqrt{2\pi}\left[\sqrt{\frac{2}{\pi}}\left(\frac{\sin \omega a}{\omega}\right)\right]^2 = 2\sqrt{\frac{2}{\pi}}\left(\frac{\sin^2 \omega a}{\omega^2}\right).$$

Confirm this result by calculating $(f * f)(x)$ and finding its Fourier transform.

Solution   In terms of the Heaviside unit step function we can write $f(t) = H(a - |t|)$ and $f(x - t) = H(a - |x - t|)$, after which consideration of the product $f(t)f(x - t)$ shows that

$$f(t)f(x - t) = \begin{cases} 1, & -a < t < x + a, (-2a < x < 0) \\ 0, & \text{otherwise} \end{cases}$$

and

$$f(t)f(x - t) = \begin{cases} 1, & x - a < t < a, (0 < x < 2a) \\ 0, & \text{otherwise.} \end{cases}$$

The required convolution is then given by

$$(f * f)(x) = \begin{cases} \int_{-a}^{x+a} dt = 2a + x, (-2a < x < 0) \\ \int_{x-a}^{a} dt = 2a - x, (0 < x < 2a) \end{cases} \quad \text{and} \quad (f * f)(x) = 0 \text{ otherwise.}$$

Taking the Fourier transform of $(f^* f)(x)$, we have

$$\mathcal{F}\{(f * f)(x)\} = \frac{1}{\sqrt{2\pi}}\left\{\int_{-2a}^{0}(2a + x)e^{-i\omega x}dx + \int_{0}^{2a}(2a - x)e^{-i\omega x}dx\right\}$$

$$= \sqrt{\frac{2}{\pi}}\left(\frac{1 - \cos 2\omega a}{\omega^2}\right),$$

but $1 - \cos 2\omega a = 2\sin^2 \omega a$, so

$$\mathcal{F}\{(f * f)(x)\} = 2\sqrt{\frac{2}{\pi}}\left(\frac{\sin^2 \omega a}{\omega^2}\right),$$

as required.   ∎

THEOREM 10.7

the Parseval relation extended to Fourier transforms

The Parseval relation for the Fourier transform   If $f(x)$ has the Fourier transform $F(\omega)$, then

$$\int_{-\infty}^{\infty}|f(x)|^2 dx = \int_{-\infty}^{\infty}|F(\omega)|^2 d\omega.$$

Proof   Setting $x = 0$ in result (b) of the convolution theorem gives

$$\int_{-\infty}^{\infty} f(t)g(-t)dt = \int_{-\infty}^{\infty} F(\omega)G(\omega)d\omega.$$

As the Fourier transform is defined for both real and complex functions, it follows from the definition of the transform that $\mathcal{F}\{\bar{f}(-x)\} = \bar{F}(\omega)$, where the bar indicates complex conjugation. If we set $g(t) = \bar{f}(-t)$, the preceding result becomes

$$\int_{-\infty}^{\infty} f(t)\bar{f}(t)dt = \int_{-\infty}^{\infty} F(\omega)\bar{F}(\omega)d(\omega),$$

or

$$\int_{-\infty}^{\infty} |f(x)|^2 dx = \int_{-\infty}^{\infty} |F(\omega)|^2 d\omega,$$

and the result is proved.   ■

EXAMPLE 10.9

Using the result of Example 10.3(a) and the Parseval relation, show that

$$\int_{-\infty}^{\infty} \frac{\sin^2 \omega a}{\omega^2} d\omega = \pi a.$$

Solution   Substituting $f(x) = \begin{cases} 1, & |x| < a \\ 0, & |x| > a \end{cases}$ and the corresponding Fourier transform $F(\omega) = \sqrt{\frac{2}{\pi}} \left(\frac{\sin \omega a}{\omega} \right)$ found in Example 10.3(a) into the Parseval relation gives

$$\int_{-a}^{a} 1^2 dx = \frac{2}{\pi} \int_{-\infty}^{\infty} \left(\frac{\sin^2 \omega a}{\omega^2} \right) d\omega, \text{ and so } 2a = \frac{2}{\pi} \int_{-\infty}^{\infty} \left(\frac{\sin^2 \omega a}{\omega^2} \right) d\omega \ (a > 0),$$

from which the required result follows.   ■

The final theorem describes the effect on the Fourier transform of $f(x)$ caused by scaling x by a factor a, shifting x by a and shifting ω by λ.

THEOREM 10.8

some useful
properties of Fourier
transforms

Fourier transforms involving scaling x by a, shifting x by a, and shifting ω by λ   If $f(x)$ has a Fourier transform $F(\omega)$, then

(i) $\mathcal{F}\{f(ax)\} = \dfrac{1}{a} F(\omega/a) \quad (a > 0)$

(ii) $\mathcal{F}\{f(x - a)\} = e^{-i\omega a} F(\omega)$

(iii) $\mathcal{F}\{e^{i\lambda x} f(x)\} = F(\omega - \lambda)$

Proof   As the results of the theorem follow immediately from the definition of the Fourier transform, only result (i) will be proved, and the derivation of results (ii) and (iii) left as exercises. Starting from the definition of $\mathcal{F}\{f(ax)\}$ and making the variable change $u = ax$ we have

$$\mathcal{F}\{f(ax)\} = \frac{1}{\sqrt{2\pi}} \int_{-\infty}^{\infty} f(ax)e^{-i\omega x} dx = \frac{1}{a\sqrt{2\pi}} \int_{-\infty}^{\infty} f(u)e^{-i\omega u/a} du$$

$$= \frac{1}{a} F(\omega/a)(a > 0).$$ ■

EXAMPLE 10.10

Using the function $f(x)$ and its Fourier transform $F(\omega)$ from Example 10.9, find (a) $\mathcal{F}\{f(2x)\}$, (b) $\mathcal{F}\{f(x - \pi)\}$, and (c) $\mathcal{F}\{e^{ix} f(x)\}$.

Solution   Using the results of Theorem 10.8 we have:

(a) $\mathcal{F}\{f(2x)\} = \dfrac{1}{2}\sqrt{\dfrac{2}{\pi}}\left(\dfrac{\sin(\omega a/2)}{(\omega/2)}\right) = \sqrt{\dfrac{2}{\pi}}\left(\dfrac{\sin(\omega a/2)}{\omega}\right)$

(b) $\mathcal{F}\{f(x-\pi)\} = e^{-i\pi\omega}\sqrt{\dfrac{2}{\pi}}\left(\dfrac{\sin\omega a}{w}\right)$

(c) $\mathcal{F}\{e^{ix}f(x)\} = \sqrt{\dfrac{2}{\pi}}\left(\dfrac{\sin(\omega-1)a}{\omega-1}\right)$ ∎

the Dirac delta function and the Fourier transform

The **Dirac delta function** $\delta(x)$ was introduced in connection with the Laplace transform, where it was recognized that it is not a function in the usual sense, but an *operation* that only has meaning when it appears in the integrand of a definite integral. Because of its many uses in connection with physical problems described by differential equations, we now extend its definition in a way that is suitable for use with Fourier transforms. This is accomplished by defining $\delta(x-a)$ in a symmetrical manner about $x = a$ in terms of the integrals

$$\int_{-\infty}^{\infty}\delta(x-a)f(x)dx = \int_{-\infty}^{\infty}\delta(a-x)f(x)dx = f(a),\qquad(30)$$

where a is any real number.

This definition allows the Fourier transform of $\delta(x-a)$ to be represented as

$$\mathcal{F}\{\delta(x-a)\} = \frac{1}{\sqrt{2\pi}}\int_{-\infty}^{\infty}\delta(x-a)e^{-i\omega x}dx = \frac{1}{\sqrt{2\pi}}e^{-i\omega a}.\qquad(31)$$

EXAMPLE 10.11   Find the Fourier transform of $f(x) = \delta(x-a)\exp[-b^2x^2]\;(b>0)$.

Solution   By definition

$$\mathcal{F}\{\delta(x-a)\exp[-b^2x^2]\} = \frac{1}{\sqrt{2\pi}}\int_{-\infty}^{\infty}\delta(x-a)\exp[-b^2x^2]e^{-i\omega x}dx$$

$$= \frac{1}{\sqrt{2\pi}}\exp[-(a^2b^2+i\omega a)].$$ ∎

Fourier Transforms of Partial Derivatives with Respect to *x* of a Function *f(x, t)* of Two Independent Variables

transforming partial derivatives

The Fourier transform with respect to x of a function $f(x,t)$ of two independent variables x and t, denoted by $F(\omega,t)$, is defined as

$$_x\mathcal{F}\{f(x,t)\} = F(\omega,t) = \frac{1}{\sqrt{2\pi}}\int_{-\infty}^{\infty}f(x,t)e^{-i\omega x}dx,\qquad(32)$$

where the prefix suffix x shows the variable that is being transformed.

In (32) the variable t is not involved in the integration with respect to x, so it follows that the integral by which $f(x, t)$ is recovered from $F(\omega, t)$ and the transform of partial derivatives of $f(x, t)$ with respect to x obey the same rules as those for the function of a single variable $f(x)$. Thus, the inversion integral is given by

$$f(x, t) = {}_x\mathcal{F}^{-1}\{F(\omega, t)\} = \frac{1}{\sqrt{2\pi}} \int_{-\infty}^{\infty} F(\omega, t)e^{i\omega x}\,d\omega, \tag{33}$$

and the Fourier transforms of the partial derivatives of $f(x, t)$ with respect to x are given by

$$\mathop{}_x\mathcal{F}\left\{\frac{\partial^n}{\partial x^n}[f(x, t)]\right\} = (i\omega)^n F(\omega, t) \tag{34}$$

$$\mathop{}_x\mathcal{F}\{x^n f(x, t)\} = i^n \frac{\partial^n}{\partial \omega^n}[F(\omega, t)] \tag{35}$$

$$\mathop{}_x\mathcal{F}\left\{x^m \frac{\partial^n}{\partial x^n}[f(x, t)]\right\} = i^{m+n} \frac{\partial^m}{\partial \omega^m}[\omega^n F(\omega, t)]. \tag{36}$$

These results are necessary when using the Fourier transform to solve partial differential equations involving a function $f(x, t)$ of two independent variables x and t where $-\infty < x < \infty$. Once the partial differential equation has been transformed, it becomes an ordinary differential equation for $F(\omega, t)$, with t as the independent variable and ω as a parameter. When $F(\omega, t)$ has been found by solving the differential equation, the solution $f(x, t)$ of the partial differential equation is recovered from $F(\omega, t)$ by means of the inversion integral (33).

an application to the heat equation

To illustrate the application of the Fourier transform to a partial differential equation we take as an example the **one-dimensional heat equation**, the derivation of which can be found in Section 18.5. This same partial differential equation was used when developing applications of the Laplace transform in Chapter 7. The heat equation that determines the one-dimensional temperature distribution $T(x, t)$ on a plane $x = $ constant at time t in an infinite block of metal with heat conduction properties characterized by the constant κ is given by

$$\frac{\partial^2 T}{\partial x^2} = \frac{1}{\kappa}\frac{\partial T}{\partial t}.$$

The problem we now consider is finding the temperature distribution throughout the metal at a time t when at $t = 0$ the one-dimensional temperature distribution throughout the block is given by

$$T(x, 0) = f(x),$$

where $f(x)$ is a prescribed function. Our objective will be to find the temperature $T(x, t)$ on a plane $x = $ constant at a time $t > 0$ caused by the redistribution of heat as time increases.

The Laplace transform cannot be used because when applied to the spatial variable x it is only valid for $x \geq 0$, so instead we must make use of the Fourier transform with respect to x because this applies for $-\infty \leq x \leq \infty$. Taking the Fourier transform of the heat equation with respect to x gives

$$\mathop{}_x\mathcal{F}\left\{\frac{\partial^2 T}{\partial x^2}\right\} = {}_x\mathcal{F}\frac{1}{\kappa}\left\{\frac{\partial T}{\partial t}\right\},$$

so if we apply (34) with $n = 2$, while regarding ω as a parameter, this becomes

$$-\omega^2 \kappa F(\omega, t) = \frac{d}{dt}[F(\omega, t)], \quad \text{where} \quad F(\omega, t) = \frac{1}{\sqrt{2\pi}} \int_{-\infty}^{\infty} T(x, t) e^{-i\omega x} dx.$$

The transform $F(\omega, t)$ satisfies the ordinary differential equation

$$F' + \omega^2 \kappa F = 0,$$

with the solution

$$F(\omega, t) = A(\omega) \exp\{-\omega^2 \kappa t\},$$

where $A(\omega)$ is to be determined (remember that ω is a constant with respect to t).

As

$$F(\omega, t) = \frac{1}{\sqrt{2\pi}} \int_{-\infty}^{\infty} T(x, t) e^{-i\omega x} dx,$$

it follows from the initial condition that

$$F(\omega, 0) = \frac{1}{\sqrt{2\pi}} \int_{-\infty}^{\infty} f(x) e^{-i\omega x} dx,$$

but $F(\omega, 0) = A(\omega)$, so

$$F(\omega, t) = \frac{1}{\sqrt{2\pi}} \int_{-\infty}^{\infty} f(x') \exp\{-i\omega x' - \omega^2 \kappa t\} dx',$$

where to avoid confusion in the next step of the calculation the dummy variable x has been replaced by x'.

Applying the inversion integral to this result gives

$$T(x, t) = \frac{1}{\sqrt{2\pi}} \int_{-\infty}^{\infty} \exp\{i\omega x\} \left[\frac{1}{\sqrt{2\pi}} \int_{-\infty}^{\infty} f(x') \exp\{-i\omega x' - \omega^2 \kappa t\} dx' \right] d\omega$$

$$= \frac{1}{2\pi} \int_{-\infty}^{\infty} f(x') \left[\int_{-\infty}^{\infty} \exp\{i\omega(x - x') - \omega^2 \kappa t\} d\omega \right] dx'.$$

We show separately that

$$\frac{1}{2\pi} \int_{-\infty}^{\infty} \exp\{i\omega(x - x') - \omega^2 \kappa t\} d\omega = \sqrt{\frac{1}{4\pi \kappa t}} \exp\left\{ -\frac{(x - x')^2}{4\kappa t} \right\},$$

so the required solution is seen to be given by

$$T(x, t) = \sqrt{\frac{1}{4\pi \kappa t}} \int_{-\infty}^{\infty} f(x') \exp\left\{ -\frac{(x - x')^2}{4\kappa t} \right\} dx'.$$

OPTIONAL   To show that

$$\frac{1}{2\pi} \int_{-\infty}^{\infty} \exp\{i\omega(x - x') - \omega^2 \kappa t\} d\omega = \sqrt{\frac{1}{4\pi \kappa t}} \exp\left\{ -\frac{(x - x')^2}{4\kappa t} \right\}$$

we need to use a complex analysis method from Chapter 15. However, before we can use this technique, the integrand of the integral on the left must be rewritten. We multiply the exponential function by $e^P e^{-P}$ (that is, by 1), where P is to be determined later, and as a result obtain

$$\exp\{i\omega(x - x') - \omega^2 \kappa t\} = e^P \exp\{-P + i\omega(x - x') - \omega^2 \kappa t\}.$$

We now choose P so that the exponent in the exponential can be expressed in the form $-(\alpha - i\beta\omega)^2$. When this is done it turns out that

$$\alpha = -\frac{i(x-x')}{2\sqrt{\kappa t}}, \quad \beta = i\sqrt{\kappa t}, \quad \text{and} \quad P = -\frac{(x-x')^2}{4\kappa t},$$

so

$$\frac{1}{2\pi}\int_{-\infty}^{\infty} \exp\{i\omega(x-x') - \omega^2\kappa t\}d\omega$$

$$= \frac{1}{2\pi}\exp\left\{-\frac{(x-x')^2}{4\kappa t}\right\}\int_{-\infty}^{\infty}\exp\left\{-\left(-\frac{i(x-x')}{2\sqrt{\kappa t}} + \omega\sqrt{\kappa t}\right)^2\right\}d\omega$$

Making the change of variable

$$u = -\frac{i(x-x')}{2\sqrt{\kappa t}} + \omega\sqrt{\kappa t},$$

we find that

$$\frac{1}{2\pi}\int_{-\infty}^{\infty} \exp\{i\omega(x-x') - \omega^2\kappa t\}d\omega$$

$$= \frac{1}{2\pi}\exp\left\{-\frac{(x-x')^2}{4\kappa t}\right\}\frac{1}{\sqrt{\kappa t}}\int_{ic-\infty}^{ic+\infty}\exp\{-u^2\}du,$$

where $c = (x-x')^2/\sqrt{(4\kappa t)}$. If we integrate $\exp\{-u^2\}$ around the rectangle with corners located at $-R$, R, $R+ic$, and $-R+ic$ in the complex plane, and proceed to the limit as $R \longrightarrow \infty$, it follows that the integrals from $-R$ to $-R+ic$ and from R to $R+ic$ vanish, so as $\exp\{-u^2\}$ has no poles inside the rectangle, we have

$$\int_{ic-\infty}^{ic+\infty}\exp\{-u^2\}du = \int_{-\infty}^{\infty}\exp\{-u^2\}du.$$

The integral on the right is related to the error function erf(v) because

$$\int_0^v \exp\{-u^2\}du = \frac{\sqrt{\pi}}{2}\text{erf}(v),$$

where $\text{erf}(-v) = -\text{erf}(v)$ and $\text{erf}(\infty) = 1$.
Thus,

$$\frac{1}{2\pi}\int_{-\infty}^{\infty} \exp\{i\omega(x-x') - \omega^2\kappa t\}d\omega$$

$$= \frac{1}{2\pi}\exp\left\{-\frac{(x-x')^2}{4\kappa t}\right\}\frac{1}{\sqrt{\kappa t}}\frac{\sqrt{\pi}}{2}[\text{erf}(\infty) - \text{erf}(-\infty)]$$

$$= \frac{1}{2\pi}\exp\left\{-\frac{(x-x')^2}{4\kappa t}\right\}\frac{1}{\sqrt{\kappa t}}\frac{\sqrt{\pi}}{2}2$$

$$= \sqrt{\frac{1}{4\pi\kappa t}}\exp\left\{-\frac{(x-x')^2}{4\kappa t}\right\},$$

so we have shown that

$$\frac{1}{2\pi}\int_{-\infty}^{\infty} \exp\{i\omega(x-x') - \omega^2\kappa t\}d\omega = \sqrt{\frac{1}{4\pi\kappa t}}\exp\left\{-\frac{(x-x')^2}{4\kappa t}\right\}. \tag{37}$$

Fourier integrals are discussed in references [4.3] and [4.4]. Tables of Fourier transform pairs are given in references [4.2] and [3.11].

Summary

The Fourier transform was introduced and its most important operational properties were established. The transforms of derivatives and partial derivatives were considered, and applications were made to functions defined by an ordinary differential equation and also to the unsteady one-dimensional heat equation. Partial differential equations such as the heat equation, and the use of integral transforms in their solution, will be considered in more detail in Chapter 18.

TABLE 10.1 Fourier Transform Pairs

$f(x)$	$F(x) = \dfrac{1}{\sqrt{2\pi}} \displaystyle\int_{-\infty}^{\infty} f(x)e^{-i\omega x}dx$
1. $af(x) + bg(x)$	$aF(\omega) + bG(\omega)$
2. $f^{(n)}(x)$	$(i\omega)^n F(\omega)$
3. $x^n f(x)$	$(i)^n \dfrac{d^n}{d\omega^n}[F(\omega)]$
4. $x^m f^{(n)}(x)$	$(i)^{m+n} \dfrac{d^m}{d\omega^m}[\omega^n F(\omega)]$
5. $f(ax)\,(a>0)$	$\dfrac{1}{a} F(\omega/a)$
6. $f(x-a)$	$e^{-i\omega a} F(\omega)$
7. $e^{i\lambda x} f(x)$	$F(\omega - \lambda)$
8. $(f * g)(x)$	$\sqrt{2\pi}\, F(\omega)G(\omega)$ (convolution theorem)
9. $\displaystyle\int_{-\infty}^{\infty}|f(x)|^2 dx$	$\displaystyle\int_{-\infty}^{\infty}|F(\omega)|^2 d\omega$ (Parseval relation)
10. $\begin{cases}1, & |x|<a \\ 0, & |x|>a\end{cases}\ (a>0)$	$\sqrt{\dfrac{2}{\pi}}\left(\dfrac{\sin a\omega}{\omega}\right)$
11. $\dfrac{\sin ax}{x}\ (a>0)$	$\begin{cases}\sqrt{\dfrac{\pi}{2}}, & |\omega|<a \\ 0, & |\omega|>a\end{cases}$
12. $\begin{cases}1, & a<x<b \\ 0, & \text{otherwise}\end{cases}\ (0<a<b)$	$\dfrac{1}{\sqrt{2\pi}}\left(\dfrac{e^{-ia\omega}-e^{-ib\omega}}{i\omega}\right)$
13. $\begin{cases}a-|x|, & |x|<a \\ 0, & |x|>a\end{cases}$	$\sqrt{\dfrac{2}{\pi}}\left(\dfrac{1-\cos\omega a}{\omega^2}\right)$
14. $\dfrac{1}{a^2+x^2}\ (a>0)$	$\sqrt{\dfrac{\pi}{2}}\dfrac{e^{-a|\omega|}}{a}$
15. $\begin{cases}e^{-ax}, & x>0 \\ 0, & x<0\end{cases}\ (a>0)$	$\dfrac{1}{\sqrt{2\pi}}\left(\dfrac{1}{a+i\omega}\right)$
16. $\begin{cases}e^{ax}, & b<x<c \\ 0, & \text{otherwise}\end{cases}\ (a>0)$	$\dfrac{1}{\sqrt{2\pi}}\left[\dfrac{e^{(a-i\omega)c}-e^{(a-i\omega)b}}{a-i\omega}\right]$

(continued)

TABLE 10.1   (*continued*)

$f(x)$	$F(x) = \dfrac{1}{\sqrt{2\pi}} \displaystyle\int_{-\infty}^{\infty} f(x)e^{-i\omega x}\,dx$
17. $e^{-a\lvert x\rvert}$ $(a > 0)$	$\sqrt{\dfrac{2}{\pi}}\left(\dfrac{a}{a^2 + \omega^2}\right)$
18. $xe^{-a\lvert x\rvert}$ $(a > 0)$	$-\sqrt{\dfrac{2}{\pi}}\,\dfrac{2ia\omega}{(a^2 + \omega^2)^2}$
19. $\begin{cases} e^{iax}, & \lvert x\rvert < b \\ 0, & \lvert x\rvert > b \end{cases}$	$\sqrt{\dfrac{2}{\pi}}\left(\dfrac{\sin b(\omega - a)}{\omega - a}\right)$
20. $\exp(-a^2 x^2)$ $(a > 0)$	$\dfrac{1}{a\sqrt{2}}\exp\left\{-\dfrac{\omega^2}{4a^2}\right\}$
21. $\begin{cases} e^{-x}x^a, & x > 0 \\ 0, & x \le 0 \end{cases}$	$\dfrac{\Gamma(a)}{\sqrt{2\pi}\,(1 + i\omega)^a}$
22. $J_0(ax)$ $(a > 0)$	$\sqrt{\dfrac{2}{\pi}}\,\dfrac{H(a - \lvert\omega\rvert)}{(a^2 - \omega^2)^{1/2}}$
23. $\delta(x - a)$ $(a$ real$)$	$\dfrac{1}{\sqrt{2\pi}}e^{-ia\omega}$

EXERCISES 10.2

In Exercises 1 through 10 establish the Fourier transform of the stated entry in Table 10.1.

1. Entry 11.
2. Entry 12.
3. Entry 13.
4. Entry 15.
5. Entry 16.
6. Entry 17.
7. Entry 18.
8. Entry 19.
9. Entry 21.

10. Entry 22, by using the fact that $f(x) = J_0(ax)$ satisfies the Bessel's differential equation of order zero

$$xf'' + f' + a^2xf = 0 \quad (a > 0),$$

together with the standard result $\int_0^\infty J_0(ax)\,dx = 1/a$.

11. Use integration by parts to show that if $f(x)$ has a finite jump discontinuity at $x = a$, then $\mathcal{F}\{f'(x)\} = i\omega F(\omega) - \frac{1}{\sqrt{2\pi}}[f(a+) - f(a-)]e^{-iwa}$.

12. (a) Use the result of Exercise 11 to find the Fourier transform of $f'(x)$ given that

$$f(x) = \begin{cases} x, & 0 \le x < 1 \\ 0, & \text{otherwise}. \end{cases}$$

(b) Calculate $f'(x)$ and use entry 12 of Table 10.1 to find $\mathcal{F}\{f'(x)\}$ directly. Hence, show that the result obtained by this direct method is in agreement with the Fourier transform found in (a). So $f'(x) = -\delta(x - 1) + \begin{cases} 1, & 0 < x < 1 \\ 0, & \text{otherwise} \end{cases}$.

10.3   Fourier Cosine and Sine Transforms

The Fourier *cosine and sine transforms* arise as special cases of the Fourier transform, according to whether $f(x)$ is even or odd. Let us start by considering the Fourier cosine transform of $f(x)$ that can be defined when $f(x)$ is an even function that is absolutely integrable over $(-\infty, \infty)$, and so possesses a Fourier transform. Result (22) of Section 10.2 can be written

$$F(\omega) = \frac{1}{\sqrt{2\pi}} \int_{-\infty}^{\infty} f(x)\{\cos \omega x - i \sin \omega x\}\,dx, \tag{38}$$

but if $f(x)$ is an even function, the product $f(x) \cos \omega x$ is also even, so its integral over $(-\infty, \infty)$ does not vanish, though the product $f(x) \sin \omega x$ is an odd function, so its integral over $(-\infty, \infty)$ vanishes, causing (38) to simplify to

$$F_C(\omega) = \frac{1}{\sqrt{2\pi}} \int_{-\infty}^{\infty} f(x) \cos \omega x dx.$$

If we use the result $f(-x) = f(x)$ to change the interval of integration to $[0, \infty)$ this last result becomes

$$F_C(\omega) = \sqrt{\frac{2}{\pi}} \int_0^{\infty} f(x) \cos \omega x dx, \tag{39}$$

Fourier sine and cosine transforms

where the integral on the right is called the **Fourier cosine transform** of $f(x)$, and to distinguish it from the ordinary Fourier transform we write $\mathcal{F}_C\{f(x)\} = F_C(\omega)$. The **Fourier cosine inversion integral** corresponding to equation (23) of Section 10.2 becomes $f(x) = \mathcal{F}_C^{-1}\{F_C(\omega)\}$, where

$$f(x) = \sqrt{\frac{2}{\pi}} \int_0^{\infty} F_C(\omega) \cos \omega x d\omega. \tag{40}$$

inversion integrals

A similar argument applied to (16) of Section 10.2 when $f(x)$ is an odd function leads to the result

$$F_S(\omega) = \sqrt{\frac{2}{\pi}} \int_0^{\infty} f(x) \sin \omega x dx, \tag{41}$$

where the integral on the right is called the **Fourier sine transform** of $f(x)$ and we write $\mathcal{F}_S\{f(x)\} = F_S(\omega)$. The corresponding **Fourier cosine inversion integral** becomes $f(x) = \mathcal{F}_S^{-1}\{F_S(\omega)\}$, where

$$f(x) = \sqrt{\frac{2}{\pi}} \int_0^{\infty} F_S(\omega) \sin \omega x d\omega. \tag{42}$$

The Fourier cosine transform of $f(x)$ in (39) only involves $f(x)$ for $x \geq 0$, though it was derived from the Fourier transform on the assumption that $f(x)$ was an even function defined for all x. Consequently, taking the Fourier cosine transform of an arbitrary function $f(x)$ defined for $x \geq 0$ is equivalent to transforming an *even* function $f_e(x)$ obtained from $f(x)$ by setting $f_e(x) = f(x)$ for $x \geq 0$ and defining $f_e(x)$ for $x < 0$ by $f_e(-x) = f(x)$. Similarly, the Fourier sine transform of $f(x)$ in (41) only involves $f(x)$ for $x \geq 0$, though it was derived on the assumption that $f(x)$ was an odd function. So, taking the Fourier sine transform of an arbitrary function $f(x)$ defined for $x \geq 0$ is equivalent to transforming *odd* function $f_o(x)$ obtained from $f(x)$ by setting $f_o(x) = f(x)$ for $x \geq 0$ and defining $f_e(x)$ for $x < 0$ by $f_e(-x) = -f(x)$.

Because (40) and (41) have been derived from (22) of Section 10.2, it follows that whenever $f(x)$ is discontinuous, the expression on the left must be replaced by $(1/2)[f(x+0) + f(x-0)]$, because the Fourier cosine and sine transforms have the same convergence properties as the Fourier transform.

EXAMPLE 10.12

Find $\mathcal{F}_C\{e^{-ax}\}$ and $\mathcal{F}_S\{e^{-ax}\}$ when $a > 0$, and use the results with the Fourier cosine and sine inversion integrals and an interchange of variables to show that

$$\mathcal{F}_C\left\{\frac{1}{x^2+a^2}\right\} = \sqrt{\frac{\pi}{2}}\frac{e^{-a\omega}}{a} \quad \text{and} \quad \mathcal{F}_S\left\{\frac{x}{x^2+a^2}\right\} = \sqrt{\frac{\pi}{2}}e^{-a\omega}.$$

Solution   By definition

$$\mathcal{F}_C\{e^{-ax}\} = \sqrt{\frac{2}{\pi}}\int_0^\infty e^{-ax}\cos\omega x\,dx$$

$$= \operatorname{Re}\sqrt{\frac{2}{\pi}}\left\{\int_0^\infty e^{-ax}e^{i\omega x}dx\right\} = \sqrt{\frac{2}{\pi}}\operatorname{Re}\left\{\frac{1}{a-i\omega}\right\} = \sqrt{\frac{2}{\pi}}\left(\frac{a}{\omega^2+a^2}\right).$$

Similarly,

$$\mathcal{F}_S\{e^{-ax}\} = \sqrt{\frac{2}{\pi}}\int_0^\infty e^{-ax}\sin\omega x\,dx$$

$$= \operatorname{Im}\sqrt{\frac{2}{\pi}}\left\{\int_0^\infty e^{-ax}e^{i\omega x}dx\right\} = \sqrt{\frac{2}{\pi}}\left(\frac{\omega}{\omega^2+a^2}\right).$$

Using these results in the Fourier cosine and sine inversion integrals gives

$$e^{-ax} = \frac{2a}{\pi}\int_0^\infty \frac{\cos\omega x}{\omega^2+a^2}d\omega = \frac{2}{\pi}\int_0^\infty \frac{\omega\sin\omega x}{\omega^2+a^2}d\omega, \text{ for } a > 0,$$

so after x and ω are interchanged, these results become

$$e^{-a\omega} = \frac{2a}{\pi}\int_0^\infty \frac{\cos\omega x}{x^2+a^2}dx = \frac{2}{\pi}\int_0^\infty \frac{x\cos\omega x}{x^2+a^2}dx.$$

However,

$$\mathcal{F}_C\left\{\frac{1}{x^2+a^2}\right\} = \sqrt{\frac{2}{\pi}}\int_0^\infty \frac{\cos\omega x}{x^2+a^2}dx \quad \text{and} \quad \mathcal{F}_S\left\{\frac{x}{x^2+a^2}\right\} = \sqrt{\frac{2}{\pi}}\int_0^\infty \frac{x\sin\omega x}{x^2+a^2}dx,$$

so combining results gives

$$\mathcal{F}_C\left\{\frac{1}{x^2+a^2}\right\} = \sqrt{\frac{\pi}{2}}\frac{e^{-a\omega}}{a} \quad \text{and} \quad \mathcal{F}_S\left\{\frac{x}{x^2+a^2}\right\} = \sqrt{\frac{\pi}{2}}e^{-a\omega}. \quad ■$$

THEOREM 10.9

Linearity of the Fourier cosine and sine transforms   Let the functions $f(x)$ and $g(x)$ have Fourier cosine and sine transforms, and let a and b be arbitrary constants. Then

$$\mathcal{F}_C\{af(x)+bg(x)\} = a\,\mathcal{F}_C\{f(x)\} + b\,\mathcal{F}_C\{g(x)\} = a\,F_C(\omega) + b\,G_C(\omega)$$

and

$$\mathcal{F}_S\{af(x)+bg(x)\} = a\,\mathcal{F}_S\{f(x)\} + b\,\mathcal{F}_S\{g(x)\} = a\,F_S(\omega) + b\,G_S(\omega).$$

Proof   The linearity properties of the Fourier cosine and sine transforms follow directly from the linearity property of the Fourier transform from which they are derived. ■

linearity of sine and cosine transforms and the transformation of derivatives

The expressions for the Fourier cosine and sine transforms of derivatives of a function $f(x)$ are slightly more complicated than those for the Fourier transform because they involve the initial values of the function and its derivatives.

THEOREM 10.10

Fourier cosine and sine transforms of derivatives Let $f(x)$ be continuous and absolutely integrable over $[0, \infty)$ and such that $\lim_{x \to \infty} f(x) = 0$. Then if $f'(x)$ and $f''(x)$ are piecewise continuous on each finite subinterval of $[0, \infty)$,

(i) $\quad \mathcal{F}_C\{f'(x)\} = \omega \mathcal{F}_S\{f(x)\} - \sqrt{\dfrac{2}{\pi}} f(0)$

(ii) $\quad \mathcal{F}_S\{f'(x)\} = -\omega \mathcal{F}_C\{f(x)\}$

(iii) $\quad \mathcal{F}_C\{f''(x)\} = -\omega^2 \mathcal{F}_C\{f(x)\} - \sqrt{\dfrac{2}{\pi}} f'(0)$

(iv) $\quad \mathcal{F}_S\{f''(x)\} = -\omega^2 \mathcal{F}_S\{f(x)\} + \sqrt{\dfrac{2}{\pi}} \omega f(0).$

Proof The proof of each result is similar, so only result (i) will be derived in detail and outlines given for the proofs of the remaining results. To obtain (i) we integrate by parts and make use of the definition of $\mathcal{F}_C\{f(x)\}$ as follows:

$$\mathcal{F}_C\{f'(x)\} = \sqrt{\frac{2}{\pi}} \int_0^\infty f'(x) \cos \omega x \, dx$$

$$= \sqrt{\frac{2}{\pi}} \left[f(x) \cos \omega x \Big|_0^\infty + \omega \int_0^\infty f(x) \sin \omega x \, dx \right]$$

$$= -\sqrt{\frac{2}{\pi}} f(0) + \omega \mathcal{F}_S\{f(x)\}.$$

Result (iii) follows from (i) by replacing f by f'. Result (ii) follows in similar fashion, and (iv) follows from (ii) by replacing f by f'. ∎

When Theorem 10.10 is used in the solution of second order differential equations, the initial conditions involved will help decide whether to use the cosine or sine transform. Thus, for example, if $f(0)$ is given but $f'(0)$ is unknown, the Fourier sine transform should be used to transform $f''(x)$ because result (iv) does not involve $f'(0)$. Conversely, if $f(0)$ is unknown but $f'(0)$ is given, then the Fourier cosine transform should be used to transform $f''(x)$, because result (iii) does not involve $f(0)$.

The Fourier cosine and sine transforms have Parseval relations that are analogous to the Parseval relation for the Fourier transform given in Theorem 10.7. To arrive at the first of these results we consider two functions $f(x)$ and $g(x)$ with the respective Fourier cosine transforms $F_C(\omega)$ and $G_C(\omega)$ and, using the definition of $G_C(\omega)$, write

$$\int_0^\infty F_C(\omega) G_C(\omega) \cos \omega x \, d\omega = \sqrt{\frac{2}{\pi}} \int_0^\infty F_C(\omega) \cos \omega x \, d\omega \int_0^\infty g(x) \cos \omega x \, dx.$$

Changing the order of integration in the expression on the right gives

$$\sqrt{\frac{2}{\pi}} \int_0^\infty F_C(\omega) \cos \omega x \, d\omega \int_0^\infty g(v) \cos \omega v \, dv$$

$$= \sqrt{\frac{2}{\pi}} \int_0^\infty g(x) dx \int_0^\infty F_C(\omega) \cos \omega x \cos \omega v \, d\omega$$

$$= \sqrt{\frac{2}{\pi}} \int_0^\infty \frac{1}{2} [\cos \omega(x+v) + \cos \omega |x-v|] F_C(\omega) d\omega$$

$$= \frac{1}{2} \int_0^\infty g(v) [f(x+v) + f(|x-v|)] dv,$$

where use has first been made of the identity $\cos u \cos v = \frac{1}{2}[\cos(u+v) + \cos(u-v)]$ and then of the Fourier cosine inversion integral.

We have established the result

$$\int_0^\infty F_C(\omega) G_C(\omega) \cos \omega x \, d\omega = \frac{1}{2} \int_0^\infty g(v) [f(x+v) + f(|x-v|)] dv.$$

Setting $x = 0$ in this last result shows that

$$\int_0^\infty F_C(\omega) G_C(\omega) d\omega = \int_0^\infty f(v) g(v) dv. \tag{43}$$

The **Parseval relation** for the **Fourier cosine transform** follows from this result by identifying $g(v)$ with $\bar{f}(v)$, for then (43) becomes

$$\int_0^\infty |F_C(\omega)|^2 d\omega = \int_0^\infty |f(x)|^2 dx, \tag{44}$$

where in the last integral the dummy variable v has been replaced by x.

A similar argument involving the Fourier sine transform establishes the corresponding results

$$\int_0^\infty F_S(\omega) G_S(\omega) d\omega = \int_0^\infty f(v) g(v) dv \tag{45}$$

and the **Parseval relation** for the **Fourier sine transform**

$$\int_0^\infty |F_S(\omega)|^2 d\omega = \int_0^\infty |f(x)|^2 dx. \tag{46}$$

We have arrived at the following theorem.

THEOREM 10.11

the Parseval relation extended to Fourier sine and cosine transforms

The Parseval relation for the Fourier cosine and sine transforms   Let $f(x)$ have the respective Fourier cosine and sine transforms $F_C(\omega)$ and $F_S(\omega)$. Then the Parseval relation for the Fourier cosine transform is

$$\int_0^\infty |F_C(\omega)|^2 d\omega = \int_0^\infty |f(x)|^2 dx,$$

and the Parseval relation for the Fourier sine transform is

$$\int_0^\infty |F_S(\omega)|^2 d\omega = \int_0^\infty |f(x)|^2 dx.$$

∎

Results (44) and (46) often provide a simple way of evaluating improper integrals, as shown by the following example.

EXAMPLE 10.13   Apply result (43) to $f(x) = xe^{-ax}$ and $g(x) = xe^{-bx}$, where $a > 0$, $b > 0$, given that

$$\mathcal{F}_C\{f(x)\} = \sqrt{\frac{2}{\pi}} \frac{(a^2 - \omega^2)}{(a^2 + \omega^2)^2} \quad \text{and} \quad \mathcal{F}_C\{g(x)\} = \sqrt{\frac{2}{\pi}} \frac{(b^2 - \omega^2)}{(b^2 + \omega^2)^2}.$$

Solution   Substituting into (43) gives

$$\frac{2}{\pi} \int_0^\infty \frac{(a^2 - \omega^2)(b^2 - \omega^2)}{(a^2 + \omega^2)^2(b^2 + \omega^2)^2} d\omega = \int_0^\infty x^2 e^{-(a+b)x} dx,$$

and after integrating the expression on the right and multiplying by $\pi/2$ we find that

$$\int_0^\infty \frac{(a^2 - \omega^2)(b^2 - \omega^2)}{(a^2 + \omega^2)^2(b^2 + \omega^2)^2} d\omega = \frac{\pi}{(a + b)^3}.$$

This integral can be evaluated by other techniques, but the preceding method is one of the simplest.

∎

The final theorem in this section is the analogue of Theorem 10.8, and it is useful when transforming known Fourier cosine and sine transforms.

THEOREM 10.12

shifting and scaling Fourier sine and cosine transforms

Shifting ω and scaling x in Fourier cosine and sine transforms   Let $f(x)$ have the respective Fourier cosine and sine transforms $F_C(\omega)$ and $F_S(\omega)$. Then

(i) $\mathcal{F}_C\{\cos(ax) f(x)\} = \frac{1}{2}\{F_C(\omega + a) + F_C(\omega - a)\}$

(ii) $\mathcal{F}_C\{\sin(ax) f(x)\} = \frac{1}{2}\{F_S(a + \omega) + F_S(a - \omega)\}$

(iii) $\mathcal{F}_S\{\cos(ax) f(x)\} = \frac{1}{2}\{F_S(\omega + a) + F_S(\omega - a)\}$

(iv) $\mathcal{F}_S\{\sin(ax) f(x)\} = \frac{1}{2}\{F_C(\omega - a) - F_C(\omega + a)\}$

(v) $\mathcal{F}_C\{f(ax)\} = \frac{1}{a} F_C(\omega/a) \quad (a > 0)$

(vi) $\mathcal{F}_S\{f(ax)\} = \frac{1}{a} F_S(\omega/a) \quad (a > 0).$

Proof   (i) $\mathcal{F}_C\{\cos(ax) f(x)\} = \sqrt{\frac{2}{\pi}} \int_0^\infty \cos(\omega x) \cos(ax) f(x) dx$, but

$$\cos(ax) \cos(\omega x) = \frac{1}{2}[\cos\{(a + \omega)x\} + \cos\{(a - \omega)x\}],$$

so

$$\mathcal{F}_C\{\cos(ax)f(x)\} = \frac{1}{2}\sqrt{\frac{2}{\pi}} \int_0^\infty \cos\{(a+\omega)x\} f(x)dx$$

$$+ \frac{1}{2}\sqrt{\frac{2}{\pi}} \int_0^\infty \cos\{(a-\omega)x\} f(x)dx$$

$$= \frac{1}{2}\{F_C(\omega+a) + F_C(\omega-a)\}.$$

Results (ii) to (iv) follow in similar fashion, whereas results (v) and (vi) follow from the definitions of the Fourier cosine and sine transforms after making the change of variable $u = ax$. ∎

EXAMPLE 10.14

Given $f(x) = e^{-ax}$ with $a > 0$, use the results of Theorem 10.12 to find (a) $\mathcal{F}_C\{\cos bx f(x)\}$ and (b) $\mathcal{F}_S\{f(bx)\}$, when $b > 0$.

Solution

(a) Using Theorem 10.12 (i) with

$$\mathcal{F}_C\{e^{-ax}\} = \sqrt{\frac{2}{\pi}}\left(\frac{a}{\omega^2 + a^2}\right),$$

gives

$$\mathcal{F}_C\{\cos bx e^{-ax}\} = \frac{1}{2}\sqrt{\frac{2}{\pi}}\left(\frac{a}{(\omega+b)^2 + a^2}\right) + \frac{1}{2}\sqrt{\frac{2}{\pi}}\left(\frac{a}{(\omega-b)^2 + a^2}\right)$$

$$= \sqrt{\frac{2}{\pi}}\frac{a(\omega^2 + a^2 + b^2)}{[(\omega+b)^2 + a^2][(\omega-b)^2 + a^2]}.$$

(b) Using Theorem 10.12 (vi) with

$$\mathcal{F}_S\{e^{-ax}\} = \sqrt{\frac{2}{\pi}}\left(\frac{\omega}{\omega^2 + a^2}\right)$$

gives

$$\mathcal{F}_S\{f(bx)\} = \mathcal{F}_S\{e^{-abx}\} = \frac{1}{b}\sqrt{\frac{2}{\pi}}\left(\frac{\omega/b}{(\omega/b)^2 + a^2}\right) = \sqrt{\frac{2}{\pi}}\left(\frac{\omega}{\omega^2 + a^2 b^2}\right).$$

This result is to be expected, as it follows directly from the original result when a is replaced by ab. ∎

When Fourier cosine and sine transforms are used in the solution of partial differential equations, the function to be transformed is a function of more than one variable. So, for example, the operation of taking the Fourier cosine transform of $f(x, y)$ with respect to x, denoted by $F_C(\omega, y)$, is given by

$$_x\mathcal{F}_C\{f(x, y)\} = F_C(\omega, y) = \sqrt{\frac{2}{\pi}} \int_0^\infty f(x, y)\cos \omega x dx. \qquad (47)$$

Similarly, the operation of taking the Fourier sine transform of $f(x, y)$ with respect to y, denoted by $F_S(x, \omega)$, is given by

$$_y\mathcal{F}_S\{f(x, y)\} = F_S(x, \omega) = \sqrt{\frac{2}{\pi}} \int_0^\infty f(x, y)\sin\omega y\, dy. \tag{48}$$

As a variable that has not been transformed only appears as a parameter in the transform, it follows immediately that the rules for transforming partial derivatives follow directly from the rules for transforming derivatives of functions of a single independent variable. As a result, when interpreted in terms of a function $f(x, y)$, the entries in Theorem 10.10 take the following form.

transform of partial derivatives by Fourier sine and cosine transforms

Fourier cosine and sine transforms of partial derivatives of a function f(x, y)

$$_x\mathcal{F}_C\{f'(x, t)\} = \omega F_S(\omega, t) - \sqrt{\frac{2}{\pi}} f(0, t) \tag{49}$$

$$_x\mathcal{F}_S\{f'(x, t)\} = -\omega F_C(\omega, t) \tag{50}$$

$$_x\mathcal{F}_C\{f''(x, t)\} = -\omega^2 F_S(\omega, t) - \sqrt{\frac{2}{\pi}} f'(0, t) \tag{51}$$

$$_x\mathcal{F}_S\{f''(x, t)\} = -\omega^2 F_S(\omega, t) + \sqrt{\frac{2}{\pi}}\omega f(0, t) \tag{52}$$

It also follows that when transforming with respect to x partial derivatives of $f(x, y)$ with respect to y, the function f is transformed and the partial derivative of $f(x, y)$ with respect to y becomes an ordinary derivative with respect to y of the transformed function. So, for example,

$$_x\mathcal{F}_C\left\{\frac{\partial^n f(x, y)}{\partial y^n}\right\} = \frac{d^n F_C(\omega, y)}{dy^n},$$

with corresponding results for mixed derivatives.

another application to the heat equation

To provide a motivation for these results we again anticipate the discussion of partial differential equations that is to follow in Chapter 18. Our objective now will be to solve the same **initial boundary value problem** for the one-dimensional **heat equation** that was solved previously by means of the Laplace transform. The one-dimensional heat equation governing the temperature $T(x, t)$ in a semi-infinite slab of metal at a distance x from its plane face at time t is

$$\frac{\partial^2 T}{\partial x^2} = \frac{1}{\kappa}\frac{\partial T}{\partial t}, \tag{53}$$

and as before we will seek a solution subject to the initial condition

$$T(x, 0) = 0 \tag{54}$$

and the boundary condition

$$T(0, t) = T_0, \quad t \geq 0. \tag{55}$$

The initial condition (54) says that at time $t = 0$ all the metal in the slab is at temperature $T = 0$, whereas the boundary condition (55) says that for $t > 0$ the plane face of the slab of metal is suddenly maintained at the constant temperature $T = T_0$.

As an initial temperature is known, but $\partial T/\partial x$ is unknown, consideration of results (49) to (52) suggests that we use the Fourier sine transform because it is valid for $x \geq 0$ and it only requires knowledge of $T(0, t) = T_0$. Accordingly, taking the Fourier sine transform of (53) with $\mathcal{F}_S\{T(x, t)\} = T_S(\omega, t)$, we have

$$\mathcal{F}_S\left\{\frac{\partial^2 T}{\partial x^2}\right\} = \frac{1}{\kappa}\ \mathcal{F}_S\left\{\frac{\partial T}{\partial t}\right\},$$

so using (52) and regarding ω as a parameter (it is independent of t), we obtain

$$\kappa\left(-\omega^2 T_S(\omega, t) + \omega T_0\sqrt{\frac{2}{\pi}}\right) = \frac{d}{dt}[T_S(\omega, t)].$$

Thus, $T_S(\omega, t)$ satisfies the linear differential equation

$$T_S' + \omega^2\kappa T_S = \omega\kappa T_0\sqrt{\frac{2}{\pi}}$$

with the solution

$$T_S(\omega, t) = \frac{T_0}{\omega}\sqrt{\frac{2}{\pi}} + A(\omega)\exp\{-\omega^2\kappa t\},$$

where the arbitrary function $A(\omega)$ enters as the integration "constant" when $T_S(\omega, t)$ is integrated with respect to t, during which ω behaves as a constant.

Applying the inverse Fourier sine transform to this last result gives

$$T(x, t) = \sqrt{\frac{2}{\pi}}\int_0^\infty\left\{\frac{T_0}{\omega}\sqrt{\frac{2}{\pi}} + A(\omega)\exp\{-\omega^2\kappa t\}\right\}\sin\omega x\,d\omega.$$

To determine $A(\omega)$ we now apply the initial condition $T(x, 0) = 0$ to the preceding result, which then becomes

$$0 = \sqrt{\frac{2}{\pi}}\int_0^\infty\left\{\frac{T_0}{\omega}\sqrt{\frac{2}{\pi}} + A(\omega)\right\}\sin\omega x\,d\omega.$$

This must be true for all ω, but this is only possible if $A(\omega) = -\frac{T_0}{\omega}\sqrt{\frac{2}{\pi}}$, and so

$$T(x, t) = T_0\sqrt{\frac{2}{\pi}}\left\{\sqrt{\frac{2}{\pi}}\int_0^\infty\left(\frac{1 - \exp(-\kappa t\omega^2)}{\omega}\right)\sin\omega x\,d\omega\right\}.$$

The bracketed term is the inverse Fourier sine transform of $\{[1 - \exp(-\kappa\omega^2)]/\omega\}$, so if we use entry 17 in Table 10.3, the solution becomes

$$T(x, t) = T_0\,\mathrm{erfc}\left\{\frac{x}{2\sqrt{\kappa t}}\right\}.$$

This is the result that was obtained in Section 7.3 (e) (ii) by means of the Laplace transform. The result agrees with physical intuition because for any fixed x we have $\lim_{t\to\infty}\mathrm{erfc}\{\frac{x}{2\sqrt{\kappa t}}\} = 1$, showing that as $t \to \infty$, so $T(x, t) \to T_0$ the constant temperature of the plane face of the metal.

Summary

The Fourier sine and cosine transforms were introduced, their inversion integrals were stated, and the main operational properties of the transforms were established. The sine and cosine transforms of ordinary and partial derivatives were derived and applications were made to the unsteady one-dimensional heat equation.

TABLE 10.2   Fourier Cosine Transform Pairs

$f(x)$	$F_C(\omega) = \sqrt{\dfrac{2}{\pi}} \displaystyle\int_0^\infty f(x)\cos\omega x\,dx$
1. $af(x) + bg(x)$	$aF(\omega) + bG(\omega)$
2. $\cos(ax)f(x)$	$\frac{1}{2}\{F_C(\omega + a) + F_C(\omega - a)\}$
3. $\sin(ax)f(x)$	$\frac{1}{2}\{F_S(a + \omega) + F_S(a - \omega)\}$
4. $f(ax)$	$\dfrac{1}{a}F_C\left(\dfrac{\omega}{a}\right)(a > 0)$
5. $f'(x)$	$\omega F_S(\omega) - \sqrt{\dfrac{2}{\pi}}\,f(0)$
6. $f''(x)$	$-\omega^2 F_C(\omega) - \sqrt{\dfrac{2}{\pi}}\,f'(0)$
7. $\displaystyle\int_0^\infty |f(x)|^2 dx$	$\displaystyle\int_0^\infty |F(\omega)|^2 d\omega$ (Parseval relation)
8. $\displaystyle\int_0^\infty f(x)g(x)dx$	$\displaystyle\int_0^\infty F_C(\omega)G_C(\omega)d\omega$
9. $\begin{cases} 1, & 0 < x < a \\ 0, & \text{otherwise} \end{cases}$	$\sqrt{\dfrac{2}{\pi}}\left(\dfrac{\sin a\omega}{\omega}\right)$
10. $\begin{cases} 1, & a < x < b \\ 0, & \text{otherwise} \end{cases}$	$\sqrt{\dfrac{2}{\pi}}\left(\dfrac{\sin b\omega - \sin a\omega}{\omega}\right)$
11. $x^{\alpha-1}(0 < \alpha < 1)$	$\sqrt{\dfrac{2}{\pi}}\dfrac{\Gamma(\alpha)}{\omega^\alpha}\cos\dfrac{\alpha\pi}{2}$
12. $\begin{cases} x, & a < x < b \\ 0, & \text{otherwise} \end{cases}$	$\sqrt{\dfrac{2}{\pi}}\left(\dfrac{\cos b\omega + b\omega\sin b\omega - \cos a\omega - a\omega\sin a\omega}{\omega^2}\right)$
13. $e^{-ax}(a > 0)$	$\sqrt{\dfrac{2}{\pi}}\left(\dfrac{a}{\omega^2 + a^2}\right)$
14. $xe^{-ax}(a > 0)$	$\sqrt{\dfrac{2}{\pi}}\dfrac{(a^2 - \omega^2)}{(a^2 + \omega^2)^2}$
15. $\exp\{-ax^2\}\,(a > 0)$	$\dfrac{1}{\sqrt{2a}}\exp\left\{-\dfrac{\omega^2}{4a}\right\}$
16. $\dfrac{1}{x^2 + a^2}(a > 0)$	$\sqrt{\dfrac{\pi}{2}}\dfrac{e^{-a\omega}}{a}$
17. $J_0(ax)(a > 0)$	$\sqrt{\dfrac{2}{\pi}}\dfrac{H(a - \omega)}{(a^2 - \omega^2)^{1/2}}$
18. $\dfrac{\sin ax}{x}(a > 0)$	$\sqrt{\dfrac{2}{\pi}}H(a - \omega)$

TABLE 10.3 Fourier Sine Transform Pairs

$f(x)$	$F_S(\omega) = \sqrt{\dfrac{2}{\pi}} \displaystyle\int_0^\infty f(x) \sin \omega x\, dx$
1. $af(x) + bg(x)$	$aF(\omega) + bG(\omega)$
2. $\cos(ax)f(x)$	$\frac{1}{2}\{F_S(\omega + a) + F_S(\omega - a)\}$
3. $\sin(ax)f(x)$	$\frac{1}{2}\{F_C(\omega - a) - F_C(\omega + a)\}$
4. $f(ax)$	$\dfrac{1}{a} F_S\left(\dfrac{\omega}{a}\right) \quad (a > 0)$
5. $f'(x)$	$-\omega F_C(\omega)$
6. $f''(x)$	$-\omega^2 F_S(\omega) + \sqrt{\dfrac{2}{\pi}}\,\omega f'(0)$
7. $\displaystyle\int_0^\infty |f(x)|^2 dx$	$\displaystyle\int_0^\infty |F(\omega)|^2 d\omega$
	(Parseval relation)
8. $\displaystyle\int_0^\infty f(x)g(x)dx$	$\displaystyle\int_0^\infty F_S(\omega)G_S(\omega)d\omega$
9. $\begin{cases} 1, & 0 < x < a \\ 0, & \text{otherwise} \end{cases}$	$\sqrt{\dfrac{2}{\pi}}\left(\dfrac{1 - \cos a\omega}{\omega}\right)$
10. $\begin{cases} 1, & a < x < b \\ 0, & \text{otherwise} \end{cases}$	$\sqrt{\dfrac{2}{\pi}}\left(\dfrac{\cos a\omega - \cos b\omega}{\omega}\right)$
11. $x^{\alpha-1} \ (0 < \alpha < 1)$	$\sqrt{\dfrac{2}{\pi}}\,\dfrac{\Gamma(\alpha)}{\omega^\alpha} \sin \dfrac{\alpha\pi}{2}$
12. $e^{-ax} \ (a > 0)$	$\sqrt{\dfrac{2}{\pi}}\,\dfrac{\omega}{(\omega^2 + a^2)}$
13. $xe^{-ax} \ (a > 0)$	$\sqrt{\dfrac{2}{\pi}}\,\dfrac{2a\omega}{(\omega^2 + a^2)^2}$
14. $x\exp\{-ax^2\} \ (a > 0)$	$\dfrac{\omega}{(2a)^{3/2}} \exp\left\{-\dfrac{\omega^2}{4a}\right\}$
15. $\dfrac{x}{x^2 + a^2} \ (a > 0)$	$\sqrt{\dfrac{\pi}{2}}\,e^{-a\omega}$
16. $\dfrac{\cos ax}{x} \ (a > 0)$	$\sqrt{\dfrac{\pi}{2}}\,H(\omega - a)$
17. $\operatorname{erfc}\left\{\dfrac{x}{2a}\right\} \ (a > 0)$	$\sqrt{\dfrac{2}{\pi}}\left\{\dfrac{1 - \exp(-a^2\omega^2)}{\omega}\right\}$

EXERCISES 10.3

In Exercises 1 through 10 establish the Fourier cosine transform of the stated entry in Table 10.2.

1. Entry 9.
2. Entry 10.
3. Entry 11.
4. Entry 12.
5. Entry 13.
6. Entry 14.
7. Entry 15.
8. Entry 16.
9. Entry 17.
10. Entry 18.

In Exercises 11 through 15 find the Fourier cosine transform of the stated function.

11. $f(x) = \begin{cases} \sin x, & 0 \le x \le \pi \\ 0, & \text{otherwise.} \end{cases}$

12. $f(x) = \begin{cases} \cos x, & 0 \le x \le \pi \\ 0, & \text{otherwise.} \end{cases}$

13. $f(x) = \begin{cases} x, & 0 \le x \le 1 \\ 2 - x, & 1 \le x \le 2 \\ 0, & \text{otherwise.} \end{cases}$

14. $f(x) = \begin{cases} 1, & 0 \le x \le 1 \\ 2 - x, & 1 \le x \le 2 \\ 0, & \text{otherwise.} \end{cases}$

15. $f(x) = \begin{cases} 1 - x^2, & 0 \le x < 1 \\ 0, & \text{otherwise.} \end{cases}$

In Exercises 16 through 23 establish the Fourier sine transform of the stated entry in Table 10.3.

16. Entry 9.

17. Entry 10.

18. Entry 11.

19. Entry 12.

20. Entry 13.

21. Entry 14.

22. Entry 15.

23. Entry 16.

In Exercises 24 through 28 find the Fourier sine transform of the stated function.

24. $f(x) = \begin{cases} \sin x, & 0 \le x \le \pi \\ 0, & \text{otherwise.} \end{cases}$

25. $f(x) = \begin{cases} \cos x, & 0 \le x \le \pi \\ 0, & \text{otherwise.} \end{cases}$

26. $f(x) = \begin{cases} x, & 0 \le x \le 1 \\ 2 - x, & 1 \le x \le 2 \\ 0, & \text{otherwise.} \end{cases}$

27. $f(x) = \begin{cases} 1, & 0 \le x \le 1 \\ 2 - x, & 1 \le x \le 2 \\ 0, & \text{otherwise.} \end{cases}$

28. $f(x) = \begin{cases} 1 - x^2, & 0 \le x < 1 \\ 0, & \text{otherwise.} \end{cases}$

PART FIVE

VECTOR CALCULUS

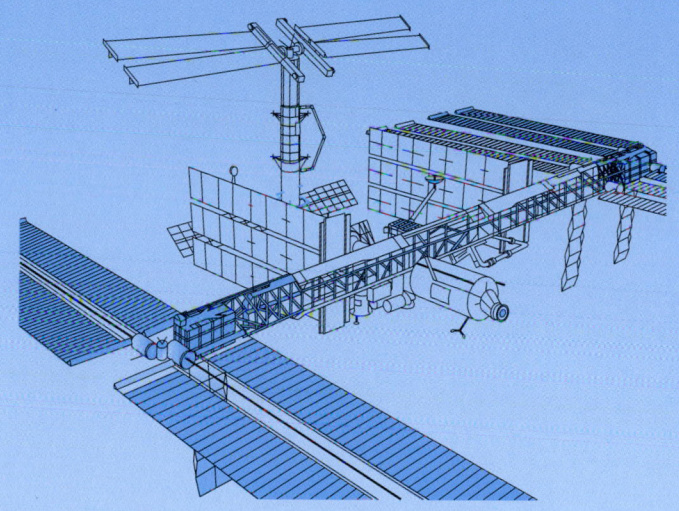

CHAPTER 11

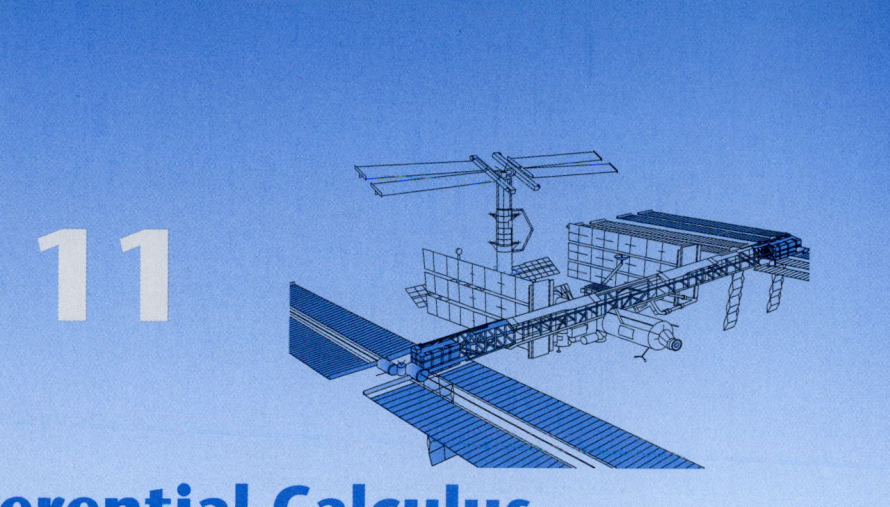

Vector Differential Calculus

Many physical quantities that occur in engineering and science require more than a single number to characterize them. When describing quantities such as force and velocity it is necessary to specify both a magnitude and a direction, and these are examples of vector quantities, whereas the air temperature, which can be specified by giving a single number, is an example of a scalar quantity. Physical problems are often best described in terms of vectors, so the objective of this chapter is to develop the most important aspects of vector differential calculus.

Scalar and vector fields are defined in Section 11.1, and these concepts are then related to the limit, continuity, and differentiability of a vector function of a single real variable. The rules for the differentiation of vector functions of a single real variable are established and used to develop the basic geometry of space curves. The definition of the derivative at a point on a space curve is used when defining the unit tangent vector **T** to such a curve, its curvature κ, its principal normal **N**, and its binormal **B**.

The integration of scalar and vector functions of a single real variable is developed in Section 11.2, after which the line integral of a vector function of position $\mathbf{F}(x, y, z)$ is defined, and by way of example it is then used to define the circulation in a fluid flow and the flux of a vector function of position.

A directional derivative of a scalar function $w = f(x, y, z)$ is defined in Section 11.3 where its most important properties are established. The directional derivative is used when developing the concept of the gradient of f, written either grad f or ∇f, after which rules for its use are developed.

The important property of path invariance of integrals in conservative fields is proved in Section 11.4. The potential function is introduced, a test for a conservative field is given, and the determination of the related potential function is discussed, all of which concepts have important applications throughout engineering and science.

The two other vector operators divergence and curl, written div **F** and curl **F**, respectively, are defined and their physical meaning is explained in Section 11.5. The properties of the divergence operator are established, and then used to prove the properties of the most important combinations of the gradient, divergence, and curl operators.

Applications involving vector operators are often simplified if an appropriate system of coordinates is adopted. The purpose of Section 11.6 is to establish the forms taken by the gradient, divergence, and curl operators in a general system of orthogonal curvilinear coordinates, with special emphasis on cylindrical and spherical polar coordinates.

11.1   Scalar and Vector Fields, Limits, Continuity, and Differentiability

A scalar function $F(x, y, z)$ defined over some region of space D is a function that assigns to each point P_0 in D with coordinates (x_0, y_0, z_0) the *number* $F(P_0) = F(x_0, y_0, z_0)$. The set of all numbers $F(P)$ for all points P in D are said to form a **scalar field** over D. If P has position vector $\mathbf{r}$, we can write the scalar field $F(x, y, z)$ in the form $F(P) = F(\mathbf{r})$ to emphasize the fact that a *scalar* value $F(\mathbf{r})$ is associated with the position vector $\mathbf{r}$ in D. In physical problems P is usually a point in space, and in addition to depending on P, the function F often also depends on the time t, so then $F(P, t) = F(x, y, z, t)$ and in this case we can write $F(P, t) = F(\mathbf{r}, t)$. A typical example of a time dependent scalar field is provided by the temperature distribution throughout a block of metal heated in such a way that the temperatures on its sides vary with time.

scalar and vector fields

More general than a scalar field $F(x, y, z)$ is a **vector field** defined by a vector function $\mathbf{F}(x, y, z)$ over some region of space D that assigns to each point P_0 in D with coordinates (x_0, y_0, z_0) the vector $\mathbf{F}(P_0) = \mathbf{F}(x_0, y_0, z_0)$ with its tail at P_0. Functions of this type are called either **vector functions** or **vector-valued functions**, and if P has position vector $\mathbf{r}$ we can write $\mathbf{F}(P) = \mathbf{F}(\mathbf{r})$ to emphasize the fact that in this case a *vector* $\mathbf{F}(P)$ is associated with each position vector $\mathbf{r}$ in D. Like scalar fields, vector fields over D often depend on both position and the time t, so then $\mathbf{F} = \mathbf{F}(x, y, z, t)$, and in this case we can write $\mathbf{F}(P, t) = \mathbf{F}(\mathbf{r}, t)$. An example of a time dependent vector field is provided by the fluid velocity vector in the unsteady flow of water around a bridge support column, because there the velocity depends on both the position vector $\mathbf{r}$ in the water and the time t at which the velocity is observed. In general, in terms of the unit vectors $\mathbf{i}$, $\mathbf{j}$, and $\mathbf{k}$, a time-dependent vector-valued function can be defined by setting

$$\mathbf{F}(\mathbf{r}, t) = f_1(\mathbf{r}, t)\mathbf{i} + f_2(\mathbf{r}, t)\mathbf{j} + f_3(\mathbf{r}, t)\mathbf{k}, \tag{1}$$

where the scalars $f_1(\mathbf{r}, t)$, $f_2(\mathbf{r}, t)$, and $f_3(\mathbf{r}, t)$ are the components of $\mathbf{F}(\mathbf{r}, t)$ that depend on both position and time and, at a point $\mathbf{r}_0$, translating the vector $\mathbf{F}(\mathbf{r}_0, t)$ until its tail is located at $\mathbf{r}_0$.

EXAMPLE 11.1

(a) The scalar function of position $F(x, y, z) = xyz^2$ for (x, y, z) inside the unit sphere $x^2 + y^2 + z^2 = 1$ defines a scalar field throughout the unit sphere.

(b) The vector-valued function $\mathbf{F}(x, y, z) = (x - y)\mathbf{i} + (y - z)\mathbf{j} + (xyz - 2)\mathbf{k}$, for (x, y, z) inside the ellipsoid $x^2/a^2 + y^2/b^2 + z^2/c^2 = 1$, defines a vector field throughout the ellipsoid. ∎

In order to perform calculus on vectors it is necessary to introduce the idea of a vector as a function. The simplest example of this kind is a vector $\mathbf{F}(t)$ of a single real variable t, which in terms of cartesian coordinates can be written

$$\mathbf{F}(t) = f_1(t)\mathbf{i} + f_2(t)\mathbf{j} + f_3(t)\mathbf{k}, \tag{2}$$

where the components $f_1(t)$, $f_2(t)$, and $f_3(t)$ of $\mathbf{F}(t)$ are functions of t defined over some interval $a \le t \le b$. Vectors of this type are called **vector functions of a single real variable**.

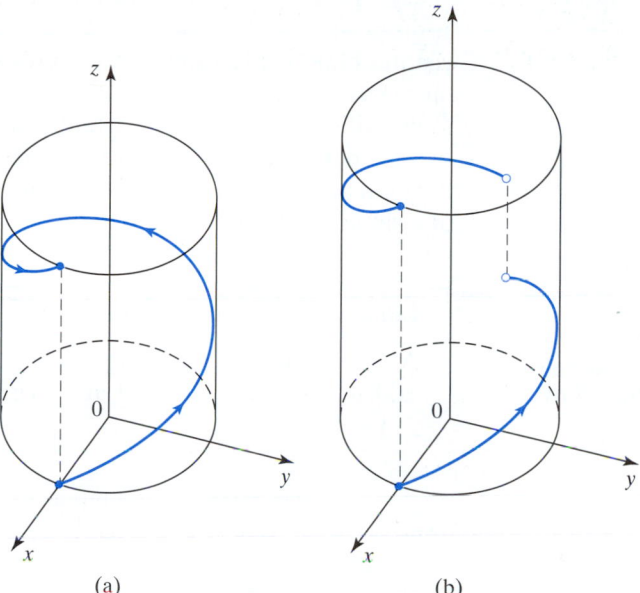

FIGURE 11.1   (a) A single turn of a helix. (b) A single turn of a broken helix.

If $\mathbf{F}(t)$ is regarded as a position vector $\mathbf{r}(t)$ in space, (2) can be interpreted as a curve in space traced out by the tip of the vector $\mathbf{r}(t)$ as t increases from a to b. Notice that a *sense* (of direction) along the curve is determined by the direction in which $\mathbf{r}(t)$ moves along the curve as t increases. When the components of $\mathbf{r}(t)$ are all continuous functions the curve, or path, traced out by the tip of $\mathbf{r}(t)$ will be an unbroken curve in space and $\mathbf{r}'(t) \neq \mathbf{0}$, though the curve will only be *smooth* if in addition to the components of $\mathbf{r}(t)$ being continuous they are also continuously differentiable for $a \leq t \leq b$, but more will be said about this later. If t is allowed to *decrease* from b to a, then the sense along the curve is *reversed*, and this fact will be important later when line integrals are considered.

EXAMPLE 11.2

(a) When interpreted as a position vector, the vector function of a single real variable $\mathbf{r}(t) = \cos t\,\mathbf{i} + \sin t\,\mathbf{j} + t\,\mathbf{k}$ for $0 \leq t \leq 2\pi$ describes a single turn of the space curve called a **helix** that is shown in Fig. 11.1(a). The fact that each component of $\mathbf{r}(t)$ is both continuous and continuously differentiable and $|d\mathbf{r}/dt| \neq 0$ ensures that the helix is a smooth curve. The form of the helix can be visualized by recognizing that, as t increases, so the projection of $\mathbf{r}(t)$ onto the (x, y)-plane given by the vector $\mathbf{r}_{(x,y)}(t) = \cos t\,\mathbf{i} + \sin t\,\mathbf{j}$ moves once in a counterclockwise direction around a unit circle centered on the origin, while the $\mathbf{k}$ component increases linearly with t.

(b) The vector function of a single real variable $\mathbf{r}(t) = \cos t\,\mathbf{i} + \sin t\,\mathbf{j} + \theta(t + H(t - \pi))\mathbf{k}$ for $0 \leq t \leq 2\pi$, where $H(t)$ is the Heaviside unit step function, has a discontinuous $\mathbf{k}$ component, and so describes the broken helix shown in Fig. 11.1(b), where the jump in the $\mathbf{k}$ component of $\mathbf{r}(t)$ occurs at $t = \pi$. ∎

It is important to recognize that because vector quantities are independent of a coordinate system, vector-valued functions and vector fields do not depend for their existence on any particular coordinate system. The choice of coordinate system used to describe vector functions is usually taken to be the one that is most appropriate for the geometry of the situation involved. So, for example, when a vector of interest depends only on distance along a straight axis and on the position on a circle centered on the axis and lying in a plane normal to the axis, it is natural to describe it in terms of the cylindrical polar coordinates (r, θ, z).

To make further progress it is necessary to generalize the related concepts of the limit and continuity of a real function of a single real variable to vector functions of a single real variable.

Limits and continuity of vector functions of a single real variable

limits and continuity of vector functions

A vector function of a single real variable $\mathbf{F}(t) = f_1(t)\mathbf{i} + f_2(t)\mathbf{j} + f_3(t)\mathbf{k}$ is said to have $\mathbf{L}$ as its **limit** at t_0, written $\lim_{t \to t_0} \mathbf{F}(t) = \mathbf{L}$, where $\mathbf{L} = L_1\mathbf{i} + L_2\mathbf{j} + L_3\mathbf{k}$, if

$$\lim_{t \to t_0} f_1(t) = L_1, \quad \lim_{t \to t_0} f_2(t) = L_2, \quad \text{and} \quad \lim_{t \to t_0} f_3(t) = L_3.$$

If, in addition, the vector function is defined at t_0 and $\lim_{t \to t_0} \mathbf{F}(t) = \mathbf{F}(t_0)$, then $\mathbf{F}(t)$ is said to be **continuous** at t_0. A vector function $\mathbf{F}(t)$ that is continuous for each t in the interval $a \leq t \leq b$ is said to be continuous over the interval. A vector function of a single real variable that is not continuous at a point t_0 is said to be **discontinuous** at t_0.

It can be seen from the preceding definitions that the limit and continuity properties of a parametrically defined vector function can be determined by examination of the behavior of its components. So, for example, the parametrically defined vector function describing the helix in Example 11.1(a) is seen to be continuous, whereas the broken helix in Example 11.1(b) is seen to be discontinuous at one point because of the behavior of its $\mathbf{k}$ component when $t = \pi$.

The notion of a limit of a vector function of a single real variable leads naturally to the definition of the differentiability of such a function. Returning to (2) we see that if t is increased to $t + \Delta t$, the change $\Delta \mathbf{F}$ produced in $\mathbf{F}$ is

$$\Delta \mathbf{F} = \mathbf{F}(t + \Delta t) - \mathbf{F}(t)$$
$$= \{f_1(t + \Delta t)\mathbf{i} + f_2(t + \Delta t)\mathbf{j} + f_3(t + \Delta t)\mathbf{k}\} - \{f_1(t)\mathbf{i} + f_2(t)\mathbf{j} + f_3(t)\mathbf{k}\},$$

so

$$\frac{\Delta \mathbf{F}}{\Delta t} = \left(\frac{f_1(t + \Delta t) - f_1(t)}{\Delta t}\right)\mathbf{i} + \left(\frac{f_2(t + \Delta t) - f_2(t)}{\Delta t}\right)\mathbf{j} + \left(\frac{f_3(t + \Delta t) - f_3(t)}{\Delta t}\right)\mathbf{k}.$$

If the functions $f_1(t)$, $f_2(t)$, and $f_3(t)$ are differentiable, by letting $\Delta t \to 0$ it follows at once that the derivative of $\mathbf{F}(t)$, denoted by $d\mathbf{F}/dt$, can be expressed in terms of the derivatives of the components of $\mathbf{F}(t)$ as

$$\frac{d\mathbf{F}}{dt} = \frac{df_1}{dt}\mathbf{i} + \frac{df_2}{dt}\mathbf{j} + \frac{df_3}{dt}\mathbf{k}. \tag{3}$$

We have arrived at the following definitions of the differentiability of $\mathbf{F}(t)$ and the derivative $d\mathbf{F}/dt$.

Differentiability and the derivative of a vector function of a single real variable

The vector function of a single real variable $\mathbf{F}(t) = f_1(t)\mathbf{i} + f_2(t)\mathbf{j} + f_3(t)\mathbf{k}$ defined over the interval $a \leq t \leq b$ is said to be **differentiable** at a point t_0 in the interval if its components are differentiable at t_0. It is said to be **differentiable over the interval** if it is differentiable at each point of the interval, and when $\mathbf{F}(t)$ is differentiable its **derivative** with respect to t is

$$\frac{d\mathbf{F}}{dt} = \frac{df_1}{dt}\mathbf{i} + \frac{df_2}{dt}\mathbf{j} + \frac{df_3}{dt}\mathbf{k}.$$

If $\mathbf{F}(t)$ is continuous over $a \leq t \leq b$, but $d\mathbf{F}/dt$ is discontinuous at a point t_0 in the interval, the derivative $d\mathbf{F}/dt$ will only be defined in the one-sided sense to the left and right of t_0 at the points $t = t_0 - 0$ and $t = t_0 + 0$.

When $d\mathbf{F}/dt$ is differentiable, the second order derivative $d^2\mathbf{F}/dt^2$ is defined as

$$\frac{d^2\mathbf{F}}{dt^2} = \frac{d}{dt}\left(\frac{d\mathbf{F}}{dt}\right)$$

and, in general, provided the derivatives exist,

$$\frac{d^n\mathbf{F}}{dt^n} = \frac{d}{dt}\left(\frac{d^{n-1}\mathbf{F}}{dt^{n-1}}\right), \quad \text{for } n \geq 2.$$

If $\mathbf{F}(t)$ is taken to be a differentiable position vector $\mathbf{r}(t)$, it follows from the definition of a derivative that $d\mathbf{r}/dt$ is a vector that is tangent to the point $\mathbf{r}(t)$ on the curve Γ traced out by the tip of the vector as t increases from $t = a$ to $t = b$. This situation, illustrated in Fig. 11.2, shows the relationship between $\mathbf{r}(t + \Delta t)$, $\mathbf{r}(t)$, and $\Delta \mathbf{r}$ before proceeding to the limit as $\Delta t \to 0$. It can be seen from this that as $\Delta t \to 0$, so $\Delta \mathbf{r}$ tends to coincidence with the tangent line T to the curve Γ at the point $\mathbf{r}(t)$. Furthermore, if $\mathbf{r}(t)$ is a position vector in space and t is the time, $d\mathbf{r}/dt$ is the **velocity** of the point with position vector $\mathbf{r}(t)$ and $d^2\mathbf{r}/dt^2$ is its **acceleration**.

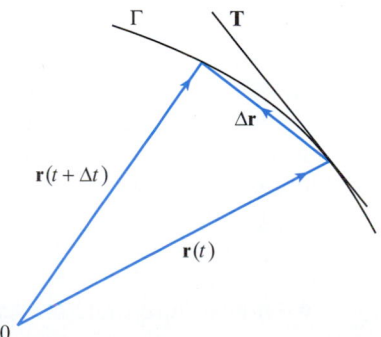

FIGURE 11.2   As $\Delta t \to 0$, so the vector $\Delta \mathbf{r}$ tends to coincidence with the tangent line T to the space curve Γ at $\mathbf{r}(t)$.

The differentiability properties of vector functions of a single real variable have been seen to be determined by the differentiability properties of the components. Consequently, as $\mathbf{F}(t)$ is a linear combination of its components in the $\mathbf{i}$, $\mathbf{j}$, and $\mathbf{k}$ directions, it follows that the rules for the differentiation of vector functions of a single real variable follow directly by applying the rules for the differentiation of a real function of a single real variable to each component in turn. The theorem that follows summarizes the basic rules for differentiation, and because vectors are independent of a coordinate system the results can be formulated without reference to a coordinate system.

THEOREM 11.1

differentiation of vector functions

Differentiation of vector functions of a single real variable   Let $\mathbf{u}(t)$ and $\mathbf{v}(t)$ be differentiable functions of t over some interval $a \le t \le b$, with $\mathbf{C}$ an arbitrary constant vector and c an arbitrary constant scalar. Then rules for differentiation of vector functions of a single real variable over the interval $a \le t \le b$ are:

(i) $\dfrac{d\mathbf{C}}{dt} = \mathbf{0}$ (differentiation of a constant vector)

(ii) $\dfrac{d}{dt}(c\mathbf{u}) = c\dfrac{d\mathbf{u}}{dt}$ (differentiation of a vector scaled by c)

(iii) $\dfrac{d}{dt}(\mathbf{u} \pm \mathbf{v}) = \dfrac{d\mathbf{u}}{dt} \pm \dfrac{d\mathbf{v}}{dt}$ (differentiation of a sum or difference)

(iv) $\dfrac{d}{dt}(\mathbf{u} \cdot \mathbf{v}) = \dfrac{d\mathbf{u}}{dt} \cdot \mathbf{v} + \mathbf{u} \cdot \dfrac{d\mathbf{v}}{dt}$ (differentiation of a dot product)

(v) $\dfrac{d}{dt}(\mathbf{u} \times \mathbf{v}) = \dfrac{d\mathbf{u}}{dt} \times \mathbf{v} + \mathbf{u} \times \dfrac{d\mathbf{v}}{dt}$ (differentiation of a cross product)

(vi) If $\mathbf{u}(t)$ is a differentiable function of t and $t = t(s)$ is a differentiable function of s, then

$$\frac{d\mathbf{u}}{ds} = \frac{d\mathbf{u}}{dt}\frac{dt}{ds}$$

or, explicitly, if $\mathbf{u}(t) = u_1(t)\mathbf{i} + u_2(t)\mathbf{j} + u_3(t)\mathbf{k}$, then

$$\frac{d\mathbf{u}}{ds} = \frac{du_1}{dt}\frac{dt}{ds}\mathbf{i} + \frac{du_2}{dt}\frac{dt}{ds}\mathbf{j} + \frac{du_3}{dt}\frac{dt}{ds}\mathbf{k}$$

(the chain rule for differentiation of $\mathbf{u}(t)$).

Proof   The proof of each result is straightforward and similar, so only the proof of result (iv) will be given, and for convenience the vectors $\mathbf{u}$ and $\mathbf{v}$ will be expressed in terms of the unit vectors $\mathbf{i}$, $\mathbf{j}$, and $\mathbf{k}$. The proofs of the remaining results will be left as exercises.

Letting $\mathbf{u} = u_1\mathbf{i} + u_2\mathbf{j} + u_3\mathbf{k}$ and $\mathbf{v} = v_1\mathbf{i} + v_2\mathbf{j} + v_3\mathbf{k}$, we have

$$\mathbf{u} \cdot \mathbf{v} = u_1v_1 + u_2v_2 + u_3v_3.$$

We now differentiate the scalar function $\mathbf{u} \cdot \mathbf{v}$ with respect to t, using the result

$$\frac{d(u_iv_i)}{dt} = \frac{du_i}{dt}v_i + u_i\frac{dv_i}{dt}, \quad \text{for } i = 1, 2, 3,$$

which when $i = 1$ can be written

$$\frac{d(u_1 v_1 \mathbf{i})}{dt} = \left(\frac{du_1}{dt}\mathbf{i}\right) \cdot (v_1\mathbf{i}) + (u_1\mathbf{i}) \cdot \left(\frac{dv_i}{dt}\mathbf{i}\right),$$

with corresponding results for $d(u_2 v_2)/dt$ and $d(u_3 v_3)/dt$. Summing the results for $d(u_i v_i)dt$ corresponding to $i = 1, 2, 3$, we arrive at result (iv), and the proof is complete. ∎

EXAMPLE 11.3   Given that $\mathbf{r}(t) = \cos t\,\mathbf{i} + \sin t\,\mathbf{j} + t\mathbf{k}$, find the first three derivatives of $\mathbf{r}$ with respect to t.

Solution $\dfrac{d\mathbf{r}}{dt} = -\sin t\,\mathbf{i} + \cos t\,\mathbf{j} + \mathbf{k},\ \dfrac{d^2\mathbf{r}}{dt^2} = -\cos t\,\mathbf{i} - \sin t\,\mathbf{j}$, and $\dfrac{d^3\mathbf{r}}{dt^3} = \sin t\,\mathbf{i} - \cos t\,\mathbf{j}.$ ∎

EXAMPLE 11.4   Given that $\mathbf{u} = t\mathbf{i} - 2t\mathbf{j} + t^2\mathbf{k}$, $\mathbf{v} = t\mathbf{j} + 3t\mathbf{k}$ and $\mathbf{w} = t\mathbf{i} - t^2\mathbf{k}$, find

$$\frac{d}{dt}[(\mathbf{u} \cdot \mathbf{v})\mathbf{w}].$$

Solution   The scalar $\mathbf{u} \cdot \mathbf{v} = -2t^2 + 3t^3$, so $(\mathbf{u} \cdot \mathbf{v})\mathbf{w} = (3t^4 - 2t^3)\mathbf{i} - (3t^5 - 2t^4)\mathbf{k}$, and so

$$\frac{d}{dt}[(\mathbf{u} \cdot \mathbf{v})\mathbf{w}] = (12t^3 - 6t^2)\mathbf{i} - (15t^4 - 8t^3)\mathbf{k}.$$ ∎

vector differential   The concept of a **vector differential** is often useful, and by analogy with the real variable calculus, if $\mathbf{F}(t) = f_1(t)\mathbf{i} + f_2(t)\mathbf{j} + f_3(t)\mathbf{k}$, the vector differential $d\mathbf{F}$ is defined as

$$d\mathbf{F} = \left(\frac{df_1}{dt}\mathbf{i} + \frac{df_2}{dt}\mathbf{j} + \frac{df_3}{dt}\mathbf{k}\right)dt. \tag{4}$$

A simple and useful application of the vector differential is to the element of arc length along a space curve Γ defined by the position vector $\mathbf{r}(t) = x_1(t)\mathbf{i} + x_2(t)\mathbf{j} + x_3(t)\mathbf{k}$ for $t \geq t_0$. If s is the arc length measured along Γ from some fixed point, then by applying Pythagoras' theorem to the differential elements

$$dx_1 = \frac{dx_1}{dt}dt, \quad dx_2 = \frac{dx_2}{dt}dt, \quad \text{and} \quad dx_3 = \frac{dx_3}{dt}dt,$$

it is seen from Fig. 11.3 that the differential element of arc length ds along Γ is given by

$$ds = \left[\left(\frac{dx_1}{dt}\right)^2 + \left(\frac{dx_2}{dt}\right)^2 + \left(\frac{dx_3}{dt}\right)^2\right]^{1/2} dt, \tag{5}$$

and so

$$\frac{ds}{dt} = \left|\frac{d\mathbf{r}}{dt}\right| = \left[\left(\frac{dx_1}{dt}\right)^2 + \left(\frac{dx_2}{dt}\right)^2 + \left(\frac{dx_3}{dt}\right)^2\right]^{1/2}. \tag{6}$$

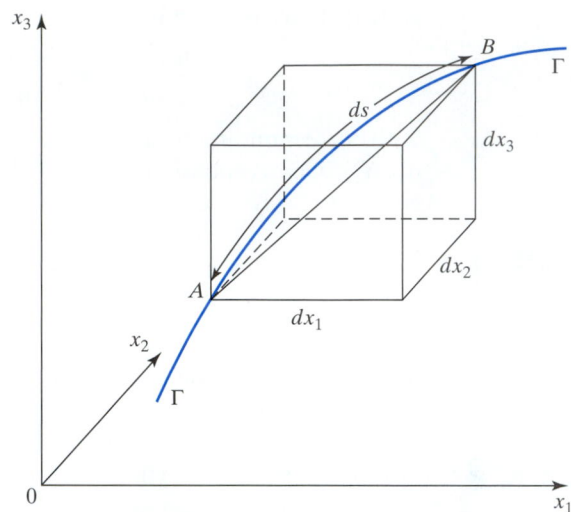

FIGURE 11.3 The geometrical relationship between the differentials ds, dx_1, dx_2, and dx_3.

This result shows that when t is the time and $\mathbf{r}(t)$ is a position vector in space, $\frac{ds}{dt} = |\frac{d\mathbf{r}}{dt}|$ is the *speed* with which the tip of position vector $\mathbf{r}(t)$ traces out a space curve Γ.

tangent vector   Examination of Fig. 11.2 and consideration of the definition of $d\mathbf{r}/dt$ shows that the *unit tangent vector* $\mathbf{T}$ along Γ as a function of t is given by

$$\mathbf{T} = \frac{d\mathbf{r}}{dt} \Big/ \left|\frac{d\mathbf{r}}{dt}\right|, \tag{7}$$

and as $ds/dt = |d\mathbf{r}/dt|$, this can be rewritten in the form

$$\frac{d\mathbf{r}}{dt} = \frac{ds}{dt}\mathbf{T}. \tag{8}$$

EXAMPLE 11.5   If $\mathbf{r}(t)$ is a position vector and t is the time, find the velocity, speed, and acceleration of a particle with position vector $\mathbf{r}(t) = a\cos\omega t\,\mathbf{i} + a\sin\omega t\,\mathbf{j}$, where a and ω are constants, and interpret the results.

Solution   We have $|\mathbf{r}(t)| = (a^2\cos^2\omega t + a^2\sin^2\omega t)^{1/2} = a$, so as the motion is two-dimensional in the plane containing $\mathbf{i}$ and $\mathbf{j}$, it takes place in a circle of radius a with its center at the origin of the coordinate system. Differentiation of $\mathbf{r}(t)$ gives

$$\frac{d\mathbf{r}}{dt} = -\omega a\sin\omega t\,\mathbf{i} + \omega a\cos\omega t\,\mathbf{j} \quad \text{and} \quad \frac{d^2\mathbf{r}}{dt^2} = -\omega^2 a\cos\omega t\,\mathbf{i} - \omega^2 a\sin\omega t\,\mathbf{j}.$$

The speed $ds/dt = |d\mathbf{r}/dt| = \omega a$ is constant, and the velocity $d\mathbf{r}/dt$ is seen to be tangential to the circular path, because $\mathbf{r}\cdot(d\mathbf{r}/dt) = 0$. The acceleration $d^2\mathbf{r}/dt^2$ is proportional to $\mathbf{r}$, but oppositely directed, so it is always directed toward the origin. Figure 11.4 illustrates the relationship between the velocity and acceleration as the particle moves around the circle at a constant speed ωa.

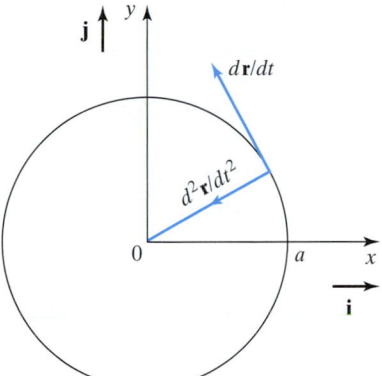

FIGURE 11.4   Uniform motion around the circle $\mathbf{r} = a\cos\omega t\mathbf{i} + a\sin\omega t\mathbf{j}$.

In dealing with the geometry of a space curve Γ, it is often convenient to specify the position vector $\mathbf{r}$ of a point on the curve in terms of the arc length s measured along the curve from some fixed point, so that then $\mathbf{r} = \mathbf{r}(s)$. When $\mathbf{r}$ is expressed in this manner the equation $\mathbf{r} = \mathbf{r}(s)$ is called the **intrinsic equation** of Γ. In addition to the unit tangent $\mathbf{T}$ at any point $\mathbf{r} = \mathbf{r}(s)$ of Γ, two other important unit vectors $\mathbf{N}$ and $\mathbf{B}$ can also be defined at that point.

intrinsic vector equation

To arrive at definitions of vectors $\mathbf{N}$ and $\mathbf{B}$, we start from the fact that as $\mathbf{T}$ is a unit vector $\mathbf{T} \cdot \mathbf{T} = 1$, so differentiating with respect to t and using Theorem 11.1(iv) we have

$$\frac{d\mathbf{T}}{ds} \cdot \mathbf{T} + \mathbf{T} \cdot \frac{d\mathbf{T}}{ds} = 0.$$

However, as the scalar product is commutative, this last result is seen to be equivalent to

$$\mathbf{T} \cdot \frac{d\mathbf{T}}{ds} = 0,$$

showing that $\mathbf{T}$ and $d\mathbf{T}/ds$ are orthogonal. The unit vector $\mathbf{N}$ in the direction of $d\mathbf{T}/ds$ at a point $\mathbf{r} = \mathbf{r}(s)$ on Γ is called the **principal normal** to Γ at $\mathbf{r}(s)$, and so

$$\mathbf{N} = \frac{d\mathbf{T}}{ds} \bigg/ \left|\frac{d\mathbf{T}}{ds}\right| \quad \text{for} \quad \left|\frac{d\mathbf{T}}{ds}\right| \neq 0. \tag{9}$$

When the connection between $d\mathbf{T}/ds$ and $\mathbf{N}$ at a point $\mathbf{r} = \mathbf{r}(s)$ on Γ is written in the form

$$\frac{d\mathbf{T}}{ds} = \kappa(s)\mathbf{N}, \tag{10}$$

curvature, normal and binormal

the nonnegative number $\kappa(s)$ is called the **curvature** of the curve Γ at $\mathbf{r} = \mathbf{r}(s)$, and $\rho(s) = 1/\kappa(s)$ is called the **radius of curvature** of the curve Γ at $\mathbf{r} = \mathbf{r}(s)$. As $\mathbf{N}$ is a unit vector, taking the modulus of (10) gives

$$\kappa(s) = \left| \frac{d\mathbf{T}}{ds} \right|. \tag{11}$$

In the case of a smooth plane curve Γ, the circle of curvature at a point P on Γ is tangent to Γ at P with radius $\rho = 1/\kappa$, and such that its center lies on the concave side of Γ.

If the curvature is required in terms of the parameter t, the relationship between $\kappa(s)$ and $\kappa(t)$ follows from the chain rule

$$\frac{d\mathbf{T}}{dt} = \frac{d\mathbf{T}}{ds} \frac{ds}{dt},$$

showing that

$$\left| \frac{d\mathbf{T}}{dt} \right| = \kappa(t) \left| \frac{ds}{dt} \right|. \tag{12}$$

As $dt/ds = 1/(ds/dt) = 1/|d\mathbf{r}/dt|$, this last result can be written in the convenient form

$$\kappa(t) = \left| \frac{d\mathbf{T}}{dt} \right| \bigg/ \left| \frac{d\mathbf{r}}{dt} \right|. \tag{13}$$

Finally, the vector $\mathbf{B}$, defined as

$$\mathbf{B} = \mathbf{T} \times \mathbf{N}, \tag{14}$$

is called the **unit binormal** to the curve Γ at $\mathbf{r} = \mathbf{r}(s)$. The three unit vectors $\mathbf{T}$, $\mathbf{N}$, and $\mathbf{B}$ at a point $\mathbf{r} = \mathbf{r}(s)$ on the space curve Γ form a *triad* of mutually orthogonal unit vectors whose orientation depends on the location of the point on Γ. When studying the geometry of space curves it proves to be more convenient to use the unit vectors $\mathbf{T}$, $\mathbf{N}$, and $\mathbf{B}$, whose orientation depends on the point on the curve under consideration, than a fixed reference system of unit vectors such as $\mathbf{i}$, $\mathbf{j}$, and $\mathbf{k}$. ∎

EXAMPLE 11.6

Show that the straight line $\mathbf{r}(t) = at\mathbf{i} + bt\mathbf{j} + ct\mathbf{k} + \mathbf{C}$, with a, b, and c scalar constants and $\mathbf{C}$ a constant vector, has an infinite radius of curvature at every point.

Solution Differentiation shows that $|d\mathbf{r}/dt| = (a^2 + b^2 + c^2)^{1/2} \neq 0$, and the tangent vector $\mathbf{T} = d\mathbf{r}/dt/|d\mathbf{r}/dt| = (a\mathbf{i} + b\mathbf{j} + c\mathbf{k})/(a^2 + b^2 + c^2)^{1/2}$, so $d\mathbf{T}/dt \equiv 0$, and $\mathbf{N}$ has to be chosen arbitrarily except for $\mathbf{T} \cdot \mathbf{N} = 0$. Consequently, from (13) $\kappa(t) \equiv 0$, and so the radius of curvature $\rho(t) = 1/\kappa(t) = \infty$ for all t. ∎

EXAMPLE 11.7

Find $\mathbf{T}$, $\mathbf{N}$, $\mathbf{B}$, and $\kappa(t)$ for the helix $\mathbf{r}(t) = a \cos t\mathbf{i} + a \sin t\mathbf{j} + bt\mathbf{k}$.

Solution From $ds/dt = |d\mathbf{r}/dt|$ we have

$$ds/dt = [(-a \sin t)^2 + (a \cos t)^2 + b^2]^{1/2} = (a^2 + b^2)^{1/2},$$

and so

$$\mathbf{T} = \frac{d\mathbf{r}}{dt} \bigg/ \frac{ds}{dt} = \frac{1}{(a^2 + b^2)^{1/2}}(-a\sin t\mathbf{i} + a\cos t\mathbf{j} + b\mathbf{k}).$$

By definition,

$$\mathbf{N} = \frac{d\mathbf{T}}{ds} \bigg/ \left|\frac{d\mathbf{T}}{ds}\right| = \frac{d\mathbf{T}}{dt}\frac{dt}{ds} \bigg/ \left|\frac{d\mathbf{T}}{dt}\frac{dt}{ds}\right| = \frac{d\mathbf{T}}{dt} \bigg/ \left|\frac{d\mathbf{T}}{dt}\right| = -\cos t\mathbf{i} - \sin t\mathbf{j}$$

and

$$\mathbf{B} = \mathbf{T} \times \mathbf{N} = \frac{1}{(a^2 + b^2)^{1/2}}(b\sin t\mathbf{i} - b\cos t\mathbf{j} + a\mathbf{k}).$$

A simple calculation shows that $|d\mathbf{T}/dt| = a/(a^2 + b^2)^{1/2}$, $|d\mathbf{r}/dt| = (a^2 + b^2)^{1/2}$, so it follows from (13) that the curvature $\kappa(t) = a/(a^2 + b^2)$ for all t. This is to be expected, because the uniform shape of the helix implies that the curvature, and hence the radius of curvature, are constant along the helix. ∎

Summary

Scalar and vector fields have been introduced, vector functions of a single real variable have been defined, and their differentiability properties have been derived. Applications to dynamics and the geometry of space curves have been made.

EXERCISES 11.1

In Exercises 1 through 6 find the first and second derivatives of the function and their values at the given value of t.

1. $\mathbf{r} = t\sin t\mathbf{i} + t\cos t\mathbf{j} + t^2\mathbf{k}$, $t = \pi/2$.

2. $\mathbf{r} = (1 + t^2)\mathbf{i} + e^{-2t}\mathbf{j} + \sqrt{t}\mathbf{k}$, $t = 1$.

3. $\mathbf{r} = (2 - \cos^2 t)\mathbf{i} + \sin^2 t\mathbf{j} + (\pi - t)\mathbf{k}$, $t = \pi/4$.

4. $\mathbf{r} = \ln(1 + t)\mathbf{i} + \ln(1 + r^2)\mathbf{j} + e^{3t}\mathbf{k}$, $t = 0$.

5. $\mathbf{r} = (t - \sin t)\mathbf{i} + (1 - \cos t)\mathbf{j}$, $t = \pi/2$ (a **cycloid**).
   Notice that $\mathbf{r}$ is arbitrarily many times differentiable, yet the cycloid has cusps for $t = n\pi$.

6. $\mathbf{r} = 4\cos t\mathbf{i} + 3\sin t\mathbf{j} + 2t\mathbf{k}$, $t = \pi/4$ (an **elliptical "helix"**).

7. Prove result (iii) in Theorem 11.1 by expressing the vectors in terms of their cartesian components.

8. Prove result (v) in Theorem 11.1 by expressing the vectors in terms of their cartesian components.

9. Given that $\mathbf{r} = t\mathbf{i} + 3t^2\mathbf{j} - (t - 1)\mathbf{k}$ and $t = \ln(1 + s^2)$, use result (vi) in Theorem 11.1 to find $d\mathbf{r}/ds$.

10. Given that $\mathbf{r} = \sin t\mathbf{i} + \cos t\mathbf{j} + \tan t\mathbf{k}$ and $t = 2 + s^2$, use result (vi) in Theorem 11.1 to find $d\mathbf{r}/ds$.

11. A particle has a position vector at time t given by

$$\mathbf{r} = t^2\mathbf{i} + 4\cos 2t\mathbf{j} + 3\sin 2t\mathbf{k}.$$

Find the component of its velocity in the direction $2\mathbf{i} + \mathbf{j} + 2\mathbf{k}$ at time t.

12. A particle has a position vector at time t given by

$$\mathbf{r} = 3\cos t\mathbf{i} + 3\sin t\mathbf{j} + (t^2 - 2)\mathbf{k}.$$

Find the component of its velocity in the direction $\mathbf{i} + 2\mathbf{j} - \mathbf{k}$ at time t.

13. If $\phi(t)$ is a differentiable function of t and $\mathbf{u}(t)$ is a differentiable parametrically defined function of t, prove that

$$\frac{d}{dt}(\phi\mathbf{u}) = \phi\frac{d\mathbf{u}}{dt} + \frac{d\phi}{dt}\mathbf{u}.$$

14. If $\mathbf{u}$, $\mathbf{v}$, and $\mathbf{w}$ are differentiable parametrically defined functions of t, prove that

$$\frac{d}{dt}(\mathbf{u} \cdot (\mathbf{v} \times \mathbf{w})) = \mathbf{u} \cdot \left(\mathbf{v} \times \frac{d\mathbf{w}}{dt}\right) + \mathbf{u} \cdot \left(\frac{d\mathbf{v}}{dt} \times \mathbf{w}\right)$$
$$+ \frac{d\mathbf{u}}{dt} \cdot (\mathbf{v} \times \mathbf{w}),$$

where the order in the products must be preserved.

15. If $\mathbf{u}$, $\mathbf{v}$, and $\mathbf{w}$ are differentiable parametrically defined functions of t, prove that

$$\frac{d}{dt}(\mathbf{u} \times (\mathbf{v} \times \mathbf{w})) = \mathbf{u} \times \left(\mathbf{v} \times \frac{d\mathbf{w}}{dt}\right) + \mathbf{u} \times \left(\frac{d\mathbf{v}}{dt} \times \mathbf{w}\right)$$
$$+ \frac{d\mathbf{u}}{dt} \times (\mathbf{v} \times \mathbf{w}),$$

where the order in the products must be preserved.

16. If **u** is a differentiable parametrically defined function of t, prove that

$$\frac{d\mathbf{u}}{dt} \times \frac{d}{dt}\left(\frac{d\mathbf{u}}{dt} \times \frac{d^2\mathbf{u}}{dt^2}\right) = \frac{d\mathbf{u}}{dt}\left(\frac{d\mathbf{u}}{dt} \cdot \frac{d^3\mathbf{u}}{dt^3}\right) - \frac{d^3\mathbf{u}}{dt^3}\left(\frac{d\mathbf{u}}{dt}\right)^2.$$

17. If **u** is a differentiable parametrically defined function of t, prove that

$$\frac{d\mathbf{u}}{dt} \times \frac{d}{dt}\left(\mathbf{u} \times \frac{d\mathbf{u}}{dt}\right) = \mathbf{u}\left(\frac{d\mathbf{u}}{dt} \cdot \frac{d^2\mathbf{u}}{dt^2}\right) - \frac{d^2\mathbf{u}}{dt^2}\left(\mathbf{u} \cdot \frac{d\mathbf{u}}{dt}\right).$$

18. Given that $\phi(t) = t^2 \cos t$ and $\mathbf{u} = \sin t\,\mathbf{i} + 2\cos t\,\mathbf{j} + (1+t^2)^{1/2}\mathbf{k}$, use the result of Exercise 13 to find $\frac{d}{dt}(\phi\mathbf{u})$, and confirm the result by direct differentiation of $\phi\mathbf{u}$ with respect to t.

19. Given that $\mathbf{u} = 2t\mathbf{i} - t^2\mathbf{j} + \mathbf{k}$, $\mathbf{v} = 2\mathbf{i} + 3t\mathbf{j} + t\mathbf{k}$, and $\mathbf{w} = t\mathbf{i} + 2t\mathbf{j} - t\mathbf{k}$, use the result of Exercise 14 to find $\frac{d}{dt}(\mathbf{u} \cdot (\mathbf{v} \times \mathbf{w}))$. Confirm the result by finding $\mathbf{u} \cdot (\mathbf{v} \times \mathbf{w})$ and differentiating the result with respect to t.

20. Given that $\mathbf{u} = t\mathbf{i} - t\mathbf{j} + t^2\mathbf{k}$, $\mathbf{v} = -t\mathbf{i} + 2t\mathbf{j} - t^2\mathbf{k}$, and $\mathbf{w} = 2t\mathbf{i} - 2t\mathbf{j} + t\mathbf{k}$, use the result of Exercise 15 to find $\frac{d}{dt}(\mathbf{u} \times (\mathbf{v} \times \mathbf{w}))$. Confirm the result by finding $\mathbf{u} \times (\mathbf{v} \times \mathbf{w})$ and differentiating the result with respect to t.

21. Find **T**, **N**, **B**, and κ as functions of t for the helix $\mathbf{r}(t) = a\cos\omega t\,\mathbf{i} + a\sin\omega t\,\mathbf{j} + bt\mathbf{k}$.

22. By differentiating $\mathbf{B} = \mathbf{T} \times \mathbf{N}$ with respect to s, show

that

$$\frac{d\mathbf{B}}{ds} = \mathbf{T} \times \frac{d\mathbf{N}}{ds},$$

and then by forming the product $\mathbf{N} \times d\mathbf{B}/ds$, show that

$$\mathbf{N} \times \frac{d\mathbf{B}}{ds} = \mathbf{0}.$$

Introduce a constant of proportionality called the **torsion** of the curve Γ at P, which by convention is denoted by $-\tau$, and deduce from this last result that

$$\frac{d\mathbf{B}}{ds} = -\tau\mathbf{N}.$$

Finally, by differentiating $\mathbf{N} = \mathbf{B} \times \mathbf{T}$ with respect to s show that

$$\frac{d\mathbf{N}}{ds} = \tau\mathbf{B} - \kappa\mathbf{T}.$$

The three equations relating the derivatives of **T**, **N**, and **B** with respect to s to **T**, **N**, **B**, κ, and τ found earlier, namely,

$$\frac{d\mathbf{T}}{ds} = \kappa\mathbf{N}, \quad \frac{d\mathbf{N}}{ds} = \tau\mathbf{B} - \kappa\mathbf{T}, \quad \text{and} \quad \frac{d\mathbf{B}}{ds} = -\tau\mathbf{N},$$

are called the **Frenet–Serret equations**, and they are fundamental to the study of the differential geometry of space curves.

11.2 Integration of Scalar and Vector Functions of a Single Real Variable

As with real functions of a single real variable, a differentiable vector function of a single real variable $\mathbf{F}(t)$ will be called an **antiderivative** of the vector function $\mathbf{f}(t)$ on some interval $a < t < b$ if at each point of the interval $d\mathbf{F}(t)/dt = \mathbf{f}(t)$. Because differentiation of a vector constant yields the null vector $\mathbf{0}$, an antiderivative of **f** is only determined up to an arbitrary additive vector constant **C**. An **indefinite integral** of **f** is any antiderivative of **f** to which has been added an arbitrary vector constant.

Indefinite and definite integrals of a vector function of a single real variable

indefinite and definite integrals of vector functions of a single real variable

If $\mathbf{F}(t)$ is any antiderivative of $\mathbf{f}(t)$, then an **indefinite integral** of the function **f** with respect to t, written $\int \mathbf{f}(t)dt$, is

$$\int \mathbf{f}(t)dt = \mathbf{F}(t) + \mathbf{C},$$

where **C** is an arbitrary vector constant.

If $\mathbf{f}(t) = f_1(t)\mathbf{i} + f_2(t)\mathbf{j} + f_3(t)\mathbf{k}$, the indefinite integral of $\mathbf{f}(t)$ is determined by integrating each component of $\mathbf{f}(t)$ with respect to t and combining the results to give

$$\int f_1(t)dt\mathbf{i} + \int f_2(t)dt\mathbf{j} + \int f_3(t)dt\mathbf{k} = \mathbf{F}(t) + \mathbf{C}.$$

The **definite integral** of $\mathbf{f}(t)$ over the interval $a \leq t \leq b$ is defined as

$$\int_a^b \mathbf{f}(t)dt = \int_a^b f_1(t)dt\mathbf{i} + \int_a^b f_2(t)dt\mathbf{j} + \int_a^b f_3(t)dt\mathbf{k}.$$

EXAMPLE 11.8   Given that $\mathbf{f}(t) = \sin t\mathbf{i} + (1 - t^2)\mathbf{j} + e^{-t}\mathbf{k}$, find

$$\text{(a)} \int \mathbf{f}(t)dt \quad \text{and} \quad \text{(b)} \int_0^2 \mathbf{f}(t)dt.$$

Solution

(a)
$$\int \mathbf{f}(t)dt = \int \sin t \, dt\mathbf{i} + \int (1 - t^2)dt\mathbf{j} + \int e^{-t} \, dt\mathbf{k}$$

$$= -\cos t\mathbf{i} + \left(t - \frac{1}{3}t^3\right)\mathbf{j} - e^{-t}\mathbf{k} + c_1\mathbf{i} + c_2\mathbf{j} + c_3\mathbf{k},$$

where c_1, c_2, and c_3 are arbitrary real constants, so

$$\int \mathbf{f}(t)dt = -\cos t\mathbf{i} + \left(t - \frac{1}{3}t^3\right)\mathbf{j} - e^{-t}\mathbf{k} + \mathbf{C},$$

where $\mathbf{C}$ is an arbitrary vector constant.

(b)
$$\int_0^2 \mathbf{f}(t)dt = \int_0^2 \sin t \, dt\mathbf{i} + \int_0^2 (1 - t^2)dt\mathbf{j} + \int_0^2 e^{-t} \, dt\mathbf{k}$$

$$= (1 - \cos 2)\mathbf{i} - \frac{2}{3}\mathbf{j} + (1 - e^{-2})\mathbf{k}. \qquad \blacksquare$$

It is sometimes necessary to find the length of arc between two points on a curve defined by a vector function of a single real variable. This can be accomplished by making use of result (6), which showed that the rate of change of distance s with respect to t along the curve Γ defined by

$$\mathbf{r}(t) = x_1(t)\mathbf{i} + x_2(t)\mathbf{j} + x_3(t)\mathbf{k}$$

is given by

$$\frac{ds}{dt} = \left[\left(\frac{dx_1}{dt}\right)^2 + \left(\frac{dx_2}{dt}\right)^2 + \left(\frac{dx_3}{dt}\right)^2\right]^{1/2}.$$

arc length along a space curve   Consequently, if the length of arc $s = s(t_2) - s(t_1)$ between the points corresponding to $t = t_1$ and $t = t_2$ is required, where $t_2 > t_1$, integration of this result gives

$$\int_{t_1}^{t_2} \frac{ds}{dt}dt = \int_{t_1}^{t_2} \left[\left(\frac{dx_1}{dt}\right)^2 + \left(\frac{dx_2}{dt}\right)^2 + \left(\frac{dx_3}{dt}\right)^2\right]^{1/2} dt,$$

so the required arc length is given by the definite integral

$$s = s(t_2) - s(t_1) = \int_{t_1}^{t_2} \left[\left(\frac{dx_1}{dt} \right)^2 + \left(\frac{dx_2}{dt} \right)^2 + \left(\frac{dx_3}{dt} \right)^2 \right]^{1/2} dt. \tag{15}$$

EXAMPLE 11.9   Find the length of arc along the helix $\mathbf{r}(t) = \cos t\mathbf{i} + \sin t\mathbf{j} + \alpha t\mathbf{k}$ between the points corresponding to $t = 0$ and $t = 2\pi$, where α is a scalar constant.

Solution   Making the identifications $x_1(t) = \cos t$, $x_2(t) = \sin t$, $x_3(t) = \alpha t$, $t_1 = 0$, and $t_2 = 2\pi$, and substituting into (15) gives

$$s = \int_0^{2\pi} [(-\sin t)^2 + (\cos t)^2 + \alpha^2]^{1/2} dt$$

$$= \sqrt{1 + \alpha^2} \int_0^{2\pi} dt = 2\pi \sqrt{1 + \alpha^2}.$$

When $\alpha = 0$ the helix reduces to a circle of unit radius, and as expected s then becomes the circumference 2π of a unit circle. ∎

Let the vector $\mathbf{F}(x, y, z)$ be defined along a piecewise smooth space curve Γ along which the arc length is s, and let Γ extend from the point $\mathbf{r}_1$ at which $s = s_1$ to the point $\mathbf{r}_2$ at which $s = s_2$. Then, if $\mathbf{T}(s)$ is the unit tangent vector to Γ at arc length s, an expression of the form

$$I = \int_{s_1}^{s_2} \mathbf{F} \cdot \mathbf{T} \, ds$$

scalar line integrals   is called a **line integral** of $\mathbf{F}$, or more precisely, the **scalar line integral** of $\mathbf{F}$ along the space curve Γ. It follows from (8) that $\mathbf{T}ds = d\mathbf{r}$, so the line integral of $\mathbf{F}$ along Γ can be written in the simpler form

$$I = \int_{s_1}^{s_2} \mathbf{F} \cdot d\mathbf{r}. \tag{16}$$

Integrals of this type have many applications, two of the most important of which are described in what follows. The first application is to mechanics, where when a constant force $\mathbf{F}$ moves its point of application a distance d along a straight line L, the **work** that is done by the force is $W = f_L d$, where f_L is the component of $\mathbf{F}$ along the line L. To find the work done by a variable force $\mathbf{F}(t)$ as it moves its point of application along a parametrically defined curve Γ, it is necessary to generalize this simple result by appealing to the notion of a line integral along the space curve Γ.

If the vector differential along Γ is denoted by $d\mathbf{r}$, its length $|d\mathbf{r}| = dr$, so the unit vector $\mathbf{T}$ in the direction $d\mathbf{r}$ will be $\mathbf{T} = d\mathbf{r}/dr$. Consequently, the component of force $\mathbf{F}$ in the direction of $d\mathbf{r}$ is given by $\mathbf{F} \cdot \mathbf{T} = (\mathbf{F} \cdot d\mathbf{r})/dr$, so the element of work dW performed by the force in moving its point of application along $d\mathbf{r}$ will be

$$dW = \mathbf{F} \cdot \left(\frac{d\mathbf{r}}{dr}\right) dr = \mathbf{F} \cdot d\mathbf{r}.$$

Integration of this result shows the work performed by the force in moving its point of application along Γ from $\mathbf{r} = \mathbf{r}_1$ to $\mathbf{r} = \mathbf{r}_2$, corresponding to $s = s_1$ and $s = s_2$, respectively, is given by the line integral

$$W = \int_{s_1}^{s_2} \mathbf{F} \cdot d\mathbf{r}. \tag{17}$$

When $\mathbf{r} = \mathbf{r}(t)$ is known as a function of t, but t is not the arc length s along Γ, and integration is between $\mathbf{r} = \mathbf{r}(t_1)$ and $\mathbf{r} = \mathbf{r}(t_2)$, $d\mathbf{r} = (d\mathbf{r}/dt)dt$ and (17) becomes

$$W = \int_{t(s_1)}^{t(s_2)} \mathbf{F}(\mathbf{r}(t)) \cdot (d\mathbf{r}/dt)dt. \tag{18}$$

Integrals of this type arise when particles move in a gravitational field or a charged particle moves in an electric field. The sign of W depends on the direction of integration, so reversing its direction changes the sign of W. Work is done by the vector field $\mathbf{F}$ when W is positive, and work is recovered from the field when W is negative.

For the second example we consider the case of fluid mechanics and identify $\mathbf{F}$ with the fluid velocity vector $\mathbf{q}$. In this case a line integral of the form (16) is called the **flow** of the fluid along Γ, because $d\mathbf{r} = (d\mathbf{r}/ds)ds = \mathbf{T}ds$, where $\mathbf{T}$ is the unit tangent along Γ, so that $\mathbf{q} \cdot \mathbf{T}$ is the component of the flow along Γ. The **circulation** k of fluid is defined as the flow around a *closed* curve Γ, so it is given by

circulation and irrotational flow

$$k = \oint_{\Gamma} \mathbf{q} \cdot d\mathbf{r} = \oint_{\Gamma} \mathbf{q} \cdot \mathbf{T} \, ds, \tag{19}$$

where the symbol $\oint_{\Gamma}$ is used to indicate that the line integral of $\mathbf{q} \cdot d\mathbf{r}$ is taken *once* around the closed curve Γ.

In fluid mechanics the circulation k describes an important characteristic of the fluid motion, and it can be seen from (19) that reversing the direction of integration around Γ reverses the sign of $\mathbf{T}$, and so leads to a reversal of the sign of the circulation. The fundamental class of fluid flow in which there is zero circulation around every simple closed curve Γ, so that $k \equiv 0$, is called **irrotational** flow.

In general, the line integral (16) depends not only on $\mathbf{F}$ and the end points of integration, but also on the path Γ along which the integral is evaluated. The method of evaluating line integrals, and the fact that they usually depend on the path, is illustrated in the next example.

EXAMPLE 11.10

Find the line integral of $\mathbf{F} = -yz^2\mathbf{i} + xz^2\mathbf{j} + yz\mathbf{k}$ (a) along the helix Γ given by $\mathbf{r}(t) = \cos t\mathbf{i} + \sin t\mathbf{j} + t\mathbf{k}$ from $t = 0$ to $t = 2\pi$, and (b) along the straight line path γ joining the points $\mathbf{r}(0)$ to $\mathbf{r}(2\pi)$.

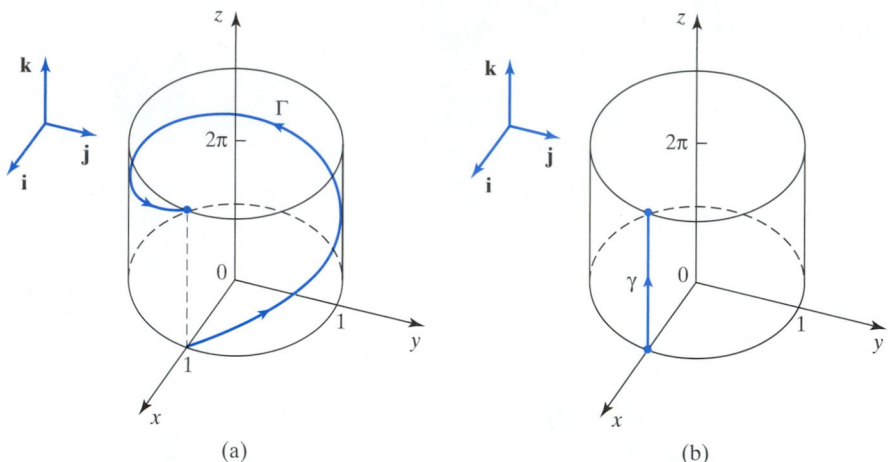

FIGURE 11.5 (a) The helix Γ. (b) The straight line path γ.

Solution

(a) The helix Γ is shown in Fig. 11.5(a).
Differentiation of $\mathbf{r}(t)$ gives

$$\frac{d\mathbf{r}}{dt} = -\sin t\,\mathbf{i} + \cos t\,\mathbf{j} + \mathbf{k},$$

but on the helix $x = \cos t$, $y = \sin t$, and $z = t$, so in the line integral along Γ the general vector-valued function $\mathbf{F}$ becomes the vector function of the single real variable t given by $\mathbf{F}(t) = -t^2 \sin t\,\mathbf{i} + t^2 \cos t\,\mathbf{j} + t \sin t\,\mathbf{k}$. As a result,

$$\mathbf{F} \cdot d\mathbf{r} = (-t^2 \sin t\,\mathbf{i} + t^2 \cos t\,\mathbf{j} + t \sin t\,\mathbf{k}) \cdot (-\sin t\,\mathbf{i} + \cos t\,\mathbf{j} + \mathbf{k})dt$$
$$= (t^2 + t \sin t)dt,$$

and so the required line integral is

$$\int_\Gamma \mathbf{F} \cdot d\mathbf{r} = \int_0^{2\pi} \mathbf{F} \cdot d\mathbf{r} = \int_0^{2\pi} (t^2 + t \sin t)dt = \frac{8}{3}\pi^3 - 2\pi.$$

(b) The straight line path γ shown in Fig. 11.5(b) joins the points $\mathbf{r}(0) = \mathbf{i}$ and $\mathbf{r}(2\pi) = \mathbf{i} + 2\pi\mathbf{k}$, so in terms of the parameter t its vector equation can be written $\mathbf{r}(t) = \mathbf{i} + t\mathbf{k}$ with $0 \leq t \leq 2\pi$. This shows that on the path γ we have $x = 1$, $y = 0$, and $z = t$, and $d\mathbf{r} = dt\,\mathbf{k}$.

Consequently, on γ the vector-valued function $\mathbf{F}$ becomes $\mathbf{F} = t^2\mathbf{j}$, and so

$$\mathbf{F} \cdot d\mathbf{r} = t^2\mathbf{j} \cdot (dt\,\mathbf{k}) = 0,$$

showing that

$$\int_\gamma \mathbf{F} \cdot d\mathbf{r} = 0. \qquad \blacksquare$$

In the next section, after the introduction of the *gradient* of a function, we will find a condition to be satisfied by $\mathbf{F}$ in order that the line integral in (16) is independent of the path Γ, and so depends only on $\mathbf{F}$ and the end points of the integration.

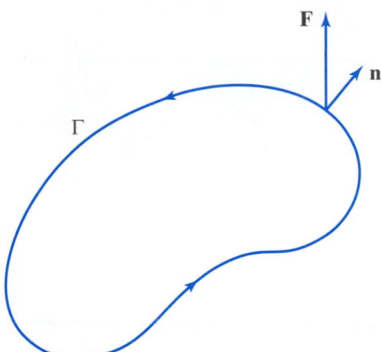

FIGURE 11.6 $\Phi_\Gamma = \oint_\Gamma \mathbf{F} \cdot \mathbf{n}\, ds$ is the flux of $\mathbf{F}$ across Γ.

As a final example of an application of line integrals we determine the *flux* of a vector $\mathbf{F}(x, y)$ across a closed two-dimensional smooth curve Γ in the (x, y)-plane. If $\mathbf{n}$ is a unit vector normal to Γ that is directed *outward* from Γ, as shown in Fig. 11.6, the **flux** Φ_Γ across the curve Γ is defined as the line integral

$$\Phi_\Gamma = \int_\Gamma \mathbf{F} \cdot \mathbf{n}\, ds,$$

the flux of a vector across a plane curve

where s is the arc length around Γ and integration is in the *counterclockwise* sense around Γ. As $\mathbf{F} \cdot \mathbf{n}$ is the component of $\mathbf{F}$ in the direction of the outward drawn normal to Γ, the flux Φ_Γ is seen to measure the total amount of the normal component of $\mathbf{F}$ that crosses the curve Γ.

For a physical illustration of the meaning of flux, let us consider a long block of metal with its axis in the z-direction in which there is a steady-state temperature distribution that is only a function of x and y. This means that the temperature distribution is the same in every plane $z =$ constant. Let us now consider a cylindrical region in the block of unit height and cross-section Γ with its axis in the z-direction. Then if $\mathbf{F}$ is identified with a heat flow vector $\mathbf{h}(x, y)$, the flux Φ_Γ is the amount of the heat that crosses the curved walls of this cylinder in Fig. 11.7 in a unit time. If $\Phi_\Gamma > 0$ there is a net *outflow* of heat from the region bounded by Γ, and if $\Phi_\Gamma < 0$ there is a net *inflow* of heat into the region. When $\Phi_\Gamma = 0$ the amount of heat in the region remains constant.

In two space dimensions it is important to recognize the difference between the circulation and flux of $\mathbf{F}$ in relation to the curve Γ. Whereas the determination of the *circulation* of $\mathbf{F}$ involves the line integral of the component of $\mathbf{F}$ *along* the tangent to curve Γ with respect to the arc length s, the *flux* of $\mathbf{F}$ involves the line integral of the component of $\mathbf{F}$ normal to (*across*) the curve Γ with respect to the arc length.

To determine the flux we proceed as follows. Let $\mathbf{F}(x, y) = f_1(x, y)\mathbf{i} + f_2(x, y)\mathbf{j}$ and Γ have the equation $\mathbf{r}(t) = x(t)\mathbf{i} + y(t)\mathbf{j}$. Then, as integration around Γ is in the counterclockwise sense, we see from Fig. 11.6 that if $\mathbf{T}$ is the unit tangent to Γ, then $\mathbf{n} = \mathbf{T} \times \mathbf{k}$. As $\mathbf{T} = (dx/ds)\mathbf{i} + (dy/ds)\mathbf{j}$, it follows that

$$\mathbf{n} = \mathbf{T} \times \mathbf{k} = [(dx/ds)\mathbf{i} + (dy/ds)\mathbf{j}] \times \mathbf{k}$$
$$= (dy/ds)\mathbf{i} - (dx/ds)\mathbf{j},$$

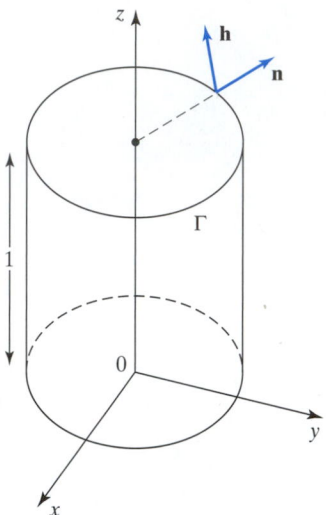

FIGURE 11.7   A cylinder of unit height and cross-section Γ with its axis in the z-direction.

and so

$$\Phi_\Gamma = \int_\Gamma \mathbf{F} \cdot \mathbf{n} \, ds = \int_\Gamma (f_1(x, y)\mathbf{i} + f_2(x, y)\mathbf{j}) \cdot ((dy/ds)\mathbf{i} - (dx/ds)\mathbf{j}) \, ds$$

$$= \int_\Gamma f_1(x, y) \, dy - f_2(x, y) \, dx.$$

EXAMPLE 11.11   Find the flux of $\mathbf{F} = (2x + y)\mathbf{i} + (y - x)\mathbf{j}$ across the ellipse with the equation $x^2/a^2 + y^2/b^2 = 1$.

Solution   By setting $x = a \cos t$ and $y = b \sin t$ and restricting t to the interval $0 \le t \le 2\pi$, the ellipse is traversed once in the counterclockwise sense as required. As $dx = -a \sin t \, dt$ and $dy = b \cos t \, dt$, substitution into the expression for Φ_Γ gives

$$\Phi_\Gamma = \int_0^{2\pi} [(2a \cos t + b \sin t)b \cos t - (b \sin t - a \cos t)(-a \sin t)] \, dt = 3ab\pi. \quad \blacksquare$$

Finally we define a different integral called a *vector line integral* of $\mathbf{F}$. To do this we let a curve Γ have the vector equation

$$\mathbf{r}(t) = x(t)\mathbf{i} + y(t)\mathbf{j} + z(t)\mathbf{k} \quad \text{for } a \le t \le b$$

and introduce a general vector function $\mathbf{F} = F_1(x, y, z)\mathbf{i} + F_2(x, y, z)\mathbf{j} + F_3(x, y, z)\mathbf{k}$ defined along the curve Γ. Then the **vector line integral** of $\mathbf{F}$ along Γ from $t = a$ to $t = b$ is defined as

$$\int_a^b \mathbf{F} \, dt = \mathbf{i} \int_a^b F_1(t) \, dt + \mathbf{j} \int_a^b F_2(t) \, dt + \mathbf{k} \int_a^b F_3(t) \, dt, \tag{20}$$

where $F_i(t) = F_i(x(t), y(t), z(t))$, for $i = 1, 2, 3$.

EXAMPLE 11.12

Find the vector line integral of the vector function $\mathbf{F} = xz\mathbf{i} + yz\mathbf{j} + z\mathbf{k}$ along the curve $\mathbf{r}(t) = a\cos t\mathbf{i} + a\sin t\mathbf{j} + t\mathbf{k}$ over the interval $0 \leq t \leq \pi$.

Solution

$$\int_0^\pi \mathbf{F}\,dt = \mathbf{i}\int_0^\pi at\cos t\,dt + \mathbf{j}\int_0^\pi at\sin t\,dt + \mathbf{k}\int_0^\pi t\,dt = -2a\mathbf{i} + \pi a\mathbf{j} + \frac{1}{2}\pi^2\mathbf{k}.$$

Summary

Indefinite and definite integrals of vector functions of a single real variable have been defined and illustrated by example. The scalar line integral of a vector $\mathbf{F}(x, y, z)$ has been defined and its application illustrated by considering the work done by a force as it moves along a space curve between two fixed points. The line integral has also been applied to fluid flow and used to define the circulation of the fluid, and the related concept of an irrotational flow for which the circulation around any closed curve in the fluid is zero. Finally, the flux of a vector across a plane curve has been defined.

EXERCISES 11.2

In Exercises 1 through 4 find the required indefinite and definite integrals.

1. (a) $\displaystyle\int (t\sin t\mathbf{i} + 3t^2\mathbf{j} - 3t\mathbf{k})\,dt$.

   (b) $\displaystyle\int_0^2 (\ln(1+3t)\mathbf{i} + (t^3 - 2t)\mathbf{j} + te^t\mathbf{k})\,dt$.

2. (a) $\displaystyle\int (\cosh^2 t\mathbf{i} + 2\sin^2 2t\mathbf{j} + \mathbf{k})\,dt$.

   (b) $\displaystyle\int_0^2 ((1+t^2)^{-1}\mathbf{i} - t\sin t\mathbf{j} - (1 - 3t^2)\mathbf{k})\,dt$.

3. (a) $\displaystyle\int (\cos^2 3t\mathbf{i} + \sin^2 t\mathbf{j} + t\mathbf{k})\,dt$.

   (b) $\displaystyle\int_0^\pi ((1+3t^2)\mathbf{i} + \cos 4t\mathbf{j} + \sin 3t\mathbf{k})\,dt$.

4. (a) $\displaystyle\int (t(1+t)^{-1}\mathbf{i} + \sec^2 3t\mathbf{j} + (t^2 - 4)\mathbf{k})\,dt$.

   (b) $\displaystyle\int_0^4 (t(1+3t^2)^{-1}\mathbf{i} + (1+t^2)^{1/2}\mathbf{j} + t^2 e^{-t}\mathbf{k})\,dt$.

5. Find the arc length along the circular helix $\mathbf{r}(t) = a\cos t\mathbf{i} + a\sin t\mathbf{j} + \alpha t\mathbf{k}$ between the points corresponding to $t = \pi$ and $t = 3\pi/2$.

6. Find the arc length along the curve $\mathbf{r}(t) = \cos t\mathbf{i} + \sin t\mathbf{j} + \frac{1}{2}t^2\mathbf{k}$ between the points corresponding to $t = 0$ and $t = 2\pi$.

7. Given the vector valued function $\mathbf{F} = -z\mathbf{i} + x\mathbf{j} - y\mathbf{k}$, find the scalar line integral of $\mathbf{F}$ along the space curve $\mathbf{r}(t) = \sin t\mathbf{i} - \cos t\mathbf{j} + e^t\mathbf{k}$ between the points on the curve corresponding to $t = 0$ and $t = \pi/2$.

8. Given the vector valued function $\mathbf{F} = 2y\mathbf{i} + x^2\mathbf{j} - 3z\mathbf{k}$, find the line integral of $\mathbf{F}$ along the space curve $\mathbf{r}(t) =$ $t\mathbf{i} + (1+2t^3)\mathbf{j} + t^2\mathbf{k}$ between the points on the curve corresponding to $t = 1$ and $t = 3$.

9. Let $\mathbf{F}$ be the vector-valued function $\mathbf{F} = -x\mathbf{i} + y\mathbf{j} + z\mathbf{k}$. Show that the line integrals of $\mathbf{F}$ along the helix $\mathbf{r}(t) = \sin t\mathbf{i} + \cos t\mathbf{j} + t\mathbf{k}$ between the points on the helix corresponding to $t = 0$ and $t = 2\pi$ and along the straight line path joining the points $\mathbf{r}(0)$ to $\mathbf{r}(2\pi)$ are the same.

10. Let $\mathbf{F}$ be the vector-valued function $\mathbf{F} = 2xy^2z\mathbf{i} + 2x^2yz\mathbf{j} + x^2y^2\mathbf{k}$. Find the line integral of $\mathbf{F}$ along the straight line Γ with the equation $\mathbf{r}(t) = t\mathbf{i} + 2t\mathbf{j} + t\mathbf{k}$ between the points corresponding to $t = 0$ and $t = 1$. Let γ be the path formed by the straight line segments joining the points $PQRS$, in this order, where P is the point $\mathbf{r} = 0$, Q is the point $\mathbf{r} = \mathbf{i}$, R is the point $\mathbf{r} = \mathbf{i} + 2\mathbf{j}$, and S is the point $\mathbf{r} = \mathbf{i} + 2\mathbf{j} + \mathbf{k}$. Find the line integral of $\mathbf{F}$ along γ from P to S, and hence show that it has the same value as the integral along Γ.

11. The velocity vector in a two-dimensional fluid flow is $\mathbf{v} = y\mathbf{i} + x^2y\mathbf{j}$. Find the circulation (a) around the ellipse $x^2 + \frac{1}{4}y^2 = 1$ and (b) around the unit circle $x^2 + y^2 = 1$, and hence show the flow is *not* irrotational.

12. The velocity vector in a two-dimensional fluid flow is $\mathbf{v} = (2x + 3y^2)\mathbf{i} + 6x\mathbf{j}$. Show that there is zero circulation around all the circles $(x - a)^2 + (y - b)^2 = c^2$, where a, b, and $c > 0$ are arbitrary real numbers. Is it correct to say this proves that the flow is irrotational? Give reasons justifying your answer.

13. Find the flux of $\mathbf{F} = (3x + 2y)\mathbf{i} + (2x - y)\mathbf{j}$ across the circle $x^2 + y^2 = 4$.

11.3 Directional Derivatives and the Gradient Operator

Consider a scalar function $w = f(x, y, z)$ with continuous first order partial derivatives with respect to x, y, and z that is defined in some region D of space, and let a space curve Γ in D have the parametric equations $x = x(t)$, $y = y(t)$, and $z = z(t)$. Then from the chain rule

$$\frac{dw}{dt} = \frac{\partial f}{\partial x}\frac{dx}{dt} + \frac{\partial f}{\partial y}\frac{dy}{dt} + \frac{\partial f}{\partial z}\frac{dz}{dt}, \tag{21}$$

and it is seen from this that dw/dt can be interpreted as the scalar product of the two vectors

$$\frac{\partial f}{\partial x}\mathbf{i} + \frac{\partial f}{\partial y}\mathbf{j} + \frac{\partial f}{\partial z}\mathbf{k} \quad \text{and} \quad \frac{dx}{dt}\mathbf{i} + \frac{dy}{dt}\mathbf{j} + \frac{dz}{dt}\mathbf{k}.$$

The first vector, denoted by

$$\text{grad } f = \frac{\partial f}{\partial x}\mathbf{i} + \frac{\partial f}{\partial y}\mathbf{j} + \frac{\partial f}{\partial z}\mathbf{k}, \tag{22}$$

the gradient of a scalar function of position

is called the **gradient** of the scalar function f expressed in terms of cartesian coordinates, whereas from Section 11.1 the second vector

$$\frac{d\mathbf{r}}{dt} = \frac{dx}{dt}\mathbf{i} + \frac{dy}{dt}\mathbf{j} + \frac{dz}{dt}\mathbf{k} \tag{23}$$

is seen to be a vector that is tangent to the space curve Γ. Consequently, dw/dt is the scalar product of grad f and $d\mathbf{r}/dt$ at the point $x = x(t)$, $y = y(t)$, and $z = z(t)$ for any given value of t.

Another notation for grad f that is also used is

$$\nabla f = \frac{\partial f}{\partial x}\mathbf{i} + \frac{\partial f}{\partial y}\mathbf{j} + \frac{\partial f}{\partial z}\mathbf{k}, \tag{24}$$

where the symbol ∇f is either read "del f" or "grad f." In this notation, the **vector operator**

$$\nabla \equiv \mathbf{i}\frac{\partial}{\partial x} + \mathbf{j}\frac{\partial}{\partial y} + \mathbf{k}\frac{\partial}{\partial z} \tag{25}$$

is the **gradient operator** expressed in terms of cartesian coordinates, and if ϕ is a suitably differentiable scalar function of x, y, and z, it is to be understood that

$$\nabla \phi = \frac{\partial \phi}{\partial x}\mathbf{i} + \frac{\partial \phi}{\partial y}\mathbf{j} + \frac{\partial \phi}{\partial z}\mathbf{k}. \tag{26}$$

Let us now introduce the unit vector $\mathbf{v}$ defined as

$$\mathbf{v} = l\mathbf{i} + m\mathbf{j} + n\mathbf{k}, \tag{27}$$

where l, m, and n are the direction cosines of the tangent to the space curve Γ in (23), so that

$$l = \frac{dx}{dt} \Big/ \left|\frac{d\mathbf{r}}{dt}\right|, \qquad m = \frac{dy}{dt} \Big/ \left|\frac{d\mathbf{r}}{dt}\right|, \qquad n = \frac{dz}{dt} \Big/ \left|\frac{d\mathbf{r}}{dt}\right|, \tag{28}$$

with

$$\left|\frac{d\mathbf{r}}{dt}\right| = \left[\left(\frac{dx}{dt}\right)^2 + \left(\frac{dy}{dt}\right)^2 + \left(\frac{dz}{dt}\right)^2\right]^{1/2}. \tag{29}$$

Then as the scalar product of a vector $\mathbf{F}$ and the unit vector $\mathbf{v}$ is the *projection* of $\mathbf{F}$ in the direction $\mathbf{v}$, it follows at once that

$$D_v f = \mathbf{v} \cdot \operatorname{grad} f = l\frac{\partial f}{\partial x} + m\frac{\partial f}{\partial y} + n\frac{\partial f}{\partial z} \tag{30}$$

the directional derivative and its properties

is the **directional derivative** of f in the direction $\mathbf{v}$. This last result has meaning irrespective of whether $\mathbf{v}$ is tangent to a space curve, so from now on $\mathbf{v}$ can be taken to be an arbitrary unit vector in space.

The directional derivative $D_v f$ can be interpreted in terms of the ordinary operation of differentiation by considering Fig. 11.8. In the diagram, a straight line T in space in the direction of a given vector $\mathbf{v}$ passes through a fixed point P, and Q is a general point on line T at a distance s from P. The directional derivative $D_v f$ is then given by

$$D_v f = \frac{df}{dv} = \lim_{s\to 0} \frac{f(Q) - f(P)}{s}. \tag{31}$$

In the two-dimensional case in the (x, y)-plane, the directional derivative defined in (30) simplifies to

$$D_v f = \mathbf{v} \cdot \operatorname{grad} f = l\frac{\partial f}{\partial x} + m\frac{\partial f}{\partial y}, \tag{32}$$

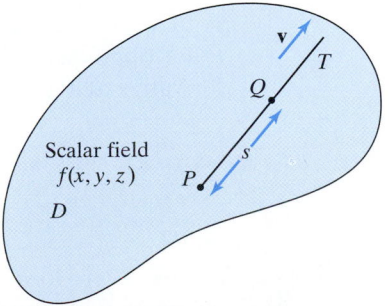

FIGURE 11.8 The directional derivative $D_v f$.

where now the unit vector $\mathbf{v} = l\mathbf{i} + m\mathbf{j}$, with $l^2 + m^2 = 1$, and the grad f in (22) simplifies to

$$\text{grad } f = \frac{\partial f}{\partial x}\mathbf{i} + \frac{\partial f}{\partial y}\mathbf{j}, \tag{33}$$

where again the unit vector $\mathbf{v} = l\mathbf{i} + m\mathbf{j}$, with $l^2 + m^2 = 1$.

EXAMPLE 11.13

Find the directional derivative of $f = x^2 + 3y^2 + 2z^2$ in the direction of the vector $2\mathbf{i} - \mathbf{j} - 2\mathbf{k}$, and determine its value at the point $(1, -3, 2)$.

Solution   grad $f = 2x\mathbf{i} + 6y\mathbf{j} + 4z\mathbf{k}$ and the unit vector in the required direction is $v = \frac{2}{3}\mathbf{i} - \frac{1}{3}\mathbf{j} - \frac{2}{3}\mathbf{k}$, and so the required directional derivative is

$$D_v f = \left(\frac{2}{3}\mathbf{i} - \frac{1}{3}\mathbf{j} - \frac{2}{3}\mathbf{k}\right) \cdot (2x\mathbf{i} + 6y\mathbf{j} + 4z\mathbf{k}),$$

and so

$$D_v f = \frac{4}{3}x - 2y - \frac{8}{3}z.$$

This shows that the directional derivative $D_v f$ at the point $(1, -3, 2)$ is

$$D_v f(1, -3, 2) = \frac{4}{3} + 6 - \frac{16}{3} = 2. \qquad \blacksquare$$

Inspection of definition (30) shows immediately that $D_v f$, which is the rate of change of f in the direction $\mathbf{v}$, must take its greatest value when $\mathbf{v}$ is in the direction of grad f, its smallest value when $\mathbf{v}$ and grad f are oppositely directed, and the value zero when $\mathbf{v}$ and grad f are orthogonal. These simple properties of a directional derivative are sufficiently important for them to be recorded separately in the following form.

Properties of directional derivatives

1. The most rapid increase of a differentiable function $f(x, y, z)$ at a point P in space occurs in the direction of the vector $\mathbf{v}_P = \text{grad } f(P)$. The directional derivative at P is then given by

$$D_v f(P) = |\text{grad } f(P)| = \left((\partial f/\partial x)^2_P + (\partial f/\partial y)^2_P + (\partial f/\partial z)^2_P\right)^{1/2}.$$

2. The most rapid decrease of a differentiable function $f(x, y, z)$ at a point P in space occurs when the vector $\mathbf{v}_P$ just defined in 1 and grad f are oppositely directed, so that $\mathbf{v}_P = -\text{grad } f(P)$. The directional derivative at P is then the *negative* of the result in 1 and so is given by

$$D_v f(P) = -|\text{grad } f(P)|$$
$$= -\left((\partial f/\partial x)^2_P + (\partial f/\partial y)^2_P + (\partial f/\partial z)^2_P\right)^{1/2}.$$

3. There is a zero local rate of change of a differentiable function $f(x, y, z)$ at a point P in space in the direction of any vector $\mathbf{v}_P$ that is orthogonal to grad f at P, so that $\mathbf{v}_P \cdot \text{grad } f(P) = 0$.

When a scalar function f defined over a region D of space is suitably differentiable, the vector-valued function grad f defines a *vector field* over D in terms of the *scalar field* defined by f. The next theorem establishes the result of performing the gradient operation on combinations of scalar functions.

THEOREM 11.2

properties of the gradient operator

Rules for the gradient operator   Let the gradients of f and g be defined over a region D. Then the gradient operator has the following properties.

(i) Gradient of a constant multiple of f:

$$\text{grad}\,(cf) = c\,\text{grad}\,f; \qquad (c \text{ a scalar constant})$$

(ii) Gradient of a sum or difference of functions:

$$\text{grad}\,(f \pm g) = \text{grad}\,f \pm \text{grad}\,g;$$

(iii) Gradient of a product of functions:

$$\text{grad}\,(fg) = f\,\text{grad}\,g + g\,\text{grad}\,f;$$

(iv) Gradient of a quotient of functions:

$$\text{grad}\left(\frac{f}{g}\right) = (g\,\text{grad}\,f - f\,\text{grad}\,g)/g^2 \qquad (g \neq 0).$$

Proof   These results all follow by applying the usual rules for partial differentiation to each component of the gradient function on the left, and then recombining the results to obtain the expression on the right. To illustrate the form of argument involved, we prove result (iii) concerning the gradient of a product of functions. By definition,

$$\text{grad}(fg) = \frac{\partial(fg)}{\partial x}\mathbf{i} + \frac{\partial(fg)}{\partial y}\mathbf{j} + \frac{\partial(fg)}{\partial z}\mathbf{k}$$

$$= \left(f\frac{\partial g}{\partial x} + g\frac{\partial f}{\partial x}\right)\mathbf{i} + \left(f\frac{\partial g}{\partial y} + g\frac{\partial f}{\partial y}\right)\mathbf{j} + \left(f\frac{\partial g}{\partial z} + g\frac{\partial f}{\partial z}\right)\mathbf{k}$$

$$= f\,\text{grad}\,g + g\,\text{grad}\,f. \qquad \blacksquare$$

A simple application of the gradient of a function involves the determination of the tangent plane to the surface S defined by the function $f(x, y, z) = \text{constant}$ at a point $P_0(x_0, y_0, z_0)$ on the surface S.

Define the function $w = f(x, y, z) - c$, where $c = \text{constant}$, so that the surface S then has the equation $w = 0$. Let any space curve Γ in the surface S have the parametric equations

$$x = x(t), \quad y = y(t), \quad \text{and} \quad z = z(t).$$

Then differentiation of $w = f(x, y, z) - c$ with respect to t gives

$$\frac{dw}{dt} = \frac{\partial f}{\partial x}\frac{dx}{dt} + \frac{\partial f}{\partial y}\frac{dy}{dt} + \frac{\partial f}{\partial z}\frac{dz}{dt},$$

but on S the function $w \equiv 0$, so this reduces to

$$\frac{\partial f}{\partial x}\frac{dx}{dt} + \frac{\partial f}{\partial y}\frac{dy}{dt} + \frac{\partial f}{\partial z}\frac{dz}{dt} = 0.$$

This result shows that any curve Γ in S must be orthogonal to grad f, and so at every point P of the surface S the vector grad f is normal to the surface. The vector equation of a plane with normal $\mathbf{n}$ containing the point P_0 with position vector $\mathbf{r}_0$ is

$$(\mathbf{r} - \mathbf{r}_0) \cdot \mathbf{n} = 0,$$

where $\mathbf{r}$ is the position vector of an arbitrary point on the plane. If we set $\mathbf{r} = x\mathbf{i} + y\mathbf{j} + z\mathbf{k}$ and $\mathbf{r}_0 = x_0\mathbf{i} + y_0\mathbf{j} + z_0\mathbf{k}$, and identify $\mathbf{n}$ with grad f at P_0, where

$$\text{grad } f(P_0) = \left(\frac{\partial f}{\partial x}\right)_{P_0} \mathbf{i} + \left(\frac{\partial f}{\partial y}\right)_{P_0} \mathbf{j} + \left(\frac{\partial f}{\partial z}\right)_{P_0} \mathbf{k},$$

the required tangent plane to the surface at $P_0(x_0, y_0, z_0)$ is seen to be given by

$$(x - x_0)\left(\frac{\partial f}{\partial x}\right)_{P_0} + (y - y_0)\left(\frac{\partial f}{\partial y}\right)_{P_0} + (z - z_0)\left(\frac{\partial f}{\partial z}\right)_{P_0} = 0 \tag{34}$$

EXAMPLE 11.14   Find the tangent plane at the point $(2, -1, 3)$ on the sphere

$$(x - 1)^2 + (y + 2)^2 + (z - 4)^2 = 3.$$

Solution   It is first necessary to check that the point $(2, -1, 3)$ does actually lie on the sphere, and this is confirmed by showing that $x = 2$, $y = -1$, and $z = 3$ satisfies the equation of the sphere. Writing $f = (x - 1)^2 + (y + 2)^2 + (z - 4)^2$, we find that $\partial f/\partial x = 2x$, $\partial f/\partial y = 2y$, and $\partial f/\partial z = 2z$, so that $(\partial f/\partial x)_{(2,-1,3)} = 4$, $(\partial f/\partial y)_{(2,-1,3)} = -2$, and $(\partial f/\partial z)_{(2,-1,3)} = 6$. Substitution into (34) shows that the equation of the tangent plane to the sphere at the point $(2, -1, 3)$ is

$$4(x - 2) - 2(y + 1) + 6(z - 3) = 0,$$

and after simplification this reduces to

$$4x - 2y + 6z = 28. \quad\blacksquare$$

In applications, the geometry of a problem often makes it necessary to express the gradient operator in terms of different coordinate systems. The coordinate systems that occur most frequently as a result of formulating problems involving either a cylindrical or a spherical geometry are the cylindrical polar coordinate system (r, θ, z) illustrated in Fig. 11.9a and the spherical polar coordinate system (r, θ, ϕ) illustrated in Fig. 11.9b, and shown in a different form in Fig. 1.15.

Consideration of the geometry of Figs. 11.9a,b establishes that the connection between these coordinate systems and the cartesian coordinates (x, y, z) is given by:

Cylindrical polar coordinates (r, θ, z)

$$x = r\cos\theta, \quad y = r\sin\theta, \quad z = z \tag{35}$$

Spherical polar coordinates (r, θ, φ)

$$x = r\sin\theta\cos\phi, \quad y = r\sin\theta\sin\phi, \quad z = r\cos\theta. \tag{36}$$

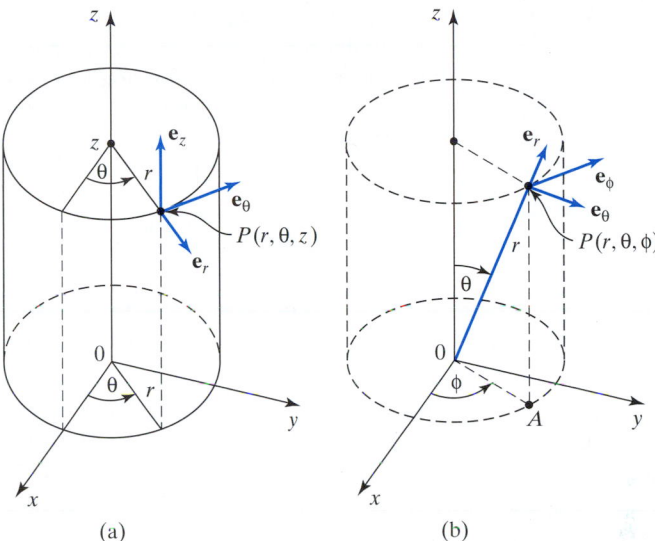

FIGURE 11.9   (a) Cylindrical polar coordinates. (b) Spherical polar coordinates.

The forms taken by grad f in cylindrical and spherical polar coordinates are given next for reference, though the derivation of these results together with related results in terms of general orthogonal curvilinear coordinates will be postponed until Section 11.6.

gradient operator in cylindrical polar coordinates

grad f in cylindrical polar coordinates (r, θ, z)

$$\text{grad } f = \nabla f = \frac{\partial f}{\partial r}\mathbf{e}_r + \frac{1}{r}\frac{\partial f}{\partial \theta}\mathbf{e}_\theta + \frac{\partial f}{\partial z}\mathbf{e}_z, \tag{37}$$

where $\mathbf{e}_r$ is a unit vector parellel to the (x, y)-plane along the radial line r, $\mathbf{e}_\theta$ is a unit vector in the (x, y)-plane normal to $\mathbf{e}_r$ in the direction of increasing θ, and $\mathbf{e}_z$ is a unit vector in the positive z-direction as shown in Fig. 11.9a, so that $\mathbf{e}_r \times \mathbf{e}_\theta = \mathbf{e}_z$.

grad f in spherical polar coordinates (r, θ, φ)

$$\text{grad } f = \nabla f = \frac{\partial f}{\partial r}\mathbf{e}_r + \frac{1}{r}\frac{\partial f}{\partial \theta}\mathbf{e}_\theta + \frac{1}{r\sin\theta}\frac{\partial f}{\partial \phi}\mathbf{e}_\phi, \tag{38}$$

where $\mathbf{e}_r$ is a unit vector along the radial line r, $\mathbf{e}_\theta$ is a unit vector in the direction of increasing θ, and $\mathbf{e}_\phi$ is a unit vector in the direction of increasing ϕ that is normal to the plane containing $\mathbf{e}_r$ and $\mathbf{e}_\theta$, as shown in Fig. 11.9b, so that $\mathbf{e}_r \times \mathbf{e}_\theta = \mathbf{e}_\phi$.

Notations for cylindrical and spherical polar coordinates are not uniform, so when consulting other works it is advisable to check the notation and conventions that are in use. This is particularly important in the case of spherical polar coordinates, where the r used here is sometimes replaced by ρ, with r then used to denote the distance OA in Fig. 11.9b; in addition, the symbols θ and ϕ are often interchanged.

Summary

The gradient of a scalar function of position is a vector, and it has been defined and used to define the concept of a directional derivative. The properties of directional derivatives have been established and the gradient operator has been used to determine the tangent plane to a sphere at a given point on its surface. For future use, the gradient operator has been expressed in terms of both cylindrical and spherical polar coordinates.

EXERCISES 11.3

In Exercises 1 through 8 find the derivative of the scalar function f in the direction of the vector v and find its value at the point P.

1. $f = x \sin y + y \cos x$, with $v = \mathbf{i} + 2\mathbf{j}$ and P the point $(\pi/4, 0)$.

2. $f = x \sinh(x + 2y)$, with $v = 3\mathbf{i} - \mathbf{j}$ and P the point $(1, -2)$.

3. $f = xe^{xy} + 2x - y$, with $v = \mathbf{i} + 4\mathbf{j}$ and P the point $(-2, 1)$.

4. $f = \ln(x + 2y^2)$, with $v = -\mathbf{i} + 2\mathbf{j}$ and P the point $(1, 3)$.

5. $f = \sin(xy) + e^{3xz}$, with $v = \mathbf{i} - 2\mathbf{j} + 2\mathbf{k}$ and P the point $(1, \pi/4, 1)$.

6. $f = (x^2y + z)^{1/2}$, with $v = \mathbf{i} + 3\mathbf{j} - 3\mathbf{k}$ and P the point $(2, -3, 1)$.

7. $f = \sinh(xy^2z + 3y)$, with $v = 2\mathbf{i} + \mathbf{k}$ and P the point $(1, -2, 2)$.

8. $f = (xz^2 + 3y)^{-1}$, with $v = -3\mathbf{i} + 2\mathbf{j} - 2\mathbf{k}$ and P the point $(1, -1, 1)$.

9. Prove result (iv) in Theorem 11.2.

10. Use result (iv) in Theorem 11.2 to find grad (f/g) given that $f = ye^{xy} + z$ and $g = xyz^2 + 1$, and confirm the result by direct calculation.

In Exercises 11 through 14 find grad f and evaluate it at the point P.

11. $f = x^2 + 3xyz - yz^2$, with P the point $(1, 3, -1)$.

12. $f = (x^2 + 2y^2 + 4z^2)^{-1}$, with P the point $(1, 2, 1)$.

13. $f = \exp(xy + 2yz - 3xz)$, with P the point $(1, 0, 2)$.

14. $f = (x^2 + yz + 3z^2)^{1/2}$, with P the point $(1, -1, 2)$.

15. Derive the cartesian form of the equation of the straight line that is normal to the curve $f(x, y) = \text{constant}$ at a point (x_0, y_0) on the curve.

16. Derive the cartesian form of the equation of the tangent line to the curve $f(x, y) = \text{constant}$ at a point (x_0, y_0) on the curve.

17. Find the equation of the tangent plane to the surface $x^3 + 3xy + z^2 = 11$ at the point on the surface $(1, 2, 2)$.

18. Find the equation of the tangent plane to the surface $\sin(xy) + 2\cos(yz) + 3x = 4$ at the point on the surface $(1, \pi/2, 1)$.

19. Derive the vector equation of the straight line that is normal to the surface $f(x, y, z) = \text{constant}$ at a point with position vector $\mathbf{r}_0$ on the surface.

20. If two surfaces $f(x, y, z) = \text{constant}$ and $g(x, y, z) = \text{Constant}$ intersect at a point with position vector $\mathbf{r}_0$, find a vector that is tangent to their curve of intersection of the two surfaces at $\mathbf{r}_0$.

21. Find grad f, given that $f(r, \theta, z) = r^2 \sin\theta + rz^2 + 1$.

22. Find grad f, given that $f(r, \phi, \theta) = r \sin\theta \cos\phi + \sin^2\phi$.

23. If $\mathbf{F} = \text{grad } f$, prove that

$$\text{grad}(f^n) = nf^{n-1}\mathbf{F}.$$

Use the result to show that when $f = r$ is the distance of a point $\mathbf{r} = x\mathbf{i} + y\mathbf{j} + z\mathbf{k}$ from the origin, then

$$\text{grad } r = \hat{\mathbf{r}} \quad \text{and} \quad \text{grad}\left(\frac{1}{r}\right) = -\frac{\mathbf{r}}{r^3},$$

where $\hat{\mathbf{r}}$ is the unit vector in the direction of $\mathbf{r}$, so $\hat{\mathbf{r}} = \mathbf{r}/r$.

11.4   Conservative Fields and Potential Functions

conservative fields and path invariance

Let us reconsider the line integral $\int_{\mathbf{r}_1}^{\mathbf{r}_2} \mathbf{F} \cdot d\mathbf{r}$ along a path Γ joining the two points $\mathbf{r}_1$ and $\mathbf{r}_2$ in a region D of space. If the value of this line integral is *independent* of the choice of path Γ in D, the vector field $\mathbf{F}$ is called a **conservative field**. The name *conservative* comes from mechanics, where it refers to the study of dynamics in which dissipative effects such as friction can be ignored, so that the sum of the kinetic and potential energy in a system remains constant (is *conserved*), though conservative fields of different types play key roles throughout physics and engineering.

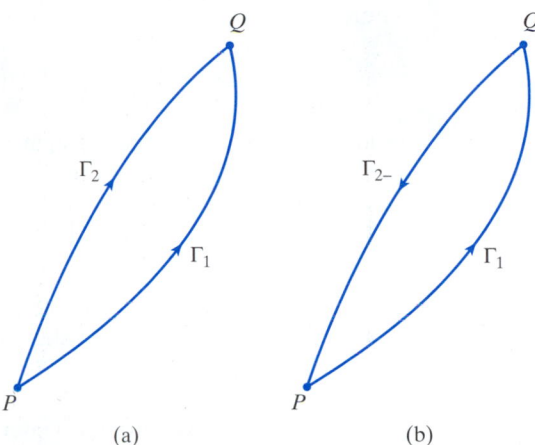

FIGURE 11.10   (a) The two paths Γ_1 and Γ_2. (b) The loop containing P and Q.

The next theorem shows that the definition of a conservative field in terms of the independence of the line integral of the path from $\mathbf{r}_1$ to $\mathbf{r}_2$ is equivalent to the vanishing of the line integral of a conservative field around any closed loop in D.

Path invariance and integrals around loops   If $\mathbf{F}$ is a conservative field in a region D, then $\oint_{\Gamma} \mathbf{F} \cdot d\mathbf{r} = 0$ around every closed loop Γ in D and, conversely, if $\oint_{\Gamma} \mathbf{F} \cdot d\mathbf{r} = 0$ around every closed loop Γ in region D, then $\mathbf{F}$ is a conservative field in D.

Proof   The proof of this result is straightforward, and it involves two steps. One is to show that if $\oint_{\Gamma} \mathbf{F} \cdot d\mathbf{r} = 0$ around every closed loop Γ in D, then the field is conservative, and the other involves showing that the converse result is true.

STEP 1   Let the points P and Q shown in Fig. 11.10(a) be any two points in a region D throughout which $\mathbf{F}$ is a conservative field, and let Γ_1 and Γ_2 be any two paths in D connecting P to Q.

As $\mathbf{F}$ is a conservative field, by definition

$$\int_{\Gamma_1} \mathbf{F} \cdot d\mathbf{r} = \int_{\Gamma_2} \mathbf{F} \cdot d\mathbf{r} \quad \text{and so} \quad \int_{\Gamma_1} \mathbf{F} \cdot d\mathbf{r} - \int_{\Gamma_2} \mathbf{F} \cdot d\mathbf{r} = 0.$$

If we reverse the direction of integration in the second integral, thereby changing its sign, and indicate the path from Q to P by Γ_{2-}, this last result becomes

$$\int_{\Gamma_1} \mathbf{F} \cdot d\mathbf{r} + \int_{\Gamma_{2-}} \mathbf{F} \cdot d\mathbf{r} = 0.$$

However, the reversal of direction of integration on path Γ_2 makes the successive paths Γ_1 and Γ_{2-} into the loop in D shown in Fig. 11.10(b). So as P and Q were any two points in D, and Γ_1 and Γ_2 were any two paths in D joining P and Q; this proves the first part of the theorem.

STEP 2   We must now prove the converse result, that if $\oint_{\Gamma} \mathbf{F} \cdot d\mathbf{r} = 0$ around every closed loop Γ in region D, then the field $\mathbf{F}$ is conservative in D. The proof involves reversing the argument used in Step 1. Let the arbitrary paths Γ_1 and Γ_{2-} in Fig. 11.10(b) form any loop in D, and let P and Q be any two points on the loop.

Then

$$\int_{\Gamma_1} \mathbf{F} \cdot d\mathbf{r} + \int_{\Gamma_{2-}} \mathbf{F} \cdot d\mathbf{r} = 0,$$

but if we reverse the direction of integration along Γ_{2-}, and compensate by reversing the sign of the integral, this becomes

$$\int_{\Gamma_1} \mathbf{F} \cdot d\mathbf{r} = \int_{\Gamma_2} \mathbf{F} \cdot d\mathbf{r}.$$

As P and Q were arbitrary points, and Γ_1 and Γ_2 are any two paths joining these points, we have succeeded in showing that the integral is path independent, so the theorem is proved. ∎

Let f be a differentiable scalar function defined over a region D and let $\mathbf{F} =$ grad f be a vector field defined in terms of f. Then f is called the **potential function** for the vector field $\mathbf{F}$. The connection between potential functions and conservative fields will become clear later.

Let us now show that if a vector field $\mathbf{F}$ has a potential function f, then the function f is unique to within an arbitrary additive constant. The proof is simple. Suppose the scalar fields f and g have the same gradient in some region D, so we can write

$$\text{grad}(f - g) \equiv 0.$$

Then if $\mathbf{v} \neq \mathbf{0}$ is an arbitrary vector in D, it follows from the preceding result that $\mathbf{v} \cdot \text{grad}(f - g) = 0$. This shows that the directional derivative of $f - g$ is equal to zero in every direction at each point of D, and this in turn implies that $f - g =$ constant, so the result is proved.

We now establish the fundamental connection between $\mathbf{F} = \text{grad } f$ and the line integral of $\mathbf{F}$ along any path Γ joining two points in a region D of space. In order to achieve this it is necessary to place some restrictions on the scalar potential function $f(x, y, z)$, the path Γ, and the region D. The function f will be assumed to have continuous first order partial derivatives in D, the path Γ in D must be continuous and piecewise smooth and comprise finitely many segments, and the region D must be open and simply connected.

The terms *open* and *simply connected* need explanation. In straightforward terms, a **simply connected** region in space can be regarded as any region that can be continuously deformed into a sphere inside of which no voids, curves, or points are missing, so it has the property that every loop in the region can be shrunk to a point that belongs to the region, without any part of the loop ever leaving the region. To understand this, consider the case of a region in space from which the points on a line are missing, and let the the loop encircle the line. Then there is no way the loop can be shrunk to a point without leaving the region, so the region is *not*

simply and multiply connected regions

simply connected (it is **multiply connected**). A region in space will be **open** if only the points on the surface of the region (its **boundary points**) are missing. A region in space is **connected** if every point in the region can be joined to every other point in the region by a piecewise continuous line that lies entirely within the region.

For example, the points between two concentric spheres, the points on the surface of each of which are missing, form an *open* region that is *connected*. The region is open because its boundary points are not included in the region, and it is connected because any two points in the region can always be joined by a space curve that lies inside the region.

As another example, consider the points inside two adjacent nonintersecting spheres, each of which is connected within itself. Then the region formed by the points inside the two spheres is *not* connected, because every path joining a point in one sphere to a point in the other sphere contains points that belong to neither sphere.

THEOREM 11.4

Condition for the path independence of a line integral   Let $\mathbf{F}$ be a vector field defined in an open connected region D of space, and let Γ be any path in D connecting two arbitrary points P at $\mathbf{r}_1$ and Q at $\mathbf{r}_2$ in D. Then:

(i) If the line integral $\oint_\Gamma \mathbf{F} \cdot d\mathbf{r}$ is independent of the path Γ joining $\mathbf{r}_1$ to $\mathbf{r}_2$, a scalar field f exists such that $\mathbf{F} = \operatorname{grad} f$.

(ii) If $\mathbf{F} = \operatorname{grad} f$ with $\mathbf{F} = F_1\mathbf{i} + F_2\mathbf{j} + F_3\mathbf{k}$ and $\mathbf{r}(t) = x(t)\mathbf{i} + y(t)\mathbf{j} + z(t)\mathbf{k}$, then

$$\oint_\Gamma \mathbf{F} \cdot d\mathbf{r} = \int_P^Q (F_1\,dx + F_2\,dy + F_3\,dz) = f(Q) - f(P).$$

a condition that ensures path invariance

Proof   Although not difficult, the proof of result (i) is a little harder than that of result (ii). To prove (i) it is necessary to show that if P and Q are any two points in an open connected region D, and the integral $f = \int_P^Q \mathbf{F} \cdot d\mathbf{r}$ is independent of the path Γ joining P to Q, then $\mathbf{F} = \operatorname{grad} f$.

Let P be an arbitrary point in D with coordinates (x_0, y_0, z_0), and Q be a point with coordinates (x, y_0, z_0), so that P and Q only differ in their x coordinates. By hypothesis f is independent of the path Γ from P to Q, so we can take it to be a straight line on which the general point can be written $\mathbf{r} = t\mathbf{i} + y_0\mathbf{j} + z_0\mathbf{k}$ for $x_0 \leq t \leq x_1$. Let $P(x)$ be any point on Γ corresponding to $\mathbf{r} = x\mathbf{i} + y_0\mathbf{j} + z_0\mathbf{k}$, so $d\mathbf{r}/dt = \mathbf{i}$, and denote by $f(x)$ the integral

$$f(x) = \int_{x_0}^x \mathbf{F} \cdot d\mathbf{r}.$$

Then, setting $\mathbf{F} = F_1\mathbf{i} + F_2\mathbf{j} + F_3\mathbf{k}$, on path Γ we can write

$$f(x) = \int_{x_0}^x \mathbf{F} \cdot \left(\frac{d\mathbf{r}}{dt}\right) dt = \int_{x_0}^x F_1(t, y_0, z_0)dt,$$

and so

$$f(x + h) - f(x) = \int_{x_0}^{x+h} F_1(t, y_0, z_0)dt - \int_{x_0}^x F_1(t, y_0, z_0)dt$$

$$= \int_x^{x+h} \mathbf{F}_1(t, y_0, z_0)dt.$$

Applying the mean value theorem for integrals (see Theorem 1.4) to the integral on the right shows that

$$f(x + h) - f(x) = hF_1(\xi, y_0, z_0),$$

where the unknown number ξ is such that $x < \xi < x + h$. The preceding expression can be rewritten in the form

$$\frac{f(x + h) - f(x)}{h} = F_1(\xi, y_0, z_0),$$

and by proceeding to the limit as $h \to 0$, when $\xi \to x$, the expression on the left reduces to $\partial f / \partial x$, because f is a function of x, y, and z, but $y = y_0$ and $z = z_0$ remain constant during the limiting process. As P was an arbitrary point in D, it follows that y_0 and z_0 are arbitrary, so we have shown that $\partial f / \partial x = F_1$. Similar arguments in which first Q is taken to be the point (x_0, y, z_0), and then to be the point (x_0, y_0, z), show that $\partial f / \partial y = F_2$ and $\partial f / \partial z = F_3$. Combining these results gives $\mathbf{F} = \text{grad } f$, and the proof of (i) is complete.

To prove (ii), let the smooth path Γ joining any two points P and Q in D have the equation $\mathbf{r} = x(t)\mathbf{i} + y(t)\mathbf{j} + z(t)\mathbf{k}$ for $a \le t \le b$. Then along Γ

$$\frac{df}{dt} = \frac{\partial f}{\partial x}\frac{dx}{dt} + \frac{\partial f}{\partial y}\frac{dy}{dt} + \frac{\partial f}{\partial z}\frac{dz}{dt}$$

$$= \text{grad} f \cdot \left(\frac{d\mathbf{r}}{dt}\right) = \mathbf{F} \cdot \left(\frac{d\mathbf{r}}{dt}\right),$$

and so

$$\int_\Gamma \mathbf{F} \cdot d\mathbf{r} = \int_a^b \mathbf{F} \cdot \left(\frac{d\mathbf{r}}{dt}\right) dt = \int_a^b \left(\frac{df}{dt}\right) dt = f(Q) - f(P),$$

and the result is proved.   ∎

To make effective use of Theorem 11.4 (ii) it is necessary to know when $\mathbf{F}$ is the gradient of a scalar function f. Theorem 11.5, which follows, provides both a test for a conservative field and a way of finding its associated potential function f.

THEOREM 11.5

a test for a conservative field

Testing for a conservative field and finding the potential function   The vector field $\mathbf{F} = F_1\mathbf{i} + F_2\mathbf{j} + F_3\mathbf{k}$ with components that are continuous and differentiable is a conservative field, and so is derivable from a scalar potential f, if

(i) $$\frac{\partial F_1}{\partial y} = \frac{\partial F_2}{\partial x}, \quad \frac{\partial F_2}{\partial z} = \frac{\partial F_3}{\partial y}, \quad \frac{\partial F_3}{\partial x} = \frac{\partial F_1}{\partial z}.$$

When $\mathbf{F}$ is a conservative field the scalar potential function f is found by integrating the equations

(ii) $$\frac{\partial f}{\partial x} = F_1, \quad \frac{\partial f}{\partial y} = F_2, \quad \frac{\partial f}{\partial z} = F_3.$$

Proof   If $\mathbf{F}$ is a conservative field, then a scalar potential f exists such that $\mathbf{F} = \text{grad } f$, and so

$$F_1\mathbf{i} + F_2\mathbf{j} + F_3\mathbf{k} = \frac{\partial f}{\partial x}\mathbf{i} + \frac{\partial f}{\partial y}\mathbf{j} + \frac{\partial f}{\partial z}\mathbf{k}.$$

Equating corresponding components gives

$$\frac{\partial f}{\partial x} = F_1, \quad \frac{\partial f}{\partial y} = F_2, \quad \frac{\partial f}{\partial z} = F_3.$$

As, by hypothesis, the components of $\mathbf{F}$ are differentiable, the equality of mixed derivatives requires that

$$\frac{\partial}{\partial y}\left(\frac{\partial f}{\partial x}\right) = \frac{\partial F_1}{\partial y} = \frac{\partial}{\partial x}\left(\frac{\partial f}{\partial y}\right) = \frac{\partial F_2}{\partial x},$$

so we have established the first result in (i). The other two results are obtained in similar fashion by equating the other two mixed derivatives, so the first part of the theorem is proved. When $\mathbf{F}$ is a conservative field the scalar potential f follows by integrating the equations in (ii), and the proof of the theorem is complete.   ∎

EXAMPLE 11.15

Show that $\mathbf{F} = y^2 z \mathbf{i} + 2xyz \mathbf{j} + (2z + xy^2)\mathbf{k}$ is a conservative field in any open connected region of space, and find the associated scalar potential f. Use the result to evaluate the line integral $I = \int_P^Q \mathbf{F} \cdot d\mathbf{r}$, where P is the point $(2, 1, 1)$ and Q is the point $(3, 2, 2)$.

Solution   In the notation of Theorem 11.5 the components of $\mathbf{F}$ are $F_1 = y^2 z$, $F_2 = 2xyz$, and $F_3 = 2z + xy^2$, and a routine calculation confirms that

$$\frac{\partial F_1}{\partial y} = \frac{\partial F_2}{\partial x}, \quad \frac{\partial F_2}{\partial z} = \frac{\partial F_3}{\partial y}, \quad \frac{\partial F_3}{\partial x} = \frac{\partial F_1}{\partial z},$$

in any region of space, so the $\mathbf{F}$ is a conservative field.

To find the scalar potential f we must integrate

$$\frac{\partial f}{\partial x} = y^2 z, \quad \frac{\partial f}{\partial y} = 2xyz, \quad \frac{\partial f}{\partial z} = 2z + xy^2.$$

Integrating the first equation with respect to x, while regarding y and z as constants, gives

$$f = xy^2 z + r(y, z),$$

where $r(y, z)$ is an arbitrary function of y and z. Combining this result with the expression for $\partial f/\partial y$ given earlier, we find that

$$\frac{\partial f}{\partial y} = 2xyz + \frac{\partial r}{\partial y} = 2xyz \quad \text{and so} \quad \frac{\partial r}{\partial y} = 0,$$

from which it follows that $r = s(z)$, with $s(z)$ an arbitrary function of z. Finally, using this result with the expression for $\partial f/\partial z$ given earlier we find that

$$\frac{\partial f}{\partial z} = xy^2 + \frac{ds}{dz} = 2z + xy^2 \quad \text{and so} \quad \frac{ds}{dz} = 2z,$$

from which it follows that $s(z) = z^2 + c$, where c is an arbitrary constant.

Combining results shows that the most general scalar potential function f associated with $\mathbf{F}$ is

$$f = xy^2 z + z^2 + c.$$

As $\mathbf{F}$ is a conservative field, the line integral between any two points in an open connected region D can be evaluated using result (ii) of Theorem 11.4. However, the arbitrary constant c in f can be omitted when evaluating a line integral using the result

$$\int_P^Q \mathbf{F} \cdot d\mathbf{r} = \int_P^Q df = f(Q) - f(P),$$

because c occurs in both $f(Q)$ and $f(P)$, and so cancels. As a result, setting $f = xy^2 z + z^2$ and using the notation $(xy^2 z + z^2)_{(p,q,r)}$ to denote $xy^2 z + z^2$ evaluated with $x = p$, $y = q$, and $z = r$, we find that

$$I = \int_P^Q \mathbf{F} \cdot d\mathbf{r} = (xy^2z + z^2)_{(3,2,2)} - (xy^2z + z^2)_{(2,1,1)}$$

$$= 28 - 3 = 25.$$ ∎

The example that follows shows the necessity of the condition in Theorem 11.4 that the region D is *simply connected*, because if this is not the case, a line integral between two arbitrary points P and Q in D will *not* be independent of the path joining them.

EXAMPLE 11.16

Show that the two-dimensional vector field $\mathbf{F} = (\frac{-y}{x^2+y^2})\mathbf{i} + (\frac{x}{x^2+y^2})\mathbf{j}$ satisfies the conditions of Theorem 11.5 (i) in any region of space that does not contain the origin. Evaluate the integral $I = \int_\Gamma \mathbf{F} \cdot d\mathbf{r}$ when (a) Γ is the circle $x^2 + y^2 = 2$ and and (b) Γ is the square with corners P at $(1, -1)$, Q at $(3, -1)$, R at $(3, 1)$, and S at $(1, 1)$, and comment on the results.

Solution The vector $\mathbf{F}$ is indeterminate at the origin, but is defined elsewhere in the plane, where it satisfies the condition

$$\frac{\partial}{\partial y}\left(\frac{-y}{x^2 + y^2}\right) = \frac{\partial}{\partial x}\left(\frac{x}{x^2 + y^2}\right).$$

This shows that $\mathbf{F}$ satisfies the two-dimensional form of Theorem 11.5 (i) in any region of the plane that does not include the origin. When the origin is excluded from the plane, vector $\mathbf{F}$ is seen to be defined in a *nonsimply connected* region.

The circle $x^2 + y^2 = 2$ and the square with its corners at $PQRS$ are shown in Fig. 11.11, from which it can be seen that the points P and S are common, so both the circle and the square represent loops in the plane containing the points P and S. The circle encloses the origin, so the points in its interior are not simply connected, while the square excludes the origin, so the points in its interior are simply connected.

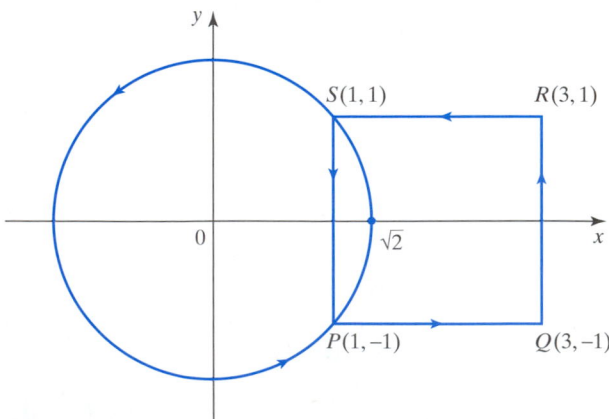

FIGURE 11.11 Two loops, each containing points P and S, in a nonsimply connected region.

Setting $x = \sqrt{2}\cos t, y = \sqrt{2}\sin t$ for $0 \le t \le 2\pi$ and evaluating the line integral I in case (a) gives

$$I = \int_\Gamma \left(\frac{-y}{x^2 + y^2}dx + \frac{x}{x^2 + y^2}dy \right) = 2\pi.$$

In case (b) we have

$$\int_P^Q \mathbf{F} \cdot d\mathbf{r} = \int_1^3 \frac{dx}{x^2 + 1}, \quad \int_Q^R \mathbf{F} \cdot d\mathbf{r} = 3\int_{-1}^1 \frac{dy}{y^2 + 9}, \quad \int_R^S \mathbf{F} \cdot d\mathbf{r} = -\int_3^1 \frac{dx}{x^2 + 1}$$

and

$$\int_S^P \mathbf{F} \cdot d\mathbf{r} = \int_1^{-1} \frac{dy}{y^2 + 1}.$$

Evaluating these integrals and adding the results shows, as expected, that in case (b) the integral $I = 0$.

These results could be used to illustrate that when a region is not simply connected, the line integral between two points (in this case P and S) of a vector $\mathbf{F}$ that satisfies the conditions of Theorem 11.5 (i) will, in general, depend on the path joining the points. ■

FURTHER RESULTS

For the sake of completeness the definitions of the terms *open*, *connected*, and *simply connected* are given below in rather more detail, and they are then illustrated diagramatically by considering regions in the plane.

Definitions of open, connected, and simply connected regions

(i) A region D in space is said to be an **open** region if every point P in D can be enclosed in a sphere centered on P whose radius can always be chosen small enough that all points inside the sphere belong to D.

(ii) A region D in space is said to be **connected** if every pair of points in D can be joined by a piecewise smooth path with finitely many segments that lies entirely inside D.

(iii) A region D in space is said to be **simply connected** if every closed non-self-intersecting loop in D can be shrunk to a point in D in such a way that during the process every point on the loop remains in D.

Figure 11.12 illustrates these definitions in the case of two-dimensional regions, where a dashed boundary is used to indicate that the points on the boundary are omitted from the region. In (a), the region D is *open*, because however close P is taken to the dashed line, a circle (the two-dimensional equivalent of the sphere referred to in (i)) can always be drawn around P in such a way that all points in the circle lie in D. In (b) the region D represented by the interior of the two circles is *not* connected, because any line joining a point in one circle to a point in the other contains points that do not belong to either circle. In (c) the region D is *connected*, because any two points can always be joined by a line that lies entirely inside D.

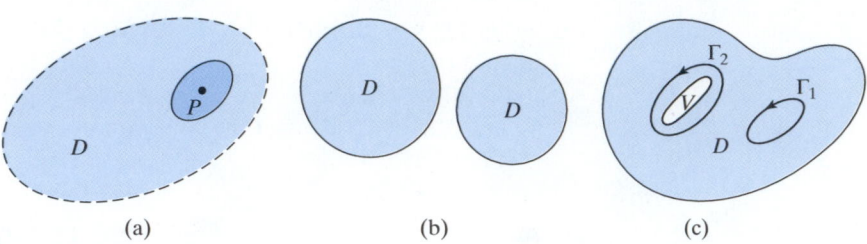

FIGURE 11.12   Regions in the plane illustrating connectivity.

However, in this case the region D is *not* simply connected, because although loop Γ_1 can be contracted to a point in such a way that every point on Γ_1 remains in D, this is not possible in the case of loop Γ_2, which encloses a void V. This last example can be visualized by considering the boundary of the void as a barrier and the loop as an elastic band. In the case of Γ_1 the elastic band can shrink to a point without hindrance, but in the case of Γ_2 this is prevented by the barrier surrounding the void.

Summary

A conservative field is one in which zero work is done when moving around a closed loop in the field and returning to the starting point. Expressed differently, a conservative field is one in which the work done when moving between two separate points is independent of the path followed between the two points. This property of conservative fields has led to this independence of a line integral on the path between two points being called the property of path invariance. The consequences of this definition have been explored and a condition has been found that ensures path invariance. A test for a conservative field has also been given.

EXERCISES 11.4

In Exercises 1 through 6 determine whether **F** is a conservative field, and if so, where.

1. $\mathbf{F} = (3x^2y^2 + yz^2)\mathbf{i} + (2x^3y + xz^2)\mathbf{j} + 2xyz\mathbf{k}$.

2. $\mathbf{F} = y\cos(xy + z^2)\mathbf{i} + x\cos(xy + z^2)\mathbf{j} + 2z\cos(xy + z^2)\mathbf{k}$.

3. $\mathbf{F} = e^x y^2\mathbf{i} + ye^x\mathbf{j} + 3xz\mathbf{k}$.

4. $\mathbf{F} = \dfrac{x}{(x^2 + y^2 + z^2)^{1/2}}\mathbf{i} - \dfrac{y}{(x^2 + y^2 + z^2)^{1/2}}\mathbf{j} + \dfrac{2z}{(x^2 + y^2 + z^2)^{1/2}}\mathbf{k}$.

5. $\mathbf{F} = \dfrac{-2xz}{(x^2 + y^2 + 2z^2)^2}\mathbf{i} + \dfrac{-2yz}{(x^2 + y^2 + 2z^2)^2}\mathbf{j} + \dfrac{x^2 + y^2 - z^2}{(x^2 + y^2 + 2z^2)^2}\mathbf{k}$.

6. $\mathbf{F} = \dfrac{z}{x^2 + y^2 + z^2}\mathbf{i} - \dfrac{y}{x^2 + y^2 + z^2}\mathbf{j} + \dfrac{x}{x^2 + y^2 + z^2}\mathbf{k}$.

In Exercises 7 to 12 show **F** is a conservative field, and by finding the scalar potential f evaluate the integral $I = \int_P^Q \mathbf{F} \cdot d\mathbf{r}$ between the given points P and Q.

7. $\mathbf{F} = (z^3 + 6xy^2)\mathbf{i} + 6x^2y\mathbf{j} + 3xz^2\mathbf{k}$ with P at $(1, 0, 1)$ and Q at $(2, 1, 0)$.

8. $\mathbf{F} = 2xz^2\cosh(x^2 + 2y^2)\mathbf{i} + 4yz^2\cosh(x^2 + 2y^2)\mathbf{j} + 2z\sinh(x^2 + 2y^2)\mathbf{k}$, with P at $(1, 1, 1)$ and Q at $(0, 2, 1)$.

9. $\mathbf{F} = e^{xyz}(1 + xyz)\mathbf{i} + x^2ze^{xyz}\mathbf{j} + x^2ye^{xyz}\mathbf{k}$, with P at $(0, 0, 0)$ and Q at $(1, 1, 2)$.

10. $\mathbf{F} = \dfrac{yz(1 - x^2)}{(1 + x^2)^2}\mathbf{i} + \dfrac{xz}{1 + x^2}\mathbf{j} + \dfrac{xy}{1 + x^2}\mathbf{k}$, with P at $(1, 1, 1)$ and Q at $(2, 2, 0)$.

11. $\mathbf{F} = 2x(1 + yz^2)\mathbf{i} + x^2z^2\mathbf{j} + 2x^2yz\mathbf{k}$, with P at $(3, 1, -1)$ and Q at $(1, 0, 2)$.

12. $\mathbf{F} = 2x(y^2 + z^2)\mathbf{i} + 2y(1 + x^2)\mathbf{j} + 2z(1 + x^2)\mathbf{k}$, with P at $(0, 1, 2)$ and Q at $(2, 0, 1)$.

13. Verify the results of Example 11.15 by performing the indicated integrations along a straight line from P to Q.

11.5   Divergence and Curl of a Vector

It is necessary to introduce two new operations involving vectors. The first operation is called the *divergence* of a vector, and it associates a *scalar function* with a differentiable vector field **F**. The second operation is called the *curl* of a vector, and it associates a *vector function* with the vector **F**. If $\mathbf{F} = F_1\mathbf{i} + F_2\mathbf{j} + F_3\mathbf{k}$ is a differentiable vector field, the **divergence** of **F**, written div **F**, is the scalar function defined in terms of cartesian coordinates as

divergence of a vector

$$\text{div } \mathbf{F} = \frac{\partial F_1}{\partial x} + \frac{\partial F_2}{\partial y} + \frac{\partial F_3}{\partial z}. \tag{39}$$

The divergence of the vector **F** can also be expressed in terms of the operator "del" defined in (25) as

$$\nabla \equiv \mathbf{i}\frac{\partial}{\partial x} + \mathbf{j}\frac{\partial}{\partial y} + \mathbf{k}\frac{\partial}{\partial z},$$

by writing

$$\text{div } \mathbf{F} = \nabla \cdot \mathbf{F} = \left(\mathbf{i}\frac{\partial}{\partial x} + \mathbf{j}\frac{\partial}{\partial y} + \mathbf{k}\frac{\partial}{\partial z}\right) \cdot (F_1\mathbf{i} + F_2\mathbf{j} + F_3\mathbf{k}), \tag{40}$$

where the mutual orthogonality of **i, j**, and **k** coupled with the fact that they are constant vectors causes the expression on the right of (40) to be reduced to the expression on the right of (39), with the operation $\nabla \cdot \mathbf{F}$ being read "del dot **F**." The form taken by div **F** in more general coordinate systems is derived in Section 11.6.

At this stage, for simplicity, the definition of div **F** is expressed in terms of cartesian coordinates, though it will be shown later that div **F** is, in fact, independent of any coordinate system. In the next chapter it will be shown that div **F** can be interpreted as the flux of the normal component of the vector **F** that crosses the surface of a unit volume in a unit time. This means that when div **F** is positive, there is a net *flow* of **F** *out* of the volume, and when div **F** is negative, there is a net *flow* of **F** *into* the volume.

In anticipation of the next chapter, we give a heuristic derivation of div **F** in terms of cartesian coordinates that shows how div **F** can be defined differently, and at the same time illustrates its physical significance. Consider the small cube of side a shown in Fig. 11.13 with faces normal to the coordinate axes, and take the positive direction of the normal to each face of the cube to be the one directed *out* of the cube. The normal component of **F** *entering* face A is $F_2(x, y_0, z)$, and the normal component of **F** *leaving* face B is $F_2(x, y_0 + a, z)$, where from Taylor's theorem for functions of several variables, to first order in a we have $F_2(x, y_0 + a, z) = F_2(x, y_0, z) + a\partial F_2(x, y_0, z)/\partial y$.

Consequently, if we average $F_2(x, y_0, z)$ over face A and denote the result by $\tilde{F}_2$, the integral of $F_2(x, y_0, z)$ over face A is approximately equal to $a^2 \tilde{F}_2$, while the integral over face B is approximately equal to $a^2[\tilde{F}_2 + a\partial \tilde{F}_2/\partial y]$, so the change of the flux of **F** from face A to face B is approximately $a^3\partial \tilde{F}_2/\partial y$. Similar results apply to the other pairs of faces, so denoting the surface of the cube by S, and letting F_n

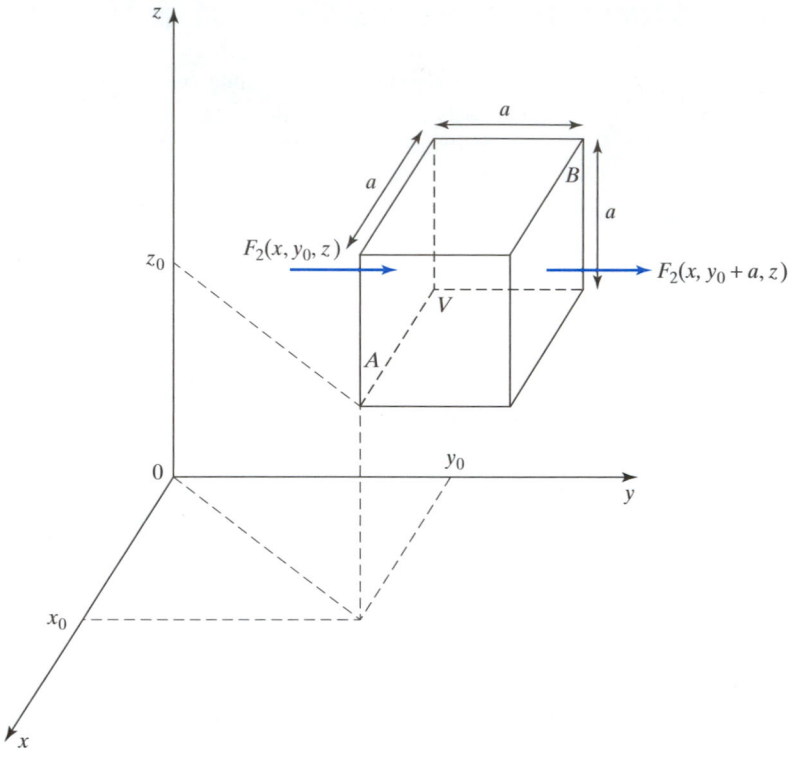

FIGURE 11.13 A representative cubic element.

denote the component of **F** normal to S, positive when outward, with dS a surface element of area of a face, we have

$$\lim_{a\to 0}\frac{1}{a^3}\iint_S F_n\,dS = \lim_{a\to 0}\frac{1}{a^3}\left(a^3\frac{\partial \tilde{F}_1}{\partial x}+a^3\frac{\partial \tilde{F}_2}{\partial y}+a^3\frac{\partial \tilde{F}_3}{\partial z}\right)=\frac{\partial F_1}{\partial x}+\frac{\partial F_2}{\partial y}+\frac{\partial F_3}{\partial z}.$$

The expression on the right is div **F**, so this result shows that the divergence of a vector field **F** in cartesian coordinates is the limit of the flux of the normal component of **F** through the surface S bounding a volume as the volume tends to zero. A different form of argument used in the next chapter will show that for *any* volume V with surface S and element of surface area dS, independently of any coordinate system

a different interpretation of div F

$$\text{div } \mathbf{F} = \lim_{V\to 0}\frac{1}{V}\iint_S F_n\,dS.$$

It is helpful to interpret this result in terms of the flow of a liquid. If we identify **q** with the liquid velocity vector, V with the volume occupied by the liquid, and S with the surface enclosing V, the product $q_n\,dS$, with q_n the component of **q** normal to dS, is seen to be the volume of liquid crossing the surface element dS in a unit time. Consequently, $\iint_S F_n\,dS$ is the total volume of liquid leaving through the surface S in a unit time. As a liquid can be considered to be incompressible, provided the volume contains neither a *source* of liquid (a point in V through which liquid enters) nor a *sink* (a point in V through which liquid is extracted), it follows that $\iint_S F_n\,dS$ will be zero for an incompressible fluid.

Thus, in an incompressible liquid free from sources and sinks, div $\mathbf{q} = 0$. If sources and sinks occur in the liquid, their strengths can be found by enclosing each in a small volume and then letting it become arbitrarily small, in which case a *positive* value of div $\mathbf{q}$ will correspond to a source and a *negative* value to a sink.

If, instead of a liquid, the flow of a gas is involved, the compressibility of a gas causes its density to vary from point to point, so then, in general, the value of div $\mathbf{q}$ will depend on position and, if the flow is unsteady, also on the time.

EXAMPLE 11.17

Find div $\mathbf{F}$ when $\mathbf{F} = xy^2\mathbf{i} + 3yz\mathbf{j} - 4xz\mathbf{k}$.

Solution   From (39) div $\mathbf{F} = \frac{\partial}{\partial x}(xy^2) + \frac{\partial}{\partial y}(3yz) + \frac{\partial}{\partial z}(-4xz) = y^2 + 3z - 4x.$ ∎

We have seen that provided f is suitably differentiable, grad f is a vector, so when f is twice differentiable it is appropriate to examine the operation div (grad f). This is usually written div grad f, because no ambiguity arises when the brackets are omitted. By definition

$$\text{div grad } f = \left(\mathbf{i}\frac{\partial}{\partial x} + \mathbf{j}\frac{\partial}{\partial y} + \mathbf{k}\frac{\partial}{\partial z}\right) \cdot \left(\mathbf{i}\frac{\partial f}{\partial x} + \mathbf{j}\frac{\partial f}{\partial y} + \mathbf{k}\frac{\partial f}{\partial z}\right)$$

$$= \frac{\partial^2 f}{\partial x^2} + \frac{\partial^2 f}{\partial y^2} + \frac{\partial^2 f}{\partial z^2} = \Delta f, \tag{41}$$

and so div grad $f = \Delta f$ is simply the **Laplacian** of f.

THEOREM 11.6

fundamental properties of the divergence operator

Properties of the divergence operator   Let the vector fields $\mathbf{F}$ and $\mathbf{G}$ and the scalar fields ϕ and ψ be a suitably differentiable, and let a and b be constants. Then the divergence operator has the following properties:

(i) $\text{div}(a\mathbf{F}) = a \, \text{div} \, \mathbf{F}$

(ii) $\text{div}(a\mathbf{F} + b\mathbf{G}) = a \, \text{div} \, \mathbf{F} + b \, \text{div} \, \mathbf{G}$

(iii) $\text{div}(\phi\mathbf{F}) = \phi \, \text{div} \, \mathbf{F} + \mathbf{F} \cdot \nabla\phi$

(iv) $\text{div}(\text{grad} \, \phi) = \Delta\phi$

(v) $\text{div}(\phi\nabla\psi) = \phi\Delta\psi + \text{grad} \, \phi \cdot \text{grad} \, \psi = \phi\Delta\psi + \nabla\phi \cdot \nabla\psi$

(vi) $\text{div}(\phi\nabla\psi) - \text{div}(\psi\nabla\phi) = \phi\Delta\psi - \psi\Delta\phi$

Proof   The derivation of these results follows directly from the definition of the divergence of a vector in (39). So, as (iv) has already been established, we will only prove (iii) and leave the other results as exercises.

If $\mathbf{F} = F_1\mathbf{i} + F_2\mathbf{j} + F_3\mathbf{k}$, it follows that $\phi\mathbf{F} = \phi F_1\mathbf{i} + \phi F_2\mathbf{j} + \phi F_3\mathbf{k}$, and so

$$\text{div}(\phi\mathbf{F}) = \frac{\partial}{\partial x}(\phi F_1) + \frac{\partial}{\partial y}(\phi F_2) + \frac{\partial}{\partial z}(\phi F_3)$$

$$= \phi\left(\frac{\partial F_1}{\partial x} + \frac{\partial F_2}{\partial y} + \frac{\partial F_3}{\partial z}\right) + F_1\frac{\partial\phi}{\partial x} + F_2\frac{\partial\phi}{\partial y} + F_3\frac{\partial\phi}{\partial z}$$

$$= \phi \, \text{div} \, \mathbf{F} + \mathbf{F} \cdot \nabla\phi. \quad ∎$$

the definition of curl F

When expressed in terms of cartesian coordinates, the **curl** of the vector $\mathbf{F} = F_1\mathbf{i} + F_2\mathbf{j} + F_3\mathbf{k}$ is defined as

$$\operatorname{curl}\mathbf{F} = \left(\frac{\partial F_3}{\partial y} - \frac{\partial F_2}{\partial z}\right)\mathbf{i} + \left(\frac{\partial F_1}{\partial z} - \frac{\partial F_3}{\partial x}\right)\mathbf{j} + \left(\frac{\partial F_2}{\partial x} - \frac{\partial F_1}{\partial y}\right)\mathbf{k}. \tag{42}$$

This form of the definition of curl $\mathbf{F}$ is more easily remembered when expressed symbolically as the determinant

$$\operatorname{curl}\mathbf{F} = \begin{vmatrix} \mathbf{i} & \mathbf{j} & \mathbf{k} \\ \dfrac{\partial}{\partial x} & \dfrac{\partial}{\partial y} & \dfrac{\partial}{\partial z} \\ F_1 & F_2 & F_3 \end{vmatrix}, \tag{43}$$

or in terms of the operator "del" as

$$\operatorname{curl}\mathbf{F} = \nabla \times \mathbf{F} = \left(\mathbf{i}\frac{\partial}{\partial x} + \mathbf{j}\frac{\partial}{\partial y} + \mathbf{k}\frac{\partial}{\partial z}\right) \times (F_1\mathbf{i} + F_2\mathbf{j} + F_3\mathbf{k}), \tag{44}$$

where it is to be understood that the differentiations are to be performed before finding the cross products, and the operation $\nabla \times \mathbf{F}$ is read as "del cross $\mathbf{F}$."

EXAMPLE 11.18

Find curl $\mathbf{F}$ given that $\mathbf{F} = xy\mathbf{i} + z\mathbf{j} + yz\mathbf{k}$.

Solution Using (43) we have

$$\operatorname{curl}\mathbf{F} = \begin{vmatrix} \mathbf{i} & \mathbf{j} & \mathbf{k} \\ \dfrac{\partial}{\partial x} & \dfrac{\partial}{\partial y} & \dfrac{\partial}{\partial z} \\ xy & z & yz \end{vmatrix}$$

$$= \left(\frac{\partial}{\partial y}(yz) - \frac{\partial}{\partial z}(z)\right)\mathbf{i} - \left(\frac{\partial}{\partial x}(yz) - \frac{\partial}{\partial z}(xy)\right)\mathbf{j} + \left(\frac{\partial}{\partial x}(z) - \frac{\partial}{\partial y}(xy)\right)\mathbf{k}$$

$$= (z - 1)\mathbf{i} - x\mathbf{k}. \qquad \blacksquare$$

EXAMPLE 11.19

Show that if ϕ is any scalar function with continuous first and second order derivatives, then $\operatorname{curl}(\operatorname{grad}\phi) \equiv \mathbf{0}$.

Solution By definition $\operatorname{grad}\phi = \phi_x\mathbf{i} + \phi_y\mathbf{j} + \phi_z\mathbf{k}$, so from (44)

$$\operatorname{curl}(\operatorname{grad}\phi) = \left(\mathbf{i}\frac{\partial}{\partial x} + \mathbf{j}\frac{\partial}{\partial y} + \mathbf{k}\frac{\partial}{\partial z}\right) \times (\phi_x\mathbf{i} + \phi_y\mathbf{j} + \phi_z\mathbf{k}).$$

After we use the properties of the vector product with the mutually orthogonal unit vectors $\mathbf{i}, \mathbf{j},$ and $\mathbf{k}$, this reduces to

$$\operatorname{curl}(\operatorname{grad}\phi) = \frac{\partial}{\partial x}(\phi_y)\mathbf{k} - \frac{\partial}{\partial x}(\phi_z)\mathbf{j} - \frac{\partial}{\partial y}(\phi_x)\mathbf{k} + \frac{\partial}{\partial y}(\phi_z)\mathbf{i} + \frac{\partial}{\partial z}(\phi_x)\mathbf{j} - \frac{\partial}{\partial z}(\phi_y)\mathbf{i}.$$

By hypothesis ϕ has continuous partial derivatives up to and including order 2, so there is equality of mixed derivatives. As a result $\phi_{xy} = \phi_{yx}$, showing that the $\mathbf{k}$ component of $\operatorname{curl}(\operatorname{grad}\phi)$ vanishes. The $\mathbf{j}$ and $\mathbf{i}$ components of $\operatorname{curl}(\operatorname{grad}\phi)$ vanish for the same reason so that $\operatorname{curl}(\operatorname{grad}\phi) \equiv \mathbf{0}$. $\blacksquare$

The operators grad, div, and curl can be combined in various ways that lead to identities, the results of which are listed in the next theorem. These identities are useful when manipulating vector operations. In some of the entries the notation $(\mathbf{F} \cdot \nabla)\mathbf{G}$ is used, and if $\mathbf{F} = F_1\mathbf{i} + F_2\mathbf{j} + F_3\mathbf{k}$ and $\mathbf{G} = G_1\mathbf{i} + G_2\mathbf{j} + G_3\mathbf{k}$ this is to be interpreted as the vector

$$(\mathbf{F} \cdot \nabla)\mathbf{G} = \left[(F_1\mathbf{i} + F_2\mathbf{j} + F_3\mathbf{k}) \cdot \left(\mathbf{i}\frac{\partial}{\partial x} + \mathbf{j}\frac{\partial}{\partial y} + \mathbf{k}\frac{\partial}{\partial z} \right) \right] (G_1\mathbf{i} + G_2\mathbf{j} + G_3\mathbf{k})$$

$$= \left(F_1\frac{\partial}{\partial x} + F_2\frac{\partial}{\partial y} + F_3\frac{\partial}{\partial z} \right) (G_1\mathbf{i} + G_2\mathbf{j} + G_3\mathbf{k}).$$

THEOREM 11.7

combining grad, div, and curl

Properties of combinations of grad, div, and curl   Let $\mathbf{F}$ and $\mathbf{G}$ be vector functions and let ϕ be a scalar function, all of which are suitably differentiable. Then the following identities hold.

(i) $\mathrm{curl}(\mathrm{grad}\,\phi) = \mathbf{0}$

(ii) $\mathrm{div}(\mathrm{curl}\,\mathbf{F}) = 0$

(iii) $\mathrm{curl}(\phi\mathbf{F}) = \phi\,\mathrm{curl}\,\mathbf{F} - \mathbf{F} \times \mathrm{grad}\,\phi$

(iv) $\mathrm{grad}(\mathbf{F} \cdot \mathbf{G}) = \mathbf{F} \times \mathrm{curl}\,\mathbf{G} + \mathbf{G} \times \mathrm{curl}\,\mathbf{F} + (\mathbf{F} \cdot \nabla)\mathbf{G} + (\mathbf{G} \cdot \nabla)\mathbf{F}$

(v) $\mathrm{div}(\mathbf{F} \times \mathbf{G}) = \mathbf{G} \cdot \mathrm{curl}\,\mathbf{F} - \mathbf{F} \cdot \mathrm{curl}\,\mathbf{G}$

(vi) $\mathrm{curl}\,(\mathbf{F} \times \mathbf{G}) = \mathbf{F}\,\mathrm{div}\,\mathbf{G} - \mathbf{G}\,\mathrm{div}\,\mathbf{F} + (\mathbf{G} \cdot \nabla)\mathbf{F} - (\mathbf{F} \cdot \nabla)\mathbf{G}$

(vii) $\mathrm{curl}(\mathrm{curl}\,\mathbf{F}) = \mathrm{grad}(\mathrm{div}\,\mathbf{F}) - \Delta\mathbf{F}$

Proof   Result (i) has already been established. As the other results follow in similar fashion from the definitions of the gradient, divergence, and curl operators, the remaining proofs are left as exercises.   ∎

The expression for curl $\mathbf{F}$ in more general coordinate systems is derived in Section 11.6, but a different definition of curl $\mathbf{F}$ together with a physical interpretation will be postponed until after the discusion of Stokes' theorem in the next chapter.

Theorem 11.7 provides a test for conservative vector fields $\mathbf{F}$. Although the test is equivalent to the test in Theorem 11.5 (i), it is in a more easily remembered form. By definition, a vector field $\mathbf{F}$ is a *conservative field* if $\mathbf{F} = \mathrm{grad}\ f$, but from (i) of Theorem 11.7, if $\mathbf{F} = \mathrm{grad}\ f$ then curl $\mathbf{F} = \mathbf{0}$, and it is this last result that provides the test. However, if after establishing that $\mathbf{F}$ is a conservative field its associated potential function f is required, it must be found by integrating the equations in Theorem 11.5 (ii), as illustrated in Example 11.14.

using curl F to test for a conservative field

Curl test for a conservative vector field

A vector field $\mathbf{F}$ is conservative, that is, it is $\mathbf{F} = \mathrm{grad}\ f$ where f is the associated scalar potential, if curl $\mathbf{F} = \mathbf{0}$.

EXAMPLE 11.20

For what values of a and b is the vector field $\mathbf{F} = (x + z)\mathbf{i} + a(y + z)\mathbf{j} + b(x + y)\mathbf{k}$ a conservative field?

Solution

$$\operatorname{curl} \mathbf{F} = \begin{vmatrix} \mathbf{i} & \mathbf{j} & \mathbf{k} \\ \dfrac{\partial}{\partial x} & \dfrac{\partial}{\partial y} & \dfrac{\partial}{\partial z} \\ x+z & a(y+z) & b(x+y) \end{vmatrix} = (b-a)\mathbf{i} + (1-b)\mathbf{j},$$

so curl $\mathbf{F} = \mathbf{0}$ if $b - a = 0$ and $1 - b = 0$. Consequently, $\mathbf{F}$ will be a conservative field if $a = b = 1$. ∎

EXAMPLE 11.21

Find curl(curl $\mathbf{F}$) given that $\mathbf{F} = x^2 y^2 \mathbf{i} + y^2 z^2 \mathbf{j} + x^2 z^2 \mathbf{k}$.

Solution To calculate curl(curl $\mathbf{F}$), we will use result (vii) of Theorem 11.7. We have

$$\operatorname{div} \mathbf{F} = 2xy^2 + 2yz^2 + 2zx^2,$$

so

$$\operatorname{grad}(\operatorname{div} \mathbf{F}) = (2y^2 + 4xz)\mathbf{i} + (2z^2 + 4xy)\mathbf{j} + (2x^2 + 4yz)\mathbf{k}.$$

Next,

$$\Delta \mathbf{F} = \left(\frac{\partial^2}{\partial x^2} + \frac{\partial^2}{\partial y^2} + \frac{\partial^2}{\partial z^2} \right) (x^2 y^2 \mathbf{i} + y^2 z^2 \mathbf{j} + x^2 z^2 \mathbf{k})$$

$$= 2(x^2 + y^2)\mathbf{i} + 2(y^2 + z^2)\mathbf{j} + 2(x^2 + z^2)\mathbf{k},$$

so combining results gives

$$\operatorname{curl}(\operatorname{curl} \mathbf{F}) = (4xz - 2x^2)\mathbf{i} + (4xy - 2y^2)\mathbf{j} + (4yz - 2z^2)\mathbf{k}. \quad \blacksquare$$

Vector fields, line integrals, the theory, application, and evaluation of multiple integrals, and the vector operators grad, div, and curl are all defined and their properties developed in standard calculus and analytic geometry texts such as those in references [1.1], [1.2], [1.5], [1.6], and [1.7]. Reference [5.6] gives a concise summary of these results together with numerous examples. More advanced and detailed accounts, where the emphasis is placed on a vector treatment, are to be found in references [5.1], [5.2], and [1.4].

Summary

The previous section introduced the gradient operator, where it was shown that it acts on a scalar function of position to produce a vector. The present section introduced two more vector operators called the divergence and curl operators. The divergence operator was seen to act on a vector to produce a scalar, while the curl operator acted on a vector to produce another vector. The general operational properties of the divergence and curl operators were developed together with the results of combining all three vector operators.

EXERCISES 11.5

In Exercises 1 through 4, find div $\mathbf{F}$ for the given vector function $\mathbf{F}$.

1. $\mathbf{F} = x^2 y \mathbf{i} + y^2 z^2 \mathbf{j} + xz^3 \mathbf{k}$.
2. $\mathbf{F} = (1 - x^2)\mathbf{i} + \sin yz \mathbf{j} + e^{xyz}\mathbf{k}$.
3. $\mathbf{F} = 3x^2 \mathbf{i} + 2x^2 y^2 \mathbf{j} + x\mathbf{k}$.

4. $\mathbf{F} = \cos x \mathbf{i} + \sin y \mathbf{j} + z^2 \mathbf{k}$.
5. Prove that $\operatorname{div}(\phi \mathbf{F}) = \phi \operatorname{div} \mathbf{F} + \mathbf{F} \cdot \nabla \phi$ (Theorem 11.6 (iii)).
6. Prove that $\operatorname{div}(\phi \nabla \psi) = \phi \Delta \psi + \nabla \phi \cdot \nabla \psi$ (Theorem 11.6 (v)).

In Exercises 7 through 10 find curl **F** for the given vector function **F**.

7. $\mathbf{F} = xyz^2\mathbf{i} + x^2yz\mathbf{j} + xy^2\mathbf{k}$.

8. $\mathbf{F} = \sinh xy\mathbf{i} + \cosh yz\mathbf{j} + xyz\mathbf{k}$.

9. $\mathbf{F} = \arctan\frac{x}{y}\mathbf{i} + \ln(x^2 + 2y^2)^{1/2}\mathbf{j} + y\mathbf{k}$.

10. $\mathbf{F} = (x^2 + y^2 + z^2)^{1/2}\mathbf{i} + (x^2 + y^2 + z^2)^{1/2}\mathbf{j} + x\mathbf{k}$.

11. Prove that div(curl **F**) $\equiv 0$ (Theorem 11.7 (ii)).

12. Prove that curl($\phi\mathbf{F}$) $\equiv \phi$ curl **F** $-$ **F** $\times$ grad ϕ (Theorem 11.7 (iii)).

13. Prove that grad(**F** $\cdot$ **G**) $\equiv$ **F** $\times$ curl **G** $+$ **G** $\times$ curl **F** $+$ (**F** $\cdot \nabla$)**G** $+$ (**G** $\cdot \nabla$)**F**   (Theorem 11.7 (iv)).

14. Prove that div(**F** $\times$ **G**) $\equiv$ **G** $\cdot$ curl **F** $-$ **F** $\cdot$ curl **G** (Theorem 11.7 (v)).

15. Prove that curl(**F** $\times$ **G**) $\equiv$ **F** div **G** $-$ **G** div **F** $+$ (**G** $\cdot \nabla$)**F** $-$ (**F** $\cdot \nabla$)**G**   (Theorem 11.7 (vi)).

16. Prove that curl(curl **F**) $=$ grad(div **F**) $-\Delta\mathbf{F}$ (Theorem 11.7 (vii)).

17. Find curl(curl **F**) given that $\mathbf{F} = 3xyz\mathbf{i} + 2y\mathbf{j} - 4z\mathbf{k}$.

In Exercises 17 and 20 use the curl test to see if or where the vector field **F** is conservative.

18. $\mathbf{F} = yz\cosh(xyz + y^2)\mathbf{i} + (xz + 2y)\cosh(xyz + y^2)\mathbf{j} + 2xy\cosh(xyz + y^2)\mathbf{k}$.

19. $\mathbf{F} = 2xy^2\mathbf{i} + (2x^2y + 6yz^3)\mathbf{j} + 9y^2z^2\mathbf{k}$.

20. $\mathbf{F} = \dfrac{1}{(x^2 + y^2 + z^2)^{1/2}}(x\mathbf{i} + y\mathbf{j} + z\mathbf{k})$.

21. $\mathbf{F} = \dfrac{1}{(1 + x^2 + 2y^2z)}(2x\mathbf{i} + 4yz\mathbf{j} + 2y^2\mathbf{k})$.

11.6   Orthogonal Curvilinear Coordinates

The geometrical configuration of a physical problem often suggests the most appropriate coordinate system that should be used when seeking its solution. For example, heat conduction in a cylindrical rod suggests the use of cylindrical polar coordinates with the z-axis aligned with the axis of the rod, whereas the distribution of an electric field inside a spherical cavity suggests the use of spherical polar coordinates. When problems of this nature are expressed in terms of vectors, and the operators grad, div, and curl are involved, it becomes necessary to find the form taken by these operators in different systems of curvilinear coordinates. The reader who wishes to omit the derivation of the main results of this section should proceed directly to Theorem 11.8 after studying the definition of an orthogonal system of curvilinear coordinates and the meaning of the scale factors h_1, h_2, and h_3.

In what follows, in order to unify notation, it is convenient to denote the usual cartesian coordinates x, y, and z by x_1, x_2, and x_3 and a general system of curvilinear coordinates by q_1, q_2, and q_3, where the two systems are related by the equations

$$x_1 = x_1(q_1, q_2, q_3), \quad x_2 = x_2(q_1, q_2, q_3), \quad x_3 = x_3(q_1, q_2, q_3). \tag{45}$$

For the curvilinear coordinates q_1, q_2, and q_3 to be equivalent to the cartesian coordinate system x_1, x_2, and x_3 it is necessary that equations (45) can be solved uniquely in the form

$$q_1 = q_1(x_1, x_2, x_3), \quad q_2 = q_2(x_1, x_2, x_3), \quad q_3 = q_3(x_1, x_2, x_3), \tag{46}$$

so that one point in cartesian coordinates corresponds to only one point in curvilinear coordinates, and conversely. As derivatives of functions occur in grad, div, and curl, it is necessary that the coordinate functions x_1, x_2, and x_3, as functions of q_1, q_2, and q_3 in (45), are all suitably differentiable with respect to their arguments. Taking the total differentials of the coordinate transformations in (45), we have

$$dx_1 = \frac{\partial x_1}{\partial q_1}dq_1 + \frac{\partial x_1}{\partial q_2}dq_2 + \frac{\partial x_1}{\partial q_3}dq_3, \quad dx_2 = \frac{\partial x_2}{\partial q_1}dq_1 + \frac{\partial x_2}{\partial q_2}dq_2 + \frac{\partial x_2}{\partial q_3}dq_3$$

$$dx_3 = \frac{\partial x_3}{\partial q_1}dq_1 + \frac{\partial x_3}{\partial q_2}dq_2 + \frac{\partial x_3}{\partial q_3}dq_3. \tag{47}$$

These results can be written in the matrix form

$$dx = J\,dq,\tag{48}$$

where

$$dx = \begin{bmatrix} dx_1 \\ dx_2 \\ dx_3 \end{bmatrix}, \quad dq = \begin{bmatrix} dq_1 \\ dq_2 \\ dq_3 \end{bmatrix}, \quad \text{and} \quad J = \begin{bmatrix} \dfrac{\partial x_1}{\partial q_1} & \dfrac{\partial x_1}{\partial q_2} & \dfrac{\partial x_1}{\partial q_3} \\[2mm] \dfrac{\partial x_2}{\partial q_1} & \dfrac{\partial x_2}{\partial q_2} & \dfrac{\partial x_2}{\partial q_3} \\[2mm] \dfrac{\partial x_3}{\partial q_1} & \dfrac{\partial x_3}{\partial q_2} & \dfrac{\partial x_3}{\partial q_3} \end{bmatrix}.\tag{49}$$

The matrix vector linear differential elements dx and dq will be uniquely related by (48) provided matrix J is nonsingular, so the coordinate transformations (45) must be such that $J = \det J \neq 0$, where

$$J = \begin{vmatrix} \dfrac{\partial x_1}{\partial q_1} & \dfrac{\partial x_2}{\partial q_1} & \dfrac{\partial x_3}{\partial q_1} \\[2mm] \dfrac{\partial x_1}{\partial q_2} & \dfrac{\partial x_2}{\partial q_2} & \dfrac{\partial x_3}{\partial q_2} \\[2mm] \dfrac{\partial x_1}{\partial q_3} & \dfrac{\partial x_2}{\partial q_3} & \dfrac{\partial x_3}{\partial q_3} \end{vmatrix}.\tag{50}$$

the Jacobian of a transformation

The determinant J is called the **Jacobian** of the transformation, and it will be shown later that the absolute value of the Jacobian occurs as a scale factor in the **volume element** in orthogonal curvilinear coordinates. Thus, the vanishing of the Jacobian signifying nonuniqueness in the transformations (45) and (46) also corresponds to the failure of the curvilinear coordinate system to define a volume element.

CARL GUSTAV JACOBI (1804–1851)
A German mathematician who studied at the University of Berlin and obtained his doctorate in 1825. In 1827 he was appointed Extraordinary Professor of Mathematics at Königsberg and, after two years, he was promoted to Ordinary Professor of Mathematics. In 1842 he moved to Berlin where he remained until his death. His most important work was in connection with elliptic functions, but he also made important contributions to number theory, ordinary and partial differential equations, and the calculus of variations. He was an outstanding teacher of mathematics.

general and orthogonal curvilinear coordinates

Keeping q_1 and $q_1 + dq_1$ constant defines two curvilinear surfaces in space, and four further curvilinear surfaces are defined by keeping q_2 and $q_2 + dq_2$ constant, and q_3 and $q_3 + dq_3$ constant. Taken together, the region between these six curvilinear surfaces defines the volume element dV in space shown in Fig. 11.14.

Allowing q_1 to vary while holding q_2 and q_3 constant in (45) will generate a *curvilinear coordinate line* in space along which only q_1 changes. Similarly, allowing q_2 to vary while holding q_1 and q_3 constant, and then q_3 to vary while holding q_1 and q_2 constant, will generate curvilinear coordinate lines in space along which, respectively, only q_2 and q_3 vary. If a general point A in space shown in Fig. 11.14 is considered, there will be three curvilinear coordinate lines passing through the point. A curvilinear coordinate system will be said to be an **orthogonal** system if at every point in space the three tangents to the coordinate lines at their point of intersection

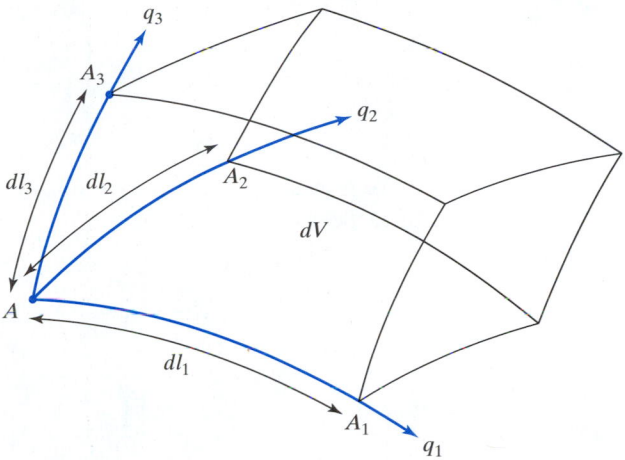

FIGURE 11.14   The curvilinear volume element dV.

are mutually orthogonal (perpendicular). Such coordinate systems are also considered to be *orthogonal* if the orthogonality condition fails at a single point or along a line. In what follows, only orthogonal coordinate systems will be considered.

With the linear differential length elements $AA_1 = dl_1$, $AA_2 = dl_2$, and $AA_3 = dl_3$, the orthogonality of the curvilinear coordinate system implies that in terms of curvilinear coordinates the linear volume element dV in Fig. 11.14 is given by

the volume element

$$dV = dl_1 \, dl_2 \, dl_3. \tag{51}$$

Now, in Fig. 11.14, let A be the point (x_1, x_2, x_3) and A_1 be the point $(x_1 + dx_1, x_2 + dx_2, x_3 + dx_3)$, where dx_1, dx_2, and dx_3 are the linear differential elements in cartesian coordinates. To find the linear differential length element dl_1 from A to A_1, we apply the Pythagoras theorem to the mutually orthogonal linear differential length elements dx_1, dx_2, and dx_3, when we obtain

$$dl_1^2 = dx_1^2 + dx_2^2 + dx_3^2, \tag{52}$$

However along AA_1 only q_1 varies, so as

$$dx_1 = \frac{\partial x_1}{\partial q_1} dq_1, \quad dx_2 = \frac{\partial x_2}{\partial q_1} dq_1, \quad dx_3 = \frac{\partial x_3}{\partial q_1} dq_1, \tag{53}$$

the square of the linear differential length element in (52) becomes

$$dl_1^2 = \left[\left(\frac{\partial x_1}{\partial q_1} \right)^2 + \left(\frac{\partial x_2}{\partial q_1} \right)^2 + \left(\frac{\partial x_3}{\partial q_1} \right)^2 \right] dq_1^2. \tag{54}$$

Similar arguments show that if dl_2 and dl_3 are the linear differential length elements along AA_2 and AA_3, then

$$dl_2^2 = \left[\left(\frac{\partial x_1}{\partial q_2} \right)^2 + \left(\frac{\partial x_2}{\partial q_2} \right)^2 + \left(\frac{\partial x_3}{\partial q_2} \right)^2 \right] dq_2^2, \tag{55}$$

and

$$dl_3^2 = \left[\left(\frac{\partial x_1}{\partial q_3} \right)^2 + \left(\frac{\partial x_2}{\partial q_3} \right)^2 + \left(\frac{\partial x_3}{\partial q_3} \right)^2 \right] dq_3^2. \tag{56}$$

the scale factors
h_1, h_2, h_3

We now adopt the standard notation and define the **scale factors** h_1, h_2, and h_3, with respect to the coordinates q_1, q_2, and q_3 in transformations (45), by

$$h_1 = \left[\left(\frac{\partial x_1}{\partial q_1} \right)^2 + \left(\frac{\partial x_2}{\partial q_1} \right)^2 + \left(\frac{\partial x_3}{\partial q_1} \right)^2 \right]^{1/2} \tag{57}$$

$$h_2 = \left[\left(\frac{\partial x_1}{\partial q_2} \right)^2 + \left(\frac{\partial x_2}{\partial q_2} \right)^2 + \left(\frac{\partial x_3}{\partial q_2} \right)^2 \right]^{1/2} \tag{58}$$

$$h_3 = \left[\left(\frac{\partial x_1}{\partial q_3} \right)^2 + \left(\frac{\partial x_2}{\partial q_3} \right)^2 + \left(\frac{\partial x_3}{\partial q_3} \right)^2 \right]^{1/2}. \tag{59}$$

In terms of h_1, h_2, and h_3 the linear differential line elements dl_1, dl_2, and dl_3 in rectangular curvilinear coordinates defined in (54) to (56) become

$$dl_1 = h_1 dq_1, \quad dl_2 = h_2 dq_2, \quad dl_3 = h_3 dl_3. \tag{60}$$

If the general linear differential length element from A to B in Fig. 11.14 is denoted by ds, then as the coordinate system is orthogonal,

$$ds^2 = dl_1^2 + dl_2^2 + dl_3^2, \tag{61}$$

so it follows from (60) that

$$ds^2 = h_1^2 dq_1^2 + h_2^2 dq_2^2 + h_3^2 dq_3^2. \tag{62}$$

In terms of the scale factors the linear differential volume element dV in (51) becomes

$$dV = h_1 h_2 h_3 \, dq_1 \, dq_2 \, dq_3. \tag{63}$$

It can be seen from this last result that the coordinate transformations (45) will fail to define a volume element in curvilinear coordinates if a scale factor vanishes. From the definitions of the scale factors, this can only happen if all of the partial derivatives in a scale factor vanish, but when this occurs the Jacobian determinant J will have a zero row, and so will also vanish. This is to be expected, because it is known from calculus that when the Jacobian vanishes, the transformation between the coordinate systems ceases to be one to one.

To understand the geometrical interpretation of the Jacobian, we make use of the elementary result from vector analysis that the scalar triple product $\mathbf{a} \cdot (\mathbf{b} \times \mathbf{c})$ can be interpreted as the volume of the parallelepiped with sides given by vectors $\mathbf{a}$, $\mathbf{b}$, and $\mathbf{c}$ that meet at a point. The value of this scalar triple product is equal to the determinant with the elements of $\mathbf{a}$, $\mathbf{b}$, and $\mathbf{c}$ as its first, second, and third rows, respectively. Considering dx_1, dx_2, and dx_3 in (47) as vectors in the curvilinear coordinate system, we see that the linear differential volume element $dV = dx_1 dx_2 dx_3$ can be written

$$\pm dV = \begin{vmatrix} \dfrac{\partial x_1}{\partial q_1}dq_1 & \dfrac{\partial x_2}{\partial q_1}dq_1 & \dfrac{\partial x_3}{\partial q_1}dq_1 \\[2mm] \dfrac{\partial x_1}{\partial q_2}dq_2 & \dfrac{\partial x_2}{\partial q_2}dq_2 & \dfrac{\partial x_3}{\partial q_2}dq_2 \\[2mm] \dfrac{\partial x_1}{\partial q_3}dq_3 & \dfrac{\partial x_2}{\partial q_3}dq_3 & \dfrac{\partial x_3}{\partial q_3}dq_3 \end{vmatrix} = \begin{vmatrix} \dfrac{\partial x_1}{\partial q_1} & \dfrac{\partial x_2}{\partial q_1} & \dfrac{\partial x_3}{\partial q_1} \\[2mm] \dfrac{\partial x_1}{\partial q_2} & \dfrac{\partial x_2}{\partial q_2} & \dfrac{\partial x_3}{\partial q_2} \\[2mm] \dfrac{\partial x_1}{\partial q_3} & \dfrac{\partial x_2}{\partial q_3} & \dfrac{\partial x_3}{\partial q_3} \end{vmatrix} dq_1\, dq_2\, dq_3. \qquad (64)$$

the Jacobian and the volume element

As a volume element is essentially nonnegative, this can be expressed in terms of the Jacobian J of the transformation as

$$dV = \pm J\, dq_1\, dq_2\, dq_3, \qquad (65)$$

where the sign in (65) is chosen to make the expression on the right positive. A comparison of (63) and (65) then shows that the absolute value of the Jacobian J is equal to the product of the scale factors forming the scale factor for the linear volume element dV, and so

$$h_1 h_2 h_3 = \pm J, \qquad (66)$$

where the sign is chosen to make the expression on the right positive.

EXAMPLE 11.22

Find the scale factors, the linear differential length elements along the curvilinear coordinate lines, the square of the general linear differential length element ds, the linear differential volume element dV, and the Jacobian for (a) cylindrical polar coordinates and (b) spherical polar coordinates.

Solution

(a) In cylindrical polar coordinates $x = r\cos\theta$, $y = r\sin\theta$, $z = z$, so to relate this system to the general one just considered, we must make the identifications $x_1 = x$, $x_2 = y$, $x_3 = z$, $q_1 = r$, $q_2 = \theta$, and $q_3 = z$. When this is done, substitution into (57) to (59) shows that

$$h_1 = 1, \quad h_2 = r, \quad h_3 = 1,$$

so from (60) the linear differential length elements along the curvilinear coordinate lines are

$$dl_1 = dr, \quad dl_2 = r\, d\theta, \quad dl_3 = dz.$$

It then follows from (62) that the square of the general linear differential length element ds is

$$ds^2 = dr^2 + r^2\, d\theta^2 + dz^2,$$

and from (63) that the linear differential volume element in terms of cylindrical polar coordinates is

$$dV = r\, dr\, d\theta\, dz.$$

The Jacobian of the transformation

$$J = \begin{vmatrix} \cos\theta & \sin\theta & 0 \\ -r\sin\theta & r\cos\theta & 0 \\ 0 & 0 & 1 \end{vmatrix} = r,$$

in agreement with (66).

The transformation ceases to be one to one when $r = 0$, because then $h_2 = 0$, though this is to be expected because $r = 0$ is the z-axis along which θ is indeterminate.

(b) In spherical polar coordinates $x = r\sin\theta\cos\phi$, $y = r\sin\theta\sin\phi$, $z = r\cos\theta$, so to relate this system to the general one just considered we must make the identifications $x_1 = x, x_2 = y, x_3 = z, q_1 = r, q_2 = \phi$, and $q_3 = \theta$. When this is done, substitution into (57) to (59) shows that

$$h_1 = 1, \quad h_2 = r, \quad h_3 = r\sin\theta,$$

so from (60) the linear differential length elements along the curvilinear coordinate lines are

$$dl_1 = dr, \quad dl_2 = r\,d\phi, \quad dl_3 = r\sin\theta\,d\theta$$

As in (a), it follows from (62) that the square of the general linear differential length element ds is

$$ds^2 = dr^2 + r^2\sin^2\theta\,d\theta^2 + r^2\,d\phi^2$$

and from (63) that the linear differential volume element in terms of spherical polar coordinates is

$$dV = r^2\sin\theta\,dr\,d\theta\,d\phi.$$

The Jacobian of the transformation

$$J = \begin{vmatrix} \sin\theta\cos\phi & \sin\theta\sin\phi & \cos\theta \\ -r\sin\theta\sin\phi & r\sin\theta\cos\phi & 0 \\ r\cos\theta\cos\phi & r\cos\theta\sin\phi & -r\sin\theta \end{vmatrix} = -r^2\sin\theta,$$

and in agreement with (66) we see that $h_1 h_2 h_3 = |J| = r^2\sin\phi$.

The Jacobian vanishes when $r = 0$, causing h_2 and h_3 to vanish, but this corresponds to the origin where θ and ϕ are indeterminate. The Jacobian also vanishes when $\phi = 0$ and $\phi = \pi$, corresponding to points on the z-axis where θ is indeterminate. ∎

To derive the form of the gradient, divergence, curl, and Laplacian operators in rectangular curvilinear coordinates, it is necessary to introduce the triad of unit vectors $\mathbf{e}_1, \mathbf{e}_2$, and $\mathbf{e}_3$ at a general point $(q_1^{(0)}, q_2^{(0)}, q_3^{(0)})$. Here, $\mathbf{e}_1$ is tangent to the q_1 coordinate line, $\mathbf{e}_2$ is tangent to the q_2 coordinate line, and $\mathbf{e}_3$ is tangent to the q_3 coordinate line at the point $(q_1^{(0)}, q_2^{(0)}, q_3^{(0)})$. If we denote a general vector in curvilinear coordinates by $\mathbf{q}(q_1, q_2, q_3)$, the vector forms of the three coordinate lines become

$$\mathbf{q} = \mathbf{q}(q_1, q_2^{(0)}, q_3^{(0)}), \quad \mathbf{q} = \mathbf{q}(q_1^{(0)}, q_2, q_3^{(0)}), \quad \text{and} \quad \mathbf{q} = \mathbf{q}(q_1^{(0)}, q_2^{(0)}, q_3).$$

$$(67)$$

As a result, the vectors $\mathbf{e}_1$, $\mathbf{e}_2$, and $\mathbf{e}_3$ are, respectively, parallel to the derivatives $\partial\mathbf{q}/\partial q_1$, $\partial\mathbf{q}/\partial q_2$, and $\partial\mathbf{q}/\partial q_3$ at the point $(q_1^{(0)}, q_2^{(0)}, q_3^{(0)})$. The scale factors along these coordinate lines are h_1, h_2, and h_3, it follows that the unit vectors at $(q_1^{(0)}, q_2^{(0)}, q_3^{(0)})$ are

$$\mathbf{e}_1 = \frac{\partial\mathbf{q}}{\partial q_1}\Big/\left|\frac{\partial\mathbf{q}}{\partial q_1}\right|, \quad \mathbf{e}_2 = \frac{\partial\mathbf{q}}{\partial q_2}\Big/\left|\frac{\partial\mathbf{q}}{\partial q_2}\right|, \quad \text{and} \quad \mathbf{e}_3 = \frac{\partial\mathbf{q}}{\partial q_3}\Big/\left|\frac{\partial\mathbf{q}}{\partial q_3}\right|,$$

where, of course, the scale factors h_1, h_2, and h_3 are given by

$$h_1 = \left|\frac{\partial\mathbf{q}}{\partial q_1}\right|, \quad h_2 = \left|\frac{\partial\mathbf{q}}{\partial q_2}\right|, \quad \text{and} \quad h_3 = \left|\frac{\partial\mathbf{q}}{\partial q_3}\right|,$$

so that

$$\mathbf{e}_1 = \frac{1}{h_1}\frac{\partial\mathbf{q}}{\partial q_1}, \quad \mathbf{e}_2 = \frac{1}{h_2}\frac{\partial\mathbf{q}}{\partial q_2}, \quad \mathbf{e}_3 = \frac{1}{h_3}\frac{\partial\mathbf{q}}{\partial q_3}. \tag{68}$$

It is important to recognize that unlike the unit vectors $\mathbf{i}$, $\mathbf{j}$, and $\mathbf{k}$, which are parallel to the fixed x-, y-, and z-axes so their derivatives are zero, the unit vectors $\mathbf{e}_1$, $\mathbf{e}_2$, and $\mathbf{e}_3$ in curvilinear coordinates are functions of position, so when finding the form of vector operators, we must take into account the derivatives of $\mathbf{e}_1$, $\mathbf{e}_2$, and $\mathbf{e}_3$.

THEOREM 11.8

grad, div, and curl in general rectangular curvilinear coordinates

Gradient, divergence, curl, and Laplacian in general rectangular curvilinear coordinates   Let the scalar function $f(q_1, q_2, q_3)$, and the vector function

$$\mathbf{F} = F_1(q_1, q_2, q_3)\mathbf{e}_1 + F_2(q_1, q_2, q_3)\mathbf{e}_2 + F_3(q_1, q_2, q_3)\mathbf{e}_3$$

be suitably differentiable functions of the rectangular curvilinear coordinates q_1, q_2, and q_3, where $\mathbf{e}_1$ is the unit vector in the direction of increasing q_1, $\mathbf{e}_2$ is the unit vector in the direction of increasing q_2, and $\mathbf{e}_3$ is the unit vector in the direction of increasing q_3 at the point (q_1, q_2, q_3). Then:

(i) $\quad \text{grad } f = \mathbf{e}_1\dfrac{1}{h_1}\dfrac{\partial f}{\partial q_1} + \mathbf{e}_2\dfrac{1}{h_2}\dfrac{\partial f}{\partial q_2} + \mathbf{e}_3\dfrac{1}{h_3}\dfrac{\partial f}{\partial q_3}$

(ii) $\quad \text{div }\mathbf{F} = \dfrac{1}{h_1h_2h_3}\left[\dfrac{\partial}{\partial q_1}(h_2h_3F_1) + \dfrac{\partial}{\partial q_2}(h_1h_3F_2) + \dfrac{\partial}{\partial q_3}(h_1h_2F_3)\right]$

(iii) $\quad \text{curl }\mathbf{F} = \dfrac{1}{h_1h_2h_3}\begin{vmatrix} h_1\mathbf{e}_1 & h_2\mathbf{e}_2 & h_3\mathbf{e}_3 \\ \dfrac{\partial}{\partial q_1} & \dfrac{\partial}{\partial q_2} & \dfrac{\partial}{\partial q_3} \\ h_1F_1 & h_2F_2 & h_3F_3 \end{vmatrix}$

(iv) $\quad \Delta \equiv \dfrac{1}{h_1h_2h_3}\left[\dfrac{\partial}{\partial q_1}\left(\dfrac{h_2h_3}{h_1}\dfrac{\partial}{\partial q_1}\right) + \dfrac{\partial}{\partial q_2}\left(\dfrac{h_1h_3}{h_2}\dfrac{\partial}{\partial q_2}\right) + \dfrac{\partial}{\partial q_3}\left(\dfrac{h_1h_2}{h_3}\dfrac{\partial}{\partial q_3}\right)\right]$

(the Laplacian operator)

Proof

(i) To find grad $f = \frac{\partial f}{\partial x_1}\mathbf{i} + \frac{\partial f}{\partial x_2}\mathbf{j} + \frac{\partial f}{\partial x_3}\mathbf{k}$ in terms of curvilinear coordinates it is necessary to find the components of this vector in the $\mathbf{e}_1$, $\mathbf{e}_2$, and $\mathbf{e}_3$ directions, and then to use them as the components of a vector expressed in terms of curvilinear coordinates. As only q_1 varies in the direction of $\mathbf{e}_1$, it follows from the first equations in (46) and (68) that

$$\mathbf{e}_1 = \frac{1}{h_1}\left(\frac{\partial x_1}{\partial q_1}\mathbf{i} + \frac{\partial x_2}{\partial q_1}\mathbf{j} + \frac{\partial x_3}{\partial q_1}\mathbf{k}\right).$$

Thus, the component of grad f in the direction of the unit vector $\mathbf{e}_1$ is

$$\mathbf{e}_1 \cdot \operatorname{grad} f = \frac{1}{h_1}\left(\frac{\partial f}{\partial x_1}\frac{\partial x_1}{\partial q_1} + \frac{\partial f}{\partial x_2}\frac{\partial x_2}{\partial q_1} + \frac{\partial f}{\partial x_3}\frac{\partial x_3}{\partial q_1}\right) = \frac{1}{h_1}\frac{\partial f}{\partial q_1},$$

where the last result follows directly from the chain rule.

Corresponding results apply for the components of grad f in the directions of the unit vectors $\mathbf{e}_2$ and $\mathbf{e}_3$, so if we use these results as the components of grad f in curvilinear coordinates, it follows that

$$\operatorname{grad} f = \mathbf{e}_1\frac{1}{h_1}\frac{\partial f}{\partial q_1} + \mathbf{e}_2\frac{1}{h_2}\frac{\partial f}{\partial q_2} + \mathbf{e}_3\frac{1}{h_3}\frac{\partial f}{\partial q_3},$$

and result (i) is established.

In what follows, for conciseness when establishing results (ii) to (iv), the operator notations $\nabla \cdot (\cdot)$ and $\nabla \times (\cdot)$ will be used to signify the divergence and curl operators.

(ii) As $\mathbf{e}_1$, $\mathbf{e}_2$, and $\mathbf{e}_3$ are orthogonal unit vectors $\mathbf{e}_1 = \mathbf{e}_2 \times \mathbf{e}_3$. By identifying f in (i) with q_1 we see that $\mathbf{e}_1 = h_1\nabla q_1$ and, similarly, by identifying f with q_2 and q_3 it follows that $\mathbf{e}_2 = h_2\nabla q_2$ and $\mathbf{e}_3 = h_3\nabla q_3$, and so $\mathbf{e}_1 = h_2 h_3 \nabla q_2 \times \nabla q_3$.

To find div $\mathbf{F}$ it is necessary to compute $\nabla \cdot (F_1\mathbf{e}_1 + F_2\mathbf{e}_2 + F_3\mathbf{e}_3)$ taking into account the dependence of $\mathbf{e}_1$, $\mathbf{e}_2$, and $\mathbf{e}_3$ on position. Because of the linearity of the divergence operator, this can be accomplished by taking the divergence of each term in $\mathbf{F} = F_1\mathbf{e}_1 + F_2\mathbf{e}_2 + F_3\mathbf{e}_3$ and then summing the results. The divergence of the first term is given by $\nabla \cdot (F_1\mathbf{e}_1) = \nabla \cdot (F_1 h_2 h_3 \nabla q_2 \times \nabla q_3)$, so using result (iii) of Theorem 11.6, this becomes

$$\nabla \cdot (F_1\mathbf{e}_1) = F_1 h_1 h_2 \nabla \cdot (\nabla q_2 \times \nabla q_3) + (\nabla q_2 \times \nabla q_3) \cdot \nabla(F_1 h_1 h_2).$$

However, applying result (v) of Theorem 11.7 to the term $\nabla \cdot (\nabla q_2 \times \nabla q_3)$ and using the fact that $\operatorname{curl}(\operatorname{grad} q_2) = \operatorname{curl}(\operatorname{grad} q_3) = 0$ simplifies this result to

$$\nabla \cdot (F_1\mathbf{e}_1) = (\nabla q_2 \times \nabla q_3) \cdot \nabla(F_1 h_1 h_2),$$

but $\mathbf{e}_1 = h_2 h_3 \nabla q_2 \times \nabla q_3$, and so

$$\nabla \cdot (F_1\mathbf{e}_1) = \frac{1}{h_2 h_3}\mathbf{e}_1 \cdot \nabla(F_1 h_2 h_3).$$

In the proof of (i) we saw that

$$\mathbf{e}_1 \cdot \operatorname{grad} f = \frac{1}{h_1}\frac{\partial f}{\partial q_1},$$

so identifying f with $F_1 h_2 h_3$ we find that

$$\nabla \cdot (F_1\mathbf{e}_1) = \frac{1}{h_1 h_2 h_3}\frac{\partial(F_1 h_2 h_3)}{\partial q_1}.$$

Corresponding results apply to $\nabla \cdot (F_2 \mathbf{e}_2)$ and $\nabla \cdot (F_3 \mathbf{e}_3)$, so summing the results we arrive at result (iii).

(iii) To find curl $\mathbf{F}$ it is necessary to compute $\nabla \times (F_1 \mathbf{e}_1 + F_2 \mathbf{e}_2 + F_3 \mathbf{e}_3)$, so as curl is a linear operator, we may compute the curl of each term in $\mathbf{F} = F_1 \mathbf{e}_1 + F_2 \mathbf{e}_2 + F_3 \mathbf{e}_3$ and then sum the results. Considering the term $\nabla \times (F_1 \mathbf{e}_1)$ and writing $\mathbf{e}_1 = h_1 \nabla q_1$, we find that $\nabla \times (F_1 \mathbf{e}_1) = \nabla \times (F_1 h_1 \nabla q_1)$. Applying result (iii) of Theorem 11.7 to this last result, we find that

$$\nabla \times (F_1 \mathbf{e}_1) = F_1 h_1 \nabla \times (\nabla q_1) - (\nabla q_1) \times (\nabla F_1 h_1),$$

but $\nabla \times (\nabla q_1) = 0$, and so

$$\nabla \times (F_1 \mathbf{e}_1) = -(\nabla q_1) \times (\nabla F_1 h_1).$$

Now $\nabla q_1 = \mathbf{e}_1 / h_1$, so if we reverse the sign in the preceding result and compensate by interchanging the order of the factors, the result becomes

$$\nabla \times (F_1 \mathbf{e}_1) = \left[\mathbf{e}_1 \frac{1}{h_1} \frac{\partial (F_1 h_1)}{\partial q_1} + \mathbf{e}_2 \frac{1}{h_2} \frac{\partial (F_1 h_1)}{\partial q_2} + \mathbf{e}_3 \frac{1}{h_3} \frac{\partial (F_1 h_1)}{\partial q_3} \right] \times \frac{\mathbf{e}_1}{h_1},$$

and so using the orthogonality of the unit vectors $\mathbf{e}_1, \mathbf{e}_2$, and $\mathbf{e}_3$, which implies $\mathbf{e}_1 \times \mathbf{e}_1 = \mathbf{0}$, $\mathbf{e}_2 \times \mathbf{e}_1 = -\mathbf{e}_3$, and $\mathbf{e}_3 \times \mathbf{e}_1 = \mathbf{e}_2$, this becomes

$$\nabla \times (F_1 \mathbf{e}_1) = \mathbf{e}_2 \frac{1}{h_1 h_3} \frac{\partial}{\partial q_3} (h_1 F_1) - \mathbf{e}_3 \frac{1}{h_1 h_2} \frac{\partial}{\partial q_2} (h_1 F_1).$$

Corresponding results exist for $\nabla \times (F_2 \mathbf{e}_2)$ and $\nabla \times (F_3 \mathbf{e}_3)$, so combining them we find that

$$\nabla \times \mathbf{F} = \mathbf{e}_2 \frac{1}{h_1 h_3} \frac{\partial}{\partial q_3} (h_1 F_1) - \mathbf{e}_3 \frac{1}{h_1 h_2} \frac{\partial}{\partial q_2} (h_1 F_1) + \mathbf{e}_3 \frac{1}{h_1 h_2} \frac{\partial}{\partial q_1} (h_2 F_2)$$

$$- \mathbf{e}_1 \frac{1}{h_2 h_3} \frac{\partial}{\partial q_3} (h_2 F_2) + \mathbf{e}_1 \frac{1}{h_2 h_3} \frac{\partial}{\partial q_1} (h_3 F_3) - \mathbf{e}_2 \frac{1}{h_1 h_3} \frac{\partial}{\partial q_2} (h_3 F_3).$$

This last result is seen to be the expansion of the determinant in (iii), so the proof is complete.

(iv) The Laplacian operator

$$\Delta = \nabla \cdot \left[\mathbf{e}_1 \frac{1}{h_1} \frac{\partial}{\partial q_1} + \mathbf{e}_2 \frac{1}{h_2} \frac{\partial}{\partial q_2} + \mathbf{e}_3 \frac{1}{h_3} \frac{\partial}{\partial q_3} \right]$$

$$= \operatorname{div} \left[\mathbf{e}_1 \frac{1}{h_1} \frac{\partial}{\partial q_1} + \mathbf{e}_2 \frac{1}{h_2} \frac{\partial}{\partial q_2} + \mathbf{e}_3 \frac{1}{h_3} \frac{\partial}{\partial q_3} \right].$$

Using result (ii) of the theorem with the operator $\frac{1}{h_1} \frac{\partial}{\partial q_1}$ in place of F_1, the operator $\frac{1}{h_2} \frac{\partial}{\partial q_2}$ in place of F_2 and the operator $\frac{1}{h_3} \frac{\partial}{\partial q_3}$ in place of F_3, we arrive at result (iv). ∎

EXAMPLE 11.23

grad, div, curl, and the Laplacian in cylindrical and spherical polar coordinates

Find the forms taken by grad, div, curl, the Laplacian, and the Laplacian operator in (a) cylindrical polar coordinates and (b) spherical polar coordinates.

Solution **(a)** Using the notation of Example 11.22 and the scale factors $h_1 = 1$, $h_2 = r$, and $h_3 = 1$ found in that example, routine calculations show that in

cylindrical polar coordinates, when $\mathbf{F} = F_r \mathbf{e}_r + F_\theta \mathbf{e}_\theta + F_z \mathbf{e}_z$,

$$\operatorname{grad} f = \frac{\partial f}{\partial r}\mathbf{e}_r + \frac{1}{r}\frac{\partial f}{\partial \theta}\mathbf{e}_\theta + \frac{\partial f}{\partial z}\mathbf{e}_z$$

$$\operatorname{div}\mathbf{F} = \frac{1}{r}\frac{\partial(r F_r)}{\partial r} + \frac{1}{r}\frac{\partial F_\theta}{\partial \theta} + \frac{\partial F_z}{\partial z}$$

$$\operatorname{curl}\mathbf{F} = \frac{1}{r}\begin{vmatrix} \mathbf{e}_r & r\mathbf{e}_\theta & \mathbf{e}_z \\ \dfrac{\partial}{\partial r} & \dfrac{\partial}{\partial \theta} & \dfrac{\partial}{\partial z} \\ F_r & r F_\theta & F_z \end{vmatrix}$$

$$\Delta f = \frac{1}{r}\frac{\partial}{\partial r}\left(r\frac{\partial f}{\partial r}\right) + \frac{1}{r^2}\frac{\partial^2 f}{\partial \theta^2} + \frac{\partial^2 f}{\partial z^2}$$

$$\Delta = \frac{1}{r}\frac{\partial}{\partial r}\left(r\frac{\partial}{\partial r}\right) + \frac{1}{r^2}\frac{\partial^2}{\partial \theta^2} + \frac{\partial^2}{\partial z^2} \qquad \text{(Laplacian operator)}.$$

(b) Again using the notation of Example 11.21 and the scale factors $h_1 = 1$, $h_2 = r\sin\phi$, $h_3 = r$ found in that example, routine calculations show that in **spherical polar coordinates**, when $\mathbf{F} = F_r \mathbf{e}_r + F_\theta \mathbf{e}_\theta + F_\phi \mathbf{e}_\phi$,

$$\operatorname{grad} f = \frac{\partial f}{\partial r}\mathbf{e}_r + \frac{1}{r}\frac{\partial f}{\partial \theta}\mathbf{e}_\theta + \frac{1}{r\sin\theta}\frac{\partial f}{\partial \phi}\mathbf{e}_\phi$$

$$\operatorname{div}\mathbf{F} = \frac{1}{r^2}\frac{\partial(r^2 F_r)}{\partial r} + \frac{1}{r\sin\theta}\frac{\partial}{\partial \theta}(F_\theta \sin\theta) + \frac{1}{r\sin\theta}\frac{\partial F_\phi}{\partial \phi}$$

$$\operatorname{curl}\mathbf{F} = \frac{1}{r^2\sin\theta}\begin{vmatrix} \mathbf{e}_r & r\mathbf{e}_\theta & r\sin\theta\,\mathbf{e}_\phi \\ \dfrac{\partial}{\partial r} & \dfrac{\partial}{\partial \theta} & \dfrac{\partial}{\partial \phi} \\ F_r & r F_\theta & r\sin\theta\, F_\phi \end{vmatrix};$$

$$\Delta f = \frac{1}{r^2}\frac{\partial}{\partial r}\left(r^2\frac{\partial f}{\partial r}\right) + \frac{1}{r^2\sin\theta}\frac{\partial}{\partial \theta}\left(\sin\theta\frac{\partial f}{\partial \theta}\right) + \frac{1}{r^2\sin^2\theta}\frac{\partial^2 f}{\partial \phi^2}$$

$$\Delta = \frac{1}{r^2}\frac{\partial}{\partial r}\left(r^2\frac{\partial}{\partial r}\right) + \frac{1}{r^2\sin\theta}\frac{\partial}{\partial \theta}\left(\sin\theta\frac{\partial}{\partial \theta}\right) + \frac{1}{r^2\sin^2\theta}\frac{\partial^2}{\partial \phi^2}$$

(Laplacian operator).   ∎

Descriptions of general orthogonal curvilinear coordinates and the form taken by vector operators in different coordinate systems are to be found in references [1.3] and [5.2], whereas applications to continuum mechanics are to be found in reference [5.4] and to hydrodynamics in reference [6.5]. Further information can also be found in Chapters 23 and 24 of reference [G.3].

Summary   After introducing the concept of general orthogonal curvilinear coordinates, this section then derived expressions for grad, div, curl, and the Laplacian operators in terms of these coordinates. Because of the importance of cylindrical and spherical polar coordinates in applications, these operators were then expressed in terms of cylindrical and spherical polar coordinates.

EXERCISES 11.6

1. Write out the results of Theorem 11.6 using the operator notation $\nabla(.)$, $\nabla \cdot (.)$, $\nabla \times (.)$ in place of grad, div, and curl.

2. Write out the results of Theorem 11.7 using the operator notation $\nabla(.)$, $\nabla \cdot (.)$, $\nabla \times (.)$ in place of grad, div, and curl.

3. Complete the calculations leading to the results of Example 11.22(a) for cylindrical polar coordinates.

4. Complete the calculations leading to the results of Example 11.22(b) for spherical polar coordinates.

5. Show the curvilinear coordinate system defined in the region $q_3 \geq 0$ by the equations $x_1 = q_1 - q_2$, $x_2 = q_1 + q_2$,

and $x_3 = \sinh q_3$ is orthogonal. Find the scale factors h_1, h_2, h_3, grad f, and div **F**.

6. Show that the **parabolic cylindrical coordinates** (u, v, z) defined by the equations $x = \frac{1}{2}(u^2 - v^2)$, $y = uv$, $z = z$ are orthogonal. Find the scale factors h_1, h_2, h_3, and $\nabla^2 f$.

7. Show that the **elliptic cylindrical coordinates** (ξ, η, z) defined by the equations $x = \cosh \xi \cos \eta$, $y = \sinh \xi \sin \eta$, $z = z$ for $0 \leq \xi < \infty$, $-\pi < \eta \leq \pi$, $-\infty < z < \infty$ are orthogonal. Find the scale factors h_1, h_2, h_3 and state the shapes of the surfaces $\xi = $ constant and $\eta = $ constant and find grad f.

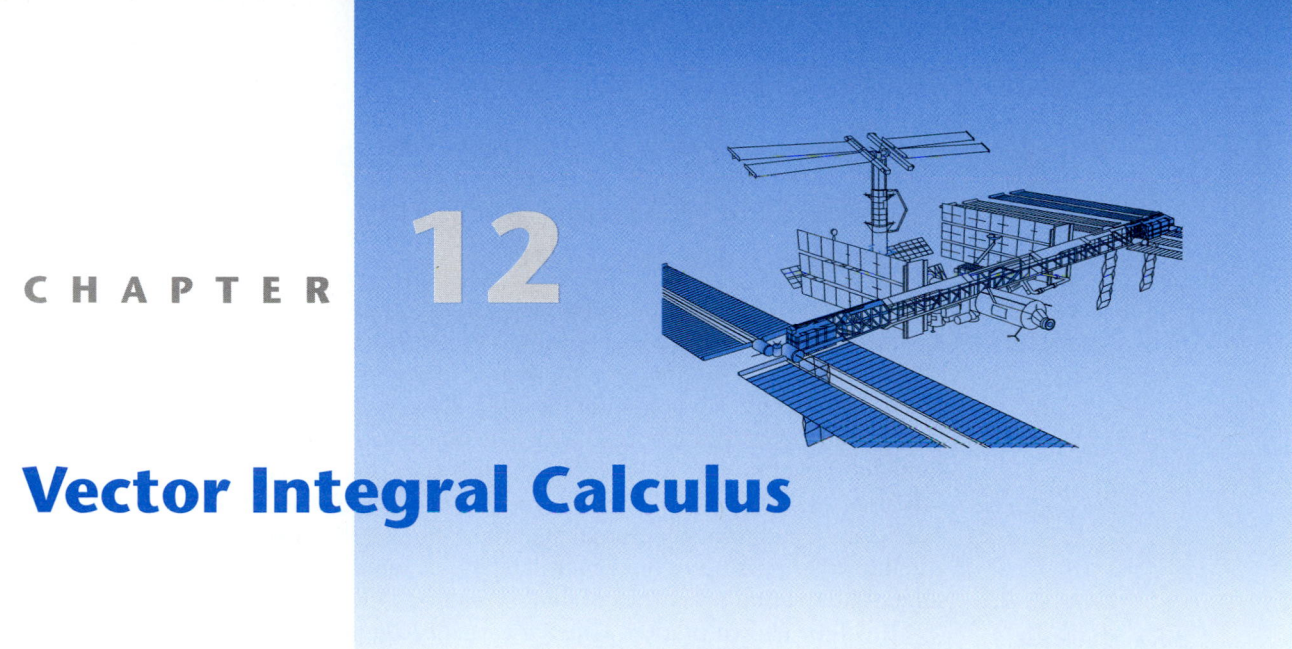

CHAPTER 12

Vector Integral Calculus

When working with the fundamental conservation laws governing engineering and physics, problems often arise that lead to the integral of the divergence of a vector function **F** over a volume V. The Gauss divergence theorem enables the integral of div **F** over volume V to be replaced by the integral of the normal component of **F** over the surface S enclosing V. This result simplifies calculations, because **F** is usually only known in general terms, whereas in physical problems the value of the normal component of **F** on S is known from the conditions of the problem.

Another vector quantity that arises naturally in engineering and physics is the vector function curl **F**, and when this occurs it is often necessary to integrate the normal component of curl **F** over an open surface S. This happens, for example, in fluid mechanics when working with the vorticity and circulation of a fluid. Stokes' theorem replaces the evaluation of the integral of the normal component of curl **F** over the open surface S by a directed line integral of **F** around the curve Γ forming the boundary of S. Here also a simplification results, because once again the vector function **F** on surface S is usually only known in general terms, whereas in physical problems its value on Γ is specified. Green's theorem in the plane is a two-dimensional form of Stokes' theorem, and it has many uses throughout engineering, physics, and mathematics.

The three most important vector integral theorems due to Gauss, Green, and Stokes are derived, followed by the derivation of two important integral transport theorems that play an essential role in mechanics, fluid mechanics, chemical engineering, electromagnetism, and elsewhere. After a review of the background of the vector integral calculus, and an introduction to the concept of an orientable surface, the Gauss divergence theorem and the theorems due to Green and Stokes are proved and applied.

The two fundamental integral transport theorems that are derived and applied are the flux transport theorem, which determines the rate of change of flux passing through an open surface bounded by a moving space curve, and Reynold's transport theorem, which concerns the rate of change of a volume integral when the volume is contained within a moving surface.

12.1   Background to Vector Integral Theorems

Information Provided by Vector Integral Theorems

Physical problems in two and three space dimensions often give rise to integrals with integrands that are determined by a vector field **F** defined over the region of integration. The most important of these integrals involves either the integration of div **F** over a finite volume V, or the integral over a finite open surface S in space of the component of curl **F** normal to S. The objective of this chapter will be to prove some fundamental integral theorems of this type due to Gauss, Stokes, and Green called, respectively, the *Gauss divergence theorem, Stokes' theorem*, and *Green's theorems*. In addition, as optional material, what is called the *flux transport theorem* and the *volume transport theorem* will be proved and, as applications, used to derive some fundamental properties of fluid mechanics.

three important theorems

It will be shown that the **Gauss divergence theorem**, often abbreviated to the **divergence theorem** or **Gauss' theorem**, relates the integral of div **F** over a volume V to the integral over the closed surface S enclosing V of the component of **F** normal to S. Thus, Gauss' theorem allows a volume integral of this type to be replaced by a simpler surface integral. **Stokes' theorem**, which will also be proved in Section 12.2, is of a different nature, in that it relates the integral of the normal component of curl **F** over an open surface S in space bounded by a closed space curve Γ to the line integral of the tangential component of **F** around Γ. So, in the case of Stokes' theorem, a surface integral of a special type over S is related to a simpler line integral around the closed space curve Γ that forms the boundary of S. **Green's theorem in the plane** is the two-dimensional form of Stokes' theorem, and a typical application is to be found in Chapter 14, where it is used in the proof of the Cauchy integral theorem for the integration of complex analytic functions.

Also proved will be two other theorems known as **Green's theorems**, though these results are also known as **Green's identities** or **Green's formulas**. They relate integrals of Laplacians of scalar functions Φ and Ψ over a volume V to the integral over the surface S enclosing V of the derivatives of these functions normal to S. Green's theorems are used extensively when working with partial differential equations involving the Laplacian operator, because they can be used to replace the integral over a volume V of a solution of Laplace's equation that is to be determined by the integral of the normal derivatives of the solution over S that occur as a prescribed boundary condition that must be satisfied by the solution.

A common feature of these theorems is that each frequently replaces an integral of a special type over a region (a volume or an open surface) by a simpler integral over the boundary of the region (a closed surface or a closed space curve), thereby reducing by one the number of dimensions involved in the integration. The integral can then be evaluated by using whichever of the two equivalent expressions is easier. When used with partial differential equations involving the Laplacian operator, Green's theorems typically allow integrals of unknown functions over a region to be replaced by simpler integrals of known functions over the boundary of the region.

The two transport theorems proved in Section 12.3 relate to the determination of the derivative with respect to time of surface and volume integrals of time-dependent integrands when the surface or volume involved moves with time. The *flux* of a vector **F** across a surface S is the integral over S of the component of

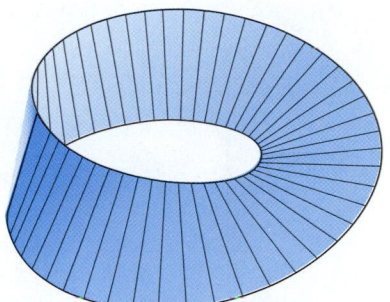

FIGURE 12.1   A Möbius strip.

F normal to S. The flux transport theorem describes the rate of change of the flux of **F** across S, taking into account the time dependence of **F** and the motion of S. A typical example of this type occurs when current is induced in a coil of wire moving in a magnetic field, because the current depends on the rate of change of magnetic flux through the moving coil.

The volume transport theorem describes the time rate of change of a volume integral due to the time dependence of the integrand and the motion of the volume over which integration takes place. A typical application of this theorem arises in fluid mechanics where the boundary of a volume of interest relating to a certain feature of the fluid flow does not move in the same way as the fluid, so that a flow takes place through the surface that encloses the volume.

Surfaces and Orientation

Section 12.2 is concerned with surfaces that have *two* sides and makes use of the normal at each point on such surfaces. It might seem unnecessary to define two-sided surfaces, but it is necessary because pathological surfaces exist that only have *one* side, and these must be excluded from the theorems of Section 12.2.

An example of a one-sided surface is provided by the **Möbius strip** shown in Fig. 12.1. This strip can be considered to be formed from a long strip of paper, the ends of which are joined after making a 180° twist in the paper about its longitudinal center line. Its one-sided nature can easily be verified by drawing a pencil line around the center line of the strip, because eventually the line will connect with the starting point, and if the strip is cut and opened out, examination will show a pencil line on both sides of the paper.

When deriving the Gauss divergence theorem, it will be necessary to work with a closed two-sided surface S, the *interior* of which contains the volume V of space that will concern us. A vector element of area of such a surface will have magnitude dS and an associated unit vector **n** normal to dS. As the normal **n** at a point on a two-sided surface S enclosing a volume V may be directed away from either side of S, it is necessary to adopt a standard convention for the direction of **n** and the vector element of area $d\mathbf{S} = \mathbf{n}\, dS$ on S. The normal **n** at a point on such a surface will always be chosen to be directed *out* of V. So if, for example, V is a sphere, the normal **n** at any point of its surface will be along a radial line drawn *outward* from the center of the sphere.

open surfaces and orientable surfaces

A two-sided **open surface** S bounded by a non-self-intersecting space curve Γ is a surface that does *not* have an interior, and so does *not* enclose a volume V. When

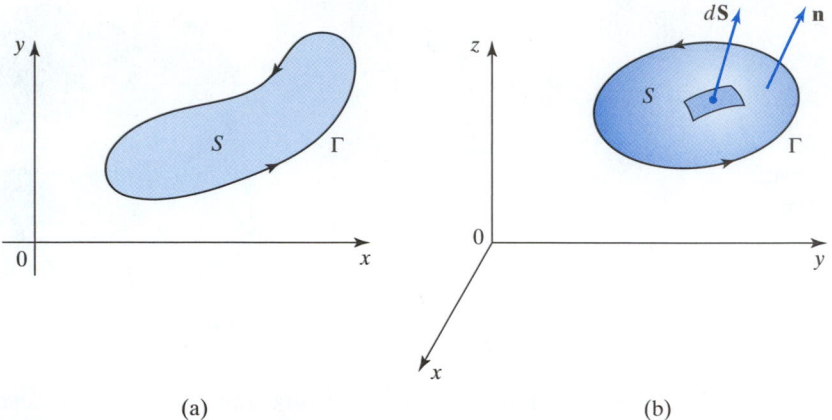

FIGURE 12.2   (a) A plane oriented surface. (b) A general oriented surface in space.

deriving Stokes' theorem it will be necessary to work with a two-sided open surface S bounded by a closed non-self-intersecting space curve Γ around which there is a given sense of direction. The normal at each point of S will be always be chosen in such a way that it points in the direction in which a right-handed screw would advance were it to be rotated in the sense of direction that is specified around the boundary curve Γ. Surfaces S of this type are called **oriented surfaces**. Pathological one-sided surfaces such as Möbius strips are said to be **nonorientable**, and they will not be considered here.

A simple but typical example of an open orientable surface S is an area in the (x, y)-plane contained within a closed curve Γ. If the sense of direction around Γ is chosen to be counterclockwise, the normal **n** to S will point in the direction of the unit vector **k**. A reversal of the sense of direction around Γ will reverse the sense of **n**, which will then point in the direction of $-$**k**. Examples of *oriented surfaces* are illustrated in Fig. 12.2, where Fig. 12.2(a) shows an open oriented surface S in the (x, y)-plane and Fig. 12.2(b) shows a general open oriented surface in space.

Let S be a two-sided surface with a boundary curve Γ around which a sense of direction is prescribed, and at each point of S let **n** be the unit normal to S pointing in the direction determined by the sense of direction around Γ, as described above. Then if dS is an element of area of S, the vector element of area on the oriented surface S is $d\mathbf{S} = \mathbf{n}dS$.

Summary

This brief section introduced the important concept of an open surface that is orientable, and established the right-handed screw convention by which the direction of the normal to an orientable surface is determined.

12.2   Integral Theorems

The first integral theorem to be established is the Gauss divergence theorem, which relates volume integrals and surface integrals. It is possible to formulate a more general statement of the theorem than the one given here, but to do so involves a lengthy argument, and Theorem 12.1 is sufficient for all practical purposes.

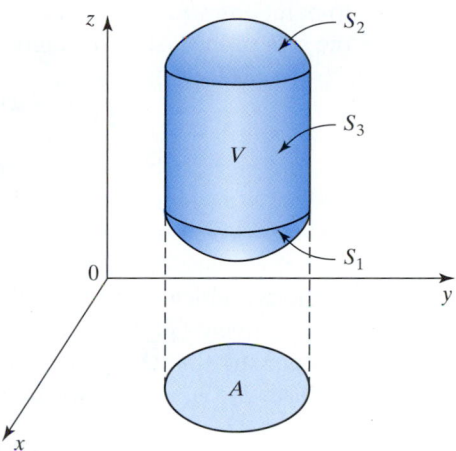

FIGURE 12.3   The volume V.

THEOREM 12.1

a theorem relating
the integral of div F
over a volume to the
integral of the
normal component
of F over the surface
bounding the volume

The Gauss divergence theorem   Let $\mathbf{F}$ be a vector field defined throughout a volume V enclosed within a piecewise smooth surface S on which the outward drawn unit normal is $\mathbf{n}$. Then, if the components of $\mathbf{F}$ and its first order partial derivatives are continuous throughout V and on S, dV is an element of volume of V, and dS is an element of area of S,

$$\iiint_V \operatorname{div} \mathbf{F} \, dV = \iint_S \mathbf{F} \cdot d\mathbf{S},$$

where $d\mathbf{S} = \mathbf{n}\, dS$ is a vector surface element of area on S.

Proof   Consider a volume V in the form of a cylinder with its sides parallel to the z-axis, a lower surface $z = z_1(x, y)$, and an upper surface $z = z_2(x, y)$, and let A be the projection of the cross-section of the cylinder onto the (x, y)-plane, as shown in Fig. 12.3.

The lower surface in Fig. 12.3 will be denoted by S_1, the upper surface by S_2, and the cylindrical side surface by S_3, so the surface S enclosing volume V is piecewise smooth and comprises these three surfaces.

Let $\mathbf{F} = F_1\mathbf{i} + F_2\mathbf{j} + F_3\mathbf{k}$, where the components of $\mathbf{F}$ and its first order partial derivatives are continuous in V and on S. The integral of $\partial F_3/\partial z$ with respect to z along a line in V drawn parallel to the z-axis is

$$\int_{z_1(x,y)}^{z_2(x,y)} \frac{\partial F_3}{\partial z} \, dz = F_3(x, y, z_2(x, y)) - F_3(x, y, z_1(x, y)).$$

The integral of this result over the area A that is the projection of V onto the (x, y)-plane is given by

$$\iiint_V \frac{\partial F_3}{\partial z} \, dV = \iint_A F_3(x, y, z_2(x, y)) \, dx \, dy - \iint_A F_3(x, y, z_1(x, y)) \, dx \, dy.$$

The first term on the right is the integral of F_3 over the *top* of the upper two-sided surface S_2, while the second term is the integral F_3 over the *top* of the lower two-sided surface S_1. As the normals to surfaces bounding the volume V are chosen to point *outward* from V, and the normal in the last term is directed *into* volume V, the sign of the last term can be reversed and the resulting equation written as

$$\iiint_V \frac{\partial F_3}{\partial z}\,dV = \iint_{S_2} F_3\,dx\,dy + \iint_{S_1} F_3\,dx\,dy.$$

To express the integrals on the right as a single integral over the complete surface S, it is necessary to take into account the integral of F_3 over the cylindrical surface S_3. The unit normal to the element of area $dx\,dy$ of A is perpendicular to the (x, y)-plane in the direction $\mathbf{k}$, but $\mathbf{k}$ is orthogonal to all outward drawn normals to the cylindrical surface, so the integral of F_3 over the cylindrical surface S_3 must vanish, giving $\iint_{S_3} F_3\,dx\,dy = 0$. Adding this integral to the preceding equation, and recognizing that the piecewise smooth surface S comprises the sum of the three surfaces S_1, S_2, and S_3, we arrive at the result

$$\iiint_V \frac{\partial F_3}{\partial z}\,dV = \iint_S F_3\,dx\,dy.$$

Corresponding results involving F_1 and F_2 that can be derived in similar fashion are

$$\iiint_V \frac{\partial F_1}{\partial x}\,dV = \iint_S F_1\,dy\,dz$$

and

$$\iiint_V \frac{\partial F_2}{\partial y}\,dV = \iint_S F_2\,dx\,dz.$$

Addition of these three integrals gives

$$\iiint_V \left(\frac{\partial F_1}{\partial x} + \frac{\partial F_2}{\partial y} + \frac{\partial F_3}{\partial z}\right)dV = \iint_S F_1\,dy\,dz + F_2\,dx\,dz + F_3\,dx\,dy,$$

or equivalently,

$$\iiint_V \operatorname{div}\mathbf{F}\,dV = \iint_S F_1\,dy\,dz + F_2\,dx\,dz + F_3\,dx\,dy.$$

Let dS with the outward drawn unit normal $\mathbf{n}$ be an element of area of the bounding surface S, and let its projection onto the (y, z)-plane be the element of area $dy\,dz$. Then if the angle between $\mathbf{n}$ and the normal to the (y, z)-plane is γ, it follows that $dy\,dz = dS\cos\gamma$. However, the unit normal to the (y, z)-plane is the vector $\mathbf{i}$, so $\cos\gamma = \mathbf{i}\cdot\mathbf{n}$, and consequently $dy\,dz = \mathbf{i}\cdot\mathbf{n}\,dS = \mathbf{i}\cdot d\mathbf{S}$. Similar arguments lead to the corresponding results $dx\,dz = \mathbf{j}\cdot d\mathbf{S}$ and $dx\,dy = \mathbf{k}\cdot d\mathbf{S}$.

Using these expressions in the preceding integral allows it to be written as

$$\iiint_V \operatorname{div}\mathbf{F}\,dV = \iint_S (F_1\mathbf{i} + F_2\mathbf{j} + F_3\mathbf{k})\cdot\mathbf{n}\,dS$$

or as

$$\iiint_V \operatorname{div}\mathbf{F}\,dV = \iint_S F\cdot d\mathbf{S},$$

and the theorem is proved for a volume V with sides parallel to the z-axis.   ∎

Modifications to the preceding form of argument that we will not detail show the theorem to be true for volumes V with boundaries formed by finitely many piecewise smooth parts, and also for boundaries on which the partial derivatives of

F_i are not differentiable at every point. The theorem remains true for domains such as a torus that have a more complicated shape. This follows because such domains can be subdivided into domains of the type covered by Theorem 12.1, and as the outward-drawn normals to each side of a dividing surface are oppositely directed, the integrals over the two sides of each such surface cancel, leaving only the integral over S of the component of **F** normal to S.

CARL FRIEDRICH GAUSS (1777–1855)
A German mathematician of truly outstanding ability who is universally regarded as the greatest mathematician of the nineteenth century. He ranks with Isaac Newton as one of the greatest mathematicians of all time. He was appointed to the directorship of the observatory in Göttingen and spent the remainder of his life there. His contributions spanned all aspects of mathematics and science, in addition to his interest in astronomy. He also made important contributions to number theory, algebra, and geometry.

The divergence theorem provides an alternative definition of div **F**, because if the result of the theorem is divided by the volume V with bounding surface S over which integration is performed, and the limit is taken as $V \to 0$ about a fixed point P in space, we obtain

$$(\text{div } \mathbf{F})_P = \lim_{V \to 0} \frac{1}{V} \iint_S \mathbf{F} \cdot d\mathbf{S}. \tag{1}$$

However, $\mathbf{F} \cdot d\mathbf{S} = \mathbf{F} \cdot \mathbf{n}\, dS$ and $\mathbf{F} \cdot \mathbf{n} = F_n$ is the component of **F** normal to dS, so $\iint_S \mathbf{F} \cdot d\mathbf{S}$ is the *flux* of **F** across S at the point P. Consequently, $(\text{div } \mathbf{F})_P$ is seen to be the flux of **F** per unit volume at P.

A physical interpretation of this last result is provided by the flow of a fluid with velocity **q**, because

$$(\text{div } \mathbf{q})_P = \lim_{V \to 0} \frac{1}{V} \iint_S \mathbf{q} \cdot d\mathbf{S} \tag{2}$$

an application to incompressible flow with sources and sinks

is seen to be the amount of fluid leaving an infinitesimal surface surrounding P in a unit time. If the fluid is **incompressible**, there can be no net flow either into or out of any volume, so in an incompressible fluid div $\mathbf{q} = 0$ throughout the fluid. If, however, there is a **source** of fluid at P causing fluid to flow into volume V and onward out of S, then $(\text{div } \mathbf{q})_P$ will be positive, whereas if there is removal of fluid from volume V at P due to the presence of a **sink** at P, then $(\text{div } \mathbf{q})_P$ will be negative. In a fluid that is **compressible**, div **q** may be either positive or negative at a point in the fluid without any source or sink being present.

Any vector **F** such that

$$\text{div } \mathbf{F} \equiv 0 \tag{3}$$

a solenoidal vector

is said to be a **solenoidal** vector. So as $\text{div}(\text{curl } \mathbf{F}) \equiv 0$, it follows that provided **F** has continuous second order partial derivatives, the vector curl **F** is a solenoidal vector.

The following examples illustrate how the divergence theorem can be used to simplify the evaluation of integrals, though more important applications arise in the formulation and solution of partial differential equations.

EXAMPLE 12.1   Evaluate

$$\iint_S 3x\,dydz + 2y\,dxdz - 5z\,dxdy$$

where S is a smooth surface bounding an arbitrary volume V.

Solution   The integral can be written

$$\iint_S 3x\,dydz + 2y\,dxdz - 5z\,dxdy = \iint_S \mathbf{F} \cdot d\mathbf{S},$$

where $\mathbf{F} = 3x\mathbf{i} + 2y\mathbf{j} - 5z\mathbf{k}$. So as the conditions of Theorem 12.1 are satisfied and div $\mathbf{F} = 0$, it follows from the divergence theorem that

$$\iint_S 3x\,dydz + 2y\,dxdz - 5z\,dxdy = \iiint_V \text{div } \mathbf{F}\, dV = 0. \qquad \blacksquare$$

EXAMPLE 12.2   Evaluate

$$\iint_S x^3\,dydz + y^3\,dxdz + z^3\,dxdy,$$

where the surface S is the boundary of the volume V occupying the region between the spheres $x^2 + y^2 + z^2 = 1$ and $x^2 + y^2 + z^2 = 4$ and above the plane $z = 0$.

Solution   The volume V is a hemispherical shell between spheres of radii 1 and 2 centered on the origin and above the plane $z = 0$, so its surface S is formed by the surfaces of two hemispheres above the $z = 0$ plane and the annulus $1 \le r \le 2$ in the plane $z = 0$. The required integral can be written

$$I = \iint_S x^3\,dydz + y^3\,dxdz + z^3\,dxdy = \iint_S \mathbf{F} \cdot d\mathbf{S},$$

where $\mathbf{F} = x^3\mathbf{i} + y^3\mathbf{j} + z^3\mathbf{k}$. As $\mathbf{F}$ is differentiable and the surface S is piecewise smooth, the divergence theorem can be used to replace the surface integral by the triple volume integral of div $\mathbf{F}$ over V, showing that

$$I = 3 \iiint_V (x^2 + y^2 + z^2)\,dxdydz.$$

The spherical symmetry of volume V suggests that integral I will be simplified if spherical polar coordinates are used. In terms of these coordinates, the volume V becomes $1 \le r \le 2, 0 \le \phi < 2\pi$, and $0 \le \theta \le \pi/2$, and the integrand becomes $x^2 + y^2 + z^2 = r^2$, so as the volume element of the transformation is given by $dV = r^2 \sin\phi\,dr\,d\theta\,d\phi$, the integral for I becomes

$$I = 3 \int_0^{2\pi} d\phi \int_0^{\pi/2} d\theta \int_1^2 r^4 \sin\theta\,dr$$

$$= 3 \int_0^{2\pi} d\phi \int_0^{\pi/2} \frac{31}{5} \sin\theta\,d\theta$$

$$= \frac{93}{5} \int_0^{2\pi} d\phi = \frac{186}{5}\pi. \qquad \blacksquare$$

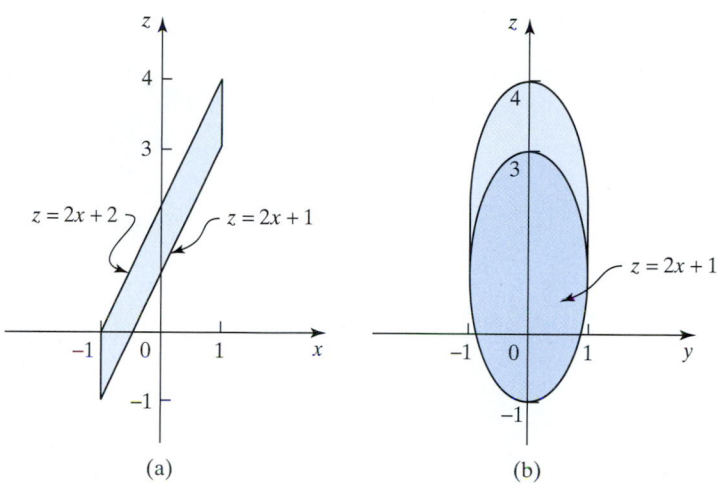

FIGURE 12.4 Cylinder with parallel oblique ends. (a) Side view; (b) front view.

EXAMPLE 12.3

Let the vector function $\mathbf{F} = (x^2 + 3y)\mathbf{i} - (3y^2 + \sin z)\mathbf{j} + 2z^2\mathbf{k}$ be defined throughout the volume V interior to the cylindrical volume with parallel oblique ends bounded by the surface S that is shown in Fig. 12.4, where the cylinder cross-section has the equation $x^2 + y^2 = 1$ and the cylinder ends are formed by the intersection of the cylinder with the planes $z = 2x + 1$ and $2x + 2$. Find the integral over S of F_n, the component of $\mathbf{F}$ normal to the surface S.

Solution The function $\mathbf{F}$ and the surface S satisfy the conditions of the divergence theorem, so as div $\mathbf{F} = 2x - 6y + 4z$, the result of applying the theorem to volume V is

$$\iint_S \mathbf{F} \cdot d\mathbf{S} = \iiint_V (2x - 6y + 4z)dV$$

$$= \iint_{x^2+y^2 \le 1} \left(\int_{1+2x}^{2+2x} (2x - 6y + 4z)dz \right) dxdy$$

$$= \iint_{x^2+y^2 \le 1} (10x - 6y + 6)dxdy.$$

To proceed further, we change to plane polar coordinates $x = r\cos\theta$, $y = r\sin\theta$ for which the Jacobian $J(r, \theta) = r$, and the area $x^2 + y^2 \le 1$ becomes $0 \le r \le 1$ with $0 \le \theta \le 2\pi$. As a result,

$$\iint_S \mathbf{F} \cdot d\mathbf{S} = \int_0^{2\pi} d\theta \int_0^1 (10r\cos\theta - 6r\sin\theta + 6)r\,dr$$

$$= \int_0^{2\pi} \left(\frac{10}{3}\cos\theta - 2\sin\theta + 3 \right) d\theta = 6\pi,$$

so the required integral over S of the component F_n of $\mathbf{F}$ normal to S is

$$\iint_S \mathrm{F}_n dS = 6\pi.$$

■

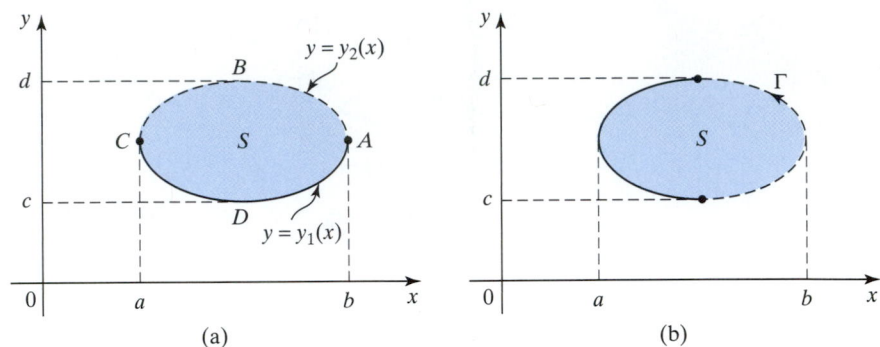

FIGURE 12.5 (a) The convex area S with lower and upper boundaries $y = y_1(x)$ and $y = y_2(x)$. (b) The convex area S with left and and right boundaries $x = x_1(y)$ and $x = x_2(y)$.

Preparatory to proving Stokes' theorem, we must prove Green's theorem in the plane that can be stated as follows.

THEOREM 12.2

a theorem relating an integral over a plane surface to an integral around its perimeter

Green's theorem in the plane   Let a finite area S in (x, y)-plane be bounded by a piecewise smooth closed non-self-intersecting plane curve Γ around which a counterclockwise sense of direction is imposed. Then if $P(x, y)$ and $Q(x, y)$ and their first order partial derivatives are continuous over S and on Γ,

$$\iint_S \left(\frac{\partial Q}{\partial x} - \frac{\partial P}{\partial y} \right) dx dy = \int_\Gamma P dx + Q dy.$$

Proof   We first prove the theorem for a plane area S that is convex, which is an area S with the property that any straight line that crosses it intersects the boundary at most twice. We then show how the theorem can be applied to more complicated areas, including those with internal boundaries. A typical area S of this type is shown in Fig. 12.5.

Let us consider the integral of $\partial P / \partial y$ over the convex area S with the lower boundary $y = y_1(x)$ and upper boundary $y = y_2(x)$, as shown in Fig. 12.5(a). The integral over S can be written as the iterated integral

$$\iint_S \frac{\partial P}{\partial y} dx dy = \int_a^b dx \int_{y_1(x)}^{y_2(x)} \frac{\partial P}{\partial y} dy$$

$$= \int_a^b P(x, y_2(x)) dx - \int_a^b P(x, y_1(x)) dx$$

or as

$$\iint_S \frac{\partial P}{\partial y} dx dy = -\int_{ABC} P(x, y) dx - \int_{CDA} P(x, y) dx,$$

where the sign of the first integral on the right has been reversed because integration from $x = a$ to $x = b$ is in the opposite sense to the counterclockwise direction of integration required along ABC. The two arcs ABC and CDA form the closed

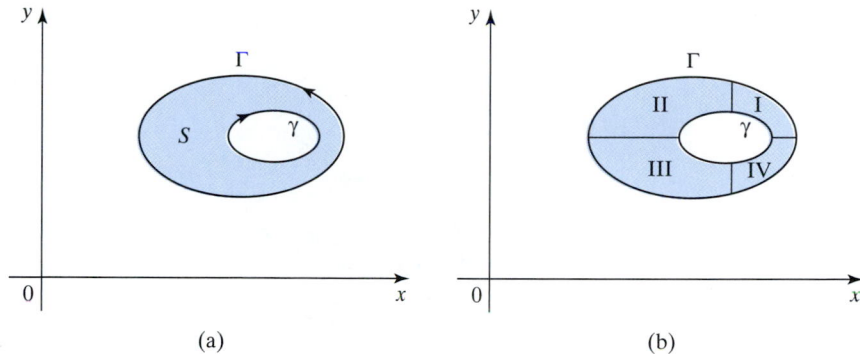

FIGURE 12.6   (a) S with an internal boundary. (b) The partitioning of S.

contour Γ, so the preceding result simplifies to

$$\iint_S \frac{\partial P}{\partial y}\,dx\,dy = -\int_\Gamma P(x,\,y)\,dx.$$

When the foregoing argument is repeated, but this time using the left and right boundaries in Fig. 12.5(b), and the integral of $\partial Q/\partial x$ over S is calculated we obtain

$$\iint_S \frac{\partial Q}{\partial x}\,dx\,dy = \int_\Gamma Q(x,\,y)\,dx.$$

However, as S is convex, each of these results is true, so subtracting them we arrive at the statement of Green's theorem

$$\iint_S \left(\frac{\partial Q}{\partial x} - \frac{\partial P}{\partial y}\right)dx\,dy = \int_\Gamma P\,dx + Q\,dy.$$

We need to show this result remains true for areas S that are not convex, and also for areas with internal boundaries. It will be sufficient to consider the area S shown in Fig. 12.6(a), in which there is a single internal boundary γ, because the argument extends immediately to arbitrary areas with finitely many internal boundaries, and to areas that are not convex.

Let S be partitioned into the four areas shown in Fig. 12.6(b), to each of which Green's theorem applies. Applying the theorem to each area and adding the integrals, we see that integrals along the adjacent straight line segments will cancel, because of the continuity of P, Q, and their first order partial derivatives in S, and the fact that the integrations take place in *opposite* directions. As a result only the integrals around the boundaries Γ and γ remain, so the theorem holds, provided the sense of integration around all boundaries (both external and internal) is such that the area S always lies to the *left* as each boundary is traversed. This argument also applies to finitely many internal boundaries, so Green's theorem in the plane is proved for this more general case.   ∎

The sense in which integration must be performed when applying Green's theorem to an area S with internal boundaries is illustrated in Fig. 12.7.

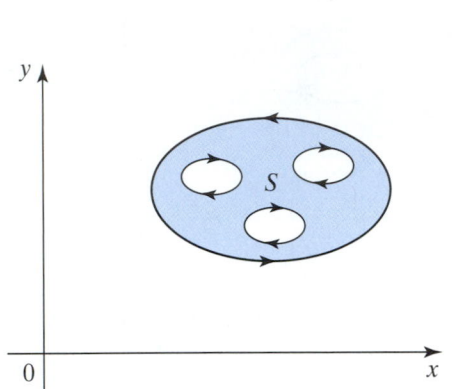

FIGURE 12.7   Direction of integration around a domain D with internal boundaries.

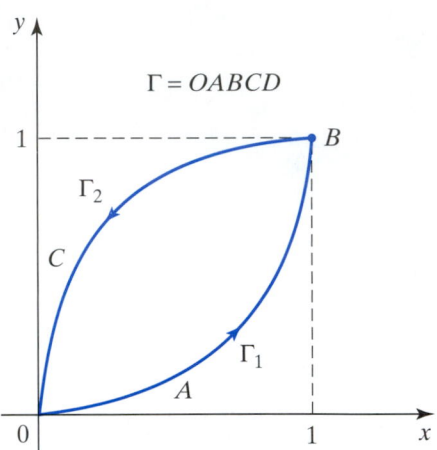

FIGURE 12.8   The curve Γ formed from two circular arcs Γ_1 and Γ.

GEORGE GREEN (1793–1841)
A self-taught English mathematical physicist who was born in Nottingham where he first worked as a baker. His contributions to electricity and magnetism, where he introduced the theorems now named after him, were first published privately in 1828, and so attracted little attention. It was not until William Thompson (Lord Kelvin) discovered his results and caused them to be republished in 1846 that their significance was recognized. Due to the limited circulation of the first published version of his work his main results were rediscovered, independently, by Lord Kelvin, Gauss, and others. He made significant contributions to the theory of optics and sound waves, and just prior to his death he was elected to a fellowship of Caius College, Cambridge.

SIR GEORGE GABRIEL STOKES (1819–1903)
A major applied mathematician and physicist who was born in County Sligo, Ireland, but spent his entire working life in Cambridge, where he was made professor of mathematics in 1849. He made fundamental contributions to the study of the flow of viscous fluids, leading to what are now called the Navier–Stokes equations, to elasticity, the propagation of sound, optics, and asymptotic series.

EXAMPLE 12.4   Evaluate

$$\int_{\Gamma} xy^2 dx - 2x^2 y \, dy$$

where Γ is the curve shown in Fig. 12.8, in which Γ_1 is an arc of a unit circle centered on the point $(0, 1)$, and Γ_2 is an arc of a unit circle centered on the point $(1, 0)$, and integration is in the counterclockwise sense around Γ.

Solution   The equation of a unit circle with its center at $(1, 0)$ is $x^2 + (y - 1)^2 = 1$, so the equation of the arc Γ_1 is $y = 1 - 1\sqrt{1 - x^2}$ for $0 \leq x \leq 1$. The equation of a unit circle with its center at $(1, 0)$ is $(x - 1)^2 + y^2 = 1$, so the equation of arc Γ_2 is $y = \sqrt{2x - x^2}$ for $0 \le x \le 1$.

Making the identifications $P = xy^2$ and $Q = -2x^2 y$ we have $\partial P/\partial y = 2xy$ and $\partial Q/\partial x = -4xy$, so substituting into Green's theorem shows that

$$\int_\Gamma xy^2 dx - 2x^2 y dy = \int_0^1 dx \int_{1-\sqrt{1-x^2}}^{\sqrt{2x-x^2}} (-6xy) dy$$

$$= \int_0^1 [-6x^2 + 6x - 6x\sqrt{1 - x^2}] dx = -1. \qquad \blacksquare$$

THEOREM 12.3

a theorem relating an integral of the normal component of curl F over an orientable surface to the line integral of F around its perimeter

Stokes' theorem   Let S be an open piecewise smooth orientable surface bounded by a closed space curve Γ around which a sense of direction is specified. At every point of the surface, let the unit normal $\mathbf{n}$ to S point in the direction specified for orientable surfaces relative to the sense around Γ. Then, if $\mathbf{F}$ is a differentiable vector function over the surface S,

$$\int_\Gamma \mathbf{F} \cdot d\mathbf{r} = \iint_S \operatorname{curl} \mathbf{F} \cdot d\mathbf{S},$$

where $\mathbf{r}$ is the position vector of a general point on Γ.

Proof   Consider Fig. 12.9, in which S is an open orientable surface $z = z(x, y)$, Γ is its bounding space curve, A is the projection of S onto the (x, y)-plane, and C is the boundary curve of A.

The proof will involve the following three steps:

(I) The line integral around Γ will be transformed into the line integral around C

(II) The line integral around C will be transformed into a double integral over A

(III) The double integral over A will be transformed into an integral over S

STEP I   Let $\mathbf{F} = F_1 \mathbf{i} + F_2 \mathbf{j} + F_3 \mathbf{k}$. Then the line integral of F_1 around Γ is

$$\int_\Gamma F_1(x, y, z) dx = \int_C F_1(x, y, z(x, y)) dx,$$

because $z = z(x, y)$ on C.

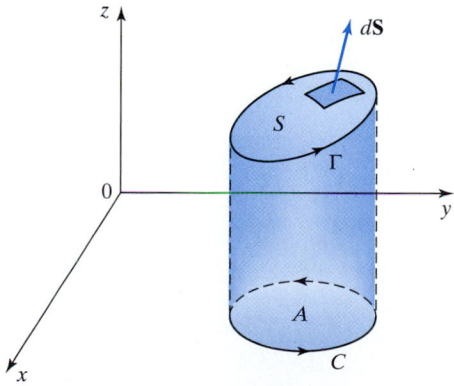

FIGURE 12.9   An orientable surface S bounded by the space curve Γ.

STEP II   In the line integral on the right $z = z(x, y)$, so

$$\frac{\partial G_1}{\partial y} = \frac{\partial F_1}{\partial y} + \frac{\partial F_1}{\partial z}\frac{\partial z}{\partial y}, \quad \text{where } G_1(x, y) \equiv F_1(x, y, z(x, y)).$$

Applying Green's theorem in the plane to the integral in Step I and using this last result gives

$$\int_C F_1(x, y, z(x, y))dx = -\iint_A \left(\frac{\partial F_1}{\partial y} + \frac{\partial F_1}{\partial z}\frac{\partial z}{\partial y}\right) dA,$$

where dA is the area element in the (x, y)-plane.

Setting $\phi = z - z(x, y)$, the surface S has the equation $\phi = 0$, so as a normal $\mathbf{N}$ to S is given by $\mathbf{N} = \text{grad } \phi$

$$\mathbf{N} = \pm\left(-\frac{\partial z}{\partial x}\mathbf{i} - \frac{\partial z}{\partial y}\mathbf{j} + \mathbf{k}\right).$$

For $\mathbf{N}$ to have the correct *upward* direction relative to S, as required by the sense of direction of integration around the oriented surface S, it is necessary that the z-component of $\mathbf{N}$ be positive. Consequently, if we take the positive sign, the unit vector $\mathbf{n}$ normal to S is

$$\mathbf{n} = n_1\mathbf{i} + n_2\mathbf{j} + n_3\mathbf{k},$$

where the direction cosines n_1, n_2, and n_3 are given by

$$n_1 = -\frac{\partial z}{\partial x}\bigg/|\mathbf{N}|, \quad n_2 = -\frac{\partial z}{\partial y}\bigg/|\mathbf{N}|, \quad n_3 = 1/|\mathbf{N}| \quad \text{with}$$

$$|\mathbf{N}| = \left(\left(\frac{\partial z}{\partial x}\right)^2 + \left(\frac{\partial z}{\partial y}\right)^2 + 1\right)^{1/2}.$$

It now follows from these results that

$$\frac{\partial z}{\partial y} = -\frac{n_2}{n_3}.$$

If we substitute this expression for $\partial z/\partial y$ in the double integral over A, it becomes

$$\int_C F_1 dx = -\iint_A \left(\frac{\partial F_1}{\partial y} - \frac{\partial F_1}{\partial z}\frac{n_2}{n_3}\right) dA.$$

STEP III   If dA is the projection of dS onto the (x, y)-plane, we have $dA = n_3\,dS$, so the last result in Step II can be written as the double integral over S

$$\int_C F_1 dx = -\iint_S \left(\frac{\partial F_1}{\partial y} - \frac{\partial F_1}{\partial z}\frac{n_2}{n_3}\right) n_3 dS$$

$$= \iint_S \left(\frac{\partial F_1}{\partial z}n_2 - \frac{\partial F_1}{\partial y}n_3\right) dS.$$

Similar arguments show that

$$\int_C F_2 dy = \iint_S \left(\frac{\partial F_2}{\partial x}n_3 - \frac{\partial F_2}{\partial z}n_1\right) dS$$

and

$$\int_C F_3 dz = \iint_S \left(\frac{\partial F_3}{\partial y}n_1 - \frac{\partial F_3}{\partial x}n_2\right) dS.$$

Finally, the addition of these three integrals gives

$$\int_C F_1 dx + F_2 dy + F_3 dz = \iint_S \left(\frac{\partial F_3}{\partial y} - \frac{\partial F_2}{\partial z} \right) n_1 dS + \left(\frac{\partial F_1}{\partial z} - \frac{\partial F_3}{\partial x} \right) n_2 dS$$
$$+ \left(\frac{\partial F_2}{\partial x} - \frac{\partial F_1}{\partial y} \right) n_3 dS,$$

or equivalently,

$$\int_\Gamma F_1 dx + F_2 dy + F_3 dz = \iint_S \left(\frac{\partial F_3}{\partial y} - \frac{\partial F_2}{\partial z} \right) dy dz + \left(\frac{\partial F_1}{\partial z} - \frac{\partial F_3}{\partial x} \right) dx dz$$
$$+ \left(\frac{\partial F_2}{\partial x} - \frac{\partial F_1}{\partial y} \right) dx dy,$$

which is one form of Stokes' theorem. To arrive at the form given in the statement of the theorem it is only necessary to write $d\mathbf{S} = \mathbf{n} dS$, and then to recognize that

$$\text{curl } \mathbf{F} = \left(\frac{\partial F_3}{\partial y} - \frac{\partial F_2}{\partial z} \right) \mathbf{i} + \left(\frac{\partial F_1}{\partial z} - \frac{\partial F_3}{\partial x} \right) \mathbf{j} + \left(\frac{\partial F_2}{\partial x} - \frac{\partial F_1}{\partial y} \right) \mathbf{k},$$

for the integral to become

$$\int_\Gamma \mathbf{F} \cdot d\mathbf{r} = \iint_S \text{curl } \mathbf{F} \cdot d\mathbf{S}. \qquad \blacksquare$$

Stokes' theorem is a generalization of Green's theorem in the plane that was used in its proof, so it is to be expected that Stokes' theorem must reduce to Green's theorem in the plane when the surface S is an area in the (x, y)-plane. That this is the case can be seen by taking $\mathbf{F}$ to be only a function of x and y, so that $\mathbf{F} = F_1(x, y)\mathbf{i} + F_2(x, y)\mathbf{j}$, because then the first form of Stokes' theorem that was proved reduces to

$$\int_\Gamma F_1 dx + F_2 dy = \iint_S \left(\frac{\partial F_2}{\partial x} - \frac{\partial F_1}{\partial y} \right) dx dy,$$

and apart from a change of notation, this is the result of Theorem 12.2.

Stokes' theorem provides a physical interpretation of curl $\mathbf{F}$ that is most easily understood in the context of a fluid flow with $\mathbf{F}$ representing the fluid velocity vector. Consider a small disc of fluid of radius ρ centered at $\mathbf{r} = \mathbf{r}_0$, as shown in Fig. 12.10,

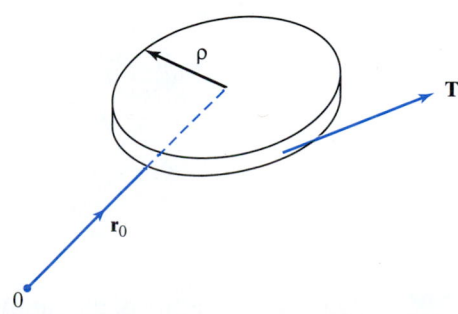

FIGURE 12.10   A disc of fluid of radius ρ with fluid velocity $\mathbf{F}$.

where S is the area of the disc and T is the unit tangent vector to the perimeter of the disc. Then $\mathbf{F} \cdot \mathbf{T}$ is the tangential component of the fluid velocity at the perimeter Γ of the disc around which the arc length is s, so the integral

$$\kappa(\mathbf{r}_0) = \int_\Gamma \mathbf{F} \cdot \mathbf{T} ds$$

is a measure of the tendency of the fluid to *rotate* around the point $\mathbf{r}_0$. This will be recognized as the *circulation* of $\mathbf{F}$ around a curve Γ introduced previously in connection with line integrals.

If the disc is small and taken on an open surface S in the fluid, and $\mathbf{N}$ is a unit normal to an element dS of the surface at $\mathbf{r} = \mathbf{r}_0$, the scalar product $(\mathrm{curl}\,\mathbf{F}) \cdot \mathbf{N}$ can be regarded as a constant over the disc, so from Stokes' theorem

$$\int_\Gamma \mathbf{F} \cdot \mathbf{T} ds = \iint_S (\mathrm{curl}\,\mathbf{F}) \cdot \mathbf{N} dS \approx [(\mathrm{curl}\,\mathbf{F}) \cdot \mathbf{N}]_{\mathbf{r}_0} (\pi \rho^2),$$

and so

$$[(\mathrm{curl}\,\mathbf{F}) \cdot \mathbf{N}]_{\mathbf{r}_0} = \lim_{\rho \to 0} \frac{1}{\pi \rho^2} \int_\Gamma \mathbf{F} \cdot \mathbf{T} ds.$$

Clearly, $(\mathrm{curl}\,\mathbf{F}) \cdot \mathbf{N}$ attains its greatest value when curl $\mathbf{F}$ is parallel to $\mathbf{N}$, and it is because curl $\mathbf{F}$ is a measure of rotation that some books use the notation rot $\mathbf{F}$ in place of curl $\mathbf{F}$. Although the circulation around Γ has been illustrated by means of a fluid flow, the general concept of the circulation of a vector $\mathbf{F}$ around a curve Γ has useful physical interpretations in other situations. Another example occurs in connection with the generation of current when a wire in the form of a closed curve Γ moves in a magnetic field. Inspection of the definition of $(\mathrm{curl}\,\mathbf{F}) \cdot \mathbf{N}$ at a point $\mathbf{r}_0$ as a limit shows it is the quotient of the circulation of $\mathbf{F}$ around Γ and the area of the disc, and so again measures the rate of circulation at $\mathbf{r}_0$.

EXAMPLE 12.5   Let $\mathbf{F} = x^2 \mathbf{i} + z^2 y \mathbf{j} + y^2 z \mathbf{k}$. Show that the line integral of $\mathbf{F}$ around any space curve Γ bounding an oriented open surface S is zero.

Solution   The conditions of Stokes' theorem apply and

$$\mathrm{curl}\,\mathbf{F} = \begin{vmatrix} \mathbf{i} & \mathbf{j} & \mathbf{k} \\ \dfrac{\partial}{\partial x} & \dfrac{\partial}{\partial y} & \dfrac{\partial}{\partial z} \\ x^2 & yz^2 & y^2 z \end{vmatrix} = \mathbf{0},$$

so

$$\int_\Gamma \mathbf{F} \cdot d\mathbf{r} = \iint_S \mathrm{curl}\,\mathbf{F} \cdot d\mathbf{S} = 0. \qquad \blacksquare$$

EXAMPLE 12.6   Let S be the surface of the paraboloid of revolution $z = 1 - x^2 - y^2$ with the domain of definition $x^2 + y^2 \leq 1$, and let Γ be the boundary of the paraboloid. Given $\mathbf{F} = x^3 \mathbf{i} + (x + y - z)\mathbf{j} + yz \mathbf{k}$, find $\iint_S \mathrm{curl}\,\mathbf{F} \cdot d\mathbf{S}$.

Solution   By Stokes' theorem

$$\iint_S \text{curl } \mathbf{F} \cdot d\mathbf{S} = \int_\Gamma \mathbf{F} \cdot d\mathbf{r},$$

so the required integral can be found by evaluating the line integral on the right. As the domain of definition of the paraboloid of revolution is $x^2 + y^2 \leq 1$, it follows that the curve Γ bounding the surface of the paraboloid is the circle $x^2 + y^2 = 1$ in the plane $z = 0$. To evaluate the line integral, we parametrize Γ as $\mathbf{r}(t) = \cos t\mathbf{i} + \sin t\mathbf{j}$, with $0 \leq t \leq 2\pi$. Then $d\mathbf{r} = (-\sin t\mathbf{i} + \cos t\mathbf{j})dt$ and on Γ the vector function

$$\mathbf{F}(t) = \cos^3 t\mathbf{i} + (\cos t + \sin t)\mathbf{j},$$

so substituting into the line integral gives

$$\iint_S \text{curl } \mathbf{F} \cdot d\mathbf{S} = \int_0^{2\pi} [\cos^3 t\mathbf{i} + (\cos t + \sin t)\mathbf{j}] \cdot [-\sin t\mathbf{i} + \cos t\mathbf{j}]dt$$

$$\int_0^{2\pi} (-\sin t \cos^3 t + \cos^2 t + \sin t \cos t)dt = \pi. \qquad \blacksquare$$

EXAMPLE 12.7   Given $\mathbf{F} = y\mathbf{i} - z^3\mathbf{j} + x^2\mathbf{k}$, use Stokes' theorem to evaluate $\int_\Gamma \mathbf{F} \cdot d\mathbf{r}$, where Γ is the boundary of the area S formed by the part of the plane $2x + 4y + z = 4$ that lies in the first octant, and integration around the boundary Γ is in the clockwise direction.

Solution   The required integral will be determined by evaluating the integral on the right of

$$\int_\Gamma \mathbf{F} \cdot d\mathbf{r} = \iint_S \text{curl } \mathbf{F} \cdot d\mathbf{S}.$$

The surface S over which integration is to be performed is the plane triangular area shown in Fig. 12.11, where the boundary of S in the plane $z = 0$ is the line $x + 2y = 2$

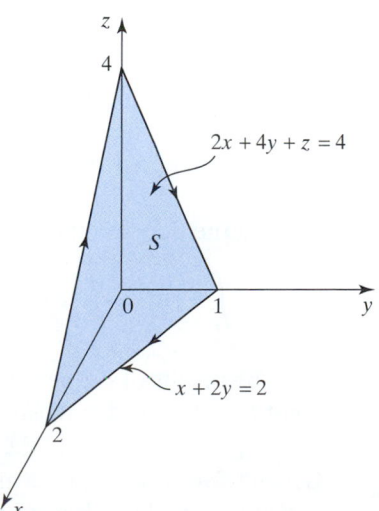

FIGURE 12.11   Plane triangular area S with clockwise direction around boundary Γ.

for $0 \leq x \leq 2$.

$$\text{curl } \mathbf{F} = \begin{vmatrix} \mathbf{i} & \mathbf{j} & \mathbf{k} \\ \dfrac{\partial}{\partial x} & \dfrac{\partial}{\partial y} & \dfrac{\partial}{\partial z} \\ y & -z^3 & x^2 \end{vmatrix} = 3z^2\mathbf{i} - 2x\mathbf{j} - \mathbf{k}.$$

If we set $\phi = 4 - 2x - 4y - z$, the equation of the plane is $\phi = 0$, so two possible normals $\mathbf{N}$ to the surface S of the plane are

$$\mathbf{N} = \pm \text{grad } \phi = \pm(-2\mathbf{i} - 4\mathbf{j} - \mathbf{k}).$$

As the direction of integration around the boundary Γ is taken to be *clockwise*, when viewed as in Fig 12.11, the normal to S must be directed away from S toward the origin, showing that the $\mathbf{k}$ component of $\mathbf{N}$ must be *negative*. Thus, the foregoing expression for $\mathbf{N}$ must be chosen with the positive sign leading to the result $\mathbf{N} = -2\mathbf{i} - 4\mathbf{j} - \mathbf{k}$, so the unit vector $\mathbf{n} = \mathbf{N}/|\mathbf{N}|$ with the required sense normal to the plane is

$$\mathbf{n} = \frac{1}{\sqrt{21}}(-2\mathbf{i} - 4\mathbf{j} - \mathbf{k}).$$

The line of intersection of the plane $2x + 4y + z = 4$ and the plane $z = 0$ is $x + 2y = 2$, so the base of the triangular plane surface S has the equation $x + 2y = 2$ for $0 \leq x \leq 2$.

We now have sufficient information to compute $\iint_S \text{curl } \mathbf{F} \cdot d\mathbf{S}$:

$$\int_\Gamma \mathbf{F} \cdot d\mathbf{r} = \iint_S \text{curl } \mathbf{F} \cdot d\mathbf{S} = \iint_S (3z^2\mathbf{i} - 2x\mathbf{j} - \mathbf{k}) \cdot d\mathbf{S},$$

but $d\mathbf{S} = \mathbf{n}\, dS$, so if A is the projection of S onto the plane $z = 0$, the integral over S can be replaced by the integral over A, giving

$$\int_\Gamma \mathbf{F} \cdot d\mathbf{r} = \iint_S (3z^2\mathbf{i} - 2x\mathbf{j} - \mathbf{k}) \cdot d\mathbf{S} = \frac{1}{\sqrt{21}} \iint_S (-6z^2 + 8x + 1)\, dS.$$

However, if n_3 is the $\mathbf{k}$ component of $\mathbf{n}$, $d A/dS = |n_3| = 1\sqrt{21}$ and so $dS = \sqrt{21}\, d A$. Using this result in the integral on the right with $z = 4 - 2x - 4y$ shows that

$$\int_\Gamma \mathbf{F} \cdot d\mathbf{r} = \iint_A [-6(4 - 2x - 4y)^2 + 8x + 1]\, d A.$$

Writing the double integral over A as a repeated integral gives

$$\int_\Gamma \mathbf{F} \cdot d\mathbf{r} = \int_0^1 dy \int_0^{-2-2y} [-6(4 - 2x - 4y)^2 + 8x + 1]\, dx = -\frac{29}{3}. \qquad \blacksquare$$

The results of the next theorem, called **Green's formulas** or sometimes **Green's identities**, are used extensively in the study of partial differential equations.

THEOREM 12.4   **Green's formulas**   Let Φ and Ψ be scalar fields such that the Laplacians $\Delta\Phi$ and $\Delta\Psi$ are defined inside a volume V enclosed in a closed piecewise smooth surface S, and if the second order partial derivatives of Φ and Ψ have any discontinuities, let them be bounded and occur only along lines on S or across finitely many surfaces in V. Then:

two useful formulas due to Green

(I) Green's first formula is

$$\iint_S \Phi \frac{\partial \Psi}{\partial n} dS = \iiint_V \{\Phi \Delta \Psi + (\text{grad } \Phi) \cdot (\text{grad } \Psi)\} dV,$$

where dV is a volume element of V.

(II) Green's second formula is

$$\iint_S \left(\Phi \frac{\partial \Psi}{\partial n} - \Psi \frac{\partial \Phi}{\partial n} \right) dS = \iiint_V (\Phi \Delta \Psi - \Psi \Delta \Phi) dV.$$

Proof　The proof is straightforward, but for simplicity it will only be offered for functions Φ and Ψ that have continuous second order partial derivatives inside a finite volume V and on its bounding surface S.

Setting $\mathbf{G} = \Phi(\text{grad } \Psi)$, it follows that

$$\text{div } \mathbf{G} = \Phi \, \text{div}(\text{grad } \Psi) + (\text{grad } \Phi) \cdot (\text{grad } \Psi),$$

so applying the divergence theorem we have

$$\iint_S \Phi(\text{grad } \Psi) \cdot d\mathbf{S} = \iiint_V \{\Phi \Delta \Psi + (\text{grad } \Phi) \cdot (\text{grad } \Psi)\} dV.$$

However, $\Phi(\text{grad } \Psi) \cdot d\mathbf{S} = \Phi \mathbf{n} \cdot (\text{grad } \Psi) dS$, but $\mathbf{n} \cdot (\text{grad } \Psi)$ is simply the directional derivative of Ψ in the direction of the unit outward normal $\mathbf{n}$ that will be denoted by $\partial \Psi / \partial n$, so

$$\Phi(\text{grad} \Psi) \cdot d\mathbf{S} = \Phi \, \partial \Psi / \partial n dS.$$

Using this in the last result gives Green's first formula,

$$\iint_S \Phi \frac{\partial \Psi}{\partial n} dS = \iiint_V \{\Phi \Delta \Psi + (\text{grad } \Phi) \cdot (\text{grad } \Psi)\} dV.$$

Green's second formula follows directly from this by interchanging Φ and Ψ and subtracting the new result from the Green's first formula.　∎

showing the uniqueness of the solution of $\Delta \phi = 0$ in a volume, on the surface of which ϕ is specified

In anticipation of Chapter 18, and as an illustration of the use of Green's first formula in the study of partial differential equations, we will prove the **uniqueness** of the solution ϕ of Laplace's equation

$$\Delta \phi = 0$$

in a volume V enclosed within a surface S on which the value of ϕ is specified at every point. Here, the *Laplacian* Δ can be considered to be expressed in terms of any system of orthogonal curvilinear coordinates, the simplest of which is, of course, the cartesian coordinate system where

$$\Delta \equiv \frac{\partial^2}{\partial x^2} + \frac{\partial^2}{\partial y^2} + \frac{\partial^2}{\partial z^2}.$$

By the *uniqueness* of the solution of Laplace's equation, we mean that when ϕ is specified over the surface S enclosing a volume V, there is only *one* function ϕ that satisfies both Laplace's equation throughout V and the specified conditions for ϕ on the surface S. A typical physical example illustrating the interpretation of this situation is provided by considering the steady state temperature distribution $T(x, y, z)$ throughout a cube of metal where the temperature is governed by the Laplace equation

$$\frac{\partial^2 T}{\partial x^2} + \frac{\partial^2 T}{\partial y^2} + \frac{\partial^2 T}{\partial z^2} = 0.$$

It is to be expected from a physical understanding of steady state heat conduction that the specification of a time-independent temperature distribution T over each face of the cube of metal will determine the temperature at each internal point of the metal, and that every time the surfaces of the same metal block are heated in the same way, the same internal temperature distribution will result. This is simply another way of saying that the solution of Laplace's equation subject to specified boundary conditions on S is expected to be *unique*.

The proof of this result is simple. Suppose, if possible, that two different solutions ϕ_1 and ϕ_2 exist that satisfy the *same* prescribed temperature conditions on S. Then, because Laplace's equation is linear, the function $\Phi = \phi_1 - \phi_2$ must also be a solution and, furthermore, $\Phi \equiv 0$ on S. Using this function Φ in Green's first formula and setting $\Psi = \Phi$ reduces it to

$$\iiint_D (\text{grad } \Phi) \cdot (\text{grad } \Phi) dV = 0.$$

The integrand is nonnegative, so this result can only be possible if grad $\Phi \equiv \mathbf{0}$, and this in turn implies that $\partial \Phi / \partial x = \partial \Phi / \partial y = \partial \Phi / \partial z = 0$, and so $\Phi = $ constant. However, as $\Phi = 0$ on the bounding surface S, this shows that $\Phi = 0$ throughout D, and so $\phi_1 \equiv \phi_2$ and the result is proved.

The theory and application of the vector integral calculus are developed in standard calculus and analytic geometry texts like those in references [1.1], [1.2], [1.5], [1.6], and [1.7]. More advanced and detailed accounts, with emphasis placed on a vector treatment, are to be found in references [5.1] to [5.3]. Extensive use of vector integral theorems in the study of hydrodynamics is made in reference [6.5].

Summary

The three fundamental integral theorems of Gauss, Green, and Stokes were proved, and in anticipation of the results of Chapter 18, a Green formula was used to establish the uniqueness of the solution of the Laplace equation $\Delta\phi = 0$ in a volume on the surface of which ϕ is specified. It will be seen later in Chapter 18 that this is called a *Dirichlet problem* for the Laplace equation, and it arises in many physical situations, such as the steady state temperature distribution in a solid, the electrostatic potential in a vacuum enclosed in a cavity, in problems of groundwater flow, and elsewhere.

EXERCISES 12.2

1. By setting $\mathbf{F} = \mathbf{a} \times \mathbf{G}$ in the divergence theorem, where $\mathbf{a}$ is an arbitrary constant vector and $\mathbf{G}$ is a differentiable vector function defined in a volume V in a closed surface S, prove by using the properties of the scalar triple product that

$$\iint_S \mathbf{G} \times d\mathbf{S} = -\iiint_V \text{curl } \mathbf{G} dV.$$

2. Given a differentiable scalar function ϕ defined in a volume V contained in a closed surface S, prove that

$$\iiint_V (\text{grad } \phi) \times d\mathbf{S} \equiv \mathbf{0}.$$

3. Given the differentiable scalar and vector functions ϕ and $\mathbf{G}$, respectively, defined in a volume V in a closed

surface S, prove that

$$\iint_S \phi\mathbf{G} \cdot d\mathbf{S} = \iiint_V (\text{grad } \phi) \cdot \mathbf{G} dV + \iiint_V \phi \text{ div } \mathbf{G} dV.$$

4. Given the differentiable vector functions $\mathbf{P}$ and $\mathbf{Q}$ defined in a volume V bounded by a closed surface S, prove that

$$\iint_S \mathbf{P} \times \mathbf{Q} \cdot d\mathbf{S} = \iiint_V \mathbf{Q} \cdot \text{curl } \mathbf{P} dV$$
$$- \iiint_V \mathbf{P} \cdot \text{curl } \mathbf{Q} dV.$$

5. The time-dependent heat equation can be written

$$\mu\rho \frac{\partial T}{\partial t} = \text{div}(\kappa \text{ grad } T),$$

where μ, ρ, and κ are material constants that may vary with position, t is the time, and T the temperature at a position $\mathbf{r}$ in a material occupying a volume V enclosed in a surface S. Prove that

$$\iint_S \kappa T(\text{grad } T) \cdot d\mathbf{S} = \iiint_V \kappa(\text{grad } T) \cdot (\text{grad } T) dV$$
$$+ \iiint_V \mu\rho T \frac{\partial T}{\partial t} dV.$$

6. Given that $\mathbf{R} = \text{curl } \mathbf{Q}$ and $\mathbf{Q} = \text{curl } \mathbf{P}$ are defined in a volume V enclosed in a surface S, prove that

$$\iiint_V \mathbf{Q} \cdot \mathbf{Q} dV = \iint_S \mathbf{P} \times \mathbf{Q} \cdot d\mathbf{S} + \iiint_V \mathbf{P} \cdot \mathbf{R} dV.$$

7. By using Stokes' theorem and considering curl $(\phi\mathbf{F})$, where ϕ and $\mathbf{F}$ are differentiable scalar and vector functions, respectively, both of which are defined over an open surface S with closed boundary curve Γ, prove that

$$\int_\Gamma \phi\mathbf{F} \cdot d\mathbf{R} = \iint_S (\text{grad } \phi) \times \mathbf{F} \cdot d\mathbf{S} + \iint_S \phi \text{ curl } \mathbf{F} \cdot d\mathbf{S}.$$

8. Given that ϕ and ψ are differentiable scalar functions defined over an open surface S with the closed boundary curve Γ, prove that

$$\int_\Gamma \phi(\text{grad } \Psi) \cdot d\mathbf{r} = \iint_S (\text{grad } \phi) \times (\text{grad } \psi) \cdot d\mathbf{S}.$$

9. Let $\mathbf{F} = -y^2\mathbf{i} + xz\mathbf{j} + z^2\mathbf{k}$ and S be the surface of the plane $x + y + 2z = 2$ lying in the first octant ($x \geq 0$, $y \geq 0$, $z \geq 0$) with a clockwise sense of direction around its triangular boundary Γ. Verify Stokes' theorem by computing $\int_\Gamma \mathbf{F} \cdot d\mathbf{r}$ and $\iint_S \text{curl } \mathbf{F} \cdot d\mathbf{S}$ and showing they are equal.

10. Given that $\mathbf{F} = yz\mathbf{i} + xy\mathbf{j} + x^2\mathbf{k}$ and S is the surface of the plane $x + 3y + z = 3$ lying in the first octant ($x \geq 0$, $y \geq 0$, $z \geq 0$) with a clockwise sense of direction around its triangular boundary Γ when seen from 0, verify Stokes' theorem by computing $\int_\Gamma \mathbf{F} \cdot d\mathbf{r}$ and $\iint_S \text{curl } \mathbf{F} \cdot d\mathbf{S}$ and showing they are equal.

12.3   Transport Theorems

In many applications the derivative with respect to time of surface and volume integrals is required where the integrand is a time-dependent field quantity and the surface or volume over which integration is to be performed moves with time. This situation arises, for example, when the rate of change of flux of a vector quantity $\mathbf{F}(\mathbf{r}, t)$ is required through an open surface $S(t)$ bounded by a moving closed space curve $\Gamma(t)$, or when the rate of change of a scalar quantity $f(\mathbf{r}, t)$ is required in a volume $V(t)$ that is enclosed in a moving surface $S(t)$. When computing the time derivative in the first case, it is necessary to take into account not only the time variation of the integrand, but also the effect of the moving boundary $\Gamma(t)$ of the surface $S(t)$ over which the time derivative of the flux is to be determined, whereas in the second case, in addition to the time dependence of $f(\mathbf{r}, t)$, the effect of the change in volume $V(t)$ must be considered.

Situations of this type occur when determining the generation of an electric current in a moving coil of wire in a magnetic field, in fluid mechanics when the energy content of a moving volume of fluid is considered and also in the study of shock waves, and in chemically reacting fluids where the chemical composition of a moving volume of fluid changes with time.

In this section two results called **transport theorems** will be derived. The first involves the rate of change of flux of a vector field across an open moving surface, whereas the second concerns the rate of change of a volume integral of a scalar quantity when the volume involved is swept out by a moving open surface.

The first result involves computing the time derivative of the flux $\Phi(t)$ of a vector function $\mathbf{F}(\mathbf{r}, t)$ through an open surface $S(t)$ bounded by a closed time-dependent space curve $\Gamma(t)$. When deriving this result it will be assumed that the points on $S(t)$ and $\Gamma(t)$ move with a specified velocity $\mathbf{v} = \mathbf{v}(\mathbf{r}, t)$ that is defined throughout the region of space involved. The **flux** $\Phi(t)$ at time t is defined as the integral of the component of $\mathbf{F}(\mathbf{r}, t)$ normal to the surface $S(t)$, and so is given by

$$\Phi(t) = \iint_{S(t)} \mathbf{F}(\mathbf{r}, t) \cdot d\mathbf{S}, \qquad (4)$$

where $d\mathbf{S}$ is an element of area of $S(t)$.

THEOREM 12.5

a transport theorem for the rate of change of flux

The flux transport theorem   Let a vector field $\mathbf{F}(\mathbf{r}, t)$ be defined and differentiable in some region of space in which the points on an open surface $S(t)$ with a closed boundary curve $\Gamma(t)$ move with a prescribed velocity $\mathbf{q}(\mathbf{r}, t)$. Then the rate of change of the flux $\Phi(t)$ of the vector field $\mathbf{F}(\mathbf{r}, t)$ through $S(t)$ is given by

$$\frac{d\Phi}{dt} = \iint_{S(t)} \left[\frac{\partial \mathbf{F}}{\partial t} + (\text{div } \mathbf{F})\mathbf{q} \right] \cdot d\mathbf{S} + \int_{\Gamma(t)} \mathbf{F} \times \mathbf{q} \cdot d\mathbf{r}.$$

Proof   Consider the surface $S(t)$ at time t and the surface $S(t + h)$ at a subsequent time $t + h$ shown in Fig. 12.12, where the points of $S(t)$ move with the given velocity $\mathbf{v}(\mathbf{r}, t)$. Then $S(t)$ sweeps out the cylindrical volume $V(t)$ shown in the diagram, where the line AB on the side surface of the cylinder shows the path followed by point A on $\Gamma(t)$ as it moves to the corresponding point B on $\Gamma(t + h)$. Correspondingly, a typical point P on $S(t)$ will move to the point Q on $S(t + h)$ along the line PQ, where for a small time increment h the vector $\underline{AB} \approx \mathbf{v}(\mathbf{r}_A, t)h$, and the vector $\underline{PQ} \approx \mathbf{v}(\mathbf{r}_P, t)h$, where $\mathbf{r}_A$ and $\mathbf{r}_P$ are the position vectors of A and P.

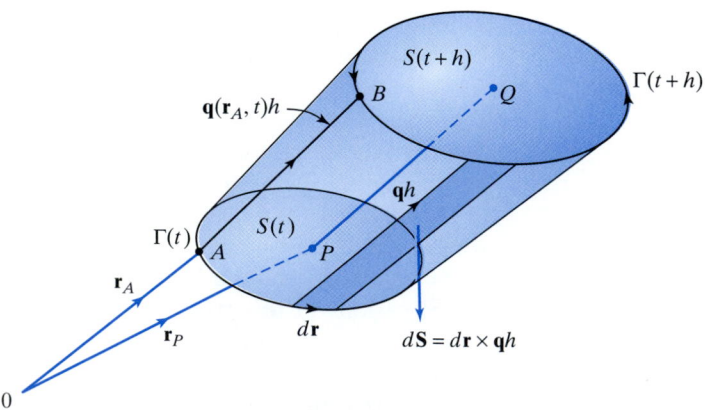

FIGURE 12.12   The surfaces S at times t and $t + h$ and the bounding curves $\Gamma(t)$ and $\Gamma(t + h)$.

It follows from the definition of a derivative that the time derivative of the flux $\Phi(t)$ is given by the limit

$$\frac{d\Phi}{dt} = \lim_{h \to 0} \left\{ \frac{1}{h} \left[\iint_{S(t+h)} \mathbf{F}(\mathbf{r}, t+h) \cdot d\mathbf{S} - \iint_{S(t)} \mathbf{F}(\mathbf{r}, t) \cdot d\mathbf{S} \right] \right\}. \tag{5}$$

In order to compute this limit, we first consider the difference

$$\iint_{S(t+h)} \mathbf{F}(\mathbf{r}, t+h) \cdot d\mathbf{S} - \iint_{S(t)} \mathbf{F}(\mathbf{r}, t) \cdot d\mathbf{S},$$

and for small h use the Taylor approximation

$$\mathbf{F}(\mathbf{r}, t+h) \approx \mathbf{F}(\mathbf{r}, t) + h\frac{\partial \mathbf{F}}{\partial t}$$

to rewrite it as

$$\iint_{S(t+h)} \mathbf{F}(\mathbf{r}, t+h) \cdot d\mathbf{S} - \iint_{S(t)} \mathbf{F}(\mathbf{r}, t) \cdot d\mathbf{S}$$

$$\approx \iint_{S(t+h)} \mathbf{F}(\mathbf{r}, t) \cdot d\mathbf{S} + h \iint_{S(t)} \frac{\partial \mathbf{F}}{\partial t} \cdot d\mathbf{S} - \iint_{S(t)} \mathbf{F}(\mathbf{r}, t) \cdot d\mathbf{S}. \tag{6}$$

To proceed further, if V is the volume swept out by $S(t)$ in time increment h, then the *outward*-drawn normal to V at $S(t+h)$ is $d\mathbf{S}$, while the *outward*-drawn normal to V at $S(t)$ is $-d\mathbf{S}$. Denoting the side of the cylindrical volume by Σ and applying the divergence theorem to $\mathbf{F}(\mathbf{r}, t)$ in V gives

$$\iiint_V \operatorname{div} \mathbf{F}(\mathbf{r}, t) dV = \iint_{S(t+h)} \mathbf{F}(\mathbf{r}, t) \cdot d\mathbf{S} - \iint_{S(t)} \mathbf{F}(\mathbf{r}, t) \cdot d\mathbf{S} + \iint_{\Sigma} \mathbf{F}(\mathbf{r}, t) \cdot d\mathbf{S}. \tag{7}$$

Using (7) to eliminate $\iint_{S(t)} \mathbf{F}(\mathbf{r}, t) \cdot d\mathbf{S}$ from (6) leads to the result

$$\iint_{S(t+h)} \mathbf{F}(\mathbf{r}, t+h) \cdot d\mathbf{S} - \iint_{S(t)} \mathbf{F}(\mathbf{r}, t) \cdot d\mathbf{S}$$

$$\approx h \iint_{S(t)} \frac{\partial \mathbf{F}}{\partial t} \cdot d\mathbf{S} + \iiint_V \operatorname{div} \mathbf{F}(\mathbf{r}, t) dV - \iint_{\Sigma} \mathbf{F}(\mathbf{r}, t) \cdot d\mathbf{S}. \tag{8}$$

Now on the side Σ of the cylindrical surface the outward-drawn surface element $d\mathbf{S} = d\mathbf{r} \times \mathbf{q}h$, where $d\mathbf{r}$ is a vector element along $\Gamma(t)$ directed in the counterclockwise direction. The volume element dV swept out by $d\mathbf{S}$ in time increment h is the product of the area $|d\mathbf{S}|$ of $d\mathbf{S}$ and the perpendicular distance l between $S(t+h)$ and $S(t)$ given by $l = |\mathbf{q}h \cdot \mathbf{n}|$, where $\mathbf{n}$ is the unit normal to $d\mathbf{S}$, so that $dV = d\mathbf{S} \cdot \mathbf{q}h$. When these results are used to simplify (8) and h is small, it becomes

$$\iint_{S(t+h)} \mathbf{F}(\mathbf{r}, t+h) \cdot d\mathbf{S} - \iint_{S(t)} \mathbf{F}(\mathbf{r}, t) \cdot d\mathbf{S}$$

$$\approx h \iint_{S(t)} \frac{\partial \mathbf{F}}{\partial t} \cdot d\mathbf{S} + h \iint_{S(t)} \operatorname{div} \mathbf{F}(\mathbf{r}, t)\mathbf{q} \cdot d\mathbf{S} + h \int_{\Gamma(t)} \mathbf{F}(\mathbf{r}, t) \times \mathbf{q} \cdot d\mathbf{r}, \tag{9}$$

where the sign of the last term has been changed by using the result $\mathbf{F} \cdot d\mathbf{r} \times \mathbf{q} = -\mathbf{F} \times \mathbf{q} \cdot d\mathbf{r}$.

Using (9) in the difference quotient (5) and proceeding to the limit as $h \to 0$ brings us to the statement of the theorem:

$$\frac{d\Phi}{dt} = \iint_{S(t)} \left[\frac{\partial \mathbf{F}}{\partial t} + (\text{div } \mathbf{F})\mathbf{q} \right] \cdot d\mathbf{S} + \int_{\Gamma(t)} \mathbf{F} \times \mathbf{q} \cdot d\mathbf{r}. \qquad \blacksquare$$

EXAMPLE 12.8

Let $S(t)$ be a plane rectangular area with its corners at the points $(0, 0, z)$, $(x, 0, z)$, $(x, 1, z)$, and $(0, 1, z)$, where $x = vt$, $z = ut$, t is the time, and u and v are constant speeds. Verify the flux transport theorem in the case that $\mathbf{F} = xz\mathbf{k}$, where $\mathbf{k}$ is the unit vector in the z-direction.

Solution   To verify Theorem 12.5 it will first be necessary to compute $\Phi(t)$ in order to find $d\Phi/dt$ directly. The theorem will be verified in this case if this expression for $d\Phi/dt$ can be shown to equal the sum of the surface and line integrals on the right of the statement of the theorem when each has been computed separately.

The geometry of the problem is shown in Fig. 12.13(a), and the projection of $S(t)$ onto the (x, y)-plane is shown in Fig. 12.13(b). It can be seen from the statement of the problem that the rectangular area remains parallel to the (x, y)-plane while moving along the z-axis with the constant speed u, and that its length increases with constant speed v in the positive x-direction.

We have $\mathbf{F} = xz\mathbf{k}$, $z = ut$, $x = vt$, so as the motion is uniform in the x- and z-directions, each point of $S(t)$ must move with the velocity $\mathbf{q} = v\mathbf{i} + u\mathbf{k}$. The flux $\Phi(t)$ is given by

$$\Phi(t) = \iint_{S(t)} \mathbf{F}(\mathbf{r}, t) \cdot d\mathbf{S} = \int_0^1 \int_0^{vt} xz\mathbf{k} \cdot \mathbf{k}\, dx\, dy = \int_0^1 \int_0^{vt} xz\, dx\, dy.$$

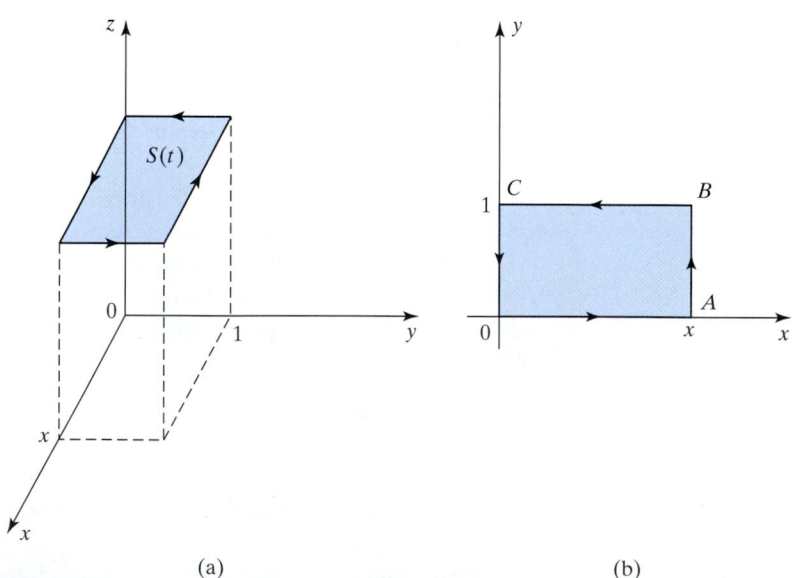

(a)   (b)

FIGURE 12.13   (a) The moving planar rectangle $S(t)$. (b) The projection of $S(t)$ onto the (x, y)-plane.

So as $z = ut$ is not involved in the integration, it can be removed as a factor to give

$$\Phi(t) = ut \int_0^1 \int_0^{vt} x\,dx\,dy = \frac{1}{2}uv^2t^3,$$

so the rate of change of flux when computed directly is given by

$$\frac{d\Phi}{dt} = \frac{3}{2}uv^2t^2.$$

Now $\partial \mathbf{F}/\partial t = \mathbf{0}$, div $\mathbf{F} = x$, and $d\mathbf{S} = dx\,dy\,\mathbf{k}$, so as

$$\left[\frac{\partial \mathbf{F}}{\partial \mathbf{t}} + (\text{div } \mathbf{F})\right]\mathbf{q} = xv\mathbf{i} + xu\mathbf{k},$$

$$\iint_{S(t)} \left[\frac{\partial \mathbf{F}}{\partial \mathbf{t}} + \text{div } \mathbf{F}\right]\mathbf{q} \cdot d\mathbf{S} = \int_0^1 \int_0^{vt} (xv\mathbf{i} + xu\mathbf{k}) \cdot \mathbf{k}\,dx\,dy$$

$$= u\int_0^1 \int_0^{vt} x\,dx\,dy = \frac{1}{2}uv^2t^2.$$

A simple calculation shows that $\mathbf{F} \times \mathbf{q} = xvz\mathbf{j}$, and so

$$\int_\Gamma \mathbf{F} \times \mathbf{q} \cdot d\mathbf{r} = \int_{\Gamma(t)} xvz\mathbf{j} \cdot d\mathbf{r} = uvt\int_{\Gamma(t)} x\mathbf{j} \cdot d\mathbf{r}.$$

Inspection of Fig. 12.13(b) shows that on OA, $d\mathbf{r} = dx\mathbf{i}$, on AB, $d\mathbf{r} = dy\mathbf{j}$, on BC, $d\mathbf{r} = -dx\mathbf{i}$, and on CO, $d\mathbf{r} = -dy\mathbf{j}$. The orthogonality of $\mathbf{i}$ and $\mathbf{j}$ means there are no contributions from the line integrals along OA and BC, and as $x = 0$ on OC there no contribution from the line integral along CO, so that

$$\int_{\Gamma(t)} \mathbf{F} \times \mathbf{q} \cdot d\mathbf{r} = uvtx\int_0^1 dy = uv^2t^2.$$

We see from this that

$$\iint_{S(t)} \left[\frac{\partial \mathbf{F}}{\partial \mathbf{t}} + (\text{div } \mathbf{F})\mathbf{q}\right] \cdot d\mathbf{S} + \int_{\Gamma(t)} \mathbf{F} \times \mathbf{q} \cdot d\mathbf{r} = \frac{1}{2}uv^2t^2 + uv^2t^2 = \frac{3}{2}uv^2t^2.$$

This result equals the expression for $d\Phi/dt$ found previously by direct computation, so the theorem has been verified in this case. ∎

a theorem determining the rate of change of an integral over a volume $V(t)$ of a function of position and time when the surface bounding $V(t)$ is moving

The second transport theorem concerns the rate of change of a volume integral of a differentiable scalar function $f(\mathbf{r}, t)$ when the volume $V(t)$ over which integration is performed is bounded by a closed moving surface $S(t)$, so for this reason it is called the **volume transport theorem**. Because of the importance of this theorem in fluid mechanics, where it was first derived by Reynolds, it is also known as the **Reynolds transport theorem**.

THEOREM 12.6

The Reynolds transport theorem   Let the scalar function $f(\mathbf{r}, t)$ be defined and differentiable in a region of space $V(t)$ through which the points inside and on a closed surface $S(t)$ move with a prescribed velocity $\mathbf{q}(\mathbf{r}, t)$. Then

$$\frac{d}{dt}\iiint_{V(t)} f(\mathbf{r}, t)dV = \iiint_{V(t)} \frac{\partial f}{\partial t}dV + \iint_{S(t)} f(\mathbf{r}, t)\mathbf{q} \cdot d\mathbf{S}.$$

Proof For simplicity we only offer an intuitive derivation of the theorem. Let a scalar function $f(\mathbf{r}, t)$ be defined and differentiable throughout some region in which a volume $V(t)$ enclosed in a closed surface $S(t)$ moves, and let the points of $V(t)$ and $S(t)$ move with a prescribed velocity $\mathbf{q}(r, t)$. Then our objective will be to compute

$$\frac{d}{dt} \iiint_{V(t)} f(\mathbf{r}, t) dV,$$

where dV is the volume element in $V(t)$. To accomplish this we start from the definition of a derivative in terms of a limit

$$\frac{d}{dt} \iiint_{V(t)} f(\mathbf{r}, t) dV = \lim_{h \to 0} \frac{1}{h} \left[\iiint_{V(t+h)} f(\mathbf{r}, t+h) dV - \iiint_{V(t)} f(\mathbf{r}, t) dV \right],$$
(10)

and write $V(t + h) = V(t) + \Delta(t, h)$, where $\Delta(t, h)$ represents the change in volume $V(t)$ in the time increment h. As a result of this (10) becomes

$$\frac{d}{dt} \iiint_{V(t)} f(\mathbf{r}, t) dV$$

$$= \lim_{h \to 0} \frac{1}{h} \left[\iiint_{V(t)} f(\mathbf{r}, t+h) dV - \iiint_{V(t)} f(\mathbf{r}, t) dV + \iiint_{\Delta(t,h)} f(\mathbf{r}, t) dV \right]$$

$$= \lim_{h \to 0} \iiint_{V(t)} \frac{1}{h} [f(\mathbf{r}, t+h) - f(\mathbf{r}, t)] dV + \lim_{h \to 0} \frac{1}{h} \left[\iiint_{\Delta(t,h)} f(\mathbf{r}, t+h) dV \right]$$

$$= \iiint_{V(t)} \frac{\partial f(\mathbf{r}, t)}{\partial t} + \lim_{h \to 0} \frac{1}{h} \left[\iiint_{\Delta(t,h)} f(\mathbf{r}, t+h) dV \right].$$
(11)

The volume $\Delta(t, h)$ is the change in volume of $V(t)$ in the time increment h, but in this time a surface element $d\mathbf{S}$ of $S(t)$ is displaced by the vector $\mathbf{q}h$, so the corresponding volume element swept out by $d\mathbf{S}$ in $\Delta(t, h)$ in this time interval is $dV \approx h\mathbf{q} \cdot d\mathbf{S}$. Consequently, (11) becomes

$$\frac{d}{dt} \iiint_{V(t)} f(\mathbf{r}, t) dV = \iiint_{V(t)} \frac{\partial f(\mathbf{r}, t)}{\partial t} dV + \lim_{h \to 0} \frac{1}{h} \left[\iint_{S(t)} h f(\mathbf{r}, t+h) \mathbf{q} \cdot d\mathbf{S} \right].$$

If we take the limit as $h \to 0$, when $f(\mathbf{r}, t+h) \to f(\mathbf{r}, t)$, this reduces to the statement of the theorem

$$\frac{d}{dt} \iiint_{V(t)} f(\mathbf{r}, t) dV = \iiint_{V(t)} \frac{\partial f}{\partial t} dV + \iint_{S(t)} f(\mathbf{r}, t) \mathbf{q} \cdot d\mathbf{S}. \qquad \blacksquare$$

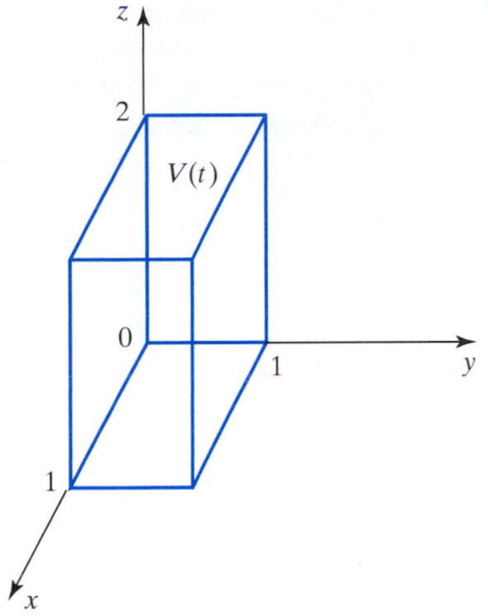

FIGURE 12.14 The rectangular parallelepiped with its top surface moving vertically with the constant speed u.

EXAMPLE 12.9

Verify the Reynolds transport theorem when $f = x^2 yzt$ and the volume $V(t)$ is the rectangular parallelepiped with the corners of its base at the points $(0, 0, 0)$, $(1, 0, 0)$, $(1, 1, 0)$, and $(0, 1, 0)$, its sides normal to the (x, y)-plane, and the corners of its upper surface at the points $(0, 0, z)$, $(1, 0, z)$, $(1, 1, z)$, and $(0, 1, z)$ when $z = ut$, with t the time and u a constant speed.

Solution The geometry of the problem is shown in Fig. 12.14. To verify the Reynolds transport theorem, it is necessary first to compute the integral $\iiint_{V(t)} f(\mathbf{r}, t) dV$, and then to find its derivative with respect to time t. The theorem will be verified if this result can be shown to equal the sum of the two integrals on the right of the theorem when they are evaluated separately:

$$\iiint_{V(t)} f(\mathbf{r}, t) dV = \int_0^1 \int_0^1 \int_0^{ut} x^2 yzt \, dz \, dy \, dx = \frac{1}{3}\frac{1}{2}\frac{1}{2} u^2 t^2 t = \frac{1}{12} u^2 t^3,$$

so

$$\frac{d}{dt} \iiint_{V(t)} f(\mathbf{r}, t) dV = \frac{1}{4} u^2 t^2.$$

We have

$$\iiint_{V(t)} \frac{\partial f}{\partial t} dV = \int_0^1 \int_0^1 \int_0^{ut} x^2 yz \, dz \, dy \, dx = \frac{1}{3}\frac{1}{2}\frac{1}{2} u^2 t^2 = \frac{1}{12} u^2 t^2,$$

and as $\mathbf{q} = u\mathbf{k}$ and $d\mathbf{S} = dx \, dy \mathbf{k}$,

$$\iint_{S(t)} f(\mathbf{r}, t) \mathbf{q} \cdot d\mathbf{S} = z \int_0^1 \int_0^1 x^2 ytu \, dy \, dx = \frac{1}{3}\frac{1}{2} u^2 t^2 = \frac{1}{6} u^2 t^2.$$

The theorem is verified, because $\frac{1}{12} u^2 t^2 + \frac{1}{6} u^2 t^2 = \frac{1}{4} u^2 t^2$. ∎

Summary   The flux transport theorem and the Reynolds' transport theorem, also known as the volume transport theorem, were proved and applied. Typical examples of the application of these theorems is the use of the first theorem to determine the rate of change of electric flux through a moving coil of wire in a generator, and the use of the second theorem when considering the continuity equation in fluid mechanics.

EXERCISES 12.3

1. Verify the rate of change of flux theorem given that $\mathbf{F} = xz\mathbf{k}$ and $S(t)$ is the plane rectangular surface with its corners at the points $(0, 0, z)$, $(x, 0, z)$, (x, y, z), and $(0, y, z)$, where $x = ut$, $y = vt$, and $z = wt$, with t the time and $u > 0$, $v > 0$, $w > 0$ a constant speed.

2. Verify the rate of change of flux theorem given that $\mathbf{F} = xz\mathbf{k}$ and $S(t)$ is the plane rectangular surface with its corners at the points $(0, 0, z)$, $(1, 0, z)$, $(1, y, z)$, and $(0, y, z)$, where $y = vt$ and $z = \alpha t^2$, with t the time and $v > 0$ a constant speed.

3.* A volume $V(t)$ in the form of a rectangular parallelepiped has the corners of its base at the points $(0, 0, z_1)$, $(1, 0, z_1)$, $(1, 1, z_1)$, and $(0, 1, z_1)$ with its sides perpendicular to the (x, y)-plane and the corners of its top surface at the points $(0, 0, z_2)$, $(1, 0, z_2)$, $(1, 1, z_2)$, and $(0, 1, z_2)$, where $z_1 = ut$ and $z_2 = vt$, with t the time and u, v constant speeds such that $u > 0$, $v > 0$. Verify the Reynolds transport theorem for the case in which $f(\mathbf{r}, t) = xyt$.

4.* A volume $V(t)$ in the form of a rectangular parallelepiped has the corners of its base at the points $(0, -\pi/2, 0)$, $(\pi, -\pi/2, 0)$, $(\pi, \pi/2, 0)$, and $(0, \pi/2, 0)$ with its sides perpendicular to the (x, y)-plane and the corners of its top surface at the points $(0, -\pi/2, z)$, $(\pi, -\pi/2, z)$, $(\pi, \pi/2, z)$, and $(0, \pi/2, z)$, where $z = ut$, with t the time and $u > 0$ a constant speed. Verify the Reynolds transport theorem for the case in which $f(\mathbf{r}, t) = \sin x \cos y e^z t^2$.

5.* A cylindrical volume $V(t)$ of height h has the center of its circular base located at the origin on the plane $z = 0$ and a radius $r = ut$, where t is the time and $u > 0$ is a constant speed. Verify the Reynolds transport theorem given that $f = r^2 t$.

6.* A hemispherical volume $V(t)$ lies in the region $z > 0$ with its center located at the origin in the plane $z = 0$ and a radius $r = ut$, where t is the time and $u > 0$ is a constant speed. Verify the Reynolds transport theorem given that $f = r^3 t$.

12.4 Fluid Mechanics Applications of Transport Theorems

When using the transport theorems, in fluid mechanics and elsewhere, two different types of time derivative occur, and for what is to follow it is important to distinguish between them. Consider a moving continuous medium, like a fluid, that has a property f associated with it, say its density, that depends on position $\mathbf{r}$ and the time t so that $f = f(\mathbf{r}, t)$. One way of finding the time derivative of f is to regard $\mathbf{r}$ as a fixed point, and then to find the time rate of change of f as seen by an observer fixed at point $\mathbf{r}$. This time derivative is denoted by $\partial f/\partial t$, and it is evaluated by differentiating f with respect to t while keeping $\mathbf{r}$ fixed. The other physically important time derivative of f involves letting the position vector $\mathbf{r}$ be a point that moves with the medium, so that $\mathbf{r} = \mathbf{r}(t)$, and then finding the time derivative of f at the moving point $\mathbf{r}$. This time derivative of f is denoted by df/dt, and in continuum mechanics it is called the **material derivative** of f, or sometimes the **convected derivative** of f, in which case it is often represented by Df/Dt.

To find the connection between the derivatives $\partial/\partial t$ and d/dt, when finding df/dt it is necessary to allow for the fact that the position vector $\mathbf{r}(t) = x(t)\mathbf{i} + y(t)\mathbf{j} + z(t)\mathbf{k}$, so that $f = f(\mathbf{r}(t), t)$. Thus, allowing for the time variation in $\mathbf{r}(t)$, we have

$$\frac{df}{dt} = \frac{\partial f}{\partial t} + \frac{\partial f}{\partial x}\frac{dx}{dt} + \frac{\partial f}{\partial y}\frac{dy}{dt} + \frac{\partial f}{\partial z}\frac{dz}{dt} \quad \text{or} \quad \frac{df}{dt} = \frac{\partial f}{\partial t} + (\mathbf{q} \cdot \nabla)f,$$

where $\mathbf{q} = (dx/dt)\mathbf{i} + (dy/dt)\mathbf{j} + (dz/dt)\mathbf{k}$ is the velocity of the moving point $\mathbf{r}(t)$. This shows that the material derivative operation can be written

$$\frac{d}{dt} = \frac{\partial}{\partial t} + (\mathbf{q} \cdot \nabla). \tag{12}$$

Before proceeding further, notice that an application of the divergence theorem to the last term in Reynolds' transport theorem (Theorem 12.6) allows it to be written in the equivalent form

$$\frac{d}{dt} \iiint_{V(t)} f\, dV = \iiint_{V(t)} \left\{ \frac{\partial f}{\partial t} + \nabla \cdot (f\mathbf{q}) \right\} dV, \tag{13}$$

but from Theorem 11.6 (iii) $\operatorname{div}(f\mathbf{q}) = f(\nabla \cdot \mathbf{q}) + (\mathbf{q} \cdot \nabla)f$, so

$$\frac{d}{dt} \iiint_{V(t)} f\, dV = \iiint_{V(t)} \left\{ \frac{\partial f}{\partial t} + (\mathbf{q} \cdot \nabla)f + f(\nabla \cdot \mathbf{q}) \right\} dV.$$

Finally, if we use (12) this becomes

$$\frac{d}{dt} \iiint_{V(t)} f\, dV = \iiint_{V(t)} \left\{ \frac{df}{dt} + f(\nabla \cdot \mathbf{q}) \right\} dV. \tag{14}$$

Let us now use this result to derive the *equation of continuity* of fluid mechanics that describes the *conservation of mass* in any volume containing fluid in which fluid is not added (by a *source*) or removed (by a *sink*). To do this we assume that $V(t)$ is an arbitrary *material* volume in a fluid, so that $V(t)$ always contains the same fluid particles and the points on the surface $S(t)$ enclosing $V(t)$ move with the fluid. If we set $f = \rho$, where $\rho(\mathbf{r}, t)$ is the density of the fluid, the mass m of fluid in $V(t)$ is

$$m = \iiint_{V(t)} \rho(\mathbf{r}, t)\, dV.$$

As $V(t)$ is a material volume, provided it contains neither sources, nor sinks, the mass m must remain constant, from which it follows that $dm/dt = 0$.

Setting $f = \rho$ in (14), we find that

$$\frac{dm}{dt} = \iiint_{V(t)} \left\{ \frac{d\rho}{dt} + \rho(\nabla \cdot \mathbf{q}) \right\} dV = 0.$$

As $V(t)$ is arbitrary, this is only possible if the integrand is identically zero, so that

$$\frac{d\rho}{dt} + \rho(\nabla \cdot \mathbf{q}) = 0, \quad \text{or} \quad \frac{\partial \rho}{\partial t} + \nabla \cdot (\rho\mathbf{q}) = 0. \tag{15}$$

These are two equivalent forms of the **equation of continuity** of a fluid, which is of fundamental importance in the study of fluid dynamics.

If the fluid velocity is such that $\nabla \cdot \mathbf{q} = 0$ (div $\mathbf{q} = 0$), setting $f = 1$ in (14) reduces it to

$$\frac{d}{dt} \iiint_{V(t)} dV = \iiint_{V(t)} \nabla \cdot \mathbf{q}\, dV.$$

If div $\mathbf{q} = 0$, then $\rho_t + \rho \nabla \cdot \mathbf{q} = 0$ simplifies to $d\rho/dt = 0$. So, if initially $\rho_0 = \rho|_{t=0}$ is constant, ρ must remain constant throughout the flow even when the fluid is compressible. As $\iiint_{V(t)} dV = V$, where V is the volume of the fluid, it follows from $d/dt \iiint_{V(t)} dV = \iiint_{V(t)} \nabla \cdot \mathbf{q}\, dV$ that $dV/dt = 0$ when $\nabla \cdot \mathbf{q} = 0$. Consequently, in this case, the fluid motion will evolve without change of volume, even though the fluid may be compressible. In fluid mechanics, a flow of a compressible fluid that takes place without a change of volume is called **isochoric** flow. Naturally this last result is true when the fluid is incompressible, because then the density ρ is an absolute constant.

Next we derive a generalization of Theorem 12.6 that allows the function $f(\mathbf{r}, t)$ to be discontinuous across some surface Σ in $V(t)$ that moves with an arbitrary velocity $\mathbf{u}$, with $f = f_1(\mathbf{r}, t)$ on one side of Σ and $f = f_2(\mathbf{r}, t)$ on the other side. Particular cases of this result are needed when a physical quantity of interest experiences a discontinuous change across a surface, as can happen, for example, in chemical engineering and fluid mechanics.

The situation is illustrated in Fig. 12.15, where a material volume $V(t)$ with bounding surface $S(t)$ is shown divided into two parts $V_1(t)$ and $V_2(t)$ by a surface Σ that moves with an arbitrary velocity $\mathbf{u}$. The volume $V_1(t)$ is bounded by the surface $S_1(t)$ that is part of $S(t)$ and Σ, where the unit normal $\mathbf{n}_1$ to Σ directed out of $V_1(t)$ is $\mathbf{n}_1 = \nu$. Similarly, volume $V_2(t)$ is bounded by the surface $S_2(t)$ that is part of $S(t)$ and Σ, where the unit normal $\mathbf{n}_2$ to Σ directed out of $V_2(t)$ is in the opposite sense to that of $\mathbf{n}_1$ so that $\mathbf{n}_2 = -\nu$.

Applying Theorem 12.6 to volume $V_1(t)$ gives

$$\frac{d}{dt} \iiint_{V_1(t)} f_1\, dV = \iiint_{V_1(t)} \frac{\partial f_1}{\partial t}\, dV + \iint_{S_1(t)} f_1 \mathbf{q} \cdot d\mathbf{S} + \iint_{\Sigma(t)} f_1 \mathbf{u} \cdot \mathbf{n}_1\, dS,$$

and an application of Theorem 12.6 to the volume $V_2(t)$ gives

$$\frac{d}{dt} \iiint_{V_2(t)} f_2\, dV = \iiint_{V_2(t)} \frac{\partial f_2}{\partial t}\, dV + \iint_{S_2(t)} f_2 \mathbf{q} \cdot d\mathbf{S} + \iint_{\Sigma(t)} f_2 \mathbf{u} \cdot \mathbf{n}_2\, dS.$$

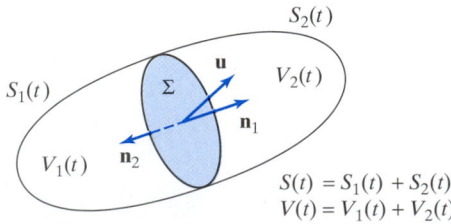

$$S(t) = S_1(t) + S_2(t)$$
$$V(t) = V_1(t) + V_2(t)$$

FIGURE 12.15   The material volume $V(t)$ and the surface Σ across which f is discontinuous.

Adding these two results and using the fact that $\mathbf{n}_1 = \boldsymbol{\nu}$ and $\mathbf{n}_2 = -\boldsymbol{\nu}$, we obtain

$$\frac{d}{dt}\iiint_{V(t)} f\,dV = \iiint_{V(t)} \frac{\partial f}{\partial t}\,dV + \iint_{S(t)} f\mathbf{q}\cdot d\mathbf{S} + \iint_{\Sigma(t)} (f_1 - f_2)\mathbf{u}\cdot d\mathbf{S}, \quad (16)$$

which is the required generalization.

Examination of the last term in (16) shows, as would be expected, that the contribution made by the jump discontinuity $f_1 - f_2$ across the surface Σ that moves with velocity $\mathbf{u}$ depends only on the component of $\mathbf{u}$ normal to Σ, so if $\mathbf{u}$ is tangential to Σ, this term will vanish.

An extension of these ideas to allow for discontinuous solutions f in a volume $V(t)$ when f satisfies an equation of the form

$$\frac{\partial f}{\partial t} + \operatorname{div}\mathbf{h}(f) = 0,$$

called a **conservation equation**, is to be found in Chapter 18, Section 18.4, where conservation equations and shock solutions are considered. It should be noticed that an equation of this type has already been encountered in (15) when deriving the continuity equation for a fluid (the *conservation of mass equation*) in the form

$$\frac{\partial \rho}{\partial t} + \operatorname{div}(\rho\mathbf{q}) = 0.$$

This is a *partial differential equation*, because it is an equation relating partial derivatives of the dependent variables ρ and $\mathbf{q}$.

Let Γ be a closed curve in a fluid flow with velocity vector $\mathbf{q}$ for which $\operatorname{div}\mathbf{q} = 0$ (an *isochoric flow*), and let S be any smooth surface with boundary Γ. Then the streamlines passing through Γ define a stream tube in the fluid flow. The integral

$$\Phi = \iint_S \mathbf{q}\cdot d\mathbf{S} \quad (17)$$

is called the **strength** of the stream tube, and it measures the flow rate through the tube. As a final application of an integral theorem, we will prove that the strength of the flow in a tube bounded by streamlines (a stream tube) remains constant along its length.

First we rewrite Theorem 12.5, which was proved in the form

$$\frac{d}{dt}\iint_{S(t)} \mathbf{F}(\mathbf{r}, t)\cdot d\mathbf{S} = \iint_{S(t)} \left[\frac{\partial \mathbf{F}}{\partial t} + (\nabla\cdot\mathbf{F})\mathbf{q}\right]\cdot d\mathbf{S} + \int_{\Gamma(t)} \mathbf{F}\times\mathbf{q}\cdot d\mathbf{r}.$$

If we apply Stokes' theorem to the last integral, this becomes

$$\frac{d}{dt}\iint_{S(t)} \mathbf{F}(\mathbf{r}, t)\cdot d\mathbf{S} = \iint_{S(t)} \left[\frac{\partial \mathbf{F}}{\partial t} + (\nabla\cdot\mathbf{F})\mathbf{q} + \nabla\times(\mathbf{F}\times\mathbf{q})\right]\cdot d\mathbf{S}. \quad (18)$$

Replacing $\mathbf{F}$ by $\mathbf{q}$, we have

$$\frac{d}{dt}\iint_{S(t)} \mathbf{q}\cdot d\mathbf{S} = \iint_{S(t)} \left[\frac{\partial \mathbf{q}}{\partial t} + (\nabla\cdot\mathbf{q})\mathbf{q} + \nabla\times(\mathbf{q}\times\mathbf{q})\right]\cdot d\mathbf{S},$$

but $\mathbf{q}\times\mathbf{q} = \mathbf{0}$, and as the flow is isochoric, $(\nabla\cdot\mathbf{q}) = 0$, this result reduces to

$$\frac{d}{dt}\iint_{S(t)} \mathbf{q}\cdot d\mathbf{S} = \iint_{S(t)} \frac{\partial \mathbf{q}}{\partial t}\cdot d\mathbf{S}.$$

An application of the divergence theorem to the integral on the right, where the closed surface $V(t)$ is formed by $S(t)$, $S(t + dt)$ and streamlines through Γ, gives

$$\frac{d}{dt}\iint_{S(t)} \mathbf{q}\cdot d\mathbf{S} = \iiint_{V(t)} \nabla\cdot(\partial\mathbf{q}/\partial t)dV = \iiint_{V(t)} \partial/\partial t(\nabla\cdot\mathbf{q})dV = 0,$$

showing that the strength $\Phi = \iint_S \mathbf{q}\cdot d\mathbf{S}$ remains constant along a stream tube.

Summary

The applications considered in this section were to fluid mechanics, and they made use of the so-called material, or convected, derivative of a function f of both position and time. The determination of this derivative was seen to involve letting a position vector move with the fluid and then finding the time derivative of f at the moving point. One result obtained by means of the transport theorems was the equation of continuity of fluid mechanics. Another result used the notion of a conservation equation to establish the invariance of the flow rate (strength) in a stream tube, the walls of which are bounded by streamlines.

EXERCISES 12.4

1. Prove the **Euler expansion formula**

$$\frac{d}{dt}\iiint_{V(t)} dV = \iint_{S(t)} \mathbf{q}\cdot d\mathbf{S}.$$

2. Show that the flux transport theorem given in (18) can also be written as

$$\frac{d}{dt}\iint_{S(t)} \mathbf{F}(\mathbf{r}, t)\cdot d\mathbf{S}$$

$$= \iint_{S(t)} \left[\frac{d\mathbf{F}}{dt} + (\nabla\cdot\mathbf{q})\mathbf{F} - (\mathbf{F}\cdot\nabla)\mathbf{q}\right]\cdot d\mathbf{S}.$$

3.* Show that if

$$\frac{\partial\mathbf{F}}{\partial t} + (\nabla\cdot\mathbf{F})\mathbf{q} + \nabla\times(\mathbf{F}\times\mathbf{q}) = \mathbf{0},$$

the strength of flow through any stream tube remains constant along its length.

PART SIX

COMPLEX ANALYSIS

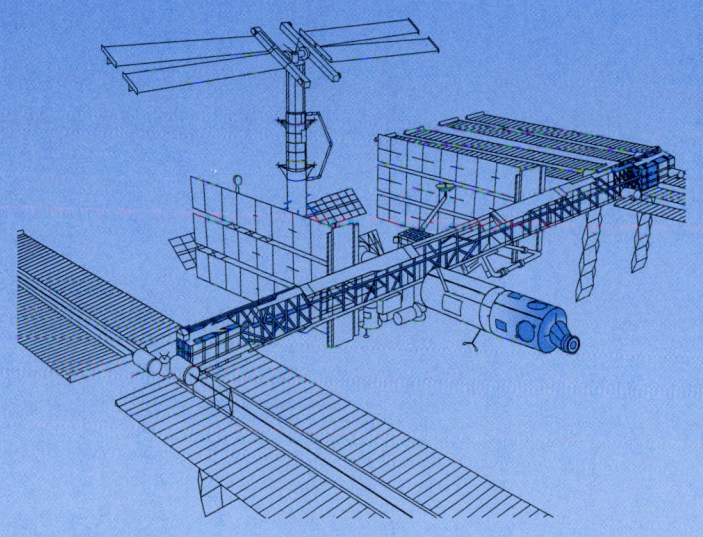

CHAPTER 13

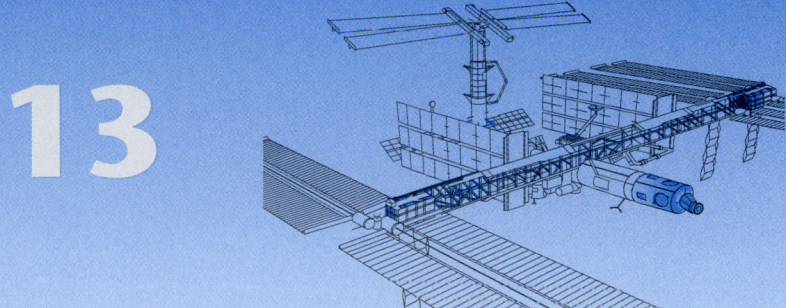

Analytic Functions

Analytic functions involve an extension of the calculus to complex functions, and they find applications throughout all of engineering and science. Examples of direct applications are to be found in two-dimensional problems in elasticity, fluid mechanics, and electrostatics, and such functions also contribute indirectly to many other applications through their use with the Laplace and Fourier transforms. The fundamental idea underlying the systematic development of analytic functions is the extension of the concept of a derivative to a function of a complex variable. The requirement that the derivative of a complex function be independent of the way the defining complex limit is evaluated is more restrictive than the definition of partial derivatives of functions of two real variables, and it leads directly to the Cauchy–Riemann equations, which are central to the development of the subject.

After a brief review of the notion of a mapping, the fundamental concepts of the limit, continuity and differentiability of a complex function are introduced, and the essential difference between derivatives of real and complex functions is explained. An analytic function is defined, and the requirement that the limiting operation in the definition of a derivative of a complex function should be independent of the direction in which it is evaluated is shown to lead to the important Cauchy–Riemann equations. These equations provide a condition that ensures that a function of a complex variable is analytic, and both the real and imaginary parts of an analytic function are shown to be harmonic functions. Some important elementary analytic functions are defined and the problem of finding their inverse is examined.

13.1 Complex Functions and Mappings

A typical example of a complex function is the nth degree polynomial

$$P(z) = a_0 z^n + a_1 z^{n-1} + a_2 z^{n-2} + \cdots + a_{n-1} z + a_n, \tag{1}$$

where the coefficients $a_0, a_1, \ldots, a_n$ are complex numbers and $z = x + iy$ is an arbitrary complex variable. Assigning to z the specific value z_1 determines a complex

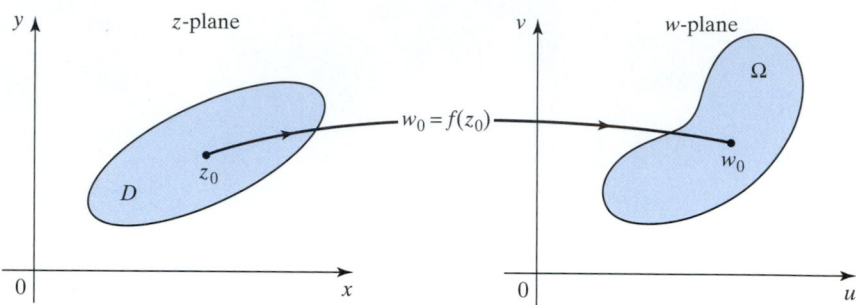

FIGURE 13.1   The function $w = f(z)$ and the w- and z-planes.

number $P(z_1)$, so to each z in the complex plane, there corresponds another complex number $P(z)$. Complex polynomials are defined for all z in the complex plane, and $P(z)$ ranges over all of the complex numbers defined by (1).

The general concept of an arbitrary complex function $w = f(z)$ can be introduced by considering two complex planes, one the z-plane containing the points $z = x + iy$ and the other the w-plane containing the points $w = u + iv$, as shown in Fig. 13.1. To develop this idea further, let a set of points D in the z-plane be such that to each point z in D there corresponds a unique complex number w belonging to another set of points Ω in the w-plane. Then the set D is said to be **mapped** onto the set Ω by a **single-valued** function of the complex variable z. A point w_0 in the w-plane corresponding to a point z_0 in the z-plane is called the **image** of z_0. The term *single-valued* is used because, by hypothesis, each point of D corresponds to one and only one point of Ω, and the name *mapping* is used because an arbitrary curve in D will correspond (be *mapped*) to a corresponding curve in Ω, with each point of the curve in Ω the image of a point in D. The notion of a mapping is important, and it will be used later in Chapter 17 when the concept of *conformal mapping* is introduced. The relationship between the points in D and the corresponding points in Ω is shown by the usual functional notation

$$w = f(z). \tag{2}$$

Set D is called the **domain of definition** of the complex function $f(z)$, and set Ω is called its **range**.

This definition of a function of a complex variable is more general than we require, because it places no restriction on the nature of the sets D and Ω. In complex analysis we will only be concerned with sets of points that possess the property of being *connected*. A set G will be said to be **connected** if every pair of points in G can be joined by an unbroken path with the property that every point of the path also belongs to G. Here, the path may be either a curve or a set of straight line segments joined end to end.

A **neighborhood** of a point z_0 in G is defined as all the points of a set contained strictly inside a circle of arbitrarily small radius with its center at z_0. A point z_0 is called an **interior** point of G if a neighborhood of z_0 only contains points of G. If a neighborhood of z_0 contains no points of G, the point z_0 is called an **exterior** point of G. When any neighborhood of z_0 contains both interior and exterior points of G, the point z_0 is called a **boundary point** of G. Collectively, the set of all boundary points is called the **boundary** of the set. In the sets to be considered later, the boundary

mappings and images

neighborhoods and boundaries

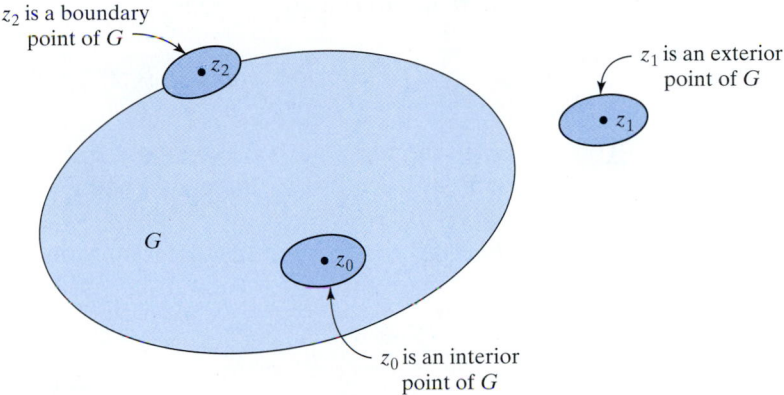

FIGURE 13.2 Interior, exterior, and boundary points of G and their associated neighborhoods.

points usually comprise a combination of straight line segments and curved arcs joined end to end to form a continuous **boundary**.

open and closed sets, and connectivity

A set G that contains no boundary points is called an **open** set. If every boundary point of set G belongs to G, then G said to be **closed**. The name **domain** is given to an open connected set, while the more general term **region** is used to describe a connected set of points that may contain none, some, or all of its boundary points. A typical open connected set G is the disc $|z| < 1$ in the z-plane. The set is *connected* because every point in G can be joined to every other point in G by a curve lying entirely inside G, and the set is *open* because however close a point in G is to the circle $|z| = 1$, a neighborhood of z_0 can always be found that only contains points of G. This becomes a closed set if the relation $|z| < 1$ is replaced by $|z| \leq 1$, because then the *boundary* of G formed by the circle $|z| = 1$ belongs to the set. These ideas are illustrated in Fig. 13.2.

In what follows we will be concerned with functions with the property that to a single element in their domain there corresponds a single element in their range and, conversely, to a single element in their range there corresponds a single element in their domain. Functions of this type are said to be **one-one**, so a function like $w = \sqrt{z}$ is to be regarded as two separate functions, each with the same domain in the z-plane, but with different ranges in the w-plane.

representing complex functions in cartesian and polar forms

The complex function (2) can be written in its **cartesian form** as

$$f(z) = u(x, y) + iv(x, y) \quad \text{for } z = x + iy \in D, \tag{3}$$

where $u(x, y)$ and $v(x, y)$ are real functions of the real variables x and y denoted by

$$u(x, y) = \text{Re}\{f(z)\} \quad \text{and} \quad v(x, y) = \text{Im}\{f(z)\}. \tag{4}$$

Similarly, when z is expressed in **modulus argument form** by setting $z = re^{i\theta}$, with $r = |z|, \theta = \text{Arg } z$ and $-\pi < \theta \leq \pi$, the complex function $f(z)$ takes the **polar form**

$$f(z) = u(r, \theta) + iv(r, \theta), \tag{5}$$

where $u(r, \theta)$ and $v(r, \theta)$ are real functions of the real variables r and θ given by

$$u(r, \theta) = \text{Re}\{f(z)\} \quad \text{and} \quad v(r, \theta) = \text{Im}\{f(z)\}. \tag{6}$$

EXAMPLE 13.1

Write the function $f(z) = z^2 - z + 2$ in both its cartesian and polar form, and in each case identify the functions u and v.

Solution   To arrive at the cartesian form we set $z = x + iy$ in $f(z)$ to obtain

$$\begin{aligned}
f(z) = u + iv &= (x + iy)^2 - (x + iy) + 2 \\
&= x^2 + 2ixy - y^2 - x - iy + 2 \\
&= x^2 - y^2 - x + 2 + i(2xy - y).
\end{aligned}$$

Equating the real and imaginary parts gives

$$u(x, y) = \text{Re}\{f(z)\} = x^2 - y^2 - x + 2 \quad \text{and} \quad v(x, y) = \text{Im}\{f(z)\} = 2xy - y.$$

The polar form is obtained by setting $z = re^{i\theta}$ in $f(z)$ to obtain

$$\begin{aligned}
f(z) = u + iv &= r^2 e^{2i\theta} - re^{i\theta} + 2 \\
&= r^2(\cos 2\theta + i \sin 2\theta) - r(\cos \theta + i \sin \theta) + 2 \\
&= r^2 \cos 2\theta - r \cos \theta + 2 + i(r^2 \sin 2\theta - r \sin \theta).
\end{aligned}$$

In this case, equating real and imaginary parts gives

$$u(r, \theta) = r^2 \cos 2\theta - r \cos \theta + 2 \quad \text{and} \quad v(r, \theta) = r^2 \sin 2\theta - r \sin \theta. \quad \blacksquare$$

EXAMPLE 13.2

Draw the straight line segment in the z-plane joining the points $z = 2 + 3i$ and $z = 4 + 5i$, and find its image in the w-plane under the mapping $w = \frac{1}{2}z + i$.

Solution   The straight line segment starts at the point A with coordinates $(2, 3)$ and ends at the point B with coordinates $(4, 5)$, so if it has the equation $y = mx + c$, its gradient $m = (5 - 3)/(4 - 2) = 1$. As the line must pass through the point $(2, 3)$, substitution into the equation $y = mx + c$ gives $3 = 2 + c$, so $c = 1$. This has established that the equation of the line to which the line segment AB belongs is $y = x + 1$.

The mapping is $w = \frac{1}{2}z + i$, so setting $w = u + iv$ and $z = x + iy$, we find that $u + iv = \frac{1}{2}x + i(\frac{1}{2}y + 1)$. Equating the real and imaginary parts of this equation gives $u = \frac{1}{2}x$, $v = \frac{1}{2}y + 1$. As the straight line segment AB in the z-plane is part of the line $y = x + 1$, substituting for x and y in terms of u and v shows that the mapping onto the w-plane of the line to which AB belongs has the equation $v = u + 3/2$. This is also the equation of a straight line, so we have established that $w = \frac{1}{2}z + i$ maps the straight line $y = x + 1$ in the z-plane onto the straight line $v = u + 3/2$ in the w-plane.

To draw the required image in the w-plane we must now determine the images A' and B' in the w-plane of A and B in the z-plane, and then join them by a straight line. As A is the point $z = 2 + 3i$ and B is the point $z = 4 + 5i$, substitution into $w = \frac{1}{2}z + i$ shows that A' is the point $w = 1 + \frac{5}{2}i$ and B' is the point $w = 2 + \frac{7}{2}i$. The line segments in the z- and w-planes are shown in Fig. 13.3. $\quad \blacksquare$

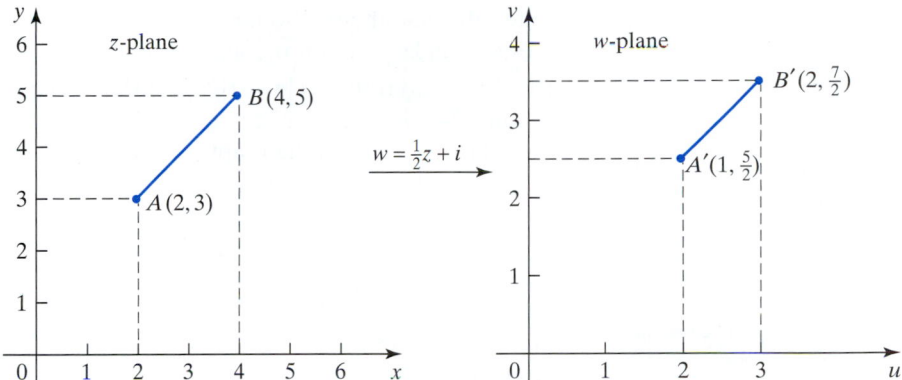

FIGURE 13.3   The image of line AB under the mapping $w = \frac{1}{2}z + i$.

EXAMPLE 13.3

(a) Draw and shade the area in the z-plane containing the points satisfying the conditions $|z - 1 + 2i| \leq 1$ and Im$\{z\} > -2$, marking a boundary that belongs to the set by a solid line and one that does not belong to it by a dashed line. (b) Draw and shade the area in the z-plane to which belong the points satisfying the conditions $r = |z - 1| \geq 2$ and $\pi/6 \leq \text{Arg}(z - 1) \leq \pi/3$.

Solution

(a) We must use the fact that the modulus of a complex number is a nonnegative real number, and $|z_1 - z_2|$ is the distance between z_1 and z_2. It follows from this that the inequality $|z - 1 + 2i| \leq 1$ is satisfied by all points z distant from the point $1 - 2i$ by an amount less than or equal to 1. So the inequality $|z - 1 + 2i| \leq 1$ is satisfied by all points inside and on a circle of radius 1 centered on the point $1 - 2i$. As Im$\{z\} = y$, the inequality Im$\{z\} > -2$ is simply $y > -2$. So the required points lie inside and on a circle of radius 1 centered on the point $1 - 2i$, and strictly above the line $y = -2$. The required area is shown in Fig. 13.4a, where the boundary of the circle has been drawn using a solid line because these boundary points belong

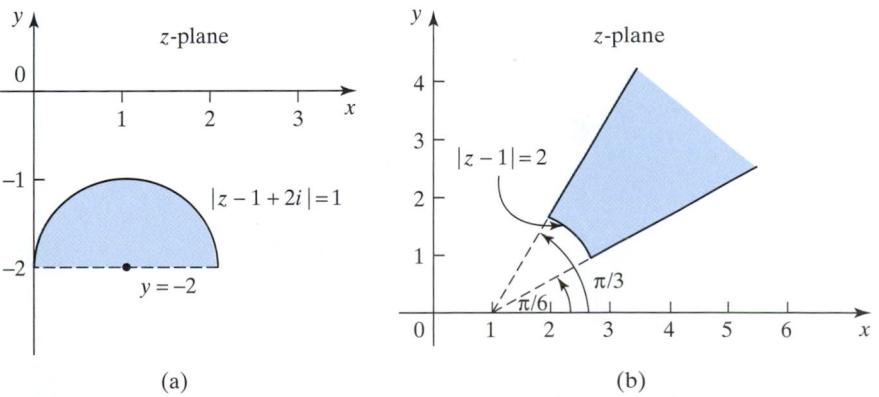

FIGURE 13.4   (a) Points satisfying $|z - 1 + 2i| \leq 1$, Im$\{z\} > -2$. (b) Points satisfying $r = |z - 1| \geq 2$ and $\pi/6 \leq \text{Arg}(z - 1) \leq \pi/3$.

to the set, while the bounding line $y = -3$ is drawn as a dashed line because points on this boundary do not belong to the set.

(b) The condition $r = |z - 1| \geq 2$ is satisfied by all points outside and on a circle of radius 2 with its center at $z = 1$, and a condition of the form $\mathrm{Arg}(z - 1) = \omega$ is a radial line drawn from the point $z = 1$ as origin making an angle ω measured counterclockwise from the positive real axis. Thus the condition $\pi/6 \leq \mathrm{Arg}(z - 1) \leq \pi/3$ gives a wedge shaped area in the upper half of the z-plane centered on the point $z = 1$ with its bounding lines making angles $\pi/6$ and $\pi/3$ with the positive real axis. The required area is shown in Fig. 13.4b. ∎

EXAMPLE 13.4

Find the image of the set of points $|\mathrm{Re}\{z\}| < 1$, $|\mathrm{Im}\{z\}| < 2$ in the z-plane under the mapping $w = 2z + 1$.

Solution   When mapping areas, the approach to be used is first to determine how the boundary transforms, and then to determine if the points in the given area in the z-plane map to points inside or outside the image of this boundary in the w-plane. As $\mathrm{Re}\{z\} = x$ and $\mathrm{Im}\{z\} = y$, the area in the z-plane lies inside the rectangle $-1 < x < 1$, $-2 < y < 2$ shown in the left of Fig. 13.5.

Setting $z = x + iy$ in $w = 2z + 1$ gives $w = u + iv = 2x + 1 + 2iy$, so $u = 2x + 1$ and $v = 2y$. The top boundary of the area in the z-plane in Fig. 13.5 is $-1 < x < 1$, $y = 2$, so using these results in the mapping shows the image of this boundary in the w-plane to be given by $u = 2x + 1$, with $-1 < x < 1$, and $v = 4$. A repetition of this form of argument applied to the other three sides of the rectangle establishes that the image in the w-plane of the rectangle in the z-plane is the one illustrated on the right of Fig. 13.5. A general point (x, y) inside the rectangle in the z-plane maps to the point $(2x + 1, 2y)$ in the w-plane with $-1 < x < 1$, $-2 < y < 2$, and this point is seen to lie inside the rectangular boundary in the w-plane. Consequently, all points inside the rectangle in the z-plane map to points inside the image rectangle in the w-plane. Inspection of Fig. 13.5 shows that the geometrical effect of this mapping is first to scale the rectangle in the z-plane uniformly by a factor 2 in both the x and y directions, and then to shift the origin parallel to the real axis. Mappings are examined in greater detail in Chapter 17 in connection with conformal mappings. ∎

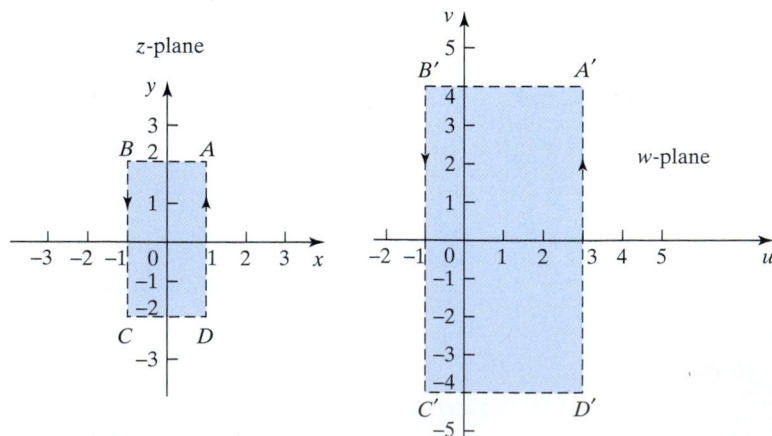

FIGURE 13.5   The effect of the mapping $w = 2z + 1$ on a rectangle.

Summary   A mapping by a single-valued complex function and the image of a point were defined, the notion of a connected set was introduced, and the definition of a neighborhood was used to define the boundary of a set in the complex plane and to identify open and closed regions in the complex plane.

EXERCISES 13.1

In Exercises 1 and 2 sketch and shade the areas in the z-plane occupied by points satisfying the given conditions. Represent a boundary that belongs to a set by a solid line and one that does not by a dashed line. Determine if the areas represent open sets, closed sets, or regions.

1. (a) $|z| \geq 1$ and $|z| \leq 2$. (b) $|z - i| \leq 1$ and $|z| < 1$.
(c) $0 < x < 1$, $0 < y < 1$.

2. (a) $1 < |z| \leq 2$.   (b) $1 < |z - 1| \leq 2$, $x > 1$, $y > 0$.
(c) $\mathrm{Re}\{z\} > 0$, $\mathrm{Im}\{z\} < 0$, $|z| \leq 2$.

3. Determine the image of the straight line segment joining the origin to the point $z = 2 + 2i$ under the mapping $w = -iz$.

4. Set $w = u + iv$ and use the fact that $z\bar{z} = 1$ on the circle $|z| = 1$ to determine the image under the mapping $w = 2z - 1$ of the part of the circular arc $|z| = 1$ that lies in the first quadrant of the z-plane.

5. Determine the image of the points satisfying $|\mathrm{Re}\{z\}| > 2$, $|\mathrm{Im}\{z\}| < 1$ in the z-plane under the mapping $w = iz + 2$.

6. Determine the image of the points satisfying $|\mathrm{Re}\{z\}| > 4$, $|\mathrm{Im}\{z\}| > 2$ in the z-plane under the mapping $w = i - 3z$.

7. By considering the lines joining the origin and the point $(2, 0)$ to a point z in the upper half of the z-plane,

show that the conditions $\mathrm{Arg}(z - 2) - \mathrm{Arg}\, z = \pi/2$ and $0 \leq \mathrm{Arg}\, z \leq \pi/2$ define a semicircular arc of radius 1 in the upper half of the z-plane with its center at $z = 1$.

8. By considering the lines joining the points $(1, 0)$ and $(3, 0)$ to a point z in the upper half of the z-plane, determine the area in the z-plane defined by the conditions $\mathrm{Arg}(z - 3) - \mathrm{Arg}(z - 1) = \pi/2$, $0 \leq |z - 2| \leq 1$, and $\pi/4 \leq \mathrm{Arg}(z - 2) \leq 3\pi/4$.

9. Use a geometrical argument to find the locus of points z such that

$$|z - 1| + |z + 1| = 4.$$

10. Use a geometrical argument to find the locus of points z such that

$$|z - 3i| = |z - i|.$$

Express the functions in Exercises 11 through 14 in both cartesian and polar form, and determine the forms taken by u and v in each case.

11. $f(z) = (2z + i)/(z + i)$.   **13.** $f(z) = ze^{iz}$.
12. $f(z) = 3z^2 - 2z + 1/z$.   **14.** $f(z) = z + 1/z$.

13.2   Limits, Derivatives, and Analytic Functions

When working with functions of a complex variable it is necessary to generalize the related concepts of a limit and continuity by extending the corresponding definitions from real analysis. These generalizations use the fact that in the complex plane the modulus $|z|$ measures the magnitude of z, so $|z_1 - z_2|$ can be considered to measure the distance between points z_1 and z_2 in the z-plane. The function

$$f(z) = u(x, y) + iv(x, y) \tag{7}$$

complex limit   will have the **complex limit** L, written

$$\lim_{z \to z_0} f(z) = L = L_1 + iL_2, \tag{8}$$

where L_1 and L_2 are real numbers, if

$$\lim_{|z-z_0|\to 0} |f(z) - L| = 0. \tag{9}$$

If $z = x + iy$ and $z_0 = x_0 + iy_0$, then z will tend to z_0, written $z \to z_0$, when $(x, y) \to (x_0, y_0)$, so (9) is equivalent to

$$\lim_{(x,y)\to(x_0,y_0)} |f(z) - L| = 0. \tag{10}$$

However, by the triangle inequality,

$$|f(z) - L| = |u(x, y) + iv(x, y) - L_1 - iL_2| = |u(x, y) - L_1 + i(v(x, y) - L_2)|$$
$$\leq |u(x, y) - L_1| + |v(x, y) - L_2|,$$

so in terms of real functions, $f(z)$ will have the limit L as $z \to z_0$ if

$$\lim_{(x,y)\to(x_0,y_0)} u(x, y) = L_1 \quad \text{and} \quad \lim_{(x,y)\to(x_0,y_0)} v(x, y) = L_2. \tag{11}$$

This shows the connection between the limit of a function $f(z)$ of a complex variable and the limits of the real functions $u(x, y)$ and $v(x, y)$. Because of this relationship, the fundamental properties of limits of functions of a real variable are transferred to functions of a complex variable, with the result that if $f(z)$ and $g(z)$ have limits as $z \to z_0$, then

$$\lim_{z\to z_0} [f(z) \pm g(z)] = \lim_{z\to z_0} f(z) \pm \lim_{z\to z_0} g(z) \tag{12}$$

$$\lim_{z\to z_0} [f(z)g(z)] = \lim_{z\to z_0} f(z) \lim_{z\to z_0} g(z) \tag{13}$$

$$\lim_{z\to z_0} [f(z)/g(z)] = \lim_{z\to z_0} f(z)/ \lim_{z\to z_0} g(z), \quad \text{when} \quad \lim_{z\to z_0} g(z) \neq 0. \tag{14}$$

continuous and discontinuous complex functions

As with real functions of a real variable, the complex function $f(z)$ will be said to be **continuous** at z_0 if it is defined in a neighborhood of z_0 and $f(z_0)$ exists and is equal to $\lim_{z\to z_0} f(z)$. When expressed in terms of real functions, it can be seen that $f(z) = u + iv$ will be continuous at $z_0 = x_0 + iy_0$ if

$$\lim_{(x,y)\to(x_0,y_0)} u(x, y) = u(x_0, y_0) \quad \text{and} \quad \lim_{(x,y)\to(x_0,y_0)} v(x, y) = v(x_0, y_0). \tag{15}$$

A function $f(z)$ that does not satisfy condition (15) at (x_0, y_0), that is, at $z = z_0$, will be said to be **discontinuous** at z_0.

It is a direct consequence of the definitions of a limit and of continuity that the sum and difference of continuous complex functions of a complex variable are themselves continuous, and the quotient of continuous functions is continuous at z_0 provided the divisor does not vanish at z_0.

EXAMPLE 13.5   Examine the continuity of the functions (a) $f(z) = z^2 + 3z - 1$ and (b) $f(z) = z/(z-1)$.

Solution

(a) Setting $z = x + iy$ in $f(z)$ and identifying the real and imaginary parts gives

$$f(z) = (x + iy)^2 + 3(x + iy) - 1$$
$$= x^2 - y^2 + 3x - 1 + i(2xy + 3y),$$

so if $f(z) = u + iv$, then

$$u(x, y) = x^2 - y^2 + 3x - 1 \quad \text{and} \quad v(x, y) = 2xy + 3y.$$

As u and v are continuous for all (x, y), that is, for all z, it follows from (15) that $f(z)$ is continuous for all z.

(b) The function $f(z)$ can be considered as the product of the functions $g(z) = z$ and $h(z) = 1/(z - 1)$, and clearly $g(z)$ is continuous for all z. To examine the behavior of $h(z)$ we set $z = x + iy$, and after separating the real and imaginary parts we have

$$h(z) = \frac{1}{x + iy - 1} = \frac{x - 1}{(x - 1)^2 + y^2} - i\frac{y}{(x - 1)^2 + y^2}.$$

So, if $h(z) = u_2 + v_2$, then

$$u_2(x, y) = \frac{x - 1}{(x - 1)^2 + y^2} \quad \text{and} \quad v_2(x, y) = -\frac{y}{(x - 1)^2 + y^2}.$$

The functions u_2 and v_2 are continuous for all (x, y) except at the point $(1, 0)$ corresponding to $z = 1$ where their divisors vanish. Thus, $h(z)$ is continuous for all z except at $z = 1$, so it follows from (13) that the product $f(z) = g(z)h(z)$ is continuous everywhere except at the point $z = 1$, where it has a discontinuity.

This same conclusion can be reached if $f(z)$ is regarded as a quotient of the functions $g(z) = z$ and $h(z) = (z - 1)$. Setting $z = x + iy$ in $f(z)$ and identifying the real and imaginary parts gives

$$f(z) = \frac{x + iy}{x + iy - 1} = \frac{x^2 + y^2 - x}{(x - 1)^2 + y^2} - i\frac{y}{(x - 1)^2 + y^2},$$

so if $f(z) = u + iv$, then

$$u(x, y) = \frac{x^2 + y^2 - x}{(x - 1)^2 + y^2} \quad \text{and} \quad v(x, y) = -\frac{y}{(x - 1)^2 + y^2}.$$

Both u and v have limits as $(x, y) \to (x_0, y_0)$ for all points (x_0, y_0) with the exception of the point $(1, 0)$, corresponding to $z = 1$, where their divisors vanish. So again we conclude that $f(z)$ is continuous for all z with the exception of the point $z = 1$, where it is discontinuous. ∎

A major difference between a real-valued function of two real variables and a single-valued function of a complex variable $w = f(z) = u(x, y) + iv(x, y)$ arises when the derivative of $f(z)$ is introduced. If a single-valued complex function $f(z)$ is defined in some domain D of the complex plane then, when it exists, its **derivative**

derivative of a complex function

$f'(z)$ is defined as

$$f'(z) = \frac{dw}{dz} = \lim_{h \to 0} \frac{f(z+h) - f(z)}{h}, \tag{16}$$

where in the limit on the right the complex variable h is allowed to tend to zero along any path in the z-plane. It is this last condition that distinguishes the derivative of a complex function from that of a real function of two real variables because, as will be seen later, the existence of a unique derivative $f'(z)$ requires a special relationship to exist between the real and imaginary parts $u(x, y)$ and $v(x, y)$ of $f(z)$.

analytic and entire functions

A function that has a continuous derivative throughout some domain D of the complex plane is said to be **analytic** in D. A function is **analytic at a point** P if there is a region containing P in which it is analytic, and a function that is analytic everywhere in the z-plane is called an **entire** function.

On account of the definition of a derivative in (16), and results (12) to (14) involving limits, it follows that the rules for the differentiation of real functions of a real variable carry over to complex functions, so for functions $f(z)$ and $g(z)$ that are analytic in D,

fundamental rules for differentiating combinations of complex functions

$$\frac{d}{dz}[f(z) \pm g(z)] = f'(z) \pm g'(z) \quad \text{is analytic in } D \tag{17}$$

$$\frac{d}{dz}[f(z)g(z)] = f'(z)g(z) + f(z)g'(z) \quad \text{is analytic in } D \tag{18}$$

$$\frac{d}{dz}\left(\frac{f(z)}{g(z)}\right) = \frac{f'(z)g(z) - f(z)g'(z)}{[g(z)]^2} \quad \text{is analytic in } D$$
$$\text{wherever } g(z) \neq 0, \tag{19}$$

and differentiation of a composite function (*function of a function*) is given by the familiar result

$$\frac{d[f(g(z))]}{dz} = g'(z)f'(g(z)), \tag{20}$$

where the expression on the right is analytic whenever the range of $g(z)$ lies within the domain of definition of $f(z)$, and $f'(g(z))$ exists.

Higher derivatives are defined in the usual manner, so that, for example,

$$\frac{d^2[f(z)]}{dz^2} = \frac{d}{dz}\left[\frac{d[f(z)]}{dz}\right] = f''(z) \quad \text{and}$$
$$\frac{d^3[f(z)]}{dz^3} = \frac{d}{dz}\left[\frac{d^2[f(z)]}{dz^2}\right] = \frac{d[f''(z)]}{dz} = f'''(z). \tag{21}$$

It follows directly that if $f(z)$ and $g(z)$ are analytic in a common domain D of the complex plane, then $f(z) \pm g(z)$ and $f(z)g(z)$ are analytic in D, and $f(z)/g(z)$ is analytic in D except for points where $g(z) = 0$ but $f(z) \neq 0$.

The formal definition of a derivative in (16) does not usually provide a convenient way of calculating $f'(z)$, though it can be used as shown by the next example

EXAMPLE 13.6

finding an important derivative from first principles

Use the definition of a derivative in (16) to show that

$$\frac{d[z^n]}{dz} = nz^{n-1} \quad \text{for } n = 0, \pm 1, \pm 2, \ldots,$$

and that z^n is analytic for all z when $n = 0, 1, 2, \ldots$, and when $n = -1, -2, \ldots$, it is analytic everywhere except at $z = 0$.

Solution   We consider the cases $n = 0, 1, 2, \ldots$, and $n = -1, -2, \ldots$, separately.

Case: n = 0. From (16) we have

$$\frac{d[1]}{dz} = \lim_{h \to 0} \left(\frac{1-1}{h} \right) = 0,$$

and this is true irrespective of how $h \to 0$, so the statement is true for $n = 0$.

Case: n a positive integer. From (16), after expanding $(z+h)^n$ by the binomial theorem, we have

$$\frac{d[z^n]}{dz} = \lim_{h \to 0} \left(\frac{(z+h)^n - z^n}{h} \right)$$

$$= \lim_{h \to 0} \left(\frac{z^n + nhz^{n-1} + \frac{n(n-1)}{2!}h^2 z^{n-2} + \cdots + h^n - z^n}{h} \right)$$

$$= nz^{n-1} + \lim_{h \to 0} h \left(\frac{n(n-1)}{2!} z^{n-2} + \frac{n(n-1)(n-2)}{3!} h z^{n-3} + \cdots + h^{n-2} \right)$$

$$= nz^{n-1}.$$

This result is also true for all z, irrespective of the path in the z-plane by which $h \to 0$, so the statement is true for all positive integers n.

Case: n a negative integer. In this case, using (19) with $f(z) = 1$ and $g(z) = z^n$ gives

$$\frac{d[z^{-n}]}{dz} = -\frac{d[z^n]/dz}{z^{2n}} = -nz^{-(n+1)}, \quad \text{for } z \neq 0,$$

so the statement in the problem is seen to be true when n is a negative integer and $z \neq 0$. We have shown that when $n = 0, 1, 2, \ldots$ the function $f(z) = z^n$ is analytic for all z, and when n is a negative integer it is analytic everywhere except at the origin. ∎

The definition of a derivative in (16) is too cumbersome to use for general purposes. A more convenient way of determining derivatives will be found as a result of arriving at conditions to be satisfied by $u(x, y)$ and $v(x, y)$ that will ensure that the function $f(z) = u(x, y) + iv(x, y)$ is analytic.

THEOREM 13.1

a fundamental condition to be satisfied if a complex function is to have a derivative

Cauchy–Riemann equations   The single-valued complex function

$$f(z) = u(x, y) + iv(x, y)$$

defined for all z in some domain D of the complex plane will have a derivative $f'(z)$ at every point of D, and so be analytic in D, if the partial derivatives $\partial u/\partial x$, $\partial u/\partial y$, $\partial v/\partial x$, and $\partial v/\partial y$ are continuous throughout D and satisfy the **Cauchy–Riemann** equations at every point of D:

$$\frac{\partial u}{\partial x} = \frac{\partial v}{\partial y} \quad \text{and} \quad \frac{\partial u}{\partial y} = -\frac{\partial v}{\partial x}.$$

Proof   To arrive at conditions to be satisfied by $f(z) = u + iv$ that will ensure that $f'(z)$ exists and is unique in D, independently of the way in which $h \to 0$ in (16), we will compute $f'(z)$ in two different ways. First we will find $f'(z)$ by letting $h \to 0$ parallel to the real axis, and then by letting $h \to 0$ parallel to the imaginary axis, as a result of which two different expressions will be obtained for $f'(z)$. If these are to be identical, their respective real and imaginary parts must be equal, and it will be this requirement that will lead to the Cauchy–Riemann equations.

First we set $h = h_1 + i0$ and let $h_1 \to 0$, so that $h \to 0$ parallel to the real axis, and as a result (16) becomes

$$f'(z) = \lim_{h_1 \to 0} \left[\frac{u(x + h_1, y) + iv(x + h_1, y) - u(x, y) - iv(x, y)}{h_1} \right]$$

$$= \lim_{h_1 \to 0} \left[\frac{u(x + h_1, y) - u(x, y)}{h_1} \right] + i \lim_{h_1 \to 0} \left[\frac{v(x + h_1, y) - v(x, y)}{h_1} \right]$$

$$= \frac{\partial u}{\partial x} + i \frac{\partial v}{\partial x}.$$

Next we set $h = 0 + ih_2$ and let $h_2 \to 0$, so that $h \to 0$ is parallel to the imaginary axis. In this case (16) becomes

$$f'(z) = \lim_{h_2 \to 0} \left[\frac{u(x, y + h_2) + iv(x, y + h_2) - u(x, y) - iv(x, y)}{ih_2} \right]$$

$$= \lim_{h_2 \to 0} \left[\frac{u(x, y + h_2) - u(x, y)}{ih_2} \right] + i \lim_{h_2 \to 0} \left[\frac{v(x, y + h_2) - v(x, y)}{ih_2} \right]$$

$$= \frac{\partial v}{\partial y} - i \frac{\partial u}{\partial y}.$$

Equating these two different expressions for $f'(z)$, whose respective real and imaginary parts must be equal, gives the Cauchy–Riemann equations

$$\frac{\partial u}{\partial x} = \frac{\partial v}{\partial y} \quad \text{and} \quad \frac{\partial u}{\partial y} = -\frac{\partial v}{\partial x},$$

that must hold throughout D if $f(z)$ is to be analytic in D.

It is somewhat harder to prove that when $u(x, y)$ and $v(x, y)$ have continuous partial derivatives u_x, u_y, v_x, and v_y in D, the function $f(z) = u(x, y) + iv(x, y)$ is analytic in D, so the details of the proof will be omitted.   ∎

AUGUSTIN-LOUIS CAUCHY (1789–1857)
A French mathematician who was born in Paris and studied and held a professorship at the Ecole Polytechnique. He was subsequently appointed to the chair of mathematical physics at the University of Turin. Cauchy published many mathematical papers, and he was responsible for introducing a rigorous definition of a limit. One of his most important contributions was to the development of complex analysis. Among his other works of a fundamental nature were contributions to number theory, differential equations, and various aspects of mathematical physics.

GEORGE FRIEDRICH BERNHARD RIEMANN (1826–1866)
A German mathematician of outstanding ability who was born in Hanover, but whose delicate health due to tuberculosis resulted in his untimely death while visiting Italy. He studied under Gauss, and after a period of time in Berlin he returned to Göttingen to study physics under Weber. He was made Professor of Mathematics in Göttingen in 1859, and he made contributions of fundamental importance to many branches of mathematics, some of which were influenced by his earlier studies in physics. Among his remarkable contributions, it was his work that led to a proper understanding of definite integrals and to the development of complex analysis and its geometrical interpretation.

The implications of the Cauchy–Riemann equations are far-reaching, because it will be shown later that if a function is analytic in D, then it possesses derivatives of *all* orders.

When $f(z) = u(x, y) + iv(x, y)$ is an analytic function in D, a convenient method for the computation of $f'(z)$ follows from the first expression found in Theorem 13.1, because then

$$f'(z) = \frac{\partial u}{\partial x} + i\frac{\partial v}{\partial x} = \frac{\partial v}{\partial y} - i\frac{\partial u}{\partial y}. \tag{22}$$

This result expresses the derivative $f'(z)$ in its cartesian form involving functions of x and y, but it is often necessary to represent $f'(z)$ as a function of z. In general, to convert the cartesian form of an analytic function $g(z) = u(x, y) + iv(x, y)$ into an expression in terms of z, it is only necessary to recognize that when z is purely real the functional forms of $g(x)$ and $g(z)$ are identical. This leads to the following general rule.

Rule for converting an analytic function $w = u + iv$ to the form $w = f(z)$

how to convert an analytic function in (x, y) form to a function of z

Let $g(z) = u(x, y) + iv(x, y)$ be an analytic function in some domain D of the complex plane. Then the cartesian representation of the function involving x and y on the right of $g(z)$ can be converted to a function of z by setting $y = 0$ and replacing x by z in $u(x, y)$ and $v(x, y)$.

EXAMPLE 13.7

Show that $f(z) = z^2$ satisfies the Cauchy–Riemann equations and is an entire function. Use result (22) and the foregoing rule to show that

$$\frac{d}{dz}[z^2] = 2z.$$

Solution If we set $f(z) = z^2 = u + iv$, it follows that $u(x, y) = x^2 - y^2$ and $v(x, y) = 2xy$. Then $\partial u/\partial x = 2x$, $\partial u/\partial y = -2y$, $\partial v/\partial x = 2y$, and $\partial v/\partial y = 2x$, so

$$\frac{\partial u}{\partial x} = \frac{\partial v}{\partial y} \quad \text{and} \quad \frac{\partial u}{\partial y} = -\frac{\partial v}{\partial x},$$

showing that the Cauchy–Riemann equations are satisfied for all (x, y), so z^2 is an entire function. From (22) the cartesian form of $f'(z)$ is

$$\frac{d}{dz}[z^2] = 2x + i2y,$$

so setting $y = 0$ and replacing x by z the above rule shows that

$$\frac{d}{dz}[z^2] = 2z,$$

in agreement with the result of Example 13.6 with $n = 2$.   ∎

Not every function of a complex variable is an analytic function, as can be seen from the next example.

EXAMPLE 13.8   Show that neither $f(z) = \bar{z}$ nor $f(z) = |z|$ is an analytic function.

Solution   Setting $f(z) = \bar{z} = x - iy$, we have $u(x, y) = x$ and $v(x, y) = -y$, so $\partial u/\partial x = 1$ and $\partial v/\partial y = -1$. As the first Cauchy–Riemann equation is not satisfied at any point in the z-plane, the function $f(z) = \bar{z}$ is not an analytic function.

Setting $f(z) = |z| = (x^2 + y^2)^{1/2}$, we find that $u(x, y) = (x^2 + y^2)^{1/2}$ and $v(x, y) \equiv 0$. As $\partial v/\partial x = \partial v/\partial y \equiv 0$, the Cauchy–Riemann equations cannot be satisfied in the z-plane, so $f(z) = |z|$ is not an analytic function. This is not surprising, because $|z|$ is a *real* function.   ∎

It should be recognized that because polynomials are sums of analytic functions, they are themselves analytic functions. As a result, derivatives of sums and products of polynomials are analytic functions, and derivatives of quotients of polynomials are analytic functions except at the zeros of their divisors. Derivatives of polynomials are obtained by repeated use of the result of Example 13.6 using the appropriate values of n.

EXAMPLE 13.9   Find $F'(z)$ given that $F(z) = z/(z^2 - 1)$.

Solution   Applying (19) with $f(z) = z$ and $g(z) = z^2 - 1$ gives

$$\frac{d}{dz}\left[\frac{z}{z^2 - 1}\right] = -\frac{(z^2 + 1)}{(z^2 - 1)^2}, \quad \text{for } z \neq \pm 1.$$ ∎

the complex exponential

It is natural to define the **complex exponential function** e^z as

$$f(z) = e^z = e^{(x+iy)} = e^x(\cos y + i \sin y), \tag{23}$$

because when $z = x + i0$ this reduces to the definition of e^x, and when $z = 0 + iy$ it becomes the Euler formula

$$e^{iy} = \cos y + i \sin y.$$

Expression (23) is compatible with the series representation

$$e^z = 1 + z + \frac{z^2}{2!} + \frac{z^3}{3!} + \cdots = \sum_{n=0}^{\infty} \frac{z^n}{n!}, \tag{24}$$

because when $z = x$ this becomes the ordinary exponential series for e^x with an infinite radius of convergence, and when $z = iy$ it becomes the Euler formula.

The form of argument used in elementary calculus to establish the ratio test for the convergence of a series in the real variable x remains true when x is replaced by the complex variable z and the absolute value of x is replaced by the modulus of z (see Section 15.1). As a result, because e^x has an infinite radius of convergence and so can be differentiated term by term, so can e^z, because it converges in a *disc* of arbitrarily large radius centered on the origin in the z-plane. Term-by-term differentiation of series (24) is permissible and shows that

$$\frac{d[e^z]}{dz} = e^z.$$

Replacing z in the series by az, with a an arbitrary complex constant, and again differentiating term by term gives the more general result

$$\frac{d[e^{az}]}{dz} = ae^{az}, \tag{25}$$

and so e^{az} is an entire function.

As with the real variable case, the **complex hyperbolic functions** $\sinh z$ and $\cosh z$ are defined by the formulas

complex hyperbolic functions

$$\sinh z = \frac{e^z - e^{-z}}{2} \quad \text{and} \quad \cosh z = \frac{e^z + e^{-z}}{2}, \tag{26}$$

and after squaring and differencing these definitions we obtain the fundamental identity

$$\cosh^2 z - \sinh^2 z = 1. \tag{27}$$

Differentiation of definitions (26) with z replaced by az shows that

$$\frac{d[\sinh az]}{dz} = a \cosh z \quad \text{and} \quad \frac{d[\cosh az]}{dz} = \sinh az, \tag{28}$$

but as e^{az} is an entire function, so also are $\sinh az$ and $\cosh az$.

By definition,

$$\tanh z = \frac{\sinh z}{\cosh z}, \tag{29}$$

so after z is replaced by az, an application of (19) together with results (27) and (28) shows that

$$\frac{d[\tanh az]}{dz} = \frac{a \cosh^2 az - a \sinh^2 az}{\cosh^2 az} = \frac{a}{\cosh^2 az} = a \operatorname{sech}^2 az, \tag{30}$$

provided $\cosh az \neq 0$. This last condition is necessary because although the real-variable hyperbolic cosine function never vanishes, the complex hyperbolic cosine function has an infinity of zeros. The complex function $\tanh az$ is seen to be analytic in any domain D that does not contain a zero of $\cosh az$, so it is *not* an entire function.

The functions $\operatorname{sech} az$, $\operatorname{csch} az$, and $\coth az$ are defined in the usual manner as

$$\operatorname{sech} az = \frac{1}{\cosh az}, \quad \operatorname{csch} az = \frac{1}{\sinh az}, \quad \text{and} \quad \coth az = \frac{1}{\tanh az}, \tag{31}$$

with the derivatives

$$\frac{d}{dz}[\operatorname{sech} az] = -a \operatorname{sech} az \tanh az \text{ for } az \neq \left(n + \frac{1}{2}\right)\pi i,$$

$$\frac{d}{dz}[\operatorname{csch} az] = -a \operatorname{csch} az \coth az \text{ for } az \neq n\pi i, \tag{32}$$

$$\frac{d}{dz}[\coth az] = -a \operatorname{csch}^2 az \text{ for } az \neq n\pi i.$$

EXAMPLE 13.10   Find the zeros of (a) $\cosh z$ and (b) $\cos z - 3$.

Solution

(a) By definition

$$\cosh z = \frac{1}{2}[e^{x+iy} + e^{-x-iy}] = \frac{1}{2}e^x x^{iy} + \frac{1}{2}e^{-x}e^{-iy}$$

$$= \frac{1}{2}e^x(\cos y + i \sin y) + \frac{1}{2}e^{-x}(\cos y - i \sin y)$$

$$= \left(\frac{e^x + e^{-x}}{2}\right)\cos y + i\left(\frac{e^x - e^{-x}}{2}\right)\sin y$$

$$= \cosh x \cos y + i \sinh x \sin y.$$

The function $\cosh z$ will vanish when

$$u(x, y) = \operatorname{Re}\{\cosh z\} = \cosh x \cos y = 0 \quad \text{and}$$
$$v(x, y) = \operatorname{Im}\{\cosh z\} = \sinh x \sin y = 0,$$

and this is only possible if $\cos y = 0$ and $\sinh x = 0$. The function $\cos y = 0$ when $y = (2n+1)\pi/2$ for $n = 0, \pm1, \pm2, \ldots$, and $\sinh x = 0$ only when $x = 0$, so the *zeros* of $\cosh z$, that is, the *roots* of $\cosh z = 0$, are $z = i(2n+1)\pi/2$ for $n = 0, \pm1, \pm2, \ldots$.

(b) A similar argument shows that $\cos z = \cos x \cosh y - i \sin x \sinh y$, so $\cos z = 3$ if $\cos x \cosh y = 3$ and $\sin x \sinh y = 0$. The first condition is true if $\cos x = 1$ and $\cosh y = 3$, from which it follows that $y = \pm\operatorname{arccosh} 3$ (remember that the inverse hyperbolic cosine function is double valued) and $x = 2n\pi$, for $n = 0, \pm1, \pm2, \ldots$. This choice of x also causes the second condition to be satisfied for all y, so the *zeros* of $\cos z - 3$, that is, the *roots* of $\cos z = 3$, are $z = 2n\pi \pm i \operatorname{arccosh} 3$, for $n = 0, \pm1, \pm2, \ldots$. ∎

EXAMPLE 13.11 Use the Cauchy–Riemann equations to show that $\cosh z$ is an entire function, and to find $d[\cosh z]/dz$.

Solution   It was shown in Example 13.10 that if $\cosh z = u(x, y) + iv(x, y)$, then

$$u(x, y) = \cosh x \cos y \quad \text{and} \quad v(x, y) = \sinh x \sin y.$$

Routine differentiation shows u and v satisfy the Cauchy–Riemann equations for all z, so $\cosh z$ is an entire function. Substituting in (22) gives

$$\frac{d}{dz}[\cosh z] = \frac{\partial u}{\partial x} + i \frac{\partial v}{\partial x} = \sinh x \cos y + i \cosh x \sin y,$$

so as $\cosh z$ is an analytic function, setting $y = 0$ and replacing x by z to express the result in terms of z, we obtain the expected result

$$\frac{d[\cosh z]}{dz} = \sinh z. \qquad \blacksquare$$

To make the complex trigonometric sine and cosine functions compatible with the definitions of the corresponding real variable trigonometric functions, we use the definitions

complex trigonometric functions

$$\sin z = \frac{e^{iz} - e^{-iz}}{2i} \quad \text{and} \quad \cos z = \frac{e^{iz} + e^{-iz}}{2} \tag{33}$$

so that, in particular, when $z = x$ is real,

$$\sin ix = i \sinh x, \quad \cos ix = \cosh x, \quad \sinh ix = i \sin x, \quad \text{and} \quad \cosh ix = \cos x. \tag{34}$$

By squaring and adding the expressions in (33), we obtain the fundamental identity

$$\sin^2 z + \cos^2 z = 1. \tag{35}$$

Replacing z by az and differentiating the definitions of $\sin az$ and $\cos az$ shows that

$$\frac{d[\sin az]}{dz} = a \cos z \quad \text{and} \quad \frac{d[\cos az]}{dz} = -a \sin az \tag{36}$$

for all z, so $\sin az$ and $\cos az$ are entire functions.

By definition

$$\tan z = \frac{\sin z}{\cos z}, \tag{37}$$

so replacing z by az followed by an application of (19) together with results (35) and (36) gives

$$\frac{d[\tan az]}{dz} = \frac{a}{\cos^2 az} = a \sec^2 az, \tag{38}$$

provided $\cos az \neq 0$, so $\tan z$ is *not* an entire function.

The functions $\sec az$, $\csc az$, and $\cot az$ are defined in the usual manner as

$$\sec az = \frac{1}{\cos az}, \quad \csc az = \frac{1}{\sin az}, \quad \text{and} \quad \cot az = \frac{1}{\tan az}, \tag{39}$$

with the derivatives

$$\frac{d}{dz}[\sec az] = a \sec az \tan az, \quad \frac{d}{dz}[\csc az] = -a \csc az \cot az, \quad \text{and}$$

$$\frac{d}{dz}[\cot az] = -a \csc^2 az. \tag{40}$$

Summary of derivatives of elementary complex functions

1. $\dfrac{d}{dz}[z^n] = nz^{n-1}$, for $n = 0, \pm 1, \pm 2, \ldots$, and $z \neq 0$ when $n < 0$.

2. $\dfrac{d}{dz}[e^{az}] = ae^{az}$, for all a and z.

3. $\dfrac{d}{dz}[\sinh az] = a \cosh az$, for all a and z.

4. $\dfrac{d}{dz}[\cosh az] = a \sinh az$, for all a and z.

5. $\dfrac{d}{dz}[\tanh az] = a \operatorname{sech}^2 az$, for $\cosh az \neq 0$.

6. $\dfrac{d}{dz}[\operatorname{sech} az] = -a \operatorname{sech} az \tanh az$, for $\cosh az \neq 0$.

7. $\dfrac{d}{dz}[\operatorname{csch} az] = -a \operatorname{csch} az \coth az$, for $\sinh az \neq 0$.

8. $\dfrac{d}{dz}[\coth az] = -a \operatorname{csch}^2 az$, for $\sinh az \neq 0$.

9. $\dfrac{d}{dz}[\sin az] = a \cos az$, for all a and z.

10. $\dfrac{d}{dz}[\cos az] = a \sin az$, for all a and z.

11. $\dfrac{d}{dz}[\tan az] = a \sec^2 az$, for $\cos az \neq 0$.

12. $\dfrac{d}{dz}[\sec az] = a \sec az \tan az$, for $\cos az \neq 0$.

13. $\dfrac{d}{dz}[\csc az] = -a \csc az \cot az$, for $\sin az \neq 0$.

14. $\dfrac{d}{dz}[\cot az] = -a \csc^2 az$, for $\sin az \neq 0$.

Summary

After the definitions of a limit and the continuity of a complex function $f(z)$, its derivative $f'(z)$ was defined. The Cauchy–Riemann conditions were shown to ensure the differentiability of a complex function, and a function that has a continuous derivative throughout some part of the complex plane was called an analytic function. Derivatives of the complex exponential, complex hyperbolic, and complex trigonometric functions were derived.

EXERCISES 13.2

In Exercises 1 through 4 find the real and imaginary parts of the functions and locate any points where they are discontinuous.

1. $f(z) = z^3 + 4z^2 - 3z + 1$.   **3.** $f(z) = z/(1 + z^2)$.

2. $f(z) = 1 + z^2 + z\bar{z}$.   **4.** $f(z) = (z - 1)/(z + 1)$.

In Exercises 5 through 8, use the definition of a derivative given in (16) to determine if the given function $f(z)$ is differentiable and, when it is, to find $f'(z)$. Locate any points where the derivative is not defined.

5. $f(z) = z^3 + z + 1$.   **8.** $f(z) = 1/(a + z)^2$, with a a complex constant.

6. $f(z) = 3 + \bar{z}$.

7. $f(z) = 1/(1 + z)$.

In Exercises 9 through 12 use the Cauchy–Riemann equations to show that the given function $f(z)$ is differentiable. Use the result to find $f'(z)$ both in its cartesian form and as a function of z, and locate any points where the derivative is not defined.

9. $f(z) = z^3$.   **11.** $f(z) = z + 1/z$.

10. $f(z) = 1/(4 + z)$.   **12.** $f(z) = 1/(z^2 + 1)$.

In Exercises 13 through 16 use the definitions of complex hyperbolic functions to establish the stated identities.

13. $\sinh(z_1 \pm z_2) = \sinh z_1 \cosh z_2 \pm \cosh z_1 \sinh z_2$, and deduce that $\sinh(x \pm iy) = \sinh x \cos y \pm i \cosh x \sin y$.

14. $\cosh(z_1 \pm z_2) = \cosh z_1 \cosh z_2 \pm \sinh z_1 \sinh z_2$, and deduce that $\cosh(x \pm iy) = \cosh x \cos y \pm i \sinh x \sin y$.

15. $\cosh^2 z - \sinh^2 z = 1$ and $\tanh^2 z = 1 - \text{sech}^2 z$.

16. $\tanh(z_1 \pm z_2) = \dfrac{\tanh z_1 \pm \tanh z_2}{1 \pm \tanh z_1 \tanh z_2}$.

In Exercises 17 through 20 use the definitions of the complex trigonometric functions to establish the stated identities.

17. $\sin(z_1 \pm z_2) = \sin z_1 \cos z_2 \pm \cos z_1 \sin z_2$, and deduce that $\sin(x \pm iy) = \sin x \cosh y \pm i \cos x \sinh y$.

18. $\cos(z_1 \pm z_2) = \cos z_1 \cos z_2 \mp \sin z_1 \sin z_2$, and deduce that $\cos(x \pm iy) = \cos x \cosh y \mp i \sin x \sinh y$.

19. $\sin^2 z + \cos^2 z = 1$ and $\tan^2 z = \sec^2 z - 1$.

20. $\tan(z_1 \pm z_2) = \dfrac{\tan z_1 \pm \tan z_2}{1 \mp \tan z_1 \tan z_2}$.

In Exercises 21 through 29 use the method of Example 13.10 to find the roots of the given equations.

21. $\sin z = 0$.   **26.** $\sin z = 7$.

22. $\cos z = 0$.   **27.** $\sinh z = i \cosh 2$.

23. $\sinh z = 0$.   **28.** $\cos z = -i \sinh 5$.

24. $\sin z = \cosh 2$.   **29.** $\tanh z = 0$.

25. $\cos z = -\cosh 3$.

In Exercises 30 and 31, locate the points where the given functions are not analytic in the specified domains.

30. (a) $\sec z$ for $|z| < 3$.   (b) $\sin z/(1 + z^2)$ for $|z| < 2$.
(c) $\cos z/(1 + z)^2$ for $|z| < \pi$.

31. (a) $\csc z/(z^2 - 3i)$ for $|z| < 4$.   (b) $1/(z^4 + 16)$ for $|z| < 3$.
(c) $|z| \tan z$ for $|z| < 2$.

32. Show that $f(z) = \cosh 2z$ satisfies the Cauchy–Riemann equations for all z. Hence, find $f'(z)$ both in its cartesian form and as a function of z.

33. Show that $f(z) = \sin 3z$ satisfies the Cauchy–Riemann equations for all z. Hence, find $f'(z)$ both in its cartesian form and as a function of z.

34. Show that $f(z) = 1/\sinh z$ satisfies the Cauchy–Riemann equations for all z other than at the zeros of $\sinh z$. Hence, find $f'(z)$ both in its cartesian form and as a function of z.

35. Use the change of variable from the cartesian coordinates (x, y) to the polar coordinates (r, θ) given by $x = r \cos \theta$ and $y = r \sin \theta$ to show that the **polar form** of the **Cauchy–Riemann equations** for a single-valued analytic function $f(z) = u(r, \theta) + iv(r, \theta)$ is

$$\frac{\partial u}{\partial r} = \frac{1}{r}\frac{\partial v}{\partial \theta} \quad \text{and} \quad \frac{1}{r}\frac{\partial u}{\partial \theta} = -\frac{\partial v}{\partial r}.$$

36. Use the change of variable from the cartesian coordinates (x, y) to the polar coordinates (r, θ) given by $x = r \cos \theta$ and $y = r \sin \theta$ to show that the derivative of a single-valued analytic function $f(z) = u(r, \theta) + iv(r, \theta)$ is given by

$$f'(z) = \left(\cos\theta \frac{\partial u}{\partial r} - \sin\theta \frac{1}{r}\frac{\partial u}{\partial \theta} \right)$$
$$+ i\left(\cos\theta \frac{\partial v}{\partial r} - \sin\theta \frac{1}{r}\frac{\partial v}{\partial \theta} \right).$$

Explain why, when $f(z)$ is a single valued analytic function, this last result can be expressed as a function of z by setting $\theta = 0$ and replacing r by z.

37. Set $z = re^{i\theta}$ in $f(z) = z + 1/z$ and use the polar form of the Cauchy–Riemann equations given in Exercise 35 to show that $f(z)$ is differentiable for $z \neq 0$. Use the result of Exercise 36 to find $f'(z)$ as a function of z.

38. Set $z = re^{i\theta}$ in $f(z) = z^2 - 1/z^2$ and use the polar form of the Cauchy–Riemann equations given in Exercise 35 to show $f(z)$ is differentiable for $z \neq 0$. Use the result of Exercise 36 to find $f'(z)$ as a function of z.

39. Use the polar form of the Cauchy–Riemann equations given in Exercise 35 to verify that

$$f(z) = (3r^3 \cos 3\theta + r \cos \theta + 1) + i(3r^3 \sin 3\theta + r \sin \theta)$$

is an entire function, and then use the result of Exercise 36 to express $f'(z)$ as a function of z. Confirm that $f(z)$ is an entire function by first expressing $f(z)$ as a function of z and then differentiating the result.

40. Repeat Exercise 39 using

$$f(z) = \left(r^2 \cos 2\theta - \frac{2}{r^2} \cos 2\theta + r \cos \theta \right)$$
$$+ i \left(r^2 \sin 2\theta + \frac{2}{r^2} \sin 2\theta + r \sin \theta \right).$$

13.3   Harmonic Functions and Laplace's Equation

Let $f(z) = u(x, y) + iv(x, y)$ be analytic in some domain D, and let functions $u(x, y)$ and $v(x, y)$ have continuous second order partial derivatives with respect to x and y. Then it is known from elementary calculus (see Theorem 1.3) that the mixed partial derivatives of $u(x, y)$ and $v(x, y)$ must be equal, so $\partial^2 u/\partial x \partial y = \partial^2 u/\partial y \partial x$ and $\partial^2 v/\partial x \partial y = \partial^2 v/\partial y \partial x$.

Differentiating the first Cauchy–Riemann equation in Theorem 13.1 partially with respect to x gives

$$\frac{\partial}{\partial x}\left[\frac{\partial u}{\partial x}\right] = \frac{\partial}{\partial x}\left[\frac{\partial v}{\partial y}\right] \quad \text{or} \quad \frac{\partial^2 u}{\partial x^2} = \frac{\partial^2 v}{\partial y \partial x},$$

and differentiating the second Cauchy–Riemann equation in Theorem 13.1 partially with respect to y gives

$$\frac{\partial}{\partial y}\left[\frac{\partial u}{\partial y}\right] = -\frac{\partial}{\partial y}\left[\frac{\partial v}{\partial x}\right] \quad \text{or} \quad \frac{\partial^2 u}{\partial y^2} = -\frac{\partial^2 v}{\partial x \partial y}.$$

Adding these two results and using the equality of mixed derivatives show that

$$\frac{\partial^2 u}{\partial x^2} + \frac{\partial^2 u}{\partial y^2} = 0. \tag{41}$$

Had the first equation been differentiated partially with respect to y and the second partially with respect to x, addition of the results would have given

$$\frac{\partial^2 v}{\partial x^2} + \frac{\partial^2 v}{\partial y^2} = 0. \tag{42}$$

Results (41) and (42) show that both the real and imaginary twice differentiable parts of an analytic function satisfy the same second order *partial differential equation*. The partial differential equation

$$\frac{\partial^2 \Phi}{\partial x^2} + \frac{\partial^2 \Phi}{\partial y^2} = 0 \tag{43}$$

the Laplace equation, harmonic functions, and the Laplacian

is called the **Laplace equation**, and any function Φ that satisfies Laplace's equation is called a **harmonic function**. Thus, both $u = \text{Re}\{f(z)\}$ and $v = \text{Im}\{f(z)\}$ are harmonic functions, and they are defined throughout the domain D. We now define the symbol

Δ, pronounced "Laplacian," as

$$\Delta \equiv \frac{\partial^2}{\partial x^2} + \frac{\partial^2}{\partial y^2}. \tag{44}$$

Then Δ is a **differential operator**, and as it stands (44) is *not* a function because it only describes a *differentiation operation*. However, when the operator Δ acts on a suitably differentiable function $\Phi(x, y)$, indicated by placing the function $\Phi(x, y)$ immediately after the symbol Δ, the result $\Delta\Phi$ becomes a *function*. As the Laplace equation in (43) can be written as $\Delta\Phi = 0$, the symbol Δ defined in (44) is called the **Laplacian operator in two dimensions**, and $\Delta\Phi$ is called the **Laplacian** of Φ. Consequently, a function Φ will be *harmonic* if its Laplacian is zero.

When $f(z)$ is an analytic function with $u(x, y) = \mathrm{Re}\{f(z)\}$ and $v(x, y) = \mathrm{Im}\{f(z)\}$, the function $v(x, y)$ is called the **harmonic conjugate** of $u(x, y)$ and, conversely, $u(x, y)$ is called the **harmonic conjugate** of $v(x, y)$. It is important to recognize that two functions $U(x, y)$ and $V(x, y)$ that are harmonic can only be *harmonic conjugates* if U and V satisfy the Cauchy–Riemann equations.

harmonic conjugates

EXAMPLE 13.12

Given $f(z) = \sin z$ and $g(z) = \cos z$, find the harmonic conjugate functions $u_1(x, y) = \mathrm{Re}\{f(z)\}$ and $v_1(x, y) = \mathrm{Im}\{f(z)\}$ associated with $f(z)$, and the harmonic conjugate functions $u_2(x, y) = \mathrm{Re}\{g(z)\}$ and $v_2(x, y) = \mathrm{Im}\{g(z)\}$ associated with $g(z)$. Verify that u_1, v_1, u_2, and v_2 are harmonic functions and show that the complex function $F(z) = u_1(x, y) + iv_2(x, y)$ is *not* analytic, and so $u_1(x, y)$ is not the harmonic conjugate of $v_2(x, y)$.

Solution   As $f(z) = \sin(x + iy) = \sin x \cosh y + i \cos x \sinh y$, writing $f(z) = u_1 + iv_1$ we see that $u_1 = \sin x \cosh y$ and $v_1 = \cos x \sinh y$. The functions u_1 and v_1 are harmonic conjugate functions because straightforward differentiation confirms that u_1 and v_1 satisfy the Cauchy–Riemann equations. To verify that u_1 and v_1 are harmonic functions, it is necessary to show that each satisfies Laplace's equation. Differentiation gives

$$\frac{\partial^2 u_1}{\partial x^2} = -\sin x \cosh y \quad \text{and} \quad \frac{\partial^2 u_1}{\partial y^2} = \sin x \cosh y,$$

so

$$\frac{\partial^2 u_1}{\partial x^2} + \frac{\partial^2 u_1}{\partial y^2} = 0, \quad \text{or} \quad \Delta u_1 = 0,$$

confirming that u_1 is a harmonic function. The fact that v_1 is harmonic follows in similar fashion.

As $g(z) = \cos z = \cos(x + iy) = \cos x \cosh y - i \sin x \sinh y$, setting $g(z) = u_2 + iv_2$ shows that $u_2 = \cos x \cosh y$ and $v_2 = -\sin x \sinh y$. These are harmonic conjugate functions because they also satisfy the Cauchy–Riemann equations.

Although the functions $u_1(x, y) = \sin x \cosh y$ and $v_2(x, y) = -\sin x \sinh y$ forming the real and imaginary parts of $F(z) = u_1(x, y) + iv_2(x, y)$ are both harmonic, $\partial u_1/\partial x \neq \partial v_2/\partial y$, and $\partial u_1/\partial y \neq -\partial v_2/\partial x$, showing that $F(z)$ does not satisfy the Cauchy–Riemann equations, and so $F(z)$ is *not* analytic and $u_1(x, y)$ and $v_2(x, y)$ are *not* harmonic conjugates. ∎

In (44) the Laplacian operator is expressed in its cartesian form, but if the cartesian coordinates (x, y) are changed to the polar coordinates (r, θ) by means of the transformation $x = r \cos \theta$ and $y = r \sin \theta$, the change of variable formulas from elementary calculus (see Theorem 1.11) shows that the Laplacian operator takes on the form

Laplacian in polar coordinates

$$\Delta \equiv \frac{\partial^2}{\partial r^2} + \frac{1}{r} \frac{\partial}{\partial r} + \frac{1}{r^2} \frac{\partial^2}{\partial \theta^2}. \tag{45}$$

This means that when polar coordinates are used to express z in the form $z = re^{i\theta}$, and a single-valued analytic function $f(z) = u(r, \theta) + iv(r, \theta)$ is considered, the functions $u(r, \theta)$ and $v(r, \theta)$ will each be harmonic, so

$$\Delta u(r, \theta) \equiv \frac{\partial^2 u}{\partial r^2} + \frac{1}{r} \frac{\partial u}{\partial r} + \frac{1}{r^2} \frac{\partial^2 u}{\partial \theta^2} = 0 \quad \text{and}$$

$$\Delta v(r, \theta) \equiv \frac{\partial^2 v}{\partial r^2} + \frac{1}{r} \frac{\partial v}{\partial r} + \frac{1}{r^2} \frac{\partial^2 v}{\partial \theta^2} = 0. \tag{46}$$

It follows that $u(r, \theta)$ will be the harmonic conjugate of $v(r, \theta)$ and, conversely, $v(r, \theta)$ will be the harmonic conjugate of $u(r, \theta)$.

EXAMPLE 13.13

Set $z = re^{i\theta}$ in $f(z) = z + 1/z$, and by showing that when $z \neq 0$ the function $f(z)$ satisfies the polar form of the Cauchy–Riemann equations given in Exercise 37 of Exercise set 13.2, confirm that $f(z)$ is analytic when $z \neq 0$. Verify that the functions $u(r, \theta) = \text{Re}\{f(z)\}$ and $v(r, \theta) = \text{Im}\{f(z)\}$ are harmonic functions.

Solution $f(z) = z + 1/z = re^{i\theta} + \frac{1}{r}e^{-i\theta} = (r + \frac{1}{r}) \cos \theta + i(r - \frac{1}{r}) \sin \theta$, and so

$$u(r, \theta) = \left(r + \frac{1}{r}\right) \cos \theta \quad \text{and} \quad v(r, \theta) = \left(r - \frac{1}{r}\right) \sin \theta.$$

Routine differentiation confirms that u and v satisfy the polar form of the Cauchy–Riemann equations

$$\frac{\partial u}{\partial r} = \frac{1}{r} \frac{\partial v}{\partial \theta} \quad \text{and} \quad \frac{1}{r} \frac{\partial u}{\partial \theta} = -\frac{\partial v}{\partial r} \quad \text{for } r \neq 0,$$

so $f(z)$ is analytic for $z \neq 0$.

Straightforward differentiation shows that u and v satisfy the polar form of Laplace's equation and so are harmonic when $z \neq 0$.   ■

In applications of complex analysis, as in Section 17.2 when solving a boundary value problem for the two-dimensional steady state temperature distribution in a solid, it can happen that a harmonic function $\Phi(x, y)$ is known, but it is required to find its harmonic conjugate $\Psi(x, y)$ so an analytic function $F(z) = \Phi(x, y) + i\Psi(x, y)$ can be constructed. The function $\Psi(x, y)$ can be found by making use of the Cauchy–Riemann equations that must be satisfied simultaneously by both $\Phi(x, y)$ and $\Psi(x, y)$.

We now show how an analytic function $f(z) = u(x, y) + iv(x, y)$ can be constructed when either one of the harmonic conjugate functions $u(x, y)$ or $v(x, y)$ is known. Let us suppose that a harmonic function $u(x, y)$ is known. Then from the first of the Cauchy–Riemann equations,

how to find an
analytic function
from one of its
harmonic conjugate
functions

$$\frac{\partial v}{\partial y} = \frac{\partial u}{\partial x}, \tag{47}$$

where the expression on the right can be found by differentiation of the known function $u(x, y)$.

If we reverse the process by which $\partial u / \partial x$ was found, by integrating (47) with respect to y while keeping x constant, we obtain

$$v(x, y) = \int \frac{\partial u}{\partial x} dy + g(x) + a, \tag{48}$$

where $g(x)$ is an arbitrary function of x and a is an arbitrary real integration constant.

The inclusion of the arbitrary function $g(x)$ in (48) in addition to the usual arbitrary integration constant a is necessary to make the expression on the right the most general antiderivative that can be obtained when (47) is integrated with respect to y while holding x constant. The result can be checked by differentiating (48) partially with respect to y to return to (47), because after differentiation the first term on the right reduces to $\partial u / \partial x$ and the remaining terms vanish because $\partial \{g(x) + a\} / \partial y \equiv 0$. It is obvious that (48) can be simplified by including the arbitrary constant a in the arbitrary function $g(x)$, but in applications it is usually better to retain it explicitly as in (48).

If we rewrite the second Cauchy–Riemann equation as

$$\frac{\partial v}{\partial x} = -\frac{\partial u}{\partial y},$$

the term on the right is again known by differentiation of $u(x, y)$. Integration of this equation with respect to x while keeping y constant gives

$$v(x, y) = -\int \frac{\partial u}{\partial y} dx + h(y) + b, \tag{49}$$

where now $h(y)$ is an arbitrary function of y and b is an arbitrary real integration constant.

Expressions (48) and (49) must be identical, so $g(x)$ in (48) must be identified with any functions on the right of (49) that only involve x, and $h(y)$ in (49) must be identified with any functions on the right of (48) that only involve y, whereas the arbitrary constants must be equal, so $b = a$. The required analytic function is then seen to be

$$f(z) = u(x, y) + iv(x, y) + ia. \tag{50}$$

An analogous argument shows how if $v(x, y)$ is known instead of $u(x, y)$, then

$$u(x, y) = \int \frac{\partial v}{\partial y} dx + H(y) + C, \tag{51}$$

and

$$u(x, y) = -\int \frac{\partial v}{\partial x} dy + G(x) + D, \tag{52}$$

with $H(y)$ an arbitrary function of y, $G(x)$ an arbitrary function of x, and C and D arbitrary real integration constants. The form of argument used to arrive at (50) then shows that the required analytic function is

$$f(z) = u(x, y) + iv(x, y) + D. \tag{53}$$

It is to be expected that the analytic function $f(z)$ can only be determined up to an arbitrary additive constant, because a constant is always a solution of Laplace's equation. In applications, either the constant occurring in (50) or (53) is unimportant, and so can be set equal to zero, or, if needed, it must be determined by some additional condition satisfied by the analytic function $f(z)$.

To understand why the introduction of an arbitrary additive constant to a solution of Laplace's equation causes no difficulties in applications, it is only necessary to consider problems like the determination of a steady state temperature distribution or an electrostatic potential distribution. In these cases, and in others of a similar type, what matters is the temperature or potential *difference*, rather than their absolute values, so the arbitrary additive constant simply represents a convenient reference level from which all other temperatures or potentials are measured.

EXAMPLE 13.14   Given $u(x, y) = x^2 - y^2 + x - y$, find its harmonic conjugate $v(x, y)$ and construct the most general analytic function $f(z)$ such that $u(x, y) = \text{Re}\{f(z)\}$.

Solution   First it is necessary to check that $u(x, y)$ is a harmonic function, and this can be seen from the fact that

$$\frac{\partial^2 u}{\partial x^2} = 2, \quad \frac{\partial^2 u}{\partial y^2} = -2, \quad \text{and so} \quad \Delta u = 0.$$

As $\partial u / \partial x = 2x + 1$, result (48) becomes

$$v(x, y) = \int (2x + 1)dy + g(x) + a,$$

so

$$v(x, y) = 2xy + y + g(x) + a.$$

Using the fact that $\partial u / \partial y = -2y - 1$, result (49) becomes

$$v(x, y) = -\int (-2y - 1)dx + h(y) + b,$$

so

$$v(x, y) = 2xy + x + h(y) + b.$$

These two expressions for $v(x, y)$ will be identical if $g(x) = x$, $h(y) = y$, and $a = b$, so

$$v(x, y) = 2xy + x + y + a,$$

with a an arbitrary real constant. The cartesian form of the required analytic function is

$$f(z) = x^2 - y^2 + x - y + i(2xy + x + y) + ia.$$

Setting $y = 0$ and replacing x by z to convert this to an analytic function in terms of z shows that

$$f(z) = z^2 + (1 + i)z + ia. \qquad \blacksquare$$

For more information and examples involving limits, continuity, differentiability, and elementary functions of a complex variable, see any one of references [6.1] to [6.4] and [6.6] to [6.9].

Summary

Harmonic functions were introduced as solutions of Laplace's equation, and in an analytic function $f(z) = u + iv$ the functions u and v were shown to be harmonic. The functions u and v in an analytic function were called harmonic conjugates, and it was shown how to reconstruct $f(z)$ when either of its harmonic conjugates u or v is known.

EXERCISES 13.3

In Exercises 1 through 10, verify that the given function is harmonic, and find its harmonic conjugate. Use the result to construct the most general analytic function $f(z)$ as a function of z.

1. $u(x, y) = x^3 - 3xy^2 + 2x + y.$

2. $u(x, y) = e^{2x}(x \cos 2y - y \sin 2y).$

3. $v(x, y) = e^{-y}(y \cos x + x \sin x) + 2x.$

4. $v(x, y) = x^3 - 3xy^2 + x + y.$

5. $v(x, y) = y \sinh 2x \cos 2y + x \cosh 2x \sin 2y.$

6. $u(x, y) = \sin 3x \cosh 3y - 2x^2 + 2y^2.$

7. $u(x, y) = x \cos 3x \cosh 3y + y \sin 3x \sinh 3y.$

8. $v(x, y) = e^{-y}(3 \cos x + 2 \sin x) - 5y.$

9. $u(r, \theta) = r \cos \theta + 2r^2 \cos 2\theta + r^2 \sin 2\theta.$

10. $v(r, \theta) = r \sin \theta + \dfrac{1}{r^2} \sin 2\theta.$

11. Show that $u(x, y) = xy$ and $v(x, y) = x^3 - 3xy^2$ are both harmonic functions, but they are not harmonic conjugates.

12. Show that $u(x, y) = -x^2 + y^2 + 2xy$ and $v(x, y) = x^3 - 3xy^2 + 3x^2y - y^3$ are both harmonic functions, but they are not harmonic conjugates.

13. Prove that if $f(z) = u(x, y) + iv(x, y)$ is analytic in a domain D, and either $u(x, y) = $ constant or $v(x, y) = $ constant, then $f(z) = $ constant in D. Does this result remain true if $f(z)$ is not analytic? If not, explain why and give an example.

14. Given that $\Phi(x, y) = a(1 - 2x^2 + 2y^2) \sin 2x \cosh 2y + 4xy \cos 2x \sinh by$, find $\Delta\Phi$, and hence determine the values of the constants a and b that make Φ a harmonic function.

15. Given that $\Phi(x, y) = (2 + ax^2 - y^2) \sinh x \cos y + bxy \cosh x \sin y$, find $\Delta\Phi$, and hence determine the values of the constants a and b that make Φ a harmonic function.

13.4 Elementary Functions, Inverse Functions, and Branches

The elementary analytic functions considered so far have been polynomials, rational functions (quotients of polynomials), the exponential function, and the trigonometric and hyperbolic functions. All of these have involved the fundamental idea that for f to be a *function*, one point in the domain of definition of f must correspond to one point in the range of f. If the domain of definition of f is D and its range is Ω and we set $w = f(z)$, then z is any point of D and w is the corresponding point in Ω.

In addition to the connection between the domain D of f and its range Ω, expressed by the functional relationship $w = f(z)$, it is also necessary to be able to proceed in the reverse direction, by starting with a point w in Ω and finding the point or points z in D to which it corresponds. This is the **inverse** relationship involving f, and it is convenient to represent it by writing $z = f^{-1}(w)$. For this inverse relationship to be a *function* it is necessary that f^{-1} has the property that to every w in Ω there corresponds only one z in D.

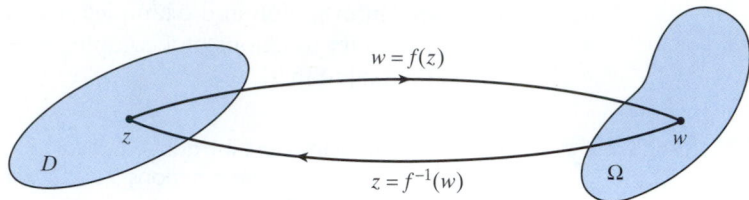

FIGURE 13.6   If f is a one-one analytic function, then $f(f^{-1}(w)) = w$ and $f^{-1}(f(z)) = z$.

In general, if the analytic function

$$w = f(z) \tag{54}$$

maps its domain of definition D onto a domain Ω and, in addition, if to each w in Ω there corresponds only one z in D given by $z = f^{-1}(w)$, the function f is **one-one**, and the function f^{-1} is called the **inverse** of the function f. This means that if an analytic function f is one-one, then

inverse function

$$f(f^{-1}(w)) = w \quad \text{and} \quad f^{-1}(f(z)) = z. \tag{55}$$

The relationship between a one-one analytic function f and its inverse f^{-1} is shown diagramatically in Fig. 13.6.

Let us now show that if f is a one-one analytic function defined for z in D, and

$$f'(z) \neq 0,$$

then the inverse function $z = f^{-1}(w)$ is analytic in Ω. This result is easily proved by using the definition of differentiability and setting $z + h = f^{-1}(w + k)$, so that $w + k = f(z + h)$. Differentiation $f^{-1}(w)$ gives

$$\frac{d}{dw}[f^{-1}(w)] = \lim_{k \to 0}\left[\frac{f^{-1}(w + k) - f^{-1}(w)}{k}\right] = \lim_{h \to 0}\left[\frac{h}{f(z + h) - f(z)}\right]$$

$$\times \lim_{h \to 0}\left\{1 \Big/ \left[\frac{f(z + h) - f(z)}{h}\right]\right\} = 1/f'(z).$$

Then, as by hypothesis $f'(z) \neq 0$, it follows that $d[f^{-1}(w)]/dw$ exists and is unique in Ω, so $f^{-1}(w)$ is analytic in Ω, and the result is proved.

One of the simplest examples of a one-one analytic function is provided by the **linear function** $w = az + b$ with $a \neq 0$, because this is analytic throughout the z-plane and maps every point of it one-one onto the w-plane, and the inverse function $z = (w - b)/a$ is also analytic throughout the w-plane.

A slightly more complicated example of a one-one analytic function is the **linear fractional function**

linear fractional function

$$w = \frac{az + b}{cz + d} \tag{56}$$

that is analytic in any domain D in the z-plane in which $z \neq -d/c$, because then dw/dz is defined throughout D. Solving the linear fractional function in (56) for z

shows the inverse function to be given by

$$z = \frac{b - wd}{wc - a}.$$

This inverse function is also analytic, and it maps any domain Ω in the w-plane where $w \neq a/c$ onto a corresponding domain D in the z-plane. The condition $w \neq a/c$ ensures the analyticity of the inverse function because then dz/dw is defined and unique throughout Ω.

Inverse functions associated with functions as simple as $w = z^2$, $w = \exp z$, and the hyperbolic and trigonometric functions require special attention because these functions exhibit periodicity in the complex plane. This periodicity has the effect that although one z corresponds to one w, the converse is not true because one w corresponds to more than one z, and often to infinitely many values of z. To overcome this difficulty it is necessary to confine z to a restricted domain in the z-plane to make the relationship between the restricted domain in the z-plane and the w-plane one-one. To illustrate this approach we will consider the function **nth root function** $w = z^n$, and its inverse the **nth root function** $z = w^{1/n}$, where n is a positive integer.

When expressed in polar form by writing $w = \rho e^{i\phi}$ and $z = r e^{i\theta}$, with $\theta = \text{Arg}\, z$, the function $w = z^n$ becomes $w = r^n e^{in\theta}$. So, as the argument of z is multiplied by n, any domain in the z-plane in the form of a sector with angle $2\pi/n$ centered on the origin will be mapped onto the entire w-plane, with the result that the function $w = z^n$ will map the entire z-plane onto the w-plane n times. Consequently, although one z corresponds to one w, the inverse operation $z = w^{1/n}$ will map n different values of w onto one point in the z-plane.

As it stands, the inverse formula $z = w^{1/n}$ represents many functions and so does *not* define a single function. To overcome this problem we divide the z-plane into n equal sectors $D_0, D_1, \ldots, D_{n-1}$, each centered on the origin, with D_k defined as the sector given by

$$(2k - 1)\frac{\pi}{n} < \theta < (2k + 1)\frac{\pi}{n}, \quad r > 0, \quad \text{for } k = 0, 1, 2, \ldots, n - 1. \tag{57}$$

If we restrict $z = r e^{i\theta}$ to any one of the sectors D_k, the function $w = z^n$ will map the sector D_k once onto the entire w-plane with the exception of points on the negative real axis up to and including the point at the origin. Conversely, when z is restricted to D_k, any point in the w-plane not on the negative real axis or at the origin will be mapped once by the function $z = w^{1/n}$ onto the sector D_k. The deletion of the points on the negative real axis up to and including the origin is called a **cut** in the w-plane.

Let ψ be such that $-\pi/n < \psi < \pi/n$, then in the kth sector $D_k, \theta = 2k\pi/n + \psi$ for $k = 0, 1, \ldots, n - 1$. Using the polar representations for w and z by setting $z = r e^{i\psi}$ and $w = \rho e^{i\phi}$ allows $w = z^n$ to be written

$$\rho e^{i\phi} = r^n \exp\left[in\left(\frac{2k\pi}{n} + \psi\right)\right],$$

so equating moduli and arguments we have

$$\rho = r^n \quad \text{and} \quad \phi = 2k\pi + n\psi,$$

showing that

$$r = \rho^{1/n} \quad \text{and} \quad \phi/n = 2k\pi/n + \psi,$$

where $\rho^{1/n}$ is the numerical value of the nth root of the positive real number ρ.

Solving for z in terms of w shows that the cut w-plane is mapped one-one onto the sector D_k by

$$z = \rho^{1/n} \left\{ \cos\left[\frac{2k\pi}{n} + \psi \right] + i \sin\left[\frac{2k\pi}{n} + \psi \right] \right\}, \quad k = 0, 1, \ldots, n-1. \qquad (58)$$

branch, principal branch, and branch cut

Each of the n different solutions in (58) is called a **branch** of the nth root function, and the branch corresponding to $n = 0$ is called the **principal branch**. The *cut* in the w-plane separating one branch from another is called a **branch cut**. So the principal branch of the nth root function $z = w^{1/n}$ is

$$z = \rho^{1/n}[\cos\psi + i\sin\psi], \quad \text{with } -\pi/n < \psi \le \pi/n. \qquad (59)$$

The mapping of the sector D_0 onto the cut w-plane by $w = z^3$ and of the cut w-plane onto the z-plane by the principal branch of the cube root function $z = w^{1/3}$ is shown in Fig. 13.7, where shading has been used to show how different areas correspond. The mapping of D_1 onto the cut w-plane by $w = z^3$ and of the cut w-plane onto the z-plane by the second branch ($k = 1$) of the cube root function is shown in Fig. 13.8, where shading has again been used to show how different areas correspond.

When it is necessary to consider the nth root function as a function of z, and not merely as the inverse of the power function $w = z^n$, all that is necessary in (59) is to interchange z and w and their associated moduli and arguments, leading to the corresponding result for the function $w = z^{1/n}$.

The complex exponential function $w = e^z$ has been defined as

$$e^z = e^x(\cos y + i\sin y),$$

so as $\sin y$ and $\cos y$ are periodic with period 2π, it can be seen that e^z is periodic with period $2\pi i$. This means that any strip of width 2π in the z-plane that is parallel to the real axis will be mapped onto the entire w-plane, with the exception of the

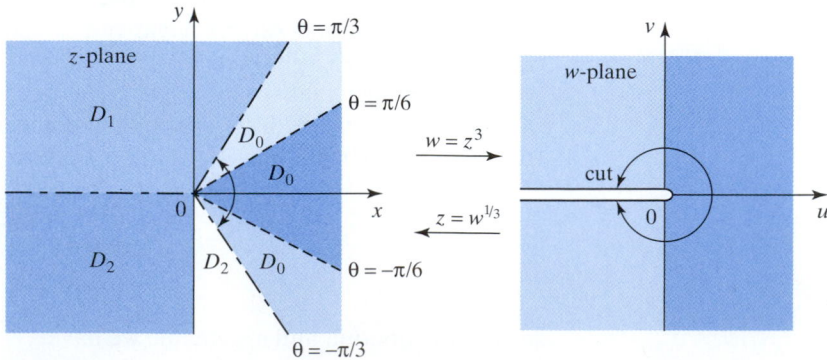

FIGURE 13.7   Mapping of sector D_0 in the z-plane onto the cut w-plane by $w = z^3$, and of the cut w-plane onto D_0 by the principal branch of $z = w^{1/3}$.

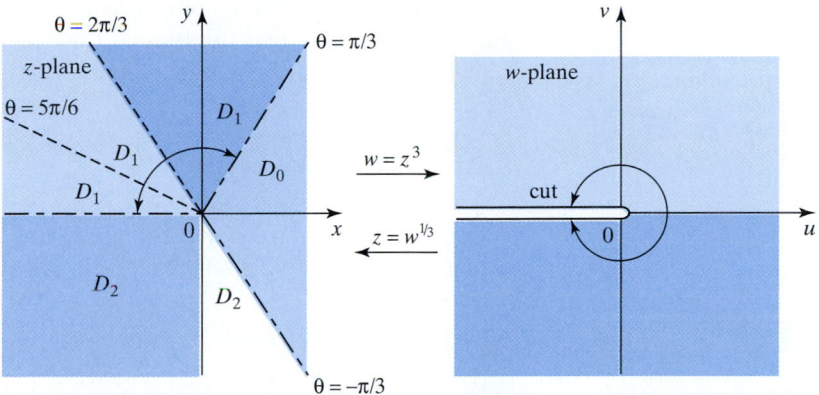

FIGURE 13.8   Mapping of sector D_1 in the z-plane onto the cut w-plane by $w = z^3$, and of the cut w-plane onto D_1 by the second branch of $z = w^{1/3}$.

origin. The origin must be excluded because $e^z \neq 0$ for any finite z, as may be seen from the fact that $|e^z| = e^x$, and e^x is never zero. The strip $-\pi < y \leq \pi$ is called the **fundamental strip** for the complex exponential function, and it is usual to refer to the complex plane from which the point at the origin has been removed as the **deleted complex plane**.

fundamental strip

Important properties of the complex exponential function are as follows:

(i) $e^{2\pi n i} = 1$ for n an integer, so $e^{z+2\pi n i} = e^z$ when n is an integer

(ii) If $w = e^z = \rho e^{i\phi}$, then $\rho = e^x$ and $\phi = \arg e^z = y \pm 2n\pi$ for all integers n

(iii) As $x = \ln \rho$, it follows that

$$z = x + iy = \ln \rho + i(\phi + 2n\pi)$$

and so

$$w = \exp[\ln |w| + i(\operatorname{Arg} w + 2n\pi)].$$

The inverse of the complex exponential function is the logarithmic function $\log z$, but the fact that any strip of width 2π parallel to the real axis in the z-plane will be mapped by $w = e^z$ onto the deleted w-plane means that the logarithmic function is infinitely many valued or, more simply, a **multivalued** function.

To make the multivalued complex logarithmic function into a one-one function, it is necessary to replace $\log z$ by a function with infinitely many branches, each corresponding to a strip of width 2π in the z-plane parallel to the real axis. The relationship between the planes then becomes one-one, because the exponential function will map a particular strip once onto the deleted w-plane and, conversely, a branch of the logarithmic function will map the deleted w-plane once onto the strip.

Using the symbol $\log z$ to denote the multifunction complex logarithmic function, and $\ln |z|$ to denote the natural logarithm of the real number $|z|$, we define the complex logarithm of the complex number z in the obvious manner as

$$\log z = \ln |z| + i \arg z, \quad \text{for } z \neq 0,$$

but $\arg z = \operatorname{Arg} z \pm 2n\pi$, with n an integer, so

$$\log z = \ln |z| + i(\operatorname{Arg} z \pm 2n\pi). \tag{60}$$

Each of the expressions in (60) is to be regarded as a **branch** of the complex logarithmic function, and the branch for which $n = 0$ is taken to be the **principal branch** of the function. To avoid confusion, the principal branch is denoted by Log z, where

principal branch of the logarithmic function and principal value

$$\text{Log } z = \ln |z| + i(\text{Arg } z), \quad \text{with } z \neq 0 \quad \text{and} \quad -\pi < \text{Arg } z \leq \pi. \tag{61}$$

For any given complex number z, the corresponding complex number defined by (61) is called the **principal value** of the logarithm of z.

EXAMPLE 13.15

Find $\log(1 + i\sqrt{3})$ and $\text{Log}(1 + i\sqrt{3})$.

Solution Setting $z = 1 + i\sqrt{3}$ we find that that $|z| = 2$ and $\text{Arg } z = \pi/3$, and so $\log(1 + i\sqrt{3}) = \ln 2 + i(\frac{\pi}{3} + 2n\pi)$, and $\text{Log}(1 + i\sqrt{3}) = \ln 2 + i\pi/3$. ∎

Applying the polar form of the Cauchy–Riemann equations to Log z shows that it is an analytic function for $z \neq 0$, and the multivalued form of the complex logarithmic function possesses all the properties of the natural logarithmic function so, for example,

$$\log(z_1 z_2) = \log z_1 + \log z_2 \quad \text{and} \quad \log(z_1/z_2) = \log z_1 - \log z_2. \tag{62}$$

However, the restriction placed on the arguments of principal values means that these results do not always remain true when the multivalued logarithm $\log z$ is replaced by Log z.

We are now in a position to generalize the power function $w = z^a$, where a is an arbitrary real number. To do this we write $w = z^a$ in the form

$$w = z^a = e^{a \, \text{Log} \, z} = e^{a[\ln |z| + i(\text{Arg } z + 2n\pi)]} \quad \text{for } n = 0, \pm 1, \pm 2, \ldots,$$

and setting $z = re^{i\theta}$ this becomes

$$w = r^a \{\cos[a(\theta + 2n\pi)] + i\sin[a(\theta + 2n\pi)]\}. \tag{63}$$

We must now consider the behavior of the complex hyperbolic and trigonometric functions that map the complex z-plane more than once onto the w-plane, causing their inverses to be multivalued. To see how suitable branches can be introduced, we consider the typical example $w = \arcsin z$, which is the inverse of the function $z = \sin w$, so $\sin(\arcsin z) = z$. From the definition of the sine function,

$$z = \sin w = \frac{e^{iw} - e^{-iw}}{2i} = \frac{e^{2iw} - 1}{2i\,e^{iw}},$$

so

$$e^{2iw} - 2i\,z e^{iw} - 1 = 0.$$

Solving this quadratic equation for e^{iw} we find

$$e^{iw} = iz + (1 - z^2)^{1/2},$$

where the $\pm$ sign usually inserted in front of the square root has been omitted because the function $w = z^{1/2}$ implies that the square root function is two-valued.

inverse trigonometric and hyperbolic functions

Taking the complex logarithm of this result, we have

$$iw = \log[iz + (1 - z^2)^{1/2}],$$

and so

$$w = \arcsin z = -i \log[iz + (1 - z^2)^{1/2}]. \tag{64}$$

Because of its branches the log function must be interpreted as many one-one functions, all with the same domain, but each branch having a different range.

Similar arguments applied to the other complex, trigonometric functions and to the complex hyperbolic functions show that

$$\arccos z = -i \log[z + i(1 - z^2)^{1/2}] \tag{65}$$

$$\arctan z = \frac{i}{2} \log \left(\frac{i + z}{i - z} \right) \tag{66}$$

$$\operatorname{arcsinh} z = \log[z + (1 + z^2)^{1/2}] \tag{67}$$

$$\operatorname{arccosh} z = \log[z + (z^2 - 1)^{1/2}] \tag{68}$$

$$\operatorname{arctanh} z = \frac{1}{2} \log \left(\frac{1 + z}{1 - z} \right). \tag{69}$$

In each of the preceding cases, the branch of the inverse function involved is determined by the choice of branch in the square root and complex logarithmic function that appears on the right.

Differentiation shows that:

derivatives of inverse trigonometric and hyperbolic functions

$$\frac{d}{dz}[\arcsin z] = \frac{1}{(1 - z^2)^{1/2}} \tag{70}$$

$$\frac{d}{dz}[\arccos z] = \frac{-1}{(1 - z^2)^{1/2}} \tag{71}$$

$$\frac{d}{dz}[\arctan z] = \frac{1}{1 + z^2} \tag{72}$$

$$\frac{d}{dz}[\operatorname{arcsinh} z] = \frac{1}{(z^2 + 1)^{1/2}} \tag{73}$$

$$\frac{d}{dz}[\operatorname{arccosh} z] = \frac{1}{(z^2 - 1)^{1/2}} \tag{74}$$

$$\frac{d}{dz}[\operatorname{arctanh} z] = \frac{1}{1 - z^2}. \tag{75}$$

EXAMPLE 13.16   Show that the result obtained from (64) with $z = 1$ is consistent with the real variable trigonometric result $\arcsin 1 = (4n + 1)\pi/2$, for $n = 0, \pm 1, \pm 2, \ldots$.

__Solution__   From (64), $\arcsin 1 = -i \log i$, but $i = \exp[i(\frac{\pi}{2} + 2n\pi)] = \exp[i(4n + 1)\frac{\pi}{2}]$, for $n = 0, \pm 1, \pm 2, \ldots$, and so

$$\arcsin 1 = -i \log i = -i \left[i(4n + 1)\frac{\pi}{2} \right] = (4n + 1)\frac{\pi}{2}, \quad \text{for } n = 0, \pm 1, \pm 2, \ldots.$$

The principal value of this result, obtained by using the principal value Log z of log z corresponding to $n = 0$, is $\arcsin 1 = \pi/2$. ∎

EXAMPLE 13.17   Find all the values of $\arcsin i$ and identify the one corresponding to the principal values of the square root and logarithmic functions.

Solution From (64), $\arcsin i = -i \log[-1 + \sqrt{2}]$, but $2 = 2e^{2m\pi i}$, for $m = 0, \pm 1, \pm 2, \dots$, so

$$\sqrt{2} = 2^{1/2} e^{m\pi i}, \quad \text{for } m = 0, \pm 1, \pm 2, \dots.$$

As $e^{m\pi i}$ is either 1 or -1, according as m is even or odd, the value corresponding to the principal branch ($m = 0$) is $\sqrt{2} = 2^{1/2}$, while the one corresponding to the second branch ($m = 1$) is $\sqrt{2} = -2^{1/2}$, where $2^{1/2}$ denotes the *positive* square root of 2.

Case $m = 0$ (The principal branch): If the principal value of $\sqrt{2}$ is used, $-1 + \sqrt{2} = 2^{1/2} - 1$ is positive and $\arcsin i = -i \log(2^{1/2} - 1)$, so writing $2^{1/2} - 1 = (2^{1/2} - 1)e^{2n\pi i}$, for $n = 0, \pm 1, \pm 2, \dots$, shows that in this case

$$\arcsin i = -i \log(2^{1/2} - 1) = 2n\pi - i \ln(2^{1/2} - 1), \quad \text{for } n = 0, \pm 1, \pm 2, \dots.$$

The value obtained for $\arcsin i$ depends on the choice of n, which in turn identifies the branch of the logarithmic function that is used to determine the value of $\log(2^{1/2} - 1)$.

Case $m = 1$ (The second branch): If the second value of $\sqrt{2}$ is used, $-1 - \sqrt{2} = -(2^{1/2} + 1)$ is negative, so now we have $\arcsin i = -i \log[-(2^{1/2} + 1)]$, but $-(2^{1/2} + 1) = (2^{1/2} + 1)e^{\pi i} = (2^{1/2} + 1)e^{\pi i} e^{2n\pi i} = (2^{1/2} + 1)e^{(2n+1)\pi i}$, for $n = 0, \pm 1, \pm 2, \dots$. So $\log[-(2^{1/2} + 1)] = \ln(2^{1/2} + 1) + (2n + 1)\pi i$, leading to the result

$$\arcsin i = (2n + 1)\pi - i \ln(2^{1/2} + 1).$$

The value of $\arcsin i$ obtained by using the principal values of the square root function ($m = 0$) and the logarithmic function ($n = 0$) is

$$\arcsin i = -i \ln(2^{1/2} - 1). \quad \blacksquare$$

More information about inverse functions and branches can be found in references [6.1] to [6.4] and [6.6] to [6.9]. In particular, reference [6.4] provides valuable insight into the nature of the inverse of elementary functions of a complex variable.

EXERCISES 13.4

In Exercises 1 through 6 find all of the values of the given inverse functions and state the value obtained by using the principal value of the function or functions involved.

1. $\arccos 2i$.
2. $\text{arccosh } 4i$.
3. $\text{arctanh } i$.
4. $\arctan 3i$
5. $\arctan\left(-\dfrac{2}{5} + \dfrac{1}{5}i\right)$.
6. $\text{arctanh}\left(\dfrac{3}{7} + i\dfrac{2\sqrt{3}}{7}\right)$.

7. Show that $\arcsin z + \arccos z = \pi/2 + 2n\pi$.
8. Show that $u(x, y) = \ln(x^2 + y^2)$ and $v(x, y) =$ $\arctan(y/x)$ are analytic throughout the (x, y)-plane with the exception of the points on the imaginary axis.

9. Use the definition of $\text{Log } z$ to show that it is discontinuous at $z = 0$, and also that it experiences a jump of πi across the negative real axis.

10. Use implicit differentiation on the function $z = \exp w$ to show that its inverse $w = \log z$ has the derivative

$$\frac{d}{dz}[\log z] = \frac{1}{z}, \quad \text{for } z \neq 0.$$

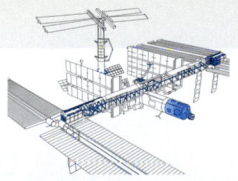

Project 1

Finding how $w = az + b$ Maps a Given Curve in the z-Plane onto the w-Plane

This project explores how the two complex constants a and b in $w = az + b$ influence the way in which a curve in the z-plane is mapped by this function onto an image curve in the w-plane. This project anticipates some of the ideas that will be examined later in more detail in the chapter on conformal mapping.

Let $z(t) = x(t) + iy(t)$, with $x(t) = t(\pi - t)$, $y(t) = \sin(2t)$; and $0 \le t \le \pi$. Then as t increases from 0 to π, so the point $(x(t), y(t))$ in the z-plane with t as a parameter will describe a curve C_z in the z-plane. If $w(t) = az(t) + b$, with a and b complex numbers, each point of the curve C_z will be mapped by this function onto an image curve C_w in the w-plane. If we set $w(t) = u(t) + iv(t) = a(x(t) + iy(t)) + b$, the image C_w in the w-plane of the curve C_z in the z-plane is obtained by plotting the parametrically defined curve $(u(t), v(t))$.

Using the same length scales on the x- and y-axes, and also on the u- and v-axes, make computer plots of C_z and the corresponding image curves C_w given that: (i) $a = 2$, $b = 0$, (ii) $a = \frac{1}{2}$, $b = 1 + i$, (iii) $a = 2e^{i\pi/4}$, $b = 0$, (iv) $a = \frac{1}{3}e^{3\pi i/4}$, $b = -1 + i$.

Repeat the preceding numerical experiments using several values of a and b of your own choosing. Comment on the effect of $|a|$, Arg a, and b on the way the curve C_z is mapped onto the curve C_w.

Project 2

Another Example of Mapping by $w = az + b$

Repeat Project 1, but this time using $x(t) = t^3 - 2t$, $y(t) = 4 - t^2$, and $-2 \le t \le 2$.

Project 3

Finding an Analytic Function from One of Its Harmonic Conjugates

This project uses computer algebra to find an analytic function $f(z)$ when only its imaginary part is known in cartesian form.

Show that the function
$$v(x, y) = 3e^{2x}(x \sin 2y + y \cos 2y) + 2 \sin x \cosh y + 6x^2 y - 2y^3 + 4x + 3$$

is harmonic. Find its harmonic conjugate $u(x, y)$ and hence find the corresponding analytic function $f(z) = u + iv$ as a function of z, given that $f(0) = 3i$.

Complex Integration

B oth derivatives and integrals of analytic functions occur extensively in applications, so this chapter extends the results of Chapter 13 to include integration. As the integral of a complex function is evaluated either along or around a curve, the chapter starts by developing the concept of integration along a parametrically defined path or curve. It is then shown why, for the result to be independent of the path, the complex function must be an analytic function, that is, it must satisfy the Cauchy–Riemann equations.

Integrals of this type are called line integrals of complex functions, and when the path of integration is a closed curve in the form of a single loop, called either a simple curve or a Jordan curve, the integral is called a contour integral. The properties of line integrals are used to define indefinite integrals of complex functions, and fundamental results concerning contour integrals are proved and illustrated by example. Various properties of analytic functions are proved in the last section, including the important fundamental theorem of algebra that asserts that every polynomial of degree n has precisely n zeros, though some may be repeated.

14.1 Complex Integrals

C omplex integration involves integrating a single valued analytic function $f(z)$ in a given direction along a curve Γ in the complex z-plane. A non-self-intersecting curve Γ whose end points are not coincident is called a **path**, and paths are usually formed by joining straight line segments and arcs end to end. A closed path Γ in the form of a simple non-self-intersecting loop is called a **contour**. Paths and contours are usually specified parametrically by defining a general point z on Γ in the form

path or contour

$$z = z(t) = x(t) + i y(t) \quad \text{for } t_0 \le t \le t_1, \tag{1}$$

where $x(t)$ and $y(t)$ are prescribed functions of the parameter t. Parametric representations are not unique, and in applications the simplest one is always used.

As t increases, so (1) determines the direction in which point z moves along Γ, and this direction is called the **sense** along the path or around the contour described

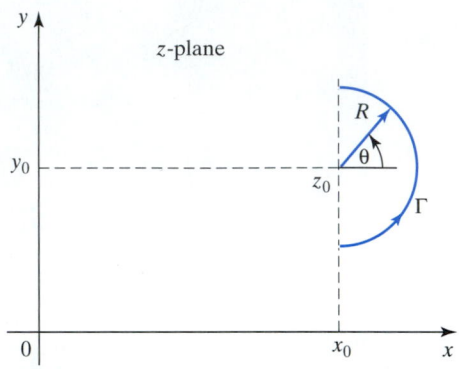

FIGURE 14.1   The semicircle Γ.

integration in
positive sense

by the parametrization. In integration around a contour, the standard convention is that integration in the **positive sense** is taken to be in the **counterclockwise direction**.

An essential feature of the parametric description of a path or contour is that, in addition to its convenience when used in complex integration, it allows the description of curves that in a cartesian representation are many-valued. This is illustrated in the following example.

EXAMPLE 14.1

parametrizing a
circular arc

Parametrize the semicircle Γ of radius R shown in Fig. 14.1 with its center at the point $z_0 = x_0 + iy_0$ in the z-plane.

Solution   The cartesian representation of the semicircle Γ is $(x - x_0)^2 + (y - y_0)^2 = R^2$, with $x_0 \leq x \leq x_0 + R$, but this is ambiguous because when it is solved for y to give $y = y_0 + [R^2 - (x - x_0)^2]^{1/2}$, the square root operation makes y double valued. One way to overcome this difficulty is to use polar coordinates to describe a point (x, y) on a semicircle of radius R located at the origin by writing

$$x = R\cos\theta \quad \text{and} \quad y = R\sin\theta \quad \text{for} \quad -\pi/2 \leq \theta \leq \pi/2.$$

Each point on Γ is now described unambiguously in terms of the parameter θ. A shift of origin to the point (x_0, y_0) shows that the required parametric representation of Γ is

$$x = x_0 + R\cos\theta \quad \text{and} \quad y = y_0 + R\sin\theta, \quad -\pi/2 \leq \theta \leq \pi/2,$$

so

$$z(\theta) = x_0 + R\cos\theta + i(y_0 + R\sin\theta), \quad -\pi/2 \leq \theta \leq \pi/2.$$

In this representation, as θ *increases*, so z moves *counterclockwise* (positively) around the semicircle Γ. The choice of symbol for the parameter is immaterial, so the result could equally well be written

$$z(t) = x_0 + R\cos t + i(y_0 + R\sin t), \quad -\pi/2 \leq t \leq \pi/2.$$

Clearly this is not the only possible parametric description of Γ in terms of sines and cosines, because the change of variable $t = 1 + s$ gives the equivalent parametric description in terms of s

$$z(s) = x_0 + R\cos(1 + s) + i[y_0 + R\sin(1 + s)], \quad -\left(\frac{1}{2}\pi + 1\right) \leq s \leq \left(\frac{1}{2}\pi - 1\right).$$

Other parametric representations of this type can be found by making different changes of variable, provided only that the new argument of the sine and cosine functions increases monotonically from $-\pi/2$ to $\pi/2$.

Differentiation of $z(t)$ shows that the differential dz along Γ as t increases is

$$dz = (-R\sin t + iR\cos t)dt,$$

so if $dz = dx + idy$, then

$$dx = -R\sin t\, dt \quad \text{and} \quad dy = R\cos t\, dt. \qquad \blacksquare$$

EXAMPLE 14.2

Let A and B be the points $(3, 1)$ and $(5, 7)$ in the z-plane. Parametrize the straight line segment AB in terms of parameter t so that (a) the sense is from A to B as t increases, and (b) the sense is from B to A as t increases.

Solution

(a) The cartesian equation of a straight line with gradient m passing through the point (x_1, y_1) is

$$\frac{y - y_1}{x - x_1} = m.$$

The gradient of the line segment AB is $m = (y_B - y_A)/(x_B - x_A) = (7 - 1)/(5 - 3) = 3$, so taking $(3, 1)$ for the point (x_1, y_1) and substituting into the foregoing result shows that the straight line through AB in Fig. 14.2 has the equation

$$y = 3x - 8.$$

The line segment AB is obtained from the equation $y = 3x - 8$ by restricting x to $3 \leq x \leq 5$. To parametrize the line segment AB in terms of t, we set

$$x = t \quad \text{and} \quad y = 3t - 8, \quad \text{with} \quad 3 \leq t \leq 5,$$

so that

$$z(t) = t + i(3t - 8), \quad 3 \leq t \leq 5.$$

It is easily seen from this parametrization that an increase in t induces a *sense* along the line segment from A to B. Differentiation shows that the differential along

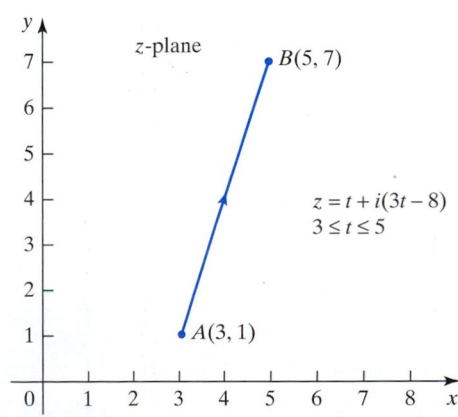

FIGURE 14.2   The line segment AB.

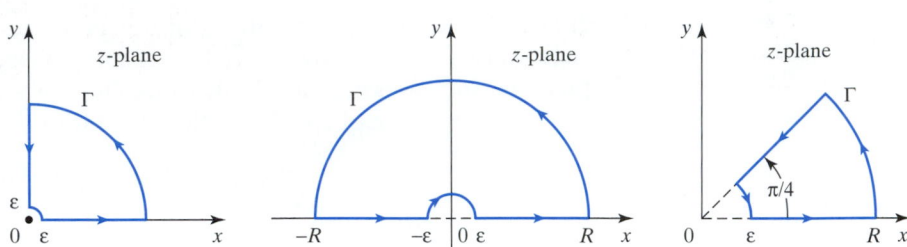

FIGURE 14.3 Some typical contours that arise in complex integration.

the line segment as t increases is

$$dz = dt + 3i\,dt \quad \text{so that } dx = dt \quad \text{and} \quad dy = 3dt.$$

(b) To reverse the sense along the line segment as t increases necessitates using a parameter that *decreases* as t *increases*. As the limits on t are $3 \le t \le 5$, this is most easily accomplished by setting $t = 5 - T$, because then $T = 0$ corresponds to $t = 5$ and $T = 2$ corresponds to $t = 3$. Substituting for t in the previous expression for $z(t)$ gives

$$z(T) = 5 - T + i(7 - 3T) \quad \text{for} \quad 0 \le T \le 2.$$

The differential dz along the line segment as T increases is now

$$dz = -dT - 3i\,dt, \quad \text{and so} \quad dx = -dT \quad \text{and} \quad dy = -3dT. \qquad \blacksquare$$

Typical examples of contours that arise in complex integration are shown in Fig. 14.3, in each of which the positive (counterclockwise) sense around the contour is shown by arrows.

The complex integral of an analytic function $f(z) = u(x, y) + iv(x, y)$ along **line integral** the path Γ from A to B shown in Fig. 14.4, called a **line integral**, is denoted by $\int_{\Gamma_{AB}} f(z)dz$, where $dz = dx + i\,dy$. This integral is defined as

$$
\begin{aligned}
\int_{\Gamma_{AB}} f(z)dz &= \int_{\Gamma_{AB}} (u + iv)(dx + i\,dy) \\
&= \int_{\Gamma_{AB}} (u\,dx - v\,dy) + i \int_{\Gamma_{AB}} (v\,dx + u\,dy).
\end{aligned}
\tag{2}
$$

When Γ is a contour, and so is a simple non-self-intersecting loop, the integral **contour integral** $\int_{\Gamma} f(z)dz$ is called a **contour integral**, and this is sometimes indicated by writing $\oint_{\Gamma} f(z)dz$, though this notation will not be used here.

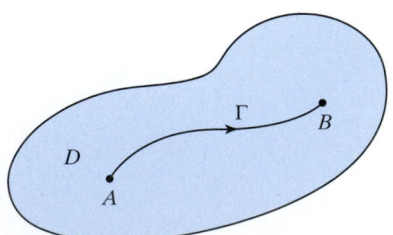

FIGURE 14.4 The path Γ for the line integral $\int_{\Gamma} f(z)dz$.

If the path Γ is parametrized as in (1), with A the point $z(t_0)$ and B the point $z(t_1)$, result (2) becomes

$$\int_{\Gamma_{AB}} f(z)dz = \int_{t_0}^{t_1} [u(x(t), y(t))x'(t) - v(x(t), y(t))y'(t)]dt$$

$$+ i \int_{t_0}^{t_1} [v(x(t), y(t))x'(t) + u(x(t), y(t))y'(t)]dt, \qquad (3)$$

where $x'(t) = dx/dt$ and $y'(t) = dy/dt$, showing that the evaluation of $\int_{\Gamma_{AB}} f(z)dz$ reduces to the calculation of two real integrals. It is usual to write (3) in the more concise form

$$\int_{\Gamma_{AB}} f(z)dz = \int_{t_0}^{t_1} f[z(t)]z'(t)dt. \qquad (4)$$

If in (4) the path Γ is constructed by joining end to end the successive paths $\Gamma_1, \Gamma_2, \ldots, \Gamma_n$, the linearity of the ordinary definite integral allows $\int_{\Gamma_{AB}} f(z)dz$ to be written

$$\int_{\Gamma_{AB}} f(z)dz = \int_{\Gamma_1} f(z)dz + \int_{\Gamma_2} f(z)dz + \cdots + \int_{\Gamma_n} f(z)dz. \qquad (5)$$

The significance of the sense along a path is apparent from (4), because reversing the sense along Γ interchanges the limits on the integral and so changes the sign of the integral. Consequently, if Γ_- denotes the path Γ with its sense reversed, then

$$\int_{\Gamma_-} f(z)dz = -\int_{\Gamma} f(z)dz. \qquad (6)$$

As a complex integral involves the sum of two real integrals, the complex integral of a linear combination $Af(z) + Bg(z)$ of two analytic functions $f(z)$ and $g(z)$ shares the same linearity property as real integrals, and so

$$\int_{\Gamma} \{Af(z) + Bg(z)\}dz = A\int_{\Gamma} f(z)dz + B\int_{\Gamma} g(z)dz, \qquad (7)$$

where A and B are arbitrary complex constants.

The following Theorems contain important results that are used when working with complex integrals.

THEOREM 14.1

A fundamental inequality for complex integrals   Let Γ be any path of finite length L, and let $f(z)$ be a complex function. Then the following inequality holds

(i) $$\left| \int_{\Gamma} f(z)dz \right| \leq \int_{\Gamma} |f(z)||dz|$$

and

(ii) $$\int_{\Gamma} |dz| = L.$$

Proof

(i) It was shown in (3) that the real and imaginary parts of a complex line integral are both real integrals, so the complex line integral $\int_\Gamma f(z)dz$ can be defined in essentially the same way as a real definite integral. Let a sequence of points $z_0, z_1, \ldots, z_n$ lie along Γ, with z_0 at one end and z_n at the other. Then if $\Delta_k = z_k - z_{k-1}$, and ζ_k is any point on the straight line segment joining z_{k-1} and z_k, generalizing the definition of a real definite integral we have

$$\int_\Gamma f(z)dz = \lim_{n \to \infty} \sum_{k=1}^{n} f(\zeta_k)\Delta z_k,$$

when $|\Delta z_k| = |z_k - z_{k-1}| \to 0$ for all k as $n \to \infty$.

Taking the modulus of $\sum_{k=1}^{n} f(\zeta_k)\Delta z_k$ and making repeated use of the triangle inequality gives

$$\left| \sum_{k=1}^{n} f(\zeta_k)\Delta_k \right| \leq \sum_{k=1}^{n} |f(\zeta_k)||\Delta_k|,$$

so proceeding to the limit as $n \to \infty$ this becomes

$$\left| \int_\Gamma f(z)dz \right| \leq \int_\Gamma |f(z)||dz|.$$

(ii) Setting $f(z) = 1$ in the result (i), and using the fact that $|dz| = [(dx)^2 + (dy)^2]^{1/2} = ds$, where ds is the element of arc length along Γ, we see that

$$\int_\Gamma |dz| = \int_\Gamma ds = L,$$

and the theorem is proved.   ■

THEOREM 14.2

a useful estimate for the modulus of an integral

Estimating the modulus of an integral   On a path Γ of finite length L, let $|f(z)|$ be bounded above by the positive real constant M, so that $|f(z)| \leq M$ when z lies on Γ. Then

$$\left| \int_\Gamma f(z)dz \right| \leq ML.$$

Proof   The result follows directly from Theorem 14.1. Using the bound $|f(z)| \leq M$ reduces (i) to

$$\left| \int_\Gamma f(z)dz \right| \leq \int_\Gamma |f(z)||dz| \leq M \int_\Gamma |dz|,$$

and using (ii) this becomes

$$\left| \int_\Gamma f(z)dz \right| \leq ML,$$

so the theorem is proved.   ■

Because an upper bound of $|f(z)|$ is denoted by M, and the length of path Γ is denoted by L, this theorem is sometimes called the *ML* theorem.

EXAMPLE 14.3

Let the points A, B, and C at $(2, 2)$, $(6, 2)$, and $(6, 3)$, respectively, form a triangle as shown in Fig. 14.5. Take Γ_1 to be the path $AB + BC$, Γ_2 to be the path AC, and Γ_3 to

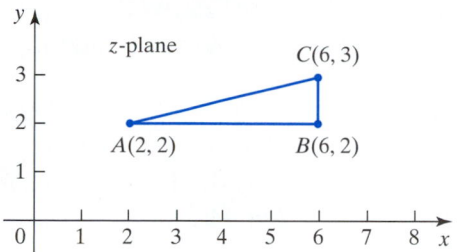

FIGURE 14.5   The points A, B, and C.

be the path $AB + BC + CA$, with the senses along the line segments indicated by the order of the letters. Set $f(z) = z$ and find the integrals $\int_{\Gamma_i} f(z)dz$, for $i = 1, 2, 3$. Verify Theorem 14.2 when $\Gamma = \Gamma_1$.

Solution

Case Γ_1:   It is necessary to parametrize the paths AB and BC before the integral can be evaluated. On $AB\ z = t + 2i$ for $2 \le t \le 6$, so an increase in t induces a sense on AB from A to B. Differentiation shows that $dz = dt$ on AB. Similarly, on $BC\ z = 6 + it$ for $2 \le t \le 3$, so an increase in t induces a sense on BC from B to C. Differentiation shows that $dz = i\,dt$ on BC. We have

$$\int_{\Gamma_1} f(z)dz = \int_{AB} f(z)dz + \int_{BC} f(z)dz$$

$$= \int_2^6 (t + 2i)dt + \int_2^3 (6 + it)i\,dt$$

$$= \left(\frac{1}{2}t^2 + 2it\right)\bigg|_{t=2}^{t=6} + \left(-\frac{1}{2}t^2 + 6it\right)\bigg|_{t=2}^{t=3} = \frac{27}{2} + 14i.$$

Case Γ_2:   Elementary coordinate geometry shows that the straight line through AC has the equation

$$y = \frac{3}{2} + \frac{x}{4},$$

so the line segment AC on this line is described by the condition $2 \le x \le 6$. This shows that a general point z on AC has the parametrization $x = t$, $y = \frac{3}{2} + \frac{t}{4}$ with $2 \le t \le 6$, and so

$$z(t) = t + i\left(\frac{3}{2} + \frac{t}{4}\right), \quad \text{for } 2 \le t \le 6.$$

Using this parametrization, an increase in t induces a sense from A to C on AC. Differentiation shows that $dz = (1 + \frac{i}{4})dt$, and so

$$\int_{\Gamma_2} f(z)dz = \int_{AC} f(z)dz = \int_2^6 \left(t + i\left(\frac{3}{2} + \frac{t}{4}\right)\right)\left(1 + \frac{i}{4}\right) dt$$

$$= \int_2^6 \left(\frac{15}{16}t - \frac{3}{8}\right) dt + i\int_2^6 \left(\frac{3}{2} + \frac{t}{2}\right) dt = \frac{27}{2} + 14i.$$

Case Γ_3:   As $\Gamma_3 = AB + BC + CA$, $\int_{\Gamma_3} z\,dz = \int_{\Gamma_1} z\,dz + \int_{CA} z\,dz$, but $\int_{\Gamma_1} z\,dz = \frac{27}{2} + 14i$, and from (6), $\int_{CA} z\,dz = -\int_{AC} z\,dz = -\frac{27}{2} - 14i$, so

$$\int_{\Gamma_3} z\,dz = \frac{27}{2} + 14i - \left(\frac{27}{2} + 14i\right) = 0.$$

To verify Theorem 14.2 for the path Γ_1 we proceed as follows. As $\int_{\Gamma_1} z\,dz = \frac{27}{2} + 14i$,

$$\left|\int_{\Gamma_1} z\,dz\right| = \left|\frac{27}{2} + 14i\right| = \frac{1}{2}\sqrt{1513} = 19.45.$$

On AB $z = t + 2i$, so $|z| = (t^2 + 4)^{1/2}$, and this assumes its largest value on AB at B when $t = 6$, so $\max_{AB}|z| = 40^{1/2} = 6.32$. On BC $z = 6 + it$, so $|z| = (t^2 + 36)^{1/2}$, and this assumes its largest value on BC at C when $t = 3$, so $\max_{BC}|z| = 45^{1/2} = 6.71$. These results show that M, the greatest value of $|z|$ on Γ_1, is $M = 6.71$. The length L of path Γ_1 is $4 + 1 = 5$, so $ML = 6.71 \times 5 = 33.55$, which is greater than $|\int_{\Gamma_1} z\,dz| = 19.45$, so the result of Theorem 14.2 is confirmed. ∎

EXAMPLE 14.4   Show that

$$\int_{\Gamma}(z - z_0)^n dz = 0, \quad \text{for } n \neq -1 \text{ a positive or negative integer;}$$

where Γ is a circle of radius R centered on the point $z = z_0$, and integration is performed around Γ in the counterclockwise sense.

Solution   It can be seen from Example 14.1 that the contour Γ in Fig. 14.6 can be parametrized by setting $z(t) = z_0 + Re^{it}$, with $0 \leq t \leq 2\pi$.

Using this parametrization, an increase in t, induces a sense of direction around contour Γ in the counterclockwise (positive) direction, and differentiation of $z(t)$ with respect to t shows that on Γ we have $dz = iRe^{it}dt$. Substituting for $z - z_0$ and dz, we obtain

$$\int_{\Gamma}(z - z_0)^n dz = \int_0^{2\pi} R^n e^{int} iRe^{it}dt = iR^{n+1}\int_0^{2\pi} e^{i(n+1)t}dt$$

$$= iR^{n+1}\left(\frac{\exp[i(n+1)t]}{i(n+1)}\right)_{t=0}^{t=2\pi} = 0, \quad \text{provided } n \neq -1. ∎$$

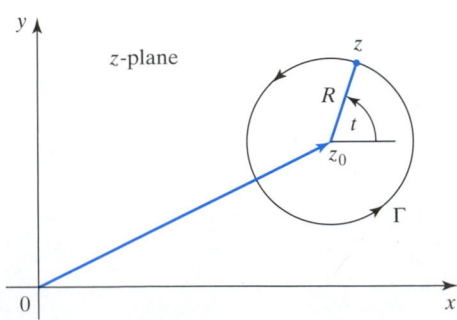

FIGURE 14.6   The circle Γ.

EXAMPLE 14.5   Show that

$$\int_\Gamma \frac{dz}{z - z_0} = 2\pi i,$$

where Γ is the circular contour used in Example 14.4.

Solution   Using the parameterization of Example 14.4 we find that the integrand becomes $dz/(z - z_0) = Rie^{it} dt / Re^{it} = i\, dt$, so

$$\int_\Gamma \frac{dz}{z - z_0} = i \int_0^{2\pi} dt = 2\pi i.$$ ∎

zeros and poles

The integrands in Examples 14.4 and 14.5 are special cases of functions which possess what are called *zeros* and *poles*. To make matters precise, a function $f(z)$ is said to have a **zero of order n** at $z = z_0$ if $n \geq 1$ is an integer and

$$f(z) = (z - z_0)^n g(z), \quad \text{with } g(z_0) \neq 0. \tag{8}$$

Expressed differently, $f(z)$ will have a zero of order n at $z = z_0$ if

$$\lim_{z \to z_0} \frac{f(z)}{(z - z_0)^n} = g(z_0), \quad \text{with } g(z_0) \neq 0.$$

A function $f(z)$ will have a **pole of order n** at $z = z_0$ if $n \geq 1$ is an integer and

$$f(z) = \frac{g(z)}{(z - z_0)^n}, \quad \text{with } g(z_0) \neq 0. \tag{9}$$

Expressed differently, $f(z)$ will have a pole of order n at $z = z_0$ if

$$\lim_{z \to z_0} (z - z_0)^n f(z) = g(z_0), \quad \text{with } g(z_0) \neq 0.$$

This shows that when $n \geq 1$ the integrand in Example 14.4 has a zero of order n at $z = z_0$ with $g(z) = 1$, and when $n \leq -1$ a pole of order $|n|$ at $z = z_0$ with $g(z) = 1$. The integrand in Example 14.5 has a pole of order 1, called a **simple pole**, at $z = z_0$ with $g(z) = 1$.

Similarly, the function

$$f(z) = \frac{(z - 2)^3}{(z - 1)(z + 5)^2}$$

has zero of order 3 at $z = 2$, a simple pole at $z = 1$ and pole of order 2 at $z = -5$.

This definition of a pole will be used first in Theorem 14.14, though later the simple poles of functions will be seen to play an essential role in complex integration.

Summary   The positive (counterclockwise) sense of direction around contours was defined and the line integral of a complex function was introduced. The useful *ML* theorem that estimates the magnitude of a complex line integral was derived and two elementary integrals around simple closed loops (contour integrals) were found.

EXERCISES 14.1

1. Given that A, B, and C are the respective points $(2, 1)$, $(4, 2)$, and $(5, 4)$ in the z-plane, find parametric representations of the straight line segments AB and BC with their respective senses from A to B and from B to C.

2. Find parametric representations for the straight line segments AB and BC illustrated in Fig. 14.7, with the senses shown by the arrows.

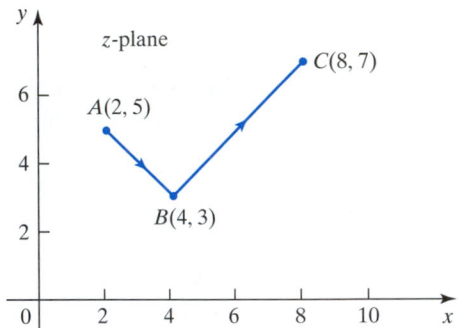

FIGURE 14.7 The straight line segments AB and BC.

3. Find parametric representations for the straight line segments AB and BC illustrated in Fig. 14.8, with the senses shown by the arrows.

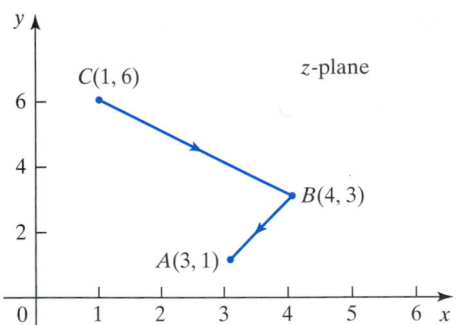

FIGURE 14.8 The line segments AB and BC.

4. Find parametric representations for the straight line segment AB, the circular arc BC, and the straight line segment CD illustrated in Fig. 14.9, with the senses shown by the arrows.

5. Integrate $f(z) = z$ in the positive sense around the square with corners at $(1, 1)$, $(2, 1)$, $(2, 2)$, and $(1, 2)$.

6. Integrate $f(z) = z$ along the consecutive straight line paths from A to B and from B to C, where A, B, and C are the respective points $(1, 1)$, $(3, 2)$, and $(5, 4)$.

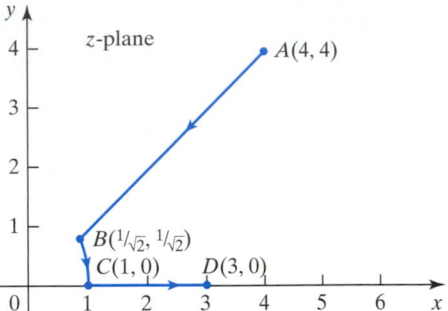

FIGURE 14.9 The straight line segments AB and CD, and the circular arc BC.

7. Integrate $f(z) = z^2 + i$ along the straight line path from point $(1, 1)$ to $(1, 4)$.

8. Integrate $f(z) = iz^2 + 1$ along the straight line path from point $(3, 1)$ to $(6, 1)$.

9. Integrate $f(z) = 2z^2 - 3i$ along the straight line path from point $(1, 1)$ to $(4, 1)$.

10. Integrate $f(z) = z^2 + z$ along the straight line path from point $(2, 3)$ to $(5, 6)$.

11. Represent $\sinh z$ in terms of its real and imaginary parts and integrate it along the straight line path from point $(3, \pi)$ to $(6, \pi)$.

12. Represent $\cosh z$ in terms of its real and imaginary parts and integrate it along the straight line path from point $(1, 2)$ to $(1, 4)$.

13. Represent $\sin z$ in terms of its real and imaginary parts and integrate it along the straight line path from point $(2, \pi)$ to $(3, \pi)$.

14. Represent $\cos z$ in terms of its real and imaginary parts and integrate it along the straight line path from point $(1, 4\pi)$ to $(1, 6\pi)$.

15. Represent $\cosh 2z$ in terms of its real and imaginary parts and integrate it along the straight line path from point $(0, 0)$ to $(4, 2)$.

16. Represent $\sin z$ in terms of its real and imaginary parts and integrate it along the straight line path from point $(0, 0)$ to $(2, 4)$.

17. Integrate e^z along the straight line path from the point $(0, 0)$ to $(4, \pi/4)$.

18. Set $f(z) = \bar{z}$, and let the corners A, B, C, and D of a square be located at the respective points $(-1, -1)$, $(1, -1)$, $(1, 1)$, and $(-1, 1)$. Integrate $f(z)$ first along the consecutive paths from A to B and from B to C, and then along the consecutive paths from A to D and from D to C, and hence show that the value of the

integral of the nonanalytic function $\bar{z}$ from A to C depends on the choice of path joining A to C.

19. Integrate $f(z) = 1/(z-1)$ in the negative sense around the semicircle with the equation $|z-1| = 1$.

20. Integrate the function $f(z) = z\bar{z}$ around the circular arc $|z-2| = 3$ in the positive sense between the points $(2,3)$ and $(5,0)$.

21. Show that $\int_{\Gamma} \frac{1}{z+i} dz = 0$, when integration is performed in the either the positive or the negative sense around the circle Γ given by $|z-2| = 2$.

22. Let A, B, and C be the respective points $(0,0), (1,0)$, and $(1,1)$, and let $f(z) = z\bar{z}$. Integrate $f(z)$ along the consecutive straight line segments AB and BC, and then along the straight line segment AC, and hence show that the value of the integral of this nonanalytic function from A to C depends on the path joining the two points.

14.2   Contours, the Cauchy–Goursat Theorem, and Contour Integrals

The definition of a complex integral of a single-valued analytic function $f(z)$ along a path introduced in Section 14.1 was for paths that were finite in length, did not intersect themselves, and had end points that were distinct. To make further progress with complex integrals it is necessary to consider integrating along general paths in the form of closed loops that are continuous, piecewise smooth, and do not

contours and simple closed curves

intersect themselves. In Section 14.1, closed paths of this type were called **contours**, though they are also often called **simple closed curves** or **Jordan curves**. A typical example of a simple closed curve is shown in Fig. 14.10a, and the self-intersecting figure-eight-shaped curve in Fig. 14.10b is a nonsimple closed curve.

Before examining contour integrals in more detail, it is necessary to introduce the notion of a *simply connected* domain in which all contour integrals are to be

simply and multiply connected domains

evaluated. A domain D is called **simply connected** if the interior points of all possible simple closed curves in D belong to D. This means that a simply connected domain is one from which no points, curves, or areas are missing. A domain D that does not satisfy this condition is said to be **multiply connected**.

An example of a simply connected domain is shown in Fig. 14.11a, and typical multiply connected domains are shown in Figs. 14.11b and c. The annular domain in Fig. 14.11b is a simple example of a multiply connected domain, and it is made multiply connected by the removal from D of the points in the disc in the center that leaves a "hole" in D. Domains containing only one "hole" are said to be **doubly**

(a)   (b)

FIGURE 14.10   (a) A simple closed curve. (b) A nonsimple closed curve.

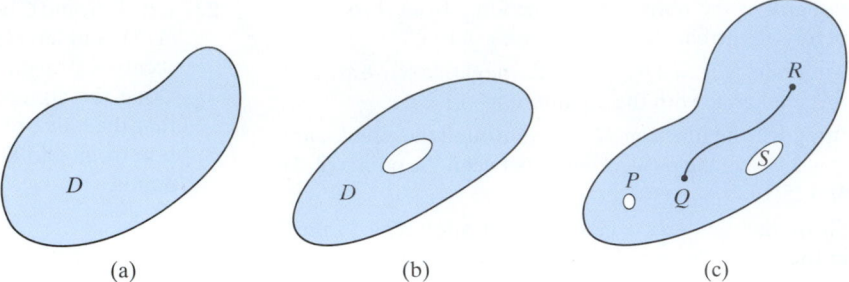

FIGURE 14.11   (a) Simply, (b) doubly, and (c) multiply connected domains.

connected. The domain in Fig. 14.11c is *multiply connected* because the point at P is missing, as are the points along the cut QR and the points in the area (hole) S.

Another way of defining a simply connected domain D is by saying it is one with the property that *every* simple closed curve connecting any two points of D can *always* be collapsed onto an arc in D that joins the two points. This definition is illustrated in Fig. 14.12a, from which it can be seen that for any two points A and B in D, all simple closed curves Γ connecting A and B can always be collapsed onto a dashed arc like the one shown joining the two points. Domain D in Fig. 14.12b is multiply connected. The reason for this can be seen by examining the curves Γ_1 and Γ_2. The simple closed curve Γ_1 joining two points A and B in D lies entirely to the side of all holes in D, and so can be collapsed onto an arc in D joining the points A and B, but this is not possible for a simple closed curve such as Γ_2 that encloses one or more of the holes in D, because the boundaries of the holes act as barriers that stop its collapse onto an arc.

In future the notation $\int_\Gamma f(z)dz$, already used to denote the line integral of a single-valued analytic function $f(z)$ along a path Γ, will be taken to include **contour integrals** around a simple closed curve Γ.

The fundamental theorem governing contour integrals is the **Cauchy–Goursat** theorem, which can be stated as follows.

THEOREM 14.3

a fundamental theorem

Cauchy–Goursat Theorem   Let f be a single-valued analytic function in a simply connected domain D. Then if Γ is any simple closed curve of finite length lying

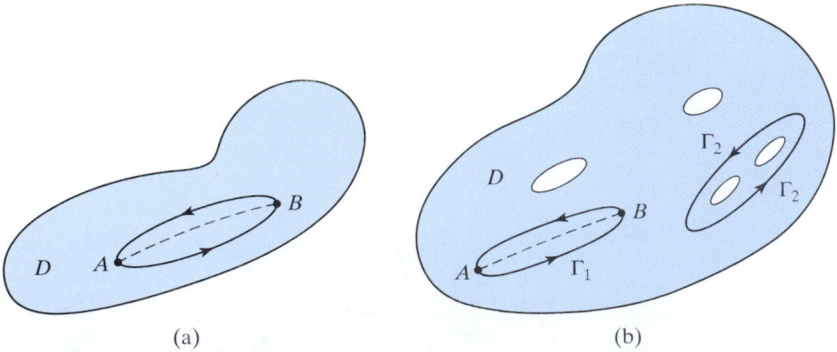

FIGURE 14.12   Illustration of the alternative definition of simply and multiply connected domains.

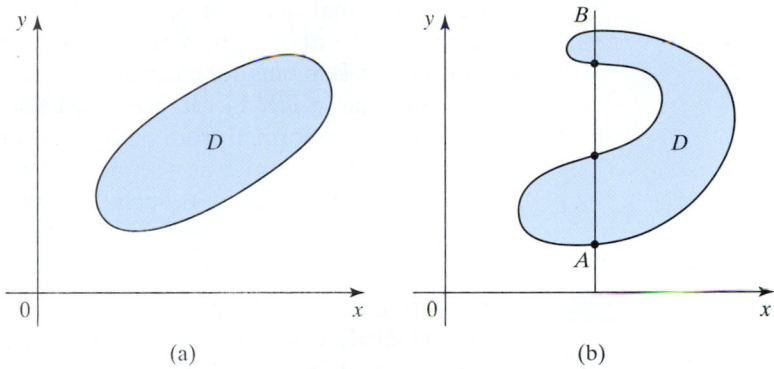

(a)   (b)

FIGURE 14.13   Standard and nonstandard domain.

entirely within D,

$$\int_\Gamma f(z)dz = 0.$$

Proof   This is the most general statement of the Cauchy–Goursat theorem that is necessary for practical purposes. We now prove it in a weaker form by requiring that in addition to f being single-valued and analytic, its derivative $f'(z)$ must be continuous in D and the contour Γ must be one for which lines passing through the interior of Γ drawn parallel to the real and imaginary axes only intersect Γ twice. Areas bounded by such closed curves Γ are called **standard domains**. A typical standard domain is shown in Fig. 14.13a, and a nonstandard one is shown in Fig. 14.13b, where lines such as AB are seen to intersect D four times.

standard and nonstandard domains

Under the stated conditions, the proof can be based on Green's theorem in the plane, which takes the form

$$\int_\Gamma Pdx + Qdy = \int_D \left(\frac{\partial Q}{\partial x} - \frac{\partial P}{\partial y} \right) dxdy,$$

where the domain D inside Γ is a simple domain and $P, Q, \partial Q/\partial x$, and $\partial P/\partial y$ are continuous in D and on Γ. If $f(z) = u + iv$, then $f'(z) = \partial u/\partial x + i\partial v/\partial x = \partial v/\partial y - i\partial u/\partial y$, so the assumption that $f'(z)$ is continuous implies the continuity of $\partial u/\partial x, \partial u/\partial y, \partial v/\partial x$, and $\partial v/\partial y$, and through them the continuity of u and v.

Applying Green's theorem to $\int_\Gamma f(z)dz$, we have

$$\int_\Gamma f(z)dz = \int_\Gamma (udx - vdy) + i \int_\Gamma (vdx + udy)$$

$$= \int_D \left(-\frac{\partial v}{\partial x} - \frac{\partial u}{\partial y} \right) dxdy + i \int_D \left(\frac{\partial u}{\partial x} - \frac{\partial v}{\partial y} \right) dxdy.$$

However, from the Cauchy–Riemann equations $\partial u/\partial x = \partial v/\partial y$ and $\partial u/\partial y = -\partial v/\partial x$, so each integrand vanishes and we obtain the statement of the theorem

$$\int_\Gamma f(z)dz = 0.$$

The form of proof given here is the one due to Cauchy. The removal of the requirements that $f'(z)$ be continuous and D be a standard domain that were necessary in the above proof allows the theorem to be used under very general circumstances. It means, for example, that the theorem remains true when domains such as the one in Fig. 14.13b arise, and also that instead of the contour Γ being smooth, it can be formed from piecewise smooth arcs joined end to end to make a simple closed curve such as a semicircle or a rectangle. The generalization of the theorem is due to Goursat, though the details of its proof will not be given here.

∎

EXAMPLE 14.6

The functions z^n with n a positive integer, $\sin z$, $\cos z$, e^z, $\sinh z$, and $\cosh z$ are analytic and single valued throughout the complex plane (they are *entire* functions), so for any simple contour Γ,

$$\int_\Gamma z^n dz = 0, \quad \int_\Gamma \sin z\, dz = 0, \quad \int_\Gamma \cos z\, dz = 0, \quad \int_\Gamma e^z dz = 0,$$

$$\int_\Gamma \sinh z\, dz = 0, \quad \int_\Gamma \cosh z\, dz = 0.$$

∎

EXAMPLE 14.7

The function $\sec z = 1/\cos z$ is analytic and single valued throughout the z-plane except at the zeros of $\cos z$ that are located at $z = (2n+1)\pi/2$, for $n = 0, \pm1, \pm2, \ldots$. Thus, $\int_\Gamma \sec z\, dz = 0$ for every contour Γ that neither contains zeros of $\cos z$ nor passes through any of its zeros.

∎

An immediate consequence of the Cauchy–Goursat theorem is that if the contour Γ in D is **deformed** into some other contour Γ_1 that is also in D, the statement in the theorem remains unchanged. When this happens the contours Γ and Γ_1 are said to be **equivalent contours**.

Examples of two equivalent contours are shown in Fig. 14.14a, and the usefulness of this result is such that we record it in the form of a theorem.

THEOREM 14.4

a suitable deformation of a contour does not change the value of a contour integral

Deformation of contours   Let f be a single-valued analytic function in a simply connected domain D, and let Γ_1 and Γ_2 be any two simple closed contours in D.

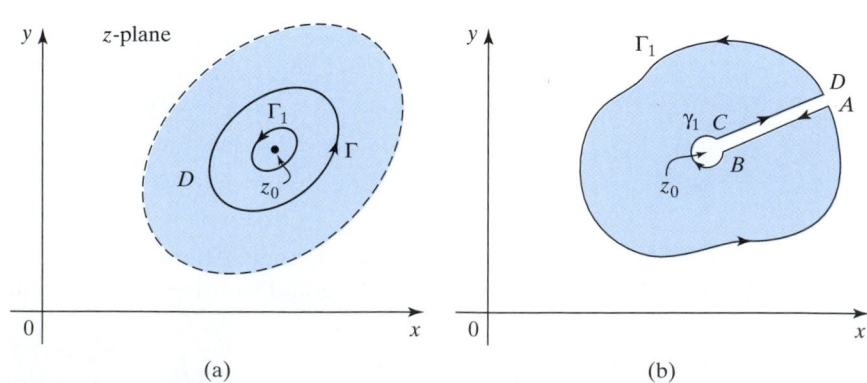

(a)   (b)

FIGURE 14.14   (a) Equivalent contours. (b) A contour that excludes a simple pole at z_0.

Then Γ_1 and Γ_2 are equivalent in the sense that

$$\int_{\Gamma_1} f(z)dz = \int_{\Gamma_2} f(z)dz.$$

If, however f has a simple pole at a point $z = z_0$ inside both Γ_1 and Γ_2, then

$$\int_{\Gamma_1} f(z)dz = \int_{\Gamma_2} f(z)dz = 2\pi i \lim_{z \to z_0}[(z - z_0)f(z)].$$

Proof   The first result has already been established, so it only remains to prove the second one. Consider Fig. 14.14b, and let there be a simple pole at $z = z_0$. Enclose the pole in a small circle γ_1 of radius r, and join the circle to the contour Γ_1 by two parallel straight lines AB and CD that are arbitrarily close together. Then, in the domain bounded by Γ_1, AB, γ_1, and CD as indicated by the arrows in Fig. 14.14b, the function f is analytic because the pole has been excluded.

Applying the Cauchy–Goursat theorem and integrating around this contour gives

$$\int_{\Gamma_{DA}} f(z)dz + \int_{AB} f(z)dz + \int_{-\gamma_1} f(z)dz + \int_{CD} f(z)dz = 0,$$

where $-\gamma_1$ indicates that integration around the circle γ_1 is in the clockwise sense.

If the radius r of circle γ_1 is now allowed to tend to zero, the second and fourth integrals vanish, because f is continuous across the lines AB and CD and f is integrated in opposite directions along each of these lines. Reversing the sense of integration around γ_1 and compensating by changing the sign of the integral, we arrive for $r \to 0$ at the result

$$\int_{\Gamma_1} f(z)dz = \lim_{r \to 0} \int_{\gamma_1} f(z)dz.$$

By definition, if f has a simple pole at $z = z_0$, then $f(z) = g(z)/(z - z_0)$ with $g(z_0) \neq 0$. So, integrating around γ_1 on which $z = z_0 + re^{i\theta}$ with $0 \leq \theta \leq 2\pi$, and using the fact that $dz = ire^{i\theta}d\theta$, gives

$$\int_{\Gamma_1} f(z)dz = \lim_{r \to 0} \int_0^{2\pi} \frac{g(z_0 + re^{i\theta})}{re^{i\theta}} ire^{i\theta}d\theta = 2\pi i g(z_0).$$

The same result would be obtained using any other contour in D that contains z_0, so the second result is proved.   ∎

EXAMPLE 14.8   Find $\int_\Gamma \frac{3}{z+i}dz$, with Γ any square of side 4 with its center at the origin.

Solution   The square Γ contains $z = -i$, which is a simple pole of the integrand, so deforming Γ into any circle centered on $z = -i$ and integrating around Γ in the positive sense using the result of Example 14.5 gives

$$\int_\Gamma \frac{3}{z+i}dz = 6\pi i. \quad ■$$

EXAMPLE 14.9

Find $\int_\Gamma(\frac{4}{z-1} - \frac{5}{z+4})dz$, where Γ is the circle $|z| = 2$.

Solution   The point $z = -4$ lies outside $|z| = 2$, so the Cauchy–Goursat theorem shows that the second term in the integrand contributes nothing to the integral. Deforming Γ into any circle centered on $z = 1$ that does not contain the point $z = -4$, and integrating around it in the positive sense using the result of Example 14.5, gives

$$\int_\Gamma \left(\frac{4}{z-1} - \frac{5}{z+4}\right) dz = 8\pi i - 0 = 8\pi i.$$ ∎

EXAMPLE 14.10

Find

$$\int_\Gamma \frac{2z-3}{z^3 - 3z^2 + 4}dz$$

simplifying integration by using partial fractions

by integrating in the positive sense around Γ when (a) Γ is the circle $|z| = 3/2$ and (b) Γ is the circle $|z - 3| = 2$.

Solution   A partial fraction decomposition of the integrand gives

$$\frac{2z-3}{z^3 - 3z^2 + 4} = \frac{5}{9}\frac{1}{z-2} - \frac{5}{9}\frac{1}{z+1} + \frac{1}{3}\frac{1}{(z-2)^2},$$

so

$$\int_\Gamma \frac{2z-3}{z^3 - 3z^2 + 4}dz = \frac{5}{9}\int_\Gamma \frac{dz}{z-2} - \frac{5}{9}\int_\Gamma \frac{dz}{z+1} + \frac{1}{3}\int_\Gamma \frac{dz}{(z-2)^2}.$$

(a) The functions $1/(z-2)$ and $1/(z-2)^2$ are analytic in and on the circle $|z| = 3/2$, so by the Cauchy–Goursat theorem the first and last integrals on the right vanish. The contour Γ is not convenient for the evaluation of the second integral on the right, so we deform the circle $|z| = 3/2$ into the circle $|z + 1| = 1$ centered on $z = -1$ and use the result of Example 14.5 to obtain

$$\int_\Gamma \frac{dz}{z+1} = 2\pi i.$$

Combining these results gives

$$\int_\Gamma \frac{2z-3}{z^3 - 3z^2 + 4}dz = -\frac{5}{9}\int_\Gamma \frac{dz}{z+1} = -\frac{10\pi i}{9}.$$

(b) The function $1/(z+1)$ is analytic in and on the circle $|z - 3| = 2$, so by the Cauchy–Goursat theorem the second integral on the right vanishes. Again the contour Γ is not convenient when determining the other two contour integrals, so deforming the circle $|z - 3| = 2$ into the circle $|z - 2| = 1$ and using the results of Examples 14.4 and 14.5 gives

$$\int_\Gamma \frac{dz}{z-2} = 2\pi i, \quad \text{and} \quad \int_\Gamma \frac{dz}{(z-2)^2} = 0.$$

Combining these results we find that

$$\int_\Gamma \frac{2z-3}{z^3 - 3z^2 + 4}dz = \frac{5}{9}\int_\Gamma \frac{dz}{z-2} = \frac{10\pi i}{9}.$$ ∎

Let f be a single-valued analytic function in some domain D in which two distinct points z_1 and z_2 are connected by two paths in D that form the simple contour Γ shown as $APBQA$ in Fig. 14.15a.

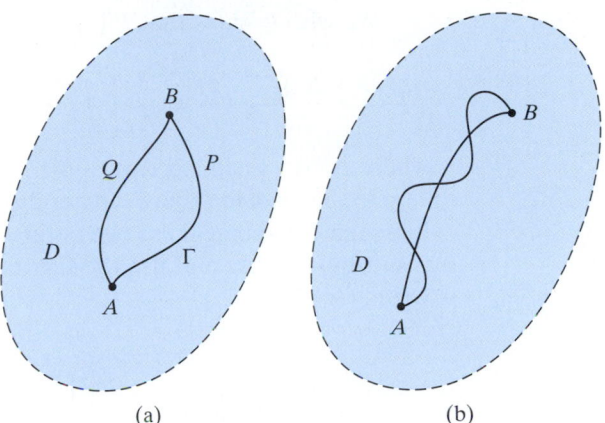

FIGURE 14.15   (a) Two paths forming a simple contour Γ.
(b) Two paths forming loops.

Using the Cauchy–Goursat theorem and dividing Γ into the two parts APB and BQA allow us to write

$$\int_{\Gamma} f(z)dz = \int_{APB} f(z)dz + \int_{BQA} f(z)dz = 0.$$

Reversing the direction of integration along BQA, and compensating by changing the sign of the integral, shows the preceding result to be equivalent to

$$\int_{APB} f(z)dz = \int_{AQB} f(z)dz. \tag{10}$$

By Theorem 14.4 the contour Γ in D through z_1 and z_2 can be deformed into any other equivalent contour in D through the two points, showing that the integral of $f(z)$ from z_1 to z_2 is independent of the path joining z_1 to z_2. The result remains true if the paths intersect finitely many times forming n loops, as shown in Fig. 14.15b. In this case the result is established by applying the preceding result to each loop in succession.

antiderivative, or indefinite integral

As in the real variable calculus, a differentiable function $F(z)$ such that $F'(z) = f(z)$ is called an **antiderivative** of $f(z)$, or an **indefinite integral**, and written

$$\int f(z)dz. \tag{11}$$

To simplify the calculation of line integrals of analytic functions, we now consider the integral of a single-valued analytic (and so *continuous*) function $f(z)$ from a fixed point z_0 in D to some other point z in D along any path in D. The result can be written

$$F(z) = \int_{z_0}^{z} f(\zeta)d\zeta, \tag{12}$$

where $F(z)$ is a function of the upper limit of integration z, and no path need be specified because the integral is independent of the path joining z_0 to z_1 in D.

We wish to show that $F'(z) = f(z)$, so let us consider the difference quotient

$$\frac{F(z + \Delta z) - F(z)}{\Delta z} = \frac{1}{\Delta z}\left[\int_{z_0}^{z+\Delta z} f(\zeta)d\zeta - \int_{z_0}^{z} f(\zeta)d\zeta \right] = \frac{1}{\Delta z}\int_{z}^{z+\Delta z} f(\zeta)d\zeta,$$

where Δz is a small increment in z.

As any path in D between z and $z + \Delta z$ can be used, we take it to be the straight line segment joining these two points. Then, as $\int_{z}^{z+\Delta z} d\zeta = \Delta z$, we can multiply this result by $f(z)/\Delta z$ and use the fact that $f(z)$ is not involved in the integration to write $f(z)$ as

$$f(z) = \frac{1}{\Delta z}\int_{z}^{z+\Delta z} f(z)d\zeta.$$

This result allows the difference quotient to be written

$$\frac{F(z + \Delta z) - F(z)}{\Delta z} - f(z) = \frac{1}{\Delta z}\int_{z}^{z+\Delta z} [f(\zeta) - f(z)]d\zeta.$$

Taking the modulus of this expression and using the fundamental integral inequality in Theorem 14.1, we obtain

$$\left| \frac{F(z + \Delta z) - F(z)}{\Delta z} - f(z) \right| \le \frac{1}{|\Delta z|}\int_{z}^{z+\Delta z} |f(\zeta) - f(z)||d\zeta|,$$

but $f(z)$ is a continuous function of z, so for any arbitrary small number $\varepsilon > 0$ we can always find a number $\delta > 0$ such that

$$|f(\zeta) - f(z)| < \varepsilon, \quad \text{when } |z - \zeta| < \delta.$$

Then, as ζ lies on the straight line segment joining z and $z + \Delta z$, we have $|z - \zeta| \le |\Delta z|$, showing that the preceding result is true if $\delta < |\Delta z|$.

It now follows that

$$\left| \frac{F(z + \Delta z) - F(z)}{\Delta z} - f(z) \right| \le \frac{1}{|\Delta z|}\varepsilon|\Delta z| = \varepsilon,$$

so in the limit as $\Delta z \to 0$ this shows that

$$\lim_{\Delta z \to 0}\left(\frac{F(z + \Delta z) - F(z)}{\Delta z} \right) = F'(z) = f(z). \tag{13}$$

As $F(z)$ has been shown to be differentiable, we have also proved the very important result that the derivative of an analytic function is itself an analytic function.

We now show how definite integrals can be evaluated. Let $F(z)$ and $G(z)$ be any two different antiderivatives of $f(z)$. Then setting $\Phi(z) = F(z) - G(z) = u + iv$, we have

$$\Phi'(z) = F'(z) - G'(z) = 0, \quad \text{for all } z \text{ in } D.$$

When this result is used with the Cauchy–Riemann equations, it shows that $\Phi(z) =$ constant, so all antiderivatives of $f(z)$ can only differ one from the other by a complex constant C, allowing us to write

$$F(z) = G(z) + C.$$

If z and z^* are any two points in D where f is defined, the antiderivative $G(z)$ of $f(z)$ can be written

$$G(z) = \int_{z^*}^{z} f(\zeta)d\zeta, \tag{14}$$

so the most general antiderivative of $f(z)$ becomes

$$F(z) = \int_{z^*}^{z} f(\zeta)d\zeta + C. \tag{15}$$

The **definite integral** $\int_{z_0}^{z_1} f(\zeta)d\zeta$ can be written

$$\int_{z^*}^{z_1} f(\zeta)d\zeta = \int_{z^*}^{z_1} f(\zeta)d\zeta - \int_{z^*}^{z_0} f(\zeta)d\zeta,$$

and after elimination of the arbitrary constant C we find that

$$\int_{z_0}^{z_1} f(\zeta)d\zeta = F(z_1) - F(z_0). \tag{16}$$

In complex analysis, this last result is the analogue of the fundamental theorem of integral calculus for real functions. We have proved the following important and useful theorem.

THEOREM 14.5

Independence of path—definite integrals Let $f(z)$ be a single-valued analytic function in some domain D to which belong the two distinct points z_1 and z_2. Then if $F(z) = \int f(z)dz$ is an antiderivative of f, the line integral of f along any path in D joining z_1 to z_2 is independent of the path, and

$$\int_{z_1}^{z_2} f(z)dz = F(z_2) - F(z_1). \quad\blacksquare$$

EXAMPLE 14.11

Find the integral of z^2 from $z_1 = 1 + i$ to $z_2 = 3 + 4i$.

Solution The function $f(z) = z^2$ is single valued and analytic in the finite z-plane, and an antiderivative of $f(z)$ is $z^3/3$, so Theorem 14.5 can be applied and gives

$$\int_{1+i}^{3+4i} z^2 dz = \left(\frac{z^3}{3}\right)_{1+i}^{3+4i} = \frac{1}{3}[(3+4i)^3 - (1+i)^3] = -\frac{115}{3} + 14i. \quad\blacksquare$$

Consider a function $f(z)$ that is analytic and single valued inside the multiply connected domain D with outer boundary Γ shown in Fig. 14.16a. The domain D can be made simply connected by inserting the n cuts $C_1, C_2, \ldots, C_n$ shown in Fig. 14.16b, and taking as the new boundary the one formed by Γ, the internal boundaries $\Gamma_1, \Gamma_2, \ldots, \Gamma_n$, and the cuts $C_1, C_2, \ldots, C_n$. In this way, as the contour is traversed in the positive sense indicated by the arrows in Fig. 14.16b, the modified domain always lies to the left and is simply connected.

The next theorem makes use of cuts to extend the Cauchy–Goursat theorem for analytic functions to multiply connected domains.

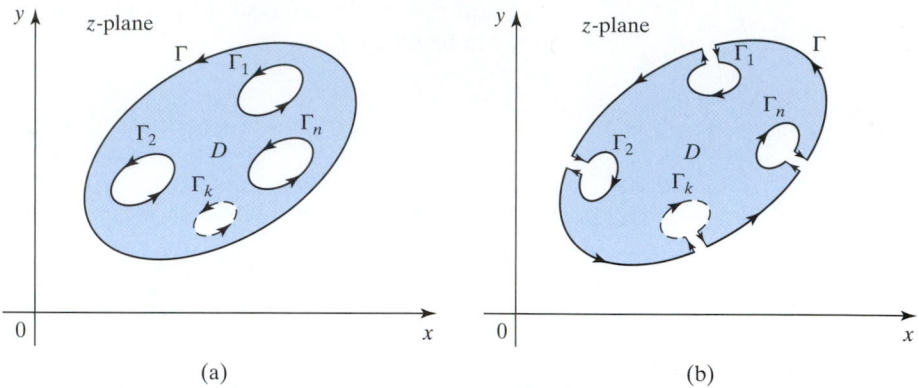

FIGURE 14.16 Cuts used to make a multiply connected domain simply connected.

THEOREM 14.6

integration in multiply connected domains

Extended Cauchy–Goursat theorem Let $f(z)$ be a single-valued analytic function in a possibly multiply connected domain D bounded externally by a simple contour Γ, and internally by the simple contours $\Gamma_1, \Gamma_2, \ldots, \Gamma_n$, as shown in Fig. 14.17, and let each of the $n + 1$ contours be traversed in the positive sense. Then

$$\int_{\Gamma} f(z)dz = \int_{\Gamma_1} f(z)dz + \int_{\Gamma_2} f(z)dz + \cdots + \int_{\Gamma_n} f(z)dz.$$

Proof Make the cuts indicated in Fig. 14.18, and integrate around the resulting composite contour using the Cauchy–Goursat theorem to obtain

$$\int_{c_1^+} f(z)dz + \int_{\Gamma_1} f(z)dz + \int_{c_1^-} f(z)dz + \int_{\Gamma(P_1^- P_2^+)} f(z)dz + \int_{c_2^+} f(z)dz$$

$$+ \int_{\Gamma_2} f(z)dz + \int_{c_2^-} f(z)dz + \cdots + \int_{c_n^+} f(z)dz + \int_{\Gamma_n} f(z)dz + \int_{c_n^-} f(z)dz$$

$$+ \int_{\Gamma(P_n^- P_1^+)} f(z)dz = 0.$$

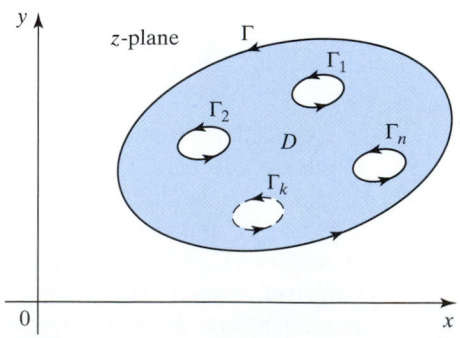

FIGURE 14.17 The multiply connected domain D.

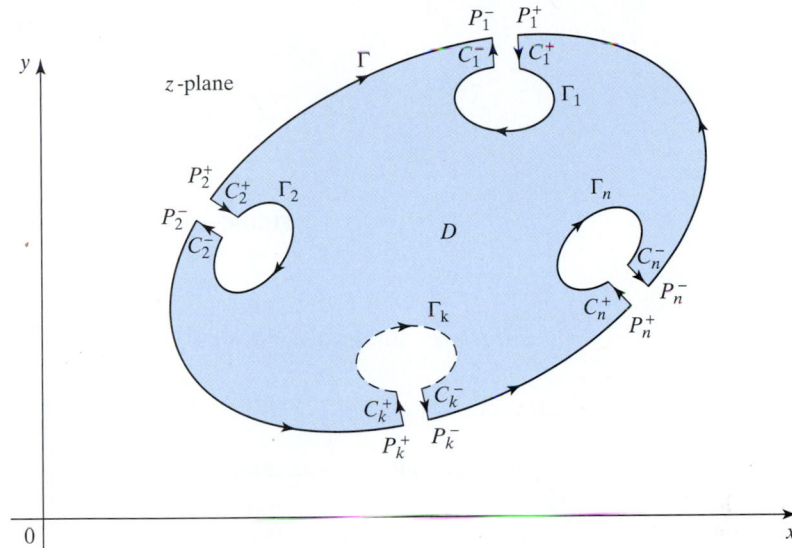

FIGURE 14.18   Composite contour for integration.

As f is analytic in D, and C_i^+ and C_i^- are opposite sides of the cut C_i, the function f is continuous across the cut. The paths C_i^+ and C_i^- are traversed in opposite directions, so the integrals along opposite sides of the cut cancel, leading to the result

$$\int_{C_i^+} f(z)dz + \int_{C_i^-} f(z)dz = 0, \quad \text{for } i = 1, 2, \ldots, n.$$

Adding the integrals around the successive segments of Γ, using the fact that $f(z)$ is continuous on Γ, cancelling the integrals along opposite sides of each cut, and denoting integration around Γ_i in the clockwise (negative) sense by Γ_{i-} reduces the preceding result to

$$\int_{\Gamma} f(z)dz + \int_{\Gamma_{1-}} f(z)dz + \int_{\Gamma_{2-}} f(z)dz + \cdots + \int_{\Gamma_{n-}} f(z)dz = 0.$$

The direction of integration around the internal contours $\Gamma_1, \Gamma_2, \ldots, \Gamma_n$ is negative (clockwise), so reversing their directions to give them a positive orientation, introducing corresponding changes of sign in the integrals, and rearranging terms, we arrive at the result

$$\int_{\Gamma} f(z)dz = \int_{\Gamma_1} f(z)dz + \int_{\Gamma_2} f(z)dz + \cdots + \int_{\Gamma_n} f(z)dz,$$

and the theorem is proved.   ■

EXAMPLE 14.12   Find the integral of $f(z) = (4z^2 + 11z - 3)/(z^3 + 2z^2 - z - 2)$ around the contour Γ shown in Fig. 14.19 with the direction of integration around the connected contours A, B, and C shown by the arrows.

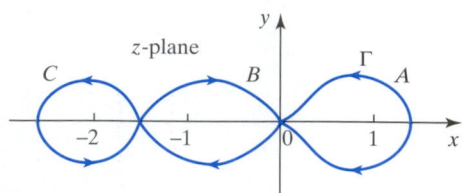

FIGURE 14.19   Connected contours A, B, and C forming Γ.

Solution   Integrating around Γ we have

$$\int_\Gamma f(z)dz = \int_A f(z)dz + \int_B f(z)dz + \int_C f(z)dz,$$

and a partial fraction expansion of $f(z)$ gives the representation

$$f(z) = \frac{2}{z-1} + \frac{5}{z+1} - \frac{3}{z+2}.$$

Inside and on contour A the functions $1/(z+1)$ and $1/(z+2)$ are analytic; inside and on contour B the functions $1/(z-1)$ and $1/(z+2)$ are analytic; and inside and on contour C the functions $1/(z-1)$ and $1/(z+1)$ are analytic. In addition, we must take account of the fact that integration around A is in the *positive* sense, integration around B is in the *negative* sense, and integration around C is in the *positive* sense. Deforming contours A, B, and C into the respective circles $|z-1| = 1$, $|z+1| = 1/2$, and $|z+2| = 1/2$ and using the Cauchy–Goursat theorem with the result of Example 14.5, we find that integration around contour A in the *positive* sense gives

$$\int_A f(z)dz = 2\int_{|z-1|=1} \frac{1}{z-1}dz = 2 \cdot 2\pi i = 4\pi i,$$

integration around contour B in the *negative* sense gives

$$\int_B f(z)dz = -5\int_{|z+1|=1/2} \frac{1}{z+1}dz = -5 \cdot 2\pi i = -10\pi i,$$

and integration around contour C in the *positive* sense gives

$$\int_C f(z)dz = -3\int_{|z+2|=1/2} \frac{1}{z+2}dz = -3 \cdot 2\pi i = -6\pi i.$$

Adding these results to find the integral around Γ we obtain

$$\int_\Gamma f(z)dz = 4\pi i - 10\pi i - 6\pi i = -12\pi i. \qquad \blacksquare$$

integrands involving quotients of trigonometric functions

By setting $z = e^{i\theta}$, expressing $\sin\theta$ and $\cos\theta$ in terms of z, and integrating around the unit circle Γ given by $|z| = 1$, the Cauchy–Goursat theorem can be used to evaluate trigonometric integrals of the form

$$\int_0^{2\pi} \frac{a\cos\theta + b\sin\theta}{c + d\cos\theta + e\sin\theta}d\theta, \tag{17}$$

where $a, b, c, d,$ and e are real numbers.

The expressions for $\sin\theta$ and $\cos\theta$ in terms of z follow by adding and subtracting

$$z = \cos\theta + i\sin\theta \quad \text{and} \quad 1/z = \cos\theta - i\sin\theta$$

to obtain

$$\sin\theta = \frac{1}{2i}\left(\frac{z^2-1}{z}\right), \quad \cos\theta = \frac{1}{2}\left(\frac{z^2+1}{z}\right), \tag{18}$$

and differentiating the result $z = e^{i\theta}$ to obtain $dz = ie^{i\theta}\,d\theta$, from which it follows that

$$d\theta = \frac{1}{iz}dz. \tag{19}$$

EXAMPLE 14.13   Find

$$\int_0^{2\pi} \frac{d\theta}{a + b\sin\theta}, \quad \text{where } a \text{ and } b \text{ are real numbers such that } |a/b| > 1.$$

Solution   The condition $|a/b| > 1$ is necessary to prevent the integrand becoming unbounded in the interval of integration.

Substituting for $d\theta$ and $\sin\theta$ in the integral, we find that

$$\int_0^{2\pi} \frac{d\theta}{a + b\sin\theta} = \frac{2}{b}\int_\Gamma \frac{dz}{z^2 + 2i(a/b)z - 1},$$

where Γ is the unit circle, and integration around Γ is in the positive sense.

As $|a/b| > 1$, the roots of the denominator $z^2 + 2i(a/b)z - 1 = 0$ can be written

$$\alpha = \frac{i}{b}(-a + \sqrt{a^2-b^2}) \quad \text{and} \quad \beta = \frac{i}{b}(-a - \sqrt{a^2-b^2}),$$

where the positive square root is taken. Then, as $|\alpha| < 1$, the point $z = \alpha$ lies inside Γ, and as $|\beta| > 1$, the point $z = \beta$ lies outside Γ. In terms of α and β the denominator can be written

$$z^2 + 2i(a/b)z - 1 = (z - \alpha)(z - \beta),$$

so when expressed in terms of z and the contour Γ, the integral becomes

$$\int_0^{2\pi} \frac{d\theta}{a + b\sin\theta} = \frac{2}{b}\int_\Gamma \frac{dz}{(z-\alpha)(z-\beta)}.$$

A partial fraction expansion of the integrand on the right gives

$$\frac{1}{(z-\alpha)(z-\beta)} = \frac{1}{(\alpha-\beta)}\left[\frac{1}{z-\alpha} - \frac{1}{z-\beta}\right],$$

showing that

$$\int_0^{2\pi} \frac{d\theta}{a + b\sin\theta} = \frac{2}{b(\alpha-\beta)}\int_\Gamma \frac{dz}{z-\alpha} - \frac{2}{b(\alpha-\beta)}\int_\Gamma \frac{dz}{z-\beta}.$$

As only $z = \alpha$ lies inside Γ, it follows from the Cauchy–Goursat theorem and Example 14.5 that

$$\int_0^{2\pi} \frac{d\theta}{a + b\sin\theta} = \frac{2}{b(\alpha-\beta)}\cdot 2\pi i - 0 = \frac{4\pi i}{b(\alpha-\beta)},$$

so as $b(\alpha - \beta) = 2i\sqrt{a^2 - b^2}$ this simplifies to

$$\int_0^{2\pi} \frac{d\theta}{a + b\sin\theta} = \frac{2\pi}{\sqrt{a^2 - b^2}}, \quad \text{for } |a/b| > 1. \quad \blacksquare$$

Summary

Simply and multiply connected domains were introduced, and the fundamental Cauchy–Goursat theorem of complex analysis for a function in a simply connected domain was proved using Green's theorem. Conditions under which contours can be deformed into more convenient shapes were given, and then used to evaluate some simple contour integrals in terms of two elementary results obtained earlier using circular contours. The Cauchy–Goursat theorem was extended to include multiply connected domains, and some simple definite integrals involving quotients of trigonometric functions were obtained.

EXERCISES 14.2

In Exercises 1 through 4 find $\int_{z_1}^{z_2} f(z)dz$ by parametrizing the given path and using the result to integrate $f(z)$ along Γ from z_1 to z_2. State when Theorem 14.5 can be used to evaluate the integral and, when appropriate, use it to check the result.

1. $f(z) = \sinh z$, and the path Γ is the straight line segment joining the points $z_1 = 1$ and $z_2 = i$.

2. $f(z) = e^{3z}$, and the path Γ is the circular arc $|z - 1| = 1$ joining the points $z_1 = 0$ and $z_2 = 1 + i$.

3. $f(z) = z + \text{Im}\{z\}$, and the path Γ is formed by the straight line segment from $z_1 = 1 + i$ to the point $z^* = 2 + i$ and the straight line segment from the point $z^* = 2 + i$ to $z_2 = 2 + 2i$.

4. $f(z) = 2 + \bar{z}$, and the path Γ is the straight line segment from the point $z_1 = 3i$ to the point $z_2 = 3 + 6i$.

In Exercises 5 through 8 find the integral $\int_\Gamma f(z)dz$, where Γ is the unit circle $|z| = 1$ and integration around Γ is taken in the positive sense, using the Cauchy–Goursat theorem whenever it is appropriate.

5. $f(z) = \tanh z$.

6. $f(z) = (z - 3)^2 + \text{Im}\{z\}$.

7. $f(z) = z + \bar{z}^2$.

8. $f(z) = e^z/(z^2 - 2)$.

9. What conditions must be satisfied by a contour Γ in order that $\int_\Gamma f(z)dz = 0$, given that (a) $f(z) = \sin z/(z^2 + 1)$, (b) $f(z) = \csc z$, (c) $f(z) = \text{sech } z$, and (d) $f(z) = \coth z$?

10. Find

$$\int_\Gamma \frac{z+1}{z^2 - 3z + 2}dz$$

where Γ is the contour $ABCADEA$ shown in Fig. 14.20, with integration in the direction indicated by the arrows.

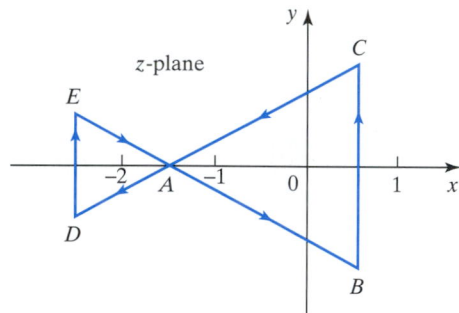

FIGURE 14.20

In Exercises 11 through 17 use analysis to find the integral of $f(z)$ when it is integrated around the given contour Γ in the positive sense. Verify the result by using computer algebra and the substitution $z = z_0 + Re^{i\theta}$, $dz = iRe^{i\theta}d\theta$, with $0 \le \theta \le 2\pi$, when Γ is the circle $|z - z_0| = R$.

11. $f(z) = \dfrac{z+5}{z^2 + 3z - 4}$ with Γ (a) the circle $|z - i| = 2$, and (b) the circle $|z + 3| = 2$.

12. $f(z) = \dfrac{3 - 4z}{z^2 + 5z + 6}$ with Γ (a) the circle $|z| = 5/2$, and (b) the rectangle with its corners at the points $(-7/2, -1), (-5/2, -1), (-5/2, 1)$, and $(-7/2, 1)$.

13. $f(z) = \dfrac{2 - 7z}{z^2 + 3z}$ with Γ (a) the circle $|z + i| = 2$, and (b) the circle $|z - 2| = 4$.

14. $f(z) = \dfrac{3z - 2}{(z + 2)^2}$ with Γ the circle $|z - 3| = 2$.

15. $f(z) = \dfrac{z^2 + 2z}{z^2 - 2z + 1}$ with Γ the circle $|z - 2| = 3$.

16. $f(z) = \dfrac{z + 4}{z^3 + 6z^2 + 9z}$ with Γ (a) the circle $|z + 4| = 2$, and (b) the square with its corners at the points $(-1, -1), (1, -1), (1, 1)$, and $(-1, 1)$.

17. $f(z) = \dfrac{2z-1}{(z+1)^3}$ with Γ the triangle with its vertices at the points $(-2, -1)$, $(0, -1)$ and $(1, 1)$.

Establish the results of Exercises 18 through 20 by using the method of Example 14.13.

18. Show that
$$\int_0^{2\pi} \frac{d\theta}{a + b\cos\theta} = \frac{2\pi}{\sqrt{a^2 - b^2}}$$
for a and b real numbers such that $|a/b| > 1$.

19. Show that
$$\int_0^{2\pi} \frac{\cos\theta}{2 + \cos\theta}\, d\theta = 2\pi - \frac{4\pi}{\sqrt{3}}.$$

20. Show that
$$\int_0^{2\pi} \frac{\sin\theta}{3 + \sin\theta}\, d\theta = 2\pi - \frac{3\pi}{\sqrt{2}}.$$

14.3   The Cauchy Integral Formulas

Two consequences of the Cauchy–Goursat theorem are the **Cauchy integral formula** and the **Cauchy integral formula for derivatives** for a function $f(z)$ that is analytic and single valued in some domain D. These results are of fundamental importance in complex analysis and the first of these formulas can be stated as follows.

THEOREM 14.7   **The Cauchy integral formula**   Let $f(z)$ be a single-valued analytic function in a simply connected domain D containing a contour Γ in the form of a simple closed curve. Then for every point z_0 inside Γ,

expressing $f(z_0)$ as an integral

$$f(z_0) = \frac{1}{2\pi i}\int_\Gamma \frac{f(z)}{z - z_0}\, dz,$$

where integration around Γ is in the positive sense.

Proof   Let z_0 be any point inside the domain D shown in Fig. 14.21, and let the contour Γ containing z_0 lie inside D. Enclose z_0 by an equivalent circular contour C of arbitrarily small radius ρ.

Let us consider the function $\varphi(z)$ defined as

$$\varphi(z) = \frac{f(z) - f(z_0)}{z - z_0} \quad \text{for } z \neq z_0,$$

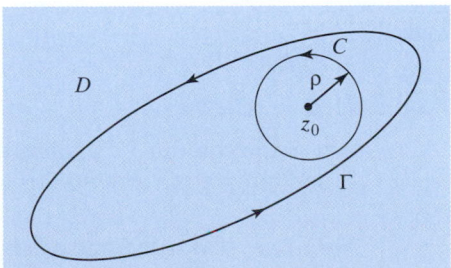

FIGURE 14.21   The equivalent contours Γ and C.

and for later use notice that

$$\lim_{z \to z_0} \varphi(z) = f'(z_0).$$

After deforming the contour Γ into an equivalent circular contour C of radius ρ with its center at z_0, we can write

$$\int_\Gamma \varphi(z)dz = \int_C \varphi(z)dz,$$

where from Example 14.5 it can be seen that the integral around C is independent of the radius ρ. The function $\varphi(z)$ is undefined at $z = z_0$, so if we define it to be $f'(z_0)$ the function $\varphi(z)$ will be continuous throughout D. This result, in turn, implies that the modulus of $\varphi(z)$ must be bounded in D, so we have $|\varphi(z)| \le M$ for some fixed M and all z in D.

It then follows from Theorem 14.2 that as the circumference of C is $2\pi\rho$,

$$\left| \int_C \varphi(z)dz \right| \le M \cdot 2\pi\rho,$$

so taking the limit as $\rho \to 0$ shows that

$$\int_C \varphi(z)dz = 0.$$

Consequently, as

$$\int_\Gamma \varphi(z)dz = \int_C \varphi(z)dz,$$

we have proved that

$$\int_\Gamma \varphi(z)dz = \int_\Gamma \frac{f(z) - f(z_0)}{z - z_0}dz = 0,$$

but this result is equivalent to

$$\int_\Gamma \frac{f(z)}{z - z_0}dz = f(z_0) \int_\Gamma \frac{dz}{z - z_0} = 2\pi i f(z_0),$$

and the theorem is proved.   ∎

Remark   The Cauchy integral formula shows how a function $f(z)$ that is defined and analytic on a contour Γ defines $f(z)$ at *every* point inside Γ.

EXAMPLE 14.14   Find

$$\int_\Gamma \frac{\sinh z}{z^2 + (\pi/2)^2}dz,$$

where the contour Γ contains the point $z = i\pi/2$ but excludes the point $z = -i\pi/2$, and integration around Γ is in the positive sense.

Solution   The integrand can be written

$$\frac{\sinh z}{z^2 + (\pi/2)^2} = \frac{\sinh z}{(z + i\pi/2)} \cdot \frac{1}{(z - i\pi/2)},$$

and because of the exclusion of the point $z = -i\pi/2$ from inside Γ, the function $\sinh z/(z + i\pi/2)$ is analytic inside Γ. Setting $f(z) = \sinh z/(z + i\pi/2)$ in the Cauchy integral formula with $z_0 = i\pi/2$, and integrating around Γ in the positive sense, gives

$$\int_\Gamma \frac{\sinh z}{z^2 + (\pi/2)^2} dz = \int_\Gamma \frac{f(z)}{z - i\pi/2} dz = 2\pi i f(i\pi/2)$$

$$= 2\pi i \cdot \frac{\sinh(i\pi/2)}{i\pi} = 2i \sin(\pi/2) = 2i. \qquad \blacksquare$$

The second Cauchy integral formula determines the derivatives of an analytic function in terms of a contour integral around a domain in which the function is analytic. The theorem can be stated as follows.

THEOREM 14.8

The Cauchy integral formula for derivatives   Let $f(z)$ be a single-valued analytic function in a simply connected domain D containing a contour Γ in the form of a simple closed curve. Then, for any point z_0 inside Γ,

expressing $f^{(n)}(z)$ as an integral

$$f^{(n)}(z) = \frac{n!}{2\pi i} \int_\Gamma \frac{f(\zeta)}{(\zeta - z)^{n+1}} d\zeta, \quad \text{for } n = 1, 2, \ldots .$$

Proof   The result follows by differentiating the statement of Theorem 14.7 with respect to z_0, and this in turn involves justifying differentiating under a contour integral sign. To simplify the proof of the Cauchy integral theorem for derivatives, this operation will be assumed to be justified, and an outline proof of its legitimacy will be postponed until the end of this section.

Let us consider the function $\varphi(\zeta, z) = f(\zeta)/(\zeta - z)$ to be a function of the two complex variables z and z_0. Differentiation of the result of Theorem 14.7 with respect to z_0 gives

$$f'(z) = \frac{1}{2\pi i} \frac{\partial}{\partial z} \int_\Gamma \frac{f(\zeta)}{\zeta - z} d\zeta,$$

so, if we assume differentiation under the integral sign is permissible, this becomes

$$f'(z) = \frac{1}{2\pi i} \int_\Gamma \frac{\partial}{\partial z} \left(\frac{f(\zeta)}{\zeta - z} \right) d\zeta = \frac{1}{2\pi i} \int_\Gamma \frac{f(\zeta)}{(\zeta - z)^2} d\zeta,$$

and the result has been established for $n = 1$. The result for $n > 1$ follows by using mathematical induction, so the theorem is proved. $\qquad \blacksquare$

EXAMPLE 14.15

Find the value of the integral

$$\int_\Gamma \frac{\cos z}{(z - \pi/4)^3} dz,$$

where integration is in the positive sense around the circle Γ given by $|z - \pi/2| = 1$.

Solution   Matching the integrand to the one in Theorem 14.8 shows that $f(z) = \cos z$, $n = 2$, and $z_0 = \pi/4$, so z_0 lies inside Γ. As $f^{(2)}(z) = -\cos z$, substitution into the Cauchy integral formula for derivatives gives

$$\frac{2!}{2\pi i} \int_\Gamma \frac{\cos z}{(z - \pi/4)^3} dz = f^{(2)}(\pi/4) = -\frac{1}{\sqrt{2}},$$

showing that

$$\int_\Gamma \frac{\cos z}{(z - \pi/4)^3} dz = -\frac{i\pi}{\sqrt{2}}.$$ ∎

The next result has far-reaching consequences, because it says that an analytic function can be differentiated arbitrarily many times and the result will still be an analytic function.

THEOREM 14.9

an analytic function can be differentiated arbitrarily many times

An analytic function has derivatives of all orders A function $f(z)$ that is analytic in a simply connected domain D has derivatives of all orders.

Proof The result follows directly from Theorem 14.8. ∎

A useful property of harmonic functions is stated in the next theorem, the proof of which makes use of the Cauchy–Riemann equations.

THEOREM 14.10

derivatives of harmonic functions are harmonic

Harmonic functions have partial derivatives that are harmonic A function $u(x, y)$ that is harmonic throughout a domain D has partial derivatives u_x, u_y, u_{xx}, u_{xy}, and u_{yy} that exist and are themselves harmonic functions.

Proof Around each point $z_0 = x_0 + iy_0$ inside D, construct a disc $|z - z_0| \leq \rho$, all points of which lie in D. The Cauchy–Riemann equations can be used to construct a conjugate harmonic function v in the disc such that $f(z) = u + iv$ is analytic throughout the disc. From the Cauchy–Riemann equations we have $f'(z) = u_x + iv_x = v_y - iu_y$, but Theorem 14.8 asserts that $f'(z)$ is analytic in the disc, so the functions u_x and u_y must themselves be harmonic in the disc.

A repetition of this argument, coupled with the fact that $f''(z)$ is also analytic in the disc, establishes that u_{xx}, u_{xy}, and u_{yy} must be harmonic functions in the disc. By selecting a suitable choice of points z_0, each as the center of a disc with an appropriate radius ρ, it is possible to include all points of D in a set of overlapping discs. The result is true in each disc, so the theorem is proved. ∎

We remark that the method used in Theorem 14.10 to extend the analytic function $f''(z)$ from the interior of disc C to the domain D, throughout which $f(z)$ is analytic, is called **analytic continuation**.

Further Results

The following is an outline proof of the legitimacy of the operation of differentiation under the integral sign with respect to a parameter. The result we obtain, known as **Leibniz' rule for analytic functions**, is a little more general than is necessary for the proof of Theorem 14.8.

THEOREM 14.11

Leibniz' rule—Differentiation under a contour integral Let $z = x + iy$ be a point on a simple closed curve Γ in a domain D, and let $z_0 = x_0 + iy_0$ be a point inside

Γ in which a function $g(z, z_0)$ is analytic with a continuous derivative $\partial g(z, z_0)/\partial z_0$ for all z and z_0. Then the function

$$G(z_0) = \int_\Gamma g(z, z_0)dz$$

is analytic in D and

$$G'(z_0) = \int_\Gamma \frac{\partial g(z, z_0)}{\partial z_0}dz.$$

Proof   Write the functions $g(z, z_0)$ and $G(z_0)$ in the cartesian form

$g(z, z_0) = u(x, y, x_0, y_0) + iv(x, y, x_0, y_0)$ and $G(z_0) = U(x_0, y_0) + iV(x_0, y_0).$

Then, as $G(z_0) = \int_\Gamma g(z, z_0)dz$, substituting for $g(z, z_0)$ in the integral we obtain

$$U(x_0, y_0) = \int_\Gamma udx - vdy \quad \text{and} \quad V(x_0, y_0) = \int_\Gamma vdx + udy.$$

As the partial derivatives of u and v are continuous with respect to all their dependent variables, it follows from real analysis that these last two real integrals can be differentiated under their integral signs with respect to x and y. Consequently,

$$\frac{\partial U}{\partial x_0} = \int_\Gamma \frac{\partial u}{\partial x_0}dx - \frac{\partial v}{\partial x_0}dy, \quad \frac{\partial V}{\partial x_0} = \int_\Gamma \frac{\partial v}{\partial x_0}dx + \frac{\partial u}{\partial x_0}dy,$$

with similar results for $\partial U/\partial y_0$ and $\partial V/\partial y_0$.

Using the Cauchy–Riemann equations, we can rewrite these results as

$$\frac{\partial U}{\partial y_0} = \int_\Gamma \frac{\partial u}{\partial y_0}dx - \frac{\partial v}{\partial y_0}dy = -\frac{\partial V}{\partial x_0} \quad \text{and, similarly,} \quad \frac{\partial U}{\partial x_0} = \frac{\partial V}{\partial y_0},$$

showing that U and V satisfy the Cauchy–Riemann equations in D. As the partial derivatives of U and V are continuous, it follows that $G(z_0)$ must be analytic in D. This proves the first part of the theorem. To prove the second part we use the fact that

$$G'(z_0) = \frac{\partial U}{\partial x_0} + i\frac{\partial V}{\partial x_0} = \int_\Gamma \left(\frac{\partial u}{\partial x_0} + i\frac{\partial v}{\partial x_0}\right)dx + \left(-\frac{\partial v}{\partial x_0} + i\frac{\partial u}{\partial x_0}\right)dy$$

$$= \int_\Gamma \left(\frac{\partial u}{\partial x_0} + i\frac{\partial v}{\partial x_0}\right)(dx + idy) = \int_\Gamma \frac{\partial g(z, z_0)}{\partial z_0}dz,$$

and the proof is complete.   ∎

GOTTFRIED WILHELM LEIBNIZ (1646–1716)
A German mathematician who studied moral philosophy and law, first at the University of Leipzig and then at the University of Altdorf, from where he obtained his degree. Declining an offer of a professorship at Altdorf, he embarked on a legal career and chose to develop his mathematical work as a personal interest. He traveled extensively, meeting distinguished people in many countries, including Isaac Newton, whom he met during a visit to the Royal Society of London. He published his work on the calculus about a decade after Newton had completed his own fundamental work on the calculus, but before its publication. It was due to Newton's cautious and suspicious nature that the publication of his work was delayed, leading to the long-standing international dispute over who should be considered to be the founder of the calculus. Shortly before his death Leibniz founded the Berlin Academy of Sciences.

Summary

The Cauchy integral formulas were derived that express $f(z_0)$ and $f^{(n)}(z_0)$ in terms of integrals involving $f(z)/(z - z_0)^{n+1}$ around a contour containing z_0. Some important properties of analytic functions were obtained, and Leibniz' rule for differentiation under a contour integral was proved.

EXERCISES 14.3

In Exercises 1 through 8 use Theorem 14.7 to evaluate the given integral when integration is around Γ in the positive sense.

1. $\int_\Gamma \dfrac{\sin 2z}{z^2 - (\pi/2)^2}\,dz$, with Γ the circle $|z - 1| = 1$.

2. $\int_\Gamma \dfrac{(1 + z)e^z}{z^2 - 3z}\,dz$, with Γ the circle $|z| = 1$.

3. $\int_\Gamma \dfrac{\sin(\pi z/4)}{z^2 - 1}\,dz$, with Γ the circle $|z - 1| = 1$.

4. $\int_\Gamma \dfrac{\cosh z}{z^2 + 1}\,dz$, with Γ the circle $|z - i| = 1$.

5. $\int_\Gamma \dfrac{e^z}{z - 4}\,dz$, with Γ the circle $|z - 6| = 3$.

6. $\int_\Gamma \dfrac{(3 + z^2)}{z\cosh z}\,dz$, with Γ the circle $|z| = 1$.

7. $\int_\Gamma \dfrac{z\sinh z}{z^2 + 1}\,dz$, with Γ the circle $|z + i/2| = 1$.

8. $\int_\Gamma \dfrac{\sin z}{z^2 + 1}\,dz$, with Γ the circle $|z - 2i| = 2$.

In Exercises 9 through 15 use Theorem 14.8 to evaluate the given integral analytically when integration is around Γ in the positive sense, and verify the result by using computer algebra.

9. $\int_\Gamma \dfrac{z\sin z}{(z - \pi/4)^5}\,dz$, with Γ the circle $|z - \pi/4| = \pi$.

10. $\int_\Gamma \dfrac{z\cosh z}{(z - i)^4}\,dz$, with Γ the circle $|z - i| = 1$.

11. $\int_\Gamma \dfrac{\sin^2 z}{(z - \pi/2)^3}\,dz$, with Γ the circle $|z| = \pi$.

12. $\int_\Gamma \dfrac{\exp z^2}{(z + i)^4}\,dz$, with Γ the circle $|z + i| = 2$.

13. $\int_\Gamma \dfrac{z^2 \sinh z}{(z - i)^4}\,dz$, with Γ the circle $|z| = 3$.

14. $\int_\Gamma \dfrac{(1 - z)\cos z}{(z + i)^5}\,dz$, with Γ the circle $|z| = 2$.

15. $\int_\Gamma \dfrac{ze^z}{(z^2 + 1)^2}\,dz$, with Γ the circle $|z + 2i| = 2$.

16. The Legendre polynomial $P_n(z)$ can be defined by the *Rodrigues formula* (Exercise 16, Section 8.2):

$$P_n(z) = \frac{1}{2^n n!}\frac{d^n}{dz^n}(z^2 - 1)^n, \quad n = 0, 1, 2, \ldots.$$

Use the Cauchy integral formula for derivatives to show that

$$P_n(z) = \frac{1}{2\pi i}\int_\Gamma \frac{(t^2 - 1)^n}{2^n(t - z)^{n+1}}\,dt,$$

where Γ is any simple closed curve containing the point $t = z$ in its interior, and integration is around Γ in the positive sense. This result is called the *Schläfli* contour integral representation of $P_n(z)$.

Further Results

The first exercise provides an upper bound for the modulus of the nth derivative of a function that is analytic in a disc, while the remaining exercises offer an introduction to the study of special functions and linear differential equations by means of contour integrals.

17.* Use the Cauchy integral formula for derivatives to prove that if $f(z)$ is an analytic function in a domain D containing a disc Γ of radius R with its center at $z = z_0$, and $|f(z)| \le M$ for all z on Γ, then

$$\left|f^{(n)}(z_0)\right| \le \frac{Mn!}{R^n}, \quad \text{for } n = 1, 2, \ldots.$$

These results are called the **Cauchy inequalities for derivatives**.

18.* Show, by considering the change in the argument of $(t^2 - 1)^{n+1}/(t - z)^{n+1}$ around a simple closed curve Γ with positive orientation that contains the point $t = z$, that

$$\int_\Gamma \frac{d}{dt}\left[\frac{(t^2 - 1)^{n+1}}{(t - z)^{n+1}}\right]dt = 0.$$

19.* Find the form taken by the result of Exercise 18 when the differentiation under the integral sign has been performed. Use the definition of $P_n(z)$ given in Exercise 16 to find $P_{n+1}(z)$, and by differentiation with respect to z find $P_n'(z)$ and $P_{n+1}'(z)$. Use these results in the first part of this exercise to derive the recurrence relation

$$P'_{n+1}(z) = zP'_n(z) + (n+1)P_n(z).$$

20.* Show, by considering the change in the argument of $t(t^2-1)^n/(t-z)^n$ around a simple closed curve Γ with positive orientation that contains the point $t=z$, that

$$\int_\Gamma \frac{d}{dt}\left[\frac{t(t^2-1)^n}{(t-z)^n}\right]dt = 0.$$

21.* Find the form taken by the result of Exercise 20 when the differentiation under the integral sign has been performed. Use the definition of $P_n(z)$ in Exercise 16 to find $P_{n-1}(z)$ and $P_{n+1}(z)$, and use them in the result of the first part of the exercise to derive the recurrence relation

$$(n+1)P_{n+1}(z) - (2n+1)zP_n(z) + nP_{n-1}(z) = 0.$$

22.* Show, by considering the change in the argument of $(t^2-1)^{n+1}/(t-z)^{n+2}$ around a simple closed curve Γ with positive orientation that contains the point $t=z$, that

$$\int_\Gamma \frac{d}{dt}\left[\frac{(t^2-1)^{n+1}}{(t-z)^{n+2}}\right]dt = 0.$$

23.* Differentiate the integral representation for $P_n(z)$ given in Exercise 16 with respect to z to find $P'_n(z)$

and $P''_n(z)$ and form the expression

$$G(z) = (1-z^2)P''_n(z) - 2zP'_n(z) + n(n+1)P_n(z).$$

Show that

$$G(z) = \frac{(n+1)}{2^{n+1}\pi i}\int_\Gamma \frac{(t^2-1)^n}{(t-z)^{n+3}}[2(n+1)t(t-z)$$
$$- (n+2)(t^2-1)]dt.$$

By comparing the integrand of $G(z)$ with the differentiated form of the integrand in Exercise 22, deduce that $G(z)=0$, and hence show that $P_n(z)$ is a solution of the Legendre differential equation

$$(1-z^2)P''_n(z) - 2zP'_n(z) + n(n+1)P_n(z) = 0.$$

24.* By integrating $\exp(-z^2)$ around the rectangle with its corners at the points $(0,0)$, $(R,0)$, (R,b) and $(0,b)$ in the complex plane, proceeding to the limit as $R\to\infty$, and using the standard result $\int_0^\infty \exp(-x)^2 dx = \frac{1}{2}\sqrt{\pi}$, show that

$$\int_0^\infty \exp(-x)^2 \cos(2ax)dx = \frac{1}{2}\sqrt{\pi}\exp(-a^2).$$

Find the value of $\int_0^\infty \exp(-x)^2 \sin(2ax)dx$ in terms of a.

14.4 Some Properties of Analytic Functions

The next group of theorems describe some of the most important properties of analytic functions that can be deduced either directly or indirectly from the Cauchy integral theorem.

The first result, known as *Morera's theorem*, is the converse of the Cauchy–Goursat theorem and it is largely of theoretical importance.

THEOREM 14.12   **Morera's theorem**   If a function $f(z)$ is continuous in a domain D and such that

$$\int_\Gamma f(z)dz = 0$$

for every simple closed contour Γ in D, then $f(z)$ is analytic in D.

Proof   The condition

$$\int_\Gamma f(z)dz = 0$$

implies that the function

$$F(z) = \int_{z_0}^{z} f(\zeta)d\zeta,$$

with z, z_0 and Γ in D, is independent of the path from z_0 to z. The continuity of $f(z)$ implies that $F(z)$ is differentiable, and from the argument preceding Theorem 14.5 it follows that $F(z)$ is analytic, with $F'(z) = f(z)$. Consequently, as $f(z)$ is the derivative of an analytic function, $f(z)$ must be analytic in D, so the theorem is proved.   ■

The next result to be established is **Liouville's theorem** and it has numerous applications, one of which will occur later in the proof of the *fundamental theorem of algebra*.

THEOREM 14.13

Liouville's theorem   If $f(z)$ is analytic in the entire z-plane, and such that $|f(z)| \leq M$ for all z, then $f(z) = $ constant.

Proof   Setting $n = 1$, $z - z_0 = Re^{i\theta}$, and $dz = iRe^{i\theta}d\theta$ in the Cauchy integral formula for derivatives and taking the modulus gives

$$|f'(z_0)| \leq \frac{1}{2\pi} \int_0^{2\pi} \frac{|f(z)|}{|z - z_0|^2} |iRe^{i\theta}|d\theta \leq \frac{1}{2\pi} \int_0^{2\pi} \frac{M}{R^2} Rd\theta,$$

and so

$$|f'(z_0)| \leq \frac{M}{R},$$

which is true for all z_0 independently of R. Taking the limit as $R \to \infty$ and dropping the suffix zero show that $|f'(z)| = 0$ for all z, but this is only possible if $f'(z) \equiv 0$, so $f(z) = $ constant and the result is proved.   ■

Liouville's theorem illustrates one of the major differences between analytic functions in compex analysis and and differentiable functions in real analysis, because the theorem has no analogue in real analysis. This is easily seen by considering the function $\sin x$, which, although differentiable, bounded, and defined for all x, is not a constant. Another important difference between analytic functions and real functions is that a real function may only be differentiable a finite number of times, whereas analytic function has derivatives of all orders.

The next theorem is used repeatedly when seeking the zeros of polynomials, and it is proved here for a general complex polynomial of degree n.

THEOREM 14.14

every polynomial of degree *n* has *n* zeros

Fundamental theorem of algebra   Every complex polynomial $P_n(z) = a_0 + a_1z + a_2z^2 + \cdots + a_nz^n$, with complex coefficients $a_0, a_1, \ldots, a_n$ with $a_n \neq 0$ and $n \geq 1$ has precisely n zeros, some of which may be repeated.

Proof   The proof will be by contradiction. Suppose, if possible, that $P_n(z)$ has no zeros. Then the function $Q_n(z) = 1/P_n(z)$ is analytic for all z (it is an *entire function*). Then, when $|z|$ is large, $|P_n(z)|$ can be approximated by $|P_n(z)| \approx |a_nz^n|$, so it follows that $\lim_{|z| \to \infty} |Q_n(z)| = \lim_{|z| \to \infty} 1/|P_n(z)| = 0$. Consequently, $|Q_n(z)|$ is bounded in the entire complex plane, so by Liouville's theorem $Q_n(z)$ must be a constant. This contradicts the definition of $Q_n(z)$, showing that $P_n(z)$ must have at least one zero.

Denoting this zero by z_1, we can remove a factor $(z - z_1)$ from $P_n(z)$ and write it as $P_n(z) = (z - z_1)P_{n-1}(z)$, where $P_{n-1}(z)$ is a polynomial of degree $n - 1$. This process of factoring out $(z - z_0)$ from $P_n(z)$ to arrive at the polynomial $P_{n-1}(z)$ of lower degree is called **deflation**. Applying the same form of argument to $P_{n-1}(z)$ proves the existence of another zero z_2, and repetition of the argument establishes the existence of precisely n zeros, not all of which need be different. ∎

THEOREM 14.15

an averaging property for analytic functions

Gauss's mean value theorem   Let $f(z)$ be analytic in a simply connected domain D containing the circle Γ of radius ρ with its center at z_0. Then,

$$f(z_0) = \frac{1}{2\pi} \int_0^{2\pi} f(z_0 + \rho e^{i\theta})d\theta.$$

Proof   From the Cauchy integral formula

$$f(z_0) = \frac{1}{2\pi i} \int_\Gamma \frac{f(z)}{z - z_0}dz,$$

but on the circle $z - z_0 = \rho e^{i\theta}$ and $dz = i\rho e^{i\theta}d\theta$, so

$$f(z_0) = \frac{1}{2\pi i} \int_0^{2\pi} \frac{f(z_0 + \rho e^{i\theta})}{\rho e^{i\theta}} i\rho e^{i\theta} d\theta$$

$$= \frac{1}{2\pi} \int_0^{2\pi} f(z_0 + \rho e^{i\theta})d\theta. \qquad ∎$$

When expressed in words, the Gauss mean value theorem says that the value of an analytic function $f(z)$ at a point z_0 in D is the average of the values of $f(z)$ around the perimeter of any circle Γ in D with its center at the point z_0. A useful consequence of this theorem is the following result for harmonic functions that we state in the form of a corollary.

COROLLARY TO THEOREM 14.15

an averaging property for harmonic functions

Mean value theorem for harmonic functions   Let $u(x, y)$ be harmonic in a domain D containing the point $z_0 = x_0 + iy_0$, and let Γ be any circle of radius ρ in D with its center at (x_0, y_0). Then $u(x_0, y_0)$ is the average of the values of $u(x, y)$ around the perimeter of Γ.

Proof   The corollary follows immediately from Theorem 14.15 by setting $f(z) = u + iv$ and equating the real parts of the statement of the theorem. ∎

THEOREM 14.16

A function with its maximum modulus at the center of a disc   Let a function $f(z)$ be analytic in a disc with its center at the point z_0, and continuous on its circular boundary Γ. Then if the modulus $|f(z)|$ attains its maximum value M at z_0, the function $f(z) =$ constant throughout the disc and on its boundary Γ.

Proof   The proof of the theorem contains two steps. The first involves showing that the conditions of the theorem lead to the result that $|f(z)| = M$ inside the disc and on its boundary Γ. The second step that completes the proof involves showing that a function with constant modulus that is analytic in a disc must, of necessity, be constant.

STEP 1   Let the function $f(z)$ be analytic inside the circle $z = z_0 + \rho e^{i\theta}$ and continuous on its boundary Γ, and let its modulus $|f(z)|$ attain its maximum value $M > 0$ at z_0. Suppose, if possible, that $|f(z)| < M$ at some point on Γ. Then because the function is continuous on Γ, it must follow that $|f(z)| < M$ over some finite part of Γ.

From the Gauss mean value theorem,

$$f(z_0) = \frac{1}{2\pi} \int_0^{2\pi} f(z_0 + \rho e^{i\theta}) d\theta,$$

so if we take the modulus, this becomes

$$|f(z_0)| \leq \frac{1}{2\pi} \int_0^{2\pi} |f(z_0 + \rho e^{i\theta})| d\theta.$$

As $|f(z_0)| = M$, this becomes

$$M \leq \frac{1}{2\pi} \int_0^{2\pi} |f(z_0 + \rho e^{i\theta})| d\theta.$$

However, the integrand is less than M over some part of Γ, so for some k such that $0 < k < 1$,

$$\int_0^{2\pi} |f(z_0 + \rho e^{i\theta})| d\theta = kM.$$

Using this result with the previous one leads to the equation

$$M \leq kM \quad (0 < k < 1),$$

but this result is impossible, so $k = 1$ and $|f(z)| = M$ on Γ. As the disc is of radius ρ the result will be true for any radius r such that $0 \leq r \leq \rho$, and we have proved that $|f(z)| = M$ inside and on the boundary of the disc.

STEP 2   Setting $f(z) = u + iv$ we can write $|f(z)|^2 = u^2 + v^2$, so from the result of Step 1 we see that $u^2 + v^2 = M^2$ throughout the disk. Differentiating this result partially with respect to x and y gives

$$2u\frac{\partial u}{\partial x} + 2v\frac{\partial v}{\partial x} = 0 \quad \text{and} \quad 2u\frac{\partial u}{\partial y} + 2v\frac{\partial v}{\partial y} = 0.$$

The Cauchy–Riemann equations then allow these equations to be rewritten as

$$u\frac{\partial u}{\partial x} - v\frac{\partial u}{\partial y} = 0 \quad \text{and} \quad v\frac{\partial u}{\partial x} + u\frac{\partial u}{\partial y} = 0.$$

Solving these equations for u_x and u_y gives $(u^2 + v^2)u_x = 0$ and $(u^2 + v^2)v_y = 0$, but $u^2 + v^2 = M^2 > 0$, so the only solution of this system of equations is $\partial u/\partial x = \partial u/\partial y = 0$, showing that $u = $ constant. Using $u = $ constant in the Cauchy–Riemann equations implies that $\partial v/\partial x = \partial v/\partial y = 0$, so $v = $ constant, and we have shown that $f(z) = $ constant throughout the disc. The proof is complete.   ∎

THEOREM 14.17

an extremum principle
for $|f(z)|$ when $f(z)$ is
analytic

The maximum/minimum modulus principle   If a nonconstant function $f(z)$ is analytic in a bounded domain D and continuous on its boundary Γ, then the maximum of $|f(z)|$ must occur on Γ. If $f(z) \neq 0$ anywhere in D, then the minimum value of $|f(z)|$ also must occur on Γ.

Proof   The conditions that $f(z)$ is analytic in D and continuous on Γ imply that $f(z)$ is continuous throughout D and on its boundary Γ. Consequently, the real function $|f(z)|$ must have both a maximum and a minimum in the closed region formed by D and its boundary Γ.

As $f(z)$ is analytic in D, so also is $[f(z)]^n$ for $n = 2, 3, \ldots$, so taking a point z_0 inside D and applying the Cauchy integral theorem to $[f(z)]^n$ gives

$$[f(z_0)]^n = \frac{1}{2\pi i} \int_\Gamma \frac{[f(z)]^n}{z - z_0} dz.$$

If $|f(z)| \le M$ on the boundary Γ of finite length L, and if d is the minimum distance from z_0 to Γ, taking the modulus of this result we obtain

$$|f(z_0)|^n \le \frac{1}{2\pi} \int_\Gamma \frac{|f(z)|^n}{|z - z_0|} |dz|$$

$$\le \frac{1}{2\pi} \frac{M^n L}{d},$$

showing that

$$|f(z_0)| \le M \left(\frac{L}{2\pi d} \right)^{1/n}.$$

Proceeding to the limit as $n \to \infty$ leads to the result $|f(z_0)| \le M$, so the value of $|f(z)|$ throughout the domain cannot exceed its maximum value on the boundary Γ.

To complete the proof suppose, if possible, that in addition to the maximum value of the modulus occurring on the boundary, it also occurs at a point z^* inside D. Construct a circle inside D with z^* as its center. Then from Theorem 14.16 the function $f(z)$ must be constant inside this circle. As the $f(z)$ is analytic in D, it is also continuous in D together with all its derivatives, and, in particular, it is continuous across the boundary of the circle. The derivatives of $f(z)$ are zero inside and on the boundary of the circle, so by continuity they must also be zero throughout the rest of D, from which it follows that $f(z) = $ constant in D. This contradicts the assumption that $f(z)$ is nonconstant, so the maximum value of $|f(z)|$ can only occur on the boundary Γ.

The minimum value of $|f(z)|$ must also occur on the boundary Γ if $f(z) \ne 0$ in D, because if the foregoing result is applied to the function $\varphi(z) = 1/f(z)$, the maximum value of $|\varphi(z)|$ must occur on the boundary of Γ, but this corresponds to the minimum value of $|f(z)|$, so the theorem is proved, because if $f(z) = 0$ in D then $1/f(z)$ is not analytic in D.   ∎

EXAMPLE 14.16

Confirm by direct calculation the maximum/minimum principle for the function $f(z) = \sin z$ in the domain D defined by $0 \le x \le \pi$ and $0 \le y \le 1$, and place bounds on $|\sin z|$ inside D.

Solution   We notice first that the function $f(z)$ is analytic for all z and the domain D is bounded. Setting $z = x + iy$ in $f(z)$ and expanding the result gives $\sin z = \sin x \cosh y + i \cos x \sinh y$, from which it follows that

$$|\sin z|^2 = \sin^2 x \cosh^2 y + \cos^2 x \sinh^2 y.$$

Differentiating this result with respect to x and y, we obtain

$$\frac{\partial}{\partial x}|\sin z|^2 = 2\sin x\cos x\cosh^2 y - 2\sin x\cos x\sinh^2 y = 2\sin x\cos x = \sin 2x$$

and

$$\frac{\partial}{\partial y}|\sin z|^2 = 2\sin^2 x\sinh y\cosh y + 2\cos^2 x\sinh y\cosh y = 2\sinh y\cosh y = \sinh 2y.$$

The maxima and minima of $|\sin z|^2$, and hence of $|\sin z|$, will occur in D if each of these derivatives vanishes simultaneously at a point or points inside D. The function $\sin 2x$ only vanishes in D on the line $x = \pi/2$, but $\sinh 2y \neq 0$ for $0 < y < 1$, so $|\sin z|^2$, and hence $|\sin z|$, has neither maxima nor minima in D. Thus, the extrema of $|\sin z|$ must occur on the straight line boundaries of D. On the boundary $x = 0$ of D, $|\sin z|$ has a minimum of 0 at $(0, 0)$ and a maximum of $\sinh 1$ at $(0, 1)$. On the boundary $x = \pi$ of D, $|\sin z|$ has a minimum of 0 at $(\pi, 0)$ and a maximum of $\sinh 1$ at $(\pi, 1)$. On the boundary $y = 0$ of D, $|\sin z|$ has two minima of 0 at $(0, 0)$ and $(0, \pi)$ and a maximum of 1 at $(\pi/2, 0)$, while on the boundary $y = 1$ of D, $|\sin z|$ has two minima equal to $\sinh 1$ at $(0, 1)$ and $(\pi, 1)$, and a maximum of $(1 + \sinh^2 1)^{1/2}$ at $(\pi/2, 1)$. This shows that the smallest value of $|\sin z|$ on the boundary of D is 0, and the largest value is $(1 + \sinh^2 1)^{1/2}$.

The results of Theorem 14.17 are confirmed, so inside the rectangle D it follows that

$$0 < |\sin z| < (1 + \sinh^2 1)^{1/2}, \quad \text{for all } z \text{ inside } D. \qquad \blacksquare$$

We now use Theorem 14.17 to prove a corresponding result for harmonic functions that has important consequences in the study of boundary value problems for Laplace's equation.

THEOREM 14.18

an extremum principle for a harmonic function u

The maximum/minimum principle for harmonic functions   The maximum and minimum values of a nonconstant function u that is harmonic in a bounded simply connected domain and continuous on its boundary must occur on the boundary.

Proof   Let u be a harmonic function satisfying the conditions of the theorem, and form the analytic function $f(z) = u + iv$, where v is the harmonic conjugate of u. Then,

$$|\exp\{f(z)\}| = |e^{u+iv}| = |e^u||e^{iv}| = e^u.$$

As e^u is a monotonic increasing function of u, this result shows that the maxima of e^u, and hence of u and those of $|\exp f(z)|$, coincide. Using this result in Theorem 14.16 shows that the maxima of u must occur on the boundary. The fact that the minima of u also occur on the boundary follows if we notice that the minima of u correspond to the maxima of the harmonic function $-u$, so the proof is complete. $\qquad \blacksquare$

Theorem 14.18 also applies to nonsimply connected bounded domains. In such a domain D, the maximum value of u is taken to be the largest of the maxima on all the internal boundaries and the external boundary of D, and the minimum value is taken to be the smallest of the minima on all the internal boundaries and the external boundary of D.

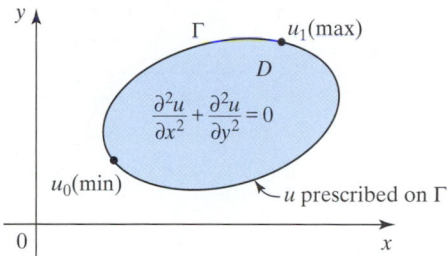

FIGURE 14.22   A two-dimensional boundary value problem for Laplace's equation.

To see how the theorem provides qualitative information about solutions of Laplace's equation $u_{xx} + u_{yy} = 0$, consider the bounded two-dimensional domain D with boundary Γ shown in Fig. 14.22 on which u assumes prescribed continuous values, and let the smallest of the values of u on Γ be u_0 and the largest be u_1. Then, for all points (x, y) in D we have

$$u_0 < u(x, y) < u_1.$$

Problems of this type are called two-dimensional **boundary value problems** for Laplace's equation. They occur, for example, when a two-dimensional steady-state temperature distribution is to be determined within a uniform heat-conducting medium on the boundary of which the temperature takes prescribed values, because the steady state temperature as a function of position in the medium is a solution of Laplace's equation.

EXAMPLE 14.17

Use Theorem 14.18 to place bounds on the function $u(x, y) = (1 + 2 \sinh^2 x) \sin 2y$ in the domain D determined by $0 \le x \le 1$ and $0 \le y \le \pi$.

Solution   Routine differentiation establishes that $u_{xx} + u_{yy} = 0$, so $u(x, y)$ is harmonic. As the domain D is bounded and $u(x, y)$ is harmonic, Theorem 14.18 applies and asserts that the smallest and largest values of $u(x, y)$ must occur on the boundary of D. Examination of the behavior of $u(x, y)$ on the straight line boundaries of D shows that the smallest value of $u(x, y)$ is $-1 - 2 \sinh^2 1$ at $(1, 3\pi/4)$, and the largest value is $1 + 2 \sinh^2 1$ at $(1, \pi/4)$, so

$$-1 - 2 \sinh^2 1 < u(x, y) < 1 + 2 \sinh^2 1 \text{ at all points inside } D. \qquad \blacksquare$$

The next example illustrates how the maximum/minimum principle may be used to place bounds on the two-dimensional temperature distribution inside a long uniform hexagonal rod of metal when an arbitrary temperature distribution is prescribed around its hexagonal faces. The bounds on the temperature distribution inside the metal can, for example, be used to estimate the thermal stress produced in the rod due to the uneven heating of its faces.

EXAMPLE 14.18

Consider the cross-section of a long hexagonal rod of metal, shown in Fig. 14.23a, where the inscribed circle that is tangent to the faces has radius $a\sqrt{3}/2$, and the circumscribed circle that passes through the vertices has radius a. Draw a ray from the origin to a point on the circumscribed circle, and let $T = f(\theta)$ be the temperature

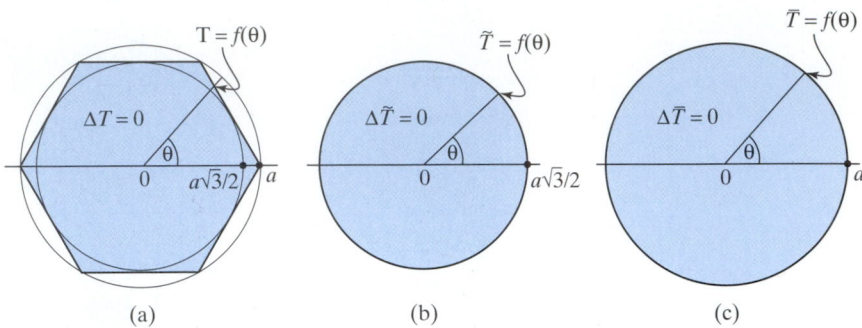

FIGURE 14.23   The hexagonal cross-section and two related cross-sections with radii $a\sqrt{3}/2$ and a.

that is imposed on the hexagonal face where the ray intersects the face. Then the function $f(\theta)$ is periodic with period 2π.

We now anticipate the result of Chapter 17 (proved in Chapter 18) that the steady state temperature distribution in a uniform heat conducting medium satisfies the Laplace equation. As the problem is two-dimensional, it follows that the temperature distribution inside the hexagonal cross-section must satisfy the two-dimensional Laplace equation $\Delta T = 0$.

Our approach will be to consider two related but far simpler problems than the problem in the hexagonal cross-section. One will be for the Laplace equation for a temperature $\tilde{T}(r, \theta)$ inside the inscribed circle, and the other for a temperature $\bar{T}(r, \theta)$ inside the circumscribed circle, when both problems satisfy the same temperature distribution at an angle θ on the perimeter of their respective circles as the temperature on the plane face at the same angle. We start by considering the problem in cylindrical polar coordinates $\Delta \tilde{T}(r, \theta) = 0$ in the disc of radius $a\sqrt{3}/2$ shown in Fig. 14.23b that is required to satisfy the temperature $\tilde{T}(a\sqrt{3}/2, \theta) = f(\theta)$ on the perimeter of the circle. Then, as the temperatures on the hexagonal faces have been transferred *inward* to corresponding points on the inscribed circle, it follows directly from the maximum/minimum principle that inside and on the inscribed circle we must have $T(r, \theta) \leq \tilde{T}(r, \theta)$. Thus, $\tilde{T}(r, \theta)$ provides an upper bound for the temperature in any cross-section of the hexagonal rod at points that lie inside the circle of radius $a\sqrt{3}/2$.

Next, we consider the corresponding problem shown in Fig. 14.23c, where this time the solution $\bar{T}(a, \theta)$ of the Laplace equation inside the circumscribed circle is required to satisfy the temperature $\bar{T}(a, \theta) = f(\theta)$ on the perimeter of the circle. Here the temperatures on the hexagonal faces have been transferred *outward* to corresponding points on the circumscribed circle, so this time by the maximum/minimum principle it follows that $\bar{T}(r, \theta) \leq T(r, \theta)$. Thus, $\bar{T}(r, \theta)$ provides a lower bound for the temperature $\bar{T}(r, \theta)$ at all points inside the hexagonal cross-section.

Consequently, we have established the following results:

(i) $\bar{T}(r, \theta) \leq T(r, \theta)$ at all points inside the hexagonal cross-section

(ii) $T(r, \theta) \leq \tilde{T}(r, \theta)$ at all points inside the hexagonal cross-section that belong to the inscribed circle

To make further progress we appeal to the Poisson integral formula for a circle that forms the result of Exercise 3 in Exercise Section 14.4. This asserts that if $u(r, \theta)$

is harmonic in a circle of radius R centered on the origin, and on the perimeter of the circle $u(R, \theta) = f(\theta)$, then

$$u(r, \theta) = \frac{1}{2\pi} \int_0^{2\pi} \frac{(R^2 - r^2) f(\psi)}{R^2 - 2rR\cos(\theta - \psi) + r^2} d\psi.$$

The bound $\tilde{T}(r, \theta)$ follows directly from this result by setting $R = a\sqrt{3}/2$ and $u(r, \theta) = \tilde{T}(r, \theta)$, while the bound for $\bar{T}(r, \theta)$ follows by setting $R = a$ and $u(r, \theta) = \bar{T}(r, \theta)$. Clearly, this approach works for any cross-section shape, though the bounds will be sharper when the radii of the inscribed and circumscribed circles are close together. ∎

The performance of an engineering system often depends on the location of the zeros of a function that may not necessarily be a polynomial. To obtain a system with satisfactory properties, the zeros are often required to lie in a particular part of the z-plane. This occurs, for example, when working with control systems governed by a system of differential equations, because the system will only be stable if the zeros of a characteristic equation all lie to the left of the imaginary axis, and so have negative real parts. However, to avoid an undesirably slow decay of any disturbances to such a system, it is usually also necessary to require that each zero have a real part that is less than some prescribed negative number, so in such cases all zeros must lie to the left of a line $z = -c$ with $c > 0$. Consequently, when such a system has parameters that can be adjusted to optimize performance, unless the zeros can be found explicitly, it is necessary to devise a practical test that determines how many zeros lie inside a given region contained within a closed curve Γ.

A powerful test of this type can be derived from the following result we will call the **restricted argument principle**, as it is a special case of what in complex analysis is known as the **argument principle**. Although this more general theorem is not difficult to establish, its proof would be out of place here and will be omitted, as it can be found in any of the references quoted at the end of this chapter.

THEOREM 14.19   **The restricted argument principle**   Let $f(z)$ be analytic and have a finite number of zeros and no poles in a bounded simply connected domain D with boundary Γ. Then, provided $f(z) \neq 0$ on Γ,

$$\frac{1}{2\pi} \Delta_\Gamma \arg f(z) = N,$$

where $\Delta_\Gamma \arg f(z)$ denotes the change in the argument of $f(z)$ when the contour Γ is traversed once in the positive (counterclockwise) sense, and N is the number of zeros in D with their multiplicity counted. ∎

The geometrical implication of this theorem is as follows. Let Γ' be the image of Γ under the mapping $w = f(z)$. Then, when a point z makes one traverse of the contour Γ in the z-plane, the number of times its image Γ' encircles the origin in the w-plane is equal to the number of zeros of $f(z)$ inside Γ. To apply this geometrical interpretation of the theorem, the contour Γ in the z-plane must first be parametrized, after which this parametrization must be used in $w = f(z)$ to construct the image Γ' in the w-plane. The number of times Γ' encircles the origin $w = 0$ can then be counted to determine the number of zeros of $f(z)$ inside Γ.

A result that can be derived from the restricted argument principle, which although weaker is both useful and simple to use, is **Rouché's theorem**.

THEOREM 14.20

Rouché's theorem Let D be a simply connected domain bounded by a contour Γ in which the functions $f(z)$ and $g(z)$ are analytic and such that $|f(z)| > |g(z)|$ for all z on Γ. Then $f(z)$ and $f(z) + g(z)$ each have the same number of zeros in D. ∎

In effect, the conditions of Rouché's theorem are such that it enables the number of zeros of a simple function $f(z)$ inside Γ to be equated to the number of zeros possessed by the more complicated function $f(z) + g(z)$ that also lie inside Γ.

EXAMPLE 14.19

Use Rouché's theorem to find the number of zeros of the polynomial $P(z) = z^4 - 8z + 10$ that lie (a) in $|z| \leq 1$ and (b) in $|z| \leq 3$. (c) Confirm results (a) and (b) by using the graphical implication of the restricted argument principle.

Solution

(a) Make the identifications $f(z) = 10$ and $g(z) = z^4 - 8z$. On $|z| = 1$ we have $|f(z)| = 10$ and $|g(z)| = |z^4 - 8z| \leq |z| + 8|z| = 9$, so $|f(z)| > |g(z)|$ on $|z| = 1$. Then by Rouché's theorem, as $f(z)$ has no zeros inside $|z| = 1$, it follows that $f(z) + g(z) = P(z)$ has no zeros inside $|z| \leq 1$.

(b) Make the identification $f(z) = z^4$ and $g(z) = -8z + 10$. On $|z| = 3$, $|f(z)| = 81$ and $|g(z)| = |-8z + 10| \leq 8|z| + 10 = 34$, so $|f(z)| > |g(z)|$ on $|z| = 3$. Then by Rouché's theorem, as $f(z)$ has four zeros inside $|z| = 3$ when their multiplicity is counted, it follows that $f(z) + g(z) = P(z)$ also has four zeros inside $|z| \leq 3$.

(c) Parametrize the circle $|z| = 1$ by setting $x = \cos t$, $y = \sin t$ with $0 \leq t \leq 2\pi$ so the unit circle is traversed once. Then setting $z = \cos t + i \sin t$ in $w = u + iv = P(z)$ and separating out the real and imaginary parts gives

$$u = \cos^4 t - 6\cos^2 t \sin^2 t + \sin^4 t - 8\cos t + 10$$
$$v = 4\cos^3 t \sin t - 4\cos t \sin^3 t - 8\sin t.$$

The image Γ' of Γ under the mapping $w = f(z)$ is obtained by plotting this parametric representation of Γ' with $0 \leq t \leq 2\pi$. This plot is shown in Fig. 14.24a, from which it can be seen that the image Γ' does not encircle the origin in the w-plane, so no zeros of $P(z)$ lie in $|z| = 1$.

Repeating this argument, but this time parametrizing the circle $|z| = 3$ by setting $z = 3(\cos t + i \sin t)$, leads to the results

$$u = 81\cos^4 t - 486\cos^2 t \sin^2 t + 81\sin^4 t - 24\cos t + 10$$
$$v = 324\cos^3 t \sin t - 324\cos t \sin^3 t - 24\sin t.$$

The plot of this image of Γ' is shown in Fig. 14.24b, from which it can be seen that Γ' encircles the origin in the w-plane four times, so $P(z)$ has four zeros inside the circle $|z| = 3$. ∎

Alternative accounts and extra information concerning the material in Sections 14.1 to 14.4 can be found in any one of references [6.1] to [6.4] and [6.6] to [6.9].

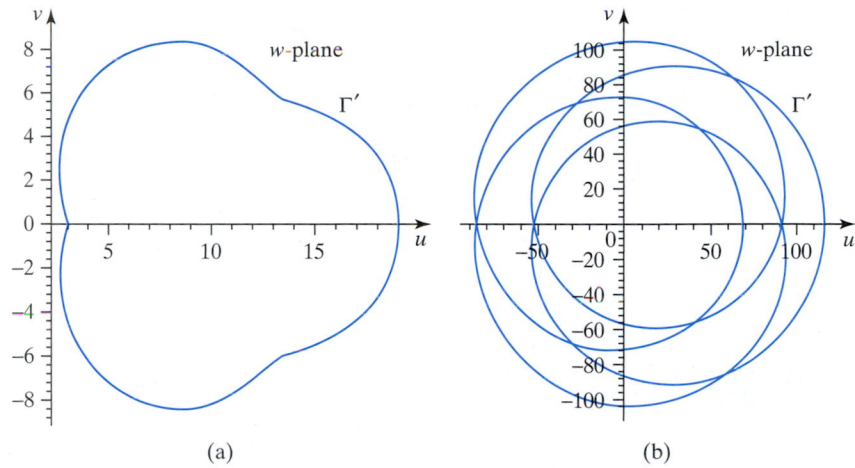

FIGURE 14.24   (a) Γ' does not encircle $w = 0$. (b) Γ' encircles $w = 0$ four times.

Summary

Some general properties of analytic functions were derived, one of which was the fundamental theorem of algebra that asserts every polynomial of degree n has precisely n zeros, though these need not all be distinct. The maximum/minimum modulus theorem for analytic functions was also proved, showing that the maximum and minimum values of the modulus of a nonconstant analytic function defined in a domain D must occur on the boundary of D. A corresponding theorem for harmonic functions was also proved.

EXERCISES 14.4

1. * Let $P_n(z) = a_0 + a_1 z + a_2 z^2 + \cdots + a_n z^n$ be a complex polynomial, and Γ be a positively oriented circle with its center at the origin. Show that

$$\frac{1}{2\pi i} \sum_{k=0}^{n} \int_{\Gamma} \frac{P_n(z)}{z^{k+1}} dz = \sum_{k=0}^{n} a_k.$$

2. * Let $f(z)$ be analytic inside and on the circle Γ defined by $|z| = R$, and let $z_0 = re^{i\theta}$, with $0 < r < R$, be a point inside the circle. Show that the point $Z = z\bar{z}/\bar{z}_0$ lies outside the circle Γ, so that

$$\frac{1}{2\pi i} \int_{\Gamma} \frac{f(z)}{z - Z} dz = 0.$$

By differencing this expression and the expression for $f(z_0)$ determined by the Cauchy integral formula, show that

$$f(re^{i\theta}) = \frac{1}{2\pi i} \int_{\Gamma} \frac{1}{z} \frac{(z\bar{z} - z_0\bar{z}_0)}{(z - z_0)(\bar{z} - \bar{z}_0)} f(z) dz.$$

3. * By setting $z_0 = re^{i\theta}$ and $z = Re^{i\psi}$ in the result of Exercise 2, show that

$$f(re^{i\theta}) = \frac{1}{2\pi} \int_{0}^{2\pi} \frac{(R^2 - r^2)}{R^2 - 2rR\cos(\psi - \theta) + r^2} f(Re^{i\psi}) d\psi.$$

Write $f(re^{i\theta}) = u(r, \theta) + iv(r, \theta)$ in the preceding result and derive the **Poisson integral formula for a disc**,

$$u(r, \theta) = \frac{1}{2\pi} \int_{0}^{2\pi} \frac{(R^2 - r^2)u(R, \psi)}{R^2 - 2rR\cos(\psi - \theta) + r^2} d\psi.$$

This formula determines the value of the harmonic function $u = \mathrm{Re}\{f(z)\}$ at any point (r, θ) inside the disc in terms of the prescribed values of u on the boundary Γ of the disc. The specification of u on the boundary of a domain in which u is harmonic constitutes what is called a **Dirichlet problem** for Laplace's equation. This formula determines, for example, the steady state electrostatic potential in a long cavity with a circular cross-section of radius R, on the walls of which the potential $u(R, \psi) = f(R, \psi)$. As the steady state two-dimensional temperature distribution in a long metal rod of circular cross-section of radius R is also a solution of Laplace's equation, this same formula determines the temperature distribution in the rod when its surface is at a temperature $u(R, \psi) = f(R, \psi)$.

4.* By setting $u(R, \psi) = M$ in the Poisson integral formula for a disc given in Exercise 3, and using the result

$$\int_0^{2\pi} \frac{dt}{1 + a\cos t} = \frac{2\pi}{\sqrt{1-a^2}} \quad \text{for } a^2 < 1$$

that can be established by the method of Example 14.13, show that when $u(R, \psi) = M$ (constant) on the boundary of the disc, it must follow that $u(r, \theta) \equiv M$ throughout the disc.

5.* Let domain D be the interior of the positively oriented contour Γ comprising the semicircle C_R of radius R in the upper half plane with its center at the origin, and the segment of the real axis from $-R$ to R. If z_0 is an interior point of D, explain why

$$f(z_0) = \frac{1}{2\pi i}\int_\Gamma \frac{f(z)}{z - z_0}dz \text{ and } 0 = \frac{1}{2\pi i}\int_\Gamma \frac{f(z)}{z - \bar z_0}dz.$$

Set $z_0 = x_0 + iy_0$ and difference these results to show that

$$f(z_0) = \frac{y_0}{\pi}\int_{-R}^R \frac{f(x)}{|x - z_0|^2}dx + \frac{y_0}{\pi}\int_{C_R} \frac{f(z)}{(z - z_0)(z - \bar z_0)}dz.$$

6.* Using the notation of Exercise 5, and writing $z = z_0 + (z - z_0)$ and $z = \bar z_0 + (z - \bar z_0)$, show that

$$(R - |z_0|)^2 \le |z - z_0| \cdot |z - \bar z_0|.$$

Deduce from this that if $|f(z)| \le K$ in the upper half plane, then

$$\left| \frac{y_0}{\pi}\int_{C_R} \frac{f(z)}{(z - z_0)(z - \bar z_0)}dz \right| \le \frac{Ky_0 R}{(R - |z_0|)^2}.$$

By taking the limit of the result of Exercise 5 as $R \to \infty$ and using the result from this exercise, deduce that

$$f(z_0) = \frac{y_0}{\pi}\int_{-\infty}^\infty \frac{f(x)}{(x - x_0)^2 + y_0^2}dx.$$

Then, by setting $f(z) = u(x, y) + iv(x, y)$ and equating the real parts of the equation, show that

$$u(x_0, y_0) = \frac{y_0}{\pi}\int_{-\infty}^\infty \frac{u(x, 0)}{(x - x_0)^2 + y_0^2}dx, \quad \text{for } y_0 > 0.$$

This result is the **Poisson integral formula for a half-plane**, and it determines the harmonic function $u(x_0, y_0)$ at points (x_0, y_0) in the upper half-plane in terms of a prescribed function $u(x, 0)$ on the real axis. The function $u(x, 0)$ is called a **Dirichlet boundary condition** for the two-dimensional boundary value problem for Laplace's equation. This formula can be used to determine the

steady state temperature distribution $u(x, y)$ in a thermally conducting half-plane when the temperature on the plane bounding surface is $u(x, 0) = T(x)$, with $T(x)$ a given function. A similar interpretation applies when the formula is used to determine the steady state electrostatic potential $u(x, y)$ in a half-space when the potential on the plane bounding surface is $u(x, 0) = T(x)$.

7. Let $P_n(z)$ be the complex polynomial $P_n(z) = a_0 + a_1 z + a_2 z^2 + \cdots + a_n z^n$ with $a_n \ne 0$, and $n \ge 1$. Justify the assertion in the proof of the fundamental theorem of algebra that if $Q_n(z) = 1/P_n(z)$, then $\lim_{|z|\to\infty}|Q_n(z)| = \lim_{|z|\to\infty} 1/|P_n(z)| = 0$.

8. Given that $z = 1 + 2i$ is a root of the polynomial $z^4 + 2z^3 + 10z^2 - 6z + 65 = 0$ with real coefficients, use the deflation method described in the proof of the fundamental theorem of algebra to find the remaining roots.

9. Verify the maximum/minimum principle for the function $f(z) = e^z$ in the domain $-1 \le x \le 1, -2 \le y \le 2$, and place bounds on $|e^z|$ inside the given domain.

10. Verify the maximum/minimum principle for the function $f(z) = \cosh z$ in the domain $-1 \le x \le 1, -1 \le y \le 1$, and place bounds on $|\cosh z|$ inside the given domain.

In Exercises 11 through 14 place bounds on the function $u(x, y)$ inside the given domain.

11. $u(x, y) = x + 2x^2 - 2y^2$ in the domain $-1 \le x \le 1, -1 \le y \le 1$.

12. $u(x, y) = e^x(y\cos y + x\sin y)$ in the domain $0 \le x \le 1, -\pi/2 \le y \le \pi/2$.

13. $u(x, y) = e^x(x\cos y - y\sin y)$ in the domain $0 \le x \le 1, -\pi/2 \le y \le \pi/2$.

14. $u(x, y) = e^x(\cos^2 y\cosh x - \sin^2 y\sinh x)$ in the domain $0 \le x \le 1, 0 \le y \le \pi/2$.

15. Show by Rouché's theorem that $P(z) = z^4 - 5z + 1$ has one zero in the disc $|z| \le 1$ and three zeros in the annulus $1 \le |z| \le 2$.

16. Use Rouché's theorem to find the number of zeros of $P(z) = 2z^3 - 4z + 1$ contained in (a) $|z| \le \frac{1}{4}$, (b) $|z| \le 1$, and (c) $|z| \le 3$.

17. Use the geometrical interpretation of the restricted argument principle to show that $f(z) = z - 2i + \exp(-z)$ has no zeros in $|z - i| \le 1$, one zero in $|z - i| \le 2$, and two zeros in $|z - i| \le 3$.

18. Given that $f(z) = z\exp(z) - 2z^5 + iz + 3i$, use the geometrical interpretation of the restricted argument principle to determine the number of zeros of $f(z)$ in (a) $|z| \le \frac{1}{4}$, (b) $|z| \le \frac{1}{2}$, (c) $|z| \le 1$, and (d) $|z| = \frac{3}{2}$.

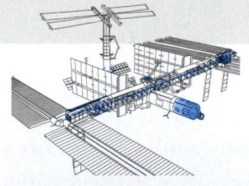

CHAPTER 14
TECHNOLOGY PROJECTS

The integral of a complex function $f(z)$ along a path Γ_{AB} from point A to point B, on which f has no singularities, is simply a line integral of $f(z)$ along Γ from A to B with respect to arc length. The complex integral can be evaluated numerically as follows. First, a general point z on the arc Γ_{AB} with its initial point A at $z = z_0$ and its final point B at $z = z_1$ is expressed parametrically as $z(t) = x(t) + iy(t)$ for the parameter t in the interval $t_0 \leq t \leq t_1$, with $z_0 = z(t_0)$ and $z_1 = z(t_1)$. Then, on Γ, $dz = (dx/dt + i\,dy/dt)dt$, so the required integral along Γ_{AB} is given by

$$\int_{\Gamma_{AB}} f(z)dz = \int_{t_0}^{t_1} f(z(t))(dx/dt + i\,dy/dt)dt.$$

If the path Γ is continuous, but defined in a piecewise manner along successive segments, each segment must be parametrized separately. The integral along Γ then follows by adding the integrals along each of the segments. A contour integral around a simple closed curve is obtained by parametrizing the curve (in segments if necessary) and integrating once around the curve in the counterclockwise direction. If f is not analytic, the integral of f from A to B will, in general, depend on the choice of path from A to B.

Project 1

The Numerical Evaluation of Integrals along Arcs

This project uses computer algebra to calculate the integrals of complex functions f along different arcs Γ from A to B to verify, in particular cases, that when f is analytic the result is independent of the path, though when f is not analytic the integral depends on the choice of path.

1. Let A be the point $z = 1 - 2i$ and B the point $z = 1 + 2i$. Parametrize the semicircular path Γ_1 from A to B that lies to the right of the line AB and has A and B as points on opposite ends of a diameter, and find dz on Γ_1. Parametrize the piecewise continuous straight line path Γ_2 joining A to C and C to B, where C is the point $z = 2 - 2i$, and find dz on the straight line segments AC and CB.

2. Given that $f_1(z) = z \sinh(2z)$, use computer algebra to show that $f_1(z)$ satisfies the Cauchy–Riemann equations for all z, and so is an entire function.

3. Evaluate $\int_{\Gamma_1} f_1(z)dz$ and $\int_{\Gamma_2} f_1(z)dz$ and hence show, as would be expected because f is an entire function, that the integrals are equal.

4. Given that $f_2(z) = z\bar{z} \sin z$, show by using computer algebra that f_2 is not analytic. By finding $\int_{\Gamma_1} f_2(z)dz$ and $\int_{\Gamma_2} f_2(z)dz$, show that $\int_{\Gamma_1} f_2(z)dz \neq \int_{\Gamma_2} f_2(z)dz$.

Project 2

Integrating around a Circular Arc Centered on a Simple Pole

This project uses computer algebra to examine the effect of integrating around a circular arc of arbitrarily small radius when its center is located at a simple pole of a complex function $f(z)$. This process is examined analytically in Chapter 15, where it is used in the determination of definite integrals of real functions $f(x)$ over the semi-infinite interval $0 \leq f < \infty$ and the infinite interval $-\infty < x < \infty$.

1. Let Γ_α be a circular arc of radius r with its center at the point $z = 1$ that subtends an angle α at $z = 1$. Denote by θ the angle from $z = 1$ to a point on the arc, with θ measured counterclockwise from the positive real axis such that $0 \leq \theta \leq \alpha$. Parametrize the arc Γ_α, and find dz on this arc.

2. Given that $f(z) = \cos z/(z - 1)$, use computer algebra to display the integral

$$\int_{\Gamma_\alpha} f(z)\,dz$$

in terms of r and the parametrization of the arc Γ_α.

3. Given that $\alpha = \pi/3$, compute the integral for $r = 0.01, 0.001,$ and 0.0001, and hence estimate its limiting value as $r \to 0$.

4. Repeat Step 3, using $\alpha = 2\pi/3$.

5. Repeat Step 3, using $\alpha = \pi$.

6. Repeat Step 3, using $\alpha = 5\pi/3$.

7. Compare the results of Steps 3 through 6 with the theoretical result $\int_{\Gamma_{2\pi}} f(z)\,dz = 2\pi i \cos(1)$, and deduce the relationship between $\int_{\Gamma_\alpha} f(z)\,dz$ and $\int_{\Gamma_{2\pi}} f(z)\,dz$ as a function of α.

Project 3

Complex Integrals around Deformed Contours

Let a function f be analytic in a region D except at a finite number of points where it has simple poles, and let Γ_1 and Γ_2 be any two contours in D both of which contain the same poles. Then contour Γ_2 can be considered to be a deformation of contour Γ_1. The purpose of the project is to use computer algebra to verify, in particular cases, that the integral around each of these contours is the same.

1. Let contour Γ_1 be the circle $|z - 1| = 4$ and contour Γ_2 be the circle $|z - 2 - i| = 3$. Parametrize the contours, and in each case find dz on the contour.

2. Given that $f(z) = (3z - 2)/(z^2 - 5z + 6)$, verify that the poles of $f(z)$ lie inside both Γ_1 and Γ_2. Use the results of 1 with computer algebra to find $\int_{\Gamma_1} f(z)\,dz$ and $\int_{\Gamma_2} f(z)\,dz$, and hence show that they are equal.

3. Use analysis to find $\int_{\Gamma_1} f(z)\,dz$, and so confirm the results obtained in 2.

4. Parametrize the contour Γ_3 given by $|z - i| = 5$, and by using computer algebra to integrate around it in the clockwise sense, show that

$$\int_{\Gamma_1} f(z)\,dz = -\int_{\Gamma_3} f(z)\,dz.$$

Project 4

The Cauchy Integral Formula for Derivatives

The purpose of this project is to use computer algebra to verify the Cauchy integral formula for derivatives.

1. Parametrize the contour Γ formed by the circle $|z| = 2$ and find dz on Γ.

2. Given that $f(z) = z^2 + 3z - 7$, use computer algebra with the Cauchy integral formula to find $f(1)$.

3. Given that $f(z) = e^z(z^3 + 2z - 1)$, use computer algebra with the Cauchy integral formula for derivatives to find $f^{(2)}(1)$, and check the result by differentiation.

Projects 5–7

The Number of Zeros of a Polynomial in Each Quadrant of the z-Plane

Let a polynomial $P(z)$ be nonvanishing on a simple closed contour Γ, and let the total number of zeros of $P(z)$ inside Γ be N when multiplicity is counted, so that if a zero $z = a$ is repeated m times it has multiplicity m. Then

$$\frac{1}{2\pi i} \int_\Gamma \frac{P'(z)}{P(z)}\,dz = N.$$

The proof is simple, because if $P(z)$ has a zero of multiplicity m at $z = a$ inside Γ, it follows that $P(z)$ can be written $P(z) = (z - a)^m h(z)$, where $h(a) \neq 0$. Thus,

$$\frac{P'(z)}{P(z)} = \frac{m}{(z - a)} + \frac{h'(z)}{h(z)},$$

and as $P(z) \neq 0$ on Γ the expression on the right remains finite on Γ, so integrating around Γ gives

$$\int_\Gamma \frac{P'(z)}{P(z)}\,dz = 2\pi i m.$$

The result now follows by applying the preceding argument to each zero inside Γ and summing the multiplicities of the zeros to obtain N.

The purpose of Projects 5 through 7 is to use the foregoing result to find the number of zeros of the given polynomial that lies in each quadrant. To accomplish this, suitable finite size contours should be chosen and, where appropriate, use should be made of the properties of the zeros of polynomials contained in Theorem 1.2.

Project 5

$P(z) = z^5 + 3z + 18.$

Project 6

$P(z) = z^4 + 2z + 6.$

Project 7

$P(z) = z^5 + 4iz + 3i.$

Project 8

Identifying Regions Where a Polynomial Has No Zeros

The location of the zeros of polynomials is important in many problems: for example, in linear differential equations, where the solution will only be stable if no zeros lie to the right of the imaginary axis. The purpose of this project is to apply a theorem (see reference [6.2], Theorem 6.4b) that identifies a disc about a point z_0, which is not a zero of a given polynomial, inside and on which the polynomial has no zeros. This means that the reciprocal of the polynomial is an analytic function inside and on the boundary of the disc. The result is then to be verified numerically by integrating the reciprocal of the polynomial around the boundary of the disc and appealing to the Cauchy±Goursat theorem that asserts the result must be zero.

Let the polynomial

$$P(z) = z^n + a_{n-1}z^{n-1} + \cdots + a_1 z + a_0$$

have real or complex coefficients, and let z_0 be any complex number that is not a zero of $P(z)$. Define the numbers $b_0, b_1, \ldots, b_n$ by

$$b_m = \frac{1}{m!} P^{(m)}(z_0), \quad b_0 = P(z_0) \neq 0,$$

where $P^{(m)}(z) = d^m P(z)/dz^m$ and $P^{(0)}(z_0) = P(z_0)$. Then if

$$\rho(z_0) = \frac{1}{2} \min_{1 \leq m \leq n} \left| \frac{b_0}{b_m} \right|^{1/m},$$

the polynomial $P(z)$ has no zeros inside or on the disc $|z - z_0| \leq \rho(z_0)$.

Given $P(z) = z^4 + (1 + i)z^3 + 2iz^2 + z + 2$, using a suitable value of z_0 apply the theorem to find a disc with boundary Γ inside and on which $P(z)$ has no zeros. Confirm this by using computer algebra to show numerically that, as expected from the Cauchy–Goursat theorem, $\int_\Gamma (1/P(z))dz = 0$.

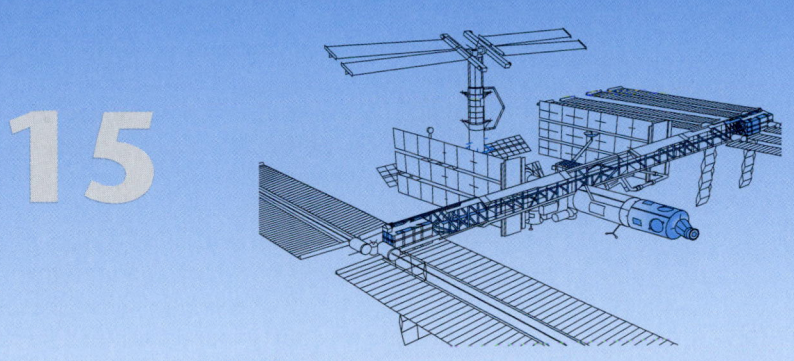

CHAPTER 15

Laurent Series, Residues, and Contour Integration

The analytical evaluation of a general contour integral with integrand $f(z)$ depends for its success on what are called the residues at the poles of $f(z)$. The residue of a function $f(z)$ at a pole is defined in terms of a special series expansion of $f(z)$ about the pole called a Laurent series. The Laurent series represents an extension of the conventional Taylor series that is no longer applicable when an expansion of $f(z)$ is required about a singular point. Various ways of obtaining Laurent series are described, and it is shown how a contour integral is related to the residues of the integrand $f(z)$ that lie either inside or on the contour of integration. Different types of contour integral are evaluated and integration around a branch point of $f(z)$ is considered.

15.1 Complex Power Series and Taylor Series

Before introducing complex power series and discussing their convergence, it is necessary to recall the definition of a sequence. A **sequence** of real or complex numbers, or of functions, is a set of such objects arranged in a specific order, so that changing the order changes the sequence. It is conventional to enclose the terms of a sequence in brackets by writing $\{\ldots\}$. Typical examples of sequences are

$$\left\{1, \frac{1}{2!}, \frac{1}{3!}, \frac{1}{4!}, \frac{1}{5!}, \frac{1}{6!}\right\}, \quad \text{a \textit{finite} sequence of real numbers}$$

$$\left\{\frac{1}{(1+i)}, \frac{1}{(1+i)^2}, \frac{1}{(1+i)^3}, \ldots, \frac{1}{(1+i)^n}, \ldots\right\}, \quad \text{an \textit{infinite} sequence of complex numbers}$$

$$\left\{z, -\frac{z^3}{3!}, \frac{z^5}{5!}, -\frac{z^7}{7!}, \frac{z^9}{9!}, \ldots, (-1)^{n+1}\frac{z^{2n-1}}{(2n-1)!}, \ldots\right\}, \quad \text{an \textit{infinite} sequence of powers of } z.$$

When working with sequences the expressions *finite* and *infinite* are used to describe to the number of terms in a sequence, and not the magnitude of any of its terms. In what follows our main concern will be with infinite sequences.

791

As the terms of a sequence occur in a specific order, they can be numbered sequentially like $u_1, u_2, u_3, \ldots$, with the suffix indicating the position of a term in the sequence. Because of this a sequence can be considered to be a function f that assigns to each positive integer n the term $u_n = f(n)$, where u_n is called the **general term** of the sequence. A convenient abbreviated notation for a sequence $\{u_1, u_2, u_3, \ldots\}$ is $\{u_n\}_{n=1}^{\infty}$ or, equivalently, $\{f(n)\}_{n=1}^{\infty}$. In an infinite sequence the behavior of the general term u_n as $n \to \infty$ is its most important property, so when numbering the terms it is usually immaterial whether the suffix of the first term is 0 or 1, so the notation for an *infinite* sequence is often simplified to $\{u_n\}$.

To illustrate how the general term of a sequence can be defined in terms of a function with a positive integer argument, we consider the function

$$f(z) = \frac{1}{3^z}\sin\left\{(2z-1)\frac{\pi}{2}\right\}.$$

Setting $z = n$ and $u_n = f(n)$, with $n = 1, 2, \ldots$, we obtain

$$u_n = (-1)^{n-1}\frac{1}{3^n},$$

so the infinite sequence with u_n as its general term becomes

$$\{u_n\}_{n=1}^{\infty} = \left\{\frac{1}{3}, -\frac{1}{3^2}, \frac{1}{3^3}, \ldots\right\} \text{ or, more simply, } \left\{(-1)^{n-1}\frac{1}{3^n}\right\}.$$

sequences, series, and nth partial sum

To understand the connection between infinite sequences and infinite series, let $s_n = u_1 + u_2 + \cdots + u_n$ be the sum of the first n terms of the infinite series $\sum_{n=1}^{\infty} u_n = u_1 + u_2 + u_3 + \cdots$. Then the sum of the series will be determined by the behavior of s_n as $n \to \infty$. The sum s_n is called the **nth partial sum** of the series, and when the terms of the series involve powers of the complex number z the nth partial sum will become a function of z, written $s_n(z)$. For any fixed z and n the function $s_n(z)$ will have a finite value. An infinite series $S(z)$ with the nth partial sum $s_n(z)$ will be said to **converge** to the value L when $z = z_0$ if, as $n \to \infty$, $\lim_{n\to\infty} s_n(z) = L$. If for some z_0 this limit is not defined, or if it is infinite, the series will be said to be **divergent**, or to **diverge** when $z = z_0$. Determining the **convergence** of an infinite power series involves finding the region in the z-plane where $\lim_{n\to\infty} s_n(z)$ is finite.

The tests for convergence that will be introduced later are applicable to the most commonly occurring types of series involving powers of z, and although they determine the *region* in the z-plane where the series converges, they do *not* determine the sum of the series.

A complex sequence $\{u_n\}$ is said to be **bounded** if some positive constant M exists such that $|u_n| < M$ for all positive integers n, and if this condition is not satisfied the sequence is said to be **unbounded**.

These ideas can be illustrated by considering the complex sequence $\{\frac{1}{6} + (-1)^n(\frac{n}{n^2+1})i\}$ that is seen to be bounded by 1 (not the sharpest bound), because the modulus of every term is less than 1. A simple example of an unbounded complex sequence is $\{ni^n\}$.

convergence, divergence, cluster points, and neighborhoods

A point α is called a **cluster point**, or a **point of accumulation**, of a sequence $\{u_n\}$ if every circle with its center at α, from which the point α itself has been deleted, contains infinitely many points of the sequence. The interior of a circle with its center at α is called a **neighborhood** of α, and a circle from which the single point α at its center has been removed is called a **deleted neighborhood** of α. A sequence

$\{u_n\}$ may have one or more cluster points, or possibly none, but when a cluster point α exists it is not necessarily a member of the sequence.

It is not difficult to see that the sequence $\{\frac{1}{6} + (-1)^n(\frac{n}{n^2+1})i\}$ only has a single cluster point at $\frac{1}{6}$, and that in this case no member of the sequence is equal to $\frac{1}{6}$. This means that however small a circle is drawn around the point $\frac{1}{6}$, infinitely many terms of the sequence will lie inside it and only a finite number will lie outside it, and no member of the sequence will lie at the center of the circle. Consequently, all but a finite number of terms of the sequence will be contained in *any* deleted neighborhood of the point $\frac{1}{6}$.

The most important type of sequence $\{u_n\}$ is one with only a single cluster point L, called the **limit** of the sequence and written

$$\lim_{n \to \infty} z_n = L.$$

A sequence with this property is said to **converge** to the limit L, and a sequence that does not converge is said to be **divergent**.

An example of a convergent infinite sequence is $\{\frac{1}{6} + (-1)^n(\frac{n}{n^2+1})i\}$, because this has a single cluster point at $\frac{1}{6}$, and so

$$\lim_{n \to \infty} \left\{ \frac{1}{6} + (-1)^n \left(\frac{n}{n^2+1} \right) i \right\} = \frac{1}{6}.$$

limits and convergence of complex series

When expressed in words, the definition of the limit of a sequence says that a sequence $\{u_n\}$ will have a limit L if, and only if, however small we take the radius of a circle with its center at L, there are infinitely many terms of the sequence inside the circle and only finitely many outside it. The limit L of a convergent sequence $\{u_n\}$ is illustrated in Fig. 15.1 where the deleted neighborhood of L is indicated by the interior of the circle centered on L with an arbitrarily small radius ε. Finitely many points of $\{u_n\}$ lie outside this circle and infinitely many lie inside it, and although in the limit as $n \to \infty$, $u_n \to L$, it is not necessary that L be a member of the sequence.

A more precise definition of the limit of a convergent complex sequence can be formulated as follows.

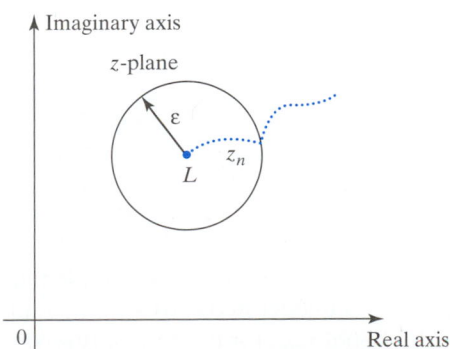

FIGURE 15.1   A convergent complex sequence $\{z_n\}$ with limit L.

A Convergent Sequence

A complex sequence $\{z_n\}$ will be said to **converge** to the **limit** L if for every arbitrarily small number $\varepsilon > 0$ a positive integer N can be found such that

$$|z_n - L| < \varepsilon \quad \text{for all } n > N.$$

As this definition of the limit of a convergent sequence applies to real and complex sequences, when $L = L_1 + iL_2$ is complex the definition implies that if $z_n = u_n + iv_n$, then

$$\lim_{n \to \infty} z_n = \lim_{n \to \infty} (u_n + iv_n) = \lim_{n \to \infty} u_n + i \lim_{n \to \infty} v_n = L_1 + iL_2,$$

and so

$$\lim_{n \to \infty} u_n = L_1 \quad \text{and} \quad \lim_{n \to \infty} v_n = L_2.$$

A formal proof of this result involves using the more precise definition of the limit of a convergent sequence given earlier, but as the proof is straightforward the details are left as an exercise.

EXAMPLE 15.1   Find any cluster points that belong to the following sequences and, where appropriate, find the limit of the sequence.

$$\text{(a)} \ \left\{ 1 + (-1)^n + \frac{1}{n!} \right\}, \quad \text{(b)} \ \{n\}, \quad \text{(c)} \ \left\{ \left(\frac{2n + 1}{n} \right) + i \left(\frac{3n^2 - 1}{n^2} \right) \right\}.$$

Solution

(a) As n increases, the first two terms combine to give either 0 or 2, according as n is odd or even, and the third term tends to zero as $n \to \infty$. Thus, as n increases, so terms of the sequence cluster ever closer around the numbers 0 and 2, showing that this sequence has two cluster points. Any small circle drawn around one of the cluster points that excludes the other will contain infinitely many points of the sequence, though infinitely many will remain outside it. This sequence is bounded but has no limit because it has more than one cluster point, and so it is divergent.

(b) It is clear by inspection that this sequence is unbounded and has no cluster points, so it is divergent.

(c) Setting

$$z_n = \left(\frac{2n + 1}{n} \right) + i \left(\frac{3n^2 - 1}{n^2} \right),$$

we see that

$$\lim_{n \to \infty} \left(\frac{2n + 1}{n} \right) = 2 \quad \text{and} \quad \lim_{n \to \infty} \left(\frac{3n^2 - 1}{n^2} \right) = 3,$$

so this sequence is bounded and only has the single cluster point $2 + 3i$. Thus, the sequence converges to the limit $2 + 3i$. This limit is *not* a member of the sequence, because for no finite n is it true that $z_n = 2 + 3i$. ∎

The foregoing definition of convergence makes use of the limit of the sequence, but this is not always easy to find, so it is desirable to have a test for convergence that does not involve the limit itself. This is made possible by introducing the concept of a Cauchy sequence. A sequence $\{z_n\}$ is called a **Cauchy sequence** if for any arbitrarily small number $\varepsilon > 0$ it is always possible to find an integer N, usually depending on ε, such that $|z_m - z_n| < \varepsilon$ for all $m > n > N$.

In effect, a Cauchy sequence is one with the property that, however small the number ε is chosen, it is always possible to find a large positive integer N such that the modulus of the difference between *any* two terms of the sequence with index greater than N will always be less than ε.

Although we omit the proof, it can be shown that a Cauchy sequence $\{z_n\}$ must converge to a limit. This result forms our next theorem.

THEOREM 15.1

Cauchy convergence principle for sequences   A sequence $\{z_n\}$ converges if, and only if, for any arbitrary small number $\varepsilon > 0$ it is possible to find an integer N depending on ε such that $|z_m - z_n| < \varepsilon$ for all $m > n > N$. ∎

EXAMPLE 15.2

Use Theorem 15.1 to prove the convergence of the sequence $\{(\cos n\pi)/n\}$.

Solution   Setting $z_n = (\cos n\pi)/n$ we have

$$|z_m - z_n| = \left| \frac{\cos m\pi}{m} - \frac{\cos n\pi}{n} \right| \leq \frac{n|\cos m\pi| + m|\cos n\pi|}{mn} = \frac{m+n}{mn}.$$

Now, if $m > n > N$, then

$$\frac{m+n}{mn} < \frac{2m}{mn} = \frac{2}{n} < \frac{2}{N},$$

so

$$|z_m - z_n| < \frac{2}{N}.$$

Consequently, for any arbitrary $\varepsilon > 0$, provided N is chosen such that $2/N < \varepsilon$, the conditions of Theorem 15.1 are satisfied and the sequence converges. In this case the convergence of the sequence to the limit 0 is obvious, because $\cos n\pi = (-1)^n$, so the general term of the sequence is simply $(-1)^n/n$. ∎

It has already been shown that the sum of an infinite series can be regarded as the limit of the operation of sequentially adding the terms of an infinite sequence. Consequently, if the sequence of partial sums has a limit L, this must be the limit of the infinite series formed in this manner. If the infinite series involves powers of z, and so is a **power series**, its convergence or divergence will depend on z. For a power series to be useful it will be necessary to determine the region in the z-plane where it converges.

The proofs of the following results for complex series closely parallel the corresponding results for real series, so the results will merely be stated.

THEOREM 15.2

Limit of complex series   Let $z_n = u_n + iv_n$, and denote the nth partial sum of the series $\sum_{n=1}^{\infty} z_n$ by

$$s_n = \sum_{m=1}^{n} u_m + i \sum_{m=1}^{n} v_m.$$

Then a necessary and sufficient condition for the series to converge is that the sequences $\{\sum_{m=1}^n u_m\}$ and $\{\sum_{m=1}^n v_m\}$ converge as $n \to \infty$. When this is true, if $\lim_{n\to\infty}\sum_{m=1}^n u_m = L_1$ and $\lim_{n\to\infty}\sum_{m=1}^n v_m = L_2$, then $\lim_{n\to\infty} s_n = L_1 + iL_2$. ∎

THEOREM 15.3

A necessary condition satisfied by convergent series   If the series $\sum_{n=1}^\infty z_n$ converges, then $\lim_{n\to\infty} z_n = 0$. ∎

The main use of this theorem is to establish the *divergence* of a series, because if $\lim_{n\to\infty} z_n \neq 0$ the series *cannot* converge. The theorem provides no information about convergence, because the condition $\lim_{n\to\infty} z_n = 0$ is *not* sufficient to ensure the convergence of a series. This is easily seen by considering the harmonic series $\sum_{n=1}^\infty \frac{1}{n}$, because setting $z_n = \frac{1}{n}$ we see that $\lim_{n\to\infty} z_n = \lim_{n\to\infty} \frac{1}{n} = 0$, but the series is known to diverge.

EXAMPLE 15.3

Show the series $\sum_{n=1}^\infty \frac{(n^2-2ni)}{3n+4}$ is divergent.

Solution   The general term is $z_n = \frac{n^2-2ni}{3n+4}$. However, $\lim_{n\to\infty} z_n = \frac{n}{3} - \frac{2i}{3} \neq 0$, so it follows from Theorem 15.3 that the series is divergent. ∎

Convergence Tests for Complex Series

The relationship that exists between sequences and series allows the Cauchy convergence principle for sequences to be reinterpreted for series in the following form.

THEOREM 15.4

Cauchy convergence principle for series   The infinite series $\sum_{n=1}^\infty z_n$ is convergent if, and only if, for every arbitrarily small number $\varepsilon > 0$ a positive integer N can be found depending on ε such that

$$|z_{n+1} + z_{n+2} + \cdots + z_{n+r}| < \varepsilon \quad \text{for every } n > N \text{ and } r = 1, 2, \ldots.$$ ∎

Expressed in words, this theorem says that if an infinite series is convergent, then, however small ε, it is always possible to find a positive integer N such that the modulus of the sum of *any* number of consecutive terms starting with index greater than N will be less than ε. If the series is written $z_1 + z_2 + \cdots + z_n + R_n$, where $R_n = z_{n+1} + z_{n+2} + z_{n+3} + \cdots = \sum_{m=n+1}^\infty z_m$ is called the **remainder** after n terms, the theorem asserts that $|R_N| < \varepsilon$. In practical terms this means that if the infinite series is approximated by the sum of the first N terms, the error involved cannot exceed ε.

A series $\sum_{n=1}^\infty z_n$ with the property that the sum of the moduli of z_n is also convergent is said to be **absolutely convergent**. Thus, the series $\sum_{n=1}^\infty z_n$ is *absolutely convergent* if the series

$$\sum_{n=1}^\infty |z_n| = |z_1| + |z_2| + \cdots$$

is convergent.

If, however, the series $\sum_{n=1}^{\infty} z_n$ is convergent but the series $\sum_{n=1}^{\infty} |z_n| = |z_1| + |z_2| + \cdots$ is divergent, the series $\sum_{n=1}^{\infty} z_n$ is said to be **conditionally convergent**. Absolute and conditional convergence are most easily illustrated by considering the real series $1 - \frac{1}{2} + \frac{1}{3} - \frac{1}{4} + \cdots + (-1)^{n+1}\frac{1}{n} + \cdots$. It is known from elementary calculus that the sum of this series is $\ln 2$, and so it is convergent. However, the sum of the absolute values is the harmonic series $1 + \frac{1}{2} + \frac{1}{3} + \frac{1}{4} + \cdots + \frac{1}{n} + \cdots$, which is known to be divergent, so the series $1 - \frac{1}{2} + \frac{1}{3} - \frac{1}{4} + \cdots + (-1)^{n+1}\frac{1}{n} + \cdots$ is conditionally convergent.

One direct consequence of Theorem 15.4 is that absolute convergence *implies* convergence. Another consequence of the theorem is the following result, which we state in the form of a theorem.

THEOREM 15.5

a simple comparison test for convergence

Comparison test for convergence   Let a series $\sum_{n=1}^{\infty} z_n = z_1 + z_2 + \cdots$ be given, and let the series $\sum_{n=1}^{\infty} b_n$ with nonnegative terms b_n be convergent and such that $|z_n| \le b_n$ for $n = 1, 2, \ldots$. Then the series $\sum_{n=1}^{\infty} z_n$ is absolutely convergent.

Proof   As the series $\sum_{n=1}^{\infty} b_n$ is convergent by hypothesis, for any $\varepsilon > 0$ there exists an integer N such that $b_{n+1} + b_{n+2} + \cdots + b_{n+r} < \varepsilon$ for all $n > N$ and $r = 1, 2, \ldots$. As $|z_n| < b_n$ for every n, it follows that

$$|z_{n+1}| + |z_{n+2}| + \cdots + |z_{n+r}| \le b_{n+1} + b_{n+2} + \cdots + b_{n+r} < \varepsilon,$$

so by Theorem 15.4 the series $\sum_{n=1}^{\infty} |z_n| = |z_1| + |z_2| + \cdots$ converges, showing the series $\sum_{n=1}^{\infty} z_n$ to be absolutely convergent.   ∎

Several tests for convergence use, for purposes of comparison, the infinite **geometric series**

$$\sum_{n=0}^{\infty} r^n = 1 + r + r^2 + \cdots,$$

which an elementary argument shows converges to the sum $1/(1-r)$ if $|r| < 1$, and diverges if $|r| \ge 1$.

Because the convergence of an infinite geometric series depends on the magnitude of $|r|$, convergence tests based on a comparison with the geometric series lead to tests for *absolute convergence*. When these tests are applied to real series with positive terms they become tests for *convergence*. The most important and useful of these tests are the ratio and nth root tests.

THEOREM 15.6

the ratio test for convergence

The ratio test   Let a series $\sum_{n=1}^{\infty} z_n = z_1 + z_2 + \cdots$, in which no term is zero, be such that

$$\lim_{n \to \infty} \left| \frac{z_{n+1}}{z_n} \right| = L.$$

Then the absolute convergence or divergence of the series is determined by the following conditions:

(i) If $L < 1$, the series converges absolutely

(ii) If $L > 1$, the series diverges

(iii) If $L = 1$, the test fails and no conclusion can be drawn about the convergence of the series.

Proof   Suppose that $|z_{n+1}/z_n| \leq \alpha < 1$ for n greater than some positive integer N. Then $|z_{n+1}| \leq |z_n|$ and we have

$$|z_{N+2}| \leq \alpha|z_{N+1}|, \quad |z_{N+3}| \leq \alpha|z_{N+2}| \leq \alpha^2|z_{N+1}|, \ldots,$$

leading to the general result $|z_{N+r}| \leq \alpha^{r-1}|z_{N+1}|$.

If R_N is the remainder of the series after N terms, this last result allows its modulus to be estimated by

$$|R_N| \leq |z_{N+1}| + |z_{N+2}| + |z_{N+3}| + \cdots \leq |z_{N+1}|(1 + \alpha + \alpha^2 + \alpha^3 + \cdots).$$

The bracketed geometric series converges when $\alpha < 1$, so as $|R_N|$ is bounded the series $\sum_{n=1}^{\infty} z_n$ is absolutely convergent. Conversely, the bracketed geometric series is divergent if $|\alpha| > 1$, showing that then the series $\sum_{n=1}^{\infty} z_n$ must be divergent. If $\alpha = 1$ the test fails in the sense that it provides no information about the convergence of the series. The statement of the theorem follows directly from these conclusions. ∎

It is important to recognize that the real constant α in the ratio test must be strictly less than 1. This is essential in order to exclude series such as the harmonic series that, although divergent, have a limiting ratio $|z_{n+1}/z_n|$ that approaches arbitrarily close to 1 as $n \to \infty$.

EXAMPLE 15.4   Apply the ratio test to the series

$$\text{(a) } \sum_{n=1}^{\infty}(-1)^{n+1}\frac{n!}{n^n}, \quad \text{(b) } \sum_{n=1}^{\infty}\frac{i^n}{(3n+2)^2}, \quad \text{and} \quad \text{(c) } \sum_{n=1}^{\infty}(-1)^{n+1}\frac{2^{n+1}}{n+2}.$$

Solution

(a) Setting $z_n = (-1)^{n+1}n!/n^n$ we find that

$$\left|\frac{z_{n+1}}{z_n}\right| = \frac{(n+1)!n^n}{(n+1)^n n!} = \left(1 + \frac{1}{n}\right)^{-n},$$

but from from Table 15.1 it is seen that $\lim_{n\to\infty}(1 + \frac{1}{n})^{-n} = 1/e$, so

$$\lim_{n\to\infty}\left|\frac{z_{n+1}}{z_n}\right| = \frac{1}{e} < 1.$$

Thus, as $L = 1/e < 1$, it follows from the ratio test that the series is absolutely convergent.

(b) Setting $z_n = i^n/(3n+2)^2$ we find that

$$\left|\frac{z_{n+1}}{z_n}\right| = \left(\frac{3n+2}{3n+5}\right)^2,$$

so

$$\lim_{n\to\infty}\left|\frac{z_{n+1}}{z_n}\right| = \lim_{n\to\infty}\left(\frac{3n+2}{3n+5}\right)^2 = 1.$$

In this case the limit $L = 1$, so the ratio test fails. In fact, the series is absolutely convergent, as may be seen by comparison with the convergent series $\sum_{n=1}^{\infty}1/n^2$ given in Table 15.1.

TABLE 15.1   Some Useful Comparison Series and Limits

1. $\displaystyle\sum_{n=0}^{\infty} \frac{1}{n!} = 1 + \frac{1}{1!} + \frac{1}{2!} + \frac{1}{3!} \cdots + \frac{1}{n!} + \cdots = e$ (convergent)

2. $\displaystyle\sum_{n=0}^{\infty} (-1)^n \frac{1}{n!} = 1 - \frac{1}{1!} + \frac{1}{2!} - \frac{1}{3!} + \cdots + (-1)^n \frac{1}{n!} + \cdots = 1/e$ (absolutely convergent)

3. $\displaystyle\sum_{n=1}^{\infty} (-1)^{n+1} \frac{1}{n} = 1 - \frac{1}{2} + \frac{1}{3} - \cdots + (-1)^{n+1} \frac{1}{n} + \cdots = \ln 2$ (conditionally convergent)

4. $\displaystyle\sum_{n=1}^{\infty} \frac{1}{n} = 1 + \frac{1}{2} + \frac{1}{3} + \cdots + \frac{1}{n} + \cdots$ (divergent; this is the *harmonic* series)

5. $\displaystyle\sum_{n=0}^{\infty} \alpha^n = 1 + \alpha + \alpha^2 + \alpha^3 + \cdots + \alpha^n + \cdots = \frac{1}{1 - \alpha}$ (convergent for $|\alpha| < 1$ and divergent for $|\alpha| \geq 1$; this is the *geometric series*)

6. $\displaystyle\sum_{n=1}^{\infty} \frac{1}{n^2} = 1 + \frac{1}{2^2} + \frac{1}{3^2} + \cdots + \frac{1}{n^2} + \cdots = \frac{\pi^2}{6}$ (convergent)

7. $\displaystyle\sum_{n=1}^{\infty} (-1)^{n+1} \frac{1}{n^2} = 1 - \frac{1}{2^2} + \frac{1}{3^2} - \cdots + (-1)^{n+1} \frac{1}{n^2} + \cdots = \frac{\pi^2}{12}$ (absolutely convergent)

8. $\displaystyle\sum_{n=1}^{\infty} \frac{1}{n^\alpha} = 1 + \frac{1}{2^\alpha} + \frac{1}{3^\alpha} + \cdots + \frac{1}{n^\alpha} + \cdots$ (convergent if $\alpha > 1$ and divergent if $0 < \alpha \leq 1$; this is the *harmonic series* of order α)

9. $\displaystyle\lim_{n\to\infty} \left(1 + \frac{\alpha}{n}\right)^n = e^\alpha$

10. $\displaystyle\lim_{n\to\infty} \sqrt[n]{n} = 1$

11. $\displaystyle\lim_{n\to\infty} \frac{n!}{n^n} = 0$

12. $\displaystyle\lim_{n\to\infty} \frac{n^n}{n!} = \infty$

(c) Setting $z_n = (-1)^{n+1} \frac{2^{n+1}}{n+2}$, we have

$$\left| \frac{z_{n+1}}{z_n} \right| = 2 \left(\frac{n+2}{n+3} \right),$$

so

$$\lim_{n\to\infty} \left| \frac{z_{n+1}}{z_n} \right| = 2 \lim_{n\to\infty} \left(\frac{n+2}{n+3} \right) = 2,$$

showing that $L = 2$, but as $L > 1$ the ratio test shows this series to be divergent.

■

The nth root test can be established in a manner similar to that of the ratio test, so the details of its proof will be omitted.

THEOREM 15.7

the nth root test for convergence

The nth root test for convergence   Let a series $\sum_{n=1}^{\infty} z_n = z_1 + z_2 + \cdots$, in which no term is zero, be such that

$$\lim_{n\to\infty} \sqrt[n]{z_n} = L.$$

Then the absolute convergence and divergence of the series is determined by the following conditions:

(i) If $L < 1$, the series converges absolutely

(ii) If $L > 1$, the series diverges

(iii) If $L = 1$ the test fails, and no conclusion can be drawn about the convergence of the series. ∎

EXAMPLE 15.5

Find conditions on the real constant α in order that the series

$$\sum_{n=1}^{\infty} \left(\frac{\alpha n}{\alpha n + 1} \right)^{n^2} i^n$$

is absolutely convergent.

Solution   Setting $z_n = \left(\frac{\alpha n}{\alpha n+1}\right)^{n^2} i^n$ we have

$$\sqrt[n]{\left| \left(\frac{\alpha n}{\alpha n + 1} \right)^{n^2} i^n \right|} = \left| \frac{\alpha n}{\alpha n + 1} \right|^n = 1 \Big/ \left(1 + \left(\frac{1}{\alpha} \right) \left(\frac{1}{n} \right) \right)^n,$$

and making use of a limit in Table 15.1 we see that

$$L = \lim_{n \to \infty} \sqrt[n]{|z_n|} = 1/e^{1/\alpha} = e^{-1/\alpha}.$$

As $L < 1$ if $\alpha > 0$ and $L > 1$ if $\alpha < 0$, the nth root test shows that the series is absolutely convergent if $\alpha > 0$ and divergent if $\alpha < 0$. ∎

Complex Power Series and Circles of Convergence

A series of the form

$$\sum_{n=0}^{\infty} a_n (z - z_0)^n = a_0 + a_1(z - z_0) + a_2(z - z_0)^2 + \cdots + a_n(z - z_0)^n + \cdots,$$

(1)

complex power series

in which the a_n, z, and z_0 are complex, is called a **complex series** in powers of $z - z_0$, or simply a **complex power series**, expanded about the point z_0. In complex power series the complex number z_0 is often called the **center** of the series. The convergence of such series depends on the **coefficients** of the series, that is, the numbers a_n, on the complex variable z, and on the point z_0 about which the series is expanded. To determine the conditions to be imposed on a_n, z, and z_0 in order to ensure convergence, we apply either the ratio test or the nth root test to the nth term $a_n(z - z_0)^n$ of the complex power series in (1).

An application of the ratio test shows that the series will be convergent if

$$L = \lim_{n \to \infty} \left| \frac{a_{n+1}(z - z_0)^{n+1}}{a_n(z - z_0)^n} \right| = \lim_{n \to \infty} \left| \frac{a_{n+1}}{a_n} \right| |z - z_0| < 1,$$

and this is equivalent to the condition

$$|z - z_0| < \lim_{n \to \infty} \left| \frac{a_n}{a_{n+1}} \right| = R, \tag{2}$$

where the number

$$R = \lim_{n \to \infty} \left| \frac{a_n}{a_{n+1}} \right| \tag{3}$$

is called the **radius of convergence** of the complex power series in (1). In terms of R the condition for absolute convergence in (2) becomes

$$|z - z_0| < R, \tag{4}$$

showing that the series is absolutely convergent for all z inside a circle of radius R with its center at the point z_0.

A similar argument applied to the complex power series in (1), but this time using the nth root test, gives

$$L = \lim_{n \to \infty} \sqrt[n]{|a_n(z - z_0)^n|} = \lim_{n \to \infty} \sqrt[n]{|a_n|}|z - z_0| < 1,$$

showing that the series will be absolutely convergent if

$$|z - z_0| < R, \quad \text{where } R = 1/\lim_{n \to \infty} \sqrt[n]{|a_n|}. \tag{5}$$

radius and circle of convergence

Summarizing these results, we see that the **radius of convergence** R of the power series in (1) and its associated **circle of convergence**, that is, the circle $|z - z_0| < R$, can be found either from

$$R = \lim_{n \to \infty} \left| \frac{a_n}{a_{n+1}} \right|, \quad \text{with } |z - z_0| < R, \tag{6}$$

or from

$$R = 1/\lim_{n \to \infty} \sqrt[n]{|a_n|}, \quad \text{with } |z - z_0| < R. \tag{7}$$

The choice of which one of these results to use in practice is determined by whichever limit is the simpler to evaluate.

EXAMPLE 15.6

Find the radius and circle of convergence of the power series

$$\text{(a) } \sum_{n=1}^{\infty} \frac{(z - i)^n}{n} \quad \text{and} \quad \text{(b) } \sum_{n=1}^{\infty} \frac{n(5 + 2i)^n}{3^n}(z - 1)^n.$$

Solution

(a) In this case result (6) is simpler to use, so setting $a_n = 1/n$ and $z_0 = i$ gives

$$R = \lim_{n \to \infty} \left| \frac{n + 1}{n} \right| = 1.$$

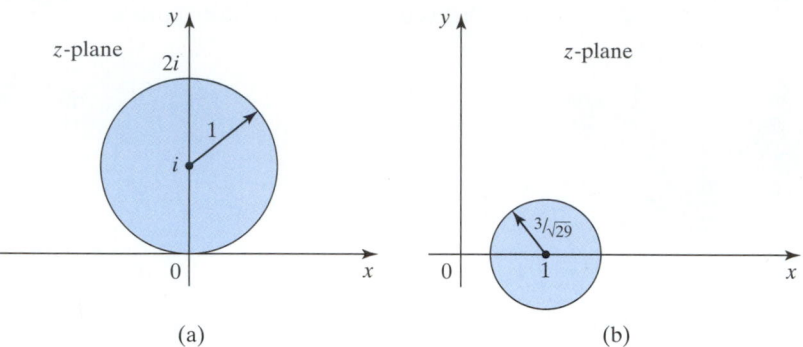

FIGURE 15.2   Circles of convergence.

So the radius of convergence is $R = 1$ and the circle of convergence is $|z - i| < 1$. This is illustrated in Fig. 15.2a.

(b) Here result (7) is simpler to use, so setting $a_n = \frac{n(5+2i)^n}{3^n}$ and $z_0 = 1$ gives

$$R = 1/\lim_{n\to\infty} \sqrt[n]{\left|\frac{n(5+2i)^n}{3^n}\right|} = \frac{3}{|5+2i|} \lim_{n\to\infty}(1/\sqrt[n]{n}) = \frac{3}{\sqrt{29}},$$

where when determining the limit use has been made of entry 10 in Table 15.1. This series converges in a circle of radius $R = 3/\sqrt{29}$ with its center at the point $z_0 = 1$, as shown in Fig. 15.2b.

Power series define functions, so it is necessary to know if they possess the property of continuity, and whether they can be added, multiplied, differentiated, and integrated. Furthermore, as each partial sum $s_n(z)$ of a power series is a polynomial in z, and so is an analytic function, it is necessary to know if the power series itself is also an analytic function. The answer to each of these questions is in the affirmative, and they form the substance of the next theorem.   ■

THEOREM 15.8

important properties
of complex power
series

Properties of power series   Power series with finite circles of convergence possesses the following properties:

(i) A power series represents a continuous function at each point inside its circle of convergence.

(ii) If two power series expanded about the same point have the same circle of convergence D and the same sum at each point of D, then they are identical.

(iii) If two power series with sums $f(z)$ and $g(z)$ and circles of convergence D_1 and D_2 are added or subtracted term by term, the result is a power series that converges to the sum $f(z) \pm g(z)$ with a circle of convergence that is at least equal to the largest circle that can be drawn in the region common to D_1 and D_2.

(iv) If two power series with sums $f(z)$ and $g(z)$ and circles of convergence D_1 and D_2 are multiplied, the result is a power series that converges to the product $f(z)g(z)$ with a circle of convergence that is at least equal to the largest circle that can be drawn in the region common to D_1 and D_2.

(v) If a power series with the sum $f(z)$ and a circle of convergence D are differentiated term by term, the result is a power series that converges to $f'(z)$ at each point in D.

(vi) If a power series with the sum $f(z)$ and a circle of convergence D is integrated term by term, the result is a sum that converges to $\int f(z)dz$ at each point in D.

(vii) A power series with a circle of convergence D is an analytic function in D.

Proof Only results with proofs that are straightforward will be outlined in order to avoid introducing unnecessary complication.

(i) It will be sufficient to prove that a power series $f(z) = \sum_{n=0}^{\infty} a_n z^n$ with a nonzero radius of convergence R and circle of convergence Γ represents a continuous function of z at every point inside Γ. This is because if the power series is expanded about a point z_0 instead of the origin, the change of variable $w = z - z_0$ will reduce it to this case. Continuity will be proved if we can show that for any point ζ inside Γ and for any given $\varepsilon > 0$, it follows that $|f(z) - f(\zeta)| < \varepsilon$ for all z inside Γ such that $|z - \zeta| < \delta$.

Set $f(z) = S_N(z) + R_N(z)$, where $S_N(z) = \sum_{n=0}^{N} a_n z^n$ and the remainder $R_N(z) = \sum_{n=N}^{\infty} a_n z^n$. Let D be the interior and boundary of any circle C with its center at the origin and its radius $r < R$. Then to proceed further it is necessary to anticipate the result of Theorem 15.11 by using the uniform convergence of the power series in C to guarantee the existence of a positive integer $N = N(\varepsilon)$ such that $|f(z) - S_N(z)| < \frac{1}{3}\varepsilon$ for all z in D. The series $S_N(z)$ is simply a polynomial in z, so it follows that it must be a continuous function of z. Consequently, with this value of N, there must be a $\delta > 0$ such that $|S_N(z) - S_N(\zeta)| < \frac{1}{3}\varepsilon$ when $|z - \zeta| < \delta$.

Then, for all z in D such that $|z - \zeta| < \delta$, we can write

$$
\begin{aligned}
|f(z) - f(\zeta)| &= |f(z) - S_N(z) + S_N(z) - S_N(\zeta) + S_N(\zeta) - f(\zeta)| \\
&\le |f(z) - S_N(z)| + |S_N(z) - S_N(\zeta)| + |S_N(\zeta) - f(\zeta)| \\
&< \tfrac{1}{3}\varepsilon + \tfrac{1}{3}\varepsilon + \tfrac{1}{3}\varepsilon = \varepsilon,
\end{aligned}
$$

so the continuity of the power series at all points of D has been established. The statement of the theorem now follows because C is any circle with its center at the origin with $r < R$.

(ii) As with (i), it will be sufficient to consider the two power series expanded about the origin $\sum_{n=0}^{\infty} a_n z^n$ and $\sum_{n=0}^{\infty} b_n z^n$, each with the same circle of convergence D throughout which each converges to the same sum. Then for all z in D we have, by hypothesis,

$$
a_0 + a_1 z + a_2 z^2 + a_3 z^3 + \cdots = b_0 + b_1 z + b_2 z^2 + b_3 z^3 + \cdots .
$$

By (i) the sums are continuous at $z = 0$, so $a_0 = b_0$. Cancelling these terms and removing a factor z, we arrive at the result

$$
a_1 + a_2 z + a_3 z^2 + \cdots = b_1 + b_2 z + b_3 z^2 + \cdots ,
$$

and a repetition of the argument shows that $a_1 = b_1$. Continuing this process by induction, we conclude that $a_n = b_n$ for $n = 0, 1, 2, \ldots$, so the uniqueness of the power series has been proved.

(iii) The result follows by adding or subtracting the two nth partial sums and proceeding to the limit as $n \to \infty$.

(iv) Though not difficult, the proof of this result is lengthy and so will be omitted.

(v) Let the circle of convergence of $f(z)$ be D. The convergence of the differentiated series to $f'(z)$, and the demonstration that the differentiated series has the same circle of convergence D, follows by using term-by-term differentiation and applying the ratio test to the result.

(vi) Let the circle of convergence of $f(z)$ be D. Then the convergence of the integrated series to $\int f(z)dz$, and the demonstration that the integrated series has the same circle of convergence D, follows by using term-by-term integration and applying the ratio test to the result.

(vii) The details of the proof of this result are complicated and so will be omitted. ∎

Complex power series arise in many different ways, the most frequent of which is in the form of Taylor series expansions of functions. The **Taylor series** expansion of an analytic function f about the point z_0 takes the same form as the Taylor series for a function of a real variable, though the derivation of the result is different.

THEOREM 15.9

the complex form of Taylor's series

Taylor's theorem Let $f(z)$ be an analytic function of z at the point z_0, and let it also be analytic inside a circle C given by $|z - z_0| = r$ that forms a neighborhood of z_0. Then there exists a power series

$$\sum_{n=0}^{\infty} a_n(z - z_0)^n$$

with coefficients a_n determined by the formula

$$a_n = \frac{f^{(n)}(z_0)}{n!} \quad \text{for } n = 0, 1, 2, \ldots,$$

which converges to $f(\zeta)$ for every ζ inside the circle C and is such that

$$f(\zeta) = \sum_{n=0}^{\infty} \frac{f^{(n)}(z_0)}{n!}(\zeta - z_0)^n.$$

Proof Without loss of generality the result will be proved for $z_0 = 0$, because a change of origin extends the result to the case $z_0 \neq 0$. The proof is based on the Cauchy integral formula for derivatives and makes use of the identity

$$\frac{f(z)}{z - \zeta} = \frac{f(z)}{z} + \zeta \frac{f(z)}{z^2} + \cdots + \zeta^{n-1} \frac{f(z)}{z^n} + \zeta^n \frac{f(z)}{(z - \zeta)z^n},$$

which is easily verified for $z \neq 0$ and $z \neq \zeta$.

As $z_0 = 0$, the circle C in the theorem becomes the circle $|z| = r$. If we multiply the preceding identity by $1/(2\pi i)$ and integrate around any positively oriented circle Γ inside C with its center at the origin and radius $\rho (0 < \rho < r)$, it follows from the analytic nature of $f(\zeta)$ and the Cauchy integral formula for derivatives that

$$f(\zeta) = f(0) + \zeta f'(0) + \zeta^2 \frac{f''(0)}{2!} + \cdots + \zeta^{n-1} \frac{f^{(n-1)}(0)}{(n-1)!}$$
$$+ \frac{\zeta^n}{2\pi i} \int_{\Gamma} \frac{f(z)}{(z - \zeta)z^n} dz.$$

This is **Taylor's theorem with a remainder**, where the last term is the remainder R_n after n terms. The proof will be complete once we have shown that $R_n \to 0$ as $n \to \infty$. From the maximum modulus theorem we know that a number $M > 0$ can be found such that $|f(z)| < M$ for all z inside the circle Γ, so on Γ

$$\left| \frac{\zeta^n f(z)}{(z - \zeta)z^n} \right| \leq \frac{M}{|z - \zeta|} \left| \frac{\zeta}{z} \right|^n.$$

Using this result in R_n leads to the estimate

$$|R_n| = \left| \frac{1}{2\pi i} \int_\Gamma \frac{\zeta^n f(z)}{(z - \zeta)z^n} dx \right| \leq \frac{1}{2\pi} \frac{M}{|z - \zeta|} \left| \frac{\zeta}{z} \right|^n 2\pi\rho = \frac{M\rho}{|z - \zeta|} \left| \frac{\zeta}{z} \right|^n.$$

Now as z lies on Γ and ζ is inside Γ, it follows that $|\zeta/z| < 1$, and so $|\zeta/z|^n \to 0$ as $n \to \infty$. The result $|z| = \rho$ allows the elementary inequality $|z - \zeta| \geq ||z| - |\zeta||$ to be written as $|z - \zeta| \geq |\rho - |\zeta||$, so that the expression for $|R_n|$ becomes

$$|R_n| \leq \frac{M\rho}{|\rho - |\zeta||} \left| \frac{\zeta}{z} \right|^n.$$

Finally, as $|\zeta|$ is a constant and $|\zeta/z|^n \to 0$ as $n \to \infty$, proceeding to the limit as $n \to \infty$ shows that $\lim_{n\to\infty} |R_n| = 0$, and hence that $\lim_{n\to\infty} R_n = 0$. Thus, we have proved that

$$f(\zeta) = \sum_{n=0}^{\infty} \frac{f^{(n)}(0)}{n!} \zeta^n$$

at all points inside Γ. As $0 < \rho < r$, this result is also true for all points inside C. The proof is complete. ∎

BROOKE TAYLOR (1685–1731)

An English mathematician educated at Cambridge and whose interests extended beyond mathematics to religion and philosophy. He was responsible for the introduction into mathematics of the method of finite differences in a work published between 1715 and 1717, that also contained what is now known as "Taylor's Theorem." Taylor did not consider the convergence of his series and it was not until a century later that Cauchy provided a satisfactory convergence proof. Taylor obtained a series solution for an initial value problem for a differential equation by repeatedly differentiating the equation to find the coefficients to substitute into his series solution.

The complex power series

$$f(\zeta) = \sum_{n=0}^{\infty} \frac{f^{(n)}(z_0)}{n!} (\zeta - z_0)^n \qquad (8)$$

is called the **Taylor series** of the analytic function $f(\zeta)$ expanded about the **point** (or **center**) z_0, and when $z_0 = 0$ this becomes the **Maclaurin series** expansion of $f(\zeta)$.

The derivation of Taylor's theorem shows that the radius of convergence of the Taylor series of a function with its center at z_0 will be the radius of the largest circle centered on z_0 inside which the function is analytic.

| EXAMPLE 15.7 | Find the Taylor series expansion of $f(z) = \cos z$ with its center at $z_0 = c$, and hence deduce its Maclaurin series expansion. |

Solution The cosine function is an entire function and so can be expanded as a power series about any center, so the resulting series will have an arbitrarily large radius of convergence.

Routine differentiation gives

$$\frac{d[\cos z]}{dz} = -\sin z, \ \frac{d^2[\cos z]}{dz^2} = -\cos z, \ \frac{d^3[\cos z]}{dz^3} = \sin z, \ \frac{d^4[\cos z]}{dz^4} = \cos z, \ldots,$$

so substituting these results in the Taylor series (8), setting $z_0 = c$, and replacing ζ by z shows the required Taylor series to be

$$\cos z = \cos c - \frac{\sin c}{1!}(z - c) - \frac{\cos c}{2!}(z - c)^2 + \frac{\sin c}{3!}(z - c)^3 + \frac{\cos c}{4!}(z - c)^4 - \cdots.$$

The cosine function is an entire function, so this series converges for all z.

The Maclaurin series for $\cos z$ is obtained from this by setting $c = 0$, when we find that

$$\cos z = \sum_{n=0}^{\infty}(-1)^n\frac{z^{2n}}{(2n)!} = 1 - \frac{z^2}{2!} + \frac{z^4}{4!} - \cdots. \qquad \blacksquare$$

It is seen from this example that the complex Maclaurin series for $\cos z$ can be obtained from the corresponding series involving the real variable x by simply replacing x by z. This result remains true in general for Taylor series of elementary functions of a real variable. Some useful results that can be obtained in this manner are listed next. Here, for completeness, the expansion of $\cos z$ has been included:

$$e^z = \sum_{n=0}^{\infty}\frac{z^n}{n!} = 1 + z + \frac{z^2}{2!} + \cdots, \qquad |z| < \infty \qquad (9)$$

$$\sin z = \sum_{n=0}^{\infty}(-1)^n\frac{z^{2n+1}}{(2n+1)!} = z - \frac{z^3}{3!} + \frac{z^5}{5!} - \cdots, \qquad |z| < \infty \qquad (10)$$

$$\cos z = \sum_{n=0}^{\infty}(-1)^n\frac{z^{2n}}{(2n)!} = 1 - \frac{z^2}{2!} + \frac{z^4}{4!} - \cdots, \qquad |z| < \infty \qquad (11)$$

$$\mathrm{Log}(1 + z) = \sum_{n=1}^{\infty}(-1)^{n+1}\frac{z^n}{n} = z - \frac{z^2}{2} + \frac{z^3}{3} - \cdots, \quad |z| < 1 \qquad (12)$$

$$\sinh z = \sum_{n=0}^{\infty}\frac{z^{2n+1}}{(2n+1)!} = z + \frac{z^3}{3!} + \frac{z^5}{5!} + \cdots, \qquad |z| < \infty \qquad (13)$$

$$\cosh z = \sum_{n=0}^{\infty}\frac{z^{2n}}{(2n)!} = 1 + \frac{z^2}{2!} + \frac{z^4}{4!} + \cdots, \qquad |z| < \infty. \qquad (14)$$

Alternative Ways of Obtaining Power Series Expansions

other ways of finding Taylor series expansions

A Taylor series is a power series, so it follows from the uniqueness of power series in Theorem 15.8 (ii) that however a power series expansion of a function $f(z)$ about a point z_0 is obtained, it must be the Taylor series expansion of the function about the same point. This property of power series is of considerable practical importance, because it is often easier to obtain a power series expansion of a function by methods that do not require the repeated differentiations needed to find the coefficients of a Taylor series. Typical ways in which power series expansions of functions can be obtained are by substitution into known simpler series, by multiplication of series, by use of the binomial theorem, or by differentiation or integration of known simpler series. Some representative examples of these ways are given below.

Expansion by the binomial theorem and a substitution

EXAMPLE 15.8

Find the Taylor series expansion of $f(z) = (8 + z)^{-1/2}$ about the point $z_0 = 1$.

Solution   To introduce powers of $(z - 1)$ into the expansion we write $f(z) = (8 + z)^{-1/2}$ as

$$f(z) = \frac{1}{3} \frac{1}{\left[1 + \frac{1}{9}(z - 1)\right]^{1/2}},$$

and after setting $u = z - 1$ we expand $\frac{1}{3}(1 + \frac{1}{9}u)^{-1/2}$ by the binomial theorem to obtain

$$\frac{1}{3(1 + \frac{1}{9}u)^{1/2}} = \frac{1}{3} - \frac{1}{54}u + \frac{1}{648}u^2 - \frac{5}{34992}u^3 + \cdots.$$

Replacing u by $z - 1$ we arrive at the required Taylor series expansion about the point $z_0 = 1$:

$$\frac{1}{(8 + z)^{1/2}} = \frac{1}{3} - \frac{1}{54}(z - 1) + \frac{1}{648}(z - 1)^2 - \frac{5}{34992}(z - 1)^3 + \cdots.$$

The binomial expansion of $(1 + \frac{1}{9}u)^{-1/2}$ converges for $|u/9| < 1$, so the required Taylor series converges for $|z - 1| < 9$.   ∎

Series obtained by integration

EXAMPLE 15.9

Find the Maclaurin series expansion of Arcsin z.

Solution   We start from the result

$$\arcsin z = \int \frac{dz}{(1 - z^2)^{1/2}}.$$

Expanding the integrand by the binomial theorem and integrating term by term gives the power series expansion for the general function arcsin z. Confining attention to the principal branch Arcsin z for which Arcsin $0 = 0$ shows that the arbitrary integration constant is zero, so

$$\text{Arcsin } z = z + \frac{1}{6}z^3 + \frac{3}{40}z^5 + \frac{5}{112}z^7 + \cdots.$$

As the principal branch is required, we must restrict z so that $\text{Re}\{\text{Arcsin } z\} < |\pi/2|$.   ∎

Series obtained by using a partial fraction representation

Find the Taylor series expansion of $f(z) = 1/[(z-2)(z-3)]$ about the point $z_0 = 1$.

Solution   To introduce powers of $(z-1)$ we write $f(z)$ as

$$f(z) = \frac{1}{[(z-1)-1][(z-1)-2]}$$

and set $u = z - 1$. A partial fraction expansion of the resulting expression in u gives

$$\frac{1}{(u-1)(u-2)} = \frac{1}{(1-u)} - \frac{1}{2}\frac{1}{\left(1 - \frac{1}{2}u\right)}.$$

Expanding each of these terms by the binomial theorem and combining the results gives

$$\frac{1}{(u-1)(u-2)} = \frac{1}{2} + \frac{3}{4}u + \frac{7}{8}u^2 + \frac{15}{16}u^3 + \cdots.$$

Replacing u by $z - 1$ shows that the required Taylor series expansion is

$$\frac{1}{(z-2)(z-3)} = \frac{1}{2} + \frac{3}{4}(z-1) + \frac{7}{8}(z-1)^2 + \frac{15}{16}(z-1)^3 + \cdots.$$

The binomial series for $(z-2)^{-1}$ converges for $|z| < 2$ and the series for $(z-3)^{-1}$ converges for $|z| < 3$, so as both will converge for $|z| < 2$, this must be the circle of convergence for the required Taylor series.   ∎

Series obtained by multiplication of series

Find up to the term in z^5 the Maclaurin series expansion of

$$f(z) = \frac{\sin z}{(1 + 3z^2)}.$$

Solution   We will obtain the result by multiplying together an appropriate number of terms of the Maclaurin series expansion of $\sin z$ and the binomial series expansion of $(1 + 3z^2)^{-1}$. To obtain a result accurate to the term in z^5 we will need to multiply the truncated series

$$\sin z = z - \frac{z^3}{6} + \frac{z^5}{120} + \cdots$$

and the truncated binomial expansion

$$\frac{1}{(1 + 3z^2)} = 1 - 3z^2 + 9z^4 - \cdots.$$

This gives

$$\frac{\sin z}{(1 + 3z^2)} = \left(z - \frac{1}{6}z^3 + \frac{1}{120}z^5 - \cdots\right)(1 - 3z^2 + 9z^4 - \cdots)$$

$$= z - \frac{19}{6}z^3 + \frac{1141}{120}z^5 - \cdots.$$

The series for $\sin z$ converges for all z, but the binomial expansion of $(1 + 3z^2)^{-1}$ only converges for $|z| < 1/\sqrt{3}$, so the required Maclaurin series converges for $|z| < 1/\sqrt{3}$. ∎

EXAMPLE 15.12

Find up to the term in z^5 the Maclaurin series expansion of $f(z) = [\log(1 - z)]^2$, using the branch of the logarithmic function for which $\log 1 = 2\pi i$.

Solution   The principal branch is the function $\text{Log}(1 - z)$ for which $\text{Log } 1 = 0$, and routine differentiation shows the Maclaurin series expansion of $\text{Log}(1 - z)$ to be

$$\text{Log}(1 - z) = -z - \frac{z^2}{2} - \frac{z^3}{3} - \cdots - \frac{z^n}{n} - \cdots.$$

Using the result $e^{2\pi i} = 1$, we can write

$$\log(1 - z) = \text{Log}[e^{2\pi i}(1 - z)]$$
$$= \text{Log } e^{2\pi i} + \text{Log}(1 - z) = 2\pi i + \text{Log}(1 - z),$$

showing that the appropriate branch of the logarithmic function has the Maclaurin series expansion

$$\log(1 - z) = 2\pi i - z - \frac{z^2}{2} - \frac{z^3}{3} - \cdots - \frac{z^n}{n} - \cdots.$$

Multiplying the series $[\log(1 - z)]^2$ term by term and collecting all terms up to and including terms in z^5, we obtain

$$[\log(1 - z)]^2 = -4\pi^2 - 4\pi i z + (1 - 2\pi i)z^2 + \left(1 - \frac{4\pi i}{3}\right)z^3$$
$$+ \left(\frac{11}{12} - \pi i\right)z^4 + \left(\frac{5}{6} - \frac{4}{5}\pi i\right)z^5 + \cdots.$$

A careful examination of the coefficients in the series shows that it can be written more systematically as

$$[\log(1 - z)]^2 = -4\pi^2 + 2\left[-2\pi i + (1 - 2\pi i)\frac{z^2}{2} + \left(1 + \frac{1}{2} - 2\pi i\right)\frac{z^3}{3}\right.$$
$$\left. + \cdots \left(1 + \frac{1}{2} + \frac{1}{3} + \cdots + \frac{1}{n - 1} - 2\pi i\right)\frac{z^n}{n} + \cdots\right].$$

The series for $\text{Log}(1 - z)$ converges for $|z| < 1$, so the series for $[\log(1 - z)]^2$ also converges for $|z| < 1$. ∎

Summary

Complex sequences and series have been defined. Tests for the convergence of complex power series were derived that gave rise to the notions of the radius and circle of convergence of the series. These tests are immediate extensions of the corresponding tests for real power series. The complex form of Taylor's theorem was derived, and alternative and often simpler methods for deriving Taylor series were illustrated by example.

EXERCISES 15.1

In Exercises 1 through 4 identify any cluster points that exist, determine whether they belong to the sequence and, where appropriate, find the limit of the sequence. State when a sequence is divergent.

1. (a) $\left\{1 + \dfrac{(-1)^n}{n}\right\}$. (b) $\left\{[1 + (-1)^n]\left(\dfrac{2n+1}{n}\right)\right\}$.

(c) $\left\{\dfrac{5n-1}{2n+6}\right\}$.

2. (a) $\{n^2\}$. (b) $\left\{\dfrac{n+1}{n} + \dfrac{(-1)^n}{n^2}\right\}$.

(c) $\left\{n \sin \dfrac{\pi}{n}\right\}$.

3. (a) $\left\{\left(1 + \dfrac{1}{n}\right)^n\right\}$. (b) $\left\{\left(\dfrac{2n^2+1}{n}\right) \tan \dfrac{\pi}{4n}\right\}$.

(c) $\{1 + \sin n\pi\}$.

4. (a) $\left\{1 + \cos n\pi + \dfrac{1}{n!}\right\}$. (b) $\left\{\left(\dfrac{n-1}{n+1}\right)^n\right\}$.

(c) $\left\{\tan\left(\dfrac{\pi}{2} - \dfrac{1}{n}\right)\right\}$.

In Exercises 5 through 22 use an appropriate test to determine the nature of the convergence of the series, stating when a series is divergent.

5. $\displaystyle\sum_{n=1}^{\infty} \dfrac{\cos n}{n^2}$.

6. $\displaystyle\sum_{n=1}^{\infty} \dfrac{2 + (-1)^n}{2^n}$.

7. $\displaystyle\sum_{n=1}^{\infty} \dfrac{1}{3+n}$.

8. $\displaystyle\sum_{n=1}^{\infty} (-1)^{n+1} \dfrac{n^3}{5^n}$.

9. $\displaystyle\sum_{n=1}^{\infty} (3\sqrt[n]{n} - 1)^n$.

10. $\displaystyle\sum_{n=1}^{\infty} \dfrac{(n!)^2}{(2n)!}$.

11. $\displaystyle\sum_{n=1}^{\infty} (-1)^{n+1} \sin \dfrac{1}{n^2}$.

12. $\displaystyle\sum_{n=1}^{\infty} \tan^2 \dfrac{1}{n}$.

13. $\displaystyle\sum_{n=1}^{\infty} \dfrac{1}{n(n+1)}$.

14. $\displaystyle\sum_{n=1}^{\infty} \dfrac{n(2+i)^n}{2^n}$.

15. $\displaystyle\sum_{n=1}^{\infty} (-1)^{n+1} \dfrac{n(2i-1)^n}{3^n}$.

16. $\displaystyle\sum_{n=1}^{\infty} \dfrac{1}{n(3+i)^n}$.

17. $\displaystyle\sum_{n=1}^{\infty} \dfrac{1}{n(n-1)}$.

18. $\displaystyle\sum_{n=1}^{\infty} \dfrac{n - (-1)^n}{3^n}$.

19. $\displaystyle\sum_{n=1}^{\infty} (-1)^{n+1} \left[\dfrac{n(2-i)+1}{n(3-2i)-3i}\right]^n$.

20. $\displaystyle\sum_{n=1}^{\infty} \left[\dfrac{2n(4+2i)-1}{n(2+i)+3}\right]^n$.

In Exercises 21 through 34 find the radius of convergence and circle of convergence of the complex power series.

21. $\displaystyle\sum_{n=1}^{\infty} \dfrac{z^n}{n \cdot 2^n}$.

22. $\displaystyle\sum_{n=1}^{\infty} \dfrac{z^{2n-1}}{2n-1}$.

23. $\displaystyle\sum_{n=1}^{\infty} \dfrac{2^{n-1} z^{2n-1}}{(4n-3)^2}$.

24. $\displaystyle\sum_{n=1}^{\infty} (-1)^{n-1} \dfrac{z^n}{n}$.

25. $\displaystyle\sum_{n=1}^{\infty} n! z^n$.

26. $\displaystyle\sum_{n=1}^{\infty} (-1)^{n+1} \dfrac{z^n}{n!}$.

27. $\displaystyle\sum_{n=1}^{\infty} \left(\dfrac{n}{n+1}\right) \left(\dfrac{z}{2}\right)^n$.

28. $\displaystyle\sum_{n=1}^{\infty} (-1)^{n-1} \dfrac{(z-5)^n}{n \cdot 3^n}$.

29. $\displaystyle\sum_{n=1}^{\infty} \dfrac{(z+3)^n}{n^2}$.

30. $\displaystyle\sum_{n=1}^{\infty} \dfrac{(z-2)^n}{2^n(2n-1)}$.

31. $\displaystyle\sum_{n=1}^{\infty} i^n z^n$.

32. $\displaystyle\sum_{n=1}^{\infty} (1+ni) z^n$.

33. $\displaystyle\sum_{n=1}^{\infty} \left(\dfrac{1+2ni}{n+2i}\right)^n z^n$.

34. $\displaystyle\sum_{n=1}^{\infty} \dfrac{(z-2i)^n}{n \cdot 3^n}$.

In Exercises 35 through 44 use Taylor's theorem to find the first four terms of the expansion of $f(z)$ about the given center.

35. $f(z) = \dfrac{\sin z}{1 + \sin z}$ with the center $\pi/4$.

36. $f(z) = \cosh(1 + 3z^2)$ with the center 1.

37. $f(z) = \sinh(2 - 3z)$ with the center 1.

38. $f(z) = \mathrm{Log}\left(\dfrac{4+z}{4-z}\right)$ with the center -1 ($\mathrm{Log}\,1 = 0$).

39. $f(z) = \dfrac{z}{(z+3i)(z-2i)}$ with center i.

40. $f(z) = \cos(2z - i)$ with center i.

41. $f(z) = [\cos z]^2$ with center 0 and $f(0) = 1$.

42. $f(z) = \exp\{z \sin z\}$ with center 0.

43. $f(z) = (z+1)^{1/2}$ with center 0 and $f(0) = (1+i)/\sqrt{2}$.

44. $f(z) = \cos^2(z - i)$ with center $-i$.

In Exercises 45 through 56 use the most appropriate alternative method to find the first four nonvanishing terms in the expansion of $f(z)$ about the given center.

45. $f(z) = \dfrac{\log(z+1)}{(1+z^2)^{1/2}}$ with the center 0 and $\sqrt{1} = 1$, $\log 1 = 4\pi i$.

46. $f(z) = \dfrac{z}{(z-3)(z+2)}$ with center -1.

47. $f(z) = \dfrac{1 - \cos z}{(1-z)^2}$ with center 0.

48. $f(z) = \dfrac{2z+5}{z^2 + z - 2}$ with center 2.

49. $f(z) = \dfrac{1+z}{1+2z^2}$ with center 0.

50. $f(z) = \log\left(\dfrac{1+z}{1-z}\right)$ with center 0 and $\log 1 = 2\pi i$.

51. $f(z) = \text{Arctan } z$ with center 0 ($\text{Arctan } 0 = 0$).

52. $f(z) = [\text{Arctan } z]^2$ with center 0 ($\text{Arctan } 0 = 0$).

53. $f(z) = \displaystyle\int_0^z \dfrac{\sin u}{u}\,du.$

54. $f(z) = (1-z)^{-3}$ with center -1.

55. $f(z) = \dfrac{\sin z}{1-z}$ with center 0.

56. $f(z) = [\cos 2z]^2$ with center $\pi/4$.

15.2   Uniform Convergence

The detailed arguments in this section may be omitted at a first reading, but before doing so, the reader should review the important properties of power series listed in Theorem 15.8.

A power series possesses a special property called *uniform convergence* in any region D of the complex plane where it is convergent. This enables power series to be manipulated as though they were ordinary functions while still retaining the property of uniform convergence.

If $\{u_0(z), u_1(z), u_2(z), \ldots\}$ is an infinite sequence of functions, a series of the form

$$\sum_{n=0}^{\infty} u_n(z) = u_0(z) + u_1(z) + u_2(z) + \cdots \tag{15}$$

is called a **functional series**, and this becomes the *power series*

$$\sum_{n=0}^{\infty} a_n(z - z_0)^n = a_0 + a_1(z - z_0) + a_2(z - z_0)^2 + \cdots, \tag{16}$$

with its center at z_0 when $u_n(z) = a_n(z - z_0)^n$. As with power series, the **nth partial sum** of the functional series (15) is denoted by $s_n(z)$, where

$$s_n(z) = u_0(z) + u_1(z) + u_2(z) + \cdots + u_{n-1}(z). \tag{17}$$

Uniform Convergence

uniform convergence

The functional series (15) is said to **converge uniformly** to the sum $U(z)$ in a region D of the complex plane if for every arbitrary number $\varepsilon > 0$ it is possible to find a number $N = N(\varepsilon)$ that depends on ε, but *not* on z, such that

$$|U(z) - s_n(z)| < \varepsilon \quad \text{for all } n > N \quad \text{and all } z \text{ in } D. \tag{18}$$

It follows from this definition that the power series (16) will be uniformly convergent in D if it can be shown that

$$\left| \sum_{k=n}^{\infty} a_k(z - z_0)^k \right| < \varepsilon \quad \text{for all } n > N \quad \text{and all } z \text{ in } D. \tag{19}$$

A comparison of the definitions of uniform convergence and convergence shows that whereas for convergence the number N depends on ε and the value of z, in the case of uniform convergence the number N depends only on ε, and *not* on z. It is because of the independence of the convergence on the value of z in D that the term *uniform* is used to describe this powerful form of convergence.

In practical terms, if a power series converges uniformly to $f(z)$ in a circle of convergence D, and it is known that when $n = N$ the Nth partial sum $s_N(z)$ at a point z_0 in D approximates $f(z_0)$ in such a way that

$$|f(z_0) - s_N(z_0)| < \varepsilon$$

for some known small number $\varepsilon > 0$, then $s_N(z)$ will approximate $f(z)$ with the *same* accuracy for *all* points z in D. This is *not* the case for series that are not uniformly convergent, because in that case the number of terms needed in the partial sum to maintain the accuracy will depend on the value of z.

The following theorem, called the **Weierstrass M-test**, provides the simplest test for uniform convergence.

THEOREM 15.10

the simplest test for uniform convergence

Weierstrass M-test Let the functional series $\sum_{n=0}^{\infty} u_n(z)$ be such that for each n, $|u_n(z)| < M_n$ for all z in a domain D. Then if the series of positive constants $\sum_{n=0}^{\infty} M_n$ converges, the series $\sum_{n=0}^{\infty} u_n(z)$ is uniformly convergent in D.

Proof Let $s_n(z)$ and $s_{n+p}(z)$ be the nth and $(n+p)$th partial sums of the series, with $p > 0$ any positive integer. Then

$$|s_{n+p}(z) - s_n(z)| = |u_n(z) + u_{n+1}(z) + \cdots + u_{n+p}(z)|$$

$$\leq |u_n(z)| + |u_{n+1}(z)| + \cdots + |u_{n+p}(z)| = \sum_{k=n}^{n+p} M_k,$$

where repeated use has been made of the triangle inequality.

By hypothesis the series $\sum_{n=0}^{\infty} M_n$ is convergent, so it follows from the Cauchy convergence principle that the sum $\sum_{k=n}^{n+p} M_k$ can be made arbitrarily small by making n sufficiently large, so a function $U(z)$ exists such that $U(z) = \lim_{n \to \infty} s_n(z)$. This has established that the conditions of the theorem ensure that the functional series is *convergent*.

To show that the convergence is *uniform* it is only necessary to notice that for any $\varepsilon > 0$, the convergence of $\sum_{n=0}^{\infty} M_n$ means that a positive integer $N(\varepsilon)$ can be found such that if $n \geq N(\varepsilon)$, then $\sum_{k=n}^{\infty} M_k < \varepsilon$. So, for $n \geq N(\varepsilon)$ and all z in D,

$$|U(z) - s_n(z)| = \left| \sum_{k=n}^{\infty} u_k(z) \right| \leq \sum_{k=n}^{\infty} M_k < \varepsilon,$$

and the theorem is proved. ∎

EXAMPLE 15.13

Prove that the power series $\sum_{n=0}^{\infty} z^n = 1 + z + z^2 + \cdots + z^n + \cdots$ is uniformly convergent inside the unit circle $|z| = 1$.

Solution   Let z^* be a point inside the unit circle $|z| = 1$ and write $|z^*| = r$, so $r < 1$ and $|(z^*)^n| = r^n < 1$. Setting $M_n = r^n$ in the Weierstrass M-test, we obtain

$$\sum_{n=0}^{\infty} M_n < \sum_{n=0}^{\infty} r^n = \frac{1}{1 - r}, \quad \text{with } r < 1.$$

As the conditions of the theorem are satisfied, the series is uniformly convergent everywhere inside the unit circle $|z| = 1$.

This result is not unexpected, because $\sum_{n=0}^{\infty} z^n$ is the Maclaurin series expansion of $1/(1 - z)$, and this is an analytic function inside the unit circle. ∎

From now on attention will be confined to power series, and the next theorem generalizes the result of the last example by proving that every power series converges uniformly inside its circle of convergence.

THEOREM 15.11

a power series converges uniformly inside its circle of convergence

Uniform convergence of power series   A power series $\sum_{n=0}^{\infty} a_n(z - z_0)^n$ with a radius of convergence $R > 0$ converges uniformly inside and on every circle $|z - z_0| = r$, where $r < R$.

Proof   The proof of the theorem makes use of the Weierstrass M-test. From the definition of the radius of convergence of a series it follows that the series $\sum_{n=1}^{\infty} a_n(z - z_0)^n$ is absolutely convergent for $|z - z_0| < r$, so for any $z = \zeta$ on the circle $|z - z_0| = r$ the series

$$\sum_{n=0}^{\infty} |a_n(\zeta - z_0)^n| = \sum_{n=0}^{\infty} |a_n| r^n$$

must also be convergent. Hence, for all z inside and on the circle $|z - z_0| = r$, the following inequality must hold:

$$|a_n(z - z_0)^n| \le |a_n| r^n.$$

The statement of the theorem now follows from this result and the convergence of the series $\sum_{n=0}^{\infty} |a_n| r^n$ if we apply the Weierstrass M-test to the power series with $M_n = |a_n| r^n$. ∎

The result of Example 15.13 is a special case of this theorem. An examination of the series $\sum_{n=0}^{\infty} z^n$ shows that its radius of convergence is $R = 1$, so as the series is a power series expansion with the origin as center it is uniformly convergent inside the circle $|z| = 1$, as was shown directly in the example.

It is useful to use Theorem 15.11 to reformulate the results of Theorem 15.8 concerning the differentiation and integration of power series.

THEOREM 15.12

a power series can be differentiated and integrated inside its circle of convergence

Differentiation and integration of power series   Let a power series with the sum $f(z)$ have a circle of convergence $|z - z_0| = R$, where $R > 0$. Then the series possesses the following properties:

   (i) The power series converges uniformly to $f(z)$ inside the circle of convergence.

   (ii) The power series obtained by term-by-term differentiation of the power series for $f(z)$ converges uniformly to $f'(z)$ and has the same circle of convergence as

$f(z)$, so if

$$f(z) = \sum_{n=0}^{\infty} a_n(z-z_0)^n, \quad \text{then} \quad f'(z) = \sum_{n=1}^{\infty} na_n(z-z_0)^{n-1}.$$

(iii) The power series obtained by term-by-term integration of the power series for $f(z)$ along any path Γ inside the circle of convergence converges uniformly to the integral of $f(z)$ along Γ, so if

$$f(z) = \sum_{n=0}^{\infty} a_n(z-z_0)^n, \quad \text{then} \quad \int_{\Gamma} f(z)dz = \sum_{n=0}^{\infty} a_n \int_{\Gamma} (z-z_0)^n dz. \quad \blacksquare$$

EXAMPLE 15.14   Use the Maclaurin series for $1/(1-2z)$ to find the Maclaurin series for $1/(1-2z)^2$, and confirm that both series have the same circle of convergence.

Solution   The Maclaurin series expansion is obtained most easily by writing the function in the form $(1-2z)^{-1}$ and expanding it by the binomial theorem, when we obtain

$$\frac{1}{1-2z} = (1-2z)^{-1} = 1 + 2z + 2^2z^2 + \cdots = \sum_{n=0}^{\infty} 2^n z^n.$$

This binomial series is convergent for $|z| < 1/2$, so this is the circle of convergence for the function. By Theorem 15.12 (ii) this series can be differentiated term by term inside its circle of convergence, so as

$$\frac{d}{dz}\left(\frac{1}{1-2z}\right) = \frac{2}{(1-2z)^2} \quad \text{and} \quad \frac{d}{dz}\sum_{n=0}^{\infty}[2^n z^n] = 2 + 2\cdot 2^2 z + 3\cdot 2^3 z^2 + \cdots$$

$$= \sum_{n=0}^{\infty}(n+1)2^{n+1}z^n,$$

equating these results and cancelling a factor 2 gives the desired expansion,

$$\frac{1}{(1-2z)^2} = \sum_{n=0}^{\infty}(n+1)2^n z^n = 1 + 4z + 12z^2 + \cdots.$$

It is easily verified that this power series has a radius of convergence $R = \frac{1}{2}$, so the differentiated series is also uniformly convergent for $|z| < \frac{1}{2}$.   ■

EXAMPLE 15.15   By integrating the Maclaurin series for $\sin \zeta$ along a suitable path, find the Maclaurin series for $\cos z$.

Solution   The Maclaurin series for $\sin \zeta$ is

$$\sin \zeta = \sum_{n=0}^{\infty}(-1)^n \frac{\zeta^{2n+1}}{(2n+1)!} = \zeta - \frac{\zeta^3}{3!} + \frac{\zeta^5}{5!} - \cdots.$$

As this power series converges for all finite ζ, it follows from Theorem 15.12 (iii) that term-by-term integration is permitted along any path in the complex plane, so integrating from the origin to an arbitrary point z gives

$$\int_0^z \sin \zeta \, d\zeta = \sum_{n=0}^{\infty}(-1)^n \frac{1}{(2n+1)!} \int_0^z \zeta^{2n+1} d\zeta.$$

After the integrations are performed this becomes

$$1 - \cos z = \sum_{n=0}^{\infty}(-1)^n \frac{z^{2n+2}}{(2n+2)!},$$

and a rearrangement of terms leads to the expected result

$$\cos z = \sum_{n=0}^{\infty}(-1)^n \frac{z^{2n}}{(2n)!},$$

where the series on the right also converges for all finite z. ∎

EXAMPLE 15.16   By integrating the Maclaurin series for $1/(1+\zeta)$ along a suitable path in its circle of convergence, show that

$$\mathrm{Log}(1+z) = \sum_{n=1}^{\infty}(-1)^{n+1}\frac{z^n}{n} \quad \text{for } |z| < 1.$$

Solution   The Maclaurin series expansion of $1/(1+\zeta)$ is most easily found by means of the binomial theorem, so we can write

$$\frac{1}{1+\zeta} = 1 - \zeta + \zeta^2 - \zeta^3 + \cdots = \sum_{n=0}^{\infty}(-1)^n \zeta^n.$$

This power series has a radius of convergence $R = 1$ and so is uniformly convergent inside the circle of convergence $|\zeta| = 1$. By Theorem 15.12 (iii), this series can be integrated term by term along any path Γ inside this circle, so

$$\int_{\Gamma}\frac{1}{1+\zeta}d\zeta = \sum_{n=0}^{\infty}(-1)^n \int_{\Gamma}\zeta^n d\zeta.$$

To obtain $\mathrm{Log}(1+z)$ we choose Γ to be the straight line path joining the origin to a point $\zeta = z$ inside the circle of convergence, and take the *principal* branch of the logarithmic function as an antiderivative of the integral on the left. As a result, on the left we obtain

$$\int_0^z \frac{1}{1+\zeta}d\zeta = \mathrm{Log}(1+z), \quad \text{where} \quad \mathrm{Log}(1+\zeta) = \ln|1+\zeta| + i\theta,$$

where θ is the argument of $\mathrm{Log}(1+\zeta)$, with $-\pi < \theta \le \pi$. Integration of the expression on the right leads to the result

$$\sum_{n=0}^{\infty}(-1)^n \int_0^z \zeta^n d\zeta = \sum_{n=1}^{\infty}(-1)^{n+1}\frac{z^n}{n},$$

so equating these expresions gives

$$\mathrm{Log}(1+z) = \sum_{n=1}^{\infty}(-1)^{n+1}\frac{z^n}{n}.$$

Care must always be exercised when working with logarithmic functions because they are multivalued. The principal branch of $\mathrm{Log}(1+z)$ used here is analytic throughout the complex plane, with the exception of the branch cut made along the negative real axis from $-\infty$ to the point $z = -1$. However, the series representation of $\mathrm{Log}(1+z)$ is only valid inside the circle $|z| = 1$ where, like the function

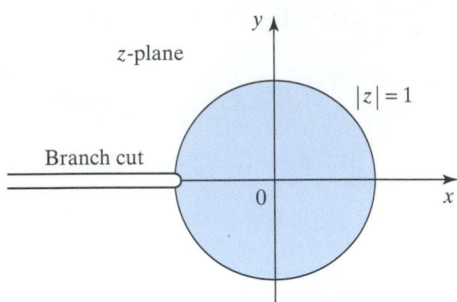

FIGURE 15.3 The circle of convergence for the series representation of $\text{Log}(1 + z)$ and the branch cut for the function $\text{Log}(1 + z)$.

$f(z) = 1/(1 + z)$, it is analytic. Figure 15.3 shows the circle of convergence for the series expansion of $\text{Log}(1 + z)$, and the branch cut used in the definition of the function $\text{Log}(1 + z)$. ■

Summary The concept of uniform convergence was defined and related to power series, and the simple Weierstrass M-test for uniform convergence was given. The importance of the uniform convergence of a power series was shown to be that it retains its uniform convergence property when it is either differentiated or integrated inside its circle of convergence, thereby allowing it to be manipulated like an ordinary function.

15.3   Laurent Series and the Classification of Singularities

We have seen how a function $f(z)$ that is analytic at a point z_0 can be expanded in a neighborhood of z_0 as a Taylor series with z_0 as its center. Although Taylor series expansions are sufficient for many purposes, the requirement that $f(z)$ must be analytic at z_0 means some other form of expansion must be used when an expansion is required about a point where $f(z)$ is not analytic. The development of a more general form of expansion that overcomes this difficulty leads to what is called a *Laurent series* expansion of a function. Arising from the study of Laurent series comes the need to classify the nature of points where a function is not analytic.

Points where a function $f(z)$ is analytic are called **regular points** of the function, and a point z_0 where $f(z)$ is analytic in every neighborhood of z_0, but not at z_0 itself, is called a **singular point** of the function. For example, the function $f(z) = 1/z$ is analytic for all finite z apart from the point $z = 0$ where its derivative is not defined, so the origin is a singular point of $f(z) = 1/z$.

A **Laurent series** $L(z)$ is a series of the form

regular and singular points and Laurent series

$$L(z) = \sum_{n=-\infty}^{\infty} a_n(z - z_0)^n \qquad (20)$$

that contains both positive and negative powers of $(z - z_0)$. It is customary to represent a Laurent series $L(z)$ as the sum of two series by setting $L(z) = L_1(z) + L_2(z)$ where

$$L_1(z) = \sum_{n=-\infty}^{-1} a_n(z - z_0)^n \quad \text{and} \quad L_2(z) = \sum_{n=0}^{\infty} a_n(z - z_0)^n. \tag{21}$$

The series $L_1(z)$ containing only negative powers of $(z - z_0)$ is called the **principal part** of the Laurent series, and the series $L_2(z)$ containing only positive powers of $(z - z_0)$ is called its **regular part**. A Laurent series is said to **converge** in a domain D when both of the series $L_1(z)$ and $L_2(z)$ are convergent in D. In general, a Laurent series converges in an annulus

$$r < |z - z_0| < R, \quad \text{where} \quad 0 < r < R.$$

A simple example of a Laurent series is obtained by considering the function $(\cos z)/z$ and expanding $\cos z$ as a Maclaurin series to arrive at the representation

$$\frac{\cos z}{z} = \frac{1}{z} \sum_{n=0}^{\infty} (-1)^n \frac{z^{2n}}{(2n)!} = \sum_{n=0}^{\infty} (-1)^n \frac{z^{2n-1}}{(2n)!} = \frac{1}{z} - \frac{z}{2!} + \frac{z^3}{4!} - \cdots.$$

The principal part of this Laurent series is the single term $L_1(z) = 1/z$, and its regular part is the power series

$$L_2(z) = \sum_{n=1}^{\infty} (-1)^n \frac{z^{2n-1}}{(2n)!} = -\frac{z}{2!} + \frac{z^3}{4!} - \frac{z^5}{6!} + \cdots.$$

In this case the principal part of the expansion is finite (converges) for all $z \neq 0$, and the regular part converges for all z, so the annulus in which this Laurent series converges becomes the complex plane from which has been deleted the single point at the origin.

The next theorem shows how a function that is analytic in an annulus with its center at the point z_0 can be expanded inside the annulus as a unique Laurent series. This theorem also provides an explicit general formula for the Laurent coefficients. The examples that follow the theorem show how simple algebraic arguments often provide easier ways of finding the Laurent coefficients than using the general formula.

THEOREM 15.13

the Laurent expansion theorem

Laurent's theorem   A function $f(z)$ that is analytic in the annulus D given by $R_1 < |z - z_0| < R_2$ can be expanded in D as a unique Laurent series

$$f(z) = \sum_{n=-\infty}^{\infty} a_n(z - z_0)^n,$$

where

$$a_n = \frac{1}{2\pi i} \int_\Gamma \frac{f(\zeta)}{(\zeta - z_0)^{n+1}} d\zeta \quad \text{with } n = 0, \pm 1, \pm 2, \ldots,$$

and Γ is any positively oriented circle in D given by $|\zeta - z_0| = \rho$, with $R_1 < \rho < R_2$.

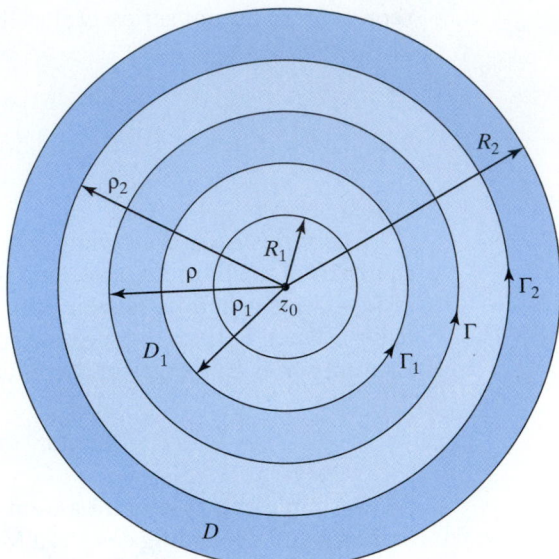

FIGURE 15.4 The annulus D determined by
$R_1 < |z - z_0| < R_2$.

Proof Let the annulus D be the one shown in Fig. 15.4 with its center at z_0, its inner boundary a circle of radius R_1, and its outer boundary a circle of radius R_2. The positively oriented circles Γ_1 and Γ_2 with the respective radii ρ_1 and ρ_2 bound the annulus D_1 contained in D, where the positively oriented circle Γ inside D_1 has radius ρ.

If z is a fixed point inside D_1, then by the extended Cauchy–Goursat theorem we can write

$$f(z) = \frac{1}{2\pi i} \int_{\Gamma_2} \frac{f(\zeta)}{\zeta - z} d\zeta + \frac{1}{2\pi i} \int_{\widetilde{\Gamma}_1} \frac{f(\zeta)}{\zeta - z} d\zeta,$$

where $\widetilde{\Gamma}_1$ denotes integration around Γ_1 in the negative (clockwise) sense.

In the integrand of the first term ζ lies on Γ_2, so we expand $1/(\zeta - z)$ as the power series in $z - \zeta$:

$$\frac{1}{\zeta - z} = \frac{1}{(\zeta - z_0)\left(1 - \frac{z - z_0}{\zeta - z_0}\right)} = \sum_{n=0}^{\infty} \frac{(z - z_0)^n}{(\zeta - z_0)^{n+1}}.$$

This is a geometric series, and because ζ lies on Γ_2 we have $|\zeta - z_0| = \rho_2$, showing that

$$\left| \frac{z - z_0}{\zeta - z_0} \right| = \frac{|z - z_0|}{\rho_2} < 1.$$

Applying the Weierstrass M-test shows that the series expansion of $1/(\zeta - z)$ is uniformly convergent. As uniform convergence allows term-by-term integration of the series, substituting the series expansion in the integral gives

$$\frac{1}{2\pi i} \int_{\Gamma_2} \frac{f(\zeta)}{\zeta - z} d\zeta = \sum_{n=0}^{\infty} a_n (z - z_0)^n,$$

where

$$a_n = \frac{1}{2\pi i} \int_{\Gamma_2} \frac{f(\zeta)}{(\zeta - z_0)^{n+1}} d\zeta.$$

A similar argument can be used to express the integrand in the second integral as

$$\frac{1}{\zeta - z} = \frac{1}{z - z_0 - (\zeta - z_0)} = \frac{-1}{(z - z_0)\left(1 - \frac{\zeta - z_0}{z - z_0}\right)} = -\sum_{k=0}^{\infty} \frac{(\zeta - z_0)^k}{(z - z_0)^{k+1}},$$

where, as ζ now lies on Γ_1,

$$\left| \frac{\zeta - z_0}{z - z_0} \right| = \frac{\rho_1}{|z - z_0|} < 1.$$

The Weierstrass M-test shows this series is also uniformly convergent. Substituting the series in the second integral and integrating term by term gives

$$\frac{1}{2\pi i} \int_{\tilde{\Gamma}_1} \frac{f(\zeta)}{\zeta - z} d\zeta = -\frac{1}{2\pi i} \int_{\Gamma_1} \frac{f(\zeta)}{\zeta - z} d\zeta = \frac{1}{2\pi i} \sum_{k=0}^{\infty} \int_{\Gamma_1} \frac{f(\zeta)(\zeta - z_0)^k}{(z - z_0)^{k+1}} d\zeta,$$

where a negative sign has been introduced to compensate for the change from contour $\tilde{\Gamma}_1$ where integration is in the clockwise sense, to contour Γ_1 where the integration is counterclockwise.

When $k + 1$ is replaced by $-n$ the summation becomes

$$\frac{1}{2\pi i} \int_{\Gamma_1} \frac{f(\zeta)}{\zeta - z} d\zeta = \sum_{n=-1}^{-\infty} a_n (z - z_0)^n,$$

with

$$a_n = \frac{1}{2\pi i} \int_{\Gamma_1} \frac{f(\zeta)}{(\zeta - z_0)^{n+1}} d\zeta.$$

Combining the two integrals, and recognizing that the positively oriented circles Γ_1 and Γ_2 bounding D_1 may both be deformed into any positively oriented contour Γ that lies in D_1 with z_0 in its interior, shows that the Laurent series coefficients a_n are *all* given by the single formula

$$a_n = \frac{1}{2\pi i} \int_{\Gamma} \frac{f(\zeta)}{(\zeta - z_0)^{n+1}} d\zeta \quad \text{with } n = 0, \pm 1, \pm 2, \dots.$$

Finally, as the fixed point z was *any* point inside the annulus D_1 that is itself contained in the annulus D, the first part of the theorem has been proved.

The uniqueness of a Laurent series expansion in a given annulus can be established as follows. Suppose, if possible, that $f(z)$ can be represented in the same annulus by the two different Laurent series

$$f(z) = \sum_{n=-\infty}^{\infty} a_n (z - z_0)^n = \sum_{n=-\infty}^{\infty} b_n (z - z_0)^n.$$

Forming the product of these series with $(z - z_0)^{-m-1}$, where m is a fixed integer, leads to the result

$$\sum_{n=-\infty}^{\infty} a_n (z - z_0)^{n-m-1} = \sum_{n=-\infty}^{\infty} b_n (z - z_0)^{n-m-1}.$$

Each of these series converges on the contour Γ inside D_1, so using the results

$$\int_\Gamma (z - z_0)^k dz = \begin{cases} 0, & k \neq -1 \\ 2\pi i, & k = -1 \end{cases} \quad (k \text{ a positive or negative integer})$$

shows that $a_k = b_k$ for each k, so the uniqueness of the Laurent series is proved. ∎

PIERRE-ALPHONSE LAURENT (1813–1854)
A French mathematician whose major contribution to complex analysis, published in 1843, was the fact that when a function is discontinuous at a single point, the Taylor series expansion of the function must be replaced by an expansion involving both increasing and decreasing powers of the variable involved. This result is the one now known as Laurent's theorem.

The uniqueness of a Laurent series expansion of an analytic function $f(z)$ in a given annulus means that any method used to generate the expansion in the annulus will produce the same series. This result can be used to considerable advantage, because instead of using the general formula given in Theorem 15.13, it frequently proves to be easier to find the coefficients of the series by using a simple algebraic approach.

If an analytic function $f(z)$ that is expanded about the point z_0 has singular points at $a_1, a_2, \ldots, a_n$, then the loss of differentiability at these points means that the radius R_2 of the outer circle of the annulus in which the expansion is valid cannot exceed the distance from z_0 to the *nearest* singular point, so that

$$R_2 = \min\{|z_0 - a_1|, |z_0 - a_2|, \ldots, |z_0 - a_n|\}.$$

how algebraic arguments can often simplify the task of finding a Laurent series

The expansion will, of course, be analytic everywhere on the outer boundary of the annulus of convergence except for any point where there is a singularity.

The next example illustrates the use of algebraic arguments to develop Laurent series, and also how the location of singularities relative to the point about which the expansion is carried out determines the outer radius of the annulus of convergence.

EXAMPLE 15.17 Find the Laurent series expansion of

$$f(z) = \frac{1}{6 - z - z^2}$$

in (a) the domain D_1 determined by $|z| < 2$, (b) the domain D_2 determined by $2 < |z| < 3$, and (c) the domain D_3 determined by $|z| > 3$.

Solution Factoring the denominator gives

$$f(z) = \frac{1}{(2 - z)(z + 3)},$$

so the function has singular points at $z = 2$ and $z = -3$, but is analytic elsewhere. As these points occur on the boundaries of the domains D_1, D_2, and D_3, the function will be analytic inside each of these domains. Consequently, $f(z)$ will have a unique though different Laurent series expansion in each of the three domains. The required expansions will now be obtained by using simple algebraic arguments that start from the partial fraction decomposition

$$f(z) = \frac{1}{5}\left(\frac{1}{2-z} + \frac{1}{z+3}\right).$$

If $|z| < 2$, by using the binomial theorem we can write

$$\frac{1}{2-z} = \frac{1}{2\left(1-\frac{z}{2}\right)} = \frac{1}{2}\left(1-\frac{z}{2}\right)^{-1} = \sum_{n=0}^{\infty}\frac{z^n}{2^{n+1}}.$$

If $|z| > 2$, it follows in similar fashion that we can write

$$\frac{1}{2-z} = \frac{1}{z\left(\frac{2}{z}-1\right)} = -\frac{1}{z}\left(1-\frac{2}{z}\right)^{-1} = -\sum_{n=1}^{\infty}\frac{2^{n-1}}{z^n}.$$

If $|z| < 3$, we can write

$$\frac{1}{z+3} = \frac{1}{3\left(1+\frac{z}{3}\right)} = \frac{1}{3}\left(1+\frac{z}{3}\right)^{-1} = \sum_{n=0}^{\infty}(-1)^n\frac{z^n}{3^{n+1}}.$$

Finally, if $|z| > 3$, we have

$$\frac{1}{z+3} = \frac{1}{z\left(1+\frac{3}{z}\right)} = \frac{1}{z}\left(1+\frac{3}{z}\right)^{-1} = \sum_{n=1}^{\infty}(-1)^{n-1}\frac{3^{n-1}}{z^n}.$$

These results can now be combined with the partial fraction decomposition to obtain the Laurent series expansions in each of the three domains.

(a) In D_1 where $|z| < 2$ we have from the first and third of the preceding expansions that

$$f(z) = \sum_{n=0}^{\infty}\frac{1}{5}\left[\frac{1}{2^{n+1}} + \frac{(-1)^n}{3^{n+1}}\right]z^n.$$

This expansion contains no principal part, and because $f(z)$ is analytic in D_1 we see that in this domain the Laurent series has degenerated into a Taylor series expansion about the origin that is, of course, just the Maclaurin series expansion of $f(z)$ in D_1.

(b) In D_2 where $2 < |z| < 3$, we have from the second and third of the preceding expansions that

$$f(z) = \sum_{n=1}^{\infty}\left(-\frac{2^{n-1}}{5}\right)\frac{1}{z^n} + \sum_{n=0}^{\infty}\left(\frac{(-1)^n}{5\cdot 3^{n+1}}\right)z^n.$$

Here the first summation represents the principal part and the second summation the regular part of the Laurent series expansion in the domain.

(c) In D_3 where $|z| > 3$, we have from the second and fourth of the preceding expansions that

$$f(z) = \sum_{n=1}^{\infty}\frac{1}{5}[-2^{n-1} + (-1)^{n-1}3^{n-1}]\frac{1}{z^n}.$$

This shows that in D_3 the Laurent series expansion has only a principal part.

Although expansions (a) and (b) are different in form, each is analytic on the circle $|z| = 2$, with the exception of the point $z = 2$ where a singularity occurs. Thus, representations (a) and (b) give different, but equivalent, representations of $f(z)$ on the circle $|z| = 2$ away from the single point $z = 2$. A similar situation occurs with representations (b) and (c) on the circle $|z| = 3$ away from the single point $z = -3$ where the other singularity is located. ∎

EXAMPLE 15.18   Expand $f(z) = \exp(z + \frac{1}{z})$ as a Laurent series about the origin.

Solution   The function $f(z)$ is analytic everywhere except at the origin, which is a singular point. Consequently, when $f(z)$ is expanded about the origin its Laurent series will converge throughout the complex plane with the exception of the single point $z = 0$, and the series will be of the form

$$\exp\left(z + \frac{1}{z}\right) = \sum_{n=1}^{\infty} a_{-n}\frac{1}{z^n} + \sum_{n=0}^{\infty} a_n z^n, \quad \text{for } |z| > 0.$$

To determine the coefficients $a_{\pm n}$, we write the function as $f(z) = (\exp z)(\exp \frac{1}{z})$ and then express this as the product of the two series

$$(\exp z)\left(\exp \frac{1}{z}\right) = \left(1 + z + \frac{z^2}{2!} + \frac{z^3}{3!} + \frac{z^4}{4!} + \frac{z^5}{5!} + \cdots\right)$$
$$\times \left(1 + \frac{1}{z} + \frac{1}{2!z^2} + \frac{1}{3!z^3} + \frac{1}{4!z^4} + \frac{1}{5!z^5} + \cdots\right).$$

The coefficient a_0 is simply the constant term in this product, so identifying this as the sum of products of the form $\frac{z^k}{k!} \cdot \frac{1}{z^k k!}$, we find that

$$a_0 = 1 + 1 + \frac{1}{(2!)^2} + \frac{1}{(3!)^2} + \frac{1}{(4!)^2} + \cdots = \sum_{k=0}^{\infty} \frac{1}{(k!)^2}.$$

Further examination of the product of the two series shows that the coefficients a_n and a_{-n} are equal, so we need only determine a_n. The coefficient a_1 in the Laurent series expansion about the origin is the coefficient of z in the preceding product, so identifying this as the sum of the products $\frac{z^{k+1}}{(k+1)!} \frac{1}{z^k k!}$ gives

$$a_1 = a_{-1} = 1 + \frac{1}{2!} + \frac{1}{2! \cdot 3!} + \frac{1}{3! \cdot 4!} + \cdots = \sum_{k=0}^{\infty} \frac{1}{k!(k+1)!}.$$

If we proceed in this manner, it is not difficult to see that

$$a_n = a_{-n} = \sum_{k=0}^{\infty} \frac{1}{k!(n+k)!}.$$

Substituting these values for a_0 and $a_{\pm n}$ into

$$\exp\left(z + \frac{1}{z}\right) = \sum_{n=1}^{\infty} a_{-n}\frac{1}{z^n} + \sum_{n=0}^{\infty} a_n z^n$$

gives the required Laurent series expansion that is convergent for $|z| > 0$. ∎

EXAMPLE 15.19   Find (a) the Laurent series expansion of $f(z) = 1/(z^2 + 1)^2$ in the largest possible circle about the point $z = i$, and (b) the expansion about the origin for $|z| > 1$.

Solution

(a) Writing $f(z)$ as

$$f(z) = \frac{1}{(z-i)^2(z+i)^2}$$

shows that the function has singularities only at $z = i$ and $z = -i$. When the function is expanded in a Laurent series about $z = i$, the radius R of the outer boundary of the largest annulus of convergence must equal the distance between $z = i$ and the singularity at $z = -i$ closest to $z = i$. As $|i - (-i)| = 2$ we see that $R = 2$, so as the point $z = i$ must be excluded from the annulus of convergence centered on $z = i$, the function $f(z)$ will be analytic in the punctured disc $0 < |z - i| < 2$, where the expansion will be in terms of powers of $z - i$.

Simplifying $f(z)$ by using partial fractions gives

$$f(z) = \frac{1}{(z^2+1)^2} = -\frac{i}{4}\frac{1}{z-i} - \frac{1}{4}\frac{1}{(z-i)^2} + \frac{i}{4}\frac{1}{z+i} - \frac{1}{4}\frac{1}{(z+i)^2}.$$

The first two terms are already expressed in terms of powers of $z - i$, so it remains to express the last two terms in this form.

The third term on the right can be written as

$$\frac{i}{4}\frac{1}{z+i} = \frac{i}{4}\frac{1}{(z-i)+2i},$$

so as $|z - i| < |2i|$ the binomial theorem can be used to expand this expression as

$$\frac{i}{4}\frac{1}{z+i} = \frac{i}{4}\frac{1}{2i}\frac{1}{\left[1+\frac{z-i}{2i}\right]} = \frac{1}{8}\left[1 - i\left(\frac{z-i}{2}\right)\right]^{-1}$$

$$= \frac{1}{8}\sum_{n=0}^{\infty}\frac{i^n(z-i)^n}{2^n} = \sum_{n=0}^{\infty}\frac{i^n(z-i)^n}{2^{n+3}}.$$

The fourth term can be written in a similar form by writing

$$-\frac{1}{4}\frac{1}{(z+i)^2} = -\frac{1}{4}\frac{1}{[(z-i)+2i]^2} = -\frac{1}{4}\frac{1}{(2i)^2}\frac{1}{\left[1+\frac{z-i}{2i}\right]^2} = \frac{1}{16}\left[1 - i\left(\frac{z-i}{2}\right)\right]^{-2}$$

$$= \frac{1}{16}\sum_{n=1}^{\infty}\frac{ni^{n-1}(z-i)^{n-1}}{2^{n-1}}.$$

The coefficients of the Laurent series expansion will be simplified if the last two results are combined. To accomplish this, we change the summation index in the last expansion to make it start from zero. This is accomplished by setting $n - 1 = m$ when we can write

$$\frac{1}{16}\sum_{n=1}^{\infty}\frac{ni^{n-1}(z-i)^{n-1}}{2^{n-1}} = \sum_{m=0}^{\infty}\frac{(1+m)i^m(z-i)^m}{2^{m+4}}.$$

As the choice of symbol for a summation index does not affect the summation, we now replace m by n to obtain the equivalent result

$$-\frac{1}{4}\frac{1}{(z+i)^2} = \sum_{n=0}^{\infty}\frac{(1+n)i^n(z-i)^n}{2^{n+4}}.$$

As a result, the last two terms of the partial fraction decomposition become

$$\frac{i}{4}\frac{1}{z+i} - \frac{1}{4}\frac{1}{(z+i)^2} = \sum_{n=0}^{\infty}\frac{i^n(z-i)^n}{2^{n+3}} + \sum_{n=0}^{\infty}\frac{(1+n)i^n(z-i)^n}{2^{n+4}}$$

$$= \sum_{n=0}^{\infty}\frac{(n+3)i^n(z-i)^n}{2^{n+4}},$$

from which the complete Laurent series expansion for $0 < |z - i| < 2$ is seen to be

$$\frac{1}{(z^2+1)^2} = -\frac{i}{4}\frac{1}{z-i} - \frac{1}{4}\frac{1}{(z-i)^2} + \sum_{n=0}^{\infty}\frac{(n+3)i^n(z-i)^n}{2^{n+4}}.$$

(b) The singularities of $f(z)$ occur on the unit circle $|z| = 1$, so outside this circle the function will be analytic. As $|1/z| < 1$ in the required domain, the binomial theorem can be used to expand the function when written in the form

$$\frac{1}{(z^2+1)^2} = \frac{1}{z^4}\frac{1}{(1+\frac{1}{z^2})^2} = \frac{1}{z^4}\left(1+\frac{1}{z^2}\right)^{-2},$$

from which it follows that

$$\frac{1}{(z^2+1)^2} = \sum_{n=1}^{\infty}(-1)^{n+1}\frac{n}{z^{2n+2}} \quad \text{for } |z| > 1. \quad \blacksquare$$

When $|z|$ is large the operations leading to a Laurent series are sometimes difficult to perform directly. In such circumstances the substitution $z = 1/u$ is made where $|u|$ is small, corresponding to $|z|$ large, and after the expansion has been developed in terms of u, the result is then transformed back to the original variable z. This approach is illustrated in the next example.

EXAMPLE 15.20   Find the Laurent series expansion of $f(z) = \text{Log}(\frac{z-1}{z-2})$ for large $|z|$.

Solution   Substituting $z = 1/u$ in $f(z)$ gives

$$f(z) = \text{Log}\left(\frac{z-1}{z-2}\right) = \text{Log}\left(\left(1-\frac{1}{z}\right)\Big/\left(1-\frac{2}{z}\right)\right) = \text{Log}\left(\frac{1-u}{1-2u}\right)$$

$$= \text{Log}(1-u) - \text{Log}(1-2u).$$

Replacing the logarithms in this last expression by their Maclaurin series expansions that will *both* be valid provided $|u| < \frac{1}{2}$ gives

$$f(u) = -\left(u + \frac{1}{2}u^2 + \frac{1}{3}u^3 + \cdots \frac{1}{n}u^n + \cdots\right) + \left(2u + 2u^2 + \frac{8}{3}u^3 + \cdots + \frac{2^n}{n}u^n + \cdots\right)$$

$$= u + \frac{3}{2}u^2 + \frac{7}{3}u^3 + \cdots + \left(\frac{2^n-1}{n}\right)u^n + \cdots, \quad \text{for } |u| < \frac{1}{2}.$$

Finally, transforming back to the variable z, and noticing that $|u| < 2$ corresponds to $|z| > 2$, we arrive at the required Laurent series expansion for large $|z|$:

$$\text{Log}\left(\frac{z-1}{z-2}\right) = \sum_{n=1}^{\infty}\frac{2^n-1}{nz^n}, \quad \text{for } |z| > 2. \quad \blacksquare$$

The expansion of functions as Laurent series makes it necessary to classify the different types of singularity that arise. The relevance of this classification, and the importance of the coefficients of a Laurent series, will become clear later once the evaluation of integrals by means of contour integration has been developed.

isolated singularities, removable singularities, poles, and essential singularities

A point z_0 is called an **isolated singularity** of a function $f(z)$ if $f(z)$ has a singularity at z_0, but is single valued and analytic in the annulus (punctured disc) $0 < |z - z_0| < R$.

Singularities are easily identified when a function is a quotient of analytic functions $f(z) = g(z)/h(z)$, because they occur at any zero z^* of $h(z)$ where the numerator $g(z^*) \neq 0$, and also at any infinity of $g(z)$ where $h(z)$ remains finite.

For example, the function $f(z) = (z + 3)/(z^2 + 4)$ has singularities at the zeros $z = \pm 2i$ of the denominator $z^2 + 4$, because the numerator $z + 3$ does not vanish at either of these points. However, the function $f(z) = (\tan z)/z^2$ has a singularity at $z = 0$ due to a zero of the denominator, because although $\tan z = z + z^3/3 + \ldots$, the function $f(z) = (\tan z)/z^2 = (1/z) + z/3 + \ldots$. So $f(z)$ has a singularity at the origin also and also at $z = (2n + 1)\pi/2$ for $n = 0, \pm 1, \pm 2, \ldots$, because of infinities of the numerator.

Consideration of the general form of the Laurent series expansion given in (15) allows three distinct cases to be identified, namely:

1. The Laurent series for $f(z)$ contains no negative powers of $(z - z_0)$.
2. The Laurent series for $f(z)$ only contains a finite number of terms involving negative powers of $(z - z_0)$, up to and including the term in $(z - z_0)^{-r}$.
3. The Laurent series for $f(z)$ contains infinitely many terms involving negative powers of $(z - z_0)$.

Case 1. Functions $f(z)$ with this property are said to have a **removable singularity** at z_0 because, irrespective of how $f(z)$ is defined at z_0 (and even if it is not defined), the Laurent series converges to the value a_0 when $z = z_0$. Consequently, by defining $f(z_0) = a_0$ the singularity (discontinuity) at z_0 is *removed*. In working with functions with removable singularities, it is always assumed that they have been removed.

Case 2. Functions $f(z)$ with this property have a principal part of the Laurent series of the form

$$\frac{a_{-1}}{(z - z_0)} + \frac{a_{-2}}{(z - z_0)^2} + \cdots + \frac{a_{-r+1}}{(z - z_0)^{r-1}} + \frac{a_{-r}}{(z - z_0)^r},$$

where some or all of the coefficients $a_{-1}, a_{-2}, \ldots, a_{-r+1}$ may be zero, but $a_{-r} \neq 0$. This type of singularity is called a **pole of order r** of the function $f(z)$ located at z_0, or sometimes a pole of **multiplicity r** located at z_0. A pole of order 1 is called a **simple pole**.

Although no further use will be made of the term, for the sake of completeness we mention that the quotient of two analytic functions is called a **meromorphic function**. Thus, a meromorphic function is analytic throughout a domain apart from points where poles arise due to a zero of the denominator where the numerator is nonvanishing.

Case 3. Functions $f(z)$ with this property are said to have an **essential singularity** located at the point z_0.

In what follows our concern will only be with Cases 1 and 2, because of the extremely erratic behavior of functions in a neighborhood of an essential singularity.

EXAMPLE 15.21   Identify the singularities of the functions

(a) $f(z) = \dfrac{\cosh z - 1}{z^2}$,   (b) $f(z) = \dfrac{2z^2 + 13z + 3}{z^3 + 3z^2 - 4}$,   (c) $f(z) = \dfrac{\sinh z}{z^5}$,

(d) $f(z) = z \exp(1/z)$.

Solution

(a) $f(z)$ is analytic everywhere apart from $z = 0$ where it is indeterminate. To examine the behavior of $f(z)$ at the origin we replace $\cosh z$ by its Maclaurin series, leading to the result

$$\frac{\cosh z - 1}{z^2} = \frac{\left(1 + \frac{z^2}{2!} + \frac{z^4}{4!} + \cdots\right) - 1}{z^2}.$$

Cancelling the 1 and dividing by z^2 gives

$$\frac{\cosh z - 1}{z^2} = \frac{1}{2} + \frac{z^2}{4!} + \cdots,$$

so taking the limit as $z \to 0$, we find that

$$\lim_{z \to 0} \frac{\cosh z - 1}{z^2} = \frac{1}{2}.$$

If we define

$$f(z) = \begin{cases} \dfrac{\cosh z - 1}{z^2}, & z \neq 0 \\[2mm] \dfrac{1}{2}, & z = 0, \end{cases}$$

the singularity at $z = 0$ has been removed, and the resulting function is analytic for all z, so this function has a *removable singularity* at $z = 0$.

(b) A partial fraction decomposition of $f(z)$ gives

$$f(z) = \frac{2}{z - 1} + \frac{5}{(z + 2)^2},$$

from which it can be seen that $f(z)$ has a *simple pole* at $z = 1$ and a *pole of order* 2 at $z = -2$.

(c) As

$$f(z) = \frac{\sinh z}{z^5} = \frac{z + \frac{z^3}{3!} + \frac{z^5}{5!} + \frac{z^7}{7!} + \cdots}{z^5} = \frac{1}{z^4} + \frac{1}{3!}\frac{1}{z^2} + \frac{1}{5!} + \frac{1}{7!}z^2 + \cdots,$$

the function is seen to have a *pole of order* 4 at the origin and to be analytic for all $z \neq 0$.

(d) Expanding the function gives

$$f(z) = z \exp(1/z) = z\left(1 + \frac{1}{z} + \frac{1}{2!z^2} + \frac{1}{3!z^3} + \cdots\right) = z + 1 + \frac{1}{2!z} + \frac{1}{3!z^2} + \cdots,$$

showing that this function has an *isolated essential singularity* at the origin. ∎

The Extended Complex Plane: The Point at Infinity

Unlike real numbers, complex numbers have no natural *order property*, so the inequality symbols $<$ and $>$ have no meaning when applied to complex numbers z_1 and z_2. However, $|z_1|$ and $|z_2|$ are real numbers that can be ordered, so this property can be used to give meaning to the "number" $z = \infty$. This is accomplished by saying that the complex sequence $\{z_n\}$ *tends to infinity*, written

$$\lim_{n \to \infty} z_n = \infty,$$

the meaning of the point at infinity in the complex plane, and the Riemann sphere

if

$$\lim_{n \to \infty} |z_n| = \infty.$$

This definition coincides with the corresponding one for real numbers, because the last result means that for any positive number L there is a positive integer N such that $|z_n| > L$ for all $n > N$. Thus, **the point at infinity** in the complex plane is taken to be the set of all points z such that $|z|$ lies outside the circle $|z| = L$ for any positive L. Accordingly, the set of all points outside a circle of arbitrarily large radius L centered on the origin is said to be a **neighborhood of infinity**. The complex plane, to which has been added the point at infinity is called the **extended complex plane**, and it is useful when performing various limiting operations.

A geometrical interpretation of $z = \infty$ that provides a justification for using the expression "point" at infinity can be obtained by making a stereographic projection of the extended complex plane onto a sphere. The concept, called the **Riemann sphere**, is illustrated in Fig. 15.5, which represents a sphere resting on the extended complex plane with its center above the origin. The point S of the sphere at the origin is called its *south pole* and the point N on its surface vertically above the origin is called its *north pole*.

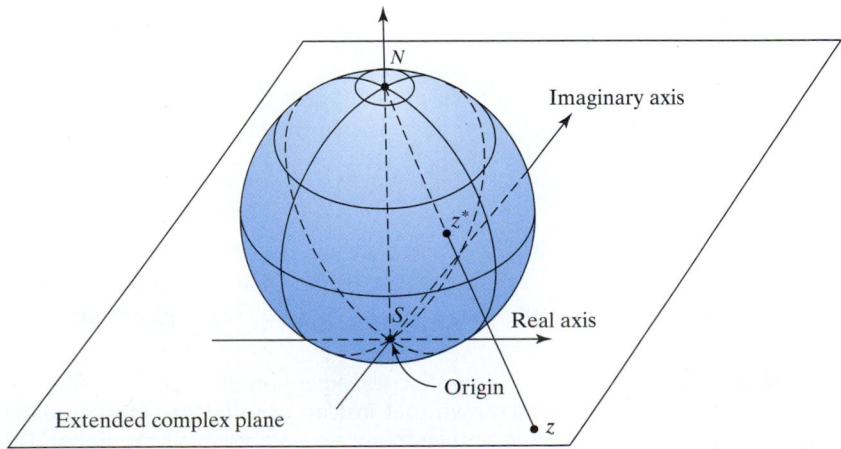

FIGURE 15.5 The Riemann sphere.

A point z on the extended complex plane is brought into correspondence with a point z^* on the sphere by taking z^* to be the point of intersection with the sphere of a straight line drawn from N to the point z. Each finite z corresponds to a unique point on the sphere, while all points in a neighborhood of $z = \infty$, which is outside a circle of arbitrarily large radius drawn in the extended complex plane with the origin as its center, correspond to an arbitrarily small neighborhood of N. Thus, the point N corresponds to the point at infinity in the extended complex plane. It is easy to see that circles in the extended complex plane with their center at the origin map to circles on the sphere (lines of latitude) while radial lines through S in the extended complex plane map to great circles (meridians) on the sphere (lines of longitude).

As already remarked, to study the behavior of a function $f(z)$ in a neighborhood of $z = \infty$, the substitution $z = 1/u$ is made, leading to an expression $F(u) = f(1/u)$. The behavior of $f(z)$ in a neighborhood of $z = \infty$ is then determined by the behavior of $F(u)$ in a neighborhood of $u = 0$.

Thus, if we consider the extended complex plane, the Laurent series for $f(z) = \frac{1}{z(1-z)}$ in a neighborhood of $z = \infty$ is obtained by setting $z = 1/u$, taking u to be arbitrarily small, and then, after expanding the result, writing $u = 1/z$. This leads to the result

$$F(u) = f(1/u) = \frac{1}{\frac{1}{u}\left(1 - \frac{1}{u}\right)} = \frac{u^2}{u - 1}$$

$$= -u^2(1 - u)^{-1} = -u^2(1 + u + u^2 + \cdots) = -\sum_{n=2}^{\infty} u^n,$$

and after substituting $u = 1/z$ this becomes the required Laurent series expansion in a neighborhood of $z = \infty$,

$$f(z) = -\sum_{n=2}^{\infty} \frac{1}{z^n} \quad \text{for } |z| > 1.$$

The same form of argument makes it possible to determine if a function $f(z)$ has a singularity at infinity and to classify such singularities. If we set $z = 1/u$ as before to obtain $F(u) = f(1/u)$, the singularity of $f(z)$ at $z = \infty$ is defined to be the same as that of $F(u)$ at $u = 0$.

For example, the function

$$f(z) = z^5 - \frac{1}{z^3}.$$

has a pole of order 3 at the origin in the ordinary complex plane, so to study its behavior at $z = \infty$ in the extended complex plane we set $z = 1/u$ when

$$F(u) = \frac{1}{u^5} - u^3,$$

showing that $F(u)$ has a pole of order 5 at $z = \infty$. Similarly, the function $f(z) = e^z$ is regular at the origin, that is, it has no singularity at the origin, but as $F(u) = f(1/u) = e^{1/u}$ we see that $f(z) = e^z$ has an essential singularity at $z = \infty$.

Summary

The Laurent series expansion of a function $f(z)$ about a singularity was defined, and it was shown that instead of using the formal definition to arrive at the expansion, it is often simpler to use a simple algebraic argument. Poles and singularities of functions were defined, and the meaning of the point at infinity in the complex plane was explained.

EXERCISES 15.3

In Exercises 1 through 12 find the Laurent series of $f(z)$ expanded about the given point and determine its annulus of convergence.

1. $f(z) = \dfrac{1}{z-2}$ expanded about $z = 0$.

2. $f(z) = \dfrac{1}{(z-a)^2}$ $(a \neq 0)$ expanded about $z = 0$.

3. $f(z) = \dfrac{1}{(z-a)(z-b)}$ $(0 < |a| < |b|)$, with $|z| < |a|$ expanded about $z = 0$.

4. $f(z) = \dfrac{1}{(z-a)(z-b)}$ $(0 < |a| < |b|)$, with $0 < |z-a| < |b-a|$ expanded about $z = a$.

5. $f(z) = \dfrac{1}{(z-a)(z-b)}$ $(0 < |a| < |b|)$, in the annulus $|a| < |z| < |b|$ when expanded about $z = 0$.

6. $f(z) = \dfrac{z^2 - 2z + 5}{(z-2)(z^2+1)}$ expanded about $z = 2$.

7. $f(z) = \exp\left(\dfrac{1}{1-z}\right)$ expanded (a) about $z = 1$, and (b) about $z = 0$ for $|z| > 1$.

8. $f(z) = \dfrac{1}{z(1-z)}$ expanded (a) about $z = 0$ and (b) about $z = 1$.

9. $f(z) = \sin\left(\dfrac{z}{1-z}\right)$ expanded about $z = 1$.

10. $f(z) = \dfrac{1}{(z-2)(z-3)}$ expanded about $z = 0$ for $|z| < 2$ and for $2 < |z| < 3$.

11. $f(z) = \dfrac{1}{(1-z)(z+2)}$ expanded about $z = 0$.

12. $f(z) = \dfrac{1}{z^2 + a^2}$ expanded about (a) $z = ia$ and (b) $z = 0$ for $|z| > |a|$.

In Exercises 13 through 16 find the first four terms of the Laurent series expansion of $f(z)$ about the given point.

13. $f(z) = \sinh\left(1 + \dfrac{1}{z}\right)$ expanded about $z = 0$.

14. $f(z) = \cosh\left(2 + \dfrac{1}{z}\right)$ expanded about $z = 0$.

15. $f(z) = \dfrac{\sin z \sin(z/3)}{z^3}$ expanded about $z = 0$.

16. $f(z) = \dfrac{\sin z \sinh(z/4)}{z^4}$ expanded about $z = 0$.

In Exercises 17 through 28 classify the nature of any singularities that occur in the finite complex plane.

17. $f(z) = \dfrac{1}{4z - z^3}$.

18. $f(z) = \dfrac{z}{1 + z^4}$.

19. $f(z) = \exp(-1/z^2)$.

20. $f(z) = \dfrac{1+z}{z(z^2+4)^2}$.

21. $f(z) = \dfrac{\sin z}{\sinh z}$.

22. $f(z) = \dfrac{1+z^2}{\cosh z}$.

23. $f(z) = \exp\left(\dfrac{z}{1-z}\right)$.

24. $f(z) = \cot(1/z)$.

25. $f(z) = \tan^2 z$.

26. $f(z) = \dfrac{\cos z}{z^2}$.

27. $f(z) = \dfrac{\cos 2z - 1}{\sin^2 z}$.

28. $f(z) = \dfrac{z^3 - 8z - 3}{z - 3}$.

Further Results

29. The integral for a_n in Theorem 15.13 defines the coefficients of the Laurent series for a function $f(z)$ expanded about the point z_0 that is convergent in the annulus $R_1 < |z - z_0| < R_2$. Use this integral to derive the **Cauchy inequalities for the coefficients of a Laurent series**

$$|a_n| \leq \frac{M}{R^n}, \quad \text{for } n = 0, \pm 1, \pm 2, \ldots,$$

where M is the greatest value of $|f(z)|$ on a circle $|z - z_0| = R$, with $R_1 < R < R_2$.

30. Use the result of Exercise 29 to show that if a function $f(z) = \sum_{n=0}^{\infty} a_n z^n$ is an entire function such that when $|z| > R_1$ (the inner radius of the annulus of convergence in Theorem 15.13), and for a given nonnegative integer N,

$$|f(z)| < M|z|^N,$$

then $f(z)$ must be a polynomial of degree no greater than N.

In Exercises 31 through 34 find the Laurent series expansion of $f(z)$ in a neighborhood of $z = \infty$.

31. $f(z) = \dfrac{1}{z + 3}$.

32. $f(z) = \dfrac{1}{(z^2 + 1)^2}$.

33. $f(z) = \text{Log}\left(\dfrac{z^2}{1 + z^2}\right)$.

34. $f(z) = \dfrac{1}{(z-a)(z-b)}$ $\times (0 < |a| < |b|)$.

In Exercises 35 through 40 determine the nature of the singularity of $f(z)$ at $z = \infty$.

35. $f(z) = \dfrac{1}{z - z^3}$.

36. $f(z) = \dfrac{z^5}{(1 + z)^2}$.

37. $f(z) = \dfrac{1}{\sin z}$.

38. $f(z) = \dfrac{\cos 3z}{z^2}$.

39. $f(z) = e^{2iz}$.

40. $f(z) = \tan z$.

15.4   Residues and the Residue Theorem

Let an analytic function $f(z)$ have an isolated singularity at z_0. Then its Laurent series expansion about the point z_0,

$$f(z) = \sum_{n=-\infty}^{\infty} a_n(z - z_0)^n = \cdots + \frac{a_{-3}}{(z - z_0)^3} + \frac{a_{-2}}{(z - z_0)^{-2}} + \frac{a_{-1}}{(z - z_0)}$$

$$+ a_0 + a_1(z - z_0) + a_2(z - z_0)^2 + \cdots, \qquad (22)$$

the residue and its connection with the Laurent expansion

will converge in some punctured disc $0 < |z - z_0| < R$. The **residue** of $f(z)$ at $z = z_0$, written Res$[f(z), z_0]$, or simply Res$[z_0]$ when there is no ambiguity about the function involved, is defined as the number a_{-1}, so that

$$\text{Res}[f(z), z_0] = a_{-1}. \qquad (23)$$

Thus, the residue of $f(z)$ at z_0 is the coefficient of the term $1/(z - z_0)$ in the principal part of its Laurent series expansion about z_0.

EXAMPLE 15.22

Find the residue of $f(z) = 1/(z^2 + 1)^2$ at the point $z = i$.

Solution   It was shown in Example 15.19 that the Laurent series of $f(z) = 1/(z^2 + 1)^2$ expanded about the point $z = i$ is

$$f(z) = -\frac{1}{4(z - i)^2} - \frac{i}{4(z - i)} + \sum_{n=0}^{\infty} \frac{(n + 3)i^n(z - i)^n}{2^{n+4}} \quad \text{for } 0 < |z - i| < 2,$$

so the residue at $z = i$ is seen to be

$$\text{Res}[i] = -\frac{i}{4}. \qquad \blacksquare$$

From now on our concern will be with residues of analytic functions $f(z)$ whose only isolated singularities are poles. Then, if z_0 is a pole of $f(z)$ of order N, its Laurent series expansion about the pole will be of the form

$$f(z) = \frac{a_{-1}}{z - z_0} + \frac{a_{-2}}{(z - z_0)^2} + \cdots + \frac{a_{-N}}{(z - z_0)^N} + \sum_{n=0}^{\infty} a_n(z - z_0)^n, \qquad (24)$$

where $a_{-N} \neq 0$, though some or all of the remaining coefficients $a_{-1}, a_{-2}, \ldots, a_{1-N}$ may vanish.

Let $f(z)$ be analytic at z_0. Then z_0 is a **zero** of the function $f(z)$ if $f(z_0) = 0$. In some neighborhood of z_0 the function will have a Taylor series expansion of the form

$$f(z) = \sum_{n=1}^{\infty} a_n(z - z_0)^n, \qquad (25)$$

where to satisfy the condition $f(z_0) = 0$ we have set the coefficient $a_0 = 0$. The zero z_0 is called a **simple zero** of $f(z)$ if $a_1 \neq 0$, and a **zero of order N** if the first nonvanishing coefficient in (25) is a_N. If the zero is of order N we can write

$$f(z) = (z - z_0)^N g(z), \qquad (26)$$

a zero of order *n* and testing for a pole of order *n*

where $g(z_0) \neq 0$ and $g(z)$ is analytic in a neighborhood of z_0, from which it follows that if $f(z)$ has a zero of order N at z_0, then $1/f(z)$ will have a pole of order N at z_0.

Inspection of (24) provides the following simple test for a pole of order N.

Test for a pole of order N

If $f(z)$ is analytic in the punctured disc $0 < |z - z_0| < R$, then a necessary and sufficient condition for it to have a pole of order N at z_0 is that

$$\lim_{z \to z_0} (z - z_0)^N f(z) = C, \quad \text{where } C \neq 0.$$

In most cases, when z_0 is a pole of $f(z)$, it is simpler to determine the residue at z_0 by one of the formulas we will now derive than to develop the Laurent series expansion of $f(z)$ about z_0 and then to identify the residue with the coefficient a_{-1}.

The simplest case occurs when a function $f(z)$ of the form

$$f(z) = \frac{g(z)}{h(z)}$$

has a simple pole at z_0, and $g(z)$ and $h(z)$ are analytic functions in a neighborhood of z_0. Suppose first that $h(z)$ contains a factor $(z - z_0)$, and so can be written $h(z) = (z - z_0)F(z)$, where $F(z_0) \neq 0$. Then

$$f(z) = \frac{1}{(z - z_0)} \frac{g(z)}{F(z)},$$

but $H(z) = g(z)/F(z)$ is analytic at z_0 and so can be expanded in a Taylor series about z_0 of the form

$$H(z) = H(z_0) + (z - z_0)H'(z_0) + \frac{1}{2!}(z - z_0)^2 H''(z_0) + \cdots.$$

Using this result in the expression for $f(z)$ and writing $H(z_0) = g(z_0)/F(z_0)$ gives

$$f(z) = \frac{1}{z - z_0} \frac{g(z_0)}{F(z_0)} + H'(z_0) + \frac{1}{2!}(z - z_0)H''(z_0) + \cdots.$$

This shows that $\text{Res}[f(z), z_0]$, the coefficient of $1/(z - z_0)$ in the Laurent series expansion of $f(z)$ about z_0, is given by

$$\text{Res}[f(z), z_0] = \frac{g(z_0)}{F(z_0)}. \tag{27}$$

Now suppose that $f(z)$ is of the form

$$f(z) = \frac{g(z)}{h(z)}$$

and has a simple pole at z_0, but that $h(z)$ does *not* contain a factor $(z - z_0)$. Then, as $f(z)$ will have a Laurent series expansion about z_0 of the form

$$f(z) = \frac{g(z)}{h(z)} = \frac{\text{Res}[f(z), z_0]}{z - z_0} + \sum_{n=0}^{\infty} a_n(z - z_0)^n,$$

we see that

$$\text{Res}[f(z), z_0] = \lim_{z \to z_0} \left\{ \frac{(z - z_0)g(z)}{h(z)} \right\}.$$

Using the fact that $h(z_0) = 0$ allows this to be written

$$\text{Res}[f(z), z_0] = \lim_{z \to z_0} \left\{ g(z) \Big/ \left(\frac{h(z) - h(z_0)}{z - z_0} \right) \right\},$$

but $h(z)$ is analytic and

$$h'(z_0) = \lim_{z \to z_0} \left(\frac{h(z) - h(z_0)}{z - z_0} \right),$$

so

$$\text{Res}[f(z), z_0] = \frac{g(z_0)}{h'(z_0)}. \tag{28}$$

Finally we consider the case where $f(z)$ has a pole of order N at z_0, and so has the Laurent series expansion about z_0 given by (24). Multiplying (24) by $(z - z_0)^N$ gives

$$(z - z_0)^N f(z) = a_{-1}(z - z_0)^{N-1} + a_{-2}(z - z_0)^{N-2} + \cdots + a_{-N} + \sum_{n=0}^{\infty} a_n(z - z_0)^{N+n},$$

and after differentiating this with respect to z we find that

$$\frac{d}{dz}\left[(z - z_0)^N f(z)\right] = (N - 1)a_{-1}(z - z_0)^{N-2} + (N - 2)a_{-2}(z - z_0)^{N-3}$$

$$+ \cdots + a_{1-N} + \sum_{n=0}^{\infty}(N + n)a_n(z - z_0)^{N+n-1}.$$

Taking the limit of this result as $z \to z_0$ reduces it to

$$\lim_{z \to z_0} \left\{ \frac{d}{dz}\left[(z - z_0)^N f(z)\right] \right\} = a_{1-N}.$$

A repetition of this process yields the formula

$$\lim_{z \to z_0} \left\{ \frac{d^2}{dz^2}\left[(z - z_0)^N f(z)\right] \right\} = a_{2-N},$$

so as $\text{Res}[f(z), z_0] = a_{-1}$, after $N - 1$ differentiations this same form of argument brings us to the final result

$$\text{Res}[f(z), z_0] = \frac{1}{(N - 1)!} \lim_{z \to z_0} \left\{ \frac{d^{N-1}}{dz^{N-1}}\left[(z - z_0)^N f(z)\right] \right\}. \tag{29}$$

formulas for finding the residue at a simple pole and at a pole of order n

Taken together, results (27) to (29) have established the following formulas for the calculation of residues.

Formulas for the residue at a pole of a function of the form $f(z) = g(z)/h(z)$

1. **(i)** Let a function $f(z)$ that is analytic in a punctured disc $0 < |z - z_0| < R$ have a simple pole at z_0. Then if $f(z) = g(z)/h(z)$, and $h(z)$ contains a factor $(z - z_0)$ and so can be written $h(z) = (z - z_0)F(z)$ where $F(z_0) \neq 0$, the residue of $f(z)$ at z_0 is given by the formula

$$\text{Res}[f(z), z_0] = \frac{g(z_0)}{F(z_0)}, \tag{30}$$

(ii) and if $h(z_0) = 0$, but $h(z)$ does not necessarily contain a factor $(z - z_0)$, the residue of $f(z)$ at z_0 is given by the formula

$$\operatorname{Res}[f(z), z_0] = \frac{g(z_0)}{h'(z_0)}. \tag{31}$$

2. Finally, if $f(z)$ has a pole of order N at z_0, the residue of $f(z)$ at z_0 is given by the formula

$$\operatorname{Res}[f(z), z_0] = \frac{1}{(N-1)!} \lim_{z \to z_0} \left\{ \frac{d^{N-1}}{dz^{N-1}} [(z - z_0)^N f(z)] \right\}. \tag{32}$$

EXAMPLE 15.23

Find the residues at the poles of the functions

(a) $f(z) = \dfrac{z^2 + 2z + 3}{z - i}$, (b) $f(z) = \dfrac{1}{(z^2 + 1)^2}$, (c) $f(z) = \dfrac{1}{z \sin z}$,

(d) $f(z) = \operatorname{sech} z$.

Solution

(a) $f(z)$ has a simple pole at $z = i$, with $g(z) = z^2 + 2z + 3$ and $h(z) = z - i$, so as the denominator contains the factor $(z - i)$, making use of (30) gives

$$\operatorname{Res}[i] = \left[(z - i) \frac{(z^2 + 2z + 3)}{(z - i)} \right]_{z=i} = [z^2 + 2z + 3]_{z=i} = 2(1 + i).$$

(b) The function has poles of order 2 at $z = \pm i$, so from (32) with $N = 2$ we see that

$$\operatorname{Res}[i] = \frac{1}{1!} \lim_{z \to i} \left\{ \frac{d}{dz} \left[(z - i)^2 \frac{1}{(z^2 + 1)^2} \right] \right\} = -\frac{i}{4},$$

and similarly

$$\operatorname{Res}[-i] = \frac{1}{1!} \lim_{z \to -i} \left\{ \frac{d}{dz} \left[(z + i)^2 \frac{1}{(z^2 + 1)^2} \right] \right\} = \frac{i}{4}.$$

This simple calculation for the determination of $\operatorname{Res}[i]$ should be compared with the extensive calculations needed to arrive at the full Laurent series for $f(z)$ expanded about the point $z = i$ in Example 15.22, where the coefficient of the term $1/(z - i)$ was, of course, equal to $-i/4$.

(c) The function has poles at the zeros of the denominator $z \sin z$. For small z

$$\sin z = z - \frac{z^3}{3!} + \frac{z^5}{5!} - \cdots = z \left(1 - \frac{z^2}{3!} + \frac{z^4}{5!} - \cdots \right),$$

so near the origin

$$f(z) = \frac{1}{z^2} \cdot \frac{1}{\left(1 - \frac{z^2}{3!} + \frac{z^4}{5!} - \cdots \right)},$$

showing that $f(z)$ has a pole of order 2 at the origin. Elsewhere, $z \neq 0$ and the factor $\sin z$ has zeros at $\pm n\pi$ for $n = 0, 1, 2, \ldots$, corresponding to simple poles of $f(z)$.

The residue at the origin, obtained from (32) with $N = 2$ and $z_0 = 0$, is

$$\text{Res}[0] = \frac{1}{1!} \lim_{z \to 0} \left\{ \frac{d}{dz} \left[z^2 \frac{1}{z \sin z} \right] \right\} = \lim_{z \to 0} \left\{ \frac{\sin z - z \cos z}{\sin^2 z} \right\}.$$

This is an indeterminate form, so applying l'Hôpital's rule we find that

$$\text{Res}[0] = \lim_{z \to 0} \left\{ \frac{z \sin z}{2 \sin z \cos z} \right\} = 0.$$

The residues at the simple zeros $\pm n\pi$, for $n = 1, 2, \ldots$, follow by setting $g(z) = 1$ and $h(z) = z \sin z$ in (31) to obtain

$$\text{Res}[\pm n\pi] = [1/(\sin z + z \cos z)]_{z = \pm n\pi} = \frac{\pm(-1)^n}{n\pi}, \quad \text{for } n = 1, 2, \ldots.$$

(d) Writing $f(z) = 1/\cosh z$ shows that poles of $f(z)$ are located at the zeros $(2n + 1)\pi i/2$ of $\cosh z$ for $n = 0, \pm 1, \pm 2, \ldots$. So $f(z)$ has simple poles at $z = (2n + 1)\pi i/2$ for $n = 0, \pm 1, \pm 2, \ldots$. From (31) using $g(z) = 1$ and $h(z) = \cosh z$ we have

$$\text{Res}[(2n + 1)\pi i] = \frac{1}{\sinh\{(2n + 1)\pi i/2\}} = \frac{1}{i \sin\{(2n + 1)\pi/2\}} = (-1)^{n+1} i,$$

for $n = 0, \pm 1, \pm 2, \ldots$. ■

When the limit in (32) is difficult to evaluate, it is necessary to determine the residue by developing the Laurent series expansion to the point where the coefficient of the term $1/(z - z_0)$ can be identified. This situation is illustrated in the next example.

EXAMPLE 15.24

Find the residue of

$$f(z) = \sin\left(\frac{z}{z + 1} \right).$$

Solution Inspection of the argument of the sine function shows that its only singularity occurs at $z = -1$, but the function is sufficiently complicated that result (32) is not useful. Accordingly, to find the coefficient of the term $1/(z + 1)$ in the Laurent series expansion about $z = -1$, we rewrite $f(z)$ as

$$f(z) = \sin\left(1 - \frac{1}{z + 1} \right),$$

and then use the familiar trigonometric identity $\sin(A - B) = \sin A \cos B - \cos A \sin B$ to expand this as

$$f(z) = \sin(1) \cos\left(\frac{1}{z + 1} \right) - \cos(1)\sin\left(\frac{1}{z + 1} \right).$$

Replacing the cosine and sine function involving z with the first few terms of their Maclaurin series gives

$$f(z) = \sin(1) \left(1 - \frac{1}{2!(z + 1)^2} + \frac{1}{4!(z + 1)^4} - \cdots \right)$$

$$- \cos(1) \left(\frac{1}{z + 1} - \frac{1}{3!(z + 1)^3} + \frac{1}{5!(z + 1)^5} - \cdots \right).$$

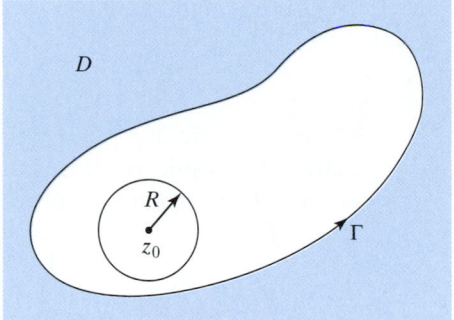

FIGURE 15.6 A contour Γ containing a point z_0 at which $f(z)$ has a pole.

Inspection then shows that the coefficient of the term $1/(z+1)$ is $-\cos(1)$, so

$$\text{Res}[\,f(z), -1] = -\cos(1).\qquad\blacksquare$$

why residues are important

The crucial importance of residues in the theory of complex integration follows from the fact that when a function $f(z)$ has a pole of any order at a point z_0 in a domain D, but is analytic elsewhere D, then the integral around any contour in D that contains the pole at z_0 depends *only* on the value of the residue at z_0. To prove this assertion, and to find the value of the integral, we consider the case in which $f(z)$ has a pole of order N at a point z_0 in a domain D but is analytic elsewhere in D.

We take a positively oriented contour Γ in D as shown in Fig. 15.6, represent $f(z)$ by its Laurent series (24) expanded about z_0, and integrate the result around Γ.

As a result we have

$$\int_{\Gamma} f(z)\,dz = \int_{\Gamma}\left(\frac{a_{-1}}{z - z_0} + \frac{a_{-2}}{(z - z_0)^2} + \cdots + \frac{a_{-N}}{(z - z_0)^N}\right)dz + \sum_{n=0}^{\infty} a_n \int_{\Gamma}(z - z_0)^n\,dz,$$

$$(33)$$

where term-by-term integration of the infinite series at the right is allowed by virtue of Theorem 15.12.

It was shown in Example 14.4 that

$$\int_{|z-z_0|=R}(z - z_0)^n\,dz = 0 \quad \text{for } n = -2, -3, \ldots, \quad \text{and} \quad n = 0, 1, 2, \ldots,$$

where the circle $|z - z_0| = R$ lies within D. The deformation of contour theorem asserts that these results are true for any contour Γ in D that contains z_0, as a result of which (33) reduces to

$$\int_{\Gamma} f(z)\,dz = \int_{\Gamma}\frac{a_{-1}}{z - z_0}\,dz,$$

and so to the equivalent result

$$\int_{\Gamma} f(z)\,dz = \text{Res}[z_0]\int_{\Gamma}\frac{dz}{z - z_0}.$$

In Example 14.5 it was shown that

$$\int_{|z-z_0|=R} \frac{dz}{z-z_0} = 2\pi i,$$

when the circle $|z - z_0| = R$ lies within D. The deformation of contour theorem allows this result to remain true when the circle $|z - z_0| = R$ is replaced by the contour Γ containing z_0, so we have proved the extremely important result that

$$\int_{\Gamma} f(z)dz = 2\pi i \, \text{Res}[f(z), z_0]. \tag{34}$$

This result is easily extended to the case of a function $f(z)$ with m poles in D located at the points $z_1, z_2, \ldots, z_m$. To see this, let the poles in D lie inside a simple positively oriented closed contour Γ contained in D, and enclose the pole at z_r in a small positively oriented circle Γ_r lying inside D, with $r = 1, 2, \ldots, m$. Integrating around Γ and using the extended Cauchy–Goursat theorem, we obtain

$$\int_{\Gamma} f(z)dz = \int_{\Gamma_1} f(z)dz + \int_{\Gamma_2} f(z)dz + \cdots + \int_{\Gamma_m} f(z)dz,$$

but from (34),

$$\int_{\Gamma_r} f(z)dz = 2\pi i \, \text{Res}[f(z), z_r], \quad \text{for } r = 1, 2, \ldots, m,$$

so

$$\int_{\Gamma} f(z)dz = 2\pi i [\text{Res}[f(z), z_1] + \text{Res}[f(z), z_2] + \cdots + \text{Res}[f(z), z_m]]. \tag{35}$$

This result contains the Cauchy–Goursat theorem as a special case, because if the contour Γ in D contains no poles of $f(z)$, the function has no residues inside Γ and so

$$\int_{\Gamma} f(z)dz = 0.$$

The fundamental result contained in (35) forms our next theorem.

THEOREM 15.14

contour integrals and the residue theorem

The residue theorem   Let $f(z)$ have poles at $z_1, z_2, \ldots, z_m$ in a domain D and be analytic elsewhere in D. Then if Γ is any simple positively oriented contour in D containing the points $z_1, z_2, \ldots, z_m$,

$$\int_{\Gamma} f(z)dz = 2\pi i \sum_{r=1}^{m} \text{Res}[f(z), z_r].$$
■

evaluating a contour integral using residues

Expressed in words, this theorem says that the integral of $f(z)$ around Γ is $2\pi i$ times the sum of the residues enclosed in Γ. The next example illustrates the application of Theorem 15.14 to a function with three poles.

EXAMPLE 15.25

Find all the residues of the function $f(z) = \frac{e^z}{(z+2i)^3(z^2-4)}$, and use them to determine $\int_{\Gamma} f(z)dz$ around the following positively oriented contours in which Γ is (a) the circle Γ_1 given by $|z + 3i| = 2$, (b) the circle Γ_2 given by $|z - 2| = 1$, and (c) the circle Γ_3 given by $|z| = 4$.

Solution   Inspection of $f(z)$ shows it has a pole of order 3 at $z = -2i$, and simple poles at $z = \pm 2$. Applying (32) to find the residue at $z = -2i$ gives

$$\text{Res}[f(z), -2i] = \frac{1}{2!} \left\{ \frac{d^2}{dz^2} \left[(z + 2i)^3 \frac{e^z}{(z + 2i)^3 (z^2 - 4)} \right] \right\}_{z=-2i}$$

$$= \frac{1}{2} \left\{ \frac{d^2}{dz^2} \left[\frac{e^z}{z^2 - 4} \right] \right\}_{z=-2i} = \frac{e^{-2i}}{16} \left(i - \frac{3}{4} \right),$$

and as the poles at $z = \pm 2$ are only simple poles, it follows from (30) that

$$\text{Res}[f(z), -2] = \left[\frac{e^z}{(z + 2i)^3 (z - 2)} \right]_{z=-2} = \frac{e^{-2}}{128} (i - 1),$$

and

$$\text{Res}[f(z), 2] = \left[\frac{e^z}{(z + 2i)^3 (z + 2)} \right]_{z=2} = -\frac{e^2}{128} (1 + i).$$

The three contours Γ_1, Γ_2, and Γ_3 and the location of the poles of $f(z)$ are shown in Fig. 15.7. Only the pole of order 3 at $z = -2i$ lies inside contour Γ_1, and only the simple pole at $z = 2$ lies inside contour Γ_2, though all three poles lie inside contour Γ_3.

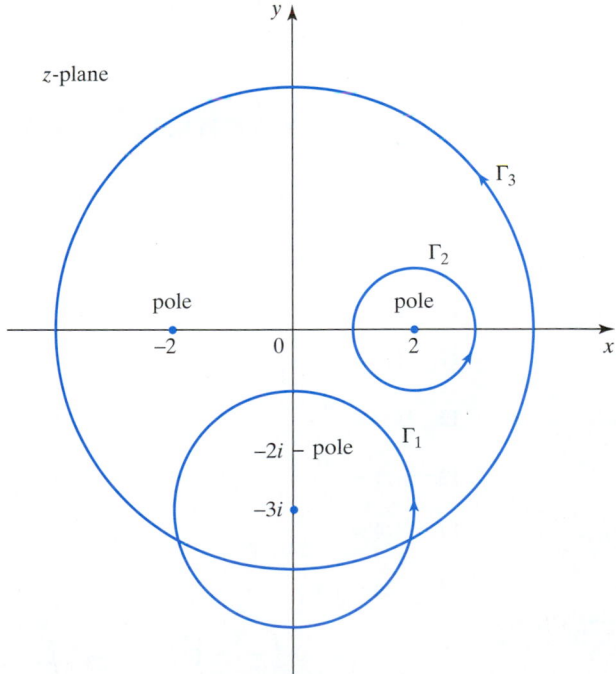

FIGURE 15.7   The contours Γ_1, Γ_2, and Γ_3 and the location of the poles of $f(z)$.

Applying Theorem 15.14 we have

$$\int_{\Gamma_1} f(z)dz = 2\pi i \left(\frac{e^{-2i}}{16} \left(i - \frac{3}{4} \right) \right) = -\frac{\pi e^{-2i}}{8} \left(1 + \frac{3i}{4} \right)$$

$$\int_{\Gamma_2} f(z)dz = 2\pi i \left(-\frac{e^2(1+i)}{128} \right) = \frac{\pi e^2(1-i)}{64}$$

and

$$\int_{\Gamma_3} f(z)dz = 2\pi i \left\{ -\frac{e^2(1+i)}{128} + \frac{e^{-2}(i-1)}{128} + \frac{e^{-2i}}{16} \left(i - \frac{3}{4} \right) \right\}$$

$$= \frac{\pi}{64}(e^2 - e^{-2}) - \frac{\pi e^{-2i}}{8} - \frac{i\pi}{64}(6e^{-2i} + e^2 + e^{-2}). \qquad \blacksquare$$

EXAMPLE 15.26   Find

$$\int_{|z+1|=1} \sin\left(\frac{z}{z+1} \right) dz.$$

Solution   We saw in Example 15.24 that the only singularity of the integrand is a simple pole at $z = -1$ with residue $-\cos(1)$. So as the circle $|z+1| = 1$ contains the pole, it follows immediately from Theorem 15.14 that

$$\int_{|z+1|=1} \sin\left(\frac{z}{z+1} \right) dz = 2\pi i \{-\cos(1)\} = -2\pi i \cos(1). \qquad \blacksquare$$

Summary   The Laurent series was used to introduce the idea of a residue, and formulas for finding the residue at a simple pole and at a pole of order n were derived. The relationship of residues to contour integrals was explained, and the fundamental residue theorem was proved.

EXERCISES 15.4

In Exercises 1 through 16 find the residues of the given functions at their poles in the finite complex plane.

1. $f(z) = \frac{z+3}{z^2-4}$.

2. $f(z) = \frac{z^2+1}{z^2(z+2)}$.

3. $f(z) = \frac{z^2+z-2}{z^2(z+1)}$.

4. $f(z) = \frac{z^2+1}{z(z+1)^3}$.

5. $f(z) = \frac{\sin z}{z^2(z-1)}$.

6. $f(z) = \frac{\cos z}{z^2-5z+6}$.

7. $f(z) = \frac{z^2+3}{\sin z}$.

8. $f(z) = \frac{\sin 3z}{(z-1)^4}$.

9. $f(z) = \tan z$.

10. $f(z) = \cot z$.

11. $f(z) = \frac{1}{e^z+1}$.

12. $f(z) = \frac{\sinh z}{\sin z}$.

13. $f(z) = \frac{\sin z}{\sinh z}$.

14. $f(z) = \frac{\pi}{z^2 \tan \pi z}$.

15. $f(z) = \cos\left(\frac{1}{z-2} \right)$.

16. $f(z) = z^3 \cos\left(\frac{1}{z-2} \right)$.

Evaluate the contour integrals in Exercises 17 through 28.

17. $\int_{|z|=1} \frac{\sin z}{z^4} dz$.

18. $\int_{|z|=2} \frac{\cos z}{z^2} dz$.

19. $\int_{|z|=1} \frac{z^2+1}{z(z-6)} dz$.

20. $\int_{|z-2|=1/2} \frac{zdz}{(z-1)(z-2)^2}$.

21. $\int_{|z-1|=1} \frac{dz}{z^4+1}$.

22. $\int_{|z|=2} \frac{dz}{(z-3)(z^5-1)}$.

23. $\int_{|z|=1} \frac{e^z}{z^2(z^2-9)} dz$.

24. $\int_{|z|=1/2} z^n e^{2/z} dz (n = 0, \pm 1, \pm 2, \ldots)$.

25. $\displaystyle\int_{|z-i|=1} \frac{1 - e^{2iz}}{z^2 + 1}\,dz.$

26. $\displaystyle\int_{|z|=2} \frac{\cos z}{z^3}\,dz.$

27. $\displaystyle\int_{|z|=2} (2z - 1)\cos\left(\frac{z}{z-1}\right)dz.$

28. $\displaystyle\int_{|z|=4} \frac{e^{1/(z-1)}}{z - 2}\,dz.$

Integrals of the form

$$\int_0^{2\pi} \text{Rational}[\cos\theta, \sin\theta]\,d\theta,$$

where $\text{Rational}[\cos\theta, \sin\theta]$ is a rational function of $\cos\theta$ and $\sin\theta$ (a quotient of polynomials in $\cos\theta$ and $\sin\theta$), can be evaluated by making the substitutions

$$\cos\theta = \frac{1}{2}\left(z + \frac{1}{z}\right), \quad \sin\theta = \frac{1}{2i}\left(z - \frac{1}{z}\right), \quad \text{and} \quad d\theta = \frac{dz}{iz},$$

which all follow from De Moivre's theorem, and then integrating around the unit circle $|z| = 1$. Use this approach to evaluate the trigonometric integrals in Exercises 29 through 33.

29. $\displaystyle\int_0^{2\pi} \frac{d\theta}{a + \cos\theta} \quad (a > 1).$

30. $\displaystyle\int_0^{2\pi} \frac{d\theta}{(a + \cos\theta)^2} \quad (a > 1).$

31. $\displaystyle\int_0^{2\pi} \frac{d\theta}{3 + \sin\theta}.$

32. $\displaystyle\int_0^{2\pi} \frac{d\theta}{3 - 2\sin\theta}.$

33. $\displaystyle\int_0^{2\pi} \frac{d\theta}{1 - 2a\cos\theta + a^2} \quad (0 < a < 1).$

34. Prove that if $f(z) = g(z)/h(z)$ is the quotient of two functions where $g(z)$ is analytic at z_0 with $g(z_0) \neq 0$, and $h(z)$ has a zero of order 2 at z_0, then

$$\text{Res}[f(z), z_0] = \frac{6g'(z_0)h''(z_0) - 2g(z_0)h'''(z_0)}{3[h''(z_0)]^2}.$$

15.5   Evaluation of Real Integrals by Means of Residues

The previous section showed how the residues at the poles of an analytic function inside a simple closed contour determine the value of integral of the function around the contour. In the present section we show how by taking some part of the contour along the real axis it is possible to use the method of residues to evaluate improper real integrals of the form

$$\int_0^\infty f(x)\,dx \quad \text{and} \quad \int_{-\infty}^\infty f(x)\,dx,$$

where $f(x)$ may become infinite at a finite number of points in the interval of integration.

(a) Convergence, Divergence, and Cauchy Principal Values of Integrals

The meaning of integration over a semi-infinite or an infinite interval obtained by complex analysis needs to be explained. It will be recalled from elementary calculus that when $f(x)$ remains finite over the interval of integration, the values of these improper integrals are defined as the limiting values

$$\int_0^\infty f(x)\,dx = \lim_{R\to\infty} \int_0^R f(x)\,dx \quad \text{and}$$

$$\int_{-\infty}^\infty f(x)\,dx = \lim_{R_1\to\infty,\, R_2\to\infty} \int_{-R_1}^{R_2} f(x)\,dx, \tag{36}$$

where in the second integral R_1 and R_2 are allowed to tend to infinity *independently* of each other. If these limiting values are finite, the improper integrals are said to *converge* to the values of their respective limits, and they are said to be *divergent* if the limits are undefined or are infinite.

If, in addition, $f(x)$ becomes infinite at a point x_0 inside the interval of integration, say in the first of these integrals, the value of the integral is to be interpreted as

$$\int_0^\infty f(x)dx = \lim_{\alpha \to 0} \int_0^{x_0 - \alpha} f(x)dx + \lim_{\beta \to 0, R \to \infty} \int_{x_0 + \beta}^R f(x)dx, \qquad (37)$$

where $\alpha > 0$ and $\beta > 0$ are allowed to tend to zero *independently* of each other. If this limit is finite, the integral is said to *converge* to the value of the limit, and it is said to be *divergent* if the limit is undefined or infinite. A corresponding interpretation applies to integrals over the interval $(-\infty, \infty)$ when $f(x)$ is infinite at a point x_0 inside the interval of integration. If $f(x)$ is infinite at several points inside the interval of integration, the limiting operation shown in (37) is extended in an obvious manner.

Improper integrals such as (37) can occur that are *divergent* if α and β are allowed to tend to zero independently of each other, but are *convergent* if $\beta = \alpha$ as $\alpha \to 0$. In convergent integrals of this type the upper limit of integration in the first integral in (37) is $x_0 - \alpha$ and the lower limit in the second integral is $x_0 + \alpha$. Similarly, improper integrals over infinite intervals such as the second integral in (36) occur that are divergent when R_1 and R_2 are allowed to tend to infinity independently of each other, but are convergent if $R_1 = R_2$, as $R_1 \to +\infty$.

The value of an improper integral when the limits of integration on either side of an infinity of the integrand at x_0 are of the form $x_0 - \alpha$ and $x_0 + \alpha$ as $\alpha \to 0$, and when the integral is over the infinite interval $(-\infty, \infty)$ the upper and lower limits of integration are of the form $R_1 = R_2$, as $R_1 \to +\infty$, is called the **Cauchy principal value** of the integral. The Cauchy principal value of an integral is indicated by inserting the symbol P.V. in front of the integral sign. So, if in the integral of $f(x)$ over the interval $[0, \infty)$, the function $f(x)$ has an infinity at x_0, its Cauchy principal value is defined as

Cauchy principal value

$$\text{P.V.} \int_0^\infty f(x)dx = \lim_{\alpha \to 0} \int_0^{x_0 - \alpha} f(x)dx + \lim_{\alpha \to 0, R \to \infty} \int_{x_0 + \alpha}^R f(x)dx \quad (\alpha > 0).$$

$$(38)$$

In some improper integrals of the type shown in the second expression in (36), allowing R_1 and R_2 to approach infinity at different rates produces the same result as the Cauchy principal value, and when this occurs the symbol P.V. can be dropped. This happens, for example, with the integral

$$\int_{-\infty}^\infty \frac{dx}{1 + x^2} = \lim_{R_1 \to \infty, R_2 \to \infty} \int_{-R_1}^{R_2} \frac{dx}{1 + x^2} = \lim_{R_1 \to \infty, R_2 \to \infty} \{\text{Arctan } R_2 - \text{Arctan}(-R_1)\}$$

$$= \left\{ \frac{\pi}{2} - \left(-\frac{\pi}{2} \right) \right\} = \pi,$$

because it is also true that

$$\int_{-\infty}^\infty \frac{dx}{1 + x^2} = \lim_{R \to \infty} \int_{-R}^R \frac{dx}{1 + x^2} = \lim_{R \to \infty} \{\text{Arctan } R - \text{Arctan}(-R)\} = \pi.$$

This is an integral for which

$$\text{P.V.}\int_{-\infty}^{\infty}\frac{dx}{1+x^2} = \int_{-\infty}^{\infty}\frac{dx}{1+x^2} = \pi.$$

As the integrand is an even function of x, these results allow us to conclude that

$$\int_{0}^{\infty}\frac{dx}{1+x^2} = \frac{1}{2}\int_{-\infty}^{\infty}\frac{dx}{1+x^2} = \frac{\pi}{2}.$$

The situation is quite different in the case of the integral

$$\int_{-\infty}^{\infty}\sin x\, dx,$$

because although $\sin x$ is continuous and bounded for all x the integral is divergent. This result follows from the fact that

$$\lim_{R_1\to\infty,\, R_2\to\infty}\int_{-R_1}^{R_2}\sin x\, dx = \lim_{R_1\to\infty,\, R_2\to\infty}\{\cos R_2 - \cos R_1\},$$

so the limit is not defined, though the Cauchy principal value of the integral is finite because

$$\text{P.V.}\int_{-\infty}^{\infty}\sin x\, dx = \lim_{R\to\infty}\int_{-R}^{R}\sin x\, dx = \lim_{R\to\infty}\{\cos R - \cos(-R)\} = 0.$$

Another example of a divergent integral for which the Cauchy principal value is finite is

$$\int_{-\infty}^{\infty}\frac{x}{1+x^2}dx.$$

The divergence of the integral follows from the fact that

$$\int_{-\infty}^{\infty}\frac{x}{1+x^2}dx = \lim_{R_1\to\infty,\, R_2\to\infty}\int_{-R_1}^{R_2}\frac{x}{1+x^2}dx$$

$$= \lim_{R_1\to\infty,\, R_2\to\infty}\frac{1}{2}\left\{\ln\left(1+R_2^2\right) - \ln\left(1+R_1^2\right)\right\},$$

because this limit is not defined if $R_1 \neq R_2$. When $R_1 = R_2$ the Cauchy principal value follows from the preceding result, from which it is seen to be zero, so we say

$$\text{P.V.}\int_{-\infty}^{\infty}\frac{x}{1+x^2}dx = 0.$$

Tests exist that enable the convergence or divergence of various types of improper integral to be established without the need for direct integration, and these are necessary because in most cases it is either difficult or impossible to evaluate the integral analytically. The simplest of these tests, called a **comparison test**, establishes the convergence (or divergence) of an improper integral by comparing its integrand with the integrand of an improper integral whose convergence or divergence properties are known. Thus, if, for example, the improper integral $\int_{-\infty}^{\infty}g(x)dx$ is known to be convergent, and $f(x)$ is such that $0 \le f(x) \le g(x)$, then the improper integral $\int_{-\infty}^{\infty}f(x)dx$ is convergent. This follows because then the integral $\int_{-\infty}^{\infty}f(x)dx$ is

a comparison test for improper integrals

bounded by

$$0 \leq \int_{-\infty}^{\infty} f(x)dx \leq \int_{-\infty}^{\infty} g(x)dx.$$

If, however, the improper integral $\int_{-\infty}^{\infty} g(x)dx$ is known to be divergent and $0 \leq g(x) \leq f(x)$, then

$$0 \leq \int_{-\infty}^{\infty} g(x)dx \leq \int_{-\infty}^{\infty} f(x)dx,$$

showing that the integral $\int_{-\infty}^{\infty} f(x)dx$ is divergent. Different forms of comparison tests exist, and corresponding tests apply to improper integrals over the interval $[0, \infty)$.

The concept of the Cauchy principal value of an integral is important when evaluating real improper integrals by means of contour integration, and especially when the integrand has an infinity at one or more points inside the interval of integration. This is because the method of evaluating such integrals gives rise automatically to the Cauchy principal value of the integral. Whether a real improper integral determined by contour integration also exists in the sense of (36) or (37), thereby allowing the symbol P.V. to be dropped from in front of the integral, must be determined separately.

(b) Improper Integrals of Rational Functions without Poles on the Real Axis

integrals of rational functions without poles on the real axis

As improper real integrals only involve integration along the real axis, in order to evaluate them by contour integration a suitable simple closed contour Γ must be introduced that includes as part of the contour the piece of the real axis that is involved. An essential feature of an analytic function $f(z)$ that is to be integrated must be that it, or its real or imaginary part, reduces to the required real improper integral on the real axis. In addition to this, in general, on the segment of the contour Γ that does not include the real axis, the modulus of $f(z)$ must tend to zero sufficiently rapidly as $|z| \to \infty$ that the integral around that segment vanishes.

When the entire real axis is involved, the contour Γ is usually taken to be the contour formed by the segment of the real axis from $-R$ to R, and the semicircle Γ_R with the equation $|z| = R$ in the upper half of the complex plane, with the sense of integration taken in the counterclockwise sense around Γ, as shown in Fig. 15.8a.

If we consider functions $f(z)$ that have no poles on the real axis, an improper integral of $f(z)$ over the interval $(-\infty, \infty)$ is evaluated by first taking R sufficiently large that all the poles of $f(z)$ in the upper-half of the complex plane lie inside Γ, applying the residue theorem to $\int_{\Gamma} f(z)dz$, and then proceeding to the limit at $R \to \infty$. It is this choice of contour that introduces the Cauchy principal value of improper integrals taken over an infinite interval.

indenting a contour

Later we will consider the situation in which a simple pole of $f(z)$ occurs on the real axis at x_0, when we will see it is necessary to exclude it from the contour of Γ by **indenting** the contour at x_0 by the addition of a small semicircle of radius

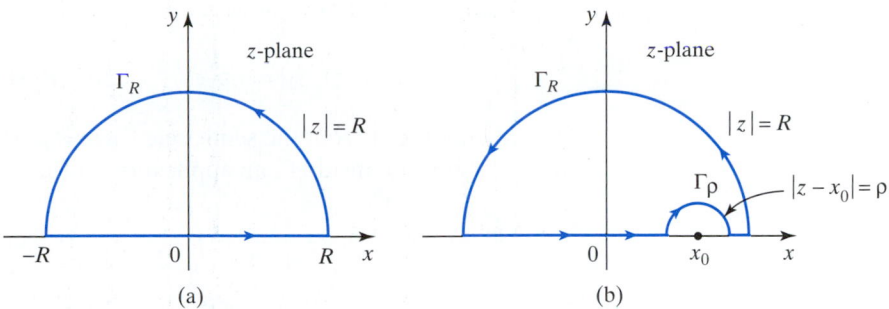

FIGURE 15.8   (a) The contour Γ in the upper half of the complex plane. (b) An indented contour Γ in the upper half of the complex plane.

ρ extending into the upper half of the complex plane, as shown in Fig. 15.8b. Then, after applying the residue theorem and giving due consideration to the effect of integration around the indentation, R is allowed to tend to infinity and ρ to tend to zero. In such a case the Cauchy principal value of the integral is due to reducing the indentation at x_0 to one of vanishingly small radius, and also to taking the limit symmetrically with respect to the origin as R tends to infinity.

This general approach to the evaluation of real integrals will be seen to work for functions $f(z)$ with the property that the integral around the semicircular part of the contour Γ_R vanishes in the limit as the radius of the semicircle $R \to \infty$. This means that for the method to succeed we must impose the condition

$$\lim_{R \to \infty} \int_{\Gamma_R} f(z)dz = 0. \tag{39}$$

Later we will find conditions to be satisfied by the most frequently occurring types of integrand for which this result is always true. First, however, to illustrate the general approach, we begin by assuming condition (39) and applying the method to a typical example.

EXAMPLE 15.27   Evaluate the integral

$$\text{P.V.} \int_{-\infty}^{\infty} \frac{dx}{1 + x^4},$$

and show that the P.V. symbol can be omitted from the result.

Solution   The function $f(z) = 1/(1 + z^4)$ reduces to $f(x) = 1/(1 + x^4)$ on the real axis, and the integrand has simple poles at the four zeros of $1 + z^4$ given by $z_k = e^{i\pi(1+2k)/4}$ with $k = 0, 1, 2, 3$, but only the two zeros at

$$z_0 = e^{\pi i/4} \quad \text{and} \quad z_1 = e^{3\pi i/4}$$

lie in the upper half of the complex plane. So, as the interval of integration extends over the entire real axis, we will consider the integral of $f(z)$ around the contour of Fig. 15.8a.

A simple calculation shows that

$$\text{Res}[\, f(z), z_0] = -\frac{1}{4}e^{i\pi/4} \quad \text{and} \quad \text{Res}[\, f(z), z_1] = \frac{1}{4}e^{-i\pi/4},$$

so when the radius R of the semicircle Γ_R in Fig. 15.8a is large enough for the poles at z_0 and z_1 to lie inside Γ, an application of the residue theorem gives

$$\int_\Gamma \frac{dz}{1+z^4} = \int_{-R}^{R} \frac{dx}{1+x^4} + \int_{\Gamma_R} \frac{dz}{1+z^4} = \{\text{Res}[\, f(z), z_0] + \text{Res}[\, f(z), z_1]\}.$$

Letting $R \to \infty$, assuming that $\lim_{R\to\infty} \int_{\Gamma_R} \frac{dz}{1+z^4} = 0$, and substituting the values of the residues reduce this to

$$\text{P.V.} \int_{-\infty}^{\infty} \frac{dx}{1+x^4} = \pi \left(\frac{e^{i\pi/4} - e^{-i\pi/4}}{2i} \right) = \pi \sin \frac{\pi}{4} = \frac{\pi}{\sqrt{2}}.$$

The symbol P.V. can only be omitted if the Cauchy principal value and the value of the improper integral are equal. This result will be true if we can show that the improper integral converges, because the Cauchy principal value is obtained as one of the possible ways in which the limits in (36) may be taken, so that then the two integrals must be equal.

We use a comparison argument to justify the removal of the P.V. symbol. The integrand $1/(1+x^4) \le 1/(1+x^2)$ for all x, so the convergence of the integral $\int_{-\infty}^{\infty} \frac{dx}{1+x^2}$ that has been established proves the convergence of $\int_{-\infty}^{\infty} \frac{dx}{1+x^4}$, and so justifies writing

$$\int_{-\infty}^{\infty} \frac{dx}{1+x^4} = \frac{\pi}{\sqrt{2}}.$$

The integrand is finite, continuous, and symmetric about the origin, so we may conclude that

$$\int_0^{\infty} \frac{dx}{1+x^4} = \frac{1}{2} \int_{-\infty}^{\infty} \frac{dx}{1+x^4} = \frac{\pi}{2\sqrt{2}}. \qquad \blacksquare$$

The theorem we now prove provides conditions that ensure the validity of the limit in (39) when the modulus of the integrand $f(z)$ decreases sufficiently rapidly as $|z|$ becomes large. The theorem is particularly useful when $f(z)$ is a quotient of two polynomials in z, that is to say when $f(z)$ is a **rational function**, which we choose to write as

$$f(z) = \frac{a_0 + a_1 z + \cdots + a_m z^m}{b_0 + b_1 z + \cdots + b_n z^n}. \qquad (40)$$

We have

$$|f(z)| = \frac{|z^m||a_0/z^m + a_1/z^{m-1} + \cdots + a_m|}{|z^n||b_0/z^n + b_1/z^{n-1} + \cdots + b_n|},$$

but as $|z|$ increases, terms such as $|c|/|z|^r$, and hence ones such as c/z^r, tend to zero, showing that when $|z|$ is large $|f(z)|$ can be overestimated by

$$|f(z)| \le \frac{K}{|z|^{n-m}}, \qquad (41)$$

for some finite positive constant K, and $n - m$ positive, zero or negative.

THEOREM 15.15

Estimation of $\int_{\Gamma_R} f(z)dz$ when $f(z)$ decays rapidly for large $|z|$ Let $f(z)$ be analytic in the upper half of the complex plane with the exception of a finite number of poles at the points $z_1, z_2, \ldots, z_N$. Then if for $|z| > R$ the function $f(z)$ is such that $|f(z)| < K/|z|^{1+\delta}$, with K and δ positive constants,

$$\lim_{R \to \infty} \int_{\Gamma_R} f(z)dz = 0,$$

where Γ_R is the part of the circle $|z| = R$ that lies in the upper half of the complex plane.

Proof   On Γ_R we have $z = Re^{i\theta}$, so from the usual integral inequality,

$$\left| \int_{\Gamma_R} f(z)dz \right| = \left| \int_0^\pi f(Re^{i\theta}) Ri\, e^{i\theta} d\theta \right| \leq \int_0^\pi |f(Re^{i\theta}) Ri\, e^{i\theta}| d\theta$$

$$< \int_0^\pi \left(\frac{K}{R^{1+\delta}} \right) R d\theta = \frac{K\pi}{R^\delta}.$$

The result of the theorem now follows directly by taking the limit as $R \to \infty$.   ■

Theorem 15.15 provides the justification for the use of property (39) that was assumed in Example 15.22, because for large $|z|$ it follows from (41) that a constant K can be found such that $|f(z)| < K/|z|^4$, showing that in this case $\delta = 3$.

EXAMPLE 15.28

Evaluate the integral

$$\int_0^\infty \frac{a + x^2}{1 + x^4} dx \quad \text{where } a \text{ is a real constant.}$$

Solution   The integrand is an even function of x, so

$$\int_0^\infty \frac{a + x^2}{1 + x^4} dx = \frac{1}{2} \int_{-\infty}^\infty \frac{a + x^2}{1 + x^4} dx.$$

The function $f(z) = (a + z^2)/(1 + z^4)$ reduces to the required integrand on the real axis, so integrating $f(z)$ around the contour in Fig. 15.8a and using the residue theorem leads to the result

$$\int_\Gamma \frac{a + z^2}{1 + z^4} dz = \int_{-R}^R \frac{a + x^2}{1 + x^4} dx + \int_{\Gamma_R} \frac{a + z^2}{1 + z^4} dz = 2\pi i \{\text{Res}[f(z), z_0] + \text{Res}[f(z), z_1]\},$$

when R is sufficiently large that Γ contains the two of the four simple poles of $f(z)$ that lie in the upper half of the complex plane at the points $z_0 = e^{\pi i/4}$ and $z_1 = e^{3\pi i/4}$. These poles occur at the same points as those of Example 15.27, though the residues are different.

We find that

$$\text{Res}[f(z), z_0] = \frac{a + i}{2\sqrt{2}(i - 1)} \quad \text{and} \quad \text{Res}[f(z), z_1] = \frac{a - i}{2\sqrt{2}(1 + i)},$$

so substituting these values in the preceding result gives

$$\int_{-R}^{R} \frac{a+x^2}{1+x^4}dx + \int_{\Gamma_R} \frac{a+z^2}{1+z^4}dz = 2\pi i \left\{ \frac{a+i}{2\sqrt{2}(i-1)} + \frac{a-i}{2\sqrt{2}(1+i)} \right\} = \frac{\pi}{\sqrt{2}}(a+1).$$

Theorem 15.15 applies, because for large $|z|$ a positive constant K can be found such that $|f(z)| < K/|z|^2$ corresponding to $\delta = 1$, so proceeding to the limit as $R \to \infty$ gives

$$\text{P.V.} \int_{-\infty}^{\infty} \frac{a+x^2}{1+x^4}dx = \frac{\pi}{\sqrt{2}}(a+1).$$

To justify removing the P.V. symbol we need to show that the improper integral is convergent. As the integrand $(a+x^2)/(1+x^4)$ is an even function of x and its integral over any finite interval is finite, it will be sufficient to show that $\int_{R}^{\infty} \frac{a+x^2}{1+x^4}dx$ is finite for any $R > 0$. This is indeed so, because for large R it is always possible to find an $M > 0$ such that $(a+x^2)/(1+x^4) \leq M/x^2$, and $\int_{R}^{\infty} M/x^2 dx = M/R$ is finite, so we are justified in writing

$$\int_{0}^{\infty} \frac{a+x^2}{1+x^4}dx = \frac{1}{2}\int_{-\infty}^{\infty} \frac{a+x^2}{1+x^4}dx = \frac{\pi}{2\sqrt{2}}(a+1). \qquad \blacksquare$$

We now combine Theorem 15.5 and the residue theorem to arrive at the following theorem that enables the rapid evaluation of a certain type of improper integral.

THEOREM 15.16

a useful theorem when $|f(z)|$ decays rapidly as $|z| \to \infty$

Integration of functions that decay rapidly as $|z|$ becomes large   Let $f(z)$ be analytic in the upper half of the complex plane with the exception of a finite number of poles at the points $z_1, z_2, \ldots, z_N$, and let no poles of $f(z)$ lie on the real axis. Then if for $|z| > R$ the function $f(z)$ is such that $|f(z)| < K/|z|^{1+\delta}$, where K and δ are positive constants,

$$\text{P.V.} \int_{-\infty}^{\infty} f(x)dx = 2\pi i \sum_{k=1}^{N} \text{Res}[f(z), z_k]. \qquad \blacksquare$$

Notice that when the function $f(z)$ in Theorem 15.16 is a rational function of the form (40), the condition $|f(z)| < K/|z|^{1+\delta}$ when $|z| > R$ becomes the condition $n - m \geq 2$.

EXAMPLE 15.29

Evaluate the integral

$$\int_{-\infty}^{\infty} \frac{x^2}{(1+x^2)^4}dx.$$

Solution   We set $f(z) = z^2/(1+z^2)^4$, because this reduces to the required integrand on the real axis, and notice that the conditions of Theorem 15.16 are satisfied by $f(z)$, because for large $|z|$ it behaves like $K/|z|^6$. Writing

$$f(z) = \frac{z^2}{(z-i)^4(z+i)^4},$$

shows that $f(z)$ only has a single pole of order 4 at $z = i$ in the upper half of the complex plane with

$$\text{Res}[\,f(z), i] = -\frac{i}{32}.$$

From Theorem 15.16 we have

$$\text{P.V.} \int_{-\infty}^{\infty} \frac{x^2 dx}{(1+x^2)^4} = 2\pi i \left(-\frac{i}{32}\right) = \frac{\pi}{16}.$$

The P.V. symbol can be omitted because the integrand is everywhere continuous and finite, and for large x the integrand behaves like $1/x^6$ showing that the improper integral is convergent, so we conclude that

$$\int_{-\infty}^{\infty} \frac{x^2 dx}{(1+x^2)^4} = \frac{\pi}{16}.$$ ∎

(c) Improper Integrals with Integrands of the Form $e^{imz}Q(z)$

Another important type of improper integral that occurs is one where the integrand is of the form $f(z) = e^{imz}Q(z)$, involving the product of an exponential factor e^{imz} with $m > 0$ and a rational function $Q(z)$. If the method of residues is to be used to evaluate improper integrals of this type it is necessary to find conditions that will ensure the validity of the limit in (39) when $f(z)$ is of this form.

The first step when seeking to establish such a condition is to prove a result known as the **Jordan inequality**, and an associated result that we will call the **Jordan integral inequality**.

LEMMA 15.1

the Jordan inequality and integral inequality

The Jordan inequality and integral inequality

(a) $\dfrac{2\theta}{\pi} \le \sin\theta \le \theta, \quad \text{for } 0 \le \theta \le \pi/2 \quad \text{(Jordan inequality)}$

(b) $\displaystyle\int_0^{\pi/2} e^{-k\sin\theta}\, d\theta \le \dfrac{\pi}{2k}(1 - e^{-k}), \quad \text{for } k > 0 \quad \text{(Jordan integral inequality)}.$

Proof

(a) Assuming the inequality to be true, division by θ allows it to be written as

$$1 \ge \frac{\sin\theta}{\theta} \ge \frac{2}{\pi}, \quad \text{for } 0 \le \theta \le \pi/2.$$

Setting $S(\theta) = \sin\theta/\theta$ we have $S(\pi/2) = 2/\pi$, and from L'Hospital's rule

$$S(0) = \lim_{\theta \to 0} S(\theta) = \lim_{\theta \to 0} \frac{\sin\theta}{\theta} = 1,$$

so the upper and lower limits of the Jordan inequality have been established. The inequality will be proved if we can show that $S'(\theta) < 0$ for $0 \le \theta \le \pi/2$, because then $S(\theta)$ will be a strictly decreasing function of θ in the interval.

Differentiation of $S(\theta)$ gives

$$S'(\theta) = \frac{\theta\cos\theta - \sin\theta}{\theta^2},$$

so the sign of $S'(\theta)$ is determined by the sign of $h(\theta) = \theta \cos\theta - \sin\theta$. Using the results $h(0) = 0$ and $h'(\theta) = -\theta \sin\theta$ shows that $h'(\theta) \leq 0$ for $0 \leq \theta \leq \pi/2$, so $h(\theta)$ and hence also $S(\theta)$ are strictly decreasing functions of θ in the given interval, and the Jordan inequality is proved.

(b) The integral form of the inequality follows by replacing $\sin\theta$ by $2\theta/\pi$ in the integrand $e^{-k\sin\theta}$ and then integrating to obtain the stated result.   ■

We now use the Jordan integral inequality to prove the next result known as **Jordan's lemma**.

THEOREM 15.17

the useful Jordan's lemma

Jordan's lemma   Let m be a positive constant and $Q(z)$ be a continuous function in the upper half of the complex plane, such that for $|z| \geq R_0$

$$M_R = \max_{z \in \Gamma_R} |Q(z)| \to 0 \quad \text{as } R \to \infty,$$

where Γ_R is the semicircle $|z| = R$ in the upper half of the complex plane. Then

$$\lim_{R \to \infty} \int_{\Gamma_R} e^{imz} Q(z) dz = 0.$$

Proof   Let z lie on the semicircle Γ_R with $R > R_0$, so $z = Re^{i\theta}$ and $dz = i Re^{i\theta} d\theta$. Then

$$|e^{imz}| = |e^{imR(\cos\theta + i\sin\theta)}| = e^{-mR\sin\theta},$$

and on Γ_R we have

$$\left| \int_{\Gamma_R} e^{imz} Q(z) dz \right| \leq \max_{z \in \Gamma_R} |Q(z)| \int_0^\pi e^{-mR\sin\theta} R d\theta = RM_R \int_0^\pi e^{-mR\sin\theta} d\theta.$$

The last integral cannot be evaluated as it stands, but because of the symmetry of $\sin\theta$ about the value $\theta = \pi/2$ the integral can be written as

$$RM_R \int_0^\pi e^{-mR\sin\theta} d\theta = 2RM_R \int_0^{\pi/2} e^{-mR\sin\theta} d\theta.$$

As the interval of integration is now $0 \leq \theta \leq \pi/2$, we can apply the Jordan integral inequality to arrive at the estimate

$$2RM_R \int_0^{\pi/2} e^{-mR\sin\theta} d\theta \leq \frac{\pi M_R}{m}(1 - e^{-mR}).$$

Thus,

$$\left| \int_{\Gamma_R} e^{imz} Q(z) dz \right| \leq \frac{\pi M_R}{m}(1 - e^{-mR}),$$

but by hypothesis $M_R \to 0$ as $R \to \infty$, so the right-hand side of this inequality vanishes and the result is proved.   ■

Coupling Jordan's lemma with the residue theorem, we arrive at the following theorem, which enables the rapid evaluation of improper integrals with integrands that involve a product of an exponential factor and a rational function.

THEOREM 15.18

Integrals with integrands of the form $e^{imz}Q(z)$

Integration of functions of the form $e^{imz}Q(z)$ Let $m > 0$ be a real constant, and $f(z) = e^{imz}Q(z)$ be analytic in the upper half of the complex plane with the exception of a finite number of poles at the points $z_1, z_2, \ldots, z_N$, and let no poles of $f(z)$ lie on the real axis. Then if for $|z| > R$ the function $Q(z)$ is such that for all z in the upper half of the complex plane

$$\lim_{|z| \to \infty} |Q(z)| \to 0,$$

it follows that

$$\text{P.V.} \int_{-\infty}^{\infty} e^{imx} Q(x)dx = 2\pi i \sum_{k=1}^{N} \text{Res}[e^{imz} Q(z), z_k].$$

∎

The following theorem is often useful in establishing the convergence of integrals obtained by using Theorem 15.19, and so justifying the omission of the P.V. symbol.

THEOREM 15.19

Convergence of integrals with integrands of the form $e^{imz}Q(z)$ Let $Q(x) > 0$ be a strictly decreasing function of x for $0 \leq x < \infty$ such that $\lim_{x \to \infty} Q(x) = 0$. Then, provided the integrands are finite at the origin, the improper integrals $\int_0^\infty Q(x)\cos mx\,dx$ and $\int_0^\infty Q(x)\sin mx\,dx$ are convergent. Furthermore, if $Q(x)$ is an even function, the improper integral $\int_{-\infty}^\infty Q(x)\cos mx\,dx$ is convergent, and if $Q(x)$ is an odd function the improper integral $\int_{-\infty}^\infty Q(x)\sin mx\,dx$ is convergent.

Proof   As $Q(x) > 0$, the sign of the integrand in $\int_0^\infty Q(x)\cos mx\,dx$ will be determined by the sign of $\cos mx$. The function $\cos mx$ changes sign in adjacent intervals of the form $(2n-1)\pi/2m < x < (2n+1)\pi/2m$, for $n = 1, 2, \ldots$, so setting

$$I_n = \int_{(2n-1)\pi/2m}^{(2n+1)\pi/2m} Q(x)|\cos mx|dx$$

allows us to write

$$\int_{(2n-1)\pi/2m}^{(2n+1)\pi/2m} Q(x)\cos mx\,dx = (-1)^n I_n.$$

This result enables the original integral to be written as

$$\int_0^\infty Q(x)\cos mx\,dx = \int_0^{\pi/2m} Q(x)\cos mx\,dx + \sum_{n=1}^\infty (-1)^n I_n.$$

By hypothesis, $Q(x)$ is a strictly decreasing function of x, so $0 < I_{n+1} < I_n$, but $\lim_{x \to \infty} Q(x) = 0$, so we also have $\lim_{x \to \infty} I_n = 0$. The series $\sum_{n=1}^\infty (-1)^n I_n$ is seen to be an *alternating series* satisfying the alternating series test for convergence, and so has a finite sum. As the integrand is assumed to be finite at the origin, the term $\int_0^{\pi/2m} Q(x)\cos mx\,dx$ is finite, showing that the integral $\int_0^\infty Q(x)\cos mx\,dx$ has a finite sum. This has proved the integral to be convergent, thus allowing the P.V. symbol to be omitted. The convergence of $\int_0^\infty Q(x)\sin mx\,dx$ can be established in similar fashion. In the case of integrals over an infinite interval, the conditions imposed on $Q(x)$ in the last part of the theorem allow the integrals to be reduced to one of the cases just considered, so the proof is complete. ∎

EXAMPLE 15.30   Evaluate the integral

$$\int_{-\infty}^{\infty} \frac{\cos x}{(1+x^2)^4} dx.$$

Solution   The real part of $f(z) = \exp(iz)/(1+z^2)^4$ reduces to the required integrand on the real axis, so we take this for our integrand. An attempt to use the more obvious choice of integrand $\cos z/(1+z^2)^4$ must be avoided because it would introduce unnecessary complications due to the behavior of $f(z)$ as $|z| \to \infty$. As in Example 15.27, the integrand only has a single pole of order 4 located at $z = i$ in the upper half of the complex plane. A routine calculation shows that the residue at $z = i$ is

$$\text{Res}[f(z), i] = -\frac{37i}{96e}.$$

The conditions of Theorem 15.19 are seen to be satisfied, so it follows that

$$\text{P.V.} \int_{-\infty}^{\infty} \frac{\exp(iz)}{(1+x^2)^4} dx = 2\pi i \left(-\frac{37i}{96e}\right) = \frac{37\pi}{48}.$$

Equating the real parts of the expressions on each side of the equation gives

$$\text{P.V.} \int_{-\infty}^{\infty} \frac{\cos x}{(1+x^2)^4} dx = \frac{37\pi}{48}.$$

The justification for the removal of the P.V. symbol follows from the form of proof used in Theorem 15.19 by setting

$$I_n = \int_{(2n+1)\pi/2}^{(2n+3)\pi/2} \frac{|\cos x|}{(1+x^2)^4} dx \quad \text{with } n = 1, 2, \ldots.$$

Consequently, we can write

$$\int_0^{\infty} \frac{\cos x}{(1+x^2)^4} dx = \frac{37\pi}{96} \quad \text{or, equivalently,} \quad \int_{-\infty}^{\infty} \frac{\cos x}{(1+x^2)^4} dx = \frac{37\pi}{48}.$$

Had the imaginary parts been equated, we would have obtained the result

$$\int_{-\infty}^{\infty} \frac{\sin x}{(1+x^2)^4} dx = 0,$$

which is to be expected because the integral is convergent and the integrand is an *odd* function. ∎

EXAMPLE 15.31   Evaluate the integral

$$\int_0^{\infty} \frac{x \sin x}{(x^2+1)^2} dx.$$

Solution   The integrand is an even function of x, so we will consider the integral

$$\int_{-\infty}^{\infty} \frac{x \sin x}{(x^2+1)^2} dx.$$

We integrate the function $f(z) = z \exp(iz)/(z^2+1)^2$ around the contour Γ in Fig. 15.8a and notice that when $|z|$ is sufficiently large, $f(z)$ only has a single pole of order 2 at the point $z = i$ inside Γ. We find that

$$\text{Res}[f(z), i] = \frac{1}{4e},$$

so as $f(z)$ satisfies the conditions of Theorem 15.18, after equating the imaginary parts we have

$$\text{P.V.}\int_{-\infty}^{\infty} \frac{x \exp(ix)}{(x^2+1)^2}dx = 2\pi i \left(\frac{1}{4e}\right) = \frac{\pi i}{2e}$$

and so

$$\text{P.V.}\int_{-\infty}^{\infty} \frac{x \sin x}{(x^2+1)^2}dx = \frac{\pi}{2e}.$$

The conditions of Theorem 15.19 are satisfied if in its proof we define

$$I_n = \int_{n\pi}^{(n+1)\pi} \frac{x|\sin x|}{(1+x^2)^2}dx \quad \text{for } n = 1, 2, \ldots,$$

so the P.V. symbol can be omitted, leading to the result

$$\int_0^{\infty} \frac{x \sin x}{(x^2+1)^2}dx = \frac{\pi}{4e}. \qquad \blacksquare$$

The last example is somewhat different, because it involves an integrand that is an entire function, so by the Cauchy–Goursat theorem its integral around any simple closed contour must be zero, though in this case the contour used is a sector of a circle and not a semicircle.

EXAMPLE 15.32   By considering the integral

$$\int_{\Gamma} \exp(iz^2)dz$$

around a suitable contour Γ, show that

$$\int_0^{\infty} \cos x^2 dx = \int_0^{\infty} \sin x^2 dx = \frac{1}{2}\sqrt{\frac{\pi}{2}}.$$

Fresnel integrals   *Solution*   These integrals, called the **Fresnel integrals**, are of importance in engineering and physics in connection with the study of diffraction phenomena. For reasons that will appear later, we take for the positively oriented contour Γ the boundary of the sector of a circle shown in Fig. 15.9 with the internal angle $\pi/4$, and the positively directed circular arc AB of radius R denoted by Γ_R.

The integrand $\exp(iz^2)$ is an entire function, so from the Cauchy–Goursat theorem

$$\int_{\Gamma} \exp(iz^2)dz = 0.$$

To derive the required improper integrals, we represent the integral around Γ as the sum of integrals along the real axis from O to A, along the arc Γ_R from A to B, and along the radial line from B to O (take note of the direction of integration

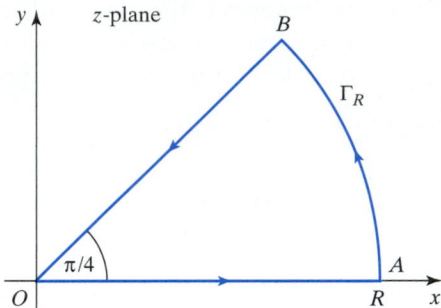

FIGURE 15.9 The sector bounded by the contour Γ.

along this line), as a result of which we find that

$$\int_\Gamma \exp(iz^2)dz = \int_{OA} \exp(iz^2)dz + \int_{\Gamma_R} \exp(iz^2)dz + \int_{BO} \exp(iz^2)dz = 0.$$

Line segment AB lies on the real axis, so on AB we have $z = x$ and hence $dz = dx$, whereas on the radial line OB inclined at an angle $\pi/4$ to the real axis $z = re^{i\pi/4}$, so here $dz = e^{i\pi/4}dr$ and $iz^2 = -r^2$. Using these results in the preceding equation reduces it to

$$\int_0^R \exp(ix^2)dx + \int_{\Gamma_R} \exp(iz^2)dz + e^{i\pi/4}\int_R^0 \exp(-r^2)dr = 0.$$

Reversing the limits of integration in the last integral and rearranging terms gives

$$\int_0^R \exp(ix^2)dx + \int_{\Gamma_R} \exp(iz^2)dz = e^{i\pi/4}\int_0^R \exp(-r^2)dr.$$

Taking the limit of this result as $R \to \infty$ gives

$$\text{P.V.} \int_0^\infty \exp(ix^2)dx + \lim_{R\to\infty} \int_{\Gamma_R} \exp(iz^2)dz = e^{i\pi/4}\frac{\sqrt{\pi}}{2},$$

where we have used the standard result from calculus that

$$\int_0^\infty \exp(-r^2)dr = \frac{1}{2}\sqrt{\pi}.$$

As neither Theorem 15.16 nor Theorem 15.18 apply to the integral around Γ_R, to make further progress we need to examine the limit

$$\lim_{R\to\infty} I_R = \lim_{R\to\infty} \int_{\Gamma_R} \exp(iz^2)dz.$$

On Γ_R we have $z = Re^{i\theta}$, with $0 \le \theta \le \pi/4$, so

$$\exp(iz^2) = \exp(iR^2\cos 2\theta)\cdot\exp(-R^2\sin 2\theta) \quad \text{and} \quad dz = iRe^{i\theta}d\theta,$$

showing that

$$I_R = \int_0^{\pi/4} \exp(iR^2\cos\theta)\cdot\exp(-R^2\sin 2\theta)iRe^{i\theta}d\theta.$$

To estimate this integral we take its modulus and use the standard integral inequality

$$|I_R| \leq \int_0^{\pi/4} |\exp(i R^2 \cos\theta) \exp(-R^2 \sin 2\theta)i\, Re^{i\theta}|d\theta,$$

together with the fact that $|i\, Re^{i\pi/4}| = R$ and $|\exp(i R^2 \cos 2\theta)| = 1$, to arrive at the inequality

$$|I_R| \leq R \int_0^{\pi/4} \exp(-R^2 \sin 2\theta)d\theta.$$

The integral on the right cannot be evaluated in terms of simple functions, but it can be estimated with the help of the Jordan inequality. The interval of integration involved is $0 \leq \theta \leq \pi/4$, so on this interval $0 \leq 2\theta \leq \pi/2$. If we replace θ by 2θ in the Jordan inequality, the result becomes

$$\sin 2\theta \leq \frac{4\theta}{\pi},$$

from which we see that

$$\exp(-R^2 \sin 2\theta) \leq \exp\left(-\frac{4R^2\theta}{\pi}\right),$$

leading to the inequality

$$|I_R| \leq R \int_0^{\pi/4} \exp\left(-\frac{4R^2\theta}{\pi}\right)d\theta = \frac{\pi}{4R}[1 - \exp(-R^2)].$$

Taking the limit of this last result as $R \to \infty$ gives $\lim_{R\to\infty} |I_R| = 0$, showing that the integral around the arc Γ_R vanishes in the limit as $R \to \infty$.

Using this result in the contour integral around Γ, we conclude that

$$\text{P.V.} \int_0^\infty \exp(ix^2)dx = e^{i\pi/4}\frac{\sqrt{\pi}}{2}.$$

The Fresnel integrals follow by omitting the P.V. symbol and equating the respective real and imaginary parts on each side of this equation to obtain

$$\int_0^\infty \cos x^2 dx = \int_0^\infty \sin x^2 dx = \frac{1}{2}\sqrt{\frac{\pi}{2}}.$$

The justification for the removal of the P.V. symbols follows by using an argument similar to the one employed in Theorem 15.19, because as x increases the integrands oscillate more frequently, causing integrals over successive periods to form convergent alternating series.

The reason for choosing the contour Γ to be the boundary of a sector with angle $\pi/4$ is now apparent, because were the angle to exceed $\pi/4$, Jordan's inequality could not be used to estimate $|I_R|$. If, on the other hand, the angle were to be less than $\pi/4$ the form of the resulting integrals would be different, and to evaluate them the values of the Fresnel integrals would need to be known.   ■

(d) Improper Integrals with Poles on the Real Axis

We now consider improper integrals where a simple pole of the integrand occurs on the real axis. Let $f(z)$ have a simple pole located at a point x_0 on the real axis

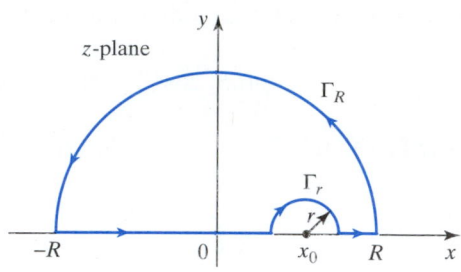

FIGURE 15.10   An indentation Γ_r at x_0 on the real axis.

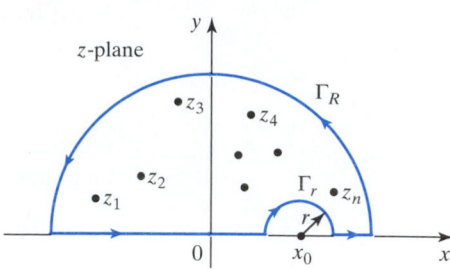

FIGURE 15.11   A contour Γ indented at x_0 on the real axis.

that forms part of the integration path in a contour integral. To prevent the contour passing through the pole, the contour is deformed in a neighborhood of x_0 by a small semicircle of radius r centered on x_0 extending into the upper half of the complex plane, as shown in Fig. 15.10, and we denote this **indentation** by Γ_r.

indentations

The Laurent series representation of $f(z)$ at x_0 is

$$f(z) = \frac{a_{-1}}{z - x_0} + \sum_{n=0}^{\infty} a_n (z - x_0)^n,$$

where $a_{-1} = \text{Res}[f(z), x_0]$. On Γ_r, $z = x_0 + re^{i\theta}$ and $dz = ire^{i\theta} d\theta$ with $0 \leq \theta \leq \pi$, so integrating around Γ_r in the positive sense gives

$$\lim_{r \to 0} \int_{\Gamma_r} f(z) dz = \lim_{r \to 0} \int_0^{\pi} \frac{a_{-1}}{re^{i\theta}} ire^{i\theta} d\theta + \lim_{r \to 0} \sum_{n=0}^{\infty} a_n \int_0^{\pi} r^n e^{in\theta} ire^{i\theta} d\theta$$

$$= ia_{-1} \int_0^{\pi} d\theta = i\pi a_{-1}$$

$$= i\pi \, \text{Res}[f(z), x_0].$$

So, in the limit as $r \to 0$, we have shown that integrating in the positive sense around the semicircular indentation Γ_r above the simple pole located at the point x_0 on the real axis yields $\pi i \, \text{Res}[f(z), x_0]$. This result is seen to be *half* the result that would have been obtained had the integration been taken around a circle with the pole at x_0 at its center.

integration around an indented simple pole

This same form of argument establishes the more general result that if a simple pole is located at z_0, then integration around the pole using a path in the form of a sector of a circle Γ_r, located at z_0 with an arbitrarily small radius r and an internal angle α yields the result

$$\int_{\Gamma_r} f(z) dz = i\alpha \, \text{Res}[f(z), z_0]. \tag{42}$$

Consider a function $f(z)$ that has a finite number of poles located at z_1, $z_2, \ldots, z_n$ in the upper half of the complex plane and a simple pole on the real axis at x_0. Let the positively oriented contour Γ be the one shown in Fig. 15.11, where the indentation above the pole at x_0 is denoted by Γ_r, and Γ_R denoting the semicircle of radius R. Then, when R is sufficiently large that all of the poles above the real axis lie inside Γ, integrating around Γ in the positive sense gives

$$\int_{\Gamma} f(z)dz = \int_{-R}^{x_0-r} f(x)dx + \int_{\Gamma_r} f(z)dz + \int_{x_0+r}^{R} f(x)dx + \int_{\Gamma_R} f(z)dz$$

$$= 2\pi i \sum_{k=1}^{n} \operatorname{Res}[f(z), z_k] \quad \text{when} \quad \lim_{R\to\infty} \int_{\Gamma_R} f(z)dz = 0.$$

Before proceeding to the limit as $R \to \infty$ and $r \to 0$, we notice that the integration around Γ_r, corresponding to $\alpha = \pi$ in (42), is in the *negative* sense, so after the limits have been taken, the result becomes

$$\int_{-\infty}^{x_0-} f(x)dx - \pi i \operatorname{Res}[f(z), x_0] + \int_{x_0+}^{\infty} f(x)dx = 2\pi i \sum_{k=1}^{n} \operatorname{Res}[f(z), z_k].$$

Combining the integrals and rearranging terms gives

$$\text{P.V.} \int_{-\infty}^{\infty} f(x)dx = \pi i \operatorname{Res}[f(z), x_0] + 2\pi i \sum_{k=1}^{n} \operatorname{Res}[f(z), z_k]. \qquad (43)$$

This result extends immediately to a function with m simple poles located on the real axis and so leads to the following theorem.

THEOREM 15.20

integrals involving functions with poles on the real axis

The residue theorem when poles are located on the real axis   Let an analytic function $f(z)$ have n poles at the points $z_1, z_2, \ldots, z_n$ in the upper half of the complex plane and m simple poles at the points $x_1, x_2, \ldots, x_m$ on the real axis. Then, provided $\lim_{R\to\infty} \int_{\Gamma_R} f(z)dz = 0$ where Γ_R is the semicircle $|z| = R$ in the upper half of the complex plane,

$$\text{P.V.} \int_{-\infty}^{\infty} f(x)dx = \pi i \sum_{k=1}^{m} \operatorname{Res}[f(z), x_k] + 2\pi i \sum_{k=1}^{n} \operatorname{Res}[f(z), z_k]. \qquad \blacksquare$$

EXAMPLE 15.33   Evaluate the integral

$$\int_{0}^{\infty} \frac{\sin x}{x} dx.$$

Solution   The integrand is an even function of x, and because $\lim_{x\to 0}(\sin x/x) = 1$ the singularity at the origin is removable, so we consider the integral

$$\int_{-\infty}^{\infty} \frac{\sin x}{x} dx.$$

To evaluate this integral we integrate the function $f(z) = \exp(iz)/z$ around a contour Γ indented at the origin, as shown in Fig. 15.12, because using the function $f(z) = \sin z/z$ would introduce unnecessary complications when z is large.

The only pole of $f(z)$ is a simple pole at the origin, where $\operatorname{Res}[f(z), 0] = 1$, so as the conditions of Jordan's lemma are satisfied, we can use Theorem 15.20 to evaluate the integral. An application of the theorem gives

$$\text{P.V.} \int_{-\infty}^{\infty} \frac{\exp(ix)}{x} dx = \pi i.$$

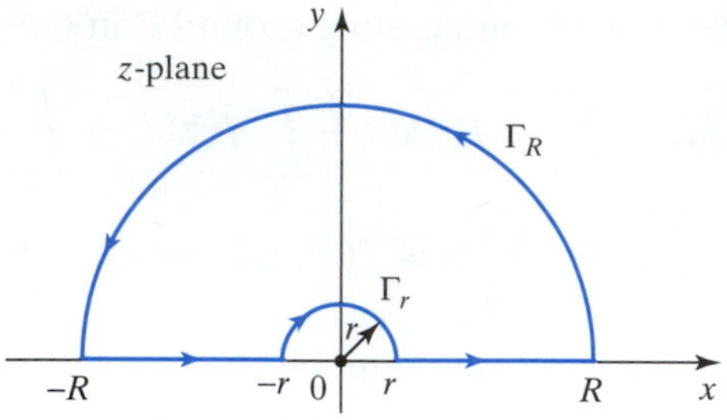

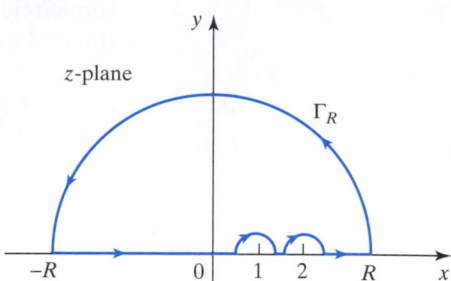

FIGURE 15.12 The contour Γ indented at the origin.

FIGURE 15.13 The contour Γ indented at $x = 1$ and $x = 2$ on the real axis.

Equating the imaginary parts of the expressions on each side of the last equation gives

$$\text{P.V.} \int_{-\infty}^{\infty} \frac{\sin x}{x} dx = \pi.$$

As $x = 0$ is a removable singularity the integrand $\sin x/x$ is finite at the origin, so this fact together with the form of argument used in Example 15.30 justifies the removal of the P.V. symbol, and we have proved that

$$\frac{1}{2} \int_{-\infty}^{\infty} \frac{\sin x}{x} dx = \int_{0}^{\infty} \frac{\sin x}{x} dx = \frac{\pi}{2}.$$ ∎

EXAMPLE 15.34   Evaluate the integral

$$\int_{-\infty}^{\infty} \frac{\cos x}{(x^2 + 1)(x^2 - 3x + 2)} dx.$$

Solution   We choose for the integrand the function $f(z) = \exp(iz)/[(z^2 + 1)(z^2 - 3z + 2)]$. This has simple poles at $z = \pm i$, $z = 1$, and $z = 2$. Modifying the contour in Fig. 15.10 to allow for the two simple poles on the real axis leads to integration around the indented contour shown in Fig. 15.13, which contains the simple pole at $z = i$. The usual calculations show that

$$\text{Res}[f(z), i] = \frac{(3 - i)}{20e}, \quad \text{Res}[f(z), 1] = -\frac{1}{2}(\cos(1) + i\sin(1)) \quad \text{and}$$

$$\text{Res}[f(z), 2] = \frac{1}{5}(\cos(2) + i\sin(2)).$$

The conditions of Theorem 15.20 are seen to be satisfied, so

$$\text{P.V.} \int_{-\infty}^{\infty} \frac{\exp(ix)}{(x^2 + 1)(x^2 - 3x + 2)} dx$$

$$= 2\pi i \, \text{Res}[f(z), i] + \pi i \{\text{Res}[f(z), 1] + \text{Res}[f(z), 2]\}$$

$$= 2\pi i \left(\frac{3 - i}{20e}\right) + \pi i \left(-\frac{[\cos(1) + i\sin(1)]}{2}\right) + \pi i \left(\frac{\cos(2) + i\sin(2)}{5}\right).$$

Equating the real parts on each side of this equation shows that

$$\text{P.V.} \int_{-\infty}^{\infty} \frac{\cos x}{(x^2 + 1)(x^2 - 3x + 2)} dx = \frac{\pi}{10}\left(\frac{1}{e} + 5\sin(1) - 2\sin(2)\right).$$

In this case, because of the complexity of the integrand, no attempt will be made to investigate whether the P.V. symbol can be omitted.

Although not required, equating imaginary parts on each side of the equation shows that

$$\text{P.V.} \int_{-\infty}^{\infty} \frac{\sin x}{(x^2+1)(x^2-3x+2)} dx = \frac{\pi}{10}\left(\frac{3}{e} + 2\cos(2) - 5\cos(1)\right).$$

This determination of two real improper integrals when only one was required is typical of the evaluation of real integrals by contour integration.   ∎

(e) Improper Integrals with Branch Points

Finally, we consider improper integrals of functions with a branch point. To evaluate these by means of contour integration it is necessary to cut the complex plane in an appropriate manner to make the integrand single valued, and to specify the branch of the integrand that is to be used. An important class of integrals of this type are of the form

$$\int_0^{\infty} x^{\alpha-1} P(x)\,dx, \tag{44}$$

where α is not an integer and $P(x)$ is a rational function of x. This integral will have a finite value if $P(x)$ has no poles on the positive real axis and it is such that

$$\lim_{z\to 0} |z|^{\alpha} P(z) = 0 \quad \text{and} \quad \lim_{|z|\to\infty} |z|^{\alpha} P(z) = 0. \tag{45}$$

Provided $z = 0$ is neither a pole nor a zero of $P(z)$, the first of these conditions implies that $\alpha > 0$. Let the rational function $P(z)$ with real coefficients $a_0, a_1, \ldots, a_m$ and $b_0, b_1, \ldots, b_n$ be written

$$P(z) = \frac{a_0 z^m + a_1 z^{m-1} + \cdots + a_m}{b_0 z^n + b_1 z^{n-1} + \cdots + b_n},$$

so that for large $|z|$ a constant K exists such that $P(z) < K/|z^{n-m}|$. Then the second condition in (45) will be satisfied when $n - m - \alpha > 0$. Taken together, these conditions show the integral will have a finite value when $0 < \alpha < n - m$, and they also imply that

$$\lim_{|z|\to\infty} |P(z)| = 0.$$

To take account of the fact that $z^{\alpha-1}$ is many valued and has a branch point at the origin, it is necessary to cut the complex plane to make $z^{\alpha-1}$ (and hence the integrand) single valued, and then to choose a branch of $z^{\alpha-1}$. The cut we will make is along the positive real axis up to and including the origin, so that arg $z = \theta + 2k\pi$, with $k = 0, \pm 1, \pm 2, \ldots$, and θ in the interval $0 \le \theta \le 2\pi$. The contour Γ that will be used is shown in Fig. 15.14 and comprises the circular contour Γ_R with equation $|z| = R$, the cut with its sides immediately above and below the positive real axis, and the circular contour Γ_ρ with equation $|z| = \rho$ around the branch point at the origin. We will work with the branch corresponding to $k = 0$, so $z = re^{i\theta}$ and $z^{\alpha-1} = r^{\alpha-1} e^{(\alpha-1)\theta i}$. The principal branch is positive on the side of the cut that lies

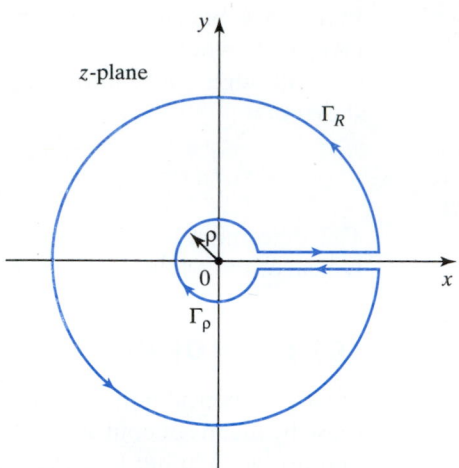

FIGURE 15.14 The contour Γ used to evaluate $\int_0^\infty x^{\alpha-1} P(x)dx$.

above the positive real axis. This branch of the function $z^{\alpha-1} P(z)$ is now single valued in the cut plane, so we can use the residue theorem to evaluate the integral.

When substituting for z in the various integrals that arise while integrating around Γ, it is necessary to express z in its modulus–argument form to take account of the different forms taken by the integrand $z^{\alpha-1} P(z)$ on either side of the cut. Setting $z = re^{i\theta}$, with $0 \leq \theta \leq 2\pi$, it follows that on AB $z = re^{0i} = r$ and $dz = dr$, so that

$$z^{\alpha-1} P(z) = r^{\alpha-1} P(r),$$

while on CD $z = re^{2\pi i}$ and $dz = e^{2\pi i} dr$, so then

$$z^{\alpha-1} P(z) = r^{\alpha-1} e^{(\alpha-1)2\pi i} P(r).$$

We now set $f(z) = z^{\alpha-1} P(z)$, and consider the case where $f(z)$ has poles at $z_1, z_2, \ldots, z_n$, none of which lies on the positive real axis. Integrating around the contour Γ in Fig. 15.14 gives

$$\int_\rho^R r^{\alpha-1} P(r)dr + \int_{\Gamma_R} z^{\alpha-1} P(z)dz + \int_R^\rho r^{\alpha-1} \exp[(\alpha-1)2\pi i] P(r)e^{2\pi i} dr$$

$$+ \int_{\Gamma_\rho} z^{\alpha-1} P(z)dz = 2\pi i \sum_{k=1}^n \text{Res}[f(z), z_k].$$

The conditions (45) with $0 < \alpha < n - m$ ensure the vanishing of both the integral around Γ_R in the limit as $R \to \infty$ and the integral around Γ_ρ as $\rho \to 0$, so taking the limit as $R \to \infty$ and $\rho \to 0$ reduces the preceding result to

$$\int_0^\infty r^{\alpha-1} P(r)dr + e^{2\pi i\alpha} \int_\infty^0 r^{\alpha-1} P(r)dr = 2\pi i \sum_{k=1}^n \text{Res}[f(z), z_k].$$

**integration around
a branch point**

Replacing the dummy variable r by x and rearranging terms, we arrive at the general result

$$\int_0^\infty x^{\alpha-1} P(x)dx = \frac{2\pi i}{1 - e^{2\pi i\alpha}} \sum_{k=1}^n \mathrm{Res}[\,f(z), z_k]. \tag{46}$$

This result forms our next theorem.

THEOREM 15.21

Evaluation of integrals of the form $\int_0^\infty x^{\alpha-1} P(x)dx$ Let $f(z) = z^{\alpha-1} P(z)$ with α not an integer and

$$P(z) = \frac{a_0 z^m + a_1 z^{m-1} + \cdots + a_m}{b_0 z^n + b_1 z^{n-1} + \cdots + b_n},$$

where the coefficients $a_0, a_1, \ldots, a_m$ and $b_0, b_1, \ldots, b_n$ are all real, $0 < \alpha < n - m$, and $P(z)$ has neither a pole nor a zero at the origin. In addition, let the poles of $P(z)$ located at $z_1, z_2, \ldots, z_n$ be such that none lies on the positive real axis. Then

$$\int_0^\infty x^{\alpha-1} P(x)dx = \frac{2\pi i}{1 - e^{2\pi i\alpha}} \sum_{k=1}^n \mathrm{Res}[\,f(z), z_k]. \qquad\blacksquare$$

EXAMPLE 15.35

Find a condition on α that ensures that the integral

$$\int_0^\infty \frac{x^{\alpha-1}}{x^2+1} dx$$

exists, and evaluate the integral subject to this condition.

Solution   In the notation of Theorem 15.21, the rational function $P(z) = 1/(1 + z^2)$, so $m = 0$ and $n = 2$. The condition on α that ensures the existence of the integral is $0 < \alpha < n - m$, so we must have $0 < \alpha < 2$. The function $P(z)$ has simple poles at $z = \pm i$, neither of which lies on the positive real axis, and $P(0) \neq 0$, so all the conditions of Theorem 15.21 are satisfied.

Using the result of the theorem with $f(z) = z^{\alpha-1}/(1 + z^2)$ we find that

$$\mathrm{Res}[\,f(z), i] = \lim_{z\to i}\left[(z-i)\frac{z^{\alpha-1}}{(z-i)(z+i)}\right] = \lim_{z\to i}\left[\frac{z^{\alpha-1}}{z+i}\right] = \frac{i^{\alpha-1}}{2i} = \frac{i^{\alpha-2}}{2},$$

but $i = e^{\pi i/2}$, so

$$\mathrm{Res}[\,f(z), i] = \frac{1}{2}e^{(\alpha-2)\pi i/2} = -\frac{1}{2}e^{\alpha\pi i/2}.$$

Similarly,

$$\mathrm{Res}[\,f(z), -i] = \lim_{z\to -i}\left[(z+i)\frac{z^{\alpha-1}}{(z-i)(z+i)}\right] = \lim_{z\to -i}\left[\frac{z^{\alpha-1}}{z-i}\right] = \frac{(-i)^{\alpha-1}}{-2i} = \frac{(-i)^{\alpha-2}}{2},$$

but $-i = e^{3\pi i/2}$, so

$$\mathrm{Res}[\,f(z), -i] = \frac{1}{2}e^{(\alpha-2)3\pi i/2} = -\frac{1}{2}e^{3\alpha\pi i/2}.$$

Using these residues in Theorem 15.21 gives

$$\int_0^\infty \frac{x^{\alpha-1}}{1+x^2}dx = \frac{2\pi i}{1-e^{2\alpha\pi i}}\left[-\frac{e^{\alpha\pi i/2}}{2}-\frac{e^{3\alpha\pi i/2}}{2}\right] = \pi i\left[\frac{e^{\alpha\pi i/2}+e^{-\alpha\pi i/2}}{e^{\alpha\pi i}-e^{-\alpha\pi i}}\right]$$

$$= \frac{\pi\cos(\alpha\pi/2)}{\sin(\alpha\pi)},$$

and we have shown that

$$\int_0^\infty \frac{x^{\alpha-1}}{x^2+1}dx = \pi\frac{\cos(\alpha\pi/2)}{\sin(\alpha\pi)}, \qquad \text{when } \alpha \text{ is not an integer and with } 0 < \alpha < 2. $$

■

Different types of function with branch points can be evaluated by means of contour integration, provided the complex plane is cut in a suitable manner to make the integrand single valued and a branch of the function is specified. The integrand in the next example involves the logarithmic function that has a branch point at the origin and infinitely many branches.

EXAMPLE 15.36   Show that

$$\int_0^\infty \frac{\log x}{x^2+a^2}dx = \frac{\pi}{2a}\ln a \quad (a > 0).$$

Solution   The function $\log z$ has infinitely many branches, so we will work with the principal branch $\text{Log }z$. The contour Γ to be used is shown in Fig. 15.15, in which the cut is made along the negative real axis, and an indentation is made around the branch point of $\text{Log }z$ located at the origin. The contour Γ_R is the semicircle with the equation $|z| = R$ and $\text{Im }z > 0$, and the contour Γ_ρ is the semicircle with the equation $|z| = \rho$ and $\text{Im }z > 0$.

With the cut as shown in Fig. 15.15, $\text{Arg }z = \theta$ is restricted to the interval $0 \le \theta \le \pi$, so $z = re^{i\theta}$ and $\text{Log }z = \ln r + i\theta$. Setting $f(z) = \text{Log }z/(z^2+a^2)$, we see that when R is large the only singularity of $f(z)$ inside the contour Γ is a simple pole at $z = ia$, where

$$\text{Res}[f(z), ia] = \lim_{z\to ia}[(z-ia)f(z)] = \text{Log }(ia)/(2ia),$$

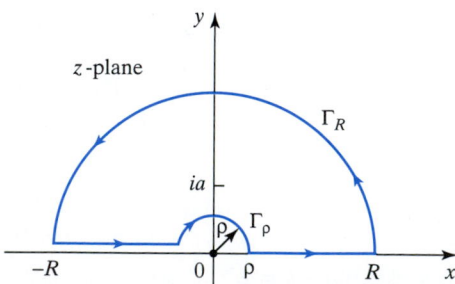

FIGURE 15.15   The contour Γ used to evaluate $\int_0^\infty \frac{\log x}{x^2+a^2}dx.$

but $i = e^{i\pi/2}$ so

$$\text{Res}[f(z), ia] = \frac{\ln a + i\pi/2}{2ia}.$$

On the positive real axis $z = re^{i0} = r$ and $dz = dr$, whereas on the negative real axis $z = re^{i\pi}$ and $dz = e^{i\pi} dr$, so as the simple pole at $z = ia$ lies inside Γ, integration around Γ leads to the result

$$\int_\rho^R \frac{\ln r}{r^2 + a^2} dr + \int_{\Gamma_R} f(z)dz + \int_R^\rho \frac{\ln r + i\pi}{r^2 e^{2\pi i} + a^2} e^{i\pi} dr + \int_{\Gamma_\rho} f(z)dz = 2\pi i \, \text{Res}[f(z), ia].$$

On $\Gamma_R \, z = Re^{i\theta}$ and $dz = iRe^{i\theta} d\theta$, so

$$\left| \int_{\Gamma_R} f(z)dz \right| = \left| \int_0^\pi \frac{\ln R + i\pi}{(R^2 e^{2i\theta} + a^2)} iR \cdot e^{i\theta} d\theta \right|$$

$$\leq \int_0^\pi \frac{R\ln R}{|R^2 e^{2i\theta} + a^2|} d\theta \leq \int_0^\pi R\frac{\ln R}{(R^2 - a^2)} d\theta \leq \pi \left(\frac{\ln R}{R} \right),$$

but as $\lim_{R\to\infty}(\ln R/R) = 0$ it follows from this that $\int_{\Gamma_R} f(z)dz \to 0$ as $R \to \infty$.

A similar argument shows $\int_{\Gamma_\rho} f(z)dz \to 0$ as $\rho \to 0$, because when ρ is small the integrand is approximated by the function $\rho \ln \rho$ that vanishes in the limit as $\rho \to 0$. Taking the limit at $R \to \infty$ and $\rho \to 0$, and using the factor $e^{i\pi} = -1$ to reverse the limits in the third integral on the left gives

$$\int_0^\infty \frac{\ln r}{r^2 + a^2} dr + \int_0^\infty \frac{\ln r + i\pi}{r^2 + a^2} dr = 2\pi i \left(\frac{\ln a + i\pi/2}{2ia} \right).$$

Equating the real parts on either side of the equation and replacing the dummy variable r by x gives the required result,

$$\int_0^\infty \frac{\log x}{x^2 + a^2} dx = \frac{\pi \ln a}{2a} \quad (a > 0).$$

Equating the imaginary parts and again replacing the dummy variable r by x gives the elementary result,

$$\int_0^\infty \frac{dx}{x^2 + a^2} = \frac{\pi}{2a}. \qquad \blacksquare$$

Alternative accounts and more information about Taylor and Laurent series, residues, the evaluation of real integrals by means of contour integrals, and the treatment of contour integrals involving branch points can be found in references [6.1] to [6.4] and [6.6] to [6.9].

Summary

After reviewing the concept of the Cauchy principal value of a definite integral, the residue theorem was used to evaluate real integrals in terms of the limit of associated contour integrals as the contour becomes arbitrarily large. The cases considered involved integrands with poles strictly inside the contour of integration, part of which was along the real axis, integrands with poles both inside and on an indented contour, and integration around an integrand with a branch point.

EXERCISES 15.5

Integrands without poles on the real axis

In Exercises 1 through 6 evaluate the integrals using the contour in Fig. 15.8a.

1. $\int_0^\infty \dfrac{x^2}{(x^2+a^2)^2}\,dx \quad (a>0)$.

5. $\int_{-\infty}^\infty \dfrac{x^2}{(x^2+1)^2(x^2+4)}\,dx$.

2. $\int_{-\infty}^\infty \dfrac{x^2}{x^4+a^4}\,dx \quad (a>0)$.

6. $\int_{-\infty}^\infty \dfrac{x^2}{(x^2+a^2)(x^2+b^2)^2}\,dx$
$\quad (a,b>0,\, a\neq b)$.

3. $\int_0^\infty \dfrac{x^2}{x^4+1}\,dx \quad (a>0)$.

4. $\int_{-\infty}^\infty \dfrac{x^2}{(x^2+a^2)(x^2+b^2)}\,dx$
$\quad (a,b>0)$.

Integrands of the form $e^{imz}Q(z)$

In Exercises 7 through 11 evaluate the integrals using the contour in Fig. 15.8a.

7. $\int_0^\infty \dfrac{\cos x}{(x^2+a^2)^2}\,dx \quad (a>0)$.

8. $\int_0^\infty \dfrac{\cos ax}{(x^2+b^2)^2}\,dx \quad (a,b>0)$.

9. $\int_0^\infty \dfrac{\cos x}{(x^2+a^2)(x^2+b^2)}\,dx \quad (a,b>0)$.

10. $\int_0^\infty \dfrac{x\sin x}{x^2+4}\,dx$.

11. $\int_0^\infty \dfrac{x^3\sin mx}{x^4+a^4}\,dx \quad (a>0)$.

12. By integrating around the contour in Fig. 15.16, show that

$$\int_{-\infty}^\infty \frac{dx}{1+x^{2n}} = \frac{\pi}{n\sin(\pi/2n)} \quad (n=1,2,\ldots).$$

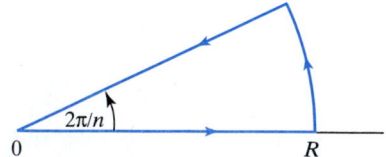

$2\pi/n$

0 R

FIGURE 15.16   The contour for Exercise 12.

Integrands with poles on the real axis

In Exercises 13 through 22 evaluate the integrals using a contour comprising the semicircle $|z|=R$ in the upper half of the complex plane and a suitably indented real axis.

13. P.V. $\displaystyle\int_0^\infty \frac{\sin \pi x}{x(1-x^2)}\,dx$.

14. P.V. $\displaystyle\int_0^\infty \frac{\sin ax}{x(x^2+b^2)^2}\,dx \quad (b>0)$.

15. P.V. $\displaystyle\int_0^\infty \frac{\cos ax - \cos bx}{x^2}\,dx \quad (a\geq 0, b\geq 0)$.

16. P.V. $\displaystyle\int_{-\infty}^\infty \frac{\sin x}{(x^2+4)(x-1)}\,dx$.

17. P.V. $\displaystyle\int_0^\infty \frac{\sin ax}{x(x^2+b^2)}\,dx \quad (a,b>0)$.

18. $\displaystyle\int_0^\infty \frac{\sin^2 x}{x^2}\,dx$ (Hint: Integrate the function $f(z) = [e^{2iz}-1]/z^2$).

19. $\displaystyle\int_0^\infty \frac{\sin^3 x}{x^3}\,dx$ (Hint: Integrate the function $f(z) = [e^{3iz}-3e^{iz}+2]/z^3$).

20. P.V. $\displaystyle\int_0^\infty \frac{x^2}{x^4-1}\,dx$.

21. P.V. $\displaystyle\int_0^\infty \frac{\cos ax}{1-x^4}\,dx \quad (a>0)$.

22. P.V. $\displaystyle\int_0^\infty \frac{x}{x^4-1}\,dx$.

Integrands with branch points

In Exercises 23 through 28 evaluate the integrals by integrating around the contour in Fig. 15.14.

23. $\displaystyle\int_0^\infty \frac{x^\alpha}{x^2+1}\,dx \quad (-1<\alpha<1)$.

24. $\displaystyle\int_0^\infty \frac{x^\alpha}{(x^2+1)^2}\,dx \quad (-1<\alpha<3, \alpha\neq 1)$.

25. $\displaystyle\int_0^\infty \frac{x^{\alpha-1}}{1+x+x^2}\,dx \quad (0<\alpha<2)$.

26. $\displaystyle\int_0^\infty \frac{dx}{x^\alpha(x+1)} \quad (0<\alpha<1)$.

27. $\displaystyle\int_0^\infty \frac{x^{1/2}}{x^3+1}\,dx$.

28. $\displaystyle\int_0^\infty \frac{x^\alpha}{(x^2+1)^2}\,dx \quad (-1<\alpha<3)$.

In Exercises 29 and 30 evaluate the integrals by integrating around the contour in Fig. 15.15.

29. Show that

$$\int_0^\infty \frac{\ln x}{(1+x^2)^2}\,dx = -\frac{\pi}{4}.$$

30. Show that

$$\int_0^\infty \frac{(\ln x)^2}{1+x^2}\,dx = \frac{\pi^3}{8}.$$

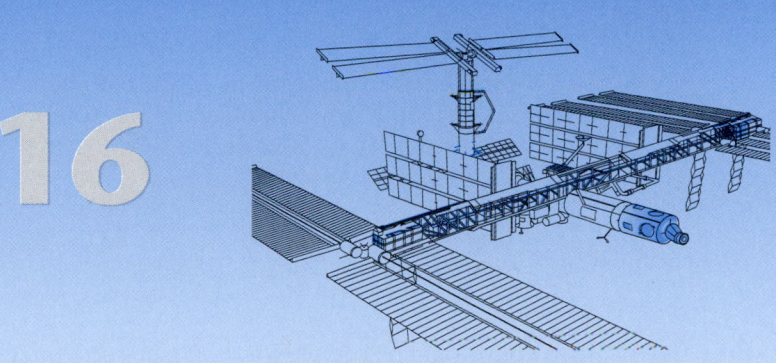

CHAPTER 16

The Laplace Inversion Integral

When applying the Laplace transform to most practical problems, and obtaining the transform $F(s)$ of the required result, it is usually possible to find the required inverse transform $f(t)$ by using tables of Laplace transform pairs together with the operational properties listed in Chapter 7. Sometimes, however, the appropriate transform pairs cannot be found, so then some other way must be developed that enables the determination of the inverse Laplace transform. This is the problem that is addressed in the present chapter, where it is shown how the inversion of a Laplace transform can be performed by means of a special contour integral called the Laplace inversion integral.

The Laplace transform $F(s)$ of a function $f(t)$ is defined as

$$F(s) = \int_0^\infty e^{-st} f(t)dt,$$

provided $f(t)$ is such that the integral exists. The inversion of the Laplace transform to find the function $f(t)$ from a given transform $F(s)$ was performed in Chapter 7 by using a table of transform pairs together with the operational properties of the Laplace transform. In that approach the fact that in general the transform variable s is a complex variable was not used. However, when more complicated transforms $F(s)$ need to be inverted, and this cannot be achieved by using a table of transform pairs, it becomes necessary to regard $F(s)$ as a function of a complex variable and to use complex analysis to find $f(t)$.

This brief chapter uses complex analysis to derive an integral called the Laplace inversion integral that expresses $f(t)$ in terms of a contour integral involving $F(s)$. The inversion integral is then applied to some typical cases, where it is shown how the residues of the transform $F(s)$ can be used to recover the original function $f(t)$.

16.1 The Inversion Integral for the Laplace Transform

When the Laplace transform was introduced in Chapter 7, a table of Laplace transform pairs was developed by considering the transform variable s to be real, and these were then used with the operational properties of the Laplace transform to recover a wide variety of functions $f(t)$ from elementary Laplace transforms $F(s)$. As tables of transform pairs do not always contain the required inverse Laplace transform and s must be allowed to be complex, some other method must be found by which to determine $f(t) = \mathcal{L}^{-1}\{F(s)\}$.

The method we now derive shows that if $f(t)$ possesses a Laplace transform $F(s)$, so that

$$F(s) = \int_0^\infty e^{-st} f(t)dt, \tag{1}$$

where s can be complex, $f(t)$ can be recovered from its Laplace transform $F(s)$ by means of the complex line integral

$$f(t) = \frac{1}{2\pi i} \int_{c-i\infty}^{c+i\infty} e^{st} F(s)ds, \tag{2}$$

the Laplace inversion integral

where $c > 0$ is a suitable real constant. The formula in (2) is called the **inversion integral** for the Laplace transform $F(s)$, and it involves an integral in the complex s-plane taken along the line $\text{Re}\{s\} = c$ from minus infinity to infinity. We show later how this inversion integral can be evaluated in terms of the residues of $e^{st}F(s)$.

To establish result (2) we use the close relationship that exists between the complex form of the Fourier integral and the Laplace transform. The nature of this relationship can be seen from the fact that if

$$f(t) = \begin{cases} e^{-ct}g(t), & t > 0 \\ 0, & t < 0, \end{cases} \tag{3}$$

where the real constant $c > 0$ is chosen to guarantee the existence of $\mathcal{F}\{f(t)\}$, then from the definition of the complex form of the Fourier transform

$$\mathcal{F}\{f(t)\} = \frac{1}{\sqrt{2\pi}} \int_{-\infty}^\infty e^{-i\omega t} f(t)dt = \frac{1}{\sqrt{2\pi}} \int_0^\infty e^{-(c+i\omega)t} g(t)dt$$

$$= \frac{1}{\sqrt{2\pi}} \int_0^\infty e^{-st} g(t)dt. \tag{4}$$

The integral on the right of (4) is simply the Laplace transform of $g(t)$, though now the Laplace transform parameter $s = c + i\omega$ is complex. If F is the Fourier transform of f, the preceding result can be written

$$F(c + i\omega) = \frac{1}{\sqrt{2\pi}} \mathcal{L}\{g(t)\}. \tag{5}$$

derivation of the inversion integral

To derive the inversion integral (2) we start from the complex form of the Fourier integral representation for $f(t)$, which for clarity in the argument that follows we write as

$$f(t) = \frac{1}{2\pi} \int_{-\infty}^\infty \left[\int_{-\infty}^\infty f(u)e^{i\omega(t-u)} du \right] d\omega.$$

If we use the expression for $f(t)$ in (3), this becomes

$$e^{-ct}g(t) = \frac{1}{2\pi} \int_{-\infty}^\infty \left[\int_0^\infty e^{i\omega t} e^{-(c+i\omega)u} g(u)du \right] d\omega.$$

However, as $e^{i\omega t}$ is not involved in the integral with respect to u, this can be rewritten as

$$e^{-ct}g(t) = \frac{1}{2\pi}\int_{-\infty}^{\infty}e^{i\omega t}\left[\int_{0}^{\infty}e^{-su}g(u)du\right]d\omega,$$

where $s = c + i\omega$, showing that the integral in brackets is simply the Laplace transform $G(s)$ of $g(t)$ that exists by hypothesis. As $s = c + i\omega$, $ds = id\omega$, so after the change of variable from ω to s in the integral with respect to ω, the limit $\omega = -\infty$ becomes $s = c - i\infty$, and the limit $\omega = \infty$ becomes $s = c + i\infty$, reducing the previous result to

$$e^{-ct}g(t) = \frac{1}{2\pi i}\int_{c-i\infty}^{c+i\infty}e^{(s-c)t}G(s)ds.$$

Finally, cancelling the factor e^{-ct} that is not involved in the integral with respect to s, we arrive at the line integral

$$g(t) = \frac{1}{2\pi i}\int_{c-i\infty}^{c+i\infty}e^{st}G(s)ds.$$

Apart from a change of notation, involving g and G in place of f and F, this is the inversion formula (2), so the derivation is complete. The function $g(t)$ will be independent of the value of c provided $\text{Re}\{s\} > c$.

An important consequence of this derivation is that $g(t)$ can be allowed to be piecewise continuous with finite jump discontinuities. This follows because of the ability of the Fourier integral representation of a function to take account of finite jump discontinuities.

For the inversion integral to be useful, the line integral involved must be capable of evaluation in a straightforward manner, so let us now find how this can be accomplished. Consider the contour C_R in Fig. 16.1, where C_{1R} is the line $\text{Re}\{s\} = c$, $-R \le \text{Im}\{s\} \le R$, and C_{2R} is the semicircle $|s - c| = R$. If the integrand $e^{st}F(s)$ in (2) has a finite number of poles, all located inside C_R, then for sufficiently large R

$$\frac{1}{2\pi i}\int_{C_R}e^{st}F(s)ds = \Sigma\text{ \{residues at each of the poles of }e^{st}F(s)\}.$$

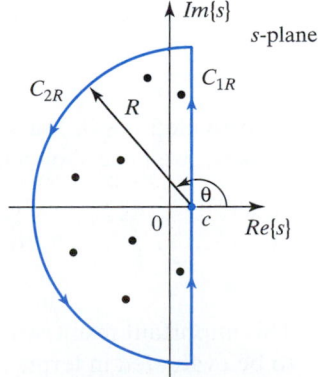

FIGURE 16.1   The contour C_R and a typical arrangement of poles inside C_R.

When expressed in terms of the contours C_{1R} and C_{2R} this result becomes

$$\frac{1}{2\pi i}\int_{c-iR}^{c+iR} e^{st}F(s)ds + \frac{1}{2\pi i}\int_{C_{2R}} e^{st}F(s)ds$$
$$= \Sigma \{\text{residues at each of the poles of } e^{st}F(s)\}.$$

Our objective will be to show that the integral around C_{2R} vanishes as $R \to \infty$. On C_{2R} we have $s = c + Re^{i\theta}$ with $\frac{\pi}{2} \le \theta \le \frac{3\pi}{2}$, so after the change of variable $\theta = \frac{\pi}{2} + \phi$ we can write $s = c + iRe^{i\phi}, 0 \le \phi \le \pi$, from which it follows that

$$ds = -Re^{i\phi}d\phi = -|s-c|e^{i\phi}d\phi.$$

Setting

$$I_R = \left|\frac{1}{2\pi i}\int_{C_{2R}} e^{st}F(s)ds\right|,$$

and transferring the modulus from outside the integral to inside, we arrive at the inequality

$$I_R \le \frac{1}{2\pi}\int_{C_{2R}} |e^{st}||F(s)||ds|$$
$$= \frac{1}{2\pi}\int_0^\pi |F(s)||\exp\{t[c+R(i\cos\phi - \sin\phi)]\}||s-c|d\phi.$$

Let us now suppose $F(s)$ is such that $|sF(s)| \le M$ on C_{2R} as $R \to \infty$. Then if we use the fact that for R sufficiently large $|s-c| \le |s| + |c| \le 2|s|$, the integral inequality becomes

$$I_R \le \frac{1}{\pi}\int_0^\pi |sF(s)|e^{ct}\exp(-Rt\sin\phi)d\phi = \frac{Me^{ct}}{\pi}\int_0^\pi \exp(-Rt\sin\phi)d\phi.$$

As $\sin\phi$ is symmetrical about the value $\frac{\pi}{2}$, this result can be rewritten as

$$I_R \le \frac{2Me^{ct}}{\pi}\int_0^{\pi/2} \exp(-Rt\sin\phi)d\phi.$$

Finally, applying the integral form of the Jordan inequality to this estimate, we find that

$$I_R \le \left(\frac{2Me^{ct}}{\pi}\right)\left(\frac{\pi}{2Rt}\right)(1-e^{-Rt})$$

so, provided $t > 0$, this shows that $\lim_{R\to\infty} I_R = 0$. Consequently, in the limit as $R \to \infty$, we have shown that when $t > 0$,

$$\frac{1}{2\pi i}\int_{c-iR}^{c+iR} e^{st}F(s)ds = \Sigma \{\text{residue at each of the poles of } e^{st}F(s)\}.$$

This important result, which forms the next theorem, enables the inversion integral to be evaluated in terms of the residues of the function $e^{st}F(s)$.

THEOREM 16.1

Inversion of a Laplace transform by means of residues   Let $F(s) = \mathcal{L}\{f(t)\}$, the Laplace transform of $f(t)$, be such that it has a finite number of poles, and choose c such that all the poles lie to the left of $\text{Re}\{s\} = c$. Then if a positive real number

the inversion integral and residues

M exists such that $|sF(s)| \leq M$ for all s to the left of $\text{Re}\{s\} = c$, the inverse Laplace transform $f(t) = \mathcal{L}^{-1}\{F(s)\}$ is given by

$$f(t) = \mathcal{L}^{-1}\{F(s)\} = \Sigma \,\{\text{residue at each of the poles of } e^{st}\, F(s)\}. \qquad \blacksquare$$

This theorem extends immediately to the case where $F(s)$ has an infinite number of poles all lying to the left of $\text{Re}\{s\} = c$ provided, as $R \to \infty$, the contour C_{2R} is allowed to expand in such a way that it never passes through a pole. The inversion of transforms of this type leads to the determination of $f(t) = \mathcal{L}^{-1}\{F(s)\}$ in the form of an infinite series of functions of t (see Example 16.4).

EXAMPLE 16.1

Use Theorem 16.1 to find $\mathcal{L}^{-1}\{(s^2 - a^2)/(s^2 + a^2)^2\}$, $a > 0$.

Solution   Before applying Theorem 16.1 it is necessary to check that its conditions are satisfied. Using the contour in Fig. 16.1 and setting $F(s) = (s^2 - a^2)/(s^2 + a^2)^2$, the poles (double) of $F(s)$ are seen to be located at $s = \pm ia$, so for suitably large R they will lie inside the contour provided $\text{Re}\{s\} < c$, with $c > 0$. In addition, $\lim_{s \to \infty} |sF(s)| = 0$ when s lies to the left of the imaginary axis, so the conditions of Theorem 16.1 are satisfied.

Routine calculations show that the residues of $e^{st}\, F(s)$ at its two double poles are

$$\text{Res}\left\{\frac{e^{st}(s^2 - a^2)}{(s^2 + a^2)^2}, s = \pm ia\right\} = \frac{t}{2}\exp(\pm iat),$$

so

$$f(t) = \mathcal{L}^{-1}\left\{\frac{(s^2 - a^2)}{(s^2 + a^2)^2}\right\} = \frac{t}{2}\{\exp(iat) + \exp(-iat)\} = t\cos at,$$

confirming entry 12 in Table 7.1 of Laplace transform pairs. ■

If a Laplace transform involves a branch point, the contour in Fig. 16.1 must be modified by inserting a branch cut to make the function single valued inside the contour, and this often involves making a cut along the negative real axis. An inversion integral requiring a branch cut of this type is given in the next example.

EXAMPLE 16.2

Find $\mathcal{L}^{-1}\{1/\sqrt{s}\}$.

some typical examples

Solution   The function $F(s) = 1/\sqrt{s}$ has a branch point at the origin of the s-plane, so instead of the contour in Fig. 16.1 we will use the contour in Fig. 16.2, where a branch cut has been made along the negative real axis with each side of the cut being connected by a small circular arc surrounding the branch point at the origin.

The semicircular contour C_{2R} in Fig. 16.1 is now replaced by the two circular arcs AB and EF of radius R together with the path BC along the top of the branch cut, the small circular arc Γ of radius ε around the branch point, and the path DE along the bottom of the branch cut. The function $F(s) = 1/\sqrt{s}$ is analytic and single valued inside this modified contour, which is bounded on the right by the vertical line C_{1R}. We will use the principal branch of the function for which the argument lies in the interval $-\pi < \theta \leq \pi$. As the branch cut along the negative real axis terminates at the origin, we must take $c > 0$.

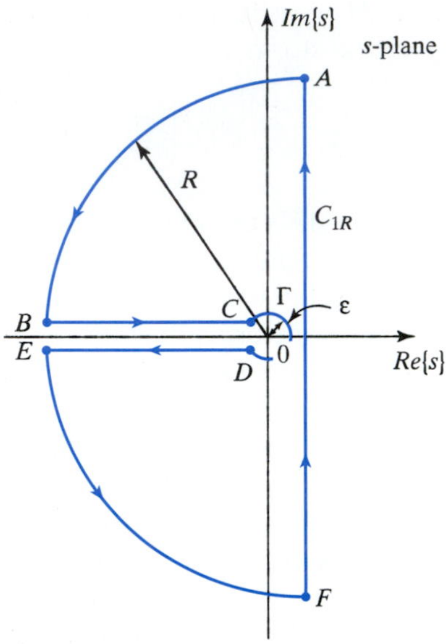

FIGURE 16.2 Modified contour with a branch cut to make $1/\sqrt{s}$ single valued.

On C_{2R} we have $s = c + Re^{i\theta}$ for $\frac{\pi}{2} \le \theta \le \frac{3\pi}{2}$. For later use we now set $\theta = \pi/2 + \phi$, so s becomes $s = c + iRe^{i\phi}$, for $0 \le \phi \le \pi$. With this change of variable $ds = -Re^{i\phi}d\phi$, so $|ds| = Rd\phi$, and provided R is sufficiently large $|s| = |c + iRe^{i\phi}| \ge ||Re^{i\phi}| - |c|| = R - c$. We will also need to use the result that $|e^{st}| = |\exp\{t[(c - R\sin\phi) + iR\cos\phi]\}| = e^{ct}\exp\{-Rt\sin\phi\}$. The integral I_R around C_{2R} can now be estimated as follows:

$$I_R = \left|\int_{ABEF} \frac{e^{st}}{\sqrt{s}}ds\right| \le \int_{ABEF} \frac{|e^{st}|}{|s|^{1/2}}|ds| \le \frac{e^{ct}R}{(R-c)^{1/2}} \int_0^{\pi} \exp[-Rt\sin\phi]d\phi.$$

The symmetry of $\sin\phi$ about $\phi = \pi/2$ allows this to be rewritten as

$$I_R \le \frac{2e^{ct}R}{(R-c)^{1/2}} \int_0^{\pi/2} \exp[-Rt\sin\phi]d\phi,$$

so applying the integral form of the Jordan inequality we find that

$$I_R \le \frac{\pi e^{ct}}{(R-c)^{1/2}t}(1 - e^{-Rt}).$$

Allowing $R \to \infty$, with $t > 0$, in this last result shows that $\lim_{R\to\infty} I_R = 0$.

The integral around the contour Γ of radius ε, on which $s = \varepsilon e^{i\varphi}, ds = i\varepsilon e^{i\varphi}d\varphi$, and $s^{1/2} = e^{i\varphi/2}\sqrt{\varepsilon}$, is given by

$$\int_{-\pi}^{\pi} \frac{1}{e^{i\varphi/2}\sqrt{\varepsilon}} \exp[\varepsilon t(\cos\varphi + i\sin\varphi)]i\varepsilon e^{i\varphi}d\varphi,$$

but this also is seen to vanish as $\varepsilon \to 0$.

Along the top BC of the branch cut $s = re^{\pi i} = -r$, so $\sqrt{s} = e^{i\pi/2}\sqrt{r} = i\sqrt{r}$, and $ds = -dr$, whereas along the bottom DE of the cut $s = re^{-i\pi} = -r$, so $\sqrt{s} = e^{-i\pi/2}\sqrt{r} = -i\sqrt{r}$, and again $ds = -dr$. As no poles lie inside the contour, it follows from the Cauchy integral theorem that

$$\frac{1}{2\pi i} \lim_{R\to\infty, \varepsilon\to 0} \left\{ \int_{C_{1R}} \frac{e^{st}}{\sqrt{s}} ds + \int_{R}^{\varepsilon} \frac{1}{i\sqrt{r}} e^{-rt}(-dr) + \int_{\Gamma} \frac{e^{st}}{\sqrt{s}} ds \right.$$

$$\left. + \int_{\varepsilon}^{R} \frac{1}{(-i)\sqrt{r}} e^{-rt}(-dr) + \int_{C_{2R}} \frac{e^{st}}{\sqrt{s}} ds \right\} = 0.$$

We have shown that when $t > 0$ the third and last terms vanish in the limit as $R \to \infty$ and $\varepsilon \to 0$, so the equation reduces to

$$\frac{1}{2\pi i} \int_{c-i\infty}^{c+i\infty} \frac{e^{st}}{\sqrt{s}} ds = \frac{1}{2\pi i} \left\{ -\int_{\infty}^{0} \frac{ie^{-rt}}{\sqrt{r}} dr + \int_{0}^{\infty} \frac{ie^{-rt}}{\sqrt{r}} dr \right\} = \frac{1}{\pi} \int_{0}^{\infty} \frac{e^{-rt}}{\sqrt{r}} dr.$$

The changes of variable $r = u^2$ followed by $v = u\sqrt{t}$ simplify this result to

$$\frac{1}{2\pi i} \int_{c-i\infty}^{c+i\infty} \frac{e^{st}}{\sqrt{s}} ds = \frac{2}{\pi\sqrt{t}} \int_{0}^{\infty} e^{-v^2} dv,$$

so using the standard result $\int_{0}^{\infty} e^{-v^2} dv = \sqrt{\pi}/2$ we find that

$$\mathcal{L}^{-1}\left\{ \frac{1}{\sqrt{s}} \right\} = \frac{1}{\sqrt{\pi t}}, \quad \text{for Re}\{s\} > 0. \qquad \blacksquare$$

In the next example we consider a Laplace transform with an exponential factor in the numerator, which is known from the operational properties of the Laplace transform to arise from a shift in t.

EXAMPLE 16.3   Find $\mathcal{L}^{-1}\{e^{-s}/(s^2 + 1)\}$.

Solution   It was shown in Chapter 7 that $\mathcal{L}^{-1}\{e^{-s}/(s^2 + 1)\} = H(t - 1)\sin(t - 1)$ for $t > 0$, where $H(t - 1)$ is the Heaviside unit step function defined as

$$H(t - a) = \begin{cases} 0, & t < a \\ 1, & t > a. \end{cases}$$

how the inversion integral generates the Heaviside step function

We now show how the result $\mathcal{L}^{-1}\{e^{-s}/(s^2 + 1)\}$ can be recovered by means of the inversion integral. It is a routine matter to establish that Theorem 16.1 applies to the function $F(s) = e^{-s}/(s^2 + 1)$, which only has simple poles at $s = \pm i$, so we proceed directly to the determination of the residues of $e^{st} F(s)$. We have

$$\text{Res}\left\{ \frac{e^{s(t-1)}}{s^2 + 1}, s = i \right\} = -\frac{i}{2} \exp[i(t - 1)]$$

and

$$\text{Res}\left\{ \frac{e^{s(t-1)}}{s^2 + 1}, s = -i \right\} = \frac{i}{2} \exp[-i(t - 1)],$$

so from Theorem 16.1

$$f(t) = \mathcal{L}^{-1}\left\{ \frac{e^{-s}}{s^2 + 1} \right\} = \left\{ -\frac{i}{2} \exp[-i(t - 1)] + \frac{i}{2} \exp[-i(t - 1)] \right\} = \sin(t - 1).$$

As the Laplace transform of a function $f(t)$ is not defined for $t < 0$, we must require $\mathcal{L}^{-1}\{e^{-s}/(s^2 + 1)\}$ to be zero for $t < 1$, so if we make use of the Heaviside unit step function this becomes

$$f(t) = \mathcal{L}^{-1}\{e^{-s}/(s^2 + 1)\} = H(t - 1)\sin(t - 1) \quad \text{for } t > 0.$$

In this example a discontinuous function has been recovered from its Laplace transform by means of the inversion integral in (2).   ∎

The extension of Theorem 16.1 to a Laplace transform $F(s)$ with an infinite number of poles is illustrated in the following example.

EXAMPLE 16.4   Find $\mathcal{L}^{-1}\{\frac{1}{s\cosh s}\}$.

Solution   Setting $F(s) = \frac{1}{s\cosh s}$, we see that $e^{st}F(s)$ has an infinite number of simple poles on the imaginary axis, with one at $s = 0$ due to the factor s in the denominator, and others at $s = (2n + 1)\pi i/2$ with $n = 0, \pm1, \pm2, \dots$, corresponding to the zeros of $\cosh s$. As all the poles lie on the imaginary axis, when applying the inversion integral we will use the contour shown in Fig. 16.3 with $c > 0$ arbitrarily small, and to prevent the contour passing through a pole we set $R = k\pi$ with $k = 1, 2, \dots$.

how the inversion integral generates a series

Routine calculations show that

$$\text{Res}\{e^{st}F(s), s = 0\} = 1$$

and

$$\text{Res}\{e^{st}F(s), s = (2n + 1)\pi i/2\} = (-1)^{n+1}\frac{2\exp[(2n + 1)\pi it/2]}{(2n + 1)\pi}.$$

Extending Theorem 16.1 in an obvious manner we have

$$f(t) = \frac{1}{2\pi i}\lim_{R\to\infty}\left\{\int_{-iR}^{iR}\frac{e^{st}}{s\cosh s}ds + \int_{ABC}\frac{e^{st}}{s\cosh s}ds\right\}$$

$$= \Sigma\{\text{residues at poles of } e^{st}F(s)\}.$$

On the semicircle ABC of radius R, $s = Re^{i\theta}$ with $\frac{\pi}{2} \le \theta \le \frac{3\pi}{2}$, so $|s| = R$ and $|ds| = Rd\theta$. Substituting for s in e^{st} gives $|e^{st}| = \exp[Rt\cos\theta]$, and

$$|\cosh s| = |\cosh(R\cos\theta)\cos(R\sin\theta) + i\sinh(R\cos\theta)\sin(R\sin\theta)|$$

$$= [\cosh^2(R\cos\theta) - \sin^2(R\sin\theta)]^{1/2}$$

The graph of $|\cosh s|$ as a function of θ is symmetrical about $\theta = \pi$ for all R, and it attains its least values at the ends of the interval $\pi/2 \le \theta \le 3\pi/2$. However, $R = k\pi$, so setting $\theta = \pi/2$ we find that on the semi-circle ABC

$$|\cosh s| \ge [1 - \sin^2(k\pi)]^{1/2} = 1.$$

Using these results to estimate the integral around ABC we find that

$$I_R = \left|\int_{ABC}\frac{e^{st}}{s\cosh s}ds\right| \le \int_{ABC}\frac{|e^{st}|}{|s||\cosh s|}|ds| \le \int_{\pi/2}^{3\pi/2}\frac{\exp[k\pi t\cos\theta]}{k\pi}k\pi d\theta$$

$$= \int_{\pi/2}^{3\pi/2}\exp[k\pi t\cos\theta]d\theta.$$

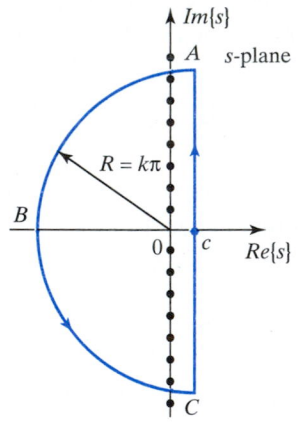

FIGURE 16.3 Contour containing poles on the imaginary axis.

After the change of variable $\theta = \pi/2 - \phi$ this becomes

$$I_R \le \int_0^\pi \exp[-k\pi t \sin\phi]d\phi,$$

but $\sin\phi$ is symmetric about $\phi = \pi/2$, so this is seen to be equivalent to

$$I_R \le 2\int_0^{\pi/2} \exp[-k\pi t \sin\phi]d\phi.$$

Applying the integral form of the Jordan inequality reduces this to

$$I_R \le \frac{1}{kt}(1 - e^{-k\pi t}),$$

so that provided $t > 0$, $\lim_{k\to\infty} I_R = 0$. Consequently we have shown that

$$f(t) = \mathcal{L}^{-1}\left\{\frac{1}{s\cosh s}\right\} = \sum \{\text{residues at the poles of } e^{st}F(s)\}.$$

Combining the residues of poles located at pairs of complex conjugate points along the imaginary axis causes the complex parts of the residues to cancel, leaving the real result

$$f(t) = \mathcal{L}^{-1}\left\{\frac{1}{s\cosh s}\right\} = 1 + \frac{4}{\pi}\sum_{n=0}^\infty (-1)^{n+1}\frac{\cos[(2n+1)\pi t/2]}{2n+1}.$$

We see that in this case the inversion integral has given rise to a function $f(t)$ in the form of a sum of an infinite series of cosine functions.

To understand why this has occurred, we need only notice that $F(s)$ is, in fact, the Laplace transform of the rectangular pulse function

$$f(t) = 2\sum_{n=0}^\infty (-1)^n H[t - (2n+1)]$$

with period 4 and amplitude 2. So what has been recovered by the inversion integral is the Fourier series representation of the piecewise continuous function

$$f(t) = \begin{cases} 0, & 0 < t < 1 \\ 2, & 1 < t < 3 \\ 0, & 3 < t < 4, \end{cases}$$

where $f(t) = 0$ for $t < 0$ and $f(t+4) = f(t)$ for $t > 0$. ∎

Although Theorem 16.1 provides a general formula for the inverse of a Laplace transform, it is not always easy to use. In certain cases the inversion integral can be avoided by employing a known transform together with one or more of the operational properties possessed by all Laplace transforms. This approach is illustrated in the next example.

EXAMPLE 16.5　Find $\mathcal{L}^{-1}\left\{\frac{1}{s\sqrt{s+1}}\right\}$.

Solution　An attempt to find this inverse transform by means of Theorem 16.1 leads to difficulties in the determination of the residues, so we will employ a different approach. The first shift theorem for Laplace transforms asserts that if

$\mathcal{L}\{f(t)\} = F(s)$, then $\mathcal{L}\{e^{at}f(t)\} = F(s-a)$, so by replacing s by $s+1$ in the result of Example 16.2 we have

$$\mathcal{L}^{-1}\left\{\frac{1}{\sqrt{s+1}}\right\} = \frac{e^{-t}}{\sqrt{\pi t}}.$$

To complete the inversion process we now make use of the Laplace transform of an integral that asserts that if $\mathcal{L}\{f(t)\} = F(s)$, then

$$\mathcal{L}\left\{\frac{F(s)}{s}\right\} = \int_0^t f(\tau)d\tau.$$

Using this result with $\mathcal{L}^{-1}\{1/\sqrt{s+1}\}$ gives

$$\mathcal{L}^{-1}\left\{\frac{1}{s\sqrt{s+1}}\right\} = \frac{1}{\sqrt{\pi}}\int_0^t \frac{e^{-u}}{\sqrt{u}}du.$$

The change of variable $u = v^2$ converts this to

$$\mathcal{L}^{-1}\left\{\frac{1}{s\sqrt{s+1}}\right\} = \frac{2}{\sqrt{\pi}}\int_0^{\sqrt{t}} \exp(-v^2)dv,$$

but the error function erf (x) is given by

$$\text{erf}\,(x) = \frac{2}{\sqrt{\pi}}\int_0^x \exp(-v^2)dv = \frac{2}{\sqrt{\pi}}\sum_{n=0}^{\infty}(-1)^n\frac{x^{2n+1}}{n!(2n+1)},$$

so

$$\mathcal{L}^{-1}\left\{\frac{1}{s\sqrt{s+1}}\right\} = \text{erf}\,(\sqrt{t}).$$ ■

EXAMPLE 16.6   Find

$$\mathcal{L}^{-1}\left\{\frac{\exp(-a\sqrt{s})}{s}\right\}, \quad \text{with } a > 0.$$

Solution   The function has a branch point at the origin, so when evaluating the Laplace inversion integral by means of a contour integral it is necessary to use a contour with a cut along the negative real axis and to enclose the origin in a small circle of radius $\varepsilon > 0$. The complete contour C is shown in Fig. 16.4, and it comprises integrals along the path AB that in the limit will become the integral from $c - i\infty$ to $c + i\infty$, and the paths $\gamma_1, \Gamma_1, C_1, \Gamma_2,$ and γ_2.

Setting

$$f(t) = \mathcal{L}^{-1}\left\{\frac{\exp(-a\sqrt{s})}{s}\right\} \quad \text{and} \quad F(s) = \frac{\exp(-a\sqrt{s})}{s}e^{st},$$

and noticing that $F(s)$ has no poles inside C, we can write

$$0 = \int_{AB} F(s)ds + \int_{\gamma_1} F(s)ds + \int_{\Gamma_1} F(s)ds + \int_{C_1} F(s)ds + \int_{\Gamma_2} F(s)ds + \int_{\gamma_2} F(s)ds,$$

and so

$$\frac{1}{2\pi i}\int_{AB} F(s)ds = \frac{1}{2\pi i}\left\{\int_{-\gamma_1} F(s)ds + \int_{-\Gamma_1} F(s)ds + \int_{-C_1} F(s)ds \right.$$
$$\left. + \int_{-\Gamma_2} F(s)ds + \int_{-\gamma_2} F(s)ds\right\},$$

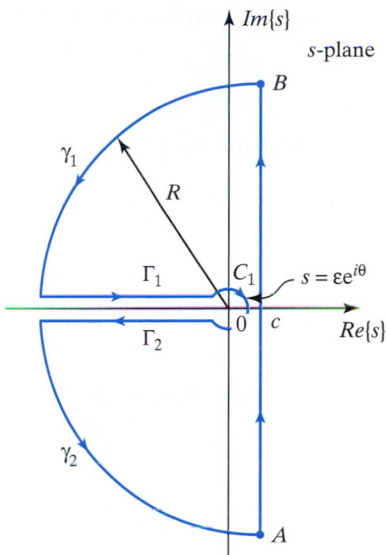

FIGURE 16.4 The contour involving a cut along the negative real axis.

where the symbols $-\gamma_1, -\Gamma, \ldots, -\gamma_2$ indicate the reversal of the direction of integration along these paths. In the limit as $A \to c - i\infty$ and $B \to c + i\infty$, the integral on the left becomes $f(t)$, and standard arguments show that as $R \to \infty$ the integrals along γ_1 and γ_2 that form part of the circle $|s| = R$ in Fig. 16.4 vanish. So, letting $R \to \infty$, the preceding result is seen to reduce to

$$\frac{1}{2\pi i} \int_{AB} F(s)ds = \frac{1}{2\pi i} \left\{ \int_{-\Gamma_1} F(s)ds + \int_{-C_1} F(s)ds + \int_{-\Gamma_2} F(s)ds \right\}.$$

The path Γ_1 lies on the upper side of the negative real axis on which $s = re^{\pi i}$, so $\sqrt{s} = \sqrt{r}e^{i\pi/2} = i\sqrt{r}$. The path Γ_2 lies on the lower side of the negative real axis on which $s = re^{-\pi i}$, so $\sqrt{s} = \sqrt{r}e^{-i\pi/2} = -i\sqrt{r}$. Using these results and allowing for the reversal of the directions of integration, we have

$$f(t) = \lim_{\varepsilon \to 0} \frac{1}{2\pi i} \left\{ \int_{\varepsilon}^{\infty} \frac{\exp(-ia\sqrt{r})}{(-r)} e^{-rt}(-dr) \right.$$

$$+ \int_{-\pi}^{\pi} \frac{\exp(-a\sqrt{\varepsilon}e^{i\theta/2})}{\varepsilon e^{i\theta}} \exp(\varepsilon t e^{i\theta}) \varepsilon i e^{i\theta} d\theta$$

$$\left. + \int_{\varepsilon}^{\infty} \frac{\exp(ia\sqrt{r})}{(-r)} e^{-rt}(-dr) \right\}.$$

Letting $\varepsilon \to 0$, the integral around the branch point becomes $\int_{-\pi}^{\pi} i\,d\theta = 2\pi i$, so after reversing the limits in the last integral the equation becomes

$$f(t) = \frac{1}{2\pi i} \left\{ \int_0^{\infty} \frac{e^{-rt}}{r}(-2i) \sin\left(a\sqrt{r}\right)dr + 2\pi i \right\},$$

or

$$f(t) = 1 - \frac{1}{\pi} \int_0^{\infty} \frac{e^{-rt}}{r} \sin(a\sqrt{r})dr.$$

This expression can be put in a more convenient form if the integral

$$I = \frac{1}{\pi} \int_0^\infty \frac{e^{-rt}}{r} \sin\left(a\sqrt{r}\right) dr$$

is transformed by setting $rt = u^2$. After this change of variable the integral becomes

$$I = \frac{2}{\pi} \int_0^\infty \frac{\exp(-u^2)}{u} \sin(\beta u) du, \quad \text{where } \beta = a/\sqrt{t}.$$

Now

$$\frac{\partial I}{\partial \beta} = \frac{2}{\pi} \int_0^\infty \exp(-u^2) \cos(\beta u) du,$$

but from Exercise 24 in Exercise Section 14.3,

$$\int_0^\infty \exp(-u^2) \cos(\beta u) du = \frac{1}{2}\sqrt{\pi}\exp(-\beta^2/4),$$

so

$$\frac{\partial I}{\partial \beta} = \frac{1}{\sqrt{\pi}}\exp(-\beta^2/4).$$

Integration of this result from 0 to β, using the fact that $I = 0$ when $\beta = 0$, gives

$$I = \frac{1}{\sqrt{\pi}} \int_0^\beta \exp(-v^2/4) dv,$$

or

$$I = \frac{1}{\sqrt{\pi}} \int_0^{a/\sqrt{t}} \exp(-v^2/4) dv.$$

In terms of the error function

$$\text{erf}(x) = \frac{2}{\pi} \int_0^x \exp(-t^2) dt,$$

integral I becomes

$$I = \text{erf}\left(\frac{a}{2\sqrt{t}}\right),$$

and so

$$f(t) = 1 - \text{erf}\left(\frac{a}{2\sqrt{t}}\right) = \text{erfc}\left(\frac{a}{2\sqrt{t}}\right).$$

We have shown that

$$f(t) = \mathcal{L}^{-1}\left\{\frac{\exp(-a/\sqrt{s})}{s}\right\} = 1 - \text{erf}\left(\frac{a}{2\sqrt{t}}\right) = \text{erfc}\left(\frac{a}{2\sqrt{t}}\right). \quad \blacksquare$$

The inversion integral for the Laplace transform is discussed in some detail in reference [3.8] together with various applications, and also in references [4.3] and [4.4]. A comprehensive account of different forms of the Laplace transform and their associated inversion integrals is given in reference [3.18]; see also reference [6.10].

Summary

A contour integral called the Laplace inversion integral was derived that allows the function $f(t)$ to be recovered from its Laplace transform $F(s)$. This more advanced method is necessary when the transform $F(s)$ is too complicated for $f(t)$ to be found by means of a table of transform pairs. The method was illustrated by being used to invert some more complicated transforms.

EXERCISES 16.1

In Exercises 1 through 13 use the inversion integral to find $\mathcal{L}^{-1}\{F(s)\}$.

1. $F(s) = \dfrac{1}{s(s^2 + a^2)}$ $(a > 0)$.

2. $F(s) = \dfrac{1}{(s + 2)(s^2 + 4)}$.

3. $F(s) = \dfrac{(s - 1)}{(s + 1)^2}$.

4. $F(s) = \dfrac{4s + 1}{s^2(s^2 + 1)}$.

5. $F(s) = \dfrac{1}{s^3(s + 1)}$.

6. $F(s) = \dfrac{s}{(s + 4)^2(s - 1)}$.

7. $F(s) = \dfrac{1}{(s^2 + a^2)^2}$ $(a > 0)$.

8. $F(s) = \dfrac{1}{s^4 - a^4}$ $(a > 0)$.

9.* $F(s) = \dfrac{1}{s^{1/3}}$ (Hint: Use the gamma function in the final result).

10.* $F(s) = \dfrac{e^{-s}}{(s^2 + 1)^2}$.

11.* $F(s) = \dfrac{(s + 1)e^{-2s}}{s^2 - 1}$.

12.* $F(s) = \dfrac{1}{\sqrt{s}(s - 1)}$.

13.* $F(s) = \dfrac{1}{s\sqrt{s + a}}$ $(a > 0)$.

14.* Find $\mathcal{L}^{-1}\{\frac{1}{s^{3/2}}\}$ without using the inversion integral by using a property of the Laplace transform that determines $\mathcal{L}^{-1}\{s^{-3/2}\}$ from the result $\mathcal{L}^{-1}\{s^{-1/2}\} = (\pi t)^{-1/2}$.

15.* Find $\mathcal{L}^{-1}\{\frac{1}{\sqrt{s+b}}\}$, and use the result with the convolution theorem for the Laplace transform to find $\mathcal{L}^{-1}\{\frac{1}{(s+a)\sqrt{s+b}}\}$ $(b > a > 0)$.

16.* Show that

$$\mathcal{L}^{-1}\left\{\frac{1}{s^3 \sinh s}\right\} = \frac{t(t^2 - 1)}{6} - \frac{2}{\pi^3}\sum_{n=1}^{\infty}(-1)^n\frac{\sin n\pi t}{n^3}.$$

17.* Show that

$$\mathcal{L}^{-1}\left\{\frac{1}{(s^2 + 1)(1 + e^{-2as})}\right\} = \frac{\sin(t + a)}{2}$$

$$+ \frac{1}{a}\sum_{n=1}^{\infty}\frac{\cos(2n - 1)\pi t/2a}{1 - (2n - 1)^2\pi^2/4a^2}\quad (a > 0).$$

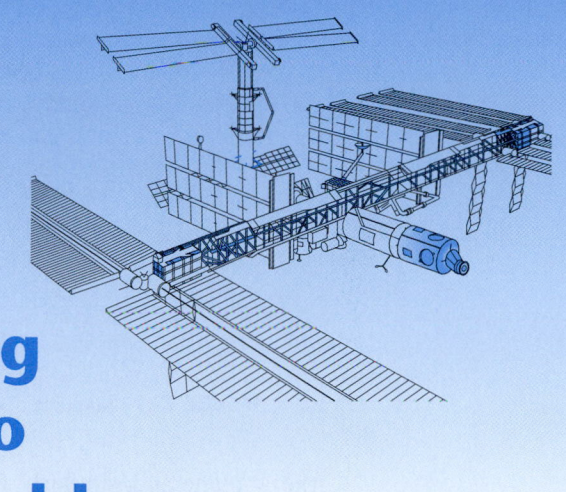

CHAPTER 17

Conformal Mapping and Applications to Boundary Value Problems

The way curves and regions in one plane are mapped by analytic functions onto another plane constitutes the study of conformal mappings. Conformal mappings concern the geometrical properties of analytic functions, and their study is closely related to the Laplace equation. This chapter defines a conformal mapping as one that preserves both the angle between intersecting curves and the sense of rotation from one curve to the other, and then proceeds to examine some of the most important examples of these mappings produced by elementary analytic functions.

Conformal mappings are shown to map a harmonic function in one plane into a harmonic function in another plane, and it is this property that is used when boundary value problems for the two-dimensional Laplace equation are solved. Applications of conformal mappings are made to two-dimensional boundary value problems involving heat flow, electrostatics, and ideal fluids.

Of particular interest is the ability of conformal mappings to map regions with a complicated boundary shape onto regions with a simple boundary shape. This is because such mappings can be used to solve two-dimensional boundary value problems for Laplace's equation in regions of complicated shape. The required solution follows directly from the fact that conformal mappings map one analytic function into another one. Consequently, if a conformal mapping can be found to map a complicated region onto one with a simple shape, once the solution of the corresponding boundary value problem in the simply shaped region has been found, it can be transformed back into the required solution in the complicated region.

17.1   Conformal Mapping

Let Γ_1 and Γ_2 be any two curves in the z-plane that radiate out from a common point of intersection P at z_0, as shown in Fig. 17.1a. Then if the curves have the respective parametric representations $z_1(t) = x_1(t) + i y_1(t)$ and $z_2(t) = x_2(t) + i y_2(t)$ for $a \leq t \leq b$, at their point of intersection P corresponding to $t = a$ we have $z_0 = z_1(a) = z_2(a)$. Now let the function $f(z)$ be a single-valued analytic function of z in some region D of the z-plane, and set $w = f(z)$. Then, as $f(z)$ is continuous, each point of Γ_1 will correspond to a unique point on some curve γ_1 in

877

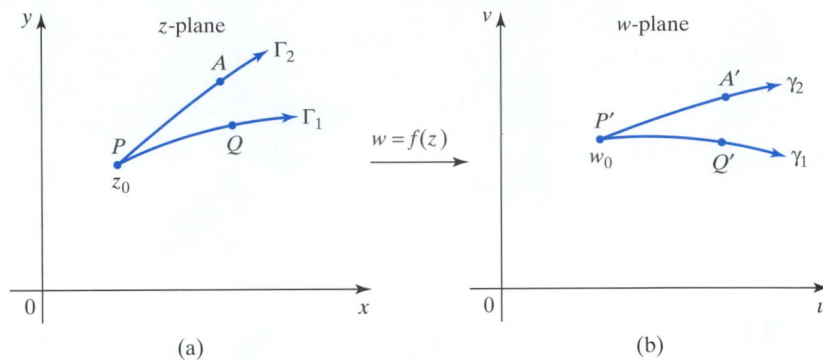

FIGURE 17.1   Mapping of curves Γ_1 and Γ_2 to γ_1 and γ_2 by $w = f(z)$.

the w-plane and, similarly, each point of Γ_2 will correspond to a unique point on some other curve γ_2 in the w-plane. As the curves Γ_1 and Γ_2 intersect at P located at z_0, the curves γ_1 and γ_2 must intersect at the point P', called the **image** of P, located at the point $w_0 = f(z_0)$ in the w-plane. In general, when points in the w-plane are identified by letters, their images in the w-plane are identified by using the same letters with the addition of a prime. So if A, B, C denote points in the z-plane, A', B', and C' will be used to denote the corresponding images in the w-plane.

As the parametrization in terms of t induces a **sense** (of direction) along the curves Γ_1 and Γ_2 as t increases, this sense is transferred to the curves γ_1 and γ_2 in the w-plane, as shown in Fig. 17.1b. Curves along which a sense of direction is defined are called **directed curves**.

> **image, directed curve, and conformal mapping**

The curves Γ_1 and Γ_2 in the z-plane are said to be **mapped** onto the respective curves γ_1 and γ_2 in the w-plane by the function $w = f(z)$. It is usual to call γ_i the **image** of Γ_i under the mapping $w = f(z)$ from the z-plane to the w-plane and, conversely, as $f(z)$ is single valued, Γ_1 is called the image of γ_1 under the inverse mapping $z = f^{-1}(w)$ from the w-plane to the z-plane. In what follows we will show that for any z_0 such that $f'(z_0) \neq 0$, the analytic nature of $f(z)$ causes the mapping to preserve the angle of intersection between the curves Γ_1 and Γ_2 at P in the z-plane, so it equals the angle between their images γ_1 and γ_2 at P' in the w-plane. In addition, and equally important, we will show that the sense of rotation is preserved, so if the tangent to Γ_2 at P is obtained by rotating the tangent to Γ_1 at P counterclockwise through an angle α, then the tangent to γ_2 at P' is obtained by rotating the tangent to γ_1 at P' counterclockwise through the same angle α. A mapping that possesses these two properties is called a **conformal mapping**, and such mappings play a useful role in connection with the solution of boundary value problems for the two-dimensional Laplace equation.

To establish the conformal nature of the mapping produced by a single valued analytic function $w = f(z)$, we now appeal to Fig. 17.2. Consider the secant PQ on curve Γ_1 in Fig. 17.2a, and the corresponding secant $P'Q'$ in Fig. 17.2b, where Q is located at z_1 and Q' at $w_1 = f(z_1)$. In the limit as $Q \to P$, the angle between the secant PQ and the real axis in the z-plane becomes the angle α_1 between the tangent to Γ_1 at P and the real axis and, correspondingly, point $Q' \to P'$, causing the angle between the secant $P'Q'$ and the real axis in the w-plane to become the angle β_1 between the tangent to γ_1 at P' and the real axis.

Consequently, as $PQ = z_1 - z_0$, we can write

$$\alpha_1 = \lim_{z_1 \to z_0} \text{Arg}(z_1 - z_0),$$

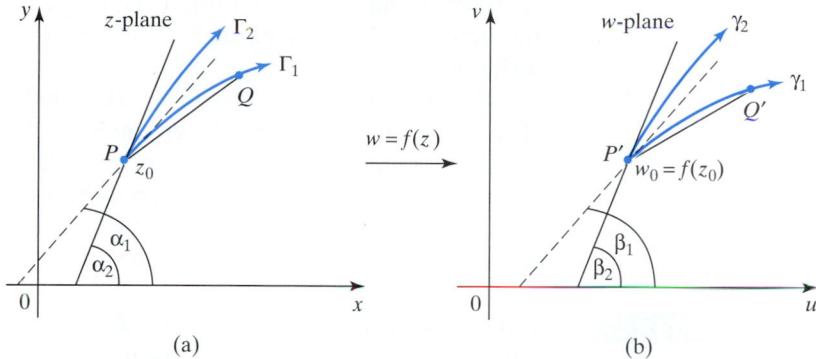

FIGURE 17.2   Secants PQ and $P'Q'$ in the z- and w-planes.

and correspondingly

$$\beta_1 = \lim_{z_1 \to z_0} \text{Arg}(w_1 - w_0).$$

Forming the difference $\beta_1 - \alpha_1$, we have

$$\beta_1 - \alpha_1 = \lim_{z_1 \to z_0} \text{Arg}(w_1 - w_0) - \lim_{z_1 \to z_0} \text{Arg}(z_1 - z_0),$$

but $\text{Arg}\, a - \text{Arg}\, b = \text{Arg}(a/b)$, so this last result can be written

$$\beta_1 - \alpha_1 = \lim_{z_1 \to z_0} \text{Arg}\left(\frac{w_1 - w_0}{z_1 - z_0}\right).$$

As $f(z)$ is an analytic function, and so has a unique derivative $f'(z)$ irrespective of the way in which $z_1 \to z_0$, the preceding result shows that when $f'(z_0) \neq 0$,

$$\beta_1 - \alpha_1 = \text{Arg}\, f'(z_0).$$

The uniqueness of the derivative $f'(z_0)$ means that the foregoing result is true for any other curve passing through P and its image curve through P', so, in particular, it is true for the curves Γ_2 and γ_2. We have shown

$$\beta_1 - \alpha_1 = \beta_2 - \alpha_2,$$

and this can be rewritten as

$$\alpha_2 - \alpha_1 = \beta_2 - \beta_1.$$

As the curves Γ_1 and Γ_2 were any two curves that intersect in the z-plane, this result has established the preservation of both the angles and their senses under the mapping $w = f(z)$, and hence the conformal nature of mappings produced by single-valued analytic functions at all points z where $f'(z) \neq 0$.

Although angles and senses of rotation are preserved by a conformal mapping, in general the length scale involved in a mapping at a point z_0 in the z-plane and at its image point $w_0 = f(z_0)$ in the w-plane is different. To find the linear scale factor $\rho(z_0)$ that is involved at $z = z_0$, we need to consider the limit of the quotient $|f(z) - f(z_0)|/|z - z_0|$ as $z \to z_0$, but this is simply $|f'(z_0)|$. So in a conformal mapping, provided $f'(z) \neq 0$, the **linear scale factor** $\rho(z)$ introduced at a point z when mapping infinitesimal line elements from the z-plane to the w-plane is $\rho(z) = |f'(z)|$ and, correspondingly, the **area scale factor** is $\rho^2(z)$. Because the scale factor and the

linear and area scale factors and critical points

rotation produced by a conformal mapping usually change throughout the w-plane, the image in the w-plane of boundaries of regions in the z-plane can look very different.

Points z_0 for which $f'(z_0) = 0$ are called **critical points** of the function $f(z)$. It can be seen from the above argument that the conformal nature of a mapping $w = f(z)$ breaks down at a critical point z_0 of $f(z)$, because at such a point the angle between intersecting curves at z_0, and between their image curves at $w_0 = f(z_0)$ are *not* preserved, and in addition the linear and area scale factors vanish at such points. We have proved the following fundamental theorem.

THEOREM 17.1

the fundamental mapping theorem

Conformal mapping   Let $f(z)$ be analytic and single valued in a region of the z-plane. Then, at every point z in the region such that $f'(z) \neq 0$, the conformal mapping $w = f(z)$ preserves angles between intersecting curves in the z-plane, and it also preserves the sense of rotation between intersecting directed curves. The linear scale factor involved in the mapping from the z-plane to the w-plane is $\rho(z) = |f'(z)|$ and the area scale factor is $\rho^2(z) = |f'(z)|^2$.   ∎

The fact that conformal mappings preserve angles between intersecting curves and their sense of rotation leads to the following rule that determines how regions in the z-plane map onto regions in the w-plane. The rule will be used in the examples that follow.

Rule for determining how a region in the z-plane is mapped onto a corresponding region in the w-plane by a conformal mapping $w = f(z)$

Let a region R in the z-plane be bounded by a continuous and piecewise smooth contour Γ, and let the z-plane be mapped conformally onto the w-plane by $w = f(z)$. Furthermore, let A and B be any two distinct points on Γ and suppose that the region R lies to the left (right) as the boundary Γ is traversed in the direction from A to B. Then if γ is the image of Γ, A' and B' are the images of A and B, and R' is the image of R, the region R' in the w-plane will lie to the left (right) as γ is traversed in the direction from A' to B'.

deciding how a region in the z-plane maps onto a region in the w-plane

---

The preceding rule implies the following simple test for the determination of regions that correspond under a one-one conformal transformation $w = f(z)$. If Z is any test point in a region of interest in the z-plane, then the corresponding region in the w-plane will be the one containing the point $w = f(Z)$.

Before examining some typical examples of conformal transformations, we will prove the important property that the curves $u = $ constant and $v = $ constant in the w-plane are mutually orthogonal at all points other than at the images of the critical points of $w = f(z)$ in the z-plane.

Setting $w = u + iv = f(z)$ and taking the total derivatives of u and v with respect to x gives

$$\frac{du}{dx} = \frac{\partial u}{\partial x} + \frac{\partial u}{\partial y}\frac{dy}{dx} \quad \text{and} \quad \frac{dv}{dx} = \frac{\partial v}{\partial x} + \frac{\partial v}{\partial y}\frac{dy}{dx}.$$

So, along the curves $u = $ constant and $v = $ constant,

$$0 = \frac{\partial u}{\partial x} + \frac{\partial u}{\partial y}\left(\frac{dy}{dx}\right)_{u=const} \quad \text{and} \quad 0 = \frac{\partial v}{\partial x} + \frac{\partial v}{\partial y}\left(\frac{dy}{dx}\right)_{v=const},$$

where $(dy/dx)_{u=const}$ and $(dy/dx)_{v=const}$ are, respectively, the gradients of $u = $ constant and $v = $ constant in the w-plane. Combining these results at an arbitrary point P that is not the image of a critical point of $w = f(z)$ in the z-plane, and writing $(dy/dx)_{u=const, P} = (dy/dx)_{P(u)}$, and $(dy/dx)_{v=const, P} = (dy/dx)_{P(v)}$, we have

$$\left(\frac{dy}{dx}\right)_{P(u)}\left(\frac{dy}{dx}\right)_{P(v)} = \left(-\frac{\partial u}{\partial x} \Big/ \frac{\partial u}{\partial y}\right)_P \left(-\frac{\partial v}{\partial x} \Big/ \frac{\partial v}{\partial y}\right)_P .$$

However, from the Cauchy–Riemann equations the product of these last factors is seen to be -1, showing that

$$\left(\frac{dy}{dx}\right)_{P(u)}\left(\frac{dy}{dx}\right)_{P(v)} = -1.$$

Thus, as P is an arbitrary point in the w-plane at which the product of the gradients of $u = $ constant and $v = $ constant equals -1, it follows directly that the curves $u = $ constant and $v = $ constant are mutually orthogonal except at points that are the images of critical points of $w = f(z)$ in the z-plane. We have proved the next theorem.

THEOREM 17.2

constant values of the real and imaginary parts of $f(z)$ map onto orthogonal trajectories

$u = $ constant and $v = $ constant are orthogonal trajectories   If $w = f(z) = u + iv$ is a single-valued analytic function, the families of curves $u = $ constant and $v = $ constant are mutually orthogonal in the w-plane except at the images of the critical points of $f(z)$.   ∎

(a) The Linear Transformation $w = az + b$

The simplest nontrivial conformal transformation is the **linear transformation**

$$w = az + b \quad \text{with } a \neq 0. \tag{1}$$

As $a \neq 0$ the transformation between the z- and w-planes is one-one, because

$$z = \left(\frac{1}{a}\right)w - \frac{b}{a}, \tag{2}$$

the geometrical properties of the linear mapping

and the transformation is conformal because $f(z) = az + b$ is an analytic function for all z. As $w' = d/dz[az + b] = a \neq 0$, the linear transformation has no critical points. To understand the geometrical interpretation of the linear transformation, notice first that we can write $a = |a| \exp[i \operatorname{Arg} a]$. As a result $w = az + b$ can be regarded as the combination of the three simple transformations,

$$w_1 = |a|z, \quad w_2 = \exp[i \operatorname{Arg} a]w_1, \quad \text{and} \quad w = w_2 + b.$$

The transformation $w_1 = |a|z$ scales z by the real constant factor $|a|$, so although the image in the w_1-plane of the boundary of an arbitrary region in the z-plane experiences neither a translation nor a rotation, it does experience a uniform *magnification* if $|a| > 1$ and uniform *contraction* if $|a| < 1$.

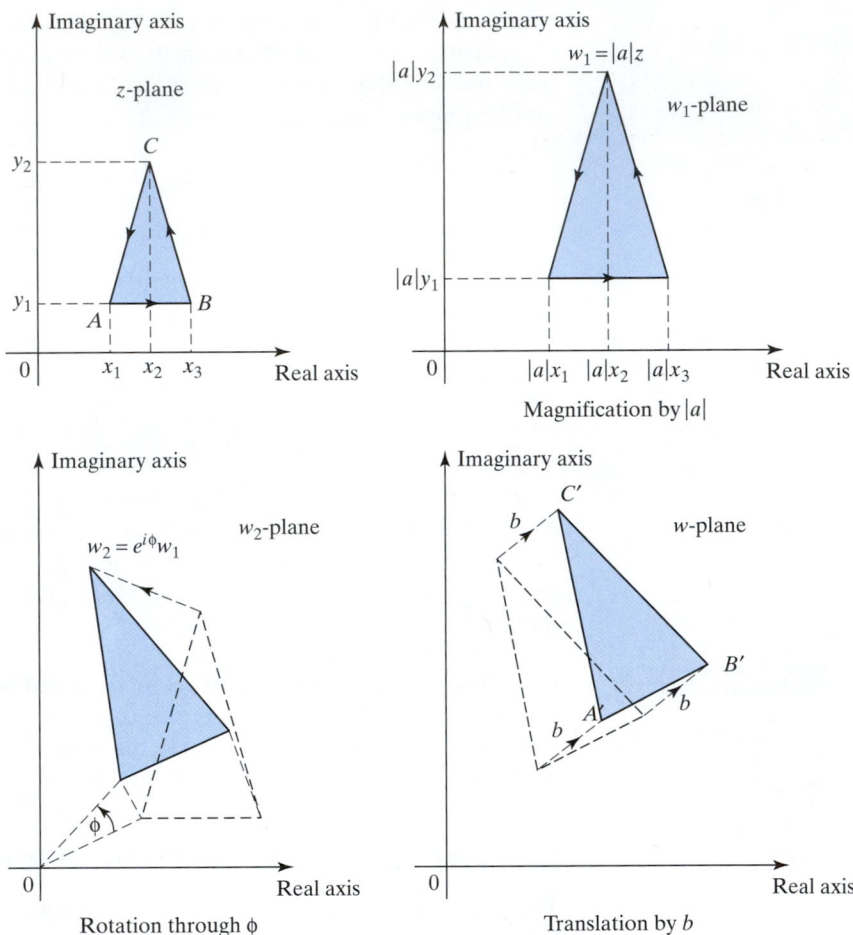

FIGURE 17.3 Successive transformations leading to $w = az + b$.

When complex numbers are multiplied their arguments are added, so setting $\text{Arg}\,a = \phi$ shows the transformation $w_2 = e^{i\phi} w_1$ produces a uniform *rotation* through an angle ϕ about the origin in the w_1-plane.

Finally, the transformation $w = w_2 + b$ is seen to involve a translation of every point in the w_2-plane by an amount b. So the combined effect of linear transformation (1) on the boundary of any region in the z-plane is to produce first a scaling by a constant factor $|a|$, then a uniform rotation through an angle $\phi = \text{Arg}\,a$, and finally a uniform translation by an amount b. Thus, a linear transformation *preserves the shapes* of boundaries of regions of interest. The sequence of diagrams in Fig. 17.3 illustrates the typical effect of these successive transformations on a triangular region in the z-plane with its vertices at A, B, and C and the image points A', B', and C' in the w-plane.

To apply the preceding rule to determine the region in the w-plane corresponding to the triangle in the z-plane, we use the fact that the interior of the triangle in the z-plane lies to the left as the boundary is traversed in the direction A, B, and C. Consequently, the corresponding region in the w-plane is the interior of the triangle A', B', and C', because this also lies to the left as the transformed boundary is traversed in the direction A', B', and C'.

It is important always to use a test point with the rule developed earlier in order to check how regions transform. This is because a conformal transformation may map the *interior* of a closed contour Γ in the z-plane onto the *exterior* of its image γ in the w-plane. An example of this type is provided by the inversion mapping that is considered next.

(b) The Inversion Mapping $w = 1/z$

The mapping

$$w = 1/z \tag{3}$$

is called the **inversion mapping**, or sometimes the **reciprocal mapping**. This provides a conformal mapping of the z-plane onto the w-plane, because $f(z) = 1/z$ is a single-valued analytic function with only the simple pole at the origin $z = 0$ where the derivative $w' = -1/z^2$ is not defined. If we set $z = re^{i\theta}$, the mapping becomes

$$w = \left(\frac{1}{r}\right)e^{-i\theta}. \tag{4}$$

This result shows that points on the unit circle $|z| = 1$ map to points on the unit circle $|w| = 1$. However, because of the reversal of the sign of θ, points on the *upper* half of the circle $|z| = 1$ are reflected in the real axis and mapped to points on the *lower* half of the circle $|w| = 1$, and conversely. Furthermore, because $|w| = 1/r$, it follows that points *inside* $|z| = 1$ are mapped to points *outside* $|w| = 1$, and conversely, as shown in Fig. 17.4. This can be confirmed by taking $z = \frac{1}{2}$ as a test point *inside* the unit circle $|z| = 1$, and noticing that it transforms to the point $w = 2$ *outside* the unit circle $|w| = 1$. Notice that the circle in the z-plane and its image in the w-plane are traversed in opposite directions.

The inversion mapping can be regarded as the composition (product) of the two simple transformations

$$Z = \frac{1}{z} \quad \text{and} \quad w = \overline{Z}. \tag{5}$$

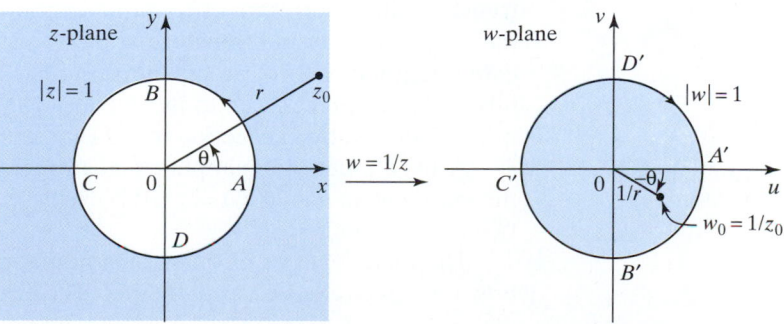

FIGURE 17.4   The inversion mapping $w = 1/z$.

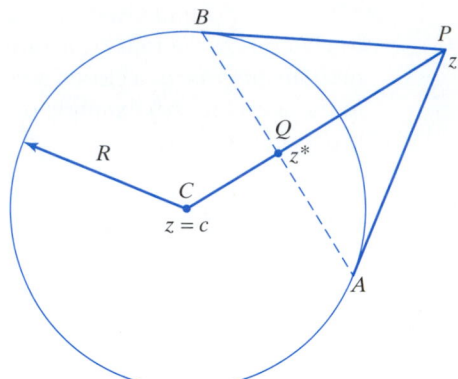

FIGURE 17.5 Inversion in a circle.

To interpret these transformations geometrically we will make use of the general concept of **inversion in a circle**. Consider the circle of radius R in Fig. 17.5, where the point P at z lies outside the circle with its center C at $z = c$, and Q at z^* lies inside it on the radial line CP at its point of intersection with the chord AB drawn from the points A and B where lines from P are tangent to the circle.

A simple argument using similar triangles shows that

the geometrical operation of inversion in a circle

$$|CP| \times |CQ| = R^2,$$

or

$$|z - c|\,|z^* - c| = R^2.$$

The points P and Q in Fig. 17.5 are said to be **symmetric** with respect to the circle with its center at C. Point Q said to be **inverse** to point P and, similarly, point P is inverse to Q. In particular, if $c = 0$ so the circle is centered on the origin, the preceding result implies that points z and z^* that are symmetric with respect to the circle $|z| = R$ are such that

$$z^* = \frac{R^2}{\bar{z}}. \tag{6}$$

Examination of the first transformation in (5) shows that $|Z||\bar{z}| = 1$, so this transformation corresponds to an inversion in the unit circle $|z| = 1$ centered on the origin. The second transformation $w = \bar{Z}$ simply involves the complex conjugate operation, and so can be interpreted as a reflection in the real axis. Thus, the inversion mapping is seen to involve a reflection in the unit circle centered on the origin followed by a reflection in the real axis.

a fixed point of a mapping

A **fixed point** of a mapping f is a point z^* that is left invariant as a result of the mapping, so that $f(z^*) = z^*$. It is easily seen that the inversion mapping has the two fixed points $z = \pm 1$.

The main features of the inversion mapping will become clear if we consider how it maps circles and straight lines. The equation

$$A(x^2 + y^2) + Bx + Cy + D = 0, \tag{7}$$

where the coefficients A, B, C, and D are real, describes a circle of radius $R = (B^2 + C^2 - 4AD)^{1/2}/2\,|A|$ with its center at $(-B/2A, -C/2A)$ provided $B^2 + C^2 > 4AD$ and $A \neq 0$, and a straight line when $A = 0$. The distance of the center of the circle from the origin is $(B^2 + C^2)^{1/2}/2\,|A|$, so the circle will *not* pass through the origin if $D \neq 0$, since then $x = 0$, $y = 0$ does not satisfy (7).

If we write $w = u + iv$, the inversion mapping $w = 1/z$ becomes

$$u + iv = \frac{1}{x + iy},$$

from which we find that

$$x = \frac{u}{u^2 + v^2}, \qquad y = -\frac{v}{u^2 + v^2}. \tag{8}$$

Substituting (8) into (7) with $A \neq 0$, $D \neq 0$ gives the equation

$$D(u^2 + v^2) + Bu - Cv + A = 0, \tag{9}$$

that describes a circle in the w-plane of radius $\rho = (B^2 + C^2 - 4AD)^{1/2}/2\,|D|$, with its center at $(-B/2D, C/2D)$. This circle will not pass through the origin in the w-plane if $A \neq 0$, since then $u = v$ does not satisfy (9). Thus, the inversion mapping transforms a circle in the z-plane that does not pass through the origin into a circle in the w-plane that does not pass through the origin.

If, however, $A = 0$ and $D \neq 0$, the straight line in the z-plane given by (7) maps to the circle

$$D(u^2 + v^2) + Bu - Cv = 0 \tag{10}$$

with radius $\rho = (B^2 + C^2)^{1/2}/2\,|D|$ and its center at $(-B/2D, C/2D)$. As the radius of this circle and the distance of its center from the origin are equal, the circle passes through the origin in the w-plane. Conversely, if $D = 0$ and $A \neq 0$, a straight line in the w-plane will map onto a circle that passes through the origin in the z-plane.

Finally, if $A = D = 0$, the straight line in the z-plane given by (7) will pass through the origin and map onto a straight line in the w-plane that passes through the origin.

In summary, the inversion mapping has the following properties:

summary of the geometrical properties of the inversion mapping

(a) A circle in one plane that does not pass through the origin will map onto a circle in the other plane that does not pass through the origin.

(b) A straight line in one plane that does not pass through the origin will map onto a circle in the other plane that passes through the origin.

(c) A straight line through the origin in one plane will map onto a straight line through the origin in the other plane.

(d) Points inside a unit circle centered on the origin in one plane will map to points outside the unit circle centered on the origin in the other plane, and conversely.

The line $x = $ constant parallel to the imaginary axis is obtained from (7) by setting $A = C = 0$, and examination of the results following (10) shows that this maps onto a circle through the origin in the w-plane with its center on the *real* axis. Similarly, the line $y = $ constant, corresponding to $A = B = 0$ in (10), is seen to map onto a circle through the origin in the w-plane with its center on the *imaginary* axis. Thus, constant coordinate lines map to families of circles through the origin, one with its centers on the real axis and the other with its centers on the imaginary

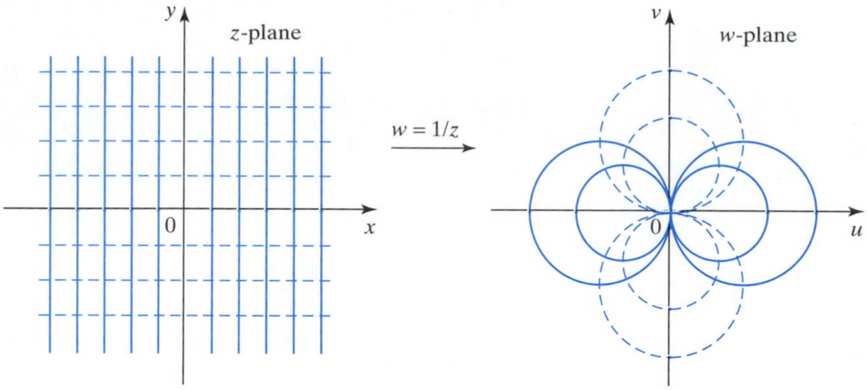

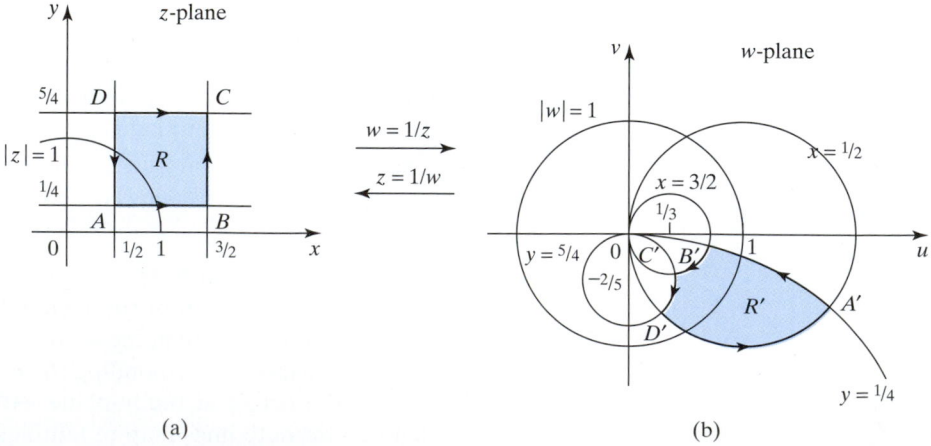

FIGURE 17.6 Mapping of coordinate lines by $w = 1/z$.

axis. As the coordinate lines $x =$ constant and $y =$ constant are orthogonal, the conformal nature of the transformation ensures that the two families of circles are themselves mutually orthogonal, as shown in Fig. 17.6.

The inversion mapping relates directly to the *extended complex plane* introduced at the end of Section 15.3. It will be recalled that the extended complex plane is formed by including in the ordinary complex plane the so-called *point at infinity*, defined as the limit as $R \to \infty$ of all points in the z-plane that lie outside the circle $|z| = R$. As a result, the inversion mapping is seen to map the origin in the z-plane to the point at infinity in the w-plane, and the point at infinity in the z-plane to the origin in the w-plane. If we set $T(z) = 1/z$, the inversion mapping becomes $w = T(z)$, and we can then write $T(0) = \infty$ and $T(\infty) = 0$.

The use of the extended complex plane unifies the treatment of the mapping of straight lines and circles by $w = 1/z$ by allowing straight lines to be regarded as circles of infinite radius.

The effect of an inversion mapping on the square in the z-plane with its sides parallel to the real and imaginary axes shown in diagram (a) on the left of Fig. 17.7 can be seen in the diagram (b) on the right. The sides of the square are seen to map

(a)

(b)

FIGURE 17.7 Inversion mapping of a square.

to four circular arcs, and the rule for determining how regions transform shows that the interior of the square maps to the interior of the region bounded by the circular arcs. For reference purposes the unit circles centered on the origin have been shown in both planes to illustrate how points B, C, and D that lie outside the unit circle in the z-plane map to points inside the unit circle in the w-plane, while point A that lies inside the unit circle in the z-plane maps to a point outside the unit circle in the w-plane. The effect of the reflection in the real axis that is involved in the inverse mapping is also apparent, because a region in the first quadrant in the z-plane has been mapped to a region in the fourth quadrant in the w-plane.

(c) The Linear Fractional Transformation

The transformation

$$w = \frac{az + b}{cz + d},$$ (11)

the linear fractional transformation, or bilinear transformation

is called either the **linear fractional transformation** or the **bilinear transformation**, and sometimes the **Möbius transformation**. It is always possible to assume that $c \neq 0$, because when $c = 0$ the transformation reduces to the linear transformation already considered. Furthermore, we may always assume that $ad - bc \neq 0$, because if $ad - bc = 0$ transformation (11) reduces to a constant.

The inverse mapping

$$z = \frac{b - dw}{cw - a}$$ (12)

is also a linear fractional mapping, and as the derivative is

$$w' = \frac{ad - bc}{(cz + d)^2},$$

the mapping is seen to be one-to-one and conformal everywhere with the exception of the point at $z = -d/c$.

Writing the linear fractional transformation in (11) in the form

$$w = \frac{az + b}{cz + d} = \frac{a}{c} + \frac{bc - ad}{c(cz + d)}$$ (13)

allows it to be regarded as the sequence of transformations

$$w_1 = cz + d, \quad w_2 = 1/w_1, \quad \text{and} \quad w = (a/c) + \frac{(bc - ad)}{c} w_2.$$ (14)

These equations show that a linear fractional transformation can be regarded as the composition of a linear transformation, an inversion mapping, and then another linear transformation.

Having interpreted a general linear fractional transformation in this manner, we can now make use of the general properties of linear transformations and inversion mappings to deduce the general properties of a linear fractional transformation. It is not difficult to see that the transformation (11) maps straight lines and circles onto straight lines and circles, though not necessarily in this order.

summary of geometrical properties of the linear fractional transformation

Furthermore, the definition of symmetry of two points with respect to a circle introduced in (b) earlier when discussing the inversion mapping enables another useful result to be proved: namely, that a pair of points that are symmetric with respect to a circle in the z-plane are mapped by a linear fractional transformation into a pair of points that are symmetric with respect to the image of the circle in the w-plane. The proof of this result is not difficult and so is left as an exercise, but the general result is important because it describes the *symmetry preserving property* of all linear fractional transformations.

When the linear fractional transformation is written in the form

$$w = \frac{(a/c)z + (b/c)}{z + d/c}, \tag{15}$$

it can be seen to be fully determined once the three numbers a/c, b/c, and d/c are specified. We now show how the transformation can be found when three distinct points z_1, z_2, and z_3 that are specified in the z-plane are required to map to three distinct points w_1, w_2, and w_3 that are specified in the w-plane. As three noncollinear points define a circle, it follows that three such points mapping to three other noncollinear points will cause the transformation to map a specific circle in one plane onto a specific circle in the other plane. Similarly, if the three points in one plane are collinear and the three in the other plane are not collinear, the transformation will map a specific straight line in one plane onto a specific circle in the other plane.

Using (11) we can write the difference $w - w_m$ as

$$w - w_m = \frac{(ad - bc)}{(cz + d)(cz_m + d)}(z - z_m), \quad \text{for } m = 1, 2, 3. \tag{16}$$

Forming the differences $w - w_1$, $w - w_2$, $w_3 - w_2$, and $w_3 - w_1$ and combining the resulting expressions leads to the result

a fundamental implicit relationship between w and z

$$\frac{w - w_1}{w - w_2} \cdot \frac{w_3 - w_2}{w_3 - w_1} = \frac{z - z_1}{z - z_2} \cdot \frac{z_3 - z_2}{z_3 - z_1}. \tag{17}$$

This is an *implicit* form of the relationship between w and z that determines the mapping between the specified points in each plane. The *explicit* transformation that produces the required mapping from the z-plane to the w-plane can be obtained from (17) by substituting the numbers z_1, z_2, z_3, w_1, w_2, and w_3 and solving for w in terms of z.

If one of the three points in either plane is the point at infinity, the factors in (17) containing it must be set equal to 1. To understand the reason for this, let us suppose for example that $z_3 = \infty$. Then from (17),

$$\lim_{z_3 \to \infty} \left[\frac{(z - z_1)(z_3 - z_2)}{(z - z_2)(z_3 - z_1)} \right] = \frac{(z - z_1)}{(z - z_2)} \lim_{z_3 \to \infty} \left[\frac{1 - z_2/z_3}{1 - z_1/z_3} \right] = \frac{z - z_1}{z - z_2},$$

using the implicit
relationship to find
a mapping

confirming that the factors containing z_3 are no longer present and so can be considered to have been set equal to 1. A corresponding result applies if either z_1 or z_2 is the point at infinity, or if any one of w_1, w_2, or w_3 is the point at infinity.

EXAMPLE 17.1

Find the linear fractional transformation that maps the points $z_1 = -1$, $z_2 = 1$, and $z_3 = i$ onto the respective points $w_1 = 0$, $w_2 = 1$, and $w_3 = -i$, and determine how the region R inside the circle through the three points in the z-plane maps onto a region R' in the w-plane.

Solution   Substitution into (17) gives

$$\frac{w}{w-1} \cdot \frac{-(1+i)}{-i} = \frac{z+1}{z-1} \cdot \frac{i-1}{i+1},$$

so solving for w shows the required linear fractional transformation to be

$$w = \frac{z+1}{(2+i)z - i}.$$

The circles in the z- and w-planes through the stated points are shown in Fig. 17.8. As the region R *inside* the circle in the z-plane lies to the left as the circle is traversed in the direction z_1, z_2, and z_3, traversing the image points in the w-plane in the order w_1, w_2, and w_3 shows that the image R' of R must lie *outside* the circle in the w-plane. This is easily confirmed by noticing that the point $z = 0$ in R maps to the point $w = i$ in R'. ∎

EXAMPLE 17.2

Find the linear fractional transformation that maps the points $z_1 = -1$, $z_2 = 0$, and $z_3 = i$ onto the three points $w_1 = 0$, $w_2 = 1$, and $w_3 = \infty$, and determine how the region R inside the circle through the three points in the z-plane maps onto a region R' in the w-plane.

Solution   Substituting z_1, z_2, z_3, w_1, and w_2 into (17), and using the fact that $w_3 = \infty$ enables the factor containing w_3 to be replaced by 1, we find that

$$\frac{w}{w-1} = \frac{z+1}{z} \cdot \frac{i}{i+1}.$$

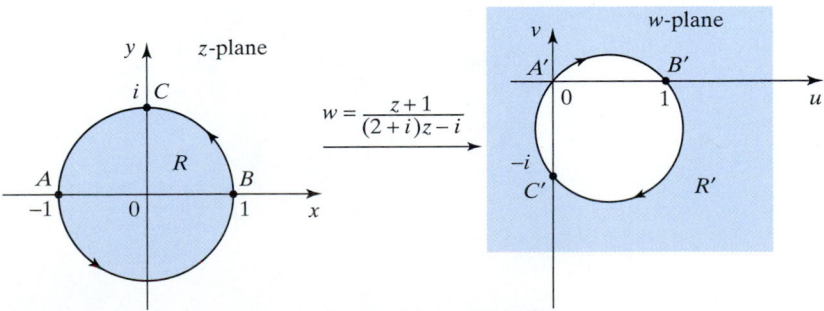

FIGURE 17.8   The mapping $\frac{z+1}{(2+i)z-i}$.

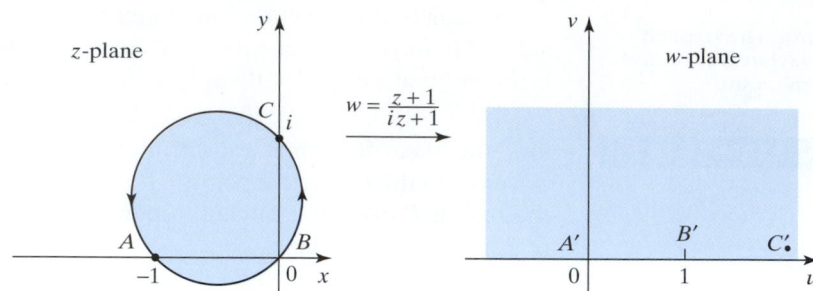

FIGURE 17.9   The mapping $w = \frac{z+1}{iz+1}$.

When solved for w the required linear fractional transformation is found to be given by

$$w = \frac{z+1}{iz+1}.$$

The circle in the z-plane and the corresponding straight line image in the w-plane are shown in Fig. 17.9. The ordering of the points in the two planes shows that as the region R *inside* the circle in the z-plane lies to the left as the circle is traversed in the direction z_1, z_2, and z_3, the image region R' in the w-plane must lie *above* (to the left) as the straight line (real axis) is traversed in the direction w_1, w_2, and w_3 in the w-plane.   ∎

(d) Mapping Eccentric Circles onto Concentric Circles

how to map eccentric circles onto concentric circles

A linear fractional transformation can map circles onto circles and, when doing so, preserves symmetry. Thus, it can be used to map the region between the eccentric circles in Fig. 17.10a onto the annular region between the concentric circles in Fig. 17.10b.

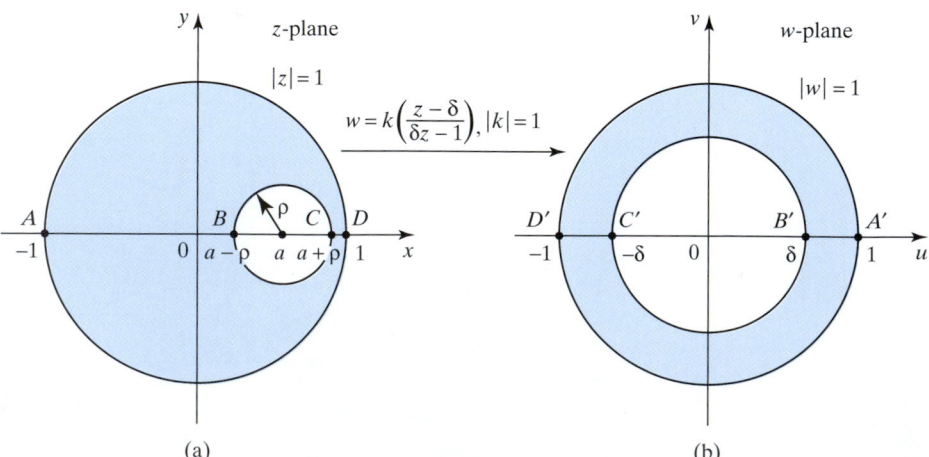

(a)

(b)

FIGURE 17.10   Mapping eccentric circles onto concentric circles.

To find the required transformation $w = T(z)$, we start from the fact that a linear fractional transformation $T(z)$ can always be written in the form

$$w = T(z) = K\left(\frac{z - \alpha}{z - \beta}\right).$$

So if the center of the inner circle of radius ρ in Fig. 17.10(a) located at $z = a$ is to map to the origin in the w-plane in Fig. 17.10b, we must set $\alpha = a$, so that $T(z)$ becomes

$$T(z) = K\left(\frac{z - a}{z - \beta}\right).$$

The circles in Fig. 17.10(a) are symmetric about the real axis, so this symmetry will be preserved by $T(z)$. In addition, a point z^* that is symmetric relative to $z = a$ with respect to the circle $|z| = 1$ will be mapped onto a point in the w-plane that is symmetric relative to the origin $w = 0$ with respect to the circle $|w| = 1$, so z^* will be mapped to the point at infinity, showing that we must set $\beta = z^*$. The mapping $T(z)$ now takes the form

$$T(z) = K\left(\frac{z - a}{z - z^*}\right).$$

As a and z^* are symmetric with respect to the circle $|z| = 1$, it follows from (6) that $az^* = 1$, but a is real, so $z^* = 1/a$ must also be real. Using this result in $T(z)$ reduces it to

$$w = T(z) = aK\left(\frac{z - a}{az - 1}\right).$$

The unit circle $|z| = 1$ maps to the unit circle $|w| = 1$, so recognizing that $|w|^2 = w\overline{w} = 1$ and $z\overline{z} = 1$, and using $w = T(z)$ to form the product $w\overline{w}$, we arrive at the equation

$$1 = w\overline{w} = a^2 K\overline{K}\left(\frac{z - a}{az - 1}\right)\left(\frac{\overline{z} - a}{a\overline{z} - 1}\right) = a^2 K\overline{K}.$$

This result shows that the factor aK must be of unit modulus, so if k is an arbitrary complex number with unit modulus, $T(z)$ can be written

$$w = T(z) = k\left(\frac{z - a}{az - 1}\right), \quad \text{with } |k| = 1.$$

The transformation $T(z)$ maps the circle $|z| = 1$ onto the circle $|w| = 1$, and it preserves symmetry about the real axis in the w-plane. As a is arbitrary, although the image of the inner circle must be symmetric about the real axis in the w-plane, the location of its center will depend on a. The two circles in the w-plane are required to be concentric, so the images of $z_1 = a + \rho$ and $z_2 = a - \rho$ must be symmetric with respect to $w = 0$ at the points $w = \pm\delta$ on the real axis in the w-plane. Thus, $T(z)$ must be such that $T(z_1) = -T(z_2)$, and so

$$\frac{a - \rho - \delta}{\delta(a - \rho) - 1} = -\left(\frac{a + \rho - \delta}{\delta(a + \rho) - 1}\right).$$

After simplification δ is found to be a solution of the quadratic equation

$$a\delta^2 - (1 + a^2 - \rho^2)\delta + a = 0.$$

Examination of the way the boundaries transform confirms that the region between the eccentric circles in the z-plane maps to the region between the concentric circles in the w-plane.

We have shown that the transformation $w = T(z)$ that maps the region between the eccentric circles in Fig. 17.10a onto the annular region between the concentric circles in Fig. 17.10b is given by

$$w = T(z) = k\left(\frac{z - \delta}{\delta z - 1}\right), \quad |k| = 1, \tag{18}$$

with δ a solution

$$a\delta^2 - (1 + a^2 - \rho^2)\delta + a = 0.$$

(e) The Mapping $w = z^2$

The function

$$w = z^2 \tag{19}$$

how $w = z^2$ maps the z-plane onto the w-plane

is analytic for all z, and so provides a conformal mapping of the z-plane onto the w-plane except at $z = 0$, which is a critical point. Setting $z = re^{i\theta}$ and $w = \rho e^{i\phi}$ in (19) gives $w = r^2 e^{2i\theta} = \rho e^{i\phi}$, so

$$\rho = r^2 \quad \text{and} \quad \phi = 2\theta. \tag{20}$$

Consequently the concentric circles $r = R$ (constant) in the z-plane map onto the concentric circles

$$u^2 + v^2 = R^2$$

in the w-plane, while the radial lines $\theta = \alpha$ (constant) radiating out from the origin in the z-plane map onto the radial lines $\phi = 2\alpha$ in the w-plane.

To make the mapping from the z-plane to the w-plane single valued, it is necessary to restrict θ to any interval of length π. It is usual to restrict z to the upper half of the z-plane so $0 < \theta \leq \pi$ and $r > 0$, because then the upper half of the z-plane maps to the entire w-plane with a cut along the positive real axis, as shown in Fig. 17.11. The image of the region R shown in the z-plane is the region R' in the w-plane. The cut is essential to keep the mapping one-one, because the same transformation also maps the lower half of the z-plane onto the same cut w-plane. Without the cut the function $w = z^2$ maps the entire z-plane *twice* onto the entire w-plane.

Setting $z = x + iy$ and $w = u + iv$ in $w = z^2$ and equating the real and imaginary parts of the equation shows that

$$u = x^2 - y^2 \quad \text{and} \quad v = 2xy. \tag{21}$$

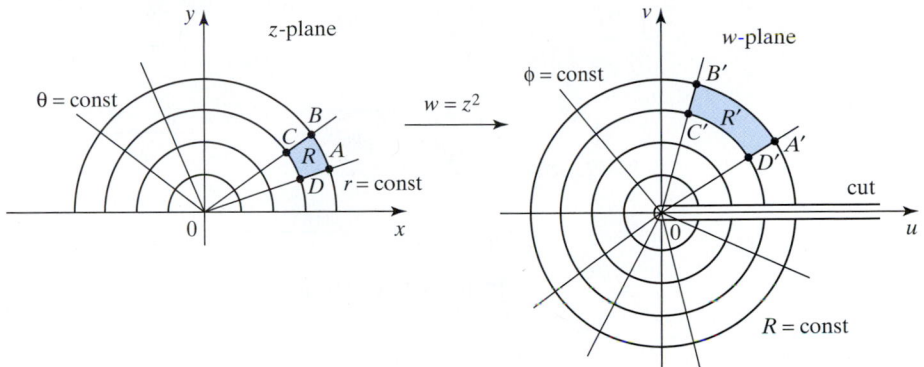

FIGURE 17.11   The mapping $w = z^2$.

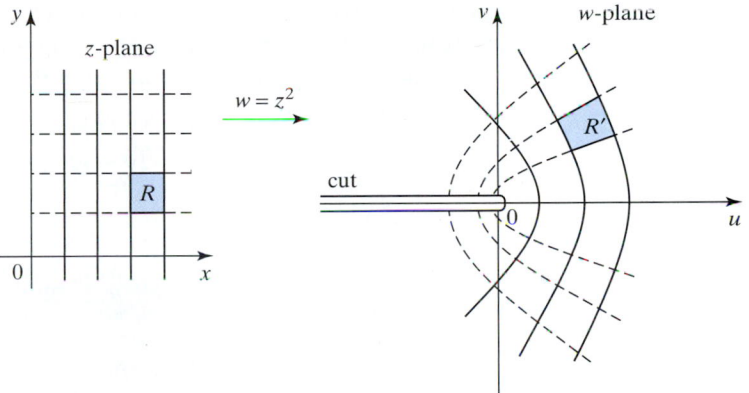

FIGURE 17.12   Mapping of cartesian coordinate lines by $w = z^2$.

So the lines $x = p$ map to the parabolas

$$v^2 = 4p^2(p^2 - u), \tag{22}$$

and the lines $y = q$ map to the parabolas

$$v^2 = 4q^2(u + q^2). \tag{23}$$

This mapping of cartesian coordinate lines in the z-plane onto parabolas in the w-plane is shown in Fig. 17.12, where region R' is the image of region R.

(f) The Function $w = z^{1/2}$

mapping by the branches of the square root function

The **square root function**

$$w = z^{1/2} \tag{24}$$

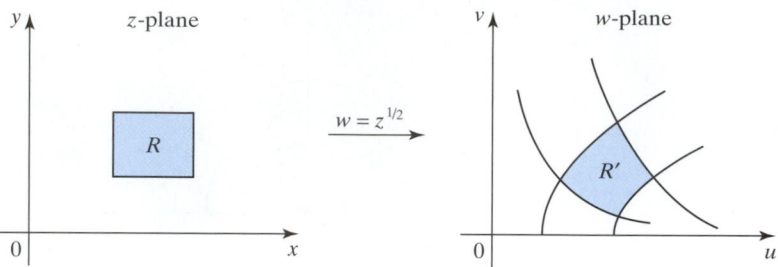

FIGURE 17.13   Mapping of a rectangle in the z-plane by the principal branch of $w = z^{1/2}$.

is the inverse of the mapping considered in (e) above. As the derivative of the square root function is $w' = \frac{1}{2}z^{-1/2}$, the square root function is seen to be an analytic function for all $z \neq 0$, so the conformal nature of the mapping from the z-plane to the w-plane will only fail at the origin. To make the function single valued, we will work with the principal branch of the square root function by setting $z = re^{i\theta}$, and then restricting θ to the interval $-\pi < \theta \leq \pi$, with $r > 0$. If we write $w = u + iv$, the mapping in (24) becomes

$$w = u + iv = r^{1/2}(\cos\theta/2 + i\sin\theta/2), \tag{25}$$

showing that

$$u = r^{1/2}\cos\theta/2 \quad \text{and} \quad v = r^{1/2}\sin\theta/2. \tag{26}$$

If the z-plane is cut along the negative real axis, results (26) show that the principal branch of the square root function maps each point of the cut z-plane once onto the right half of the w-plane, as illustrated in Fig. 17.13. Had the other branch of the square root function been used, where w is determined by

$$w = z^{1/2} = r^{1/2}\left(\cos\frac{(\theta + 2\pi)}{2} + i\sin\frac{(\theta + 2\pi)}{2}\right), \tag{27}$$

each point of the same cut z-plane would have been mapped once onto the left half of the w-plane.

To see how the square root function maps the cartesian coordinate lines in the z-plane onto the w-plane, we set $z = x + iy$ and $w = u + iv$ in (24) and square the result. Equating the real and imaginary parts then shows that

$$x = u^2 - v^2 \quad \text{and} \quad y = 2uv. \tag{28}$$

Thus, the cartesian coordinate lines $x = $ constant and $y = $ constant each map to families of rectangular hyperbolas. The conformal nature of the transformation ensures that the two families of hyperbolas are mutually orthogonal everywhere except at the origin where the critical point of the mapping is located. Figure 17.13 illustrates how the principal branch of the square root function maps a rectangular region in the z-plane onto a curvilinear region in the w-plane.

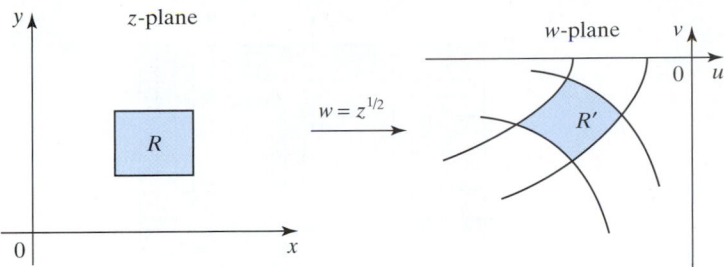

FIGURE 17.14 Mapping of a rectangle in the z-plane by the second branch of $w = z^{1/2}$.

The mapping of the same rectangular region by the second branch of the square root function given in (27) is shown in Fig. 17.14, obtained by rotating the first branch by an angle π.

(g) The Joukowski Transformation $w = z + 1/z$

the Joukowski transformation

The mapping

$$w = z + \frac{1}{z} \tag{29}$$

is called the **Joukowski transformation**, and as $w' = 1 - 1/z^2$ it is seen that w is analytic everywhere except at $z = 0$, and conformal everywhere except at the critical points located at $z = \pm 1$ that map to the points $w = \pm 2$. Setting $z = re^{i\theta}$ in (29), with $-\pi < \theta \leq \pi$ and $w = u + iv$, gives

$$w = u + iv = \left(r + \frac{1}{r}\right)\cos\theta + i\left(r - \frac{1}{r}\right)\sin\theta,$$

so that

$$u = \left(r + \frac{1}{r}\right)\cos\theta \quad \text{and} \quad v = \left(r - \frac{1}{r}\right)\sin\theta. \tag{30}$$

Examination of these results shows that the unit circle $|z| = 1$ maps onto the segment $-2 < u < 2$, $v = 0$, of the real axis in the w-plane, and that its *exterior* maps to the w-plane from which the cut represented by this segment has been removed. The mapping of the z-plane onto the w-plane by the Joukowski transformation is double valued, because the *interior* of the unit circle is also mapped onto this same cut w-plane. The mapping (29) will be single valued if z is restricted to either the interior or the exterior of the unit circle $|z| = 1$.

Setting $z = x + iy$ and $w = u + iv$ in (29) and equating the real and imaginary parts of the equation give

$$u = \frac{x(x^2 + y^2 + 1)}{x^2 + y^2} \quad \text{and} \quad v = \frac{y(x^2 + y^2 - 1)}{x^2 + y^2}. \tag{31}$$

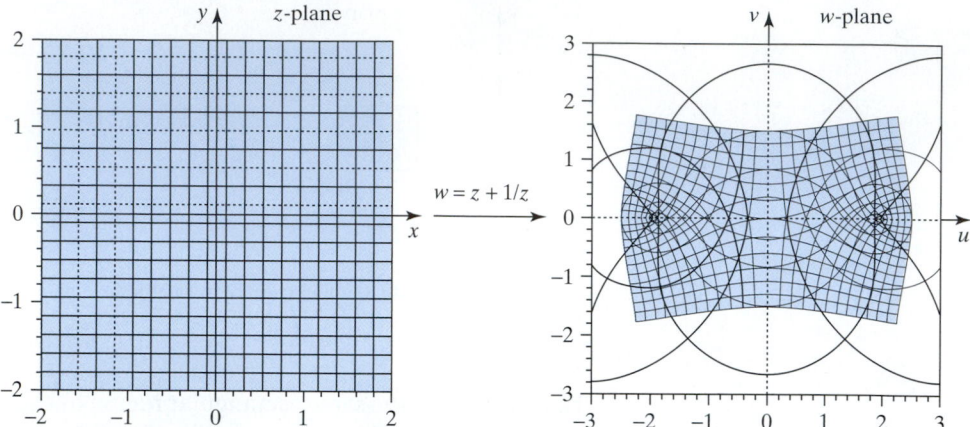

FIGURE 17.15 Mapping of cartesian coordinate lines by $w = z + 1/z$.

These equations determine the way the cartesian coordinate lines $x = \text{constant}$ and $y = \text{constant}$ map onto the w-plane. Figure 17.15 shows a representative set of mutually orthogonal curves in the w-plane corresponding to a set of cartesian coordinate lines in the z-plane.

Interest in this transformation, which was introduced by the Russian aerodynamicist N. J. Joukowski (1847–1921), first arose because of the way it maps a circle of radius R passing through the point $z = -1$ with its center at a point in the first quadrant of the z-plane onto the w-plane. A typical result of the mapping, called a **Joukowski airfoil profile**, is illustrated in Fig. 17.16. The mapping was used by Joukowski in early studies of the subsonic airflow when calculating the aerodynamic lift of wings with a cross-section in the form of a Joukowski profile.

The inverse mapping from the w-plane to the z-plane is obtained by multiplying the Joukowski transformation in (29) by z and solving the resulting quadratic equation for z in terms of w to obtain

$$z = \frac{1}{2}(w + \sqrt{w^2 - 4}). \tag{32}$$

The square root function is double valued, so this inverse transformation maps both the *exterior* and *interior* of $|z| = 1$ onto the w-plane, with a cut along the real axis from $w = -2$ to $w = 2$. Because of this it is necessary to use the branch of

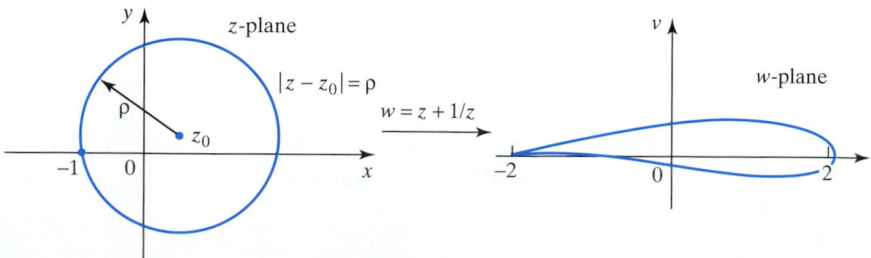

FIGURE 17.16 A typical Joukowski airfoil.

the square root function that is appropriate for the region to be mapped. So, for example, if the exterior of $|z| = 1$ is to be mapped onto the cut w-plane it is necessary to use the branch of the square root function for which

$$|w + \sqrt{w^2 - 4}| > 2.$$

This branch will give a one-one mapping of the upper half of the cut w-plane onto the exterior of the circle $|z| = 1$ in the upper half of the z-plane, with a corresponding mapping of the lower half of the cut w-plane onto the exterior of the circle $|z| = 1$ in the lower half of the z-plane.

(h) The Mappings $w = \sin z$ and Arcsin z

mapping by the sine function and its inverse

The next mapping to be considered is

$$w = \sin z \tag{33}$$

and its inverse Arcsin z.

The function $f(z) = \sin z$ is an entire function, and its critical points are determined by the zeros of $f'(z) = \cos z$ that occur when $z = (k + \frac{1}{2})\pi$ for $k = 0, \pm 1, \pm 2, \ldots$. This means that the mapping $w = \sin z$ will be conformal everywhere except at this infinite set of critical points along the real axis in the z-plane.

Setting $z = x + iy$ and $w = u + iv$ in (33), we have

$$w = \sin z = u + iv = \sin x \cosh y + i \cos x \sinh y,$$

so

$$u = \sin x \cosh y \quad \text{and} \quad v = \cos x \sinh y. \tag{34}$$

As $\sin x$ and $\cos x$ are periodic functions of x, equations (34) show that $w = \sin z$ maps the z-plane infinitely many times onto the w-plane. To make the mapping between the z- and w-planes conformal and one-one, it is necessary to restrict x to lie between any two successive critical points. We choose to require x to lie in the interval $-\frac{\pi}{2} \le x \le \frac{\pi}{2}$ and y to be such that $y \ge 0$, so z lies inside or on the boundary of the semi-infinite strip shown in Fig. 17.17.

As on the side $A_\infty B$ of the semi-infinite strip $x = -\frac{\pi}{2}$ and $y \ge 0$, it follows from (34) that this side must map onto the semi-infinite line segment $A'_\infty B'$ in the w-plane given by $u = -\cosh y$, $y > 0$ and $v = 0$, which lies along the real axis in the w-plane from $-\infty$ to the point $w = -1$. On the line BC, $y = 0$ and $-\frac{\pi}{2} \le x \le \frac{\pi}{2}$,

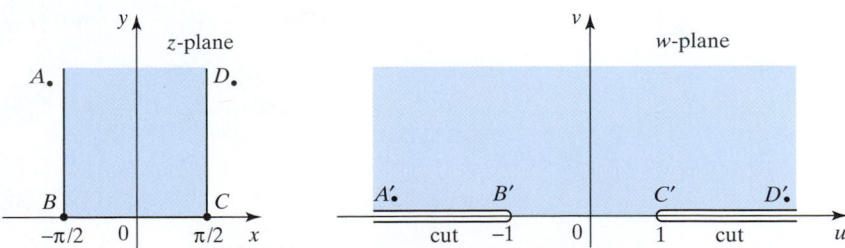

FIGURE 17.17   The mapping of a semi-infinite strip by $w = \sin z$.

so from (34) this line segment is seen to map onto the line segment $B'C'$ given by $-1 \leq u \leq 1$, which is simply the line segment of the real axis in the w-plane extending from $w = -1$ to $w = 1$. Similarly, the side CD_∞ is seen to map to the semi-infinite line segment $C'D'_\infty$ of the real axis in the w-plane extending from $w = 1$ to ∞.

As the interior of the semi-infinite strip lies to the left as the region is traversed in the direction $A_\infty BCD_\infty$, it follows that the interior of the strip must map to the upper half of the w-plane. A similar argument shows that the semi-infinite strip $-\frac{\pi}{2} \leq x \leq \frac{\pi}{2}$, $y \leq 0$, is mapped by $w = \sin z$ onto the lower half of the w-plane, so that $w = \sin z$ maps the infinite strip $-\frac{\pi}{2} \leq x \leq \frac{\pi}{2}$ one-one and conformally onto the w-plane cut along the real axis from -1 to $-\infty$ and from 1 to ∞, with the exception of the points $w = \pm 1$ at B' and C' that are the images of the critical points of the mapping located at B and C. These cuts are necessary, because the multivalued nature of $\sin z$ causes the boundaries of each of the semi-infinite strips between successive critical points to map onto the cuts.

The inverse mapping from w to z, denoted by $z = \arcsin w$, is many valued. The mapping can be made one-one by cutting the w-plane along the real axis from $-\frac{\pi}{2}$ to $-\infty$ and from $\frac{\pi}{2}$ to ∞, and then restricting z to any strip of width π that is parallel to the imaginary axis in the z-plane and lies between two adjacent critical points of $\sin z$. When the strip is taken to be $-\frac{\pi}{2} \leq x \leq \frac{\pi}{2}$, the inverse function is written $z = \text{Arcsin } w$, and this is called the **principal branch of the inverse sine function**.

principal branch of the inverse sine function

If the inverse sine function is considered as a function in its own right, it is usual to interchange w and z and to consider the function $w = \text{Arcsin } z$. The principal branch of the inverse sine function $w = \text{Arcsin } z$ is defined in the z-plane where the cuts along $x < -\frac{\pi}{2}$, $y = 0$, and $x > \frac{\pi}{2}$, $y = 0$, have been made, and $w = \text{Arcsin } z$ is restricted to the strip $-\frac{\pi}{2} \leq \text{Re } w \leq \frac{\pi}{2}$ in the w-plane.

It follows from (34) that the cartesian coordinate lines $x = a$ and $y = b$ map, respectively, to the mutually orthogonal families of hyperbolas and ellipses

$$\frac{u^2}{\sin^2 a} - \frac{v^2}{\cos^2 a} = 1 \quad \text{and} \quad \frac{u^2}{\cosh^2 b} + \frac{v^2}{\sinh^2 b} = 1.$$

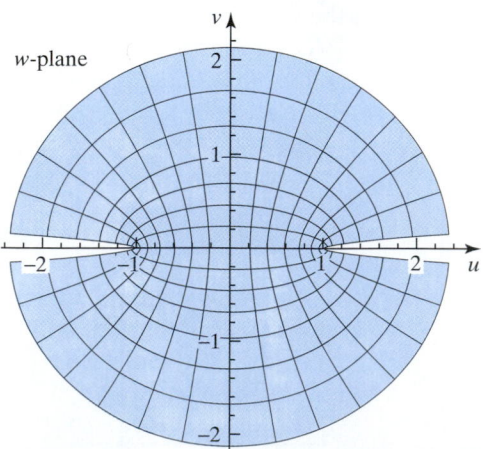

FIGURE 17.18 The mapping of cartesian coordinate lines by $w = \sin z$.

Figure 17.18 illustrates the mapping of these coordinate lines in the z-plane onto the hyperbolas and ellipses in the w-plane by the function $w = \sin z$. The inverse mapping from the w-plane to the z-plane is given by $z = \text{Arcsin } w$.

(i) The Mappings $w = \exp z$ and $w = \text{Log } z$

The function $\exp z$ is an entire function, so writing it in the form

$$w = \exp(z) = e^x(\cos y + i \sin y) \tag{35}$$

shows that $\exp z$ is periodic in y with period 2π. Thus, $w = \exp z$ will map any strip of width 2π parallel to the imaginary axis one-one and conformally onto the w-plane from which the point $w = 0$ has been deleted. The deletion of the point $w = 0$ is necessary because for no finite z is it true that $\exp z = 0$. The strip $-\pi < y \le \pi$ is called the **fundamental strip** of the $\exp z$, and from now on y will be restricted to this strip.

the exponential and logarithmic mappings and fundamental strips

Setting $w = u + iv$ in (35) and equating real and imaginary parts give

$$u = e^x \cos y \quad \text{and} \quad v = e^x \sin y. \tag{36}$$

Eliminating y from (36) shows that the cartesian coordinate lines $x = a$ map to the concentric circles $u^2 + v^2 = e^{2a}$. Setting $y = b$ in (36) and eliminating x shows that the cartesian coordinate lines $y = b$ map to the to radial lines (rays) $v = u \tan b$ emanating from the origin. Because of the restriction on y, the strip in the z-plane maps to the w-plane with a cut along the real axis from the origin to $-\infty$, as shown in Fig. 17.19.

In working with the fundamental strip, the inverse function is the principal branch of the logarithmic function Log w, and it will provide a one-one and conformal mapping of the w-plane onto the z-plane. If the logarithmic function is considered as a function in its own right, w and z are interchanged and we obtain the function

$$\text{Log } z = \ln |z| + i \text{ Arg } z, \quad \text{with } |z| > 0 \text{ and } -\pi < \text{Arg } z \le \pi. \tag{37}$$

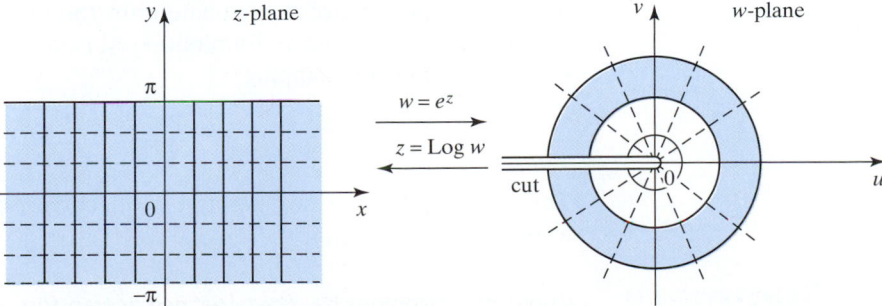

FIGURE 17.19   The mappings $w = \exp z$ and $z = \text{Log } w$.

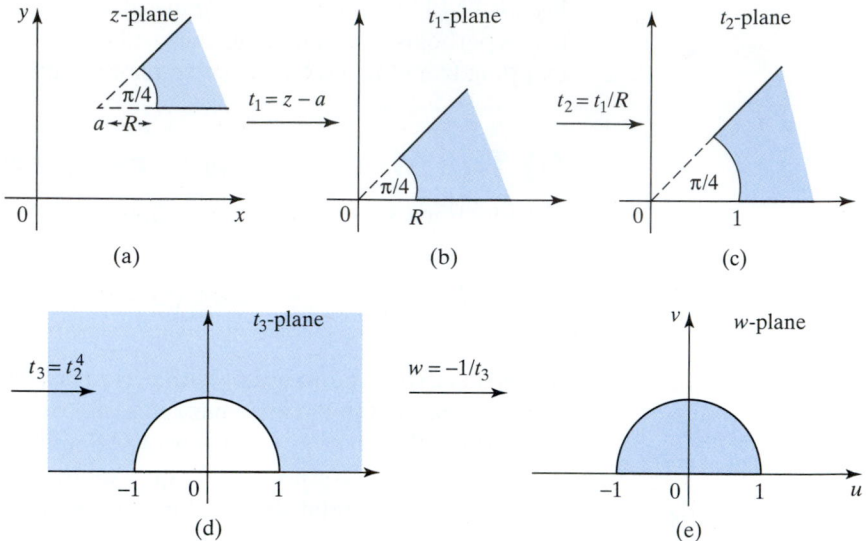

FIGURE 17.20   Mapping an indented semi-infinite sector onto a semicircle.

(j) Composite Mappings

When considering fundamental mappings such as the inversion mapping and the linear fractional transformation, we have seen how they can be interpreted as a sequence of very simple mappings. The combination of mappings in this manner is called the **composition** of mappings, by analogy with the real variable case where if $w = f(u)$ and $u = g(x)$, the "function of a function" $w = f(g(x))$ is called the composition of the functions g and f.

This approach is also used to build up more complicated mappings when it is required to map a given region onto a more conveniently shaped one. We illustrate this by showing how the interior of the semi-infinite indented wedge-shaped region shown in Fig. 17.20a can be mapped onto the interior of the semicircle $|w| \leq 1$, Im $w \geq 0$, shown in Fig. 17.20e.

The linear mapping $t_1 = z - a$ shifts the vertex of the indented wedge to the origin in Fig. 17.20b without change of scale or rotation. In Fig. 17.20c the mapping $t_2 = t_1/R$ scales the indented wedge so the radius of the circular boundary is 1, again without rotation. In Fig. 17.20d the mapping $t_3 = t_2^4$ opens out the indented wedge so the required region lies in the upper half of the t_3-plane above the unit circle. In Fig. 17.20e the final mapping $w = -1/t_3$ is the inversion mapping, so it maps the indented upper half of the t_3-plane onto the interior of the unit semicircle in the upper half of the w-plane. Eliminating t_1, t_2, and t_3 from these mappings gives the required composite mapping

$$w = \frac{-R^4}{(z-a)^4}.$$

This mapping has a critical point at $z = a$, corresponding to the point $w = \infty$ in the w-plane.

Summary

Conformal mappings have been defined as transformations that preserve both the angle between intersecting curves and the sense of rotation between the curves, when they are mapped from one plane to another. The scale factors determining the stretching of curves and areas at any point have been derived, and a critical point has been defined as one where the conformal nature of a mapping breaks down. The simple but important linear mapping and its inverse were introduced and their properties combined to give the linear fractional transformation that was then applied to various examples. The quadratic mapping was introduced and shown to map the z-plane twice onto the w-plane and, correspondingly, its inverse mapping by the square-root function was seen to be double valued. The exponential and logarithmic mappings were introduced and composite mappings were defined.

EXERCISES 17.1

1. Describe the effect of the linear transformation $w = 2iz + 3$ when mapping geometrical shapes from the z-plane onto the w-plane. Sketch the image of the rectangle in the z-plane with its corners at $(1, 1), (3, 1), (3, 2)$, and $(1, 2)$, and show the correspondence between corners in the two planes.

2. Describe the effect of the linear transformation $w = (1 + i)z - i$ when mapping geometrical shapes from the z-plane to the w-plane. Sketch (a) the image of the unit circle $|z| = 1$ and (b) the image of the ellipse $(x - 3)^2/9 + y^2/4 = 1$. In each case show how four points on the curve in the z-plane map to the w-plane.

3. Find a linear transformation that maps the triangle with its vertices A, B, and C at points $0, 1 + i$, and $2 - i$ in the z-plane onto the similar triangle with vertices A', B', and C' at $1 - i, 5 - i$, and $3 - 7i$ in the w-plane.

4. Find the linear transformation with the fixed point $2 - i$ that maps $z = -i$ to $w = 2 - 3i$.

5. Find the linear transformation with the fixed point $3 + 2i$ that maps $z = 1$ to $w = -7$.

6. In the following transformations find the fixed point z^* when one exists, the angle of rotation α about z^* that is introduced, and the magnification factor ρ:

    (a) $w = 2z + 1 - 3i$.   (b) $w = iz + 4$.   (c) $w = z + 1 - 2i$.

7. Find a linear transformation $w = az + b$ that maps the infinite strip $k < y < k + h$ in the z-plane onto the strip $0 < u < 1$ in the w-plane in such a way that $w(ik) = 0$.

8. Find a linear transformation $w = az + b$ that maps the infinite strip $k < x < k + h$ in the z-plane onto the strip $0 < u < 1$ in the w-plane in such a way that $w(k) = 0$.

9. Given that $w = 1/z$, find the image in the w-plane of the family of parallel straight lines $y = x + c$ in the z-plane.

10. By using the symmetry properties of linear fractional mappings, or otherwise, find how $w = z/(z - 1)$ maps the annulus $1 \le |z| \le 2$ in the z-plane onto the w-plane.

In Exercises 11 through 14 find the linear fractional transformation that maps the three given points in the z-plane onto the three given points in the w-plane. Determine the region in the w-plane that corresponds to the region to the left of the given points in the z-plane when the points are traversed in the order z_1, z_2, and z_3.

11. Map points $z_1 = i$, $z_2 = -i$, and $z_3 = 1$ onto the points $w_1 = -1$, $w_2 = 1$, and $w_3 = \infty$.

12. Map the points $z_1 = -1$, $z_2 = -i$, and $z_3 = 1$ onto the points $w_1 = -3 + i$, $w_2 = (2 - 4i)/5$, and $w_3 = 1 + i/3$.

13. Map the points $z_1 = 1$, $z_2 = 2 + i$, and $z_3 = i$ onto the points $w_1 = i$, $w_2 = (-1 + 2i)/5$, and $w_3 = 1/3$.

14. Map the points $z_1 = -1$, $z_2 = 1$, and $z_3 = \infty$ onto the points $w_1 = i$, $w_2 = -i$, and $w_3 = 1$.

15. Prove that the function $w = \exp(\pi z/a)$ maps the infinite strip of width a in the z-plane shown in the diagram on the left of Fig. 17.21 onto the upper half of the w-plane in the manner shown in the diagram on the right. Determine the images in the w-plane of the lines $x = c$ and $y = k$.

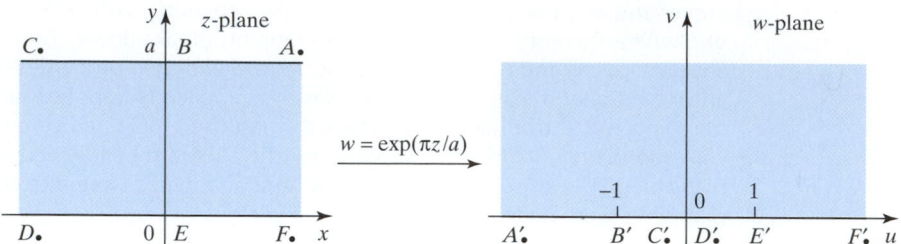

FIGURE 17.21 Mapping by $w = \exp(\pi z/a)$.

16. Prove that the function $w = \sin(\pi z/a)$ maps the semi-infinite strip of width a in the z-plane shown in the diagram on the left of Fig. 17.22 onto the upper half of the w-plane in the manner shown in the diagram on the right. Determine the images in the w-plane of the lines $x = c$ and $y = k$.

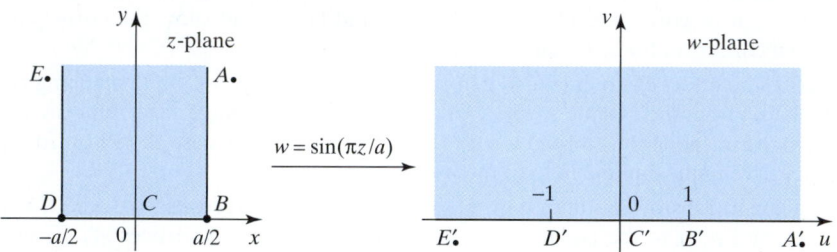

FIGURE 17.22 Mapping by $w = \sin(\pi z/a)$.

17. Prove that the function $w = \cos(\pi z/a)$ maps the semi-infinite strip of width a in the z-plane shown in the diagram on the left of Fig. 17.23 onto the upper-half of the w-plane in the manner shown in the diagram on the right. Determine the images in the w-plane of the lines $x = c$ and $y = k$.

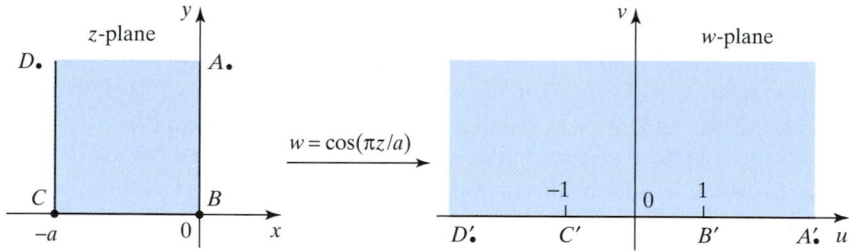

FIGURE 17.23 Mapping by $w = \cos(\pi z/a)$.

18. Prove that the function $w = \cosh(\pi z/a)$ maps the semi-infinite strip of width a in the z-plane shown in the diagram on the left of Fig. 17.24 onto the upper half of the w-plane in the manner shown in the diagram on the right. Determine the images in the w-plane of the lines $x = c$ and $y = k$.

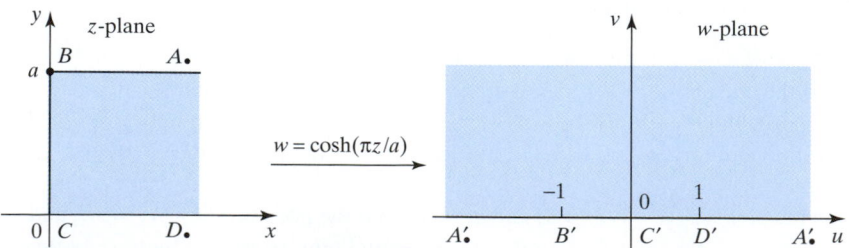

FIGURE 17.24 Mapping by $w = \cosh(\pi z/a)$.

19. Prove that the function $w = (\frac{1+z}{1-z})^2$ maps the interior of the unit semicircle in the z-plane in the diagram on the left of Fig. 17.25 onto the upper half of the w-plane in the manner shown in the diagram on the right.

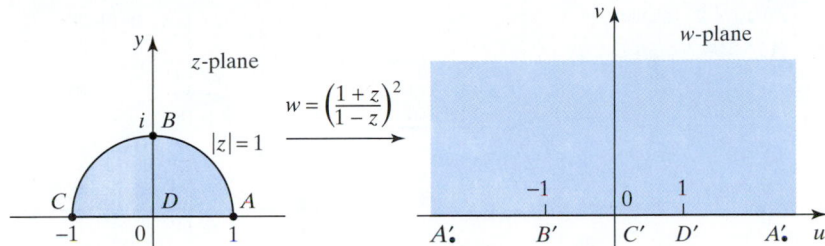

FIGURE 17.25   Mapping by $w = (\frac{1+z}{1-z})^2$.

(Hint: First find the image of $(1 + z)/(1 - z)$ in the unit circle $|z| = 1$.)

20. Given that $w = z + k/z$, with k real, find the image in the w-plane of the lines $x = c$ and $y = d$. Find the values of k and R such that for given real a and b the transformation will map the circle $|z| = R$ onto the ellipse

$$\frac{u^2}{a^2} + \frac{v^2}{b^2} = 1$$

in the w-plane.

21. Verify that $w = k(\frac{z-z_0}{z-\bar{z}_0})$, with $|k| = 1$ and z_0 an arbitrary point in the upper half of the z-plane, maps the upper half of the z-plane onto $|w| < 1$ and z_0 to the point $w = 0$.

22. Verify that $w = k(\frac{z-z_0}{\bar{z}_0 z-1})$, with $|k| = 1$ and z_0 an arbitrary point such that $|z_0| < 1$, maps $|z| < 1$ onto $|w| < 1$ and z_0 to the point $w = 0$.

23. Show that $w = \tanh z$ maps the semi-infinite strip $0 < y < \pi/2a$ in the diagram on the left of Fig. 17.26 onto the upper half of the w-plane in the manner shown in the diagram on the right.

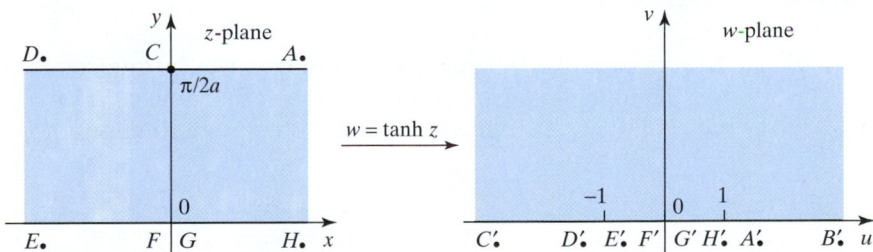

FIGURE 17.26   The mapping $w = \tanh z$.

24. Show that $w = [(1 + z^n)/(1 - z^n)]^2$ maps the sector in the diagram on the left of Fig. 17.27 onto the upper half of the w-plane in the manner shown in the diagram on the right in the w-plane.

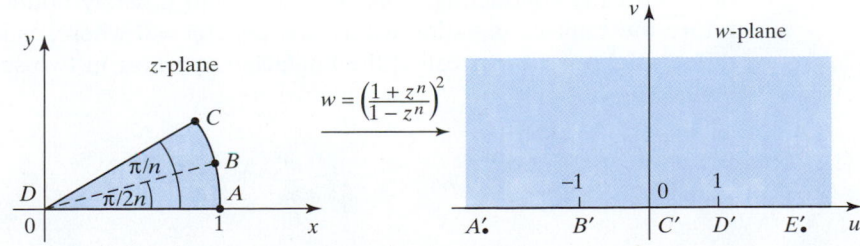

FIGURE 17.27   The mapping $w = [(1 + z^n)/(1 - z^n)]^2$.

25. Show that $w = (1 - \cos z)/(1 + \cos z)$ maps the semi-infinite strip $0 < x < \pi/2$, $y > 0$ in the diagram on the left of Fig. 17.28 onto the interior of the unit semicircle $|w| = 1$ in the upper half of the w-plane in the manner shown in the diagram on the right.

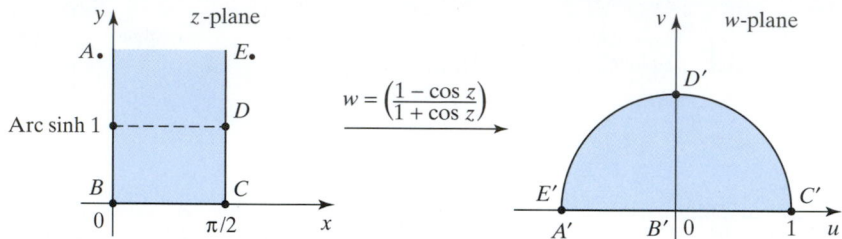

FIGURE 17.28   The mapping $w = (1 - \cos z)/(1 + \cos z)$.

26. Show that $w = \mathrm{Log}(\frac{z-1}{z+1})$ maps the upper half of the z-plane in the diagram on the left of Fig. 17.29 onto the infinite strip $0 < v < \pi$ in the w-plane in the manner shown in the diagram on the right.

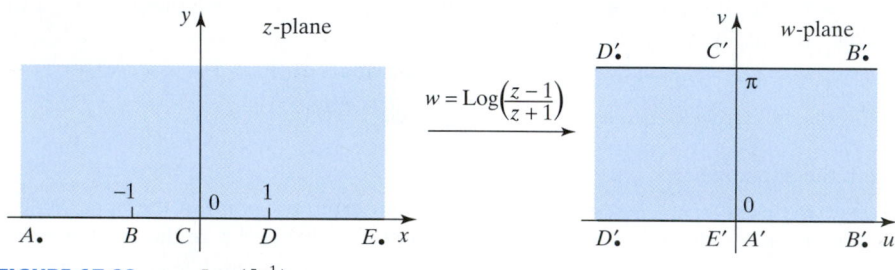

FIGURE 17.29 $w = \mathrm{Log}(\frac{z-1}{z+1})$.

17.2   Conformal Mapping and Boundary Value Problems

boundary value problem for the Laplace equation

The concept of a **boundary value problem** was introduced in connection with the maximum/minimum property of harmonic functions $\phi(x, y)$ (see Theorem 14.17), in which the two independent variables x and y are solutions of the **Laplace equation**

$$\frac{\partial^2 \phi}{\partial x^2} + \frac{\partial^2 \phi}{\partial y^2} = 0. \tag{38}$$

Solutions of Laplace's equation are also called **potential functions** because of the role played by the *gravitational potential* that determines the gravitational force acting on a body and the *electric potential* in space caused by a potential distribution on electrically conducting walls present in, and possibly bounding, the space. In future the Laplace equation will be written $\Delta\phi = 0$ where, as in Chapter 13, the **differential operator** Δ called the **Laplacian operator** in two space dimensions is defined as

$$\Delta \equiv \frac{\partial^2}{\partial x^2} + \frac{\partial^2}{\partial y^2},$$

and $\Delta\phi$ is read "Laplacian ϕ."

In complex analysis only the two-dimensional Laplacian is involved, but in other branches of mathematics both two- and three-dimensional Laplacians occur. To avoid confusion, the two-dimensional Laplacian of ϕ is often denoted by $\Delta_2\phi$ and the three-dimensional Laplacian by $\Delta_3\phi$.

The simplest boundary value problems for the Laplace equation involve specifying either ϕ on the boundary Γ of a region R in which ϕ is harmonic, or the derivative of ϕ normal to the boundary Γ, usually denoted by $\partial\phi/\partial n$. The specification of ϕ on the boundary Γ is called a **Dirichlet boundary condition**, and the requirement that ϕ satisfy both (38) and a Dirichlet boundary condition is called a

Dirichlet and Neumann boundary conditions

Dirichlet boundary value problem for the harmonic function ϕ. The specification of $\partial\phi/\partial n$ on the boundary Γ of R is called a **Neumann boundary condition**, and the requirement that ϕ satisfy both (38) and a Neumann boundary condition is called a **Neumann boundary value problem** for the harmonic function ϕ. Dirichlet and Neumann boundary value problems are also known as **boundary value problems of the first and second kind**, respectively.

CARL NEUMANN (1832–1925)
A German mathematician and physicist who in 1868 was appointed Professor of Mathematics at the University of Leipzig. His main contributions were to the study of potential theory and to integral equations.

It is not difficult to show that a Dirichlet boundary value problem for a harmonic function ϕ determines ϕ uniquely at every point of R, and that a Neumann boundary value problem for ϕ determines it uniquely apart from an arbitrary additive constant.

A useful application of conformal mapping is to the solution of two-dimensional boundary value problems for harmonic functions. Various quite different methods of solution exist for such problems, but conformal mapping provides a method that offers valuable geometrical insight into the nature of the solution. The approach comes from the fact that if $w = f(z) = u + iv$ is a single-valued analytic function that maps a region R in the z-plane onto a region R' in the w-plane and $\phi(x, y)$ is harmonic in R, the change of variable from (x, y) to (u, v) transforms $\phi(x, y)$ to a function $\Phi(u, v)$ that is harmonic in R'. Furthermore, either a Dirichlet or a Neumann boundary condition at a point P on the boundary Γ of region R is mapped without change to a point P' on the boundary γ of R', where γ is the image of Γ and P' is the image of P under the mapping $w = f(z)$. In some problems Dirichlet and Neumann boundary conditions apply on different parts of a continuous piecewise smooth boundary Γ, and when this occurs these boundary conditions are transferred to the appropriate parts of the transformed boundary γ. Problems of this type are

mixed boundary value problems

called **mixed boundary value problems**. In applications to steady state temperature distributions, the temperature satisfies Laplace's equation and a Dirichlet condition on a boundary corresponds to the specification of the temperature on the boundary, whereas the specification of a Neumann condition corresponds to the specification of the temperature gradient across a boundary, and hence the heat flow across the boundary because the heat flow is proportional to the temperature gradient.

The idea behind a conformal mapping approach to the solution of a boundary value problem for the two-dimensional Laplace equation is to use a conformal transformation $w = f(z)$ to transform a region R in the z-plane with a complicated boundary shape, into a region R' in the w-plane with a more simply shaped boundary. Then, if the solution of the simpler boundary value problem can be found, the conformal mapping can be used in reverse to transform this simpler solution back into the solution for the more complicated region. As the choice of mapping $w = f(z)$ determines the way in which the boundary of a region R with a simple shape is mapped to a region R' with a more complicated boundary shape, a knowledge of the fundamental mapping properties of elementary functions is necessary when using conformal mapping to solve boundary value problems.

We now give a direct proof that a function $\phi(x, y)$ remains harmonic under the change of variable from (x, y) to (u, v) that transforms $\phi(x, y)$ to $\Phi(u, v)$, where $w = f(z) = u + iv$, and $f(z)$ is a single-valued analytic function. From the chain rule, if $u = u(x, y)$, $v = v(x, y)$ and all functions involved are suitably differentiable,

showing that a harmonic function remains harmonic under a conformal mapping

$$\frac{\partial \phi}{\partial x} = \frac{\partial \Phi}{\partial u}\frac{\partial u}{\partial x} + \frac{\partial \Phi}{\partial v}\frac{\partial v}{\partial x}, \tag{39}$$

and

$$\frac{\partial^2 \phi}{\partial x^2} = \left[\frac{\partial}{\partial x}\left(\frac{\partial \Phi}{\partial u}\right)\right]\left(\frac{\partial u}{\partial x}\right) + \left(\frac{\partial \Phi}{\partial u}\right)\left(\frac{\partial^2 u}{\partial x^2}\right)$$
$$+ \left[\frac{\partial}{\partial x}\left(\frac{\partial \Phi}{\partial v}\right)\right]\left(\frac{\partial v}{\partial x}\right) + \left(\frac{\partial \Phi}{\partial v}\right)\left(\frac{\partial^2 v}{\partial x^2}\right). \tag{40}$$

Examination of (39) shows that the differentiation operation $\partial/\partial x$ is related to the differentiation operations $\partial/\partial u$ and $\partial/\partial v$ by

$$\frac{\partial}{\partial x} \equiv \frac{\partial u}{\partial x}\frac{\partial}{\partial u} + \frac{\partial v}{\partial x}\frac{\partial}{\partial v}.$$

Using this result in the terms involving $\frac{\partial}{\partial x}(\frac{\partial \Phi}{\partial u})$ and $\frac{\partial}{\partial x}(\frac{\partial \Phi}{\partial v})$ in (40) changes it to

$$\frac{\partial^2 \phi}{\partial x^2} = \frac{\partial^2 \Phi}{\partial u^2}\left(\frac{\partial u}{\partial x}\right)^2 + \frac{\partial^2 \Phi}{\partial v^2}\left(\frac{\partial v}{\partial x}\right)^2 + \left(\frac{\partial^2 \Phi}{\partial u \partial v} + \frac{\partial^2 \Phi}{\partial v \partial u}\right)\left(\frac{\partial u}{\partial x}\right)\left(\frac{\partial v}{\partial x}\right)$$
$$+ \frac{\partial \Phi}{\partial u}\left(\frac{\partial^2 u}{\partial x^2}\right) + \frac{\partial \Phi}{\partial v}\left(\frac{\partial^2 v}{\partial x^2}\right).$$

A corresponding expression exists for $\frac{\partial^2 \phi}{\partial y^2}$, so combining the two results and using the equality of the mixed derivatives $\frac{\partial^2 \Phi}{\partial u \partial v} = \frac{\partial^2 \Phi}{\partial v \partial u}$, which is justified when Φ is continuous and twice differentiable, leads to the result

$$\frac{\partial^2 \phi}{\partial x^2} + \frac{\partial^2 \phi}{\partial y^2} = \frac{\partial^2 \Phi}{\partial u^2}\left[\left(\frac{\partial u}{\partial x}\right)^2 + \left(\frac{\partial u}{\partial y}\right)^2\right] + \frac{\partial^2 \Phi}{\partial v^2}\left[\left(\frac{\partial v}{\partial x}\right)^2 + \left(\frac{\partial v}{\partial y}\right)^2\right]$$
$$+ 2\frac{\partial^2 \Phi}{\partial u \partial v}\left(\frac{\partial u}{\partial x}\frac{\partial v}{\partial x} + \frac{\partial u}{\partial y}\frac{\partial v}{\partial y}\right) + \frac{\partial \Phi}{\partial u}\left(\frac{\partial^2 u}{\partial x^2} + \frac{\partial^2 u}{\partial y^2}\right) + \frac{\partial \Phi}{\partial v}\left(\frac{\partial^2 v}{\partial x^2} + \frac{\partial^2 v}{\partial y^2}\right). \tag{41}$$

Examination of (41) shows that the last two terms vanish because u and v are harmonic, while the Cauchy–Riemann equations cause the factor multiplying $\partial^2 \Phi/\partial u \partial v$ to vanish. To simplify the equation further, we now make use of result (21) in Section 13.1 where it was shown that

$$f'(z) = \frac{\partial u}{\partial x} + i\frac{\partial v}{\partial x},$$

and notice that the Cauchy–Riemann equations allow it to be written in either of the following ways:

$$f'(z) = \frac{\partial u}{\partial x} - i\frac{\partial u}{\partial y} \quad \text{or} \quad f'(z) = \frac{\partial v}{\partial y} + i\frac{\partial v}{\partial x}. \tag{42}$$

When the results of (42) are used in the two nonvanishing terms that remain in (41), the equation is seen to reduce to

$$\frac{\partial^2 \phi}{\partial x^2} + \frac{\partial^2 \phi}{\partial y^2} = |f'(z)|^2 \left(\frac{\partial^2 \Phi}{\partial u^2} + \frac{\partial^2 \Phi}{\partial v^2} \right) \tag{43}$$

or, equivalently, to $\Delta\phi = |f'(z)|^2 \Delta\Phi$.

This last result shows that if $\phi(x, y)$ is harmonic in the z-plane, then $\Phi(u, v)$ is harmonic in the w-plane, with the exception of points in the w-plane that are images of the critical points of the mapping $w = f(z)$ in the z-plane. We have proved the following important result.

THEOREM 17.3

Harmonic functions remain harmonic under a conformal transformation   Let $w = u + iv = f(z)$ be a single-valued analytic function and $\phi(x, y)$ be harmonic in a region R. Then if $\phi(x, y)$ becomes the function $\Phi(u, v)$ under the change of variables $u = u(x, y)$ and $v = v(x, y)$, and R' is the image of R under the transformation, the function $\Phi(u, v)$ is harmonic in R'.   ■

To see how the boundary conditions transform, notice first that if P' is the image in the w-plane of a point P on the boundary in the z-plane, then as $\Phi(u, v)$ is simply the function $\phi(x, y)$ expressed in terms of the variables u and v, it follows that $\Phi(P') = \phi(P)$. Also, if $(\partial\phi/\partial n)_P = k(P)$ at a point P on the boundary in the z-plane, then because the mapping is conformal it follows that $(\partial\Phi/\partial n)_{P'}$ will still be normal to the transformed boundary curve in the w-plane at P', so that $(\partial\Phi/\partial n)_{P'} = k(P')$. Thus, Dirichlet and Neumann conditions at P on the boundary in the z-plane are transferred directly to the image of P at P' on the boundary in the w-plane.

A fundamental Dirichlet boundary value problem that has many applications involves finding the harmonic function ϕ at an arbitrary point P in the upper half of the (x, y)-plane that satisfies piecewise constant Dirichlet conditions on the x-axis. As the result generalizes in an obvious manner, we will only consider the Dirichlet boundary value problem for the Laplace equation when the solution ϕ is required to assume the three piecewise constant values ϕ_1, ϕ_2, and ϕ_3 on the x-axis. That is, we will solve the Laplace equation

solving a fundamental boundary value problem

$$\Delta\phi = 0, \quad -\infty < x < \infty, y > 0$$

subject to the boundary conditions

$$\phi(x, 0) = \phi_1 \quad \text{for } x < x_1, y = 0$$
$$\phi(x, 0) = \phi_2 \quad \text{for } x < x_1 < x_2, y = 0 \tag{44}$$
$$\phi(x, 0) = \phi_3 \quad \text{for } x > x_2, y = 0.$$

This boundary value problem is illustrated in Fig. 17.30.

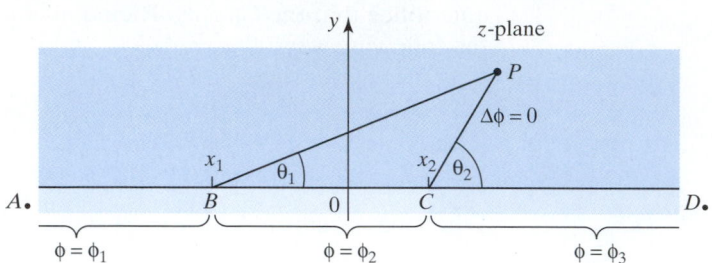

FIGURE 17.30 A piecewise constant Dirichlet boundary value problem.

Inspection shows that the following function ϕ satisfies these boundary conditions:

$$\phi(P) = \phi_3 + \frac{1}{\pi}[(\phi_1 - \phi_2)\theta_1 + (\phi_2 - \phi_3)\theta_2]. \qquad (45)$$

To check this it is only necessary to notice that when P in Fig. 17.30 is on the line segment CD_∞ the angles $\theta_1 = \theta_2 = 0$, so $\phi(P) = \phi_3$. Similarly, when P is on the line segment BC, $\theta_1 = 0$ and $\theta_2 = \pi$, so $\phi(P) = \phi_2$, whereas when P is on the line segment $A_\infty B$, $\theta_1 = \theta_2 = \pi$, so $\phi(P) = \phi_1$. The uniqueness of a Dirichlet problem for Laplace's equation then guarantees that (45) is the only solution for this simple boundary value problem, once it has been verified that it is a solution of the Laplace equation.

If Fig. 17.30 is regarded as the complex z-plane, we can write $\theta_1 = \text{Arg}(z - x_1)$ and $\theta_2 = \text{Arg}(z - x_2)$, allowing ϕ to be written

$$\phi(x, y) = \phi_3 + \frac{1}{\pi}(\phi_1 - \phi_2)\text{Arg}(z - x_1) + \frac{1}{\pi}(\phi_2 - \phi_3)\,\text{Arg}(z - x_2).$$

This expression for $\phi(x, y)$ is simply the imaginary part of the complex function

$$w = i\phi_3 + \frac{1}{\pi}(\phi_1 - \phi_2)\,\text{Log}(z - x_1) + \frac{1}{\pi}(\phi_2 - \phi_3)\,\text{Log}(z - x_2).$$

As the function w is analytic for $z \neq x_1, x_2$, its real and imaginary parts are harmonic for $z \neq x_1, x_2$ so, in particular, ϕ must be harmonic for $z \neq x_1, x_2$. The uniqueness of solutions of Dirichlet boundary value problems for harmonic functions then implies that the solution of the boundary value problem in (44) is given by

$$\phi(x, y) = \phi_3 + \frac{1}{\pi}(\phi_1 - \phi_2)\text{Arg}(z - x_1) + \frac{1}{\pi}(\phi_2 - \phi_3)\text{Arg}(z - x_2). \qquad (46)$$

Care must be exercised when determining Arg z in terms of the inverse tangent function arctan t. To understand why this is, let point $P(x, y)$ be located at $z = x + iy$ in the upper half of the z-plane, and define θ to be the angle measured counterclockwise from the positive real axis to the line OP drawn from the origin to P, so that $\tan \theta = y/x$. Then, to use (46), an inverse tangent function must be constructed that defines an angle θ that increases *continuously* from 0 to π as P moves counterclockwise around an arc in the upper half of the z-plane, from a point on the positive real axis to one on the negative real axis.

To accomplish this, notice first that the function $\tan t$ is defined over the interval $-\pi/2 < t < \pi/2$, and by periodicity elsewhere, so the standard inverse tangent function arctan t cannot be used in (46) when determining θ because it is defined over the wrong interval. However, consideration of the behavior of the function arctan t over the interval $0 < t < \pi$ shows an Arctan function defined as follows has the required properties:

$$\text{Arctan } t = \begin{cases} \arctan t, & t > 0 \\ \pi/2, & t = \pm\infty \\ \pi + \arctan t, & t < 0 \end{cases}. \tag{47}$$

It is this function that must be used in conjunction with (46) when determining ϕ.

The solution of the simplest boundary value problem in which ϕ only assumes two different constant values on the x-axis, with $\phi(x, 0) = \phi_1$ for $x < x_1$, $y = 0$ and $\phi(x, 0) = \phi_2$ for $x > x_1$, $y = 0$, follows directly from the preceding result if we omit the last term (i.e., set $\phi_3 = \phi_2$). If ϕ is required to assume more than three different constant values on the x-axis, result (46) can be extended in an obvious manner. So, for example, if the four constant values $\phi_1, \phi_2, \phi_3,$ and ϕ_4 are involved, and the points separating them on the x-axis are $x_1, x_2,$ and x_3, then in place of (46) we would use

$$\phi(x, y) = \phi_4 + \frac{1}{\pi}(\phi_1 - \phi_2)\operatorname{Arg}(z - x_1) + \frac{1}{\pi}(\phi_2 - \phi_3)\operatorname{Arg}(z - x_2)$$

$$+ \frac{1}{\pi}(\phi_3 - \phi_4)\operatorname{Arg}(z - x_3).$$

EXAMPLE 17.3

equipotentials

Find the lines of constant electric potential, called either **equipotential lines** or **equipotentials**, in the region between two perpendicular infinitely long electrically conducting walls, when parts of the surfaces are maintained at the constant potentials $\phi_1 = 60$, $\phi_2 = 0$, and $\phi_3 = 20$, as shown in Fig. 17.31.

Solution   In space an electric potential ϕ satisfies Laplace's equation so as the conducting walls in Fig. 17.31 are assumed to be infinitely long in the direction perpendicular to the plane of the diagram, and the potentials on the sections of the walls are constant, it follows that ϕ must satisfy the two-dimensional Laplace equation

$$\frac{\partial^2 \phi}{\partial x^2} + \frac{\partial^2 \phi}{\partial y^2} = 0.$$

The mapping $w = z^2$ will open up the right angle between the walls in Fig. 17.32a to the half-plane shown in Fig. 17.32b.

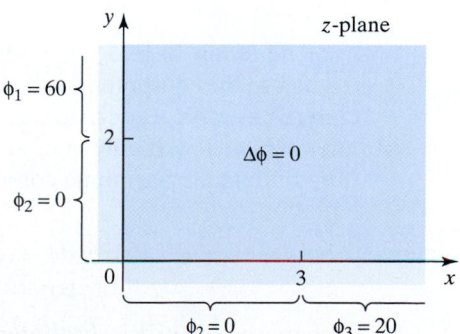

FIGURE 17.31   A Dirichlet problem for the electric potential between two conducting walls.

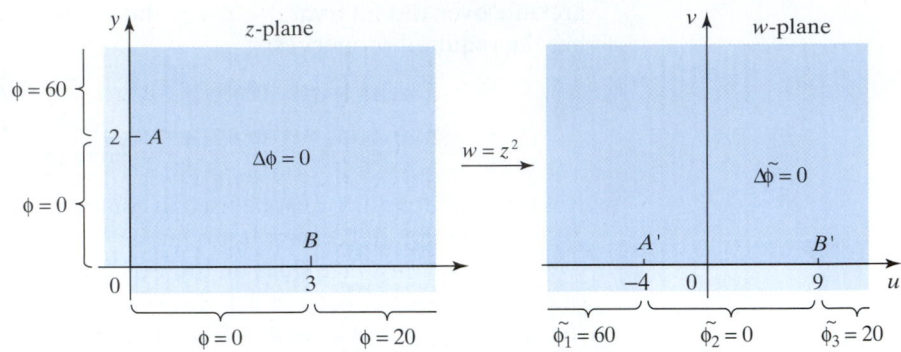

FIGURE 17.32   The effect of the mapping $w = z^2$ on the perpendicular conducting walls.

Setting $w = u + iv$ and and changing from the variables x and y to u and v will cause the potential function $\phi(x, y)$ to become the function $\widetilde{\phi}(u, v)$, and the boundary conditions transform as shown in Fig. 17.32b. The solution of the boundary value problem for $\widetilde{\phi}(u, v)$ follows directly from (46) by replacing z by w, and x_1 and x_2 by $u_1 = -4$ and $u_2 = 9$, respectively, and by setting $\widetilde{\phi}_1 = 60$, $\widetilde{\phi}_2 = 0$, and $\widetilde{\phi}_3 = 20$ to obtain

$$\widetilde{\phi}(u, v) = 60 + \frac{20}{\pi}\text{Arg}(w + 4) - \frac{60}{\pi}\text{Arg}(w - 9).$$

To return to the z-plane we now use the definition of Arctan t in (47), set $z = x + iy$ in $w = z^2$, and write $w = u + iv$ so that $u = x^2 - v^2$ and $v = 2xy$. Then, as $w + 4 = x^2 - y^2 + 4 + i2xy$, we have

$$\text{Arg}(w + 4) = \text{Arctan}\left(\frac{2xy}{x^2 - y^2 + 4}\right)$$

and, similarly,

$$\text{Arg}(w - 9) = \text{Arctan}\left(\frac{2xy}{x^2 - y^2 - 9}\right).$$

So the electric potential at the point (x, y) is seen to be given by

$$\phi(x, y) = 60 + \frac{20}{\pi}\text{Arctan}\left(\frac{2xy}{x^2 - y^2 + 4}\right) - \frac{60}{\pi}\text{Arctan}\left(\frac{2xy}{x^2 - y^2 - 9}\right),$$

for (x, y) in the first quadrant.   ∎

flux lines   The family of lines $\psi(x, y) = $ constant that form orthogonal trajectories with respect to the equipotentials are called **flux lines**. In electrostatics these are lines of electrostatic force, and in a steady state temperature distribution they correspond to lines of heat flow. If only $\phi(x, y)$ is known, the function $\psi(x, y)$ can be obtained from it by finding the harmonic conjugate function $\psi(x, y)$ using the Cauchy–Riemann equations

$$\frac{\partial \phi}{\partial x} = \frac{\partial \psi}{\partial y} \quad \text{and} \quad \frac{\partial \phi}{\partial y} = -\frac{\partial \psi}{\partial x}.$$

This method is precisely the one given in Section 13.3, by which $\psi(x, y)$ can be recovered from $\phi(x, y)$.

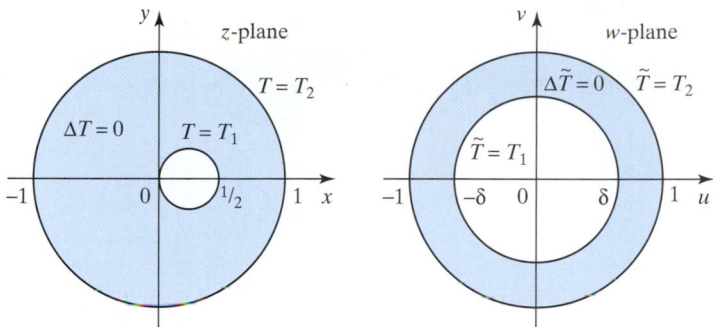

FIGURE 17.33   Equivalent problems in the z-plane and the w-plane.

EXAMPLE 17.4

isothermal lines between eccentric circles

By mapping the region between the eccentric circles on the left of Fig. 17.33 onto the annulus shown on the right, find the lines of constant temperature, called **isothermal lines** or simply **isothermals**, in the region between the eccentric circles when the constant temperature on the inner boundary is T_1 and that on the outer boundary is T_2.

Solution   It is shown in Section 18.5 that the two-dimensional steady-state temperature distribution T in a uniform solid is determined by the solution of the two-dimensional Laplace equation $\Delta T = 0$, subject to suitable boundary conditions on the surface of the solid. The two-dimensional formulation of a three-dimensional problem is satisfactory if the solid is in the form of a long uniform bar of constant cross-section and the boundary conditions are constant along the length of the bar, because then the variation of temperature along the length of the bar close to its end faces can be neglected. Under such circumstances the problem reduces to finding the two-dimensional temperature distribution in a lamina in the form of a cross-section of the bar.

When cartesian coordinates are used, the Laplace equation $\Delta T = 0$ satisfied by T is

$$\frac{\partial^2 T}{\partial x^2} + \frac{\partial^2 T}{\partial y^2} = 0.$$

As T is harmonic, and the problem involves Dirichlet boundary conditions, a conformal transformation $w = f(z)$ with $w = u + iv$ that maps the eccentric circles on the left of Fig. 17.33 onto the concentric circles on the right will lead to an equivalent problem for the temperature $\widetilde{T}$ in the annulus. In what follows the notation $\widetilde{T}(u, v)$ is used to represent $T(x, y)$ after the change of variables from (x, y) to (u, v).

The transformation $w = T(z)$ that maps the eccentric circles onto concentric circles can be found from (18) in Section 17.1. Inspection of the diagram on the left of Fig. 17.10 and a comparison with the geometry of Fig. 17.33 shows that $a = \frac{1}{4}$ and $\rho = \frac{1}{4}$. A simple calculation gives $\delta = 2 - \sqrt{3}$, from which it follows that the required transformation is

$$w = \frac{z - 2 + \sqrt{3}}{(2 - \sqrt{3})z - 1}.$$

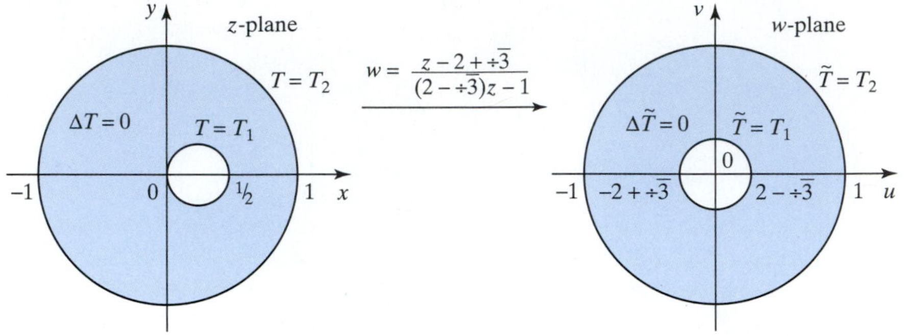

FIGURE 17.34   The mapping $w = \frac{z-2+\sqrt{3}}{(2-\sqrt{3})z-1}$.

The mapping by this function of the region between the eccentric circles onto the annular region is illustrated in Fig. 17.34.

The concentric circular boundaries in the w-plane suggest that $\Delta \widetilde{T} = 0$ should be expressed in terms of cylindrical polar coordinates, leading to the equation

$$\frac{\partial^2 \widetilde{T}}{\partial r^2} + \frac{1}{r}\frac{\partial \widetilde{T}}{\partial r} + \frac{1}{r^2}\frac{\partial^2 \widetilde{T}}{\partial \theta^2} = 0.$$

The radial symmetry of the problem in the w-plane shows that the solution must be independent of θ, as a result of which all derivatives with respect to θ vanish, causing Laplace's equation to reduce to the ordinary second order differential equation

$$\frac{d^2 \widetilde{T}}{dr^2} + \frac{1}{r}\frac{d\widetilde{T}}{dr} = 0.$$

Setting $d\widetilde{T}/dr = u$ and integrating gives $u = A/r$, and a further integration then shows the general solution to be

$$\widetilde{T}(r) = A \ln r + B.$$

Matching the integration constants A and B to the boundary conditions $\widetilde{T}(\delta) = T_1$ and $\widetilde{T}(1) = T_2$ gives the solutions in the annulus

$$\widetilde{T}(r) = T_2 - \left(\frac{T_2 - T_1}{\ln(2 - \sqrt{3})}\right) \ln r.$$

To return to the (x, y)-plane, it is necessary to express r in terms of x and y, but $r = |w|$, so setting $z = x + iy$ in the expression for w we arrive at the solution

$$T(x, y) = T_2 - \left(\frac{T_2 - T_1}{\ln(2 - \sqrt{3})}\right) \ln \left| \frac{x + iy - 2 + \sqrt{3}}{(2 - \sqrt{3})(x + iy) - 1} \right|.$$

This solution is complicated, but its typical behavior can be seen by considering the temperature variation along the x-axis, where it reduces to

$$T(x, 0) = T_2 - \left(\frac{T_2 - T_1}{\ln(2 - \sqrt{3})}\right) \ln \left| \frac{x - 2 + \sqrt{3}}{(2 - \sqrt{3})x - 1} \right|,$$

for $-1 \leq x \leq 0$ and $1/2 \leq x \leq 1$.

In Example 17.4 the family of lines $\psi(x, y) = $ constant that form orthogonal
trajectories with respect to the isothermals are called **heat flux lines**, and these
are lines along which heat flows. When required, the function $\psi(x, y)$ determining
the heat flux lines can be obtained from the temperature $T(x, y)$ by finding the
harmonic conjugate function $\psi(x, y)$ from the Cauchy–Riemann equations

heat flux lines

$$\frac{\partial T}{\partial x} = \frac{\partial \psi}{\partial y} \quad \text{and} \quad \frac{\partial T}{\partial y} = -\frac{\partial \psi}{\partial x},$$

using the method described in Section 13.3.

Before discussing the next examples it is necessary to preface them with an
introduction to the two-dimensional steady flow of an ideal fluid, and its relationship to conformal mapping. An **ideal fluid** is defined as one that is **incompressible**,
inviscid (free from viscosity), and **irrotational** (its velocity vector $\mathbf{q}$ is such that curl
$\mathbf{q} = \mathbf{0}$). The flow of water at low speeds and even of air at subsonic speeds is well
approximated by the flow of an ideal fluid.

ideal fluids

If in the steady (time-independent) two-dimensional flow of an ideal fluid the
velocity vector is $\mathbf{q} = q_1 \mathbf{i} + q_2 \mathbf{j}$, it is shown in introductory accounts of fluid mechanics that the incompressibility condition follows from the equation of conservation
of mass in the form

$$\frac{\partial q_1}{\partial x} + \frac{\partial q_2}{\partial y} = 0 \quad \text{or, equivalently, as div } \mathbf{q} = 0. \tag{48}$$

A simple calculation shows that the irrotational condition curl $\mathbf{q} = \mathbf{0}$ leads to the
equation

$$\frac{\partial q_2}{\partial x} - \frac{\partial q_1}{\partial y} = 0, \tag{49}$$

so equations (48) and (49) are seen to take the form of the Cauchy–Riemann
equations for the analytic function

$$f(z) = q_1 - iq_2, \tag{50}$$

where the harmonic functions q_1 and q_2 are the components of the fluid velocity
vector $\mathbf{q} = q_1 \mathbf{i} + q_2 \mathbf{j}$.

From vector analysis it is known that if curl $\mathbf{q} = \mathbf{0}$, a scalar function ϕ can always
be found with the property that

$$\mathbf{q} = \text{grad } \phi, \tag{51}$$

so

$$q_1 = \frac{\partial \phi}{\partial x} \quad \text{and} \quad q_2 = \frac{\partial \phi}{\partial y}. \tag{52}$$

Combining (48) and (52) shows that the real function ϕ satisfies the Laplace equation

$$\Delta\phi = 0, \tag{53}$$

and hence that ϕ is harmonic. Because of (52) the function ϕ is called the **velocity potential** of the fluid flow. Associated with the velocity potential $\phi(x, y)$ is its harmonic conjugate $\psi(x, y)$, called the **stream function** of the flow, so an analytic function

velocity potential, stream function, streamlines, and the complex potential

$$w(z) = \phi(x, y) + i\psi(x, y) \tag{54}$$

can always be defined, called the **complex potential** of the flow, with the property that the curves $\phi(x, y) = $ constant and $\psi(x, y) = $ constant are mutually orthogonal trajectories. The lines along which the stream function is constant are called the **streamlines** of the flow, because the velocity vector is tangent to each point on a streamline. Drawing streamlines enables a flow to be visualized, because any particle of fluid that lies on a streamline will remain on it as it moves steadily across the (x, y)-plane.

We mention here that in many applications the vector $\mathbf{q}$ is often defined in terms of the scalar potential ϕ by writing $\mathbf{q} = -\text{grad } \phi$, because it still remains true that curl $\mathbf{q} = \mathbf{0}$. For example, when studying the flow of heat in a steady-state temperature distribution, where ϕ is identified with the temperature T and $\mathbf{q}$ is the heat flow vector, as would be expected the heat then flows in the direction of decreasing temperature. A similar situation also applies in electrostatics.

When required, a stream function can always be found from a given velocity potential $\phi(x, y)$ by the method described in Section 13.3. Result (54) shows that any analytic function can be interpreted as a complex potential, and the streamlines of the flow are then described by the lines along which the stream function is constant. As already mentioned, the functions $\phi(x, y)$ and $\psi(x, y)$ are harmonic conjugates, so the streamlines and lines of constant velocity potential are mutually orthogonal.

Using (52) and (54) together with the fact that ϕ and ψ satisfy the Cauchy–Riemann equations, we can easily show that

$$w'(z) = q_1 - iq_2 \quad \text{and the speed } q = |\mathbf{q}| = \left[\left(\frac{\partial\phi}{\partial x}\right)^2 + \left(\frac{\partial\phi}{\partial y}\right)^2 \right]^{1/2}. \tag{55}$$

The connection between the two-dimensional steady flow of an ideal fluid and conformal mapping arises because the complex potential representing the flow in a given region can be mapped conformally onto a different region. This enables the flow in a simple region to be used to determine the flow in a more complicated one.

EXAMPLE 17.5

Interpret the flow of an ideal fluid with the complex potential $w = z^2$, when z is restricted to the first quadrant.

Solution   The transformation $w = z^2$ maps the first quadrant in the z-plane onto the upper half of the w-plane. Setting $z = x + iy$ and $w = \phi + i\psi$ and equating real and imaginary parts shows the velocity potential in the w-plane to be $\phi = x^2 - y^2$ and the stream function to be $\psi = 2xy$. The streamlines $\psi = $ constant in the w-plane

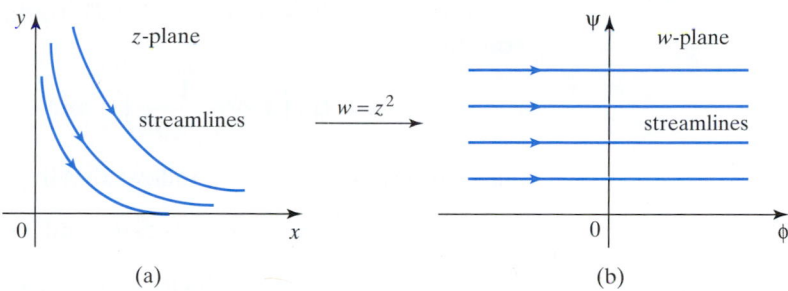

FIGURE 17.35   Flow around two perpendicular walls.

are straight lines parallel to the real axis, so they represent a uniform flow parallel to the real axis as shown in Fig. 17.35b. As no flow crosses the real axis in the w-plane, the axis can be regarded as a rigid wall bounding the flow. The map of this uniform parallel flow in the w-plane onto the z-plane is the family of streamlines $xy =$ constant that form the rectangular hyperbolas shown in Fig. 17.35a. So the complex potential $w = z^2$ describes the flow between two perpendicular walls where, far from the corner, the flow is parallel to a wall.

The velocity components at any point (x, y) in the first quadrant found from from (52) are $q_1 = \partial\phi/\partial x = 2x$ and $q_2 = \partial\phi/\partial y = -2y$, so the flow in the z-plane is in the direction indicated by the arrows in Fig. 17.35a. The speed $q = 2(x^2 + y^2)^{1/2}$ at the point (x, y) follows from (55). It should be recognized that because fluid cannot cross a streamline, in an ideal fluid it is always possible to replace a streamline by a rigid boundary without disturbing the remainder of the flow.   ∎

EXAMPLE 17.6   Interpret the flow of an ideal fluid with the complex potential

$$w = U\left(z + \frac{1}{z}\right), \quad \text{where } U \text{ is real.}$$

Describe the flow that results when the additional transformation $z = e^{-i\alpha}\zeta$ is made, with α real.

Solution   We have seen that the Joukowski transformation maps the exterior of the unit circle $|z| = 1$ in the z-plane onto the w-plane cut along the real axis from $w = -2$ to $w = 2$, as shown in Fig. 17.36.

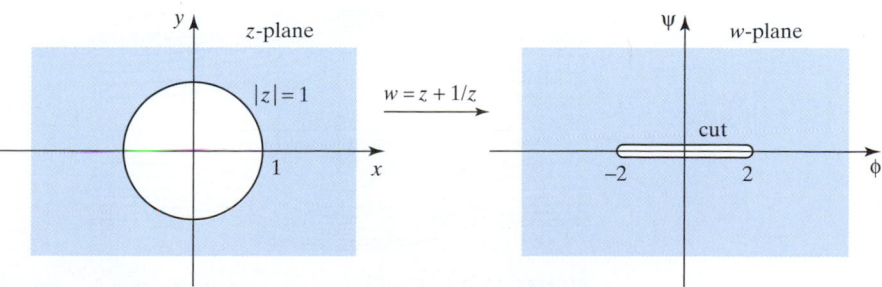

FIGURE 17.36   The effect of the mapping $w = z + 1/z$ on $|z| = 1$.

If we set $w = u + iv$ and $z = x + iy$, routine calculation shows that in cartesian coordinates

$$u = Ux\left(1 + \frac{1}{x^2 + y^2}\right) \quad \text{and} \quad v = Uy\left(1 - \frac{1}{x^2 + y^2}\right),$$

whereas if we set $z = re^{i\theta}$, it follows that in polar coordinates

$$u = (r + 1/r)\cos\theta \quad \text{and} \quad v = (r - 1/r)\sin\theta.$$

When $|x|$ is large the velocity potential $u \approx Ux$, so (52) shows that far from the origin in the z-plane the fluid velocity tends to $\mathbf{q} = U\mathbf{i}$, corresponding to a uniform flow parallel to the x-axis with speed U at infinity. On the unit circle $|z| = 1$ the stream function $\psi = 0$, so this is a *streamline*. Thus, fluid will flow around the unit circle as though it is a solid cylinder of unit radius centered on the origin with its axis perpendicular to the z-plane.

The **streamlines** around the unit circle are described by either

$$Uy\left(1 - \frac{1}{x^2 + y^2}\right) = \text{constant} \quad \text{or} \quad \left(r - \frac{1}{r}\right)\sin\theta = \text{constant},$$

whereas the **equipotentials** around the unit circle (lines of constant velocity potential) are described by either

$$Ux\left(1 + \frac{1}{x^2 + y^2}\right) = \text{constant} \quad \text{or} \quad \left(r + \frac{1}{r}\right)\cos\theta = \text{constant}.$$

Figure 17.37 shows some representative streamlines in the z-plane and their images in the w-plane. As no fluid crosses the streamline around the unit circle $|z| = 1$, none will flow across the cut in the w-plane, so the cut can be taken to represent the cross-section of flat plate normal to the z-plane that forms an impenetrable barrier.

The inverse of this transformation can be used to determine the flow past a flat plate when the flow at infinity is incident from the left at an angle α to the plate. From (55) it follows that in the ζ-plane $w_1 = \zeta e^{-i\alpha}$ represents the complex potential of a uniform parallel flow at infinity that is incident from the left at an angle α to the real axis. Consequently, if we use the Joukowski transformation,

$$w = \left(w_1 + \frac{1}{w_1}\right) = \left(\zeta e^{-i\alpha} + \frac{e^{i\alpha}}{\zeta}\right)$$

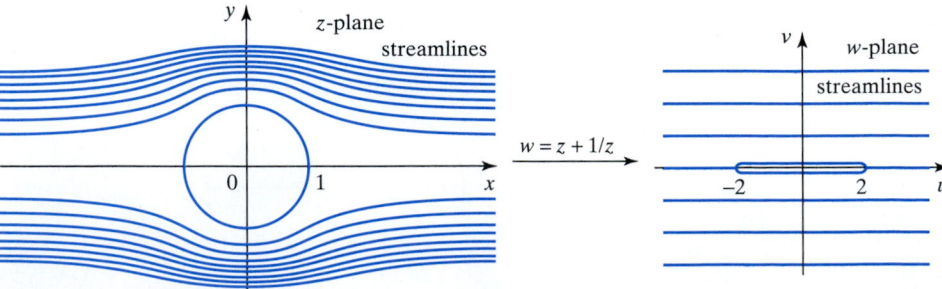

FIGURE 17.37   Flow past a cylinder mapping onto flow parallel to a flat plate.

is the complex potential of a uniform parallel flow that at infinity is incident from the left on the unit circle in the ζ-plane, with the flow at infinity making an angle α with the real axis.

Solving the transformation $\zeta = z + 1/z$ for z, and then interchanging ζ and z, we find that inverse mapping back from the unit circle in the ζ-plane to the z-plane cut from $z = -2$ to $z = 2$ is given by $\zeta = \frac{1}{2}(z + \sqrt{z^2 - 1})$. If we substitute for ζ in the previous result, the required complex potential in the z-plane for flow with speed U past a flat plate formed by the cut from $z = -2$ to $z = 2$, when the flow is incident from the left of the plate and at an angle α, is seen to be given by

$$w = U\left(\frac{1}{2}e^{-i\alpha}(z + \sqrt{z^2 - 4}) + \frac{2e^{i\alpha}}{z + \sqrt{z^2 - 4}}\right).$$

When simplified using the result $z + \sqrt{z^2 - 4} = \frac{1}{4}(z - \sqrt{z^2 - 4})$, this reduces to

$$w = U(z\cos\alpha - i\sqrt{z^2 - 4}\sin\alpha).$$

stagnation point in a flow

In this complex potential, as the square root function has a branch point, we must interpret the square root as $\sqrt{z^2 - 4} = |z^2 - 4|e^{(i/2)(\theta_1 + \theta_2)}$, where $z - 2 = |z - 2|e^{i\theta_1}$ and $z + 2 = |z + 2|e^{i\theta_2}$, with $0 \le \theta_1 \le 2\pi$ and $0 \le \theta_2 \le 2\pi$ measured as shown in the cut plane in Fig. 17.38c.

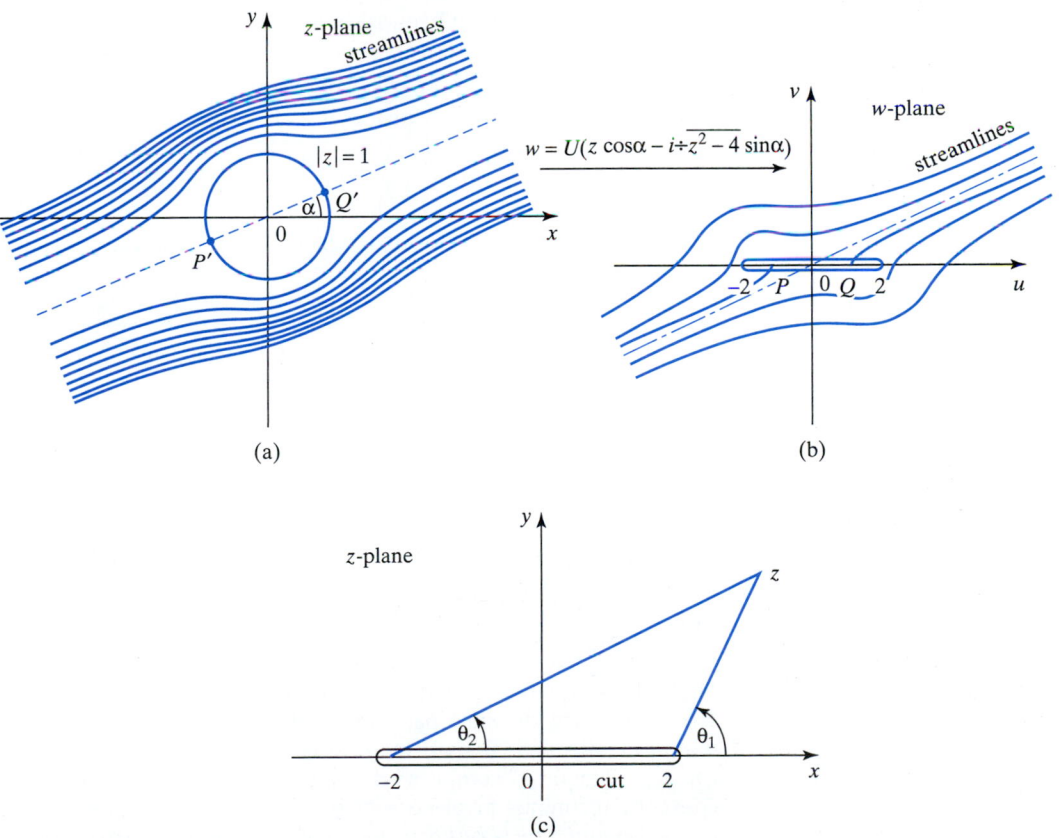

(a)   (b)

(c)

FIGURE 17.38   Inclined flow past a flat plate.

Representative streamlines around the unit circle in the z-plane with the flow at infinity inclined to the x-axis at an angle α are shown in Fig. 17.38a, where the points P' and Q' are streamlines that terminate on the unit circle. These points are called **stagnation points**, because the fluid velocity is zero at such points. Fig. 17.38b shows the inverse mapping of this flow, corresponding to inclined flow around a flat plate in the w-plane, where the stagnation points at P and Q on the plate are the images of the stagnation points P' and Q' in Fig. 17.38b.

The pressure p at any point on a streamline can be found from a result called the **Bernoulli equation**, and for the steady two-dimensional flow of an ideal fluid this takes the form

$$\frac{1}{2}\rho \left(q_1^2 + q_2^2\right) + p = \text{constant},$$

where ρ is the density of the fluid. This shows that the pressures in the vicinity of the stagnation points on either side of the plate apply turning moments to the plate that both act in the same sense. When the plate is broadside to the flow, points P and Q are opposite one another at the center of the plate, about which the flow is then symmetrical. Such a flow provides a good approximation to the actual flow of fluid past a flat plate, and it only fails at the ends of the plate where in the real world the speed of flow is finite, whereas in an ideal fluid it is infinite. The existence of a turning moment about the center line of the plate, which vanishes when the plate is perpendicular to the flow, explains why a boat allowed to drift from rest down a stream will always turn broadside to the direction of flow. ∎

The Laplace equation arises in many other steady-state physical situations, the most important of which are in the description of gravitational fields, diffusion, electric current flow, magnetism, and elasticity. When restricted to two space dimensions the real and imaginary parts of an analytic function $w = \phi + i\psi$ can be interpreted as follows:

Application of Laplace's Equation	$\phi(x, y) = \text{Constant}$	$\psi(x, y) = \text{Constant}$
Gravitational fields	Gravitational equipotentials	Lines of force
Diffusion phenomena	Concentration	Lines of flow
Electric current flow	Potential	Lines of current flow
Magnetism	Magnetic potential	Lines of force
Elasticity	Strain function	Stress lines

The development of conformal transformations together with various applications is to be found in references [6.1], [6.2], [6.4], [6.6], [6.8], and [6.9]. A systematic application of conformal transformations is made to hydrodynamics in reference [6.5].

Summary

The Laplace equation is fundamental to the study of heat flow, electricity and magnetism, fluid mechanics, gravitational fields, and elsewhere. This section has shown how conformal mappings can be used to solve certain types of boundary value problems for the Laplace equation in complicated two-dimensional regions bounded by arcs and straight lines. The technique involved first solving a boundary value problem in a simply shaped region bounded by coordinate lines in one plane, and then mapping the region onto one in another plane with a more complicated shape that is of interest. The approach was seen to work because conformal mappings transform harmonic functions in one plane into harmonic functions in another plane, while the boundary conditions are mapped without change onto the corresponding boundaries. Consequently, the solution of a simple boundary value problem in one plane can be transformed into the solution of a corresponding boundary value problem in a region of more complicated shape in another plane. Applications to various boundary value problems of physical interest were made, including ones to the flow of ideal fluids.

EXERCISES 17.2

1. Let the function $\phi(x, y)$ be harmonic in some region of the (x, y)-plane. If $\phi(x, y)$ becomes $\Phi(u, v)$ under the change of variable $u = x^2 - y^2$ and $v = 2xy$, confirm by direct calculation that the transformation $w = z^2$ leaves Φ harmonic.

2. Using the definition of Arctan t in (47) and setting $t = y/x$, confirm that if (a) P is the point $(\sqrt{3}, 1)$ then Arctan $t = \frac{\pi}{6}$, (b) P is the point $(-2, 2)$ then Arctan $t = \frac{3\pi}{4}$, and (c) if P is the point $(\pm\varepsilon, 2)$, then $\lim_{\varepsilon \to 0}$ Arctan $t = \frac{\pi}{2}$. Find Arctan t when (d) P is the point $(4, 1)$ and (e) when P is the point $(-3, 2)$.

3. Derive the function $\phi(x, y)$ that is harmonic in the upper half of the (x, y)-plane and satisfies the piecewise constant Dirichlet boundary value problem

$$\phi = \phi_1 \quad \text{on } x < x_1, y = 0$$
$$\phi = \phi_2 \quad \text{on } x_1 < x < x_2, y = 0$$
$$\phi = \phi_3 \quad \text{on } x_2 < x < x_3, y = 0$$
$$\phi = \phi_4 \quad \text{on } x > x_3, y = 0.$$

4. Derive the function $\phi(x, y)$ that is harmonic in the right half of the (x, y)-plane and satisfies the piecewise constant Dirichlet boundary value problem

$$\phi = \phi_1 \quad \text{on } y > y_1, x = 0$$
$$\phi = \phi_2 \quad \text{on } y_2 < y < y_1, x = 0$$
$$\phi = \phi_3 \quad \text{on } y < y_2, x = 0.$$

Is there a simple way of finding $\phi(x, y)$ from (46)?

5. Prove that the transformation $w = \left(\frac{1+z}{1-z}\right)^2$ maps the interior of the semicircle of radius 1 on the left of Fig. 17.39 onto a half-plane in the manner shown in the diagram on the right. If the semicircle represents a cross-section of a long heat-conducting bar, find the temperature distribution and the isothermals in a cross-section of the bar when the flat boundary AB is maintained at the constant temperature $T = 30$ and the semicircular boundary ACB is maintained at the constant temperature $T = 150$.

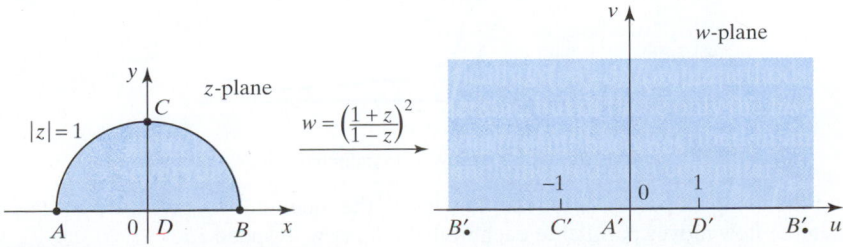

FIGURE 17.39   The mapping $w = \left(\frac{1+z}{1-z}\right)^2$.

6. Repeat Exercise 5 assuming that the semicircle on the left represents a cross-section of an electrically conducting wall of a cavity. Find the electric potential inside the cavity and the

equipotentials when the flat section of the wall AO is maintained at the constant electric potential $\phi = 20$, the flat section of the wall OB at the constant electric potential $\phi = 100$, and the curved wall ACB at the constant electric potential $\phi = 50$.

7. Prove that the transformation $w = i\left(\frac{1-z}{1+z}\right)$ maps the inside of the circle on the left in Fig. 17.40 onto the upper half-plane in the manner shown in the diagram on the right. If the circle is considered to be the electrically conducting wall of a cavity, find the electric potential and electric force lines inside a cross-section of the cavity if the upper semicircular boundary ABC is maintained at the constant electric potential $\phi = 320$ and the lower semicircular boundary CDA at the constant electric potential $\phi = 100$.

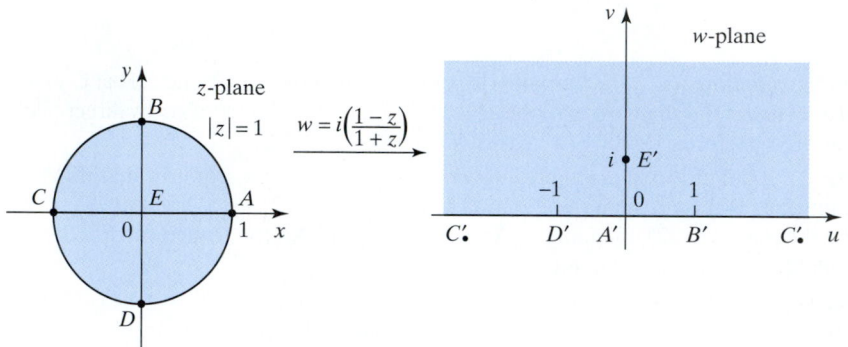

FIGURE 17.40 The mapping $w = i\left(\frac{1-z}{1+z}\right)$.

8. Repeat Exercise 7 assuming the circle to be the cross-section of a long solid heat-conducting cylinder. Find the temperature distribution and the isothermals in a cross-section of the cylinder if the circular boundary CD is maintained at a temperature $T = 50$, the circular boundary DAB is maintained at a constant temperature $T = 200$, and the circular boundary BC is maintained at a constant temperature $T = 0$.

9. Explain why $w = U\left(z^3 + \frac{1}{z^3}\right)$ is the complex potential of the flow inside the indented wedge shown in Fig. 17.41, in which the flow moves parallel to each wall at infinity with speed U.

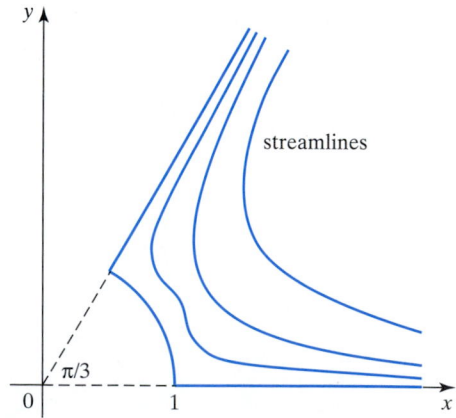

FIGURE 17.41 Flow in an indented wedge.

10. Find the complex potential for the flow inside the indented wedge shown in Fig. 17.42 when the flow moves parallel to each wall at infinity with speed U.

11. The Joukowski transformation $w = z + 1/z$ maps the upper half of the z-plane from which has been deleted a unit semicircle centered on the origin onto the upper half of the w-plane with a cut along the real axis from $w = -2$ to $w = 2$, as shown in Fig. 17.43. If w is the

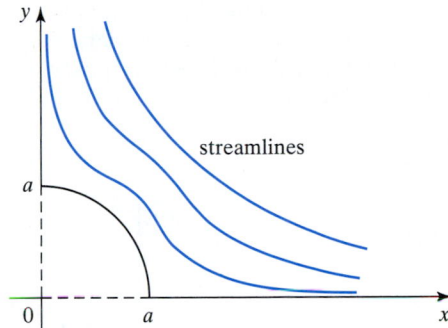

FIGURE 17.42 Flow in an indented right-angled wedge.

complex potential of a fluid flow, by setting $z = x + iy$ and $w = u + iv$, find the implicit equation of the streamlines in the z-plane corresponding to the flow lines $v = c (c \geq 0)$ in the w-plane. By examining the qualitative properties of the implicit equation of the streamlines, confirm they have the properties shown in Fig. 17.43, which can be interpreted as the flow of very deep water over a semicircular obstacle resting on the bottom. State how the diagram on the left can be used to describe the flow of a stream of water of finite depth over a submerged obstacle, when the surface of the stream is a *free surface* (a fluid–air interface).

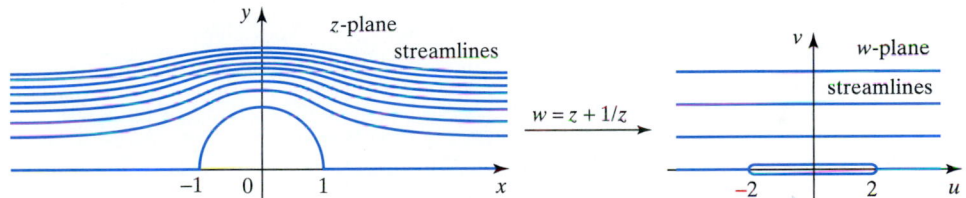

FIGURE 17.43 Flow over a semicircular obstacle.

12. The transformation $w = z + \exp z$ maps the strip $-\pi \leq y \leq \pi$ in the z-plane onto the w-plane with cuts along the lines $u \leq -1, v = \pm \pi$, as shown in Fig. 17.44. If w is the complex potential of a fluid flow, by setting $z = x + iy$ and $w = u + iv$ find the equation of the streamline $y = c$ in parametric form. As the cuts are bounded by streamlines, and fluid cannot cross a streamline, the cuts can be interpreted as parallel barriers, allowing the diagram on the right to be interpreted as flow emerging from a parallel channel into an unrestricted region. How can this problem be interpreted in terms of an electrostatic potential?

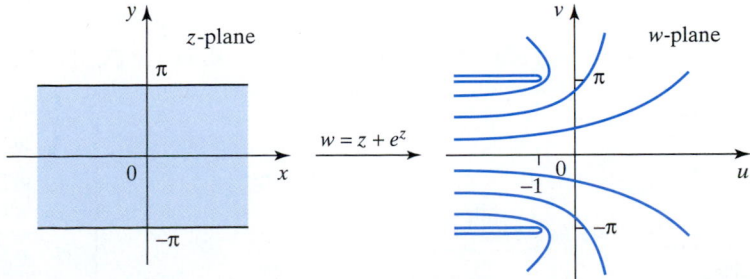

FIGURE 17.44 Flow from a parallel channel into an unrestricted region.

13. The transformation $w = \text{Arcsin}\, z$ maps the upper half of the z-plane with a cut along the real axis from $z = -1$ to $z = 1$, onto the semi-infinite strip $-\pi \leq u \leq \pi, v \geq 0$ in the

w-plane, as shown in Fig. 17.45. Use this result to find the equipotentials and flux lines if $A_\infty B$ is an electrically conducting plate at the constant electric potential $\phi = 200$, CD_∞ is an electrically conducting plate at the constant potential $\phi = 100$, and BC is an insulator (no flux lines can cross it). How can this problem be interpreted in terms of steady state heat conduction?

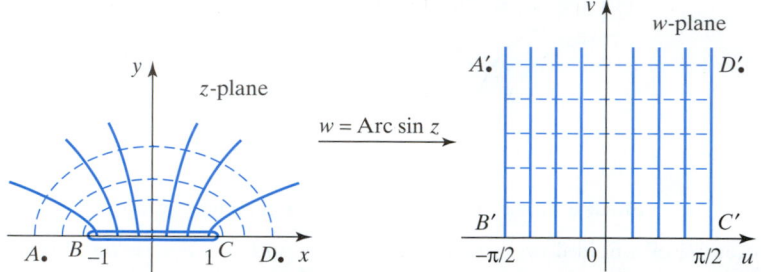

FIGURE 17.45 Electrically conducting plates separated by an insulator.

14. The diagram in Fig. 17.46 represents a metal lamina occupying the first quadrant of the (x, y)-plane with the edge $x = 0$, $y > 1$ maintained at the constant temperature $T = 200$, the edge $x > 1$, $y = 0$ maintained at the constant temperature $T = 50$, and the edges $x = 0, 0 < y < 1$ and $0 < x < 1, y = 0$, maintained at the constant temperature $T = 0$. Find the temperature $T(x, y)$ at any point (x, y) in the lamina.

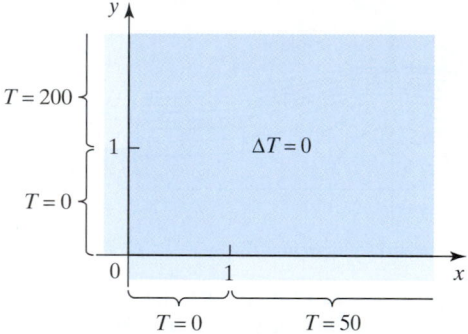

FIGURE 17.46 Mapping the exterior of a hole onto a half-plane.

15. The diagram on the left of Fig. 17.47 shows a cross-section of an infinite metal block pierced by a hole of unit diameter, the boundary DAB of which is maintained at the constant temperature $T = 450$, while the boundary DCB is maintained at the constant temperature $T = 100$. Use the fact that the transformation

$$ w = i \left(\frac{z+1}{(1-i)z - 1 - i} \right) $$

maps $|z| \geq 1$ onto the upper half of the w-plane in the manner shown in Fig. 17.47 to find the temperature and isothermals in the plate.

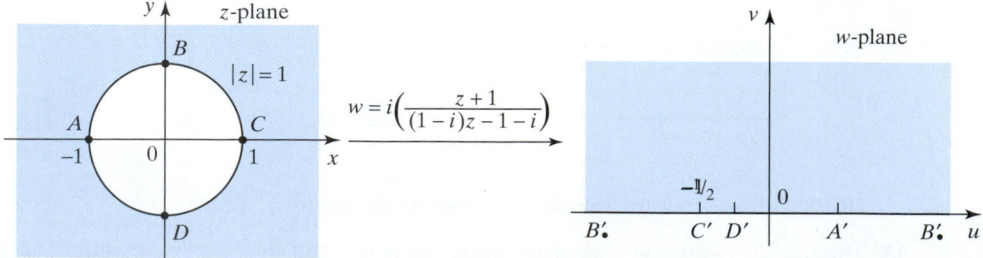

FIGURE 17.47 A metal block pierced by a hole.

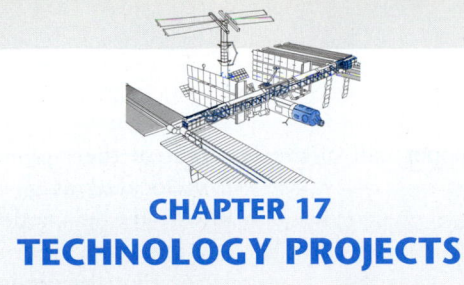

CHAPTER 17
TECHNOLOGY PROJECTS

Project 1

Examining the Mapping of Lines and Circles by the Linear Fractional Transformation

The purpose of this project is to apply computer algebra and graphics to the linear fractional transformation

$$w = \frac{2z - i}{z + i}$$

to explore how it maps various straight lines and circles in the z-plane onto lines and circles in the w-plane, though not necessarily in this order.

Find the maps of (a) $y = 0$, (b) $y = 2$, (c) $y = x$, (d) the circle of radius $\frac{1}{2}$ centered on the origin, (e) the unit circle centered on the origin, and (f) the unit circle centered on the point $z = 1 + i$.

Project 2

This project examines the way the Joukowski transformation

$$w = z + \frac{1}{z}$$

maps a circle of radius R passing through the point $z = -1$ with its center in the first quadrant.

Experiment by choosing different positions for the center of the circle and then mapping the boundary of the circle onto the w-plane.

Project 3

Verify the results of Example 17.4 by using the function

$$w = \frac{z - 2 + \sqrt{3}}{(2 - \sqrt{3})z - 1}$$

to plot the map of the circles $|z - \frac{1}{4}| = \frac{1}{4}$ and $|z| = 1$ in the z-plane onto the w-plane. Hence, show that

the circles map onto the concentric circles shown in Fig. 17.34.

Project 4

By considering the way $w = z + \exp(z)$ maps the infinite strip $-\pi \le y \le \pi$ in the z-plane onto the w-plane, show how this mapping can be interpreted as the two-dimensional discharge of fluid from between parallel semi-infinite planes into a surrounding infinite volume of fluid. Find the slope of the fluid flow lines far from the place of discharge, and plot some representative flow lines.

Explain how this same mapping can describe equipotentials inside and outside a parallel plate capacitor in a vacuum when the lower plate is at a potential V_1 and the upper plate is at a potential V_2, and determine the potential associated with each equipotential.

Project 5

In two-dimensional fluid mechanics, a **line source** of strength m is a line normal to the plane of the flow from which fluid enters the surrounding medium symmetrically at a steady rate of m volume units per unit line length per unit time. Similarly, when m is negative, this becomes a **line sink** that removes fluid from the surrounding medium symmetrically at a steady rate of m volume units per unit line length per unit time.

By considering the fluid complex potential

$$w = \phi + i\psi = m \operatorname{Log}(z - z_0), \quad (m > 0)$$

find the curves $\phi = $ constant and $\psi = $ constant that are, respectively, the equipotentials and streamlines of the flow. Hence, explain why w is the fluid complex potential of a line source located at a point z_0, with the line source perpendicular to the z-plane.

If attention is confined to the upper half of the z-plane, explain why the function

$$w = m \operatorname{Log}\left(\frac{z - z_0}{z - \bar{z}_0}\right)$$

is the complex potential for fluid flow in the upper half of the z-plane due to a line source of strength m located at z_0 when the region is bounded below by a fixed impenetrable barrier along the x-axis.

Plot the equipotentials and streamlines for such a flow for $-3 \le x \le 3, 0 \le y \le 3$ when $m = 1, z_0 = i$.

PART SEVEN

PARTIAL DIFFERENTIAL EQUATIONS

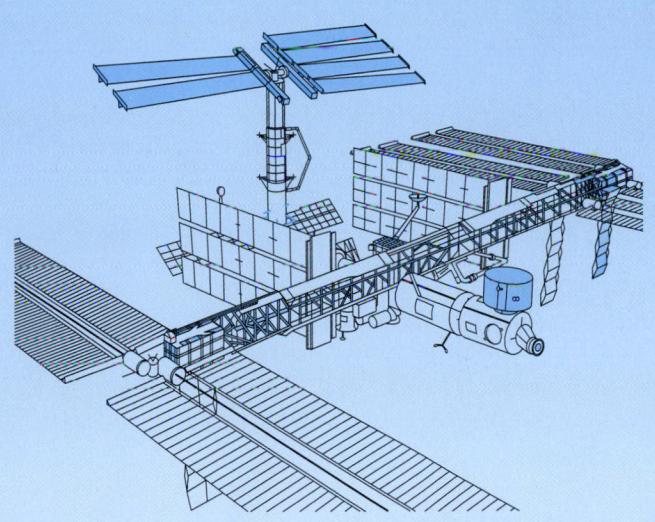

Chapter 18   **Partial Differential Equations**

CHAPTER 18

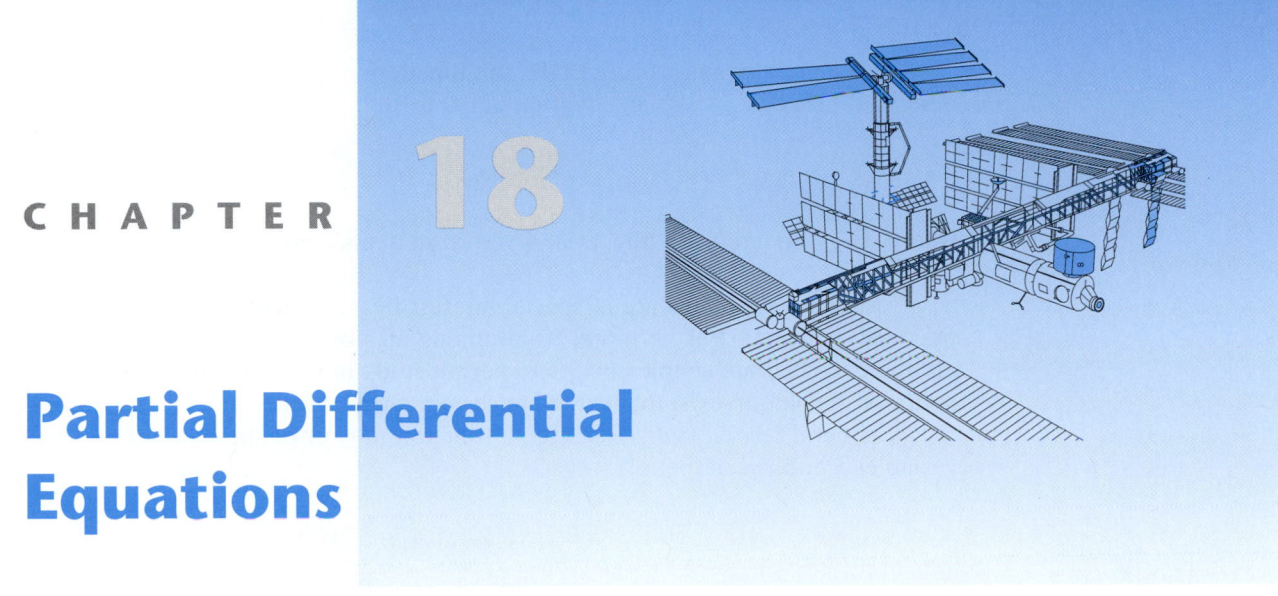

Partial Differential Equations

Partial differential equations (PDEs) are equations satisfied by partial derivatives of functions of two or more independent variables. They describe all types of physical phenomena in engineering and science, ranging from transient heat conduction through vibrations of strings and plates to fluid flow and the behavior of electric and magnetic fields. The solution of first order equations is developed using the method of characteristics, and the three fundamentally different types of second order PDE are derived from first principles using typical physical examples. After classifying second order equations and describing suitable boundary and initial conditions, it is shown how the PDEs can be reduced to their standard forms to simplify the task of finding a solution. The wave equation is interpreted in terms of two disturbances propagating with equal speed, but in opposite directions, and the D'Alembert solution is derived.

The separation of variables method of solution is developed and related to the Sturm–Liouville systems, eigenvalues, and eigenfunctions already discussed in connection with ordinary differential equations. The method is then applied to various physical problems involving cartesian, cylindrical, and spherical polar coordinates. Some results of general importance to the study of PDEs are derived, and the chapter ends with an introduction to Laplace and Fourier transform methods of solution for PDEs.

18.1 What Is a Partial Differential Equation?

The simplest form of partial differential equation (PDE) involving a suitably differentiable unknown function (**dependent variable**) $u(x, y)$ of the two independent variables x and y is an equation that relates x, y, u, and some partial derivatives of u with respect to x and y. The **order** of the PDE is the order of the highest partial derivative of u that occurs in the equation, so a general *first order* PDE for the function $u(x, y)$ is of the form

$$F(x, y, u, u_x, u_y) = 0, \tag{1}$$

order of a PDE

where F is an arbitrary function of its arguments.

927

More generally, a first order PDE for a function $u(x_1, \ldots, x_n)$ of the n independent variables $x_1, \ldots, x_n$ is an equation of the form

$$G(x_1, \ldots, x_n, u, u_{x_1}, \ldots, u_{x_n}) = 0, \tag{2}$$

where G is an arbitrary function of its arguments and $u_{x_i} = \partial u / \partial x_i$, for $i = 1, 2, \ldots, n$.

First order equations are of special interest because they occur frequently in practical problems. Furthermore, from among all possible classes of PDE, they are the ones that are simple enough to permit study in great detail, and for which methods of solution exist that extend to certain types of second order equation.

A general *second order* PDE for a function $u(x, y)$ of the two independent variables x and y is of the form

$$H(x, y, u, u_x, u_y, u_{xx}, u_{xy}, u_{yy}) = 0, \tag{3}$$

where H is an arbitrary function of its arguments, and for conciseness the suffix notation $u_x = \partial u / \partial x, u_y = \partial u / \partial y, u_{xx} = \partial^2 u / \partial x^2, u_{yx} = \partial^2 u / \partial x \partial y$, and $u_{yy} = \partial^2 u / \partial y^2$ has been used.

classical and generalized solutions

A **classical solution** of a PDE defined in some region D of the (x, y)-plane is a *real* function u with the property that all of its partial derivatives that occur in the PDE are defined and *continuous* throughout D, and when the function is substituted into the PDE it satisfies the equation identically. We will see later that in certain cases a slightly more general class of solution is also possible where a derivative may be discontinuous. Solutions of this type are called **generalized solutions**, and they are often used in connection with wave propagation problems.

The expressions in (1) and (2) are too general to be directly useful, so only some important special cases will be examined. In the case of (1) the three special cases to be considered are called, respectively, first order PDEs of *linear*, *semilinear*, and *quasilinear* type.

The Linear First Order PDE for $u(x, y)$

A *linear first order* PDE for the unknown function $u(x, y)$ can always be written as

$$p(x, y)u_x + q(x, y)u_y = r(x, y)u + s(x, y), \tag{4}$$

where $p(x, y)$, $q(x, y)$, $r(x, y)$, and $s(x, y)$ are arbitrary functions of x and y, and the term $s(x, y)$ that does not multiply u, u_x, or u_y is called the **nonhomogeneous** term. The PDE is called **homogeneous** when $s(x, y) = 0$. When, as often happens, the functions p, q, and r are constants, the PDE becomes a **constant coefficient** equation. The equation in (4) is called *linear* because u, u_x, and u_y all occur linearly (with degree 1) in each term. The following is a typical linear first order PDE:

$$u_x + xu_y = u + 2.$$

The solution $u = u(x, y)$ of (4) in a region D of the (x, y)-plane where the PDE is defined can be represented in the form of a surface above D called an **integral surface**. For most PDEs it is impossible to find a general solution so instead, when solving a PDE, it is usual to consider a specific problem by requiring that as well as the solution satisfying the PDE, it also satisfies some auxiliary (additional) conditions that identify the particular problem.

In the case of a linear first order PDE it will be seen later that in principle a general solution can be found, though usually only the solution of a specific problem is required. In order to specify such a problem for a first order PDE, the auxiliary condition that identifies the problem uniquely involves prescribing the value the solution u is required to attain along a line in D. An auxiliary condition of this **Cauchy conditions** nature is called a **Cauchy condition**, and the problem of finding the solution of a PDE in D that satisfies a Cauchy condition is called a **Cauchy problem** for the PDE. More will be said about the Cauchy problems in the next section.

The Semilinear First Order PDE for $u(x, y)$

A *semilinear first order* PDE is slightly more complicated than a linear first order equation because it is of the form

$$p(x, y)u_x + q(x, y)u_y = f(x, y, u), \tag{5}$$

where f is an arbitrary *nonlinear* function of u. The left sides of the PDEs in (4) and (5) are identical, but the right side of the semilinear PDE in (5) depends *nonlinearly* **linear,** on u instead of linearly as in (4). A typical example of a semilinear first order PDE is **semilinear, and quasilinear first order PDEs**

$$u_x + (1 + x)u_y = (1 + x + y)u^2,$$

where the term $f(x, y, u) = (1 + x + y)u^2$ is nonlinear because of the term u^2.

The Quasilinear First Order PDE

A *quasilinear first order* PDE is one that can be written in the form

$$p(x, y, u)u_x + q(x, y, u)u_y = f(x, y, u) \tag{6}$$

where the functions p and q may or may not depend on x and y, but at least one of them depends on the undifferentiated function u. When f is present in (6) it may or may not depend on all of x, y, and u, though the presence or absence of f does not alter the quasilinear nature of the equation. A typical quasilinear first order PDE is

$$u_x + uu_y = u,$$

where in this case the quasilinearity is due to the presence of the term uu_y.

Both linear and quasilinear first order PDEs often occur in systems involving several dependent variables, and on occasion it is possible for all but one of the dependent variables to be eliminated, leading to a single higher order equation in the remaining dependent variable. The following is an example of a simple linear system of first order equations involving the variables $v(x, t)$ and $w(x, t)$:

$$v_t - c^2 w_x = 0 \quad \text{and} \quad w_t - v_x = 0. \tag{7}$$

Here c is a constant. In these equations the independent variables are denoted by x and t, because in physical problems governed by these equations x is usually a space variable (a length) and t is the time.

When v and w are twice differentiable functions, partial differentiation of the first equation with respect to t gives

$$v_{tt} - c^2 w_{xt} = 0,$$

and partial differentiation of the second equation with respect to x gives

$$w_{tx} - v_{xx} = 0.$$

Provided the second derivatives are continuous, the mixed derivatives are equal, so that $w_{xt} = w_{tx}$. After the elimination of w_{xt} between these two equations, the following linear second order equation for v is obtained:

$$v_{tt} - c^2 v_{xx} = 0. \tag{8}$$

Had the first equation in (7) been differentiated partially with respect to x and the second equation partially with respect to t, this same argument would have given

$$w_{tt} - c^2 w_{xx} = 0,$$

showing that v and w both satisfy the same PDE.

Later this equation will be seen to describe an important form of wave propagation in one space dimension and time, and for this reason it is called the *one-dimensional wave equation*. In the wave equation the constant c is the speed with which waves (disturbances) are propagated. Another linear example is provided by the *Cauchy–Riemann equations* (see Section 13.2)

$$u_x = v_y \quad \text{and} \quad u_y = -v_x,$$

where u and v are the real and imaginary parts of an analytic function $f(z) = u + iv$, with $z = x + iy$. In this case an argument similar to the one just used shows that both u and v are harmonic functions, so as each is a solution of Laplace's equation,

$$u_{xx} + u_{yy} = 0 \quad \text{and} \quad v_{xx} + v_{yy} = 0.$$

A more complicated system of quasilinear equations is provided by the equations of *unsteady* (time dependent) *gas dynamics*. In their simplest form these equations relate the gas density ρ, its pressure $p = k\rho^\gamma$ with k and γ constants, and the gas velocity $\mathbf{u}$, all at time t and at some position vector $\mathbf{r}$ in space, through the system of equations

$$\rho_t + \operatorname{div}(\rho \mathbf{u}) = 0 \quad \text{and} \quad \mathbf{u}_t + \mathbf{u} \cdot \nabla \mathbf{u} + (1/\rho)\nabla p = 0. \tag{9}$$

The first equation is a scalar equation that describes the conservation of mass, and the second is a vector equation with three scalar components that is related to the equation that describes the conservation of momentum. The system in (9) couples the density ρ and the three scalar components of $\mathbf{u}$ through a system of four scalar quasilinear equations. In this case the structure of the system is such that it cannot be replaced by a single higher order equation for one of the unknowns.

When introducing the linear first order PDE, mention was made of the fact that the complexity of PDEs is such that general solutions can only be found in very special cases. As a result, when dealing with higher order PDEs, instead of seeking general solutions, methods are developed that enable solutions of specific problems to be found. As already mentioned, to find the solution of a particular problem involving a PDE it is necessary to require that the solution satisfy some auxiliary conditions that identify the problem. The additional conditions may be imposed on spatial boundaries belonging to a region D where the solution is required, and when this is done the conditions are called **boundary conditions**. A typical boundary condition for a second order PDE defined in a rectangle could be that the solution is required to assume specified values on the sides of the rectangle. If time is involved, it is necessary to specify how the solution starts, and a condition of this type is called an **initial condition**. Problems requiring initial and boundary conditions are called **initial boundary value problems** (IBVPs).

boundary and initial conditions

The definitions of linearity and quasilinearity extend quite naturally to PDEs of all orders. A PDE of any order is **linear** if the unknown function u and all its derivatives only appear linearly (to degree 1), so a general linear second order PDE for the unknown function $u(x, y)$ can be written

$$a(x, y)u_{xx} + b(x, y)u_{xy} + c(x, y)u_{yy} + d(x, y)u_x + e(x, y)u_y + f(x, y)u = h(x, y).$$
(10)

Analogously, a PDE of order n is said to be **quasilinear** when its partial derivatives of order n occur linearly in the equation, but combinations of u and some of its derivatives up to order $n - 1$ occur as coefficients of the nth order partial derivatives. A general quasilinear second order PDE for the unknown function $u(x, y)$ can be written

linear, quasilinear, and nonlinear higher order PDEs

$$a(x, y, u, u_x, u_y)u_{xx} + b(x, y, u, u_x, u_y)u_{xy}$$
$$+ c(x, y, u, u_x, u_y)u_{yy} + h(x, y, u, u_x, u_y) = 0,$$
(11)

where a, b, c, and h are arbitrary functions of their arguments, with at least one of the functions a, b, and c depending on u and/or one or more of its first order partial derivatives.

A PDE of any order that is not linear, semilinear, or quasilinear is said to be **nonlinear**. The following is an example of a nonlinear second order PDE:

$$uu_{xx} + \sin(u_{yy}) + xu_x + u_y + u = 0.$$

Here the nonlinearity is caused by the term $\sin(u_{yy})$.

Although in principle a general solution of a linear first order PDE can be found, unlike the general solution of a linear first order ordinary differential equation (ODE) that contains an arbitrary *constant*, the general solution of a linear first order PDE contains an arbitrary *function*. This situation is illustrated by the first order PDE

$$u_x + xu_y = u + 2,$$
(12)

which can be shown to have the general solution

$$u(x, y) = C \exp\{x + \phi(\xi)\} - 2,$$
(13)

where $\xi^2 = x^2 - 2y$, ϕ is an arbitrary differentiable function of its argument ξ and C is a constant.

To find a specific solution suppose, for example, that a solution of (12) is required to satisfy the auxiliary condition $u(x, 0) = -1$. Setting $y = 0$ in the general solution, and noticing that as $\xi^2 = x^2 - 2y$ it follows that on the x-axis $\xi = x$, we find from the condition $u(x, 0) = -1$ that the arbitrary function ϕ must be chosen such that

$$-1 = C \exp\{x + \phi(x)\} - 2, \quad \text{and so } 1 = C \exp\{x + \phi(x)\}.$$

This is only possible if $C = 1$ and $\phi(x) = -x$, so replacing x in $\phi(x)$ by $\xi = (x^2 - 2y)^{1/2}$ gives $\phi(\xi) = -(x^2 - 2y)^{1/2}$, so the solution becomes

$$u(x, y) = C \exp\{x - (x^2 - 2y)^{1/2}\} - 2.$$

Differentiation confirms that this expression satisfies the PDE, so as it also satisfies the additional condition $u(x, 0) = -1$ it is the required classical solution. The solution will be real provided $x^2 \geq 2y$, so the line $y = 0$ on which the Cauchy condition is specified is seen to bound the region of the (x, y)-plane where the classical solution is defined.

existence and uniqueness

Two important questions that must be answered when working with PDEs are (i) the **existence** question (does the PDE have a solution?) and (ii) the **uniqueness** question (if a solution exists, is it the only possible one?).

These questions can be answered in some detail for first order PDEs and higher order linear equations, and to a lesser extent for other types of PDEs, but it will suffice to say here that a solution of a linear PDE exists, and when the additional condition in the form of a Cauchy condition is specified in a manner to be described later, the corresponding solution will be unique.

To see that not every first order PDE has a solution, it is only necessary to consider the nonlinear equation

$$u_x^2 + u_y^2 = -1.$$

The expression on the left is nonnegative, so clearly this equation cannot be satisfied by any *real* function $u(x, y)$.

derivation of the first order PDE involving a transient heat balance

To illustrate one of the ways in which first order PDEs arise from physical situations, we will derive the equation governing the transient heat balance between a pipe transporting a hot fluid and the air surrounding the pipe at a constant temperature T_0. Let the length of the pipe be L, the constant speed of the fluid through the pipe be u, and the temperature of the fluid be $T(x, t)$, where x is the distance along the pipe and t is the time measured from the moment a particle of fluid enters the pipe. The physical situation is represented in Fig. 18.1, and in order to arrive at the transient heat balance equation we will consider the situation in an element of the pipe of length Δx.

The instantaneous energy balance that is to be modeled in the element of pipe of length Δx can be represented as follows:

{energy entering with fluid}−{energy leaving with fluid}−{heat transferred to air}
={energy stored in fluid}.

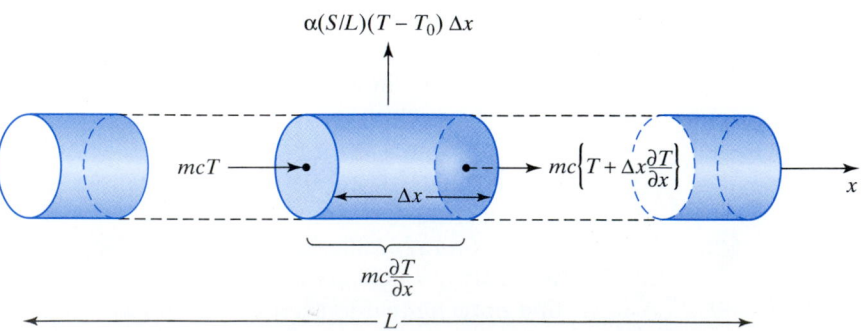

FIGURE 18.1   Transient heat distribution in an element of the pipe of length Δx.

If Δt is the time taken for a particle of fluid to travel through an element of the pipe of length Δx, the fluid speed $u \approx \Delta x/\Delta t$. If we denote the mass of fluid present in this element by M and the mass flow rate by m, the quantities M and m are related by $M = m\Delta x/u$.

If the fluid enters the element at the temperature $T(x, t)$, its temperature when leaving it can be approximated by $T + \Delta x(\partial T/\partial x)$. If we assume that the transfer of heat from the surface of the pipe to the air is proportional to the temperature difference $T(x, t) - T_0$, and denote the heat transfer coefficient by α, the heat transferred from the surface of the pipe to the air will be $(\alpha S\Delta x/L)(T - T_0)$, where S is the surface area of the pipe. The heat energy entering the element due to the fluid is mcT, where c is the specific heat of the fluid, and the heat energy leaving with the fluid is $mc(T + \Delta x\partial T/\partial x)$, whereas the stored energy in the fluid occupying the element is $Mc(\partial T/\partial t)$. Substituting these quantities into the energy balance equation gives

$$mcT - mc\left(T + \Delta x\frac{\partial T}{\partial x}\right) - \alpha\left(\frac{S\Delta x}{L}\right)(T - T_0) = Mc\frac{\partial T}{\partial t}. \tag{14}$$

Cancelling terms, and dividing (14) by $Mc = cm\Delta x/u$, this balance equation becomes the PDE for transient heat transfer:

$$\frac{\partial T}{\partial t} + u\frac{\partial T}{\partial x} = -\frac{\alpha u S}{mcL}(T - T_0). \tag{15}$$

Other examples of the derivation of PDEs that govern the behavior of important but very different physical situations are to be found in Section 18.5 where the three fundamental types of linear second order PDE are derived.

Summary

First and second order partial differential equations (PDEs) of linear, quasilinear, and nonlinear type have been defined. The Cauchy problem has been introduced and the questions of the existence and uniqueness of solutions raised. A typical first order PDE has been derived from a physical problem involving the transient heat balance between a pipe carrying hot water and the surrounding air.

EXERCISES 18.1

Classify the PDEs in Exercises 1 and 2 as linear, semilinear, quasilinear, or nonlinear.

1. (a) $u_x + u^2 u_y = x + 2y$.
   (b) $3u_x + 4u_y = \sin x$.
   (c) $u_x + xu_y^2 = u + 1$.
   (d) $u_x + 2u_y = \cos u$.
   (e) $(x + 1)u_x + yu_y = 2u + e^x$.
   (f) $u_x + (1 + u_x)u_y = u^2$.
   (g) $(x^2 + 1)u_{xx} - yu_{yy} = 1 + \cos x$.
   (h) $u_{xx} + (1 + u_x^{3/2})u_{yy} = \sin u$.

2. (a) $u_x \sin y + u_y \cos x = 1 + x^2 + y$.
   (b) $u_x + (1 + u)u_y = 2xy$.

   (c) $(x^2 + 1)u_x + u_y^2 = 2x + 3$.
   (d) $(1 + x + x^2)u_x + (2y + 1)u_y = 1$.
   (e) $(xy + 2)u_x + (1 + y + u)u_y = u$.
   (f) $u_x \sin x + u_y \cos y = x + y + 3u$.
   (g) $u_{xx} - u_{yy} = \sin u$.
   (h) $u_{xx} - 2xu_{xy} + (1 + \cos u)u_{yy} = 4$.

In Exercises 3 through 6 use the general solution of the PDE in (12) given in (13) to find the solution that satisfies the given condition, stating any restriction that is required for the solution to be valid.

3. $u(x, 0) = 2$, $y > 0$.   **4.** $u(x, 0) = e^{2x} - 2$, $y > 0$.
5. $u(x, 1) = -1$, $y > 0$.   **6.** $u(x, 2) = x - 2$, $y > 2$.

The method of solution of a quasilinear first order PDE involving the unknown function $u(x, y)$ contains within it as special cases the solution of linear and semilinear first order PDEs. Consequently it is only necessary to discuss the solution of a Cauchy problem for a quasilinear equation that we will write in the form

$$p(x, y, u)u_x + q(x, y, u)u_y = f(x, y, u), \tag{16}$$

where p, q, and f are assumed to be continuous functions of their arguments. The Cauchy condition for u will be imposed on a curve Γ in the (x, y)-plane on which u will be required to assume a prescribed functional form, with the function depending on the position on Γ.

Cauchy data curve, initial line

When the independent variables x and y are space variables, the curve Γ will be called the **Cauchy data curve**. If, however, one independent variable is a space variable and the other is the time, and Γ coincides with the x-axis, it is natural to refer to Γ as the **initial line** and to the Cauchy condition itself as the **initial condition** (or the **initial data**) for the PDE. It is then understood that as time increases the solution will evolve away from the initial condition.

If the Cauchy data curve Γ is complicated, it is usually necessary to define it parametrically by writing

$$x = x_0(s), \quad y = y_0(s), \tag{17}$$

for all values of a parameter s in some appropriate interval I. So, for example, if Γ is the straight line through the origin $ax - by = 0$, one possible parametrization of the line involves setting $x = bs$ and $y = as$ for $-\infty < s < \infty$.

In (17) the functions $x_0(s)$ and $y_0(s)$ are assumed to be continuous with piecewise continuous derivatives $x_0'(s)$ and $y_0'(s)$ such that $(x_0'(s))^2 + (y_0'(s))^2 \neq 0$. This last condition ensures that the length element $dl = \sqrt{\{(x_0'(s))^2 + (y_0'(s))^2\}}\,ds$ along Γ increases steadily with s. We will see later that the Cauchy data curve Γ cannot be specified in a completely arbitrary manner, and the nature of the restriction that must be placed on it will become clear when the method of solution has been developed.

When Γ has been defined parametrically in terms of s, the initial condition $u = u_\Gamma$ on Γ can also be defined in terms of s by setting

$$u_\Gamma(s) = u_0(s), \tag{18}$$

where $u_0(s) = u_0(x_0(s), y_0(s))$ is a prescribed function.

The total derivative of a function $u(x, y)$ along an arbitrary curve defined parametrically in terms of a parametric variable σ by the differentiable functions $x = x(\sigma)$, $y = y(\sigma)$ is

$$\frac{du}{d\sigma} = \frac{\partial u}{\partial x}\frac{dx}{d\sigma} + \frac{\partial u}{\partial y}\frac{dy}{d\sigma}. \tag{19}$$

A comparison of (16) and (19) shows that by setting

$$\frac{dx}{d\sigma} = p(x, y, u) \quad \text{and} \quad \frac{dy}{d\sigma} = q(x, y, u), \tag{20}$$

the PDE in (16) can be expressed as the ODE

$$\frac{du}{d\sigma} = f(x, y, u), \tag{21}$$

provided x and y satisfy (20).

characteristic equations, characteristics, and the compatibility condition

The two ODEs in (20) are called the parametric form of the **characteristic equations** of the PDE in (16), and when they are integrated to obtain an expression of the form

$$\Phi(x, y, k) = 0, \tag{22}$$

where k is a constant of integration, they define a family of curves C in the (x, y)-plane called the **characteristic curves** of the PDE, each of which is identified by a different value of k. Notice that in quasilinear PDEs the characteristics depend on the solution u, so in such cases it is necessary to solve (20) and (21) simultaneously. For conciseness, the curves belonging to the family C are usually called the **characteristics** of the PDE. The ODE in (21) is called the **compatibility condition** along the characteristic.

If required, the parameter σ can be eliminated from the characteristic equations and the compatibility condition by dividing the second ODE in (20), and the ODE in (21), by $dx/d\sigma$ given in the first of the equations in (20). This leads to the equation for the *characteristic curves*

$$\frac{dy}{dx} = \frac{q(x, y, u)}{p(x, y, u)} \tag{23}$$

and to the *compatibility condition*

$$\frac{du}{dx} = \frac{f(x, y, u)}{p(x, y, u)}. \tag{24}$$

Although the equations (23) and (24) appear simpler than the equivalent ones in (20) and (21), in many cases the equations in terms of the parameter σ are easier to integrate.

The representation of the PDE in (16) as the set of ODEs in (20) and (21) or, equivalently, as the ODEs in (23) and (24) forms the basis of a method of solution for a first order PDE for $u(x, y)$ called the **method of characteristics**.

method of characteristics

The significance of the characteristic curves and the compatibility condition is most easily understood by considering the intersection of a representative characteristic curve and the Cauchy data curve Γ. Consider the characteristic curve C^* in Fig. 18.2 that intersects Γ at a point P corresponding to $s = s^*$ in the parametrization of Γ. As P is the point $(x_0(s^*), y_0(s^*))$, in the (x, y)-plane, the Cauchy condition at P is $u = u_0(s^*)$. The solution $u(x, y)$ of the PDE will then be determined along the characteristic curve C^* by integration of the compatibility condition (21) subject to the initial condition $u = u_0(s^*)$, with similar interpretations when (23) and (24) are used.

It can be seen from this argument that when the PDE in (16) is either linear or semilinear, the characteristic curves can be determined *independently* of the solution by integrating either (20) or (23), because in these two cases the solution u does not enter into the functions p and q. Consequently, in these two cases, solving the PDE in (16) reduces to the integration of the ODEs that determine the family of characteristic curves C, followed by the integration of the compatibility condition along the characteristic curves subject to an appropriate initial condition. Figure 18.2 illustrates the application of the method of characteristics to linear and semilinear PDEs written in the form

$$p(x, y)u_x + q(x, y)u_y = f(x, y, u), \tag{25}$$

where f depends linearly on u when (25) is linear, and nonlinearly on u when it is semilinear.

If the PDE is quasilinear, the solution u enters into the equations determining the characteristics, so when this occurs the integrations can only be performed analytically when the equations involved are simple. In general, when working with quasilinear first order PDEs, and also with linear and semilinear PDEs with complicated coefficients, the system of ODEs comprising the characteristic equations and the compatibility condition must be solved simultaneously using a numerical integration technique such as the Runge–Kutta method described in Chapter 19.

The **uniqueness** of the solution $u(x, y)$ in (25) follows directly from the way in which the method of characteristics produces the solution, and the fact that integration along a typical characteristic C^* of the compatibility condition (see Fig. 18.2) leads to a solution for $u(x, y)$ that depends *uniquely* on the initial condition $u = u_0(s^*)$ associated with the characteristic. The solution will cease to be unique if intersection of characteristics occurs at a point Q in the (x, y)-plane. This is because, in general, the value of u at Q determined by integration of the compatibility condition along each of the characteristics that meet there cannot be expected to be in agreement.

The restriction that must be placed on the initial curve Γ can be seen by considering Fig. 18.3. Provided Γ is nowhere tangent to a characteristic, as is the case for the characteristic C_P through point P, the solution along C_P will evolve according to

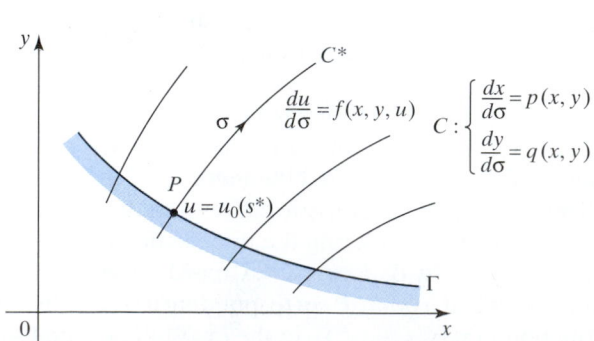

FIGURE 18.2 The solution of a linear or semilinear PDE by the method of characteristics.

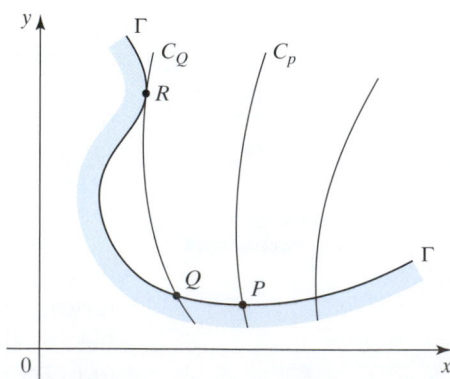

FIGURE 18.3 Tangency and nontangency of characteristic curves and the initial line Γ.

the solution of the compatibility condition subject to the initial condition $u = u_0(P)$. The situation is different, however, in the case of the characteristic curve C_Q through the point Q that becomes tangent to the Cauchy data curve Γ at point R.

In this case the Cauchy condition $u = u_0(R)$ specified at R where the Cauchy data curve Γ is tangent to C_Q cannot be expected to be in agreement with the solution obtained by integrating the compatibility condition along C_Q from Q to R subject to the initial condition $u = u_0(Q)$ at Q. This shows that when specifying a Cauchy problem for the PDE in (16) it is necessary that the initial curve Γ be nowhere tangent to a characteristic curve. As the characteristics can be determined independently of the solution u when the PDE is linear or semilinear, for such equations it is always possible to determine in advance that the nontangency condition is satisfied. If, however, the equation is quasilinear, then although the nontangency condition for Γ may be satisfied in neighborhood of Γ, this may not remain true as the solution evolves.

A special case of the Cauchy problem for the PDE in (16) arises when the Cauchy data curve Γ coincides with a characteristic curve of the equation. The determination of a solution for such a problem, when it exists, is called the **characteristic Cauchy problem**.

characteristic Cauchy problem

The following examples illustrate the application of the method of characteristics to linear, semilinear, and quasilinear first order PDEs, and also to a simple characteristic Cauchy problem. In general, equations (23) and (24) are the simplest to use when the Cauchy condition is prescribed on any straight line, and the parametric representation of the characteristic equations is only necessary when Cauchy data is prescribed on a curve. However, to illustrate the parametric approach, the second example makes use of equations (20) and (21) for the case where the Cauchy data is prescribed on a straight line through the origin.

Once a solution has been found it must always be checked to see that it satisfies both the prescribed Cauchy condition and the original PDE. The solution should also be examined to identify any restrictions that need to be placed on it in order to ensure that it remains real and finite.

EXAMPLE 18.1   Solve the Cauchy problem

$$u_x + 3u_y = 2u, \quad \text{given that } u(x, 0) = e^x.$$

Solution   This is a linear equation, and as the Cauchy data curve is the x-axis, we will use the characteristic equations given in (23) and (24).

From (23) the characteristic curves of the PDE are determined by $dy/dx = 3$, so integration shows their equation to be $y = 3x + \xi$, where ξ is a constant of integration that corresponds to the point of intersection $(0, \xi)$ of the characteristic and the x-axis.

The compatibility condition is $du/dx = 2u$, so integration shows that

$$\ln u = 2x + f(\xi),$$

where $f(\xi)$ represents the arbitrary constant introduced as a result of the integration. This constant depends on the characteristic involved, but as a characteristic depends on ξ because of its point of intersection $(\xi, 0)$ with the x-axis, it is necessary to introduce the constant (on a particular characteristic) as $f(\xi)$, where f is an arbitrary function.

Substituting $\xi = y - 3x$ into the solution for u gives

$$u(x, y) = \exp\{2x + f(y - 3x)\}.$$

To find the form of the arbitrary function f, we now make use of the Cauchy condition, which in this case is $u(x, 0) = e^x$. Setting $y = 0$ in the expression for $u(x, y)$ and imposing the Cauchy condition gives

$$e^x = \exp\{2x + f(-3x)\},$$

and after taking logarithms this becomes

$$-x = f(-3x), \text{ that is equivalent to } f(x) = \frac{1}{3}x.$$

Replacing x by $y - 3x$ in $f(x)$, we have $f(y - 3x) = \frac{1}{3}y - x$, so substituting $f(y - 3x)$ into the expression for $u(x, y)$ gives

$$u(x, y) = \exp\left\{x + \frac{1}{3}y\right\}.$$

This function satisfies the Cauchy condition and differentiation confirms that it is a solution of the original PDE, so it is a classical solution of the equation. Inspection shows the solution to be valid throughout the entire (x, y)-plane. A solution such as this that is valid without restriction on its independent variables is called a **global** solution. ∎

EXAMPLE 18.2

Cauchy problems for linear, semilinear, and quasilinear PDEs

Solve the Cauchy problem

$3u_x + 2u_y = x$, given that $u(x, y) = 1$ on the line Γ with the equation $ax = by$.

Solution   This is a linear equation, with the Cauchy data curve Γ a straight line through the origin, so to illustrate the parametric approach we will use the characteristic equations given in (20) and (21).

We parametrize Γ by setting $x = bs$, $y = as$, where $-\infty < s < \infty$. The characteristic curves (lines in this case) are determined by (20), which when integrated become

$$x = 3\sigma + k_1, \quad y = 2\sigma + k_2.$$

When $\sigma = 0$ we know that x and y lie on Γ, but then $x = bs$ and $y = as$, so it follows that $k_1 = bs$, $k_2 = as$, showing that

$$x = 3\sigma + bs, \quad y = 2\sigma + as.$$

Solving these expressions for s and σ gives

$$s = \frac{3y - 2x}{3a - 2b}, \quad \sigma = \frac{ax - by}{3a - 2b}, \quad \text{for } 3a \neq 2b.$$

The compatibility equation (21) becomes

$$\frac{du}{d\sigma} = x, \text{ but } x = 3\sigma + bs,$$

so after integration

$$u(s, \sigma) = \frac{3}{2}\sigma^2 + b\sigma s + f(s),$$

where $f(s)$ represents the usual arbitrary additive integration constant. As the characteristic depends on the parameter s, the integration constant $f(s)$ is shown as a function of s. The Cauchy condition $u(x, y) = 1$ is imposed on Γ, corresponding to $\sigma = 0$ in the preceding expression, so setting $\sigma = 0$ and replacing $u(s, 0)$ by 1 we find that $1 = f(s)$ for all s, and so in terms of s and σ the solution is seen to be given by

$$u(s, \sigma) = \frac{3}{2}\sigma^2 + b\sigma s + 1.$$

Replacing s and σ by their expressions in terms of x and y, we arrive at the explicit solution in terms of x and y

$$u(x, y) = \frac{3}{2}\left(\frac{ax - by}{3a - 2b}\right)^2 + b\left(\frac{ax - by}{3a - 2b}\right)\left(\frac{3y - 2x}{3a - 2b}\right) + 1, \quad \text{for } 3a \neq 2b.$$

This function satisfies the Cauchy condition on Γ, and differentiation confirms that it satisfies the original PDE, so it is a classical solution of the equation. Inspection shows the solution to be valid in the entire (x, y)-plane provided $3a \neq 2b$, so it is a global solution if this condition is satisfied.

When $3a = 2b$, the preceding solution fails because the Cauchy data line Γ with the equation $2x - 3y = 0$ coincides with the characteristic through the origin, causing the problem to become a *characteristic Cauchy problem*.

To examine the solution in this case, we must allow for the fact that although both Γ and the characteristic through the origin coincide, they are each parametrized differently. From the equations defining the characteristics we have $dx/d\sigma = 3$ on the Cauchy data line $x = bs$, so $dx/ds = b$.

The compatibility condition is

$$\frac{du}{d\sigma} = x, \quad \text{so in terms of } s \text{ this can be written } \frac{du}{d\sigma} = bs.$$

To express the derivative on the left of this last result in terms of s we use the chain rule

$$\frac{du}{ds} = \frac{du}{d\sigma}\frac{d\sigma}{ds} = \frac{d\sigma}{dx}\frac{dx}{ds}\frac{du}{d\sigma}, \quad \text{and so} \quad \frac{du}{ds} = \frac{b}{3}\frac{du}{d\sigma}.$$

Combining this result with $du/d\sigma = bs$ gives

$$\frac{du}{ds} = \frac{b^2}{3}s, \quad \text{and after integration this becomes } u = \frac{b^2}{6}s^2 + c, (c = \text{constant}).$$

Substituting $x = bs$ into this result, we arrive at the solution

$$u(x, y) = \frac{1}{6}x^2 + c.$$

This expression for $u(x, y)$ is a degenerate solution of the original PDE along the characteristic through the origin that coincides with the Cauchy data line. However, this is *not* a solution of the characteristic Cauchy problem, because it does not satisfy the Cauchy condition $u(x, y) = 1$ along the line Γ.

This shows that this characteristic Cauchy problem with the stated Cauchy condition along Γ has *no* solution. A solution for the characteristic Cauchy problem could only exist if the Cauchy condition on Γ is changed to $u(x, y) = \frac{1}{6}x^2 + c$.

This solution is not the most general one, because the fact that Γ has the equation $3y - 2x = 0$ allows us to add to the preceding solution any arbitrary differentiable function $f(3y - 2x)$ that is a solution of the *homogeneous* form of the PDE $3u_x + 2u_y = 0$, since the result will still be a solution. This shows that the most general solution of this characteristic Cauchy problem is

$$u(x, y) = \frac{1}{6}x^2 + f(3y - 2x),$$

provided this expression also satisfies the Cauchy condition on Γ. In this result the constant c that appeared earlier has been absorbed into the arbitrary function f.

This example demonstrates the fact that, in general, the characteristic Cauchy problem has no solution, but when it does the solution is not unique, because it contains an arbitrary function. ∎

EXAMPLE 18.3

Solve the Cauchy problem

$$u_x + u_y = e^u, \text{ given that } u(0, y) = y.$$

Solution   This is a semilinear equation, but this time the Cauchy condition is specified on the y-axis so it will be simplest to use the nonparametric form of the characteristic equations.

The characteristics are determined by

$$\frac{dy}{dx} = 1, \text{ and integration gives } y = x + \xi,$$

where ξ is the point $(0, \xi)$ on the y-axis through which the characteristic passes. The compatibility condition is

$$\frac{du}{dx} = e^u, \text{ and after integration this becomes } -e^{-u} + f(\xi) = x.$$

Here f, an arbitrary function of its argument ξ that identifies the characteristic as the one passing through the point $(0, \xi)$, again represents the arbitrary constant that enters as a result of the integration. Substituting $\xi = y - x$ into this last result gives

$$-e^{-u} + f(y - x) = x, \quad \text{or } u(x, y) = 1/\ln\{f(y - x) - x\}.$$

To find f we must now make use of the Cauchy condition $u(0, y) = y$. Setting $x = 0$, and replacing u by y, the preceding expression becomes

$$-e^y + f(y) = 0, \quad \text{so } f(y) = e^y,$$

from which it follows that $f(y - x) = e^{x-y}$. Substituting for $f(y - x)$ in the expression for $u(x, y)$, we find that

$$u(x, y) = \ln\left(\frac{1}{e^{x-y} - x}\right) \quad \text{for } e^{x-y} > x.$$

This expression satisfies the Cauchy condition specified, and differentiation confirms that it is a solution of the original PDE, so it is a classical solution. The restriction $e^{x-y} > x$ that ensures $u(x, y)$ is real shows that the solution is not defined over all of the (x, y)-plane, and so it is not a global solution. ∎

EXAMPLE 18.4   Solve the Cauchy problem

$$u_x + uu_y + u = 0, \text{ given that } u(0, y) = 1 + y.$$

Solution   This equation is quasilinear because of the presence of the term uu_y, and again the Cauchy condition is specified on an axis so the nonparametric form of the characteristic equations will be used.

The characteristic curves follow by integration of the equation

$$\frac{dy}{dx} = u,$$

on which the compatibility condition that determines u is

$$\frac{du}{dx} = -u.$$

Let the solution along the characteristic through the point $(0, \xi)$ on the y-axis be $u = g(\xi)$. Then integration of the compatibility condition along the characteristic with respect to x gives

$$\ln u = -x + \ln g(\xi), \quad \text{so} \quad u = g(\xi)e^{-x}.$$

It follows from the Cauchy condition that $u = 1 + \xi$ at the point $(0, \xi)$, so setting $x = 0$ and replacing u by $1 + \xi$ in this last result, we find that

$$g(\xi) = 1 + \xi,$$

and so

$$u = (1 + \xi)e^{-x}.$$

The equation determining the characteristic curves now follows if we use this last result in the equation $dy/dx = u$, to obtain

$$\frac{dy}{dx} = (1 + \xi)e^{-x}.$$

Integration of this result using the fact that the characteristic passes through the point $(0, \xi)$ leads to the result

$$y = \xi + (1 + \xi) \int_0^x e^{-\eta} d\eta,$$

so

$$y = (1 - e^{-x}) + \xi(2 - e^{-x}).$$

When ξ is eliminated the solution becomes

$$u = \left(\frac{1 + y}{2 - e^{-x}} \right) e^{-x}, \text{ provided } x \neq -\ln 2.$$

This function satisfies the Cauchy condition, and differentiation shows that it satisfies the original PDE, so it is a classical solution. The solution is not defined everywhere because it becomes infinite when $x = -\ln 2$.   ∎

Summary

This section introduced the method of characteristics for first order PDEs involving a scalar function of two independent variables. The method was seen to involve replacing the single PDE by two coupled ordinary differential equations (ODEs), one of which determined the family of characteristic curves, while the other determined the variation of the solution along the characteristic curves. The method was seen to apply to linear, semilinear, and quasilinear equations, and in the linear and semilinear cases the characteristic curves could be determined independently of the solution. However, in the quasilinear case, the equations for the characteristics and for the variation of the solution along the characteristics had to be solved simultaneously.

EXERCISES 18.2

In Exercises 1 through 18 solve the given Cauchy problem. Verify that the result obtained is a solution, and comment on any restrictions that need to be placed on it.

1. $u_x + 2u_y = 2$, $u(x, 0) = x$.
2. $3u_x + 2u_y = x$, $u(x, 0) = 1$.
3. $4u_x + 3u_y = 1$, $u(x, y) = 3$ on $y = x$.
4. $yu_x + 3u_y = y(1 + u)$, $u(0, y) = y^2$.
5. $2u_x + u_y = \cos x$, $u(x, 0) = \dfrac{3}{2}\sin x$.
6. $u_x + u_y = u - 1$, $u(x, 0) = 2x$.
7. $u_x + 2u_y = 2x$, $u(x, y) = 2$ on $y = 3x + 1$.
8. $u_x + xu_u = u + 2$, $u(0, y) = 3y$.

9. $u_x + 4xu_y = 3 + 2x\sec^2 x^2$, $u(x, 0) = 3x$.
10. $yu_x + u_y = y(u + 4)$, $u(x, 0) = e^{2x}$.
11. $u_x + u_y = u^2$, $u(0, y) = y$.
12. $u_x + 2u_y = (1 + 2x)e^{-u}$, $u(x, y) = 1$ on $y = x$.
13. $u_x + uu_y + u = 0$, $u(0, y) = \sin y$.
14. Obtain the solution in Example 18.2 without parameterizing the line $ax = by$.
15. $u_x + 2uu_y + 3u = 0$, $u(0, y) = 4y$.
16. $u_x + uu_y + u = 0$, $u(0, y) = e^y$.
17. $u_x + uu_y + u = 0$, $u(0, y) = 3 + 2y$.
18. $u_x + 2uu_y + 3u = 0$, $u(0, y) = 4y$.

18.3   Wave Propagation and First Order PDEs

A first order PDE for the unknown function $u(x, t)$ of two independent variables x and t of the form

$$u_t + p(x, t, u)u_x + q(x, t, u) = 0, \tag{26}$$

wave propagation and hyperbolic PDEs

where x has the dimensions of length and t is the time, can be considered to describe **wave propagation**. Here the term *wave* is used to describe an identifiable disturbance such as a sound or water wave that propagates at a finite speed through space as time increases. The PDE in (26) is called a first order **hyperbolic equation** because, like the second order wave equation to be considered later, it describes wave propagation. A typical equation of this type characterizing a physical problem was derived in Section 18.1, where a linear first order PDE was shown to model the transient heat flow from a pipe transporting a hot fluid.

To understand the different types of wave propagation that can be described by hyperbolic equations such as (26), it will be necessary to examine some typical cases. The method of solution that will be used is the method of characteristics described in Section 18.2. However, this time the variable x will be replaced by t, as it represents the time, and the variable y will be replaced by x, which represents a length. A Cauchy condition for (26) specified at some fixed time, typically $t = 0$, is an **initial condition**, and the line on which the initial condition is specified is then the **initial line**.

The Traveling Wave Equation

The simplest possible form of wave propagation described by the PDE in (26) occurs when $p(x, t, u) = c$ and $q(x, t, u) = 0$, causing the equation to simplify to

traveling wave equation or the advection equation

$$u_t + cu_x = 0 \quad (c = \text{constant}). \tag{27}$$

This is a linear homogeneous constant coefficient first order PDE that is often known as the **advection equation**.

The classical general solution of (27) can be found by inspection, but for what is to follow it will be more useful if it is obtained by the method of characteristics. Using the characteristic equations (23) and (24) with the new independent variables x and t, we find that the characteristic curves are determined by integrating the equation

$$\frac{dx}{dt} = c \quad \text{to obtain} \quad x = ct + \xi,$$

where the characteristic curve passes through the point $(\xi, 0)$ on the x-axis (the initial line). As $c = $ constant, the characteristics are all parallel straight lines, and the equation of the characteristic through the point $(\xi, 0)$ has the equation

$$x - ct = \xi. \tag{28}$$

The solution $u(x, t)$ along the characteristic curve (line) through the point $(\xi, 0)$ follows by integrating the compatibility equation

$$\frac{du}{dt} = 0 \quad \text{to obtain} \quad u(x, t) = f(\xi).$$

As $u(x, t)$ is constant on a characteristic, the constant value must be equal to the value assigned by the initial condition at the point where the characteristic intersects the x-axis. It follows from this that along the characteristic $x - ct = \xi$ that passes through the point $(\xi, 0)$ we must have $u(x, t) = f(\xi)$. Substituting for ξ shows that the general solution of (27) is

$$u(x, t) = f(x - ct). \tag{29}$$

The derivative $dx/dt = c$ has the dimensions of a speed, so (29) shows that the **profile** of the initial disturbance determined by the function $f(x)$ at the time $t = 0$ is propagated with speed c, without change of shape or scale (size), in the positive x-direction when $c > 0$, and in the negative x-direction when $c < 0$. A wave of this type is called a **traveling wave**, and sometimes a **wave of constant form**. Figure 18.4 shows a typical traveling wave with an initial wave profile in the form of a symmetrical pulse and a propagation speed $c = 2$. The plot illustrates the steady propagation to the right of the initial profile in such a way that at a time $t = t_1$ each point has moved to the right through a distance $2t_1$.

wave profile and a traveling wave

A Typical Linear Constant Coefficient Nonhomogeneous Equation

Let us consider the initial value problem

$$u_t + 3u_x - u = kx, \quad \text{with} \quad u(x, 0) = \sin x \quad (k = \text{constant}).$$

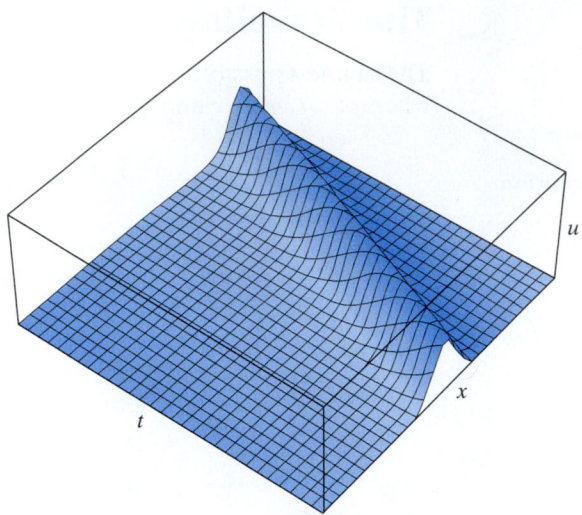

FIGURE 18.4 A traveling wave moving in the positive
x-direction with $c = 2$.

**traveling wave
problems involving
linear, semilinear,
and quasilinear PDEs**

The characteristics determined by integrating $dx/dt = 3$ are $x = 3t + \xi$, where
the characteristic intersects the initial line at $(\xi, 0)$. The compatibility condition
is $du/dt = u + kx$, but $x = 3t + \xi$ along the characteristic through $(\xi, 0)$, so along
this characteristic u is determined by the solution of the ODE

$$\frac{du}{dt} = u + 3kt + k\xi.$$

Solving this linear first order ODE shows that

$$u(x, t) = e^t f(\xi) - k(3t + 3 + \xi),$$

where $f(\xi)$ with f an arbitrary function represents the arbitrary additive integration
constant introduced by the integration.

As $\xi = x - 3t$ this solution becomes

$$u(x, t) = e^t f(x - 3t) - k(3 + x).$$

To determine the form of the function f, we now make use of the initial condition $u(x, 0) = \sin x$. Setting $t = 0$ in the expression for $u(x, t)$ and using the initial
condition we have

$$\sin x = f(x) - k(3 + x),$$

and so

$$f(x) = \sin x + k(3 + x).$$

Finally, replacing x in $f(x)$ by $x - 3t$ and substituting the result in $u(x, t)$ we arrive
at the result

$$u(x, t) = e^t \{\sin(x - 3t) + k(3 + x - 3t)\} - k(3 + x).$$

This expression satisfies the initial condition and the PDE, so it is the required
classical solution. Although the speed of propagation of the wave is constant, because $dx/dt = 3$, the wave shape changes from the initial sinusoid as it propagates.

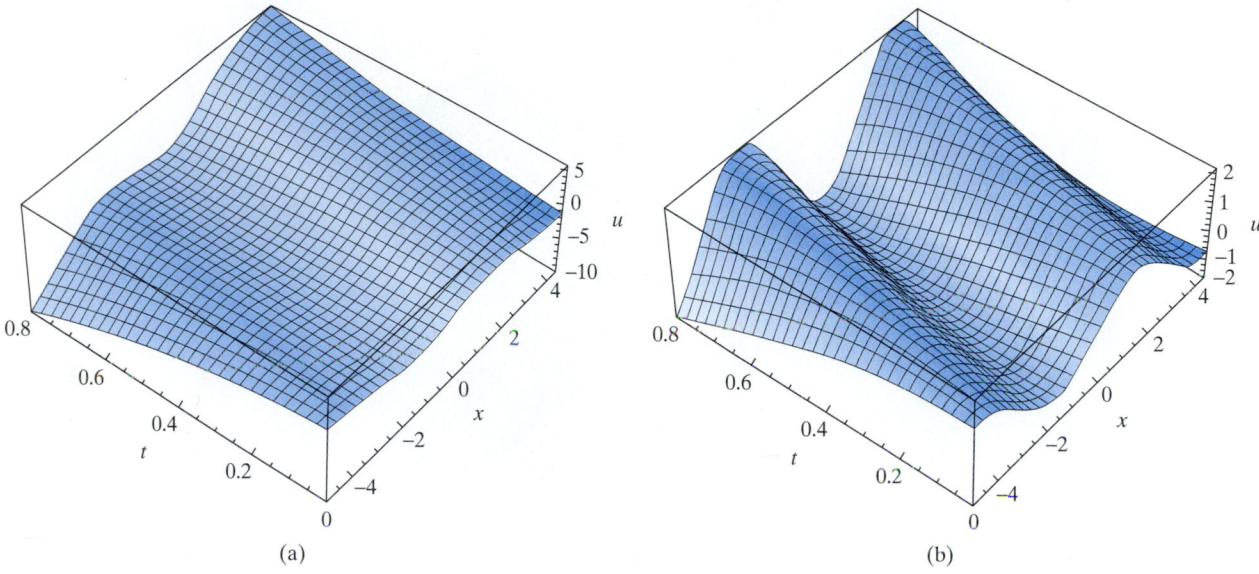

FIGURE 18.5   (a) The solution when $k = 1$; (b) the solution when $k = 0$.

Only when $k = 0$ is the shape of the wave preserved, though *not* its scale, because of the presence of the multiplicative scale factor e^t. Figure 18.5a shows a plot of the solution when $k = 1$ and a plot when $k = 0$ is shown in Fig. 18.5b, in each case for $-5 \le x \le 5$ and $0 \le t \le 0.8$. ∎

A Typical Linear Variable Coefficient Nonhomogeneous Equation

The following PDE illustrates the wave propagation properties of a typical linear variable coefficient nonhomogeneous equation. Consider the initial value problem

$$u_t + xu_x + u = 1, \quad \text{with } u(x, 0) = \tanh x.$$

The characteristic curves are determined by integrating the equation

$$\frac{dx}{dt} = x \quad \text{to obtain } x = \xi e^t,$$

where the characteristic curve passes through the point $(\xi, 0)$ on the initial line $t = 0$.

The compatibility condition is

$$\frac{du}{dt} = 1 - u,$$

so when this is integrated along a characteristic curve we find that

$$u = 1 + e^{-t} f(\xi),$$

where f is an arbitrary function of ξ. Substituting for ξ we have

$$u = 1 + e^{-t} f(xe^{-t}).$$

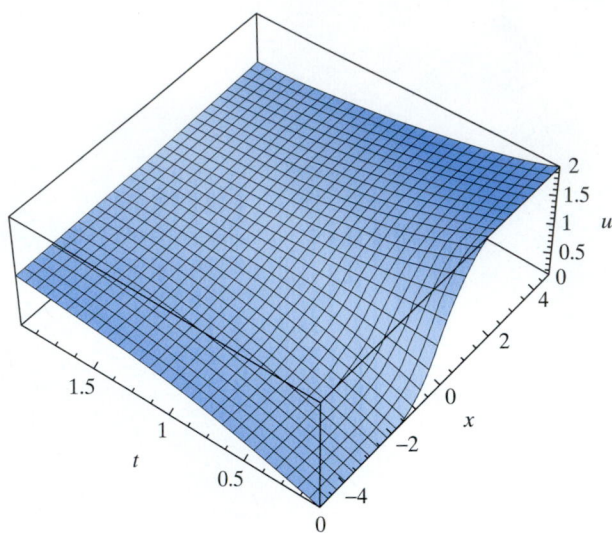

FIGURE 18.6 Decay of the initial condition $u(x, 0) = \tanh x$ to the constant value $u = 1$.

The arbitrary function f must be determined by using the initial condition $u(x, 0) = \tanh x$. Setting $t = 0$ in the preceding expression for u and imposing the initial condition gives

$$\tanh x = 1 + f(x), \quad \text{so that} \quad f(x) = \tanh x - 1.$$

Replacing x in $f(x)$ by xe^{-t} and using the result in the expression for u gives

$$u(x, t) = 1 + e^{-t} \tanh(xe^{-t}).$$

Wave propagation described by this PDE is not at a constant speed, because $dx/dt = x$, nor is its initial shape preserved. Examination of the solution shows that the wave profile changes shape as it propagates, and that after a suitable period of time the profile decays to the constant solution $u(x, t) = 1$, as illustrated in Fig. 18.6. ∎

The last examples show that, in general, wave propagation described by first order linear equations that are *not* of the form of (27) describe wave propagation that may or may not preserve the shape of the initial wave profile, but will not preserve the scale as time evolves, so their solutions are not traveling waves.

A Typical Semilinear Equation

The properties of semilinear PDEs can be illustrated by considering the initial value problem

$$u_t + u_x = u^2, \quad \text{with } u(x, 0) = \sin x.$$

The characteristic passing through the point $(\xi, 0)$ in the (x, t)-plane obtained by integrating $dx/dt = 1$ is $x = t + \xi$, and the compatibility condition along this characteristic is $du/dt = u^2$. Integrating the compatibility condition along the characteristic gives

$$-\frac{1}{u} = t + f(\xi),$$

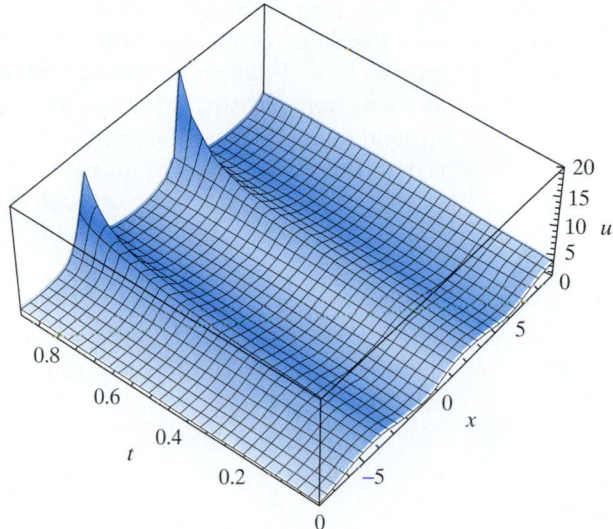

FIGURE 18.7   The evolution of infinite values of $u(x, t)$ as $t \to 1$.

where f is an arbitrary function of ξ. Substituting $\xi = x - t$ into this result, we have

$$u(x, t) = \frac{-1}{t + f(x - t)}.$$

As $u(x, 0) = \sin x$, setting $t = 0$ in $u(x, t)$ and using the initial condition shows that $f(x) = -1/\sin x$, from which it follows that $f(x - t) = -1/\sin(x - t)$. Substituting for $f(x - t)$ in the expression for $u(x, t)$ then gives

$$u(x, t) = \frac{\sin(x - t)}{1 - t \sin(x - t)}.$$

This function satisfies both the initial condition and the PDE, so it is the required classical solution.

Examination of this solution shows that it is only defined in the strip $0 < t < 1$, because only in this strip is the denominator of $u(x, t)$ nonzero. So, unlike linear equations, this semilinear equation has a classical solution for only a finite time, after which for some x the solution becomes infinite. The plot of $u(x, t)$ in Fig. 18.7 shows the development of infinite values of the solution as $t \to 1$.   ■

A Typical Quasilinear Equation

The general properties of solutions of the first order quasilinear PDE

$$u_t + p(x, t, u)u_x + q(x, t, u) = 0 \tag{30}$$

can all be illustrated by considering the typical initial value problem

$$u_t + f(u)u_x = 0, \quad \text{with } u(x, 0) = g(x), \tag{31}$$

where f and g are arbitrary functions of their arguments.

The characteristics of (31) are determined by integrating $dx/dt = f(u)$, while the compatibility condition determining the solution u that is valid along a characteristic is seen to be $du/dt = 0$.

The compatibility condition shows that $u = $ constant along a characteristic, with the value of the constant determined by the initial condition at the point of intersection of the characteristic and the initial line. Furthermore, as $u = $ constant along a characteristic, it follows from $dx/dt = f(u)$ that all characteristics will be straight lines, and that the propagation speed $f(u)$ associated with a characteristic is determined by the constant value of u that is transported along it.

Thus, the characteristic through the point $(\xi, 0)$ on the initial line (the x-axis) where the initial condition is $u = g(\xi)$ will have the equation

$$x = \xi + f(g(\xi))t, \quad \text{and along this characteristic } u = g(\xi). \tag{32}$$

Elimination of ξ between these equations, where it appears as a parameter, shows that the solution u of the initial value problem in (31) is determined by the *implicit* relationship

$$u = g\{x - f(u)t\}. \tag{33}$$

To examine the nature of solutions of (31) we must consider the behavior of the characteristic curves (lines in this case), and when doing so we follow the usual convention that the x-axis is taken to be horizontal and the t-axis vertical. Consequently, when drawn in the (x, t)-plane, the gradient of a characteristic curve is $dt/dx = 1/f(u)$.

Let us now suppose that the function $f(u)$ in (31) is a steadily *increasing* function of u. Then the characteristics radiating out from points on the initial line will all fan out, as illustrated in Fig. 18.8a. This shows that the initial value problem (31) will have a unique solution throughout the upper half of the (x, t)-plane, because the solution at any point will be the value of u associated with the characteristic that passes through the point, and the characteristics never intersect. However, if $f(u)$ is a steadily *decreasing* function of u, the characteristics radiating out from points on the initial line will converge, leading to the intersection of characteristics as shown in Fig. 18.8b. When this happens the nature of the solution changes dramatically, because different characteristics transport *different* constant values of u into the upper half of the (x, t)-plane, so the intersection of characteristics corresponds

how solutions of quasilinear PDEs can break down

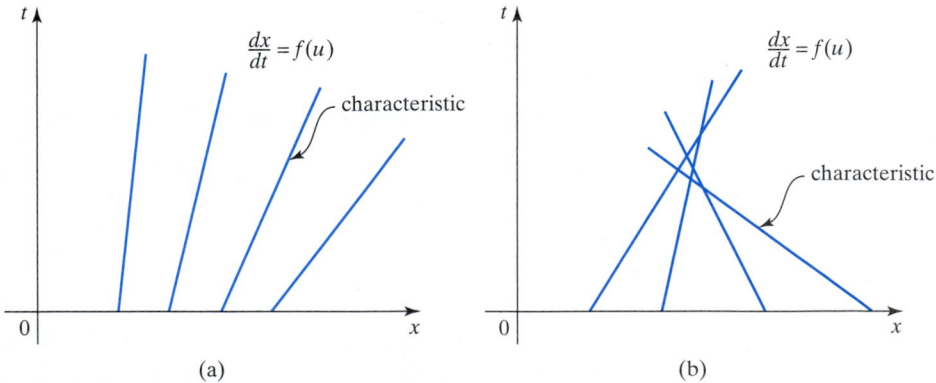

FIGURE 18.8   The influence of $f(u)$ on the behavior of characteristics. (a) $f(u)$ an increasing function of u; (b) $f(u)$ a decreasing function of u.

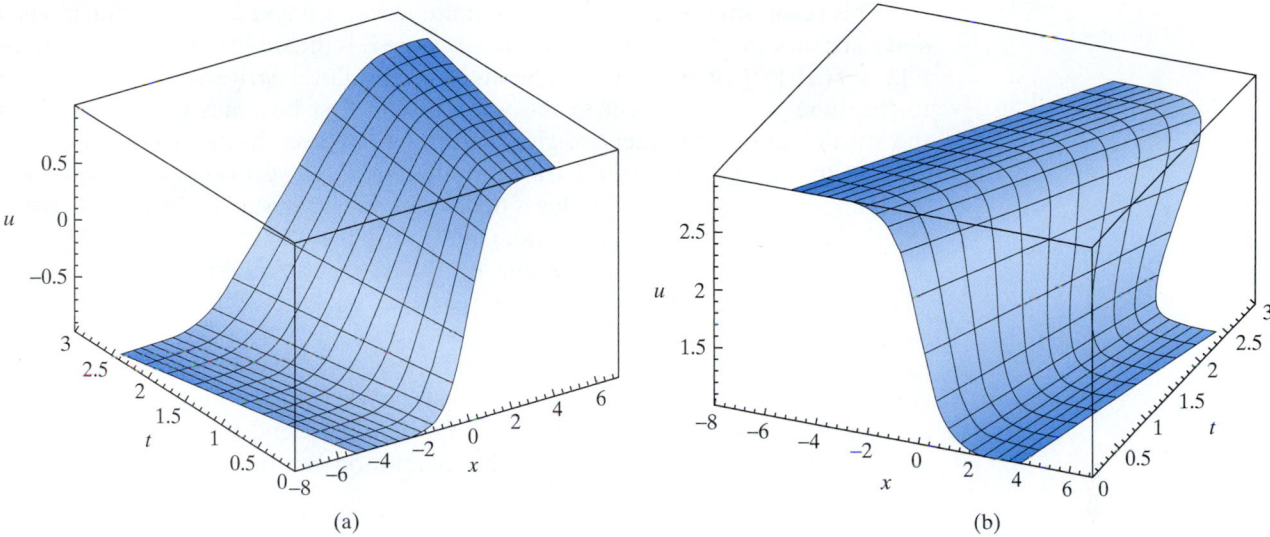

FIGURE 18.9 (a) $f(u)$ an increasing function leading to smoothing because the top of the wave then moves slower than the bottom; (b) $f(u)$ a decreasing function leading to steepening due to the top of the wave moving faster than the bottom.

to the nonuniqueness of the solution of the initial value problem (31) wherever intersection of characteristics occurs. This conclusion is implied by the implicit form of the solution found in (33), because it is known from analysis that a function determined by an implicit relationship need not be unique.

The qualitative properties of waves propagated by a PDE of the form

$$u_t + f(u)u_x = 0$$

can be deduced from the equation $dx/dt = f(u)$ determining the characteristics along which constant initial values of u are transported. To see this, suppose $f(u)$ is an increasing function of u, and consider the wave profile $u(x, t)$. Then if P and Q are adjacent points on a wave profile, with Q to the right of P and $u(Q) > u(P)$, it follows that point Q will propagate faster than point P, causing the wave to become *smoother* as it evolves, as illustrated in Fig. 18.9a. When the converse is true, and $f(u)$ is a decreasing function of u, point P will propagate faster than point Q, causing the wavefront to *steepen*, and eventually this will cause the solution to become nonunique because of the intersection of characteristics, as illustrated in Fig. 18.9b.

Partial differentiation of (33) with respect to x gives

$$\frac{\partial u}{\partial x} = g'\{x - f(u)t\}\left\{1 - f'(u)\frac{\partial u}{\partial x}t\right\},$$

so

$$\frac{\partial u}{\partial x} = \frac{g'\{x - f(u)t\}}{1 + g'\{x - f(u)t\}f'(u)t}. \tag{34}$$

This result shows u_x can become infinite at finite time $t = t_c$ if the functions f and g are such that $g'\{x - f(u)t\} f'(u) < 0$ where t_c is the *smallest* time for which $1 + g'\{x - f(u)t\} f'(u)t = 0$. The development of an infinite derivative u_x corresponds to the time when a tangent to the wave profile first becomes vertical, marking the start of the nonuniqueness. This feature can be seen in Fig. 18.9b, where the tangent to the mid-point of the wave profile tends to a vertical position as $t \to 1$. An immediate consequence of this is that when characteristics converge, a classical solution can only exist for a finite time in the strip $0 < t < t_c$ in the (x, t)-plane.

Solutions of initial value problems for the more general first order quasilinear PDE in (30) exhibit the same general properties as those of (31). As typical functions $p(x, t, u)$ in (30) and $f(u)$ in (31) will have domains where they are increasing functions of u and others where they are decreasing functions, in general classical solutions of first order quasilinear PDEs can only exist for a finite time. The next section examines how the concept of a solution can be extended to allow the solution of some PDEs to be generalized so that a solution can be extended beyond the time t_c.

EXAMPLE 18.5

Solve the initial value problem

$$u_t + (1+u)u_x + u = 0, \quad \text{given that} \quad u(x, 0) = 1 + x.$$

Solution   This PDE is quasilinear because of the product term uu_x. The characteristic curves are obtained by integrating $dx/dt = 1 + u$, and the compatibility condition determining u along a characteristic is $du/dt = -u$.

Let the solution along the characteristic through the point $(\xi, 0)$ on the initial line be $u = g(\xi)$, then integration of the compatibility condition gives

$$\ln u = -t + \ln g(\xi), \quad \text{and so} \quad u = g(\xi)e^{-t}.$$

From this result and the initial condition at $(\xi, 0)$ we have $g(\xi) = 1 + \xi$, so the solution can be written

$$u = (1 + \xi)e^{-t}.$$

Substitution of this result into the equation determining the characteristic curves gives

$$\frac{dx}{dt} = 1 + (1 + \xi)e^{-t},$$

and so

$$\int_\xi^x ds = \int_0^t [1 + (1 + \xi)e^{-\tau}]d\tau,$$

where s and τ are dummy variables. After integration this becomes

$$x = \xi(2 - e^{-t}) + t + 1 - e^{-t},$$

from which it follows that

$$1 + \xi = \frac{1 + x - t}{2 - e^{-t}}.$$

Finally, using this result to eliminate ξ from the expression for u, we find that

$$u(x, t) = \left(\frac{1 + x - t}{2 - e^{-t}}\right)e^{-t}.$$

This function satisfies the initial condition and the original PDE, so it is the required classical solution. As the denominator does not vanish for $t > 0$, this is the classical solution for the initial value problem for $t > 0$. ∎

More information on the method of characteristics, including applications, can be found in references [7.1], [7.4], [7.6], [7.8], [7.11], [7.12], and [7.20].

Summary

The concept of wave propagation was introduced and related to the method of characteristics. Each characteristic curve was seen to transport the initial condition appropriate to the characteristic according to the ODE determining the evolution of the solution along the curve. It was shown how homogeneous linear first order PDEs can have traveling wave solutions where the shape of the wave remains unchanged as it propagates with time. However, the introduction of nonlinearity was seen to make traveling wave solutions impossible, and in certain cases to lead to the solution becoming nonunique after a finite time.

EXERCISES 18.3

Solve the following initial value problems.

1. $2u_t + 4u_x = 3u$, given that $u(x, 0) = \sin 2x$.
2. $u_t - 2u_x = x$, given that $u(x, 0) = x^2$.
3. $u_t - 3u_x = 2u + 1$, given that $u(x, 0) = \frac{1}{2}\cos x$.
4. $u_t - u_x = u + \sin x$, given that $u(x, 0) = 1$.
5. $u_t - 4u_x = 3x$, given that $u(x, 0) = e^x$.
6. $u_t + 2u_x = 2u + x$, given that $u(x, 0) = x$.
7. $u_t - 3xu_x + 2u = x$, given that $u(x, 0) = x$.
8. $u_t + 3xu_x - 2u = 4$, given that $u(x, 0) = x$.
9. $u_t - 3xu_x + 2u = x$, given that $u(x, 0) = 3x$.
10. $(1 + t^2)u_t + u_x = (1 + t^2)(u - 1)$, given that $u(x, 0) = \sinh x$.

11. $3u_t - 9xu_x + 6u = x$, given that $u(x, 0) = x$.
12. $u_t + e^{2t}u_x = u + x$, given that $u(x, 0) = 1$.
13. $u_t + u_x = 2u^2$, given that $u(x, 0) = \cos x$.
14. $u_t + 4xu_x = u^2$, given that $u(x, 0) = \sinh x$.
15. $u_t + 2uu_x - u = 0$, given that $u(x, 0) = -2x$.
16. $u_t + 2uu_x + 2u = 0$, given that $u(x, 0) = 3x$.
17. $u_t - 3uu_x + 4u = 0$, given that $u(x, 0) = 1 + x$.
18. $u_t + tuu_x - u = 0$, given that $u(x, 0) = 2x$.
19. $u_t + (1 + t)uu_x - \left(\dfrac{1}{1+t}\right)u = 0$, given that $u(x, 0) = 3x - 1$.
20. $u_t + uu_x - \left(\dfrac{1}{1+t}\right)u = 0$, given that $u(x, 0) = 1 - x$.

18.4   Generalizing Solutions: Conservation Laws and Shocks

In many physical situations a commonly occurring feature of wave propagation is the evolution of smooth solutions of PDEs to a point where their nature changes, and jump discontinuities occur and propagate in a manner quite different from the smooth solution. This happens in fluid and solid mechanics, in magneto-hydrodynamics, and elsewhere when the governing PDEs are quasilinear and describe wave propagation.

generalizing solutions

The propagation of discontinuities in otherwise continuous and differentiable solutions represents an extension of the concept of a solution that has been used thus far. This is because although the solution on either side of the discontinuity satisfies the original PDE, the solution is not a classical solution since it is not differentiable at a jump discontinuity. In high-speed gas dynamics, and in elastic materials that behave nonlinearly, discontinuous solutions of this type are called **shock waves**.

Jump discontinuities can also develop and propagate in water, as can be seen in estuaries subject to suitable tidal conditions, where a mass of water across which there is a large and abrupt change of level can propagate in a stable manner for a considerable distance. A steplike disturbance of this type in water is called a tidal **bore**, and when the effects of viscosity and turbulence are neglected the situation can be approximated mathematically by a jump discontinuity in the water height.

Behavior of this type was suggested in the last section where it was seen that classical solutions of initial value problems for first order quasilinear equations may only exist for a finite time until the solution becomes nondifferentiable. This suggests that a possible generalization of a classical solution $u(x, t)$ could involve a function that is differentiable and satisfies a PDE on either side of a moving point $x = \sigma(t)$ inside a fixed interval $x_1 \le x \le x_2$, but that across the moving point the solution is discontinuous and experiences a finite jump. Let us see how such a generalization of a solution can be obtained, and in the process examine some of its properties and how it depends fundamentally on the notion of a conservation law.

The fundamental idea that will be used to extend the notion of a classical solution is most easily understood by considering the simple PDE

$$u_t + uu_x = 0, \tag{35}$$

which is a special case of (31) with $f(u) = u$. As $uu_x = \frac{\partial}{\partial x}(\frac{1}{2}u^2)$, the PDE in (35) can be written

$$u_t + \left(\frac{1}{2}u^2\right)_x = 0. \tag{36}$$

To allow for a discontinuity we will use an integral representation of (36), because although the derivative of $u(x, t)$ is not defined at a point where the function is discontinuous, its integral over an interval $x_1 \le x \le x_2$ containing the discontinuity is well defined. Let us now attempt to generalize the concept of a solution of (36) to allow for a situation where $u(x, t)$ satisfies the PDE to the left and right of a moving interior point $x = \sigma(t)$ in the interval $x_1 \le x \le x_2$, but across which it is discontinuous, with $u = u_L$ at the point $x = \sigma(t)_L$ to the immediate left of $x = \sigma(t)$ and $u = u_R$ at the point $x = \sigma(t)_R$ to the immediate right, with $u_L \ne u_R$.

Integrating (36) over the interval $x_1 \le x \le x_2$ gives

$$\int_{x_1}^{x_2} \frac{\partial u}{\partial t} dx + \int_{x_1}^{x_2} \left(\frac{1}{2}u^2\right)_x dx = 0. \tag{37}$$

Provided u is differentiable with respect to t, the time derivative can be taken outside the first integral in (37), which then becomes

$$\frac{d}{dt} \int_{x_1}^{x_2} u(x, t) dx + \int_{x_1}^{x_2} \frac{\partial}{\partial x}\left(\frac{1}{2}u^2\right) dx = 0. \tag{38}$$

An application of the fundamental theorem of integral calculus to the second integral leads to the result

$$\frac{d}{dt}\int_{x_1}^{x_2} u(x,t)dx + \frac{1}{2}\{u^2(x_2,t) - u^2(x_1,t)\} = 0. \tag{39}$$

To develop this result further by allowing for the discontinuity in $u(x,t)$ across $x = \sigma(t)$, we now rewrite (39) as

$$\frac{d}{dt}\int_{x_1}^{\sigma(t)_L} u(x,t)dx + \frac{d}{dt}\int_{\sigma(t)_R}^{x_2} u(x,t)dx = \frac{1}{2}\{u^2(x_1,t) - u^2(x_2,t)\}. \tag{40}$$

conservation law in integral form

This result is a **conservation law** in **integral form** for the quantity represented by $u(x,t)$. The term on the left is the rate of change of the amount of $u(x,t)$ in the interval $x_1 \leq x \leq x_2$, and the term on the right represents the difference between the amount of $u(x,t)$ entering through $x = x_1$ and leaving through $x = x_2$.

If Leibniz' theorem (Theorem 1.5) for the differentiation of a definite integral with respect to a parameter is applied to the term on the left of this equation, we find that

$$\int_{x_1}^{\sigma(t)_L} u_t(x,t)dx + \int_{\sigma(t)_L}^{x_2} u_t(x,t)dx + \frac{d\sigma}{dt}(u_L - u_R) = \frac{1}{2}\{u^2(x_1,t) - u^2(x_2,t)\}. \tag{41}$$

Letting $x_1 \to \sigma(t)_L$ and $x_2 \to \sigma(t)_R$, when $u(x_1,t) \to u_L$ and $u(x_2,t) \to u_R$, simplifies this result to

$$\frac{d\sigma}{dt}(u_L - u_R) = \frac{1}{2}(u_L^2 - u_R^2), \tag{42}$$

because the boundedness of u_t causes the two integrals to vanish in the limit as their intervals of integration tend to zero.

If we set $s = d\sigma/dt$, and introduce the notation $[[\alpha]] = \alpha_L - \alpha_R$, the **jump condition** experienced by a discontinuous solution of this PDE across the discontinuity at $x = \sigma(t)$ becomes

$$s[[u]] = \frac{1}{2}[[u^2]]. \tag{43}$$

In terms of u_L and u_R this can be written

$$s(u_L - u_R) = \frac{1}{2}(u_L^2 - u_R^2), \tag{44}$$

so the speed of propagation of the discontinuity

$$s = \frac{1}{2}(u_L + u_R). \tag{45}$$

shock waves and the Riemann problem

A discontinuity across $x = \sigma(t)$ is called a **shock wave**, or simply a **shock**, when it arises because of the intersection of characteristics.

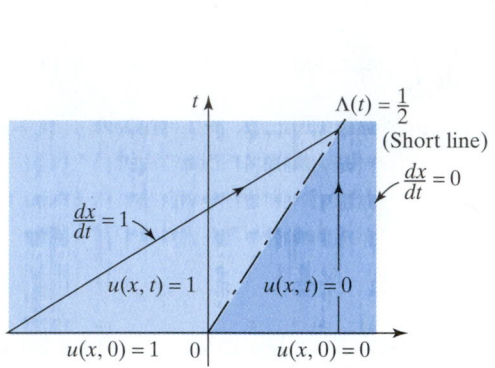

FIGURE 18.10   Characteristics in Riemann problem (I) converge to produce a discontinuous generalized solution that forms a propagating shock wave.

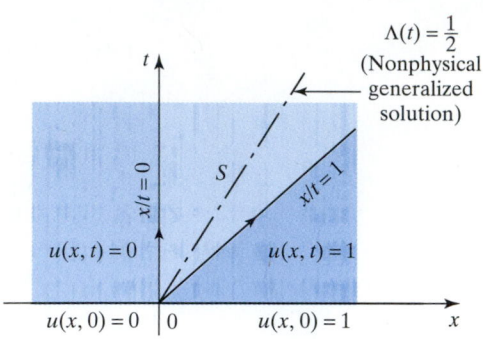

FIGURE 18.11   A mathematically permissible but nonphysical discontinuous solution in S for Riemann problem (II) that is *not* produced by the intersection of characteristics. The two constant solutions to the left and right of region S are joined continuously in a physically realistic manner by a centered simple wave in S.

To illustrate some of the properties of this extension of a classical solution, we now consider two special piecewise continuous initial value problems for (35) called **Riemann problems**.

Riemann problem (I): Solve the initial value problem

$$u_t + uu_x = 0, \quad \text{with} \quad u(x, 0) = \begin{cases} 1, & x < 0 \\ 0, & x > 0, \end{cases} \qquad (46)$$

where the initial condition is piecewise constant and decreases as x increases.

From (45) the speed of propagation of the discontinuity initiated by the discontinuity in the initial data is seen to be $s = \frac{1}{2}$. Figure 18.10 shows that this propagating discontinuity is a *shock*, because characteristics converge onto the discontinuity line from both the left and right. In gas dynamics a discontinuity of this type models an ideal shock wave in supersonic flow across which there is a sudden change of pressure, which in supersonic flight causes the sonic boom as an aircraft flies past.

Riemann problem (II): Solve the initial value problem

$$u_t + uu_x = 0, \quad \text{with} \quad u(x, 0) = \begin{cases} 0, & x < 0 \\ 1, & x > 0, \end{cases} \qquad (47)$$

where the initial condition is piecewise constant and increases as x increases.

In this problem the speed of propagation of the discontinuity is again $s = \frac{1}{2}$, but Fig. 18.11 shows that the discontinuity cannot be a shock because no characteristics converge onto the line along which the discontinuity is propagated. In applications, a discontinuous solution of this type is a mathematical solution but *not* a physically realizable one, as was the case in Riemann problem (I). This illustrates the fact that a consequence of extending a classical solution to permit discontinuous solutions can be to introduce nonphysical solutions that must be rejected when they do not arise because of the intersection of characteristics.

a mathematical solution that is nonphysical

To examine Riemann problem (II) in more detail, having rejected its discontinuous generalized solution as not physically realizable, we need to consider how a differentiable solution can be found in the wedge-shaped region S in Fig. 18.11. For a differentiable solution to exist in S it is necessary that the region be covered by a family of characteristics that at the left and right extremes of S coincide with the characteristics bounding the adjacent regions where $u(x, t)$ is constant. This can be achieved by straight line characteristics (rays) emanating from the origin O, the equation of which can be written $\zeta = x/t$ with $0 \leq \zeta \leq 1$, because then the rays at the edges of S coincide with the characteristics bounding the constant state regions.

Let us now try to find a solution of (47) in region S of the form $u(x, t) = U(\zeta)$, where $\zeta = x/t$. Then, as $u_t = U'(\zeta)\partial\zeta/\partial t = -(x/t^2)U'(\zeta)$ and $u_x = U'(\zeta)\partial\zeta/\partial x = (1/t)U'(\zeta)$, substitution into (35) followed by the cancellation of t and $U'(\zeta)$, neither of which is zero, shows that

$$U(\zeta) = u(x, t) = x/t. \tag{48}$$

This is the required solution of Riemann problem (II) in S. The solution $u(x, t)$ in S is constant along every characteristic issuing out from the origin, and at the extremes of S these characteristics coincide with the characteristics bounding the constant solutions to the left and right of S. This solution in S resolves the initial discontinuity immediately and joins the constant solutions to the left and right of S in a continuous manner. A solution of this type is called a **centered simple wave** with its center located at the origin 0. This is a *generalized solution* because of the discontinuity in derivatives across the characteristics that bound S. In applications, a centered simple wave resolves discontinuous initial conditions that do not give rise to the intersection of characteristics, and in Riemann problem (II) the nonphysical discontinuous generalized solution that is also possible must be rejected and replaced by the physically realizable centered simple wave.

centered simple wave

A proper examination of shock waves, centered simple waves, and simple waves of a more general type is beyond this brief introduction, as is a discussion of a different form of generalization of a solution called a **weak solution**. Nevertheless, the extension of a classical solution outlined here to include shock solutions has many important practical consequences, as, for example, in fluid mechanics, solid mechanics, and electromagnetic theory. In three space dimensions and time, these ideas are used to examine shock waves produced by aircraft in supersonic flight, and the bow shock wave produced by the Space Shuttle during its reentry into the atmosphere.

A classical account of shock waves in gases can be found in reference [7.4]. References [7.9] and [7.13] consider the generalization of differentiable solutions of PDEs to allow for discontinuous solutions; see also reference [7.20]. Reference [7.13] also covers in considerable detail various types of reaction–diffusion problems. A useful and elementary introduction to the mathematical theory of waves of several different types is to be found in reference [7.8]; reference [7.10] develops the mathematical theory of PDEs in considerable detail. A standard reference to various types of wave propagation problem is to be found in reference [7.18].

Summary

It was shown how, when a first order PDE describing a conservation law is written in integral form, it is possible to extend the classical concept of a differentiable solution by incorporating discontinuous solutions called shocks. This becomes necessary in order to extend the concept of a solution to take into account the situation when the classical solution becomes nonunique because of the intersection of characteristics, causing the solution to become nondifferentiable. It was seen that this generalization of a solution can give rise to more than one shock solution. In physical situations, such as gas dynamics, only one of these shock solutions is possible, so some selection principle must be introduced to allow the physically realizable solution to be distinguished from among the mathematically possible ones.

EXERCISES 18.4

1. Find the jump condition that must be satisfied by a shock solution of

$$u_t + u^n u_x = 0 \quad \text{for } n = 1, 2, \ldots .$$

2. Given that the differential equation

$$u_t + f(u)u_x = 0$$

has a discontinuous solution and that $f(u)$ is a continuous function of u, find the jump condition that must be satisfied by its shock solution.

3. Given the two Riemann problems for the equation

$$u_t + u^2 u_x = 0,$$

determined by (a) $u(x, 0) = \begin{cases} 1, & x < 0 \\ 2, & x > 0 \end{cases}$ and (b) $u(x, 0) = \begin{cases} 3, & x < 0 \\ 1, & x > 0 \end{cases}$, find which problem has a shock solution and determine its speed of propagation.

4. Show that the Riemann problem

$$u_t + u^3 u_x = 0 \quad \text{with} \quad u(x, 0) = \begin{cases} 1, & x < 0 \\ 2, & x > 0 \end{cases}$$

has a centered simple wave solution located at the origin. By setting $\zeta = x/t, u(x, t) = U(\zeta)$ and substituting

into the differential equation, find the analytical solution for the centered simple wave and determine the region in the (x, t)-plane occupied by the simple wave solution.

5.* Show that the Riemann problem

$$u_t + uu_x = 0 \quad \text{with} \quad u(x, 0) = \begin{cases} 0, & x < 2 \\ 1, & x > 2 \end{cases}$$

has a centered simple wave solution. Generalize the approach suggested in Exercise 4 to find the analytical solution for the centered simple wave, stating the region in the (x, t)-plane occupied by the centered simple wave solution.

6.* The compound Riemann problem

$$u_t + uu_x = 0 \quad \text{with} \quad u(x, 0) = \begin{cases} 1, & x < 0 \\ 2, & 0 < x < 2 \\ 0, & x > 2 \end{cases}$$

describes a solution that starts with both a centered simple wave and a shock located at different points on the initial line. By considering the path of the shock and the boundary of the centered simple wave, determine the time at which the simple wave and the shock first meet.

18.5   The Three Fundamental Types of Linear Second Order PDE

We now show how the three most important types of linear second order PDEs can be derived from some representative physical problems. The equations are classified as being of hyperbolic, parabolic, or elliptic type, and the basis of this system of classification will be developed in the next section.

Vibrating Strings and Plates

vibrating strings and plates and the wave equation

Let us consider a uniform stretched linearly elastic string under a tension T that is displaced from its equilibrium position and then released. This could, for example, represent the response of a plucked violin string. To derive the PDE governing the motion of the string after its release we must examine the forces acting on an element PQ of the string at time t when it has been displaced through a small

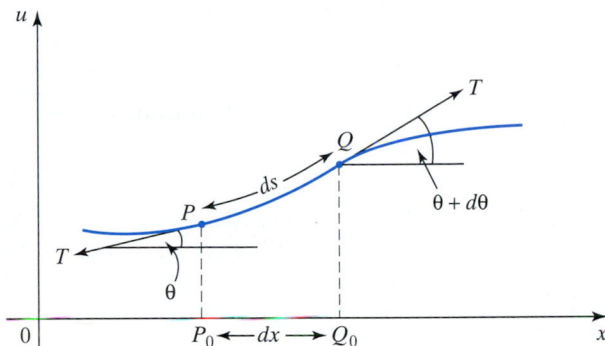

FIGURE 18.12   A transverse displacement of element PQ of a stretched string.

distance in the u-direction transverse to its equilibrium position along the x-axis. Figure 18.12 shows a typical element PQ when in its displaced position.

The element of arc length ds along the string when the displacement is $u(x, t)$ is given by $ds = \sqrt{(1 + u_x^2)}dx$. As the displacement u is small, the term u_x^2 is small relative to 1, so to this order of approximation $ds \approx dx$. In a linearly elastic string the tension is proportional to the extension of the string, so as $ds \approx dx$ we may assume that the string tension T remains constant as long as the transverse displacement is small.

In the equilibrium condition let the element PQ lie along the x-axis between the points P_0 and Q_0, where the length PQ is ds and the length $P_0 Q_0$ is dx. The equation of motion of the element is obtained by equating the forces acting on the element due to the tension T (gravity is neglected) and the rate of change of momentum of the element in the u-direction. As the string is uniform, the mass of the element PQ is $dm = \rho ds$, where ρ, called the **line density** of the string, is the mass per unit length of the string. The momentum of the element in the u-direction is $\rho ds\, u_t$, so its rate of change of momentum in the direction is $\rho ds\, u_{tt}$. As T is considered to be constant, the force acting on the element is simply the difference in the components of the tension normal to the x-axis at each of its ends due to the change in inclination of the string from an angle θ at P to an angle $\theta + d\theta$ at Q. The resultant force acting on the element is thus

$$T \sin(\theta + d\theta) - T \sin\theta = T\sin\theta\cos d\theta + T\cos\theta\sin d\theta - T\sin\theta.$$

As $d\theta$ is small we may replace $\cos d\theta$ by 1 and $\sin d\theta$ by $d\theta$, as a result of which the transverse force acting on the string can be approximated by $T\cos\theta d\theta$. Finally, equating the resultant force and the rate of change of momentum in the u-direction shows that when $d\theta$ and the transverse displacements are small the equation of motion is

$$T\cos\theta d\theta = \rho ds\, u_{tt}.$$

To eliminate θ we now use the fact that $\tan\theta = u_x$, from which it follows by differentiation with respect to x that $\sec^2\theta\, d\theta/dx = u_{xx}$, and so $\sec^2\theta\, d\theta = u_{xx}dx$. Multiplying this by $T\cos^3\theta$, substituting into the preceding result, and using the fact that in the limit as $dx \to 0$ we have $dx/ds = \cos\theta$ leads to the result

$$\rho u_{tt} = T\cos^4\theta\, u_{xx}.$$

As $\tan \theta = u_x$ and $\sec^2 \theta = 1 + \tan^2 \theta$, we see that

$$\cos^2 \theta = 1/\{1 + (u_x)^2\},$$

so the equation of motion becomes

$$u_{tt} = c^2\{1 + (u_x)^2\}^{-2}u_{xx},$$

the wave equation is the prototype second order hyperbolic PDE

where $c^2 = T/\rho$. This second order partial differential equation governing the motion of the string is quasilinear, but if the transverse displacement is sufficiently small the term $(u_x)^2$ can be neglected, the *linearized* one-dimensional form of the equation of motion becomes

$$u_{tt} = c^2 u_{xx}. \tag{49}$$

This is a linear second order PDE of *hyperbolic* type called the **one-dimensional wave equation**, and it is one of the three fundamentally different classes of second order PDE.

Vibrations of membranes can be treated in a similar fashion to vibrating strings. Figure 18.13 shows a vibrating rectangular element $ABCD$ of a thin uniform membrane with its sides of lengths dx and dy parallel to the x- and y-axes displaced a small amount in the u-direction normal to its equilibrium position in the (x, y)-plane (the plane $u = 0$). If L is a line of unit length drawn in the membrane, the **tension** T in the membrane is defined as the force exerted on L by the material on one side of the line. The tension will be said to be *uniform* when T is independent of the direction of L and its location in the membrane.

Reasoning as in the case of the vibrating string, and considering a membrane with a *uniform* tension T, we see that the resultant of the forces Tdx normal to the boundaries AB and CD of the element is $(Tdx)(u_{yy}dy)$ and, similarly, the resultant of the forces Tdy normal to the boundaries AD and BC of the element is $(Tdy)(u_{xx}dx)$. If the mass per unit area of the membrane ρ, called its **area density**, is constant, the momentum of the element in the u direction is $\rho dx dy u_t$, so its rate of change of momentum in that direction is $\rho dx dy u_{tt}$. Equating the forces acting to the rate of change of momentum and proceeding to the limit as $dx \to 0$ and

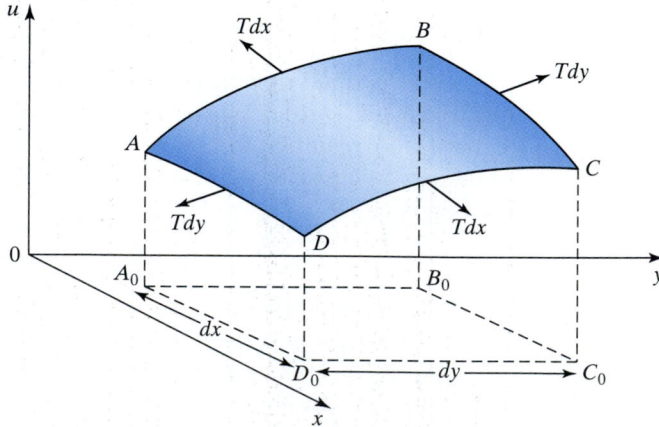

FIGURE 18.13   An element of a uniform vibrating membrane with tension T.

$dy \to 0$, we find that the PDE describing the vibrations is

$$\rho u_{tt} = T\{u_{xx} + u_{yy}\},$$

and after we set $c^2 = T/\rho$ this becomes

$$u_{tt} = c^2\{u_{xx} + u_{yy}\}. \tag{50}$$

This linear second order PDE, which is also of *hyperbolic* type, is called the **two-dimensional wave equation**. Notice that the one-dimensional and two-dimensional wave equations have *second order* partial derivatives with respect to both the *time* and the *space variables* involved.

The Heat (Diffusion) Equation

We now derive the **heat equation**, also known as the **diffusion equation**, that describes the flow of heat through a heat-conducting solid material. The derivation is based on the experimentally observed fact that heat flows in the direction of decreasing temperature, and on the assumption that the rate of heat flow $\mathbf{j}$ at any point P in the body is given by **Fourier's law**

$$\mathbf{j} = -K \operatorname{grad} T, \tag{51}$$

where $T(x, y, z, t)$ is the temperature at any point P in the material at time t, and K, called the **thermal conductivity** of the material, is a physical property that is usually taken to be a constant.

If V is an arbitrary volume in the solid bounded by a surface S, the quantity of heat leaving V in unit time is given by the surface integral

$$\int_S \mathbf{j} \cdot \mathbf{n} dS, \tag{52}$$

where $\mathbf{n}$ is the *outward* drawn unit normal to S. If we substitute for $\mathbf{j}$ in (52) and allow K to be a function of position, an application of the divergence theorem to this integral gives

$$\int_S \mathbf{j} \cdot \mathbf{n} dS = -\int_V \operatorname{div}(K \operatorname{grad} T) dV.$$

However, $\operatorname{div}(K \operatorname{grad} T) = K\Delta T + \operatorname{grad} K \cdot \operatorname{grad} T$, so the preceding expression becomes

$$\int_S \mathbf{j} \cdot \mathbf{n} dS = -\int_V (K\Delta T + \operatorname{grad} K \cdot \operatorname{grad} T) dV. \tag{53}$$

If the density of the material is ρ and its specific heat is c, the amount of heat in an element of volume dV is given by $c\rho T dV$, where both ρ and c can be functions of position. Integration of $c\rho T dV$ over V shows that the total amount of heat Q in V must be

$$Q = \int_V \rho c T dV.$$

As V is a fixed arbitrary volume in the solid, differentiating this result with respect to the time t shows that the rate at which Q decreases with respect to time is

$$-Q_t = -\int_V \frac{\partial}{\partial t}(\rho c T)dV. \tag{54}$$

Equating (53) and (54) and combining the integrals gives

$$\int_V \left\{ \frac{\partial}{\partial t}(\rho c T) - K\Delta T - \operatorname{grad} K \cdot \operatorname{grad} T \right\} dV = 0. \tag{55}$$

This result must be true for all arbitrary volumes V, but this can only be possible if the integrand of (55) is identically zero, so the PDE determining the flow of heat when expressed in terms of the temperature T is

the heat or diffusion equation is the prototype second order parabolic PDE

$$\frac{\partial}{\partial t}(\rho c T) = K\Delta T + \operatorname{grad} K \cdot \operatorname{grad} T. \tag{56}$$

This PDE is a linear variable coefficient second order PDE for the temperature distribution throughout the solid, and in general its independent variables are three space variables and time. When, as is usually the case, the conductivity K, the density ρ, and the specific heat c are taken to be constants, the linear second order PDE in (56), which is an equation of *parabolic* type, reduces to

$$\rho c T_t = K\Delta T, \tag{57}$$

heat conduction and diffusion

called the **heat conduction equation**, or simply the **heat equation**. The constant κ^2, where $\kappa^2 = K/(\rho c)$, is called the **diffusivity** of the material, so in terms of the diffusivity, (57) becomes

$$T_t = \kappa^2 \Delta T. \tag{58}$$

Values of the diffusivity κ^2 for some common materials, in c.g.s. units and degrees Celsius, are steel 0.12, copper 1.14, aluminum 0.86, silver 1.71, glass 0.006, and concrete 0.004.

Notice that the heat equation that is of *parabolic type* involves a *first order* partial derivative with respect to time and *second order* partial derivatives with respect to the space variables involved.

An equation of the form (58) also describes the diffusion process caused by an imbalance of concentration of a substance diffusing through material, and for this reason (58) is also known as the **diffusion equation**. A typical diffusion process involves the passage of a chemical with concentration k_1 present in a liquid or gas through a membrane to a liquid or gas on the other side of the membrane where the concentration is k_2 with $k_1 > k_2$.

Diffusion is used in many ways for the concentration of chemicals, and it occurs naturally in plants where nutrients obtained from the soil are passed through the plant by diffusion through plant membranes.

The Laplace Equation

the Laplace equation
is the prototype
second order elliptic
PDE

The **Laplace equation** characterizes a large group of physical problems that are *independent* of the time, and for this reason they are usually called **steady state problems**. An obvious example is provided by the heat equation in (58), because if a heat transfer process attains a steady state the time derivative T_t vanishes and the heat equation reduces to the Laplace equation $\Delta T = 0$ that is the simplest PDE of *elliptic* type. Some typical two-dimensional steady state temperature distributions have already been obtained in Section 17.2 as applications of conformal transformation techniques, where it was also shown that Laplace's equation governs the velocity potential of the steady fluid flow of an incompressible, irrotational, and inviscid fluid.

Other physical situations that give rise to Laplace's equation can be found in the study of steady state electromagnetic fields. When the field exists in an isotropic medium with dielectric constant ε, permeability μ, and charge distribution density ρ, the electric vector **E**, the magnetic vector **H**, and the current **j** are related by the **Maxwell equations**

the Maxwell
equations of
electromagnetic
theory

$$\operatorname{curl} \mathbf{H} = \mathbf{j} + \varepsilon \frac{\partial}{\partial t} \mathbf{E}$$

$$\operatorname{curl} \mathbf{E} = -\mu \frac{\partial}{\partial t} \mathbf{H}$$

$$\operatorname{div} \mathbf{H} = 0$$

$$\operatorname{div} \mathbf{E} = \rho/\varepsilon. \tag{59}$$

In **electrostatics** there is no change with respect to time of the electric vector **E**, so the time derivative $\mathbf{E}_t$ vanishes, and in an uncharged region ($\rho = 0$) Maxwell's equations reduce to

$$\operatorname{div} \mathbf{E} = 0 \quad \text{and} \quad \operatorname{curl} \mathbf{E} = \mathbf{0}.$$

This pair of equations can be satisfied by introducing a scalar **electric potential** ϕ such that $\mathbf{E} = \operatorname{grad} \phi$, because then curl $\mathbf{E} = \operatorname{curl}(\operatorname{grad} \phi) = \mathbf{0}$, so

$$\operatorname{div} \mathbf{E} = \operatorname{div}(\operatorname{grad} \phi) = 0 \quad \text{and so} \quad \Delta \phi = 0. \tag{60}$$

This has shown that the electrostatic potential distribution ϕ is a solution of the Laplace equation, and that the electric field vector can be found from ϕ by using $\mathbf{E} = \operatorname{grad} \phi$. Various electrostatic potential distributions were found in Section 17.2 by means of conformal transformations.

electrostatics and
magnetostatics

A similar situation occurs in **magnetostatics**, because if the medium is non-conducting $\mathbf{j} = \mathbf{0}$, so the Maxwell equations reduce to

$$\operatorname{div} \mathbf{H} = 0 \quad \text{and} \quad \operatorname{curl} \mathbf{H} = \mathbf{0}.$$

This time a **magnetic potential** ϕ can be introduced by setting $\mathbf{H} = \operatorname{grad} \phi$, and then the magnetic potential is seen to be a solution of the Laplace equation $\Delta \phi = 0$.

An important physical problem that gives rise to the Laplace equation in three dimensions is the gravitational potential $\phi(x, y, z)$. The mathematics of gravitational potentials is closely related to the cases considered above, but before we proceed further, some definitions are necessary.

A **force field** in a region D of space exerts a force $\mathbf{F}$ on a material solid particle at a point (x, y, z) in D, where

$$\mathbf{F} = F_1(x, y, z)\mathbf{i} + F_2(x, y, z)\mathbf{j} + F_3(x, y, z)\mathbf{k}.$$

It may happen that the force is proportional to the mass m of the particle, as occurs in the earth's gravitational field, where the constant of proportionality between the mass of the particle and its weight is g, the acceleration due to gravity.

force fields and lines of force

A curve in a force field with the property that at each point on the curve the tangent to the curve is parallel to the direction of the force is called a **line of force**. If the vector element along the line of force is $d\mathbf{r} = dx\mathbf{i} + dy\mathbf{j} + dz\mathbf{k}$, the lines of force are determined by the equations

$$\frac{dx}{F_1} = \frac{dy}{F_2} = \frac{dz}{F_3}. \tag{61}$$

When a particle moves in a force field from A to B along a path AB, the work W done by the action of the field on the particle is given by the line integral

$$W_{AB} = \int_{AB} (F_1 dx + F_2 dy + F_3 dz).$$

In general, the work W_{AB} will depend not only on A and B, but also on the path taken from A to B. A **potential field** is a force field in which the work done by the force depends only on the points A and B, and not on the path joining them.

potentials and conservative fields

Consequently, a field is a *potential field* if the work done along every loop joining A to itself is zero. It is for this reason that potential fields are also called **conservative fields**, because work done by the force on a particle moving away from a point is returned if the particle arrives back at its starting point.

Consider the two arbitrary paths APB and AQB shown in Fig. 18.14a. Then in a potential field $W_{APB} + W_{BQA} = 0$, so $W_{BQA} = -W_{APB}$. Now let A in Fig. 18.14b be a fixed point (x_0, y_0, z_0), B be a general point (x, y, z), and C be a point (x^*, y^*, z^*). Then if $W_{AB} = \phi(x, y, z)$ is the work done moving from A to B, in a

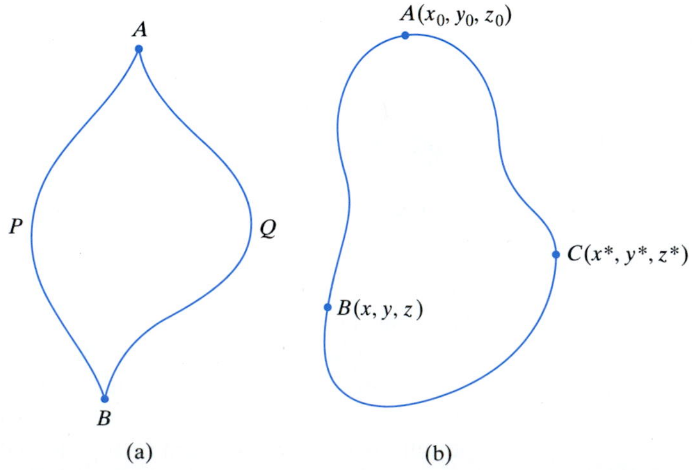

(a)   (b)

FIGURE 18.14   (a) Two paths joining A to B. (b) A loop containing a fixed point A.

potential field $W_{AB} + W_{BC} + W_{CA} = 0$, so $W_{CA} = -W_{AC} = -\phi(x^*, y^*, z^*)$ and so $W_{BC} = \phi(x^*, y^*, z^*) - \phi(x, y, z)$. This shows that the work done by the force moving between any two points in a potential field is equal to the difference of the potential between the two points.

A gravitational field is due to the presence of matter, so in free space between the matter producing a gravitational force field there can be no sources, and so div $\mathbf{F} = 0$. This means that in a potential field div grad $\phi = 0$ or $\Delta\phi = 0$, so a gravitational potential ϕ is seen to be a solution of the Laplace equation.

The linear second order PDE called the **Poisson equation** is

the Poisson equation and its connection with the Laplace equation

$$\Delta\phi = F(x, y, z), \tag{62}$$

and it is also a PDE of *elliptic* type. The Poisson equation arises in a variety of ways, one of which is in electrostatics when a charge distribution is present in a dielectric medium so that div $\mathbf{E} = \rho/\varepsilon$. If we set $F(x, y, z) = \rho/\varepsilon$, and again introduce an electric potential through $\mathbf{E} = \text{grad } \phi$, the equation div $\mathbf{E} = \rho/\varepsilon$ becomes the three-dimensional Poisson equation in (62).

Electromagnetic Waves

Finally, we use Maxwell's equations to show how the wave equation in three space dimensions and time determines electromagnetic wave propagation though space. Returning to the equations in (59), and considering the situation in a dielectric medium where no current can flow so $\mathbf{j} = \mathbf{0}$ and where there is no charge distribution so $\rho = 0$, the equations reduce to

electromagnetic waves in space

$$\text{curl } \mathbf{H} = \varepsilon\frac{\partial \mathbf{E}}{\partial t} \quad \text{and} \quad \text{curl } \mathbf{E} = -\mu\frac{\partial \mathbf{H}}{\partial t}.$$

Differentiating the first equation with respect to t and substituting for $\partial\mathbf{H}/\partial t$ from the second equation gives

$$-\text{curl curl } \mathbf{E} = \varepsilon\mu\,\mathbf{E}_{tt},$$

but curl curl $\mathbf{E} = \text{grad div } \mathbf{E} - \Delta\mathbf{E}$, but div $\mathbf{E} = 0$, and so

$$\mathbf{E}_{tt} = (1/\varepsilon\mu)\Delta\mathbf{E}. \tag{63}$$

We have shown that the electric vector $\mathbf{E}$ is a solution of the three-dimensional wave equation.

A similar argument shows that the magnetic vector $\mathbf{H}$ is also a solution of the same three-dimensional wave equation

$$\mathbf{H}_{tt} = (1/\varepsilon\mu)\Delta\mathbf{H}, \tag{64}$$

so that waves involving both the electric and the magnetic vector propagate with the same speed c that is determined by $c^2 = 1/(\varepsilon\mu)$. In free space the speed of propagation c of these electromagnetic waves is the velocity of light.

Summary Using typical physical examples, the three fundamental types of linear constant coefficient second order PDEs have been derived from first principles. These are the wave equation that is of hyperbolic type, the heat or diffusion equation that is of parabolic type, and the Laplace equation that is of elliptic type. Potential functions and conservative fields were also defined and interpreted in terms of a force acting on a particle moving in the field.

18.6 Classification and Reduction to Standard Form of a Second Order Constant Coefficient Partial Differential Equation for $u(x, y)$

In the previous section the three fundamental types of PDE were derived from typical physical situations, and they were then classed as being of hyperbolic, parabolic, or elliptic type. The purpose of the present section is to explain the basis of this classification where, for simplicity, in the main the discussion will be limited to linear second order partial differential equations whose coefficients are either constants or functions of the independent variables involved.

We have already seen that in the case of two dimensions, examples of these equations involving a function u are as follows:

The one-dimensional wave equation

$$u_{tt} = c^2 u_{xx} \tag{65}$$

for the function $u(x, t)$, where x is a space variable, t is the time, and c is a constant.

the three
fundamental types
of second order PDE

The one-dimensional heat equation

$$u_t = \kappa^2 u_{xx} \tag{66}$$

for the function $u(x, t)$, where x is a space variable, t is the time, and κ is a constant.

The two-dimensional Laplace equation

$$u_{xx} + u_{yy} = 0 \tag{67}$$

for the function $u(x, y)$, where x and y are both space variables.

These three equations are all special cases of the general linear PDE for an unknown twice differentiable classical solution $u(x, y)$ of the two independent variables x and y, or sometimes t and x, which is defined in some region D and can be written

$$Au_{xx} + 2Bu_{xy} + Cu_{yy} + Pu_x + Qu_y + Ru = F(x, y), \tag{68}$$

where A, B, C, P, Q, and R are functions of x and y.

In equation (68) the factor 2 multiplying B has been introduced for convenience as it simplifies the calculations that are to follow. The functions $A, B, \ldots, R$ multiplying u and its derivatives are called the **coefficients** of the PDE, and $F(x, y)$ is called the **nonhomogeneous term**. Equation (68) is called **homogeneous** if $F(x, y)$ is identically zero.

The two-dimensional Laplace equation (67) is an example of a homogeneous constant coefficient PDE that can be derived from (68) by setting $A = C = 1$, $B = 0$, and $F(x, y) = 0$. The corresponding nonhomogeneous equation

$$u_{xx} + u_{yy} = F(x, y) \tag{69}$$

is the **two-dimensional Poisson equation**.

The operations of partial differentiation with respect to x and y are linear when performed on $u(x, y)$, so if two functions $u_1(x, y)$ and $u_2(x, y)$ are solutions of the nonhomogeneous equation (68), it follows that their difference $v(x, y) = u_1(x, y) - u_2(x, y)$ will be a solution of the *homogeneous* equation

$$Av_{xx} + 2Bv_{xy} + Cv_{yy} + Pv_x + Qv_y + Rv = 0. \tag{70}$$

An immediate extension of this result that will be needed later is that if $u_i(x, y)$ with $i = 1, 2, \ldots, k$ are solutions of the homogeneous equation and $c_1, c_2, \ldots, c_k$ are constants, then

$$u(x, y) = \sum_{i=1}^{k} c_i u_i(x, y) \tag{71}$$

is also a solution of the homogeneous equation.

To understand why the three PDEs in (65) to (67) have fundamentally different mathematical properties, it is necessary to examine their mathematical **classification** according to type.

To arrive at the method of classification of second order linear constant coefficient PDEs, we consider the group of second order terms $L[u]$ in (68) given by

$$L[u] = Au_{xx} + 2Bu_{xy} + Cu_{yy}, \tag{72}$$

the quadratic form used to classify a PDE

called the **principal part** of (68), and at some point (x_0, y_0) in a region D where the equation is defined associate with it the *quadratic form*

$$Q(\alpha, \beta) = A(x_0, y_0)\alpha^2 + 2B(x_0, y_0)\alpha\beta + C(x_0, y_0)\beta^2, \tag{73}$$

where α and β are real variables. The differential equation in (68) is then classified according to the following criteria:

classification of PDEs according to type

(a) the PDE is of **hyperbolic** type in D if $B^2 - AC > 0$
(b) the PDE is of **parabolic** type in D if $B^2 - AC = 0$ (74)
(c) the PDE is of **elliptic** type in D if $B^2 - AC < 0$

The expression

$$d = B^2 - AC \tag{75}$$

is called the **discriminant** of the PDE, so it is hyperbolic if $d > 0$, parabolic if $d = 0$, and elliptic if $d < 0$.

When this system of classification is applied to equations (65) to (67), it is seen that the wave equation (65) is of *hyperbolic* type, the heat equation (66) is of *parabolic* type, and the Laplace equation (67) is of *elliptic* type, as is the Poisson equation in (62) because the nonhomogeneous term does not enter into the classification.

This apparently arbitrary classification of the PDEs in (68) is of fundamental importance for the following reasons:

(a) The classification of a PDE is *independent* of the choice of coordinate system used when formulating the equation. Expressed differently, the classification is such that it does not depend on the choice of independent variables. So, for example, if a PDE is of elliptic type when expressed in terms of the cartesian coordinates x and y, it will still be of elliptic type when expressed in terms of any other coordinate system like the cylindrical polar coordinates r, θ, and z.

why the
classification of
PDEs is important

(b) The nature of an appropriate domain D and the associated auxiliary conditions (initial and/or boundary conditions) that must be imposed on the PDE in order to ensure a unique solution throughout D differ according to the classification.

We will only justify statement (a), as the significance of (b) will become apparent when boundary and initial conditions are considered. Let us make a transformation of the independent coordinate variables x and y to ξ and η in such a way that one point in the domain D in the (x, y)-plane corresponds to one point in the corresponding domain in the (ξ, η)-plane, and conversely (the transformation is one-one between the two domains), by setting

$$\xi = \xi(x, y) \quad \text{and} \quad \eta = \eta(x, y), \tag{76}$$

where the functions ξ and η are assumed to be twice continuously differentiable. The transformation will be one-one if its Jacobian $J(x, y)$ is nonvanishing throughout D, where

$$J(x, y) = \frac{\partial(\xi, \eta)}{\partial(x, y)} = \begin{vmatrix} \xi_x & \xi_y \\ \eta_x & \eta_y \end{vmatrix} \neq 0. \tag{77}$$

Using the rules from the calculus for a change of variables to express the partial derivatives of u with respect to x and y in terms of those with respect to ξ and η, we find that

$$u_x = \xi_x u_\xi + \eta_x u_\eta, \tag{78}$$

so dropping the variable u we obtain the operator relationship

$$\frac{\partial}{\partial x} = \xi_x \frac{\partial}{\partial \xi} + \eta_x \frac{\partial}{\partial \eta} \tag{79}$$

with the corresponding result

$$u_y = \xi_y u_\xi + \eta_y u_\eta \tag{80}$$

and the associated operator relationship

$$\frac{\partial}{\partial y} = \xi_y \frac{\partial}{\partial \xi} + \eta_y \frac{\partial}{\partial \eta}. \tag{81}$$

To find u_{xx} we start from its definition and proceed as follows:

$$u_{xx} = \frac{\partial(u_x)}{\partial x} = \frac{\partial}{\partial x}(\xi_x u_\xi + \eta_x u_\eta) = \xi_{xx} u_\xi + \eta_{xx} u_\eta + \xi_x \frac{\partial(u_\xi)}{\partial x} + \eta_x \frac{\partial(u_\eta)}{\partial x}.$$

Next, replacing the operator $\partial/\partial x$ by the result in (79), simplifying the result, and using the equality of mixed derivatives $u_{\xi\eta} = u_{\eta\xi}$, which is justified because we are considering classical solutions that are continuously twice differentiable, we find that

$$u_{xx} = \xi_x^2 u_{\xi\xi} + 2\xi_x \eta_x u_{\xi\eta} + \eta_x^2 u_{\eta\eta} + \xi_{xx} u_\xi + \eta_{xx} u_\eta. \tag{82}$$

Similar arguments show that

$$u_{xy} = \xi_x \xi_y u_{\xi\xi} + (\xi_x \eta_y + \xi_y \eta_x) u_{\xi\eta} + \eta_x \eta_y u_{\eta\eta} + \xi_{xy} u_\xi + \eta_{xy} u_\eta \tag{83}$$

and

$$u_{yy} = \xi_y^2 u_{\xi\xi} + 2\xi_y \eta_y u_{\xi\eta} + \eta_y^2 u_{\eta\eta} + \xi_{yy} u_\xi + \eta_{yy} u_\eta. \tag{84}$$

When working with transformations of derivatives, the use of the suffixes x and y with u denoting partial differentiation with respect to x and y is to be understood to imply that u is to be regarded as the *original* function of x and y, but that when the suffixes ξ and η are used it is to be understood that u is then to be regarded as the transformed function $u = u(\xi, \eta)$. The expressions for x and y follow, if the coordinate transformations (76) are solved to obtain $x = \sigma(\xi, \eta)$ and $y = \mu(\xi, \eta)$, which because the transformation is one-one will always enable x and y to be expressed uniquely as functions of ξ and η.

After substituting these results into (68) and collecting terms, we obtain

$$\widetilde{A} u_{\xi\xi} + 2\widetilde{B} u_{\xi\eta} + \widetilde{C} u_{\eta\eta} + \widetilde{P} u_\xi + \widetilde{Q} u_\eta + \widetilde{R} u = \widetilde{F}(\xi, \eta), \tag{85}$$

where

$$\widetilde{A} = A(\xi_x)^2 + 2B\xi_x \xi_y + C(\xi_y)^2 \tag{86}$$
$$\widetilde{B} = A\xi_x \eta_x + B(\xi_x \eta_y + \xi_y \eta_x) + C\xi_y \eta_y \tag{87}$$
$$\widetilde{C} = A(\eta_x)^2 + 2B\eta_x \eta_y + C(\eta_y)^2, \tag{88}$$

with $\widetilde{P}, \widetilde{Q}$, and $\widetilde{R}$ defined in similar fashion, and $\widetilde{F}(\xi, \eta) = F(\sigma(\xi, \eta), \mu(\xi, \eta))$.

A routine calculation establishes the important result that

$$\widetilde{B}^2 - \widetilde{A}\widetilde{C} = (\xi_x \eta_y - \xi_y \eta_x)^2 (B^2 - AC) = J^2(x, y)(B^2 - AC).$$

why a change of variables does not alter the classification of a PDE

As the Jacobian is nonvanishing and $J^2(x, y)$ is positive, the classification of the equation is seen to be unchanged by the transformation of the independent variables in (76), so statement (a) has been proved.

The transformed PDE in (85) will be simplified if the coordinate transformation can be chosen so that at the point (x_0, y_0):

(a) $\widetilde{A} = \widetilde{C} = 0$, or $\widetilde{A} = -\widetilde{C}, \widetilde{B} = 0$, if the PDE is of **hyperbolic** type
(b) $\widetilde{A} = \widetilde{B} = 0$, if the PDE is of **parabolic** type
(c) $\widetilde{A} = \widetilde{C}$ and $\widetilde{B} = 0$, if the PDE is of **elliptic** type.

Clearly this classification depends on the functions $\widetilde{A}, \widetilde{B}, \widetilde{C}$, and the point (x_0, y_0), though if the original PDE has constant coefficients this classification will be the same for all points in the region D where the PDE is defined.

To see how to accomplish these reductions we again consider the quadratic form $Q(\alpha, \beta)$ in (73) and make the substitutions

$$\alpha = p\xi_x + q\eta_x \quad \text{and} \quad \beta = p\xi_y + q\eta_y,$$

when we find that

$$Q(\alpha, \beta) = \widetilde{A}\lambda^2 + 2\widetilde{B}\lambda\mu + \widetilde{C}\mu^2.$$

This is seen to be of exactly the same algebraic form as the transformation of the principal term $L[u]$ of (68).

So far the functions $\xi(x, y)$ and $\eta(x, y)$ have been arbitrary, so they can now be used to achieve the simplifications in (a), (b), or (c). The **standard forms**, also called **canonical forms**, of the hyperbolic, parabolic, and elliptic PDEs associated with (68) that correspond to cases (a), (b), and (c) are as follows:

Hyperbolic standard forms

$$u_{\xi\eta} = F_1(\xi, \eta, u, u_\xi, u_\eta) \quad \text{or} \quad u_{\xi\xi} = u_{\eta\eta} + F_2(\xi, \eta, u, u_\xi, u_\eta); \tag{89}$$

Parabolic standard form

$$u_{\eta\eta} = G(\xi, \eta, u, u_\xi, u_\eta); \tag{90}$$

Elliptic standard form

$$u_{\xi\xi} + u_{\eta\eta} = H(\xi, \eta, u, u_\xi, u_\eta), \tag{91}$$

where F_1, F_2, G, and H are linear combinations of u, u_ξ, and u_η.

The equivalence of the two different standard forms in the hyperbolic case (89) will be shown later.

Reduction of a Hyperbolic Equation to Standard Form

how to reduce a hyperbolic PDE to standard form

To arrive at the first standard form in (89), ξ and η must be chosen such that $\widetilde{A} = \widetilde{C} = 0$. We see from this that ξ and η must be solutions of the first order PDE

$$A(\varphi_x)^2 + 2B\varphi_x\varphi_y + C(\varphi_y)^2 = 0, \tag{92}$$

which can be factored into the product

$$(A\varphi_x + \{B + \sqrt{B^2 - AC}\}\varphi_y)(A\varphi_x + \{B - \sqrt{B^2 - AC}\}\varphi_y) = 0.$$

Now if φ_1 and φ_2 are solutions of

$$A\varphi_{1x} + (B + \sqrt{B^2 - AC})\varphi_{1y} = 0 \quad \text{and} \quad A\varphi_{2x} + (B - \sqrt{B^2 - AC})\varphi_{2y} = 0, \tag{93}$$

characteristic equations and characteristic curves

they are also solutions of (92). These are called the **characteristic equations** associated with PDE (68), and as the discriminant $d = B^2 - AC > 0$ it follows from Section 18.2 that each defines a family of **characteristic curves** of PDE (68) determined by the solutions of

$$\frac{dy}{dx} = \frac{B + \sqrt{B^2 - AC}}{A} \quad \text{and} \quad \frac{dy}{dx} = \frac{B - \sqrt{B^2 - AC}}{A}. \tag{94}$$

These solutions can be written

$$\varphi_1(x, y) = \text{constant} \quad \text{and} \quad \varphi_2(x, y) = \text{constant}, \tag{95}$$

so we now define the functions ξ and η in (76) as

$$\xi = \varphi_1(x, y) \quad \text{and} \quad \eta = \varphi_2(x, y). \tag{96}$$

With this change of variables (68), and hence (85), reduces to

$$2\widetilde{B}u_{\xi\eta} + \widetilde{P}u_\xi + \widetilde{Q}u_\eta + Ru = \widetilde{F}(\xi, \eta), \tag{97}$$

so

$$u_{\xi\eta} = \frac{1}{2\widetilde{B}}[\widetilde{F}(\xi, \eta) - \widetilde{P}u_\xi - \widetilde{Q}u_\eta - \widetilde{R}u], \tag{98}$$

from which the first result in (89) follows by setting $\widetilde{F}_1(\xi, \eta, u, u_\xi, u_\eta) = [\widetilde{F}(\xi, \eta) - \widetilde{P}u_\xi - \widetilde{Q}u_\eta - \widetilde{R}u]/(2\widetilde{B})$.

The equivalence of the two different standard forms in (90) is established by making the substitution $\xi = X + Y$, $\eta = X - Y$ in $u_{\xi\eta} = F_1(\xi, \eta, u, u_\xi, u_\eta)$. This transforms the equation into $u_{XX} - u_{YY} = F_2(X, Y, u, u_X, u_Y)$, and apart from a change of notation the two results are the same, because F_2 is simply the transformation of F_1.

In the hyperbolic case the discriminant d is positive, so the two families of characteristic curves associated with (68) are two separate families of *real* curves in the (x, y)-plane.

EXAMPLE 18.6   Reduce to standard form

$$u_{xx} + 8u_{xy} + 7u_{yy} + u_x + 2u_y + 3u + y = 0,$$

and find its characteristic equations and curves.

Solution   Identifying the PDE with (68) shows $A = 1$, $B = 4$, and $C = 7$, so as the discriminant $d = 4^2 - (1)(7) = 9 > 0$, the equation is hyperbolic. It is unconditionally hyperbolic because the coefficients of the PDE do not depend on position. From (94) the characteristic equations are

$$\frac{dy}{dx} = 1 \quad \text{and} \quad \frac{dy}{dx} = 7.$$

Integrating these equations shows the characteristic curves to be given by the two families of parallel straight lines

$$y = x + \alpha \quad \text{and} \quad y = 7x + \beta,$$

where α and β are arbitrary constants of integration.

Setting $\xi = \alpha = y - x$ and $\eta = \beta = y - 7x$ allows the principal terms in the PDE to be replaced by $2\widetilde{B}u_{\xi\eta}$, and simple calculations establish that $\widetilde{B} = -18$, $u_x = -(u_\xi + 7u_\eta)$, $u_y = u_\xi + u_\eta$, and $y = \frac{1}{6}(7\xi - \eta)$. Substituting for u_x, u_y, and y in the PDE and rearranging terms leads to its being expressed in the standard form

$$u_{\xi\eta} = \frac{1}{36}\left[u_\xi - 5u_\eta + 3u + \frac{1}{6}(7\xi - \eta)\right]. \qquad \blacksquare$$

Reduction of a Parabolic Equation to Standard Form

The standard form in (90) arises when the discriminant $d = B^2 - AC = 0$, in which case the two characteristic equations in (94) coincide and so determine only one family of characteristic curves given by

how to reduce a parabolic PDE to standard form

$$\frac{dy}{dx} = \frac{B}{A} \quad \text{with the characteristics } y = (B/A)x + \alpha, \tag{99}$$

where α is an arbitrary constant of integration.

The required reduction is accomplished by equating ξ and α and choosing for η *any* function of x and y that is independent of ξ, so in general we can set $\eta = x$. Then with the change of variables

$$\xi = y - (B/A)x, \quad \eta = x, \tag{100}$$

the principal terms of PDE (68) can be replaced by $Au_{\eta\eta}$, so that (85) becomes

$$\widetilde{A}u_{\eta\eta} + \widetilde{P}u_\xi + \widetilde{Q}u_\eta + \widetilde{R}u = \widetilde{F}(\xi, \eta),$$

and so

$$u_{\eta\eta} = \frac{1}{\widetilde{A}}\left[\widetilde{F}(\xi, \eta) - \widetilde{P}u_\xi - \widetilde{Q}u_\eta - \widetilde{R}u\right], \tag{101}$$

from which (90) follows by setting $G(\xi, \eta, u, u_\xi, u_\eta) = (1/\widetilde{A})[\widetilde{F}(\xi, \eta) - \widetilde{P}u_\xi - \widetilde{Q}u_\eta - \widetilde{R}u]$.

EXAMPLE 18.7   Reduce to standard form

$$u_{xx} + 4u_{xy} + 4u_{yy} + u_x + 3x = 0.$$

Solution   Here $A = 1$, $B = 2$, and $C = 4$, so the discriminant $d = B^2 - AC = 0$, showing that the PDE is unconditionally parabolic. In this case the transformation $\xi = y - (B/A)x$ and $\eta = x$ becomes $\xi = y - 2x$, $\eta = x$, and this change of variables allows the principal terms to be replaced by $Au_{\eta\eta}$, so as $u_x = -2u_\xi + u_\eta$ and $x = \eta$, substitution into the PDE leads to the required reduction to standard form

$$u_{\eta\eta} = 2u_\xi - u_\eta - 3\eta. \qquad \blacksquare$$

Reduction of an Elliptic Equation to Standard Form

When PDE (68) is elliptic, its discriminant $d = B^2 - AC < 0$, so the right-hand sides of the characteristic equations in (94) become complex, showing that an elliptic PDE has no real characteristic curves. However, in the elliptic case the transformation

how to reduce an elliptic PDE to standard form

$$\xi = \frac{Ay - Bx}{\sqrt{AC - B^2}}, \quad \eta = x, \tag{102}$$

reduces (68) to

$$A(u_{\xi\xi} + u_{\eta\eta}) + \widetilde{P}u_\xi + \widetilde{Q}u_\eta + \widetilde{R}u = \widetilde{F}(\xi, \eta), \tag{103}$$

so as $A \neq 0$ an elliptic equation can always be written in the standard form

$$u_{\xi\xi} + u_{\eta\eta} = \frac{1}{A}[\tilde{F}(\xi, \eta) - \tilde{P}u_{\xi} - \tilde{Q}u_{\eta} - \tilde{R}u], \qquad (104)$$

that is, of the form in (91) with $H(\xi, \eta, u, u_{\xi}, u_{\eta}) = (1/A)[\tilde{F}(\xi, \eta) - \tilde{P}u_{\xi} - \tilde{Q}u_{\eta} - \tilde{R}u]$.

EXAMPLE 18.8

Reduce to standard form

$$5u_{xx} - 2u_{xy} + 2u_{yy} + 2u_y + 4y = 0.$$

Solution   Here $A = 5$, $B = -1$, and $C = 2$, so the discriminant $d = B^2 - AC = -9$, showing that the PDE is unconditionally elliptic. From (102) the transformation to be used is $\xi = \frac{1}{3}(5y - x)$ and $\eta = x$, and when this change of variables has been made the principal terms can be replaced by $A(u_{\xi\xi} + u_{\eta\eta})$, so substituting into (103) and using the results $u_y = \frac{5}{3}u_{\eta}$ and $y = \frac{1}{5}(3\xi - \eta)$ gives the required reduction

$$u_{\xi\xi} + u_{\eta\eta} = \frac{1}{75}[12\eta - 36\xi - 50u_{\xi}]. \qquad \blacksquare$$

EXAMPLE 18.9

Classify and reduce to standard form the PDE

$$u_{xx} + yu_{yy} + \frac{1}{2}u_y + 4yu_x = 0.$$

Solution   This is now a variable coefficient PDE with $A = 1$, $B = 0$, and $C = y$, so the discriminant $d = B^2 - AC = -y$. This shows the equation to be elliptic when $y > 0$, hyperbolic when $y < 0$, and degenerately parabolic on the x-axis.

Elliptic Case $y > 0$

The characteristic equations become

$$\frac{dy}{dx} = -\sqrt{-y} \quad \text{or} \quad \frac{dy}{dx} = -i\sqrt{y}, \quad \text{and} \quad \frac{dy}{dx} = \sqrt{-y} \quad \text{or} \quad \frac{dy}{dx} = i\sqrt{y}.$$

Integrating these complex characteristic equations gives

$$2\sqrt{y} = -ix + \xi - i\eta \quad \text{and} \quad 2\sqrt{y} = ix + \xi + i\eta,$$

and solving for ξ and η we find that $\xi = 2\sqrt{y}$ and $\eta = -x$.
Substituting into (78), (80), (82), and (84) gives

$$u_x = -u_{\eta}, \quad u_y = \frac{1}{\sqrt{y}}u_{\xi}, \quad u_{xx} = u_{\eta\eta}, \quad \text{and} \quad u_{yy} = \frac{1}{y}u_{\xi\xi} - \frac{1}{2y^{3/2}}.$$

Using these results to transform the original PDE gives the standard form

$$u_{\xi\xi} + u_{\eta\eta} = \xi^2 u_{\eta} - \frac{1}{2}(1 - 2/\xi)u_{\xi}.$$

Hyperbolic Case $y < 0$

The characteristic equations become

$$\frac{dy}{dx} = -\sqrt{-y} \quad \text{and} \quad \frac{dy}{dx} = \sqrt{-y},$$

with the respective solutions

$$-2\sqrt{-y} = -x + \xi \quad \text{and} \quad -2\sqrt{-y} = x + \eta,$$

so

$$\xi = x - 2\sqrt{-y} \quad \text{and} \quad \eta = -x - 2\sqrt{-y}.$$

Substituting into (78), (80), (82), and (84) gives

$$u_x = u_\xi - u_\eta, \quad u_y = (1/\sqrt{-y})(u_\xi + u_\eta), \quad u_{xx} = u_{\xi\xi} - 2u_{\xi\eta} + u_{\eta\eta},$$

$$u_{yy} = -(1/y)u_{\xi\xi} - (2/y)u_{\xi\eta} - (1/y)u_{\eta\eta} + \frac{1}{2}(-y)^{-3/2}(u_\xi + u_\eta).$$

When these are substituted into the original PDE it becomes

$$u_{\xi\eta} = \frac{1}{16}(\xi + \eta)^2 (u_\eta - u_\xi) - \left(\frac{1}{\xi + \mu}\right)(u_\xi + u_\eta). \qquad \blacksquare$$

This classification of PDEs can be extended to equations with n independent variables by using the property of orthogonal matrices, which were introduced in Chapter 4. Let the second order constant coefficient PDE for an unknown function $u(x_1, x_2, \ldots, x_n)$ in the n independent variables $x_1, x_2, \ldots, x_n$ be written

classification of PDEs in n independent variables

$$\sum_{i,j=1}^n a_{ij} u_{x_i x_j} + \sum_{i=1}^n b_i u_{x_i} + cu = F(x_1, x_2, \ldots, x_n), \qquad (105)$$

where the coefficients a_{ij}, b_i, and c are real constants and F is a real function of its n arguments. Then, because of the equivalence of mixed partial derivatives, it is always possible to assume the a_{ij} to be symmetric and to write $a_{ij} = a_{ji}$.

We now define an n element column vector $\mathbf{x} = [x_1, x_2, \ldots, x_n]^T$ involving the independent variables, and make a linear transformation of $\mathbf{x}$ to a new set of variables $\xi_1, \xi_2, \ldots, \xi_n$ that can be written as the column vector $\boldsymbol{\xi} = [\xi_1, \xi_2, \ldots, \xi_n]^T$. The linear transformation can be expressed in terms of an $n \times n$ matrix $\mathbf{B} = [b_{ij}]$ with real elements by writing

$$\boldsymbol{\xi} = \mathbf{Bx}. \qquad (106)$$

As with second order PDEs in two independent variables, the classification of the second order PDE (105) is determined by the way in which

$$L[u] = \sum_{i,j=1}^n a_{ij} u_{x_i x_j} \qquad (107)$$

transforms into a standard form that is free from mixed derivatives, so we need only consider the effect of this linear transformation on its principal part $L[u]$, the result of which can be written

$$L[u] = \sum_{i,j=1}^n a_{ij} u_{x_i x_j} = \sum_{p,q=1}^n \left(\sum_{i,j=1}^n b_{pi} a_{ij} b_{qj} \right) u_{\xi_p \xi_q}. \qquad (108)$$

In matrix form this transformation of the leading terms is seen to have the coefficient matrix $\mathbf{BAB}^T$. As $\mathbf{A}$ is symmetric, its eigenvalues $\lambda_1, \lambda_2, \ldots, \lambda_n$ will all be *real*, and it follows from Theorem 4.10 that an orthogonal matrix $\mathbf{Q}$ can always be associated with $\mathbf{A}$ in such a way that $\mathbf{Q}^{\mathrm{T}}\mathbf{A}\mathbf{Q} = \mathbf{D}$, where $\mathbf{D}$ is a diagonal matrix with the eigenvalues of $\mathbf{A}$ as the elements along its leading diagonal. Consequently, if we set $\mathbf{B} = \mathbf{Q}^{\mathrm{T}}$, and

$$
\mathbf{Q} = \begin{bmatrix}
\lambda_1 & 0 & 0 & \cdots & 0 \\
0 & \lambda_2 & 0 & \cdots & 0 \\
0 & 0 & \lambda_3 & \cdots & 0 \\
\cdots & \cdots & \cdots & \cdots & \cdots \\
0 & 0 & 0 & \cdots & \lambda_n
\end{bmatrix},
$$

the principal terms of PDE (105) become

$$
L(u) = \lambda_1 u_{\xi_1\xi_1} + \lambda_2 u_{\xi_2\xi_2} + \cdots + \lambda_n u_{\xi_n\xi_n}. \tag{109}
$$

A simple scaling of the variables $\xi_1, \xi_2, \ldots, \xi_n$ will always reduce the principal terms in $L[u]$ to the form

$$
L(u) = \varepsilon_1 u_{\xi_1\xi_1} + \varepsilon_2 u_{\xi_2\xi_2} + \cdots + \varepsilon_n u_{\xi_n\xi_n}, \tag{110}
$$

where ε_i is $+1$ when $\lambda_i > 0$, -1 when $\lambda_i < 0$, and 0 when $\lambda_i = 0$.

The classification of PDE (105) involves a generalization of the case of two independent variables to n independent variables as follows:

(a) PDE (105) is of **hyperbolic type** if none of the eigenvalues $\lambda_1, \lambda_2, \ldots, \lambda_n$ of $\mathbf{A}$ vanishes and only one eigenvalue has a sign opposite to that of the remaining $n - 1$ eigenvalues. So, if the eigenvalues are ordered such that $\lambda_1 > 0$, after scaling the independent variables $\xi_1, \xi_2, \ldots, \xi_n$ a hyperbolic PDE of type (105) will have the **standard form**

$$
u_{\xi_1\xi_1} = u_{\xi_2\xi_2} + u_{\xi_3\xi_3} \cdots + u_{\xi_n\xi_n} + F(\xi_1, \ldots, \xi_n, u, u_{\xi_1}, \ldots, u_{\xi_n}), \tag{111}
$$

where F is a linear combination of $u, u_{\xi_1}, \ldots, u_{\xi_n}$.

(b) PDE (105) is of **parabolic type** if one of the eigenvalues $\lambda_1, \lambda_2, \ldots, \lambda_n$ of $\mathbf{A}$ vanishes and the remaining $n - 1$ eigenvalues are all of the same sign. So, if the eigenvalues are ordered so that $\lambda_1 = 0$, after scaling the independent variables $\xi_1, \xi_2, \ldots, \xi_n$ a parabolic PDE of type (105) will have the **standard form**

$$
u_{\xi_2\xi_2} + u_{\xi_3\xi_3} \cdots + u_{\xi_n\xi_n} = G(\xi_1, \ldots, \xi_n, u, u_{\xi_1}, \ldots, u_{\xi_n}), \tag{112}
$$

where G is a linear combination of $u, u_{\xi_1}, \ldots, u_{\xi_n}$.

(c) PDE (105) is of **elliptic type** if none of the eigenvalues $\lambda_1, \lambda_2, \ldots, \lambda_n$ of $\mathbf{A}$ vanishes and all have the same sign that may be either positive or negative. So after scaling the independent variables $\xi_1, \xi_2, \ldots, \xi_n$ an elliptic PDE of type (105) will have the **standard form**

$$
u_{\xi_1\xi_1} + u_{\xi_2\xi_2} + u_{\xi_3\xi_3} \cdots + u_{\xi_n\xi_n} = H(\xi_1, \ldots, \xi_n, u, u_{\xi_1}, \ldots, u_{\xi_n}), \tag{113}
$$

where H is a linear combination of $u, u_{\xi_1}, \ldots, u_{\xi_n}$.

EXAMPLE 18.10   Classify the PDE

$$4u_{x_1 x_1} + 4u_{x_2 x_2} + u_{x_3 x_3} - 2u_{x_1 x_2} = 0,$$

and find the form to which it is reduced by an orthogonal transformation that converts its coefficient matrix to a diagonal matrix.

Solution   Because of the equality of mixed derivatives, the matrix form of the PDE can be written $\mathbf{AU} = \mathbf{0}$, where

$$\mathbf{A} = \begin{bmatrix} 4 & -1 & 0 \\ -1 & 4 & 0 \\ 0 & 0 & 1 \end{bmatrix} \quad \text{and} \quad \mathbf{U} = \begin{bmatrix} u_{x_1 x_1} \\ u_{x_2 x_2} \\ u_{x_3 x_3} \end{bmatrix}.$$

The eigenvalues of $\mathbf{A}$ are $\lambda_1 = 1$, $\lambda_2 = 5$, and $\lambda_3 = 3$, so from (a) just shown, the PDE is seen to be of *elliptic* type. As the PDE only contains principal terms, an orthogonal transformation that transforms $\mathbf{A}$ into a diagonal matrix will transform the PDE into

$$u_{\xi_1 \xi_1} + 5u_{\xi_2 \xi_2} + 3u_{\xi_3 \xi_3} = 0.$$

The actual change of variables from x_1, x_2, and x_3 to ξ_1, ξ_2, and ξ_3 necessary to accomplish this was shown in Example 4.18 to be given by $\xi = \mathbf{Qx}$, where $\mathbf{x} = [x_1, x_2, x_3]^T$ and $\xi = [\xi_1, \xi_2, \xi_3]^T$, with the orthogonal matrix $\mathbf{Q}$ and the diagonal matrix $\mathbf{D}$ given by

$$\mathbf{Q} = \begin{bmatrix} 0 & -1/\sqrt{2} & 1/\sqrt{2} \\ 0 & 1/\sqrt{2} & 1/\sqrt{2} \\ 1 & 0 & 0 \end{bmatrix} \quad \text{and} \quad \mathbf{D} = \begin{bmatrix} 1 & 0 & 0 \\ 0 & 5 & 0 \\ 0 & 0 & 3 \end{bmatrix}.$$

So the necessary change of variables determined by $\xi = \mathbf{Qx}$ becomes

$$\xi_1 = -\frac{1}{\sqrt{2}}x_2 + \frac{1}{\sqrt{2}}x_3, \quad \xi_2 = \frac{1}{\sqrt{2}}x_2 + \frac{1}{\sqrt{2}}x_3 \quad \text{and} \quad \xi_3 = x_1.$$

If necessary, the PDE can be further simplified by scaling the variables ξ_1, ξ_2, ξ_3 to arrive at the new variables ζ_1, ζ_2, and ζ_3 by writing

$$\zeta_1 = \xi_1, \quad \zeta_2 = \frac{1}{\sqrt{5}}\xi_2 \quad \text{and} \quad \zeta_3 = \frac{1}{\sqrt{3}}\xi_3,$$

because then the PDE reduces to the standard form

$$u_{\zeta_1 \zeta_1} + u_{\zeta_2 \zeta_2} + u_{\zeta_3 \zeta_3} = 0,$$

which is Laplace's equation in three independent variables.   ∎

    For more information on the classification of PDEs, see references [7.6] and [7.19].

Summary

Linear second order PDEs in two independent variables have been classified and shown to belong to one of three distinct types, namely, PDEs of hyperbolic, parabolic, and elliptic type. Changes of variable were introduced that simplified the structure of each type of equation by reducing it to one of three standard forms. In each case, the method of reduction to standard form was illustrated by an example, and the classification was then extended to linear second order PDEs in n independent variables.

EXERCISES 18.6

In Exercises 1 through 6 classify the given PDE.

1. $4u_{xx} - 6u_{xy} + 3u_{yy} + 2u_x + 6 = 0$.

2. $u_{xx} + 8u_{xy} - 2u_{yy} + u_x + 3u_y + 2u - 3 = 0$.

3. $2u_{xx} - 2u_{xy} + u_{yy} + 4u_x + 2u + 1 = 0$.

4. $4u_{xx} - 4u_{xy} + u_{yy} + 6u_x - u_y + (1 + x)u + 2 = 0$.

5. $3u_{xx} + 6u_{xy} + 3u_{yy} + (1 + \sin x)u = 0$.

6. $2u_{xx} + 2u_{xy} - u_{yy} + 3u_y + u + 5 = 0$.

In Exercises 7 through 12 classify and reduce to standard form the given PDE.

7. $u_{xx} - 2u_{xy} + 5u_{yy} + 3u_x + 1 = 0$.

8. $4u_{xx} + 4u_{xy} + u_{yy} + 4u_y + u = 0$.

9. $u_{xx} - 10u_{xy} + 9u_{yy} + u_x = 0$.

10. $u_{xx} - 4u_{xy} - 5u_{yy} + 3u_y + u + 4 = 0$.

11. $u_{xx} + 6u_{xy} + 9u_{yy} - u + 5 = 0$.

12. $2u_{xx} - 6u_{xy} + 5u_{yy} + 4u_x + u_y - 2 = 0$.

In Exercises 13 through 16 classify the PDE, and by using a suitable orthogonal matrix **Q** followed, if necessary, by a scaling of the independent variables, reduce it to standard form.

13.* $5u_{x_1 x_1} + 2u_{x_2 x_2} + 8u_{x_2 x_3} + 2u_{x_2} + 4u + 1 = 0$.

14.* $2u_{x_2 x_2} - 4u_{x_1 x_3} + u_{x_3} + 1 = 0$.

15.* $3u_{x_1 x_1} + 2u_{x_2 x_2} - 2u_{x_2 x_3} + 2u_{x_3 x_3} + 4u - 7 = 0$.

16.* $u_{x_1 x_1} + 2u_{x_2 x_3} + u_{x_2} + 5u + 2 = 0$.

18.7   Boundary Conditions and Initial Conditions

The PDEs derived in Section 18.5, and classified in Section 18.6, are special cases of the general linear PDE for an unknown function $u(x, y)$ of the two independent variables x and y

$$Au_{xx} + 2Bu_{xy} + Cu_{yy} + Pu_x + Qu_y + Ru = F(x, y), \qquad (114)$$

though sometimes with y replaced by t.

Physical problems whose solution is governed by a PDE of this type are formulated in some region D of the (x, y)-plane on the boundary Γ of which suitable auxiliary conditions, called **boundary conditions**, are imposed that serve to identify a particular problem. The most important types of boundary conditions are as follows:

(a) The specification of the functional form to be taken by the solution $u(x, y)$ on the boundary Γ, by requiring that

$$u(x, y) = \Phi(x, y) \quad \text{for } (x, y) \text{ on } \Gamma, \qquad (115)$$

Dirichlet boundary condition

where $\Phi(x, y)$ is a given function. A boundary condition of this type is called a **Dirichlet condition**.

(b) The specification of the functional form to be taken by the derivative of the solution $u(x, y)$ normal to the boundary Γ, by requiring that

$$\frac{\partial u}{\partial n}(x, y) = \Psi(x, y) \quad \text{for } (x, y) \text{ on } \Gamma, \qquad (116)$$

Neumann boundary condition

where $\Psi(x, y)$ is a given function and $\partial/\partial n$ is the directional derivative normal to the boundary Γ. A boundary condition of this type is called a **Neumann condition**.

(c) The specification of the functional form to be taken by a linear combination of a Dirichlet condition and a Neumann condition by the solution $u(x, y)$ on the boundary Γ, by requiring that

$$a(x, y)u(x, y) + b\frac{\partial u}{\partial n}(x, y) = c(x, y) \quad \text{for } (x, y) \text{ on } \Gamma, \qquad (117)$$

mixed or Robin boundary condition

where $a(x, y)$, $b(x, y)$, and $c(x, y)$ are given functions. A boundary condition of this type is called a **mixed condition**, and sometimes either a **Robin condition** or a **boundary condition of the third kind**. When $c(x, y) = 0$, this condition is called a **homogeneous mixed condition**.

(d) The specification on Γ of the functional form to be taken by both the solution $u(x, y)$ and its derivative normal to the boundary, by requiring that

$$u(x, y) = \Phi(x, y) \quad \text{and} \quad \frac{\partial u}{\partial n}(x, y) = \Psi(x, y) \quad \text{for } (x, y) \text{ on } \Gamma, \qquad (118)$$

Cauchy conditions

where $\Phi(x, y)$ and $\Psi(x, y)$ are given functions and $\partial/\partial n$ is the directional derivative normal to the boundary Γ. Boundary conditions of this type are called **Cauchy conditions** for a second order PDE.

When the solution u is a function of a space variable x and the time t, and Cauchy conditions are specified when $t = 0$, so that Γ becomes the x-axis and

$$u(x, 0) = \Phi(x) \quad \text{and} \quad \frac{\partial u}{\partial t}(x, 0) = \Psi(x), \qquad (119)$$

initial conditions

the Cauchy conditions are usually called **initial conditions** for a second order PDE.

The types of boundary condition that can be imposed on PDE (114) depend on its classification and the nature of the region D that is involved. Some typical examples of boundary conditions and their associated regions for PDEs of hyperbolic, parabolic, and elliptic type were seen in Section 18.5 when the three types of equation were derived from physical problems.

open and closed regions

A region D is classified as being **closed** when it is enclosed by a boundary and every point on the boundary belongs to D, and as being **open** when either the region D extends to infinity or, although D is contained within a boundary, not all of the points of the boundary belong to D. Typical closed regions are the rectangle $a \leq x \leq b, c \leq y \leq d$, and the annular region $R_1 \leq r \leq R_2$ centered on the origin. Examples of open regions are the semi-infinite strip $a \leq x \leq b, y \geq 0$, where the boundary points on three sides of the strip belong to the region but there is no upper boundary because y extends to infinity, and the annular region $R_1 < r \leq R_2$, where the points on the outer rim of the annulus belong to the region but the points on the inner rim do not.

well-posed and improperly posed problems

When the boundary conditions and the region D are such that a unique solution exists, and small changes in the boundary conditions only produce small changes in the solution, the boundary value problem is said to be **well posed**, and the solution is said to be **stable**. If, however, the boundary conditions and/or region are such that although a solution exists, a small change in the boundary conditions causes a large change in the solution, the boundary value problem is said to be **improperly**

posed and the solution is then said to be **unstable**. In what follows our concern will only be with well-posed problems.

Listed next are the most frequently occurring combinations of boundary conditions and regions that lead to properly posed problems for hyperbolic, parabolic, and elliptic PDEs.

appropriate conditions and regions for the three types of PDE

Type of PDE	Conditions	Type of Region
Hyperbolic	Cauchy conditions	Open
Parabolic	Dirichlet, Neumann, or mixed	Open
Elliptic	Dirichlet, Neumann, or mixed	Closed

The effect of imposing inappropriate boundary conditions on a PDE can lead to one of the following situations: (a) no solution exists, (b) a solution exists, but it is either trivial (identically zero) or not unique, and (c) a solution exists, but it is not stable.

To demonstrate that appropriateness of the preceding conditions, by way of example we prove that the Dirichlet problem for the Laplace equation in a closed region is a properly posed problem. To do this we will make use of Theorem 14.17, which showed that a harmonic function defined in a closed region D with boundary Γ must attain its greatest and least values on the boundary Γ. A trivial corollary of this theorem that will be needed, and that is almost immediately obvious, is that if $u(x, y)$ is harmonic in D and is equal to the constant k on the boundary of Γ of D, then $u(x, y) = k$ throughout D.

Let us use this theorem to prove the uniqueness of a function u that is harmonic in D and satisfies a Dirichlet condition $u|_{\Gamma} = f(s)$ on Γ, where the parameter s can be taken to be the arc length measured around Γ from some fixed point on the boundary. Suppose, if possible, that this Dirichlet problem has two *different* solutions u and v that satisfy the *same* Dirichlet condition, and set $w = u - v$. Then because the Laplace equation is linear, w is also a solution of Laplace's equation, and on Γ it satisfies the homogeneous boundary condition $w|_{\Gamma} = 0$. Using the corollary of the maximum–minimum theorem mentioned earlier, it follows at once that $w \equiv 0$ throughout D, and so $u \equiv v$, and the uniqueness of the solution has been established.

As a further demonstration of the appropriateness of Dirichlet conditions for the Laplace equation, we now prove that small changes in the boundary conditions produce small changes in the solution, because this shows the continuous dependence of the solution on the boundary data (Dirichlet condition). Let u_1 and u_2 be solutions of two different Dirichlet problems for the Laplace equation in a closed region D with boundary Γ on which u_1 satisfies the Dirichlet condition $u_1|_{\Gamma} = f_1(s)$ and u_2 satisfies the Dirichlet condition $u_2|_{\Gamma} = f_2(s)$, where s is defined as before and $|f_1(s) - f_2(s)| < \varepsilon$ on Γ, with $\varepsilon > 0$ arbitrarily small. The difference $u_1 - u_2$ is also harmonic in D, so the condition on $f_1 - f_2$ is equivalent to $-\varepsilon < u_1 - u_2 < \varepsilon$ on Γ. It follows directly from the maximum–minimum theorem that throughout D we must have $-\varepsilon < u_1 - u_2 < \varepsilon$, so $|u_1 - u_2| < \varepsilon$. This has established that when the Dirichlet data is only changed by a small amount, the same is true of the solution, so the continuous dependence of the solution on the Dirichlet data has been established. This result, combined with the uniqueness of the solution, shows that the Dirichlet problem for the Laplace equation is well posed.

Summary

The main types of boundary condition suitable for second order PDEs were described, and the notion of a well-posed problem was introduced. Open and closed regions were defined and a short table was given listing suitable combinations of boundary condition and region according to the type of PDE involved.

18.8   Waves and the One-Dimensional Wave Equation

The general solution of the one-dimensional wave equation

$$\frac{\partial^2 u}{\partial t^2} = c^2 \frac{\partial^2 u}{\partial x^2} \tag{120}$$

has a useful interpretation in terms disturbances that move with speed c in opposite directions along the x-axis.

It is known from Section 18.6 that the characteristic equations for the wave equation are

$$\frac{dx}{dt} = c \quad \text{and} \quad \frac{dx}{dt} = -c.$$

Integrating the first of these equations to find the characteristic through the point $(\xi, 0)$ on the x-axis (the initial line) gives

$$\xi = x - ct, \tag{121}$$

and integrating the second equation to find the characteristic through the point $(\eta, 0)$ on the x-axis gives

$$\eta = x + ct. \tag{122}$$

Changing the independent variables in (120) from x and t to ξ and η reduces it to the standard form $u_{\xi\eta} = 0$. Integrating this result partially with respect to η, while regarding ξ as a constant in order to reverse the process of partial differentiation, gives $u_\xi = f(\xi)$, where F is an arbitrary differentiable function of ξ. Next, integrating this result partially with respect to ξ, where η is now regarded as a constant, leads to the result $u(\xi, \eta) = f(\xi) + g(\eta)$, where $f(\xi) = \int F(\xi)d\xi$ and g is an arbitrary differentiable function of η. Notice that as $F(\xi)$ is an arbitrary function, so also is $f(\xi)$. Finally, if we revert to the original variables x and t, the general solution of (120) becomes

general solution of wave equation as the sum of two waves moving in opposite directions

$$u(x, t) = f(x - ct) + g(x + ct). \tag{123}$$

The function $f(x - ct)$ will be constant along the characteristic $x - ct = $ constant, so considering all possible characteristics of this type the term $f(x - ct)$, as in Section 18.3, in (123) is seen to transport the initial shape of the function f to the right along the x-axis with constant speed c without change of either shape or scale. Similarly, the term $g(x + ct)$ will transport the initial shape of the function g to the left along the x-axis, also with the constant speed c and without change of shape or scale.

Disturbances that propagate through space at a finite speed as time increases are called **waves**, so the general solution in (123) represents two traveling waves moving in opposite directions, each with the constant speed c. The initial disturbances $f(x)$ and $g(x)$ are called the **wave profiles**, so the shape of each wave profile in (123) is preserved as it propagates.

The interpretation of the general solution (123) of the wave equation in (120) is now clear, because it shows that an initial disturbance $u(x, 0)$ is resolved into two traveling waves, one moving to the right and the other to the left, each with the same speed c and without change of shape or scale. The general solution also shows that the disturbance (wave) propagated by the wave equation at any time t is the *sum* of the disturbances caused by the traveling waves as they move to the left and right, so that (123) describes the *interaction* of the two wave profiles. This very important property of the wave equation is due to its *linearity*, which allows the sum of any two solutions to be another solution.

To make effective use of this result when seeking the solution of a Cauchy problem for the wave equation, it is necessary to know how the initial disturbance $u(x, 0)$ is resolved into the functions $f(x)$ and $g(x)$. We will see how to find $f(x)$ and $g(x)$ in terms of the Cauchy conditions when the D'Alembert solution of the one-dimensional wave equation is derived in the next section.

In Section 18.5 the wave equation was derived under the assumption that $u(x, t)$ is a continuous and twice differentiable function of its arguments. We will now use result (123) to show how these conditions can be relaxed to allow for initial wave profiles that have a discontinuity in their derivative, or even a finite jump discontinuity in the functions themselves.

Suppose that the wave profile $f(x)$ has a discontinuity either in its derivative or in the function itself at some point $x = x^*$. The characteristic through $x = x^*$ does not depend on the solution, so the propagating wave profile can be separated into two distinct parts, one to the left of the characteristic $x - ct = x^*$, and the other to the right.

The characteristics to the immediate left and right of $x - ct = x^*$ are both parallel to it. This means that the wave profile to the left of this characteristic will propagate in the manner just described, independently of the wave to the right, but bounded on the right by $x - ct = x^*$. Similarly, the wave profile to the right of this characteristic will propagate in the manner described, independently of the solution to the left, but bounded on the left by $x - ct = x^*$. As the solutions to the immediate left and right move along the same characteristic $x - ct = x^*$, any initial discontinuity in f will be propagated along this characteristic without change.

The same result is true for the initial wave profile $g(x)$, so the interpretation of the general solution (123) as the sum of disturbances due to the two wave profiles propagating to the right and left remains valid even when a discontinuity in the derivative or in the initial disturbance $u(x, 0)$ itself is present. This generalizes the concept of a solution of the wave equation, because it permits initial disturbances with discontinuities in either a derivative or the function itself. This situation is quite different from the quasilinear case considered in Section 18.4, because there a discontinuity in the solution was seen to propagate at the shock speed, which is quite different from that of the adjacent characteristic speeds.

We now use this generalization to examine the resolution of an initial disturbance that is localized in a finite part $a < x < b$ of the x-axis, and zero outside it. The purpose of this is to make clear how the two wave profiles interact until, after a suitable lapse of time, they move clear of one another, after which all interaction

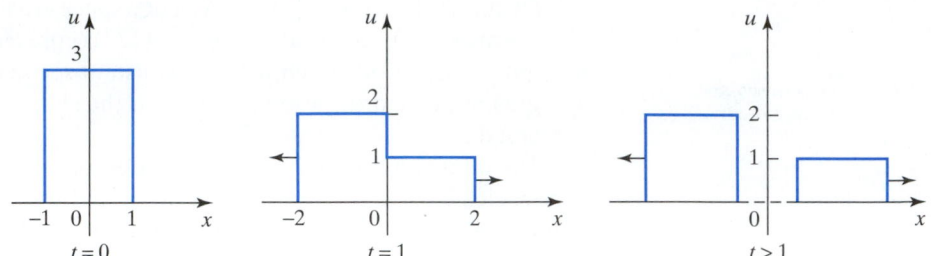

FIGURE 18.15 The resolution of a finite-width initial disturbance into two separate waves that move apart.

ceases. We consider the special case of two initial wave profiles of rectangular shape and the same width, but different heights, that at time $t = 0$ are given by

$$f(x) = \begin{cases} 0, & x < -1 \\ 1, & -1 < x < 1 \quad \text{and } g(x) = \begin{cases} 0, & x < -1 \\ 2, & -1 < x < 1 \\ 0, & x > 1. \end{cases} \\ 0, & x > 1 \end{cases}$$

The evolution of this initial disturbance is shown in Fig. 18.15 for the case $c = 1$. Wave interaction continues until the two disturbances have separated, after which the initial disturbance is represented by two distinct traveling waves.

This result can be explained differently if the wave equation is written in either of the two equivalent forms

$$\left(\frac{\partial}{\partial t} - c\frac{\partial}{\partial x}\right)\left(\frac{\partial u}{\partial t} + c\frac{\partial u}{\partial x}\right) = 0 \quad \text{or} \quad \left(\frac{\partial}{\partial t} + c\frac{\partial}{\partial x}\right)\left(\frac{\partial u}{\partial t} - c\frac{\partial u}{\partial x}\right) = 0. \quad (124)$$

An examination of the first of these representations shows that a solution of the *first order* PDE obtained by equating to zero the second group of bracketed terms, namely $u_t + cu_x = 0$, is also a solution of the wave equation. The first order PDE describes a traveling wave of constant shape that propagates to the right at a constant speed c. This special solution is a **degenerate solution** of the wave equation, because it is a solution of a *first order* PDE that is also a *special* solution of a *second order* PDE. Furthermore, unlike the general solution of the wave equation, it is a wave that only moves in *one* direction. When interpreted in terms of the initial conditions used in Fig. 18.15, this degenerate solution is seen to describe the initial wave profile $f(x)$ that after all interaction has ceased becomes the part of the solution of the wave equation that moves to the right.

degenerate solutions

A corresponding argument applied to the other form of the wave equation when the second bracketed term is set equal to zero, so that $u_t + cu_x = 0$, describes a similar degenerate solution that this time moves to the left.

Summary

The wave equation was shown to have a general solution that can be interpreted as the sum of two independent waves moving with the same speed, but in opposite directions. The nature of the solution was used to explain how in wave propagation involving an initial wave profile with discontinuities, the discontinuities propagate along the characteristic curves of the wave equation. A factorization of the wave equation operator was then used to show how special degenerate solutions can arise.

In Exercises 1 through 4, the functions $f(x)$ and $g(x)$ refer to the functions in the general solution of the wave equation given in (123). Taking $c = 1$, plot the form of the solution $u(x, t)$ at two different stages during the interaction of the waves, and plot the form of the solution after the waves have separated and all interaction has ceased.

1. $f(x) = \begin{cases} 0, & x < -1 \\ 1+x, & -1 < x < 1 \\ 0, & x > 1 \end{cases}$,

$g(x) = \begin{cases} 0, & x < -1 \\ 1, & -1 < x < 1 \\ 0, & x > 1. \end{cases}$

2. $f(x) = \begin{cases} 0, & x < -\pi/2 \\ \cos x, & -\pi/2 < x < \pi/2 \\ 0, & x > \pi/2 \end{cases}$,

$g(x) = \begin{cases} 0, & x < -\pi/2 \\ 1 + \dfrac{2}{\pi}x, & -\pi/2 < x < \pi/2 \\ 0, & x > \pi/2. \end{cases}$

3.* $f(x) = \begin{cases} 0, & x < -1 \\ 1 - x^2, & -1 < x < 1 \\ 0, & x > 1 \end{cases}$,

$g(x) = \begin{cases} 0, & x < -1 \\ 1, & -1 < x < 1 \\ 0, & x > 1. \end{cases}$

4.* $f(x) = \begin{cases} 0, & x < -1 \\ 2, & -1 < x < 1 \\ 0, & x > 1 \end{cases}$,

$g(x) = \begin{cases} 0, & x < -1 \\ 1 - x^2, & -1 < x < 1 \\ 0, & x > 1. \end{cases}$

18.9   The D'Alembert Solution of the Wave Equation and Applications

We now derive the promised representation of the solution of the one-dimensional wave equation in terms of its Cauchy conditions that shows explicitly the way in which each of the initial conditions influences the solution. This form of solution is called the **D'Alembert solution**, and the starting point for its derivation is the one-dimensional wave equation for the unknown function $u(x, t)$ where x is a space variable and t is the time.

Let us consider the initial value problem for the homogeneous one-dimensional wave equation

$$\frac{\partial^2 u}{\partial t^2} = c^2 \frac{\partial^2 u}{\partial x^2} \quad (c = \text{constant}), \tag{125}$$

subject to the Cauchy conditions

$$u(x, 0) = h(x) \quad \text{and} \quad u_t(x, 0) = k(x), \tag{126}$$

where h and k are suitably differentiable functions defined on the initial line $-\infty < x < \infty$.

It is known from (123) that the general solution of (125) is

$$u(x, t) = f(x - ct) + g(x + ct), \tag{127}$$

where f and g are arbitrary functions of their arguments. Our task will be to find the functions f and g so the solution of the wave equation satisfies the Cauchy conditions in (126). One equation relating f and g follows immediately by setting $t = 0$ in (127) and using the first condition in (126), which gives

$$f(x) + g(x) = h(x). \tag{128}$$

To find another equation we differentiate (127) once partially with respect to t, set $t = 0$, and use the second condition in (126), when we obtain

$$-cf'(x) + cg'(x) = k(x). \tag{129}$$

Integration of (129) from an arbitrary fixed point a on the initial line to a general point x gives

$$-f(x) + g(x) = \frac{1}{c}\int_a^x k(\sigma)d\sigma + g(a) - f(a). \tag{130}$$

Eliminating first $f(x)$ and then $g(x)$ between (128) and (130) gives

$$f(x) = \frac{1}{2}h(x) - \frac{1}{2c}\int_a^x k(\sigma)d\sigma - \frac{1}{2}(g(a) - f(a))$$

and

$$g(x) = \frac{1}{2}h(x) + \frac{1}{2c}\int_a^x k(\sigma)d\sigma + \frac{1}{2}(g(a) - f(a)).$$

If in the expression for $f(x)$ we now replace x by $x - ct$, and in the expression for $g(x)$ we replace x by $x + ct$ and add the results, it follows from (127) that the solution $u(x, t)$ becomes

$$u(x, t) = \frac{1}{2}\left\{h(x - ct) + h(x + ct) - \frac{1}{c}\int_a^{x-ct} k(\sigma)d\sigma + \frac{1}{c}\int_a^{x+ct} k(\sigma)d\sigma\right\}.$$

the D'Alembert solution of the wave equation

Reversing the limits on the first integral, and compensating by changing its sign, allows the two integrals to be combined to give the **D'Alembert solution** of the wave equation:

$$u(x, t) = \frac{h(x - ct) + h(x + ct)}{2} + \frac{1}{2c}\int_{x-ct}^{x+ct} k(\sigma)d\sigma. \tag{131}$$

The structure of this solution gives important information about the way the Cauchy conditions enter into the solution of the initial value problem. The implications of (131) can best be understood by interpreting the D'Alembert solution in terms of Fig. 18.16. Consider a representative point P located at (x_0, t_0) in the upper half of the (x, t)-plane, and trace back to the initial line the two characteristics that pass through P with slopes $\pm c$ until they meet the line at points A at $x_0 - ct_0$ and B at $x_0 + ct_0$.

The D'Alembert solution in (131) then shows that the solution at P only depends on the Cauchy conditions over the interval AB on the initial line. Specifically, the solution $u(x_0, t_0)$ only depends on the function $h(x)$ through the two values $h(x_0 - ct_0)$ and $h(x_0 + ct_0)$ at the *ends* of the interval AB, and on $k(x)$ through its integral over the same interval. Because of this, the interval $x_0 - ct_0 \leq x \leq x_0 + ct_0$ on the initial line is called the **domain of dependence** of the solution at the point (x_0, t_0), and points inside the triangle ABP are said to belong to the **domain of**

domain of dependence and determinacy

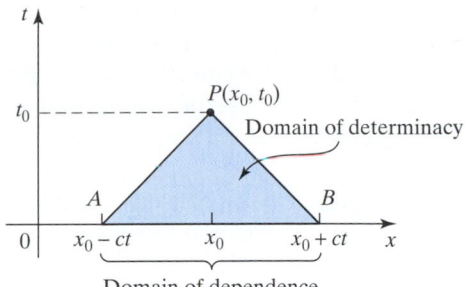

FIGURE 18.16 Domain of dependence and the D'Alembert solution.

determinacy of the interval, because the solution at every point inside this triangle is completely determined by the Cauchy conditions on this interval.

The D'Alembert solution also shows the suitability of Cauchy conditions for the wave equation because they lead to a solution.

The solution is unique, because from the linearity of the wave equation, if two *different* solutions $u(x, t)$ and $v(x, t)$ exist, both satisfying the *same* Cauchy conditions, then the difference between the two solutions $w(x, t) = u(x, t) - v(x, t)$ must also be a solution. The Cauchy conditions for w are $w(x, 0) = 0$ and $w_t(x, 0) = 0$, corresponding to $h(x) \equiv 0$ and $k(x) \equiv 0$, so we conclude from the D'Alembert solution that $w \equiv 0$, and hence that $u \equiv v$.

We can also use the D'Alembert solution to show the stability of the solution of the wave equation subject to Cauchy conditions, in the sense that a small change in the Cauchy conditions only produces a correspondingly small change in the solution. To show this, let us suppose that $u_1(x, t)$ and $u_2(x, t)$ are two *different* solutions of the wave equation that correspond to the respective *different* Cauchy conditions

showing the stability of the solution of a Cauchy problem for the wave equation

$$u_1(x, 0) = h_1(x), \quad u_{1t}(x, 0) = k_1(x), \quad u_2(x, 0) = h_2(x), \quad \text{and}$$

$$u_{2t}(x, 0) = k_2(x).$$

Now let these two sets of Cauchy conditions be close together in the sense that

$$|h_1(x) - h_2(x)| < \varepsilon_1 \quad \text{and} \quad |k_1(x) - k_2(x)| < \varepsilon_2,$$

where $\varepsilon_1 > 0$ and $\varepsilon_2 > 0$ are two arbitrarily small numbers. Applying the elementary integral inequality $|\int_a^b p(x)dx| \leq \int_a^b |p(x)|dx$ to this last result gives

$$|u_1(x, t) - u_2(x, t)| < \frac{1}{2}|h_1(x - ct) - h_2(x - ct)| + \frac{1}{2}|h_1(x + ct) - h_2(x + ct)|$$

$$+ \frac{1}{2c} \int_{x-ct}^{x+ct} |k_1(\sigma) - k_2(\sigma)|d\sigma,$$

so as $|k_1(x) - k_2(x)| < \varepsilon_2$ this last result becomes

$$|u_1(x, t) - u_2(x, t)| < \frac{1}{2}\varepsilon_1 + \frac{1}{2}\varepsilon_1 + \frac{\varepsilon_2}{2c} \int_{x-ct}^{x+ct} d\sigma.$$

Finally, after evaluating the integral, we arrive at the result

$$|u_1(x, t) - u_2(x, t)| < \varepsilon_1 + \varepsilon_2 t.$$

This shows that for any time $\tau \leq t$, and arbitrary fixed t, when the two sets of Cauchy data are close together, the corresponding solutions of the wave equation will also be close together, confirming the stability of the solution. The existence of a unique stable solution of the wave equation subject to Cauchy conditions has established that the problem is properly posed.

JEAN-LE-ROND D'ALEMBERT (1717–1783)
A French mathematician born in Paris who was abandoned as a baby near the church of Saint Jean-le-Ronde where he was found by a gendarme who had him christened with the name of the church where he was found. Later, for an unknown reason, he added the name D'Alembert. He was brought up by the wife of a poor glazier, and when he showed early brilliance, his education in law was paid for by his natural father, but his fascination with mathematics was such that he soon abandoned law and devoted himself to the study of mathematics. At the age of 24 he was admitted to the French Academy, and in 1743 he published his great work on mechanics based on what is now known as D'Alembert's principle. He made important contributions to the study of fluid flow, to the study of waves on vibrating strings and elsewhere, and in 1754 made the important suggestion, not to be acted upon until much later, that the then theory of limits needed to be placed on a sound basis. His last years were spent working on the great French encyclopedia.

For reference purposes we state without proof (see, for example, reference [7.20]) that a modification of the preceding argument shows the solution of the **nonhomogeneous wave equation**

the solution of the nonhomogeneous wave equation

$$\frac{\partial^2 u}{\partial t^2} = c^2 \frac{\partial^2 u}{\partial x^2} + f(x, t) \tag{132}$$

is given by

$$u(x, t) = \frac{h(x - ct) + h(x + ct)}{2} + \frac{1}{2c} \int_{x-ct}^{x+ct} k(\sigma)\,d\sigma$$
$$+ \frac{1}{2c} \int_0^t \int_{x-c(t-\tau)}^{x+c(t-\tau)} f(\sigma, \tau)\,d\sigma\,d\tau. \tag{133}$$

An important and useful result can be derived directly from the general solution of the wave equation in (123), and the fact that its characteristics are $x - ct = $ constant and $x + ct = $ constant. Consider Fig. 18.17, where the four points A at (x_A, t_A), B at (x_B, t_B), C at (x_C, t_C), and D at (x_D, t_D) lie at the corners of a parallelogram, the sides of which are characteristics.

Using the equations of the characteristics, the coordinates of the points A, B, C, and D are seen to be related by

$$x_B - ct_B = x_C - ct_C, \quad x_A - ct_A = x_D - ct_D$$
$$x_A + ct_A = x_B + ct_B, \quad x_D + ct_D = x_C + ct_C. \tag{134}$$

a useful functional relationship connecting solutions at the corners of a parallelogram formed by characteristic lines

The sums $u(A) + u(C)$ and $u(B) + u(D)$ of the solutions at A, B, C, and D can be written

$$u(A) + u(C) = f(x_A - ct_A) + g(x_A + ct_A) + f(x_C - ct_C) + g(x_C + ct_C)$$

and

$$u(B) + u(D) = f(x_B - ct_B) + g(x_B + ct_B) + f(x_D - ct_D) + g(x_D + ct_D).$$

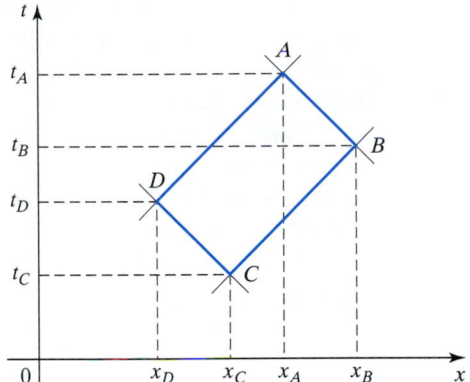

FIGURE 18.17   A parallelogram with sides that coincide with characteristics.

Using the results in (134), we see that these two results are equal, so we have proved that

$$u(A) + u(C) = u(B) + u(D). \tag{135}$$

This result can be used in various ways, one of which is in conjunction with the D'Alembert solution to solve an initial boundary value problem for the wave equation. Let us now find the solution of the wave equation

$$\frac{\partial^2 u}{\partial t^2} = c^2 \frac{\partial^2 u}{\partial x^2} \tag{136}$$

in the quarter-plane $x \geq 0$, $t > 0$ shown in Fig. 18.18, where the solution $u(x, t)$ is required to satisfy the Cauchy conditions $u(x, 0) = h(x)$ and $u_t(x, 0) = k(x)$ on the positive x-axis $x \geq 0$, and the boundary condition $u(0, t) = U(t)$ on the line $x = 0$.

solving an initial boundary value problem

The D'Alembert solution (131) gives the solution in the lower triangular region in Fig. 18.18, but not in the upper triangular region. To find the solution in the upper triangular region we will make use of the D'Alembert solution and result (135).

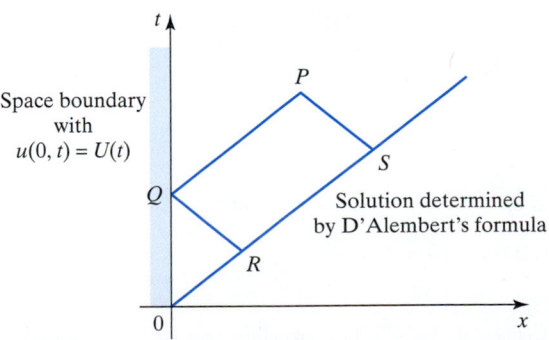

FIGURE 18.18   An initial boundary value problem.

Let P be any point in the upper triangular region, and draw the two character-istics of the wave equation with slopes c and $-c$ that pass through it. Let Q be the point where the characteristic with slope c meets the boundary $x = 0$, and S be the point where the characteristic with slope $-c$ meets the upper boundary of the lower triangular region. Let R be the point where the characteristic through Q with slope $-c$ meets the upper boundary of the lower triangular region. Then, as the sides of the parallelogram $PQRS$ are characteristics, result (135) can be used to relate the solutions at P, Q, R, and S.

The solution $u(x, t)$ at any point in the upper triangular region is now known, because from (135)

$$u(P) = u(Q) + u(S) - u(R),$$

and the solutions at $u(R)$ and $u(S)$ are determined by the D'Alembert solution, while the solution at $u(Q)$ is determined by the given boundary condition $u(0, t) = U(t)$.

This method of solution of an initial boundary value problem in the first quad-rant of the (x, t)-plane can be extended to include the case of a semi-infinite strip $a \leq x \leq b, t > 0$ in a straightforward manner, though the details are left as an exercise.

A special case of an initial boundary value problem can be solved by means of the D'Alembert solution without appeal to result (135). To see how this can be done we consider the pure initial value problem for the wave equation

$$u(x, 0) = h(x) \quad \text{and} \quad u_t(x, 0) = k(x), \tag{137}$$

where h and k are bounded odd functions, so that $h(-x) = -h(x)$ and $k(-x) = -k(x)$. Notice that as h and k are odd functions, this implies that $h(0) = k(0) = 0$. The D'Alembert solution applies for all x and $t > 0$, so

$$u(x, t) = \frac{h(x - ct) + h(x + ct)}{2} + \frac{1}{2c} \int_{x-ct}^{x+ct} k(\sigma) d\sigma, \tag{138}$$

but as $h(0) = k(0) = 0$, (138) shows that $u(0, t) = 0$.

When in (131) the sign of x is reversed the result becomes

$$u(-x, t) = \frac{h(-x - ct) + h(-x + ct)}{2} + \frac{1}{2c} \int_{-x-ct}^{-x+ct} k(\sigma) d\sigma. \tag{139}$$

However, as h is an odd function, $h(-x - ct) = -h(x + ct)$ and $h(-x + ct) = -h(x - ct)$, so the change of variable $s = -\sigma$ coupled with the fact that k is also an odd function shows that

$$\frac{1}{2c} \int_{-x-ct}^{-x+ct} k(\sigma) d\sigma = -\frac{1}{2c} \int_{x-ct}^{x+ct} k(s) ds.$$

Using these results in (139) and comparing the result with (138) shows that

$$u(-x, t) = -u(x, t). \tag{140}$$

The implication of this result is that if in the D'Alembert solution the Cauchy conditions imposed on the initial line $t = 0$ are such that h and k are *odd* functions, then if attention is restricted to the *first* quadrant $x \geq 0$, $t > 0$, the D'Alembert solution of this initial value problem solves the initial boundary value problem in which

$$u(x,0) = h(x), \quad u_t(x,0) = k(x) \quad \text{and} \quad u(0,t) = 0, \text{ for } x > 0. \tag{141}$$

reflecting boundary

A useful physical interpretation of this result can be obtained by considering the boundary $x = 0$ to be a **reflecting boundary**, with the property that when a wave moving to the left encounters the boundary it is reflected back in the positive x-direction with a change of sign.

A corresponding result can be derived by assuming h and k to be *even* functions, for then a similar argument shows that the boundary condition imposed on $x = 0$ is the condition

$$u_x(0,t) = 0, \tag{142}$$

but this time when a reflection occurs at the boundary $x = 0$, a wave moving to the left is reflected back in the positive x-direction *without* a change of sign. The details of the proof of this result are left as an exercise.

One-dimensional wave propagation governed by the wave equation is discussed in some detail in references [7.3], [7.10], [7.11], and [7.17] to [7.20].

EXERCISES 18.9

1. Show by differentiation that if f and g are twice differentiable functions of their arguments, $u(x,t) = f(x - ct) + g(x + ct)$ is a solution of the wave equation $u_{tt} = c^2 u_{xx}$.

2. For what value of c is

$$u(x,t) = \frac{1}{2}(x - 4t + 1)e^{-(x-4t)} + \frac{1}{2}(x + 4t - 1)e^{-(x+4t)}$$

a solution of the wave equation $u_{tt} = c^2 u_{xx}$? Find the Cauchy conditions that, when applied to this wave equation, give rise to this solution.

In Exercises 3 through 6 use the D'Alembert solution to solve the given Cauchy problem for the wave equation $u_{tt} = c^2 u_{xx}$.

3. $u(x,0) = \sin x$, $u_t(x,0) = 1/(1 + x^2)$.

4. $u(x,0) = 1$, $u_t(x,0) = \cos x$.

5. $u(x,0) = \tanh x$, $u_t(x,0) = \text{sech}^2 x$.

6. $u(x,0) = e^x$, $u_t(x,0) = e^{-x}$.

7. Suggest how the D'Alembert solution and result (135) can be used to solve the initial boundary value problem for the wave equation $u_{tt} = c^2 u_{xx}$ in the semi-infinite strip $a \leq x \leq b$, $t > 0$ when $u(x,0) = h(x)$ with $h(a) = h(b) = 0$, $u_t(x,0) = k(x)$ and $u(a,t) = u(b,0) = 0$. Does this method provide a practical way of solving this initial boundary value problem?

8. By using the form of argument that led to the notion of a reflecting boundary, show that by taking $h(x)$ and

$k(x)$ to be *even* functions, the solution given by the D'Alembert formula in the first quadrant solves the initial boundary value problem in that quadrant when

$$u(x,0) = f(x), \quad u_t(x,0) = g(x) \text{ for } x \leq 0,$$

and u satisfies the boundary condition $u_x(0,t) = 0$.

9. Suggest how the D'Alembert solution may be used together with a reflecting boundary to solve the initial boundary value problem in the semi-infinite strip $-a \leq x \leq a$, $t > 0$, subject to the initial and boundary conditions

$$u(x,0) = f(x), u_t(x,0) = g(x) \quad \text{and}$$
$$u(-a,t) = u(a,t) = 0.$$

10. Repeat Exercise 9 with the same initial conditions but with the boundary conditions changed to

$$u(-a,t) = 0 \quad \text{and} \quad u_x(a,t) = 0.$$

11. Write down the D'Alembert solution for the wave equation $u_{tt} = c^2 u_{xx}$ given that the Cauchy conditions are $u(x,0) = f(x)$ and $u_t(x,t) = 0$. Sketch the solution at the times $t = 0$, $1/(2c)$, $1/c$, and $3/(2c)$ using the foregoing initial conditions with

$$f(x) = \begin{cases} 0, & x < -1 \\ -1 - x, & -1 \leq x < 0 \\ 1 - x, & 0 \leq x < 1 \\ 0, & x \geq 1. \end{cases}$$

12. Repeat Exercise 11, but with

$$f(x) = \begin{cases} 0, & x < -1 \\ 1+x, & -1 \le x < 0 \\ 1-x, & 0 \le x < 1 \\ 0, & x \ge 1. \end{cases}$$

conditions are $u(x, 0) = 0$ and $u_t(x, 0) = g(x)$, where

$$g(x) = \begin{cases} 0, & x < -1 \\ 1-x^2, & -1 \le x \le 1 \\ 0, & x > 1. \end{cases}$$

13. Write down the D'Alembert solution at the time $t = \frac{1}{4}$ for the wave equation $u_{tt} = u_{xx}$, given that the Cauchy

14. Repeat Exercise 13 with the same Cauchy conditions, but at time $t = \frac{1}{2}$.

18.10 Separation of Variables

The method of solution described in this section applies to homogeneous second and higher order constant coefficient linear PDEs defined in regions D whose spatial boundaries coincide with constant values of the coordinate variables involved. For example, D may be a rectangle with sides parallel to the x-, and y-coordinate axes, a semi-infinite strip parallel to the x-axis, the wedge $r > 0, 0 \le \theta \le \frac{\pi}{4}$ in cylindrical polar coordinates, or the exterior of a sphere of radius R, where it is natural to use spherical polar coordinates with their origin located at the center of the sphere. The success of the method of separation of variables rests on the following results:

1. If u_1 and u_2 are two linearly independent solutions of a homogeneous linear PDE of first or higher order, then the **linear superposition** of the two solutions to give $u = c_1 u_1 + c_2 u_2$ is also a solution of the PDE, where c_1 and c_2 are arbitrary constants.

2. Under conditions that are satisfied in all ordinary applications, Property 1 extends to the fact that if $u_1, u_2, \ldots$, is an infinite sequence of linearly independent solutions of a homogeneous linear PDE of second or higher order, then the **linear superposition** of an infinite number of the solutions to give $u = c_1 u_1 + c_2 u_2 + \cdots$ is also a solution of the PDE, where $c_1, c_2, \ldots$ are arbitrary constants.

3. The orthogonality properties of the eigenfunctions associated with the PDE, special cases of which were developed in Chapter 8, can be used to determine the coefficients $c_1, c_2, \ldots$ in the linear superposition $u = c_1 u_1 + c_2 u_2 + \cdots$ to make it satisfy the boundary conditions imposed on the PDE, and so become the solution of the boundary value problem.

To illustrate the method we will solve some typical boundary value problems for each of the three fundamental types of second order linear PDE.

Vibrations of a Clamped String

It was shown in Section 18.5 that if a uniform stretched string vibrates in a fixed plane containing its equilibrium position, and the transverse displacement u of the string in this plane remains small, then u must be a solution of the one-dimensional wave equation. If the equilibrium position of the string is taken to coincide with the x-axis and t is the time, the transverse displacement of the string $u(x, t)$ will satisfy the hyperbolic PDE

$$(1/c^2)u_{tt} = u_{xx},$$

where the propagation speed $c = \sqrt{T/\rho}$, with T the tension in the string and ρ the line density of the string.

Let a string of finite length L be clamped rigidly at each end, and choose the origin of the x-axis to coincide with the left end of the string, so its right end will be at the point $x = L$. The *boundary conditions* for the problem then become

$$u(0, t) = u(L, t) = 0, \quad t \geq 0,$$

because these conditions ensure that the ends of the string remain motionless for all time. The Cauchy conditions

$$u(x, 0) = g(x) \quad \text{and} \quad u_t(x, 0) = h(x)$$

determine how the vibration starts at time $t = 0$, with the initial transverse displacement of the string defined by $g(x)$ and its initial transverse speed by $h(x)$. In general the functions g and h are arbitrary, apart from the fact that as the ends of the string are clamped they must be such that $g(0) = g(L) = 0$ and $h(0) = h(L) = 0$.

EXAMPLE 18.11

Consider the vibrations of a stretched string of length L that is clamped at each end and starts from rest with the initial shape $u(x, 0) = kx(L - x)$. Here $k > 0$ is a positive constant chosen such that the maximum transverse displacement is small, in agreement with the approximations made when deriving the wave equation. As the string starts from rest, the Cauchy conditions to be imposed on the wave equation in (143) are

$$u(x, 0) = kx(L - x) \quad \text{and} \quad u_t(x, 0) \equiv 0.$$

The approach to be adopted involves seeking elementary solutions of the wave equation of the form $u(x, t) = X(x)T(t)$, and then using the linearity of the PDE to express the required solution, subject to the boundary and Cauchy conditions, as a linear combination of these elementary solutions. The name **separation of variables** comes from the way the independent variables are *separated* in each elementary solution. In this case the separation involves the product of a function $X(x)$ only of x and a function $T(t)$ only of t.

the method of separation of variables

Partial differentiation of $u(x, t) = X(x)T(t)$ with respect to x only acts on the function $X(x)$, and partial differentiation with respect to t only acts on $T(t)$, so $u_{xx} = X''(x)T(t)$ and $u_{tt} = X(x)T''(t)$, where primes indicate differentiation of the associated function with respect to the appropriate single independent variable.

Substituting these results into the wave equation and dividing by $X(x)T(t)$ gives

$$\frac{1}{c^2} \frac{T''}{T} = \frac{X''}{X}.$$

Inspection of this result shows that the expression on the left is independent of x and so is only a function of t, while the expression on the right is independent of t and so is only a function of x. As x and t are independent variables, the only way a function of t can equal a function of x is if they are each equal to some constant p, so that

$$\frac{1}{c^2} \frac{T''}{T} = \frac{X''}{X} = p,$$

where p is a constant. So T and X must be solutions of the two ordinary differential equations

$$T'' = pc^2 T \quad \text{and} \quad X'' = pX.$$

separation constant

The constant p is called a **separation constant**, and before we proceed further it is necessary to determine its *sign*.

Examination of the first equation for $T(t)$ shows that the *time variation* is determined by $T'' = pc^2 T$, where $c^2 > 0$, so this equation can only describe *oscillatory behavior* with respect to the time if $p < 0$. Setting $p = -\lambda^2$, with λ a positive real constant, we see that the time variation of the solution is determined by

$$T'' + c^2 \lambda^2 T = 0.$$

Our next task will be to find the permissible values of λ, and to do this we must consider the x-variation of the solution that is described by the Sturm–Liouville equation

a Sturm–Liouville problem

$$X'' + \lambda^2 X = 0.$$

The function $X(x)$ determined by this equation must satisfy the boundary conditions on $u(x, t)$ that require $u(0, t) = u(L, t) = 0$. However, as $u(x, t) = X(x)T(t)$ and x and t are independent variables, these boundary conditions on $u(x, t)$ can only hold for all t if $X(0) = X(L) = 0$. This requires that we choose λ so X satisfies the two-point boundary value problem

$$X'' + \lambda^2 X = 0, \text{ with } X(0) = X(L) = 0.$$

This has the general solution

$$X(x) = \widetilde{A}\cos \lambda x + \widetilde{B}\sin \lambda x,$$

where $\widetilde{A}$ and $\widetilde{B}$ are arbitrary constants. Imposing the two-point boundary conditions $X(0) = X(L) = 0$, we have

$$
\begin{aligned}
(\text{condition } X(0) = 0) && 0 &= \widetilde{A}, \\
(\text{condition } X(L) = 0) && 0 &= \widetilde{B}\sin \lambda L.
\end{aligned}
$$

The last condition is satisfied if either $\widetilde{B} = 0$, or λL is a zero of the sine function. The condition $\widetilde{B} = 0$ is unacceptable because it makes $X(x)$ identically zero, in which case $u(x, t)$ will also vanish identically, so there can be no vibration of the string. The only alternative is to make λL a zero of the sine function by setting $\lambda L = n\pi$ for $n = 0, 1, 2, \ldots$, where the case $n = 0$ must be omitted because it corresponds to $u(x, t) \equiv 0$. The permissible values of λ, called the **eigenvalues** of the differential equation for $X(x)$, are

$$\lambda_n = \frac{n\pi}{L}, \quad n = 1, 2, \ldots.$$

eigenvalues and eigenfunctions of the Sturm–Liouville problem

The x variation is now seen to be given by

$$X_n(x) = \widetilde{B}\sin \frac{n\pi x}{L}, \quad n = 1, 2, \ldots,$$

where the functions $X_n(x)$ are called the **eigenfunctions** of the differential equation for $X(x)$, and as the equation for X is homogeneous, the value of the constant $\widetilde{B}$ is unimportant.

Once we have determined the permissible values of the eigenvalues λ, the time variation follows by integrating the equation $T'' + c^2\lambda^2 T = 0$, when we find that

$$T_n(t) = \widetilde{C}\cos\frac{nc\pi t}{L} + \widetilde{D}\sin\frac{nc\pi t}{L}, \quad n = 1, 2, \ldots,$$

where the constants $\widetilde{C}$ and $\widetilde{D}$ still remain to be determined. If we substitute for the functions $X_n(x)$ and $T_n(t)$, the permissible elementary solutions become $u_n(x, t) = X_n(x)T_n(t)$ for $n = 1, 2, \ldots$, and these are called the **eigensolutions** of the wave equation. As the constants in $u_n(x, t)$ depend on n, if we replace $\widetilde{B}\widetilde{C}$ by C_n and $\widetilde{B}\widetilde{D}$ by D_n, the eigensolutions of the wave equation become

eigensolutions

$$u_n(x, t) = \sin\frac{n\pi x}{L}\left\{C_n\cos\frac{nc\pi t}{L} + D_n\sin\frac{nc\pi t}{L}\right\},$$

with $n = 1, 2, \ldots$.

Each eigensolution is an elementary solution of the wave equation that satisfies the boundary conditions $u(0, t) = u(L, t) = 0$ for $t \geq 0$, but not the Cauchy conditions. As the wave equation is linear, a linear combination of eigensolutions will also satisfy these same boundary conditions, so we now seek a solution of the initial boundary value problem of the form

$$u(x, t) = \sum_{n=1}^{\infty} u_n(x, t) = \sum_{n=1}^{\infty}\sin\frac{n\pi x}{L}\left\{C_n\cos\frac{nc\pi t}{L} + D_n\sin\frac{nc\pi t}{L}\right\},$$

where the coefficients C_n and D_n are to be chosen so that $u(x, t)$ satisfies the Cauchy conditions

$$u(x, 0) = kx(L - x) \quad \text{and} \quad u_t(x, 0) \equiv 0.$$

To find the coefficients C_n and D_n we need to make use of Fourier series. First setting $t = 0$ in the expression for $u(x, t)$ and using the first initial condition gives

$$kx(L - x) = \sum_{n=1}^{\infty} C_n\sin\frac{n\pi x}{L}.$$

Then, assuming that differentiation of the series for $u(x, t)$ with respect to t is permissible, setting $t = 0$ in the result, and using the second initial condition gives

$$0 = \frac{c\pi}{L}\sum_{n=1}^{\infty} nD_n\sin\frac{n\pi x}{L}.$$

The series involving the coefficients C_n and D_n are simply the Fourier sine series expansion of the functions on the left, so it follows immediately that $D_n = 0$ for $n = 1, 2, \ldots$. To find the coefficients C_n we multiply series for $kx(L - x)$ by $\sin m\pi x/L$ and integrate from $x = 0$ to $x = L$, when we obtain

$$\int_0^L kx(L - x)\sin\frac{m\pi x}{L}dx = \int_0^L\sum_{n=1}^{\infty}C_n\sin\frac{n\pi x}{L}\sin\frac{m\pi x}{L}dx$$

$$= \sum_{n=1}^{\infty}\int_0^L C_n\sin\frac{n\pi x}{L}\sin\frac{m\pi x}{L}dx,$$

where the justification for the interchange of the summation and integral signs has been omitted. As the set of functions $\{\sin(m\pi x/L)\}_{m=1}^{\infty}$ is orthogonal on the interval

$0 \le x \le L$, the preceding result reduces to

$$\int_0^L kx(L-x) \sin \frac{m\pi x}{L} dx = C_m \int_0^L \sin^2 \frac{m\pi x}{L} dx.$$

After the integrations are performed, this becomes

$$\left(-\frac{2kL^3}{m^3 \pi^3} \cos m\pi + \frac{2kL^3}{m^3 \pi^3} \right) = \frac{L}{2} C_m \quad \text{for} \quad m = 1, 2, \ldots.$$

Using the result $\cos m\pi = (-1)^m$, we see that the expression on the left vanishes when m is even, so setting $m = 2r$ with $r = 1, 2, \ldots$, we have $C_{2r} = 0$. However, when m is odd the expression on the left no longer vanishes, and setting $m = 2r+1$ with $r = 0, 1, \ldots$ simplifies the result to

$$\frac{4kL^3}{(2r+1)^3 \pi^3} = \frac{L}{2} C_{2r+1}.$$

The coefficients C_r are now all known and are given by

$$C_{2r} = 0 \quad \text{and} \quad C_{2r+1} = \frac{8kL^2}{(2r+1)^3 \pi^3} \quad \text{for } r = 0, 1, \ldots.$$

Substituting for the coefficients C_n in the series for $u(x, t)$, and setting the coefficients $D_n = 0$, we arrive at the required solution

$$u(x, t) = \frac{8kL^2}{\pi^3} \sum_{r=0}^{\infty} \frac{1}{(2r+1)^3} \sin \frac{(2r+1)\pi x}{L} \cos \frac{(2r+1)c\pi t}{L},$$

for $0 \le x \le L$ and $t \ge 0$.

The justification for differentiating the functional series $u(x, t)$ term by term with respect to t and for interchanging the summation and integral signs requires arguments involving uniform convergence and so will be omitted.

It is instructive to interpret the eigenfunctions $X_n(x)$ and the eigensolutions $u_n(x, t)$ in physical terms. Inspection of the solution shows that the eigenfunction $X_n(x)$ defines the nth **mode** of the vibration, in the sense that however $X_n(x)$ is scaled, it always specifies the *shape* of the string corresponding to a given value of n. The nth eigensolution $u_n(x, t)$ is seen to be the time modulation of the nth mode.

modes of vibration

This describes how the nth mode vibrates with time and shows that it experiences a periodic variation of amplitude and a change of sign. The solution is a linear combination of all of the possible modes of vibration, chosen such that when $t = 0$ the shape of the string is $u(x, 0) = kx(L-x)$.

If the initial shape of the string is changed, but the second Cauchy condition $u_t(x, 0) \equiv 0$ is retained, the new solution will simply be a *different* linear combination of these *same* eigensolutions.

Figure 18.19 shows the initial shape of the string at time $t = 0$, and its shape at three subsequent times where, for convenience, we have set $L = \pi$, $c = 1$ and graphed the approximation to the function $\hat{u} = (\frac{\pi}{8k})u(x, t)$ using only the first 10 terms of the series solution.   ∎

Vibrations of a Circular Membrane

To illustrate the method of separation of variables when applied to the wave equation in more than one space variable, we will examine the vibrations of a uniform

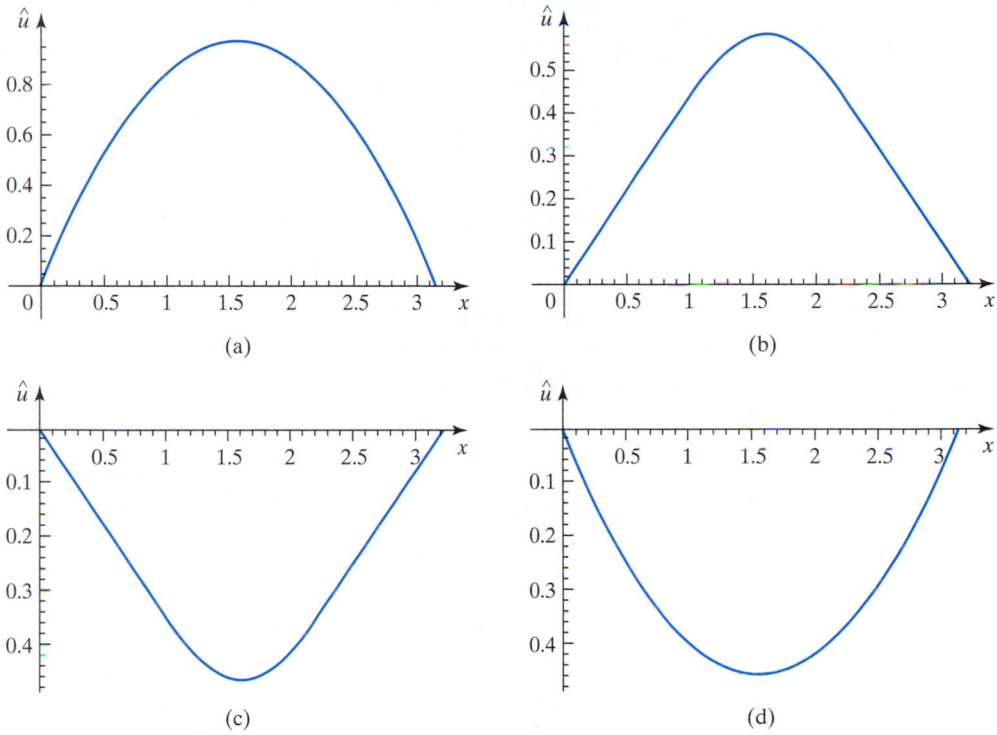

FIGURE 18.19   The shape of the string at different times (a) $t = 0$, (b) $t = 1$, (c) $t = 2$, (d) $t = 3$.

circular membrane of unit radius clamped around its rim. Because of the circular boundary, when we solve this problem the two space variables will be taken to be the cylindrical polar coordinates (r, θ), with their origin at the center of the membrane when in its equilibrium position, and the third independent variable will be the time t. The displacement of the membrane normal to its equilibrium position will be denoted by $u(r, \theta, t)$.

This problem can be considered to be a mathematical description of the vibrations of a circular membrane covering a drum that is subjected to Cauchy conditions at an initial time $t = 0$ that describe the vertical displacement $u(r, \theta, t)$ and the speed $u_t(r, \theta, t)$ of the membrane in a direction normal to its equilibrium position. It will be shown that the response to arbitrary Cauchy conditions is expressible as a sum of eigensolutions in a manner analogous to that of the vibrating string.

EXAMPLE 18.12

a vibration problem involving cylindrical polar coordinates

The geometry of the circular membrane problem suggests that cylindrical polar coordinates should be used. When the wave equation is expressed in terms of cylindrical polar coordinates it becomes

$$u_{tt} = c^2 \left(u_{rr} + \frac{1}{r} u_r + \frac{1}{r^2} u_{\theta\theta} \right) \quad \text{or} \quad u_{tt} = c^2 \Delta u,$$

where in cylindrical polar coordinates the Laplacian $\Delta = \frac{\partial^2}{\partial r^2} + \frac{1}{r} \frac{\partial}{\partial r} + \frac{1}{r^2} \frac{\partial^2}{\partial \theta^2}$.

The boundary conditions are

$$u(1, \theta, t) = 0 \text{ for } 0 \le \theta \le 2\pi \text{ and } t > 0 \quad \text{(the rim is clamped)}$$

and

$$u(r, \theta, t) \text{ is finite for } 0 \le r \le 1 \text{ and } t > 0 \quad \text{(the displacement is finite),}$$

while the initial, or Cauchy, conditions describing the initial shape of the membrane and its initial speed normal to its equilibrium position are

$$u(r, \theta, 0) = f(r, \theta) \quad \text{and} \quad u_t(r, \theta, 0) = g(r, \theta).$$

It will be simplest if the variables are separated in two stages, so first we separate out the time t by setting $u(r, \theta, t) = H(r, \theta)T(t)$, and then substitute into the differential equation to obtain

$$HT'' = c^2 T \nabla^2 H,$$

where primes denote differentiation with respect to the independent variable t occurring in $T(t)$. Dividing by HT, we have

$$\frac{1}{c^2}\frac{T''}{T} = \frac{\nabla^2 H}{H},$$

but as the expression on the left is a function of the independent time variable t, and the one on the right is a function of the independent space variables r and θ, this can only be true if

$$\frac{1}{c^2}\frac{T''}{T} = \frac{\nabla^2 H}{H} = k,$$

where k is a constant. The time variation is determined by $T'' - c^2 kT = 0$, so for the solution to be periodic in time, as is necessary if it is to describe vibrations, it is necessary that $k < 0$. Accordingly, if we set the separation constant $k = -\lambda^2$, the equations for T and H become

$$T'' + \lambda^2 c^2 T = 0$$

and

$$\Delta^2 H + \lambda^2 H = 0.$$

Helmholtz equation

The partial differential equation for H is called the **Helmholtz equation**, and it plays a fundamental role in studies of the wave equation. To find the permissible values of the eigenvalues λ we must now solve the Helmholtz equation, because the eigenvalues will be determined by the boundary conditions that must be imposed on H.

To this end we set $H(r, \theta) = R(r)\Theta(\theta)$, and after substituting for H in the Helmholtz equation we obtain

$$\Theta\left(R'' + \frac{1}{r}R'\right) + \frac{R}{r^2}\Theta'' + \lambda^2 R\Theta = 0.$$

Dividing this result by $R\Theta$ and rearranging terms gives

$$\frac{r^2}{R}\left(R'' + \frac{1}{r}R'\right) + \lambda^2 r^2 = -\frac{\Theta''}{\Theta}.$$

The expression on the left is only a function of the independent variable r, and the one on the right is only a function of the independent variable θ, so this can only be possible if

$$\frac{r^2}{R}\left(R'' + \frac{1}{r}R'\right) + \lambda^2 r^2 = -\frac{\Theta''}{\Theta} = m,$$

where m is another separation constant. The preceding result can now be decoupled to give the two Sturm–Liouville equations for $R(r)$ and $\Theta(\theta)$

$$r^2 R'' + rR' + (\lambda^2 r^2 - m)R = 0 \quad \text{and} \quad \Theta'' + m\Theta = 0.$$

To solve these equations it is necessary to supply boundary conditions for both R and Θ. As the variables are separable, these conditions follow if we interpret the boundary conditions for $u(r, \theta, t)$ in terms of $H(r, \theta) = R(r)\Theta(\theta)$. The boundary conditions give rise to two conditions, the first of which corresponds to the clamping of the rim that can be expressed by the requirement $R(1) = 0$, which ensures that the rim of the membrane remains fixed at all times. The second condition, which at first sight appears a little strange, is the requirement that $R(r)$ be *finite* for $0 \le r \le 1$. The need for this seemingly obvious requirement will become clear later.

The condition to be imposed on θ follows from the fact that the membrane is circular, so for the solution to have circular symmetry θ must be periodic with period 2π. The equation for Θ can only give rise to solutions that are periodic if $\sqrt{m} > 0$, in which case the solution becomes

$$\Theta(\theta) = \widetilde{A}\cos(\sqrt{m}\theta + \phi),$$

where $\widetilde{A}$ and ϕ are arbitrary constants. This solution will only be periodic with period 2π, as is required by the nature of the problem, if $\sqrt{m} = n$ for $n = 0, 1, \ldots$, so setting $m = n^2$, we see that the angular variation is determined by

$$\Theta(\theta) = \widetilde{A}\cos(n\theta + \phi).$$

The choice of reference line through the origin relative to which the polar angle θ is measured is immaterial, so without loss of generality it will be chosen to make the constant $\phi = 0$, because then the angular variation is determined by

$$\Theta(\theta) = \widetilde{A}\cos(n\theta).$$

If we substitute $m = n^2$ for the separation constant, the radial variation is seen to be governed by *Bessel's equation*

$$r^2 R'' + rR' + (\lambda^2 r^2 - n^2)R = 0 \quad \text{for } 0 < r < 1, n = 0, 1, 2, \ldots.$$

how Bessel's equation and its zeros enter into this solution of the wave equation

The general solution of this form of Bessel's equation (see Sections 8.6 and 8.7) is

$$R(r) = \widetilde{B}J_n(\lambda r) + \widetilde{C}Y_n(\lambda r),$$

and to determine the two arbitrary constants $\widetilde{B}$ and $\widetilde{C}$ we now make use of the two boundary conditions for $R(r)$ that were found earlier. The need for the condition that $R(r)$ remains finite for $0 \leq r \leq 1$ will be used first. This boundary condition shows that the term $Y_n(\lambda r)$ must be omitted from the solution $R(r)$ if u is to remain finite when $r = 0$, because $Y_n(x)$ is infinite at the origin. So we must set $\widetilde{C} = 0$, when the radial variation becomes

$$R(r) = \widetilde{B} J_n(\lambda r).$$

The permissible values of λ now follow by using the remaining boundary condition $R(1) = 0$. This condition shows that we must set $J_n(\lambda) = 0$, so λ must be one of the infinite number of nonvanishing zeros of $J_n(x)$. If we denote these by $j_{n,s}$ for $s = 1, 2, \ldots$, the eigenvalues λ must be

$$\lambda = j_{n,s}.$$

A listing of the first few of these zeros is given in Section 8.6.

Combining the foregoing results shows that the eigenfunction determining the (n, s)-mode of vibration is

$$H_{ns}(r, \theta) = \widetilde{A}\,\widetilde{B} J_n(j_{n,s}r)\cos(n\theta),$$

where the product of the arbitrary constants, itself another arbitrary constant, will depend on n and s.

The time variation follows by integrating $T'' + \lambda^2 c^2 T = 0$, when we find that

$$T(t) = \widetilde{D}\cos(j_{n,s}ct) + \widetilde{E}\sin(j_{n,s}ct),$$

where here also the two arbitrary constants depend on n and s.

Finally, combining results to obtain a general eigensolution gives

$$u_{ns}(r, \theta, t) = J_n(j_{n,s}r)\cos(n\theta)\{P_{ns}\cos(j_{n,s}ct) + Q_{ns}\sin(j_{n,s}ct)\}.$$

Here, because the arbitrary constants depend on n and s, and a product of arbitrary constants is also an arbitrary constant, we have set $P_{ns} = \widetilde{A}\widetilde{B}\widetilde{D}$ and $Q_{ns} = \widetilde{A}\widetilde{B}\widetilde{E}$.

Before we solve the initial value problem, let us first examine the nature of the eigenfunctions $H_{ns}(r, \theta)$ that determine the *shape* of each mode of vibration. As $H_{ns}(r, \theta)$ is modulated by the time variation $T(t)$, the general shape of the (n, s)-mode can be seen by setting the product of arbitrary constants equal to 1 and taking the eigenfunction to be $H_{ns}(r, \theta) = J_n(j_{n,s}r)\cos(n\theta)$. The diagrams in Fig. 18.20 illustrate the first few vibrational modes. The shaded and unshaded areas in the diagrams indicate where displacement occurs in *opposite* directions. The modulation of an eigenfunction by the time variation $T(t)$ simply alters the amplitude of the displacement, and periodically reverses its direction. The lines bordering the shaded and unshaded areas are called **nodal lines**, and these represent lines on the surface of the membrane that are never displaced from their equilibrium position. As n and s increase, so also does the complexity of the pattern of the nodal lines. Figure 18.21a illustrates the membrane displacement in the eigenmode corresponding to $n = 2$ and $s = 1$, and Fig. 18.21b shows the corresponding contour lines.

typical vibrational modes and nodal lines

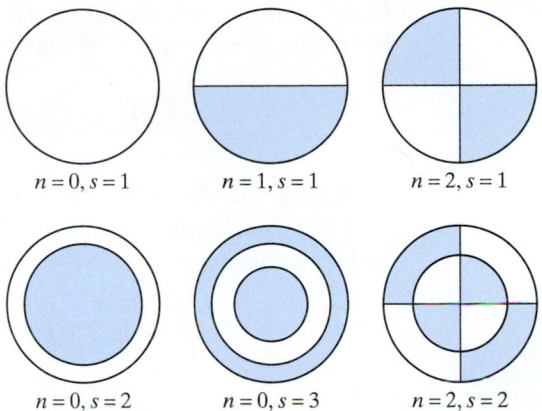

$n = 0, s = 1$ $n = 1, s = 1$ $n = 2, s = 1$

$n = 0, s = 2$ $n = 0, s = 3$ $n = 2, s = 2$

FIGURE 18.20   Some typical vibrational modes.

As with the stretched string, we now express the required solution that satisfies the Cauchy conditions as the linear combination of eigensolutions

$$u(r, \theta, t) = \sum_{n=0, s=1}^{\infty} u_{ns}(r, \theta, t).$$

Substituting for $u_{ns}(r, \theta, t)$ gives

$$u(r, \theta, t) = \sum_{n=0, s=1}^{\infty} J_n(j_{n,s} r) \cos(n\theta)\{P_{ns} \cos(j_{n,s} ct) + Q_{ns} \sin(j_{n,s} ct)\}.$$

To satisfy the Cauchy conditions it is necessary to set $u(r, \theta, 0) = f(r, \theta)$ and $u_t(r, \theta, 0) = g(r, \theta)$, and then to solve for the coefficients P_{ns} and Q_{ns}. To do this we will make use of the orthogonality of the set of cosine functions $\{\cos(n\theta)\}|_{n=0}^{\infty}$ over the interval $0 \leq \theta \leq 2\pi$ and the orthogonality of the set of Bessel functions

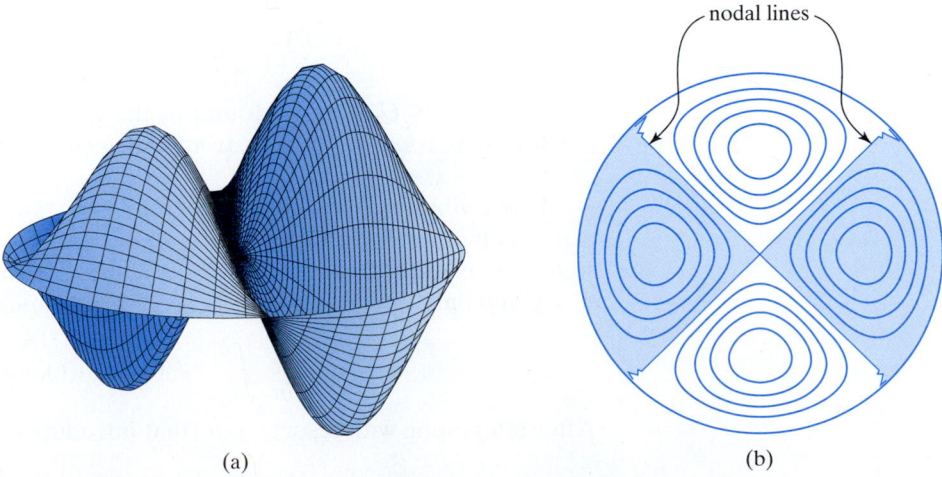

nodal lines

(a)   (b)

FIGURE 18.21   (a) The membrane displacement and (b) a contour plot for $H_{21}(r, \theta)$.

$\{J_m(j_{m,q}r)\}|_{q=1}^{\infty}$ over the interval $0 \le r \le 1$, and when doing so we will make use of the results of Example 8.25, where it was shown that

$$\int_0^1 r J_m(j_{m,p}r) J_m(j_{m,q}r) dr = \begin{cases} 0, & p \ne q \\ \frac{1}{2}[J_{m+1}(j_{m,q})]^2, & p = q. \end{cases}$$

Using the first Cauchy condition, setting $t = 0$, multiplying the result by $r J_m(j_{m,q}r) \cos(m\theta)$, and integrating with respect to r over the interval $0 \le r \le 1$, and then with respect to θ over the interval $0 \le \theta \le 2\pi$ gives

using the orthogonality of Bessel functions to determine the coefficients

$$\int_0^1 \int_0^{2\pi} r J_m(j_{m,q}r) \cos(m\theta) f(r, \theta) d\theta dr$$

$$= \sum_{n=0,s=1}^{\infty} P_{ns} \int_0^1 \int_0^{2\pi} r J_m(j_{m,q}r) J_n(j_{n,s}r) \cos(m\theta) \cos(n\theta) d\theta dr.$$

The orthogonality properties of the Bessel and cosine functions in the series on the right cause all but the term in P_{mq} to vanish, so that the result reduces to the single term

$$\int_0^1 \int_0^{2\pi} r J_m(j_{m,q}r) \cos(m\theta) f(r, \theta) d\theta dr$$

$$= P_{mq} \left\{ \int_0^1 r[J_m(j_{m,q}r)]^2 dr \right\} \left\{ \int_0^{2\pi} \cos^2(m\theta) d\theta \right\}.$$

Evaluating the integrals and solving for P_{mq}, we find that

$$P_{0q} = \frac{1}{\pi} \int_0^1 \int_0^{2\pi} r J_0(j_{0,q}r) f(r, \theta) d\theta dr / [J_1(j_{0,q})]^2 \quad \text{for } m = 0, q = 1, 2, \ldots,$$

and

$$P_{mq} = \frac{2}{\pi} \int_0^1 \int_0^{2\pi} r J_m(j_{m,q}r) \cos(m\theta) f(r, \theta) d\theta dr / [J_{m+1}(j_{m,q})]^2 \text{ for } m, q = 1, 2, \ldots.$$

Differentiation of $u(r, \theta, t)$ with respect to t, followed by setting $t = 0$, shows that after setting $u_t(r, \theta, 0) = g(r, \theta)$ we obtain

$$g(r, \theta) = \sum_{n=0,s=1}^{\infty} Q_{ns} j_{n,s} J_n(j_{n,s}r) \cos(n\theta).$$

The coefficients Q_{ns} can be found in the same way as the coefficients P_{ns}, and the formulas for them follow from the results for P_{ns} by replacing $f(r, \theta)$ by $g(r, \theta)$.

If the vibrations are circularly symmetric, and so do not depend on θ, the expression $u(r, \theta, 0)$ simplifies to $u(r, \theta, 0) = h(r)$, say. If, in addition, the vibrations start from rest, so $u_t(r, \theta, 0) = 0$, the solution simplifies still further, because then $m = 0$ and only the coefficients P_{0q} are nonvanishing, so that

$$P_{0q} = \frac{1}{\pi} \int_0^1 \int_0^{2\pi} r J_0(j_{0,q}r) h(r) d\theta dr / [J_1(j_{0,q})]^2.$$

After integrating with respect to θ (that introduces a factor 2π), we find that

$$P_{0q} = 2 \int_0^1 r J_0(j_{0,q}r) h(r) dr / [J_1(j_{0,q})]^2 \quad \text{for } q = 1, 2, \ldots.$$

In terms of these coefficients the solution then takes the particularly simple form

$$u(r, t) = \sum_{q=1}^{\infty} J_0(j_{0,q} r) P_{0q} \cos(j_{0,q} ct) \quad \text{for } 0 \le r \le 1, t > 0.$$

■

This same method of analysis can be used when the membrane is in the form of an annulus $r_1 \le r \le r_2$, with Dirichlet and/or Neumann conditions imposed on its inner and outer boundaries. In this case the solution is not required at the origin $r = 0$, so the term $Y_n(r)$ must be retained in the solution for $R(r)$, which then becomes $R(r) = \tilde{B} J_n(\lambda r) + \tilde{C} Y_n(\lambda r)$. The eigenvalues λ_n follow by applying the appropriate boundary conditions to $R(r)$ at $r = r_1$ and $r = r_2$, but depending on the boundary conditions the determination of the numerical values of the eigenvalues can be difficult, so it is usually necessary to obtain them by numerical methods.

Time Variation of Temperature in a Long Thin Metal Plate or Rod

The following example illustrates how the method of separation of variables can be applied to a time-dependent heat flow problem in a long thin metal plate of width L.

EXAMPLE 18.13

We consider the long thin metal plate of width L in the x-direction illustrated in Fig. 18.22, with negligible thickness in the y-direction and a length in the z-direction that is much greater than L. The edge $x = 0$ is kept at zero temperature and the edge $x = L$ is thermally insulated, so no heat can pass through it. The temperature distribution across the width of the plate will be assumed to be independent of z, so as the thickness in the y-direction is negligible, the temperature distribution will depend only on x and t. The initial temperature distribution across the width of the plate applied at $t = 0$ will be taken to be $u(x, 0) = u_0(1 + x/L)$.

As the temperature distribution across the plate will be the same in any plane $z = $ constant, this situation also models a rod of length L in the plane $z = 0$, along the x-axis, when its faces above and below the plane $z = 0$ are thermally insulated. In each case the temperature distribution $u(x, t)$ will be determined by the one-dimensional heat equation

$$u_t = \kappa^2 u_{xx}.$$

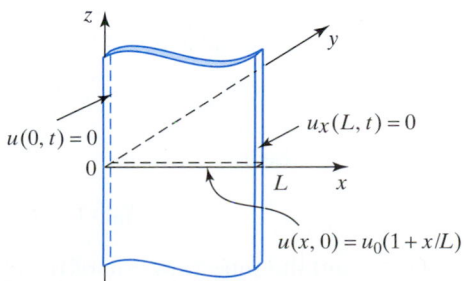

FIGURE 18.22   The plate of width L and the boundary and initial conditions.

Solution   The boundary conditions on the plate are

$$u(0, t) = 0 \quad \text{and} \quad u_x(L, t) = 0 \quad \text{for } t > 0,$$

where the first condition says that the left edge of the plate is maintained at zero temperature, and the second says that there is no heat flux across the edge $x = L$. The initial condition to be imposed across the plate is

$$u(x, 0) = u_0(1 + x/L).$$

Setting $u(x, t) = X(x)T(t)$, substituting into the heat equation, and dividing by XT gives

$$\frac{T'}{T} = \kappa^2 \frac{X''}{X}.$$

As the expression on the left is only a function of t and the one on the right is only a function of x, this can only be possible if

$$\frac{T'}{T} = \kappa^2 \frac{X''}{X} = k,$$

where k is a separation constant. To determine the sign of k, we appeal to the physical condition that the temperature cannot become infinite with the increase of time, so as $T' = kT$, this can only be possible if $k < 0$, so we set $k = -\lambda$ with $\lambda > 0$. The differential equations governing T and X now become

$$X'' + \frac{\lambda}{\kappa^2} X = 0 \quad \text{and} \quad T' + \lambda T = 0,$$

so the X variation is given by

$$X(x) = \widetilde{A}\cos(\sqrt{\lambda}x/\kappa) + \widetilde{B}\sin(\sqrt{\lambda}x/\kappa).$$

The boundary conditions for X follow from the boundary conditions for the temperature, so as the variables are separable, we require that $X(0)T(t) = 0$ and $X'(L)T(t) = 0$ for $t > 0$. Thus, the boundary conditions on X must be $X(0) = 0$ and $X'(L) = 0$. The equation for $X(x)$ is a Sturm–Liouville problem, so applying these boundary conditions gives

$$\text{(the condition } X(0) = 0) \qquad 0 = \widetilde{A}$$
$$\text{(the condition } X'(L) = 0) \qquad 0 = \frac{\sqrt{\lambda}}{\kappa}\widetilde{B}\cos(\sqrt{\lambda}L/\kappa).$$

If $\widetilde{B} = 0$, the solution vanishes identically, so as this is impossible, the eigenvalues λ must be solutions of

$$\cos(\sqrt{\lambda}L/\kappa) = 0,$$

which are the zeros of the cosine function

$$\frac{\sqrt{\lambda_n}}{\kappa}L = (2n + 1)\frac{\pi}{2}, \quad \text{or} \quad \lambda_n = \frac{(2n+1)^2\pi^2\kappa^2}{4L^2} \quad \text{for } n = 0, 1, \ldots.$$

The eigenfunctions are thus

$$X_n(x) = \widetilde{B}\sin(2n+1)\frac{\pi x}{2L} \quad \text{for } n = 0, 1, \ldots,$$

and the time variation of the eigenfunctions follows if we integrate

$$T' + \frac{(2n+1)^2\pi^2\kappa^2}{4L^2}T = 0$$

to obtain

$$T_n(t) = \tilde{C} \exp\left[-\frac{(2n+1)^2 \pi^2 \kappa^2 t}{4L^2}\right].$$

If we set $C_n = \tilde{B}\tilde{C}$, because both coefficients depend on n, the nth eigensolution becomes

$$u_n(x, t) = C_n \sin(2n+1)\frac{\pi x}{2L} \exp\left[-\frac{(2n+1)^2 \pi^2 \kappa^2 t}{4L^2}\right], \quad \text{for } n = 0, 1, \ldots.$$

We now seek a solution in the form of the linear combination of eigensolutions

$$u(x, t) = \sum_{n=0}^{\infty} u_n(x, t) = \sum_{n=0}^{\infty} C_n \sin(2n+1)\frac{\pi x}{2L} \exp\left[-\frac{(2n+1)^2 \pi^2 \kappa^2 t}{4L^2}\right].$$

To determine the coefficients C_n it is necessary to make use of the initial condition $u(x, 0) = u_0(1 + x/L)$. Setting $t = 0$ in this expression and using the initial condition gives

$$u_0(1 + x/L) = \sum_{n=0}^{\infty} C_n \sin(2n+1)\frac{\pi x}{2L}.$$

Multiplying this result by $\sin(2m+1)\frac{\pi x}{2L}$, integrating with respect to x over the interval $0 \le x \le L$, and using the orthogonality properties of the set of functions $\left\{\sin(2n+1)\frac{\pi x}{2L}\right\}$ leads to the equation for C_n

$$u_0 \int_0^L (1 + x/L) \sin(2n+1)\frac{\pi x}{2L} dx = C_n \int_0^L \left[\sin(2n+1)\frac{\pi x}{2L}\right]^2 dx.$$

Evaluating the integrals and then solving for C_n, we have

$$C_n = \frac{4u_0}{\pi(2n+1)}\left[1 + (-1)^n \frac{2}{(2n+1)\pi}\right] \quad \text{for } n = 0, 1, \ldots,$$

and the solution now follows if we substitute this expression for C_n into the series solution for $u(x, t)$.

A computer plot of $\hat{u}(x, t)/u_0$, obtained by using the first 50 terms in the series solution with $L = 1$ and $\kappa = 1$, is shown in Fig. 18.23. This confirms, as expected, that the solution decays to zero as t increases. The scale of the plot is too small to show the Gibbs phenomenon near $x = 0, t = 0$ where there is a discontinuity. ∎

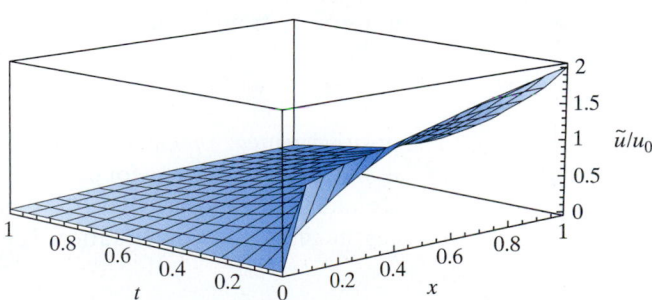

FIGURE 18.23   A plot of $\hat{u}(x, t)/u_0$ for $0 \le x \le 1, t > 0$.

The boundary conditions used so far have been particularly simple, but in physical situations they are often more complicated, and instead of being either a Dirichlet or Neumann condition they may involve a linear combination of both of these conditions. For example, a condition of the form $(\partial u/\partial x + Ku)|_{x=a} = f(t)$ describes how a combination of the temperature and heat flux is required to vary as a function of time t at the boundary $x = a$. The next example involves a boundary condition of this type, and it demonstrates how under such conditions the eigenvalues can become the zeros of a transcendental equation, and so must be found numerically.

EXAMPLE 18.14   Solve the heat equation

$$\frac{\partial u}{\partial t} = \kappa^2 \frac{\partial^2 u}{\partial x^2}, \quad 0 \le x \le L, \ t > 0,$$

subject to the boundary conditions

$$u(0, t) = 0 \quad \text{and} \quad \left(\frac{\partial u}{\partial x} + Ku\right)\Bigg|_{x=L} = 0, \ K > 0,$$

and the initial condition

$$u(x, 0) = \sin(\pi x/L).$$

Solution   Separating variables by seeking elementary solutions of the form $u(x, t) = X(x)T(t)$, substituting into the heat equation, and dividing by $X(x)T(t)$, we obtain

$$\frac{X''}{X} = \frac{1}{\kappa^2}\frac{T'}{T} = -\lambda^2,$$

where λ^2 is a positive real separation constant. So, as usual, we arrive at the two ordinary differential equations

$$X'' + \lambda^2 X = 0 \quad \text{and} \quad T' + \lambda^2 \kappa^2 T = 0,$$

the first of which is a Sturm–Liouville problem.

The general solution for $X(x)$ is $X(x) = A\cos \lambda x + B\sin \lambda x$. The boundary condition $u(0, t) = 0$ shows that $A = 0$, so $X(x) = B\sin \lambda x$, while the boundary condition $(\partial u/\partial x + Ku)|_{x=L} = 0$ leads to the condition

$$\lambda B\cos \lambda L + KB\sin \lambda L = 0, \quad \text{and so} \quad \tan \lambda L = -\lambda/K.$$

a transcendental equation for the eigenvalues

Setting $\mu = \lambda L$ and $p = KL > 0$, we find that the eigenvalues μ are determined by the zeros of the transcendental equation

$$\tan \mu = -\frac{\mu}{p}.$$

The positive values of μ can be estimated from the points of intersection the graphs of $y = \tan \mu$ and $y = -\mu/p$ for $\mu > 0$. Figure 18.24 shows a typical case when $p = 1$.

Denoting the positive roots (the eigenvalues) of this equation by $\mu_1, \mu_2, \ldots$ and solving the time variation equation $T_n' + \lambda_n^2 \kappa^2 T_n = 0$ gives

$$T_n(t) = C_n \exp\left[-\left(\frac{\mu_n \kappa}{L}\right)^2 t\right].$$

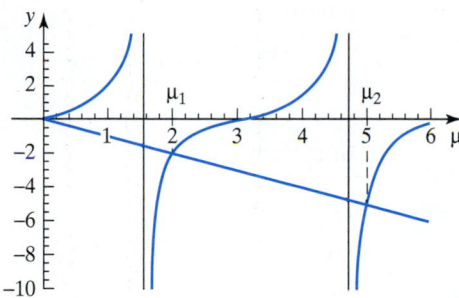

FIGURE 18.24   Graphs of $y = \tan \mu$ and $y = -\mu$
for $\mu > 0$.

The eigenfunction $X_n(x)$ becomes $X_n(x) = B_n \sin\left(\frac{\mu_n x}{L}\right)$, so the eigensolution $X_n T_n$ becomes

$$X_n T_n = D_n \exp\left[-\left(\frac{\mu_n \kappa}{L}\right)^2 t\right] \sin\left(\frac{\mu_n x}{L}\right).$$

We now seek a solution involving the linear combination of eigensolutions

$$u(x, t) = \sum_{n=1}^{\infty} D_n \exp\left[-\left(\frac{\mu_n \kappa}{L}\right)^2 t\right] \sin\left(\frac{\mu_n x}{L}\right),$$

where the constants $D_n = B_n C_n$ are to be determined by use of the initial condition $u(x, 0) = \sin(\pi x / L)$. Setting $t = 0$ and using this condition gives

$$\sin\left(\frac{\pi x}{L}\right) = \sum_{n=1}^{\infty} D_n \sin\left(\frac{\mu_n x}{L}\right).$$

Multiplying by $\sin(\mu_m x / L)$ and integrating over $0 \le x \le L$ gives

$$D_n = \left(\frac{2\pi \sin \mu_n}{\pi^2 - \mu_n^2}\right)\left(\frac{p^2 + \mu_n^2}{p(p+1) + \mu_n^2}\right),$$

and so

$$u(x, t) = \sum_{n=1}^{\infty} \left(\frac{2\pi \sin \mu_n}{\pi^2 - \mu_n^2}\right)\left(\frac{p^2 + \mu_n^2}{p(p+1) + \mu_n^2}\right) \exp\left[-\left(\frac{\mu_n \kappa}{L}\right)^2 t\right] \sin\left(\frac{\mu_n x}{L}\right).$$

When obtaining this solution we have used the result

$$\int_0^L \sin(\pi x / L) \sin(\mu_m x / L) dx = \frac{\pi L \sin \mu_m}{\pi^2 - \mu_m^2},$$

and the orthogonality of the eigenfunctions $X_n(x)$ of the associated Sturm–Liouville problem over the interval $0 \le x \le L$ with respect to the weight function $w(x) \equiv 1$, that after integration gives

$$\int_0^L \sin(\mu_m x / L) \sin(\mu_n x / L) dx = \begin{cases} 0, & m \ne n \\ \dfrac{L}{2}\left(\dfrac{p(p+1) + \mu_n^2}{p^2 + \mu_n^2}\right), & m = n, \end{cases}$$

where

$$\sin \mu_n = -\mu_n/\left(p^2 + \mu_n^2\right)^{1/2}, \quad \cos \mu_n = p/\left(p^2 + \mu_n^2\right)^{1/2}. \qquad \blacksquare$$

In the next heat conduction example we consider a problem that requires the use of cylindrical polar coordinates.

EXAMPLE 18.15

a heat problem involving plane polar coordinates

Find the time-dependent temperature distribution $u(r, \theta, t)$ in a thin semicircular metal plate $0 \leq r \leq 1, 0 \leq \theta \leq \pi$, given that its plane faces are insulated to prevent heat loss through them, the straight edge of the plate formed by the diameter $0 \leq r \leq 1, \theta = 0$ and $\theta = \pi$ is insulated, the semicircular boundary is maintained at zero temperature, and the initial temperature distribution is $u(r, \theta, 0) = (1 - r) \cos \theta$.

Solution   The geometry of this problem requires the use of plane polar coordinates, in terms of which the temperature $u(r, \theta, t)$ must satisfy the two-dimensional time-dependent heat equation (see Section 11.6)

$$\frac{\partial u}{\partial t} = \kappa^2 \left(\frac{\partial^2 u}{\partial r^2} + \frac{1}{r} \frac{\partial u}{\partial r} + \frac{1}{r^2} \frac{\partial^2 u}{\partial \theta^2} \right).$$

The bounding diameter $0 \leq r \leq 1$, $\theta = 0$, and $\theta = \pi$ is thermally insulated, so as the derivative normal to the diameter is u_θ, the boundary condition on this line becomes $u_\theta(r, 0, t) = 0$ and $u_\theta(r, \pi, t) = 0$. The semicircular boundary is maintained at zero temperature, so the boundary condition there is $u(1, \theta, t) = 0$. A routine check shows the initial condition to be appropriate, because it satisfies both the boundary condition on the diameter and the one on the semicircular boundary.

We now separate the variables by seeking elementary solution of the form $u(r, \theta, t) = R(r)\Theta(\theta)T(t)$. Substituting into the heat equation and dividing by $R\Theta T$ gives

$$\frac{T'}{T} = \kappa^2 \left(\frac{R''}{R} + \frac{1}{r} \frac{R'}{R} + \frac{1}{r^2} \frac{\Theta''}{\Theta} \right).$$

The expression on the left is only a function of t, and the one on the right is a function of r and θ, so each must be equal to a separation constant. As the temperature must decrease with time, it follows that the separation constant must be negative, so setting it equal to $-\lambda^2$ with $\lambda > 0$, we arrive at the two equations

$$T' + \kappa^2 \lambda^2 T = 0 \quad \text{and} \quad r^2 \frac{R''}{R} + r \frac{R'}{R} + \lambda^2 r^2 = -\frac{\Theta''}{\Theta}.$$

In the second equation the expression on the left is only a function of r and the one on the right is only a function of θ, so each must be equal to another separation constant μ, so we obtain the two Sturm–Liouville equations

$$\Theta'' + \mu\Theta = 0 \quad \text{and} \quad r^2 R'' + r R' + (\lambda^2 r^2 - \mu)R = 0.$$

The general solution for Θ is

$$\Theta(\theta) = A \cos \sqrt{\mu}\, \theta + B \sin \sqrt{\mu}\, \theta,$$

so as the boundary conditions on the diameter are $u_\theta(r, 0, t) = 0$ and $u_\theta(r, \pi, t) = 0$, it follows that we must have $\Theta'(\theta)|_{\theta=0} = 0$ and $\Theta'(\theta)|_{\theta=\pi} = 0$. The first of these conditions gives $B = 0$, and the second gives $\sin \sqrt{\mu}\pi = 0$, so $\sqrt{\mu} = 0, 1, \ldots$. Setting $\sqrt{\mu} = m$, and using the fact that the equation for Θ is homogeneous, we may set the arbitrary constant $A = 1$ when

$$\Theta_m(\theta) = \cos m\theta, \quad \text{for} \quad m = 0, 1, \ldots.$$

how Bessel's equation and its zeros enter into this time-dependent heat equation

The equation for $R(r)$ now becomes Bessel's equation

$$r^2 R'' + r R' + (\lambda^2 r^2 - m^2) R = 0,$$

with the general solution

$$R_m(r) = P J_m(\lambda r) + Q Y_m(\lambda r).$$

The temperature must remain finite throughout the plate, so as $Y_m(\lambda r)$ becomes infinite when $r = 0$, we must set $Q = 0$, reducing the equation to $R_m(r) = J_m(\lambda r)$, where because the equation is homogeneous we have set the arbitrary constant $P = 1$.

To satisfy the boundary condition on the semicircular boundary $u(1, \theta, t) = 0$, we must have $R(1) = 0$, and so λ must satisfy the eigenvalue equation $J_m(\lambda) = 0$, showing that the eigenvalues λ must be the positive zeros $j_{m,n}$ of the Bessel function $J_m(r) = 0$, where $j_{m,n}$ is the nth positive zero of $J_m(r)$. A short list of these zeros can be found in Table 8.1 of Chapter 8.

Using these eigenvalues in the equation for the time variation $T' + \kappa^2 \lambda^2 T = 0$ shows that $T_{m,n}(t) = C_{m,n} \exp\{-j_{m,n}^2 \kappa^2 t\}$, so combining the results for $R(r)$, $\Theta(\theta)$, and $T(t)$, we now seek a solution in the form of the following linear combination of elementary solutions:

$$u(r, \theta, t) = \sum_{m=0,n=1}^{\infty} C_{m,n} J_m(j_{m,n} r) \cos m\theta \exp\{-j_{m,n}^2 \kappa^2 t\}.$$

To find the coefficients $C_{m,n}$ we now make use of the initial condition, the orthogonality of the cosine functions over the interval $0 \le \theta \le \pi$, and the orthogonality of the Bessel functions over the interval $0 \le r \le 1$. Setting $t = 0$ in the preceding series solution and equating the result to the initial condition $u(r, \theta, 0) = (1 - r)\cos\theta$ gives

$$(1 - r)\cos\theta = \sum_{m=0,n=1}^{\infty} C_{m,n} J_m(j_{m,n} r) \cos m\theta.$$

Multiplying this by $\cos\theta$ and integrating over the interval $0 \le \theta \le \pi$ causes every term on the right to vanish, with the exception of the one involving $\cos\theta$ corresponding to $m = 1$. Thus, the required series representation simplifies to

$$(1 - r)\cos\theta = \sum_{n=1}^{\infty} C_{1,n} J_1(j_{1,n} r) \cos\theta,$$

and so after cancellation of the factor $\cos\theta$ to

$$(1 - r) = \sum_{n=1}^{\infty} C_{1,n} J_1(j_{1,n} r).$$

This same result could have been obtained by noticing that as only $\cos\theta$ occurs on the left, the linear independence of cosines of multiple angles requires that all terms involving $\cos m\theta$ on the right must vanish for $m \ne 1$.

To find the coefficients $C_{1,n}$ we multiply the last result by $r J_1(j_{1,s} r)$, integrate over the interval $0 \le r \le 1$, and after using the orthogonality of Bessel functions derived in (148) of Appendix 2 in Chapter 8, we obtain

$$\int_0^1 r(1-r)J_1(j_{1,s}r)dr = C_{1,s}\frac{1}{2}[J_2(j_{1,s})]^2.$$

Replacing s by n gives

$$C_{1,n} = \frac{2\int_0^1 (r-r^2)J_1(j_{1,n}r)dr}{[J_2(j_{1,n})]^2} \qquad \text{for } n = 1, 2, \ldots.$$

In terms of these coefficients $C_{1,n}$, the required solution becomes

$$u(r,\theta,t) = \sum_{n=1}^{\infty} C_{1,n}J_1(j_{1,n}r)\cos\theta \exp\{-j_{1,n}^2\kappa^2 t\}.$$

Evaluating the first few coefficients numerically gives

$$C_{1,1} = 0.917184, \quad C_{1,2} = 0.432800, \quad C_{1,3} = 0.317323, \quad C_{1,4} = 0.232474,$$
$$C_{1,5} = 0.193256, \quad C_{1,6} = 0.158851, \quad C_{1,7} = 0.139139, \quad C_{1,8} = 0.120617.$$

On the diameter bounding the semicircle, when $\theta = 0$ the initial condition is $u(r, 0, t) = 1 - r$, and when $\theta = \pi$ it is $u(r, \pi, t) = r - 1$, so the solution u is discontinuous at $r = 0$ on the bounding diameter.

Figure 18.25 shows a plot of the solution along the insulated diameter as a function of time, using the eight terms in the series solution for $u(r, \theta, t)$ with $\kappa^2 = 0.1$. The plot shows the development of the Gibbs phenomenon at $t = 0$ due to the discontinuity in u at $r = 0$, and the way the temperature along the diameter relaxes to zero as $t \to \infty$. ∎

Separation of Variables in the Elliptic Case

Laplace's equation describes many different physical situations, from among which we choose to solve three problems. The first two involve steady-state temperature

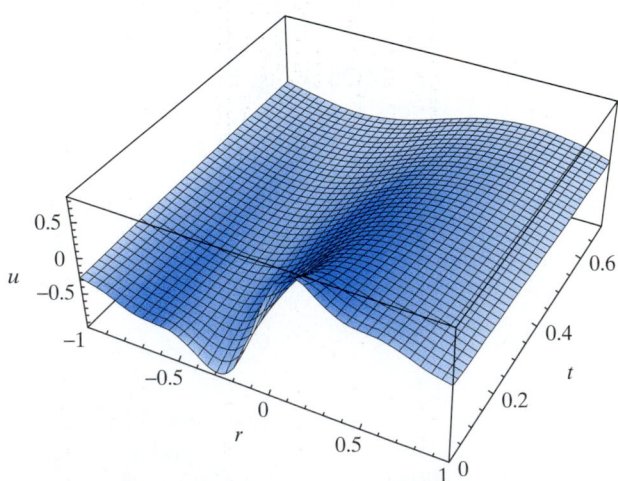

FIGURE 18.25   The relaxation of the initial temperature distribution with time along the diameter bounding the plate.

distributions in two-dimensional regions, and the third involves finding the electro-static potential distribution inside a spherical cavity. The equation determining the steady-state temperature u in a heat-conducting material is the Laplace equation $\Delta u = 0$, and the first problem to be considered is as follows.

EXAMPLE 18.16

The diagram in Fig. 18.26 shows a rectangular region $0 \leq x \leq \pi, 0 \leq y \leq 2$, in which the steady state temperature distribution $u(x, y)$ is required subject to the temperature on the side $0 \leq x \leq \pi, y = 0$, being $u(x, 0) = x \sin x$, and the temperature on the other three sides being maintained at $u = 0$. This can either be considered to represent a cross-section of a long metal bar extending in the z-direction with the boundary conditions on its sides independent of z, or as a thin metal plate with its faces parallel to the (x, y)-plane thermally insulated.

Solution   The domain is rectangular with its sides parallel to the coordinate axes, so it is appropriate to express the Laplace equation in terms of the cartesian coordinates x and y so the temperature must satisfy

$$u_{xx} + u_{yy} = 0.$$

Separating variables by setting $u(x, y) = X(x)Y(y)$, substituting into the Laplace equation, dividing by XY, and rearranging terms gives

$$\frac{X''}{X} = -\frac{Y''}{Y}.$$

As the expression on the left is a function of only x and the one on the right is a function of only y, these expressions must be equal to a separation constant k, so that

$$\frac{X''}{X} = -\frac{Y''}{Y} = k.$$

The sign of the separation constant must be chosen so the boundary conditions are satisfied. As $u(x, y) = X(x)Y(y)$, and neither $X(x)$ nor $Y(y)$ can be identically zero, the boundary conditions $u(0, y) = 0$ and $u(\pi, y) = 0$ imply that $X(0) = X(\pi) = 0$. When $k > 0$, the general solution for $X(x)$ is $X(x) = A \cosh x\sqrt{k} + B \sinh x\sqrt{k}$, and the boundary conditions can only be satisfied if $A = B = 0$, which is impossible. Consequently, k must be negative, so we set $k = -\lambda^2$, where λ is positive and real. The separated equations give the following Sturm–Liouville equation

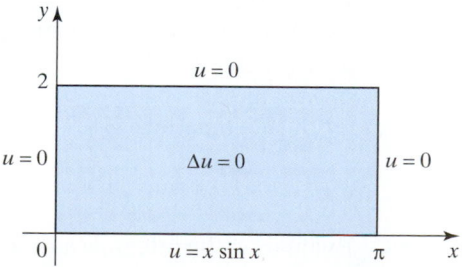

FIGURE 18.26   The rectangular region and its boundary conditions.

for $X(x)$ and the equation for $Y(y)$:

$$X'' + \lambda^2 X = 0 \quad \text{and} \quad Y'' - \lambda^2 Y = 0.$$

Solving for X gives

$$X(x) = \widetilde{A} \cos \lambda x + \widetilde{B} \sin \lambda x,$$

and imposing the left boundary condition $X(0) = 0$ shows that $\widetilde{A} = 0$. The imposition of the right boundary condition $X(\pi) = 0$ gives $\widetilde{B} \sin \lambda \pi = 0$, so as $\widetilde{B} \neq 0$, it follows that the eigenvalues are the zeros of $\sin \pi x$, and so

$$\lambda_n = n, \quad \text{for } n = 1, 2, \ldots.$$

Thus, the eigenfunctions are proportional to

$$X_n(x) = \widetilde{B} \sin nx, \quad \text{for } n = 1, 2, \ldots,$$

where, as the equation for $X_n(x)$ is homogeneous, the value of $\widetilde{B}$ is unimportant. Solving the differential equation for $Y(y)$ gives

$$Y_n(y) = \widetilde{C} \cosh ny + \widetilde{D} \sinh ny.$$

The boundary condition $u(x, 2) = 0$ is equivalent to $X(x)Y(2) = 0$, but as $X(x)$ is not identically zero, we must have $Y(2) = 0$. Applying this condition to $Y_n(y)$ gives

$$0 = \widetilde{C} \cosh 2n + \widetilde{D} \sinh 2n,$$

but only the ratio is important, so we can set $\widetilde{D} = 1$ when

$$\widetilde{C} = -\frac{\sinh 2n}{\cosh 2n}.$$

Using this result in the expression for $Y_n(y)$ gives

$$Y_n(y) = \frac{\widetilde{C}}{\cosh 2n}(\sinh ny \cosh 2n - \cosh ny \sinh 2n) = \frac{\widetilde{C}}{\cosh 2n} \sinh n(y - 2).$$

If we replace the product $\widetilde{B}\widetilde{C}/\cosh 2n$ by C_n, the eigensolution $u_n(x, y) = X_n(x) Y_n(y)$ becomes

$$u_n(x, y) = C_n \sin nx \sinh n(y - 2), \quad \text{for } n = 1, 2, \ldots.$$

We now seek a solution of the boundary value problem in the form of the linear combination of the eigensolutions

$$u(x, y) = \sum_{n=1}^{\infty} u_n(x, y) = \sum_{n=1}^{\infty} C_n \sin nx \sinh n(y - 2).$$

To determine the coefficients C_n we must use the boundary condition $u(x, 0) = x \sin x$ together with the orthogonality properties of the set of functions $\{\sin nx\}|_1^{\infty}$ over the interval $0 \leq x \leq \pi$. Setting $y = 0$ in $u(x, y)$ and multiplying the result by $\sin mx$, integrating over $0 \leq x \leq \pi$, and using the orthogonality properties of the set of sine functions gives

$$\int_0^{\pi} x \sin x \sin nx \, dx = -C_n \sinh 2n \int_0^{\pi} \sin^2 nx \, dx, \quad n = 1, 2, \ldots.$$

Evaluating the integrals and solving for C_n we find that

$$C_1 = -\frac{\pi}{2 \sinh 2} \quad \text{and} \quad C_n = \frac{4n(1 + (-1)^n)}{(n^2 - 1)^2 \pi \sinh 2n} \quad \text{for } n = 2, 3, \ldots.$$

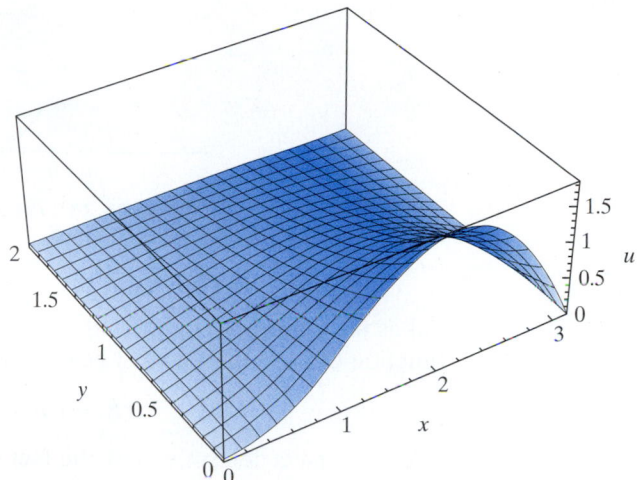

FIGURE 18.27   A plot of the temperature distribution $u(x, y)$ using five terms.

The problem is solved by substituting these values of C_n into

$$u(x, y) = \sum_{n=1}^{\infty} C_n \sin nx \sinh n(y - 2).$$

Figure 18.27 shows a computer plot of the temperature distribution $u(x, y)$ in the region $0 \le x \le \pi$, $0 \le y \le 2$ obtained by using the preceding result with five terms. ∎

The following is another example of the application of the method of separation of variables to the Laplace equation when finding the steady state temperature distribution.

EXAMPLE 18.17

Find the steady state temperature distribution in the semicircular region of radius ρ lying in the upper half-plane and centered on the origin, as shown in Fig. 18.28. The temperature on the straight boundary is $u = 0$, and that on the semicircular boundary is $u = u_0 \theta(\pi - \theta)$.

Solution   The geometry of the problem suggests that the Laplace equation for the steady state temperature distribution u should be expressed in terms of the polar coordinates r and θ. In terms of these variables the Laplace equation $\Delta u = 0$ becomes

$$u_{rr} + \frac{1}{r} u_r + \frac{1}{r^2} u_{\theta\theta} = 0.$$

To separate the variables we now set $u(r, \theta) = R(r)\Theta(\theta)$ and substitute into the equation. After dividing by $R\Theta$ and rearranging terms, we find that

$$r^2 \frac{R''}{R} + r \frac{R'}{r} = -\frac{\Theta''}{\Theta},$$

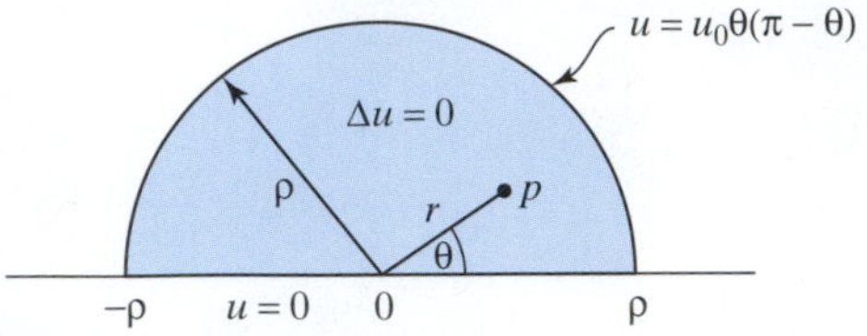

FIGURE 18.28   The semicircular domain and its boundary conditions.

but as the expression on the left is a function of only r and the one on the right is a function of only θ, both must be equal to a separation constant k, so we have

$$r^2 R'' + rR' - kR = 0 \quad \text{and} \quad \Theta'' + k\Theta = 0.$$

The sign of k is determined by the fact that only when $k > 0$ will the θ variation be periodic in nature, as would be expected, because increasing θ by a multiple of 2π will simply reproduce the original problem. If we set $k = \lambda^2$, the functions R and Θ are seen to satisfy the two equations

$$r^2 R'' + rR' - \lambda^2 R = 0 \quad \text{and} \quad \Theta'' + \lambda^2 \Theta = 0.$$

how the Cauchy–Euler equation arises

The first of these equations is a **Cauchy–Euler equation**, which was seen in Section 6.5 to have the general solution

$$R(r) = \widetilde{A} r^\lambda + \widetilde{B} \frac{1}{r^\lambda}.$$

As the solution must be *finite* at the origin, we must set $\widetilde{B} = 0$, so $R(r)$ must be of the form $R(r) = \widetilde{A} r^\lambda$. Now, as $u(r, \theta) = R(r)\Theta(\theta)$ and $u(r, 0) = u(r, \pi) = 0$ (in polar coordinates these two conditions represent the boundary condition on the straight line boundary), it follows that the boundary conditions for Θ are $\Theta(0) = \Theta(\pi) = 0$.

The general solution for Θ is

$$\Theta(\theta) = \widetilde{C} \cos \lambda\theta + \widetilde{D} \sin \lambda\theta,$$

so imposing the first of the boundary conditions gives $\widetilde{C} = 0$, and when the second one is imposed we find that λ must satisfy

$$0 = \widetilde{D} \sin \lambda\pi,$$

so the eigenvalues λ_n are

$$\lambda_n = n, \quad \text{for } n = 1, 2, \ldots.$$

The eigenfunctions $R_n(r)$ become

$$R_n(x) = A_n r^n, \quad \text{for } n = 1, 2, \ldots,$$

and the eigensolutions $u_n(r, \theta) = A_n r^n \sin n\theta$, where the product of the arbitrary constants $\widetilde{A}\widetilde{D}$, each of which depends on n, has been denoted by A_n.

We now seek a solution in the form of the linear combination of the eigensolutions

$$u(r, \theta) = \sum_{n=1}^{\infty} u_n(r, \theta) = \sum_{n=1}^{\infty} A_n r^n \sin n\theta.$$

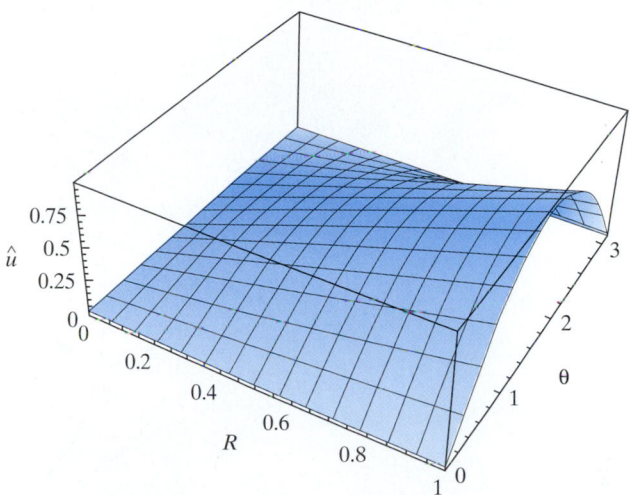

FIGURE 18.29   A plot of the normalized solution $\hat{u} = (\pi/8u_0)u(r, \theta)$.

Substituting the boundary condition $u(\rho, \theta) = u_0\theta(\pi - \theta)$ on the left of this series and setting $r = \rho$ in the expression on the right gives

$$u_0\theta(\pi - \theta) = \sum_{n=1}^{\infty} A_n\rho^n \sin n\theta.$$

The coefficients A_n now follow from the orthogonality properties of the sine function over the interval $0 \le \theta \le \pi$. Multiplying the last result by $\sin m\theta$ and integrating over the interval $0 \le \theta \le \pi$, we find that

$$2u_0\left(\frac{1 - (-1)^n}{n^3}\right) = \frac{1}{2}A_n\rho^n\pi \quad \text{and so} \quad A_n = \frac{4u_0}{\pi}\frac{(1 - (-1)^n)}{n^3\rho^n}.$$

Substituting these coefficients into the series now gives the required solution,

$$u(r, \theta) = \frac{8u_0}{\pi}\sum_{n=1}^{\infty}\left(\frac{r}{\rho}\right)^{2n-1}\frac{\sin(2n - 1)\theta}{(2n - 1)^3}.$$

Figure 18.29 shows a plot of $\hat{u} = (\pi/8u_0)u(r, \theta)$ as a function of $R = r/\rho$ for $0 \le R \le 1$ and $0 \le \theta \le \pi$ using 10 terms of the series.   ∎

The next example involving Laplace's equation is a three-dimensional problem for which spherical polar coordinates form the natural coordinate system to be used. This example also shows how Legendre polynomials arise naturally when we work with Laplace's equation in spherical polar coordinates.

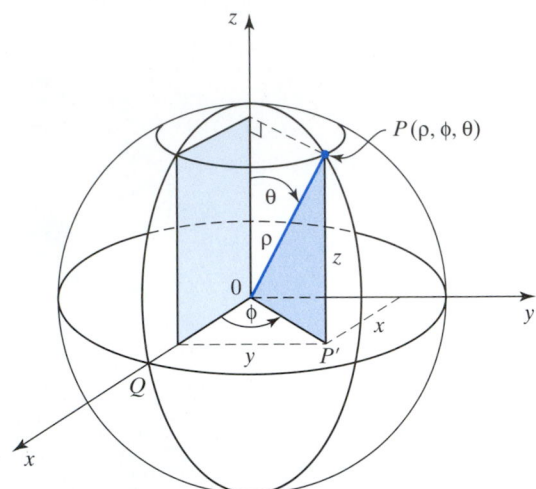

FIGURE 18.30   The spherical polar coordinate system.

Find the electrostatic potential inside a spherical cavity of radius ρ when the bottom half of the spherical boundary is maintained at a potential U_0 and the upper half is maintained at a potential U_1.

Solution   The geometry of the problem indicates that for simplicity spherical polar coordinates should be used, because the boundary of the region involved is a sphere of radius ρ. Figure 18.30 shows the standard system of spherical coordinates. As the potential on the boundary assumes a different constant value on each of two hemispheres, the problem will be simplified if the origin is taken to be at the center of the sphere with the z-axis chosen so the potential is $u = U_1$ on the upper hemisphere where $z > 0$, corresponding to $r = \rho$ and $0 \leq \theta < \frac{\pi}{2}$, and $u = U_0$ on the lower hemisphere where $z < 0$, corresponding to $r = \rho$ and $\frac{\pi}{2} < \theta \leq \pi$.

In this case the boundary conditions are such that there is no variation with respect to the angle ϕ (called the *azimuthal* angle), so as the potential inside the spherical cavity will depend only on r and θ we set $u = u(r, \theta)$. Making use of the expression for the Laplacian in spherical polar coordinates found in Example 11.23(b) of Chapter 11, and setting the partial derivative with respect to ϕ equal to zero, because there is no variation with respect to ϕ, gives

$$\Delta u = \frac{1}{r^2 \sin \theta} \left[\frac{\partial}{\partial r} \left(r^2 \sin \theta \frac{\partial u}{\partial r} \right) + \frac{\partial}{\partial \theta} \left(\sin \theta \frac{\partial u}{\partial \theta} \right) \right] = 0,$$

or

$$r^2 \frac{\partial^2 u}{\partial r^2} + 2r \frac{\partial u}{\partial r} + \cot \theta \frac{\partial u}{\partial \theta} + \frac{\partial^2 u}{\partial \theta^2} = 0.$$

For what is to follow, derivatives with respect to θ need to be transformed into derivatives with respect to ξ, where $\xi = \cos \theta$. Using the results obtained from the chain rule

$$\frac{\partial u}{\partial \theta} = -\sin \theta \frac{\partial u}{\partial \xi} \quad \text{and} \quad \frac{\partial^2 u}{\partial \theta^2} = \sin^2 \theta \frac{\partial^2 u}{\partial \xi^2} - \cos \theta \frac{\partial u}{\partial \xi},$$

we find that

$$\Delta u = r^2 \frac{\partial^2 u}{\partial r^2} + 2r \frac{\partial u}{\partial r} - 2\xi \frac{\partial u}{\partial \xi} + (1 - \xi^2) \frac{\partial^2 u}{\partial \xi^2} = 0.$$

Separating variables by seeking elementary solutions of the form $u(r, \xi) = R(r)Q(\xi)$, substituting into the preceding equation, and then dividing by RQ gives

$$\frac{r^2 R'' + 2r R'}{R} = \frac{2\xi Q' - (1 - \xi^2)Q''}{Q} = k,$$

where, as $R = R(r)$ and $Q = Q(\xi)$, these expressions must both be equal to a separation constant k whose value will be assigned later. Now that the variables have been separated, the two differential equations that follow from this are

$$r^2 R'' + 2rR' - kR = 0 \quad \text{and} \quad (1 - \xi^2)Q'' - 2\xi Q' + kQ = 0.$$

If we now choose the separation constant to be $k = n(n + 1)$ with $n = 0, 1, \ldots,$ the second equation becomes

$$(1 - \xi^2)Q'' - 2\xi Q' + n(n + 1)Q = 0,$$

and from Section 8.2 of Chapter 8 its solution is seen to be $Q(\xi) = P_n(\xi)$, where $P_n(\xi)$ is the Legendre polynomial of degree n. The equation for R now becomes the Cauchy–Euler equation

$$r^2 R'' + 2rR' - n(n + 1)R = 0.$$

The solution of this equation is found by setting $R = r^\alpha$ and solving for α. As a result we find $\alpha = n$ or $\alpha = -(n + 1)$, so the general solution for $R(r)$ is

$$R(r) = Ar^n + Br^{-(n+1)}.$$

The potential $u(r, \xi)$ must remain finite at the origin, so we must set $B = 0$. Thus, the required elementary eigensolution $u_n(r, \xi) = R(r)Q(\xi)$ becomes

$$u_n(r, \xi) = A_n r^n P_n(\xi).$$

how a Fourier–Legendre expansion arises

We now use this result to find the potential inside the sphere in the form of the linear combination of eigensolutions

$$u(r, \xi) = \sum_{n=0}^{\infty} A_n r^n P_n(\xi),$$

which form a Fourier–Legendre expansion of $u(r, \xi)$.

In terms of the new variable ξ, the boundary conditions on the spherical boundary $r = \rho$ become $u(\rho, \xi) = U_0$ for $-1 \le \xi < 0$ and $u(\rho, \xi) = U_1$ for $0 < \xi \le 1$. The coefficients A_n now follow by setting $r = \rho$ in the Fourier–Legendre expansion for $u(r, \xi)$, substituting the boundary conditions, multiplying by $P_m(\xi)$, and integrating the result with respect to ξ over the interval $-1 \le \xi \le 1$, followed by use of the orthogonality property of Legendre polynomials (see Chapter 8),

$$\int_{-1}^{1} P_m(\xi) P_n(\xi) d\xi = \begin{cases} \frac{2}{2n+1}, & m = n \\ 0, & m \ne n. \end{cases}$$

When this is done the coefficients A_n are found to be given by

$$A_n = \left(\frac{2n + 1}{2\rho^n}\right)\left(\int_{-1}^{0} U_0 P_n(\xi) d\xi + \int_{0}^{1} U_1 P_n(\xi) d\xi\right), \quad \text{for } n = 0, 1, \ldots,$$

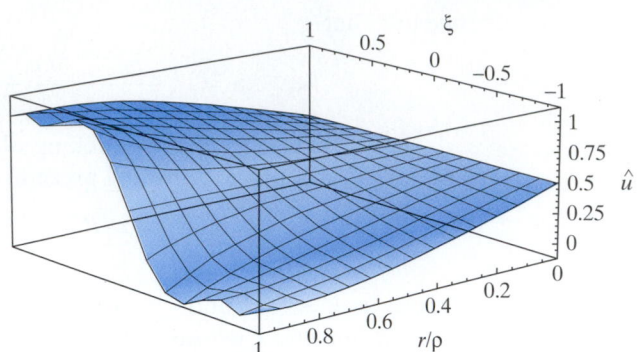

FIGURE 18.31   A plot of the normalized solution $\hat{u}(r, \xi)$.

and so

$$A_0 = \frac{1}{2}(U_0 + U_1), \ A_1 = \frac{3}{4\rho}(U_1 - U_0), \ A_2 = 0, \ A_3 = -\frac{7}{16\rho^3}(U_1 - U_0), \ A_4 = 0$$

$$A_5 = \frac{11}{32\rho^5}(U_1 - U_0), \ A_6 = 0, \ A_7 = -\frac{55}{256\rho^7}(U_1 - U_0), \ A_8 = 0, \ldots.$$

Substituting for the A_n in the Fourier–Legendre series for $u(r, \xi)$ shows the solution to be

$$\frac{u(r, \xi) - U_0}{U_1 - U_0} =$$

$$\left\{ \frac{1}{2} + \frac{3}{4}\left(\frac{r}{\rho}\right)P_1(\xi) - \frac{7}{16}\left(\frac{r}{\rho}\right)^3 P_3(\xi) + \frac{11}{32}\left(\frac{r}{\rho}\right)^5 P_5(\xi) - \frac{55}{256}\left(\frac{r}{\rho}\right)^7 P_7(\xi) + \cdots \right\},$$

for $-1 \le \xi \le 1$ with $\xi = \cos\theta$.

Figure 18.31 shows a plot of $\hat{u}(r, \xi) = [u(r, \xi) - U_0]/(U_1 - U_0)$ obtained using the preceding approximation with $0 \le r/\rho \le 1$ and $-1 \le \xi \le 1$. The plot exhibits the start of the Gibbs phenomenon in this Fourier–Legendre expansion due to the discontinuity in the boundary condition across $r = \rho$ when $\theta = \pi/2$. ∎

So far the method of separation of variables has only been applied to homogeneous equations. The next example illustrates a way in which the nonhomogeneous one-dimensional heat equation may be solved by using variation of parameters in the method of separation of variables.

EXAMPLE 18.19

The temperature $u(x, t)$ in a slab of metal $0 < x < L$ with heat generated in it at time t and position x at a rate $H(x, t)$ is determined by the nonhomogeneous heat equation

$$\frac{\partial u}{\partial t} = \kappa \frac{\partial^2 u}{\partial x^2} + H(x, t),$$

subject to the initial condition

$$u(x, 0) = U(x)$$

and the boundary conditions

$$u(0, t) = u(L, t) = 0 \text{ for } t > 0.$$

Find the temperature distribution $u(x, t)$ in the slab by combining method of variation of parameters with separation of the variables.

Solution   The nonhomogeneous term does not allow separation of variables to be used directly, so a modified approach must be adopted. Let us consider first the solution of the problem when $H(x, t) \equiv 0$.

Separating variables by setting $u(x, t) = X(x)T(t)$ and proceeding in the usual manner leads to the separated equations

$$\frac{T'(t)}{\kappa T(t)} = \frac{X''(x)}{X(x)}.$$

Introducing a separation constant $-\lambda$ with $\lambda > 0$, where the negative sign is chosen to make the solution satisfy the physical requirement that it decays with time, we arrive at the two separated ordinary differential equations

$$\frac{dT}{dt} = -\lambda\kappa T \quad \text{and} \quad \frac{d^2 X}{dx^2} + \lambda X = 0.$$

To satisfy the boundary conditions on the temperature $u(x, t)$, the function $X(x)$ must satisfy the boundary conditions $X(0) = X(L) = 0$. The equation for $X(x)$ together with these boundary conditions is a Sturm–Liouville problem that determines the eigenvalues λ_n and the associated eigenfunctions $X_n(x)$. As the general solution for $X(x)$ is

$$X(x) = A\cos(\sqrt{\lambda}x) + B\sin(\sqrt{\lambda}x),$$

the boundary conditions will only be satisfied when $\lambda = (n\pi/L)^2$ and $A = 0$, so the eigenvalues are $\lambda_n = (n\pi/L)^2$ and the associated eigenfunctions can be taken to be $X_n(x) = \sin(n\pi x/L)$, with $n = 1, 2 \ldots$.

Integrating the equation for the time variation $T(t)$ with $\lambda = \lambda_n$ gives $T_n(t) = \exp(-\lambda_n\kappa t)$, so the elementary solutions for this problem are

$$u_n(x, t) = \exp\left(-\left(\frac{n\pi}{L}\right)^2 \kappa t\right) \sin\left(\frac{n\pi x}{L}\right), \quad \text{with } n = 1, 2, \ldots.$$

It follows from this that the solution for the temperature distribution will be of the form

$$u(x, t) = \sum_{n=1}^{\infty} a_n u_n(x, t) = \sum_{n=1}^{\infty} a_n \exp\left(-\left(\frac{n\pi}{L}\right)^2 \kappa t\right) \sin\left(\frac{n\pi x}{L}\right).$$

The coefficients a_n follow in the usual manner by setting $t = 0$ and using the initial condition that $n(x, 0) = U(x)$, when we find that

$$U(x) = \sum_{n=1}^{\infty} a_n \sin\left(\frac{n\pi x}{L}\right).$$

Multiplying this result by $\sin(n\pi x/L)$ and integrating over the interval $0 \leq x \leq L$ gives

$$a_n = \frac{2}{L} \int_0^L U(x)\sin\left(\frac{n\pi x}{L}\right) dx, \quad \text{for } n = 1, 2, \ldots.$$

This completes the solution for the temperature $u(x, t)$ when the heat equation is homogeneous, because

$$u(x, t) = \sum_{n=1}^{\infty} a_n u_n(x, t) = \sum_{n=1}^{\infty} a_n \exp\left(-\left(\frac{n\pi}{L}\right)^2 \kappa t\right) \sin\left(\frac{n\pi x}{L}\right).$$

To make use of this solution in the nonhomogeneous case, we start by seeking a solution of the form

$$u(x, t) = \sum_{n=1}^{\infty} \Psi_n(t) \sin\left(\frac{n\pi x}{L}\right),$$

where the functions $\Psi_n(t)$ are still to be determined. We then expand $H(x, t)$ in terms of x as

$$H(x, t) = \sum_{n=1}^{\infty} H_n(t) \sin\left(\frac{n\pi x}{L}\right),$$

where the time-dependent coefficients $H_n(t)$ are obtained from $H(x, t)$ by multiplying this last result by $\sin(n\pi x/L)$ and integrating over the interval $0 \leq x \leq L$.

The initial condition $u(x, 0) = U(x)$ has already been expanded as

$$U(x) = \sum_{n=1}^{\infty} a_n \sin\left(\frac{n\pi x}{L}\right), \quad \text{with } a_n = \frac{2}{L} \int_0^L U(x) \sin\left(\frac{n\pi x}{L}\right) dx, \quad \text{for } n = 1, 2, \ldots$$

so after substituting these results in the PDE and combining terms in $\sin(n\pi x/L)$, we obtain

$$\sum_{n=1}^{\infty} \left[\frac{d\Psi_n(t)}{dt} + \kappa \left(\frac{n\pi}{L}\right)^2 \Psi_n(t) - H_n(t) \right] \sin\left(\frac{n\pi x}{L}\right) = 0.$$

As the right-hand side of this equation is zero, multiplying the series by $\sin(n\pi x/L)$ and integrating the result over the interval $0 \leq x \leq L$ shows that the unknown functions $\Psi_n(t)$ are solutions of the linear first order equation

$$\frac{d\Psi_n(t)}{dt} + \kappa \left(\frac{n\pi}{L}\right)^2 \Psi_n(t) = H_n(t), \quad \text{with } n = 1, 2, \ldots.$$

The initial conditions for these equations follow from the two different expressions for $u(x, 0)$, namely,

$$u(x, 0) = \sum_{n=1}^{\infty} \Psi_n(0) \sin\left(\frac{n\pi x}{L}\right) \quad \text{and} \quad u(x, 0) = \sum_{n=1}^{\infty} a_n \sin\left(\frac{n\pi x}{L}\right).$$

These must be true for all x, so when equated they give $\Psi_n(0) = a_n$, for $n = 1, 2, \ldots$. A straightforward integration of the linear first order differential equations for $\Psi_n(t)$

shows the solutions, subject to these initial conditions, to be

$$\Psi_n(t) = a_n \exp\left(-\left(\frac{n\pi}{L}\right)^2 \kappa t\right)$$

$$+ \int_0^t \exp\left(-\left(\frac{n\pi}{L}\right)^2 \kappa(t-s)\right) H_n(s)\,ds, \text{ for } n = 1, 2, \ldots.$$

Finally, after substituting for $\Psi_n(t)$ in

$$u(x, t) = \sum_{n=1}^{\infty} \Psi_n(t)\sin\left(\frac{n\pi x}{L}\right),$$

we arrive at the required solution

$$u(x, t) = \sum_{n=1}^{\infty} a_n \exp\left(-\left(\frac{n\pi}{L}\right)^2 \kappa t\right) \sin\left(\frac{n\pi x}{L}\right)$$

$$+ \sum_{n=1}^{\infty} \left(\int_0^t \exp\left(-\left(\frac{n\pi}{L}\right)^2 \kappa(t-s)\right) H_n(s)\,ds\right) \sin\left(\frac{n\pi x}{L}\right).$$

The first summation on the right is seen to be the solution of the homogeneous equation, whereas the second summation represents the contribution made to the solution by the nonhomogeneous term. ∎

The following example shows how the wave equation can be solved by separation of variables when the boundary conditions are dependent on the time.

EXAMPLE 18.20   Solve the wave equation

$$\frac{\partial^2 u}{\partial t^2} = c^2 \frac{\partial^2 u}{\partial x^2}$$

in the interval $0 \le x \le L$, subject to the initial conditions

$$u(x, 0) = f(x) \quad \text{and} \quad u_t(x, 0) = g(x)$$

and the time-dependent boundary conditions

$$u(0, t) = h(t) \quad \text{and} \quad u(L, t) = k(t).$$

Solution   To take account of the time-dependent boundary conditions, we define an auxiliary function

$$v(x, t) = \left(\frac{L-x}{L}\right) h(t) + \left(\frac{x}{L}\right) k(t)$$

that agrees with the boundary conditions at $x = 0$ and $x = L$. Next we seek a solution $u(x, t)$ of the form

$$u(x, t) = v(x, t) + w(x, t).$$

With this choice of $u(x, t)$, it is seen that $w(x, t)$ must be a solution of

$$\frac{\partial^2 w}{\partial t^2} = c^2 \frac{\partial^2 w}{\partial x^2} + \left(\frac{x-L}{L}\right) \frac{d^2 h}{dt^2} - \left(\frac{x}{L}\right) \frac{d^2 k}{dt^2},$$

with

$$w(x, 0) = f(x) + \left(\frac{x-L}{L}\right)h(0) - \left(\frac{x}{L}\right)k(0) = F(x), \text{ say,}$$

$$w_t(x, 0) = g(x) + \left(\frac{x-L}{L}\right)h'(0) - \left(\frac{x}{L}\right)k'(0) = G(x), \text{ say,}$$

and

$$w(0, t) = w(L, t) = 0.$$

The trick now is to write $w(x, t) = P(x, t) + Q(x, t)$, with $P(x, t)$ the solution of the homogeneous boundary value problem

$$\frac{\partial^2 P}{\partial t^2} = c^2 \frac{\partial^2 P}{\partial x^2},$$

with the initial conditions

$$P(x, 0) = F(x), \ P_t(x, 0) = G(x)$$

and the homogeneous boundary conditions

$$P(0, t) = P(L, t) = 0.$$

Arguments similar to those used with Example 18.11 then show that

$$P(x, t) = \sum_{n=1}^{\infty} \left[A_n \cos\left(\frac{n\pi ct}{L}\right) + B_n \sin\left(\frac{n\pi ct}{L}\right) \right] \sin\left(\frac{n\pi x}{L}\right),$$

where

$$A_n = \frac{2}{L} \int_0^L F(x)\sin\left(\frac{n\pi x}{L}\right)dx \text{ and } B_n = \frac{2}{n\pi c} \int_0^L G(x)\sin\left(\frac{n\pi x}{L}\right)dx, \ n = 1, 2, \ldots.$$

The function $Q(x, t)$ is then a solution of the nonhomogeneous problem

$$\frac{\partial^2 Q}{\partial t^2} = c^2 \frac{\partial^2 Q}{\partial x^2} + \left(\frac{x-L}{L}\right)\frac{d^2 h}{dt^2} - \left(\frac{x}{L}\right)\frac{d^2 k}{dt^2}.$$

If we use the method of Example 18.19, the solution $Q(x, t)$ becomes

$$Q(x, t) = \sum_{n=1}^{\infty} \Psi_n(t) \sin\left(\frac{n\pi x}{L}\right),$$

where

$$\Psi_n(t) = \frac{L}{n\pi} \int_0^t \sin\left(\frac{n\pi}{L}(t - \tau)\right) S_n(\tau)d\tau,$$

with

$$S_n(t) = \frac{2}{L} \int_0^L \left[\left(\frac{x-L}{L}\right)\frac{d^2 h}{dt^2} - \left(\frac{x}{L}\right)\frac{d^2 k}{dt^2} \right] \sin\left(\frac{n\pi x}{L}\right)dx. \qquad \blacksquare$$

The next example concerns the Laplace equation subject to Dirichlet conditions that are imposed on the boundaries of an annulus, and it demonstrates how a logarithmic term can appear in the solution.

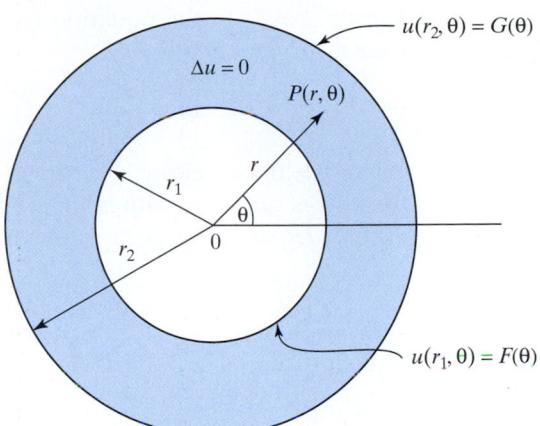

FIGURE 18.32   The Laplace equation in the annulus
$r_1 \leq r \leq r_2$.

EXAMPLE 18.21

Find solution $u(r, \theta)$ of the Laplace equation in cylindrical polar coordinates

$$\frac{\partial^2 u}{\partial r^2} + \frac{1}{r}\frac{\partial u}{\partial r} + \frac{1}{r^2}\frac{\partial^2 u}{\partial \theta^2} = 0,$$

in the annulus $r_1 \leq r \leq r_2$ shown in Fig. 18.32, where $u(r, \theta)$ is periodic in θ with period 2π and subject to the general Dirichlet boundary conditions

$$u(r_1, \theta) = F(\theta) \quad \text{and} \quad u(r_2, \theta) = G(\theta),$$

where $F(\theta)$ and $G(\theta)$ are continuous functions of θ that are periodic with period 2π. Apply the result to find $u(r, \theta)$ in the annulus $2 \leq r \leq 3$, when $F(\theta) = 1 + \sin\theta$ and $G(\theta) = \cos\theta + \frac{1}{3}\cos 2\theta$.

Solution   First, it is necessary to remember that in polar coordinates the polar angle θ is indeterminate to within a multiple of 2π, so for $u(r, \theta)$ to be a continuous function of θ it is necessary that the Dirichlet (boundary) conditions should be periodic with period 2π. This can be expressed analytically by requiring that $F(\theta) = F(\theta + 2\pi)$ and $G(\theta) = G(\theta + 2\pi)$.

   Separating variables by writing $u(r, \theta) = R(r)\Theta(\theta)$, substituting $u(r, \theta)$ into the Laplace equation, dividing by $R(r)\Theta(\theta)$, and separating the terms in $R(r)$ and $\Theta(\theta)$ gives

$$\frac{r^2 R''(r) + r R'(r)}{R(r)} = -\frac{\Theta''(\theta)}{\Theta(\theta)} = \lambda,$$

where λ is a separation constant whose values and sign remain to be determined.

   The equation for $\Theta(\theta)$, namely $\Theta'' + \lambda\Theta = 0$, will only be periodic in θ when $\lambda > 0$, and it will only be periodic with period 2π if $\lambda = n^2$ with $n = 0, 1, \ldots$. Thus, the eigenvalues of the problem are $\lambda_n = n^2$ and the associated eigenfunctions are

$$\Theta_n(\theta) = A_n \cos(n\theta) + B_n \sin(n\theta), \quad \text{for } n = 0, 1, \ldots.$$

Setting $\lambda = \lambda_n$ in the equation for $R(r)$ shows that it must be a solution of the Cauchy–Euler equation

$$r^2 \frac{d^2 R}{dr^2} + r \frac{dR}{dr} + n^2 R = 0.$$

When $n = 0$, cancelling r, setting $dR/dr = v(r)$, separating variables, and solving for v gives $v = b_0/r$, with b_0 an arbitrary constant of integration. After we replace $v(r)$ by dR/dr in this last result, a further integration gives

$$R_0(r) = a_0 + b_0 \ln r,$$

with a_0 as a second arbitrary constant of integration. When $n = 1, 2, \ldots$, the Cauchy–Euler equation has the usual solution

$$R_n(r) = a_n r^n + \frac{b_n}{r^n},$$

with a_n and b_n arbitrary constants.

Adding these results, which is permissible because Laplace's equation is linear and homogeneous, shows that we must now seek a solution for $u(r, \theta)$ of the form

$$u(r, \theta) = a_0 + b_0 \ln r + \sum_{n=1}^{\infty} \left[\left(a_n r^n + \frac{b_n}{r^n} \right) \cos(n\theta) + \left(c_n r^n + \frac{d_n}{r^n} \right) \sin(n\theta) \right],$$

though at present it is unclear how the coefficients a_n, b_n, c_n, and d_n are to be determined.

The approach we now use to find these coefficients in the series for $u(r, \theta)$ involves first expanding the Dirichlet condition $F(\theta)$ as Fourier series in θ over the interval $0 \le \theta \le 2\pi$ (remember that $F(\theta)$ is periodic in θ with period 2π). Then, after setting $r = r_1$ in the expression for $u(r, \theta)$ and using the Dirichlet boundary condition $u(r_1, \theta) = F(\theta)$, we will equate the known coefficients of $\cos(n\theta)$ and $\sin(n\theta)$ in the expansion of $F(\theta)$ and the unknown coefficients of the corresponding terms in $\cos(n\theta)$ and $\sin(n\theta)$ in the representation of $u(r_1, \theta)$. A further set of equations will then be obtained in similar fashion by expanding $G(\theta)$ as a Fourier series, setting $r = r_2$ in $u(r, \theta)$, and using the second Dirichlet boundary condition, which gives $u(r_2, \theta) = G(\theta)$. Taken together, these equations will determine all of the coefficients a_n, b_n, c_n, and d_n.

Accordingly, let us represent the Fourier series expansions of $F(\theta)$ and $G(\theta)$ as follows:

$$F(\theta) = \tfrac{1}{2} P_0 + \sum_{n=1}^{\infty} [P_n \cos(n\theta) + Q_n \sin(n\theta)]$$

and

$$G(\theta) = \tfrac{1}{2} S_0 + \sum_{n=1}^{\infty} [S_n \cos(n\theta) + T_n \sin(n\theta)].$$

Equating the coefficients of corresponding terms in $\cos(n\theta)$ and $\sin(n\theta)$ gives

$$\tfrac{1}{2} P_0 = a_0 + b_0 \ln r_1 \qquad \tfrac{1}{2} S_0 = a_0 + b_0 \ln r_2$$

$$P_n = a_n r_1^n + \frac{b_n}{r_1^n} \qquad Q_n = c_n r_1^n + \frac{d_n}{r_1^n}$$

$$S_n = a_n r_2^n + \frac{b_n}{r_2^n} \qquad T_n = c_n r_2^n + \frac{d_n}{r_2^n}.$$

Once these equations have been solved for a_n, b_n, c_n, and d_n, the expansion of $u(r, \theta)$ can be determined, so the general approach to the solution of the Dirichlet problem for Laplace's equation in an annulus has been established.

When $F(\theta) = 1 + \sin\theta$ and $G(\theta) = \cos\theta + \frac{1}{3}\cos 2\theta$, the solution simplifies, because the functions $F(\theta)$ and $G(\theta)$ are already their own Fourier series. The only nonzero coefficients in the Fourier expansion of $F(\theta)$ and $P_0 = 2$ and $Q_1 = 1$, whereas the only nonzero coefficients in the Fourier expansion of $G(\theta)$ are $S_1 = 1$ and $S_2 = \frac{1}{3}$. Consequently, we only need equate coefficients of terms up to the multiple 2θ, so that when $r = 2$ we obtain

$$1 = a_0 + b_0 \ln 2, \quad 0 = 2a_1 + \tfrac{1}{2}b_1, \quad 0 = 4a_2 + \tfrac{1}{4}b_2,$$
$$1 = 2c_1 + \tfrac{1}{2}d_1, \quad 0 = 4c_2 + \tfrac{1}{4}d_2,$$

and when $r = 3$ we obtain

$$0 = a_0 + b_0 \ln 3, \quad 1 = 3a_1 + \tfrac{1}{3}b_1, \quad \tfrac{1}{3} = 9a_2 + \tfrac{1}{9}b_2,$$
$$0 = 3c_1 + \tfrac{1}{3}d_1, \quad 0 = 9c_2 + \tfrac{1}{9}d_2.$$

These have the solutions

$$a_0 = -\frac{\ln 3}{\ln(2/3)}, \ b_0 = \frac{1}{\ln(2/3)}, \ a_1 = \frac{3}{5}, b_1 = -\frac{12}{5}, a_2 = \frac{3}{65},$$
$$b_2 = -\frac{48}{65}, \ c_1 = -\frac{2}{5}, d_1 = \frac{18}{5}, c_2 = d_2 = 0,$$

and so

$$u(r, \theta) = \frac{\ln r - \ln 3}{\ln(2/3)} + \frac{3}{5}\left(r - \frac{4}{r}\right)\cos\theta + \frac{3}{65}\left(r^2 - \frac{16}{r^2}\right)\cos 2\theta - \frac{2}{5}\left(r - \frac{9}{r}\right)\sin\theta,$$

and the solution is complete. ∎

The next example is of a different type again, in that it involves the solution of Laplace's equation in a region that is *unbounded* in one direction.

EXAMPLE 18.22

Find the steady state temperature distribution $T(x, y)$ in the uniform slab of metal shown in Fig. 18.33, given that no heat sources are present in the slab and the temperatures on the boundaries are

$$T(x, 0) = T(x, a) = 0 \quad \text{for } 0 < x < \infty, \text{ and } T(0, y) = f(y),$$

where $f(y)$ is a bounded function. State any additional condition that must be imposed on $T(x, y)$ for the solution to be physically possible.

Solution As the metal is uniform and there are no heat sources present, it follows that the steady state temperature must be a solution of the Laplace equation

$$\frac{\partial^2 T}{\partial x^2} + \frac{\partial^2 T}{\partial y^2} = 0.$$

The sides of the slab are parallel to the coordinate axes, and the equation is homogeneous, so we may separate variables by setting

$$T(x) = X(x)Y(y).$$

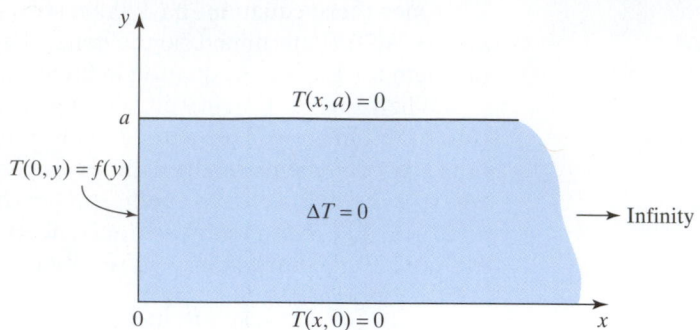

FIGURE 18.33 A semi-infinite slab of metal.

Substituting this expression into Laplace's equation and proceeding in the normal manner, we arrive at the separated form of the equation

$$\frac{Y''}{Y} = -\frac{X''}{X} = -\lambda,$$

where $\lambda > 0$ is a separation constant.

This last result separates Laplace's equation into the two differential equations

$$Y'' + \lambda Y = 0 \quad \text{and} \quad X'' - \lambda X = 0,$$

where the boundary conditions for $Y(y)$ are easily seen to be $Y(0) = Y(a) = 0$. Thus, we have arrived at the following Sturm–Liouville problem for $Y(y)$:

$$Y'' + \lambda Y = 0 \quad \text{with} \quad Y(0) = Y(a) = 0.$$

The general solution for $Y(y)$ is

$$Y(y) = A\cos\left(\sqrt{\lambda}\,y\right) + B\sin\left(\sqrt{\lambda}\,y\right).$$

Imposing these boundary conditions on the general solution for $Y(y)$ shows that the eigenvalues are $\lambda_n = n^2\pi^2/a^2$ and the corresponding eigenfunctions are $Y_n(y) = \sin(n\pi y/a)$, for $n = 1, 2, \ldots$.

Setting $\lambda = \lambda_n$ in the equation for $X(x)$ and integrating gives

$$X_n(x) = C_n \exp(-n\pi x/a) + D_n \exp(n\pi x/a).$$

To make further progress it is now necessary to recognize that when no sources are present in the metal, and a finite temperature is imposed along the boundary $x = 0, 0 < y < a$, a physically possible temperature distribution is one that must be bounded throughout the metal. This being so, we must set the coefficients $D_n = 0$ to remove the terms $\exp(n\pi x/a)$ that would otherwise become infinite as $x \to \infty$, thereby causing the functions $X_n(x)$ to simplify to $X_n(x) = \exp(-n\pi x/a)$. Notice that for convenience we have set all scale factors $C_n = 1$, since in what is to follow they will be absorbed into the new arbitrary constants d_n.

Writing $T_n(x, y) = X_n(x)Y_n(y) = \exp(-n\pi x/a)\sin(n\pi y/a)$, we now seek a solution of the form

$$T(x, y) = \sum_{n=1}^{\infty} d_n X_n(x)Y_n(y) = \sum_{n=1}^{\infty} d_n \exp(-n\pi x/a)\sin(n\pi y/a).$$

If we set $x = 0$ in this summation and use the boundary condition $T(0, y) = f(y)$, this reduces to

$$f(y) = \sum_{n=1}^{\infty} d_n \sin(n\pi y/a),$$

from which it follows in the usual manner that

$$d_n = \frac{2}{a} \int_0^a f(y) \sin\left(\frac{n\pi y}{a}\right) dy, \text{ for } n = 1, 2, \ldots.$$

The solution has been found by imposing the extra condition that $T(x, y)$ remains *bounded* in the (open) semi-infinite strip, which compensates for the normal requirement for elliptic equations that the region is closed (see page 977). ■

Other accounts of the method of separation of variables are to be found in references [3.7], [7.5], [7.7], [7.10], [7.15], [7.17], [7.19], and [7.20].

Summary

An application of the separation of variables method of solution to a PDE was seen to lead to a Sturm–Liouville problem with its parameter formed by a separation constant. When time was involved, the eigenvalues and eigenfunctions of the Sturm–Liouville problem were seen to be determined by the boundary conditions of the problem. This, in turn, was seen to determine the general structure of the solution as a series of functions of space and time, but with the multiplicative coefficients of these functions undetermined. The unknown coefficients were obtained by requiring the general series solution to satisfy the initial conditions, and by using the orthogonality properties of the functions involved. An exception was the solution of a Dirichlet problem for the Laplace equation in an annular region, where the coefficients in the series solution were obtained by matching the coefficients of corresponding sines and cosines of multiple angles. The examples given required the use of cartesian, cylindrical, and spherical polar coordinates.

EXERCISES 18.10

In Exercises 1 through 9 solve the stated boundary value problems for the wave equation in two independent variables $u_{tt} = c^2 u_{xx}$ on the interval $0 \le x \le L$.

1. A stretched string of length L, clamped at each end, starts from rest at time $t = 0$ with the initial shape $u(x, 0) = kx^2(1 - x/L)$. Find its transverse displacement $u(x, t)$ at any subsequent time $t > 0$.

2. A stretched string of length L, clamped at each end, starts from rest at time $t = 0$ with the initial shape $u(x, 0) = kx(1 - x^2/L^2)$. Find its transverse displacement $u(x, t)$ at any subsequent time $t > 0$.

3. A stretched string, clamped at each end, is displaced from its equilibrium position by having its mid-point given a small transverse displacement k, so that its initial shape is given by

$$u(x, 0) = \begin{cases} 2kx/L, & 0 \le x \le L/2 \\ 2k(1 - x/L), & L/2 \le x \le L. \end{cases}$$

If, while in this position, the string is released from rest at time $t = 0$, find its transverse displacement $u(x, t)$ at any subsequent time $t > 0$.

4. A stretched string, clamped at each end, is displaced from its equilibrium position by having a point on the string at $x = L/3$ given a small transverse displacement k, so that its initial shape is given by

$$u(x, 0) = \begin{cases} 3kx/L, & 0 \le x \le L/3 \\ \dfrac{3}{2}k(1 - x/L), & L/3 \le x \le L. \end{cases}$$

If, while in this position, the string is released from rest at time $t = 0$, find its transverse displacement $u(x, t)$ at any subsequent time $t > 0$.

5. A stretched string of length L, clamped at each end, starts from rest at time $t = 0$ with the initial shape $u(x, 0) = k \sin(\pi x/L)$. Use a simple argument to find its transverse displacement $u(x, t)$ at any subsequent time $t > 0$.

6. At time $t = 0$ a stretched string of length L, clamped at each end, starts from its equilibrium position $u(x, 0) = 0$ with the transverse speed $u_t(x, 0) = k\sin(2\pi x/L)$. Use simple arguments to find its transverse displacement $u(x, t)$ at any subsequent time $t > 0$.

7. At time $t = 0$ a stretched string of length L, clamped at both ends, starts from its equilibrium position $u(x, 0) = 0$ with the transverse speed $u_t(x, 0) = kx(1 - x/L)$. Find its transverse displacement $u(x, t)$ at any subsequent time $t > 0$.

8. At time $t = 0$ a stretched string of length L, clamped at both ends, starts from its equilibrium position $u(x, 0) = 0$ with the transverse speed $u_t(x, 0) = kx^2(1 - x/L)$. Find its transverse displacement $u(x, t)$ at any subsequent time $t > 0$.

9. A string of length L is clamped at the end $x = 0$, and its other end is allowed to move along the line $x = L$ in such a way that its slope at $x = L$ remains horizontal, so that $u_x(L, t) = 0$. If the string starts from rest at the time $t = 0$ with the initial shape $u(x, 0) = kx/L$ with $0 \le x \le L$, find its transverse displacement at any subsequent time $t > 0$.

10. An approximate description of the oscillations of air caused by blowing across the end of a tube is provided by the wave equation $p_{tt} = c^2 p_{xx}$, where c is the speed of sound in air and p is the air pressure in the tube. The velocity v of the air transverse to the axis of the tube is given by $\rho v_t = -p_x$, where ρ is the density of the air. When the tube is closed at the end $x = 0$ and open at the end $x = L$, the boundary conditions are $p_x(0, t) = p_x(L, t) = 0$. Find the eigenvalues determining the possible frequencies of oscillation, the associated eigensolutions, and the transverse speed $v(x, t)$ associated with each mode.

11. Solve the initial boundary value problem $u_{xx} = u_{yy} + 5u_y$ when $u(x, 0) = e^{-6x}$ and $u_y(x, 0) = 0$. Find the approximate form of the solution when y is large and positive.

12. A rectangular membrane with its corners at $(0, 0), (a, 0), (a, b),$ and $(0, b)$ has its edges clamped. Show that the eigenvalues λ_{mn} determining the vibrational frequencies $\lambda_{mn}c/2\pi$ are given by

$$\lambda_{mn}^2 = \sqrt{(n\pi/a)^2 + (m\pi/b)^2},$$

and that the corresponding eigensolutions determining the modes of vibration are proportional to

$$u_{mn}(x, y) = \sin(n\pi x/a)\sin(m\pi y/b)\cos(\lambda_{mn}ct).$$

13. The temperature $u(x, t)$ in a strip of metal of width L is governed by the heat equation $ku_{xx} = u_t$ for $0 \le x \le L$ and $t > 0$. Find the temperature in the strip given that the initial condition is $u(x, 0) = x$ and the boundary

conditions, corresponding to insulated ends of the strip, are $u_x(0, t) = u_x(L, t) = 0$ for $t > 0$.

14. The electric potential $u(x, y)$ in the semi-infinite strip $x > 0, 0 < y < a$ satisfies the Laplace equation $u_{xx} + u_{yy} = 0$. Find the potential in the strip if $u(x, y)$ is finite throughout the strip and it satisfies the boundary conditions on the top and bottom of the strip

$$u_y(x, 0) = u_y(x, a) = 0,$$

corresponding to insulator sides of the strip, and the potential

$$u(0, y) = \begin{cases} 1, & 0 \le y \le a/2 \\ 0, & a/2 < y \le a \end{cases}$$

at $x = 0$ on the y-axis at the end of the strip.

15. Find the potential inside the spherical cavity in Example 18.17 when the potential on the spherical boundary $r = \rho$ is zero for $0 \le \theta < \frac{\pi}{4}$, U for $\frac{\pi}{4} < \theta < \frac{3\pi}{4}$, and zero for $\frac{3\pi}{4} < \theta \le \pi$.

16. Explain why when in spherical coordinates the solution $u(r, \theta)$ of the Laplace equation does not depend on ϕ, the solution outside a sphere on which the potential u is given can be written as a linear combination of the eigensolutions

$$u_n(r, \theta) = \frac{1}{r^{n+1}} P_n(\xi),$$

for $n = 0, 1, \ldots$, where the $P_n(\xi)$ with $\xi = \cos\theta$ are Legendre polynomials of degree n. Use this result to find the first four terms in the Fourier–Legendre expansion of the potential $u(r, \xi)$ outside a sphere of radius ρ when the potential on the surface $r = \rho$ of the sphere is zero for $0 \le \theta < \frac{\pi}{4}$, U for $\frac{\pi}{4} < \theta < \frac{\pi}{2}$, and zero for $\frac{\pi}{2} < \theta \le \pi$.

17. A uniform rectangular membrane $0 \le x \le c, 0 \le y \le d$ is clamped around its edges and performs small oscillations governed by the equation $c^2(u_{xx} + u_{yy}) = u_{tt}$, where $u(x, y, t)$ is the displacement of the membrane normal to the (x, y)-plane at time t and position (x, y), and c is a constant. Derive a general series expansion for $u(x, y, t)$ when the membrane satisfies the boundary conditions

$$u(0, y, t) = u(c, y, t) = 0 \text{ for } 0 \le y \le d$$
$$\text{and} \quad u(x, 0, t) = u(x, d, t) = 0 \quad \text{for } 0 \le x \le c$$

and the initial conditions

$$u(x, y, 0) = f(x, y) \quad \text{and} \quad u_t(x, y, 0) = g(x, y).$$

Use the result to find the form of the solution when

$$f(x, y) = 2\sin\left(\frac{3\pi x}{c}\right)\sin\left(\frac{\pi y}{d}\right) \quad \text{and } g(x, y) = 0.$$

Explain why the solution is so simple.

18. Show that the solution of $\Delta u = 0$ in the rectangle $0 \le x \le l, 0 \le y \le L$ subject to the boundary conditions $u(0, y) = u(l, y) = 0$ and $u(x, 0) = \sin(\pi x/l)$ and $u(x, L) = \sin(2\pi x/l)$ is given by

$$u(x, y) = \frac{\sinh(2\pi y/l)}{\sinh(2\pi L/l)} \sin\left(\frac{2\pi x}{l}\right)$$

$$- \frac{\sinh(\pi(y - L)/l)}{\sinh(\pi L/l)} \sin\left(\frac{\pi x}{l}\right).$$

19. Show that the solution of the diffusion equation $u_t = \kappa^2 u_{xx}$ for $0 \le x \le L, t > 0$ subject to the boundary conditions

$$u(0, t) = u(L, t) = 0, \quad t > 0,$$

and the initial condition

$$u(x, 0) = \begin{cases} x, & 0 \le x \le L/2 \\ L - x, & L/2 \le x \le L \end{cases}$$

is

$$u(x, t) = \frac{4L}{\pi^2} \sum_{n=0}^{\infty} \frac{(-1)^n}{(2n + 1)^2}$$

$$\times \exp\left[-\frac{(2n + 1)^2 \pi^2 \kappa^2}{L^2} t\right] \sin \frac{(2n + 1)\pi x}{L}.$$

20. Solve the Laplace equation

$$\frac{\partial^2 u}{\partial r^2} + \frac{1}{r} \frac{\partial u}{\partial r} + \frac{1}{r^2} \frac{\partial^2 u}{\partial \theta^2} = 0$$

in the annulus $3 \le r \le 5$, subject to the Dirichlet conditions

$$u(3, \theta) = 2 + \cos\theta \quad \text{and} \quad u(5, \theta) = 1 - \sin 2\theta.$$

21. Find the steady state temperature distribution determined by the Laplace equation

$$\frac{\partial^2 T}{\partial x^2} + \frac{\partial^2 T}{\partial y^2} = 0$$

in the semi-infinite block of metal $x \ge 0, 0 \le y \le \pi$ subject to the boundary conditions

$$T(x, 0) = T(x, \pi) = 0 \quad \text{for} \quad 0 \le x < \infty$$

$$\text{and} \quad T(0, y) = y \cos(y - \pi/2).$$

18.11 Some General Results for the Heat and Laplace Equations

(a) Equations Reducible to the Heat Equation

The simplest form of the heat equation for the function $u(x, t)$ occurs when the thermal conductivity κ is a constant and $\kappa = 1$, so the equation becomes

$$\frac{\partial u}{\partial t} = \frac{\partial^2 u}{\partial x^2}. \tag{143}$$

The following transformations reduce the given form of parabolic equation to the form given in (143).

(i) The transformation $\tau = \kappa^2 t$ reduces the equation

$$\frac{\partial u}{\partial t} = \kappa^2 \frac{\partial^2 u}{\partial x^2} \quad \text{to} \quad \frac{\partial u}{\partial \tau} = \frac{\partial^2 u}{\partial x^2}.$$

PDEs that can be reduced to the heat equation

(ii) The transformation $v(x, t) = \exp(-at)u(x, t)$ reduces the equation

$$\frac{\partial v}{\partial t} = \kappa^2 \frac{\partial^2 v}{\partial x^2} - ae^{at} v \quad \text{to} \quad \frac{\partial u}{\partial t} = \kappa^2 \frac{\partial^2 u}{\partial x^2}.$$

(iii) The transformation $v(x,t) = \exp[b(x - \frac{1}{2}bt)/(2\kappa^2)]u(x,t)$ reduces the equation

$$\frac{\partial v}{\partial t} = \kappa^2 \frac{\partial^2 v}{\partial x^2} - bv_x \quad \text{to} \quad \frac{\partial u}{\partial t} = \kappa^2 \frac{\partial^2 u}{\partial x^2}.$$

(iv) Successive applications of transformations (i), (ii), and (iii) reduce the equation

$$\frac{\partial w}{\partial t} = \kappa^2 \frac{\partial^2 w}{\partial x^2} - bw_x - aw \quad \text{to} \quad \frac{\partial u}{\partial t} = \frac{\partial^2 u}{\partial x^2}.$$

(b) The Weak Maximum/Minimum Principle for the Heat Equation

Physical intuition suggests that because heat flows from a region of high temperature to one of lower temperature, the temperature $u(x,t)$ at any interior point of the interval $0 \leq x \leq L$ at a time $t_0 > 0$ must be less than the maximum of the initial temperature distribution on the interval when $t = 0$, or the maximum at the ends $x = 0$ and $x = L$ during the time $0 < t < t_0$. Conversely, the temperature $u(x,t)$ in the time interval $0 < t < t_0$ will be greater than the least of the minima of the temperature distributions over the interval at the initial time, and at the ends $x = 0$ and $x = L$. These observations form the substance of Theorem 18.1, which is called the **weak maximum/minimum principle** for the heat equation. The theorem is useful when proving general properties of the heat equation, and also for finding bounds on the solution without the need to solve the equation. The proof of the theorem that follows is based on the approach used by Petrovsky.

THEOREM 18.1

The maximum/minimum principle for the heat equation Let $u(x,t)$ be the solution of the heat equation

$$\frac{\partial u}{\partial t} = \frac{\partial^2 u}{\partial x^2}$$

in the rectangular region D formed by $0 \leq x \leq L, 0 \leq t \leq t_0$, and subject to the boundary conditions

the form taken by the max/min principle for the heat equation

$$u(0,t) = h_1(t) \quad \text{and} \quad u(L,t) = h_2(t) \text{ for } 0 \leq t \leq t_0$$

and the initial condition

$$u(x,0) = \Phi(x).$$

Let m and M, respectively, be the smallest and greatest values assumed by u on the partial boundary Γ of the rectangle D formed by the interval $0 \leq x \leq L$ on the x-axis and the two vertical lines $x = 0, 0 \leq t \leq t_0$ and $x = L, 0 \leq t \leq t_0$, the line forming the top of the rectangle being omitted. Then the solution $u(x,t)$ is such that

$$m \leq u(x,t) \leq M.$$

Proof Let M be the maximum of $u(x,t)$ in D and Γ, and m be the minimum of u on Γ. Assume, if possible, that the statement of the theorem is false and there exists a solution $u(x, t)$ such that $M > m$ at some point (ξ, τ) strictly inside D. Now consider the function

$$v(x, t) = u(x, t) + \frac{M - m}{4L^2}(x - \xi)^2.$$

Then on Γ we have

$$v(\xi, \tau) \leq m + \frac{1}{4}(M - m) < \frac{1}{4}M + \frac{3}{4}m = kM,$$

where $0 < k < 1$ and $v(\xi, \tau) = M$.

This shows that v does not assume its maximum value on Γ, so it must occur at some point (ξ_1, τ_1) inside D. From the elementary calculus of maxima of twice continuously differentiable functions of two variables, we must have $\partial^2 v / \partial x^2 \leq 0$ and $\partial v / \partial t \geq 0$ at (ξ_1, τ_1). Consequently, at the point (ξ_1, τ_1) we have shown that

$$\frac{\partial v}{\partial t} - \frac{\partial^2 v}{\partial x^2} \geq 0,$$

but direct calculation shows that

$$\frac{\partial v}{\partial t} - \frac{\partial^2 v}{\partial x^2} = -\frac{M - m}{2L^2} < 0.$$

This is a contradiction, so the assumption that the maximum of $u(x, t)$ can occur inside D is false. The result concerning the minimum of $u(x, t)$ follows by applying the preceding result to $-u(x, t)$, so the theorem is proved. ∎

An almost immediate consequence to this theorem is the continuous dependence of the solution of the heat equation on the boundary and initial conditions, showing that it is a properly posed problem.

THEOREM 18.2

showing the continuous dependence of the solution of the heat equation on the initial and boundary conditions

The continuous dependence of $u(x, t)$ on the boundary and initial conditions Consider the two problems

(I) $$\frac{\partial u}{\partial t} = \frac{\partial^2 u}{\partial x^2}$$

in the rectangular region D formed by $0 \leq x \leq L, 0 \leq t \leq t_0$, and subject to the boundary conditions

$$u(0, t) = h_1(t) \quad \text{and} \quad u(L, t) = h_2(t) \text{ for } 0 \leq t \leq t_0$$

and the initial condition

$$u(x, 0) = \Phi(x),$$

and

(II) $$\frac{\partial v}{\partial t} = \frac{\partial^2 v}{\partial x^2}$$

in the rectangular region D formed by $0 \leq x \leq L, 0 \leq t \leq t_0$, and subject to the boundary conditions

$$v(0, t) = H_1(t) \quad \text{and} \quad v(L, t) = H_2(t) \text{ for } 0 \leq t \leq t_0$$

and the initial condition

$$v(x, 0) = \Omega(x).$$

Then, if for some arbitrarily small number $\varepsilon > 0$

$$|h_1(t) - H_1(t)| \leq \varepsilon \quad \text{and} \quad |h_2(t) - H_2(t)| \leq \varepsilon \quad \text{for } 0 \leq t \leq t_0,$$

and

$$|\Phi(x) - \Omega(x)| \leq \varepsilon \quad \text{for } 0 \leq x \leq L,$$

it follows that $|u(x, t) - v(x, t)| \leq \varepsilon$ for $0 \leq x \leq L$ and $0 \leq t \leq t_0$.

Proof   Set $w(x, t) = u(x, t) - v(x, t)$, and notice that as the heat equation is linear, $w(x, t)$ will also be a solution of the heat equation. It then follows from the boundary conditions that

$$|w(0, t)| = |h_1(t) - H_1(t)| \leq \varepsilon \quad \text{and} \quad |w(L, t)| = |h_2(t) - H_2(t)| \leq \varepsilon \quad \text{for } 0 \leq t \leq t_0,$$

and from the initial conditions that

$$|w(x, 0)| = |\Phi(x) - \Omega(x)| \leq \varepsilon \quad \text{for } 0 \leq x \leq L.$$

From Theorem 18.1, the maximum of $w(x, t)$ on the partial boundary Γ defined in the theorem cannot exceed ε and it cannot be less than $-\varepsilon$, so $-\varepsilon \leq w(x, t) \leq \varepsilon$. This is equivalent to

$$|u(x, t) - v(x, t)| \leq \varepsilon,$$

so the theorem is proved.   ∎

To see how Theorem 18.1 can be used to place bounds on solutions of the heat equation $u_t = u_{xx}$, consider the problem corresponding to $h_1(t) = t \sin t$ and $h_2(t) = 0$ for $0 \leq t \leq \frac{\pi}{2}$ and $\Phi(x) = \sin(3x/2) - \sin x$ for $0 \leq x \leq \pi$.

The maximum and minimum values of $h_1(t)$ for $0 \leq t \leq \frac{\pi}{2}$ are $\frac{\pi}{2}$ and 0, respectively, and $h_2(t)$ is identically zero, whereas on the interval $0 \leq x \leq \pi$ a plot of $\Phi(x)$ shows it has a maximum of 0.2233 at $x = 0.6858$ and a minimum of -1.2160 at $x = 2.7084$. The partial boundary Γ in Theorem 18.1 comprises the interval $0 \leq x \leq \pi$ on the x-axis, and the two vertical lines $x = 0$ and $x = \pi$ for $0 \leq t \leq \frac{\pi}{2}$, so from Theorem 18.1

$$-1.2160 \leq u(x, t) \leq \pi/2 \quad \text{for } 0 \leq x \leq \pi \quad \text{and} \quad 0 \leq t \leq \frac{\pi}{2}.$$

(c) The Fundamental Solution of the Heat Equation

It was proved in Section 10.2, using the Fourier transform, that when the heat equation defined in the infinite interval $-\infty < x < \infty$ is written in the form

$$\frac{\partial^2 u}{\partial x^2} = \frac{1}{k} \frac{\partial u}{\partial t} \qquad (k = \kappa^2), \tag{144}$$

the fundamental solution of the heat equation and the delta function

its solution subject to the initial condition $u(x, 0) = f(x)$ is given by

$$u(x, t) = \sqrt{\frac{1}{4\pi kt}} \int_{-\infty}^{\infty} f(x') \exp\left\{-\frac{(x - x')^2}{4kt}\right\} dx'. \tag{145}$$

Setting $f(x) = \delta(x)$, where $\delta(x)$ is the Dirac delta function, simplifies this result to

$$u(x, t) = \sqrt{\frac{1}{4\pi kt}} \exp\left\{-\frac{x^2}{4kt}\right\}.$$

This elementary solution, which corresponds to an initial condition in the form of a single delta function located at the origin, is called the **fundamental solution** of the heat equation, and it is often denoted by $K(x, t)$, so that

$$K(x, t) = \sqrt{\frac{1}{4\pi kt}} \exp\left\{-\frac{x^2}{4kt}\right\}. \tag{146}$$

In terms of $K(x, t)$, the solution of

$$\frac{\partial^2 u}{\partial x^2} = \frac{1}{k}\frac{\partial u}{\partial t}$$

subject to the initial condition $u(x, 0) = f(x)$ can be written

$$u(x, t) = \int_{-\infty}^{\infty} f(x')K(x - x', t)dx',$$

showing that $u(x, t)$ is the convolution of the initial condition $f(x)$ and $K(x, t)$.

The fundamental solution plays an important role in more advanced studies of the heat/diffusion equation (see, for example, references [7.14] and [7.20]).

(d) The Maximum/Minimum Principle for Solutions of the Laplace Equation

For the sake of completeness we restate the maximum–minimum theorem for harmonic functions (solutions of the Laplace equation) that was established in Theorem 14.17 of Chapter 14.

THEOREM 18.3

The maximum/minimum theorem for harmonic functions   If the function $u(x, y)$ satisfies the Laplace equation (is harmonic)

$$\frac{\partial^2 u}{\partial x^2} + \frac{\partial^2 u}{\partial y^2} = 0$$

again the max/min theorem for harmonic functions and continuous dependence on Dirichlet conditions

in some open bounded region D and continuous on its boundary Γ, then the maximum and minimum values of u occur on Γ.   ∎

An argument similar to the one used in Theorem 18.2 establishes the continuous dependence of solutions of the Laplace equation on Dirichlet conditions imposed on Γ, showing that the problem is well posed.

Summary   Substitutions were given that reduce certain types of parabolic equation to the standard heat equation. A maximum/minimum theorem was proved for the heat equation, and used to show the continuous dependence of the solution on the initial and boundary conditions. The delta function was then employed to derive the fundamental solution of the heat equation that enables the solution to be found subject to an arbitrary initial condition for a problem defined in the infinite interval $-\infty < x < \infty$.

18.12 An Introduction to Laplace and Fourier Transform Methods for PDEs

The solution of partial differential equations by means of Laplace and Fourier transforms has already been illustrated in Section 7.3(e)(ii) and Section 10.2. In the examples just mentioned, the application of the Fourier transform, the Fourier sine transform, and the Laplace transform to the one-dimensional heat equation all involved the same three fundamental steps that are typical of transform methods, so these are summarized below in terms of a function $u(x, t)$ that satisfies a linear constant coefficient PDE.

Steps in the solution of a PDE by means of an integral transform

the basic steps to be followed when solving a PDE using an integral transform

STEP 1   Let the solution of a PDE be the function $u(x, t)$ of the two independent variables x and t. Transform $u(x, t)$ with respect to one of its independent variables by means of an integral transform suited to the problem. If, for example, the transform is with respect to x, then a transformed variable $U(\alpha, t)$ is obtained, where α is the transform variable. If a Laplace transform is appropriate, the transform variable α will be s, and when a Fourier transform is appropriate, α will be ω. Rearrange the result to obtain an *ordinary* differential equation for the transformed variable $U(\alpha, t)$ where t is the single independent variable and α is a parameter.

STEP 2   Find the general solution of the ODE for $U(\alpha, t)$ as a function of t, with the transform variable α still appearing as a parameter in the solution, and use the boundary and/or initial conditions of the original problem to determine the precise form of the transform $U(\alpha, t)$.

STEP 3   Invert the transform $U(\alpha, t)$ to find the required solution $u(x, t)$. In simple cases the inversion can be performed with the help of a table of transform pairs, but in general $U(\alpha, t)$ must be inverted using the appropriate inversion integral.

The type of transform to be used, and the independent variable in $u(x, t)$ that is to be transformed, depends on the region in which the solution is required, and also on the boundary and initial conditions of the original problem. In general, the Laplace and the Fourier sine and cosine transforms can be used when the variable to be transformed is defined over the semi-infinite interval $[0, \infty)$, and a Fourier transform is used when the variable to be transformed is defined over the entire real line $(-\infty, \infty)$. If the transformed variable is defined over the semi-infinite interval $[0, \infty)$, the appropriate choice of transform is determined by the partial derivatives that are to be transformed and the nature of the boundary and/or initial conditions of the original problem.

The following summary of the way in which derivatives transform illustrates what must be known about $u(x, t)$ in order that the necessary transforms of partial derivatives can be determined and, consequently, which transform should be used.

The transform of derivatives by different transforms

how partial derivatives transform when using different transforms

The **Laplace transforms** of $u(x, t)$ and its partial derivatives:

$$_t\mathcal{L}\{u(x, t)\} = U(x, s) = \int_0^\infty e^{-st} u(x, t)\,dt$$

$$_t\mathcal{L}\left\{\frac{\partial u(x, t)}{\partial t}\right\} = sU(x, s) - u(x, 0)$$

$$_x\mathcal{L}\left\{\frac{\partial u(x, t)}{\partial x}\right\} = sU(s, t) - u(0, t)$$

$$_t\mathcal{L}\left\{\frac{\partial^2 u(x, t)}{\partial t^2}\right\} = s^2 U(x, s) - su(x, 0) - u_t(x, 0)$$

$$_x\mathcal{L}\left\{\frac{\partial^2 u(x, t)}{\partial x^2}\right\} = s^2 U(s, t) - su(0, t) - u_x(0, t)$$

$$_t\mathcal{L}\left\{\frac{\partial^n u(x, t)}{\partial x^n}\right\} = \frac{d^n U(x, s)}{dx^n}, \quad n = 1, 2, \dots.$$

Corresponding results are easily written down for mixed and higher order derivatives using the results for the ordinary Laplace transform given in Theorem 7.3, so, for example,

$$_t\mathcal{L}\left\{\frac{\partial^2 u(x, t)}{\partial x \partial t}\right\} = \frac{\partial}{\partial x}\int_0^\infty e^{-st}\frac{\partial u(x, t)}{\partial t}\,dt = \frac{\partial}{\partial x}(sU(x, s) - u(x, 0))$$

$$= s\frac{dU(x, s)}{dx} - u_x(x, 0).$$

The **Fourier transform** of $u(x, t)$ and its partial derivatives:

$$_t\mathcal{F}\{u(x, t)\} = U(x, \omega) = \frac{1}{\sqrt{2\pi}}\int_{-\infty}^\infty u(x, t)\exp\{-i\omega t\}\,dt$$

$$_x\mathcal{F}\{u(x, t)\} = U(\omega, t) = \frac{1}{\sqrt{2\pi}}\int_{-\infty}^\infty u(x, t)\exp\{-i\omega x\}\,dx.$$

Here the replacement of an independent variable by ω in the transformed function U indicates that the Fourier transform has been performed with respect to that variable:

$$_t\mathcal{F}\left\{\frac{\partial^n u(x, t)}{\partial t^n}\right\} = (i\omega)^n U(x, \omega), \quad n = 1, 2, \dots$$

$$_x\mathcal{F}\left\{\frac{\partial^n u(x, t)}{\partial t^n}\right\} = \frac{\partial^n U(\omega, t)}{\partial t^n}, \quad n = 1, 2, \dots$$

$$_t\mathcal{F}\left\{\frac{\partial^n u(x, t)}{\partial x^n}\right\} = \frac{\partial^n U(x, \omega)}{\partial x^n}, \quad n = 1, 2, \dots.$$

Corresponding results apply when mixed partial derivatives are involved so, for example,

$$_t\mathcal{F}\left\{\frac{\partial^2 u(x,t)}{\partial x \partial t}\right\} = \frac{\partial}{\partial x}\,_t\mathcal{F}\left\{\frac{\partial u(x,t)}{\partial t}\right\} = i\omega\frac{\partial U(x,\omega)}{\partial x}.$$

The **Fourier sine and cosine transforms** of $u(x,t)$ and its partial derivatives:

$$_x\mathcal{F}_C\left\{\frac{\partial f(x,t)}{\partial x}\right\} = \omega F_S(\omega,t) - \sqrt{\frac{2}{\pi}}\,f(0,t)$$

$$_x\mathcal{F}_S\left\{\frac{\partial f(x,t)}{\partial x}\right\} = -\omega F_C(\omega,t)$$

$$_x\mathcal{F}_C\left\{\frac{\partial^2 f(x,t)}{\partial x^2}\right\} = -\omega^2 F_S(\omega,t) - \sqrt{\frac{2}{\pi}}\,f_x(0,t)$$

$$_x\mathcal{F}_S\left\{\frac{\partial^2 f(x,t)}{\partial x^2}\right\} = -\omega^2 F_S(\omega,t) + \omega\sqrt{\frac{2}{\pi}}\,f(0,t)$$

$$_x\mathcal{F}_C\left\{\frac{\partial^n f(x,t)}{\partial t^n}\right\} = \frac{\partial^n F_C(\omega,t)}{\partial t^n}.$$

Corresponding results can be written down for the transform of higher order partial derivatives and also when the transform is with respect to t instead of x. The transforms of mixed partial derivatives are obtained straightforwardly from the preceding results so that, for example,

$$_x\mathcal{F}_C\left\{\frac{\partial^2 f(x,t)}{\partial x \partial t}\right\} = \frac{\partial}{\partial t}\,_x\mathcal{F}_C\left\{\frac{\partial f(x,t)}{\partial x}\right\} = \omega\frac{\partial F_S(\omega,t)}{\partial t} - \sqrt{\frac{2}{\pi}}\,f_t(0,t).$$

The examples that follow illustrate the use of different integral transforms when solving some simple but typical problems.

EXAMPLE 18.23

Use a transform method to obtain the Poisson integral formula

$$u(x,y) = \frac{1}{\pi}\int_{-\infty}^{\infty}\frac{yf(\xi)}{(x-\xi)^2 + y^2}\,d\xi,$$

finding some solutions using integral transforms

which solves the boundary value problem for the Laplace equation $u_{xx} + u_{yy} = 0$ in the half-plane $-\infty < x < \infty$, $y > 0$ subject to the boundary condition $u(x,0) = f(x)$.

Solution   As x belongs to the entire real line $-\infty < x < \infty$, only the Fourier transform with respect to x can be used. Setting $_x\mathcal{F}\{u(x,y)\} = U(\omega,y)$ and transforming the Laplace equation with respect to x gives

$$(i\omega)^2 U(\omega,y) + \frac{d^2}{dy^2}U(\omega,y) = 0.$$

This has the general solution

$$U(\omega,y) = A(\omega)e^{\omega y} + B(\omega)e^{-\omega y},$$

where $A(\omega)$ and $B(\omega)$ are functions of ω that are to be determined. As $y > 0$, and the solution must be bounded for both positive and negative ω, this can only be possible if $A(\omega) = 0$ when $\omega > 0$ and $B(\omega) = 0$ when $\omega < 0$. Defining $C(\omega) = A(\omega) + B(\omega)$ allows the transform $U(\omega, y)$ to be written

$$U(\omega, y) = C(\omega)e^{-y|\omega|}, \text{ for } -\infty < \omega < \infty \text{ and } y > 0.$$

Provided $f(x)$ has a Fourier transform $\mathcal{F}\{f(x)\} = F(\omega)$, the result of transforming $u(x, 0) = f(x)$ is $U(\omega) = F(\omega)$. Setting $y = 0$ in $U(\omega, y)$ and using this last result shows that $C(\omega) = F(\omega)$, and so

$$U(\omega, y) = F(\omega)e^{-y|\omega|}.$$

The result of Example 10.3(c) can be rewritten as

$$_x\mathcal{F}\left\{\sqrt{\frac{2}{\pi}}\left(\frac{y}{x^2 + y^2}\right)\right\} = e^{-y|\omega|},$$

so applying the convolution theorem to $U(\omega, y)$ and using the foregoing result yields the Poisson integral formula

$$u(x, y) = \frac{1}{\pi}\int_{-\infty}^{\infty}\frac{yf(\xi)}{(x - \xi)^2 + y^2}d\xi. \qquad \blacksquare$$

EXAMPLE 18.24

Use a transform method to derive the D'Alembert formula

$$u(x, t) = \frac{h(x - ct) + h(x + ct)}{2} + \frac{1}{2c}\int_{x-ct}^{x+ct}k(\sigma)d\sigma,$$

which solves the initial value problem for the wave equation $u_{tt} = c^2 u_{xx}$ with $u(x, 0) = h(x)$ and $u_t(x, 0) = k(x)$, where $-\infty < x < \infty, t > 0$.

Solution   As x belongs to the entire real line $-\infty < x < \infty$, only the Fourier transform with respect to x can be used. Setting $_x\mathcal{F}\{u(x, t)\} = U(\omega, t)$ and transforming the wave equation with respect to x gives

$$\frac{d^2 U(\omega, t)}{dt^2} = c^2(i\omega)^2 U(\omega, t).$$

This ordinary differential equation in which ω appears as a parameter has the general solution

$$U(\omega, t) = A(\omega)\cos(\omega ct) + B(\omega)\sin(\omega ct),$$

where the functions $A(\omega)$ and $B(\omega)$ of ω are to be determined.

Provided $h(x)$ has the Fourier transform $\mathcal{F}\{h(x)\} = H(\omega)$, the result of transforming the first initial condition $u(x, 0) = h(x)$ with respect to x is

$$_x\mathcal{F}\{u(x, 0)\} = H(\omega).$$

Differentiation of $U(\omega, t)$ with respect to t gives

$$\frac{\partial U(\omega, t)}{\partial t} = -\omega c A(\omega)\sin(\omega ct) + \omega c B(\omega)\cos(\omega ct),$$

and so

$$U_t(\omega, 0) = \omega c B(\omega).$$

Provided $k(x)$ has the Fourier transform $\mathcal{F}\{k(x)\} = K(\omega)$, as $_x\mathcal{F}\{u_t(x,t)\} = U_t(\omega, t)$ and $u_t(x, 0) = k(x)$, we see that $U_t(\omega, 0) = K(\omega)$. Using these results in the expression for $U(\omega, t)$ we find that the Fourier transform of the solution is

$$U(\omega, t) = H(\omega)\cos(\omega ct) + K(\omega)\frac{\sin(\omega ct)}{\omega c}.$$

If we replace $\cos(\omega ct)$ by $\frac{1}{2}(e^{i\omega ct} + e^{-i\omega ct})$, this becomes

$$U(\omega, t) = \frac{1}{2}H(\omega)(e^{i\omega ct} + e^{-i\omega ct}) + K(\omega)\frac{\sin(\omega ct)}{\omega c}.$$

The solution is now obtained by finding $_x\mathcal{F}^{-1}\{U(\omega, t)\}$. The transform $U(\omega, t)$ is sufficiently simple that the inversion of the first group of terms can be performed using Fourier transform pairs and Theorem 10.8, while the inversion of the last term can be obtained with the help of Example 10.3(a) and the convolution theorem. From Theorem 10.8(ii) the inverse transform of the first group of terms is seen to be

$$_x\mathcal{F}^{-1}\left\{\frac{1}{2}H(\omega)(e^{i\omega ct} + e^{-i\omega ct})\right\} = \frac{1}{2}[h(x + ct) + h(x + ct)],$$

while appeal to Example 10.3(a) and the convolution theorem shows that

$$_x\mathcal{F}^{-1}\left\{K(\omega)\frac{\sin(\omega ct)}{\omega c}\right\} = \frac{1}{2c}\int_{x-ct}^{x+ct} k(\sigma)d\sigma.$$

The D'Alembert formula now follows by addition of these results.   ■

EXAMPLE 18.25

Use a transform method to find the solution of the modified wave equation

$$v_{xx} = c^2 v_{tt} + 2ckv_t + k^2 v$$

that remains finite for $t > 0$ and satisfies the initial conditions $v(x, 0) = 0$ and $v_t(x, 0) = 0$ and the boundary condition $v(0, t) = \sin t$ for $t > 0$.

Solution   Although both x and t lie in semi-infinite intervals, only the initial conditions imposed on $v(x, t)$ are sufficient to allow the Laplace transform of the PDE to be taken with respect to t. Defining $_t\mathcal{L}\{v(x, t)\} = V(x, s)$, using the initial conditions $v(x, 0) = 0$ and $v_t(x, 0) = 0$, and taking the Laplace transform of the PDE with respect to t gives

$$\frac{d^2 V(x, s)}{dx^2} = c^2 s^2 V(x, s) + 2cks V(x, s) + k^2 V(x, s),$$

so

$$\frac{d^2 V(x, s)}{dx^2} - (cs + k)^2 V(x, s) = 0.$$

This ordinary differential equation with s appearing as a parameter has the solution

$$V(x, s) = A(s)\exp[(cs + k)x] + B(s)\exp[-(cs + k)x],$$

where the functions $A(s)$ and $B(s)$ of s are to be determined. For the solution to remain bounded for all t it is necessary that $A(s) = 0$, and so when $x = 0$

$$V(0, s) = B(s)\exp[-(cs + k)x].$$

Taking the Laplace transform of the boundary condition gives

$$_t\mathcal{L}\{v(0,t)\} = \mathcal{L}\{\sin t\} = 1/(s^2+1),$$

and so $B(s) = 1/(s^2+1)$ and

$$V(x,s) = \frac{1}{s^2+1}\exp[-(cs+k)x] = e^{-kx}\frac{e^{-cxs}}{s^2+1}.$$

Using the table of transform pairs and the second shift theorem to invert the Laplace transform $V(x,s)$, we arrive at the solution

$$v(x,t) = e^{-kx}\sin(t-cx)H(t-cx),$$

where H is the Heaviside unit step function.

Examination of the form of the solution shows it to be a traveling wave that decays exponentially with distance, and because of the delay introduced by the Heaviside unit step function, the periodic disturbance at $x=0$ will have no effect at a position $x = x_0$ until a time t such that $t > cx_0$. ∎

EXAMPLE 18.26

Use an integral transform to find the solution of the two-dimensional Laplace equation $u_{xx} + u_{yy} = 0$ in the infinite strip $0 \le y \le a$, given that $u(x,0) = 0$ and $u(x,a) = f(x)$, and interpret the result in terms of two different physical problems.

Solution   As $-\infty < x < \infty$, it is necessary to use the Fourier transform with respect to x, so transforming the Laplace equation we find that

$$(i\omega)^2 U(\omega,y) + \frac{d^2 U(\omega,y)}{dy^2} = 0.$$

The solution of this ODE for the Fourier transform $U(\omega,y)$ of the solution $u(x,y)$ is

$$U(\omega,y) = A(\omega)e^{\omega y} + B(\omega)e^{-\omega y},$$

where the functions $A(\omega)$ and $B(\omega)$ of ω are to be determined. Assuming that $f(x)$ has the Fourier transform $F(\omega)$, the Fourier transform of the boundary conditions becomes

$$_x\mathcal{F}\{u(x,0)\} = U(\omega,0) = 0 \quad \text{and} \quad _x\mathcal{F}\{u(x,a)\} = U(\omega,a) = F(\omega).$$

The transform $U(\omega,y)$ is required to satisfy these two-point boundary conditions, and a routine calculation shows that

$$U(\omega,y) = F(\omega)\frac{\sinh(\omega y)}{\sinh(\omega a)}.$$

Applying the Fourier inversion integral to $U(\omega,y)$ gives

$$u(x,y) = \frac{1}{\sqrt{2\pi}}\int_{-\infty}^{\infty} U(\omega,y)e^{i\omega x}d\omega.$$

If $G(\omega,y)$ is defined as

$$G(\omega,y) = \frac{\sinh(\omega y)}{\sinh(\omega a)},$$

we can write

$$U(\omega,y) = F(\omega)G(\omega,y),$$

and so

$$u(x, y) = \frac{1}{\sqrt{2\pi}} \int_{-\infty}^{\infty} F(\omega)G(\omega, y)e^{i\omega x}\,d\omega.$$

If $g(x, y) = {}_x\mathcal{F}^{-1}\{G(\omega, y)\}$, an application of the Fourier convolution theorem to the expression on the right gives

$$u(x, y) = \frac{1}{\sqrt{2\pi}}(f * g).$$

By definition

$$g(x, y) = \frac{1}{\sqrt{2\pi}} \int_{-\infty}^{\infty} \frac{\sinh(\omega y)}{\sinh(\omega a)} e^{i\omega x}\,d\omega,$$

so after expansion of the factor $e^{i\omega x}$ this becomes

$$g(x, y) = \frac{1}{\sqrt{2\pi}} \int_{-\infty}^{\infty} \frac{\sinh(\omega y)}{\sinh(\omega a)} \cos(\omega x)\,d\omega + \frac{i}{\sqrt{2\pi}} \int_{-\infty}^{\infty} \frac{\sinh(\omega y)}{\sinh(\omega a)} \sin(\omega x)\,d\omega.$$

The last integral is zero because its integrand is an odd function of ω, but the integrand of the first integral is an even function of ω, so

$$g(x, y) = \frac{1}{\sqrt{2\pi}} \int_{-\infty}^{\infty} \frac{\sinh(\omega y)}{\sinh(\omega a)} \cos(\omega x)\,d\omega = \sqrt{\frac{2}{\pi}} \int_{0}^{\infty} \frac{\sinh(\omega y)}{\sinh(\omega a)} \cos(\omega x)\,d\omega.$$

Using these results in the convolution theorem now gives

$$u(x, y) = \frac{1}{\sqrt{2\pi}}(f * g) = \frac{1}{\sqrt{2\pi}} \sqrt{\frac{2}{\pi}} \int_{\omega=0}^{\infty} \int_{-\infty}^{\infty} f(\tau) \frac{\sinh(\omega y)}{\sinh(\omega a)} \cos[(\omega - \tau)x]\,d\tau\,d\omega,$$

and so

$$u(x, y) = \frac{1}{\pi} \int_{\omega=0}^{\infty} \int_{-\infty}^{\infty} f(\tau) \frac{\sinh(\omega y)}{\sinh(\omega a)} \cos[(\omega - \tau)x]\,d\tau\,d\omega.$$

One physical interpretation of this problem is that it provides the steady state temperature distribution in a slab of metal of thickness a when the lower face is maintained at a temperature $u(x, 0) = 0$ and the upper face is maintained at the temperature $u(x, a) = f(x)$. Another interpretation is that it provides the potential distribution in air between two parallel conducting plates a distance a apart, when the lower plate is maintained at zero potential and the upper one is maintained at the potential $u(x, a) = f(x)$.   ■

Fourier and Laplace transform methods for the solution of PDEs are also discussed in references [3.8] and [7.14].

Summary

The basic steps to be followed when attempting to solve a PDE by means of an integral transform were outlined, and the way in which partial derivatives are transformed by different integral transforms was listed. The examples that followed showed how the nature of the problem to be solved, together with the boundary and initial conditions, serves to determine the appropriate form of transform that is to be used.

EXERCISES 18.12

1. Find the solution $T(x, t)$ that is finite for all $x > 0$, $t > 0$ and such that $T_t = kT_{xx}$ subject to the conditions $T(x, 0) = T_0$ for $x > 0$ and $T(0, t) = 0$ for $t > 0$.

2. Find the solution $T(x, t)$ that is finite for all $x > 0$, $t > 0$ and such that $T_t = kT_{xx}$ subject to the conditions $T(x, 0) = 0$ for $x > 0$ and $T(0, t) = e^{-t}$ for $t > 0$.

3. Find the solution $T(x, t)$ that is finite for all $x > 0$, $t > 0$ and such that $T_t = kT_{xx}$ subject to the conditions $T(x, 0) = T_0$ for $x > 0$ and $T(0, t) = T_0 \cos at$ for $t > 0$.

4. Use the Fourier transform to solve the problem $T_t = kT_{xx}$ subject to the condition $T(x, 0) = T_0/(1 + x^2)$.

5. Solve $u_{tt} = c^2 u_{xx} - ku$ for $-\infty < x < \infty$, $t > 0$ subject to the conditions

$$u(x, 0) = \begin{cases} U, & |x| \le 1 \\ 0, & |x| > 1 \end{cases} \quad \text{and} \quad u_t(x, 0) = 0.$$

6. Find the bounded solution of $u_t = \kappa u_{xx} + Q\delta(x)$ subject to the initial condition $u(x, 0) = 0$ for $t > 0$, where $\delta(x)$ is the Dirac delta function.

7. Find the bounded solution of $u_{xx} + u_{yy} = 0$ in the upper half-plane $-\infty < x < \infty$, $y > 0$ subject to the condition that $u(x, 0) = f(x)$.

8. Find the bounded solution of $u_{xx} + u_{yy} = 0$ in the strip $-\infty < x < \infty$, $0 < y < a$ subject to the conditions $u(x, 0) = f(x)$ and $u(x, a) = 0$.

9. It was shown in Section 10.2 that

$$\frac{1}{2\pi} \int_{-\infty}^{\infty} \exp\{i\omega x - \omega^2 \kappa t\} d\omega = \sqrt{\frac{1}{4\pi \kappa t}} \exp\left\{-\frac{x^2}{4\kappa t}\right\}.$$

By differentiating this result with respect to x, show that

$$_x\mathcal{F}_s^{-1}\{\omega \exp(-\omega^2 \kappa t)\} = \frac{x}{2\sqrt{2}(\kappa t)^{3/2}} \exp\left\{-\frac{x^2}{4\kappa t}\right\}.$$

10.* Find the Fourier sine transform with respect to x of the bounded solution of the heat equation $u_t = ku_{xx}$ defined for $x > 0$, $t > 0$ that is subject to the initial condition $u(x, 0) = 0$ and the boundary condition $u(0, t) = u_0 e^{-t}$. Use the result of Exercise 9 to show the solution $u(x, t)$ is given by

$$u(x, t) = \frac{u_0 x}{\sqrt{4\pi k}} \int_0^t \exp\left\{-\left(\tau + \frac{x^2}{4k(t - \tau)}\right)\right\}$$

$$\frac{d\tau}{(t - \tau)^{3/2}}, \quad \text{for } x > 0 \text{ and } t > 0.$$

11.* Find the Fourier transform with respect to x of the bounded solution of the heat equation $T_t = kT_{xx}$ that is defined for $-\infty < x < \infty$ and $t > 0$ and such that it satisfies the initial condition

$$T(x, 0) = \begin{cases} T_0, & |x| \le a \\ 0, & |x| > a. \end{cases}$$

Use result (36) of Section 10.2 to invert the Fourier transform, and express the solution in terms of the error function. Verify the solution by substituting $f(x) = T(x, 0)$ in the solution for $T(x, t)$ derived in the heat conduction problem in Section 10.2.

12.* Find the Fourier transform with respect to x of the bounded solution of the heat equation $T_t = kT_{xx}$ that is defined for $-\infty < x < \infty$ and $t > 0$ and is such that it satisfies the the initial condition

$$T(x, 0) = \begin{cases} T_0, & x > a \\ 0, & x < a. \end{cases}$$

Use result (36) of Section 10.2 to invert the Fourier transform, and express the solution in terms of the error function. Verify the solution by substituting $f(x) = T(x, 0)$ in the solution for $T(x, t)$ derived in the heat conduction problem in Section 10.2.

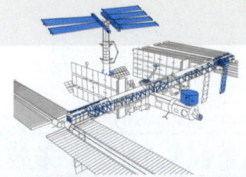

CHAPTER 18
TECHNOLOGY PROJECTS

Project 1

Linear Wave Interaction

The linear wave equation $u_{tt} = c^2 u_{xx}$ with the propagation speed c has been shown to have the general solution

$$u(x, t) = f(x - ct) + g(x + ct),$$

where the functions f and g are arbitrary. The aim of this project is first to use this general solution to obtain a 3D plot showing the resolution of an initial pulse into two waves propagating in opposite directions. Then computer algebra is to be used with the general D'Alembert solution for the wave equation to make a 3D plot of the solution to a Cauchy problem with localized initial conditions.

1. Make a 3D plot showing the interaction of two waves, each with the propagation speed $c = 1$, when

$$f(x) = \begin{cases} 0, & x < -\pi/2 \\ \cos x, & -\pi/2 < x < \pi/2 \\ 0, & x > \pi/2 \end{cases}$$

$$\text{and} \quad g(x) = \begin{cases} 0, & x < -\pi/2 \\ 1, & -\pi/2 < x < \pi/2 \\ 0, & x > \pi/2. \end{cases}$$

2. Use computer algebra to find the D'Alembert solution of the wave equation $u_{tt} = u_{xx}$ when

$$u(x, 0) = \begin{cases} 0, & x < -\pi/2 \\ 2\cos x, & -\pi/2 < x < \pi/2 \\ 0, & x > \pi/2 \end{cases}$$

$$\text{and} \quad u_t(x, 0) = \begin{cases} 0, & x < -\pi/2 \\ x, & -\pi/2 < x < \pi/2 \\ 0, & x > \pi/2. \end{cases}$$

Make a 3D plot of the result for $-5 \le x \le 5$ and $0 \le t \le 3$ to show how the initial condition is resolved into waves propagating in opposite directions.

Project 2

Vibrating Membranes

The aim of this project is to plot the shapes of some of the eigenmodes in vibrating membranes, and to identify the nodal lines in each of these modes

1. Using the information in Example 18.12, write procedures to make 3D plots and contour plots of the eigenmodes H_{31}, H_{13}, H_{22}, and H_{23}, and in each case identify the nodal lines.

2. The eigenfunctions of a square vibrating membrane with $0 \le x \le \pi$ and $0 \le y \le \pi$ are defined by $u(m, n, x, y) = \sin(mx)\sin(ny)\cos(m^2 + n^2)$. Make a 3D plot and a contour plot of the mode u in which $m = 4, n = 3$, and identify the nodal lines.

Project 3

A Vibrating String Problem

The objective of this project is to write a procedure that reproduces the steps in the vibrating string problem at the start of Section 18.10, and then to make a 3D plot of the solution showing how the shape of the string changes with time.

1. Write a procedure that mimics the steps leading to the solution

$$u(x, t) = \frac{8kL^2}{\pi^3} \sum_{r=0}^{\infty} \frac{1}{(2r + 1)^3}$$

$$\times \sin \frac{(2r + 1)\pi x}{L} \cos \frac{(2r + 1)c\pi t}{L}$$

of the wave equation $u_{tt} = c^2 u_{xx}$ subject to the initial condition $u(x, 0) = kx(L - x)$ and $u_t(x, 0) = 0$, and the boundary conditions $u(0, t) = u(L, t) = 0$.

2. By making 3D plots of the solution with $L = \pi, c = 1$ using 5, 10, and 20 terms in the summation approximating $u(x, t)$, show that a satisfactory result is obtained by using only five terms.

Project 4

The Korteweg–de Vries Equation

The motion of long waves in shallow water is governed by the nonlinear partial differential equation

$$u_t - 6uu_x + u_{xxx} = 0,$$

called the *Korteweg–de Vries equation*, usually abbreviated to the KdV equation, where $u(x, t)$ can be considered to describe the profile of the surface wave as a function of distance x and time t. This equation, which was first derived by Korteweg and de Vries in 1895, has been shown to be of fundamental importance to various types of nonlinear wave propagation.

When the term u_{xxx} is absent from the KdV equation, it reduces to a quasilinear hyperbolic equation. It is known from Section 18.3 that the solution of a Cauchy problem for such an equation may become nonunique, and from Section 18.4 that the solution can develop into a shock wave. However, the term u_{xxx}, called a *dispersive term*, smooths the effect of the terms $u_t - 6uu_x$ in the KdV equation and balances their steepening effect and leads to the existence of smooth traveling wave solutions.

One form of smooth motion described by the KdV equation involves what is called a *solitary wave*. This is a localized disturbance in the form of the square of a hyperbolic secant function that propagates without change of shape with a speed proportional to its amplitude relative to the equilibrium water level on either side of the solitary wave. The KdV equation is first order in time, and so describes unidirectional wave propagation (propagation in one direction). Thus, if propagation is to the right, and a solitary wave of large amplitude starts well to the left of a solitary wave of smaller amplitude, the larger wave will overtake the smaller one.

The nonlinear nature of the KdV equation might be expected to cause the solution to cease to describe the propagation of such waves once interaction occurs. However, this is not the case, and after a nonlinear interaction during which the amplitudes are *not* additive, the waves reappear with their identity preserved, though with their positions slightly altered because of the interaction. This remarkable property, which occurs however many times these solitary waves interact, led to these solitary waves being called *solitons* by Zabusky and Kruskal, who were the first to observe this phenomenon as a result of numerical experiments. The interaction process is now understood analytically, but

the purpose of this project is to observe this interaction and to confirm some of its qualitative features.

1. Use computer algebra to confirm by differentiation that

$$u_1(x, t) = -2 \operatorname{sech}^2(x - 4t)$$
$$\text{and} \quad u_2(x, t) = -8 \operatorname{sech}^2(2x - 32t)$$

are both solutions of the KdV equation $u_t - 6uu_x + u_{xxx} = 0$. Make 3D plots of the negative of $u_1(x, t)$ and $u_2(x, t)$ to show their shape and amplitude, and that their respective speeds of propagation are $dx/dt = 4$ and $dx/dt = 16$.

2. An analytical solution exhibiting soliton interaction for the KdV equation is

$$u(x, t) =$$
$$- 12 \frac{3 + 4\cosh(2x - 8t) + \cosh(4x - 64t)}{[3\cosh(x - 28t) + \cosh(3x - 36t)]^2}.$$

Using computer algebra, substitute $u(x, t)$ into $F(x, t) = u_t - 6uu_x + u_{xxx}$, and after simplification by grouping terms show that $F(x, t) \equiv 0$, confirming that $u(x, t)$ is a solution of the KdV equation. If simplification by grouping of terms proves difficult, substitute various pairs of values of x and t into $F(x, t)$ to show that $F(x, t) = 0$, to verify that in these particular cases $u(x, t)$ is indeed a solution of the KdV equation.

3. Make a 3D plot of the negative of $u(x, t)$ for $-10 \leq x \leq 10$ and $-0.5 \leq t \leq 0.5$, using sufficient points for the plot to be relatively smooth. Choose a suitable orientation for the plot so that the crests of the propagating solitary waves are easy to follow. Notice (a) that during the interaction process around the time $t = 0$ the amplitudes are not additive, (b) that the solitons preserve their shapes after interaction, and (c) that after interaction, the path followed by the slow soliton has been slightly delayed while the path followed by the faster soliton has been slightly advanced.

4. Compare the shapes of $u_1(x, t)$ and $u_2(x, t)$ with the slow and fast solitons, respectively, both well before and after their interaction, to confirm that their shapes have been preserved.

Project 5

The Sine–Gordon Equation

This project illustrates a different type of soliton that is a solution of the nonlinear *Sine–Gordon equation*

$$u_{xx} - u_{tt} = \sin u.$$

The Sine-Gordon equation is second order in time and so describes bi-directional wave propagation (propagation in both directions).

1. Confirm by computer algebra that the function

$$u(x,t) = 4\arctan\left[\exp\left(\frac{1}{3}(5x - 4t)\right)\right]$$

   is a solution of the Sine–Gordon equation and, using sufficient points, make a smooth 3D plot of $u(x,t)$ for $-25 < x < 25$ and $-5 < t < 5$. This steplike function is called a *kink soliton*, and when the step changes in the opposite sense the result is called an *antikink soliton*.

2. Confirm by computer algebra that the function

$$u(x,t) = 4\arctan\left[\frac{2}{\sqrt{3}}\frac{\sinh(\sqrt{3}t)}{\cosh(2x)}\right]$$

   is a solution of the Sine–Gordon equation and, using sufficient points, make a smooth 3D plot of $u(x,t)$ for $-15 < x < 15$ and $-8 < t < 8$. This shows the collision of a kink soliton and an antikink soliton.

Project 6

Dispersive Wave Propagation and the Telegraph Equation

This project demonstrates how linear equations that describe wave propagation can distort a propagating disturbance because of an effect called *dispersion*. The *telegraph equation*

$$u_{tt} - c^2 u_{xx} + a u_t + b u = 0,$$

with c, a, and b positive constants describes bidirectional wave propagation, and it was first derived to model telephonic communication along land lines. To see how a harmonic plane wave (a sinusoid) moving along the x-axis and governed by this equation is propagated, we consider the function $u(x,t)$ that is the real part of

$$\hat{u}(x,t) = A\exp[im(x - ct)] \quad (A \text{ real}),$$

and start by substituting $\hat{u}(x,t)$ into the telegraph equation. (This is equivalent to substituting $u(x,t) = A\cos[m(x - ct)]$ into the equation.)

Defining the *wavelength* $\lambda = 2\pi/m$, the *wave number* $k = 2\pi/\lambda$, and the frequency $\omega = 2\pi c/\lambda$ of the harmonic wave allows $\hat{u}(x,t)$ to be written

$$\hat{u}(x,t) = A\exp[i(kx - \omega t)].$$

When this expression is substituted into the telegraph equation, the following compatibility condition is found between k and ω in order that the harmonic wave is a solution of the equation:

$$\omega^2 + ia\omega - (b + c^2 k^2) = 0.$$

This result is called the *dispersion relation* for the telegraph equation, and for real k it shows that ω is complex, with

$$\frac{\omega}{k} = -i\frac{a}{2k} \pm \frac{1}{2k}(4c^2 k^2 + 4b - a^2)^{1/2}.$$

The quantity $kx - \omega t$ determines the *phase* of the wave, so that a wave of constant phase propagates with $kx - \omega t =$ constant, showing that the *phase velocity* of the wave is $v_P = \omega/k$. However, the dispersion relation shows that ω/k is a function of ω, so it follows that waves with different frequencies ω will propagate with different phase speeds v_p. Consequently, with the use of Fourier series, any periodic initial disturbance at time $t = 0$ can be decomposed into a sum of harmonic components, so because each component propagates with a different phase speed, when they are recombined to form the solution at later times $t_1, t_2, \ldots$, it follows that the wave shape will have changed with time. This change of shape of the wave is said to be due to *dispersion*.

When the dispersion relation is used in $\hat{u}(x,t)$, it turns out that

$$u(x,t) = \mathrm{Re}\left\{ A\exp\left(-\frac{at}{2}\right) \right.$$
$$\left. \times \exp\left[ik\left[x \mp \frac{t}{2k}(4c^2 k^2 + 4b - a^2)^{1/2}\right]\right]\right\}. \quad \text{(I)}$$

This confirms the dispersive nature of the telegraph equation, and when $a > 0$ it shows that the magnitude of the wave decays exponentially with time. If, however, $4b = a^2$ the dispersive effect vanishes and the wave propagates without change of shape, but with an exponential decay called *dissipation*. Such waves are said to be *relatively undistorted*. It was this condition that was first used to adjust the parameters in a telephone land line to remove distortion of the transmitted message due to dispersion. The decay, or dissipation, was corrected by the insertion of amplifiers at regular points along the line.

1. Let the initial wave profile be $u(x, 0) = x(\pi - x)$ in the interval $0 \leq x \leq \pi$, and let this profile be repeated periodically along the x-axis with period π. Use computer algebra to find the coefficients $a_0, a_1, \ldots, a_6$ of the Fourier cosine series expansion of $u(x, 0)$ on the interval $0 \leq x \leq \pi$.

2. Set $a = 0.2$, $b = 0.4$, and $c = 1$ in (I), and take the negative sign to describe a wave moving

to the right with speed $c = 1$. Let $u_k(x, t)$ denote the solution corresponding to $A = a_k$ for $k = 0, 1, \ldots, 6$, and use computer algebra to form the approximate solution of (I) given by $u_A(x, t) = \sum_{k=0}^{6} u_k(x, t)$.

3. The combined effects of dispersion and dissipation on the initial wave profile can be seen by making 2D plots of $u_A(x, t)$ at the times $t = n$ for $n = 0, 1, 2, 3$, and 4 over the respective intervals $n \leq x \leq n + \pi$, where the x-interval moves with speed $c = 1$ to follow the initial wave profile.

4. Repeat the calculations using $a = 0.2$, $b = 0.01$, and $c = 1$, and by again making the 2D plot in Step 3 confirm that in this case the wave decays, but is relatively undistorted (it preserves its shape as it propagates, but not its amplitude).

5. A special case of the telegraph equation is the **Klein–Gordon equation**

$$u_{tt} = au_{xx} - bu, \quad \text{with} \quad a > 0, b > 0.$$

Relate this equation to the dispersion relation in (I), and hence show that the Klein–Gordon equation is purely dispersive and so does not decay as time increases.

Project 7

Development of a Nonunique Solution

This project involves the construction of the envelope of characteristics for the first order quasilinear equation

$u_t + uu_x = 0$ subject to the initial condition $u(x, 0) = \sin x$,

to demonstrate where and when the solution first becomes nonunique because of the intersection of characteristics. It also examines the shape of the nonlinear wave as it propagates.

1. Plot the envelope of the characteristics together with their asymptotes for the preceding problem for $0 \leq x \leq 2\pi$ and $0 \leq t \leq 4$, and confirm that its cusp forms at $x = \pi$ and $t = 1$.

2. Make 2D implicit plots of the solution $u = \sin(x - ut)$ in the interval $-5 \leq x \leq 5$ for the times $t = 0$, 0.5, 0.75, 1, and 2 to demonstrate how the nonuniqueness of the solution develops, using sufficient points for the plots to be smooth.

3. Make a 3D plot of the solution $u = \sin(x - ut)$ for $-2\pi \leq x \leq 3\pi, 0 \leq t \leq 3$, and $-1 \leq u \leq 1$ to show the global development of the nonunique solution, using sufficient points for the plot to be smooth. Compare the result with the 2D plots made in Step 2. (Hint: In the program MAPLE V, this 3D plot can be made with PDEtools and PDEplot).

PART EIGHT

NUMERICAL MATHEMATICS

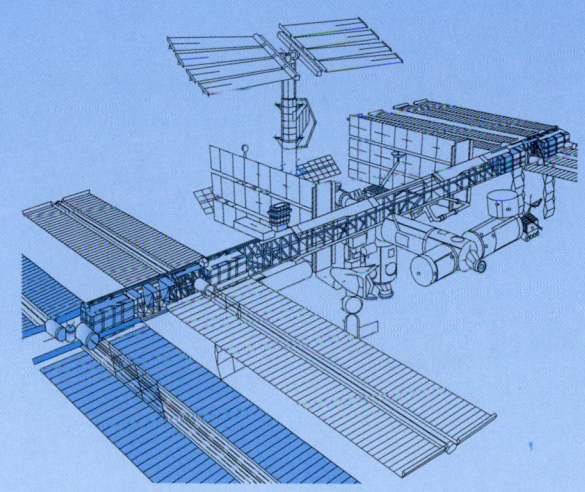

Chapter 19   **Numerical Mathematics**

CHAPTER 19

Numerical Mathematics

Unlike theoretical solutions to problems that give rise to general results that can then be related to specific problems, numerical methods only yield answers to specific problems. Because of this, numerical methods are used in the analysis of specific mathematical problems, where numerical solutions can become necessary for many different reasons. It may, for example, happen that a theoretical solution is available but is inconvenient to use, possibly because a system of linear equations arises requiring a solution that is so complex the theoretical solution is not useful. When studying a specific problem it can also happen that a definite integral occurs with no known closed form solution, or a nonlinear differential equation arises that cannot be solved theoretically. Yet another reason might be that a solution to a group of interrelated problems is so complicated that no theoretical solution is possible. In all such cases, when solving specific problems, it becomes necessary to use efficient numerical methods.

This chapter describes how to deal with the most frequently occurring types of numerical problem. These are interpolation, root finding, numerical integration, the numerical solution of large systems of linear equations, the numerical determination of eigenvalues and eigenvectors, and the numerical solution of initial value problems for linear and nonlinear differential equations and systems.

The methods described here are the classical ones, so they are neither as efficient nor as sophisticated as the methods used in currently available numerical symbolic algebra packages, though they are practical and can be used for straightforward calculations. The reason for their inclusion is because they illustrate in a concise way some of the most important general principles that are involved, while at the same time showing both the shortcomings and advantages of different methods. One essential difference between the classical methods described in this chapter, and many of the codes used in practice, is that modern codes are adaptive, so they can switch between methods of solution to speed up convergence, or adjust step size when integrating differential equations to maintain a predetermined accuracy.

Decimal Places and Significant Figures

Many of the problems that occur in engineering and physics have no analytical solution, and even when one can be found it is frequently the case that the form in which it arises is difficult to use directly if numerical results are required. There are many reasons for such limitations, some typical ones being that the zeros of a function involved in the solution cannot be found analytically, a definite integral that arises cannot be evaluated analytically, an analytical solution of a nonlinear differential equation cannot be found, or a large system of linear simultaneous equations must be solved. A situation of a different type arises when an analytical solution is known, but its application in specific cases leads to a prohibitive amount of calculation, so a more efficient numerical method becomes necessary.

As most numerical results can only be approximate, such as calculations involving $\sqrt{2}$, e, or π, it is necessary to have a simple way of indicating their accuracy. This is accomplished either by stating that a result is accurate to ***n* decimal places**, **decimal places** or that it is accurate to a given number of **significant digits (figures)**. For example, when approximating a number such as

$$17.213622,$$

to *three* decimal places, the *fourth* digit after the decimal point is examined, and if the digit is 5 or more the preceding digit is increased by one and the result truncated to three places after the decimal point. However, if the *fourth* digit is 4 or less, the previous digit is left unchanged and the result is truncated to the existing three digits that follow the decimal point. When this process is applied to the above number to approximate it to an accuracy of three decimal places it becomes

$$17.214,$$

whereas if it is approximated to an accuracy of four decimal places it becomes

$$17.2136.$$

rounding up and down This process of approximating a number to n decimal places by increasing the nth digit by 1, if the $(n + 1)$th digit is a 5 or more, and then truncating the result after n decimal places is called **rounding up** to an accuracy of n decimal places. Similarly, the process of leaving the nth digit unchanged when the $(n + 1)$th digit is a 4 or less, and truncating the result after n decimal places is called **rounding down** to an accuracy of n decimal places.

To express a number accurately to n significant figures involves a somewhat different argument from the one just described. The first nonzero digit that occurs in a number, irrespective of where the decimal point is located, is called the *first* (and most) *significant digit*, so in a number such as 3.496221 the first significant digit is 3, and in a number such as 0.004713 the first significant digit is 4. Starting from the first significant digit and counting $n + 1$ digits to the right, the nth digit is rounded up or down, according as the $(n + 1)$th digit is 5 or more, or 4 or less, as previously described. The number is then truncated after the group of n digits obtained in this way, with zeros being entered in place of any other digits that appear *before* the decimal point. This process is called expressing the number accurately **significant digits** to ***n* significant digits (figures)**. So, to three significant digits, the number

$$315,814$$

becomes 316,000, while to four significant digits the number

$$0.004723217$$

becomes 0.004723.

    Accuracy can be lost if the (approximate) result of one numerical calculation is used in a subsequent numerical calculation, and certainly if this process is repeated many times. To avoid loss of accuracy it is necessary to work to a fixed number of digits that is sufficiently large. Calculators and computers use a fixed number of digits, but symbolic algebra computer packages allow the user to choose the number so that high accuracy can be maintained throughout a sequence of calculations.

    The form in which numbers have been represented so far is called a **fixed point** decimal representation, because the numbers are displayed relative to the decimal point that is involved. The **floating point** representation used in most computer calculations involves writing a number x in the form

fixed and floating point numbers

$$x = r \cdot N^s,$$

where the number N is called the **base** of the representation, the number r is called the **mantissa**, and s is called the **exponent**. The mantissa is usually chosen to have one digit in front of the decimal point. So, to the base 10, the number 453.7 has the floating point representation 4.537×10^2, while the number 0.000369 has the representation 3.69×10^{-4}. A notation used for floating point representations in machine computation to the base 10 involves representing x in floating point form by writing the mantissa r first, then the symbol E followed by the exponent s, which may be positive or negative. Most computers normalize so that the mantissa is between 0 and 1, so when using this convention the number 453.7 becomes 0.4537E3, and the number 0.000369 becomes 0.369E–3.

Summary

Accuracy in terms of decimal places and significant figures was defined, and the convention for rounding numbers up or down was explained. Floating point calculations were introduced, and the importance of expressing accuracy in terms of significant digits when working with floating point numbers was stressed.

19.2   Roots of Nonlinear Functions

Let $f(x)$ be a real valued function defined for $a \le x \le b$. A number ξ is called a **root** of the function $f(x)$ in this interval if $f(\xi) = 0$ and, correspondingly, a number $x = \xi$ that makes $f(x)$ vanish is called a **zero** of $f(x)$. The need to find roots of functions is fundamental to the development and application of mathematics, and only in simple cases can the roots be determined analytically, so in all other cases it is necessary to find them numerically. Many different methods exist for the numerical determination of roots of functions, but of these only the *bisection method*, the *fixed point method*, and *Newton's method* will be described in any detail, as they are in everyday use and are easily implemented on a computer.

(a) The Bisection Method

Apart from graphing $f(x)$ and finding by inspection those values of x for which $f(x) = 0$, the simplest systematic method for finding the roots of a function $f(x)$

is the **bisection method**. The method is easily programmed, and it applies to roots of functions $f(x)$ with the property that $f(x)$ changes sign when x crosses a root. The determination of a root accurately by this method depends on the ability to evaluate the function with sufficient accuracy that its sign change can be determined correctly.

To understand how the method works, consider a continuous function $f(x)$ and numbers $\alpha < \beta$ such that $f(\alpha)$ and $f(\beta)$ have opposite signs. Then from the intermediate value theorem the function $f(x)$ must vanish at least once (have at least one root) ξ between α and β, as shown in Figs. 19.1a,b. However, if $f(\alpha)$ and $f(\beta)$ have the same sign, nothing can be deduced about the existence of roots in the interval, as can be seen from Figs. 19.1c–e, which illustrate situations in which

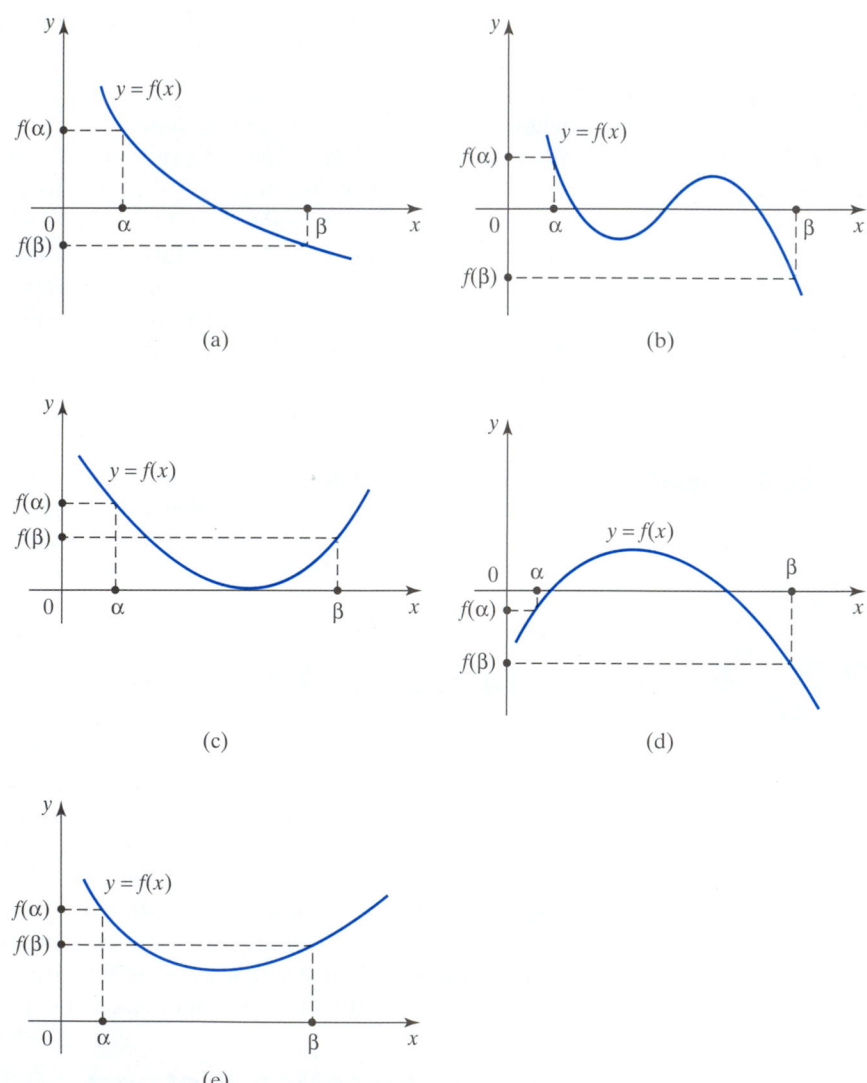

FIGURE 19.1   Roots and the product $f(\alpha)f(\beta)$ in the interval $\alpha \le x \le \beta$. (a) $f(\alpha)f(\beta) < 0$, one root. (b) $f(\alpha)f(\beta) < 0$, three roots. (c) $f(\alpha)f(\beta) > 0$, double root. (d) $f(\alpha)f(\beta) > 0$, two roots. (e) $f(\alpha)f(\beta) > 0$, no roots.

there are a double root, two roots, and no root, respectively. In what follows we will assume that $f(x)$ experiences a change of sign across the interval, and that α and β are chosen sufficiently close that there is only *one* root in the interval, as illustrated in Fig. 19.1a. When $f(x)$ is sufficiently simple this can usually be achieved by graphing $f(x)$ and selecting suitable values for α and β.

geometrical interpretation of the bisection method

To implement the bisection method, a simple test is needed to see if a function $f(x)$ has opposite signs at the ends of an interval $\alpha < x < \beta$. Such a test is provided by examining the product $f(\alpha)f(\beta)$, because when this is negative a sign change occurs, but when it is positive there is no such sign change. When, as may happen during a computation, a computer finds that $f(\alpha)f(\beta) = 0$, the value of $f(\alpha)$ must be examined to avoid interpreting as a true zero an approximate number α that causes the computer arithmetic system to regard this product function as zero.

The first step in the bisection method involves dividing (bisecting) the interval $\alpha \leq x \leq \beta$ into the two subintervals $\alpha < x < x_1$ and $x_1 < x < \beta$, where $x_1 = \frac{1}{2}(\alpha + \beta)$. The subinterval to be considered next is obtained by replacing α by x_1 if $f(\alpha)f(x_1) > 0$, because in this case $f(x)$ changes sign in the subinterval $x_1 < x < \beta$ so this interval must contain a root of $f(x)$. Conversely, if $f(\alpha)f(x_1) < 0$, the subinterval to be considered is obtained by replacing β by x_1, because in this case $f(x)$ experiences a change of sign in the subinterval $\alpha < x < x_1$, and so this interval must contain a root ξ. The task of finding the root has now been refined from considering the interval $\alpha \leq x \leq \beta$ and replaced by the task of finding the root in an interval half the size.

The bisection process involves a repetition of this procedure, each time using the smaller subinterval found at the previous stage of the calculation, so that after m steps the root ξ will be contained in an interval of length $|\alpha - \beta|/2^m$. If the root is required to be accurate to within an error of ε, where $\varepsilon > 0$ is a preassigned small quantity, machine computation that works with a fixed number of digits proceeds until the first time successive iterates x_m and x_{m+1} satisfy the condition $|x_m - x_{m+1}| < \varepsilon$. The required approximation to the root ξ is then taken to be $x_m \pm \varepsilon$.

The bisection method has the property that the bound placed on the error involved is halved at each iteration. Unlike some other methods, provided the bisection method is applicable it *always* converges to a root, though if more than one root occurs in the initial interval $\alpha \leq x \leq \beta$ it is not known in advance to which root the method will converge.

The bisection method has the advantage of being simple and using the minimum amount of information, because it only depends on the functional values of $f(x)$ at the end points of an interval and not on the calculation of derivatives, though other methods may converge faster. The practical implementation of the method on a computer suffers from the fact that when the product $f(\alpha)f(\beta)$ is determined, underflow of this floating point number becomes inevitable as the upper and lower bounds approach the root. However, this is easily overcome by determining the sign of $f(\alpha)f(\beta)$ by examining the signs of $f(\alpha)$ and $f(\beta)$. Because the bisection method is affected less by limiting precision, a different and faster method is often used to start the calculation, and a switch is made to the bisection method once a very accurate approximation to the root has been obtained.

The bisection method cannot be used to find a root $x = \xi$ of a function that is either convex or concave at $x = \xi$, as illustrated in Fig. 19.1e, because such functions do not change sign as x crosses ξ. This can happen, for example, when seeking the roots of polynomials of even order, the simplest case of which is $f(x) = (x - a)^2$ with a double root at $x = a$.

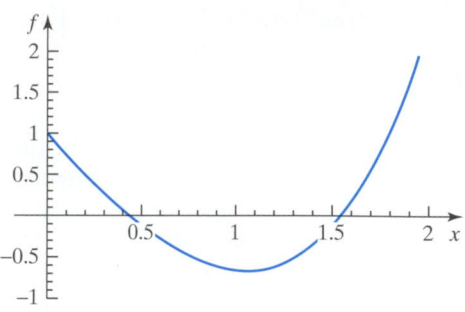

FIGURE 19.2   The function $f(x) = 1 - 3x + \frac{1}{2}xe^x$.

The numerical determination of multiple (repeated) roots is difficult, so only an outline of a possible approach will be given here for a polynomial of degree n with real roots, one of which is a double root. The difficulty that arises when seeking multiple roots is because the calculation always leads to an ill-conditioned problem—that is, to a problem in which an extremely small error in part of the calculation leads to a very large error in the result.

deflation of a polynomial

The approach we now describe involves what is called the **deflation** of the polynomial. First a single root of the polynomial is found, and the polynomial is then divided by the corresponding factor to obtain a polynomial of degree $n - 1$. A repetition of this process involving each of the $n - 2$ single roots will lead to a quadratic whose double root can then be found from the quadratic formula. When it is necessary, deflation must always be carried out with care to avoid the compounding of errors.

It is important to remember that the bisection method cannot be used to compute roots of even order, because in such cases no sign change is involved, but it works well for roots of odd order irrespective of their multiplicity. One approach to the multiple root problem involves using the bisection method with different starting intervals, and another involves using other methods with different guesses.

EXAMPLE 19.1

Use the bisection method to find the smallest root of the function $f(x) = 1 - 3x + \frac{1}{2}xe^x$.

Solution   Examination of Fig. 19.2 shows that an approximation to the smallest root of $f(x) = 0$ is $x = 0.45$, and that suitable values for α and β are $\alpha = 0.43$ and $\beta = 0.47$, because $f(\alpha) = 0.0405$ and $f(\beta) = -0.0340$, and the graph shows that there is only one root between α and β.

If at each stage of the calculation the left end point of an interval containing the root ξ is denoted by x_l and the right end point by x_r, the calculation can be arranged as follows.

n	Left End Point x_l	Right End Point x_r	x_n	$f(x_l)$	$f(x_n)$	$f(x_l)f(x_n)$	New Interval	Approximate Root
1	$\alpha = 0.43$	$\beta = 0.47$	0.45	0.0405	0.0029	>0	$0.45 < \xi < 0.47$	0.45
2	0.45	0.47	0.46	0.0029	−0.0157	<0	$0.45 < \xi < 0.46$	0.46
3	0.45	0.46	0.455	0.0029	−0.0064	<0	$0.45 < \xi < 0.455$	0.455
4	0.45	0.455	0.4525	0.0029	−0.0018	<0	$0.45 < \xi < 0.4525$	0.4525

Continuing this process shows that to an accuracy of five decimal places the required value of the root is $x = 0.45154$. ∎

(b) Fixed Point Iteration

This method is well suited to machine computation provided numerical values of the function involved are easily calculated, and a good approximation to the root is used to start the iteration process. The idea is straightforward, and its success depends on rewriting the given function $f(x)$ whose root is required in the form

$$f(x) = x - g(x). \tag{1}$$

Then if $x = \xi$ makes the expression on the right of (1) vanish, it follows that ξ is a root of $f(x)$. The representation of $f(x)$ in the form given in (1) is not unique, because as will be seen in the examples that follow, $g(x)$ can be written in more than one way. Later we will derive a simple condition on the form of $g(x)$ that must be satisfied together with the value $x_0 = \alpha$ used to start the iteration process in order that the calculations are likely to converge to the root ξ.

If we now consider the function $g(x)$ to *map* a point x into a point $g(x)$, then a root $x = \xi$ of equation (1) has the property that $g(x)$ maps the point ξ into itself, and for this reason ξ is called a **fixed point** of the equation

$$x = g(x). \tag{2}$$

fixed points and iteration

The fixed point iterative scheme follows from (2) by writing it as

$$x_{n+1} = g(x_n), \tag{3}$$

and starting the iteration process by setting $x_0 = \alpha$. The iteration will be said to **converge** if the sequence of iterates x_n approaches a limit as $n \to \infty$, and to **diverge** if no such limit exists. Suppose that when the iterations converge the result is required to be accurate to within an error of ε, where $\varepsilon > 0$ is a preassigned small quantity. Then the calculation proceeds until the first time successive iterates x_m and x_{m+1} satisfy the condition $|x_m - x_{m+1}| < \varepsilon$. The required approximation to the root ξ is then taken to be $x_n \pm \varepsilon$.

EXAMPLE 19.2   Find a fixed point iterative scheme for determining $\sqrt{a}$ when $a > 0$, and use it to calculate $\sqrt{2}$ to an accuracy of six decimal places.

Solution   The required number $\sqrt{a}$ is a solution of the equation $x^2 = a$, so to express this in the form given in (2) we write it as $2x^2 = x^2 + a$, and then divide the result by $2x$ to arrive at the result

$$x = \frac{1}{2}\left(x + \frac{a}{x}\right),$$

so in the notation of (2) the function $g(x) = \frac{1}{2}(x + \frac{a}{x})$.

The fixed point iterative scheme follows from this, as in (2), by replacing x on the left by x_{n+1} and x on the right by x_n to obtain

$$x_{n+1} = \frac{1}{2}\left(x_n + \frac{a}{x_n}\right).$$

The iteration is started by setting $n = 0$ and $x_0 = k$, where k is an approximation to $\sqrt{a}$.

To illustrate the scheme we will calculate $\sqrt{2}$, so as $a = 2$ the scheme becomes

$$x_{n+1} = \frac{1}{2}\left(x_n + \frac{2}{x_n}\right),$$

and for simplicity we start by setting $x_0 = 1$. The results of the calculation are

$$x_0 = 1$$
$$x_1 = 1.5$$
$$x_2 = 1.41666667$$
$$x_3 = 1.41421569$$
$$x_4 = 1.41421356$$
$$x_5 = 1.41421356.$$

As the x_4 and x_5 iterates are identical, rounding the result of x_5 to six decimal places gives $\sqrt{2} = 1.414214$. ∎

The fixed point iterative scheme in Example 19.2 converged rapidly, and it is this scheme that is used in computers to determine the square root of any positive number to an accuracy that is within the capability of the computing system and software being used. Experimentation will show that this iterative scheme is stable with respect to the choice of the starting approximation, because it will always converge to $\sqrt{2}$, though a starting approximation close to $\sqrt{2}$ will, of course, lead to the most rapid convergence.

To examine iterative schemes a little further, and to show that convergence does not always occur, we consider the next example.

EXAMPLE 19.3

examination of two fixed point iterative schemes

Devise fixed point iterative schemes to find the roots of the quadratic equation

$$2x^2 - 24x + 41 = 0,$$

and test them numerically.

Solution Two obvious fixed point iterative schemes that can be obtained directly from the equation follow by first writing it in either of the forms

$$x = \frac{1}{24}(2x^2 + 41) \quad \text{or} \quad x = 12 - \frac{41}{2x}.$$

Replacing x by x_{n+1} on the left and by x_n on the right we obtain the following two schemes:

Scheme A: $x_{n+1} = \frac{1}{24}\left(2x_n^2 + 41\right)$, and Scheme B: $x_{n+1} = 12 - \frac{41}{2x_n}$.

An application of the quadratic formula shows the two roots to be $x = 6 - \frac{1}{2}\sqrt{62} = 2.0630$ and $x = 6 + \frac{1}{2}\sqrt{62} = 9.9370$, so starting approximations close to these values are $x_0 = 2$ and $x_0 = 10$. Scheme A leads to the results

$x_0 = 2$	$x_0 = 10$
$x_1 = 2.0417$	$x_1 = 10.0417$
$x_2 = 2.0557$	$x_2 = 10.1113$
$x_3 = 2.0605$	$\vdots$
$x_4 = 2.0621$	$x_8 = 12.4801$
$x_5 = 2.0627$	$x_9 = 14.6877$
$x_6 = 2.0630$	$\vdots$
$\vdots$	$x_\infty = \infty$
$x_\infty = 2.0630.$	

Clearly Scheme A is only partially successful, because although when started with $x_0 = 2$ it converges to the zero close to 2, it diverges when started with $x_0 = 10$.

Scheme B produces the following results:

$x_0 =$	2	$x_0 = 10$
$x_1 =$	1.75	$x_1 =$ 9.7222
$x_2 =$	0.2857	$x_2 =$ 9.8914
$x_3 =$	-59.7500	$x_3 =$ 9.9275
$x_4 =$	12.3431	$x_4 =$ 9.9350
$x_5 =$	10.3392	$x_5 =$ 9.9370
$x_6 =$	10.0172	$x_6 =$ 9.9370
$x_7 =$	9.9535	$\vdots$
$\vdots$		$x_\infty =$ 9.9370
$x_\infty =$	9.9370.	

Here also scheme B is also only partially successful, though this time for a different reason. Although, as required, the iterates converge to the root close to 10 when started with $x_0 = 9$, when started with $x_0 = 2$ they fail to converge to the root close to 2 and again converge to the root close to 10. ∎

To understand this behavior of iterative schemes we need the following theorem that gives conditions for the choice of $g(x)$ and the starting approximation x_0 that will ensure the convergence of the scheme.

THEOREM 19.1

Convergence of a fixed point iterative scheme   Let $g(x)$ be defined in the interval $a \leq x \leq b$ in which it has a fixed point ξ, and let $g(x)$ be continuous throughout this interval with a continuous derivative $g'(x)$ such that $|g'(x)| \leq k < 1$. Then the equation $x = g(x)$ has a unique fixed point ξ in the interval, and if x_0 is such that $a \leq x_0 \leq b$ the iterative scheme

condition for convergence of a fixed point iterative scheme

$$x_{n+1} = g(x_n)$$

will converge to ξ.

Proof   The proof involves two steps, in the first of which a fixed point ξ is assumed and shown to be unique, whereas in the second we go on to prove the convergence of the scheme and to justify the assumption of the existence of a fixed point. To show that the fixed point is unique let us assume, if possible, that two *different* fixed points ξ_1 and ξ_2 occur inside the interval, so that $\xi_1 = g(\xi_1)$ and $\xi_2 = g(\xi_2)$. Considering the expression $|\xi_1 - \xi_2|$, applying the mean value theorem, and using the condition $|g'(x)| \leq x_0 < 1$, we find that for some number η inside the interval $a \leq x \leq b$

$$|\xi_1 - \xi_2| = |g(\xi_1) - g(\xi_2)| = |g'(\eta)(\xi_1 - \xi_2)| \leq x_0|\xi_1 - \xi_2| < |\xi_1 - \xi_2|,$$

but this is impossible, so the contradiction implies the uniqueness of the fixed point.

Next, to prove the convergence of the scheme, we again make use of the mean value theorem that asserts there is some point ζ_n between x_{n-1} and ξ such that

$$|\xi - x_n| = |g(\xi) - g(x_{n-1})| = |g'(\zeta_n)(\xi - x_{n-1})| = |g'(\zeta_n)||\xi - x_{n-1}| \leq x_0|\xi - x_{n-1}|.$$

Repeated application of this inequality leads to the result $|\xi - x_n| \leq x_0^n|\xi - x_0|$, but as $0 \leq x_0 < 1$ we have $\lim_{n\to\infty} x_0^n = 0$, so that

$$\lim_{n\to\infty} |\xi - x_n| = 0, \quad \text{and hence} \quad \lim_{n\to\infty} x_n = \xi.$$

With a little more trouble, the iterates can be shown to form a Cauchy sequence, and an appeal to the completeness of real numbers then guarantees that the sequence has a limit ξ, so the theorem is proved.   ■

This theorem explains the results of Example 19.2. In Scheme A the function $g(x) = \frac{1}{24}(2x^2 + 41)$, so $|g'(x)| = \frac{1}{6}|x|$ and $|g'(x)| < 1$ when $0 < x < 6$, showing the scheme to be convergent to the root close to 2 when an initial approximation close to 2 is used. However, when $x = 10$ the conditions of the theorem are not satisfied so the scheme cannot be expected to converge to the root close to 10, though it does not assert that it will diverge.

In the case of Scheme B we have $g(x) = 12 - \frac{41}{2x}$ so that $|g'(x)| = \frac{41}{2x^2}$. This shows that the scheme will converge to the root close to 10 for an x_0 close to 10, because then $|g'(x)| < 1$, but that it cannot be expected to converge to the root close to 2 where the condition is violated, though again the theorem does not assert that in this case it will diverge. It is possible to show that if $|g'(\xi)| > 1$, the iteration will not converge, except by accident.

The reason for the convergence or divergence of iterative schemes is most easily understood by using a graphical representation of a fixed point iteration process. Typical cases are illustrated in Fig. 19.3, where diagrams (a) and (b) show how the mapping $x_{n+1} = g(x_n)$, using the lines $y = x$ and $y = g(x)$, can lead to *convergent* processes, while diagrams (c) and (d) show how *divergent* processes can arise.

convergent and divergent iterations

(c) Newton's Method

Our starting point for the derivation of **Newton's method** for the determination of a zero of a differentiable function $f(x)$, also known as the **Newton–Raphson** method, is the mean value theorem representation of $f(x)$ about a point $x = x_0$ that can be written

$$f(x) = f(x_0) + (x - x_0)f'(\xi), \tag{4}$$

where ξ is a point between x_0 and x.

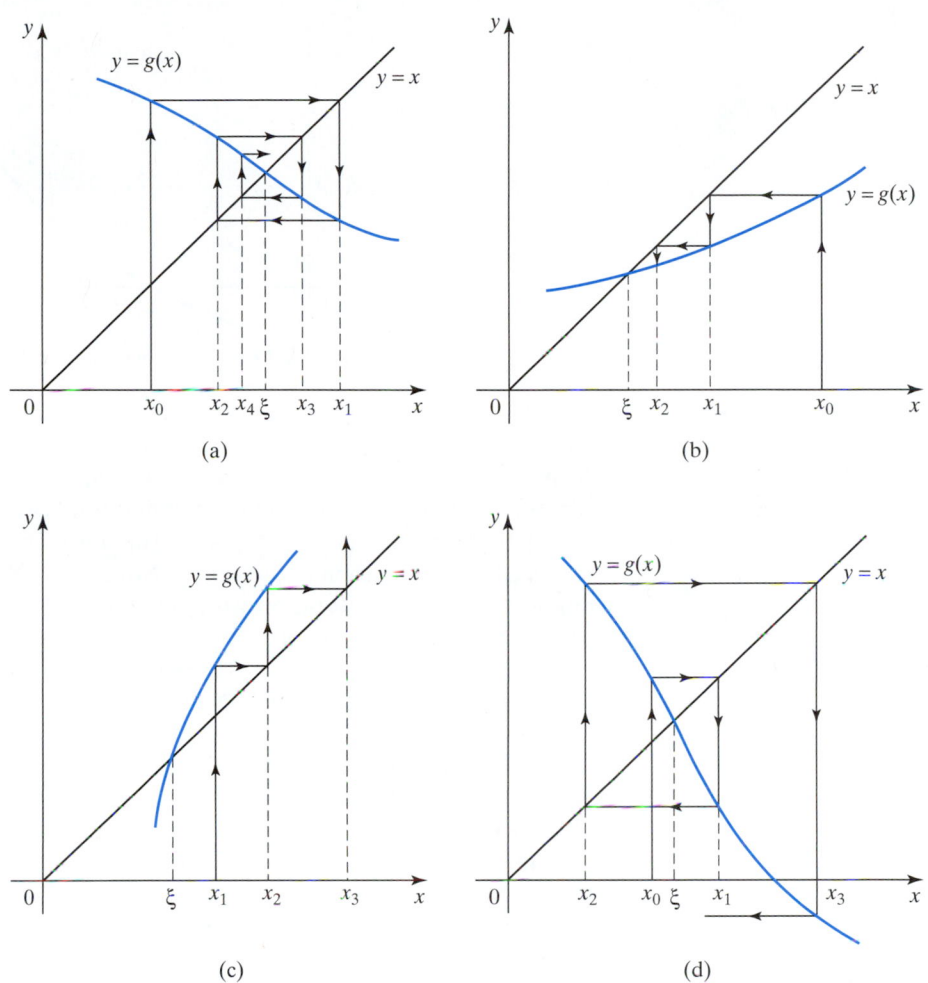

FIGURE 19.3 Typical convergent iterative processes in (a) and (b), and typical divergent iterative processes in (c) and (d).

If we set $h_0 = x - x_0$, and choose h_0 so that $x_0 + h_0$ is a zero of $f(x)$, result (4) becomes

$$h_0 = -\frac{f(\xi)}{f'(\xi)},$$

so the zero $x = x_0 + h_0$ of $f(x)$ is given by

$$x = x_0 - f(\xi)/f'(\xi). \tag{5}$$

As ξ is unknown, replacing it by x_0 produces the approximation x_1 given by

$$x_1 = x_0 - f(x_0)/f'(x_0).$$

Newton's method   Iterating this result leads to the **Newton's method**

$$x_{n+1} = x_n - f(x_n)/f'(x_n), \tag{6}$$

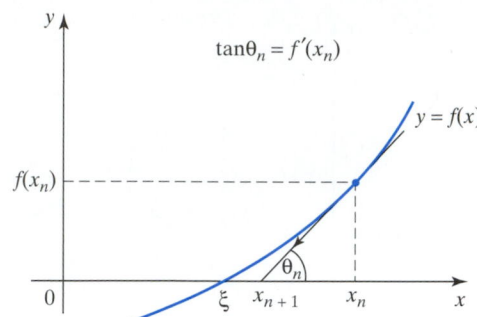

FIGURE 19.4 The tangent approximation used in Newton's method.

with $n = 0, 1, 2, \ldots$. If a tolerance ε is set, where $\varepsilon > 0$ is a preassigned small quantity, the calculations proceed until the first time the successive iterates x_m and x_{m+1} satisfy the condition $|x_m - x_{m+1}| < \varepsilon$. The number $x_{m+1} \pm \varepsilon$ is then taken to be the required approximation to the root ξ. Notice that Newton's method is a special example of fixed point iteration with $g(x) = x - f(x)/f'(x)$ and, in connection with Theorem 19.1, that the expression $|\xi - x_n| = |g'(\zeta_n)||\xi - x_{n-1}|$ tells us that $|\xi - x_n|$ approximates $|g'(\xi)||\xi - x_{n-1}|$ as the iterations converge. Clearly, the smaller $|g'(\xi)|$, the faster the convergence. For Newton's method and a simple root, this quantity is zero. So the argument suggests that for Newton's method the iterations converge faster than linearly, as is indeed the case. Typically, both fixed point iteration and Newton's method converge to the root nearest to the initial guess, though as has already been remarked, this is not true of the bisection method. Newton's method is generally much faster than the bisection method for simple roots, though not for multiple roots.

The geometrical interpretation of Newton's method is illustrated in Fig. 19.4, where the $(n + 1)$th approximation x_{n+1} is obtained from the nth approximation x_n by tracing back the tangent to the curve $y = f(x)$ at the point $(x_n, f(x_n))$ to the point x_{n+1} where it intersects the x-axis.

how Newton's method uses the tangent line approximation

EXAMPLE 19.4

Use Newton's method to find the zeros of $f(x) = 1 - 3x + \frac{1}{2}xe^x$ accurate to five decimal places.

Solution A graph of $f(x)$ shows that it has zeros close to 0.5 and 1.6, so we will use these as our starting approximations. As $f'(x) = \frac{1}{2}(1 + x)e^x - 3$, Newton's method becomes

$$x_{n+1} = x_n - \left(1 - 3x_n + \frac{1}{2}x_n e^{x_n}\right) \bigg/ \left(\frac{1}{2}(1 + x_n)e^{x_n} - 3\right) \quad \text{for } n = 0, 1, 2, \ldots.$$

Starting the calculation with $x_0 = 0.5$ gives

$$
\begin{aligned}
x_0 &= 0.5 & x_3 &= 0.451542 \\
x_1 &= 0.450200 & x_4 &= 0.451542 \\
x_2 &= 0.451541,
\end{aligned}
$$

so to an accuracy of five decimal places the smallest zero of $f(x)$ is 0.45154.

Similarly, when the calculation is started with $x_0 = 1.6$, we find that

$$
\begin{aligned}
x_0 &= 1.6 & x_3 &= 1.549538 \\
x_1 &= 1.552769 & x_4 &= 1.549538 \\
x_2 &= 1.549552 &
\end{aligned}
$$

so to an accuracy of five decimal places the largest zero of $f(x)$ is 1.54954.   ■

divergent and repeated cycle Newton iterations

This example illustrates the speed with which Newton's method can converge to a zero when a good starting approximation is used and the tangent to the graph $y = f(x)$ at a zero is not inclined at a small angle to the x-axis making high accuracy difficult to obtain. A poor starting approximation can cause Newton's method to diverge from the required zero, as illustrated in Fig. 19.5a where successive approximations move further away from the zero. Sometimes an unfortunate choice of starting approximation can lead to the situation illustrated in Fig. 19.5b where the iteration cycles indefinitely. To avoid situations like these, machine computations place a limit on the number of iterations to be performed to achieve the required accuracy, after which a new starting approximation must be used.

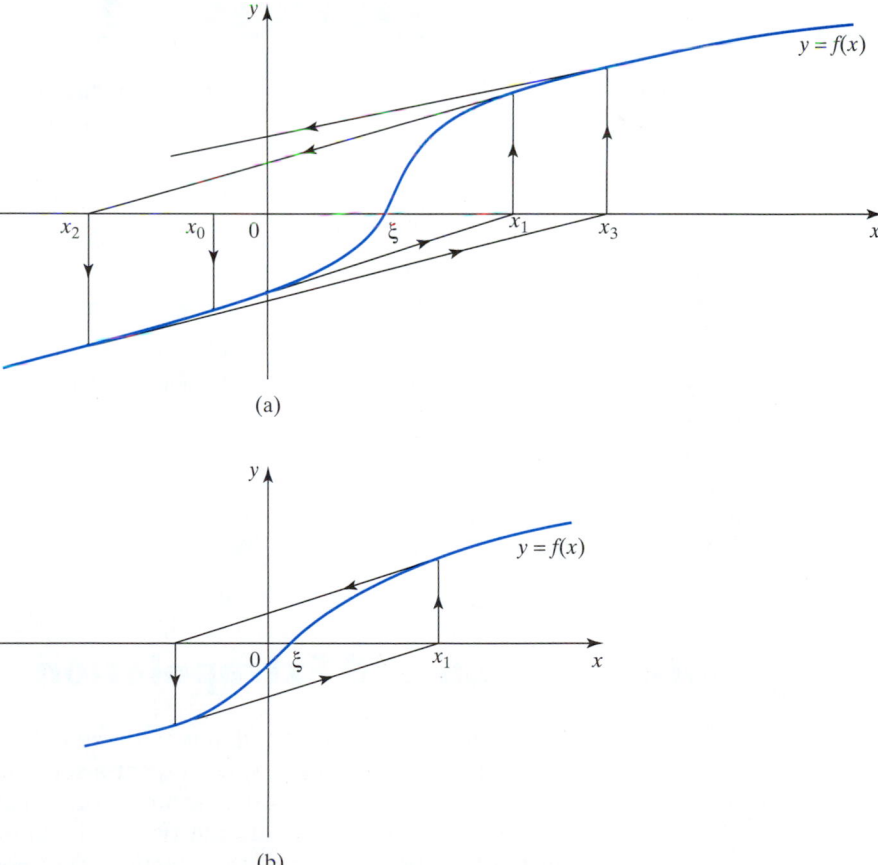

(a)

(b)

FIGURE 19.5   (a) Divergent process. (b) Repeated cycle.

Summary

The need for the determination of roots of nonlinear functions arises in many ways. The methods for the determination of roots discussed in this section were the bisection method, fixed point iteration methods, and Newton's method, which can be considered as a special fixed point iteration method. It was stressed that the bisection method only works for functions that change sign across a root, that its rate of convergence to a root is slow, and that if more than one root occurs in an interval it is not known in advance to which one the method will converge. The relative speeds of convergence of these methods were mentioned.

EXERCISES 19.2

In Exercises 1 through 6 use the bisection method to find the required root.

1. The root of $\sin x - \frac{1}{3}x = 0$ close to $x = 2.2$.
2. The root of $e^{x/3} - x^2 = 0$ close to $x = 1.1$.
3. The root of $3 \ln x + x^2 - 3 = 0$ close to $x = 1.3$.
4. The largest positive root of $x^3 - 1.9x^2 - 2.3x + 3.7 = 0$.
5. The smallest root of $x^3 - 4.5x^2 + 1.3x + 8 = 0$.
6. The root of $\frac{1}{2}\sqrt{1 - x^2} - x^2 = 0$.

In Exercises 7 through 12 use a fixed point iteration scheme to find the required roots.

7. Determine $a^{1/n}$ where $a > 0$ and n is an integer. Check the result by finding $4^{1/3}$.
8. Find the roots of $x^2 + 4x + 1 = 0$ and check the results by using the quadratic formula.

9. Find all three roots of $x^3 - 4.3x^2 + 1.4x + 7.8 = 0$.
10. Find the positive root of $\sin x - \frac{1}{2}x = 0$.
11. Find the positive root of $x^2 - 2\sinh x + 1 = 0$.
12. Find the positive root of $x^2 + 2 \ln x - 4 = 0$.

In Exercises 13 through 18 use Newton's method to find the required root.

13. Find $23^{1/3}$ by solving for the zero of $f(x) = 23 - x^3$.
14. Find the smallest positive root of $\tan x + 2\tanh x = 0$.
15. Find the largest root of $x^4 - 4x^3 + x^2 + 1.2 = 0$.
16. Find the smallest root of $x^4 - 3x^3 + 2x^2 - 3x - 1.6 = 0$.
17. Find the root of $3x - e^{-x} = 0$.
18. Find the root of $1 + \tanh x - 2\tan x = 0$.

19.3 Interpolation and Extrapolation

Sometimes a function $f(x)$ that is assumed to be smooth is only known in the form of a set of discrete values $y_i = f(x_i)$ at a set of arguments $x_1, x_2, \ldots, x_n$ such that $x_1 < x_1 < \cdots < x_n$. When this occurs it often becomes necessary to estimate the value $f(\alpha)$ when α lies between two of the known arguments x_i. This process is called the **interpolation** of the function $f(x)$ between its known values, and the interpolated value $f(\alpha)$ is estimated using some or all of the known values y_i. Various methods are available for interpolation, but nothing can be said about the error involved unless some assumptions are made about the function. As a general rule the error is best reduced by selecting a method that reflects the apparent variation of $f(x)$. Some of the factors to be taken into account when choosing an interpolation method are whether $f(x)$ appears to be convex or concave for $x_1 < x < x_n$, whether it is oscillatory, and whether it exhibits sharp curvature at a point or points belonging to the interval.

The estimation of $f(\alpha)$ when α lies outside the interval, either to the left of x_1 or to the right of x_n, is called **extrapolation** of the function $f(x)$, and as the process can be liable to considerable error it should be used with care. As with interpolation, nothing can be said about errors produced by extrapolation unless some general properties of the function involved either are known or are assumed. The use of extrapolation is more frequent than might be expected. It is, for example, used in Newton's method when the curve at a point is replaced by its tangent that is then extended (extrapolated) until it intersects the x-axis, again in the numerical solution of ordinary differential equations to be discussed later, and elsewhere.

Linear Interpolation

Let the data points $(x_1, y_1), (x_2, y_2), \ldots, (x_n, y_n)$ belonging to an unknown smooth function $y = f(x)$ be plotted on a graph. Then the simplest way to estimate the value of $y(x)$ when x lies in the interval $x_i < x < x_{i+1}$ is to join the points (x_i, y_i) and (x_{i+1}, y_{i+1}) by a straight line segment, and then to use the point on the line segment with argument x as the approximation to $y(x)$. This process is called **linear interpolation**, and it is illustrated in Fig. 19.6, where A is the point (x_i, y_i), B is the point (x_{i+1}, y_{i+1}), and the straight line segment AB has the equation $y = \widetilde{y}(x)$. Then, in linear interpolation, point P on the line segment AB is used as the approximation to Q on the curve $y = f(x)$.

graphical interpretation of linear interpolation

A simple calculation shows that the straight line segment $y = \widetilde{y}(x)$ representing the **linear interpolation function** between the two points (x_i, y_i) and (x_{i+1}, y_{i+1}) is given by

$$\widetilde{y}(x) = \left(\frac{y_{i+1} - y_i}{x_{i+1} - x_i} \right)(x - x_i) + y_i, \quad \text{for } x_i < x < x_{i+1}. \tag{7}$$

linear extrapolation

If x is chosen so that either $x < x_1$ or $x > x_n$, result (7) becomes a **linear extrapolation formula** for $y = f(x)$ outside the interval $x_1 < x < x_n$.

Result (7) is useful for interpolation when the variation of x_i and y_i between adjacent data points is small, but as the formula introduces an error due to its failure

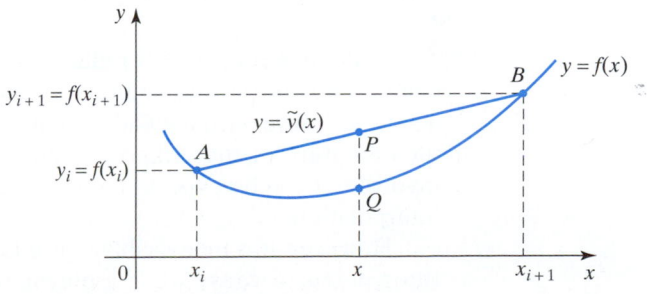

FIGURE 19.6 Linear interpolation.

to take account of the curvature of the curve, the error can become large when the result is used for extrapolation.

Lagrange Interpolation

Instead of using linear interpolation to join successive pairs of data points (x_1, y_1), $(x_2, y_2), \ldots, (x_n, y_n)$, it is possible that a better result can be obtained by constructing a polynomial $y = P(x)$ that passes through each data point. As a polynomial is a smooth curve, it is to be hoped that it will take some account of the curvature of the function to which the data points belong, as exhibited by a set of data points, and so provide a better interpolation formula.

In Lagrange interpolation the polynomial $P(x)$ that is used is taken to be the one with the lowest possible degree that passes through each of the data points, so that when there are n data points the polynomial will be at most of degree $n - 1$. The polynomial is unique, because n equations for its n coefficients can be found by requiring it to pass through each of the n data points. The graph of this polynomial over the interval $x_1 \leq x \leq x_n$ is then used as an approximation to the unknown function $y = f(x)$ from which the data points are presumed to have been derived, on the assumption that $y = f(x)$ does not exhibit large variations as its argument x moves between the successive arguments $x_1, x_2, \ldots, x_n$ of the data points.

The polynomial $y = P(x)$ given by

$$P(x) = \sum_{k=1}^{n} L_k(x) y_k,$$

where

$$L_k(x) = \frac{(x - x_1)(x - x_2) \cdots (x - x_{k-1})(x - x_{k+1}) \cdots (x - x_n)}{(x_k - x_1)(x_k - x_2) \cdots (x_k - x_{k-1})(x_k - x_{k+1}) \cdots (x_k - x_n)}, \tag{8}$$

has the property we require, because it is of degree at most $n - 1$, and it passes through each data point, so it defines an interpolation formula over the interval $x_1 \leq x \leq x_n$. The polynomials $L_k(x)$, called **fundamental Lagrangian interpolation polynomials**, are all of degree $n - 1$, but the linear combination forming the function $P(x)$ involving the set of data points can have a lower degree. That the $L_k(x)$ have the required property is easily seen from the fact that when $x = x_k$ each $L_r(x_k)$ with $r \neq k$ contains a zero factor in its numerator so that $L_r(x_k) = 0$, but when $r = k$ we have $L_k(x_k) = 1$, showing that $P(x_k) = y_k$. The polynomial $P(x)$ provides the required **Lagrange interpolation formula** for the set of n data points $(x_1, y_1), (x_2, y_2), \ldots, (x_n, y_n)$.

fundamental Lagrangian interpolation polynomials

When $n = 2$ result (8) reduces to linear interpolation, and when $n = 3$ it becomes a quadratic, and so fits a parabola through the three points. A parabola is a smooth curve with a steadily changing gradient, so as it takes some account of the curvature of the unknown function $y = f(x)$ over the three points that are involved, it can be expected to provide a better approximation than simple linear interpolation.

However, it is inadvisable to use Lagrange interpolation over many more than three points, because when a polynomial of degree $(n - 1) \gg 1$ is forced to pass

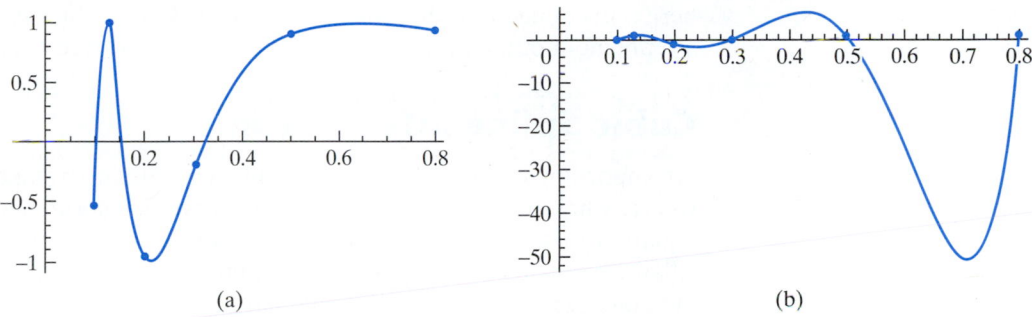

(a)   (b)

FIGURE 19.7   The function $y = f(x)$ and its Lagrange interpolation approximation $y = P(x)$ using six points.

through a set of n fixed points it usually produces a polynomial that introduces large oscillations between adjacent pairs of data points, even though the points themselves indicate no such behavior of the original function.

This undesirable characteristic of high degree Lagrange interpolation polynomials can be illustrated by constructing a fifth degree interpolation polynomial for the function

$$y(x) = \sin(1/x),$$

in the interval $0.1 \leq x \leq 0.8$ shown in Fig. 19.7a. When constructing an interpolation function, the precise extrema of the function are seldom known, so to reflect this uncertainty the six data points used will be the two end points and four internal points, two of which are close to, though not at, the extrema of $y(x) = \sin(1/x)$ in the interval $0.1 \leq x \leq 0.8$. These six data points are shown as dots on the graph of $y(x)$, and they have the following (x, y)-coordinates:

$$(0.1, -0.544021), (0.13, 0.986959), (0.2, -0.958924), (0.3, -0.190568),$$
$$(0.5, 0.909297), (0.8, 0.948985).$$

The Lagrangian interpolation polynomial that passes through these six points is

$$p(x) = -47.953442 + 1039.947347x - 7963.493901x^2 + 26828.578780x^3$$
$$-39901.683910x^4 + 21121.453960x^5.$$

The extreme oscillations that occur between the interpolation data points can be seen by inspection of Fig. 19.7b that shows the graph of $P(x)$ in the interval $0.1 \leq x \leq 0.8$, on which the data points are marked as dots.

In this case, as only six data points are involved, it would have been better to use three consecutive three point Lagrangian interpolation polynomials over the intervals $0.1 \leq x \leq 0.2$, $0.2 \leq x \leq 0.5$, and $0.3 \leq x \leq 0.8$, with the last interpolation polynomial used only in the interval $0.5 \leq x \leq 0.8$. However, although such a composite interpolation scheme would provide a *continuous* approximation to $y(x) = \sin(1/x)$ over the entire interval, the curve would not be smooth because of discontinuities in its derivative at $x = 0.2$ and $x = 0.5$ where the parabolic approximations meet.

We conclude this brief introduction to Lagrange interpolation by mentioning that its main use is of a theoretical nature in connection with the derivation of effective numerical techniques of various kinds. The only one of which to be developed here is in connection with *cubic spline interpolation*, which can be considered to be a refinement of the fitting of a polynomial of low degree over two points.

Cubic Spline Interpolation

An important use of an interpolation function arises in engineering design, and elsewhere, when it becomes necessary to generate a smooth curve with an unknown equation that passes through a set of data points, without the introduction of oscillations between these points. The approach to be outlined is motivated by the old engineering drafting technique that produced such a curve by tracing along a thin flexible metal strip, called a *spline*, that by the application of pressure at points along its length was constrained to pass through each data point.

Clearly a Lagrange interpolation polynomial is unsuitable because of the oscillations it can introduce, and because in practice there may be many data points. The approach we will use instead will be to approximate the curve in a piecewise manner by a polynomial of degree 3 over each interval $x_i \leq x \leq x_{i+1}$ in such a way that both the first and second derivatives of the curve at the ends of the interval match those of the approximations to the immediate left at x_i and those of the approximation to the immediate right at x_{i+1}. Composite approximations of this type are called **cubic spline function** approximations. In the mathematical approach to the determination of the spline function approximation through the n data points $(x_1, y_1), (x_2, y_2), \ldots, (x_n, y_n)$, the x_i are called the **nodes** of the approximation, and the corresponding points y_i where adjacent curves meet are called the **knots** of the approximation.

The mathematical requirements to be satisfied by a spline function approximation are seen to be:

(a) Each curve through the adjacent points (x_i, y_i) and (x_{i+1}, y_{i+1}) is a cubic.

(b) The composite curve over the entire interval must interpolate the data by passing through each knot.

(c) The curve itself and the first and second derivatives of the composite curve must be continuous at the nodes x_i.

(d) Conditions must be prescribed at the end points x_1 and x_n of the interval, depending on whether the data points indicate that *beyond* these points the extrapolation curve is required to approach a straight line or a parabola, or to exhibit some other behavior such as periodicity over the interval $x_1 \leq x \leq x_n$.

Because of conditions (i) to (iii) the second derivative $f''(x)$ must vary linearly over each interval $x_i \leq x \leq x_{i+1}$ and be continuous across each node, so using the Lagrange interpolation formula we can write

$$f''(x) = \left(\frac{x_{i+1} - x}{x_{i+1} - x_i} \right) f''(x_i) + \left(\frac{x - x_i}{x_{i+1} - x_i} \right) f''(x_{i+1}) \quad \text{for } x_i \leq x \leq x_{i+1}. \quad (9)$$

Integrating this result twice with respect to x gives

$$f(x) = \frac{1}{6} \left(\frac{3x_{i+1}x^2 - x^3}{x_{i+1} - x_i} \right) f''(x_i) + \frac{1}{6} \left(\frac{x^3 - 3x_i x^2}{x_{i+1} - x_i} \right) f''(x_{i+1}) + ax + b,$$
$$\text{for } x_i \leq x \leq x_{i+1}, \quad (10)$$

where a and b are arbitrary constants of integration. As $f(x)$ is required to pass through the points (x_i, y_i) and (x_{i+1}, y_{i+1}), substituting these two conditions into

(10) determines a and b, and after setting $d_i = x_{i+1} - x_i$ we find that

$$
\begin{aligned}
f(x) = \frac{1}{6d_i} & \left[(x_{i+1} - x)^3 f''(x_i) + (x - x_i)^3 f''(x_i) \right] \\
& + \frac{1}{6d_i} \left[6y_i - d_i^2 f''(x_i) \right] (x_{i+1} - x) \\
& + \frac{1}{6d_i} \left[6y_{i+1} - d_i^2 f''(x_{i+1}) \right] (x - x_i), \quad \text{for } x_i \leq x \leq x_{i+1}.
\end{aligned}
\tag{11}
$$

To proceed further we must now find conditions determining the derivatives $f''(x_i)$ and $f''(x_{i+1})$, and this can be accomplished by using the as yet unused condition that the first derivative $f'(x)$ must be continuous across each node. To apply this condition we differentiate (11) once with respect to x, and require the derivative when $x = x_{i+1}$ in the ith interval, that is, at its *right-hand* end point, to equal the derivative when $x = x_{i+1}$ in the $(i+1)$th interval, corresponding to its *left-hand* end point, as a result of which we find that

$$
d_{i-1} f''(x_{i-1}) + 2(d_{i-1} + d_i) f''(x_i) + d_i f''(x_{i+1}) = Y_i,
\tag{12}
$$

where

$$
Y_i = 6 \left(\frac{y_{i+1} - y_i}{d_i} - \frac{y_i - y_{i-1}}{d_{i-1}} \right).
\tag{13}
$$

Result (12) is a set of $n - 2$ linear simultaneous equations for the n derivatives $f''(x_i)$, and when these are known the spline function approximation formed by the set of functions in (11) defined over the consecutive intervals $x_i \leq x \leq x_{i+1}$ with $i = 1, 2, \ldots, n - 1$ can be constructed. It is crucial to the practical use of splines that this linear system of equations be nonsingular, and that an extremely efficient algorithm be available for solving it.

As the values of $f''(x_1)$ and $f''(x_n)$ cannot be found from the condition that $f'(x)$ is continuous across the nodes x_1 and x_n these values must be specified as additional conditions.

The choice of values for $f''(x_1)$ and $f''(x_n)$ prescribed as end conditions must be made intuitively, based on the way the data points indicate that the interpolated curve is most likely to behave (be extrapolated) *beyond* the end points of the interval $x_1 \leq x \leq x_n$. Three typical choices are the **natural** or **linear spline** end condition, the **parabolic spline** end condition, and **periodic spline** end conditions.

Natural or linear spline end conditions

spline end conditions

This choice of end conditions involves setting

$$
f''(x_1) = f''(x_n) = 0.
\tag{14}
$$

These conditions are also called the *linear spline end conditions* because although the polynomial used over the intervals is a cubic, the vanishing of the second derivative at $x = x_1$ and $x = x_n$ causes the approximation to become linear beyond the ends of the interval.

Parabolic spline end conditions

This choice of end conditions involves setting

$$f''(x_1) = f''(x_2) \quad \text{and} \quad f''(x_{n-1}) = f''(x_n). \tag{15}$$

These conditions are called the *parabolic spline end conditions* because the consequence of this choice is that $f''(x)$ is constant in each of the end intervals, causing the cubic interpolation formula to reduce to a quadratic or *parabolic* approximation.

Periodic spline end conditions

If there is reason to believe that the data is periodic over the interval $x_1 \leq x \leq x_n$, then the following are the appropriate end conditions

$$f(x_1) = f(x_{n-1}) \quad \text{and} \quad f'(x_n) = f'(x_2). \tag{16}$$

Other end conditions can be used and, of course, a linear spline end condition may be applied at one end of an interval and a parabolic spline end condition at the other if this is appropriate. An end condition that is more important than the parabolic end condition is the condition that leads to the complete cubic spline, namely the spline that interpolates $f'(x)$ as well as $f(x)$ at both x_1 and x_n. This spline has a higher rate of convergence as the maximum step size tends to zero, and it is often implemented using a local approximation to the derivatives that preserves the higher rate of convergence.

an example of a spline approximation

The function $y(x) = \sin(1/x)$ is shown in Fig. 19.8 as the dashed curve, on which is superimposed the cubic spline approximation with natural end conditions. The six interpolation data points are shown as dots.

For more information about topics in this section see references [2.14] through [2.20].

Summary

Linear and Lagrange interpolation were defined, and the desirability of using low degree Lagrange interpolation in order to avoid the introduction of excessive oscillations between interpolation data points was illustrated by example. Extrapolation was then defined, and its attendant dangers were stressed unless something is known about the nature of the function being extrapolated. Finally, spline interpolation was introduced, which produces

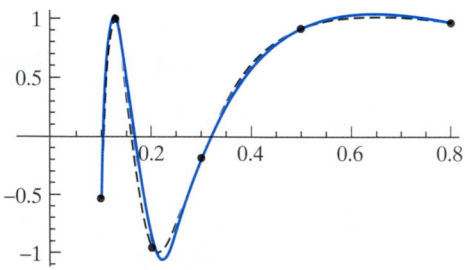

FIGURE 19.8 The function $y(x) = \sin(1/x)$, the cubic spline approximation and the data interpolation points.

a smooth interpolated curve through each data point, and the different end conditions that can be applied were explained together with their effects.

EXERCISES 19.3

Exercises in this set require the use of a computer.

1. Graph the function $f(x) = x/(1+x^2)$ in the interval $0 \le x \le 3$. Select four points on the graph, and after constructing a polynomial that passes through each of the points graph the polynomial and compare the result with the original function.

2. Graph the function $f(x) = \sin x/(1+x^2)$ in the interval $0 \le x \le \pi$. Select four points on the graph, and after constructing a polynomial that passes through each of the points graph the polynomial and compare the result with the original function.

3. Graph the function $f(x) = 1 + x\sin x$ in the interval $0 \le x \le 2\pi$. Select seven points on the graph, and after constructing a polynomial that passes through each of the points graph the polynomial and compare the result with the original function. Try to improve the approximation by choosing the seven points differently.

4. Graph the function $f(x) = (1 - x^5)^{1/5}$ in the interval $0 \le x \le 1$. Select seven points on the graph, and after constructing a polynomial that passes through each of

the points graph the polynomial and compare the result with the original function. Try to improve the approximation by choosing the seven points differently.

5. Graph the function $f(x) = 1 - 2x\cos x$ in the interval $0 \le x \le 2\pi$. Select seven points on the graph and construct a spline function approximation to the function in the interval $0 \le x \le 2\pi$ using parabolic spline function end conditions. Graph the spline function and compare it with the graph of the original function. Repeat the calculation using linear spline function end conditions and compare the result with the previous graph.

6. Graph the function $f(x) = (1 - x^7)^{1/7}$ in the interval $0 \le x \le 1$. Select seven points on the graph, and construct a spline function approximation to the function in the interval $0 \le x \le 1$ using linear spline function end conditions. Graph the spline function and compare the result with the original function. Repeat the calculation using parabolic spline function end conditions and compare the result with the previous graph.

19.4   Numerical Integration

The need for **numerical integration**, also called **numerical quadrature**, arises when either a definite integral that is required cannot be evaluated analytically, or when special functions involved in an analytical solution are too complicated to be of direct use. A typical definite integral that can only be evaluated numerically is

$$I = \int_0^5 \frac{\sin 3x}{\sqrt{x^2 + x + 1}} dx,$$

the value of which can be shown to be $I = 0.364\,873$. In what follows, three different numerical integration schemes for the evaluation of definite integrals will be derived called, respectively, the *trapezoidal rule*, *Simpson's rule*, and *Gaussian integration*. Of these three methods the first is the least accurate, whereas the last provides high accuracy with far fewer computational steps than the frequently used Simpson's rule.

The Trapezoidal Rule

The basis of this very simple rule can be understood from Fig. 19.9 in which the integral $I = \int_a^b f(x)dx$ is approximated by the area of the *trapezoid PQRS* shown as the shaded area in the interval $a \le x \le b$ associated with the graph of $y = f(x)$.

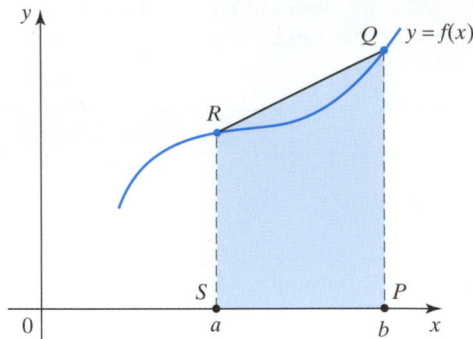

FIGURE 19.9 A trapezoidal approximation to $I = \int_a^b f(x)\,dx.$

As the area $PQRS = \frac{1}{2}(b - a)[f(a) + f(b)]$, the approximation to the definite integral in Fig. 19.9 is given by

$$\int_a^b f(x)\,dx \approx \frac{1}{2}(b - a)[f(a) + f(b)]. \tag{17}$$

Setting $b - a = h$, and denoting by $E(h)$ the error made when approximating the definite integral by a single trapezoid with base h, we have

$$E(h) = \frac{1}{2}(b - a)[f(a) + f(b)] - \int_a^b f(x)\,dx,$$

so in terms of $E(h)$ the approximation (17) can be replaced by the exact result

$$\int_a^b f(x)\,dx = \frac{1}{2}(b - a)[f(a) + f(b)] - E(h). \tag{18}$$

A different way of deriving result (17) is to use linear interpolation to represent $y(x)$ between $x = a$ and $x = b$, and then to integrate the result.

Although the exact error $E(h)$ is not known, an expression for the error can be derived on the assumption that $f(x)$ is suitably differentiable in the range of integration $a \leq x \leq b$. The error term for the trapezoidal rule will be stated without proof because its derivation is similar to that for the more accurate Simpson's rule, and this will be given later. When a definite integral is approximated by a single trapezoid, as in Fig. 19.9, the error term in (18) is given by $E(h) = \frac{1}{12}h^3 f''(\xi)$, for some ξ such that $a \leq \xi \leq b$. If we use this error term (18) becomes

$$\int_a^b f(x)\,dx = \frac{1}{2}(b - a)[f(a) + f(b)] - \frac{1}{12}h^3 f''(\xi), \tag{19}$$

for some ξ such that $a \leq \xi \leq b$.

A better estimate of the definite integral $\int_a^b f(x)\,dx$ can by obtained by dividing $a \leq x \leq b$ into n subintervals, applying (19) to each of the n subintervals, and then summing the result. Although not necessary, it is usual to choose all n subintervals to be of equal length $h = (b - a)/n$, where h is usually called the **step size**. Consequently, setting $x_i = a + ih$ for $i = 0, 1, \ldots, n$, we arrive at what is called the **composite trapezoidal rule**

$$\int_a^b f(x)dx = \frac{1}{2}h\left[f(a) + 2\sum_{i=1}^{n-1} f(x_i) + f(b)\right] - \frac{1}{12}(b-a)h^2 f''(\eta),\qquad (20)$$

where the unknown number η in the error term is such that $a \leq \eta \leq b$.

The error term in the composite trapezoidal rule is obtained from the error term in (19) by addition of the error terms in each subinterval. The details of the derivation will be left as an exercise, because they parallel those for the corresponding case in the composite Simpson's rule that will shortly be discussed in detail.

Although η is not known, whenever it is possible to estimate the greatest and least values of $f''(x)$ in the interval $a \leq x \leq b$, bounds can be placed on the composite trapezoidal rule result by assigning these values of $f''(x)$ to $f''(\eta)$.

In practical applications of the composite trapezoidal rule the error term is usually only used to show that as the number n of subintervals increases, so the error decreases as $(b-a)h^2/12$, where $h = (b-a)/n$. The error is often approximated by forming two approximations with different h and using the asymptotic behavior to estimate the error of the result corresponding to the smaller h. Another approach is to compare the result with the one obtained with Simpson's method.

EXAMPLE 19.5   Use the composite trapezoidal rule with $n = 10, 30,$ and 50 subintervals to evaluate

$$I = \int_0^5 \frac{\sin 3x}{\sqrt{x^2 + x + 1}}dx,$$

and approximate the error when 50 subintervals are used.

Solution   The following results were obtained by computer:

n	10	30	50
$I_{\text{trap}(n)}$	0.290422	0.356897	0.362010

The result for $I_{\text{trap}(50)}$ should be compared with the result $I = 0.364873$ obtained by a higher order method that is known to be correct to six decimal places.

Instead of using $f''(\eta)$ when approximating the error with $n = 50$, where η is unknown, we will use the easily computed average f''_{av} of $f''(x)$ over the interval, where

$$f''_{av} = \frac{1}{b-a}\int_a^b f''(x)dx = \frac{1}{b-a}[f'(b) - f'(a)].$$

We have $b - a = 5$ and the step size $h = 5/50 = 0.1$, so

$$f''_{av} = \frac{1}{5}\int_0^5 f''(x)dx = \frac{1}{5}[f'(5) - f'(0)] = -0.686.$$

Using f''_{av} in the error term instead of $f''(\eta)$ leads to $\frac{1}{12}\cdot 5\cdot(0.1)^2\cdot(-0.686) = -0.002858$ as the approximation to the error. Consequently, allowing for this error, the estimate of the integral is $0.362010 - (-0.002858) = 0.364868$. When this is compared with the result $I = 0.364873$ we see that in this case the error approximation is good.   ■

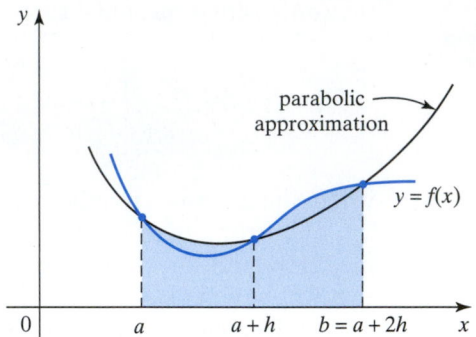

FIGURE 19.10 A parabolic approximation to $I = \int_a^b f(x)dx$.

Simpson's Rule

In its simplest form, the trapezoidal rule applied to $I = \int_a^b f(x)dx$ represents $f(x)$ by the single trapezoidal area $PQRS$ shown in Fig. 19.9, where in the interval $a \le x \le b$ the function $y = f(x)$ is approximated by the straight line segment QR. A more accurate result would be expected if a point on the curve $y = f(x)$ is chosen inside the interval $a \le x \le b$ and $f(x)$ is approximated by a parabola that passes through the two end points and the single internal point, as shown in Fig. 19.10.

Setting $b = a + 2h$, where h is the step size, and taking the additional point in the interval of integration to be $x = a + h$, so it is *midway* between the ends of the interval, the parabola to be fitted must pass through the three consecutive points $(a, f(a))$, $(a + h, f(a + h))$, and $(a + 2h, f(a + 2h))$. The Lagrange interpolation formula that fits a quadratic through these three points is

$$L(x) = \frac{1}{2}\frac{(x - a - h)(x - a - 2h)}{h^2}f(a) - \frac{(x - a)(x - a - 2h)}{h^2}f(a + h)$$

$$+ \frac{1}{2}\frac{(x - a)(x - a - h)}{h^2}f(a + 2h). \tag{21}$$

Integrating $L(x)$ over the interval $a \le x \le a + 2h$ and simplifying the result gives

$$\int_a^{a+2h} f(x)dx \approx \frac{1}{3}h[f(a) + 4f(a + h) + f(a + 2h)], \tag{22}$$

which is the result known as **Simpson's rule**, or sometimes **Simpson's 1/3 rule**. Result (22) can also be written in terms of the limits of integration a and $b = a + 2h$ as

$$\int_a^{a+2h} f(x)dx \approx \frac{1}{6}(b - a)\left[f(a) + 4f\left(\frac{a + h}{2}\right) + f(b)\right]. \tag{23}$$

If the error in Simpson's rule is denoted by $E(h)$, the approximation in (22) can be replaced by the exact result

$$\int_a^{a+2h} f(x)dx = \frac{1}{3}h[f(a) + 4f(a+h) + f(a+2h)] - E(h). \tag{24}$$

We will now derive an expression for $E(h)$ but before doing so, in order to simplify the manipulation, it will be convenient to write the limits of integration in the more symmetrical form $a = c - h$ and $b = c + h$. In terms of c and h (24) becomes

$$E(h) = \frac{1}{3}h[f(c-h) + 4f(c) + f(c+h)] - \int_{c-h}^{c+h} f(x)dx.$$

We now differentiate this result with respect to h to obtain

$$E'(h) = \frac{1}{3}[f(c-h) + 4f(c) + f(c+h)]$$
$$+ \frac{1}{3}h[-f'(c-h) + f'(c+h)] - [f(c+h) + f(c-h)],$$

where the last group of terms on the right follow from differentiating the definite integral using Leibniz's theorem (Theorem 1.5). If we set $h = 0$, this result shows that $E'(0) = 0$.

Differentiation of $E'(h)$ gives

$$E''(h) = \frac{1}{3}[f'(c-h) - f'(c+h)] + \frac{1}{3}h[f''(c+h) + f''(c-h)],$$

so setting $h = 0$ we find that $E''(0) = 0$. One final differentiation gives

$$E'''(h) = \frac{1}{3}h[f'''(c+h) - f'''(c-h)],$$

but this can be simplified by using the Taylor expansion of $f'''(c+h)$ with a remainder after the first term, where the expansion is about the point $c - h$, so that

$$f'''(c+h) = f'''(c-h) + 2hf^{(4)}(\xi),$$

where ξ is unknown but lies in the interval $c - h < \xi < c + h$.

The error term can now be found by integrating this last result three times using the results $E'(0) = E''(0) = 0$.

We have

$$\int_0^h E'''(t)dt = E''(h) - E''(0) = E''(h),$$

so

$$E''(h) = \frac{2}{3}f^{(4)}(\xi)\int_0^h t^2 dt = \frac{2}{9}h^3 f^{(4)}(\xi).$$

A further integration using the result

$$\int_0^h E''(t)dt = E'(h) - E'(0) = E'(h)$$

gives

$$E'(h) = \frac{2}{9} f^{(4)}(\xi) \int_0^h t^3 dt = \frac{1}{18} h^4 f^{(4)}(\xi).$$

Finally, after another integration we arrive at the result

$$E(h) = \frac{1}{18} f^{(4)}(\xi) \int_0^h t^4 dt = \frac{1}{90} h^5 f^{(4)}(\xi),$$ (25)

which is the required expression for the error term. Using this result in (24) gives

$$\int_a^{a+2h} f(x)dx = \frac{1}{3} h[f(a) + 4f(a+h) + f(a+2h)] - \frac{1}{90} h^5 f^{(4)}(\xi).$$ (26)

As $f^{(4)}(\xi)$ enters as a factor in $E(h)$, this shows the rather surprising result that although Simpson's rule was derived by requiring a quadratic polynomial to pass through three points, the rule is actually exact for cubic polynomials.

As with the trapezoidal rule, the accuracy of Simpson's rule can be improved by increasing the number of subintervals, but as the rule is equivalent to constructing parabolas through three consecutive equispaced points, to use the rule over more than three points the number of points chosen for the interval $a \le x \le b$ must be *odd*, so that the number of intervals must be *even*. Dividing the interval $a \le x \le b$ into $2n$ equal subintervals each of length $h = (b-a)/2n$, and adding the results, gives the **composite Simpson's rule**

composite Simpson's rule with error term

$$\int_a^b f(x)dx = \frac{1}{3} h \left[f(a) + 4 \sum_{i=1}^{n} f(a + (2i-1)h) + 2 \sum_{i=1}^{n-1} f(a + 2ih) + f(b) \right]$$
$$- \frac{1}{180}(b-a)h^4 f^{(4)}(\eta)$$

(27)

where η is unknown but is such that $a < \eta < b$.

The error term in the composite rule (27) is obtained as follows. Let $x_i = a + 2ih$, with $i = 0, 1, \ldots, n$, and let ξ_i be the value of ξ in the interval $x_i \le x \le x_{i+1}$ appropriate to the Simpson's rule applied to that interval. Consequently, when the composite Simpson's rule is formed, the error term in each of these intervals will be added. Now, each derivative $f^{(4)}(\xi_i)$ must satisfy the inequality

$$\min_{a \le x \le b} f^{(4)}(x) \le f^{(4)}(\xi_i) \le \max_{a \le x \le b} f^{(4)}(x),$$

so the addition of these n results followed by division by n gives

$$\min_{a \le x \le b} f^{(4)}(x) \le \frac{1}{n} \sum_{i=1}^{n} f^{(4)}(\xi_i) \le \max_{a \le x \le b} f^{(4)}(x).$$

Finally, assuming $f^{(4)}(x)$ is continuous, it follows from the intermediate value theorem that some number η exists, with $a < \eta < b$, such that

$$f^{(4)}(\eta) = \frac{1}{n} \sum_{i=1}^{n} f^{(4)}(\xi_i).$$

If we use the result $h = (b - a)/2n$, the error term in the composite Simpson's rule is seen to be given by

$$-\frac{1}{90} h^5 \sum_{i=1}^{n} f^{(4)}(\xi_i) = -\frac{1}{180}(b - a)h^4 f^{(4)}(\eta).$$

EXAMPLE 19.6   Use the composite Simpson's rule with $n = 10, 30$, and 50 subintervals to evaluate

$$I = \int_0^5 \frac{\sin 3x}{\sqrt{x^2 + x + 1}} dx$$

and compare the results obtained with the result $I = 0.364873$, which is accurate to six decimal places. Compare the result of integrating this definite integral by the trapezoidal rule and Simpson's rule.

Solution   The following results were obtained by computer:

n	10	30	50
$I_{\text{simp}(n)}$	0.376738	0.365019	0.364892

Comparison of the result $I = 0.364873$, known to be correct to six decimal places, with $I_{\text{simp}(50)} = 0.364892$, shows that $I_{\text{simp}(50)}$ only overestimates the true result by 0.000025.

When comparing the composite Simpson's rule with the composite trapezoidal rule it should be remembered that Simpson's rule subdivides the interval of integration into $2n$ subintervals, whereas the composite trapezoidal rule only uses n subintervals. The following computer results provide a comparison on this basis:

n	20	40	60	80	100
$I_{\text{trap}(n)}$	0.346825	0.360395	0.362886	0.363756	0.364158
$I_{\text{simp}(n/2)}$	0.376738	0.365626	0.365019	0.364919	0.364892

Gaussian Quadrature

Many more numerical integration methods exist than have been outlined so far, but the only other important one to be mentioned here is due to C. F. Gauss. He showed that if, when evaluating numerically an integral in the standard form

$$\int_{-1}^{1} f(x)dx,$$

the points x_i at which the values of the integrand $f(x)$ are sampled are chosen in a special way, then when n sample points are used the result can be made exact in the case that $f(x)$ is an arbitrary polynomial of degree $2n - 1$ or less. Unlike Simpson's rule, in this method the n sample points x_i are *nonuniformly* spaced throughout the interval of integration $-1 \leq x \leq 1$, and they are all contained *inside* the interval.

The sample points, or **nodes** as they are called, are chosen to get a formula that will integrate exactly polynomials of as high degree as possible. It turns out that the n sample points are real and lie in the open interval $(-1, 1)$, and polynomials of degree $2n - 1$ are integrated exactly.

A somewhat different approach to integration involves specifying some of the sample points to be used, and then trying to find the remaining ones so as to integrate polynomials of as high degree as possible. Formulas of this type that evaluate function values at the two ends of the interval of integration are called **Lobatto** formulas, and the trapezoidal rule and Simpson's rule are formulas of the lowest order that belong to this family.

The point is that if it is useful to specify sample points at the end points of an interval of integration it is possible to proceed in this way. However, as would be expected, if this approach is adopted it is not possible to get a formula that is as accurate as one in which no constraint is placed on the sample points.

The previous arguments are all based on the assumption that functions are approximated by (algebraic) polynomials, though sometimes it is more natural to approximate them by trigonometric polynomials (finite Fourier series).

The composite trapezoidal rule is, in fact, the optimal formula of Gaussian integration type based on trigonometric approximation. As a result it converges faster than any power of h when applied to a periodic analytic function over a multiple of a period, so for this reason it is used to compute Fourier coefficients.

To illustrate the approach used to obtain this type of integration formula, we consider the simplest situation in which $n = 2$, so as only the two sample points x_1 and x_2 are involved, with $-1 < x_1 < x_2 < 1$, the integration formula becomes

$$\int_{-1}^{1} f(x)dx \approx w_1 f(x_1) + w_2 f(x_2).$$

weights in integration formula

At this stage the values of the two sample points x_1 and x_2 are unknown, as are the numbers w_1 and w_2, called the **weights** for the integration formula at these sample points. To determine these four numbers we impose the requirement that this formula should be exact when $f(x)$ is an arbitrary polynomial of degree $2n - 1 = 3$ or less.

Let $f(x)$ be the cubic polynomial

$$f(x) = c_0 + c_1 x + c_2 x^2 + c_3 x^3,$$

in which the coefficients c_0, c_1, c_2, and c_3 are arbitrary. Then for the integration to be exact, the numbers x_1, x_2, w_1, and w_2 must be such that

$$\int_{-1}^{1} (c_0 + c_1 x + c_2 x^2 + c_3 x^3)dx = w_1 \left(c_0 + c_1 x_1 + c_2 x_1^2 + c_3 x_1^3\right)$$
$$+ w_2 \left(c_0 + c_1 x_2 + c_2 x_2^2 + c_3 x_2^3\right).$$

Evaluating the integral on the left, and equating the respective multipliers of the arbitrary coefficients c_0, c_1, c_2, and c_3 to make this result an identity, leads to the results

$$
\begin{aligned}
&\text{(coefficient } c_0) & w_1 + w_2 &= \int_{-1}^{1} dx = 2 \\
&\text{(coefficient } c_1) & w_1 x_1 + w_2 x_2 &= \int_{-1}^{1} x\, dx = 0 \\
&\text{(coefficient } c_2) & w_1 x_1^2 + w_2 x_2^2 &= \int_{-1}^{1} x^2\, dx = \tfrac{2}{3} \\
&\text{(coefficient } c_3) & w_1 x_1^3 + w_2 x_2^3 &= \int_{-1}^{1} x^3\, dx = 0.
\end{aligned}
$$

This set of equations has the unique solution $x_1 = -1/\sqrt{3}$, $x_2 = 1/\sqrt{3}$, $w_1 = 1$, and $w_2 = 1$. Consequently, when $n = 2$, we have

The sample points

$$
x_1 = -1/\sqrt{3}, \quad x_2 = 1/\sqrt{3},
$$

The weights

$$
w_1 = 1, \quad w_2 = 1,
$$

so the extremely simple two-point integration formula that gives exact results when $f(x)$ is a polynomial of degree 3 or less is seen to be given by

$$
\int_{-1}^{1} f(x)\,dx = f\left(-\frac{1}{\sqrt{3}}\right) + f\left(\frac{1}{\sqrt{3}}\right).
$$

When this approach is extended to n points, an examination of the derivation of the formula shows that the sample points $x_1, x_2, \ldots, x_n$ are simply the n roots of the Legendre polynomial $P_n(x) = 0$ of degree n, with the corresponding weight w_i at x_i given by $w_i = 2[P'(x_i)]^2/(1 - x_i^2)$, for $i = 1, 2, \ldots, n$. The general integration formula involving n points becomes

$$
\int_{-1}^{1} f(x)\,dx \approx \sum_{i=1}^{n} w_i\, f(x_i)
$$

Gauss–Legendre integration formulas

and, collectively, these results are called **Gaussian** integration formulas or, sometimes, **Gauss–Legendre integration formulas**. It can be shown that the remainder term that must be added to the right-hand side of this last result for it to be exact for *any* function $f(x)$ with a continuous derivative $f^{(2n)}(x)$ is $R_n = \frac{2^{2n+1}(n!)^4}{(2n+1)[(2n)!]^3} f^{(2n)}(\xi)$, for some unknown ξ such that $-1 < \xi < 1$. A list of Gaussian sampling points x_i and their associated weights w_i is given in Table 19.1 for $n = 2, 3, 4, 5, 10$, and 16.

error term in Gaussian integration

As would be expected, if $f(x)$ is an arbitrary polynomial of degree $2n - 1$ or less, it follows directly that $R_n \equiv 0$, confirming that in this case the result is exact.

TABLE 19.1 Gaussian Sampling Points and Weights

n	i	x_i	w_i
2	1	−0.57735 02692	1.00000 00000
	2	0.57735 02692	1.00000 00000
3	1	−0.77459 66692	0.55555 55556
	2	0.00000 00000	0.88888 88889
	3	0.77459 66692	0.55555 55556
4	1	−0.86113 63115	0.34785 48451
	2	−0.33998 10436	0.65214 51548
	3	0.33998 10436	0.65214 51548
	4	0.86113 63115	0.34785 48451
5	1	−0.90617 98459	0.23692 68851
	2	−0.53846 93101	0.47862 86705
	3	0.00000 00000	0.56888 88889
	4	0.53846 93101	0.47862 86705
	5	0.90617 98459	0.23692 68851
10	1	−0.97390 65285	0.06667 13443
	2	−0.86506 33667	0.14945 13492
	3	−0.67940 95683	0.21908 63625
	4	−0.43339 53941	0.26926 67193
	5	−0.14887 43390	0.29552 42247
	6	0.14887 43390	0.29552 42247
	7	0.43339 53941	0.26926 67193
	8	0.67940 95683	0.21908 63625
	9	0.86506 33667	0.14945 13492
	10	0.97390 65285	0.06667 13443
16	1	−0.98940 09350	0.02715 24594
	2	−0.94457 50231	0.06225 35230
	3	−0.86563 12024	0.09515 85117
	4	−0.75540 44084	0.12462 89713
	5	−0.61787 62444	0.14959 59888
	6	−0.45801 67777	0.16915 65194
	7	−0.28160 35508	0.18260 34150
	8	−0.09501 25098	0.18945 06105
	9	0.09501 25098	0.18945 06105
	10	0.28160 35508	0.18260 34150
	11	0.45801 67777	0.16915 65194
	12	0.61787 62444	0.14959 59888
	13	0.75540 44084	0.12462 89713
	14	0.86563 12024	0.09515 85117
	15	0.94457 50231	0.06225 35239
	16	0.98940 09350	0.02715 24594

The apparent restriction of the integration to the standard interval $-1 \leq x \leq 1$ is unimportant, because if the integral involved is

$$I = \int_a^b f(x)dx,$$

where a and b are finite, the simple change of variable

$$x = \frac{1}{2}(b+a) + \frac{1}{2}(b-a)u$$

converts the integral to

$$I = \frac{b-a}{2} \int_{-1}^{1} F(u)\,du,$$

where $F(u)$ is the function $f(x)$ after the change of variable.

The accuracy obtained when using an n-point Gaussian integration formula depends on the extent to which the integrand can be approximated by a polynomial of degree $2n - 1$. To illustrate matters, we apply the five-point formula to the following integral for which there is an analytical solution that can be used for comparison:

$$I = \int_{0}^{1/2} \frac{dx}{(1-x^2)^{1/2}} = \text{Arcsin}(1/2) = \frac{\pi}{6} = 0.523599.$$

The change of variable $x = \frac{1}{4}(1+u)$ maps the interval $0 \le x \le \frac{1}{2}$ onto the interval $-1 \le u \le 1$, so as $dx/du = \frac{1}{4}$, after changing the variable

$$I = \int_{-1}^{1} \frac{du}{(15 - 2u - u^2)^{1/2}}.$$

Setting $f(u) = 1/(15 - 2u - u^2)^{1/2}$ and applying the five-point Gaussian formula gives

$$I \approx 0.236927\,f(-0.906180) + 0.478629\,f(-0.538469) + 0.568889\,f(0)$$

$$+ 0.478629\,f(0.538469) + 0.236927\,f(0.906180) = 0.523599.$$

In this case the numerical approximation is seen to be correct to six decimal places.

modern adaptive integration codes

The key idea used in modern integration codes involves the use of an adaptive algorithm. In such codes the error of an integral evaluated over an interval is approximated by comparing it to the result obtained by a higher order formula. Thus, the error of the trapezoidal rule can be estimated by comparing the result to the one obtained using Simpson's rule. If the result is not sufficiently accurate, the interval is split in half and the two intervals are then treated separately. Reducing the length of an interval produces a significant reduction in the error. This can be seen by considering the low-order trapezoidal rule. The effect of halving the interval is to reduce the error in a half interval by a factor of approximately an eighth, so as the operation of integration is linear, the error over the full interval is reduced by a factor of approximately a fourth. When this argument is extended, we see that if the interval of integration is divided into many pieces, accurate values of the integral over all the pieces can be added together to get an accurate value over the whole interval, with the same being true of the error estimates.

In this approach two formulas are applied to an interval using as many values of f as possible in both formulas. That the method is computationally efficient when the combination of the trapezoidal rule is used and Simpson's rule is used can be seen from the fact that only one extra evaluation of f is necessary in order to estimate the error. Modern codes use a Gaussian formula of high order as the basic formula, and a special formula of much higher order that makes use of as many function evaluations as possible for estimating the error.

As Gaussian integration formulas make no use of the values of the integrand at the end points of the interval $-1 \le x \le 1$, they can be used to approximate a convergent improper integral of the type $\int_a^b f(x)dx$, where the integrand becomes infinite at either end point.

For more information about numerical integration, see references [2.14] through [2.20].

Summary

The methods for numerical integration, also called numerical quadrature, introduced in this section were the trapezoidal and composite trapezoidal rule, Simpson's rule and the composite Simpson's rule, and Gaussian quadrature. The relative accuracies of the methods were explained; the high accuracies of Gaussian quadrature was stressed. The suitability of the composite trapezoidal rule for the computation of Fourier coefficients was mentioned.

EXERCISES 19.4

The following exercises require the use of a computer.

1. Use the composite Simpson's rule with step length $h = 0.5$ to determine

$$I = \int_1^3 (2x^3 - 3x^2 + 4x - 1)dx,$$

and hence verify that the rule integrates cubics exactly.

2. Use the composite trapezoidal rule with the step length $h = 0.1$ to evaluate

$$I = \int_0^1 \frac{dx}{1 + x^2},$$

and estimate the error term involved. Compare your results with the exact value $I = \frac{\pi}{4}$. Repeat the calculation using the composite Simpson's rule with the same step length, but without estimating the error.

3. Use the composite trapezoidal rule and Simpson's rule, each with 10 subintervals, to estimate

$$I = \int_0^\pi \frac{\sin x}{x}dx,$$

and compare your results with $I = 1.851937$, which is exact to six decimal places.

4. Use the composite trapezoidal and Simpson's rule, each with step length $h = 0.2$, to estimate

$$I = \int_0^2 x^2 e^{-x}dx,$$

and compare your results with the analytical solution $I = \frac{1}{13} + \frac{1}{2} \text{Arctan } 5 - \frac{1}{8}\pi$.

5. Use the composite Simpson's rule with step length $h = 0.4$ to estimate

$$I = \int_2^6 \frac{\ln(2 + 3\sqrt{x})}{1 + x^2}dx,$$

and compare your result with the result $I = 0.596545$, which is correct to six decimal places.

6. Use the composite trapezoidal and Simpson's rule, each with step length $h = 0.4$, to estimate

$$\int_0^4 \left(1 - \frac{x}{4}\right)^4 x^{1/2}dx.$$

Compare your results with the exact solution that follows from the general result

$$I(z, n) = \int_0^n \left(1 - \frac{x}{n}\right)^n x^{z-1}dx$$

$$= \frac{1 \cdot 2 \cdot 3 \cdots n}{z(z+1)(z+2)\cdots(z+n)}n^z.$$

It follows from the definition of the gamma function that $\lim_{n \to \infty} I(z, n) = \Gamma(z)$. Explain why replacing 4 by 50 in the original integral and evaluating the result using the composite Simpson's rule with more subdivisions is not likely to lead to much improvement of the poor estimate it provides of $\Gamma(3/2) = \frac{1}{2}\sqrt{\pi}$.

7. The Bessel function $J_1(x)$ has the integral representation

$$J_1(x) = \frac{2}{\pi} \int_0^{\pi/2} \sin(x \cos \theta) \cos \theta d\theta.$$

Use the composite Simpson's rule with step length $h = \pi/20$ to estimate $J_1(2)$, and compare your result with the result $J_1(2) = 0.576725$, which is accurate to six decimal places.

In Exercises 8 through 10 use the integral representation

$$J_n(x) = \frac{1}{\pi} \int_0^\pi \cos(x \sin \theta - n\theta)d\theta.$$

8. Estimate $J_2(2)$ using the composite Simpson's rule with step length $h = \pi/8$, and compare your result with $J_2(2) = 0.352834$, which is accurate to six decimal places.

9. Estimate $J_1(4)$ using the composite Simpson's rule with step length $h = \pi/10$, and compare your result with $J_1(4) = -0.066043$, which is accurate to six decimal places.

10. Estimate $J_3(4)$ using the composite Simpson's rule with step length $h = \pi/10$, and compare your result with $J_3(4) = 0.430171$, which is accurate to six decimal places.

11. The modified Bessel function $I_0(x)$ has the integral representation

$$I_0(x) = \frac{2}{\pi} \int_0^{\pi/2} \cosh(x \sin\theta)\, d\theta.$$

Use the composite Simpson's rule with step length $h = \pi/16$ to estimate $I_0(3.5)$, and compare your result with $I_0(3.5) = 7.378203$, which is correct to six decimal places.

12. The modified Bessel function $I_1(x)$ has the integral representation

$$I_1(x) = \frac{2x}{\pi} \int_0^{\pi/2} \cosh(x \sin\theta)(\cos\theta)^2\, d\theta.$$

Use the composite Simpson's rule with step size $h =$

$\pi/16$ to estimate $I_1(3)$, and compare your result with $I_1(3) = 3.953370$, which is correct to six decimal places.

In Exercises 13 and 16 use the 3-, 5-, and 10-point Gaussian formulas to estimate the given integral and compare the results with the exact value.

13. $I = \displaystyle\int_0^{3\pi/2} \cos x\, dx.$

The exact value is $I = -1$.

14. $I = \displaystyle\int_0^{\pi/2} e^{-x}\cos x.$

The exact value to six decimal places is $I = \frac{1}{2}[1 + \exp(-\frac{1}{2}\pi)] = 0.603940$.

15. Use the 10-point Gaussian formula to estimate the value of the convergent improper integral

$$I = \int_0^{1/2} \frac{dx}{(1 - 4x^2)^{1/2}}.$$

Compare the result with the exact value to six decimal places $I = \frac{\pi}{4} = 0.785398$.

18. Use the 10-point Gaussian formula to estimate the value of the convergent improper integral

$$I = \int_0^{\pi/2} \frac{\sqrt{x}}{\sin x}\, dx.$$

Compare the result with the exact value to six decimal places $I = 2.753142$.

19.5   Numerical Solution of Linear Systems of Equations

This section describes two approaches to the solution of systems of n nonhomogeneous linear equations in the n unknowns $x_1, x_2, \ldots, x_n$, both of which are important. These methods, with various refinements, are all found in major linear algebra software packages.

The first method, involving the successive elimination of unknowns, is of the *direct type*, in which the solution is obtained after systematically eliminating $n - 1$ of the n unknowns to find x_n. The process of back-substitution is then used to find the remaining unknowns in the reverse order $x_{n-1}, x_{n-2}, \ldots, x_1$. This method can also be used when the number of equations is not equal to the number of the unknowns, when it also shows automatically if a system of equations is inconsistent.

A related method is essentially the same as the first, apart from the way in which details of the elimination process are recorded to permit solving conveniently more than one system of equations with the same coefficient matrix. It applies to systems in which the number of equations equals the number of unknowns. The approach is to attempt to factorize the coefficient matrix $\mathbf{A}$ in the system $\mathbf{Ax} = \mathbf{b}$ into the product $\mathbf{PA} = \mathbf{LU}$, where $\mathbf{L}$ is a *lower-triangular* matrix with 1's on its leading diagonal, $\mathbf{U}$ is an upper-triangular matrix, and $\mathbf{P}$ is a permutation matrix, the reason for which will be explained later. The method uses this factorization to determine the solution vector **x**. A failure of the method to achieve this factorization indicates that **A** is singular, so then one or more of its rows is linearly dependent on other rows.

The second type of approach is an *iterative* one, and it only applies to a system of n nonhomogeneous equations in the n unknowns $x_1, x_2, \ldots, x_n$. The methods start with an arbitrary approximation $\mathbf{x}^{(0)}$ to the solution vector **x**, and this is iterated in such a way that it leads to successive improved approximations $\mathbf{x}^{(1)}, \ldots, \mathbf{x}^{(i)}, \mathbf{x}^{(i+1)}$ to **x**. The iterative process is terminated after N iterations as soon as the two successive

tolerance in iterations

iterates $\mathbf{x}^{(N-1)}$ and $\mathbf{x}^{(N)}$ yield approximations $x_i^{(N-1)}$ and $x_i^{(N)}$ to x_i, for $i = 1, 2, \ldots, n$, that differ by less than a small preassigned number $\varepsilon > 0$, called the **tolerance**. The final iterate is taken to be the solution of the system of equations to within the chosen tolerance. The number of iterations necessary to arrive at this approximation to the solution vector is indeterminate, because it depends on the structure of the equations, the iterative scheme involved, and the tolerance.

As all methods of the direct type are, in a sense, derived from the standard **Gaussian elimination** process, it will be sufficient to describe this process in some detail. Later a comment will be offered concerning a modification that must be made to the process to ensure that the elimination procedure does not fail unnecessarily, and that round-off errors are minimized. The second direct method retains information contained in the Gaussian elimination process and uses it to derive the factorization $\mathbf{PA} = \mathbf{LU}$, after which the result is used to solve the system $\mathbf{Ax} = \mathbf{b}$. This method is useful when solutions are required to a system $\mathbf{Ax} = \mathbf{b}$ for a sequence of nonhomogeneous vectors **b** while leaving the coefficient matrix **A** unchanged. This can happen, for example, in the analysis of forces in a structure due to changes in loading, where the matrix **A** representing the structure stays the same, while the loading represented by the vector **b** is altered repeatedly.

Of the various iterative schemes that are available, we describe only the **Jacobi** and **Gauss–Seidel** schemes. These are widely used, though for somewhat different purposes, and they are applicable to systems of equations that possess a property called **diagonal dominance** that will be described later. Iterative methods are used when working with large matrices, where it frequently happens that many zero elements are present, often occurring in diagonal bands parallel to the leading diagonal of matrix **A**. Matrices of this type are called **sparse matrices**, and they arise when solving partial differential equations, in spline interpolation, and in many other applications of matrices. More information about refinements to the Gaussian elimination process and about iterative methods in general can be found in the references cited at the end of the section.

The Gaussian Elimination Process

Let us assume that the system of equations to be solved is of the form

$$\mathbf{Ax} = \mathbf{b}, \tag{28}$$

where $\mathbf{A} = [a_{ij}]$ is an $n \times n$ matrix with constant coefficients, the column vector $\mathbf{x} = [x_1, x_2, \ldots, x_n]^{\mathrm{T}}$ is the required solution vector, and the column vector $\mathbf{b} = [b_1, b_2, \ldots, b_n]^{\mathrm{T}}$ contains the constant nonhomogeneous terms, not every one of which is zero.

When written out explicitly, (28) becomes

$$
\begin{bmatrix}
a_{11} & a_{12} & \cdots & a_{1n} \\
a_{21} & a_{22} & \cdots & a_{2n} \\
\vdots & \vdots & \vdots & \vdots \\
a_{n1} & a_{n2} & \cdots & a_{nn}
\end{bmatrix}
\begin{bmatrix}
x_1 \\ x_2 \\ \vdots \\ x_n
\end{bmatrix}
=
\begin{bmatrix}
b_1 \\ b_2 \\ \vdots \\ b_n
\end{bmatrix}.
\tag{29}
$$

It was shown in Chapter 3 that (29), equivalently (28), possesses a unique solution provided the rank of matrix $\mathbf{A}$ and the rank of the augmented matrix $[\mathbf{A}|\mathbf{b}]$ are both equal to n, in which case the formal solution of (28) can be written $\mathbf{x} = \mathbf{A}^{-1}\mathbf{b}$. However, the need to find different ways of calculating $\mathbf{x}$ arises from the fact that solutions in terms of the inverse matrix are *not* practicable when n is large, because of the magnitude of the task of calculating $\mathbf{A}^{-1}$ when n is large.

In both machine and hand computation, the foregoing full matrix form of the system in (29) is abbreviated to the augmented matrix, and the calculations are then performed on its entries. The augmented array corresponding to (29) is

$$
\begin{bmatrix}
a_{11} & a_{12} & \cdots & a_{1n} & \vdots & b_1 \\
a_{21} & a_{22} & \cdots & a_{2n} & \vdots & b_2 \\
\vdots & \vdots & \vdots & \vdots & \vdots & \vdots \\
a_{n1} & a_{n2} & \cdots & a_{nn} & \vdots & b_n
\end{bmatrix}.
\tag{30}
$$

In this abbreviated notation the coefficients of $x_1, x_2, \ldots, x_n$ in each equation are identified by their position in the array, so the coefficient of x_1 in the second equation is a_{12}, while the coefficient of x_2 in the nth equation is a_{n2}.

As individual equations can be scaled by a number k, and a multiple of an equation can be added to another equation, all without altering the solution, it follows that these same operations can be performed on the array in (30). The basic **Gaussian elimination process** makes use of these properties. The first stage of the elimination process involves assuming $a_{11} \neq 0$, multiplying the first row by a_{21}/a_{11}, and subtracting the result from the second row, when its first entry becomes zero. The next step is to multiply the first row by a_{31}/a_{11} and subtract the result from the third row, to make its first entry zero. A repetition of this process $n - 1$ times completes the first stage of the process, after which all entries below a_{11} are zero, causing (30) to become

Gaussian elimination

$$
\begin{bmatrix}
a_{11} & a_{12} & \cdots & a_{1n} & b_1 \\
0 & a_{22}^{(1)} & \cdots & a_{2n}^{(1)} & b_2^{(1)} \\
\vdots & \vdots & \vdots & \vdots & \vdots \\
0 & a_{n2}^{(1)} & \cdots & a_{nn}^{(1)} & b_n^{(1)}
\end{bmatrix},
\tag{31}
$$

where the $a_{ij}^{(1)}$ and $b_i^{(1)}$ represent the modified elements a_{ij} and b_i after subtraction of the multiple of the corresponding elements in the first row.

The second stage of the elimination process involves assuming $a_{22}^{(1)} \neq 0$, subtracting suitable multiples of the modified second row in (31) from the $n - 2$ rows below it to make all entries in the column below $a_{22}^{(1)}$ zero. A continuation of this process, assuming no element used to eliminate those below it is zero, leads in the end to all elements below the leading diagonal of the first n columns of the modified augmented array becoming zero, so the final array becomes

$$
\begin{bmatrix}
a_{11} & a_{12} & \cdots & a_{1n} & b_1 \\
0 & a_{22}^{(1)} & \cdots & a_{2n}^{(1)} & b_2^{(1)} \\
\vdots & \vdots & \vdots & \vdots & \vdots \\
0 & 0 & \cdots & a_{nn}^{(n-1)} & b_n^{(n-1)}
\end{bmatrix}.
\tag{32}
$$

The solution is then found by the process called **back-substitution**, which starts with the last row in (32) that is equivalent to the equation $a_{nn}^{(n-1)} x_n = b_n^{(n-1)}$, from which it follows that

$$
x_n = b_n^{(n-1)} / a_{nn}^{(n-1)}.
\tag{33}
$$

The second row from the bottom in (32) is equivalent to the equation

$$
a_{n-1,n-1}^{(n-2)} x_{n-1} + a_{n-1,n}^{(n-2)} x_n = b_{n-1}^{(n-2)},
\tag{34}
$$

from which x_{n-1} can be found after substituting the value of x_n found in (33). Continuing in this manner, all elements $x_1, x_2, \ldots, x_n$ of the required solution vector $\mathbf{x}$ can be found in the reverse order $x_n, x_{n-1}, \ldots, x_1$.

The elements $a_{11}, a_{22}^{(1)}, \ldots, a_{nn}^{(n-1)}$ used to reduce the coefficient matrix $\mathbf{A}$ to the upper triangular form shown in the first n columns of (32) are called the **pivotal elements** in the Gaussian elimination process, and the row containing a pivotal element is called the **pivotal row**. This completes the basic Gaussian elimination process.

pivotal elements

Clearly, if at the rth stage in the process a row of zeros is obtained in the modified coefficient matrix $\mathbf{A}$, but the modified rth element in the nonhomogeneous vector $\mathbf{b}$ is nonzero, the system of equations is incompatible and *no* solution exists. If, however, at the rth stage in the elimination process a row of zeros is obtained in the modified coefficient matrix $\mathbf{A}$, and the modified rth element in the nonhomogeneous vector $\mathbf{b}$ is also zero, then the rth equation is linearly dependent on the first $r - 1$ equations, so the solution cannot be unique.

A difficulty arises if at any stage of the process the pivotal element in the mth position on the leading diagonal of the modified matrix $\mathbf{A}$ becomes zero, as would happen at the start if $a_{11} = 0$. Should this occur, the difficulty is overcome by interchanging the order of the rows to bring a nonzero element into the pivotal position. Errors can be introduced during the elimination process if a very small pivotal element is used to reduce to zero entries in the column below it that are significantly larger, so this must be avoided. As the order of equations can be changed without altering the solution, these disadvantages can both be avoided as follows. At the mth stage, from among rows m to n, a row is selected that contains one of the elements of largest magnitude in its mth column. This row is then moved upward to form the new mth row, after which the elimination process continues as before. This process is called **Gaussian elimination with partial pivoting**, and it is a standard feature of software codes.

Gaussian elimination with partial pivoting

It is easy to see this same method can be used when the number of equations is not equal to the number of unknowns. The form of the modified augmented matrix will then, as just described, indicate whether the system has no solution, a unique solution that can be found, or a nonunique solution depending on some arbitrary parameters because of linear independence of rows.

Although det $\mathbf{A}$ is not required when using the Gaussian elimination process, because the process reduces the original coefficient matrix $\mathbf{A}$ in an efficient manner to the upper-triangular form shown in the first n columns of (32), it follows at once that

Gaussian elimination and det A

$$\det \mathbf{A} = a_{11}a_{22}^{(1)}a_{33}^{(2)}\ldots a_{nn}^{(n-1)}, \tag{35}$$

and it is this method that is used by software programs when finding det $\mathbf{A}$, thereby avoiding the many time-consuming multiplications involved when computing cofactors.

EXAMPLE 19.7

Solve the following system of equations by Gaussian elimination:

$$2x_1 - 2x_2 + 3x_3 + 4x_4 = -18$$
$$4x_1 + x_2 - x_3 + 2x_4 = -11$$
$$x_1 - x_2 - x_3 + 5x_4 = -26$$
$$2x_1 - 3x_2 + 2x_3 - x_4 = -3.$$

Use (35) to find the determinant of the coefficient matrix $\mathbf{A}$.

Solution   The array to be considered is

$$\begin{bmatrix} 2 & -2 & 3 & 4 & -18 \\ 4 & 1 & -1 & 2 & -11 \\ 1 & -1 & -1 & 5 & -26 \\ 2 & -3 & 2 & -1 & -3 \end{bmatrix},$$

in which the first four columns represent the coefficient matrix $\mathbf{A}$ and the last column the nonhomogeneous vector $\mathbf{b}$. As no element in the first column is small, there is no need to interchange rows, so we will use the entry $a_{11} = 2$ as the initial pivotal element. Subtracting twice the first row from the second row, half the first row from the third row, and the first row from the last row shows that at the end of the first stage of the Gaussian elimination process the modified array becomes

$$\begin{bmatrix} 2 & -2 & 3 & 4 & -18 \\ 0 & 5 & -7 & -6 & 25 \\ 0 & 0 & -\frac{5}{2} & 3 & -17 \\ 0 & -1 & -1 & -5 & 15 \end{bmatrix}.$$

The next element in the pivotal position is 5, so as this element is not small, the order of the rows can be left unchanged and the element 5 used as the next pivotal element. Adding one-fifth of row 2 to row 4 gives

$$\begin{bmatrix} 2 & -2 & 3 & 4 & -18 \\ 0 & 5 & -7 & -6 & 25 \\ 0 & 0 & -\frac{5}{2} & 3 & -17 \\ 0 & 0 & -\frac{12}{5} & -\frac{31}{5} & 20 \end{bmatrix}.$$

In the last stage of the elimination process we use $-5/2$ as the pivotal element and subtract $24/25$ times row 3 from row 4 to obtain

$$\begin{bmatrix} 2 & -2 & 3 & 4 & -18 \\ 0 & 5 & -7 & -6 & 25 \\ 0 & 0 & -\frac{5}{2} & 3 & -17 \\ 0 & 0 & 0 & -\frac{227}{25} & \frac{908}{25} \end{bmatrix}.$$

Back substitution now gives the solution, because if we reinsert the unknown quantities $x_1, x_2, \ldots, x_n$ it follows from the last row that

$$-\frac{227}{25}x_4 = \frac{908}{25}, \quad \text{so } x_4 = -4,$$

while the second row from the bottom becomes

$$-\frac{5}{2}x_3 + 3x_4 = -17, \quad \text{so using } x_4 = -4 \text{ we find that } x_3 = 2.$$

Continuing in this manner and using the remaining two rows leads first to the result $x_2 = 3$ and then to $x_1 = -1$, so the solution is seen to be

$$x_1 = -1, \quad x_2 = 3, \quad x_3 = 2, \quad x_4 = -4.$$

Notice that in this case no pivotal element was small enough to necessitate an interchange of rows, so the solution was obtained without the need for partial pivoting.

The value of det $\mathbf{A}$ follows immediately from (35) as the product of the diagonal entries in the upper-triangular array to which the matrix $\mathbf{A}$ has been reduced at the end of the Gaussian elimination process, so

$$\det \mathbf{A} = 2 \cdot 5 \cdot \left(-\frac{5}{2}\right) \cdot \left(-\frac{277}{25}\right) = 277. \qquad \blacksquare$$

The LU Factorization Method

Suppose the $n \times n$ nonsingular matrix $\mathbf{A}$ in the system $\mathbf{A}\mathbf{x} = \mathbf{b}$ can be factored as the product $\mathbf{A} = \mathbf{L}\mathbf{U}$, where $\mathbf{L}$ is an $n \times n$ lower-triangular matrix with 1's along its leading diagonal and $\mathbf{U}$ is an $n \times n$ upper-triangular matrix.

The method of solution of the system of equations $\mathbf{A}\mathbf{x} = \mathbf{b}$ reduces to finding the column vector $\mathbf{y}$ that is the solution of $\mathbf{L}\mathbf{y} = \mathbf{b}$, and then determining $\mathbf{x}$ from the system of equations $\mathbf{U}\mathbf{x} = \mathbf{y}$. The advantage of this approach is that once $\mathbf{L}$ and $\mathbf{U}$ have been found, the elements of the vector $\mathbf{y}$ can be obtained by *forward substitution*, after which the elements of the vector $\mathbf{x}$ then follow by *backward substitution*. As already remarked, this approach is very efficient when the system $\mathbf{A}\mathbf{x} = \mathbf{b}$ has to be solved repeatedly with the same coefficient matrix $\mathbf{A}$, but different nonhomogeneous vectors $\mathbf{b}$. This is because $\mathbf{L}$ and $\mathbf{U}$ remain unchanged, so the solution vector $\mathbf{x}$ can be found using only multiplications, the vector $\mathbf{b}$, and the known factorization of $\mathbf{A}$. We remark here that, without introducing row permutations, it may not be possible to factor a nonsingular matrix.

All the information necessary for the factorization of $\mathbf{A}$ into the product $\mathbf{A} = \mathbf{L}\mathbf{U}$ is already contained in the Gaussian elimination method, so the most straightforward form of $\mathbf{L}\mathbf{U}$ factorization in which partial pivoting is not necessary will be illustrated by means of an example. We will factor the matrix $\mathbf{A}$ in Example 19.7, and then use the result to solve the system of equations in that example.

When the first stage of the Gaussian elimination process was applied to matrix **A** in the example, 2 times row 1 was *subtracted* from row 2, $\frac{1}{2}$ row 1 was *subtracted* from row 3, and 1 times row 1 was *subtracted* from row 4, causing matrix

$$\mathbf{A} = \begin{bmatrix} 2 & -2 & 3 & 4 \\ 4 & 1 & -1 & 2 \\ 1 & -1 & -1 & 5 \\ 2 & -3 & 2 & -1 \end{bmatrix} \quad \text{to become the matrix } \mathbf{A}_1 = \begin{bmatrix} 2 & -2 & 3 & 4 \\ 0 & 5 & -7 & -6 \\ 0 & 0 & -\frac{5}{2} & 3 \\ 0 & -1 & -1 & -5 \end{bmatrix}.$$

If we represent the elementary row operations involved in terms of premultiplication of **A** by a matrix $\mathbf{M}_1$, this can be written $\mathbf{M}_1 \mathbf{A} = \mathbf{A}_1$, where

$$\mathbf{M}_1 = \begin{bmatrix} 1 & 0 & 0 & 0 \\ -2 & 1 & 0 & 0 \\ -\frac{1}{2} & 0 & 1 & 0 \\ -1 & 0 & 0 & 1 \end{bmatrix}.$$

When the second stage of the Gaussian elimination process was applied to the matrix $\mathbf{A}_1$, $-\frac{1}{5}$ times row 2 was *subtracted* from row 4, causing $\mathbf{A}_2$ to become the matrix

$$\mathbf{A}_2 = \begin{bmatrix} 2 & -2 & 3 & 4 \\ 0 & 5 & -7 & -6 \\ 0 & 0 & -\frac{5}{2} & 3 \\ 0 & 0 & -\frac{12}{5} & -\frac{31}{5} \end{bmatrix},$$

so in terms of matrix multiplication this becomes $\mathbf{M}_2 \mathbf{A}_1 = \mathbf{A}_2$, or $\mathbf{M}_2 \mathbf{M}_1 \mathbf{A} = \mathbf{A}_2$, where

$$\mathbf{M}_2 = \begin{bmatrix} 1 & 0 & 0 & 0 \\ 0 & 1 & 0 & 0 \\ 0 & 0 & 1 & 0 \\ 0 & \frac{1}{5} & 0 & 1 \end{bmatrix}.$$

Finally, when the last stage of the Gaussian elimination process was applied to matrix $\mathbf{A}_2$, 24/25 times row 3 was *subtracted* from row 4 to give the upper-triangular matrix

$$\mathbf{A}_3 = \begin{bmatrix} 2 & -2 & 3 & 4 \\ 0 & 5 & -7 & -6 \\ 0 & 0 & -\frac{5}{2} & 3 \\ 0 & 0 & 0 & -\frac{227}{25} \end{bmatrix},$$

so in terms of matrix multiplication $\mathbf{M}_3 \mathbf{A}_2 = \mathbf{A}_3$, or $\mathbf{M}_3 \mathbf{M}_2 \mathbf{M}_1 \mathbf{A} = \mathbf{A}_3$, where

$$\mathbf{M}_3 = \begin{bmatrix} 1 & 0 & 0 & 0 \\ 0 & 1 & 0 & 0 \\ 0 & 0 & 1 & 0 \\ 0 & 0 & -\frac{24}{25} & 1 \end{bmatrix}.$$

However, $\mathbf{A}_3 = \mathbf{U}$ is an upper-triangular matrix, and we have shown that

$$\mathbf{M}_3 \mathbf{M}_2 \mathbf{M}_1 \mathbf{A} = \mathbf{U}, \quad \text{with } \mathbf{U} = \begin{bmatrix} 2 & -2 & 3 & 4 \\ 0 & 5 & -7 & -6 \\ 0 & 0 & -\frac{5}{2} & 3 \\ 0 & 0 & 0 & -\frac{227}{25} \end{bmatrix},$$

and so

$$\mathbf{A} = \mathbf{M}_1^{-1}\mathbf{M}_2^{-1}\mathbf{M}_3^{-1}\mathbf{U}.$$

We will have succeeded in factoring $\mathbf{A}$ if we can show that $\mathbf{M}_1^{-1}\mathbf{M}_2^{-1}\mathbf{M}_3^{-1}$ is a lower-triangular matrix of the required type.

To accomplish this last step notice that the special structure of the matrices $\mathbf{M}_i$, for $i = 1, 2, 3$ is such that from the definition of the inverse matrix in terms of its cofactors, the inverse matrix $\mathbf{M}_i^{-1}$ can be obtained directly from $\mathbf{M}_i$ by reversing the signs of the elements in its ith column that lie below the element 1, so without further computation we can write

$$\mathbf{M}_1^{-1}\mathbf{M}_2^{-1}\mathbf{M}_3^{-1} = \begin{bmatrix} 1 & 0 & 0 & 0 \\ 2 & 1 & 0 & 0 \\ \frac{1}{2} & 0 & 1 & 0 \\ 1 & 0 & 0 & 1 \end{bmatrix} \begin{bmatrix} 1 & 0 & 0 & 0 \\ 0 & 1 & 0 & 0 \\ 0 & 0 & 1 & 0 \\ 0 & -\frac{1}{5} & 0 & 1 \end{bmatrix} \begin{bmatrix} 1 & 0 & 0 & 0 \\ 0 & 1 & 0 & 0 \\ 0 & 0 & 1 & 0 \\ 0 & 0 & \frac{24}{25} & 1 \end{bmatrix}.$$

The structure of these matrices allows their product to be written down on sight, because the ith column of the product matrix is simply the ith column of the matrix $\mathbf{M}_i$, so that

$$\mathbf{M}_1^{-1}\mathbf{M}_2^{-1}\mathbf{M}_3^{-1} = \begin{bmatrix} 1 & 0 & 0 & 0 \\ 2 & 1 & 0 & 0 \\ \frac{1}{2} & 0 & 1 & 0 \\ 1 & -\frac{1}{5} & \frac{24}{25} & 1 \end{bmatrix}.$$

This is a lower-triangular matrix of the required form, so

$$\mathbf{L} = \begin{bmatrix} 1 & 0 & 0 & 0 \\ 2 & 1 & 0 & 0 \\ \frac{1}{2} & 0 & 1 & 0 \\ 1 & -\frac{1}{5} & \frac{24}{25} & 1 \end{bmatrix},$$

and the factored form of $\mathbf{A}$ is

$$\mathbf{A} = \mathbf{LU} = \begin{bmatrix} 1 & 0 & 0 & 0 \\ 2 & 1 & 0 & 0 \\ \frac{1}{2} & 0 & 1 & 0 \\ 1 & -\frac{1}{5} & \frac{24}{25} & 1 \end{bmatrix} \begin{bmatrix} 2 & -2 & 3 & 4 \\ 0 & 5 & -7 & -6 \\ 0 & 0 & -\frac{5}{2} & 3 \\ 0 & 0 & 0 & -\frac{227}{25} \end{bmatrix}.$$

To use $\mathbf{L}$ and $\mathbf{U}$ to solve the system of equations in Example 19.7, we must first solve the system $\mathbf{Ly} = \mathbf{b}$, where $\mathbf{b} = [-18, -11, -26, -3]^\mathrm{T}$. This is the system

$$\begin{bmatrix} 1 & 0 & 0 & 0 \\ 2 & 1 & 0 & 0 \\ \frac{1}{2} & 0 & 1 & 0 \\ 1 & -\frac{1}{5} & \frac{24}{25} & 1 \end{bmatrix} \begin{bmatrix} y_1 \\ y_2 \\ y_3 \\ y_4 \end{bmatrix} = \begin{bmatrix} -18 \\ -11 \\ -26 \\ -3 \end{bmatrix},$$

from which we see that $y_1 = -18$, and *forward substitution* then shows $y_2 = 25$, $y_3 = -17$, and $y_4 = 908/25$.

The elements x_1, x_2, x_3, and x_4 of the required solution vector $\mathbf{x}$ now follow by solving $\mathbf{Ux} = \mathbf{y}$, that is, the system

$$\begin{bmatrix} 2 & -2 & 3 & 4 \\ 0 & 5 & -7 & -6 \\ 0 & 0 & -\frac{5}{2} & 3 \\ 0 & 0 & 0 & -\frac{227}{25} \end{bmatrix} \begin{bmatrix} x_1 \\ x_2 \\ x_3 \\ x_4 \end{bmatrix} = \begin{bmatrix} -18 \\ 25 \\ -17 \\ \frac{908}{25} \end{bmatrix}.$$

This shows $x_4 = -4$, so using *back substitution* we find that $x_3 = 2, x_2 = -3$, and $x_1 = -1$, so the system is solved.

This method has been described in its simplest form where straightforward Gaussian elimination is used without partial pivoting. The modification that is necessary to allow for row interchanges simply involves premultiplication at the appropriate stage by a permutation matrix. It will be recalled that a **permutation matrix** $\mathbf{P}$ is a matrix obtained from a unit matrix by interchanging its rows. If, for example, rows i and j of a unit matrix are interchanged to give the permutation matrix $\mathbf{P}$, then $\mathbf{PA}$ is the matrix obtained from $\mathbf{A}$ by interchanging its ith and jth rows. Use is then made of the result $\mathbf{PA} = \mathbf{LU}$.

An analysis of the steps involved in the foregoing approach leads to the following algorithm for the $\mathbf{LU}$ factorization of a nonsingular matrix $\mathbf{A}$ when no row interchanges are involved.

The LU factorization algorithm

the steps in LU factorization

The factorization of an $n \times n$ nonsingular matrix $\mathbf{A}$ into the product $\mathbf{A} = \mathbf{LU}$, where $\mathbf{L}$ is a lower-triangular matrix with 1's on its leading diagonal and $\mathbf{U}$ is an upper-triangular matrix, can be accomplished as follows.

1. The matrix $\mathbf{U}$ is obtained by applying the Gaussian elimination process to the rows of $\mathbf{A}$ to reduce it to an upper-triangular matrix.

2. At the ith stage of the Gaussian elimination process in Step 1, and in the ith column, let m_{ij} be the multiple of the ith element that must be *subtracted* from the jth element to reduce the jth element to zero. Then the matrix $\mathbf{L}$ is given by

$$\mathbf{L} = \begin{bmatrix} 1 & 0 & 0 & \cdots & 0 & 0 \\ m_{21} & 1 & 0 & \cdots & 0 & 0 \\ m_{31} & m_{32} & 1 & \cdots & 0 & 0 \\ \vdots & \vdots & \vdots & \cdots & 1 & 0 \\ m_{n1} & m_{n2} & m_{n3} & \cdots & m_{nn-1} & 1 \end{bmatrix}.$$

EXAMPLE 19.8

Apply the $\mathbf{LU}$ factorization algorithm to determine the matrix $\mathbf{L}$ in Example 19.7.

Solution   An examination of the Gaussian elimination process described in the example used to derive the algorithm shows that in the first step $m_{21} = 2, m_{31} = \frac{1}{2}$, and $m_{41} = 1$, and in the second step $m_{32} = 0$ and $m_{42} = -\frac{1}{5}$, while in the last step

$m_{43} = 24/25$, so from the algorithm

$$\mathbf{L} = \begin{bmatrix} 1 & 0 & 0 & 0 \\ 2 & 1 & 0 & 0 \\ \frac{1}{2} & 0 & 1 & 0 \\ 1 & -\frac{1}{5} & \frac{24}{25} & 1 \end{bmatrix}.$$ ∎

The Jacobi Iterative Process

To derive the Jacobi iterative process, the individual equations in (29) are rearranged so the first expresses x_1 in terms of the remaining unknowns and b_1, the second expresses x_2 in terms of the remaining unknowns and b_2, and so on until the last is rearranged to express x_n in terms of the other unknowns and b_n, leading to the result

$$
\begin{aligned}
x_1 &= (b_1 - a_{12}x_2 - a_{13}x_3 - \cdots - a_{1n}x_n)/a_{11} \\
x_2 &= (b_2 - a_{21}x_1 - a_{23}x_3 - \cdots - a_{2n}x_n)/a_{22} \\
&\quad \cdot \quad \cdot \quad \cdot \quad \cdot \quad \cdot \quad \cdot \quad \cdot \quad \cdot \\
x_n &= (b_n - a_{n1}x_1 - a_{n2}x_2 - \cdots - a_{nn-1}x_{n-1})/a_{nn}.
\end{aligned}
\tag{36}
$$

Jacobi iterative method

The **Jacobi iterative process** follows from this by defining the rth approximation to the solution denoted by $x_1^{(r)}, x_2^{(r)}, \ldots, x_n^{(r)}$, in terms of the $(r-1)$th approximation denoted by $x_1^{(r-1)}, x_2^{(r-1)}, \ldots, x_1^{(r-1)}$, by means of the equations

$$
\begin{aligned}
x_1^{(r)} &= \left(b_1 - a_{12}x_2^{(r-1)} - a_{13}x_3^{(r-1)} - \cdots - a_{1n}x_n^{(r-1)} \right) / a_{11} \\
x_2^{(r)} &= \left(b_2 - a_{21}x_1^{(r-1)} - a_{23}x_3^{(r-1)} - \cdots - a_{2n}x_n^{(r-1)} \right) / a_{22} \\
x_3^{(r)} &= \left(b_3 - a_{31}x_1^{(r-1)} - a_{32}x_2^{(r-1)} - \cdots - a_{3n}x_n^{(r-1)} \right) / a_{33} \\
&\quad \cdot \quad \cdot \quad \cdot \quad \cdot \quad \cdot \quad \cdot \quad \cdot \quad \cdot \quad \cdot \\
x_n^{(r)} &= \left(b_n - a_{n1}x_1^{(r-1)} - a_{n2}x_2^{(r-1)} - \cdots - a_{nn-1}x_{n-1}^{(r-1)} \right) / a_{nn}.
\end{aligned}
\tag{37}
$$

The iteration is started with any initial choice for $x_1^{(0)}, x_2^{(0)}, \ldots, x_n^{(0)}$, typically $x_1^{(0)} = 1, x_2^{(0)} = 1, \ldots, x_n^{(0)} = 1$. The iterative process is continued until for some r the magnitude of the difference between corresponding elements of the $(r-1)$th and the rth iterates given by $|x_i^{(r)} - x_i^{(r-1)}|$ for $i = 1, 2, \ldots, n$ is less than some preassigned tolerance $\varepsilon > 0$, so that

$$
\left| x_i^{(r)} - x_i^{(r-1)} \right| < \varepsilon, \quad \text{for } i = 1, 2, \ldots, n.
\tag{38}
$$

This is the simplest of many possible conditions for the **convergence** of an iterative process. The values $x_1^{(r)}, x_2^{(r)}, \ldots, x_n^{(r)}$ obtained from the rth iteration at which conditions (38) are first satisfied are taken to be the required solution $x_1, x_2, \ldots, x_n$, to within the tolerance ε. It should be noticed that the Jacobi iteration process is a fixed point iteration process for a system of linear equations.

Although it will not be proved here, a *sufficient* condition for the convergence of the Jacobi iterative process for any initial choice of $x_1^{(0)}, x_2^{(0)}, \ldots, x_n^{(0)}$ is that the

diagonal dominance

system (29) is **diagonally dominant**. This means that in each row of the coefficient matrix **A**, the magnitude of the element lying on the leading diagonal exceeds the sum of the magnitudes of all the other elements in the row. Thus, matrix **A** will be diagonally dominant if

$$|a_{ii}| > |a_{i1}| + |a_{i2}| + \cdots + |a_{ii-1}| + |a_{ii+1}| + \cdots + |a_{in}|, \quad \text{for } i = 1, 2, \ldots, n$$

(39)

An examination of the equations in (37) shows the Jacobi method fails to make use of current (improved) approximations as they are generated. This can be seen in the second equation where the better estimate $x_1^{(r)}$ could be used in place of the estimate $x_1^{(r-1)}$, as it has already been found from the first equation. Proceeding in this manner, and in each equation always using the currently available estimates, leads to the **Gauss–Seidel iterative process** defined by

Gauss–Seidel iterative method

$$
\begin{aligned}
x_1^{(r)} &= \left(b_1 - a_{12}x_2^{(r-1)} - a_{13}x_3^{(r-1)} - \cdots - a_{1n}x_n^{(r-1)} \right) / a_{11} \\
x_2^{(r)} &= \left(b_2 - a_{21}x_1^{(r)} - a_{23}x_3^{(r-1)} - \cdots - a_{2n}x_n^{(r-1)} \right) / a_{22} \\
x_3^{(r)} &= \left(b_3 - a_{31}x_1^{(r)} - a_{32}x_2^{(r)} - \cdots - a_{3n}x_n^{(r-1)} \right) / a_{33} \\
& \quad \cdot \quad \cdot \quad \cdot \quad \cdot \quad \cdot \quad \cdot \quad \cdot \quad \cdot \\
x_n^{(r)} &= \left(b_n - a_{n1}x_1^{(r)} - a_{n2}x_2^{(r)} - \cdots - a_{nn-1}x_{n-1}^{(r)} \right) / a_{nn}.
\end{aligned}
$$

(40)

A sufficient condition for the convergence of the Gauss–Seidel process is the same as that for the Jacobi process, namely that **A** is diagonally dominant.

Other conditions for the convergence of iterative processes can be derived in terms of the magnitude of the largest eigenvalue of **A**, called its **spectral radius**, but as this eigenvalue is difficult to compute when the number of equations n is large, such results are mainly of theoretical importance.

spectral radius

When an iterative process **diverges**, successive iterates usually alternate in sign and their magnitude grows without bound. In software programs a check is made on the behavior of successive iterates, and if divergence is detected the computer produces a message to this effect and terminates the computation.

EXAMPLE 19.9

Use the Gauss–Seidel iterative process to find the solution of the following system of equations

$$
\begin{aligned}
1.2x_1 + 4.4x_2 - 1.9x_3 &= -4.2 \\
5.1x_1 - 1.3x_2 + 2.4x_3 &= 2.7 \\
-2.6x_1 + 1.7x_2 - 6.3x_3 &= 9.6.
\end{aligned}
$$

Solution   Applying the test for diagonal dominance in (39) shows that only the third equation satisfies the condition, because

$$|-6.3| > |-2.6| + |1.7| \quad \text{but} \quad |1.2| < |4.4| + |-1.9| \quad \text{and} \quad |-1.3| < |5.1| + |2.4|.$$

However, if the first two equations are interchanged the system becomes diagonally dominant, so when setting up the Gauss–Seidel iterative process in this case the equations must be used in the order

$$5.1x_1 - 1.3x_2 + 2.4x_3 = 2.7$$
$$1.2x_1 + 4.4x_2 - 1.9x_3 = -4.2$$
$$-2.6x_1 + 1.7x_2 - 6.3x_3 = 9.6.$$

From (40) the Gauss–Seidel iterative process for this system of equations becomes

$$x_1^{(r)} = \frac{1}{5.1}\left(1.3x_2^{(r-1)} - 2.4x_3^{(r-1)} + 2.7\right)$$

$$x_2^{(r)} = \frac{1}{4.4}\left(-1.2x_1^{(r)} + 1.9x_3^{(r-1)} - 4.2\right)$$

$$x_3^{(r)} = \frac{1}{6.3}\left(-2.6x_1^{(r)} + 1.7x_2^{(r)} - 9.6\right).$$

The result of starting the iterations with $x_1^{(0)} = x_2^{(0)} = x_3^{(0)} = 1$ is shown in the following tables, and the values obtained in the 10th iteration should be compared with the solution $x_1 = 1.162946$, $x_2 = -2.418817$, and $x_3 = -2.656452$ obtained by Gaussian elimination.

Iteration Number					
	0	1	2	3	4	5
x_1	1	0.313726	1.229617	1.219913	1.175631	1.162857
x_2	1	−0.608289	−2.074693	−2.406137	−2.430951	−2.422740
x_3	1	−1.817425	−2.591108	−2.676541	−2.664962	−2.657887

	6	7	8	9	10
x_1	1.162621	1.162815	1.162924	1.162946	1.162947
x_2	−2.419348	−2.418785	−2.418784	−2.418809	−2.418816
x_3	−2.656461	−2.656389	−2.656434	−2.656450	−2.656452

These results demonstrate the *convergence* of the iterations obtained from a diagonally dominant scheme to the solution obtained by the direct method.

If, instead, an iterative scheme had been derived from the original system of equations without first rearranging them to make the system diagonally dominant, we would have obtained

how nondiagonal dominance can lead to divergence

$$x_1^{(r)} = \frac{1}{1.2}\left(-4.4x_2^{(r-1)} + 1.9x_3^{(r-1)} - 4.2\right)$$

$$x_2^{(r)} = \frac{1}{1.3}\left(5.1x_1^{(r)} + 2.4x_3^{(r-1)} - 2.7\right)$$

$$x_3^{(r)} = \frac{1}{6.3}\left(-2.6x_1^{(r)} + 1.7x_2^{(r)} - 9.6\right).$$

Using this scheme, and starting the iterations as before with $x_1^{(0)} = x_2^{(0)} = x_3^{(0)} = 1$, gives the results

$$x_1^{(1)} = -5.58333, \quad x_2^{(1)} = -22.13462, \quad x_3^{(1)} = -5.19241$$
$$x_1^{(2)} = 69.43894, \quad x_2^{(2)} = 260.75140, \quad x_3^{(2)} = 40.18034$$

that demonstrate very clearly the *divergence* of the nondiagonally dominant scheme.　∎

Something must be said about how these two iterative methods are used. The Gauss–Seidel method is used in computer codes mainly as a preconditioner for more advanced schemes, where its use of the current approximation at each stage requires only half as much storage as the Jacobi method. The Jacobi schemes are used extensively as building blocks in much more complicated and efficient iterative procedures, such as preconditioned conjugate gradient and multigrid methods.

For more information about numerical linear algebra, see references [2.15], [2.16], [2.17], [2.19], and [2.20].

Summary

Various examples were given, and it was seen that the **LU** factorization of an $n \times n$ matrix **A** is only possible if det $\mathbf{A} \neq 0$.

Two essentially different types of methods have been derived for the solution of systems of nonhomogeneous linear equations, one of a direct type and the other based on iteration. The two direct methods were Gaussian elimination and the **LU** factorization method that is derived from it. The necessity to interchange rows when a pivotal element was either zero or small was shown to lead to Gaussian elimination with partial pivoting. The **LU** factorization method was shown to make use of the information produced by the Gaussian elimination process at each step in a different manner, and it may also involve partial pivoting.

The other method, involving iteration, started from an arbitrary initial approximation and converged to the required solution to within a prescribed tolerance, provided the system of equations was diagonally dominant.

EXERCISES 19.5

In Exercises 1 through 4, (a) solve the system of equations using Gaussian elimination, and (b) compare the results obtained in (a) with those found by solving the system using Gauss–Seidel iteration starting from the initial iterates $x_1^{(0)} = x_2^{(0)} = x_3^{(0)} = 1$ and performing 10 iterations.

1. $4.7x_1 + 1.3x_2 - 1.6x_3 = 1.3$
   $x_1 - 4.1x_2 + 1.1x_3 = 4.6$
   $2.1x_1 + 1.4x_2 + 6.2x_3 = 5.2.$

2. $1.7x_1 - 4.6x_2 - 1.2x_3 = 3.4$
   $-3.1x_1 + 2.3x_2 + 7.2x_3 = 2.7$
   $3.2x_1 + 1.2x_2 + 1.4x_3 = -4.2.$

3. $2.1x_1 + 6.5x_2 - 3.1x_3 = -6.4$
   $-5.2x_1 + 2.1x_2 - 1.5x_3 = 3.7$
   $1.8x_1 - 2.9x_2 + 6.2x_3 = -4.2.$

4. $6.2x_1 - 2.2x_2 + 3.1x_3 = -2.6$
   $-1.6x_1 + 1.9x_2 + 8.4x_3 = -2.6$
   $2.3x_1 - 8.4x_2 + 3.2x_3 = 6.5.$

5. The $n \times n$ real symmetric matrix $\mathbf{H}_n$ with the element $h_{ij} = 1/(i + j - 1)$ in its ith row and jth column is

called the **Hilbert matrix**, and its determinant rapidly becomes vanishingly small as n increases. Matrices of this type are said to be **ill-conditioned**, and when ill-conditioned matrices occur as coefficient matrices in systems of linear equations, large errors arise unless the calculations are performed using very high precision. The development of a vanishingly small determinant of a Hilbert matrix can be seen, for example, even when $n = 4$, because

$$\mathbf{H}_4 = \begin{bmatrix} 1 & \frac{1}{2} & \frac{1}{3} & \frac{1}{4} \\ \frac{1}{2} & \frac{1}{3} & \frac{1}{4} & \frac{1}{5} \\ \frac{1}{3} & \frac{1}{4} & \frac{1}{5} & \frac{1}{6} \\ \frac{1}{4} & \frac{1}{5} & \frac{1}{6} & \frac{1}{7} \end{bmatrix}, \text{ and det } \mathbf{H}_4 = 1/6{,}048{,}000.$$

When the fractions involved are not approximated, the exact solution of the system of equations

$$\begin{bmatrix} 1 & \frac{1}{2} & \frac{1}{3} & \frac{1}{4} \\ \frac{1}{2} & \frac{1}{3} & \frac{1}{4} & \frac{1}{5} \\ \frac{1}{3} & \frac{1}{4} & \frac{1}{5} & \frac{1}{6} \\ \frac{1}{4} & \frac{1}{5} & \frac{1}{6} & \frac{1}{7} \end{bmatrix} \begin{bmatrix} x_1 \\ x_2 \\ x_3 \\ x_4 \end{bmatrix} = \begin{bmatrix} 1 \\ 2 \\ 3 \\ 4 \end{bmatrix}$$

can be shown to be $x_1 = -64$, $x_2 = 900$, $x_3 = -2520$, and $x_4 = 1820$. Typically, ill-conditioned matrices arise in least squares approximations and orthogonalization.

Demonstrate the errors that arise when Gaussian elimination is used to solve this system of equations and the calculations are rounded to five decimal places. Use the Gaussian elimination to calculate det $\mathbf{H}_4$ working to five decimal places and compare the value obtained with the true result.

6. Use Jacobi and Gauss–Seidel iteration to solve the system

$$-4.2x_1 + 1.1x_2 - 2.1x_3 = 1.4$$

$$3.6x_1 + 9.2x_2 - 3.1x_3 = -3.2$$

$$1.4x_1 + 2.9x_2 - 6.4x_3 = -1.2,$$

starting from the initial iterates $x_1^{(0)} = x_2^{(0)} = x_3^{(0)} = 0$ and performing six iterations. Compare the results with the exact solution $x_1 = -0.39101$, $x_2 = -0.18938$, $x_3 = 0.01615$. Derive an iterative scheme when the equations are arranged in a *nondiagonally dominant* form, and using the initial iterates $x_1^{(0)} = x_2^{(0)} = x_3^{(0)} = 0$ perform three iterations to demonstrate the divergence of the scheme.

In Exercises 7 through 12 use **LU** factorization to solve the system of equations $\mathbf{Ax} = \mathbf{b}$ for the given matrices $\mathbf{A}$ and $\mathbf{b}$.

7. $\mathbf{A} = \begin{bmatrix} -4 & 1 & -1 \\ 12 & -1 & 5 \\ -12 & 5 & -4 \end{bmatrix}$, $\mathbf{b} = \begin{bmatrix} 3 \\ -2 \\ 2 \end{bmatrix}$.

8. $\mathbf{A} = \begin{bmatrix} -1 & 2 & 3 \\ -5 & 7 & 16 \\ 2 & -10 & -2 \end{bmatrix}$, $\mathbf{b} = \begin{bmatrix} -5 \\ 2 \\ 6 \end{bmatrix}$.

9. $\mathbf{A} = \begin{bmatrix} 4 & -1 & -1 \\ -16 & 6 & 1 \\ -4 & 7 & -9 \end{bmatrix}$, $\mathbf{b} = \begin{bmatrix} 0 \\ 6 \\ -7 \end{bmatrix}$.

10. $\mathbf{A} = \begin{bmatrix} -5 & -2 & 0 \\ -15 & -9 & 2 \\ 0 & -6 & 8 \end{bmatrix}$, $\mathbf{b} = \begin{bmatrix} 1 \\ -2 \\ 3 \end{bmatrix}$.

11. $\mathbf{A} = \begin{bmatrix} 2 & 1 & 0 & 2 \\ -1 & 0 & 1 & 0 \\ 4 & \frac{3}{2} & 2 & 3 \\ -2 & 0 & 8 & 1 \end{bmatrix}$, $\mathbf{b} = \begin{bmatrix} 1 \\ 2 \\ 1 \\ 4 \end{bmatrix}$.

12. $\mathbf{A} = \begin{bmatrix} 3 & 0 & 1 & -1 \\ 6 & -1 & 3 & -3 \\ -3 & 1 & 0 & 1 \\ -3 & 0 & -5 & 4 \end{bmatrix}$, $\mathbf{b} = \begin{bmatrix} -2 \\ 3 \\ -1 \\ 5 \end{bmatrix}$.

19.6   Eigenvalues and Eigenvectors

In Chapter 4 an **eigenvalue** associated with an $n \times n$ matrix $\mathbf{A}$ was defined as a number λ satisfying the matrix equation

$$\mathbf{Ax} = \lambda\mathbf{x}, \tag{41}$$

and the corresponding $n \times 1$ vector $\mathbf{x}$ was defined as the associated **eigenvector**. It follows directly from (41) that an eigenvector $\mathbf{x}$ of $\mathbf{A}$ corresponding to an eigenvalue λ can be multiplied (scaled) by a nonzero number k and still remain an eigenvector, because

$$\mathbf{A}(k\mathbf{x}) = \lambda(k\mathbf{x}) \quad \text{is equivalent to} \quad k\mathbf{Ax} = k\lambda\mathbf{x},$$

and cancellation of the scalar k reduces this last result to (41).

When eigenvalues and eigenvectors were determined in Chapter 4, result (41) was rewritten as the homogeneous system $(\mathbf{A} - \lambda\mathbf{I})\mathbf{x} = \mathbf{0}$, and the eigenvalues were found by requiring the determinant of the coefficient matrix $\det(\mathbf{A} - \lambda\mathbf{I})$ to vanish, leading to a polynomial in λ of degree n of the general form

$$P(\lambda) = \lambda^n + a_1\lambda^{n-1} + a_2\lambda^{n-2} + \cdots + a_n,$$

called the **characteristic polynomial** associated with $\mathbf{A}$. Once the zeros of $P(\lambda)$ had been found, that is, the eigenvalues $\lambda_1, \lambda_2, \ldots, \lambda_n$ of $\mathbf{A}$, the associated eigenvectors $\mathbf{x}_1, \mathbf{x}_2, \ldots, \mathbf{x}_n$ were then obtained by solving the matrix equation

$$\mathbf{Ax}_i = \lambda_i\mathbf{x}_i \quad \text{for } i = 1, 2, \ldots, n. \tag{42}$$

This theoretical approach is only useful when $n \leq 3$, because then the zeros of $P(\lambda)$ can be determined analytically. In all other cases the task of finding the zeros is difficult, and unless they are known accurately, significant errors can be introduced when using them in (42) to compute the associated eigenvectors.

Computationally efficient numerical methods are available in computer algebra packages for the determination of eigenvalues and eigenvectors that do not involve first solving the characteristic equation for the eigenvalues. These are capable of finding real and complex eigenvalues, including repeated eigenvalues, and the corresponding eigenvectors. Because of this the only method that will be described here will be the **power method**, as it is easy to apply and its derivation is straightforward. However, this is not the method that is used in practice, except in certain special situations.

The derivation requires all of the eigenvalues of $\mathbf{A}$ to be ordered according to absolute magnitude so that

$$|\lambda_1| > |\lambda_2| \geq |\lambda_3| \geq \cdots \geq |\lambda_n|. \tag{43}$$

When this ordering is adopted, the eigenvalue λ_1 with the greatest magnitude is called the **dominant** eigenvalue of matrix $\mathbf{A}$, and the remaining eigenvalues $\lambda_2, \lambda_3, \ldots, \lambda_n$ are then called the **subdominant** eigenvalues of $\mathbf{A}$.

dominant and subdominant eigenvalues

It was seen in Chapter 4 that an arbitrary n element column vector $\mathbf{v}_0$ can always be expressed as the linear combination of eigenvectors $\mathbf{x}_1, \mathbf{x}_2, \ldots, \mathbf{x}_n$,

$$\mathbf{v}_0 = c_1\mathbf{x}_1 + c_2\mathbf{x}_2 + \cdots + c_n\mathbf{x}_n, \tag{44}$$

the power method for the dominant eigenvalue and its eigenvector

for some suitable choice of constants $c_1, c_2, \ldots, c_n$. The **power method** for the simultaneous determination of the eigenvalues and eigenvectors of $\mathbf{A}$ is an *iterative* method, and it involves setting $\mathbf{v}_r = \mathbf{A}^r\mathbf{v}_0$, multiplying (44) by $\mathbf{A}^r$, and making use of results (42) and (43). For $r = 0, 1, 2, \ldots$, we have

$$\begin{aligned}\mathbf{v}_r &= \mathbf{A}^r(c_1\mathbf{x}_1 + c_2\mathbf{x}_2 + \cdots + c_n\mathbf{x}_n) \\ &= c_1\lambda_1^r\mathbf{x}_1 + c_2\lambda_2^r\mathbf{x}_2 + \cdots + c_n\lambda_n^r\mathbf{x}_n \\ &= \lambda_1^r\{c_1\mathbf{x}_1 + c_2(\lambda_2/\lambda_1)^r\mathbf{x}_2 + \cdots + c_n(\lambda_n/\lambda_1)^r\mathbf{x}_n\}. \end{aligned} \tag{45}$$

The ordering of the eigenvalues in (43) causes the factors $(\lambda_2/\lambda_1)^r$, $(\lambda_3/\lambda_1)^r$, $\ldots$, $(\lambda_n/\lambda_1)^r$ in (45) all to tend to zero as r increases, so assuming that $c_1 \neq 0$, for suitably large r equation (45) can be approximated by

$$\mathbf{x}_r \approx \lambda_1^r c_1\mathbf{x}_1. \tag{46}$$

The assumption that $c_1 \neq 0$ is not restrictive, because if this is true, roundoff can be expected to introduce a component in the direction of $\mathbf{x}_1$, so that although convergence will be delayed, it will still take place in practice.

Result (46) shows that when r is large, $\mathbf{v}_r$ can be taken to be proportional to the eigenvector $\mathbf{x}_1$ associated with the dominant eigenvalue λ_1. As $\mathbf{v}_r = \mathbf{A}^r\mathbf{v}_0 = \mathbf{A}(\mathbf{A}^{r-1}\mathbf{v}_0) = \mathbf{A}\mathbf{v}_{r-1}$, it follows that the ratio (quotient) of corresponding elements in $\mathbf{v}_r$ and $\mathbf{v}_{r-1}$ approximate the dominant eigenvalue λ_1.

When the power method is implemented, the elements in $\mathbf{v}_r$ can become very large or very small, so to keep the exponent range of the machine from being exceeded, the fact that an eigenvector can be scaled and still remain an eigenvector is used to redefine the vector $\mathbf{v}_r$ as $\mathbf{v}_r = \mathbf{A}\widetilde{\mathbf{v}}_{r-1}$, where $\widetilde{\mathbf{v}}_{r-1}$ is a *normalized* vector $\mathbf{v}_{r-1}$. Many normalizations are possible, but the most convenient one involves obtaining $\widetilde{\mathbf{v}}_{r-1}$ from $\mathbf{v}_{r-1}$ by dividing each element of $\mathbf{v}_{r-1}$ by α_{r-1}, where α_{r-1} is its element of greatest magnitude. As a result of this normalization $\mathbf{v}_{r-1} = \alpha_{r-1}\widetilde{\mathbf{v}}_{r-1}$, and the element in $\widetilde{\mathbf{v}}_{r-1}$ with greatest magnitude becomes 1.

The iteration equation $\mathbf{v}_r = \mathbf{A}\mathbf{v}_{r-1}$ must now be replaced by $\mathbf{v}_r = \mathbf{A}\widetilde{\mathbf{v}}_{r-1}$ for $r = 1, 2, \ldots$, or, equivalently, by

$$\mathbf{v}_{r+1} = \mathbf{A}\widetilde{\mathbf{v}}_r \quad \text{for } r = 0, 1, \ldots. \tag{47}$$

normalization of vectors

Substituting $\mathbf{v}_{r+1} = \alpha_{r+1}\widetilde{\mathbf{v}}_{r+1}$ in the preceding result gives $\mathbf{A}\widetilde{\mathbf{v}}_r = \alpha_{r+1}\widetilde{\mathbf{v}}_{r+1}$, so as r becomes large and $\widetilde{\mathbf{v}}_r \to \widetilde{\mathbf{v}}_{r+1}$, it follows that $\alpha_{r+1} \to \lambda_1$ and $\widetilde{\mathbf{v}}_{r+1} \to \tilde{\mathbf{x}}_1$, the normalized eigenvector associated with λ_1. The iteration process in (47) can be started with any constant vector $\mathbf{v}_0 = [v_1, v_2, \ldots, v_n]^T$ that is often taken to be $\mathbf{v}_0 = [1, 1, \ldots, 1]^T$. The rate of convergence of the iterations is fastest when $|\lambda_1| \gg |\lambda_2|$, but the convergence becomes very slow when $|\lambda_1|$ and $|\lambda_2|$ are close together.

Various methods exist for the determination of the subdominant eigenvalues once the dominant eigenvalue is known, though these will not be discussed here.

EXAMPLE 19.10

Use the power method to find the dominant eigenvalue λ_1 and the normalized eigenvector $\mathbf{x}_1$ when

$$\mathbf{A} = \begin{bmatrix} 1 & 4 & 1 & 2 \\ 4 & 0 & 3 & 1 \\ 1 & 3 & 1 & 2 \\ 2 & 1 & 2 & 1 \end{bmatrix}.$$

Solution   As the matrix $\mathbf{A}$ is symmetric, its eigenvalues will all be real, so it is appropriate to use the power method to determine its eigenvalues and eigenvectors. In order to determine the dominant eigenvalue and its associated eigenvector, the iterative process $\mathbf{v}_r = \mathbf{A}\widetilde{\mathbf{v}}_{r-1}$ will be started by setting $\mathbf{v}_0 = [1, 1, 1, 1]^T$, and in the table that follows the ith element of $\mathbf{v}_r$ is denoted by $v_r^{(i)}$ while the corresponding normalized ith element of $\widetilde{\mathbf{v}}_r$ is denoted by $\widetilde{v}_r^{(i)}$.

Iterations Using $\mathbf{v}_{r+1} = \mathbf{A}\widetilde{\mathbf{v}}_r$

Iteration r	0	1	2	3	4	5	6	7	8	9	10
$v_r^{(1)}$	1	8	7.375	7.35593	7.34334	7.35018	7.34608	7.34881	7.34748	7.34770	7.34756
$v_r^{(2)}$	1	8	7.375	7.33899	7.33642	7.33732	7.33674	7.33797	7.33695	7.33696	7.33695
$v_r^{(3)}$	1	7	6.375	6.35593	6.34569	6.35112	6.34783	6.35008	6.34896	6.34913	6.34902
$v_r^{(4)}$	1	6	5.5	5.47458	5.47006	5.47224	5.47091	5.47229	5.47137	5.47143	5.47135
α_r	1	8	7.375	7.35593	7.34334	7.35018	7.34608	7.34881	7.34748	7.34770	7.34756
$\widetilde{v}_r^{(1)}$	1	1	1	1	1	1	1	1	1	1	1
$\widetilde{v}_r^{(2)}$	1	1	1	0.99770	0.99906	0.99825	0.99873	0.99852	0.99857	0.99854	0.99856
$\widetilde{v}_r^{(3)}$	1	0.87500	0.86441	0.86406	0.86414	0.86408	0.86411	0.86410	0.86410	0.86410	0.86410
$\widetilde{v}_r^{(4)}$	1	0.75000	0.74576	0.74424	0.74490	0.74450	0.74474	0.74465	0.74466	0.74465	0.74465

This shows that after 10 iterations the approximation to λ_1 provided by α_1 is $\lambda_1 \approx 7.34756$, and the associated normalized eigenvector $\widetilde{\mathbf{x}}_1$ is

$$\widetilde{\mathbf{x}}_1 \approx [1, 0.99856, 0.86410, 0.74465]^{\mathrm{T}}.$$

A calculation using a software package shows that when approximated to five decimal places $\lambda_1 = 7.34760$ and $\widetilde{\mathbf{v}}_1 = [1, 0.99855, 0.86410, 0.74465]^{\mathrm{T}}$.   ■

Euclidean norm of a vector

A different normalization that is often used involves dividing a vector $\mathbf{u}$ by $\|\mathbf{u}\| = (u_1^2 + u_2^2 + \cdots + u_n^2)^{1/2}$, where $u_1, u_2, \ldots, u_n$ are the n elements of $\mathbf{u}$. $\|\mathbf{u}\|$ is called the **Euclidean norm**, and it is useful when working with eigenvalues and eigenvectors of symmetric matrices, because then the quotient of corresponding terms in successive iterations provides a higher order approximation to the eigenvalue.

the inverse power method and finding the eigenvalue closest to a given number

The power method can also be used to find the eigenvalue λ_n of an $n \times n$ matrix $\mathbf{A}$ with the *smallest* magnitude, together with its associated eigenvector. The idea is simple, and it starts from the fact that if $\mathbf{A}$ is a nonsingular $n \times n$ matrix with the real eigenvalues $\lambda_1, \lambda_2, \ldots, \lambda_n$, then these are solutions of $\mathbf{A}\mathbf{x} = \lambda\mathbf{x}$. As $\mathbf{A}$ is nonsingular, it has an inverse $\mathbf{A}^{-1}$, and premultiplication of $\mathbf{A}\mathbf{x} = \lambda\mathbf{x}$ by $\mathbf{A}^{-1}$ gives $\mathbf{A}^{-1}\mathbf{A}\mathbf{x} = \lambda\mathbf{A}^{-1}\mathbf{x}$, or $\mathbf{A}^{-1}\mathbf{x} = (1/\lambda)\mathbf{x}$, showing that $1/\lambda_1, 1/\lambda_2, \ldots, 1/\lambda_n$ are the eigenvalues of $\mathbf{A}^{-1}$ and that the eigenvectors associated with λ_i and $1/\lambda_i$ are identical. Consequently, if the eigenvalues are ordered so that $|\lambda_1| > |\lambda_2| \geq |\lambda_3| \geq \cdots \geq |\lambda_n|$, the eigenvalue of $\mathbf{A}$ with the *smallest* magnitude will be the *dominant* eigenvalue of $\mathbf{A}^{-1}$. Thus, an application of the power method to $\mathbf{A}^{-1}$ will generate its dominant eigenvalue $\mu_1 = 1/\lambda_n$, so that $\lambda_n = 1/\mu_1$.

When using this method the inverse matrix $\mathbf{A}^{-1}$ is *not* constructed, and instead the equation

$$\mathbf{A}\mathbf{v}_{r+1} = \mathbf{v}_r \tag{48}$$

is iterated, having first used **LU** decomposition to solve for $\mathbf{v}_{r+1}$ in terms of $\mathbf{v}_r$. The decomposition only needs to be performed once because afterwards, at each stage of the iteration, the elements of $\mathbf{v}_{r+1}$ can be found by back-substitution using the elements of $\mathbf{v}_r$. This is just the situation where an **LU** decomposition is needed, because the right-hand sides are not available in advance, so it is necessary to solve a sequence of problems with the same matrix. Without the **LU** decomposition this process is not really practical.

As with the previous iteration procedure it is again necessary to normalize $\mathbf{v}_r$ by dividing each of its elements by its element of greatest magnitude α_r, or to use some other form of normalization, to keep calculations within the exponent range of the machine. This is because, unlike the previous case where the nonnormalized elements of $\mathbf{v}_r$ increased in magnitude as r increased, in this case they will decrease, causing accuracy to be lost if normalization is not performed. This method is called the **inverse power method** because it is equivalent to iterating the inverse matrix $\mathbf{A}^{-1}$. If we denote the normalized column vector $\mathbf{v}_r$ by $\widetilde{\mathbf{v}}_r$, the iteration scheme to be used analogous to (47) becomes

$$\mathbf{A}\mathbf{v}_{r+1} = \widetilde{\mathbf{v}}_r \quad \text{for } r = 0, 1, \ldots. \tag{49}$$

EXAMPLE 19.11   Use the inverse power method to find the eigenvalue of $\mathbf{A}$ with the smallest magnitude, given that

$$\mathbf{A} = \begin{bmatrix} 4 & 2 & 4 \\ 3 & 9 & 2 \\ 5 & 6 & 9 \end{bmatrix}.$$

Solution   The required eigenvalue will be obtained by iterating $\mathbf{A}\tilde{\mathbf{v}}_{r+1} = \mathbf{v}_r$ with the given matrix $\mathbf{A}$, so the system to be considered is

$$\begin{bmatrix} 4 & 2 & 4 \\ 3 & 9 & 2 \\ 5 & 6 & 9 \end{bmatrix} \begin{bmatrix} v_1^{(r+1)} \\ v_2^{(r+1)} \\ v_3^{(r+1)} \end{bmatrix} = \begin{bmatrix} \tilde{v}_1^{(r)} \\ \tilde{v}_2^{(r)} \\ \tilde{v}_3^{(r)} \end{bmatrix} \quad \text{with } r = 0, 1, \ldots \text{ and } \begin{bmatrix} v_1^{(0)} \\ v_2^{(0)} \\ v_3^{(0)} \end{bmatrix} = \begin{bmatrix} 1 \\ 1 \\ 1 \end{bmatrix}.$$

Using **LU** decomposition the system becomes

$$4v_1^{(r+1)} + 2v_2^{(r+1)} + 4v_3^{(r+1)} = \tilde{v}_1^{(r)}$$

$$\frac{15}{2} v_2^{(r+1)} - v_3^{(r)} = \tilde{v}_2^{(r)}$$

$$\frac{67}{15} v_3^{(r)} = \tilde{v}_3^{(r)}$$

and $\mathbf{v}_{r+1}$ now follows from $\tilde{\mathbf{v}}_r$ by back-substitution. As r increases, so the ratio of corresponding components of $\tilde{\mathbf{v}}_{r+1}$ and $\tilde{\mathbf{v}}_r$ will tend to the eigenvalue μ_1 of $\mathbf{A}^{-1}$ of greatest magnitude, so that the eigenvalue of $\mathbf{A}$ of *smallest* magnitude will be $\lambda_3 = 1/\mu_1$. The results of eight iterations are listed below, as in Example 19.10.

Iterations Using $\mathbf{A}\mathbf{v}_{r+1} = \tilde{\mathbf{v}}_r$								
Iteration 0	1	2	3	4	5	6	7	8
$v_r^{(1)}$ 1	0.32090	0.57914	0.61488	0.61215	0.60984	0.60898	0.60871	0.60862
$v_r^{(2)}$ 1	0.02239	−0.12617	−0.16659	−0.17289	−0.17403	−0.17429	−0.17436	−0.17438
$v_r^{(3)}$ 1	−0.08209	−0.26606	−0.28158	−0.27571	−0.27282	−0.27183	−0.27152	−0.27143
α_r 1	0.32090	0.57914	0.61488	0.61215	0.60984	0.60898	0.60871	0.60862
$\tilde{v}_r^{(1)}$ 1	1	1	1	1	1	1	1	1
$\tilde{v}_r^{(2)}$ 1	0.06977	−0.21786	−0.27093	−0.28243	−0.28637	−0.28620	−0.28644	−0.28652
$\tilde{v}_r^{(3)}$ 1	−0.25582	−0.45941	−0.45794	−0.45040	−0.44736	−0.44637	−0.44606	−0.44598

This shows that the approximate value of the *largest* eigenvalue of $\mathbf{A}^{-1}$ given by α_8 is $\mu_1 \approx 0.60862$, so the approximate value of the *smallest* eigenvalue of $\mathbf{A}$ is $\lambda_3 = 1/\mu_1 = 1.64306$, and the corresponding approximation to the associated normalized eigenvector $\mathbf{x}_3$ provided by $\mathbf{v}_8$ is

$$\mathbf{x}_3 \approx [1, -0.28652, -0.44598]^\mathrm{T}.$$

The results accurate to five decimal places found by using a software package are $\lambda_3 = 1.64315$ and $\mathbf{x}_3 = [1, -0.28656, -0.44592]^\mathrm{T}$. ∎

As an extension of the previous argument, let k be a specified constant, and consider the matrix $\mathbf{B} = \mathbf{A} - k\mathbf{I}$. Then, in terms of matrix $\mathbf{B}$, the eigenvalue equation

$\mathbf{A}\mathbf{x}_i = \lambda_i \mathbf{x}_i$ becomes

$$\mathbf{B}\mathbf{x}_i = (\lambda_i - k)\mathbf{x}_i, \tag{50}$$

showing the eigenvectors of $\mathbf{A}$ and $\mathbf{B}$ are identical, but the eigenvalues $\lambda_i - k$ of $\mathbf{B}$ are those of $\mathbf{A}$ reduced by k. This means that the eigenvalues of $(\mathbf{A} - k\mathbf{I})^{-1}$ for $k \neq \lambda_i$, with $i = 1, 2, \dots, n$, are $1/(\lambda_1 - k), 1/(\lambda_2 - k), \dots, 1/(\lambda_n - k)$. An application of the inverse power method to $(\mathbf{A} - k\mathbf{I})^{-1}$ then determines the eigenvalue of $\mathbf{A}$ closest to the specified constant k. This can be used as a basis for computing an eigenvector once an eigenvalue has been found. In terms of this approach, the initial application of the inverse power method is seen to involve the determination of the eigenvalue of $\mathbf{A}$ closest to 0.

For more information about the numerical computation of eigenvalues and eigenvectors see references [2.15], [2.16], [2.17], [2.19], and [2.20].

Summary   The power method for the calculation of the eigenvalue of greatest magnitude of a matrix together with its associated eigenvector was described. It was then shown how the inverse power method can be used to find the eigenvalue of smallest magnitude, and by making a small modification to the inverse power method, how the eigenvalue closest to a given number k can be found.

EXERCISES 19.6

In Exercises 1 through 4 use the power method to find the approximate value of the dominant eigenvalue and the associated normalized eigenvector of the given matrix, starting with $\mathbf{x}_0 = [1, 1, 1]^T$ and performing 10 iterations.

1. $\mathbf{A} = \begin{bmatrix} 18 & 3 & -1 \\ 3 & 12 & 2 \\ -1 & 2 & 4 \end{bmatrix}.$ **3.** $\mathbf{A} = \begin{bmatrix} 2 & -3 & 2 \\ -3 & 12 & 1 \\ 2 & 1 & 28 \end{bmatrix}.$

2. $\mathbf{A} = \begin{bmatrix} 20 & -2 & 1 \\ -2 & 3 & 4 \\ 1 & 4 & 0 \end{bmatrix}.$ **4.** $\mathbf{A} = \begin{bmatrix} -31 & -1 & 2 \\ -1 & -10 & 4 \\ 2 & 4 & -2 \end{bmatrix}.$

In Exercises 5 and 6 use the power method to find approximations to the dominant eigenvalue λ_1, and the associated normalized eigenvector, starting with $\mathbf{x}_0 = [1, -1, 1]^T$ and performing 10 iterations.

5. $\mathbf{A} = \begin{bmatrix} 26 & 3 & 1 \\ 3 & 20 & 2 \\ 1 & 2 & 1 \end{bmatrix}.$ **6.** $\mathbf{A} = \begin{bmatrix} 19 & 2 & 2 \\ 2 & 14 & 1 \\ 2 & 1 & 2 \end{bmatrix}.$

In Exercises 8 through 10 use the inverse power method to find approximations to the eigenvalue of smallest magnitude of the given matrix $\mathbf{A}$ and its associated eigenvector, starting with $\mathbf{x}_0 = [1, 1, 1]^T$ and performing six iterations.

7. $\mathbf{A} = \begin{bmatrix} 6 & 1 & -4 \\ 1 & 4 & 0 \\ -1 & -1 & 3 \end{bmatrix}.$ **9.** $\mathbf{A} = \begin{bmatrix} 2 & 5 & -2 \\ 4 & 2 & 4 \\ -3 & 1 & 0 \end{bmatrix}.$

8. $\mathbf{A} = \begin{bmatrix} 3 & 3 & -4 \\ 3 & 5 & 0 \\ -5 & -1 & 1 \end{bmatrix}.$ **10.** $\mathbf{A} = \begin{bmatrix} -3 & 5 & -3 \\ 3 & 1 & 1 \\ -2 & 1 & 2 \end{bmatrix}.$

19.7 Numerical Solution of Differential Equations

Most differential equations have no known analytical solution, and even when one can be found it is often difficult to use. As a result, when solutions are required and an analytical solution either is not known or is inconvenient to use, it becomes necessary to use methods that produce a numerical solution directly. However, unlike the general analytical solution of an initial value problem that can be adapted to any appropriate initial conditions, a numerical solution is the solution of a specific initial value problem, so the calculation must be repeated if the initial conditions are changed.

Many different techniques are available for the generation of a numerical solution of an initial value problem, the most powerful of which are implemented in the various numerical analysis software packages that are available. These include extrapolation methods, codes based on a family of Adams–Moulton methods, and others that use predictor–corrector methods with an Adams–Basforth method as the predictor and an Adams–Moulton method as a corrector. References for these methods are given later. In this section attention will be confined to the popular family of Runge–Kutta methods.

Predictor–corrector methods first use an explicit formula and previously computed solutions to predict a new solution. This prediction is then refined by using it in an implicit corrector formula. The Runge–Kutta methods are *one-step* methods, in the sense that the solution of a differential equation at the next step is determined solely by the solution at the previous step.

To illustrate how numerical solutions can be obtained by Runge–Kutta type methods, and to show the varying degrees of accuracy that can be attained by different approaches, a few of the simpler methods of this type will be described.

Euler's Method

The basis of this method has already been encountered in Section 5.3 when considering the **direction field** that can be associated with the first order differential equation

$$\frac{dy}{dx} = f(x, y). \tag{51}$$

Preparatory to developing Euler's method let us first recall the definition of the direction field associated with (51). At any point (x_0, y_0) in the (x, y)-plane at which $f(x, y)$ is defined, (51) shows that the slope (gradient) of the solution curve through the point is $f(x_0, y_0)$. If a short line segment is drawn through the point (x_0, y_0), making an angle θ with the positive x-axis, where $\tan \theta = f(x_0, y_0)$, the line segment will be *tangent* to the solution curve through (x_0, y_0). This line segment will define a *direction* of change of the solution at the point (x_0, y_0) if an arrow is added to the line segment indicating the sense in which y changes at that point as x increases. A repetition of this construction at a mesh of points over the region of the (x, y)-plane in which differential equation (51) is defined will then generate a **direction field** associated with the equation. Examples of direction fields have already been given in Chapter 5, and another for the linear differential equation

a typical direction field

$$\frac{dy}{dx} = \sin x - y$$

is shown in Fig. 19.11.

It is a short step from the notion of the direction field for differential equation (51) to **Euler's algorithm** for the solution of an initial value problem for the differential equation. An approximate numerical solution by Euler's method for the initial value problem

$$\frac{dy}{dx} = f(x, y), \quad \text{subject to the initial condition } y(x_0) = y_0, \tag{52}$$

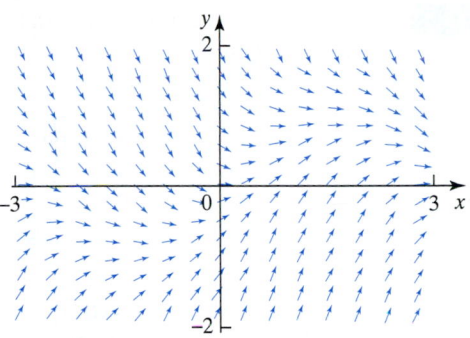

FIGURE 19.11   The direction field for $\frac{dy}{dx} = \sin x - y$.

is obtained as follows. A step size h in x is chosen, and the line segment through (x_0, y_0) is extended from x_0 to $x_0 + h$, and the y-coordinate $y_0 + \Delta y$ of the end point of the line segment is taken as the approximation to y at $x_0 + h$.

An increase in x of h from x_0 will cause the point on the tangent line approximation to the solution curve through (x_0, y_0) to increase from y_0 to $y_0 + \Delta y$, where $\Delta y = h \tan \theta$, but $\tan \theta = f(x_0, y_0)$, so $\Delta y = h f(x_0, y_0)$. It then follows that if P is the point $(x_0 + h, y_1)$ on the tangent line approximation (cf Fig. 19.4),

$$y_1 = y_0 + h f(x_0, y_0). \tag{53}$$

A repetition of this process produces a sequence of points $(x_0, y_0), (x_1, y_1), \ldots,$ $(x_n, y_n), \ldots,$ where $x_n = x_0 + nh$ and $n = 0, 1, 2, \ldots.$ When these points are joined by straight line segments, a polygonal line approximation to the solution of the initial value problem in (52) is generated, called an **Euler polygonal approximation** to the solution. The algorithm for generating such an approximate solution is easily seen to be as follows.

The Euler algorithm

finding an approximate solution by the Euler method

The approximate numerical solution of the initial value problem

$$\frac{dy}{dx} = f(x, y) \quad \text{subject to the initial condition } y(x_0) = y_0$$

generated by the Euler method with step size h is obtained from the algorithm

$$y_n = y_{n-1} + h f(x_{n-1}, y_{n-1}) \quad \text{for } n = 1, 2, \ldots,$$

where $x_n = x_0 + nh$.

This is the simplest example of a one-step method, and an obvious modification involves varying the step size from point to point, reducing it when the solution changes rapidly and lengthening it when it changes slowly. However, it is not possible to make such changes in a systematic manner without first having a way of estimating the error. This is usually done by comparing the result at each step to the result obtained by using a formula of higher order.

Use the Euler algorithm with a step size $h = 0.2$ to find an approximate solution of the linear first order initial value problem

$$\frac{dy}{dx} = \sin x - y \quad \text{with } y(0) = 1$$

in the interval $0 \le x \le 2$, and compare it with the exact solution

$$y = \frac{1}{2}(\sin x - \cos x) + \frac{3}{2}e^{-x}.$$

Solution   This is an initial value problem for the differential equation whose direction field is shown in Fig. 19.11. Setting $h = 0.2$, $n = 10$, and $f(x, y) = \sin x - y$ in the Euler algorithm leads to the following results. The column y_{exact} contains the analytical solution.

n	x_n	y_n	$0.2\,f(x_n, y_n)$	$y_{n+1} = y_n + 0.2\,f(x_n, y_n)$	y_{exact}
0	0	1	−0.2	0.8	1
1	0.2	0.8	−0.1203	0.6797	0.8374
2	0.4	0.6797	−0.0581	0.6217	0.7397
3	0.6	0.6217	−0.0114	0.6103	0.6929
4	0.8	0.6103	0.0214	0.6317	0.6843
5	1	0.6317	0.0420	0.6736	0.7024
6	1.2	0.6736	0.0517	0.7253	0.7366
7	1.4	0.7253	0.0520	0.7773	0.7776
8	1.6	0.7773	0.0444	0.8218	0.8172
9	1.8	0.8218	0.0304	0.8522	0.8485
10	2	0.8522	0.0114	0.8636	0.8657

The error between y_{n+1} and y_{exact} can be reduced, but not eliminated, by choosing a smaller step size, though for significantly greater accuracy it is necessary to make use of a different method.   ■

Modified Euler's Method

A source of error in Euler's method is its failure to take account of the curvature of the solution curve at a point (x_i, y_i) when using the tangent line approximation to the curve to estimate y_{i+1}. An improvement can be obtained by using a two-stage process to arrive at a modified gradient $\tilde{f}(x_i, y_i)$ that can be used in Euler's method in place of $f(x_i, y_i)$.

The first step when finding the modified gradient involves computing the gradient $f(x_i, y_i)$ and then using it in Euler's method to compute the gradient $f(x_{i+1}, y_{i+1})$ at the point (x_{i+1}, y_{i+1}). The second and final step involves averaging these two gradients, to obtain the new gradient

$$\tilde{f}(x_i, y_i) = \frac{1}{2}\{f(x_i, y_i) + f(x_{i+1}, y_{i+1})\}, \tag{54}$$

and then using $\tilde{f}(x_i, y_i)$ in place of $f(x_i, y_i)$ in Euler's method at (x_i, y_i) to find an improved estimate $\tilde{y}_{i+1}$ at the point (x_{i+1}, y_{i+1}). This way of computing the gradient is known as **Heun's method**, and it takes some account of the curvature of the solution curve at (x_i, y_i). The following is an algorithm for the modified Euler method.

The modified Euler algorithm

The approximate numerical solution of the initial value problem

$$\frac{dy}{dx} = f(x, y) \quad \text{subject to the initial condition } y(x_0) = y_0$$

generated by the modified Euler method with step size h is obtained from the algorithm

$$y_{n+1} = y_n + \frac{1}{2}h[\,f(x_n, y_n) + f(x_n + h,\ y_n + hf(x_n, y_n))\,]$$
$$\text{for } n = 1, 2, \ldots,$$

where $x_n = x_0 + nh$.

finding an approximate solution by the modified Euler method

EXAMPLE 19.13

Repeat Example 19.12 using the modified Euler method with $n = 10$ and $h = 0.2$, and compare the results obtained with both the Euler method and the exact solution.

Solution   The results of the calculations together with the comparisons are shown in the following table, in which results obtained using Euler's method are denoted by $y_n^{(e)}$, results obtained using Euler's modified method are denoted by $y_n^{(mod)}$, and the analytical result is denoted by y_{exact}. As the calculations are straightforward, the details have been omitted.

n	0	1	2	3	4	5	6	7	8	9	10
x_n	0	0.2	0.4	0.6	0.8	1	1.2	1.4	1.6	1.8	2
$y_n^{(e)}$	1	0.8	0.6797	0.6217	0.6103	0.6317	0.6736	0.7253	0.7773	0.8212	0.8522
$y_n^{(mod)}$	1	0.8399	0.7435	0.6973	0.6887	0.7063	0.7397	0.7796	0.8181	0.8482	0.8643
y_{exact}	1	0.8374	0.7397	0.6929	0.6843	0.7024	0.7366	0.7776	0.8172	0.8485	0.8657

A comparison of the results in last three rows of the table shows the improvement in accuracy obtained when the modified Euler method is used.   ∎

Euler's method is effectively a Taylor series expansion of the solution $y(x)$, in which $y(x_n + h)$ is predicted from $y(x_n)$ using only the first two terms of the Taylor series expansion of $y(x)$ about the point x_n. An often-used general purpose numerical method for the integration of initial value problems for first order differential equations is the Runge–Kutta fourth order method.

There are several families of four-stage, fourth order Runge–Kutta formulas in which the error after a step size h is of the order h^5, but as their derivation involves tedious algebra we will simply describe the most familiar one. However, before quoting this method, we first demonstrate the general approach to the derivation of Runge–Kutta methods by finding the modified Euler method.

In essence, all Runge–Kutta methods are one-step methods that can be considered to be of the form

$$y_{i+1} = y_i + hF(x_i, y_i, h), \tag{55}$$

where $F(x_i, y_i, h)$ represents some form of *averaged* value of $f(x, y)$ over the interval $x_i \leq x \leq x_{i+1}$. All of these methods can be obtained by adopting a particular form of F that contains some undetermined constants, and then finding the equations determining the constants by requiring that F agree with the Taylor series expansion of f up to a certain power of h.

In the case where F contains terms up to order h, so the error at each step will be of order h^2, using the chain rule and the fact that $f(x, y) = f(x, y(x))$, the function F in (55) is approximated by the truncated Taylor series expansion

a Runge–Kutta type derivation of the modified Euler method

$$F(x, y, h) = f(x, y) + \frac{1}{2}h\left\{ \frac{\partial f}{\partial x} + \frac{\partial f}{\partial y}\frac{dy}{dx} \right\},$$

but $dy/dx = f(x, y)$, so

$$F(x, y, h) = f(x, y) + \frac{1}{2}h\{ f_x(x, y) + f_y(x, y) f(x, y) \}. \tag{56}$$

We now seek a representation of the function F of the form

$$F(x, y, h) = w_1 f(x, y) + w_2 f(x + w_3 h, y + w_4 h f(x, y)), \tag{57}$$

where as yet the constants w_1 to w_4 are unknown. Expanding $f(x + w_3 h, y + w_4 h f(x, y))$ about the point (x, y) as a two-variable Taylor series with a remainder after the first derivative terms gives

$$\begin{aligned} f(x + w_3 h, y + w_4 h(f(x, y))) &= f(x, y) + w_3 h f_x(x, y) \\ &\quad + w_4 h f_y(x, y) f(x, y) + R(h), \end{aligned} \tag{58}$$

where the error term $R(h)$ is of order h^2.

Substituting (58) into (57) and combining terms gives

$$F(x, y, h) = (w_1 + w_2) f(x, y) + h(w_2 w_3 f_x(x, y) + w_2 w_4 f_y(x, y) f(x, y)). \tag{59}$$

If (57) and (59) are required to agree up to terms in h, by equating terms with corresponding powers of h we find that

$$w_1 + w_2 = 1, \quad w_2 w_3 = \frac{1}{2}, \quad \text{and} \quad w_2 w_4 = \frac{1}{2}.$$

These three equations relate the four arbitrary constants w_1 to w_4, so if one of these constants, say w_2, is assigned arbitrarily, the others will be determined in terms of w_2. From (57) we then have

$$F(x, y, h) = (1 - w_2) f(x, y) + w_2 f\left(x + \frac{1}{2}h/w_2, y + \frac{1}{2}hf(x, y)/w_2 \right). \tag{60}$$

Making the choice $w_2 = \frac{1}{2}$ in (60), and using it in (55), gives the modified Euler method

$$y_{i+1} = y_i + \frac{1}{2}h\{ f(x_i, y_i) + f(x_i + h, y_i + h \ f(x_i, y_i)) \}. \tag{61}$$

The fourth order Runge–Kutta method for a first order differential equation

The approximate numerical solution of the initial value problem

$$\frac{dy}{dx} = f(x, y) \text{ subject to the initial condition } y(x_0) = y_0$$

with step length h is obtained from the following fourth order Runge–Kutta
algorithm, with $x_n = x_0 + nh$ and $y_n = y(x_n)$.

STEP 1   Calculate

$$k_{1n} = hf(x_n, y_n)$$

$$k_{2n} = hf\left(x_n + \frac{1}{2}h, y_n + \frac{1}{2}k_{1n}\right)$$

$$k_{3n} = hf\left(x_n + \frac{1}{2}h, y_n + \frac{1}{2}k_{2n}\right)$$

$$k_{4n} = hf(x_n + h, y_n + k_{3n}).$$

STEP 2   Calculate

$$d_n = \frac{1}{6}(k_{1n} + 2k_{2n} + 2k_{3n} + k_{4n}).$$

STEP 3   The numerical approximation y_{n+1} of the solution $y = y(x_{n+1})$ is
given by

$$y_{n+1} = y_n + d_n,$$

for $n = 1, 2, \ldots.$

EXAMPLE 19.14   Use the fourth order Runge–Kutta algorithm with a step size $h = 0.2$ to solve the
initial value problem

$$\frac{dy}{dx} + 2y = \sin 3x \quad \text{with } y(0) = 1$$

in the interval $0 \leq x \leq 2.4$. Compare the results obtained with the results found by the modified Euler method and the analytical solution

$$y = \frac{1}{13}[9\cos x - 2\sin x + 4\sin 2x \cos x - 12\cos^3 x + 16e^{-2x}].$$

Solution   In the following calculations $f(x, y) = \sin 3x - 2y$, the step length $h = 0.2$, so as the solution is required in the interval $0 \le x \le 2.4$ it follows that $n = 0, 1, \ldots, 12$. The details of the intermediate calculations for $x = 0, 0.2, 0.4$, and 0.6 are listed in the first of the following tables. Under the heading y_{rk}, the second table lists all of the results obtained by the Runge–Kutta algorithm up to $x = 2.4$, and for purposes of comparison the columns with headings y_{mod} and y_{exact} show the results obtained by using the modified Euler method and the analytical solution, respectively.

Detailed Calculations for $x = 0, 0.2$, and 0.4

n	x_n	y_n	$f(x_n, y_n)$	k_{1n}	k_{2n}	k_{3n}	k_{4n}	y_{n+1}
0	0	1	-2	-0.4	-0.2609	-0.28872	-0.17158	0.72153
1	0.2	0.72153	-0.87842	-0.17568	-0.09681	-0.11258	-0.05717	0.61292
2	0.4	0.61292	-0.29380	-0.05876	-0.03392	-0.03889	-0.03484	0.57305
3	0.6	0.57305	—	—	—	—	—	—

Comparison of Results in the Interval $0 \le x \le 2.4$

n	x_n	y_{rk}	y_{mod}	y_{exact}	n	x_n	y_{rk}	y_{mod}	y_{exact}
0	0	1.0	1.0	1.0	7	1.4	0.05390	0.05090	0.05389
1	0.2	0.72153	0.73646	0.72142	8	1.6	-0.12324	-0.11730	-0.12328
2	0.4	0.61292	0.62788	0.61279	9	1.8	-0.23165	-0.21681	-0.23173
3	0.6	0.57305	0.58026	0.57295	10	2.0	-0.24192	-0.22174	-0.24202
4	0.8	0.52262	0.52056	0.52257	11	2.2	-0.15615	-0.13639	-0.15624
5	1.0	0.41675	0.40862	0.41674	12	2.4	-0.00809	-0.00531	-0.00816
6	1.2	0.25051	0.24208	0.25052	—	—	—	—	—

The fourth order Runge–Kutta algorithm is easily adapted to solve two simultaneous first order differential equations or, as a special case, a single second order differential equation as follows.

The fourth order Runge–Kutta algorithm for two first order simultaneous equations

The approximate numerical solution of the initial value problem for the simultaneous first order initial value problem

$$\frac{dy}{dx} = f(x, y, z) \quad \text{and} \quad \frac{dz}{dx} = g(x, y, z)$$

subject to the initial conditions

$$y(x_0) = y_0 \quad \text{and} \quad z(x_0) = z_0$$

generated by the fourth order Runge–Kutta method with step size h is obtained from the following algorithm in which $x_n = x_0 + nh$, $y_n = y(x_n)$, and $z_n = z(x_n)$.

STEP 1   Calculate in the following order

$$k_{1n} = hf(x_n, y_n, z_n) \qquad\qquad K_{1n} = hg(x_n, y_n, z_n)$$

$$k_{2n} = hf\left(x_n + \tfrac{1}{2}h, y_n + \tfrac{1}{2}k_{1n}, z_n + \tfrac{1}{2}K_{1n}\right) \qquad K_{2n} = hg\left(x_n + \tfrac{1}{2}h, y_n + \tfrac{1}{2}k_{1n}, z_n + \tfrac{1}{2}K_{1n}\right)$$

$$k_{3n} = hf\left(x_n + \tfrac{1}{2}h, y_n + \tfrac{1}{2}k_{2n}, z_n + \tfrac{1}{2}K_{2n}\right) \qquad K_{3n} = hg\left(x_n + \tfrac{1}{2}h, y_n + \tfrac{1}{2}k_{2n}, z_n + \tfrac{1}{2}K_{2n}\right)$$

$$k_{4n} = hf(x_n + h, y_n + k_{3n}, z_n + K_{3n}) \qquad K_{4n} = hg(x_n + h, y_n + k_{3n}, z_n + K_{3n}).$$

STEP 2   Calculate

$$d_n = \frac{1}{6}(k_{1n} + 2k_{2n} + 2k_{3n} + k_{4n}) \quad \text{and}$$

$$D_n = \frac{1}{6}(K_{1n} + 2K_{2n} + 2K_{3n} + K_{4n}).$$

STEP 3   The numerical approximations of the solutions $y = y(x_{n+1})$ and $z = z(x_{n+1})$ are given by

$$y_{n+1} = y_n + d_n \quad \text{and} \quad z_{n+1} = z_n + D_n,$$

for $n = 1, 2, \ldots$.

This fourth order Runge–Kutta algorithm with step size h is easily modified to find the solution of the following initial value problem for the single second order differential equation written in the standard form

adapting the Runge–Kutta method to solve second order equations

$$\frac{d^2 y}{dx^2} = g\left(x, y, \frac{dy}{dx}\right) \quad \text{with } y(x_0) = y_0 \text{ and } z(x_0) = z_0. \tag{62}$$

All that is necessary is to reduce the second order equation to a system of two simultaneous first order equations by setting

$$\frac{dy}{dx} = z \quad \text{and} \quad \frac{dz}{dx} = g(x, y, z) \tag{63}$$

in the preceding fourth order Runge–Kutta algorithm, and then to use the initial conditions

$$y(x_0) = y_0 \quad \text{and} \quad z(x_0) = y'(x_0) = z_0. \tag{64}$$

EXAMPLE 19.15   Use the fourth order Runge–Kutta algorithm with step length 0.1 to find a numerical approximation to the solution of the initial value problem for the Hermite equation

$$y'' - 2xy' + 8y = 0 \quad \text{with } y(0) = 12 \quad \text{and} \quad y'(0) = 0$$

in the interval $0 \le x \le 1$. Compare the results of the calculations with the analytical solution $y(x) = 16x^4 - 48x^2 + 12$.

Solution   This is the Hermite equation with $n = 4$, and it has the analytical solution $H_4(x) = 16x^4 - 48x^2 + 12$. Using (62) and (63) we set $z = dy/dx$ and $g(x, y, z) = 2xz - 8y$, and use the step size $h = 0.1$. The initial conditions are imposed at the origin, so $x_0 = 0$, $y(x_0) = 12$, and $z(x_0) = y'(x_0) = 0$, corresponding to $y_0 = 12$ and $z_0 = 0$. The details of the intermediate calculations for $x = 0$ and 0.1 are set out below; the table that follows lists the results for the interval $0 \le x \le 1$, with the second order Runge–Kutta solution denoted by y_{rk} and the analytical solution by y_{exact}.

$\underline{x_0 = 0}$

$f(x_0, y_0, z_0) = 0$, $g(x_0, y_0, z_0) = -96$, $k_1 = 0$, $K_1 = -9.6$, $k_2 = -0.48$

$K_2 = -9.648$, $k_3 = -0.4824$, $K_3 = -9.45624$, $k_4 = -0.945624$

$K_4 = -9.403205$, $d = -0.478404$, $D = -9.535281$ so that

$y_1 = 11.521596$ and $z_1 = -9.535281$,   where $z_1 = y'(x_1)$.

$\underline{x_1 = 0.2}$

$f(x_1, y_1, z_1) = -9.535281$, $g(x_1, y_1, z_1) = -94.079824$, $k_1 = -0.953528$,

$K_1 = -9.407982$, $k_2 = -1.423927$, $K_2 = -9.263044$, $k_3 = -1.416680$,

$K_3 = -9.072710$, $k_4 = -1.860799$, $K_4 = -8.828252$, $d = -1.415924$,

$D = -9.151290$ so that $y_2 = 10.105672$ and

$z_2 = -18.686571$,   where $z_2 = y'(x_2)$.

Comparison of Solutions for $0 \le x \le 1$

n	x_n	y_{rk}	y_{exact}	n	x_n	y_{rk}	y_{exact}
0	0	12	12	6	0.6	−3.205311	−3.2064
1	0.1	11.521596	11.5216	7	0.7	−7.676938	−7.6784
2	0.2	10.105672	10.1056	8	0.8	−12.164555	−12.1664
3	0.3	7.809827	7.8096	9	0.9	−16.380188	−16.3824
4	0.4	4.730055	4.7296	10	1.0	−19.997470	−20.0
5	0.5	1.000747	1.0	—	—	—	—

the F(4,5) adaptive step size algorithm

When the solution of a differential equation changes rapidly in some intervals, and slowly in others, it becomes necessary to vary the step size as the calculation progresses if accuracy is to be maintained. The **F(4,5) Runge–Kutta–Fehlberg algorithm**, based on a form of the fourth order Runge–Kutta scheme, is implemented in many numerical analysis software programs that are readily available, and it determines the step size at each stage of the calculation. The increase in complexity of the calculation is indicated by the fact that the **F(4,5) algorithm** uses six stages in the calculation in place of the four used by the classical fourth order Runge–Kutta algorithm. As the calculation proceeds, numerical estimates of the solution after a given step size h are made using a form of the fourth order Runge–Kutta method and an efficient fifth order formula. The difference of these two estimates is compared with a preassigned tolerance, and the result is then used to either reduce or increase the step size until the difference lies within the required tolerance. The resulting step size is then used to advance the calculation to the next stage.

More detailed information about the numerical integration schemes for ordinary differential equations can be found in references [2.19], [2.20], and [3.20] through [3.26].

Summary

Of the many methods available for the numerical integration of ordinary differential equations, at an elementary level only the Euler and modified Euler methods have been described. For greater accuracy the classical fourth order Runge–Kutta algorithm, which belongs to a family of similar algorithms, was presented without derivation, though the form of argument used was illustrated by deriving the modified Euler method. Finally, the important adaptive F(4,5) Runge–Kutta–Fehlberg algorithm was mentioned that adjusts the step size automatically as the calculation progresses in order to preserve a preassigned accuracy.

EXERCISES 19.7

Solve the following initial value problems by computer using the fourth order Runge–Kutta algorithm.

1. $y' = (3x^2 + y^2)^{1/2} - y$ with $y(2) = 0$ and $h = 0.2$ over the interval $2 \le x \le 3$.

2. $y' = xy/(x^2 + y^2)^{1/2}$ with $y(1) = 1$ and $h = 0.2$ over the interval $1 \le x \le 2$.

3. $y' = (x^2 + y^2)^{1/2} - xy$ with $y(1) = 2$ and $h = 0.2$ over the interval $1 \le x \le 2$.

4. $y' = \frac{1}{2}(x^2 + 2y^2) - xy$ with $y(1) = 0$ and $h = 0.1$ over the interval $1 \le x \le 1.5$.

5. $y' = \cos(2x + y) - 3y$ with $y(1) = 1$ and $h = 0.2$ over the interval $1 \le x \le 2$.

6. $y' = \sin(x + y) - 2y$ with $y(0) = 1$ and $h = 0.2$ over the interval $0 \le x \le 1$.

7. $y'' - xyy' + 2y = 0$ with $y(0) = 2$, $y'(0) = -1$, and $h = 0.1$ over the interval $0 \le x \le 0.5$.

8. $y'' + (3 + x)y' + y^2 = 0$ with $y(1) = 1$, $y'(1) = 2$, and $h = 0.1$ over the interval $1 \le x \le 1.5$.

9. $y'' + (1 + \sin 2x)y' + 3y = 0$ with $y(0) = 1$, $y'(0) = 1$, and $h = 0.1$ over the interval $0 \le x \le 0.5$.

10. $y'' + (1 + y^2)^{1/2}y' + y = 0$ with $y(2) = 0$, $y'(2) = 1$, and $h = 0.1$ over the interval $2 \le x \le 2.5$.

11. $y'' + 2y' - y^2 = 0$ with $y(0) = 2$, $y'(0) = 1$, and $h = 0.2$ over the interval $0 \le x \le 1$.

12. $y'' - xy' - y^2 = 0$ with $y(0) = -1$, $y'(0) = 2$, and $h = 0.2$ over the interval $0 \le x \le 1$.

13. $y'' + yy' - 3y = 0$ with $y(1) = 1$, $y'(1) = 1$, and $h = 0.2$ over the interval $1 \le x \le 2$.

14. $y'' + x^2 \sin y' - 2y = 0$ with $y(1) = 0$, $y'(1) = -1$, and $h = 0.2$ over the interval $1 \le x \le 2$.

15. $y'' - xy' - y^2 = 2x$ with $y(0) = -2$, $y'(0) = 1$, and $h = 0.2$ over the interval $0 \le x \le 1$.

16. $y'' + 2yy' - 3y = 1 - x^2$ with $y(0) = 3$, $y'(0) = 2$, and $h = 0.2$ over the interval $0 \le x \le 1$.

17. $\frac{dx}{dt} = tx + (x + y)y$ and $\frac{dy}{dt} = ty - (x + y)x$ with $x(0) = 1$, $y(0) = 0$, and $h = 0.2$ over the interval $0 \le t \le 1$.

18. $\frac{dx}{dt} = (1 + t)y^2 - 2x$ and $\frac{dy}{dt} = y^2 + tx$ with $x(0) = -1$, $y(0) = -3$, and $h = 0.2$ over the interval $0 \le t \le 1$.

19. $\frac{dx}{dt} = \sin(x + 4y)$ and $\frac{dy}{dt} = 2\cos(x - 3y)$ with $x(0) = 1$, $y(0) = 1$, and $h = 0.2$ over the interval $0 \le t \le 1$.

20. $\frac{dx}{dt} = \sin x + 4\cos y$ and $\frac{dy}{dt} = \sin y - 3\sin x$ with $x(0) = 1$, $y(0) = -2$, and $h = 0.2$ over the interval $0 \le t \le 1$.

CHAPTER 19
TECHNOLOGY PROJECTS

Project 1

Spline Function Approximation

This project uses a spline function computer package to generate a natural spline approximation to a given data set. The data provided can be considered to be the scaled set of nine points through which the profile of the side elevation of a yacht hull complete with its keel must pass.

1. Make and plot a natural cubic spline function approximation to the following set of data points, where in each number pair the first number represents the x-coordinate and the second the y-coordinate:

   $(0, 0), (4.5, -2.3), (10, -3.7), (12.3, -6.8),$
   $(16.7, -6.8), (18.4, -3.4), (21.2, -2.3), (23, 0).$

2. Design a different profile of your own involving at least nine number pairs. Construct and plot a corresponding spline function approximation, and compare the result with the original profile. If necessary, reposition the data points to make the approximation a better fit.

Project 2

Newton's Method

The purpose of this project is to construct a procedure for Newton's method, and then to use it to determine the zeros of two expressions involving Bessel functions.

1. Plot $f(x) = J_2(x)$ for $0 \leq x \leq 35$ and use the graph to determine the approximate zeros of $J_2(x)$ in this interval, the first six of which are listed in Table 8.1. Construct a procedure for Newton's method involving 10 iterations and use it with the approximate values found from the graph to determine the zeros of $f(x)$ to 10 decimal places. Print out the values of these

zeros together with the value of $f(x)$ at each zero to confirm the accuracy.

2. Repeat some of the previous calculations using poorer initial approximations to experience how sometimes the calculation does not converge to the expected zero and sometimes it diverges. Notice that this numerical method only works when $f'(x)$ can be found analytically.

3. The eigenvalues of a certain problem are determined by the zeros of the expression

   $$J_0(x)J_1(1.5x) - J_0(1.5x)J_1(x) = 0.$$

   Plot $f(x) = J_0(x)J_1(1.5x) - J_0(1.5x)J_1(x)$ in the interval $0 \leq x \leq 20$ and determine from the graph the approximate values of the first three positive zeros of $f(x)$. Use the procedure developed in part 1 with these approximate zeros to find their values to 10 decimal places.

Project 3

Modified Euler and Runge–Kutta Methods

The purpose of this project is to construct procedures for the modified Euler and the fourth order Runge–Kutta method and then to compare the results obtained when they are applied first to a simple linear initial value problem and then to a nonlinear initial value problem.

1. Construct a procedure for the modified Euler method derived in Section 19.7.

2. Construct a procedure for the fourth order Runge–Kutta method defined as follows:

   Consider the differential equation $dy/dx = f(x, y)$, and let the initial condition at $x = x_0$ be $y(x_0) = y_0$. Let the step size be h and $y_1, y_2, \ldots, y_r$ be the approximations to $y(x)$ at the respective points $x_1 = x_0 + h, x_2 = x_0 + 2h, \ldots, x_r = x_0 + rh$. Then, for $n = 0, 1, \ldots,$ the values $y_1, y_2, \ldots$ are determined from the algorithm

$$k_1 = hf(x_n, y_n)$$

$$k_2 = hf\left(x_n + \frac{1}{2}h, y_n + \frac{1}{2}k_1\right)$$

$$k_3 = hf\left(x_n + \frac{1}{2}h, y_n + \frac{1}{2}k_2\right)$$

$$k_4 = hf(x_n + h, y_n + k_3)$$

with $x_{n+1} = x_n + h$ and

$$y_{n+1} = y_n + \frac{1}{6}(k_1 + 2k_2 + 2k_3 + k_4).$$

3. Apply both methods to the linear initial value problem $dy/dx = y$ with $y(0) = 1$ and $h = 0.1$. Print out the results for the interval $0 \le x \le 1$ and compare them with the exact solution $y(x) = e^x$.

4. Apply both methods to the nonlinear initial value problem

$$dy/dx = \sin(xy)\sin(3x), \quad \text{with}$$
$$y(0) = 1 \text{ and } h = 0.1,$$

and compare the results over the interval $0 \le x \le 2$.

Project 4

The Shooting Method

This project provides an introduction to the **shooting method** when used to solve a two-point boundary value problem for a linear second order differential equation. The underlying idea of the method can be likened to the problem of projecting a particle from a fixed point at different angles to the horizontal, and finding the angle of projection at which the particle attains a prescribed altitude when at a fixed horizontal distance from its point of origin.

Consider the two-point boundary value problem

$$\frac{d^2y}{dx^2} + P(x)\frac{dy}{dx} + Q(x)y = R(x),$$

with $y(a) = k$ and $y(b) = K$ $(b > a)$,

where a, b, k, and K are given numbers.

Now consider two initial value problems with the different initial conditions

(I) $y(a) = k$ and $y'(a) = K_1$

and

(II) $y(a) = k$ and $y'(a) = K_2,$

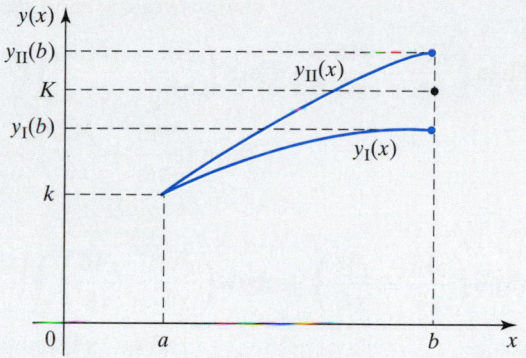

FIGURE 19.12 The two solutions $y_I(x)$ and $y_{II}(x)$.

where for the moment numbers $K_1 \ne K_2$ are specified arbitrarily. If the corresponding solutions are $y_I(x)$ and $y_{II}(x)$, their typical behavior is shown in Fig. 19.12, where the value $y(b) = K$ necessary to satisfy the original two-point boundary value problem is shown as the point (b, K).

Now set $y(x) = K_1 y_I(x) + K_2 y_{II}(x)$, with $K_1 + K_2 = 1$. Then substituting this result into the differential equation shows that it is a solution and, in addition, that $y(x)$ satisfies the boundary condition $y(a) = k$. Setting $x = b$ and $y(b) = K$ in $y(x)$ gives

$$K = K_1 y_I(b) + K_2 y_{II}(b),$$

so using the condition $K_1 + K_2 = 1$, solving for K_1 and K_2, and substituting the results into $y(x)$ shows that the solution of the two-point boundary value problem is given by

$$y(x) = \left[\frac{K - y_{II}(b)}{y_I(b) - y_{II}(b)}\right] y_I(x)$$
$$+ \left[\frac{y_I(b) - K}{y_I(b) - y_{II}(b)}\right] y_{II}(x).$$

Using the fourth order Runge-Kutta method to find $y_I(x)$ and $y_{II}(x)$, apply this method to the two-point boundary value problem

$$2x^2\frac{d^2y}{dx^2} - 7x\frac{dy}{dx} + 10y = 3x,$$
$$\text{with } y(1) = 1 \text{ and } y(2) = 4,$$

and find the solution for $1 \le x \le 2$ at step increments of 0.2. Compare your result with the analytical solution

$$y(x) = x + \frac{x^2(1 - \sqrt{x})}{2 - 2\sqrt{2}}, \quad \text{for } 1 \le x \le 2.$$

Least Squares Fitting of Data

Instead of Lagrange or spline interpolation between known data points, it is sometimes better to fit an expression of the form

$$P(x) = a_0\varphi_0(x) + a_1\varphi_1(x) + \cdots + \varphi_m(x),$$

where the $\varphi_0(x), \varphi_1(x), \ldots, \varphi_m(x)$ is some convenient set of functions. In the method of least squares, the function $P(x)$ is fitted to the set of data points $(x_0, y_0), (x_1, y_1), \ldots, (x_n, y_n)$ by setting

$$S(a_0, a_1, \ldots, a_m) = \sum_{i=0}^{n}[P(x_j) - y_i]^2,$$

and requiring this sum of squares of errors between $P(x)$ at the points x_i and the corresponding numbers y_i to be minimized.

A typical case involves fitting a quadratic in x to the data points, so $\varphi_r(x) = x^r$ and

$$P(x) = a_0 + a_1x + a_2x^2.$$

The method of least squares then requires the sum $S(a_0, a_1, a_2)$ to be minimized, where

$$S(a_0, a_1, a_2) = \sum_{i=0}^{n}\left[a_0 + a_1x_i + a_2x_i^2 - y_i\right]^2.$$

Regarding the coefficients a_0, a_1, a_2 as parameters, the extremum of the square error S will be found by taking the coefficients to be the solutions of the three equations $\partial S/\partial a_j = 0$; that is, by finding a_0, a_1, and a_2 from the three linear nonhomogeneous equations

$$a_0\sum_{r=0}^{n}x_r^j + a_1\sum_{r=0}^{n}x_r^{j+1} + a_2\sum_{r=0}^{n}x_r^{j+2} = \sum_{r=0}^{n}x_r^j y_r,$$
$$\text{for } j = 0, 1, 2.$$

Substituting the coefficients a_0, a_1, and a_2 into $P(x)$ then gives the required least squares fit.

(a) Define a function $f(x)$ that can reasonably be approximated by $P(x) = a_0 + a_1x + a_2x^2$ over an interval $x_0 \leq x \leq x_n$. For some arbitrary increasing set of points $x_0, x_1, \ldots, x_n$, typically with $n = 20$, set $y_j = f(x_j)$. Using the points $(x_0, y_0), (x_1, y_1), \ldots, (x_n, y_n)$ as data points, make a least squares fit of $P(x)$. Plot $P(x)$ and the data points together to show the nature of fit that is obtained. Examine how changing the set of points $x_0, x_1, \ldots, x_n$ alters the nature of the fit.

(b) Extend the preceding analysis using $P(x) = a_0 = a_1x + a_2x^2 + a_3x^3$. Repeat the calculations in (a), but this time using a function $f(x)$ that can reasonably be approximated by a cubic. Again plot $P(x)$ and the original set of data points together to show the nature of the fit. Again examine how changing the set of points $x_0, x_1, \ldots, x_n$ alters the nature of the fit.

ANSWERS

Exercise Set 1.1

1. Consider $a/\sqrt{b}+b/\sqrt{a}-\sqrt{a}-\sqrt{b}=[a-\sqrt{(ab)}]/$
$\sqrt{b}+[b-\sqrt{(ab)}]/\sqrt{a}=(a-b)(\sqrt{a}-\sqrt{b})/$
$\sqrt{(ab)}\geq 0$. Numerator and denominator have the same sign, so the result follows.

3. $P(n)$ is the stated proposition and $P(1)$ is true. $(1-r^n)/(1-r)+r^n=(1-r^{n+1})/(1-r)$ so $P(n)$ implies $P(n+1)$, but $P(1)$ is true so $P(n)$ is true for $n \geq 1$.

5. Use the same form of argument as in Example 1.1. A quick noninductive proof follows from Example 1.1 by replacing ax by $ax + \pi/2$.

7. $81 + 216x + 216x^2 + 96x^3 + 16x^4$

9. $\frac{1}{9} - \frac{4}{27}x + \frac{4}{27}x^2 - \frac{32}{243}x^3 + \cdots, |x| < \frac{3}{2}$

11. $\frac{1}{2} - \frac{1}{8}x^2 + \frac{3}{64}x^4 - \frac{5}{256}x^6 + \cdots, |x| < \sqrt{2}$

Exercise Set 1.2

1. $-\frac{1}{2} \pm i\frac{\sqrt{3}}{2}$

3. $-\frac{1}{2} \pm i\frac{\sqrt{23}}{2}$

5. $-\frac{1}{2} \pm i\frac{\sqrt{3}}{6}$

7. $-\frac{1}{4} \pm i\frac{\sqrt{31}}{4}$

9. $a = 5, b = -40$

11. $\sqrt{10}, 4 - i, -7 - 3i, 8 - i,$
$-30 - 45i, \sqrt{65}/5, \frac{1}{5} + i\frac{4}{15}$

Exercise Set 1.3

1. $u + v = 3 + i, u - v = 1 + 5i$

3. $u + v = -6 - 4i, u - v = 4i$

5. $u + v = -1 + 8i, u - v = 7 + 4i$

7. $u + v = -8 - 8i, u - v = 12i$

Exercise Set 1.4

1. Straightforward

3. Expand the left side of the identity $(\cos\theta + i\sin\theta)^5 = \cos 5\theta + i\sin 5\theta$ and then equate real and imaginary parts to obtain $\cos 5\theta = \cos^5\theta - 10\cos^3\theta\sin^2\theta + 5\cos\theta\sin^4\theta$ and $\sin 5\theta = 5\cos^4\theta\sin\theta - 10\cos^2\theta\sin^3\theta + \sin^5\theta$.

5. Straightforward

7. $z^n + 1/z^n = \exp(in\theta) + \exp(-in\theta) = 2\cos n\theta$, so $\cos n\theta = \frac{1}{2}(z^n + 1/z^n)$ and, similarly, $\sin n\theta = \frac{1}{2i}(z^n - 1/z^n)$, and with $n = 1$, $\cos\theta = \frac{1}{2}(z + 1/z)$ and $\sin\theta = \frac{1}{2i}(z - 1/z)$. Thus $\cos^3\theta\sin^3\theta = (1/2)^3(z + 1/z)^3(1/2i)^3(z - 1/z)^3$. Expanding, grouping terms, and using the above results gives $\cos^3\theta\sin^3\theta = \frac{3}{32}\sin 2\theta - \frac{1}{32}\sin 6\theta$.

9. Proceed as in Exercise 7.

11. Proceed as in Exercise 7.

13. $8\sqrt{2}\exp(i\pi/12), \sqrt{2}/4\exp(7i\pi/12),$
$128\exp(-2\pi i/3)$

15. $24\exp(-i\pi/3), 2/3\exp(i\pi/3), \sqrt{2}/32\exp(-i\pi/4)$

17. $64, \pi/2$

19. Multiply numerator and denominator on the right of Exercise 18 by $e^{i\theta/2}$ to obtain $\sum_{k=1}^{n}\exp(ik\theta) = \frac{\exp[i(n+\frac{1}{2})\theta]-\exp(\frac{1}{2}i\theta)}{\exp(\frac{1}{2}i\theta)-\exp(-\frac{1}{2}i\theta)}$. The Lagrange identity follows by equating the real parts of this identity.

Exercise Set 1.5

1. $\pm\left\{\left(\frac{\sqrt{2}-1}{2}\right)^{1/2} + i\left(\frac{\sqrt{2}+1}{2}\right)^{1/2}\right\}$

3. $\pm\frac{1}{\sqrt{2}}(1 + i)$

5. $\pm\left\{\left(\frac{\sqrt{13}+2}{2}\right)^{1/2} - i\left(\frac{\sqrt{13}-2}{2}\right)^{1/2}\right\}$

7. $\pm(1/\sqrt{2})(3 - i)$

9. $2^{1/3}\exp(\pi i/9), 2^{1/3}\exp(7\pi i/9), 2^{1/3}\exp(13\pi i/9)$

11. $-(1/\sqrt{2})(1 + i), (1/\sqrt{2})(1 - i), (1/\sqrt{2})(-1 + i),$
$(1/\sqrt{2})(1 + i)$

13. $i, -(1/2)(\sqrt{3} + i), (1/2)(\sqrt{3} - i)$

15. $0, [(\sqrt{2}+1)/2]^{1/2} - i[(\sqrt{2}-1)/2]^{1/2},$
$-[(\sqrt{2}+1)/2]^{1/2} + i[(\sqrt{2}-1)]^{1/2}$

17. ω may be any nth root of unity. Choose $\omega = \exp(2\pi i/n)$ and substitute for ω. The first result follows by equating the real parts of the expression and the second by equating the imaginary parts.

19. $1, 2 - 3i, 2 + 3i$

21. The polynomial has complex coefficients, so its roots do not occur in complex conjugate pairs. $z_{\pm} = \pm[(1/\sqrt{2} + 1/2)^{1/2} - i(1/\sqrt{2} - 1/2)^{1/2}]$

Exercise Set 1.6

1. $\dfrac{5}{3}\dfrac{1}{2x+1} + \dfrac{2}{3}\dfrac{1}{x+2}$

3. $\dfrac{29}{2x+5} - \dfrac{13}{x+2}$

5. $\dfrac{1}{x+2} - \dfrac{1}{(x+2)^2}$

7. $1 - \dfrac{4}{x+2} + \dfrac{5}{(x+2)^2}$

9. $(x+2)^2 + 1$

11. $2(x+3/4)^2 - 57/8$

13. $9(x-1/9)^2 + 17/9$

Exercise Set 1.7

1. 18

3. 21

5. 0

7. 1

11. $x_1 = 10/23,$
$x_2 = 15/23,$
$x_3 = -6/23$

Exercise Set 1.13

1. In Theorem 1.10 set $n = 3$ and make the identifications $x_1 = x$, $x_2 = y$, $x_3 = z$, $u_1 = r$, $u_2 = \theta$, $u_3 = z$, $X_1 = r\cos\theta$, $X_2 = r\sin\theta$, $X_3 = z$ and substitute into the theorem.

3. In Theorem 1.10 set $n = 3$ and make the identifications $x_1 = x$, $x_2 = y$, $x_3 = z$, $u_1 = r$, $u_2 = \theta$, $u_3 = \phi$, $X_1 = r\sin\theta\cos\phi$, $X_2 = r\sin\theta\sin\phi$, $X_3 = r\cos\theta$ and substitute the results into the theorem.

Exercise Set 2.1

3. (a) $-(3/2)\mathbf{i} - \mathbf{j} - 3\mathbf{k}$ (b) $2\mathbf{i} - 9\mathbf{j} - 9\mathbf{k}$

5. $\underline{AB} = -\mathbf{i} - \mathbf{j} + 5\mathbf{k}$, unit vector is $(1/\sqrt{27})$ $(-\mathbf{i} - \mathbf{j} + 5\mathbf{k})$

7. $\underline{AB} = \mathbf{b} - \mathbf{a}$, so the unit vector in this direction is $\hat{\mathbf{v}} = (\mathbf{b} - \mathbf{a})/|\mathbf{b} - \mathbf{a}|$. Divide AB into $m + n$ parts of length $|\mathbf{b} - \mathbf{a}|/(m+n)$, then AP $= m|\mathbf{b} - \mathbf{a}|/(m+n)$, so $\underline{AP} = AP\,\hat{\mathbf{v}} = m(\mathbf{b} - \mathbf{a})/ (m+n)$. As $\underline{OP} = \underline{OA} + \underline{AP}$ we have $\mathbf{r} = \mathbf{a} + m(\mathbf{b} - \mathbf{a})/(m+n) = (n\mathbf{a} + m\mathbf{b})/(m+n)$.

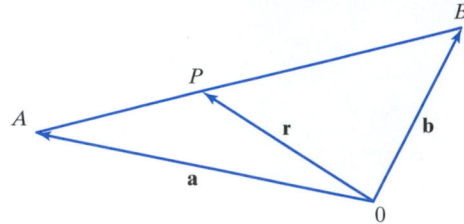

9. Use the same form of argument as in Exercise 7 with M the mid-point of AC. Hence, show that $\underline{AM} = (\mathbf{c} - \mathbf{a})/2$, $\underline{OM} = \underline{OA} + \underline{AM} = (\mathbf{a} + \mathbf{c})/2$ and $\underline{MB} = \underline{OB} - \underline{OM} = \mathbf{b} - (\mathbf{a} + \mathbf{c})/2$. If P is $1/3$ the distance along MB from M, $\underline{MP} = \underline{MB}/3$. Position vector $\underline{OP} = \underline{OM} + \underline{MP} = (1/3)(\mathbf{a} + \mathbf{b} + \mathbf{c})$. A similar argument yields the identical result using the other two mid-points of sides of the triangle, so the result is proved.

11. Let the forces along the x, y, and z axes be $\mathbf{F}_1$, $\mathbf{F}_2$, and $\mathbf{F}_3$. Then $\mathbf{F}_1 = 2\mathbf{i}$, $\mathbf{F}_2 = \mathbf{j}$, and $\mathbf{F}_3 = 4\mathbf{k}$, so $\mathbf{S} = \mathbf{F}_1 + \mathbf{F}_2 + \mathbf{F}_3 = 2\mathbf{i} + \mathbf{j} + 4\mathbf{k}$, $\|\mathbf{S}\| = \sqrt{21}$, and $\hat{\mathbf{S}} = (1/\sqrt{21})(2\mathbf{i} + \mathbf{j} + 4\mathbf{k})$.

13. The standard form of the equation of L is $\frac{x+1/2}{3/2} = \frac{y+2/3}{4/3} = \frac{z-1/2}{1/4}$, so the position vector of a point on the line is $\mathbf{a} = -(1/2)\mathbf{i} - (2/3)\mathbf{j} + (1/2)\mathbf{k}$. A vector along the line is $\mathbf{b} = (3/2)\mathbf{i} + (4/3)\mathbf{j} - (1/4)\mathbf{k}$, so a unit vector along L is $\mathbf{b}/\|\mathbf{b}\|$ where $\|\mathbf{b}\| = \sqrt{589}/12$. To find the position vector of another point on L choose an arbitrary value for x and use it in the equation for L to find the corresponding values of y and z.

15. (a) As $(3, 2, 4)$ lies on L_1 its position vector is $\mathbf{a} = 3\mathbf{i} + 2\mathbf{j} + 4\mathbf{k}$. As $(3, 2, 4)$ and $(2, 1, 6)$ also lie on L_1 a vector $\mathbf{b}$ along L_1 is $\mathbf{b} = (2\mathbf{i} + \mathbf{j} + 6\mathbf{k}) - (3\mathbf{i} + 2\mathbf{j} + 4\mathbf{k}) = -\mathbf{i} - \mathbf{j} + 2\mathbf{k}$. (b) The line L_2 is also parallel to $\mathbf{b}$, but it passes through $\mathbf{a} = -2\mathbf{i} + \mathbf{j} + 2\mathbf{k}$, so L_2 has the equation

$$\frac{x+2}{-1} = \frac{y-1}{-1} = \frac{z-2}{2}.$$

17. The position vector of a point on the line is $\mathbf{a} = 3\mathbf{i} + 2\mathbf{j} - 3\mathbf{k}$ and a vector parallel to the line is $\mathbf{b} = 2\mathbf{i} + 3\mathbf{j} - 3\mathbf{k}$. If we set $\mathbf{r} = x\mathbf{i} + y\mathbf{j} + z\mathbf{k}$ the vector equation of the line $\mathbf{r} = \mathbf{a} + \lambda\mathbf{b}$ becomes $x\mathbf{i} + y\mathbf{j} + z\mathbf{k} = 3\mathbf{i} + 2\mathbf{j} - 3\mathbf{k} + \lambda(2\mathbf{i} + 3\mathbf{j} - 3\mathbf{k})$, so the cartesian form of the equation is

$$\frac{x-3}{2} = \frac{y-2}{3} = \frac{z+3}{-3} = \lambda.$$

The coordinates of three arbitrary points on the line follow by assigning λ three different numerical values and then solving for x, y and z.

Exercise Set 2.2

1. (a) 2 (b) 4 (c) 0 **3.** (a) No (b) No (c) No (d) Yes

5. (a) 16 (b) -15 (c) 17 (d) 1

7. (a) $\cos\theta = (\sqrt{2}/3), \theta = 61.9°$ (b) $\cos\theta = 6/7$, $\theta = 31°$ (c) $\cos\theta = 8/\sqrt{154}, \theta = 49.9°$

9. $F_C = \mathbf{F}\cdot\hat{\mathbf{n}} = \mathbf{F}\cdot(\mathbf{i}+\mathbf{j}+\mathbf{k})/\sqrt{3}$ so $F_C = 9/\sqrt{3}$

11. $\mathbf{a}\cdot\hat{\mathbf{b}} = -2/\sqrt{14}, \mathbf{b}\cdot\hat{\mathbf{a}} = -2/3$

13. (a) $l = m = n = 1/\sqrt{3}, \theta = 54.7°$

(b) $l = 1/3, \theta = 70.5°,\quad m = -2/3,\quad \theta = 131.8°$, $n = 2/3, \theta = 48.2°$

(c) $l = 4/\sqrt{29},\qquad \theta = 42°,\qquad m = -2/\sqrt{29}$, $\theta = 111.8°, n = 3/\sqrt{29}, \theta = 56.1°$

15. $\|\mathbf{a}\| = \sqrt{14}, \|\mathbf{b}\| = \sqrt{54}, \|\mathbf{a}+\mathbf{b}\| = \sqrt{118}, \sqrt{118} < \sqrt{14}+\sqrt{54}$

17. $2x - 3y + z = 3$

19. $2x + z = -1$

21. $\mathbf{r}\cdot\mathbf{n}/\|\mathbf{n}\|$ is the projection of the position vector of a point on the plane onto the unit normal to the plane, and so is the perpendicular distance of the plane from the origin. If $\mathbf{a}\cdot\mathbf{n} > 0$ the perpendicular distance of the plane from the origin is positive, so the plane then lies on the side of O toward which $\mathbf{n}$ is directed, and conversely.

23. $\mathbf{n}_1 = \mathbf{i}+3\mathbf{j}+2\mathbf{k}, \mathbf{n}_2 = 2\mathbf{i}-5\mathbf{j}+\mathbf{k}$ so $\cos\theta = \mathbf{n}_1\cdot\mathbf{n}_2/\|\mathbf{n}_1\|\|\mathbf{n}_2\| = -11/(\sqrt{14}\sqrt{30}), \theta = 122°$

25. Component of $\mathbf{a}$ in direction of $\mathbf{b}$ is $a_b = \mathbf{a}\cdot\hat{\mathbf{b}}$ so $\mathbf{a}_b = (\mathbf{a}\cdot\mathbf{b})\mathbf{b}/\|\mathbf{b}\|^2$, but $\mathbf{a} = \mathbf{a}_b+\mathbf{a}_p$, so $\mathbf{a}_p = \mathbf{a}-(\mathbf{a}\cdot\mathbf{b})\mathbf{b}/\|\mathbf{b}\|^2$

27. $W = \mathbf{F}\cdot\hat{\mathbf{a}}d = (\mathbf{F}\cdot\mathbf{a}/\|\mathbf{a}\|)d$

29. $\|\mathbf{a}\|^2 = 26, \|\mathbf{b}\|^2 = 14, |\mathbf{a}\cdot\mathbf{b}| = 5, \|\lambda\mathbf{a}+\mu\mathbf{b}\|^2 = 170$. $170 \le (4)\cdot(26)-(12)\cdot(5)+(9)\cdot(14) = 170$, so in this case the equality holds.

Exercise Set 2.3

1. $5\mathbf{i}-14\mathbf{j}+\mathbf{k}$

3. $-18\mathbf{i}-7\mathbf{j}+21\mathbf{k}$

5. $2\mathbf{i}-4\mathbf{k}$

7. $-5\mathbf{i}+8\mathbf{j}-\mathbf{k}$

9. $-2\mathbf{i}-11\mathbf{j}-5\mathbf{k}$

11. $(\mathbf{b}+\mathbf{c})\times\mathbf{a} = -24\mathbf{i}-12\mathbf{j}+18\mathbf{k}$

13. $(\mathbf{b}+\mathbf{c})\times\mathbf{a} = -7\mathbf{j}$

15. $(-\mathbf{i}-2\mathbf{j}+5\mathbf{k})/\sqrt{30}$

17. $(-4\mathbf{i}+3\mathbf{j}+\mathbf{k})/\sqrt{26}$

19. $(\mathbf{i}-\mathbf{j})/\sqrt{z}$

21. $3x + 3y - z = 10$

23. $4x + 2z = 10$

25. No

27. Yes

29. $\mathbf{N} = \alpha\mathbf{i}+\beta\mathbf{j}+\gamma\mathbf{k}, \mathbf{a} = \mathbf{i}+\mathbf{j}+3\mathbf{k}, \mathbf{b} = 3\mathbf{i}+2\mathbf{j}+\mathbf{k}$ $\mathbf{a}\cdot\mathbf{N} = 0$ gives $\alpha+\beta+3\gamma = 0$ and $\mathbf{b}\cdot\mathbf{N} = 0$ gives $3\alpha+2\beta+\gamma = 0$. Set $\alpha = c$ (arbitrary). Then $\beta = -(8/5)c$ and $\gamma = (1/5)c$, so $\mathbf{N} = c(\mathbf{i}-(8/5)\mathbf{j}+(1/5)\mathbf{k})$ and $\hat{\mathbf{N}} = (5\mathbf{i}-8\mathbf{j}+\mathbf{k})/\sqrt{90}$. Next $\mathbf{a}\times\mathbf{b} = -5\mathbf{i}+8\mathbf{j}-\mathbf{k}$, so $\hat{\mathbf{n}} = (-5\mathbf{i}+8\mathbf{j}-\mathbf{k})/\sqrt{90}$, showing that $\hat{\mathbf{N}} = -\hat{\mathbf{n}}$. The difference in sign is due to the fact that $\mathbf{a}, \mathbf{b}$, and $\mathbf{N}$ do not necessarily form a right-handed set of vectors.

Exercise Set 2.4

1. $\mathbf{a}\cdot(\mathbf{b}\times\mathbf{c}) = -15, V = 15$

3. $\mathbf{a}\cdot(\mathbf{b}\times\mathbf{c}) = 25, V = 25$

5. Yes

7. No

9. Yes

11. $[\mathbf{a},\mathbf{b},\mathbf{c}] = -10$

13. $[\mathbf{a},\mathbf{b},\mathbf{c}] = 0$

15. $[\lambda\mathbf{a}+\mu\mathbf{b},\mathbf{c},\mathbf{d}] = (\lambda\mathbf{a}+\mu\mathbf{b})\cdot(\mathbf{c}\times\mathbf{d}) = \lambda\mathbf{a}\cdot(\mathbf{c}\times\mathbf{d}) + \mu\mathbf{b}\cdot(\mathbf{c}\times\mathbf{d}) = \lambda[\mathbf{a},\mathbf{c},\mathbf{d}] + \mu[\mathbf{b},\mathbf{c},\mathbf{d}]$

17. $7x + 2y - 4z = 2, \hat{\mathbf{n}} = (7\mathbf{i}+2\mathbf{j}-4\mathbf{k})/\sqrt{69}$

19. $5x - 10y - z = -20, \hat{\mathbf{n}} = (5\mathbf{i}-10\mathbf{j}-\mathbf{k})/\sqrt{126}$

21. From Theorem 2.4(a) $\mathbf{a}\times(\mathbf{b}\times\mathbf{c}) = (\mathbf{a}\cdot\mathbf{c})\mathbf{b}-(\mathbf{a}\cdot\mathbf{b})\mathbf{c}$. Make the substitutions $\mathbf{a}\to\mathbf{b}, \mathbf{b}\to\mathbf{c}$, and $\mathbf{c}\to\mathbf{a}$ to get $\mathbf{b}\times(\mathbf{c}\times\mathbf{a}) = (\mathbf{a}\cdot\mathbf{b})\mathbf{c}-(\mathbf{b}\cdot\mathbf{c})\mathbf{a}$. Now make the substitutions $\mathbf{b}\to\mathbf{c}, \mathbf{c}\to\mathbf{a}$, and $\mathbf{a}\to\mathbf{b}$ to get $\mathbf{c}\times(\mathbf{a}\times\mathbf{b}) = (\mathbf{b}\cdot\mathbf{c})\mathbf{a}-(\mathbf{a}\cdot\mathbf{c})\mathbf{b}$. The result follows by adding these results.

23. Yes **25.** Yes

27. Area of base $= 1/2$ area of parallelogram with sides $\mathbf{b}$ and $\mathbf{c}$, so $S = (1/2)\|\mathbf{b}\times\mathbf{c}\|$. Vertical height $h = \mathbf{a}\cdot\hat{\mathbf{n}}$, so volume of tetrahedron is $V = (1/3)hS = (1/6)|\mathbf{a}\cdot(\mathbf{b}\times\mathbf{c})|$.

29. Take the dot product with $\mathbf{b}\times\mathbf{c}$ to get $\lambda\mathbf{a}\cdot(\mathbf{b}\times\mathbf{c}) + \mu\mathbf{b}\cdot(\mathbf{b}\times\mathbf{c}) + \nu\mathbf{c}\cdot(\mathbf{b}\times\mathbf{c}) + \mathbf{d}\cdot(\mathbf{b}\times\mathbf{c}) = 0$. The two middle terms are zero, so $\lambda\mathbf{a}\cdot(\mathbf{b}\times\mathbf{c}) + \mathbf{d}\cdot(\mathbf{b}\times\mathbf{c}) = 0$. So, provided $\mathbf{a}, \mathbf{b}$, and $\mathbf{c}$ are linearly independent, $\mathbf{a}\cdot(\mathbf{b}\times\mathbf{c}) \neq 0$, so then $\lambda = -\mathbf{d}\cdot(\mathbf{b}\times\mathbf{c})/[\mathbf{a}\cdot(\mathbf{b}\times\mathbf{c})]$, and the other result follows in similar fashion.

31. Write Theorem 2.4 in the form $\mathbf{b}\times(\mathbf{c}\times\mathbf{d}) = (\mathbf{b}\cdot\mathbf{d})\mathbf{c}-(\mathbf{b}\cdot\mathbf{c})\mathbf{d}$ and form the dot product with $\mathbf{a}$ to obtain $\mathbf{a}\cdot[\mathbf{b}\times(\mathbf{c}\times\mathbf{d})] = (\mathbf{a}\cdot\mathbf{c})(\mathbf{b}\cdot\mathbf{d})-(\mathbf{a}\cdot\mathbf{d})(\mathbf{b}\cdot\mathbf{c})$. Interchanging the dot and cross on the left gives the result.

Exercise Set 2.5

1. Sum $(3, 0, 2, 4, 6)$, norms $\sqrt{13}, \sqrt{26}$, dot product 13

3. Sum $(0, 0, 0, 0, 0)$, norms $\sqrt{11}$, $\sqrt{11}$, dot product -11

5. Sum $(3, 2, 1, 4)$, norms $\sqrt{10}$, $\sqrt{20}$, dot product 0

7. Sum $(1, 1, -3, 0, 3)$, norms $\sqrt{22}$, $\sqrt{10}$, dot product -6

9. 0.859 rad, unit n-vectors $(1/\sqrt{15})(3, 1, 2, 1)$, $(1/\sqrt{10})(1, -1, 2, 2)$

11. 0 rad, unit n-vectors $(1/\sqrt{7})(1, -1, -1, 2)$, $(1/\sqrt{7})$ $(1, -1, -1, 2)$

13. No. Null vector belongs to S but the summation and scaling laws fail.

15. No. The null vector is not contained in S and both the scaling and summation laws are not satisfied.

17. Yes.

19. Yes, since a linear equation and a constant are special cases of quadratic functions.

21. Yes.

23. No. As $f'(x) > 0$ the zero function does not belong to S, and the scaling law is not satisfied when $\lambda < 0$, for then $f'(x) < 0$.

25. The null vector $(0, 0, 0)$ in $\mathbf{R}^3$ does not belong to S.

27. $\|x + \lambda y\|^2 = (x + \lambda y) \cdot (x + \lambda y) = \|x\|^2 + 2\lambda(x \cdot y) + \lambda^2 \|y\|^2$ and $\|x - \lambda y\|^2 = (x - \lambda y) \cdot (x - \lambda y) = \|x\|^2 - 2\lambda(x \cdot y) + \lambda^2 \|y\|^2$. The result follows by addition of these equalities.

29. Corresponding components must be equal, or $\mathbf{x} = c\mathbf{y}, c > 0$.

Exercise Set 2.6

1. Linearly independent   **9.** Linearly independent

3. Linearly independent   **11.** Linearly dependent

5. Linearly independent   **13.** Linearly independent

7. Linearly dependent   **15.** Linearly dependent

17. $e_1 = (1, 1, 0, 0, 0)$, $e_2 = (1, 1, 1, 0, 0)$, $e_3 = (1, 1, 0, 1, 0)$, $e_4 = (1, 1, 0, 0, 1)$; dimension 4

19. $e_1 = (2, 2, 1, 0, 0)$, $e_2 = (2, 2, 1, 1, 0)$, $e_3 = (2, 2, 1, 0, 1)$; dimension 3

21. (a) $2 = 2(u + v)$ lies in V (b) No, because $\sin 2x = 2 \sin x \cos x$ does not lie in V (c) $0 = 0u + 0v$ lies in V (d) $\cos 2x = \cos^2 x - \sin^2 x = u - v$ lies in V (e) $2 + 3x$ does not lie in V (f) Yes, because 3 and $-4 \cos 2x$ both lie in V

Exercise Set 2.7

1. $\mathbf{i} + 2\mathbf{j} + \mathbf{k}$, $(7/6)\mathbf{i} - (2/3)\mathbf{j} + (1/6)\mathbf{k}$, $(5/11)\mathbf{i} + (5/11)\mathbf{j} - (15/11)\mathbf{k}$

3. $2\mathbf{i} + \mathbf{j}$, $-(4/5)\mathbf{i} + (8/5)\mathbf{j} + \mathbf{k}$, $(4/21)\mathbf{i} - (8/21)\mathbf{j} + (16/21)\mathbf{k}$

5. $-\mathbf{i} + \mathbf{k}$, $(1/2)\mathbf{i} + 2\mathbf{j} + (1/2)\mathbf{k}$, $(2/3)\mathbf{i} - (1/3)\mathbf{j} + (2/3)\mathbf{k}$

7. $\mathbf{a}_1 = 3\mathbf{j} - \mathbf{k}$, $\mathbf{a}_2 = \mathbf{i} + \mathbf{j}$, $\mathbf{a}_3 = \mathbf{i} + 2\mathbf{k}$. Starting with $\mathbf{u}_1 = \mathbf{a}_1$; $3\mathbf{j} - \mathbf{k}$, $\mathbf{i} + (1/10)\mathbf{j} + (3/10)\mathbf{k}$, $-(5/11)\mathbf{i} + (5/11)\mathbf{j} + (15/11)\mathbf{k}$. Rearrange with $\mathbf{a}_1 = \mathbf{i} + \mathbf{j}$, $\mathbf{a}_2 = 3\mathbf{j} - \mathbf{k}$, $\mathbf{a}_3 = \mathbf{i} + 2\mathbf{k}$; $\mathbf{i} + \mathbf{j}$, $-(3/2)\mathbf{i} + (3/2)\mathbf{j} - \mathbf{k}$, $-(5/11)\mathbf{i} + (5/11)\mathbf{j} + (15/11)\mathbf{k}$

Exercise Set 3.1

1. $a = -1, b = 3, c = 4$ **3.** $a = 1, b = 3, c = 2$

5. $\mathbf{A} + \mathbf{B} = \begin{bmatrix} 3 & 4 & 4 & 4 \\ 3 & 2 & -3 & 3 \\ 1 & 0 & 1 & 1 \end{bmatrix}$,

$\mathbf{A} - \mathbf{B} = \begin{bmatrix} -1 & 4 & 2 & 8 \\ 1 & 0 & 3 & 1 \\ 1 & -2 & -1 & 1 \end{bmatrix}$

7. $\mathbf{A} + \mathbf{B} = \begin{bmatrix} 1 & 4 & 7 \\ 6 & 0 & 1 \\ 1 & 2 & 1 \\ 3 & 5 & 6 \end{bmatrix}$, $\mathbf{A} - \mathbf{B} = \begin{bmatrix} 1 & 0 & 1 \\ 0 & 2 & -1 \\ 1 & 0 & -1 \\ 1 & -1 & 2 \end{bmatrix}$

9. $\mathbf{A} + 3\mathbf{B} = \begin{bmatrix} 7 & 13 & -1 \\ 5 & 7 & 16 \\ 6 & 2 & 11 \end{bmatrix}$

11. $4\mathbf{A} - 2\mathbf{B} = \begin{bmatrix} 4 & 10 & 4 \\ 4 & -4 & 0 \\ 2 & 6 & 0 \end{bmatrix}$

13. 14   **15.** 15   **17.** $\mathbf{BA} = \begin{bmatrix} 6 & 17 & 7 \\ 4 & 2 & 6 \end{bmatrix}$

19. $\mathbf{AB} = \mathbf{BA} = \mathbf{B}$

21. $\mathbf{AB} = \begin{bmatrix} 17 & 8 & 25 \\ 24 & 16 & 30 \\ 20 & 28 & 56 \\ 17 & 10 & 37 \end{bmatrix}$

25. $\mathbf{A} = \begin{bmatrix} 4 & 5 & -1 & 7 \\ 3 & 2 & 0 & 3 \\ 0 & 1 & 6 & -7 \end{bmatrix}$, $\mathbf{x} = \begin{bmatrix} u \\ v \\ w \\ z \end{bmatrix}$, $\mathbf{b} = \begin{bmatrix} 25 \\ 6 \\ 0 \end{bmatrix}$

27. $\mathbf{A} = \begin{bmatrix} 3 - \lambda & 4 & -2 \\ 2 & -7 - \lambda & 6 \\ 8 & 3 & 5 - \lambda \end{bmatrix}$, $\mathbf{x} = \begin{bmatrix} x \\ y \\ z \end{bmatrix}$, $\mathbf{b} = \mathbf{0}$

29. $\mathbf{X} = \dfrac{1}{4}\begin{bmatrix} 11 & -1 & 1 \\ 25 & 7 & 19 \\ 12 & -7 & 9 \end{bmatrix}$

43. Use $\mathbf{A}^4 = \mathbf{A}^2\mathbf{A}^2$ and $\mathbf{A}^6 = \mathbf{A}^2\mathbf{A}^4$ to show that $\mathbf{A}^6 = \mathbf{I}$.

45. $(p = 0, q = 0, r = 1), (p = 0, q = 1, r = 0),$ $(p = 1, q = 0, r = 0)$

46. $n = 3$

47. The structure of $\mathbf{x}^{\mathrm{T}}\mathbf{A}\mathbf{x}$ is such that it is a sum of products of the form $x_i x_j$ with $i, j = 1, 2, 3$. $\mathbf{x}^{\mathrm{T}}\mathbf{A}\mathbf{x} = 3x_1^2 + 8x_1x_2 + 6x_1x_4 + 2x_2^2 + 4x_2x_3 + 12x_2x_4 + 5x_3^2 + 2x_3x_4 + 7x_4^2$.

51. $\begin{bmatrix} 2 & 2 & 7/2 \\ 2 & 6 & 0 \\ 7/2 & 0 & -9 \end{bmatrix}$

53. Use the fact that $\mathbf{PE} = \mathbf{E}$, so $\mathbf{P}^2\mathbf{E} = \mathbf{PE} = \mathbf{E}$, etc.

Exercise Set 3.2

1. (a) Yes (b) No, there is one negative entry (c) No, second row sum >1 (d) Yes

3. $\begin{bmatrix} 0 & 1 & 0 & 0 & 0 & 1 \\ 1 & 0 & 1 & 1 & 0 & 1 \\ 0 & 1 & 0 & 1 & 0 & 0 \\ 0 & 1 & 1 & 0 & 1 & 0 \\ 0 & 0 & 0 & 1 & 0 & 1 \\ 1 & 1 & 0 & 0 & 1 & 0 \end{bmatrix}$ **5.** $\begin{bmatrix} 0 & 1 & 0 & 0 & 0 \\ 0 & 0 & 1 & 0 & 0 \\ 0 & 0 & 0 & 1 & 1 \\ 0 & 0 & 0 & 1 & 1 \\ 1 & 0 & 0 & 0 & 0 \end{bmatrix}$

Exercise Set 3.3

1. $\det\mathbf{A} = -7$ **3.** $\det\mathbf{A} = 43$

13. $x_1 = -7, x_2 = -11, x_3 = -15$

15. $P(\lambda) = -\lambda^3 + 6\lambda^2 - 3\lambda - 10; P(\lambda) = 0$ when $\lambda = -1, 2, 5$

17. $\det\mathbf{A} = -14, \det\mathbf{B} = -18, \det(\mathbf{AB}) = 252$

Exercise Set 3.5

5. $\mathbf{E}_{12} = \begin{bmatrix} 0 & 1 & 0 \\ 1 & 0 & 0 \\ 0 & 0 & 1 \end{bmatrix}$, $\mathbf{E}_2(3) = \begin{bmatrix} 1 & 0 & 0 \\ 0 & 3 & 0 \\ 0 & 0 & 1 \end{bmatrix}$,

$\mathbf{E}_{12}(6) = \begin{bmatrix} 1 & 0 & 0 \\ 6 & 1 & 0 \\ 0 & 0 & 1 \end{bmatrix}$

7. $\begin{bmatrix} 1 & 0 & 0 & 1/2 \\ 0 & 1 & 0 & 0 \\ 0 & 0 & 1 & 1/4 \end{bmatrix}$ **9.** $\begin{bmatrix} 1 & 0 & 0 & 3/2 & 1 \\ 0 & 1 & 0 & 2/3 & 0 \\ 0 & 0 & 1 & -5/6 & -3/2 \end{bmatrix}$

11. $\begin{bmatrix} 1 & 0 & 0 & 0 & -2 \\ 0 & 1 & 0 & 0 & 1 \\ 0 & 0 & 1 & 0 & 2 \\ 0 & 0 & 0 & 1 & -2 \end{bmatrix}$

13. $\begin{bmatrix} 1 & 0 & 0 & -4 \\ 0 & 1 & 0 & 0 \\ 0 & 0 & 1 & 8 \end{bmatrix}$, $x_1 = -4, x_2 = 0, x_3 = 8$

15. $\begin{bmatrix} 1 & 0 & 0 & -1 & -2 \\ 0 & 1 & 0 & 2 & 2 \\ 0 & 0 & 1 & -3 & -3 \end{bmatrix}$, $x_1 = k - 2, x_2 = 2 - 2k,$ $x_3 = -3 + 3k, x_4 = k$

17. $\begin{bmatrix} 1 & 0 & 1 & 0 & 2 & 0 \\ 0 & 1 & 2 & 0 & 1 & 0 \\ 0 & 0 & 0 & 1 & 4 & 0 \\ 0 & 0 & 0 & 0 & 0 & 1 \end{bmatrix}$, no solution

Exercise Set 3.6

1. $\begin{bmatrix} 1 & 0 & 0 & -2 & -1/2 & -1 \\ 0 & 1 & 0 & -2 & 1/2 & -1 \\ 0 & 0 & 1 & 8 & 0 & 5 \end{bmatrix}$,
rank $= 3$, row space $\{[1, 0, 0, -2, -1/2, -1],$ $[0, 1, 0, -2, 1/2, -1], [0, 0, 1, 8, 0, 5]\}$
column space $\{[0, 0, 1]^{\mathrm{T}}, [1, 0, 0]^{\mathrm{T}}, [0, 1, 0]^{\mathrm{T}}\}$

3. $\begin{bmatrix} 1 & 0 & 0 & 2 & 0 \\ 0 & 1 & 0 & 3 & 0 \\ 0 & 0 & 1 & 0 & 0 \\ 0 & 0 & 0 & 0 & 1 \end{bmatrix}$
rank $= 4$, row space $\{[0, 0, 0, 0, 1], [1, 0, 0, 2, 0],$ $[0, 0, 1, 0, 0], [0, 1, 0, 3, 0]\}$
column space $\{[0, 0, 0, 1]^{\mathrm{T}}, [0, 0, 1, 0]^{\mathrm{T}},$ $[0, 1, 0, 0]^{\mathrm{T}}, [1, 0, 0, 0]^{\mathrm{T}}\}$

5. $\begin{bmatrix} 1 & 0 & 0 \\ 0 & 1 & 0 \\ 0 & 0 & 1 \end{bmatrix}$
rank $= 3$, row space $\{[1, 0, 0], [0, 1, 0], [0, 0, 1]\}$
column space $\{[1, 0, 0]^{\mathrm{T}}, [0, 1, 0]^{\mathrm{T}}, [0, 0, 1]^{\mathrm{T}}\}$

7. $\begin{bmatrix} 1 & 0 & 0 \\ 0 & 1 & 0 \\ 0 & 0 & 1 \\ 0 & 0 & 0 \end{bmatrix}$
rank $= 3$, row space $\{[1, 0, 0], [0, 1, 0], [0, 0, 1]\}$,
column space $\{[0, 1, 0, -2/13]^{\mathrm{T}}, [0, 0, 1, -7/13]^{\mathrm{T}},$ $[1, 0, 0, 20/13]^{\mathrm{T}}\}$

9. $\begin{bmatrix} 1 & 0 & -1/3 & -2/3 & -1/3 & -5/3 \\ 0 & 1 & 2/3 & 7/3 & 8/3 & 13/3 \\ 0 & 0 & 0 & 0 & 0 & 0 \end{bmatrix}$
rank $= 2$, row space $\{[0, 1, 2/3, 7/3, 8/3, 13/3],$

$[1, 0, -1/3, -2/3, -1/3, -5/3]\}$,
column space $\{[0, 1, 1]^T, [1, 0, 1]^T\}$

11. $\begin{bmatrix} 1 & 0 & 0 & 0 \\ 0 & 1 & 0 & 0 \\ 0 & 0 & 1 & 0 \\ 0 & 0 & 0 & 1 \end{bmatrix}$

rank = 4, row space $\{[1, 0, 0, 0], [0, 1, 0, 0],$
$[0, 0, 1, 0], [0, 0, 0, 1]\}$,
column space $\{[1, 0, 0, 0]^T, [0, 1, 0, 0]^T,$
$[0, 0, 1, 0]^T, [0, 0, 0, 1]^T\}$

13. $\begin{bmatrix} 1 & 7 & 0 & 0 \\ 0 & 0 & 1 & 0 \\ 0 & 0 & 0 & 1 \end{bmatrix}$

rank = 3, row space $\{[0, 0, 1, 0], [0, 0, 0, 1],$
$[1, 7, 0, 0]\}$,
column space $\{[0, 1, 0]^T, [1, 0, 0]^T, [0, 0, 1]^T\}$

Exercise Set 3.7

1. $x_1 = -2a - 6b, \quad x_2 = a + 4b, \quad x_3 = -a - (7/2)b,$
$x_4 = a, x_5 = b$

3. $x_1 = k, x_2 = -k, x_3 = 0, x_4 = k$

5. $x_1 = x_2 = x_3 = 0$

7. $x_1 = -(1/4)a + (5/4)b - (3/4)c,$
$x_2 = (1/20)a - (29/20)b + (7/20)c,$
$x_3 = (3/20)a - (7/20)b + (1/20)c,$
$x_4 = -(13/20)a + (37/20)b - (31/20)c,$
$x_5 = a, x_7 = b, x_7 = c$

9. $x_1 = (4/9)a + (37/9)b - (14/9)c, \quad x_2 = -a - 3b,$
$x_3 = (1/9)a - (2/9)b + (1/9)c, \quad x_4 = a, \quad x_5 = b,$
$x_6 = c$

Exercise Set 3.8

1. $x_1 = 3, x_2 = 1, x_3 = -2, x_4 = 4$

3. $x_1 = -5/12, x_2 = -1/12, x_3 = 1/6, x_4 = 1/2$

5. Inconsistent; no solution

7. $x_1 = -15/11, x_2 = 1/11, x_3 = 8/11, x_4 = 5/11$

9. Inconsistent: no solution

Exercise Set 3.9

1. $\begin{bmatrix} -1/5 & 4/15 & 1/3 \\ 2/5 & 3/10 & -1/2 \\ 0 & -1/6 & 1/6 \end{bmatrix}$

3. $\begin{bmatrix} -2/73 & 16/73 & -5/73 \\ -9/73 & -1/73 & 14/73 \\ 28/73 & -5/73 & -3/73 \end{bmatrix}$

5. $\begin{bmatrix} 2 & 1 & -2 \\ -1 & 0 & 1 \\ 0 & -2 & 1 \end{bmatrix}$

7. $\begin{bmatrix} 37/131 & 8/131 & -31/131 & 43/131 \\ 52/131 & -10/131 & 6/131 & -21/131 \\ -1/131 & -25/131 & 15/131 & 13/131 \\ -10/131 & 12/131 & 19/131 & -1/131 \end{bmatrix}$

9. $(\mathbf{AB})^{-1} = \mathbf{B}^{-1}\mathbf{A}^{-1}$
$= \begin{bmatrix} 31/276 & 1/207 & -7/69 \\ -19/276 & 1/414 & -7/138 \\ -4/69 & 19/414 & 5/138 \end{bmatrix}$

11. $\begin{bmatrix} 25/27 & -31/27 & 13/9 \\ -7/27 & 13/27 & -4/9 \\ -1/27 & -2/27 & 2/9 \end{bmatrix}$

13. $\begin{bmatrix} -2/27 & -1/9 & 16/27 \\ 28/27 & 5/9 & -89/27 \\ -11/27 & -1/9 & 34/27 \end{bmatrix}$

15. $\begin{bmatrix} 27/29 & -7/29 & -1/58 & -7/58 \\ -28/29 & 18/29 & 15/58 & -11/58 \\ -3/29 & 4/29 & -8/29 & 2/29 \\ -11/29 & 5/29 & 9/58 & 5/58 \end{bmatrix}$

17. Elementary row operations require far less computational effort.

Exercise Set 3.10

1. $\dfrac{d\mathbf{C}}{dt} = \begin{bmatrix} 3t^2 & 1 + 4t & \sin t + t\cos t + \sinh t \\ 1 + 2t & -\sin t & 2\cos 2t - \sin t \end{bmatrix}$

$\dfrac{d^2\mathbf{C}}{dt} = \begin{bmatrix} 6t & 4 & 2\cos t - t\sin t + \cosh t \\ 2 & -\cos t & -4\sin 2t - \cos t \end{bmatrix}$

3. $\dfrac{d\mathbf{C}}{dt} = \begin{bmatrix} 1 - 4e^{2t} & 0 & -3t^2 \\ 0 & 3 - 4t & 2e^{2t} - 2\cosh t \end{bmatrix}$

$\dfrac{d^2\mathbf{C}}{dt^2} = \begin{bmatrix} -8e^{2t} & 0 & -6t \\ 0 & -4 & 4e^{2t} - 2\sinh t \end{bmatrix}$

7. $\dfrac{d\mathbf{A}^{-1}}{dt} =$

$\begin{bmatrix} -\sin t & -\cos t & 0 \\ \cos t & -\sin t & 0 \\ -\sin t - 3t\cos t + t^2\sin t & -\cos t + 3t\sin t + t^2\cos t & 0 \end{bmatrix}$

9. As $\mathbf{AA}^{-1} = \mathbf{I}$, $\dfrac{d}{dt}(\mathbf{AA}^{-1}) = \dfrac{d\mathbf{A}}{dt}\mathbf{A}^{-1} + \mathbf{A}\dfrac{d\mathbf{A}^{-1}}{dt} = \mathbf{0}$, so another differentiation gives $\dfrac{d^2\mathbf{A}}{dt^2}\mathbf{A}^{-1} + 2\dfrac{d\mathbf{A}}{dt}\dfrac{d\mathbf{A}^{-1}}{dt} + \mathbf{A}\dfrac{d^2\mathbf{A}^{-1}}{dt^2} = \mathbf{0}$. Now substitute for $\dfrac{d\mathbf{A}^{-1}}{dt}$ to find $\dfrac{d^2\mathbf{A}^{-1}}{dt^2}$.

Exercise Set 4.1

1. $P(\lambda) = \lambda^3 - 3\lambda^2$

3. $P(\lambda) = \lambda^3 - 3\lambda^2 + 5\lambda + 1$

5. $P(\lambda) = \lambda^3 - 4\lambda^2 - 2\lambda$

7. $P(\lambda) = \lambda(\lambda - 1)(\lambda^2 - \lambda - 2)$

9. $1, \begin{bmatrix} -1 \\ 0 \\ 1 \end{bmatrix}; \ 2, \begin{bmatrix} 0 \\ 1 \\ 1 \end{bmatrix}; \ -1, \begin{bmatrix} 1 \\ 2 \\ 0 \end{bmatrix}$

11. $-1, \begin{bmatrix} -1 \\ 0 \\ 1 \end{bmatrix}; \ 1, \begin{bmatrix} 1 \\ 2 \\ 0 \end{bmatrix}; \ 3, \begin{bmatrix} 0 \\ 1 \\ 1 \end{bmatrix}$

13. $-2, \begin{bmatrix} 2 \\ 1 \\ -2 \end{bmatrix}; \ 1, \begin{bmatrix} 1 \\ 1 \\ -2 \end{bmatrix}; \ 0, \begin{bmatrix} 1 \\ 1 \\ -1 \end{bmatrix}$

15. $1, \begin{bmatrix} 1 \\ 1 \\ -2 \end{bmatrix}; \ 2, \begin{bmatrix} 1 \\ 1 \\ -1 \end{bmatrix}; \ -2, \begin{bmatrix} 2 \\ 1 \\ -2 \end{bmatrix}$

17. $1, \begin{bmatrix} 1 \\ 1 \\ 1 \end{bmatrix}; \ 2, \begin{bmatrix} 0 \\ 1 \\ 0 \end{bmatrix}; \ 0, \begin{bmatrix} 2 \\ 1 \\ 1 \end{bmatrix}$

19. $2, \begin{bmatrix} 1 \\ 1 \\ 1 \end{bmatrix}; \ 1, \begin{bmatrix} 0 \\ 1 \\ 0 \end{bmatrix}; \ 1, \begin{bmatrix} 2 \\ 0 \\ 1 \end{bmatrix}$

21. $0, \begin{bmatrix} 1 \\ 1 \\ 1 \end{bmatrix}; \ 2, \begin{bmatrix} 0 \\ 1 \\ 0 \end{bmatrix}; \ 2, \begin{bmatrix} 2 \\ 0 \\ 1 \end{bmatrix}$

23. $P(\lambda) = (\lambda + 1)(\lambda^3 - \lambda^2 - 4\lambda + 4)$;

$1, \begin{bmatrix} 0 \\ 1 \\ 1 \\ 1 \end{bmatrix}; \ 2, \begin{bmatrix} 0 \\ 1 \\ 1 \\ 0 \end{bmatrix}; \ -2, \begin{bmatrix} -1 \\ 0 \\ 1 \\ 0 \end{bmatrix}; \ -1, \begin{bmatrix} 1 \\ 0 \\ 0 \\ 1 \end{bmatrix}$

25. To obtain the first result expand the characteristic determinant in terms of elements of the first column. The second part of the problem is illustrated by $\mathbf{A} = \begin{bmatrix} 1 & -2 & 1 \\ 0 & 1 & 2 \\ 0 & 0 & 2 \end{bmatrix}$ with eigenvalues $\lambda_1 = \lambda_2 = 1$ and $\lambda_3 = 2$ and eigenvectors $\mathbf{x}_{1,2} = [1, \ 0, \ 0]^T$ and $\mathbf{x}_3 = [-3, \ 2, \ 1]^T$.

31. Premultiplication of a matrix by $\mathbf{E}$ interchanges its ith and jth rows, while premultiplication by $\mathbf{E}^T$ reverses the process. Thus as $\mathbf{E}$ is obtained from $\mathbf{I}$, it follows that $\mathbf{E}^T\mathbf{E} = \mathbf{I}$. This shows that $\mathbf{E}^T = \mathbf{E}^{-1}$, and so $\mathbf{E}$ is an orthogonal matrix. As the product of two orthogonal matrices is an orthogonal matrix, if $\mathbf{Q}$ is an orthogonal matrix, so also is the matrix $\mathbf{EQ}$ obtained from $\mathbf{Q}$ by a row interchange. Multiplication of $\mathbf{Q}$ by a sequence of elementary matrices $\mathbf{E}_1, \mathbf{E}_2, \ldots, \mathbf{E}_t$ will interchange the rows of $\mathbf{Q}$ in any desired order while leaving the result still an orthogonal matrix.

Exercise Set 4.2

In solutions 1 through 12 a diagonalizing matrix $\mathbf{P}$ is formed by using the given eigenvectors in any order as the columns of $\mathbf{P}$. The elements on the leading diagonal of the corresponding diagonal matrix are then arranged in the same order as the eigenvectors to which they belong.

1. $1, \begin{bmatrix} -1 \\ 1 \\ 0 \end{bmatrix}; \ -1, \begin{bmatrix} 1 \\ 0 \\ -1 \end{bmatrix}; \ 2, \begin{bmatrix} -1 \\ 1 \\ 1 \end{bmatrix}$

3. $2, \begin{bmatrix} 1 \\ 1 \\ 1 \end{bmatrix}; \ -1, \begin{bmatrix} 0 \\ 2 \\ 1 \end{bmatrix}; \ 1, \begin{bmatrix} 1 \\ 0 \\ 1 \end{bmatrix}$

5. $1, \begin{bmatrix} -1 \\ 1 \\ 2 \end{bmatrix}; \ 1, \begin{bmatrix} 0 \\ 1 \\ 1 \end{bmatrix}; \ -1, \begin{bmatrix} 1 \\ -1 \\ -1 \end{bmatrix}$

7. $1, \begin{bmatrix} 1 \\ 1 \\ -1 \end{bmatrix}; \ 3, \begin{bmatrix} 1 \\ 0 \\ -1 \end{bmatrix}; \ 3, \begin{bmatrix} -1 \\ 1 \\ 2 \end{bmatrix}$

9. $1, \begin{bmatrix} 1 \\ 0 \\ 1 \end{bmatrix}; \ 2, \begin{bmatrix} 1 \\ 1 \\ 1 \end{bmatrix}; \ -1, \begin{bmatrix} 0 \\ 2 \\ 1 \end{bmatrix}$

11. $0, \begin{bmatrix} -1 \\ -1 \\ 1 \end{bmatrix}; \ -2, \begin{bmatrix} 1 \\ 2 \\ 0 \end{bmatrix}; \ -2, \begin{bmatrix} 0 \\ 2 \\ 1 \end{bmatrix}$

13. $\begin{bmatrix} 1 \\ 1 \\ 1 \end{bmatrix}, \begin{bmatrix} -1/3 \\ -1/3 \\ 2/3 \end{bmatrix}, \begin{bmatrix} -1/2 \\ 1/2 \\ 0 \end{bmatrix}$

15. $\begin{bmatrix} -1 \\ 1 \\ 0 \end{bmatrix}, \begin{bmatrix} 3/\sqrt{18} \\ 3/\sqrt{18} \\ 0 \end{bmatrix}, \begin{bmatrix} 0 \\ 0 \\ 1 \end{bmatrix}$

17. $3, \begin{bmatrix} 1 \\ 0 \\ 0 \end{bmatrix}; \ 2, \begin{bmatrix} 0 \\ -1/\sqrt{2} \\ 1/\sqrt{2} \end{bmatrix}; \ 4, \begin{bmatrix} 0 \\ 1/\sqrt{2} \\ 1/\sqrt{2} \end{bmatrix}$

19. Two equal eigenvalues, but the corresponding eigenvectors are orthogonal:

$$5, \begin{bmatrix} 1/\sqrt{2} \\ 1/\sqrt{2} \\ 0 \end{bmatrix}; \ 3, \begin{bmatrix} -1/\sqrt{2} \\ 1/\sqrt{2} \\ 0 \end{bmatrix}; \ 3, \begin{bmatrix} 0 \\ 0 \\ 1 \end{bmatrix}$$

21. Two equal eigenvalues, but the corresponding eigenvectors are orthogonal:

$$6, \begin{bmatrix} 1/\sqrt{2} \\ 1/\sqrt{2} \\ 0 \end{bmatrix}; \ 2, \begin{bmatrix} 0 \\ 0 \\ 1 \end{bmatrix}; \ 2, \begin{bmatrix} 1/\sqrt{2} \\ -1/\sqrt{2} \\ 0 \end{bmatrix}$$

25. $P(\lambda) = \lambda^2 - 6\lambda + 11$, $\quad \mathbf{A}^{-1} = \begin{bmatrix} 4/11 & -3/11 \\ 1/11 & 2/11 \end{bmatrix}$

27. $P(\lambda) = -\lambda^3 + \lambda^2 - 4$,

$$\mathbf{A}^{-1} = \begin{bmatrix} 0 & -1/2 & -1/2 \\ 1 & 1 & 1 \\ -1/2 & -1/2 & -1 \end{bmatrix}$$

Exercise Set 4.3

1. $\begin{bmatrix} 1 & 1-i & 2i \\ 1+i & 2 & 3-i \\ -2i & 3+i & 4 \end{bmatrix} + \begin{bmatrix} i & 2+2i & 3 \\ -2+2i & 0 & 1+2i \\ -3 & -1+2i & 2i \end{bmatrix}$

$\qquad\qquad$ Hermitian $\qquad\qquad\qquad$ skew-Hermitian

3. $\begin{bmatrix} 4 & 2i & 1+i \\ -2i & 1 & 3 \\ 1-i & 3 & 0 \end{bmatrix} + \begin{bmatrix} -2i & 1-i & 1+i \\ -1-i & 2i & 1 \\ -1+i & -1 & 0 \end{bmatrix}$

$\qquad\qquad$ Hermitian $\qquad\qquad\qquad$ skew-Hermitian

5. $\frac{3}{2} \pm \frac{\sqrt{21}}{2}$ $\qquad\qquad$ **9.** $\frac{1}{2}i(3 \pm \sqrt{41})$

7. $2 \pm \sqrt{14}$ $\qquad\qquad$ **11.** $\pm i\sqrt{13}$

13. $(1+i)/\sqrt{2}, \begin{bmatrix} 1 \\ 1 \end{bmatrix}; \ (1-i)/\sqrt{2}, \begin{bmatrix} -1 \\ 1 \end{bmatrix}$

15. $i, \begin{bmatrix} i \\ 1 \end{bmatrix}; \ 1, \begin{bmatrix} -i \\ 1 \end{bmatrix}$

Exercise Set 4.4

1. $\begin{bmatrix} 1 & 0 & 2 \\ 0 & 3 & -3 \\ 2 & -3 & -2 \end{bmatrix}$ $\quad$ **3.** $\begin{bmatrix} -2 & 0 & -1 \\ 0 & 3 & 2 \\ -1 & 2 & 0 \end{bmatrix}$

5. $\begin{bmatrix} 3 & -2 & 0 & 0 \\ -2 & 0 & -3 & -1 \\ -3 & -3 & 2 & 0 \\ 0 & -1 & 0 & 8 \end{bmatrix}$

7. $2x_1^2 + x_2^2 - x_3^2 + 3x_4^2 + 8x_1x_2 + 8x_1x_3 + 4x_2x_3 + 2x_2x_4 + 4x_3x_4$

9. $3x_2^2 + 2x_3^2 + 7x_4^2 + 4x_1x_2 - 8x_1x_3 + 4x_1x_4 + 2x_2x_3 + 2x_3x_4$

11. $\mathbf{Q} = \begin{bmatrix} 0 & 1/\sqrt{2} & -1/\sqrt{2} \\ 1 & 0 & 0 \\ 0 & 1/\sqrt{2} & 1/\sqrt{2} \end{bmatrix}$, $\mathbf{D} = \begin{bmatrix} 1 & 0 & 0 \\ 0 & 3 & 0 \\ 0 & 0 & 2 \end{bmatrix}$,

$\mathbf{P} = y_1^2 + 3y_2^2 + y_3^2$, $\mathbf{y} = \mathbf{Q}^T\mathbf{x}$, positive definite

13. $\mathbf{Q} = \begin{bmatrix} 0 & 1 & 0 \\ -1/\sqrt{2} & 0 & 1/\sqrt{2} \\ 1/\sqrt{2} & 0 & 1/\sqrt{2} \end{bmatrix}$, $\mathbf{D} = \begin{bmatrix} 3 & 0 & 0 \\ 0 & 4 & 0 \\ 0 & 0 & 5 \end{bmatrix}$,

$\mathbf{P} = 3y_1^2 + 4y_2^2 + 5y_3^2$, $\mathbf{y} = \mathbf{Q}^T\mathbf{x}$, positive definite

15. $\mathbf{Q} = \begin{bmatrix} 0 & -1/\sqrt{2} & 1/\sqrt{2} \\ 1 & 0 & 0 \\ 0 & 1/\sqrt{2} & 1/\sqrt{2} \end{bmatrix}$, $\mathbf{D} = \begin{bmatrix} -1 & 0 & 0 \\ 0 & 1 & 0 \\ 0 & 0 & 2 \end{bmatrix}$,

$\mathbf{P} = -y_1^2 + y_2^2 + 2y_3^2$, $\mathbf{y} = \mathbf{Q}^T\mathbf{x}$, indefinite

17. $\mathbf{Q} = \begin{bmatrix} 1 & 0 & 0 \\ 0 & -1/\sqrt{2} & 1/\sqrt{2} \\ 0 & 1/\sqrt{2} & 1/\sqrt{2} \end{bmatrix}$, $\mathbf{D} = \begin{bmatrix} 2 & 0 & 0 \\ 0 & 3 & 0 \\ 0 & 0 & -1 \end{bmatrix}$,

$\mathbf{P} = 2y_1^2 + 3y_2^2 - y_3^2$, $\mathbf{y} = \mathbf{Q}^T\mathbf{x}$, indefinite

19. Ellipse $\quad$ **21.** Hyperbolic $\quad$ **23.** Ellipse

25. $\mathbf{A} = \begin{bmatrix} 1 & 4 \\ 4 & 1 \end{bmatrix}$, $\mathbf{Q} = \begin{bmatrix} -1/\sqrt{2} & 1/\sqrt{2} \\ 1/\sqrt{2} & 1/\sqrt{2} \end{bmatrix}$,

$\mathbf{D} = \begin{bmatrix} -3 & 0 \\ 0 & 5 \end{bmatrix}$, $\mathbf{x} = \mathbf{Q}\mathbf{y}$, $x_1 = (-y_1 + y_2)/\sqrt{2}$,

$x_2 = (y_1 + y_2)/\sqrt{2}, -3y_1^2 + 5y_2^2$

27. $\mathbf{A} = \begin{bmatrix} -2 & 2 \\ 2 & 1 \end{bmatrix}$, $\mathbf{Q} = \begin{bmatrix} -2/\sqrt{5} & 1/\sqrt{5} \\ 1/\sqrt{5} & 2/\sqrt{5} \end{bmatrix}$,

$\mathbf{D} = \begin{bmatrix} -3 & 0 \\ 0 & 2 \end{bmatrix}$, $\mathbf{x} = \mathbf{Q}\mathbf{y}$, $x_1 = -(2/\sqrt{5})y_1 + (1/\sqrt{5})y_2$, $\quad x_2 = (1/\sqrt{5})y_1 + (2/\sqrt{5})y_2, -3y_1^2 + 2y_2^2 - (9/\sqrt{5})y_1 + (2/\sqrt{5})y_2$

29. $\mathbf{A} = \begin{bmatrix} 35/17 & 4/17 \\ 4/17 & 50/17 \end{bmatrix}$, $\mathbf{Q} = \begin{bmatrix} -4/17 & 1/\sqrt{17} \\ 1/\sqrt{17} & 4/\sqrt{17} \end{bmatrix}$,

$\mathbf{D} = \begin{bmatrix} 2 & 0 \\ 0 & 3 \end{bmatrix}$, $\mathbf{x} = \mathbf{Q}\mathbf{y}$, $x_1 = (-4y_1 + y_2)/\sqrt{17}$ $x_2 = (y_1 + 4y_2)/\sqrt{17}$, $\quad 2y_1^2 + 3y_2^2 + (4/\sqrt{17})y_1 + (16/\sqrt{17})y_2$

Exercise Set 4.5

1. $n = 4$ $\quad$ **5.** $e^{\mathbf{A}t} = \begin{bmatrix} e^{mt} & 0 \\ 0 & e^{nt} \end{bmatrix}$

7. $e^{\mathbf{A}t} = \begin{bmatrix} \frac{4}{5}e^{-3t} + \frac{1}{5}e^{2t} & -\frac{2}{5}e^{-3t} + \frac{2}{5}e^{2t} \\ -\frac{2}{5}e^{-3t} + \frac{2}{5}e^{2t} & \frac{1}{5}e^{-3t} + \frac{4}{5}e^{2t} \end{bmatrix}$

9. $e^{At} = \begin{bmatrix} 2e^t - e^{2t} & -e^t - e^{2t} & 2e^t - 2e^{2t} \\ 2e^t - 2 & 2 - e^t & 2e^t - 2 \\ e^{2t} - 1 & 1 - e^{2t} & 2e^{2t} - 1 \end{bmatrix}$

11. Follows from the definitions.

Exercise Set 5.1

1. Homogeneous linear of order 3 and degree 1
3. Nonlinear of order 2 and degree 1
5. Nonlinear of order 2 and degree 1
7. Nonhomogeneous linear of order 1 and degree 1
9. Nonlinear of order 1

Exercise Set 5.2

1. $\left(y - x\frac{dy}{dx}\right)^2 = 2xy\left(1 + \left(\frac{dy}{dx}\right)^2\right)$

3. $x\frac{d^2y}{dx^2} = \frac{U}{V}\left(1 - \left(\frac{dy}{dx}\right)^2\right)^{1/2}$

Exercise Set 5.3

1.

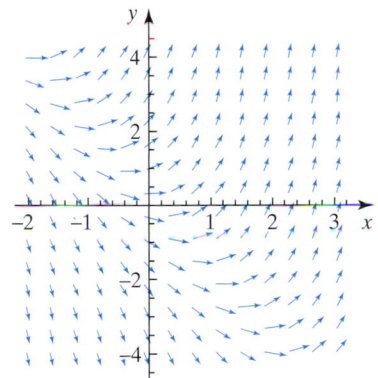

3.

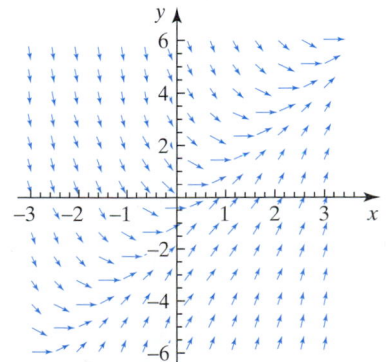

5.

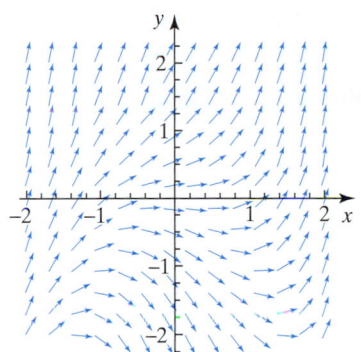

Exercise Set 5.4

1. $x^2 + 2y + \ln|2y - 1| = 3$;   Singular solution $y = 1/2$ does not satisfy $y(1) = 1$

3. $y = (x^2 - 3)/[2(x^2 - 4)]$

5. $\ln|y + \sqrt{(y^2 - 1)}| = 3(1 + x^2)^{1/2} + C$

7. $2\ln|y + 2| + 2/(y + 2) = C - \ln|x + 1|$

9. $2\ln|y| + 3y^2 = 4x - 4(x + 1)\ln|x + 1| + C$

11. $\ln[(1 + x^2)/(y^2 + y + 1)] + (2/\sqrt{3})\text{Arctan}[(2y + 1)/\sqrt{3}] = C$

13. $y = 2 + C\cos^2 x$

15. Eliminate k between the original equation and $dy/dx = -1/k$ to obtain the differential equation of the orthogonal trajectories $dy/dx = -(x - a)/(y - b)$, with the solution $x^2 + y^2 - 2ax - 2by = C$, the equation of a family of concentric circles with their center at (a, b).

17. Eliminate C between the original equation and $dy/dx = -1/\{2Cxe^{2x}(1 + x)\}$ to obtain the differential equation of the orthogonal trajectories $dy/dx = -x/\{2y(1 + x)\}$, with the solution $y^2 = -x + \ln|1 + x| + C$.

19. $\lambda = \ln(N_2/N_1)/(t_2 - t_1)$; predicts infinite growth

21. Approximately 50,200 years

Exercise Set 5.5

1. $x/y^2 + 1/y = C$ **3.** $y = x(4\ln|x| + C)^{1/2}$

5. $-(1/2)x^2 + xy + y^2 = C$

7. $-2\ln|x| + (1/2)\cos(y/x)\sin(y/x) + (1/2)y/x = C$

9. $-\ln|x| - (1/2)\cos(y/x)\sin(y/x) + (1/2)y/x = C$

11. $x/(y + 2) - \ln|y + 2| = C$

13. $x + 1 = [C(1 + x)\exp\{\text{Arctan}[y/(1 + x)]\}]/[y^2 + (1 + x)^2]^{1/2}$

Exercise Set 5.6

1. (a) Not exact
   (b) $f(x, y) = x^4 + \sin x + 3xy^2 + 2y = C$
3. (a) $f(x, y) = x \sin x + y^3 + \sinh(x + 2y) = C$
   (b) Not exact
5. (a) $f(x, y) = (x^3 + y^2)^{1/2} + 3y^2 = C$
   (b) $f(x, y) = y \ln x + x^2 \sinh(y^2) = C$
7. (a) $x^2 y + 6 \ln x + 4 \ln y = C$
   (b) $f(x, y) = x^2/(2x + 3y^2) + 2x = C$

Exercise Set 5.7

1. $y = 1/2 + Ce^{-2x}$
3. $y = (1/3)(2x^3 + 3x^2 + 3C)/(x + 1)$
5. $y = (1/6)(6Cx^3 - 3x - 2)/x$
7. $y = (1/4)(4C + x^4)/x^2$
9. $y = \sin x\{C + 2\ln(\cos x - 1)\}/(1 + \cos x)$
11. $y = x \sin x + x$ 13. $y = 2x^2 - x - 1$
14. $y = x^4/3 + 2/(3x^2)$
15. $y = x/\sin x - \pi/(2\sin x) - \cos x$
17. Approximately 173 seconds
19. $dv/dt + kv + kt = 0$; $v(t) = \frac{(v_0 k - 1)}{k}e^{-kt} + \frac{1}{k} - t$;
    $k = \frac{4}{(4-e)v_0}$

Exercise Set 5.8

1. $y^{1/2} = x - 1 + Ce^{-x}$
3. $y^{1/2} = 1/(4 - 2x + Ce^{-x/2})$
5. $y = 1/(1 + Ce^{-2\cos x})$
7. $y^{1/2} = 4x/(4C - x^2)$
9. $n(t) = \frac{n_0 a}{n_0 b + (a - n_0 b)e^{-at}}$. If $a/b = n_0$, then $n(t) = n_0$
   (constant); otherwise $n(t)$ approaches the value
   a/b. Thus, if $a/b > n_0$ the stock level increases to a
   value greater than n_0, and if $a/b < n_0$ it decreases
   to a value less than n_0.

Exercise Set 5.9

3. $y = x + \exp(2x^2/3)/\{C - 2\int x \exp(2x^3/3)dx\}$
5. $y = 1 + 1/(Ce^{-x} - 2)$

Exercise Set 5.10

1. Initial conditions can be imposed anywhere in the
   part of the plane $x < 1$ other than on the line $x = 1$,
   where $\partial f/\partial x$ is infinite.
3. Initial conditions can be imposed anywhere in the
   (x, y)-plane.

5. Initial conditions can be imposed anywhere in the
   (x, y)-plane other than on the y-axis.

Exercise Set 6.1

1. (a) Linearly independent   (b) Linearly independent   (c) Linearly independent
3. (a) Linearly independent   (b) Linearly independent   (c) Linearly dependent
5. $y = c_1 e^x + c_2 e^{-4x}$ 9. $y = c_1 e^x + c_2 e^{-3x}$
7. $y = e^x(c_1 \cos x + c_2 \sin x)$ 11. $y = (c_1 + c_2 x)e^{-3x}$
13. $y = e^{2x}(c_1 \cos x + c_2 \sin x)$
15. $y = e^{-3x}(c_1 \cos 4x + c_2 \sin 4x)$
17. $y = c_1 e^{-4x} + c_2 e^{-x}$
19. $y = e^{3x/2}\{c_1 \cos(x\sqrt{3}/2) + c_2 \sin(x\sqrt{3}/2)\}$
21. $y = 5e^{-2x} - 4e^{-3x}$
23. $y = e^{-x}(3\cos x + 4\sin x)$ 25. $y = 5e^{2x} - 3e^{3x}$
27. $y = e^{4x}/5 - 6e^{-x}/5$
29. $y = 3e^{-x}/(3 - e^2) - e^{-3x}/(3e^{-2} - 1)$
31. $y = (1/5)e^{-3(1+x)}(2 - 3x)$
33. $y = e^{-x}\{\cos 5x + (3/2)\sin 5x\}$
35. $y = e^{-2x}/(3e^{-3} - 2e^{-2}) - e^{-3x}/(3e^{-3} - 2e^{-2})$
37. (a) Not unique   (b) No solution   (c) Unique
39. $y = b \sin \lambda x$, b arbitrary and $\lambda = 0, \pm 1, \pm 2, \dots$.
41. $\theta(t) = (\alpha/p)\exp(-kt)\sin pt$, and so the angular
    velocity is $d\theta/dt = -(ak/p)\exp(-kt)\sin(pt) + a\exp(-kt)\cos pt$. The pendulum comes to rest
    for the first time when $d\theta/dt$ first becomes
    zero. This occurs at the smallest positive value
    $t = t_C$, say, such that $\tan pt_C = p/k$. The angu-
    lar displacement at $t = t_C$ is given by $\theta(t_C) = a\exp(-kt_C)/(k^2 + p^2)^{1/2}$.

Exercise Set 6.2

1. $y_p = (2/5)\sin x - (1/5)\cos x$,
   $y_c = (2/5)(3\cos 2x + \sin 2x)e^{-x}$
3. $y_p = -(1/2)\cos x$, $y_c = (3/2)(1 + x)e^{-x}$
5. $y_p = -(1/130)(9\cos 3x + 7\sin 3x)$,
   $y_c = (13/10)e^{-x} - (16/13)e^{-2x}$
7. $y_p = (A/10)(\sin x - \cos x)$,
   $y_c = (A/5 + 10)e^{-2x} - (A/10 + 7)e^{-3x}$
9. $y_p = (1/5)(\cos x + \sin x)$,
   $y_c = (4/5)(4e^{-2x} - 3e^{-3x})$, $\tan \phi = -1$
11. $y_p = (1/9)\sin 3x$,
    $y_c = \{2 + (23/3)x\}e^{-3x}$, $\phi = 0$

13. $y_p = (3/40)(\cos 2x + 3\sin 2x)$,
$y_c = (65/8)e^{-2x} - (21/5)e^{-4x}$, $\tan \phi = -1/3$

15. $y(t) = 2000 - \dfrac{m}{10} - \dfrac{mt}{10} + \dfrac{m^2}{3200}$

$$+ \left(\dfrac{m}{10} - \dfrac{m^2}{3200}\right)\exp\left(-\dfrac{320t}{m}\right),$$

$$\dfrac{dy}{dt} = -\dfrac{m}{10} - \dfrac{320}{m}\left(\dfrac{m}{10} - \dfrac{m^2}{3200}\right)\exp\left(-\dfrac{320t}{m}\right).$$

After a long fall the terminal speed is $|dy/dt| = m/10$, so setting $|dy/dt| = 24$ shows that $M = 240$ lbs.

17. $x(t) = \dfrac{2}{9}\dfrac{g(\rho_2 - \rho_1)}{\eta}a^2 t + \dfrac{4}{81}\dfrac{g(\rho_2 - \rho)a^4\rho_1}{\eta^2}$

$$\times \left[\exp\left(-\dfrac{9}{2}\dfrac{\eta t}{a^2\rho_1}\right) - 1\right]$$

The container reaches the surface at a time $t = T$ given by $x(T) = h$. As it will reach its terminal speed soon after release, the exponential term can be ignored so $T \approx 9\eta h/[2g(\rho_2 - \rho_1)a^2]$.

19. Try, for example, $\omega_1 = 1$ and $\omega_2 = 1.05$ with $0 \le t \le 20$. Use the result $\cos\omega_1 t + \cos\omega_2 t = 2\cos\{\frac{1}{2}(\omega_1 + \omega_2)t\}\cos\{\frac{1}{2}(\omega_1 - \omega_2)t\}$. The high frequency component is the term with argument $\frac{1}{2}(\omega_1 + \omega_2)t$, and this is modulated by the term with argument $\frac{1}{2}(\omega_1 - \omega_2)t$.

Exercise Set 6.3

3. Not linearly independent; $(1 + 2x)^2$ is a linear combination of 3, $-x$ and x^2

5. $y = c_1 \cosh x^2 + c_2 \sinh x^2$ (for all x)

7. General solution: $y = c_1 e^x + (c_2 \cos 3x + c_3 \sin 3x)e^{-2x}$ (for all x); solution of i. v. p. is $y = (13/18)e^x + (5/18)e^{-2x}\cos 3x - (1/18)e^{-2x}\sin 3x$

9. $y = c_1 x + c_2(8x^2 - 1)$ (for all x)

11. $y = c_1 x + c_2 \sin(x/2)$ (for all x)

13. $3/4 + (1/68)[9\sqrt{17}\sinh(x\sqrt{17}/2) + 17\cosh(x\sqrt{17}/2)]e^{-x/2}$

15. $y = ((5/4) + (1/2)\sin 2x - (1/4)\cos 2x)e^{-x}$

17. $y = (1/3)\cosh(x\sqrt{2}) + (2/3)\cos x$

19. $x(t) = A\cos(\omega_1 t - \phi) + B\cos(\omega_2 t - \psi)$, $y(t) = A\sin(\omega_1 t - \phi) - B\sin(\omega_2 t - \psi)$, with $\omega_1 = \frac{1}{2}(\sqrt{4c^2 + a} + a)$, $\omega_2 = \frac{1}{2}(\sqrt{4c^2 + a} - a)$. If the initial conditions make $B = 0$, the motion is in a circle with angular speed ω_1, whereas if they

make $A = 0$ the motion is also in a circle, but this time in the opposite sense with angular speed ω_2.

Exercise Set 6.4

1. $y = -(14/9) - (1/3)x + (4/5)e^{2x} + c_1 e^x + c_2 e^{-3x}$

3. $y = 5 + (3/8)e^x - (1/2)xe^x + (1/4)x^2 e^x + c_1 e^{-x} + c_2 xe^{-x}$

5. $y = -(2/5)\cos x - (1/5)\sin x + c_1 e^{-2x} + c_2 xe^{-2x}$

7. $y = (1/2)x + e^{-x} + c_1 e^{-x}\cos x + c_2 e^{-x}\sin x$

9. $y = -x + 2x^2 - (2/3)x^3 + c_1 + c_2 e^{-x}\cos x + c_3 e^{-x}\sin x$

11. $y = (7/144) + (1/12)x + (1/2)e^{2x} - e^{3x} - xe^{3x} + c_1 e^{3x} + c_2 e^{4x}$

13. $y = -(9/80)x\cos 4x + (3/80)x\sin 4x + (57/1600)\sin 4x + (3/200)\cos 4x + c_1 e^{-4x} + c_2 e^{2x}$

15. $y = (1/18)\cos 3x + (1/36)\sin 3x - (1/6)x\cos 3x + (1/3)x\sin 3x + c_1 \cos 3x + c_2 \sin 3x$

17. $y = (7/4) - (3/2)x + (1/2)x^2 - 3e^{-2x} - 3xe^{-2x} + c_1 e^{-x} + c_2 e^{-2x}$

19. $y = -(1/2)xe^{-2x}\cos x + c_1 e^{-2x}\cos x + c_2 e^{-2x}\sin x$

21. $y = (1/3)e^{-3x}\cos x + (5/3)e^{-3x}\cos 2x + (7/2)e^{-3x}\sin 2x$

23. $y = (7/9) - (16/9)\cos 3x + (4/9)\sin 3x - (1/3)x\cos 3x - (2/3)x\sin 3x$

25. $y = (1/5) + (1/8)e^{-x} + (67/40)e^x\cos 2x - (11/40)e^x\sin 2x$

27. $y = -(3/2) - (3/5)\cos x - (1/5)\sin x + e^x + (11/10)e^{-x}\cos x + (13/10)e^{-x}\sin x$

Exercise Set 6.5

1. $y = c_1 x + c_2/x^3$

3. $y = (c_1/x^2)\cos(\sqrt{5}\ln|x|) + (c_2/x^2)\sin(\sqrt{5}\ln|x|)$

5. $y = c_1 x^2 + c_2/x^4$

7. $y = c_1/x + c_2/x^4$

9. $y = (c_1/x^{2/3})\cos(\frac{1}{2}\sqrt{7}\ln|x|) + (c_2/x^{2/3})\sin(\frac{1}{2}\sqrt{7}\ln|x|)$

11. The general solution is given in Solution 3.

13. $y = c_1 x + c_2 x^2 + c_3 x^3$

Exercise Set 6.6

1. $y = c_1 e^x + c_2 e^{-2x} + (1/27)e^x - (1/9)xe^x + (1/6)x^2 e^x$

3. $y = c_1 e^{-2x} + c_2 e^{-3x} - 2e^{-2x} + 2xe^{-2x} - x^2 e^{-2x} + (1/3)x^3 e^{-2x}$

5. $y = c_1 e^x + c_2 x e^x - 2x e^x + 2x e^x \ln|x|$

7. $y = c_1 e^{-2x} \cos x + c_2 e^{-2x} \sin x + (1/4) x e^{-2x}$
$\cos x + (1/4) x^2 e^{-2x} \sin x$

9. $y = c_1 \cos 4x + c_2 \sin 4x - (26/4913) e^x - (4/289)$
$x e^x + (1/17) x^2 e^x$

11. $y = c_1 e^{-2x} + c_2 e^{-x} + 3e^{-2x} \ln(1 + e^x) + 3e^{-x}$
$\ln(1 + e^x)$

13. $y = c_1 \cos x + c_2 \sin x - 1 - \cos x +$
$2 \text{Arctanh}[\sin x/(1 + \cos x)] \sin x$

15. $y = c_1 x + c_2/x^3 - (4/7)\sqrt{x}$

17. $y = c_1(2x^2 - 1) + c_2 x(x^2 - 1)^{1/2} + x/3$

19. $y = x^3 - 2x^2 \ln x - x$

21. $y = 2\cos x - 2 + 4\text{Arctanh}\left(\dfrac{\sin x}{1 + \cos x}\right) \sin x$

23. $y_1(x) = x$, $y_2(x) = 1 - x$, $W(t) = -1$,
$$G(x, t) = \begin{cases} t(x - 1), & 0 \le t < x \\ x(t - x), & x < t \le 1 \end{cases}$$

25. $y_1(x) = \sin \lambda x$, $y_2(x) = \dfrac{\sin \lambda(1 - x)}{\sin \lambda}$, $W(t) = -\lambda$,
$$G(x, t) = \begin{cases} \dfrac{\sin \lambda t \sin \lambda (x - 1)}{\lambda \sin \lambda}, & 0 \le t < x \\ \dfrac{\sin \lambda x \sin \lambda (t - 1)}{\lambda \sin \lambda}, & x < t \le 1 \end{cases}$$

27. $y_1(x) = x - 1/x$, $y_2(x) = x - 4/x$, $W(t) = 6/t$,
$$y(x) = \frac{e^{-x}}{x}(1 + x) + \frac{1}{x e^2}(1 - x^2) + \frac{2}{3 x e}(x^2 - 4)$$

29. $y_1(x) = 3x - x^3$, $y_2(x) = 4x - x^3$, $W(t) = 2t^3$,
$y(x) = x(4 - x \ln x - x^2) - 2x \ln 2(3 - x^2)$

Exercise Set 6.7

1. $y_2 = e^{-2x}$

3. $y_2 = e^{-x} \sin x$

5. $y_2 = x \ln|x|$

7. $y_2 = (1/x) \cos x$

9. $y_2 = \ln|x|$

Exercise Set 6.8

1. $u'' + \left(\dfrac{9}{x} - \dfrac{1}{4x^2}\right) u = 0$ **3.** $y = c_1 e^x + c_2 x e^{-x}$

5. $y = e^{2x}(c_1 \cos x + c_2 \sin x)$

7. $y = c_1(1/x)\sin x + c_2(1/x)\cos x$

Exercise Set 6.9

1. $x_1 = c_1 e^{2t} - c_2 e^t$, $x_2 = -3c_1 e^{2t} + c_2 e^t$

3. $x_1 = -6e^{2t} + 6e^{-t}$, $x_2 = 4e^{2t} - 3e^{-t}$

5. $x_1 = (5/3) - 4e^t + 9e^{2t} - (25/6)e^{3t} - (3/2)e^{-t}$,
$x_2 = -(4/3) + 2e^t - 3e^{2t} + (25/12)e^{3t} + (1/4)e^{-t}$,
$x_3 = -(1/2)e^{-t} + 2/3 - 2e^t + 6e^{2t} - (25/6)e^{3t}$

Exercise Set 6.10

1. $\Phi(t) = \begin{bmatrix} e^t \cos t & e^t \sin t \\ -e^t \sin t & e^t \cos t \end{bmatrix}$

3. $\Phi(t) = \begin{bmatrix} \cos t & \sin t \\ \frac{1}{2}(\cos t + \sin t) & \frac{1}{2}(\sin t - \cos t) \end{bmatrix}$

5. $\Phi(t) =$
$\begin{bmatrix} e^{3t/2} \cos \frac{1}{2}t & e^{3t/2} \sin \frac{1}{2}t \\ e^{3t/2}\left(\sin \frac{1}{2}t - \cos \frac{1}{2}t\right) & -e^{3t/2}\left(\cos \frac{1}{2}t + \sin \frac{1}{2}t\right) \end{bmatrix}$

Exercise Set 6.11

1. $\Phi(t) = \begin{bmatrix} \sin t\sqrt{2} & \cos t\sqrt{2} \\ -\sqrt{2}\cos t\sqrt{2} & \sqrt{2}\sin t\sqrt{2} \end{bmatrix}$;
$x_1(t) = C_1 \sin t\sqrt{2} + C_2 \cos t\sqrt{2}$,
$x_2(t) = C_2 \sqrt{2} \sin t\sqrt{2} - C_1 \sqrt{2} \cos t\sqrt{2}$

3. $\Phi(t) =$
$\begin{bmatrix} -2e^{-t} \sin 2t & e^{-t}(\cos 2t - 2\sin 2t) \\ e^{-t}(\sin 2t + \cos 2t) & e^{-t} \sin 2t \end{bmatrix}$;
$x_1(t) = -(2C_1 + C_2)e^{-t} \sin 2t + C_2 e^{-t} \cos 2t$,
$x_2(t) = (C_1 + C_2)e^{-t} \sin 2t + C_1 e^{-t} \cos 2t$

5. $\Phi(t) = \begin{bmatrix} 1 & \sin 2t & \cos 2t \\ 0 & \cos 2t & -\sin 2t \\ 0 & \sin 2t & \cos 2t \end{bmatrix}$;
$x_1(t) = C_1 + C_2 \sin 2t + C_3 \cos 2t$,
$x_2(t) = -C_3 \sin 2t + C_2 \cos 2t$,
$x_3(t) = C_2 \sin 2t + C_3 \cos 2t$

7. $x_1(t) = \frac{95}{4} + \frac{11}{2}t - \frac{3}{2}C_1 e^{2t} - 2C_2 e^{-t}$,
$x_2(t) = -\frac{27}{2} - 3t + C_1 e^{2t} + C_2 e^{-t}$

9. $x_1(t) = -(1/5)\cos t + (3/5)\sin t - (1/3)e^{3t}$
$+ C_1 + C_2 e^{2t}$
$x_2(t) = (2/3)e^{3t} + (2/5)\sin t + (1/5)\cos t$
$+ C_1 - C_2 e^{2t}$

11. $x_1(t) = (1/8)\cos t + (1/4)\sin t + C_1 e^{t\sqrt{7}}$
$+ C_2 e^{-t\sqrt{7}}$
$x_2(t) = (1/8)\sin t + (1/3)C_1(\sqrt{7} - 2)e^{t\sqrt{7}}$
$- (1/3)C_2(\sqrt{7} + 2)e^{-t\sqrt{7}}$

13. $x_1(t) = -3 - (3/5)\cos t + (4/5)\sin t + C_2 e^{2t}$
$+ 2(C_1 + C_3)e^{-t}$,
$x_2(t) = 3 + C_1 e^{-t}$ $x_3(t) = -6 - (1/5)\cos t$
$+ (3/5)\sin t + 2C_2 e^{2t} + C_3 e^{-t}$

15. $x_1(t) = -3/5 - t - 2C_1 e^{-t} + (C_3 - C_2)e^{2t} \sin t$
$+ C_2 e^{2t} \cos t$
$x_2(t) = -4/5 + C_1 e^{-t} - C_2 e^{2t} \sin t + (C_3 - C_2)$
$e^{2t} \cos t$
$x_3(t) = 6/5 + 2C_1 e^{-t} + C_3 e^{2t} \sin t + (2C_2 - C_3)$
$e^{2t} \cos t$

17. $x_1(t) = -4/3 - e^{-t} + 2C_1 e^{3t} + (C_2 - C_3)\sin t$
$\quad + C_2 \cos t$
$\quad x_2(t) = 1/3 + t - (1/2)e^{-t} + C_1 e^{3t} - C_2 \sin t$
$\quad + (C_2 - C_3)\cos t$
$\quad x_3(t) = 2/3 - 2t + e^{-t} + 2C_1 e^{3t} + C_3 \sin t$
$\quad + (C_3 - 2C_2)\cos t$

19. $x_1(t) = C_1 e^{2t} + C_2 e^t, \quad x_2(t) = -3C_1 e^{2t} - C_2 e^t$

21. $x_1(t) = C_1 \sin t\sqrt{2} + C_2 \cos t\sqrt{2},$
$\quad x_2(t) = C_2\sqrt{2}\sin t\sqrt{2} - C_1\sqrt{2}\cos t\sqrt{2}$

23. $x_1(t) = (4C_2 - 17C_1)e^{-t}\sin 2t + C_2 e^{-t}\cos 2t$
$\quad x_2(t) = (C_2 - 4C_1)e^{-t}\sin 2t + C_1 e^{-t}\cos 2t$

25. $x_1(t) = -(2C_1 + C_2)e^{-t}\sin 2t + C_2 e^{-t}\cos 2t,$
$\quad x_2(t) = (C_1 + C_2)e^{-t}\sin 2t + C_1 e^{-t}\cos 2t$

27. $x_1(t) = -(7/5)\cos t - (16/5)\sin t - 9t - 9/2$
$\quad - 2C_1 e^t - (3/2)C_2 e^{-2t}$
$\quad x_2(t) = (3/5)\cos t + (9/5)\sin t + 2 + 5t + C_1 e^t$
$\quad + C_2 e^{-2t}$

29. $x_1(t) = -(4/5)t^2 - (16/25)t + 8/125 + (2C_1 + C_2)$
$\quad e^t \sin 2t + C_2 e^t \cos 2t$
$\quad x_2(t) = (2/25)t - 26/125 + (3/5)t^2 - (C_1 + C_2)$
$\quad e^t \sin 2t + C_1 e^t \cos 2t$

31. $x_1(t) = -(3/4) - (1/2)t + (5/3)e^t + (1/12)e^{-2t},$
$\quad x_2(t) = -3/2 + (5/3)e^t - (1/6)e^{-2t}$

33. $x_1(t) = 3t - e^t + 1 - 2te^t,$
$\quad x_2(t) = -6t + 1 + 2te^t$

35. $x_1(t) = -5/2 + (1/10)\cos t + (3/10)\sin t + 2e^{2t}$
$\quad - (61/10)e^{-2t} + (15/2)e^{-t}$
$\quad x_2(t) = -15/2 + 6t - (1/5)\sin t + (3/5)\cos t$
$\quad - 2e^{2t} - (61/10)e^{-2t} + 15e^{-t}$
$\quad x_3(t) = 15/4 - (5/2)t + (1/10)\sin t - (3/10)$
$\quad \cos t + (61/20)e^{-2t} - (15/2)e^{-t}$

Exercise Set 6.12

1. Saddle point at $(0, 0)$

3. Stable focus at $(0, 0)$

5. Stable focus at $\left(\frac{46}{13}, \frac{2}{13}\right)$

7. Saddle point at $(-2, 0)$ and an unstable node at $(2, 0)$

9. Saddle point at $(0, 0)$ and linear theory predicts a center at $\left(\frac{1}{4}, -\frac{1}{2}\right)$. An examination of the phase portrait shows that the point $\left(\frac{1}{4}, -\frac{1}{2}\right)$ is also a center of the nonlinear system.

11. For $\varepsilon \leq -2$, the point $(0, 0)$ is a stable node.
For $-2 < \varepsilon < 0$, the point $(0, 0)$ is a stable focus.
For $0 < \varepsilon < 2$, the point $(0, 0)$ is an unstable focus.
For $\varepsilon \geq 2$, the point $(0, 0)$ is an unstable node.

Exercise Set 7.1

5. $1/(s - 2)^2$

9. $s/(s^2 + 4s + 8)$

7. $1/s^2 - 2/s^3 + 6/s^4$

11. $1/(5 + 3)^2$

13. $(s^3 - 2s - 5)/[s^2(s^2 + 2s + 5)]$

15. $e^{\pi/2}e^{-\pi s/2}(s - 1)/(s^2 - 2s + 2)$

17. $\pi e^{-\pi s/2}/(2s) + e^{-\pi s/2}/s^2 - \pi e^{-\pi s}/s - e^{-\pi s}/s^2$

19. $-e^{-\pi/2}e^{-\pi s/2}/(s^2 + 2s + 2)$

21. $-1/4 + (5/4)\cos 2t$

23. $5/9 + \sin 3t - (5/9)\cos 3t$

25. $(9/5)te^{-2t} - (96/25)e^{-2t} + (13/75)e^{3t} + (14/3)e^{-3t}$

27. $(1/4)e^{-t} + (1/2)te^t + (3/4)e^t$

29. $-(5/8)e^t + (13/12)e^{3t} + (13/24)e^{-3t}$

31. $F(s) = 1/s + (e^{-2as} - 2e^{-as})/s$

33. $F(s) = k/s^2 - ke^{-s}/s^2$

35. $F(s) = k(1 + e^{-2as} - 2e^{-as})/as^2$

Exercise Set 7.2

3. $s^3 F(s) - s^2 - 1$ **5.** $(1 - se^{-\pi s/2})/(s^2 + 1)$

7. $(1/10)\cos t - (3/10)\sin t + (5/2)e^t - (8/5)e^{2t}$

9. $-(8/81) - (1/9)t + 2e^t + (8/81)e^{-9t}$

11. $2/(s + 2) + 6/(s + 2)^4$

13. $(4s + 4)/(s^2 + 2s + 5)^2$

15. $3/[s^2 - 4s + 13]$ **17.** $(1/3)e^{2t}\sin 3t$

19. $e^{-t}(2\sin 2t - 3\cos 2t)$

21. $-(1/18)e^{-t} + e^{2t}[(1/18)\cos 3t + (5/18)\sin 3t]$

23. $3/2 + e^{-2t}[(3/2)\cos 2t - (9/2)\sin 2t]$

25.

27.

29.

31.

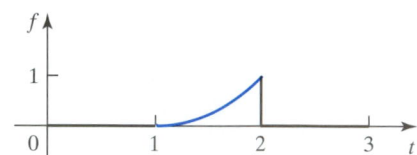

33. $6e^{-3s}/s^4$ **37.** $3e^{-4s}/(s^2 - 9)$

35. $2e^{-3\pi s/2}/(4 + s^2)$ **39.** $H(t - 2)\cos(2t - 4)$

41. $H(t - \pi/2)e^{\pi - 2t}(\cos t + \sin t)$

43. $H(t - 4)e^{8 - 2t}\{(1/3)\sin(3t - 12) + \cos(3t - 12)\}$

45. $y(t) = 3e^{-2t} - 2e^{-3t} + (1/10)(3e^{3(\pi - t)} - 4e^{2(\pi - 1)} - \cos t - \sin t)H(t - \pi)$

47. $y(t) = -(3/2)e^{2t} + (4/3)e^{3t} + 1/6 + (1/36)(5 + 6t - 27e^{2t - 4} + 28e^{3t - 6})H(t - 2)$

49. $y(t) = -e^{-t}\cos 3t - (1/3)e^{-t}\sin 3t + (1/9)e^{-t}(1 - \cos(3t - 3))H(t - 1)$

51. $2(3s^2 - 18s + 26)/(s^2 - 6s + 10)^3$

53. $48s(s - 2)(s - 4)/(s^2 - 4s + 8)^4$

55. $2e^{-3s/2}(s^2 - 4)/(s^4 - 16a^4)$

57. $1/(27s^4 + 12s^2)$ **59.** $1/(s + se^{-ks})$

61. $(1/s^2)\tanh ks$ **63.** $k/(as^2) - ke^{-as}/(s - se^{-as})$

65. $k(1 - 2ase^{-as} - e^{-2as})/[as^2(1 - e^{-2as})]$

67. $e^{-t} - e^{-2t}$

69. $t^2 + 2\cos t - 2$

71. $(1/2)(\sin t + t\cos t)$

73. $1/[s^2(s + 2)]$

75. $1/[s^2(s^2 + 2s + 2)]$

77. $(1/4)t - (1/8)\sin 2t$

79. $(1/2)t\cosh t + (1/2)\sinh t$

81. $y(t) = t$

83. $y(t) = t^2 + 2t + 2 - e^{t/2}\{(2/\sqrt{3})\sin(t\sqrt{3}/2) + 2\cos(t\sqrt{3}/2)\}$

85. $y(t) = 1 - (4/\sqrt{3})e^{-2t}\sinh t\sqrt{3}$

87. $y(t) = (1/2)(1 + \cosh t\sqrt{2})$

89. $(12s^2 - 16)/[s(s^2 + 4)^3]$

91. $(\sin at - at\cos at)/(2a^3)$

93. $(1/2)\ln\{(s + 2)/(s - 2)\}$

95. $(2/t)(1 - \cosh at)$

97. $f(t) = 3/2 + (1/2)\cos 3t$; $f(0) = 1$, $f'(0) = 0$, $f''(0) = -3$

99. $f(t) = e^{2t}(1 + t)$; $f(0) = 1$, $f'(0) = 3$, $f''(0) = 8$

101. $-4/\pi$

103. $-8/(21\pi)$

105. $y(t) = (2/9)\sin^2(3t/2) + (1/3)\sin(3t - 3)H(t - 1)$

107. $y(t) = (1/2)e^{-t}(1 + t) - (1/2)\cos t + (t - \pi)e^{\pi - t}H(t - \pi)$

109. $y(t) = 1/2 + \cos 2t - (1/2)\cos^2 t - (1/2)(1 - \cos^2(t - 1))H(t - 1) + (1/2)\sin(2t - 4)H(t - 2)$

Exercise Set 7.3

1. $x(t) = 27/49 + (8/7)t - (27/49)e^{-7t}$,
$y(t) = 71/49 + (20/7)t + (27/49)e^{-7t}$

3. $x(t) = 3/2 + \sqrt{2}\sinh t\sqrt{2} - (5/2)\cosh t\sqrt{2}$,
$y(t) = 1/2 + (3/2)\sinh t\sqrt{2} + (1/2)\cosh t\sqrt{2}$

5. $x(t) = 5/2 + (1/2)t + e^t\{3^{3/2}\sinh t\sqrt{3} - (1/2)\cosh t\sqrt{3}\}$
$y(t) = 1 + (1/2)t + e^t\{(1/6)\sqrt{3}\sinh t\sqrt{3} - 3\cosh t\sqrt{3}\}$

7. $x(t) = 7/8 + (5/4)t - (1/4)t^2 + (1/8)e^{-2t}$,
$y(t) = 1/8 + (7/4)t - (1/4)t^2 - (1/8)e^{-2t}$
$z(t) = 9/8 + (3/4)t - (1/4)t^2 - (1/8)e^{-2t}$

9. $x(t) = -1 + (1/4)e^{-t} + (1/4)e^t(3 + 2t)$,
$y(t) = 1 + 2t + (1/4)e^{-t} + (1/4)e^t(2t - 1)$,
$z(t) = -(1/4)e^{-t} + (1/4)e^t(1 + 2t)$

11. $\begin{bmatrix} \frac{1}{4}e^{-2t} + \frac{3}{4}e^{2t} & \frac{3}{4}e^{2t} - \frac{3}{4}e^{-2t} \\ \frac{1}{4}e^{2t} - \frac{1}{4}e^{-2t} & \frac{3}{4}e^{-2t} + \frac{1}{4}e^{2t} \end{bmatrix}$

13. $\begin{bmatrix} \frac{1}{4}e^{-3t} + \frac{3}{4}e^{5t} & \frac{3}{4}e^{5t} - \frac{3}{4}e^{-3t} \\ \frac{1}{4}e^{5t} - \frac{1}{4}e^{-3t} & \frac{3}{4}e^{-3t} + \frac{1}{4}e^{5t} \end{bmatrix}$

15. $\begin{bmatrix} e^{2t}\cos 4t + \frac{1}{2}e^{2t}\sin 4t & -\frac{5}{4}e^{2t}\sin 4t \\ e^{2t}\sin 4t & e^{2t}\cos 4t - \frac{1}{2}e^{2t}\sin 4t \end{bmatrix}$

17. $\begin{bmatrix} e^{2t}\cos 2t & -2e^{2t}\sin 2t \\ \frac{1}{2}e^{2t}\sin 2t & e^{2t}\cos 2t \end{bmatrix}$ **19.** $\begin{bmatrix} e^{6t} & -te^{6t} \\ 0 & e^{6t} \end{bmatrix}$

21. $\begin{bmatrix} e^{-2t} & 4te^{-2t} \\ 0 & e^{-2t} \end{bmatrix}$

23. $\begin{bmatrix} e^{5t} & \frac{11}{5}e^{5t} - e^t - \frac{6}{5} & \frac{8}{5}e^{5t} - e^t - \frac{3}{5} \\ 0 & 2 - e^t & 1 - e^t \\ 0 & 2e^t - 2 & 2e^t - 1 \end{bmatrix}$

27. $y(t) = (1/10)e^{2t} - (3/10)e^{-t}\sin t - (1/10)e^{-t}\cos t$,
$W(t) = e^{-t}\sin t$

29. $y(t) = (1/16)e^{-5t} + (1/4)te^{-t} - (1/16)e^{-t}$,
$W(t) = (1/4)(e^{-t} - e^{-5t})$

31. $x(t) = -1 - (5/14)e^{-t} - (1/7)e^{6t} + (3/2)e^t$
$y(t) = -3/2 - (3/14)e^{-t} + (3/14)e^{6t} + (3/2)e^t$

39. $x(t) = \sin t - (1/3)\sin 2t$

41. $y(x) = \dfrac{M}{24a\,EI}x^4 + \dfrac{Q}{6EI}(x - 3a/4)^3\,H(x -$
$3a/4) + \dfrac{a}{384\,EI}(16M+9Q)x^2 - \dfrac{Q}{192\,EI}$
$(16M + 5Q)x^3;\, w(x) = M/a + Q\delta(x - 3a/4)$

43. $i(t) = \dfrac{E_0 C}{\sqrt{R^2C^2 - 4LC}}\exp\left(-\dfrac{Rt}{2L}\right)$
$\times \left\{\exp\left(\dfrac{t}{2}\dfrac{\sqrt{R^2C^2 - 4LC}}{LC}\right)\right.$
$\left. - \exp\left(-\dfrac{t}{2}\dfrac{\sqrt{R^2C^2 - 4LC}}{LC}\right)\right\}$

The solution is oscillatory if $4L > R^2C$; otherwise it behaves exponentially.

45. $x(t) = Qe^{-2t},\ y(t) = 2Q(e^{-2t} - e^{-3t}),\ z(t) = 6Q(e^{-2t} - e^{-3t} - te^{-3t})$ so $w(t) = Q(1 - 9e^{-2t} + 8e^{-3t} + 6te^{-3t})$. After 1, 2, and 3 time units $w(t)/Q = 48\%, 88\%$, and 98%, respectively.

Exercise Set 7.4

1. (a) Order 3, roots $s = 1, s = -2 \pm 4i$, unstable
  (b) Order 3, roots $s = -2, s = -1 \pm 3i$, stable
  (c) Order 2, roots $s = -\frac{1}{3} \pm i$, stable

Exercise Set 8.1

1. $y(x) = 1 - x + (1/2)x^2 - (1/6)x^3 + (7/24)x^4 - (19/120)x^5 + \cdots$

3. $y(x) = -1 + x - x^2 + x^3 - (3/4)x^4 + (11/20)x^5 + \cdots$

5. $y(x) = 1 + x - (1/2)x^2 + (1/3)x^3 + (5/8)x^4 - (4/15)x^5 + \cdots$

7. $y(x) = 2 - (1/3)x + (1/18)x^2 + (35/162)x^3 - (89/1944)x^4 + (197/29160)x^5 + \cdots$

9. $y(x) = a + bx + (1/3)bx^3 - (1/12)ax^4 + (1/20)bx^5 - (1/45)ax^6 + (1/252)bx^7 + \cdots$

11. $y(x) = a + bx + \{-(1/2)a + 1/2\}x^2 + \{-(2/3)b + 1/6\}x^3 + \{(11/24)a - 3/8\}x^4 + \cdots$

13. $y(x) = a + bx - (1/6)ax^3 - (1/12)bx^4 + (1/180)ax^6 + (1/504)bx^7 + \cdots$

15. $y(x) = a + bx - (1/4)ax^2 + (1/12)(2 - b)x^3 + (1/96)(5a - 12b)x^4 + \cdots$

Exercise Set 8.2

1. $y(x) = 2 - 3x - x^2 + x^3 - (3/10)x^5 + (1/10)x^6 + \cdots$

3. $y(x) = 1 - 3x + (3/2)x^2 - (2/3)x^3 + (2/3)x^4 - (43/120)x^5 + \cdots$

5. $y(x) = 2 - x + x^2 + (1/12)x^4 + (1/40)x^6 + \cdots$

7. $y(x) = 1 - x + x^2 - (1/2)x^3 + (1/3)x^4 - (2/15)x^5 + \cdots$

9. $y(x) = 1 - x - (1/2)x^2 + (5/6)x^3 - (11/24)x^4 + (67/120)x^5 + \cdots$

11. $y(x) = 1 + 4x + 3x^2 + 3x^3 + (11/4)x^4 + (31/10)x^5 + \cdots$

13. $y(x) = 2 - 3(x - 1) + (7/3)(x - 1)^2 - (53/54)(x - 1)^3 + (11/81)(x - 1)^4 + (319/3240)(x - 1)^5 + \cdots$

15. $y(x) = 1 + 5(x - 2) + 8(x - 2)^2 + 6(x - 2)^3 + (13/6)(x - 2)^4 + (7/30)(x - 2)^5 + \cdots$

17. Proceed as outlined in the exercise

19. Proceed as outlined in the exercise

Exercise Set 8.3

1. Regular singular point at $x = 1$
3. Irregular singular point at $x = -1$
5. Irregular singular point at $x = -4$
7. Irregular singular point at $x = 3$

Exercise Set 8.4

1. (a) $a_0 x^{c-2} + (a_0 + a_1)x^{c-1} + \sum_{n=0}^{\infty}(2a_n + a_{n+1} + a_{n+2})x^{n+c}$
  (b) $3a_0 x^c + \sum_{n=0}^{\infty}(2a_n + 3a_{n+1})x^{n+c+1}$

3. (a) $1 + (1/2)x - (1/12)x^2 + (1/24)x^3 - (9/720)x^4 + \cdots$
  (b) $1 - (1/4)x^2 - (5/24)x^3 - (1/16)x^4 - (11/480)x^5 - \cdots$
  (c) $1 - (3/2)x + (4/3)x^2 - (7/6)x^3 + (31/30)x^4 + \cdots$

5. (a) $\ln x - 2x - (1/4)x^2 - (4/9)x^3 - (15/32)x^4 + \cdots + \text{constant}$
  (b) $\ln x - (1/4)x^2 + (2/9)x^3 - (1/32)x^4 - (8/75)x^5 + \cdots + \text{constant}$
    (Hint: write the integrand as $\frac{1}{x}\frac{e^x}{(1+x+x^2)}$)

7. $c = 1, y_1(x) = x\{1 - (1/10)x + (1/280)x^2 - (1/15120)x^3 + \cdots\}$;
  $c = -1/2, y_2(x) = x^{-1/2}\{1 + (1/2)x - (1/8)x^2 + (1/144)x^3 + \cdots\}$

9. $c = 1$, $y_1(x) = x\{1 + (2/5)x + (2/35)x^2 + (4/945)x^3 + \cdots\}$;
$c = -1/2$, $y_2(x) = x^{-1/2}\{1 - 2x - 2x^2 - (4/9)x^3 - (2/45)x^4 + \cdots\}$

11. $y_1(x) = 1 + \frac{1}{2!}x^2 + \frac{7}{4!}x^4 + \frac{49}{240}x^6 + \cdots$, $y_2(x) = x + \frac{1}{2}x^2 + \frac{13}{40}x^5 + \frac{403}{1680}x^7 + \cdots$

13. $y_1(x) = 1 + x + \frac{2}{4}x^2 + \frac{2\cdot5}{4\cdot9}x^3 + \frac{2\cdot5\cdot10}{4\cdot9\cdot16}x^4 + \cdots$
$y_2(x) = y_1(x)\ln x - 2x - x^2 - (14/27)x^3 - \cdots$

15. $c = 1$ (twice), $y_1(x) = xe^{-2x}$; $c = 1$, $y_2(x) = y_1(x)\{\ln x + 2x + x^2 + (4/9)x^3 + \cdots\}$

17. $c = 2$, $y_1(x) = x^2 e^{-x}$; $c = 1$, $y_2(x) = y_1(x)\{\ln x - 1/x + (1/2)x + (1/12)x^2 + \cdots\}$

19. $c = 1/4$ (twice), $y_1(x)$
$= x^{1/4}\left\{1 - x + \frac{1}{2^2}x^2 - \frac{1}{2^2 3^2}x^3 + \frac{1}{2^2 3^2 4^2}x^4 + \cdots\right\}$
$c = 1/4$, $y_2(x) = y_1(x)\{\ln x + 2x + (5/4)x^2 + (23/27)x^3 + \cdots\}$

21. $c = 3$, $y_1(x) = x^3\{1 - (3/5)x + (1/5)x^2 - (1/21)x^3 + (1/112)x^4 + \cdots\}$;
$c = -1$, $y_2(x) = x^{-1}\{1 - (1/3)x\}$

23. $c = 2$, $y_1(x) = x^2(1 - (2/5)x + (1/10)x^2 - (2/105)x^3 + \cdots)$; $c = -2$, $y_2(x) = y_1(x)$
$\left[\frac{1}{168}\ln x - \frac{1}{4x^4} + \frac{1}{15x^3} + \frac{1}{100x^2} - \frac{13}{1750x} + \cdots\right]$

25. $c = 2 \pm 4i$, $y_1(x) = x^2 \cos(4\ln|x|)$; $y_2(x) = x^2 \sin(4\ln|x|)$

27. Shift the critical point at $x = -1$ to the origin by setting $X = x + 1$ and solve the resulting equation to get

$$c = 1, \quad y_1(X) = 1 + \frac{1}{2\cdot3}X^2 + \frac{1}{(2\cdot4)(3\cdot7)}X^4$$
$$+ \frac{1}{(2\cdot4\cdot6)(3\cdot7\cdot9)}X^6 + \cdots$$

and

$$c = 1/2, \quad y_2(X)$$
$$= X^{1/2}\left(1 + \frac{1}{2\cdot5}X^2 + \frac{1}{(2\cdot4)(5\cdot9)}X^4 + \cdots\right)$$

The required results follows by substituting $X = x + 1$. The results converge in an interval of the form $0 < x + 1 < d$ for some suitable d.

Exercise Set 8.5

1. $\Gamma(5/2) = (3/4)\sqrt{\pi}$, $\Gamma(-5/2) = -(8/15)\sqrt{\pi}$, $\Gamma(9/2) = (105/16)\sqrt{\pi}$

3. $\Gamma(5/4) = (1/4)\Gamma(1/4)$, $\Gamma(-5/4) = -(4/5)\Gamma(-1/4)$, $\Gamma(7/4) = -(3/16)\Gamma(-1/4)$

5. $5^{n+1}\Gamma(6/5 + n + 1)/\Gamma(6/5)$

7. $3^{n+1}\Gamma(8/3 + n)/\Gamma(5/3)$

9. $(\frac{1}{2} - n)\Gamma(\frac{1}{2} - n) = \Gamma(\frac{3}{2} - n)$, so $\Gamma(\frac{1}{2} - n) = -\Gamma(\frac{3}{2} - n)/(n - \frac{1}{2})$, similarly, $(\frac{3}{2} - n)\Gamma(\frac{3}{2} - n) = \Gamma(\frac{5}{2} - n)$, so $\Gamma(\frac{3}{2} - n) = -\Gamma(\frac{5}{2} - n)/(n - \frac{3}{2})$ giving $\Gamma(\frac{1}{2} - n) = (-1)^2\Gamma(\frac{5}{2} - n)/(n - \frac{1}{2}) \times (n - \frac{3}{2})$. Continuing this process leads to
$\Gamma(\frac{1}{2} - n) = (-1)^n\Gamma(\frac{1}{2})/(n - \frac{1}{2})(n - \frac{3}{2})\ldots(\frac{1}{2}) = (-1)^n\sqrt{\pi}/(n - \frac{1}{2})(n - \frac{3}{2})\ldots(\frac{1}{2})$

11. $\Gamma(2n)$
$= (2n - 1)! = (2n - 1)(2n - 2)\ldots3\cdot2\cdot1$
$= 2^{2n-1}(n - \frac{1}{2})(n - 1)(n - \frac{3}{2})\ldots(\frac{3}{2})\cdot1$
$= 2^{2n-1}\{(n - \frac{1}{2})(n - \frac{3}{2})\ldots(\frac{1}{2})\}$
$\quad \times\{(n - 1)(n - 2)\ldots2\cdot1\}$
$= 2^{n-1}\{(n - \frac{1}{2})(n - \frac{3}{2})\ldots(\frac{1}{2})\}\Gamma(n)$
$= 2^{n-1}\{(n - \frac{1}{2})(n - \frac{3}{2})\ldots(\frac{1}{2})\Gamma(\frac{1}{2})\}\Gamma(n)/\Gamma(\frac{1}{2})$
$= 2^{n-1}\Gamma(n + \frac{1}{2})\Gamma(n)/\sqrt{\pi}$

13. Make the substitution $t = u^2$ in the definition of $\Gamma(x)$ in (32).

15. $\psi(x+n) = d/dx\{\ln\Gamma(x+n)\} = d/dx\{\ln[(x + n - 1)\Gamma(x+n-1)]\} = 1/(x+n-1) + d/dx\{\ln\Gamma(x+n-1)\}$ a repetition of this argument leads to $\psi(x+n) = 1/(x+n-1) + 1/(x+n-2) + \cdots 1/x + \psi(x) = \Sigma_{k=0}^{n-1}1/(x+k) + \psi(x)$

17. The result follows directly after integrating by parts.

Exercise Set 8.6

1. $J_2(x) = (1/8)x^2 - (1/96)x^4 + (1/3072)x^6 - (1/184320)x^8 + (1/17694720)x^{10} - (1/2477260800)x^{12} + \cdots$

5. 6 terms **7.** 6 terms **9.** 6 terms

11. $(1/4)x^2 - (1/64)x^4 + (1/2304)x^6 - (1/147456)x^8$; max magnitude of error is $a^{10}/14745600$

12 to 17. If $x = \lambda X$, then $d/dx = (dX/dx)d/dX = (1/X)d/dx$. Substitute $x = \lambda X$ and use results (64)–(67).

19. The first two limits follow from the series for $J_v(x)$ in (54). The third follows by taking the limit as $x \to \infty$ in result (70):

$$\int_0^\infty J_1(x)dx = -\int_0^\infty J_0'(x)dx = [-J_0(x)]_0^\infty = 1.$$

21. $\mathcal{L}\{J_0(x)\} = \int_0^\infty e^{-xs}J_0(x)dx = 1/(s^2 + 1)^{1/2}$. Setting $s = 0$ gives $\int_0^\infty J_0(x)dx = 1$. From (67)

with $v = 2n + 1$ we have $\int_0^\infty J_{2n}(x)dx -$
$\int_0^\infty J_{2n+2}(x)dx = 2[J_{2n+1}(x)]_0^\infty = 0$.
As $\int_0^\infty J_0(x)dx = 1$ we have $1 = \int_0^\infty J_0(x)dx = \int_0^\infty J_1(x)dx = \int_0^\infty J_2(x)dx = \cdots$

23. $\displaystyle\int J_4(x)dx = -2J_1(x) - 2J_3(x) + \int J_0(x)dx$

25. $\displaystyle\int xJ_1(x)dx = -xJ_0(x) + \int J_0(x)dx$

Exercise Set 8.7

1. $y(x) = C_1 J_2(x) + C_2 Y_2(x)$
3. $y(x) = C_1 J_0(x) + C_2 Y_0(x)$
5. $y(x) = C_1 J_0(x^2) + C_2 Y_0(x^2)$
7. $y(x) = C_1 J_2(2x) + C_2 Y_2(2x)$
9. $y(x) = x^{1/2}\{C_1 J_0(2x) + C_2 Y_0(2x)\}$
11. $a = 1, b = 1, c = 2, n = 1; y(x) = x Z_1(x^2)$
13. $a = 1, b = 3, c = 1, n = 0; y(x) = x Z_0(3x)$
15. $a = 2, b = 2, c = 4, n = 1; y(x) = x^3 Z_1(2x^4)$
17. For u to depend on J_0 and Y_0, we must set $a = 3$ and $v = 1$. Thus the general solution for u is $u(x) = AJ_0(kx) + BY_0(kx)$, so the general solution for y is $y(x) = (1/x)(AJ_0(kx) + BY_0(kx))$.

Exercise Set 8.8

5. Replacing $\sinh x$ and $\cosh x$ by their definitions in terms of exponentials and comparing with (106) shows that $C_1 = C_2 = \sqrt{(2/\pi)}$, so

$$I_{1/2}(x) = \sqrt{2/\pi x}\,\sinh x \quad \text{and}$$
$$I_{-1/2}(x) = \sqrt{2/\pi x}\,\cosh x.$$

Using this with the result of Exercise 2 gives

$$I_{3/2}(x) = -\sqrt{\frac{2}{\pi x}}\left(\frac{\sinh x}{x} - \cosh x\right) \text{ and}$$
$$I_{-3/2}(x) = -\sqrt{\frac{2}{\pi x}}\left(\frac{\cosh x}{x} - \sinh x\right).$$

7. Replace x by ix in $J_{\pm 1/2}(x)$ and $J_{\pm 3/2}(x)$ and remove any multiplicative factors i to obtain the results of Exercise 5.

9. Substituting the series for $I_\nu(x)$ and $I_{-\nu}(x)$ into the expression on the left of Exercise 8 shows that C, the coefficient of the term in $(1/x)$, is given by $C = -2\nu/\{\Gamma(1 + \nu)\Gamma(1 - \nu)\}$. Using $\Gamma(1 + \nu) = \nu\Gamma(\nu)$ and the result $\Gamma(\nu)\Gamma(1 - \nu) = \pi/\sin \pi\nu$ then gives $C = -(2/\pi)\sin \pi\nu$.

11. The expression $(\frac{d^2}{dr^2} + \frac{1}{r}\frac{d}{dr} + 1)(\frac{d^2}{dr^2} + \frac{1}{r} - 1)R$ is equal to the left-hand side of the governing equation, so $\frac{d^2R}{dr^2} + \frac{1}{r}\frac{dR}{dr} + R = 0$ and $\frac{d^2R}{dr^2} + \frac{1}{r}\frac{dR}{dr} - R = 0$ are both special solutions of the original fourth order equation. They have the respective solutions $R_1(r) = AJ_0(r) + BY_0(r)$ and $R_2(r) = CI_0(r) + DK_0(r)$, so the general solution of the original equation is $R(r) = R_1(r) + R_2(r)$. In a particular problem the initial conditions will determine the arbitrary constants $A, B, C,$ and D.

Exercise Set 8.10

1. $\dfrac{d}{dx}[xe^{-x}y'] + \lambda e^{-x}y = 0$ (Laguerre's equation)

3. $\dfrac{d}{dx}[(1 - x^2)^{1/2}y'] + \lambda(1 - x^2)^{-1/2}y = 0$
(Chebyshev's equation)

5. $\lambda_n = n^2\pi^2/L^2, n = 1, 2, \ldots, \varphi_n = \sin\dfrac{n\pi x}{L}$

7. $\lambda_n = (2n - 1)^2\pi^2/4, n = 1, 2, \ldots,$
$\varphi_n = \cos\dfrac{(2n - 1)\pi x}{2}$

9. $\lambda_n = k_n^2$ where k_n are the roots of $\tan x = 2x$, $\varphi_n = \sin k_n x, \lambda_1 = k_1^2 \approx (1.166)^2 = 1.340, \lambda_2 = k_2^2 \approx (4.604)^2 = 21.197$

11. $\lambda_n = n^2\pi^2, n = 0, 1, \ldots,$
$\varphi_n = \{1, \cos n\pi x, \sin n\pi x\}$

13. General solution $y = C_1\cos(k \ln x) + C_2\sin(k \ln x)$, Eigenvalues $\lambda_n = k_n^2 = \left(\dfrac{n\pi}{2\ln 2}\right)^2,$
$\varphi_n = \sin\left(\dfrac{n\pi \ln x}{2\ln 2}\right)$

15. $\|\varphi_n\| = \sqrt{L/2}$ **16.** $\|\varphi_n\| = 1/\sqrt{2}$
17. $\|\varphi_0\| = \sqrt{L}, \|\varphi_n\| = \sqrt{L/2}, n = 1, 2, \ldots.$
19. An upper bound to λ_1 is

$$\int_0^\pi 4(\pi - x)^2 dx \Big/ \int_0^\pi x^2(2\pi - x)^2 dx = 5/2\pi^2$$
$$= 0.2533.$$

When Φ is substituted into the Rayleigh quotient, the constant C cancels.

21. An upper bound to λ_1 is

$$\left\{\left(\int_0^1 x(1 - 2x)^2 dx + \int_0^1 x(1 - x)^2 dx\right)\Big/ \int_0^1 x^3(1 - x)^2 dx\right\} = 15,$$

so $j_{1,1} \approx \sqrt{15} = 3.87$.

Exercise Set 8.11

1. $(1/3)\,P_0(x) + (12/5)\,P_1(x) - (4/3)\,P_2(x)$
   $+ (8/5)\,P_3(x)$

3. $(42/35)\,P_0(x) + 2\,P_1(x) + (18/7)\,P_2(x)$
   $+ (8/35)\,P_4(x)$

5. $f(x) = (3/4)\,P_0(x) - (1/4)\,P_1(x) + (5/16)\,P_2(x)$
   $+ (7/16)\,P_3(x) + \cdots$

7. $f(x) = (5/8)\,P_0(x) + (9/32)\,P_1(x) - (45/64)$
   $P_2(x) - (133/512)\,P_3(x) + \cdots$

9. $f(x) = (1/2)(e - 1/e)\,P_0(x) + (3/e)\,P_1(x) + (5/2)$
   $(e - 7/e)\,P_2(x) - (1/2)(35e - 259/e)\,P_3(x) + \cdots$

11. $-(7/8)\,T_0(x) - T_1(x) - (1/2)\,T_2(x) + (3/8)\,T_3(x)$

13. $(15/4)\,T_0(x) + (1/4)\,T_1(x) + T_2(x) - (1/4)\,T_3(x)$
    $+ (1/4)\,T_4(x)$

15. $f(x) = (1/2\pi)(5\pi - 2)\,T_0(x) + (1/2\pi)(\pi + 4)$
    $T_1(x) - (2/3\pi)\,T_2(x) - (2/3\pi)\,T_3(x) + \cdots$

Exercise Set 9.1

1. 2π 3. π 5. 12π

7. $f(x)$ is not periodic   9. $f(x)$ is not periodic

11. (a) $(1/2)\sin 2x$ (b) $\cos 2x$ (c) $(1/2)\sin 2x +$
    $(1/2)\sin 4x$

17. If $f(-x) = f(x)$ and $g(-x) = g(x)$ then $f(-x) +$
    $g(-x) = f(x) + g(x)$, so the sum is an even func-
    tion. If $f(-x) = -f(x)$ and $g(-x) = -g(x)$, then
    $f(-x) + g(-x) = -f(x) - g(x)$, so the sum is an
    odd function.

19. (a) $2L^2/\pi$ (b) $-L^2/\pi$ (c) $2L^2/3\pi$

23. $f(x) = \dfrac{a + b}{2} - \dfrac{2(a - b)}{\pi} \displaystyle\sum_{n=0}^{\infty} \dfrac{\sin(2n + 1)x}{2n + 1}$.
    Graph for $a = 1$, $b = 3$.

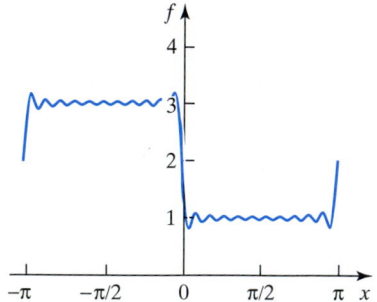

25. $f(x) = \dfrac{1}{2} + \dfrac{4}{\pi^2} \displaystyle\sum_{n=1}^{\infty} \dfrac{\cos(2n - 1)\pi x}{(2n - 1)^2}$

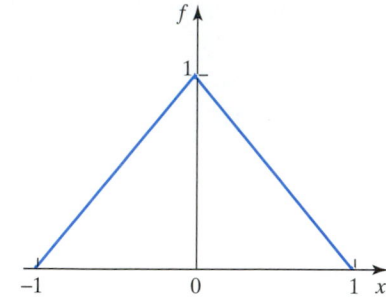

27. $f(x) = \dfrac{2}{\pi} - \dfrac{4}{\pi} \displaystyle\sum_{n=1}^{\infty} \dfrac{\cos 2nx}{4n^2 - 1}$

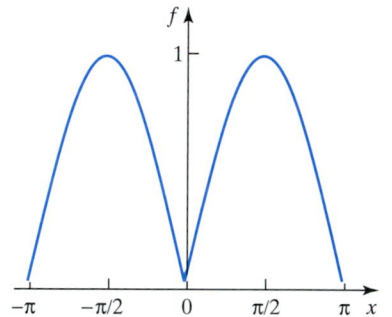

29. $f(x) = \dfrac{1}{\pi} + \dfrac{\sin x}{2} - \dfrac{2}{\pi} \displaystyle\sum_{n=1}^{\infty} \dfrac{\cos 2nx}{(4n^2 - 1)}$

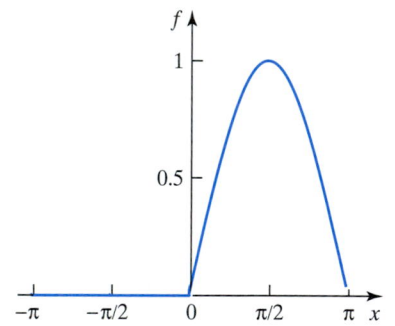

31. $f(x) = \dfrac{4\pi^2}{3} + 16 \displaystyle\sum_{n=1}^{\infty} \dfrac{(-1)^n \cos \frac{1}{2}nx}{n^2}$

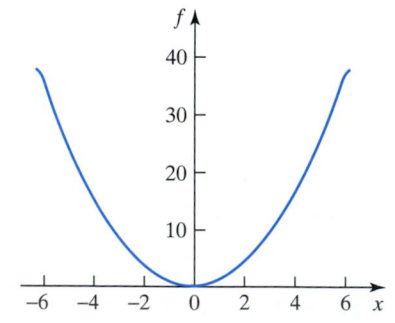

33. $f(x) = \dfrac{2 \sin a\pi}{\pi} \left\{ \dfrac{1}{2a} + \displaystyle\sum_{n=1}^{\infty} \dfrac{(-1)^n a \cos nx}{a^2 - n^2} \right\}.$

Graph for $a = 0.7, n = 10$.

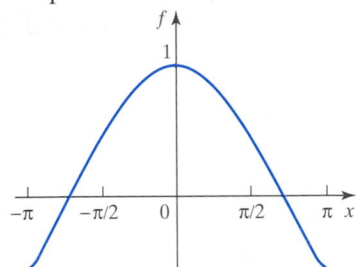

35. $f(x) = \dfrac{4}{3\pi} \sin \dfrac{1}{2} x + \dfrac{1}{2} \sin x$

$\qquad + \dfrac{4}{\pi} \displaystyle\sum_{n=1}^{\infty} (-1)^{n+1} \dfrac{\sin \frac{1}{2}(2n+1)x}{(2n+1)^2 - 4}$

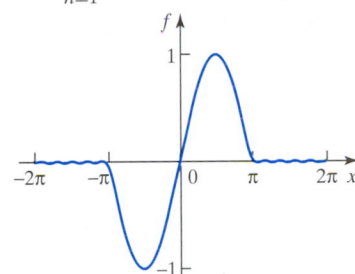

Exercise Set 9.2

1. $\dfrac{\pi^2}{8} = \displaystyle\sum_{n=1}^{\infty} \dfrac{1}{(2n-1)^2}$ **3.** $\dfrac{\pi^4}{90} = \displaystyle\sum_{n=1}^{\infty} \dfrac{1}{n^4}$

5. Proceed as in the derivation of the Parseval relation (27), but starting from the Fourier series representation of $f(x)$ on $-L \le x \le L$.

7. Set $x = 0$ with $f(0) = 0$ to get $\dfrac{\pi^2}{12} = \sum_{n=1}^{\infty} \dfrac{(-1)^{n+1}}{n^2}$

9. $f(x) = \dfrac{1}{2} - \dfrac{2}{\pi} \displaystyle\sum_{n=1}^{\infty} \dfrac{(-1)^n}{n} \sin \dfrac{1}{2} n\pi x$

$\qquad - \dfrac{4}{\pi^2} \displaystyle\sum_{n=1}^{\infty} \dfrac{\cos \frac{1}{2}(2n-1)\pi x}{(2n-1)^2}.$

Set $x = 0$ with $f(0) = 0$ to get $\dfrac{\pi^2}{8} = \sum_{n=1}^{\infty} \dfrac{1}{(2n-1)^2}$, or set $x = 2$ with $f(2) = 1$ for the same result.

11. The Fourier series for $f(x) = \pi^2 - x^2$ is $f(x) = \dfrac{2\pi^2}{3} + 4\sum_{n=1}^{\infty} (-1)^{n+1} \dfrac{\cos nx}{n^2}$. As $f(-\pi) = f(\pi)$, Theorem 9.3 can be used to find the Fourier series for $f'(x)$ by differentiating term by term to obtain

$$x = 2 \sum_{n=1}^{\infty} (-1)^{n+1} \frac{\sin nx}{n}.$$

Theorem 9.2 can also be applied to obtain

$$x(\pi^2 - x^2) = 12 \sum_{n=1}^{\infty} (-1)^{n+1} \frac{\sin nx}{n^3}.$$

13. Transform the result to

$$S_n(x) = \frac{1}{\pi} \int_{-\pi}^{\pi} f(u) \left[\frac{1}{2} + \sum_{r=1}^{\infty} \cos[r(x-u)] \right] du.$$

Now set $t = x - u$ to obtain

$$S_n(x) = \frac{1}{\pi} \int_{x-\pi}^{x+\pi} f(x-t) \frac{\sin \left[(n + \frac{1}{2}) t \right]}{2 \sin \frac{1}{2} t} dt.$$

17. $S_n(x) = \frac{1}{\pi} \int_0^{\pi} [f(x-t) + f(x+t)] D_n(t) dt.$

When n is large $D_n(t)$ can be replaced by $\Delta(t)$ to give

$$S_n(x) \approx \frac{(2n+1)}{4\pi} \int_0^{2\pi/(2n+1)} \times [f(x-t) + f(x+t)] dt,$$

and for large n the interval of integration is very small so the integrand is almost constant over the interval of integration, as a result of which integral can be replaced by

$$S_n(x) \approx \frac{(2n+1)}{4\pi} [f(x-t) + f(x+t)] \times$$

$$\int_0^{2\pi/(2n+1)} dt = \frac{1}{2}[f(x-t) + f(x+t)],$$

and in the limit as $n \to \infty$ this becomes an equality. So when f is continuous at x the Fourier series converges to $f(x_0)$, and when it is discontinuous it converges to the mid-point of the jump $\frac{1}{2}[f(x_{0-} - t) + f(x_{0+} + t)].$

Exercise Set 9.3

1. $b_1 = \dfrac{2}{\pi}(\pi^2 - 4)$, $b_2 = -\pi$, $b_3 = \dfrac{2}{27\pi}(9\pi^2 - 4)$,

$b_4 = -\dfrac{\pi}{2}$, $b_5 = \dfrac{2}{125\pi}(25\pi^2 - 4)$

3. $b_1 = 1/\pi$, $b_2 = 4/(3\pi)$, $b_3 = 1/\pi$, $b_4 = 8/(15\pi)$, $b_5 = 1/(3\pi)$

5. $\dfrac{1}{\pi} + \dfrac{1}{2} \cos x + \dfrac{2}{\pi} \displaystyle\sum_{n=1}^{\infty} (-1)^{n+1} \dfrac{\cos 2nx}{(4n^2 - 1)}$

7. $\dfrac{1}{\pi} + \dfrac{1}{\pi} \cos x - \dfrac{2}{3\pi} \cos 2x - \dfrac{1}{\pi} \cos 3x - \dfrac{2}{15\pi}$

$\cos 4x + \dfrac{1}{3\pi} \cos 5x - \dfrac{2}{35\pi} \cos 6x + \cdots$

11. $\dfrac{2}{\pi} \displaystyle\sum_{n=1}^{\infty} [1 + (-1)^{n+1} e^{-\pi}] \dfrac{n \sin nx}{(n^2 + 1)}$

13. The linearity of the integral used in the derivation of the Fourier series coefficients allows the
Fourier series of $f(x) \pm g(x)$ to be added or subtracted term by term. The Parseval relation gives

$$\frac{1}{\pi} \int_{-\pi}^{\pi} [f(x) \pm g(x)]^2 dx$$

$$= a_0 \pm A_0 + \sum_{n=1}^{\infty} [(a_n \pm A_n)^2 + (b_n \pm B_n)^2].$$

The result follows by subtracting the result with
the negative sign from the corresponding result
with the positive sign.

Exercise Set 9.4

1. $\dfrac{1}{2} - \dfrac{2}{\pi} \displaystyle\sum_{n=1}^{\infty} \dfrac{\sin(2n-1)x}{(2n-1)}$

3. $\dfrac{\pi}{2} - \displaystyle\sum_{n=1}^{\infty} \dfrac{\sin n\pi x}{n}$

5. $c_n = (-1)^n \dfrac{\sinh 1(1 - in\pi)}{1 + n^2 \pi^2}, \quad n = 0, \pm 1, \pm 2, \dots$

7. $c_n = \dfrac{e-1}{1 - 2n\pi i}, \quad n = 0, \pm 1, \pm 2, \dots$

9. $c_n = (-1)^n \dfrac{\sinh \pi}{\pi(1 - in)}, \quad n = 0, \pm 1, \pm 2, \dots$

Exercise Set 9.5

1. $\omega_0 = 1/2, \quad f(x) = \dfrac{\pi}{2} - \dfrac{4}{\pi} \displaystyle\sum_{n=1}^{\infty} \dfrac{\cos\frac{1}{2}(2n-1)x}{(2n-1)^2}$

$+ 2\displaystyle\sum_{n=1}^{\infty}(-1)^{n+1}\dfrac{\sin\frac{1}{2}nx}{n} \quad A_0 = \dfrac{\pi}{2},$

$A_1 = \left[\left(\dfrac{4}{\pi}\right)^2 + 2\right]^{1/2}, \quad A_2 = 1,$

$A_3 = \left[\left(\dfrac{4}{5\pi}\right)^2 + \left(\dfrac{2}{3}\right)^2\right]^{1/2}, \quad A_4 = \dfrac{1}{2}, \dots$

3. $\omega_0 = 1, \quad f(x) = -2 - \dfrac{8}{\pi}\displaystyle\sum_{n=1}^{\infty}\dfrac{\sin(2n-1)x}{(2n-1)},$

$A_0 = 2, \quad A_{2n-1} = \dfrac{8}{\pi(2n-1)},$

$A_{2n} = 0, \quad n = 1, 2, \dots$

5. $\omega_0 = 4, \quad f(x) = \dfrac{\pi^2}{48} + \dfrac{1}{4}\displaystyle\sum_{n=1}^{\infty}(-1)^n\dfrac{\cos 4nx}{n^2},$

$A_0 = \dfrac{\pi^2}{48}, \quad A_n = \dfrac{1}{4n^2}, \quad n = 1, 2, \dots$

Exercise Set 9.6

3. Case (d); $\quad d_{mn} = (-1)^{m+n}\dfrac{4}{m^3 n}[m^2\pi^2 - 6]$

5. Case (d); $\quad d_{mn} = \dfrac{16}{mn\pi}$ for m, n odd and $d_{mn} = 0$ for m, n even

7. Case (d); $\quad d_{mn} = (-1)^{m+n}\dfrac{32}{\pi^2 mn}$

9. Case (d); $\quad (-1)^{m+1}\dfrac{4}{mn^3\pi}\{2[(-1)^n - 1]$
$+ (-1)^{n+1}n^2\pi^2\}$

Exercise Set 10.1

1. $A(\omega) = \dfrac{2\sin\omega}{\omega\pi}, B(\omega) \equiv 0,$

$f(x) = \dfrac{2}{\pi}\displaystyle\int_0^{\infty}\dfrac{\cos\omega x \sin\omega}{\omega}d\omega$

3. $A(\omega) \equiv 0, B(\omega) = \dfrac{2b}{\omega^2 a\pi}(\sin\omega a - \omega a \cos\omega a),$

$f(x) = \dfrac{2b}{a\omega}\displaystyle\int_0^{\infty}\dfrac{\sin\omega x(\sin\omega a - \omega a\cos\omega a)}{\omega^2}d\omega$

When $x = a$, $\frac{1}{2}[f(a+0) + f(a-0)] = b/2$, so
this result also shows that

$$\int_0^{\infty}\dfrac{\sin\omega a(\sin\omega a - \omega a\cos\omega a)}{\omega^2}d\omega = \dfrac{\pi a}{4}$$

5. $f(x) = \displaystyle\int_0^{\infty}\dfrac{\cos\frac{1}{2}\omega\pi\cos\omega x}{1 - \omega^2}d\omega$

7. $f(x) = \dfrac{1}{\pi}\displaystyle\int_0^{\infty}\dfrac{\omega[\sin\omega x - \sin\omega(\pi + x)]}{\omega^2 - 1}d\omega$

Exercise Set 10.3

11. $F_C\{f(x)\} = \sqrt{\dfrac{2}{\pi}}\left(\dfrac{1 + \cos\omega\pi}{1 - \omega^2}\right)$

13. $F_C\{f(x)\} = \sqrt{\dfrac{2}{\pi}}\left(\dfrac{2\cos\omega - 1 - \cos 2\omega}{\omega^2}\right)$

15. $F_C\{f(x)\} = 2\sqrt{\dfrac{2}{\pi}}\left(\dfrac{\sin\omega - \omega\cos\omega}{\omega^3}\right)$

25. $F_S\{f(x)\} = -\sqrt{\dfrac{2}{\pi}}\left(\dfrac{\omega(1 + \cos\omega\pi)}{1 - \omega^2}\right)$

27. $F_S\{f(x)\} = \sqrt{\dfrac{2}{\pi}}\left(\dfrac{\omega - \sin 2\omega + \sin\omega}{\omega^2}\right)$

Exercise Set 11.1

1. $d\mathbf{r}/dt = (\sin t + t\cos t)\mathbf{i} + (\cos t - t\sin t)\mathbf{j} + 2t\mathbf{k},$
$(d\mathbf{r}/dt)_{t=\pi/2} = \mathbf{i} - (\pi/2)\mathbf{j} + \pi\mathbf{k}$

$d^2\mathbf{r}/dt^2 = (2\cos t - t\sin t)\mathbf{i} - (2\sin t + t\cos t)\mathbf{j}$
$+ 2\mathbf{k}$, $(d^2\mathbf{r}/dt^2)_{t=\pi/2} = -(\pi/2)\mathbf{i} - 2\mathbf{j} + 2\mathbf{k}$

3. $d\mathbf{r}/dt = 2\sin t\cos t\,\mathbf{i} + 2\sin t\cos t\,\mathbf{j} - \mathbf{k}$,
$(d\mathbf{r}/dt)_{t=\pi/4} = \mathbf{i} + \mathbf{j} - \mathbf{k}$
$d^2\mathbf{r}/dt^2 = 2(\cos^2 t - \sin^2 t)\mathbf{i} + 2(\cos^2 t - \sin^2 t)\mathbf{j}$,
$(d^2\mathbf{r}/dt^2)_{t=\pi/4} = \mathbf{0}$

5. $d\mathbf{r}/dt = (1 - \cos t)\mathbf{i} + \sin t\,\mathbf{j}$, $(d\mathbf{r}/dt)_{t=\pi/2} = \mathbf{i} + \mathbf{j}$
$d^2\mathbf{r}/dt^2 = \sin t\,\mathbf{i} + \cos t\,\mathbf{j}$, $(d^2\mathbf{r}/dt^2)_{t=\pi/2} = \mathbf{i}$

9. $d\mathbf{r}/ds = 2s\mathbf{i}/(1 + s^2) + 12s\ln(1 + s^2)\mathbf{j}/(1 + s^2)$
$- 2s\mathbf{k}/(1 + s^2)$

11. $d\mathbf{r}/dt = 2t\mathbf{i} - 8\sin 2t\mathbf{j} + 6\cos 2t\mathbf{k}$. A unit vector in the given direction is $\hat{\mathbf{a}} = \frac{2}{3}\mathbf{i} + \frac{1}{3}\mathbf{j} + \frac{2}{3}\mathbf{k}$ so the component in the required direction is $\hat{\mathbf{a}} \cdot d\mathbf{r}/dt = \frac{4}{3}t - \frac{8}{3}\sin 2t + 4\cos 2t$

19. $\dfrac{d}{dt}\{\mathbf{u} \cdot (\mathbf{v} \times \mathbf{w})\} = -4t^3 - 36t^2 - 6t + 4$

21. $\mathbf{T} = \dfrac{1}{(a^2\omega^2 + b^2)^{1/2}}[-a\omega\sin\omega t\mathbf{i} + a\omega\cos\omega t\mathbf{j} + b\mathbf{k}]$

$\mathbf{N} = -\cos\omega t\mathbf{i} - \sin\omega t\mathbf{j}$

$\mathbf{B} = \dfrac{1}{(a^2\omega^2 + b^2)^{1/2}}[b\sin\omega t\mathbf{i} - b\cos\omega t\mathbf{j} + a\omega\mathbf{k}]$

$\kappa = \dfrac{a\omega^2}{(a^2\omega^2 + b^2)}$

Exercise Set 11.2

1. (a) $((1/4)\sin 2t - (1/2)t\cos 2t)\mathbf{i} + t^3\mathbf{j} - (3/2)t^2\mathbf{k}$
(b) $[(7/3)\ln 7 - 2]\mathbf{i} + (1 + e^2)\mathbf{k}$

3. (a) $[(1/6)\cos 3t\sin 3t + t(1/2)]\mathbf{i} + (1/2)$
$[t - \cos t\sin t]\mathbf{j} + (1/2)t^2\mathbf{k}$
(b) $(\pi + \pi^3)\mathbf{i} + (1/3)\mathbf{k}$ **5.** $(\pi/2)(a^2 + \alpha^2)^{1/2}$

7. Integrate $\mathbf{F} \cdot d\mathbf{r}$ between the limits $t = 0$ and $t = \pi/2$ to obtain $\pi/4$

9. $2\pi^2$ **10.** 4   **11.** (a) 0, (b) $-3\pi/4$ **13.** 8π

Exercise Set 11.3

1. $\sqrt{5}(\pi + 2\sqrt{2})/10$ **3.** $(15e^{-2} - 2)/\sqrt{17}$
5. $\sqrt{2}[(\pi/8) - 1]/3 + 2e^3$
7. $4\sqrt{5}\cosh 2$
11. $(2x + 3yz)\mathbf{i} + (3xz - z^2)\mathbf{j} + (3xy - 2yz)\mathbf{k}$
13. $[(y - 3z)\mathbf{i} + (x + 2z)\mathbf{j} + (2y - 3x)\mathbf{k}]$
$\exp(xy + 2yz - 3xz)$
15. A normal $\mathbf{n}$ to $f(x, y) = $ constant is $\mathbf{n} = \text{grad } f$, so at point $P(x_0, y_0)$, $\mathbf{n} = (\text{grad } f)_P$, so $\mathbf{n} = (f_x)_P\mathbf{i} + (f_y)_P\mathbf{j}$. The vector equation of a line normal to f at P is $\mathbf{r} = \mathbf{r}_0 + \lambda(\text{grad } f)_P$ with $\mathbf{r}_0 = x_0\mathbf{i} + y_0\mathbf{j}$.

The cartesian equation is found by eliminating λ between $x = x_0 + \lambda(f_x)_P$ and $y = y_0 + \lambda(f_y)_P$ to obtain $y = y_0 + (x - x_0)(f_x/f_y)_P$.

17. A normal to the surface is grad f, so at $(1, 2, 2)$ the normal $\mathbf{n} = 9\mathbf{i} + 3\mathbf{j} + 4\mathbf{k}$. The tangent plane through $\mathbf{r}_0 = \mathbf{i} + 2\mathbf{j} + 2\mathbf{k}$ is $(\mathbf{r} - \mathbf{r}_0) \cdot \mathbf{n} = 0$, so the plane has the equation $9x + 3y + 4z = 23$.

19. The normal to the surface at $\mathbf{r}_0$ is $(\text{grad } f)_{\mathbf{r}_0}$ so the required equation is $(\mathbf{r} - \mathbf{r}_0) \cdot (\text{grad } f)_{\mathbf{r}_0} = 0$.

21. $(2r\sin\theta + z^2)\mathbf{e}_r + r\cos\theta\mathbf{e}_\theta + 2rz\mathbf{e}_z$

23. grad $(f^n) = nf^{n-1}(f_x\mathbf{i} + f_y\mathbf{j} + f_z\mathbf{k}) = nf^{n-1}\mathbf{F}$
If $f = r$ then $f = (x^2 + y^2 + z^2)^{1/2}$ and grad $r = (x\mathbf{i} + y\mathbf{j} + z\mathbf{k})/(x^2 + y^2 + z^2)^{1/2} = \hat{\mathbf{r}}$.
If $f = 1/r$ then grad $f = -(1/r^2)\text{grad } r = -(1/r^2)\hat{\mathbf{r}} = -\mathbf{r}/r^3$.

Exercise Set 11.4

1. Yes   **3.** No   **5.** No
7. $f = xz^3 + 3x^2y^2 + $ constant; $I = f(Q) - f(P) = 11$
9. $f = x\exp(xyz) + $ constant; $I = f(Q) - f(P) = e^2$
11. $f = x^2 + x^2yz^2 + $ constant; $I = f(Q) - f(P) = -17$

Exercise Set 11.5

1. div $\mathbf{F} = 2xy + 2yz^2 + 3xz^2$
3. div $\mathbf{F} = 6x + 4x^2y$
5. Substitute $\phi\mathbf{F}$ into the definition of divergence and expand the result.
7. curl $\mathbf{F} = (2xy - x^2y)\mathbf{i} + (2xyz - y^2)\mathbf{j} + (2xyz - xz^2)\mathbf{k}$
9. curl $\mathbf{F} = \mathbf{i} + \dfrac{x(3y^2 + 2x^2)}{(x^2 + 2y^2)(x^2 + y^2)}\mathbf{k}$
11. Expand curl $\mathbf{F}$, substitute into the definition of divergence, and make use of the equality of mixed derivatives.
13. Substitute $\mathbf{F} \cdot \mathbf{G}$ into the definition of grad and expand the result.
15. Substitute $\mathbf{F} \times \mathbf{G}$ into the definition of curl and expand the result.
17. $\nabla^2\mathbf{F} = \mathbf{0}$, so curl(curl$\mathbf{F}$) = grad div $\mathbf{F} - \nabla^2\mathbf{F} = $ grad div $\mathbf{F} = 3(z\mathbf{i} + y\mathbf{k})$
21. Yes; $f = \ln(1 + x^2 + 2y^2z) = $ constant

Exercise Set 11.6

1. $\nabla \cdot (a\mathbf{F}) = a\nabla \cdot \mathbf{F}; \quad \nabla \cdot (a\mathbf{F} + b\mathbf{G}) = a\nabla \cdot \mathbf{F} + b\nabla \cdot \mathbf{G};$

$\nabla \cdot (\phi\mathbf{F}) = \phi\nabla \cdot \mathbf{G} + \mathbf{F} \cdot \nabla\phi;$

$\nabla \cdot (\nabla\phi) = \nabla^2\phi; \; \nabla \cdot (\phi\nabla\psi) = \phi\nabla^2\psi + \nabla\phi \cdot \nabla\psi;$

$\nabla \cdot (\phi\nabla\psi) - \nabla \cdot (\psi\nabla\phi) = \phi\nabla^2\psi - \psi\nabla^2\phi$

5. $h_1 = h_2 = \sqrt{2}, \; h_3 = \cosh q_3; \; \mathbf{q} = (q_1 - q_2)\mathbf{i} + (q_1 + q_2)\mathbf{j} + \sinh q_3\mathbf{k}; \; \mathbf{e}_1 = \dfrac{1}{h_1}\dfrac{\partial \mathbf{q}}{\partial q_1} = \dfrac{1}{\sqrt{2}}(\mathbf{i} + \mathbf{j}),$

$\mathbf{e}_2 = \dfrac{1}{h_2}\dfrac{\partial \mathbf{q}}{\partial q_2} = \dfrac{1}{\sqrt{2}}(-\mathbf{i} + \mathbf{j}), \mathbf{e}_3 = \mathbf{k},$ so $\mathbf{e}_1, \mathbf{e}_2,$ and $\mathbf{e}_3$ form an orthonormal set.

$$\text{grad } f = \mathbf{e}_1\frac{1}{\sqrt{2}}\frac{\partial f}{\partial q_1} + \mathbf{e}_2\frac{1}{\sqrt{2}}\frac{\partial f}{\partial q_2} + \mathbf{e}_3\frac{1}{\cosh q_3}\frac{\partial f}{\partial q_3}$$

$$\text{div }\mathbf{F} = \frac{1}{\sqrt{2}}\frac{\partial F_1}{\partial q_1} + \frac{1}{\sqrt{2}}\frac{\partial F_2}{\partial q_2} + \frac{1}{\cosh q_3}\frac{\partial F_3}{\partial q_3}$$

7. $h_1 = h_2 = \sinh^2\xi + \sin^2\eta, h_3 = 1$

$\mathbf{q} = \cosh\xi\cos\eta\mathbf{i} + \sinh\xi\sin\eta\mathbf{j} + z\mathbf{k}$

$\mathbf{e}_\xi = \dfrac{1}{h_1}\dfrac{\partial \mathbf{q}}{\partial \xi} = \dfrac{1}{\sinh^2\xi + \sin^2\eta}(\sinh\xi\cos\eta\mathbf{i} + \cosh\xi\sin\eta\mathbf{j})$

$\mathbf{e}_\eta = \dfrac{1}{h_1}\dfrac{\partial \mathbf{q}}{\partial \eta} = \dfrac{1}{\sinh^2\xi + \sin^2\eta}(-\cosh\xi\sin\eta\mathbf{i} + \sinh\xi\cos\eta\mathbf{j})$

$\mathbf{e}_z = \mathbf{k},$ so $\mathbf{e}_\xi, \mathbf{e}_\eta,$ and $\mathbf{e}_z$ form an orthonormal set. $\xi = $ constant are ellipses and $\eta = $ constant are hyperbolas

$$\text{grad } f = \frac{1}{\sinh^2\xi + \sin^2\eta}\frac{\partial f}{\partial \xi}\mathbf{e}_\xi + \frac{1}{\sinh^2\xi + \sin^2\eta}$$
$$\times \frac{\partial f}{\partial \eta}\mathbf{e}_\eta + \mathbf{e}_z\frac{\partial f}{\partial z}$$

Exercise Set 12.2

1. Set $\mathbf{F} = \mathbf{a} \times \mathbf{G}$ in the divergence theorem to obtain

$$\iint_S (\mathbf{a} \times \mathbf{G}) \cdot d\mathbf{S} = \iiint_D \text{div}(\mathbf{a} \times \mathbf{G})dV \text{ but}$$
$$\text{div}(\mathbf{a} \times \mathbf{G}) = -\mathbf{a} \cdot \text{curl } \mathbf{G}, \text{ so}$$
$$\iint_S (\mathbf{a} \times \mathbf{G}) \cdot d\mathbf{S} = -\iiint_D \mathbf{a} \cdot \text{curl}\mathbf{G}dV \text{ or}$$
$$\iint_S (\mathbf{a} \times \mathbf{G}).\mathbf{n}d\mathbf{S} = -\iiint_D \mathbf{a} \cdot \text{curl}\mathbf{G}dV$$

The properties of the scalar triple product allow the interchange of the dot and the cross to give

(because $\mathbf{a}$ is a constant vector)

$$\mathbf{a} \cdot \iint_S \mathbf{G} \times d\mathbf{S} = -\mathbf{a} \cdot \iiint_D \text{curl}\mathbf{G}dV.$$ As $\mathbf{a}$ is arbitrary this last result implies that $\displaystyle\iint_S \mathbf{G} \times d\mathbf{S}$

$$= -\iiint_D \text{curl}\mathbf{G}dV.$$

3. Set $\mathbf{F} = \phi\mathbf{G}$ in the divergence theorem and use the result that $\text{div}(\phi\mathbf{G}) = (\text{grad }\phi) \cdot \mathbf{G} + \phi \text{ div } \mathbf{G}$

5. Write $\text{div}(\kappa T\text{grad } T) = \text{div}(T[\kappa\text{grad } T])$ and expand the expression to get $\text{div}(\kappa T\text{grad } T) = (\text{grad } T) \cdot (\kappa\text{grad } T) + T \text{ div}$
$(\kappa\text{grad } T)$, so the heat equation becomes
$\text{div}(\kappa T \text{ grad } T) = \kappa(\text{grad } T) \cdot (\text{grad } T) + \mu\rho T\partial T/\partial t$. Now integrate over D and use the divergence theorem to get

$$\iint_S \kappa T(\text{grad } T) \cdot d\mathbf{S} = \iiint_D \kappa(\text{grad } T) \cdot (\text{grad } T)dV$$
$$+ \iiint_D \mu\rho T\frac{\partial T}{\partial t}dV$$

7. Replace $\mathbf{F}$ in Stokes's theorem by $\phi \mathbf{F}$ and use curl $(\phi\mathbf{F}) = (\text{grad }\phi) \times \mathbf{F} + \phi \text{ curl } \mathbf{F}$

Exercise Set 12.3

1. Reason as in Example 12.16 with $\mathbf{q} = u\mathbf{i} + v\mathbf{j} + w\mathbf{k}$

3. $\dfrac{d}{dt}\displaystyle\iiint_{D(t)} f(\mathbf{r}, t)dV$

$$= \frac{d}{dt}\left[\int_0^1 \int_0^1 \int_{ut}^{vt} xyt\,dz\,dy\,dx\right]$$
$$= \frac{d}{dt}\left[\frac{1}{4}(v - u)t^2\right] = \frac{1}{2}(v - u)t$$

Here, on the upper surface $\mathbf{q} = v\mathbf{k}$ so $d\mathbf{S} = dx\,dy\mathbf{k}$, while on the lower surface $\mathbf{q} = u\mathbf{k}$ and $d\mathbf{S} = -dx\,dy\mathbf{k}$, so

$$\iiint_{D(t)} \frac{\partial f(\mathbf{r}, t)}{\partial t}dV + \iint_{S(t)} f\mathbf{q} \cdot d\mathbf{S}$$
$$= \int_0^1 \int_0^1 \int_{ut}^{vt} xy\,dx\,dy\,dz + \int_0^1 \int_0^1 xytv\,dy\,dx$$
$$- \int_0^1 \int_0^1 xytu\,dy\,dx = \frac{1}{2}(v - u)t,$$

so the two results are in agreement.

5. Use cylindrical symmetry when evaluating the integrals with $dV = 2\pi r h dr$ and $d\mathbf{S} = hr\,d\theta$.

$$\frac{d}{dt}\iiint_{D(t)} f(\mathbf{r}, t)dV = \frac{d}{dt}\left[\int_0^{ut} r^2 t 2\pi r h dr\right]$$
$$= \frac{5}{2}\pi h u^4 t^4 \quad \text{and}$$

$$\iiint_{D(t)} \frac{\partial f(\mathbf{r}, t)}{\partial t}dV + \int_{S(t)} f\mathbf{q}\cdot d\mathbf{S}$$
$$= \int_0^{ut} r^2 2\pi r h dr + h u^4 t^4 \int_0^{2\pi} d\theta = \frac{5}{2}\pi h u^4 t^4,$$

so the two results are in agreement.

Exercise Set 13.1

1.

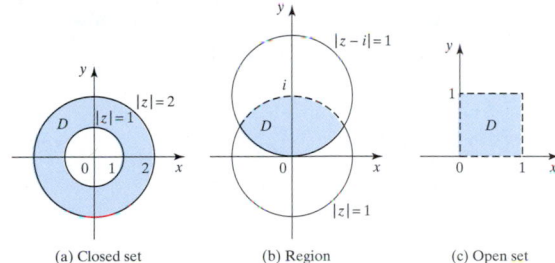

(a) Closed set   (b) Region   (c) Open set

3. line $y = -x$ from the origin to the point $(-2, -2)$

5.

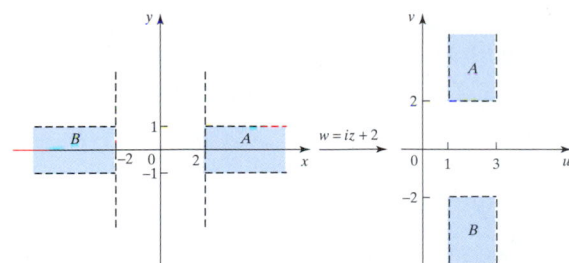

7.

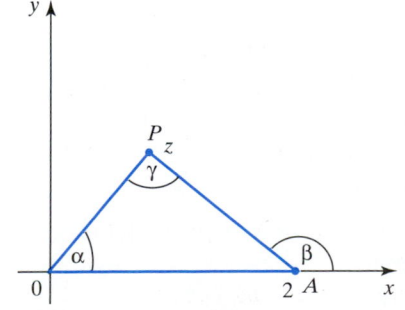

Angle $OAP = \pi - \beta$, but $\alpha +$ angle $OAP + \gamma = \pi$, so $\gamma = \beta - \alpha$. As $\alpha = \text{Arg}\, z$, $\beta = \text{Arg}\,(z-2)$, so $\text{Arg}\,(z-2) - \text{Arg}\, z = \gamma = \pi/2$. From Euclidean geometry point P must lie on a circle with its diameter from the point $(0, 0)$ to $(2, 0)$. The condition $0 \le \text{Arg}\, z \le \pi/2$ defines the part of the circle that lies in the upper half of the z-plane.

9. An ellipse with the foci at $z = \pm 1$ and eccentricity $e = 1/2$

11. $f(z)$
$$= \left(\frac{2x^2 + 2y^2 + 3y + 1}{x^2 + (1+y)^2}\right) - i\left(\frac{x}{x^2 + (1+y)^2}\right)$$
$$= \left(\frac{2r^2 + 3r\sin\theta + 1}{r^2 + 2r\sin\theta + 1}\right)$$
$$- i\left(\frac{r\cos\theta}{r^2 + 2r\sin\theta + 1}\right)(z \ne 0)$$
$$u = \text{Re}\{f(z)\},\ v = \text{Im}\{f(z)\}$$

13. $f(z) = e^{-y}(x\cos x - y\sin x)$
$$+ ie^{-y}(y\cos x + x\sin x)$$
$$= r\exp(-r\sin\theta)\{\cos\theta\cos(r\cos\theta)$$
$$- \sin\theta\sin(r\cos\theta)\}$$
$$+ ir\exp(-r\sin\theta)\{\sin\theta\cos(r\cos\theta)$$
$$+ \cos\theta\sin(r\cos\theta)\}$$
$$u = \text{Re}\{f(z)\},\ v = \text{Im}\{f(z)\}$$

Exercise Set 13.2

1. $\text{Re}\{f(x)\} = x^3 - 3xy^2 + 4x^2 - 4y^2 - 3x + 1$,
$\text{Im}\{f(x)\} = 3x^2 y - y^3 + 8xy - 3y$;   continuous for all z

3. $\text{Re}\{f(z)\} = \dfrac{2xy^2 + x(1 + x^2 - y^2)}{(1 + x^2 - y^2)^2 + 4x^2 y^2}$,

$\text{Im}\{f(z)\} = \dfrac{y(1 + x^2 - y^2) - 2x^2 y}{(1 + x^2 - y^2)^2 + 4x^2 y^2}$;

discontinuous at $z = \pm i$

5. $f'(z) = 3z^2 + 1$ for all z

7. $f'(z) = -1/(1 + z)^2$ for $z \ne -1$

9. $f'(z) = 3z^2$ for all z

11. $f'(z) = 1 - 1/z^2$ for $z \ne 0$

13. Substitute in the definitions of the functions on the right and show they simplify to the function on the left. The second result follows by setting $z_1 = x$ and $z_2 = iy$ and using $\cosh(iy) = \cos y$ and $\sinh(iy) = i\sin y$.

15. To establish the first identity substitute in the definitions of the functions on the left and show they simplify to unity. The second identity follows from the first one after division by $\cosh^2 z$ and rearrangement of the result.

17. In the first identity substitute in the definitions of the functions on the right and show they simplify to the function on the left. The second result follows from the first by setting $z_1 = x$ and $z_2 = iy$ and using $\cos(iy) = \cosh y$ and $\sin(iy) = i \sinh y$.

19. Establish the first identity by substituting into the definitions of the functions on the left and showing the result simplifies to unity. The second result follows from the first after division by $\cos^2 z$.

21. $z = n\pi$, $n = 0, \pm1, \pm2, \ldots$

23. $z = n\pi i$, $n = 0, \pm1, \pm2, \ldots$

25. $z = (2n+1)\pi \pm 3i$, $n = 0, \pm1, \pm2, \ldots$

27. $z = \pm2 + (4n+1)\pi i/2$, $n = 0, \pm1, \pm2, \ldots$

29. $z = n\pi i$, $n = 0, \pm1, \pm2, \ldots$ (the zeros of $\sinh z$)

31. (a) $0, \pm\pi, \sqrt{3}e^{i\pi/4}, \sqrt{3}e^{5i\pi/4}$ (b) $z = 2\{\cos(2k+1)\pi/4 + i\sin(2k+1)\pi/4\}$, $k = 0, 1, \ldots$ (c) Nowhere analytic because $|z|$ is not an analytic function

33. $3\cos 3x \cosh 3y - i3\sin 3x \sinh 3y = 3\cos 3z$

35. Using the change of variables from cartesian to polar coordinates $x = r\cos\theta$, $y = r\sin\theta$, substitute in the change of variable formulas $u_x = r_x u_r + \theta_x u_\theta$ etc. to find u_x, u_y, v_x and v_y. Use these results in the cartesian form of the Cauchy–Riemann equations to obtain their polar form.

37. $f'(z) = 1 - 1/z^2$

39. $f(z) = 3z^3 + z + 1$, $f'(z) = 9z^2 + 1$

Exercise Set 13.3

1. $f(z) = z^3 + (2-i)z + ic$

3. $f(z) = ze^{iz} + 2iz + a$

5. $f(z) = z\sinh 2z + a$ 7. $f(z) = z\cos 3z + ic$

9. $f(z) = z + (2-i)z^2 + ic$

11. Show that the functions do not satisfy the Cauchy–Riemann equations.

13. Say $u \equiv$ constant. Then from the Cauchy–Riemann equations $v_x = v_y = 0$, so $v =$ constant, and hence $f(z) = u + iv \equiv$ constant in D. If $f(z)$ is not analytic there is no connection between u and v, so if $u \equiv 0$ it is not necessary that $v = 0$. A simple example is $f(z) = |z| + i$ constant.

15. Combine similar terms and chose a and b to make $\Delta\Phi = 0$ to get $a = 1, b = -2$.

Exercise Set 13.4

1. $(4n+1)\pi/2 - i\ln(\sqrt{5}+2)$ using the principal value of the square root function. $(4n-1)\pi/2 - i\ln(\sqrt{5}-2)$ using the value from the second branch of the square root function. $\pi/2 - i\ln(\sqrt{5}+2)$ using the principal values of the square root and logarithmic functions.

3. $(4n+1)\pi i/4$, $\pi i/4$ using the principal value of the logarithmic function.

5. $-(1/8)(8n+1)\pi + (1/4)i\ln 2$, $-\pi/8 + (1/4)i\ln 2$ using the principal branch of the logarithmic function.

7. $\arcsin z + \arccos z = -i\log[iz + (1-z^2)^{1/2}] - i\log[z + i(1-z^2)^{1/2}] = -i\log\{[iz + (1-z^2)^{1/2}][z + i(1-z^2)^{1/2}]\} = -i\log i$. However, as $i = e^{i\pi/2} \cdot e^{2n\pi i}$, so $-i\log i = \pi/2 + 2n\pi$.

9. From (59) $\log z = \ln|z| + i\,\text{Arg}\,z$ so immediately above the negative real axis $\text{Arg}\,a = \pi$ and immediately below it $\text{Arg}\,z = -\pi$, so there is a jump of $2\pi i$ across the negative real axis.

Exercise Set 14.1

1. $AB: z = t + it/2, 2 \le t \le 4$
   $BC: z = t + i(2t-6), 4 \le t \le 5$

3. $AB: z = t + i(2t-5), 3 \le t \le 4$
   $BC: z = 4 - t + i(3+t), 0 \le t \le 3$

5. 0   7. $-18 - 18i$

9. $36 + 21i$ 11. $\cosh 3 - \cosh 6$

13. $\cosh\pi(\cos 2 - \cos 3) + i\sinh\pi(\sin 3 - \sin 2)$

15. $(1/2)(\sinh 8\cos 4 + i\cosh 8\sin 4)$

17. $e^4/\sqrt{2} - 1 + ie^4/\sqrt{2}$

19. On the semicircle $\Gamma: z = 1 + e^{it}$, from $t = \pi$ to $t = 0$ (in the negative sense)
$$\int_\Gamma \frac{dz}{z-1} = \int_\pi^0 \frac{1}{e^{it}} ie^{it}\,dt = -\pi i$$

21. $\Gamma: z = 2 + 2e^{it}$ and as integration is in the positive sense $\int_\Gamma \frac{1}{z+i}dz = \int_0^{2\pi} \frac{1}{2 + 2e^{it} + i} 2ie^{it}\,dt = [\log(2 + 2e^{it} + i]_0^{2\pi} = 0$. Reversal of the direction of integration gives the same result.

Exercise Set 14.2

1. $\cos 1 - (1/2)(e + 1/e)$; $f(z)$ is analytic, so Theorem 14.4 applies.

3. $5/2 + 3i$; $f(z)$ is not analytic, so Theorem 14.4 cannot be used.

5. 0; $f(z)$ is analytic in $|z| \leq 1$, so the Cauchy–Goursat theorem applies.

7. 0; z is analytic but $\bar{z}^2$ is not, so $\int_\Gamma f(z)dz = \int_\Gamma z\,dz + \int_\Gamma \bar{z}^2 dz = 0 + 0 = 0$.

9. (a) The points $\pm i$ must not lie inside Γ. (b) The points $z = n\pi, n = 0, \pm 1, \ldots$ (the zeros of $\sin z$) must not lie inside Γ. (c) The points $z = (2n+1)i\pi/2, n = 0, \pm 1 \ldots$ (the zeros of $\cosh z$) must not lie inside Γ. (d) The points $z = n\pi i, n = 0, \pm 1, \ldots$ must not lie inside Γ.

11. $f(z) = \dfrac{z+5}{z^2 + 3z - 4} = \dfrac{6}{5}\dfrac{1}{z-1} - \dfrac{1}{5}\dfrac{1}{z+4}$ so

(a) $\displaystyle\int_\Gamma f(z)dz = \dfrac{6}{5}\int_\Gamma \dfrac{dz}{z-1} + 0 = \dfrac{12\pi i}{5}$

(b) $\displaystyle\int_\Gamma f(z)dz = 0 - \dfrac{1}{5}\int_\Gamma \dfrac{dz}{z+1} + 0 = -\dfrac{2\pi i}{5}$

13. $f(z) = \dfrac{2-7z}{z^2 + 3z} = \dfrac{2}{3}\dfrac{1}{z} - \dfrac{23}{3}\dfrac{1}{z+3}$ so

(a) $\displaystyle\int_\Gamma f(z)dz = \dfrac{2}{3}\int_\Gamma \dfrac{dz}{z} + 0 = \dfrac{4\pi i}{3}$

(b) $\displaystyle\int_\Gamma f(z)dz = 0 - \dfrac{23}{3}\int_\Gamma \dfrac{dz}{z+3} = -\dfrac{46\pi i}{3}$

15. $f(z) = \dfrac{z^2 + 2z}{z^2 - 2z + 1} = 1 + \dfrac{3}{(z-1)^2} + \dfrac{4}{z-1}$;

$\displaystyle\int_\Gamma f(z)dz = 0 + 0 + 4\int_\Gamma \dfrac{dz}{z-1} = 8\pi i$

17. $f(z) = \dfrac{2z-1}{(z+1)^3} = \dfrac{2}{(z+1)^2} - \dfrac{3}{(z+1)^3}$;

$\displaystyle\int_\Gamma f(z)dz = 2\int_\Gamma \dfrac{dz}{(z+1)^2} - 3\int_\Gamma \dfrac{dz}{(z+1)^3} = 0 - 0 = 0$

Exercise Set 14.3

1. 0

3. $\pi i/\sqrt{2}$

5. $2\pi e^4 i$

7. $\pi \sin 1$

9. $\pi i\sqrt{2}\left(\dfrac{\pi}{86} - \dfrac{1}{6}\right)$

11. $-2\pi i$

13. $\dfrac{\pi i}{3}(5\cos 1 - 6\sin 1)$

15. $-\dfrac{\pi}{2}e^{-i}$

17. Set $z - z_0 = Re^{i\theta}$ in the Cauchy integral formula for derivatives, take the absolute value, and use

the integral inequality in Theorem 14.1 to obtain

$$|f^n(z_0)| \leq \left|\dfrac{n!}{2\pi i}\int_0^{2\pi} \dfrac{f(z)Rie^{i\theta}}{R^{n+1}e^{i(n+1)\theta}}d\theta\right|$$

$$\leq \dfrac{n!M}{2\pi R^n}\int_0^{2\pi} d\theta = \dfrac{n!M}{R^n}.$$

19. $\displaystyle\int_\Gamma \dfrac{d}{dt}\left[\dfrac{(t^2-1)^{n+1}}{(t-z)^{n1}}\right]dt$

$= \displaystyle\int_\Gamma \dfrac{(n+1)(t^2-1)^n(t^2 - 2tz + 1)}{(t-z)^{n+2}}dt = 0.$

Express $P'_{n+1}(z) - zP'_n(z) - (n+1)P_n(z)$ in terms of the integral definition of $P_n(z)$ to show that apart from a constant factor it is given by the contour integral in Exercise 18, so $P'_{n+1}(z) - zP'_n(z) - (n+1)P_n(z) = 0$.

21. $\displaystyle\int_\Gamma \dfrac{d}{dt}\left[\dfrac{t(t^2-1)}{(t-z)^n}\right]dt = \int_\Gamma \left[\dfrac{(t^2-1)^n}{(t-z)^n}\right.$

$\left. + 2\dfrac{nt^2(t^2-1)^{n-1}}{(t-z)^n} - \dfrac{nt(t^2-1)^n}{(t-z)^{n+1}}\right]dt = 0.$

Express $(n+1)P_{n+1}(z) - (2n+1)zP_n(z) + nP_{n-1}(z)$ in terms of the integral definition of $P_n(z)$ to show that apart from a constant factor it is given by the contour integral in Exercise 18, so $(n+1)P_{n+1}(z) - (2n+1)zP_n(z) + nP_{n-1}(z) = 0$.

23. Perform the indicated differentiation in Exercise 22 to obtain an equivalent expression for that result. Construct $G(z)$ using the integral representation for $P_n(z)$ and show that after simplification it reduces to $G(z)$. As Exercise 22 establishes that $G(z) = 0$ it follows that the Legendre differential equation is $(1 - z^2)P''_n(z) - 2zP'(z) + n(n+1)P_n(z) = 0$.

Exercise Set 14.4

1. If $0 \leq k \leq n$,

$$\dfrac{1}{2\pi i}\dfrac{P_k}{z^{k+1}} = \dfrac{1}{2\pi i}\left[\dfrac{a_0}{z^{k+1}} + \dfrac{a_1}{z^k} + \cdots + \dfrac{a_k}{z} + a_{k+1}\right.$$

$$\left. + \cdots a_n z^{n-k+1}\right].$$

Integrating around Γ shows that all integrals but that of a_k/z vanish, while $\dfrac{1}{2\pi i}\displaystyle\int_\Gamma \dfrac{a_k}{z}dz = a_k$, so

$$\dfrac{1}{2\pi i}\sum_{k=0}^n \int_\Gamma \dfrac{P_n(z)}{z^{k+1}}dz = \sum_{k=0}^n a_k.$$

3. In terms of the given substitutions

$$f(re^{i\theta}) = \frac{1}{2\pi i} \int_0^{2\pi} \frac{(R^2 - r^2) f(e^{i\psi} R)}{e^{i\psi} R(z\bar{z} - z\bar{z}_0 - z_0\bar{z} + z_0\bar{z}_0)}$$
$$ie^{i\psi} R d\psi, \quad \text{but} \quad z\bar{z} = R^2, z_0\bar{z}_0 = r^2, z\bar{z}_0 + z_0\bar{z} = r R\cos(\psi - \theta), \text{ so}$$

$$f(re^{i\theta}) = \frac{1}{2\pi i} \int_0^{2\pi} \frac{(R^2 - r^2) f(Re^{i\psi})}{R^2 - 2r R\cos(\psi - \theta) + r^2} d\psi.$$

The Poisson integral formula follows from this by writing $f(re^{i\theta}) = u(r, \theta) + iv(r, \theta)$ and equating the real parts.

5. If z_0 lies inside the semicircle, then $\bar{z}_0$ lies outside it, so from the Cauchy integral formula $f(z_0) = \frac{1}{2\pi i} \int_\Gamma \frac{f(z)}{z - z_0} dz$ and $0 = \frac{1}{2\pi i} \int_\Gamma \frac{f(z)}{z - \bar{z}_0} dz$. Subtracting these results and combining the integrands gives

$$f(z_0) = \frac{1}{2\pi i} \int_\Gamma \frac{f(z)(z_0 - \bar{z}_0)}{(z - z_0)(z - \bar{z}_0)} dz$$
$$= \frac{1}{2\pi i} \int_\Gamma \frac{f(z)2iy_0}{(z - z_0)(z - \bar{z}_0)} dz \quad \text{where}$$
$$z_0 = x_0 + iy_0$$

On the real axis

$z = x$ so $(z - z_0)(z - \bar{z}_0) = x^2 - 2xx_0 + x_0^2 + y_0^2 = |x - z_0|^2$ so

$$f(z_0) = \frac{1}{2\pi i} \int_{-R}^{R} \frac{f(x)2iy_0}{|x - z_0|^2} dx$$
$$+ \frac{1}{2\pi i} \int_{C_R} \frac{f(z)2iy_0}{(z - z_0)(z - \bar{z}_0)} dz,$$

which after cancellation of the factors i and removal of the constant y_0 from the integrand gives the required result.

7. $P_n(z) = a_n z^n \left(1 + \frac{a_{n-1}}{a_n z} + \frac{a_{n-2}}{a_n z^2} + \cdots + \frac{a_0}{a_n z^n}\right)$,
so as $|z| \to \infty$ the bracketed term tends to 1, showing that $|P_n(z)| \to |a_n z^n|$ as $|z| \to \infty$. Thus, as $|z| \to \infty$, $|Q_n(z)| \to 1/|a_n z^n| = 1/(|a_n| r^n)$, showing that $|Q_n(z)| \to 0$ as $|z| \to \infty$.

9. $f(z) = e^z = e^{x+iy} = e^x(\cos y + i \sin y)$, so $|e^z| = e^x$. In $-1 \le x \le 1, -2 \le y \le 2$, $|e^z| = e^x$ has its greatest value e on $x = 1$ for all y and its least value $1/e$ on $x = -1$ for all y, and thus $1/e < |e^z| < e$ for $-1 \le x \le 1, -2 \le y \le 2$.

11. $u = x + 2x^2 - 2y^2$ is harmonic so the max/min of u occur on the boundary of the domain. Examination of u on the boundary shows Max $u = 3$ at $x =$

1, $y = 0$, and Min $u = -17/8$ at $x = -1/4, y = \pm 1$, so $-17/8 < u < 3$ inside the domain.

13. $u = e^x(x \cos y - y \sin y)$ is harmonic so the max/min of u occur on the boundary of the domain. Examination of u on the boundary shows Max $u = e$ at $x = 1$ on $y = 0$ and Min $u = -e\pi/2$ at $x = 1, y = \pm \pi/2$, so $-e\pi/2 < u < e$ in the domain.

Exercise Set 15.1

1. (a) Only cluster point is at 1, so the sequence converges to the limit 1, but the limit is not a member of the series.
(b) Cluster points at 0 and 4. The point 0 belongs to the sequence but the point 4 does not. The sequence has no limit.
(c) Only cluster point is at 5/2, so the sequence converges to the limit 5/2, but the limit is not a member of the sequence.

3. (a) This is one definition of the Euler number e, so the sequence converges to the limit e, but the limit is not a member of the sequence.
(b) Only cluster point is at $\pi/2$, so the sequence converges to $\pi/2$, but the limit is not a member of the sequence.
(c) Every member of the sequence is 1, so the sequence converges to the limit 1 that is a member of the sequence.

5. Convergent by comparison with $\Sigma 1/n^2$.

7. Divergent by comparison with $\Sigma 1/n$.

9. Divergent by nth root test as $L = 2$.

11. Absolutely convergent by comparison with $\Sigma 1/n^2$ because for large $n \sin(1/n^2) \approx 1/n^2$.

13. Write $\frac{1}{r(r+1)} = \frac{1}{r} - \frac{1}{r+1}$ so

$$\sum_{r=1}^{n} \frac{1}{r(r+1)} = \left(\frac{1}{1} - \frac{1}{2}\right) + \left(\frac{1}{2} - \frac{1}{3}\right) + \cdots$$
$$+ \left(\frac{1}{n} - \frac{1}{n+1}\right) = 1 - \frac{1}{n+1}.$$

So in the limit at $n \to \infty$ the series converges to 1. This cancellation of terms is called the *telescoping* of the series.

15. Convergent by nth root test because $L = (1/3)|2i - 1| \lim \sqrt[n]{n} = \sqrt{5}/2 > 1$.

17. Use the approach in Exercise 13 to show that the series converges to 1.

19. Absolutely convergent by the nth root test.

21. $R = 2$; convergence for $|z| < 2$.

23. Alternate powers are missing so set $u = z^2$ and write as $2z\Sigma 2^n u^n/(4n+1)^2$. This has a radius of convergence $R = 1/2$, and so it converges for $|u| < 1/2$, and so for $|z| \ll 1/\sqrt{2}$.

25. $R = 0$; convergence only for $z = 0$.

27. $R = 2$; convergence for $|z| < 2$.

29. $R = 1$; convergence for $|z + 3| < 1$.

30. $R = 2$; convergence for $|z - 2| < 2$.

31. $R = 1$; convergence for $|z| < 1$.

33. $R = 1/2$; convergence for $|z| < 1/2$.

35. $\dfrac{\sqrt{2}}{2 + \sqrt{2}} + \dfrac{2\sqrt{2}}{(2 + \sqrt{2})^2}(z - \pi/4) - \dfrac{2\sqrt{2} + 6}{(2 + \sqrt{2})^3}$ $(z - \pi/4)^2 + \cdots$

37. $\dfrac{1}{2}\left(\dfrac{1}{e} - e\right) - \dfrac{3}{2}\left(\dfrac{1}{e} + e\right)(z - 1) + \dfrac{9}{4}\left(\dfrac{1}{e} - e\right)$ $(z - 1)^2 - \cdots$

39. $\dfrac{i}{4} + \dfrac{7}{16}(z - i) - \dfrac{25}{64}i(z - i)^2 - \dfrac{103}{256}(z - i)^3 + \cdots$

41. $1 - \dfrac{1}{4}x^2 - \dfrac{1}{96}x^4 - \dfrac{19}{5760}x^6 - \cdots$

43. $\dfrac{(1 + i)}{\sqrt{2}}\left[1 - \dfrac{i}{2}z + \displaystyle\sum_{n=2}^{\infty}(-1)^{n-1}\dfrac{1 \cdot 3 \ldots (2n-3)}{2 \cdot 4 \ldots 2n}z^n\right]$

45. $4\pi i + z - \left(\dfrac{1}{2} + 2\pi i\right)z^2 - \dfrac{1}{6}z^3 + \cdots$

47. $\dfrac{1}{2}z^2 + z^3 + \dfrac{35}{24}z^4 + \cdots$

49. $1 + z - 2z^2 - 2z^3 + \cdots$

51. $z - \dfrac{z^3}{3} + \dfrac{z^5}{5} - \dfrac{z^7}{7} + \cdots$

53. $\displaystyle\int_0^z \dfrac{\sin u}{u}\,du = z - \dfrac{1}{18}z^3 + \dfrac{1}{600}z^5 - \cdots$ (divide the series for $\sin u$ by u and integrate the result term by term)

55. $z + z^2 + \dfrac{5}{6}z^3 + \dfrac{5}{6}z^4 + \cdots$

Exercise Set 15.3

1. $-\dfrac{1}{2}\displaystyle\sum_{n=0}^{\infty}\left(\dfrac{z}{2}\right)^n$, $|z| < 2$

3. $\dfrac{1}{b - a}\displaystyle\sum_{n=0}^{\infty}\dfrac{b^{n+1} - a^{n+1}}{a^{n+1} + b^{n+1}}z^n$, $\quad |z| < |a|$

5. $\dfrac{1}{a - b}\displaystyle\sum_{n=0}^{\infty}\left(\dfrac{z^n}{b^{n+1}} + \dfrac{a^n}{z^{n+1}}\right)$, $\quad |a| < |z| < |b|$

7. For $z = 0$;
$f(z) = \exp[1/(1 - z)] = \exp[-1/(z - 1)]$
$$= 1 - \dfrac{1}{(z - 1)} + \dfrac{1}{2!(z - 1)^2} - \dfrac{1}{3!(z - 1)^3}$$
$$+ \cdots = \sum_{n=0}^{\infty}(-1)^n\dfrac{1}{n!(z - 1)^n},$$
$$0 < |z - 1| < \infty.$$
For $|z| > 1$; $f(z) = \exp[-1/(z - 1)] = \exp[-\frac{1}{z}(1 - \frac{1}{z})^{-1}]$. Now expand $(1 - \frac{1}{z})^{-1}$ by the binomial theorem and multiply the result by $-1/z$ to obtain
$$f(z) = \exp\left[-\dfrac{1}{z} - \dfrac{1}{z^2} - \cdots\right]$$
$$= 1 - \left(\dfrac{1}{z} + \dfrac{1}{z^2} + \cdots\right) + \dfrac{1}{2!}\left(\dfrac{1}{z} + \dfrac{1}{z^2} + \cdots\right)^2$$
$$- \cdots = 1 - \dfrac{1}{z} - \dfrac{1}{z^2} + \cdots.$$

9. $\sin\left(\dfrac{z}{1 - z}\right) = -\sin\left(1 + \dfrac{1}{z - 1}\right)$
$$= -\sin 1\cos\left(\dfrac{1}{z - 1}\right) - \cos 1\sin\left(\dfrac{1}{z - 1}\right)$$
Now substitute $1/(z - 1)$ into the series for sine and cosine to obtain
$$\sin\left(\dfrac{z}{1 - z}\right)$$
$$= -\sin 1\left(1 - \dfrac{1}{2!(z - 1)^2} + \dfrac{1}{4!(z - 1)^4} - \cdots\right)$$
$$- \cos 1\left(\dfrac{1}{z - 1} - \dfrac{1}{3!(z - 1)^3} + \cdots\right)$$
$$= -\sum_{n=0}^{\infty}\dfrac{\sin\left(1 + \frac{1}{2}n\pi\right)}{n!(z - 1)^n}, 0 < |z - 1| < \infty.$$

11. $\dfrac{1}{3}\displaystyle\sum_{n=1}^{\infty}\dfrac{(-1)^{n-1}2^{n-1} - 1}{z^n}$, $\quad |z| > 2$

13. Expand $\sinh(1 + u)$ as a Maclaurin series and then set $u = 1/z$ to obtain
$$\dfrac{1}{2}\left(e - \dfrac{1}{e}\right) + \dfrac{1}{2}\left(e + \dfrac{1}{e}\right)\dfrac{1}{z} + \dfrac{1}{4}\left(e - \dfrac{1}{e}\right)\dfrac{1}{z^2} - \cdots,$$
$$|z| > 0$$

15. Multiply the series for $\sin z$ and $\sin z/3$ and divide the result by z^3 to obtain
$$\dfrac{1}{3}\dfrac{1}{z} - \dfrac{5}{81}z + \dfrac{14}{3645}z^3 + \cdots, \quad |z| > 0$$

17. Simple poles at $z = 0$ and $z = \pm 2$

19. $z = 0$ is an essential singularity
21. Removable singularity at $z = 0$ obtained by defining $f(0) = 1$
23. $z = 1$ is an essential singularity
25. $z = (1 \pm 2k)\pi/2, k = 0, 1, \ldots$ are second order poles
27. Removable singularity at $z = 0$ obtained by defining $f(0) = -2$
29. $a_n = \frac{1}{2\pi i} \int_\Gamma \frac{f(\varsigma)}{(\varsigma - z)^{n+1}} d\varsigma, n = 0, \pm 1, \pm 2, \ldots$ where Γ is the circle $|z - z_0| = R$ with
$$R_1 < R < R_2. \text{ So } |a_n| \leq \frac{1}{2\pi} \int_0^{2\pi} \frac{|f(\varsigma)|}{|\varsigma - z^{n+1}|} R d\theta$$
$$\leq \frac{1}{2\pi} \frac{M}{R^n} \int_0^{2\pi} d\theta = \frac{M}{R^n}.$$
31. $\frac{1}{z} \sum_{n=0}^{\infty} \left(\frac{a}{z}\right)^n, \quad |z| > 3$ **33.** $-\sum_{n=1}^{\infty} \frac{(-1)^n}{n} \frac{1}{z^{2n}}, |z| > 1$
35. $z = \infty$ is a regular point
37. $z = \infty$ is a limit point of poles
39. There is an essential singularity at $z = \infty$

Exercise Set 15.4

1. $\text{Res}[z = 2] = 5/4; \text{Res}[z = -2] = -1/4$
3. $\text{Res}[z = 0] = 3; \text{Res}[z = -1] = -2$
5. $\text{Res}[z = 0] = -1; \text{Res}[z = -1] = 0$
7. $\text{Res}[z = n\pi] = (-1)^n(n^2\pi + 3), n = 0, \pm 1, \pm 2, \ldots$
9. $\text{Res}[z = (2n + 1)\pi/2] = -1, n = 0, \pm 1, \pm 2, \ldots$
11. $\text{Res}[z = (2n + 1)\pi i] = -1, n = 0, \pm 1, \pm 2, \ldots$
13. $z = 0$ is a removable singularity so $\text{Res}[z = 0] = 0$;
$\text{Res}[z = n\pi i] = (-1)^n i \sinh n\pi, n = \pm 1, \pm 2, \ldots$
15. $\text{Res}[z = 2] = 0$
17. $-\pi i/3$ **19.** $12\pi i$ **21.** $-\pi i/\sqrt{2}$
23. $-2\pi i/9$ **25.** $\pi(1 - e^{-2})$
27. $-2\pi i\{\cos 1 + i \sin 1\}$
29. $2\pi/(a^2 - 1)^{1/2}$ **31.** $\pi/\sqrt{2}$ **33.** $2\pi/(1 - a^2)$

Exercise Set 15.5

1. $\pi/(4a)$
3. $\pi/(2\sqrt{2})$
5. $\pi/18$
7. $\dfrac{\pi(1 + a)}{4a^3 e^a}$
9. $\dfrac{\pi}{(a^2 - b^2)}\left(\dfrac{e^{-b}}{b} - \dfrac{e^{-a}}{a}\right)$
11. $\dfrac{\pi}{2} \dfrac{\exp[-ma/\sqrt{2}]}{\cos(ma/\sqrt{2})}$

13. π
15. $\dfrac{\pi}{2}(b - a)$
17. $\dfrac{\pi}{2b^2}(1 - e^{-ab})$
19. $3\pi/8$
21. $\dfrac{\pi}{4}[e^{-a} + \sin a]$
23. $\pi/\sqrt{2}$
25. $\pi/\sqrt{3}$
27. $\pi/3$

Exercise Set 16.1

1. $f(t) = (1/a^2)(1 - \cos at)$
3. $f(t) = (1/2)(t \cos t + t \sin t - \sin t)$
5. $f(t) = t^2/2 - t + 1 - e^{-t}$
7. $f(t) = \dfrac{1}{2a^2}\left(\dfrac{\sin at}{a} - t \cos at\right)$
9. $f(t) = \dfrac{\sqrt{3}}{2} \dfrac{\Gamma(2/3)}{\pi t^{2/3}}$
11. $f(t) = H(t - 2)[\cosh(t - 2) + \sinh(t - 2)]$
13. $f(t) = \dfrac{1}{\sqrt{a}} \operatorname{erf}(\sqrt{at})$
15. $f(t) = \dfrac{e^{-at}}{\sqrt{b-a}} \operatorname{erf}(\sqrt{(b - a)t})$. Set $\mathcal{L}^{-1}\{1/\sqrt{s + b}\}$ $= e^{-bt}/\sqrt{\pi t}$ and $\mathcal{L}^{-1}\{1/(s + a)\} = e^{-at}$ and use the convolution theorem followed by a change of variable)

Exercise Set 17.1

1. A $\pi/2$ counterclockwise rotation, a uniform magnification by a factor 2, and a shift of origin causing the point $z = 1 + i$ to map to the point $w = 1 + 2i$
3. $w = (1 - i)(1 + 2z)$
5. $w = (3 - 2i)z + 2i - 10$
7. As the transformation is linear it preserves shape, so a mapping of one strip onto the other is obtained by mapping a point on one side of the strip in the z-plane onto a point on one side of the strip in the w-plane, and then repeating the process by mapping a point on the other side of the strip in the z-plane onto a point on the other side of the strip in the w-plane. Only the correspondence between one pair of points is specified, namely the point $z = ik$ in the z-plane maps to the point $w = 0$ in the w-plane, so the transformation will not be unique. If we choose to map the point $z = i(k + h)$ on the top of the strip in the z-plane to the point $w = 1$ on the other side of the strip in the w-plane, we must solve the equations $0 = iak + b$ and $1 = ia(k + h) + b$, leading to the transformation $w = -(iz + k)/h$. A different choice of points will lead to a different

transformation between the two strips that still preserves the condition $w(ik) = 0$.

9. Family of circles $c(u^2 + v^2) + u + v = 0$ tangent to the straight line $v = -u$ at the origin

11. $w = i(1 + z)/(1 - z)$; interior of circle maps to upper-half of the w-plane

13. $w = (2i - z)/(2z + i)$; interior of circle maps to the interior of a circle

15. $x = c$ maps to circle $u^2 + v^2 = \exp(2\pi c/a)$; $y = k$ maps to radial line $v = u \tan \pi k/a$

17. $x = c$ maps to hyperbola $\dfrac{u^2}{\cos^2 \pi c/a} - \dfrac{v^2}{\sin^2 \pi c/a} = 1$;

$y = k$ maps to ellipse $\dfrac{u^2}{\cosh^2 \pi k/a} + \dfrac{v^2}{\sinh^2 \pi k/a} = 1$

19. Write transformation as $w = \left(\dfrac{(1+z)(1-\bar{z})}{(1-z)(1-\bar{z})}\right)^2$ and use the fact that on the circle $z\bar{z} = |\bar{z}|^2 = |z|^2 = 1$. Then find how the semicircular boundary and the strip CA map and, finally, show that a point inside the semicircle maps to a point in the upper half of the w-plane.

Exercise Set 17.2

3. $\phi(x, y) = \phi_4 + \dfrac{1}{\pi}\left[(\phi_1 - \phi_2)\text{Arctan}\left(\dfrac{y}{x - x_1}\right)\right.$

$\left. + (\phi_2 - \phi_3)\text{Arctan}\left(\dfrac{y}{x - x_2}\right) \right.$

$\left. + (\phi_3 - \phi_4)\text{Arctan}\left(\dfrac{y}{x - x_3}\right)\right]$

5. $T(x, y) = 30 + \dfrac{240}{\pi}\text{Arctan}\left(\dfrac{2y}{1 - x^2 - y^2}\right)$

7. $\phi(x, y) = 320 - \dfrac{220}{\pi}\text{Arctan}\left(\dfrac{1 - x^2 - y^2}{2y}\right)$

9. $U\left(x^2 y - y^3 - \dfrac{3x^2 y - y^3}{(x^3 - 3xy^2)^2 + (3x^2 y - y^3)^2}\right)$
$= \text{constant}$

11. The equation of the streamline is $y\left(1 - \dfrac{1}{x^2 + y^2}\right) = $ constant. As this equation is an even function of x, the streamlines are symmetric about the y-axis and $y' = 0$ for $x = 0$, $y \geq 1$. Far from the origin the streamlines are parallel to the x-axis. A bounding streamline lies along the x-axis and around the unit semicircle. Routine calculations show $y' > 0$ for $x < 0$ and $y' < 0$ for $x > 0$. Any streamline can be replaced by a boundary, so as the flow is steady any streamline $\psi = $ constant can represent a free surface.

13. The equipotentials $u = c$ in the w-plane are the hyperbolas $\dfrac{x^2}{\sin^2 c} - \dfrac{y^2}{\cos^2 c} = 1$ and the flux lines $v = k$ are the ellipses $\dfrac{x^2}{\cosh^2 k} + \dfrac{y^2}{\sinh^2 k} = 1$. In steady state heat conditions this represents a semi-infinite metal lamina with edge $A_\infty B$ at $T = 200$, edge CD_∞ at $T = 100$, with the edge BC insulated. The equipotentials become isotherms and flux lines become heat flow lines.

15. $T(x, y) = 450 - \dfrac{350}{\pi}\text{Arctan}\left(\dfrac{1 - x^2 - y^2}{2x}\right)$

Exercise Set 18.1

1. (a) Quasilinear first order
   (b) Linear first order
   (c) Nonlinear first order
   (d) Semilinear first order
   (e) Linear first order
   (f) Nonlinear first order
   (g) Linear second order
   (h) Nonlinear second order

3. $u(x, y) = 4\exp[x - (x^2 - 2y)^{1/2}] - 2$, $x^2 \geq 2y$

5. $u(x, y) = \exp[x - (x^2 - 2y + 2)^{1/2}] - 2$, $x^2 \geq 2y - 2$

Exercise Set 18.2

1. $u(x, y) = x + \frac{1}{2}y$; global

3. $u(x, y) = x - y + 3$; global

5. $u(x, y) = \frac{1}{2}\sin x - \sin(2y - x)$; global

7. $u(x, y) = 1 + 2y - 4x - y^2 + 4xy - 3x^2$; global

9. $u(x, y) = 3x + \tan x^2 + \tan(\frac{1}{2}y - x^2)$ for (x, y) such that $\tan x^2$ and $\tan(\frac{1}{2}y - x^2)$ are both finite

11. $u(x, y) = (y - x)/(x^2 - xy + 1)$ for (x, y) such that $x^2 - xy + 1 \neq 0$

13. The solution in parametric form is $u = e^{-x} \sin \xi$, $y = \xi + (1 - e^{-x})\sin y$. An attempt to eliminate the parameter ξ leads to an implicit solution, so it is best to use the parametric form.

15. The parametric form of the solution is $u = 4\xi e^{-3x}$, $y = \xi + \frac{8}{3}\xi(1 - e^{-3x})$. In this case the parameter ξ can be eliminated to give the simple explicit solution $u(x, y) = 12y/(11e^{3x} - 8)$, for x such that the denominator does not vanish.

17. The solution in parametric form is $u = (3 + 2\xi)e^{-x}$, $y = \xi + (3 + 2\xi)(1 - e^{-x})$. In this case the parameter ξ can be eliminated to give the simple explicit solution $u(x, y) = (2y + 3)/(3e^x - 2)$, for x such that the denominator does not vanish.

Exercise Set 18.3

1. $u(x, t) = e^{3t/2} \sin(2x - 4t)$
3. $u(x, t) = \frac{1}{2}e^{2t}\{\cos(x + 3t) + 1\} - \frac{1}{2}$
5. $u(x, t) = e^{x+4t} + 6t^2 + 3xt$
7. $u(x, t) = x(2e^t - 1)$
9. $u(x, t) = x(4e^t - 1)$
11. $u(x, t) = \frac{1}{3}x(4e^t - 1)$
13. $u(x, t) = \dfrac{\cos(x - t)}{1 - 2t\cos(x - t)}$; provided the denominator does not vanish
15. $u(x, t) = \dfrac{-2xe^t}{5 - 4e^t}$;   for $0 \le t < \ln\frac{5}{4}$
17. $u(x, t) = \dfrac{4(1 + x)e^{-4t}}{1 + 3e^{-4t}}$
19. $u(x, t) = \dfrac{(3x - 1)(1 + t)}{1 + 3t + 3t^2 + t^3}$;   for $t > -1$

Exercise Set 18.4

1. Write the equation in the conservation form $\frac{\partial u}{\partial t} + \frac{\partial}{\partial x}\left(\frac{u^{n+1}}{n+1}\right) = 0$. The shock condition is $\Lambda(t)[[u]] = \frac{1}{n+1}[[u^{n+1}]]$.
3. Riemann problem (b) has a shock solution because of the intersection of its characteristics. The conservation form of the equation is $\frac{\partial u}{\partial t} + \frac{\partial}{\partial x}\left(\frac{1}{3}u^3\right) = 0$, so the shock condition is $\Lambda(t)[[u]] = \frac{1}{3}[[u^3]]$, and hence the shock speed is seen to be given by $\Lambda(t) = \frac{(27-1)}{3(3-1)} = \frac{13}{3}$.
5. A similar problem was solved in Section 18.4 with the initial condition $u(x, 0) = \begin{cases} 0, x < 0 \\ 1, x > 0 \end{cases}$. The solution of Exercise 5 follows from the solution given in Section 18.4 by replacing x by $x - 2$ to obtain $u = (x - 2)/t$. The solution lies in the region $t > 0$ bounded by the characteristic $x = 2$ and the characteristic $x = t + 2$.

Exercise Set 18.6

1. Elliptic 3. Elliptic 5. Parabolic
7. Elliptic; $\xi = \frac{1}{2}(x + y)$, $\eta = x : u_{\xi\xi} + u_{\eta\eta} + \frac{3}{2}u_\xi + 3u_\eta + 1 = 0$
9. Hyperbolic; $\xi = 9x + y$, $\eta = x + y : u_{\xi\eta} = -\frac{1}{64}(9u_\xi + u_\eta)$
11. Parabolic; $\xi = y - 3x$, $\eta = x : u_{\eta\eta} = u - 5$

13. $\mathbf{A} = \begin{bmatrix} 1 & 0 & 0 \\ 0 & \frac{2}{5} & \frac{4}{5} \\ 0 & \frac{4}{5} & \frac{8}{5} \end{bmatrix}$, $\lambda_1 = 2, \lambda_2 = 1, \lambda_3 = 0$, so the PDE is parabolic

$\mathbf{Q} = \begin{bmatrix} 0 & 1/\sqrt{5} & 2/\sqrt{5} \\ 1 & 0 & 0 \\ 0 & 2/\sqrt{5} & -1/\sqrt{5} \end{bmatrix}$, $\mathbf{D} = \begin{bmatrix} 2 & 0 & 0 \\ 0 & 1 & 0 \\ 0 & 0 & 0 \end{bmatrix}$,

so as $\xi = \mathbf{Q}\mathbf{x}$,
$\xi_1 = (1/\sqrt{5})x_2 + (2/\sqrt{5})x_3$, $\xi_2 = x_2$, $\xi_3 = (2/\sqrt{5})x_2 - (1/\sqrt{5})x_3$.
The PDE becomes $\frac{\partial^2 u}{\partial \xi_1^2} + \frac{1}{\sqrt{5}}\frac{\partial u}{\partial \xi_1} + \frac{2}{\sqrt{5}}\frac{\partial u}{\partial \xi_3} + 2u + \frac{1}{2} = 0$.

15. $\mathbf{A} = \begin{bmatrix} 3 & 0 & 0 \\ 0 & 2 & -1 \\ 0 & -1 & 2 \end{bmatrix}$, $\lambda_1 = 3, \lambda_2 = 3, \lambda_3 = 1$, so the PDE is elliptic

$\mathbf{Q} = \begin{bmatrix} 1 & 0 & 0 \\ 0 & -1/\sqrt{2} & 1/\sqrt{2} \\ 0 & 1/\sqrt{2} & 1/\sqrt{2} \end{bmatrix}$, $\mathbf{D} = \begin{bmatrix} 3 & 0 & 0 \\ 0 & 3 & 0 \\ 0 & 0 & 1 \end{bmatrix}$,

so as $\xi = \mathbf{Q}\mathbf{x}$,
$\xi_1 = x_1$, $\xi_2 = -(1/\sqrt{2})x_2 + (1/\sqrt{2})x_3$, $\xi_3 = (1/\sqrt{2})x_2 + (1/\sqrt{2})x_3$. The PDE becomes $3u_{\xi_1\xi_1} + 3u_{\xi_2\xi_2} + u_{\xi_3\xi_3} + 4u - 7 = 0$. The further scaling $\zeta_1 = (1/\sqrt{3})\xi_1$, $\zeta_2 = (1/\sqrt{3})\xi_2$, $\zeta_3 = \xi_3$ reduces the PDE to the still simpler form $u_{\zeta_1\zeta_1} + u_{\zeta_2\zeta_2} + u_{\zeta_3\zeta_3} + 4u - 7 = 0$.

Exercise Set 18.8

In each case the solutions are given in the form of computer generated plots at the respective times $t = 0, t = 0.5, t = 1$ and $t = 3$. The 3D plot shown at the end of each solution illustrates how the waves evolve away from the initial condition.

 1.

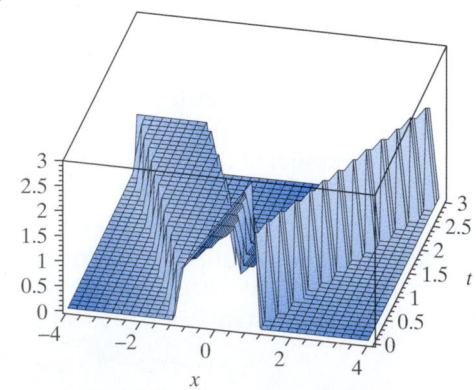

3.

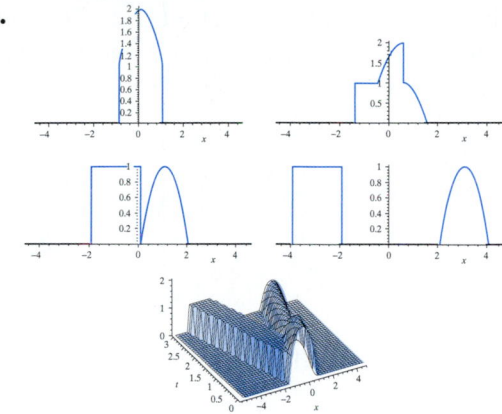

Exercise Set 18.9

1. $f_{xx} = f''(x - ct)$, $g_{xx} = g''(x - ct)$,
$f_{tt} = c^2 f''(x - ct)$, $g_{tt} = c^2 g''(x - ct)$, so
$u = f + g$ satisfies $u_{tt} = c^2 u_{xx}$

3. $u(x, t) = \dfrac{1}{2}\{\sin(x - ct) + \sin(x + ct)\}$
$+ \dfrac{1}{2c}\displaystyle\int_{x-ct}^{x+ct} \dfrac{ds}{1 + s^2}$, and so $u(x, t) = \sin x \cos ct$
$+ \dfrac{1}{2c}\{\text{Arctan}(x + ct) - \text{Arctan}(x - ct)\}$

4. $u(x, t) = 1 + \dfrac{1}{2c}\displaystyle\int_{x-ct}^{x+ct} \cos s\, ds = 1 +$
$\dfrac{1}{c}\cos c \sin ct$

5. $u(x, t) = \dfrac{1}{2}\{\tanh(x - ct) + \tanh(x + ct)\} +$
$\dfrac{1}{2c}\{\tanh(x + ct) - \tanh(x - ct)\}$, and so $u(x, t) =$
$\left(\dfrac{c + 1}{2c}\right)\tanh(x + ct) + \left(\dfrac{c - 1}{2c}\right)\tanh(x - ct)$

6. $u(x, t) = \dfrac{1}{2}\{e^{x-ct} + e^{x+ct}\} + \dfrac{1}{2c}\displaystyle\int_{x-ct}^{x+ct} e^{-s}\, ds =$
$e^x \cosh ct + \dfrac{1}{c}\sinh ct$

7.

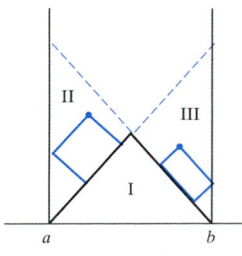

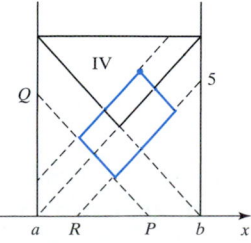

Use D'Alembert in (I), then
(128) to find u in (II) and
(III)

Use solutions in (I), (II) and
(III) and (128) with characteristics PQ and RS to find
solution in (IV)

The situation is now back to the original problem and so can be continued as long as necessary.
This is a *theoretical* rather than a practical way of
solving the problem.

9.

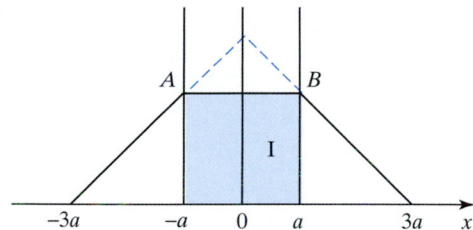

Reflect the initial conditions as odd functions
about $x = -a$ and $x = a$. Then the initial conditions are known for $-3a \le x \le 3a$. D'Alembert's
formula can now be used to find the solution in
(I). The solution is then known along AB, so the
argument can be repeated using the conditions
along AB as new initial conditions, etc.

11. From D'Alembert's formula with $g(x) \equiv 0$ we
have

$$u(x, t) = \dfrac{1}{2}\{f(x - ct) + f(x + ct)\}.$$

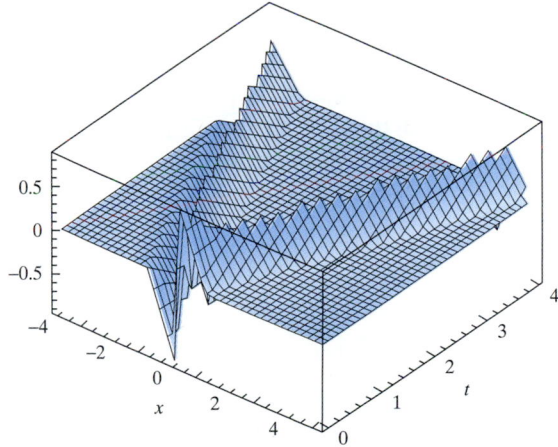

13. $u(x, 1/4) = \dfrac{1}{2}\displaystyle\int_{x-1/4}^{x+1/4} g(s)\, ds$, so

$u\left(x, \dfrac{1}{4}\right)$

$= \begin{cases} 0, & x < -5/4 \\ \dfrac{1}{2}\int_{-1}^{x+1/4}(1 - s^2)\, ds, & -5/4 \le x \le -3/4 \\ \dfrac{1}{2}\int_{x-1/4}^{x+1/4}(1 - s^2)\, ds, & -3/4 \le x \le 3/4 \\ \dfrac{1}{2}\int_{x-1/4}^{1}(1 - s^2)\, ds, & 3/4 \le x \le 5/4 \\ 0, & x > 5/4 \end{cases}$

1140 Answers

Exercise Set 18.10

1. $u(x,t) = \dfrac{4kL^2}{\pi^3} \displaystyle\sum_{n=1}^{\infty} \dfrac{1}{n^3}[2(-1)^{n+1} - 1]$
$\times \sin \dfrac{n\pi x}{L} \cos \dfrac{n\pi ct}{L}$

3. $u(x,t) = \dfrac{8k}{\pi^2} \displaystyle\sum_{n=1}^{\infty} \dfrac{1}{n^2} \sin \dfrac{n\pi}{2} \sin \dfrac{n\pi x}{L} \cos \dfrac{n\pi ct}{L}$

5. $u(x,t) = k \sin \dfrac{\pi x}{L} \cos \dfrac{2\pi ct}{L}$

7. $u(x,t) = \dfrac{4kL^3}{\pi^2} \displaystyle\sum_{n=1}^{\infty} \dfrac{1}{n^4}[1 + (-1)^{n+1}]$
$\times \sin \dfrac{n\pi x}{L} \sin \dfrac{n\pi ct}{L}$

9. $u(x,t) = \dfrac{8k}{\pi^2} \displaystyle\sum_{n=0}^{\infty} \dfrac{(-1)^n}{(2n+1)^2} \sin \dfrac{(2n+1)\pi x}{2L}$
$\times \cos \dfrac{(2n+1)\pi ct}{2L}$

11. $u(x,y) = \dfrac{e^{-6x}}{13}(4e^{-9y} + 9e^{4y})$. When $y \gg 0$,
$u(x,y) \approx \dfrac{9}{13} \exp(4y - 6x)$.

13. $u(x,t) = \dfrac{L}{2} - \dfrac{4L}{\pi^2} \displaystyle\sum_{n=1}^{\infty} \exp\left[-\dfrac{(2n-1)^2\pi^2 kt}{L^2}\right]$
$\times \dfrac{1}{(2n-1)^2} \cos \dfrac{(2n-1)\pi x}{L}$

17. $u(x,y,t) = 2 \sin \dfrac{3\pi x}{c} \sin \left(\dfrac{\pi y}{d}\right) \cos(\pi t \sqrt{(3/c)^2 + (1/d)^2})$.
The initial condition is an eigenfunction.

21. $T(x,y) = 1 - \dfrac{1}{2}e^{-x} \cos y + 2 \displaystyle\sum_{n=2}^{\infty}(-1)^n$
$\times \dfrac{(1-n^2)}{(1 - 2n^2 + n^4)} \exp(-nx) \cos(ny)$

Exercise Set 18.12

1. Taking the Laplace transform of the PDE with respect to t gives $s\bar{T} - T_0 = \kappa \dfrac{d^2\bar{T}}{dx^2}$ and $\bar{T}(0,s) = 0$ with the general solution $\bar{T}(x,s) = A\exp(\sqrt{\frac{s}{\kappa}}x) + B\exp(-\sqrt{\frac{s}{\kappa}}x) + \frac{T_0}{s}$. The solution can only be finite for all x if $A = 0$, so $\bar{T}(x,s) = T_0\left\{\dfrac{1 - \exp\left(-x\sqrt{s/\kappa}\right)}{s}\right\}$. Finding the inverse of this transform then gives $T(x,t) = T_0\text{erf}\left[\dfrac{x}{2\sqrt{\kappa t}}\right]$.

3. Taking the Laplace transform of the PDE with respect to t gives $s\bar{T} - T_0 = \kappa \dfrac{d^2\bar{T}}{dx^2}$ so the general solution is $\bar{T}(x,s) = A\exp(x\sqrt{s/\kappa}) + B\exp(-x\sqrt{s/\kappa}) + T_0/s$. The solution can only

be bounded for all x if $A = 0$, so $\bar{T}(x,s) = B\exp(-x\sqrt{s/\kappa}) + T_0/s$. Now $\mathcal{L}\{T(0,t)\} = T_0 s/(s^2 + a^2)$ so setting $x = 0$ in the above result gives $T_0 s/(s^2 + a^2) = B + T_0/s$ so that

$$\bar{T}(x,s) = T_0\left(\dfrac{s}{s^2 + a^2}\right)\exp\left(-x\sqrt{\dfrac{s}{\kappa}}\right)$$
$$- T_0\dfrac{1}{s}\exp\left(-x\sqrt{\dfrac{s}{\kappa}}\right) + \dfrac{T_0}{s}.$$

Using the convolution theorem to invert the transform gives

$$T(x,t) = \dfrac{T_0}{2\sqrt{\pi\kappa}}\int_0^t \dfrac{\cos a\tau}{(t-\tau)^{3/2}}\exp\left(\dfrac{-x^2}{4\kappa(t-\tau)}\right)d\tau$$
$$- \dfrac{T_0 x}{2\sqrt{\pi\kappa}}\int_0^t \dfrac{1}{\tau^3}\exp\left(\dfrac{-x^2}{4\kappa\tau}\right)d\tau + T_0.$$

5. Taking the Fourier transform of the PDE with respect to x gives $\bar{u}_{tt}(\omega,t) + (k + c^2\omega^2)\bar{u}(\omega,t) = 0$, and so $\bar{u}(\omega,t) = a(\omega)\cos(t\sqrt{k + c^2\omega^2}) + b(\omega) \times \sin(t\sqrt{k + c^2\omega^2})$, showing that $\bar{u}_t(\omega,t) = \sqrt{k + c^2\omega^2}\left\{-a(\omega)\sin(t\sqrt{k + c^2\omega^2}) + b(\omega) \times \cos(t\sqrt{k + c^2\omega^2})\right\}$.

From the initial conditions

$$a(\omega) = \bar{u}(\omega,0) = \dfrac{U}{\sqrt{2\pi}}\int_{-1}^{1}e^{-i\omega x}dx$$
$$= U\sqrt{\dfrac{2}{\pi}}\dfrac{\sin\omega}{\omega}, \quad \sqrt{k + c^2\omega^2}b(\omega)$$
$$= \bar{u}_t(\omega,0) = 0, \text{ so that } \bar{u}(\omega,t)$$
$$= U\sqrt{\dfrac{2}{\pi}}\dfrac{\sin\omega}{\omega}\cos(t\sqrt{k + c^2\omega^2}).$$

Taking the inverse transform then gives

$$u(x,t) = \dfrac{1}{\sqrt{2\pi}}\int_{-\infty}^{\infty}\bar{u}(\omega,t)e^{i\omega x}d\omega$$
$$= \dfrac{2U}{\pi}\int_0^{\infty}\dfrac{\sin\omega}{\omega}\cos(t\sqrt{k + c^2\omega^2})\cos\omega x\,d\omega.$$

7. Take the Fourier transform with respect to x of the PDE to obtain $-\omega\bar{u}(\omega,y) + \dfrac{d^2\bar{u}}{dy^2} = 0$ for $y > 0$, where $\bar{u}(\omega,0) = F(\omega)$, the transform of $f(x)$. For the solution to remain bounded when y is large it then follows that $\bar{u}(\omega,y) = F(\omega)e^{-|\omega|}y$. Taking the inverse transform then gives

$$u(x, y) = \frac{y}{\pi} \int_{-\infty}^{\infty} \frac{f(\tau)}{y^2 + (x - \tau)^2} d\tau.$$

9. Differentiate the result with respect to x and expand $e^{i\omega x}$ by de Moivre's theorem. The integral containing $\omega \cos \omega x$ vanishes because this is an odd function of ω and the remaining integral containing the function $\omega \sin \omega x$ is an even function of ω, so the result follows from the definition of the sine transform after changing the interval of integration to $[0, \infty)$.

11. Proceed as in the heat conduction example in Section 10.2 using the given form of $T(x, 0)$. The solution reduces to $T(x, t) = \frac{1}{2} T_0 \{ \text{erf}[(x + a)/(2\sqrt{\kappa t})] - \text{erf}[(x - a)/(2\sqrt{\kappa t})] \}$.

Exercise Set 19.2

1. 2.27886
3. 1.40619
5. −1.08601
7. $x_{r+1} = \frac{1}{2}(x_r + a/x_r^{n-1})$
9. −1.08090, 2.54109, 2.83981
11. 0.67567
13. 2.84387
15. 3.70665
17. 0.25763

Exercise Set 19.4

1. $I = 28$ 3. $I_{\text{trap}} = 1.849317$, $I_{\text{simp}} = 1.851944$, $I_{\text{exact}} = 1.851937$
5. 0.596584
7. $J_1(2) = 0.576725$ (the result obtained by Simpson's rule agrees with the exact result to six decimal places)
9. $J_1(4) = -0.065743$ (using Simpson's rule)
11. $I_0(3.5) = 7.378203$ (the result obtained by Simpson's rule agrees with the exact result to six decimal places)

Exercise Set 19.5

1. $x_1 = 0.73826$, $x_2 = -0.73918$, $x_3 = 0.75556$ (Gaussian elimination)
3. $x_1 = -0.90034$, $x_2 = -1.14831$, $x_3 = -0.95315$ (Gaussian elimination) Not diagonally dominant: interchange first and second equations
5. $x_1 = -66.51395$, $x_2 = 927.64721$, $x_3 = -2585.93671$, $x_4 = 1862.64259$

When calculations are rounded to five decimal places $\det \mathbf{H}_4 = 1.6111 \times 10^{-7}$. Exact value $\det \mathbf{H}_4 = 1/6048000 \approx 1.65344 \times 10^{-7}$

7. $\mathbf{L} = \begin{bmatrix} 1 & 0 & 0 \\ -3 & 1 & 0 \\ 3 & 1 & 1 \end{bmatrix}$, $\mathbf{U} = \begin{bmatrix} -4 & 1 & -1 \\ 0 & 2 & 2 \\ 0 & 0 & -3 \end{bmatrix}$,

$x_1 = -53/24$, $x_2 = -7/6$, $x_3 = 14/3$

9. $\mathbf{L} = \begin{bmatrix} 1 & 0 & 0 \\ -4 & 1 & 0 \\ -1 & 3 & 1 \end{bmatrix}$, $\mathbf{U} = \begin{bmatrix} 4 & -1 & -1 \\ 0 & 2 & -3 \\ 0 & 0 & -1 \end{bmatrix}$,

$x_1 = 131/8$, $x_2 = 81/2$, $x_3 = 25$

11. $\mathbf{L} = \begin{bmatrix} 1 & 0 & 0 & 0 \\ -1/2 & 1 & 0 & 0 \\ 2 & -1 & 1 & 0 \\ -1 & 2 & 2 & 1 \end{bmatrix}$, $\mathbf{U} = \begin{bmatrix} 2 & 1 & 0 & 2 \\ 0 & 1/2 & 1 & 1 \\ 0 & 0 & 3 & 0 \\ 0 & 0 & 0 & 1 \end{bmatrix}$

$x_1 = -3/2$, $x_2 = 10$, $x_3 = 1/2$, $x_4 = -3$

Exercise Set 19.6

1. $\lambda = 19.24435$ (exact), $\tilde{\mathbf{x}} = [1, 0.41089, -0.01169]^T$
3. $\lambda = 28.19020$ (exact), $\tilde{\mathbf{x}} = [0.07079, 0.04865, 1]^T$
5. $\lambda = 27.35196$ (exact), $\tilde{\mathbf{x}} = [1, 0.42720, 0.07037]^T$
7. $\lambda = 2.55051$ (exact), $\mathbf{x} = [1.44949, -1, 1]^T$ (not normalized)
9. $\lambda = -3.04390$ (exact), $\mathbf{x} = [-4.68367, 1, 4.94464]^T$ (not normalized)

Exercise Set 19.7

1.

x_n	2.0	2.2	2.4	2.6	2.8	3.0
y_n	0	0.66419	1.28937	1.89393	2.48875	3.08063

3.

x_n	1.0	1.2	1.4	1.6	1.8	2.0
y_n	2.0	2.17043	2.27255	2.29924	2.25314	2.14619

5.

x_n	1.0	1.2	1.4	1.6	1.8	2.0
y_n	1.0	0.40577	0.08015	−0.10414	−0.20593	−0.24801

7.

x_n	0	0.1	0.2	0.3	0.4	0.5
y_n	2.0	1.87998	1.71971	1.51888	1.27772	0.99787

9.

x_n	0	0.1	0.2	0.3	0.4	0.5
y_n	1.0	1.07995	1.12053	1.12465	1.09709	1.04377

11.

x_n	0	0.2	0.4	0.6	0.8	1.0
y_n	2.0	2.24068	2.57043	3.01382	3.61800	4.46785

13.

x_n	1.0	1.2	1.4	1.6	1.8	2.0
y_n	1.0	1.23999	1.55909	1.95332	2.41386	2.92755

15.

x_n	0	0.2	0.4	0.6	0.8	1.0
y_n	−2.0	−1.72167	−1.25453	−0.72717	−0.01088	0.90446

17.

t_n	0	0.2	0.4	0.6	0.8	1.0
x_n	1.0	1.00348	1.02480	1.07075	1.17222	1.32949
y_n	0	−0.18397	−0.35117	−0.52343	−0.72280	−0.97510

19.

t_n	0	0.2	0.4	0.6	0.8	1.0
x_n	1.0	0.80588	0.64974	0.55084	0.49643	0.46921
y_n	1.0	0.87511	0.79475	0.74186	0.70938	0.69102

REFERENCES

General References

[G.1] M. Abrabowitz and I. A. Stegun (Eds.), *Handbook of Mathematical Functions*, Dover (reprint), New York, 1970

[G.2] I. S. Gradshteyn and I. M. Ryzhik (Ed. A. Jeffrey), *Tables of Integrals, Series, and Products*, 6th ed., Academic Press, Boston, 2000

[G.3] A. Jeffrey, *Handbook of Mathematical Formulas and Integrals*, Academic Press, New York, 1995

Part One   Review Material

[1.1] I. D. Faires and B. T. Faires, *Calculus*, 2nd ed., Random House, New York, 1988

[1.2] R. L. Finney and G. B. Thomas Jr., *Calculus and Analytic Geometry*, 9th ed., Addison-Wesley, Reading, MA, 1996

[1.3] R. E. Larson, R. P. Hostetler, and B. E. Edwards, *Calculus with Analytic Geometry*, 4th ed., D. C. Heath, Lexington, MA, 1990

[1.4] J. E. Marsden and A. J. Tromba, *Vector Calculus*, 2nd ed., W. H. Freeman, San Francisco, 1981

[1.5] M. H. Protter and P. E. Protter, *Calculus with Analytic Geometry*, 4th ed., Jones and Bartlett, Boston, 1988

[1.6] M. H. Protter and C. B. Morrey, Jr., *Modern Mathematical Analysis*, Addison-Wesley, Reading, MA, 1964

[1.7] D. G. Zill, *Calculus with Analytic Geometry*, 2nd ed., PWS-Kent, Boston, 1988

Part Two   Vectors and Matrices

[2.1] H. Anton and C. Rorres, *Elementary Linear Algebra: Applications Version*, 6th ed., Wiley, New York, 1991

[2.2] K. P. Bogart, *Introductory Combinatorics*, Pitman, London, 1983

[2.3] D. E. Bourne and P. C. Kendall, *Vector Analysis and Cartesian Tensors*, 3rd ed., Chapman and Hall, London, 1992

[2.4] A. B. Clarke and R. L. Disney, *Probability and Random Processes for Engineers and Scientists*, Wiley, New York, 1970

[2.5] C. G. Cullen, *Linear Algebra with Applications*, 2nd ed., Addison-Wesley, Reading, MA, 1997

[2.6] R. L. Finney and G. B. Thomas Jr., *Calculus*, 2nd ed., Addison-Wesley, Reading, MA, 1994

[2.7] S. I. Grossman, *Elementary Linear Algebra*, 3rd ed., Wadsworth, Belmont, CA, 1987

[2.8] L. Mirsky, *An Introduction to Linear Algebra*, Oxford University Press, Oxford, 1963

[2.9] P. V. O'Neal, *Introduction to Linear Algebra: Theory and Applications*, Wadsworth, Belmont, CA, 1979

[2.10] E. D. Nering, *Linear Algebra and Matrix Theory*, Wiley, New York, 1970

[2.11] B. Nobel and J. W. Daniel, *Applied Linear Algebra*, 3rd ed., Prentice-Hall, Englewood Cliffs, NJ, 1988

[2.12] G. Strang, *Linear Algebra and its Applications*, 2nd ed., Academic Press, New York, 1980

[2.13] S. A. Wiitala, *Discrete Mathematics: A Unified Approach*, McGraw-Hill, New York, 1987

[2.14] K. E. Atkinson, *An Introduction to Numerical Analysis*, 2nd ed., Wiley, New York, 1989

[2.15] G. J. Borse, *Numerical Mathematics with MATLAB*, PWS, Boston, 1997

[2.16] W. Cheney and D. Kincaid, *Numerical Mathematics and Computing*, Brooks/Cole, San Francisco, 1994

[2.17] C. E. Froberg, *Numerical Mathematics*, Benjamin Cummings, Menlo Park, CA, 1985

[2.18] L. W. Johnson and R. D. Riess, *Numerical Analysis*, 2nd ed., Addison-Wesley, Reading, MA, 1982

[2.19] W. H. Press, B. P. Flannen, S. A. Teukolsky, and W. T. Vetterling, *Numerical Recipes: The Art of Scientific Computing*, Cambridge University Press, Cambridge, UK, 1987

[2.20] J. Stoer and R. Bulirsch, *Introduction to Numerical Analysis*, Springer-Verlag, New York, 1980

Part Three   Ordinary Differential Equations

[3.1] V. I. Arnold, *Ordinary Differential Equations*, Springer-Verlag, New York, 1992

[3.2] D. K. Arrowsmith and C. M. Place, *Dynamical Systems and Nonlinear Ordinary Differential Equations*, Chapman and Hall, London, 1995

[3.3] G. Birkhoff and Gian-Carlo Rota, *Ordinary Differential Equations*, 4th ed., Wiley, New York, 1989

[3.4] W. E. Boyce and R. C. DiPrima, *Elementary Differential Equations and Boundary Value Problems*, 3rd ed., Wiley, New York, 1977

[3.5] F. Brauer and J. A. Nohel, *Introduction to Differential Equations with Applications*, Harper and Row, New York, 1986

[3.6] M. Braun, *Differential Equations and Their Applications*, Springer-Verlag, New York, 1975

[3.7] J. W. Brown and R. V. Churchill, *Fourier Series and Boundary Value Problems*, 5th ed., McGraw-Hill, 1993

[3.8] H. S. Carslaw and J. C. Jaeger, *Operational Methods in Applied Mathematics*, 2nd ed., Oxford University Press, London, 1949

[3.9] R. V. Churchill, *Operational Methods*, 3rd ed., McGraw-Hill, New York, 1972

[3.10] E. A. Coddington and N. Levinson, *Theory of Ordinary Differential Equations*, McGraw-Hill, New York, 1955

[3.11] A. Erdelyi, W. Magnus, F. Oberhettinger and F. Tricomi, *Tables of Integral Transforms*, Vols. I and II, McGraw-Hill, New York, 1954

[3.12] E. L. Ince, *Ordinary Differential Equations*, Dover (reprint), New York, 1956

[3.13] D. W. Jordan and P. Smith, *Nonlinear Ordinary Differential Equations*, 3rd ed., Clarendon Press, Oxford, 1999

[3.14] F. Oberhettinger and L. Bandii, *Tables of Laplace Transforms*, Springer-Verlag, New York, 1973

[3.15] M. Krusemeyer, *Differential Equations*, Macmillan, New York, 1994

[3.16] S. L. Ross, *Differential Equations*, 3rd ed., Wiley, New York, 1984

[3.17] G. N. Watson, *A Treatise on the Theory of Bessel Functions*, 2nd ed., Cambridge University Press, Cambridge, UK, 1966

[3.18] D. V. Widder, *The Laplace Transform*, Princeton University Press, Princeton, NJ, 1941

[3.19] D. G. Zill, *A First Course in Differential Equations with Applications*, 3rd ed., PWS, Boston, 1986

[3.20] K. E. Atkinson, *An Introduction to Numerical Analysis*, 2nd ed., Wiley, New York, 1989

[3.21] G. J. Borse, *Numerical Mathematics with MATLAB*, PWS, Boston, 1997

[3.22] W. Cheney and D. Kincaid, *Numerical Mathematics and Computing*, Brooks/Cole, San Francisco, 1994

[3.23] C. E. Froberg, *Numerical Mathematics*, Benjamin Cummings, Menlo Park, California, 1985

[3.24] L. W. Johnson and R. D. Riess, *Numerical Analysis*, 2nd ed., Addison-Wesley, Reading, MA, 1982

[3.25] J. L. Morris, *Computational Methods in Elementary Numerical Analysis*, Wiley, New York, 1983

[3.26] A. Ralston and P. Rabinowitz, *A First Course in Numerical Analysis*, 2nd ed., McGraw-Hill, New York, 1978

Part Four   Fourier Series, Integrals, and the Fourier Transform

[4.1] W. Brown and R. V. Churchill, *Fourier Series and Boundary Value Problems*, 5th ed., McGraw-Hill, New York, 1993

[4.2] A. Erdelyi, W. Magnus, F. Oberhettinger, and F. Tricomi, *Tables of Integral Transforms*, Vols. I and II, McGraw-Hill, New York, 1954

[4.3] I. N. Sneddon, *Fourier Transforms*, McGraw-Hill, New York, 1951

[4.4] I. N. Sneddon, *The Use of Integral Transforms*, McGraw-Hill, New York, 1972

[4.5] A. Zygmund, *Trigonometric Series*, 2nd ed. (Volumes I and II combined), Cambridge University Press, Cambridge, UK, 1988

Part Five Vector Calculus

[5.1] P. Baxandall and H. Liebeck, *Vector Calculus*, Oxford University Press, Oxford, 1986

[5.2] D. E. Bourne and P. C. Kendall, *Vector Analysis and Cartesian Tensors*, 3rd ed., Chapman and Hall, London, 1992

[5.3] J. E. Marsden and A. J. Tromba, *Vector Calculus*, 2nd ed., W. H. Freeman, San Francisco, 1981

[5.4] G. E. Mase and G. T. Mase, *Continuum Mechanics for Engineers*, CRC Press, Boca Raton, FL, 1992

[5.5] M. H. Protter and C. B. Morrey, Jr., *Modern Mathematical Analysis*, Addison-Wesley, Reading, MA, 1964

[5.6] M. R. Spiegel, *Vector Analysis*, Schaum Outline Series, McGraw-Hill, New York, 1974

Part Six Complex Analysis

[6.1] R. V. Churchill and J. W. Brown, *Complex Variables and Applications*, 5th ed., McGraw-Hill, New York, 1990

[6.2] P. Henrici, *Applied Computational Complex Analysis* (3 volumes), Wiley, New York, 1977, 1988, 1991

[6.3] J. E. Marsden, *Basic Complex Analysis*, Freeman, San Francisco, 1973

[6.4] J. H. Mathews and R.W. Howell, *Complex Analysis for Mathematics and Engineering*, 3rd ed., Jones and Bartlett, Boston, 1997

[6.5] L. M. Milne-Thompson, *Theoretical Hydrodynamics*, 5th ed., Macmillan, London, 1972

[6.6] J. D. Paliouras and D. S. Meadows, *Complex Variables for Scientists and Engineers*, 2nd ed., Macmillan, New York, 1975

[6.7] L. Pennisi, *Elements of Complex Variables*, 2nd ed., Holt, Rinehart and Winston, New York, 1976

[6.8] L. R. Rubenfeld, *A First Course in Applied Complex Variables*, Wiley, New York, 1985

[6.9] E. B. Saff and A. D. Snider, *Fundamentals of Complex Analysis for Mathematicians, Scientists and Engineers*, 2nd ed., Prentice Hall, Englewood Cliffs, NJ, 1993

[6.10] J. L. Schiff, *The Laplace Transform: Theory and Applications*, Springer-Verlag, New York, 1999

Part Seven Partial Differential Equations

[7.1] D. R. Bland, *Wave Theory and Applications*, Clarendon Press, Oxford, 1988

[7.2] J. W. Brown and R. V. Churchill, *Fourier Series and Boundary Value Problems*, 5th ed., McGraw-Hill, New York, 1993

[7.3] C. A. Coulson and A. Jeffrey, *Waves: A Mathematical Approach to the Common Types of Wave Motion*, 2nd ed., Longman, London, 1977

[7.4] R. Courant and K. O. Friedrichs, *Supersonic Flow and Shock Waves*, Wiley-Interscience, New York, 1956

[7.5] G. F. D. Duff and D. Naylor, *Differential Equations of Applied Mathematics*, Wiley, New York, 1966

[7.6] P. R. Garabedian, *Partial Differential Equations*, Wiley, New York, 1964

[7.7] R. Haberman, *Elementary Applied Partial Differential Equations*, 2nd ed., Prentice Hall, Englewood Cliffs, NJ, 1983

[7.8] R. Knobel, *An Introduction to the Mathematical Theory of Waves*, Student Mathematical Library Volume 3, American Mathematical Society, Rhode Island, 1999

[7.9] R. J. LeVeque, *Numerical Methods for Conservation Laws*, Birkhauser, Boston, 1990

[7.10] H. Levine, *Partial Differential Equations, Studies in Advanced Mathematics*, Vol. 6, American Mathematical Society, Rhode Island, 1991

[7.11] J. D. Logan, *Applied Partial Differential Equations*, Springer-Verlag, Berlin, 1998

[7.12] P. V. O'Neal, *Beginning Partial Differential Equations*, Wiley, New York, 1999

[7.13] J. Smoller, *Shock Waves and Reaction–Diffusion Equations*, Springer-Verlag, Berlin, 1983

[7.14] I. N. Sneddon, *The Use of Integral Transforms*, McGraw-Hill, New York, 1972

[7.15] W. A. Strauss, *Partial Differential Equations: An Introduction*, Wiley, New York, 1992

[7.16] M. E. Taylor, *Partial Differential Equations: Basic Theory*, Springer-Verlag, New York, 1996

[7.17] J. L. Troutman, *Boundary Value Problems of Applied Mathematics*, PWS, Boston, 1994

[7.18] G. B. Whitham, *Linear and Nonlinear Waves*, Wiley, New York, 1974; reprinted by Wiley, New York, 1999

[7.19] E. C. Zachmanoglou and D. W. Thoe, *Introduction to Partial Differential Equations with Applications*, Williams and Wilkins, Baltimore, 1976

[7.20] E. Zauderer, *Partial Differential Equations of Applied Mathematics*, 2nd ed., Wiley, New York, 1989

Part Eight   Numerical Mathematics

[8.1] K. E. Atkinson, *An Introduction to Numerical Analysis*, 2nd ed., Wiley, New York, 1989

[8.2] G. J. Borse, *Numerical Mathematics with MATLAB*, PWS, Boston, 1997

[8.3] W. Cheney and D. Kincaid, *Numerical Mathematics and Computing*, Brooks/Cole, San Francisco, 1994

[8.4] C. E. Fröberg, *Numerical Mathematics*, Benjamin Cummings, Menlo Park, CA, 1985

[8.5] L. W. Johnson and R. D. Riess, *Numerical Analysis*, 2nd ed., Addison-Wesley, Reading, MA, 1982

[8.6] R. J. LeVeque, *Numerical Methods for Conservation Laws*, Birkhauser, Boston, 1990

[8.7] J. Ll. Morris, *Computational Methods in Elementary Numerical Analysis*, Wiley, New York, 1983

[8.8] J. M. Ortega and W. G. Poole, Jr., *An Introduction to Numerical Methods for Differential Equations*, Pitman, London, 1981

[8.9] A. Ralston and P. Rabinowitz, *A First Course in Numerical Analysis*, 2nd ed., McGraw-Hill, New York, 1978

Suggested Further Reading

Linear Algebra

F. Chatelin, *Eigenvalues of Matrices*, Wiley-Interscience, New York, 1993

G. H. Golub and C. F. Van Loan, *Matrix Computations*, 3rd ed., Baltimore, Johns Hopkins University Press, 1996

J. H. Wilkinson, *The Algebraic Eigenvalue Problem*, Clarendon Press, Oxford, 1988

Analytic Functions

E. C. Titchmarsh, *The Theory of Functions*, 2nd ed., Oxford University Press, 1975 (reprint)

Applied Mathematics and Differential Equations

H. S. Carslaw and J. C. Jaeger, *Conduction of Heat in Solids*, 2nd ed., Clarendon Press, Oxford, 1986 (reprint)

J. P. Keener, *Principle of Applied Mathematics: Transformation and Approximation*, Addison-Wesley, Reading, MA, 1988

J. D. Logan, *Applied Mathematics: A Contemporary Approach*, Wiley-Interscience, New York, 1981

J. D. Logan, *An Introduction to Nonlinear Partial Differential Equations*, Wiley-Interscience, New York, 1994

D. Zwillinger, *Handbook of Differential Equations*, 2nd ed., Academic Press, Boston, 1992

Numerical Analysis

L. F. Shampine, *Numerical Solution of Ordinary Differential Equations*, Chapman and Hall, New York, 1994

L. F. Shampine et al., *Fundamentals of Numerical Computation*, Wiley-Interscience, New York, 1996

INDEX